# HANDBOOK OF
# MATHEMATICS FOR ENGINEERS AND SCIENTISTS

# HANDBOOK OF
# MATHEMATICS FOR ENGINEERS AND SCIENTISTS

### Andrei D. Polyanin
### Alexander V. Manzhirov

Chapman & Hall/CRC
Taylor & Francis Group
Boca Raton  London  New York

Chapman & Hall/CRC is an imprint of the
Taylor & Francis Group, an informa business

Chapman & Hall/CRC
Taylor & Francis Group
6000 Broken Sound Parkway NW, Suite 300
Boca Raton, FL 33487-2742

© 2007 by Taylor & Francis Group, LLC
Chapman & Hall/CRC is an imprint of Taylor & Francis Group, an Informa business

No claim to original U.S. Government works
Printed in the United States of America on acid-free paper
10 9 8 7 6 5 4 3 2 1

International Standard Book Number-10: 1-58488-502-5 (Hardcover)
International Standard Book Number-13: 978-1-58488-502-3 (Hardcover)

This book contains information obtained from authentic and highly regarded sources. Reprinted material is quoted with permission, and sources are indicated. A wide variety of references are listed. Reasonable efforts have been made to publish reliable data and information, but the author and the publisher cannot assume responsibility for the validity of all materials or for the consequences of their use.

No part of this book may be reprinted, reproduced, transmitted, or utilized in any form by any electronic, mechanical, or other means, now known or hereafter invented, including photocopying, microfilming, and recording, or in any information storage or retrieval system, without written permission from the publishers.

For permission to photocopy or use material electronically from this work, please access www.copyright.com (http://www.copyright.com/) or contact the Copyright Clearance Center, Inc. (CCC) 222 Rosewood Drive, Danvers, MA 01923, 978-750-8400. CCC is a not-for-profit organization that provides licenses and registration for a variety of users. For organizations that have been granted a photocopy license by the CCC, a separate system of payment has been arranged.

**Trademark Notice:** Product or corporate names may be trademarks or registered trademarks, and are used only for identification and explanation without intent to infringe.

**Visit the Taylor & Francis Web site at**
http://www.taylorandfrancis.com

**and the CRC Press Web site at**
http://www.crcpress.com

# CONTENTS

Authors .................................................................... xxv
Preface ................................................................... xxvii
Main Notation ............................................................. xxix

## Part I. Definitions, Formulas, Methods, and Theorems    1

### 1. Arithmetic and Elementary Algebra    3
1.1. Real Numbers ......................................................... 3
    1.1.1. Integer Numbers ............................................... 3
    1.1.2. Real, Rational, and Irrational Numbers ........................ 4
1.2. Equalities and Inequalities. Arithmetic Operations. Absolute Value ... 5
    1.2.1. Equalities and Inequalities ................................... 5
    1.2.2. Addition and Multiplication of Numbers ........................ 6
    1.2.3. Ratios and Proportions ........................................ 6
    1.2.4. Percentage .................................................... 7
    1.2.5. Absolute Value of a Number (Modulus of a Number) .............. 8
1.3. Powers and Logarithms ................................................ 8
    1.3.1. Powers and Roots .............................................. 8
    1.3.2. Logarithms .................................................... 9
1.4. Binomial Theorem and Related Formulas ............................... 10
    1.4.1. Factorials. Binomial Coefficients. Binomial Theorem .......... 10
    1.4.2. Related Formulas ............................................. 10
1.5. Arithmetic and Geometric Progressions. Finite Sums and Products ..... 11
    1.5.1. Arithmetic and Geometric Progressions ........................ 11
    1.5.2. Finite Series and Products ................................... 12
1.6. Mean Values and Inequalities of General Form ........................ 13
    1.6.1. Arithmetic Mean, Geometric Mean, and Other Mean Values. Inequalities for Mean Values ............................................. 13
    1.6.2. Inequalities of General Form ................................. 14
1.7. Some Mathematical Methods ........................................... 15
    1.7.1. Proof by Contradiction ....................................... 15
    1.7.2. Mathematical Induction ....................................... 16
    1.7.3. Proof by Counterexample ...................................... 17
    1.7.4. Method of Undetermined Coefficients .......................... 17
References for Chapter 1 ................................................. 18

### 2. Elementary Functions    19
2.1. Power, Exponential, and Logarithmic Functions ....................... 19
    2.1.1. Power Function: $y = x^\alpha$ ............................... 19
    2.1.2. Exponential Function: $y = a^x$ .............................. 21
    2.1.3. Logarithmic Function: $y = \log_a x$ ......................... 22
2.2. Trigonometric Functions ............................................. 24
    2.2.1. Trigonometric Circle. Definition of Trigonometric Functions .. 24
    2.2.2. Graphs of Trigonometric Functions ............................ 25
    2.2.3. Properties of Trigonometric Functions ........................ 27

## 2.3. Inverse Trigonometric Functions ... 30
### 2.3.1. Definitions. Graphs of Inverse Trigonometric Functions ... 30
### 2.3.2. Properties of Inverse Trigonometric Functions ... 33
## 2.4. Hyperbolic Functions ... 34
### 2.4.1. Definitions. Graphs of Hyperbolic Functions ... 34
### 2.4.2. Properties of Hyperbolic Functions ... 36
## 2.5. Inverse Hyperbolic Functions ... 39
### 2.5.1. Definitions. Graphs of Inverse Hyperbolic Functions ... 39
### 2.5.2. Properties of Inverse Hyperbolic Functions ... 41
## References for Chapter 2 ... 42

# 3. Elementary Geometry ... 43
## 3.1. Plane Geometry ... 43
### 3.1.1. Triangles ... 43
### 3.1.2. Polygons ... 51
### 3.1.3. Circle ... 56
## 3.2. Solid Geometry ... 59
### 3.2.1. Straight Lines, Planes, and Angles in Space ... 59
### 3.2.2. Polyhedra ... 61
### 3.2.3. Solids Formed by Revolution of Lines ... 65
## 3.3. Spherical Trigonometry ... 70
### 3.3.1. Spherical Geometry ... 70
### 3.3.2. Spherical Triangles ... 71
## References for Chapter 3 ... 75

# 4. Analytic Geometry ... 77
## 4.1. Points, Segments, and Coordinates on Line and Plane ... 77
### 4.1.1. Coordinates on Line ... 77
### 4.1.2. Coordinates on Plane ... 78
### 4.1.3. Points and Segments on Plane ... 81
## 4.2. Curves on Plane ... 84
### 4.2.1. Curves and Their Equations ... 84
### 4.2.2. Main Problems of Analytic Geometry for Curves ... 88
## 4.3. Straight Lines and Points on Plane ... 89
### 4.3.1. Equations of Straight Lines on Plane ... 89
### 4.3.2. Mutual Arrangement of Points and Straight Lines ... 93
## 4.4. Second-Order Curves ... 97
### 4.4.1. Circle ... 97
### 4.4.2. Ellipse ... 98
### 4.4.3. Hyperbola ... 101
### 4.4.4. Parabola ... 104
### 4.4.5. Transformation of Second-Order Curves to Canonical Form ... 107
## 4.5. Coordinates, Vectors, Curves, and Surfaces in Space ... 113
### 4.5.1. Vectors. Cartesian Coordinate System ... 113
### 4.5.2. Coordinate Systems ... 114
### 4.5.3. Vectors. Products of Vectors ... 120
### 4.5.4. Curves and Surfaces in Space ... 123

| 4.6. Line and Plane in Space | 124 |
|---|---|
|    4.6.1. Plane in Space | 124 |
|    4.6.2. Line in Space | 131 |
|    4.6.3. Mutual Arrangement of Points, Lines, and Planes | 135 |
| 4.7. Quadric Surfaces (Quadrics) | 143 |
|    4.7.1. Quadrics (Canonical Equations) | 143 |
|    4.7.2. Quadrics (General Theory) | 148 |
| References for Chapter 4 | 153 |
| **5. Algebra** | **155** |
| 5.1. Polynomials and Algebraic Equations | 155 |
|    5.1.1. Polynomials and Their Properties | 155 |
|    5.1.2. Linear and Quadratic Equations | 157 |
|    5.1.3. Cubic Equations | 158 |
|    5.1.4. Fourth-Degree Equation | 159 |
|    5.1.5. Algebraic Equations of Arbitrary Degree and Their Properties | 161 |
| 5.2. Matrices and Determinants | 167 |
|    5.2.1. Matrices | 167 |
|    5.2.2. Determinants | 175 |
|    5.2.3. Equivalent Matrices. Eigenvalues | 180 |
| 5.3. Linear Spaces | 187 |
|    5.3.1. Concept of a Linear Space. Its Basis and Dimension | 187 |
|    5.3.2. Subspaces of Linear Spaces | 190 |
|    5.3.3. Coordinate Transformations Corresponding to Basis Transformations in a Linear Space | 191 |
| 5.4. Euclidean Spaces | 192 |
|    5.4.1. Real Euclidean Space | 192 |
|    5.4.2. Complex Euclidean Space (Unitary Space) | 195 |
|    5.4.3. Banach Spaces and Hilbert Spaces | 196 |
| 5.5. Systems of Linear Algebraic Equations | 197 |
|    5.5.1. Consistency Condition for a Linear System | 197 |
|    5.5.2. Finding Solutions of a System of Linear Equations | 198 |
| 5.6. Linear Operators | 204 |
|    5.6.1. Notion of a Linear Operator. Its Properties | 204 |
|    5.6.2. Linear Operators in Matrix Form | 208 |
|    5.6.3. Eigenvectors and Eigenvalues of Linear Operators | 209 |
| 5.7. Bilinear and Quadratic Forms | 213 |
|    5.7.1. Linear and Sesquilinear Forms | 213 |
|    5.7.2. Bilinear Forms | 214 |
|    5.7.3. Quadratic Forms | 216 |
|    5.7.4. Bilinear and Quadratic Forms in Euclidean Space | 219 |
|    5.7.5. Second-Order Hypersurfaces | 220 |
| 5.8. Some Facts from Group Theory | 225 |
|    5.8.1. Groups and Their Basic Properties | 225 |
|    5.8.2. Transformation Groups | 228 |
|    5.8.3. Group Representations | 230 |
| References for Chapter 5 | 233 |

## 6. Limits and Derivatives .................................................. 235

### 6.1. Basic Concepts of Mathematical Analysis ............................... 235
- 6.1.1. Number Sets. Functions of Real Variable ............................. 235
- 6.1.2. Limit of a Sequence ................................................. 237
- 6.1.3. Limit of a Function. Asymptotes ..................................... 240
- 6.1.4. Infinitely Small and Infinitely Large Functions ..................... 242
- 6.1.5. Continuous Functions. Discontinuities of the First and the Second Kind ....... 243
- 6.1.6. Convex and Concave Functions ........................................ 245
- 6.1.7. Functions of Bounded Variation ...................................... 246
- 6.1.8. Convergence of Functions ............................................ 249

### 6.2. Differential Calculus for Functions of a Single Variable .............. 250
- 6.2.1. Derivative and Differential, Their Geometrical and Physical Meaning .. 250
- 6.2.2. Table of Derivatives and Differentiation Rules ...................... 252
- 6.2.3. Theorems about Differentiable Functions. L'Hospital Rule ............ 254
- 6.2.4. Higher-Order Derivatives and Differentials. Taylor's Formula ........ 255
- 6.2.5. Extremal Points. Points of Inflection ............................... 257
- 6.2.6. Qualitative Analysis of Functions and Construction of Graphs ........ 259
- 6.2.7. Approximate Solution of Equations (Root-Finding Algorithms for Continuous Functions) .................................................... 260

### 6.3. Functions of Several Variables. Partial Derivatives ................... 263
- 6.3.1. Point Sets. Functions. Limits and Continuity ........................ 263
- 6.3.2. Differentiation of Functions of Several Variables ................... 264
- 6.3.3. Directional Derivative. Gradient. Geometrical Applications .......... 267
- 6.3.4. Extremal Points of Functions of Several Variables ................... 269
- 6.3.5. Differential Operators of the Field Theory .......................... 272

References for Chapter 6 ..................................................... 272

## 7. Integrals .............................................................. 273

### 7.1. Indefinite Integral ................................................... 273
- 7.1.1. Antiderivative. Indefinite Integral and Its Properties .............. 273
- 7.1.2. Table of Basic Integrals. Properties of the Indefinite Integral. Integration Examples ....................................................... 274
- 7.1.3. Integration of Rational Functions ................................... 276
- 7.1.4. Integration of Irrational Functions ................................. 279
- 7.1.5. Integration of Exponential and Trigonometric Functions .............. 281
- 7.1.6. Integration of Polynomials Multiplied by Elementary Functions ....... 283

### 7.2. Definite Integral ..................................................... 286
- 7.2.1. Basic Definitions. Classes of Integrable Functions. Geometrical Meaning of the Definite Integral ................................................ 286
- 7.2.2. Properties of Definite Integrals and Useful Formulas ................ 287
- 7.2.3. General Reduction Formulas for the Evaluation of Integrals .......... 289
- 7.2.4. General Asymptotic Formulas for the Calculation of Integrals ........ 290
- 7.2.5. Mean Value Theorems. Properties of Integrals in Terms of Inequalities. Arithmetic Mean and Geometric Mean of Functions ..................... 295
- 7.2.6. Geometric and Physical Applications of the Definite Integral ........ 299
- 7.2.7. Improper Integrals with Infinite Integration Limit .................. 301
- 7.2.8. General Reduction Formulas for the Calculation of Improper Integrals ........ 304
- 7.2.9. General Asymptotic Formulas for the Calculation of Improper Integrals ....... 307
- 7.2.10. Improper Integrals of Unbounded Functions .......................... 308
- 7.2.11. Cauchy-Type Singular Integrals ..................................... 310

|  |  |  |
|---|---|---|
| | 7.2.12. Stieltjes Integral | 312 |
| | 7.2.13. Square Integrable Functions | 314 |
| | 7.2.14. Approximate (Numerical) Methods for Computation of Definite Integrals | 315 |
| 7.3. | Double and Triple Integrals | 317 |
| | 7.3.1. Definition and Properties of the Double Integral | 317 |
| | 7.3.2. Computation of the Double Integral | 319 |
| | 7.3.3. Geometric and Physical Applications of the Double Integral | 323 |
| | 7.3.4. Definition and Properties of the Triple Integral | 324 |
| | 7.3.5. Computation of the Triple Integral. Some Applications. Iterated Integrals and Asymptotic Formulas | 325 |
| 7.4. | Line and Surface Integrals | 329 |
| | 7.4.1. Line Integral of the First Kind | 329 |
| | 7.4.2. Line Integral of the Second Kind | 330 |
| | 7.4.3. Surface Integral of the First Kind | 332 |
| | 7.4.4. Surface Integral of the Second Kind | 333 |
| | 7.4.5. Integral Formulas of Vector Calculus | 334 |
| References for Chapter 7 | | 335 |

# 8. Series — 337

| | | |
|---|---|---|
| 8.1. | Numerical Series and Infinite Products | 337 |
| | 8.1.1. Convergent Numerical Series and Their Properties. Cauchy's Criterion | 337 |
| | 8.1.2. Convergence Criteria for Series with Positive (Nonnegative) Terms | 338 |
| | 8.1.3. Convergence Criteria for Arbitrary Numerical Series. Absolute and Conditional Convergence | 341 |
| | 8.1.4. Multiplication of Series. Some Inequalities | 343 |
| | 8.1.5. Summation Methods. Convergence Acceleration | 344 |
| | 8.1.6. Infinite Products | 346 |
| 8.2. | Functional Series | 348 |
| | 8.2.1. Pointwise and Uniform Convergence of Functional Series | 348 |
| | 8.2.2. Basic Criteria of Uniform Convergence. Properties of Uniformly Convergent Series | 349 |
| 8.3. | Power Series | 350 |
| | 8.3.1. Radius of Convergence of Power Series. Properties of Power Series | 350 |
| | 8.3.2. Taylor and Maclaurin Power Series | 352 |
| | 8.3.3. Operations with Power Series. Summation Formulas for Power Series | 354 |
| 8.4. | Fourier Series | 357 |
| | 8.4.1. Representation of $2\pi$-Periodic Functions by Fourier Series. Main Results | 357 |
| | 8.4.2. Fourier Expansions of Periodic, Nonperiodic, Odd, and Even Functions | 359 |
| | 8.4.3. Criteria of Uniform and Mean-Square Convergence of Fourier Series | 361 |
| | 8.4.4. Summation Formulas for Trigonometric Series | 362 |
| 8.5. | Asymptotic Series | 363 |
| | 8.5.1. Asymptotic Series of Poincaré Type. Formulas for the Coefficients | 363 |
| | 8.5.2. Operations with Asymptotic Series | 364 |
| References for Chapter 8 | | 366 |

# 9. Differential Geometry — 367

| | | |
|---|---|---|
| 9.1. | Theory of Curves | 367 |
| | 9.1.1. Plane Curves | 367 |
| | 9.1.2. Space Curves | 379 |

| | |
|---|---|
| 9.2. Theory of Surfaces | 386 |
|     9.2.1. Elementary Notions in Theory of Surfaces | 386 |
|     9.2.2. Curvature of Curves on Surface | 392 |
|     9.2.3. Intrinsic Geometry of Surface | 395 |
| References for Chapter 9 | 397 |

## 10. Functions of Complex Variable ... 399

| | |
|---|---|
| 10.1. Basic Notions | 399 |
|     10.1.1. Complex Numbers. Functions of Complex Variable | 399 |
|     10.1.2. Functions of Complex Variable | 401 |
| 10.2. Main Applications | 419 |
|     10.2.1. Conformal Mappings | 419 |
|     10.2.2. Boundary Value Problems | 427 |
| References for Chapter 10 | 433 |

## 11. Integral Transforms ... 435

| | |
|---|---|
| 11.1. General Form of Integral Transforms. Some Formulas | 435 |
|     11.1.1. Integral Transforms and Inversion Formulas | 435 |
|     11.1.2. Residues. Jordan Lemma | 435 |
| 11.2. Laplace Transform | 436 |
|     11.2.1. Laplace Transform and the Inverse Laplace Transform | 436 |
|     11.2.2. Main Properties of the Laplace Transform. Inversion Formulas for Some Functions | 437 |
|     11.2.3. Limit Theorems. Representation of Inverse Transforms as Convergent Series and Asymptotic Expansions | 440 |
| 11.3. Mellin Transform | 441 |
|     11.3.1. Mellin Transform and the Inversion Formula | 441 |
|     11.3.2. Main Properties of the Mellin Transform. Relation Among the Mellin, Laplace, and Fourier Transforms | 442 |
| 11.4. Various Forms of the Fourier Transform | 443 |
|     11.4.1. Fourier Transform and the Inverse Fourier Transform | 443 |
|     11.4.2. Fourier Cosine and Sine Transforms | 445 |
| 11.5. Other Integral Transforms | 446 |
|     11.5.1. Integral Transforms Whose Kernels Contain Bessel Functions and Modified Bessel Functions | 446 |
|     11.5.2. Summary Table of Integral Transforms. Areas of Application of Integral Transforms | 448 |
| References for Chapter 11 | 451 |

## 12. Ordinary Differential Equations ... 453

| | |
|---|---|
| 12.1. First-Order Differential Equations | 453 |
|     12.1.1. General Concepts. The Cauchy Problem. Uniqueness and Existence Theorems | 453 |
|     12.1.2. Equations Solved for the Derivative. Simplest Techniques of Integration | 456 |
|     12.1.3. Exact Differential Equations. Integrating Factor | 458 |
|     12.1.4. Riccati Equation | 460 |
|     12.1.5. Abel Equations of the First Kind | 462 |
|     12.1.6. Abel Equations of the Second Kind | 464 |
|     12.1.7. Equations Not Solved for the Derivative | 465 |
|     12.1.8. Contact Transformations | 468 |
|     12.1.9. Approximate Analytic Methods for Solution of Equations | 469 |
|     12.1.10. Numerical Integration of Differential Equations | 471 |

12.2. Second-Order Linear Differential Equations ........................... 472
    12.2.1. Formulas for the General Solution. Some Transformations ....... 472
    12.2.2. Representation of Solutions as a Series in the Independent Variable ........ 475
    12.2.3. Asymptotic Solutions ........................................ 477
    12.2.4. Boundary Value Problems .................................... 480
    12.2.5. Eigenvalue Problems ........................................ 482
    12.2.6. Theorems on Estimates and Zeros of Solutions .................. 487

12.3. Second-Order Nonlinear Differential Equations ......................... 488
    12.3.1. Form of the General Solution. Cauchy Problem ................. 488
    12.3.2. Equations Admitting Reduction of Order ....................... 489
    12.3.3. Methods of Regular Series Expansions with Respect to the Independent Variable ........................................... 492
    12.3.4. Movable Singularities of Solutions of Ordinary Differential Equations. Painlevé Transcendents ........................................ 494
    12.3.5. Perturbation Methods of Mechanics and Physics ................. 499
    12.3.6. Galerkin Method and Its Modifications (Projection Methods) ...... 508
    12.3.7. Iteration and Numerical Methods .............................. 511

12.4. Linear Equations of Arbitrary Order .................................. 514
    12.4.1. Linear Equations with Constant Coefficients ................... 514
    12.4.2. Linear Equations with Variable Coefficients ................... 518
    12.4.3. Asymptotic Solutions of Linear Equations ...................... 522
    12.4.4. Collocation Method and Its Convergence ....................... 523

12.5. Nonlinear Equations of Arbitrary Order ............................... 524
    12.5.1. Structure of the General Solution. Cauchy Problem .............. 524
    12.5.2. Equations Admitting Reduction of Order ....................... 525

12.6. Linear Systems of Ordinary Differential Equations ..................... 528
    12.6.1. Systems of Linear Constant-Coefficient Equations ............... 528
    12.6.2. Systems of Linear Variable-Coefficient Equations ............... 539

12.7. Nonlinear Systems of Ordinary Differential Equations ................... 542
    12.7.1. Solutions and First Integrals. Uniqueness and Existence Theorems .. 542
    12.7.2. Integrable Combinations. Autonomous Systems of Equations ...... 545
    12.7.3. Elements of Stability Theory .................................. 546

References for Chapter 12 ..................................................... 550

## 13. First-Order Partial Differential Equations ........................... 553

13.1. Linear and Quasilinear Equations .................................... 553
    13.1.1. Characteristic System. General Solution ........................ 553
    13.1.2. Cauchy Problem. Existence and Uniqueness Theorem ............. 556
    13.1.3. Qualitative Features and Discontinuous Solutions of Quasilinear Equations .. 558
    13.1.4. Quasilinear Equations of General Form. Generalized Solution, Jump Condition, and Stability Condition ............................. 567

13.2. Nonlinear Equations ................................................ 570
    13.2.1. Solution Methods ............................................. 570
    13.2.2. Cauchy Problem. Existence and Uniqueness Theorem ............. 576
    13.2.3. Generalized Viscosity Solutions and Their Applications .......... 579

References for Chapter 13 ..................................................... 584

## 14. Linear Partial Differential Equations . . . . . . . . . . . . . . . . . . . . . . . . . . . . . . . . . . . . 585

14.1. Classification of Second-Order Partial Differential Equations . . . . . . . . . . . . . . . . . . . . 585
    14.1.1. Equations with Two Independent Variables . . . . . . . . . . . . . . . . . . . . . . . . . . 585
    14.1.2. Equations with Many Independent Variables . . . . . . . . . . . . . . . . . . . . . . . . 589

14.2. Basic Problems of Mathematical Physics . . . . . . . . . . . . . . . . . . . . . . . . . . . . . . . . . . 590
    14.2.1. Initial and Boundary Conditions. Cauchy Problem. Boundary Value Problems 590
    14.2.2. First, Second, Third, and Mixed Boundary Value Problems . . . . . . . . . . . . . . 593

14.3. Properties and Exact Solutions of Linear Equations . . . . . . . . . . . . . . . . . . . . . . . . . . . 594
    14.3.1. Homogeneous Linear Equations and Their Particular Solutions . . . . . . . . . . . . 594
    14.3.2. Nonhomogeneous Linear Equations and Their Particular Solutions . . . . . . . . . 598
    14.3.3. General Solutions of Some Hyperbolic Equations . . . . . . . . . . . . . . . . . . . . . . 600

14.4. Method of Separation of Variables (Fourier Method) . . . . . . . . . . . . . . . . . . . . . . . . . . 602
    14.4.1. Description of the Method of Separation of Variables. General Stage of Solution . . . . . . . . . . . . . . . . . . . . . . . . . . . . . . . . . . . . . . . . . . . . . . . . . . . . . 602
    14.4.2. Problems for Parabolic Equations: Final Stage of Solution . . . . . . . . . . . . . . . 605
    14.4.3. Problems for Hyperbolic Equations: Final Stage of Solution . . . . . . . . . . . . . . 607
    14.4.4. Solution of Boundary Value Problems for Elliptic Equations . . . . . . . . . . . . . 609

14.5. Integral Transforms Method . . . . . . . . . . . . . . . . . . . . . . . . . . . . . . . . . . . . . . . . . . . . 611
    14.5.1. Laplace Transform and Its Application in Mathematical Physics . . . . . . . . . . . 611
    14.5.2. Fourier Transform and Its Application in Mathematical Physics . . . . . . . . . . . 614

14.6. Representation of the Solution of the Cauchy Problem via the Fundamental Solution . . 615
    14.6.1. Cauchy Problem for Parabolic Equations . . . . . . . . . . . . . . . . . . . . . . . . . . . . 615
    14.6.2. Cauchy Problem for Hyperbolic Equations . . . . . . . . . . . . . . . . . . . . . . . . . . . 617

14.7. Boundary Value Problems for Parabolic Equations with One Space Variable. Green's Function . . . . . . . . . . . . . . . . . . . . . . . . . . . . . . . . . . . . . . . . . . . . . . . . . . . . . . . . . . 618
    14.7.1. Representation of Solutions via the Green's Function . . . . . . . . . . . . . . . . . . . 618
    14.7.2. Problems for Equation $s(x)\frac{\partial w}{\partial t} = \frac{\partial}{\partial x}\left[p(x)\frac{\partial w}{\partial x}\right] - q(x)w + \Phi(x,t)$ . . . . . . . . . . . 620

14.8. Boundary Value Problems for Hyperbolic Equations with One Space Variable. Green's Function. Goursat Problem . . . . . . . . . . . . . . . . . . . . . . . . . . . . . . . . . . . . . . . . . . . . . 623
    14.8.1. Representation of Solutions via the Green's Function . . . . . . . . . . . . . . . . . . . 623
    14.8.2. Problems for Equation $s(x)\frac{\partial^2 w}{\partial t^2} = \frac{\partial}{\partial x}\left[p(x)\frac{\partial w}{\partial x}\right] - q(x)w + \Phi(x,t)$ . . . . . . . . . 624
    14.8.3. Problems for Equation $\frac{\partial^2 w}{\partial t^2} + a(t)\frac{\partial w}{\partial t} = b(t)\left\{\frac{\partial}{\partial x}\left[p(x)\frac{\partial w}{\partial x}\right] - q(x)w\right\} + \Phi(x,t)$ 626
    14.8.4. Generalized Cauchy Problem with Initial Conditions Set Along a Curve . . . . . 627
    14.8.5. Goursat Problem (a Problem with Initial Data of Characteristics) . . . . . . . . . . 629

14.9. Boundary Value Problems for Elliptic Equations with Two Space Variables . . . . . . . . . 631
    14.9.1. Problems and the Green's Functions for Equation
$a(x)\frac{\partial^2 w}{\partial x^2} + \frac{\partial^2 w}{\partial y^2} + b(x)\frac{\partial w}{\partial x} + c(x)w = -\Phi(x,y)$ . . . . . . . . . . . . . . . . . . . . . . . 631
    14.9.2. Representation of Solutions to Boundary Value Problems via the Green's Functions . . . . . . . . . . . . . . . . . . . . . . . . . . . . . . . . . . . . . . . . . . . . . . . . . . . . 633

14.10. Boundary Value Problems with Many Space Variables. Representation of Solutions via the Green's Function . . . . . . . . . . . . . . . . . . . . . . . . . . . . . . . . . . . . . . . . . . . . . . 634
    14.10.1. Problems for Parabolic Equations . . . . . . . . . . . . . . . . . . . . . . . . . . . . . . . . 634
    14.10.2. Problems for Hyperbolic Equations . . . . . . . . . . . . . . . . . . . . . . . . . . . . . . . 636
    14.10.3. Problems for Elliptic Equations . . . . . . . . . . . . . . . . . . . . . . . . . . . . . . . . . . 637
    14.10.4. Comparison of the Solution Structures for Boundary Value Problems for Equations of Various Types . . . . . . . . . . . . . . . . . . . . . . . . . . . . . . . . . . . . . 638

14.11. Construction of the Green's Functions. General Formulas and Relations .......... 639
    14.11.1. Green's Functions of Boundary Value Problems for Equations of Various Types in Bounded Domains .......... 639
    14.11.2. Green's Functions Admitting Incomplete Separation of Variables .......... 640
    14.11.3. Construction of Green's Functions via Fundamental Solutions .......... 642
14.12. Duhamel's Principles in Nonstationary Problems .......... 646
    14.12.1. Problems for Homogeneous Linear Equations .......... 646
    14.12.2. Problems for Nonhomogeneous Linear Equations .......... 648
14.13. Transformations Simplifying Initial and Boundary Conditions .......... 649
    14.13.1. Transformations That Lead to Homogeneous Boundary Conditions .......... 649
    14.13.2. Transformations That Lead to Homogeneous Initial and Boundary Conditions .......... 650

References for Chapter 14 .......... 650

## 15. Nonlinear Partial Differential Equations .......... **653**

15.1. Classification of Second-Order Nonlinear Equations .......... 653
    15.1.1. Classification of Semilinear Equations in Two Independent Variables .......... 653
    15.1.2. Classification of Nonlinear Equations in Two Independent Variables .......... 653
15.2. Transformations of Equations of Mathematical Physics .......... 655
    15.2.1. Point Transformations: Overview and Examples .......... 655
    15.2.2. Hodograph Transformations (Special Point Transformations) .......... 657
    15.2.3. Contact Transformations. Legendre and Euler Transformations .......... 660
    15.2.4. Bäcklund Transformations. Differential Substitutions .......... 663
    15.2.5. Differential Substitutions .......... 666
15.3. Traveling-Wave Solutions, Self-Similar Solutions, and Some Other Simple Solutions. Similarity Method .......... 667
    15.3.1. Preliminary Remarks .......... 667
    15.3.2. Traveling-Wave Solutions. Invariance of Equations Under Translations .......... 667
    15.3.3. Self-Similar Solutions. Invariance of Equations Under Scaling Transformations .......... 669
    15.3.4. Equations Invariant Under Combinations of Translation and Scaling Transformations, and Their Solutions .......... 674
    15.3.5. Generalized Self-Similar Solutions .......... 677
15.4. Exact Solutions with Simple Separation of Variables .......... 678
    15.4.1. Multiplicative and Additive Separable Solutions .......... 678
    15.4.2. Simple Separation of Variables in Nonlinear Partial Differential Equations .......... 678
    15.4.3. Complex Separation of Variables in Nonlinear Partial Differential Equations .......... 679
15.5. Method of Generalized Separation of Variables .......... 681
    15.5.1. Structure of Generalized Separable Solutions .......... 681
    15.5.2. Simplified Scheme for Constructing Solutions Based on Presetting One System of Coordinate Functions .......... 683
    15.5.3. Solution of Functional Differential Equations by Differentiation .......... 684
    15.5.4. Solution of Functional-Differential Equations by Splitting .......... 688
    15.5.5. Titov–Galaktionov Method .......... 693
15.6. Method of Functional Separation of Variables .......... 697
    15.6.1. Structure of Functional Separable Solutions. Solution by Reduction to Equations with Quadratic Nonlinearities .......... 697
    15.6.2. Special Functional Separable Solutions. Generalized Traveling-Wave Solutions .......... 697

- 15.6.3. Differentiation Method .................................................. 700
- 15.6.4. Splitting Method. Solutions of Some Nonlinear Functional Equations and Their Applications .................................................. 704
- 15.7. Direct Method of Symmetry Reductions of Nonlinear Equations .................. 708
  - 15.7.1. Clarkson–Kruskal Direct Method ...................................... 708
  - 15.7.2. Some Modifications and Generalizations ............................... 712
- 15.8. Classical Method of Studying Symmetries of Differential Equations ............ 716
  - 15.8.1. One-Parameter Transformations and Their Local Properties ............. 716
  - 15.8.2. Symmetries of Nonlinear Second-Order Equations. Invariance Condition .... 719
  - 15.8.3. Using Symmetries of Equations for Finding Exact Solutions. Invariant Solutions .................................................. 724
  - 15.8.4. Some Generalizations. Higher-Order Equations ......................... 730
- 15.9. Nonclassical Method of Symmetry Reductions .................................. 732
  - 15.9.1. Description of the Method. Invariant Surface Condition ............... 732
  - 15.9.2. Examples: The Newell–Whitehead Equation and a Nonlinear Wave Equation 733
- 15.10. Differential Constraints Method .......................................... 737
  - 15.10.1. Description of the Method .......................................... 737
  - 15.10.2. First-Order Differential Constraints ................................ 739
  - 15.10.3. Second- and Higher-Order Differential Constraints ................... 744
  - 15.10.4. Connection Between the Differential Constraints Method and Other Methods .................................................. 746
- 15.11. Painlevé Test for Nonlinear Equations of Mathematical Physics .............. 748
  - 15.11.1. Solutions of Partial Differential Equations with a Movable Pole. Method Description .................................................. 748
  - 15.11.2. Examples of Performing the Painlevé Test and Truncated Expansions for Studying Nonlinear Equations .................................... 750
  - 15.11.3. Construction of Solutions of Nonlinear Equations That Fail the Painlevé Test, Using Truncated Expansions ...................................... 753
- 15.12. Methods of the Inverse Scattering Problem (Soliton Theory) .................. 755
  - 15.12.1. Method Based on Using Lax Pairs .................................... 755
  - 15.12.2. Method Based on a Compatibility Condition for Systems of Linear Equations .................................................. 757
  - 15.12.3. Solution of the Cauchy Problem by the Inverse Scattering Problem Method 760
- 15.13. Conservation Laws and Integrals of Motion .................................. 766
  - 15.13.1. Basic Definitions and Examples ...................................... 766
  - 15.13.2. Equations Admitting Variational Formulation. Noetherian Symmetries ... 767
- 15.14. Nonlinear Systems of Partial Differential Equations ........................ 770
  - 15.14.1. Overdetermined Systems of Two Equations ............................. 770
  - 15.14.2. Pfaffian Equations and Their Solutions. Connection with Overdetermined Systems .................................................. 772
  - 15.14.3. Systems of First-Order Equations Describing Convective Mass Transfer with Volume Reaction .................................................. 775
  - 15.14.4. First-Order Hyperbolic Systems of Quasilinear Equations. Systems of Conservation Laws of Gas Dynamic Type ................................ 780
  - 15.14.5. Systems of Second-Order Equations of Reaction-Diffusion Type ........ 796
- References for Chapter 15 .................................................. 798

## 16. Integral Equations . . . . . . . . . . . . . . . . . . . . . . . . . . . . . . . . . . . . . . . . . . . . . . . . . . . . . . . . . . . . . . . . 801

16.1. Linear Integral Equations of the First Kind with Variable Integration Limit . . . . . . . . . 801
    16.1.1. Volterra Equations of the First Kind . . . . . . . . . . . . . . . . . . . . . . . . . . . . . . . . . . . . 801
    16.1.2. Equations with Degenerate Kernel: $K(x,t) = g_1(x)h_1(t) + \cdots + g_n(x)h_n(t)$ . . 802
    16.1.3. Equations with Difference Kernel: $K(x,t) = K(x-t)$ . . . . . . . . . . . . . . . . . . . . . 804
    16.1.4. Reduction of Volterra Equations of the First Kind to Volterra Equations of the Second Kind . . . . . . . . . . . . . . . . . . . . . . . . . . . . . . . . . . . . . . . . . . . . . . . . . . . . . . 807
    16.1.5. Method of Quadratures . . . . . . . . . . . . . . . . . . . . . . . . . . . . . . . . . . . . . . . . . . . . . . 808

16.2. Linear Integral Equations of the Second Kind with Variable Integration Limit . . . . . . . 810
    16.2.1. Volterra Equations of the Second Kind . . . . . . . . . . . . . . . . . . . . . . . . . . . . . . . . . 810
    16.2.2. Equations with Degenerate Kernel: $K(x,t) = g_1(x)h_1(t) + \cdots + g_n(x)h_n(t)$ . . 811
    16.2.3. Equations with Difference Kernel: $K(x,t) = K(x-t)$ . . . . . . . . . . . . . . . . . . . . . 813
    16.2.4. Construction of Solutions of Integral Equations with Special Right-Hand Side 815
    16.2.5. Method of Model Solutions . . . . . . . . . . . . . . . . . . . . . . . . . . . . . . . . . . . . . . . . . . 818
    16.2.6. Successive Approximation Method . . . . . . . . . . . . . . . . . . . . . . . . . . . . . . . . . . . 822
    16.2.7. Method of Quadratures . . . . . . . . . . . . . . . . . . . . . . . . . . . . . . . . . . . . . . . . . . . . . . 823

16.3. Linear Integral Equations of the First Kind with Constant Limits of Integration . . . . . . 824
    16.3.1. Fredholm Integral Equations of the First Kind . . . . . . . . . . . . . . . . . . . . . . . . . . . 824
    16.3.2. Method of Integral Transforms . . . . . . . . . . . . . . . . . . . . . . . . . . . . . . . . . . . . . . . 825
    16.3.3. Regularization Methods . . . . . . . . . . . . . . . . . . . . . . . . . . . . . . . . . . . . . . . . . . . . . 827

16.4. Linear Integral Equations of the Second Kind with Constant Limits of Integration . . . . 829
    16.4.1. Fredholm Integral Equations of the Second Kind. Resolvent . . . . . . . . . . . . . . . 829
    16.4.2. Fredholm Equations of the Second Kind with Degenerate Kernel . . . . . . . . . . 830
    16.4.3. Solution as a Power Series in the Parameter. Method of Successive Approximations . . . . . . . . . . . . . . . . . . . . . . . . . . . . . . . . . . . . . . . . . . . . . . . . . . . . 832
    16.4.4. Fredholm Theorems and the Fredholm Alternative . . . . . . . . . . . . . . . . . . . . . . . 834
    16.4.5. Fredholm Integral Equations of the Second Kind with Symmetric Kernel . . . . 835
    16.4.6. Methods of Integral Transforms . . . . . . . . . . . . . . . . . . . . . . . . . . . . . . . . . . . . . . 841
    16.4.7. Method of Approximating a Kernel by a Degenerate One . . . . . . . . . . . . . . . . . 844
    16.4.8. Collocation Method . . . . . . . . . . . . . . . . . . . . . . . . . . . . . . . . . . . . . . . . . . . . . . . . 847
    16.4.9. Method of Least Squares . . . . . . . . . . . . . . . . . . . . . . . . . . . . . . . . . . . . . . . . . . . . 849
    16.4.10. Bubnov–Galerkin Method . . . . . . . . . . . . . . . . . . . . . . . . . . . . . . . . . . . . . . . . . . 850
    16.4.11. Quadrature Method . . . . . . . . . . . . . . . . . . . . . . . . . . . . . . . . . . . . . . . . . . . . . . . . 852
    16.4.12. Systems of Fredholm Integral Equations of the Second Kind . . . . . . . . . . . . . . 854

16.5. Nonlinear Integral Equations . . . . . . . . . . . . . . . . . . . . . . . . . . . . . . . . . . . . . . . . . . . . . . . 856
    16.5.1. Nonlinear Volterra and Urysohn Integral Equations . . . . . . . . . . . . . . . . . . . . . 856
    16.5.2. Nonlinear Volterra Integral Equations . . . . . . . . . . . . . . . . . . . . . . . . . . . . . . . . . 856
    16.5.3. Equations with Constant Integration Limits . . . . . . . . . . . . . . . . . . . . . . . . . . . . 863

References for Chapter 16 . . . . . . . . . . . . . . . . . . . . . . . . . . . . . . . . . . . . . . . . . . . . . . . . . . . . . . 871

## 17. Difference Equations and Other Functional Equations . . . . . . . . . . . . . . . . . . . . . . . . 873

17.1. Difference Equations of Integer Argument . . . . . . . . . . . . . . . . . . . . . . . . . . . . . . . . . . . 873
    17.1.1. First-Order Linear Difference Equations of Integer Argument . . . . . . . . . . . . . 873
    17.1.2. First-Order Nonlinear Difference Equations of Integer Argument . . . . . . . . . . 874
    17.1.3. Second-Order Linear Difference Equations with Constant Coefficients . . . . . . 877
    17.1.4. Second-Order Linear Difference Equations with Variable Coefficients . . . . . . 879
    17.1.5. Linear Difference Equations of Arbitrary Order with Constant Coefficients . . 881
    17.1.6. Linear Difference Equations of Arbitrary Order with Variable Coefficients . . . 882
    17.1.7. Nonlinear Difference Equations of Arbitrary Order . . . . . . . . . . . . . . . . . . . . . . 884

17.2. Linear Difference Equations with a Single Continuous Variable .................. 885
    17.2.1. First-Order Linear Difference Equations .......................... 885
    17.2.2. Second-Order Linear Difference Equations with Integer Differences ........ 894
    17.2.3. Linear $m$th-Order Difference Equations with Integer Differences ........ 898
    17.2.4. Linear $m$th-Order Difference Equations with Arbitrary Differences ........ 904

17.3. Linear Functional Equations ............................................ 907
    17.3.1. Iterations of Functions and Their Properties ..................... 907
    17.3.2. Linear Homogeneous Functional Equations ...................... 910
    17.3.3. Linear Nonhomogeneous Functional Equations ................... 912
    17.3.4. Linear Functional Equations Reducible to Linear Difference Equations with Constant Coefficients ......................................... 916

17.4. Nonlinear Difference and Functional Equations with a Single Variable ........... 918
    17.4.1. Nonlinear Difference Equations with a Single Variable .............. 918
    17.4.2. Reciprocal (Cyclic) Functional Equations ....................... 919
    17.4.3. Nonlinear Functional Equations Reducible to Difference Equations ........ 921
    17.4.4. Power Series Solution of Nonlinear Functional Equations ............. 922

17.5. Functional Equations with Several Variables .............................. 922
    17.5.1. Method of Differentiation in a Parameter ...................... 922
    17.5.2. Method of Differentiation in Independent Variables ................. 925
    17.5.3. Method of Substituting Particular Values of Independent Arguments ....... 926
    17.5.4. Method of Argument Elimination by Test Functions ................. 928
    17.5.5. Bilinear Functional Equations and Nonlinear Functional Equations Reducible to Bilinear Equations ........................................ 930

References for Chapter 17 ................................................. 935

# 18. Special Functions and Their Properties ............................... 937

18.1. Some Coefficients, Symbols, and Numbers ............................... 937
    18.1.1. Binomial Coefficients ................................... 937
    18.1.2. Pochhammer Symbol .................................... 938
    18.1.3. Bernoulli Numbers ..................................... 938
    18.1.4. Euler Numbers ........................................ 939

18.2. Error Functions. Exponential and Logarithmic Integrals ..................... 939
    18.2.1. Error Function and Complementary Error Function ................. 939
    18.2.2. Exponential Integral .................................... 940
    18.2.3. Logarithmic Integral .................................... 941

18.3. Sine Integral and Cosine Integral. Fresnel Integrals ......................... 941
    18.3.1. Sine Integral ......................................... 941
    18.3.2. Cosine Integral ....................................... 942
    18.3.3. Fresnel Integrals ...................................... 942

18.4. Gamma Function, Psi Function, and Beta Function ........................ 943
    18.4.1. Gamma Function ...................................... 943
    18.4.2. Psi Function (Digamma Function) ........................... 944
    18.4.3. Beta Function ........................................ 945

18.5. Incomplete Gamma and Beta Functions .................................. 946
    18.5.1. Incomplete Gamma Function ............................... 946
    18.5.2. Incomplete Beta Function ................................. 947

- 18.6. Bessel Functions (Cylindrical Functions) ... 947
  - 18.6.1. Definitions and Basic Formulas ... 947
  - 18.6.2. Integral Representations and Asymptotic Expansions ... 949
  - 18.6.3. Zeros and Orthogonality Properties of Bessel Functions ... 951
  - 18.6.4. Hankel Functions (Bessel Functions of the Third Kind) ... 952
- 18.7. Modified Bessel Functions ... 953
  - 18.7.1. Definitions. Basic Formulas ... 953
  - 18.7.2. Integral Representations and Asymptotic Expansions ... 954
- 18.8. Airy Functions ... 955
  - 18.8.1. Definition and Basic Formulas ... 955
  - 18.8.2. Power Series and Asymptotic Expansions ... 956
- 18.9. Degenerate Hypergeometric Functions (Kummer Functions) ... 956
  - 18.9.1. Definitions and Basic Formulas ... 956
  - 18.9.2. Integral Representations and Asymptotic Expansions ... 959
  - 18.9.3. Whittaker Functions ... 960
- 18.10. Hypergeometric Functions ... 960
  - 18.10.1. Various Representations of the Hypergeometric Function ... 960
  - 18.10.2. Basic Properties ... 960
- 18.11. Legendre Polynomials, Legendre Functions, and Associated Legendre Functions ... 962
  - 18.11.1. Legendre Polynomials and Legendre Functions ... 962
  - 18.11.2. Associated Legendre Functions with Integer Indices and Real Argument ... 964
  - 18.11.3. Associated Legendre Functions. General Case ... 965
- 18.12. Parabolic Cylinder Functions ... 967
  - 18.12.1. Definitions. Basic Formulas ... 967
  - 18.12.2. Integral Representations, Asymptotic Expansions, and Linear Relations ... 968
- 18.13. Elliptic Integrals ... 969
  - 18.13.1. Complete Elliptic Integrals ... 969
  - 18.13.2. Incomplete Elliptic Integrals (Elliptic Integrals) ... 970
- 18.14. Elliptic Functions ... 972
  - 18.14.1. Jacobi Elliptic Functions ... 972
  - 18.14.2. Weierstrass Elliptic Function ... 976
- 18.15. Jacobi Theta Functions ... 978
  - 18.15.1. Series Representation of the Jacobi Theta Functions. Simplest Properties ... 978
  - 18.15.2. Various Relations and Formulas. Connection with Jacobi Elliptic Functions ... 978
- 18.16. Mathieu Functions and Modified Mathieu Functions ... 980
  - 18.16.1. Mathieu Functions ... 980
  - 18.16.2. Modified Mathieu Functions ... 982
- 18.17. Orthogonal Polynomials ... 982
  - 18.17.1. Laguerre Polynomials and Generalized Laguerre Polynomials ... 982
  - 18.17.2. Chebyshev Polynomials and Functions ... 983
  - 18.17.3. Hermite Polynomials ... 985
  - 18.17.4. Jacobi Polynomials and Gegenbauer Polynomials ... 986
- 18.18. Nonorthogonal Polynomials ... 988
  - 18.18.1. Bernoulli Polynomials ... 988
  - 18.18.2. Euler Polynomials ... 989

References for Chapter 18 ... 990

## 19. Calculus of Variations and Optimization ... 991

19.1. Calculus of Variations and Optimal Control ... 991
    19.1.1. Some Definitions and Formulas ... 991
    19.1.2. Simplest Problem of Calculus of Variations ... 993
    19.1.3. Isoperimetric Problem ... 1002
    19.1.4. Problems with Higher Derivatives ... 1006
    19.1.5. Lagrange Problem ... 1008
    19.1.6. Pontryagin Maximum Principle ... 1010

19.2. Mathematical Programming ... 1012
    19.2.1. Linear Programming ... 1012
    19.2.2. Nonlinear Programming ... 1027

References for Chapter 19 ... 1028

## 20. Probability Theory ... 1031

20.1. Simplest Probabilistic Models ... 1031
    20.1.1. Probabilities of Random Events ... 1031
    20.1.2. Conditional Probability and Simplest Formulas ... 1035
    20.1.3. Sequences of Trials ... 1037

20.2. Random Variables and Their Characteristics ... 1039
    20.2.1. One-Dimensional Random Variables ... 1039
    20.2.2. Characteristics of One-Dimensional Random Variables ... 1042
    20.2.3. Main Discrete Distributions ... 1047
    20.2.4. Continuous Distributions ... 1051
    20.2.5. Multivariate Random Variables ... 1057

20.3. Limit Theorems ... 1068
    20.3.1. Convergence of Random Variables ... 1068
    20.3.2. Limit Theorems ... 1069

20.4. Stochastic Processes ... 1071
    20.4.1. Theory of Stochastic Processes ... 1071
    20.4.2. Models of Stochastic Processes ... 1074

References for Chapter 20 ... 1079

## 21. Mathematical Statistics ... 1081

21.1. Introduction to Mathematical Statistics ... 1081
    21.1.1. Basic Notions and Problems of Mathematical Statistics ... 1081
    21.1.2. Simplest Statistical Transformations ... 1082
    21.1.3. Numerical Characteristics of Statistical Distribution ... 1087

21.2. Statistical Estimation ... 1088
    21.2.1. Estimators and Their Properties ... 1088
    21.2.2. Estimation Methods for Unknown Parameters ... 1091
    21.2.3. Interval Estimators (Confidence Intervals) ... 1093

21.3. Statistical Hypothesis Testing ... 1094
    21.3.1. Statistical Hypothesis. Test ... 1094
    21.3.2. Goodness-of-Fit Tests ... 1098
    21.3.3. Problems Related to Normal Samples ... 1101

References for Chapter 21 ... 1109

## Part II. Mathematical Tables — 1111

### T1. Finite Sums and Infinite Series — 1113

#### T1.1. Finite Sums — 1113
- T1.1.1. Numerical Sum — 1113
- T1.1.2. Functional Sums — 1116

#### T1.2. Infinite Series — 1118
- T1.2.1. Numerical Series — 1118
- T1.2.2. Functional Series — 1120

References for Chapter T1 — 1127

### T2. Integrals — 1129

#### T2.1. Indefinite Integrals — 1129
- T2.1.1. Integrals Involving Rational Functions — 1129
- T2.1.2. Integrals Involving Irrational Functions — 1134
- T2.1.3. Integrals Involving Exponential Functions — 1137
- T2.1.4. Integrals Involving Hyperbolic Functions — 1137
- T2.1.5. Integrals Involving Logarithmic Functions — 1140
- T2.1.6. Integrals Involving Trigonometric Functions — 1142
- T2.1.7. Integrals Involving Inverse Trigonometric Functions — 1147

#### T2.2. Tables of Definite Integrals — 1147
- T2.2.1. Integrals Involving Power-Law Functions — 1147
- T2.2.2. Integrals Involving Exponential Functions — 1150
- T2.2.3. Integrals Involving Hyperbolic Functions — 1152
- T2.2.4. Integrals Involving Logarithmic Functions — 1152
- T2.2.5. Integrals Involving Trigonometric Functions — 1153

References for Chapter T2 — 1155

### T3. Integral Transforms — 1157

#### T3.1. Tables of Laplace Transforms — 1157
- T3.1.1. General Formulas — 1157
- T3.1.2. Expressions with Power-Law Functions — 1159
- T3.1.3. Expressions with Exponential Functions — 1159
- T3.1.4. Expressions with Hyperbolic Functions — 1160
- T3.1.5. Expressions with Logarithmic Functions — 1161
- T3.1.6. Expressions with Trigonometric Functions — 1161
- T3.1.7. Expressions with Special Functions — 1163

#### T3.2. Tables of Inverse Laplace Transforms — 1164
- T3.2.1. General Formulas — 1164
- T3.2.2. Expressions with Rational Functions — 1166
- T3.2.3. Expressions with Square Roots — 1170
- T3.2.4. Expressions with Arbitrary Powers — 1172
- T3.2.5. Expressions with Exponential Functions — 1172
- T3.2.6. Expressions with Hyperbolic Functions — 1174
- T3.2.7. Expressions with Logarithmic Functions — 1174
- T3.2.8. Expressions with Trigonometric Functions — 1175
- T3.2.9. Expressions with Special Functions — 1176

T3.3. Tables of Fourier Cosine Transforms .................................... 1177
    T3.3.1. General Formulas ................................................ 1177
    T3.3.2. Expressions with Power-Law Functions ........................... 1177
    T3.3.3. Expressions with Exponential Functions .......................... 1178
    T3.3.4. Expressions with Hyperbolic Functions ........................... 1179
    T3.3.5. Expressions with Logarithmic Functions .......................... 1179
    T3.3.6. Expressions with Trigonometric Functions ........................ 1180
    T3.3.7. Expressions with Special Functions .............................. 1181

T3.4. Tables of Fourier Sine Transforms ...................................... 1182
    T3.4.1. General Formulas ................................................ 1182
    T3.4.2. Expressions with Power-Law Functions ........................... 1182
    T3.4.3. Expressions with Exponential Functions .......................... 1183
    T3.4.4. Expressions with Hyperbolic Functions ........................... 1184
    T3.4.5. Expressions with Logarithmic Functions .......................... 1184
    T3.4.6. Expressions with Trigonometric Functions ........................ 1185
    T3.4.7. Expressions with Special Functions .............................. 1186

T3.5. Tables of Mellin Transforms ............................................ 1187
    T3.5.1. General Formulas ................................................ 1187
    T3.5.2. Expressions with Power-Law Functions ........................... 1188
    T3.5.3. Expressions with Exponential Functions .......................... 1188
    T3.5.4. Expressions with Logarithmic Functions .......................... 1189
    T3.5.5. Expressions with Trigonometric Functions ........................ 1189
    T3.5.6. Expressions with Special Functions .............................. 1190

T3.6. Tables of Inverse Mellin Transforms .................................... 1190
    T3.6.1. Expressions with Power-Law Functions ........................... 1190
    T3.6.2. Expressions with Exponential and Logarithmic Functions ......... 1191
    T3.6.3. Expressions with Trigonometric Functions ........................ 1192
    T3.6.4. Expressions with Special Functions .............................. 1193

References for Chapter T3 ...................................................... 1194

**T4. Orthogonal Curvilinear Systems of Coordinate** .......................... **1195**

T4.1. Arbitrary Curvilinear Coordinate Systems ............................... 1195
    T4.1.1. General Nonorthogonal Curvilinear Coordinates ................... 1195
    T4.1.2. General Orthogonal Curvilinear Coordinates ...................... 1196

T4.2. Special Curvilinear Coordinate Systems ................................. 1198
    T4.2.1. Cylindrical Coordinates .......................................... 1198
    T4.2.2. Spherical Coordinates ............................................ 1199
    T4.2.3. Coordinates of a Prolate Ellipsoid of Revolution ................. 1200
    T4.2.4. Coordinates of an Oblate Ellipsoid of Revolution ................. 1201
    T4.2.5. Coordinates of an Elliptic Cylinder .............................. 1202
    T4.2.6. Conical Coordinates .............................................. 1202
    T4.2.7. Parabolic Cylinder Coordinates ................................... 1203
    T4.2.8. Parabolic Coordinates ............................................ 1203
    T4.2.9. Bicylindrical Coordinates ........................................ 1204
    T4.2.10. Bipolar Coordinates (in Space) .................................. 1204
    T4.2.11. Toroidal Coordinates ............................................ 1205

References for Chapter T4 ...................................................... 1205

## T5. Ordinary Differential Equations ..... 1207
T5.1. First-Order Equations ..... 1207
T5.2. Second-Order Linear Equations ..... 1212
    T5.2.1. Equations Involving Power Functions ..... 1213
    T5.2.2. Equations Involving Exponential and Other Functions ..... 1220
    T5.2.3. Equations Involving Arbitrary Functions ..... 1222
T5.3. Second-Order Nonlinear Equations ..... 1223
    T5.3.1. Equations of the Form $y''_{xx} = f(x,y)$ ..... 1223
    T5.3.2. Equations of the Form $f(x,y)y''_{xx} = g(x,y,y'_x)$ ..... 1225
References for Chapter T5 ..... 1228

## T6. Systems of Ordinary Differential Equations ..... 1229
T6.1. Linear Systems of Two Equations ..... 1229
    T6.1.1. Systems of First-Order Equations ..... 1229
    T6.1.2. Systems of Second-Order Equations ..... 1232
T6.2. Linear Systems of Three and More Equations ..... 1237
T6.3. Nonlinear Systems of Two Equations ..... 1239
    T6.3.1. Systems of First-Order Equations ..... 1239
    T6.3.2. Systems of Second-Order Equations ..... 1240
T6.4. Nonlinear Systems of Three or More Equations ..... 1244
References for Chapter T6 ..... 1246

## T7. First-Order Partial Differential Equations ..... 1247
T7.1. Linear Equations ..... 1247
    T7.1.1. Equations of the Form $f(x,y)\frac{\partial w}{\partial x} + g(x,y)\frac{\partial w}{\partial y} = 0$ ..... 1247
    T7.1.2. Equations of the Form $f(x,y)\frac{\partial w}{\partial x} + g(x,y)\frac{\partial w}{\partial y} = h(x,y)$ ..... 1248
    T7.1.3. Equations of the Form $f(x,y)\frac{\partial w}{\partial x} + g(x,y)\frac{\partial w}{\partial y} = h(x,y)w + r(x,y)$ ..... 1250
T7.2. Quasilinear Equations ..... 1252
    T7.2.1. Equations of the Form $f(x,y)\frac{\partial w}{\partial x} + g(x,y)\frac{\partial w}{\partial y} = h(x,y,w)$ ..... 1252
    T7.2.2. Equations of the Form $\frac{\partial w}{\partial x} + f(x,y,w)\frac{\partial w}{\partial y} = 0$ ..... 1254
    T7.2.3. Equations of the Form $\frac{\partial w}{\partial x} + f(x,y,w)\frac{\partial w}{\partial y} = g(x,y,w)$ ..... 1256
T7.3. Nonlinear Equations ..... 1258
    T7.3.1. Equations Quadratic in One Derivative ..... 1258
    T7.3.2. Equations Quadratic in Two Derivatives ..... 1259
    T7.3.3. Equations with Arbitrary Nonlinearities in Derivatives ..... 1261
References for Chapter T7 ..... 1265

## T8. Linear Equations and Problems of Mathematical Physics ..... 1267
T8.1. Parabolic Equations ..... 1267
    T8.1.1. Heat Equation $\frac{\partial w}{\partial t} = a\frac{\partial^2 w}{\partial x^2}$ ..... 1267
    T8.1.2. Nonhomogeneous Heat Equation $\frac{\partial w}{\partial t} = a\frac{\partial^2 w}{\partial x^2} + \Phi(x,t)$ ..... 1268
    T8.1.3. Equation of the Form $\frac{\partial w}{\partial t} = a\frac{\partial^2 w}{\partial x^2} + b\frac{\partial w}{\partial x} + cw + \Phi(x,t)$ ..... 1270
    T8.1.4. Heat Equation with Axial Symmetry $\frac{\partial w}{\partial t} = a\left(\frac{\partial^2 w}{\partial r^2} + \frac{1}{r}\frac{\partial w}{\partial r}\right)$ ..... 1270
    T8.1.5. Equation of the Form $\frac{\partial w}{\partial t} = a\left(\frac{\partial^2 w}{\partial r^2} + \frac{1}{r}\frac{\partial w}{\partial r}\right) + \Phi(r,t)$ ..... 1271
    T8.1.6. Heat Equation with Central Symmetry $\frac{\partial w}{\partial t} = a\left(\frac{\partial^2 w}{\partial r^2} + \frac{2}{r}\frac{\partial w}{\partial r}\right)$ ..... 1272
    T8.1.7. Equation of the Form $\frac{\partial w}{\partial t} = a\left(\frac{\partial^2 w}{\partial r^2} + \frac{2}{r}\frac{\partial w}{\partial r}\right) + \Phi(r,t)$ ..... 1273
    T8.1.8. Equation of the Form $\frac{\partial w}{\partial t} = \frac{\partial^2 w}{\partial x^2} + \frac{1-2\beta}{x}\frac{\partial w}{\partial x}$ ..... 1274

T8.1.9. Equations of the Diffusion (Thermal) Boundary Layer .............. 1276
T8.1.10. Schrödinger Equation $i\hbar \frac{\partial w}{\partial t} = -\frac{\hbar^2}{2m} \frac{\partial^2 w}{\partial x^2} + U(x)w$ .............. 1276

T8.2. Hyperbolic Equations .............................................. 1278
T8.2.1. Wave Equation $\frac{\partial^2 w}{\partial t^2} = a^2 \frac{\partial^2 w}{\partial x^2}$ .............................. 1278
T8.2.2. Equation of the Form $\frac{\partial^2 w}{\partial t^2} = a^2 \frac{\partial^2 w}{\partial x^2} + \Phi(x,t)$ ............ 1279
T8.2.3. Klein–Gordon Equation $\frac{\partial^2 w}{\partial t^2} = a^2 \frac{\partial^2 w}{\partial x^2} - bw$ ............ 1280
T8.2.4. Equation of the Form $\frac{\partial^2 w}{\partial t^2} = a^2 \frac{\partial^2 w}{\partial x^2} - bw + \Phi(x,t)$ ........ 1281
T8.2.5. Equation of the Form $\frac{\partial^2 w}{\partial t^2} = a^2 \big(\frac{\partial^2 w}{\partial r^2} + \frac{1}{r}\frac{\partial w}{\partial r}\big) + \Phi(r,t)$ ... 1282
T8.2.6. Equation of the Form $\frac{\partial^2 w}{\partial t^2} = a^2 \big(\frac{\partial^2 w}{\partial r^2} + \frac{2}{r}\frac{\partial w}{\partial r}\big) + \Phi(r,t)$ ... 1283
T8.2.7. Equations of the Form $\frac{\partial^2 w}{\partial t^2} + k \frac{\partial w}{\partial t} = a^2 \frac{\partial^2 w}{\partial x^2} + b \frac{\partial w}{\partial x} + cw + \Phi(x,t)$ ...... 1284

T8.3. Elliptic Equations ................................................. 1284
T8.3.1. Laplace Equation $\Delta w = 0$ .................................. 1284
T8.3.2. Poisson Equation $\Delta w + \Phi(\mathbf{x}) = 0$ .............. 1287
T8.3.3. Helmholtz Equation $\Delta w + \lambda w = -\Phi(\mathbf{x})$ .. 1289

T8.4. Fourth-Order Linear Equations ..................................... 1294
T8.4.1. Equation of the Form $\frac{\partial^2 w}{\partial t^2} + a^2 \frac{\partial^4 w}{\partial x^4} = 0$ ................ 1294
T8.4.2. Equation of the Form $\frac{\partial^2 w}{\partial t^2} + a^2 \frac{\partial^4 w}{\partial x^4} = \Phi(x,t)$ ........... 1295
T8.4.3. Biharmonic Equation $\Delta\Delta w = 0$ ....................... 1297
T8.4.4. Nonhomogeneous Biharmonic Equation $\Delta\Delta w = \Phi(x,y)$ ... 1298

References for Chapter T8 ............................................... 1299

# T9. Nonlinear Mathematical Physics Equations .......................... 1301

T9.1. Parabolic Equations ............................................... 1301
T9.1.1. Nonlinear Heat Equations of the Form $\frac{\partial w}{\partial t} = \frac{\partial^2 w}{\partial x^2} + f(w)$ .......... 1301
T9.1.2. Equations of the Form $\frac{\partial w}{\partial t} = \frac{\partial}{\partial x}\big[f(w)\frac{\partial w}{\partial x}\big] + g(w)$ ........ 1303
T9.1.3. Burgers Equation and Nonlinear Heat Equation in Radial Symmetric Cases .. 1307
T9.1.4. Nonlinear Schrödinger Equations ............................... 1309

T9.2. Hyperbolic Equations .............................................. 1312
T9.2.1. Nonlinear Wave Equations of the Form $\frac{\partial^2 w}{\partial t^2} = a\frac{\partial^2 w}{\partial x^2} + f(w)$ ........ 1312
T9.2.2. Other Nonlinear Wave Equations ................................ 1316

T9.3. Elliptic Equations ................................................ 1318
T9.3.1. Nonlinear Heat Equations of the Form $\frac{\partial^2 w}{\partial x^2} + \frac{\partial^2 w}{\partial y^2} = f(w)$ .......... 1318
T9.3.2. Equations of the Form $\frac{\partial}{\partial x}\big[f(x)\frac{\partial w}{\partial x}\big] + \frac{\partial}{\partial y}\big[g(y)\frac{\partial w}{\partial y}\big] = f(w)$ ........ 1321
T9.3.3. Equations of the Form $\frac{\partial}{\partial x}\big[f(w)\frac{\partial w}{\partial x}\big] + \frac{\partial}{\partial y}\big[g(w)\frac{\partial w}{\partial y}\big] = h(w)$ ...... 1322

T9.4. Other Second-Order Equations ...................................... 1324
T9.4.1. Equations of Transonic Gas Flow ................................ 1324
T9.4.2. Monge–Ampère Equations ......................................... 1326

T9.5. Higher-Order Equations ............................................ 1327
T9.5.1. Third-Order Equations .......................................... 1327
T9.5.2. Fourth-Order Equations ......................................... 1332

References for Chapter T9 ............................................... 1335

# T10. Systems of Partial Differential Equations ........................ 1337

T10.1. Nonlinear Systems of Two First-Order Equations ................... 1337
T10.2. Linear Systems of Two Second-Order Equations ..................... 1341

T10.3. Nonlinear Systems of Two Second-Order Equations .......................... 1343
    T10.3.1. Systems of the Form $\frac{\partial u}{\partial t} = a\frac{\partial^2 u}{\partial x^2} + F(u,w)$, $\frac{\partial w}{\partial t} = b\frac{\partial^2 w}{\partial x^2} + G(u,w)$ ...... 1343
    T10.3.2. Systems of the Form $\frac{\partial u}{\partial t} = \frac{a}{x^n}\frac{\partial}{\partial x}\left(x^n \frac{\partial u}{\partial x}\right) + F(u,w)$,
          $\frac{\partial w}{\partial t} = \frac{b}{x^n}\frac{\partial}{\partial x}\left(x^n \frac{\partial w}{\partial x}\right) + G(u,w)$ ........................ 1357
    T10.3.3. Systems of the Form $\Delta u = F(u,w)$, $\Delta w = G(u,w)$ .................. 1364
    T10.3.4. Systems of the Form $\frac{\partial^2 u}{\partial t^2} = \frac{a}{x^n}\frac{\partial}{\partial x}\left(x^n \frac{\partial u}{\partial x}\right) + F(u,w)$,
          $\frac{\partial^2 w}{\partial t^2} = \frac{b}{x^n}\frac{\partial}{\partial x}\left(x^n \frac{\partial w}{\partial x}\right) + G(u,w)$ ........................ 1368
    T10.3.5. Other Systems ......................................................... 1373
T10.4. Systems of General Form ................................................................ 1374
    T10.4.1. Linear Systems ....................................................... 1374
    T10.4.2. Nonlinear Systems of Two Equations Involving the First Derivatives in $t$ .. 1374
    T10.4.3. Nonlinear Systems of Two Equations Involving the Second Derivatives in $t$ 1378
    T10.4.4. Nonlinear Systems of Many Equations Involving the First Derivatives in $t$ . 1381
References for Chapter T10 ..................................................................... 1382

## T11. Integral Equations ........................................................................... 1385

T11.1. Linear Equations of the First Kind with Variable Limit of Integration ............. 1385
T11.2. Linear Equations of the Second Kind with Variable Limit of Integration .......... 1391
T11.3. Linear Equations of the First Kind with Constant Limits of Integration ........... 1396
T11.4. Linear Equations of the Second Kind with Constant Limits of Integration ........ 1401
References for Chapter T11 ..................................................................... 1406

## T12. Functional Equations ....................................................................... 1409

T12.1. Linear Functional Equations in One Independent Variable ......................... 1409
    T12.1.1. Linear Difference and Functional Equations Involving Unknown Function
            with Two Different Arguments ............................................ 1409
    T12.1.2. Other Linear Functional Equations .................................... 1421
T12.2. Nonlinear Functional Equations in One Independent Variable ...................... 1428
    T12.2.1. Functional Equations with Quadratic Nonlinearity ..................... 1428
    T12.2.2. Functional Equations with Power Nonlinearity ........................ 1433
    T12.2.3. Nonlinear Functional Equation of General Form ...................... 1434
T12.3. Functional Equations in Several Independent Variables ............................ 1438
    T12.3.1. Linear Functional Equations ........................................... 1438
    T12.3.2. Nonlinear Functional Equations ....................................... 1443
References for Chapter T12 ..................................................................... 1450

**Supplement. Some Useful Electronic Mathematical Resources** ........................ 1451

**Index** ............................................................................................. 1453

# AUTHORS

**Andrei D. Polyanin, D.Sc., Ph.D.,** is a well-known scientist of broad interests who is active in various areas of mathematics, mechanics, and chemical engineering sciences. He is one of the most prominent authors in the field of reference literature on mathematics and physics.

Professor Polyanin graduated with honors from the Department of Mechanics and Mathematics of Moscow State University in 1974. He received his Ph.D. in 1981 and his D.Sc. in 1986 from the Institute for Problems in Mechanics of the Russian (former USSR) Academy of Sciences. Since 1975, Professor Polyanin has been working at the Institute for Problems in Mechanics of the Russian Academy of Sciences; he is also Professor of Mathematics at Bauman Moscow State Technical University. He is a member of the Russian National Committee on Theoretical and Applied Mechanics and of the Mathematics and Mechanics Expert Council of the Higher Certification Committee of the Russian Federation.

Professor Polyanin has made important contributions to exact and approximate analytical methods in the theory of differential equations, mathematical physics, integral equations, engineering mathematics, theory of heat and mass transfer, and chemical hydrodynamics. He has obtained exact solutions for several thousand ordinary differential, partial differential, and integral equations.

Professor Polyanin is an author of more than 30 books in English, Russian, German, and Bulgarian as well as more than 120 research papers and three patents. He has written a number of fundamental handbooks, including A. D. Polyanin and V. F. Zaitsev, *Handbook of Exact Solutions for Ordinary Differential Equations*, CRC Press, 1995 and 2003; A. D. Polyanin and A. V. Manzhirov, *Handbook of Integral Equations*, CRC Press, 1998; A. D. Polyanin, *Handbook of Linear Partial Differential Equations for Engineers and Scientists*, Chapman & Hall/CRC Press, 2002; A. D. Polyanin, V. F. Zaitsev, and A. Moussiaux, *Handbook of First Order Partial Differential Equations*, Taylor & Francis, 2002; and A. D. Polyanin and V. F. Zaitsev, *Handbook of Nonlinear Partial Differential Equation*, Chapman & Hall/CRC Press, 2004.

Professor Polyanin is editor of the book series Differential and Integral Equations and Their Applications, Chapman & Hall/CRC Press, London/Boca Raton, and Physical and Mathematical Reference Literature, Fizmatlit, Moscow. He is also Editor-in-Chief of the international scientific-educational Website *EqWorld — The World of Mathematical Equations* (http://eqworld.ipmnet.ru), which is visited by over 1000 users a day worldwide. Professor Polyanin is a member of the Editorial Board of the journal *Theoretical Foundations of Chemical Engineering*.

In 1991, Professor Polyanin was awarded a Chaplygin Prize of the Russian Academy of Sciences for his research in mechanics. In 2001, he received an award from the Ministry of Education of the Russian Federation.

**Address:** Institute for Problems in Mechanics, Vernadsky Ave. 101 Bldg 1, 119526 Moscow, Russia
**Home page:** http://eqworld.ipmnet.ru/polyanin-ew.htm

**Alexander V. Manzhirov, D.Sc., Ph.D.,** is a noted scientist in the fields of mechanics and applied mathematics, integral equations, and their applications.

After graduating with honors from the Department of Mechanics and Mathematics of Rostov State University in 1979, Professor Manzhirov attended postgraduate courses at Moscow Institute of Civil Engineering. He received his Ph.D. in 1983 from Moscow Institute of Electronic Engineering Industry and his D.Sc. in 1993 from the Institute for Problems in Mechanics of the Russian (former USSR) Academy of Sciences. Since 1983, Professor Manzhirov has been working at the Institute for Problems in Mechanics of the Russian Academy of Sciences, where he is currently head of the Laboratory for Modeling in Solid Mechanics.

Professor Manzhirov is also head of a branch of the Department of Applied Mathematics at Bauman Moscow State Technical University, professor of mathematics at Moscow State University of Engineering and Computer Science, vice-chairman of Mathematics and Mechanics Expert Council of the Higher Certification Committee of the Russian Federation, executive secretary of Solid Mechanics Scientific Council of the Russian Academy of Sciences, and an expert in mathematics, mechanics, and computer science of the Russian Foundation for Basic Research. He is a member of the Russian National Committee on Theoretical and Applied Mechanics and the European Mechanics Society (EUROMECH), and a member of the editorial board of the journal *Mechanics of Solids* and the international scientific-educational Website *EqWorld — The World of Mathematical Equations* (http://eqworld.ipmnet.ru).

Professor Manzhirov has made important contributions to new mathematical methods for solving problems in the fields of integral equations and their applications, mechanics of growing solids, contact mechanics, tribology, viscoelasticity, and creep theory. He is the author of ten books (including *Contact Problems in Mechanics of Growing Solids* [in Russian], Nauka, Moscow, 1991; *Handbook of Integral Equations*, CRC Press, Boca Raton, 1998; *Handbuch der Integralgleichungen: Exacte Lösungen*, Spektrum Akad. Verlag, Heidelberg, 1999; *Contact Problems in the Theory of Creep* [in Russian], National Academy of Sciences of Armenia, Erevan, 1999), more than 70 research papers, and two patents.

Professor Manzhirov is a winner of the First Competition of the Science Support Foundation 2001, Moscow.

**Address:** Institute for Problems in Mechanics, Vernadsky Ave. 101 Bldg 1, 119526 Moscow, Russia
**Home page:** http://eqworld.ipmnet.ru/en/board/manzhirov.htm

# PREFACE

This book can be viewed as a reasonably comprehensive compendium of mathematical definitions, formulas, and theorems intended for researchers, university teachers, engineers, and students of various backgrounds in mathematics. The absence of proofs and a concise presentation has permitted combining a substantial amount of reference material in a single volume.

When selecting the material, the authors have given a pronounced preference to practical aspects, namely, to formulas, methods, equations, and solutions that are most frequently used in scientific and engineering applications. Hence some abstract concepts and their corollaries are not contained in this book.

- This book contains chapters on arithmetics, elementary geometry, analytic geometry, algebra, differential and integral calculus, differential geometry, elementary and special functions, functions of one complex variable, calculus of variations, probability theory, mathematical statistics, etc. Special attention is paid to formulas (exact, asymptotical, and approximate), functions, methods, equations, solutions, and transformations that are of frequent use in various areas of physics, mechanics, and engineering sciences.

- The main distinction of this reference book from other general (nonspecialized) mathematical reference books is a significantly wider and more detailed description of methods for solving equations and obtaining their exact solutions for various classes of mathematical equations (ordinary differential equations, partial differential equations, integral equations, difference equations, etc.) that underlie mathematical modeling of numerous phenomena and processes in science and technology. In addition to well-known methods, some new methods that have been developing intensively in recent years are described.

- For the convenience of a wider audience with different mathematical backgrounds, the authors tried to avoid special terminology whenever possible. Therefore, some of the methods and theorems are outlined in a schematic and somewhat simplified manner, which is sufficient for them to be used successfully in most cases. Many sections were written so that they could be read independently. The material within subsections is arranged in increasing order of complexity. This allows the reader to get to the heart of the matter quickly.

The material in the first part of the reference book can be roughly categorized into the following three groups according to meaning:

1. The main text containing a concise, coherent survey of the most important definitions, formulas, equations, methods, and theorems.

2. Numerous specific examples clarifying the essence of the topics and methods for solving problems and equations.

3. Discussion of additional issues of interest, given in the form of remarks in small print.

For the reader's convenience, several long mathematical tables — finite sums, series, indefinite and definite integrals, direct and inverse integral transforms (Laplace, Mellin, and Fourier transforms), and exact solutions of differential, integral, functional, and other mathematical equations — which contain a large amount of information, are presented in the second part of the book.

This handbook consists of chapters, sections, subsections, and paragraphs (the titles of the latter are not included in the table of contents). Figures and tables are numbered separately in each section, while formulas (equations) and examples are numbered separately in each subsection. When citing a formula, we use notation like (3.1.2.5), which means formula 5 in Subsection 3.1.2. At the end of each chapter, we present a list of main and additional literature sources containing more detailed information about topics of interest to the reader.

Special font highlighting in the text, cross-references, an extensive table of contents, and an index help the reader to find the desired information.

We would like to express our deep gratitude to Alexei Zhurov for fruitful discussions and valuable remarks. We also appreciate the help of Vladimir Nazaikinskii and Grigorii Yosifian for translating several chapters of this book and are thankful to Kirill Kazakov and Mikhail Mikhin for their assistance in preparing the camera-ready copy of the book.

The authors hope that this book will be helpful for a wide range of scientists, university teachers, engineers, and students engaged in the fields of mathematics, physics, mechanics, control, chemistry, biology, engineering sciences, and social and economical sciences. Some sections and examples can be used in lectures and practical studies in basic and special mathematical courses.

*Andrei D. Polyanin*
*Alexander V. Manzhirov*

# Main Notation

## Special symbols

| | |
|---|---|
| $=$ | equal to |
| $\equiv$ | identically equal to |
| $\neq$ | not equal to |
| $\approx$ | approximately equal to |
| $\sim$ | of same order as (used in comparisons of infinitesimals or infinites) |
| $<$ | less than; "$a$ less than $b$" is written as $a < b$ (or, equivalently, $b > a$) |
| $\leq$ | less than or equal to; $a$ less than or equal to $b$ is written as $a \leq b$ |
| $\ll$ | much less than; $a$ much less than $b$ is written as $a \ll b$ |
| $>$ | greater than; $a$ greater than $b$ is written as $a > b$ (or, equivalently, $b < a$) |
| $\geq$ | greater than or equal to; $a$ greater than or equal to $b$ is written as $a \geq b$ |
| $\gg$ | much greater than; $a$ much greater than $b$ is written as $a \gg b$ |
| $+$ | plus sign; the sum of numbers $a$ and $b$ is denoted by $a + b$ and has the property $a + b = b + a$ |
| $-$ | minus sign; the difference of numbers $a$ and $b$ is denoted by $a - b$ |
| $\cdot$ | multiplication sign; the product of numbers $a$ and $b$ is denoted by either $ab$ or $a \cdot b$ (sometimes $a \times b$) and has the property $ab = ba$; the inner product of vectors $\mathbf{a}$ and $\mathbf{b}$ is denoted by $\mathbf{a} \cdot \mathbf{b}$ |
| $\times$ | multiplication sign; the product of numbers $a$ and $b$ is sometimes denoted by $a \times b$; the cross-product of vectors $\mathbf{a}$ and $\mathbf{b}$ is denoted by $\mathbf{a} \times \mathbf{b}$ |
| $:$ | division sign; the ratio of numbers $a$ and $b$ is denoted by $a:b$ or $a/b$ |
| $!$ | factorial sign: $0! = 1! = 1$, $n! = 1 \cdot 2 \cdot 3 \ldots (n-1)n$, $n = 2, 3, 4, \ldots$ |
| $!!$ | double factorial sign: $0!! = 1!! = 1$, $(2n)!! = 2 \cdot 4 \cdot 6 \ldots (2n)$, $(2n+1)!! = 1 \cdot 3 \cdot 5 \ldots (2n+1)$, where $n = 1, 2, 3, \ldots$ |
| $\%$ | percent sign; 1% is one hundredth of the entire quantity |
| $\infty$ | infinity |
| $\to$ | tends (infinitely approaches) to; $x \to a$ means that $x$ tends to $a$ |
| $\Longrightarrow$ | implies; consequently |
| $\Longleftrightarrow$ | is equivalent to (if and only if ...) |
| $\forall$ | for all, for any |
| $\exists$ | there exists |
| $\in$ | belongs to; $a \in A$ means that $a$ is an element of the set $A$ |
| $\notin$ | does not belong to; $a \notin A$ means that $a$ is not an element of the set $A$ |
| $\cup$ | union (Boolean addition); $A \cup B$ stands for the union of sets $A$ and $B$ |
| $\cap$ | intersection (Boolean multiplication); $A \cap B$ stands for the intersection (common part) of sets $A$ and $B$ |
| $\subset$ | inclusion; $A \subset B$ means that the set $A$ is part of the set $B$ |
| $\subseteq$ | nonstrict inclusion; $A \subseteq B$ means that the set $A$ is part of the set $B$ or coincides with $B$ |
| $\varnothing$ | empty set |
| $\sum$ | sum, $\sum_{k=1}^{n} a_k = a_1 + a_2 + \cdots + a_n$ |
| $\prod$ | product, $\prod_{k=1}^{n} a_k = a_1 \cdot a_2 \cdot \ldots \cdot a_n$ |
| $\partial$ | symbol used to denote partial derivatives and differential operators; $\partial_x$ is the operator of differentiation with respect to $x$ |

| | |
|---|---|
| $\nabla$ | vector differential operator "nabla"; $\nabla a$ is the gradient of a scalar $a$ |
| $\int$ | integral; $\int_a^b f(x)\,dx$ is the integral of a function $f(x)$ over the interval $[a, b]$ |
| $\oint$ | contour integral (denotes an integral over a closed contour) |
| $\perp$ | perpendicular |
| $\parallel$ | parallel |

## Roman alphabet

| | |
|---|---|
| Arg $z$ | argument of a complex number $z = x + iy$; by definition, $\tan(\text{Arg } z) = y/x$ |
| arg $z$ | principal value of Arg $z$; by definition, arg $z$ = Arg $z$, where $-\pi < \text{Arg } z \leq \pi$ |
| $\sqrt{a}$ | square root of a number $a$, defined by the property $(\sqrt{a})^2 = a$ |
| $\sqrt[n]{a}$ | $n$th root of a number $a$ ($n = 2, 3, \ldots, a \geq 0$), defined by the property $(\sqrt[n]{a})^n = a$ |
| $\|a\|$ | absolute value (modulus) of a real number $a$, $\|a\| = \begin{cases} a & \text{if } a \geq 0 \\ -a & \text{if } a < 0 \end{cases}$ |
| $\mathbf{a}$ | vector, $\mathbf{a} = \{a_1, a_2, a_3\}$, where $a_1, a_2, a_3$ are the vector components |
| $\|\mathbf{a}\|$ | modulus of a vector $\mathbf{a}$, $\|\mathbf{a}\| = \sqrt{\mathbf{a} \cdot \mathbf{a}}$ |
| $\mathbf{a} \cdot \mathbf{b}$ | inner product of vectors $\mathbf{a}$ and $\mathbf{b}$, denoted also by $(\mathbf{a} \cdot \mathbf{b})$ |
| $\mathbf{a} \times \mathbf{b}$ | cross-product of vectors $\mathbf{a}$ and $\mathbf{b}$ |
| $[\mathbf{abc}]$ | triple product of vectors $\mathbf{a}, \mathbf{b}, \mathbf{c}$ |
| $(a, b)$ | interval (open interval) $a < x < b$ |
| $(a, b]$ | half-open interval $a < x \leq b$ |
| $[a, b)$ | half-open interval $a \leq x < b$ |
| $[a, b]$ | interval (closed interval) $a \leq x \leq b$ |
| arccos $x$ | arccosine, the inverse function of cosine: $\cos(\arccos x) = x$, $\|x\| \leq 1$ |
| arccot $x$ | arccotangent, the inverse function of cotangent: $\cot(\text{arccot } x) = x$ |
| arcsin $x$ | arcsine, the inverse function of sine: $\sin(\arcsin x) = x$, $\|x\| \leq 1$ |
| arctan $x$ | arctangent, the inverse function of tangent: $\tan(\arctan x) = x$ |
| arccosh $x$ | hyperbolic arccosine, the inverse function of hyperbolic cosine; also denoted by arccosh $x = \cosh^{-1} x$; arccosh $x = \ln\left(x + \sqrt{x^2 - 1}\right)$ ($x \geq 1$) |
| arccoth $x$ | hyperbolic arccotangent, the inverse function of hyperbolic cotangent; also denoted by arccoth $x = \coth^{-1} x$; arccoth $x = \dfrac{1}{2} \ln \dfrac{x+1}{x-1}$ ($\|x\| > 1$) |
| arcsinh $x$ | hyperbolic arcsine, the inverse function of hyperbolic sine; also denoted by arcsinh $x = \sinh^{-1} x$; arcsinh $x = \ln\left(x + \sqrt{x^2 + 1}\right)$ |
| arctanh $x$ | hyperbolic arctangent, the inverse function of hyperbolic tangent; also denoted by arctanh $x = \tanh^{-1} x$; arctanh $x = \dfrac{1}{2} \ln \dfrac{1+x}{1-x}$ ($\|x\| < 1$) |
| $C_n^k$ | binomial coefficients, also denoted by $\binom{n}{k}$, $C_n^k = \dfrac{n!}{k!(n-k)!}$, $k = 1, 2, \ldots, n$ |
| $\mathcal{C}$ | Euler constant, $\mathcal{C} = \lim\limits_{n \to \infty} \left(1 + \dfrac{1}{2} + \dfrac{1}{3} + \cdots + \dfrac{1}{n} - \ln n\right) = 0.5772156\ldots$ |
| cos $x$ | cosine, even trigonometric function of period $2\pi$ |
| cosec $x$ | cosecant, odd trigonometric function of period $2\pi$: cosec $x = \dfrac{1}{\sin x}$ |
| cosh $x$ | hyperbolic cosine, $\cosh x = \tfrac{1}{2}(e^x + e^{-x})$ |
| cot $x$ | cotangent, odd trigonometric function of period $\pi$, $\cot x = \cos x / \sin x$ |
| coth $x$ | hyperbolic cotangent, $\coth x = \cosh x / \sinh x$ |
| det $A$ | determinant of a matrix $A = (a_{ij})$ |

# MAIN NOTATION

| | |
|---|---|
| div **a** | divergence of a vector **a** |
| $e$ | the number "e" (base of natural logarithms), $e = 2.718281\ldots$; definition: $e = \lim_{n\to\infty}\left(1+\dfrac{1}{n}\right)^n$ |
| erf $x$ | Gauss error function, $\operatorname{erf} x = \dfrac{2}{\sqrt{\pi}}\int_0^x \exp(-\xi^2)\,d\xi$ |
| erfc $x$ | complementary error function, $\operatorname{erfc} x = \dfrac{2}{\sqrt{\pi}}\int_x^\infty \exp(-\xi^2)\,d\xi$ |
| exp $x$ | exponential (exponential function), denoted also by $\exp x = e^x$ |
| grad $a$ | gradient of a scalar $a$, denoted also by $\nabla a$ |
| $H_n(x)$ | Hermite polynomial, $H_n(x) = (-1)^n e^{x^2} \dfrac{d^n}{dx^n}\left(e^{-x^2}\right)$ |
| $I_\nu(x)$ | modified Bessel function of the first kind, $I_\nu(x) = \displaystyle\sum_{n=0}^{\infty}\dfrac{(x/2)^{\nu+2n}}{n!\,\Gamma(\nu+n+1)}$ |
| Im $z$ | imaginary part of a complex number; if $z = x + iy$, then $\operatorname{Im} z = y$ |
| inf $A$ | infimum of a (numerical) set $A$; if $A = (a,b)$ or $A = [a,b)$, then $\inf A = a$ |
| $J_\nu(x)$ | Bessel function of the first kind, $J_\nu(x) = \displaystyle\sum_{n=0}^{\infty}\dfrac{(-1)^n(x/2)^{\nu+2n}}{n!\,\Gamma(\nu+n+1)}$ |
| $K_\nu(x)$ | modified Bessel function of the second kind, $K_\nu(x) = \dfrac{\pi}{2}\dfrac{I_{-\nu}(x) - I_\nu(x)}{\sin(\pi\nu)}$ |
| $\lim\limits_{x\to a} f(x)$ | limit of a function $f(x)$ as $x \to a$ |
| ln $x$ | natural logarithm (logarithm to base $e$) |
| $\log_a x$ | logarithm to base $a$ |
| $\max\limits_{a\le x\le b} f(x)$ | maximum of a function $f(x)$ on the interval $a \le x \le b$ |
| $\min\limits_{a\le x\le b} f(x)$ | minimum of a function $f(x)$ on the interval $a \le x \le b$ |
| $n!$ | factorial: $0! = 1! = 1$, $n! = 1\cdot 2\cdot 3\ldots(n-1)n$, $n = 2, 3, 4, \ldots$ |
| $P_n(x)$ | Legendre polynomial, $P_n(x) = \dfrac{1}{n!\,2^n}\dfrac{d^n}{dx^n}(x^2-1)^n$ |
| $\mathbb{R}$ | set of real numbers, $\mathbb{R} = \{-\infty < x < \infty\}$ |
| Re $z$ | real part of a complex number; if $z = x + iy$, then $\operatorname{Re} z = x$ |
| $r, \varphi, z$ | cylindrical coordinates, $r = \sqrt{x^2+y^2}$ and $x = r\cos\varphi$, $y = r\sin\varphi$ |
| $r, \theta, \varphi$ | spherical coordinates, $r = \sqrt{x^2+y^2+z^2}$ and $x = r\sin\theta\cos\varphi$, $y = \sin\theta\sin\varphi$, $z = r\cos\theta$ |
| rank $A$ | rank of a matrix $A$ |
| curl **a** | curl of a vector **a**, also denoted by rot **a** |
| sec $x$ | secant, even trigonometric function of period $2\pi$: $\sec x = \dfrac{1}{\cos x}$ |
| sign $x$ | "sign" function: it is equal to 1 if $x > 0$, $-1$ if $x < 0$, and 0 if $x = 0$ |
| sin $x$ | sine, odd trigonometric function of period $2\pi$ |
| sinh $x$ | hyperbolic sine, $\sinh x = \tfrac{1}{2}(e^x - e^{-x})$ |
| sup $A$ | supremum of a (numerical) set $A$; if $A = (a,b)$ or $A = (a,b]$, then $\sup A = b$ |
| tan $x$ | tangent, odd trigonometric function of period $\pi$, $\tan x = \sin x/\cos x$ |
| tanh $x$ | hyperbolic tangent, $\tanh x = \sinh x/\cosh x$ |
| $x$ | independent variable, argument |
| $x, y, z$ | spatial variables (Cartesian coordinates) |

| | |
|---|---|
| $Y_\nu(x)$ | Bessel function of the second kind; $Y_\nu(x) = \dfrac{J_\nu(x)\cos(\pi\nu) - J_{-\nu}(x)}{\sin(\pi\nu)}$ |
| $y$ | dependent variable, function; one often writes $y = y(x)$ or $y = f(x)$ |
| $y'_x$ | first derivative of a function $y = f(x)$, also denoted by $y'$, $\dfrac{dy}{dx}$, $f'(x)$ |
| $y''_{xx}$ | second derivative of a function $y = f(x)$, also denoted by $y''$, $\dfrac{d^2y}{dx^2}$, $f''(x)$ |
| $y_x^{(n)}$ | $n$th derivative of a function $y = f(x)$, also denoted by $\dfrac{d^ny}{dx^n}$ |
| $z = x + iy$ | complex number; $x$ is the real part of $z$, $y$ is the imaginary part of $z$, $i^2 = -1$ |
| $\bar{z} = x - iy$ | complex conjugate number, $i^2 = -1$ |
| $\|z\|$ | modulus of a complex number; if $z = x + iy$, then $\|z\| = \sqrt{x^2 + y^2}$. |

## Greek alphabet

| | |
|---|---|
| $\Gamma(\alpha)$ | gamma function, $\Gamma(\alpha) = \int_0^\infty e^{-t} t^{\alpha-1}\, dt$ |
| $\gamma(\alpha, x)$ | incomplete gamma function, $\gamma(\alpha, x) = \int_0^x e^{-t} t^{\alpha-1}\, dt$ |
| $\Phi(a, b; x)$ | degenerate hypergeometric function, $\Phi(a, b; x) = 1 + \sum_{n=1}^\infty \dfrac{a(a+1)\ldots(a+n-1)}{b(b+1)\ldots(b+n-1)} \dfrac{x^n}{n!}$ |
| $\Delta$ | Laplace operator; in the two-dimensional case, $\Delta w = \dfrac{\partial^2 w}{\partial x^2} + \dfrac{\partial^2 w}{\partial y^2}$, where $x$ and $y$ are Cartesian coordinates |
| $\Delta x$ | increment of the argument |
| $\Delta y$ | increment of the function; if $y = f(x)$, then $\Delta y = f(x + \Delta x) - f(x)$ |
| $\delta_{nm}$ | Kronecker delta, $\delta_{nm} = \begin{cases} 1 & \text{if } n = m \\ 0 & \text{if } n \ne m \end{cases}$ |
| $\pi$ | the number "pi" (ratio of the circumference to the diameter), $\pi = 3.141592\ldots$ |

## Remarks

**1.** If a formula or a solution contains an expression like $\dfrac{f(x)}{a-2}$, it is often not stated explicitly that the assumption $a \ne 2$ is implied.

**2.** If a formula or a solution contains derivatives of some functions, then the functions are assumed to be differentiable.

**3.** If a formula or a solution contains definite integrals, then the integrals are supposed to be convergent.

**4.** ODE and PDE are conventional abbreviations for ordinary differential equation and partial differential equation, respectively.

# Part I
# Definitions, Formulas, Methods, and Theorems

# Chapter 1
# Arithmetic and Elementary Algebra

## 1.1. Real Numbers
### 1.1.1. Integer Numbers

**1.1.1-1. Natural, integer, even, and odd numbers.**

*Natural numbers*: 1, 2, 3, ... (all positive whole numbers).
*Integer numbers* (or simply *integers*): 0, ±1, ±2, ±3, ...
*Even numbers*: 0, 2, 4, ... (all nonnegative integers that can be divided evenly by 2). An even number can generally be represented as $n = 2k$, where $k = 0, 1, 2, ...$

Remark 1. Sometimes all integers that are multiples of 2, such as 0, ±2, ±4, ..., are considered to be even numbers.

*Odd numbers*: 1, 3, 5, ... (all natural numbers that cannot be divided evenly by 2). An odd number can generally be represented as $n = 2k + 1$, where $k = 0, 1, 2, ...$

Remark 2. Sometimes all integers that are not multiples of 2, such as ±1, ±3, ±5, ..., are considered to be odd numbers.

All integers as well as even numbers and odd numbers form *infinite countable sets*, which means that the elements of these sets can be enumerated using the natural numbers 1, 2, 3, ...

**1.1.1-2. Prime and composite numbers.**

A *prime number* is a positive integer that is greater than 1 and has no positive integer divisors other than 1 and itself. The prime numbers form an infinite countable set. The first ten prime numbers are: 2, 3, 5, 7, 11, 13, 17, 19, 23, 29, ...

A *composite number* is a positive integer that is greater than 1 and is not prime, i.e., has factors other than 1 and itself. Any composite number can be uniquely factored into a product of prime numbers. The following numbers are composite: $4 = 2 \times 2$, $6 = 2 \times 3$, $8 = 2^3$, $9 = 3^2$, $10 = 2 \times 5$, $12 = 2^2 \times 3$, ...

The number 1 is a special case that is considered to be neither composite nor prime.

**1.1.1-3. Divisibility tests.**

Below are some simple rules helping to determine if an integer is divisible by another integer.

All integers are divisible by 1.
Divisibility by 2: last digit is divisible by 2.
Divisibility by 3: sum of digits is divisible by 3.
Divisibility by 4: two last digits form a number divisible by 4.
Divisibility by 5: last digit is either 0 or 5.

Divisibility by 6: divisible by both 2 and 3.
Divisibility by 9: sum of digits is divisible by 9.
Divisibility by 10: last digit is 0.
Divisibility by 11: the difference between the sum of the odd numbered digits (1st, 3rd, 5th, etc.) and the sum of the even numbered digits (2nd, 4th, etc.) is divisible by 11.

**Example 1.** Let us show that the number 80729 is divisible by 11.
The sum of the odd numbered digits is $\Sigma_1 = 8 + 7 + 9 = 24$. The sum of the even numbered digits is $\Sigma_2 = 0 + 2 = 2$. The difference between them is $\Sigma_1 - \Sigma_2 = 22$ and is divisible by 11. Consequently, the original number is also divisible by 11.

### 1.1.1-4. Greatest common divisor and least common multiple.

$1°$. The *greatest common divisor* of natural numbers $a_1, a_2, \ldots, a_n$ is the largest natural number, $b$, which is a common divisor to $a_1, \ldots, a_n$.

Suppose some positive numbers $a_1, a_2, \ldots, a_n$ are factored into products of primes so that

$$a_1 = p_1^{k_{11}} p_2^{k_{12}} \ldots p_m^{k_{1m}}, \quad a_2 = p_1^{k_{21}} p_2^{k_{22}} \ldots p_m^{k_{2m}}, \quad \ldots, \quad a_n = p_1^{k_{n1}} p_2^{k_{n2}} \ldots p_m^{k_{nm}},$$

where $p_1, p_2, \ldots, p_n$ are different prime numbers, the $k_{ij}$ are positive integers ($i = 1, 2, \ldots, n$; $j = 1, 2, \ldots, m$). Then the greatest common divisor $b$ of $a_1, a_2, \ldots, a_n$ is calculated as

$$b = p_1^{\sigma_1} p_2^{\sigma_2} \ldots p_m^{\sigma_m}, \qquad \sigma_j = \min_{1 \le i \le n} k_{ij}.$$

**Example 2.** The greatest common divisor of 180 and 280 is $2^2 \times 5 = 20$ due to the following factorization:

$$180 = 2^2 \times 3^2 \times 5 = 2^2 \times 3^2 \times 5^1 \times 7^0,$$
$$280 = 2^3 \times 5 \times 7 = 2^3 \times 3^0 \times 5^1 \times 7^1.$$

$2°$. The *least common multiple* of $n$ natural numbers $a_1, a_2, \ldots, a_n$ is the smallest natural number, $A$, that is a multiple of all the $a_k$.

Suppose some natural numbers $a_1, \ldots, a_n$ are factored into products of primes just as in Item $1°$. Then the least common multiple of all the $a_k$ is calculated as

$$A = p_1^{\nu_1} p_2^{\nu_2} \ldots p_m^{\nu_m}, \qquad \nu_j = \max_{1 \le i \le n} k_{ij}.$$

**Example 3.** The least common multiple of 180 and 280 is equal to $2^3 \times 3^2 \times 5^1 \times 7^1 = 2520$ due to their factorization given in Example 2.

## 1.1.2. Real, Rational, and Irrational Numbers

### 1.1.2-1. Real numbers.

The *real numbers* are all the positive numbers, negative numbers, and zero. Any real number can be represented by a *decimal fraction* (or simply *decimal*), finite or infinite. The set of all real numbers is denoted by $\mathbb{R}$.

All real numbers are categorized into two classes: the *rational* numbers and *irrational* numbers.

### 1.1.2-2. Rational numbers.

A *rational number* is a real number that can be written as a fraction (ratio) $p/q$ with integer $p$ and $q$ ($q \neq 0$). It is only the rational numbers that can be written in the form of finite (terminating) or periodic (recurring) decimals (e.g., $1/8 = 0.125$ and $1/6 = 0.16666\ldots$). Any integer is a rational number.

The rational numbers form an infinite countable set. The set of all rational numbers is everywhere dense. This means that, for any two distinct rational numbers $a$ and $b$ such that $a < b$, there exists at least one more rational number $c$ such that $a < c < b$, and hence there are infinitely many rational numbers between $a$ and $b$. (Between any two rational numbers, there always exist irrational numbers.)

### 1.1.2-3. Irrational numbers.

An *irrational number* is a real number that is not rational; no irrational number can be written as a fraction $p/q$ with integer $p$ and $q$ ($q \neq 0$). To the irrational numbers there correspond nonperiodic (nonrepeating) decimals. Examples of irrational numbers: $\sqrt{3} = 1.73205\ldots$, $\pi = 3.14159\ldots$

The set of irrational numbers is everywhere dense, which means that between any two distinct irrational numbers, there are both rational and irrational numbers. The set of irrational numbers is uncountable.

## 1.2. Equalities and Inequalities. Arithmetic Operations. Absolute Value

### 1.2.1. Equalities and Inequalities

#### 1.2.1-1. Basic properties of equalities.

▶ *Throughout Paragraphs 1.2.1-1 and 1.2.1-2, it is assumed that $a$, $b$, $c$, $d$ are real numbers.*

1. If $a = b$, then $b = a$.
2. If $a = b$, then $a + c = b + c$, where $c$ is any real number; furthermore, if $a + c = b + c$, then $a = b$.
3. If $a = b$, then $ac = bc$, where $c$ is any real number; furthermore, if $ac = bc$ and $c \neq 0$, then $a = b$.
4. If $a = b$ and $b = c$, then $a = c$.
5. If $ab = 0$, then either $a = 0$ or $b = 0$; furthermore, if $ab \neq 0$, then $a \neq 0$ and $b \neq 0$.

#### 1.2.1-2. Basic properties of inequalities.

1. If $a < b$, then $b > a$.
2. If $a \leq b$ and $b \leq a$, then $a = b$.
3. If $a \leq b$ and $b \leq c$, then $a \leq c$.
4. If $a < b$ and $b \leq c$ (or $a \leq b$ and $b < c$), then $a < c$.
5. If $a < b$ and $c < d$ (or $c = d$), then $a + c < b + d$.
6. If $a \leq b$ and $c > 0$, then $ac \leq bc$.
7. If $a \leq b$ and $c < 0$, then $ac \geq bc$.
8. If $0 < a \leq b$ (or $a \leq b < 0$), then $1/a \geq 1/b$.

## 1.2.2. Addition and Multiplication of Numbers

**1.2.2-1. Addition of real numbers.**

The sum of real numbers is a real number.
Properties of addition:

$$a + 0 = a \qquad \text{(property of zero),}$$
$$a + b = b + a \qquad \text{(addition is commutative),}$$
$$a + (b + c) = (a + b) + c = a + b + c \qquad \text{(addition is associative),}$$

where $a, b, c$ are arbitrary real numbers.

For any real number $a$, there exists its unique *additive inverse*, or its *opposite*, denoted by $-a$, such that
$$a + (-a) = a - a = 0.$$

**1.2.2-2. Multiplication of real numbers.**

The product of real numbers is a real number.
Properties of multiplication:

$$a \times 0 = 0 \qquad \text{(property of zero),}$$
$$ab = ba \qquad \text{(multiplication is commutative),}$$
$$a(bc) = (ab)c = abc \qquad \text{(multiplication is associative),}$$
$$a \times 1 = 1 \times a = a \qquad \text{(multiplication by unity),}$$
$$a(b + c) = ab + ac \qquad \text{(multiplication is distributive),}$$

where $a, b, c$ are arbitrary real numbers.

For any nonzero real number $a$, there exists its unique *multiplicative inverse*, or its *reciprocal*, denoted by $a^{-1}$ or $1/a$, such that
$$aa^{-1} = 1 \qquad (a \neq 0).$$

## 1.2.3. Ratios and Proportions

**1.2.3-1. Operations with fractions and properties of fractions.**

Ratios are written as fractions: $a : b = a/b$. The number $a$ is called the *numerator* and the number $b$ ($b \neq 0$) is called the *denominator* of a fraction.

Properties of fractions and operations with fractions:

$$\frac{a}{1} = a, \qquad \frac{a}{b} = \frac{ab}{bc} = \frac{a:c}{b:c} \qquad \text{(simplest properties of fractions);}$$

$$\frac{a}{b} \pm \frac{c}{b} = \frac{a \pm c}{b}, \qquad \frac{a}{b} \pm \frac{c}{d} = \frac{ad \pm bc}{bd} \qquad \text{(addition and subtraction of fractions);}$$

$$\frac{a}{b} \times c = \frac{ac}{b}, \qquad \frac{a}{b} \times \frac{c}{d} = \frac{ac}{bd} \qquad \text{(multiplication by a number and by a fraction);}$$

$$\frac{a}{b} : c = \frac{a}{bc}, \qquad \frac{a}{b} : \frac{c}{d} = \frac{ad}{bc} \qquad \text{(division by a number and by a fraction).}$$

### 1.2.3-2. Proportions. Simplest relations. Derivative proportions.

A proportion is an equation with a ratio on each side. A proportion is denoted by $a/b = c/d$ or $a : b = c : d$.

$1°$. The following simplest relations follow from $a/b = c/d$:

$$ad = bc, \quad \frac{a}{c} = \frac{b}{d}, \quad a = \frac{bc}{d}, \quad b = \frac{ad}{c}.$$

$2°$. The following derivative proportions follow from $a/b = c/d$:

$$\frac{ma + nb}{pa + qb} = \frac{mc + nd}{pc + qd},$$

$$\frac{ma + nc}{pa + qc} = \frac{mb + nd}{pb + qd},$$

where $m, n, p, q$ are arbitrary real numbers.

Some special cases of the above formulas:

$$\frac{a \pm b}{b} = \frac{c \pm d}{d}, \quad \frac{a - b}{a + b} = \frac{c - d}{c + d}.$$

## 1.2.4. Percentage

### 1.2.4-1. Definition. Main percentage problems.

A *percentage* is a way of expressing a ratio or a fraction as a whole number, by using 100 as the denominator. One *percent* is one per one hundred, or one hundredth of a whole number; notation: 1%.

Below are the statements of main percentage problems and their solutions.

$1°$. Find the number $b$ that makes up $p\%$ of a number $a$. Answer: $b = \frac{ap}{100}$.

$2°$. Find the number $a$ whose $p\%$ is equal to a number $b$. Answer: $a = \frac{100b}{p}$.

$3°$. What percentage does a number $b$ make up of a number $a$? Answer: $p = \frac{100b}{a}\%$.

### 1.2.4-2. Simple and compound percentage.

$1°$. *Simple percentage.* Suppose a cash deposit is increased yearly by the same amount defined as a percentage, $p\%$, of the initial deposit, $a$. Then the amount accumulated after $t$ years is calculated by the simple percentage formula

$$x = a\left(1 + \frac{pt}{100}\right).$$

$2°$. *Compound percentage.* Suppose a cash deposit is increased yearly by an amount defined as a percentage, $p\%$, of the deposit in the previous year. Then the amount accumulated after $t$ years is calculated by the compound percentage formula

$$x = a\left(1 + \frac{p}{100}\right)^t,$$

where $a$ is the initial deposit.

## 1.2.5. Absolute Value of a Number (Modulus of a Number)

**1.2.5-1. Definition.**

The absolute value of a real number $a$, denoted by $|a|$, is defined by the formula

$$|a| = \begin{cases} a & \text{if } a \geq 0, \\ -a & \text{if } a < 0. \end{cases}$$

An important property: $|a| \geq 0$.

**1.2.5-2. Some formulas and inequalities.**

1°. The following relations hold true:

$$|a| = |-a| = \sqrt{a^2}, \quad a \leq |a|,$$
$$\big||a| - |b|\big| \leq |a+b| \leq |a| + |b|,$$
$$\big||a| - |b|\big| \leq |a-b| \leq |a| + |b|,$$
$$|ab| = |a||b|, \quad |a/b| = |a|/|b|.$$

2°. From the inequalities $|a| \leq A$ and $|b| \leq B$ it follows that $|a+b| \leq A+B$ and $|ab| \leq AB$.

## 1.3. Powers and Logarithms

### 1.3.1. Powers and Roots

**1.3.1-1. Powers and roots: the main definitions.**

Given a positive real number $a$ and a positive integer $n$, the $n$th power of $a$, written as $a^n$, is defined as the multiplication of $a$ by itself repeated $n$ times:

$$a^n = \underbrace{a \times a \times a \times \cdots \times a}_{n \text{ multipliers}}.$$

The number $a$ is called the *base* and $n$ is called the *exponent*.

Obvious properties: $0^n = 0$, $1^n = 1$, $a^1 = a$.

Raising to the zeroth power: $a^0 = 1$, where $a \neq 0$. Sometimes $0^0$ is taken as undefined, but it is often sensibly defined as 1.

Raising to a negative power: $a^{-n} = \dfrac{1}{a^n}$, where $n$ is a positive integer.

If $a$ is a positive real number and $n$ is a positive integer, then the *$n$th arithmetic root* or *radical* of $a$, written as $\sqrt[n]{a}$, is the unique positive real number $b$ such that $b^n = a$. In the case of $n = 2$, the brief notation $\sqrt{a}$ is used to denote $\sqrt[2]{a}$.

The following relations hold:

$$\sqrt[n]{0} = 0, \quad \sqrt[n]{1} = 1, \quad \left(\sqrt[n]{a}\right)^n = a.$$

Raising to a fractional power $p = m/n$, where $m$ and $n$ are natural numbers:

$$a^p = a^{m/n} = \sqrt[n]{a^m}, \quad a \geq 0.$$

### 1.3.1-2. Operations with powers and roots.

The properties given below are valid for any rational and real exponents $p$ and $q$ ($a > 0$, $b > 0$):

$$a^{-p} = \frac{1}{a^p}, \quad a^p a^q = a^{p+q}, \quad \frac{a^p}{a^q} = a^{p-q},$$

$$(ab)^p = a^p b^p, \quad \left(\frac{a}{b}\right)^q = \frac{a^q}{b^q}, \quad (a^p)^q = a^{pq}.$$

In operations with roots (radicals) the following properties are used:

$$\sqrt[n]{ab} = \sqrt[n]{a}\sqrt[n]{b}, \quad \sqrt[n]{\frac{a}{b}} = \frac{\sqrt[n]{a}}{\sqrt[n]{b}}, \quad \sqrt[n]{a^m} = \left(\sqrt[n]{a}\right)^m, \quad \sqrt[n]{\sqrt[m]{a}} = \sqrt[mn]{a}.$$

**Remark.** It often pays to represent roots as powers with rational exponents and apply the properties of operations with powers.

## 1.3.2. Logarithms

### 1.3.2-1. Definition. The main logarithmic identity.

The *logarithm of a positive number $b$ to a given base $a$* is the exponent of the power $c$ to which the base $a$ must be raised to produce $b$. It is written as $\log_a b = c$.

Equivalent representations:

$$\log_a b = c \quad \Longleftrightarrow \quad a^c = b,$$

where $a > 0$, $a \neq 1$, and $b > 0$.

Main logarithmic identity:

$$a^{\log_a b} = b.$$

Simple properties:

$$\log_a 1 = 0, \quad \log_a a = 1.$$

### 1.3.2-2. Properties of logarithms. The common and natural logarithms.

Properties of logarithms:

$$\log_a(bc) = \log_a b + \log_a c, \quad \log_a\left(\frac{b}{c}\right) = \log_a b - \log_a c,$$

$$\log_a(b^k) = k \log_a b, \quad \log_{a^k} b = \frac{1}{k}\log_a b \quad (k \neq 0),$$

$$\log_a b = \frac{1}{\log_b a} \quad (b \neq 1), \quad \log_a b = \frac{\log_c b}{\log_c a} \quad (c \neq 1),$$

where $a > 0$, $a \neq 1$, $b > 0$, $c > 0$, and $k$ is any number.

The logarithm to the base 10 is called the *common* or *decadic logarithm* and written as

$$\log_{10} b = \log b \quad \text{or sometimes} \quad \log_{10} b = \lg b.$$

The logarithm to the base $e$ (the *base of natural logarithms*) is called the *natural logarithm* and written as

$$\log_e b = \ln b,$$

where $e = \lim\limits_{n \to \infty} \left(1 + \frac{1}{n}\right)^n = 2.718281\ldots$

## 1.4. Binomial Theorem and Related Formulas
### 1.4.1. Factorials. Binomial Coefficients. Binomial Theorem

**1.4.1-1. Factorials. Binomial coefficients.**

*Factorial*:
$$0! = 1! = 1,$$
$$n! = 1 \times 2 \times 3 \times \ldots \times (n-1)n, \quad n = 2, 3, 4, \ldots$$

*Double factorial*:
$$0!! = 1!! = 1,$$
$$n!! = \begin{cases} (2k)!! & \text{if } n = 2k, \\ (2k+1)!! & \text{if } n = 2k+1, \end{cases}$$
$$(2k)!! = 2 \times 4 \times 6 \times \ldots \times (2k-2)(2k) = 2^k k!,$$
$$(2k+1)!! = 1 \times 3 \times 5 \times \ldots \times (2k-1)(2k+1),$$

where $n$ and $k$ are natural numbers.

*Binomial coefficients*:
$$C_n^k = \binom{n}{k} = \frac{n!}{k!(n-k)!} = \frac{n(n-1)\ldots(n-k+1)}{k!}, \quad k = 1, 2, 3, \ldots, n;$$
$$C_a^k = \frac{a(a-1)\ldots(a-k+1)}{k!}, \quad \text{where} \quad k = 1, 2, 3, \ldots,$$

where $n$ is a natural number and $a$ is any number.

**1.4.1-2. Binomial theorem.**

Let $a$, $b$, and $c$ be real (or complex) numbers. The following formulas hold true:
$$(a \pm b)^2 = a^2 \pm 2ab + b^2,$$
$$(a \pm b)^3 = a^3 \pm 3a^2b + 3ab^2 \pm b^3,$$
$$(a \pm b)^4 = a^4 \pm 4a^3b + 6a^2b^2 \pm 4ab^3 + b^4,$$
$$\ldots\ldots\ldots\ldots\ldots\ldots\ldots\ldots\ldots\ldots\ldots\ldots\ldots\ldots$$
$$(a+b)^n = \sum_{k=0}^{n} C_n^k a^{n-k} b^k, \quad n = 1, 2, \ldots$$

The last formula is known as the *binomial theorem*, where the $C_n^k$ are binomial coefficients.

### 1.4.2. Related Formulas

**1.4.2-1. Formulas involving powers ≤ 4.**

$$a^2 - b^2 = (a-b)(a+b),$$
$$a^3 + b^3 = (a+b)(a^2 - ab + b^2),$$
$$a^3 - b^3 = (a-b)(a^2 + ab + b^2),$$
$$a^4 - b^4 = (a-b)(a+b)(a^2 + b^2),$$
$$(a+b+c)^2 = a^2 + b^2 + c^2 + 2ab + 2ac + 2bc,$$
$$a^4 + a^2 b^2 + b^4 = (a^2 + ab + b^2)(a^2 - ab + b^2).$$

### 1.4.2-2. Formulas involving arbitrary powers.

Let $n$ be any positive integer. Then
$$a^n - b^n = (a-b)(a^{n-1} + a^{n-2}b + \cdots + ab^{n-2} + b^{n-1}).$$
If $n$ is a positive even number, then
$$a^n - b^n = (a+b)(a^{n-1} - a^{n-2}b + \cdots + ab^{n-2} - b^{n-1})$$
$$= (a-b)(a+b)(a^{n-2} + a^{n-4}b^2 + \cdots + a^2 b^{n-4} + b^{n-2}).$$
If $n$ is a positive odd number, then
$$a^n + b^n = (a+b)(a^{n-1} - a^{n-2}b + \cdots - ab^{n-2} + b^{n-1}).$$

## 1.5. Arithmetic and Geometric Progressions. Finite Sums and Products

### 1.5.1. Arithmetic and Geometric Progressions

#### 1.5.1-1. Arithmetic progression.

$1°$. An *arithmetic progression*, or *arithmetic sequence*, is a sequence of real numbers for which each term, starting from the second, is the previous term plus a constant $d$, called the *common difference*. In general, the terms of an arithmetic progression are expressed as
$$a_n = a_1 + (n-1)d, \quad n = 1, 2, 3, \ldots,$$
where $a_1$ is the first term of the progression. An arithmetic progression is called *increasing* if $d > 0$ and *decreasing* if $d < 0$.

$2°$. An arithmetic progression has the property
$$a_n = \tfrac{1}{2}(a_{n-1} + a_{n+1}).$$

$3°$. The sum of $n$ first terms of an arithmetic progression is called an *arithmetic series* and is calculated as
$$S_n = a_1 + \cdots + a_n = \tfrac{1}{2}(a_1 + a_n)n = \tfrac{1}{2}[2a_1 + (n-1)d]n.$$

#### 1.5.1-2. Geometric progression.

$1°$. A *geometric progression*, or *geometric sequence*, is a sequence of real numbers for which each term, starting from the second, is the previous term multiplied by a constant $q$, called the *common ratio*. In general, the terms of a geometric progression are expressed as
$$a_n = a_1 q^{n-1}, \quad n = 1, 2, 3, \ldots,$$
where $a_1$ is the first term of the progression.

$2°$. A geometric progression with positive terms has the property
$$a_n = \sqrt{a_{n-1} a_{n+1}}.$$

$3°$. The sum of $n$ first terms of an arithmetic progression is called a *geometric series* and is calculated as ($q \neq 1$)
$$S_n = a_1 + \cdots + a_n = a_1 \frac{1-q^n}{1-q}.$$

## 1.5.2. Finite Series and Products

**1.5.2-1. Notations for finite series and products.**

A finite series is just the sum of a finite number of terms and a finite product is the product of a finite number of terms. These are written as

$$a_1 + a_2 + \cdots + a_n = \sum_{k=1}^{n} a_k, \qquad a_m + a_{m+1} + \cdots + a_n = \sum_{k=m}^{n} a_k;$$

$$a_1 a_2 \ldots a_n = \prod_{k=1}^{n} a_k, \qquad a_m a_{m+1} \ldots a_n = \prod_{k=m}^{n} a_k,$$

where $m$ is a nonnegative integer ($m \leq n$). The variable $k$ appearing on the right-hand sides of the above formulas is called the *index of summation* (for series) or the *index of multiplication* (for products). The 1 and $n$ (or the $m$ and $n$) are the *upper* and *lower limits of summation (multiplication)*.

The values of sums (products) are independent of the names used to denote the index of summation (multiplication): $\sum_{k=1}^{n} a_k = \sum_{j=1}^{n} a_j$, $\prod_{k=1}^{n} a_k = \prod_{i=1}^{n} a_i$. Such indices are called dummy indices.

**1.5.2-2. Formulas for summation of some finite series.**

$$\sum_{k=1}^{n} k = \frac{n(n+1)}{2};$$

$$\sum_{k=1}^{n} (-1)^k k = (-1)^n \left[\frac{n-1}{2}\right], \qquad [m] \text{ is the integer part of } m;$$

$$\sum_{k=0}^{n} (2k+1) = (n+1)^2;$$

$$\sum_{k=0}^{n} (-1)^k (2k+1) = (-1)^n (n+1);$$

$$\sum_{k=1}^{n} k^2 = \frac{1}{6} n(n+1)(2n+1);$$

$$\sum_{k=1}^{n} (-1)^k k^2 = (-1)^n \frac{n(n+1)}{2};$$

$$\sum_{k=0}^{n} (2k+1)^2 = \frac{1}{3}(n+1)(2n+1)(2n+3);$$

$$\sum_{k=0}^{n} (-1)^k (2k+1)^2 = 2(-1)^n (n+1)^2 - \frac{1}{2}\left[1 + (-1)^n\right];$$

$$\sum_{k=1}^{n} (k+a)(k+b) = \frac{1}{6} n(n+1)(2n+1+3a+3b) + nab.$$

▶ *A large number of formulas for the summation of various finite series can be found in Section T1.1.*

## 1.6. Mean Values and Inequalities of General Form

### 1.6.1. Arithmetic Mean, Geometric Mean, and Other Mean Values. Inequalities for Mean Values

**1.6.1-1. Arithmetic mean, geometric mean, and other mean values.**

The *arithmetic mean* of a set of $n$ real numbers $a_1, a_2, \ldots, a_n$ is defined as

$$m_\text{a} = \frac{a_1 + a_2 + \cdots + a_n}{n}. \tag{1.6.1.1}$$

*Geometric mean* of $n$ positive numbers $a_1, a_2, \ldots, a_n$:

$$m_\text{g} = (a_1 a_2 \ldots a_n)^{1/n}. \tag{1.6.1.2}$$

*Harmonic mean* of $n$ real numbers $a_1, a_2, \ldots, a_n$:

$$m_\text{h} = \frac{n}{(1/a_1) + (1/a_2) + \cdots + (1/a_n)}, \qquad a_k \neq 0. \tag{1.6.1.3}$$

*Quadratic mean* (or *root mean square*) of $n$ real numbers $a_1, a_2, \ldots, a_n$:

$$m_\text{q} = \sqrt{\frac{a_1^2 + a_2^2 + \cdots + a_n^2}{n}}. \tag{1.6.1.4}$$

**1.6.1-2. Basic inequalities for mean values.**

Given $n$ positive numbers $a_1, a_2, \ldots, a_n$, the following inequalities hold true:

$$m_\text{h} \leq m_\text{g} \leq m_\text{a} \leq m_\text{q}, \tag{1.6.1.5}$$

where the mean values are defined above by (1.6.1.1)–(1.6.1.4). The equalities in (1.6.1.5) are attained only if $a_1 = a_2 = \cdots = a_n$.

To make it easier to remember, let us rewrite inequalities (1.6.1.5) in words as

$$\boxed{\text{harmonic mean}} \leq \boxed{\text{geometric mean}} \leq \boxed{\text{arithmetic mean}} \leq \boxed{\text{quadratic mean}}.$$

**1.6.1-3. General approach to defining mean values.**

Let $f(x)$ be a continuous monotonic function defined on the interval $0 \leq x < \infty$.

The *functional mean* with respect to the function $f(x)$ for $n$ positive real numbers $a_1, a_2, \ldots, a_n$ is introduced as follows:

$$m_f = f^{-1}\left(\frac{f(a_1) + f(a_2) + \cdots + f(a_n)}{n}\right), \tag{1.6.1.6}$$

where $f^{-1}(y)$ is the inverse of $f(x)$.

The mean values defined by (1.6.1.1)–(1.6.1.4) in Paragraph 1.6.1-1 are all special cases of the functional mean (1.6.1.6), provided the real numbers $a_1, a_2, \ldots, a_n$ are all positive. Specifically,

the arithmetic mean is the functional mean with respect to $f(x) = x$,
the geometric mean is the functional mean with respect to $f(x) = \ln x$,
the harmonic mean is the functional mean with respect to $f(x) = 1/x$,
the quadratic mean is the functional mean with respect to $f(x) = x^2$.

### 1.6.2. Inequalities of General Form

**1.6.2-1. Triangle inequality, Cauchy inequality, and related inequalities.**

Let $a_k$ and $b_k$ be real numbers with $k = 1, 2, \ldots, n$.
*Generalized triangle inequality*:

$$\left|\sum_{k=1}^{n} a_k\right| \le \sum_{k=1}^{n} |a_k|.$$

*Cauchy's inequality* (also known as the *Cauchy–Bunyakovski inequality* or *Cauchy–Schwarz–Bunyakovski inequality*):

$$\left(\sum_{k=1}^{n} a_k b_k\right)^2 \le \left(\sum_{k=1}^{n} a_k^2\right)\left(\sum_{k=1}^{n} b_k^2\right).$$

*Minkowski's inequality*:

$$\left(\sum_{k=1}^{n} |a_k + b_k|^p\right)^{\frac{1}{p}} \le \left(\sum_{k=1}^{n} |a_k|^p\right)^{\frac{1}{p}} + \left(\sum_{k=1}^{n} |b_k|^p\right)^{\frac{1}{p}}, \qquad p \ge 1.$$

*Hölder's inequality* (reduces to Cauchy's inequality at $p = 2$):

$$\left|\sum_{k=1}^{n} a_k b_k\right| \le \left(\sum_{k=1}^{n} |a_k|^p\right)^{\frac{1}{p}} \left(\sum_{k=1}^{n} |b_k|^{\frac{p}{p-1}}\right)^{\frac{p-1}{p}}, \qquad p > 1.$$

**1.6.2-2. Chebyshev's inequalities.**

*Chebyshev's inequalities*:

$$\left(\sum_{k=1}^{n} a_k\right)\left(\sum_{k=1}^{n} b_k\right) \le n\left(\sum_{k=1}^{n} a_k b_k\right) \quad \text{if } 0 < a_1 \le a_2 \le \cdots < a_n,\ 0 < b_1 \le b_2 \le \cdots < b_n;$$

$$\left(\sum_{k=1}^{n} a_k\right)\left(\sum_{k=1}^{n} b_k\right) \ge n\left(\sum_{k=1}^{n} a_k b_k\right) \quad \text{if } 0 < a_1 \le a_2 \le \cdots < a_n,\ b_1 \ge b_2 \ge \cdots \ge b_n > 0;$$

*Generalized Chebyshev inequalities*:

$$\left(\frac{1}{n}\sum_{k=1}^{n}a_k^p\right)^{1/p}\left(\frac{1}{n}\sum_{k=1}^{n}b_k^p\right)^{1/p} \leq \left(\frac{1}{n}\sum_{k=1}^{n}a_k^p b_k^p\right)^{1/p}$$

if $0 < a_1 \leq a_2 \leq \cdots < a_n$, $0 < b_1 \leq b_2 \leq \cdots < b_n$;

$$\left(\frac{1}{n}\sum_{k=1}^{n}a_k^p\right)^{1/p}\left(\frac{1}{n}\sum_{k=1}^{n}b_k^p\right)^{1/p} \geq \left(\frac{1}{n}\sum_{k=1}^{n}a_k^p b_k^p\right)^{1/p}$$

if $0 < a_1 \leq a_2 \leq \cdots < a_n$, $b_1 \geq b_2 \geq \cdots \geq b_n > 0$.

### 1.6.2-3. Generalizations of inequalities for means.

1°. The following inequality holds:

$$\left(a_1^{p_1} a_2^{p_2} \ldots a_n^{p_n}\right)^{\frac{1}{p_1+p_2+\cdots+p_n}} \leq \frac{a_1 p_1 + a_2 p_2 + \cdots + a_n p_n}{p_1 + p_2 + \cdots + p_n},$$

where the $a_k$ and $p_k$ are all positive. In the special case $p_1 = p_2 = \cdots = p_n = 1$, we have the well-known inequality stating that the geometric mean of a series of positive numbers does not exceed their arithmetic mean (see Paragraph 1.6.1-2).

2°. The following inequality holds:

$$\frac{p_1 + p_2 + \cdots + p_n}{(p_1/a_1) + (p_2/a_2) + \cdots + (p_n/a_n)} \leq \left(a_1^{p_1} a_2^{p_2} \ldots a_n^{p_n}\right)^{\frac{1}{p_1+p_2+\cdots+p_n}},$$

where the $a_k$ and $p_k$ are all positive. In the special case $p_1 = p_2 = \cdots = p_n = 1$, we have the well-known inequality stating that the harmonic mean of a series of positive numbers does not exceed their geometric mean (see Paragraph 1.6.1-2).

### 1.6.2-4. Jensen's inequality.

If $f(x)$ is a convex function (in particular, with $f'' > 0$), then the following *Hölder–Jensen inequality* holds:

$$f\left(\frac{\sum p_k x_k}{\sum p_k}\right) \leq \frac{\sum p_k f(x_k)}{\sum p_k}, \tag{1.6.2.1}$$

where the $x_k$ are any numbers and the $p_k$ are any positive numbers; the summation is performed over all $k$ (the limits are omitted for simplicity). The equality is attained if and only if either $x_1 = x_2 = \cdots = x_n$ or $f(x)$ is a linear function. If $f(x)$ is a concave function ($f'' < 0$), inequality (1.6.2.1) is the other way around.

The Hölder–Jensen inequality is often used to obtain various inequalities; in particular, the previous two inequalities as well as the Hölder inequality follow from it.

## 1.7. Some Mathematical Methods

### 1.7.1. Proof by Contradiction

*Proof by contradiction* (also known as *reductio ad absurdum*) is an indirect method of mathematical proof. It is based on the following reasoning:

1. Suppose one has to prove some statement $S$.
2. One assumes that the opposite of $S$ is true.
3. Based on known axioms, definitions, theorems, formulas, and the assumption of Item 2, one arrives at a contradiction (deduces some obviously false statement).
4. One concludes that the assumption of Item 2 is false and hence the original statement $S$ is true, which was to be proved.

**Example.** (Euclid's proof of the irrationality of the square root of 2 by contradiction.)
1. It is required to prove that $\sqrt{2}$ is an irrational number, that is, a real number that cannot be represented as a fraction $p/q$, where $p$ and $q$ are both integers.
2. Assume the opposite: $\sqrt{2}$ is a rational number. This means that $\sqrt{2}$ can be represented as a fraction
$$\sqrt{2} = p/q. \tag{1.7.1.1}$$
Without loss of generality the fraction $p/q$ is assumed to be irreducible, implying that $p$ and $q$ are mutually prime (have no common factor other than 1).
3. Square both sides of (1.7.1.1) and then multiply by $q^2$ to obtain
$$2q^2 = p^2. \tag{1.7.1.2}$$
The left-hand side is divisible by 2. Then the right-hand side, $p^2$, and hence $p$ is also divisible by 2. Consequently, $p$ is an even number so that
$$p = 2n, \tag{1.7.1.3}$$
where $n$ is an integer. Substituting (1.7.1.3) into (1.7.1.2) and then dividing by 2 yields
$$q^2 = 2p^2. \tag{1.7.1.4}$$
Now it can be concluded, just as above, that $q^2$ and hence $q$ must be divisible by 2. Consequently, $q$ is an even number so that
$$q = 2m, \tag{1.7.1.5}$$
where $m$ is an integer.

It is now apparent from (1.7.1.3) and (1.7.1.5) that the fraction $p/q$ is not simple, since $p$ and $q$ have a common factor 2. This contradicts the assumption made in Item 2.
4. It follows from the results of Item 3 that the representation of $\sqrt{2}$ in the form of a fraction (1.7.1.1) is false, which means that $\sqrt{2}$ is irrational.

## 1.7.2. Mathematical Induction

The method of proof by (complete) *mathematical induction* is based on the following reasoning:

1. Let $A(n)$ be a statement dependent on $n$ with $n = 1, 2, \ldots$ ($A$ is a hypothesis at this stage).
2. Base case. Suppose the initial statement $A(1)$ is true. This is usually established by direct substitution $n = 1$.
3. Induction step. Assume that $A(n)$ is true for any $n$ and then, based on this assumption, prove that $A(n+1)$ is also true.
4. Principle of mathematical induction. From the results of Items 2–3 it is concluded that the statement $A(n)$ is true for any $n$.

**Example.**
1. Prove the formula for the sum of odd numbers,
$$1 + 3 + 5 + \cdots + (2n-1) = n^2, \tag{1.7.2.1}$$
for any natural $n$.
2. For $n = 1$, we have an obvious identity: $1 = 1$.
3. Let us assume that formula (1.7.2.1) holds for any $n$. To consider the case of $n+1$, let us add the next term, $(2n+1)$, to both sides of (1.7.2.1) to obtain
$$1 + 3 + 5 + \cdots + (2n-1) + (2n+1) = n^2 + (2n+1) = (n+1)^2.$$
Thus, from the assumption of the validity of formula (1.7.2.1) for any $n$ it follows that (1.7.2.1) is also valid for $n+1$.
4. According to the principle of mathematical induction, this proves formula (1.7.2.1).

**Remark.** The first step, the formulation of an original hypothesis, is the most difficult part of the method of mathematical induction. This step is often omitted from the method.

### 1.7.3. Proof by Counterexample

A *counterexample* is an example which is used to prove that a statement (proposition) is false. Counterexamples play an important role in mathematics. Whereas a complicated proof may be the only way to demonstrate the validity of a particular theorem, a single counter example is all that is need to refute the validity of a proposed theorem.

In general, the scheme of a proof by counterexample is as follows:

1. Given a proposition: all elements $a$ that belong to a set $A$ also belong to a set (possess a property) $B$.

2. Refutation of the proposition: one specifies an element $a_*$ (counterexample) that belongs to $A$ but does not belong to $B$.

**Example.** *Proposition:* Numbers in the form $2^{2^n} + 1$, where $n$ is a positive integer, were once thought to be prime.

These numbers are prime for $n = 1, 2, 3, 4$. But for $n = 5$, we have a counterexample, since

$$2^{2^5} + 1 = 4294967297 = 641 \times 6700417;$$

it is a composite number.

*Conclusion*: When faced with a number in the form $2^{2^n} + 1$, we are not allowed to assume it is either prime or composite, unless we know for sure for some other reason.

### 1.7.4. Method of Undetermined Coefficients

The *method of undetermined coefficients* is employed to find coefficients in expressions (such as formulas, series expansions, solutions to mathematical equations), the form of which is either known in advance or assigned based on intuitive judgment.

**Example.** The fractional function

$$\frac{x+2}{x(x^2-1)}, \qquad (1.7.4.1)$$

whose denominator can be rewritten in the factored form $x(x+1)(x-1)$, can be represented as the sum of partial fractions

$$\frac{A}{x} + \frac{B}{x+1} + \frac{C}{x-1}, \qquad (1.7.4.2)$$

where $A, B, C$ are (undetermined) coefficients whose values are to be found. Equating (1.7.4.1) with (1.7.4.2), multiplying by $x(x^2-1)$, and rearranging, one obtains

$$(A+B+C)x^2 + (-B+C-1)x - A - 2 = 0.$$

For this equation to be valid for any $x$, the coefficients of the different powers of $x$ must be set equal to zero. This results in the system of linear algebraic equations

$$A+B+C = 0, \quad C-B-1 = 0, \quad -A-2 = 0.$$

On solving this system, one determines the coefficients in (1.7.4.2):

$$A = -2, \quad B = \tfrac{1}{2}, \quad C = \tfrac{3}{2}.$$

# References for Chapter 1

**Bronshtein, I. N. and Semendyayev, K. A.**, *Handbook of Mathematics, 4th Edition*, Springer-Verlag, Berlin, 2004.

**Courant, R. and Robbins, H.**, *What Is Mathematics?: An Elementary Approach to Ideas and Methods, 2nd Edition*, Oxford University Press, Oxford, 1996.

**Franklin, J. and Daoud, A.**, *Introduction to Proofs in Mathematics*, Prentice Hall, New York, 1988.

**Garnier, R. and Taylor, J.**, *100% Mathematical Proof*, John Wiley & Sons, New York, 1996.

**Gelbaum, B. R. and Olmsted, J. M. H.**, *Theorems and Counterexamples in Mathematics*, Springer-Verlag, New York, 1990.

**Jordan, B. E. and Palow, W. P.**, *Integrated Arithmetic & Algebra*, Addison-Wesley, Boston, 1999.

**Krantz, S. G.**, *Dictionary of Algebra, Arithmetic, and Trigonometry*, CRC Press, Boca Raton, 2001.

**Pólya, G.**, *Mathematics and Plausible Reasoning, Vol. 1: Induction and Analogy in Mathematics*, Princeton University Press, Princeton, 1990.

**Rossi, R. J.**, *Theorems, Corollaries, Lemmas, and Methods of Proof*, Wiley-Interscience, Hoboken, N.J., 2006.

**Thompson, J. E.**, *Arithmetic for the Practical Man*, Van Nostrand Reinhold, New York, 1973.

**Weisstein, E. W.**, *CRC Concise Encyclopedia of Mathematics, 2nd Edition*, CRC Press, Boca Raton, 2003.

**Zwillinger, D.**, *CRC Standard Mathematical Tables and Formulae, 31st Edition*, CRC Press, Boca Raton, 2002.

# Chapter 2
# Elementary Functions

*Basic elementary functions*: power, exponential, logarithmic, trigonometric, and inverse trigonometric (arc-trigonometric or antitrigonometric) functions. All other elementary functions are obtained from the basic elementary functions and constants by means of the four arithmetic operations (addition, subtraction, multiplication, and division) and the operation of composition (composite functions).

The graphs and the main properties of the basic as well as some other frequently occurring elementary functions of the real variable are described below.

## 2.1. Power, Exponential, and Logarithmic Functions

### 2.1.1. Power Function: $y = x^\alpha$ ($\alpha$ is an Arbitrary Real Number)

2.1.1-1. Graphs of the power function.

General properties of the graphs: the point $(1, 1)$ belongs to all the graphs, and $y > 0$ for $x > 0$. For $\alpha > 0$, the graphs pass through the origin $(0, 0)$; for $\alpha < 0$, the graphs have the vertical asymptote $x = 0$ ($y \to +\infty$ as $x \to 0$, $x > 0$). For $\alpha = 0$, the graph is a straight line parallel to the $x$-axis.

Consider more closely the following cases.

*Case 1*: $y = x^{2n}$, where $n$ is a positive integer ($n = 1, 2, \ldots$). This function is defined for all real $x$ and its range consists of all $y \geq 0$. This function is even, nonperiodic, and unbounded. It crosses the axis $Oy$ and is tangential to the axis $Ox$ at the origin $x = 0$, $y = 0$. On the interval $(-\infty, 0)$ this function decreases, and it increases on the interval $(0, +\infty)$. It attains its minimum $y = 0$ at $x = 0$. The graph of the function $y = x^2$ (parabola) is given in Fig. 2.1 $a$.

*Case 2*: $y = x^{2n+1}$, where $n$ is a positive integer. This function is defined on the entire $x$-axis and its range coincides with the $y$-axis. This function is odd, nonperiodic, and unbounded. It crosses the $x$-axis and the $y$-axis at the origin $x = 0$, $y = 0$. It is an increasing function on the entire real axis with no points of extremum, the origin being its inflection point. The graph of the function $y = x^3$ (cubic parabola) is shown in Fig. 2.1 $a$.

*Case 3*: $y = x^{-2n}$, where $n$ is a positive integer. This function is defined for all $x \neq 0$, and its range is the semiaxis $y > 0$. It is an even, nonperiodic, unbounded function having no intersection with the coordinate axes. It increases on the interval $(-\infty, 0)$, decreases on the interval $(0, +\infty)$, and has no points of extremum. This function has a vertical asymptote $x = 0$. The graph of the function $y = x^{-2}$ is given in Fig. 2.1 $b$.

*Case 4*: $y = x^{-2n+1}$, where $n$ is a positive integer. This function is defined for all $x \neq 0$, and its range is the entire $y$-axis. It is an odd, nonperiodic, unbounded function with no intersections with the coordinate axes. This is a decreasing function on the entire real axis with no points of extremum. It has a vertical asymptote $x = 0$. The graph of the function $y = x^{-1}$ is given in Fig. 2.1 $b$.

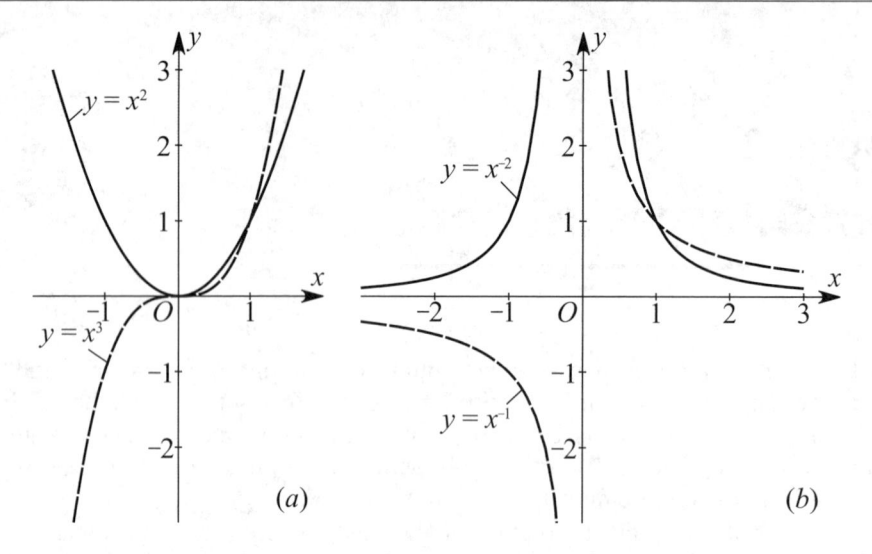

**Figure 2.1.** Graphs of the power function $y = x^n$, where $n$ is an integer.

*Case 5*: $y = x^\alpha$ with a noninteger $\alpha > 0$. This function is defined for all* $x \geq 0$ and its range is the semiaxis $y \geq 0$. This function is neither odd nor even and it is nonperiodic and unbounded. It crosses the axes $Ox$ and $Oy$ at the origin $x = 0$, $y = 0$ and increases everywhere in its domain, taking its smallest value at the limit point $x = 0$, $y = 0$. The graph of the function $y = x^{1/2}$ is given in Fig. 2.2.

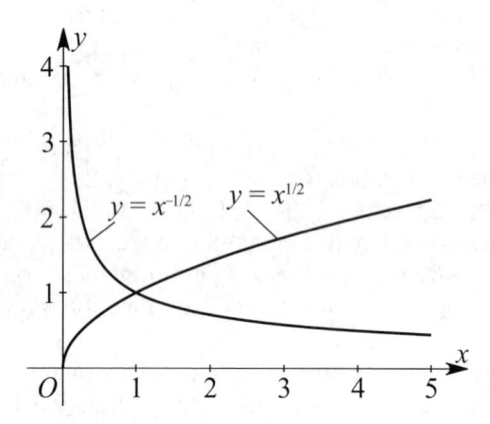

**Figure 2.2.** Graphs of the power function $y = x^\alpha$, where $\alpha$ is a noninteger.

*Case 6*: $y = x^\alpha$ with a noninteger $\alpha < 0$. This function is defined for all $x \geq 0$ and its range is the semiaxis $y \geq 0$. This function is neither odd nor even, it is nonperiodic and unbounded, and it has no intersections with the coordinate axes, which coincide with its horizontal and vertical asymptotes. This function is decreasing on its entire domain and has no points of extremum. The graph of the function $y = x^{-1/2}$ is given in Fig. 2.2.

---

* In fact, the power function $y = x^{1/n}$ with an odd integer $n$ is defined and negative for all $x < 0$. Here, however, it is always assumed that $x \geq 0$. A similar assumption is made with regard to the functions of the form $y = x^{m/n}$, where $m$ is a positive integer and $m/n$ is an irreducible fraction.

### 2.1.1-2. Properties of the power function.

Basic properties of the power function:

$$x^\alpha x^\beta = x^{\alpha+\beta}, \quad (x_1 x_2)^\alpha = x_1^\alpha x_2^\alpha, \quad (x^\alpha)^\beta = x^{\alpha\beta},$$

for any $\alpha$ and $\beta$, where $x > 0$, $x_1 > 0$, $x_2 > 0$.

Differentiation and integration formulas:

$$(x^\alpha)' = \alpha x^{\alpha-1}, \quad \int x^\alpha \, dx = \begin{cases} \dfrac{x^{\alpha+1}}{\alpha+1} + C & \text{if } \alpha \neq -1, \\ \ln|x| + C & \text{if } \alpha = -1. \end{cases}$$

The Taylor series expansion in a neighborhood of an arbitrary point:

$$x^\alpha = \sum_{n=0}^{\infty} C_\alpha^n x_0^{\alpha-n}(x - x_0)^n \quad \text{for} \quad |x - x_0| < |x_0|,$$

where $C_\alpha^n = \dfrac{\alpha(\alpha-1)\dots(\alpha-n+1)}{n!}$ are binomial coefficients.

## 2.1.2. Exponential Function: $y = a^x$ ($a > 0$, $a \neq 1$)

### 2.1.2-1. Graphs of the exponential function.

This function is defined for all $x$ and its range is the semiaxis $y > 0$. This function is neither odd nor even, it is nonperiodic and unbounded, and it crosses the axis $Oy$ at $y = 1$ and does not cross the axis $Ox$. For $a > 1$, it is an increasing function on the entire real axis; for $0 < a < 1$, it is a decreasing function. This function has no extremal points; the axis $Ox$ is its horizontal asymptote. The graphs of these functions have the following common property: they pass through the point $(0, 1)$. The graph of $y = a^x$ is symmetrical to the graph of $y = (1/a)^x$ with respect to the $y$-axis. For $a > 1$, the function $a^x$ grows faster than any power of $x$ as $x \to +\infty$, and it decays faster than any power of $1/x$ as $x \to -\infty$. The graphs of the functions $y = 2^x$ and $y = (1/2)^x$ are given in Fig. 2.3.

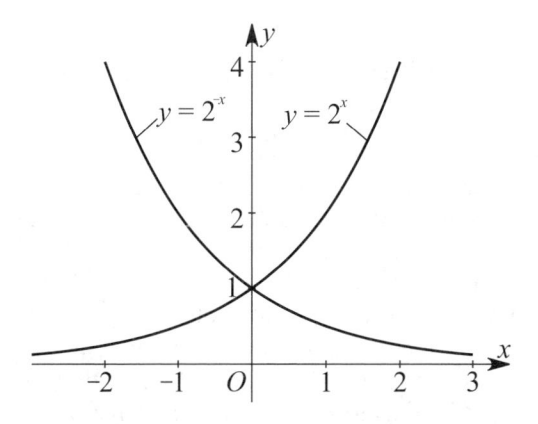

**Figure 2.3.** Graphs of the exponential function.

### 2.1.2-2. Properties of the exponential function.

Basic properties of the exponential function:

$$a^{x_1} a^{x_2} = a^{x_1+x_2}, \qquad a^x b^x = (ab)^x, \qquad (a^{x_1})^{x_2} = a^{x_1 x_2}.$$

Number $e$, *base of natural (Napierian) logarithms*, and the function $e^x$:

$$e = \lim_{n \to \infty} \left(1 + \frac{1}{n}\right)^n = 2.718281\ldots, \qquad e^x = \lim_{n \to \infty} \left(1 + \frac{x}{n}\right)^n.$$

The formula for passing from an arbitrary base $a$ to the base $e$ of natural logarithms:

$$a^x = e^{x \ln a}.$$

The inequality

$$a^{x_1} > a^{x_2} \iff \begin{cases} x_1 > x_2 & \text{if } a > 1, \\ x_1 < x_2 & \text{if } 0 < a < 1. \end{cases}$$

The limit relations for any $a > 1$ and $b > 0$:

$$\lim_{x \to +\infty} \frac{a^x}{|x|^b} = \infty, \qquad \lim_{x \to -\infty} a^x |x|^b = 0.$$

Differentiation and integration formulas:

$$(e^x)' = e^x, \qquad \int e^x \, dx = e^x + C;$$

$$(a^x)' = a^x \ln a, \qquad \int a^x \, dx = \frac{a^x}{\ln a} + C.$$

The expansion in power series:

$$e^x = 1 + \frac{x}{1!} + \frac{x^2}{2!} + \frac{x^3}{3!} + \cdots + \frac{x^n}{n!} + \cdots = \sum_{k=0}^{\infty} \frac{x^k}{k!}.$$

## 2.1.3. Logarithmic Function: $y = \log_a x$ $(a > 0, a \neq 1)$

### 2.1.3-1. Graphs of the logarithmic function.

This function is defined for all $x > 0$ and its range is the entire $y$-axis. The function is neither odd nor even; it is nonperiodic and unbounded; it crosses the axis $Ox$ at $x = 1$ and does not cross the axis $Oy$. For $a > 1$, this function is increasing, and for $0 < a < 1$, it is a decreasing function; it has no extremal points, and the axis $Oy$ is its vertical asymptote. The common property of the graphs of such functions is that they all pass through the point $(1, 0)$. The graph of the function $y = \log_a x$ is symmetric to that of $y = \log_{1/a} x$ with respect to the $x$-axis. The modulus of the logarithmic function tends to infinity slower than any power of $x$ as $x \to +\infty$; and it tends to zero slower than any power of $1/x$ as $x \to +0$. The graphs of the functions $y = \log_2 x$ and $y = \log_{1/2} x$ are shown in Fig. 2.4.

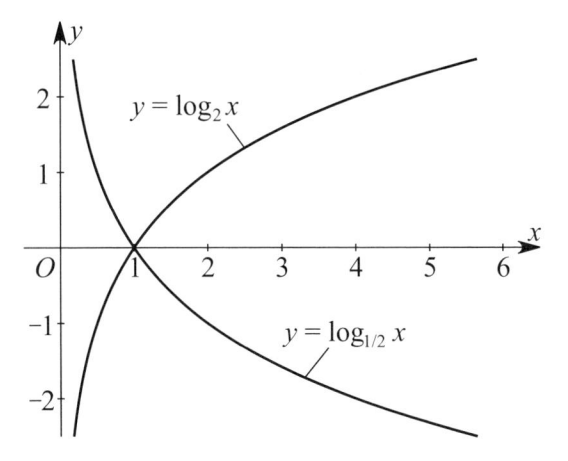

**Figure 2.4.** Graphs of the logarithmic function.

2.1.3-2. **Properties of the logarithmic function.**

By definition, the logarithmic function is the inverse of the exponential function. The following equivalence relation holds:

$$y = \log_a x \quad \Longleftrightarrow \quad x = a^y,$$

where $a > 0$, $a \neq 1$.

Basic properties of the logarithmic function:

$$a^{\log_a x} = x, \qquad \log_a(x_1 x_2) = \log_a x_1 + \log_a x_2,$$

$$\log_a(x^k) = k \log_a x, \quad \log_a x = \frac{\log_b x}{\log_b a},$$

where $x > 0$, $x_1 > 0$, $x_2 > 0$, $a > 0$, $a \neq 1$, $b > 0$, $b \neq 1$.

The simplest inequality:

$$\log_a x_1 > \log_a x_2 \quad \Longleftrightarrow \quad \begin{cases} x_1 > x_2 & \text{if } a > 1, \\ x_1 < x_2 & \text{if } 0 < a < 1. \end{cases}$$

For any $b > 0$, the following limit relations hold:

$$\lim_{x \to +\infty} \frac{\log_a x}{x^b} = 0, \quad \lim_{x \to +0} x^b \log_a x = 0.$$

The logarithmic function with the base $e$ (*base of natural logarithms, Napierian base*) is denoted by

$$\log_e x = \ln x,$$

where $e = \lim_{n \to \infty} \left(1 + \frac{1}{n}\right)^n = 2.718281\ldots$

Formulas for passing from an arbitrary base $a$ to the Napierian base $e$:

$$\log_a x = \frac{\ln x}{\ln a}.$$

Differentiation and integration formulas:

$$(\ln x)' = \frac{1}{x}, \quad \int \ln x \, dx = x \ln x - x + C.$$

Expansion in power series:

$$\ln(1+x) = x - \frac{x^2}{2} + \frac{x^3}{3} - \cdots + (-1)^{n-1} \frac{x^n}{n} + \cdots = \sum_{k=1}^{\infty} (-1)^{k-1} \frac{x^k}{k}, \quad -1 < x < 1.$$

## 2.2. Trigonometric Functions

### 2.2.1. Trigonometric Circle. Definition of Trigonometric Functions

**2.2.1-1. Trigonometric circle. Degrees and radians.**

*Trigonometric circle* is the circle of unit radius with center at the origin of an orthogonal coordinate system $Oxy$. The coordinate axes divide the circle into four quarters (*quadrants*); see Fig. 2.5. Consider rotation of the polar radius issuing from the origin $O$ and ending at a point $M$ of the trigonometric circle. Let $\alpha$ be the angle between the $x$-axis and the polar radius $OM$ measured from the positive direction of the $x$-axis. This angle is assumed positive in the case of counterclockwise rotation and negative in the case of clockwise rotation.

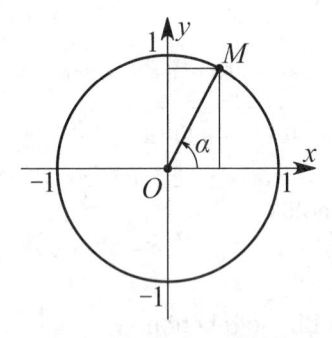

**Figure 2.5.** Trigonometric circle.

Angles are measured either in radians or in degrees. One radian is the angle at the vertex of the sector of the trigonometric circle supported by its arc of unit length. One degree is the angle at the vertex of the sector of the trigonometric circle supported by its arc of length $\pi/180$. The radians are related to the degrees by the formulas

$$1 \text{ radian} = \frac{180°}{\pi}; \quad 1° = \frac{\pi}{180}.$$

**2.2.1-2. Definition of trigonometric functions.**

The *sine* of $\alpha$ is the ordinate (the projection to the axis $Oy$) of the point on the trigonometric circle corresponding to the angle of $\alpha$ radians. The *cosine* of $\alpha$ is the abscissa (projection to the axis $Ox$) of that point (see Fig. 2.5). The sine and the cosine are basic trigonometric functions and are denoted, respectively, by $\sin \alpha$ and $\cos \alpha$.

Other trigonometric functions are *tangent*, *cotangent*, *secant*, and *cosecant*. These are derived from the basic trigonometric functions, sine and cosine, as follows:

$$\tan \alpha = \frac{\sin \alpha}{\cos \alpha}, \quad \cot \alpha = \frac{\cos \alpha}{\sin \alpha}, \quad \sec \alpha = \frac{1}{\cos \alpha}, \quad \operatorname{cosec} \alpha = \frac{1}{\sin \alpha}.$$

Table 2.1 gives the signs of the trigonometric functions in different quadrants. The signs and the values of $\sin \alpha$ and $\cos \alpha$ do not change if the argument $\alpha$ is incremented by $\pm 2\pi n$, where $n = 1, 2, \ldots$ The signs and the values of $\tan \alpha$ and $\cot \alpha$ do not change if the argument $\alpha$ is incremented by $\pm \pi n$, where $n = 1, 2, \ldots$

Table 2.2 gives the values of trigonometric functions for some values of their argument (the symbol $\infty$ means that the function is undefined for the corresponding value of its argument).

**TABLE 2.1**
Signs of trigonometric functions in different quarters

| Quarter | Angle in radians | $\sin \alpha$ | $\cos \alpha$ | $\tan \alpha$ | $\cot \alpha$ | $\sec \alpha$ | $\csc \alpha$ |
|---------|------------------|---------------|---------------|---------------|---------------|---------------|---------------|
| I | $0 < \alpha < \frac{\pi}{2}$ | + | + | + | + | + | + |
| II | $\frac{\pi}{2} < \alpha < \pi$ | + | − | − | − | − | + |
| III | $\pi < \alpha < \frac{3\pi}{2}$ | − | − | + | + | − | − |
| IV | $\frac{3\pi}{2} < \alpha < 2\pi$ | − | + | − | − | + | − |

**TABLE 2.2**
Numerical values of trigonometric functions for some angles $\alpha$ (in radians)

| Angle $\alpha$ | 0 | $\frac{\pi}{6}$ | $\frac{\pi}{4}$ | $\frac{\pi}{3}$ | $\frac{\pi}{2}$ | $\frac{2\pi}{3}$ | $\frac{3\pi}{4}$ | $\frac{5\pi}{6}$ | $\pi$ |
|----------------|---|-----------------|-----------------|-----------------|-----------------|------------------|------------------|------------------|-------|
| $\sin \alpha$ | 0 | $\frac{1}{2}$ | $\frac{\sqrt{2}}{2}$ | $\frac{\sqrt{3}}{2}$ | 1 | $\frac{\sqrt{3}}{2}$ | $\frac{\sqrt{2}}{2}$ | $\frac{1}{2}$ | 0 |
| $\cos \alpha$ | 1 | $\frac{\sqrt{3}}{2}$ | $\frac{\sqrt{2}}{2}$ | $\frac{1}{2}$ | 0 | $-\frac{1}{2}$ | $-\frac{\sqrt{2}}{2}$ | $-\frac{\sqrt{3}}{2}$ | −1 |
| $\tan \alpha$ | 0 | $\frac{\sqrt{3}}{3}$ | 1 | $\sqrt{3}$ | $\infty$ | $-\sqrt{3}$ | −1 | $-\frac{\sqrt{3}}{3}$ | 0 |
| $\cot \alpha$ | $\infty$ | $\sqrt{3}$ | 1 | $\frac{\sqrt{3}}{3}$ | 0 | $-\frac{\sqrt{3}}{3}$ | −1 | $-\sqrt{3}$ | $\infty$ |

## 2.2.2. Graphs of Trigonometric Functions

### 2.2.2-1. Sine: $y = \sin x$.

This function is defined for all $x$ and its range is $y \in [-1, 1]$. The sine is an odd, bounded, periodic function (with period $2\pi$). It crosses the axis $Oy$ at the point $y = 0$ and crosses the axis $Ox$ at the points $x = \pi n$, $n = 0, \pm 1, \pm 2, \ldots$ The sine is an increasing function on every segment $[-\frac{\pi}{2} + 2\pi n, \frac{\pi}{2} + 2\pi n]$ and is a decreasing function on every segment $[\frac{\pi}{2} + 2\pi n, \frac{3}{2}\pi + 2\pi n]$. For $x = \frac{\pi}{2} + 2\pi n$, it attains its maximal value ($y = 1$), and for $x = -\frac{\pi}{2} + 2\pi n$ it attains its minimal value ($y = -1$). The graph of the function $y = \sin x$ is called the *sinusoid* or *sine curve* and is shown in Fig. 2.6.

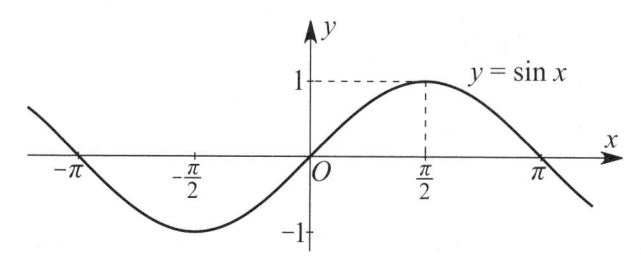

**Figure 2.6.** The graph of the function $y = \sin x$.

### 2.2.2-2. Cosine: $y = \cos x$.

This function is defined for all $x$ and its range is $y \in [-1, 1]$. The cosine is a bounded, even, periodic function (with period $2\pi$). It crosses the axis $Oy$ at the point $y = 1$, and crosses the axis $Ox$ at the points $x = \frac{\pi}{2} + \pi n$. The cosine is an increasing function on every segment $[-\pi + 2\pi n, 2\pi n]$ and is a decreasing function on every segment $[2\pi n, \pi + 2\pi n]$, $n = 0, \pm 1, \pm 2, \ldots$ For $x = 2\pi n$ it attains its maximal value ($y = 1$), and for $x = \pi + 2\pi n$ it attains its minimal value ($y = -1$). The graph of the function $y = \cos x$ is a sinusoid obtained by shifting the graph of the function $y = \sin x$ by $\frac{\pi}{2}$ to the left along the axis $Ox$ (see Fig. 2.7).

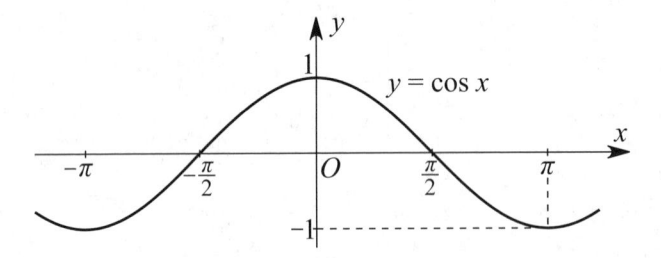

**Figure 2.7.** The graph of the function $y = \cos x$.

**2.2.2-3. Tangent:** $y = \tan x$.

This function is defined for all $x \neq \frac{\pi}{2} + \pi n$, $n = 0, \pm 1, \pm 2, \ldots$, and its range is the entire $y$-axis. The tangent is an unbounded, odd, periodic function (with period $\pi$). It crosses the axis $Oy$ at the point $y = 0$ and crosses the axis $Ox$ at the points $x = \pi n$. This is an increasing function on every interval $(-\frac{\pi}{2} + \pi n, \frac{\pi}{2} + \pi n)$. This function has no points of extremum and has vertical asymptotes at $x = \frac{\pi}{2} + \pi n$, $n = 0, \pm 1, \pm 2, \ldots$ The graph of the function $y = \tan x$ is given in Fig. 2.8.

**2.2.2-4. Cotangent:** $y = \cot x$.

This function is defined for all $x \neq \pi n$, $n = 0, \pm 1, \pm 2, \ldots$, and its range is the entire $y$-axis. The cotangent is an unbounded, odd, periodic function (with period $\pi$). It crosses the axis $Ox$ at the points $x = \frac{\pi}{2} + \pi n$, and does not cross the axis $Oy$. This is a decreasing function on every interval $(\pi n, \pi + \pi n)$. This function has no extremal points and has vertical asymptotes at $x = \pi n$, $n = 0, \pm 1, \pm 2, \ldots$ The graph of the function $y = \cot x$ is given in Fig. 2.9.

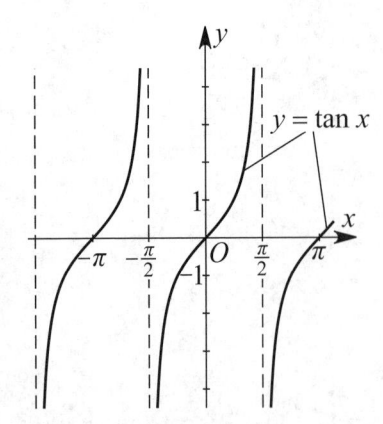

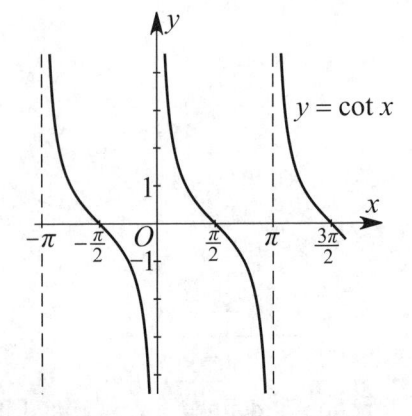

**Figure 2.8.** The graph of the function $y = \tan x$.

**Figure 2.9.** The graph of the function $y = \cot x$.

## 2.2.3. Properties of Trigonometric Functions

**2.2.3-1. Simplest relations.**

$$\sin^2 x + \cos^2 x = 1, \qquad \tan x \cot x = 1,$$
$$\sin(-x) = -\sin x, \qquad \cos(-x) = \cos x,$$
$$\tan x = \frac{\sin x}{\cos x}, \qquad \cot x = \frac{\cos x}{\sin x},$$
$$\tan(-x) = -\tan x, \qquad \cot(-x) = -\cot x,$$
$$1 + \tan^2 x = \frac{1}{\cos^2 x}, \qquad 1 + \cot^2 x = \frac{1}{\sin^2 x}.$$

**2.2.3-2. Reduction formulas.**

$$\sin(x \pm 2n\pi) = \sin x, \qquad \cos(x \pm 2n\pi) = \cos x,$$
$$\sin(x \pm n\pi) = (-1)^n \sin x, \qquad \cos(x \pm n\pi) = (-1)^n \cos x,$$
$$\sin\left(x \pm \frac{2n+1}{2}\pi\right) = \pm(-1)^n \cos x, \qquad \cos\left(x \pm \frac{2n+1}{2}\pi\right) = \mp(-1)^n \sin x,$$
$$\sin\left(x \pm \frac{\pi}{4}\right) = \frac{\sqrt{2}}{2}(\sin x \pm \cos x), \qquad \cos\left(x \pm \frac{\pi}{4}\right) = \frac{\sqrt{2}}{2}(\cos x \mp \sin x),$$
$$\tan(x \pm n\pi) = \tan x, \qquad \cot(x \pm n\pi) = \cot x,$$
$$\tan\left(x \pm \frac{2n+1}{2}\pi\right) = -\cot x, \qquad \cot\left(x \pm \frac{2n+1}{2}\pi\right) = -\tan x,$$
$$\tan\left(x \pm \frac{\pi}{4}\right) = \frac{\tan x \pm 1}{1 \mp \tan x}, \qquad \cot\left(x \pm \frac{\pi}{4}\right) = \frac{\cot x \mp 1}{1 \pm \cot x},$$

where $n = 1, 2, \ldots$

**2.2.3-3. Relations between trigonometric functions of single argument.**

$$\sin x = \pm\sqrt{1 - \cos^2 x} = \pm\frac{\tan x}{\sqrt{1 + \tan^2 x}} = \pm\frac{1}{\sqrt{1 + \cot^2 x}},$$
$$\cos x = \pm\sqrt{1 - \sin^2 x} = \pm\frac{1}{\sqrt{1 + \tan^2 x}} = \pm\frac{\cot x}{\sqrt{1 + \cot^2 x}},$$
$$\tan x = \pm\frac{\sin x}{\sqrt{1 - \sin^2 x}} = \pm\frac{\sqrt{1 - \cos^2 x}}{\cos x} = \frac{1}{\cot x},$$
$$\cot x = \pm\frac{\sqrt{1 - \sin^2 x}}{\sin x} = \pm\frac{\cos x}{\sqrt{1 - \cos^2 x}} = \frac{1}{\tan x}.$$

The sign before the radical is determined by the quarter in which the argument takes its values.

## 2.2.3-4. Addition and subtraction of trigonometric functions.

$$\sin x + \sin y = 2\sin\left(\frac{x+y}{2}\right)\cos\left(\frac{x-y}{2}\right),$$
$$\sin x - \sin y = 2\sin\left(\frac{x-y}{2}\right)\cos\left(\frac{x+y}{2}\right),$$
$$\cos x + \cos y = 2\cos\left(\frac{x+y}{2}\right)\cos\left(\frac{x-y}{2}\right),$$
$$\cos x - \cos y = -2\sin\left(\frac{x+y}{2}\right)\sin\left(\frac{x-y}{2}\right),$$
$$\sin^2 x - \sin^2 y = \cos^2 y - \cos^2 x = \sin(x+y)\sin(x-y),$$
$$\sin^2 x - \cos^2 y = -\cos(x+y)\cos(x-y),$$
$$\tan x \pm \tan y = \frac{\sin(x\pm y)}{\cos x \cos y}, \qquad \cot x \pm \cot y = \frac{\sin(y\pm x)}{\sin x \sin y},$$
$$a\cos x + b\sin x = r\sin(x+\varphi) = r\cos(x-\psi).$$

Here, $r = \sqrt{a^2+b^2}$, $\sin\varphi = a/r$, $\cos\varphi = b/r$, $\sin\psi = b/r$, and $\cos\psi = a/r$.

## 2.2.3-5. Products of trigonometric functions.

$$\sin x \sin y = \tfrac{1}{2}[\cos(x-y) - \cos(x+y)],$$
$$\cos x \cos y = \tfrac{1}{2}[\cos(x-y) + \cos(x+y)],$$
$$\sin x \cos y = \tfrac{1}{2}[\sin(x-y) + \sin(x+y)].$$

## 2.2.3-6. Powers of trigonometric functions.

$$\cos^2 x = \tfrac{1}{2}\cos 2x + \tfrac{1}{2}, \qquad \sin^2 x = -\tfrac{1}{2}\cos 2x + \tfrac{1}{2},$$
$$\cos^3 x = \tfrac{1}{4}\cos 3x + \tfrac{3}{4}\cos x, \qquad \sin^3 x = -\tfrac{1}{4}\sin 3x + \tfrac{3}{4}\sin x,$$
$$\cos^4 x = \tfrac{1}{8}\cos 4x + \tfrac{1}{2}\cos 2x + \tfrac{3}{8}, \qquad \sin^4 x = \tfrac{1}{8}\cos 4x - \tfrac{1}{2}\cos 2x + \tfrac{3}{8},$$
$$\cos^5 x = \tfrac{1}{16}\cos 5x + \tfrac{5}{16}\cos 3x + \tfrac{5}{8}\cos x, \qquad \sin^5 x = \tfrac{1}{16}\sin 5x - \tfrac{5}{16}\sin 3x + \tfrac{5}{8}\sin x,$$

$$\cos^{2n} x = \frac{1}{2^{2n-1}} \sum_{k=0}^{n-1} C_{2n}^k \cos[2(n-k)x] + \frac{1}{2^{2n}} C_{2n}^n,$$

$$\cos^{2n+1} x = \frac{1}{2^{2n}} \sum_{k=0}^{n} C_{2n+1}^k \cos[(2n-2k+1)x],$$

$$\sin^{2n} x = \frac{1}{2^{2n-1}} \sum_{k=0}^{n-1} (-1)^{n-k} C_{2n}^k \cos[2(n-k)x] + \frac{1}{2^{2n}} C_{2n}^n,$$

$$\sin^{2n+1} x = \frac{1}{2^{2n}} \sum_{k=0}^{n} (-1)^{n-k} C_{2n+1}^k \sin[(2n-2k+1)x].$$

Here, $n = 1, 2, \ldots$ and $C_m^k = \dfrac{m!}{k!\,(m-k)!}$ are binomial coefficients ($0! = 1$).

## 2.2.3-7. Addition formulas.

$$\sin(x \pm y) = \sin x \cos y \pm \cos x \sin y, \quad \cos(x \pm y) = \cos x \cos y \mp \sin x \sin y,$$
$$\tan(x \pm y) = \frac{\tan x \pm \tan y}{1 \mp \tan x \tan y}, \quad \cot(x \pm y) = \frac{1 \mp \tan x \tan y}{\tan x \pm \tan y}.$$

## 2.2.3-8. Trigonometric functions of multiple arguments.

$$\cos 2x = 2\cos^2 x - 1 = 1 - 2\sin^2 x, \quad \sin 2x = 2 \sin x \cos x,$$
$$\cos 3x = -3 \cos x + 4 \cos^3 x, \quad \sin 3x = 3 \sin x - 4 \sin^3 x,$$
$$\cos 4x = 1 - 8\cos^2 x + 8\cos^4 x, \quad \sin 4x = 4 \cos x (\sin x - 2 \sin^3 x),$$
$$\cos 5x = 5 \cos x - 20 \cos^3 x + 16 \cos^5 x, \quad \sin 5x = 5 \sin x - 20 \sin^3 x + 16 \sin^5 x,$$
$$\cos(2nx) = 1 + \sum_{k=1}^{n} (-1)^k \frac{n^2(n^2-1)\dots[n^2-(k-1)^2]}{(2k)!} 4^k \sin^{2k} x,$$
$$\cos[(2n+1)x] = \cos x \left\{ 1 + \sum_{k=1}^{n} (-1)^k \frac{[(2n+1)^2-1][(2n+1)^2-3^2]\dots[(2n+1)^2-(2k-1)^2]}{(2k)!} \sin^{2k} x \right\},$$
$$\sin(2nx) = 2n \cos x \left[ \sin x + \sum_{k=1}^{n} (-4)^k \frac{(n^2-1)(n^2-2^2)\dots(n^2-k^2)}{(2k-1)!} \sin^{2k-1} x \right],$$
$$\sin[(2n+1)x] = (2n+1) \left\{ \sin x + \sum_{k=1}^{n} (-1)^k \frac{[(2n+1)^2-1][(2n+1)^2-3^2]\dots[(2n+1)^2-(2k-1)^2]}{(2k+1)!} \sin^{2k+1} x \right\},$$
$$\tan 2x = \frac{2 \tan x}{1 - \tan^2 x}, \quad \tan 3x = \frac{3 \tan x - \tan^3 x}{1 - 3 \tan^2 x}, \quad \tan 4x = \frac{4 \tan x - 4 \tan^3 x}{1 - 6 \tan^2 x + \tan^4 x},$$

where $n = 1, 2, \dots$

## 2.2.3-9. Trigonometric functions of half argument.

$$\sin^2 \frac{x}{2} = \frac{1 - \cos x}{2}, \quad \cos^2 \frac{x}{2} = \frac{1 + \cos x}{2},$$
$$\tan \frac{x}{2} = \frac{\sin x}{1 + \cos x} = \frac{1 - \cos x}{\sin x}, \quad \cot \frac{x}{2} = \frac{\sin x}{1 - \cos x} = \frac{1 + \cos x}{\sin x},$$
$$\sin x = \frac{2 \tan \frac{x}{2}}{1 + \tan^2 \frac{x}{2}}, \quad \cos x = \frac{1 - \tan^2 \frac{x}{2}}{1 + \tan^2 \frac{x}{2}}, \quad \tan x = \frac{2 \tan \frac{x}{2}}{1 - \tan^2 \frac{x}{2}}.$$

## 2.2.3-10. Differentiation formulas.

$$\frac{d \sin x}{dx} = \cos x, \quad \frac{d \cos x}{dx} = -\sin x, \quad \frac{d \tan x}{dx} = \frac{1}{\cos^2 x}, \quad \frac{d \cot x}{dx} = -\frac{1}{\sin^2 x}.$$

### 2.2.3-11. Integration formulas.

$$\int \sin x \, dx = -\cos x + C, \qquad \int \cos x \, dx = \sin x + C,$$

$$\int \tan x \, dx = -\ln|\cos x| + C, \qquad \int \cot x \, dx = \ln|\sin x| + C,$$

where $C$ is an arbitrary constant.

### 2.2.3-12. Expansion in power series.

$$\cos x = 1 - \frac{x^2}{2!} + \frac{x^4}{4!} - \frac{x^6}{6!} + \cdots + (-1)^n \frac{x^{2n}}{(2n)!} + \cdots \qquad (|x| < \infty),$$

$$\sin x = x - \frac{x^3}{3!} + \frac{x^5}{5!} - \frac{x^7}{7!} + \cdots + (-1)^n \frac{x^{2n+1}}{(2n+1)!} + \cdots \qquad (|x| < \infty),$$

$$\tan x = x + \frac{x^3}{3} + \frac{2x^5}{15} + \frac{17x^7}{315} + \cdots + \frac{2^{2n}(2^{2n}-1)|B_{2n}|}{(2n)!} x^{2n-1} + \cdots \qquad (|x| < \pi/2),$$

$$\cot x = \frac{1}{x} - \left( \frac{x}{3} + \frac{x^3}{45} + \frac{2x^5}{945} + \cdots + \frac{2^{2n}|B_{2n}|}{(2n)!} x^{2n-1} + \cdots \right) \qquad (0 < |x| < \pi),$$

where $B_n$ are Bernoulli numbers (see Subsection 18.1.3).

### 2.2.3-13. Representation in the form of infinite products.

$$\sin x = x \left(1 - \frac{x^2}{\pi^2}\right)\left(1 - \frac{x^2}{4\pi^2}\right)\left(1 - \frac{x^2}{9\pi^2}\right) \cdots \left(1 - \frac{x^2}{n^2\pi^2}\right) \cdots$$

$$\cos x = \left(1 - \frac{4x^2}{\pi^2}\right)\left(1 - \frac{4x^2}{9\pi^2}\right)\left(1 - \frac{4x^2}{25\pi^2}\right) \cdots \left(1 - \frac{4x^2}{(2n+1)^2\pi^2}\right) \cdots$$

### 2.2.3-14. Euler and de Moivre formulas. Relationship with hyperbolic functions.

$$e^{y+ix} = e^y(\cos x + i \sin x), \quad (\cos x + i \sin x)^n = \cos(nx) + i \sin(nx), \quad i^2 = -1,$$
$$\sin(ix) = i \sinh x, \quad \cos(ix) = \cosh x, \quad \tan(ix) = i \tanh x, \quad \cot(ix) = -i \coth x.$$

## 2.3. Inverse Trigonometric Functions

### 2.3.1. Definitions. Graphs of Inverse Trigonometric Functions

#### 2.3.1-1. Definitions of inverse trigonometric functions.

*Inverse trigonometric functions* (*arc functions*) are the functions that are inverse to the trigonometric functions. Since the trigonometric functions $\sin x$, $\cos x$, $\tan x$, $\cot x$ are periodic, the corresponding inverse functions, denoted by $\operatorname{Arcsin} x$, $\operatorname{Arccos} x$, $\operatorname{Arctan} x$,

Arccot $x$, are multi-valued. The following relations define the multi-valued inverse trigonometric functions:
$$\sin(\operatorname{Arcsin} x) = x, \quad \cos(\operatorname{Arccos} x) = x,$$
$$\tan(\operatorname{Arctan} x) = x, \quad \cot(\operatorname{Arccot} x) = x.$$

These functions admit the following verbal definitions: Arcsin $x$ is the angle whose sine is equal to $x$; Arccos $x$ is the angle whose cosine is equal to $x$; Arctan $x$ is the angle whose tangent is equal to $x$; Arccot $x$ is the angle whose cotangent is equal to $x$.

The principal (single-valued) branches of the inverse trigonometric functions are denoted by

$$\arcsin x \equiv \sin^{-1} x \quad \text{(arcsine is the inverse of sine)},$$
$$\arccos x \equiv \cos^{-1} x \quad \text{(arccosine is the inverse of cosine)},$$
$$\arctan x \equiv \tan^{-1} x \quad \text{(arctangent is the inverse of tangent)},$$
$$\operatorname{arccot} x \equiv \cot^{-1} x \quad \text{(arccotangent is the inverse of cotangent)}$$

and are determined by the inequalities

$$-\tfrac{\pi}{2} \leq \arcsin x \leq \tfrac{\pi}{2}, \quad 0 \leq \arccos x \leq \pi \quad (-1 \leq x \leq 1);$$
$$-\tfrac{\pi}{2} < \arctan x < \tfrac{\pi}{2}, \quad 0 < \operatorname{arccot} x < \pi \quad (-\infty < x < \infty).$$

The following equivalent relations can be taken as definitions of single-valued inverse trigonometric functions:

$$y = \arcsin x, \quad -1 \leq x \leq 1 \iff x = \sin y, \quad -\tfrac{\pi}{2} \leq y \leq \tfrac{\pi}{2};$$
$$y = \arccos x, \quad -1 \leq x \leq 1 \iff x = \cos y, \quad 0 \leq y \leq \pi;$$
$$y = \arctan x, \quad -\infty < x < +\infty \iff x = \tan y, \quad -\tfrac{\pi}{2} < y < \tfrac{\pi}{2};$$
$$y = \operatorname{arccot} x, \quad -\infty < x < +\infty \iff x = \cot y, \quad 0 < y < \pi.$$

The multi-valued and the single-valued inverse trigonometric functions are related by the formulas
$$\operatorname{Arcsin} x = (-1)^n \arcsin x + \pi n,$$
$$\operatorname{Arccos} x = \pm \arccos x + 2\pi n,$$
$$\operatorname{Arctan} x = \arctan x + \pi n,$$
$$\operatorname{Arccot} x = \operatorname{arccot} x + \pi n,$$

where $n = 0, \pm 1, \pm 2, \ldots$

The graphs of inverse trigonometric functions are obtained from the graphs of the corresponding trigonometric functions by mirror reflection with respect to the straight line $y = x$ (with the domain of each function being taken into account).

2.3.1-2. Arcsine: $y = \arcsin x$.

This function is defined for all $x \in [-1, 1]$ and its range is $y \in [-\tfrac{\pi}{2}, \tfrac{\pi}{2}]$. The arcsine is an odd, nonperiodic, bounded function that crosses the axes $Ox$ and $Oy$ at the origin $x = 0$, $y = 0$. This is an increasing function in its domain, and it takes its smallest value $y = -\tfrac{\pi}{2}$ at the point $x = -1$; it takes its largest value $y = \tfrac{\pi}{2}$ at the point $x = 1$. The graph of the function $y = \arcsin x$ is given in Fig. 2.10.

## 2.3.1-3. Arccosine: $y = \arccos x$.

This function is defined for all $x \in [-1, 1]$, and its range is $y \in [0, \pi]$. It is neither odd nor even. It is a nonperiodic, bounded function that crosses the axis $Oy$ at the point $y = \frac{\pi}{2}$ and crosses the axis $Ox$ at the point $x = 1$. This is a decreasing function in its domain, and at the point $x = -1$ it takes its largest value $y = \pi$; at the point $x = 1$ it takes its smallest value $y = 0$. For all $x$ in its domain, the following relation holds: $\arccos x = \frac{\pi}{2} - \arcsin x$. The graph of the function $y = \arccos x$ is given in Fig. 2.11.

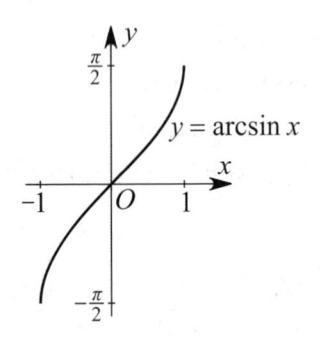

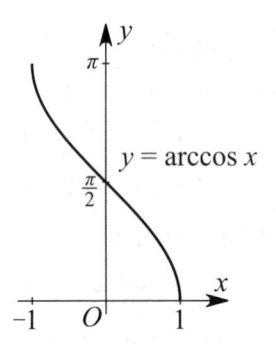

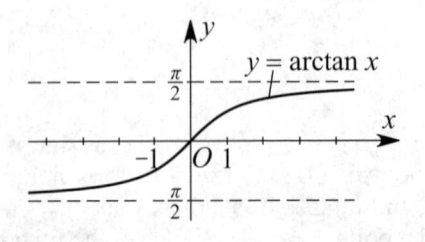

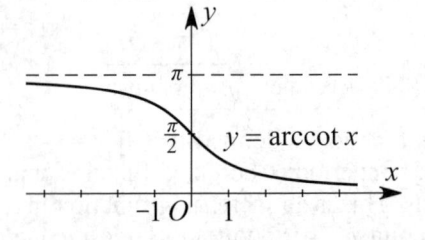

**Figure 2.10.** The graph of the function $y = \arcsin x$.

**Figure 2.11.** The graph of the function $y = \arccos x$.

## 2.3.1-4. Arctangent: $y = \arctan x$.

This function is defined for all $x$, and its range is $y \in (-\frac{\pi}{2}, \frac{\pi}{2})$. The arctangent is an odd, nonperiodic, bounded function that crosses the coordinate axes at the origin $x = 0$, $y = 0$. This is an increasing function on the real axis with no points of extremum. It has two horizontal asymptotes: $y = -\frac{\pi}{2}$ (as $x \to -\infty$) and $y = \frac{\pi}{2}$ (as $x \to +\infty$). The graph of the function $y = \arctan x$ is given in Fig. 2.12.

## 2.3.1-5. Arccotangent: $y = \text{arccot } x$.

This function is defined for all $x$, and its range is $y \in (0, \pi)$. The arccotangent is neither odd nor even. It is a nonperiodic, bounded function that crosses the axis $Oy$ at the point $y = \frac{\pi}{2}$ and does not cross the axis $Ox$. This is a decreasing function on the entire real axis with no points of extremum. It has two horizontal asymptotes $y = 0$ (as $x \to +\infty$) and $y = \pi$ (as $x \to -\infty$). For all $x$, the following relation holds: $\text{arccot } x = \frac{\pi}{2} - \arctan x$. The graph of the function $y = \text{arccot } x$ is given in Fig. 2.13.

**Figure 2.12.** The graph of the function $y = \arctan x$.

**Figure 2.13.** The graph of the function $y = \text{arccot } x$.

## 2.3.2. Properties of Inverse Trigonometric Functions

**2.3.2-1. Simplest formulas.**

$$\sin(\arcsin x) = x, \quad \cos(\arccos x) = x,$$
$$\tan(\arctan x) = x, \quad \cot(\arccot x) = x.$$

**2.3.2-2. Some properties.**

$$\arcsin(-x) = -\arcsin x, \quad \arccos(-x) = \pi - \arccos x,$$
$$\arctan(-x) = -\arctan x, \quad \arccot(-x) = \pi - \arccot x,$$

$$\arcsin(\sin x) = \begin{cases} x - 2n\pi & \text{if } 2n\pi - \tfrac{\pi}{2} \le x \le 2n\pi + \tfrac{\pi}{2}, \\ -x + 2(n+1)\pi & \text{if } (2n+1)\pi - \tfrac{\pi}{2} \le x \le 2(n+1)\pi + \tfrac{\pi}{2}, \end{cases}$$

$$\arccos(\cos x) = \begin{cases} x - 2n\pi & \text{if } 2n\pi \le x \le (2n+1)\pi, \\ -x + 2(n+1)\pi & \text{if } (2n+1)\pi \le x \le 2(n+1)\pi, \end{cases}$$

$$\arctan(\tan x) = x - n\pi \quad \text{if } n\pi - \tfrac{\pi}{2} < x < n\pi + \tfrac{\pi}{2},$$
$$\arccot(\cot x) = x - n\pi \quad \text{if } n\pi < x < (n+1)\pi.$$

**2.3.2-3. Relations between inverse trigonometric functions.**

$$\arcsin x + \arccos x = \tfrac{\pi}{2}, \qquad \arctan x + \arccot x = \tfrac{\pi}{2};$$

$$\arcsin x = \begin{cases} \arccos \sqrt{1-x^2} & \text{if } 0 \le x \le 1, \\ -\arccos \sqrt{1-x^2} & \text{if } -1 \le x \le 0, \\ \arctan \dfrac{x}{\sqrt{1-x^2}} & \text{if } -1 < x < 1, \\ \arccot \dfrac{\sqrt{1-x^2}}{x} - \pi & \text{if } -1 \le x < 0; \end{cases} \quad \arccos x = \begin{cases} \arcsin \sqrt{1-x^2} & \text{if } 0 \le x \le 1, \\ \pi - \arcsin \sqrt{1-x^2} & \text{if } -1 \le x \le 0, \\ \arctan \dfrac{\sqrt{1-x^2}}{x} & \text{if } 0 < x \le 1, \\ \arccot \dfrac{x}{\sqrt{1-x^2}} & \text{if } -1 < x < 1; \end{cases}$$

$$\arctan x = \begin{cases} \arcsin \dfrac{x}{\sqrt{1+x^2}} & \text{for any } x, \\ \arccos \dfrac{1}{\sqrt{1+x^2}} & \text{if } x \ge 0, \\ -\arccos \dfrac{1}{\sqrt{1+x^2}} & \text{if } x \le 0, \\ \arccot \dfrac{1}{x} & \text{if } x > 0; \end{cases} \quad \arccot x = \begin{cases} \arcsin \dfrac{1}{\sqrt{1+x^2}} & \text{if } x > 0, \\ \pi - \arcsin \dfrac{1}{\sqrt{1+x^2}} & \text{if } x < 0, \\ \arctan \dfrac{1}{x} & \text{if } x > 0, \\ \pi + \arctan \dfrac{1}{x} & \text{if } x < 0. \end{cases}$$

**2.3.2-4. Addition and subtraction of inverse trigonometric functions.**

$$\arcsin x + \arcsin y = \arcsin\left(x\sqrt{1-y^2} + y\sqrt{1-x^2}\right) \quad \text{for } x^2 + y^2 \le 1,$$
$$\arccos x \pm \arccos y = \pm \arccos\left[xy \mp \sqrt{(1-x^2)(1-y^2)}\right] \quad \text{for } x \pm y \ge 0,$$

$$\arctan x + \arctan y = \arctan \frac{x+y}{1-xy} \quad \text{for} \quad xy < 1,$$
$$\arctan x - \arctan y = \arctan \frac{x-y}{1+xy} \quad \text{for} \quad xy > -1.$$

### 2.3.2-5. Differentiation formulas.

$$\frac{d}{dx}\arcsin x = \frac{1}{\sqrt{1-x^2}}, \qquad \frac{d}{dx}\arccos x = -\frac{1}{\sqrt{1-x^2}},$$
$$\frac{d}{dx}\arctan x = \frac{1}{1+x^2}, \qquad \frac{d}{dx}\operatorname{arccot} x = -\frac{1}{1+x^2}.$$

### 2.3.2-6. Integration formulas.

$$\int \arcsin x\, dx = x\arcsin x + \sqrt{1-x^2} + C, \qquad \int \arccos x\, dx = x\arccos x - \sqrt{1-x^2} + C,$$
$$\int \arctan x\, dx = x\arctan x - \frac{1}{2}\ln(1+x^2) + C, \qquad \int \operatorname{arccot} x\, dx = x\operatorname{arccot} x + \frac{1}{2}\ln(1+x^2) + C,$$

where $C$ is an arbitrary constant.

### 2.3.2-7. Expansion in power series.

$$\arcsin x = x + \frac{1}{2}\frac{x^3}{3} + \frac{1\times 3}{2\times 4}\frac{x^5}{5} + \frac{1\times 3\times 5}{2\times 4\times 6}\frac{x^7}{7} + \cdots + \frac{1\times 3\times\cdots\times(2n-1)}{2\times 4\times\cdots\times(2n)}\frac{x^{2n+1}}{2n+1} + \cdots \quad (|x|<1),$$
$$\arctan x = x - \frac{x^3}{3} + \frac{x^5}{5} - \frac{x^7}{7} + \cdots + (-1)^{n-1}\frac{x^{2n-1}}{2n-1} + \cdots \quad (|x|\leq 1),$$
$$\arctan x = \frac{\pi}{2} - \frac{1}{x} + \frac{1}{3x^3} - \frac{1}{5x^5} + \cdots + (-1)^n\frac{1}{(2n-1)x^{2n-1}} + \cdots \quad (|x|>1).$$

The expansions for $\arccos x$ and $\operatorname{arccot} x$ can be obtained from the relations $\arccos x = \frac{\pi}{2} - \arcsin x$ and $\operatorname{arccot} x = \frac{\pi}{2} - \arctan x$.

## 2.4. Hyperbolic Functions

### 2.4.1. Definitions. Graphs of Hyperbolic Functions

#### 2.4.1-1. Definitions of hyperbolic functions.

Hyperbolic functions are defined in terms of the exponential functions as follows:

$$\sinh x = \frac{e^x - e^{-x}}{2}, \quad \cosh x = \frac{e^x + e^{-x}}{2}, \quad \tanh x = \frac{e^x - e^{-x}}{e^x + e^{-x}}, \quad \coth x = \frac{e^x + e^{-x}}{e^x - e^{-x}}.$$

The graphs of hyperbolic functions are given below.

### 2.4.1-2. Hyperbolic sine: $y = \sinh x$.

This function is defined for all $x$ and its range is the entire $y$-axis. The hyperbolic sine is an odd, nonperiodic, unbounded function that crosses the axes $Ox$ and $Oy$ at the origin $x = 0$, $y = 0$. This is an increasing function in its domain with no points of extremum. The graph of the function $y = \sinh x$ is given in Fig. 2.14.

### 2.4.1-3. Hyperbolic cosine: $y = \cosh x$.

This function is defined for all $x$, and its range is $y \in [1, +\infty)$. The hyperbolic cosine is a nonperiodic, unbounded function that crosses the axis $Oy$ at the point 1 and does not cross the axis $Ox$. This function is decreasing on the interval $(-\infty, 0)$ and is increasing on the interval $(0, +\infty)$; it takes its smallest value $y = 1$ at $x = 0$. The graph of the function $y = \cosh x$ is given in Fig. 2.15.

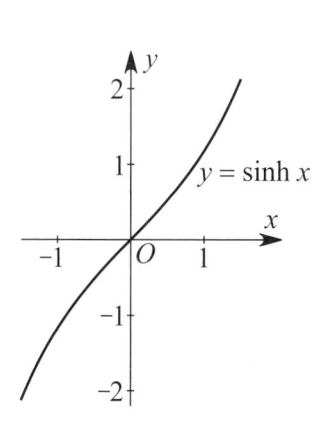

**Figure 2.14.** The graph of the function $y = \sinh x$.

**Figure 2.15.** The graph of the function $y = \cosh x$.

### 2.4.1-4. Hyperbolic tangent: $y = \tanh x$.

This function is defined for all $x$, and its range is $y \in (-1, 1)$. The hyperbolic tangent is an odd, nonperiodic, bounded function that crosses the coordinate axes at the origin $x = 0$, $y = 0$. This is an increasing function on the entire real axis and has two horizontal asymptotes: $y = -1$ (as $x \to -\infty$) and $y = 1$ (as $x \to +\infty$). The graph of the function $y = \tanh x$ is given in Fig. 2.16.

### 2.4.1-5. Hyperbolic cotangent: $y = \coth x$.

This function is defined for all $x \neq 0$, and its range consists of all $y \in (-\infty, -1)$ and $y \in (1, +\infty)$. The hyperbolic cotangent is an odd, nonperiodic, unbounded function that does not cross the coordinate axes. This is a decreasing function on each of the semiaxes of its domain; it has no points of extremum and does not cross the coordinate axes. It has two horizontal asymptotes: $y = -1$ (as $x \to -\infty$) and $y = 1$ (as $x \to +\infty$). The graph of the function $y = \coth x$ is given in Fig. 2.17.

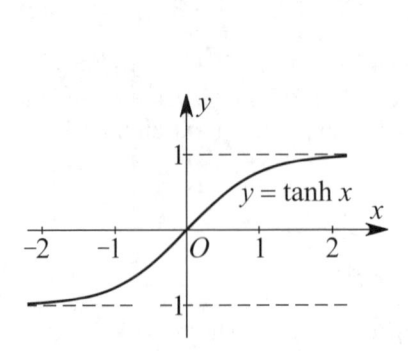

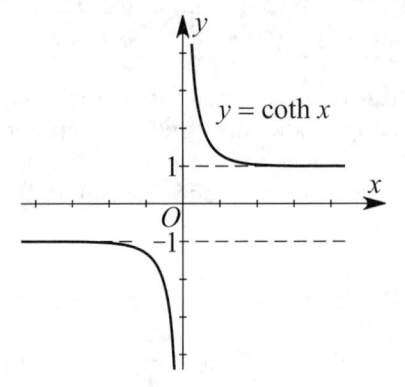

**Figure 2.16.** The graph of the function $y = \tanh x$.

**Figure 2.17.** The graph of the function $y = \coth x$.

## 2.4.2. Properties of Hyperbolic Functions

**2.4.2-1. Simplest relations.**

$$\cosh^2 x - \sinh^2 x = 1, \qquad \tanh x \coth x = 1,$$
$$\sinh(-x) = -\sinh x, \qquad \cosh(-x) = \cosh x,$$
$$\tanh x = \frac{\sinh x}{\cosh x}, \qquad \coth x = \frac{\cosh x}{\sinh x},$$
$$\tanh(-x) = -\tanh x, \qquad \coth(-x) = -\coth x,$$
$$1 - \tanh^2 x = \frac{1}{\cosh^2 x}, \qquad \coth^2 x - 1 = \frac{1}{\sinh^2 x}.$$

**2.4.2-2. Relations between hyperbolic functions of single argument ($x \geq 0$).**

$$\sinh x = \sqrt{\cosh^2 x - 1} = \frac{\tanh x}{\sqrt{1 - \tanh^2 x}} = \frac{1}{\sqrt{\coth^2 x - 1}},$$
$$\cosh x = \sqrt{\sinh^2 x + 1} = \frac{1}{\sqrt{1 - \tanh^2 x}} = \frac{\coth x}{\sqrt{\coth^2 x - 1}},$$
$$\tanh x = \frac{\sinh x}{\sqrt{\sinh^2 x + 1}} = \frac{\sqrt{\cosh^2 x - 1}}{\cosh x} = \frac{1}{\coth x},$$
$$\coth x = \frac{\sqrt{\sinh^2 x + 1}}{\sinh x} = \frac{\cosh x}{\sqrt{\cosh^2 x - 1}} = \frac{1}{\tanh x}.$$

**2.4.2-3. Addition formulas.**

$$\sinh(x \pm y) = \sinh x \cosh y \pm \sinh y \cosh x, \qquad \cosh(x \pm y) = \cosh x \cosh y \pm \sinh x \sinh y,$$
$$\tanh(x \pm y) = \frac{\tanh x \pm \tanh y}{1 \pm \tanh x \tanh y}, \qquad \coth(x \pm y) = \frac{\coth x \coth y \pm 1}{\coth y \pm \coth x}.$$

## 2.4. HYPERBOLIC FUNCTIONS

### 2.4.2-4. Addition and subtraction of hyperbolic functions.

$$\sinh x \pm \sinh y = 2 \sinh\left(\frac{x \pm y}{2}\right) \cosh\left(\frac{x \mp y}{2}\right),$$
$$\cosh x + \cosh y = 2 \cosh\left(\frac{x + y}{2}\right) \cosh\left(\frac{x - y}{2}\right),$$
$$\cosh x - \cosh y = 2 \sinh\left(\frac{x + y}{2}\right) \sinh\left(\frac{x - y}{2}\right),$$
$$\sinh^2 x - \sinh^2 y = \cosh^2 x - \cosh^2 y = \sinh(x+y)\sinh(x-y),$$
$$\sinh^2 x + \cosh^2 y = \cosh(x+y)\cosh(x-y),$$
$$(\cosh x \pm \sinh x)^n = \cosh(nx) \pm \sinh(nx),$$
$$\tanh x \pm \tanh y = \frac{\sinh(x \pm y)}{\cosh x \cosh y}, \quad \coth x \pm \coth y = \pm\frac{\sinh(x \pm y)}{\sinh x \sinh y},$$

where $n = 0, \pm 1, \pm 2, \ldots$

### 2.4.2-5. Products of hyperbolic functions.

$$\sinh x \sinh y = \tfrac{1}{2}[\cosh(x+y) - \cosh(x-y)],$$
$$\cosh x \cosh y = \tfrac{1}{2}[\cosh(x+y) + \cosh(x-y)],$$
$$\sinh x \cosh y = \tfrac{1}{2}[\sinh(x+y) + \sinh(x-y)].$$

### 2.4.2-6. Powers of hyperbolic functions.

$\cosh^2 x = \tfrac{1}{2}\cosh 2x + \tfrac{1}{2},$      $\sinh^2 x = \tfrac{1}{2}\cosh 2x - \tfrac{1}{2},$

$\cosh^3 x = \tfrac{1}{4}\cosh 3x + \tfrac{3}{4}\cosh x,$      $\sinh^3 x = \tfrac{1}{4}\sinh 3x - \tfrac{3}{4}\sinh x,$

$\cosh^4 x = \tfrac{1}{8}\cosh 4x + \tfrac{1}{2}\cosh 2x + \tfrac{3}{8},$      $\sinh^4 x = \tfrac{1}{8}\cosh 4x - \tfrac{1}{2}\cosh 2x + \tfrac{3}{8},$

$\cosh^5 x = \tfrac{1}{16}\cosh 5x + \tfrac{5}{16}\cosh 3x + \tfrac{5}{8}\cosh x,$    $\sinh^5 x = \tfrac{1}{16}\sinh 5x - \tfrac{5}{16}\sinh 3x + \tfrac{5}{8}\sinh x,$

$$\cosh^{2n} x = \frac{1}{2^{2n-1}} \sum_{k=0}^{n-1} C_{2n}^k \cosh[2(n-k)x] + \frac{1}{2^{2n}} C_{2n}^n,$$

$$\cosh^{2n+1} x = \frac{1}{2^{2n}} \sum_{k=0}^{n} C_{2n+1}^k \cosh[(2n-2k+1)x],$$

$$\sinh^{2n} x = \frac{1}{2^{2n-1}} \sum_{k=0}^{n-1} (-1)^k C_{2n}^k \cosh[2(n-k)x] + \frac{(-1)^n}{2^{2n}} C_{2n}^n,$$

$$\sinh^{2n+1} x = \frac{1}{2^{2n}} \sum_{k=0}^{n} (-1)^k C_{2n+1}^k \sinh[(2n-2k+1)x].$$

Here, $n = 1, 2, \ldots$; and $C_m^k$ are binomial coefficients.

### 2.4.2-7. Hyperbolic functions of multiple argument.

$\cosh 2x = 2\cosh^2 x - 1,$  $\qquad\qquad$ $\sinh 2x = 2\sinh x \cosh x,$

$\cosh 3x = -3\cosh x + 4\cosh^3 x,$ $\qquad$ $\sinh 3x = 3\sinh x + 4\sinh^3 x,$

$\cosh 4x = 1 - 8\cosh^2 x + 8\cosh^4 x,$ $\qquad$ $\sinh 4x = 4\cosh x(\sinh x + 2\sinh^3 x),$

$\cosh 5x = 5\cosh x - 20\cosh^3 x + 16\cosh^5 x,$ $\qquad$ $\sinh 5x = 5\sinh x + 20\sinh^3 x + 16\sinh^5 x.$

$$\cosh(nx) = 2^{n-1}\cosh^n x + \frac{n}{2}\sum_{k=0}^{[n/2]}\frac{(-1)^{k+1}}{k+1}C_{n-k-2}^{k-2}2^{n-2k-2}(\cosh x)^{n-2k-2},$$

$$\sinh(nx) = \sinh x \sum_{k=0}^{[(n-1)/2]} 2^{n-k-1}C_{n-k-1}^{k}(\cosh x)^{n-2k-1}.$$

Here, $C_m^k$ are binomial coefficients and $[A]$ stands for the integer part of the number $A$.

### 2.4.2-8. Hyperbolic functions of half argument.

$$\sinh\frac{x}{2} = \operatorname{sign} x\sqrt{\frac{\cosh x - 1}{2}}, \qquad \cosh\frac{x}{2} = \sqrt{\frac{\cosh x + 1}{2}},$$

$$\tanh\frac{x}{2} = \frac{\sinh x}{\cosh x + 1} = \frac{\cosh x - 1}{\sinh x}, \qquad \coth\frac{x}{2} = \frac{\sinh x}{\cosh x - 1} = \frac{\cosh x + 1}{\sinh x}.$$

### 2.4.2-9. Differentiation formulas.

$$\frac{d\sinh x}{dx} = \cosh x, \qquad \frac{d\cosh x}{dx} = \sinh x,$$

$$\frac{d\tanh x}{dx} = \frac{1}{\cosh^2 x}, \qquad \frac{d\coth x}{dx} = -\frac{1}{\sinh^2 x}.$$

### 2.4.2-10. Integration formulas.

$$\int \sinh x\, dx = \cosh x + C, \qquad \int \cosh x\, dx = \sinh x + C,$$

$$\int \tanh x\, dx = \ln\cosh x + C, \qquad \int \coth x\, dx = \ln|\sinh x| + C,$$

where $C$ is an arbitrary constant.

### 2.4.2-11. Expansion in power series.

$$\cosh x = 1 + \frac{x^2}{2!} + \frac{x^4}{4!} + \frac{x^6}{6!} + \cdots + \frac{x^{2n}}{(2n)!} + \cdots \qquad (|x| < \infty),$$

$$\sinh x = x + \frac{x^3}{3!} + \frac{x^5}{5!} + \frac{x^7}{7!} + \cdots + \frac{x^{2n+1}}{(2n+1)!} + \cdots \qquad (|x| < \infty),$$

$$\tanh x = x - \frac{x^3}{3} + \frac{2x^5}{15} - \frac{17x^7}{315} + \cdots + (-1)^{n-1}\frac{2^{2n}(2^{2n}-1)|B_{2n}|x^{2n-1}}{(2n)!} + \cdots \qquad (|x| < \pi/2),$$

$$\coth x = \frac{1}{x} + \frac{x}{3} - \frac{x^3}{45} + \frac{2x^5}{945} - \cdots + (-1)^{n-1}\frac{2^{2n}|B_{2n}|x^{2n-1}}{(2n)!} + \cdots \qquad (|x| < \pi),$$

where $B_n$ are Bernoulli numbers (see Subsection 18.1.3).

### 2.4.2-12. Relationship with trigonometric functions.

$$\sinh(ix) = i\sin x, \quad \cosh(ix) = \cos x, \quad \tanh(ix) = i\tan x, \quad \coth(ix) = -i\cot x, \quad i^2 = -1.$$

## 2.5. Inverse Hyperbolic Functions
### 2.5.1. Definitions. Graphs of Inverse Hyperbolic Functions

#### 2.5.1-1. Definitions of inverse hyperbolic functions.

*Inverse hyperbolic functions* are the functions that are inverse to hyperbolic functions. The following notation is used for inverse hyperbolic functions:

$$\operatorname{arcsinh} x \equiv \sinh^{-1} x \quad \text{(inverse of hyperbolic sine)},$$
$$\operatorname{arccosh} x \equiv \cosh^{-1} x \quad \text{(inverse of hyperbolic cosine)},$$
$$\operatorname{arctanh} x \equiv \tanh^{-1} x \quad \text{(inverse of hyperbolic tangent)},$$
$$\operatorname{arccoth} x \equiv \coth^{-1} x \quad \text{(inverse of hyperbolic cotangent)}.$$

Inverse hyperbolic functions can be expressed in terms of logarithmic functions:

$$\operatorname{arcsinh} x = \ln\left(x + \sqrt{x^2 + 1}\right) \quad (x \text{ is any}); \qquad \operatorname{arccosh} x = \ln\left(x + \sqrt{x^2 - 1}\right) \quad (x \geq 1);$$

$$\operatorname{arctanh} x = \frac{1}{2}\ln\frac{1+x}{1-x} \quad (|x| < 1); \qquad \operatorname{arccoth} x = \frac{1}{2}\ln\frac{x+1}{x-1} \quad (|x| > 1).$$

Here, only one (principal) branch of the function $\operatorname{arccosh} x$ is listed, the function itself being double-valued. In order to write out both branches of $\operatorname{arccosh} x$, the symbol $\pm$ should be placed before the logarithm on the right-hand side of the formula.

Below, the graphs of the inverse hyperbolic functions are given. These are obtained from the graphs of the corresponding hyperbolic functions by mirror reflection with respect to the straight line $y = x$ (with the domain of each function being taken into account).

#### 2.5.1-2. Inverse hyperbolic sine: $y = \operatorname{arcsinh} x$.

This function is defined for all $x$, and its range coincides with the $y$-axis. The $\operatorname{arcsinh} x$ is an odd, nonperiodic, unbounded function that crosses the axes $Ox$ and $Oy$ at the origin $x = 0$, $y = 0$. This is an increasing function on the entire real axis with no points of extremum. The graph of the function $y = \operatorname{arcsinh} x$ is given in Fig. 2.18.

### 2.5.1-3. Inverse hyperbolic cosine: $y = \text{arccosh}\, x$.

This function is defined for all $x \in [1, +\infty)$, and its range consists of $y \in [0, +\infty)$. The arccosh $x$ is neither odd nor even; it is nonperiodic and unbounded. It does not cross the axis $Oy$ and crosses the axis $Ox$ at the point $x = 1$. It is an increasing function in its domain with the minimal value $y = 0$ at $x = 1$. The graph of the function $y = \text{arccosh}\, x$ is given in Fig. 2.19.

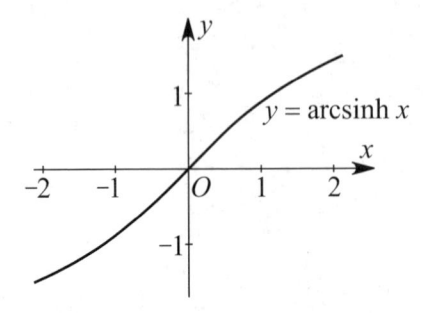

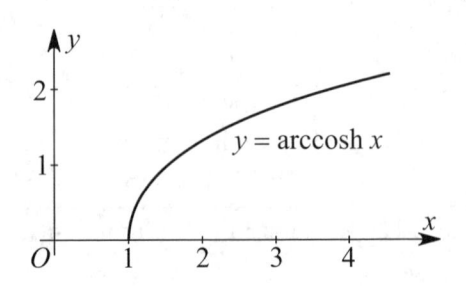

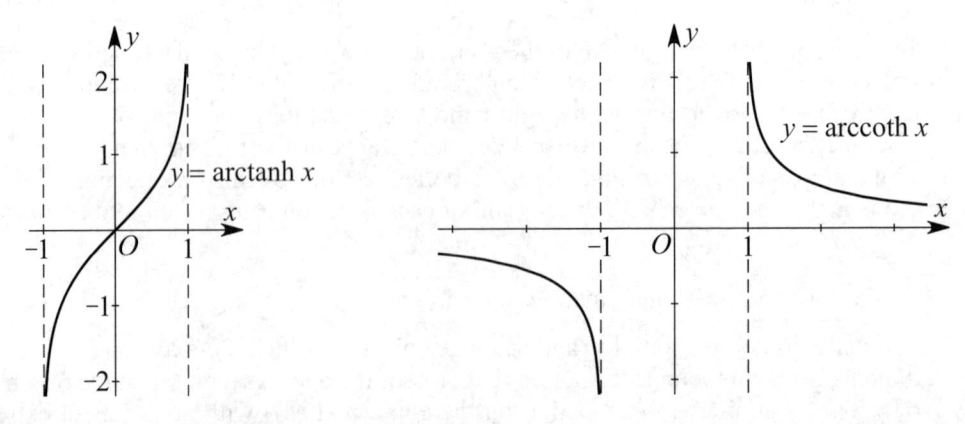

**Figure 2.18.** The graph of the function $y = \text{arcsinh}\, x$.

**Figure 2.19.** The graph of the function $y = \text{arccosh}\, x$.

### 2.5.1-4. Inverse hyperbolic tangent: $y = \text{arctanh}\, x$.

This function is defined for all $x \in (-1, 1)$, and its range consists of all $y$. The arctanh $x$ is an odd, nonperiodic, unbounded function that crosses the coordinate axes at the origin $x = 0$, $y = 0$. This is an increasing function in its domain with no points of extremum and an inflection point at the origin. It has two vertical asymptotes: $x = \pm 1$. The graph of the function $y = \text{arctanh}\, x$ is given in Fig. 2.20.

### 2.5.1-5. Inverse hyperbolic cotangent: $y = \text{arccoth}\, x$.

This function is defined for $x \in (-\infty, -1)$ and $x \in (1, +\infty)$. Its range consists of all $y \neq 0$. The arccoth $x$ is an odd, nonperiodic, unbounded function that does not cross the coordinate axes. It is a decreasing function on each of the semiaxes of its domain. This function has no points of extremum and has one horizontal asymptote $y = 0$ and two vertical asymptotes $x = \pm 1$. The graph of the function $y = \text{arccoth}\, x$ is given in Fig. 2.21.

**Figure 2.20.** The graph of the function $y = \text{arctanh}\, x$.

**Figure 2.21.** The graph of the function $y = \text{arccoth}\, x$.

## 2.5.2. Properties of Inverse Hyperbolic Functions

**2.5.2-1. Simplest relations.**

$$\operatorname{arcsinh}(-x) = -\operatorname{arcsinh} x, \quad \operatorname{arctanh}(-x) = -\operatorname{arctanh} x, \quad \operatorname{arccoth}(-x) = -\operatorname{arccoth} x.$$

**2.5.2-2. Relations between inverse hyperbolic functions.**

$$\operatorname{arcsinh} x = \operatorname{arccosh} \sqrt{x^2 + 1} = \operatorname{arctanh} \frac{x}{\sqrt{x^2 + 1}},$$

$$\operatorname{arccosh} x = \operatorname{arcsinh} \sqrt{x^2 - 1} = \operatorname{arctanh} \frac{\sqrt{x^2 - 1}}{x},$$

$$\operatorname{arctanh} x = \operatorname{arcsinh} \frac{x}{\sqrt{1 - x^2}} = \operatorname{arccosh} \frac{1}{\sqrt{1 - x^2}} = \operatorname{arccoth} \frac{1}{x}.$$

**2.5.2-3. Addition and subtraction of inverse hyperbolic functions.**

$$\operatorname{arcsinh} x \pm \operatorname{arcsinh} y = \operatorname{arcsinh} \left( x\sqrt{1 + y^2} \pm y\sqrt{1 + x^2} \right),$$

$$\operatorname{arccosh} x \pm \operatorname{arccosh} y = \operatorname{arccosh} \left[ xy \pm \sqrt{(x^2 - 1)(y^2 - 1)} \right],$$

$$\operatorname{arcsinh} x \pm \operatorname{arccosh} y = \operatorname{arcsinh} \left[ xy \pm \sqrt{(x^2 + 1)(y^2 - 1)} \right],$$

$$\operatorname{arctanh} x \pm \operatorname{arctanh} y = \operatorname{arctanh} \frac{x \pm y}{1 \pm xy}, \quad \operatorname{arctanh} x \pm \operatorname{arccoth} y = \operatorname{arctanh} \frac{xy \pm 1}{y \pm x}.$$

**2.5.2-4. Differentiation formulas.**

$$\frac{d}{dx} \operatorname{arcsinh} x = \frac{1}{\sqrt{x^2 + 1}}, \qquad \frac{d}{dx} \operatorname{arccosh} x = \frac{1}{\sqrt{x^2 - 1}},$$

$$\frac{d}{dx} \operatorname{arctanh} x = \frac{1}{1 - x^2} \quad (x^2 < 1), \qquad \frac{d}{dx} \operatorname{arccoth} x = \frac{1}{1 - x^2} \quad (x^2 > 1).$$

**2.5.2-5. Integration formulas.**

$$\int \operatorname{arcsinh} x \, dx = x \operatorname{arcsinh} x - \sqrt{1 + x^2} + C,$$

$$\int \operatorname{arccosh} x \, dx = x \operatorname{arccosh} x - \sqrt{x^2 - 1} + C,$$

$$\int \operatorname{arctanh} x \, dx = x \operatorname{arctanh} x + \frac{1}{2} \ln(1 - x^2) + C,$$

$$\int \operatorname{arccoth} x \, dx = x \operatorname{arccoth} x + \frac{1}{2} \ln(x^2 - 1) + C,$$

where $C$ is an arbitrary constant.

**2.5.2-6. Expansion in power series.**

$$\operatorname{arcsinh} x = x - \frac{1}{2}\frac{x^3}{3} + \frac{1 \times 3}{2 \times 4}\frac{x^5}{5} - \cdots + (-1)^n \frac{1 \times 3 \times \cdots \times (2n-1)}{2 \times 4 \times \cdots \times (2n)} \frac{x^{2n+1}}{2n+1} + \cdots \qquad (|x|<1),$$

$$\operatorname{arcsinh} x = \ln(2x) + \frac{1}{2}\frac{1}{2x^2} + \frac{1 \times 3}{2 \times 4}\frac{1}{4x^4} + \cdots + \frac{1 \times 3 \times \cdots \times (2n-1)}{2 \times 4 \times \cdots \times (2n)} \frac{1}{2nx^{2n}} + \cdots \qquad (|x|>1),$$

$$\operatorname{arccosh} x = \ln(2x) - \frac{1}{2}\frac{1}{2x^2} - \frac{1 \times 3}{2 \times 4}\frac{1}{4x^4} - \cdots - \frac{1 \times 3 \times \cdots \times (2n-1)}{2 \times 4 \times \cdots \times (2n)} \frac{1}{2nx^{2n}} - \cdots \qquad (|x|>1),$$

$$\operatorname{arctanh} x = x + \frac{x^3}{3} + \frac{x^5}{5} + \frac{x^7}{7} + \cdots + \frac{x^{2n+1}}{2n+1} + \cdots \qquad (|x|<1),$$

$$\operatorname{arccoth} x = \frac{1}{x} + \frac{1}{3x^3} + \frac{1}{5x^5} + \frac{1}{7x^7} + \cdots + \frac{1}{(2n+1)x^{2n+1}} + \cdots \qquad (|x|>1).$$

## References for Chapter 2

**Abramowitz, M. and Stegun, I. A. (Editors)**, *Handbook of Mathematical Functions with Formulas, Graphs and Mathematical Tables*, National Bureau of Standards Applied Mathematics, Washington, D. C., 1964.
**Adams, R.**, *Calculus: A Complete Course, 6th Edition*, Pearson Education, Toronto, 2006.
**Anton, H., Bivens, I., and Davis, S.**, *Calculus: Early Transcendental Single Variable, 8th Edition*, John Wiley & Sons, New York, 2005.
**Bronshtein, I. N. and Semendyayev, K. A.**, *Handbook of Mathematics, 4th Edition*, Springer-Verlag, Berlin, 2004.
**Courant, R. and John, F.**, *Introduction to Calculus and Analysis, Vol. 1*, Springer-Verlag, New York, 1999.
**Edwards, C. H., and Penney, D.**, *Calculus, 6th Edition*, Pearson Education, Toronto, 2002.
**Gradshteyn, I. S. and Ryzhik, I. M.**, *Tables of Integrals, Series, and Products, 6th Edition*, Academic Press, New York, 2000.
**Kline, M.**, *Calculus: An Intuitive and Physical Approach, 2nd Edition*, Dover Publications, New York, 1998.
**Korn, G. A. and Korn, T. M.**, *Mathematical Handbook for Scientists and Engineers, 2nd Edition*, Dover Publications, New York, 2000.
**Prudnikov, A. P., Brychkov, Yu. A., and Marichev, O. I.**, *Integrals and Series, Vol. 1, Elementary Functions*, Gordon & Breach, New York, 1986.
**Sullivan, M.**, *Trigonometry, 7th Edition*, Prentice Hall, Englewood Cliffs, 2004.
**Thomas, G. B. and Finney, R. L.**, *Calculus and Analytic Geometry, 9th Edition*, Addison Wesley, Reading, Massachusetts, 1996.
**Weisstein, E. W.**, *CRC Concise Encyclopedia of Mathematics, 2nd Edition*, CRC Press, Boca Raton, 2003.
**Zill, D. G. and Dewar, J. M.**, *Trigonometry, 2nd Edition*, McGraw-Hill, New York, 1990.
**Zwillinger, D.**, *CRC Standard Mathematical Tables and Formulae, 31st Edition*, CRC Press, Boca Raton, 2002.

# Chapter 3
# Elementary Geometry

## 3.1. Plane Geometry
### 3.1.1. Triangles

**3.1.1-1. Plane triangle and its properties.**

1°. A *plane triangle*, or simply a *triangle*, is a plane figure bounded by three straight line segments (*sides*) connecting three noncollinear points (*vertices*) (Fig. 3.1a). The smaller angle between the two rays issuing from a vertex and passing through the other two vertices is called an (*interior*) *angle* of the triangle. The angle adjacent to an interior angle is called an *external angle* of the triangle. An external angle is equal to the sum of the two interior angles to which it is not adjacent.

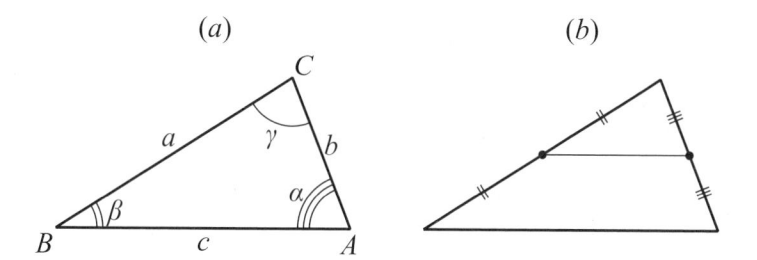

**Figure 3.1.** Plane triangle (*a*). Midline of a triangle (*b*).

*A triangle is uniquely determined by any of the following sets of its parts:*
1. Two angles and their included side.
2. Two sides and their included angle.
3. Three sides.

*Depending on the angles, a triangle is said to be:*
1. *Acute* if all three angles are acute.
2. *Right* (or *right-angled*) if one of the angles is right.
3. *Obtuse* if one of the angles is obtuse.

*Depending on the relation between the side lengths, a triangle is said to be:*
1. *Regular* (or *equilateral*) if all sides have the same length.
2. *Isosceles* if two of the sides are of equal length.
3. *Scalene* if all sides have different lengths.

2°. *Congruence tests for triangles:*
1. If two sides of a triangle and their included angle are congruent to the corresponding parts of another triangle, then the triangles are congruent.
2. If two angles of a triangle and their included side are congruent to the corresponding parts of another triangle, then the triangles are congruent.

3. If three sides of a triangle are congruent to the corresponding sides of another triangle, then the triangles are congruent.

3°. Triangles are said to be *similar* if their corresponding angles are equal and their corresponding sides are proportional.

*Similarity tests for triangles:*
1. If all three pairs of corresponding sides in a pair of triangles are in proportion, then the triangles are similar.
2. If two pairs of corresponding angles in a pair of triangles are congruent, then the triangles are similar.
3. If two pairs of corresponding sides in a pair of triangles are in proportion and the included angles are congruent, then the triangles are similar.

The areas of similar triangles are proportional to the squares of the corresponding linear parts (such as sides, altitudes, diagonals, etc.).

4°. The line connecting the midpoints of two sides of a triangle is called a *midline* of the triangle. The midline is parallel to and half as long as the third side (Fig. 3.1b).

Let $a$, $b$, and $c$ be the lengths of the sides of a triangle; let $\alpha$, $\beta$, and $\gamma$ be the respective opposite angles (Fig. 3.1a); let $R$ and $r$ be the circumradius and the inradius, respectively; and let $p = \frac{1}{2}(a + b + c)$ be the semiperimeter.

Table 3.1 represents the basic properties and relations characterizing triangles.

TABLE 3.1
Basic properties and relations characterizing plane triangles

| No. | The name of property | Properties and relations |
|---|---|---|
| 1 | Triangle inequality | The length of any side of a triangle does not exceed the sum of lengths of the other two sides |
| 2 | Sum of angles of a triangle | $\alpha + \beta + \gamma = 180°$ |
| 3 | Law of sines | $\dfrac{a}{\sin \alpha} = \dfrac{b}{\sin \beta} = \dfrac{c}{\sin \gamma} = 2R$ |
| 4 | Law of cosines | $c^2 = a^2 + b^2 - 2ab\cos\gamma$ |
| 5 | Law of tangents | $\dfrac{a+b}{a-b} = \dfrac{\tan[\frac{1}{2}(\alpha+\beta)]}{\tan[\frac{1}{2}(\alpha-\beta)]} = \dfrac{\cot(\frac{1}{2}\gamma)}{\tan[\frac{1}{2}(\alpha-\beta)]}$ |
| 6 | Theorem on projections (law of cosines) | $c = a\cos\beta + b\cos\alpha$ |
| 7 | Trigonometric angle formulas | $\sin\dfrac{\gamma}{2} = \sqrt{\dfrac{(p-a)(p-b)}{ab}}$, $\cos\dfrac{\gamma}{2} = \sqrt{\dfrac{p(p-c)}{ab}}$, $\tan\dfrac{\gamma}{2} = \sqrt{\dfrac{(p-a)(p-b)}{p(p-c)}}$, $\sin\gamma = \dfrac{2}{ab}\sqrt{p(p-a)(p-b)(p-c)}$ |
| 8 | Law of tangents | $\tan\gamma = \dfrac{c\sin\alpha}{b - c\cos\alpha} = \dfrac{c\sin\beta}{a - c\cos\beta}$ |
| 9 | Mollweide's formulas | $\dfrac{a+b}{c} = \dfrac{\cos[\frac{1}{2}(\alpha-\beta)]}{\sin(\frac{1}{2}\gamma)} = \dfrac{\cos[\frac{1}{2}(\alpha-\beta)]}{\cos[\frac{1}{2}(\alpha+\beta)]}$, $\dfrac{a-b}{c} = \dfrac{\sin[\frac{1}{2}(\alpha-\beta)]}{\cos(\frac{1}{2}\gamma)} = \dfrac{\sin[\frac{1}{2}(\alpha-\beta)]}{\sin[\frac{1}{2}(\alpha+\beta)]}$ |

Table 3.2 permits one to find the sides and angles of an arbitrary triangle if three appropriately chosen sides and/or angles are given. From the relations given in Tables 3.1 and 3.2, one can derive all missing relations by cyclic permutations of the sides $a$, $b$, and $c$ and the angles $\alpha$, $\beta$, and $\gamma$.

TABLE 3.2
Solution of plane triangles

| No. | Three parts specified | Formulas for the remaining parts |
|---|---|---|
| 1 | Three sides $a, b, c$ | *First method.* <br> One of the angles is determined by the law of cosines, $\cos \alpha = \dfrac{b^2 + c^2 - a^2}{2bc}$. <br> Then either the law of sines or the law of cosines is applied. <br> *Second method.* <br> One of the angles is determined by trigonometric angle formulas. Further proceed in a similar way. <br> **Remark.** The sum of lengths of any two sides must be greater than the length of the third side. |
| 2 | Two sides $a, b$ and the included angle $\gamma$ | *First method.* <br> The side $c$ is determined by the law of cosines, $c = \sqrt{a^2 + b^2 - 2ab\cos\gamma}$. <br> The angle $\alpha$ is determined by either the law of cosines or the law of sines. The angle $\beta$ is determined from the sum of angles in triangle, $\beta = 180° - \alpha - \gamma$. <br> *Second method.* <br> $\alpha + \beta$ is found from the sum of angles in triangle, $\alpha + \beta = 180° - \gamma$; <br> $\alpha - \beta$ is found from the law of tangents, $\tan\dfrac{\alpha-\beta}{2} = \dfrac{a-b}{a+b}\cot\dfrac{\gamma}{2}$. <br> Then $\alpha$ and $\beta$ can be found. The third side $c$ is determined by either the law of cosines or the law of sines. |
| 3 | A side $c$ and the two angles $\alpha, \beta$ adjacent to it | The third angle $\gamma$ is found from the sum of angles in triangle, $\gamma = 180° - \alpha - \beta$. Sides $a$ and $b$ are determined by the law of sines. |
| 4 | Two sides $a, b$ and the angle $\alpha$ opposite one of them | The second angle is determined by the law of sines, $\sin\beta = \dfrac{b}{a}\sin\alpha$. <br> The third angle is $\gamma = 180° - \alpha - \beta$. <br> The third side is determined by the law of sines, $c = a\dfrac{\sin\gamma}{\sin\alpha}$. <br> **Remark.** Five cases are possible: <br> 1. $a > b$; i.e., the angle is opposite the greater side. Then $\alpha > \beta$, $\beta < 90°$ (the larger angle is opposite the larger side), and the triangle is determined uniquely. <br> 2. $a = b$; i.e., the triangle is isosceles and is determined uniquely. <br> 3. $a < b$ and $b\sin\alpha < a$. Then there are two solutions, $\beta_1 + \beta_2 = 180°$. <br> 4. $a < b$ and $b\sin\alpha = a$. Then the solution is unique, $\beta = 90°$. <br> 5. $a < b$ and $b\sin\alpha > a$. Then there are no solutions. |

### 3.1.1-2. Medians, angle bisectors, and altitudes of triangle.

A straight line through a vertex of a triangle and the midpoint of the opposite side is called a *median* of the triangle (Fig. 3.2a). The three medians of a triangle intersect in a single point lying strictly inside the triangle, which is called the *centroid* or *center of gravity* of the triangle. This point cuts the medians in the ratio 2:1 (counting from the corresponding vertices).

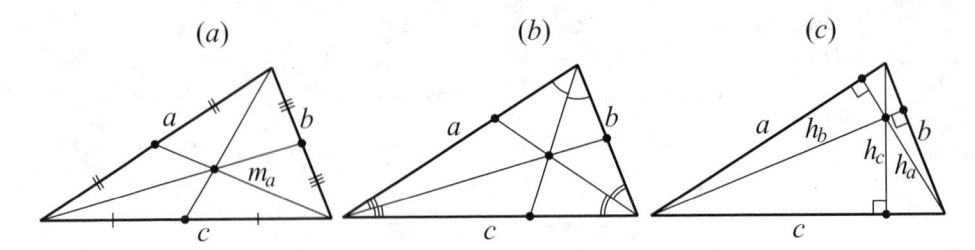

**Figure 3.2.** Medians (*a*), angle bisectors (*b*), and altitudes (*c*) of a triangle.

The length of the median $m_a$ to the side $a$ opposite the angle $\alpha$ is equal to

$$m_a = \frac{1}{2}\sqrt{2(b^2+c^2)-a^2} = \frac{1}{2}\sqrt{a^2+b^2+2ab\cos\gamma}. \qquad (3.1.1.1)$$

An *angle bisector* of a triangle is a line segment between a vertex and a point of the opposite side and dividing the angle at that vertex into two equal parts (Fig. 3.2*b*). The three angle bisectors intersect in a single point lying strictly inside the triangle. This point is equidistant from all sides and is called the *incenter* (the center of the *incircle* of the triangle). Concerning the radius $r$ of the incircle, see Paragraph 3.1.1-3. The angle bisector through a vertex cuts the opposite side in ratio proportional to the adjacent sides of the triangle.

The length of the angle bisector $l_a$ drawn to the side $a$ is given by the formulas

$$\begin{aligned} l_a &= \sqrt{bc - b_1 c_1} = \frac{\sqrt{bc[(b+c)^2 - a^2]}}{b+c} = \frac{\sqrt{4p(p-a)bc}}{b+c}, \\ l_a &= \frac{2cb\cos(\tfrac{1}{2}\alpha)}{b+c} = 2R\frac{\sin\beta \sin\gamma}{\cos[\tfrac{1}{2}(\beta-\gamma)]} = 2p\frac{\sin(\tfrac{1}{2}\beta)\sin(\tfrac{1}{2}\gamma)}{\sin\beta + \sin\gamma}, \end{aligned} \qquad (3.1.1.2)$$

where $b_1$ and $c_1$ are the segments of the side $a$ cut by bisector $l_a$ and adjacent to the sides $b$ and $c$, respectively, and $R$ is the circumradius (see Paragraph 3.1.1-3).

An *altitude* of a triangle is a straight line passing through a vertex and perpendicular to the straight line containing the opposite side (Fig. 3.2*c*). The three altitudes of a triangle intersect in a single point, called the *orthocenter* of the triangle.

The length of the altitude $h_a$ to the side $a$ is given by the formulas

$$\begin{aligned} h_a &= b\sin\gamma = c\sin\beta = \frac{bc}{2R}, \\ h_a &= 2(p-a)\cos\frac{\alpha}{2}\cos\frac{\beta}{2}\cos\frac{\gamma}{2} = 2(p-b)\sin\frac{\alpha}{2}\sin\frac{\beta}{2}\cos\frac{\gamma}{2}. \end{aligned} \qquad (3.1.1.3)$$

The lengths of the altitude, the angle bisector, and the median through the same vertex satisfy the inequality $h_a \leq l_a \leq m_a$. If $h_a = l_a = m_a$, then the triangle is isosceles; moreover, the first equality implies the second, and vice versa.

### 3.1.1-3. Circumcircle, incircle, and excircles.

A straight line passing through the midpoint of a segment and perpendicular to it is called the *perpendicular bisector* of the segment. The circle passing through the vertices of a triangle is called the *circumcircle* of the triangle. The center $O_1$ of the circumcircle, called the *circumcenter*, is the point where the perpendicular bisectors of the sides of the triangle

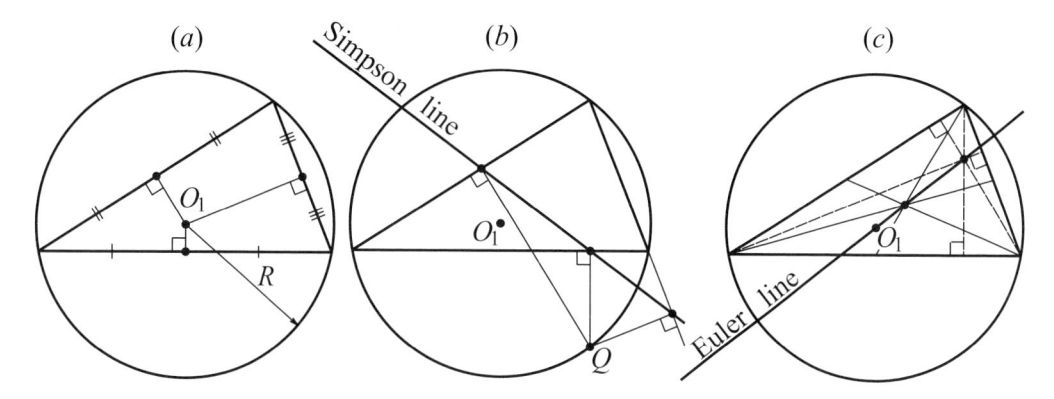

**Figure 3.3.** The circumcircle of a triangle. The circumcenter (*a*), the Simpson line (*b*), and the Euler line (*c*).

meet (Fig. 3.3*a*). The feet of the perpendiculars drawn from a point $Q$ on the circumcircle to the three sides of the triangle lie on the same straight line called the *Simpson line* of $Q$ with respect to the triangle (Fig. 3.3*b*). The circumcenter, the orthocenter, and the centroid lie on a single line, called the *Euler line* (Fig. 3.3*c*).

The circle tangent to the three sides of a triangle and lying inside the triangle is called the *incircle* of the triangle. The center $O_2$ of the incircle (the *incenter*) is the point where the angle bisectors meet (Fig. 3.4*a*). The straight lines connecting the vertices of a triangle with the points at which the incircle is tangent to the respective opposite sides intersect in a single point $G$ called the *Gergonne point* (Fig. 3.4*b*).

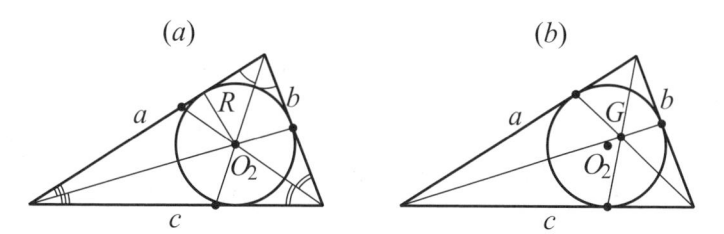

**Figure 3.4.** The incircle of a triangle (*a*). The incenter and the Gergonne point (*b*).

The circle tangent to one side of a triangle and to the extensions of the other two sides is called an *excircle* of the triangle. Each triangle has three excircles. The center of an excircle (an *excenter*) is the point of concurrency of two external angle bisectors and an interior angle bisector. The straight lines connecting the vertices of a triangle with the points at which the respective opposite sides are tangent to the excircles intersect in a single point $N$, called the *Nagel point* (Fig. 3.5).

The inradius $r$, the circumradius $R$, and the exradii $\rho_a$, $\rho_b$, and $\rho_c$ satisfy the relations

$$r = \sqrt{\frac{(p-a)(p-b)(p-c)}{p}} = p \tan\frac{\alpha}{2} \tan\frac{\beta}{2} \tan\frac{\gamma}{2}$$

$$= 4R \sin\frac{\alpha}{2} \sin\frac{\beta}{2} \sin\frac{\gamma}{2} = (p-c)\tan\frac{\gamma}{2} = \frac{S}{p}, \quad (3.1.1.4)$$

$$R = \frac{a}{2\sin\alpha} = \frac{b}{2\sin\beta} = \frac{c}{2\sin\gamma} = \frac{abc}{4S} = \frac{p}{4\cos\left(\frac{1}{2}\alpha\right)\cos\left(\frac{1}{2}\beta\right)\cos\left(\frac{1}{2}\gamma\right)}, \quad (3.1.1.5)$$

$$\frac{1}{r} = \frac{1}{\rho_a} + \frac{1}{\rho_b} + \frac{1}{\rho_c}. \quad (3.1.1.6)$$

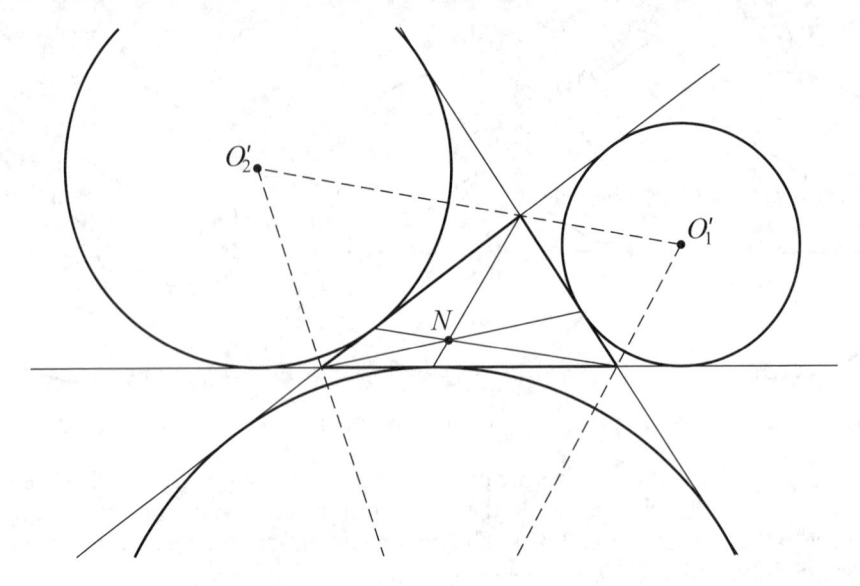

**Figure 3.5.** Excircles of a triangle. The Nagel point.

The distance $d_1$ between the circumcenter and the incenter and the distance $d_2$ between the circumcenter and the excenter are given by the expressions

$$d_1 = \sqrt{R^2 - 2Rr}, \qquad (3.1.1.7)$$
$$d_2 = \sqrt{R^2 + 2R\rho_a}. \qquad (3.1.1.8)$$

**3.1.1-4. Area of a triangle.**

The area $S$ of a triangle is given by the formulas

$$\begin{aligned} S &= ah_a = \frac{1}{2}ab\sin\gamma = rp, \\ S &= \sqrt{p(p-a)(p-b)(p-c)} \quad \text{(Heron's formula)}, \\ S &= \frac{abc}{4R} = 2R^2\sin\alpha\sin\beta\sin\gamma, \\ S &= c^2\frac{\sin\alpha\sin\beta}{2\sin\gamma} = c^2\frac{\sin\alpha\sin\beta}{2\sin(\alpha+\beta)}. \end{aligned} \qquad (3.1.1.9)$$

**3.1.1-5. Theorems about points and lines related to triangle.**

CEVA'S THEOREM. *Let points $C_1$, $A_1$, and $B_1$ lie on the sides $AB$, $BC$, and $CA$, respectively, of a triangle (Fig. 3.6). The straight lines $AA_1$, $BB_1$, and $CC_1$ are concurrent or parallel if and only if*

$$\frac{AC_1}{C_1B} \cdot \frac{BA_1}{A_1C} \cdot \frac{CB_1}{B_1A} = 1. \qquad (3.1.1.10)$$

STEWART'S THEOREM. *If a straight line through a vertex of a triangle divides the opposite side into segments of lengths $m$ and $n$ (Fig. 3.7), then*

$$(m+n)(p^2+mn) = b^2m + c^2n. \qquad (3.1.1.11)$$

**Figure 3.6.** Ceva's theorem.

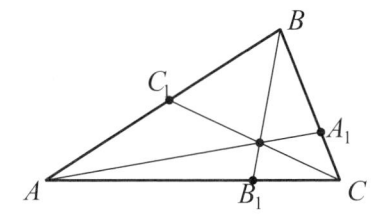

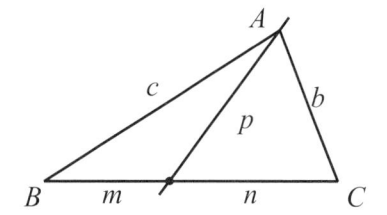

**Figure 3.7.** Stewart's theorem.

MENELAUS'S THEOREM. *If a straight line intersects sides $AB$, $BC$, and $CA$ of a triangle (Fig. 3.8) or their extensions at points $C_1$, $A_1$, and $B_1$, respectively, then*

$$\frac{AC_1}{C_1B} \cdot \frac{BA_1}{A_1C} \cdot \frac{CB_1}{B_1A} = -1. \qquad (3.1.1.12)$$

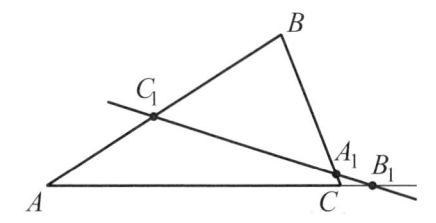

**Figure 3.8.** Menelaus's theorem.

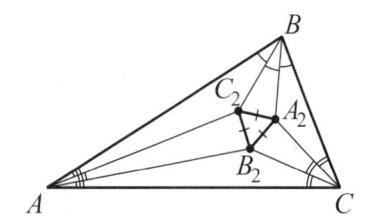

**Figure 3.9.** Morley's theorem.

Straight lines dividing the interior angles of a triangle into three equal parts are called *angle trisectors*.

MORLEY'S THEOREM. *The three points of intersection of adjacent angle trisectors of a triangle form an equilateral triangle (Fig. 3.9).*

In a triangle, the midpoints of the three sides, the feet of the three altitudes, and the midpoints of the segments of the altitudes between the orthocenter and the vertices all lie on a single circle, the *nine-point circle* (Fig. 3.10).

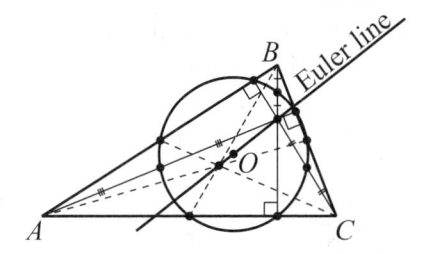

**Figure 3.10.** Nine-point circle.

FEUERBACH'S THEOREM. *The nine-point circle is tangent to the incircle and the three excircles. The points of tangency are called the Feuerbach points. The center of the nine-point circle lies on the Euler line (see Paragraph 3.1.1-3).*

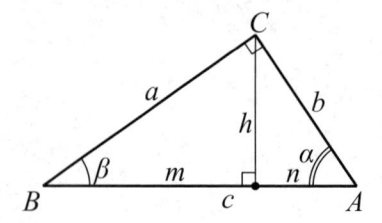

**Figure 3.11.** A right triangle.

### 3.1.1-6. Right (right-angled) triangles.

A *right* triangle is a triangle with a right angle. The side opposite the right angle is called the *hypotenuse*, and the other two sides are called the *legs* (Fig. 3.11).

The hypotenuse $c$, the legs $a$ and $b$, and the angles $\alpha$ and $\beta$ opposite the legs satisfy the following relations:

$$\alpha + \beta = 90°;$$

$$\sin\alpha = \cos\beta = \frac{a}{c}, \quad \sin\beta = \cos\alpha = \frac{b}{c}, \qquad (3.1.1.13)$$

$$\tan\alpha = \cot\beta = \frac{a}{b}, \quad \tan\beta = \cot\alpha = \frac{b}{a}.$$

One also has

$$a^2 + b^2 = c^2 \quad \text{(Pythagorean theorem)}, \qquad (3.1.1.14)$$

$$h^2 = mn, \quad a^2 = mc, \quad b^2 = nc, \qquad (3.1.1.15)$$

where $h$ is the length of the altitude drawn to the hypotenuse; moreover, the altitude cuts the hypotenuse into segments of lengths $m$ and $n$.

In a right triangle, the length of the median $m_c$ drawn from the vertex of the right angle coincides with the circumradius $R$ and is equal to half the length of the hypotenuse $c$, $m_c = R = \frac{1}{2}c$. The inradius is given by the formula $r = \frac{1}{2}(a+b-c)$. The area of the right triangle is $S = ah_a = \frac{1}{2}ab$ (see also Paragraphs 3.1.1-2 to 3.1.1-4).

### 3.1.1-7. Isosceles and equilateral triangles.

1°. An *isosceles* triangle is a triangle with two equal sides. These sides are called the *legs*, and the third side is called the *base* (Fig. 3.12a).

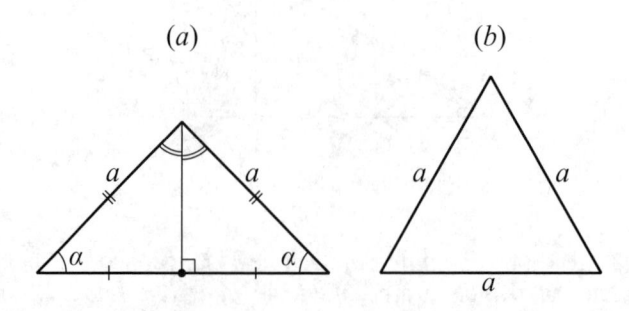

**Figure 3.12.** An isosceles triangle (*a*). An equilateral triangle (*b*).

*Properties of isosceles triangles:*

1. In an isosceles triangle, the angles adjacent to the base are equal.
2. In an isosceles triangle, the median drawn to the base is the angle bisector and the altitude.
3. In an isosceles triangle, the sum of distances from a point of the base to the legs is constant.

*Criteria for a triangle to be isosceles:*

1. If two angles in a triangle are equal, then the triangle is isosceles.
2. If a median in a triangle is also an altitude, then the triangle is isosceles.
3. If a bisector in a triangle is also an altitude, then the triangle is isosceles.

$2°$. An *equilateral* (or *regular*) triangle is a triangle with all three sides equal (Fig. 3.12b). All angles of an equilateral triangle are equal to $60°$. In an equilateral triangle, the circumradius $R$ and the inradius $r$ satisfy the relation $R = 2r$.

For an equilateral triangle with side length $a$, the circumradius and the inradius are given by the formulas $R = \frac{\sqrt{3}}{3}a$ and $r = \frac{\sqrt{3}}{6}a$, and the area is equal to $S = \frac{\sqrt{3}}{4}a^2$.

## 3.1.2. Polygons

### 3.1.2-1. Polygons. Basic information.

A *polygon* is a plane figure bounded by a closed broken line, i.e., a line obtained if one takes $n$ distinct points such that no three successive points are collinear and draws a straight line segment between each of these points and its successor as well as between the last point and the first point (Fig. 3.13a). The segments forming a polygon are called the *sides* (or *edges*), and the points themselves are called the *vertices* of the polygon. Two sides sharing a vertex, as well as two successive vertices (the endpoints of the same edge), are said to be adjacent. A polygon can be self-intersecting, but the points of self-intersection should not be vertices (Fig. 3.13b). A polygon is said to be *plane* if its vertices are coplanar. A polygon is said to be *simple* if its nonadjacent sides do not have common interior or endpoints. A polygon is said to be *convex* if it lies on one side of any straight line passing through two neighboring vertices (Fig. 3.13c). In what follows, we consider only plane simple convex polygons.

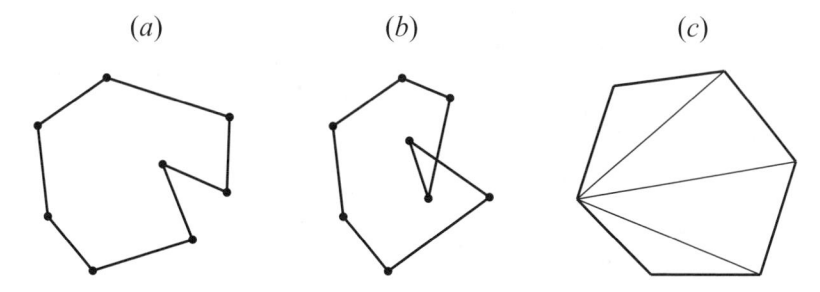

(a)  (b)  (c)

**Figure 3.13.** Polygons. Nonself-intersecting (*a*), self-intersecting (*b*), and convex (*c*) polygon.

An *(interior) angle* of a convex polygon is the angle between two sides meeting in a vertex. An angle adjacent to an interior angle is called an *external angle* of the convex polygon. A convex polygon is said to be *inscribed* in a circle if all of its vertices lie on the circle. A polygon is said to be *circumscribed* about a circle if all of its sides are tangent to the circle.

For a convex polygon with $n$ sides, the sum of interior angles is equal to $180°(n-2)$, and the sum of external angles is equal to $360°$.

One can find the area of an arbitrary polygon by dividing it into triangles.

**3.1.2-2. Properties of quadrilaterals.**

1. The diagonals of a convex quadrilateral meet.
2. The sum of interior angles of a convex quadrilateral is equal to 360° (Fig. 3.14a and b).
3. The lengths of the sides $a$, $b$, $c$, and $d$, the diagonals $d_1$ and $d_2$, and the segment $m$ connecting the midpoints of the diagonals satisfy the relation $a^2 + b^2 + c^2 + d^2 = d_1^2 + d_2^2 + 4m^2$.
4. A convex quadrilateral is circumscribed if and only if $a + c = b + d$.
5. A convex quadrilateral is inscribed if and only if $\alpha + \gamma = \beta + \delta$.
6. The relation $ac + bd = d_1 d_2$ holds for inscribed quadrilaterals (PTOLEMY'S THEOREM).

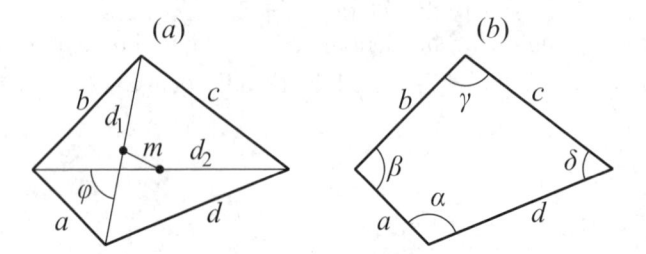

**Figure 3.14.** Quadrilaterals.

**3.1.2-3. Areas of quadrilaterals.**

The area of a convex quadrilateral is equal to

$$S = \frac{1}{2} d_1 d_2 \sin \varphi = \sqrt{p(p-a)(p-b)(p-c)(p-d) - abcd \cos^2 \frac{\beta + \delta}{2}}, \qquad (3.1.2.1)$$

where $\varphi$ is the angle between the diagonals $d_1$ and $d_2$ and $p = \frac{1}{2}(a + b + c + d)$.

The area of an inscribed quadrilateral is

$$S = \sqrt{p(p-a)(p-b)(p-c)(p-d)}. \qquad (3.1.2.2)$$

The area of a circumscribed quadrilateral is

$$S = \sqrt{abcd \sin^2 \frac{\beta + \delta}{2}}. \qquad (3.1.2.3)$$

If a quadrilateral is simultaneously inscribed and circumscribed, then

$$S = \sqrt{abcd}. \qquad (3.1.2.4)$$

**3.1.2-4. Basic quadrilaterals.**

1°. A *parallelogram* is a quadrilateral such that both pairs of opposite sides are parallel (Fig. 3.15a).

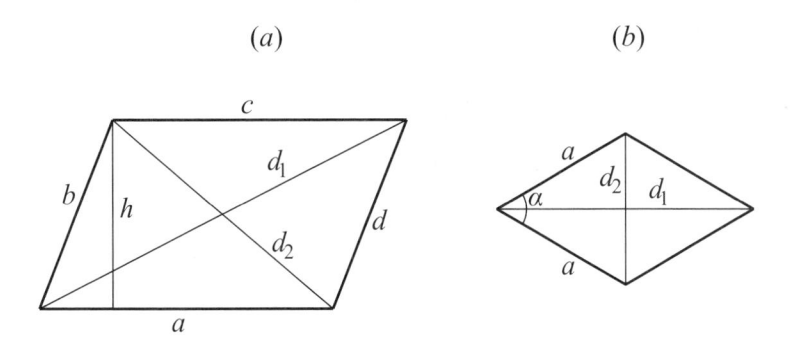

**Figure 3.15.** A parallelogram (*a*) and a rhombus (*b*).

*Attributes of parallelograms (a quadrilateral is a parallelogram if):*

1. Both pairs of opposite sides have equal length.
2. Both pairs of opposite angles are equal.
3. Two opposite sides are parallel and have equal length.
4. The diagonals meet and bisect each other.

*Properties of parallelograms:*

1. The diagonals meet and bisect each other.
2. Opposite sides have equal length, and opposite angles are equal.
3. The diagonals and the sides satisfy the relation $d_1^2 + d_2^2 = 2(a^2 + b^2)$.
4. The area of a parallelogram is $S = ah$, where $h$ is the altitude (see also Paragraph 3.1.2-3).

2°. A *rhombus* is a parallelogram in which all sides are of equal length (Fig. 3.15*b*).

*Properties of rhombi:*

1. The diagonals are perpendicular.
2. The diagonals are angle bisectors.
3. The area of a rhombus is $S = ah = a^2 \sin \alpha = \frac{1}{2} d_1 d_2$.

3°. A *rectangle* is a parallelogram in which all angles are right angles (Fig. 3.16*a*).

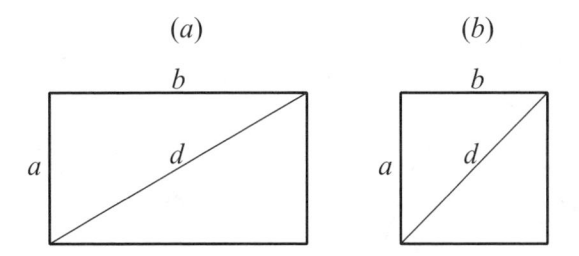

**Figure 3.16.** A rectangle (*a*) and a square (*b*).

*Properties of rectangles:*

1. The diagonals have equal lengths.
2. The area of a rectangle is $S = ab$.

4°. A *square* is a rectangle in which all sides have equal lengths (Fig. 3.16*b*). A square is also a special case of a rhombus (all angles are right angles).

*Properties of squares:*

1. All angles are right angles.
2. The diagonals are equal to $d = a\sqrt{2}$.
3. The diagonals meet at a right angle and are angle bisectors.
4. The area of a square is equal to $S = a^2 = \frac{1}{2}d^2$.

5°. A *trapezoid* is a quadrilateral in which two sides are parallel and the other two sides are nonparallel (Fig. 3.17). The parallel sides $a$ and $b$ are called the *bases* of the trapezoid, and the other two sides are called the *legs*. In an *isosceles* trapezoid, the legs are of equal length. The line segment connecting the midpoints of the legs is called the *median* of the trapezoid. The length of the median is equal to half the sum of the lengths of the bases, $m = \frac{1}{2}(a + b)$.

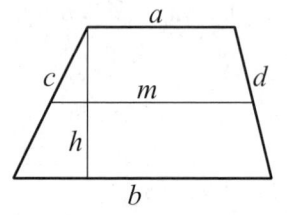

**Figure 3.17.** A trapezoid.

The perpendicular distance between the bases is called the *altitude* of a trapezoid.

*Properties of trapezoids:*

1. A trapezoid is circumscribed if and only if $a + b = c + d$.
2. A trapezoid is inscribed if and only if it is isosceles.
3. The area of a trapezoid is $S = \frac{1}{2}(a+b)h = mh = \frac{1}{2}d_1 d_2 \sin\varphi$, where $\varphi$ is the angle between the diagonals $d_1$ and $d_2$.
4. The segment connecting the midpoints of the diagonals is parallel to the bases and has the length $\frac{1}{2}(b-a)$.

**Example 1.** Consider an application of plane geometry to measuring distances in geodesy. Suppose that the angles $\alpha$, $\beta$, $\gamma$, and $\delta$ between a straight line $AB$ and the directions to points $D$ and $C$ are known at points $A$ and $B$ (Fig. 3.18a). Suppose also that the distance $a = AB$ (or $b = DC$) is known and the task is to find the distance $b = DC$ (or $a = AB$).

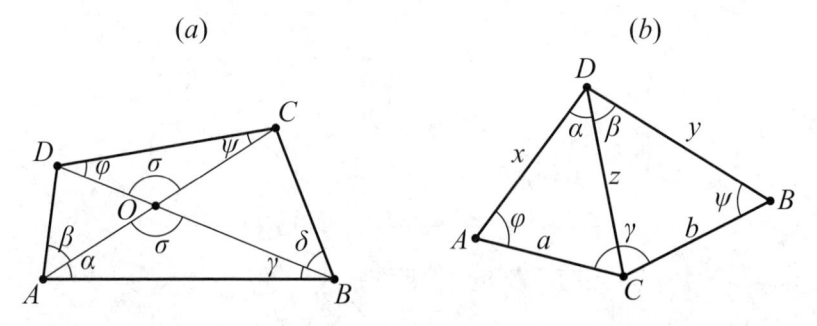

**Figure 3.18.** Applications of plane geometry in geodesy.

Let us find the angles $\varphi$ and $\psi$. Since $\sigma$ is the angle at the vertex $O$ in both triangles $AOB$ and $DOC$, it follows that $\alpha + \gamma = \varphi + \psi$. Let $\varepsilon_1 = \frac{1}{2}(\varphi + \psi)$. We twice apply the law of sines (Table 3.1) and find the half-difference of the desired angles. The main formulas read

$$\frac{AD}{a} = \frac{\sin\gamma}{\sin(\pi - \alpha - \beta - \gamma)} = \frac{\sin\gamma}{\sin(\alpha + \beta + \gamma)}, \quad \frac{BC}{a} = \frac{\sin\alpha}{\sin(\alpha + \gamma + \delta)},$$

$$\frac{b}{AD} = \frac{\sin\beta}{\sin\gamma}, \quad \frac{b}{BC} = \frac{\sin\delta}{\sin\varphi}.$$

These relations imply that

$$\frac{b}{a} = \frac{\sin\beta\sin\gamma}{\sin\psi\sin(\alpha+\beta+\gamma)} = \frac{\sin\delta\sin\alpha}{\sin\varphi\sin(\alpha+\gamma+\delta)} \tag{3.1.2.5}$$

and hence

$$\frac{\sin\varphi}{\sin\psi} = \frac{\sin\delta\sin\alpha\sin(\alpha+\beta+\gamma)}{\sin\beta\sin\gamma\sin(\alpha+\gamma+\delta)} = \cot\eta,$$

where $\eta$ is an auxiliary angle. By adding and subtracting, we obtain

$$\frac{\sin\varphi - \sin\psi}{\sin\varphi + \sin\psi} = \frac{\cot\eta - 1}{\cot\eta + 1}, \quad \frac{2\cos[\tfrac{1}{2}(\varphi+\psi)]\sin[\tfrac{1}{2}(\varphi-\psi)]}{2\sin[\tfrac{1}{2}(\varphi+\psi)]\cos[\tfrac{1}{2}(\varphi-\psi)]} = \frac{\cot(\tfrac{1}{4}\pi)\cot\eta - 1}{\cot\eta + \cot(\tfrac{1}{4}\pi)},$$

$$\tan\frac{\varphi-\psi}{2} = \tan\frac{\varphi+\psi}{2}\cot\left(\frac{\pi}{4}+\eta\right) = \tan\frac{\alpha+\gamma}{2}\cot\left(\frac{\pi}{4}+\eta\right).$$

From this we find $\varepsilon_2 = \tfrac{1}{2}(\varphi-\psi)$ and, substituting $\varphi = \varepsilon_1 + \varepsilon_2$ and $\psi = \varepsilon_1 - \varepsilon_2$ into (3.1.2.5), obtain the desired distance.

**Example 2.** Suppose that the mutual position of three points $A$, $B$, and $C$ is determined by the segments $AC = a$ and $BC = b$, and the angle $\angle ACB = \gamma$. Suppose that the following angles have been measured at some point $D$: $\angle CDA = \alpha$ and $\angle CDB = \beta$.

In the general case, one can find the position of point $D$ with respect to $A$, $B$, and $C$, i.e., uniquely determine the segments $x$, $y$, and $z$ (Fig. 3.18*b*). For this to be possible, it is necessary that $D$ does not lie on the circumcircle of the triangle $ABC$. We have

$$\varphi + \psi = 2\pi - (\alpha+\beta+\gamma) = 2\varepsilon_1. \tag{3.1.2.6}$$

By the law of sines (Table 3.1), we obtain

$$\sin\varphi = \frac{z}{a}\sin\alpha, \quad \sin\psi = \frac{z}{b}\sin\beta, \tag{3.1.2.7}$$

which implies that

$$\frac{\sin\varphi}{\sin\psi} = \frac{b\sin\alpha}{a\sin\beta} = \cot\eta, \tag{3.1.2.8}$$

where $\eta$ is an auxiliary angle. We find the angles $\varphi$ and $\psi$ from (3.1.2.6) and (3.1.2.8), substitute them into (3.1.2.7) to determine $z$, and finally apply the law of sines to obtain $x$ and $y$.

### 3.1.2-5. Regular polygons.

A convex polygon is said to be *regular* if all of its sides have the same length and all of its interior angles are equal. A convex $n$-gon is regular if and only if it is taken to itself by the rotation by an angle of $2\pi/n$ about some point $O$. The point $O$ is called the *center* of the regular polygon. The angle between two rays issuing from the center and passing through two neighboring vertices is called the *central angle* (Fig. 3.19).

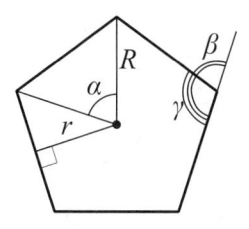

**Figure 3.19.** A regular polygon.

*Properties of regular polygons:*

1. The center is equidistant from all vertices as well as from all sides of a regular polygon.
2. A regular polygon is simultaneously inscribed and circumscribed; the centers of the circumcircle and the incircle coincide with the center of the polygon itself.
3. In a regular polygon, the central angle is $\alpha = 360°/n$, the external angle is $\beta = 360°/n$, and the interior angle is $\gamma = 180° - \beta$.
4. The circumradius $R$, the inradius $r$, and the side length $a$ of a regular polygon satisfy the relations

$$a = 2\sqrt{R^2 - r^2} = 2R \sin \frac{\alpha}{2} = 2r \tan \frac{\alpha}{2}. \tag{3.1.2.9}$$

5. The area $S$ of a regular $n$-gon is given by the formula

$$S = \frac{arn}{2} = nr^2 \tan \frac{\alpha}{2} = nR^2 \sin \frac{\alpha}{2} = \frac{1}{4} na^2 \cot \frac{\alpha}{2}. \tag{3.1.2.10}$$

Table 3.3 presents several useful formulas for regular polygons.

TABLE 3.3
Regular polygons ($a$ is the side length)

| No. | Name | Inradius $r$ | Circumradius $R$ | Area $S$ |
|---|---|---|---|---|
| 1 | Regular polygon | $\dfrac{a}{2 \tan \frac{\pi}{n}}$ | $\dfrac{a}{2 \sin \frac{\pi}{n}}$ | $\dfrac{1}{2} arn$ |
| 2 | Triangle | $\dfrac{\sqrt{3}}{6} a$ | $\dfrac{\sqrt{3}}{3} a$ | $\dfrac{\sqrt{3}}{4} a^2$ |
| 3 | Square | $\dfrac{1}{2} a$ | $\dfrac{1}{\sqrt{2}} a$ | $a^2$ |
| 4 | Pentagon | $\sqrt{\dfrac{5 + 2\sqrt{5}}{20}}\, a$ | $\sqrt{\dfrac{5 + \sqrt{5}}{10}}\, a$ | $\dfrac{\sqrt{25 + 10\sqrt{5}}}{4} a^2$ |
| 5 | Hexagon | $\dfrac{\sqrt{3}}{2} a$ | $a$ | $\dfrac{3\sqrt{3}}{2} a^2$ |
| 6 | Octagon | $\dfrac{1 + \sqrt{2}}{2} a$ | $\dfrac{\sqrt{2 + \sqrt{2}}}{2} a$ | $2(1 + \sqrt{2}) a^2$ |
| 7 | Enneagon | $\dfrac{5 + 2\sqrt{5}}{2} a$ | $\dfrac{1 + \sqrt{5}}{2} a$ | $\dfrac{\sqrt{5 + 2\sqrt{5}}}{2} a^2$ |
| 8 | Dodecagon | $\dfrac{2 + \sqrt{3}}{2} a$ | $\dfrac{3 + \sqrt{3}}{\sqrt{6}} a$ | $3(2 + \sqrt{3}) a^2$ |

## 3.1.3. Circle

### 3.1.3-1. Some definitions and formulas.

The *circle of radius $R$ centered at $O$* is the set of all points of the plane at a fixed distance $R$ from a fixed point $O$ (Fig. 3.20a). A plane figure bounded by a circle is called a *disk*. A segment connecting two points on a circle is called a *chord*. A chord passing through the center of a circle is called a *diameter* of the circle (Fig. 3.20b). The diameter length is $d = 2R$. A straight line that meets a circle at a single point is called a *tangent*, and the common point is called the *point of tangency* (Fig. 3.20c). An angle formed by two radii is called a *central angle*. An angle formed by two chords with a common endpoint is called an *inscribed angle*.

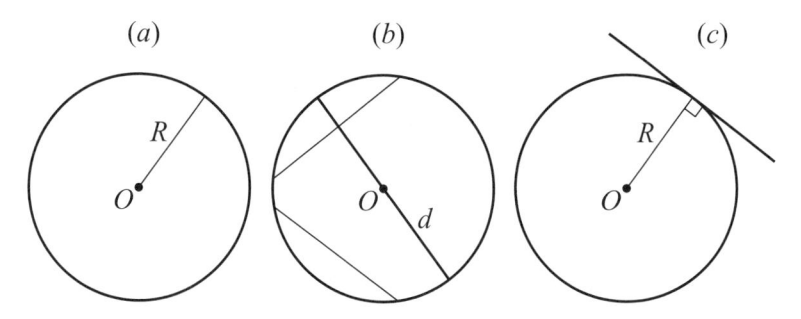

**Figure 3.20.** A circle (*a*). A diameter (*b*) and a tangent (*c*) of a circle.

*Properties of circles and disks:*

1. The circumference is $L = 2\pi R = \pi d = 2\sqrt{\pi S}$.
2. The area of a disk is $S = \pi R^2 = \frac{1}{4}\pi d^2 = \frac{1}{4} L d$.
3. The diameter of a circle is a longest chord.
4. The diameter passing through the midpoint of the chord is perpendicular to the chord.
5. The radius drawn to the point of tangency is perpendicular to the tangent.
6. An inscribed angle is half the central angle subtended by the same chord, $\alpha = \frac{1}{2}\angle BOC$ (Fig. 3.21*a*).
7. The angle between a chord and the tangent to the circle at an endpoint of the chord is $\beta = \frac{1}{2}\angle AOC$ (Fig. 3.21*a*).
8. The angle between two chords is $\gamma = \frac{1}{2}(\breve{BC} + \breve{ED})$ (Fig. 3.21*b*).
9. The angle between two secants is $\alpha = \frac{1}{2}(\breve{DE} - \breve{BC})$ (Fig. 3.21*c*).

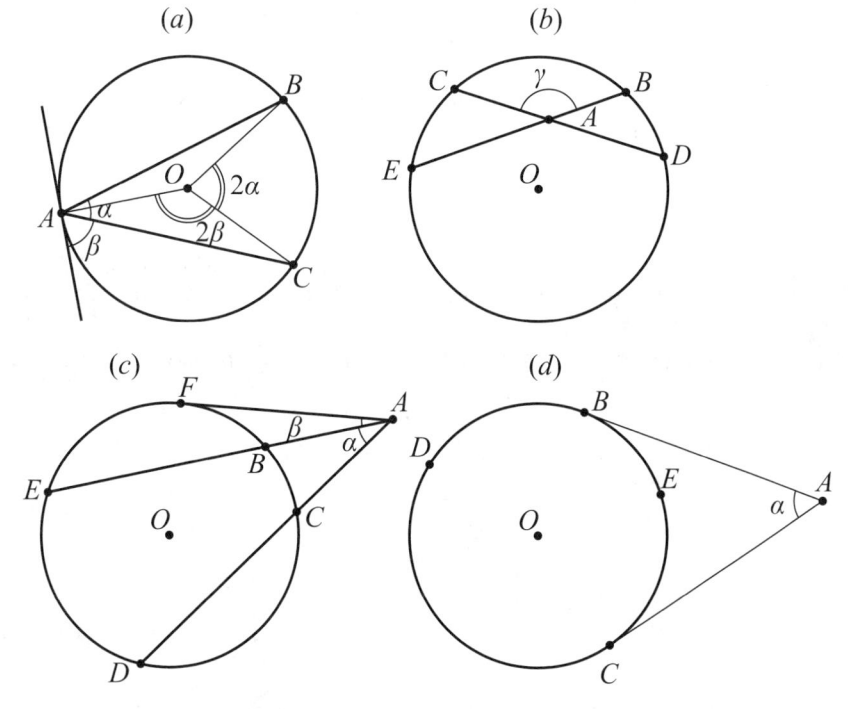

**Figure 3.21.** Properties of circles and disks.

10. The angle between a secant and the tangent to the circle at an endpoint of the secant is

$\beta = \frac{1}{2}(\breve{FE} - \breve{BF})$ (Fig. 3.21c).
11. The angle between two tangents is $\alpha = \frac{1}{2}(\breve{BDC} - \breve{BEC})$ (Fig. 3.21d).
12. If two chords meet, then $AC \cdot AD = AB \cdot AE = R^2 - m^2$ (Fig. 3.21b).
13. For secants, $AC \cdot AD = AB \cdot AE = m^2 - R^2$ (Fig. 3.21c).
14. For a tangent and a secant, $AF \cdot AF = AC \cdot AD$ (Fig. 3.21c).

### 3.1.3-2. Segment and sector.

A plane figure bounded by two radii and one of the subtending arcs is called a *(circular) sector*. A plane figure bounded by an arc and the corresponding chord is called a *segment* (Fig. 3.22a). If $R$ is the radius of the circle, $l$ is the arc length, $a$ is the chord length, $\alpha$ is the central angle (in degrees), and $h$ is the height of the segment, then the following formulas hold:

$$a = 2\sqrt{2hR - h^2} = 2R\sin\frac{\alpha}{2},$$

$$h = R - \sqrt{R^2 - \frac{a^2}{4}} = R\left(1 - \cos\frac{\alpha}{2}\right) = \frac{a}{2}\tan\frac{\alpha}{4}, \qquad (3.1.3.1)$$

$$l = \frac{2\pi R\alpha}{360} \approx 0.01745\,R\alpha.$$

The area of a circular sector is given by the formula

$$S = \frac{lR}{2} = \frac{\pi R^2 \alpha}{360} \approx 0.00873\,R^2\alpha, \qquad (3.1.3.2)$$

and the area of a segment not equal to a half-disk is given by the expression

$$S_1 = \frac{\pi R^2 \alpha}{360} \pm S_\triangle, \qquad (3.1.3.3)$$

where $S_\triangle$ is the area of the triangle with vertices at the center of the disk and at the endpoints of the radii bounding the corresponding sector. One takes the minus sign for $\alpha < 180$ and the plus sign for $\alpha > 180$.

The arc length and the area of a segment can be found by the approximate formulas

$$l \approx \frac{8b - a}{3}, \quad l \approx \sqrt{a^2 + \frac{16h^2}{3}},$$
$$S_1 \approx \frac{h(6a + 8b)}{15}, \qquad (3.1.3.4)$$

where $b$ is the chord of the half-segment (see Fig. 3.22a).

### 3.1.3-3. Annulus.

An *annulus* is a plane figure bounded by two concentric circles of distinct radii (Fig. 3.22b). Let $R$ be the outer radius of an annulus (the radius of the outer bounding circle), and let $r$

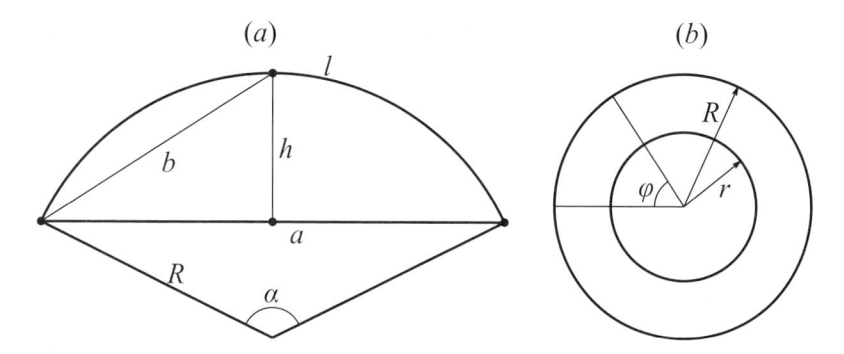

**Figure 3.22.** A segment (*a*) and an annulus (*b*).

be the inner radius (the radius of the inner bounding circle). Then the area of the annulus is given by the formula

$$S = \pi(R^2 - r^2) = \frac{\pi}{4}(D^2 - d^2) = 2\pi\rho\delta, \tag{3.1.3.5}$$

where $D = 2R$ and $d = 2r$ are the outer and inner diameters, $\rho = \frac{1}{2}(R + r)$ is the midradius, and $\delta = R - r$ is the width of the annulus.

The area of the part of the annulus contained in a sector of central angle $\varphi$, given in degrees (see Fig. 3.22*b*), is given by the formula

$$S = \frac{\pi\varphi}{360}(R^2 - r^2) = \frac{\pi\varphi}{1440}(D^2 - d^2) = \frac{\pi\varphi}{180}\rho\delta. \tag{3.1.3.6}$$

## 3.2. Solid Geometry

### 3.2.1. Straight Lines, Planes, and Angles in Space

3.2.1-1. Mutual arrangement of straight lines and planes.

1°. Two distinct straight lines lying in a single plane either have exactly one point of intersection or do not meet at all. In the latter case, they are said to be *parallel*. If two straight lines do not lie in a single plane, then they are called *skew lines*.

The angle between skew lines is determined as the angle between lines parallel to them and lying in a single plane (Fig. 3.23*a*). The distance between skew lines is the length of the straight line segment that meets both lines and is perpendicular to them.

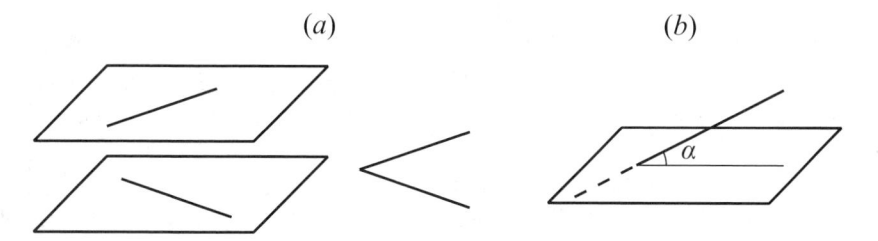

**Figure 3.23.** The angle between skew lines (*a*). The angle between a line and a plane (*b*).

2°. Two distinct planes either intersect in a straight line or do not have common points. In the latter case, they are said to be *parallel*. Coinciding planes are also assumed to be parallel. If two planes are perpendicular to a single straight line or each of them contains a pair of intersecting straight lines parallel to the corresponding lines in the other pair, then the planes are parallel.

3°. A straight line either entirely lies in the plane, meets the plane at a single point, or has no common points with the plane. In the last case, the line is said to be *parallel* to the plane.

The angle between a straight line and a plane is equal to the angle between the line and its projection onto the plane (Fig. 3.23b). If a straight line is perpendicular to two intersecting straight lines on a plane, then it is perpendicular to each line on the plane, i.e., *perpendicular to the plane*.

3.2.1-2. Polyhedral angles.

1°. A *dihedral angle* is a figure in space formed by two half-planes issuing from a single straight line as well as the part of space bounded by these half-planes. The half-planes are called the *faces* of the dihedral angle, and their common straight line is called the *edge*. A dihedral angle is measured by its linear angle $ABC$ (Fig. 3.24a), i.e., by the angle between the perpendiculars raised to the edge $DE$ of the dihedral angle in both planes (*faces*) at the same point.

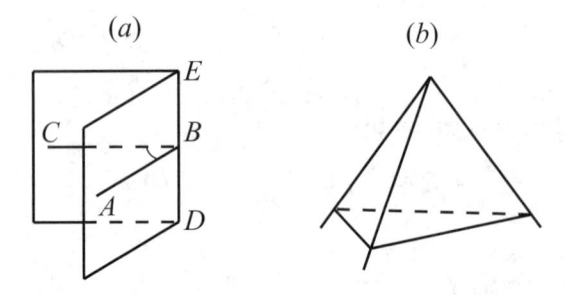

**Figure 3.24.** A dihedral (*a*) and a trihedral (*b*) angle.

2°. A part of space bounded by an infinite triangular pyramid is called a *trihedral angle* (Fig. 3.24b). The faces of this pyramid are called the *faces* of the trihedral angle, and the vertex of the pyramid is called the *vertex of a trihedral angle*. The rays in which the faces intersect are called the *edges* of a trihedral angle. The edges form *face angles*, and the faces form the dihedral angles of the trihedral angle. As a rule, one considers trihedral angles with dihedral angles less than $\pi$ (or $180°$), i.e., convex trihedral angles. Each face angle of a convex trihedral angle is less than the sum of the other two face angles and greater than their difference.

*Two trihedral angles are equal if one of the following conditions is satisfied:*

1. Two face angles, together with the included dihedral angle, of the first trihedral angle are equal to the respective parts (arranged in the same order) of the second trihedral angle.
2. Two dihedral angles, together with the included face angle, of the first trihedral angle are equal to the respective parts (arranged in the same order) of the second trihedral angle.
3. The three face angles of the first trihedral angle are equal to the respective face angles (arranged in the same order) of the second trihedral angle.
4. The three dihedral angles of the first trihedral angle are equal to the respective dihedral angles (arranged in the same order) of the second trihedral angle.

3°. A *polyhedral angle* $OABCDE$ (Fig. 3.25a) is formed by several planes (*faces*) having a single common point (the *vertex*) and successively intersecting along straight lines $OA$,

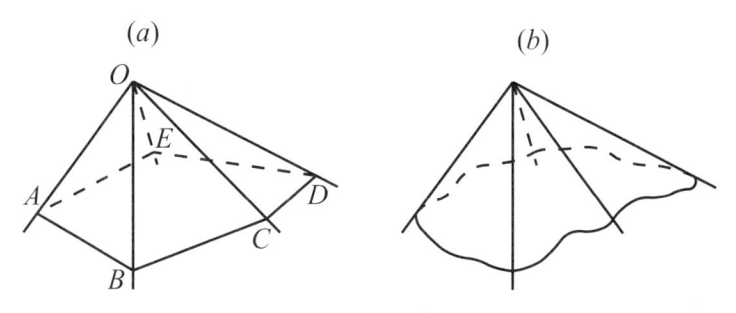

**Figure 3.25.** A polyhedral (*a*) and a solid (*b*) angle.

$OB, \ldots, OE$ (the *edges*). Two edges belonging to the same face form a *face angle* of the polyhedral angle, and two neighboring faces form a *dihedral angle*.

Polyhedral angles are equal (*congruent*) if one can be transformed into the other by translations and rotations. For polyhedral angles to be congruent, the corresponding parts (face and dihedral angles) must be equal. However, if the corresponding equal parts are arranged in reverse order, then the polyhedral angles cannot be transformed into each other by translations and rotations. In this case, they are said to be *symmetric*.

A convex polyhedral angle lies entirely on one side of each of its faces. The sum $\angle AOB + \angle BOC + \cdots + \angle EOA$ of face angles (Fig. 3.25*a*) of any convex polyhedral angle is less that $2\pi$ (or $360°$).

4°. A *solid angle* is a part of space bounded by straight lines issuing from a single point (vertex) to all points of some closed curve (Fig. 3.25*b*). Trihedral and polyhedral angles are special cases of solid angles. A solid angle is measured by the area cut by the solid angle on the sphere of unit radius centered at the vertex. Solid angles are measured in steradians. The entire sphere forms a solid angle of $4\pi$ steradians.

### 3.2.2. Polyhedra

#### 3.2.2-1. General concepts.

A *polyhedron* is a solid bounded by planes. In other words, a polyhedron is a set of finitely many plane polygons satisfying the following conditions:

1. Each side of each polygon is simultaneously a side of a unique other polygon, which is said to be adjacent to the first polygon (via this side).
2. From each of the polygons forming a polyhedron, one can reach any other polygon by successively passing to adjacent polygons.

These polygons are called the *faces*, their sides are called the *edges*, and their vertices are called the *vertices* of a polyhedron.

A polyhedron is said to be *convex* if it lies entirely on one side of the plane of any of its faces; if a polyhedron is convex, then so are its faces.

EULER'S THEOREM. *If the number of vertices in a convex polyhedron is $e$, the number of edges is $f$, and the number of faces is $g$, then $e + f - g = 2$.*

#### 3.2.2-2. Prism. Parallelepiped.

1°. A *prism* is a polyhedron in which two faces are $n$-gons (the *base faces* of the prism) and the remaining $n$ faces (*joining faces*) are parallelograms. The base faces of a prism are

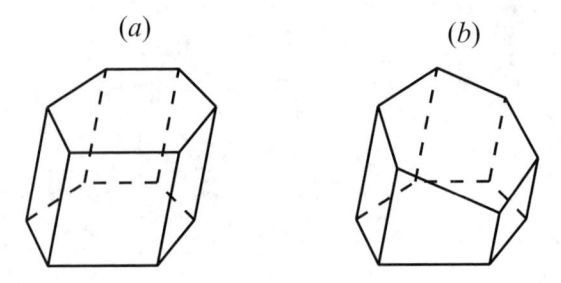

Figure 3.26. A prism (a) and a truncated prism (b).

equal (congruent) and lie in parallel planes (Fig. 3.26a). A *right* prism is a prism in which the joining faces are perpendicular to the base faces. A right prism is said to be *regular* if its base face is a regular polygon.

If $l$ is the joining edge length, $S$ is the area of the base face, $H$ is the altitude of the prism, $P_{\sec}$ is the perimeter of a perpendicular section, and $S_{\sec}$ is the area of the perpendicular section, then the area of the lateral surface $S_{\text{lat}}$ and the volume $V$ of the prism can be determined by the formulas

$$S_{\text{lat}} = P_{\sec} l$$
$$V = SH = S_{\sec} l. \tag{3.2.2.1}$$

The portion of a prism cut by a plane nonparallel to the base face is called a *truncated prism* (Fig. 3.26b). The volume of a truncated prism is

$$V = L P_1, \tag{3.2.2.2}$$

where $L$ is the length of the segment connecting the centers of the base faces and $P_1$ is the area of the section of the prism by a plane perpendicular to this segment.

2°. A prism whose bases are parallelograms is called a *parallelepiped*. All four diagonals in a parallelepiped intersect at a single point and bisect each other (Fig. 3.27a). A parallelepiped is said to be *rectangular* if it is a right prism and its base faces are rectangles. In a rectangular parallelepiped, all diagonals are equal (Fig. 3.27b).

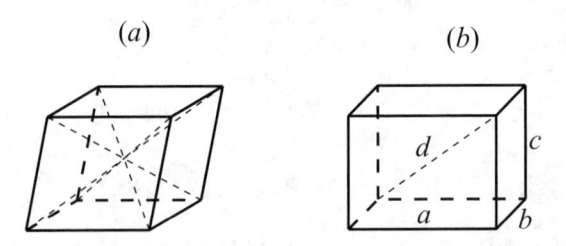

Figure 3.27. A parallelepiped (a) and a rectangular parallelepiped (b).

If $a$, $b$, and $c$ are the lengths of the edges of a rectangular parallelepiped, then the diagonal $d$ can be determined by the formula $d^2 = a^2 + b^2 + c^2$. The volume of a rectangular parallelepiped is given by the formula $V = abc$, and the lateral surface area is $S_{\text{lat}} = PH$, where $P$ is the perimeter of the base face.

3°. A rectangular parallelepiped all of whose edges are equal ($a = b = c$) is called a *cube*. The diagonal of a cube is given by the formula $d^2 = 3a^2$. The volume of the cube is $V = a^3$, and the lateral surface area is $S_{lat} = 4a^2$.

### 3.2.2-3. Pyramid, obelisk, and wedge.

1°. A *pyramid* is a polyhedron in which one face (the *base* of the pyramid) is an arbitrary polygon and the other (*lateral*) faces are triangles with a common vertex, called the *apex* of the pyramid (Fig. 3.28a). The base of an $n$-sided pyramid is an $n$-gon. The perpendicular through the apex to the base of a pyramid is called the *altitude* of the pyramid.

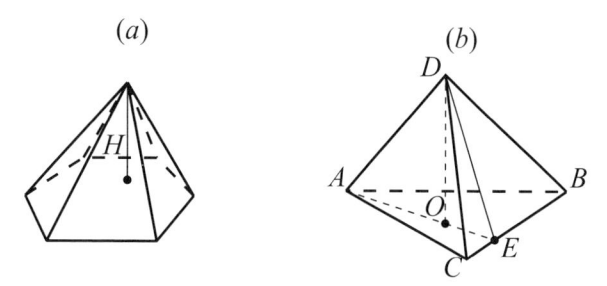

**Figure 3.28.** A pyramid (a). The attitude $DO$, the plane $DAE$, and the side $BC$ in a triangular pyramid (b).

The volume of a pyramid is given by the formula

$$V = \frac{1}{3}SH, \qquad (3.2.2.3)$$

where $S$ is the area of the base and $H$ is the altitude of the pyramid.

The apex of a pyramid is projected onto the circumcenter of the base if one of the following conditions is satisfied:
1. The lengths of all lateral edges are equal.
2. All lateral edges make equal angles with the base plane.

The apex of a pyramid is projected onto the incenter of the base if one of the following conditions is satisfied:
3. All lateral faces have equal apothems.
4. The angles between all lateral faces and the base are the same.

If $DO$ is the altitude of the pyramid $ABCD$ and $DA \perp BC$, then the plane $DAE$ is perpendicular to $BC$ (Fig. 3.28b).

If the pyramid is cut by a plane (Fig. 3.29a) parallel to the base, then

$$\begin{aligned} \frac{SA_1}{A_1A} &= \frac{SB_1}{B_1B} = \cdots = \frac{SO_1}{O_1O}, \\ \frac{S_{ABCDEF}}{S_{A_1B_1C_1D_1E_1F_1}} &= \left(\frac{SO}{SO_1}\right)^2, \end{aligned} \qquad (3.2.2.4)$$

where $SO$ is the altitude of the pyramid, i.e., the segment of the perpendicular through the vertex to the base.

The altitude of a *triangular pyramid* passes through the orthocenter of the base if and only if all pairs of opposite edges of the pyramid are perpendicular. The volume of a triangular pyramid (Fig. 3.29b), where $DA = a$, $DB = b$, $DC = c$, $BC = p$, $AC = q$, and $AB = r$, is given by the formula

$$V^2 = \frac{1}{288} \begin{vmatrix} 0 & r^2 & q^2 & a^2 & 1 \\ r^2 & 0 & p^2 & b^2 & 1 \\ q^2 & p^2 & 0 & c^2 & 1 \\ a^2 & b^2 & c^2 & 0 & 1 \\ 1 & 1 & 1 & 1 & 0 \end{vmatrix}, \qquad (3.2.2.5)$$

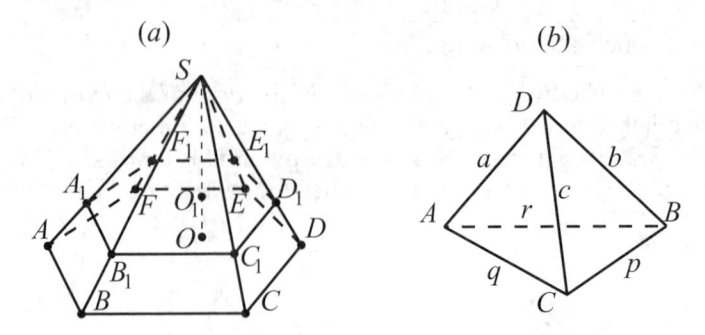

**Figure 3.29.** The pyramid cut by a plane and the original pyramid (*a*). A triangular pyramid (*b*).

where the right-hand side contains a determinant.

A pyramid is said to be *regular* if its base is a regular $n$-gon and the altitude passes through the center of the base. The altitude (issuing from the apex) of a lateral face is called the *apothem* of a regular pyramid. For a regular pyramid, the lateral surface area is

$$S_{\text{lat}} = \frac{1}{2} Pl, \qquad (3.2.2.6)$$

where $P$ is the perimeter of the base and $l$ is the apothem.

2°. If a pyramid is cut by a plane parallel to the base, then it splits into two parts, a pyramid similar to the original pyramid and the *frustum* (Fig. 3.30a). The volume of the frustum is

$$V = \frac{1}{3} h (S_1 + S_2 + \sqrt{S_1 S_2}) = \frac{1}{3} h S_2 \left[ 1 + \frac{a}{A} + \frac{a^2}{A^2} \right], \qquad (3.2.2.7)$$

where $S_1$ and $S_2$ are the areas of the bases, $a$ and $A$ are two respective sides of the bases, and $h$ is the altitude (the perpendicular distance between the bases).

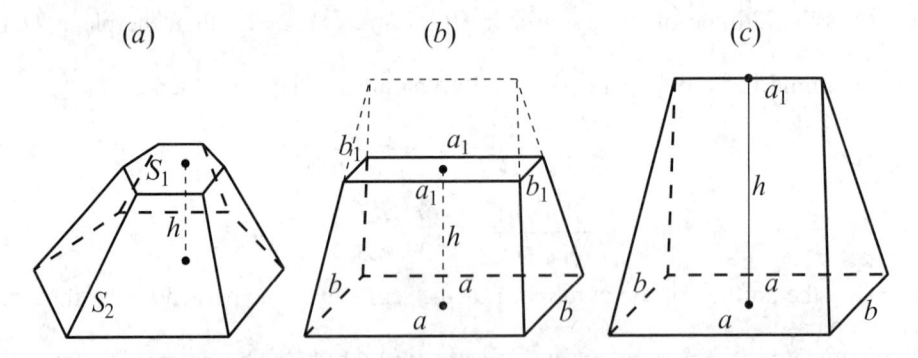

**Figure 3.30.** A frustum of a pyramid (*a*), an obelisk (*b*), and a wedge (*c*).

For a regular frustum, the lateral surface area is

$$S_{\text{lat}} = \frac{1}{2} (P_1 + P_2) l, \qquad (3.2.2.8)$$

where $P_1$ and $P_2$ are the perimeters of the bases and $l$ is the altitude of the lateral face.

3°. A hexahedron whose bases are rectangles lying in parallel planes and whose lateral faces form equal angles with the base, but do not meet at a single point, is called an *obelisk* (Fig. 3.30*b*). If $a$, $b$ and $a_1$, $b_1$ are the sides of the bases and $h$ is the altitude, then the volume of the hexahedron is

$$V = \frac{h}{6}[(2a + a_1)b + (2a_1 + a)b_1]. \qquad (3.2.2.9)$$

4°. A pentahedron whose base is a rectangle and whose lateral faces are isosceles triangles and isosceles trapezoids is called a *wedge* (Fig. 3.30*c*). The volume of the wedge is

$$V = \frac{h}{6}(2a + a_1)b. \qquad (3.2.2.10)$$

3.2.2-4. Regular polyhedra.

A polyhedron is said to be *regular* if all of its faces are equal regular polygons and all polyhedral angles are equal to each other. There exist five regular polyhedra (Fig. 3.31), whose properties are given in Table 3.4.

Tetrahedron  Cube  Octahedron

Dodecahedron  Icosahedron

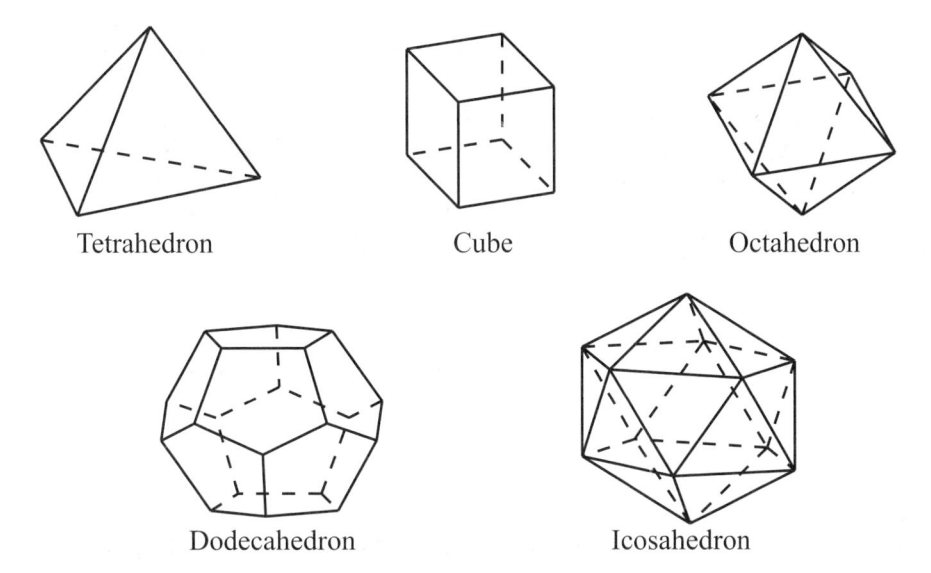

**Figure 3.31.** Five regular polyhedra.

## 3.2.3. Solids Formed by Revolution of Lines

3.2.3-1. Cylinder.

A *cylindrical surface* is a surface in space swept by a straight line (the *generator*) moving parallel to a given direction along some curve (the *directrix*) (Fig. 3.32*a*).

1°. A solid bounded by a closed cylindrical surface and two planes is called a *cylinder*; the planes are called the *bases* of the cylinder (Fig. 3.32*b*).

If $P$ is the perimeter of the base, $P_{\text{sec}}$ is the perimeter of the section perpendicular to the generator, $S_{\text{sec}}$ is the area of this section, $S_{\text{bas}}$ is the area of the base, and $l$ is the length of

TABLE 3.4
Regular polyhedra ($a$ is the edge length)

| No. | Name | Number of faces and its shape | Number of vertices | Number of edges | Total surface area | Volume |
|---|---|---|---|---|---|---|
| 1 | Tetrahedron | 4 triangles | 4 | 6 | $a^2\sqrt{3}$ | $\dfrac{a^3\sqrt{2}}{12}$ |
| 2 | Cube | 6 squares | 8 | 12 | $6a^2$ | $a^3$ |
| 3 | Octahedron | 8 triangles | 6 | 12 | $2a^2\sqrt{3}$ | $\dfrac{a^3\sqrt{2}}{3}$ |
| 4 | Dodecahedron | 12 pentagons | 20 | 30 | $3a^2\sqrt{25+10\sqrt{5}}$ | $\dfrac{a^3}{4}(15+7\sqrt{5})$ |
| 5 | Icosahedron | 20 triangles | 12 | 30 | $3a^2\sqrt{3}$ | $\dfrac{5a^3}{12}(3+\sqrt{5})$ |

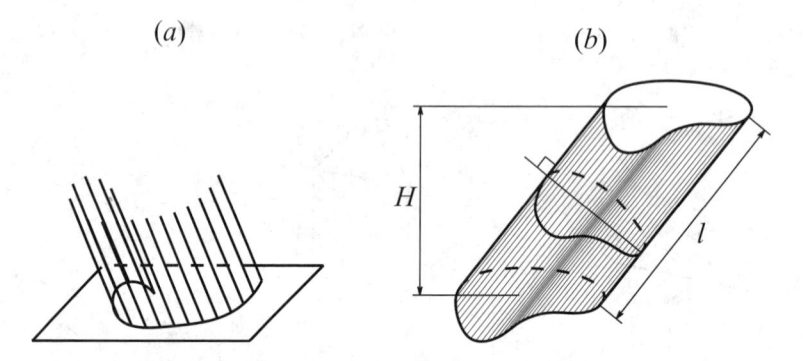

**Figure 3.32.** A cylindrical surface ($a$). A cylinder ($b$).

the generator, then the lateral surface area $S_{\text{lat}}$ and the volume $V$ of the cylinder are given by the formulas

$$S_{\text{lat}} = PH = P_{\text{sec}}l,$$
$$V = S_{\text{bas}}H = S_{\text{sec}}l. \tag{3.2.3.1}$$

In a *right* cylinder, the bases are perpendicular to the generator. In particular, if the bases are disks, then one speaks of a *right circular cylinder*. The volume, the lateral surface area, and the total surface area of a right circular cylinder are given by the formulas

$$V = \pi R^2 H,$$
$$S_{\text{lat}} = 2\pi RH,$$
$$S = 2\pi R(R+H), \tag{3.2.3.2}$$

where $R$ is the radius of the base.

A right circular cylinder is also called a *round cylinder*, or simply a *cylinder*.

2°. The part of a cylinder cut by a plane nonparallel to the base is called a *truncated cylinder* (Fig. 3.33a).

The volume, the lateral surface area, and the total surface area of a truncated cylinder

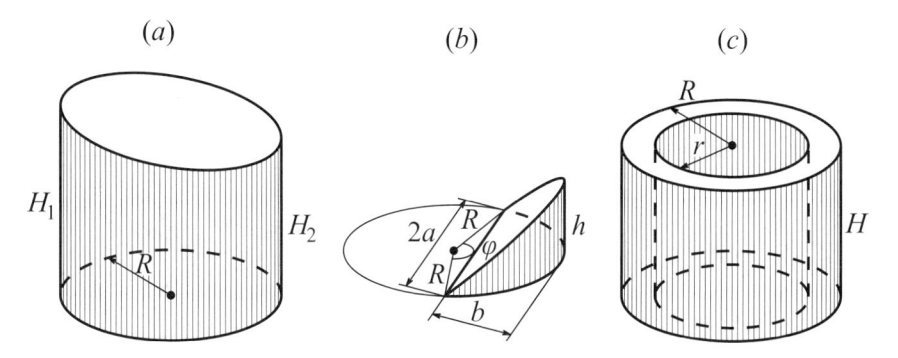

**Figure 3.33.** A truncated cylinder (*a*), a "hoof" (*b*), and a cylindrical tube (*c*).

are given by the formulas

$$V = \pi R^2 \frac{H_1 + H_2}{2},$$
$$S_{\text{lat}} = \pi R(H_1 + H_2), \qquad (3.2.3.3)$$
$$S = \pi R \left[ H_1 + H_2 + R + \sqrt{R^2 + \left(\frac{H_2 - H_1}{2}\right)^2} \right],$$

where $H_1$ and $H_2$ are the maximal and minimal generators.

3°. A *segment of a round cylinder (a "hoof")* is a portion of the cylinder cut by a plane that is nonparallel to the base and intersects it. If $R$ is the radius of the cylindrical segment, $h$ is the height of the "hoof," and $b$ is its width (for the other notation, see Fig. 3.33*b*), then the volume $V$ and the lateral surface area $S_{\text{lat}}$ of the "hoof" can be determined by the formulas

$$V = \frac{h}{3b}\left[a(3R^2 - a^2) + 3R^2(b-R)\alpha\right] = \frac{hR^3}{b}\left(\sin\alpha - \frac{\sin^3\alpha}{3} - \alpha\cos\alpha\right),$$
$$S_{\text{lat}} = \frac{2\pi R}{b}[(b-R)\alpha + a], \qquad (3.2.3.4)$$

where $\alpha = \frac{1}{2}\varphi$ is measured in radians.

4°. A solid bounded by two closed cylindrical surfaces and two planes is called a *cylindrical tube*; the planes are called the bases of the tube. The volume of a round cylindrical tube (Fig. 3.33*c*) is

$$V = \pi H(R^2 - r^2) = \pi H\delta(2R - r) = \pi H\delta(2r + \delta) = 2\pi H\delta\rho, \qquad (3.2.3.5)$$

where $R$ and $r$ are the outer and inner radii, $\delta = R - r$ is the thickness, $\rho = \frac{1}{2}(R + r)$ is the midradius, and $H$ is the height of the pipe.

### 3.2.3-2. Conical surface. Cone. Frustum of cone.

A *conical surface* is the union of straight lines (*generators*) passing through a fixed point (the *apex*) in space and any point of some space curve (the *directrix*) (Fig. 3.34*a*).

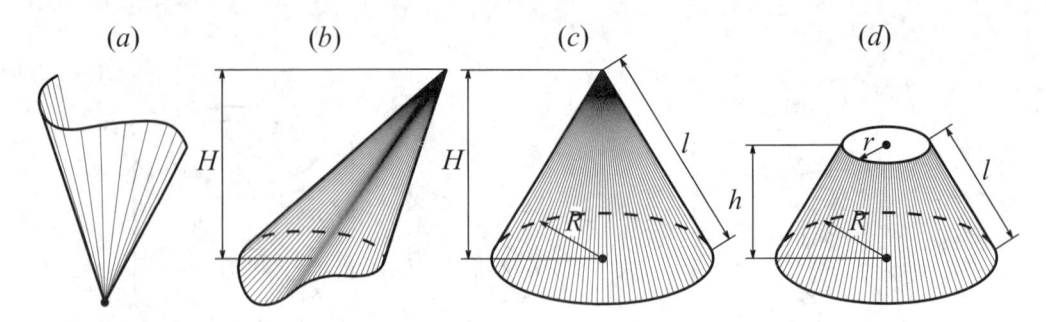

**Figure 3.34.** Conical surface (*a*). A cone (*b*), a right circular cone (*c*), and a frustum of a cone (*d*).

1°. A solid bounded by a conical surface with closed directrix and a plane is called a *cone*; the plane is the base of the cone (Fig. 3.34*b*). The volume of an arbitrary cone is given by the formula

$$V = \frac{1}{3} H S_{\text{bas}}, \qquad (3.2.3.6)$$

where $H$ is the altitude of the cone and $S_{\text{bas}}$ is the area of the base.

A *right circular cone* (Fig. 3.34*c*) has a disk as the base, and its vertex is projected onto the center of the disk. If $l$ is the length of the generator and $R$ is the radius of the base, then the volume, the lateral surface area, and the total surface area of the right circular cone are given by the formulas

$$\begin{aligned} V &= \frac{1}{3}\pi R^2 H, \\ S_{\text{lat}} &= \pi R l = \pi R \sqrt{R^2 + H^2}, \\ S &= \pi R(R + l). \end{aligned} \qquad (3.2.3.7)$$

2°. If a cone is cut by a plane parallel to the base, then we obtain a *frustum of a cone* (Fig. 3.34*d*). The length $l$ of the generator, the volume $V$, the lateral surface area $S_{\text{lat}}$, and the total surface area $S$ of the frustum of a right circular cone are given by the formulas

$$\begin{aligned} l &= \sqrt{h^2 + (R-r)^2}, \\ V &= \frac{\pi h}{3}(R^2 + r^2 + Rr), \\ S_{\text{lat}} &= \pi l (R + r), \\ S &= \pi [l(R+r) + R^2 + r^2], \end{aligned} \qquad (3.2.3.8)$$

where $r$ is the radius of the upper base and $h$ is the altitude of the frustum of a cone.

### 3.2.3-3. Sphere. Spherical parts. Torus.

1°. The *sphere of radius $R$ centered at $O$* is the set of points in space at the distance $R$ from the point $O$ (Fig. 3.35*a*). A solid bounded by a sphere is called a *ball*. Any section of the sphere by a plane is a circle. The section of the sphere by a plane passing through its center is called a *great circle* of radius $R$. There exists exactly one great circle passing through two arbitrary points on the sphere that are not antipodal (i.e., are not the opposite endpoints of a diameter). The smaller arc of this great circle is the shortest distance on the sphere between these points. Concerning the geometry of the sphere, see Section 3.3. The surface area $S$ of the sphere and the volume $V$ of the ball bounded by the sphere are given by the formulas

$$S = 4\pi R = \pi D^2 = \sqrt[3]{36\pi V^2},$$
$$V = \frac{4\pi R^3}{3} = \frac{\pi D^3}{6} = \frac{1}{6}\sqrt{\frac{S^3}{\pi}},$$ (3.2.3.9)

where $D = 2R$ is the *diameter* of the sphere.

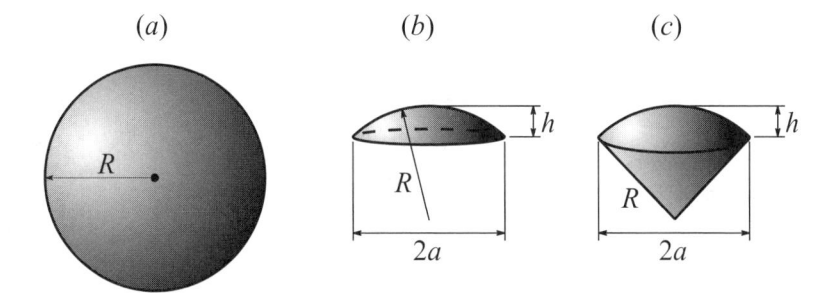

**Figure 3.35.** A sphere (*a*), a spherical cap (*b*), and a spherical sector (*c*).

2°. A portion of a ball cut from it by a plane is called a *spherical cap* (Fig. 3.35*b*). The width $a$, the area $S_{\text{lat}}$ of the curved surface, the total surface area $S$, and the volume $V$ of a spherical cap can be found from the formulas

$$a^2 = h(2R - h),$$
$$S_{\text{lat}} = 2\pi R h = \pi(a^2 + h^2),$$
$$S = S_{\text{lat}} + \pi a^2 = \pi(2Rh + a^2) = \pi(h^2 + 2a^2),$$ (3.2.3.10)
$$V = \frac{\pi h}{6}(3a^2 + h^2) = \frac{\pi h^2}{3}(3R - h),$$

where $R$ and $h$ are the radius and the height of the spherical cap.

3°. A portion of a ball bounded by the curved surface of a spherical cap and the conical surface whose base is the base of the cap and whose vertex is the center of the ball is called a *spherical sector* (Fig. 3.35*c*). The total surface area $S$ and the volume $V$ of a spherical sector are given by the formulas

$$S = \pi R(2h + a),$$
$$V = \frac{2}{3}\pi R^2 h,$$ (3.2.3.11)

where $a$ is the width, $h$ is the height, and $R$ is the radius of the sector.

4°. A portion of a ball contained between two parallel plane secants is called a *spherical segment* (Fig. 3.36*a*). The curved surface of a spherical segment is called a *spherical zone*, and the plane circular surfaces are the *bases* of a spherical segment. The radius $R$ of the ball, the radii $a$ and $b$ of the bases, and the height $h$ of a spherical segment satisfy the relation

$$R^2 = a^2 + \left(\frac{a^2 - b^2 - h^2}{2h}\right)^2.$$ (3.2.3.12)

The curved surface area $S_{\text{lat}}$, the total surface area $S$, and the volume $V$ of a spherical segment are given by the formulas

$$S_{\text{lat}} = 2\pi R h,$$
$$S = S_{\text{lat}} + \pi(a^2 + b^2) = \pi(2Rh + a^2 + b^2), \qquad (3.2.3.13)$$
$$V = \frac{\pi h}{6}(3a^2 + 3b^2 + h^2).$$

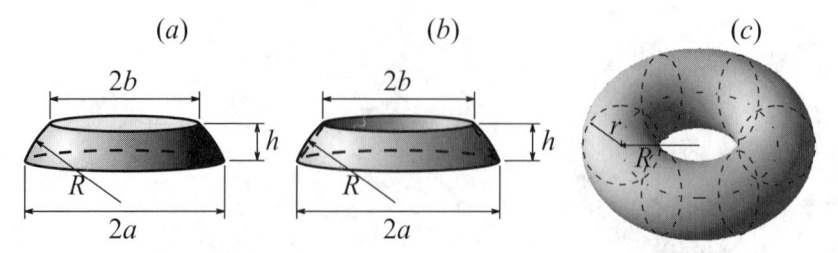

**Figure 3.36.** A spherical segment (*a*) and a spherical segment without the truncated cone inscribed in it (*b*). A torus (*c*).

If $V_1$ is the volume of the truncated cone inscribed in a spherical segment (Fig. 3.36*b*) and $l$ is the length of its generator, then

$$V - V_1 = \frac{\pi h l^2}{6}. \qquad (3.2.3.14)$$

4°. A *torus* is a surface generated by revolving a circle about an axis coplanar with the circle but not intersecting it. If the directrix is a circle (Fig. 3.36*c*), the radius $R$ of the directrix is not less than the radius $r$ of the generating circle ($R \geq r$), and the center of the generator moves along the directrix, then the surface area and the volume of the torus are given by the formulas

$$S = 4\pi^2 Rr = \pi^2 Dd,$$
$$V = 2\pi^2 Rr^2 = \frac{\pi^2 Dd^2}{4}, \qquad (3.2.3.15)$$

where $D = 2R$ and $d = 2r$ are the diameters of the generator and the directrix.

## 3.3. Spherical Trigonometry
### 3.3.1. Spherical Geometry

3.3.1-1. Great circle.

A *great circle* is a section of a sphere by a plane passing through the center.

*Properties of great circles:*
1. The radius of a great circle is equal to the radius of the sphere.
2. There is only one great circle through two arbitrary points that are not the opposite endpoints of a diameter.

The smaller arc of the great circle through two given points is called a *geodesic*, and the length of this arc is the *shortest distance* on the sphere between the two points. The great circles on the sphere play a role similar to the role of straight lines on the plane.

Any two points on the sphere determine a pencil of planes. The intersection of each plane in the pencil with the sphere is a circle. If two points are not the opposite endpoints of a diameter, then the plane passing through the center of the sphere determines the largest circle in the pencil, which is a great circle. The other circles are called *small circles*; the intersection with the sphere of the plane perpendicular to the plane containing the great circle is the smallest circle.

3.3.1-2. Measurement of arcs and angles on sphere. Spherical biangles.

The distances on the sphere are measured along great circle arcs. The great circle arc length between points $A$ and $B$ is given by the relation

$$\overset{\frown}{AB} = R\alpha, \qquad (3.3.1.1)$$

where $R$ is the radius of the sphere and $\alpha$ is the corresponding central angle (in radians). If only the unit sphere (the radius $R = 1$) is considered, then each great circle arc can be characterized by the corresponding central angle (in radians). The angle between two intersecting great circle arcs is measured by the linear angle between the tangents to the great circles at the point of intersection or, which is the same, by the dihedral angle between the planes of the great circles.

Two intersecting great circles on the sphere form four *spherical biangles*. The area of a spherical biangle with the angle $\alpha$ is given by the formula

$$S = 2R^2\alpha. \qquad (3.3.1.2)$$

## 3.3.2. Spherical Triangles

3.3.2-1. Basic notions and properties.

A figure formed by three great circle arcs pairwise connecting three arbitrary points on the sphere is called a *spherical triangle* (Fig. 3.37a). The *vertices* of a spherical triangle are the points of intersection of three rays issuing from the center of the sphere with the sphere. The angles less than $\pi$ between the rays are called the sides $a$, $b$, and $c$ of a spherical triangle. Such spherical triangles are called *Euler triangles*. To each side of a triangle there corresponds a great circle arc on the sphere. The angles $\alpha$, $\beta$, and $\gamma$ opposite the sides $a$, $b$, and $c$ of a spherical triangle are the angles between the great circle arcs corresponding to the sides of the triangle, or, equivalently, the angles between the planes determined by these rays.

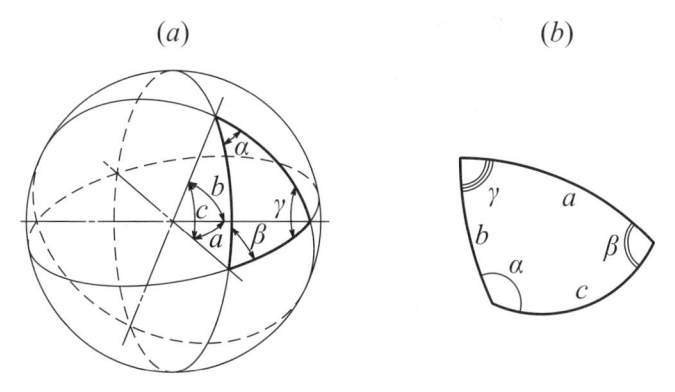

**Figure 3.37.** A spherical triangle.

By analogy with the circumcircle of a plane triangle, there is a "circumscribed cone of revolution" that contains the three straight lines determining the triangle; the axis of this cone is the intersection of the planes perpendicular to the sides at their midpoints. There also exists an "inscribed cone of revolution" that is tangent to the three planes corresponding to the spherical triangle; the axis of this cone is the intersection of the angle bisector planes. The "circumradius" $\overline{R}$ and the "inradius" $r$ are defined as the angles equal to half the angles at the vertices of the first and the second cone, respectively.

If $R$ is the radius of the sphere, then the area $S$ of the spherical triangle is given by the formula
$$S = R^2 \varepsilon, \qquad (3.3.2.1)$$
where $\varepsilon$ is the *spherical excess* defined as
$$\varepsilon = \alpha + \beta + \gamma - \pi \qquad (3.3.2.2)$$
and measured in radians.

*A spherical triangle is uniquely determined (up to a symmetry transformation) by:*
1. Three sides.
2. Three angles.
3. Two sides and their included angle.
4. Two angles and their included side.

Let $\alpha$, $\beta$, and $\gamma$ be the angles and $a$, $b$, and $c$ the sides opposite these angles in a spherical triangle (Fig. 3.37b). Table 3.5 presents the basic properties and relations characterizing spherical triangles (with the notation $2p = a + b + c$ and $2P = \alpha + \beta + \gamma - \pi$). From the relations given in Table 3.5, one can derive all missing relations by cyclically permuting the sides $a$, $b$, and $c$ and the angles $\alpha$, $\beta$, and $\gamma$.

LEGENDRE'S THEOREM. *The area of a spherical triangle with small sides (i.e., with sides that are small compared with the radius of the sphere) is approximately equal to the area of a plane triangle with the same sides; the difference between each angle of the plane triangle and the corresponding angle of the spherical triangle is approximately equal to one-third of the spherical excess.*

The law of sines, the law of cosines, and the half-angle theorem in spherical trigonometry for small sides become the corresponding theorems of the linear (plane) trigonometry.

Table 3.6 allows one to find the sides and angles of an arbitrary spherical triangle if three appropriately chosen sides and/or angles are given.

### 3.3.2-2. Rectangular spherical triangle.

A spherical triangle is said to be *rectangular* if at least one of its angles, for example, $\gamma$, is equal to $\frac{1}{2}\pi$ (Fig. 3.38a); the opposite side $c$ is called the *hypotenuse*.

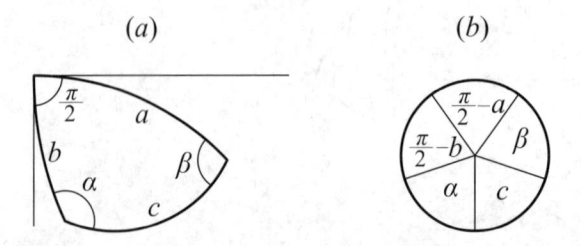

**Figure 3.38.** A rectangular spherical triangle (a). The Neper rules (b).

TABLE 3.5
Basic properties and relations characterizing spherical triangles

| No. | The name of property | Properties and relations |
|---|---|---|
| 1 | Triangle inequality | The sum of lengths of two sides is greater than the length of the third side. The absolute value of the difference between the lengths of two sides is less than the length of the third side, $a + b > c$,  $\|a - b\| < c$ |
| 2 | Sum of two angles of a triangle | The sum of two angles of a triangle is greater than the third angle increased by $\pi$, $\alpha + \beta < \pi + \gamma$ |
| 3 | The greatest side and the greatest angle | The greatest side is opposite the greatest angle, $a < b$ if $\alpha < \beta$; $a = b$ if $\alpha = \beta$ |
| 4 | Sum of angles of a triangle | The sum of the angles lies between $\pi$ and $3\pi$, $\pi < \alpha + \beta + \gamma < 3\pi$ |
| 5 | Sum of sides of a triangle | The sum of sides lies between 0 and $2\pi$ $0 < a + b + c < 2\pi$ |
| 6 | The law of sines | $\dfrac{\sin a}{\sin \alpha} = \dfrac{\sin b}{\sin \beta} = \dfrac{\sin c}{\sin \gamma}$ |
| 7 | The law of cosines of sides | $\cos c = \cos a \cos b + \sin a \sin b \cos \gamma$ |
| 8 | The law of cosines of angles | $\cos \gamma = -\cos \alpha \cos \beta + \sin \alpha \sin \beta \cos c$ |
| 9 | Half-angle formulas | $\sin \dfrac{\gamma}{2} = \sqrt{\dfrac{\sin(p-a)\sin(p-b)}{\sin a \sin b}}$, $\cos \dfrac{\gamma}{2} = \sqrt{\dfrac{\sin p \sin(p-c)}{\sin a \sin b}}$, $\tan \dfrac{\gamma}{2} = \sqrt{\dfrac{\sin(p-a)\sin(p-b)}{\sin p \sin(p-c)}}$ |
| 10 | Half-side theorem | $\sin \dfrac{c}{2} = \sqrt{\dfrac{-\sin P \sin(P-\gamma)}{\sin \alpha \sin \beta}}$, $\cos \dfrac{c}{2} = \sqrt{\dfrac{\sin(P-\alpha)\sin(P-\beta)}{\sin \alpha \sin \beta}}$, $\tan \dfrac{c}{2} = \sqrt{\dfrac{-\sin P \sin(P-\gamma)}{\sin(P-\alpha)\sin(P-\beta)}}$ |
| 11 | Neper's analogs | $\tan \dfrac{c}{2} \cos \dfrac{\alpha - \beta}{2} = \tan \dfrac{a+b}{2} \cos \dfrac{\alpha + \beta}{2}$, $\tan \dfrac{c}{2} \sin \dfrac{\alpha - \beta}{2} = \tan \dfrac{a-b}{2} \sin \dfrac{\alpha + \beta}{2}$, $\cot \dfrac{\gamma}{2} \cos \dfrac{a-b}{2} = \tan \dfrac{\alpha + \beta}{2} \cos \dfrac{a+b}{2}$, $\cot \dfrac{\gamma}{2} \sin \dfrac{a-b}{2} = \tan \dfrac{\alpha - \beta}{2} \sin \dfrac{a+b}{2}$ |
| 12 | D'Alembert (Gauss) formulas | $\sin \dfrac{\gamma}{2} \sin \dfrac{a+b}{2} = \sin \dfrac{c}{2} \cos \dfrac{\alpha - \beta}{2}$,  $\sin \dfrac{\gamma}{2} \sin \dfrac{a+b}{2} = \cos \dfrac{c}{2} \cos \dfrac{\alpha + \beta}{2}$, $\cos \dfrac{\gamma}{2} \sin \dfrac{a-b}{2} = \sin \dfrac{c}{2} \sin \dfrac{\alpha - \beta}{2}$,  $\cos \dfrac{\gamma}{2} \cos \dfrac{a-b}{2} = \cos \dfrac{c}{2} \sin \dfrac{\alpha + \beta}{2}$ |
| 13 | Product formulas | $\sin a \cos \beta = \cos b \sin c - \cos \alpha \sin b \cos c$, $\sin a \cos b = \cos \beta \sin c - \cos a \sin \beta \cos \gamma$ |
| 14 | The "circumradius" $\overline{R}$ | $\cot \overline{R} = \sqrt{\dfrac{\sin(P-\alpha)\sin(P-\beta)\sin(P-\gamma)}{\sin P}} = \cot \dfrac{\alpha}{2} \sin(\alpha - P)$ |
| 15 | The "inradius" $r$ | $\tan r = \sqrt{\dfrac{\sin(p-\alpha)\sin(p-\beta)\sin(p-\gamma)}{\sin p}} = \tan \dfrac{\alpha}{2} \sin(p - \alpha)$ |

## TABLE 3.5 (continued)
### Basic properties and relations characterizing spherical triangles

| No. | The name of property | Properties and relations |
|---|---|---|
| 16 | Willier's formula for the spherical excess $\varepsilon$ | $\tan \dfrac{P}{2} = \tan \dfrac{\varepsilon}{4} = \sqrt{\tan \dfrac{p}{2} \tan \dfrac{p-a}{2} \tan \dfrac{p-b}{2} \tan \dfrac{p-c}{2}}$ |
| 17 | L'Huiller equation | $\tan\left(\dfrac{\gamma}{2} - \dfrac{\varepsilon}{4}\right) = \sqrt{\dfrac{\tan \dfrac{p-a}{2} \tan \dfrac{p-b}{2}}{\tan \dfrac{p}{2} \tan \dfrac{p-c}{2}}}$ |

## TABLE 3.6.
### Solution of spherical triangles

| No. | Three parts specified | Formulas for the remaining parts |
|---|---|---|
| 1 | Three sides $a, b, c$ | The angles $\alpha$, $\beta$, and $\gamma$ are determined by the half-angle formulas and the cyclic permutation.<br>**Remark.** $0 < a+b+c < 2\pi$. The sum and difference of two sides are greater than the third. |
| 2 | Three angles $\alpha, \beta, \gamma$ | The sides $a$, $b$, and $c$ are determined by the half-side theorems and the cyclic permutation.<br>**Remark.** $\pi < \alpha+\beta+\gamma < 3\pi$. The sum of two angles is less than $\pi$ plus the third angle. |
| 3 | Two sides $a, b$ and the included angle $\gamma$ | *First method.*<br>$\alpha + \beta$ and $\alpha - \beta$ are determined from Neper's analogs, then $\alpha$ and $\beta$ can be found; side $a$ is determined from the law of cosines, $\sin c = \sin \gamma \dfrac{\sin a}{\sin \alpha}$.<br>*Second method.*<br>The law of cosines of sides is applied, $\cos c = \cos a \cos b + \sin a \sin b \cos \gamma$, $\cos \beta = \dfrac{\cos b - \sin a \sin c}{\sin a \sin c}$, $\cos \alpha = \dfrac{\cos a - \sin b \sin c}{\sin b \sin c}$.<br>**Remark 1.** If $\gamma > \beta$ ($\gamma < \beta$), then $c$ must be chosen so that $c > b$ ($c < b$).<br>**Remark 2.** The quantities $c$, $\alpha$, and $\beta$ are determined uniquely. |
| 4 | A side $c$ and the two angles $\alpha, \beta$ adjacent to it | *First method.*<br>$a + b$ and $a - b$ are determined from Neper's analogs, then $a$ and $b$ can be found; angle $\gamma$ is determined from the law of sines, $\sin \gamma = \sin c \dfrac{\sin \alpha}{\sin a}$.<br>*Second method.*<br>The law of cosines of angles is applied, $\cos \gamma = -\cos \alpha \cos \beta + \sin \alpha \sin \beta \cos c$, $\cos a = \dfrac{\cos \alpha + \cos \beta \cos \gamma}{\sin \beta \sin \gamma}$, $\cos b = \dfrac{\cos \beta + \cos \alpha \cos \gamma}{\sin \alpha \sin \gamma}$.<br>**Remark 1.** If $c > b$ ($c < b$), then $\gamma$ must be chosen so that $\gamma > \beta$ ($\gamma < \beta$).<br>**Remark 2.** The quantities $\gamma$, $a$, and $b$ are determined uniquely. |
| 5 | Two sides $a, b$ and the angle $\alpha$ opposite one of them | $\beta$ is determined by the law of sines, $\sin \beta = \sin \alpha \dfrac{\sin b}{\sin a}$.<br>The elements $c$ and $\gamma$ can be found from Neper's analogs.<br>**Remark 1.** The problem has a solution for $\sin b \sin \alpha \leq \sin a$.<br>**Remark 2.** Different cases are possible:<br>1. If $\sin a \geq \sin b$, then the solution is determined uniquely.<br>2. If $\sin b \sin \alpha < \sin a$, then there are two solutions $\beta_1$ and $\beta_2$, $\beta_1 + \beta_2 = \pi$.<br>3. If $\sin b \sin \alpha = \sin a$, then the solution is unique: $\beta = \tfrac{1}{2}\pi$. |
| 6 | Two angles $\alpha, \beta$ and the side $a$ opposite one of them | $b$ is determined by the law of sines, $\sin b = \sin a \dfrac{\sin \beta}{\sin \alpha}$.<br>The elements $c$ and $\gamma$ can be found from Neper's analogs.<br>**Remark 1.** The problem has a solution for $\sin a \sin \beta \leq \sin \alpha$.<br>**Remark 2.** Different cases are possible:<br>1. If $\sin \alpha \geq \sin \beta$, then the solution is determined uniquely.<br>2. If $\sin \beta \sin \alpha < \sin a$, then there are two solutions $b_1$ and $b_2$, $b_1 + b_2 = \pi$.<br>3. If $\sin \beta \sin \alpha = \sin a$, then the solution is unique: $b = \tfrac{1}{2}\pi$. |

The following basic relations hold for spherical triangles:

$$\sin a = \cos\left(\tfrac{\pi}{2} - a\right) = \sin\alpha \sin c = \cot\left(\tfrac{\pi}{2} - b\right) \cot\beta = \tan b \cot\beta,$$
$$\sin b = \cos\left(\tfrac{\pi}{2} - b\right) = \sin\beta \sin c = \cot\left(\tfrac{\pi}{2} - a\right) \cot\alpha = \tan a \cot\alpha,$$
$$\cos c = \sin\left(\tfrac{\pi}{2} - a\right) \sin\left(\tfrac{\pi}{2} - b\right) = \cos a \cos b = \cot\alpha \cot\beta, \qquad (3.3.2.3)$$
$$\cos\alpha = \sin\left(\tfrac{\pi}{2} - a\right) \sin\beta = \cos a \sin\beta = \cot\left(\tfrac{\pi}{2} - b\right) \cot c = \tan b \cot c,$$
$$\cos\beta = \sin\left(\tfrac{\pi}{2} - b\right) \sin\alpha = \cos b \sin\alpha = \cot\left(\tfrac{\pi}{2} - a\right) \cot c = \tan a \cot c,$$

which can be obtained from the *Neper rules*: if the five parts of a spherical triangle (the right angle being omitted) are written in the form of a circle in the order in which they appear in the triangle and the legs $a$ and $b$ are replaced by their complements to $\tfrac{1}{2}\pi$ (Fig. 3.38$b$), then the cosine of each part is equal to the product of sines of the two parts not adjacent to it, as well as to the product of the cotangents of the two parts adjacent to it.

## References for Chapter 3

**Alexander, D. C. and Koeberlein, G. M.**, *Elementary Geometry for College Students, 3rd Edition*, Houghton Mifflin Company, Boston, 2002.
**Alexandrov, A. D., Verner, A. L., and Ryzhik, B. I.**, *Solid Geometry [in Russian]*, Alpha, Moscow, 1998.
**Chauvenet, W.**, *A Treatise on Elementary Geometry*, Adamant Media Corporation, Boston, 2001.
**Fogiel, M. (Editor)**, *High School Geometry Tutor, 2nd Edition*, Research & Education Association, Englewood Cliffs, New Jersey, 2003.
**Gustafson, R. D. and Frisk, P. D.**, *Elementary Geometry, 3rd Edition*, Wiley, New York, 1991.
**Hadamard, J.**, *Leçons de géométrie élémentaire, Vols 1 and 2, Rep. Edition*, H.-C. Hege and K. Polthier (Editors), Editions J. Gabay, Paris, 1999.
**Hartshorne, R.**, *Geometry: Euclid and Beyond*, Springer, New York, 2005.
**Jacobs, H. R.**, *Geometry, 2nd Edition*, W. H. Freeman & Company, New York, 1987.
**Jacobs, H. R.**, *Geometry: Seeing, Doing, Understanding, 3rd Edition*, W. H. Freeman & Company, New York, 2003.
**Jurgensen, R. and Brown, R. G.**, *Geometry*, McDougal Littell/Houghton Mifflin, Boston, 2000.
**Kay, D.**, *College Geometry: A Discovery Approach, 2nd Edition*, Addison Wesley, Boston, 2000.
**Kiselev, A. P.**, *Plain and Solid Geometry [in Russian]*, Fizmatlit Publishers, Moscow, 2004.
**Leff, L. S.**, *Geometry the Easy Way, 3rd Edition*, Barron's Educational Series, Hauppauge, New York, 1997.
**Moise, E.**, *Elementary Geometry from an Advanced Standpoint, 3rd Edition*, Addison Wesley, Boston, 1990.
**Musser, G. L., Burger, W. F., and Peterson, B. E.**, *Mathematics for Elementary Teachers: A Contemporary Approach, 6th Edition*, Wiley, New York, 2002.
**Musser, G. L., Burger, W. F., and Peterson, B. E.**, *Essentials of Mathematics for Elementary Teachers: A Contemporary Approach, 6th Edition*, Wiley, New York, 2003.
**Musser, G. L. and Trimpe, L. E.**, *College Geometry: A Problem Solving Approach with Applications*, Prentice Hall, Englewood Cliffs, New Jersey, 1994.
**Pogorelov, A. V.**, *Elementary Geometry [in Russian]*, Nauka Publishers, Moscow, 1977.
**Pogorelov, A.**, *Geometry*, Mir Publishers, Moscow, 1987.
**Prasolov, V. V.**, *Problems in Plane Geometry [in Russian]*, MTsNMO, Moscow, 2001.
**Prasolov, V. V. and Tikhomirov, V. M.**, *Geometry, Translations of Mathematical Monographs, Vol. 200*, American Mathematical Society, Providence, Rhode Island, 2001.
**Roe, J.**, *Elementary Geometry*, Oxford University Press, Oxford, 1993.
**Schultze, A. and Sevenoak, F. L.**, *Plane and Solid Geometry*, Adamant Media Corporation, Boston, 2004.
**Tussy, A. S.**, *Basic Geometry for College Students: An Overview of the Fundamental Concepts of Geometry, 2nd Edition*, Brooks Cole, Stamford, 2002.
**Vygodskii, M. Ya.**, *Mathematical Handbook: Elementary Mathematics, Rev. Edition*, Mir Publishers, Moscow, 1972.

# Chapter 4
# Analytic Geometry

## 4.1. Points, Segments, and Coordinates on Line and Plane

### 4.1.1. Coordinates on Line

**4.1.1-1. Axis and segments on axis.**

A straight line on which a sense is chosen is called an *axis*. If an axis is given and a *scale segment*, i.e., a linear unit used to measure any segment of the axis, is indicated, then the *segment length* is defined (see Fig. 4.1).

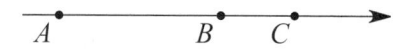

**Figure 4.1.** Axis.

A segment bounded by points $A$ and $B$ is called a *directed segment* if its initial point and endpoint are chosen. Such a segment with initial point $A$ and endpoint $B$ is denoted by $\overrightarrow{AB}$. Directed segments are usually called simply "segments" for brevity.

The *value of a segment* $\overrightarrow{AB}$ of some axis is defined as the number $AB$ equal to its length taken with the plus sign if the senses of the interval and the axis coincide, and with the minus sign if the senses are opposite. Obviously, the length of a segment is its absolute value. The segment length is usually denoted by the symbol $|AB|$. It follows from the above that

$$AB = -BA, \qquad |AB| = |BA|. \tag{4.1.1.1}$$

*Main identity.* For any arbitrary arrangement of points $A$, $B$, and $C$ on the axis, the values of the segments $\overrightarrow{AB}$, $\overrightarrow{BC}$, and $\overrightarrow{AC}$ satisfy the relation

$$AB + BC = AC. \tag{4.1.1.2}$$

**4.1.1-2. Coordinates on line. Number axis.**

One says that a *coordinate system* is introduced on an axis if there is a one-to-one correspondence between points of the axis and numbers.

Suppose that a sense, a scale segment, and a point $O$ called the *origin* are chosen on a line. The value of a segment $\overrightarrow{OA}$ is called the *coordinate of the point* $A$ on the axis. It is usually denoted by the letter $x$. The coordinates of different points are usually denoted by subscripts; for example, the coordinates of points $A_1, \ldots, A_n$ are $x_1, \ldots, x_n$. The point $A_n$ with coordinate $x_n$ is denoted by $A_n(x_n)$. An axis with a coordinate system on it is called a *number axis*.

### 4.1.1-3. Distance between points on axis.

The value $A_1 A_2$ of the segment $\overrightarrow{A_1 A_2}$ on an axis is equal to the difference between the coordinate $x_2$ of the endpoint and the coordinate $x_1$ of the initial point:

$$A_1 A_2 = x_2 - x_1. \tag{4.1.1.3}$$

The distance $d$ between two arbitrary points $A_1(x_1)$ and $A_2(x_2)$ on the line is given by the relation

$$d = |A_1 A_2| = |x_2 - x_1|. \tag{4.1.1.4}$$

Remark. If segments do not lie on some axis but are treated as arbitrary segments on the plane or in space, then there is no reason to assign any sign to their lengths. In such cases, the symbol of absolute value is usually omitted in the notation of lengths of segments. We adopt this convention in the sequel.

## 4.1.2. Coordinates on Plane

### 4.1.2-1. Rectangular Cartesian coordinates on plane.

If a one-to-one correspondence between points on the plane and numbers (pairs of numbers) is specified, then one says that a *coordinate system* is introduced on the plane.

A *rectangular Cartesian coordinate system* is determined by a scale segment for measuring lengths and two mutually perpendicular axes. The point of intersection of the axes is usually denoted by the letter $O$ and is called the *origin*, while the axes themselves are called the *coordinate axes*. As a rule, one of the coordinate axes is horizontal and the right sense is positive. This axis is called the *abscissa axis* and is denoted by the letter $X$ or by $OX$. On the vertical axis, which is called the *ordinate axis* and is denoted by $Y$ or $OY$, the upward sense is usually positive (see Fig. 4.2a). The coordinate system introduced above is often denoted by $XY$ or $OXY$.

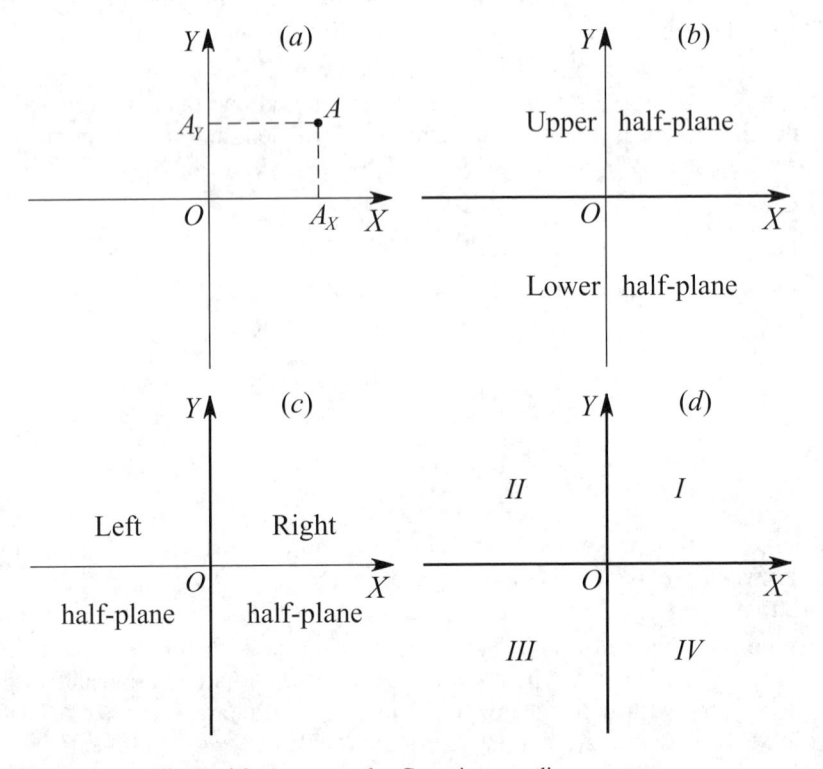

**Figure 4.2.** A rectangular Cartesian coordinate system.

The abscissa axis divides the plane into the *upper* and *lower* half-planes (see Fig. 4.2b), while the ordinate axis divides the plane into the *right* and *left* half-planes (see Fig. 4.2c). The two coordinate axes divide the plane into four parts, which are called *quadrants* and numbered as shown in Fig. 4.2d.

Take an arbitrary point $A$ on the plane and project it onto the coordinate axes, i.e., draw perpendiculars to the axes $OX$ and $OY$ through $A$. The points of intersection of the perpendiculars with the axes are denoted by $A_X$ and $A_Y$, respectively (see Fig. 4.2a). The numbers

$$x = OA_X, \qquad y = OA_Y, \qquad (4.1.2.1)$$

where $OA_X$ and $OA_Y$ are the respective values of the segments $\overrightarrow{OA}_X$ and $\overrightarrow{OA}_Y$ on the abscissa and ordinate axes, are called the *coordinates of the point* $A$ in the rectangular Cartesian coordinate system. The number $x$ is the first coordinate, or the *abscissa*, of the point $A$, and $y$ is the second coordinate, or the *ordinate*, of the point $A$. One says that the point $A$ has the coordinates $(x, y)$ and uses the notation $A(x, y)$.

**Example 1.** Let $A$ be an arbitrary point in the right half-plane. Then the segment $\overrightarrow{OA}_X$ has the positive sense on the axis $OX$, and hence the abscissa $x = OA_X$ of $A$ is positive. But if $A$ lies in the left half-plane, then the segment $A_X$ has the negative sense on the axis $OX$, and the number $x = OA_X$ is negative. If the point $A$ lies on the axis $OY$, then its projection on the axis $OX$ coincides with the point $O$ and $x = OA_X = 0$.

Thus all points in the right half-plane have positive abscissas ($x > 0$), all points in the left half-plane have negative abscissas ($x < 0$), and the abscissas of points lying on the axis $OY$ are zero ($x = 0$).

Similarly, all points in the upper half-plane have positive ordinates ($y > 0$), all points in the lower half-plane have negative ordinates ($y < 0$), and the ordinates of points lying on the axis $OX$ are zero ($y = 0$).

**Remark 1.** Strictly speaking, the coordinate system introduced above is a *right rectangular Cartesian coordinate system*. A *left rectangular Cartesian coordinate system* can, for example, be obtained by changing the sense of one of the axes. There also exist right and left *oblique Cartesian coordinate systems*, where the coordinate axes intersect at an arbitrary angle.

**Remark 2.** A right rectangular Cartesian coordinate system is usually called simply a Cartesian coordinate system.

### 4.1.2-2. Transformation of Cartesian coordinates under parallel translation of axes.

Suppose that two rectangular Cartesian coordinate systems $OXY$ and $\widehat{O}\widehat{X}\widehat{Y}$ are given and the first system is taken to the second by the translation of the origin $O$ of the first system to the origin $\widehat{O}$ of the second system. Under this translation, the axes preserve their directions (the respective axes of the systems are parallel), and the origin moves by $x_0$ in the direction of the $OX$-axis and by $y_0$ in the direction of the $OY$-axis (see Fig. 4.3a). Obviously, the point $\widehat{O}$ has the coordinates $(x_0, y_0)$ in the coordinate system $OXY$.

Let an arbitrary point $A$ have coordinates $(x, y)$ in the system $OXY$ and coordinates $(\hat{x}, \hat{y})$ in the system $\widehat{O}\widehat{X}\widehat{Y}$. The transformation of rectangular Cartesian coordinates by the parallel translation of the axes is given by the formulas

$$\begin{array}{ccc} x = \hat{x} + x_0, & & \hat{x} = x - x_0, \\ & \text{or} & \\ y = \hat{y} + y_0 & & \hat{y} = y - y_0. \end{array} \qquad (4.1.2.2)$$

### 4.1.2-3. Transformation of Cartesian coordinates under rotation of axes.

Suppose that two rectangular Cartesian coordinate systems $OXY$ and $O\widehat{X}\widehat{Y}$ are given and the first system is taken to the second by the rotation around the point $O$ by an angle $\alpha$ (see Fig. 4.3b).

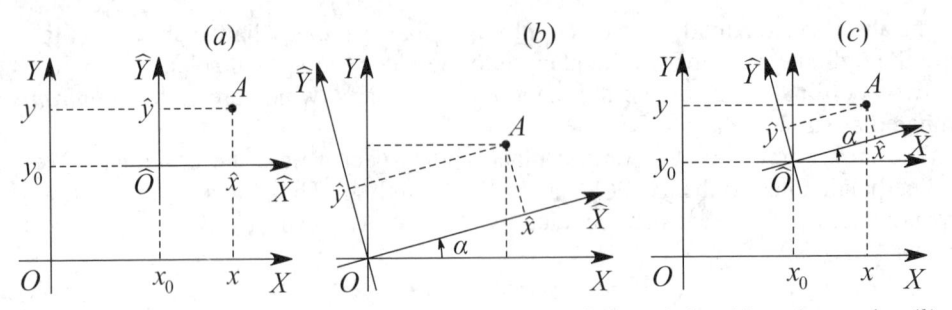

**Figure 4.3.** Transformation of Cartesian coordinates under parallel translation (a), under rotation (b), and under translation and rotation (c) of axes.

Let an arbitrary point $A$ have coordinates $(x, y)$ in the system $OXY$ and coordinates $(\hat{x}, \hat{y})$ in the system $O\hat{X}\hat{Y}$. The transformation of rectangular Cartesian coordinates by the rotation of axes is given by the formulas

$$\begin{array}{ll} x = \hat{x}\cos\alpha - \hat{y}\sin\alpha, & \hat{x} = x\cos\alpha + y\sin\alpha, \\ y = \hat{x}\sin\alpha + \hat{y}\cos\alpha & \text{or} \quad \hat{y} = -x\sin\alpha + y\cos\alpha. \end{array} \qquad (4.1.2.3)$$

### 4.1.2-4. Transformation of coordinates under translation and rotation of axes.

Suppose that two rectangular Cartesian coordinate systems $OXY$ and $\hat{O}\hat{X}\hat{Y}$ are given and the first system is taken to the second by the translation of the origin $O(0,0)$ of the first system to the origin $\hat{O}(x_0, y_0)$ of the second system followed by the rotation of the system around the point $\hat{O}$ by an angle $\alpha$ (see Fig. 4.3c and Paragraphs 4.1.2-2 and 4.1.2-3).

Let an arbitrary point $A$ have coordinates $(x, y)$ in the system $OXY$ and coordinates $(\hat{x}, \hat{y})$ in the system $\hat{O}\hat{X}\hat{Y}$. The transformation of rectangular Cartesian coordinates by the parallel translation and rotation of axes is given by the formulas

$$\begin{array}{ll} x = \hat{x}\cos\alpha - \hat{y}\sin\alpha + x_0, & \hat{x} = (x-x_0)\cos\alpha + (y-y_0)\sin\alpha, \\ y = \hat{x}\sin\alpha + \hat{y}\cos\alpha + y_0, & \text{or} \quad \hat{y} = -(x-x_0)\sin\alpha + (y-y_0)\cos\alpha. \end{array} \qquad (4.1.2.4)$$

### 4.1.2-5. Polar coordinates.

A *polar coordinate system* is determined by a point $O$ called the *pole*, a ray $OA$ issuing from this point, which is called the *polar axis*, a scale segment for measuring lengths, and the positive sense of rotation around the pole. Usually, the anticlockwise sense is assumed to be positive (see Fig. 4.4a).

The position of each point $B$ on the plane is determined by two *polar coordinates*, the *polar radius* $\rho = |OB|$ and the *polar angle* $\theta = \angle AOB$ (the values of the angle $\theta$ are defined up to the addition of $\pm 2\pi n$, where $n$ is an integer). To be definite, one usually assumes that $0 \leq \theta \leq 2\pi$ or $-\pi \leq \theta \leq \pi$. The polar radius of the pole is zero, and its polar angle does not have any definite value.

### 4.1.2-6. Relationship between Cartesian and polar coordinates.

Suppose that $B$ is an arbitrary point on the plane, $(x, y)$ are its rectangular Cartesian coordinates, and $(\rho, \theta)$ are its polar coordinates (see Fig. 4.4b). The formulas of transformation

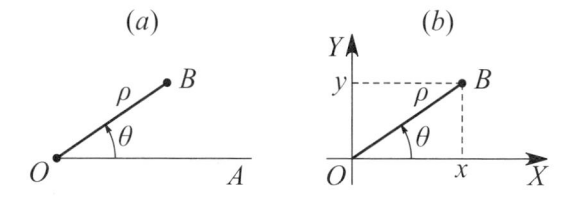

**Figure 4.4.** A polar coordinate system (*a*). Relationship between Cartesian and polar coordinates (*b*).

from one coordinate system to the other have the form

$$x = \rho \cos \theta, \quad \text{or} \quad \rho = \sqrt{x^2 + y^2},$$
$$y = \rho \sin \theta \qquad \tan \theta = y/x, \qquad (4.1.2.5)$$

where the polar angle $\theta$ is determined with regard to the quadrant where the point $B$ lies.

**Example 2.** Let us find the polar coordinates $\rho, \theta$ ($0 \leq \theta \leq 2\pi$) of the point $B$ whose Cartesian coordinates are $x = -3$, $y = -3$.

From formulas (4.1.2.5), we obtain $\rho = \sqrt{(-3)^2 + (-3)^2} = 3\sqrt{2}$ and $\tan \theta = \frac{-3}{-3} = 1$. Since the point $B$ lies in the third quadrant, we have $\theta = \arctan 1 + \pi = \frac{5}{4}\pi$.

## 4.1.3. Points and Segments on Plane

### 4.1.3-1. Distance between points on plane.

The distance $d$ between two arbitrary points $A_1$ and $A_2$ on the plane is given by the formula

$$d = \sqrt{(x_2 - x_1)^2 + (y_2 - y_1)^2}, \qquad (4.1.3.1)$$

where $x$ and $y$ with the corresponding subscripts are the Cartesian coordinates of these points, and by the formula

$$d = \sqrt{\rho_1^2 + \rho_2^2 - 2\rho_1\rho_2 \cos(\theta_2 - \theta_1)}, \qquad (4.1.3.2)$$

where $\rho$ and $\theta$ with the corresponding subscripts are the polar coordinates of these points.

### 4.1.3-2. Segment and its projections.

Suppose that an axis $u$ and an arbitrary segment $\overrightarrow{A_1 A_2}$ are given on the plane (see Fig. 4.5*a*). From the points $A_1$ and $A_2$, we draw the perpendiculars to $u$ and denote the points of intersection of the perpendiculars with the axis by $P_1$ and $P_2$. The value $P_1 P_2$ of the segment $\overrightarrow{P_1 P_2}$ of the axis $u$ is called the *projection of the segment* $\overrightarrow{A_1 A_2}$ *onto the axis* $u$. Usually one writes $\text{pr}_u \overrightarrow{A_1 A_2} = P_1 P_2$. If $\varphi$ ($0 \leq \varphi \leq \pi$) is the angle between the segment $\overrightarrow{A_1 A_2}$ and the axis $u$, then

$$\text{pr}_u \overrightarrow{A_1 A_2} = d \cos \varphi. \qquad (4.1.3.3)$$

For two arbitrary points $A_1(x_1, y_1)$ and $A_2(x_2, y_2)$, the projections $x$ and $y$ of the segment $\overrightarrow{A_1 A_2}$ onto the coordinate $X$- and $Y$-axes are given by the formulas (see Fig. 4.5*b*)

$$\text{pr}_X \overrightarrow{A_1 A_2} = x_2 - x_1, \quad \text{pr}_Y \overrightarrow{A_1 A_2} = y_2 - y_1. \qquad (4.1.3.4)$$

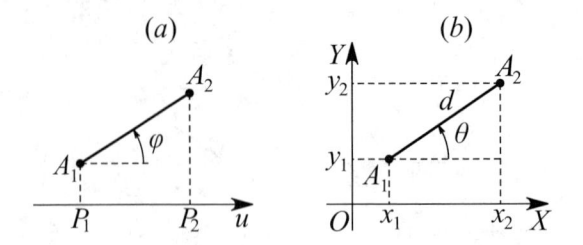

**Figure 4.5.** Projection of the segment onto the axis $u$ (*a*) and onto the coordinate $X$- and $Y$-axes (*b*).

Thus, to obtain the projections of a segment onto the coordinate axes, one subtracts the coordinates of its initial point from the respective coordinates of its endpoint.

The projections of the segment $\overrightarrow{A_1 A_2}$ onto the coordinate axes can be found if its length $d$ (see (4.1.3.1)) and *polar angle* $\theta$ are known (see Fig. 4.5*b*). The corresponding formulas are

$$\mathrm{pr}_X \overrightarrow{A_1 A_2} = d\cos\theta, \quad \mathrm{pr}_Y \overrightarrow{A_1 A_2} = d\sin\theta, \quad \tan\theta = \frac{y_2 - y_1}{x_2 - x_1}. \tag{4.1.3.5}$$

### 4.1.3-3. Angles between coordinate axes and segments.

The angles $\alpha_x \equiv \theta$ and $\alpha_y$ between the segment $\overrightarrow{A_1 A_2}$ and the coordinate $x$- and $y$-axes are determined by the expressions

$$\cos\alpha_x = \frac{x_2 - x_1}{\sqrt{(x_2 - x_1)^2 + (y_2 - y_1)^2}}, \quad \cos\alpha_y = \frac{y_2 - y_1}{\sqrt{(x_2 - x_1)^2 + (y_2 - y_1)^2}}, \tag{4.1.3.6}$$

and $\alpha_y = \pi - \alpha_x$.

The angle $\beta$ between arbitrary segments $\overrightarrow{A_1 A_2}$ and $\overrightarrow{A_3 A_4}$ joining the points $A_1(x_1, y_1)$, $A_2(x_2, y_2)$ and $A_3(x_3, y_3)$, $A_4(x_4, y_4)$, respectively, can be found from the relation

$$\cos\beta = \frac{(x_2 - x_1)(x_4 - x_3) + (y_2 - y_1)(y_4 - y_3)}{\sqrt{(x_2 - x_1)^2 + (y_2 - y_1)^2}\sqrt{(x_4 - x_3)^2 + (y_4 - y_3)^2}}. \tag{4.1.3.7}$$

### 4.1.3-4. Division of segment in given ratio.

The number $\lambda = p/q$, where $p = A_1 A$ and $q = A A_2$ are the values of the directed segments $\overrightarrow{A_1 A}$ and $\overrightarrow{A A_2}$, is called the ratio in which point $A$ divides the segment $\overrightarrow{A_1 A_2}$. It is independent of the sense of the segment (i.e., one could use the segment $\overrightarrow{A_2 A_1}$) and the scale segment. The coordinates of the point $A$ dividing the segment $\overrightarrow{A_1 A_2}$ in a ratio $\lambda$ are given by the formulas

$$x = \frac{x_1 + \lambda x_2}{1 + \lambda} = \frac{q x_1 + p x_2}{q + p}, \quad y = \frac{y_1 + \lambda y_2}{1 + \lambda} = \frac{q y_1 + p y_2}{q + p}, \tag{4.1.3.8}$$

where $-\infty \leq \lambda \leq \infty$.

For the coordinates of the midpoint of the segment $\overrightarrow{A_1 A_2}$, we have

$$x = \frac{x_1 + x_2}{2}, \quad y = \frac{y_1 + y_2}{2}; \tag{4.1.3.9}$$

i.e., each coordinate of the midpoint of a segment is equal to the half-sum of the respective coordinates of its endpoints.

### 4.1.3-5. Area of triangle area.

The area $S_3$ of the triangle with vertices $A_1$, $A_2$, and $A_3$ is given by the formula

$$\pm S_3 = \frac{1}{2}[(x_2 - x_1)(y_3 - y_1) - (x_3 - x_1)(y_2 - y_1)]$$

$$= \frac{1}{2}\begin{vmatrix} x_2 - x_1 & y_2 - y_1 \\ x_3 - x_1 & y_3 - y_1 \end{vmatrix} = \frac{1}{2}\begin{vmatrix} x_1 & y_1 & 1 \\ x_2 & y_2 & 1 \\ x_3 & y_3 & 1 \end{vmatrix}, \qquad (4.1.3.10)$$

where $x$ and $y$ with respective subscripts are the Cartesian coordinates of the vertices, and by the formula

$$\pm S_3 = \frac{1}{2}[\rho_1\rho_2 \sin(\theta_2 - \theta_1) + \rho_2\rho_3 \sin(\theta_3 - \theta_2) + \rho_3\rho_1 \sin(\theta_1 - \theta_3)], \qquad (4.1.3.11)$$

where $\rho$ and $\theta$ with respective subscripts are the polar coordinates of the vertices. In formulas (4.1.3.10) and (4.1.3.11), one takes the plus sign if the vertices are numbered anticlockwise (see Fig. 4.6$a$) and the minus sign otherwise.

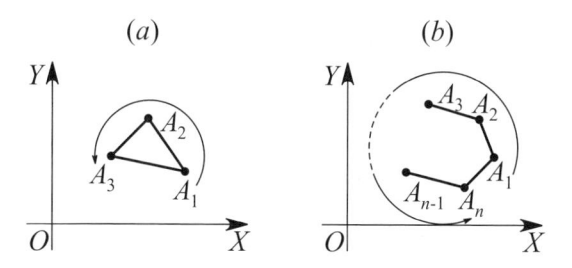

**Figure 4.6.** Area of triangle ($a$) and of a polygon ($b$).

### 4.1.3-6. Area of a polygon.

The area $S_n$ of the polygon with vertices $A_1, \ldots, A_n$ is given by the formula

$$\pm S_n = \frac{1}{2}[(x_1 - x_2)(y_1 + y_2) + (x_2 - x_3)(y_2 + y_3) + \cdots + (x_n - x_1)(y_n + y_1)], \qquad (4.1.3.12)$$

where $x$ and $y$ with respective subscripts are the Cartesian coordinates of the vertices, and by the formula

$$\pm S_n = \frac{1}{2}[\rho_1\rho_2 \sin(\theta_2 - \theta_1) + \rho_2\rho_3 \sin(\theta_3 - \theta_2) + \cdots + \rho_n\rho_1 \sin(\theta_1 - \theta_n)], \qquad (4.1.3.13)$$

where $\rho$ and $\theta$ with respective subscripts are the polar coordinates of the vertices. In formulas (4.1.3.12) and (4.1.3.13), one takes the plus sign if the vertices are numbered anticlockwise (see Fig. 4.6$b$) and the minus sign otherwise.

*Remark.* One often says that formulas (4.1.3.10)–(4.1.3.13) express the *oriented area* of the corresponding figures.

## 4.2. Curves on Plane

### 4.2.1. Curves and Their Equations

**4.2.1-1. Basic definitions.**

A *curve on the plane* determined by an equation in some coordinate system is the geometric locus of points of the plane whose coordinates satisfy this equation.

An *equation of a curve on the plane* in a given coordinate system is an equation with two variables such that the coordinates of the points lying on the curve satisfy the equation and the coordinates of the points that do not lie on the curve do not satisfy it.

The coordinates of an arbitrary point of a curve occurring in an equation of the curve are called *current coordinates*.

**4.2.1-2. Equation of curve in Cartesian coordinate system.**

An *equation of a curve in the Cartesian coordinate system* $OXY$ can be written as

$$F(x, y) = 0. \qquad (4.2.1.1)$$

The image of a curve determined by an equation of the form

$$y = f(x) \qquad (4.2.1.2)$$

is called the *graph of the function* $f(x)$.

**Example 1.** Let us plot the curve determined by the equation $x^2 - y = 0$. We express one coordinate via the other (e.g., $y$ via $x$) from this equation: $y = x^2$. Specifying various values of $x$, we find the corresponding values of $y$ and thus construct a sequence of points of the desired curve. By joining these points, we obtain the curve itself (see Fig. 4.7a).

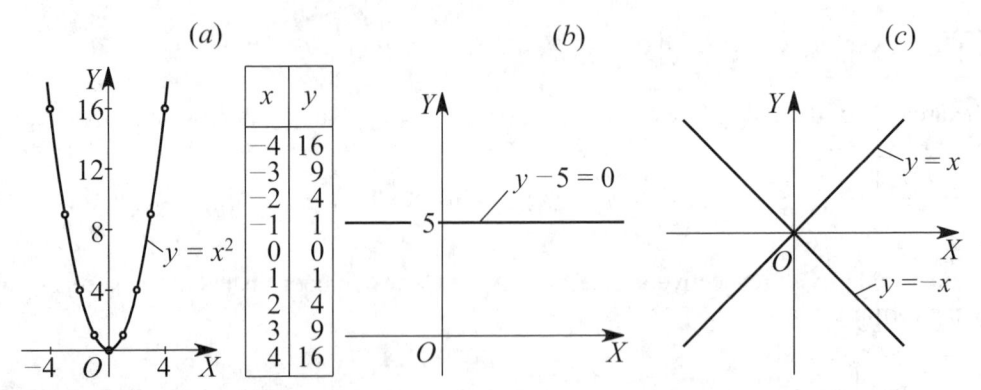

**Figure 4.7.** Cartesian coordinate system. Loci of points for equations $x^2 - y = 0$ (a), $y - 5 = 0$ (b), and $x^2 - y^2 = 0$ (c).

**4.2.1-3. Special kinds of equations.**

1. The equation of a curve on the plane may contain only one of the current coordinates but still determine a certain curve.

**Example 2.** Suppose that the equation $y - 5 = 0$ (or $y = 5$) is given. The locus of points whose coordinates are equal to five is the straight line parallel to the axis $OX$ and passing through the point $y = 5$ of the axis $OY$ (see Fig. 4.7b).

Similarly, the equation $x + 7 = 0$ determines a straight line parallel to the axis $OY$.

2. If the left-hand side of equation (4.2.1.1) can be factorized, then, equating each factor separately with zero, we obtain several new equations, each of which can determine a certain curve.

**Example 3.** Consider the equation $x^2 - y^2 = 0$. Factorizing the left-hand side of this equation, we obtain $(x + y)(x - y) = 0$. Obviously, the latter equation determines the pair of straight lines $x + y = 0$ and $x - y = 0$, which are the bisectors of the coordinate angles (see Fig. 4.7c).

3. Equation (4.2.1.1) may determine a locus consisting of one or several isolated points.

**Example 4.** The equation $x^2 + y^2 = 0$ determines the single point with coordinates $(0, 0)$.

**Example 5.** The equation $(x^2 - 9)^2 + (y^2 - 25)^2 = 0$ defines the locus consisting of the four points $(3, 5)$, $(3, -5)$, $(-3, 5)$, and $(-3, -5)$.

4. There exist equations that do not determine any locus.

**Example 6.** The equation $x^2 + y^2 + 5 = 0$ does not have solutions for any real $x$ and $y$.

### 4.2.1-4. Equation of curve in polar coordinate system.

An *equation of a curve in a polar coordinate system* can be written as

$$\Phi(\rho, \theta) = 0, \tag{4.2.1.3}$$

where $\rho$ is the polar radius and $\theta$ is the polar angle. This equation is satisfied by the polar coordinates of any point lying on the curve and is not satisfied by the coordinates of the points that do not lie on the curve.

**Example 7.** Consider the equation $\rho - a \cos \theta = 0$ (or $\rho = a \cos \theta$), where $a$ is a positive number. By $B$ we denote the point with polar coordinates $(\rho, \theta)$, and by $A$ we denote the point with coordinates $(a, 0)$. If $\rho = a \cos \theta$, then the angle $OBA$ is a right angle, and vice versa. Therefore, the locus of points whose coordinates satisfy this equation is a circle with diameter $a$ (see Fig. 4.8).

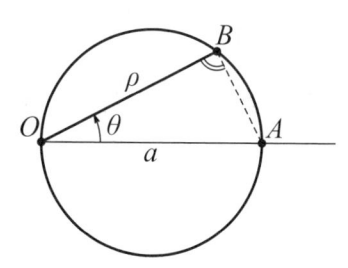

**Figure 4.8.** Polar coordinate system. Locus of points for equations $\rho - a \cos \theta = 0$.

**Example 8.** Consider the equation $\rho - a\theta = 0$ (or $\rho = a\theta$), where $a$ is a positive constant. The curve determined by this equation is called a *spiral of Archimedes*.

As $\theta$ increases starting from zero, the point $B(\rho, \theta)$ issues from the pole and moves around it in the positive sense, simultaneously moving away from it. For each point of this curve with positive coordinates $(\rho, \theta)$, one has the corresponding point $(-\rho, -\theta)$ on the same curve. Figures 4.9a and b show the branches of the spiral of Archimedes corresponding to the positive and negative values of $\theta$, respectively.

Note that the spiral of Archimedes divides each polar ray into equal segments (except for the segment nearest to the pole).

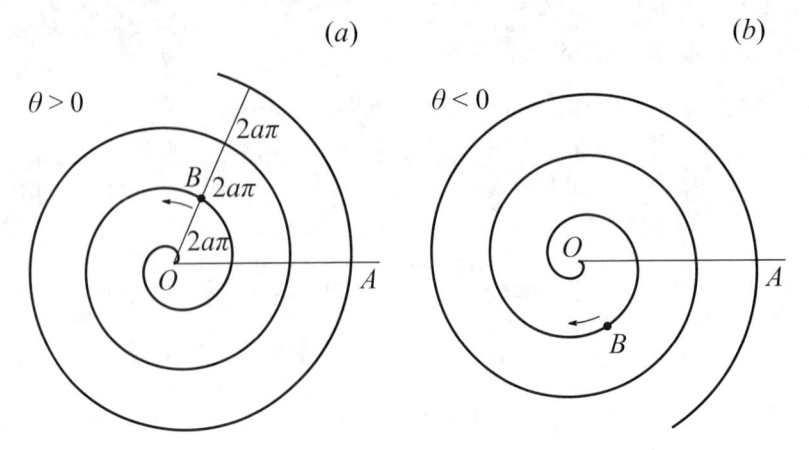

**Figure 4.9.** A spiral of Archimedes $\rho = a\theta$ corresponding to the positive (a) and negative (b) values of $\theta$.

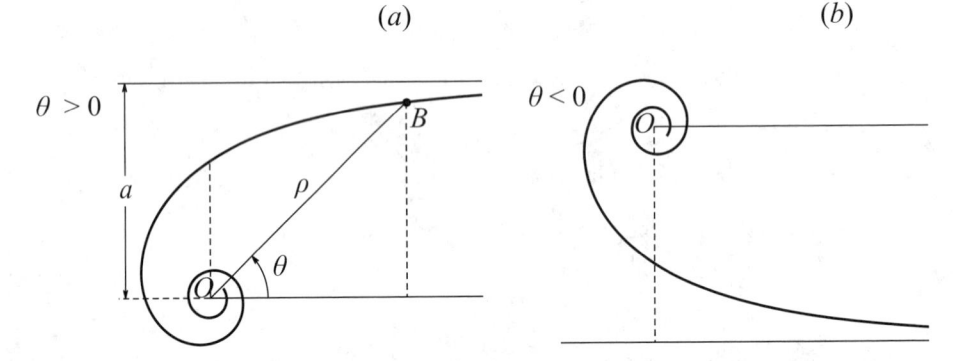

**Figure 4.10.** A hyperbolic spiral $\rho = a/\theta$ corresponding to the positive (a) and negative (b) values of $\theta$.

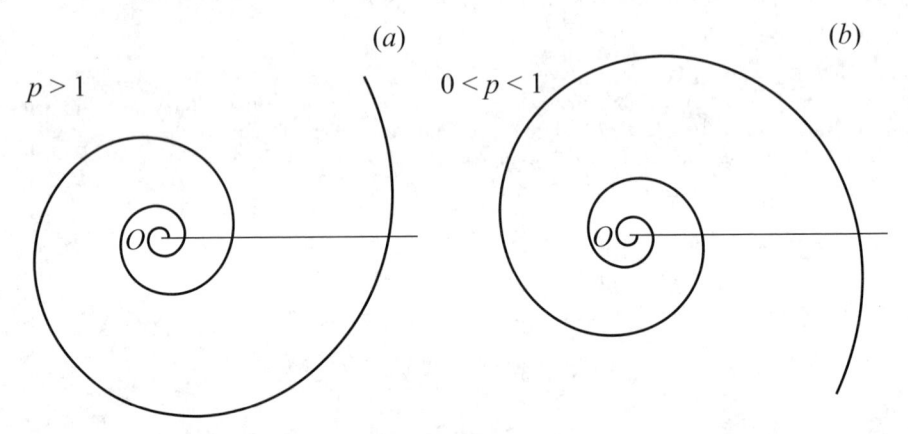

**Figure 4.11.** A logarithmic spiral $\rho = a^\theta$ corresponding to the values $a > 1$ (a) and $0 < a < 1$ (b).

**Example 9.** Consider the equation $\rho - a/\theta = 0$ (or $\rho = a/\theta$), where $a$ is a positive number. The curve determined by this equation is called a *hyperbolic spiral*.

As $\theta$ increases, the point $B(\rho, \theta)$ moves around the pole in the positive sense while approaching it endlessly. As $\theta$ tends to zero, the point $B$ approaches the line $y = a$ while moving to infinity. Figures 4.10a and b show the branches of the hyperbolic spiral corresponding to the positive and negative values of $\theta$, respectively.

**Example 10.** Consider the equation $\rho - a^\theta = 0$ (or $\rho = a^\theta$), where $a$ is a positive number. The curve determined by this equation is called the *logarithmic spiral*.

Figures 4.11a and b show the branches of the logarithmic spiral corresponding to the values $a > 1$ and $0 < a < 1$, respectively. For $a = 1$, we obtain the equation of a circle.

### 4.2.1-5. Parametric equations of a curve.

*Parametric equations of a curve on the plane* have the form

$$x = \varphi(t), \qquad y = \psi(t), \qquad (4.2.1.4)$$

where $x$ and $y$ are treated as the coordinates of some point $A$ for each value of the *variable parameter t*. In general, the variables $x$ and $y$ vary with $t$, and the point $A$ moves on the plane.

Parametric equations play an important role in applied mathematics and mechanics, where they are called the *equations of motion of a mass point*. The parameter $t$ has the meaning of time.

Remark 1. Eliminating the parameter $t$ from equations (4.2.1.4), we obtain an equation of the curve in the form (4.2.1.1).

Remark 2. In different problems, the variable parameter in equations (4.2.1.4) may have different meanings.

**Example 11.** The circle of radius $a$ centered at the origin is described by the following parametric equations for the Cartesian system:

$$x = a \cos t, \quad y = a \sin t.$$

By squaring these equations and by adding them, we obtain the equation of the circle in the form

$$x^2 + y^2 = a^2.$$

**Example 12.** The spiral of Archimedes (*a*), the hyperbolic spiral (*b*), and the logarithmic spiral (*c*) are described by the following equations in the polar coordinate system:

$$(a) \ \rho = a\theta; \qquad (b) \ \rho = \frac{a}{\theta}; \qquad (c) \ \rho = a^\theta.$$

The parametric equations for the Cartesian coordinates of these curves have the form

$$(a) \quad x = a\theta \cos\theta, \quad y = a\theta \sin\theta;$$
$$(b) \quad x = \frac{a \cos\theta}{\theta}, \quad y = \frac{a \sin\theta}{\theta};$$
$$(c) \quad x = a^\theta \cos\theta, \quad y = a^\theta \sin\theta.$$

In all three cases, the variable parameter is the polar angle $\theta$.

### 4.2.1-6. Algebraic curves.

The curves given by algebraic equations of the form

$$\begin{aligned} &Ax + By + C = 0, \\ &Ax^2 + Bxy + Cy^2 + Dx + Ey + F = 0, \\ &Ax^3 + Bx^2y + Cxy^2 + Dy^3 + Ex^2 + Fxy + Gy^2 + Hx + Iy + K = 0, \\ &\dots\dots\dots\dots\dots\dots\dots\dots\dots\dots\dots\dots\dots\dots\dots\dots\dots\dots\dots\dots\dots\dots\dots\dots\dots \end{aligned} \qquad (4.2.1.5)$$

in a rectangular Cartesian coordinate system are called *algebraic curves on the plane*.

A curve given by an algebraic equation of degree $n$ in a rectangular Cartesian coordinate system is called an *nth-order algebraic curve*.

When passing from one rectangular Cartesian coordinate system to another, the degree of the equation of an algebraic curve does not change; i.e., any $n$th-order algebraic curve remains such in any rectangular Cartesian coordinate system.

## 4.2.2. Main Problems of Analytic Geometry for Curves

### 4.2.2-1. Construction of equation for given curve.

Suppose that a curve is defined on the plane as a locus of points and one needs to construct an equation of this curve. This is the first main problem of analytic geometry for curves.

**Example 1.** Suppose that a circle of radius $R$ is given on the plane. In the rectangular Cartesian coordinate system $OXY$, its origin is point $A(x_0, y_0)$. It is required to derive an equation of this circle (see Fig. 4.12a).

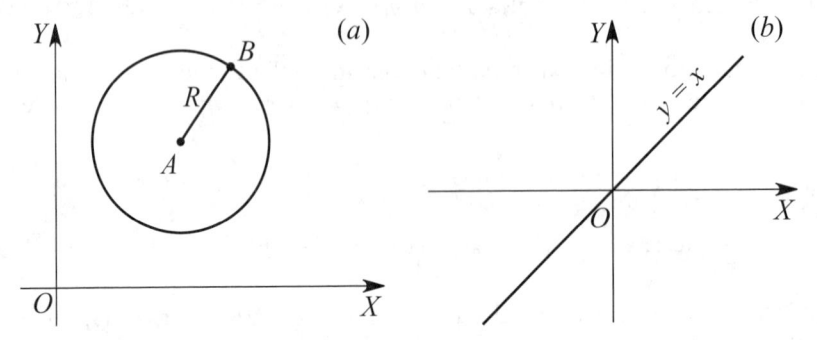

**Figure 4.12.** Construction of equation for given curve (a) and construction of curve with given equation (b).

By $B(x, y)$ we denote the variable point of the circle. Obviously, the circle is the locus of points lying at the distance $R$ from the point $A(x_0, y_0)$; i.e., $AB = R$, or
$$\sqrt{(x - x_0)^2 + (y - y_0)^2} = R.$$
This is the desired equation of the circle of radius $R$ centered at the point $A(x_0, y_0)$, which is usually written as
$$(x - x_0)^2 + (y - y_0)^2 = R^2.$$

### 4.2.2-2. Construction of curve with given equation.

Suppose that an equation with two variables is given and one needs to construct the curve determined by this equation on the plane. This is the second main problem of analytic geometry for curves.

**Example 2.** Consider the equation $x - y = 0$ or $y = x$. The points whose coordinates satisfy this equation lie in the first and third quadrants at equal distances from the coordinate axes. Thus the locus of points whose coordinates satisfy this equation is the bisector of the first and third coordinate angles (see Fig. 4.12b).

### 4.2.2-3. Intersection of two curves.

Consider two curves determined by the equations
$$F(x, y) = 0, \qquad G(x, y) = 0. \tag{4.2.2.1}$$
To find the points of intersection of these curves, one solves system (4.2.2.1). Each real solution of this system gives a point of intersection. If the system is inconsistent or does not have real solutions, then the curves do not meet.

**Example 3.** Let us find the points of intersection of two curves (a circle and a straight line) with the equations
$$x^2 + y^2 - 25 = 0, \qquad 2x - y + 5 = 0.$$
These equations form a system of two equations with two unknowns. For example, if we express $y$ from the second equation and substitute the resulting expression into the first equation, then we obtain $x^2 + 4x = 0$, whence we find $x_1 = 0$, $x_2 = -4$ and $y_1 = 5$, $y_2 = -3$. This means that the curves have two points of intersection, $(0, 5)$ and $(-4, -3)$.

**Remark.** For equations (4.2.2.1) written, for example, in a polar coordinate system, it is natural to use $\rho$ and $\theta$ instead of the variables $x$ and $y$.

## 4.3. Straight Lines and Points on Plane

### 4.3.1. Equations of Straight Lines on Plane

**4.3.1-1. Slope-intercept equation of straight line.**

The tangent of the angle of inclination of a straight line to the axis $OX$ is called the *slope* of the straight line. The slope characterizes the direction of the line. For straight lines perpendicular to the $OX$-axis, slope does not make sense, although one often says that the slope of such straight lines is equal to infinity.

The *slope-intercept equation of a straight line* in the rectangular Cartesian coordinate system $OXY$ has the form

$$y = kx + b, \qquad (4.3.1.1)$$

where $k = \tan \varphi = (y - b)/x$ is the slope of the line and $b$ is the $y$-intercept of the line, i.e., the signed distance from the point of intersection of the line with the ordinate axis to the origin. Equation (4.3.1.1) is meaningful for any straight line that is not perpendicular to the abscissa axis (see Fig. 4.13a).

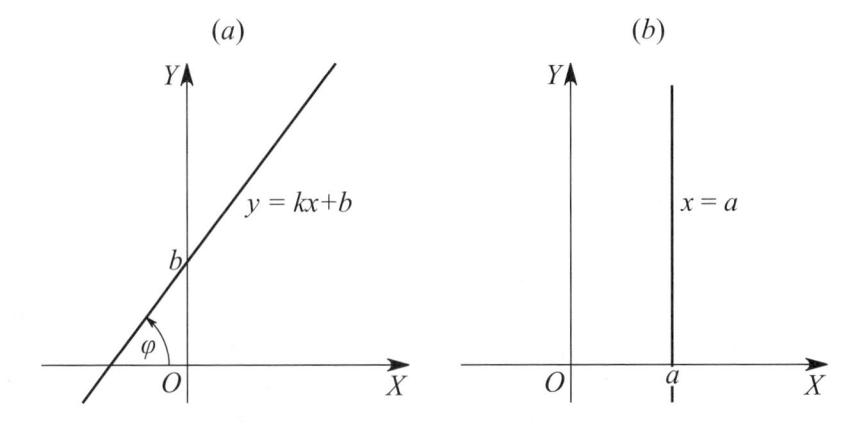

**Figure 4.13.** Straight lines on plane.

If a straight line is not perpendicular to the $OX$-axis, then its equation can be written as (4.3.1.1), but if a straight line is perpendicular to the $OX$-axis, then its equation can be written as

$$x = a, \qquad (4.3.1.2)$$

where $a$ is the abscissa of the point of intersection of this line with the $OX$-axis (see Fig. 4.13b).

For the slope of a straight line, we also have the formula

$$k = \frac{y_2 - y_1}{x_2 - x_1}, \qquad (4.3.1.3)$$

where $A_1(x_1, y_1)$ and $A_2(x_2, y_2)$ are two arbitrary points of the line.

**4.3.1-2. Point-slope equation of straight line.**

In the rectangular Cartesian coordinate system $OXY$, the equation of a straight line with slope $k$ passing through a point $A(x_1, y_1)$ has the form

$$y - y_1 = k(x - x_1). \qquad (4.3.1.4)$$

**Remark.** If we set $x_1 = 0$ and $y_1 = b$ in equation (4.3.1.4), then we obtain equation (4.3.1.1).

### 4.3.1-3. Equation of straight line passing through two given points.

In the rectangular Cartesian coordinate system $OXY$, the equation of a straight line with slope $k$ passing through points $A_1(x_1, y_1)$ and $A_2(x_2, y_2)$ has the form (4.3.1.4), where $k$ is given by the expression (4.3.1.3):

$$y - y_1 = \frac{y_2 - y_1}{x_2 - x_1}(x - x_1). \qquad (4.3.1.5)$$

This equation is usually written as

$$\frac{x - x_1}{x_2 - x_1} = \frac{y - y_1}{y_2 - y_1}. \qquad (4.3.1.6)$$

Equation (4.3.1.6) is also called the *canonical equation of the straight line passing through two given points on the plane*.

Sometimes one writes this equation in terms of a third-order determinant as follows:

$$\begin{vmatrix} x & y & 1 \\ x_1 & y_1 & 1 \\ x_2 & y_2 & 1 \end{vmatrix} = 0. \qquad (4.3.1.7)$$

**Example 1.** Let us derive the equation of the straight line passing through the points $A_1(5, 1)$ and $A_2(7, 3)$. Substituting the coordinates of these points into formula (4.3.1.5), we obtain

$$\frac{x - 5}{2} = \frac{y - 1}{2} \quad \text{or} \quad y = x - 4.$$

### 4.3.1-4. General equation of straight line on plane.

An equation of the form

$$Ax + By + C = 0 \qquad (4.3.1.8)$$

is called the *general equation of a straight line* in the rectangular Cartesian coordinate system $OXY$. In rectangular Cartesian coordinates, each straight line is determined by an equation of degree 1, and, conversely, each equation of degree 1 determines a straight line.

If $B \ne 0$, then equation (4.3.1.8) can be written as (4.3.1.1), where $k = -A/B$ and $b = -C/B$. If $B = 0$, then equation (4.3.1.8) can be written as (4.3.1.2), where $a = -C/A$.

If $C = 0$, then the equation of a straight line becomes $Ax + By = 0$ and determines a straight line passing through the origin.

If $B = 0$ and $A \ne 0$, then the equation of a straight line becomes $Ax + C = 0$ and determines a straight line parallel to the axis $OY$.

If $A = 0$ and $B \ne 0$, then the equation of a straight line becomes $By + C = 0$ and determines a straight line parallel to the axis $OX$.

### 4.3.1-5. General equation of straight line passing through given points on plane.

In the rectangular Cartesian coordinate system $OXY$, the general equation of a straight line passing through the point $A(x_1, y_1)$ on the plane has the form

$$A(x - x_1) + B(y - y_1) = 0. \qquad (4.3.1.9)$$

If this equation is written in the form

$$\frac{x - x_1}{B} = \frac{y - y_1}{-A}, \qquad (4.3.1.10)$$

then it is called the *canonical equation of a straight line passing through a given point on the plane*. If $B = 0$, then one sets $x - x_0 = 0$, and if $A = 0$, then one sets $y - y_0 = 0$.

**Remark.** The general equation of the straight line passing through two given points on the plane has the form (4.3.1.6).

### 4.3.1-6. Parametric equations of straight line on plane.

The *parametric equations of a straight line on the plane* through the point $A(x_1, y_1)$ in the rectangular Cartesian coordinate system $OXY$ have the form

$$x = x_1 + Bt, \quad y = y_1 - At, \quad (4.3.1.11)$$

where $A$ and $B$ are the coefficients of the general equation (4.3.1.8) or (4.3.1.9) of a straight line and $t$ is a variable parameter.

In the rectangular Cartesian coordinate system $OXY$, the parametric equations of the straight line passing through two points $A(x_1, y_1)$ and $A(x_2, y_2)$ on the plane can be written as

$$\begin{aligned} x &= x_1(1-t) + x_2 t, \\ y &= y_1(1-t) - y_2 t. \end{aligned} \quad (4.3.1.12)$$

**Remark.** Eliminating the parameter $t$ from equations (4.3.1.11) and (4.3.1.12), we obtain equations (4.3.1.9) and (4.3.1.6), respectively.

### 4.3.1-7. Intercept-intercept equation of straight line.

The *intercept-intercept equation of a straight line* in the rectangular Cartesian coordinate system $OXY$ has the form

$$\frac{x}{a} + \frac{y}{b} = 1, \quad (4.3.1.13)$$

where $a$ and $b$ are the $x$- and $y$-intercepts of the line, i.e., the signed distances from the points of intersection of the line with the coordinate axes to the origin (see Fig. 4.14).

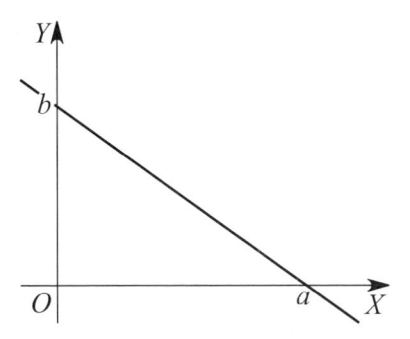

**Figure 4.14.** A straight line with intercept-intercept equation.

### 4.3.1-8. Normalized equation of straight line.

Suppose that a rectangular Cartesian coordinate system $OXY$ and a straight line are given on the plane. We draw the perpendicular to the straight line through the origin. This perpendicular is called the *normal* to the line. By $P$ we denote the point of intersection of the normal with the line.

The equation

$$x \cos \alpha + y \sin \alpha - p = 0, \quad (4.3.1.14)$$

where $\alpha$ is the polar angle of the normal and $p$ is the length of the segment $OP$ (the distance from the origin to the straight line) (see Fig. 4.15), is called the *normalized equation of*

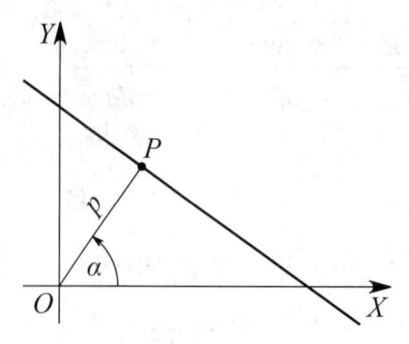

**Figure 4.15.** A straight line with normalized equation.

the straight line in the rectangular Cartesian coordinate system $OXY$. In the normalized equation of a straight line, $p \geq 0$ and $\cos^2 \alpha + \sin^2 \alpha = 1$.

For all positions of the straight line with respect to the coordinate axes, its equation can always be written in normalized form.

The general equation of a straight line (4.3.1.8) can be reduced to a normalized form (4.3.1.14) by setting

$$\cos \alpha = \pm \frac{A}{\sqrt{A^2 + B^2}}, \quad \sin \alpha = \pm \frac{B}{\sqrt{A^2 + B^2}}, \quad p = \mp \frac{C}{\sqrt{A^2 + B^2}}, \qquad (4.3.1.15)$$

where the upper sign is taken for $C < 0$ and the lower sign for $C > 0$. For $C = 0$, either sign can be taken.

### 4.3.1-9. Equation of straight line in polar coordinates.

The *equation of a straight line in polar coordinates* can be written as

$$\rho \cos(\theta - \alpha) = p, \qquad (4.3.1.16)$$

where $\rho$ and $\theta$ are the polar radius and the polar angle of the current point $B$ of the straight line, $\alpha$ is the polar angle of the normal, and $p$ is the distance from the pole to the line (see Fig. 4.16).

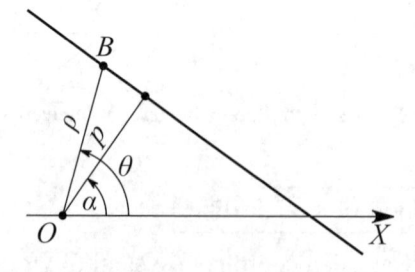

**Figure 4.16.** A straight line with equation in polar coordinates.

### 4.3.1-10. Equation of a pencil of straight lines.

The set of all straight lines passing through a point $A$ on the plane is called a *pencil of straight lines*, and the point $A$ itself is called the *center of the pencil*. The equations determining all straight lines in the pencil are called *equations of the pencil*.

$1°$. If the Cartesian coordinates of the pencil center $A(x_0, y_0)$ are known, then the equation of any straight line in the pencil that is not parallel to the $OY$-axis has the form

$$y - y_0 = k(x - x_0), \qquad (4.3.1.17)$$

where $k$ is the slope of the straight line. For the straight line in the pencil parallel to the $OY$-axis,

$$x = x_0. \qquad (4.3.1.18)$$

$2°$. If the equations of two straight lines in the pencil are known, for example, $A_1 x + B_1 y + C_1 = 0$ and $A_2 x + B_2 y + C_2 = 0$, then the equation of the pencil can be written as

$$\alpha(A_1 x + B_1 y + C_1) + \beta(A_2 x + B_2 y + C_2) = 0, \qquad (4.3.1.19)$$

where $\alpha$ and $\beta$ are any numbers that are not simultaneously zero.

## 4.3.2. Mutual Arrangement of Points and Straight Lines

### 4.3.2-1. Condition for three points to be collinear.

Suppose that points $A_1(x_1, y_1)$, $A_2(x_2, y_2)$, and $A_3(x_3, y_3)$ are given in the Cartesian coordinate system $OXY$ on the planes. They are collinear (lie on the same straight line) if and only if

$$\frac{x_3 - x_1}{x_2 - x_1} = \frac{y_3 - y_1}{y_2 - y_1}. \qquad (4.3.2.1)$$

### 4.3.2-2. Deviation of point from straight line.

The number $\delta$, equal to the length of the perpendicular to a straight line through a point taken with the plus sign if the point and the origin lie on opposite sides of the line and with the minus sign otherwise, is called the *deviation of the point from the line*. Obviously, the deviation is zero for points lying on the line.

To find the deviation of a point $A(x_0, y_0)$ from a straight line, one substitutes the coordinates of $A$ into the left-hand side of the normalized equation of the line (see Paragraph 4.3.1-8) for the current Cartesian coordinates:

$$\delta = x_0 \cos \alpha + y_0 \sin \alpha - p. \qquad (4.3.2.2)$$

### 4.3.2-3. Distance from point to straight line.

The *distance $d$ from a point to a straight line* is the absolute value of the deviation. It can be calculated by the formula

$$d = |x_0 \cos \alpha + y_0 \sin \alpha - p|. \qquad (4.3.2.3)$$

The distance from a point $A(x_0, y_0)$ to a straight line given by the general equation $Ax + By + C = 0$ can be calculated by the formula

$$d = \frac{|A x_0 + B y_0 + C|}{\sqrt{A^2 + B^2}}. \qquad (4.3.2.4)$$

**Example 1.** Let us find the distance from the point $A(2, 1)$ to the straight line $3x + 4y + 5 = 0$. We use formula (4.3.2.4) to obtain

$$d = \frac{|A x_0 + B y_0 + C|}{\sqrt{A^2 + B^2}} = \frac{|3 \cdot 2 + 4 \cdot 1 + 5|}{\sqrt{3^2 + 4^2}} = \frac{15}{5} = 3.$$

### 4.3.2-4. Angle between two straight lines.

We consider two straight lines given by the equations

$$y = k_1 x + b_1 \quad \text{and} \quad y = k_2 x + b_2, \tag{4.3.2.5}$$

where $k_1 = \tan \varphi_1$ and $k_2 = \tan \varphi_2$ are the slopes of the respective lines (see Fig. 4.17). The angle $\alpha$ between these lines can be obtained by the formula

$$\tan \alpha = \frac{k_2 - k_1}{1 + k_1 k_2}, \tag{4.3.2.6}$$

where $k_1 k_2 \neq -1$. If $k_1 k_2 = -1$, then $\alpha = \frac{1}{2}\pi$.

**Remark.** If at least one of the lines is perpendicular to the axis $OX$, then formula (4.3.2.6) does not make sense. In this case, the angle between the lines can be calculated by the formula

$$\alpha = \varphi_2 - \varphi_1. \tag{4.3.2.7}$$

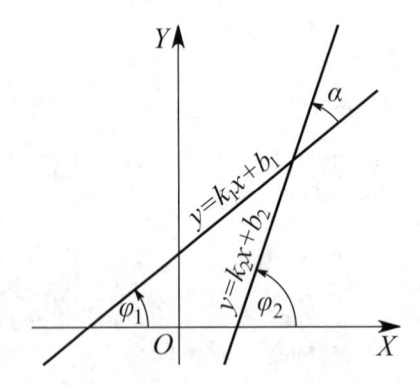

**Figure 4.17.** Angle between two straight lines.

The angle $\alpha$ between the two straight lines given by the general equations

$$A_1 x + B_1 y + C_1 = 0 \quad \text{and} \quad A_2 x + B_2 y + C_2 = 0 \tag{4.3.2.8}$$

can be calculated using the expression

$$\tan \alpha = \frac{A_1 B_2 - A_2 B_1}{A_1 A_2 + B_1 B_2}, \tag{4.3.2.9}$$

where $A_1 A_2 + B_1 B_2 \neq 0$. If $A_1 A_2 + B_1 B_2 = 0$, then $\alpha = \frac{1}{2}\pi$.

**Remark.** If one needs to find the angle between straight lines and the order in which they are considered is not defined, then this order can be chosen arbitrarily. Obviously, a change in the order results in a change in the sign of the tangent of the angle.

### 4.3.2-5. Point of intersection of straight lines.

Suppose that two straight lines are defined by general equations in the form (4.3.2.8). Consider the system of two first-order algebraic equations (4.3.2.8):

$$\begin{aligned} A_1 x + B_1 y + C_1 &= 0, \\ A_2 x + B_2 y + C_2 &= 0. \end{aligned} \quad (4.3.2.10)$$

Each common solution of equations (4.3.2.10) determines a common point of the tow lines.

If the determinant of system (4.3.2.10) is not zero, i.e.,

$$\begin{vmatrix} A_1 & B_1 \\ A_2 & B_2 \end{vmatrix} = A_1 B_2 - A_2 B_1 \neq 0, \quad (4.3.2.11)$$

then the system is consistent and has a unique solution; hence these straight lines are distinct and nonparallel and meet at the point $A(x_0, y_0)$, where

$$x_0 = \frac{B_1 C_2 - B_2 C_1}{A_1 B_2 - A_2 B_1}, \quad y_0 = \frac{C_1 A_2 - C_2 A_1}{A_1 B_2 - A_2 B_1}. \quad (4.3.2.12)$$

Condition (4.3.2.11) is often written as

$$\frac{A_1}{A_2} \neq \frac{B_1}{B_2}. \quad (4.3.2.13)$$

**Example 2.** To find the point of intersection of the straight lines $y = 2x - 1$ and $y = -4x + 5$, we solve system (4.3.2.10):

$$\begin{aligned} 2x - y - 1 &= 0, \\ -4x - y + 5 &= 0, \end{aligned}$$

and obtain $x = 1$, $y = 1$. Thus the intersection point has the coordinates $(1, 1)$.

### 4.3.2-6. Condition for straight lines to be perpendicular.

For two straight lines determined by slope-intercept equations (4.3.2.5) to be perpendicular, it is necessary and sufficient that

$$k_1 k_2 = -1. \quad (4.3.2.14)$$

Relation (4.3.2.14) is usually written as

$$k_1 = -\frac{1}{k_2}, \quad (4.3.2.15)$$

and one also says that the slopes of perpendicular straight lines are inversely proportional in absolute value and opposite in sign.

If the straight lines are given by general equations (4.3.2.8), then a necessary and sufficient condition for them to be perpendicular can be written as (see Paragraph 4.3.2-4)

$$A_1 A_2 + B_1 B_2 = 0. \quad (4.3.2.16)$$

**Example 3.** The lines $3x+y-3=0$ and $x-3y+8=0$ are perpendicular since they satisfy condition (4.3.2.16):

$$A_1 A_2 + B_1 B_2 = 3 \cdot 1 + 1 \cdot (-3) = 0.$$

### 4.3.2-7. Condition for straight lines to be parallel.

For two straight lines defined by slope-intercept equations (4.3.2.5) to be parallel and not to coincide, it is necessary and sufficient that

$$k_1 = k_2, \quad b_1 \neq b_2. \tag{4.3.2.17}$$

If the straight lines are given by general equations (4.3.2.8), then a necessary and sufficient condition for them to be parallel can be written as

$$\frac{A_1}{A_2} = \frac{B_1}{B_2} \neq \frac{C_1}{C_2}; \tag{4.3.2.18}$$

in this case, the straight lines do not coincide (see Fig. 4.18).

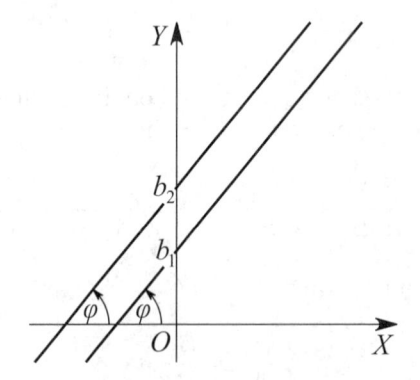

**Figure 4.18.** Parallel straight lines.

**Example 4.** The straight lines $3x + 4y + 5 = 0$ and $3/2\,x + 2y + 6 = 0$ are parallel since the following condition (4.3.2.18) is satisfied:

$$\frac{3}{3/2} = \frac{4}{2} \neq \frac{5}{6}.$$

### 4.3.2-8. Condition for straight lines to coincide.

For two straight lines given by slope-intercept equations (4.3.2.5) to coincide, it is necessary and sufficient that

$$k_1 = k_2, \quad b_1 = b_2. \tag{4.3.2.19}$$

If the straight lines are given by general equations (4.3.2.8), then a necessary and sufficient condition for them to coincide has the form

$$\frac{A_1}{A_2} = \frac{B_1}{B_2} = \frac{C_1}{C_2}. \tag{4.3.2.20}$$

**Remark.** Sometimes the case of coinciding straight lines is considered as a special case of parallel straight lines and it not distinguished as an exception.

### 4.3.2-9. Distance between parallel lines.

The distance between the parallel lines given by equations (see Paragraph 4.3.2-7)

$$A_1 x + B_1 y + C_1 = 0 \quad \text{and} \quad A_1 x + B_1 y + C_2 = 0 \tag{4.3.2.21}$$

can be found using the formula (see Paragraph 4.3.2-3)

$$d = \frac{|C_1 - C_2|}{\sqrt{A_1^2 + B_1^2}}. \tag{4.3.2.22}$$

**4.3.2-10. Condition for a straight line to separate points of plane.**

Suppose that a straight line in the Cartesian coordinate system $OXY$ is given by an equation of the form
$$Ax + By + C = 0. \qquad (4.3.2.23)$$
Obviously, this straight line divides the plane into two half-planes. We consider two arbitrary points $A_1(x_1, y_1)$ and $A_2(x_2, y_2)$ of the plane that do not lie on the line. The points are said to be *nonseparated by the straight line* if they belong to the same half-plane (lie on the same side of the straight line and possibly coincide). The points are said to be *separated by the straight line* if they belong to different half-planes (lie on opposite sides of the straight line).

Two points $A_1(x_1, y_1)$ and $A_2(x_2, y_2)$ that do not belong to the straight line (4.3.2.23) are separated by this line if and only if the numbers $Ax_1 + By_1 + C$ and $Ax_2 + By_2 + C$ have opposite signs.

## 4.4. Second-Order Curves
### 4.4.1. Circle

**4.4.1-1. Definition and canonical equation of circle.**

A curve on the plane is called a *circle* if there exists a rectangular Cartesian coordinate system $OXY$ in which the equation of this curve has the form (see Fig. 4.19a)
$$x^2 + y^2 = a^2, \qquad (4.4.1.1)$$
where the point $O(0,0)$ is the center of the circle and $a > 0$ is its radius. Equation (4.4.1.1) is called the *canonical equation of a circle*.

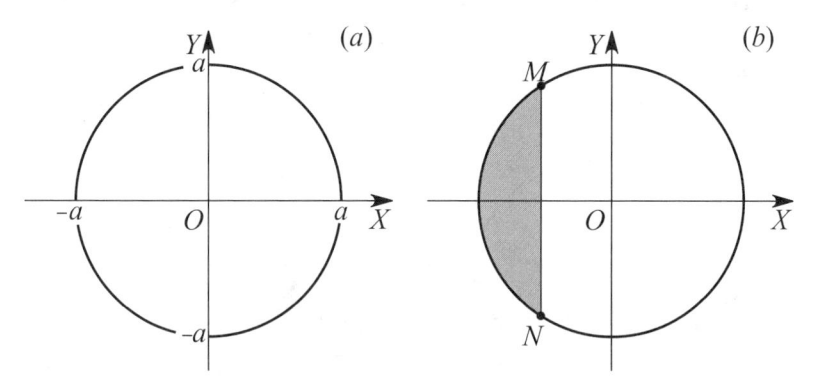

**Figure 4.19.** Circle.

The circle defined by equation (4.4.1.1) is the locus of points equidistant (lying at the distance $a$) from its center. If a circle of radius $a$ is centered at a point $C(x_0, y_0)$, then its equation can be written as
$$(x - x_0)^2 + (y - y_0)^2 = a^2. \qquad (4.4.1.2)$$
The area of the *disk* bounded by a circle of radius $a$ is given by the formula $S = \pi a^2$. The length of this circle is $L = 2\pi a$. The area of the figure bounded by the circle and the chord passing through the points $M(x_0, y_0)$ and $N(x_0, -y_0)$ is (see Fig. 19b)
$$S = \frac{\pi a^2}{2} + x_0 \sqrt{a^2 - x_0^2} + a^2 \arcsin \frac{x_0}{a}. \qquad (4.4.1.3)$$

### 4.4.1-2. Parametric and other equations of circle.

The parametric equations of the circle (4.4.1.1) have the form

$$x = a\cos\theta, \quad y = a\sin\theta, \qquad (4.4.1.4)$$

where the angle in the polar coordinate system plays the role of the variable parameter (see Paragraphs 4.2.1-4 and 4.2.1-5).

The equation of the circle (4.4.1.1) in the polar coordinate system has the form

$$\rho = a \qquad (4.4.1.5)$$

and does not contain the polar angle $\theta$.

*Remark.* The form of the equation of a circle in a polar coordinate system depends on the choice of the pole and the polar axis (see Example 7 in Subsection 4.2.1).

## 4.4.2. Ellipse

### 4.4.2-1. Definition and canonical equation of ellipse.

A curve on the plane is called an *ellipse* if there exists a rectangular Cartesian coordinate system $OXY$ in which the equation of the curve has the form

$$\frac{x^2}{a^2} + \frac{y^2}{b^2} = 1, \qquad (4.4.2.1)$$

where $a \geq b > 0$ (see Fig. 4.20). The coordinates in which the equation of an ellipse has the form (4.4.2.1) are called the *canonical coordinates* for this ellipse, and equation (4.4.2.1) itself is called the *canonical equation of the ellipse*.

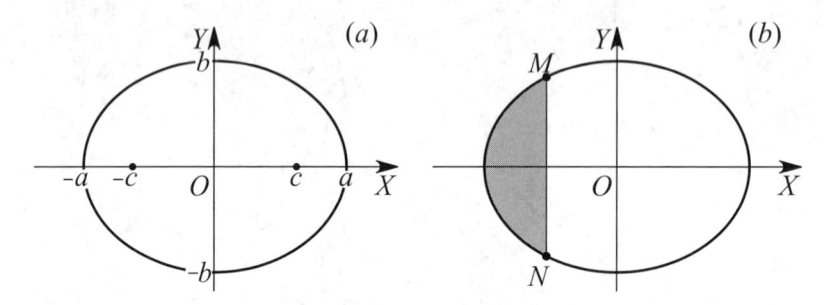

**Figure 4.20.** Ellipse.

The segments $A_1 A_2$ and $B_1 B_2$ joining the opposite vertices of an ellipse, as well as their lengths $2a$ and $2b$, are called the *major* and *minor axes*, respectively, *of the ellipse*. The axes of an ellipse are its axes of symmetry. In Fig. 4.20a, the axes of symmetry of the ellipse coincide with the axes of the rectangular Cartesian coordinate system $OXY$. The numbers $a$ and $b$ are called the *semimajor* and *semiminor axes of the ellipse*. The number $c = \sqrt{a^2 - b^2}$ is called the *linear eccentricity*, and the number $2c$ is called the *focal distance*. The number $e = c/a = \sqrt{1 - a^2/b^2}$, where, obviously, $0 \leq e < 1$, is called the *eccentricity* or the *numerical eccentricity*. The number $p = b^2/a$ is called the *focal parameter* or simply the *parameter of the ellipse*.

The point $O(0,0)$ is called the *center of the ellipse*. The points of intersection $A_1(-a,0)$, $A_2(a,0)$ and $B_1(0,-b)$, $B_2(0,b)$ of the ellipse with the axes of symmetry are called its *vertices*. The points $F_1(-c,0)$ and $F_2(c,0)$ are the *focus of the ellipse*. This explains why the major axis of an ellipse is sometimes called its *focal axis*. The straight lines $x = \pm a/e$ ($e \neq 0$) are called the *directrices*. The focus $F_2(c,0)$ and the directrix $x = a/e$ are said to be *right*, and the focus $F_1(-c,0)$ and the directrix $x = -a/e$ are said to be *left*. A focus and a directrix are said to be *like* if both of them are right or left simultaneously.

The segments joining a point $M(x,y)$ of an ellipse with the foci $F_1(-c,0)$ and $F_2(c,0)$ are called the *left* and *right focal radii* of this point. We denote the lengths of the left and right focal radii by $r_1 = |F_1 M_0|$ and $r_2 = |F_2 M_0|$, respectively.

**Remark.** For $a = b$ ($c = 0$), equation (4.4.2.1) becomes $x^2 + y^2 = a^2$ and determines a circle; hence a circle can be considered as an ellipse for which $b = a$, $c = 0$, $e = 0$, and $p = 0$, i.e., the semiaxes are equal to each other, the foci coincide with the center, the eccentricity is zero (the directrices are not defined), and the focal parameter is zero.

The area of the figure bounded by the ellipse is given by the formula $S = \pi ab$. The length of the ellipse can be calculated approximately by the formula $L \approx \pi \left[1.5(a+b) - \sqrt{ab}\right]$. The area of the figure bounded by the ellipse and the chord passing through the points $M(x_0, y_0)$ and $N(x_0, -y_0)$ is equal to (see Fig. 20b)

$$S = \frac{\pi ab}{2} + \frac{b}{a}\left(x_0 \sqrt{a^2 - x_0^2} + a^2 \arcsin \frac{x_0}{a}\right). \tag{4.4.2.2}$$

### 4.4.2-2. Focal property of ellipse.

The ellipse defined by equation (4.4.2.1) is the locus of points on the plane for which the sum of distances to the foci $F_1$ and $F_2$ is equal to $2a$ (see Fig. 4.21). We write this property as

$$r_1 + r_2 = 2a, \tag{4.4.2.3}$$

where $r_1$ and $r_2$ satisfy the relations

$$\begin{aligned} r_1 &= \sqrt{(x+c)^2 + y^2} = a + ex, \\ r_2 &= \sqrt{(x-c)^2 + y^2} = a - ex. \end{aligned} \tag{4.4.2.4}$$

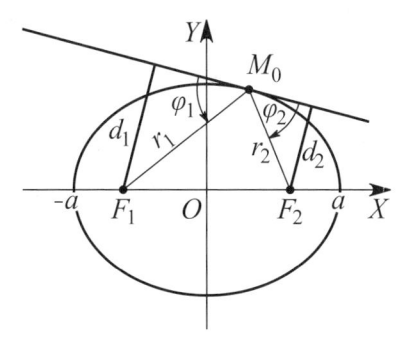

**Figure 4.21.** Focal property of ellipse.

**Remark.** One can show that equation (4.4.2.1) implies equation (4.4.2.3) and vice versa; hence the focal property of an ellipse is often used as its definition.

### 4.4.2-3. Focus-directrix property of ellipse.

The ellipse determined by equation (4.4.2.1) on the plane is the locus of points for which the ratio of distances to a focus and the like directrix is equal to $e$:

$$r_1 \left| x + \frac{a}{e} \right|^{-1} = e, \qquad r_2 \left| x - \frac{a}{e} \right|^{-1} = e. \qquad (4.4.2.5)$$

### 4.4.2-4. Equation of tangent and optical property of ellipse.

The tangent to the ellipse (4.4.2.1) at an arbitrary point $M_0(x_0, y_0)$ is described by the equation

$$\frac{x_0 x}{a^2} + \frac{y_0 y}{b^2} = 1. \qquad (4.4.2.6)$$

The distances $d_1$ and $d_2$ from the foci $F_1(-c, 0)$ and $F_2(c, 0)$ to the tangent to the ellipse at the point $M_0(x_0, y_0)$ are given by the formulas (see Paragraph 4.3.2-4)

$$d_1 = \frac{1}{Na} |x_0 e + a| = \frac{r_1(M_0)}{Na},$$
$$d_2 = \frac{1}{Na} |x_0 e - a| = \frac{r_2(M_0)}{Na}, \qquad N = \sqrt{\left(\frac{x_0}{a^2}\right)^2 + \left(\frac{y_0}{b^2}\right)^2}, \qquad (4.4.2.7)$$

where $r_1(M_0)$ and $r_2(M_0)$ are the lengths of the focal radii of $M_0$.

The tangent at an arbitrary point $M_0(x_0, y_0)$ of an ellipse forms acute angles $\varphi_1$ and $\varphi_2$ with the focal radii of the point of tangency, and

$$\sin \varphi_1 = \frac{d_1}{r_1} = \frac{1}{Na}, \qquad \sin \varphi_2 = \frac{d_2}{r_2} = \frac{1}{Na}. \qquad (4.4.2.8)$$

This implies the *optical property of the ellipse*:

$$\varphi_1 = \varphi_2, \qquad (4.4.2.9)$$

which means that all light rays issuing from one focus of the ellipse converge at the other focus after the reflection in the ellipse.

### 4.4.2-5. Diameters of ellipse.

A straight line passing through the midpoints of parallel chords of an ellipse is called a *diameter of the ellipse*. All diameters of an ellipse pass through its center. Two diameters of an ellipse are said to be *conjugate* if their slopes satisfy the relation

$$k_1 k_2 = -\frac{b^2}{a^2}. \qquad (4.4.2.10)$$

Two perpendicular conjugate diameters are called the *principal diameters of the ellipse*.

**Remark.** If $a = b$, i.e., the ellipse is a circle, then condition (4.4.2.10) becomes the perpendicularity condition: $k_1 k_2 = -1$. Thus any two conjugate diameters of a circle are perpendicular to each other, and each of the diameters is a principal diameter.

**4.4.2-6. Ellipse in polar coordinate system.**

In polar coordinates $(\rho, \varphi)$, the equation of an ellipse becomes

$$\rho = \frac{p}{1 - e \cos \varphi}, \qquad (4.4.2.11)$$

where $0 \leq \varphi \leq 2\pi$.

### 4.4.3. Hyperbola

**4.4.3-1. Definition and canonical equation of hyperbola.**

A curve on the plane is called a *hyperbola* if there exists a rectangular Cartesian coordinate system $OXY$ in which the equation of this curve has the form

$$\frac{x^2}{a^2} - \frac{y^2}{b^2} = 1, \qquad (4.4.3.1)$$

where $a > 0$ and $b > 0$ (see Fig. 4.22a). The coordinates in which the equation of a hyperbola has the form (4.4.3.1) are called the *canonical coordinates* for the hyperbola, and equation (4.4.3.1) itself is called the *canonical equation of the hyperbola*.

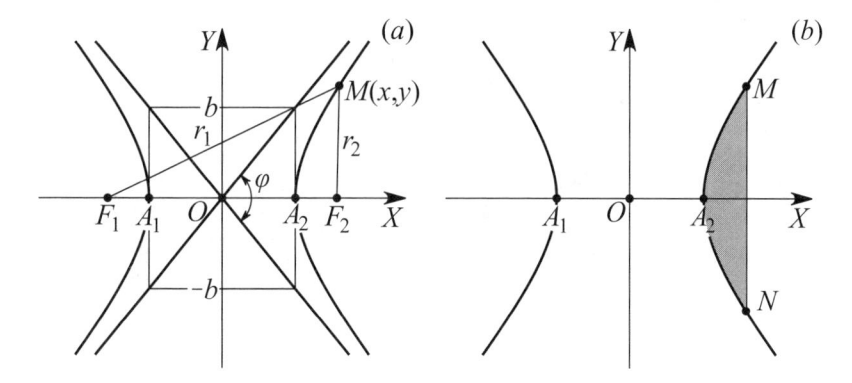

**Figure 4.22.** Hyperbola.

The hyperbola is a central curve of second order. It is described by equation (4.4.3.1) and consists of two connected parts (*arms*) lying in the domains $x > a$ and $x < -a$. The hyperbola has two *asymptotes* given by the equations

$$y = \frac{b}{a}x \quad \text{and} \quad y = -\frac{b}{a}x. \qquad (4.4.3.2)$$

More precisely, its arms lie in the two vertical angles formed by the asymptotes and are called the *left* and *right arms* of the hyperbola. A hyperbola is symmetric about the axes $OX$ and $OY$, which are called the *principal (real, or focal, and imaginary) axes*.

The angle between the asymptotes of a hyperbola is determined by the equation

$$\tan \frac{\varphi}{2} = \frac{b}{a}, \qquad (4.4.3.3)$$

and if $a = b$, then $\varphi = \frac{1}{2}\pi$ (an *equilateral hyperbola*).

The number $a$ is called the *real semiaxis*, and the number $b$ is called the *imaginary semiaxis*. The number $c = \sqrt{a^2 + b^2}$ is called the *linear eccentricity*, and $2c$ is called the *focal distance*. The number $e = c/a = \sqrt{a^2 + b^2}/a$, where, obviously, $e > 1$, is called the *eccentricity*, or the *numerical eccentricity*. The number $p = b^2/a$ is called the *focal parameter* or simply the *parameter of the hyperbola*.

The point $O(0,0)$ is called the *center of the hyperbola*. The points $A_1(-a, 0)$ and $A_2(a, 0)$ of intersection of the hyperbola with the real axis are called the *vertices of the hyperbola*. Points $F_1(-c, 0)$ and $F_2(c, 0)$ are called the *foci of the hyperbola*. This is why the real axis of a hyperbola is sometimes called the focal axis. The straight lines $x = \pm a/e$ ($y \neq 0$) are called the *directrices of the hyperbola* corresponding to the foci $F_2$ and $F_1$. The focus $F_2(c, 0)$ and the directrix $x = a/e$ are said to be *right*, and the focus $F_1(-c, 0)$ and the directrix $x = -a/e$ are said to be *left*. A focus and a directrix are said to be *like* if both of them are right or left simultaneously.

The segments joining a point $M(x, y)$ of the hyperbola with the foci $F_1(-c, 0)$ and $F_2(c, 0)$ are called the *left* and *right focal radii* of this point. We denote the lengths of the left and right focal radii by $r_1 = |F_1M|$ and $r_2 = |F_2M|$, respectively.

**Remark.** For $a = b$, the hyperbola is said to be *equilateral*, and its asymptotes are mutually perpendicular. The equation of an equilateral hyperbola has the form $x^2 - y^2 = a^2$. If we take the asymptotes to be the coordinate axes, then the equation of the hyperbola becomes $xy = a^2/2$; i.e., an equilateral hyperbola is the graph of inverse proportionality.

The curvature radius of a hyperbola at a point $M(x, y)$ is

$$R = a^2 b^2 \left( \frac{x^2}{a^4} + \frac{y^2}{b^4} \right)^{3/2} = \frac{\sqrt{(r_1 r_2)^3}}{ab}. \qquad (4.4.3.4)$$

The area of the figure bounded by the right arm of the hyperbola and the chord passing through the points $M(x_1, y_1)$ and $N(x_1, -y_1)$ is equal to (see Fig. 4.22b)

$$S = x_1 y_1 - ab \ln\left( \frac{x_1}{a} + \frac{y_1}{b} \right). \qquad (4.4.3.5)$$

### 4.4.3-2. Focal properties of hyperbola.

The hyperbola determined by equation (4.4.3.1) is the locus of points on the plane for which the difference of the distances to the foci $F_1$ and $F_2$ has the same absolute value $2a$ (see Fig. 4.22a). We write this property as

$$|r_1 - r_2| = 2a, \qquad (4.4.3.6)$$

where $r_1$ and $r_2$ satisfy the relations

$$\begin{aligned} r_1 &= \sqrt{(x+c)^2 + y^2} = \begin{cases} a + ex & \text{for } x > 0, \\ -a - ex & \text{for } x < 0, \end{cases} \\ r_2 &= \sqrt{(x-c)^2 + y^2} = \begin{cases} -a + ex & \text{for } x > 0, \\ a - ex & \text{for } x < 0. \end{cases} \end{aligned} \qquad (4.4.3.7)$$

**Remark.** One can show that equation (4.4.3.1) implies equation (4.4.3.6) and vice versa; hence the focal property of a hyperbola is often used as the definition.

### 4.4.3-3. Focus-directrix property of hyperbola.

The hyperbola defined by equation (4.4.3.1) on the plane is the locus of points for which the ratio of distances to a focus and the like directrix is equal to $e$:

$$r_1 \left| x + \frac{a}{e} \right|^{-1} = e, \qquad r_2 \left| x - \frac{a}{e} \right|^{-1} = e. \tag{4.4.3.8}$$

### 4.4.3-4. Equation of tangent and optical property of hyperbola.

The tangent to the hyperbola (4.4.3.1) at an arbitrary point $M_0(x_0, y_0)$ is described by the equation

$$\frac{x_0 x}{a^2} - \frac{y_0 y}{b^2} = 1. \tag{4.4.3.9}$$

The distances $d_1$ and $d_2$ from the foci $F_1(-c, 0)$ and $F_2(0, c)$ to the tangent to the hyperbola at the point $M_0(x_0, y_0)$ are given by the formulas (see Paragraph 4.3.2-4)

$$d_1 = \frac{Na}{|x_0 e + a|} = \frac{r_1}{Na},$$
$$d_2 = \frac{Na}{|x_0 e - a|} = \frac{r_2}{Na}, \qquad N = \sqrt{\left(\frac{x_0}{a^2}\right)^2 + \left(\frac{y_0}{b^2}\right)^2}, \tag{4.4.3.10}$$

where $r_1$ and $r_2$ are the lengths of the focal radii of the point $M_0$.

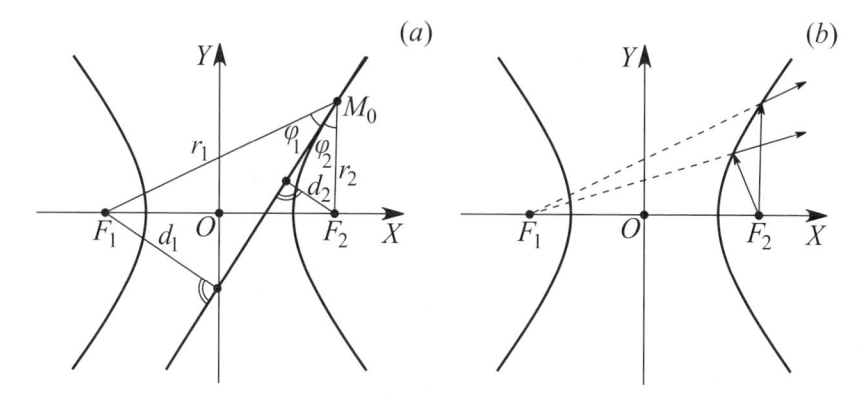

**Figure 4.23.** The tangent to the hyperbola (*a*). Optical property of a hyperbola (*b*).

The tangent at any point $M_0(x_0, y_0)$ of the hyperbola forms acute angles $\varphi_1$ and $\varphi_2$ with the focal radii of the point of tangency (see Fig. 4.23*a*), and

$$\sin \varphi_1 = \frac{d_1}{r_1} = \frac{1}{Na}, \qquad \sin \varphi_2 = \frac{d_2}{r_2} = \frac{1}{Na}. \tag{4.4.3.11}$$

This implies the *optical property of a hyperbola*:

$$\varphi_1 = \varphi_2, \tag{4.4.3.12}$$

which means that all light rays issuing from a focus appear to be issuing from the other focus after the mirror reflection in the hyperbola (see Fig. 4.23*b*).

The tangent and normal to a hyperbola at any point bisect the angles between the straight lines joining this point with the foci. The tangent to a hyperbola at either of its vertices intersects the asymptotes at two points such that the distance between them is equal to $2b$.

### 4.4.3-5. Diameters of hyperbola.

A straight line passing through the midpoints of parallel chords of a hyperbola is called a *diameter of the hyperbola*. Two diameters of a hyperbola are said to be *conjugate* if their slopes satisfy the relation

$$k_1 k_2 = \frac{b^2}{a^2}. \tag{4.4.3.13}$$

A hyperbola meets the diameter $y = kx$ if and only if

$$k^2 < \frac{b^2}{a^2}. \tag{4.4.3.14}$$

The lengths $l_1$ and $l_2$ of the conjugate diameters with slopes $k_1$ and $k_2$ satisfy the relation

$$l_1 l_2 \sin(\arctan k_2 - \arctan k_1) = 4ab. \tag{4.4.3.15}$$

Two perpendicular conjugate diameters are called the *principal diameters of a hyperbola*; they are its principal axes.

### 4.4.3-6. Hyperbola in polar coordinate system.

In polar coordinates $(\rho, \varphi)$, the equation for two connected parts of a hyperbola becomes

$$\rho = \frac{\pm p}{1 \mp e \cos \varphi}, \tag{4.4.3.16}$$

where upper and lower signs correspond to right and left parts of a hyperbola, respectively.

## 4.4.4. Parabola

### 4.4.4-1. Definition and canonical equation of parabola.

A curve on the plane is called a *parabola* if there exists a rectangular Cartesian coordinate system $OXY$, in which the equation of this curve has the form

$$y^2 = 2px, \tag{4.4.4.1}$$

where $p > 0$ (see Fig. 4.24a). The coordinates in which the equation of a parabola has the form (4.4.4.1) are called the *canonical coordinates* for the parabola, and equation (4.4.4.1) itself is called the *canonical equation of the parabola*.

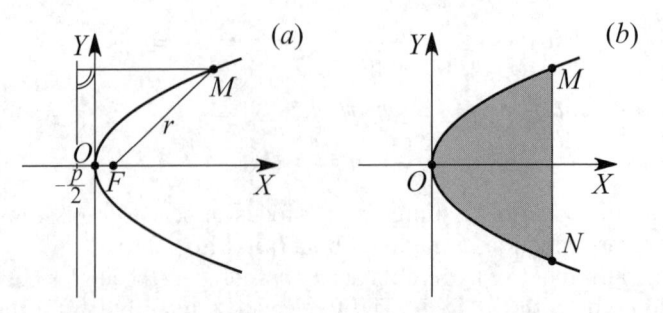

**Figure 4.24.** Parabola.

A parabola is a noncentral line of second order. It consists of an infinite branch symmetric about the $OX$-axis. The point $O(0,0)$ is called the *vertex of the parabola*. The point $F(p/2, 0)$ is called the *focus of the parabola*. The straight line $x = -p/2$ is called the *directrix*. The focal parameter $p$ is the distance from the focus to the directrix. The number $p/2$ is called the *focal distance*.

The segment joining a point $M(x, y)$ of the parabola with the focus $F(p/2, 0)$ is called the *focal radius* of the point. The curvature radius of the parabola at a point $M(x, y)$ can be found from the formula

$$R = \frac{(p + 2x)^{3/2}}{\sqrt{p}}. \tag{4.4.4.2}$$

The area of the figure bounded by the parabola and the chord passing through the points $M(x_1, y_1)$ and $N(x_1, -y_1)$ is equal to (see Fig. 4.24b)

$$S = \frac{4}{3} x_1 y_1. \tag{4.4.4.3}$$

### 4.4.4-2. Focal properties of parabola.

The parabola defined by equation (4.4.4.1) on the plane is the locus of points equidistant from the focus $F(p/2, 0)$ and the directrix $x = -p/2$ (see Fig. 4.24a).

We denote the length of the focal radius by $r$ and write this property as

$$r = x + \frac{p}{2}, \tag{4.4.4.4}$$

where $r$ satisfies the relation

$$r = \sqrt{\left(x - \frac{p}{2}\right)^2 + y^2}. \tag{4.4.4.5}$$

**Remark.** One can show that equation (4.4.4.1) implies equation (4.4.4.5) and vice versa; hence the focal property of a parabola is often used as the definition.

### 4.4.4-3. Focus-directrix property of parabola.

The parabola defined by equation (4.4.4.1) on the plane is the locus of points for which the ratio of distances to the focus and the directrix is equal to 1:

$$\frac{r}{|x + p/2|} = 1. \tag{4.4.4.6}$$

### 4.4.4-4. Equation of tangent and optical property of parabola.

The tangent to the parabola (4.4.4.1) at an arbitrary point $M_0(x_0, y_0)$ is described by the equation

$$y y_0 = p(x + x_0). \tag{4.4.4.7}$$

The direction vector of the tangent (4.4.4.7) has the coordinates $(y_0, p)$, and the direction vector of the line passing through the points $M_0(x_0, y_0)$ and $F(p/2, 0)$ has the coordinates

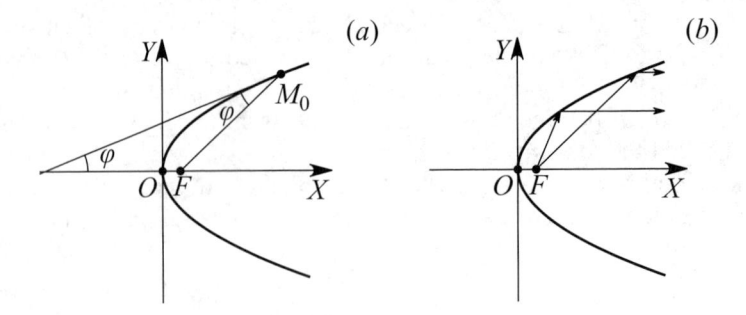

**Figure 4.25.** The tangent to the parabola (*a*). Optical property of a parabola (*b*).

$(x_0 - p/2, y_0)$ (see Fig. 4.25*a*). Thus, in view of the focus-directrix property, the angle $\varphi$ between these lines satisfies the relation

$$\cos \varphi = \frac{y_0(x_0 - p/2) + py_0}{\sqrt{y_0^2 + p^2}\sqrt{(x_0 - p/2)^2 + y_0^2}} = \frac{y_0}{\sqrt{y_0^2 + p^2}}. \qquad (4.4.4.8)$$

But the same relation also holds for the angle between the tangent (4.4.4.7) and the $OX$-axis.

This property of a parabola is called the *optical property*: all light rays issuing from the focus of a parabola form a pencil parallel to the axis of the parabola after the mirror reflection in the parabola (see Fig. 4.25*b*).

The tangent and normal to a parabola at any point bisect the angles between the focal radius and the diameter.

### 4.4.4-5. Diameters of parabola.

A straight line passing through the midpoints of parallel chords of a parabola is called a *diameter of the parabola*. The diameter corresponding to the chords perpendicular to the axis of the parabola is the axis itself. The diameter of the parabola $y^2 = 2px$ corresponding to the chords with slope $k$ ($k > 0$) is given by the equation

$$y = \frac{p}{k}. \qquad (4.4.4.9)$$

The $OX$-axis (the axis of symmetry of a parabola), in contrast to the other diameters of the parabola, is the diameter perpendicular to the chords conjugate to it. This diameter is called the *principal diameter of the parabola*. The slope of any diameter of a parabola is zero. A parabola does not have mutually conjugate diameters.

### 4.4.4-6. Parabola with vertical axis.

The equation of a parabola with vertical axis has the form

$$y = ax^2 + bx + c \quad (a \neq 0). \qquad (4.4.4.10)$$

For $a > 0$, the vertex of the parabola is directed downward, and for $a < 0$, the vertex is directed upward. The vertex of a parabola has the coordinates

$$x_0 = \frac{b}{2}, \quad y_0 = \frac{4ac - b^2}{4a}. \qquad (4.4.4.11)$$

#### 4.4.4-7. Parabola in polar coordinates.

In the polar coordinates $(\rho, \varphi)$ (the pole lies at the focus of the parabola, and the polar axis is directed along the parabola axis), the equation of the parabola has the form

$$\rho = \frac{p}{1 - \cos \varphi}, \qquad (4.4.4.12)$$

where $-\frac{1}{2}\pi \leq \varphi \leq \frac{1}{2}\pi$.

### 4.4.5. Transformation of Second-Order Curves to Canonical Form

#### 4.4.5-1. General equation of second-order curve.

The set of points on the plane whose coordinates in the rectangular Cartesian coordinate system satisfy the second-order algebraic equation

$$\begin{aligned} & a_{11}x^2 + 2a_{12}xy + a_{22}y^2 + 2a_{13}x + 2a_{23}y + a_{33} = 0 \quad \text{or} \\ & (a_{11}x + a_{12}y + a_{13})x + (a_{21}x + a_{22}y + a_{23})y + a_{31}x + a_{32}y + a_{33} = 0, \quad (4.4.5.1) \\ & a_{ij} = a_{ji} \quad (i, j = 1, 2, 3) \end{aligned}$$

is called a *second-order curve*.

#### 4.4.5-2. Nine canonical second-order curves.

There exists a rectangular Cartesian coordinate system in which equations (4.4.5.1) can be reduced to one of the following *nine canonical forms*:

1. $\dfrac{x^2}{a^2} + \dfrac{y^2}{b^2} = 1$, an ellipse;
2. $\dfrac{x^2}{a^2} - \dfrac{y^2}{b^2} = 1$, a hyperbola;
3. $y^2 = 2px$, a parabola;
4. $\dfrac{x^2}{a^2} + \dfrac{y^2}{b^2} = -1$, an imaginary ellipse;
5. $\dfrac{x^2}{a^2} - \dfrac{y^2}{b^2} = 0$, a pair of intersecting straight lines;
6. $\dfrac{x^2}{a^2} + \dfrac{y^2}{b^2} = 0$, a pair of imaginary intersecting straight lines;
7. $x^2 - a^2 = 0$, a pair of parallel straight lines;
8. $x^2 + a^2 = 0$, a pair of imaginary parallel straight lines;
9. $x^2 = 0$, a pair of coinciding straight lines.

#### 4.4.5-3. Invariants of second-order curves.

Second-order curves can be studied with the use of the three *invariants*

$$I = a_{11} + a_{22}, \quad \delta = \begin{vmatrix} a_{22} & a_{23} \\ a_{32} & a_{33} \end{vmatrix}, \quad \Delta = \begin{vmatrix} a_{11} & a_{12} & a_{13} \\ a_{21} & a_{22} & a_{23} \\ a_{31} & a_{32} & a_{33} \end{vmatrix}, \qquad (4.4.5.2)$$

whose values do not change under parallel translation and rotation of the coordinate axes, and the *semi-invariant*

$$\sigma = \begin{vmatrix} a_{22} & a_{23} \\ a_{32} & a_{33} \end{vmatrix} + \begin{vmatrix} a_{11} & a_{13} \\ a_{31} & a_{33} \end{vmatrix}, \qquad (4.4.5.3)$$

whose value does not change under rotation of the coordinate axes.

The invariant $\Delta$ is called the *large discriminant* of equation (4.4.2.1). The invariant $\delta$ is called the *small discriminant*.

Table 1 presents the classification of second-order curves based on invariants.

TABLE 4.1
Classification of second-order curves

| Type | | $\Delta \neq 0$ | $\Delta = 0$ | |
|---|---|---|---|---|
| Elliptic $\delta > 0$ | $\dfrac{\Delta}{I} > 0$ | Real ellipse (for $I^2 = 4\delta$ or $a_{11} = a_{22}$ and $a_{12} = 0$, this is a circle) | Pair of imaginary straight lines intersecting at a real point (ellipse degenerating into a point) | |
| | $\dfrac{\Delta}{I} < 0$ | Imaginary ellipse (no real points) | | |
| Hyperbolic $\delta < 0$ | | Hyperbola | Pair of real intersecting straight lines (degenerate hyperbola) | |
| Parabolic $\delta = 0$ | | Parabola | $\sigma > 0$ | Pair of imaginary parallel straight lines |
| | | | $\sigma < 0$ | Pair of real parallel straight lines |
| | | | $\sigma = 0$ | Pair of coinciding straight lines (a single straight line) |

### 4.4.5-4. Characteristic equation of second-order curves.

The properties of second-order curves can be studied with the use of the characteristic equation

$$\begin{vmatrix} a_{11} - \lambda & a_{12} \\ a_{21} & a_{22} - \lambda \end{vmatrix} = 0 \quad \text{or} \quad \lambda^2 - I\lambda + \delta = 0. \qquad (4.4.5.4)$$

The roots $\lambda_1$ and $\lambda_2$ of the characteristic equation (4.4.5.4) are the eigenvalues of the real symmetric matrix $[a_{ij}]$ and, as a consequence, are real.

Obviously, the invariants $I$ and $\delta$ of second-order curves are expressed as follows in terms of the roots $\lambda_1$ and $\lambda_2$ of the characteristic equation (4.4.5.4):

$$I = \lambda_1 + \lambda_2, \quad \delta = \lambda_1 \lambda_2. \qquad (4.4.5.5)$$

### 4.4.5-5. Centers and diameters of second-order curves.

A straight line passing through the midpoints of parallel chords of a second-order curve is called a *diameter* of this curve. A diameter is said to be *conjugate* to the chords (or to the direction of chords) which it divides into two parts. The diameter conjugate to chords forms an angle $\varphi$ with the positive direction of the $OX$-axis and is determined by the equation

$$(a_{11}x + a_{12}y + a_{13})\cos\varphi + (a_{21}x + a_{22}y + a_{23})\sin\varphi = 0. \qquad (4.4.5.6)$$

## 4.4. SECOND-ORDER CURVES

All diameters of a second-order curve with $\delta \neq 0$ meet at a single point called the *center of the curve*, and in this case the curve is said to be *central*. The center coordinates $(x_0, y_0)$ satisfy the system of equations

$$a_{11}x_0 + a_{12}y_0 + a_{13} = 0,$$
$$a_{21}x_0 + a_{22}y_0 + a_{23} = 0, \quad (4.4.5.7)$$

which implies that

$$x_0 = -\frac{1}{\delta}\begin{vmatrix} a_{13} & a_{12} \\ a_{23} & a_{22} \end{vmatrix}, \quad y_0 = -\frac{1}{\delta}\begin{vmatrix} a_{11} & a_{13} \\ a_{21} & a_{23} \end{vmatrix}. \quad (4.4.5.8)$$

All diameters of a second-order curve with $\delta = 0$ are parallel or coincide. A second-order curve does not have a center if and only if $\delta = 0$ and $\Delta \neq 0$. A second-order curve has a center line if and only if $\delta = 0$ and $\Delta = 0$.

**Example 1.** The centers of nine canonical second-order curves are as follows:
1. Ellipse, $\delta = 1$ and $\Delta = -1$: the single center $O(0,0)$;
2. Hyperbola, $\delta = -1$ and $\Delta = 1$: the single center $O(0,0)$;
3. Parabola, $\delta = 0$ and $\Delta = -1$: no centers;
4. Imaginary ellipse, $\delta = 1$ and $\Delta = -1$: the single center $O(0,0)$;
5. Pair of intersecting straight lines, $\delta = -1$ and $\Delta = 0$: the single center $O(0,0)$;
6. Pair of imaginary intersecting straight lines, $\delta = 1$ and $\Delta = 0$: the single center $O(0,0)$;
7. Pair of imaginary parallel straight lines, $\delta = 0$ and $\Delta = 0$: the center line $y = 0$;
8. Pair of coinciding straight lines, $\delta = 0$ and $\Delta = 0$: the center line $y = 0$.

Each of the two conjugate diameters of a central second-order curve bisects the chords parallel to the other diameter.

### 4.4.5-6. Principal axes.

A diameter perpendicular to the chords conjugate to it is called a *principal axis*. A principal axis is a symmetry axis of a second-order curve. For each central second-order curve ($\delta \neq 0$), either there are two perpendicular principal axes or each of its diameters is a principal axis (the circle). A second-order curve with $\delta = 0$ has a unique principal axis. The points of intersection of a second-order curve with its principal axes are called its *vertices*.

The directions of principal axes coincide with the directions of the eigenvectors of the symmetric matrix $[a_{ij}]$ ($i, j = 1, 2$); i.e., the direction cosines $\cos\theta$, $\sin\theta$ of the normals to the principal axes are determined from the system of equations

$$(a_{11} - \lambda)\cos\theta + a_{12}\sin\theta = 0,$$
$$a_{21}\cos\theta + (a_{22} - \lambda)\sin\theta = 0, \quad (4.4.5.9)$$

where $\lambda$ is a nonzero root of characteristic equation (4.4.5.4).

The directions of principal axes and of their conjugate chords are called the *principal directions* of a second-order curve. The angle between the positive direction of the axis $OX$ and each of the two principal directions of a second-order curve is given by the formula

$$\tan 2\varphi = \tan 2\theta = \frac{2a_{12}}{a_{11} - a_{22}}. \quad (4.4.5.10)$$

**Remark.** The circle has undetermined principal directions.

## 4.4.5-7. Reduction of central second-order curves to canonical form.

A second-order curve with $\delta \neq 0$ has a center. By shifting the origin to the center $O_1(x_0, y_0)$ whose coordinates are determined by formula (4.4.5.8), we can reduce equation (4.4.5.1) to the form

$$a_{11}x_1^2 + 2a_{12}x_1y_1 + a_{22}y_1^2 + \frac{\Delta}{\delta} = 0, \qquad (4.4.5.11)$$

where $x_1$ and $y_1$ are the new coordinates.

By rotating the axes $O_1X_1$ and $O_1Y_1$ by the angle $\theta$ determined by (4.4.5.10), we transform equation (4.4.5.11) as follows:

$$A_{11}\hat{x}^2 + A_{22}\hat{y}^2 + \frac{\Delta}{\delta} = 0. \qquad (4.4.5.12)$$

The coefficients $A_{11}$ and $A_{22}$ are the roots of the characteristic equation (4.4.5.4).

We note the following formulas for the ellipse:

$$a^2 = -\frac{1}{\lambda_2}\frac{\Delta}{\delta} = -\frac{\Delta}{\lambda_1\lambda_2^2}, \quad b^2 = -\frac{1}{\lambda_1}\frac{\Delta}{\delta} = -\frac{\Delta}{\lambda_1^2\lambda_2}, \qquad (4.4.5.13)$$

where $a$ and $b$ are the parameters of the canonical equation, $\delta$ and $\Delta$ are the invariants, and $\lambda_1$ and $\lambda_2$ are the roots of the characteristic equation (4.4.5.4).

Similarly, for the hyperbola one has

$$a^2 = -\frac{1}{\lambda_1}\frac{\Delta}{\delta} = -\frac{\Delta}{\lambda_1^2\lambda_2}, \quad b^2 = \frac{1}{\lambda_1}\frac{\Delta}{\delta} = \frac{\Delta}{\lambda_1^2\lambda_2}. \qquad (4.4.5.14)$$

## 4.4.5-8. Reduction of noncentral second-order curves to canonical form.

If $\delta = 0$, then the curve does not have any center or does not have a definite center, and its equation can be written as

$$(\alpha x + \beta y)^2 + 2a_{13}x + 2a_{23}y + a_{33} = 0. \qquad (4.4.5.15)$$

If the coefficients $a_{13}$ and $a_{23}$ are proportional to the coefficients $\alpha$ and $\beta$, i.e., $a_{13} = k\alpha$ and $a_{23} = k\beta$, then equation (4.4.5.15) becomes $(\alpha x + \beta y)^2 + 2k(\alpha x + \beta y) + a_{33} = 0$, and hence

$$\alpha x + \beta y = -k \pm \sqrt{k^2 - a_{33}} \qquad (4.4.5.16)$$

is a pair of real parallel straight lines.

If the coefficients $a_{13}$ and $a_{23}$ are not proportional to the coefficients $\alpha$ and $\beta$, then equation (4.4.5.15) can be written as

$$(\alpha x + \beta y + \gamma)^2 + 2k(\beta x - \alpha x + q) = 0. \qquad (4.4.5.17)$$

The parameters $k$, $\gamma$, and $q$ can be determined by comparing the coefficients in equations (4.4.5.15) and (4.4.5.17). For the axis $O_1X$ one should take the line $\alpha x + \beta y + \gamma = 0$, and for the axis $O_1Y$, the line $\beta x - \alpha x + q = 0$. We denote

$$\hat{x} = \frac{\beta x - \alpha x + q}{\pm\sqrt{\alpha^2 + \beta^2}}, \quad \hat{y} = \frac{\alpha x + \beta y + \gamma}{\pm\sqrt{\alpha^2 + \beta^2}}; \qquad (4.4.5.18)$$

then equation (4.4.5.17) acquires the form

$$\hat{y}^2 = 2p\hat{x}, \tag{4.4.5.19}$$

where $p = |k|/\sqrt{\alpha^2 + \beta^2}$. The axis $O_1X$ points into the half-plane where the sign of $\beta x - \alpha x + q$ is opposite to that of $k$.

If one only needs to find the canonical equation of a parabola and it is not necessary to construct the graph of the parabola in the coordinate system $OXY$, then the parameter $p$ is determined via the invariants $I$, $\delta$, and $\Delta$ and the roots $\lambda_1$ and $\lambda_2$ ($\lambda_1 \geq \lambda_2$) of the characteristic equation (4.4.5.4) by the formulas

$$p = \frac{1}{I}\sqrt{-\frac{\Delta}{I}} = \frac{1}{\lambda_1}\sqrt{-\frac{\Delta}{\lambda_1}} > 0, \quad \lambda_2 = 0. \tag{4.4.5.20}$$

### 4.4.5-9. Geometric definition of nondegenerate second-order curve.

There exists a coordinate system in which equation (4.4.5.1) has the form

$$y^2 = 2px - (1-e^2)x^2, \tag{4.4.5.21}$$

where $p > 0$ is a parameter and $e$ is the eccentricity. Obviously, the curve (4.4.5.21) passes through the origin of the new coordinate system. The axis $OX$ is a symmetry axis of the curve.

The equation of the directrix of the curve (4.4.5.21) is

$$x = -\frac{p}{(1+e)e}. \tag{4.4.5.22}$$

The coordinates of the focus are

$$x = \frac{p}{1+e}, \quad y = 0. \tag{4.4.5.23}$$

The distance from the focus to the directrix is equal to $p/e$. For a central second-order curve, the line

$$x = \frac{p}{1-e^2} = a \tag{4.4.5.24}$$

is a symmetry axis.

Remark. All types of second-order curves can be obtained as plane sections of a right circular cone for various positions of the secant plane with respect to the cone.

### 4.4.5-10. Tangents and normals to second-order curves.

The *equation of the tangent* to a second-order curve at a point $M_0(x_0, y_0)$ has the form

$$(a_{11}x_0 + a_{12}y_0 + a_{13})x + (a_{21}x_0 + a_{22}y_0 + a_{23})y + a_{31}x_0 + a_{32}y_0 + a_{33} = 0. \tag{4.4.5.25}$$

The *equation of the normal* to a second-order curve at the point $M_0(x_0, y_0)$ has the form

$$\frac{x - x_0}{a_{11}x_0 + a_{12}y_0 + a_{13}} = \frac{y - y_0}{a_{21}x_0 + a_{22}y_0 + a_{23}}. \tag{4.4.5.26}$$

TABLE 4.2
Ellipse, hyperbola, parabola. Main formulas

| | Ellipse | Hyperbola | Parabola |
|---|---|---|---|
| Canonical equation | $\dfrac{x^2}{a^2} + \dfrac{y^2}{b^2} = 1$ | $\dfrac{x^2}{a^2} - \dfrac{y^2}{b^2} = 1$ | $y^2 = 2px$ |
| Equation in polar coordinates | $\rho = \dfrac{p}{1 - e\cos\varphi}$ | $\rho = \dfrac{\pm p}{1 \mp e\cos\varphi}$ | $\rho = \dfrac{p}{1 - \cos\varphi}$ |
| Eccentricity | $e = \dfrac{c}{a} = \sqrt{1 - \dfrac{b^2}{a^2}} < 1$ | $e = \dfrac{c}{a} = \sqrt{1 + \dfrac{b^2}{a^2}} > 1$ | $e = \dfrac{c}{a} = 1$ |
| Foci | $(\pm ae, 0)$ | $(\pm ae, 0)$ | $\left(\dfrac{p}{2}, 0\right)$ |
| Focal radii (distance from the foci to an arbitrary point $(x,y)$ of curve) | $r_1 = a + ex$<br>$r_2 = a - ex$ | $r_1 = \begin{cases} a + ex & \text{for } x > 0 \\ -a - ex & \text{for } x < 0 \end{cases}$<br>$r_2 = \begin{cases} -a + ex & \text{for } x > 0 \\ a - ex & \text{for } x < 0 \end{cases}$ | $r = x + \dfrac{p}{2}$ |
| Focal parameter | $p = \dfrac{b^2}{a}$ | $p = \dfrac{b^2}{a}$ | $p$ |
| Equation of directrices | $x = \pm \dfrac{a}{e}$ | $x = \pm \dfrac{a}{e}$ | $x = -\dfrac{p}{2}$ |
| Equation of diameter conjugate to chords with slope $k$ | $y = -\dfrac{b^2}{a^2 k} x$ | $y = \dfrac{b^2}{a^2 k} x$ | $y = \dfrac{p}{k}$ |
| Area of segment bounded by an arc convex to the left and the chord joining points $(x_0, y_0)$ and $(x_0, -y_0)$ | $\dfrac{\pi ab}{2} + \dfrac{b}{a}\left(x_0\sqrt{a^2 - x_0^2} + a^2 \arcsin\dfrac{x_0}{a}\right)$ | $x_0 y_0 - ab\ln\left(\dfrac{x_0}{a} + \dfrac{y_0}{b}\right)$ $= x_0 y_0 - ab\operatorname{arccosh}\dfrac{x_0}{a}$ | $\dfrac{4}{3} x_0 y_0$ |
| Curvature radius at point $(x,y)$ | $a^2 b^2 \left(\dfrac{x^2}{a^4} + \dfrac{y^2}{b^4}\right)^{3/2} = \dfrac{\sqrt{(r_1 r_2)^3}}{ab}$ | $a^2 b^2 \left(\dfrac{x^2}{a^4} + \dfrac{y^2}{b^4}\right)^{3/2} = \dfrac{\sqrt{(r_1 r_2)^3}}{ab}$ | $\dfrac{(p + 2x)^{3/2}}{\sqrt{p}}$ |
| Equations of tangents to a curve which pass through an arbitrary point $(x_0, y_0)$ | $\dfrac{y - y_0}{x - x_0} = \dfrac{-x_0 y_0 \pm \sqrt{a^2 y_0^2 + b^2 x_0^2 - a^2 b^2}}{a^2 - x_0^2}$ | $\dfrac{y - y_0}{x - x_0} = \dfrac{-x_0 y_0 \pm \sqrt{a^2 y_0^2 - b^2 x_0^2 + a^2 b^2}}{a^2 - x_0^2}$ | $\dfrac{y - y_0}{x - x_0} = \dfrac{y_0 \pm \sqrt{y_0^2 - 2px_0}}{2x_0}$ |
| Equation of tangent at point $(x_0, y_0)$ | $\dfrac{x_0 x}{a^2} + \dfrac{y_0 y}{b^2} = 1$ | $\dfrac{x_0 x}{a^2} - \dfrac{y_0 y}{b^2} = 1$ | $y y_0 = p(x + x_0)$ |
| Equation of tangent with slope $k$ | $y = kx \pm \sqrt{k^2 a^2 + b^2}$ | $y = kx \pm \sqrt{k^2 a^2 - b^2}$ | $y = kx + \dfrac{p}{2k}$ |
| Equation of normal at point $(x_0, y_0)$ | $\dfrac{y - y_0}{x - x_0} = \dfrac{a^2 y_0}{b^2 x_0}$ | $\dfrac{y - y_0}{x - x_0} = -\dfrac{a^2 y_0}{b^2 x_0}$ | $\dfrac{y - y_0}{x - x_0} = -\dfrac{y_0}{p}$ |
| Coordinates of pole $(x_0, y_0)$ of straight line $Ax + By + C = 0$ w.r.t. a curve | $x_0 = -\dfrac{a^2 A}{C}, \quad y_0 = -\dfrac{b^2 B}{C}$ | $x_0 = -\dfrac{a^2 A}{C}, \quad y_0 = \dfrac{b^2 B}{C}$ | $x_0 = \dfrac{A}{C}, \quad y_0 = -\dfrac{pB}{C}$ |

The *polar* is the set of points $Q$ that are harmonically conjugate to a point $P$, called the pole of a second-order curve, with respect to points $R_1$ and $R_2$ of intersection of the second-order curve with secants passing through $P$; i.e.,

$$\frac{R_1 P}{P R_2} = -\frac{R_1 Q}{Q R_2}.$$

If the point $P$ lies outside the second-order curve (one can draw two tangents through $P$), then the polar line passes through the points at which this curve is tangent to straight lines drawn through point $P$. If the point lies on a second-order curve, then the polar line is a straight line tangent to this curve at this point. If the polar line of a point $P$ passes through a point $Q$, then the polar line of $Q$ passes through $P$.

## 4.5. Coordinates, Vectors, Curves, and Surfaces in Space

### 4.5.1. Vectors. Cartesian Coordinate System

#### 4.5.1-1. Notion of vector.

A directed segment with initial point $A$ and endpoint $B$ (see Fig. 4.26) is called the *vector* $\overrightarrow{AB}$. A nonnegative number equal to the length of the segment $AB$ joining the points $A$ and $B$ is called the *length* $|\overrightarrow{AB}|$ of the *vector* $\overrightarrow{AB}$. The vector $\overrightarrow{BA}$ is said to be *opposite to the vector* $\overrightarrow{AB}$.

**Figure 4.26.** Vector.

Two directed segments $\overrightarrow{AB}$ and $\overrightarrow{CD}$ of the same length and the same direction determine the same vector $\mathbf{a}$; i.e., $\mathbf{a} = \overrightarrow{AB} = \overrightarrow{CD}$.

Two vectors are said to be *collinear (parallel)* if they lie on the same straight line or on parallel lines. Three vectors are said to be *coplanar* if they lie in the same plane or in parallel planes. A vector $0$ whose initial point and endpoint coincide is called the *zero vector*. The length of the zero vector is equal to zero ($|0| = 0$), and the direction of the zero vector is assumed to be arbitrary. A vector $\mathbf{e}$ of unit length is called a *unit vector*.

The *sum* $\mathbf{a} + \mathbf{b}$ *of vectors* $\mathbf{a}$ *and* $\mathbf{b}$ is defined as the vector directed from the initial point of $\mathbf{a}$ to the endpoint of $\mathbf{b}$ under the condition that $\mathbf{b}$ is applied at the endpoint of $\mathbf{a}$. The rule for addition of vectors, which is contained in this definition, is called the *triangle rule* or the *rule of closing a chain of vectors* (see Fig. 4.27a). The sum $\mathbf{a} + \mathbf{b}$ can also be found using the *parallelogram rule* (see Fig. 4.27b). The *difference* $\mathbf{a} - \mathbf{b}$ *of vectors* $\mathbf{a}$ *and* $\mathbf{b}$ is defined as follows: $\mathbf{b} + (\mathbf{a} - \mathbf{b}) = \mathbf{a}$ (see Fig. 4.27c).

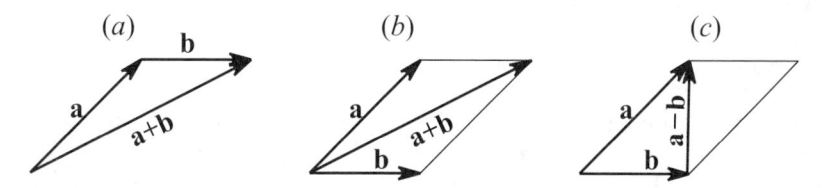

**Figure 4.27.** The sum of vectors: triangle rule (*a*) and parallelogram rule (*b*). The difference of vectors (*c*).

The *product* $\lambda\mathbf{a}$ *of a vector* $\mathbf{a}$ *by a number* $\lambda$ is defined as the vector whose length is equal to $|\lambda\mathbf{a}| = |\lambda||\mathbf{a}|$ and whose direction coincides with that of the vector $\mathbf{a}$ if $\lambda > 0$ or is opposite to the direction of the vector $\mathbf{a}$ if $\lambda < 0$.

**Remark.** If $\mathbf{a} = 0$ or $\lambda = 0$, then the absolute value of the product is zero, i.e., it is the zero vector. In this case, the direction of the product $\lambda\mathbf{a}$ is undetermined.

*Main properties of operations with vectors:*

1. $\mathbf{a} + \mathbf{b} = \mathbf{b} + \mathbf{a}$ (commutativity).
2. $\mathbf{a} + (\mathbf{b} + \mathbf{c}) = \mathbf{a} + (\mathbf{b} + \mathbf{c})$ (associativity of addition).
3. $\mathbf{a} + 0 = \mathbf{a}$ (existence of the zero vector).
4. $\mathbf{a} + (-\mathbf{a}) = 0$ (existence of the opposite vector).
5. $\lambda(\mathbf{a} + \mathbf{b}) = \lambda\mathbf{a} + \lambda\mathbf{b}$ (distributivity with respect to addition of vectors).
6. $(\lambda + \mu)\mathbf{a} = \lambda\mathbf{a} + \mu\mathbf{a}$ (distributivity with respect to addition of constants).
7. $\lambda(\mu\mathbf{a}) = (\lambda\mu)\mathbf{a}$ (associativity of product).
8. $1\mathbf{a} = \mathbf{a}$ (multiplication by unity).

**4.5.1-2. Projection of vector onto axis.**

A straight line with a unit vector $\mathbf{e}$ lying on it determining the positive sense of the line is called an *axis*. The *projection* $\mathrm{pr}_\mathbf{e}\,\mathbf{a}$ of a vector $\mathbf{a}$ onto the axis (see Fig. 4.28) is defined as the directed segment on the axis whose signed length is equal to the scalar product of $\mathbf{a}$ by the unit vector $\mathbf{e}$, i.e., is determined by the formula

$$\mathrm{pr}_\mathbf{e}\,\mathbf{a} = |\mathbf{a}|\cos\varphi, \qquad (4.5.1.1)$$

where $\varphi$ is the angle between the vectors $\mathbf{a}$ and $\mathbf{e}$.

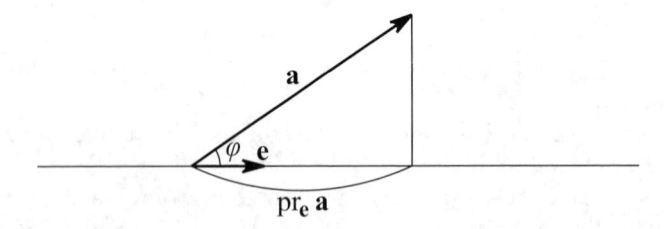

**Figure 4.28.** Projection of a vector onto the axes.

*Properties of projections:*

1. $\mathrm{pr}_\mathbf{e}(\mathbf{a} + \mathbf{b}) = \mathrm{pr}_\mathbf{e}\,\mathbf{a} + \mathrm{pr}_\mathbf{e}\,\mathbf{b}$ (additivity).
2. $\mathrm{pr}_\mathbf{e}(\lambda\mathbf{a}) = \lambda\,\mathrm{pr}_\mathbf{e}\,\mathbf{a}$ (homogeneity).

## 4.5.2. Coordinate Systems

**4.5.2-1. Cartesian coordinate system.**

If a one-to-one correspondence between points in space and numbers (triples of numbers) is given, then one says that a *coordinate system* is introduced in space.

A *rectangular Cartesian coordinate system* is determined by a scale segment for measuring lengths and three pairwise perpendicular directed straight lines $OX$, $OY$, and $OZ$ (the *coordinate axes*) concurrent at a single point $O$ (the *origin*). The three coordinate axes divide the space into eight parts called *octants*.

We choose an arbitrary point $M$ in space and project it onto the coordinate axes, i.e., draw the perpendiculars to the axes $OX$, $OY$, and $OZ$ through $M$. We denote the points

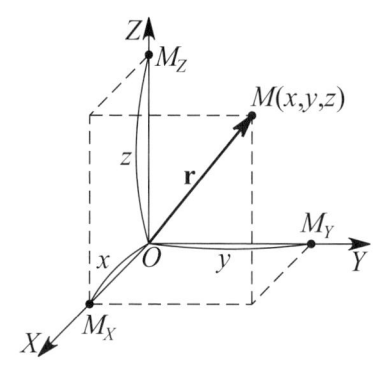

## 4.5. COORDINATES, VECTORS, CURVES, AND SURFACES IN SPACE

**Figure 4.29.** Point in rectangular Cartesian coordinate system.

of intersection of the perpendiculars with the axes by $M_X$, $M_Y$, and $M_Z$, respectively. The numbers (see Fig. 4.29)

$$x = OM_X, \quad y = OM_Y, \quad z = OM_Z, \qquad (4.5.2.1)$$

where $OM_X$, $OM_Y$, and $OM_Z$ are the signed lengths of the directed segments $\overrightarrow{OM}_X$, $\overrightarrow{OM}_Y$, and $\overrightarrow{OM}_Z$ of the axes $OX$, $OY$, and $OZ$, respectively, are called the *coordinates of the point* $M$ in the rectangular Cartesian coordinate system. The number $x$ is called the first coordinate or the *abscissa* of the point $M$, the number $y$ is called the second coordinate or the *ordinate* of the point $M$, and the number $z$ is called the third coordinate or the *applicate* of the point $M$. Usually one says that the point $M$ has the coordinates $(x, y, z)$, and the notation $M(x, y, z)$ is used.

To each point $M$ of three-dimensional space one can assign its position vector. The directed segment $\overrightarrow{OM}$ is called the *position vector of the point* $M$. The position vector determines the vector $\mathbf{r}$ ($\mathbf{r} = \overrightarrow{OM}$) whose coordinates are its projections on the axes $OX$, $OY$, and $OZ$, respectively. Obviously, the triple $(x, y, z)$ of numbers can be called the point $M$ whose coordinates are these numbers, or the position vector $\overrightarrow{OM}$ whose projections are these numbers. An arbitrary vector $(x, y, z)$ can be represented as

$$(x, y, z) = x\mathbf{i} + y\mathbf{j} + z\mathbf{k}, \qquad (4.5.2.2)$$

where $\mathbf{i} = (1, 0, 0)$, $\mathbf{j} = (0, 1, 0)$, and $\mathbf{k} = (0, 0, 1)$ are the unit vectors with the same directions as the coordinate axes $OX$, $OY$, and $OZ$. The distance between points $M_1(x_1, y_1, z_1)$ and $M_2(x_2, y_2, z_2)$ is determined by the formula

$$d = \sqrt{(x_2 - x_1)^2 + (y_2 - y_1)^2 + (z_2 - z_1)^2} = |\mathbf{r}_2 - \mathbf{r}_1|, \qquad (4.5.2.3)$$

where $\mathbf{r}_2 = \overrightarrow{OM}_2$ and $\mathbf{r}_1 = \overrightarrow{OM}_1$ are the position vectors of the points $M_1$ and $M_2$, respectively (see Fig. 4.30).

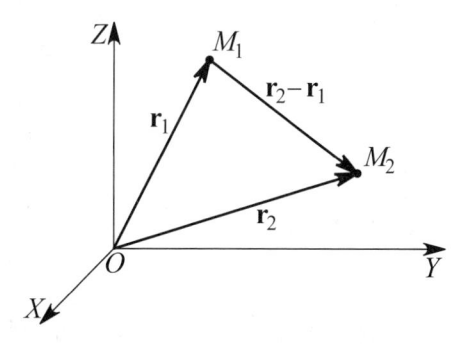

**Figure 4.30.** Distance between points.

Two vectors $\mathbf{a} = (x_1, y_1, z_1)$ and $\mathbf{b} = (x_2, y_2, z_2)$ are equal to each other if and only if the following relations hold simultaneously:

$$x_1 = x_2, \quad y_1 = y_2, \quad z_1 = z_2.$$

For arbitrary vectors, one has the relations

$$(x_1, y_1, z_1) \pm (x_2, y_2, z_2) = (x_1 \pm x_2, y_1 \pm y_2, z_1 \pm z_2),$$
$$\alpha(x, y, z) = (\alpha x, \alpha y, \alpha z). \tag{4.5.2.4}$$

If a point $M$ divides the directed segment $\overrightarrow{M_1 M_2}$ in the ratio $p/q = \lambda$, then the coordinates of this point are given by the formulas

$$x = \frac{x_1 + \lambda x_2}{1 + \lambda} = \frac{q x_1 + p x_2}{q + p},$$
$$y = \frac{y_1 + \lambda y_2}{1 + \lambda} = \frac{q y_1 + p y_2}{q + p}, \quad \text{or} \quad \mathbf{r} = \frac{\mathbf{r}_1 + \lambda \mathbf{r}_2}{1 + \lambda} = \frac{q \mathbf{r}_1 + p \mathbf{r}_2}{q + p}, \tag{4.5.2.5}$$
$$z = \frac{z_1 + \lambda z_2}{1 + \lambda} = \frac{q z_1 + p z_2}{q + p},$$

where $-\infty \leq \lambda \leq \infty$. In the special case of the midpoint of the segment $\overrightarrow{M_1 M_2}$ ($p = q$, $\lambda = 1$), the coordinates are

$$x = \frac{x_1 + x_2}{2}, \quad y = \frac{y_1 + y_2}{2}, \quad z = \frac{z_1 + z_2}{2} \quad \text{or} \quad \mathbf{r} = \frac{\mathbf{r}_1 + \mathbf{r}_2}{2};$$

i.e., each coordinate of the midpoint of a segment is equal to the half-sum of the corresponding coordinates of its endpoints.

The angles $\alpha$, $\beta$, and $\gamma$ between the segment $\overrightarrow{M_1 M_2}$ and the coordinate axes $OX$, $OY$, and $OZ$ are determined by the expressions

$$\cos \alpha = \frac{x_2 - x_1}{|\mathbf{r}_2 - \mathbf{r}_1|}, \quad \cos \beta = \frac{y_2 - y_1}{|\mathbf{r}_2 - \mathbf{r}_1|}, \quad \cos \gamma = \frac{z_2 - z_1}{|\mathbf{r}_2 - \mathbf{r}_1|}, \tag{4.5.2.6}$$

and

$$\cos^2 \alpha + \cos^2 \beta + \cos^2 \gamma = 1.$$

The numbers $\cos \alpha$, $\cos \beta$, and $\cos \gamma$ are called the *direction cosines of the segment* $\overrightarrow{M_1 M_2}$.

The angle $\varphi$ between arbitrary directed segments $\overrightarrow{M_1 M_2}$ and $\overrightarrow{M_3 M_4}$ joining the points $M_1(x_1, y_1, z_1)$, $M_2(x_2, y_2, z_2)$ and $M_3(x_3, y_3, z_3)$, $M_4(x_4, y_4, z_4)$, respectively, can be found from the relation

$$\cos \varphi = \frac{(x_2 - x_1)(x_4 - x_3) + (y_2 - y_1)(y_4 - y_3) + (z_2 - z_1)(z_4 - z_3)}{|\mathbf{r}_2 - \mathbf{r}_1| \, |\mathbf{r}_4 - \mathbf{r}_3|}. \tag{4.5.2.7}$$

The area of the triangle with vertices $M_1$, $M_2$, and $M_3$ is given by the formula

$$S = \frac{1}{4} \sqrt{ \begin{vmatrix} y_1 & z_1 & 1 \\ y_2 & z_2 & 1 \\ y_3 & z_3 & 1 \end{vmatrix}^2 + \begin{vmatrix} z_1 & x_1 & 1 \\ z_2 & x_2 & 1 \\ z_3 & x_3 & 1 \end{vmatrix}^2 + \begin{vmatrix} x_1 & y_1 & 1 \\ x_2 & y_2 & 1 \\ x_3 & y_3 & 1 \end{vmatrix}^2 }. \tag{4.5.2.8}$$

The volume of the pyramid with vertices $M_1$, $M_2$, $M_3$, and $M_4$ is equal to

$$V = \pm \frac{1}{6} \begin{vmatrix} x_2 - x_1 & y_2 - y_1 & z_2 - z_1 \\ x_3 - x_1 & y_3 - y_1 & z_3 - z_1 \\ x_4 - x_1 & y_4 - y_1 & z_4 - z_1 \end{vmatrix} = \pm \frac{1}{6} \begin{vmatrix} 1 & x_1 & y_1 & z_1 \\ 1 & x_2 & y_2 & z_2 \\ 1 & x_3 & y_3 & z_3 \\ 1 & x_4 & y_4 & z_4 \end{vmatrix}, \qquad (4.5.2.9)$$

and the volume of the parallelepiped spanned by vectors $\overrightarrow{M_1M_2}$, $\overrightarrow{M_1M_3}$, and $\overrightarrow{M_1M_4}$ is equal to

$$V = \pm \begin{vmatrix} x_2 - x_1 & y_2 - y_1 & z_2 - z_1 \\ x_3 - x_1 & y_3 - y_1 & z_3 - z_1 \\ x_4 - x_1 & y_4 - y_1 & z_4 - z_1 \end{vmatrix} = \pm \begin{vmatrix} 1 & x_1 & y_1 & z_1 \\ 1 & x_2 & y_2 & z_2 \\ 1 & x_3 & y_3 & z_3 \\ 1 & x_4 & y_4 & z_4 \end{vmatrix}. \qquad (4.5.2.10)$$

The coordinate surfaces, on which one of the coordinates is constant, are planes parallel to the coordinate planes, and the coordinate lines, along which only one coordinate varies, are straight lines parallel to the coordinate axes. Coordinate surfaces meet in the coordinate lines.

**4.5.2-2. Transformation of Cartesian coordinates under parallel translation of axes.**

Suppose that two rectangular Cartesian coordinate systems $OXYZ$ and $\widehat{O}\widehat{X}\widehat{Y}\widehat{Z}$ are given and the first system can be made to coincide with the second system by translating the origin $O$ of the first system to the origin $\widehat{O}$ of the second system. Under this translation the axes preserve their direction (the respective axes of the systems are parallel), and the origin moves by $x_0$ in the direction of the axis $OX$, by $y_0$ in the direction of the axis $OY$, and by $z_0$ in the direction of the axis $OZ$. Obviously, the point $\widehat{O}$ in the coordinate system $OXYZ$ has the coordinates $(x_0, y_0, z_0)$.

An arbitrary point $M$ has coordinates $(x, y, z)$ in the system $OXYZ$ and coordinates $(\hat{x}, \hat{y}, \hat{z})$ in the system $\widehat{O}\widehat{X}\widehat{Y}\widehat{Z}$. The transformation of rectangular Cartesian coordinates by the parallel translation of axes is determined by the formulas

$$\begin{aligned} x &= \hat{x} + x_0, & \hat{x} &= x - x_0, \\ y &= \hat{y} + y_0, \quad \text{or} \quad & \hat{y} &= y - y_0, \\ z &= \hat{z} + z_0 & \hat{z} &= z - z_0. \end{aligned} \qquad (4.5.2.11)$$

**4.5.2-3. Transformation of Cartesian coordinates under rotation of axes.**

Suppose that two rectangular Cartesian coordinate systems $OXYZ$ and $O\widehat{X}\widehat{Y}\widehat{Z}$ are given and the first system can be made to coincide with the second system by rotating the first system around the point $O$.

An arbitrary point $M$ has coordinates $(x, y, z)$ in the system $OXYZ$ and coordinates $(\hat{x}, \hat{y}, \hat{z})$ in the system $O\widehat{X}\widehat{Y}\widehat{Z}$. If the axis $O\widehat{X}$ has the direction cosines $e_{11}, e_{21}, e_{31}$, the axis $O\widehat{Y}$ has the direction cosines $e_{12}, e_{22}, e_{32}$, and the axis $O\widehat{Z}$ has the direction cosines $e_{13}, e_{23}, e_{33}$ in the coordinate system $OXYZ$, then the axis $OX$ has the direction cosines $e_{11}, e_{12}, e_{13}$, the axis $OY$ has the direction cosines $e_{21}, e_{22}, e_{23}$, and the axis $OZ$ has the direction cosines $e_{31}, e_{32}, e_{33}$ in the coordinate system $O\widehat{X}\widehat{Y}\widehat{Z}$. The transformation of rectangular Cartesian coordinates by the rotation of axes is determined by the formulas

$$
\begin{aligned}
x &= e_{11}\hat{x} + e_{12}\hat{y} + e_{13}\hat{z}, \\
y &= e_{21}\hat{x} + e_{22}\hat{y} + e_{23}\hat{z}, \quad \text{or} \quad \\
z &= e_{31}\hat{x} + e_{32}\hat{y} + e_{33}\hat{z},
\end{aligned}
\qquad
\begin{aligned}
\hat{x} &= e_{11}x + e_{21}y + e_{31}z, \\
\hat{y} &= e_{12}x + e_{22}y + e_{32}z, \\
\hat{z} &= e_{13}x + e_{23}y + e_{33}z.
\end{aligned}
\qquad (4.5.2.12)
$$

### 4.5.2-4. Transformation of coordinates under translation and rotation of axes.

Suppose that two rectangular Cartesian coordinate systems $OXYZ$ and $\widehat{O}\widehat{X}\widehat{Y}\widehat{Z}$ are given and the first system can be made to coincide with the second system by translating the origin $O(0,0,0)$ of the first system to the origin $\widehat{O}(x_0, y_0, z_0)$ of the second system, and then by rotating the first system around the point $\widehat{O}$ (see Paragraphs 4.5.1-2 and 4.5.1-3).

An arbitrary point $M$ has coordinates $(x, y, z)$ in the system $OXYZ$ and coordinates $(\hat{x}, \hat{y}, \hat{z})$ in the system $\widehat{O}\widehat{X}\widehat{Y}\widehat{Z}$. The transformation of rectangular Cartesian coordinates by the parallel translation and the rotation of axes is determined by the formulas

$$
\begin{aligned}
x &= e_{11}\hat{x} + e_{12}\hat{y} + e_{13}\hat{z} + x_0, \\
y &= e_{21}\hat{x} + e_{22}\hat{y} + e_{23}\hat{z} + y_0, \quad \text{or} \quad \\
z &= e_{31}\hat{x} + e_{32}\hat{y} + e_{33}\hat{z} + z_0,
\end{aligned}
\qquad
\begin{aligned}
\hat{x} &= e_{11}(x - x_0) + e_{21}(y - y_0) + e_{31}(z - z_0), \\
\hat{y} &= e_{12}(x - x_0) + e_{22}(y - y_0) + e_{32}(z - z_0), \\
\hat{z} &= e_{13}(x - x_0) + e_{23}(y - y_0) + e_{33}(z - z_0).
\end{aligned}
\qquad (4.5.2.13)
$$

### 4.5.2-5. Cylindrical and spherical coordinates.

A more general curvilinear coordinate system is obtained if one introduces three families of coordinate surfaces such that exactly one surface of each family passes through each point of space. The position of a point in such a system is determined by the values of the parameters of the coordinate surfaces passing through this point. The most commonly used curvilinear coordinate systems (cylindrical and spherical) are described below.

The *cylindrical coordinates* of a point $M$ are defined as the polar coordinates $\rho$ and $\varphi$ (see Paragraph 4.1.2-5) of the projection of $M$ onto the base plane (usually $OXY$) and the distance (usually $z$) from $M$ to the base plane, which is called the *applicate* (see Fig. 4.31a). To be definite, one usually assumes that $0 < \varphi \leq 2\pi$ or $-\pi < \varphi \leq \pi$. For cylindrical coordinates, the coordinate surfaces are the planes $z = \text{const}$ perpendicular to the axis $OZ$, the half-planes $\varphi = \text{const}$ bounded by the axis $OZ$, and the cylindrical surfaces $\rho = \text{const}$ with axis $OZ$. The coordinate surfaces intersect in the coordinate lines.

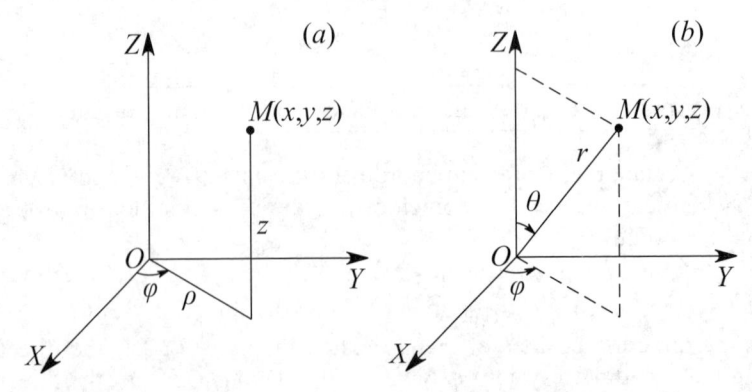

**Figure 4.31.** Point in cylindrical (*a*) and spherical (*b*) coordinates.

The *spherical (polar) coordinates* are defined as the length $r = |\overrightarrow{OM}|$ of the radius vector, the *longitude* $\varphi$, and the polar distance $\theta$, called the *latitude* (see Fig. 4.31b). To be definite, one usually assumes that $0 < \varphi \leq 2\pi$, $0 \leq \theta \leq \pi$ or $-\pi < \varphi \leq \pi$, and $0 \leq \theta \leq \pi$. For spherical coordinates, the coordinate surfaces are the spheres $r = $ const centered at the origin, the half-planes $\varphi = $ const bounded by the axis $OZ$, and the cones $\theta = $ const with vertex $O$ and axis $OZ$. The coordinate surfaces intersect in the coordinate lines.

### 4.5.2-6. Relationship between Cartesian, cylindrical, and spherical coordinates.

Let $M$ be an arbitrary point in space with rectangular Cartesian coordinates $(x, y, z)$, cylindrical coordinates $(\rho, \varphi, z)$, and spherical coordinates $(r, \varphi, \theta)$. The formulas of transition from the cylindrical coordinate system to the Cartesian coordinate system and vice versa have the form

$$x = \rho \cos \varphi, \qquad \rho = \sqrt{x^2 + y^2},$$
$$y = \rho \sin \varphi, \qquad \text{or} \qquad \tan \varphi = y/x, \qquad (4.5.2.14)$$
$$z = z, \qquad z = z,$$

where the polar angle $\varphi$ is taken with regard to the quadrant in which the projection of the point $M$ onto the base plane lies. The formulas of transition from the spherical coordinate system to the Cartesian coordinate system and vice versa have the form

$$x = r \sin \theta \cos \varphi, \qquad r = \sqrt{x^2 + y^2 + z^2},$$
$$y = r \sin \theta \sin \varphi, \qquad \text{or} \qquad \tan \varphi = y/x, \qquad (4.5.2.15)$$
$$z = r \cos \theta, \qquad \tan \theta = \sqrt{x^2 + y^2}/z,$$

where the angle $\varphi$ is determined from the same considerations as in the case of cylindrical coordinates.

### 4.5.2-7. Surfaces and curves in space.

A *surface in space* determined by an equation in some coordinate system is the locus of points in space whose coordinates satisfy this equation.

An *equation of a surface in space* in a given coordinate system is an equation with three variables satisfied by the coordinates of points lying on the surface and not satisfied by the coordinates of points that do not lie on the surface.

The coordinates of an arbitrary point of the surface occurring in the equation of the surface are called the *current coordinates*.

**Example 1.** The equation
$$(x - x_0)^2 + (y - y_0)^2 + (z - z_0)^2 = r^2$$
defines the sphere of radius $r$ centered at the point $(x_0, y_0, z_0)$, i.e., the locus of points lying at the distance $r$ from the point $(x_0, y_0, z_0)$.

**Example 2.** The equation $x^2 + y^2 + (z - 1)^2 = 0$ determines the single point with coordinates $(0, 0, 1)$.

**Example 3.** The equation $x^2 + y^2 + z^2 + 1 = 0$ does not have solutions for any real $x$, $y$, and $z$.

In the general case, the *equation of a surface in the Cartesian coordinate system* $OXYZ$ can be written as

$$F(x, y, z) = 0. \qquad (4.5.2.16)$$

On the other hand, a surface (continuous surface) can be defined parametrically; i.e., a *surface* is defined as the set of points whose coordinates satisfy the system of parametric equations

$$x = x(u, v), \quad y = y(u, v), \quad z = z(u, v) \qquad (4.5.2.17)$$

for appropriate values of the parameters $u$ and $v$.

In spatial analytic geometry, each curve is treated as the intersection of two surfaces and hence is defined by a system of two equations

$$F(x, y, z) = 0, \qquad G(x, y, z) = 0. \qquad (4.5.2.18)$$

On the other hand, each curve (continuous curve) can be defined parametrically; i.e., a *curve* is defined as the set of points whose coordinates satisfy the system of parametric equations

$$x = x(t), \quad y = y(t), \quad z = z(t) \quad \text{or} \quad \mathbf{r} = \mathbf{r}(t) \qquad (-\infty \le t_1 \le t \le t_2 \le \infty). \qquad (4.5.2.19)$$

### 4.5.3. Vectors. Products of Vectors

**4.5.3-1. Scalar product of two vectors.**

The *scalar product* of two vectors is defined as the product of their absolute values times the cosine of the angle between the vectors (see Fig. 4.32),

$$\mathbf{a} \cdot \mathbf{b} = |\mathbf{a}||\mathbf{b}| \cos \varphi. \qquad (4.5.3.1)$$

If the angle between vectors $\mathbf{a}$ and $\mathbf{b}$ is acute, then $\mathbf{a} \cdot \mathbf{b} > 0$; if the angle is obtuse, then $\mathbf{a} \cdot \mathbf{b} < 0$; if the angle is right, then $\mathbf{a} \cdot \mathbf{b} = 0$. Taking into account (4.5.1.1), we can write the scalar product as

$$\mathbf{a} \cdot \mathbf{b} = |\mathbf{a}||\mathbf{b}| \cos \varphi = |\mathbf{a}| \operatorname{pr}_\mathbf{a} \mathbf{b} = |\mathbf{b}| \operatorname{pr}_\mathbf{b} \mathbf{a}. \qquad (4.5.3.2)$$

Remark. The scalar product of a vector $\mathbf{a}$ by a vector $\mathbf{b}$ is also denoted by $(\mathbf{a}, \mathbf{b})$ or $\mathbf{ab}$.

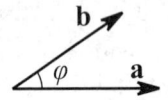

**Figure 4.32.** Scalar product of two vectors.

The *angle $\varphi$ between vectors* is determined by the formula

$$\cos \varphi = \frac{\mathbf{a} \cdot \mathbf{b}}{|\mathbf{a}||\mathbf{b}|} = \frac{a_x b_x + a_y b_y + a_z b_z}{\sqrt{a_x^2 + a_y^2 + a_z^2} \sqrt{b_x^2 + b_y^2 + b_z^2}}. \qquad (4.5.3.3)$$

*Properties of scalar product:*

1. $\mathbf{a} \cdot \mathbf{b} = \mathbf{b} \cdot \mathbf{a}$ (commutativity).
2. $\mathbf{a} \cdot (\mathbf{b} + \mathbf{c}) = \mathbf{a} \cdot \mathbf{b} + \mathbf{a} \cdot \mathbf{c}$ (distributivity with respect to addition of vectors). This property holds for any number of summands.
3. If vectors $\mathbf{a}$ and $\mathbf{b}$ are collinear, then $\mathbf{a} \cdot \mathbf{b} = \pm|\mathbf{a}||\mathbf{b}|$. (The sign + is taken if the vectors $\mathbf{a}$ and $\mathbf{b}$ have the same sense, and the sign − is taken if the senses are opposite.)
4. $(\lambda \mathbf{a}) \cdot \mathbf{b} = \lambda(\mathbf{a} \cdot \mathbf{b})$ (associativity with respect to a scalar factor).
5. $\mathbf{a} \cdot \mathbf{a} = |\mathbf{a}|^2$. The scalar product $\mathbf{a} \cdot \mathbf{a}$ is denoted by $\mathbf{a}^2$ (the *scalar square* of the vector $\mathbf{a}$).

6. The length of a vector is expressed via the scalar product by

$$|\mathbf{a}| = \sqrt{\mathbf{a} \cdot \mathbf{a}} = \sqrt{\mathbf{a}^2}.$$

7. Two nonzero vectors **a** and **b** are perpendicular if and only if **ab** = 0.
8. The scalar products of basis vectors are

$$\mathbf{i} \cdot \mathbf{j} = \mathbf{i} \cdot \mathbf{k} = \mathbf{j} \cdot \mathbf{k} = 0, \quad \mathbf{i} \cdot \mathbf{i} = \mathbf{j} \cdot \mathbf{j} = \mathbf{k} \cdot \mathbf{k} = 1.$$

9. If vectors are given by their coordinates, $\mathbf{a} = (a_x, a_y, a_z)$ and $\mathbf{b} = (b_x, b_y, b_z)$, then

$$\mathbf{a} \cdot \mathbf{b} = (a_x\mathbf{i} + a_y\mathbf{j} + a_z\mathbf{j})(b_x\mathbf{i} + b_y\mathbf{j} + b_z\mathbf{j}) = a_x b_x + a_y b_y + a_z b_z. \qquad (4.5.3.4)$$

10. The *Cauchy–Schwarz inequality*

$$|\mathbf{a} \cdot \mathbf{b}| \leq |\mathbf{a}||\mathbf{b}|.$$

11. The *Minkowski inequality*

$$|\mathbf{a} + \mathbf{b}| \leq |\mathbf{a}| + |\mathbf{b}|.$$

### 4.5.3-2. Cross product of two vectors.

The *cross product* of a vector **a** by a vector **b** is defined as the vector **c** (see Fig. 4.33) satisfying the following three conditions:

1. Its absolute value is equal to the area of the parallelogram spanned by the vectors **a** and **b**; i.e.,

$$|\mathbf{c}| = |\mathbf{a} \times \mathbf{b}| = |\mathbf{a}||\mathbf{b}| \sin \varphi. \qquad (4.5.3.5)$$

2. It is perpendicular to the plane of the parallelogram; i.e., $\mathbf{c} \perp \mathbf{a}$ and $\mathbf{c} \perp \mathbf{b}$.
3. The vectors **a**, **b**, and **c** form a *right-handed trihedral*; i.e., the vector **c** points to the side from which the sense of the shortest rotation from **a** to **b** is anticlockwise.

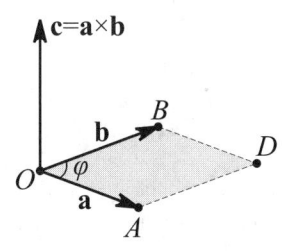

**Figure 4.33.** Cross product of two vectors.

Remark 1. The cross product of a vector **a** by a vector **b** is also denoted by **c** = [**a**, **b**].

Remark 2. If vectors **a** and **b** are collinear, then the parallelogram $OADB$ is degenerate and should be assigned the zero area. Hence the cross product of collinear vectors is defined to be the zero vector whose direction is arbitrary.

*Properties of cross product:*
1. $\mathbf{a} \times \mathbf{b} = -\mathbf{b} \times \mathbf{a}$ (anticommutativity).
2. $\mathbf{a} \times (\mathbf{b} + \mathbf{c}) = \mathbf{a} \times \mathbf{b} + \mathbf{a} \times \mathbf{c}$ (distributivity with respect to the addition of vectors). This property holds for any number of summands.

3. Vectors **a** and **b** are collinear if and only if $\mathbf{a} \times \mathbf{b} = 0$. In particular, $\mathbf{a} \times \mathbf{a} = 0$ and $\mathbf{a}(\mathbf{a} \times \mathbf{b}) = \mathbf{b}(\mathbf{a} \times \mathbf{b}) = 0$.
4. $(\lambda \mathbf{a}) \times \mathbf{b} = \mathbf{a} \times (\lambda \mathbf{b}) = \lambda(\mathbf{a} \times \mathbf{b})$ (associativity with respect to a scalar factor).
5. The cross product of basis vectors is

$$\mathbf{i} \times \mathbf{i} = \mathbf{j} \times \mathbf{j} = \mathbf{k} \times \mathbf{k} = 0, \quad \mathbf{i} \times \mathbf{j} = \mathbf{k}, \quad \mathbf{j} \times \mathbf{k} = \mathbf{i}, \quad \mathbf{k} \times \mathbf{i} = \mathbf{j}.$$

6. If the vectors are given by their coordinates $\mathbf{a} = (a_x, a_y, a_z)$ and $\mathbf{b} = (b_x, b_y, b_z)$, then

$$\mathbf{a} \times \mathbf{b} = \begin{vmatrix} \mathbf{i} & \mathbf{j} & \mathbf{k} \\ a_x & a_y & a_z \\ b_x & b_y & b_z \end{vmatrix} = (a_y b_z - a_z b_y)\mathbf{i} + (a_z b_x - a_x b_z)\mathbf{j} + (a_x b_y - a_y b_x)\mathbf{k}. \quad (4.5.3.6)$$

7. The area of the parallelogram spanned by vectors **a** and **b** is equal to

$$S = |\mathbf{a} \times \mathbf{b}| = \sqrt{\begin{vmatrix} a_y & a_z \\ b_y & b_z \end{vmatrix}^2 + \begin{vmatrix} a_x & a_z \\ b_x & b_z \end{vmatrix}^2 + \begin{vmatrix} a_x & a_y \\ b_x & b_y \end{vmatrix}^2}. \quad (4.5.3.7)$$

8. The area of the triangle spanned by vectors **a** and **b** is equal to

$$S = \frac{1}{2}|\mathbf{a} \times \mathbf{b}| = \frac{1}{2}\sqrt{\begin{vmatrix} a_y & a_z \\ b_y & b_z \end{vmatrix}^2 + \begin{vmatrix} a_x & a_z \\ b_x & b_z \end{vmatrix}^2 + \begin{vmatrix} a_x & a_y \\ b_x & b_y \end{vmatrix}^2}. \quad (4.5.3.8)$$

**Example 1.** The moment with respect to the point $O$ of a force **F** applied at a point $M$ is the cross product of the position vector $\overrightarrow{OM}$ by the force **F**; i.e., $\mathbf{M} = \overrightarrow{OM} \times \mathbf{F}$.

### 4.5.3-3. Conditions for vectors to be parallel or perpendicular.

A vector **a** is collinear to a vector **b** if

$$\mathbf{b} = \lambda \mathbf{a} \quad \text{or} \quad \mathbf{a} \times \mathbf{b} = 0. \quad (4.5.3.9)$$

A vector **a** is perpendicular to a vector **b** if

$$\mathbf{a} \cdot \mathbf{b} = 0. \quad (4.5.3.10)$$

Remark. In general, the condition $\mathbf{a} \cdot \mathbf{b} = 0$ implies that the vectors **a** and **b** are perpendicular or one of them is the zero vector. The zero vector can be viewed to be perpendicular to any other vector.

### 4.5.3-4. Triple cross product.

The *triple cross product* of vectors **a**, **b**, and **c** is defined as the vector

$$\mathbf{d} = \mathbf{a} \times (\mathbf{b} \times \mathbf{c}). \quad (4.5.3.11)$$

The triple cross product is coplanar to the vectors **b** and **c**; it can be expressed via **b** and **c** as follows:

$$\mathbf{a} \times (\mathbf{b} \times \mathbf{c}) = \mathbf{b} \cdot (\mathbf{a} \cdot \mathbf{c}) - \mathbf{c} \cdot (\mathbf{a} \cdot \mathbf{b}). \quad (4.5.3.12)$$

### 4.5.3-5. Scalar triple product of three vectors.

The *scalar triple product* of vectors **a**, **b**, and **c** is defined as the scalar product of **a** by the cross product of **b** and **c**:

$$[\mathbf{abc}] = \mathbf{a} \cdot (\mathbf{b} \times \mathbf{c}). \tag{4.5.3.13}$$

*Remark.* The scalar triple product of three vectors **a**, **b**, and **c** is also denoted by **abc**.

*Properties of scalar triple product:*

1. [**abc**] = [**bca**] = [**cab**] = –[**bac**] = –[**cba**] = –[**acb**].
2. [(**a** + **b**)**cd**] = [**acd**] + [**bcd**] (distributivity with respect to addition of vectors). This property holds for any number of summands.
3. [λ**abc**] = λ[**abc**] (associativity with respect to a scalar factor).
4. If the vectors are given by their coordinates $\mathbf{a} = (a_x, a_y, a_z)$, $\mathbf{b} = (b_x, b_y, b_z)$, and $\mathbf{c} = (c_x, c_y, c_z)$, then

$$[\mathbf{abc}] = \begin{vmatrix} a_x & a_y & a_z \\ b_x & b_y & b_z \\ c_x & c_y & c_z \end{vmatrix}. \tag{4.5.3.14}$$

5. The scalar triple product [**abc**] is equal to the volume $V$ of the parallelepiped spanned by the vectors **a**, **b**, and **c** taken with the sign + if the vectors **a**, **b**, and **c** form a right-handed trihedral and the sign – if the vectors form a left-handed trihedral,

$$[\mathbf{abc}] = \pm V. \tag{4.5.3.15}$$

6. Three nonzero vectors **a**, **b**, and **c** are coplanar if and only if [**abc**] = 0. In this case, the vectors **a**, **b**, and **c** are linearly dependent; they satisfy a relation of the form $\mathbf{c} = \alpha\mathbf{a} + \beta\mathbf{b}$.

## 4.5.4. Curves and Surfaces in Space

### 4.5.4-1. Methods for defining curves.

A *continuous curve* in three-dimensional space is the set of points whose coordinates satisfy a system of parametric equations (4.5.2.19). This method for defining a curve is referred to as *parametric*. The curve can also be defined by an equivalent system of equations (4.5.2.18), i.e., described as the intersection of two surfaces (see Paragraph 4.5.4-2).

*Remark 1.* A curve may have more than one branch.

*Remark 2.* One can obtain the equation of the projection of the curve (4.5.2.19) onto the plane $OXY$ by eliminating the variable $x$ from equations (4.5.2.19)

### 4.5.4-2. Methods for defining surfaces.

A *continuous surface* in three-dimensional space is the set of points whose coordinates satisfy a system of parametric equations (4.5.2.17). This method for defining a surface is referred to as *parametric*. The surface can also be determined by an equation (4.5.2.16) or $z = f(x, y)$.

*Remark 1.* A surface may have more than one sheet.

*Remark 2.* The surfaces determined by the equations

$$F(x, y, z) = 0 \quad \text{and} \quad \lambda F(x, y, z) = 0$$

coincide provided that the constant $\lambda$ is nonzero.

**Remark 3.** For any real number $\lambda$, the equation
$$F_1(x, y, z) + \lambda F_2(x, y, z) = 0$$
describes a surface passing through the line of intersection of the surfaces (4.5.2.18), provided that this line exists.

**Remark 4.** The equation
$$F_1(x, y, z) \cdot F_2(x, y, z) = 0$$
describes the surface that is formed by points of both surfaces in (4.5.2.18) and does not contain any other points.

## 4.6. Line and Plane in Space
### 4.6.1. Plane in Space

**4.6.1-1. Equation of plane passing through point $M_0$ and perpendicular to vector N.**

A *plane* is a first-order algebraic surface. In a Cartesian coordinate system, a plane is given by a first-order equation.

The equation of the plane passing through a point $M_0(x_0, y_0, z_0)$ and perpendicularly to a vector $\mathbf{N} = (A, B, C)$ has the form

$$A(x - x_0) + B(y - y_0) + C(z - z_0) = 0, \quad \text{or} \quad (\mathbf{r} - \mathbf{r}_0) \cdot \mathbf{N} = 0, \qquad (4.6.1.1)$$

where $\mathbf{r}$ and $\mathbf{r}_0$ are the position vectors of the point $M(x, y, z)$ and $M_0(x_0, y_0, z_0)$, respectively (see Fig. 4.34). The vector $\mathbf{N}$ is called a *normal vector*. Its direction cosines are

$$\cos\alpha = \frac{A}{\sqrt{A^2 + B^2 + C^2}}, \quad \cos\beta = \frac{B}{\sqrt{A^2 + B^2 + C^2}}, \quad \cos\gamma = \frac{C}{\sqrt{A^2 + B^2 + C^2}}. \qquad (4.6.1.2)$$

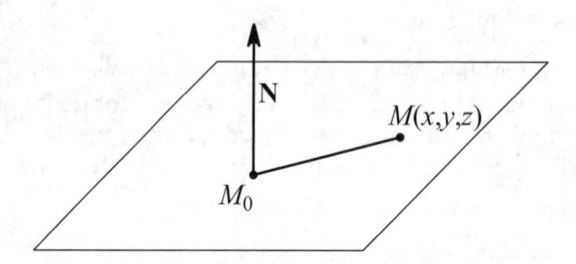

**Figure 4.34.** Plane passing through a point $M_0$ and perpendicularly to a vector $\mathbf{N}$.

**Example 1.** Let us write out the equation of the plane that passes through the point $M_0(1, 2, 1)$ and is perpendicular to the vector $\mathbf{N} = (3, 2, 3)$.
According to (4.6.1.1), the desired equation is $3(x - 1) + 2(y - 2) + 3(z - 1) = 0$ or $3x + 2y + 3z - 10 = 0$.

**4.6.1-2. General equation of plane.**

The *general (complete) equation of a plane* has the form

$$Ax + By + Cz + D = 0, \quad \text{or} \quad \mathbf{r} \cdot \mathbf{N} + D = 0. \qquad (4.6.1.3)$$

It follows from (4.6.1.1) that $D = -Ax_0 - By_0 - Cz_0$. If one of the coefficients in the equation of a plane is zero, then the equation is said to be *incomplete*:

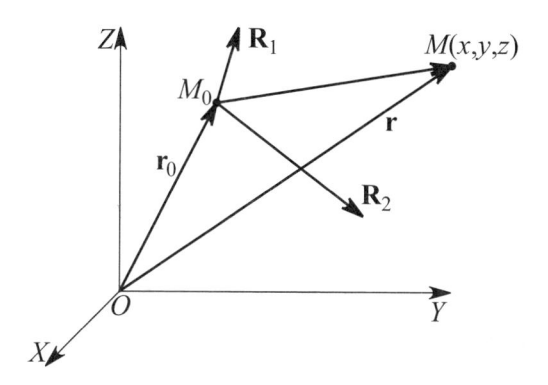

**Figure 4.35.** Basis in plane.

1. For $D = 0$, the equation has the form $Ax + By + Cz = 0$ and defines a plane passing through the origin.
2. For $A = 0$ (respectively, $B = 0$ or $C = 0$), the equation has the form $By + Cz + D = 0$ and defines a plane parallel to the axis $OX$ (respectively, $OY$ or $OZ$).
3. For $A = D = 0$ (respectively, $B = D = 0$ or $B = D = 0$), the equation has the form $By + Cz = 0$ and defines a plane passing through the axis $OX$ (respectively, $OY$ or $OZ$).
4. For $A = B = 0$ (respectively, $A = C = 0$ or $B = C = 0$), the equation has the form $Cz + D = 0$ and defines a plane parallel to the plane $OXY$ (respectively, $OXZ$ or $OYZ$).

### 4.6.1-3. Parametric equation of plane.

Each vector $\overrightarrow{M_0M} = \mathbf{r} - \mathbf{r}_0$ lying in a plane (where $\mathbf{r}$ and $\mathbf{r}_0$ are the position vectors of the points $M$ and $M_0$, respectively) can be represented as (see Fig. 4.35)

$$\overrightarrow{M_0M} = t\mathbf{R}_1 + s\mathbf{R}_2, \tag{4.6.1.4}$$

where $\mathbf{R}_1 = (l_1, m_1, n_1)$ and $\mathbf{R}_2 = (l_2, m_2, n_2)$ are two arbitrary noncollinear vectors lying in the plane. Obviously, these two vectors form a basis in this plane. The *parametric equation of a plane* passing through the point $M_0(x_0, y_0, z_0)$ has the form

$$\mathbf{r} = \mathbf{r}_0 + t\mathbf{R}_1 + s\mathbf{R}_2, \quad \text{or} \quad \begin{aligned} x &= x_0 + tl_1 + sl_2, \\ y &= y_0 + tm_1 + sm_2, \\ z &= z_0 + tn_1 + sn_2. \end{aligned} \tag{4.6.1.5}$$

### 4.6.1-4. Intercept equation of plane.

A plane $Ax + By + Cz + D = 0$ that is not parallel to the axis $Ox$ (i.e., $A \neq 0$) meets this axis at a (signed) distance $a = -D/A$ from the origin (see Fig. 4.36). The number $a$ is called the $x$-intercept of the plane. Similarly, one defines the $y$-intercepts $b = -D/B$ (for $B \neq 0$) and the $z$-intercept $c = -D/C$ (for $C \neq 0$). Then such a plane can be defined by the equation

$$\frac{x}{a} + \frac{y}{b} + \frac{z}{c} = 1, \tag{4.6.1.6}$$

which is called the *intercept equation of the plane*.

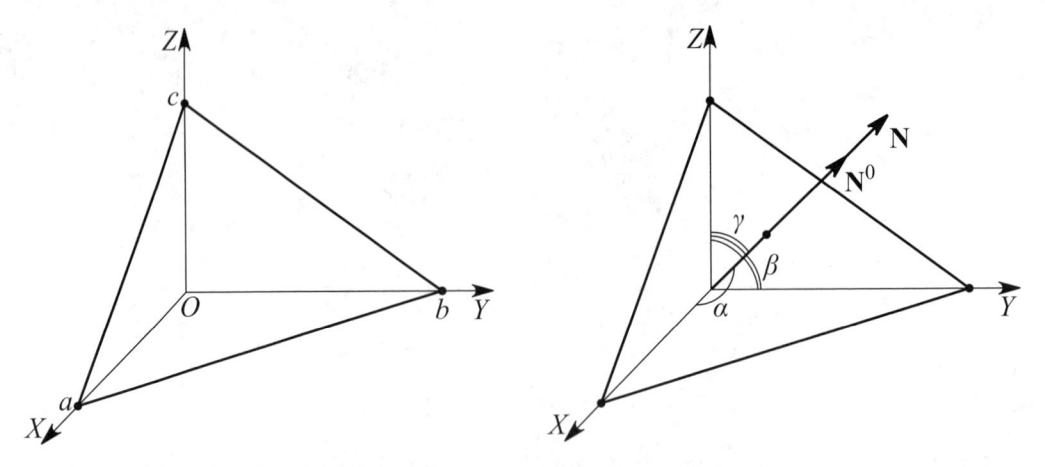

**Figure 4.36.** A plane with intercept equation.

**Figure 4.37.** A plane with normalized equation.

Remark 1. Equation (4.6.1.6) can be obtained as the equation of the plane passing through three given points.

Remark 2. A plane parallel to the axis $OX$ but nonparallel to the other two axes is defined by the equation $y/b + z/c = 1$, where $b$ and $c$ are the $y$- and $z$-intercepts of the plane. A plane simultaneously parallel to the axes $OY$ and $OZ$ can be represented in the form $z/c = 1$.

**Example 2.** Consider the plane given by the general equation $2x + 3y - z + 6 = 0$. Let us rewrite it in intercept form.

The $x$-, $y$-, and $z$-intercepts of this plane are

$$a = -\frac{D}{A} = -\frac{6}{2} = -3, \quad b = -\frac{D}{B} = -\frac{6}{3} = -2, \quad \text{and} \quad c = -\frac{D}{C} = -\frac{6}{-1} = 6.$$

Thus the intercept equation of the plane reads

$$\frac{x}{-3} + \frac{y}{-2} + \frac{z}{6} = 0.$$

### 4.6.1-5. Normalized equation of plane.

The *normalized equation of a plane* has the form

$$\mathbf{r} \cdot \mathbf{N}^0 - p = 0, \quad \text{or} \quad x\cos\alpha + y\cos\beta + z\cos\gamma - p = 0, \qquad (4.6.1.7)$$

where $\mathbf{N}^0 = (\cos\alpha, \cos\beta, \cos\gamma)$ is a unit vector and $p$ is the distance from the plane to the origin; here $\cos\alpha$, $\cos\beta$, and $\cos\gamma$ are the direction cosines of the normal to the plane (see Fig. 4.37). The numbers $\cos\alpha$, $\cos\beta$, $\cos\gamma$, and $p$ can be expressed via the coefficients $A$, $B$, $C$ as follows:

$$\cos\alpha = \pm\frac{A}{\sqrt{A^2 + B^2 + C^2}}, \quad \cos\beta = \pm\frac{B}{\sqrt{A^2 + B^2 + C^2}},$$
$$\cos\gamma = \pm\frac{C}{\sqrt{A^2 + B^2 + C^2}}, \quad p = \mp\frac{D}{\sqrt{A^2 + B^2 + C^2}}, \qquad (4.6.1.8)$$

where the upper sign is taken if $D < 0$ and the lower sign is taken if $D > 0$. For $D = 0$, either sign can be taken.

The normalized equation (4.6.1.7) can be obtained from a general equation (4.6.1.3) by multiplication by the normalizing factor

$$\mu = \pm\frac{1}{\sqrt{A^2 + B^2 + C^2}}, \qquad (4.6.1.9)$$

where the sign of $\mu$ must be opposite to that of $D$.

**Example 3.** Let us reduce the equation of the plane $-2x + 2y - z - 6 = 0$ to normalized form. Since $D = -6 < 0$, we see that the normalizing factor is

$$\mu = \frac{1}{\sqrt{(-2)^2 + 2^2 + (-1)^2}} = \frac{1}{3}.$$

We multiply the equation by this factor and obtain

$$-\frac{2}{3}x + \frac{2}{3}y - \frac{1}{3}z - 2 = 0.$$

Hence for this plane we have

$$\cos\alpha = -\frac{2}{3}, \quad \cos\beta = \frac{2}{3}, \quad \cos\gamma = -\frac{1}{3}, \quad p = 2.$$

**Remark.** The numbers $\cos\alpha$, $\cos\beta$, $\cos\gamma$, and $p$ are also called the *polar parameters of a plane*.

### 4.6.1-6. Equation of plane passing through point and parallel to another plane.

The plane that passes through a point $M_1(x_1, y_1, z_1)$ and is parallel to a plane $Ax + By + Cz + D = 0$ is given by the equation

$$A(x - x_1) + B(y - y_1) + C(z - z_1) + D = 0. \quad (4.6.1.10)$$

**Example 4.** Let us derive the equation of the plane that passes through the point $M_1(1, 2, -1)$ and is parallel to the plane $x + 2y + z + 2 = 0$.
According to (4.6.1.1), the desired equation is $(x - 1) + 2(y - 2) + (z + 1) + 2 = 0$ or

$$x + 2y + z - 2 = 0.$$

### 4.6.1-7. Equation of plane passing through three points.

The plane passing through three points $M_1(x_1, y_1, z_1)$, $M_2(x_2, y_2, z_2)$, and $M_3(x_3, y_3, z_3)$ (see Fig. 4.38) is described by the equation

$$\begin{vmatrix} x - x_1 & y - y_1 & z - z_1 \\ x_2 - x_1 & y_2 - y_1 & z_2 - z_1 \\ x_3 - x_1 & y_3 - y_1 & z_3 - z_1 \end{vmatrix} = 0, \quad \text{or} \quad \big[(\mathbf{r} - \mathbf{r}_1)(\mathbf{r}_2 - \mathbf{r}_1)(\mathbf{r}_3 - \mathbf{r}_1)\big] = 0, \quad (4.6.1.11)$$

where $\mathbf{r}$, $\mathbf{r}_1$, $\mathbf{r}_2$, and $\mathbf{r}_3$ are the position vectors of the points $M(x, y, z)$, $M_1(x_1, y_1, z_1)$, $M_2(x_2, y_2, z_2)$, and $M_3(x_3, y_3, z_3)$, respectively.

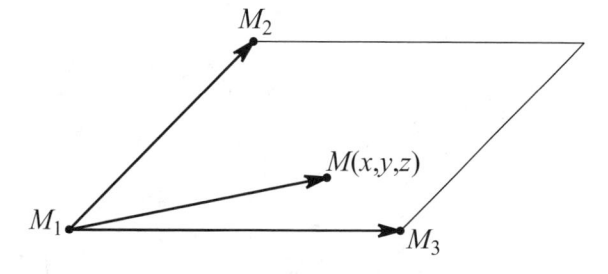

**Figure 4.38.** Plane passing through three points.

**Remark 1.** Equation (4.6.1.11) means that the vectors $\overrightarrow{M_1M}$, $\overrightarrow{M_1M_2}$, and $\overrightarrow{M_1M_3}$ are coplanar.

**Remark 2.** Equation (4.6.1.11) of the plane passing through three given points can be represented via a fourth-order determinant as follows:
$$\begin{vmatrix} x & y & z & 1 \\ x_1 & y_1 & z_1 & 1 \\ x_2 & y_2 & z_2 & 1 \\ x_3 & y_3 & z_3 & 1 \end{vmatrix} = 0. \qquad (4.6.1.11a)$$

**Remark 3.** If the three points $M_1(x_1, y_1, z_1)$, $M_2(x_2, y_2, z_2)$, and $M_3(x_3, y_3, z_3)$ are collinear, then equations (4.6.1.11) and (4.6.1.11a) become identities.

**Example 5.** Let us construct an equation of the plane passing through the three points $M_1(1,1,1)$, $M_2(2,2,1)$, and $M_3(1,2,2)$.

Obviously, the points $M_1$, $M_2$, and $M_3$ are not collinear, since the vectors $\overrightarrow{M_1M_2} = (1,1,0)$ and $\overrightarrow{M_1M_3} = (0,1,1)$ are not collinear. According to (4.6.1.11), the desired equation is
$$\begin{vmatrix} x-1 & y-1 & z-1 \\ 1 & 1 & 0 \\ 0 & 1 & 1 \end{vmatrix} = 0,$$
whence
$$x - y + z - 1 = 0.$$

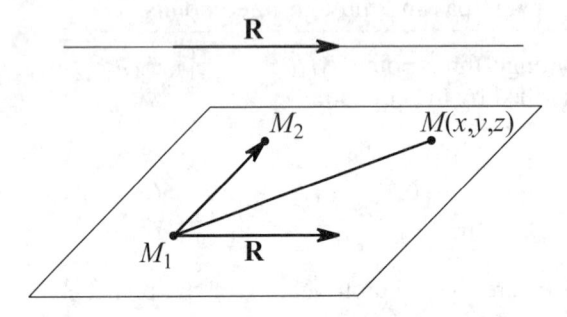

### 4.6.1-8. Equation of plane passing through two points and parallel to line.

The equation of the plane passing through two points $M_1(x_1, y_1, z_1)$ and $M_2(x_2, y_2, z_2)$ and parallel to a straight line with direction vector $\mathbf{R} = (l, m, n)$ (see Fig. 4.39) is

$$\begin{vmatrix} x-x_1 & y-y_1 & z-z_1 \\ x_2-x_1 & y_2-y_1 & z_2-z_1 \\ l & m & n \end{vmatrix} = 0, \qquad \text{or} \qquad [(\mathbf{r}-\mathbf{r}_1)(\mathbf{r}_2-\mathbf{r}_1)\mathbf{R}] = 0, \qquad (4.6.1.12)$$

where $\mathbf{r}$, $\mathbf{r}_1$, and $\mathbf{r}_2$ are the position vectors of the points $M(x,y,z)$, $M_1(x_1,y_1,z_1)$, and $M_2(x_2,y_2,z_2)$, respectively.

**Figure 4.39.** Plane passing through two points and parallel to line.

**Remark.** If the vectors $\overrightarrow{M_1M_2}$ and $\mathbf{R}$ are collinear, then equations (4.6.1.12) become identities.

**Example 6.** Let us construct an equation of the plane passing through the points $M_1(0,1,0)$ and $M_2(1,1,1)$ and parallel to the straight line with direction vector $\mathbf{R} = (0,1,1)$.

According to (4.6.1.12), the desired equation is
$$\begin{vmatrix} x-0 & y-1 & z-0 \\ 1-0 & 1-1 & 1-0 \\ 0 & 1 & 1 \end{vmatrix} = 0,$$
whence
$$-x - y + z + 1 = 0.$$

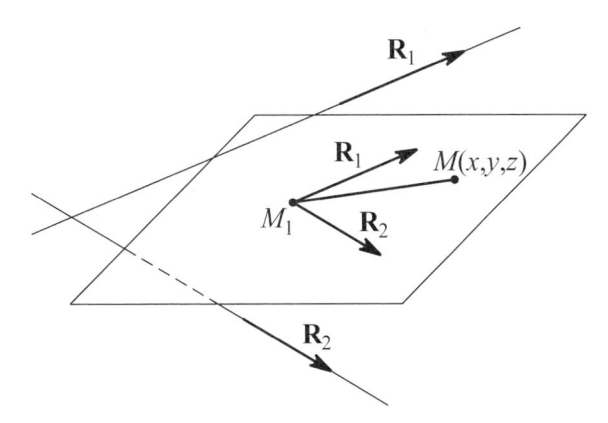

**Figure 4.40.** Plane passing through a point and parallel to two straight lines.

### 4.6.1-9. Equation of plane passing through point and parallel to two straight lines.

The equation of the plane passing through a point $M_1(x_1, y_1, z_1)$ and parallel to two straight lines with direction vectors $\mathbf{R}_1 = (l_1, m_1, n_1)$ and $\mathbf{R}_2 = (l_2, m_2, n_2)$ (see Fig. 4.40) is

$$\begin{vmatrix} x - x_1 & y - y_1 & z - z_1 \\ l_1 & m_1 & n_1 \\ l_2 & m_2 & n_2 \end{vmatrix} = 0, \quad \text{or} \quad \big[(\mathbf{r} - \mathbf{r}_1)\mathbf{R}_1\mathbf{R}_2\big] = 0, \qquad (4.6.1.13)$$

where $\mathbf{r}$ and $\mathbf{r}_1$ are the position vectors of the points $M(x, y, z)$ and $M_1(x_1, y_1, z_1)$, respectively.

**Example 7.** Let us find the equation of the plane passing through the point $M_1(0, 1, 0)$ and parallel to the straight lines with direction vectors $\mathbf{R}_1 = (1, 0, 1)$ and $\mathbf{R}_2 = (0, 1, 2)$.

According to (4.6.1.13), the desired equation is

$$\begin{vmatrix} x - 0 & y - 1 & z - 0 \\ 1 & 0 & 1 \\ 0 & 1 & 2 \end{vmatrix} = 0,$$

whence

$$-x - 2y + z + 2 = 0.$$

### 4.6.1-10. Plane passing through two points and perpendicular to given plane.

The plane (see Fig. 4.41) passing through two points $M_1(x_1, y_1, z_1)$ and $M_2(x_2, y_2, z_2)$ and perpendicular to the plane given by the equation $Ax + By + Cz + D = 0$ is determined by the equation

$$\begin{vmatrix} x - x_1 & y - y_1 & z - z_1 \\ x_2 - x_1 & y_2 - y_1 & z_2 - z_1 \\ A & B & C \end{vmatrix} = 0, \quad \text{or} \quad \big[(\mathbf{r} - \mathbf{r}_1)(\mathbf{r}_2 - \mathbf{r}_1)\mathbf{N}\big] = 0, \qquad (4.6.1.14)$$

where $\mathbf{r}$, $\mathbf{r}_1$, and $\mathbf{r}_2$ are the position vectors of the points $M(x, y, z)$, $M_1(x_1, y_1, z_1)$, and $M_2(x_2, y_2, z_2)$, respectively.

**Remark.** If the straight line passing through points $M_1(x_1, y_1, z_1)$ and $M_2(x_2, y_2, z_2)$ is perpendicular to the original plane, then the desired plane is undetermined and equations (4.6.1.14) become identities.

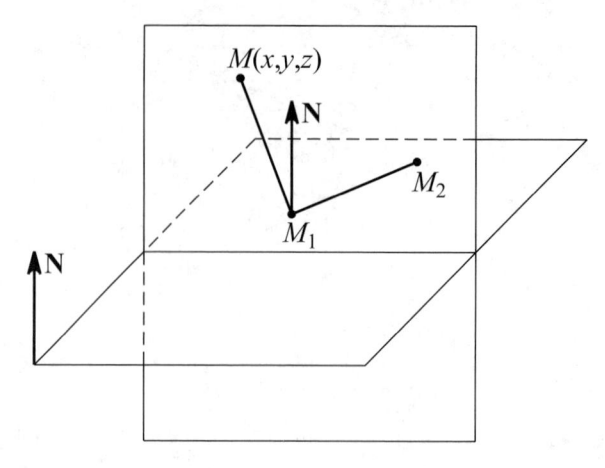

**Figure 4.41.** Plane passing through two points and perpendicular to given plane.

**Example 8.** Let us find an equation of the plane passing through the points $M_1(0, 1, 2)$ and $M_2(2, 2, 3)$ and perpendicular to the plane $x - y + z + 5 = 0$.

According to (4.6.1.14), the desired equation is

$$\begin{vmatrix} x-0 & y-1 & z-2 \\ 2-0 & 2-1 & 3-2 \\ 1 & -1 & 1 \end{vmatrix} = 0,$$

whence

$$2x - y - 3z + 7 = 0.$$

### 4.6.1-11. Plane passing through point and perpendicular to two planes.

The plane (see Fig. 4.42) passing through a point $M_1(x_1, y_1, z_1)$ and perpendicular to two (nonparallel) planes $A_1 x + B_1 y + C_1 z + D_1 = 0$ and $A_2 x + B_2 y + C_2 z + D_2 = 0$ is given by the equation

$$\begin{vmatrix} x-x_1 & y-y_1 & z-z_1 \\ A_1 & B_1 & C_1 \\ A_2 & B_2 & C_2 \end{vmatrix} = 0, \quad \text{or} \quad \big[(\mathbf{r} - \mathbf{r}_1)\mathbf{N}_1 \mathbf{N}_2\big] = 0, \qquad (4.6.1.15)$$

where $\mathbf{N}_1 = (A_1, B_1, C_1)$ and $\mathbf{N}_2 = (A_2, B_2, C_2)$ are the normals to the given planes and $\mathbf{r}$ and $\mathbf{r}_1$ are the position vectors of the points $M(x, y, z)$ and $M_1(x_1, y_1, z_1)$, respectively.

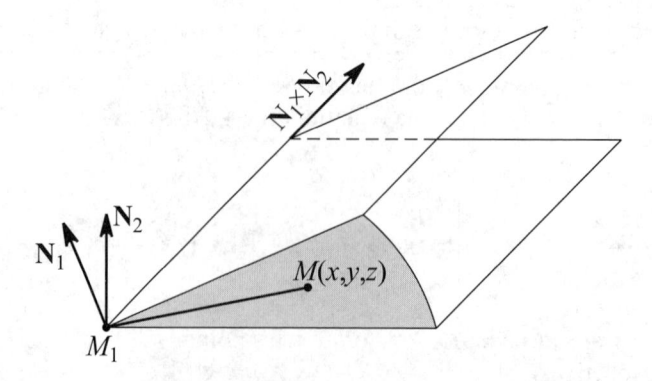

**Figure 4.42.** Plane passing through a point and perpendicular to two planes.

**Remark 1.** Equations (4.6.1.15) mean that the vectors $\overrightarrow{M_1 M}$, $\mathbf{N}_1$, and $\mathbf{N}_2$ are coplanar.

**Remark 2.** If the original planes are parallel, then the desired plane is undetermined. In this case, equations (4.6.1.15) become identities.

**Example 9.** Let us find an equation of the plane passing through the point $M_1(0, 1, 2)$ and perpendicular to the planes $x - y + z + 5 = 0$ and $-x + y + z - 1 = 0$.

According to (4.6.1.15), the desired equation is

$$\begin{vmatrix} x-0 & y-1 & z-2 \\ 1 & -1 & 1 \\ -1 & 1 & 1 \end{vmatrix} = 0,$$

whence

$$x + y - 1 = 0.$$

#### 4.6.1-12. Equation of plane passing through line of intersection of planes.

The planes passing through the line of intersection of the planes $A_1x + B_1y + C_1z + D_1 = 0$ and $A_2x + B_2y + C_2z + D_2 = 0$ are given by the equation

$$\alpha(A_1x + B_1y + C_1z + D_1) + \beta(A_2x + B_2y + C_2z + D_2) = 0, \qquad (4.6.1.16)$$

which is called the *equation of a pencil of planes*. Here $\alpha$ and $\beta$ are arbitrary parameters.

Let $\alpha \neq 0$. Set $\beta/\alpha = \lambda$; then equation (4.6.1.16) becomes

$$A_1x + B_1y + C_1z + D_1 + \lambda(A_2x + B_2y + C_2z + D_2) = 0. \qquad (4.6.1.17)$$

By varying the parameter $\lambda$ from $-\infty$ to $+\infty$, we obtain all the planes in the pencil. For $\lambda = \pm 1$, we obtain equations of the planes that bisect the angles between the given planes provided that the equations of the latter are given in normalized form.

**Remark.** The passage from equation (4.6.1.16) to equation (4.6.1.17) excludes the case $\alpha = 0$. Equation (4.6.1.17) does not define the plane $A_2x + B_2y + C_2z + D_2 = 0$; i.e., equation (4.6.1.17) for various $\lambda$ defines all the planes in the pencil but one (the second of the two given planes).

### 4.6.2. Line in Space

#### 4.6.2-1. Parametric equation of straight line.

The *parametric equation of the line* that passes through a point $M_1(x_1, y_1, z_1)$ and is parallel to a direction vector $\mathbf{R} = (l, m, n)$ (see Fig. 4.43) is

$$x = x_1 + lt, \quad y = y_1 + mt, \quad z = z_1 + nt, \quad \text{or} \quad \mathbf{r} = \mathbf{r}_1 + t\mathbf{R}, \qquad (4.6.2.1)$$

where $\mathbf{r} = \overrightarrow{OM}$ and $\mathbf{r}_1 = \overrightarrow{M_1M}$. As the parameter $t$ varies from $-\infty$ to $+\infty$, the point $M$ with position vector $\mathbf{r} = (x, y, z)$ determined by formula (4.6.2.1) runs over the entire straight line in question. It is convenient to use the parametric equation (4.6.2.1) if one needs to find the point of intersection of a straight line with a plane.

The numbers $l$, $m$, and $n$ characterize the direction of the straight line in space; they are called the *direction coefficients* of the straight line. For a unit vector $\mathbf{R} = \mathbf{R}^0$, the coefficients $l, m, n$ are the cosines of the angles $\alpha$, $\beta$, and $\gamma$ formed by this straight line (the direction of the vector $\mathbf{R}^0$) with the coordinate axes $OX, OY$, and $OZ$. These cosines can be expressed via the coordinates of the direction vector $\mathbf{R}$ as

$$\cos\alpha = \frac{l}{\sqrt{l^2 + m^2 + n^2}}, \quad \cos\beta = \frac{m}{\sqrt{l^2 + m^2 + n^2}}, \quad \cos\gamma = \frac{n}{\sqrt{l^2 + m^2 + n^2}}. \qquad (4.6.2.2)$$

**Example 1.** Let us find the equation of the straight line that passes through the point $M_1(2, -3, 1)$ and is parallel to the direction vector $\mathbf{R} = (1, 2, -3)$.

According to (4.6.2.1), the desired equation is

$$x = 2 + t, \quad y = -3 + 2t, \quad z = 1 - 3t.$$

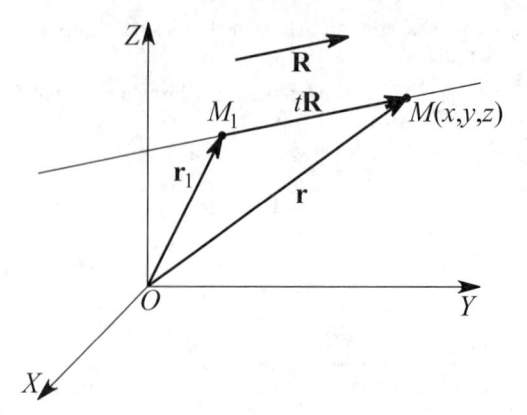

**Figure 4.43.** Straight line passing through a point and parallel to direction vector.

#### 4.6.2-2. Canonical equation of straight line.

The equation

$$\frac{x-x_1}{l} = \frac{y-y_1}{m} = \frac{z-z_1}{n}, \quad \text{or} \quad (\mathbf{r}-\mathbf{r}_1) \times \mathbf{R} = 0, \qquad (4.6.2.3)$$

is called the *canonical equation of the straight line* passing through the point $M_1(x_1, y_1, z_1)$ with position vector $\mathbf{r}_1 = (x_1, y_1, z_1)$ and parallel to the direction vector $\mathbf{R} = (l, m, n)$.

**Remark 1.** One can obtain the canonical equation (4.6.2.3) from the parametric equations (4.6.2.1) by eliminating the parameter $t$.

**Remark 2.** In the canonical equation, all coefficients $l$, $m$, and $n$ cannot be zero simultaneously, since $|\mathbf{R}| \neq 0$. But some of them may be zero. If one of the denominators in equations (4.6.2.3) is zero, this means that the corresponding numerator is also zero.

**Example 2.** The equations $(x-1)/1 = (y-3)/4 = (z-3)/0$ determine the straight line passing through the point $M_1(1, 3, 3)$ and perpendicular to the axis $OZ$. This means that the line lies in the plane $z = 3$, and hence $z - 3 = 0$ for all points of the line.

**Example 3.** Let us find the equation of the straight line passing through the point $M_1(2, -3, 1)$ and parallel to the direction vector $\mathbf{R} = (1, 2, -3)$.

According to (4.6.2.3), the desired equation is

$$\frac{x-2}{1} = \frac{y+3}{2} = \frac{z-1}{-3}.$$

#### 4.6.2-3. General equation of straight line.

The *general equation of a straight line in space* defines it as the line of intersection of two planes (see Fig. 4.44) and is given analytically by a system of two linear equations

$$\begin{aligned} A_1 x + B_1 y + C_1 z + D_1 &= 0, \\ A_2 x + B_2 y + C_2 z + D_2 &= 0, \end{aligned} \quad \text{or} \quad \begin{aligned} \mathbf{r} \cdot \mathbf{N}_1 + D_1 &= 0, \\ \mathbf{r} \cdot \mathbf{N}_2 + D_2 &= 0, \end{aligned} \qquad (4.6.2.4)$$

where $\mathbf{N}_1 = (A_1, B_1, C_1)$ and $\mathbf{N}_2 = (A_2, B_2, C_2)$ are the normals to the planes and $\mathbf{r}$ is the position vector of the point $(x, y, z)$.

The direction vector $\mathbf{R}$ is equal to the cross product of the normals $\mathbf{N}_1$ and $\mathbf{N}_2$; i.e.,

$$\mathbf{R} = \mathbf{N}_1 \times \mathbf{N}_2, \qquad (4.6.2.5)$$

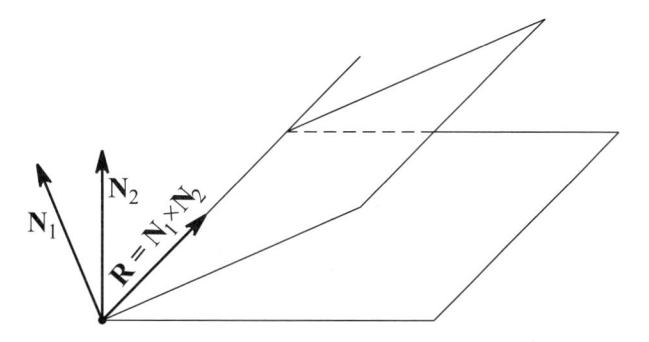

Figure 4.44. Straight line as intersection of two planes.

and its coordinates $l$, $m$, and $n$ can be obtained by the formulas

$$l = \begin{vmatrix} B_1 & C_1 \\ B_2 & C_2 \end{vmatrix}, \quad m = \begin{vmatrix} C_1 & A_1 \\ C_2 & A_2 \end{vmatrix}, \quad n = \begin{vmatrix} A_1 & B_1 \\ A_2 & B_2 \end{vmatrix}. \tag{4.6.2.6}$$

**Remark 1.** Simultaneous equations of the form (4.6.2.4) define a straight line if and only if the coefficients $A_1$, $B_1$, and $C_1$ in one of them are not proportional to the respective coefficients $A_2$, $B_2$, and $C_2$ in the other.

**Remark 2.** For $D_1 = D_2 = 0$ (and only in this case), the line passes through the origin.

**Example 4.** Let us reduce the equation of the straight line

$$x + 2y - z + 1 = 0, \quad x - y + z + 3 = 0$$

to canonical form.

We choose one of the coordinates arbitrarily; say, $x = 0$. Then

$$2y - z + 1 = 0, \quad -y + z + 3 = 0,$$

and hence $y = -4$, $z = -7$. Thus the desired line contains the point $M(0, -4, -7)$. We find the cross product of the vectors $\mathbf{N}_1 = (1, 2, -1)$ and $\mathbf{N}_2 = (1, -1, 1)$ and, according to (4.6.2.5), obtain the direction vector $\mathbf{R} = (1, -2, -3)$ of the desired line. Therefore, with (4.6.2.3) taken into account, the equation of the line becomes

$$\frac{x}{1} = \frac{y+4}{-2} = \frac{z+7}{-3}.$$

### 4.6.2-4. Equation of line in projections.

The *equation of a line in projections* can be obtained by eliminating first $z$ and then $y$ from the general equations (4.6.2.4):

$$y = kx + a, \quad z = hx + b. \tag{4.6.2.7}$$

Each of two equations (4.6.2.7) defines a plane projecting the straight line onto the planes $OXY$ and $OXZ$ (see Fig. 4.45).

**Remark 1.** For straight lines parallel to the plane $OYZ$, this form of the equations cannot be used; one should take the projections onto some other pair of coordinate planes.

**Remark 2.** Equations (4.6.2.7) can be represented in the canonical form

$$\frac{x-0}{1} = \frac{y-a}{k} = \frac{z-b}{h}. \tag{4.6.2.8}$$

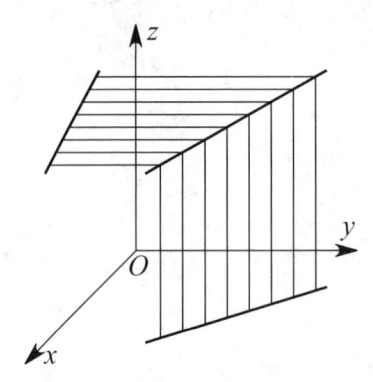

**Figure 4.45.** Straight line with equation in projections.

### 4.6.2-5. Equation of straight line passing through two points.

The canonical equation of the straight line (see Fig. 4.46) passing through two points $M_1(x_1, y_1, z_1)$ and $M_2(x_2, y_2, z_2)$ is

$$\frac{x - x_1}{x_2 - x_1} = \frac{y - y_1}{y_2 - y_1} = \frac{z - z_1}{z_2 - z_1}, \quad \text{or} \quad (\mathbf{r} - \mathbf{r}_1) \times (\mathbf{r}_2 - \mathbf{r}_1) = 0, \qquad (4.6.2.9)$$

where $\mathbf{r}$, $\mathbf{r}_1$, and $\mathbf{r}_2$ are the position vectors of the points $M(x, y, z)$, $M_1(x_1, y_1, z_1)$, and $M_2(x_2, y_2, z_2)$, respectively.

The parametric equations of the straight line passing through two points $M_1(x_1, y_1, z_1)$ and $M_2(x_2, y_2, z_2)$ in the rectangular Cartesian coordinate system $OXYZ$ can be written as

$$\begin{aligned} x &= x_1(1-t) + x_2 t, \\ y &= y_1(1-t) + y_2 t, \quad \text{or} \quad \mathbf{r} = (1-t)\mathbf{r}_1 + t\mathbf{r}_2. \\ z &= z_1(1-t) + z_2 t, \end{aligned} \qquad (4.6.2.10)$$

**Remark.** Eliminating the parameter $t$ from equations (4.6.2.10), we obtain equations (4.6.2.9).

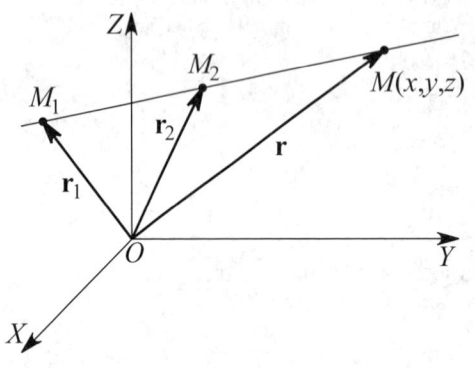

**Figure 4.46.** Straight line passing through two points.

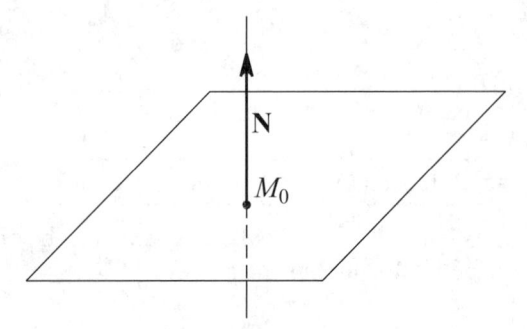

**Figure 4.47.** Straight line passing through point and perpendicular to plane.

### 4.6.2-6. Equation of straight line passing through point and perpendicular to plane.

The equation of the straight line passing through a point $M_0(x_0, y_0, z_0)$ and perpendicular to the plane given by the equation $Ax + By + Cz + D = 0$, or $\mathbf{r} \cdot \mathbf{N} + D = 0$ (see Fig. 4.47), is

$$\frac{x - x_0}{A} = \frac{y - y_0}{B} = \frac{z - z_0}{C}, \qquad (4.6.2.11)$$

where $\mathbf{N} = (A, B, C)$ is the normal to the plane.

## 4.6.3. Mutual Arrangement of Points, Lines, and Planes

**4.6.3-1. Angles between lines in space.**

Consider two straight lines determined by vector parametric equations $\mathbf{r} = \mathbf{r}_1 + t\mathbf{R}_1$ and $\mathbf{r} = \mathbf{r}_2 + t\mathbf{R}_2$. The angle $\varphi$ between these lines (see Fig. 4.48) can be obtained from the formulas

$$\cos\varphi = \frac{\mathbf{R}_1 \cdot \mathbf{R}_2}{|\mathbf{R}_1||\mathbf{R}_2|}, \quad \sin\varphi = \frac{|\mathbf{R}_1 \times \mathbf{R}_2|}{|\mathbf{R}_1||\mathbf{R}_2|}.$$

If the lines are given by the canonical equations

$$\frac{x-x_1}{l_1} = \frac{y-y_1}{m_1} = \frac{z-z_1}{n_1} \quad \text{and} \quad \frac{x-x_2}{l_2} = \frac{y-y_2}{m_2} = \frac{z-z_2}{n_2},$$

then the angle $\varphi$ between the lines can be found from the formulas

$$\cos\varphi = \frac{l_1 l_2 + m_1 m_2 + n_1 n_2}{\sqrt{l_1^2 + m_1^2 + n_1^2}\sqrt{l_2^2 + m_2^2 + n_2^2}},$$

$$\sin\varphi = \frac{\sqrt{\left|\begin{matrix} m_1 & n_1 \\ m_2 & n_2 \end{matrix}\right|^2 + \left|\begin{matrix} n_1 & l_1 \\ n_2 & l_2 \end{matrix}\right|^2 + \left|\begin{matrix} l_1 & m_1 \\ l_2 & m_2 \end{matrix}\right|^2}}{\sqrt{l_1^2 + m_1^2 + n_1^2}\sqrt{l_2^2 + m_2^2 + n_2^2}}, \quad (4.6.3.1)$$

which coincide with formulas (4.6.3.1) written in coordinate form.

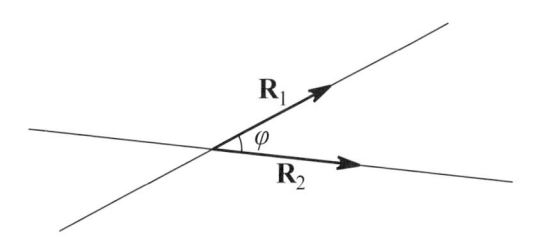

**Figure 4.48.** Angles between lines in space.

**Example 1.** Let us find the angle between the lines

$$\frac{x}{1} = \frac{y-2}{2} = \frac{z+1}{2} \quad \text{and} \quad \frac{x}{0} = \frac{y-2}{3} = \frac{z+1}{4}.$$

Using the first formula in (4.6.3.1), we obtain

$$\cos\varphi = \frac{1\cdot 0 + 3\cdot 2 + 4\cdot 2}{\sqrt{0^2 + 3^2 + 4^2}\sqrt{1^2 + 2^2 + 2^2}} = \frac{14}{15},$$

and hence $\varphi \approx 0.3672$ rad.

**4.6.3-2. Conditions for two lines to be parallel.**

Two straight lines given by vector parametric equations $\mathbf{r} = \mathbf{r}_1 + t\mathbf{R}_1$ and $\mathbf{r} = \mathbf{r}_2 + t\mathbf{R}_2$ are parallel if

$$\mathbf{R}_2 = \lambda \mathbf{R}_1 \quad \text{or} \quad \mathbf{R}_2 \times \mathbf{R}_1 = 0,$$

i.e., if their direction vectors $\mathbf{R}_1$ and $\mathbf{R}_2$ are collinear. If the straight lines are given by canonical equations, then the condition that they are parallel can be written as

$$\frac{l_1}{l_2} = \frac{m_1}{m_2} = \frac{n_1}{n_2}. \qquad (4.6.3.2)$$

Remark. If parallel lines have a common point (i.e., $\mathbf{r}_1 = \mathbf{r}_2$ in parametric equations), then they coincide.

**Example 2.** Let us show that the lines

$$\frac{x-1}{2} = \frac{y-3}{1} = \frac{z}{2} \quad \text{and} \quad \frac{x-3}{4} = \frac{y+1}{2} = \frac{z}{4}$$

are parallel to each other.
Indeed, condition (4.6.3.2) is satisfied,

$$\frac{2}{4} = \frac{1}{2} = \frac{2}{4},$$

and hence the lines are parallel.

### 4.6.3-3. Conditions for two lines to be perpendicular.

Two straight lines given by vector parametric equation $\mathbf{r} = \mathbf{r}_1 + t\mathbf{R}_1$ and $\mathbf{r} = \mathbf{r}_2 + t\mathbf{R}_2$ are perpendicular if

$$\mathbf{R}_1 \cdot \mathbf{R}_2 = 0. \qquad (4.6.3.3)$$

If the lines are given by canonical equations, then the condition that they are perpendicular can be written as

$$l_1 l_2 + m_1 m_2 + n_1 n_2 = 0, \qquad (4.6.3.3a)$$

which coincides with formula (4.6.3.3) written in coordinate form.

**Example 3.** Let us show that the lines

$$\frac{x-1}{2} = \frac{y-3}{1} = \frac{z}{2} \quad \text{and} \quad \frac{x-2}{1} = \frac{y+1}{2} = \frac{z}{-2}$$

are perpendicular.
Indeed, condition (4.6.3.3a) is satisfied,

$$2 \cdot 1 + 1 \cdot 2 + 2 \cdot (-2) = 0,$$

and hence the lines are perpendicular.

### 4.6.3-4. Theorem on the arrangement of two lines in space.

THEOREM ON THE ARRANGEMENT OF TWO LINES IN SPACE. *Two lines in space can:*
a) *be skew;*
b) *lie in the same plane and not meet each other, i.e., be parallel;*
c) *meet at a point;*
d) *coincide.*

A general characteristic of all four cases is the determinant of the matrix

$$\begin{pmatrix} x_2 - x_1 & y_2 - y_1 & z_2 - z_1 \\ l_1 & m_1 & n_1 \\ l_2 & m_2 & n_2 \end{pmatrix}, \qquad (4.6.3.4)$$

whose entries are taken from the canonical equations

$$\frac{x-x_1}{l_1} = \frac{y-y_1}{m_1} = \frac{z-z_1}{n_1} \quad \text{and} \quad \frac{x-x_2}{l_2} = \frac{y-y_2}{m_2} = \frac{z-z_2}{n_2}$$

of the lines.

In cases a–d of the theorem, for the matrix (4.6.3.4) we have, respectively:
a) the determinant is nonzero;
b) the last two rows are proportional to each other but are not proportional to the first row;
c) the last two rows are not proportional, and the first row is their linear combination;
d) all rows are proportional.

### 4.6.3-5. Angles between planes.

Consider two planes given by the general equations

$$\begin{array}{ll} A_1x + B_1y + C_1z + D_1 = 0, & \mathbf{r} \cdot \mathbf{N}_1 + D_1 = 0, \\ A_2x + B_2y + C_2z + D_2 = 0, & \mathbf{r} \cdot \mathbf{N}_2 + D_2 = 0, \end{array} \quad \text{or} \quad \quad (4.6.3.5)$$

where $\mathbf{N}_1 = (A_1, B_1, C_1)$ and $\mathbf{N}_2 = (A_2, B_2, C_2)$ are the normals to the planes and $\mathbf{r}$ is the position vector of the point $(x, y, z)$.

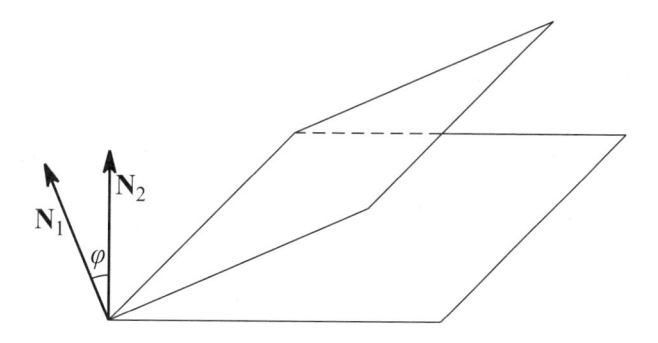

**Figure 4.49.** Angles between planes.

The angle between two planes (see Fig. 4.49) is defined as any of the two adjacent dihedral angles formed by the planes (if the planes are parallel, then the angle between them is by definition equal to 0 or $\pi$). One of these dihedral angles is equal to the angle $\varphi$ between the normal vectors $\mathbf{N}_1 = (A_1, B_1, C_1)$ and $\mathbf{N}_2 = (A_2, B_2, C_2)$ to the planes, which can be determined by the formula

$$\cos\varphi = \frac{A_1A_2 + B_1B_2 + C_1C_2}{\sqrt{A_1^2 + B_1^2 + C_1^2}\sqrt{A_2^2 + B_2^2 + C_2^2}} = \frac{\mathbf{N}_1 \cdot \mathbf{N}_2}{|\mathbf{N}_1||\mathbf{N}_2|}. \quad (4.6.3.6)$$

If the planes are given by vector parametric equations

$$\mathbf{r} = \mathbf{r}_1 + \mathbf{R}_1 t + \mathbf{R}_2 s \quad \text{or} \quad \mathbf{r}' = \mathbf{r}'_1 + \mathbf{R}'_1 t + \mathbf{R}'_2 s, \quad (4.6.3.7)$$

then the angle between the planes is given by the formula

$$\cos\varphi = \frac{(\mathbf{R}_1 \times \mathbf{R}_2) \cdot (\mathbf{R}'_1 \times \mathbf{R}'_2)}{|\mathbf{R}_1 \times \mathbf{R}_2||\mathbf{R}'_1 \times \mathbf{R}'_2|}. \quad (4.6.3.8)$$

### 4.6.3-6. Conditions for two planes to be parallel.

Two planes given by the general equations (4.6.3.5) in coordinate form are parallel if and only if the following *condition for the planes to be parallel* is satisfied:

$$\frac{A_1}{A_2} = \frac{B_1}{B_2} = \frac{C_1}{C_2} \neq \frac{D_1}{D_2}; \tag{4.6.3.9}$$

in this case, the planes do not coincide. For planes given by the general equations (4.6.3.5) in vector form, the condition becomes

$$\mathbf{N}_2 = \lambda \mathbf{N}_1 \quad \text{or} \quad \mathbf{N}_2 \times \mathbf{N}_1 = 0; \tag{4.6.3.10}$$

i.e., the planes are parallel if their normals are parallel.

**Example 4.** Let us show that the planes $x - y + z = 0$ and $2x - 2y + 2z + 5 = 0$ are parallel. Since condition (4.6.3.9) is satisfied,

$$\frac{1}{2} = \frac{-1}{-2} = \frac{1}{2},$$

we see that the planes are parallel to each other.

### 4.6.3-7. Conditions for planes to coincide.

Two planes coincide if they are parallel and have a common point.

Two planes given by the general equations (4.6.3.5) coincide if and only if the following *condition for the planes to coincide* is satisfied:

$$\frac{A_1}{A_2} = \frac{B_1}{B_2} = \frac{C_1}{C_2} = \frac{D_1}{D_2}. \tag{4.6.3.11}$$

**Remark.** Sometimes the case in which the planes coincide is treated as a special case of parallel straight lines and is not distinguished as an exceptional case.

### 4.6.3-8. Conditions for two planes to be perpendicular.

Planes are perpendicular if their normals are perpendicular.

Two planes determined by the general equations (4.6.3.5) are perpendicular if and only if the following *condition for the planes to be perpendicular* is satisfied:

$$A_1 A_2 + B_1 B_2 + C_1 C_2 = 0 \quad \text{or} \quad \mathbf{N}_1 \cdot \mathbf{N}_2 = 0, \tag{4.6.3.12}$$

where $\mathbf{N}_1 = (A_1, B_1, C_1)$ and $\mathbf{N}_2 = (A_2, B_2, C_2)$ are the normals to the planes.

**Example 5.** Let us show that the planes $x - y + z = 0$ and $x - y - 2z + 5 = 0$ are perpendicular. Since condition (4.6.3.12) is satisfied,

$$1 \cdot 1 + (-1) \cdot (-1) + 1 \cdot (-2) = 0,$$

we see that the planes are perpendicular to each other.

## 4.6.3-9. Intersection of three planes.

The point of intersection of three planes given by the equations

$$A_1x + B_1y + C_1z + D_1 = 0, \quad A_2x + B_2y + C_2z + D_2 = 0, \quad A_3x + B_3y + C_3z + D_3 = 0,$$

has the following coordinates:

$$x_0 = -\frac{1}{\Delta}\begin{vmatrix} D_1 & B_1 & C_1 \\ D_2 & B_2 & C_2 \\ D_3 & B_3 & C_3 \end{vmatrix}, \quad y_0 = -\frac{1}{\Delta}\begin{vmatrix} A_1 & D_1 & C_1 \\ A_2 & D_2 & C_2 \\ A_3 & D_3 & C_3 \end{vmatrix}, \quad z_0 = -\frac{1}{\Delta}\begin{vmatrix} A_1 & B_1 & D_1 \\ A_2 & B_2 & D_2 \\ A_3 & B_3 & D_3 \end{vmatrix}, \quad (4.6.3.13)$$

where $\Delta$ is given by the formula

$$\Delta = \begin{vmatrix} A_1 & B_1 & C_1 \\ A_2 & B_2 & C_2 \\ A_3 & B_3 & C_3 \end{vmatrix}. \quad (4.6.3.14)$$

**Remark.** Three planes are concurrent at a single point if $\Delta \neq 0$. If $\Delta = 0$ and at least one of the second-order minors is nonzero, then all planes are parallel to a single line. If all minors are zero, then the planes are concurrent in a single line.

## 4.6.3-10. Intersection of four planes.

If four planes given by the equations

$$A_1x + B_1y + C_1z + D_1 = 0, \qquad A_2x + B_2y + C_2z + D_2 = 0,$$
$$A_3x + B_3y + C_3z + D_3 = 0, \qquad A_4x + B_4y + C_4z + D_4 = 0,$$

are concurrent at a single point, then

$$\begin{vmatrix} x & y & z & 1 \\ x_1 & y_1 & z_1 & 1 \\ x_2 & y_2 & z_2 & 1 \\ x_3 & y_3 & z_3 & 1 \end{vmatrix} = 0. \quad (4.6.3.15)$$

To find the points of intersection, it suffices to find the point of intersection of any three of them (see Paragraph 4.6.3-9). The remaining equation follows from the three other equations.

## 4.6.3-11. Angle between straight line and plane.

Consider a plane given by the general equation

$$Ax + By + Cz + D = 0, \qquad \text{or} \qquad \mathbf{r} \cdot \mathbf{N} + D = 0 \quad (4.6.3.16)$$

and a line given by the canonical equation

$$\frac{x - x_1}{l} = \frac{y - y_1}{m} = \frac{z - z_1}{n}, \qquad \text{or} \qquad (\mathbf{r} - \mathbf{r}_1) \times \mathbf{R} = 0, \quad (4.6.3.17)$$

where $\mathbf{N} = (A, B, C)$ is the normal to the plane, $\mathbf{r}$ and $\mathbf{r}_1$ are the respective position vectors of the points $(x, y, z)$ and $(x_1, y_1, z_1)$, and $\mathbf{R} = (l, m, n)$ is the direction vector of the line.

The angle between the line and the plane (see Fig. 4.50) is defined as the complementary angle $\theta$ of the angle $\varphi$ between the direction vector $\mathbf{R}$ of the line and the normal $\mathbf{N}$ to the plane. For this angle, one has the formula

$$\sin\theta = \cos\varphi = \frac{Al + Bm + Cn}{\sqrt{A^2 + B^2 + C^2}\sqrt{l^2 + m^2 + n^2}} = \frac{\mathbf{N} \cdot \mathbf{R}}{|\mathbf{N}||\mathbf{R}|}. \quad (4.6.3.18)$$

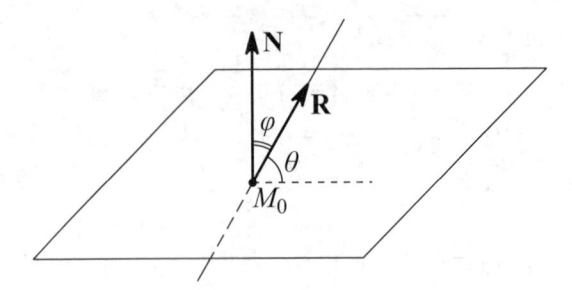

**Figure 4.50.** Angle between straight line and plane.

### 4.6.3-12. Conditions for line and plane to be parallel.

A plane given by the general equation (4.6.3.16) and a line given by the canonical equation (4.6.3.17) are parallel if

$$Al + Bm + Cn = 0, \qquad \mathbf{N} \cdot \mathbf{R} = 0,$$
$$Ax_1 + By_1 + Cz_1 + D \neq 0, \quad \text{or} \quad \mathbf{N} \cdot \mathbf{r}_1 + D \neq 0; \qquad (4.6.3.19)$$

i.e., a line is parallel to a plane if the direction vector of the line is perpendicular to the normal to the plane. Conditions (4.6.3.19) include the condition under which the line is not contained in the plane.

### 4.6.3-13. Condition for line to be entirely contained in plane.

A straight line given by the canonical equation (4.6.3.17) is entirely contained in a plane given by the general equation (4.6.3.16) if

$$Al + Bm + Cn = 0, \qquad \mathbf{N} \cdot \mathbf{R} = 0,$$
$$Ax_1 + By_1 + Cz_1 + D = 0, \quad \text{or} \quad \mathbf{N} \cdot \mathbf{r}_1 + D = 0. \qquad (4.6.3.20)$$

*Remark.* Sometimes the case in which a line is entirely contained in a plane is treated as a special case of parallel straight lines and is not distinguished as an exception.

### 4.6.3-14. Condition for line and plane to be perpendicular.

A line given by the canonical equation (4.6.3.17) and a plane given by the general equation (4.6.3.16) are perpendicular if the line is collinear to the normal to the plane (is a normal itself), i.e., if

$$\frac{A}{l} = \frac{B}{m} = \frac{C}{n}, \quad \text{or} \quad \mathbf{N} = \lambda \mathbf{R}, \quad \text{or} \quad \mathbf{N} \times \mathbf{R} = 0. \qquad (4.6.3.21)$$

### 4.6.3-15. Intersection of line and plane.

Consider a plane given by the general equation (4.6.3.16) and a straight line given by the parametric equation

$$x = x_1 + lt, \quad y = y_1 + mt, \quad z = z_1 + nt, \quad \text{or} \quad \mathbf{r} = \mathbf{r}_1 + t\mathbf{R}.$$

The coordinates of the point $M_0(x_0, y_0, z_0)$ of intersection of the line with the plane (see Fig. 4.50), if the point exists at all, are determined by the formulas

$$x_0 = x_1 + lt_0, \quad y_0 = y_1 + mt_0, \quad z_0 = z_1 + nt_0, \quad \text{or} \quad \mathbf{r} = \mathbf{r}_1 + t_0 \mathbf{R}, \qquad (4.6.3.22)$$

where the parameter $t_0$ is determined from the relation

$$t_0 = -\frac{Ax_1 + By_1 + Cz_1 + D}{Al + Bm + Cn} = -\frac{\mathbf{N} \cdot \mathbf{r}_1 + D}{\mathbf{N} \cdot \mathbf{R}}. \qquad (4.6.3.23)$$

**Remark.** To obtain formulas (4.6.3.22) and (4.6.3.23), one should rewrite the equation of the straight line in parametric form and replace $x$, $y$, and $z$ in equation (4.6.3.16) of the plane by their expressions via $t$. From the resulting expression, one finds the parameter $t_0$ and then the coordinates $x_0$, $y_0$, and $z_0$ themselves.

**Example 6.** Let us find the point of intersection of the line $x/2 = (y-1)/1 = (z+1)/2$ with the plane $x + 2y + 3z - 29 = 0$.

We use formula (4.6.3.23) to find the value of the parameter $t_0$:

$$t_0 = -\frac{1 \cdot 0 + 2 \cdot 1 + 3 \cdot (-1) - 29}{1 \cdot 2 + 2 \cdot 1 + 3 \cdot 2} = -\frac{-30}{10} = 3.$$

Then, according to (4.6.3.22), we finally obtain the coordinates of the point of intersection in the form

$$x_0 = 0 - 2 \cdot 3 = 6, \quad y_0 = 1 - 1 \cdot 3 = 4, \quad z_0 = -1 - 2 \cdot 3 = 5.$$

### 4.6.3-16. Distance from point to plane.

The *deviation of a point from a plane* is defined as the number $\delta$ equal to the length of the perpendicular drawn from this point to the plane and taken with sign + if the point and the origin lie on opposite sides of the plane and with sign − if they lie on the same side of the plane. Obviously, the deviation is zero for the points lying on the plane.

To obtain the deviation of a point $M_1(x_1, y_1, z_1)$ from a given plane, one should replace the current Cartesian coordinates $(x, y, z)$ on the left-hand side in the normal equation (4.6.1.7) of this plane by the coordinates of the point $M_1$:

$$\delta = x_1 \cos\alpha + y_1 \cos\beta + z_1 \cos\gamma - p = \mathbf{r}_1 \cdot \mathbf{N}^0 - p, \qquad (4.6.3.24)$$

where $\mathbf{N}^0 = (\cos\alpha, \cos\beta, \cos\gamma)$ is a unit vector and $\mathbf{r}_1$ is the position vector of the point $M_1(x_1, y_1, z_1)$. If the plane is given by the parametric equation (4.6.1.5), then the deviation of the point $M_1$ from the plane is equal to

$$\delta = \frac{[(\mathbf{r}_1 - \mathbf{r}_0)\mathbf{R}_1\mathbf{R}_2]}{|\mathbf{R}_1 \times \mathbf{R}_2|}. \qquad (4.6.3.25)$$

The *distance from a point to a plane* is defined as the nonnegative number $d$ equal to the absolute value of the deviation; i.e.,

$$d = |\delta| = |x_0 \cos\alpha + y_0 \cos\beta + z_0 \cos\gamma - p|. \qquad (4.6.3.26)$$

Let us write out some more representations of the distance for the cases in which the plane is given by the general equation (4.6.1.3) and the parametric equation (4.6.1.5):

$$d = \frac{|Ax_0 + By_0 + Cz_0 + D|}{\sqrt{A^2 + B^2 + C^2}} = \frac{|[(\mathbf{r}_1 - \mathbf{r}_0)\mathbf{R}_1\mathbf{R}_2]|}{|\mathbf{R}_1 \times \mathbf{R}_2|}. \qquad (4.6.3.27)$$

**Example 7.** Let us find the distance from the point $M(5, 1, -1)$ to the plane $x - 2y - 2z + 1 = 0$.
Using formula (4.6.3.27), we obtain the desired distance

$$d = \frac{|1 \cdot 5 + (-2) \cdot 1 + (-2) \cdot (-1) + 1|}{\sqrt{1^2 + (-2)^2 + (-2)^2}} = \frac{6}{3} = 2.$$

### 4.6.3-17. Distance between two parallel planes.

We consider two parallel planes given by the general equations $Ax + By + Cz + D_1 = 0$ and $Ax + By + Cz + D_2 = 0$. The distance between them is

$$d = \frac{|D_1 - D_2|}{\sqrt{A^2 + B^2 + C^2}}. \qquad (4.6.3.28)$$

### 4.6.3-18. Distance from point to line.

The distance from a point $M_0(x_0, y_0, z_0)$ to a line given by the canonical equation (4.6.2.3) is determined by the formula

$$d = \frac{|\mathbf{R} \times (\mathbf{r}_0 - \mathbf{r}_1)|}{|\mathbf{R}|} = \frac{\sqrt{\begin{vmatrix} m & n \\ y_1 - y_0 & z_1 - z_0 \end{vmatrix}^2 + \begin{vmatrix} n & l \\ z_1 - z_0 & x_1 - x_0 \end{vmatrix}^2 + \begin{vmatrix} l & m \\ x_1 - x_0 & y_1 - y_0 \end{vmatrix}^2}}{\sqrt{l^2 + m^2 + n^2}}. \qquad (4.6.3.29)$$

Note that the last formulas are significantly simplified if $\mathbf{R}$ is the unit vector ($l^2 + m^2 + n^2 = 1$).

**Remark.** The numerator of the fraction (4.6.3.29) is the area of the triangle spanned by the vectors $\mathbf{r}_0 - \mathbf{r}_1$ and $\mathbf{R}$, while the denominator of this fraction is the length of the base of the triangle. Hence the fraction itself is the altitude $d$ of this triangle.

**Example 8.** Let us find the distance from the point $M_0(3, 0, 4)$ to the line $x/1 = (y-1)/2 = z/2$. We use formula (4.6.3.29) to obtain the desired distance

$$d = \frac{\sqrt{\begin{vmatrix} 2 & 2 \\ 1-0 & 0-4 \end{vmatrix}^2 + \begin{vmatrix} 2 & 1 \\ 0-4 & 0-3 \end{vmatrix}^2 + \begin{vmatrix} 1 & 2 \\ 0-3 & 1-0 \end{vmatrix}^2}}{\sqrt{1^2 + 2^2 + 2^2}} = \frac{\sqrt{153}}{3}.$$

### 4.6.3-19. Distance between lines.

Consider two nonparallel lines (see Fig. 4.51) given in the canonical form

$$\frac{x - x_1}{l_1} = \frac{y - y_1}{m_1} = \frac{z - z_1}{n_1}, \qquad \text{or} \qquad (\mathbf{r} - \mathbf{r}_1) \times \mathbf{R}_1 = 0$$

and

$$\frac{x - x_2}{l_2} = \frac{y - y_2}{m_2} = \frac{z - z_2}{n_2}, \qquad \text{or} \qquad (\mathbf{r} - \mathbf{r}_2) \times \mathbf{R}_2 = 0.$$

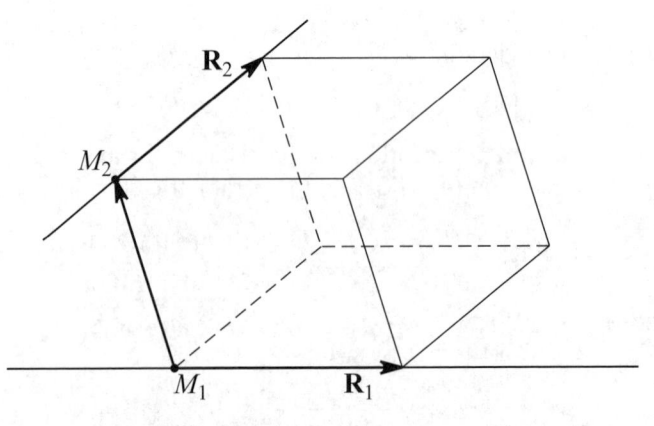

**Figure 4.51.** Distance between lines.

The distance between them can be calculated by the formula

$$d = \frac{|[(\mathbf{r}_1 - \mathbf{r}_2)\mathbf{R}_1\mathbf{R}_2]|}{|\mathbf{R}_1 \times \mathbf{R}_2|} = \frac{\pm \begin{vmatrix} x_1 - x_2 & y_1 - y_2 & z_1 - z_2 \\ l_1 & m_1 & n_1 \\ l_2 & m_2 & n_2 \end{vmatrix}}{\sqrt{\begin{vmatrix} l_1 & m_1 \\ l_2 & m_2 \end{vmatrix}^2 + \begin{vmatrix} m_1 & n_1 \\ m_2 & n_2 \end{vmatrix}^2 + \begin{vmatrix} n_1 & l_1 \\ n_2 & l_2 \end{vmatrix}^2}}. \quad (4.6.3.30)$$

The condition that the determinant in the numerator in (4.6.3.30) is zero is the *condition for the two lines in space to meet*.

Remark 1. The numerator of the fraction in (4.6.3.30) is the volume of the parallelepiped spanned by the vectors $\mathbf{r}_1 - \mathbf{r}_2$, $\mathbf{R}_1$, and $\mathbf{R}_2$, while the denominator of the fraction is the area of its base. Hence the fraction itself is the altitude $d$ of this parallelepiped.

Remark 2. If the lines are parallel (i.e., $l_1 = l_2 = l$, $m_1 = m_2 = m$, and $n_1 = n_2 = n$, or $\mathbf{R}_1 = \mathbf{R}_2 = \mathbf{R}$), then the distance between them should be calculated by formula (4.6.3.29) with $\mathbf{r}_0$ replaced by $\mathbf{r}_2$.

## 4.7. Quadric Surfaces (Quadrics)

### 4.7.1. Quadrics (Canonical Equations)

**4.7.1-1. Central surfaces.**

A segment joining two points of a surface is called a *chord*. If there exists a point in space, not necessarily lying on the surface, that bisects all chords passing through it, then the surface is said to be *central* and the point is called the *center* of the surface.

The equations listed below in Paragraphs 4.7.1-2 to 4.7.1-4 for central surfaces are given in *canonical form*; i.e., the center of a surface is at the origin, and the surface symmetry axes are the coordinate axes. Moreover, the coordinate planes are symmetry planes.

**4.7.1-2. Ellipsoid.**

An *ellipsoid* is a surface defined by the equation

$$\frac{x^2}{a^2} + \frac{y^2}{b^2} + \frac{z^2}{c^2} = 1, \quad (4.7.1.1)$$

where the numbers $a$, $b$, and $c$ are the lengths of the segments called the *semiaxes* of the ellipsoid (see Fig 4.52a).

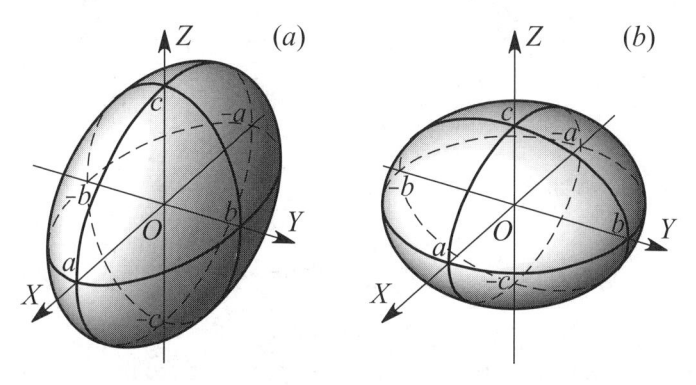

**Figure 4.52.** Triaxial ellipsoid (*a*) and spheroid (*b*).

If $a \neq b \neq c$, then the ellipsoid is said to be *triaxial*, or *scalene*. If $a = b \neq c$, then the ellipsoid is called a *spheroid*; it can be obtained by rotating the ellipse $x^2/a^2 + z^2/c^2 = 1$, $y = 0$ lying in the plane $OXZ$ about the axis $OZ$ (see Fig. 4.52b). If $a = b > c$, then the ellipsoid is an *oblate spheroid*, and if $a = b < c$, then the ellipsoid is a *prolate spheroid*. If $a = b = c$, then the ellipsoid is the sphere of radius $a$ given by the equation $x^2 + y^2 + z^2 = a^2$.

An arbitrary plane section of an ellipsoid is an ellipse (in a special case, a circle). The volume of an ellipsoid is equal to $V = \frac{4}{3}\pi abc$.

Remark. About the sphere, see also Paragraph 3.2.3-3.

### 4.7.1-3. Hyperboloids.

A *one-sheeted hyperboloid* is a surface defined by the equation

$$\frac{x^2}{a^2} + \frac{y^2}{b^2} - \frac{z^2}{c^2} = 1, \qquad (4.7.1.2)$$

where $a$ and $b$ are the *real semiaxes* and $c$ is the *imaginary semiaxis* (see Fig. 4.53a).

A *two-sheeted hyperboloid* is a surface defined by the equation

$$\frac{x^2}{a^2} + \frac{y^2}{b^2} - \frac{z^2}{c^2} = -1, \qquad (4.7.1.3)$$

where $c$ is the *real semiaxis* and $a$ and $b$ are the *imaginary semiaxes* (see Fig 4.53b).

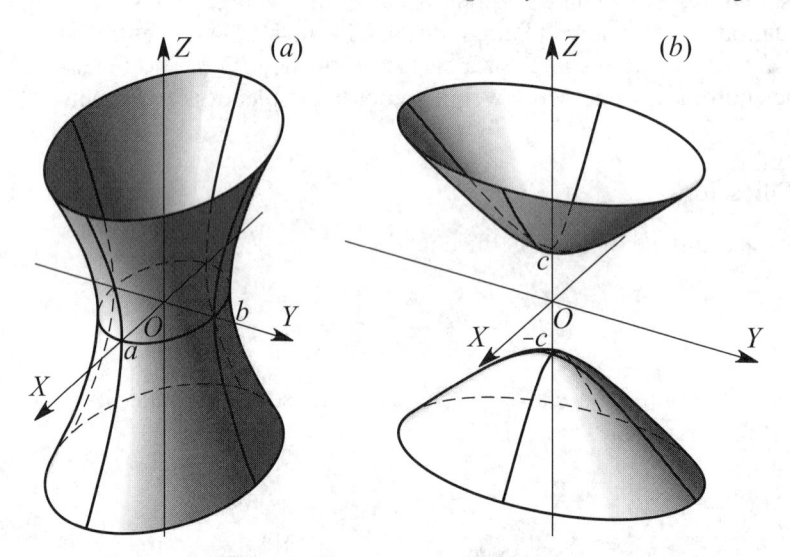

**Figure 4.53.** One-sheeted (a) and two-sheeted (b) hyperboloids.

A hyperboloid approaches the surface

$$\frac{x^2}{a^2} + \frac{y^2}{b^2} - \frac{z^2}{c^2} = 0,$$

which is called an *asymptotic cone*, infinitely closely.

A plane passing through the axis $OZ$ intersects each of the hyperboloids (4.7.1.2) and (4.7.1.3) in two hyperbolas and the asymptotic cone in two straight lines, which are the

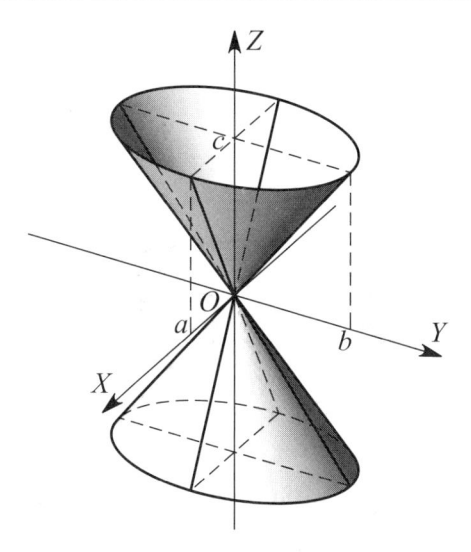

**Figure 4.54.** A cone.

asymptotes of these hyperbolas. The section of a hyperboloid by a plane parallel to $OXY$ is an ellipse. The section of a one-sheeted hyperboloid by the plane $z = 0$ is an ellipse, which is called the *gorge* or *throat* ellipse.

For $a = b$, we deal with the *hyperboloid of revolution* obtained by rotating a hyperbola with semiaxes $a$ and $c$ about its focal axis $2c$ (which is an imaginary axis for a one-sheeted hyperboloid and a real axis for a two-sheeted hyperboloid). If $a = b = c$, then the hyperboloid of revolution is said to be *right*, and its sections by the planes $OXZ$ and $OYZ$ are equilateral hyperbolas.

A one-sheeted hyperboloid is a ruled surface (see Paragraph 4.7.1-6).

### 4.7.1-4. Cone.

A *cone* is a surface defined by the equation

$$\frac{x^2}{a^2} + \frac{y^2}{b^2} - \frac{z^2}{c^2} = 0. \tag{4.7.1.4}$$

The cone (see Fig. 4.54) defined by (4.7.1.4) has vertex at the origin, and for its base we can take the ellipse with semiaxes $a$ and $b$ in the plane perpendicular to the axis $OZ$ at the distance $c$ from the origin. This cone is the asymptotic cone for the hyperboloids (4.7.1.2) and (4.7.1.3). For $a = b$, we obtain a *right circular cone*.

A cone is a ruled surface (see Paragraph 4.7.1-6).

Remark. About the cone, see also Paragraph 3.2.3-2.

### 4.7.1-5. Paraboloids.

In contrast to the surfaces considered above, paraboloids are not central surfaces. For the equations listed below, the vertex of a paraboloid lies at the origin, the axis $OZ$ is the symmetry axis, and the planes $OXZ$ and $OYZ$ are symmetry planes.

An *elliptic paraboloid* (see Fig 4.55a) is a surface defined by the equation

$$\frac{x^2}{p} + \frac{y^2}{q} = 2z, \tag{4.7.1.5}$$

where $p > 0$ and $q > 0$ are parameters.

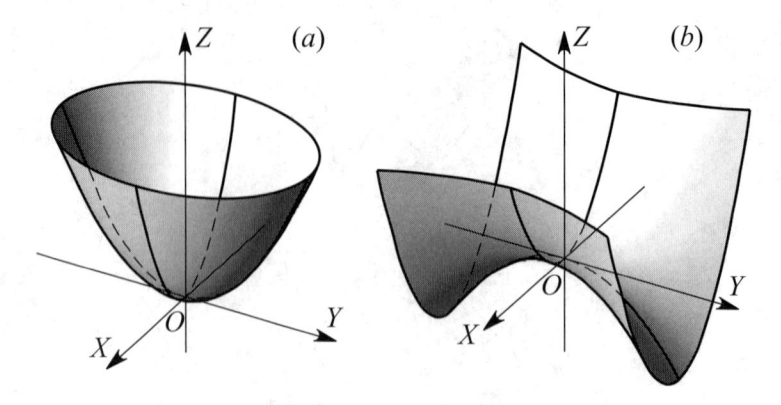

**Figure 4.55.** Elliptic (*a*) and hyperbolic (*b*) paraboloids.

The sections of an elliptic paraboloid by planes parallel to the axis $OZ$ are parabolas, and the sections by planes parallel to the plane $OXY$ are ellipses. For example, let the parabola $x^2 = 2pz$, $y = 0$, obtained by the section of an elliptic paraboloid by the plane $OXZ$ be fixed and used as the directrix, and let the parabola $x^2 = 2qz$, $x = 0$, obtained by the section of the elliptic paraboloid by the plane $OYZ$ be movable and used as the generator. Then the paraboloid can be obtained by parallel translation of the movable parabola (the generator) in a given direction along the fixed parabola (the directrix).

If $p = q$, then we have a *paraboloid of revolution*, which is obtained by rotating the parabola $2pz = x^2$ lying in the plane $OXZ$ about its axis.

The volume of the part of an elliptic paraboloid cut by the plane perpendicular to its axis at a height $h$ is equal to $V = \frac{1}{2}\pi abh$, i.e., half the volume of the elliptic cylinder with the same base and altitude.

A *hyperbolic paraboloid* (see Fig 4.55*b*) is a surface defined by the equation

$$\frac{x^2}{p} - \frac{y^2}{q} = 2z, \qquad (4.7.1.6)$$

where $p > 0$ and $q > 0$ are parameters.

The sections of a hyperbolic paraboloid by planes parallel to the axis $OZ$ are parabolas, and the sections by planes parallel to the plane $OXY$ are hyperbolas. For example, let the parabola $x^2 = 2pz$, $y = 0$, obtained by the section of the hyperbolic paraboloid by the plane $OXZ$ be fixed and used as the directrix, and let the parabola $x^2 = -2qz$, $x = 0$, obtained by the section of the hyperbolic paraboloid by the plane $OYZ$ be movable and used as the generator. Then the paraboloid can be obtained by parallel translation of the movable parabola (the generator) in a given direction along the fixed parabola (the directrix).

A hyperbolic paraboloid is a ruled surface (see Paragraph 4.7.1-6).

### 4.7.1-6. Rulings of ruled surfaces.

A *ruled surface* is a surface swept out by a moving line in space. The straight lines forming a ruled surface are called *rulings*. Examples of ruled surfaces include the cone (see Paragraph 3.2.3-2 and 4.7.1-4), the cylinder (see Paragraph 3.2.3-1), the one-sheeted hyperboloid (see Paragraph 4.7.1-3), and the hyperbolic paraboloid (see Paragraph 4.7.1-5).

The cone (4.7.1.4) has one family of rulings,

$$\alpha x = \beta y, \quad \frac{\sqrt{\alpha^2 a^2 + \beta^2 b^2}}{ab} x = \frac{\beta}{c} z.$$

*Properties of rulings of the cone:*
1. There is a unique ruling through each point of the cone.
2. Two arbitrary distinct rulings of the cone meet at the point $O(0,0,0)$.
3. Three pairwise distinct rulings of the cone are not parallel to any plane.

The one-sheeted hyperboloid (4.7.1.2) has two families of rulings:

$$\alpha\left(\frac{x}{a} + \frac{z}{c}\right) = \beta\left(1 + \frac{y}{b}\right), \quad \beta\left(\frac{x}{a} - \frac{z}{c}\right) = \alpha\left(1 - \frac{y}{b}\right);$$
$$\gamma\left(\frac{x}{a} + \frac{z}{c}\right) = \delta\left(1 - \frac{y}{b}\right), \quad \delta\left(\frac{x}{a} - \frac{z}{c}\right) = \gamma\left(1 + \frac{y}{b}\right).$$
(4.7.1.7)

One of these families is shown in Fig. 4.56a.

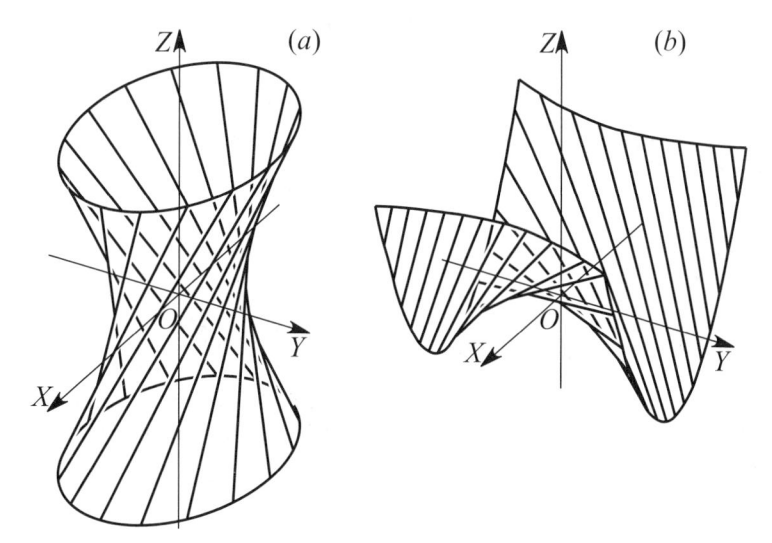

**Figure 4.56.** Families of rulings for one-sheeted hyperboloid (*a*) and for hyperbolic paraboloid (*b*).

*Properties of rulings of the one-sheeted hyperboloid:*
1. In either family, there is a unique ruling through each point of the one-sheeted hyperboloid.
2. Any two rulings in different families lie in a single plane.
3. Any two distinct rulings in the same family are skew.
4. Three distinct rulings in the same family are not parallel to any plane.

The hyperbolic paraboloid (4.7.1.6) has two families of rulings:

$$\alpha\left(\frac{x}{\sqrt{p}} + \frac{y}{\sqrt{q}}\right) = 2\beta, \quad \beta\left(\frac{x}{\sqrt{p}} - \frac{y}{\sqrt{q}}\right) = \alpha z;$$
$$\gamma\left(\frac{x}{\sqrt{p}} + \frac{y}{\sqrt{q}}\right) = \delta z, \quad \delta\left(\frac{x}{\sqrt{p}} - \frac{y}{\sqrt{q}}\right) = 2\gamma.$$
(4.7.1.8)

One of these families is shown in Fig. 4.56b.

*Properties of rulings of a hyperbolic paraboloid:*
1. In either family, there is a unique ruling through each point of the hyperbolic paraboloid.
2. Any two rulings in different families lie in a single plane and meet.
3. Any two distinct rulings in the same family are skew.
4. All rulings in either family are parallel to a single plane.

### 4.7.2. Quadrics (General Theory)

**4.7.2-1. General equation of quadric.**

A *quadric* is a set of points in three-dimensional space whose coordinates in the rectangular Cartesian coordinate system satisfy a second-order algebraic equation

$$a_{11}x^2 + a_{22}y^2 + a_{33}z^2 + 2a_{12}xy + 2a_{13}xz + 2a_{23}yz$$
$$+ 2a_{14}x + 2a_{24}y + 2a_{34}z + a_{44} = 0, \qquad (4.7.2.1)$$

or

$$(a_{11}x + a_{12}y + a_{13}z + a_{14})x + (a_{21}x + a_{22}y + a_{23}z + a_{24})y$$
$$+ (a_{31}x + a_{32}y + a_{33}z + a_{34})z + a_{41}x + a_{42}y + a_{43}z + a_{44} = 0,$$

where $a_{ij} = a_{ji}$ ($i, j = 1, 2, 3, 4$). If equation (4.7.2.1) does not define a real geometric object, then one says that this equation defines an *imaginary quadric*. Equation (4.7.2.1) in vector form reads

$$(\mathbf{A}\mathbf{r}) \cdot \mathbf{r} + 2\mathbf{a} \cdot \mathbf{r} + a_{44} = 0, \qquad (4.7.2.2)$$

where $\mathbf{A}$ is the affinor with coordinates $A^i_j = a_{ij}$ and $\mathbf{a}$ is the vector with coordinates $a_i = a_{i4}$.

**4.7.2-2. Classification of quadrics.**

There exists a rectangular Cartesian coordinate system in which equation (4.7.2.1), depending on the coefficients, has 1 of 17 *canonical forms*, each of which is associated with a certain class of quadrics (see Table 4.3).

**4.7.2-3. Invariants of quadrics.**

The shape of a quadric can be determined by using four *invariants* and two *semi-invariants* without reducing equation (4.7.2.1) to canonical form.

The main invariants are the quantities

$$S = a_{11} + a_{22} + a_{33}, \qquad (4.7.2.3)$$

$$T = \begin{vmatrix} a_{11} & a_{12} \\ a_{21} & a_{22} \end{vmatrix} + \begin{vmatrix} a_{11} & a_{13} \\ a_{31} & a_{33} \end{vmatrix} + \begin{vmatrix} a_{22} & a_{23} \\ a_{23} & a_{33} \end{vmatrix}, \qquad (4.7.2.4)$$

$$\delta = \begin{vmatrix} a_{11} & a_{12} & a_{13} \\ a_{12} & a_{22} & a_{23} \\ a_{13} & a_{23} & a_{33} \end{vmatrix}, \qquad (4.7.2.5)$$

$$\Delta = \begin{vmatrix} a_{11} & a_{12} & a_{13} & a_{14} \\ a_{12} & a_{22} & a_{23} & a_{24} \\ a_{13} & a_{23} & a_{33} & a_{34} \\ a_{14} & a_{24} & a_{34} & a_{44} \end{vmatrix}, \qquad (4.7.2.6)$$

whose values are preserved under parallel translations and rotations of the coordinate axes.

TABLE 4.3
Canonical equations and classes of quadrics

| No. | Surface | Canonical equation | Type | Class |
|---|---|---|---|---|
| | **Irreducible surfaces** | | | |
| 1 | Ellipsoid | $\dfrac{x^2}{a^2}+\dfrac{y^2}{b^2}+\dfrac{z^2}{c^2}=1$ | Elliptic | Nondegenerate |
| 2 | Imaginary ellipsoid | $\dfrac{x^2}{a^2}+\dfrac{y^2}{b^2}+\dfrac{z^2}{c^2}=-1$ | | |
| 3 | One-sheeted hyperboloid | $\dfrac{x^2}{a^2}+\dfrac{y^2}{b^2}-\dfrac{z^2}{c^2}=1$ | Hyperbolic | |
| 4 | Two-sheeted hyperboloid | $\dfrac{x^2}{a^2}+\dfrac{y^2}{b^2}-\dfrac{z^2}{c^2}=-1$ | | |
| 5 | Elliptic paraboloid | $\dfrac{x^2}{p}+\dfrac{y^2}{q}=2z$ | Parabolic $(p>0,\,q>0)$ | |
| 6 | Hyperbolic paraboloid | $\dfrac{x^2}{p}-\dfrac{y^2}{q}=2z$ | | |
| 7 | Elliptic cylinder | $\dfrac{x^2}{a^2}+\dfrac{y^2}{b^2}=1$ | Cylindrical | Degenerate |
| 8 | Imaginary elliptic cylinder | $\dfrac{x^2}{a^2}+\dfrac{y^2}{b^2}=-1$ | | |
| 9 | Hyperbolic cylinder | $\dfrac{x^2}{a^2}-\dfrac{y^2}{b^2}=1$ | | |
| 10 | Parabolic cylinder | $y^2=2px$ | | |
| 11 | Real cone | $\dfrac{x^2}{a^2}+\dfrac{y^2}{b^2}-\dfrac{z^2}{c^2}=0$ | Conic | |
| 12 | Imaginary cone with real vertex | $\dfrac{x^2}{a^2}+\dfrac{y^2}{b^2}+\dfrac{z^2}{c^2}=0$ | | |
| | **Reducible surfaces** | | | |
| 13 | Pair of real intersecting planes | $\dfrac{x^2}{a^2}-\dfrac{y^2}{b^2}=0$ | Pairs of planes | Degenerate |
| 14 | Pair of imaginary planes intersecting in a real straight line | $\dfrac{x^2}{a^2}+\dfrac{y^2}{b^2}=0$ | | |
| 15 | Pair of real parallel planes | $x^2=a^2$ | | |
| 16 | Pair of imaginary parallel planes | $x^2=-a^2$ | | |
| 17 | Pair of real coinciding planes | $x^2=0$ | | |

The semi-invariants are the quantities

$$\sigma = \Delta_{11} + \Delta_{22} + \Delta_{33}, \tag{4.7.2.7}$$

$$\Sigma = \begin{vmatrix} a_{11} & a_{14} \\ a_{41} & a_{44} \end{vmatrix} + \begin{vmatrix} a_{22} & a_{24} \\ a_{42} & a_{44} \end{vmatrix} + \begin{vmatrix} a_{33} & a_{34} \\ a_{44} & a_{44} \end{vmatrix}, \tag{4.7.2.8}$$

whose values are preserved only under rotations of the coordinate axes. Here $\Delta_{ij}$ is the cofactor of the entry $a_{ij}$ in $\Delta$.

The classification of quadrics based on the invariants $S$, $T$, $\delta$, and $\Delta$ and the semi-invariants $\sigma$ and $\Sigma$ is given in Tables 4.4 and 4.5.

TABLE 4.4
Classification of quadrics (central surfaces $\delta \neq 0$)

| Class | | $S\delta > 0$ or $T > 0$ | $S\delta > 0$ and $T > 0$ but not both |
|---|---|---|---|
| Nondegenerate surfaces ($\Delta \neq 0$) | $\Delta < 0$ | Ellipsoid $\dfrac{x^2}{a^2} + \dfrac{y^2}{b^2} + \dfrac{z^2}{c^2} = 1$ | Two-sheeted hyperboloid $\dfrac{x^2}{a^2} + \dfrac{y^2}{b^2} - \dfrac{z^2}{c^2} = -1$ |
| | $\Delta > 0$ | Imaginary ellipsoid $\dfrac{x^2}{a^2} + \dfrac{y^2}{b^2} + \dfrac{z^2}{c^2} = -1$ | One-sheeted hyperboloid $\dfrac{x^2}{a^2} + \dfrac{y^2}{b^2} - \dfrac{z^2}{c^2} = 1$ |
| Degenerate surfaces ($\Delta = 0$) | | Imaginary cone with real vertex $\dfrac{x^2}{a^2} + \dfrac{y^2}{b^2} + \dfrac{z^2}{c^2} = 0$ | Real cone $\dfrac{x^2}{a^2} + \dfrac{y^2}{b^2} - \dfrac{z^2}{c^2} = 0$ |

TABLE 4.5
Classification of quadrics (central surfaces $\delta = 0$)

| Class and type | | | $T > 0$ | | $T < 0$ | $T = 0$ | |
|---|---|---|---|---|---|---|---|
| Nondegenerate surfaces ($\Delta \neq 0$) | $\Delta > 0$ | | | | Hyperbolic paraboloid $\dfrac{x^2}{p} - \dfrac{y^2}{q} = 2z$ | | |
| | $\Delta < 0$ | | Elliptic paraboloid $\dfrac{x^2}{p} + \dfrac{y^2}{q} = 2z$ | | | | |
| Degenerate surfaces ($\Delta = 0$) | Cylindrical surfaces ($\sigma \neq 0$) | | Elliptic cylinder | | Hyperbolic cylinder $\dfrac{x^2}{a^2} - \dfrac{y^2}{b^2} = 1$ | Parabolic cylinder $y^2 = 2px$ | |
| | | | Imaginary ($\sigma S > 0$) $\dfrac{x^2}{a^2} + \dfrac{y^2}{b^2} = -1$ | Real ($\sigma S < 0$) $\dfrac{x^2}{a^2} + \dfrac{y^2}{b^2} = 1$ | | | |
| | Reducible surfaces ($\sigma = 0$) | | Pair of imaginary planes, intersecting in a real straight line $\dfrac{x^2}{a^2} + \dfrac{y^2}{b^2} = 0$ | | Pair of real intersecting planes $\dfrac{x^2}{a^2} - \dfrac{y^2}{b^2} = 0$ | Pair of real reducible planes ($\Sigma = 0$) $x^2 = 0$ | |
| | | | | | | Pair of imaginary parallel planes ($\Sigma > 0$) $x^2 = -a^2$ | Pair of real parallel planes ($\Sigma > 0$) $x^2 = a^2$ |

### 4.7.2-4. Characteristic quadratic form of quadric.

The *characteristic quadratic form*

$$F(x, y, z) = a_{11}x^2 + a_{22}y^2 + a_{33}z^2 + 2a_{12}xy + 2a_{13}xz + a_{23}yz \qquad (4.7.2.9)$$

corresponding to equation (4.7.2.1) and its *characteristic equation*

$$\begin{vmatrix} a_{11}-\lambda & a_{12} & a_{13} \\ a_{12} & a_{22}-\lambda & a_{23} \\ a_{13} & a_{23} & a_{33}-\lambda \end{vmatrix} = 0, \quad \text{or} \quad \lambda^3 - S\lambda^2 + T\lambda - \delta = 0 \qquad (4.7.2.10)$$

permit studying the main properties of quadrics.

The roots $\lambda_1$, $\lambda_2$, and $\lambda_3$ of the characteristic equation (4.7.2.10) are the eigenvalues of the real symmetric matrix $[a_{ij}]$ and hence are always real. The invariants $S$, $T$, and $\delta$ can be expressed in terms of the roots $\lambda_1$, $\lambda_2$, and $\lambda_3$ as follows:

$$S = \lambda_1 + \lambda_2 + \lambda_3, \quad T = \lambda_1\lambda_2 + \lambda_1\lambda_3 + \lambda_2\lambda_3, \quad \delta = \lambda_1\lambda_2\lambda_3. \tag{4.7.2.11}$$

### 4.7.2-5. Diameters and diameter plane.

The locus of midpoints of parallel chords of a quadric is the *diameter plane* conjugate to these chords (or the direction of these chords). The diameter plane conjugate to the chords with direction cosines $\cos\alpha$, $\cos\beta$, and $\cos\gamma$ is determined by the equation

$$(a_{11}x + a_{12}y + a_{13}z + a_{14})\cos\alpha + (a_{21}x + a_{22}y + a_{23}z + a_{24})\cos\beta$$
$$+ (a_{31}x + a_{32}y + a_{33}z + a_{34})\cos\gamma = 0. \tag{4.7.2.12}$$

The line in which two diameter planes meet is called the *diameter* conjugate to the family of planes parallel to the conjugate chords of these diameter planes. The equation of the diameter conjugate to the family of planes with given direction cosines $\cos l$, $\cos m$, and $\cos n$ of the normals has the form

$$\frac{a_{11}x+a_{12}y+a_{13}z+a_{14}}{\cos l} = \frac{a_{21}x+a_{22}y+a_{23}z+a_{24}}{\cos m} = \frac{a_{31}x+a_{32}y+a_{33}z+a_{34}}{\cos n}. \tag{4.7.2.13}$$

If a surface is central (for $\delta \neq 0$), then all diameters are concurrent at a single point, called the *center* of the surface. For $\delta = 0$, all diameters are parallel or lie in a single plane.

For central quadrics, the coordinates $x_0$, $y_0$, $z_0$ of the center are determined by the system of equations

$$a_{11}x_0 + a_{12}y_0 + a_{13}z_0 + a_{14} = 0,$$
$$a_{21}x_0 + a_{22}y_0 + a_{23}z_0 + a_{24} = 0,$$
$$a_{31}x_0 + a_{32}y_0 + a_{33}z_0 + a_{34} = 0,$$

whence it follows that

$$x_0 = -\frac{1}{\delta}\begin{vmatrix} a_{14} & a_{12} & a_{13} \\ a_{24} & a_{22} & a_{23} \\ a_{34} & a_{32} & a_{33} \end{vmatrix}, \quad y_0 = -\frac{1}{\delta}\begin{vmatrix} a_{11} & a_{14} & a_{13} \\ a_{21} & a_{24} & a_{23} \\ a_{31} & a_{34} & a_{33} \end{vmatrix}, \quad z_0 = -\frac{1}{\delta}\begin{vmatrix} a_{11} & a_{12} & a_{14} \\ a_{21} & a_{22} & a_{24} \\ a_{31} & a_{32} & a_{34} \end{vmatrix}. \tag{4.7.2.14}$$

For a central surface, the translation of the origin to its center (4.7.2.14) transforms the equation of the surface to the form

$$a_{11}x_1^2 + a_{22}y_1^2 + a_{33}z_1^2 + 2a_{12}x_1y_1 + 2a_{13}x_1z_1 + a_{23}y_1z_1 + \frac{\Delta}{\delta} = 0, \tag{4.7.2.15}$$

where $x_1$, $y_1$, and $z_1$ are the coordinates in the new system.

Three diameters of a central quadric are said to be *conjugate* if each of them is conjugate to the plane of the other two diameters.

### 4.7.2-6. Principal planes and principal axes.

A *principal plane* of a quadric is a diameter plane perpendicular to the chords conjugate to it. A principal plane is a plane of symmetry of the quadric. Any quadric has at least one principal plane and at least two symmetry planes. Cylinders have symmetry planes perpendicular to the generators; these are not principal planes. Each nondegenerate quadric has at least two principal planes perpendicular to each other. Any central quadric has at least three principal planes, three of which are pairwise perpendicular.

A *principal axis* is a diameter that is the line of intersection of two principal planes. A principal axis is a symmetry axis of the quadric. If a quadric has two principal axes, then it also has the third axis perpendicular to the first two principal axes. Any nondegenerate quadric has at least one principal axis. A central quadric has at least three pairwise perpendicular principal axes, which are normal to its principal diameter planes; three of the principal axes of a central surface are pairwise perpendicular.

### 4.7.2-7. Transformation of equation of quadric to canonical form.

Translating the origin and rotating the coordinate axes so that the normal to each of the pairwise perpendicular symmetry planes of a surface becomes parallel to one of the new coordinate axes, we can transform equation (4.7.2.1) of any nondegenerate quadric to one of the forms listed in Table 4.6.

TABLE 4.6
Expression of the parameters of the main quadrics via the invariants $T$, $\delta$, and $\Delta$ and the roots $\lambda_1$, $\lambda_2$, and $\lambda_3$ of the characteristic equation

| Surface | Canonical equation | The parameters $a, b, c, p, q$ expressed via the invariants $T, \delta, \Delta$ and the roots $\lambda_1, \lambda_2, \lambda_3$ of the characteristic equation | Remarks |
|---|---|---|---|
| Ellipsoid | $\dfrac{x^2}{a^2} + \dfrac{y^2}{b^2} + \dfrac{z^2}{c^2} = 1$ | $a^2 = -\dfrac{1}{\lambda_3}\dfrac{\Delta}{\delta}$, $\quad b^2 = -\dfrac{1}{\lambda_2}\dfrac{\Delta}{\delta}$, $\quad c^2 = -\dfrac{1}{\lambda_1}\dfrac{\Delta}{\delta}$, $\quad \delta = \lambda_1\lambda_2\lambda_3$ | $a \geq b \geq c$, $\lambda_1 \geq \lambda_2 \geq \lambda_3 > 0$ |
| One-sheeted hyperboloid | $\dfrac{x^2}{a^2} + \dfrac{y^2}{b^2} - \dfrac{z^2}{c^2} = 1$ | $a^2 = -\dfrac{1}{\lambda_2}\dfrac{\Delta}{\delta}$, $\quad b^2 = -\dfrac{1}{\lambda_1}\dfrac{\Delta}{\delta}$, $\quad c^2 = \dfrac{1}{\lambda_3}\dfrac{\Delta}{\delta}$, $\quad \delta = \lambda_1\lambda_2\lambda_3$ | $a \geq b$, $\lambda_1 \geq \lambda_2 > 0 > \lambda_3$ |
| Two-sheeted hyperboloid | $\dfrac{x^2}{a^2} + \dfrac{y^2}{b^2} - \dfrac{z^2}{c^2} = -1$ | $a^2 = \dfrac{1}{\lambda_3}\dfrac{\Delta}{\delta}$, $\quad b^2 = \dfrac{1}{\lambda_2}\dfrac{\Delta}{\delta}$, $\quad c^2 = -\dfrac{1}{\lambda_1}\dfrac{\Delta}{\delta}$, $\quad \delta = \lambda_1\lambda_2\lambda_3$ | $a \geq b$, $\lambda_1 > 0 > \lambda_2 \geq \lambda_3$ |
| Elliptic paraboloid | $\dfrac{x^2}{p} + \dfrac{y^2}{q} = 2z$ | $p = \dfrac{1}{\lambda_2}\sqrt{-\dfrac{\Delta}{T}}$, $\quad q = \dfrac{1}{\lambda_1}\sqrt{-\dfrac{\Delta}{T}}$, $\quad T = \lambda_1\lambda_2$ | $p > 0$, $q > 0$, $\lambda_1 \geq \lambda_2 > \lambda_3 = 0$ |
| Hyperbolic paraboloid | $\dfrac{x^2}{p} - \dfrac{y^2}{q} = 2z$ | $p = \dfrac{1}{\lambda_1}\sqrt{-\dfrac{\Delta}{T}}$, $\quad q = -\dfrac{1}{\lambda_3}\sqrt{-\dfrac{\Delta}{T}}$, $\quad T = \lambda_1\lambda_3$ | $p > 0$, $q > 0$, $\lambda_1 > \lambda_2 = 0 > \lambda_3$ |

### 4.7.2-8. Tangent planes and normals to quadric.

The *equation of the tangent plane to the surface* (4.7.2.1) at a point $M(x_1, y_1, z_1)$ has the form

$$a_{11}x_1 x + a_{22}y_1 y + a_{33}z_1 z + a_{12}(y_1 x + x_1 y) + a_{13}(z_1 x + x_1 z)$$
$$+ a_{23}(y_1 z + z_1 y) + a_{14}(x_1 + x) + a_{24}(y_1 + y) + a_{34}(z_1 + z) + a_{44} = 0.$$

or

$$(a_{11}x_1 + a_{12}y_1 + a_{13}z_1 + a_{14})x + (a_{21}x_1 + a_{22}y_1 + a_{23}z_1 + a_{24})y$$
$$+ (a_{31}x_1 + a_{32}y_1 + a_{33}z_1 + a_{34})z + a_{41}x_1 + a_{42}y_1 + a_{43}z_1 + a_{44} = 0.$$

The *equation of the normal to the surface* (4.7.2.1) at a point $M(x_1, y_1, z_1)$ has the form

$$\frac{x - x_1}{a_{11}x_1 + a_{12}y_1 + a_{13}z_1 + a_{14}} = \frac{y - y_1}{a_{21}x_1 + a_{22}y_1 + a_{23}z_1 + a_{24}} = \frac{z - z_1}{a_{31}x_1 + a_{32}y_1 + a_{33}z_1 + a_{34}}.$$

## References for Chapter 4

**Alexander, D. C. and Koeberlein, G. M.**, *Elementary Geometry for College Students, 3rd Edition*, Houghton Mifflin Company, Boston, 2002.
**Blau, H. I.**, *Foundations of Plane Geometry*, Prentice Hall, Englewood Cliffs, New Jersey, 2002.
**Chauvenet, W.**, *A Treatise on Elementary Geometry*, Adamant Media Corporation, Boston, 2001.
**Efimov, N. V.**, *Higher Geometry*, Mir Publishers, Moscow, 1980.
**Fogiel, M. (Editor)**,*High School Geometry Tutor, 2nd Edition*, Research & Education Association, Englewood Cliffs, New Jersey, 2003.
**Fuller, G. and Tarwater, D.**, *Analytic Geometry, 7th Edition*, Addison Wesley, Boston, 1993.
**Gustafson, R. D. and Frisk, P. D.**, *Elementary Geometry, 3rd Edition*, Wiley, New York, 1991.
**Hartshorne, R.**, *Geometry: Euclid and Beyond*, Springer, New York, 2005.
**Jacobs, H. R.**, *Geometry, 2nd Edition*, W. H. Freeman & Company, New York, 1987.
**Jacobs, H. R.**, *Geometry: Seeing, Doing, Understanding, 3rd Edition*, W. H. Freeman & Company, New York, 2003.
**Jurgensen, R. and Brown, R. G.**, *Geometry*, McDougal Littell/Houghton Mifflin, Boston, 2000.
**Kay, D.**, *College Geometry: A Discovery Approach, 2nd Edition*, Addison Wesley, Boston, 2000.
**Kletenik, D. V.**, *Problems in Analytic Geometry*, University Press of the Pacific, Honolulu, Hawaii, 2002.
**Kostrikin, A. I. (Editor)**, *Exercises in Algebra: A Collection of Exercises in Algebra, Linear Algebra and Geometry*, Gordon & Breach, New York, 1996.
**Kostrikin, A. I. and Shafarevich, I. R. (Editors)**, *Algebra I: Basic Notions of Algebra*, Springer-Verlag, Berlin, 1990.
**Leff, L. S.**, *Geometry the Easy Way, 3rd Edition*, Barron's Educational Series, Hauppauge, New York, 1997.
**Moise, E.**, *Elementary Geometry from an Advanced Standpoint, 3rd Edition*, Addison Wesley, Boston, 1990.
**Musser, G. L., Burger, W. F., and Peterson, B. E.**, *Mathematics for Elementary Teachers: A Contemporary Approach, 6th Edition*, Wiley, New York, 2002.
**Musser, G. L., Burger, W. F., and Peterson, B. E.**, *Essentials of Mathematics for Elementary Teachers: A Contemporary Approach, 6th Edition*, Wiley, New York, 2003.
**Musser, G. L. and Trimpe, L. E.**, *College Geometry: A Problem Solving Approach with Applications*, Prentice Hall, Englewood Cliffs, New Jersey, 1994.
**Postnikov, M. M.**, *Lectures in Geometry, Semester I, Analytic Geometry*, Mir Publishers, Moscow, 1982.
**Privalov, I. I.**, *Analytic Geometry, 32nd Edition [in Russian]*, Lan, Moscow, 2003.
**Riddle, D. R.**, *Analytic Geometry, 6th Edition*, Brooks Cole, Stamford, 1995.
**Roe, J.**, *Elementary Geometry*, Oxford University Press, Oxford, 1993.
**Schultze, A. and Sevenoak, F. L.**, *Plane and Solid Geometry*, Adamant Media Corporation, Boston, 2004.
**Suetin, P. K., Kostrikin, A. I., and Manin, Yu. I.**, *Linear Algebra and Geometry*, Gordon & Breach, New York, 1997.
**Tussy, A. S.**, *Basic Geometry for College Students: An Overview of the Fundamental Concepts of Geometry, 2nd Edition*, Brooks Cole, Stamford, 2002.
**Vygodskii, M. Ya.**, *Mathematical Handbook: Higher Mathematics*, Mir Publishers, Moscow, 1971.
**Weeks, A. W. and Adkins, J. B.**, *A Course in Geometry: Plane and Solid*, Bates Publishing, Sandwich, 1982.
**Woods, F. S.**, *Higher Geometry: An Introduction to Advanced Methods in Analytic Geometry, Phoenix Edition*, Dover Publications, New York, 2005.

# Chapter 5
# Algebra

## 5.1. Polynomials and Algebraic Equations
### 5.1.1. Polynomials and Their Properties

**5.1.1-1. Definition of polynomial.**

A *polynomial of degree* $n$ of a scalar variable $x$ is an expression of the form

$$f(x) \equiv a_n x^n + a_{n-1} x^{n-1} + \cdots + a_1 x + a_0 \quad (a_n \neq 0), \tag{5.1.1.1}$$

where $a_0, \ldots, a_n$ are real or complex numbers ($n = 0, 1, 2, \ldots$). Polynomials of degree zero are nonzero numbers.

Two polynomials are *equal* if they have the same coefficients of like powers of the variable.

**5.1.1-2. Main operations over polynomials.**

1°. The *sum (difference)* of two polynomials $f(x)$ of degree $n$ and $g(x)$ of degree $m$ is the polynomial of degree $l \leq \max\{n, m\}$ whose coefficient of each power of $x$ is equal to the sum (difference) of the coefficients of the same power of $x$ in $f(x)$ and $g(x)$, i.e. if

$$g(x) \equiv b_m x^m + b_{m-1} x^{m-1} + \cdots + b_1 x + b_0, \tag{5.1.1.2}$$

then the sum (difference) of polynomials (5.1.1.1) and (5.1.1.2) is

$$f(x) \pm g(x) = c_l x^l + c_{l-1} x^{l-1} + \cdots + c_1 x + c_0, \quad \text{where} \quad c_k = a_k \pm b_k \quad (k = 0, 1, \ldots, l).$$

If $n > m$ then $b_{m+1} = \ldots = b_n = 0$; if $n < m$ then $a_{n+1} = \ldots = a_m = 0$.

2°. To *multiply* a polynomial $f(x)$ of degree $n$ by a polynomial $g(x)$ of degree $m$, one should multiply each term in $f(x)$ by each term in $g(x)$, add the products, and collect similar terms. The degree of the resulting polynomial is $n+m$. The product of polynomials (5.1.1.1) and (5.1.1.2) is

$$f(x)g(x) = c_{n+m} x^{n+m} + c_{n+m-1} x^{n+m-1} + \cdots + c_1 x + c_0, \quad c_k = \sum_{i,j=0}^{i+j=k} a_i b_j,$$

where $k = 0, 1, \ldots, n+m$.

3°. Each polynomial $f(x)$ of degree $n$ can be *divided* by any other polynomial $p(x)$ of degree $m$ ($p(x) \neq 0$) *with remainder*, i.e., uniquely represented in the form $f(x) = p(x)q(x) + r(x)$, where $q(x)$ is a polynomial of degree $n - m$ (for $m \leq n$) or $q(x) = 0$ (for $m > n$), referred to as the *quotient*, and $r(x)$ is a polynomial of degree $l < m$ or $r(x) = 0$, referred to as the *remainder*.

If $r(x) = 0$, then $f(x)$ is said to be *divisible* by $p(x)$ (without remainder).
If $m > n$, then $q(x) = 0$ and $r(x) \equiv f(x)$.

### 5.1.1-3. Methods for finding quotient and remainder.

**1°. *Horner's scheme*.** To divide a polynomial $f(x)$ of degree $n$ (see (5.1.1.1)) by the polynomial $p(x) = x - b$, one uses Horner's scheme: the coefficients of $f(x)$ are written out in a row, starting from $a_n$; $b$ is written on the left; then one writes the number $a_n$ under $a_n$, the number $a_n b + a_{n-1} = b_{n-1}$ under $a_{n-1}$, the number $b_{n-1} b + a_{n-2} = b_{n-2}$ under $a_{n-2}$, ..., the number $b_1 b + a_0 = b_0$ under $a_0$. The number $b_0$ is the remainder in the division of $f(x)$ by $p(x)$, and $a_n, b_{n-1}, \ldots, b_1$ are the coefficients of the quotient.

**Remark.** To divide $f(x)$ by $p(x) = ax + b$ ($a \neq 0$) with remainder, one first uses Horner's scheme to divide by $p_1(x) = x - (-\frac{b}{a})$; now if $q_1(x)$ and $r_1$ are the quotient and remainder in the division of $f(x)$ by $p_1(x)$, then $q(x) = \frac{1}{a} q_1(x)$ and $r = r_1$ are the quotient and remainder in the division of $f(x)$ by $p(x)$.

**Example 1.** Let us divide $f(x) = x^3 - 2x^2 - 10x + 3$ by $p(x) = 2x + 5$.
We use Horner's scheme to divide $f(x)$ by $p_1(x) = x + 5/2$:

$$\begin{array}{c|cccc} & 1 & -2 & -10 & 3 \\ \hline -\dfrac{5}{2} & 1 & -\dfrac{9}{2} & \dfrac{5}{4} & -\dfrac{1}{8} \end{array}$$

Thus $f(x) = p(x) q(x) + r(x)$, where

$$q(x) = \frac{1}{2}\left(x^2 - \frac{9}{2}x + \frac{5}{4}\right) = \frac{1}{2}x^2 - \frac{9}{4}x + \frac{5}{8}, \quad r = -\frac{1}{8}.$$

**POLYNOMIAL REMAINDER THEOREM.** *The remainder in the division of a polynomial $f(x)$ by the polynomial $p(x) = x - b$ is the number equal to the value of the polynomial $f(x)$ at $x = b$.*

**2°. *Long division*.** To divide a polynomial $f(x)$ of degree $n$ by a polynomial $p(x)$ of degree $m \leq n$, one can use *long division*.

**Example 2.** Let us divide $f(x) = x^3 + 8x^2 + 14x - 5$ by $p(x) = x^2 + 3x - 1$.
We use long division:

$$\begin{array}{r|l} x^3 + 8x^2 + 14x - 5 & x^2 + 3x - 1 \\ \underline{x^3 + 3x^2 \quad\;\;\; - x} & x + 5 \\ 5x^2 + 15x - 5 & \\ \underline{5x^2 + 15x - 5} & \\ 0 & \end{array}$$

Thus $f(x) = p(x) q(x) + r(x)$, where $q(x) = x + 5$ and $r(x) = 0$; i.e., $f(x)$ is divisible by $p(x)$.

**Example 3.** Let us divide $f(x) = x^3 - 4x^2 + x + 1$ by $p(x) = x^2 + 1$.
We use long division:

$$\begin{array}{r|l} x^3 - 4x^2 + x + 1 & x^2 + 1 \\ \underline{x^3 \quad\quad\;\; + x} & x - 4 \\ -4x^2 \quad\;\; + 1 & \\ \underline{-4x^2 \quad\;\; - 4} & \\ 5 & \end{array}$$

Thus $f(x) = p(x) q(x) + r(x)$, where $q(x) = x - 4$ and $r(x) = 5$.

### 5.1.1-4. Expansion of polynomials in powers of linear binomial.

For each polynomial $f(x)$ given by equation (5.1.1.1) and any number $c$, one can write out the *expansion of $f(x)$ in powers of $x - c$*:

$$f(x) = b_n(x - c)^n + b_{n-1}(x - c)^{n-1} + \cdots + b_1(x - c) + b_0.$$

To find the coefficients $b_0, \ldots, b_n$ of this expansion, one first divides $f(x)$ by $x - c$ with remainder. The remainder is $b_0$, and the quotient is some polynomial $g_0(x)$. Then one divides $g_0(x)$ by $x - c$ with remainder. The remainder is $b_1$, and the quotient is some polynomial $g_1(x)$. Then one divides $g_1(x)$ by $x - c$, obtaining the coefficient $b_2$ as the remainder, etc. It is convenient to perform the computations by Horner's scheme (see Paragraph 5.1.1-3).

**Example 4.** Expand the polynomial $f(x) = x^4 - 5x^3 - 3x^2 + 9$ in powers of the difference $x - 3$ ($c = 3$).

We write out Horner's scheme, where the first row contains the coefficients of the polynomial $f(x)$, the second row contains the coefficients of the quotient $g_0(x)$ and the remainder $b_0$ obtained when dividing $f(x)$ by $x - 3$, the third row contains the coefficients of the quotient $g_1(x)$ and the remainder $b_1$ obtained when dividing $g_0(x)$ by $x - 3$, etc.:

$$\begin{array}{r|rrrrr} & 1 & -5 & -3 & 0 & 9 \\ \hline 3 & 1 & -2 & -9 & -27 & -72 \\ & 1 & 1 & -6 & -45 & \\ & 1 & 4 & 6 & & \\ & 1 & 7 & & & \\ & 1 & & & & \end{array}$$

Thus the expansion of $f(x)$ in powers of $x - 3$ has the form

$$f(x) = (x-3)^4 + 7(x-3)^3 + 6(x-3)^2 - 45(x-3) - 72.$$

The coefficients in the expansion of a polynomial $f(x)$ in powers of the difference $x - c$ are related to the values of the polynomial and its derivatives at $x = c$ by the formulas

$$b_0 = f(c), \quad b_1 = \frac{f'_x(c)}{1!}, \quad b_2 = \frac{f''_{xx}(c)}{2!}, \quad \ldots, \quad b_n = \frac{f^{(n)}_x(c)}{n!},$$

where the *derivative of a polynomial* $f(x) = a_n x^n + a_{n-1} x^{n-1} + \cdots + a_1 x + x_0$ with real or complex coefficients $a_0, \ldots, a_n$ is the polynomial $f'_x(x) = n a_n x^{n-1} + (n-1) a_{n-1} x^{n-2} + \cdots + a_1$, $f''_{xx}(x) = [f'_x(x)]'_x$, etc. Thus Horner's scheme permits one to find the values of the derivatives of the polynomial $f(x)$ at $x = c$.

**Example 5.** In Example 4, the values of the derivatives of the polynomial $f(x)$ at $x = 3$ are

$$f(3) = -72, \quad f'(3) = -45 \times 1! = -45, \quad f''(3) = 6 \times 2! = 12, \quad f'''(3) = 7 \times 3! = 42, \quad f^{\text{IV}}(3) = 1 \times 4! = 24.$$

The expansion of a polynomial in powers of $x - c$ can be used to compute the partial fraction decomposition of a rational function whose denominator is a power of a linear binomial.

**Example 6.** Find the partial fraction decomposition of the rational function $\Phi(x) = \dfrac{x^2 + x + 1}{(x-2)^4}$.

First, we expand the polynomial $f(x) = x^2 + x + 1$ in powers of the binomial $x - (-2) = x + 2$:

$$\begin{array}{r|rrr} & 1 & 1 & 1 \\ \hline -2 & 1 & -1 & 3 \\ & 1 & -3 & \\ & 1 & & \end{array}$$

Thus $f(x) = (x+2)^2 - 3(x+2) + 3$. As a result, we obtain $\Phi(x) = \dfrac{f(x)}{(x+2)^4} = \dfrac{1}{(x+2)^2} - \dfrac{3}{(x+2)^3} + \dfrac{3}{(x+2)^4}$.

## 5.1.2. Linear and Quadratic Equations

**5.1.2-1. Linear equations.**

The linear equation

$$ax + b = 0 \quad (a \neq 0)$$

has the solution

$$x = -\frac{b}{a}.$$

5.1.2-2. Quadratic equations.

The quadratic equation
$$ax^2 + bx + c = 0 \quad (a \neq 0) \tag{5.1.2.1}$$
has the roots
$$x_{1,2} = \frac{-b \pm \sqrt{b^2 - 4ac}}{2a}.$$

The existence of real or complex roots is determined by the sign of the discriminant $D = b^2 - 4ac$:

Case $D > 0$. There are two distinct real roots.
Case $D < 0$. There are two distinct complex conjugate roots.
Case $D = 0$. There are two equal real roots.

VIÈTE THEOREM. The roots of a quadratic equation (5.1.2.1) satisfy the following relations:
$$x_1 + x_2 = -\frac{b}{a},$$
$$x_1 x_2 = \frac{c}{a}.$$

## 5.1.3. Cubic Equations

5.1.3-1. Incomplete cubic equation.

$1°$. *Cardano's solution.* The roots of the incomplete cubic equation
$$y^3 + py + q = 0 \tag{5.1.3.1}$$
have the form
$$y_1 = A + B, \quad y_{2,3} = -\frac{1}{2}(A + B) \pm i\frac{\sqrt{3}}{2}(A - B),$$
where
$$A = \left(-\frac{q}{2} + \sqrt{D}\right)^{1/3}, \quad B = \left(-\frac{q}{2} - \sqrt{D}\right)^{1/3}, \quad D = \left(\frac{p}{3}\right)^3 + \left(\frac{q}{2}\right)^2, \quad i^2 = -1,$$
and $A$, $B$ are arbitrary values of the cubic roots such that $AB = -\frac{1}{3}p$.

The number of real roots of a cubic equation depends on the sign of the discriminant $D$:

Case $D > 0$. There is one real and two complex conjugate roots.
Case $D < 0$. There are three real roots.
Case $D = 0$. There is one real root and another real root of double multiplicity (this case is realized for $p = q = 0$).

$2°$. *Trigonometric solution.* If an incomplete cubic equation (5.1.3.1) has real coefficients $p$ and $q$, then its solutions can be found with the help of the trigonometric formulas given below.

(a) Let $p < 0$ and $D < 0$. Then
$$y_1 = 2\sqrt{-\frac{p}{3}} \cos\frac{\alpha}{3}, \quad y_{2,3} = -2\sqrt{-\frac{p}{3}} \cos\left(\frac{\alpha}{3} \pm \frac{\pi}{3}\right),$$
where the values of the trigonometric functions are calculated from the relation
$$\cos\alpha = -\frac{q}{2\sqrt{-(p/3)^3}}.$$

(b) Let $p > 0$ and $D \geq 0$. Then
$$y_1 = 2\sqrt{\frac{p}{3}} \cot(2\alpha), \quad y_{2,3} = \sqrt{\frac{p}{3}} \left[\cot(2\alpha) \pm i\frac{\sqrt{3}}{\sin(2\alpha)}\right],$$
where the values of the trigonometric functions are calculated from the relations
$$\tan \alpha = \left(\tan \frac{\beta}{2}\right)^{1/3}, \quad \tan \beta = \frac{2}{q}\left(\frac{p}{3}\right)^{3/2}, \quad |\alpha| \leq \frac{\pi}{4}, \quad |\beta| \leq \frac{\pi}{2}.$$

(c) Let $p < 0$ and $D \geq 0$. Then
$$y_1 = -2\sqrt{-\frac{p}{3}}\frac{1}{\sin(2\alpha)}, \quad y_{2,3} = \sqrt{-\frac{p}{3}}\left[\frac{1}{\sin(2\alpha)} \pm i\sqrt{3}\cot(2\alpha)\right],$$
where the values of the trigonometric functions are calculated from the relations
$$\tan \alpha = \left(\tan \frac{\beta}{2}\right)^{1/3}, \quad \sin \beta = \frac{2}{q}\left(-\frac{p}{3}\right)^{3/2}, \quad |\alpha| \leq \frac{\pi}{4}, \quad |\beta| \leq \frac{\pi}{2}.$$

In the above three cases, the real value of the cubic root should be taken.

**5.1.3-2. Complete cubic equation.**

The roots of a complete cubic equation
$$ax^3 + bx^2 + cx + d = 0 \quad (a \neq 0) \tag{5.1.3.2}$$
are calculated by the formulas
$$x_k = y_k - \frac{b}{3a}, \quad k = 1, 2, 3,$$
where $y_k$ are the roots of the incomplete cubic equation (5.1.3.1) with the coefficients
$$p = -\frac{1}{3}\left(\frac{b}{a}\right)^2 + \frac{c}{a}, \quad q = \frac{2}{27}\left(\frac{b}{a}\right)^3 - \frac{bc}{3a^2} + \frac{d}{a}.$$

VIÈTE THEOREM. *The roots of a complete cubic equation (5.1.3.2) satisfy the following relations:*
$$x_1 + x_2 + x_3 = -\frac{b}{a},$$
$$x_1 x_2 + x_1 x_3 + x_2 x_3 = \frac{c}{a},$$
$$x_1 x_2 x_3 = -\frac{d}{a}.$$

## 5.1.4. Fourth-Degree Equation

**5.1.4-1. Special cases of fourth-degree equations.**

1°. The *biquadratic equation*
$$ax^4 + bx^2 + c = 0$$
can be reduced to a quadratic equation (5.1.2.1) by the substitution $\xi = x^2$. Therefore, the roots of the biquadratic equations are given by
$$x_{1,2} = \pm\sqrt{\frac{-b + \sqrt{b^2 - 4ac}}{2a}}, \quad x_{3,4} = \pm\sqrt{\frac{-b - \sqrt{b^2 - 4ac}}{2a}}.$$

2°. The *reciprocal (algebraic) equation*
$$ax^4 + bx^3 + cx^2 + bx + a = 0$$
can be reduced to a quadratic equation by the substitution
$$y = x + \frac{1}{x}.$$
The resulting quadratic equation has the form
$$ay^2 + by + c - 2a = 0.$$
3°. The *modified reciprocal equation*
$$ax^4 + bx^3 + cx^2 - bx + a = 0$$
can be reduced to a quadratic equation by the substitution
$$y = x - \frac{1}{x}.$$
The resulting quadratic equation has the form
$$ay^2 + by + 2a + c = 0.$$
4°. The *generalized reciprocal equation*
$$ab^2 x^4 + bx^3 + cx^2 + dx + ad^2 = 0$$
can be reduced to a quadratic equation by the substitution
$$y = bx + \frac{d}{x}.$$
The resulting quadratic equation has the form
$$ay^2 + y + c - 2abd = 0.$$

### 5.1.4-2. General fourth-degree equation.

1°. *Reduction of a general equation of fourth-degree to an incomplete equation.* The general equation of fourth-degree
$$ax^4 + bx^3 + cx^2 + dx + e = 0 \quad (a \neq 0)$$
can be reduced to an incomplete equation of the form
$$y^4 + py^2 + qy + r = 0 \tag{5.1.4.1}$$
by the substitution
$$x = y - \frac{b}{4a}.$$

2°. *Descartes–Euler solution.* The roots of the incomplete equation (5.1.4.1) are given by the formulas
$$y_1 = \tfrac{1}{2}(\sqrt{z_1} + \sqrt{z_2} + \sqrt{z_3}), \quad y_2 = \tfrac{1}{2}(\sqrt{z_1} - \sqrt{z_2} - \sqrt{z_3}),$$
$$y_3 = \tfrac{1}{2}(-\sqrt{z_1} + \sqrt{z_2} - \sqrt{z_3}), \quad y_4 = \tfrac{1}{2}(-\sqrt{z_1} - \sqrt{z_2} + \sqrt{z_3}), \tag{5.1.4.2}$$
where $z_1$, $z_2$, $z_3$ are the roots of the cubic equation (cubic resolvent of equation (5.1.4.1))
$$z^3 + 2pz^2 + (p^2 - 4r)z - q^2 = 0. \tag{5.1.4.3}$$
The signs of the roots in (5.1.4.2) are chosen from the condition
$$\sqrt{z_1}\sqrt{z_2}\sqrt{z_3} = -q.$$

The roots of the fourth-degree equation (5.1.4.1) are determined by the roots of the cubic resolvent (5.1.4.3); see Table 5.1.

**3°. *Ferrari solution.*** Let $z_0$ be any of the roots of the auxiliary cubic equation (5.1.4.3). Then the four roots of the incomplete equation (5.1.4.1) are found by solving the following two quadratic equations:

$$y^2 - \sqrt{z_0}\, y + \frac{p + z_0}{2} + \frac{q}{2\sqrt{z_0}} = 0,$$

$$y^2 + \sqrt{z_0}\, y + \frac{p + z_0}{2} - \frac{q}{2\sqrt{z_0}} = 0.$$

TABLE 5.1
Relations between the roots of an incomplete equation of fourth-degree and the roots of its cubic resolvent

| Cubic resolvent (5.1.4.3) | Fourth-degree equation (5.1.4.1) |
|---|---|
| All roots are real and positive* | Four real roots |
| All roots are real: one is positive and two are negative* | Two pairs of complex conjugate roots |
| One real root and two complex conjugate roots | Two real roots and two complex conjugate roots |

## 5.1.5. Algebraic Equations of Arbitrary Degree and Their Properties

**5.1.5-1. Simplest equations of degree $n$ and their solutions.**

1°. The *binomial algebraic equation*

$$x^n - a = 0 \quad (a \neq 0)$$

has the solutions

$$x_{k+1} = \begin{cases} a^{1/n}\left(\cos \dfrac{2k\pi}{n} + i \sin \dfrac{2k\pi}{n}\right) & \text{for } a > 0, \\ |a|^{1/n}\left(\cos \dfrac{(2k+1)\pi}{n} + i \sin \dfrac{(2k+1)\pi}{n}\right) & \text{for } a < 0, \end{cases}$$

where $k = 0, 1, \ldots, n-1$ and $i^2 = -1$.

2°. Equations of the form

$$x^{2n} + ax^n + b = 0,$$
$$x^{3n} + ax^{2n} + bx^n + c = 0,$$
$$x^{4n} + ax^{3n} + bx^{2n} + cx^n + d = 0$$

are reduced by the substitution $y = x^n$ to a quadratic, cubic, and fourth-degree equation, respectively, whose solution can be expressed by radicals.

*Remark.* In the above equations, $n$ can be noninteger.

3°. The *reciprocal (algebraic) equation*

$$a_0 x^{2n} + a_1 x^{2n-1} + a_2 x^{2n-2} + \cdots + a_2 x^2 + a_1 x + a_0 = 0 \quad (a_0 \neq 0)$$

can be reduced to an equation of degree $n$ by the substitution

$$y = x + \frac{1}{x}.$$

---

* By the Viète theorem, the product of the roots $z_1$, $z_2$, $z_3$ is equal to $q^2 \geq 0$.

**Example 1.** The equation

$$ax^6 + bx^5 + cx^4 + dx^3 + cx^2 + bx + a = 0$$

can be reduced to the cubic equation

$$ay^3 + by^2 + (c - 3a)y + d - 2b = 0$$

by the substitution $y = x + 1/x$.

### 5.1.5-2. Equations of general form and their properties.

An *algebraic equation of degree* $n$ has the form

$$a_n x^n + a_{n-1} x^{n-1} + \cdots + a_1 x + a_0 = 0 \quad (a_n \neq 0), \tag{5.1.5.1}$$

where $a_k$ are real or complex coefficients. Denote the polynomial of degree $n$ on the right-hand side in equation (5.1.5.1) by

$$P_n(x) \equiv a_n x^n + a_{n-1} x^{n-1} + \cdots + a_1 x + a_0 \quad (a_n \neq 0). \tag{5.1.5.2}$$

A value $x = x_1$ such that $P_n(x_1) = 0$ is called a *root* of equation (5.1.5.1) (and also a root of the polynomial $P_n(x)$). A value $x = x_1$ is called a *root of multiplicity* $m$ if $P_n(x) = (x - x_1)^m Q_{n-m}(x)$, where $m$ is an integer ($1 \leq m \leq n$), and $Q_{n-m}(x)$ is a polynomial of degree $n - m$ such that $Q_{n-m}(x_1) \neq 0$.

THEOREM 1 (FUNDAMENTAL THEOREM OF ALGEBRA). *Any algebraic equation of degree* $n$ *has exactly* $n$ *roots (real or complex), each root counted according to its multiplicity.*

Thus, the left-hand side of equation (5.1.5.1) with roots $x_1, x_2, \ldots, x_s$ of the respective multiplicities $k_1, k_2, \ldots, k_s$ ($k_1 + k_2 + \cdots + k_s = n$) can be factorized as follows:

$$P_n(x) = a_n (x - x_1)^{k_1} (x - x_2)^{k_2} \ldots (x - x_s)^{k_s}.$$

THEOREM 2. *Any algebraic equation of an odd degree with real coefficients has at least one real root.*

THEOREM 3. *Suppose that equation (5.1.5.1) with real coefficients has a complex root* $x_1 = \alpha + i\beta$. *Then this equation has the complex conjugate root* $x_2 = \alpha - i\beta$, *and the roots* $x_1, x_2$ *have the same multiplicity.*

THEOREM 4. *Any rational root of equation (5.1.5.1) with integer coefficients* $a_k$ *is an irreducible fraction of the form* $p/q$, *where* $p$ *is a divisor of* $a_0$ *and* $q$ *is a divisor of* $a_n$. *If* $a_n = 1$, *then all rational roots of equation (5.1.5.1) (if they exist) are integer divisors of the free term.*

THEOREM 5 (ABEL–RUFFINI THEOREM). *Any equation (5.1.5.1) of degree* $n \leq 4$ *is solvable by radicals, i.e., its roots can be expressed via its coefficients by the operations of addition, subtraction, multiplication, division, and taking roots (see Subsections 5.1.2–5.1.4). In general, equation (5.1.5.1) of degree* $n > 4$ *cannot be solved by radicals.*

### 5.1.5-3. Relations between roots and coefficients. Discriminant of an equation.

VIÈTE THEOREM. *The roots of equation (5.1.5.1) (counted according to their multiplicity) and its coefficients satisfy the following relations:*

$$(-1)^k \frac{a_{n-k}}{a_n} = S_k \quad (k = 1, 2, \ldots, n),$$

where $S_k$ are *elementary symmetric functions* of $x_1, x_2, \ldots, x_n$:

$$S_1 = \sum_{i=1}^{n} x_i, \quad S_2 = \sum_{1 \le i < j}^{n} x_i x_j, \quad S_3 = \sum_{1 \le i < j < k}^{n} x_i x_j x_k, \quad \ldots, \quad S_n = x_1 x_2 \ldots x_n.$$

Note also the following relations:

$$(n-k)a_{n-k} + \sum_{j=1}^{k} a_{n-(k-j)} s_j = 0 \quad (k = 1, 2, \ldots, n)$$

with symmetric functions $s_j = \sum_{i=1}^{n} x_i^j$.

The *discriminant $D$ of an algebraic equation* is the product of $a_n^{2n-2}$ and the squared Vandermonde determinant $\Delta(x_1, x_2, \ldots, x_n)$ of its roots:

$$D = a_n^{2n-2}[\Delta(x_1, x_2, \ldots, x_n)]^2 = a_n^{2n-2} \prod_{1 \le j < i \le n} (x_i - x_j)^2.$$

The discriminant $D$ is a symmetric function of the roots $x_1, x_2, \ldots, x_n$, and is equal to zero if and only if the polynomial $P_n(x)$ has at least one multiple root.

### 5.1.5-4. Bounds for the roots of algebraic equations with real coefficients.

1°. All roots of equation (5.1.5.1) in absolute value do not exceed

$$N = 1 + \frac{A}{|a_n|}, \tag{5.1.5.3}$$

where $A$ is the largest of $|a_0|, |a_1|, \ldots, |a_{n-1}|$.

The last result admits the following generalization: all roots of equation (5.1.5.1) in absolute value do not exceed

$$N_1 = \rho + \frac{A_1}{|a_n|}, \tag{5.1.5.4}$$

where $\rho > 0$ is arbitrary and $A_1$ is the largest of

$$|a_{n-1}|, \quad \frac{|a_{n-2}|}{\rho}, \quad \frac{|a_{n-3}|}{\rho^2}, \quad \ldots, \quad \frac{|a_0|}{\rho^{n-1}}.$$

For $\rho = 1$, formula (5.1.5.4) turns into (5.1.5.3).

Remark. Formulas (5.1.5.3) and (5.1.5.4) can also be used for equations with complex coefficients.

Example 2. Consider the following equation of degree 4:

$$P_4(x) = 9x^4 - 9x^2 - 36x + 1.$$

Formula (5.1.5.3) for $n = 4$, $|a_n| = 9$, $A = 36$ yields a fairly rough estimate $N = 5$, i.e., the roots of the equation belong to the interval $[-5, 5]$. Formula (5.1.5.4) for $\rho = 2$, $n = 4$, $|a_n| = 9$, $A_1 = 9$ yields a better estimate for the bounds of the roots of this polynomial, $N_1 = 3$.

2°. A constant $K$ is called an upper bound for the real roots of equation (5.1.5.1) or the polynomial $P_n(x)$ if equation (5.1.5.1) has no real roots greater than or equal to $K$; in a similar way, one defines a lower and an upper bound for positive and negative roots of an equation or the corresponding polynomial.

Let

$K_1$ be an upper bound for the positive roots of the polynomial $P_n(x)$,
$K_2$ be an upper bound for the positive roots of the polynomial $P_n(-x)$,
$K_3 > 0$ be an upper bound for the positive roots of the polynomial $x^n P_n(1/x)$,
$K_4 > 0$ be an upper bound for the positive roots of the polynomial $x^n P_n(-1/x)$.

Then all nonzero real roots of the polynomial $P_n(x)$ (if they exist) belong to the intervals $(-K_2, -1/K_4)$ and $(1/K_3, K_1)$.

Next, we describe three methods for finding upper bounds for positive roots of a polynomial.

*Maclaurin method.* Suppose that the first $m$ leading coefficients of the polynomial (5.1.5.2) are nonnegative, i.e., $a_n > 0$, $a_{n-1} \geq 0$, ..., $a_{n-m+1} \geq 0$, and the next coefficient is negative, $a_{n-m} < 0$. Then

$$K = 1 + \left(\frac{B}{a_n}\right)^{1/m} \tag{5.1.5.5}$$

is an upper bound for the positive roots of this polynomial, where $B$ is the largest of the absolute values of negative coefficients of $P_n(x)$.

**Example 3.** Consider the fourth-degree equation from Example 2. In this case, $m = 2$, $B = 36$ and formula (5.1.5.5) yields $K = K_1 = 1 + (36/9)^{1/2} = 3$. Now, consider the polynomial $P_4(-x) = 9x^4 - 9x^2 + 36x + 1$. Its positive roots has the upper bound $K_2 = 1 + (9/9)^{1/2} = 2$. For the polynomial $x^4 P_4(1/x) = x^4 - 36x^3 - 9x^2 + 9$, we have $m = 1$, $K_3 = 1 + 36 = 37$. Finally, for the polynomial $x^4 P_4(-1/x) = x^4 + 36x^3 - 9x^2 + 9$, we have $m = 2$, $k_4 = 1 + 9^{1/2} = 4$. Thus if $P_4(x)$ has real roots, they must belong to the intervals $(-2, -1/4)$ and $(1/37, 3)$.

*Newton method.* Suppose that for $x = c$, the polynomial $P_n(x)$ and all its derivatives $P'_n(x)$, ..., $P_n^{(n)}(x)$ take positive values. Then $c$ is an upper bound for the positive roots of $P_n(x)$.

**Example 4.** Consider the polynomial from Example 2 and calculate the derivatives

$$P_4(x) = 9x^4 - 9x^2 - 36x + 1,$$
$$P'_4(x) = 36x^3 - 18x - 36,$$
$$P''_4(x) = 108x^2 - 18,$$
$$P'''_4(x) = 216x,$$
$$P''''_4(x) = 216.$$

It is easy to check that for $x = 2$ this polynomial and all its derivatives take positive values, and therefore $c = 2$ is an upper bound for its positive roots.

*A method based on the representation of a polynomial as a sum of polynomials.* Assuming $a_n > 0$, let us represent the polynomial (5.1.5.4) (without rearranging its terms) as the sum $P_n(x) = f_1(x) + \ldots + f_m(x)$, where each polynomial $f_k(x)$ ($k = 1, 2, \ldots, m$) has a positive leading coefficient and the sequence of its coefficients does not change sign more than once. Suppose that for $c > 0$ all these polynomials are positive, $f_1(c) > 0, \ldots, f_m(c) > 0$. Then $c$ is an upper bound for the positive roots of $P_n(x)$.

**Example 5.** The polynomial

$$P_7(x) = x^7 + 2x^6 - 4x^5 - 7x^4 + 2x^3 - 3x^2 + ax + b \quad (a > 0, \ b > 0)$$

can be represented as a sum of three polynomials

$$f_1(x) = x^7 + 2x^6 - 4x^5 - 7x^4 = x^4(x^3 + 2x^2 - 4x - 7), \quad f_2(x) = 2x^3 - 3x^2 = x^2(2x - 3), \quad f_3(x) = ax + b$$

(in the first two polynomials the sign of the sequence of coefficients changes once, and in the last polynomial the coefficients do not change sign). It is easy to see that all these polynomials are positive for $x = 2$. Therefore, $c = 2$ is an upper bound for the positive roots of the given polynomial.

### 5.1.5-5. Theorems on the number of real roots of polynomials.

The number all negative roots of a polynomial $P_n(x)$ is equal to the number of all positive roots of the polynomial $P_n(-x)$.

1°. The exact number of positive roots of a polynomial whose coefficients form a sequence that does not change sign or changes sign only once can be found with the help of the Descartes theorem.

DESCARTES THEOREM. *The number of positive roots (counted according to their multiplicity) of a polynomial $P_n(x)$ with real coefficients is either equal to the number of sign alterations in the sequence of its coefficients or is by an even number less.*

Applying the Descartes theorem to $P_n(-x)$, we obtain a similar theorem for the negative roots of the polynomial $P_n(x)$.

**Example 6.** Consider the cubic polynomial

$$P_3(x) = x^3 - 3x + 4.$$

Its coefficients have the signs $+ - +$, and therefore we have two alterations of sign. Therefore, the number of positive roots of $P_3(x)$ is equal either to 2 or to 0. Now, consider the polynomial $P_3(-x) = -x^3 + 2x + 1$. The sequence of its coefficients changes sign only once. Therefore, the original equation has one negative root.

2°. *A stronger version of the Descartes theorem.* Suppose that all roots of a polynomial $P_n(x)$ are real*; then the number of positive roots of $P_n(x)$ is equal to the number of sign alterations in the sequence of its coefficients, and the number of its negative roots is equal to the number of sign alterations in the sequence of coefficients of the polynomial $P_n(-x)$.

**Example 7.** Consider the characteristic polynomial of the symmetric matrix

$$P_3(x) = \begin{vmatrix} -2-x & 1 & 1 \\ 1 & 1-x & 3 \\ 1 & 3 & 1-x \end{vmatrix} = -x^3 + 14x + 20,$$

which has only real roots. The sequence of its coefficients changes sign only once, and therefore it has a single positive root. The number of its negative roots is equal to two, since this polynomial has three nonzero real roots and only one of them can be positive.

3°. If two neighboring coefficients of a polynomial $P_n(x)$ are equal to zero, then the roots of the polynomial cannot be all real (in this case, the stronger version of the Descartes theorem cannot be used).

4°. The number of real roots of a polynomial $P_n(x)$ greater than a fixed $c$ is either equal to the number of sign alterations in the sequence $P_n(c), \ldots, P_n^{(n)}(c)$ or is by an even number less. If all roots of $P_n(x)$ are real, then the number of its roots greater than $c$ coincides with the number of sign alterations in the sequence $P_n(c), \ldots, P_n^{(n)}(c)$.

**Example 8.** Consider the polynomial

$$P_4(x) = x^4 - 3x^3 + 2x^2 - 2a^2 x + a^2.$$

For $x = 1$, we have $P_4(1) = -a^2$, $P_4'(1) = -1 - 2a^2$, $P_4''(1) = -2$, $P_4'''(1) = 6$, $P_4''''(1) = 24$. Thus, there is a single sign alteration, and therefore the polynomial has a single real root greater than unity.

---

* This is the case, for instance, if we are dealing with the characteristic polynomial of a symmetric matrix.

5°. *Budan–Fourier method.* Let $N(x)$ be the number of sign alterations in the sequence $P_n(x), \ldots, P_n^{(n)}(x)$ consisting of the values of the polynomial (5.1.5.2) and its derivatives. Then the number of real roots of equation (5.1.5.1) on the interval $[a,b]$ with $P_n(a) \neq 0$, $P_n(b) \neq 0$ is either equal to $N(a) - N(b)$ or is less by an even number. When calculating $N(a)$, zero terms of the sequence are dropped. When calculating $N(b)$, it may happen that $P_n^{(i)}(b) = 0$ for $k \leq i \leq m$ and $P_n^{(k-1)}(b) \neq 0$, $P_n^{(m+1)}(b) \neq 0$; then $P_n^{(i)}(b)$ should be replaced by $(-1)^{m+1-i}$ sign $P_n^{(m+1)}(b)$.

6°. *Sturm method for finding the number of real roots.* Consider a polynomial $P_n(x)$ with no multiple roots and denote by $N(x)$ the number of sign alterations in the sequence of values of the polynomials (zero terms of the sequence are not taken into account):

$$f_0(x) = g_0(x)f_1(x) - f_2(x),$$
$$f_1(x) = g_1(x)f_2(x) - f_3(x),$$
$$\ldots\ldots\ldots\ldots\ldots\ldots\ldots,$$

where $f_0(x) = P_n(x)$, $f_1(x) = P_n'(x)$; for $k > 1$, every polynomial $-f_k(x)$ is the residue after dividing the polynomial $f_{k-2}(x)$ by $f_{k-1}(x)$; the last polynomial $f_n(x)$ is a nonzero constant. Then the number of all real roots of equation (5.1.5.1) on the segment $[a,b]$ for $P_n(a) \neq 0$, $P_n(b) \neq 0$ is equal to $N(a) - N(b)$.

**Remark 1.** Taking $a = -L$ and $b = L$ and passing to the limit as $L \to \infty$, we obtain the overall number of real roots of the algebraic equation.

**Example 9.** Consider the following cubic equation with the parameter $a$:

$$P_3(x) = x^3 + 3x^2 - a = 0.$$

The Sturm system for this equation has the form

$$P_3(x) = f_0(x) = x^3 + 3x^2 - a,$$
$$[P_3(x)]_x' = f_1(x) = 3x^2 + 6x,$$
$$f_2(x) = 2x + a,$$
$$f_3(x) = \tfrac{3}{4}a(4 - a).$$

Case $0 < a < 4$. Let us find the number of sign alterations in the Sturm system for $x = -\infty$ and $x = \infty$:

| $x$ | $f_0(x)$ | $f_1(x)$ | $f_2(x)$ | $f_3(x)$ | number of sign alterations |
|---|---|---|---|---|---|
| $-\infty$ | $-$ | $+$ | $-$ | $+$ | 3 |
| $\infty$ | $+$ | $+$ | $+$ | $+$ | 0 |

It follows that $N(-\infty) - N(\infty) = 3$. Therefore, for $0 < a < 4$, the given polynomial has three real roots.

Case $a < 0$ or $a > 4$. Let us find the number of sign alterations in the Sturm system:

| $x$ | $f_0(x)$ | $f_1(x)$ | $f_2(x)$ | $f_3(x)$ | number of sign alterations |
|---|---|---|---|---|---|
| $-\infty$ | $-$ | $+$ | $-$ | $-$ | 2 |
| $\infty$ | $+$ | $+$ | $+$ | $-$ | 1 |

It follows that $N(-\infty) - N(\infty) = 1$, and therefore for $a < 0$ or $a > 4$, the given polynomial has one real root.

**Remark 2.** If equation $P_n(x) = 0$ has multiple roots, then $P_n(x)$ and $P_n'(x)$ have a common divisor and the multiple roots are found by equating to zero this divisor. In this case, $f_n(x)$ is nonconstant and $N(a) - N(b)$ is the number of roots between $a$ and $b$, each multiple root counted only once.

**5.1.5-6. Bounds for complex roots of polynomials with real coefficients.**

$1°$. *Routh–Hurwitz criterion.* For an algebraic equation (5.1.5.1) with real coefficients, the number of roots with positive real parts is equal to the number of sign alterations in any of the two sequences

$$T_0, \ T_1, \ T_2/T_1, \ \ldots, \ T_n/T_{n-1};$$
$$T_0, \ T_1, \ T_1T_2, \ \ldots, \ T_{n-2}T_{n-1}, \ a_0;$$

where $T_m$ (it is assumed that $T_m \neq 0$ for all $m$) are defined by

$$T_0 = a_n > 0, \quad T_1 = a_{n-1}, \quad T_2 = \begin{vmatrix} a_{n-1} & a_n \\ a_{n-3} & a_{n-2} \end{vmatrix}, \quad T_3 = \begin{vmatrix} a_{n-1} & a_n & 0 \\ a_{n-3} & a_{n-2} & a_{n-1} \\ a_{n-5} & a_{n-4} & a_{n-3} \end{vmatrix},$$

$$T_4 = \begin{vmatrix} a_{n-1} & a_n & 0 & 0 \\ a_{n-3} & a_{n-2} & a_{n-1} & a_n \\ a_{n-5} & a_{n-4} & a_{n-3} & a_{n-2} \\ a_{n-7} & a_{n-6} & a_{n-5} & a_{n-4} \end{vmatrix}, \quad T_5 = \begin{vmatrix} a_{n-1} & a_n & 0 & 0 & 0 \\ a_{n-3} & a_{n-2} & a_{n-1} & a_n & 0 \\ a_{n-5} & a_{n-4} & a_{n-3} & a_{n-2} & a_{n-1} \\ a_{n-7} & a_{n-6} & a_{n-5} & a_{n-4} & a_{n-3} \\ a_{n-9} & a_{n-8} & a_{n-7} & a_{n-6} & a_{n-5} \end{vmatrix}, \quad \ldots$$

$2°$. All roots of equation (5.1.5.1) have negative real parts if and only if all $T_0, T_1, \ldots, T_n$ are positive.

$3°$. All roots of an $n$th-degree equation (5.1.5.1) have negative real parts if and only if this is true for the following $(n-1)$st-degree equation:

$$a_{n-1}x^{n-1} + \left(a_{n-2} - \frac{a_n}{a_{n-1}}a_{n-3}\right)x^{n-2} + a_{n-3}x^{n-3} + \left(a_{n-4} - \frac{a_n}{a_{n-1}}a_{n-5}\right)x^{n-2} + \cdots = 0.$$

## 5.2. Matrices and Determinants

### 5.2.1. Matrices

**5.2.1-1. Definition of a matrix. Types of matrices.**

A *matrix* of *size* (or *dimension*) $m \times n$ is a rectangular table with *entries* $a_{ij}$ ($i = 1, 2, \ldots, m$; $j = 1, 2, \ldots, n$) arranged in $m$ rows and $n$ columns:

$$A \equiv \begin{pmatrix} a_{11} & a_{12} & \cdots & a_{1n} \\ a_{21} & a_{22} & \cdots & a_{2n} \\ \vdots & \vdots & \ddots & \vdots \\ a_{m1} & a_{m2} & \cdots & a_{mn} \end{pmatrix}.$$

Note that, for each entry $a_{ij}$, the index $i$ refers to the $i$th row and the index $j$ to the $j$th column. Matrices are briefly denoted by uppercase letters (for instance, $A$, as here), or by the symbol $[a_{ij}]$, sometimes with more details: $A \equiv [a_{ij}]$ ($i = 1, 2, \ldots, m$; $j = 1, 2, \ldots, n$). The numbers $m$ and $n$ are called the *dimensions* of the matrix. A matrix is said to be *finite* if it has finitely many rows and columns; otherwise, the matrix is said to be *infinite*. In what follows, only finite matrices are considered.

The *null* or *zero matrix* is a matrix whose entries are all equal to zero: $a_{ij} = 0$ ($i = 1, 2, \ldots, m$, $j = 1, 2, \ldots, n$).

A *column vector* or *column* is a matrix of size $m \times 1$. A *row vector* or *row* is a matrix of size $1 \times n$. Both column and row vectors are often simply called *vectors*.

TABLE 5.2
Types of square matrices ($\bar{a}_{ij}$ is the complex conjugate of a number $a_{ij}$)

| Type of square matrix $[a_{ij}]$ | Entries |
|---|---|
| Unit (identity) $I = [\delta_{ij}]$ | $a_{ij} = \delta_{ij} = \begin{cases} 1, & i = j, \\ 0, & i \neq j, \end{cases}$  ($\delta_{ij}$ is the Kronecker delta) |
| Diagonal | $a_{ij} = \begin{cases} \text{any}, & i = j, \\ 0, & i \neq j \end{cases}$ |
| Upper triangular (superdiagonal) | $a_{ij} = \begin{cases} \text{any}, & i \leq j, \\ 0, & i > j \end{cases}$ |
| Strictly upper triangular | $a_{ij} = \begin{cases} \text{any}, & i < j, \\ 0, & i \geq j \end{cases}$ |
| Lower triangular (subdiagonal) | $a_{ij} = \begin{cases} \text{any}, & i \geq j, \\ 0, & i < j \end{cases}$ |
| Strictly lower triangular | $a_{ij} = \begin{cases} \text{any}, & i > j, \\ 0, & i \leq j \end{cases}$ |
| Symmetric | $a_{ij} = a_{ji}$ (see also Paragraph 5.2.1-3) |
| Skew-symmetric (antisymmetric) | $a_{ij} = -a_{ji}$ (see also Paragraph 5.2.1-3) |
| Hermitian (self-adjoint) | $a_{ij} = \bar{a}_{ji}$ (see also Paragraph 5.2.1-3) |
| Skew-Hermitian (antihermitian) | $a_{ij} = -\bar{a}_{ji}$ (see also Paragraph 5.2.1-3) |
| Monomial (generalized permutation) | Each column and each row contain exactly one nonzero entry |

A *square matrix* is a matrix of size $n \times n$, and $n$ is called the dimension of this square matrix. The *main diagonal* of a square matrix is its diagonal from the top left corner to the bottom right corner with the entries $a_{11}\ a_{22}\ \ldots\ a_{nn}$. The *secondary diagonal* of a square matrix is the diagonal from the bottom left corner to the top right corner with the entries $a_{n1}\ a_{(n-1)2}\ \ldots\ a_{1n}$. Table 5.2 lists the main types of square matrices (see also Paragraph 5.2.1-3).

5.2.1-2. Basic operations with matrices.

Two matrices are *equal* if they are of the same size and their respective entries are equal.

The *sum* of two matrices $A \equiv [a_{ij}]$ and $B \equiv [b_{ij}]$ of the same size $m \times n$ is the matrix $C \equiv [c_{ij}]$ of size $m \times n$ with the entries

$$c_{ij} = a_{ij} + b_{ij} \quad (i = 1, 2, \ldots, m;\ j = 1, 2, \ldots, n).$$

The sum of two matrices is denoted by $C = A + B$, and the operation is called *addition of matrices*.

Properties of addition of matrices:

$$A + O = A \qquad \text{(property of zero)},$$
$$A + B = B + A \qquad \text{(commutativity)},$$
$$(A + B) + C = A + (B + C) \qquad \text{(associativity)},$$

where matrices $A, B, C$, and zero matrix $O$ have the same size.

The *difference* of two matrices $A \equiv [a_{ij}]$ and $B \equiv [b_{ij}]$ of the same size $m \times n$ is the matrix $C \equiv [c_{ij}]$ of size $m \times n$ with entries

$$c_{ij} = a_{ij} - b_{ij} \quad (i = 1, 2, \ldots, m;\ j = 1, 2, \ldots, n).$$

The difference of two matrices is denoted by $C = A - B$, and the operation is called *subtraction of matrices*.

The *product* of a matrix $A \equiv [a_{ij}]$ of size $m \times n$ by a scalar $\lambda$ is the matrix $C \equiv [c_{ij}]$ of size $m \times n$ with entries

$$c_{ij} = \lambda a_{ij} \quad (i = 1, 2, \ldots, m;\ j = 1, 2, \ldots, n).$$

The product of a matrix by a scalar is denoted by $C = \lambda A$, and the operation is called *multiplication of a matrix by a scalar*.

Properties of multiplication of a matrix by a scalar:

$0A = O$ (property of zero),
$(\lambda\mu)A = \lambda(\mu A)$ (associativity with respect to a scalar factor),
$\lambda(A + B) = \lambda A + \lambda B$ (distributivity with respect to addition of matrices),
$(\lambda + \mu)A = \lambda A + \mu A$ (distributivity with respect to addition of scalars),

where $\lambda$ and $\mu$ are scalars, matrices $A$, $B$, $C$, and zero matrix $O$ have the same size.

The *additively inverse (opposite) matrix* for a matrix $A \equiv [a_{ij}]$ of size $m \times n$ is the matrix $C \equiv [c_{ij}]$ of size $m \times n$ with entries

$$c_{ij} = -a_{ij} \quad (i = 1, 2, \ldots, m;\ j = 1, 2, \ldots, n),$$

or, in matrix form,

$$C = (-1)A.$$

**Remark.** The difference $C$ of two matrices $A$ and $B$ can be expressed as $C = A + (-1)B$.

The *product* of a matrix $A \equiv [a_{ij}]$ of size $m \times p$ and a matrix $B \equiv [b_{ij}]$ of size $p \times n$ is the matrix $C \equiv [c_{ij}]$ of size $m \times n$ with entries

$$c_{ij} = \sum_{k=1}^{p} a_{ik} b_{kj} \quad (i = 1, 2, \ldots, m;\ j = 1, 2, \ldots, n);$$

i.e., the entry $c_{ij}$ in the $i$th row and $j$th column of the matrix $C$ is equal to the sum of products of the respective entries in the $i$th row of $A$ and the $j$th column of $B$. Note that the product is defined for matrices of *compatible size*; i.e., the number of the columns in the first matrix should be equal to the number of rows in the second matrix. The product of two matrices $A$ and $B$ is denoted by $C = AB$, and the operation is called *multiplication of matrices*.

**Example 1.** Consider two matrices

$$A = \begin{pmatrix} 1 & 2 \\ 6 & -3 \end{pmatrix} \quad \text{and} \quad B = \begin{pmatrix} 0 & 10 & 1 \\ -6 & -0.5 & 20 \end{pmatrix}.$$

The product of the matrix $A$ and the matrix $B$ is the matrix

$$C = AB = \begin{pmatrix} 1 & 2 \\ 6 & -3 \end{pmatrix} \begin{pmatrix} 0 & 10 & 1 \\ -6 & -0.5 & 20 \end{pmatrix}$$
$$= \begin{pmatrix} 1 \times 0 + 2 \times (-6) & 1 \times 10 + 2 \times (-0.5) & 1 \times 1 + 2 \times 20 \\ 6 \times 0 + (-3) \times (-6) & 6 \times 10 + (-3) \times (-0.5) & 6 \times 1 + (-3) \times 20 \end{pmatrix} = \begin{pmatrix} -12 & 9 & 41 \\ 18 & 61.5 & -54 \end{pmatrix}.$$

Two square matrices $A$ and $B$ are said to *commute* if $AB = BA$, i.e., if their multiplication is subject to the commutative law.

Properties of multiplication of matrices:

$AO = O_1$, $\quad A + O = A$ $\quad$ (property of zero matrix),
$(AB)C = A(BC)$ $\quad$ (associativity of the product of three matrices),
$AI = A$ $\quad$ (multiplication by unit matrix),
$A(B + C) = AB + AC$ $\quad$ (distributivity with respect to a sum of two matrices),
$\lambda(AB) = (\lambda A)B = A(\lambda B)$ $\quad$ (associativity of the product of a scalar and two matrices),
$SD = DS$ $\quad$ (commutativity for any square and any diagonal matrices),

where $\lambda$ is a scalar, matrices $A$, $B$, $C$, square matrix $S$, diagonal matrix $D$, zero matrices $O$ and $O_1$, and unit matrix $I$ have the compatible sizes.

**5.2.1-3. Transpose, complex conjugate matrix, adjoint matrix.**

The *transpose* of a matrix $A \equiv [a_{ij}]$ of size $m \times n$ is the matrix $C \equiv [c_{ij}]$ of size $n \times m$ with entries

$$c_{ij} = a_{ji} \quad (i = 1, 2, \ldots, n; \; j = 1, 2, \ldots, m).$$

The transpose is denoted by $C = A^T$.

**Example 2.** If $A = (a_1, a_2)$ then $A^T = \begin{pmatrix} a_1 \\ a_2 \end{pmatrix}$.

Properties of transposes:

$$(A + B)^T = A^T + B^T, \quad (\lambda A)^T = \lambda A^T, \quad (A^T)^T = A,$$
$$(AC)^T = C^T A^T, \quad O^T = O_1, \quad I^T = I,$$

where $\lambda$ is a scalar; matrices $A$, $B$, and zero matrix $O$ have size $m \times n$; matrix $C$ has size $n \times l$; zero matrix $O_1$ has size $n \times m$.

A square matrix $A$ is said to be *orthogonal* if $A^T A = AA^T = I$, i.e., $A^T = A^{-1}$ (see Paragraph 5.2.1-6).

Properties of orthogonal matrices:
1. If $A$ is an orthogonal matrix, then $A^T$ is also orthogonal.
2. The product of two orthogonal matrices is an orthogonal matrix.
3. Any symmetric orthogonal matrix is involutive (see Paragraph 5.2.1-7).

The *complex conjugate* of a matrix $A \equiv [a_{ij}]$ of size $m \times n$ is the matrix $C \equiv [c_{ij}]$ of size $m \times n$ with entries

$$c_{ij} = \bar{a}_{ij} \quad (i = 1, 2, \ldots, m; \; j = 1, 2, \ldots, n),$$

where $\bar{a}_{ij}$ is the complex conjugate of $a_{ij}$. The complex conjugate matrix is denoted by $C = \overline{A}$.

The *adjoint matrix* of a matrix $A \equiv [a_{ij}]$ of size $m \times n$ is the matrix $C \equiv [c_{ij}]$ of size $n \times m$ with entries

$$c_{ij} = \bar{a}_{ji} \quad (i = 1, 2, \ldots, n; \; j = 1, 2, \ldots, m).$$

The adjoint matrix is denoted by $C = A^*$.

Properties of adjoint matrices:

$$(A + B)^* = A^* + B^*, \quad (\lambda A)^* = \bar{\lambda} A^*, \quad (A^*)^* = A,$$
$$(AC)^* = C^* A^*, \quad O^* = O_1, \quad I^* = I,$$

where $\lambda$ is a scalar; matrices $A$, $B$, and zero matrix $O$ have size $m \times n$; matrix $C$ has size $n \times l$; zero matrix $O_1$ has a size $n \times m$.

Remark. If a matrix is *real* (i.e., all its entries are real), then the corresponding transpose and the adjoint matrix coincide.

A square matrix $A$ is said to be *normal* if $A^*A = AA^*$. A normal matrix $A$ is said to be *unitary* if $A^*A = AA^* = I$, i.e., $A^* = A^{-1}$ (see Paragraph 5.2.1-6).

### 5.2.1-4. Trace of a matrix.

The *trace* of a square matrix $A \equiv [a_{ij}]$ of size $n \times n$ is the sum $S$ of its diagonal entries,

$$S = \text{Tr}(A) = \sum_{i=1}^{n} a_{ii}.$$

If $\lambda$ is a scalar and square matrices $A$ and $B$ has the same size, then

$$\text{Tr}(A + B) = \text{Tr}(A) + \text{Tr}(B), \quad \text{Tr}(\lambda A) = \lambda \text{Tr}(A), \quad \text{Tr}(AB) = \text{Tr}(BA),$$

### 5.2.1-5. Linear dependence of row vectors (column vectors).

A row vector (column vector) $B$ is a *linear combination* of row vectors (column vectors) $A_1, \ldots, A_k$ if there exist scalars $\alpha_1, \ldots, \alpha_k$ such that

$$B = \alpha_1 A_1 + \cdots + \alpha_k A_k.$$

Row vectors (column vectors) $A_1, \ldots, A_k$ are said to be *linearly dependent* if there exist scalars $\alpha_1, \ldots, \alpha_k$ ($\alpha_1^2 + \cdots + \alpha_k^2 \neq 0$) such that

$$\alpha_1 A_1 + \cdots + \alpha_k A_k = O,$$

where $O$ is the zero row vector (column vector).

Row vectors (column vectors) $A_1, \ldots, A_k$ are said to be *linearly independent* if, for any $\alpha_1, \ldots, \alpha_k$ ($\alpha_1^2 + \cdots + \alpha_k^2 \neq 0$) we have

$$\alpha_1 A_1 + \cdots + \alpha_k A_k \neq O.$$

THEOREM. *Row vectors (column vectors) $A_1, \ldots, A_k$ are linearly dependent if and only if one of them is a linear combination of the others.*

### 5.2.1-6. Inverse matrices.

Let $A$ be a square matrix of size $n \times n$, and let $I$ be the unit matrix of the same size.

A square matrix $B$ of size $n \times n$ is called a *right inverse* of $A$ if $AB = I$. A square matrix $C$ of size $n \times n$ is called a *left inverse* of $A$ if $CA = I$. If one of the matrices $B$ or $C$ exists, then the other exists, too, and these two matrices coincide. In such a case, the matrix $A$ is said to be *nondegenerate (nonsingular)*.

THEOREM. *A square matrix is nondegenerate if and only if its rows (columns) are linearly independent.*

Remark. Generally, instead of the terms "left inverse matrix" and "right inverse matrix", the term "*inverse matrix*" is used with regard to the matrix $B = A^{-1}$ for a nondegenerate matrix $A$, since $AB = BA = I$.

UNIQUENESS THEOREM. *The matrix $A^{-1}$ is the unique matrix satisfying the condition $AA^{-1} = A^{-1}A = I$ for a given nondegenerate matrix $A$.*

Remark. For the existence theorem, see Paragraph 5.2.2-7.

Properties of inverse matrices:

$$(AB)^{-1} = B^{-1}A^{-1}, \quad (\lambda A)^{-1} = \lambda^{-1}A^{-1},$$
$$(A^{-1})^{-1} = A, \quad (A^{-1})^T = (A^T)^{-1}, \quad (A^{-1})^* = (A^*)^{-1},$$

where square matrices $A$ and $B$ are assumed to be nondegenerate and scalar $\lambda \neq 0$.

The problem of finding the inverse matrix is considered in Paragraphs 5.2.2-7, 5.2.4-5, and 5.5.2-3.

### 5.2.1-7. Powers of matrices.

A product of several matrices equal to one and the same matrix $A$ can be written as a *positive integer power* of the matrix $A$: $AA = A^2$, $AAA = A^2A = A^3$, etc. For a positive integer $k$, one defines $A^k = A^{k-1}A$ as the $k$th power of $A$. For a nondegenerate matrix $A$, one defines $A^0 = AA^{-1} = I$, $A^{-k} = (A^{-1})^k$. Powers of a matrix have the following properties:

$$A^p A^q = A^{p+q}, \quad (A^p)^q = A^{pq},$$

where $p$ and $q$ are arbitrary positive integers and $A$ is an arbitrary square matrix; or $p$ and $q$ are arbitrary integers and $A$ is an arbitrary nondegenerate matrix.

There exist matrices $A^k$ whose positive integer power is equal to the zero matrix, even if $A \neq O$. If $A^k = O$ for some integer $k > 1$, then $A$ is called a *nilpotent matrix*.

A matrix $A$ is said to be *involutive* if it coincides with its inverse: $A = A^{-1}$ or $A^2 = I$.

### 5.2.1-8. Polynomials and matrices. Basic functions of matrices.

A *polynomial with matrix argument* is the expression obtained from a scalar polynomial $f(x)$ by replacing the scalar argument $x$ with a square matrix $X$:

$$f(X) = a_0 I + a_1 X + a_2 X^2 + \cdots,$$

where $a_i$ ($i = 0, 1, 2, \ldots$) are real or complex coefficients. The polynomial $f(X)$ is a square matrix of the same size as $X$.

A *polynomial with matrix coefficients* is an expression obtained from a polynomial $f(x)$ by replacing its coefficients $a_i$ ($i = 0, 1, 2, \ldots$) with matrices $A_i$ ($i = 0, 1, 2, \ldots$) of the same size:

$$F(x) = A_0 + A_1 x + A_2 x^2 + \cdots.$$

**Example 3.** For the matrix

$$A = \begin{pmatrix} 4 & -8 & 1 \\ 5 & -9 & 1 \\ 4 & -6 & -1 \end{pmatrix},$$

the characteristic matrix (see Paragraph 5.2.3-2) is a polynomial with matrix coefficients and argument $\lambda$:

$$F(\lambda) \equiv A - \lambda I = A_0 + A_1 \lambda = \begin{pmatrix} 4-\lambda & -8 & 1 \\ 5 & -9-\lambda & 1 \\ 4 & -6 & -1-\lambda \end{pmatrix},$$

where

$$A_0 = A = \begin{pmatrix} 4 & -8 & 1 \\ 5 & -9 & 1 \\ 4 & -6 & -1 \end{pmatrix}, \quad A_1 = -I = \begin{pmatrix} -1 & 0 & 0 \\ 0 & -1 & 0 \\ 0 & 0 & -1 \end{pmatrix}.$$

The corresponding adjugate matrix (see Paragraph 5.2.2-7) can also be represented as a polynomial with matrix coefficients:

$$G(\lambda) = \begin{pmatrix} \lambda^2 + 10\lambda + 15 & -8\lambda - 14 & \lambda + 1 \\ 5\lambda + 9 & \lambda^2 - 3\lambda - 8 & \lambda + 1 \\ 4\lambda + 6 & -6\lambda - 8 & \lambda^2 + 5\lambda + 4 \end{pmatrix} = A_0 + A_1 \lambda + A_2 \lambda^2,$$

where
$$A_0 = \begin{pmatrix} 15 & -14 & 1 \\ 9 & -8 & 1 \\ 6 & -8 & 4 \end{pmatrix}, \quad A_1 = \begin{pmatrix} 10 & -8 & 1 \\ 5 & -3 & 1 \\ 4 & -6 & 5 \end{pmatrix}, \quad A_2 = I = \begin{pmatrix} 1 & 0 & 0 \\ 0 & 1 & 0 \\ 0 & 0 & 1 \end{pmatrix}.$$

The variable $x$ in a polynomial with matrix coefficients can be replaced by a matrix $X$, which yields a polynomial of matrix argument with matrix coefficients. In this situation, one distinguishes between the "left" and the "right" values:

$$F(X) = A_0 + A_1 X + A_2 X^2 + \cdots,$$
$$\widehat{F}(X) = A_0 + X A_1 + X^2 A_2 + \cdots.$$

The exponential function of a square matrix $X$ can be represented as the following convergent series:

$$e^X = 1 + X + \frac{X^2}{2!} + \frac{X^3}{3!} + \cdots = \sum_{k=0}^{\infty} \frac{X^k}{k!}.$$

The inverse matrix has the form

$$(e^X)^{-1} = e^{-X} = 1 - X + \frac{X^2}{2!} - \frac{X^3}{3!} + \cdots = \sum_{k=0}^{\infty} (-1)^k \frac{X^k}{k!}.$$

**Remark.** Note that $e^X e^Y \neq e^Y e^X$, in general. The relation $e^X e^Y = e^{X+Y}$ holds only for commuting matrices $X$ and $Y$.

Some other functions of matrices can be expressed in terms of the exponential function:

$$\sin X = \frac{1}{2i}(e^{iX} - e^{-iX}), \quad \cos X = \frac{1}{2}(e^{iX} + e^{-iX}),$$
$$\sinh X = \frac{1}{2}(e^X - e^{-X}), \quad \cosh X = \frac{1}{2}(e^X + e^{-X}).$$

5.2.1-9. Decomposition of matrices.

THEOREM 1. *For any square matrix $A$, the matrix $S_1 = \frac{1}{2}(A + A^T)$ is symmetric and the matrix $S_2 = \frac{1}{2}(A - A^T)$ is skew-symmetric. The representation of $A$ as the sum of symmetric and skew-symmetric matrices is unique:* $A = S_1 + S_2$.

THEOREM 2. *For any square matrix $A$, the matrices $H_1 = \frac{1}{2}(A + A^*)$ and $H_2 = \frac{1}{2i}(A - A^*)$ are Hermitian, and the matrix $iH_2$ is skew-Hermitian. The representation of $A$ as the sum of Hermitian and skew-Hermitian matrices is unique:* $A = H_1 + iH_2$.

THEOREM 3. *For any square matrix $A$, the matrices $AA^*$ and $A^*A$ are nonnegative Hermitian matrices (see Paragraph 5.7.3-1).*

THEOREM 4. *Any square matrix $A$ admits a polar decomposition*

$$A = QU \quad \text{and} \quad A = U_1 Q_1,$$

*where $Q$ and $Q_1$ are nonnegative Hermitian matrices, $Q^2 = AA^*$ and $Q_1^2 = A^*A$, and $U$ and $U_1$ are unitary matrices. The matrices $Q$ and $Q_1$ are always unique, while the matrices $U$ and $U_1$ are unique only in the case of a nondegenerate $A$.*

## 5.2.1-10. Block matrices.

Let us split a given matrix $A \equiv [a_{ij}]$ ($i = 1, 2, \ldots, m$; $j = 1, 2, \ldots, n$) of size $m \times n$ into separate rectangular cells with the help of $(M-1)$ horizontal and $(N-1)$ vertical lines. Each cell is a matrix $A_{\alpha\beta} \equiv [a_{ij}]$ ($i = i_\alpha, i_\alpha+1, \ldots, i_\alpha+m_\alpha-1$; $j = j_\beta, j_\beta+1, \ldots, j_\beta+n_\beta-1$) of size $m_\alpha \times n_\beta$ and is called a *block* of the matrix $A$. Here $i_\alpha = m_{\alpha-1} + i_{\alpha-1}$, $j_\beta = n_{\beta-1} + j_{\beta-1}$. Then the given matrix $A$ can be regarded as a new matrix whose entries are the blocks: $A \equiv [A_{\alpha\beta}]$ ($\alpha = 1, 2, \ldots, M$; $\beta = 1, 2, \ldots, N$). This matrix is called a *block matrix*.

**Example 4.** The matrix

$$A \equiv \begin{pmatrix} a_{11} & a_{12} & a_{13} & a_{14} & a_{15} \\ a_{21} & a_{22} & a_{23} & a_{24} & a_{25} \\ \hline a_{31} & a_{32} & a_{33} & a_{34} & a_{35} \\ a_{41} & a_{42} & a_{43} & a_{44} & a_{45} \\ a_{51} & a_{52} & a_{53} & a_{54} & a_{55} \end{pmatrix}$$

can be regarded as the block matrix

$$A \equiv \begin{pmatrix} A_{11} & A_{12} \\ A_{21} & A_{22} \end{pmatrix}$$

of size $2 \times 2$ with the entries being the blocks

$$A_{11} \equiv \begin{pmatrix} a_{11} & a_{12} & a_{13} \\ a_{21} & a_{22} & a_{23} \end{pmatrix}, \quad A_{12} \equiv \begin{pmatrix} a_{14} & a_{15} \\ a_{24} & a_{25} \end{pmatrix},$$

$$A_{21} \equiv \begin{pmatrix} a_{31} & a_{32} & a_{33} \\ a_{41} & a_{42} & a_{43} \\ a_{51} & a_{52} & a_{53} \end{pmatrix}, \quad A_{22} \equiv \begin{pmatrix} a_{34} & a_{35} \\ a_{44} & a_{45} \\ a_{54} & a_{55} \end{pmatrix}$$

of size $2 \times 3$, $2 \times 2$, $3 \times 3$, $3 \times 2$, respectively.

Basic operations with block matrices are practically the same as those with common matrices, the role of the entries being played by blocks:

1. For matrices $A \equiv [a_{ij}] \equiv [A_{\alpha\beta}]$ and $B \equiv [b_{ij}] \equiv [B_{\alpha\beta}]$ of the same size and the same block structure, their sum $C \equiv [C_{\alpha\beta}] = [A_{\alpha\beta} + B_{\alpha\beta}]$ is a matrix of the same size and the same block structure.
2. For a matrix $A \equiv [a_{ij}]$ of size $m \times n$ regarded as a block matrix $A \equiv [A_{\alpha\beta}]$ of size $M \times N$, the multiplication by a scalar is defined by $\lambda A = [\lambda A_{\alpha\beta}] = [\lambda a_{ij}]$.
3. Let $A \equiv [a_{ik}] \equiv [A_{\alpha\gamma}]$ and $B \equiv [b_{kj}] \equiv [B_{\gamma\beta}]$ be two block matrices such that the number of columns of each block $A_{\alpha\gamma}$ is equal to the number of the rows of the block $B_{\gamma\beta}$. Then the product of the matrices $A$ and $B$ can be regarded as the block matrix $C \equiv [C_{\alpha\beta}] = [\sum_\gamma A_{\alpha\gamma} B_{\gamma\beta}]$.
4. For a matrix $A \equiv [a_{ij}]$ of size $m \times n$ regarded as a block matrix $A \equiv [A_{\alpha\beta}]$ of size $M \times N$, the transpose has the form $A^T = [A_{\beta\alpha}^T]$.
5. For a matrix $A \equiv [a_{ij}]$ of size $m \times n$ regarded as a block matrix $A \equiv [A_{\alpha\beta}]$ of size $M \times N$, the adjoint matrix has the form $A^* = [A_{\beta\alpha}^*]$.

Let $A$ be a nondegenerate matrix of size $n \times n$ represented as the block matrix

$$A \equiv \begin{pmatrix} A_{11} & A_{12} \\ A_{21} & A_{22} \end{pmatrix},$$

where $A_{11}$ and $A_{22}$ are square matrices of size $p \times p$ and $q \times q$, respectively ($p + q = n$). Then the following relations, called the *Frobenius formulas*, hold:

$$A^{-1} = \begin{pmatrix} A_{11}^{-1} + A_{11}^{-1} A_{12} N A_{21} A_{11}^{-1} & -A_{11}^{-1} A_{12} N \\ -N A_{21} A_{11}^{-1} & N \end{pmatrix},$$

$$A^{-1} = \begin{pmatrix} K & -K A_{12} A_{22}^{-1} \\ -A_{22}^{-1} A_{21} K & A_{22}^{-1} + A_{22}^{-1} A_{21} K A_{12} A_{22}^{-1} \end{pmatrix}.$$

Here, $N = (A_{22} - A_{21} A_{11}^{-1} A_{12})^{-1}$, $K = (A_{11} - A_{12} A_{22}^{-1} A_{21})^{-1}$; in the first formula, the matrix $A_{11}$ is assumed nondegenerate, and in the second formula, $A_{22}$ is assumed nondegenerate.

The *direct sum* of two square matrices $A$ and $B$ of size $m \times m$ and $n \times n$, respectively, is the block matrix $C = A \oplus B = \begin{pmatrix} A & 0 \\ 0 & B \end{pmatrix}$ of size $m + n$.

Properties of the direct sum of matrices:

1. For any square matrices $A$, $B$, and $C$ the following relations hold:

$$(A \oplus B) \oplus C = A \oplus (B \oplus C) \quad \text{(associativity)},$$
$$\text{Tr}(A \oplus B) = \text{Tr}(A) + \text{Tr}(B) \quad \text{(trace property)}.$$

2. For nondegenerate square matrices $A$ and $B$, the following relation holds:

$$(A \oplus B)^{-1} = A^{-1} \oplus B^{-1}.$$

3. For square matrices $A_m$, $B_m$ of size $m \times m$ and square matrices $A_n$, $B_n$ of size $n \times n$, the following relations hold:

$$(A_m \oplus A_n) + (B_m \oplus B_n) = (A_m + B_m) \oplus (A_n + B_n);$$
$$(A_m \oplus A_n)(B_m \oplus B_n) = A_m B_m \oplus A_n B_n.$$

**5.2.1-11. Kronecker product of matrices.**

The *Kronecker product* of two matrices $A \equiv [a_{i_a j_a}]$ and $B \equiv [b_{i_b j_b}]$ of size $m_a \times n_a$ and $m_b \times n_b$, respectively, is the matrix $C \equiv [c_{kh}]$ of size $m_a m_b \times n_a n_b$ with entries

$$c_{kh} = a_{i_a j_a} b_{i_b j_b} \quad (k = 1, 2, \ldots, m_a m_b;\ h = 1, 2, \ldots, n_a n_b),$$

where the index $k$ is the serial number of a pair $(i_a, i_b)$ in the sequence $(1, 1), (1, 2), \ldots, (1, m_b), (2, 1), (2, 2), \ldots (m_a, m_b)$, and the index $h$ is the serial number of a pair $(j_a, j_b)$ in a similar sequence. This Kronecker product can be represented as the block matrix $C \equiv [a_{i_a j_a} B]$.

Note that if $A$ and $B$ are square matrices and the number of rows in $C$ is equal to the number of rows in $A$, and the number of rows in $D$ is equal to the number of rows in $B$, then

$$(A \otimes B)(C \otimes D) = AC \otimes BD.$$

The following relations hold:

$$(A \otimes B)^T = A^T \otimes B^T, \qquad \text{Tr}(A \otimes B) = \text{Tr}(A)\text{Tr}(B).$$

## 5.2.2. Determinants

**5.2.2-1. Notion of determinant.**

With each square matrix $A \equiv [a_{ij}]$ of size $n \times n$ one can associate a numerical characteristic, called its *determinant*. The determinant of such a matrix can be defined by induction with respect to the size $n$.

For a matrix of size $1 \times 1$ ($n = 1$), the *first-order determinant* is equal to its only entry,

$$\Delta \equiv \det A = a_{11}.$$

For a matrix of size $2 \times 2$ ($n = 2$), the *second-order determinant*, is equal to the difference of the product of its entries on the main diagonal and the product of its entries on the secondary diagonal:

$$\Delta \equiv \det A \equiv \begin{vmatrix} a_{11} & a_{12} \\ a_{21} & a_{22} \end{vmatrix} = a_{11}a_{22} - a_{12}a_{21}.$$

For a matrix of size $3 \times 3$ ($n = 3$), the *third-order determinant*,

$$\Delta \equiv \det A \equiv \begin{vmatrix} a_{11} & a_{12} & a_{13} \\ a_{21} & a_{22} & a_{23} \\ a_{31} & a_{32} & a_{33} \end{vmatrix}$$

$$= a_{11}a_{22}a_{33} + a_{12}a_{23}a_{31} + a_{21}a_{32}a_{13} - a_{13}a_{22}a_{31} - a_{12}a_{21}a_{33} - a_{23}a_{32}a_{11}.$$

This expression is obtained by the *triangle rule* (*Sarrus scheme*), illustrated by the following diagrams, where entries occurring in the same product with a given sign are joined by segments:

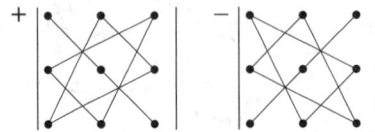

For a matrix of size $n \times n$ ($n > 2$), the *nth-order determinant* is defined as follows under the assumption that the $(n-1)$st-order determinant has already been defined for a matrix of size $(n-1) \times (n-1)$.

Consider a matrix $A = [a_{ij}]$ of size $n \times n$. The *minor* $M_j^i$ corresponding to an entry $a_{ij}$ is defined as the $(n-1)$st-order determinant of the matrix of size $(n-1) \times (n-1)$ obtained from the original matrix $A$ by removing the $i$th row and the $j$th column (i.e., the row and the column whose intersection contains the entry $a_{ij}$). The *cofactor* $A_j^i$ of the entry $a_{ij}$ is defined by $A_j^i = (-1)^{i+j} M_j^i$ (i.e., it coincides with the corresponding minor if $i + j$ is even, and is the opposite of the minor if $i + j$ is odd).

The *nth-order determinant* of the matrix $A$ is defined by

$$\Delta \equiv \det A \equiv \begin{vmatrix} a_{11} & a_{12} & \cdots & a_{1n} \\ a_{21} & a_{22} & \cdots & a_{2n} \\ \vdots & \vdots & \ddots & \vdots \\ a_{n1} & a_{n2} & \cdots & a_{nn} \end{vmatrix} = \sum_{k=1}^{n} a_{ik} A_k^i = \sum_{k=1}^{n} a_{kj} A_j^k.$$

This formula is also called the *$i$th row expansion of the determinant of* $A$ and also the *$j$th column expansion of the determinant of* $A$.

**Example 1.** Let us find the third-order determinant of the matrix

$$A = \begin{pmatrix} 1 & -1 & 2 \\ 6 & 1 & 5 \\ 2 & -1 & -4 \end{pmatrix}.$$

To this end, we use the second-column expansion of the determinant:

$$\det A = \sum_{i=1}^{3} (-1)^{i+2} a_{i2} M_2^i = (-1)^{1+2} \times (-1) \times \begin{vmatrix} 6 & 5 \\ 2 & -4 \end{vmatrix} + (-1)^{2+2} \times 1 \times \begin{vmatrix} 1 & 2 \\ 2 & -4 \end{vmatrix} + (-1)^{3+2} \times (-1) \times \begin{vmatrix} 1 & 2 \\ 6 & 5 \end{vmatrix}$$

$$= 1 \times [6 \times (-4) - 5 \times 2] + 1 \times [1 \times (-4) - 2 \times 2] + 1 \times [1 \times 5 - 2 \times 6] = -49.$$

5.2.2-2. Properties of determinants.

Basic properties:

1. Invariance with respect to transposition of matrices:

$$\det A = \det A^T.$$

2. Antisymmetry with respect to the permutation of two rows (or columns): if two rows (columns) of a matrix are interchanged, its determinant preserves its absolute value, but changes its sign.

3. Linearity with respect to a row (or column) of the corresponding matrix: suppose that the $i$th row of a matrix $A \equiv [a_{ij}]$ is a linear combination of two row vectors, $(a_{i1}, \ldots, a_{i3}) = \lambda(b_1, \ldots, b_n) + \mu(c_1, \ldots, c_n)$; then
$$\det A = \lambda \det A_b + \mu \det A_c,$$
where $A_b$ and $A_c$ are the matrices obtained from $A$ by replacing its $i$th row with $(b_1, \ldots, b_n)$ and $(c_1, \ldots, c_n)$. This fact, together with the first property, implies that a similar linearity relation holds if a column of the matrix $A$ is a linear combination of two column vectors.

Some useful corollaries from the basic properties:
1. The determinant of a matrix with two equal rows (columns) is equal to zero.
2. If all entries of a row are multiplied by $\lambda$, the determinant of the resulting matrix is multiplied by $\lambda$.
3. If a matrix contains a row (columns) consisting of zeroes, then its determinant is equal to zero.
4. If a matrix has two proportional rows (columns), its determinant is equal to zero.
5. If a matrix has a row (column) that is a linear combination of its other rows (columns), its determinant is equal to zero.
6. The determinant of a matrix does not change if a linear combination of some of its rows is added to another row of that matrix.

THEOREM (NECESSARY AND SUFFICIENT CONDITION FOR A MATRIX TO BE DEGENERATE). *The determinant of a square matrix is equal to zero if and only if its rows (columns) are linearly dependent.*

### 5.2.2-3. Minors. Basic minors. Rank and defect of a matrix.

Let $A \equiv [a_{ij}]$ be a matrix of size $n \times n$. Its *$m$th-order* ($m \leq n$) *minor of the first kind*, denoted by $M^{i_1 i_2 \ldots i_m}_{j_1 j_2 \ldots j_m}$, is the $m$th-order determinant of a submatrix obtained from $A$ by removing some of its $n - m$ rows and $n - m$ columns. Here, $i_1, i_2, \ldots, i_m$ are the indices of the rows and $j_1, j_2, \ldots, j_m$ are the indices of the columns involved in that submatrix. The $(n-m)$th-order *determinant of the second kind*, denoted by $\overline{M}^{i_1 i_2 \ldots i_m}_{j_1 j_2 \ldots j_m}$, is the $(n-m)$th-order determinant of the submatrix obtained from $A$ by removing the rows and the columns involved in $M^{i_1 i_2 \ldots i_m}_{j_1 j_2 \ldots j_m}$. The *cofactor* of the minor $M^{i_1 i_2 \ldots i_m}_{j_1 j_2 \ldots j_m}$ is defined by
$$A^{i_1 i_2 \ldots i_m}_{j_1 j_2 \ldots j_m} = (-1)^{i_1 + i_2 + \cdots + i_m + j_1 + j_2 + \cdots + j_m} \overline{M}^{i_1 i_2 \ldots i_m}_{j_1 j_2 \ldots j_m}.$$

*Remark.* minors of the first kind can be introduced for any rectangular matrix $A \equiv [a_{ij}]$ of size $m \times n$. Its $k$th-order ($k \leq \min\{m, n\}$) minor $M^{i_1 i_2 \ldots i_k}_{j_1 j_2 \ldots j_k}$ is the determinant of the submatrix obtained from $A$ by removing some of its $m - k$ rows and $n - k$ columns.

LAPLACE THEOREM. *Given $m$ rows with indices $i_1, \ldots, i_m$ (or $m$ columns with indices $j_1, \ldots, j_m$) of a square matrix $A$, its determinant $\Delta$ is equal to the sum of products of all $m$th-order minors $M^{i_1 i_2 \ldots i_m}_{j_1 j_2 \ldots j_m}$ in those rows (resp., columns) and their cofactors $A^{i_1 i_2 \ldots i_m}_{j_1 j_2 \ldots j_m}$, i.e.,*
$$\Delta \equiv \det A = \sum_{j_1, j_2, \ldots, j_m} M^{i_1 i_2 \ldots i_m}_{j_1 j_2 \ldots j_m} A^{i_1 i_2 \ldots i_m}_{j_1 j_2 \ldots j_m} = \sum_{i_1, i_2, \ldots, i_m} M^{i_1 i_2 \ldots i_m}_{j_1 j_2 \ldots j_m} A^{i_1 i_2 \ldots i_m}_{j_1 j_2 \ldots j_m}.$$

Here, in the first sum $i_1, \ldots, i_m$ are fixed, and in the second sum $j_1, \ldots, j_m$ are fixed.

Let $A \equiv [a_{ij}]$ be a matrix of size $m \times n$ with at least one nonzero entry. Then there is a positive integer $r \leq n$ for which the following conditions hold:
i) the matrix $A$ has an $r$th-order nonzero minor, and
ii) any minor of $A$ of order $(r + 1)$ and higher (of it exists) is equal to zero.

The integer $r$ satisfying these two conditions is called the *rank* of the matrix $A$ and is denoted by $r = \text{rank}(A)$. Any nonzero $r$th-order minor of the matrix $A$ is called its *basic minor*. The rows and the columns whose intersection yields its basic minor are called *basic rows* and *basic columns* of the matrix. The rank of a matrix is equal to the maximal number of its linearly independent rows (columns). This implies that for any matrix, the number of its linearly independent rows is equal to the number of its linearly independent columns.

When calculating the rank of a matrix $A$, one should pass from submatrices of a smaller size to those of a larger size. If, at some step, one finds a submatrix $A_k$ of size $k \times k$ such that it has a nonzero $k$th-order determinant and the $(k+1)$st-order determinants of all submatrices of size $(k+1) \times (k+1)$ containing $A_k$ are equal to zero, then it can be concluded that $k$ is the rank of the matrix $A$.

Properties of the rank of a matrix:

1. For any matrices $A$ and $B$ of the same size the following inequality holds:
$$\text{rank}(A + B) \leq \text{rank}(A) + \text{rank}(B).$$

2. For a matrix $A$ of size $m \times n$ and a matrix $B$ of size $n \times k$, the *Sylvester inequality* holds:
$$\text{rank}(A) + \text{rank}(B) - n \leq \text{rank}(AB) \leq \min\{\text{rank}(A), \text{rank}(B)\}.$$

For a square matrix $A$ of size $n \times n$, the value $d = n - \text{rank}(A)$ is called the *defect* of the matrix $A$, and $A$ is called a *$d$-fold degenerate matrix*. The rank of a nondegenerate square matrix $A \equiv [a_{ij}]$ of size $n \times n$ is equal to $n$.

THEOREM ON BASIC MINOR. *Basic rows (resp., basic columns) of a matrix are linearly independent. Any row (resp., any column) of a matrix is a linear combination of its basic rows (resp., columns).*

5.2.2-4. Expression of the determinant in terms of matrix entries.

1°. Consider a system of mutually distinct $\beta_1, \beta_2, \ldots, \beta_n$, with each $\beta_i$ taking one of the values $1, 2, \ldots, n$. In this case, the system $\beta_1, \beta_2, \ldots, \beta_n$ is called a *permutation* of the set $1, 2, \ldots, n$. If we interchange two elements in a given permutation $\beta_1, \beta_2, \ldots, \beta_n$, leaving the remaining $n-2$ elements intact, we obtain another permutation, and this transformation of $\beta_1, \beta_2, \ldots, \beta_n$ is called *transposition*. All permutations can be arranged in such an order that the next is obtained from the previous by a single transposition, and one can start from an arbitrary permutation.

**Example 2.** Let us demonstrate this statement in the case of $n = 3$ (there are $n! = 6$ permutations).
If we start from the permutation 1 2 3, then we can order all permutations, for instance, like this (we underline the numbers to be interchanged):
$$\underline{1}\,2\,\underline{3} \longrightarrow 2\,1\,\underline{3} \longrightarrow \underline{3}\,1\,2 \longrightarrow 1\,3\,\underline{2} \longrightarrow 2\,\underline{3}\,1 \longrightarrow 321.$$

Thus, from any given permutation of $n$ symbols, one can pass to any other permutation by finitely many transpositions.

One says that in a given permutation, the elements $\beta_i$ and $\beta_j$ form an *inversion* if $\beta_i > \beta_j$ for $i < j$. The total number of inversions in a permutation $\beta_1, \beta_2, \ldots, \beta_n$ is denoted by $N(\beta_1, \beta_2, \ldots, \beta_n)$. A permutation is said to be *even* if it contains an even number of inversions; otherwise, the permutation is said to be *odd*.

**Example 3.** The permutation 4 5 1 3 2 ($n = 5$) contains $N(4\,5\,1\,3\,2) = 7$ inversions and is, therefore, odd. Any of its transposition (for instance, that resulting in the permutation 4 3 1 5 2) yields an even permutation.

The $n$th-order determinant of a matrix $A \equiv [a_{ij}]$ of size $n \times n$ can be defined as follows:
$$\Delta \equiv \det A = \sum_{\beta_1, \beta_2, \ldots, \beta_n} (-1)^{N(\beta_1, \beta_2, \ldots, \beta_n)} a_{\beta_1 1} a_{\beta_2 2} \cdots a_{\beta_n n},$$

where the sum is over all possible permutations $\beta_1, \beta_2, \ldots, \beta_n$ of the set $1, 2, \ldots, n$.

**Example 4.** Using the last formula, let us calculate the third-order determinant of the matrix from Example 1. The numbers $\beta_1, \beta_2, \beta_3$ represent permutations of the set 1, 2, 3. We have

$$\Delta \equiv \det A = (-1)^{N(1,2,3)} a_{11} a_{22} a_{33} + (-1)^{N(1,3,2)} a_{11} a_{32} a_{23} + (-1)^{N(2,1,3)} a_{21} a_{12} a_{33}$$
$$+ (-1)^{N(2,3,1)} a_{21} a_{32} a_{13} + (-1)^{N(3,1,2)} a_{31} a_{12} a_{23} + (-1)^{N(3,2,1)} a_{31} a_{22} a_{13}$$
$$= (-1)^0 \times 1 \times 1 \times (-4) + (-1)^1 \times 1 \times (-1) \times 5 + (-1)^1 \times 6 \times (-1) \times (-4)$$
$$+ (-1)^2 \times 6 \times (-1) \times 2 + (-1)^2 \times 2 \times (-1) \times 5 + (-1)^3 \times 2 \times 1 \times 2 = -49,$$

which coincides with the result of Example 1.

2°. The $n$th-order determinant can also be defined as follows:

$$\Delta \equiv \det A = \sum_{\beta_1=1}^{n} \sum_{\beta_2=1}^{n} \cdots \sum_{\beta_n=1}^{n} \delta_{\beta_1 \beta_2 \ldots \beta_n} a_{\beta_1 1} a_{\beta_2 2} \cdots a_{\beta_n n},$$

where $\delta_{\beta_1 \beta_2 \ldots \beta_n}$ is the *Levi-Civita symbol*:

$$\delta_{\beta_1 \beta_2 \ldots \beta_n} = \begin{cases} 0, & \text{if some of } \beta_1, \beta_1, \ldots, \beta_n \text{ coincide,} \\ 1, & \text{if } \beta_1, \beta_1, \ldots, \beta_n \text{ form an even permutation,} \\ -1, & \text{if } \beta_1, \beta_1, \ldots, \beta_n \text{ form an odd permutation.} \end{cases}$$

### 5.2.2-5. Calculation of determinants.

1°. Determinants of specific matrices are often calculated with the help of the formulas for row expansion or column expansion (see Paragraph 5.2.2-1). For this purpose, its is convenient to take rows or columns containing many zero entries.

2°. The determinant of a triangular (upper or lower) and a diagonal matrices is equal to the product of its entries on the main diagonal. In particular, the determinant of the unit matrix is equal to 1.

3°. The determinant of a strictly triangular (upper or lower) matrix is equal to zero.

4°. For block matrices, the following formula can be used:

$$\begin{vmatrix} A & O \\ B & C \end{vmatrix} = \begin{vmatrix} A & B \\ O & C \end{vmatrix} = \det A \det C,$$

where $A, B, C$ are square matrices of size $n \times n$ and $O$ is the zero matrix of size $n \times n$.

5°. The *Vandermonde determinant* is the determinant of the *Vandermonde matrix*:

$$\Delta(x_1, x_2, \ldots, x_n) \equiv \begin{vmatrix} 1 & 1 & \cdots & 1 \\ x_1 & x_2 & \cdots & x_n \\ x_1^2 & x_2^2 & \cdots & x_n^2 \\ \vdots & \vdots & \ddots & \vdots \\ x_1^{n-1} & x_2^{n-1} & \cdots & x_n^{n-1} \end{vmatrix} = \prod_{1 \le j < i \le n} (x_i - x_j).$$

### 5.2.2-6. Determinant of a sum and a product of matrices.

The determinant of the product of two matrices $A$ and $B$ of the same size is equal to the product of their determinants,

$$\det(AB) = \det A \det B.$$

The determinant of the direct sum of a matrix $A$ of size $m \times m$ and $B$ of size $n \times n$ is equal to the product of their determinants,

$$\det(A \oplus B) = \det A \det B.$$

The determinant of the direct product of a matrix $A$ of size $m \times m$ and $B$ of size $n \times n$ is calculated by the formula

$$\det(A \otimes B) = (\det A)^n (\det B)^m.$$

**5.2.2-7. Relation between the determinant and the inverse matrix.**

EXISTENCE THEOREM. *A square matrix $A$ is invertible if and only if its determinant is different from zero.*

Remark. A square matrix $A$ is nondegenerate if its determinant is different from zero.

The *adjugate (classical adjoint)* for a matrix $A \equiv [a_{ij}]$ of size $n \times n$ is a matrix $C \equiv [c_{ij}]$ of size $n \times n$ whose entries coincide with the cofactors of the entries of the transpose $A^T$, i.e.,

$$c_{ij} = A_{ji} \quad (i, j = 1, 2, \ldots, n). \tag{5.2.2.9}$$

The inverse matrix of a square matrix $A \equiv [a_{ij}]$ of size $n \times n$ is the matrix of size $n \times n$ obtained from the adjugate matrix by dividing all its entries by $\det A$, i.e.,

$$A^{-1} = \begin{pmatrix} \frac{A_{11}}{\det A} & \frac{A_{21}}{\det A} & \cdots & \frac{A_{n1}}{\det A} \\ \frac{A_{12}}{\det A} & \frac{A_{22}}{\det A} & \cdots & \frac{A_{n2}}{\det A} \\ \vdots & \vdots & \ddots & \vdots \\ \frac{A_{1n}}{\det A} & \frac{A_{2n}}{\det A} & \cdots & \frac{A_{nn}}{\det A} \end{pmatrix}. \tag{5.2.2.10}$$

JACOBI THEOREM. *For minors of the matrix of cofactors of a matrix $A$, the following relations hold:*

$$\begin{vmatrix} A_{j_1}^{i_1} & A_{j_2}^{i_1} & \cdots & A_{j_k}^{i_1} \\ A_{j_1}^{i_2} & A_{j_2}^{i_2} & \cdots & A_{j_k}^{i_2} \\ \vdots & \vdots & \ddots & \vdots \\ A_{j_1}^{i_k} & A_{j_2}^{i_k} & \cdots & A_{j_k}^{i_k} \end{vmatrix} = (\det A)^{k-1} A_{j_1 j_2 \ldots j_k}^{i_1 i_2 \ldots i_k}.$$

## 5.2.3. Equivalent Matrices. Eigenvalues

**5.2.3-1. Equivalence transformation.**

Matrices $A$ and $\widetilde{A}$ of size $m \times n$ are said to be *equivalent* if there exist nondegenerate matrices $S$ and $T$ of size $m \times m$ and $n \times n$, respectively, such that $A$ and $\widetilde{A}$ are related by the *equivalence transformation*

$$\widetilde{A} = SAT.$$

THEOREM. *Two matrices of the same size are equivalent if and only if they are of the same rank.*

Remark 1. One of the square matrices $S$ and $T$ may coincide with the unit matrix. Thus, we have equivalent matrices $A$ and $B$ if there is a nondegenerate square matrix $S$ such that $\widetilde{A} = SA$ or $\widetilde{A} = AS$.

Remark 2. *Triangular decomposition* of a matrix $A$ corresponds to its equivalence transformation with $\widetilde{A} \equiv I$, so that $A = S^{-1}T^{-1} = LU$, where $L = S^{-1}$ and $P = T^{-1}$ are an upper and lower triangular matrix. This representation is also called the *LU-decomposition*.

Any equivalence transformation can be reduced to a sequence of *elementary transformations* of the following types:
1. Interchange of two rows (columns).
2. Multiplication of a row (column) by a nonzero scalar.
3. Addition to some row (column) of another row (column) multiplied by a scalar.

These elementary transformations are accomplished with the help of *elementary matrices* obtained from the unit matrix by the corresponding operations with its rows (columns). With the help of elementary transformations, an arbitrary matrix $A$ of rank $r > 0$ can be reduced to *normal (canonical) form*, which has a block structure with the unit matrix $I$ of size $r \times r$ in the top left corner.

**Example 1.** The $LU$-decomposition of a matrix

$$A = \begin{pmatrix} 2 & 1 & 4 \\ 3 & 2 & 1 \\ 1 & 3 & 3 \end{pmatrix}$$

can be obtained with the help of the following sequence of elementary transformations:

$$\overbrace{\begin{pmatrix} 1/2 & 0 & 0 \\ 0 & 1 & 0 \\ 0 & 0 & 1 \end{pmatrix}}^{S_1}\begin{pmatrix} 2 & 1 & 4 \\ 3 & 2 & 1 \\ 1 & 3 & 3 \end{pmatrix} \to \overbrace{\begin{pmatrix} 1 & 0 & 0 \\ -3 & 1 & 0 \\ 0 & 0 & 1 \end{pmatrix}}^{S_2}\begin{pmatrix} 1 & 1/2 & 2 \\ 3 & 2 & 1 \\ 1 & 3 & 3 \end{pmatrix} \to \overbrace{\begin{pmatrix} 1 & 0 & 0 \\ 0 & 1 & 0 \\ -1 & 0 & 1 \end{pmatrix}}^{S_3}\begin{pmatrix} 1 & 1/2 & 2 \\ 0 & 1/2 & -5 \\ 1 & 3 & 3 \end{pmatrix} \to$$

$$\to \begin{pmatrix} 1 & 1/2 & 2 \\ 0 & 1/2 & -5 \\ 0 & 5/2 & 1 \end{pmatrix}\overbrace{\begin{pmatrix} 1 & -1/2 & 0 \\ 0 & 1 & 0 \\ 0 & 0 & 1 \end{pmatrix}}^{T_1} \to \begin{pmatrix} 1 & 0 & 2 \\ 0 & 1/2 & -5 \\ 0 & 5/2 & 1 \end{pmatrix}\overbrace{\begin{pmatrix} 1 & 0 & -2 \\ 0 & 1 & 0 \\ 0 & 0 & 1 \end{pmatrix}}^{T_2} \to$$

$$\to \overbrace{\begin{pmatrix} 1 & 0 & 0 \\ 0 & 2 & 0 \\ 0 & 0 & 1 \end{pmatrix}}^{S_4}\begin{pmatrix} 1 & 0 & 0 \\ 0 & 1/2 & -5 \\ 0 & 5/2 & 1 \end{pmatrix} \to \overbrace{\begin{pmatrix} 1 & 0 & 0 \\ 0 & 1 & 0 \\ 0 & -5/2 & 1 \end{pmatrix}}^{S_5}\begin{pmatrix} 1 & 0 & 0 \\ 0 & 1 & -10 \\ 0 & 5/2 & 1 \end{pmatrix} \to$$

$$\to \begin{pmatrix} 1 & 0 & 0 \\ 0 & 1 & -10 \\ 0 & 0 & 26 \end{pmatrix}\overbrace{\begin{pmatrix} 1 & 0 & 0 \\ 0 & 1 & 10 \\ 0 & 0 & 1 \end{pmatrix}}^{T_3} \to \overbrace{\begin{pmatrix} 1 & 0 & 0 \\ 0 & 1 & 0 \\ 0 & 0 & 1/26 \end{pmatrix}}^{S_6}\begin{pmatrix} 1 & 0 & 0 \\ 0 & 1 & 0 \\ 0 & 0 & 26 \end{pmatrix} \to \begin{pmatrix} 1 & 0 & 0 \\ 0 & 1 & 0 \\ 0 & 0 & 1 \end{pmatrix}.$$

These transformations amount to the equivalence transformation $I = SAT$, where $T = T_1 T_2 T_3$:

$$S = S_6 S_5 S_4 S_3 S_2 S_1 = \begin{pmatrix} 1/2 & 0 & 0 \\ -3 & 2 & 0 \\ 7/26 & -5/26 & 1/26 \end{pmatrix} \quad \text{and} \quad T = T_1 T_2 T_3 = \begin{pmatrix} 1 & -1/2 & -7 \\ 0 & 1 & 10 \\ 0 & 0 & 1 \end{pmatrix}.$$

Hence, we obtain

$$L = S^{-1} = \begin{pmatrix} 2 & 0 & 0 \\ 3 & 1/2 & 0 \\ 1 & 5/2 & 26 \end{pmatrix} \quad \text{and} \quad U = T^{-1} = \begin{pmatrix} 1 & 1/2 & 2 \\ 0 & 1 & -10 \\ 0 & 0 & 1 \end{pmatrix}.$$

### 5.2.3-2. Similarity transformation.

Two square matrices $A$ and $\widetilde{A}$ of the same size are said to be *similar* if there exists a square nondegenerate matrix $S$ of the same size, the so-called *transforming matrix*, such that $A$ and $\widetilde{A}$ are related by the *similarity transformation*

$$\widetilde{A} = S^{-1} A S \quad \text{or} \quad A = S \widetilde{A} S^{-1}.$$

Properties of similar matrices:
1. If $A$ and $B$ are square matrices of the same size and $C = A + B$, then

$$\widetilde{C} = \widetilde{A} + \widetilde{B} \quad \text{or} \quad S^{-1}(A + B)S = S^{-1}AS + S^{-1}BS.$$

2. If $A$ and $B$ are square matrices of the same size and $C = AB$, then
$$\widetilde{C} = \widetilde{A}\widetilde{B} \quad \text{or} \quad S^{-1}(AB)S = (S^{-1}AS)(S^{-1}BS).$$
3. If $A$ is a square matrix and $C = \lambda A$ where $\lambda$ is a scalar, then
$$\widetilde{C} = \lambda \widetilde{B} \quad \text{or} \quad S^{-1}(\lambda B)T = \lambda S^{-1}BS.$$
4. Two similar matrices have the same rank, the same trace, and the same determinant.

Under some additional conditions, there exists a similarity transformation that turns a square matrix $A$ into a diagonal matrix with the eigenvalues of $A$ (see Paragraph 5.2.3-5) on the main diagonal. There are three cases in which a matrix can be reduced to diagonal form:
1. All eigenvalues of $A$ are mutually distinct (see Paragraph 5.2.3-5).
2. The defects of the matrices $A - \lambda_i I$ are equal to the multiplicities $m'_i$ of the corresponding eigenvalues $\lambda_i$ (see Paragraph 5.2.3-6). In this case, one says that the matrix has a *simple structure*.
3. Symmetric matrices.

For a matrix of general structure, one can only find a similarity transformation that reduces the matrix to the so-called *quasidiagonal canonical form* or the *canonical Jordan form* with a quasidiagonal structure. The main diagonal of the latter matrix consists of the eigenvalues of $A$, each repeated according to its multiplicity. The entries just above the main diagonal are equal either to 1 or 0. The other entries of the matrix are all equal to zero. The matrix in canonical Jordan form is a diagonal block matrix whose blocks form its main diagonal, each block being either a diagonal matrix or a so-called *Jordan cell* of the form

$$\Lambda_k \equiv \begin{pmatrix} \lambda_k & 1 & 0 & \cdots & 0 \\ 0 & \lambda_k & 1 & \cdots & 0 \\ 0 & 0 & \lambda_k & \cdots & 0 \\ \vdots & \vdots & \vdots & \ddots & \vdots \\ 0 & 0 & 0 & \cdots & \lambda_k \end{pmatrix}.$$

### 5.2.3-3. Congruent and orthogonal transformations.

Square matrices $A$ and $\widetilde{A}$ of the same size are said to be *congruent* if there is a nondegenerate square matrix $S$ such that $A$ and $\widetilde{A}$ are related by the so-called *congruent* or *congruence transformation*
$$\widetilde{A} = S^T A S \quad \text{or} \quad A = S\widetilde{A}S^T.$$

This transformation is characterized by the fact that it preserves the symmetry of the original matrix.

For any symmetric matrix $A$ of rank $r$ there is a congruent transformation that reduces $A$ to a canonical form which is a diagonal matrix of the form,
$$\widetilde{A} = S^T A S = \begin{pmatrix} I_p & & \\ & -I_{r-p} & \\ & & O \end{pmatrix},$$

where $I_p$ and $I_{r-p}$ are unit matrices of size $p \times p$ and $(r-p) \times (r-p)$. The number $p$ is called the *index* of the matrix $A$, and $s = p - (r-p) = 2p - r$ is called its *signature*.

THEOREM. *Two symmetric matrices are congruent if they are of the same rank and have the same index (or signature).*

A similarity transformation defined by an orthogonal matrix $S$ (i.e., $S^T = S^{-1}$) is said to be *orthogonal*. In this case
$$\widetilde{A} = S^{-1} A S = S^T A S.$$

**Example 2.** Consider a three-dimensional orthogonal coordinate system with the axes $OX_1, OX_2, OX_3$ and a new coordinate system obtained from this one by its rotation by the angle $\varphi$ around the axis $OX_3$, i.e.,

$$\widetilde{x}_1 = x_1 \cos\varphi - x_2 \sin\varphi, \quad \widetilde{x}_2 = x_1 \sin\varphi + x_2 \cos\varphi, \quad \widetilde{x}_3 = x_3.$$

The matrix of this coordinate transformation has the form

$$S_3 = \begin{pmatrix} \cos\varphi & -\sin\varphi & 0 \\ \sin\varphi & \cos\varphi & 0 \\ 0 & 0 & 1 \end{pmatrix}.$$

Rotations of the given coordinate system by the angles $\psi$ and $\theta$ around the axes $OX_1$ and $OX_2$, respectively, correspond to the matrices

$$S_1 = \begin{pmatrix} 1 & 0 & 0 \\ 0 & \cos\psi & -\sin\psi \\ 0 & \sin\psi & \cos\psi \end{pmatrix}, \quad S_2 = \begin{pmatrix} \cos\theta & 0 & \sin\theta \\ 0 & 1 & 0 \\ -\sin\theta & 0 & \cos\theta \end{pmatrix}.$$

The matrices $S_1, S_2, S_3$ are orthogonal ($S_j^{-1} = S_j^T$).

The transformation that consists of simultaneous rotations around of the coordinate axes by the angles $\psi, \theta, \varphi$ is defined by the matrix

$$S = S_3 S_2 S_1.$$

### 5.2.3-4. Conjunctive and unitary transformations.

$1°$. Square matrices $A$ and $\widetilde{A}$ of the same size are said to be *conjunctive* if there is a nondegenerate matrix $S$ such that $A$ and $\widetilde{A}$ are related by the *conjunctive transformation*

$$\widetilde{A} = S^* A S \quad \text{or} \quad A = S\widetilde{A}S^*,$$

where $S^*$ is the adjoint of $S$.

$2°$. A similarity transformation of a matrix $A$ is said to be *unitary* if it is defined by a unitary matrix $S$ (i.e., $S^* = S^{-1}$). In this case,

$$\widetilde{A} = S^{-1} A S = S^* A S.$$

Some basic properties of the above matrix transformations are listed in Table 5.3.

TABLE 5.3
Matrix transformations

| Transformation | $\widetilde{A}$ | Invariants |
|---|---|---|
| Equivalence | $SAT$ | Rank |
| Similarity | $S^{-1}AS$ | Rank, determinant, eigenvalues |
| Congruent | $S^T AS$ | Rank and symmetry |
| Orthogonal | $S^{-1}AS = S^T AS$ | Rank, determinant, eigenvalues, and symmetry |
| Conjunctive | $S^* AS$ | Rank and self-adjointness |
| Unitary | $S^{-1}AS = S^* AS$ | Rank, determinant, eigenvalues, and self-adjointness |

### 5.2.3-5. Eigenvalues and spectra of square matrices.

An *eigenvalue* of a square matrix $A$ is any real or complex $\lambda$ for which the matrix $F(\lambda) \equiv A - \lambda I$ is degenerate. The set of all eigenvalues of a matrix $A$ is called its *spectrum*, and $F(\lambda)$ is called its *characteristic matrix*. The inverse of an eigenvalue, $\mu = 1/\lambda$, is called a *characteristic value*.

Properties of spectrum of a matrices:
1. Similar matrices have the same spectrum.
2. If $\lambda$ is an eigenvalue of a normal matrix $A$ (see Paragraph 5.2.1-3), then $\bar{\lambda}$ is an eigenvalue of the matrix $A^*$; Re $\lambda$ is an eigenvalue of the matrix $H_1 = \frac{1}{2}(A + A^*)$; and Im $\lambda$ is an eigenvalue of the matrix $H_2 = \frac{1}{2i}(A - A^*)$.
3. All eigenvalues of a normal matrix are real if and only if this matrix is similar to a Hermitian matrix.
4. All eigenvalues of a unitary matrix have absolute values equal to 1.
5. A square matrix is nondegenerate if and only if all its eigenvalues are different from zero.

A nonzero (column) vector $X$ (see Paragraphs 5.2.1-1 and 5.2.1-2) satisfying the condition
$$AX = \lambda X$$
is called an *eigenvector* of the matrix $A$ corresponding to the eigenvalue $\lambda$. Eigenvectors corresponding to distinct eigenvalues of $A$ are linearly independent.

### 5.2.3-6. Reduction of a square matrix to triangular form.

THEOREM. *For any square matrix $A$ there exists a similarity transformation $\widetilde{A} = S^{-1}AS$ such that $\widetilde{A}$ is a triangular matrix.*

The diagonal entries of any triangular matrix similar to a square matrix $A$ of size $n \times n$ coincide with the eigenvalues of $A$; each eigenvalue $\lambda_i$ of $A$ occurs $m'_i \geq 1$ times on the diagonal. The positive integer $m'_i$ is called the *algebraic multiplicity* of the eigenvalue $\lambda_i$. Note that $\sum_i m'_i = n$.

The trace Tr($A$) is equal to the sum of all eigenvalues of $A$, each eigenvalue counted according to its multiplicity, i.e.,
$$\mathrm{Tr}(A) = \sum_i m'_i \lambda_i.$$

The determinant det $A$ is equal to the product of all eigenvalues of $A$, each eigenvalue counted according to its multiplicity, i.e.,
$$\det A = \prod_i \lambda_i^{m'_i}.$$

### 5.2.3-7. Reduction of a square matrix to diagonal form.

THEOREM 1. *If $A$ is a square matrix similar to some normal matrix, then there is a similarity transformation $\widetilde{A} = S^{-1}AS$ such that the matrix $\widetilde{A}$ is diagonal.*

THEOREM 2. *Two Hermitian matrices $A$ and $B$ can be reduced to diagonal form by the same similarity transformation if and only if $AB = BA$.*

THEOREM 3. *For any Hermitian matrix $A$, there is a nondegenerate matrix $S$ such that $\widetilde{A} = S^*AS$ is a diagonal matrix. The entries of $\widetilde{A}$ are real.*

THEOREM 4. *For any real symmetric matrix $A$, there is a real nondegenerate matrix $T$ such that $\widetilde{A} = S^T AS$ is a diagonal matrix.*

**Example 3.** Consider the real symmetric matrix

$$A = \begin{pmatrix} 11 & -6 & 2 \\ -6 & 10 & -4 \\ 2 & -4 & 6 \end{pmatrix}.$$

Its eigenvalues are $\lambda_1 = 18$, $\lambda_2 = 6$, $\lambda_3 = 3$ and the respective eigenvectors are

$$X_1 = \begin{pmatrix} 1 \\ 2 \\ 2 \end{pmatrix}, \quad X_2 = \begin{pmatrix} 2 \\ 1 \\ -2 \end{pmatrix}, \quad X_3 = \begin{pmatrix} 2 \\ -2 \\ 1 \end{pmatrix}.$$

Consider the matrix $S$ with the columns $X_1$, $X_2$, and $X_3$:

$$S = \begin{pmatrix} 1 & 2 & 2 \\ 2 & 1 & -2 \\ 2 & -2 & 1 \end{pmatrix}.$$

Taking $\widetilde{A}_1 = S^T A S$, we obtain a diagonal matrix:

$$\widetilde{A}_1 = S^T A S = \begin{pmatrix} 1 & 2 & 2 \\ 2 & 1 & -2 \\ 2 & -2 & 1 \end{pmatrix} \begin{pmatrix} 11 & -6 & 2 \\ -6 & 10 & -4 \\ 2 & -4 & 6 \end{pmatrix} \begin{pmatrix} 1 & 2 & 2 \\ 2 & 1 & -2 \\ 2 & -2 & 1 \end{pmatrix} = \begin{pmatrix} 27 & 0 & 0 \\ 0 & 54 & 0 \\ 0 & 0 & 162 \end{pmatrix}.$$

Taking $\widetilde{A}_2 = S^{-1} A S$, we obtain a diagonal matrix with the eigenvalues on the main diagonal:

$$\widetilde{A}_2 = S^{-1} A S = -\frac{1}{27} \begin{pmatrix} -3 & -6 & -6 \\ -6 & -3 & 6 \\ -6 & 6 & -3 \end{pmatrix} \begin{pmatrix} 11 & -6 & 2 \\ -6 & 10 & -4 \\ 2 & -4 & 6 \end{pmatrix} \begin{pmatrix} 1 & 2 & 2 \\ 2 & 1 & -2 \\ 2 & -2 & 1 \end{pmatrix} = \begin{pmatrix} 3 & 0 & 0 \\ 0 & 6 & 0 \\ 0 & 0 & 18 \end{pmatrix}.$$

We note that $\widetilde{A}_1 = 9\widetilde{A}_2$.

### 5.2.3-8. Characteristic equation of a matrix.

The algebraic equation of degree $n$

$$f_A(\lambda) \equiv \det(A - \lambda I) \equiv \det[a_{ij} - \lambda \delta_{ij}] \equiv \begin{vmatrix} a_{11} - \lambda & a_{12} & \cdots & a_{1n} \\ a_{21} & a_{22} - \lambda & \cdots & a_{2n} \\ \vdots & \vdots & \ddots & \vdots \\ a_{n1} & a_{n2} & \cdots & a_{nn} - \lambda \end{vmatrix} = 0$$

is called the *characteristic equation* of the matrix $A$ of size $n \times n$, and $f_A(\lambda)$ is called its *characteristic polynomial*. The spectrum of the matrix $A$ (i.e., the set of all its eigenvalues) coincides with the set of all roots of its characteristic equation. The multiplicity of every root $\lambda_i$ of the characteristic equation is equal to the multiplicity $m'_i$ of the eigenvalue $\lambda_i$.

**Example 4.** The characteristic equation of the matrix

$$A = \begin{pmatrix} 4 & -8 & 1 \\ 5 & -9 & 1 \\ 4 & -6 & -1 \end{pmatrix}$$

has the form

$$f_A(\lambda) \equiv \det \begin{pmatrix} 4-\lambda & -8 & 1 \\ 5 & -9-\lambda & 1 \\ 4 & -6 & -1-\lambda \end{pmatrix} = -\lambda^3 - 6\lambda^2 - 11\lambda - 6 = -(\lambda+1)(\lambda+2)(\lambda+3).$$

Similar matrices have the same characteristic equation.

Let $\lambda_j$ be an eigenvalue of a square matrix $A$. Then
1) $\alpha \lambda_j$ is an eigenvalue of the matrix $\alpha A$ for any scalar $\alpha$;
2) $\lambda_j^p$ is an eigenvalue of the matrix $A^p$ ($p = 0, \pm 1, \ldots, \pm N$ for a nondegenerate $A$; otherwise, $p = 0, 1, \ldots, N$), where $N$ is a natural number;
3) a polynomial $f(A)$ of the matrix $A$ has the eigenvalue $f(\lambda)$.

Suppose that the spectra of matrices $A$ and $B$ consist of eigenvalues $\lambda_j$ and $\mu_k$, respectively. Then the spectrum of the Kronecker product $A \otimes B$ is the set of all products $\lambda_j \mu_k$. The spectrum of the direct sum of matrices $A = A_1 \oplus \ldots \oplus A_n$ is the union of the spectra of the matrices $A_1, \ldots, A_n$. The algebraic multiplicities of the same eigenvalues of matrices $A_1, \ldots, A_n$ are summed.

Regarding bounds for eigenvalues see Paragraph 5.6.3-4.

## 5.2.3-9. Cayley–Hamilton theorem. Sylvester theorem.

CAYLEY–HAMILTON THEOREM. *Each square matrix $A$ satisfies its own characteristic equation; i.e., $f_A(A) = 0$.*

**Example 5.** Let us illustrate the Cayley–Hamilton theorem by the matrix in Example 4:

$$f_A(A) = -A^3 - 6A^2 - 11A - 6I$$

$$= -\begin{pmatrix} 70 & -116 & 19 \\ 71 & -117 & 19 \\ 64 & -102 & 11 \end{pmatrix} - 6\begin{pmatrix} -20 & 34 & -5 \\ -21 & 35 & -5 \\ -18 & 28 & -1 \end{pmatrix} - 11\begin{pmatrix} 4 & -8 & 1 \\ 5 & -9 & 1 \\ 4 & -6 & -1 \end{pmatrix} - 6\begin{pmatrix} 1 & 0 & 0 \\ 0 & 1 & 0 \\ 0 & 0 & 1 \end{pmatrix} = 0.$$

A scalar polynomial $p(\lambda)$ is called an *annihilating polynomial* of a square matrix $A$ if $p(A) = 0$. For example, the characteristic polynomial $f_A(\lambda)$ is an annihilating polynomial of $A$. The unique monic annihilating polynomial of least degree is called the *minimal polynomial* of $A$ and is denoted by $\psi(\lambda)$. The minimal polynomial is a divisor of every annihilating polynomial.

By dividing an arbitrary polynomial $f(\lambda)$ of degree $n$ by an annihilating polynomial $p(\lambda)$ of degree $m$ ($p(\lambda) \neq 0$), one obtains the representation

$$f(\lambda) = p(\lambda)q(\lambda) + r(\lambda),$$

where $q(\lambda)$ is a polynomial of degree $n - m$ (if $m \leq n$) or $q(\lambda) = 0$ (if $m > n$) and $r(\lambda)$ is a polynomial of degree $l < m$ or $r(\lambda) = 0$. Hence

$$f(A) = p(A)q(A) + r(A),$$

where $p(A) = 0$ and $f(A) = r(A)$. The polynomial $r(\lambda)$ in this representation is called the *interpolation polynomial* of $A$.

**Example 6.** Let

$$f(A) = A^4 + 4A^3 + 2A^2 - 12A - 10I,$$

where the matrix $A$ is defined in Example 4. Dividing $f(\lambda)$ by the characteristic polynomial $f_A(\lambda) = -\lambda^3 - 6\lambda^2 - 11\lambda - 6$, we obtain the remainder $r(\lambda) = 3\lambda^2 + 4\lambda + 2$. Consequently,

$$f(A) = r(A) = 3A^2 + 4A + 2I.$$

THEOREM. *Every analytic function of a square $n \times n$ matrix $A$ can be represented as a polynomial of the same matrix,*

$$f(A) = \frac{1}{\Delta(\lambda_1, \lambda_2, \ldots, \lambda_n)} \sum_{k=1}^{n} \Delta_{n-k} A^{n-k},$$

where $\Delta(\lambda_1, \lambda_2, \ldots, \lambda_n)$ is the Vandermonde determinant and $\Delta_i$ is obtained from $\Delta$ by replacing the $(i + 1)$st row by $(f(\lambda_1), f(\lambda_2), \ldots, f(\lambda_n))$.

**Example 7.** Let us find $r(A)$ by this formula for the polynomial in Example 6.

We find the eigenvalues of $A$ from the characteristic equation $f_A(\lambda) = 0$: $\lambda_1 = -1$, $\lambda_2 = -2$, and $\lambda_3 = -3$. Then the Vandermonde determinant is equal to $\Delta(\lambda_1, \lambda_2, \lambda_3) = -2$, and the other determinants are $\Delta_1 = -4$, $\Delta_2 = -8$, and $\Delta_3 = -6$. It follows that

$$f(A) = \frac{1}{-2}[(-6)A^2 + (-8)A + (-4)I] = 3A^2 + 4A + 2I.$$

The Cayley–Hamilton theorem can also be used to find the powers and the inverse of a matrix $A$ (since if $f_A(A) = 0$, then $A^k f_A(A) = 0$ for any positive integer $k$).

**Example 8.** For the matrix in Examples 4–7, one has
$$f_A(A) = -A^3 - 6A^2 - 11A - 6I = 0.$$
Hence we obtain
$$A^3 = -6A^2 - 11A - 6I.$$
By multiplying this expression by $A$, we obtain
$$A^4 = -6A^3 - 11A^2 - 6A.$$
Now we use the representation of the cube of $A$ via lower powers of $A$ and eventually arrive at the formula
$$A^4 = 25A^2 + 60A + 36I.$$
For the inverse matrix, by analogy with the preceding, we obtain
$$A^{-1} f_A(A) = A^{-1}(-A^3 - 6A^2 - 11A - 6I) = -A^2 - 6A - 11I - 6A^{-1} = 0.$$
The definitive result is
$$A^{-1} = -\frac{1}{6}(A^2 + 6A + 11I).$$

In some cases, an analytic function of a matrix $A$ can be computed by a formula in the following theorem.

SYLVESTER'S THEOREM. *If all eigenvalues of a matrix $A$ are distinct, then*
$$f(A) = \sum_{k=1}^{n} f(\lambda_k) Z_k, \quad Z_k = \frac{\prod_{i \neq k}(A - \lambda_i I)}{\prod_{i \neq k}(\lambda_k - \lambda_i)},$$
*and, moreover,* $Z_k = Z_k^m$ ($m = 1, 2, 3, \ldots$).

## 5.3. Linear Spaces

### 5.3.1. Concept of a Linear Space. Its Basis and Dimension

#### 5.3.1-1. Definition of a linear space.

A *linear space* or a *vector space* over a field of *scalars* (usually, the field of real numbers or the field of complex numbers) is a set $\mathcal{V}$ of elements $\mathbf{x}$, $\mathbf{y}$, $\mathbf{z}$, ... (also called *vectors*) of any nature for which the following conditions hold:
I. There is a rule that establishes correspondence between any pair of elements $\mathbf{x}, \mathbf{y} \in \mathcal{V}$ and a third element $\mathbf{z} \in \mathcal{V}$, called the *sum* of the elements $\mathbf{x}$, $\mathbf{y}$ and denoted by $\mathbf{z} = \mathbf{x} + \mathbf{y}$.
II. There is a rule that establishes correspondence between any pair $\mathbf{x}$, $\lambda$, where $\mathbf{x}$ is an element of $\mathcal{V}$ and $\lambda$ is a scalar, and an element $\mathbf{u} \in \mathcal{V}$, called the *product of a scalar $\lambda$ and a vector $\mathbf{x}$* and denoted by $\mathbf{u} = \lambda \mathbf{x}$.
III. The following eight axioms are assumed for the above two operations:
   1. Commutativity of the sum: $\mathbf{x} + \mathbf{y} = \mathbf{y} + \mathbf{x}$.
   2. Associativity of the sum: $(\mathbf{x} + \mathbf{y}) + \mathbf{z} = \mathbf{x} + (\mathbf{y} + \mathbf{z})$.
   3. There is a zero element $\mathbf{0}$ such that $\mathbf{x} + \mathbf{0} = \mathbf{x}$ for any $\mathbf{x}$.
   4. For any element $\mathbf{x}$ there is an opposite element $\mathbf{x}'$ such that $\mathbf{x} + \mathbf{x}' = \mathbf{0}$.
   5. A special role of the unit scalar 1: $1 \cdot \mathbf{x} = \mathbf{x}$ for any element $\mathbf{x}$.
   6. Associativity of the multiplication by scalars: $\lambda(\mu \mathbf{x}) = (\lambda \mu) \mathbf{x}$.
   7. Distributivity with respect to the addition of scalars: $(\lambda + \mu)\mathbf{x} = \lambda \mathbf{x} + \mu \mathbf{x}$.
   8. Distributivity with respect to a sum of vectors: $\lambda(\mathbf{x} + \mathbf{y}) = \lambda \mathbf{x} + \lambda \mathbf{y}$.

This is the definition of an abstract linear space. We obtain a *specific linear space* if the nature of the elements and the operations of addition and multiplication by scalars are concretized.

**Example 1.** Consider the set of all free vectors in three-dimensional space. If addition of these vectors and their multiplication by scalars are defined as in analytic geometry (see Paragraph 4.5.1-1), this set becomes a linear space denoted by $B_3$.

**Example 2.** Consider the set $\{x\}$ whose elements are all positive real numbers. Let us define the sum of two elements $x$ and $y$ as the product of $x$ and $y$, and define the product of a real scalar $\lambda$ and an element $x$ as the $\lambda$th power of the positive real $x$. The number 1 is taken as the zero element of the space $\{x\}$, and the opposite of $x$ is taken equal to $1/x$. It is easy to see that the set $\{x\}$ with these operations of addition and multiplication by scalars is a linear space.

**Example 3.** Consider the *n-dimensional coordinate space* $\mathbb{R}^n$, whose elements are ordered sets of $n$ arbitrary real numbers $(x_1, \ldots, x_n)$. The generic element of this space is denoted by $\mathbf{x}$, i.e., $\mathbf{x} = (x_1, \ldots, x_n)$, and the reals $x_1, \ldots, x_n$ are called the *coordinates* of the element $\mathbf{x}$. From the algebraic standpoint, the set $\mathbb{R}^n$ may be regarded as the set of all row vectors with $n$ real components.

The operations of addition of element of $\mathbb{R}^n$ and their multiplication by scalars are defined by the following rules:
$$(x_1, \ldots, x_n) + (y_1, \ldots, y_n) = (x_1 + y_1, \ldots, x_n + y_n),$$
$$\lambda(x_1, \ldots, x_n) = (\lambda x_1, \ldots, \lambda x_n).$$

**Remark.** If the field of scalars $\lambda, \mu, \ldots$ in the above definition is the field of all real numbers, the corresponding linear spaces are called *real linear spaces*. If the field of scalars is that of all complex numbers, the corresponding space is called a *complex linear space*. In many situations, it is clear from the context which field of scalars is meant.

The above axioms imply the following properties of an arbitrary linear space:

1. The zero vector is unique, and for any element $\mathbf{x}$ the opposite element is unique.
2. The zero vector 0 is equal to the product of any element $\mathbf{x}$ by the scalar 0.
3. For any element $\mathbf{x}$, the opposite element is equal to the product of $\mathbf{x}$ by the scalar $-1$.
4. The *difference* of two elements $\mathbf{x}$ and $\mathbf{y}$, i.e., the element $\mathbf{z}$ such that $\mathbf{z} + \mathbf{y} = \mathbf{x}$, is unique.

---

**5.3.1-2. Basis and dimension of a linear space. Isomorphisms of linear spaces.**

An element $\mathbf{y}$ is called a *linear combination* of elements $\mathbf{x}_1, \ldots, \mathbf{x}_k$ of a linear space $\mathcal{V}$ if there exist scalars $\alpha_1, \ldots, \alpha_k$ such that

$$\mathbf{y} = \alpha_1 \mathbf{x}_1 + \cdots + \alpha_k \mathbf{x}_k.$$

Elements $\mathbf{x}_1, \ldots, \mathbf{x}_k$ of the space $\mathcal{V}$ are said to be *linearly dependent* if there exist scalars $\alpha_1, \ldots, \alpha_k$ such that $|\alpha_1|^2 + \cdots + |\alpha_k|^2 \neq 0$ and

$$\alpha_1 \mathbf{x}_1 + \cdots + \alpha_k \mathbf{x}_k = 0,$$

where 0 is the zero element of $\mathcal{V}$.

Elements $\mathbf{x}_1, \ldots, \mathbf{x}_k$ of the space $\mathcal{V}$ are said to be *linearly independent* if for any scalars $\alpha_1, \ldots, \alpha_k$ such that $|\alpha_1|^2 + \cdots + |\alpha_k|^2 \neq 0$, we have

$$\alpha_1 \mathbf{x}_1 + \cdots + \alpha_k \mathbf{x}_k \neq 0.$$

**THEOREM.** *Elements $\mathbf{x}_1, \ldots, \mathbf{x}_k$ of a linear space $\mathcal{V}$ are linearly dependent if and only if one of them is a linear combination of the others.*

**Remark.** If at least one of the elements $\mathbf{x}_1, \ldots, \mathbf{x}_k$ is equal to zero, then these elements are linearly dependent. If some of the elements $\mathbf{x}_1, \ldots, \mathbf{x}_k$ are linearly dependent, then all these elements are linearly dependent.

**Example 4.** The elements $\mathbf{i}_1 = (1, 0, \ldots, 0)$, $\mathbf{i}_2 = (0, 1, \ldots, 0), \ldots, \mathbf{i}_n = (0, 0, \ldots, 1)$ of the space $\mathbb{R}^n$ (see Example 3) are linearly independent. For any $\mathbf{x} = (x_1, \ldots, x_n) \in \mathbb{R}^n$, the vectors $\mathbf{x}, \mathbf{i}_1, \ldots, \mathbf{i}_n$ are linearly dependent.

A *basis* of a linear space $\mathcal{V}$ is defined as any system of linearly independent vectors $\mathbf{e}_1, \ldots, \mathbf{e}_n$ such that for any element $\mathbf{x}$ of the space $\mathcal{V}$ there exist scalars $x_1, \ldots, x_n$ such that

$$\mathbf{x} = x_1 \mathbf{e}_1 + \cdots + x_n \mathbf{e}_n.$$

This relation is called the *representation of an element* $\mathbf{x}$ *in terms of the basis* $\mathbf{e}_1, \ldots, \mathbf{e}_n$, and the scalars $x_1, \ldots, x_n$ are called the *coordinates* of the element $\mathbf{x}$ in that basis.

UNIQUENESS THEOREM. *The representation of any element* $\mathbf{x} \in \mathcal{V}$ *in terms of a given basis* $\mathbf{e}_1, \ldots, \mathbf{e}_n$ *is unique.*

Let $\mathbf{e}_1, \ldots, \mathbf{e}_n$ be any basis in $\mathcal{V}$ and vectors $\mathbf{x}$ and $\mathbf{y}$ have the coordinates $x_1, \ldots, x_n$ and $y_1, \ldots, y_n$ in that basis. Then the coordinates of the vector $\mathbf{x} + \mathbf{y}$ in that basis are $x_1 + y_1, \ldots, x_n + y_n$, and the coordinates of the vector $\lambda \mathbf{x}$ are $\lambda x_1, \ldots, \lambda x_n$ for any scalar $\lambda$.

**Example 5.** Any three noncoplanar vectors form a basis in the linear space $B_3$ of all free vectors. The $n$ elements $\mathbf{i}_1 = (1, 0, \ldots, 0)$, $\mathbf{i}_2 = (0, 1, \ldots, 0), \ldots, \mathbf{i}_n = (0, 0, \ldots, 1)$ form a basis in the linear space $\mathbb{R}^n$. Any basis of the linear space $\{x\}$ from Example 2 consists of a single element. This element can be arbitrarily chosen of nonzero elements of this space.

A linear space $\mathcal{V}$ is said to be *$n$-dimensional* if it contains $n$ linearly independent elements and any $n + 1$ elements are linearly dependent. The number $n$ is called the *dimension* of that space, $n = \dim \mathcal{V}$.

A linear space $\mathcal{V}$ is said to be *infinite-dimensional* ($\dim \mathcal{V} = \infty$) if for any positive integer $N$ it contains $N$ linearly independent elements.

THEOREM 1. *If $\mathcal{V}$ is a linear space of dimension $n$, then any $n$ linearly independent elements of that space form its basis.*

THEOREM 2. *If a linear space $\mathcal{V}$ has a basis consisting of $n$ elements, then $\dim \mathcal{V} = n$.*

**Example 6.** The dimension of the space $B_3$ of all vectors is equal to 3. The dimension of the space $\mathbb{R}^n$ is equal to $n$. The dimension of the space $\{x\}$ is equal to 1.

Two linear spaces $\mathcal{V}$ and $\mathcal{V}'$ over the same field of scalars are said to be *isomorphic* if there is a one-to-one correspondence between the elements of these spaces such that if elements $\mathbf{x}$ and $\mathbf{y}$ from $\mathcal{V}$ correspond to elements $\mathbf{x}'$ and $\mathbf{y}'$ from $\mathcal{V}'$, then the element $\mathbf{x} + \mathbf{y}$ corresponds to $\mathbf{x}' + \mathbf{y}'$ and the element $\lambda \mathbf{x}$ corresponds to $\lambda \mathbf{x}'$ for any scalar $\lambda$.

**Remark.** If linear spaces $\mathcal{V}$ and $\mathcal{V}'$ are isomorphic, then the zero element of one space corresponds to the zero element of the other.

THEOREM. *Any two $n$-dimensional real (or complex) spaces $\mathcal{V}$ and $\mathcal{V}'$ are isomorphic.*

5.3.1-3. Affine space.

An *affine space* is a nonempty set $\mathcal{A}$ that consists of elements of any nature, called *points*, for which the following conditions hold:
 I. There is a given linear (vector) space $\mathcal{V}$, called the *associated linear space*.
 II. There is a rule by which any ordered pair of points $A, B \in \mathcal{A}$ is associated with an element (vector) from $\mathcal{V}$; this vector is denoted by $\overrightarrow{AB}$ and is called the *vector issuing from the point $A$ with endpoint at $B$*.
 III. The following conditions (called *axioms of affine space*) hold:
  1. For any point $A \in \mathcal{A}$ and any vector $\mathbf{a} \in \mathcal{V}$, there is a unique point $B \in \mathcal{A}$ such that $\overrightarrow{AB} = \mathbf{a}$.
  2. $\overrightarrow{AB} + \overrightarrow{BC} = \overrightarrow{AC}$ for any three points $A, B, C \in \mathcal{A}$.

By definition, the *dimension of an affine space* $\mathcal{A}$ is the dimension of the associated linear space $\mathcal{V}$, $\dim \mathcal{A} = \dim \mathcal{V}$.

Any linear space may be regarded as an affine space.

In particular, the space $\mathbb{R}^n$ can be naturally considered as an affine space. Thus if $A = (a_1, \ldots, a_n)$ and $B = (b_1, \ldots, b_n)$ are points of the affine space $\mathbb{R}^n$, then the corresponding vector $\overrightarrow{AB}$ from the linear space $\mathbb{R}^n$ is defined by $\overrightarrow{AB} = (b_1 - a_1, \ldots, b_n - a_n)$.

Let $\mathcal{A}$ be an $n$-dimensional affine space with the associated linear space $\mathcal{V}$. A *coordinate system* in the affine space $\mathcal{A}$ is a fixed point $O \in \mathcal{A}$, together with a fixed basis $\mathbf{e}_1, \ldots, \mathbf{e}_n \in \mathcal{V}$. The point $O$ is called the *origin* of this coordinate system.

Let $M$ be a point of an affine space $\mathcal{A}$ with a coordinate system $O\mathbf{e}_1 \ldots \mathbf{e}_n$. One says that the point $M$ has *affine coordinates* (or simply coordinates) $x_1, \ldots, x_n$ in this coordinate system, and one writes $M = (x_1, \ldots, x_n)$ if $x_1, \ldots x_n$ are the coordinates of the radius-vector $\overrightarrow{OM}$ in the basis $\mathbf{e}_1, \ldots, \mathbf{e}_n$, i.e., $\overrightarrow{OM} = x_1 \mathbf{e}_1 + \cdots + x_n \mathbf{e}_n$.

## 5.3.2. Subspaces of Linear Spaces

**5.3.2-1. Concept of a linear subspace and a linear span.**

A subset $\mathcal{L}$ of a linear space $\mathcal{V}$ is called a *linear subspace* of $\mathcal{V}$ if the following conditions hold:

1. If $\mathbf{x}$ and $\mathbf{y}$ belong to $\mathcal{L}$, then the sum $\mathbf{x} + \mathbf{y}$ belongs to $\mathcal{L}$.
2. If $\mathbf{x}$ belongs to $\mathcal{L}$ and $\lambda$ is an arbitrary scalar, then the element $\lambda \mathbf{x}$ belongs to $\mathcal{L}$.

The *null subspace* in a linear space $\mathcal{V}$ is its subset consisting of the single element zero. The space $\mathcal{V}$ itself can be regarded as its own subspace. These two subspaces are called *improper subspaces*. All other subspaces are called *proper subspaces*.

**Example 1.** A subset $B_2$ consisting of all free vectors parallel to a given plane is a subspace in the linear space $B_3$ of all free vectors.

The *linear span* $L(\mathbf{x}_1, \ldots, \mathbf{x}_m)$ of vectors $\mathbf{x}_1, \ldots, \mathbf{x}_m$ in a linear space $\mathcal{V}$ is, by definition, the set of all linear combinations of these vectors, i.e., the set of all vectors of the form

$$\alpha_1 \mathbf{x}_1 + \cdots + \alpha_m \mathbf{x}_m,$$

where $\alpha_1, \ldots, \alpha_m$ are arbitrary scalars. The linear span $L(\mathbf{x}_1, \ldots, \mathbf{x}_m)$ is the least subspace of $\mathcal{V}$ containing the elements $\mathbf{x}_1, \ldots, \mathbf{x}_m$.

If a subspace $\mathcal{L}$ of an $n$-dimensional space $\mathcal{V}$ does not coincide with $\mathcal{V}$, then $\dim \mathcal{L} < n = \dim \mathcal{V}$.

Let elements $\mathbf{e}_1, \ldots, \mathbf{e}_k$ form a basis in a $k$-dimensional subspace of an $n$-dimensional linear space $\mathcal{V}$. Then this basis can be supplemented by elements $\mathbf{e}_{k+1}, \ldots, \mathbf{e}_n$ of the space $\mathcal{V}$, so that the system $\mathbf{e}_1, \ldots, \mathbf{e}_k, \mathbf{e}_{k+1}, \ldots, \mathbf{e}_n$ forms a basis in the space $\mathcal{V}$.

THEOREM OF THE DIMENSION OF A LINEAR SPAN. *The dimension of a linear span $L(\mathbf{x}_1, \ldots, \mathbf{x}_m)$ of elements $\mathbf{x}_m, \ldots, \mathbf{x}_m$ is equal to the maximal number of linearly independent vectors in the system $\mathbf{x}_1, \ldots, \mathbf{x}_m$.*

**5.3.2-2. Sum and intersection of subspaces.**

The *intersection* of subspaces $\mathcal{L}_1$ and $\mathcal{L}_2$ of one and the same linear space $\mathcal{V}$ is, by definition, the set of all elements $\mathbf{x}$ of $\mathcal{V}$ that belong simultaneously to both spaces $\mathcal{L}_1$ and $\mathcal{L}_2$. Such elements form a subspace of $\mathcal{V}$.

The *sum* of subspaces $\mathcal{L}_1$ and $\mathcal{L}_2$ of one and the same linear space $\mathcal{V}$ is, by definition, the set of all elements of $\mathcal{V}$ that can be represented in the form $\mathbf{y} + \mathbf{z}$, where $\mathbf{y}$ is an element of $\mathcal{V}_1$ and $\mathbf{z}$ is an element of $\mathcal{L}_2$. The sum of subspaces is also a subspace of $\mathcal{V}$.

THEOREM. *The sum of dimensions of arbitrary subspaces $\mathcal{L}_1$ and $\mathcal{L}_2$ of a finite-dimensional space $\mathcal{V}$ is equal to the sum of the dimension of their intersection and the dimension of their sum.*

**Example 2.** Let $\mathcal{V}$ be the linear space of all free vectors (in three-dimensional space). Denote by $\mathcal{L}_1$ the subspace of all free vectors parallel to the plane $OXY$, and by $\mathcal{L}_2$ the subspace of all free vectors parallel to the plane $OXZ$. Then the sum of the subspaces $\mathcal{L}_1$ and $\mathcal{L}_2$ coincides with $\mathcal{V}$, and their intersection consists of all free vectors parallel to the axis $OX$.

The dimension of each space $\mathcal{L}_1$ and $\mathcal{L}_2$ is equal to two, the dimension of their sum is equal to three, and the dimension of their intersection is equal to unity.

**5.3.2-3. Representation of a linear space as a direct sum of its subspaces.**

A linear space $\mathcal{V}$ can be represented as a *direct sum* of its subspaces, $\mathcal{V}_1$ and $\mathcal{V}_2$ if each element $\mathbf{x} \in \mathcal{V}$ admits the unique representation $\mathbf{x} = \mathbf{x}_1 + \mathbf{x}_2$, where $\mathbf{x}_1 \in \mathcal{V}_1$ and $\mathbf{x}_2 \in \mathcal{V}_2$. In this case, one writes $\mathcal{V} = \mathcal{V}_1 \oplus \mathcal{V}_2$.

**Example 3.** The space $\mathcal{V}$ of all free vectors (in three-dimensional space) can be represented as the direct sum of the subspace $\mathcal{V}_1$ formed by all free vectors parallel to the plane $OXY$ and the subspace $\mathcal{V}_2$ formed by all free vectors parallel to the axis $OZ$.

THEOREM. *An $n$-dimensional space $\mathcal{V}$ is a direct sum of its subspaces $\mathcal{V}_1$ and $\mathcal{V}_2$ if and only if the intersection of $\mathcal{V}_1$ and $\mathcal{V}_2$ is the null subspace and $\dim \mathcal{V} = \dim \mathcal{V}_1 + \dim \mathcal{V}_2$.*

**Remark.** If $R$ is the sum of its subspaces $R_1$ and $R_2$, but not the direct sum, then the representation $\mathbf{x} = \mathbf{x}_1 + \mathbf{x}_2$ is nonunique, in general.

## 5.3.3. Coordinate Transformations Corresponding to Basis Transformations in a Linear Space

**5.3.3-1. Basis transformation and its inverse.**

Let $\mathbf{e}_1, \ldots, \mathbf{e}_n$ and $\widetilde{\mathbf{e}}_1, \ldots, \widetilde{\mathbf{e}}_n$ be two arbitrary bases of an $n$-dimensional linear space $\mathcal{V}$. Suppose that the elements $\widetilde{\mathbf{e}}_1, \ldots, \widetilde{\mathbf{e}}_n$ are expressed via $\mathbf{e}_1, \ldots, \mathbf{e}_n$ by the formulas

$$\widetilde{\mathbf{e}}_1 = a_{11}\mathbf{e}_1 + a_{12}\mathbf{e}_2 + \cdots + a_{1n}\mathbf{e}_n,$$
$$\widetilde{\mathbf{e}}_2 = a_{21}\mathbf{e}_1 + a_{22}\mathbf{e}_2 + \cdots + a_{2n}\mathbf{e}_n,$$
$$\cdots\cdots\cdots\cdots\cdots\cdots\cdots\cdots\cdots\cdots$$
$$\widetilde{\mathbf{e}}_n = a_{n1}\mathbf{e}_1 + a_{n2}\mathbf{e}_2 + \cdots + a_{nn}\mathbf{e}_n.$$

Thus, the transition from the basis $\mathbf{e}_1, \ldots, \mathbf{e}_n$ to the basis $\widetilde{\mathbf{e}}_1, \ldots, \widetilde{\mathbf{e}}_n$ is determined by the matrix

$$A \equiv \begin{pmatrix} a_{11} & a_{12} & \cdots & a_{1n} \\ a_{21} & a_{22} & \cdots & a_{2n} \\ \vdots & \vdots & \ddots & \vdots \\ a_{n1} & a_{n2} & \cdots & a_{nn} \end{pmatrix}.$$

Note that $\det A \neq 0$, i.e., the matrix $A$ is nondegenerate.

The transition from the basis $\widetilde{\mathbf{e}}_1, \ldots, \widetilde{\mathbf{e}}_n$ to the basis $\mathbf{e}_1, \ldots, \mathbf{e}_n$ is determined by the matrix $B \equiv [b_{ij}] = A^{-1}$. Thus, we can write

$$\widetilde{\mathbf{e}}_i = \sum_{j=1}^{n} a_{ij}\mathbf{e}_j, \qquad \mathbf{e}_k = \sum_{j=1}^{n} b_{kj}\widetilde{\mathbf{e}}_j \qquad (i, k = 1, 2, \ldots, n). \tag{5.3.3.1}$$

**5.3.3-2. Relations between coordinate transformations and basis transformations.**

Suppose that in a linear $n$-dimensional space $\mathcal{V}$, the transition from its basis $\mathbf{e}_1, \ldots, \mathbf{e}_n$ to another basis $\widetilde{\mathbf{e}}_1, \ldots, \widetilde{\mathbf{e}}_n$ is determined by the matrix $A$ (see Paragraph 5.3.3-1). Let $\mathbf{x}$ be any element of the space $\mathcal{V}$ with the coordinates $(x_1, \ldots, x_n)$ in the basis $\mathbf{e}_1, \ldots, \mathbf{e}_n$ and the coordinates $(\widetilde{x}_1, \ldots, \widetilde{x}_n)$ in the basis $\widetilde{\mathbf{e}}_1, \ldots, \widetilde{\mathbf{e}}_n$, i.e.,

$$\mathbf{x} = x_1 \mathbf{e}_1 + \cdots + x_n \mathbf{e}_n = \widetilde{x}_1 \widetilde{\mathbf{e}}_1 + \cdots + \widetilde{x}_n \widetilde{\mathbf{e}}_n.$$

Then using formulas (5.3.3.1), we obtain the following relations between these coordinates:

$$x_j = \sum_{i=1}^n \widetilde{x}_i a_{ij}, \qquad \widetilde{x}_k = \sum_{l=1}^n x_l b_{lk}, \qquad j, k = 1, \ldots, n.$$

In terms of matrices and row vectors, these relations can be written as follows:

$$(x_1, \ldots, x_n) = (\widetilde{x}_1, \ldots, \widetilde{x}_n) A, \qquad (\widetilde{x}_1, \ldots, \widetilde{x}_n) = (x_1, \ldots, x_n) A^{-1}$$

or, in terms of column vectors,

$$(x_1, \ldots, x_n)^T = A^T (\widetilde{x}_1, \ldots, \widetilde{x}_n)^T, \qquad (\widetilde{x}_1, \ldots, \widetilde{x}_n)^T = (A^{-1})^T (x_1, \ldots, x_n)^T,$$

where the superscript $T$ indicates the transpose of a matrix.

## 5.4. Euclidean Spaces

### 5.4.1. Real Euclidean Space

**5.4.1-1. Definition and properties of a real Euclidean space.**

A *real Euclidean space* (or simply, *Euclidean space*) is a real linear space $\mathcal{V}$ endowed with a *scalar product* (also called *inner product*), which is a real-valued function of two arguments $\mathbf{x} \in \mathcal{V}, \mathbf{y} \in \mathcal{V}$ called the scalar product of these elements, denoted by $\mathbf{x} \cdot \mathbf{y}$, and satisfying the following conditions (axioms of the scalar product):

1. Symmetry: $\mathbf{x} \cdot \mathbf{y} = \mathbf{y} \cdot \mathbf{x}$.
2. Distributivity: $(\mathbf{x}_1 + \mathbf{x}_2) \cdot \mathbf{y} = \mathbf{x}_1 \cdot \mathbf{y} + \mathbf{x}_2 \cdot \mathbf{y}$.
3. Homogeneity: $(\lambda \mathbf{x}) \cdot \mathbf{y} = \lambda (\mathbf{x} \cdot \mathbf{y})$ for any real $\lambda$.
4. Positive definiteness: $\mathbf{x} \cdot \mathbf{x} \geq 0$ for any $\mathbf{x}$, and $\mathbf{x} \cdot \mathbf{x} = 0$ if and only if $\mathbf{x} = 0$.

If the nature of the elements and the scalar product is concretized, one obtains a *specific Euclidean space*.

**Example 1.** Consider the linear space $B_3$ of all free vectors in three-dimensional space. The space $B_3$ becomes a Euclidean space if the scalar product is introduced as in analytic geometry (see Paragraph 4.5.3-1):

$$\mathbf{x} \cdot \mathbf{y} = |\mathbf{x}| |\mathbf{y}| \cos \varphi,$$

where $\varphi$ is the angle between the vectors $\mathbf{x}$ and $\mathbf{y}$.

**Example 2.** Consider the $n$-dimensional coordinate space $\mathbb{R}^n$ whose elements are ordered systems of $n$ arbitrary real numbers, $\mathbf{x} = (x_1, \ldots, x_n)$. Endowing this space with the scalar product

$$\mathbf{x} \cdot \mathbf{y} = x_1 y_1 + \cdots + x_n y_n,$$

we obtain a Euclidean space.

THEOREM. *For any two elements* **x** *and* **y** *of a Euclidean space, the Cauchy–Schwarz inequality holds:*
$$(\mathbf{x} \cdot \mathbf{y})^2 \le (\mathbf{x} \cdot \mathbf{x})(\mathbf{y} \cdot \mathbf{y}).$$

A linear space $\mathcal{V}$ is called a *normed space* if it is endowed with a *norm*, which is a real-valued function of $\mathbf{x} \in \mathcal{V}$, denoted by $\|\mathbf{x}\|$ and satisfying the following conditions:
1. Homogeneity: $\|\lambda \mathbf{x}\| = |\lambda| \|\mathbf{x}\|$ for any real $\lambda$.
2. Positive definiteness: $\|\mathbf{x}\| \ge 0$ and $\|\mathbf{x}\| = 0$ if and only if $\mathbf{x} = 0$.
3. The *triangle inequality* (also called the *Minkowski inequality*) holds for all elements **x** and **y**:
$$\|\mathbf{x} + \mathbf{y}\| \le \|\mathbf{x}\| + \|\mathbf{y}\|. \tag{5.4.1.1}$$

The value $\|\mathbf{x}\|$ is called the *norm of an element* **x** or its *length*.

THEOREM. *Any Euclidean space becomes a normed space if the norm is introduced by*
$$\|\mathbf{x}\| = \sqrt{\mathbf{x} \cdot \mathbf{x}}. \tag{5.4.1.2}$$

COROLLARY. *In any Euclidean space with the norm (5.4.1.2), the triangle inequality (5.4.1.1) holds for all its elements* **x** *and* **y**.

The *distance between elements* **x** and **y** of a Euclidean space is defined by
$$d(\mathbf{x}, \mathbf{y}) = \|\mathbf{x} - \mathbf{y}\|.$$

One says that $\varphi \in [0, 2\pi]$ is the *angle* between two elements **x** and **y** of a Euclidean space if
$$\cos \varphi = \frac{\mathbf{x} \cdot \mathbf{y}}{\|\mathbf{x}\| \|\mathbf{y}\|}.$$

Two elements **x** and **y** of a Euclidean space are said to be *orthogonal* if their scalar product is equal to zero, $\mathbf{x} \cdot \mathbf{y} = 0$.

PYTHAGOREAN THEOREM. *Let* $\mathbf{x}_1, \ldots \mathbf{x}_m$ *be mutually orthogonal elements of a Euclidean space, i.e.,* $\mathbf{x}_j \cdot \mathbf{x}_j = 0$ *for* $i \ne j$. *Then*
$$\|\mathbf{x}_1 + \cdots + \mathbf{x}_m\|^2 = \|\mathbf{x}_1\|^2 + \cdots + \|\mathbf{x}_m\|^2.$$

**Example 3.** In the Euclidean space $B_3$ of free vectors with the usual scalar product (see Example 1), the following relations hold:
$$\|\mathbf{a}\| = |\mathbf{a}|, \quad (\mathbf{a} \cdot \mathbf{b})^2 \le |\mathbf{a}|^2 |\mathbf{b}|^2, \quad |\mathbf{a} + \mathbf{b}| \le |\mathbf{a}| + |\mathbf{b}|.$$

In the Euclidean space $\mathbb{R}^n$ of ordered systems of $n$ numbers with the scalar product defined in Example 2, the following relations hold:
$$\|\mathbf{x}\| = \sqrt{x_1^2 + \cdots + x_n^2},$$
$$(x_1 y_1 + \cdots + x_n y_n)^2 \le (x_1^2 + \cdots + x_n^2)(y_1^2 + \cdots + y_n^2),$$
$$\sqrt{(x_1 + y_1)^2 + \cdots + (x_n + y_n)^2} \le \sqrt{x_1^2 + \cdots + x_n^2} \sqrt{y_1^2 + \cdots + y_n^2}.$$

### 5.4.1-2. Orthonormal basis in a finite-dimensional Euclidean space.

For elements $\mathbf{x}_1, \ldots, \mathbf{x}_m$ of a Euclidean space, the $m$th-order determinant $\det[\mathbf{x}_i \cdot \mathbf{x}_j]$ is called their *Gram determinant*. These elements are linearly independent if and only if their Gram determinant is different from zero.

One says that $n$ elements $\mathbf{i}_1, \ldots, \mathbf{i}_n$ of an $n$-dimensional Euclidean space $\mathcal{V}$ form its *orthonormal basis* if these elements have unit norm and are mutually orthogonal, i.e.,

$$\mathbf{i}_i \cdot \mathbf{i}_j = \begin{cases} 1 & \text{for } i = j, \\ 0 & \text{for } i \neq j. \end{cases}$$

THEOREM. *In any $n$-dimensional Euclidean space $\mathcal{V}$, there exists an orthonormal basis.*

*Orthogonalization of linearly independent elements:*

Let $\mathbf{e}_1, \ldots, \mathbf{e}_n$ be $n$ linearly independent vectors of an $n$-dimensional Euclidean space $\mathcal{V}$. From these vectors, one can construct an orthonormal basis of $\mathcal{V}$ using the following algorithm (called *Gram–Schmidt orthogonalization*):

$$\mathbf{i}_i = \frac{\mathbf{g}_i}{\sqrt{\mathbf{g}_i \cdot \mathbf{g}_i}}, \quad \text{where} \quad \mathbf{g}_i = \mathbf{e}_i - \sum_{j=1}^{i}(\mathbf{e}_i \cdot \mathbf{i}_j)\mathbf{i}_j \quad (i = 1, 2, \ldots, n). \tag{5.4.1.3}$$

**Remark.** In any $n$-dimensional ($n > 1$) Euclidean space $\mathcal{V}$, there exist infinitely many orthonormal bases.

Properties of an orthonormal basis of a Euclidean space:
1. Let $\mathbf{i}_1, \ldots, \mathbf{i}_n$ be an orthonormal basis of a Euclidean space $\mathcal{V}$. Then the scalar product of two elements $\mathbf{x} = x_1\mathbf{i}_1 + \cdots + x_n\mathbf{i}_n$ and $\mathbf{y} = y_1\mathbf{i}_1 + \cdots + y_n\mathbf{i}_n$ is equal to the sum of products of their respective coordinates:

$$\mathbf{x} \cdot \mathbf{y} = x_1 y_1 + \cdots + x_n y_n.$$

2. The coordinates of any vector $\mathbf{x}$ in an orthonormal basis $\mathbf{i}_1, \ldots, \mathbf{i}_n$ are equal to the scalar product of $\mathbf{x}$ and the corresponding vector of the basis (or the projection of the element $\mathbf{x}$ on the axis in the direction of the corresponding vector of the basis):

$$x_k = \mathbf{x} \cdot \mathbf{i}_k \quad (k = 1, 2, \ldots, n).$$

**Remark.** In an arbitrary basis $\mathbf{e}_1, \ldots, \mathbf{e}_n$ of a Euclidean space, the scalar product of two elements $\mathbf{x} = x_1\mathbf{e}_1 + \cdots + x_n\mathbf{e}_n$ and $\mathbf{y} = y_1\mathbf{e}_1 + \cdots + y_n\mathbf{e}_n$ has the form

$$\mathbf{x} \cdot \mathbf{y} = \sum_{i=1}^{n}\sum_{j=1}^{n} a_{ij} x_i y_j,$$

where $a_{ij} = \mathbf{e}_i \cdot \mathbf{e}_j$ ($i, j = 1, 2, \ldots, n$).

Let $\mathcal{X}, \mathcal{Y}$ be subspaces of a Euclidean space $\mathcal{V}$. The subspace $\mathcal{X}$ is called the *orthogonal complement* of the subspace $\mathcal{Y}$ in $\mathcal{V}$ if any element $\mathbf{x}$ of $\mathcal{X}$ is orthogonal to any element $\mathbf{y}$ of $\mathcal{Y}$ and $\mathcal{X} \oplus \mathcal{Y} = \mathcal{V}$.

THEOREM. *Any $n$-dimensional Euclidean space $\mathcal{V}$ can be represented as the direct sum of its arbitrary subspace $\mathcal{Y}$ and its orthogonal complement $\mathcal{X}$.*

Two Euclidean spaces $\mathcal{V}$ and $\widetilde{\mathcal{V}}$ are said to be *isomorphic* if one can establish a one-to-one correspondence between the elements of these spaces satisfying the following conditions: if elements $\mathbf{x}$ and $\mathbf{y}$ of $\mathcal{V}$ correspond to elements $\widetilde{\mathbf{x}}$ and $\widetilde{\mathbf{y}}$ of $\widetilde{\mathcal{V}}$, then the element $\mathbf{x} + \mathbf{y}$ corresponds to $\widetilde{\mathbf{x}} + \widetilde{\mathbf{y}}$; the element $\lambda\mathbf{x}$ corresponds to $\lambda\widetilde{\mathbf{x}}$ for any $\lambda$; the scalar product $(\mathbf{x} \cdot \mathbf{y})_{\mathcal{V}}$ is equal to the scalar product $(\widetilde{\mathbf{x}} \cdot \widetilde{\mathbf{y}})_{\widetilde{\mathcal{V}}}$.

THEOREM. *Any two $n$-dimensional Euclidean spaces $\mathcal{V}$ and $\widetilde{\mathcal{V}}$ are isomorphic.*

## 5.4.2. Complex Euclidean Space (Unitary Space)

**5.4.2-1. Definition and properties of complex Euclidean space (unitary space).**

A *complex Euclidean space* (or *unitary space*) is a complex linear space $\mathcal{V}$ endowed with a *scalar product* (also called *inner product*), which is a complex-valued function of two arguments $\mathbf{x} \in \mathcal{V}$ and $\mathbf{y} \in \mathcal{V}$ called their scalar product, denoted by $\mathbf{x} \cdot \mathbf{y}$, satisfying the following conditions (called axioms of the scalar product):

1. Commutativity: $\mathbf{x} \cdot \mathbf{y} = \overline{\mathbf{y} \cdot \mathbf{x}}$.
2. Distributivity: $(\mathbf{x}_1 + \mathbf{x}_2) \cdot \mathbf{y} = \mathbf{x}_1 \cdot \mathbf{y} + \mathbf{x}_2 \cdot \mathbf{y}$.
3. Homogeneity: $(\lambda \mathbf{x}) \cdot \mathbf{y} = \lambda(\mathbf{x} \cdot \mathbf{y})$ for any complex $\lambda$.
4. Positive definiteness: $\mathbf{x} \cdot \mathbf{x} \geq 0$; and $\mathbf{x} \cdot \mathbf{x} = 0$ if and only if $\mathbf{x} = 0$.

Here $\overline{\mathbf{y} \cdot \mathbf{x}}$ is the complex conjugate of a number $\mathbf{y} \cdot \mathbf{x}$.

**Example 1.** Consider the $n$-dimensional complex linear space $\mathbb{R}^n_*$ whose elements are ordered systems of $n$ complex numbers, $\mathbf{x} = (x_1, \dots, x_n)$. We obtain a unitary space if the scalar product of two elements $\mathbf{x} = (x_1, \dots, x_n)$ and $\mathbf{y} = (y_1, \dots, y_n)$ is introduced by

$$\mathbf{x} \cdot \mathbf{y} = x_1 \bar{y}_1 + \cdots + x_n \bar{y}_n,$$

where $\bar{y}_j$ is the complex conjugate of $y_j$.

THEOREM. *For any two elements $\mathbf{x}$ and $\mathbf{y}$ of an arbitrary unitary space, the Cauchy–Schwarz inequality holds:*

$$|\mathbf{x} \cdot \mathbf{y}|^2 \leq (\mathbf{x} \cdot \mathbf{x})(\mathbf{y} \cdot \mathbf{y}).$$

THEOREM. *Any unitary space becomes a normed space if the norm of its element $\mathbf{x}$ is introduced by*

$$\|\mathbf{x}\| = \sqrt{\mathbf{x} \cdot \mathbf{x}}. \qquad (5.4.2.1)$$

COROLLARY. *For any two elements $\mathbf{x}$ and $\mathbf{y}$ of a normed Euclidean space with the norm (5.4.2.1), the triangle inequality (5.4.1.1) holds.*

The *distance between elements* $\mathbf{x}$ and $\mathbf{y}$ of a unitary space is defined by

$$d(\mathbf{x}, \mathbf{y}) = \|\mathbf{x} - \mathbf{y}\|. \qquad (5.4.2.2)$$

Two elements $\mathbf{x}$ and $\mathbf{y}$ of a unitary space are said to be *orthogonal* if their scalar product is equal to zero, $\mathbf{x} \cdot \mathbf{y} = 0$.

**5.4.2-2. Orthonormal basis in a finite-dimensional unitary space.**

Elements $\mathbf{x}_1, \dots, \mathbf{x}_m$ of a unitary space $\mathcal{V}$ are linearly independent if and only if their *Gram determinant* is different from zero, $\det[\mathbf{x}_i \cdot \mathbf{x}_j] \neq 0$.

One says that elements $\mathbf{i}_1, \dots, \mathbf{i}_n$ of an $n$-dimensional unitary space $\mathcal{V}$ form an *orthonormal basis* of that space if these elements are mutually orthogonal and have unit norm, i.e.,

$$\mathbf{i}_i \cdot \mathbf{i}_j = \begin{cases} 1 & \text{for } i = j, \\ 0 & \text{for } i \neq j. \end{cases}$$

Given any $n$ linearly independent elements of a unitary space, one can construct an orthonormal basis of that space using the procedure described in Paragraph 5.4.1-2 (see formulas (5.4.1.3)).

Properties of an orthonormal basis of a unitary space:
1. Let $\mathbf{i}_1, \ldots, \mathbf{i}_n$ be an orthonormal basis in a unitary space. Then the scalar product of two elements $\mathbf{x} = x_1\mathbf{i}_1 + \cdots + x_n\mathbf{i}_n$ and $\mathbf{y} = y_1\mathbf{i}_1 + \cdots + y_n\mathbf{i}_n$ is equal to the sum

$$\mathbf{x} \cdot \mathbf{y} = x_1\bar{y}_1 + \cdots + x_n\bar{y}_n.$$

2. The coordinates of any vector in an orthonormal basis $\mathbf{i}_1, \ldots, \mathbf{i_n}$ are equal to the scalar products of this vector and the vectors of the bases (or the projections of this element on the axes in the direction of the corresponding basis vectors):

$$x_k = \mathbf{x} \cdot \mathbf{i}_k \qquad (k = 1, 2, \ldots, n).$$

Two unitary spaces $\mathcal{V}$ and $\tilde{\mathcal{V}}$ are said to be *isomorphic* if there is a one-to-one correspondence between their elements satisfying the following conditions: if elements $\mathbf{x}$ and $\mathbf{y}$ of $\mathcal{V}$ correspond to elements $\tilde{\mathbf{x}}$ and $\tilde{\mathbf{y}}$ of $\tilde{\mathcal{V}}$, then $\mathbf{x} + \mathbf{y}$ corresponds to $\tilde{\mathbf{x}} + \tilde{\mathbf{y}}$; the element $\lambda\mathbf{x}$ corresponds to $\lambda\tilde{\mathbf{x}}$ for any complex $\lambda$; the scalar product $(\mathbf{x} \cdot \mathbf{y})_{\mathcal{V}}$ is equal to the scalar product $(\tilde{\mathbf{x}} \cdot \tilde{\mathbf{y}})_{\tilde{\mathcal{V}}}$.

THEOREM. *Any two $n$-dimensional unitary spaces $\mathcal{V}$ and $\tilde{\mathcal{V}}$ are isomorphic.*

### 5.4.3. Banach Spaces and Hilbert Spaces

**5.4.3-1. Convergence in unitary spaces. Banach space.**

Any normed linear space is a *metric space* with the metric (5.4.2.2).

A sequence $\{\mathbf{b}_s\}$ of elements of a normed space $\mathcal{V}$ is said to be convergent to an element $\mathbf{b} \in \mathcal{V}$ as $s \to \infty$ if $\lim\limits_{s \to \infty} \|\mathbf{b}_s - \mathbf{b}\| = 0$.

A series $\mathbf{x}_0 + \mathbf{x}_1 + \cdots$ with terms in a normed space is said to be convergent to an element $\mathbf{x}$ (called its sum; one writes $\mathbf{x} = \lim\limits_{n \to \infty} \sum\limits_{k=0}^{n} \mathbf{x}_k = \sum\limits_{k=0}^{\infty} \mathbf{x}_k$) if the sequence of its partial sums forms a sequence convergent to $\mathbf{x}$, i.e., $\lim\limits_{n \to \infty} \left\| \mathbf{x} - \sum\limits_{k=0}^{n} \mathbf{x}_k \right\| = 0$.

A normed linear space $\mathcal{V}$ is said to be *complete* if any sequence of its elements $\mathbf{s}_0, \mathbf{s}_1, \ldots$ satisfying the condition

$$\lim_{n,m \to \infty} \|\mathbf{s}_n - \mathbf{s}_m\| = 0$$

is convergent to some element $\mathbf{s}$ of the space $\mathcal{V}$.

A complete normed linear space is called a *Banach space*.

Remark. Any finite-dimensional normed linear space is complete.

**5.4.3-2. Hilbert space.**

A complete unitary space is called a *Hilbert space*.

Any complete subspace of a Hilbert space is itself a Hilbert space.

PROJECTION THEOREM. *Let $\mathcal{V}_1$ be a complete subspace of a unitary space $\mathcal{V}$. Then for any $\mathbf{x} \in \mathcal{V}$, there is a unique vector $\mathbf{x}_p \in \mathcal{V}_1$ such that*

$$\min_{\mathbf{y} \in \mathcal{V}_1} \|\mathbf{x} - \mathbf{y}\| = \|\mathbf{x} - \mathbf{x}_p\|.$$

Moreover, the vector $\mathbf{x}_p$ is the unique element of $\mathcal{V}_1$ for which the difference $\mathbf{x} - \mathbf{x}_p$ is orthogonal to any element $\mathbf{x}_1$ of $\mathcal{V}_1$, i.e.,

$$(\mathbf{x} - \mathbf{x}_p) \cdot \mathbf{x}_1 = 0 \quad \text{for all} \quad \mathbf{x}_1 \in \mathcal{V}_1.$$

The mapping $\mathbf{x} \to \mathbf{x}_p$ is a bounded linear operator from $\mathcal{V}$ to $\mathcal{V}_1$ called the *orthogonal projection of the space* $\mathcal{V}$ *to its subspace* $\mathcal{V}_1$.

## 5.5. Systems of Linear Algebraic Equations

### 5.5.1. Consistency Condition for a Linear System

5.5.1-1. Notion of a system of linear algebraic equations.

A *system of* $m$ *linear equations with* $n$ *unknown quantities* has the form

$$\begin{aligned} a_{11}x_1 + a_{12}x_2 + \cdots + a_{1k}x_k + \cdots + a_{1n}x_n &= b_1, \\ a_{21}x_1 + a_{22}x_2 + \cdots + a_{2k}x_k + \cdots + a_{2n}x_n &= b_2, \\ &\cdots\cdots\cdots\cdots\cdots\cdots\cdots\cdots\cdots\cdots\cdots\cdots\cdots \\ a_{m1}x_1 + a_{m2}x_2 + \cdots + a_{mk}x_k + \cdots + a_{mn}x_n &= b_m, \end{aligned} \qquad (5.5.1.1)$$

where $a_{11}, a_{12}, \ldots, a_{mn}$ are the *coefficients of the system*; $b_1, b_2, \ldots, b_m$ are its *free terms*; and $x_1, x_2, \ldots, x_n$ are the unknown quantities.

System (5.5.1.1) is said to be *homogeneous* if all its free terms are equal to zero. Otherwise (i.e., if there is at least one nonzero free term) the system is called *nonhomogeneous*.

If the number of equations is equal to that of the unknown quantities ($m = n$), system (5.5.1.1) is called a *square system*.

A *solution* of system (5.5.1.1) is a set of $n$ numbers $x_1, x_2, \ldots, x_n$ satisfying the equations of the system. A system is said to be *consistent* if it admits at least one solution. If a system has no solutions, it is said to be *inconsistent*. A consistent system of the form (5.5.1.1) is called a *determined system* — it has a unique solution. A consistent system with more than one solution is said to be *underdetermined*.

It is convenient to use matrix notation for systems of the form (5.5.1.1),

$$AX = B, \qquad (5.5.1.2)$$

where $A \equiv [a_{ij}]$ is a matrix of size $m \times n$ called the *basic matrix* of the system; $X \equiv [x_i]$ is a column vector of size $n$; $B \equiv [b_i]$ is a column vector of size $m$.

5.5.1-2. Existence of nontrivial solutions of a homogeneous system.

Consider a homogeneous system

$$AX = O_m, \qquad (5.5.1.3)$$

where $A \equiv [a_{ij}]$ is its basic matrix of size $m \times n$, $X \equiv [x_i]$ is a column vector of size $n$, and $O_m \equiv [0]$ is a column vector of size $m$. System (5.5.1.3) is always consistent since it always has the so-called *trivial solution* $X \equiv O_n$.

THEOREM. *A homogeneous system (5.5.1.3) has a nontrivial solution if and only if the rank of the matrix $A$ is less than the number of the unknown quantities $n$.*

It follows that a square homogeneous system has a nontrivial solution if and only if the determinant of its matrix of coefficients is equal to zero, $\det A = 0$.

Properties of the set of all solutions of a homogeneous system:
1. All solutions of a homogeneous system (5.5.1.3) form a linear space.
2. The linear space of all solutions of a homogeneous system (5.5.1.3) with $n$ unknown quantities and a basic matrix of rank $r$ is isomorphic to the space $A^{n-r}$ of all ordered systems of $(n-r)$ numbers. The dimension of the space of solutions is equal to $n-r$.
3. Any system of $(n-r)$ linearly independent solutions of the homogeneous system (5.5.1.3) forms a basis in the space of all its solutions and is called a *fundamental system of solutions* of that system. The fundamental system of solutions corresponding to the basis $\mathbf{i}_1 = (1, 0, \ldots, 0)$, $\mathbf{i}_2 = (0, 1, \ldots, 0)$, $\ldots$, $\mathbf{i}_{n-r} = (0, 0, \ldots, 1)$ of the space $A^{n-r}$ is said to be *normal*.

5.5.1-3. Consistency condition for a general linear system.

System (5.5.1.1) or (5.5.1.2) is associated with two matrices: the basic matrix $A$ of size $m \times n$ and the *augmented matrix* $A_1$ of size $m \times (n+1)$ formed by the matrix $A$ supplemented with the column of the free terms, i.e.,

$$A_1 \equiv \begin{pmatrix} a_{11} & a_{12} & \cdots & a_{1n} & b_1 \\ a_{21} & a_{22} & \cdots & a_{2n} & b_2 \\ \vdots & \vdots & \ddots & \vdots & \vdots \\ a_{m1} & a_{m2} & \cdots & a_{mn} & b_m \end{pmatrix}. \tag{5.5.1.4}$$

KRONECKER–CAPELLI THEOREM. *A linear system (5.5.1.1) [or (5.5.1.2)] is consistent if and only if its basic matrix and its augmented matrix (5.5.1.4) have the same rank, i.e.* rank $(A_1) =$ rank $(A)$.

## 5.5.2. Finding Solutions of a System of Linear Equations

5.5.2-1. System of two equations with two unknown quantities.

A system of two equations with two unknown quantities has the form

$$\begin{aligned} a_1 x + b_1 y &= c_1, \\ a_2 x + b_2 y &= c_2. \end{aligned} \tag{5.5.2.1}$$

Depending on the coefficients $a_k$, $b_k$, $c_k$, the following three cases are possible:

1°. If $\Delta = a_1 b_2 - a_2 b_1 \neq 0$, then system (5.5.2.1) has a unique solution,

$$x = \frac{c_1 b_2 - c_2 b_1}{a_1 b_2 - a_2 b_1}, \quad y = \frac{a_1 c_2 - a_2 c_1}{a_1 b_2 - a_2 b_1}.$$

2°. If $\Delta = a_1 b_2 - a_2 b_1 = 0$ and $a_1 c_2 - a_2 c_1 = 0$ (the case of proportional coefficients), then system (5.5.2.1) has infinitely many solutions described by the formulas

$$x = t, \quad y = \frac{c_1 - a_1 t}{b_1} \quad (b_1 \neq 0),$$

where $t$ is arbitrary.

3°. If $\Delta = a_1 b_2 - a_2 b_1 = 0$ and $a_1 c_2 - a_2 c_1 \neq 0$, then system (5.5.2.1) has no solutions.

## 5.5.2-2. General square system of linear equations.

A square system of linear equations has the form
$$AX = B, \qquad (5.5.2.2)$$
where $A$ is a square matrix.

$1°$. If the determinant of system (5.5.2.2) is different from zero, i.e. $\det A \neq 0$, then the system has a unique solution,
$$X = A^{-1}B.$$

$2°$. *Cramer rule.* If the determinant of the matrix of system (5.5.2.2) is different from zero, i.e. $\Delta = \det A \neq 0$, then the system admits a unique solution, which is expressed by
$$x_1 = \frac{\Delta_1}{\Delta}, \quad x_2 = \frac{\Delta_2}{\Delta}, \quad \ldots, \quad x_n = \frac{\Delta_n}{\Delta}, \qquad (5.5.2.3)$$
where $\Delta_k$ ($k = 1, 2, \ldots, n$) is the determinant of the matrix obtained from $A$ by replacing its $k$th column with the column of free terms:
$$\Delta_k = \begin{vmatrix} a_{11} & a_{12} & \ldots & b_1 & \ldots & a_{1n} \\ a_{21} & a_{22} & \ldots & b_2 & \ldots & a_{2n} \\ \vdots & \vdots & \vdots & \vdots & \ddots & \vdots \\ a_{n1} & a_{n2} & \ldots & b_n & \ldots & a_{nn} \end{vmatrix}.$$

**Example 1.** Using the Cramer rule, let us find the solution of the system of linear equations
$$2x_1 + x_2 + 4x_3 = 16,$$
$$3x_1 + 2x_2 + x_3 = 10,$$
$$x_1 + 3x_2 + 3x_3 = 16.$$

The determinant of its basic matrix is different from zero,
$$\Delta = \begin{vmatrix} 2 & 1 & 4 \\ 3 & 2 & 1 \\ 1 & 3 & 3 \end{vmatrix} = 26 \neq 0,$$
and we have
$$\Delta_1 = \begin{vmatrix} 16 & 1 & 4 \\ 10 & 2 & 1 \\ 16 & 3 & 3 \end{vmatrix} = 26, \quad \Delta_2 = \begin{vmatrix} 2 & 16 & 4 \\ 3 & 10 & 1 \\ 1 & 16 & 3 \end{vmatrix} = 52, \quad \Delta_3 = \begin{vmatrix} 2 & 1 & 16 \\ 3 & 2 & 10 \\ 1 & 3 & 16 \end{vmatrix} = 78.$$

Therefore, by the Cramer rule (5.5.2.3), the only solution of the system has the form
$$x_1 = \frac{\Delta_1}{\Delta} = \frac{26}{26} = 1, \quad x_2 = \frac{\Delta_2}{\Delta} = \frac{52}{26} = 2, \quad x_3 = \frac{\Delta_3}{\Delta} = \frac{78}{26} = 3.$$

$3°$. *Gaussian elimination of unknown quantities.*

Two systems are said to be *equivalent* if their sets of solutions coincide.

The method of Gaussian elimination consists in the reduction of a given system to an equivalent system with an upper triangular basic matrix. The latter system can be easily solved. This reduction is carried out in finitely many steps. On every step, one performs an *elementary transformation* of the system (or the corresponding augmented matrix) and obtains an equivalent system. The elementary transformations are of the following three types:

1. Interchange of two equations (or the corresponding rows of the augmented matrix).
2. Multiplication of both sides of one equation (or the corresponding row of the augmented matrix) by a nonzero constant.
3. Adding to both sides of one equation both sides of another equation multiplied by a nonzero constant (adding to some row of the augmented matrix its another row multiplied by a nonzero constant).

Suppose that det $A \neq 0$. Then by consecutive elementary transformations, the augmented matrix of the system $A_1$ [see (5.5.1.4)] of size $n \times (n+1)$ can be reduced to the form

$$U_1 \equiv \begin{pmatrix} 1 & u_{12} & \cdots & u_{1n} & \bigm| & y_1 \\ 0 & 1 & \cdots & u_{2n} & \bigm| & y_2 \\ \vdots & \vdots & \ddots & \vdots & \bigm| & \vdots \\ 0 & 0 & \cdots & 1 & \bigm| & y_n \end{pmatrix}$$

and one obtains an equivalent system with an upper triangular basic matrix,

$$x_1 + u_{12}x_2 + u_{13}x_3 + \cdots + u_{1n}x_n = y_1,$$
$$x_2 + u_{23}x_3 + \cdots + u_{2n}x_n = y_2,$$
$$\dots\dots\dots\dots\dots\dots\dots\dots\dots\dots$$
$$x_n = y_n.$$

This system is solved by the so-called "backward substitution": inserting $x_n = y_n$ (obtained from the last equation) into the preceding $(n-1)$st equation, one finds $x_{n-1}$. Then inserting the values obtained for $x_n$, $x_{n-1}$ into the $(n-2)$nd equation, one finds $x_{n-2}$. Proceeding in this way, one finally finds $x_1$. This back substitution process is described by the formulas

$$x_k = y_k - \sum_{s=k+1}^{n} u_{ks} x_s \quad (k = n-1, n-2, \dots, 1).$$

Suppose that det $A = 0$ and rank$(A) = r$, $0 < r < n$. In this case, the system is either inconsistent (i.e., has no solutions) or has infinitely many solutions. By elementary transformations and, possibly, reindexing the unknown quantities (i.e., introducing new unknown quantities $y_1 = x_{\sigma(1)}, \dots, y_n = x_{\sigma(n)}$, where $\sigma(1), \dots, \sigma(n)$ is a permutation of the indices $1, 2, \dots, n$), one obtains a system of the form (for the sake of brevity, we retain the notation $x_j$ for the reindexed unknown quantities)

$$c_{11}x_1 + \cdots + c_{1r}x_r + c_{1,r+1}x_{r+1} + \cdots + c_{1n}x_n = d_1,$$
$$\dots\dots\dots\dots\dots\dots\dots\dots\dots\dots$$
$$c_{rr}x_r + c_{r,r+1}x_{r+1} + \cdots + c_{rn}x_n = d_r,$$
$$0 = d_{r+1},$$
$$\dots$$
$$0 = d_n,$$

where the matrix $[c_{ij}]$ $(i, j = 1, 2, \dots, r)$ of size $r \times r$ is nondegenerate. If at least one of the right-hand sides $d_{r+1}, \dots, d_n$ is different from zero, then the system is inconsistent. If $d_{r+1} = \dots = d_n = 0$, then the last $n-r$ equations can be dropped and it remains to find all solutions of the first $r$ equations. Transposing all terms containing the variables $x_{r+1}, \dots, x_n$ to the right-hand sides and regarding these variables as arbitrary free parameters, we obtain a linear system for the unknown quantities $x_1, \dots, x_r$ with the nondegenerate basic matrix $[c_{ij}]$ $(j, j = 1, 2, \dots, r)$.

**Example 2.** Let us find a solution of the system from Example 1 by the Gaussian elimination method. By elementary transformations of the augmented matrix, we obtain

$$\begin{pmatrix} 2 & 1 & 4 & \bigm| & 16 \\ 3 & 2 & 1 & \bigm| & 10 \\ 1 & 3 & 3 & \bigm| & 16 \end{pmatrix} \to \begin{pmatrix} 1 & 1/2 & 2 & \bigm| & 8 \\ 0 & 1/2 & -5 & \bigm| & -14 \\ 0 & 5/2 & 1 & \bigm| & 8 \end{pmatrix} \to \begin{pmatrix} 1 & 1/2 & 2 & \bigm| & 8 \\ 0 & 1 & -10 & \bigm| & -28 \\ 0 & 0 & 26 & \bigm| & 78 \end{pmatrix} \to \begin{pmatrix} 1 & 1/2 & 2 & \bigm| & 8 \\ 0 & 1 & -10 & \bigm| & -28 \\ 0 & 0 & 1 & \bigm| & 3 \end{pmatrix}.$$

The transformed system has the form

$$x_1 + \tfrac{1}{2}x_2 + 2x_3 = 8,$$
$$x_2 - 10x_3 = -28,$$
$$x_3 = 3.$$

Hence, we find that

$$x_3 = 3, \quad x_2 = -28 + 10x_3 = 2, \quad x_1 = 8 - \tfrac{1}{2}x_2 - 2x_3 = 1.$$

4°. *Gauss–Jordan elimination of unknown quantities.*

This method consists of applying elementary transformations for reducing a system with a nondegenerate basic matrix to an equivalent system with the identity matrix. On the $k$th step ($k = 1, 2, \ldots, n$) the rows of the augmented matrix $A'_1$ obtained on the preceding step can be transformed as follows:

$$a''_{kj} = \frac{a'_{kj}}{a'_{kk}}, \qquad b''_k = \frac{b'_k}{a'_{kk}} \qquad (j = k, k+1, \ldots, n),$$

$$a''_{ij} = a'_{ij} - a'_{ik}\frac{a'_{kj}}{a'_{kk}}, \quad b''_i = b'_i - a_{ik}\frac{b'_k}{a'_{kk}} \quad (i = 1, 2, \ldots, n,\ i \neq k,\ j = k, k+1, \ldots, n),$$

provided that the diagonal element obtained on each step is not equal to zero. After $n$ steps, the basic matrix is transformed to the identity matrix and the right-hand side turns into the desired solution.

**Example 3.** For the linear system from Examples 1 and 2 we have

$$\begin{pmatrix} 2 & 1 & 4 & | & 16 \\ 3 & 2 & 1 & | & 10 \\ 1 & 3 & 3 & | & 16 \end{pmatrix} \to \begin{pmatrix} 1 & 1/2 & 2 & | & 8 \\ 0 & 1/2 & -5 & | & -14 \\ 0 & 5/2 & 1 & | & 8 \end{pmatrix} \to \begin{pmatrix} 1 & 0 & 7 & | & 22 \\ 0 & 1 & -10 & | & -28 \\ 0 & 0 & 26 & | & 78 \end{pmatrix} \to \begin{pmatrix} 1 & 0 & 0 & | & 1 \\ 0 & 1 & 0 & | & 2 \\ 0 & 0 & 1 & | & 3 \end{pmatrix},$$

and therefore $x_1 = 1$, $x_2 = 2$, $x_3 = 3$.

The diagonal element obtained on some step of the above elimination procedure may happen to be equal to zero. In this case, the formulas become more complicated and reindexing of the unknown quantities may be required.

5°. *Method of LU-decomposition.*

This method is based on the representation of the basic matrix $A$ as the product of a lower triangular matrix $L$ and an upper triangular matrix $U$, i.e., in the form $A = LU$. This factorization is called a *triangular representation* or the *LU-representation* of a matrix (see also Paragraph 5.2.3-1).

Given such an $LU$-representation of the matrix $A$, the system $AX = B$ can be represented in the form $LUX = B$, and its solution can be obtained by solving the following two systems:

$$LY = B, \quad UX = Y.$$

Due to the triangular structure of the matrices $L \equiv [l_{ij}]$ and $U \equiv [u_{ij}]$, these systems can be solved with the help of the formulas

$$y_i = \frac{1}{l_{ii}}\left(b_i - \sum_{j=i}^{i-n} l_{ij} y_j\right) \quad (i = 1, 2, \ldots, n),$$

$$x_k = y_k - \sum_{s=k+1}^{n} u_{ks} x_s \quad (k = n, n-1, \ldots, 1),$$

provided that $l_{ii} \neq 0$.

There exist various methods for the construction of $LU$-decompositions. In particular if the following conditions hold:

$$a_{11} \neq 0, \quad \begin{vmatrix} a_{11} & a_{12} \\ a_{21} & a_{22} \end{vmatrix} \neq 0, \quad \ldots, \quad \det A \neq 0,$$

then the elements of the desired matrices $L$ and $U$ can be calculated by the formulas

$$l_{ij} = \begin{cases} a_{ij} - \sum_{s=1}^{j-1} l_{is} u_{sj} & \text{for } i \geq j, \\ 0 & \text{for } i < j, \end{cases}$$

$$u_{ij} = \begin{cases} \dfrac{1}{l_{ii}}\left(a_{ij} - \sum_{s=1}^{i-1} l_{is} u_{sj}\right) & \text{for } i < j \\ 1 & \text{for } i = j, \\ 0 & \text{for } i > j. \end{cases}$$

### 5.5.2-3. Solutions of a square system with different right-hand sides.

1°. One often has to solve a system of linear equations with a given basic matrix $A$ and different right-hand sides. For instance, consider the systems $AX^{(1)} = B^{(1)}, \ldots, AX^{(m)} = B^{(m)}$. These $m$ systems can be regarded as a single matrix equation $AX = B$, where $X$ and $B$ are matrices of size $n \times m$ whose columns coincide with $X^{(j)}$ and $B^{(j)}$ ($j = 1, 2, \ldots, m$).

**Example 5.** Suppose that we have to solve the equation $AX = B$ with the given basic matrix $A$ and different right-hand sides:

$$A = \begin{pmatrix} 1 & 2 & -3 \\ 3 & -2 & 1 \\ -2 & 1 & 3 \end{pmatrix}, \quad B^{(1)} = \begin{pmatrix} 7 \\ 1 \\ 5 \end{pmatrix}, \quad B^{(2)} = \begin{pmatrix} 10 \\ 6 \\ -5 \end{pmatrix}.$$

Using the Gauss-Jordan procedure, we obtain

$$\begin{pmatrix} 1 & 2 & -3 & | & 7 & 10 \\ 3 & -2 & 1 & | & 1 & 6 \\ -2 & 1 & 3 & | & 5 & -5 \end{pmatrix} \to \begin{pmatrix} 1 & 2 & -3 & | & 7 & 10 \\ 0 & -8 & 10 & | & -20 & -24 \\ 0 & 5 & -3 & | & 19 & 15 \end{pmatrix} \to \begin{pmatrix} 1 & 0 & -1/2 & | & 2 & 4 \\ 0 & 1 & -5/4 & | & 5/2 & 3 \\ 0 & 0 & 13/4 & | & 1/32 & 0 \end{pmatrix} \to \begin{pmatrix} 1 & 0 & 0 & | & 3 & 4 \\ 0 & 1 & 0 & | & 5 & 3 \\ 0 & 0 & 1 & | & 2 & 0 \end{pmatrix}.$$

Therefore,

$$X^{(1)} = \begin{pmatrix} 3 \\ 5 \\ 2 \end{pmatrix}, \quad X^{(2)} = \begin{pmatrix} 4 \\ 3 \\ 0 \end{pmatrix}.$$

2°. If $B = I$, where $I$ is the identity matrix of size $n \times n$, then the solution of the matrix equation $AX = I$ coincides with the matrix $X = A^{-1}$.

**Example 6.** Find the inverse of the matrix

$$A = \begin{pmatrix} 2 & 1 & 0 \\ -3 & 0 & 7 \\ -5 & 4 & 1 \end{pmatrix}.$$

Let us transform the augmented matrix of the system, using the Gauss-Jordan method. We get

$$\begin{pmatrix} 2 & 1 & 0 & | & 1 & 0 & 0 \\ -3 & 0 & 7 & | & 0 & 1 & 0 \\ -5 & 4 & -1 & | & 0 & 0 & 1 \end{pmatrix} \to \begin{pmatrix} 1 & 1/2 & 0 & | & 1/2 & 0 & 0 \\ 0 & 3/2 & 7 & | & 3/2 & 1 & 0 \\ 0 & 13/2 & -1 & | & 5/2 & 0 & 1 \end{pmatrix} \to$$

$$\to \begin{pmatrix} 1 & 0 & -7/3 & | & 0 & -1/3 & 0 \\ 0 & 1 & 14/3 & | & 1 & 2/3 & 0 \\ 0 & 0 & -94/3 & | & -4 & -13/3 & 1 \end{pmatrix} \to \begin{pmatrix} 1 & 0 & 0 & | & 14/47 & -1/94 & -7/94 \\ 0 & 1 & 0 & | & 19/47 & 1/47 & 7/47 \\ 0 & 0 & 1 & | & 6/47 & 13/94 & -3/94 \end{pmatrix}.$$

### 5.5.2-4. General system of $m$ linear equations with $n$ unknown quantities.

Suppose that system (5.5.1.1) is consistent and its basic matrix $A$ has rank $r$. First, in the matrix $A$, one finds a submatrix of size $r \times r$ with nonzero $r$th-order determinant and drops the $m - r$ equations whose coefficients do not belong to this submatrix (the dropped equations follow from the remaining ones and can, therefore, be neglected). In the remaining equations, the $n - r$ unknown quantities (free unknown quantities) that are not involved in the said submatrix should be transferred to the right-hand sides. Thus, one obtains a system of $r$ equations with $r$ unknown quantities, which can be solved by any of the methods described in Paragraph 5.5.2-2.

*Remark.* If the rank $r$ of the basic matrix and the rank of the augmented matrix of system (5.5.1.1) are equal to the number of the unknown quantities $n$, then the system has a unique solution.

### 5.5.2-5. Solutions of homogeneous and corresponding nonhomogeneous systems.

$1°$. Suppose that the basic matrix $A$ of the homogeneous system (5.5.1.3) has rank $r$ and its submatrix in the left top corner, $B = [a_{ij}]$ ($i, j = 1, \ldots, r$), is nondegenerate. Let $M = \det B \neq 0$ be the determinant of that submatrix. Any solution $x_1, \ldots, x_n$ has $n - r$ free components $x_{r+1}, \ldots, x_n$ and its first components $x_1, \ldots, x_r$ are expressed via the free components as follows:

$$
\begin{aligned}
x_1 &= -\frac{1}{M}[x_{r+1}M_1(a_{i(r+1)}) + x_{r+2}M_1(a_{i(r+2)}) + \cdots + x_n M_1(a_{in})], \\
x_2 &= -\frac{1}{M}[x_{r+1}M_2(a_{i(r+1)}) + x_{r+2}M_2(a_{i(r+2)}) + \cdots + x_n M_2(a_{in})], \\
&\quad \cdots\cdots\cdots\cdots\cdots\cdots\cdots\cdots\cdots\cdots\cdots\cdots\cdots\cdots\cdots\cdots \\
x_r &= -\frac{1}{M}[x_{r+1}M_r(a_{i(r+1)}) + x_{r+2}M_r(a_{i(r+2)}) + \cdots + x_n M_r(a_{in})],
\end{aligned}
\tag{5.5.2.4}
$$

where $M_j(a_{ik})$ is the determinant of the matrix obtained from $B$ by replacing its $j$th column with the column whose components are $a_{1k}, a_{2k}, \ldots, a_{rk}$:

$$
M_j(a_{ik}) = \begin{vmatrix} a_{11} & a_{12} & \cdots & a_{1k} & \cdots & a_{1r} \\ a_{21} & a_{22} & \cdots & a_{2k} & \cdots & a_{2r} \\ \vdots & \vdots & \vdots & \vdots & \ddots & \vdots \\ a_{r1} & a_{r2} & \cdots & a_{rk} & \cdots & a_{rr} \end{vmatrix}.
$$

$2°$. Using (5.5.2.4), we obtain the following $n - r$ linearly independent solutions of the original system (5.5.1.3):

$$
\begin{aligned}
X_1 &= \left(-\frac{M_1(a_{i(r+1)})}{M} \quad -\frac{M_2(a_{i(r+1)})}{M} \quad \cdots \quad -\frac{M_r(a_{i(r+1)})}{M} \quad 1 \quad 0 \quad \cdots \quad 0\right), \\
X_2 &= \left(-\frac{M_1(a_{i(r+2)})}{M} \quad -\frac{M_2(a_{i(r+2)})}{M} \quad \cdots \quad -\frac{M_r(a_{i(r+2)})}{M} \quad 0 \quad 1 \quad \cdots \quad 0\right), \\
X_{n-r} &= \left(-\frac{M_1(a_{in})}{M} \quad -\frac{M_2(a_{in})}{M} \quad \cdots \quad -\frac{M_r(a_{in})}{M} \quad 0 \quad 0 \quad \cdots \quad 1\right).
\end{aligned}
$$

Any solution of system (5.5.1.3) can be represented as their linear combination

$$
X = C_1 X_1 + C_2 X_2 + \cdots + C_{n-r} X_{n-r},
\tag{5.5.2.5}
$$

where $C_1, C_2, \ldots, C_{n-r}$ are arbitrary constants. This formula gives the *general solution* of the homogeneous system.

3°. Relations between solutions of the nonhomogeneous system (5.5.1.1) and solutions of the corresponding homogeneous system (5.5.1.3).
1. The sum of any solution of the nonhomogeneous system (5.5.1.1) and any solution of the corresponding homogeneous system (5.5.1.3) is a solution of system (5.5.1.1).
2. The difference of any two solutions of the nonhomogeneous system (5.5.1.1) is a solution of the homogeneous system (5.5.1.3).
3. The sum of a particular solution $X_0$ of the nonhomogeneous system (5.5.1.1) and the general solution (5.5.2.5) of the corresponding homogeneous system (5.5.1.3) yields the general solution $X$ of the nonhomogeneous system (5.5.1.1).

## 5.6. Linear Operators
### 5.6.1. Notion of a Linear Operator. Its Properties

5.6.1-1. Definition of a linear operator.

An *operator* **A** acting from a linear space $\mathcal{V}$ of dimension $n$ to a linear space $\mathcal{W}$ of dimension $m$ is a mapping $\mathbf{A} : \mathcal{V} \to \mathcal{W}$ that establishes correspondence between each element **x** of the space $\mathcal{V}$ and some element **y** of the space $\mathcal{W}$. This fact is denoted by $\mathbf{y} = \mathbf{Ax}$ or $\mathbf{y} = \mathbf{A}(\mathbf{x})$.

An operator $\mathbf{A} : \mathcal{V} \to \mathcal{W}$ is said to be *linear* if for any elements $\mathbf{x}_1$ and $\mathbf{x}_2$ of the space $\mathcal{V}$ and any scalar $\lambda$, the following relations hold:

$$\mathbf{A}(\mathbf{x}_1 + \mathbf{x}_2) = \mathbf{A}\mathbf{x}_1 + \mathbf{A}\mathbf{x}_2 \quad \text{(additivity of the operator)},$$
$$\mathbf{A}(\lambda \mathbf{x}) = \lambda \mathbf{A}\mathbf{x} \quad \text{(homogeneity of the operator)}.$$

A linear operator $\mathbf{A} : \mathcal{V} \to \mathcal{W}$ is said to be *bounded* if it has a finite *norm*, which is defined as follows:

$$\|\mathbf{A}\| = \sup_{\substack{\mathbf{x} \in \mathcal{V} \\ \|\mathbf{x}\| \neq 0}} \frac{\|\mathbf{A}\mathbf{x}\|}{\|\mathbf{x}\|} = \sup_{\|\mathbf{x}\|=1} \|\mathbf{A}\mathbf{x}\| \geq 0.$$

**Remark.** If **A** is a linear operator from a Hilbert space $\mathcal{V}$ into itself, then

$$\|\mathbf{A}\| = \sup_{\substack{\mathbf{x} \in \mathcal{V} \\ \|\mathbf{x}\| \neq 0}} \frac{\|\mathbf{A}\mathbf{x}\|}{\|\mathbf{x}\|} = \sup_{\|\mathbf{x}\|=1} \|\mathbf{A}\mathbf{x}\| = \sup_{\mathbf{x},\mathbf{y} \neq 0} \frac{|(\mathbf{x}, \mathbf{A}\mathbf{y})|}{\|\mathbf{x}\| \|\mathbf{y}\|} = \sup_{\|\mathbf{x}\|=\|\mathbf{y}\|=1} |(\mathbf{x}, \mathbf{A}\mathbf{y})|.$$

THEOREM. *Any linear operator in a finite-dimensional normed space is bounded.*

The set of all linear operators $\mathbf{A} : \mathcal{V} \to \mathcal{W}$ is denoted by $L(\mathcal{V}, \mathcal{W})$.

A linear operator **O** in $L(\mathcal{V}, \mathcal{W})$ is called the *zero operator* if it maps any element **x** of $\mathcal{V}$ to the zero element of the space $\mathcal{W}$: $\mathbf{Ox} = \mathbf{0}$.

A linear operator **A** in $L(\mathcal{V}, \mathcal{V})$ is also called a *linear transformation of the space* $\mathcal{V}$.

A linear operator **I** in $L(\mathcal{V}, \mathcal{V})$ is called the *identity operator* if it maps each element **x** of $\mathcal{V}$ into itself: $\mathbf{Ix} = \mathbf{x}$.

5.6.1-2. Basic operations with linear operators.

The *sum* of two linear operators **A** and **B** in $L(\mathcal{V}, \mathcal{W})$ is a linear operator denoted by $\mathbf{A} + \mathbf{B}$ and defined by

$$(\mathbf{A} + \mathbf{B})\mathbf{x} = \mathbf{A}\mathbf{x} + \mathbf{B}\mathbf{x} \quad \text{for any} \quad \mathbf{x} \in \mathcal{V}.$$

The *product* of a scalar $\lambda$ and a linear operator **A** in $L(\mathcal{V}, \mathcal{W})$ is a linear operator denoted by $\lambda \mathbf{A}$ and defined by

$$(\lambda \mathbf{A})\mathbf{x} = \lambda \mathbf{A}\mathbf{x} \quad \text{for any} \quad \mathbf{x} \in \mathcal{V}.$$

The *opposite* operator for an operator $\mathbf{A} \in L(\mathcal{V}, \mathcal{W})$ is an operator denoted by $-\mathbf{A}$ and defined by
$$-\mathbf{A} = (-1)\mathbf{A}.$$

The *product of two linear operators* $\mathbf{A}$ *and* $\mathbf{B}$ in $L(\mathcal{V}, \mathcal{V})$ is a linear operator denoted by $\mathbf{AB}$ and defined by
$$(\mathbf{AB})\mathbf{x} = \mathbf{A}(\mathbf{Bx}) \qquad \text{for any} \quad \mathbf{x} \in \mathcal{V}.$$

Properties of linear operators in $L(\mathcal{V}, \mathcal{V})$:

$(\mathbf{AB})\mathbf{C} = \mathbf{A}(\mathbf{BC})$           (associativity of the product of three operators),

$\lambda(\mathbf{AB}) = (\lambda\mathbf{A})\mathbf{B}$           (associativity of multiplication of a scalar and two operators),

$(\mathbf{A} + \mathbf{B})\mathbf{C} = \mathbf{AC} + \mathbf{BC}$     (distributivity with respect to the sum of operators),

where $\lambda$ is a scalar; $\mathbf{A}$, $\mathbf{B}$, and $\mathbf{C}$ are linear operators in $L(\mathcal{V}, \mathcal{V})$.

**Remark.** Property 1 allows us to define the product $\mathbf{A}_1 \mathbf{A}_2 \dots \mathbf{A}_k$ of finitely many operators in $L(\mathcal{V}, \mathcal{V})$ and the $k$th power of an operator $\mathbf{A}$,
$$\mathbf{A}^k = \underbrace{\mathbf{A}\mathbf{A}\dots\mathbf{A}}_{k \text{ times}}.$$

The following relations hold:
$$\mathbf{A}^{p+q} = \mathbf{A}^p \mathbf{A}^q, \quad (\mathbf{A}^p)^q = \mathbf{A}^{pq}. \tag{5.6.1.1}$$

### 5.6.1-3. Inverse operators.

A linear operator $\mathbf{B}$ is called the inverse of an operator $\mathbf{A}$ in $L(\mathcal{V}, \mathcal{V})$ if $\mathbf{AB} = \mathbf{BA} = \mathbf{I}$. The *inverse* operator is denoted by $\mathbf{B} = \mathbf{A}^{-1}$. If the inverse operator exists, the operator $\mathbf{A}$ is said to be *invertible* or *nondegenerate*.

**Remark.** If $\mathbf{A}$ is an invertible operator, then $\mathbf{A}^{-k} = (\mathbf{A}^{-1})^k = (\mathbf{A}^k)^{-1}$ and relations (5.6.1.1) still hold.

A linear operator $\mathbf{A}$ from $\mathcal{V}$ to $\mathcal{W}$ is said to be *injective* if it maps any two different elements of $\mathcal{V}$ into different elements of $\mathcal{W}$, i.e., for $\mathbf{x}_1 \neq \mathbf{x}_2$, we have $\mathbf{A}\mathbf{x}_1 \neq \mathbf{A}\mathbf{x}_2$.

If $\mathbf{A}$ is an injective linear operator from $\mathcal{V}$ to $\mathcal{V}$, then each element $\mathbf{y} \in \mathcal{V}$ is an image of some element $\mathbf{x} \in \mathcal{V}$: $\mathbf{y} = \mathbf{Ax}$.

**THEOREM.** *A linear operator* $\mathbf{A} : \mathcal{V} \to \mathcal{V}$ *is invertible if and only if it is injective.*

### 5.6.1-4. Kernel, range, and rank of a linear operator.

The *kernel* of a linear operator $\mathbf{A} : \mathcal{V} \to \mathcal{V}$ is the set of all $\mathbf{x}$ in $\mathcal{V}$ such that $\mathbf{Ax} = 0$. The kernel of an operator $\mathbf{A}$ is denoted by $\ker \mathbf{A}$ and is a linear subspace of $\mathcal{V}$.

The *range* of a linear operator $\mathbf{A} : \mathcal{V} \to \mathcal{V}$ is the set of all $\mathbf{y}$ in $\mathcal{V}$ such that $\mathbf{y} = \mathbf{Ax}$. The range of a linear operator $\mathbf{A}$ is denoted by $\text{im}\,\mathbf{A}$ and is a subspace of $\mathcal{V}$.

Properties of the kernel, the range, and their dimensions:

1. For a linear operator $\mathbf{A} : \mathcal{V} \to \mathcal{V}$ in $n$-dimensional space $\mathcal{V}$, the following relation holds:
$$\dim(\text{im}\,\mathbf{A}) + \dim(\ker \mathbf{A}) = n.$$

2. Let $\mathcal{V}_1$ and $\mathcal{V}_2$ be two subspaces of a linear space $\mathcal{V}$ and $\dim \mathcal{V}_1 + \dim \mathcal{V}_2 = \dim \mathcal{V}$. Then there exists a linear operator $\mathbf{A} : \mathcal{V} \to \mathcal{V}$ such that $\mathcal{V}_1 = \text{im}\,\mathbf{A}$ and $\mathcal{V}_2 = \ker \mathbf{A}$.

A subspace $\mathcal{V}_1$ of the space $\mathcal{V}$ is called an *invariant subspace* of a linear operator $\mathbf{A} : \mathcal{V} \to \mathcal{V}$ if for any $\mathbf{x}$ in $\mathcal{V}_1$, the element $\mathbf{Ax}$ also belongs to $\mathcal{V}_1$. A linear operator $\mathbf{A} : \mathcal{V} \to \mathcal{V}$ is said to be *reducible* if $\mathcal{V}$ can be represented as a direct sum $\mathcal{V} = \mathcal{V}_1 \oplus \dots \oplus \mathcal{V}_N$ of two or more invariant subspaces $\mathcal{V}_1, \dots, \mathcal{V}_N$ of the operator $\mathbf{A}$, where $N$ is a natural number.

**Example 1.** $\ker \mathbf{A}$ and $\text{im}\,\mathbf{A}$ are invariant subspaces of any linear operator $\mathbf{A} : \mathcal{V} \to \mathcal{V}$.

The *rank* of a linear operator **A** is the dimension of its range: rank (**A**) = dim (im **A**).
Properties of the rank of a linear operator:

$$\text{rank}(\mathbf{AB}) \leq \min\{\text{rank}(\mathbf{A}), \text{rank}(\mathbf{B})\},$$
$$\text{rank}(\mathbf{A}) + \text{rank}(\mathbf{B}) - n \leq \text{rank}(\mathbf{AB}),$$

where **A** and **B** are linear operators in $L(\mathcal{V}, \mathcal{V})$ and $n = \dim \mathcal{V}$.

Remark. If rank (**A**) = $n$ then rank (**AB**) = rank (**BA**) = rank (**B**).

THEOREM. *Let* $\mathbf{A} : \mathcal{V} \to \mathcal{V}$ *be a linear operator. Then the following statements are equivalent:*
1. **A** *is invertible (i.e., there exists* $\mathbf{A}^{-1}$*).*
2. ker **A** = 0.
3. im **A** = $\mathcal{V}$.
4. rank (**A**) = dim $\mathcal{V}$.

5.6.1-5. Notion of a adjoint operator. Hermitian operators.

Let $\mathbf{A} \in L(\mathcal{V}, \mathcal{V})$ be a bounded linear operator in a Hilbert space $\mathcal{V}$. The operator $\mathbf{A}^*$ in $L(\mathcal{V}, \mathcal{V})$ is called its *adjoint operator* if

$$(\mathbf{Ax}) \cdot \mathbf{y} = \mathbf{x} \cdot (\mathbf{A}^*\mathbf{y})$$

for all **x** and **y** in $\mathcal{V}$.

THEOREM. *Any bounded linear operator* **A** *in a Hilbert space has a unique adjoint operator.*

Properties of adjoint operators:

$$(\mathbf{A} + \mathbf{B})^* = \mathbf{A}^* + \mathbf{B}^*, \quad (\lambda \mathbf{A})^* = \bar{\lambda}\mathbf{A}^*, \quad (\mathbf{A}^*)^* = \mathbf{A},$$
$$(\mathbf{AB})^* = \mathbf{B}^*\mathbf{A}^*, \quad \mathbf{O}^* = \mathbf{O}, \quad \mathbf{I}^* = \mathbf{I},$$
$$(\mathbf{A}^{-1})^* = (\mathbf{A}^*)^{-1}, \quad \|\mathbf{A}^*\| = \|\mathbf{A}\|, \quad \|\mathbf{A}^*\mathbf{A}\| = \|\mathbf{A}\|^2,$$
$$(\mathbf{Ax}) \cdot (\mathbf{By}) \equiv \mathbf{x} \cdot (\mathbf{A}^*\mathbf{By}) \equiv (\mathbf{B}^*\mathbf{Ax}) \cdot \mathbf{y} \quad \text{for all } \mathbf{x} \text{ and } \mathbf{y} \text{ in } \mathcal{V},$$

where **A** and **B** are bounded linear operators in a Hilbert space $\mathcal{V}$, $\bar{\lambda}$ is the complex conjugate of a number $\lambda$.

A linear operator $\mathbf{A} \in L(\mathcal{V}, \mathcal{V})$ in a Hilbert space $\mathcal{V}$ is said to be *Hermitian (self-adjoint)* if

$$\mathbf{A}^* = \mathbf{A} \quad \text{or} \quad (\mathbf{Ax}) \cdot \mathbf{y} = \mathbf{x} \cdot (\mathbf{Ay}).$$

A linear operator $\mathbf{A} \in (\mathcal{V}, \mathcal{V})$ in a Hilbert space $\mathcal{V}$ is said to be *skew-Hermitian* if

$$\mathbf{A}^* = -\mathbf{A} \quad \text{or} \quad (\mathbf{Ax}) \cdot \mathbf{y} = -\mathbf{x} \cdot (\mathbf{Ay}).$$

5.6.1-6. Unitary and normal operators.

A linear operator $\mathbf{U} \in L(\mathcal{V}, \mathcal{V})$ in a Hilbert space $\mathcal{V}$ is called a *unitary operator* if for all **x** and **y** in $\mathcal{V}$, the following relation holds:

$$(\mathbf{Ux}) \cdot (\mathbf{Uy}) = \mathbf{x} \cdot \mathbf{y}.$$

This relation is called the *unitarity condition*.

Properties of a unitary operator **U**:

$$\mathbf{U}^* = \mathbf{U}^{-1} \quad \text{or} \quad \mathbf{U}^*\mathbf{U} = \mathbf{U}\mathbf{U}^* = \mathbf{I},$$
$$\|\mathbf{U}\mathbf{x}\| = \|\mathbf{x}\| \quad \text{for all } \mathbf{x} \text{ in } \mathcal{V}.$$

A linear operator **A** in $L(\mathcal{V}, \mathcal{V})$ is said to be *normal* if

$$\mathbf{A}^*\mathbf{A} = \mathbf{A}\mathbf{A}^*.$$

THEOREM. *A bounded linear operator* **A** *is normal if and only if* $\|\mathbf{A}\mathbf{x}\| = \|\mathbf{A}\| \, \|\mathbf{x}\|$.

Remark. Any unitary or Hermitian operator is normal.

### 5.6.1-7. Transpose, symmetric, and orthogonal operators.

The *transpose operator* of a bounded linear operator $\mathbf{A} \in L(\mathcal{V}, \mathcal{V})$ in a real Hilbert space $\mathcal{V}$ is the operator $\mathbf{A}^T \in L(\mathcal{V}, \mathcal{V})$ such that for all **x**, **y** in $\mathcal{V}$, the following relation holds:

$$(\mathbf{A}\mathbf{x}) \cdot \mathbf{y} = \mathbf{x} \cdot (\mathbf{A}^T \mathbf{y}).$$

THEOREM. *Any bounded linear operator* **A** *in a real Hilbert space has a unique transpose operator.*

The properties of transpose operators in a real Hilbert space are similar to the properties of adjoint operators considered in Paragraph 5.6.1-5 if one takes $\mathbf{A}^T$ instead of $\mathbf{A}^*$.

A linear operator $\mathbf{A} \in L(\mathcal{V}, \mathcal{V})$ in a real Hilbert space $\mathcal{V}$ is said to be *symmetric* if

$$\mathbf{A}^T = \mathbf{A} \quad \text{or} \quad (\mathbf{A}\mathbf{x}) \cdot \mathbf{y} = \mathbf{x} \cdot (\mathbf{A}\mathbf{y}).$$

A linear operator $\mathbf{A} \in L(\mathcal{V}, \mathcal{V})$ in a real Hilbert space $\mathcal{V}$ is said to be *skew-symmetric* if

$$\mathbf{A}^T = -\mathbf{A} \quad \text{or} \quad (\mathbf{A}\mathbf{x}) \cdot \mathbf{y} = -\mathbf{x} \cdot (\mathbf{A}\mathbf{y}).$$

The properties of symmetric linear operators in a real Hilbert space are similar to the properties of Hermitian operators considered in Paragraph 5.6.1-5 if one takes $\mathbf{A}^T$ instead of $\mathbf{A}^*$.

A linear operator $\mathbf{P} \in L(\mathcal{V}, \mathcal{V})$ in a real Hilbert space $\mathcal{V}$ is said to be *orthogonal* if for any **x** and **y** in $\mathcal{V}$, the following relations hold:

$$(\mathbf{P}\mathbf{x}) \cdot (\mathbf{P}\mathbf{y}) = \mathbf{x} \cdot \mathbf{y}.$$

This relation is called the *orthogonality condition*.

Properties of orthogonal operator **P**:

$$\mathbf{P}^T = \mathbf{P}^{-1} \quad \text{or} \quad \mathbf{P}^T\mathbf{P} = \mathbf{P}\mathbf{P}^T = \mathbf{I},$$
$$\|\mathbf{P}\mathbf{x}\| = \|\mathbf{x}\| \quad \text{for all } \mathbf{x} \text{ in } \mathcal{V}.$$

### 5.6.1-8. Positive operators. Roots of an operator.

A Hermitian (symmetric, in the case of a real space) operator **A** is said to be
a) *nonnegative* (resp., *nonpositive*), and one writes $\mathbf{A} \geq 0$ (resp., $\mathbf{A} \leq 0$) if $(\mathbf{A}\mathbf{x}) \cdot \mathbf{x} \geq 0$ (resp., $(\mathbf{A}\mathbf{x}) \cdot \mathbf{x} \leq 0$) for any **x** in $\mathcal{V}$.
b) *positive* or *positive definite* (resp., *negative* or *negative definite*) and one writes $\mathbf{A} > 0$ ($\mathbf{A} < 0$) if $(\mathbf{A}\mathbf{x}) \cdot \mathbf{x} > 0$ (resp., $(\mathbf{A}\mathbf{x}) \cdot \mathbf{x} < 0$) for any $\mathbf{x} \neq 0$.

An *mth root* of an operator **A** is an operator **B** such that $\mathbf{B}^m = \mathbf{A}$.

THEOREM. *If* **A** *is a nonnegative Hermitian (symmetric) operator, then for any positive integer* $m$ *there exists a unique nonnegative Hermitian (symmetric) operator* $\mathbf{A}^{1/m}$.

5.6.1-9. Decomposition theorems.

THEOREM 1. *For any bounded linear operator* $\mathbf{A}$ *in a Hilbert space* $\mathcal{V}$, *the operator* $\mathbf{H}_1 = \frac{1}{2}(\mathbf{A} + \mathbf{A}^*)$ *is Hermitian and the operator* $\mathbf{H}_2 = \frac{1}{2}(\mathbf{A} - \mathbf{A}^*)$ *is skew-Hermitian. The representation of* $\mathbf{A}$ *as a sum of Hermitian and skew-Hermitian operators is unique:* $\mathbf{A} = \mathbf{H}_1 + \mathbf{H}_2$.

THEOREM 2. *For any bounded linear operator* $\mathbf{A}$ *in a real Hilbert space, the operator* $\mathbf{S}_1 = \frac{1}{2}(\mathbf{A} + \mathbf{A}^T)$ *is symmetric and the operator* $\mathbf{S}_2 = \frac{1}{2}(\mathbf{A} - \mathbf{A}^T)$ *is skew-symmetric. The representation of* $\mathbf{A}$ *as a sum of symmetric and skew-symmetric operators is unique:* $\mathbf{A} = \mathbf{S}_1 + \mathbf{S}_2$.

THEOREM 3. *For any bounded linear operator* $\mathbf{A}$ *in a Hilbert space,* $\mathbf{A}\mathbf{A}^*$ *and* $\mathbf{A}^*\mathbf{A}$ *are nonnegative Hermitian operators.*

THEOREM 4. *For any linear operator* $\mathbf{A}$ *in a Hilbert space* $\mathcal{V}$, *there exist polar decompositions*
$$\mathbf{A} = \mathbf{Q}\mathbf{U} \quad \text{and} \quad \mathbf{A} = \mathbf{U}_1 \mathbf{Q}_1,$$
*where* $\mathbf{Q}$ *and* $\mathbf{Q}_1$ *are nonnegative Hermitian operators,* $\mathbf{Q}^2 = \mathbf{A}\mathbf{A}^*$, $\mathbf{Q}_1^2 = \mathbf{A}^*\mathbf{A}$, *and* $\mathbf{U}$, $\mathbf{U}_1$ *are unitary operators. The operators* $\mathbf{Q}$ *and* $\mathbf{Q}_1$ *are always unique, while the operators* $\mathbf{U}$ *and* $\mathbf{U}_1$ *are unique only if* $\mathbf{A}$ *is nondegenerate.*

## 5.6.2. Linear Operators in Matrix Form

5.6.2-1. Matrices associated with linear operators.

Let $\mathbf{A}$ be a linear operator in an $n$-dimensional linear space $\mathcal{V}$ with a basis $\mathbf{e}_1, \ldots, \mathbf{e}_n$. Then there is a matrix $[a_j^i]$ such that
$$\mathbf{A}\mathbf{e}_j = \sum_{i=1}^n a_j^i \mathbf{e}_i.$$

The coordinates $y^j$ of the vector $\mathbf{y} = \mathbf{A}\mathbf{x}$ in that basis can be represented in the form
$$y^i = \sum_{j=1}^n a_j^i x^j \quad (i = 1, 2, \ldots, n), \tag{5.6.2.1}$$

where $x^j$ are the coordinates of $\mathbf{x}$ in the same basis $\mathbf{e}_1, \ldots, \mathbf{e}_n$. The matrix $A \equiv [a_j^i]$ of size $n \times n$ is called the *matrix of the linear operator* $\mathbf{A}$ in a given basis $\mathbf{e}_1, \ldots, \mathbf{e}_n$.

Thus, given a basis $\mathbf{e}_1, \ldots, \mathbf{e}_n$, any linear operator $\mathbf{y} = \mathbf{A}\mathbf{x}$ can be associated with its matrix in that basis with the help of (5.6.2.1).

If $\mathbf{A}$ is the zero operator, then its matrix is the zero matrix in any basis. If $\mathbf{A}$ is the unit operator, then its matrix is the unit matrix in any basis.

THEOREM 1. *Let* $\mathbf{e}_1, \ldots, \mathbf{e}_n$ *be a given basis in a linear space* $\mathcal{V}$ *and let* $A \equiv [a_j^i]$ *be a given square matrix of size* $n \times n$. *Then there exists a unique linear operator* $\mathbf{A} : \mathcal{V} \to \mathcal{V}$ *whose matrix in that basis coincides with the matrix* $A$.

THEOREM 2. *The rank of a linear operator* $\mathbf{A}$ *is equal to the rank of its matrix* $A$ *in any basis:* rank $(\mathbf{A})$ = rank $(A)$.

THEOREM 3. *A linear operator* $\mathbf{A} : \mathcal{V} \to \mathcal{V}$ *is invertible if and only if* rank $(A)$ = dim $V$. *In this case, the matrix of the operator* $\mathbf{A}$ *is invertible.*

**5.6.2-2. Transformation of the matrix of a linear operator.**

Suppose that the transition from the basis $\mathbf{e}_1, \ldots, \mathbf{e}_n$ to another basis $\widetilde{\mathbf{e}}_1, \ldots, \widetilde{\mathbf{e}}_n$ is determined by a matrix $U \equiv [u_{ij}]$ of size $n \times n$, i.e.

$$\widetilde{\mathbf{e}}_i = \sum_{j=1}^{n} u_{ij} \mathbf{e}_j \qquad (i = 1, 2, \ldots, n).$$

THEOREM. *Let $A$ and $\widetilde{A}$ be the matrices of a linear operator $\mathbf{A}$ in the basis $\mathbf{e}_1, \ldots, \mathbf{e}_n$ and the basis $\widetilde{\mathbf{e}}_1, \ldots, \widetilde{\mathbf{e}}_n$, respectively. Then*

$$A = U^{-1} \widetilde{A} U \quad \text{or} \quad \widetilde{A} = U A U^{-1}.$$

Note that the determinant of the matrix of a linear operator does not depend on the basis: $\det A = \det \widetilde{A}$. Therefore, one can correctly define the *determinant* $\det \mathbf{A}$ *of a linear operator* as the determinant of its matrix in any basis:

$$\det \mathbf{A} = \det A.$$

The trace of the matrix of a linear operator, $\mathrm{Tr}(A)$, is also independent of the basis. Therefore, one can correctly define the *trace* $\mathrm{Tr}(\mathbf{A})$ *of a linear operator* as the trace of its matrix in any basis:

$$\mathrm{Tr}(\mathbf{A}) = \mathrm{Tr}(A).$$

In the case of an orthonormal basis, a Hermitian, skew-Hermitian, normal, or unitary operator in a Hilbert space corresponds to a Hermitian, skew-Hermitian, normal, or unitary matrix; and a symmetric, skew-symmetric, or transpose operator in a real Hilbert space corresponds to a symmetric, skew-symmetric, or transpose matrix.

### 5.6.3. Eigenvectors and Eigenvalues of Linear Operators

**5.6.3-1. Basic definitions.**

1°. A scalar $\lambda$ is called an *eigenvalue* of a linear operator $\mathbf{A}$ in a vector space $\mathcal{V}$ if there is a nonzero element $\mathbf{x}$ in $\mathcal{V}$ such that

$$\mathbf{A}\mathbf{x} = \lambda \mathbf{x}. \qquad (5.6.3.1)$$

A nonzero element $\mathbf{x}$ for which (5.6.3.1) holds is called an *eigenvector* of the operator $\mathbf{A}$ corresponding to the eigenvalue $\lambda$. Eigenvectors corresponding to distinct eigenvalues are linearly independent. For an eigenvalue $\lambda \neq 0$, the inverse $\mu = 1/\lambda$ is called a *characteristic value* of the operator $A$.

THEOREM. *If $\mathbf{x}_1, \ldots, \mathbf{x}_k$ are eigenvectors of an operator $\mathbf{A}$ corresponding to its eigenvalue $\lambda$, then $\alpha_1 \mathbf{x}_1 + \cdots + \alpha_k \mathbf{x}_k$ ($\alpha_1^2 + \cdots + \alpha_k^2 \neq 0$) is also an eigenvector of the operator $\mathbf{A}$ corresponding to the eigenvalue $\lambda$.*

The *geometric multiplicity* $m_i$ of an eigenvalue $\lambda_i$ is the maximal number of linearly independent eigenvectors corresponding to the eigenvalue $\lambda_i$. Thus, the geometric multiplicity of $\lambda_i$ is the dimension of the subspace formed by all eigenvectors corresponding to the eigenvalue $\lambda_i$.

The *algebraic multiplicity* $m'_i$ of an eigenvalue $\lambda_i$ of an operator $\mathbf{A}$ is equal to the algebraic multiplicity of $\lambda_i$ regarded as an eigenvalue of the corresponding matrix $A$.

The algebraic multiplicity $m'_i$ of an eigenvalue $\lambda_i$ is always not less than the geometric multiplicity $m_i$ of this eigenvalue.

The trace Tr(**A**) is equal to the sum of all eigenvalues of the operator **A**, each eigenvalue counted according to its multiplicity, i.e.,

$$\text{Tr}(\mathbf{A}) = \sum_i m'_i \lambda_i.$$

The determinant det **A** is equal to the product of all eigenvalues of the operator **A**, each eigenvalue entering the product according to its multiplicity,

$$\det \mathbf{A} = \prod_i \lambda_i^{m'_i}.$$

**5.6.3-2. Eigenvectors and eigenvalues of normal and Hermitian operators.**

Properties of eigenvalues and eigenvectors of a normal operator:
1. A normal operator **A** in a Hilbert space $\mathcal{V}$ and its adjoint operator $\mathbf{A}^*$ have the same eigenvectors and their eigenvalues are complex conjugate.
2. For a normal operator **A** in a Hilbert space $\mathcal{V}$, there is a basis $\{\mathbf{e}_k\}$ formed by eigenvectors of the operators **A** and $\mathbf{A}^*$. Therefore, there is a basis in $\mathcal{V}$ in which the operator **A** has a diagonal matrix.
3. Eigenvectors corresponding to distinct eigenvalues of a normal operator are mutually orthogonal.
4. Any bounded normal operator **A** in a Hilbert space $\mathcal{V}$ is reducible. The space $\mathcal{V}$ can be represented as a direct sum of the subspace spanned by an orthonormal system of eigenvectors of **A** and the subspace consisting of vectors orthogonal to all eigenvectors of **A**. In the finite-dimensional case, an orthonormal system of eigenvectors of **A** is a basis of $\mathcal{V}$.
5. The algebraic multiplicity of any eigenvalue $\lambda$ of a normal operator is equal to its geometric multiplicity.

Properties of eigenvalues and eigenvectors of a Hermitian operator:
1. Since any Hermitian operator is normal, all properties of normal operators hold for Hermitian operators.
2. All eigenvalues of a Hermitian operator are real.
3. Any Hermitian operator **A** in an $n$-dimensional unitary space has $n$ mutually orthogonal eigenvectors of unit length.
4. Any eigenvalue of a nonnegative (positive) operator is nonnegative (positive).
5. *Minimax property.* Let **A** be a Hermitian operator in an $n$-dimensional unitary space $\mathcal{V}$, and let $\mathcal{E}_m$ be the set of all $m$-dimensional subspaces of $\mathcal{V}$ ($m < n$). Then the eigenvalues $\lambda_1, \ldots, \lambda_n$ of the operator **A** ($\lambda_1 \geq \ldots \geq \lambda_n$) can be defined by the formulas

$$\lambda_{m+1} = \min_{\mathcal{Y} \in \mathcal{E}_m} \max_{\mathbf{x} \perp \mathcal{Y}} \frac{(\mathbf{A}\mathbf{x}) \cdot \mathbf{x}}{\mathbf{x} \cdot \mathbf{x}}.$$

6. Let $\mathbf{i}_1, \ldots, \mathbf{i}_n$ be an orthonormal basis in an $n$-dimensional space $\mathcal{V}$, and let all $\mathbf{i}_k$ are eigenvectors of a Hermitian operator **A**, i.e., $\mathbf{A}\mathbf{i}_k = \lambda_k \mathbf{i}_k$. Then the matrix of the operator **A** in the basis $\mathbf{i}_1, \ldots, \mathbf{i}_n$ is diagonal and its diagonal elements have the form $a_k^k = \lambda_k$.

7. Let $\mathbf{i}_1, \dots, \mathbf{i}_n$ be an arbitrary orthonormal basis in an $n$-dimensional Euclidean space $\mathcal{V}$. Then the matrix of an operator $\mathbf{A}$ in the basis $\mathbf{i}_1, \dots, \mathbf{i}_n$ is symmetric if and only if the operator $\mathbf{A}$ is Hermitian.
8. In an orthonormal basis $\mathbf{i}_1, \dots, \mathbf{i}_n$ formed by eigenvectors of a nonnegative Hermitian operator $\mathbf{A}$, the matrix of the operator $\mathbf{A}^{1/m}$ has the form

$$\begin{pmatrix} \lambda_1^{1/m} & 0 & \cdots & 0 \\ 0 & \lambda_2^{1/m} & \cdots & 0 \\ \vdots & \vdots & \ddots & \vdots \\ 0 & 0 & \cdots & \lambda_n^{1/m} \end{pmatrix}.$$

### 5.6.3-3. Characteristic polynomial of a linear operator.

Consider the finite-dimensional case. The algebraic equation

$$f_{\mathbf{A}}(\lambda) \equiv \det(\mathbf{A} - \lambda \mathbf{I}) = 0 \qquad (5.6.3.2)$$

of degree $n$ is called the *characteristic equation* of the operator $\mathbf{A}$ and $f_{\mathbf{A}}(\lambda)$ is called the *characteristic polynomial* of the operator $\mathbf{A}$.

Since the value of the determinant $\det(\mathbf{A} - \lambda \mathbf{I})$ does not depend on the basis, the coefficients of $\lambda^k$ ($k = 0, 1, \dots, n$) in the characteristic polynomial $f_{\mathbf{A}}(\lambda)$ are *invariants* (i.e., quantities whose values do not depend on the basis). In particular, the coefficient of $\lambda^{k-1}$ is equal to the trace of the operator $\mathbf{A}$.

In the finite-dimensional case, $\lambda$ is an eigenvalue of a linear operator $\mathbf{A}$ if and only if $\lambda$ is a root of the characteristic equation (5.6.3.2) of the operator $\mathbf{A}$. Therefore, a linear operator always has eigenvalues.

In the case of a real space, a root of the characteristic equation can be an eigenvalue of a linear operator only if this root is real. In this connection, it would be natural to find a class of linear operators in a real Euclidean space for which all roots of the corresponding characteristic equations are real.

THEOREM. *The matrix $A$ of a linear operator $\mathbf{A}$ in a given basis $\mathbf{i}_1, \dots, \mathbf{i}_n$ is diagonal if and only if all $\mathbf{i}_i$ are eigenvectors of this operator.*

### 5.6.3-4. Bounds for eigenvalues of linear operators.

The modulus of any eigenvalue $\lambda$ of a linear operator $\mathbf{A}$ in an $n$-dimensional unitary space satisfies the estimate:

$$|\lambda| \leq \min(M_1, M_2), \qquad M_1 = \max_{1 \leq i \leq n} \sum_{j=1}^{n} |a_{ij}|, \quad M_2 = \max_{1 \leq j \leq n} \sum_{i=1}^{n} |a_{ij}|,$$

where $A \equiv [a_{ij}]$ is the matrix of the operator $\mathbf{A}$. The real and the imaginary parts of eigenvalues satisfy the estimates:

$$\min_{1 \leq i \leq n}(\operatorname{Re} a_{ii} - P_i) \leq \operatorname{Re} \lambda \leq \max_{1 \leq i \leq n}(\operatorname{Re} a_{ii} + P_i),$$
$$\min_{1 \leq i \leq n}(\operatorname{Im} a_{ii} - P_i) \leq \operatorname{Im} \lambda \leq \max_{1 \leq i \leq n}(\operatorname{Im} a_{ii} + P_i),$$

where $P_i = \sum_{j=1, j\neq i}^{n} |a_{ij}|$, and $P_i$ can be replaced by $Q_i = \sum_{j=1, i\neq i}^{n} |a_{ji}|$.

The modulus of any eigenvalue $\lambda$ of a Hermitian operator $\mathbf{A}$ in an $n$-dimensional unitary space satisfies the inequalities

$$|\lambda|^2 \leq \sum_i \sum_j |a_{ij}|^2, \quad |\lambda| \leq \|\mathbf{A}\| = \sup_{\|\mathbf{x}\|=1} [(\mathbf{Ax}) \cdot \mathbf{x}],$$

and its smallest and its largest eigenvalues, denoted, respectively, by $m$ and $M$, can be found from the relations

$$m = \inf_{\|\mathbf{x}\|=1} [(\mathbf{Ax}) \cdot \mathbf{x}], \quad M = \sup_{\|\mathbf{x}\|=1} [(\mathbf{Ax}) \cdot \mathbf{x}].$$

### 5.6.3-5. Spectral decomposition of Hermitian operators.

Let $\mathbf{i}_1, \ldots, \mathbf{i}_n$ be a fixed orthonormal basis in an $n$-dimensional unitary space $\mathcal{V}$. Then any element of $\mathcal{V}$ can be represented in the form (see Paragraph 5.4.2-2)

$$\mathbf{x} = \sum_{j=1}^{n} (\mathbf{x} \cdot \mathbf{i}_j)\mathbf{i}_j.$$

The operator $\mathbf{P}_k$ ($k = 1, 2, \ldots, n$) defined by

$$\mathbf{P}_k \mathbf{x} = (\mathbf{x} \cdot \mathbf{i}_k)\mathbf{i}_k$$

is called the *projection* onto the one-dimensional subspace generated by the vector $\mathbf{i}_k$. The projection $\mathbf{P}_k$ is a Hermitian operator.

Properties of the projection $\mathbf{P}_k$:

$$\mathbf{P}_k \mathbf{P}_l = \begin{cases} \mathbf{P}_k & \text{for } k = l, \\ \mathbf{O} & \text{for } k \neq l, \end{cases} \quad \mathbf{P}_k^m = \mathbf{P}_k \quad (m = 1, 2, 3, \ldots),$$

$$\sum_{j=1}^{n} \mathbf{P}_j = \mathbf{I}, \quad \text{where } \mathbf{I} \text{ is the identity operator.}$$

For a normal operator $\mathbf{A}$, there is an orthonormal basis consisting of its eigenvectors, $\mathbf{A}\mathbf{i}_k = \lambda \mathbf{i}_k$. Then one obtains the *spectral decomposition of a normal operator*:

$$\mathbf{A}^k = \sum_{j=1}^{n} \lambda_j^k \mathbf{P}_j \quad (k = 1, 2, 3, \ldots). \tag{5.6.3.3}$$

Consider an arbitrary polynomial $p(\lambda) = \sum_{j=1}^{m} c_j \lambda^j$. By definition, $p(\mathbf{A}) = \sum_{j=1}^{m} c_j \mathbf{A}^j$. Then, using (5.6.3.3), we get

$$p(\mathbf{A}) = \sum_{i=1}^{m} p(\lambda_i) \mathbf{P}_i.$$

CAYLEY-HAMILTON THEOREM. *Every normal operator satisfies its own characteristic equation, i.e.,* $f_{\mathbf{A}}(\mathbf{A}) = \mathbf{O}$.

**5.6.3-6. Canonical form of linear operators.**

An element $\mathbf{x}$ is called an *associated vector* of an operator $\mathbf{A}$ corresponding to its eigenvalue $\lambda$ if for some $m \geq 1$, we have

$$(\mathbf{A} - \lambda\mathbf{I})^m \mathbf{x} \neq 0, \quad (\mathbf{A} - \lambda\mathbf{I})^{m+1} \mathbf{x} = 0.$$

The number $m$ is called the *order of the associated vector* $\mathbf{x}$.

THEOREM. *Let $\mathbf{A}$ be a linear operator in an $n$-dimensional unitary space $\mathcal{V}$. Then there is a basis $\{\mathbf{i}_k^m\}$ ($k = 1, 2, \ldots, l$, $m = 1, 2, \ldots, n_k$, $n_1 + n_2 + \cdots + n_l = n$) in $\mathcal{V}$ consisting of eigenvectors and associated vectors of the operator $\mathbf{A}$ such that the action of the operator $\mathbf{A}$ is determined by the relations*

$$A\mathbf{i}_k^1 = \lambda_k \mathbf{i}_k^1 \quad (k = 1, 2, \ldots, l),$$
$$A\mathbf{i}_k^m = \lambda_k \mathbf{i}_k^m + \mathbf{i}_k^{m-1} \quad (k = 1, 2, \ldots, l, \ m = 2, 3, \ldots, n_k).$$

Remark 1. The vectors $\mathbf{i}_k^1$ ($k = 1, 2, \ldots, l$) are eigenvectors of the operator $\mathbf{A}$ corresponding to the eigenvalues $\lambda_k$.

Remark 2. The matrix $A$ of the linear operator $\mathbf{A}$ in the basis $\{\mathbf{i}_k^m\}$ has canonical Jordan form, and the above theorem is also called the theorem on the reduction of a matrix to canonical Jordan form.

## 5.7. Bilinear and Quadratic Forms

### 5.7.1. Linear and Sesquilinear Forms

**5.7.1-1. Linear forms in a unitary space.**

A *linear form* or *linear functional* on $\mathcal{V}$ is a linear operator $\mathbf{A}$ in $L(\mathcal{V}, \mathcal{C})$, where $\mathcal{C}$ is the complex plane.

THEOREM. *For any linear form $f$ in a finite-dimensional unitary space $\mathcal{V}$, there is a unique element $\mathbf{h}$ in $\mathcal{V}$ such that*

$$f(\mathbf{x}) = \mathbf{x} \cdot \mathbf{h} \quad \text{for all} \quad \mathbf{x} \in \mathcal{V}.$$

Remark. This statement is true also for a Euclidean space $\mathcal{V}$ and a real-valued linear functional.

**5.7.1-2. Sesquilinear forms in unitary space.**

A *sesquilinear form* on a unitary space $\mathcal{V}$ is a complex-valued function $B(\mathbf{x}, \mathbf{y})$ of two arguments $\mathbf{x}, \mathbf{y} \in \mathcal{V}$ such that for any $\mathbf{x}, \mathbf{y}, \mathbf{z}$ in $\mathcal{V}$ and any complex scalar $\lambda$, the following relations hold:
1. $B(\mathbf{x} + \mathbf{y}, \mathbf{z}) = B(\mathbf{x}, \mathbf{z}) + B(\mathbf{y}, \mathbf{z})$.
2. $B(\mathbf{x}, \mathbf{y} + \mathbf{z}) = B(\mathbf{x}, \mathbf{y}) + B(\mathbf{x}, \mathbf{z})$.
3. $B(\lambda\mathbf{x}, \mathbf{y}) = \lambda B(\mathbf{x}, \mathbf{y})$.
4. $B(\mathbf{x}, \lambda\mathbf{y}) = \bar{\lambda} B(\mathbf{x}, \mathbf{y})$.

**Remark.** Thus, $B(\mathbf{x},\mathbf{y})$ is a scalar function that is linear with respect to its first argument and antilinear with respect to its second argument. For a real space $\mathcal{V}$, sesquilinear forms turn into bilinear forms (see Paragraph 5.7.2).

THEOREM. *Let $B(\mathbf{x},\mathbf{y})$ be a sesquilinear form in a unitary space $\mathcal{V}$. Then there is a unique linear operator $\mathbf{A}$ in $L(\mathcal{V},\mathcal{V})$ such that*

$$B(\mathbf{x},\mathbf{y}) = \mathbf{x} \cdot (\mathbf{A}\mathbf{y}).$$

COROLLARY. *If $B(\mathbf{x},\mathbf{y})$ is a sesquilinear form in a unitary space $V$, then there is a unique linear operator $\mathbf{A}$ in $L(\mathcal{V},\mathcal{V})$ such that*

$$B(\mathbf{x},\mathbf{y}) = (\mathbf{A}\mathbf{x}) \cdot \mathbf{y}.$$

### 5.7.1-3. Matrix of a sesquilinear form.

Any sesquilinear form $B(\mathbf{x},\mathbf{y})$ on an $n$-dimensional linear space with a given basis $\mathbf{e}_1, \ldots, \mathbf{e}_n$ can be uniquely represented as

$$B(\mathbf{x},\mathbf{y}) = \sum_{i,j=1}^{n} b_{ij}\xi_i\bar{\eta}_j, \qquad b_{ij} = B(\mathbf{e}_i,\mathbf{e}_j),$$

and $\xi_i$, $\eta_j$ are the coordinates of $\mathbf{x}$ and $\mathbf{y}$ in the given basis. The matrix $B \equiv [b_{ij}]$ of size $n \times n$ is called the *matrix of the sesquilinear form* $B(\mathbf{x},\mathbf{y})$ in the given basis $\mathbf{e}_1, \ldots, \mathbf{e}_n$. This sesquilinear form can also be represented as

$$B(\mathbf{x},\mathbf{y}) = X^T B Y, \qquad X^T \equiv (\xi_1,\ldots,\xi_n), \quad Y^T \equiv (\bar{\eta}_1,\ldots,\bar{\eta}_n).$$

## 5.7.2. Bilinear Forms

### 5.7.2-1. Definition of a bilinear form.

A *bilinear form* on a real linear space $\mathcal{V}$ is a real-valued function $B(\mathbf{x},\mathbf{y})$ of two arguments $\mathbf{x} \in L$, $\mathbf{y} \in \mathcal{V}$ satisfying the following conditions for any vectors $\mathbf{x}$, $\mathbf{y}$, and $\mathbf{z}$ in $\mathcal{V}$ and any real $\lambda$:
1. $B(\mathbf{x}+\mathbf{y},\mathbf{z}) = B(\mathbf{x},\mathbf{z}) + B(\mathbf{y},\mathbf{z})$.
2. $B(\mathbf{x},\mathbf{y}+\mathbf{z}) = B(\mathbf{x},\mathbf{y}) + B(\mathbf{x},\mathbf{z})$.
3. $B(\lambda\mathbf{x},\mathbf{y}) = B(\mathbf{x},\lambda\mathbf{y}) = \lambda B(\mathbf{x},\mathbf{y})$.

THEOREM. *Let $B(\mathbf{x},\mathbf{y})$ be a bilinear form in a Euclidean space $\mathcal{V}$. Then there is a unique linear operator $\mathbf{A}$ in $L(\mathcal{V},\mathcal{V})$ such that*

$$B(\mathbf{x},\mathbf{y}) = (\mathbf{A}\mathbf{x}) \cdot \mathbf{y}.$$

A bilinear form $B(\mathbf{x},\mathbf{y})$ is said to be *symmetric* if for any $\mathbf{x}$ and $\mathbf{y}$, we have

$$B(\mathbf{x},\mathbf{y}) = B(\mathbf{y},\mathbf{x}).$$

A bilinear form $B(\mathbf{x},\mathbf{y})$ is said to be *skew-symmetric* if for any $\mathbf{x}$ and $\mathbf{y}$, we have

$$B(\mathbf{x},\mathbf{y}) = -B(\mathbf{y},\mathbf{x}).$$

Any bilinear form can be represented as the sum of symmetric and skew-symmetric bilinear forms.

THEOREM. *A bilinear form $B(\mathbf{x},\mathbf{y})$ on a Euclidean space $\mathcal{V}$ is symmetric if and only if the linear operator $\mathbf{A}$ in the representation (5.6.6.1) is Hermitian ($\mathbf{A} = \mathbf{A}^*$).*

### 5.7.2-2. Bilinear forms in finite-dimensional spaces.

Any bilinear form $B(\mathbf{x}, \mathbf{y})$ on an $n$-dimensional linear space with a given basis $\mathbf{e}_1, \ldots, \mathbf{e}_n$ can be uniquely represented as

$$B(\mathbf{x}, \mathbf{y}) = \sum_{i,j=1}^{n} b_{ij} \xi_i \eta_j, \qquad b_{ij} = B(\mathbf{e}_i, \mathbf{e}_j),$$

and $\xi_i, \eta_j$ are the coordinates of the vectors $\mathbf{x}$ and $\mathbf{y}$ in the given basis. The matrix $B \equiv [b_{ij}]$ of size $n \times n$ is called the *matrix of the bilinear form* in the given basis $\mathbf{e}_1, \ldots, \mathbf{e}_n$. The bilinear form can also be represented as

$$B(\mathbf{x}, \mathbf{y}) = X^T B Y, \qquad X^T \equiv (\xi_1, \ldots, \xi_n), \quad Y^T \equiv (\eta_1, \ldots, \eta_n).$$

**Remark.** Any square matrix $B \equiv [b_{ij}]$ can be regarded as a matrix of some bilinear form in a given basis $\mathbf{e}_1, \ldots, \mathbf{e}_n$. If this matrix is symmetric (skew-symmetric), then the bilinear form is symmetric (skew-symmetric).

The *rank of a bilinear form* $B(\mathbf{x}, \mathbf{y})$ on a finite-dimensional linear space $L$ is defined as the rank of the matrix $B$ of this form in any basis: $\operatorname{rank} B(\mathbf{x}, \mathbf{y}) = \operatorname{rank}(B)$.

A bilinear form on a finite dimensional space $\mathcal{V}$ is said to be *nondegenerate* (*degenerate*) if its rank is equal to (is less than) the dimension of the space $\mathcal{V}$, i.e., $\operatorname{rank} B(\mathbf{x}, \mathbf{y}) = \dim \mathcal{V}$ ($\operatorname{rank} B(\mathbf{x}, \mathbf{y}) < \dim \mathcal{V}$).

### 5.7.2-3. Transformation of the matrix of a bilinear form in another basis.

Suppose that the transition from a basis $\mathbf{e}_1, \ldots, \mathbf{e}_n$ to a basis $\widetilde{\mathbf{e}}_1, \ldots, \widetilde{\mathbf{e}}_n$ is determined by the matrix $U \equiv [u_{ij}]$ of size $n \times n$, i.e.

$$\widetilde{\mathbf{e}}_i = \sum_{j=1}^{n} u_{ij} \mathbf{e}_j \qquad (i = 1, 2, \ldots, n).$$

THEOREM. *The matrices $B$ and $\widetilde{B}$ of a bilinear form $B(\mathbf{x}, \mathbf{y})$ in the bases $\mathbf{e}_1, \ldots, \mathbf{e}_n$ and $\widetilde{\mathbf{e}}_1, \ldots, \widetilde{\mathbf{e}}_n$, respectively, are related by*

$$\widetilde{B} = U^T B U.$$

### 5.7.2-4. Multilinear forms.

A *multilinear form* on a linear space $\mathcal{V}$ is a scalar function $B(\mathbf{x}_1, \ldots, \mathbf{x}_p)$ of $p$ arguments $\mathbf{x}_1, \ldots, \mathbf{x}_p \in \mathcal{V}$, which is linear in each argument for fixed values of the other arguments.

A multilinear form $B(\mathbf{x}, \mathbf{y})$ is said to be *symmetric* if for any two arguments $\mathbf{x}_l$ and $\mathbf{x}_l$, we have

$$B(\mathbf{x}_1, \ldots, \mathbf{x}_k, \ldots, \mathbf{x}_l, \ldots, \mathbf{x}_p) = B(\mathbf{x}_1, \ldots, \mathbf{x}_l, \ldots, \mathbf{x}_k, \ldots, \mathbf{x}_p).$$

A multilinear form $B(\mathbf{x}, \mathbf{y})$ is said to be *skew-symmetric* if for any two arguments $\mathbf{x}_l$ and $\mathbf{x}_l$, we have

$$B(\mathbf{x}_1, \ldots, \mathbf{x}_k, \ldots, \mathbf{x}_l, \ldots, \mathbf{x}_p) = -B(\mathbf{x}_1, \ldots, \mathbf{x}_l, \ldots, \mathbf{x}_k, \ldots, \mathbf{x}_p).$$

## 5.7.3. Quadratic Forms

### 5.7.3-1. Definition of a quadratic form.

A *quadratic form* on a real linear space is a scalar function $B(\mathbf{x}, \mathbf{x})$ obtained from a bilinear form $B(\mathbf{x}, \mathbf{y})$ for $\mathbf{x} = \mathbf{y}$.

Any symmetric bilinear form $B(\mathbf{x}, \mathbf{y})$ is *polar* with respect to the quadratic form $B(\mathbf{x}, \mathbf{x})$. These forms are related by

$$B(\mathbf{x}, \mathbf{y}) = \tfrac{1}{2}[B(\mathbf{x}+\mathbf{y}, \mathbf{x}+\mathbf{y}) - B(\mathbf{x}, \mathbf{x}) - B(\mathbf{y}, \mathbf{y})].$$

### 5.7.3-2. Quadratic forms in a finite-dimensional linear space.

Any quadratic form $B(\mathbf{x}, \mathbf{x})$ in an $n$-dimensional linear space with a given basis $\mathbf{e}_1, \ldots, \mathbf{e}_n$ can be uniquely represented in the form

$$B(\mathbf{x}, \mathbf{x}) = \sum_{i,j=1}^{n} b_{ij} \xi_i \xi_j, \qquad (5.7.3.1)$$

where $\xi_i$ are the coordinates of the vector $\mathbf{x}$ in the given basis, and $B \equiv [b_{ij}]$ is a symmetric matrix of size $n \times n$, called the *matrix of the bilinear form* $B(\mathbf{x}, \mathbf{x})$ in the given basis. This quadratic form can also be represented as

$$B(\mathbf{x}, \mathbf{x}) = X^T B X, \qquad X^T \equiv (\xi_1, \ldots, \xi_n).$$

**Remark.** Any quadratic form can be represented in the form (5.7.3.1) with infinitely many matrices $B$ such that $B(\mathbf{x}, \mathbf{x}) = X^T B X$. In what follows, we consider only one of such matrices, namely, the symmetric matrix. A quadratic form is real-valued if its symmetric matrix is real.

A real-valued quadratic form $B(\mathbf{x}, \mathbf{x})$ is said to be:
a) *positive definite* (*negative definite*) if $B(\mathbf{x}, \mathbf{x}) > 0$ ($B(\mathbf{x}, \mathbf{x}) < 0$) for any $\mathbf{x} \neq 0$;
b) *alternating* if there exist $\mathbf{x}$ and $\mathbf{y}$ such that $B(\mathbf{x}, \mathbf{x}) > 0$ and $B(\mathbf{y}, \mathbf{y}) < 0$;
c) *nonnegative* (*nonpositive*) if $B(\mathbf{x}, \mathbf{x}) \geq 0$ ($B(\mathbf{x}, \mathbf{x}) \leq 0$) for all $\mathbf{x}$.

If $B(\mathbf{x}, \mathbf{y})$ is a polar bilinear form with respect to some positive definite quadratic form $B(\mathbf{x}, \mathbf{x})$, then $B(\mathbf{x}, \mathbf{y})$ satisfies all axioms of the scalar product in a Euclidean space.

**Remark.** The axioms of the scalar product can be regarded as the conditions that determine a bilinear form that is polar to some positive definite quadratic form.

The *rank of a quadratic form* on a finite-dimensional linear space $\mathcal{V}$ is, by definition, the rank of the matrix of that form in any basis of $\mathcal{V}$, rank $B(\mathbf{x}, \mathbf{x}) = $ rank $(B)$.

A quadratic form on a finite-dimensional linear space $\mathcal{V}$ is said to be *nondegenerate* (*degenerate*) if its rank is equal to (is less than) the dimension of $\mathcal{V}$, i.e., rank $B(\mathbf{x}, \mathbf{x}) = \dim \mathcal{V}$ (rank $B(\mathbf{x}, \mathbf{x}) < \dim \mathcal{V}$).

### 5.7.3-3. Transformation of a bilinear form in another basis.

Suppose that the transition from the basis $\mathbf{e}_1, \ldots, \mathbf{e}_n$ to the basis $\widetilde{\mathbf{e}}_1, \ldots, \widetilde{\mathbf{e}}_n$ is given by the matrix $U \equiv [u_{ij}]$ of size $n \times n$, i.e.

$$\widetilde{\mathbf{e}}_i = \sum_{j=1}^{n} u_{ij} \mathbf{e}_j \qquad (i = 1, 2, \ldots, n).$$

Then the matrices $B$ and $\widetilde{B}$ of the quadratic form $B(\mathbf{x}, \mathbf{x})$ in the bases $\mathbf{e}_1, \ldots, \mathbf{e}_n$ and $\widetilde{\mathbf{e}}_1, \ldots, \widetilde{\mathbf{e}}_n$, respectively, are related by

$$\widetilde{B} = U^T B U.$$

## 5.7.3-4. Canonical representation of a real quadratic form.

Let $\mathbf{g}_1, \ldots, \mathbf{g}_n$ be a basis in which the real quadratic form $B(\mathbf{x}, \mathbf{x})$ in a linear space $\mathcal{V}$ admits the representation

$$B(\mathbf{x}, \mathbf{x}) = \sum_{i=1}^{n} \lambda_i \eta_i^2, \tag{5.7.3.2}$$

where $\eta_1, \ldots, \eta_n$ are the coordinates of $\mathbf{x}$ in that basis. This representation is called a *canonical representation* of the quadratic form, the real coefficients $\lambda_1, \ldots, \lambda_n$ are called the *canonical coefficients*, and the basis $\mathbf{g}_1, \ldots, \mathbf{g}_n$ is called the *canonical basis*.

The number of nonzero canonical coefficients is equal to the rank of the quadratic form.

THEOREM. *Any real quadratic form on an $n$-dimensional real linear space $\mathcal{V}$ admits a canonical representation (5.7.3.2).*

$1°$. *Lagrange method.* The basic idea of the method consists of consecutive transformations of the quadratic form: on every step, one should single out the perfect square of some linear form.

Consider a quadratic form

$$B(\mathbf{x}, \mathbf{x}) = \sum_{i,j=1}^{n} b_{ij} \xi_i \xi_j.$$

*Case 1.* Suppose that for some $m$ ($1 \leq m \leq n$), we have $b_{mm} \neq 0$. Then, letting

$$B(\mathbf{x}, \mathbf{x}) = \frac{1}{b_{mm}} \left( \sum_{k=1}^{n} b_{mk} \xi_k \right)^2 + B_2(\mathbf{x}, \mathbf{x}),$$

one can easily verify that the quadratic form $B_2(\mathbf{x}, \mathbf{x})$ does not contain the variable $\xi_m$. This method of separating a perfect square in a quadratic form can always be applied if the matrix $[b_{ij}]$ ($i, j = 1, 2, \ldots, n$) contains nonzero diagonal elements.

*Case 2.* Suppose that $b_{mm} = 0$, $b_{ss} = 0$, but $b_{ms} \neq 0$. In this case, the quadratic form can be represented as

$$B(\mathbf{x}, \mathbf{x}) = \frac{1}{2b_{sm}} \left[ \sum_{k=1}^{n} (b_{mk} + b_{sk}) \xi_k \right]^2 - \frac{1}{2b_{sm}} \left[ \sum_{k=1}^{n} (b_{mk} - b_{sk}) \xi_k \right]^2 + B_2(\mathbf{x}, \mathbf{x}),$$

where $B_2(\mathbf{x}, \mathbf{x})$ does not contain the variables $\xi_m, \xi_s$, and the linear forms in square brackets are linearly independent (and therefore can be taken as new independent variables or coordinates).

By consecutive combination of the above two procedures, the quadratic form $B(\mathbf{x}, \mathbf{x})$ can always be represented in terms of squared linear forms; these forms are linearly independent, since each contains a variable which is absent in the other linear forms.

$2°$. *Jacobi method.* Suppose that

$$\Delta_1 \equiv b_{11} \neq 0, \quad \Delta_2 \equiv \begin{vmatrix} b_{11} & b_{12} \\ b_{21} & b_{22} \end{vmatrix} \neq 0, \quad \ldots, \quad \Delta_n \equiv \det B \neq 0,$$

where $B \equiv [b_{ij}]$ is the matrix of the quadratic form $B(\mathbf{x}, \mathbf{x})$ in some basis $\mathbf{e}_1, \ldots, \mathbf{e}_n$. One can obtain a canonical representation of this form using the formulas

$$\lambda_1 = \Delta_1, \quad \lambda_i = \frac{\Delta_i}{\Delta_{i-1}} \quad (i = 2, 3, \ldots, n).$$

The basis $\mathbf{e}_1, \ldots, \mathbf{e}_n$ is transformed to the canonical basis $\mathbf{g}_1, \ldots, \mathbf{g}_n$ by the formulas

$$\mathbf{g}_i = \sum_{j=1}^{n} \alpha_{ij} \mathbf{e}_j \quad (i = 1, 2, \ldots, n),$$

$$\alpha_{ij} = (-1)^{i+j} \frac{\Delta_{i-1,j}}{\Delta_{i-1}},$$

where $\Delta_{i-1,j}$ is the minor of the submatrix of $B \equiv [b_{ij}]$ formed by the elements on the intersection of its rows with indices $1, 2, \ldots, i-1$ and columns with indices $1, 2, \ldots, j-1, j+1, i$.

### 5.7.3-5. Normal representation of a real quadratic form.

Let $\mathbf{g}_1, \ldots, \mathbf{g}_n$ be a basis of a linear space $\mathcal{V}$ in which the quadratic form $B(\mathbf{x}, \mathbf{x})$ is written as

$$B(\mathbf{x}, \mathbf{x}) = \sum_{i=1}^{n} \varepsilon_i \eta_i^2, \tag{5.7.3.3}$$

where $\eta_1, \ldots, \eta_n$ are the coordinates of $\mathbf{x}$ in that basis, and $\varepsilon_1, \ldots, \varepsilon_n$ are coefficients taking the values $-1$, $0$, or $1$. Such a representation of a quadratic form is called its *normal representation*.

Any real quadratic form $B(\mathbf{x}, \mathbf{x})$ in an $n$-dimensional real linear space $\mathcal{V}$ admits a normal representation (5.7.3.3). Such a representation can be obtained by the following transformations:

1. One obtains its canonical representation (see Paragraph 5.7.3-4):

$$B(\mathbf{x}, \mathbf{x}) = \sum_{i=1}^{n} \lambda_i \mu_i^2.$$

2. By the nondegenerate coordinate transformation

$$\eta_i = \begin{cases} \frac{1}{\sqrt{\lambda_i}} \mu_i & \text{for } \lambda_i > 0, \\ \frac{1}{\sqrt{-\lambda_i}} \mu_i & \text{for } \lambda_i < 0, \\ \mu_i & \text{for } \lambda_i = 0, \end{cases}$$

the canonical representation turns into a normal representation.

LAW OF INERTIA OF QUADRATIC FORMS. *The number of terms with positive coefficients and the number of terms with negative coefficients in any normal representation of a real quadratic form does not depend on the method used to obtain such a representation.*

The *index of inertia* of a real quadratic form is the integer $k$ equal to the number of nonzero coefficients in its canonical representation (this number coincides with the rank of the quadratic form). Its *positive index of inertia* is the integer $p$ equal to the number of positive coefficients in the canonical representation of the form, and its *negative index of inertia* is the integer $q$ equal to the number of its negative canonical coefficients. The integer $s = p - q$ is called the *signature* of the quadratic form.

A real quadratic form $B(\mathbf{x}, \mathbf{x})$ on an $n$-dimensional real linear space $\mathcal{V}$ is
a) positive definite (resp., negative definite) if $p = n$ (resp., $q = n$);
b) alternating if $p \neq 0$, $q \neq 0$;
c) nonnegative (resp., nonpositive) if $q = 0$, $p < n$ (resp., $p = 0$, $q < n$).

**5.7.3-6. Criteria of positive and negative definiteness of a quadratic form.**

1°. A real quadratic form $B(\mathbf{x}, \mathbf{x})$ is positive definite, negative definite, alternating, nonnegative, nonpositive if the eigenvalues $\lambda_i$ of its matrix $B \equiv [b_{ij}]$ are all positive, are all negative, some are positive and some negative, are all nonnegative, are all nonpositive, respectively.

2°. *Sylvester criterion.* A real quadratic form $B(\mathbf{x}, \mathbf{x})$ is positive definite if and only if the matrix of $B(\mathbf{x}, \mathbf{x})$ in some basis $\mathbf{e}_1, \ldots, \mathbf{e}_n$ satisfies the conditions

$$\Delta_1 \equiv b_{11} > 0, \quad \Delta_2 \equiv \begin{vmatrix} b_{11} & b_{12} \\ b_{21} & b_{22} \end{vmatrix} > 0, \ldots, \quad \Delta_n \equiv \det B > 0.$$

If the signs of the minor determinants alternate,

$$\Delta_1 < 0, \quad \Delta_2 > 0, \quad \Delta_3 < 0, \ldots,$$

then the quadratic form is negative definite.

3°. A real matrix $B$ is nonnegative and symmetric if and only if there is a real matrix $C$ such that $B = C^T C$.

### 5.7.4. Bilinear and Quadratic Forms in Euclidean Space

**5.7.4-1. Reduction of a quadratic form to a sum of squares.**

THEOREM 1. *Let $B(\mathbf{x}, \mathbf{y})$ be a symmetric bilinear form on a $n$-dimensional Euclidean space $\mathcal{V}$. Then there is an orthonormal basis $\mathbf{i}_1, \ldots, \mathbf{i}_n$ in $\mathcal{V}$ and there are real numbers $\lambda_k$ such that for any $\mathbf{x} \in \mathcal{V}$ the real quadratic form $B(\mathbf{x}, \mathbf{x})$ can be represented as the sum of squares of the coordinates $\xi_k$ of $\mathbf{x}$ in the basis $\mathbf{i}_1, \ldots, \mathbf{i}_n$:*

$$B(\mathbf{x}, \mathbf{x}) = \sum_{k=1}^{n} \lambda_k \xi_k^2.$$

THEOREM 2. *Let $A(\mathbf{x}, \mathbf{y})$ and $B(\mathbf{x}, \mathbf{y})$ be symmetric bilinear forms in a $n$-dimensional real linear space $\mathcal{V}$, and suppose that the quadratic form $A(\mathbf{x}, \mathbf{x})$ is positive definite. Then there is a basis $\mathbf{i}_1, \ldots, \mathbf{i}_n$ of $\mathcal{V}$ such that the quadratic forms $A(\mathbf{x}, \mathbf{x})$ and $B(\mathbf{x}, \mathbf{x})$ can be represented in the form*

$$A(\mathbf{x}, \mathbf{x}) = \sum_{k=1}^{n} \lambda_k \xi_k^2, \quad B(\mathbf{x}, \mathbf{x}) = \sum_{k=1}^{n} \xi_k^2,$$

*where $\xi_k$ are the coordinates of $\mathbf{x}$ in the basis $\mathbf{i}_1, \ldots, \mathbf{i}_n$. The set of real $\lambda_1, \ldots, \lambda_n$ coincides with the spectrum of eigenvalues of the matrix $B^{-1}A$ (the matrices $A$ and $B$ can be taken in any basis), and this set consists of the roots of the algebraic equation*

$$\det(A - \lambda B) = 0.$$

### 5.7.4-2. Extremal properties of quadratic forms.

A point $\mathbf{x}_0$ on a smooth surface $S$ is called a *stationary point* of a differentiable function $f$ defined on $S$ if the derivative of $f$ at the point $\mathbf{x}_0$ in any direction on $S$ is equal to zero. The value $f(\mathbf{x}_0)$ of the function $f$ at a stationary point $\mathbf{x}_0$ is called its *stationary value*.

The *unit sphere* in a Euclidean space $\mathcal{V}$ is the set of all $\mathbf{x} \in \mathcal{V}$ such that

$$\mathbf{x} \cdot \mathbf{x} = 1 \quad (\|\mathbf{x}\| = 1). \tag{5.7.4.1}$$

THEOREM. *Let $B(\mathbf{x}, \mathbf{x})$ be a real quadratic form and let $B(\mathbf{x}, \mathbf{y}) = (\mathbf{A}\mathbf{x}) \cdot \mathbf{y}$ be the corresponding polar bilinear form, where $\mathbf{A}$ is a Hermitian operator. The stationary values of the quadratic form $B(\mathbf{x}, \mathbf{x})$ on the unit sphere (5.7.4.1) coincide with eigenvalues of the operator $\mathbf{A}$. These stationary values are attained, in particular, on the unit eigenvectors $\mathbf{e}_k$ of the operator $\mathbf{A}$.*

Remark. If the eigenvalues of the operator $\mathbf{A}$ satisfy the inequalities $\lambda_1 \geq \ldots \geq \lambda_n$, then $\lambda_1$ and $\lambda_n$ are the largest and the smallest values of $B(\mathbf{x}, \mathbf{x})$ on the sphere $\mathbf{x} \cdot \mathbf{x} = 1$.

## 5.7.5. Second-Order Hypersurfaces

### 5.7.5-1. Definition of a second-order hypersurface.

A *second-order hypersurface* in an $n$-dimensional Euclidean space $\mathcal{V}$ is the set of all points $\mathbf{x} \in \mathcal{V}$ satisfying an equation of the form

$$A(\mathbf{x}, \mathbf{x}) + 2B(\mathbf{x}) + c = 0, \tag{5.7.5.1}$$

where $A(\mathbf{x}, \mathbf{x})$ is a real quadratic form different from identical zero, $B(\mathbf{x})$ is a linear form, and $c$ is a real constant. Equation (5.7.5.1) is called the *general equation of a second-order hypersurface*.

Suppose that in some orthonormal basis $\mathbf{i}_1, \ldots, \mathbf{i}_n$, we have

$$A(\mathbf{x}, \mathbf{x}) = X^T A X = \sum_{i,j=1}^{n} a_{ij} x_i x_k, \quad B(\mathbf{x}) = BX = \sum_{i=1}^{n} b_i x_i,$$

$$X^T = (x_1, \ldots, x_n), \quad A \equiv [a_{ij}], \quad B = (b_1, \ldots, b_n).$$

Then the general equation (5.7.5.1) of a second-order hypersurface in the Euclidean space $\mathcal{V}$ with the given orthonormal basis $\mathbf{i}_1, \ldots, \mathbf{i}_n$ can be written as

$$X^T A X + 2BX + c = 0.$$

The term $A(\mathbf{x}, \mathbf{x}) = X^T A X$ is called the *group of the leading terms* of equation (5.7.5.1), and the terms $B(\mathbf{x}) + c = BX + c$ are called the *linear part of the equation*.

### 5.7.5-2. Parallel translation.

A *parallel translation* in a Euclidean space $\mathcal{V}$ is a transformation defined by the formulas

$$X = X' + \overset{\circ}{X}, \tag{5.7.5.2}$$

where $\overset{\circ}{X}$ is a fixed point, called the *new origin*.

In terms of coordinates, (5.7.5.2) takes the form

$$x_k = x'_k + \overset{\circ}{x}_k \quad (k = 1, 2, \ldots, n),$$

where $X^T = (x_1, \ldots, x_n)$, $X'^T = (x'_1, \ldots, x'_n)$, $\overset{\circ}{X}^T = (\overset{\circ}{x}_1, \ldots, \overset{\circ}{x}_n)$.

Under parallel translations any basis remains unchanged.

The transformation of the space $\mathcal{V}$ defined by (5.7.5.2) reduces the hypersurface equation (5.7.5.1) to

$$A(\mathbf{x}', \mathbf{x}') + 2B'(\mathbf{x}') + c' = 0,$$

where the linear form $B'(\mathbf{x}')$ and the constant $c'$ are defined by

$$B'(\mathbf{x}') = A(\mathbf{x}', \overset{\circ}{\mathbf{x}}) + B(\mathbf{x}'), \quad c' = A(\overset{\circ}{\mathbf{x}}, \overset{\circ}{\mathbf{x}}) + 2B(\overset{\circ}{\mathbf{x}}) + c,$$

or, in coordinate notation,

$$B'(\mathbf{x}') \equiv B'X' = \sum_{i=1}^{n} b'_i x'_i, \quad c' = \sum_{i=1}^{n}(b'_i + b_i)\overset{\circ}{x}_i + c, \quad b'_i = \sum_{j=1}^{n} a_{ij} \overset{\circ}{x}_j + b_i.$$

Under parallel translation the group of the leading terms preserves its form.

### 5.7.5-3. Transformation of one orthonormal basis into another.

The transition from one orthonormal basis $\mathbf{i}_1, \ldots, \mathbf{i}_n$ to another orthonormal basis $\mathbf{i}'_1, \ldots, \mathbf{i}'_n$ is defined by an orthogonal matrix $P \equiv [p_{ij}]$ of size $n \times n$, i.e.,

$$\mathbf{i}'_i = \sum_{j=1}^{n} p_{ij} \mathbf{i}_j \quad (i = 1, 2, \ldots, n).$$

Under this orthogonal transformation, the coordinates of points are transformed as follows:

$$X' = PX,$$

or, in coordinate notation,

$$x'_k = \sum_{i=1}^{n} p_{ki} x_i \quad (k = 1, 2, \ldots, n), \tag{5.7.5.3}$$

where $X^T = (x_1, \ldots, x_n)$, $X'^T = (x'_1, \ldots, x'_n)$.

If the transition from the orthonormal basis $\mathbf{i}_1, \ldots, \mathbf{i}_n$ to the orthonormal basis $\mathbf{i}'_1, \ldots, \mathbf{i}'_n$ is defined by an orthogonal matrix $P$, then the hypersurface equation (5.7.5.1) in the new basis takes the form

$$A'(\mathbf{x}', \mathbf{x}') + 2B'(\mathbf{x}') + c' = 0.$$

The matrix $A' \equiv [a'_{ij}]$ ($A'(\mathbf{x}', \mathbf{x}') = X'^T A' X'$) is found from the relation

$$A' = P^{-1} A P.$$

Thus, when passing from one orthonormal basis to another orthonormal basis, the matrix of a quadratic form is transformed similarly to the matrix of some linear operator. Note that the operator $\mathbf{A}$ whose matrix in an orthonormal basis coincides with the matrix of the quadratic form $A(\mathbf{x}, \mathbf{x})$ is Hermitian.

The coefficients $b'_i$ of the linear form $B'(\mathbf{x}') = \sum_{i=1}^{n} b'_i x'_i$ are found from the relations [to this end, one should use (5.7.5.3)]

$$\sum_{i=1}^{n} b'_i x'_i = \sum_{i=1}^{n} b_i x_i,$$

and the constant is $c' = c$.

### 5.7.5-4. Invariants of the general equation of a second-order hypersurface.

An *invariant* of the general second-order hypersurface equation (5.7.5.1) with respect to parallel translations and orthogonal transformations of an orthogonal basis is, by definition, any function $f(a_{11}, \ldots, a_{nn}, b_1, \ldots, b_n, c)$ of the coefficients of this equation that does not change under such transformations of the space.

THEOREM. *The coefficients of the characteristic polynomial of the matrix $A$ of the quadratic form $A(\mathbf{x}, \mathbf{x})$ and the determinant $\det \widetilde{A}$ of the block matrix $\widetilde{A} = \begin{pmatrix} A & B \\ B^T & c \end{pmatrix}$ are invariants of the general second-order hypersurface equation (5.7.5.1).*

Remark. The quantities $\det A$, $\mathrm{Tr}(A)$, rank $(A)$, and rank $(\widetilde{A})$ are invariants of equation (5.7.5.1).

### 5.7.5-5. Center of a second-order hypersurface.

The *center* of a second-order hypersurface is a point $\overset{\circ}{\mathbf{x}}$ such that the linear form $B'(\mathbf{x}')$ becomes identically equal to zero after the parallel translation that makes $\overset{\circ}{\mathbf{x}}$ the new origin. Thus, the coordinates of the center can be found from the system of *equations of the center of a second-order hypersurface*

$$\sum_{j=1}^{n} a_{ij} \overset{\circ}{x}_j + b_i = 0 \qquad (i = 1, 2, \ldots, n).$$

If the center equations for a hypersurface $S$ have a unique solution, then $S$ is called a *central hypersurface*. If a hypersurface $S$ has a center, then $S$ consists of pairs of points, each pair being symmetric with respect to the center.

Remark 1. For a second-order hypersurface $S$ with a center, the invariants $\det A$, $\det \widetilde{A}$, and the free term $c'$ are related by

$$\det \widetilde{A} = c' \det A.$$

Remark 2. If the origin is shifted to the center of a central hypersurface $S$, then the equation of that hypersurface in new coordinates has the form

$$A(\mathbf{x}, \mathbf{x}) + \frac{\det \widetilde{A}}{\det A} = 0.$$

### 5.7.5-6. Simplification of a second-order hypersurface equation.

Let $\mathbf{A}$ be the operator whose matrix in an orthonormal basis $\mathbf{i}_1, \ldots, \mathbf{i}_n$ coincides with the matrix of a quadratic form $A(\mathbf{x}, \mathbf{x})$. Suppose that the transition from the orthonormal basis $\mathbf{i}_1, \ldots, \mathbf{i}_n$ to the orthonormal basis $\mathbf{i}'_1, \ldots, \mathbf{i}'_n$ is defined by an orthogonal matrix $P$, and $A' = P^{-1}AP$ is a diagonal matrix with the eigenvalues of the operator $\mathbf{A}$ on the main diagonal. Then the equation of the hypersurface (5.7.5.1) in the new basis takes the form

$$\sum_{i=1}^{n} \lambda_i x_i'^2 + 2\sum_{i=1}^{n} b_i' x_i' + c = 0, \tag{5.7.5.4}$$

where the coefficients $b_i'$ are determined by the relations

$$\sum_{i=1}^{n} b_i' x_i' = \sum_{i=1}^{n} b_i x_i.$$

The reduction of any equation of a second-order hypersurface $S$ to the form (5.7.5.4) is called the *standard simplification of this equation (by an orthogonal transformation of the basis)*.

### 5.7.5-7. Classification of central second-order hypersurfaces.

$1^\circ$. Let $\mathbf{i}_2, \ldots, \mathbf{i}_n$ be an orthonormal basis in which a second-order central hypersurface is defined by the equation (called its *canonical equation*)

$$\sum_{i=1}^{n} \varepsilon_i \frac{x_i^2}{a_i^2} + \operatorname{sign} \frac{\det \widetilde{A}}{\det A} = 0, \tag{5.7.5.5}$$

where $x_1, \ldots, x_n$ are the coordinates of $\mathbf{x}$ in that basis, and the coefficients $\varepsilon_1, \ldots, \varepsilon_n$ take the values $-1$, $0$, or $1$. The constants $a_k > 0$ are called the *semiaxes* of the hypersurface.

The equation of any central hypersurface $S$ can be reduced to the canonical equation (5.7.5.5) by the following transformations:

1. By the parallel translation that shifts the origin to the center of the hypersurface, its equation is transformed to (see Paragraph 5.7.5-5):

$$A(\mathbf{x}, \mathbf{x}) + \frac{\det \widetilde{A}}{\det A} = 0.$$

2. By the standard simplification of the last equation, one obtains an equation of the hypersurface in the form

$$\sum_{i=1}^{n} \lambda_i x_i^2 + \frac{\det \widetilde{A}}{\det A} = 0.$$

3. Letting

$$\frac{1}{a_k^2} = \begin{cases} |\lambda_k| \dfrac{|\det A|}{|\det \widetilde{A}|} & \text{if } \det \widetilde{A} \neq 0, \\ |\lambda_k| & \text{if } \det \widetilde{A} = 0, \end{cases} \qquad \varepsilon_k = \operatorname{sign} \lambda_k \qquad (k = 1, 2, \ldots, n),$$

one passes to the canonical equation (5.7.5.5) of the central second-order hypersurface.

2°. Let $p$ be the number of positive eigenvalues of the matrix $A$ and $q$ the number of negative ones. Central second-order hypersurfaces admit the following classification. A hypersurface is called:

a) an $(n-1)$-*dimensional ellipsoid* if $p = n$ and sign $\frac{\det \widetilde{A}}{\det A} = -1$, or $q = n$ and sign $\frac{\det \widetilde{A}}{\det A} = 1$*;

b) an *imaginary ellipsoid* if $p = n$ and sign $\frac{\det \widetilde{A}}{\det A} = 1$, or $q = n$ and sign $\frac{\det \widetilde{A}}{\det A} = -1$;

c) a *hyperboloid* if $0 < p < n$ ($0 < q < n$) and sign $\frac{\det \widetilde{A}}{\det A} \neq 0$;

d) *degenerate* if sign $\frac{\det \widetilde{A}}{\det A} = 0$.

### 5.7.5-8. Classification of noncentral second-order hypersurfaces.

1°. Let $\mathbf{i}_2, \ldots, \mathbf{i}_n$ be an orthonormal basis in which a noncentral second-order hypersurface is defined by the equation (called its *canonical equation*)

$$\sum_{i=1}^{p} \lambda_i x_i^2 + 2\mu x_n + c' = 0, \tag{5.7.5.6}$$

where $x_1, \ldots, x_n$ are the coordinates of $\mathbf{x}$ in that basis: $p = \text{rank}(A)$.

The equation of any noncentral second-order hypersurface $S$ can be reduced to the canonical form (5.7.5.6) by the following transformations:

1. If $p = \text{rank}(A)$, then after the standard simplification and renumbering the basis vectors, equation (5.7.5.1) turns into

$$\sum_{i=1}^{p} \lambda_i x_i'^2 + 2 \sum_{i=1}^{p} b_i' x_i' + 2 \sum_{i=p+1}^{n} b_i' x_i' + c = 0.$$

2. After the parallel translation

$$x_k'' = \begin{cases} x_k' + \dfrac{b_k'}{\lambda_k} & \text{for } k = 1, 2, \ldots, p, \\ x_k' & \text{for } k = p+1, p+2, \ldots, n, \end{cases}$$

the last equation can be represented in the form

$$\sum_{i=1}^{p} \lambda_i x_i''^2 + 2 \sum_{i=p+1}^{n} b_i' x_i'' + c' = 0, \qquad c' = c - \sum_{i=1}^{p} \frac{b_i'}{\lambda_i}.$$

3. Leaving intact the first $p$ basis vectors and transforming the last basis vectors so that the term $\sum_{i=p+1}^{n} b_i' x_i''$ turns into $\mu x_n'''$, one reduces the hypersurface equation to the canonical form (5.7.5.6).

2°. Noncentral second-order hypersurfaces admit the following classification. A hypersurface is called:

a) a *paraboloid* if $\mu \neq 0$ and $p = n-1$; in this case, the parallel translation in the direction of the $x_n$-axis by $-\frac{c'}{2\mu}$ yields the canonical equation of a paraboloid

$$\sum_{i=1}^{n-1} \lambda_i x_i^2 + 2\mu x_n = 0;$$

---

* If $a_1 = \ldots = a_n = R$, then the hypersurface is a *sphere* of radius $R$ in $n$-dimensional space.

b) a *central cylinder* if $\mu = 0$, $p < n$; its canonical equation has the form

$$\sum_{i=1}^{p} \lambda_i x_i^2 + c' = 0;$$

c) a *paraboloidal cylinder* if $\mu \neq 0$, $p < n-1$; in this case, the parallel translation along the $x_n$-axis by $-\frac{c'}{2\mu}$ yields the canonical equation of a paraboloidal cylinder

$$\sum_{i=1}^{p} \lambda_i x_i^2 + 2\mu x_n = 0.$$

## 5.8. Some Facts from Group Theory

### 5.8.1. Groups and Their Basic Properties

#### 5.8.1-1. Composition laws.

Let $\mathbf{T}$ be a mapping defined on ordered pairs $a, b$ of elements of a set $A$ and mapping each pair $a, b$ to an element $c$ of $A$. In this case, one says that a *composition law* is defined on the set $A$. The element $c \in A$ is called the *composition* of the elements $a, b \in A$ and is denoted by $c = a\mathbf{T}b$.

A composition law is commonly expressed in one of the two forms:

1. *Additive form*: $c = a + b$; the corresponding composition law is called *addition* and $c$ is called the *sum* of $a$ and $b$.
2. *Multiplicative form*: $c = ab$; the corresponding composition law is called *multiplication* and $c$ is called the *product* of $a$ and $b$.

A composition law is said to be *associative* if

$$a\mathbf{T}(b\mathbf{T}c) = (a\mathbf{T}b)\mathbf{T}c \qquad \text{for all} \quad a, b, c \in A.$$

In additive form, this relation reads $a + (b + c) = (a + b) + c$; and in multiplicative form, $a(bc) = (ab)c$.

A composition law is said to be *commutative* if

$$a\mathbf{T}b = b\mathbf{T}a \qquad \text{for all} \quad a, b \in A.$$

In additive form, this relation reads $a + b = b + a$; and in multiplicative form, $ab = ba$.

An element $e$ of the set $A$ is said to be *neutral* with respect to the composition law $\mathbf{T}$ if $a\mathbf{T}e = a$ for any $a \in A$.

In the additive case, a neutral element is called a *zero element*, and in the multiplicative case, an *identity element*.

An element $b$ is called an *inverse* of $a \in A$ if $a\mathbf{T}b = e$. The inverse element is denoted by $b = a^{-1}$.

In the additive case, the inverse element of $a$ is called the *negative* of $a$ and it is denoted by $-a$.

**Example 1.** Addition and multiplication of real numbers are composition laws on the set of real numbers. Both these laws are commutative. The neutral element for the addition is zero. The neutral element for the multiplication is unity.

## 5.8.1-2. Notion of a group.

A *group* is a set $G$ with a composition law **T** satisfying the conditions:
1. The law **T** is associative.
2. There is a neutral element $e \in G$.
3. For any $a \in G$, there is an inverse element $a^{-1}$.

A group $G$ is said to be *commutative* or *abelian* if its composition law **T** is commutative.

**Example 2.** The set $Z$ of all integer numbers is an abelian group with respect to addition. The set of all positive real numbers is an abelian group with respect to multiplication. Any linear space is an abelian group with respect to the addition of its elements.

**Example 3.** *Permutation groups.* Let $E$ be a set consisting of finitely many elements $a, b, c, \ldots$. A *permutation* of $E$ is a one-to-one mapping of $E$ onto itself. A permutation $f$ of the set $E$ can be expressed in the form
$$f = \begin{pmatrix} a & b & c & \ldots \\ f(a) & f(b) & f(c) & \ldots \end{pmatrix}.$$
On the set $P$ of all permutations of $E$, the composition law is introduced as follows: if $f_1$ and $f_2$ are two permutations of $E$, then their composition $f_2 \circ f_1$ is the permutation obtained by consecutive application of $f_1$ and $f_2$. This composition law is associative. The set of all permutations of $E$ with this composition law is a group.

**Example 4.** The group $Z_2$ that consists of two elements 0 and 1 with the multiplication defined by
$$0 \cdot 0 = 0, \quad 0 \cdot 1 = 1, \quad 1 \cdot 0 = 1, \quad 1 \cdot 1 = 1$$
and the neutral element 0 is called the *group of modulo 2 residues*.

Properties of groups:
1. If $a\mathbf{T}a^{-1} = e$, then $a^{-1}\mathbf{T}a = e$.
2. $e\mathbf{T}a = a$ for any $a$.
3. If $a\mathbf{T}x = e$ and $a\mathbf{T}y = e$, then $x = y$.
4. The neutral element $e$ is unique.

## 5.8.1-3. Homomorphisms and isomorphisms.

Recall that a mapping $f : A \to B$ of a set $A$ into a set $B$ is a correspondence that associates each element of $A$ with an element of $B$. The *range* of the mapping $f$ is the set of all $b \in B$ such that $b = f(a)$. One says that $f$ is a *one-to-one mapping* if it maps different elements of $A$ into different elements of $B$, i.e., for any $a_1, a_2 \in A$ such that $a_1 \neq a_2$, we have $f(a_1) \neq f(a_2)$.

A mapping $f : A \to B$ is called a *mapping of the set $A$ onto the set $B$* if each element of $B$ is an image of some element of $A$, i.e., for any $b \in B$, there is $a \in A$ such that $b = f(a)$.

A mapping $f$ of $A$ onto $B$ is said to be *invertible* if there is a mapping $g : B \to A$ such that $g(f(a)) = a$ for any $a \in A$. The mapping $g$ is called the *inverse* of the mapping $f$ and is denoted by $g = f^{-1}$.

For definiteness, we use the multiplicative notation for composition laws in what follows, unless indicated otherwise.

Let $G$ be a group and let $\widetilde{G}$ be a set with a composition law. A mapping $f : G \to \widetilde{G}$ is called a *homomorphism* if
$$f(ab) = f(a)f(b) \quad \text{for all} \quad a, b \in G;$$
and the subset of $\widetilde{G}$ consisting of all elements of the form $f(a)$, $a \in G$, is called a *homomorphic image* of the group $G$ and is denoted by $f(G)$. Note that here the set $\widetilde{G}$ with a composition law is not necessarily a group. However, the following result holds.

THEOREM. *The homomorphic image $f(G)$ is a group. The image $f(e)$ of the identity element $e \in G$ is the identity element of the group $f(G)$. Mutually inverse elements of $G$ correspond to mutually inverse images in $f(G)$.*

Two groups $G_1$ and $G_2$ are said to be *isomorphic* if there exists a one-to-one mapping $f$ of $G_1$ onto $G_2$ such that $f(ab) = f(a)f(b)$ for all $a, b \in G_1$. Such a mapping is called *an isomorphism* or *isomorphic mapping* of the group $G_1$ onto the group $G_2$.

THEOREM. *Any isomorphism of groups is invertible, and the inverse mapping is also an isomorphism.*

An isomorphic mapping of a group $G$ onto itself is called an *automorphism* of $G$. If $f_1 : G \to G$ and $f_2 : G \to G$ are two automorphisms of a group $G$, one can define another automorphism $f_1 \circ f_2 : G \to G$ by letting $(f_1 \circ f_2)(g) = f_1(f_2(g))$ for all $g \in G$. This automorphism is called the *composition* of $f_1$ and $f_2$, and with this composition law, the set of all automorphisms of $G$ becomes a group called the *automorphism group* of $G$.

### 5.8.1-4. Subgroups. Cosets. Normal subgroups.

Let $G$ be a group. A subset $G_1$ of the group $G$ is called a *subgroup* if the following conditions hold:
1. For any $a$ and $b$ belonging to $G_1$, the product $ab$ belongs to $G_1$.
2. For any $a$ belonging to $G_1$, its inverse $a^{-1}$ belongs to $G_1$.

These conditions ensure that any subgroup of a group is itself a group.

**Example 5.** The identity element of a group is a subgroup. The subset of all even numbers is a subgroup of the additive group of all integers.

The *product* of two subsets $H_1$ and $H_2$ of a group $G$ is a set $H_3$ that consists of all elements of the form $h_1 h_2$, where $h_1 \in H_1$, $h_2 \in H_2$. In this case, one writes $H_3 = H_1 H_2$.

Let $H$ be a subgroup of a group $G$ and $a$ some fixed element of $G$. The set $aH$ is called a *left coset*, and the set $Ha$ is called a *right coset* of the subgroup $H$ in $G$.

Properties of left cosets (right cosets have similar properties):
1. If $a \in H$, then $aH \equiv H$.
2. Cosets $aH$ and $bH$ coincide if $a^{-1}b \in H$.
3. Two cosets of the same subgroup $H$ either coincide or have no common elements.
4. If $aH$ is a coset, then $a \in aH$.

A subgroup $H$ of a group $G$ is called a *normal subgroup* of $G$ if $H = a^{-1}Ha$ for any $a \in G$. This is equivalent to the condition that $aH = Ha$ for any $a \in G$, i.e., every right coset is a left coset.

### 5.8.1-5. Factor groups.

Let $H$ be a normal subgroup of a group $G$. Then the product of two cosets $aH$ and $bH$ (as subsets of $G$) is the coset $abH$. Consider the set $Q$ whose elements are cosets of the subgroup $H$ in $G$, and define the product of the elements of $Q$ as the product of cosets. Endowed with this product, $Q$ becomes a group, denoted by $Q = G/H$ and called the *quotient group* of $G$ with respect to the normal subgroup $H$.

The mapping $f : G \to G/H$ that maps each $a \in G$ to the corresponding coset $aH$ is a homomorphism of $G$ onto $G/H$.

If $f : G \to \overline{G}$ is a homomorphism of groups, the set of all elements of $G$ mapped into the identity element of $\overline{G}$ is called the *kernel* of $f$ and is denoted by $\ker f = \{g \in G : f(g) = f(e)\}$.

THEOREM 1. *If $f$ is a homomorphism of a group $G$ onto a group $\overline{G}$ and $H$ is the set of all elements of $G$ that are mapped to $f(e)$ ($e$ is the identity element of $G$), then $H$ is a normal subgroup in $G$.*

THEOREM 2 (ON GROUP HOMOMORPHISMS). *If $f$ is a homomorphism of a group $G$ onto a group $\overline{G}$ and $H$ is the normal subgroup of $G$ consisting of the elements mapped to the identity element of $\overline{G}$, then the group $\overline{G}$ and the quotient group $G/H$ are isomorphic.*

Thus, given a homomorphism $f$ of a group $G$ onto a group $\widetilde{G}$, the kernel $H$ of the homomorphism is a normal subgroup of $G$, and conversely any normal subgroup $H$ of $G$ is the kernel of the homomorphism of $G$ onto the quotient group $G/H$.

**Remark.** Given a homomorphism of a group $G$ onto a set $\overline{G}$, all elements of the group $G$ are divided into mutually disjoint classes, each class containing all elements of $G$ that are mapped into the same element of $\overline{G}$.

**Example 6.** Let $\mathbb{R}^n$ be the $n$-dimensional linear coordinate space, which is an abelian group with respect to addition of its elements. This space is the direct product of one-dimensional spaces:
$$\mathbb{R}^n = \mathbb{R}^1_{(1)} \otimes \cdots \otimes \mathbb{R}^1_{(n)}.$$

Since $\mathbb{R}^1_{(n)}$ is an abelian subgroup, the set $\mathbb{R}^1_{(n)}$ is a normal subgroup of the group $\mathbb{R}^n$. The coset corresponding to an element $a \in \mathbb{R}^n$ is the straight line passing through $a$ in the direction parallel to the straight line $\mathbb{R}^1_{(n)}$, and the quotient group $\mathbb{R}^n/\mathbb{R}^1_{(n)}$ is isomorphic to the $(n-1)$-dimensional space $\mathbb{R}^{n-1}$:
$$\mathbb{R}^{n-1} = \mathbb{R}^n/\mathbb{R}^1_{(n)} = \mathbb{R}^1_{(1)} \otimes \cdots \otimes \mathbb{R}^1_{(n-1)}.$$

## 5.8.2. Transformation Groups

**5.8.2-1. Group of linear transformations. Its subgroups.**

Let $V$ be a real finite-dimensional linear space and let $\mathbf{A}: V \to V$ be a nondegenerate linear operator. This operator can be regarded as a nondegenerate linear transformation of the space $V$, since $A$ maps different elements of $V$ into different elements, and for any $\mathbf{y} \in V$ there is a unique $\mathbf{x} \in V$ such that $\mathbf{Ax} = \mathbf{y}$.

The set of all nondegenerate linear transformations $\mathbf{A}$ of the $n$-dimensional real linear space $V$ is denoted by $GL(n)$.

The *product* $\mathbf{AB}$ of linear transformations $\mathbf{A}$ and $\mathbf{B}$ in $GL(n)$ is defined by the relation

$$(\mathbf{AB})\mathbf{x} = \mathbf{A}(\mathbf{Bx}) \qquad \text{for all} \quad \mathbf{x} \in V.$$

This product is a composition law on $GL(n)$.

THEOREM. *The set $GL(n)$ of nondegenerate linear transformations of an $n$-dimensional real linear space $V$ with the above product is a group.*

The group $GL(n)$ is called the *general linear group of dimension $n$*.

A subset of $GL(n)$ consisting of all linear transformations $\mathbf{A}$ such that $\det \mathbf{A} = 1$ is a subgroup of $GL(n)$ called the *special linear group of dimension $n$* and denoted by $SL(n)$.

A sequence $\{\mathbf{A}_k\}$ of elements of $GL(n)$ is said to be *convergent* to an element $\mathbf{A} \in GL(n)$ as $k \to \infty$ if the sequence $\{\mathbf{A}_k \mathbf{x}\}$ converges to $\mathbf{Ax}$ for any $\mathbf{x} \in V$.

Types of subgroups of $GL(n)$:
1. *Finite subgroups* are subgroups with finitely many elements.
2. *Discrete subgroups* are subgroups with countably many elements.
3. *Continuous subgroups* are subgroups with uncountably many elements.

**Example 1.** The subgroup of reflections with respect to the origin is finite and consists of two elements: the identity transformation and the reflection $\mathbf{x} \to -\mathbf{x}$.

The subgroup of rotations of a plane with respect to the origin by the angles $k\varphi$ ($k = 0, \pm 1, \pm 2, \ldots$ and $\varphi$ is a fixed angle incommensurable with $\pi$) is a discrete subgroup.

The subgroups of all rotations of a three-dimensional space about a fixed axis are a continuous subgroup.

A continuous subgroup of $GL(n)$ is said to be *compact* if from any infinite sequence of its elements one can extract a subsequence convergent to some element of the subgroup.

### 5.8.2-2. Group of orthogonal transformations. Its subgroups.

Consider the set $O(n)$ that consists of all orthogonal transformations $\mathbf{P}$ of the $n$-dimensional Euclidean space $V$, i.e., $\mathbf{P}^T\mathbf{P} = \mathbf{P}\mathbf{P}^T = \mathbf{I}$ (see Paragraph 5.2.3-3 and Section 5.4). This set is a subgroup of $GL(n)$ called the *orthogonal group* of dimension $n$.

All orthogonal transformations are divided into two classes:
1. *Proper orthogonal transformations*, for which $\det \mathbf{P} = +1$.
2. *Improper orthogonal transformations*, for which $\det \mathbf{P} = -1$.

The set of proper orthogonal transformations forms a group called the *special orthogonal group* of dimension $n$ and denoted by $SO(n)$.

In the two-dimensional orthogonal group $O(2)$ there is a subgroup of rotations by the angles $k\varphi$, where $k = 0, \pm 1, \pm 2, \ldots$ and $\varphi$ is fixed. If $a_k$ is its element corresponding to $k$ and $a = a_1$, then the element $a_k$ ($k > 0$) has the form

$$a_k = \underbrace{a \cdot a \cdot \ldots \cdot a}_{k \text{ times}} = a^k \qquad (k = 1, 2, 3, \ldots).$$

Denoting by $a^{-1}$ the inverse of $a = a_1$, and the identity element by $a^0$, we see that each element of this group has the form

$$a_k = a^k \qquad (k = 0, \pm 1, \pm 2, \ldots).$$

Groups whose elements admit such a representation in terms of a single element are said to be *cyclic*. Such groups are discrete.

There are two cyclic groups of rotations ($p$ and $q$ are coprime numbers):
1. If $\varphi \neq 2\pi p/q$ (i.e., the angle $\varphi$ is incommensurable with $\pi$), then all elements are distinct.
2. If $\varphi = 2\pi p/q$, then $a_{k+q} = a_k$ ($a^q = a^0$). Such groups are called *cyclic groups of order q*.

Consider *groups of mirror symmetry*. Each of them consists of two elements: the identity element and a reflection with respect to the origin.

Let $\{\mathbf{I}, \mathbf{P}\}$ be a subgroup of $O(3)$ consisting of the identity $\mathbf{I}$ and the reflection $\mathbf{P}$ of the three-dimensional space with respect to the origin, $\mathbf{Px} = -\mathbf{x}$. This is an improper subgroup. It is isomorphic to the group $Z_2$ of residues modulo 2. The subgroup $\{\mathbf{I}, \mathbf{P}\}$ is a normal subgroup in $O(3)$, and the subgroup $SO(3)$ (consisting of proper orthogonal transformations) is isomorphic to the quotient group $O(3)/\{\mathbf{I}, \mathbf{P}\}$.

### 5.8.2-3. Unitary groups.

By analogy with Paragraph 5.8.2-2, one can consider groups of linear transformations of a complex linear space.

In the general linear group of transformations of a unitary space, one considers *unitary groups* $U(n)$, which are analogues of orthogonal groups. In the group $U(n)$ of unitary transformations, one considers the subgroup $SU(n)$ that consists of unitary transformations whose determinant is equal to 1.

## 5.8.3. Group Representations

**5.8.3-1. Linear representations of groups. Terminology.**

A *linear representation* of a group $G$ in the finite-dimensional Euclidean space $V^n$ is a homomorphism of $G$ to the group of nondegenerate linear transformations of $V^n$; in other words, a linear representation of $G$ is a mapping $D$ that associates each element $a \in G$ with a nondegenerate linear transformation $D(a)$ of the space $V^n$, so that for any $a_1$ and $a_2$ in $G$, we have $D(a_1 a_2) = D(a_1)D(a_2)$.

Thus, for any $g \in G$, its image $D(g)$ is an element of the group $GL(n)$, and the set $D(G)$ consisting of all transformations $D(g)$, $g \in G$, is a subgroup of $GL(n)$ isomorphic to the quotient group $G/\ker D$, where $\ker D$ is the kernel of the homomorphism $D$, i.e., the set of all $g$ such that $D(g)$ is the identity element of the group $GL(n)$.

The subgroup $D(G)$ is often also called a *representation* of the group $G$.

The space $V^n$ is called the *representation space*; $n$ is called the *dimension of the representation*; and the basis in $V^n$ is called the *representation basis*.

The *trivial representation* of a group is its homomorphic mapping onto the identity element of the group $GL(n)$.

A *faithful representation* of a group $G$ is an isomorphism of $G$ onto a subgroup of $GL(n)$.

**5.8.3-2. Matrices of linear representations. Equivalent representations.**

If $D^{(\mu)}(G)$ is a representation of a group $G$, each $g \in G$ corresponds to a linear transformation $D^{(\mu)}(g)$, whose matrix in the basis of the representation $D^{(\mu)}(G)$ is denoted by $[D_{ij}^{(\mu)}(g)]$.

Two representations $D^{(\mu_1)}(G)$ and $D^{(\mu_2)}(G)$ of a group $G$ in the same space $E^n$ are said to be *equivalent* if there exists a nondegenerate linear transformation $C$ of the space $E^n$ such that $D^{(\mu_1)}(g) = C^{-1} D^{(\mu_2)}(g) C$ for each $g \in G$.

The choice of a basis in the representation space is important, since the matrices corresponding to the group elements may have some standard fairly simple form in that basis, and this allows one to make important conclusions with regards to a given representation.

**5.8.3-3. Reducible and irreducible representations.**

A subspace $V'$ of $V^n$ is called *invariant* for a representation $D(G)$ if it is invariant with respect to each linear operator in $D(G)$.

Suppose that all matrices of some three-dimensional representation $D(G)$ have the form

$$\begin{pmatrix} A_1 & A_2 \\ O & A_3 \end{pmatrix}, \quad A_1 \equiv \begin{pmatrix} a_{11} & a_{12} \\ a_{21} & a_{22} \end{pmatrix}, \quad A_2 \equiv \begin{pmatrix} a_{13} \\ a_{23} \end{pmatrix}, \quad A_3 \equiv (a_{33}), \quad O \equiv (0\ 0).$$

The product of such matrices has the form (see Paragraph 5.2.1-10)

$$\begin{pmatrix} A'_1 & A'_2 \\ O & A'_3 \end{pmatrix} \begin{pmatrix} A''_1 & A''_2 \\ O & A''_3 \end{pmatrix} = \begin{pmatrix} A'_1 A''_1 & A'''_2 \\ O & A'_3 A''_3 \end{pmatrix},$$

and therefore the structure of the matrices is preserved. Thus, the matrices $A_1$ form a two-dimensional representation of the given group $G$ and the matrices $A_3$ form its one-dimensional representation. In such cases, one says that $D(G)$ is a *reducible representation*.

If all matrices of a representation have the form of size $n \times n$

$$\begin{pmatrix} A_1 & O \\ O & A_2 \end{pmatrix},$$

then the square matrices $A_1$ and $A_2$ form representations, the sum of their dimensions being equal to $n$. In this case, the representation is said to be *completely reducible*. The representation induced on an invariant space by a given representation $D(G)$ is called a *part* of the representation $D(G)$.

A representation $D(G)$ of a group $G$ is said to be *irreducible* if it has only two invariant subspaces, $V^n$ and $O$. Otherwise, it is said to be *reducible*. Any representation can be expressed in terms of irreducible representations.

5.8.3-4. Characters.

Let $D(G)$ be an $n$-dimensional representation of a group $G$, and let $[D_{ij}(g)]$ be the matrix of the operator corresponding to the element $g \in G$. The *character* of an element $g \in G$ in the representation $D(G)$ is defined by

$$\chi(g) = \sum_{i=1}^{n} D_{ii}(g) = \text{Tr}([D_{ij}(g)]).$$

Thus, the character of an element does not depend on the representation basis and is, therefore, an invariant quantity.

An element $b \in G$ is said to be *conjugate* to the element $a \in G$ if there exists $u \in G$ such that

$$uau^{-1} = b.$$

Properties of conjugate elements:
1. Any element is conjugate to itself.
2. If $b$ is conjugate to $a$, then $a$ is conjugate to $b$.
3. If $b$ is conjugate to $a$ and $c$ is conjugate to $b$, then $c$ is conjugate to $a$.

The characters of all elements belonging to one and the same class of conjugate elements coincide. The characters of elements for equivalent representations coincide.

5.8.3-5. Examples of group representations.

1°. Let $G$ be a group of symmetry of three-dimensional space consisting of two elements: the identity transformation **I** and the reflection **P** with respect to the origin, $G = \{\mathbf{I}, \mathbf{P}\}$.

The multiplication of elements of this group is described by the table

|   | **I** | **P** |
|---|---|---|
| **I** | **I** | **P** |
| **P** | **P** | **I** |

1. *One-dimensional representation of the group $G$.*

In the space $E^1$, we chose a basis $\mathbf{e}_1$ and consider the matrix $A^{(1)}$ of the nondegenerate transformation $\mathbf{A}^1$ of this space: $A^{(1)} = (1)$. The transformation $\mathbf{A}^1$ forms a subgroup in the group $GL(1)$ of all linear transformations of $E^1$, and the multiplication in this subgroup is described by the table

|        | $\mathbf{A}^{(1)}$ |
|--------|--------------------|
| $\mathbf{A}^{(1)}$ | $\mathbf{A}^{(1)}$ |

We obtain a one-dimensional representation $D^{(1)}(G)$ of the group $G$ by letting $D^{(1)}(\mathbf{I}) = \mathbf{A}^{(1)}$, $D^{(1)}(\mathbf{P}) = \mathbf{A}^{(1)}$. These relations define a homomorphism of the group $G$ to $GL(1)$ and thus define its representation.

2. *A two-dimensional representation of the group $G$.*

In $E^2$, we choose a basis $\mathbf{e}_1$, $\mathbf{e}_2$ and consider the matrices $A^{(2)}$, $B^{(2)}$ of linear transformations $\mathbf{A}^2$, $\mathbf{B}^2$ of this space: $A^{(2)} = \begin{pmatrix} 1 & 0 \\ 0 & 1 \end{pmatrix}$, $B^{(2)} = \begin{pmatrix} 0 & 1 \\ 1 & 0 \end{pmatrix}$. The transformations $\mathbf{A}^2$, $\mathbf{B}^2$ form a subgroup in the group $GL(2)$ of linear transformations of $E^2$. The multiplication in this subgroup is defined by the table

|        | $\mathbf{A}^{(2)}$ | $\mathbf{B}^{(2)}$ |
|--------|--------------------|--------------------|
| $\mathbf{A}^{(2)}$ | $\mathbf{A}^{(2)}$ | $\mathbf{B}^{(2)}$ |
| $\mathbf{B}^{(2)}$ | $\mathbf{B}^{(2)}$ | $\mathbf{A}^{(2)}$ |

We obtain a two-dimensional representation $D^{(2)}(G)$ of the group $G$ by letting $D^{(2)}(\mathbf{I}) = \mathbf{A}^{(2)}$, $D^{(2)}(\mathbf{P}) = \mathbf{B}^{(2)}$. These relations define an isomorphism of $G$ onto the subgroup $\{\mathbf{A}^{(2)}, \mathbf{B}^{(2)}\}$ of $GL(2)$, and therefore define its representation.

3. *A three-dimensional representation of the group $G$.*

Consider the linear transformation $\mathbf{A}^{(3)}$ of $E^3$ defined by the matrix

$$A^{(3)} = \begin{pmatrix} 1 & 0 & 0 \\ 0 & 1 & 0 \\ 0 & 0 & 1 \end{pmatrix}.$$

This transformation forms a subgroup in $GL(3)$ with the multiplication law $\mathbf{A}^{(3)}\mathbf{A}^{(3)} = \mathbf{A}^{(3)}$. One obtains a three-dimensional representation $D^{(3)}(G)$ of the group $G$ by letting $D^{(3)}(\mathbf{I}) = \mathbf{A}^{(3)}$, $D^{(3)}(\mathbf{P}) = \mathbf{A}^{(3)}$.

4. *A four-dimensional representation of the group $G$.*

Consider linear transformations $\mathbf{A}^{(4)}$ and $\mathbf{B}^{(4)}$ of $E^4$ defined by the matrices

$$A^{(4)} = \begin{pmatrix} 1 & 0 & 0 & 0 \\ 0 & 1 & 0 & 0 \\ 0 & 0 & 1 & 0 \\ 0 & 0 & 0 & 1 \end{pmatrix}, \quad B^{(4)} = \begin{pmatrix} 0 & 1 & 0 & 0 \\ 1 & 0 & 0 & 0 \\ 0 & 0 & 0 & 1 \\ 0 & 0 & 1 & 0 \end{pmatrix}.$$

The transformations $\mathbf{A}^{(4)}$ and $\mathbf{B}^{(4)}$ form a subgroup in $GL(4)$ with the multiplication defined by a table similar to that in the two-dimensional case. One obtains a four-dimensional representation $D^{(4)}(G)$ of the group $G$ by letting $D^{(4)}(\mathbf{I}) = \mathbf{A}^{(4)}$, $D^{(4)}(\mathbf{P}) = \mathbf{A}^{(B)}$.

**Remark.** The matrices $A^{(4)}$ and $B^{(4)}$ may be written in the form $A^{(4)} = \begin{pmatrix} A^{(2)} & 0 \\ 0 & A^{(2)} \end{pmatrix}$, $B^{(4)} = \begin{pmatrix} B^{(2)} & 0 \\ 0 & B^{(2)} \end{pmatrix}$, and therefore the representation $D^{(4)}(G)$ is sometimes denoted by $D^{(4)}(G) = D^{(2)}(G) + D^{(2)}(G) = 2D^{(2)}(G)$. In a similar way, one may use the notation $D^{(3)}(G) = 3D^{(1)}(G)$. In this way, one can construct representations of the group $G$ of arbitrary dimension.

2°. The symmetry group $G = \{\mathbf{I}, \mathbf{P}\}$ for the three-dimensional space is a normal subgroup of the group $O(3)$. The subgroup $SO(3) \subset O(3)$ formed by proper orthogonal transformations is isomorphic to the quotient group $O(3)/\{\mathbf{I}, \mathbf{P}\}$.

Since any group admits a homomorphic mapping onto its quotient group, there is a homomorphism of the group $O(3)$ onto $SO(3)$. This homomorphism is defined as follows: if $a$ is a proper orthogonal transformation in $O(3)$, its image in $SO(3)$ coincides with $a$; and if $a'$ is an improper orthogonal transformation, its image is the proper orthogonal transformation $Pa'$.

In this way, one obtains a three-dimensional representation $DO(3)$ of the group of orthogonal transformations $O(3)$ in terms of the group $SO(3)$ of proper orthogonal transformations.

## References for Chapter 5

**Anton, H.**, *Elementary Linear Algebra, 8th Edition*, Wiley, New York, 2000.
**Barnett, S.**, *Matrices in Control Theory, Rev. Edition*, Krieger Publishing, Malabar, 1985.
**Beecher, J. A., Penna, J. A., and Bittinger, M. L.**, *College Algebra, 2nd Edition*, Addison Wesley, Boston, 2004.
**Bellman, R. E.**, *Introduction to Matrix Analysis* (McGraw-Hill Series in Matrix Theory), McGraw-Hill, New York, 1960.
**Bernstein, D. S.**, *Matrix Mathematics: Theory, Facts, and Formulas with Application to Linear Systems Theory*, Princeton University Press, Princeton, 2005.
**Blitzer, R. F.**, *College Algebra, 3rd Edition*, Prentice Hall, Englewood Cliffs, New Jersey, 2003.
**Bronson, R.**, *Schaum's Outline of Matrix Operations*, McGraw-Hill, New York, 1988.
**Cullen, C. G.**, *Matrices and Linear Transformations, 2nd Edition*, Dover Publications, New York, 1990.
**Davis, P. J. (Editor)**, *The Mathematics of Matrices: A First Book of Matrix Theory and Linear Algebra, 2nd Edition*, Xerox College Publications, Lexington, 1998.
**Demmel, J. W.**, *Applied Numerical Linear Algebra*, Society for Industrial & Applied Mathematics, University City Science Center, Philadelphia, 1997.
**Dixon, J. D.**, *Problems in Group Theory*, Dover Publications, New York, 1973.
**Dugopolski, M.**, *College Algebra, 3rd Edition*, Addison Wesley, Boston, 2002.
**Eves, H.**, *Elementary Matrix Theory*, Dover Publications, New York, 1980.
**Franklin, J. N.**, *Matrix Theory*, Dover Publications, New York, 2000.
**Gantmacher, F. R.**, *Matrix Theory, Vol. 1, 2nd Edition*, American Mathematical Society, Providence, Rhode Island, 1990.
**Gantmacher, F. R.**, *Applications of the Theory of Matrices*, Dover Publications, New York, 2005.
**Gelfand, I. M.**, *Lectures on Linear Algebra*, Dover Publications, New York, 1989.
**Gelfand, I. M. and Shen, A.**, *Algebra*, Birkhauser, Boston, 2003.
**Gilbert, J. and Gilbert, L.**, *Linear Algebra and Matrix Theory, 2nd Edition*, Brooks Cole, Stamford, 2004.
**Golub, G. H. and Van Loan, C. F.**, *The Matrix Computations, 3nd Edition* (Johns Hopkins Studies in Mathematical Sciences), The Johns Hopkins University Press, Baltimore, Maryland, 1996.
**Hazewinkel, M. (Editor)**, *Handbook of Algebra, Vol. 1*, North Holland, Amsterdam, 1996.
**Hazewinkel, M. (Editor)**, *Handbook of Algebra, Vol. 2*, North Holland, Amsterdam, 2000.
**Hazewinkel, M. (Editor)**, *Handbook of Algebra, Vol. 3*, North Holland, Amsterdam, 2003.
**Horn, R. A. and Johnson, C. R.**, *Matrix Analysis, Rep. Edition*, Cambridge University Press, Cambridge, 1990.
**Horn, R. A. and Johnson, C. R.**, *Topics in Matrix Analysis, New Ed. Edition*, Cambridge University Press, Cambridge, 1994.
**Householder, A. S.**, *The Theory of Matrices in Numerical Analysis* (Dover Books on Mathematics), Dover Publications, New York, 2006.
**Jacob, B.**, *Linear Functions and Matrix Theory* (Textbooks in Mathematical Sciences), Springer, New York, 1995.
**Kostrikin, A. I. and Artamonov, V. A. (Editors)**, *Exercises in Algebra: A Collection of Exercises in Algebra, Linear Algebra and Geometry (Expanded)*, Gordon & Breach, New York, 1996.
**Kostrikin, A. I. and Manin, Yu. I.**, *Linear Algebra and Geometry*, Gordon & Breach, New York, 1997.
**Kostrikin, A. I. and Shafarevich, I. R. (Editor)**, *Algebra I: Basic Notions of Algebra*, Springer-Verlag, Berlin, 1990.
**Kurosh, A. G.**, *Lectures on General Algebra*, Chelsea Publishing, New York, 1965.
**Kurosh, A. G.**, *Algebraic Equations of Arbitrary Degrees*, Firebird Publishing, New York, 1977.
**Kurosh, A. G.**, *Theory of Groups*, Chelsea Publishing, New York, 1979.
**Lancaster, P. and Tismenetsky, M.**, *The Theory of Matrices, Second Edition: With Applications* (Computer Science and Scientific Computing), Academic Press, Boston, 1985.

**Lay, D. C.**, *Linear Algebra and Its Applications, 3rd Edition*, Addison Wesley, Boston, 2002.
**Lial, M. L.**, *Student's Solutions Manual for College Algebra, 3rd Sol. Mn Edition*, Addison Wesley, Boston, 2004.
**Lial, M. L., Hornsby, J., and Schneider, D. I.**, *College Algebra, 9th Edition*, Addison Wesley, Boston, 2004.
**Lipschutz, S.**, *3,000 Solved Problems in Linear Algebra*, McGraw-Hill, New York, 1989.
**Lipschutz, S. and Lipson, M.**, *Schaum's Outline of Linear Algebra, 3rd Edition*, McGraw-Hill, New York, 2000.
**MacDuffee, C. C.**, *The Theory of Matrices, Phoenix Edition*, Dover Publications, New York, 2004.
**McWeeny, R.**, *Symmetry: An Introduction to Group Theory and Its Applications, Unabridged edition*, Dover Publications, New York, 2002.
**Meyer, C. D.**, *Matrix Analysis and Applied Linear Algebra, Package Edition*, Society for Industrial & Applied Mathematics, University City Science Center, Philadelphia, 2001.
**Mikhalev, A. V. and Pilz, G.**, *The Concise Handbook of Algebra*, Kluwer Academic, Dordrecht, Boston, 2002.
**Perlis, S.**, *Theory of Matrices, Dover Ed. Edition*, Dover Publications, New York, 1991.
**Poole, D.**, *Linear Algebra: A Modern Introduction*, Brooks Cole, Stamford, 2002.
**Poole, D.**, *Student Solutions Manual for Poole's Linear Algebra: A Modern Introduction, 2nd Edition*, Brooks Cole, Stamford, 2005.
**Rose, J. S.**, *A Course on Group Theory* (Dover Books on Advanced Mathematics), Dover Publications, New York, 1994.
**Schneider, H. and Barker, G. Ph.**, *Matrices and Linear Algebra, 2nd Edition* (Dover Books on Advanced Mathematics), Dover Publications, New York, 1989.
**Scott, W. R.**, *Group Theory*, Dover Publications, New York, 1987.
**Shilov, G. E.**, *Linear Algebra, Rev. English Ed. Edition*, Dover Publications, New York, 1977.
**Strang, G.**, *Introduction to Linear Algebra, 3rd Edition*, Wellesley Cambridge Pr., Wellesley, 2003.
**Strang, G.**, *Linear Algebra and Its Applications, 4th Edition*, Brooks Cole, Stamford, 2005.
**Sullivan, M.**, *College Algebra, 7th Edition*, Prentice Hall, Englewood Cliffs, New Jersey, 2004.
**Tobey, J. and Slater, J.**, *Beginning Algebra, 6th Edition*, Prentice Hall, Englewood Cliffs, New Jersey, 2004.
**Tobey, J. and Slater, J.**, *Intermediate Algebra, 5th Edition*, Prentice Hall, Englewood Cliffs, New Jersey, 2005.
**Trefethen, L. N. and Bau, D.**, *Numerical Linear Algebra*, Society for Industrial & Applied Mathematics, University City Science Center, Philadelphia, 1997.
**Turnbull, H. W. and Aitken, A. C.**, *An Introduction to the Theory of Canonical Matrices, Phoenix Edition*, Dover Publications, New York, 2004.
**Vygodskii, M. Ya.**, *Mathematical Handbook: Higher Mathematics*, Mir Publishers, Moscow, 1971.
**Zassenhaus, H. J.**, *The Theory of Groups, 2nd Ed. Edition*, Dover Publications, New York, 1999.
**Zhang, F.**, *Matrix Theory*, Springer, New York, 1999.

# Chapter 6
# Limits and Derivatives

## 6.1. Basic Concepts of Mathematical Analysis

### 6.1.1. Number Sets. Functions of Real Variable

#### 6.1.1-1. Real axis, intervals, and segments.

The *real axis* is a straight line with a point $O$ chosen as the origin, a positive direction, and a scale unit.

There is a one-to-one correspondence between the set of all real numbers $\mathbb{R}$ and the set of all points of the real axis, with each real $x$ being represented by a point on the real axis separated from $O$ by the distance $|x|$ and lying to the right of $O$ for $x > 0$, or to the left of $O$ for $x < 0$.

One often has to deal with the following number sets (sets of real numbers or sets on the real axis).

1. Sets of the form $(a, b)$, $(-\infty, b)$, $(a, +\infty)$, and $(-\infty, +\infty)$ consisting, respectively, of all $x \in \mathbb{R}$ such that $a < x < b$, $x < b$, $x > a$, and $x$ is arbitrary are called *open intervals* (sometimes simply *intervals*).

2. Sets of the form $[a, b]$ consisting of all $x \in \mathbb{R}$ such that $a \leq x \leq b$ are called *closed intervals* or *segments*.

3. Sets of the form $(a, b]$, $[a, b)$, $(-\infty, b]$, $[a, +\infty)$ consisting of all $x$ such that $a < x \leq b$, $a \leq x < b$, $x \leq b$, $x \geq a$ are called *half-open intervals*.

A *neighborhood of a point* $x_\circ \in \mathbb{R}$ is defined as any open interval $(a, b)$ containing $x_\circ$ ($a < x_\circ < b$). A neighborhood of the "point" $+\infty$, $-\infty$, or $\infty$ is defined, respectively, as any set of the form $(b, +\infty)$, $(-\infty, c)$ or $(-\infty, -a) \cup (a, +\infty)$ (here, $a \geq 0$).

#### 6.1.1-2. Lower and upper bound of a set on a straight line.

The *upper bound* of a set of real numbers is the least number that bounds the set from above. The *lower bound* of a set of real numbers is the largest number that bounds the set from below.

In more details: let a set of real numbers $X \in \mathbb{R}$ be given. A number $\beta$ is called its upper bound and denoted $\sup X$ if for any $x \in X$ the inequality $x \leq \beta$ holds and for any $\beta_1 < \beta$ there exists an $x_1 \in X$ such that $x_1 > \beta_1$. A number $\alpha$ is called the lower bound of $X$ and denoted $\inf X$ if for any $x \in X$ the inequality $x \geq \alpha$ holds and for any $\alpha_1 > \alpha$ there exists an $x_1 \in X$ such that $x_1 < \alpha_1$.

**Example 1.** For a set $X$ consisting of two numbers $a$ and $b$ ($a < b$), we have
$$\inf X = a, \quad \sup X = b.$$

**Example 2.** For intervals (open, closed, and half-open), we have
$$\inf(a, b) = \inf[a, b] = \inf(a, b] = \inf[a, b) = a,$$
$$\sup(a, b) = \sup[a, b] = \sup(a, b] = \sup[a, b) = b.$$

One can see that the upper and lower bounds may belong to a given set (e.g., for closed intervals) and may not (e.g., for open intervals).

The symbol $+\infty$ (resp., $-\infty$) is called the upper (resp., lower) bound of a set unbounded from above (resp., from below).

### 6.1.1-3. Real-valued functions of real variable. Methods of defining a function.

$1°$. Let $D$ and $E$ be two sets of real numbers. Suppose that there is a relation between the points of $D$ and $E$ such that to each $x \in D$ there corresponds some $y \in E$, denoted by $y = f(x)$. In this case, one speaks of a *function* $f$ defined on the set $D$ and taking its values in the set $E$. The set $D$ is called the *domain of the function* $f$, and the subset of $E$ consisting of all elements $f(x)$ is called the *range of the function* $f$. This functional relation is often denoted by $y = f(x)$, $f : D \to E$, $f : x \mapsto y$.

The following terms are also used: $x$ is the *independent variable* or the *argument*; $y$ is the *dependent variable*.

$2°$. The most common and convenient way to define a function is the *analytic method*: the function is defined explicitly by means of a formula (or several formulas) depending on the argument $x$; for instance, $y = 2 \sin x + 1$.

*Implicit definition* of a function consists of using an equation of the form $F(x, y) = 0$, from which one calculates the value $y$ for any fixed value of the argument $x$.

*Parametric definition* of a function consists of defining the values of the independent variable $x$ and the dependent variable $y$ by a pair of formulas depending on an auxiliary variable $t$ (parameter): $x = p(t)$, $y = q(t)$.

Quite often functions are defined in terms of convergent series or by means of tables or graphs. There are some other methods of defining functions.

$3°$. The *graph of a function* is the representation of a function $y = f(x)$ as a line on the plane with orthogonal coordinates $x, y$, the points of the line having the coordinates $x, y = f(x)$, where $x$ is an arbitrary point from the domain of the function.

### 6.1.1-4. Single-valued, periodic, odd and even functions.

$1°$. A function is *single-valued* if each value of its argument corresponds to a unique value of the function. A function is *multi-valued* if there is at least one value of its argument corresponding to two or more values of the function. In what follows, we consider only single-valued functions, unless indicated otherwise.

$2°$. A function $f(x)$ is called *periodic* with period $T$ (or $T$-*periodic*) if $f(x + T) = f(x)$ for any $x$.

$3°$. A function $f(x)$ is called *even* if it satisfies the condition $f(x) = f(-x)$ for any $x$. A function $f(x)$ is called *odd* if it satisfies the condition $f(x) = -f(-x)$ for any $x$.

### 6.1.1-5. Decreasing, increasing, monotone, and bounded functions.

$1°$. A function $f(x)$ is called *increasing or strictly increasing* (resp., *nondecreasing*) on a set $D \subset \mathbb{R}$ if for any $x_1, x_2 \in D$ such that $x_1 > x_2$, we have $f(x_1) > f(x_2)$ (resp., $f(x_1) \geq f(x_2)$). A function $f(x)$ is called *decreasing or strictly decreasing* (resp., *nonincreasing*) on a set $D$ if for all $x_1, x_2 \in D$ such that $x_1 > x_2$, we have $f(x_1) < f(x_2)$ (resp., $f(x_1) \leq f(x_2)$). All such functions are called *monotone functions*. Strictly increasing or decreasing functions are called *strictly monotone*.

$2°$. A function $f(x)$ is called *bounded* on a set $D$ if $|f(x)| < M$ for all $x \in D$, where $M$ is a finite constant. A function $f(x)$ is called *bounded from above* (*bounded from below*) on a set $D$ if $f(x) < M$ ($M < f(x)$) for all $x \in D$, where $M$ is a real constant.

### 6.1.1-6. Composite and inverse functions.

$1°$. Consider a function $u = u(x)$, $x \in D$, with values $u \in E$, and let $y = f(u)$ be a function defined on $E$. Then the function $y = f(u(x))$, $x \in D$, is called a *composite function* or the *superposition* of the functions $f$ and $u$.

$2°$. Consider a function $y = f(x)$ that maps $x \in D$ into $y \in E$. The *inverse function* of $y = f(x)$ is a function $x = g(y)$ defined on $E$ and such that $x = g(f(x))$ for all $x \in D$. The inverse function is often denoted by $g = f^{-1}$.

For strictly monotone functions $f(x)$, the inverse function always exists. In order to construct the inverse function $g(y)$, one should use the relation $y = f(x)$ to express $x$ through $y$. The function $g(y)$ is monotonically increasing or decreasing together with $f(x)$.

## 6.1.2. Limit of a Sequence

### 6.1.2-1. Some definitions.

Suppose that there is a correspondence between each positive integer $n$ and some (real or complex) number denoted, for instance, by $x_n$. In this case, one says that a *numerical sequence* (or, simply, a *sequence*) $x_1, x_2, \ldots, x_n, \ldots$ is defined. Such a sequence is often denoted by $\{x_n\}$; $x_n$ is called the *generic term* of the sequence.

**Example 1.** For the sequence $\{n^2 - 2\}$, we have $x_1 = -1$, $x_2 = 2$, $x_3 = 7$, $x_4 = 14$, etc.

A sequence is called *bounded* (bounded from above, bounded from below) if there is a constant $M$ such that $|x_n| < M$ (respectively, $x_n < M$, $x_n > M$) for all $n = 1, 2, \ldots$

### 6.1.2-2. Limit of a sequence.

A number $b$ is called the *limit of a sequence* $x_1, x_2, \ldots, x_n, \ldots$ if for any $\varepsilon > 0$ there is $N = N(\varepsilon)$ such that $|x_n - b| < \varepsilon$ for all $n > N$.

If $b$ is the limit of the sequence $\{x_n\}$, one writes $\lim\limits_{n \to \infty} x_n = b$ or $x_n \to b$ as $n \to \infty$.

The limit of a constant sequence $\{x_n = c\}$ exists and is equal to $c$, i.e., $\lim\limits_{n \to \infty} = c$. In this case, the inequality $|x_n - c| < \varepsilon$ takes the form $0 < \varepsilon$ and holds for all $n$.

**Example 2.** Let us show that $\lim\limits_{n \to \infty} \dfrac{n}{n+1} = 1$.
Consider the difference $\left|\dfrac{n}{n+1} - 1\right| = \dfrac{1}{n+1}$. The inequality $\dfrac{1}{n+1} < \varepsilon$ holds for all $n > \dfrac{1}{\varepsilon} - 1 = N(\varepsilon)$. Therefore, for any positive $\varepsilon$ there is $N = \dfrac{1}{\varepsilon} - 1$ such that for $n > N$ we have $\left|\dfrac{n}{n+1} - 1\right| < \varepsilon$.

It may happen that a sequence $\{x_n\}$ has no limit at all, for instance, the sequence $\{x_n\} = \{(-1)^n\}$. A sequence that has a finite limit is called *convergent*.

THEOREM (BOLZANO–CAUCHY). *A sequence $x_n$ has a finite limit if and only if for any $\varepsilon > 0$, there is $N$ such that the inequality*

$$|x_n - x_m| < \varepsilon$$

*holds for all $n > N$ and $m > N$.*

### 6.1.2-3. Properties of convergent sequences.

1. Any convergent sequence can have only one limit.
2. Any convergent sequence is bounded. From any bounded sequence one can extract a convergent subsequence.*
3. If a sequence converges to $b$, then any of its subsequence also converges to $b$.
4. If $\{x_n\}$, $\{y_n\}$ are two convergent sequences, then the sequences $\{x_n \pm y_n\}$, $\{x_n \cdot y_n\}$, and $\{x_n/y_n\}$ (in this ratio, it is assumed that $y_n \neq 0$ and $\lim\limits_{n\to\infty} y_n \neq 0$) are also convergent and

$$\lim_{n\to\infty} (x_n \pm y_n) = \lim_{n\to\infty} x_n \pm \lim_{n\to\infty} y_n;$$

$$\lim_{n\to\infty} (cx_n) = c \lim_{n\to\infty} x_n \quad (c = \text{const});$$

$$\lim_{n\to\infty} (x_n \cdot y_n) = \lim_{n\to\infty} x_n \cdot \lim_{n\to\infty} y_n;$$

$$\lim_{n\to\infty} \frac{x_n}{y_n} = \frac{\lim\limits_{n\to\infty} x_n}{\lim\limits_{n\to\infty} y_n}.$$

5. If $\{x_n\}$, $\{y_n\}$ are convergent sequences and the inequality $x_n \leq y_n$ holds for all $n$, then $\lim\limits_{n\to\infty} x_n \leq \lim\limits_{n\to\infty} y_n$.
6. If the inequalities $x_n \leq y_n \leq z_n$ hold for all $n$ and $\lim\limits_{n\to\infty} x_n = \lim\limits_{n\to\infty} z_n = b$, then $\lim\limits_{n\to\infty} y_n = b$.

### 6.1.2-4. Increasing, decreasing, and monotone sequences.

A sequence $\{x_n\}$ is called *increasing or strictly increasing* (resp., *nondecreasing*) if the inequality $x_{n+1} > x_n$ (resp., $x_{n+1} \geq x_n$) holds for all $n$. A sequence $\{x_n\}$ is called *decreasing or strictly decreasing* (resp., *nonincreasing*) if the inequality $x_{n+1} < x_n$ (resp., $x_{n+1} \leq x_n$) holds for all $n$. All such sequences are called *monotone sequences*. Strictly increasing or decreasing sequences are called *strictly monotone*.

THEOREM. *Any monotone bounded sequence has a finite limit.*

**Example 3.** It can be shown that the sequence $\left\{\left(1+\dfrac{1}{n}\right)^n\right\}$ is bounded and increasing. Therefore, it is convergent. Its limit is denoted by the letter $e$:

$$e = \lim_{n\to\infty} \left(1 + \frac{1}{n}\right)^n \quad (e \approx 2.71828).$$

Logarithms with the base $e$ are called *natural* or *Napierian*, and $\log_e x$ is denoted by $\ln x$.

### 6.1.2-5. Properties of positive sequences.

1°. If a sequence $x_n$ ($x_n > 0$) has a limit (finite or infinite), then the sequence

$$y_n = \sqrt[n]{x_1 \cdot x_2 \ldots x_n}$$

has the same limit.

---

* Let $\{x_n\}$ be a given sequence and let $\{n_k\}$ be a strictly increasing sequence with $k$ and $n_k$ being natural numbers. The sequence $\{x_{n_k}\}$ is called a *subsequence* of the sequence $\{x_n\}$.

$2°$. From property $1°$ for the sequence

$$x_1, \frac{x_2}{x_1}, \frac{x_3}{x_2}, \ldots, \frac{x_n}{x_{n-1}}, \frac{x_{n+1}}{x_n}, \ldots,$$

we obtain a useful corollary

$$\lim_{n\to\infty} \sqrt[n]{x_n} = \lim_{n\to\infty} \frac{x_{n+1}}{x_n},$$

under the assumption that the second limit exists.

**Example 4.** Let us show that $\lim_{n\to\infty} \dfrac{n}{\sqrt[n]{n!}} = e$.

Taking $x_n = \dfrac{n^n}{n!}$ and using property $2°$, we get

$$\lim_{n\to\infty} \frac{n}{\sqrt[n]{n!}} = \lim_{n\to\infty} \frac{x_{n+1}}{x_n} = \lim_{n\to\infty} \left(1 + \frac{1}{n}\right)^n = e.$$

### 6.1.2-6. Infinitely small and infinitely large quantities.

A sequence $x_n$ converging to zero as $n \to \infty$ is called *infinitely small or infinitesimal.*

A sequence $x_n$ whose terms infinitely grow in absolute values with the growth of $n$ is called *infinitely large* or *"tending to infinity."* In this case, the following notation is used: $\lim_{n\to\infty} x_n = \infty$. If, in addition, all terms of the sequence starting from some number are positive (negative), then one says that the sequence $x_n$ converges to "plus (minus) infinity," and one writes $\lim_{n\to\infty} x_n = +\infty$ ($\lim_{n\to\infty} x_n = -\infty$). For instance, $\lim_{n\to\infty} (-1)^n n^2 = \infty$, $\lim_{n\to\infty} \sqrt{n} = +\infty$, $\lim_{n\to\infty} (-n) = -\infty$.

**THEOREM (STOLZ).** *Let $x_n$ and $y_n$ be two infinitely large sequences, $y_n \to +\infty$, and $y_n$ increases with the growth of $n$ (at least for sufficiently large $n$): $y_{n+1} > y_n$. Then*

$$\lim_{n\to\infty} \frac{x_n}{y_n} = \lim_{n\to\infty} \frac{x_n - x_{n-1}}{y_n - y_{n-1}},$$

*provided that the right limit exists (finite or infinite).*

**Example 5.** Let us find the limit of the sequence

$$z_n = \frac{1^k + 2^k + \cdots + n^k}{n^{k+1}}.$$

Taking $x_n = 1^k + 2^k + \cdots + n^k$ and $y_n = n^{k+1}$ in the Stolz theorem, we get

$$\lim_{n\to\infty} z_n = \lim_{n\to\infty} \frac{n^k}{n^{k+1} - (n-1)^{k+1}}.$$

Since $(n-1)^{k+1} = n^{k+1} - (k+1)n^k + \cdots$, we have $n^{k+1} - (n-1)^{k+1} = (k+1)n^k + \cdots$, and therefore

$$\lim_{n\to\infty} z_n = \lim_{n\to\infty} \frac{n^k}{(k+1)n^k + \cdots} = \frac{1}{k+1}.$$

### 6.1.2-7. Upper and lower limits of a sequence.

The limit (finite or infinite) of a subsequence of a given sequence $x_n$ is called a partial limit of $x_n$. In the set of all partial limits of any sequence of real numbers, there always exists the largest and the least (finite or infinite). The largest (resp., least) partial limit of a sequence is called its *upper* (resp., *lower*) *limit*. The upper and lower limits of a sequence $x_n$ are denoted, respectively,

$$\overline{\lim_{n\to\infty}} x_n, \qquad \underline{\lim_{n\to\infty}} x_n.$$

**Example 6.** The upper and lower limits of the sequence $x_n = (-1)^n$ are, respectively,

$$\overline{\lim_{n\to\infty}} x_n = 1, \qquad \underline{\lim_{n\to\infty}} x_n = -1.$$

A sequence $x_n$ has a limit (finite or infinite) if and only if its upper limit coincides with its lower limit:

$$\lim_{n\to\infty} x_n = \overline{\lim_{n\to\infty}} x_n = \underline{\lim_{n\to\infty}} x_n.$$

## 6.1.3. Limit of a Function. Asymptotes

### 6.1.3-1. Definition of the limit of a function. One-sided limits.

1°. One says that $b$ is the *limit of a function* $f(x)$ as $x$ tends to $a$ if for any $\varepsilon > 0$ there is $\delta = \delta(\varepsilon) > 0$ such that $|f(x) - b| < \varepsilon$ for all $x$ such that $0 < |x - a| < \delta$.

Notation: $\lim_{x\to a} f(x) = b$ or $f(x) \to b$ as $x \to a$.

One says that $b$ is the limit of a function $f(x)$ as $x$ tends to $+\infty$ if for any $\varepsilon > 0$ there is $N = N(\varepsilon) > 0$ such that $|f(x) - b| < \varepsilon$ for all $x > N$.

Notation: $\lim_{x\to +\infty} f(x) = b$ or $f(x) \to b$ as $x \to +\infty$.

In a similar way, one defines the limits for $x \to -\infty$ or $x \to \infty$.

THEOREM (BOLZANO–CAUCHY 1). *A function $f(x)$ has a finite limit as $x$ tends to $a$ ($a$ is assumed finite) if and only if for any $\varepsilon > 0$ there is $\delta > 0$ such that the inequality*

$$|f(x_1) - f(x_2)| < \varepsilon \tag{6.1.3.1}$$

holds for all $x_1, x_2$ such that $|x_1 - a| < \delta$ and $|x_2 - a| < \delta$.

THEOREM (BOLZANO–CAUCHY 2). *A function $f(x)$ has a finite limit as $x$ tends to $+\infty$ if and only if for any $\varepsilon > 0$ there is $\Delta > 0$ such that the inequality (6.1.3.1) holds for all $x_1 > \Delta$ and $x_2 > \Delta$.*

2°. One says that $b$ is the *left-hand limit* (resp., *right-hand limit*) of a function $f(x)$ as $x$ tends to $a$ if for any $\varepsilon > 0$ there is $\delta = \delta(\varepsilon) > 0$ such that $|f(x) - b| < \varepsilon$ for $a - \delta < x < a$ (resp., for $a < x < a + \delta$).

Notation: $\lim_{x\to a-0} f(x) = b$ or $f(a-0) = b$ (resp., $\lim_{x\to a+0} f(x) = b$ or $f(a+0) = b$).

### 6.1.3-2. Properties of limits.

Let $a$ be a number or any of the symbols $\infty, +\infty, -\infty$.
1. If a function has a limit at some point, this limit is unique.
2. If $c$ is a constant function of $x$, then $\lim_{x\to a} c = c$.

3. If there exist $\lim\limits_{x\to a} f(x)$ and $\lim\limits_{x\to a} g(x)$, then

$$\lim_{x\to a}\bigl[f(x)\pm g(x)\bigr] = \lim_{x\to a} f(x)\pm \lim_{x\to a} g(x);$$

$$\lim_{x\to a} cf(x) = c\lim_{x\to a} f(x) \quad (c=\text{const});$$

$$\lim_{x\to a} f(x)\cdot g(x) = \lim_{x\to a} f(x)\cdot \lim_{x\to a} g(x);$$

$$\lim_{x\to a}\frac{f(x)}{g(x)} = \frac{\lim\limits_{x\to a} f(x)}{\lim\limits_{x\to a} g(x)} \quad \left(\text{if } g(x)\neq 0,\ \lim_{x\to a} g(x)\neq 0\right).$$

4. Let $f(x)\leq g(x)$ in a neighborhood of a point $a$ ($x\neq a$). Then $\lim\limits_{x\to a} f(x)\leq \lim\limits_{x\to a} g(x)$, provided that these limits exist.

5. If $f(x)\leq g(x)\leq h(x)$ in a neighborhood of a point $a$ and $\lim\limits_{x\to a} f(x)=\lim\limits_{x\to a} h(x)=b$, then $\lim\limits_{x\to a} g(x)=b$.

These properties hold also for one-sided limits.

### 6.1.3-3. Limits of some functions.

*First noteworthy limit:* $\quad \lim\limits_{x\to 0}\dfrac{\sin x}{x} = 1.$

*Second noteworthy limit:* $\quad \lim\limits_{x\to\infty}\left(1+\dfrac{1}{x}\right)^x = e.$

Some other frequently used limits:

$$\lim_{x\to 0}\frac{(1+x)^n-1}{x} = n, \quad \lim_{x\to\infty}\frac{a_n x^n + a_{n-1}x^{n-1}+\cdots+a_1 x + a_0}{b_n x^n + b_{n-1}x^{n-1}+\cdots+b_1 x + b_0} = \frac{a_n}{b_n},$$

$$\lim_{x\to 0}\frac{1-\cos x}{x^2} = \frac{1}{2}, \quad \lim_{x\to 0}\frac{\tan x}{x}=1, \quad \lim_{x\to 0}\frac{\arcsin x}{x}=1, \quad \lim_{x\to 0}\frac{\arctan x}{x}=1,$$

$$\lim_{x\to 0}\frac{e^x-1}{x}=1, \quad \lim_{x\to 0}\frac{a^x-1}{x}=\ln a, \quad \lim_{x\to 0}\frac{\ln(1+x)}{x}=1, \quad \lim_{x\to 0}\frac{\log_a(1+x)}{x}=\log_a e,$$

$$\lim_{x\to 0}\frac{\sinh x}{x}=1, \quad \lim_{x\to 0}\frac{\tanh x}{x}=1, \quad \lim_{x\to 0}\frac{\operatorname{arcsinh} x}{x}=1, \quad \lim_{x\to 0}\frac{\operatorname{arctanh} x}{x}=1,$$

$$\lim_{x\to +0} x^a \ln x = 0, \quad \lim_{x\to +\infty} x^{-a}\ln x = 0, \quad \lim_{x\to +\infty} x^a e^{-x}=0, \quad \lim_{x\to +0} x^x = 1,$$

where $a>0$ and $b_n\neq 0$.

▶ See Paragraph 6.2.3-2, where L'Hospital rules for calculating limits with the help of derivatives are given.

### 6.1.3-4. Asymptotes of the graph of a function.

An *asymptote* of the graph of a function $y=f(x)$ is a straight line whose distance from the point $(x,y)$ on the graph of $y=f(x)$ tends to zero if at least one of the coordinates $(x,y)$ tends to zero.

The line $x = a$ is a *vertical asymptote* of the graph of the function $y = f(x)$ if at least one of the one-sided limits of $f(x)$ as $x \to a \pm 0$ is equal to $+\infty$ or $-\infty$.

The line $y = kx + b$ is an *oblique asymptote* of the graph of $y = f(x)$ if at least one of the limit relations holds: $\lim\limits_{x \to +\infty} [f(x) - kx - b] = 0$ or $\lim\limits_{x \to -\infty} [f(x) - kx - b] = 0$.

If there exist finite limits

$$\lim_{x \to +\infty} \frac{f(x)}{x} = k, \quad \lim_{x \to +\infty} [f(x) - kx] = b, \qquad (6.1.3.2)$$

then the line $y = kx + b$ is an oblique asymptote of the graph for $x \to +\infty$ (in a similar way, one defines an asymptote for $x \to -\infty$).

**Example.** Let us find the asymptotes of the graph of the function $y = \dfrac{x^2}{x-1}$.

1°. The graph has a vertical asymptote $x = 1$, since $\lim\limits_{x \to 1} \dfrac{x^2}{x-1} = \infty$.

2°. Moreover, for $x \to \pm\infty$, there is an oblique asymptote $y = kx + b$ whose coefficients are determined by the formulas (6.1.3.2):

$$k = \lim_{x \to \pm\infty} \frac{x}{x-1} = 1, \quad b = \lim_{x \to \pm\infty}\left(\frac{x^2}{x-1} - x\right) = \lim_{x \to \pm\infty} \frac{x}{x-1} = 1.$$

Thus, the equation of the oblique asymptote has the form $y = x + 1$. Fig. 6.1 shows the graph of the function under consideration and its asymptotes.

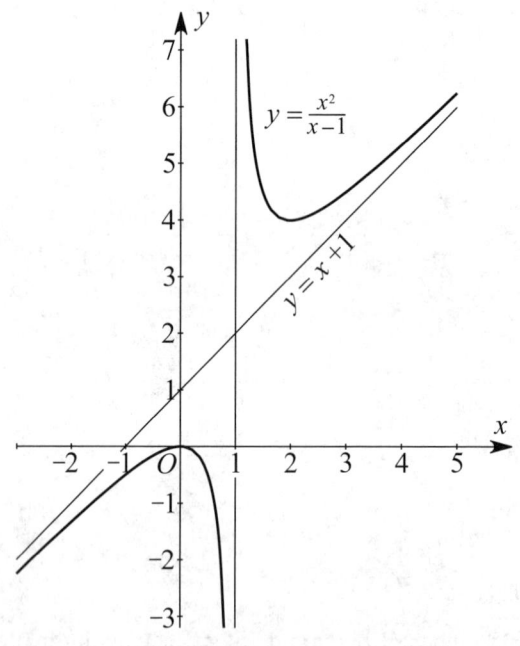

**Figure 6.1.** The graph and asymptotes of the function $y = \dfrac{x^2}{x-1}$.

## 6.1.4. Infinitely Small and Infinitely Large Functions

**6.1.4-1. Definitions.**

A function $f(x)$ is called *infinitely small* for $x \to a$ if $\lim\limits_{x \to a} f(x) = 0$.

A function $f(x)$ is said to be *infinitely large* for $x \to a$ if for any $K > 0$ the inequality $|f(x)| > K$ holds for all $x \neq a$ in a small neighborhood of the point $a$. In this case, one writes $f(x) \to \infty$ as $x \to a$ or $\lim_{x \to a} f(x) = \infty$. (In these definitions, $a$ is a finite number or any of the symbols $\infty$, $+\infty$, $-\infty$.) If $f(x)$ is infinitely large for $x \to a$ and $f(x) > 0$ ($f(x) < 0$) in a neighborhood of $a$ (for $x \neq a$), one writes $\lim_{x \to a} f(x) = +\infty$ (resp., $\lim_{x \to a} f(x) = -\infty$).

### 6.1.4-2. Properties of infinitely small and infinitely large functions.

1. The sum and the product of finitely many infinitely small functions for $x \to a$ is an infinitely small function.

2. The product of an infinitely small function $f(x)$ for $x \to a$ and a function $g(x)$ which is bounded in a neighborhood $U$ of the point $a$ (i.e., $|g(x)| < M$ for all $x \in U$, where $M > 0$ is a constant) is an infinitely small function.

3. $\lim_{x \to a} f(x) = b$ if and only if $f(x) = b + g(x)$, where $g(x)$ is infinitely small for $x \to a$.

4. A function $f(x)$ is infinitely large at some point if and only if the function $g(x) = \dfrac{1}{f(x)}$ is infinitely small at the same point.

### 6.1.4-3. Comparison of infinitely large quantities. Symbols of the order: $O$ and $o$.

Functions $f(x)$ and $g(x)$ that are infinitely small for $x \to a$ are called *equivalent* near $a$ if $\lim_{x \to a} \dfrac{f(x)}{g(x)} = 1$. In this case one writes $f(x) \sim g(x)$.

Examples of equivalent infinitely small functions:

$$(1+\varepsilon)^n - 1 \sim n\varepsilon, \qquad a^\varepsilon - 1 \sim \varepsilon \ln a, \qquad \log_a(1+\varepsilon) \sim \varepsilon \log_a e,$$

$$\sin \varepsilon \sim \varepsilon, \qquad \tan \varepsilon \sim \varepsilon, \qquad 1 - \cos \varepsilon \sim \tfrac{1}{2}\varepsilon^2, \qquad \arcsin \varepsilon \sim \varepsilon, \qquad \arctan \varepsilon \sim \varepsilon,$$

where $\varepsilon = \varepsilon(x)$ is infinitely small for $x \to a$.

Functions $f(x)$ and $g(x)$ are said to be of the *same order* for $x \to a$, and one writes $f(x) = O(g(x))$ if $\lim_{x \to a} \dfrac{f(x)}{g(x)} = K$, $0 < |K| < \infty$.*

A function $f(x)$ is of a *higher order of smallness* compared with $g(x)$ for $x \to a$ if $\lim_{x \to a} \dfrac{f(x)}{g(x)} = 0$, and in this case, one writes $f(x) = o(g(x))$.

## 6.1.5. Continuous Functions. Discontinuities of the First and the Second Kind

### 6.1.5-1. Continuous functions.

A function $f(x)$ is called *continuous* at a point $x = a$ if it is defined in that point and its neighborhood and $\lim_{x \to a} f(x) = f(a)$.

For continuous functions, a small variation of their argument $\Delta x = x - a$ corresponds to a small variation of the function $\Delta y = f(x) - f(a)$, i.e., $\Delta y \to 0$ as $\Delta x \to 0$. (This property is often used as a definition of continuity.)

---

* There is another definition of the symbol $O$. Namely, $f(x) = O(g(x))$ for $x \to a$ if the inequality $|f(x)| \leq K|g(x)|$, $K = \text{const}$, holds in some neighborhood of the point $a$ (for $x \neq a$).

A function $f(x)$ is called *right-continuous* at a point $x = a$ if it is defined in that point (and to its right) and $\lim_{x \to a+0} f(x) = f(a)$. A function $f(x)$ is called *left-continuous* at a point $x = a$ if it is defined in that point (and to its left) and $\lim_{x \to a-0} f(x) = f(a)$.

### 6.1.5-2. Properties of continuous functions.

1. Suppose that functions $f(x)$ and $g(x)$ are continuous at some point $a$. Then the functions $f(x) \pm g(x)$, $cf(x)$, $f(x)g(x)$, $\dfrac{f(x)}{g(x)}$ $(g(a) \neq 0)$ are also continuous at $a$.

2. Suppose that a function $f(x)$ is continuous on the segment $[a, b]$ and takes values of different signs at its endpoints, i.e., $f(a)f(b) < 0$. Then there is a point $c$ between $a$ and $b$ at which $f(x)$ vanishes:
$$f(c) = 0 \quad (a < c < b).$$

3. If $f(x)$ is continuous at a point $a$ and $f(a) > 0$ (resp., $f(a) < 0$), then there is $\delta > 0$ such that $f(x) > 0$ (resp., $f(x) < 0$) for all $x \in (a - \delta, a + \delta)$.

4. Any function $f(x)$ that is continuous at each point of a segment $[a, b]$ attains its largest and its smallest values, $M$ and $m$, on that segment.

5. A function $f(x)$ that is continuous on a segment $[a, b]$ takes any value $c \in [m, M]$ on that segment, where $m$ and $M$ are, respectively, its smallest and its largest values on $[a, b]$.

6. If $f(x)$ is continuous and increasing (resp., decreasing) on a segment $[a, b]$, then on the segment $[f(a), f(b)]$ (resp., $[f(b), f(a)]$) the inverse function $x = g(y)$ exists, and is continuous and increasing (resp., decreasing).

7. If $u(x)$ is continuous at a point $a$ and $f(u)$ is continuous at $b = u(a)$, then the composite function $f(u(x))$ is continuous at $a$.

**Remark.** Any elementary function is continuous at each point of its domain.

### 6.1.5-3. Points of discontinuity of a function.

A point $a$ is called a *point of discontinuity of the first kind* for a function $f(x)$ if there exist finite one-sided limits $f(a+0)$ and $f(a-0)$, but the relations $\lim_{x \to a+0} f(x) = \lim_{x \to a-0} f(x) = f(a)$ do not hold. The value $|f(a+0) - f(a-0)|$ is called the *jump* of the function at the point $a$. In particular, if $f(a+0) = f(a-0) \neq f(a)$, then $a$ is called a *point of removable discontinuity*.

**Examples of function with discontinuities of the first kind.**

1. The function $f(x) = \begin{cases} 0 & \text{for } x < 0, \\ 1 & \text{for } x \geq 0 \end{cases}$ has a jump equal to 1 at the discontinuity point $x = 0$.

2. The function $f(x) = \begin{cases} 0 & \text{for } x \neq 0, \\ 1 & \text{for } x = 0 \end{cases}$ has a removable discontinuity at the point $x = 0$.

A point $a$ is called a *point of discontinuity of the second kind* if at least one of the one-sided limits $f(a+0)$ or $f(a-0)$ does not exist or is equal to infinity.

**Examples of functions with discontinuities of the second kind.**

1. The function $f(x) = \sin \dfrac{1}{x}$ has a second-kind discontinuity at the point $x = 0$ (since this function has no one-sided limits as $x \to \pm 0$).

2. The function $f(x) = 1/x$ has an infinite jump at the point $x = 0$.

**6.1.5-4. Properties of monotone functions at points of discontinuity.**

Any monotone function $f(x)$ always has a left-hand limit and a right-hand limit at its discontinuity point $x = x_0$; moreover, if $f(x)$ is a nonincreasing function, then

$$f(x_0 - 0) \geq f(x_0) \geq f(x_0 + 0);$$

if $f(x)$ is a nondecreasing function, then

$$f(x_0 - 0) \leq f(x_0) \leq f(x_0 + 0).$$

## 6.1.6. Convex and Concave Functions

**6.1.6-1. Definition of convex and concave functions.**

1°. A function $f(x)$ defined and continuous on a segment $[a, b]$ is called *convex* (or *convex downward*) if for any $x_1, x_2$ in $[a, b]$, the *Jensen inequality* holds:

$$f\left(\frac{x_1 + x_2}{2}\right) \leq \frac{f(x_1) + f(x_2)}{2}. \tag{6.1.6.1}$$

The geometrical meaning of convexity is that all points of the graph curve between two graph points lie below or on the rectilinear segment joining the two graph points (see Fig. 6.2 *a*).

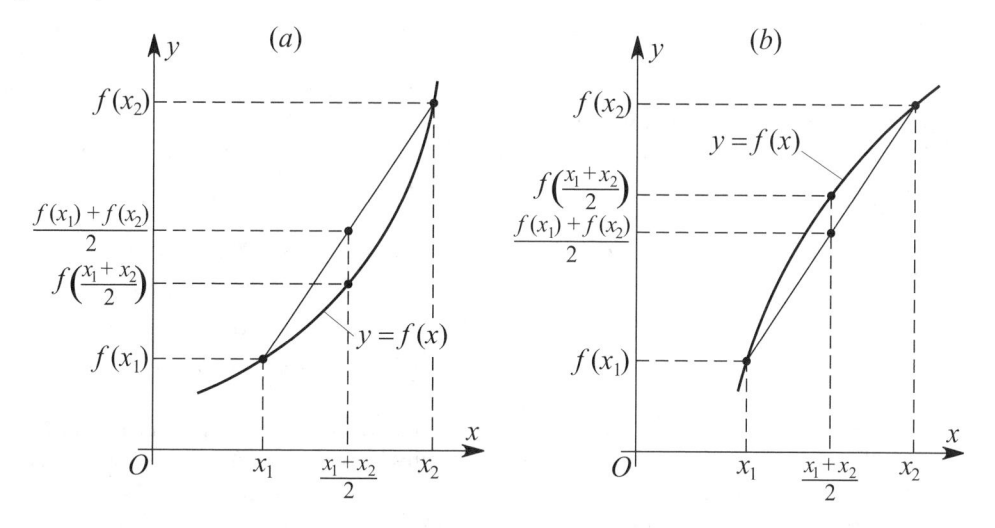

**Figure 6.2.** Graphs of convex (*a*) and concave (*b*) functions.

If for $x_1 \neq x_2$, condition (6.1.6.1) holds with $<$ instead of $\leq$, then the function $f(x)$ is called *strictly convex*.

2°. A function $f(x)$ defined and continuous on a segment $[a, b]$ is called *concave* (or *convex upward*) if for any $x_1, x_2$ in $[a, b]$ the following inequality holds:

$$f\left(\frac{x_1 + x_2}{2}\right) \geq \frac{f(x_1) + f(x_2)}{2}. \tag{6.1.6.2}$$

The geometrical meaning of concavity is that all points of the graph curve between two graph points lie above or on the rectilinear segment joining the two graph points (see Fig. 6.2 *b*).

If for $x_1 \neq x_2$, condition (6.1.6.2) holds with $>$ instead of $\geq$, then the function $f(x)$ is called *strictly concave*.

**6.1.6-2. Generalized Jensen inequalities.**

The inequalities (6.1.6.1) and (6.1.6.2) admit the following generalizations:

$$f(q_1 x_1 + \cdots + q_n x_n) \leq q_1 f(x_1) + \cdots + q_n x_n \quad \text{for a convex function,}$$
$$f(q_1 x_1 + \cdots + q_n x_n) \geq q_1 f(x_1) + \cdots + q_n x_n \quad \text{for a concave function,}$$

where $q_1, \ldots, q_n$ are arbitrary positive numbers such that $q_1 + \cdots + q_n = 1$, and $x_1, \ldots, x_n$ are arbitrary points of the segment $[a, b]$.

**6.1.6-3. Properties of convex and concave functions.**

1. The product of a convex (concave) function and a positive constant is a convex (concave) function.
2. The sum of two or more convex (concave) functions is a convex (concave) function.
3. If $\varphi(u)$ is a convex increasing function and $u = f(x)$ is a convex function, then the composite function $\varphi(f(x))$ is convex. Some other properties of composite functions:

$\varphi(u)$ *is convex and decreasing,* $\quad u = f(x)$ *is concave* $\quad \Longrightarrow \quad \varphi(f(x))$ *is convex,*
$\varphi(u)$ *is concave and increasing,* $\quad u = f(x)$ *is concave* $\quad \Longrightarrow \quad \varphi(f(x))$ *is concave,*
$\varphi(u)$ *is concave and decreasing,* $\quad u = f(x)$ *is convex* $\quad \Longrightarrow \quad \varphi(f(x))$ *is concave.*

4. A non-constant convex (resp., concave) function $f(x)$ on a segment $[a, b]$ cannot attain its largest (resp., smallest) value inside the segment.
5. If $y = f(x)$ and $x = g(y)$ are single-valued mutually inverse functions (on the corresponding intervals), then the following properties hold:

$f(x)$ *is convex and increasing* $\quad \Longleftrightarrow \quad g(y)$ *is concave and increasing,*
$f(x)$ *is convex and decreasing* $\quad \Longleftrightarrow \quad g(y)$ *is convex and decreasing,*
$f(x)$ *is concave and increasing* $\quad \Longleftrightarrow \quad g(y)$ *is convex and increasing,*
$f(x)$ *is concave and decreasing* $\quad \Longleftrightarrow \quad g(y)$ *is concave and decreasing.*

6. A function $f(x)$ that is continuous on a segment $[a, b]$ and twice differentiable on the interval $(a, b)$ is convex downward (resp., convex upward) if and only if $f''(x) \geq 0$ (resp., $f''(x) \leq 0$) on that interval.
7. Any convex function $f(x)$ satisfying the condition $f(x_0) = 0$ can be represented as the integral

$$f(x) = \int_{x_0}^{x} h(t)\, dt,$$

where $h(t)$ is a nondecreasing right-continuous function.

## 6.1.7. Functions of Bounded Variation

**6.1.7-1. Definition of a function of bounded variation.**

$1°$. Let $f(x)$ be a function defined on a finite segment $[a, b]$. Consider an arbitrary partition of the segment by the points

$$a = x_0 < x_1 < x_2 < \cdots < x_{n-1} < x_n = b$$

and construct the sum
$$v = \sum_{k=0}^{n-1} |f(x_{k+1}) - f(x_k)| \tag{6.1.7.1}$$
whose terms are absolute values of the increments of $f(x)$ on each segment of the partition.

If, for all partitions, the sums (6.1.7.1) are bounded by a constant independent of the partition, one says that the function $f(x)$ has *bounded variation* on the segment $[a, b]$. The supremum of all such sums over all partitions is called the *total variation* of the function $f(x)$ on the segment $[a, b]$. The total variation is denoted by
$$\bigvee_a^b f(x) = \sup\{v\}.$$

A function $f(x)$ is said to have bounded variation on the infinite interval $[a, \infty)$ if it is a function of bounded variation on any finite segment $[a, b]$ and its total variation on $[a, b]$ is bounded by a constant independent of $b$. By definition,
$$\bigvee_a^\infty f(x) = \sup_{b>a}\left\{\bigvee_a^b f(x)\right\}.$$

2°. In the above definitions, the continuity of the function $f(x)$ is not mentioned. A continuous function (without additional conditions) may have bounded or unbounded variation.

**Example.** Consider the continuous function
$$f(x) = \begin{cases} x \cos \dfrac{\pi}{2x} & \text{if } x \neq 0, \\ 0 & \text{if } x = 0 \end{cases}$$
and the partition of the segment $[0, 1]$ by the points
$$0 < \frac{1}{2n} < \frac{1}{2n-1} < \cdots < \frac{1}{3} < \frac{1}{2} < 1.$$
Then the sums (6.1.7.1) corresponding to this partition have the form
$$v_n = 1 + \frac{1}{2} + \cdots + \frac{1}{n} \to \infty \quad \text{as} \quad n \to \infty.$$
Therefore, $\bigvee_0^1 f(x) = \infty$.

### 6.1.7-2. Classes of functions of bounded variation.

Next, we list some common classes of functions of bounded variation.

1. Any bounded monotone function has bounded variation. Its total variation on the segment $[a, b]$ is defined by $\bigvee_a^b f(x) = |f(b) - f(a)|$.

**Remark.** The last statement is true for infinite intervals $(-\infty, a]$ and $[a, \infty)$; in the latter case, the total variation is equal to $\bigvee_a^\infty f(x) = |f(\infty) - f(a)|$.

2. Suppose that $f(x)$ is a bounded function on $[a, b]$ and this segment can be divided into finitely many parts
$$[a_k, a_{k+1}] \quad (k = 0, 1, \ldots, m-1; \quad a_0 = a, \quad a_m = b),$$
so that the function $f(x)$ is monotone on each part. Then $f(x)$ has bounded variation on $[a, b]$.

**Remark.** This statement is also true for infinite segments.

3. Let $f(x)$ be a function on a finite segment $[a, b]$ satisfying the *Lipschitz condition*

$$|f(x_1) - f(x_2)| \le L|x_1 - x_2|,$$

for any $x_1$ and $x_2$ in $[a, b]$, where $L$ is a constant. Then $f(x)$ has bounded variation and $\bigvee_a^b f(x) \le L(b-a)$.

4. Let $f(x)$ be a function on a finite segment $[a, b]$ with a bounded derivative $|f'(x)| \le L$, where $L = \text{const}$. Then, $f(x)$ is of bounded variation and $\bigvee_a^b f(x) \le L(b-a)$.

5. Let $f(x)$ be a function on $[a, b]$ or $[a, \infty)$ and suppose that $f(x)$ can be represented as an integral with variable upper limit,

$$f(x) = c + \int_a^x \varphi(t)\, dt,$$

where $\varphi(t)$ is an absolutely continuous function on the interval under consideration. Then $f(x)$ has bounded variation and

$$\bigvee_a^b f(x) = \int_a^b |\varphi(x)|\, dx.$$

COROLLARY. *Suppose that $\varphi(t)$ on a finite segment $[a, b]$ or $[a, \infty)$ is integrable, but not absolutely integrable. Then the total variation of $f(x)$ is infinite.*

6.1.7-3. Properties of functions of bounded variation.

Here, all functions are considered on a finite segment $[a, b]$.

1. Any function of bounded variation is bounded.

2. The sum, difference, or product of finitely many functions of bounded variation is a function of bounded variation.

3. Let $f(x)$ and $g(x)$ be two functions of bounded variation and $|g(x)| \ge K > 0$. Then the ratio $f(x)/g(x)$ is a function of bounded variation.

4. Let $a < c < b$. If $f(x)$ has bounded variation on the segment $[a, b]$, then it has bounded variation on each segment $[a, c]$ and $[c, b]$; and the converse statement is true. In this case, the following additivity condition holds:

$$\bigvee_a^b f(x) = \bigvee_a^c f(x) + \bigvee_c^b f(x).$$

5. Let $f(x)$ be a function of bounded variation of the segment $[a, b]$. Then, for $a \le x \le b$, the variation of $f(x)$ with variable upper limit

$$F(x) = \bigvee_a^x f(x)$$

is a monotonically increasing bounded function of $x$.

6. Any function $f(x)$ of bounded variation on the segment $[a, b]$ has a left-hand limit $\lim_{x \to x_0 - 0} f(x)$ and a right-hand limit $\lim_{x \to x_0 + 0} f(x)$ at any point $x_0 \in [a, b]$.

### 6.1.7-4. Criteria for functions to have bounded variation.

1. A function $f(x)$ has bounded variation on a finite segment $[a, b]$ if and only if there is a monotonically increasing bounded function $\Phi(x)$ such that for all $x_1, x_2 \in [a, b]$ ($x_1 < x_2$), the following inequality holds:

$$|f(x_2) - f(x_1)| \leq \Phi(x_2) - \Phi(x_1).$$

2. A function $f(x)$ has bounded variation on a finite segment $[a, b]$ if and only if $f(x)$ can be represented as the difference of two monotonically increasing bounded functions on that segment: $f(x) = g_2(x) - g_1(x)$.

**Remark.** The above criteria are valid also for infinite intervals $(-\infty, a]$, $[a, \infty)$, and $(-\infty, \infty)$.

### 6.1.7-5. Properties of continuous functions of bounded variation.

1. Let $f(x)$ be a function of bounded variation on the segment $[a, b]$. If $f(x)$ is continuous at a point $x_0$ ($a < x_0 < b$), then the function $F(x) = \bigvee_a^x f(x)$ is also continuous at that point.

2. A continuous function of bounded variation can be represented as the difference of two continuous increasing functions.

3. Let $f(x)$ be a continuous function on the segment $[a, b]$. Consider a partition of the segment

$$a = x_0 < x_1 < x_2 < \cdots < x_{n-1} < x_n = b$$

and the sum $v = \sum_{k=0}^{n-1} |f(x_{k+1}) - f(x_k)|$. Letting $\lambda = \max |x_{k+1} - x_k|$ and passing to the limit as $\lambda \to 0$, we get

$$\lim_{\lambda \to 0} v = \bigvee_a^b f(x).$$

## 6.1.8. Convergence of Functions

### 6.1.8-1. Pointwise, uniform, and nonuniform convergence of functions.

Let $\{f_n(x)\}$ be a sequence of functions defined on a set $X \subset \mathbb{R}$. The sequence $\{f_n(x)\}$ is said to be *pointwise convergent* to $f(x)$ as $n \to \infty$ if for any fixed $x \in X$, the numerical sequence $\{f_n(x)\}$ converges to $f(x)$. The sequence $\{f_n(x)\}$ is said to be *uniformly convergent* to a function $f(x)$ on $X$ as $n \to \infty$ if for any $\varepsilon > 0$ there is an integer $N = N(\varepsilon)$ and such that for all $n > N$ and all $x \in X$, the following inequality holds:

$$|f_n(x) - f(x)| < \varepsilon. \tag{6.1.8.1}$$

Note that in this definition, $N$ is independent of $x$. For a sequence $\{f_n(x)\}$ pointwise convergent to $f(x)$ as $n \to \infty$, by definition, for any $\varepsilon > 0$ and any $x \in X$, there is $N = N(\varepsilon, x)$ such that (6.1.8.1) holds for all $n > N(\varepsilon, x)$. If one cannot find such $N$ independent of $x$ and depending only on $\varepsilon$ (i.e., one cannot ensure (6.1.8.1) uniformly; to be more precise, there is $\delta > 0$ such that for any $N > 0$ there is $k_N > N$ and $x_N \in X$ such that $|f_{k_N}(x_N) - f(x_N)| \geq \delta$), then one says that the sequence $\{f_n(x)\}$ *converges nonuniformly* to $f(x)$ on the set $X$.

### 6.1.8-2. Basic theorems.

Let $X$ be an interval on the real axis.

THEOREM. *Let $f_n(x)$ be a sequence of continuous functions uniformly convergent to $f(x)$ on $X$. Then $f(x)$ is continuous on $X$.*

COROLLARY. *If the limit function $f(x)$ of a pointwise convergent sequence of continuous functions $\{f_n(x)\}$ is discontinuous, then the convergence of the sequence $\{f_n(x)\}$ is nonuniform.*

**Example.** The sequence $\{f_n(x)\} = \{x^n\}$ converges to $f(x) \equiv 0$ as $n \to \infty$ uniformly on each segment $[0, a]$, $0 < a < 1$. However, on the segment $[0, 1]$ this sequence converges nonuniformly to the discontinuous function $f(x) = \begin{cases} 0 & \text{for } 0 \leq x < 1, \\ 1 & \text{for } x = 1. \end{cases}$

CAUCHY CRITERION. *A sequence of functions $\{f_n(x)\}$ defined on a set $X \in \mathbb{R}$ uniformly converges to $f(x)$ as $n \to \infty$ if and only if for any $\varepsilon > 0$ there is an integer $N = N(\varepsilon) > 0$ such that for all $n > N$ and $m > N$, the inequality $|f_n(x) - f_m(x)| < \varepsilon$ holds for all $x \in X$.*

### 6.1.8-3. Geometrical meaning of uniform convergence.

Let $f_n(x)$ be continuous functions on the segment $[a, b]$ and suppose that $\{f_n(x)\}$ uniformly converges to a continuous function $f(x)$ as $n \to \infty$. Then all curves $y = f_n(x)$, for sufficiently large $n > N$, belong to the strip between the two curves $y = f(x) - \varepsilon$ and $y = f(x) + \varepsilon$ (see Fig. 6.3).

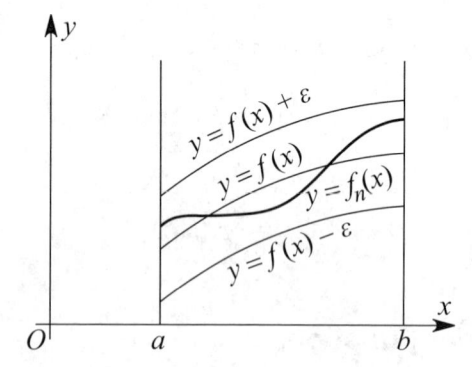

**Figure 6.3.** Geometrical meaning of uniform convergence of a sequence of functions $\{f_n(x)\}$ to a continuous function $f(x)$.

## 6.2. Differential Calculus for Functions of a Single Variable

### 6.2.1. Derivative and Differential, Their Geometrical and Physical Meaning

#### 6.2.1-1. Definition of derivative and differential.

The *derivative* of a function $y = f(x)$ at a point $x$ is the limit of the ratio

$$y' = \lim_{\Delta x \to 0} \frac{\Delta y}{\Delta x} = \lim_{\Delta x \to 0} \frac{f(x + \Delta x) - f(x)}{\Delta x},$$

where $\Delta y = f(x + \Delta x) - f(x)$ is the increment of the function corresponding to the increment of the argument $\Delta x$. The derivative $y'$ is also denoted by $y'_x$, $\dot{y}$, $\dfrac{dy}{dx}$, $f'(x)$, $\dfrac{df(x)}{dx}$.

## 6.2. Differential Calculus for Functions of a Single Variable

**Example 1.** Let us calculate the derivative of the function $f(x) = x^2$.
By definition, we have

$$f'(x) = \lim_{\Delta x \to 0} \frac{(x + \Delta x)^2 - x^2}{\Delta x} = \lim_{\Delta x \to 0} (2x + \Delta x) = 2x.$$

The increment $\Delta x$ is also called the differential of the independent variable $x$ and is denoted by $dx$.

A function $f(x)$ that has a derivative at a point $x$ is called *differentiable* at that point. The differentiability of $f(x)$ at a point $x$ is equivalent to the condition that the increment of the function, $\Delta y = f(x + dx) - f(x)$, at that point can be represented in the form $\Delta y = f'(x)\, dx + o(dx)$ (the second term is an infinitely small quantity compared with $dx$ as $dx \to 0$).

A function differentiable at some point $x$ is continuous at that point. The converse is not true, in general; continuity does not always imply differentiability.

A function $f(x)$ is called *differentiable* on a set $D$ (interval, segment, etc.) if for any $x \in D$ there exists the derivative $f'(x)$. A function $f(x)$ is called *continuously differentiable* on $D$ if it has the derivative $f'(x)$ at each point $x \in D$ and $f'(x)$ is a continuous function on $D$.

The *differential* $dy$ of a function $y = f(x)$ is the principal part of its increment $\Delta y$ at the point $x$, so that $dy = f'(x)dx$, $\Delta y = dy + o(dx)$.

The approximate relation $\Delta y \approx dy$ or $f(x + \Delta x) \approx f(x) + f'(x)\Delta x$ (for small $\Delta x$) is often used in numerical analysis.

---

**6.2.1-2. Physical and geometrical meaning of the derivative. Tangent line.**

1°. Let $y = f(x)$ be the function describing the path $y$ traversed by a body by the time $x$. Then the derivative $f'(x)$ is the velocity of the body at the instant $x$.

2°. The *tangent line* or simply the *tangent* to the graph of the function $y = f(x)$ at a point $M(x_0, y_0)$, where $y_0 = f(x_0)$, is defined as the straight line determined by the limit position of the secant $MN$ as the point $N$ tends to $M$ along the graph. If $\alpha$ is the angle between the $x$-axis and the tangent line, then $f'(x_0) = \tan \alpha$ is the slope ratio of the tangential line (Fig. 6.4).

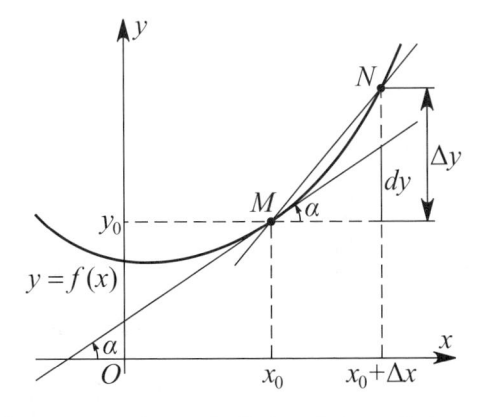

**Figure 6.4.** The tangent to the graph of a function $y = f(x)$ at a point $(x_0, y_0)$.

*Equation of the tangent line to the graph of a function $y = f(x)$ at a point $(x_0, y_0)$:*

$$y - y_0 = f'(x_0)(x - x_0).$$

*Equation of the normal to the graph of a function $y = f(x)$ at a point $(x_0, y_0)$:*

$$y - y_0 = -\frac{1}{f'(x_0)}(x - x_0).$$

### 6.2.1-3. One-sided derivatives.

One-sided derivatives are defined as follows:

$$f'_+(x) = \lim_{\Delta x \to +0} \frac{\Delta y}{\Delta x} = \lim_{\Delta x \to +0} \frac{f(x + \Delta x) - f(x)}{\Delta x} \quad \text{right-hand derivative,}$$

$$f'_-(x) = \lim_{\Delta x \to -0} \frac{\Delta y}{\Delta x} = \lim_{\Delta x \to -0} \frac{f(x + \Delta x) - f(x)}{\Delta x} \quad \text{left-hand derivative.}$$

**Example 2.** The function $y = |x|$ at the point $x = 0$ has different one-sided derivatives: $y'_+(0) = 1$, $y'_-(0) = -1$, but has no derivative at that point. Such points are called *angular points*.

Suppose that a function $y = f(x)$ is continuous at $x = x_0$ and has equal one-sided derivatives at that point, $y'_+(x_0) = y'_-(x_0) = a$. Then this function has a derivative at $x = x_0$ and $y'(x_0) = a$.

## 6.2.2. Table of Derivatives and Differentiation Rules

The derivative of any elementary function can be calculated with the help of derivatives of basic elementary functions and differentiation rules.

### 6.2.2-1. Table of derivatives of basic elementary functions ($a$ = const).

$$(a)' = 0, \qquad (x^a)' = ax^{a-1},$$
$$(e^x)' = e^x, \qquad (a^x)' = a^x \ln a,$$
$$(\ln x)' = \frac{1}{x}, \qquad (\log_a x)' = \frac{1}{x \ln a},$$
$$(\sin x)' = \cos x, \qquad (\cos x)' = -\sin x,$$
$$(\tan x)' = \frac{1}{\cos^2 x}, \qquad (\cot x)' = -\frac{1}{\sin^2 x},$$
$$(\arcsin x)' = \frac{1}{\sqrt{1-x^2}}, \qquad (\arccos x)' = -\frac{1}{\sqrt{1-x^2}},$$
$$(\arctan x)' = \frac{1}{1+x^2}, \qquad (\text{arccot } x)' = -\frac{1}{1+x^2},$$
$$(\sinh x)' = \cosh x, \qquad (\cosh x)' = \sinh x,$$
$$(\tanh x)' = \frac{1}{\cosh^2 x}, \qquad (\coth x)' = -\frac{1}{\sinh^2 x},$$
$$(\text{arcsinh } x)' = \frac{1}{\sqrt{1+x^2}}, \qquad (\text{arccosh } x)' = \frac{1}{\sqrt{x^2-1}},$$
$$(\text{arctanh } x)' = \frac{1}{1-x^2}, \qquad (\text{arccoth } x)' = \frac{1}{x^2-1}.$$

### 6.2.2-2. Differentiation rules.

1. *Derivative of a sum (difference) of functions*:
$$[u(x) \pm v(x)]' = u'(x) \pm v'(x).$$

2. *Derivative of the product of a function and a constant*:
$$[au(x)]' = au'(x) \qquad (a = \text{const}).$$

3. *Derivative of a product of functions*:
$$[u(x)v(x)]' = u'(x)v(x) + u(x)v'(x).$$

4. *Derivative of a ratio of functions*:
$$\left[\frac{u(x)}{v(x)}\right]' = \frac{u'(x)v(x) - u(x)v'(x)}{v^2(x)}.$$

5. *Derivative of a composite function*:
$$[f(u(x))]' = f'_u(u)u'(x).$$

6. *Derivative of a parametrically defined function* $x = x(t)$, $y = y(t)$:
$$y'_x = \frac{y'_t}{x'_t}.$$

7. *Derivative of an implicit function defined by the equation* $F(x, y) = 0$:
$$y'_x = -\frac{F_x}{F_y} \qquad (F_x \text{ and } F_y \text{ are partial derivatives}).$$

8. *Derivative of the inverse function* $x = x(y)$ (for details see footnote*):
$$x'_y = \frac{1}{y'_x}.$$

9. *Derivative of a composite exponential function*:
$$[u(x)^{v(x)}]' = u^v \ln u \cdot v' + vu^{v-1}u'.$$

10. *Derivative of a composite function of two arguments*:
$$[f(u(x), v(x))]' = f_u(u, v)u' + f_v(u, v)v' \qquad (f_u \text{ and } f_v \text{ are partial derivatives}).$$

**Example 1.** Let us calculate the derivative of the function $\dfrac{x^2}{2x+1}$.
Using the rule of differentiating the ratio of two functions, we obtain
$$\left(\frac{x^2}{2x+1}\right)' = \frac{(x^2)'(2x+1) - x^2(2x+1)'}{(2x+1)^2} = \frac{2x(2x+1) - 2x^2}{(2x+1)^2} = \frac{2x^2 + 2x}{(2x+1)^2}.$$

**Example 2.** Let us calculate the derivative of the function $\ln \cos x$.
Using the rule of differentiating composite functions and the formula for the logarithmic derivative from Paragraph 6.2.2-1, we get
$$(\ln \cos x)' = \frac{1}{\cos x}(\cos x)' = -\tan x.$$

**Example 3.** Let us calculate the derivative of the function $x^x$. Using the rule of differentiating the composite exponential function with $u(x) = v(x) = x$, we have
$$(x^x)' = x^x \ln x + xx^{x-1} = x^x(\ln x + 1).$$

---

* Let $y = f(x)$ be a differentiable monotone function on the interval $(a, b)$ and $f'(x_0) \neq 0$, where $x_0 \in (a, b)$. Then the inverse function $x = g(y)$ is differentiable at the point $y_0 = f(x_0)$ and $g'(y_0) = \dfrac{1}{f'(x_0)}$.

## 6.2.3. Theorems about Differentiable Functions. L'Hospital Rule

**6.2.3-1. Main theorems about differentiable functions.**

ROLLE THEOREM. *If the function $y = f(x)$ is continuous on the segment $[a, b]$, differentiable on the interval $(a, b)$, and $f(a) = f(b)$, then there is a point $c \in (a, b)$ such that $f'(c) = 0$.*

LAGRANGE THEOREM. *If the function $y = f(x)$ is continuous on the segment $[a, b]$ and differentiable on the interval $(a, b)$, then there is a point $c \in (a, b)$ such that*

$$f(b) - f(a) = f'(c)(b - a).$$

This relation is called the *formula of finite increments*.

CAUCHY THEOREM. *Let $f(x)$ and $g(x)$ be two functions that are continuous on the segment $[a, b]$, differentiable on the interval $(a, b)$, and $g'(x) \neq 0$ for all $x \in (a, b)$. Then there is a point $c \in (a, b)$ such that*

$$\frac{f(b) - f(a)}{g(b) - g(a)} = \frac{f'(c)}{g'(c)}.$$

**6.2.3-2. L'Hospital's rules on indeterminate expressions of the form $0/0$ and $\infty/\infty$.**

THEOREM 1. *Let $f(x)$ and $g(x)$ be two functions defined in a neighborhood of a point $a$, vanishing at this point, $f(a) = g(a) = 0$, and having the derivatives $f'(a)$ and $g'(a)$, with $g'(a) \neq 0$. Then*

$$\lim_{x \to a} \frac{f(x)}{g(x)} = \frac{f'(a)}{g'(a)}.$$

**Example 1.** Let us calculate the limit $\lim_{x \to 0} \frac{\sin x}{1 - e^{-2x}}$.

Here, both the numerator and the denominator vanish for $x = 0$. Let us calculate the derivatives

$$f'(x) = (\sin x)' = \cos x \quad \Longrightarrow \quad f'(0) = 1,$$
$$g'(x) = (1 - e^{-2x})' = 2e^{-2x} \quad \Longrightarrow \quad g'(0) = 2 \neq 0.$$

By the L'Hospital rule, we find that

$$\lim_{x \to 0} \frac{\sin x}{1 - e^{-2x}} = \frac{f'(0)}{g'(0)} = \frac{1}{2}.$$

THEOREM 2. *Let $f(x)$ and $g(x)$ be two functions defined in a neighborhood of a point $a$, vanishing at $a$, together with their derivatives up to the order $n - 1$ inclusively. Suppose also that the derivatives $f^{(n)}(a)$ and $g^{(n)}(a)$ exist and are finite, $g^{(n)}(a) \neq 0$. Then*

$$\lim_{x \to a} \frac{f(x)}{g(x)} = \frac{f^{(n)}(a)}{g^{(n)}(a)}.$$

THEOREM 3. *Let $f(x)$ and $g(x)$ be differentiable functions and $g'(x) \neq 0$ in a neighborhood of a point $a$ ($x \neq a$). If $f(x)$ and $g(x)$ are infinitely small or infinitely large functions for $x \to a$, i.e., the ratio $\frac{f(x)}{g(x)}$ at the point $a$ is an indeterminate expression of the form $\frac{0}{0}$ or $\frac{\infty}{\infty}$, then*

$$\lim_{x \to a} \frac{f(x)}{g(x)} = \lim_{x \to a} \frac{f'(x)}{g'(x)}$$

*(provided that there exists a finite or infinite limit of the ratio of the derivatives).*

**Remark.** The L'Hospital rule 3 is applicable also in the case of $a$ being one of the symbols $\infty$, $+\infty$, $-\infty$.

#### 6.2.3-3. Methods for interpreting other indeterminate expressions.

1°. Expressions of the form $0 \cdot \infty$ and $\infty - \infty$ can be reduced to indeterminate expressions $\dfrac{0}{0}$ or $\dfrac{\infty}{\infty}$ by means of algebraic transformations, for instance:

$$u(x) \cdot v(x) = \frac{u(x)}{1/v(x)} \qquad \text{transformation rule} \quad 0 \cdot \infty \Longrightarrow \frac{0}{0},$$

$$u(x) - v(x) = \left(\frac{1}{u(x)} - \frac{1}{v(x)}\right) : \frac{1}{u(x)v(x)} \qquad \text{transformation rule} \quad \infty - \infty \Longrightarrow \frac{0}{0}.$$

2°. Indeterminate expressions of the form $1^\infty$, $\infty^0$, $0^0$ can be reduced to expressions of the form $\dfrac{0}{0}$ or $\dfrac{\infty}{\infty}$ by taking logarithm and using the formulas $\ln u^v = v \ln u = \dfrac{\ln u}{1/v}$.

**Example 2.** Let us calculate the limit $\lim\limits_{x \to 0} (\cos x)^{1/x^2}$.
We have the indeterminate expression $1^\infty$. We find that

$$\ln \lim_{x \to 0} (\cos x)^{1/x^2} = \lim_{x \to 0} \ln(\cos x)^{1/x^2} = \lim_{x \to 0} \frac{\ln \cos x}{x^2} = \lim_{x \to 0} \frac{(\ln \cos x)'}{(x^2)'} = \lim_{x \to 0} \frac{(-\tan x)}{2x} = -\frac{1}{2}.$$

Therefore, $\lim\limits_{x \to 0} (\cos x)^{1/x^2} = e^{-1/2} = \dfrac{1}{\sqrt{e}}$.

### 6.2.4. Higher-Order Derivatives and Differentials. Taylor's Formula

#### 6.2.4-1. Derivatives and differentials of higher orders.

The *second-order derivative* or the second derivative of a function $y = f(x)$ is the derivative of the derivative $f'(x)$. The second derivative is denoted by $y''$ and also by $y''_{xx}$, $\dfrac{d^2 y}{dx^2}$, $f''(x)$.

The derivative of the second derivative of a function $y = f(x)$ is called the third-order derivative, $y''' = (y'')'$. The $n$th-order derivative of the function $y = f(x)$ is defined as the derivative of its $(n-1)$th derivative:

$$y^{(n)} = (y^{(n-1)})'.$$

The $n$th-order derivative is also denoted by $y_x^{(n)}$, $\dfrac{d^n y}{dx^n}$, $f^{(n)}(x)$.

The *second-order differential* is the differential of the first-order differential, $d^2 y = d(dy)$. If $x$ is the independent variable, then $d^2 y = y'' \cdot (dx)^2$. In a similar way, one defines differentials of higher orders.

#### 6.2.4-2. Table of higher-order derivatives of some elementary functions.

$$(x^a)^{(n)} = a(a-1)\ldots(a-n+1)x^{a-n}, \qquad (a^x)^{(n)} = (\ln a)^n a^x,$$

$$(\ln x)^{(n)} = (-1)^{n-1}(n-1)! \frac{1}{x^n}, \qquad (\log_a x)^{(n)} = (-1)^{n-1} \frac{(n-1)!}{\ln a} \frac{1}{x^n},$$

$$(\sin x)^{(n)} = \sin\left(x + \frac{\pi n}{2}\right), \qquad (\cos x)^{(n)} = \cos\left(x + \frac{\pi n}{2}\right),$$

$$(\sinh x)^{(n)} = \begin{cases} \cosh x & \text{if } n \text{ is odd,} \\ \sinh x & \text{if } n \text{ is even,} \end{cases} \qquad (\cosh x)^{(n)} = \begin{cases} \cosh x & \text{if } n \text{ is even,} \\ \sinh x & \text{if } n \text{ is odd.} \end{cases}$$

## 6.2.4-3. Rules for calculating higher-order derivatives.

1. *Derivative of a sum (difference) of functions*:

$$[u(x) \pm v(x)]^{(n)} = u^{(n)}(x) \pm v^{(n)}(x).$$

2. *Derivatives of a function multiplied by a constant*:

$$[au(x)]^{(n)} = au^{(n)}(x) \qquad (a = \text{const}).$$

3. *Derivatives of a product*:

$$[u(x)v(x)]'' = u''(x)v(x) + 2u'(x)v'(x) + u(x)v''(x),$$
$$[u(x)v(x)]''' = u'''(x)v(x) + 3u''(x)v'(x) + 3u'(x)v''(x) + u(x)v'''(x),$$
$$[u(x)v(x)]^{(n)} = \sum_{k=0}^{n} C_n^k u^{(k)}(x) v^{(n-k)}(x) \qquad \text{(Leibnitz formula)},$$

where $C_n^k$ are binomial coefficients, $u^{(0)}(x) = u(x)$, $v^{(0)}(x) = v(x)$.

4. *Derivatives of a composite function*:

$$\bigl[f(u(x))\bigr]'' = f''_{uu}(u'_x)^2 + f'_u u''_{xx},$$
$$\bigl[f(u(x))\bigr]''' = f'''_{uuu}(u'_x)^3 + 3f''_{uu} u'_x u''_{xx} + f'_u u'''_{xxx}.$$

5. *Derivatives of a parametrically defined function* $x = x(t)$, $y = y(t)$:

$$y'' = \frac{x'_t y''_{tt} - y'_t x''_{tt}}{(x'_t)^3}, \qquad y''' = \frac{(x'_t)^2 y'''_{ttt} - 3x'_t x''_{tt} y''_{tt} + 3y'_t (x''_{tt})^2 - x'_t y'_t x'''_{ttt}}{(x'_t)^5}, \qquad y^{(n)} = \frac{(y^{(n-1)})'_t}{x'_t}.$$

6. *Derivatives of an implicit function defined by the equation* $F(x, y) = 0$:

$$y'' = \frac{1}{F_y^3}\bigl(-F_y^2 F_{xx} + 2F_x F_y F_{xy} - F_x^2 F_{yy}\bigr),$$

$$y''' = \frac{1}{F_y^5}\bigl(-F_y^4 F_{xxx} + 3F_x F_y^3 F_{xxy} - 3F_x^2 F_y^2 F_{xyy} + F_x^3 F_y F_{yyy} + 3F_y^3 F_{xx} F_{xy}$$
$$- 3F_x F_y^2 F_{xx} F_{yy} - 6F_x F_y^2 F_{xy}^2 - 3F_x^3 F_{yy}^2 + 9F_x^2 F_y F_{xy} F_{yy}\bigr),$$

where the subscripts denote the corresponding partial derivatives.

7. *Derivatives of the inverse function* $x = x(y)$:

$$x''_{yy} = -\frac{y''_{xx}}{(y'_x)^3}, \qquad x'''_{yyy} = -\frac{y'''_{xxx}}{(y'_x)^4} + 3\frac{(y''_{xx})^2}{(y'_x)^5}, \qquad x_y^{(n)} = \frac{1}{y'_x}[x_y^{(n-1)}]'_x.$$

### 6.2.4-4. Taylor's formula.

Suppose that in a neighborhood of a point $x = a$, the function $y = f(x)$ has derivatives up to the order $(n + 1)$ inclusively. Then for all $x$ in that neighborhood, the following representation holds:

$$f(x) = f(a) + \frac{f'(a)}{1!}(x-a) + \frac{f''(a)}{2!}(x-a)^2 + \cdots + \frac{f^{(n)}(a)}{n!}(x-a)^n + R_n(x), \quad (6.2.4.1)$$

where $R_n(x)$ is the *remainder term* in the Taylor formula.

The remainder term can be represented in different forms (6.2.4.1):

$$R_n(x) = o[(x-a)^n] \quad \text{(Peano)},$$

$$R_n(x) = \frac{f^{(n+1)}(a + k(x-a))}{(n+1)!}(x-a)^{n+1} \quad \text{(Lagrange)},$$

$$R_n(x) = \frac{f^{(n+1)}(a + k(x-a))}{n!}(1-k)^n(x-a)^{n+1} \quad \text{(Cauchy)},$$

$$R_n(x) = \frac{f^{(n+1)}(a + k(x-a))}{n!\,p}(1-k)^{n+1-p}(x-a)^{n+1} \quad \text{(Schlömilch and Roche)},$$

$$R_n(x) = \frac{1}{n!}\int_a^x f^{(n+1)}(t)(x-t)^n\,dt \quad \text{(integral form)},$$

where $0 < k < 1$ and $p > 0$; $k$ depends on $x$, $n$, and the structure of the remainder term. The remainders in the form of Lagrange and Cauchy can be obtained as special cases of the Schlömilch formula with $p = n + 1$ and $p = 1$, respectively.

For $a = 0$, the Taylor formula (6.2.4.1) turns into

$$f(x) = f(0) + \frac{f'(0)}{1!}x + \frac{f''(0)}{2!}x^2 + \cdots + \frac{f^{(n)}(0)}{n!}x^n + R_n(x)$$

and is called the *Maclaurin formula*.

The Maclaurin formula for some functions:

$$e^x = 1 + \frac{x}{1!} + \frac{x^2}{2!} + \frac{x^3}{3!} + \cdots + \frac{x^n}{n!} + R_n(x),$$

$$\sin x = x - \frac{x^3}{3!} + \frac{x^5}{5!} - \frac{x^7}{7!} + \cdots + (-1)^n\frac{x^{2n+1}}{(2n+1)!} + R_{2n+1}(x),$$

$$\cos x = 1 - \frac{x^2}{2!} + \frac{x^4}{4!} - \frac{x^6}{6!} + \cdots + (-1)^n\frac{x^{2n}}{(2n)!} + R_{2n}(x).$$

## 6.2.5. Extremal Points. Points of Inflection

### 6.2.5-1. Maximum and minimum. Points of extremum.

Let $f(x)$ be a differentiable function on the interval $(a, b)$ and $f'(x) > 0$ (resp., $f'(x) < 0$) on $(a, b)$. Then $f(x)$ is an *increasing* (resp., *decreasing*) function on that interval*.

Suppose that there is a neighborhood of a point $x_0$ such that for all $x \neq x_0$ in that neighborhood we have $f(x) > f(x_0)$ (resp., $f(x) < f(x_0)$). Then $x_0$ is called a point of *local minimum* (resp., *local maximum*) of the function $f(x)$.

Points of local minimum or maximum are called points of *extremum*.

---

* At some points of the interval, the derivative may vanish.

### 6.2.5-2. Necessary and sufficient conditions for the existence of extremum.

NECESSARY CONDITION OF EXTREMUM. *A function $f(x)$ can have an extremum only at points in which its derivative either vanishes or does not exist (or is infinite).*

FIRST SUFFICIENT CONDITION OF EXTREMUM. *Suppose that $f(x)$ is continuous in some neighborhood $(x_0-\delta, x_0+\delta)$ of a point $x_0$ and differentiable at all points of the neighborhood except, possibly, $x_0$. If $f'(x) > 0$ for $x \in (x_0 - \delta, x_0)$ and $f'(x) < 0$ for $x \in (x_0, x_0 + \delta)$, then $x_0$ is a point of local maximum of this function. If $f'(x) < 0$ for $x \in (x_0 - \delta, x_0)$ and $f'(x) > 0$ for $x \in (x_0, x_0 + \delta)$, then $x_0$ is a point of local minimum of this function.*

*If $f'(x)$ is of the same sign for all $x \neq x_0$, $x \in (x_0-\delta, x_0+\delta)$, then $x_0$ cannot be a point of extremum.*

SECOND SUFFICIENT CONDITION OF EXTREMUM. *Let $f(x)$ be a twice differentiable function in a neighborhood of $x_0$. Then the following implications hold:*

(i)    $f'(x_0) = 0$ and $f''(x_0) < 0$    $\Longrightarrow$    $f(x)$ has a local maximum at the point $x_0$;

(ii)    $f'(x_0) = 0$ and $f''(x_0) > 0$    $\Longrightarrow$    $f(x)$ has a local minimum at the point $x_0$.

THIRD SUFFICIENT CONDITION OF EXTREMUM. *Let $f(x)$ be a function that is $n$ times differentiable in a neighborhood of a point $x_0$ and $f'(x_0) = f''(x_0) = \cdots = f^{(n-1)}(x_0) = 0$, but $f^{(n)}(x_0) \neq 0$. Then the following implications hold:*

(i)    $n$ is even and $f^{(n)}(x_0) < 0$    $\Longrightarrow$    $f(x)$ has a local maximum at the point $x_0$;

(ii)    $n$ is even and $f^{(n)}(x_0) > 0$    $\Longrightarrow$    $f(x)$ has a local minimum at the point $x_0$.

*If $n$ is odd, then $x_0$ cannot be a point of extremum.*

### 6.2.5-3. Largest and the smallest values of a function.

Let $y = f(x)$ be continuous on the segment $[a, b]$ and differentiable at all points of this segment except, possibly, finitely many points. Then the largest and the smallest values of $f(x)$ on $[a, b]$ belong to the set consisting of $f(a)$, $f(b)$, and the values $f(x_i)$, where $x_i \in (a, b)$ are the points at which $f'(x)$ is either equal to zero or does not exist (is infinite).

### 6.2.5-4. Direction of the convexity of the graph of a function.

The graph of a differentiable function $y = f(x)$ is said to be *convex upward* (resp., *convex downward*) on the interval $(a, b)$ if for each point of this interval, the graph lies below (resp., above) the tangent line at that point.

If the function $y = f(x)$ is twice differentiable on the interval $(a, b)$ and $f''(x) < 0$ (resp., $f''(x) > 0$), then its graph is convex upward (resp., downward) on that interval. (At some points of the interval, the second derivative may vanish.)

Thus, in order to find the intervals on which the graph of a twice differentiable function $f(x)$ is convex upward (resp., downward), one should solve the inequality $f''(x) < 0$ (resp., $f''(x) > 0$).

### 6.2.5-5. Inflection points.

An *inflection point* on the graph of a function $y = f(x)$ is defined as a point $(x_0, f(x_0))$ at which the graph passes from one side of its tangent line to another. At an inflection point, the graph changes the direction of its convexity.

Suppose that the function $y = f(x)$ has a continuous second derivative $f''(x)$ in some neighborhood of a point $x_0$. If $f''(x_0) = 0$ and $f''(x)$ changes sign as $x$ passes through the point $x_0$, then $(x_0, f(x_0))$ is an inflection point.

## 6.2.6. Qualitative Analysis of Functions and Construction of Graphs

**6.2.6-1. General scheme of analysis of a function and construction of its graph.**

1. Determine the domain in which the function is defined.
2. Find the asymptotes of the graph.
3. Find extremal points and intervals of monotonicity.
4. Determine the directions of convexity of the graph and its inflection points.
5. Determine whether the function is odd or even and whether it is periodic.
6. Find the points at which the graph crosses the coordinate axes.
7. Draw the graph, using the properties 1 to 6.

**Example.** Let us examine the function $y = \dfrac{\ln x}{x}$ and construct its graph.

We use the above general scheme.
1. This function is defined for all $x$ such that $0 < x < +\infty$.
2. The straight line $x = 0$ is a vertical asymptote, since $\lim\limits_{x \to +0} \dfrac{\ln x}{x} = -\infty$. We find the oblique asymptotes:

$$k = \lim_{x \to +\infty} \frac{y}{x} = 0, \quad b = \lim_{x \to +\infty} (y - kx) = 0.$$

Therefore, the line $y = 0$ is a horizontal asymptote of the graph.

3. The derivative $y' = \dfrac{1 - \ln x}{x^2}$ vanishes for $\ln x = 1$. Therefore, the function may have an extremum at $x = e$. For $x \in (0, e)$, we have $y' > 0$, i.e., the function is increasing on this interval. For $x \in (e, +\infty)$, we have $y' < 0$, and therefore the function is decreasing on this interval. At $x = e$ the function attains its maximal value $y_{\max} = \dfrac{1}{e}$.

One should also examine the points at which the derivative does not exist. There is only one such point, $x = 0$, and it corresponds to the vertical asymptote (see Item 1).

4. The second derivative $y'' = \dfrac{2 \ln x - 3}{x^3}$ vanishes for $x = e^{3/2}$. On the interval $(0, e^{3/2})$, we have $y'' < 0$, and therefore the graph is convex upward on this interval. For $x \in (e^{3/2}, +\infty)$, we have $y'' > 0$, and therefore the graph is convex downward on this interval. The value $x = e^{3/2}$ corresponds to an inflection point of the graph, with the ordinate $y = \tfrac{3}{2} e^{-3/2}$.

5. This function is neither odd nor even, since it is defined only for $x > 0$ and the relations $f(-x) = f(x)$ or $f(-x) = -f(x)$ cannot hold. Obviously, this function is nonperiodic.

6. The graph of this function does not cross the $y$-axis, since for $x = 0$ the function is undefined. Further, $y = 0$ only if $x = 1$, i.e., the graph crosses the $x$-axis only at the point $(1, 0)$.

7. Using the above results, we construct the graph (Fig. 6.5).

**6.2.6-2. Transformations of graphs of functions.**

Let us describe some methods which in many cases allow us to construct the graph of a function if we have the graph of a simpler function.

1. The graph of the function $y = f(x) + a$ is obtained from that of $y = f(x)$ by shifting the latter along the axis $Oy$ by the distance $|a|$. For $a > 0$ the shift is upward, and for $a < 0$ downward (see Fig. 6.6 a).

2. The graph of the function $y = f(x + a)$ is obtained from that of $y = f(x)$ by shifting the latter along the $Ox$ by the distance $|a|$. For $a > 0$ the shift is to the left, and for $a < 0$ to the right (see Fig. 6.6 b).

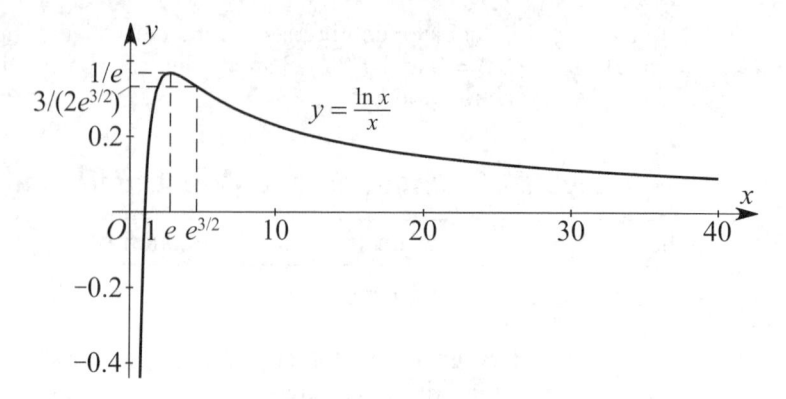

**Figure 6.5.** Graph of the function $y = \frac{\ln x}{x}$.

3. The graph of the function $y = -f(x)$ is obtained from that of $y = f(x)$ by symmetric reflection with respect to the axis $Ox$ (see Fig. 6.6 c).

4. The graph of the function $y = f(-x)$ is obtained from that of $y = f(x)$ by symmetric reflection with respect to the axis $Oy$ (see Fig. 6.6 d).

5. The graph of the function $y = kf(x)$ for $k > 1$ is obtained from that of $y = f(x)$ by extending the latter $k$ times from the axis $Ox$, and for $0 < k < 1$ by contracting the latter $1/k$ times to the axis $Ox$. The points at which the graph crosses the axis $Ox$ remain unchanged (see Fig. 6.6 e).

6. The graph of the function $y = f(kx)$ for $k > 1$ is obtained from that of $y = f(x)$ by contracting the latter $k$ times to the axis $Oy$, and for $0 < k < 1$ by extending the latter $1/k$ times from the axis $Oy$. The points at which the graph crosses the axis $Oy$ remain unchanged (see Fig. 6.6 f).

7. The graph of the function $y = |f(x)|$ is obtained from that of $y = f(x)$ by preserving the parts of the latter for which $f(x) \geq 0$ and symmetric reflection, with respect to the axis $Ox$, of the parts for which $f(x) < 0$ (see Fig. 6.6 g).

8. The graph of the inverse function $y = f^{-1}(x)$ is obtained from that of $y = f(x)$ by symmetric reflection with respect to the straight line $y = x$ (see Fig. 6.6 h).

## 6.2.7. Approximate Solution of Equations (Root-Finding Algorithms for Continuous Functions)

### 6.2.7-1. Preliminaries.

For a vast majority of algebraic (transcendental) equations of the form

$$f(x) = 0, \qquad (6.2.7.1)$$

where $f(x)$ is a continuous function, there are no exact closed-form expressions for the roots.

When solving the equation approximately, the first step is to bracket the roots, i.e., find sufficiently small intervals containing exactly one root each. Such an interval $[a, b]$, where the numbers $a$ and $b$ satisfy the condition $f(a)f(b) < 0$ (which is assumed to hold in what follows), can be found, say, graphically.

The second step is to compute successive approximations $x_n \in [a, b]$ ($n = 1, 2, \ldots$) to the desired root $c = \lim_{n \to \infty} x_n$, usually by one of the following methods.

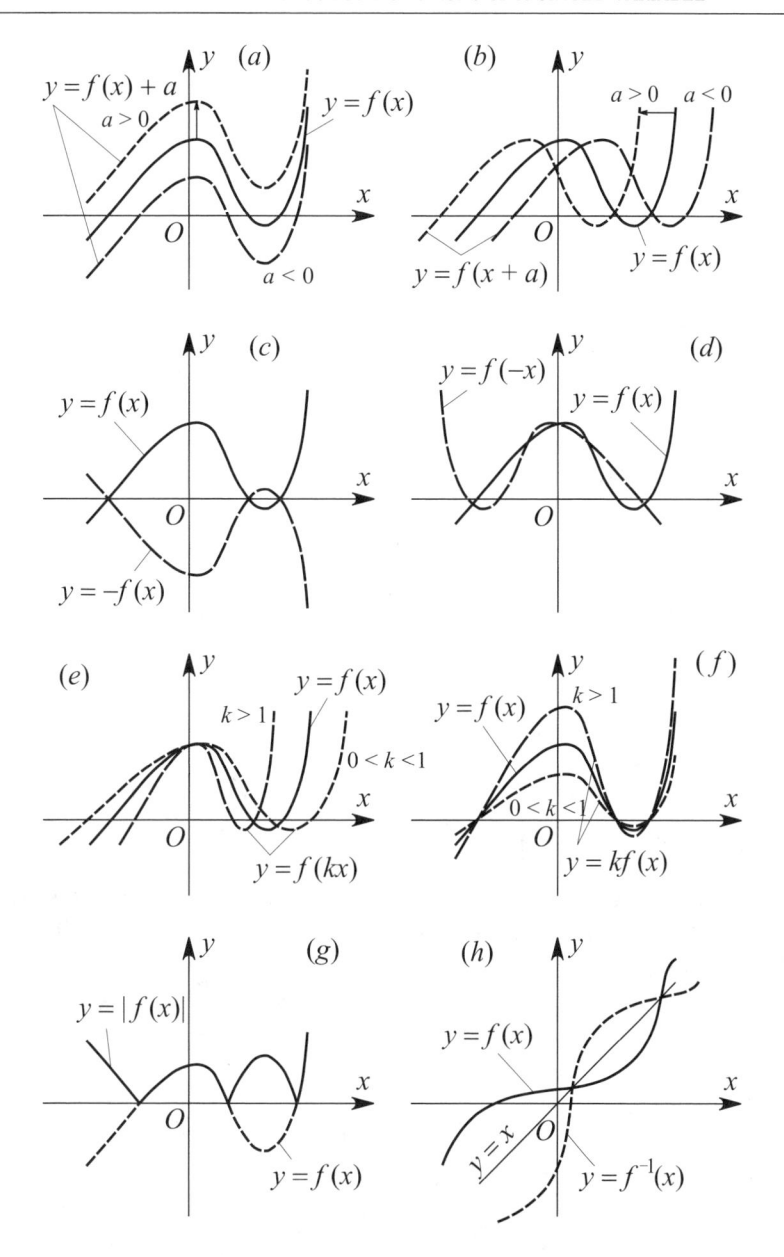

**Figure 6.6.** Transformations of graphs of functions.

6.2.7-2. **Bisection method.**

To find the root of equation (6.2.7.1) on the interval $[a, b]$, we bisect the interval. If $f\left(\frac{a+b}{2}\right) = 0$, then $c = \frac{a+b}{2}$ is the desired root. If $f\left(\frac{a+b}{2}\right) \neq 0$, then of the two intervals $\left[a, \frac{a+b}{2}\right]$ and $\left[\frac{a+b}{2}, b\right]$ we take the one at whose endpoints the function $f(x)$ has opposite signs. Now we bisect the new, smaller interval, etc. As a result, we obtain either an exact root of equation (6.2.7.1) at some step or an infinite sequence of nested intervals $[a_1, b_1], [a_2, b_2], \ldots$ such that $f(a_n)f(b_n) < 0$. The root is given by the formula

$c = \lim_{n \to \infty} a_n = \lim_{n \to \infty} b_n$, and the estimate

$$0 \le c - a_n \le \frac{1}{2^n}(b-a)$$

is valid.

The following two methods are more efficient.

### 6.2.7-3. Regula falsi method (false position method).

Suppose that the derivatives $f'(x)$ and $f''(x)$ exist on the interval $[a,b]$ and the inequalities $f'(x) \ne 0$ and $f''(x) \ne 0$ hold for all $x \in [a,b]$.

If $f'(a)f''(a) > 0$, then we take $x_0 = a$ for the zero approximation; the subsequent approximations are given by the formulas

$$x_{n+1} = x_n - \frac{f(x_n)}{f(b) - f(x_n)}(b - x_n), \qquad n = 0, 1, \ldots$$

If $f'(a)f''(a) < 0$, then we take $x_0 = b$ for the zero approximation; the subsequent approximations are given by the formulas

$$x_{n+1} = x_n - \frac{f(x_n)}{f(a) - f(x_n)}(a - x_n), \qquad n = 0, 1, \ldots$$

The regula falsi method has the first order of local convergence as $n \to \infty$:

$$|x_{n+1} - c| \le k|x_n - c|,$$

where $k$ is a constant depending on $f(x)$ and $c$ is the root of equation (6.2.7.1).

The regula falsi method has a simple geometric interpretation. The straight line (secant) passing through the points $(a, f(a))$ and $(b, f(b))$ of the curve $y = f(x)$ meets the abscissa axis at the point $x_1$; the value $x_{n+1}$ is the abscissa of the point where the line passing through the points $(x_0, f(x_0))$ and $(x_n, f(x_n))$ meets the $x$-axis (see Fig. 6.7 a).

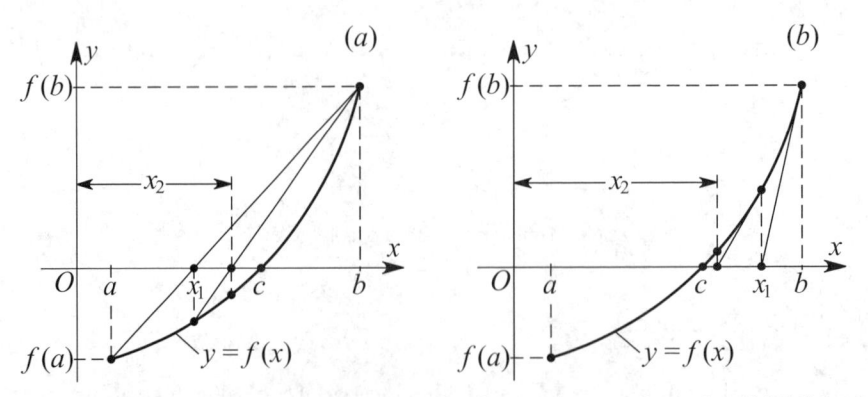

**Figure 6.7.** Graphical construction of successive approximations to the root of equation (6.2.7.1) by the regula falsi method (a) and the Newton–Raphson method (b).

### 6.2.7-4. Newton–Raphson method.

Suppose that the derivatives $f'(x)$ and $f''(x)$ exist on the interval $[a,b]$ and the inequalities $f'(x) \ne 0$ and $f''(x) \ne 0$ hold for all $x \in [a,b]$.

If $f(a)f''(a) > 0$, then we take $x_0 = a$ for the zero approximation; if $f(b)f''(b) > 0$, then $x_0 = b$. The subsequent approximations are computed by the formulas

$$x_{n+1} = x_n - \frac{f(x_n)}{f'(x_n)}, \qquad n = 0, 1, \ldots$$

If the initial approximation $x_0$ is sufficiently close to the desired root $c$, then the Newton–Raphson method exhibits quadratic convergence:

$$|x_{n+1} - c| \leq \frac{M}{2m} |x_n - c|^2,$$

where $M = \max\limits_{a \leq x \leq b} |f''(x)|$ and $m = \min\limits_{a \leq x \leq b} |f'(x)|$.

The Newton–Raphson method has a simple geometric interpretation. The tangent to the curve $y = f(x)$ through the point $(x_n, f(x_n))$ meets the abscissa axis at the point $x_{n+1}$ (see Fig. 6.7 b).

The Newton–Raphson method has a higher order of convergence than the regula falsi method. Hence the former is more often used in practice.

## 6.3. Functions of Several Variables. Partial Derivatives
### 6.3.1. Point Sets. Functions. Limits and Continuity

**6.3.1-1. Sets on the plane and in space.**

The distance between two points $A$ and $B$ on the plane and in space can be defined as follows:

$$\rho(A, B) = \sqrt{(x_A - x_B)^2 + (y_A - y_B)^2} \qquad \text{(on the plane)},$$

$$\rho(A, B) = \sqrt{(x_A - x_B)^2 + (y_A - y_B)^2 + (z_A - z_B)^2} \qquad \text{(in three-dimensional space)},$$

$$\rho(A, B) = \sqrt{(x_{1A} - x_{1B})^2 + \cdots + (x_{nA} - x_{nB})^2} \qquad \text{(in $n$-dimensional space)}.$$

where $x_A$, $y_A$ and $x_B$, $y_B$, and $x_A$, $y_A$, $z_A$ and $x_B$, $y_B$, $z_B$, and $x_{1A}, \ldots, x_{nA}$ and $x_{1B}, \ldots, x_{nB}$ are Cartesian coordinates of the corresponding points.

An *$\varepsilon$-neighborhood of a point* $M_0$ (on the plane or in space) is the set consisting of all points $M$ (resp., on the plane or in space) such that $\rho(M, M_0) < \varepsilon$, where it is assumed that $\varepsilon > 0$. An *$\varepsilon$-neighborhood of a set* $K$ (on the plane or in space) is the set consisting of all points $M$ (resp., on the plane or in space) such that $\inf\limits_{M_0 \in K} \rho(M, M_0) < \varepsilon$, where it is assumed that $\varepsilon > 0$.

An *interior point* of a set $D$ is a point belonging to $D$, together with some neighborhood of that point. An *open set* is a set containing only interior points. A *boundary point* of a set $D$ is a point such that any of its neighborhoods contains points outside $D$. A *closed set* is a set containing all its boundary points. A set $D$ is called a *bounded set* if $\rho(A, B) < C$ for any points $A, B \in D$, where $C$ is a constant independent of $A, B$. Otherwise (i.e., if there is no such constant), the set $D$ is called *unbounded*.

**6.3.1-2. Functions of two or three variables.**

A (numerical) *function* on a set $D$ is, by definition, a relation that sets up a correspondence between each point $M \in D$ and a unique numerical value. If $D$ is a plane set, then each point $M \in D$ is determined by two coordinates $x$, $y$, and a function $z = f(M) = f(x,y)$ is called a *function of two variables*. If $D$ belongs to a three-dimensional space, then one speaks of a *function of three variables*. The set $D$ on which the function is defined is called the *domain* of the function. For instance, the function $z = \sqrt{1 - x^2 - y^2}$ is defined on the closed circle $x^2 + y^2 \leq 1$, which is its domain.

The *graph* of a function $z = f(x,y)$ is the surface formed by the points $(x, y, f(x,y))$ in three-dimensional space. For instance, the graph of the function $z = ax + by + c$ is a plane, and the graph of the function $z = \sqrt{1 - x^2 - y^2}$ is a half-sphere.

A *level line* of a function $z = f(x,y)$ is a line on the plane $x$, $y$ with the following property: the function takes one and the same value $z = c$ at all points of that line. Thus, the equation of a level line has the form $f(x,y) = c$. A *level surface* of a function $u = f(x,y,z)$ is a surface on which the function takes a constant value, $u = c$; the equation of a level surface has the form $f(x,y,z) = c$.

A function $f(M)$ is called *bounded* on a set $D$ if there is a constant $C$ such that $|f(M)| \leq C$ for all $M \in D$.

### 6.3.1-3. Limit of a function at a point and its continuity.

Let $M$ be a point that comes infinitely close to some point $M_0$, i.e., $\rho = \rho(M_0, M) \to 0$. It is possible that the values $f(M)$ come close to some constant $b$.

One says that $b$ is the *limit of the function* $f(M)$ at the point $M_0$ if for any (arbitrarily small) $\varepsilon > 0$, there is $\delta > 0$ such that for all points $M$ belonging to the domain of the function and satisfying the inequality $0 < \rho(M_0, M) < \delta$, we have $|f(M) - b| < \varepsilon$. In this case, one writes $\lim\limits_{\rho(M,M_0) \to 0} f(M) = b$.

A function $f(M)$ is called *continuous* at a point $M_0$ if $\lim\limits_{\rho(M,M_0) \to 0} f(M) = f(M_0)$. A function is called *continuous on a set* $D$ if it is continuous at each point of $D$. Any continuous function $f(M)$ on a closed bounded set is bounded on that set and attains its smallest and its largest values on that set.

## 6.3.2. Differentiation of Functions of Several Variables

For the sake of brevity, we consider the case of a function of two variables. However, all statements can be easily extended to the case of $n$ variables.

### 6.3.2-1. Total and partial increments of a function. Partial derivatives.

A *total increment* of a function $z = f(x,y)$ at a point $(x,y)$ is
$$\Delta z = f(x + \Delta x, y + \Delta y) - f(x,y),$$
where $\Delta x$, $\Delta y$ are increments of the independent variables. *Partial increments* in $x$ and in $y$ are, respectively,
$$\Delta_x z = f(x + \Delta x, y) - f(x,y),$$
$$\Delta_y z = f(x, y + \Delta y) - f(x,y).$$

*Partial derivatives* of a function $z$ in $x$ and in $y$ at a point $(x,y)$ are defined as follows:
$$\frac{\partial z}{\partial x} = \lim_{\Delta x \to 0} \frac{\Delta_x z}{\Delta x}, \quad \frac{\partial z}{\partial y} = \lim_{\Delta y \to 0} \frac{\Delta_y z}{\Delta y}$$

(provided that these limits exist). Partial derivatives are also denoted by $z_x$ and $z_y$, $\partial_x z$ and $\partial_y z$, or $f_x(x,y)$ and $f_y(x,y)$.

### 6.3.2-2. Differentiable functions. Differential.

A function $z = f(x, y)$ is called *differentiable* at a point $(x, y)$ if its increment at that point can be represented in the form

$$\Delta z = A(x, y)\Delta x + B(x, y)\Delta y + o(\rho), \qquad \rho = \sqrt{(\Delta x)^2 + (\Delta y)^2},$$

where $o(\rho)$ is a quantity of a higher order of smallness compared with $\rho$ as $\rho \to 0$ (i.e., $o(\rho)/\rho \to 0$ as $\rho \to 0$). In this case, there exist partial derivatives at the point $(x, y)$, and $z'_x = A(x, y)$, $z'_y = B(x, y)$.

A function that has continuous partial derivatives at a point $(x, y)$ is differentiable at that point.

The *differential of a function* $z = f(x, y)$ is defined as follows:

$$dz = f'_x(x, y)\Delta x + f'_y(x, y)\Delta y.$$

Taking the differentials $dx$ and $dy$ of the independent variables equal to $\Delta x$ and $\Delta y$, respectively, one can also write $dz = f'_x(x, y)\, dx + f'_y(x, y)\, dy$.

The relation $\Delta z = dz + o(\rho)$ for small $\Delta x$ and $\Delta y$ is widely used for approximate calculations, in particular, for finding errors in numerical calculations of values of a function.

**Example 1.** Suppose that the values of the arguments of the function $z = x^2 y^5$ are known with the error $x = 2 \pm 0.01$, $y = 1 \pm 0.01$. Let us calculate the approximate value of the function.

We find the increment of the function $z$ at the point $x = 2$, $y = 1$ for $\Delta x = \Delta y = 0.01$, using the formula $\Delta z \approx dz = 2 \cdot 2 \cdot 1^5 \cdot 0.01 + 5 \cdot 2^2 \cdot 1^4 \cdot 0.01 = 0.24$. Therefore, we can accept the approximation $z = 4 \pm 0.24$.

If a function $z = f(x, y)$ is differentiable at a point $(x_0, y_0)$, then

$$f(x, y) = f(x_0, y_0) + f'_x(x_0, y_0)(x - x_0) + f'_y(x_0, y_0)(y - y_0) + o(\rho).$$

Hence, for small $\rho$ (i.e., for $x \approx x_0$, $y \approx y_0$), we obtain the approximate formula

$$f(x, y) \approx f(x_0, y_0) + f'_x(x_0, y_0)(x - x_0) + f'_y(x_0, y_0)(y - y_0).$$

The replacement of a function by this linear expression near a given point is called *linearization*.

### 6.3.2-3. Composite function.

Consider a function $z = f(x, y)$ and let $x = x(u, v)$, $y = y(u, v)$. Suppose that for $(u, v) \in D$, the functions $x(u, v)$, $y(u, v)$ take values for which the function $z = f(x, y)$ is defined. In this way, one defines a *composite function* on the set $D$, namely, $z(u, v) = f\bigl(x(u, v), y(u, v)\bigr)$. In this situation, $f(x, y)$ is called the outer function and $x(u, v)$, $y(u, v)$ are called the inner functions.

Partial derivatives of a composite function are expressed by

$$\frac{\partial z}{\partial u} = \frac{\partial f}{\partial x}\frac{\partial x}{\partial u} + \frac{\partial f}{\partial y}\frac{\partial y}{\partial u},$$

$$\frac{\partial z}{\partial v} = \frac{\partial f}{\partial x}\frac{\partial x}{\partial v} + \frac{\partial f}{\partial y}\frac{\partial y}{\partial v}.$$

For $z = z(t, x, y)$, let $x = x(t)$, $y = y(t)$. Thus, $z$ is actually a function of only one variable $t$. The derivative $\frac{dz}{dt}$ is calculated by

$$\frac{dz}{dt} = \frac{\partial z}{\partial t} + \frac{\partial z}{\partial x}\frac{dx}{dt} + \frac{\partial z}{\partial y}\frac{dy}{dt}.$$

This derivative, in contrast to the partial derivative $\frac{\partial z}{\partial t}$, is called a *total derivative*.

### 6.3.2-4. Second partial derivatives and second differentials.

The *second partial derivatives* of a function $z = f(x, y)$ are defined as the derivatives of its first partial derivatives and are denoted as follows:

$$\frac{\partial^2 z}{\partial x^2} = z_{xx} \equiv (z_x)_x, \qquad \frac{\partial^2 z}{\partial x\, \partial y} = z_{xy} \equiv (z_x)_y,$$

$$\frac{\partial^2 z}{\partial y\, \partial x} = z_{yx} \equiv (z_y)_x, \qquad \frac{\partial^2 z}{\partial y^2} = z_{yy} \equiv (z_y)_y.$$

The derivatives $z_{xy}$ and $z_{yx}$ are called *mixed derivatives*. If the mixed derivatives are continuous at some point, then they coincide at that point, $z_{xy} = z_{yx}$.

In a similar way, one defines higher-order partial derivatives.

The *second differential* of a function $z = f(x, y)$ is the expression

$$d^2 z = d(dz) = (dz)_x \Delta x + (dz)_y \Delta y = z_{xx}(\Delta x)^2 + 2 z_{xy}\Delta x \Delta y + z_{yy}(\Delta y)^2.$$

In a similar way, one defines $d^3 z$, $d^4 z$, etc.

### 6.3.2-5. Taylor's formula.

If at some point $(x, y)$ the function $z = f(x, y)$ possesses partial derivatives up to the order $n$ inclusively, then its increment $\Delta z$ at that point can be expressed by

$$\Delta z = dz + \frac{d^2 z}{2!} + \frac{d^3 z}{3!} + \cdots + \frac{d^n z}{n!} + o(\rho^n),$$

where $\rho = \sqrt{(\Delta x)^2 + (\Delta y)^2}$.

### 6.3.2-6. Implicit functions and their differentiation.

Consider the equation $F(x, y) = 0$ with a solution $(x_0, y_0)$. Suppose that the derivative $F_y(x, y)$ is continuous in a neighborhood of the point $(x_0, y_0)$ and $F_y(x, y) \neq 0$ in that neighborhood. Then the equation $F(x, y) = 0$ defines a continuous function $y = y(x)$ (called an *implicit function*) of the variable $x$ in a neighborhood of the point $x_0$. Moreover, if in a neighborhood of $(x_0, y_0)$ there exists a continuous derivative $F_x$, then the implicit function $y = y(x)$ has a continuous derivative expressed by $\dfrac{dy}{dx} = -\dfrac{F_x}{F_y}$.

Consider the equation $F(x, y, z) = 0$ that establishes a relation between the variables $x$, $y$, $z$. If $F(x_0, y_0, z_0) = 0$ and in a neighborhood of the point $(x_0, y_0, z_0)$ there exist continuous partial derivatives $F_x$, $F_y$, $F_z$ such that $F_z(x_0, y_0, z_0) \neq 0$, then equation $F(x, y, z) = 0$, in a neighborhood of $(x_0, y_0)$, has a unique solution $z = \varphi(x, y)$ such that $\varphi(x_0, y_0) = z_0$; moreover, the function $z = \varphi(x, y)$ is continuous and has continuous partial derivatives expressed by

$$\frac{\partial z}{\partial x} = -\frac{F_x}{F_z}, \qquad \frac{\partial z}{\partial y} = -\frac{F_y}{F_z}.$$

**Example 2.** For the equation $x \sin y + z + e^z = 0$ we have $F_z = 1 + e^z \neq 0$. Therefore, this equation defines a function $z = \varphi(x, y)$ on the entire plane, and its derivatives have the form $\dfrac{\partial z}{\partial x} = -\dfrac{\sin y}{1 + e^z}$, $\dfrac{\partial z}{\partial y} = -\dfrac{x \cos y}{1 + e^z}$.

### 6.3.2-7. Jacobian. Dependent and independent functions. Invertible transformations.

1°. Two functions $f(x, y)$ and $g(x, y)$ are called *dependent* if there is a function $\Phi(z)$ such that $g(x, y) = \Phi(f(x, y))$; otherwise, the functions $f(x, y)$ and $g(x, y)$ are called *independent*.

The *Jacobian* is the determinant of the matrix whose elements are the first partial derivatives of the functions $f(x, y)$ and $g(x, y)$:

$$\frac{\partial(f, g)}{\partial(x, y)} \equiv \begin{vmatrix} \frac{\partial f}{\partial x} & \frac{\partial f}{\partial y} \\ \frac{\partial g}{\partial x} & \frac{\partial g}{\partial y} \end{vmatrix}. \tag{6.3.2.1}$$

1) If the Jacobian (6.3.2.1) in a domain $D$ is identically equal to zero, then the functions $f(x, y)$ and $g(x, y)$ are dependent in $D$.

2) If the Jacobian (6.3.2.1) is separated from zero in $D$, then the functions $f(x, y)$ and $g(x, y)$ are independent in $D$.

2°. Functions $f_k(x_1, x_2, \ldots, x_n)$, $k = 1, 2, \ldots, n$, are called dependent in a domain $D$ if there is a function $\Phi(z_1, z_2, \ldots, z_n)$ such that

$$\Phi\bigl(f_1(x_1, x_2, \ldots, x_n), f_2(x_1, x_2, \ldots, x_n), \ldots, f_n(x_1, x_2, \ldots, x_n)\bigr) = 0 \qquad \text{(in } D\text{)};$$

otherwise, these functions are called independent.

The *Jacobian* is the determinant of the matrix whose elements are the first partial derivatives:

$$\frac{\partial(f_1, f_2, \ldots, f_n)}{\partial(x_1, x_2, \ldots, x_n)} \equiv \det\left(\frac{\partial f_i}{\partial x_j}\right). \tag{6.3.2.2}$$

The functions $f_k(x_1, x_2, \ldots, x_n)$ are dependent in a domain $D$ if the Jacobian (6.3.2.2) is identically equal to zero in $D$. The functions $f_k(x_1, x_2, \ldots, x_n)$ are independent in $D$ if the Jacobian (6.3.2.2) does not vanish in $D$.

3°. Consider the transformation

$$y_k = f_k(x_1, x_2, \ldots, x_n), \qquad k = 1, 2, \ldots, n. \tag{6.3.2.3}$$

Suppose that the functions $f_k$ are continuously differentiable and the Jacobian (6.3.2.2) differs from zero at the point $(x_1^\circ, x_2^\circ, \ldots, x_n^\circ)$. Then, in a sufficiently small neighborhood of this point, equations (6.3.2.3) specify a one-to-one correspondence between the points of that neighborhood and the set of points $(y_1, y_2, \ldots, y_n)$ consisting of the values of the functions (6.3.2.3) in the corresponding neighborhood of the point $(y_1^\circ, y_2^\circ, \ldots, y_n^\circ)$. This means that the system (6.3.2.3) is locally solvable in a neighborhood of the point $(x_1^\circ, x_2^\circ, \ldots, x_n^\circ)$, i.e., the following representation holds:

$$x_k = g_k(y_1, y_2, \ldots, y_n), \qquad k = 1, 2, \ldots, n,$$

where $g_k$ are continuously differentiable functions in the corresponding neighborhood of the point $(y_1^\circ, y_2^\circ, \ldots, y_n^\circ)$.

## 6.3.3. Directional Derivative. Gradient. Geometrical Applications

### 6.3.3-1. Directional derivative.

One says that a *scalar field* is defined in a domain $D$ if any point $M(x, y)$ of that domain is associated with a certain value $z = f(M) = f(x, y)$. Thus, a thermal field and a pressure field are examples of scalar fields. A *level line* of a scalar field is a level line of the function that specifies the field (see Subsection 6.3.1). Thus, isothermal and isobaric curves are, respectively, level lines of thermal and pressure fields.

In order to examine the behavior of a field $z = f(x,y)$ at a point $M_0(x_0, y_0)$ in the direction of a vector $\vec{a} = \{a_1, a_2\}$, one should construct a straight line passing through $M_0$ in the direction of the vector $\vec{a}$ (this line can be specified in terms of the parametric equations $x = x_0 + a_1 t$, $y = y_0 + a_2 t$) and study the function $z(t) = f(x_0 + a_1 t, y_0 + a_2 t)$. The derivative of the function $z(t)$ at the point $M_0$ (i.e., for $t = 0$) characterizes the change rate of the field at that point in the direction $\vec{a}$. Dividing $z'(0)$ by $|\vec{a}| = \sqrt{a_1^2 + a_2^2}$, we obtain the so-called *derivative in the direction* $\vec{a}$ of the given field at the given point:

$$\frac{\partial f}{\partial \vec{a}} = \frac{1}{|\vec{a}|}\left[ a_1 f'_x(x_0, y_0) + a_2 f'_y(x_0, y_0) \right].$$

The *gradient* of the scalar field $z = f(x,y)$ is, by definition, the vector-valued function

$$\operatorname{grad} f = f'_x(x,y)\vec{i} + f'_y(x,y)\vec{j},$$

where $\vec{i}$ and $\vec{j}$ are unit vectors along the coordinate axes $x$ and $y$. At each point, the gradient of a scalar field is orthogonal to the level line passing through that point. The gradient indicates the direction of maximal growth of the field. In terms of the gradient, the directional derivative can be expressed as follows:

$$\frac{\partial f}{\partial \vec{a}} = \frac{\vec{a}}{|\vec{a}|} \operatorname{grad} f.$$

The gradient is also denoted by $\nabla f = \operatorname{grad} f$.

**Remark.** The above facts for a plane scalar field obviously can be extended to the case of a spatial scalar field.

---

**6.3.3-2. Geometrical applications of the theory of functions of several variables.**

The *equation of the tangent plane* to the surface $z = f(x,y)$ at a point $(x_0, y_0, z_0)$, where $z_0 = f(x_0, y_0)$, has the form

$$z = f(x_0, y_0) + f_x(x_0, y_0)(x - x_0) + f_y(x_0, y_0)(y - y_0).$$

The vector of the normal to the surface at that point is

$$\vec{n} = \{-f_x(x_0, y_0), -f_y(x_0, y_0), 1\}.$$

If a surface is defined implicitly by the equation $\Phi(x, y, z) = 0$, then the equation of its tangent plane at the point $(x_0, y_0, z_0)$ has the form

$$\Phi_x(x_0, y_0, z_0)(x - x_0) + \Phi_y(x_0, y_0, z_0)(y - y_0) + \Phi_z(x_0, y_0, z_0)(z - z_0) = 0.$$

The vector of the normal to the surface at that point is

$$\vec{n} = \{\Phi_x(x_0, y_0, z_0), \Phi_y(x_0, y_0, z_0), \Phi_z(x_0, y_0, z_0)\}.$$

Consider a surface defined by the parametric equations

$$x = x(u, v), \quad y = y(u, v), \quad z = z(u, v)$$

or, in vector form, $\vec{r} = \vec{r}(u, v)$, where $\vec{r} = \{x, y, z\}$, and let $M_0\big(x(u_0, v_0), y(u_0, v_0), z(u_0, v_0)\big)$ be the point of the surface corresponding to the parameter values $u = u_0$, $v = v_0$. Then the vector of the normal to the surface at the point $M_0$ can be expressed by

$$\vec{n}(u, v) = \frac{\partial \vec{r}}{\partial u} \times \frac{\partial \vec{r}}{\partial v} = \begin{vmatrix} \vec{i} & \vec{j} & \vec{k} \\ x_u & y_u & z_u \\ x_v & y_v & z_v \end{vmatrix},$$

where all partial derivatives are calculated at the point $M_0$.

## 6.3.4. Extremal Points of Functions of Several Variables

**6.3.4-1. Conditions of extremum of a function of two variables.**

$1°$. *Points of minimum, maximum, or extremum.* A point $(x_0, y_0)$ is called a *point of local minimum* (resp., *maximum*) of a function $z = f(x, y)$ if there is a neighborhood of $(x_0, y_0)$ in which the function is defined and satisfies the inequality $f(x, y) > f(x_0, y_0)$ (resp., $f(x, y) < f(x_0, y_0)$). Points of maximum or minimum are called *points of extremum*.

$2°$. *A necessary condition of extremum.* If a function has the first partial derivatives at a point of its extremum, these derivatives must vanish at that point. It follows that in order to find points of extremum of such a function $z = f(x, y)$, one should find solutions of the system of equations

$$f_x(x, y) = 0, \quad f_y(x, y) = 0.$$

The points whose coordinates satisfy this system are called *stationary points*. Any point of extremum of a differentiable function is its stationary point, but not every stationary point is a point of its extremum.

$3°$. *Sufficient conditions of extremum* are used for the identification of points of extremum among stationary points. Some conditions of this type are given below.

Suppose that the function $z = f(x, y)$ has continuous second derivatives at a stationary point. Let us calculate the value of the determinant at that point:

$$\Delta = f_{xx} f_{yy} - f_{xy}^2.$$

The following implications hold:

1) If $\Delta > 0$, $f_{xx} > 0$, then the stationary point is a point of local minimum;
2) If $\Delta > 0$, $f_{xx} < 0$, then the stationary point is a point of local maximum;
3) If $\Delta < 0$, then the stationary point cannot be a point of extremum.

In the degenerate case, $\Delta = 0$, a more delicate analysis of a stationary point is required. In this case, a stationary point may happen to be a point of extremum and maybe not.

*Remark.* In order to find points of extremum, one should check not only stationary points, but also points at which the first derivatives do not exist or are infinite.

$4°$. *The smallest and the largest values of a function.* Let $f(x, y)$ be a continuous function in a closed bounded domain $D$. Any such function takes its smallest and its largest values in $D$.

If the function has partial derivatives in $D$, except at some points, then the following method can be helpful for determining the coordinates of the points $(x_{\min}, y_{\min})$ and $(x_{\max}, y_{\max})$ at which the function attains its minimum and maximum, respectively. One should find all internal stationary points and all points at which the derivatives are infinite or do not exist. Then one should calculate the values of the function at these points and compare these with its values at the boundary points of the domain, and then choose the largest and the smallest values.

**6.3.4-2. Extremal points of functions of three variables.**

For functions of three variables, points of extremum are defined in exactly the same way as for functions of two variables. Let us briefly describe the scheme of finding extremal points of a function $u = \Phi(x, y, z)$. Finding solutions of the system of equations

$$\Phi_x(x, y, z) = 0, \quad \Phi_y(x, y, z) = 0, \quad \Phi_z(x, y, z) = 0,$$

we determine stationary points. For each stationary point, we calculate the values of

$$\Delta_1 = \Phi_{xx}, \quad \Delta_2 = \begin{vmatrix} \Phi_{xx} & \Phi_{xy} \\ \Phi_{xy} & \Phi_{yy} \end{vmatrix}, \quad \Delta_3 = \begin{vmatrix} \Phi_{xx} & \Phi_{xy} & \Phi_{xz} \\ \Phi_{xy} & \Phi_{yy} & \Phi_{yz} \\ \Phi_{xz} & \Phi_{yz} & \Phi_{zz} \end{vmatrix}.$$

The following implications hold:

1) If $\Delta_1 > 0$, $\Delta_2 > 0$, $\Delta_3 > 0$, then the stationary point is a point of local minimum;
2) If $\Delta_1 < 0$, $\Delta_2 > 0$, $\Delta_3 < 0$, then the stationary point is a point of local maximum.

### 6.3.4-3. Conditional extremum of a function of two variables. Lagrange function.

A point $(x_0, y_0)$ is called a *point of conditional or constrained minimum* (resp., *maximum*) of a function
$$z = f(x, y) \tag{6.3.4.1}$$
under the additional condition*
$$\varphi(x, y) = 0 \tag{6.3.4.2}$$
if there is a neighborhood of the point $(x_0, y_0)$ in which $f(x, y) > f(x_0, y_0)$ (resp., $f(x, y) < f(x_0, y_0)$) for all points $(x, y)$ satisfying the condition (6.3.4.2).

For the determination of points of conditional extremum, it is common to use the *Lagrange function*
$$\Phi(x, y, \lambda) = f(x, y) + \lambda \varphi(x, y),$$
where $\lambda$ is the so-called *Lagrange multiplier*. Solving the system of three equations (the last equation coincides with the condition (6.3.4.2))
$$\frac{\partial \Phi}{\partial x} = 0, \quad \frac{\partial \Phi}{\partial y} = 0, \quad \frac{\partial \Phi}{\partial \lambda} = 0,$$
one finds stationary points of the Lagrange function (and also the value of the coefficient $\lambda$). The stationary points may happen to be points of extremum. The above system yields only necessary conditions of extremum, but these conditions may be insufficient; it may happen that there is no extremum at some stationary points. However, with the help of other properties of the function under consideration, it is often possible to establish the character of a critical point.

**Example 1.** Let us find an extremum of the function
$$z = x^n y, \tag{6.3.4.3}$$
under the condition
$$x + y = a \quad (a > 0, \; n > 0, \; x \geq 0, \; y \geq 0). \tag{6.3.4.4}$$
Taking $F(x, y) = x^n y$ and $\varphi(x, y) = x + y - a$, we construct the Lagrange function
$$\Phi(x, y, \lambda) = x^n y + \lambda(x + y - a).$$
Solving the system of equations
$$\Phi_x \equiv nx^{n-1} y + \lambda = 0,$$
$$\Phi_y \equiv x^n + \lambda = 0,$$
$$\Phi_\lambda \equiv x + y - a = 0,$$
we find the coordinates of a unique stationary point,
$$x_\circ = \frac{an}{n+1}, \quad y_\circ = \frac{a}{n+1}, \quad \lambda_\circ = -\left(\frac{an}{n+1}\right)^n,$$
which corresponds to the maximum of the given function, $z_{\max} = \dfrac{a^{n+1} n^n}{(n+1)^{n+1}}.$

---

\* This condition is also called a *constraint*.

**Remark.** In order to find points of conditional extremum of functions of two variables, it is often convenient to express the variable $y$ through $x$ (or vice versa) from the additional equation (6.3.4.2) and substitute the resulting expression into the right-hand side of (6.3.4.1). In this way, the original problem is reduced to the problem of extremum for a function of a single variable.

**Example 2.** Consider again the extremum problem of Example 1 for the function of two variables (6.3.4.3) with the constraint (6.3.4.4). After the elimination of the variable $y$ from (6.3.4.3)–(6.3.4.4), the original problem is reduced to the extremum problem for the function $z = x^n(a-x)$ of one variable.

### 6.3.4-4. Conditional extremum of functions of several variables.

Consider a function $u = f(x_1, \ldots, x_n)$ of $n$ variables under the condition that $x_1, \ldots, x_n$ satisfy $m$ equations ($m < n$):

$$\begin{cases} \varphi_1(x_1, \ldots, x_n) = 0, \\ \varphi_2(x_1, \ldots, x_n) = 0, \\ \cdots\cdots\cdots\cdots\cdots\cdots, \\ \varphi_m(x_1, \ldots, x_n) = 0. \end{cases}$$

In order to find the values of $x_1, \ldots, x_n$ for which $f$ may have a conditional maximum or minimum, one should construct the Lagrange function

$$\Phi(x_1, \ldots, x_n; \lambda_1, \ldots, \lambda_m) = f + \lambda_1 \varphi_1 + \lambda_2 \varphi_2 + \cdots + \lambda_m \varphi_m$$

and equate to zero its first partial derivatives in the variables $x_1, \ldots, x_n$ and the parameters $\lambda_1, \ldots, \lambda_m$. From the resulting $n + m$ equations, one finds $x_1, \ldots, x_n$ (and also the values of the unknown Lagrange multipliers $\lambda_1, \ldots, \lambda_m$). As in the case of functions of two variables, the question whether the given function has points of conditional extremum can be answered on the basis of additional investigation.

**Example 3.** Consider the problem of finding the shortest distance from the point $(x_0, y_0, z_0)$ to the plane

$$Ax + By + Cz + D = 0. \tag{6.3.4.5}$$

The squared distance between the points $(x_0, y_0, z_0)$ and $(x, y, z)$ is equal to

$$R^2 = (x - x_0)^2 + (y - y_0)^2 + (z - z_0)^2. \tag{6.3.4.6}$$

In our case, the coordinates $(x, y, z)$ should satisfy equation (6.3.4.5) (this point should belong to the plane). Thus, our problem is to find the minimum of the expression (6.3.4.6) under the condition (6.3.4.5). The Lagrange function has the form

$$\Phi = (x - x_0)^2 + (y - y_0)^2 + (z - z_0)^2 + \lambda(Ax + By + Cz + D).$$

Equating to zero the derivatives of $\Phi$ in $x, y, z, \lambda$, we obtain the following system of algebraic equations:

$$2(x - x_0) + A\lambda = 0, \quad 2(y - y_0) + B\lambda = 0, \quad 2(z - z_0) + C\lambda = 0, \quad Ax + By + Cz + D = 0.$$

Its solution has the form

$$x = x_0 - \frac{1}{2}A\lambda, \quad y = y_0 - \frac{1}{2}B\lambda, \quad z = z_0 - \frac{1}{2}C\lambda, \quad \lambda = \frac{2(Ax_0 + By_0 + Cz_0 + D)}{A^2 + B^2 + C^2}. \tag{6.3.4.7}$$

Thus we have a unique answer, and since the distance between a given point and the plane can be realized at a single point $(x, y, z)$, the values obtained should correspond to that distance. Substituting the values (6.3.4.7) into (6.3.4.6), we find the squared distance

$$R^2 = \frac{(Ax_0 + By_0 + Cz_0 + D)^2}{A^2 + B^2 + C^2}.$$

## 6.3.5. Differential Operators of the Field Theory

**6.3.5-1. Hamilton's operator and first-order differential operators.**

The Hamilton's operator or the *nabla vector* is the symbolic vector

$$\nabla = \vec{i}\,\frac{\partial}{\partial x} + \vec{j}\,\frac{\partial}{\partial y} + \vec{k}\,\frac{\partial}{\partial z}.$$

This vector can be used for expressing the following differential operators:

1) gradient of a scalar function $u(x, y, z)$:

$$\operatorname{grad} u = \vec{i}\,\frac{\partial u}{\partial x} + \vec{j}\,\frac{\partial u}{\partial y} + \vec{k}\,\frac{\partial u}{\partial z} = \nabla u;$$

2) divergence of a vector field $\vec{a} = P\vec{i} + Q\vec{j} + R\vec{k}$:

$$\operatorname{div} \vec{a} = \frac{\partial P}{\partial x} + \frac{\partial Q}{\partial y} + \frac{\partial R}{\partial z} = \nabla \cdot \vec{a}$$

(scalar product of the nabla vector and the vector $\vec{a}$);

3) rotation of a vector field $\vec{a} = P\vec{i} + Q\vec{j} + R\vec{k}$:

$$\operatorname{curl} \vec{a} = \begin{vmatrix} \vec{i} & \vec{j} & \vec{k} \\ \frac{\partial}{\partial x} & \frac{\partial}{\partial y} & \frac{\partial}{\partial z} \\ P & Q & R \end{vmatrix} = \nabla \times \vec{a}$$

(vector product of the nabla vector and the vector $\vec{a}$).

Each scalar field $u(x, y, z)$ generates a vector field $\operatorname{grad} u$. A vector field $\vec{a}(x, y, z)$ generates two fields: the scalar field $\operatorname{div} \vec{a}$ and the vector field $\operatorname{curl} \vec{a}$.

**6.3.5-2. Second-order differential operators.**

The following differential identities hold:

1) $\operatorname{curl} \operatorname{grad} u = \vec{0}$ or $(\nabla \times \nabla)u = \vec{0}$,
2) $\operatorname{div} \operatorname{curl} \vec{a} = 0$ or $\nabla \cdot (\nabla \times \vec{a}) = 0$.

The following differential relations hold:

1) $\operatorname{div} \operatorname{grad} u = \Delta u = \dfrac{\partial^2 u}{\partial x^2} + \dfrac{\partial^2 u}{\partial y^2} + \dfrac{\partial^2 u}{\partial z^2}$,

2) $\operatorname{curl} \operatorname{curl} \vec{a} = \operatorname{grad} \operatorname{div} \vec{a} - \Delta \vec{a}$,

where $\Delta$ is the *Laplace operator*, $\Delta u = \nabla \cdot (\nabla u) = \nabla^2 u$.

# References for Chapter 6

Adams, R., *Calculus: A Complete Course, 6th Edition*, Pearson Education, Toronto, 2006.
Boyer, C. B., *The History of the Calculus and Its Conceptual Development*, Dover Publications, New York, 1989.
Brannan, D., *A First Course in Mathematical Analysis*, Cambridge University Press, Cambridge, 2006.
Browder, A., *Mathematical Analysis: An Introduction*, Springer-Verlag, New York, 1996.
Courant, R. and John, F., *Introduction to Calculus and Analysis, Vol. 1*, Springer-Verlag, New York, 1999.
Edwards, C. H., and Penney, D., *Calculus, 6th Edition*, Pearson Education, Toronto, 2002.
Kline, M., *Calculus: An Intuitive and Physical Approach, 2nd Edition*, Dover Publications, New York, 1998.
Landau, E., *Differential and Integral Calculus*, American Mathematical Society, Providence, Rhode Island, 2001.
Silverman, R. A., *Essential Calculus with Applications*, Dover Publications, New York, 1989.
Zorich, V. A., *Mathematical Analysis*, Springer-Verlag, Berlin, 2004.

# Chapter 7
# Integrals

## 7.1. Indefinite Integral
### 7.1.1. Antiderivative. Indefinite Integral and Its Properties

**7.1.1-1. Antiderivative.**

An *antiderivative* (or *primitive function*) of a given function $f(x)$ on an interval $(a, b)$ is a differentiable function $F(x)$ such that its derivative is equal to $f(x)$ for all $x \in (a, b)$:
$$F'(x) = f(x).$$

**Example 1.** Let $f(x) = 2x$. Then the functions $F(x) = x^2$ and $F_1(x) = x^2 - 1$ are antiderivatives of $f(x)$, since $(x^2)' = 2x$ and $(x^2 - 1)' = 2x$.

THEOREM. *Any function $f(x)$ continuous on an interval $(a, b)$ has infinitely many continuous antiderivatives on $(a, b)$. If $F(x)$ is one of them, then any other antiderivative has the form $F(x) + C$, where $C$ is a constant.*

**7.1.1-2. Indefinite integral.**

The *indefinite integral* of a function $f(x)$ is the set, $F(x) + C$, of all its antiderivatives. This fact is conventionally written as
$$\int f(x)\,dx = F(x) + C.$$

Here, $f(x)$ is called the *integrand* (or the *integrand function*). The process of finding an integral is called integration. The differential $dx$ indicates that the integration is carried out with respect to $x$.

**Example 2.** $\int 6x^2\,dx = 2x^3 + C$, since $(2x^3)' = 6x^2$.

**7.1.1-3. Most important corollaries of the definition of the indefinite integral.**

*Differentiation is the inverse of integration*:
$$\frac{d}{dx}\left(\int f(x)\,dx\right) = f(x).$$

*Integration is the inverse of differentiation*:*
$$\int f'(x)\,dx = f(x) + C.$$

The latter formula serves to make up tables of indefinite integrals. The procedure is often reverse here: an integral is first given in explicit form (i.e., the function $f(x)$ on the right-hand side is prescribed), and then the integrand is obtained by differentiation.

---
\* Integration recovers the function from its derivative, to an additive constant.

## 7.1.2. Table of Basic Integrals. Properties of the Indefinite Integral. Integration Examples

**7.1.2-1. Table of basic integrals.**

Listed below are most common indefinite integrals, which are important for the integration of more complicated expressions:

$$\int x^a \, dx = \frac{x^{a+1}}{a+1} + C \quad (a \neq -1),$$

$$\int \frac{dx}{x} = \ln|x| + C,$$

$$\int \frac{dx}{x^2 + a^2} = \frac{1}{a} \arctan \frac{x}{a} + C,$$

$$\int \frac{dx}{x^2 - a^2} = \frac{1}{2a} \ln \left| \frac{x-a}{x+a} \right| + C,$$

$$\int \frac{dx}{\sqrt{a^2 - x^2}} = \arcsin \frac{x}{a} + C,$$

$$\int \frac{dx}{\sqrt{x^2 + a}} = \ln \left| x + \sqrt{x^2 + a} \right| + C,$$

$$\int e^x \, dx = e^x + C,$$

$$\int a^x \, dx = \frac{a^x}{\ln a} + C,$$

$$\int \ln x \, dx = x \ln x - x + C,$$

$$\int \ln ax \, dx = x \ln ax - x + C,$$

$$\int \sin x \, dx = -\cos x + C,$$

$$\int \cos x \, dx = \sin x + C,$$

$$\int \tan x \, dx = -\ln|\cos x| + C,$$

$$\int \cot x \, dx = \ln|\sin x| + C,$$

$$\int \frac{dx}{\sin x} = \ln \left| \tan \frac{x}{2} \right| + C,$$

$$\int \frac{dx}{\cos x} = \ln \left| \tan \left( \frac{x}{2} + \frac{\pi}{4} \right) \right| + C,$$

$$\int \frac{dx}{\sin^2 x} = -\cot x + C,$$

$$\int \frac{dx}{\cos^2 x} = \tan x + C,$$

$$\int \arcsin x \, dx = x \arcsin x + \sqrt{1 - x^2} + C,$$

$$\int \arccos x \, dx = x \arccos x - \sqrt{1 - x^2} + C,$$

$$\int \arctan x \, dx = x \arctan x - \frac{1}{2} \ln(1 + x^2) + C,$$

$$\int \operatorname{arccot} x \, dx = x \operatorname{arccot} x + \frac{1}{2} \ln(1 + x^2) + C,$$

$$\int \sinh x \, dx = \cosh x + C,$$

$$\int \cosh x \, dx = \sinh x + C,$$

$$\int \tanh x \, dx = \ln \cosh x + C,$$

$$\int \coth x \, dx = \ln |\sinh x| + C,$$

$$\int \frac{dx}{\sinh x} = \ln \left| \tanh \frac{x}{2} \right| + C,$$

$$\int \frac{dx}{\cosh x} = 2 \arctan e^x + C,$$

$$\int \frac{dx}{\sinh^2 x} = -\coth x + C,$$

$$\int \frac{dx}{\cosh^2 x} = \tanh x + C,$$

$$\int \operatorname{arcsinh} x \, dx = x \operatorname{arcsinh} x - \sqrt{1 + x^2} + C,$$

$$\int \operatorname{arccosh} x \, dx = x \operatorname{arccosh} x - \sqrt{x^2 - 1} + C,$$

$$\int \operatorname{arctanh} x \, dx = x \operatorname{arctanh} x + \frac{1}{2} \ln(1 - x^2) + C,$$

$$\int \operatorname{arccoth} x \, dx = x \operatorname{arccoth} x + \frac{1}{2} \ln(x^2 - 1) + C,$$

where $C$ is an arbitrary constant.

▶ A more extensive table of indefinite integrals can be found in Section T2.1.

**7.1.2-2. Properties of the indefinite integral.**

1. *A constant factor can be taken outside the integral sign*:

$$\int a f(x) \, dx = a \int f(x) \, dx \quad (a = \text{const}).$$

2. *Integral of the sum or difference of functions* (*additivity*):
$$\int [f(x) \pm g(x)] \, dx = \int f(x) \, dx \pm \int g(x) \, dx.$$

3. *Integration by parts*:
$$\int f(x) g'(x) \, dx = f(x) g(x) - \int f'(x) g(x) \, dx.$$

4. *Repeated integration by parts* (generalization of the previous formula):
$$\int f(x) g^{(n+1)}(x) \, dx = f(x) g^{(n)}(x) - f'(x) g^{(n-1)}(x) + \cdots + (-1)^n f^{(n)}(x) g(x)$$
$$+ (-1)^{n+1} \int f^{(n+1)}(x) g(x) \, dx, \qquad n = 0, 1, \ldots$$

5. *Change of variable* (*integration by substitution*):
$$\int f(x) \, dx = \int f(\varphi(t)) \, \varphi'_t(t) \, dt, \qquad x = \varphi(t).$$

On computing the integral using the change of variable $x = \varphi(t)$, one should rewrite the resulting expression in terms of the original variable $x$ using the back substitution $t = \varphi^{-1}(x)$.

### 7.1.2-3. Examples of direct integration of elementary functions.

$1°$. With simple algebraic manipulation and the properties listed in Paragraph 7.1.2-2, the integration may often be reduced to tabulated integrals.

**Example 1.** $\int \dfrac{2x-1}{\sqrt{x}} \, dx = \int \left( 2\sqrt{x} - \dfrac{1}{\sqrt{x}} \right) dx = 2 \int x^{1/2} \, dx - \int x^{-1/2} \, dx = \dfrac{4}{3} x^{3/2} - 2 x^{1/2} + C.$

$2°$. Tabulated integrals can also be used where any function $\varphi(x)$ appears in place of $x$; for example,
$$\int e^x \, dx = e^x + C \quad \Longrightarrow \quad \int e^{\varphi(x)} \, d\varphi(x) = e^{\varphi(x)} + C;$$
$$\int \frac{dx}{x} = \ln |x| + C \quad \Longrightarrow \quad \int \frac{d \sin x}{\sin x} = \ln |\sin x| + C.$$

The reduction of an integral to a tabulated one may often be achieved by taking some function inside the differential sign.

**Example 2.** $\int \tan x \, dx = \int \dfrac{\sin x \, dx}{\cos x} = \int \dfrac{-d \cos x}{\cos x} = -\int \dfrac{d \cos x}{\cos x} = -\ln |\cos x| + C.$

$3°$. Integrals of the form $\int \dfrac{dx}{ax^2 + bx + c}, \int \dfrac{dx}{\sqrt{ax^2 + bx + c}}$ can be computed by making a perfect square:
$$ax^2 + bx + c = a \left( x + \frac{b}{2a} \right)^2 - \frac{b^2}{4a} + c.$$

Then one should replace $dx$ with the equal differential $d \left( x + \dfrac{b}{2a} \right)$ and use one of the four formulas in the second and third rows in the table of integrals given in Paragraph 7.1.2-1.

**Example 3.** $\int \dfrac{dx}{\sqrt{2x-x^2}} = \int \dfrac{dx}{\sqrt{1-(x-1)^2}} = \int \dfrac{d(x-1)}{\sqrt{1-(x-1)^2}} = \arcsin(x-1) + C.$

4°. The integration of a polynomial multiplied by an exponential function can be accomplished by using the formula of integration by parts (or repeated integration by parts) given in Paragraph 7.1.2-2.

**Example 4.** Compute the integral $\int (3x+1) e^{2x}\, dx$.

Taking $f(x) = 3x + 1$ and $g'(x) = e^{2x}$, one finds that $f'(x) = 3$ and $g(x) = \tfrac{1}{2} e^{2x}$. On substituting these expressions into the formula of integration by pars, one obtains

$$\int (3x+1) e^{2x}\, dx = \frac{1}{2}(3x+1) e^{2x} - \frac{3}{2}\int e^{2x}\, dx = \frac{1}{2}(3x+1) e^{2x} - \frac{3}{4} e^{2x} + C = \left(\frac{3}{2}x - \frac{1}{4}\right) e^{2x} + C.$$

**Remark 1.** More complex examples of the application of integration by parts or repeated integration by parts can be found in Subsection 7.1.6.

**Remark 2.** Examples of using a change of variables (see Item 5 in Paragraph 7.1.2-2) for the computation of integrals can be found in Subsections 7.1.4 and 7.1.5.

**7.1.2-4. Remark on uncomputable integrals.**

The differentiation of elementary functions is known to always result in elementary functions. However, this is not the case with integration, which is the reverse of differentiation. The integrals of elementary functions are often impossible to express in terms elementary functions using finitely many arithmetic operations and compositions.

Here are examples of integrals that cannot be expressed via elementary functions:

$$\int \frac{dx}{\sqrt{x^3+1}}, \quad \int \exp(-x^2)\, dx, \quad \int \frac{e^x}{x}\, dx, \quad \int \frac{dx}{\ln x}, \quad \int \frac{\cos x}{x}\, dx, \quad \int \sin(x^2)\, dx.$$

Such integrals are sometimes called intractable. It is significant that all these integrals exist; they generate nonelementary (special) functions.

## 7.1.3. Integration of Rational Functions

**7.1.3-1. Partial fraction decomposition of a rational function.**

A *rational function* (also know as a *rational polynomial function*) is a quotient of polynomials:

$$R(x) = \frac{P_n(x)}{Q_m(x)}, \qquad (7.1.3.1)$$

where

$$P_n(x) = a_n x^n + \cdots + a_1 x + a_0,$$
$$Q_m(x) = b_m x^m + \cdots + b_1 x + b_0.$$

The fraction (7.1.3.1) is called *proper* if $m > n$ and *improper* if $m \le n$.

Every proper fraction (7.1.3.1) can be decomposed into a sum of partial fractions. To this end, one should factorize the denominator $Q_m(x)$ into irreducible multipliers of the form

$$(x - \alpha_i)^{p_i}, \qquad i = 1, 2, \ldots, k; \qquad (7.1.3.2a)$$
$$(x^2 + \beta_j x + \gamma_j)^{q_j}, \qquad j = 1, 2, \ldots, s, \qquad (7.1.3.2b)$$

where the $p_i$ and $q_j$ are positive integers satisfying the condition $p_1+\cdots+p_k+2(q_1+\cdots+q_s) = m$; $\beta_j^2 - 4\gamma_j < 0$. The rational function (7.1.3.1) can be represented as a sum of irreducibles and to each irreducible of the form (7.1.3.2) there correspond as many terms as the power $p_i$ or $q_i$:

$$\frac{A_{i,1}}{x-\alpha_i} + \frac{A_{i,2}}{(x-\alpha_i)^2} + \cdots + \frac{A_{i,p_i}}{(x-\alpha_i)^{p_i}}; \tag{7.1.3.3a}$$

$$\frac{B_{j,1}x + D_{j,1}}{x^2+\beta_j x+\gamma_j} + \frac{B_{j,2}x + D_{j,2}}{(x^2+\beta_j x+\gamma_j)^2} + \cdots + \frac{B_{j,q_j}x + D_{j,q_j}}{(x^2+\beta_j x+\gamma_j)^{q_j}}. \tag{7.1.3.3b}$$

The constants $A_{i,l}$, $B_{j,r}$, $D_{j,r}$ are found by the *method of undetermined coefficients*. To that end, one should equate the original rational fraction (7.1.3.1) with the sum of the above partial fractions (7.1.3.3) and reduce both sides of the resulting equation to a common denominator. Then, one collects the coefficients of like powers of $x$ and equates them with zero, thus arriving at a system of linear algebraic equations for the $A_{i,l}$, $B_{j,r}$, and $D_{j,r}$.

**Example 1.** This is an illustration of how a proper fraction can be decomposed into partial fractions:

$$\frac{b_5 x^5 + b_4 x^4 + b_3 x^3 + b_2 x^2 + b_1 x + b_0}{(x+a)(x+c)^3(x^2+k^2)} = \frac{A_{1,1}}{x+a} + \frac{A_{2,1}}{x+c} + \frac{A_{2,2}}{(x+c)^2} + \frac{A_{2,3}}{(x+c)^3} + \frac{Bx+D}{x^2+k^2}.$$

### 7.1.3-2. Integration of a proper fraction.

1°. To integrate a proper fraction, one should first rewrite the integrand (7.1.3.1) in the form of a sum of partial fractions. Below are the integrals of most common partial fractions (7.1.3.3a) and (7.1.3.3b) (with $q_j = 1$):

$$\int \frac{A}{x-\alpha}\,dx = A\ln|x-\alpha|, \qquad \int \frac{A}{(x-\alpha)^p}\,dx = -\frac{A}{(p-1)(x-\alpha)^{p-1}},$$
$$\int \frac{Bx+D}{x^2+\beta x+\gamma}\,dx = \frac{B}{2}\ln(x^2+\beta x+\gamma) + \frac{2D-B\beta}{\sqrt{4\gamma-\beta^2}}\arctan\frac{2x+\beta}{\sqrt{4\gamma-\beta^2}}. \tag{7.1.3.4}$$

The constant of integration $C$ has been omitted here. More complex integrals of partial fractions (7.1.3.3b) with $q_j > 1$ can be computed using the formula

$$\int \frac{Bx+D}{(x^2+\beta x+\gamma)^q}\,dx = \frac{P(x)}{(x^2+\beta x+\gamma)^{q-1}} + \lambda\int \frac{dx}{x^2+\beta x+\gamma}, \tag{7.1.3.5}$$

where $P(x)$ is a polynomial of degree $2q-3$. The coefficients of $P(x)$ and the constant $\lambda$ can be found by the method of undetermined coefficients by differentiating formula (7.1.3.5).

**Remark.** The following recurrence relation may be used in order to compute the integrals on the left-hand side in (7.1.3.5):

$$\int \frac{Bx+D}{(x^2+\beta x+\gamma)^q}\,dx = \frac{(2D-B\beta)x + D\beta - 2B\gamma}{(q-1)(4\gamma-\beta^2)(x^2+\beta x+\gamma)^{q-1}} + \frac{(2q-3)(2D-B\beta)}{(q-1)(4\gamma-\beta^2)}\int \frac{dx}{(x^2+\beta x+\gamma)^{q-1}}.$$

**Example 2.** Compute the integral $\displaystyle\int \frac{3x^2-x-2}{x^3+8}\,dx$.

Let us factor the denominator of the integrand, $x^3+8 = (x+2)(x^2-2x+4)$, and perform the partial fraction decomposition:

$$\frac{3x^2-x-2}{(x+2)(x^2-2x+4)} = \frac{A}{x+2} + \frac{Bx+D}{x^2-2x+4}.$$

Multiplying both sides by the common denominator and collecting the coefficients of like powers of $x$, we obtain

$$(A + B - 3)x^2 + (-2A + 2B + D + 1)x + 4A + 2D + 2 = 0.$$

Now equating the coefficients of the different powers of $x$ with zero, we arrive at a system of algebraic equations for $A$, $B$, and $D$:

$$A + B - 3 = 0, \quad -2A + 2B + D + 1 = 0, \quad 4A + 2D + 2 = 0.$$

Its solution is: $A = 1$, $B = 2$, $D = -3$. Hence, we have

$$\int \frac{3x^2 - x - 2}{x^3 + 8} dx = \int \frac{1}{x+2} dx + \int \frac{2x-3}{x^2 - 2x + 4} dx$$

$$= \ln|x+2| + \ln(x^2 - 2x + 4) - \frac{1}{\sqrt{3}} \arctan \frac{x-1}{\sqrt{3}} + C.$$

Here, the last integral of (7.1.3.4) has been used.

$2°$. The integrals of proper rational functions defined as the ratio of a polynomial to a power function $(x - \alpha)^m$ are given by the formulas

$$\int \frac{P_n(x)}{(x-\alpha)^m} dx = -\sum_{k=0}^{n} \frac{P_n^{(k)}(\alpha)}{k!(m-k-1)(x-\alpha)^{m-k-1}} + C, \quad m > n+1;$$

$$\int \frac{P_n(x)}{(x-\alpha)^{n+1}} dx = -\sum_{k=0}^{n-1} \frac{P_n^{(k)}(\alpha)}{k!(n-k)(x-\alpha)^{n-k}} + \frac{P_n^{(n)}(\alpha)}{n!} \ln|x-\alpha| + C,$$

where $P_n(x)$ is a polynomial of degree $n$ and $P_n^{(k)}(\alpha)$ is its $k$th derivative at $x = \alpha$.

$3°$. Suppose the roots in the factorization of the denominator of the fraction (7.1.3.1) are all real and distinct:

$$Q_m(x) = b_m x^m + \cdots + b_1 x + b_0 = b_m(x - \alpha_1)(x - \alpha_2)\ldots(x - \alpha_m), \quad \alpha_i \neq \alpha_j.$$

Then the following formula holds:

$$\int \frac{P_n(x)}{Q_m(x)} dx = \sum_{k=1}^{m} \frac{P_n(\alpha_k)}{Q'_m(\alpha_k)} \ln|x - \alpha_k| + C,$$

where $m > n$ and the prime denotes a derivative.

### 7.1.3-3. Integration of improper fractions.

$1°$. In order to integrate an improper fraction, one should first isolate a proper fraction by division with remainder. As a result, the improper fraction is represented as the sum of a polynomial and a proper fraction,

$$\frac{a_n x^n + \cdots + a_1 x + a_0}{b_m x^m + \cdots + b_1 x + b_0} = c_m x^{n-m} + \cdots + c_1 x + c_0 + \frac{s_{m-1} x^{m-1} + \cdots + s_1 x + s_0}{b_m x^m + \cdots + b_1 x + b_0} \quad (n \geq m),$$

which are then integrated separately.

**Example 3.** Evaluate the integral $I = \int \dfrac{x^2}{x-1}\,dx$.

Let us rewrite the integrand (improper fraction) as the sum of a polynomial and a proper fraction: $\dfrac{x^2}{x-1} = x+1+\dfrac{1}{x-1}$. Hence, $I = \int \left(x+1+\dfrac{1}{x-1}\right)dx = \tfrac{1}{2}x^2 + x + \ln|x-1| + C$.

2°. The integrals of improper rational functions defined as the ratio of a polynomial to a simple power function $(x-\alpha)^m$ are evaluated by the formula

$$\int \frac{P_n(x)}{(x-\alpha)^m}\,dx = \sum_{k=m}^{n} \frac{P_n^{(k)}(\alpha)}{k!\,(k-m+1)}(x-\alpha)^{k-m+1} + \frac{P_n^{(m-1)}(\alpha)}{(m-1)!}\ln|x-\alpha|$$
$$- \sum_{k=0}^{m-2} \frac{P_n^{(k)}(\alpha)}{k!\,(m-k-1)(x-\alpha)^{m-k-1}} + C,$$

where $n \geq m$.

**Remark 1.** The indefinite integrals of rational functions are always expressed in terms of elementary functions.

**Remark 2.** Some of the integrals reducible to integrals of rational functions are considered in Subsections 7.1.5 and 7.1.6.

### 7.1.4. Integration of Irrational Functions

The integration of some irrational functions can be reduced to that of rational functions using a suitable change of variables. In what follows, the functions $R(x,y)$ and $R(x_1,\ldots,x_k)$ are assumed to be rational functions in each of the arguments.

**7.1.4-1. Integration of expressions involving radicals of linear-fractional functions.**

1°. The integrals with roots of linear functions

$$\int R\!\left(x, \sqrt[n]{ax+b}\right) dx$$

are reduced to integrals of rational functions by the change of variable $z = \sqrt[n]{ax+b}$.

**Example 1.** Evaluate the integral $I = \int x\sqrt{1-x}\,dx$.

With the change of variable $\sqrt{1-x} = z$, we have $x = 1-z^2$ and $dx = -2z\,dz$. Substituting these expressions into the integral yields

$$I = -2\int (1-z^2)z^2\,dz = -\tfrac{2}{3}z^3 + \tfrac{2}{5}z^5 + C = -\tfrac{2}{3}\sqrt{(1-x)^3} + \tfrac{2}{5}\sqrt{(1-x)^5} + C.$$

2°. The integrals with roots of linear-fractional functions

$$\int R\!\left(x, \sqrt[n]{\frac{ax+b}{cx+d}}\right) dx$$

are reduced to integrals of rational functions by the substitution $z = \sqrt[n]{\dfrac{ax+b}{cx+d}}$.

3°. The integrals of the more general form

$$\int R\left(x, \left(\frac{ax+b}{cx+d}\right)^{q_1}, \ldots, \left(\frac{ax+b}{cx+d}\right)^{q_k}\right) dx,$$

where $q_1, \ldots, q_k$ are rational numbers, are reduced to integrals of rational functions using the change of variable $z^m = \dfrac{ax+b}{cx+d}$, where $m$ is the common denominator of the fractions $q_1, \ldots, q_k$.

4°. Integrals containing the product of a polynomial by a simple power function of the form $(x-a)^\beta$ are evaluated by the formula

$$\int P_n(x)(x-a)^\beta \, dx = \sum_{k=0}^{n} \frac{P_n^{(k)}(a)}{k!\,(k+\beta+1)}(x-a)^{k+\beta+1},$$

where $P_n(x)$ is a polynomial of degree $n$, $P_n^{(k)}(a)$ is its $k$th derivative at $x = a$, and $\beta$ is any positive or negative proper fraction (to be more precise, $\beta \neq -1, -2, \ldots, -n-1$).

---

**7.1.4-2. Euler substitutions. Trigonometric substitutions.**

We will be considering integrals involving the radical of a quadratic trinomial:

$$\int R\left(x, \sqrt{ax^2+bx+c}\right) dx,$$

where $b^2 \neq 4ac$. Such integrals are expressible in terms of elementary functions.

1°. *Euler substitutions.* The given integral is reduced to the integral of a rational fraction by one of the following three Euler substitutions:

1) $\sqrt{ax^2+bx+c} = t \mp x\sqrt{a}$ if $a > 0$;
2) $\sqrt{ax^2+bx+c} = xt \pm \sqrt{c}$ if $c > 0$;
3) $\sqrt{ax^2+bx+c} = t(x - x_1)$ if $4ac - b^2 < 0$,

where $x_1$ is a root of the quadratic equation $ax^2+bx+c = 0$. In all three cases, the variable $x$ and the radical $\sqrt{ax^2+bx+c}$ are expressible in terms of the new variable $t$ as (the formulas correspond to the upper signs in the substitutions):

1) $x = \dfrac{t^2 - c}{2\sqrt{a}\,t + b}$, $\sqrt{ax^2+bx+c} = \dfrac{\sqrt{a}\,t^2 + bt + c\sqrt{a}}{2\sqrt{a}\,t + b}$, $dx = 2\dfrac{\sqrt{a}\,t^2 + bt + c\sqrt{a}}{(2\sqrt{a}\,t + b)^2}\,dt$;

2) $x = \dfrac{2\sqrt{c}\,t - b}{a - t^2}$, $\sqrt{ax^2+bx+c} = \dfrac{\sqrt{c}\,t^2 - bt + c\sqrt{a}}{a - t^2}$, $dx = 2\dfrac{\sqrt{c}\,t^2 - bt + c\sqrt{a}}{(a - t^2)^2}\,dt$;

3) $x = \dfrac{(t^2 + a)x_1 + b}{t^2 - a}$, $\sqrt{ax^2+bx+c} = \dfrac{(2ax_1 + b)t}{t^2 - a}$, $dx = -2\dfrac{(2ax_1 + b)t}{(t^2 - a)^2}\,dt$.

2°. *Trigonometric substitutions.* The function $\sqrt{ax^2+bx+c}$ can be reduced, by making a perfect square in the radicand, to one of the three forms:

1) $\sqrt{a}\sqrt{(x-p)^2 + q^2}$ if $a > 0$;
2) $\sqrt{a}\sqrt{(x-p)^2 - q^2}$ if $a > 0$;
3) $\sqrt{-a}\sqrt{q^2 - (x-p)^2}$ if $a < 0$,

where $p = -\frac{1}{2}b/a$. Different trigonometric substitutions are further used in each case to evaluate the integral:

1) $x - p = q\tan t$, $\sqrt{(x-p)^2 + q^2} = \dfrac{q}{\cos t}$, $dx = \dfrac{q\,dt}{\cos^2 t}$;

2) $x - p = \dfrac{q}{\cos t}$, $\sqrt{(x-p)^2 - q^2} = q\tan t$, $dx = \dfrac{q\sin t\,dt}{\cos^2 t}$;

3) $x - p = q\sin t$, $\sqrt{q^2 - (x-p)^2} = q\cos t$, $dx = q\cos t\,dt$.

**Example 2.** Evaluate the integral $\int \sqrt{6 + 4x - 2x^2}\,dx$.

This integral corresponds to case 3 with $a = -2$, $p = 1$, and $q = 2$. The integrand can be rewritten in the form:
$$\sqrt{6 + 4x - 2x^2} = \sqrt{2}\sqrt{3 + 2x - x^2} = \sqrt{2}\sqrt{4 - (x-1)^2}.$$

Using the trigonometric substitution $x - 1 = 2\sin t$ and the formulas $\sqrt{3 + 2x - x^2} = 2\cos t$ and $dx = 2\cos t\,dt$, we obtain

$$\int \sqrt{6 + 4x - 2x^2}\,dx = 4\sqrt{2}\int \cos^2 t\,dt = 2\sqrt{2}\int (1 + \cos 2t)\,dt$$

$$= 2\sqrt{2}t + \sqrt{2}\sin 2t + C = 2\sqrt{2}\arcsin\frac{x-1}{2} + \sqrt{2}\sin\left(2\arcsin\frac{x-1}{2}\right) + C$$

$$= 2\sqrt{2}\arcsin\frac{x-1}{2} + \frac{\sqrt{2}}{2}(x-1)\sqrt{4 - (x-1)^2} + C.$$

### 7.1.4-3. Integral of a differential binomial.

The *integral of a differential binomial*,
$$\int x^m(a + bx^n)^p\,dx,$$

where $a$ and $b$ are constants, and $n$, $m$, $p$ are rational numbers, is expressible in terms of elementary functions in the following three cases only:

1) If $p$ is an integer. For $p \geq 0$, removing the brackets gives the sum of power functions. For $p < 0$, the substitution $x = t^r$, where $r$ is the common denominator of the fractions $m$ and $n$, leads to the integral of a rational function.

2) If $\dfrac{m+1}{n}$ is an integer. One uses the substitution $a + bx^n = t^k$, where $k$ is the denominator of the fraction $p$.

3) If $\dfrac{m+1}{n} + p$ is an integer. One uses the substitution $ax^{-n} + b = t^k$, where $k$ is the denominator of the fraction $p$.

**Remark.** In cases 2 and 3, the substitution $z = x^n$ leads to integrals of the form 3° from Paragraph 7.1.4-1.

## 7.1.5. Integration of Exponential and Trigonometric Functions

### 7.1.5-1. Integration of exponential and hyperbolic functions.

1. Integrals of the form $\int R(e^{px}, e^{qx})\,dx$, where $R(x, y)$ is a rational function of its arguments and $p$, $q$ are rational numbers, may be evaluated using the substitution $z^m = e^x$, where $m$ is the common denominator of the fractions $p$ and $q$. In the special case of integer $p$ and $q$, we have $m = 1$, and the substitution becomes $z = e^x$.

**Example 1.** Evaluate the integral $\int \dfrac{e^{3x}\,dx}{e^x+2}$.

This integral corresponds to integer $p$ and $q$: $p=1$ and $q=3$. So we use the substitution $z=e^x$. Then $x=\ln z$ and $dx=\dfrac{dz}{z}$. Therefore,

$$\int \frac{e^{3x}\,dx}{e^x+2} = \int \frac{z^2\,dz}{z+2} = \int\left(z-2+\frac{4}{z+2}\right)dz = \tfrac{1}{2}z^2 - 2z + 4\ln|z+2| + C = \tfrac{1}{2}e^{2x} - 2e^x + 4\ln(e^x+2) + C.$$

2. Integrals of the form $\int R(\sinh ax, \cosh ax)\,dx$ are evaluated by converting the hyperbolic functions to exponentials, using the formulas $\sinh ax = \tfrac{1}{2}(e^{ax}-e^{-ax})$ and $\cosh ax = \tfrac{1}{2}(e^{ax}+e^{-ax})$, and performing the substitution $z=e^{ax}$. Then

$$\int R(\sinh ax, \cosh ax)\,dx = \frac{1}{a}\int R\!\left(\frac{z^2-1}{2z},\, \frac{z^2+1}{2z}\right)\frac{dz}{z}.$$

Alternatively, the substitution $t = \tanh\!\left(\dfrac{ax}{2}\right)$ can also be used to evaluate integrals of the above form. Then

$$\int R(\sinh ax, \cosh ax)\,dx = \frac{2}{a}\int R\!\left(\frac{2t}{1-t^2},\, \frac{1+t^2}{1-t^2}\right)\frac{dt}{1-t^2}.$$

### 7.1.5-2. Integration of trigonometric functions.

1. Integrals of the form $\int R(\sin ax, \cos ax)\,dx$ can be converted to integrals of rational functions using the versatile trigonometric substitution $t = \tan\!\left(\dfrac{ax}{2}\right)$:

$$\int R(\sin ax, \cos ax)\,dx = \frac{2}{a}\int R\!\left(\frac{2t}{1+t^2},\, \frac{1-t^2}{1+t^2}\right)\frac{dt}{1+t^2}.$$

**Example 2.** Evaluate the integral $\int \dfrac{dx}{2+\sin x}$.

Using the versatile trigonometric substitution $t = \tan\dfrac{x}{2}$, we have

$$\int \frac{dx}{2+\sin x} = 2\int \frac{dt}{\left(2+\dfrac{2t}{1+t^2}\right)(1+t^2)} = \int \frac{dt}{t^2+t+1} = 2\int \frac{d(2t+1)}{(2t+1)^2+3}$$

$$= \frac{2}{\sqrt{3}}\arctan\frac{2t+1}{\sqrt{3}} + C = \frac{2}{\sqrt{3}}\arctan\!\left(\frac{2}{\sqrt{3}}\tan\frac{x}{2}+\frac{1}{\sqrt{3}}\right)+C.$$

2. Integrals of the form $\int R(\sin^2 ax, \cos^2 ax, \tan ax)\,dx$ are converted to integrals of rational functions with the change of variable $z = \tan ax$:

$$\int R(\sin^2 ax, \cos^2 ax, \tan ax)\,dx = \frac{1}{a}\int R\!\left(\frac{z^2}{1+z^2},\, \frac{1}{1+z^2},\, z\right)\frac{dz}{1+z^2}.$$

3. Integrals of the form

$$\int \sin ax \cos bx\,dx,\quad \int \cos ax \cos bx\,dx,\quad \int \sin ax \sin bx\,dx$$

are evaluated using the formulas

$$\sin\alpha\cos\beta = \tfrac{1}{2}[\sin(\alpha+\beta) + \sin(\alpha-\beta)],$$
$$\cos\alpha\cos\beta = \tfrac{1}{2}[\cos(\alpha+\beta) + \cos(\alpha-\beta)],$$
$$\sin\alpha\sin\beta = \tfrac{1}{2}[\cos(\alpha-\beta) - \cos(\alpha+\beta)].$$

4. Integrals of the form $\int \sin^m x \cos^n x\, dx$, where $m$ and $n$ are integers, are evaluated as follows:
   (a) if $m$ is odd, one uses the change of variable $\cos x = z$, with $\sin x\, dx = -dz$;
   (b) if $n$ is odd, one uses the change of variable $\sin x = z$, with $\cos x\, dx = dz$;
   (c) if $m$ and $n$ are both even nonnegative integers, one should use the degree reduction formulas

$$\sin^2 x = \tfrac{1}{2}(1 - \cos 2x), \quad \cos^2 x = \tfrac{1}{2}(1 + \cos 2x), \quad \sin x \cos x = \tfrac{1}{2}\sin 2x.$$

**Example 3.** Evaluate the integral $\int \sin^5 x\, dx$.

This integral corresponds to odd $m$: $m = 5$. With simple rearrangement and the change of variable $\cos x = z$, we have

$$\int \sin^5 x\, dx = \int (\sin^2 x)^2 \sin x\, dx = -\int (1 - \cos^2 x)^2\, d\cos x = -\int (1 - z^2)^2\, dz$$
$$= \tfrac{2}{3}z^3 - \tfrac{1}{5}z^5 - z + C = \tfrac{2}{3}\cos^3 x - \tfrac{1}{5}\cos^5 x - \cos x + C.$$

**Remark.** In general, the integrals $\int \sin^p x \cos^q x\, dx$ are reduced to the integral of a differential binomial by the substitution $y = \sin x$.

## 7.1.6. Integration of Polynomials Multiplied by Elementary Functions

▶ Throughout this section, $P_n(x)$ designates a polynomial of degree $n$.

### 7.1.6-1. Integration of the product of a polynomial by exponential functions.

General formulas:

$$\int P_n(x) e^{ax}\, dx = e^{ax}\left[\frac{P_n(x)}{a} - \frac{P_n'(x)}{a^2} + \cdots + (-1)^n \frac{P_n^{(n)}(x)}{a^{n+1}}\right] + C,$$

$$\int P_n(x) \cosh(ax)\, dx = \sinh(ax)\left[\frac{P_n(x)}{a} + \frac{P_n''(x)}{a^3} + \cdots\right] - \cosh(ax)\left[\frac{P_n'(x)}{a^2} + \frac{P_n'''(x)}{a^4} + \cdots\right] + C,$$

$$\int P_n(x) \sinh(ax)\, dx = \cosh(ax)\left[\frac{P_n(x)}{a} + \frac{P_n''(x)}{a^3} + \cdots\right] - \sinh(ax)\left[\frac{P_n'(x)}{a^2} + \frac{P_n'''(x)}{a^4} + \cdots\right] + C.$$

These formulas are obtained by repeated integration by parts; see formula 4 from Paragraph 7.1.2-2 with $f(x) = P_n(x)$ for $g^{(n+1)}(x) = e^{ax}$, $g^{(n+1)}(x) = \cosh(ax)$, and $g^{(n+1)}(x) = \sinh(ax)$, respectively.

In the special case $P_n(x) = x^n$, the first formula gives

$$\int x^n e^{ax}\, dx = e^{ax} \sum_{k=0}^{n} \frac{(-1)^{n-k}}{a^{n+1-k}} \frac{n!}{k!} x^k + C.$$

**7.1.6-2. Integration of the product of a polynomial by a trigonometric function.**

1°. General formulas:

$$\int P_n(x)\cos(ax)\,dx = \sin(ax)\left[\frac{P_n(x)}{a} - \frac{P_n''(x)}{a^3} + \cdots\right] + \cos(ax)\left[\frac{P_n'(x)}{a^2} - \frac{P_n'''(x)}{a^4} + \cdots\right] + C,$$

$$\int P_n(x)\sin(ax)\,dx = \sin(ax)\left[\frac{P_n'(x)}{a^2} - \frac{P_n'''(x)}{a^4} + \cdots\right] - \cos(ax)\left[\frac{P_n(x)}{a} - \frac{P_n''(x)}{a^3} + \cdots\right] + C.$$

These formulas are obtained by repeated integration by parts; see formula 4 from Paragraph 7.1.2-2 with $f(x) = P_n(x)$ for $g^{(n+1)}(x) = \cos(ax)$ and $g^{(n+1)}(x) = \sin(ax)$, respectively.

2°. To evaluate integrals of the form

$$\int P_n(x)\cos^m(ax)\,dx, \qquad \int P_n(x)\sin^m(ax)\,dx,$$

with $m = 2, 3, \ldots$, one should first use the trigonometric formulas

$$\cos^{2k}(ax) = \frac{1}{2^{2k-1}}\sum_{i=0}^{k-1} C_{2k}^i \cos[2(k-i)ax] + \frac{1}{2^{2k}}C_{2k}^k \qquad (m = 2k),$$

$$\cos^{2k+1}(ax) = \frac{1}{2^{2k}}\sum_{i=0}^{k} C_{2k+1}^i \cos[(2k-2i+1)ax] \qquad (m = 2k+1),$$

$$\sin^{2k}(ax) = \frac{1}{2^{2k-1}}\sum_{i=0}^{k-1} (-1)^{k-i} C_{2k}^i \cos[2(k-i)ax] + \frac{1}{2^{2k}}C_{2k}^k \qquad (m = 2k),$$

$$\sin^{2k+1}(ax) = \frac{1}{2^{2k}}\sum_{i=0}^{k} (-1)^{k-i} C_{2k+1}^i \sin[(2k-2i+1)ax] \qquad (m = 2k+1),$$

thus reducing the above integrals to those considered in Item 1°.

3°. Integrals of the form

$$\int P_n(x)\,e^{ax}\sin(bx)\,dx, \qquad \int P_n(x)\,e^{ax}\cos(bx)\,dx$$

can be evaluated by repeated integration by parts.

In particular,

$$\int x^n e^{ax}\sin(bx) = e^{ax}\sum_{k=1}^{n+1}\frac{(-1)^{k+1}n!}{(n-k+1)!\,(a^2+b^2)^{k/2}}x^{n-k+1}\sin(bx + k\theta) + C,$$

$$\int x^n e^{ax}\cos(bx) = e^{ax}\sum_{k=1}^{n+1}\frac{(-1)^{k+1}n!}{(n-k+1)!\,(a^2+b^2)^{k/2}}x^{n-k+1}\cos(bx + k\theta) + C,$$

where

$$\sin\theta = -\frac{b}{\sqrt{a^2+b^2}}, \qquad \cos\theta = \frac{a}{\sqrt{a^2+b^2}}.$$

### 7.1.6-3. Integrals involving power and logarithmic functions.

$1°$. The formula of integration by parts with $g'(x) = P_n(x)$ is effective in the evaluation of integrals of the form

$$\int P_n(x) \ln(ax)\, dx = Q_{n+1}(x) \ln(ax) - a \int \frac{Q_{n+1}(x)}{x}\, dx,$$

where $Q_{n+1}(x) = \int P_n(x)\, dx$ is a polynomial of degree $n+1$. The integral on the right-hand side is easy to take, since the integrand is the sum of power functions.

**Example.** Evaluate the integral $\int \ln x\, dx$.

Setting $f(x) = \ln x$ and $g'(x) = 1$, we find $f'(x) = \dfrac{1}{x}$ and $g(x) = x$. Substituting these expressions into the formula of integration by parts, we obtain $\int \ln x\, dx = x \ln x - \int dx = x \ln x - x + C$.

$2°$. The easiest way to evaluate integrals of the more general form

$$I = \int \sum_{i=0}^{n} \ln^i(ax) \left( \sum_{j=0}^{m} b_{ij} x^{\beta_{ij}} \right) dx,$$

where the $\beta_{ij}$ are arbitrary numbers, is to use the substitution $z = \ln(ax)$, so that

$$I = \int \sum_{i=0}^{n} z^i \left( \sum_{j=0}^{m} \frac{b_{ij}}{a^{\beta_{ij}+1}} e^{(\beta_{ij}+1)z} \right) dz.$$

By removing the brackets, one obtains a sum of integrals like $\int x^n e^{ax}\, dx$, which are easy to evaluate by the last formula in Paragraph 7.1.6-1.

### 7.1.6-4. Integrals involving inverse trigonometric functions.

$1°$. The formula of integration by parts with $g'(x) = P_n(x)$ also allows the evaluation of the following integrals involving inverse trigonometric functions:

$$\int P_n(x) \arcsin(ax)\, dx = Q_{n+1}(x) \arcsin(ax) - a \int \frac{Q_{n+1}(x)}{\sqrt{1 - a^2 x^2}}\, dx,$$

$$\int P_n(x) \arccos(ax)\, dx = Q_{n+1}(x) \arccos(ax) + a \int \frac{Q_{n+1}(x)}{\sqrt{1 - a^2 x^2}}\, dx,$$

$$\int P_n(x) \arctan(ax)\, dx = Q_{n+1}(x) \arctan(ax) - a \int \frac{Q_{n+1}(x)}{a^2 x^2 + 1}\, dx,$$

$$\int P_n(x) \operatorname{arccot}(ax)\, dx = Q_{n+1}(x) \operatorname{arccot}(ax) + a \int \frac{Q_{n+1}(x)}{a^2 x^2 + 1}\, dx,$$

where $Q_{n+1}(x) = \int P_n(x)\, dx$ is a polynomial of degree $n+1$. The integrals with radicals on the right-hand side in the first two formulas can be evaluated using the techniques described in Paragraph 7.1.4-2. The integrals of rational functions on the right-hand side in the last two formulas can be evaluated using the techniques described in Subsection 7.1.3.

**Remark.** The above formulas can be generalized to contain any rational functions $R'(x)$ and $R(x)$ instead of the polynomials $P_n(x)$ and $Q_{n+1}(x)$, respectively.

2°. The following integrals are taken using a change of variable:

$$\int P_n(x) \arcsin^m(ax)\, dx = \frac{1}{a} \int \cos z\, P_n\left(\frac{\sin z}{a}\right) z^m\, dz, \quad \text{substitution } z = \arcsin(ax);$$

$$\int P_n(x) \arccos^m(ax)\, dx = -\frac{1}{a} \int \sin z\, P_n\left(\frac{\cos z}{a}\right) z^m\, dz, \quad \text{substitution } z = \arccos(ax),$$

where $m = 2, 3, \ldots$ The expressions $\cos z \sin^k z$ and $\sin z \cos^k z$ ($k = 1, \ldots, n$) in the integrals on the right-hand sides should be expressed as sums of sines and cosines with appropriate arguments. Then it remains to evaluate integrals considered in Paragraph 7.1.6-2.

## 7.2. Definite Integral

### 7.2.1. Basic Definitions. Classes of Integrable Functions. Geometrical Meaning of the Definite Integral

**7.2.1-1. Basic definitions.**

Let $y = f(x)$ be a bounded function defined on a finite closed interval $[a, b]$. Let us partition this interval into $n$ elementary subintervals defined by a set of points $\{x_0, x_1, \ldots, x_n\}$ such that $a = x_0 < x_1 < \cdots < x_n = b$. Each subinterval $[x_{k-1}, x_k]$ will be characterized by its length $\Delta x_k = x_k - x_{k-1}$ and an arbitrarily chosen point $\xi_k \in [x_{k-1}, x_k]$. Let us make up an *integral sum* (a *Cauchy–Riemann sum*, also known as a *Riemann sum*)

$$s_n = \sum_{k=1}^{n} f(\xi_k) \Delta x_k \qquad (x_{k-1} \leq \xi_k \leq x_k).$$

If, as $n \to \infty$ and, accordingly, $\Delta x_k \to 0$ for all $k$, there exists a finite limit of the integral sums $s_n$ and it depends on neither the way the interval $[a, b]$ was split up, nor the selection of the points $\xi_k$, then this limit is denoted $\int_a^b f(x)\, dx$ and is called the *definite integral* (also the *Riemann integral*) of the function $y = f(x)$ over the interval $[a, b]$:

$$\int_a^b f(x)\, dx = \lim_{n \to \infty} s_n \qquad \left(\max_{1 \leq k \leq n} \Delta x_k \to 0 \text{ as } n \to \infty\right).$$

In this case, the function $f(x)$ is called *integrable* on the interval $[a, b]$.

**7.2.1-2. Classes of integrable functions.**

1. If a function $f(x)$ is continuous on an interval $[a, b]$, then it is integrable on this interval.
2. If a bounded function $f(x)$ has finitely many jump discontinuities on $[a, b]$, then it is integrable on $[a, b]$.
3. A monotonic bounded function $f(x)$ is always integrable.

### 7.2.1-3. Geometric meaning of the definite integral.

If $f(x) \geq 0$ on $[a, b]$, then the integral $\int_a^b f(x)\,dx$ is equal to the area of the domain $D = \{a \leq x \leq b,\ 0 \leq y \leq f(x)\}$ (the area of the curvilinear trapezoid shown in Fig. 7.1).

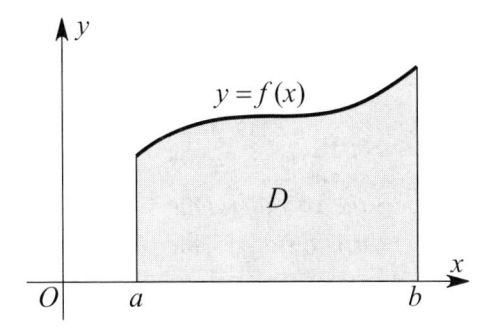

**Figure 7.1.** The integral of a nonnegative function $f(x)$ on an interval $[a, b]$ is equal to the area of the shaded region.

## 7.2.2. Properties of Definite Integrals and Useful Formulas

### 7.2.2-1. Qualitative properties of integrals.

1. If a function $f(x)$ is integrable on $[a, b]$, then the functions $cf(x)$, with $c = \text{const}$, and $|f(x)|$ are also integrable on $[a, b]$.

2. If two functions $f(x)$ and $g(x)$ are integrable on $[a, b]$, then their sum, difference, and product are also integrable on $[a, b]$.

3. If a function $f(x)$ is integrable on $[a, b]$ and its values lie within an interval $[c, d]$, where a function $g(y)$ is defined and continuous, then the composite function $g(f(x))$ is also integrable on $[a, b]$.

4. If a function $f(x)$ is integrable on $[a, b]$, then it is also integrable and on any subinterval $[\alpha, \beta] \subset [a, b]$. Conversely, if an interval $[a, b]$ is partitioned into a number of subintervals and $f(x)$ is integrable on each of the subintervals, then it is integrable on the whole interval $[a, b]$.

5. If the values of a function are changed at finitely many points, this will not affect the integrability of the function and will not change the value of the integral.

### 7.2.2-2. Properties of integrals in terms of identities.

1. *The integral over a zero-length interval is zero*:

$$\int_a^a f(x)\,dx = 0.$$

2. *Antisymmetry under the swap of the integration limits*:

$$\int_a^b f(x)\,dx = -\int_b^a f(x)\,dx.$$

This property can be taken as the definition of a definite integral with $a > b$.

3. *Linearity.* If functions $f(x)$ and $g(x)$ are integrable on an interval $[a, b]$, then

$$\int_a^b \left[Af(x) \pm Bg(x)\right] dx = A \int_a^b f(x)\, dx \pm B \int_a^b g(x)\, dx$$

for any numbers $A$ and $B$.

4. *Additivity.* If $c \in [a, b]$ and $f(x)$ is integrable on $[a, b]$, then

$$\int_a^b f(x)\, dx = \int_a^c f(x)\, dx + \int_c^b f(x)\, dx.$$

5. *Differentiation with respect the variable upper limit.* If $f(x)$ is continuous on $[a, b]$, then the function $\Phi(x) = \int_a^x f(t)\, dt$ is differentiable on $[a, b]$, and $\Phi'(x) = f(x)$. This can be written in one relation:

$$\frac{d}{dx}\left(\int_a^x f(t)\, dt\right) = f(x).$$

6. *Newton–Leibniz formula*:

$$\int_a^b f(x)\, dx = F(x)\Big|_a^b = F(b) - F(a),$$

where $F(x)$ is an antiderivative of $f(x)$ on $[a, b]$.

7. *Integration by parts.* If functions $f(x)$ and $g(x)$ have continuous derivatives on $[a, b]$, then

$$\int_a^b f(x)g'(x)\, dx = \left[f(x)g(x)\right]\Big|_a^b - \int_a^b f'(x)g(x)\, dx.$$

8. *Repeated integration by parts*:

$$\int_a^b f(x)g^{(n+1)}(x)\, dx = \left[f(x)g^{(n)}(x) - f'(x)g^{(n-1)}(x) + \cdots + (-1)^n f^{(n)}(x)g(x)\right]_a^b$$

$$+ (-1)^{n+1} \int_a^b f^{(n+1)}(x)g(x)\, dx, \qquad n = 0, 1, \ldots$$

9. *Change of variable (substitution) in a definite integral.* Let $f(x)$ be a continuous function on $[a, b]$ and let $x(t)$ be a continuously differentiable function on $[\alpha, \beta]$. Suppose also that the range of values of $x(t)$ coincides with $[a, b]$, with $x(\alpha) = a$ and $x(\beta) = b$. Then

$$\int_a^b f(x)\, dx = \int_\alpha^\beta f\bigl(x(t)\bigr) x'(t)\, dt.$$

**Example.** Evaluate the integral $\int_0^3 \frac{dx}{(x-8)\sqrt{x+1}}$.

Perform the substitution $x + 1 = t^2$, with $dx = 2t\, dt$. We have $t = 1$ at $x = 0$ and $t = 2$ at $x = 3$. Therefore

$$\int_0^3 \frac{dx}{(x-8)\sqrt{x+1}} = \int_1^2 \frac{2t\, dt}{(t^2-9)t} = 2\int_1^2 \frac{dt}{t^2-9} = \frac{1}{3}\ln\left|\frac{t-3}{t+3}\right|\Big|_1^2 = \frac{1}{3}\ln\frac{2}{5}.$$

10. *Differentiation with respect to a parameter.* Let $f(x, \lambda)$ be a continuous function in a domain $a \le x \le b$, $\lambda_1 \le \lambda \le \lambda_2$ and let it has a continuous partial derivative $\frac{\partial}{\partial \lambda} f(x, \lambda)$ in the

same domain. Also let $u(\lambda)$ and $v(\lambda)$ be differentiable functions on the interval $\lambda_1 \leq \lambda \leq \lambda_2$ such that $a \leq u(\lambda) \leq b$ and $a \leq v(\lambda) \leq b$. Then

$$\frac{d}{d\lambda} \int_{v(\lambda)}^{u(\lambda)} f(x, \lambda)\, dx = f(u(\lambda), \lambda) \frac{du(\lambda)}{d\lambda} - f(v(\lambda), \lambda) \frac{dv(\lambda)}{d\lambda} + \int_{v(\lambda)}^{u(\lambda)} \frac{\partial}{\partial \lambda} f(x, \lambda)\, dx.$$

11. *Cauchy's formula for multiple integration*:

$$\int_a^x dx_1 \int_a^{x_1} dx_2 \ldots \int_a^{x_{n-1}} f(x_n)\, dx_n = \frac{1}{(n-1)!} \int_a^x (x-t)^{n-1} f(t)\, dt.$$

## 7.2.3. General Reduction Formulas for the Evaluation of Integrals

Below are some general formulas, involving arbitrary functions and parameters, that could facilitate the evaluation of integrals.

7.2.3-1. Integrals involving functions of a linear or rational argument.

$$\int_a^b f(a+b-x)\, dx = \int_a^b f(x)\, dx;$$

$$\int_0^a [f(x) + f(a-x)]\, dx = 2 \int_0^a f(x)\, dx;$$

$$\int_0^a [f(x) - f(a-x)]\, dx = 0;$$

$$\int_{-a}^a f(x)\, dx = 0 \quad \text{if } f(x) \text{ is odd};$$

$$\int_{-a}^a f(x)\, dx = 2 \int_0^a f(x)\, dx \quad \text{if } f(x) \text{ is even};$$

$$\int_a^b f(x, a+b-x)\, dx = 0 \quad \text{if } f(x, y) = -f(y, x);$$

$$\int_0^1 f\left(2x\sqrt{1-x^2}\right) dx = \int_0^1 f(1-x^2)\, dx.$$

7.2.3-2. Integrals involving functions with trigonometric argument.

$$\int_0^\pi f(\sin x)\, dx = 2 \int_0^{\pi/2} f(\sin x)\, dx;$$

$$\int_0^{\pi/2} f(\sin x)\, dx = \int_0^{\pi/2} f(\cos x)\, dx;$$

$$\int_0^{\pi/2} f(\sin x, \cos x)\, dx = 0 \quad \text{if } f(x, y) = -f(y, x);$$

$$\int_0^{\pi/2} f(\sin 2x) \cos x\, dx = \int_0^{\pi/2} f(\cos^2 x) \cos x\, dx;$$

$$\int_0^{n\pi} xf(\sin x)\,dx = \pi n^2 \int_0^{\pi/2} f(\sin x)\,dx \quad \text{if } f(x) = f(-x);$$

$$\int_0^{n\pi} xf(\sin x)\,dx = (-1)^{n-1}\pi n \int_0^{\pi/2} f(\sin x)\,dx \quad \text{if } f(-x) = -f(x);$$

$$\int_0^{2\pi} f(a\sin x + b\cos x)\,dx = \int_0^{2\pi} f\left(\sqrt{a^2+b^2}\,\sin x\right)dx = 2\int_0^{\pi} f\left(\sqrt{a^2+b^2}\,\cos x\right)dx;$$

$$\int_0^{\pi} f\left(\frac{\sin^2 x}{1+2a\cos x + a^2}\right)dx = \int_0^{\pi} f(\sin^2 x)\,dx \quad \text{if } |a| \geq 1;$$

$$\int_0^{\pi} f\left(\frac{\sin^2 x}{1+2a\cos x + a^2}\right)dx = \int_0^{\pi} f\left(\frac{\sin^2 x}{a^2}\right)dx \quad \text{if } 0 < |a| < 1.$$

7.2.3-3. **Integrals involving logarithmic functions.**

$$\int_a^b f(x)\ln^n x\,dx = \left[\left(\frac{d}{d\lambda}\right)^n \int_a^b x^\lambda f(x)\,dx\right]_{\lambda=0},$$

$$\int_a^b f(x)\ln^n g(x)\,dx = \left[\left(\frac{d}{d\lambda}\right)^n \int_a^b f(x)[g(x)]^\lambda\,dx\right]_{\lambda=0},$$

$$\int_a^b f(x)[g(x)]^\lambda \ln^n g(x)\,dx = \left(\frac{d}{d\lambda}\right)^n \int_a^b f(x)[g(x)]^\lambda\,dx.$$

## 7.2.4. General Asymptotic Formulas for the Calculation of Integrals

Below are some general formulas, involving arbitrary functions and parameters, that may be helpful for obtaining asymptotics of integrals.

7.2.4-1. **Asymptotic formulas for integrals with weak singularity as $\varepsilon \to 0$.**

1°. We will consider integrals of the form

$$I(\varepsilon) = \int_0^a \frac{x^{\beta-1} f(x)\,dx}{(x+\varepsilon)^\alpha},$$

where $0 < a < \infty$, $\beta > 0$, $f(0) \neq 0$, and $\varepsilon > 0$ is a small parameter.

The integral diverges as $\varepsilon \to 0$ for $\alpha \geq \beta$, that is, $\lim_{\varepsilon \to 0} I(\varepsilon) = \infty$. In this case, the leading term of the asymptotic expansion of the integral $I(\varepsilon)$ is given by

$$I(\varepsilon) = \frac{\Gamma(\beta)\Gamma(\alpha-\beta)}{\Gamma(\alpha)} f(0)\varepsilon^{\beta-\alpha} + O(\varepsilon^\sigma) \quad \text{if } \alpha > \beta,$$

$$I(\varepsilon) = -f(0)\ln\varepsilon + O(1) \quad \text{if } \alpha = \beta,$$

where $\Gamma(\beta)$ is the gamma function and $\sigma = \min[\beta - \alpha + 1, 0]$.

2°. The leading term of the asymptotic expansion, as $\varepsilon \to 0$, of the more general integral

$$I(\varepsilon) = \int_0^a \frac{x^{\beta-1} f(x)\, dx}{(x^k + \varepsilon^k)^\alpha}$$

with $0 < a < \infty$, $\beta > 0$, $k > 0$, $\varepsilon > 0$, and $f(0) \neq 0$ is expressed as

$$I(\varepsilon) = \frac{f(0)}{k\Gamma(\alpha)} \Gamma\left(\frac{\beta}{k}\right) \Gamma\left(\alpha - \frac{\beta}{k}\right) \varepsilon^{\beta - \alpha k} + O(\varepsilon^\sigma) \quad \text{if} \quad \alpha k > \beta,$$

$$I(\varepsilon) = -f(0) \ln \varepsilon + O(1) \quad \text{if} \quad \alpha k = \beta,$$

where $\sigma = \min[\beta - \alpha k + 1, 0]$.

3°. The leading terms of the asymptotic expansion, as $\varepsilon \to 0$, of the integral

$$I(\varepsilon) = \int_a^\infty x^\alpha \exp(-\varepsilon x^\beta) f(x)\, dx$$

with $a > 0$, $\beta > 0$, $\varepsilon > 0$, and $f(0) \neq 0$ has the form

$$I(\varepsilon) = \frac{1}{\beta} f(0) \Gamma\left(\frac{\alpha+1}{\beta}\right) \varepsilon^{-\frac{\alpha+1}{\beta}} \quad \text{if} \quad \alpha > -1,$$

$$I(\varepsilon) = -\frac{1}{\beta} f(0) \ln \varepsilon \quad \text{if} \quad \alpha = -1.$$

4°. Now consider *potential-type integrals*

$$\Pi(f) = \int_{-1}^1 \frac{f(\xi)\, d\xi}{\sqrt{(\xi - z)^2 + r^2}},$$

with $z, r, \varphi$ being cylindrical coordinates in the three-dimensional space. The function $\Pi(f)$ is simple layer potential concentrated on the interval $z \in [-1, 1]$ with density $f(z)$. If the density is continuous, then $\Pi(f)$ satisfies the Laplace equation $\Delta\Pi = 0$ outside $z \in [-1, 1]$ and vanishes at infinity.

Asymptotics of the integral as $r \to 0$:

$$\Pi(f) = -2f(z) \ln r + O(1),$$

where $|z| \leq 1 - \delta$ with $0 < \delta < 1$.

**7.2.4-2. Asymptotic formulas for Laplace integrals of special form as $\lambda \to \infty$.**

1°. Consider a *Laplace integral of the special form*

$$I(\lambda) = \int_0^a x^{\beta-1} \exp(-\lambda x^\alpha) f(x)\, dx,$$

where $0 < a < \infty$, $\alpha > 0$, and $\beta > 0$.

The following formula, called *Watson's asymptotic formula*, holds as $\lambda \to \infty$:

$$I(\lambda) = \frac{1}{\alpha} \sum_{k=0}^{n} \frac{f^{(k)}(0)}{k!} \Gamma\left(\frac{k+\beta}{\alpha}\right) \lambda^{-(k+\beta)/\alpha} + O\left(\lambda^{-(n+\beta+1)/\alpha}\right).$$

**Remark 1.** Watson's formula also holds for improper integrals with $a = \infty$ if the original integral converges absolutely for some $\lambda_0 > 0$.

**Remark 2.** Watson's formula remains valid in the case of complex parameter $\lambda$ as $|\lambda| \to \infty$, where $|\arg \lambda| \leq \frac{\pi}{2} - \varepsilon < \frac{\pi}{2}$ ($\varepsilon > 0$ can be chosen arbitrarily small but independent of $\lambda$).

**Remark 3.** The Laplace transform corresponds to the above integral with $a = \infty$ and $\alpha = \beta = 1$.

2°. The leading term of the asymptotic expansion, as $\lambda \to \infty$, of the integral

$$I(\lambda) = \int_0^a x^{\beta-1} |\ln x|^\gamma e^{-\lambda x} f(x)\, dx$$

with $0 < a < \infty$, $\beta > 0$, and $f(0) \neq 0$ is expressed as

$$I(\lambda) = \Gamma(\beta) f(0) \lambda^{-\beta} (\ln \lambda)^\gamma.$$

**7.2.4-3. Asymptotic formulas for Laplace integrals of general form as $\lambda \to \infty$.**

Consider a *Laplace integral of the general form*

$$I(\lambda) = \int_a^b f(x) \exp[\lambda g(x)]\, dx, \tag{7.2.4.1}$$

where $[a, b]$ is a finite interval and $f(x)$, $g(x)$ are continuous functions.

1°. *Leading term of the asymptotic expansion of the integral (7.2.4.1) as $\lambda \to \infty$.* Suppose the function $g(x)$ attains a maximum on $[a, b]$ at only one point $x_0 \in [a, b]$ and is differentiable in a neighborhood of it, with $g'(x_0) = 0$, $g''(x_0) \neq 0$, and $f(x_0) \neq 0$. Then the leading term of the asymptotic expansion of the integral (7.2.4.1), as $\lambda \to \infty$, is expressed as

$$\begin{aligned} I(\lambda) &= f(x_0) \sqrt{-\frac{2\pi}{\lambda g''(x_0)}} \exp[\lambda g(x_0)] &&\text{if } a < x_0 < b, \\ I(\lambda) &= \frac{1}{2} f(x_0) \sqrt{-\frac{2\pi}{\lambda g''(x_0)}} \exp[\lambda g(x_0)] &&\text{if } x_0 = a \text{ or } x_0 = b. \end{aligned} \tag{7.2.4.2}$$

Note that the latter formula differs from the former by the factor $1/2$ only.

Under the same conditions, if $g(x)$ attains a maximum at either endpoint, $x_0 = a$ or $x_0 = b$, but $g'(x_0) \neq 0$, then the leading asymptotic term of the integral, as $\lambda \to \infty$, is

$$I(\lambda) = \frac{f(x_0)}{|g'(x_0)|} \frac{1}{\lambda} \exp[\lambda g(x_0)], \quad \text{where } x_0 = a \text{ or } x_0 = b. \tag{7.2.4.3}$$

For more accurate asymptotic estimates for the Laplace integral (7.2.4.1), see below.

**2°.** *Leading and subsequent asymptotic terms of the integral (7.2.4.1) as $\lambda \to \infty$.* Let $g(x)$ attain a maximum at only one internal point of the interval, $x_0 \in (a < x_0 < b)$, with $g'(x_0) = 0$ and $g''(x_0) \neq 0$, and let the functions $f(x)$ and $g(x)$ be, respectively, $n$ and $n+1$ times differentiable in a neighborhood of $x = x_0$. Then the asymptotic formula

$$I(\lambda) = \exp[\lambda g(x_0)] \left( \sum_{k=0}^{n-1} c_k \lambda^{-k-1} + O(\lambda^{-n}) \right) \qquad (7.2.4.4)$$

holds as $\lambda \to \infty$, with

$$c_k = \frac{1}{(2k)!\, 2^{k+1/2}} \Gamma\left(k + \frac{1}{2}\right) \lim_{x \to x_0} \left(\frac{d}{dx}\right)^k \left[ f(x) \left( \frac{g(x_0) - g(x)}{(x - x_0)^2} \right)^{-k-1/2} \right].$$

Suppose $g(x)$ attains a maximum at the endpoint $x = a$ only, with $g'(a) \neq 0$. Suppose also that $f(x)$ and $g(x)$ are, respectively, $n$ and $n+1$ times differentiable in a neighborhood of $x = a$. Then we have, as $\lambda \to \infty$,

$$I(\lambda) = \exp[\lambda g(a)] \left( \sum_{k=0}^{n-1} c_k \lambda^{-k-1} + O(\lambda^{-n}) \right), \qquad (7.2.4.5)$$

where

$$c_0 = -\frac{f(a)}{g'(a)}; \quad c_k = (-1)^{k+1} \left[ \left(\frac{1}{g'(x)} \frac{d}{dx}\right)^k \frac{f(x)}{g'(x)} \right]_{x=a}, \quad k = 1, 2, \ldots$$

**Remark 1.** The asymptotic formulas (7.2.4.2)–(7.2.4.5) hold also for improper integrals with $b = \infty$ if the original integral (7.2.4.1) converges absolutely at some $\lambda_0 > 0$.

**Remark 2.** The asymptotic formulas (7.2.4.2)–(7.2.4.5) remain valid also in the case of complex $\lambda$ as $|\lambda| \to \infty$, where $|\arg \lambda| \leq \frac{\pi}{2} - \varepsilon < \frac{\pi}{2}$ ($\varepsilon > 0$ can be chosen arbitrarily small but independent of $\lambda$).

**3°.** *Some generalizations.* Let $g(x)$ attain a maximum at only one internal point of the interval, $x_0 \in (a < x_0 < b)$, with $g'(x_0) = \cdots = g^{(2m-1)}(x_0) = 0$ and $g^{(2m)}(x_0) \neq 0$, $m \geq 1$ and $f(x_0) \neq 0$. Then the leading asymptotic term of the integral (7.2.4.1), as $\lambda \to \infty$, is expressed as

$$I(\lambda) = \frac{1}{m} \Gamma\left(\frac{1}{2m}\right) f(x_0) \left[ -\frac{(2m)!}{g^{(2m)}(x_0)} \right]^{\frac{1}{2m}} \lambda^{-\frac{1}{2m}} \exp[\lambda g(x_0)].$$

Let $g(x)$ attain a maximum at the endpoint $x = a$ only, with $g'(a) = \cdots = g^{(m-1)}(a) = 0$ and $g^{(m)}(a) \neq 0$, where $m \geq 1$ and $f(a) \neq 0$. Then the leading asymptotic term of the integral (7.2.4.1), as $\lambda \to \infty$, has the form

$$I(\lambda) = \frac{1}{m} \Gamma\left(\frac{1}{m}\right) f(a) \left[ -\frac{m!}{g^{(m)}(a)} \right]^{\frac{1}{m}} \lambda^{-\frac{1}{m}} \exp[\lambda g(a)].$$

### 7.2.4-4. Asymptotic formulas for a power Laplace integral.

Consider the *power Laplace integral*, which is obtained from the exponential Laplace integral (7.2.4.1) by substituting $\ln g(x)$ for $g(x)$:

$$I(\lambda) = \int_a^b f(x)[g(x)]^\lambda \, dx, \tag{7.2.4.6}$$

where $[a, b]$ is a finite closed interval and $g(x) > 0$. It is assumed that the functions $f(x)$ and $g(x)$ appearing in the integral (7.2.4.6) are continuous; $g(x)$ is assumed to attain a maximum at only one point $x_0 = [a, b]$ and to be differentiable in a neighborhood of $x = x_0$, with $g'(x_0) = 0$, $g''(x_0) \neq 0$, and $f(x_0) \neq 0$. Then the leading asymptotic term of the integral, as $\lambda \to \infty$, is expressed as

$$I(\lambda) = f(x_0)\sqrt{-\frac{2\pi}{\lambda g''(x_0)}}\,[g(x_0)]^{\lambda+1/2} \quad \text{if } a < x_0 < b,$$

$$I(\lambda) = \frac{1}{2}f(x_0)\sqrt{-\frac{2\pi}{\lambda g''(x_0)}}\,[g(x_0)]^{\lambda+1/2} \quad \text{if } x_0 = a \text{ or } x_0 = b.$$

Note that the latter formula differs from the former by the factor $1/2$ only.

Under the same conditions, if $g(x)$ attains a maximum at either endpoint, $x_0 = a$ or $x_0 = b$, but $g'(x_0) \neq 0$, then the leading asymptotic term of the integral, as $\lambda \to \infty$, is

$$I(\lambda) = \frac{f(x_0)}{|g'(x_0)|}\frac{1}{\lambda}[g(x_0)]^{\lambda+1/2}, \quad \text{where } x_0 = a \text{ or } x_0 = b.$$

### 7.2.4-5. Asymptotic behavior of integrals with variable integration limit as $x \to \infty$.

Let $f(t)$ be a continuously differentiable function and let $g(t)$ be a twice continuously differentiable function. Also let the following conditions hold:

$$f(t) > 0, \quad g'(t) > 0; \quad g(t) \to \infty \text{ as } t \to \infty;$$
$$f'(t)/f(t) = o\big(g'(t)\big) \text{ as } t \to \infty; \quad g''(t) = o\big(g'^2(t)\big) \text{ as } t \to \infty.$$

Then the following asymptotic formula holds, as $x \to \infty$:

$$\int_0^x f(t)\exp[g(t)]\, dt \simeq \frac{f(x)}{g'(x)}\exp[g(x)].$$

### 7.2.4-6. Limiting properties of integrals involving periodic functions with parameter.

$1^\circ$. *Riemann property of integrals involving periodic functions.* Let $f(x)$ be a continuous function on a finite interval $[a, b]$. Then the following limiting relations hold:

$$\lim_{\lambda \to \infty} \int_a^b f(x)\sin(\lambda x)\, dx = 0,$$

$$\lim_{\lambda \to \infty} \int_a^b f(x)\cos(\lambda x)\, dx = 0.$$

*Remark.* The condition of continuity of $f(x)$ can be replaced by the more general condition of absolute integrability of $f(x)$ on a finite interval $[a, b]$.

2°. *Dirichlet's formula.* Let $f(x)$ be a monotonically increasing and bounded function on a finite interval $[0, a]$, with $a > 0$. Then the following limiting formula holds:

$$\lim_{\lambda \to \infty} \int_0^a f(x) \frac{\sin(\lambda x)}{x}\, dx = \frac{\pi}{2} f(+0).$$

7.2.4-7. Limiting properties of other integrals with parameter.

Let $f(x)$ and $g(x)$ be continuous and positive functions on $[a, b]$. Then the following limiting relations hold:

$$\lim_{n \to \infty} \sqrt[n]{I_n} = \max_{x \in [a,b]} f(x),$$

$$\lim_{n \to \infty} \frac{I_{n+1}}{I_n} = \max_{x \in [a,b]} f(x),$$

where $I_n = \int_a^b g(x)[f(x)]^n\, dx$.

## 7.2.5. Mean Value Theorems. Properties of Integrals in Terms of Inequalities. Arithmetic Mean and Geometric Mean of Functions

7.2.5-1. Mean value theorems.

THEOREM 1. If $f(x)$ is a continuous function on $[a, b]$, there exists at least one point $c \in (a, b)$ such that

$$\int_a^b f(x)\, dx = f(c)(b - a).$$

The number $f(c)$ is called the mean value of the function $f(x)$ on $[a, b]$.

THEOREM 2. If $f(x)$ is a continuous function on $[a, b]$, and $g(x)$ is integrable and of constant sign ($g(x) \geq 0$ or $g(x) \leq 0$) on $[a, b]$, then there exists at least one point $c \in (a, b)$ such that

$$\int_a^b f(x) g(x)\, dx = f(c) \int_a^b g(x)\, dx.$$

THEOREM 3. If $f(x)$ is a monotonic and nonnegative function on an interval $(a, b)$, with $a \geq b$, and $g(x)$ is integrable, then there exists at least one point $c \in (a, b)$ such that

$$\int_a^b f(x) g(x)\, dx = f(a) \int_a^c g(x)\, dx \qquad \text{if } f(x) \text{ is nonincreasing};$$

$$\int_a^b f(x) g(x)\, dx = f(b) \int_c^b g(x)\, dx \qquad \text{if } f(x) \text{ is nondecreasing}.$$

THEOREM 4. If $f(x)$ and $g(x)$ are bounded and integrable functions on an interval $[a, b]$, with $a < b$, and $g(x)$ satisfies inequalities $A \leq g(x) \leq B$, then there exists a point $c \in [a, b]$

such that

$$\int_a^b f(x)g(x)\,dx = A\int_a^c f(x)\,dx + B\int_c^b f(x)\,dx \qquad \text{if } g(x) \text{ is nondecreasing;}$$

$$\int_a^b f(x)g(x)\,dx = B\int_a^c f(x)\,dx + A\int_c^b f(x)\,dx \qquad \text{if } g(x) \text{ is nonincreasing;}$$

$$\int_a^b f(x)g(x)\,dx = g(a)\int_a^c f(x)\,dx + g(b)\int_c^b f(x)\,dx \qquad \text{if } g(x) \text{ is strictly monotonic.}$$

**7.2.5-2. Properties of integrals in terms of inequalities.**

1. *Estimation theorem.* If $m \le f(x) \le M$ on $[a, b]$, then

$$m(b-a) \le \int_a^b f(x)\,dx \le M(b-a).$$

2. *Inequality integration theorem.* If $\varphi(x) \le f(x) \le g(x)$ on $[a, b]$, then

$$\int_a^b \varphi(x)\,dx \le \int_a^b f(x)\,dx \le \int_a^b g(x)\,dx.$$

In particular, if $f(x) \ge 0$ on $[a, b]$, then $\int_a^b f(x)\,dx \ge 0$.

▶ Further on, it is assumed that the integrals on the right-hand sides of the inequalities of Items 3–8 exist.

3. *Absolute value theorem (integral analogue of the triangle inequality):*

$$\left|\int_a^b f(x)\,dx\right| \le \int_a^b |f(x)|\,dx.$$

4. *Bunyakovsky's inequality (Cauchy–Bunyakovsky inequality):*

$$\left(\int_a^b f(x)g(x)\,dx\right)^2 \le \int_a^b f^2(x)\,dx \int_a^b g^2(x)\,dx.$$

5. *Cauchy's inequality:*

$$\left(\int_a^b [f(x)+g(x)]^2\,dx\right)^{1/2} \le \left(\int_a^b f^2(x)\,dx\right)^{1/2} + \left(\int_a^b g^2(x)\,dx\right)^{1/2}.$$

6. *Minkowski's inequality (generalization of Cauchy's inequality):*

$$\left(\int_a^b |f(x)+g(x)|^p\,dx\right)^{\frac{1}{p}} \le \left(\int_a^b |f(x)|^p\,dx\right)^{\frac{1}{p}} + \left(\int_a^b |g(x)|^p\,dx\right)^{\frac{1}{p}}, \qquad p \ge 1.$$

7. *Hölder's inequality* (at $p = 2$, it translates into Bunyakovsky's inequality):

$$\left| \int_a^b f(x)g(x)\,dx \right| \leq \left( \int_a^b |f(x)|^p\,dx \right)^{\frac{1}{p}} \left( \int_a^b |g(x)|^{\frac{p}{p-1}}\,dx \right)^{\frac{p-1}{p}}, \qquad p > 1.$$

8. *Chebyshev's inequality*:

$$\left( \int_a^b f(x)h(x)\,dx \right) \left( \int_a^b g(x)h(x)\,dx \right) \leq \left( \int_a^b h(x)\,dx \right) \left( \int_a^b f(x)g(x)h(x)\,dx \right),$$

where $f(x)$ and $g(x)$ are monotonically increasing functions and $h(x)$ is a positive integrable function on $[a, b]$.

9. *Jensen's inequality*:

$$f\left( \frac{\int_a^b g(t)x(t)\,dt}{\int_a^b g(t)\,dt} \right) \leq \frac{\int_a^b g(t)f(x(t))\,dt}{\int_a^b g(t)\,dt} \qquad \text{if } f(x) \text{ is convex } (f'' > 0);$$

$$f\left( \frac{\int_a^b g(t)x(t)\,dt}{\int_a^b g(t)\,dt} \right) \geq \frac{\int_a^b g(t)f(x(t))\,dt}{\int_a^b g(t)\,dt} \qquad \text{if } f(x) \text{ is concave } (f'' < 0),$$

where $x(t)$ is a continuous function ($a \leq x \leq b$) and $g(t) \geq 0$. The equality is attained if and only if either $x(t) = \text{const}$ or $f(x)$ is a linear function. Jensen's inequality serves as a general source for deriving various integral inequalities.

10. *Steklov's inequality*. Let $f(x)$ be a continuous function on $[0, \pi]$ and let it have everywhere on $[0, \pi]$, except maybe at finitely many points, a square integrable derivative $f'(x)$. If either of the conditions

(a) $\quad f(0) = f(\pi) = 0,$

(b) $\quad \int_0^\pi f(x)\,dx = 0$

is satisfied, then the following inequality holds:

$$\int_0^\pi [f'(x)]^2\,dx \geq \int_0^\pi [f(x)]^2\,dx.$$

The equality is only attained for functions $f(x) = A \sin x$ in case (a) and functions $f(x) = B \cos x$ in case (b).

11. *A $\pi$-related inequality*. If $a > 0$ and $f(x) \geq 0$ on $[0, a]$, then

$$\left( \int_0^a f(x)\,dx \right)^4 \leq \pi^2 \left( \int_0^a f^2(x)\,dx \right) \left( \int_0^a x^2 f^2(x)\,dx \right).$$

### 7.2.5-3. Arithmetic, geometric, harmonic, and quadratic means of functions.

Let $f(x)$ be a positive function integrable on $[a, b]$. Consider the values of $f(x)$ on a discrete set of points:

$$f_{kn} = f(a + k\delta_n), \qquad \delta_n = \frac{b - a}{n} \qquad (k = 1, \ldots, n).$$

The arithmetic mean, geometric mean, harmonic mean, and quadratic mean of a function $f(x)$ on an interval $[a, b]$ are introduced using the definitions of the respective mean values for finitely many numbers (see Subsection 1.6.1) and going to the limit as $n \to \infty$.

1. *Arithmetic mean of a function $f(x)$ on $[a, b]$*:

$$\lim_{n \to \infty} \frac{1}{n} \sum_{k=1}^{n} f_{kn} = \frac{1}{b-a} \int_{a}^{b} f(x)\, dx.$$

This definition is in agreement with another definition of the mean value of a function $f(x)$ on $[a, b]$ given in Theorem 1 from Paragraph 7.2.5-1.

2. *Geometric mean of a function $f(x)$ on $[a, b]$*:

$$\lim_{n \to \infty} \left( \prod_{k=1}^{n} f_{kn} \right)^{1/n} = \exp\left( \frac{1}{b-a} \int_{a}^{b} \ln f(x)\, dx \right).$$

3. *Harmonic mean of a function $f(x)$ on $[a, b]$*:

$$\lim_{n \to \infty} n \left( \sum_{k=1}^{n} \frac{1}{f_{kn}} \right)^{-1} = (b-a) \left( \int_{a}^{b} \frac{dx}{f(x)} \right)^{-1}.$$

4. *Quadratic mean of a function $f(x)$ on $[a, b]$*:

$$\lim_{n \to \infty} \left( \frac{1}{n} \sum_{k=1}^{n} f_{kn}^2 \right)^{1/2} = \left( \frac{1}{b-a} \int_{a}^{b} f^2(x)\, dx \right)^{1/2}.$$

This definition differs from the common definition of the norm of a square integrable function given in Paragraph 7.2.13-2 by the constant factor $1/\sqrt{b-a}$.

The following inequalities hold:

$$(b-a)\left( \int_{a}^{b} \frac{dx}{f(x)} \right)^{-1} \leq \exp\left( \frac{1}{b-a} \int_{a}^{b} \ln f(x)\, dx \right) \leq \frac{1}{b-a} \int_{a}^{b} f(x)\, dx \leq \left( \frac{1}{b-a} \int_{a}^{b} f^2(x)\, dx \right)^{1/2}.$$

To make it easier to remember, let us rewrite these inequalities in words as

$$\boxed{\text{harmonic mean}} \leq \boxed{\text{geometric mean}} \leq \boxed{\text{arithmetic mean}} \leq \boxed{\text{quadratic mean}}.$$

The equality is attained for $f(x) = \text{const}$ only.

**7.2.5-4. General approach to defining mean values.**

Let $g(y)$ be a continuous monotonic function defined in the range $0 \leq y < \infty$.

The *mean of a function $f(x)$ with respect to a function $g(x)$ on an interval $[a, b]$* is defined as

$$\lim_{n \to \infty} g^{-1}\left( \frac{1}{n} \sum_{k=1}^{n} g(f_{kn}) \right) = g^{-1}\left( \frac{1}{b-a} \int_{a}^{b} g(f(x))\, dx \right),$$

where $g^{-1}(z)$ is the inverse of $g(y)$.

The means presented in Paragraph 7.2.5-3 are special cases of the mean with respect to a function:

arithmetic mean of $f(x)$ = mean of $f(x)$ with respect to $g(y) = y$,
geometric mean of $f(x)$ = mean of $f(x)$ with respect to $g(y) = \ln y$,
harmonic mean of $f(x)$ = mean of $f(x)$ with respect to $g(y) = 1/y$,
quadratic mean of $f(x)$ = mean of $f(x)$ with respect to $g(y) = y^2$.

## 7.2.6. Geometric and Physical Applications of the Definite Integral

**7.2.6-1. Geometric applications of the definite integral.**

1. The *area of a domain* $D$ bounded by curves $y = f(x)$ and $y = g(x)$ and straight lines $x = a$ and $x = b$ in the $x, y$ plane (see Fig. 7.2 a) is calculated by the formula

$$S = \int_a^b \left[f(x) - g(x)\right] dx.$$

If $g(x) \equiv 0$, this formula gives the area of a curvilinear trapezoid bounded by the $x$-axis, the curve $y = f(x)$, and the straight lines $x = a$ and $x = b$.

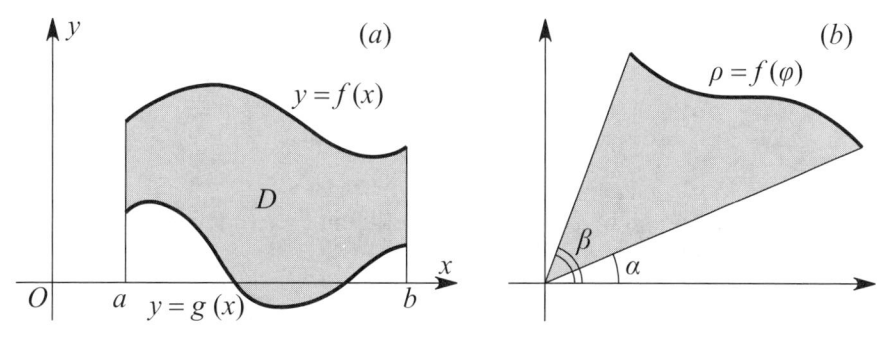

**Figure 7.2.** (a) A domain $D$ bounded by two curves $y = f(x)$ and $y = g(x)$ on an interval $[a, b]$; (b) a curvilinear sector.

2. *Area of a domain* $D$. Let $x = x(t)$ and $y = y(t)$, with $t_1 \le t \le t_2$, be parametric equations of a piecewise-smooth simple closed curve bounding on its left (traced counterclockwise) a domain $D$ with area $S$. Then

$$S = -\int_{t_1}^{t_2} y(t)x'(t)\,dt = \int_{t_1}^{t_2} x(t)y'(t)\,dt = \frac{1}{2}\int_{t_1}^{t_2} \left[x(t)y'(t) - y(t)x'(t)\right] dt.$$

3. *Area of a curvilinear sector.* Let a curve $\rho = f(\varphi)$, with $\varphi \in [\alpha, \beta]$, be defined in the polar coordinates $\rho, \varphi$. Then the area of the curvilinear sector $\{\alpha \le \varphi \le \beta;\ 0 \le \rho \le f(\varphi)\}$ (see Fig. 7.2 b) is calculated by the formula

$$S = \frac{1}{2}\int_\alpha^\beta [f(\varphi)]^2\,d\varphi.$$

4. *Area of a surface of revolution.* Let a surface of revolution be generated by rotating a curve $y = f(x) \ge 0$, $x \in [a, b]$, about the $x$-axis; see Fig. 7.3. The area of this surface is calculated as

$$S = 2\pi \int_a^b f(x)\sqrt{1 + [f'(x)]^2}\,dx.$$

5. *Volume of a body of revolution.* Let a body of revolution be obtained by rotating about the $x$-axis a curvilinear trapezoid bounded by a curve $y = f(x)$, the $x$-axis, and straight lines $x = a$ and $x = b$; see Fig. 7.3. Then the volume of this body is calculated as

$$V = \pi \int_a^b [f(x)]^2\,dx.$$

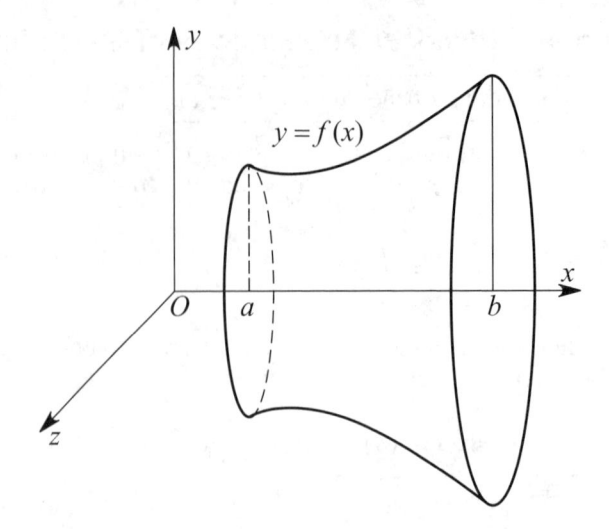

**Figure 7.3.** A surface of revolution generated by rotating a curve $y = f(x)$.

6. *Arc length of a plane curve defined in different ways.*

(a) If a curve is the graph of a continuously differentiable function $y = f(x)$, $x \in [a, b]$, then its length is determined as

$$L = \int_a^b \sqrt{1 + [f'(x)]^2}\, dx.$$

(b) If a plane curve is defined parametrically by equations $x = x(t)$ and $y = y(t)$, with $t \in [\alpha, \beta]$ and $x(t)$ and $y(t)$ being continuously differentiable functions, then its length is calculated by

$$L = \int_\alpha^\beta \sqrt{[x'(t)]^2 + [y'(t)]^2}\, dt.$$

(c) If a curve is defined in the polar coordinates $\rho$, $\varphi$ by an equation $\rho = \rho(\varphi)$, with $\varphi \in [\alpha, \beta]$, then its length is found as

$$L = \int_\alpha^\beta \sqrt{\rho^2(\varphi) + [\rho'(\varphi)]^2}\, d\varphi.$$

7. The *arc length of a spatial curve* defined parametrically by equations $x = x(t)$, $y = y(t)$, and $z = z(t)$, with $t \in [\alpha, \beta]$ and $x(t)$, $y(t)$, and $z(t)$ being continuously differentiable functions, is calculated by

$$L = \int_\alpha^\beta \sqrt{[x'(t)]^2 + [y'(t)]^2 + [z'(t)]^2}\, dt.$$

7.2.6-2. Physical application of the integral.

1. *Work of a variable force.* Suppose a point mass moves along the $x$-axis from a point $x = a$ to a point $x = b$ under the action of a variable force $F(x)$ directed along the $x$-axis. The mechanical work of this force is equal to

$$A = \int_a^b F(x)\, dx.$$

2. *Mass of a rectilinear rod of variable density.* Suppose a rod with a constant cross-sectional area $S$ occupies an interval $[0, l]$ on the $x$-axis and the density of the rod material is a function of $x$: $\rho = \rho(x)$. The mass of this rod is calculated as

$$m = S \int_0^l \rho(x)\, dx.$$

3. *Mass of a curvilinear rod of variable density.* Let the shape of a plane curvilinear rod with a constant cross-sectional area $S$ be defined by an equation $y = f(x)$, with $a \leq x \leq b$, and let the density of the material be coordinate dependent: $\rho = \rho(x, y)$. The mass of this rod is calculated as

$$m = S \int_a^b \rho(x, f(x)) \sqrt{1 + [y'(x)]^2}\, dx.$$

If the shape of the rod is defined parametrically by $x = x(t)$ and $y = y(t)$, then its mass is found as

$$m = S \int_a^b \rho(x(t), y(t)) \sqrt{[x'(t)]^2 + [y'(t)]^2}\, dt.$$

4. The *coordinates of the center of mass of a plane homogeneous material curve* whose shape is defined by an equation $y = f(x)$, with $a \leq x \leq b$, are calculated by the formulas

$$x_c = \frac{1}{L} \int_a^b x\sqrt{1 + [y'(x)]^2}\, dx, \qquad y_c = \frac{1}{L} \int_a^b f(x)\sqrt{1 + [y'(x)]^2}\, dx,$$

where $L$ is the length of the curve.

If the shape of a plane homogeneous material curve is defined parametrically by $x = x(t)$ and $y = y(t)$, then the coordinates of its center of mass are obtained as

$$x_c = \frac{1}{L} \int_a^b x(t) \sqrt{[x'(t)]^2 + [y'(t)]^2}\, dt, \qquad y_c = \frac{1}{L} \int_a^b y(t) \sqrt{[x'(t)]^2 + [y'(t)]^2}\, dt.$$

### 7.2.7. Improper Integrals with Infinite Integration Limit

An improper integral is an integral with an infinite limit (limits) of integration or an integral of an unbounded function.

**7.2.7-1. Integrals with infinite limits.**

1°. Let $y = f(x)$ be a function defined and continuous on an infinite interval $a \leq x < \infty$. If there exists a finite limit $\lim\limits_{b \to \infty} \int_a^b f(x)\, dx$, then it is called a (convergent) *improper integral* of $f(x)$ on the interval $[a, \infty)$ and is denoted $\int_a^\infty f(x)\, dx$. Thus, by definition

$$\int_a^\infty f(x)\, dx = \lim_{b \to \infty} \int_a^b f(x)\, dx. \qquad (7.2.7.1)$$

If the limit is infinite or does not exist, the improper integral is called *divergent*.

The *geometric meaning of an improper integral* is that the integral $\int_a^\infty f(x)\, dx$, with $f(x) \geq 0$, is equal to the area of the unbounded domain between the curve $y = f(x)$, its asymptote $y = 0$, and the straight line $x = a$ on the left.

2°. Suppose an antiderivative $F(x)$ of the integrand function $f(x)$ is known. Then the improper integral (7.2.7.1) is

(i) *convergent* if there exists a finite limit $\lim\limits_{x\to\infty} F(x) = F(\infty)$;

(ii) *divergent* if the limit is infinite or does not exist.

In case (i), we have
$$\int_a^\infty f(x)\,dx = F(x)\Big|_a^\infty = F(\infty) - F(a).$$

**Example 1.** Let us investigate the issue of convergence of the improper integral $I = \int_a^\infty \dfrac{dx}{x^\lambda}$, $a > 0$.

The integrand $f(x) = x^{-\lambda}$ has an antiderivative $F(x) = \dfrac{1}{1-\lambda} x^{1-\lambda}$. Depending on the value of the parameter $\lambda$, we have
$$\lim_{x\to\infty} F(x) = \frac{1}{1-\lambda} \lim_{x\to\infty} x^{1-\lambda} = \begin{cases} 0 & \text{if } \lambda > 1, \\ \infty & \text{if } \lambda \leq 1. \end{cases}$$

Therefore, if $\lambda > 1$, the integral is convergent and is equal to $I = F(\infty) - F(a) = \dfrac{a^{1-\lambda}}{\lambda - 1}$, and if $\lambda \leq 1$, the integral is divergent.

3°. Improper integrals for other infinite intervals are defined in a similar way:
$$\int_{-\infty}^b f(x)\,dx = \lim_{a\to-\infty} \int_a^b f(x)\,dx,$$
$$\int_{-\infty}^\infty f(x)\,dx = \int_{-\infty}^c f(x)\,dx + \int_c^\infty f(x)\,dx.$$

Note that if either improper integral on the right-hand side of the latter relation is convergent, then, by definition, the integral on the left-hand side is also convergent.

4°. Properties 2–4 and 6–9 from Paragraph 7.2.2-2, where $a$ can be equal to $-\infty$ and $b$ can be $\infty$, apply to improper integrals as well; it is assumed that all quantities on the right-hand sides exist (the integrals are convergent).

7.2.7-2. **Sufficient conditions for convergence of improper integrals.**

In many problems, it suffices to establish whether a given improper integral is convergent or not and, if yes, evaluate it. The theorems presented below can be useful in doing so.

THEOREM 1 (CAUCHY'S CONVERGENCE CRITERION). *For the integral (7.2.7.1) to be convergent it is necessary and sufficient that for any $\varepsilon > 0$ there exists a number $R$ such that the inequality*
$$\left| \int_\alpha^\beta f(x)\,dx \right| < \varepsilon$$
*holds for any $\beta > \alpha > R$.*

THEOREM 2. *If $0 \leq f(x) \leq g(x)$ for $x \geq a$, then the convergence of the integral $\int_a^\infty g(x)\,dx$ implies the convergence of the integral $\int_a^\infty f(x)\,dx$; moreover, $\int_a^\infty f(x)\,dx \leq \int_a^\infty g(x)\,dx$. If the integral $\int_a^\infty f(x)\,dx$ is divergent, then the integral $\int_a^\infty g(x)\,dx$ is also divergent.*

THEOREM 3. *If the integral $\int_a^\infty |f(x)|\,dx$ is convergent, then the integral $\int_a^\infty f(x)\,dx$ is also convergent; in this case, the latter integral is called absolutely convergent.*

**Example 2.** The improper integral $\int_1^\infty \frac{\sin x}{x^2}\,dx$ is absolutely convergent, since $\left|\frac{\sin x}{x^2}\right| \le \frac{1}{x^2}$ and the integral $\int_1^\infty \frac{1}{x^2}\,dx$ is convergent (see Example 1).

THEOREM 4. *Let $f(x)$ and $g(x)$ be integrable functions on any finite interval $a \le x \le b$ and let there exist a limit, finite or infinite,*

$$\lim_{x \to \infty} \frac{f(x)}{g(x)} = K.$$

*Then the following assertions hold:*
*1. If $0 < K < \infty$, both integrals*

$$\int_a^\infty f(x)\,dx, \qquad \int_a^\infty g(x)\,dx \tag{7.2.7.2}$$

*are convergent or divergent simultaneously.*
*2. If $0 \le K < \infty$, the convergence of the latter integral in (7.2.7.2) implies the convergence of the former integral.*
*3. If $0 < K \le \infty$, the divergence of the latter integral in (7.2.7.2) implies the divergence of the former integral.*

THEOREM 5 (COROLLARY OF THEOREM 4). *Given a function $f(x)$, let its asymptotics for sufficiently large $x$ have the form*

$$f(x) = \frac{\varphi(x)}{x^\lambda} \qquad (\lambda > 0).$$

*Then: (i) if $\lambda > 1$ and $\varphi(x) \le c < \infty$, then the integral $\int_a^\infty f(x)\,dx$ is convergent; (ii) if $\lambda \le 1$ and $\varphi(x) \ge c > 0$, then the integral is divergent.*

THEOREM 6. *Let $f(x)$ be an absolutely integrable function on an interval $[a, \infty)$ and let $g(x)$ be a bounded function on $[a, \infty)$. Then the product $f(x)g(x)$ is an absolutely integrable function on $[a, \infty)$.*

THEOREM 7 (ANALOGUE OF ABEL'S TEST FOR CONVERGENCE OF INFINITE SERIES). *Let $f(x)$ be an integrable function on an interval $[a, \infty)$ such that the integral (7.2.7.1) is convergent (maybe not absolutely) and let $g(x)$ be a monotonic and bounded function on $[a, \infty)$. Then the integral*

$$\int_a^\infty f(x)g(x)\,dx \tag{7.2.7.3}$$

*is convergent.*

THEOREM 8 (ANALOGUE OF DIRICHLET'S TEST FOR CONVERGENCE OF INFINITE SERIES). *Let (i) $f(x)$ be an integrable function on any finite interval $[a, A]$ and*

$$\left|\int_a^A f(x)\,dx\right| \le K < \infty \qquad (a \le A < \infty);$$

*(ii) $g(x)$ be a function tending to zero monotonically as $x \to \infty$: $\lim_{x \to \infty} g(x) = 0$. Then the integral (7.2.7.3) is convergent.*

**Example 3.** Let us show that the improper integral $\int_a^\infty \frac{\sin x}{x^\lambda}\,dx$ is convergent for $a > 0$ and $\lambda > 0$.
Set $f(x) = \sin x$ and $g(x) = x^{-\lambda}$ and verify conditions (i) and (ii) of Theorem 8. We have

(i) $\left|\int_a^A \sin x\,dx\right| = |\cos a - \cos A| \leq 2;$

(ii) since $\lambda > 0$, the function $x^{-\lambda}$ is monotonically decreasing and goes to zero as $x \to \infty$.

So both conditions of Theorem 8 are met, and therefore the given improper integral is convergent.

### 7.2.7-3. Some remarks.

1°. If an improper integral is convergent and the integrand function tends to a limit as $x \to \infty$, then this limit can only be zero (it is such situations that were dealt with in Examples 1–3). However, the property $\lim_{x\to\infty} f(x) = 0$ is not a *necessary condition for convergence of the integral* (7.2.7.1).

An integral can also be convergent if the integrand function does not have a limit as $x \to \infty$. For example, this is the case for *Fresnel's integrals*:

$$\int_0^\infty \sin(x^2)\,dx = \int_0^\infty \cos(x^2)\,dx = \frac{1}{2}\sqrt{\frac{\pi}{2}}.$$

Furthermore, it can be shown that the integral $\int \frac{x}{1+x^6 \sin^2 x}\,dx$ is convergent regardless of the fact that the integrand function, being everywhere positive, is not even bounded ($f(\pi k) = \pi k$, $k = 1, 2, \ldots$). The graph of this function has infinitely many spikes with heights increasing indefinitely and base widths vanishing. At the points lying outside the spike bases, the function rapidly goes to zero.

2°. If $f(x)$ is a monotonic function for $x \geq 0$ and the improper integral $\int_0^\infty f(x)\,dx$ is convergent, then the following limiting relation holds:

$$\int_0^\infty f(x)\,dx = \lim_{\varepsilon \to 0} \varepsilon \sum_{n=1}^\infty f(\varepsilon n).$$

## 7.2.8. General Reduction Formulas for the Calculation of Improper Integrals

Below are some general formulas, involving arbitrary functions and parameters, that may facilitate the calculation of improper integrals.

### 7.2.8-1. Improper integrals involving power functions.

$$\int_0^\infty f\left(\frac{a+bx}{1+x}\right)\frac{dx}{(1+x)^2} = \frac{1}{b-a}\int_a^b f(x)\,dx;$$

$$\int_0^\infty \frac{f(ax) - f(bx)}{x}\,dx = [f(0) - f(\infty)]\ln\frac{b}{a} \quad \text{if } a > 0,\ b > 0,\ f(x) \text{ is continuous}$$

on $[0, \infty)$, and $f(\infty) = \lim_{x\to\infty} f(x)$ is a finite quantity;

$$\int_0^\infty \frac{f(ax)-f(bx)}{x}\,dx = f(0)\ln\frac{b}{a} \quad \text{if } a>0,\ b>0,\ f(x) \text{ is continuous on } [0,\infty),$$

and the integral $\int_c^\infty \frac{f(x)}{x}\,dx$ exists; $c>0$;

$$\int_0^\infty f\left(\left|ax-\frac{b}{x}\right|\right)dx = \frac{1}{a}\int_0^\infty f(|x|)\,dx \quad \text{if } a>0,\ b>0;$$

$$\int_0^\infty x^2 f\left(\left|ax-\frac{b}{x}\right|\right)dx = \frac{1}{a^3}\int_0^\infty (x^2+ab)f(|x|)\,dx \quad \text{if } a>0,\ b>0;$$

$$\int_0^\infty f\left(\left|ax-\frac{b}{x}\right|\right)\frac{dx}{x^2} = \frac{1}{b}\int_0^\infty f(|x|)\,dx \quad \text{if } a>0,\ b>0;$$

$$\int_0^\infty f\left(x,\frac{1}{x}\right)\frac{dx}{x} = 2\int_0^1 f\left(x,\frac{1}{x}\right)\frac{dx}{x} \quad \text{if } f(x,y)=f(y,x);$$

$$\int_0^\infty f\left(x,\frac{a}{x}\right)\frac{dx}{x} = 0 \quad \text{if } f(x,y)=-f(y,x),\ a>0 \text{ (the integral is assumed to exist).}$$

7.2.8-2. **Improper integrals involving logarithmic functions.**

$$\int_0^\infty f\left(\frac{x}{a}+\frac{a}{x}\right)\frac{\ln x}{x}\,dx = \ln a \int_0^\infty f\left(\frac{x}{a}+\frac{a}{x}\right)\frac{dx}{x} \quad \text{if } a>0;$$

$$\int_0^\infty f\left(\frac{x^p}{a}+\frac{a}{x^p}\right)\frac{\ln x}{x}\,dx = \frac{\ln a}{p}\int_0^\infty f\left(\frac{x^p}{a}+\frac{a}{x^p}\right)\frac{dx}{x} \quad \text{if } a>0,\ p>0;$$

$$\int_0^\infty f(x^a + x^{-a})\frac{\ln x}{1+x^2}\,dx = 0 \quad \text{(a special case of the integral below);}$$

$$\int_0^\infty f\left(x,\frac{1}{x}\right)\frac{\ln x}{1+x^2}\,dx = 0 \quad \text{if } f(x,y)=f(y,x) \text{ (the integral is assumed to exist);}$$

$$\int_0^\infty f\left(x,\frac{1}{x}\right)\frac{\ln x}{x}\,dx = 0 \quad \text{if } f(x,y)=f(y,x) \text{ (the integral is assumed to exist).}$$

7.2.8-3. **Improper integrals involving trigonometric functions.**

$$\int_0^\infty f(x)\frac{\sin x}{x}\,dx = \int_0^{\pi/2} f(x)\,dx \quad \text{if } f(x)=f(-x) \text{ and } f(x+\pi)=f(x);$$

$$\int_0^\infty f(x)\frac{\sin x}{x}\,dx = \int_0^{\pi/2} f(x)\cos x\,dx \quad \text{if } f(x)=f(-x) \text{ and } f(x+\pi)=-f(x);$$

$$\int_0^\infty \frac{f(\sin x)}{x}\,dx = \int_0^{\pi/2}\frac{f(\sin x)}{\sin x}\,dx \quad \text{if } f(-x)=-f(x);$$

$$\int_0^\infty \frac{f(\sin x)}{x^2}\,dx = \int_0^{\pi/2}\frac{f(\sin x)}{\sin^2 x}\,dx \quad \text{if } f(x)=f(-x);$$

$$\int_0^\infty \frac{f(\sin x)}{x}\cos x\,dx = \int_0^{\pi/2}\frac{f(\sin x)}{\sin x}\cos^2 x\,dx \quad \text{if } f(-x)=-f(x);$$

$$\int_0^\infty \frac{f(\sin x)}{x} \tan x \, dx = \int_0^{\pi/2} f(\sin x) \, dx \quad \text{if } f(-x) = f(x);$$

$$\int_0^\infty \frac{f(\sin x)}{x^2 + a^2} \, dx = \frac{\sinh(2a)}{2a} \int_0^{\pi/2} \frac{f(\sin x) \, dx}{\cosh^2 a - \cos^2 x} \quad \text{if } f(-x) = f(x);$$

$$\int_0^\infty f\left(x + \frac{1}{x}\right) \frac{\arctan x}{x} \, dx = \frac{\pi}{4} \int_0^\infty f\left(x + \frac{1}{x}\right) \frac{dx}{x}.$$

### 7.2.8-4. Calculation of improper integrals using analytic functions.

Suppose
$$F(z) = f(r, x) + ig(r, x), \quad z = r(\cos x + i \sin x), \quad i^2 = -1,$$
where $F(z)$ is a function analytic in a circle of radius $r$. Then the following formulas hold:

$$\int_0^\infty \frac{f(x, r)}{x^2 + a^2} \, dx = \frac{\pi}{2a} F(re^{-a});$$

$$\int_0^\infty \frac{xg(x, r)}{x^2 + a^2} \, dx = \frac{\pi}{2}[F(re^{-a}) - F(0)];$$

$$\int_0^\infty \frac{g(x, r)}{x} \, dx = \frac{\pi}{2}[F(r) - F(0)];$$

$$\int_0^\infty \frac{g(x, r)}{x(x^2 + a^2)} \, dx = \frac{\pi}{2a^2}[F(r) - F(re^{-a})].$$

▶ Paragraph 10.1.2-8 presents a method for the calculation of improper integrals using the theory of functions of a complex variable.

### 7.2.8-5. Calculation of improper integrals using the Laplace transform.

The following classes of improper integrals may be evaluated using the Laplace transform:

$$\int_0^\infty \frac{f(x)}{x} \, dx = \int_0^\infty \widetilde{f}(p) \, dp,$$

$$\int_0^\infty x^n f(x) \, dx = (-1)^{n+1} \left[\frac{d^n}{dp^n} \widetilde{f}(p)\right]_0^\infty, \quad n = 1, 2, \ldots,$$
(7.2.8.1)

where $\widetilde{f}(p)$ is the *Laplace transform* of the function $f(x)$, which is defined as

$$\widetilde{f}(p) = \int_0^\infty e^{-px} f(x) \, dx.$$

Short notation for the Laplace transform: $\widetilde{f}(p) = \mathcal{L}\{f(x)\}$.

Section 11.2 presents properties and methods for determining the *Laplace transform*, and Section T3.1 gives tables of the Laplace transforms of various functions.

**Example 1.** Evaluate the integral $\int_0^\infty \frac{\sin(ax)}{x}\,dx$.

Using Table 11.2 from Subsection 11.2.2 (or the table from Subsection T3.1.6), we find the Laplace transform of the function $\sin(ax)$: $\mathcal{L}\{\sin(ax)\} = \frac{a}{a^2 + p^2}$. Substitute this expression into the first formula in (7.2.8.1) and integrate to obtain

$$\int_0^\infty \frac{\sin(ax)}{x}\,dx = \int_0^\infty \frac{a\,dp}{a^2 + p^2} = \arctan\frac{p}{a}\bigg|_0^\infty = \frac{\pi}{2}.$$

**Example 2.** Evaluate *Frullani's integral* $\int_0^\infty \frac{e^{-ax} - e^{-bx}}{x}\,dx$, where $a > 0$ and $b > 0$.

Using the first formula in (7.2.8.1) and Table 11.2 from Subsection 11.2.2 (or the table from Subsection T3.1.3), we obtain $\mathcal{L}\{e^{-ax}\} = \frac{1}{p + a}$. Integrating yields

$$\int_0^\infty \frac{e^{-ax} - e^{-bx}}{x}\,dx = \int_0^\infty \left(\frac{1}{p+a} - \frac{1}{p+b}\right) dp = \ln\frac{p+a}{p+b}\bigg|_0^\infty = -\ln\frac{a}{b} = \ln\frac{b}{a}.$$

### 7.2.9. General Asymptotic Formulas for the Calculation of Improper Integrals

Below are some general formulas, involving arbitrary functions and parameters, that may be useful for determining the asymptotic behavior of improper integrals.

#### 7.2.9-1. Asymptotic formulas for some improper integrals with parameter.

1°. For asymptotics of improper Laplace integrals

$$I(\lambda) = \int_a^\infty f(x)\exp[\lambda g(x)]\,dx$$

as $\lambda \to \infty$, see Remark 1 in Paragraph 7.2.4-3.

2°. For $\lambda \to \infty$, the following asymptotic expansions of improper integrals involving trigonometric functions and a Bessel function hold:

$$\int_0^\infty \cos(\lambda x)f(x)\,dx = \sum_{k=1}^n (-1)^k f^{(2k-1)}(0)\lambda^{-2k} + O(\lambda^{-2n-1}),$$

$$\int_0^\infty \sin(\lambda x)f(x)\,dx = \sum_{k=0}^{n-1} (-1)^k f^{(2k)}(0)\lambda^{-2k-1} + O(\lambda^{-2n-1}),$$

$$\int_0^\infty J_0(\lambda x)f(x)\,dx = \frac{1}{\sqrt{\pi}}\sum_{k=0}^{n-1} \frac{(-1)^k}{k!}\Gamma\left(k+\frac{1}{2}\right)f^{(2k)}(0)\lambda^{-2k-1} + O(\lambda^{-2n-1}).$$

The function $f(x)$ is assumed to have $2n + 1$ partial derivatives with respect to $x$ for $x \geq 0$ that monotonically go to zero as $x \to \infty$.

3°. For $\lambda \to \infty$, the following asymptotic expansions hold:

$$\int_0^\infty f(x)g\left(\frac{x}{\lambda}\right) dx = \sum_{k=0}^n \frac{(-1)^k}{k!} F^{(k)}(0)g^{(k)}(0)\lambda^{-k} + O(\lambda^{-n-1}),$$

$$\int_0^\infty f(\lambda x)g(x) dx = \sum_{k=0}^n \frac{(-1)^k}{k!} F^{(k)}(0)g^{(k)}(0)\lambda^{-k-1} + O(\lambda^{-n-2}),$$

where $F(t) = \int_0^\infty f(x)e^{-xt} dx$ is the Laplace transform of the function $f(x)$.

4°. For $\lambda \to \infty$, the following asymptotic expansions hold:

$$\int_0^\infty \frac{f(x) dx}{x + \lambda} = \sum_{k=0}^n \frac{F^{(k)}(0)}{\lambda^{k+1}} + O(\lambda^{-n-2}),$$

where $F(t) = \int_0^\infty f(x)e^{-xt} dx$ is the Laplace transform of the function $f(x)$.

**7.2.9-2. Behavior of integrals with variable limit of integration as $x \to \infty$.**

Let $f(t)$ be a continuously differentiable function, let $g(t)$ be a twice continuously differentiable function, and let the following conditions hold:

$$f(t) > 0, \quad g'(t) > 0; \quad g(t) \to \infty \text{ as } t \to \infty;$$
$$f'(t)/f(t) = o(g'(t)) \text{ as } t \to \infty; \quad g''(t) = o(g'^2(t)) \text{ as } t \to \infty.$$

Then the following asymptotic relation holds as $x \to \infty$:

$$\int_x^\infty f(t) \exp[-g(t)] dt \simeq \frac{f(x)}{g'(x)} \exp[-g(x)].$$

**7.2.9-3. $\pi$-related inequality.**

If $f(x) \geq 0$, the inequality

$$\left(\int_0^\infty f(x) dx\right)^4 \leq \pi^2 \left(\int_0^\infty f^2(x) dx\right)\left(\int_0^\infty x^2 f^2(x) dx\right)$$

holds, provided the integral on the left-hand side exists. The constant $\pi^2$ is best in the sense that there exist functions $f(x) \not\equiv 0$ for which the equality is attained.

## 7.2.10. Improper Integrals of Unbounded Functions

**7.2.10-1. Basic definitions.**

1°. Let a function $f(x)$ be defined and continuous for $a \leq x < b$, but $\lim_{x \to b-0} f(x) = \infty$. If there exists a finite limit $\lim_{\lambda \to b-0} \int_a^\lambda f(x) dx$, it is called the (convergent) *improper integral of the unbounded function* $f(x)$ over the interval $[a, b]$. Thus, by definition

$$\int_a^b f(x) dx = \lim_{\lambda \to b-0} \int_a^\lambda f(x) dx. \quad (7.2.10.1)$$

If no finite limit exists, the integral is called *divergent*.

If $\lim_{x \to a+0} f(x) = \infty$, then, by definition, it is assumed that

$$\int_a^b f(x)\,dx = \lim_{\gamma \to a+0} \int_\gamma^b f(x)\,dx.$$

Finally, if $f(x)$ is unbounded near a point $c \in (a,b)$ and both integrals $\int_a^c f(x)\,dx$ and $\int_c^b f(x)\,dx$ are convergent, then, by definition,

$$\int_a^b f(x)\,dx = \int_a^c f(x)\,dx + \int_c^b f(x)\,dx.$$

2°. The geometric meaning of an improper integral of an unbounded function and also sufficient conditions for convergence of such integrals are similar to those for improper integrals with infinite limit(s).

### 7.2.10-2. Convergence tests for improper integrals of unbounded functions.

Presented below are theorems for the case where the only singular point of the integrand function is the right endpoint of the interval $[a,b]$.

THEOREM 1 (CAUCHY'S CONVERGENCE CRITERION). *For the integral (7.2.10.1) to be convergent is it necessary and sufficient that for any $\varepsilon > 0$ there exists a number $\delta > 0$ such that for any $\delta_1$ and $\delta_2$ satisfying $0 < \delta_1 < \delta$ and $0 < \delta_2 < \delta$ the following inequality holds:*

$$\left| \int_{b-\delta_1}^{b-\delta_2} f(x)\,dx \right| < \varepsilon.$$

THEOREM 2. *If $0 \le f(x) \le g(x)$ for $a \le x < b$, then the convergence of the integral $\int_a^b g(x)\,dx$ implies the convergence of the integral $\int_a^b f(x)\,dx$, with $\int_a^b f(x)\,dx \le \int_a^b g(x)\,dx$. If the integral $\int_a^b f(x)\,dx$ is divergent, then the integral $\int_a^b g(x)\,dx$ is also divergent.*

**Example.** For any continuous function $\varphi(x)$ such that $\varphi(1) = 0$, the improper integral $\int_0^1 \dfrac{dx}{\varphi^2(x) + \sqrt{1-x}}$ is convergent and does not exceed 2, since $\dfrac{1}{\varphi^2(x) + \sqrt{1-x}} < \dfrac{1}{\sqrt{1-x}}$, while the integral $\int_0^1 \dfrac{dx}{\sqrt{1-x}}$ is convergent and is equal to 2.

THEOREM 3. *Let $f(x)$ and $g(x)$ be continuous functions on $[a,b)$ and let the following limit exist:*

$$\lim_{x \to b} \frac{f(x)}{g(x)} = K \qquad (0 < K < \infty).$$

*Then both integrals*

$$\int_a^b f(x)\,dx, \quad \int_a^b g(x)\,dx$$

*are either convergent or divergent simultaneously.*

THEOREM 4. *Let a function $f(x)$ be representable in the form*

$$f(x) = \frac{\varphi(x)}{(b-x)^\lambda} \qquad (\lambda > 0),$$

*where $\varphi(x)$ is continuous on $[a, b]$ and the condition $\varphi(b) \neq 0$ holds.*

*Then: (i) if $\lambda < 1$ and $\varphi(x) \leq c < \infty$, then the integral $\int_a^b f(x)\,dx$ is convergent; (ii) if $\lambda \geq 1$ and $\varphi(x) \geq c > 0$, this integral is divergent.*

Remark. The issue of convergence of the integral of an unbounded function (7.2.10.1) at $x = b$ can be reduced by a simple change of variable to the issue of convergence of an improper integral with an infinite limit:

$$\int_a^b f(x)\,dx = (b-a)\int_1^\infty f\left(\frac{a-b}{z} + b\right)\frac{dz}{z^2}, \qquad z = \frac{b-a}{b-x}.$$

### 7.2.10-3. Calculation of integrals using infinite sums of special form.

Let a function $f(x)$ be continuous and monotonic on the interval $(0,1)$, whose endpoints can be singular. If the integral (proper or improper) $\int_0^1 f(x)\,dx$ exists, then the following limiting relations hold:

$$\int_0^1 f(x)\,dx = \lim_{n\to\infty} \frac{1}{n}\sum_{k=1}^{n-1} f\left(\frac{k}{n}\right) = \lim_{n\to\infty} \frac{1}{n}\sum_{k=1}^{n} f\left(\frac{2k-1}{2n}\right).$$

## 7.2.11. Cauchy-Type Singular Integrals

### 7.2.11-1. Hölder and Lipschitz conditions.

We say that $f(x)$ satisfies the *Hölder condition* on $[a,b]$ if for any two points $x_1 \in [a,b]$ and $x_2 \in [a,b]$ we have

$$|f(x_2) - f(x_1)| < A|x_2 - x_1|^\lambda, \tag{7.2.11.1}$$

where $A$ and $\lambda$ are positive constants. The number $A$ is called the *Hölder constant* and $\lambda$ is called the *Hölder exponent*. If $\lambda > 1$, then by condition (7.2.11.1) the derivative $f'_x(x)$ vanishes everywhere, and $f(x)$ must be constant. Therefore, we assume that $0 < \lambda \leq 1$. For $\lambda = 1$, the Hölder condition is often called the *Lipschitz condition*. Sometimes the Hölder condition is called the *Lipschitz condition of order $\lambda$*.

If $x_1$ and $x_2$ are sufficiently close to each other and if the Hölder condition holds for some exponent $\lambda_1$, then this condition certainly holds for each exponent $\lambda < \lambda_1$. In general, the converse assertion fails. The smaller $\lambda$, the broader is the class of Hölder continuous functions. The narrowest class is that of functions satisfying the Lipschitz condition.

It follows from the last property that if functions $f_1(x)$ and $f_2(x)$ satisfy the Hölder condition with exponents $\lambda_1$ and $\lambda_2$, respectively, then their sum and the product, as well as their ratio provided that the denominator is nonzero, satisfy the Hölder condition with exponent $\lambda = \min(\lambda_1, \lambda_2)$.

If $f(x)$ is differentiable and has a bounded derivative, then $f(x)$ satisfies the Lipschitz condition. In general, the converse assertion fails.

### 7.2.11-2. Principal value of a singular integral.

**1°.** Consider the integral
$$\int_a^b \frac{dx}{x-c}, \qquad a < c < b.$$

Evaluating this integral as an improper integral, we obtain

$$\int_a^b \frac{dx}{x-c} = \lim_{\substack{\varepsilon_1 \to 0 \\ \varepsilon_2 \to 0}} \left( -\int_a^{c-\varepsilon_1} \frac{dx}{c-x} + \int_{c+\varepsilon_2}^b \frac{dx}{x-c} \right) = \ln \frac{b-c}{c-a} + \lim_{\substack{\varepsilon_1 \to 0 \\ \varepsilon_2 \to 0}} \ln \frac{\varepsilon_1}{\varepsilon_2}. \qquad (7.2.11.2)$$

The limit of the last expression obviously depends on the way in which $\varepsilon_1$ and $\varepsilon_2$ tend to zero. Hence, the improper integral does not exist. This integral is called a *singular integral*. However, this integral can be assigned a meaning if we assume that there is some relationship between $\varepsilon_1$ and $\varepsilon_2$. For example, if the deleted interval is symmetric with respect to the point $c$, i.e.,

$$\varepsilon_1 = \varepsilon_2 = \varepsilon, \qquad (7.2.11.3)$$

we arrive at the notion of the Cauchy principal value of a singular integral.

The *Cauchy principal value* of the singular integral

$$\int_a^b \frac{dx}{x-c}, \qquad a < c < b$$

is the number

$$\lim_{\varepsilon \to 0} \left( \int_a^{c-\varepsilon} \frac{dx}{x-c} + \int_{c+\varepsilon}^b \frac{dx}{x-c} \right).$$

With regard to formula (7.2.11.2), we have

$$\int_a^b \frac{dx}{x-c} = \ln \frac{b-c}{c-a}. \qquad (7.2.11.4)$$

**2°.** Consider the more general integral

$$\int_a^b \frac{f(x)}{x-c} dx, \qquad (7.2.11.5)$$

where $f(x) \in [a, b]$ is a function satisfying the Hölder condition. Let us understand this integral in the sense of the Cauchy principal value, which we define as follows:

$$\int_a^b \frac{f(x)}{x-c} dx = \lim_{\varepsilon \to 0} \left( \int_a^{c-\varepsilon} \frac{f(x)}{x-c} dx + \int_{c+\varepsilon}^b \frac{f(x)}{x-c} dx \right).$$

We have the identity

$$\int_a^b \frac{f(x)}{x-c} dx = \int_a^b \frac{f(x) - f(c)}{x-c} dx + f(c) \int_a^b \frac{dx}{x-c};$$

moreover, the first integral on the right-hand side is convergent as an improper integral, because it follows from the Hölder condition that

$$\left| \frac{f(x) - f(c)}{x-c} \right| < \frac{A}{|x-c|^{1-\lambda}}, \qquad 0 < \lambda \le 1,$$

and the second integral coincides with (7.2.11.4).

Thus, we see that the singular integral (7.2.11.5), where $f(x)$ satisfies the Hölder condition, exists in the sense of the Cauchy principal value and is equal to

$$\int_a^b \frac{f(x)}{x-c} \, dx = \int_a^b \frac{f(x) - f(c)}{x-c} \, dx + f(c) \ln \frac{b-c}{c-a}.$$

Some authors denote singular integrals by special symbols like v.p. $\int$ (*valeur principale*). However, this is not necessary because, on the one hand, if an integral of the form (7.2.11.5) exists as a proper or an improper integral, then it exists in the sense of the Cauchy principal value, and their values coincide; on the other hand, we shall always understand a singular integral in the sense of the Cauchy principal value. For this reason, we denote a singular integral by the usual integral sign.

### 7.2.12. Stieltjes Integral

#### 7.2.12-1. Basic definitions.

Let $f(x)$ and $\varphi(x)$ be functions defined on an interval $[a,b]$. Let us partition this interval into $n$ elementary subintervals defined by a set of points $\{x_0, x_1, \ldots, x_n\}$ such that $a = x_0 < x_1 < \cdots < x_n = b$. Each subinterval $[x_{k-1}, x_k]$ will be characterized by its length $\Delta x_k = x_k - x_{k-1}$ and an arbitrarily chosen point $\xi_k \in [x_{k-1}, x_k]$. Let us make up a *Stieltjes integral sum*

$$s_n = \sum_{k=1}^{n} f(\xi_k) \Delta_k \varphi(x),$$

where $\Delta_k \varphi(x) = \varphi(x_k) - \varphi(x_{k-1})$ is the increment of the function $\varphi(x)$ on the $k$th elementary subinterval.

If there exists a limit of the integral sums $s_n$, as the number of subintervals $n$ increases indefinitely so that the length of every subinterval $\Delta x_k$ vanishes, and this limit depends on neither the way the interval $[a,b]$ was partitioned nor the way the points $\xi_k$ were selected, then this limit is called the *Stieltjes integral* of the function $f(x)$ with respect to the function $\varphi(x)$ over the interval $[a,b]$:

$$\int_a^b f(x) \, d\varphi(x) = \lim_{\lambda \to 0} s_n \quad \left( \max_{1 \le k \le n} \Delta x_k \to 0 \text{ as } n \to \infty \right).$$

Then $f(x)$ is called an *integrable function* with respect to $\varphi(x)$, and $\varphi(x)$ is called an *integrating function*.

The Stieltjes integral is a generalization of the Riemann integral; the latter corresponds to the special case $\varphi(x) = x + \text{const}$.

#### 7.2.12-2. Properties of the Stieltjes integral.

The Stieltjes integral has properties analogous to those of the definite Riemann integral:

1) $\int_a^b d\varphi(x) = \varphi(b) - \varphi(a);$

2) $\int_a^b [Af(x) \pm Bg(x)] \, d\varphi(x) = A \int_a^b f(x) \, d\varphi(x) \pm B \int_a^b g(x) \, d\varphi(x);$

3) $\displaystyle\int_a^b f(x)\,d[A\varphi(x) \pm B\psi(x)] = A\int_a^b f(x)\,d\varphi(x) \pm B\int_a^b f(x)\,d\psi(x);$

4) $\displaystyle\int_a^b f(x)\,d\varphi(x) = \int_a^c f(x)\,d\varphi(x) + \int_c^b f(x)\,d\varphi(x) \quad (a < c < b).$

It is assumed that all integrals on the left- and right-hand sides exist.

THEOREM (MEAN VALUE). *If a function $f(x)$ satisfies inequalities $m \le f(x) \le M$ on an interval $[a, b]$ and is integrable with respect to an increasing function $\varphi(x)$, then*

$$\int_a^b f(x)\,d\varphi(x) = \mu[\varphi(b) - \varphi(a)],$$

where $m < \mu < M$.

### 7.2.12-3. Existence theorems for the Stieltjes integral.

The existence of the Stieltjes integral and its reduction to the Riemann integral is established by the following theorem.

THEOREM 1. *If $f(x)$ is continuous on $[a, b]$ and $\varphi(x)$ has a bounded variation\* on $[a, b]$, then the integral $\int_a^b f(x)\,d\varphi(x)$ exists.*

THEOREM 2. *Let $f(x)$ be integrable on $[a, b]$ in the sense of Riemann and let $\varphi(x)$ satisfy the Lipschitz condition*

$$|\varphi(x_2) - \varphi(x_1)| < K|x_2 - x_1|,$$

*where $x_1$ and $x_2$ are arbitrary points of the interval $[a, b]$ and $K$ is a fixed positive constant. Then the function $f(x)$ is integrable with respect to the function $\varphi(x)$.*

THEOREM 3. *Let $f(x)$ be integrable on $[a, b]$ in the sense of Riemann and let $\varphi(x)$ be differentiable and have an integrable derivative on $[a, b]$. Then the function $f(x)$ is integrable with respect to the function $\varphi(x)$ and, moreover,*

$$\int_a^b f(x)\,d\varphi(x) = \int_a^b f(x)\varphi'(x)\,dx,$$

*where the integral on the right-hand side is understood in the sense of Riemann.*

Remark. If a function $f(x)$ is integrable on an interval $[a, b]$ with respect to a function $\varphi(x)$, then, vice versa, the function $\varphi(x)$ is also integrable with respect to the function $f(x)$ on $[a, b]$. Owing to this property, the functions $f(x)$ and $\varphi(x)$ are interchangeable in Theorems 1 and 2.

THEOREM 4. *Let $f(x)$ be continuous on $[a, b]$ and let $\varphi(x)$ have an absolutely integrable derivative $\varphi'(x)$ everywhere on $[a, b]$, except, perhaps, finitely many points. Let, in addition, the function $\varphi(x)$ undergo a jump discontinuity at finitely many points*

$$a = c_0 < c_1 < \cdots < c_m = b.$$

---

\* A function $\varphi(x)$ is said to have a *bounded variation* on an interval $[a, b]$ if there exists a number $M > 0$ such that for any set of points $a = x_0 < x_1 < \cdots < x_n = b$ the inequality $\sum_{k=1}^{n} |\varphi(x_{k+1}) - \varphi(x_k)| < M$ holds (see also Subsection 6.1.7).

Then the Stieltjes integral exists and is calculated as

$$\int_a^b f(x)\,d\varphi(x) = \int_a^b f(x)\varphi'(x)\,dx + f(a)[\varphi(a+0) - \varphi(a)]$$
$$+ \sum_{k=1}^{m-1} f(c_k)[\varphi(c_k+0) - \varphi(c_k-0)] + f(b)[\varphi(b) - \varphi(b-0)],$$

where the right-hand side contains a Riemann integral. Note the presence of terms outside the integral on the right-hand side, where, apart from the ordinary jumps of the function $\varphi(x)$ at the internal points of discontinuity, there are terms with one-sided jumps at the endpoints (if there is no jump at either endpoint, the corresponding term vanishes).

The Stieltjes integral is useful for finding static moments, moments of inertia, and some other distributed quantities on an interval $[a, b]$, where, apart from continuous distributions, there are concentrated quantities like point masses that correspond to a discontinuous function $\varphi(x)$ with finite jumps.

### 7.2.13. Square Integrable Functions

#### 7.2.13-1. Definitions.

A function $f(x)$ is said to be *square integrable* on an interval $[a, b]$ if $f^2(x)$ is integrable on $[a, b]$. The set of all square integrable functions is denoted by $L_2(a, b)$ or, briefly, $L_2$.* Likewise, the set of all integrable functions on $[a, b]$ is denoted by $L_1(a, b)$ or, briefly, $L_1$.

#### 7.2.13-2. Basic properties of functions from $L_2$.

1°. The sum of two square integrable functions is a square integrable function.

2°. The product of a square integrable function by a constant is a square integrable function.

3°. The product of two square integrable functions is an integrable function.

4°. If $f(x) \in L_2$ and $g(x) \in L_2$, then the following Cauchy–Schwarz–Bunyakovsky inequality holds:

$$(f,g)^2 \leq \|f\|^2 \|g\|^2,$$
$$(f,g) = \int_a^b f(x)g(x)\,dx, \quad \|f\|^2 = (f,f) = \int_a^b f^2(x)\,dx.$$

The number $(f, g)$ is called the *inner product* of the functions $f(x)$ and $g(x)$ and the number $\|f\|$ is called the $L_2$-*norm* of $f(x)$.

5°. For $f(x) \in L_2$ and $g(x) \in L_2$, the following *triangle inequality* holds:

$$\|f + g\| \leq \|f\| + \|g\|.$$

6°. Let functions $f(x)$ and $f_1(x), f_2(x), \ldots, f_n(x), \ldots$ be square integrable on an interval $[a, b]$. If

$$\lim_{n \to \infty} \int_a^b [f_n(x) - f(x)]^2\,dx = 0,$$

---

* In the most general case the integral is understood as the Lebesgue integral of measurable functions. As usual, two equivalent functions — i.e., equal everywhere, or distinct on a negligible set (of zero measure) — are regarded as one and the same element of $L_2$.

then the sequence $f_1(x), f_2(x), \ldots$ is said to be *mean-square convergent* to $f(x)$.

Note that if a sequence of functions $\{f_n(x)\}$ from $L_2$ converges uniformly to $f(x)$, then $f(x) \in L_2$ and $\{f_n(x)\}$ is mean-square convergent to $f(x)$.

### 7.2.14. Approximate (Numerical) Methods for Computation of Definite Integrals

**7.2.14-1. Rectangle, trapezoidal, and Simpson's rules.**

For approximate computation of an integral like $\int_a^b f(x)\,dx$, let us break up the interval $[a, b]$ into $n$ equal subintervals with length $h = \dfrac{b-a}{n}$. Introduce the notation: $x_0 = a$, $x_1$, $\ldots$, $x_n = b$ (the partition points), $y_i = f(x_i)$, $i = 0, 1, \ldots, n$.

1°. *Rectangle rules*:
$$\int_a^b f(x)\,dx \approx h(y_0 + y_1 + \cdots + y_{n-1}),$$
$$\int_a^b f(x)\,dx \approx h(y_1 + y_2 + \cdots + y_n).$$

The error of these formulas, $R_n$, is proportional to $h$ and is estimated using the inequality
$$|R_n| \leq \tfrac{1}{2} h(b-a) M_1, \qquad M_1 = \max_{a \leq x \leq b} |f'(x)|.$$

2°. *Trapezoidal rule*:
$$\int_a^b f(x)\,dx \approx h\left( \frac{y_0 + y_n}{2} + y_1 + y_2 + \cdots + y_{n-1} \right).$$

The error of this formula is proportional to $h^2$ and is estimated as
$$|R_n| \leq \tfrac{1}{12} h^2 (b-a) M_2, \qquad M_2 = \max_{a \leq x \leq b} |f''(x)|.$$

3°. *Simpson's rule*:
$$\int_a^b f(x)\,dx \approx \tfrac{1}{3} h [y_0 + y_n + 4(y_1 + y_3 + \cdots + y_{n-1}) + 2(y_2 + y_4 + \cdots + y_{n-2})],$$

where $n$ is even. The error of approximation by Simpson's rule is proportional to $h^4$:
$$|R_n| \leq \tfrac{1}{180} h^4 (b-a) M_4, \qquad M_4 = \max_{a \leq x \leq b} |f^{(4)}(x)|.$$

Simpson's rule yields exact results for the case where the integrand function is a polynomial of degree two or three.

*Remark.* The above approximation formulas are often used for numerical computation of definite integrals; to achieve a higher accuracy, large $n$ are normally taken.

**7.2.14-2. Computation of integrals using uniformly distributed sequences.**

1°. Consider a sequence of real numbers $\{x_n\}$ whose members all belong to the closed interval [0, 1]. Let $\nu_n(a,b)$ be the number of the members of the sequence with subscript $< n$ that belong to $[a,b] \in [0,1]$. The sequence $\{x_n\}$ is called uniformly distributed on the interval [0, 1] if

$$\lim_{n\to\infty} \frac{\nu_n(a,b)}{n} = b - a.$$

In the language of probability theory, this definition means that the probability of a randomly selected element of the sequence to fall into a subinterval $[a, b]$ is equal to the length of this subinterval.

A simple example of a uniformly distributed sequence is the sequence of all proper fractions $m/n$ with $n = 2, 3, \ldots$ and $m = 1, 2, \ldots n-1$:

$$\frac{1}{2}, \frac{1}{3}, \frac{2}{3}, \frac{1}{4}, \frac{2}{4}, \frac{3}{4}, \ldots$$

The following theorem serves as an unlimited source for constructing uniformly distributed sequences.

THEOREM 1. *Let $\vartheta$ be an arbitrary irrational number. Then the sequence $x_n = n\vartheta - [n\vartheta]$, $n = 1, 2, \ldots$, is uniformly distributed on the interval* [0, 1]; *the symbol $[z]$ stands for the integer part of $z$ (the maximum integer not exceeding $z$).*

2°. Uniformly distributed sequences can be used for the calculation of integrals on the basis of the following two theorems.

THEOREM 2. *If a number sequence $\{x_n\}$ is uniformly distributed on the interval* [0, 1], *then the following limiting relation holds for any function $f(x)$ integrable on* [0, 1]:

$$\int_0^1 f(x)\,dx = \lim_{n\to\infty} \frac{1}{n} \sum_{k=1}^n f(x_k). \tag{7.2.14.1}$$

*Conversely, if this relation holds, it follows that the sequence $\{x_n\}$ is uniformly distributed on* [0, 1].

Remark. Formula (7.2.14.1) can be used for approximate calculation of integrals by defining a uniformly distributed sequence $\{x_n\}$ and computing the sum with a large $n$. The integrals defined on an arbitrary interval $[a, b]$ are first reduced by the change of variable $z = \dfrac{x-a}{b-a}$ to an integral over the interval [0, 1].

THEOREM 3. *If a number sequence $\{x_n\}$ is uniformly distributed on* [0, 1], *then the following limiting relation holds for any function $f(x)$ integrable on* [0, 1] *and any monotonically decreasing sequence of positive numbers $\{\alpha_n\}$ with a divergent sum:*

$$\int_0^1 f(x)\,dx = \lim_{n\to\infty} \frac{\alpha_1 f(x_1) + \alpha_2 f(x_2) + \cdots + \alpha_n f(x_n)}{\alpha_1 + \alpha_2 + \cdots + \alpha_n}.$$

**7.2.14-3. Computation of integrals by series expansion of the integrand function.**

THEOREM 4. *If a sequence of functions, $\{f_n(x)\}$, continuous on an interval $[a,b]$ uniformly converges on $[a,b]$ to a function $f(x)$, then the sequence of functions $\left\{\int_a^b f_n(x)\,dx\right\}$ uniformly converges on $[a,b]$ to the function $\left\{\int_a^b f(x)\,dx\right\}$.*

THEOREM 5. *If the functions of a sequence $\{u_n(x)\}$ are continuous on an interval $[a,b]$ and the series $\sum_{n=1}^{\infty} u_n(x)$ uniformly converges on $[a,b]$ to a function $f(x)$, then the series $\sum_{n=1}^{\infty} \int_a^x u_n(t)\,dt$ is also uniformly convergent on $[a,b]$ and*

$$\sum_{n=1}^{\infty} \int_a^b u_n(x)\,dx = \int_a^b \left(\sum_{n=1}^{\infty} u_n(x)\,dx\right) = \int_a^b f(x)\,dx;$$

$$\sum_{n=1}^{\infty} \int_a^x u_n(t)\,dt = \int_a^x \left(\sum_{n=1}^{\infty} u_n(t)\,dt\right) = \int_a^x f(t)\,dt \quad (a \leq x \leq b).$$

This theorem is used for the calculation of integrals by expanding the integrand functions $f(x)$ into a uniformly convergent series.

**Remark 1.** Since a series of positive continuous functions with a continuous sum is always uniformly convergent, then this series can be integrated termwise.

**Remark 2.** In general, the convergence of a sequence of continuous integrand functions, $\{f_n(x)\}$, at each point of the integration interval does not guarantee that it is permitted to proceed to the limit under the sign of integral.

## 7.3. Double and Triple Integrals
### 7.3.1. Definition and Properties of the Double Integral

**7.3.1-1. Definition and properties of the double integral.**

Suppose there is a bounded set of points defined on the plane, so that it can be placed in a minimal enclosing circle. The diameter of this circle is called the *diameter of the set*. Consider a domain $D$ in the $x,y$ plane. Let us partition $D$ into $n$ nonintersecting subdomains (cells). The largest of the cell diameters is called the *partition diameter* and is denoted $\lambda = \lambda(\mathcal{D}_n)$, where $\mathcal{D}_n$ stands for the partition of the domain $D$ into cells. Let a function $z = f(x,y)$ be defined in $D$. Select an arbitrary point in each cell $(x_i, y_i)$, $i = 1, 2, \ldots, n$, and make up an *integral sum*,

$$s_n = \sum_{i=1}^{n} f(x_i, y_i)\,\Delta S_i,$$

where $\Delta S_i$ is the area of the $i$th subdomain.

If there exists a finite limit, $\mathcal{J}$, of the sums $s_n$ as $\lambda \to 0$ and it depends on neither the partition $\mathcal{D}_n$ nor the selection of the points $(x_i, y_i)$, this limit is denoted $\iint_D f(x,y)\,dx\,dy$ and is called the double integral of the function $f(x,y)$ over the domain $D$:

$$\iint_D f(x,y)\,dx\,dy = \lim_{\lambda \to 0} s_n.$$

This means that for any $\varepsilon > 0$ there exists a $\delta > 0$ such that for all partitions $\mathcal{D}_n$ such that $\lambda(\mathcal{D}_n) < \delta$ and for any selection of the points $(x_i, y_i)$, the inequality $|s_n - \mathcal{J}| < \varepsilon$ holds.

## 7.3.1-2. Classes of integrable functions.

Further on, it is assumed that $D$ is a closed bounded domain.

1. If $f(x, y)$ is continuous in $D$, then the double integral $\iint_D f(x, y)\, dx\, dy$ exists.

2. If $f(x, y)$ is bounded and the set of points of discontinuity of $f(x, y)$ has a zero area (e.g., the points of discontinuity lie on finitely many continuous curves in the $x, y$ plane), then the double integral of $f(x, y)$ over the domain $D$ exists.

## 7.3.1-3. Properties of the double integral.

1. *Linearity.* If functions $f(x, y)$ and $g(x, y)$ are integrable in $D$, then

$$\iint_D [af(x, y) \pm bg(x, y)]\, dx\, dy = a \iint_D f(x, y)\, dx\, dy \pm b \iint_D g(x, y)\, dx\, dy,$$

where $a$ and $b$ are any numbers.

2. *Additivity.* If the domain $D$ is split into two subdomains $D_1$ and $D_2$ that do not have common internal points and if the function $f(x, y)$ is integrable in either subdomain, then

$$\iint_D f(x, y)\, dx\, dy = \iint_{D_1} f(x, y)\, dx\, dy + \iint_{D_2} f(x, y)\, dx\, dy.$$

3. *Estimation theorem.* If $m \leq f(x, y) \leq M$ in $D$, then

$$mS \leq \iint_D f(x, y)\, dx\, dy \leq MS,$$

where $S$ is the area of the domain $D$.

4. *Mean value theorem.* If $f(x, y)$ is continuous in $D$, then there exists at least one internal point $(\bar{x}, \bar{y}) \in D$ such that

$$\iint_D f(x, y)\, dx\, dy = f(\bar{x}, \bar{y})\, S.$$

The number $f(\bar{x}, \bar{y})$ is called the *mean value of the function* $f(x, y)$ in $D$.

5. *Integration of inequalities.* If $\varphi(x, y) \leq f(x, y) \leq g(x, y)$ in $D$, then

$$\iint_D \varphi(x, y)\, dx\, dy \leq \iint_D f(x, y)\, dx\, dy \leq \iint_D g(x, y)\, dx\, dy.$$

In particular, if $f(x, y) \geq 0$ in $D$, then $\iint_D f(x, y)\, dx\, dy \geq 0$.

6. *Absolute value theorem*

$$\left| \iint_D f(x, y)\, dx\, dy \right| \leq \iint_D |f(x, y)|\, dx\, dy.$$

## 7.3. DOUBLE AND TRIPLE INTEGRALS

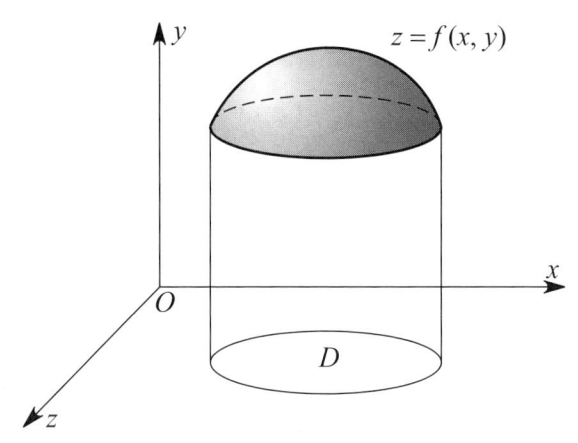

**Figure 7.4.** A double integral of a nonnegative function $f(x, y)$ over a domain $D$ is equal to the volume of a cylindrical body with base $D$ in the plane $z = 0$ and bounded from above by the surface $z = f(x, y)$.

**7.3.1-4. Geometric meaning of the double integral.**

Let a function $f(x, y)$ be nonnegative in $D$. Then the double integral $\iint_D f(x, y)\, dx\, dy$ is equal to the volume of a cylindrical body with base $D$ in the plane $z = 0$ and bounded from above by the surface $z = f(x, y)$; see Fig. 7.4.

### 7.3.2. Computation of the Double Integral

**7.3.2-1. Use of iterated integrals.**

1°. If a domain $D$ is defined in the $x, y$ plane by the inequalities $a \le x \le b$ and $y_1(x) \le y \le y_2(x)$ (see Fig. 7.5 $a$), then*

$$\iint_D f(x, y)\, dx\, dy = \int_a^b dx \int_{y_1(x)}^{y_2(x)} f(x, y)\, dy. \tag{7.3.2.1}$$

The integral on the right-hand side is called an *iterated integral*.

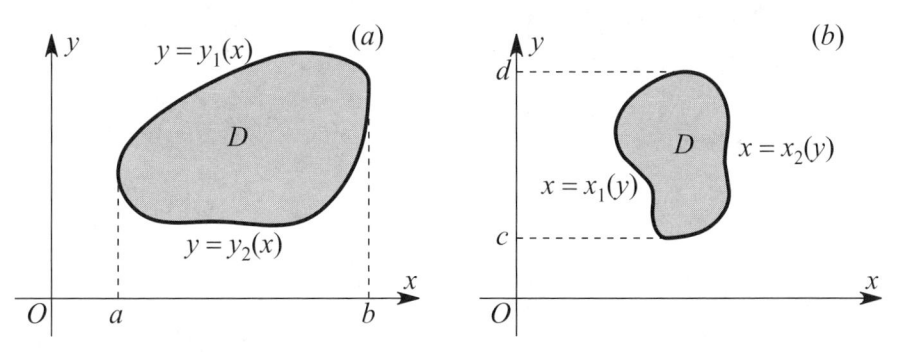

**Figure 7.5.** Computation of a double integral using iterated integrals: ($a$) illustration to formula (7.3.2.1), ($b$) illustration to formula (7.3.2.2).

---

* It is assumed that in (7.3.2.1) and (7.3.2.2) the double integral on the right-hand side and the inner integral on the right-hand side exist.

2°. If $D = \{c \leq y \leq d,\ x_1(y) \leq x \leq x_2(y)\}$ (see Fig. 7.5 b), then

$$\iint_D f(x,y)\,dx\,dy = \int_c^d dy \int_{x_1(y)}^{x_2(y)} f(x,y)\,dx. \qquad (7.3.2.2)$$

**Example 1.** Compute the integral

$$I = \iint_D \frac{dx\,dy}{(ax+by)^2},$$

where $D = \{0 \leq x \leq 1,\ 1 \leq y \leq 3\}$ is a rectangle, $a > 0$, and $b > 0$.

Using formula (7.3.2.2), we get

$$\iint_D \frac{dx\,dy}{(ax+by)^2} = \int_1^3 dy \int_0^1 \frac{dx}{(ax+by)^2}.$$

Compute the inner integral:

$$\int_0^1 \frac{dx}{(ax+by)^2} = -\frac{1}{a(ax+by)}\bigg|_{x=0}^{x=1} = \frac{1}{a}\left(\frac{1}{by} - \frac{1}{by+a}\right).$$

It follows that

$$I = \frac{1}{a}\int_1^3 \left(\frac{1}{by} - \frac{1}{by+a}\right) dy = \frac{1}{ab}\ln\frac{3(a+b)}{a+3b}.$$

3°. Consider a domain $D$ inscribed in a rectangle $\{a \leq x \leq b,\ c \leq y \leq d\}$. Let the boundary of $D$, within the rectangle, be intersected by straight lines parallel to the coordinate axes at two points only, as shown in Fig. 7.6 a. Then, by comparing formulas (7.3.2.1) and (7.3.2.2), we arrive at the relation

$$\int_a^b dx \int_{y_1(x)}^{y_2(x)} f(x,y)\,dy = \int_c^d dy \int_{x_1(y)}^{x_2(y)} f(x,y)\,dx,$$

which shows how the order of integration can be changed.

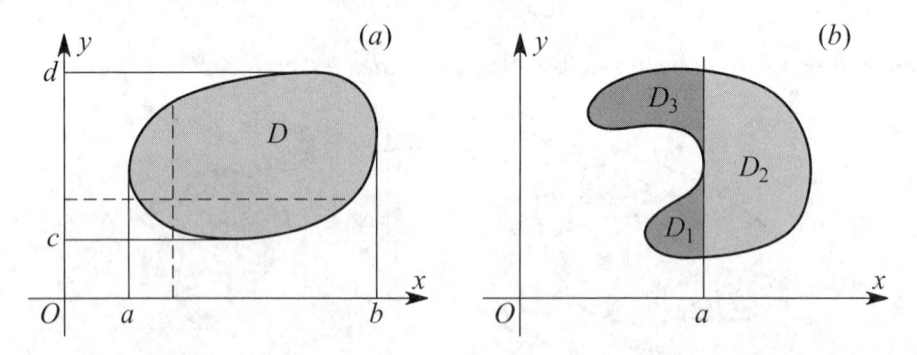

**Figure 7.6.** Illustrations to the computation of a double integral in a simple (a) and a complex (b) domain.

4°. In the general case, the domain $D$ is first split into subdomains considered in Items 1° and 2°, and then the property of additivity of the double integral is used. For example, the domain $D$ shown in Fig. 7.6 b is divided by the straight line $x = a$ into three subdomains $D_1$, $D_2$, and $D_3$. Then the integral over $D$ is represented as the sum of three integrals over the resulting subdomains.

### 7.3.2-2. Change of variables in the double integral.

1°. Let $x = x(u, v)$ and $y = y(u, v)$ be continuously differentiable functions that map one-to-one a domain $D_1$ in the $u, v$ plane onto a domain $D$ in the $x, y$ plane, and let $f(x, y)$ be a continuous function in $D$. Then

$$\iint_D f(x, y)\, dx\, dy = \iint_{D_1} f(x(u, v), y(u, v)) |J(u, v)|\, du\, dv,$$

where $J(u, v)$ is the *Jacobian* (or *Jacobian determinant*) of the mapping of $D_1$ onto $D$:

$$J(u, v) = \frac{\partial(x, y)}{\partial(u, v)} = \begin{vmatrix} \frac{\partial x}{\partial u} & \frac{\partial x}{\partial v} \\ \frac{\partial y}{\partial u} & \frac{\partial y}{\partial v} \end{vmatrix} = \frac{\partial x}{\partial u}\frac{\partial y}{\partial v} - \frac{\partial x}{\partial v}\frac{\partial y}{\partial u}.$$

The fraction before the determinant is a common notation for a Jacobian.

The absolute value of the Jacobian characterizes the extension (contraction) of an infinitesimal area element when passing from $x, y$ to $u, v$.

2°. The Jacobian of the mapping defining the change from the Cartesian coordinates $x, y$ to the polar coordinates $\rho, \varphi$,

$$x = \rho \cos \varphi, \quad y = \rho \sin \varphi, \tag{7.3.2.3}$$

is equal to

$$J(\rho, \varphi) = \rho. \tag{7.3.2.4}$$

**Example 2.** Given a sphere of radius $R$ and a right circular cylinder of radius $a < R$ whose axis passes through the sphere center, find the volume of the figure the cylinder cuts out of the sphere.

The volume of this figure is calculated as

$$V = \iint_{x^2+y^2 \leq a^2} \sqrt{R^2 - x^2 - y^2}\, dx\, dy.$$

Passing in the integral from $x, y$ to the polar coordinates (7.3.2.3) and taking into account (7.3.2.4), we obtain

$$V = \int_0^{2\pi} \int_0^a \sqrt{R^2 - \rho^2}\, \rho\, d\rho\, d\varphi = \frac{4\pi}{3}\left[R^3 - (R^2 - a^2)^{3/2}\right].$$

3°. The Jacobians of some common transformations in the plane are listed in Table 7.1.

TABLE 7.1
Some common curvilinear coordinates in the plane and the respective Jacobians

| Name of coordinates | Transformation | Jacobian, $J$ |
|---|---|---|
| Polar coordinates $\rho, \varphi$ | $x = \rho \cos \varphi$, $y = \rho \sin \varphi$ | $\rho$ |
| Generalized polar coordinates $\rho, \varphi$ | $x = a\rho \cos \varphi$, $y = b\rho \sin \varphi$ | $ab\rho$ |
| Elliptic coordinates $u, v$ (special system; $u \geq 0$, $0 \leq v \leq \pi$) | $x = a \cosh u \cos v$, $y = a \sinh u \sin v$ | $a^2(\sinh^2 u + \sin^2 v)$ |
| Parabolic coordinates $\sigma, \tau$ | $x = a\sigma\tau$, $y = \frac{1}{2}a(\tau^2 - \sigma^2)$ | $a^2(\sigma^2 + \tau^2)$ |
| Bipolar coordinates $\sigma, \tau$ | $x = \dfrac{a \sinh \tau}{\cosh \tau - \cos \sigma}$, $y = \dfrac{a \sin \sigma}{\cosh \tau - \cos \sigma}$ | $\dfrac{a^2}{(\cosh \tau - \cos \sigma)^2}$ |

**7.3.2-3. Differentiation of the double integral with respect to a parameter.**

1°. In various applications, situations may arise where the integrand function and the integration domain depend on a parameter, $t$. The derivative of such a double integral with respect to $t$ is expressed as

$$\frac{d}{dt}\iint_{D(t)} f(x,y,t)\,dx\,dy = \iint_{D(t)} \frac{\partial}{\partial t} f(x,y,t)\,dx\,dy + \int_{L(t)} (\vec{n}\cdot\vec{v})f(x,y,t)\,dl, \quad (7.3.2.5)$$

where $L(t)$ is the boundary of the domain $D(t)$, $\vec{n}$ is outer unit normal to $L(t)$, and $\vec{v}$ is the velocity of motion of the points of $L(t)$.

2°. If the boundary $L(t)$ is specified by equations in parametric form

$$x = X(\lambda, t), \quad y = Y(\lambda, t), \quad \alpha \leq \lambda \leq \beta, \quad (7.3.2.6)$$

then

$$\vec{n} = \left\{ \frac{Y_\lambda}{\sqrt{X_\lambda^2 + Y_\lambda^2}}, -\frac{X_\lambda}{\sqrt{X_\lambda^2 + Y_\lambda^2}} \right\}, \quad \vec{v} = \{X_t, Y_t\}, \quad (\vec{n}\cdot\vec{v}) = \frac{Y_\lambda X_t - X_\lambda Y_t}{\sqrt{X_\lambda^2 + Y_\lambda^2}}, \quad dl = \sqrt{X_\lambda^2 + Y_\lambda^2};$$

the subscripts $\lambda$ and $t$ denote the respective partial derivatives. The last integral in (7.3.2.5) becomes

$$\int_{L(t)} (\vec{n}\cdot\vec{v})f(x,y,t)\,dl = \int_\alpha^\beta f\big(X(\lambda,t), Y(\lambda,t), t\big)(Y_\lambda X_t - X_\lambda Y_t)\,d\lambda. \quad (7.3.2.7)$$

**Example 3.** Let $D(t)$ be a deformable plane domain bounded by an ellipse

$$\frac{x^2}{a^2(t)} + \frac{y^2}{b^2(t)} = 1. \quad (7.3.2.8)$$

Let us rewrite the equation of the ellipse (7.3.2.8) in parametric form as

$$x = a(t)\cos\lambda, \quad y = b(t)\sin\lambda, \quad 0 \leq \lambda \leq 2\pi,$$

which corresponds to the special case of (7.3.2.6) with $X(\lambda,t) = a(t)\cos\lambda$ and $Y(\lambda,t) = b(t)\sin\lambda$. Taking into account the aforesaid and using formula (7.3.2.7), we obtain

$$\int_{L(t)} (\vec{n}\cdot\vec{v})f(x,y,t)\,dl = \int_0^{2\pi} f\big(a(t)\cos\lambda, b(t)\sin\lambda, t\big)(a_t' b\cos^2\lambda + ab_t'\sin^2\lambda)\,d\lambda. \quad (7.3.2.9)$$

Let us dwell on the simple special case of $f(x,y,t) = 1$, when the first integral on the right-hand side of (7.3.2.5) vanishes. Evaluate the second integral by formula (7.3.2.9) to obtain

$$\int_{L(t)} (\vec{n}\cdot\vec{v})\,dl = \int_0^{2\pi} (a_t' b\cos^2\lambda + ab_t'\sin^2\lambda)\,d\lambda = a_t' b\int_0^{2\pi}\cos^2\lambda\,d\lambda + ab_t'\int_0^{2\pi}\sin^2\lambda\,d\lambda$$

$$= \pi[a'(t)b(t) + a(t)b'(t)]. \quad (7.3.2.10)$$

**Remark.** Formula (7.3.2.10) is easy to derive directly from (7.3.2.5) noting that, for $f(x,y,t) = 1$, the integral on the left-hand side of (7.3.2.5) gives the area of the ellipse, $S = \pi a(t)b(t)$. Differentiating this formula yields (7.3.2.10).

3°. If the boundary $L(t)$ is specified by an equation in implicit form,

$$F(x, y, t) = 0,$$

then one should take into account in (7.3.2.5) that

$$\vec{n} = \left\{ \frac{F_x}{\sqrt{F_x^2 + F_y^2}}, \frac{F_y}{\sqrt{F_x^2 + F_y^2}} \right\}, \quad \vec{v} = \left\{ -\frac{F_t}{F_x}, -\frac{F_t}{F_y} \right\}, \quad (\vec{n}\cdot\vec{v}) = -\frac{2F_t}{\sqrt{F_x^2 + F_y^2}}.$$

For an elliptic domain, specified by equation (7.3.2.8), we have

$$F(x,y,t) = \frac{x^2}{a^2(t)} + \frac{y^2}{b^2(t)} - 1.$$

## 7.3.3. Geometric and Physical Applications of the Double Integral

7.3.3-1. Geometric applications of the double integral.

1. *Area of a domain $D$ in the $x, y$ plane*:

$$S = \iint_D dx\, dy.$$

2. *Area of a surface defined by an equation $z = f(x, y)$ with $(x, y) \in D$ (the surface is projected onto a domain $D$ in the $x, y$ plane)*:

$$S = \iint_D \sqrt{\left(\frac{\partial f}{\partial x}\right)^2 + \left(\frac{\partial f}{\partial y}\right)^2 + 1}\, dx\, dy.$$

3. *Area of a surface defined parametrically by equations $x = x(u, v)$, $y = y(u, v)$, $z = z(u, v)$, with $(u, v) \in D_1$*:

$$S = \iint_{D_1} \sqrt{EG - F^2}\, du\, dv.$$

Notation used:

$$E = \left(\frac{\partial x}{\partial u}\right)^2 + \left(\frac{\partial y}{\partial u}\right)^2 + \left(\frac{\partial z}{\partial u}\right)^2,$$

$$G = \left(\frac{\partial x}{\partial v}\right)^2 + \left(\frac{\partial y}{\partial v}\right)^2 + \left(\frac{\partial z}{\partial v}\right)^2,$$

$$F = \frac{\partial x}{\partial u}\frac{\partial x}{\partial v} + \frac{\partial y}{\partial u}\frac{\partial y}{\partial v} + \frac{\partial z}{\partial u}\frac{\partial z}{\partial v}.$$

4. *Area of a surface defined by a vector equation $\vec{r} = \vec{r}(u, v) = x(u, v)\vec{i} + y(u, v)\vec{j} + z(u, v)\vec{k}$, with $(u, v) \in D_1$*:

$$S = \iint_{D_1} |\vec{n}(u, v)|\, du\, dv.$$

Here, the unit normal is calculated as $\vec{n}(u, v) = \vec{r}'_u \times \vec{r}'_v$.

**Remark.** The formulas from Items 3 and 4 are equivalent — they define one and the same surface in two forms, scalar and vector, respectively.

5. *Calculation of volumes.* If a domain $U$ of the three-dimensional space is defined by $\{(x, y) \in D,\ f(x, y) \leq z \leq g(x, y)\}$, where $D$ is a domain in the $x, y$ plane, the volume of $U$ is calculated as

$$V = \iint_D [g(x, y) - f(x, y)]\, dx\, dy.$$

The three-dimensional domain $U$ is a cylinder with base $D$ bounded by the surface $z = f(x, y)$ from below and the surface $z = g(x, y)$ from above. The lateral surface of this body consists of segments of straight lines parallel to the $z$-axis.

### 7.3.3-2. Physical applications of the double integral.

Consider a flat plate that occupies a domain $D$ in the $x, y$ plane. Let $\gamma(x, y)$ be the surface density of the plate material (the case $\gamma =$ const corresponds to a homogeneous plate).

1. *Mass of a flat plate*:
$$m = \iint_D \gamma(x, y)\, dx\, dy.$$

2. *Coordinates of the center of mass of a flat plate*:
$$x_c = \frac{1}{m} \iint_D x\gamma(x, y)\, dx\, dy, \quad y_c = \frac{1}{m} \iint_D y\gamma(x, y)\, dx\, dy,$$
where $m$ is the mass of the plate.

3. *Moments of inertia of a flat plate* about the coordinate axes:
$$I_x = \iint_D y^2 \gamma(x, y)\, dx\, dy, \quad I_y = \iint_D x^2 \gamma(x, y)\, dx\, dy.$$

The moment of inertia of the plate about the origin of coordinates is calculated as $I_0 = I_x + I_y$.

## 7.3.4. Definition and Properties of the Triple Integral

### 7.3.4-1. Definition of the triple integral.

Let a function $f(x, y, z)$ be defined in a domain $U$ of the three-dimensional space. Let us break up $U$ into $n$ subdomains (cells) that do not have common internal points. Denote by $\lambda = \lambda(\mathcal{U}_n)$ the *diameter* of the resulting partition $\mathcal{U}_n$, i.e., the maximum of the cell diameters (the diameter of a domain in space is the diameter of the minimal sphere enclosing the domain). Select an arbitrary point, $(x_i, y_i, z_i)$, $i = 1, 2, \ldots, n$, in each cell and make up an *integral sum*
$$s_n = \sum_{i=1}^{n} f(x_i, y_i, z_i)\, \Delta V_i,$$
where $\Delta V_i$ is the volume of the $i$th cell. If there exists a finite limit of the sums $s_n$ as $\lambda(\mathcal{U}_n) \to 0$ that depends on neither the partition $\mathcal{U}_n$ nor the selection of the points $(x_i, y_i, z_i)$, then it is called the triple integral of the function $f(x, y, z)$ over the domain $U$ and is denoted
$$\iiint_U f(x, y, z)\, dx\, dy\, dz = \lim_{\lambda \to 0} s_n.$$

### 7.3.4-2. Properties of the triple integral.

The properties of triple integrals are similar to those of double integrals.

1. *Linearity*. If functions $f(x, y, z)$ and $g(x, y, z)$ are integrable in a domain $U$, then
$$\iiint_U [af(x, y, z) \pm bg(x, y, z)]\, dx\, dy\, dz$$
$$= a \iiint_U f(x, y, z)\, dx\, dy\, dz \pm b \iiint_U g(x, y, z)\, dx\, dy\, dz,$$
where $a$ and $b$ are any numbers.

**2. *Additivity*.** If a domain $U$ is split into two subdomains, $U_1$ and $U_2$, that do not have common internal points and if a function $f(x,y,z)$ is integrable in either subdomain, then

$$\iiint_U f(x,y,z)\,dx\,dy\,dz = \iiint_{U_1} f(x,y,z)\,dx\,dy\,dz + \iiint_{U_2} f(x,y,z)\,dx\,dy\,dz.$$

**3. *Estimation theorem*.** If $m \leq f(x,y,z) \leq M$ in a domain $U$, then

$$mV \leq \iiint_U f(x,y,z)\,dx\,dy\,dz \leq MV,$$

where $V$ is the volume of $U$.

**4. *Mean value theorem*.** If $f(x,y,z)$ is continuous in $U$, then there exists at least one internal point $(\bar{x},\bar{y},\bar{z}) \in U$ such that

$$\iiint_U f(x,y,z)\,dx\,dy\,dz = f(\bar{x},\bar{y},\bar{z})\,V.$$

The number $f(\bar{x},\bar{y},\bar{z})$ is called the *mean value of the function* $f$ in the domain $U$.

**5. *Integration of inequalities*.** If $\varphi(x,y,z) \leq f(x,y,z) \leq g(x,y,z)$ in a domain $U$, then

$$\iiint_U \varphi(x,y,z)\,dx\,dy\,dz \leq \iiint_U f(x,y,z)\,dx\,dy\,dz \leq \iiint_U g(x,y,z)\,dx\,dy\,dz.$$

**6. *Absolute value theorem*:**

$$\left|\iiint_U f(x,y,z)\,dx\,dy\,dz\right| \leq \iiint_U |f(x,y,z)|\,dx\,dy\,dz.$$

### 7.3.5. Computation of the Triple Integral. Some Applications. Iterated Integrals and Asymptotic Formulas

**7.3.5-1. Use of iterated integrals.**

$1°$. Consider a three-dimensional body $U$ bounded by a surface $z = g(x,y)$ from above and a surface $z = h(x,y)$ from below, with a domain $D$ being the projection of it onto the $x,y$ plane. In other words, the domain $U$ is defined as $\{(x,y) \in D : h(x,y) \leq z \leq g(x,y)\}$. Then

$$\iiint_U f(x,y,z)\,dx\,dy\,dz = \iint_D dx\,dy \int_{h(x,y)}^{g(x,y)} f(x,y,z)\,dz.$$

$2°$. If, under the same conditions as in Item $1°$, the domain $D$ of the $x,y$ plane is defined as $\{a \leq x \leq b,\ y_1(x) \leq y \leq y_2(x)\}$, then

$$\iiint_U f(x,y,z)\,dx\,dy\,dz = \int_a^b dx \int_{y_1(x)}^{y_2(x)} dy \int_{h(x,y)}^{g(x,y)} f(x,y,z)\,dz.$$

**7.3.5-2. Change of variables in the triple integral.**

1°. Let $x = x(u, v, w)$, $y = y(u, v, w)$, and $z = z(u, v, w)$ be continuously differentiable functions that map, one to one, a domain $\Omega$ of the $u, v, w$ space into a domain $U$ of the $x, y, z$ space, and let a function $f(x, y, z)$ be continuous in $U$. Then

$$\iiint_U f(x,y,z)\,dx\,dy\,dz = \iiint_\Omega f\bigl(x(u,v,w),\, y(u,v,w),\, z(u,v,w)\bigr)\,|J(u,v,w)|\,du\,dv\,dw,$$

where $J(u, v, w)$ is the *Jacobian* of the mapping of $\Omega$ into $U$:

$$J(u,v,w) = \frac{\partial(x,y,z)}{\partial(u,v,w)} = \begin{vmatrix} \dfrac{\partial x}{\partial u} & \dfrac{\partial x}{\partial v} & \dfrac{\partial x}{\partial w} \\ \dfrac{\partial y}{\partial u} & \dfrac{\partial y}{\partial v} & \dfrac{\partial y}{\partial w} \\ \dfrac{\partial z}{\partial u} & \dfrac{\partial z}{\partial v} & \dfrac{\partial z}{\partial w} \end{vmatrix}.$$

The expression in the middle is a very common notation for a Jacobian.

The absolute value of the Jacobian characterizes the expansion (or contraction) of an infinitesimal volume element when passing from $x, y, z$ to $u, v, w$.

2°. The Jacobians of most common transformations in space are listed in Table 7.2.

**TABLE 7.2**
Common curvilinear coordinates in space and the respective Jacobians

| Name of coordinates | Transformation | Jacobian, $J$ |
|---|---|---|
| Cylindrical coordinates $\rho, \varphi, z$ | $x = \rho\cos\varphi,\ y = \rho\sin\varphi,\ z = z$ | $\rho$ |
| Generalized cylindrical coordinates $\rho, \varphi, z$ | $x = a\rho\cos\varphi,\ y = b\rho\sin\varphi,\ z = z$ | $ab\rho$ |
| Spherical coordinates $r, \varphi, \theta$ | $x = r\cos\varphi\sin\theta,\ y = r\sin\varphi\sin\theta,\ z = r\cos\theta$ | $r^2\sin\theta$ |
| Generalized spherical coordinates $r, \varphi, \theta$ | $x = ar\cos\varphi\sin\theta,\ y = br\sin\varphi\sin\theta,$ $z = cr\cos\theta$ | $abcr^2\sin\theta$ |
| Coordinates of prolate ellipsoid of revolution $\sigma, \tau, \varphi$ ($\sigma \geq 1 \geq \tau \geq -1$) | $x = a\sqrt{(\sigma^2-1)(1-\tau^2)}\cos\varphi,$ $y = a\sqrt{(\sigma^2-1)(1-\tau^2)}\sin\varphi,$ $z = a\sigma\tau$ | $a^3(\sigma^2 - \tau^2)$ |
| Coordinates of oblate ellipsoid of revolution $\sigma, \tau, \varphi$ ($\sigma \geq 0,\ -1 \leq \tau \leq 1$) | $x = a\sqrt{(1+\sigma^2)(1-\tau^2)}\cos\varphi,$ $y = a\sqrt{(1+\sigma^2)(1-\tau^2)}\sin\varphi,$ $z = a\sigma\tau$ | $a^3(\sigma^2 + \tau^2)$ |
| Parabolic coordinates $\sigma, \tau, \varphi$ | $x = \sigma\tau\cos\varphi,\ y = \sigma\tau\sin\varphi,\ z = \tfrac{1}{2}(\tau^2 - \sigma^2)$ | $\sigma\tau(\sigma^2 + \tau^2)$ |
| Parabolic cylinder coordinates $\sigma, \tau, z$ | $x = \sigma\tau,\ y = \tfrac{1}{2}(\tau^2 - \sigma^2),\ z = z$ | $\sigma^2 + \tau^2$ |
| Bicylindrical coordinates $\sigma, \tau, z$ | $x = \dfrac{a\sinh\tau}{\cosh\tau - \cos\sigma},\ y = \dfrac{a\sin\sigma}{\cosh\tau - \cos\sigma},\ z = z$ | $\dfrac{a^2}{(\cosh\tau - \cos\sigma)^2}$ |
| Toroidal coordinates $\sigma, \tau, \varphi$ ($-\pi \leq \sigma \leq \pi,\ 0 \leq \tau < \infty,\ 0 \leq \varphi < 2\pi$) | $x = \dfrac{a\sinh\tau\cos\varphi}{\cosh\tau - \cos\sigma},\ y = \dfrac{a\sinh\tau\sin\varphi}{\cosh\tau - \cos\sigma},$ $z = \dfrac{a\sin\sigma}{\cosh\tau - \cos\sigma}$ | $\dfrac{a^3\sinh\tau}{(\cosh\tau - \cos\sigma)^2}$ |
| Bipolar coordinates $\sigma, \tau, \varphi$ ($\sigma$ is any, $0 \leq \tau < \pi,\ 0 \leq \varphi < 2\pi$) | $x = \dfrac{a\sin\tau\cos\varphi}{\cosh\sigma - \cos\tau},\ y = \dfrac{a\sin\tau\sin\varphi}{\cosh\sigma - \cos\tau},$ $z = \dfrac{a\sinh\sigma}{\cosh\sigma - \cos\tau}$ | $\dfrac{a^3\sin\tau}{(\cosh\sigma - \cos\tau)^2}$ |

### 7.3.5-3. Differentiation of the triple integral with respect to a parameter.

Let the integrand function and the integration domain of a triple integral depend on a parameter, $t$. The derivative of this integral with respect to $t$ is expressed as

$$\frac{d}{dt} \iiint_{U(t)} f(x, y, z, t)\, dx\, dy\, dz$$
$$= \iiint_{U(t)} \frac{\partial}{\partial t} f(x, y, z, t)\, dx\, dy\, dz + \iint_{S(t)} (\vec{n} \cdot \vec{v}) f(x, y, z, t)\, ds,$$

where $S(t)$ is the boundary of the domain $U(t)$, $\vec{n}$ is the unit normal to $S(t)$, and $\vec{v}$ is the velocity of motion of the points of $S(t)$.

### 7.3.5-4. Some geometric and physical applications of the triple integral.

1. *Volume of a domain $U$*:

$$V = \iiint_U dx\, dy\, dz.$$

2. *Mass of a body of variable density $\gamma = \gamma(x, y, z)$ occupying a domain $U$*:

$$m = \iiint_U \gamma\, dx\, dy\, dz.$$

3. *Coordinates of the center of mass*:

$$x_c = \frac{1}{m} \iiint_U x\gamma\, dx\, dy\, dz, \quad y_c = \frac{1}{m} \iiint_U y\gamma\, dx\, dy\, dz, \quad z_c = \frac{1}{m} \iiint_U z\gamma\, dx\, dy\, dz.$$

4. *Moments of inertia* about the coordinate axes:

$$I_x = \iiint_U \rho_{yz}^2 \gamma\, dx\, dy\, dz, \quad I_y = \iiint_U \rho_{xz}^2 \gamma\, dx\, dy\, dz, \quad I_z = \iiint_U \rho_{xy}^2 \gamma\, dx\, dy\, dz,$$

where $\rho_{yz}^2 = y^2 + z^2$, $\rho_{xz}^2 = x^2 + z^2$, and $\rho_{xy}^2 = x^2 + y^2$.
If the body is homogeneous, then $\gamma = \text{const}$.

**Example.** Given a bounded homogeneous elliptic cylinder,

$$\frac{x^2}{a^2} + \frac{y^2}{b^2} = 1, \quad 0 \le z \le h,$$

find its moment of inertia about the $z$-axis.
Using the generalized cylindrical coordinates (see the second row in Table 7.2), we obtain

$$I_z = \gamma \iiint_U (x^2 + y^2)\, dx\, dy\, dz = \gamma \int_0^h \int_0^{2\pi} \int_0^1 \rho^2(a^2 \cos^2 \varphi + b^2 \sin^2 \varphi) ab\rho\, d\rho\, d\varphi\, dz$$
$$= \frac{1}{4} ab\gamma \int_0^h \int_0^{2\pi} (a^2 \cos^2 \varphi + b^2 \sin^2 \varphi)\, d\varphi\, dz = \frac{1}{4} ab\gamma \int_0^{2\pi} \int_0^h (a^2 \cos^2 \varphi + b^2 \sin^2 \varphi)\, dz\, d\varphi$$
$$= \frac{1}{4} abh\gamma \int_0^{2\pi} (a^2 \cos^2 \varphi + b^2 \sin^2 \varphi)\, d\varphi = \frac{1}{4} ab(a^2 + b^2) h\gamma.$$

**5. *Potential of the gravitational field*** of a body $U$ at a point $(x, y, z)$:

$$\Phi(x, y, z) = \iiint_U \gamma(\xi, \eta, \zeta) \frac{d\xi\, d\eta\, d\zeta}{r}, \quad r = \sqrt{(x-\xi)^2 + (y-\eta)^2 + (z-\zeta)^2},$$

where $\gamma = \gamma(\xi, \eta, \zeta)$ is the body density. A material point of mass $m$ is pulled by the gravitating body $U$ with a force $\vec{F}$. The projections of $\vec{F}$ onto the $x$-, $y$-, and $z$-axes are given by

$$F_x = km \frac{\partial \Phi}{\partial x} = km \iiint_U \gamma(\xi, \eta, \zeta) \frac{\xi - x}{r^3} d\xi\, d\eta\, d\zeta,$$

$$F_y = km \frac{\partial \Phi}{\partial y} = km \iiint_U \gamma(\xi, \eta, \zeta) \frac{\eta - y}{r^3} d\xi\, d\eta\, d\zeta,$$

$$F_z = km \frac{\partial \Phi}{\partial z} = km \iiint_U \gamma(\xi, \eta, \zeta) \frac{\zeta - z}{r^3} d\xi\, d\eta\, d\zeta,$$

where $k$ is the gravitational constant.

### 7.3.5-5. Multiple integrals. Asymptotic formulas.

Multiple integrals in $n$ variables of integration are an obvious generalization of double and triple integrals.

$1°$. Consider the *Laplace-type multiple integral*

$$I(\lambda) = \int_\Omega f(\mathbf{x}) \exp[\lambda g(\mathbf{x})]\, d\mathbf{x},$$

where $\mathbf{x} = \{x_1, \ldots, x_n\}$, $d\mathbf{x} = dx_1 \ldots dx_n$, $\Omega$ is a bounded domain in $\mathbb{R}^n$, $f(\mathbf{x})$ and $g(\mathbf{x})$ are real-valued functions of $n$ variable, and $\lambda$ is a real or complex parameter.

Denote by

$$S_\varepsilon = \left\{\lambda : \arg|\lambda| \le \frac{\pi}{2} - \varepsilon\right\}, \quad 0 < \varepsilon < \frac{\pi}{2},$$

a sector in the complex plane of $\lambda$.

THEOREM 1. *Let the following conditions hold:*
(1) *the functions $f(\mathbf{x})$ and $g(\mathbf{x})$ are continuous in $\Omega$,*
(2) *the maximum of $g(\mathbf{x})$ is attained at only one point $\mathbf{x}_0 \in \Omega$ ($\mathbf{x}_0$ is a nondegenerate maximum point), and*
(3) *the function $g(\mathbf{x})$ has continuous third derivatives in a neighborhood of $\mathbf{x}_0$.*

*Then the following asymptotic formula holds as $\lambda \to \infty$, $\lambda \in S_\varepsilon$:*

$$I(\lambda) = (2\pi)^{n/2} \exp[\lambda g(\mathbf{x}_0)] \frac{f(\mathbf{x}_0) + O(\lambda^{-1})}{\sqrt{\lambda^n \det[g_{x_i x_j}(\mathbf{x}_0)]}},$$

where the $g_{x_i x_j}(\mathbf{x})$ are entries of the matrix of the second derivatives of $g(\mathbf{x})$.

$2°$. Consider the *power Laplace multiple integral*

$$I(\lambda) = \int_\Omega f(\mathbf{x})[g(\mathbf{x})]^\lambda\, d\mathbf{x}.$$

THEOREM 2. *Let $g(\mathbf{x}) > 0$ and let the conditions of Theorem 1 hold. Then the following asymptotic formula holds as $\lambda \to \infty$, $\lambda \in S_\varepsilon$:*

$$I(\lambda) = (2\pi)^{n/2} [g(\mathbf{x}_0)]^{(2\lambda+n)/2} \frac{f(\mathbf{x}_0) + o(1)}{\sqrt{\lambda^n \det[g_{x_i x_j}(\mathbf{x}_0)]}}.$$

## 7.4. Line and Surface Integrals
### 7.4.1. Line Integral of the First Kind

**7.4.1-1. Definition of the line integral of the first kind.**

Let a function $f(x, y, z)$ be defined on a piecewise smooth curve $\overset{\frown}{AB}$ in the three-dimensional space $\mathbb{R}^3$. Let the curve $\overset{\frown}{AB}$ be divided into $n$ subcurves by points $A = M_0, M_1, M_2, \ldots, M_n = B$, thus defining a partition $\mathcal{L}_n$. The longest of the chords $M_0 M_1, M_1 M_2, \ldots, M_{n-1} M_n$ is called the *diameter of the partition* $\mathcal{L}_n$ and is denoted $\lambda = \lambda(\mathcal{L}_n)$. Let us select on each arc $\overset{\frown}{M_{i-1} M_i}$ an arbitrary point $(x_i, y_i, z_i)$, $i = 1, 2, \ldots, n$, and make up an *integral sum*

$$s_n = \sum_{i=1}^{n} f(x_i, y_i, z_i) \Delta l_i,$$

where $\Delta l_i$ is the length of $\overset{\frown}{M_{i-1} M_i}$.

If there exists a finite limit of the sums $s_n$ as $\lambda(\mathcal{L}_n) \to 0$ that depends on neither the partition $\mathcal{L}_n$ nor the selection of the points $(x_i, y_i, z_i)$, then it is called the *line integral of the first kind* of the function $f(x, y, z)$ over the curve $\overset{\frown}{AB}$ and is denoted

$$\int_{AB} f(x, y, z) \, dl = \lim_{\lambda \to 0} s_n.$$

A line integral is also called a *curvilinear integral* or a *path integral*.

If the function $f(x, y, z)$ is continuous, then the line integral exists. The line integral of the first kind does not depend of the direction the path $\overset{\frown}{AB}$ is traced; its properties are similar to those of the definite integral.

**7.4.1-2. Computation of the line integral of the first kind.**

1. If a plane curve is defined in the form $y = y(x)$, with $x \in [a, b]$, then

$$\int_{AB} f(x, y) \, dl = \int_a^b f\bigl(x, y(x)\bigr) \sqrt{1 + (y'_x)^2} \, dx.$$

2. If a curve $\overset{\frown}{AB}$ is defined in parametric form by equations $x = x(t)$, $y = y(t)$, and $z = z(t)$, with $t \in [\alpha, \beta]$, then

$$\int_{AB} f(x, y, z) \, dl = \int_\alpha^\beta f\bigl(x(t), y(t), z(t)\bigr) \sqrt{(x'_t)^2 + (y'_t)^2 + (z'_t)^2} \, dt. \qquad (7.4.1.1)$$

If a function $f(x, y)$ is defined on a plane curve $x = x(t)$, $y = y(t)$, with $t \in [\alpha, \beta]$, one should set $z'_t = 0$ in (7.4.1.1).

**Example.** Evaluate the integral $\int_{AB} xy \, dl$, where $\overset{\frown}{AB}$ is a quarter of an ellipse with semiaxes $a$ and $b$.
Let us write out the equations of the ellipse for the first quadrant in parametric form:

$$x = a \cos t, \quad y = b \sin t \quad (0 \le t \le \pi/2).$$

We have $\sqrt{(x'_t)^2 + (y'_t)^2} = \sqrt{a^2 \sin^2 t + b^2 \cos^2 t}$. To evaluate the integral, we use formula (7.4.1.1) with $z'_t = 0$:

$$\int_{AB} xy\, dl = \int_0^{\pi/2} (a \cos t)(b \sin t)\sqrt{a^2 \sin^2 t + b^2 \cos^2 t}\, dt$$

$$= \frac{ab}{2} \int_0^{\pi/2} \sin 2t \sqrt{\frac{a^2}{2}(1 - \cos 2t) + \frac{b^2}{2}(1 + \cos 2t)}\, dt = \frac{ab}{4} \int_{-1}^{1} \sqrt{\frac{a^2+b^2}{2} + \frac{b^2-a^2}{2}z}\, dz$$

$$= \frac{ab}{4} \frac{2}{b^2-a^2} \frac{2}{3} \left(\frac{a^2+b^2}{2} + \frac{b^2-a^2}{2}z\right)^{3/2}\bigg|_{-1}^{1} = \frac{ab}{3} \frac{a^2 + ab + b^2}{a+b}.$$

7.4.1-3. **Applications of the line integral of the first kind.**

1. *Length of a curve $\overset{\frown}{AB}$*:
$$L = \int_{AB} dl.$$

2. *Mass of a material curve $\overset{\frown}{AB}$* with a given line density $\gamma = \gamma(x, y, z)$:
$$m = \int_{AB} \gamma\, dl.$$

3. *Coordinates of the center of mass* of a material curve $\overset{\frown}{AB}$:
$$x_c = \frac{1}{m} \int_{AB} x\gamma\, dl, \quad y_c = \frac{1}{m} \int_{AB} y\gamma\, dl, \quad z_c = \frac{1}{m} \int_{AB} z\gamma\, dl.$$

To a material line with uniform density there corresponds $\gamma$ = const.

## 7.4.2. Line Integral of the Second Kind

7.4.2-1. **Definition of the line integral of the second kind.**

Let a vector field
$$\vec{a}(x, y, z) = P(x, y, z)\vec{i} + Q(x, y, z)\vec{j} + R(x, y, z)\vec{k}$$

and a piecewise smooth curve $\overset{\frown}{AB}$ be defined in some domain in $\mathbb{R}^3$. By dividing the curve by points $A = M_0, M_1, M_2, \ldots, M_n = B$ into $n$ subcurves, we obtain a partition $\mathcal{L}_n$. Let us select on each arc $M_{i-1}M_i$ an arbitrary point $(x_i, y_i, z_i)$, $i = 1, 2, \ldots, n$, and make up a sum of dot products
$$s_n = \sum_{i=1}^{n} \vec{a}(x_i, y_i, z_i) \cdot \overrightarrow{M_{i-1}M_i},$$

called an *integral sum*.

If there exists a finite limit of the sums $s_n$ as $\lambda(\mathcal{L}_n) \to 0$ that depends on neither the partition $\mathcal{L}_n$ nor the selection of the points $(x_i, y_i, z_i)$, then it is called the *line integral of the second kind* of the vector field $\vec{a}(x, y, z)$ along the curve $\overset{\frown}{AB}$ and is denoted

$$\int_{AB} \vec{a} \cdot d\vec{r}, \quad \text{or} \quad \int_{AB} P\, dx + Q\, dy + R\, dz.$$

The line integral of the second kind depends on the direction the path is traced, so that

$$\int_{AB} \vec{a} \cdot d\vec{r} = -\int_{BA} \vec{a} \cdot d\vec{r}.$$

A line integral over a closed contour $\mathcal{C}$ is called a *closed path integral* (or a *circulation*) of a vector field $\vec{a}$ around $\mathcal{C}$ and is denoted

$$\oint_{\mathcal{C}} \vec{a} \cdot d\vec{r}.$$

*Physical meaning of the line integral of the second kind:* $\int_{AB} \vec{a} \cdot d\vec{r}$ determines the work done by the vector field $\vec{a}(x,y,z)$ on a particle of unit mass when it travels along the arc $\widehat{AB}$.

**7.4.2-2. Computation of the line integral of the second kind.**

1°. For a plane curve $\widehat{AB}$ defined as $y = y(x)$, with $x \in [a,b]$, and a plane vector field $\vec{a}$, we have

$$\int_{AB} \vec{a} \cdot d\vec{r} = \int_a^b \left[ P(x, y(x)) + Q(x, y(x)) y'_x(x) \right] dx.$$

2°. Let $\widehat{AB}$ be defined by a vector equation $\vec{r} = \vec{r}(t) = x(t)\vec{i} + y(t)\vec{j} + z(t)\vec{k}$, with $t \in [\alpha, \beta]$. Then

$$\int_{AB} \vec{a} \cdot d\vec{r} = \int_{AB} P\,dx + Q\,dy + R\,dz$$
$$= \int_\alpha^\beta \left[ P(x(t), y(t), z(t)) x'_t(t) + Q(x(t), y(t), z(t)) y'_t(t) + R(x(t), y(t), z(t)) z'_t(t) \right] dt. \quad (7.4.2.1)$$

For a plane curve $\widehat{AB}$ and a plane vector field $\vec{a}$, one should set $z'(t) = 0$ in (7.4.2.1).

**7.4.2-3. Potential and curl of a vector field.**

1°. A vector field $\vec{a} = \vec{a}(x,y,z)$ is called *potential* if there exists a function $\Phi(x,y,z)$ such that

$$\vec{a} = \operatorname{grad} \Phi, \quad \text{or} \quad \vec{a} = \frac{\partial \Phi}{\partial x}\vec{i} + \frac{\partial \Phi}{\partial y}\vec{j} + \frac{\partial \Phi}{\partial z}\vec{k}.$$

The function $\Phi(x,y,z)$ is called a *potential* of the vector field $\vec{a}$. The line integral of the second kind of a potential vector field along a path $\widehat{AB}$ is equal to the increment of the potential along the path:

$$\int_{AB} \vec{a} \cdot d\vec{r} = \Phi\big|_B - \Phi\big|_A.$$

2°. The *curl* of a vector field $\vec{a}(x,y,z) = P\vec{i} + Q\vec{j} + R\vec{k}$ is the vector defined as

$$\operatorname{curl} \vec{a} = \left( \frac{\partial R}{\partial y} - \frac{\partial Q}{\partial z} \right)\vec{i} + \left( \frac{\partial P}{\partial z} - \frac{\partial R}{\partial x} \right)\vec{j} + \left( \frac{\partial Q}{\partial x} - \frac{\partial P}{\partial y} \right)\vec{k} = \begin{vmatrix} \vec{i} & \vec{j} & \vec{k} \\ \frac{\partial}{\partial x} & \frac{\partial}{\partial y} & \frac{\partial}{\partial z} \\ P & Q & R \end{vmatrix}.$$

The vector curl $\vec{a}$ characterizes the rate of rotation of $\vec{a}$ and can also be described as the circulation density of $\vec{a}$. Alternative notations: $\operatorname{curl} \vec{a} \equiv \nabla \times \vec{a} \equiv \operatorname{rot} \vec{a}$.

**7.4.2-4. Necessary and sufficient conditions for a vector field to be potential.**

Let $U$ be a simply connected domain in $\mathbb{R}^3$ (i.e., a domain in which any closed contour can be deformed to a point without leaving $U$) and let $\vec{a}(x, y, z)$ be a vector field in $U$. Then the following four assertions are equivalent to each other:
  (1) the vector field $\vec{a}$ is potential;
  (2) curl $\vec{a} \equiv \vec{0}$;
  (3) the circulation of $\vec{a}$ around any closed contour $\mathcal{C} \in U$ is zero, or, equivalently, $\oint_\mathcal{C} \vec{a} \cdot d\vec{r} = 0$;
  (4) the integral $\int_{AB} \vec{a} \cdot d\vec{r}$ is independent of the shape of $\breve{AB} \in U$ (it depends on the starting and the finishing point only).

## 7.4.3. Surface Integral of the First Kind

**7.4.3-1. Definition of the surface integral of the first kind.**

Let a function $f(x, y, z)$ be defined on a smooth surface $D$. Let us break up this surface into $n$ elements (cells) that do not have common internal points and let us denote this partition by $\mathcal{D}_n$. The *diameter*, $\lambda(\mathcal{D}_n)$, of a partition $\mathcal{D}_n$ is the largest of the diameters of the cells (see Paragraph 7.3.4-1). Let us select in each cell an arbitrary point $(x_i, y_i, z_i)$, $i = 1, 2, \ldots, n$, and make up an *integral sum*

$$s_n = \sum_{i=1}^n f(x_i, y_i, z_i) \Delta S_i,$$

where $\Delta S_i$ is the area of the $i$th element.

If there exists a finite limit of the sums $s_n$ as $\lambda(\mathcal{D}_n) \to 0$ that depends on neither the partition $\mathcal{D}_n$ nor the selection of the points $(x_i, y_i, z_i)$, then it is called the *surface integral of the first kind* of the function $f(x, y, z)$ and is denoted $\iint_D f(x, y, z)\, dS$.

**7.4.3-2. Computation of the surface integral of the first kind.**

1°. If a surface $D$ is defined by an equation $z = z(x, y)$, with $(x, y) \in D_1$, then

$$\iint_D f(x, y, z)\, dS = \iint_{D_1} f\big(x, y, z(x, y)\big) \sqrt{1 + (z'_x)^2 + (z'_y)^2}\, dx\, dy.$$

2°. If a surface $D$ is defined by a vector equation $\vec{r} = \vec{r}(x, y, z) = x(u, v)\vec{i} + y(u, v)\vec{j} + z(u, v)\vec{k}$, where $(u, v) \in D_2$, then

$$\iint_D f(x, y, z)\, dS = \iint_{D_2} f\big(x(u, v), y(u, v), z(u, v)\big) |\vec{n}(u, v)|\, du\, dv,$$

where $\vec{n}(u, v) = \vec{r}_u \times \vec{r}_v$ is the unit normal to the surface $D$; the subscripts $u$ and $v$ denote the respective partial derivatives.

### 7.4.3-3. Applications of the surface integral of the first kind.

1°. Area of a surface $D$:
$$S_D = \iint_D dS.$$

2°. Mass of a material surface $D$ with a surface density $\gamma = \gamma(x, y, z)$:
$$m = \iint_D \gamma(x, y, z)\, dS.$$

3°. Coordinates of the center of mass of a material surface $D$:
$$x_c = \frac{1}{m} \iint_D x\gamma\, dS, \quad y_c = \frac{1}{m} \iint_D y\gamma\, dS, \quad z_c = \frac{1}{m} \iint_D z\gamma\, dS.$$

To the uniform surface density there corresponds $\gamma = \text{const}$.

## 7.4.4. Surface Integral of the Second Kind

### 7.4.4-1. Definition of the surface integral of the second kind.

Let $D$ be an *oriented surface* defined by an equation
$$\vec{r} = \vec{r}(u, v) = x(u,v)\,\vec{i} + y(u,v)\,\vec{j} + z(u,v)\,\vec{k},$$

where $u$ and $v$ are parameters. The fact that $D$ is oriented means that every point $M \in D$ has an associated unit normal $\vec{n}(M) = \vec{n}(u, v)$ continuously dependent on $M$. Two cases are possible: (i) the associated unit normal is $\vec{n}(u, v) = \vec{r}_u \times \vec{r}_v$ or (ii) the associated unit normal is opposite, $\vec{n}(u, v) = \vec{r}_v \times \vec{r}_u = -\vec{r}_u \times \vec{r}_v$.

**Remark.** If a surface is defined traditionally by an equation $z = z(x, y)$, its representation in vector form is as follows: $\vec{r} = \vec{r}(x, y) = x\,\vec{i} + y\,\vec{j} + z(x,y)\,\vec{k}$.

Let a vector field $\vec{a}(x, y, z) = P\vec{i} + Q\vec{j} + R\vec{k}$ be defined on a smooth oriented surface $D$. Let us perform a partition, $\mathcal{D}_n$, of the surface $D$ into $n$ elements (cells) that do not have common internal points. Also select an arbitrary point $M_i(x_i, y_i, z_i)$, $i = 1, 2, \ldots, n$, for each cell and make up an *integral sum* $s_n = \sum_{i=1}^{n} \vec{a}(x_i, y_i, z_i) \cdot \vec{n}_i^\circ\, \Delta S_i$, where $\Delta S_i$ is area of the $i$th cell and $\vec{n}_i^\circ$ is the unit normal to the surface at the point $M_i$, the orientation of which coincides with that of the surface.

If there exists a finite limit of the sums $s_n$ as $\lambda(\mathcal{D}_n) \to 0$ that depends on neither the partition $\mathcal{D}_n$ nor the selection of the points $M_i(x_i, y_i, z_i)$, then it is called the *surface integral of the second kind* (or the *flux of the vector field $\vec{a}$ across the oriented surface $D$*) and is denoted
$$\iint_D \vec{a}(x, y, z) \cdot \vec{dS}, \quad \text{or} \quad \iint_D P\, dy\, dz + Q\, dx\, dz + R\, dx\, dy.$$

Note that the surface integral of the second kind changes its sign when the orientation of the surface is reversed.

**7.4.4-2. Computation of the surface integral of the second kind.**

1°. If a surface $D$ is defined by a vector equation $\vec{r} = \vec{r}(u, v)$, where $(u, v) \in D_1$, then

$$\iint_D \vec{a}(x, y, z) \cdot \overrightarrow{dS} = \pm \iint_{D_1} \vec{a}\big(x(u, v), y(u, v), z(u, v)\big) \cdot \vec{n}(u, v)\, du\, dv.$$

The plus sign is taken if the unit normal associated with the surface is $\vec{n}(u, v) = \vec{r}_u \times \vec{r}_v$, and the minus sign is taken in the opposite case.

2°. If a surface $D$ is defined by an equation $z = z(x, y)$, with $(x, y) \in D_2$, then the normal $\vec{n}(x, y) = \vec{r}_x \times \vec{r}_y = -z_x \vec{i} - z_y \vec{j} + \vec{k}$ orients the surface $D$ "upward," in the positive direction of the $z$-axis; the subscripts $x$ and $y$ denote the respective partial derivatives. Then

$$\iint_D \vec{a} \cdot \overrightarrow{dS} = \pm \iint_{D_2} \big(-z_x P - z_y Q + R\big)\, dx\, dy,$$

where $P = P\big(x, y, z(x, y)\big)$, $Q = Q\big(x, y, z(x, y)\big)$, and $R = R\big(x, y, z(x, y)\big)$. The plus sign is taken if the surface has the "upward" orientation, and the minus sign is chosen in the opposite case.

### 7.4.5. Integral Formulas of Vector Calculus

**7.4.5-1. Ostrogradsky–Gauss theorem (divergence theorem).**

Let a vector field $\vec{a}(x, y, z) = P(x, y, z)\vec{i} + Q(x, y, z)\vec{j} + R(x, y, z)\vec{k}$ be continuously differentiable in a finite simply connected domain $V \subset \mathbb{R}^3$ oriented by the outward normal and let $S$ denote the boundary of $V$. Then the *Ostrogradsky–Gauss theorem* (or the *divergence theorem*) holds:

$$\iint_S \vec{a} \cdot \overrightarrow{dS} = \iiint_V \operatorname{div} \vec{a}\, dx\, dy\, dz,$$

where $\operatorname{div} \vec{a}$ is the *divergence* of the vector $\vec{a}$, which is defined as follows:

$$\operatorname{div} \vec{a} = \frac{\partial P}{\partial x} + \frac{\partial Q}{\partial y} + \frac{\partial R}{\partial z}.$$

Thus, the flux of a vector field across a closed surface in the outward direction is equal to the triple integral of the divergence of the vector field over the volume bounded by the surface. In coordinate form, the Ostrogradsky–Gauss theorem reads

$$\iint_S P\, dy\, dz + Q\, dx\, dz + R\, dx\, dy = \iiint_V \left(\frac{\partial P}{\partial x} + \frac{\partial Q}{\partial y} + \frac{\partial R}{\partial z}\right) dx\, dy\, dz.$$

**7.4.5-2. Stokes's theorem (curl theorem).**

1°. Let a vector field $\vec{a}(x, y, z)$ be continuously differentiable in a domain of the three-dimensional space $\mathbb{R}^3$ that contains an oriented surface $D$. The orientation of a surface uniquely defines the direction in which the boundary of the surface is traced; specifically, the boundary is traced counterclockwise when looked at from the direction of the normal to the surface. Then the circulation of the vector field around the boundary $\mathcal{C}$ of the surface $D$ is equal to the flux of the vector curl $\vec{a}$ across $D$:

$$\oint_{\mathcal{C}} \vec{a} \cdot d\vec{r} = \iint_{D} \operatorname{curl} \vec{a} \cdot \vec{dS}.$$

In coordinate notation, *Stokes's theorem* reads

$$\oint_{\mathcal{C}} P\,dx + Q\,dy + R\,dz = \iint_{D} \left(\frac{\partial R}{\partial y} - \frac{\partial Q}{\partial z}\right) dy\,dz + \left(\frac{\partial P}{\partial z} - \frac{\partial R}{\partial x}\right) dx\,dz + \left(\frac{\partial Q}{\partial x} - \frac{\partial P}{\partial y}\right) dx\,dy.$$

2°. For a plane vector field $\vec{a}(x,y) = P(x,y)\vec{i} + Q(x,y)\vec{j}$, Stokes's theorem reduces to *Green's theorem*:

$$\oint_{\mathcal{C}} P\,dx + Q\,dy = \iint_{D} \left(\frac{\partial Q}{\partial x} - \frac{\partial P}{\partial y}\right) dx\,dy,$$

where the contour $\mathcal{C}$ of the domain $D$ on the $x, y$ plane is traced counterclockwise.

### 7.4.5-3. Green's first and second identities. Gauss's theorem.

1°. Let $\Phi = \Phi(x,y,z)$ and $\Psi = \Psi(x,y,z)$ be twice continuously differentiable functions defined in a finite simply connected domain $V \subset \mathbb{R}^3$ bounded by a piecewise smooth boundary $S$.

Then the following formulas hold:

$$\iiint_{V} \Psi \Delta \Phi \, dV + \iiint_{V} \nabla \Phi \cdot \nabla \Psi \, dV = \iint_{S} \Psi \frac{\partial \Phi}{\partial n}\, dS \quad \text{(Green's first identity)},$$

$$\iiint_{V} (\Psi \Delta \Phi - \Phi \Delta \Psi) \, dV = \iint_{S} \left(\Psi \frac{\partial \Phi}{\partial n} - \Phi \frac{\partial \Psi}{\partial n}\right) dS \quad \text{(Green's second identity)},$$

where $\frac{\partial}{\partial n}$ denotes a derivative along the (outward) normal to the surface $S$, and $\Delta$ is the Laplace operator.

2°. In applications, the following special cases of the above formulas are most common:

$$\iiint_{V} \Phi \Delta \Phi \, dV + \iiint_{V} |\nabla \Phi|^2 \, dV = \iint_{S} \Phi \frac{\partial \Phi}{\partial n}\, dS \quad \text{(first identity with } \Psi = \Phi),$$

$$\iiint_{V} \Delta \Phi \, dV = \iint_{S} \frac{\partial \Phi}{\partial n}\, dS \quad \text{(second identity with } \Psi = 1).$$

## References for Chapter 7

**Adams, R.**, *Calculus: A Complete Course, 6th Edition*, Pearson Education, Toronto, 2006.
**Anton, H.**, *Calculus: A New Horizon, 6th Edition*, Wiley, New York, 1999.
**Anton, H., Bivens, I., and Davis, S.**, *Calculus: Early Transcendental Single Variable, 8th Edition*, John Wiley & Sons, New York, 2005.
**Aramanovich, I. G., Guter, R. S., et al.**, *Mathematical Analysis (Differentiation and Integration)*, Fizmatlit Publishers, Moscow, 1961.
**Borden, R. S.**, *A Course in Advanced Calculus*, Dover Publications, New York, 1998.
**Brannan, D.**, *A First Course in Mathematical Analysis*, Cambridge University Press, Cambridge, 2006.
**Bronshtein, I. N. and Semendyayev, K. A.**, *Handbook of Mathematics, 4th Edition*, Springer-Verlag, Berlin, 2004.

Browder, A., *Mathematical Analysis: An Introduction*, Springer-Verlag, New York, 1996.
Clark, D. N., *Dictionary of Analysis, Calculus, and Differential Equations*, CRC Press, Boca Raton, 2000.
Courant, R. and John, F., *Introduction to Calculus and Analysis, Vol. 1*, Springer-Verlag, New York, 1999.
Danilov, V. L., Ivanova, A. N., et al., *Mathematical Analysis (Functions, Limits, Series, Continued Fractions)* [in Russian], Fizmatlit Publishers, Moscow, 1961.
Dwight, H. B., *Tables of Integrals and Other Mathematical Data*, Macmillan, New York, 1961.
Edwards, C. H., and Penney, D., *Calculus, 6th Edition*, Pearson Education, Toronto, 2002.
Fedoryuk, M. V., *Asymptotics, Integrals and Series* [in Russian], Nauka Publishers, Moscow, 1987.
Fikhtengol'ts, G. M., *Fundamentals of Mathematical Analysis, Vol. 2*, Pergamon Press, London, 1965.
Fikhtengol'ts, G. M., *A Course of Differential and Integral Calculus, Vol. 2* [in Russian], Nauka Publishers, Moscow, 1969.
Gradshteyn, I. S. and Ryzhik, I. M., *Tables of Integrals, Series and Products, 6th Edition*, Academic Press, New York, 2000.
Kaplan, W., *Advanced Calculus, 5th Edition*, Addison Wesley, Reading, Massachusetts, 2002.
Kline, M., *Calculus: An Intuitive and Physical Approach, 2nd Edition*, Dover Publications, New York, 1998.
Landau, E., *Differential and Integral Calculus*, American Mathematical Society, Providence, 2001.
Marsden, J. E. and Weinstein, A., *Calculus, 2nd Edition*, Springer-Verlag, New York, 1985.
Mendelson, E., *3000 Solved Problems in Calculus*, McGraw-Hill, New York, 1988.
Polyanin, A. D., Polyanin, V. D., et al., *Handbook for Engineers and Students. Higher Mathematics. Physics. Theoretical Mechanics. Strength of Materials, 3rd Edition* [in Russian], AST/Astrel, Moscow, 2005.
Prudnikov, A. P., Brychkov, Yu. A., and Marichev, O. I., *Integrals and Series, Vol. 1, Elementary Functions*, Gordon & Breach, New York, 1986.
Silverman, R. A., *Essential Calculus with Applications*, Dover Publications, New York, 1989.
Strang, G., *Calculus*, Wellesley-Cambridge Press, Massachusetts, 1991.
Taylor, A. E. and Mann, W. R., *Advanced Calculus, 3rd Edition*, John Wiley, New York, 1983.
Thomas, G. B. and Finney, R. L., *Calculus and Analytic Geometry, 9th Edition*, Addison Wesley, Reading, Massachusetts, 1996.
Widder, D. V., *Advanced Calculus, 2nd Edition*, Dover Publications, New York, 1989.
Zorich, V. A., *Mathematical Analysis*, Springer-Verlag, Berlin, 2004.
Zwillinger, D., *CRC Standard Mathematical Tables and Formulae, 31st Edition*, CRC Press, Boca Raton, 2002.

# Chapter 8
# Series

## 8.1. Numerical Series and Infinite Products
### 8.1.1. Convergent Numerical Series and Their Properties. Cauchy's Criterion

**8.1.1-1. Basic definitions.**

Let $\{a_n\}$ be a numerical sequence. The expression

$$a_1 + a_2 + \cdots + a_n + \cdots = \sum_{n=1}^{\infty} a_n$$

is called a *numerical series* (*infinite sum*, *infinite numerical series*), $a_n$ is the *generic term of the series*, and

$$s_n = a_1 + a_2 + \cdots + a_n = \sum_{k=1}^{n} a_k$$

is the *nth partial sum of the series*. If there exists a finite limit $\lim\limits_{n\to\infty} s_n = S$, the series is called *convergent*, and $S$ is called the *sum of the series*. In this case, one writes $\sum_{n=1}^{\infty} a_n = S$. If $\lim\limits_{n\to\infty} s_n$ does not exist (or is infinite), the series is called *divergent*. The series $a_{n+1} + a_{n+2} + a_{n+3} + \cdots$ is called the *nth remainder of the series*.

**Example 1.** Consider the series $\sum_{n=1}^{\infty} aq^{n-1} = a + aq + aq^2 + \cdots$ whose terms form a *geometric progression* with ratio $q$. This series is convergent for $|q| < 1$ (its sum has the form $S = \frac{a}{1-q}$) and is divergent for $|q| \geq 1$.

**8.1.1-2. Necessary condition for a series to be convergent. Cauchy's criterion.**

1. *A necessary condition for a series to be convergent.* For a convergent series $\sum_{n=1}^{\infty} a_n$, the generic term must tend to zero, $\lim\limits_{n\to\infty} a_n = 0$. If $\lim\limits_{n\to\infty} a_n \neq 0$, then the series is divergent.

**Example 2.** The series $\sum_{n=1}^{\infty} \cos\frac{1}{n}$ is divergent, since its generic term $a_n = \cos\frac{1}{n}$ does not tend to zero as $n \to \infty$.

The above necessary condition is insufficient for the convergence of a series.

**Example 3.** Consider the series $\sum_{n=1}^{\infty} \frac{1}{\sqrt{n}}$. Its generic term tends to zero, $\lim\limits_{n\to\infty} \frac{1}{\sqrt{n}} = 0$, but the series $\sum_{n=1}^{\infty} \frac{1}{\sqrt{n}}$ is divergent because its partial sums are unbounded,

$$s_n = \frac{1}{\sqrt{1}} + \frac{1}{\sqrt{2}} + \cdots + \frac{1}{\sqrt{n}} > n\frac{1}{\sqrt{n}} = \sqrt{n} \to \infty \quad \text{as} \quad n \to \infty.$$

2. *Cauchy's criterion of convergence of a series.* A series $\sum_{n=1}^{\infty} a_n$ is convergent if and only if for any $\varepsilon > 0$ there is $N = N(\varepsilon)$ such that for all $n > N$ and any positive integer $k$, the following inequality holds: $|a_{n+1} + \cdots + a_{n+k}| < \varepsilon$.

### 8.1.1-3. Properties of convergent series.

1. If a series is convergent, then any of its remainders is convergent. Removal or addition of finitely many terms does not affect the convergence of a series.

2. If all terms of a series are multiplied by a nonzero constant, the resulting series preserves the property of convergence or divergence (its sum is multiplied by that constant).

3. If the series $\sum_{n=1}^{\infty} a_n$ and $\sum_{n=1}^{\infty} b_n$ are convergent and their sums are equal to $S_1$ and $S_2$, respectively, then the series $\sum_{n=1}^{\infty} (a_n \pm b_n)$ are convergent and their sums are equal to $S_1 \pm S_2$.

4. Terms of a convergent series can be grouped in successive order; the resulting series has the same sum. In other words, one can insert brackets inside a series in an arbitrary order. The inverse operation of opening brackets is not always admissible. Thus, the series $(1-1)+(1-1)+\cdots$ is convergent (its sum is equal to zero), but, after removing the brackets, we obtain the divergent series $1 - 1 + 1 - 1 + \cdots$ (its generic term does not tend to zero).

## 8.1.2. Convergence Criteria for Series with Positive (Nonnegative) Terms

### 8.1.2-1. Basic convergence (divergence) criteria for series with positive terms.

1. *The first comparison criterion.* If $0 \leq a_n \leq b_n$ (starting from some $n$), then convergence of the series $\sum_{n=1}^{\infty} b_n$ implies convergence of $\sum_{n=1}^{\infty} a_n$; and divergence of the series $\sum_{n=1}^{\infty} a_n$ implies divergence of $\sum_{n=1}^{\infty} b_n$.

2. *The second convergence criterion.* Suppose that there is a finite limit

$$\lim_{n \to \infty} \frac{a_n}{b_n} = \sigma,$$

where $0 < \sigma < \infty$. Then $\sum_{n=1}^{\infty} a_n$ is convergent (resp., divergent) if and only if $\sum_{n=1}^{\infty} b_n$ is convergent (resp., divergent).

*Corollary.* Suppose that $a_{n+1}/a_n \leq b_{n+1}/b_n$ starting from some $N$ (i.e., for $n > N$). Then convergence of the series $\sum_{n=1}^{\infty} b_n$ implies convergence of $\sum_{n=1}^{\infty} a_n$, and divergence of $\sum_{n=1}^{\infty} a_n$ implies divergence of $\sum_{n=1}^{\infty} b_n$.

3. *D'Alembert criterion.* Suppose that there exists the limit (finite or infinite)

$$\lim_{n \to \infty} \frac{a_{n+1}}{a_n} = D.$$

If $D < 1$, then the series $\sum_{n=1}^{\infty} a_n$ is convergent. If $D > 1$, then the series is divergent. For $D = 1$, the D'Alembert criterion cannot be used for deciding whether the series is convergent or divergent.

**Example 1.** Let us examine convergence of the series $\sum_{n=1}^{\infty} n^k x^n$ with $x > 0$, using the D'Alembert criterion. Taking $a_n = n^k x^n$, we get
$$\frac{a_{n+1}}{a_n} = \left(1 + \frac{1}{n}\right)^k x \to x \quad \text{as} \quad n \to \infty.$$
Therefore, $D = x$. It follows that the series is convergent for $x < 1$ and divergent for $x > 1$.

4. *Cauchy's criterion.* Suppose that there exists the limit (finite or infinite)
$$\lim_{n \to \infty} \sqrt[n]{a_n} = K.$$

For $K < 1$, the series $\sum_{n=1}^{\infty} a_n$ is convergent; for $K > 1$, the series is divergent. For $K = 1$, the Cauchy criterion cannot be used to establish convergence of a series.

Remark. The Cauchy criterion is stronger than the D'Alembert criterion, but the latter is simpler than the former.

5. *Gauss' criterion.* Suppose that the ratio of two consecutive terms of a series can be represented in the form
$$\frac{a_n}{a_{n+1}} = \lambda + \frac{\mu}{n} + o\left(\frac{1}{n}\right) \quad \text{as} \quad n \to \infty.$$

The series $\sum_{n=1}^{\infty} a_n$ is convergent if $\lambda > 1$ or $\lambda = 1$, $\mu > 1$. The series is divergent if $\lambda < 1$ or $\lambda = 1$, $\mu \le 1$.

6. *Maclaurin–Cauchy integral criterion.* Let $f(x)$ be a nonnegative nonincreasing continuous function on the interval $1 \le x < \infty$. Let $f(1) = a_1$, $f(2) = a_2, \ldots, f(n) = a_n, \ldots$ Then the series $\sum_{n=1}^{\infty} a_n$ is convergent if and only if the improper integral $\int_1^{\infty} f(x)\,dx$ is convergent.

**Example 2.** The *harmonic series* $\sum_{n=1}^{\infty} \frac{1}{n} = 1 + \frac{1}{2} + \frac{1}{3} + \cdots$ is divergent, since the integral $\int_1^{\infty} \frac{1}{x}\,dx$ is divergent. In a similar way, one finds that the series $\sum_{n=1}^{\infty} \frac{1}{n^\alpha}$ is convergent for $\alpha > 1$ and divergent for $\alpha \le 1$.

### 8.1.2-2. Other criteria of convergence (divergence) of series with positive terms.

1. *Raabe criterion.* Suppose that there exists the limit (finite or infinite)
$$\lim_{n \to \infty} n\left(\frac{a_n}{a_{n+1}} - 1\right) = \mathcal{R}.$$

Then, for $\mathcal{R} > 1$ the series $\sum_{n=1}^{\infty} a_n$ is convergent, and for $\mathcal{R} < 1$ it is divergent. For $\mathcal{R} = 1$, the Raabe criterion is inapplicable.

**2. Bertrand criterion.** Suppose that there exists the limit (finite or infinite)

$$\lim_{n\to\infty} \ln n \left[ n\left( \frac{a_n}{a_{n+1}} - 1 \right) - 1 \right] = \mathcal{B}.$$

Then, for $\mathcal{B} > 1$ the series $\sum_{n=1}^{\infty} a_n$ is convergent, and for $\mathcal{B} < 1$ it is divergent. For $\mathcal{R} = 1$, the Bertrand criterion is inapplicable.

**3. Kummer criterion.** Let $b_n$ be an arbitrary sequence with positive terms. Suppose that there exists the limit (finite or infinite)

$$\lim_{n\to\infty} \left( b_n \frac{a_n}{a_{n+1}} - b_{n+1} \right) = \mathcal{K}.$$

Then, for $\mathcal{K} > 0$, the series $\sum_{n=1}^{\infty} a_n$ is convergent. If $\mathcal{K} < 0$ and the additional condition $\sum_{n=1}^{\infty} \frac{1}{b_n} = \infty$ holds, then the series is divergent.

**Remark.** From the Kummer criterion, we obtain the D'Alembert criterion (taking $b_n = 1$), the Raabe criterion (taking $b_n = n$), and the Bertrand criterion (taking $b_n = n \ln n$).

**4. Ermakov criterion.** Let $f(x)$ be a positive monotonically decreasing continuous function on the interval $1 \leq x < \infty$. Let $f(1) = a_1, f(2) = a_2, \ldots, f(n) = a_n, \ldots$ Then the following implications hold:

1) If $\dfrac{e^x f(e^x)}{f(x)} \leq q < 1,$ then the series $\sum_{n=1}^{\infty} a_n$ is convergent.

2) If $\dfrac{e^x f(e^x)}{f(x)} \geq 1,$ then the series $\sum_{n=1}^{\infty} a_n$ is divergent.

Here, it suffices to have the inequalities on the left for sufficiently large $x \geq x_0$.

**5. Generalized Ermakov criterion.** Let $f(x)$ be the function involved in the Ermakov criterion, and let $\varphi(x)$ be an arbitrary positive monotonically increasing function that has a continuous derivative and satisfies the inequality $\varphi(x) > x$. Then the following implications hold:

1) If $\dfrac{\varphi(x) f(\varphi(x))}{f(x)} \leq q < 1,$ then the series $\sum_{n=1}^{\infty} a_n$ is convergent.

2) If $\dfrac{\varphi(x) f(\varphi(x))}{f(x)} \geq 1,$ then the series $\sum_{n=1}^{\infty} a_n$ is divergent.

Here, it suffices to have the inequalities on the left for sufficiently large $x \geq x_0$.

**6. Sapogov criterion.** Let $a_1, a_2, \ldots$ be a monotonically increasing sequence. Then the series

$$\sum_{n=1}^{\infty} \left( 1 - \frac{a_n}{a_{n+1}} \right) \quad \left[ \text{as well as } \sum_{n=1}^{\infty} \left( \frac{a_n}{a_{n+1}} - 1 \right) \right]$$

is convergent, provided that the sequence $a_n$ is bounded ($a_n \leq L$). Otherwise, this series is divergent.

**7. *Special Cauchy criterion.*** Suppose that $a_1, a_2, \ldots$ is a monotonically decreasing sequence. Then the series $\sum_{n=1}^{\infty} a_n$ is convergent (resp., divergent) if and only if the series $\sum_{n=1}^{\infty} 2^n a_{2^n}$ is convergent (resp., divergent).

**8. *Abel–Dini criterion.*** If the series $\sum_{n=1}^{\infty} a_n$ is divergent and $s_n$ denotes its partial sum, then the series $\sum_{n=1}^{\infty} \frac{a_n}{s_n}$ is also divergent, while the series $\sum_{n=1}^{\infty} \frac{a_n}{s_n^{1+\sigma}}$ ($\sigma > 0$) is convergent.

**9. *Dini criterion.*** If the series $\sum_{n=1}^{\infty} a_n$ is convergent and $\gamma_n$ denotes its remainder after the $n$th term, then the series $\sum_{n=1}^{\infty} \frac{a_n}{\gamma_{n-1}}$ is divergent, while the series $\sum_{n=1}^{\infty} \frac{a_n}{\gamma_{n-1}^{\sigma}}$ ($0 < \sigma < 1$) is convergent.

**10. *Bugaev criterion.*** If the function $\varphi'(x) u(\varphi(x))$ is monotone for large enough $x$, then the series $\sum_{n=1}^{\infty} u(n)$ and $\sum_{n=1}^{\infty} \varphi'(n) u(\varphi(n))$ are convergent or divergent simultaneously.

**11. *Lobachevsky criterion.*** Let $u(x)$ be a monotonically decreasing function defined for all $x$. Then the series $\sum_{n=1}^{\infty} u(n)$ is convergent or divergent if and only if the series $\sum_{k=1}^{\infty} p_k 2^{-k}$ is convergent or divergent, where $p_k$ is defined from the relation $u(p_k) = 2^{-k}$.

## 8.1.3. Convergence Criteria for Arbitrary Numerical Series. Absolute and Conditional Convergence

8.1.3-1. Arbitrary series. Leibnitz, Abel, and Dirichlet convergence criteria.

**1. *Leibnitz criterion.*** Suppose that the terms $a_n$ of a series $\sum_{n=1}^{\infty} a_n$ have alternating signs, their absolute values form a nonincreasing sequence, and $a_n \to 0$ as $n \to \infty$. Then this "alternating" series is convergent. If $S$ is the sum of the series and $s_n$ is its $n$th partial sum, then the following inequality holds for the error $|S - s_n| \leq |a_{n+1}|$.

**Example 1.** The series $1 - \frac{1}{2^2} + \frac{1}{3^3} - \frac{1}{4^4} + \frac{1}{5^5} - \cdots$ is convergent by the Leibnitz criterion. Taking $S \approx s_4 = 1 - \frac{1}{2^2} + \frac{1}{3^3} - \frac{1}{4^4}$, we obtain the error less than $a_5 = \frac{1}{5^5} = 0.00032$.

**2. *Abel criterion.*** Consider the series

$$\sum_{n=1}^{\infty} a_n b_n = a_1 b_1 + a_2 b_2 + \cdots + a_n b_n + \cdots, \tag{8.1.3.1}$$

where $a_n$ and $b_n$ are two sequences or real numbers.

Series (8.1.3.1) is convergent if the series

$$\sum_{n=1}^{\infty} b_n = b_1 + b_2 + \cdots + b_n + \cdots \tag{8.1.3.2}$$

is convergent and the $a_n$ form a bounded monotone sequence ($|a_n| < K$).

3. *Dirichlet criterion.* Series (8.1.3.1) is convergent if partial sums of series (8.1.3.2) are bounded uniformly in $n$,

$$\left|\sum_{k=1}^{n} b_k\right| \leq M \qquad (n = 1, 2, \ldots),$$

and the sequence $a_n \to 0$ is monotone.

**Example 2.** Consider the series $\sum_{n=1}^{\infty} a_n \sin(nx)$, where $a_n \to 0$ is a monotonically decreasing sequence. Taking $b_n = \sin(nx)$ and using a well-known identity, we find the partial sum

$$s_n = \sum_{k=1}^{n} \sin(kx) = \frac{\cos\left(\frac{1}{2}x\right) - \cos\left[\left(n + \frac{1}{2}\right)x\right]}{2\sin\left(\frac{1}{2}x\right)} \qquad (x \neq 2m\pi; \ m = 0, \pm 1, \pm 2, \ldots).$$

This sum is bounded for $x \neq 2m\pi$:

$$|s_n| \leq \frac{1}{\left|\sin\left(\frac{1}{2}x\right)\right|}.$$

Therefore, by the Dirichlet criterion, the series $\sum_{n=1}^{\infty} a_n \sin(nx)$ is convergent for any $x \neq 2m\pi$. Direct verification shows that this series is also convergent for $x = 2m\pi$ (since all its terms at these points are equal to zero).

**Remark.** The Leibnitz and the Able criteria can be deduced from the Dirichlet criterion.

---

**8.1.3-2. Absolute and conditional convergence.**

1. *Absolutely convergent series.* A series $\sum_{n=1}^{\infty} a_n$ (with terms of arbitrary sign) is called *absolutely convergent* if the series $\sum_{n=1}^{\infty} |a_n|$ is convergent.

Any absolutely convergent series is convergent. In order to establish absolute convergence of a series, one can use all convergence criteria for series with nonnegative terms given in Subsection 8.1.2 (in these criteria, $a_n$ should be replaced by $|a_n|$).

**Example 3.** The series $1 + \frac{1}{2^2} - \frac{1}{3^2} - \frac{1}{4^2} + \frac{1}{5^2} + \frac{1}{6^2} - \cdots$ is absolutely convergent, since the series with the absolute values of its terms, $\sum_{n=1}^{\infty} \frac{1}{n^2}$, is convergent (see the second series in Example 2 of Subsection 8.1.2 for $\alpha = 2$).

2. *Conditionally convergent series.* A convergent series $\sum_{n=1}^{\infty} a_n$ is called *conditionally convergent* if the series $\sum_{n=1}^{\infty} |a_n|$ is divergent.

**Example 4.** The series $1 - \frac{1}{2} + \frac{1}{3} - \frac{1}{4} + \cdots$ is conditionally convergent, since it is convergent (by the Leibnitz criterion), but the series with absolute values of its terms is divergent (it is a harmonic series; see Example 2 in Subsection 8.1.2).

Any rearrangement of the terms of an absolutely convergent series (in particular, a convergent series with nonnegative terms) neither violates its absolute convergence nor changes its sum. Conditionally convergent series do not possess this property: the terms of a conditionally convergent series can be rearranged in such order that the sum of the new series becomes equal to any given value; its terms can also be rearranged so as to result in a divergent series.

## 8.1.4. Multiplication of Series. Some Inequalities

**8.1.4-1. Multiplication of series. Cauchy, Mertens, and Abel theorems.**

A product of two infinite series $\sum_{n=0}^{\infty} a_n$ and $\sum_{n=0}^{\infty} b_n$ is understood as a series whose terms have the form $a_n b_m$ ($n, m = 0, 1, \ldots$). The products $a_n b_m$ can be ordered to form a series in many different ways. The following theorems allow us to decide whether it is possible to multiply series.

CAUCHY THEOREM. *Suppose that the series $\sum_{n=0}^{\infty} a_n$ and $\sum_{n=0}^{\infty} b_n$ are absolutely convergent and their sums are equal to $A$ and $B$, respectively. Then any product of these series is an absolutely convergent series and its sum is equal to $AB$. The following Cauchy multiplication formula holds:*

$$\left(\sum_{n=0}^{\infty} a_n\right)\left(\sum_{n=0}^{\infty} b_n\right) = \sum_{n=0}^{\infty}\left(\sum_{m=0}^{n} a_m b_{n-m}\right). \tag{8.1.4.1}$$

MERTENS THEOREM. *The Cauchy multiplication formula (8.1.4.1) is also valid if one of the series, $\sum_{n=0}^{\infty} a_n$ or $\sum_{n=0}^{\infty} b_n$, is absolutely convergent and the other is (conditionally) convergent. In this case, the product is a convergent series, possibly, not absolutely convergent.*

ABEL THEOREM. *Consider two convergent series with sums $A$ and $B$. Suppose that the product of these series in the form of Cauchy (8.1.4.1) is a convergent series with sum $C$. Then $C = AB$.*

**8.1.4-2. Inequalities.**

1. *Generalized triangle inequality*:

$$\left|\sum_{n=1}^{\infty} a_n\right| \leq \sum_{n=1}^{\infty} |a_n|.$$

2. *Cauchy inequality (Cauchy–Bunyakovsky inequality)*:

$$\left(\sum_{n=1}^{\infty} a_n b_n\right)^2 \leq \left(\sum_{n=1}^{\infty} a_n^2\right)\left(\sum_{n=1}^{\infty} b_n^2\right).$$

3. *Minkowski inequality*:

$$\left(\sum_{n=1}^{\infty} |a_n + b_n|^p\right)^{\frac{1}{p}} \leq \left(\sum_{n=1}^{\infty} |a_n|^p\right)^{\frac{1}{p}} + \left(\sum_{n=1}^{\infty} |b_n|^p\right)^{\frac{1}{p}}, \qquad p \geq 1.$$

4. *Hölder inequality* (for $p = 2$ coincides with the Cauchy inequality):

$$\left|\sum_{n=1}^{\infty} a_n b_n\right| \leq \left(\sum_{n=1}^{\infty} |a_n|^p\right)^{\frac{1}{p}} \left(\sum_{n=1}^{\infty} |b_n|^{\frac{p}{p-1}}\right)^{\frac{p-1}{p}}, \qquad p > 1.$$

5. *An inequality with $\pi$*:

$$\left(\sum_{n=1}^{\infty} a_n\right)^4 \leq \pi^2 \left(\sum_{n=1}^{\infty} a_n^2\right)\left(\sum_{n=1}^{\infty} n^2 a_n^2\right).$$

In all these inequalities it is assumed that the series in the right-hand sides are convergent.

## 8.1.5. Summation Methods. Convergence Acceleration

**8.1.5-1. Some simple methods for calculating the sum of a series.**

THEOREM 1. *Suppose that the terms of a series $\sum_{n=1}^{\infty} a_n$ can be represented in the form $a_n = b_n - b_{n+1}$, where $b_n$ is a sequence with a finite limit $b_\infty$. Then*

$$\sum_{n=1}^{\infty} a_n = b_1 - b_\infty.$$

**Example 1.** Let us find the sum of the infinite series $S = \sum_{n=1}^{\infty} \frac{1}{n(n+1)}$.

We have
$$a_n = \frac{1}{n(n+a)} = \frac{1}{n} - \frac{1}{n+1} \implies b_n = \frac{1}{n}.$$

Since $b_1 = 1$ and $\lim_{n \to \infty} b_n = 0$, we get $S = 1$.

THEOREM 2. *Suppose that the terms of a series $\sum_{n=0}^{\infty} a_n$ can be represented in the form*

$$a_n = b_n - b_{n+m},$$

*where $m$ is a positive integer and the sequence $b_n$ has a finite limit $b_\infty$. Then*

$$\sum_{n=1}^{\infty} a_n = b_0 + b_1 + \cdots + b_{m-1} - m b_\infty.$$

THEOREM 3. *Suppose that the terms of a series $\sum_{n=0}^{\infty} a_n$ can be represented in the form*

$$a_n = \alpha_1 b_{n+1} + \alpha_2 b_{n+2} + \cdots + \alpha_m b_{n+m}, \qquad (8.1.5.1)$$

*where $m \geq 2$ is a fixed positive integer, the sequence $b_n$ has a finite limit $b_\infty$, and $\alpha_k$ satisfy the condition*

$$\alpha_1 + \alpha_2 + \cdots + \alpha_m = 0. \qquad (8.1.5.2)$$

*Then the series is convergent and*

$$\sum_{n=0}^{\infty} a_n = \alpha_1 b_1 + (\alpha_1+\alpha_2) b_2 + \cdots + (\alpha_1+\alpha_2+\cdots+\alpha_{m-1}) b_{m-1} + [\alpha_2 + 2\alpha_3 + \cdots + (m-1)\alpha_m] b_\infty.$$

$$(8.1.5.3)$$

**Example 2.** For the series $\sum_{n=0}^{\infty} \frac{4n+6}{(n+1)(n+2)(n+3)}$, we have

$$a_n = \frac{4n+6}{(n+1)(n+2)(n+3)} = \frac{1}{n+1} + \frac{2}{n+2} - \frac{3}{n+3},$$

which corresponds to the following values in (8.1.5.1):

$$\alpha_1 = 1, \quad \alpha_2 = 2, \quad \alpha_3 = -3, \quad b_n = \frac{1}{n}.$$

Thus, condition (8.1.5.2) holds: $1 + 2 - 3 = 0$, and the sequence $b_n$ tends to zero: $b_\infty = 0$. Using (8.1.5.3), we find the sum of the series

$$\sum_{n=0}^{\infty} \frac{4n+6}{(n+1)(n+2)(n+3)} = 1 \cdot 1 + (1+2)\frac{1}{2} = \frac{5}{2}.$$

### 8.1.5-2. Summation of series with the help of Laplace transforms.

Let $a(k)$ be the Laplace transform of a given function $f(x)$,

$$a(k) = \int_0^\infty e^{-kx} f(x)\, dx.$$

Then the following summation formulas hold:

$$\sum_{k=1}^\infty a(k) = \int_0^\infty \frac{f(x)\,dx}{e^x - 1},$$
$$\sum_{k=1}^\infty (-1)^k a(k) = -\int_0^\infty \frac{f(x)\,dx}{e^x + 1}, \tag{8.1.5.4}$$

provided that the series are convergent.

**Example 3.** It is easy to check that

$$\frac{a}{k^2 + a^2} = \int_0^\infty e^{-kx} \sin(ax)\, dx.$$

Therefore, using the first formula in (8.1.5.4), we get

$$\sum_{k=1}^\infty \frac{1}{k^2 + a^2} = \frac{1}{a}\int_0^\infty \frac{\sin(ax)\,dx}{e^x - 1} = \frac{\pi}{2a}\coth(\pi a) - \frac{1}{2a^2}.$$

### 8.1.5-3. Kummer and Abel transformations. Acceleration of convergence of series.

**1°.** *Kummer transformation.* Consider a series with positive (nonnegative) terms

$$\sum_{n=1}^\infty a_n \tag{8.1.5.5}$$

and an auxiliary series with a finite sum

$$B = \sum_{n=1}^\infty b_n. \tag{8.1.5.6}$$

Suppose that there is a finite limit

$$K = \lim_{n\to\infty} \frac{b_n}{a_n} \neq 0.$$

Under these conditions, the series (8.1.5.5) is convergent and the following identity holds:

$$\sum_{n=1}^\infty a_n = \frac{B}{K} + \sum_{n=1}^\infty \left(1 - \frac{1}{K}\frac{b_n}{a_n}\right) a_n. \tag{8.1.5.7}$$

The right-hand side of (8.1.5.7) is called the Kummer transformation of the given series.

**Remark.** The Kummer transformation is used for accelerating convergence of series, since the generic term $a_n^{(1)} = \left(1 - \dfrac{1}{K}\dfrac{b_n}{a_n}\right)a_n$ of the transformed series (8.1.5.7) tends to zero faster that the generic term $\{a_n\}$ of the original series (8.1.5.5): $\lim\limits_{n\to\infty} a_n^{(1)}/a_n = 0$. The auxiliary sequence $\{b_n\}$ is chosen in such a way that the sum (8.1.5.6) is known beforehand.

2°. *Abel transformation*:

$$\sum_{n=1}^{\infty} a_n b_n = \sum_{n=1}^{\infty}(a_n - a_{n+1})B_n, \qquad B_n = \sum_{k=1}^{n} b_k.$$

Here, it is assumed that the sequence $\{B_n\}$ is bounded and $\lim\limits_{n\to\infty} a_n = 0$ (for example, these conditions are satisfied if the sequence $\{a_n\}$ has finitely many nonzero terms).

**Remark.** The Abel transformation is used for accelerating convergence of series whose convergence is slow (if $\lim\limits_{n\to\infty} a_{n+1}/a_n = 1$ and $b_n = O(1)$).

## 8.1.6. Infinite Products

### 8.1.6-1. Convergent and divergent infinite products.

An *infinite product* is an expression of the form

$$a_1 a_2 a_3 \cdots = \prod_{n=1}^{\infty} a_n \qquad (a_n \neq 0),$$

where $a_n$ are real (in the general case, complex) numbers. The expression

$$p_n = a_1 a_2 \cdots a_n = \prod_{k=1}^{n} a_k \tag{8.1.6.1}$$

is called a *finite product*.

One says that an *infinite product is convergent to* $p$ if there exists a finite nonzero limit of the partial products:

$$\lim_{n\to\infty} \prod_{k=1}^{n} a_k = p. \tag{8.1.6.2}$$

If there is no such limit, or $p = \infty$, or $p = 0$, one says that the *infinite product is divergent*. In the last case one says that the *infinite product diverges to zero* ($p = 0$).

The simplest examples of infinite products $\prod\limits_{n=1}^{\infty} a_n$:

| | | |
|---|---|---|
| for $a_n = 1$ | the infinite product is convergent, | $p_n \to 1$; |
| for $a_n = (-1)^n$ | the infinite product is divergent, | $p_n$ has no limit; |
| for $a_n = n$ | the infinite product is divergent, | $p_n \to \infty$; |
| for $a_n = 1/n$ | the infinite product is divergent to zero, | $p_n \to 0$. |

### 8.1.6-2. Infinite products with positive factors.

Taking logarithm of relation (8.1.6.1), one can reduce the problem of convergence of infinite products with positive factors $a_n$ to the problem of convergence of partial sums of numerical series:

$$s_n \equiv \ln p_n = \sum_{k=1}^{n} \ln a_k.$$

Therefore, in order to examine convergence of infinite products, one can use the convergence criteria for infinite series considered in Subsections 8.1.2–8.1.3 (with $a_n$ replaced by $\ln a_n$).

In a similar way, one can examine convergence of infinite products containing finitely many negative factors (infinite products with infinitely many negative factors are divergent; see Corollary in Paragraph 8.1.6-3). To that end, one should consider the part of the infinite product with positive factors.

### 8.1.6-3. Necessary condition for an infinite product to be convergent.

If an infinite product $\prod_{n=1}^{\infty} a_n$ is convergent, then $\lim_{n \to \infty} a_n = 1$; if $\lim_{n \to \infty} a_n \neq 1$, then the infinite product is divergent. (This necessary condition is insufficient to ensure convergence of a product.)

COROLLARY. *For a convergent infinite product, there is $N$ such that $a_n > 0$ for all $n > N$ (i.e., a convergent infinite product can have only finitely many negative factors).*

### 8.1.6-4. Convergence criteria for infinite products.

In the criteria given below, $\Delta_n$ denotes the difference of the factor and its limit, which is equal to 1 (see the above necessary condition of convergence):

$$\Delta_n = a_n - 1.$$

THEOREM 1. *Suppose that for sufficiently large $n$, the difference $\Delta_n$ does not change sign. Then the infinite product $\prod_{n=1}^{\infty} a_n$ is convergent if and only if the series*

$$\sum_{n=1}^{\infty} \Delta_n \qquad (8.1.6.3)$$

*is convergent.*

THEOREM 2. *If series (8.1.6.3) and the series*

$$\sum_{n=1}^{\infty} \Delta_n^2$$

*are convergent, then the infinite product $\prod_{n=1}^{\infty} a_n$ is convergent. (Here, the difference $\Delta_n$ may change sign.)*

THEOREM 3 (AN ANALOGUE OF THE D'ALEMBERT THEOREM). *Suppose that the terms of an infinite product satisfy the necessary condition of convergence, i.e., $a_n \to 1$ as $n \to \infty$, and there exists the limit (finite or infinite)*

$$\lim_{n \to \infty} \frac{\Delta_{n+1}}{\Delta_n} = \lim_{n \to \infty} \frac{a_{n+1} - 1}{a_n - 1} = D.$$

*Then, for $D < 1$ the infinite product $\prod_{n=1}^{\infty} a_n$ is convergent, and for $D > 1$ it is divergent. (For $D = 1$, this criterion is inapplicable.)*

**8.1.6-5. Absolute convergence of infinite products.**

An infinite product $\prod_{n=1}^{\infty} a_n$ is said to be *absolutely convergent* if the product $\prod_{n=1}^{\infty} (1 + |\Delta_n|)$ is convergent.

THEOREM 4. *An infinite product $\prod_{n=1}^{\infty} a_n$ is absolutely convergent if the series $\sum_{n=1}^{\infty} |\Delta_n|$ is convergent.*

An infinite product is *commutation convergent* (i.e., its value does not depend on the order of its factors) if and only if it is absolutely convergent.

## 8.2. Functional Series

### 8.2.1. Pointwise and Uniform Convergence of Functional Series

**8.2.1-1. Convergence of a functional series at a point. Convergence domain.**

A *functional series* is a series of the form

$$u_1(x) + u_2(x) + \cdots + u_n(x) + \cdots = \sum_{n=1}^{\infty} u_n(x),$$

where $u_n(x)$ are functions defined on a set $X$ on the real axis. The series $\sum_{n=1}^{\infty} u_n(x)$ is called *convergent at a point* $x_0 \in X$ if the numerical series $\sum_{n=1}^{\infty} u_n(x_0)$ is convergent. The set of all $x \in X$ for which the functional series is convergent is called its *convergence domain*. The sum of the series is a function of $x$ defined on its convergence domain.

In order to find the convergence domain for a functional series, one can use the convergence criteria for numerical series described in Subsections 8.1.2 and 8.1.3 (with the variable $x$ regarded as a parameter).

A series $\sum_{n=1}^{\infty} u_n(x)$ is called *absolutely convergent on a set* $X$ if the series $\sum_{n=1}^{\infty} |u_n(x)|$ is convergent on that set.

**Example.** The functional series $1 + x + x^2 + x^3 + \cdots$ is convergent for $-1 < x < 1$ (see Example 1 in Subsection 8.1.1). Its sum is defined on this interval, $S = \dfrac{1}{1-x}$.

The series $\sum_{k=n+1}^{\infty} u_k(x)$ is called the *remainder of a functional series* $\sum_{n=1}^{\infty} u_n(x)$. For a series convergent on a set $X$, the relation $S(x) = s_n(x) + r_n(x)$ ($s_n(x)$ is the partial sum of the series, $r_n(x)$ is the sum of its remainder) implies that $\lim_{n \to \infty} r_n(x) = 0$ for $x \in X$.

### 8.2.1-2. Uniformly convergent series. Condition of uniform convergence.

A functional series is called *uniformly convergent* on a set $X$ if for any $\varepsilon > 0$ there is $N$ (depending on $\varepsilon$ but not on $x$) such that for all $n > N$, the inequality $\left| \sum_{k=n+1}^{\infty} u_k(x) \right| < \varepsilon$ holds for all $x \in X$.

A *necessary and sufficient condition of uniform convergence of a series.* A series $\sum_{n=1}^{\infty} u_n(x)$ is uniformly convergent on a set $X$ if and only if for any $\varepsilon > 0$ there is $N$ (independent of $x$) such that for all $n > N$ and all $m = 1, 2, \ldots$, the inequality

$$\left| \sum_{k=n+1}^{n+m} u_k(x) \right| < \varepsilon$$

holds for all $x \in X$.

## 8.2.2. Basic Criteria of Uniform Convergence. Properties of Uniformly Convergent Series

### 8.2.2-1. Criteria of uniform convergence of series.

1. *Weierstrass criterion of uniform convergence.* A functional series $\sum_{n=1}^{\infty} u_n(x)$ is uniformly convergent on a set $X$ if there is a convergent series $\sum_{n=1}^{\infty} a_n$ with nonnegative terms such that $|u_n(x)| \leq a_n$ for all sufficiently large $n$ and all $x \in X$. The series $\sum_{n=1}^{\infty} a_n$ is called a *majorant series* for $\sum_{n=1}^{\infty} u_n(x)$.

**Example.** The series $\sum_{n=1}^{\infty} (-1)^n \dfrac{\sin nx}{n^2}$ is uniformly convergent for $-\infty < x < \infty$, since $\left| (-1)^n \dfrac{\sin nx}{n^2} \right| \leq \dfrac{1}{n^2}$, and the numerical series $\sum_{n=1}^{\infty} \dfrac{1}{n^2}$ is convergent (see the second series in Example 2 in Subsection 8.1.2).

2. *Abel criterion of uniform convergence of functional series.* Consider a functional series

$$\sum_{n=1}^{\infty} u_n(x) v_n(x) = u_1(x) v_1(x) + u_2(x) v_2(x) + \cdots + u_n(x) v_n(x) + \cdots, \qquad (8.2.2.1)$$

where $u_n(x)$ and $v_n(x)$ are sequences of functions of the real variable $x \in [a, b]$.

Series (8.2.2.1) is uniformly convergent on the interval $[a, b]$ if the series

$$\sum_{n=1}^{\infty} v_n(x) = v_1(x) + v_2(x) + \cdots + v_n(x) + \cdots \qquad (8.2.2.2)$$

is uniformly convergent on $[a, b]$ and the functions $u_n(x)$ form a monotone sequence for each $x$ and are uniformly bounded (i.e., $|u_n(x)| \leq K$ with a constant $K$ independent of $n, x$).

3. *Dirichlet criterion of uniform convergence of functional series.* Series (8.2.2.1) is uniformly convergent on the interval $[a, b]$ if the partial sums of the series (8.2.2.2) are uniformly bounded, i.e.,

$$\left| \sum_{k=1}^{n} v_k(x) \right| \leq M = \text{const} \qquad (x \in [a, b], \quad n = 1, 2, \dots),$$

and the functions $u_n(x)$ form a monotone sequence (for each $x$) that uniformly converges to zero on $[a, b]$ as $n \to \infty$.

### 8.2.2-2. Properties of uniformly convergent series.

Let $\sum_{n=1}^{\infty} u_n(x)$ be a functional series that is uniformly convergent on a segment $[a, b]$, and let $S(x)$ be its sum. Then the following statements hold.

THEOREM 1. *If all terms $u_n(x)$ of the series are continuous at a point $x_0 \in [a, b]$, then the sum $S(x)$ is continuous at that point.*

THEOREM 2. *If the terms $u_n(x)$ are continuous on $[a, b]$, then the series admits term-by-term integration:*

$$\int_a^b S(x)\,dx = \int_a^b \left( \sum_{n=1}^{\infty} u_n(x) \right) dx = \sum_{n=1}^{\infty} \int_a^b u_n(x)\,dx.$$

*Remark.* The condition of continuity of the functions $u_n(x)$ on $[a, b]$ can be replaced by a weaker condition of their integrability on $[a, b]$.

THEOREM 3. *If all terms of the series have continuous derivatives and the functional series $\sum_{n=1}^{\infty} u'_n(x)$ is uniformly convergent on $[a, b]$, then the sum $S(x)$ is continuously differentiable on $[a, b]$ and*

$$S'(x) = \left( \sum_{n=1}^{\infty} u_n(x) \right)' = \sum_{n=1}^{\infty} u'_n(x)$$

(i.e., the series admits term-by-term differentiation).

## 8.3. Power Series
### 8.3.1. Radius of Convergence of Power Series. Properties of Power Series

#### 8.3.1-1. Abel theorem. Convergence radius of a power series.

A *power series* is a functional series of the form

$$\sum_{n=0}^{\infty} a_n x^n = a_0 + a_1 x + a_2 x^2 + a_3 x^3 + \cdots \qquad (8.3.1.1)$$

(the constants $a_0, a_1, \ldots$ are called the *coefficients* of the power series), and also a series of a more general form

$$\sum_{n=0}^{\infty} a_n(x-x_0)^n = a_0 + a_1(x-x_0) + a_2(x-x_0)^2 + a_3(x-x_0)^3 + \cdots,$$

where $x_0$ is a fixed point. Below, we consider power series of the first form, since the second series can be transformed into the first by the replacement $\bar{x} = x - x_0$.

ABEL THEOREM. *A power series $\sum_{n=0}^{\infty} a_n x^n$ that is convergent for some $x = x_1$ is absolutely convergent for all $x$ such that $|x| < |x_1|$. A power series that is divergent for some $x = x_2$ is divergent for all $x$ such that $|x| > |x_2|$.*

Remark. There exist series convergent for all $x$, for instance, $\sum_{n=1}^{\infty} \frac{x^n}{n!}$. There are series convergent only for $x = 0$, for instance, $\sum_{n=1}^{\infty} n! \, x^n$.

For a given power series (8.3.1.1), let $R$ be the least upper bound of all $|x|$ such that the series (8.3.1.1) is convergent at point $x$. Thus, by the Abel theorem, the series is (absolutely) convergent for all $|x| < R$, and the series is divergent for all $|x| > R$. The constant $R$ is called the *radius of convergence* of the power series, and the interval $(-R, R)$ is called its *interval of convergence*. The problem of convergence of a power series at the endpoints of its convergence interval has to be studied separately in each specific case. If a series is convergent only for $x = 0$, the convergence interval degenerates into a point (and $R = 0$); if a series is convergent for all $x$, then, obviously, $R = \infty$.

8.3.1-2. Formulas for the radius of convergence of power series.

1°. The radius of convergence of a power series (8.3.1.1) with finitely many zero terms can be calculated by the formulas

$$R = \lim_{n \to \infty} \left| \frac{a_n}{a_{n+1}} \right| \quad \text{(obtained from the D'Alembert criterion for numerical series),}$$

$$R = \lim_{n \to \infty} \frac{1}{\sqrt[n]{|a_n|}} \quad \text{(obtained from the Cauchy criterion for numerical series).}$$

**Example 1.** For the power series $\sum_{n=1}^{\infty} \frac{3^n}{n} x^n$, using the first formula for the radius of convergence, we get

$$R = \lim_{n \to \infty} \left| \frac{a_n}{a_{n+1}} \right| = \lim_{n \to \infty} \left| \frac{n+1}{3n} \right| = \frac{1}{3}.$$

Therefore, the series is absolutely convergent on the interval $-\frac{1}{3} < x < \frac{1}{3}$ and is divergent outside that interval. At the left endpoint of the interval, for $x = -\frac{1}{3}$, we have the conditionally convergent series $\sum_{n=1}^{\infty} \frac{(-1)^n}{n}$, and at the right endpoint, for $x = \frac{1}{3}$, we have the divergent numerical series $\sum_{n=1}^{\infty} \frac{1}{n}$. Thus, the series under consideration is convergent on the semi-open interval $\left[-\frac{1}{3}, \frac{1}{3}\right)$.

2°. Suppose that a power series (8.3.1.1) is convergent at a boundary point of its convergence interval, say, for $x = R$. Then its sum is left-hand continuous at that point,

$$\lim_{x \to R-0} \sum_{n=0}^{\infty} a_n x^n = \sum_{n=0}^{\infty} a_n R^n.$$

**Example 2.** Having the expansion

$$\ln(1+x) = x - \frac{x^2}{2} + \frac{x^3}{3} - \cdots + (-1)^{n+1}\frac{x^n}{n} + \cdots \qquad (R = 1)$$

in the domain $-1 < x < 1$ and knowing that the series

$$1 - \frac{1}{2} + \frac{1}{3} - \cdots + (-1)^{n+1}\frac{1}{n} + \cdots$$

is convergent (by the Leibnitz criterion for series with terms of alternating sign), we conclude that the sum of the last series is equal to $\ln 2$.

### 8.3.1-3. Properties of power series.

On any closed segment belonging to the (open) convergence interval of a power series, the series is uniformly convergent. Therefore, on any such segment, the series has all the properties of uniformly convergent series described in Subsection 8.2.2. Therefore, the following statements hold:

1. A power series (8.3.1.1) admits term-by-term integration on any segment $[0, x]$ for $|x| < R$,

$$\int_0^x \left( \sum_{n=0}^{\infty} a_n x^n \right) dx = \sum_{n=0}^{\infty} \frac{a_n}{n+1} x^{n+1}$$

$$= a_0 x + \frac{a_1}{2} x^2 + \frac{a_2}{3} x^3 + \cdots + \frac{a_n}{n+1} x^{n+1} + \cdots.$$

*Remark 1.* The value of $x$ in this formula may coincide with an endpoint of the convergence interval ($x = -R$ and/or $x = R$), provided that series (8.3.1.1) is convergent at that point.

*Remark 2.* The convergence radii of the original series and the series obtained by its term-by-term integration on the segment $[0, x]$ coincide.

2. Inside the convergence interval (for $|x| < R$), the series admits term-by-term differentiation of any order, in particular,

$$\frac{d}{dx}\left( \sum_{n=0}^{\infty} a_n x^n \right) = \sum_{n=1}^{\infty} n a_n x^{n-1}$$

$$= a_1 + 2a_2 x + 3a_3 x^2 + \cdots + n a_n x^{n-1} + \cdots.$$

*Remark 1.* This statement remains valid for an endpoint of the convergence interval if the series (8.3.1.1) is convergent at that point.

*Remark 2.* The convergence radii of the original series and the series obtained by its term-by-term differentiation coincide.

## 8.3.2. Taylor and Maclaurin Power Series

### 8.3.2-1. Basic definitions.

Let $f(x)$ is an infinitely differentiable function at a point $x_0$. The *Taylor series* for this function is the power series

$$\sum_{n=0}^{\infty} \frac{1}{n!} f^{(n)}(x_0)(x - x_0)^n = f(x_0) + f'(x_0)(x - x_0) + \frac{1}{2} f''(x_0)(x - x_0)^2 + \cdots,$$

where $0! = 1$ and $f^{(0)}(x_0) = f(x_0)$.

A special case of the Taylor series (for $x_0 = 0$) is the *Maclaurin series*:

$$\sum_{n=0}^{\infty} \frac{1}{n!} f^{(n)}(0) x^n = f(0) + f'(0)x + \frac{1}{2} f''(0) x^2 + \cdots.$$

A formal Taylor series (Maclaurin series) for a function $f(x)$ may be:

1) divergent for $x \neq x_0$,
2) convergent in a neighborhood of $x_0$ to a function different from $f(x)$,
3) convergent in a neighborhood of $x_0$ to the function $f(x)$.

In the last case, one says that $f(x)$ admits *expansion in Taylor series* in the said neighborhood, and one writes

$$f(x) = \sum_{n=0}^{\infty} \frac{1}{n!} f^{(n)}(x_0)(x - x_0)^n.$$

### 8.3.2-2. Conditions of expansion in Taylor series.

A *necessary and sufficient condition for a function $f(x)$ to be represented by its Taylor series* in a neighborhood of a point $x_0$ is that the remainder term in the Taylor formula* should tend to zero as $n \to \infty$ in this neighborhood of $x_0$.

In order that $f(x)$ could be represented by its Taylor series in a neighborhood of $x_0$, it suffices that all its derivatives in that neighborhood be bounded by the same constant, $|f^{(n)}(x)| \leq M$ for all $n$.

*Uniqueness of the Taylor series expansion.* For a function $f(x)$ that can be represented as the sum of a power series, the coefficients of this series are determined uniquely (since this series is the Taylor series of $f(x)$ and its coefficients have the form $\dfrac{f^{(n)}(x_0)}{n!}$, where $n = 0, 1, 2, \dots$). Therefore, in problems of representing a function by a power series, the answer does not depend on the method adopted for this purpose.

### 8.3.2-3. Representation of some functions by the Maclaurin series.

The following representations of elementary functions by Maclaurin series are often used in applications:

$$e^x = 1 + x + \frac{x^2}{2!} + \frac{x^3}{3!} + \cdots + \frac{x^n}{n!} + \cdots;$$

$$\sin x = x - \frac{x^3}{3!} + \frac{x^5}{5!} - \cdots + (-1)^{n-1} \frac{x^{2n-1}}{(2n-1)!} + \cdots;$$

$$\cos x = 1 - \frac{x^2}{2!} + \frac{x^4}{4!} - \cdots + (-1)^n \frac{x^{2n}}{(2n)!} + \cdots;$$

$$\sinh x = x + \frac{x^3}{3!} + \frac{x^5}{5!} + \cdots + \frac{x^{2n-1}}{(2n-1)!} + \cdots;$$

---

* Different representations of the remainder in the Taylor formula are given in Paragraph 6.2.4-4.

$$\cosh x = 1 + \frac{x^2}{2!} + \frac{x^4}{4!} + \cdots + \frac{x^{2n}}{(2n)!} + \cdots;$$

$$(1+x)^\alpha = 1 + \alpha x + \frac{\alpha(\alpha-1)}{2!}x^2 + \cdots + \frac{\alpha(\alpha-1)\ldots(\alpha-n+1)}{n!}x^n + \cdots;$$

$$\ln(1+x) = x - \frac{x^2}{2} + \frac{x^3}{3} - \cdots + (-1)^{n+1}\frac{x^n}{n} + \cdots;$$

$$\arctan x = x - \frac{x^3}{3} + \frac{x^5}{5} - \cdots + (-1)^{n+1}\frac{x^{2n-1}}{2n-1} + \cdots.$$

The first five series are convergent for $-\infty < x < \infty$ ($R = \infty$), and the other series have unit radius of convergence, $R = 1$.

### 8.3.3. Operations with Power Series. Summation Formulas for Power Series

**8.3.3-1. Addition, subtraction, multiplication, and division of power series.**

1. *Addition and subtraction of power series.* Two series $\sum_{n=0}^{\infty} a_n x^n$ and $\sum_{n=0}^{\infty} b_n x^n$ with convergence radii $R_a$ and $R_b$, respectively, admit term-by-term addition and subtraction on the intersection of their convergence intervals:

$$\sum_{n=0}^{\infty} a_n x^n \pm \sum_{n=0}^{\infty} b_n x^n = \sum_{n=0}^{\infty} c_n x^n, \quad c_n = a_n \pm b_n.$$

The radius of convergence of the resulting series satisfies the inequality $R_c \geq \min[R_a, R_b]$.

2. *Multiplication of power series.* Two series $\sum_{n=0}^{\infty} a_n x^n$ and $\sum_{n=0}^{\infty} b_n x^n$, with the respective convergence radii $R_a$ and $R_b$, can be multiplied on the intersection of their convergence intervals, and their product has the form

$$\left(\sum_{n=0}^{\infty} a_n x^n\right)\left(\sum_{n=0}^{\infty} b_n x^n\right) = \sum_{n=0}^{\infty} c_n x^n, \quad c_n = \sum_{k=0}^{n} a_k b_{n-k}.$$

The convergence radius of the product satisfies the inequality $R_c \geq \min[R_a, R_b]$.

3. *Division of power series.* The ratio of two power series $\sum_{n=0}^{\infty} a_n x^n$ and $\sum_{n=0}^{\infty} b_n x^n$, $b_0 \neq 0$, with convergence radii $R_a$ and $R_b$ can be represented as a power series

$$\frac{\sum_{n=0}^{\infty} a_n x^n}{\sum_{n=0}^{\infty} b_n x^n} = c_0 + c_1 x + c_2 x^2 + \cdots = \sum_{n=0}^{\infty} c_n x^n, \tag{8.3.3.1}$$

whose coefficients can be found, by the method of indefinite coefficients, from the relation

$$(a_0 + a_1 x + a_2 x^2 + \cdots) = (b_0 + b_1 x + b_2 x^2 + \cdots)(c_0 + c_1 x + c_2 x^2 + \cdots).$$

Thus, for the unknown $c_n$, we obtain a triangular system of linear algebraic equations

$$a_n = \sum_{k=0}^{n} b_k c_{n-k}, \qquad n = 0, 1, \ldots,$$

which is solved consecutively, starting from the first equation:

$$c_0 = \frac{a_0}{b_0}, \quad c_1 = \frac{a_1 b_0 - a_0 b_1}{b_0^2}, \quad c_n = \frac{a_n}{b_0} - \frac{1}{b_0} \sum_{k=1}^{n} b_k c_{n-k}, \quad n = 2, 3, \ldots$$

The convergence radius of the series (8.3.3.1) is determined by the formula

$$R_1 = \min\left[R_a, \frac{\rho}{M+1}\right],$$

where $\rho$ is any constant such that $0 < \rho < R_b$; $\rho$ can be chosen arbitrarily close to $R_b$; and $M$ is the least upper bound of the quantities $|b_m/b_0|\rho^m$ ($m = 1, 2, \ldots$), so that $|b_m/b_0|\rho^m \leq M$ for all $m$.

**8.3.3-2. Composition of functions representable by power series.**

Consider a power series

$$z = f(y) = a_0 + a_1 y + a_2 y^2 + \cdots = \sum_{n=0}^{\infty} a_n y^n \qquad (8.3.3.2)$$

with convergence radius $R$. Let the variable $y$ be a function of $x$ that can be represented by a power series

$$y = \varphi(x) = b_0 + b_1 x + a_2 x^2 + \cdots = \sum_{n=0}^{\infty} b_n x^n \qquad (8.3.3.3)$$

with convergence radius $r$. It is required to represent $z$ as a power series of $x$ and find the convergence radius of this series.

Formal substitution of (8.3.3.3) into (8.3.3.2) yields

$$z = f(\varphi(x)) = \sum_{n=0}^{\infty} a_n \left(\sum_{k=0}^{\infty} b_k x^k\right)^n = A_0 + A_1 x + A_2 x^2 + \cdots = \sum_{n=0}^{\infty} A_n x^n, \qquad (8.3.3.4)$$

where

$$A_0 = a_0 + a_1 b_0 + a_2 b_0^2 + \cdots,$$
$$A_1 = a_1 b_1 + 2a_2 b_0 b_1 + 3a_3 b_0^2 b_1 + \cdots,$$
$$A_2 = a_1 b_2 + a_2(b_1^2 + 2b_0 b_2) + 3a_3(b_0 b_1^2 + b_0^2 b_2) + \cdots,$$
$$\ldots\ldots\ldots\ldots\ldots\ldots\ldots\ldots\ldots\ldots\ldots\ldots\ldots\ldots\ldots\ldots\ldots$$

**THEOREM ON CONVERGENCE OF SERIES (8.3.3.4).**
(i) If series (8.3.3.2) is convergent for all $y$ (i.e., $R = \infty$), then the convergence radius of series (8.3.3.4) coincides with the convergence radius $r$ of series (8.3.3.3).
(ii) If $0 \le |b_0| < R$, then series (8.3.3.4) is convergent on the interval $(-R_1, R_1)$, where

$$R_1 = \frac{(R - |b_0|)\rho}{M + R - |b_0|},$$

and $\rho$ is an arbitrary constant such that $0 < \rho < r$; $\rho$ can be chosen arbitrarily close to $r$; and $M$ is the least upper bound of the quantities $|b_m|\rho^m$ ($m = 1, 2, \ldots$), so that $|b_m|\rho^m \le M$ for all $m$.
(iii) If $|b_0| > R$, then series (8.3.3.4) is divergent.

**Remark.** Case (i) is realized, for instance if (8.3.3.2) has finitely many terms.

**8.3.3-3. Local inversion of a function represented by power series.**

1. Suppose that $y = y(x)$ is a function that can be represented, in a neighborhood of a point $x = x_0$, by the power series

$$y = y_0 + a(x - x_0) + b(x - x_0)^2 + c(x - x_0)^3 + d(x - x_0)^4 + \cdots, \qquad a \ne 0.$$

Then the inverse function $x = x(y)$, in a neighborhood of $y = y_0$, can be represented by the series

$$x = x_0 + \frac{1}{a}(y - y_0) - \frac{b}{a^3}(y - y_0)^2 + \frac{2b^2 - ac}{a^5}(y - y_0)^3 + \frac{5abc - 5b^3 - a^2 d}{a^7}(y - y_0)^4 + \cdots.$$

2. **Bürman–Lagrange formula.** Suppose that for a given function

$$y = f(x), \qquad (8.3.3.5)$$

the auxiliary function

$$\varphi(x) = \frac{x}{f(x)}$$

is holomorphic in a neighborhood of the point $x = 0$ (i.e., it can be represented by a convergent power series in a neighborhood of that point). Then there is $\varepsilon > 0$ such that on the interval $|y| < \varepsilon$, the function (8.3.3.5) is invertible and its inverse $x = g(y)$ is holomorphic on that interval,

$$x = \sum_{n=1}^{\infty} b_n y^n, \qquad b_n = \frac{1}{n!}\left[\frac{d^{n-1}}{dx^{n-1}}\varphi^n(x)\right]_{x=0}.$$

The expression for the coefficients $b_n$ is called the *Bürman–Lagrange formula*.

**Example 1.** Consider the function

$$y = x(x + b) \qquad (b \ne 0),$$

for which the auxiliary function has the form $\varphi(x) = (x + a)^{-1}$. Using the Bürman–Lagrange formula and the relation

$$\frac{d^{n-1}}{dx^{n-1}}\frac{1}{(x + a)^n} = \frac{(-1)^{n-1} n(n + 1)\cdots(2n - 2)}{(x + a)^{2n-1}},$$

we find the representation of the given function by power series:

$$x = \frac{y}{a} - \frac{y^2}{a^3} + \cdots + (-1)^{n-1}\frac{(2n - 2)!}{(n - 1)! \, n!}\frac{y^n}{a^{2n-1}} + \cdots.$$

### 8.3.3-4. Simplest summation formulas for power series.

Suppose that the sum of a power series is known,

$$\sum_{k=0}^{\infty} a_k x^k = S(x). \tag{8.3.3.6}$$

Then, using term-by-term integration (on the convergence interval), one can find the following sums:

$$\sum_{k=0}^{\infty} a_k k^m x^k = \left(x \frac{d}{dx}\right)^m S(x);$$

$$\sum_{k=0}^{\infty} a_k (nk+m) x^{nk+m-1} = \frac{d}{dx}\left[x^m S(x^n)\right];$$

$$\sum_{k=0}^{\infty} \frac{a_k}{nk+m} x^{nk+m} = \int_0^x x^{m-1} S(x^n)\, dx, \qquad n > 0,\ m > 0; \tag{8.3.3.7}$$

$$\sum_{k=0}^{\infty} a_k \frac{nk+s}{nk+m} x^{nk+s} = x \frac{d}{dx}\left[x^{s-m}\int_0^x x^{m-1} S(x^n)\, dx\right], \qquad n > 0,\ m > 0;$$

$$\sum_{k=0}^{\infty} a_k \frac{nk+m}{nk+s} x^{nk+s} = \int_0^x x^{s-m} \frac{d}{dx}\left[x^m S(x^n)\right] dx, \qquad n > 0,\ s > 0.$$

**Example 2.** Let us find the sum of the series $\sum_{n=0}^{\infty} k x^{k-1}$.

We start with the well-known formula for the sum of an infinite geometrical progression:

$$\sum_{k=0}^{\infty} x^k = \frac{1}{1-x} \qquad (|x| < 1).$$

This series is a special case of (8.3.3.6) with $a_k = 1$, $S(x) = 1/(1-x)$. The series $\sum_{n=0}^{\infty} k x^{k-1}$ can be obtained from the left-hand side of the second formula in (8.3.3.7) for $m = 0$ and $n = 1$. Substituting $S(x) = 1/(1-x)$ into the right-hand side of that formula, we get

$$\sum_{k=0}^{\infty} k x^{k-1} = \frac{d}{dx}\frac{1}{1-x} = \frac{1}{(1-x)^2} \qquad (|x| < 1).$$

## 8.4. Fourier Series

### 8.4.1. Representation of $2\pi$-Periodic Functions by Fourier Series. Main Results

#### 8.4.1-1. Dirichlet theorem on representation of a function by Fourier series.

A function $f(x)$ is said to satisfy the *Dirichlet conditions* on an interval $(a, b)$ if:
   1) this interval can be divided into finitely many intervals on which $f(x)$ is monotone and continuous;
   2) at any discontinuity point $x_0$ of the function, there exist finite one-sided limits $f(x_0 + 0)$ and $f(x_0 - 0)$.

DIRICHLET THEOREM. *Any $2\pi$-periodic function that satisfies the Dirichlet conditions on the interval $(-\pi, \pi)$ can be represented by its Fourier series*

$$f(x) = \frac{a_0}{2} + \sum_{n=1}^{\infty} \left( a_n \cos nx + b_n \sin nx \right) \qquad (8.4.1.1)$$

whose coefficients are defined by the *Euler–Fourier formulas*

$$\begin{aligned} a_n &= \frac{1}{\pi} \int_{-\pi}^{\pi} f(x) \cos nx \, dx, & n &= 0, 1, 2, \ldots, \\ b_n &= \frac{1}{\pi} \int_{-\pi}^{\pi} f(x) \sin nx \, dx, & n &= 1, 2, 3, \ldots \end{aligned} \qquad (8.4.1.2)$$

At the points of continuity of $f(x)$, the Fourier series converges to $f(x)$, and at any discontinuity point $x_0$, the series converges to $\frac{1}{2}[f(x_0 + 0) + f(x_0 - 0)]$.

The coefficients $a_n$ and $b_n$ of the series (8.4.1.1) are called the *Fourier coefficients*.

*Remark.* Instead of the integration limits $-\pi$ and $\pi$ in (8.4.1.2), one can take $c$ and $c + 2\pi$, where $c$ is an arbitrary constant.

### 8.4.1-2. Lipschitz and Dirichlet–Jordan convergence criteria for Fourier series.

LIPSCHITZ CRITERION. *Suppose that $f(x)$ is continuous at a point $x_0$ and for sufficiently small $\varepsilon > 0$ satisfies the inequality $|f(x_0 \pm \varepsilon) - f(x_0)| \leq K\varepsilon^{\sigma}$, where $L$ and $\sigma$ are constants, $0 < \sigma \leq 1$. Then the representation (8.4.1.1)–(8.4.1.2) holds at $x = x_0$.*

In particular, the conditions of the Lipschitz criterion hold for continuous piecewise differentiable functions.

*Remark.* The Fourier series of a continuous periodic function with no additional conditions (for instance, of its regularity) may happen to be divergent at infinitely many (even uncountably many) points.

DIRICHLET–JORDAN CRITERION. *Suppose that $f(x)$ is a function of bounded variation on some interval $(x_0 - h, x_0 + h) \in (-\pi, \pi)$ (i.e., $f(x)$ can be represented as a difference of two monotonically increasing functions). Then the Fourier series (8.4.1.1)–(8.4.1.2) of the function $f(x)$ at the point $x_0$ converges to the value $\frac{1}{2}[f(x_0 + 0) + f(x_0 - 0)]$.*

### 8.4.1-3. Riemann localization principle.

RIEMANN LOCALIZATION PRINCIPLE. *The behavior of the Fourier series of a function $f(x)$ at a point $x_0$\* depends only on its values near that point, i.e., values in an arbitrarily small neighborhood of that point.*

Thus, for two functions that coincide in a neighborhood of a point $x_0$, but differ outside that neighborhood, the corresponding Fourier series at $x_0$ are either both convergent or divergent and have the same sum in the case of convergence, although their Fourier coefficients may be different, being dependent on all values of the functions.

---

\* What is meant here is the fact of convergence or divergence of the Fourier series at $x_0$, and also the numerical value of its sum in the case of convergence.

**8.4.1-4. Asymptotic properties of Fourier coefficients.**

1°. Fourier coefficients of an absolutely integrable function tend to zero as $n$ goes to infinity: $a_n \to 0$ and $b_n \to 0$ as $n \to \infty$.

2°. Fourier coefficients of a continuous $2\pi$-periodic function have the following limit properties:
$$\lim_{n \to \infty} (na_n) = 0, \quad \lim_{n \to \infty} (nb_n) = 0,$$
i.e., $a_n = o(1/n)$, $b_n = o(1/n)$.

3°. If a continuous periodic function is continuously differentiable up to the order $m-1$ inclusively, then its Fourier coefficients have the following limit properties:
$$\lim_{n \to \infty} (n^m a_n) = 0, \quad \lim_{n \to \infty} (n^m b_n) = 0,$$
i.e., $a_n = o(n^{-m})$, $b_n = o(n^{-m})$.

**8.4.1-5. Integration and differentiation of Fourier series.**

1°. The Fourier series of a continuous periodic function of bounded variation admits term-by-term integration, and the resulting series is uniformly convergent.

2°. The Fourier series of a $k$ times continuously differentiable function admits term-by-term differentiation $(k-1)$ times, the resulting series still being uniformly convergent (the $k$th differentiation yields the $k$th derivative of the function, but the resulting series may have only mean-square convergence, not necessarily pointwise convergence).

## 8.4.2. Fourier Expansions of Periodic, Nonperiodic, Odd, and Even Functions

**8.4.2-1. Expansion of $2l$-periodic and nonperiodic functions in Fourier series.**

1°. The case of $2l$-periodic functions can be easily reduced to that of $2\pi$-periodic functions by changing the variable $x$ to $z = \dfrac{\pi x}{l}$. In this way, all the results described above for $2\pi$-periodic functions can be easily extended to $2l$-periodic functions.

The Fourier expansion of a $2l$-periodic function $f(x)$ has the form
$$f(x) = \frac{a_0}{2} + \sum_{n=1}^{\infty} \left( a_n \cos \frac{n\pi x}{l} + b_n \sin \frac{n\pi x}{l} \right), \tag{8.4.2.1}$$

where
$$a_n = \frac{1}{l} \int_{-l}^{l} f(x) \cos \frac{n\pi x}{l} \, dx, \quad b_n = \frac{1}{l} \int_{-l}^{l} f(x) \sin \frac{n\pi x}{l} \, dx. \tag{8.4.2.2}$$

2°. A nonperiodic (aperiodic) function $f(x)$ defined on the interval $(-l, l)$ can also be represented by a Fourier series (8.4.2.1)–(8.4.2.2); however, outside that interval, the sum of that series $S(x)$ may differ from $f(x)$*.

---

* The sum $S(x)$ is a $2l$-periodic function defined for all $x$, but $f(x)$ may happen to be nonperiodic, or even undefined outside the interval $(-l, l)$.

### 8.4.2-2. Fourier expansion of odd and even functions.

1°. Let $f(x)$ be an even function, i.e., $f(x) = f(-x)$. Then the Fourier expansion of $f(x)$ on the interval $(-l, l)$ has the form of the *cosine Fourier series*:

$$f(x) = \frac{a_0}{2} + \sum_{n=1}^{\infty} a_n \cos \frac{n\pi x}{l},$$

where the Fourier coefficients have the form

$$a_n = \frac{2}{l} \int_0^l f(x) \cos \frac{n\pi x}{l} \, dx \qquad (b_n = 0).$$

2°. Let $f(x)$ be an odd function, i.e., $f(x) = -f(-x)$. Then the Fourier expansion of $f(x)$ on the interval $(-l, l)$ has the form of the *sine Fourier series*:

$$f(x) = \sum_{n=1}^{\infty} b_n \sin \frac{n\pi x}{l},$$

where the Fourier coefficients have the form

$$b_n = \frac{2}{l} \int_0^l f(x) \sin \frac{n\pi x}{l} \, dx \qquad (a_n = 0).$$

**Example.** Let us find the Fourier expansion of the function $f(x) = x$ on the interval $(-\pi, \pi)$. Taking $l = \pi$ and $f(x) = x$ in the formula for the Fourier coefficients and integrating by parts, we obtain

$$b_n = \frac{2}{\pi} \int_0^\pi x \sin(nx) \, dx = \frac{2}{\pi} \left( -\frac{1}{n} x \cos(nx) \Big|_0^\pi + \frac{1}{n} \int_0^\pi \cos(nx) \, dx \right) = -\frac{2}{n} \cos(n\pi) = (-1)^{n+1} \frac{2}{n}.$$

Therefore, the Fourier expansion of $f(x) = x$ has the form

$$f(x) = 2 \sum_{n=1}^{\infty} (-1)^{n+1} \frac{\sin(nx)}{n} \qquad (-\pi < x < \pi).$$

3°. If $f(x)$ is defined on the interval $(0, l)$ and satisfies the Dirichlet conditions, it can be represented by the cosine Fourier series, as well as the sine Fourier series (with the help of the above formulas).* Both series on the interval $(0, l)$ give the values of $f(x)$ at points of its continuity and the value $\frac{1}{2}[f(x_0 + 0) + f(x_0 - 0)]$ at points of its discontinuity; outside the interval $(0, l)$, these two series represent different functions.

### 8.4.2-3. Fourier series in complex form.

The complex Fourier expansion of a function $f(x)$ on an interval $(-l, l)$ has the form

$$f(x) = \sum_{n=-\infty}^{\infty} c_n e^{i\omega_n x},$$

where

$$\omega_n = \frac{n\pi}{l}, \quad c_n = \frac{1}{2l} \int_{-l}^{l} f(x) e^{-i\omega_n x} \, dx; \quad n = 0, \pm 1, \pm 2, \ldots$$

The expressions $e^{i\omega_n x}$ are called *harmonics*, the coefficients $c_n$ are *complex amplitudes*, $\omega_n$ are *wave numbers* of the function $f(x)$, and the set of all wave numbers $\{\omega_n\}$ is called the *discrete spectrum* of the function.

---

\* The cosine Fourier expansion of $f(x)$ on the interval $(0, l)$ corresponds to the extension of $f(x)$ to the interval $(-l, 0)$ as an even function: $f(-x) = f(x)$. The sine Fourier expansion of $f(x)$ on $(0, l)$ corresponds to the extension of $f(x)$ to the interval $(-l, 0)$ as an odd function: $f(-x) = -f(x)$.

### 8.4.3. Criteria of Uniform and Mean-Square Convergence of Fourier Series

**8.4.3-1. Criteria of uniform convergence of Fourier series.**

LIPSCHITZ CRITERION. *The Fourier series of a function $f(x)$ converges uniformly to that function on an interval $[-l, l]$ if on a wider interval $[-L, L]$ ($-L < -l < l < L$) the following inequality holds:*

$$|f(x_1) - f(x_2)| \leq K|x_1 - x_2|^\sigma \quad \text{for all} \quad x_1, x_2 \in [L, L],$$

*where $K$ and $\sigma$ are constants, $0 < \sigma \leq 1$.*

*Corollary.* The Fourier series of a continuous function $f(x)$ uniformly converges to that function on an interval $[-l, l]$ if on a wider interval the function $f(x)$ has a bounded derivative $f'(x)$.

DIRICHLET–JORDAN CRITERION. *Suppose that on an interval $(-l, l) \in [-L, L]$, a continuous function $f(x)$ has bounded variation (i.e., can be represented as the difference of two monotonically increasing functions). Then its Fourier series is uniformly convergent to that function on the interval $(-l, l)$.*

For any continuously differentiable $2l$-periodic function $f(x)$, its Fourier series [defined by formulas (8.4.2.1)–(8.4.2.2)] is uniformly convergent to $f(x)$.

**8.4.3-2. Fourier series of square-integrable functions. Parseval identity.**

$1°$. For a continuous $2\pi$-periodic function $f(x)$, its Fourier series (8.4.1.1)–(8.4.1.2) converges to $f(x)$ in mean square, i.e.,

$$\int_{-\pi}^{\pi} |f(x) - f_n(x)|^2 \, dx \to 0 \quad \text{as} \quad n \to \infty,$$

where $f_n(x) = \frac{1}{2}a_0 + \sum_{k=1}^{n}(a_k \cos kx + b_k \sin kx)$ is a partial sum of the Fourier series.

$2°$. If $f(x)$ is integrable on the segment $[-\pi, \pi]$ and the integral $\int_{-\pi}^{\pi} |f(x)|^2 \, dx$ exists as an improper integral with finitely many singularities, then the Fourier series (8.4.1.1)–(8.4.1.2) is mean-square convergent to $f(x)$.

$3°$. Let $f(x) \in L_2[-\pi, \pi]$ be a square-integrable function on the segment $[-\pi, \pi]$. Then its Fourier series (8.4.1.1)–(8.4.1.2) is mean-square convergent to $f(x)$, and the *Parseval identity* holds:

$$\frac{a_0^2}{2} + \sum_{n=1}^{\infty}(a_n^2 + b_n^2) = \frac{1}{\pi}\int_{-\pi}^{\pi} f^2(x) \, dx,$$

where $a_n, b_n$ are defined by (8.4.1.2). Note that the functions considered in Items $1°$ and $2°$ belong to $L^2[-\pi, \pi]$.

## 8.4.4. Summation Formulas for Trigonometric Series

**8.4.4-1. Summation of trigonometric series with the help of Laplace transforms.**

When finding sums of trigonometric series, the following formulas may be useful:

$$\sum_{n=1}^{\infty} F(n)\sin(nx) = \frac{1}{2}\sin x \int_0^{\infty} \frac{f(t)\,dt}{\cosh t - \cos x}, \qquad 0 \le x \le \pi;$$

$$\sum_{n=1}^{\infty} F(n)\cos(nx) = \frac{1}{2}\int_0^{\infty} \frac{\cos x - e^{-t}}{\cosh t - \cos x} f(t)\,dt, \qquad 0 \le x \le \pi,$$

where

$$F(x) = \int_0^{\infty} f(t) e^{-xt}\,dt.$$

For specific $F(x)$, the corresponding functions $f(t)$ can be found in tables of inverse Laplace transforms (see Section T3.2).

**8.4.4-2. Summation of series with the help of functions of complex variable.**

Suppose that trigonometric series

$$\frac{a_0}{2} + \sum_{n=1}^{\infty} a_n \cos(nx) = f_1(x), \tag{8.4.4.1}$$

$$\sum_{n=1}^{\infty} a_n \sin(nx) = f_2(x) \tag{8.4.4.2}$$

have positive (nonnegative) coefficients $a_n$ and the auxiliary series $\sum_{n=1}^{\infty} \frac{a_n}{n}$ is convergent. Then the series (8.4.4.1) and (8.4.4.2) are Fourier series representing continuous functions.

In order to find the sums $f_1(x)$ and $f_2(x)$ of the series (8.4.4.1) and (8.4.4.2), it is sometimes possible to use functions of complex variable. Let

$$\varphi(z) = \frac{a_0}{2} + \sum_{n=1}^{\infty} a_n z^n$$

be the sum of a series that is convergent in the circle $|z| < 1$. If $\lim_{|z|\to 1} \varphi(z) = \lim_{r\to 1} \varphi(re^{ix}) = \varphi(e^{ix})$, then $f_1(x) + if_2(x) = \varphi(e^{ix})$. Thus, after separating the real and the imaginary parts of the function $\varphi(e^{ix})$, we get

$$f_1(x) = \operatorname{Re}[\varphi(e^{ix})], \qquad f_2(x) = \operatorname{Im}[\varphi(e^{ix})].$$

**Example.** Let us sum up the series

$$1 + \sum_{n=1}^{\infty} \frac{\cos(nx)}{n!} = f_1(x),$$

$$\sum_{n=1}^{\infty} \frac{\sin(nx)}{n!} = f_2(x).$$

Taking
$$\varphi(z) = 1 + \sum_{n=1}^{\infty} \frac{z^n}{n!} = e^z$$
and representing $z$ in exponential form $z = re^{ix}$, let us pass to the limit as $r \to 1$. We find that
$$\exp(e^{ix}) = e^{\cos x + i \sin x} = e^{\cos x}[\cos(\sin x) + i \sin(\sin x)] = 1 + \sum_{n=1}^{\infty} \frac{(\cos x + i \sin x)^n}{n!}.$$
Using the Moivre formula $(\cos x + i \sin x)^n = \cos(nx) + i \sin(nx)$, we obtain
$$e^{\cos x} \cos(\sin x) + i e^{\cos x} \sin(\sin x) = 1 + \sum_{n=1}^{\infty} \frac{\cos(nx)}{n!} + i \sum_{n=1}^{\infty} \frac{\sin(nx)}{n!}.$$
It follows that
$$1 + \sum_{n=1}^{\infty} \frac{\cos(nx)}{n!} = e^{\cos x} \cos(\sin x),$$
$$\sum_{n=1}^{\infty} \frac{\sin(nx)}{n!} = e^{\cos x} \sin(\sin x).$$

## 8.5. Asymptotic Series

### 8.5.1. Asymptotic Series of Poincaré Type. Formulas for the Coefficients

8.5.1-1. Definition of asymptotic series. Illustrative example.

Suppose that for large $x$ the function $f(x)$ can be represented in the form
$$f(x) = \sum_{k=0}^{n} \frac{a_k}{x^k} + o(x^{-n}) \quad \text{as} \quad x \to \infty. \tag{8.5.1.1}$$
Then one writes
$$f(x) \sim \sum_{k=0}^{\infty} \frac{a_k}{x^k}. \tag{8.5.1.2}$$
The infinite series on the right-hand side is called an *asymptotic series of Poincaré's type* (or an *asymptotic expansion*) of the function $f(x)$.

Asymptotic series may happen to be convergent or divergent.

**Example.** Consider the function defined by the integral
$$f(x) = \int_x^{\infty} e^{x-t} \frac{dt}{t}.$$
Repeated integration by parts yields
$$f(x) = \frac{1}{x} - \frac{1}{x^2} + \frac{2!}{x^3} - \cdots + (-1)^{n-1} \frac{(n-1)!}{x^n} + R_n(x), \tag{8.5.1.3}$$
where the remainder is defined by
$$R_n(x) = (-1)^n n! \int_x^{\infty} \frac{e^{x-t}}{t^{n+1}} dt.$$
The following estimate holds:
$$|R_n(x)| = n! \int_x^{\infty} \frac{e^{x-t}}{t^{n+1}} dt < n! \frac{1}{x^{n+1}} \int_x^{\infty} e^{x-t} dt = -\frac{n!}{x^{n+1}} e^{x-t}\Big|_{t=x}^{t=\infty} = \frac{n!}{x^{n+1}}.$$

Therefore, $R_n(x) = o(x^{-n})$ and we can write

$$f(x) \sim \frac{1}{x} - \frac{1}{x^2} + \frac{2!}{x^3} - \cdots + (-1)^{n-1} \frac{(n-1)!}{x^n} + \cdots.$$

This asymptotic series is divergent, since for any fixed $x$ we have

$$\lim_{n \to \infty} |a_n| = \lim_{n \to \infty} \frac{(n-1)!}{x^n} = \infty.$$

The motivation of using divergent asymptotic series of the form (8.5.1.2) is that a finite sum of the series (8.5.1.1) provides a good approximation for the given function for $x \to \infty$. For a fixed $n$, the accuracy of the approximations increases with the growth of $x$.

8.5.1-2. **Uniqueness of an asymptotic series representing a function.**

1. For a function $f(x)$ that admits the representation by asymptotic series (8.5.1.2), such a representation is unique and the coefficients $a_n$ of the series are determined by

$$a_n = \lim_{x \to \infty} x^n \left[ f(x) - \sum_{k=0}^{n-1} a_k x^{-k} \right].$$

2. There exist functions that are not identically equal to zero, but having all coefficients $a_n$ in their asymptotic expansion (8.5.1.2) equal to zero. Such a function is called *asymptotic zero*. The class of asymptotic zeroes includes any function $f(x)$ such that for large $x$ and any positive integer $n$ the following estimate holds:

$$f(x) = o(x^{-n}).$$

For instance, the functions $f(x) = e^{-x}$, $f(x) = xe^{-2x}$, $f(x) = \exp(-x^2)$ belong to this class. Adding such functions to the left-hand side of the asymptotic expansion (8.5.1.2) does not change its right-hand side.

COROLLARY. *The coefficients of the asymptotic expansion of a function $f(x)$ do not uniquely determine the function.*

Remark. Addition of a suitable asymptotic zero can be used, in some cases, for the construction of approximations that not only correctly describe the behavior of a function for large $x$, but give fairly accurate numerical results for moderate values of $x$ (i.e., $x = O(1)$; sometimes even for small $x$). This method is often used in applications, for instance in engineering, to obtain formulas applicable in a wide range of some characteristic parameter.

## 8.5.2. Operations with Asymptotic Series

8.5.2-1. **Addition, subtraction, multiplication, and division of asymptotic series.**

1. Asymptotic expansions of functions $f(x) \sim \sum_{n=0}^{\infty} a_n x^{-n}$ and $g(x) \sim \sum_{n=0}^{\infty} b_n x^{-n}$ admit term-by-term addition and subtraction:

$$f(x) \pm g(x) \sim \sum_{n=0}^{\infty} a_n x^{-n} \pm \sum_{n=0}^{\infty} b_n x^{-n} = \sum_{n=0}^{\infty} (a_n \pm b_n) x^{-n}.$$

2. Asymptotic expansions $f(x) \sim \sum_{n=0}^{\infty} a_n x^{-n}$ and $g(x) \sim \sum_{n=0}^{\infty} b_n x^{-n}$ can be formally multiplied, according to the Cauchy rule:

$$f(x)g(x) \sim \left(\sum_{n=0}^{\infty} a_n x^{-n}\right)\left(\sum_{n=0}^{\infty} b_n x^{-n}\right) = \sum_{n=0}^{\infty} c_n x^{-n}, \qquad c_n = \sum_{k=0}^{n} a_k b_{n-k}.$$

3. Asymptotic expansion $f(x) \sim \sum_{n=0}^{\infty} a_n x^{-n}$ can be formally divided by $g(x) \sim \sum_{n=0}^{\infty} b_n x^{-n}$, provided that $b_0 \neq 0$. The coefficients of the resulting asymptotic expansion $h(x) \sim \sum_{n=0}^{\infty} b_n x^{-n}$ are found by the method of indefinite coefficients with the help of the relation $f(x) = g(x)h(x)$: the corresponding expansions should be inserted into this relation and then similar terms gathered.

### 8.5.2-2. Composition of series. Integration of asymptotic series.

1. Suppose that in a neighborhood of $y = 0$, the function $g(y)$ can be represented by a convergent power series, $g(y) = \sum_{m=0}^{\infty} b_m y^m$, and for $x \to \infty$, the function $f(x)$ can be represented as an asymptotic series $f(x) \sim \sum_{n=1}^{\infty} a_n x^{-n}$ with $a_0 = 0$. Then, for large enough $x$, the composite function

$$g(f(x)) = \sum_{m=0}^{\infty} b_m [f(x)]^m \qquad (8.5.2.1)$$

makes sense and admits an asymptotic expansion, which can be obtained from (8.5.2.1) if $f(x)$ is replaced by its asymptotic expansion, after which similar terms should be gathered.

2. Suppose that the asymptotic expansion of a function $f(x)$ starts with $a_2 x^{-2}$ [i.e., $a_0 = a_1 = 0$ in (8.5.1.2)]. Then this asymptotic expansion admits term-by-term integration from $x$ to $\infty$,

$$\int_x^{\infty} f(x)\,dx \sim \int_x^{\infty} \left(\sum_{n=2}^{\infty} \frac{a_n}{x^n}\right) dx = \sum_{n=2}^{\infty} \int_x^{\infty} \frac{a_n}{x^n}\,dx = \sum_{n=2}^{\infty} \frac{a_n}{(n-1)x^{n-1}}.$$

3. Term-by-term differentiation of asymptotic series is impossible, in general.

**Example.** To illustrate the last statement, consider the function $f(x) = e^{-x} \sin(e^x)$. For all $n$, we have

$$\lim_{x \to \infty} x^n f(x) = 0,$$

and therefore the asymptotic expansion of this function has zero coefficients, $f(x) \sim 0$. Term-by-term differentiation of this relation also yields zero, $f'(x) \sim 0$.

On the other hand, the derivative of this function, $f'(x) = -e^{-x} \sin(e^x) + \cos(e^x)$, admits no asymptotic expansion at all, since there is no limit $\lim_{x \to \infty} f'(x)$!

## References for Chapter 8

**Brannan, D.**, *A First Course in Mathematical Analysis*, Cambridge University Press, Cambridge, 2006.
**Bromwich, T. J. I.**, *Introduction to the Theory of Infinite Series*, American Mathematical Society, Providence, Rhode Island, 2005.
**Bronshtein, I. N. and Semendyayev, K. A.**, *Handbook of Mathematics, 4th Edition*, Springer-Verlag, Berlin, 2004.
**Danilov, V. L., Ivanova, A. N., et al.**, *Mathematical Analysis (Functions, Limits, Series, Continued Fractions)* [in Russian], Fizmatlit Publishers, Moscow, 1961.
**Davis, H. F.**, *Fourier Series and Orthogonal Functions*, Dover Publications, New York, 1989.
**Fedoryuk, M. V.** *Asymptotics, Integrals, Series* [in Russian], Nauka, Moscow, 1987.
**Fichtenholz, G. M.**, *Functional Series*, Taylor & Francis, London, 1970.
**Fikhtengol'ts (Fichtenholz), G. M.**, *A Course in Differential and Integral Calculus, Vols. 2 and 3* [in Russian], Nauka Publishers, Moscow, 1969.
**Gradshteyn, I. S. and Ryzhik, I. M.**, *Tables of Integrals, Series, and Products, 6th Edition*, Academic Press, New York, 2000.
**Hansen, E. R.**, *A Table of Series and Products*, Printice Hall, Englewood Cliffs, London, 1975.
**Hirschman, I.**, *Infinite Series*, Greenwood Press, New York, 1978.
**Hyslop, J. M.**, *Infinite Series*, Dover Publications, New York, 2006.
**Jolley, L. B. W.**, *Summation of Series*, Dover Publications, New York, 1961.
**Knopp, K.**, *Infinite Sequences and Series*, Dover Publications, New York, 1956.
**Knopp, K.**, *Theory and Application of Infinite Series*, Dover Publications, New York, 1990.
**Mangulis, V.**, *Handbook of Series for Scientists and Engineers*, Academic Press, New York, 1965.
**Pinkus, A. and Zafrany, S.**, *Fourier Series and Integral Transforms*, Cambridge University Press, Cambridge, 1997.
**Prudnikov, A. P., Brychkov, Yu. A., and Marichev, O. I.**, *Integrals and Series, Vol. 1, Elementary Functions*, Gordon & Breach, New York, 1986.
**Zhizhiashvili, L.V.**, *Trigonometric Fourier Series and Their Conjugates*, Kluwer Academic, Dordrecht, 1996.
**Zygmund, A.**, *Trigonometric Series, 3rd Edition*, Cambridge University Press, Cambridge, 2003.

# Chapter 9
# Differential Geometry

## 9.1. Theory of Curves

### 9.1.1. Plane Curves

#### 9.1.1-1. Regular points of plane curve.

A plane curve $\Gamma$ in a Cartesian coordinate system can be defined by equations in the following form:

*Explicitly,*
$$y = f(x). \tag{9.1.1.1}$$

*Implicitly,*
$$F(x, y) = 0.$$

*Parametrically,*
$$x = x(t), \quad y = y(t). \tag{9.1.1.2}$$

*In vector form,*
$$\mathbf{r} = \mathbf{r}(t), \quad \text{where} \quad \mathbf{r}(t) = x(t)\mathbf{i} + y(t)\mathbf{j}.$$

In a polar coordinate system, the curve is usually given by the equation

$$r = r(\varphi),$$

where the relationship between Cartesian and polar coordinates is given by formulas $x = r \cos \varphi$ and $y = r \sin \varphi$.

**Remark.** The explicit equation (9.1.1.1) can be obtained from the parametric equations (9.1.1.2) if the abscissa is taken for the parameter: $x = t, y = f(t)$.

A point $M(x(t), y(t))$ is said to be *regular* if the functions $x(t)$ and $y(t)$ have continuous first derivatives not simultaneously equal to zero in a sufficiently small neighborhood of this point. For implicitly defined functions, a point $M(x, y)$ is said to be *regular* if grad $F = \nabla F \neq 0$ at this point.

If a curve is given parametrically (9.1.1.2), then the *positive sense* is defined on this curve, i.e., the direction in which the point $M(x(t), y(t))$ of the curve moves as the parameter $t$ increases. If the curve is given explicitly by (9.1.1.1), then the positive sense corresponds to the direction in which the abscissa increases (i.e., moves from left to right). In a polar coordinate system, the positive sense corresponds to the direction in which the angle $\varphi$ increases (i.e., the positive sense is counterclockwise).

If $s$ is the curve length from some constant point $M_0$ to $M$, then the infinitesimal length increment of the arc $M_0 M$ is approximately expressed by the formula for the arc length differential $ds$; i.e., the following formulas hold:

$$\Delta s \approx ds = \sqrt{1 + (y'_x)^2}\, dx, \quad \text{if the curve is given explicitly,}$$

$$\Delta s \approx ds = \sqrt{(x'_t)^2 + (y'_t)^2}\, dt, \quad \text{if the curve is given parametrically,}$$

$$\Delta s \approx ds = \sqrt{r^2 + (r'_\varphi)^2}\, d\varphi, \quad \text{for a curve in the polar coordinate system.}$$

**Example 1.** The arc length differential of the curve $y = \cos x$ has the form $ds = \sqrt{1 + \sin x}\, dx$.

**Example 2.** For the semicubical parabola $x = t^2$, $y = t^3$, the arc length differential is equal to $ds = t\sqrt{4 + 9t^2}\, dt$.

**Example 3.** For the *hyperbolic spiral* $r = a/\varphi$ for $r > 0$, the arc length differential is equal to $ds = a\sqrt{1 + \varphi^2}/\varphi^2\, d\varphi$.

### 9.1.1-2. Tangent and normal.

The *tangent* to a curve $\Gamma$ at a regular point $M_0$ is defined to be the straight line that is the limit position of the secant $M_0 M_1$ as the point $M_1$ approaches the point $M_0$; the *normal* is defined to be the straight line passing through $M_1$ and perpendicular to the tangent (Fig. 9.1).

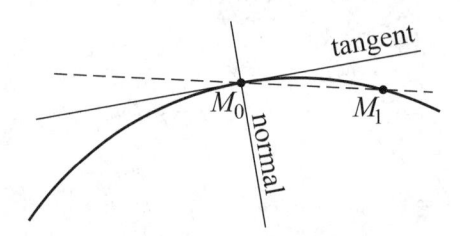

**Figure 9.1.** Tangent and normal.

At each regular point $M(x_0, y_0) = M(x(t_0), y(t_0))$, the curve $\Gamma$ has a unique tangent given by one of the equations (depending on how the curve is defined)

$$y - y_0 = y'_x(x - x_0), \quad \text{if the curve is given explicitly,}$$

$$F_x(x - x_0) + F_y(y - y_0) = 0, \quad \text{if the curve is given implicitly,}$$

$$\frac{y - y_0}{y'_t} = \frac{x - x_0}{x'_t}, \quad \text{if the curve is given parametrically,}$$

$$\mathbf{r} = \mathbf{r}_0 + \lambda \mathbf{r}_t, \quad \text{if the curve is given in vector form,}$$

where $\mathbf{r}_0$ is the position vector of the point $M_0$, $\lambda$ is an arbitrary parameter, and all derivatives are evaluated at $x = x_0$, $y = y_0$, and $t = t_0$.

The *slope of the tangent* is determined by the angle $\alpha$ between the positive direction of the $OX$-axis and the positive direction of the tangent (Fig. 9.2a). The slope of the tangent (and the angle $\alpha$) is determined by the formulas

$$\tan \alpha = y'_x = -\frac{F_x}{F_y} = \frac{y'_t}{x'_t}.$$

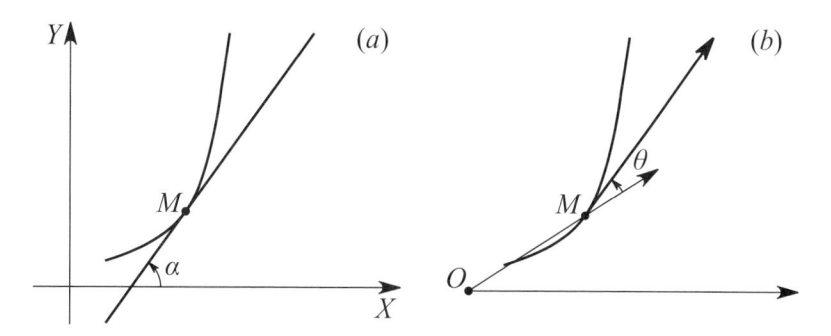

**Figure 9.2.** Slope of the tangent.

If a curve is given in the polar coordinate system, then the slope of the tangent is determined by the angle $\theta$ between the direction of the position vector $r = OM$ and the positive direction of the tangent (Fig. 9.2b). The angle $\theta$ is determined by the formula

$$\tan \theta = \frac{r}{\varphi'_r}.$$

The normal at each regular point $M(x_0, y_0) = M(x(t_0), y(t_0))$ is given, depending on the method for defining the curve, by the equations

$$y - y_0 = -\frac{x - x_0}{y'_x}, \qquad \text{if the curve is given explicitly,}$$

$$\frac{x - x_0}{F_x} = \frac{y - y_0}{F_y}, \qquad \text{if the curve is given implicitly,}$$

$$(y - y_0)y'_t + (x - x_0)x'_t = 0, \quad \text{if the curve is given parametrically,}$$

where all the derivatives are evaluated at $x = x_0$, $y = y_0$, and $t = t_0$.

The *positive sense* of the tangent coincides with the positive sense of the curve at the point of tangency; and the *positive sense* of the normal can in some way be made consistent with the positive sense of the tangent; for example, it can be obtained from the positive sense of the tangent by counterclockwise rotation around $M$ by an angle of 90° (Fig. 9.3). The point $M$ divides the tangent and the normal into positive and negative half-lines.

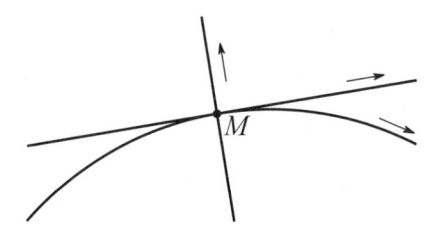

**Figure 9.3.** Positive sense of tangent.

**Example 4.** Let us find the equations of the tangent and the normal to the parametrically given semicubical parabola $x = t^2$, $y = t^3$ at the point $M_0(1, 1)$, $t = 1$.

The equation of the tangent,

$$\frac{y - t^3}{3t^2} = \frac{x - t^2}{2t} \quad \text{or} \quad y = \frac{3}{2}tx - \frac{1}{2}t^3,$$

at the point $M_0(1, 1)$ is

$$y = \frac{3}{2}x - \frac{1}{2}.$$

The equation of the normal,

$$2t(x - t^2) + 3t^2(y - t^3) = 0 \quad \text{or} \quad 2x + 3ty = t^2(2 + 3t^2),$$

at the point $M_0(1, 1)$ is (Fig. 9.4a)

$$2x + 3y = 5.$$

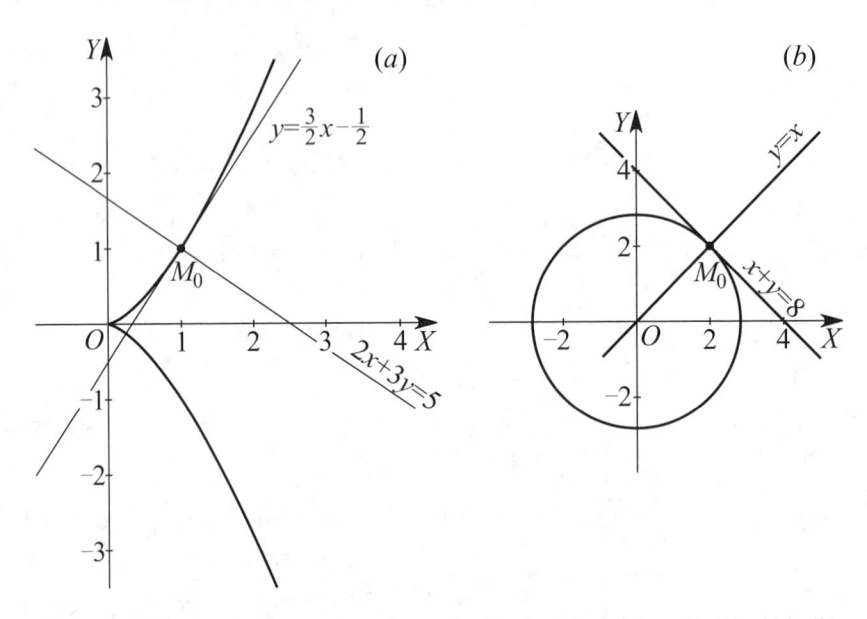

**Figure 9.4.** Tangents and normals to the semicubical parabola (*a*) and to the circle (*b*).

**Example 5.** Let us find the equation of the tangent and the normal to the circle $x^2 + y^2 = 8$ at the point $M_0(2, 2)$.

We write the equation of the circle as $F(x, y) = 0$:

$$x^2 + y^2 - 8 = 0,$$

i.e., $F(x, y) = x^2 + y^2 - 8$. Obviously, we obtain

$$F_x = 2x, \quad F_y = 2y.$$

The equation of the tangent is

$$2x_0(x - x_0) + 2y_0(y - y_0) = 0,$$

or, taking into account the original equation of the circle,

$$xx_0 + yy_0 = 8.$$

At the point $M_0(2, 2)$, we have

$$x + y = 4.$$

The equation of the normal is

$$\frac{x - x_0}{2x_0} = \frac{y - y_0}{2y_0},$$

or

$$y = \frac{y_0}{x_0} x.$$

At the point $M_0(2, 2)$ (Fig. 9.4b), we have

$$y = x.$$

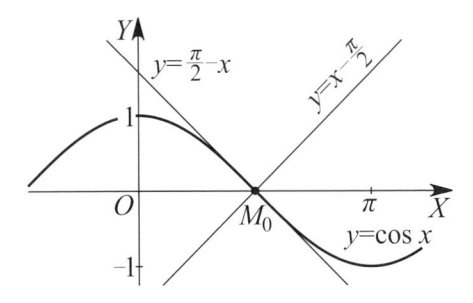

**Figure 9.5.** The tangent and the normal to the curve $y = \cos x$.

**Example 6.** Let us find the equations of the tangent and the normal to the curve $y = \cos x$ at the point $M_0(\pi/2, 0)$.

The equation of the tangent is
$$y - \cos x_0 = -\sin x_0 (x - x_0) \quad \text{or} \quad y = \cos x_0 - \sin x_0 (x - x_0).$$

At the point $M_0(\pi/2, 0)$, we have
$$y = \frac{\pi}{2} - x.$$

The equation of the normal is
$$y - \cos x_0 = -\frac{x - x_0}{-\sin x_0} \quad \text{or} \quad y = \cos x_0 + \frac{x - x_0}{\sin x_0}.$$

At the point $M_0(\pi/2, 0)$ (Fig. 9.5), we have
$$y = x - \frac{\pi}{2}.$$

### 9.1.1-3. Singular points.

A point is said to be *singular* if it is not regular.

Implicit equations of the form $F(x, y) = 0$ are used as a rule to find singular points of a curve and analyze their character. At any singular point $M_0(x_0, y_0)$, both partial derivatives of the function $F(x, y)$ are zero:
$$F_x(x_0, y_0) = 0 \quad \text{and} \quad F_y(x_0, y_0) = 0.$$

If both first partial derivatives are zero at $M_0$ and simultaneously at least one of the second derivatives $F_{xx}$, $F_{xy}$, and $F_{yy}$ is nonzero, then $M_0$ is called a *double point*. This is the most widely known case of singular points. If both first partial derivatives and simultaneously all second partial derivatives are zero at $M_0$ but not all third partial derivatives are zero at $M_0$, then the point $M_0$ is said to be *triple*. In general, if all partial derivatives of $F(x, y)$ up to order $n - 1$ inclusive are zero at $M_0$ but at least one of the $n$th derivatives is nonzero at $M_0$, then the point $M_0$ is called an *$n$-fold singular point*. At an $n$-fold singular point $M_0$, the curve has $n$ tangents, some of which may coincide or be imaginary. For example, for a double singular point $M_0$, the slopes $\lambda = y'_x$ of the two tangents at this point are the roots of the quadratic equation
$$F_{yy}(x_0, y_0)\lambda^2 + 2F_{xy}(x_0, y_0)\lambda + F_{xx}(x_0, y_0) = 0. \tag{9.1.1.3}$$

The roots of equation (9.1.1.3) depend on the sign of the expression
$$\Delta = \begin{vmatrix} F_{xx} & F_{xy} \\ F_{yx} & F_{yy} \end{vmatrix} = F_{xx} F_{yy} - F_{xy}^2,$$

where the second derivatives are evaluated at the point $M_0(x_0, y_0)$.

If $\Delta > 0$, then the roots of the quadratic equation (9.1.1.3) are complex conjugate. In this case, a sufficiently small neighborhood of $M_0(x_0, y_0)$ does not contain any other points of the curve except for $M_0$ itself. Such a point is called an *isolated point*.

**Example 7.** The curve
$$y^2 + 4x^2 - x^4 = 0$$
has the isolated point $(0,0)$ (Fig. 9.6a).

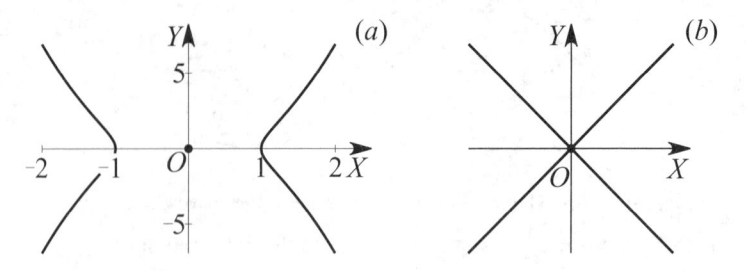

Figure 9.6. Examples of the isolated point (a) and the node (b).

If $\Delta < 0$, then the quadratic equation (9.1.1.3) has two distinct real roots. In this case, there are two branches of the curve passing through the point $M_0(x_0, y_0)$; these branches have distinct tangents whose directions are just determined by equation (9.1.1.3). Such a point is called a *node (a point of self-intersection)*.

**Example 8.** The point $(0,0)$ of the curve
$$y^2 - x^2 = 0$$
is the node $(0,0)$ (Fig. 9.6b).

If $\Delta = 0$, then the roots of the quadratic equation (9.1.1.3) coincide. In this case, the singular point of the curve is either isolated or characterized by the fact that all branches approaching the singular point $M_0$ have a common tangent at this point:

1. *Cusps of the first kind* are points approached by two branches of the curve that have a common tangent at this point and lie on the same side of the common normal and on opposite sides of the common tangent.
2. *Cusps of the second kind* are points approached by two branches of the curve that have a common tangent at this point and lie on the same side of the common normal and on the same side of the common tangent.
3. *Points of osculation* are points at which the curve is tangential to itself.

**Example 9.** For the curve (*cissoid of Diocles*)
$$(2a - x)y^2 - x^3 = 0,$$
shown in Fig. 9.7a, the origin is a cusp of the first kind, which is clear from the explicit equation
$$y = \pm\sqrt{\frac{x^3}{2a - x}}$$
of the curve. For $x < 0$, no $y$ satisfy the equation, and for $x > 0$ the values $\pm y$ lie on opposite sides of the tangent $x = 0$ at the origin.

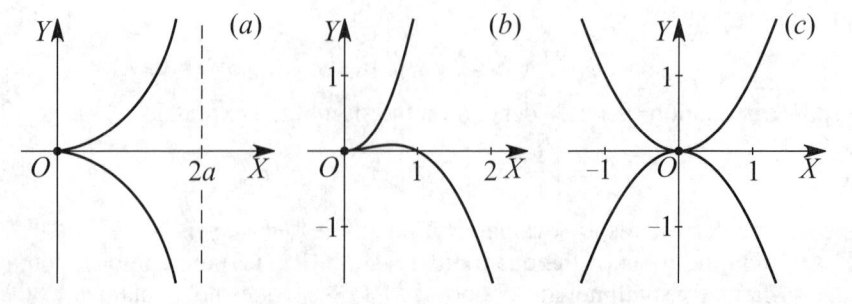

Figure 9.7. Examples of a cusp of the first kind (a), a cusp of the second kind (b), and an osculation point (c).

## 9.1. THEORY OF CURVES

**Example 10.** For the curve (Fig. 9.7b)
$$(y^2 - x^2)^2 - x^5 = 0,$$
the origin is a cusp of the second kind, which easily follows from equation
$$y = x^2 \pm x^{5/2}$$
of the curve. It is also obvious that the curve consists of two branches tangent to the axis $OX$ at the origin, and for $0 < x < 1$ the value of $y$ is positive for both branches.

**Example 11.** The curve (Fig. 9.7c)
$$y^2 - x^4 = 0$$
has an osculation point at the origin.

**Remark 1.** If all the second partial derivatives are zero at the point $M_0$, i.e., $F_{xx} = F_{xy} = F_{yy} = 0$, then more than two branches of the curve can pass through this point. For example, for the *trefoil*
$$(x^2 + y^2)^2 - ax(x^2 - y^2) = 0,$$
three branches with tangents $x = 0$ and $x \pm y = 0$ pass through the origin (Fig. 9.8).

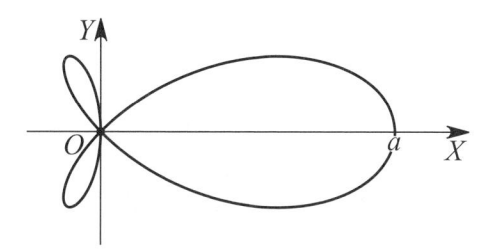

**Figure 9.8.** The trefoil.

**Remark 2.** If the equation $F(x, y) = 0$ does not contain constant terms and terms of degree 1, then the origin is a double point. The equation of the tangent at a double point can readily be obtained by equating all terms of degree 2 with zero. For example, for the cissoid of Diocles (Example 9), the equation of the tangent $x = 0$ follows from the equation $-xy^2 - x^3 = 0$. If the equation $F(x, y) = 0$ does not contain constant terms and terms of degrees 1 and 2, then the origin is a triple point, etc.

Along with the singular points listed above, there are many other singular points with specific names:

1. *Break points* are points at which the curve changes its direction by a "jump" and, in contrast to cusps, the tangents to both parts of the curve are distinct (Fig. 9.9a).
2. *Termination points* are points at which the curve terminates (Fig. 9.9b).
3. *Asymptotic points* are points around which the curve winds infinitely many times while infinitely approaching them (Fig. 9.9c).

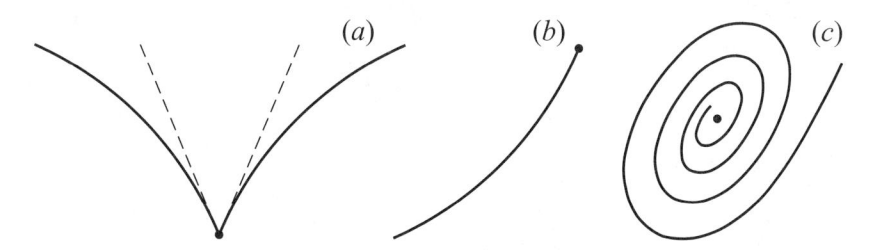

**Figure 9.9.** The break point (*a*), the termination point (*b*), and the asymptotic point (*c*).

**Remark.** A break point corresponds to a jump discontinuity of the derivative $dy/dx$. Termination points correspond to either a jump discontinuity or termination of the function $y = f(x)$. Asymptotic points can be found most easily in curves given in the polar coordinate system.

### 9.1.1-4. Asymptotes.

A straight line is called an *asymptote* of a curve $\Gamma$ if the distance from a point $M(x,y)$ of the curve to this straight line tends to zero as $x^2 + y^2 \to \infty$. The limit position of the tangent to a regular point of the curve is an asymptote; the converse assertion is generally not true.

For a curve given explicitly as $y = f(x)$, vertical asymptotes are determined as points of discontinuity of the function $y = f(x)$, while horizontal and skew asymptotes have the form $y = kx + b$, where

$$k = \lim_{t \to \infty} \frac{f(x)}{x}, \quad b = \lim_{t \to \infty} [f(x) - kx];$$

both limits must be finite.

**Example 12.** Let us find the asymptotes of the curve $y = x^3/(x^2 + 1)$. Since the limits

$$k = \lim_{t \to \infty} \frac{f(x)}{x} = \lim_{t \to \infty} \frac{x^3}{x(x^2 + 1)} = 1,$$

$$b = \lim_{t \to \infty} [f(x) - kx] = \lim_{t \to \infty} \left(\frac{x^3}{x^2 + 1} - x\right) = \lim_{t \to \infty} \left(\frac{-x}{x^2 + 1}\right) = 0$$

exist, the asymptote is given by the equation $y = x$.

To find an asymptote of a parametrically defined curve $x = x(t)$, $y = y(t)$, one should find the values $t = t_i$ for which $x(t) \to \infty$ or $y(t) \to \infty$.

If

$$x(t_i) = \infty \quad \text{but} \quad y(t_i) = c \neq \infty,$$

then the straight line $y = c$ is a horizontal asymptote. If

$$y(t_i) = \infty \quad \text{but} \quad x(t_i) = a \neq \infty,$$

then the straight line $x = a$ is a vertical asymptote. If

$$x(t_i) = \infty \quad \text{and} \quad y(t_i) = \infty,$$

then one should calculate the following two limits:

$$k = \lim_{t \to t_i} \frac{y(t)}{x(t)} \quad \text{and} \quad b = \lim_{t \to t_i} [y(t) - kx(t)]. \tag{9.1.1.4}$$

If both limits exist, then the curve has the asymptote $y = kx + b$.

**Example 13.** Let us find the asymptote of the *Folium of Descartes*

$$x = \frac{3at}{t^3 + 1}, \quad y = \frac{3at^2}{t^3 + 1} \quad (-\infty \leq t \leq \infty).$$

Since $x(-1) = \infty$ and $y(-1) = \infty$, one should use formulas (9.1.1.4),

$$k = \lim_{t \to -1} \frac{y(t)}{x(t)} = \lim_{t \to -1} \frac{\frac{3at^2}{t^3 + 1}}{\frac{3at}{t^3 + 1}} = \lim_{t \to -1} t = -1,$$

$$b = \lim_{t \to -1} [y(t) - kx(t)] = \lim_{t \to -1} \left(\frac{3at^2}{t^3 + 1} + \frac{3at}{t^3 + 1}\right) = \lim_{t \to -1} \frac{3at(t + 1)}{(t + 1)(t^2 - t + 1)} = -a,$$

which imply that the asymptote is given by the equation $y = -x - a$.

Suppose that the function $F(x, y)$ in the equation $F(x, y) = 0$ is a polynomial in the variables $x$ and $y$. We choose the terms of the highest order in $F(x, y)$. By $\Phi(x, y)$ we denote the set of highest-order terms and solve the equation for the variables $x$ and $y$:

$$x = \varphi(y), \quad y = \psi(x).$$

The values $y_i = c$ for which $x = \infty$ give the horizontal asymptotes $y = c$; the values $x_i = a$ for which $y = \infty$ give the horizontal asymptotes $x = a$.

To find skew asymptotes, one should substitute the expression $y = kx + b$ into $F(x, y)$. We write the resulting polynomial $F(x, kx + b)$ as

$$F(x, kx + b) = f_1(k, b)x^n + f_2(k, b)x^{n-1} + \ldots$$

If the system of equations

$$f_1(k, b) = 0, \quad f_2(k, b) = 0$$

is consistent, then its solutions $k, b$ are the parameters of the asymptotes $y = kx + b$.

### 9.1.1-5. Osculating circle.

The *osculating circle (circle of curvature)* of a curve $\Gamma$ at a point $M_0$ is defined to be the limit position of the circle passing through $M_0$ and two neighboring points $M_1$ and $M_2$ of the curve as $M_1 \to M_0$ and $M_2 \to M_0$ (Fig. 9.10).

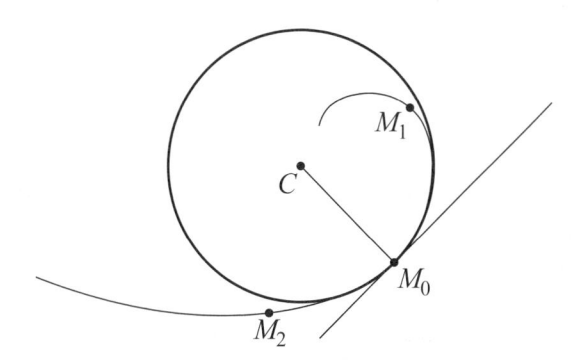

**Figure 9.10.** The osculating circle.

The center of this circle (the *center of curvature* of the curve $\Gamma$ at the point $M_1$) is called the center of the osculating circle and lies on the normal to this curve (Fig. 9.10). The coordinates of the center of curvature can be found by the following formulas:

*for a curve defined explicitly,*

$$x_c = x_0 - \frac{y'_x[1 + (y'_x)^2]}{y''_{xx}}, \quad y_c = y_0 + \frac{1 + (y'_x)^2}{y''_{xx}};$$

*for a curve defined implicitly,*

$$x_c = x_0 - \frac{F_x(F_x^2 + F_y^2)}{2F_{xy}F_xF_y - F_x^2F_{yy} - F_y^2F_{xx}}, \quad y_c = y_0 + \frac{F_y(F_x^2 + F_y^2)}{2F_{xy}F_xF_y - F_x^2F_{yy} - F_y^2F_{xx}};$$

for a curve defined parametrically,

$$x_c = x_0 - \frac{y'_t\left[(x'_t)^2 + (y'_t)^2\right]}{x'_t y''_{tt} - y'_t x''_{tt}}, \quad y_c = y_0 + \frac{x'_t\left[(x'_t)^2 + (y'_t)^2\right]}{x'_t y''_{tt} - y'_t x''_{tt}};$$

for a curve in polar coordinates,

$$x_c = r_0 \cos\varphi_0 - \frac{\left[r_0^2 + (r'_\varphi)^2\right](r_0 \cos\varphi_0 + r'_\varphi \sin\varphi_0)}{r_0^2 + 2(r'_\varphi)^2 - r_0 r''_{\varphi\varphi}},$$

$$y_c = r_0 \sin\varphi_0 - \frac{\left[r_0^2 + (r'_\varphi)^2\right](r_0 \sin\varphi_0 - r'_\varphi \cos\varphi_0)}{r_0^2 + 2(r'_\varphi)^2 - r_0 r''_{\varphi\varphi}},$$

$$x_0 = r_0 \cos\varphi_0, \quad y_0 = r_0 \sin\varphi_0,$$

where all derivatives are evaluated at $x = x_0$, $y = y_0$, $t = t_0$, and $\varphi = \varphi_0$.

The radius of the osculating circle is called the *radius of curvature* of the curve at the point $M_0(x_0, y_0)$; its length in a Cartesian coordinate system is

$$\rho = \frac{\left[1 + (y'_x)^2\right]^{3/2}}{|y''_{xx}|} = \frac{\left(F_x^2 + F_y^2\right)^{3/2}}{|2F_{xy}F_x F_y - F_x^2 F_{yy} - F_y^2 F_{xx}|} = \frac{\left[(x'_t)^2 + (y'_t)^2\right]^{3/2}}{|x'_t y''_{tt} - y'_t x''_{tt}|},$$

and in the polar coordinate system it is

$$\rho = \frac{\left[r_0^2 + (r'_\varphi)^2\right]^{3/2}}{|r_0^2 + 2(r'_\varphi)^2 - r_0 r''_{\varphi\varphi}|},$$

where all derivatives are evaluated at $x = x_0$, $y = y_0$, $t = t_0$, and $\varphi = \varphi_0$.

### 9.1.1-6. Curvature of plane curves.

The limit ratio of the tangent rotation angle $\Delta\varphi$ to the corresponding arc length $\Delta s$ of the curve $\Gamma$ as $\Delta s \to 0$ (Fig. 9.11),

$$k = \lim_{\Delta s \to 0} \left|\frac{\Delta\varphi}{\Delta s}\right|,$$

is called the *curvature* of $\Gamma$ at the point $M_1$.

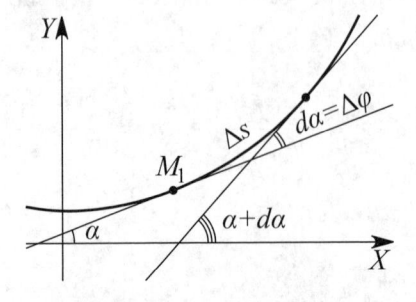

**Figure 9.11.** The curvature of the curve.

The curvature and the radius of curvature are reciprocal quantities,

$$k = \frac{1}{\rho}.$$

The more bent a curve is near a point, the larger $k$ is and the smaller $\rho$ is at this point. For a circle of radius $a$, the radius of curvature is $\rho = a$ and the curvature is $k = 1/a$ (they are constant at all points of the circle); for a straight line, $\rho = \infty$ and $k = 0$; for all other curves, the curvature varies from point to point.

**Remark.** All points of inflection are points of zero curvature.

### 9.1.1-7. Frénet formulas.

To each point $M$ of a plane curve, one can naturally assign a local coordinate system. The role of the origin $O$ is played by point $M$ itself, and the role of the axes $OX$ and $OY$ are played by the tangent and normal at this point. The unit tangent and normal vectors to the curve are usually denoted by **t** and **n**, respectively (Fig. 9.12).

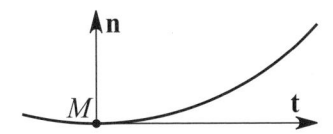

**Figure 9.12.** The unit tangent **t** and normal **n** vectors to the curve.

Suppose that the arc length is taken as a (natural) parameter on the curve:

$$\mathbf{r} = \mathbf{r}(s);$$

then the *Frénet formulas*

$$\mathbf{t}'_s = k\mathbf{n}, \quad \mathbf{n}'_s = -k\mathbf{t}$$

hold, where $k$ is the curvature of the curve.

With first-order accuracy, the Frénet formulas determine the rotation of the vectors **t** and **n** when translated along the curve to a close point, $s \to s + \Delta s$.

### 9.1.1-8. Envelope of a family of curves.

A *one-parameter family of curves* is the set of curves defined by the equation

$$F(x, y, C) = 0, \qquad (9.1.1.5)$$

which is called the *equation of the family*. Here $C$ is a parameter varying in a certain range, $C_1 \leq C \leq C_2$; in particular, the range can be $-\infty \leq C + \infty$.

A curve that is tangent at each point to some curve of a one-parameter family of curves (9.1.1.5) is called the *envelope* of the family. The point of tangency of the envelope to a curve of the family is called a *characteristic point* of the curve of the family (Fig. 9.13).

**Figure 9.13.** The envelope of the family of curves.

The equation of the envelope of a one-parameter family is obtained by elimination of the parameter $C$ from the system of equations

$$F(x, y, C) = 0,$$
$$\frac{\partial F(x, y, C)}{\partial C} = 0, \qquad (9.1.1.6)$$

which determines the *discriminant curve* of this family.

The discriminant curve (9.1.1.6) of a one-parameter family is an envelope if it does not consist of singular points of the curves.

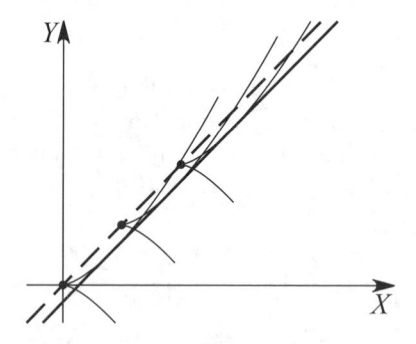

**Figure 9.14.** The envelope of the family of semicubical parabolas.

**Example 14.** Consider the family of semicubical parabolas (Fig. 9.14)

$$3(y - C)^2 - 2(x - C)^3 = 0.$$

Differentiating with respect to the parameter $C$, we obtain

$$y - C - (x - C)^2 = 0.$$

Solving the system

$$3(y - C)^2 - 2(x - C)^3 = 0,$$
$$y - C - (x - C)^2 = 0,$$

we obtain

$$x - C = 0, \quad y - C = 0;$$
$$x - C = \frac{2}{3}, \quad y - C = \frac{4}{9}.$$

Eliminating the parameter $C$, we see that the discriminant curve splits into the pair of straight lines $x = y$ and $x - y = 2/9$. Only the second of these two straight lines is an envelope, because the first straight line is the locus of singular points.

### 9.1.1-9. Evolute and evolvent.

The locus of centers of curvature of a curve is called its *evolute*. If a curve is defined via a natural parameter $s$, then the position vector of a point of the evolute of a plane curve $\mathbf{p}$ can be expressed in terms of the radius vector $\mathbf{r}$ of a point of this curve, the normal vector $\mathbf{n}$, and its radius of curvature $\rho$ as follows:

$$\mathbf{p} = \mathbf{r} + \rho \mathbf{n}.$$

To obtain the equation of the evolute, it also suffices to treat $x_c$ and $y_c$ in the equations for the coordinates of the center of curvature as the current coordinates of the evolute.

Geometric properties of the evolute:

1. The normals to the original curve coincide with the tangents to the evolute at the corresponding points.
2. If the radius of curvature $\rho$ varies monotonically on a given part of the curve, then its increment is equal to the distance passed by the center of curvature along the evolute.
3. At the points of extremum of the radius of curvature $\rho$, the evolute has a cusp of the first kind.
4. Since the radius of curvature $\rho$ is always positive, any point of the evolute lies on a normal to the curve on the concave side.
5. The evolute is the envelope of the family of normals to the original curve.

A curve that intersects all curves of a family at the right angle is called an *orthogonal trajectory* of a one-parameter family of curves. A trajectory orthogonal to tangents to a given curve is called an *evolvent* of this curve.

If a curve is defined via its natural parameter $s$, then the vector equation of its evolvent has the form

$$\mathbf{p} = \mathbf{r} + (s_0 - s)\mathbf{t},$$

where $s_0$ is an arbitrary constant.

Basic properties of the evolvent:

1. The tangent to the original curve at each point is the normal to the evolvent at the corresponding point.
2. The distance between the corresponding points of two evolvents of a given curve is constant.
3. For $s = s_0$, the evolvent has cusps of the first kind.

The evolute and the evolvent are related to each other. The original curve is the evolvent of its evolute. The converse assertion is also true; i.e., the original curve is the evolute of its evolvent. The normal to the evolvent is tangent to the evolute.

## 9.1.2. Space Curves

### 9.1.2-1. Regular points of space curve.

A space curve $\Gamma$ is in general determined parametrically or in vector form by the equations

$$x = x(t), \quad y = y(t), \quad z = z(t) \quad \text{or} \quad \mathbf{r} = \mathbf{r}(t) = x(t)\mathbf{i} + y(t)\mathbf{j} + z(t)\mathbf{k},$$

where $\mathbf{i}$, $\mathbf{j}$, and $\mathbf{k}$ are the unit vectors (see (4.5.2.2)), $t$ is an arbitrary parameter ($t \in [t_1, t_2]$), and $t_1$ and $t_2$ can be $-\infty$ and $+\infty$, respectively.

A point $M(x(t), y(t), z(t))$ is said to be *regular* if the functions $x(t)$, $y(t)$, and $z(t)$ have continuous first derivatives in a sufficiently small neighborhood of this point and these derivatives are not simultaneously zero, i.e., if $d\mathbf{r}/dt \neq 0$.

If the functions $x(t)$, $y(t)$, and $z(t)$ have continuous derivatives with respect to $t$ and $d\mathbf{r}/dt \neq 0$ for all $t \in [t_1, t_2]$, then $\Gamma$ is a *regular arc*.

For the parameter $t$ it is convenient to take the arc length $s$, that is, the length of the arc from a point $M_0(x(t_0), y(t_0), z(t_0))$ to $M_1(x(t_1), y(t_1), z(t_1))$,

$$s = \int_\Gamma ds = \int_\Gamma \sqrt{d\mathbf{r} \cdot d\mathbf{r}} = \int_{t_0}^{t_1} \sqrt{(x'_t)^2 + (y'_t)^2 + (z'_t)^2}\, dt. \tag{9.1.2.1}$$

The sign of $ds$ is chosen arbitrarily, and it determines the positive sense of the curve and the tangent.

A space curve can also be defined as the intersection of two surfaces

$$F_1(x,y,z) = 0, \quad F_2(x,y,z) = 0. \tag{9.1.2.2}$$

For a curve determined as the intersection of two planes, a point $M_0$ is regular if the vectors $\nabla F_1$ and $\nabla F_2$ are not linearly dependent at this point.

### 9.1.2-2. Tangents and normals.

A straight line is called the *tangent* to a curve $\Gamma$ at a regular point $M_0$ if it is the limit position of the secant passing through $M_0$ and a point $M_1$ infinitely approaching the point $M_0$.

At a regular point $M_0$, the equation of the tangent has the form

$$\mathbf{r} = \mathbf{r}_0 + \lambda \mathbf{r}'_t(t_0), \tag{9.1.2.3}$$

where $\lambda$ is a variable parameter.

Eliminating the parameter $\lambda$ from (9.1.2.3), we obtain the canonical equation

$$\frac{x - x_0}{x'_t(t_0)} = \frac{y - y_0}{y'_t(t_0)} = \frac{z - z_0}{z'_t(t_0)}$$

of the tangent.

The equation of the tangent at a point $M_0$ of the curve $\Gamma$ obtained as the intersection (9.1.2.2) of two planes is

$$\frac{x - x_0}{(F_1)_y(F_2)_z - (F_1)_z(F_2)_y} = \frac{y - y_0}{(F_1)_z(F_2)_x - (F_1)_x(F_2)_z} = \frac{z - z_0}{(F_1)_x(F_2)_y - (F_1)_y(F_2)_x},$$

where all derivatives are evaluated at $x = x_0$ and $y = y_0$.

A perpendicular to the tangent at the point of tangency is called a *normal* to a space curve. Obviously, at each point of the curve, there are infinitely many normals that form the plane perpendicular to the tangent.

The plane passing through the point of tangency and perpendicular to the tangent is called the *normal plane* (Fig. 9.15).

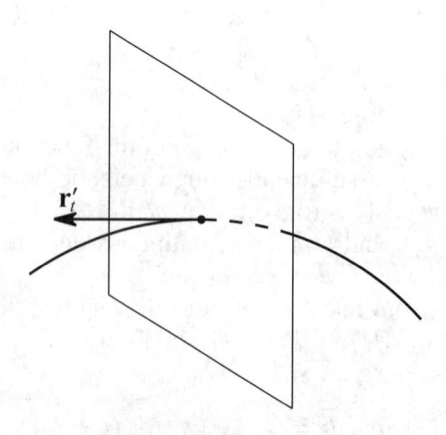

**Figure 9.15.** The normal plane.

At a regular point $M_0$, the equation of the normal plane has the form
$$(x - x_0)x'_t(t_0) + (y - y_0)y'_t(t_0) + (z - z_0)z'_t(t_0) = 0 \quad \text{or} \quad (\mathbf{r} - \mathbf{r}_0) \cdot \mathbf{r}'_t(t_0) = 0, \quad (9.1.2.4)$$
and for the curve $\Gamma$ obtained as the intersection (9.1.2.2) of two planes, we have
$$\begin{vmatrix} x - x_0 & y - y_0 & z - z_0 \\ (F_1)_x & (F_1)_y & (F_1)_z \\ (F_2)_x & (F_2)_y & (F_2)_z \end{vmatrix} = 0, \quad (9.1.2.5)$$
where all derivatives are evaluated at $x = x_0$ and $y = y_0$.

1°. Consider a curve defined via its natural parameter $s$, $\mathbf{r} = \mathbf{r}(s)$. Unlike the derivatives with respect to an arbitrary parameter, the derivatives with respect to $s$ will be denoted by primes. Consider the vectors $\mathbf{r}'_s$ and $\mathbf{r}''_{ss}$. It follows from (9.1.2.1) that $|\mathbf{r}'_s| = 1$. Thus the first derivative with respect to the natural parameter $s$ of the position vector of a point on a curve is the unit vector tangent to the curve.

2°. The second derivative with respect to the natural parameter $s$ of the position vector of a point of a curve is equal to the first derivative of the unit vector $\mathbf{r}'_s$, i.e., of a vector of constant length, and hence it is perpendicular to this vector. But since the vector of the first derivative is tangent to the curve, the vector of the second derivative with respect to the natural parameter $s$ is normal to the curve. This normal is called the *principal normal* to the curve.

At a regular point $M_0$, the equation of the principal normal has the form
$$\frac{x - x_0}{y'_t n - z'_t m} = \frac{y - y_0}{z'_t l - x'_t n} = \frac{z - z_0}{x'_t m - y'_t l} \quad \text{or} \quad \mathbf{r} = \mathbf{r}_0 + \lambda \mathbf{r}'_t \times (\mathbf{r}'_t \times \mathbf{r}''_{tt}),$$
$$l = y'_t z''_{tt} - y''_{tt} z'_t, \quad m = z'_t x''_{tt} - z''_{tt} x'_t, \quad n = x'_t y''_{tt} - x''_{tt} y'_t, \quad (9.1.2.6)$$
where all derivatives are evaluated at $t = t_0$.

### 9.1.2-3. Osculating plane.

Any plane passing through the tangent line to a curve is called a *tangent plane*. A tangent plane passing through a principal normal to the curve is called the *osculating plane*.

A curve has exactly one osculating plane at each of its points (assuming that the vectors $\mathbf{r}'_t$ and $\mathbf{r}''_{tt}$ are linearly independent). This plane passes through the vectors $\mathbf{r}'_t$ and $\mathbf{r}''_{tt}$ drawn from the point of tangency. The osculating plane is independent of the choice of the parameter $t$ on the curve.

*Remark.* If $t$ is time and $\mathbf{r} = \mathbf{r}(t)$ is an equation of motion, then the vector $\mathbf{r}''_{tt}$ is called the *acceleration vector* of a moving point. The acceleration vector always lies in the osculating plane of the trajectory of a moving point.

Since the family of all planes in space depends on three parameters and the position of a plane is determined by three noncollinear points, the osculating plane can be determined as the limit position of the plane passing through three points of the curve that infinitely approach one another (Fig. 9.16).

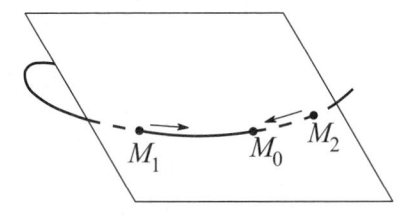

**Figure 9.16.** The osculating plane.

The equation of the osculating plane at the point $M_0$ has the form

$$\begin{vmatrix} x - x_0 & y - y_0 & z - z_0 \\ x'_t(t_0) & y'_t(t_0) & z'_t(t_0) \\ x''_{tt}(t_0) & y''_{tt}(t_0) & z''_{tt}(t_0) \end{vmatrix} = 0 \quad \text{or} \quad \left[(\mathbf{r} - \mathbf{r}_0)\mathbf{r}'_t(t_0)\mathbf{r}''_{tt}(t_0)\right] = 0.$$

This equation becomes meaningless for the points of the curve at which

$$\mathbf{r}'_t(t_0) \times \mathbf{r}''_{tt}(t_0) = \mathbf{0}. \tag{9.1.2.7}$$

The points satisfying (9.1.2.7) are called *points of rectification*. The osculating plane at these points is undefined and, in what follows, we exclude them from consideration together with singular points of the curve.

**Remark.** This is inadmissible for a curve entirely consisting of points of rectification, because a curve consisting of points of rectification is a straight line.

### 9.1.2-4. Moving trihedral of curve.

The normal perpendicular to the osculating plane is called the *binormal*.

At each point of a curve, the tangent, the principal normal, and the binormal determine a trihedral with three right angles at the vertex, which lies on the curve. This trihedral is called the *moving* or *natural trihedral* of a curve. The faces of the moving trihedral are three mutually perpendicular planes:

1. The *normal plane* is the plane containing the principal normal and the binormal.
2. The *osculating plane* is the plane containing the tangent and the principal normal.
3. The *rectifying plane* is the plane containing the tangent and the binormal.

The direction vector of the tangent is equal to the first derivative

$$\mathbf{T} = \mathbf{r}'_t.$$

The direction vector of the binormal is equal to the vector product of the vectors of the first and second derivatives (see Fig. 9.17):

$$\mathbf{B} = \mathbf{r}'_t \times \mathbf{r}''_{tt}.$$

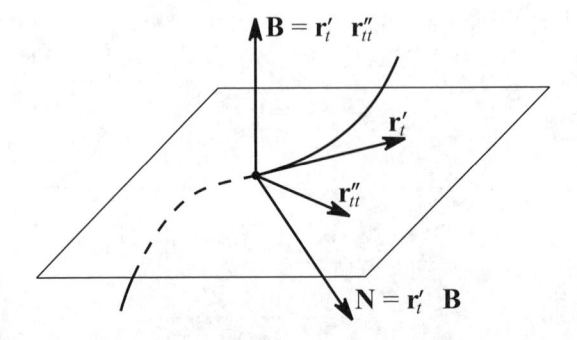

**Figure 9.17.** Direction vectors of the tangent, of the binormal, and of the principal normal.

The direction vector of the principal normal is perpendicular to the tangent and binormal vectors. Hence it can be set equal to their vector product

$$\mathbf{N} = \mathbf{r}'_t \times \mathbf{B}.$$

The moving trihedral permits one to assign a rectangular coordinate system to each point of the curve; the axes of this coordinate system coincide with the tangent, the principal normal, and the binormal. To determine the sense of these axes, one introduces positive unit vectors of these axes.

Consider a curve defined via the natural parameter $s$.

1. The unit tangent vector is defined to coincide with the first derivative of the position vector with respect to the natural parameter,

$$\mathbf{t} = \mathbf{r}'_s. \tag{9.1.2.8}$$

2. The unit principal normal vector $\mathbf{n}$ is defined in such a way that its sense coincides with that of the vector of the second derivative with respect to the parameter,

$$\mathbf{n} = \frac{\mathbf{r}''_{ss}}{|\mathbf{r}''_{ss}|} = \frac{1}{k}\mathbf{r}''_{ss}.$$

The vector $k\mathbf{n} = \mathbf{r}''_{ss}$ is called the *curvature vector*; here $k$ is the curvature, studied in Paragraph 9.1.2-6.

3. The unit binormal vector $\mathbf{b}$ is chosen from the condition that $\mathbf{t}$, $\mathbf{n}$, and $\mathbf{b}$ is a right triple of vectors,

$$\mathbf{b} = \mathbf{t} \times \mathbf{n}.$$

If the direction of arc length increase is changed, then the tangent and binormal vectors also change their sense, but the sense of the principal normal vector remains the same.

9.1.2-5. Equations for elements of trihedral.

At a regular point $M_0$, the equation of the tangent has the form [see also (9.1.2.3)]

$$\mathbf{r} = \mathbf{r}_0 + \lambda \mathbf{t},$$

where $\lambda$ is a variable parameter.

At a regular point $M_0$, the equation of the principal normal has the form [see also (9.1.2.5)]

$$\mathbf{r} = \mathbf{r}_0 + \lambda \mathbf{n},$$

where $\lambda$ is a variable parameter.

At a regular point $M_0$, the equation of the binormal has the form

$$\mathbf{r} = \mathbf{r}_0 + \lambda \mathbf{b},$$

where $\lambda$ is a variable parameter.

The vector equation of the osculating plane at a regular point $M_0$ is

$$(\mathbf{r} - \mathbf{r}_1) \cdot \mathbf{b} = 0$$

[see also (9.1.2.6)].

The vector equation of the normal plane at a regular point $M_0$ is

$$(\mathbf{r} - \mathbf{r}_0) \cdot \mathbf{t} = 0$$

[see also (9.1.2.4)].

The vector equation of the rectifying plane at a regular point $M_0$ is

$$(\mathbf{r} - \mathbf{r}_0) \cdot \mathbf{n} = 0.$$

### 9.1.2-6. Curvature of space curves.

The curvature of a space curve at a point $M_0$ is determined by analogy with the curvature of a plane curve (see Paragraph 9.1.1-7); i.e., the *curvature* of a curve at a point $M_0$ is the limit ratio of the tangent rotation angle along an arc shrinking to $M_0$ to the arc length (Fig. 9.18),

$$k = \lim_{\Delta s \to 0} \left| \frac{\Delta \mathbf{t}}{\Delta s} \right| = \left| \frac{d\mathbf{t}}{ds} \right|. \tag{9.1.2.9}$$

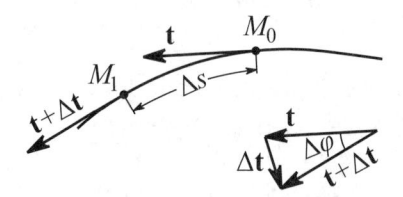

**Figure 9.18.** The curvature of a space curve.

**Remark.** The curvature $k$ can be determined as the angular velocity of the vector $\mathbf{t}$ (or, which is the same, the angular velocity of the tangent) at a given point of the curve with respect to the distance $s$ passed along the curve.

It follows from (9.1.2.8) and (9.1.2.9) that

$$k = |\mathbf{r}''_{ss}| = \sqrt{(x''_{ss})^2 + (y''_{ss})^2 + (z''_{ss})^2}. \tag{9.1.2.10}$$

For an arbitrary choice of the parameter, the curvature $k$ is calculated by the formulas

$$k = \frac{|\mathbf{r}'_t \times \mathbf{r}''_{tt}|}{|\mathbf{r}'_t|^3},$$

$$k = \frac{\sqrt{\left[(x'_t)^2 + (y'_t)^2 + (z'_t)^2\right]\left[(x''_{tt})^2 + (y''_{tt})^2 + (z''_{tt})^2\right] - \left(x'_t x''_{tt} + y'_t y''_{tt} + z'_t z''_{tt}\right)^2}}{\left[(x'_t)^2 + (y'_t)^2 + (z'_t)^2\right]^{3/2}}.$$

The *radius of curvature* and the curvature are reciprocal quantities; i.e., $\rho = 1/k$. For space curves, $\rho$ and $k$ are always positive.

**Example 1.** Let us find the curvature of the helix $x = a \cos t$, $y = a \sin t$, $z = bt$. Expressing the parameter $t$ in terms of the arc length $s$ as

$$s = \sqrt{a^2 + b^2}\, t,$$

we obtain

$$x = a \cos \frac{s}{\sqrt{a^2 + b^2}}, \quad y = a \sin \frac{s}{\sqrt{a^2 + b^2}}, \quad z = \frac{bs}{\sqrt{a^2 + b^2}}.$$

Formula (9.1.2.10) implies

$$k = \frac{a}{a^2 + b^2}, \quad \rho = \frac{a^2 + b^2}{a}.$$

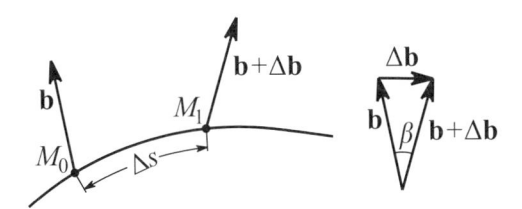

**Figure 9.19.** The torsion of a space curve.

### 9.1.2-7. Torsion of space curves.

In addition to curvature, space curves are also characterized by torsion. The *torsion* of a space curve is defined to be a quantity whose absolute value is equal to the limit ratio of the rotation angle of the binormal along an arc, shrinking to a given point to the arc length (Fig. 9.19):

$$|\tau| = \lim_{\Delta s \to 0} \frac{\beta}{\Delta s} = \lim_{\Delta s \to 0} \left|\frac{\Delta \mathbf{b}}{\Delta s}\right| = \left|\frac{d\mathbf{b}}{ds}\right|.$$

**Remark 1.** The torsion $\tau$ does not change (including the sign) if the direction of increase of the parameter $s$ is changed to the opposite.

**Remark 2.** The torsion $\tau$ can be defined as the angular velocity of the binormal $\mathbf{b}$ at the corresponding point with respect to the distance $s$ passed along the curve.

In the special case of a plane curve, the torsion is zero at all points; conversely, if the torsion is zero at all points of a curve, then the entire curve lies in a single plane, which is the common osculating plane for all of its points.

For an arbitrary choice of the parameter $t$, the torsion $\tau$ is calculated by the formulas

$$\tau = \rho^2 \frac{\mathbf{r}'_t \mathbf{r}''_{tt} \mathbf{r}'''_{ttt}}{|\mathbf{r}'_t|^3} = \frac{\begin{vmatrix} x'_t & y'_t & z'_t \\ x''_{tt} & y''_{tt} & z''_{tt} \\ x'''_{ttt} & y'''_{ttt} & z'''_{ttt} \end{vmatrix}}{[(x'_t)^2 + (y'_t)^2 + (z'_t)^2][(x''_{tt})^2 + (y''_{tt})^2 + (z''_{tt})^2] - (x'_t x''_{tt} + y'_t y''_{tt} + z'_t z''_{tt})^2}. \tag{9.1.2.11}$$

But if the arc length is taken as the parameter, then $\tau$ is calculated by the formulas

$$\tau = \rho^2 \frac{(\mathbf{r}'_s \mathbf{r}''_{ss} \mathbf{r}'''_{sss})}{|\mathbf{r}'_s|^3} = \frac{\begin{vmatrix} x'_s & y'_s & z'_s \\ x''_{ss} & y''_{ss} & z''_{ss} \\ x'''_{sss} & y'''_{sss} & z'''_{sss} \end{vmatrix}}{\left[(x''_{ss})^2 + (y''_{ss})^2 + (z''_{ss})^2\right]^2}.$$

The quantity $\rho_\tau = 1/\tau$ inverse to the torsion is called the *radius of torsion*.

**Example 2.** Let us find the torsion of the helix $x = a\cos t$, $y = a\sin t$, $z = bt$. Using formula (9.1.2.11), we obtain

$$\tau = \frac{\begin{vmatrix} -a\sin t & a\cos t & b \\ -a\cos t & -a\sin t & 0 \\ a\sin t & -a\cos t & 0 \end{vmatrix}}{(a^2\sin^2 t + a^2\cos^2 t + b^2)(a^2\cos^2 t + a^2\sin^2 t) - (a^2\sin t\cos t - a^2\cos t\sin t)^2} = \frac{b}{a^2 + b^2}.$$

### 9.1.2-8. Serret–Frénet formulas.

The derivatives of the vectors **t**, **n**, and **b** with respect to the parameter $s$ at each point of the curve satisfy the *Serret–Frénet formulas*

$$\mathbf{t}'_s = k\mathbf{n}, \quad \mathbf{n}'_s = -k\mathbf{t} + \tau\mathbf{b}, \quad \mathbf{b}'_s = -\tau\mathbf{n},$$

where $k$ is the curvature of the curve and $\tau$ is the torsion of the curve.

As the parameter $s$ increases, the point $M$ moves along $\Gamma$; in this case

1. The tangent rotates around the instantaneous position of the binormal **b** at a positive angular velocity equal to the curvature $k$ of $\Gamma$ at $M$.
2. The binormal **b** rotates around the instantaneous position of the tangent at an angular velocity equal to the torsion $\tau$ of $\Gamma$ at $M$; here the torsion is positive if the shape of the curve resembles a right helix.
3. The trihedral rotates as a solid around an instantaneous axis whose direction is determined by the *Darboux vector* $\Omega = \tau\mathbf{t} + k\mathbf{b}$ at a positive angular velocity equal to the total curvature of $\Gamma$ $M$, i.e., to $|\Omega| = \sqrt{\tau^2 + k^2}$.

The Serret–Frénet formulas permit one to find the coefficients in the decomposition of the derivatives of the position vector in the vectors of the moving trihedral. The following recursion formulas hold for the decomposition of the derivatives of the position vector of a point on the curve with respect to the arc length:

$$\mathbf{r}^{(n)}_s = A_n\mathbf{t} + B_n\mathbf{n} + C_n\mathbf{b},$$

where

$$A_{n+1} = (A_n)'_s - B_nk, \quad B_{n+1} = (B_n)'_s + A_nk - C_n\tau, \quad C_{n+1} = (C_n)'_s + B_n\tau.$$

For example, using the relation $\mathbf{r}''_{ss} = k\mathbf{n}$, we can decompose $\mathbf{r}'''_{sss}$ as

$$\mathbf{r}'''_{sss} = -k^2\mathbf{t} + k'_s\mathbf{n} + k\tau\mathbf{b}.$$

## 9.2. Theory of Surfaces

### 9.2.1. Elementary Notions in Theory of Surfaces

#### 9.2.1-1. Equation of surface.

Any surface can be determined by equations in one of the following forms:

*Explicitly,*
$$z = f(x, y).$$

*Implicitly,*
$$F(x, y, z) = 0. \tag{9.2.1.1}$$

*Parametrically,*
$$x = x(u, v), \quad y = y(u, v), \quad z = z(u, v).$$

*In vector form,*
$$\mathbf{r} = \mathbf{r}(u, v), \quad \text{or} \quad \mathbf{r} = x(u, v)\mathbf{i} + y(u, v)\mathbf{j} + z(u, v)\mathbf{k}.$$

Varying the parameters $u$ and $v$ arbitrarily, we obtain the position vector and the coordinates of various points on the surface. Eliminating the parameters $u$ and $v$ from a parametric equation of a surface, we obtain an implicit equation of the surface. The explicit equation is a special case ($u = x$ and $v = y$) of the parametric equation.

It is assumed that the vectors $\mathbf{r}_u \equiv \partial \mathbf{r}/\partial u$ and $\mathbf{r}_v \equiv \partial \mathbf{r}/\partial v$ are nonparallel, i.e.,

$$\mathbf{r}_u \times \mathbf{r}_v \neq 0. \tag{9.2.1.2}$$

The points at which condition (9.2.1.2) holds are said to be *regular*. This condition satisfied at a point $M$ on the surface guarantees that the equation of the surface near this point can be solved for one of the coordinates. Condition (9.2.1.2) also guarantees a one-to-one correspondence between points on the surface and pairs of values of $u$ and $v$ in the corresponding range of $u$ and $v$.

Condition (9.2.1.2) for a surface defined implicitly becomes

$$\operatorname{grad} F \neq 0.$$

For a surface defined parametrically, all functions have continuous first partial derivatives, and the rank of the matrix

$$\begin{pmatrix} x_u & y_u & z_u \\ x_v & y_v & z_v \end{pmatrix}$$

is equal to 2.

**Example 3.** The sphere defined by the implicit equation

$$x^2 + y^2 + z^2 - a^2 = 0$$

is described in parametric form as

$$x = a \cos u \sin v, \quad y = a \sin u \sin v, \quad z = a \cos v$$

and in vector form as

$$\mathbf{r} = a \cos u \sin v \, \mathbf{i} + a \sin u \sin v \, \mathbf{j} + a \cos v \, \mathbf{k}.$$

### 9.2.1-2. Curvilinear coordinates on surface.

We consider a surface or part of the surface such that it can be topologically (i.e., bijectively and continuously) mapped onto a plane domain and assume that a point $M$ of this surface is taken to a point $M_0$ with rectangular coordinates $u$ and $v$ on the plane (see Fig. 9.20). If such a mapping is given, then the surface is said to be *parametrized*, and $u$ and $v$ are called *curvilinear (Gaussian) coordinates* of the point $M$ on the surface. Since the mapping is continuous, each curve on the plane gives rise to a curve on the surface. In particular, the straight lines $u = \operatorname{const}$ and $v = \operatorname{const}$ are associated with the curves on the surface, which are called *parametric* or *curvilinear lines* of the surface. Since the mapping is one-to-one, there is a single curve of the family $u = \operatorname{const}$ and a single curve of the family $v = \operatorname{const}$ that pass through each point of the parametrized surface. Both families together form a regular net, which is called the *coordinate net*.

In the case of rectangular Cartesian coordinates, the coordinate net on the plane is formed by all possible straight lines parallel to coordinate axes; in the case of polar coordinates, the coordinate net is formed by circles centered at the pole and half-lines issuing from the pole.

**Example 4.** In the parametric equations

$$x = a \cos u \sin v, \quad y = a \sin u \sin v, \quad z = a \cos v$$

of the sphere, $u$ is the longitude and $v$ the polar distance of a point.

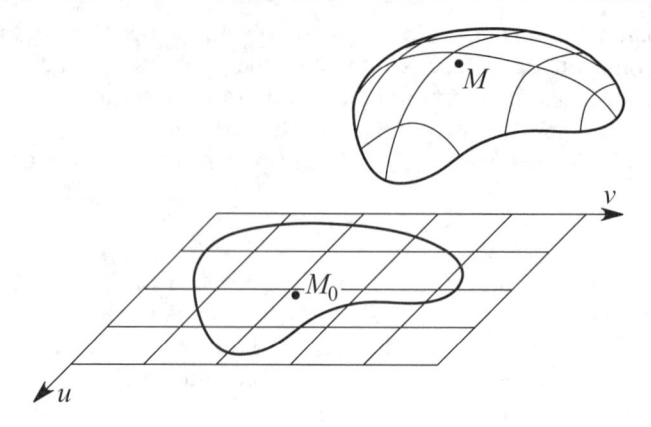

**Figure 9.20.** The parametrized surface.

9.2.1-3. Tangent line to surface.

A straight line is said to be *tangent* to a surface if it is tangent to a curve lying in this surface. Suppose that a surface is given in vector form (9.2.1.1) and a curve lying on it is parametrized by the parameter $t$. Then to each parameter value there corresponds a point of the curve, and the position of this point on the surface is specified by some values of the curvilinear coordinates $u$ and $v$. Thus the curvilinear coordinates of points of a curve lying on a surface are functions of the parameter $t$.

The system of equations

$$u = u(t), \quad v = v(t) \tag{9.2.1.3}$$

is called the *intrinsic* equations of the curve on the surface.

The intrinsic equations completely characterize the curve if the vector equation of the surface is given, since the substitution of (9.2.1.3) into (9.2.1.1) results in the equation

$$\mathbf{r} = \mathbf{r}[u(t), v(t)], \tag{9.2.1.4}$$

which is called the *parametric equation* of the curve.

The differential of the position vector is equal to

$$d\mathbf{r} = \mathbf{r}_u \, du + \mathbf{r}_v \, dv, \tag{9.2.1.5}$$

where $du = u'_t(t) \, dt$ and $dv = v'_t(t) \, dt$. The vectors $\mathbf{r}_u$ and $\mathbf{r}_v$ are called the *coordinate vectors* corresponding to the point whose curvilinear coordinates have been used in the computations. The coordinate vectors are tangent vectors to coordinate curves (Fig. 9.21).

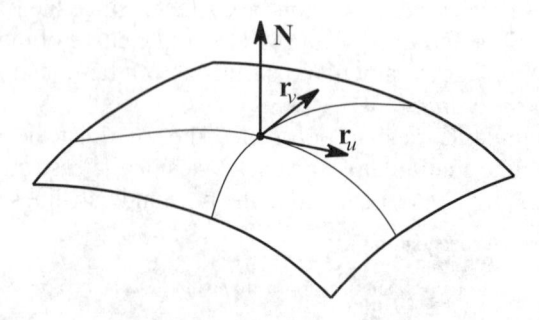

**Figure 9.21.** Coordinate vectors.

Formula (9.2.1.5) shows that the direction vector of any tangent line to a surface at a given point is a linear combination of the coordinate vectors corresponding to this point; i.e., the tangent to a curve lies in the plane spanned by the vectors $\mathbf{r}_u$ and $\mathbf{r}_v$ at this point.

The direction of the tangent to a curve on a surface at a point $M$ is completely characterized by the ratio $dv:du$ of differentials taken along this curve.

### 9.2.1-4. Tangent plane and normal.

If all possible curves are drawn on a surface through a given regular point $M_0(r_0) = M_0(x_0, y_0, z_0) = M(u_0, v_0)$ of the surface, then their tangents at $M_0$ lie in the same plane, which is called the *tangent plane* to the surface at $M_0$. The tangent plane can be defined as the limit position of the plane passing through three distinct points $M_0$, $M_1$, and $M_2$ on the surface as $M_1 \to M_0$ and $M_2 \to M_0$; here $M_1$ and $M_2$ should move along curves with distinct tangents at $M_0$.

The tangent plane at $M_0$ can be viewed as the plane passing through $M_0$ and perpendicular to the vector

$$\mathbf{r}_u \times \mathbf{r}_v;$$

i.e., it passes through the vectors $\mathbf{r}_u$ and $\mathbf{r}_v$. Thus the tangent plane at $M_0$, depending on the method for defining the surface, is given by one of the equations

$$\begin{vmatrix} x-x_0 & y-y_0 & z-z_0 \\ x_u & y_u & z_u \\ x_v & y_v & z_v \end{vmatrix} = 0, \quad [(\mathbf{r}-\mathbf{r}_0)\mathbf{r}_u\mathbf{r}_v] = 0,$$

$$F_x(x-x_0) + F_y(y-y_0) + F_z(z-z_0) = 0, \quad z_x(x-x_0) + z_y(y-y_0) = z - z_0,$$

where all the derivatives are evaluated at the point $M_0(r_0) = M_0(x_0, y_0, z_0) = M(u_0, v_0)$.

The straight line passing through $M_0$ and perpendicular to the tangent plane is called the *normal* to the surface at $M_0$. The vector $\mathbf{N} = \mathbf{r}_u \times \mathbf{r}_v / |\mathbf{r}_u \times \mathbf{r}_v|$ is called the *unit normal vector*. The sense of the vector $\mathbf{N}$ is called the *positive normal direction*; the vector $\mathbf{r}_u$, the vector $\mathbf{r}_v$, and the positive normal form a right triple.

The equation of the normal, depending on the method for defining the surface, has one of the forms

$$\frac{x-x_0}{\begin{vmatrix} y_u & z_u \\ y_v & z_v \end{vmatrix}} = \frac{y-y_0}{\begin{vmatrix} z_u & x_u \\ z_v & x_v \end{vmatrix}} = \frac{z-z_0}{\begin{vmatrix} x_u & y_u \\ x_v & y_v \end{vmatrix}}, \quad \mathbf{r} = \mathbf{r}_0 + \lambda(\mathbf{r}_u \times \mathbf{r}_v) \quad \text{or} \quad \mathbf{r} = \mathbf{r}_0 + \lambda \mathbf{N},$$

$$\frac{x-x_0}{F_x} = \frac{y-y_0}{F_y} = \frac{z-z_0}{F_z}, \quad \frac{x-x_0}{z_x} = \frac{y-y_0}{z_y} = \frac{z-z_0}{-1},$$

where all derivatives are evaluated at the point $M_0(r_0) = M_0(x_0, y_0, z_0) = M(u_0, v_0)$.

**Example 5.** For the sphere given by the implicit equation

$$x^2 + y^2 + z^2 - a^2 = 0,$$

the tangent plane at the point $M_0(x_0, y_0, z_0)$ is given by the equation

$$2x_0(x-x_0) + 2y_0(y-y_0) + 2z_0(z-z_0) = 0 \quad \text{or} \quad xx_0 + yy_0 + zz_0 = 0,$$

and the normal is given by the equation

$$\frac{x-x_0}{2x_0} = \frac{y-y_0}{2y_0} = \frac{z-z_0}{2z_0} \quad \text{or} \quad \frac{x}{x_0} = \frac{y}{y_0} = \frac{z}{z_0}.$$

### 9.2.1-5. First quadratic form.

If a surface is given parametrically or in vector form, $M_1(u, v)$ is an arbitrary point, and $M_2(u + du, v + dv)$ is a nearby point on the surface, then the length of the arc $M_1 M_2$ on the surface is approximately expressed in terms of the *arc length differential* or in terms of the *linear surface element* by the formula

$$ds^2 = E\,du^2 + 2F\,du\,dv + G\,dv^2, \tag{9.2.1.6}$$

where the coefficients $E$, $F$, and $G$ are given by the formulas

$$E = \mathbf{r}_u^2 = x_u^2 + y_u^2 + z_u^2,$$
$$F = \mathbf{r}_u \cdot \mathbf{r}_v = x_u x_v + y_u y_v + z_u z_v,$$
$$G = \mathbf{r}_v^2 = x_v^2 + y_v^2 + z_v^2.$$

The right-hand side of formula (9.2.1.6) is also called the *first quadratic form of the surface* given parametrically or in vector form; its coefficients $E$, $F$, and $G$ depend on the point on the surface. At each regular point on the surface corresponding to (real) coordinates $u$ and $v$, the first quadratic form (0.2.1.12) is positive definite; i.e.,

$$E > 0, \quad G > 0, \quad EG - F^2 > 0.$$

**Example 6.** For the sphere given by the equation

$$\mathbf{r} = a\cos u \sin v\,\mathbf{i} + a\sin u \sin v\,\mathbf{j} + a\cos v\,\mathbf{k},$$

the coefficients $E$, $F$, and $G$ are equal to

$$E = a^2 \sin^2 v, \quad F = 0, \quad G = a^2,$$

and the first quadratic form is

$$ds^2 = a^2(\sin^2 v\,du^2 + dv^2).$$

For a surface given explicitly, the coefficients $E$, $F$, and $G$ are given by the formulas

$$E = 1 + z_x^2, \quad F = z_x z_y, \quad G = 1 + z_y^2.$$

The *arc length* of the curve $u = u(t)$, $v = v(t)$, $t \in [t_0, t_1]$, on the surface can be calculated by the formula

$$L = \int_{t_0}^{t_1} ds = \int_{t_0}^{t_1} \sqrt{E u_t^2 + 2F u_t v_t + G v_t^2}\,dt. \tag{9.2.1.7}$$

The angle $\gamma$ between two curves (i.e., between their tangents) intersecting in a point $M$ and having the direction vectors $d\mathbf{r} = (du, dv)$ and $\delta\mathbf{r} = (\delta u, \delta v)$ at this point (Fig. 9.22) can be calculated by the formula

$$\cos\gamma = \frac{d\mathbf{r}\,\delta\mathbf{r}}{|d\mathbf{r}||\delta\mathbf{r}|} = \frac{E\,du\,\delta u + F(du\,\delta v + dv\,\delta u) + G\,dv\,\delta v}{\sqrt{E\,du^2 + 2F\,du\,dv + G\,dv^2}\sqrt{E\,\delta u^2 + 2F\,\delta u\,\delta v + G\,\delta v^2}}. \tag{9.2.1.8}$$

(The coefficients $E$, $F$, and $G$ are evaluated at point $M$.)

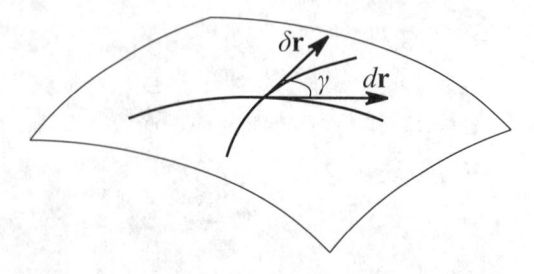

**Figure 9.22.** The angle between two space curves.

In particular, the angle $\gamma_1$ between the coordinate curves $u = \text{const}$ and $v = \text{const}$ passing through a point $M(u, v)$ is determined by the formulas

$$\cos\gamma_1 = \frac{F}{\sqrt{EG}}, \quad \sin\gamma_1 = \frac{\sqrt{EG - F^2}}{\sqrt{EG}}.$$

The coordinate lines are perpendicular if $F = 0$.

The area of a domain $U$ bounded by some curve on the surface can be calculated as the double integral

$$S = \int_U dS = \int_U \sqrt{EG - F^2}\, du\, dv. \tag{9.2.1.9}$$

Thus if the coefficients $E$, $F$, and $G$ of the first quadratic form are known, then one can measure lengths, angles, and areas on the surface according to formulas (9.2.1.7), (9.2.1.8), and (9.2.1.9); i.e., the first quadratic form completely determines the *intrinsic geometry of the surface* (see Subsection 9.2.3 for details).

To calculate surface areas in three-dimensional space, one can use the following theorems.

THEOREM 1. *If a surface is given in the explicit form $z = f(x, y)$ and a domain $U$ on the surface is projected onto a domain $V$ on the plane $(x, y)$, then*

$$S = \int_V \sqrt{1 + f_x^2 + f_y^2}\, dx\, dy.$$

THEOREM 2. *If the surface is given implicitly ($F(x, y, z) = 0$) and a domain $U$ on the surface is projected bijectively onto a domain $V$ on the plane $(x, y)$, then*

$$S = \int_V \frac{|\operatorname{grad} F|}{|F_z|}\, dx\, dy,$$

*where $|F_z| = \partial F/\partial z \ne 0$ for $(x, y, z)$ lying in the domain $U$.*

THEOREM 3. *If a surface is the parametric form $\mathbf{r} = \mathbf{r}(u, v)$ or $x = x(u, v)$, $y = y(u, v)$, $z = z(u, v)$, then*

$$S = \int_V |\mathbf{r}_u \times \mathbf{r}_v|\, du\, dv.$$

9.2.1-6. Singular (conic) points of surface.

A point $M_0(x_0, y_0, z_0)$ on a surface given implicitly, i.e., determined by the equation $F(x, y) = 0$, is said to be *singular (conic)* if its coordinates satisfy the system of equations

$$F_x(x_0, y_0, z_0) = 0, \quad F_y(x_0, y_0, z_0) = 0, \quad F_z(x_0, y_0, z_0) = 0, \quad F(x_0, y_0, z_0) = 0.$$

All tangents passing through a singular point $M_0(x_0, y_0, z_0)$ do not lie in the same plane but form a second-order cone defined by the equation

$$F_{xx}(x - x_0) + F_{yy}(y - y_0) + F_{zz}(z - z_0) + 2F_{xy}(x - x_0)(y - y_0)$$
$$+ 2F_{yz}(y - y_0)(z - z_0) + 2F_{zx}(z - z_0)(x - x_0) = 0.$$

The derivatives are evaluated at the point $M_0(x_0, y_0, z_0)$; if all six second partial derivatives are simultaneously zero, then the singular point is of a more complicated type (the tangents form a cone of third or higher order).

## 9.2.2. Curvature of Curves on Surface

### 9.2.2-1. Normal curvature. Meusnier's theorem.

Of the plane sections of a surface, the planes containing the normal to the surface at a given point are said to be *normal*. In this case, there exists a unique normal section $\Gamma_0$ containing a given tangent to the curve $\Gamma$.

MEUSNIER THEOREM. *The radius of curvature at a given point of a curve $\Gamma$ lying on a surface is equal to the radius of curvature of the normal section $\Gamma_0$ taken at the same point with the same tangent, multiplied by the cosine of the angle $\alpha$ between the osculating plane of the curve at this point and the plane of the normal section $\Gamma_0$; i.e.,*

$$\rho = \rho_N \cos \alpha \ddagger$$

The *normal curvature* of a curve $\Gamma$ at a point $M(u,v)$ is defined as

$$k_N = k\mathbf{n} \cdot \mathbf{N} = \mathbf{r}''_{ss} \cdot \mathbf{N} = -\mathbf{r}'_s \cdot \mathbf{N}'_s.$$

The normal curvature is the curvature of the *normal section*.

The *geodesic curvature* of a curve $\Gamma$ at a point $M(u,v)$ is defined as

$$k_G = k\mathbf{r}'_s \mathbf{n} \mathbf{N} = \mathbf{r}'_s \mathbf{r}''_{ss} \mathbf{N}.$$

The geodesic curvature is the angular velocity of the tangent to the curve around the normal. The geodesic curvature is the curvature of the projection of the curve $\Gamma$ onto the tangent plane.

For any point $u, v$ of the curve $\Gamma$ given by equation (9.2.1.4) and lying on the surface, the curvature vector can be represented as a sum of two vectors,

$$\mathbf{r}''_{ss} = k\mathbf{n} = k_G \mathbf{N} \times \mathbf{r}'_s + k_N \mathbf{N}, \tag{9.2.2.1}$$

where $\mathbf{N}$ is the unit normal vector to the surface. The first term on the right-hand side in (9.2.2.1) is called the *geodesic (tangential) curvature vector*, and the second term is called the *normal curvature vector*. The geodesic curvature vector lies in the tangent plane, and the normal curvature vector is normal to the surface.

### 9.2.2-2. Second quadratic form. Curvature of curve on surface.

The quadratic differential form

$$-d\mathbf{r} \cdot d\mathbf{N} = L\,du^2 + 2M\,du\,dv + N\,dv^2,$$

$$L = -\mathbf{r}_u \cdot \mathbf{N}_u = \frac{\mathbf{r}_{uu}\mathbf{r}_u\mathbf{r}_v}{\sqrt{EG - F^2}}, \quad N = -\mathbf{r}_v \cdot \mathbf{N}_v = \frac{\mathbf{r}_{vv}\mathbf{r}_u\mathbf{r}_v}{\sqrt{EG - F^2}},$$

$$M = -\mathbf{r}_u \cdot \mathbf{N}_v = -\mathbf{r}_v \cdot \mathbf{N}_u = \frac{\mathbf{r}_{uv}\mathbf{r}_u\mathbf{r}_v}{\sqrt{EG - F^2}}$$

(all derivatives are evaluated at the point $M(u,v)$) is called the *second quadratic form of the surface*.

The coefficients $L$, $N$, and $M$ for surfaces given parametrically or implicitly can be calculated by the formulas

$$L = \frac{1}{\sqrt{EG - F^2}} \begin{vmatrix} x_{uu} & y_{uu} & z_{uu} \\ x_u & y_u & z_u \\ x_v & y_v & z_v \end{vmatrix} = z_{xx}\sqrt{1 + z_x^2 + z_y^2},$$

$$M = \frac{1}{\sqrt{EG - F^2}} \begin{vmatrix} x_{uv} & y_{uv} & z_{uv} \\ x_u & y_u & z_u \\ x_v & y_v & z_v \end{vmatrix} = z_{xy}\sqrt{1 + z_x^2 + z_y^2},$$

$$N = \frac{1}{\sqrt{EG - F^2}} \begin{vmatrix} x_{vv} & y_{vv} & z_{vv} \\ x_u & y_u & z_u \\ x_v & y_v & z_v \end{vmatrix} = z_{yy}\sqrt{1 + z_x^2 + z_y^2}.$$

The curvature $k_N$ of a normal section can be calculated by the formula

$$k_N = -\frac{d\mathbf{r} \cdot d\mathbf{N}}{ds^2} = \frac{L\,du^2 + 2M\,du\,dv + N\,dv^2}{E\,du^2 + 2F\,du\,dv + G\,dv^2}.$$

A point on the surface at which the curvature $\rho_N$ of a normal section takes the same value for any normal section ($L : M : N = E : F : G$) is said to be *umbilical (circular)*. At each nonumbilical point, there are two normal sections called the *principal normal sections*. They are characterized by the maximum and minimum values $k_1$ and $k_2$ of the curvature $\rho_N$, which are called the *principal curvatures* of the surface $U$ at the point $M(u, v)$. The planes of principal normal sections are mutually perpendicular.

EULER THEOREM. *For a normal section at $M(u, v)$ whose plane forms an angle $\theta$ with the plane of one of the principal normal sections, one has*

$$k_N = k_1 \cos^2 \theta + k_2 \sin^2 \theta \quad \text{or} \quad k_N = k_1 + (k_2 - k_1) \sin^2 \theta.$$

The quantities $k_1$ and $k_2$ are the roots of the characteristic equation

$$\begin{vmatrix} L - kE & M - kF \\ M - kF & N - kG \end{vmatrix} = 0.$$

The curves on a surface whose directions at each point coincide with the directions of the principal normal sections are called the *curvature lines*; their differential equation is

$$\begin{vmatrix} dv^2 & -dv\,du & du^2 \\ E & F & G \\ L & M & N \end{vmatrix} = 0.$$

*Asymptotic lines* are defined to be the curves for which $\rho_N = 0$ at each point. The asymptotic lines are determined by the differential equation

$$L\,du^2 + 2M\,du\,dv + N\,dv^2 = 0.$$

### 9.2.2-3. Mean and Gaussian curvatures.

The symmetric functions

$$H(u, v) = \frac{k_1 + k_2}{2} \quad \text{and} \quad K(u, v) = k_1 k_2$$

are called the *mean and Gaussian (extrinsic) curvature*, respectively, of the surface $U$ at the point $M(u, v)$. They are given by the formulas

$$H(u, v) = \frac{EN - 2FM + LG}{2(EG - F^2)},$$

$$K(u, v) = \frac{LN - M^2}{EG - F^2}. \quad (9.2.2.2)$$

The mean and Gaussian curvatures are related by the inequality

$$H^2 - K = \frac{(k_1 + k_2)^2}{4} - k_1 k_2 = \frac{(k_1 - k_2)^2}{4} \geq 0.$$

The mean and Gaussian curvatures can be used to characterize the deviation of the surface from a plane. In particular, if $H = 0$ and $K = 0$ at all points of the surface, then the surface is a plane.

**Example 1.** For a circular cylinder (of radius $a$),

$$H = \frac{a}{2}, \quad K = 0.$$

If a surface is represented by the equation $z = f(x, y)$, then the mean and Gaussian curvature can be determined by the formulas

$$H = \frac{r(1 + q^2) - 2pqs + t(1 + p^2)}{2\sqrt{(1 + p^2 + q^2)^3}}, \quad K = \frac{rt - s^2}{(1 + p^2 + q^2)^3},$$

where the following notation is used:

$$p = z_x, \quad q = z_y, \quad r = z_{xx}, \quad s = z_{xy}, \quad t = z_{yy}, \quad h = \sqrt{1 + p^2 + q^2}.$$

The surfaces for which the mean curvature $H$ is zero at all points are said to be *minimal*. The surfaces for which the Gaussian curvature $K$ is constant at all points are called *surfaces of constant curvature*.

### 9.2.2-4. Classification of points on surface.

The points of a surface can be classified according to the values of the Gaussian curvature:

1. A point $M$ at which $K = k_1 k_2 > 0$ (the principal normal sections are convex in the same direction from the tangent plane; example: any point of an ellipsoid) is called an *elliptic point*; the analytic criterion for this case is $LN - M^2 > 0$. In the special case $k_1 = k_2$, the point is umbilical (circular): $R = \text{const}$ for all normal sections at this point.
2. A point $M$ at which $K = k_1 k_2 < 0$ (the principal normal sections are convex in opposite directions; the surface intersects the tangent plane and has a saddle character; example: any point of a one-sheeted hyperboloid) is called a *hyperbolic (saddle) point*; the analytic criterion for this case is $LN - M^2 < 0$.
3. A point $M$ at which $K = k_1 k_2 = 0$ (one principal normal section has an inflection point or is a straight line; example: any point of a cylinder) is called a *parabolic point*; the analytic criterion for this case is $LN - M^2 = 0$.

Any umbilical point ($k_1 = k_2$) is either elliptic or parabolic.

### 9.2.3. Intrinsic Geometry of Surface

**9.2.3-1. Intrinsic geometry and bending of surface.**

Suppose that two surfaces $U$ and $U^*$ are given and there is one-to-one correspondence between their points such that the length of each curve on $U$ is equal to the length of the corresponding curve on $U^*$. Such a one-to-one mapping of $U$ into $U^*$ is called a *bending* of the surface $U$ into the surface $U^*$, and the surfaces $U$ and $U^*$ are said to be *applicable*.

The correspondence between $U$ and $U^*$ is said to be *isometric*.

The intrinsic geometry of a surface studies geometric constructions and quantities related to the surface that can be determined solely from the first quadratic form. The notions of length of a segment, angle between two curves, and area of part of a surface all belong in intrinsic geometry.

On the opposite, the curvature of a curve given on the surface by the equations

$$u = u(t), \quad v = v(t)$$

cannot be found using only the first quadratic form, and hence it does not belong in intrinsic geometry.

**9.2.3-2. Index notation. Surface as Riemannian space.**

From now on in this chapter, the following notation related to tensor analysis is used:

$$u^1 = u, \quad u^2 = v;$$
$$g_{11} = E, \quad g_{12} = g_{21} = F, \quad g_{22} = G;$$
$$b_{11} = L, \quad b_{12} = b_{21} = M, \quad b_{22} = N;$$
$$\mathbf{r}_1 = \mathbf{r}_u, \quad \mathbf{r}_2 = \mathbf{r}_v, \quad \mathbf{r}_{12} = \mathbf{r}_{21} = \mathbf{r}_{vu} = \mathbf{r}_{uv}, \quad \mathbf{r}_{11} = \mathbf{r}_{uu}, \quad \mathbf{r}_{22} = \mathbf{r}_{vv}.$$

In the new notation, the first fundamental quadratic form becomes

$$E\,du^2 + 2F\,du\,dv + G\,dv^2 = \sum_{\alpha=1}^{2}\sum_{\beta=1}^{2} g_{\alpha\beta}\,du^\alpha\,du^\beta = g_{\alpha\beta}\,du^\alpha\,du^\beta,$$

and the second fundamental quadratic form is

$$L\,du^2 + 2M\,du\,dv + N\,dv^2 = \sum_{\alpha=1}^{2}\sum_{\beta=1}^{2} b_{\alpha\beta}\,du^\alpha\,du^\beta = b_{\alpha\beta}\,du^\alpha\,du^\beta.$$

The expressions for the coefficients of the first and second quadratic forms in the new notation become

$$g_{ij} = \mathbf{r}_i \cdot \mathbf{r}_j, \quad b_{ij} = \mathbf{r}_{ij} \cdot \mathbf{N},$$

where $\mathbf{N}$ is the unit normal vector to the surface; $i$ and $j$ are equal to either 1 or 2.

### 9.2.3-3. Derivation formula.

The *Christoffel symbols* $\Gamma_{k,ij}$ *of the first kind* are defined to be the scalar products of the vectors $\mathbf{r}_k$ and $\mathbf{r}_{ij}$; i.e.,

$$\mathbf{r}_k \cdot \mathbf{r}_{ij} = \Gamma_{k,ij}.$$

The Christoffel symbols of the first kind satisfy the formula

$$\Gamma_{k,ij} = \frac{1}{2}\left(\frac{\partial g_{ki}}{\partial u^j} + \frac{\partial g_{kj}}{\partial u^i} - \frac{\partial g_{ij}}{\partial u^k}\right).$$

This is one of the basic formulas in the theory of surfaces; this formula means that the scalar products of the second partial derivatives of the position vector $\mathbf{r}(u^i, u^j)$ by its partial derivatives can be expressed in terms of the coefficients of the first quadratic form (more precisely, in terms of their derivatives).

The *Christoffel symbols* $\Gamma_{ij}^k$ *of the second kind* are defined by the relations

$$\Gamma_{ij}^k = \sum_{t=1}^{n} g^{kt}\Gamma_{k,ij},$$

where $g^{kt}$ is given by

$$\sum_{k=1}^{n} g^{kt}g_{ks} = \begin{cases} 1, & \text{if } t = s, \\ 0, & \text{if } t \neq s. \end{cases}$$

The Christoffel symbols of the second kind are the coefficients in the decomposition of the vector $\mathbf{r}_{ij}$ in two noncollinear vectors $\mathbf{r}_1$ and $\mathbf{r}_2$ and the unit normal vector $\mathbf{N}$:

$$\mathbf{r}_{ij} = \Gamma_{ij}^1 \mathbf{r}_1 + \Gamma_{ij}^2 \mathbf{r}_2 + b_{ij}\mathbf{N} \quad (i, j = 1, 2). \tag{9.2.3.1}$$

Formulas (9.2.3.1) are called the *first group of derivation formulas (the Gauss derivation formulas)*.

The formulas

$$\mathbf{N}_1 = -b_1^1 \mathbf{r}_1 - b_1^2 \mathbf{r}_2, \quad \mathbf{N}_1 = -b_2^1 \mathbf{r}_1 - b_2^2 \mathbf{r}_2, \tag{9.2.3.2}$$

where $b_i^j = b_{i\alpha}g^{\alpha j}$, are called the *second group of derivation formulas (Weingarten formulas)*. The formulas in the second group of derivation formulas express the partial derivatives of the unit normal vector $\mathbf{N}$ in terms of the variables $u^1$ and $u^2$ in the decomposition in the basis vectors $\mathbf{r}_1$, $\mathbf{r}_2$, and $\mathbf{N}$.

Formulas (9.2.3.1) and (9.2.3.2) express the partial derivatives with respect to $u^1$ and $u^2$ of the basis vectors, i.e., of the two tangent vectors $\mathbf{r}_1$ and $\mathbf{r}_2$ and the normal vector $\mathbf{N}$, at a given point on a surface. These partial derivatives of $\mathbf{r}_1$, $\mathbf{r}_2$ and $\mathbf{N}$ are obtained as a decomposition in the vectors $\mathbf{r}_1$, $\mathbf{r}_2$ and $\mathbf{N}$ themselves.

### 9.2.3-4. Gauss formulas. Peterson–Codazzi formulas.

If the first quadratic form of the surface is given, then the second quadratic form cannot be chosen arbitrarily, since its discriminant $(LN - M^2)$ is completely determined by the *Gauss formula*

$$b_{11}b_{22} - b_{12}^2 = \frac{\partial^2 g_{12}}{\partial u^1 \partial u^2} - \frac{1}{2}\frac{\partial^2 g_{11}}{\partial u^2 \partial u^2} - \frac{1}{2}\frac{\partial^2 g_{22}}{\partial u^1 \partial u^1} + \Gamma_{12}^\gamma \Gamma_{12}^\delta g_{\gamma\delta} - \Gamma_{11}^\alpha \Gamma_{22}^\beta g_{\alpha\beta}, \tag{9.2.3.3}$$

where $\gamma$, $\delta$, $\alpha$, and $\beta$ are independent summation indices equal to either 1 or 2.

We use (9.2.3.3) to reduce relation (9.2.2.2) to the form

$$K(u,v) = k_1 k_2 = \frac{LN - M^2}{EG - F^2} = \frac{b_{11}b_{22} - b_{12}^2}{g_{11}g_{22} - g_{12}^2}. \qquad (9.2.3.4)$$

In view of (9.2.3.4), the Gaussian curvature of the surface is completely determined by the coefficients of the first quadratic form and by their first and second derivatives with respect to $u^1$ and $u^2$.

The coefficients $b_{11}$, $b_{12}$, $b_{22}$ of the second quadratic form and their first derivatives are related to the coefficients $g_{11}$, $g_{12}$, $g_{22}$ of the first quadratic form and their first derivatives (contained only in $\Gamma_{ij}^k$) by differential equations known as the *Peterson–Codazzi formulas*:

$$\frac{\partial b_{i1}}{\partial u^2} - \Gamma_{i2}^1 b_{11} - \Gamma_{i2}^2 b_{21} = \frac{\partial b_{i2}}{\partial u^1} - \Gamma_{i1}^1 b_{12} - \Gamma_{i1}^2 b_{22} \quad (i = 1, 2).$$

The Gauss formula and the Peterson–Codazzi formulas are necessary and sufficient conditions for two analytically determined quadratic differential forms to be the fist and second quadratic forms of some surface.

## References for Chapter 9

**Aubin, T.,** *A Course in Differential Geometry*, American Mathematical Society, Providendce, Rhoad Island, 2000.
**Burke, W. L.,** *Applied Differential Geometry*, Cambridge University Press, Cambridge, 1985.
**Byushgens, S. S.,** *Differential Geometry* [in Russian], Komkniga, Moscow, 2006.
**Chern, S.-S., Chen, W.-H., and Lam, K. S.,** *Lectures on Differential Geometry*, World Scientific Publishing Co., Hackensack, New Jersey, 2000.
**Danielson, D. A.,** *Vectors and Tensors in Engineering and Physics, 2nd Rep Edition*, Westview Press, Boulder, Colorado, 2003.
**Dillen, F. J. E. and Verstraelen, L. C. A.,** *Handbook of Differential Geometry, Vol. 1*, North Holland, Amsterdam, 2000.
**Dillen, F. J. E. and Verstraelen, L. C. A.,** *Handbook of Differential Geometry, Vol. 2*, North Holland, Amsterdam, 2006.
**Guggenheimer, H. W.,** *Differential Geometry*, Dover Publications, New York, 1977.
**Kay, D. C.,** *Schaum's Outline of Tensor Calculus*, McGraw-Hill, New York, 1988.
**Kobayashi, S. and Nomizu, K.,** *Foundations of Differential Geometry, Vol. 1*, Wiley, New York, 1996.
**Kreyszig, E.,** *Differential Geometry*, Dover Publications, New York, 1991.
**Lang, S.,** *Fundamentals of Differential Geometry*, Springer, New York, 2001.
**Lebedev, L. P. and Cloud, M. J.,** *Tensor Analysis*, World Scientific Publishing Co., Hackensack, New Jersey, 2003.
**Lovelock, D. and Rund, H.,** *Tensors, Differential Forms, and Variational Principles*, Dover Publications, New York, 1989.
**O'Neill, V.,** *Elementary Differential Geometry, Rev. 2nd Edition*, Academic Press, New York, 2006.
**Oprea, J.,** *Differential Geometry and Its Applications, 2nd Edition*, Prentice Hall, Englewood Cliffs, New Jersey, 2003.
**Pogorelov, A. V.,** *Differential Geometry*, P. Noordhoff, Groningen, 1967.
**Postnikov, M. M.,** *Linear Algebra and Differential Geometry* (Lectures in Geometry), Mir Publishers, Moscow, 1982.
**Pressley, A.,** *Elementary Differential Geometry*, Springer, New York, 2002.
**Rashevsky, P. K.,** *A Course in Differential Geometry, 4th Edition* [in Russian], URSS, Moscow, 2003.
**Simmonds, J. G.,** *A Brief on Tensor Analysis, 2nd Edition*, Springer, New York, 1997.
**Somasundaram, D.,** *Differential Geometry: A First Course*, Alpha Science International, Oxford, 2004.
**Spain, B.,** *Tensor Calculus: A Concise Course*, Dover Publications, New York, 2003.
**Spivak, M.,** *A Comprehensive Introduction to Differential Geometry. Vols 1–5, 3rd Edition*, Publish or Perish, Houston, 1999.
**Struik, D. J.,** *Lectures on Classical Differential Geometry, 2nd Edition*, Dover Publications, New York, 1988.
**Temple, G.,** *Cartesian Tensors: An Introduction*, Dover Publications, New York, 2004.

# Chapter 10
# Functions of Complex Variable

## 10.1. Basic Notions
### 10.1.1. Complex Numbers. Functions of Complex Variable

**10.1.1-1. Complex numbers.**

The set of complex numbers is an extension of the set of real numbers. An expression of the form $z = x + iy$, where $x$ and $y$ are real numbers, is called a *complex number*, and the symbol $i$ is called the *imaginary unit*: $i^2 = -1$. The numbers $x$ and $y$ are called, respectively, the *real* and *imaginary parts* of $z$ and denoted by

$$x = \operatorname{Re} z \quad \text{and} \quad y = \operatorname{Im} z. \tag{10.1.1.1}$$

The complex number $x + i0$ is identified with real number $x$, and the number $0 + iy$ is denoted by $iy$ and is said to be *pure imaginary*. Two complex numbers $z_1 = x_1 + iy_1$ and $z_2 = x_2 + iy_2$ are assumed to be *equal* if $x_1 = x_2$ and $y_1 = y_2$.

The complex number $\bar{z} = x - iy$ is said to be *conjugate* to the number $z$.

The *sum or difference* of complex numbers $z_1 = x_1 + iy_1$ and $z_2 = x_2 + iy_2$ is defined to be the number

$$z_1 \pm z_2 = x_1 \pm x_2 + i(y_1 \pm y_2). \tag{10.1.1.2}$$

*Addition laws:*
1. $z_1 + z_2 = z_2 + z_1$ (commutativity).
2. $z_1 + (z_2 + z_3) = (z_1 + z_2) + z_3$ (associativity).

The *product* $z_1 z_2$ of complex numbers $z_1 = x_1 + iy_1$ and $z_2 = x_2 + iy_2$ is defined to be the number

$$z_1 z_2 = (x_1 x_2 - y_1 y_2) + i(x_1 y_2 - x_2 y_1). \tag{10.1.1.3}$$

*Multiplication laws:*
1. $z_1 z_2 = z_2 z_1$ (commutativity).
2. $z_1(z_2 z_3) = (z_1 z_2)z_3$ (associativity).
3. $(z_1 + z_2)z_3 = z_1 z_3 + z_2 z_3$ (distributivity with respect to addition).

The product of a complex number $z = x + iy$ by its conjugate is always nonnegative:

$$z\bar{z} = x^2 + y^2. \tag{10.1.1.4}$$

For a positive integer $n$, the $n$-fold product of $z$ by itself is called the $n$th *power* of the number $z$ and is denoted by $z^n$. A number $w$ is called an $n$th *root* of a number $z$ and is denoted by $w = \sqrt[n]{z}$ if $w^n = z$.

If $z_2 \neq 0$, then the *quotient* of $z_1$ and $z_2$ is defined as

$$\frac{z_1}{z_2} = \frac{x_1 x_2 + y_1 y_2}{x_2^2 + y_2^2} + i\frac{x_2 y_1 - x_1 y_2}{x_2^2 + y_2^2}. \tag{10.1.1.5}$$

Relation (10.1.1.5) can be obtained by multiplying the numerator and the denominator of the fraction $z_1/z_2$ by $\bar{z}_2$.

### 10.1.1-2. Geometric interpretation of complex number.

There is a one-to-one correspondence between complex numbers $z = x + iy$ and points $M$ with coordinates $(x, y)$ on the plane with a Cartesian rectangular coordinate system $OXY$ or with vectors $\overrightarrow{OM}$ connecting the origin $O$ with $M$ (Fig. 10.1). The length $r$ of the vector $\overrightarrow{OM}$ is called the *modulus* of the number $z$ and is denoted by $r = |z|$, and the angle $\varphi$ formed by the vector $\overrightarrow{OM}$ and the positive direction of the $OX$-axis is called the *argument* of the number $z$ and is denoted by $\varphi = \text{Arg } z$.

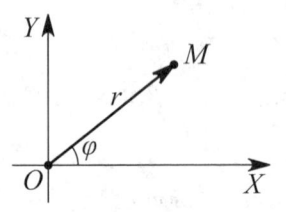

**Figure 10.1.** Geometric interpretation of complex number.

The modulus of a complex number is determined by the formula

$$|z| = \sqrt{z\bar{z}} = \sqrt{x^2 + y^2}. \tag{10.1.1.6}$$

The argument $\text{Arg } z$ is determined up to a multiple of $2\pi$, $\text{Arg } z = \arg z + 2k\pi$, where $k$ is an arbitrary integer and $\arg z$ is the *principal value* of $\text{Arg } z$ determined by the condition $-\pi < \arg z \leq \pi$. The principal value $\arg z$ is given by the formula

$$\arg z = \begin{cases} \arctan(y/x) & \text{for } x > 0, \\ \pi + \arctan(y/x) & \text{for } x < 0,\, y \geq 0, \\ -\pi + \arctan(y/x) & \text{for } x < 0,\, y < 0, \\ \pi/2 & \text{for } x = 0,\, y > 0, \\ -\pi/2 & \text{for } x = 0,\, y < 0. \end{cases} \tag{10.1.1.7}$$

For $z = 0$, $\text{Arg } z$ is undefined.

Since $x = r \cos \varphi$ and $y = r \sin \varphi$, it follows that the complex number can be written in the *trigonometric form*

$$z = x + iy = r(\cos \varphi + i \sin \varphi). \tag{10.1.1.8}$$

For numbers $z_1 = r_1(\cos \varphi_1 + i \sin \varphi_1)$ and $z_2 = r_2(\cos \varphi_2 + i \sin \varphi_2)$, written in trigonometric form, the following rules of algebraic operations are valid:

$$\begin{aligned} z_1 z_2 &= r_1 r_2 \big[\cos(\varphi_1 + \varphi_2) + i \sin(\varphi_1 + \varphi_2)\big], \\ \frac{z_1}{z_2} &= \frac{r_1}{r_2} \big[\cos(\varphi_1 - \varphi_2) + i \sin(\varphi_1 - \varphi_2)\big]. \end{aligned} \tag{10.1.1.9}$$

In the latter formula, it is assumed that $z \neq 0$. For any positive integer $n$, this implies the *de Moivre formula*

$$z^n = r^n(\cos n\varphi + i \sin n\varphi), \tag{10.1.1.10}$$

as well as the *formula for extracting the root of a complex number*. For $z \neq 0$, there are exactly $n$ distinct values of the $n$th root of the number $z = r(\cos \varphi + i \sin \varphi)$. They are determined by the formulas

$$\sqrt[n]{z} = \sqrt[n]{r}\left(\cos \frac{\varphi + 2k\pi}{n} + i \sin \frac{\varphi + 2k\pi}{n}\right) \quad (k = 0, 1, 2, \ldots, n-1). \tag{10.1.1.11}$$

**Example.** Let us find all values of $\sqrt[3]{i}$.

We represent the complex number $z = i$ in trigonometric form. We have $r = 1$ and $\varphi = \arg z = \frac{1}{2}\pi$. The distinct values of the third root are calculated by the formula

$$\omega_k = \sqrt[3]{i}\left(\cos\frac{\frac{\pi}{2} + 2\pi k}{3} + i\sin\frac{\frac{\pi}{2} + 2\pi k}{3}\right) \quad (k = 0, 1, 2),$$

so that

$$\omega_0 = \cos\frac{\pi}{6} + i\sin\frac{\pi}{6} = \frac{\sqrt{3}}{2} + i\frac{1}{2},$$

$$\omega_1 = \cos\frac{5\pi}{6} + i\sin\frac{5\pi}{6} = -\frac{\sqrt{3}}{2} + i\frac{1}{2},$$

$$\omega_2 = \cos\frac{3\pi}{2} + i\sin\frac{3\pi}{2} = -i.$$

The roots are shown in (Fig. 10.2).

**Figure 10.2.** The roots of $\sqrt[3]{i}$.

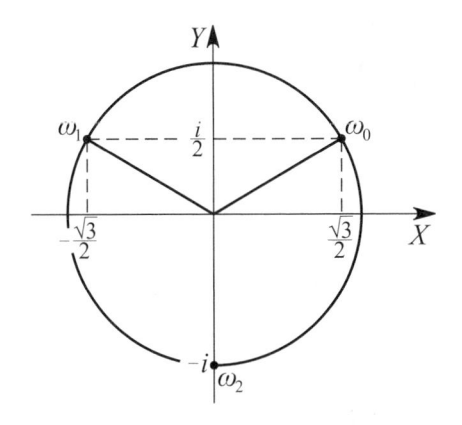

**Figure 10.3.** The sum and difference of complex numbers.

The plane $OXY$ is called the *complex plane*, the axis $OX$ is called the *real axis*, and the axis $OY$ is called the *imaginary axis*. The notions of complex number and point on the complex plane are identical.

The *geometric meaning of the operations of addition and subtraction of complex numbers* is as follows: the sum and the difference of complex numbers $z_1$ and $z_2$ are the vectors equal to the directed diagonals of the parallelogram spanned by the vectors $z_1$ and $z_2$ (Fig. 10.3). The following inequalities hold (Fig. 10.3):

$$|z_1 + z_2| \leq |z_1| + |z_2|, \quad |z_1 - z_2| \geq \big||z_1| - |z_2|\big|. \tag{10.1.1.12}$$

Inequalities (10.1.1.12) become equalities if and only if the arguments of the complex numbers $z_1$ and $z_2$ coincide (i.e., $\arg z_1 = \arg z_2$) or one of the numbers is zero.

## 10.1.2. Functions of Complex Variable

**10.1.2-1. Notion of function of complex variable.**

A subset $D$ of the complex plane such that each point of $D$ has a neighborhood contained in $D$ (i.e., $D$ is open) and two arbitrary points of $D$ can be connected by a broken line lying in $D$ (i.e., $D$ is connected) is called a *domain* on the complex plane. A point that does not itself lie in $D$ but whose arbitrary neighborhood contains points of $D$ is called a *boundary point* of $D$. The set of all boundary points of $D$ is called the *boundary* of $D$, and the union of $D$ with its boundary is called a *closed domain* and is denoted by $\overline{D}$. The *positive sense* of the boundary is defined to be the sense for which the domain lies to the left of the boundary. The boundary of a domain can consist of finitely many closed curves, segments, and points; the curves and cuts are assumed to be piecewise smooth.

The simplest examples of domains are neighborhoods of points on the complex plane. A *neighborhood* of a point $a$ on the complex plane is understood as the set of points $z$ such that $|z - a| < R$, i.e., the interior of the disk of radius $R > 0$ centered at the point $a$. The *extended complex plane* is obtained by augmenting the complex plane with the fictitious *point at infinity*. A *neighborhood of the point at infinity* is understood as the set of points $z$ such that $|z| > R$ (including the point at infinity itself).

If to each point $z$ of a domain $D$ there corresponds a point $w$ (resp., a set of points $w$), then one says that there is a *single-valued* (resp., *multivalued*) *function* $w = f(z)$ defined on the domain $D$. If we set $z = x + iy$ and $w = u + iv$, then defining a function $w = f(z)$ of the complex variable is equivalent to defining two functions $\operatorname{Re} f = u = u(x, y)$ and $\operatorname{Im} f = v = v(x, y)$ of two real variables. If the function $w = f(z)$ is single-valued on $D$ and the images of distinct points of $D$ are distinct, then the mapping determined by this function is said to be *schlicht*. The notions of boundedness, limit, and continuity for single-valued functions of the complex variable do not differ from the corresponding notions for functions of two real variables.

### 10.1.2-2. Differentiability and analyticity.

Let a single-valued function $w = f(z)$ be defined in a neighborhood of a point $z$. If there exists a limit
$$\lim_{h \to 0} \frac{f(z+h) - f(z)}{h} = f'_z(z), \qquad (10.1.2.1)$$
then the function $w = f(z)$ is said to be *differentiable* at the point $z$ and $f'_z(z)$ is called its *derivative* at the point $z$.

*Cauchy–Riemann conditions.* If the functions $u(x, y) = \operatorname{Re} f(z)$ and $v(x, y) = \operatorname{Im} f(z)$ are differentiable at a point $(x, y)$, then the *Cauchy–Riemann conditions*
$$\frac{\partial u}{\partial x} = \frac{\partial v}{\partial y}, \quad \frac{\partial u}{\partial y} = -\frac{\partial v}{\partial x} \qquad (10.1.2.2)$$
are necessary and sufficient for the function $w = f(z)$ to be differentiable at the point $z = x+iy$.

If the function $w = f(z)$ is differentiable, then
$$w'_z = u_x + iv_x = v_y - iu_y = u_x + iu_y = v_y + iv_x, \qquad (10.1.2.3)$$
where the subscripts $x$ and $y$ indicate the corresponding partial derivatives.

*Remark.* The Cauchy–Riemann conditions are sometimes also called the d'Alembert–Euler conditions.

The rules for algebraic operations on the derivatives and for calculating the derivative of the composite function and the inverse function (if it exists) have exactly the same form as in the real case:

1. $\big[\alpha f_1(z) \pm \beta f_2(z)\big]'_z = \alpha [f_1(z)]'_z \pm \beta [f_2(z)]'_z$, where $\alpha$ and $\beta$ are arbitrary complex constants.
2. $\big[f_1(z) f_2(z)\big]'_z = [f_1(z)]'_z f_2(z) + f_1(z)[f_2(z)]'_z$.
3. $\left[\dfrac{f_1(z)}{f_2(z)}\right]'_z = \dfrac{[f_1(z)]'_z f_2(z) - f_1(z)[f_2(z)]'_z}{f_2^2(z)}$ $\quad (f_2(z) \neq 0)$.
4. If a function $w = f(z)$ is differentiable at a point $z$ and a function $W = F(w)$ is differentiable at the point $w = f(z)$, then the composite function $W = F(f(z))$ is differentiable at the point $z$ and $W'_z = [F(f(z))]'_z = F'_f(f) f'_z(z)$.

5. If a function $w=f(z)$ is differentiable at a point $z$ and the inverse function $z=g(w)\equiv f^{-1}(w)$ exists and is differentiable at the point $w$ and satisfies $[f^{-1}(w)]'_w \neq 0$, then

$$f'_z(z) = \frac{1}{[f^{-1}(w)]'_w}.$$

A single-valued function differentiable in some neighborhood of a point $z_0$ is said to be *analytic (regular, holomorphic)* at this point.

A function $w = f(z)$ is analytic at a point $z_0$ if and only if it can be represented by a power series

$$f(z) = \sum_{k=0}^{\infty} c_k(z - z_0)^k \qquad (10.1.2.4)$$

converging in some neighborhood of $z_0$.

A function analytic at each point of the domain $D$ is said to be *analytic* in $D$.

A function $w = f(z)$ is said to be *analytic at the point at infinity* if the function $F(z) = f(1/z)$ is analytic at the point $z = 0$. In this case, $f'_z(\infty) = (-z^2 F'_z)\big|_{z=0}$ by definition.

A function $w = f(z)$ is analytic at the point at infinity if and only if this function can be represented by a power series

$$f(z) = \sum_{k=0}^{\infty} b_k z^{-k} \qquad (10.1.2.5)$$

converging for sufficiently large $|z|$.

If a function $w = f(z)$ is analytic at a point $z_0$ and $f'_z(z_0) \neq 0$, then $f(z)$ has an analytic inverse function $z(w)$ defined in a neighborhood of the point $w_0 = f(z_0)$. If a function $w = f(z)$ is analytic at a point $z_0$ and the function $W = F(w)$ is analytic at the point $w_0 = f(z_0)$, then the composite function $W = F[f(z)]$ is analytic at the point $z_0$. If a function is analytic in a domain $D$ and continuous in $\overline{D}$, then its value at any interior point of the domain is uniquely determined by its values on the boundary of the domain. The analyticity of a function at a point implies the existence and analyticity of its derivatives of arbitrary order at this point.

MAXIMUM MODULUS PRINCIPLE. *If a function $w = f(z)$ that is not identically constant is analytic in a domain $D$ and continuous in $\overline{D}$, then its modulus cannot attain a maximum at an interior point of $D$.*

LIOUVILLE'S THEOREM. *If a function $w = f(z)$ is analytic and bounded in the entire complex plane, then it is constant.*

Remark. The Liouville theorem can be stated in the following form:
*if a function $w = f(z)$ is analytic in the extended complex plane, then it is constant.*

*Geometric meaning of the absolute value of the derivative.* Suppose that a function $w = f(z)$ is analytic at a point $z_0$ and $f'_z(z_0) \neq 0$. Then the value $|f'_z(z_0)|$ determines the dilatation (similarity) coefficient at the point $z_0$ under the mapping $w = f(z)$. The value $|f'_z(z_0)|$ is called the *dilatation ratio* if $|f'_z(z_0)| > 1$ and the *contraction ratio* if $|f'_z(z_0)| < 1$.

*Geometric meaning of the argument of the derivative.* The argument of the derivative $f'_z(z_0)$ is equal to the angle by which the tangent at the point $z_0$ to any curve passing through $z_0$ should be rotated to give the tangent to the image of the curve at the point $w_0 = f(z_0)$. For $\varphi = \arg f'_z(z) > 0$, the rotation is anticlockwise, and for $\varphi = \arg f'_z(z) < 0$, the rotation is clockwise.

Single-valued functions, as well as single-valued branches of multi-valued functions, are analytic everywhere on the domains where they are defined. It follows from (10.1.2.2) that the real and imaginary parts $u(x,y)$ and $v(x,y)$ of a function analytic in a domain are *harmonic* in this domain, i.e., satisfy the Laplace equation

$$\Delta f = f_{xx} + f_{yy} = 0 \qquad (10.1.2.6)$$

in this domain.

**Remark.** If $u(x,y)$ and $v(x,y)$ are two arbitrary harmonic functions, then the function $f(z) = u(x,y) + iv(x,y)$ is not necessarily analytic, since for the analyticity of $f(z)$ the functions $u(x,y)$ and $v(x,y)$ must satisfy the Cauchy–Riemann conditions.

**Example 1.** The function $w = z^2$ is analytic.
Indeed, since $z = x + iy$, we have $w = (x+iy)^2 = x^2 - y^2 + i2xy$, $u(x,y) = x^2 - y^2$, and $v(x,y) = 2xy$. The Cauchy–Riemann conditions

$$u_x = v_y = 2x, \qquad u_y = v_x = -2y$$

are satisfied at all points of the complex plane, and the function $w = z^2$ is analytic.

**Example 2.** The function $w = \bar{z}$ is not analytic.
Indeed, since $z = x + iy$, we have $w = x - iy$, $u(x,y) = x$, $v(x,y) = -y$. The Cauchy–Riemann conditions are not satisfied,

$$u_x = 1 \neq -1 = v_y, \qquad u_y = v_x = 0,$$

and the function $w = \bar{z}$ is not analytic.

### 10.1.2-3. Elementary functions.

1°. The *functions* $w = z^n$ *and* $w = \sqrt[n]{z}$ for positive integer $n$ are defined in Paragraph 10.1.1-2. The function

$$w = z^n \qquad (10.1.2.7)$$

is single-valued. It is schlicht in the sectors $2\pi k/n < \varphi < 2\pi(k+1)/n$, $k = 0, 1, 2, \ldots$, each of which is transformed by the mapping $w = z^n$ into the plane $w$ with a cut on the positive semiaxis.

The function

$$w = \sqrt[n]{z} \qquad (10.1.2.8)$$

is an $n$-valued function for $z \neq 0$, and its value is determined by the value of the argument chosen for the point $z$. If a closed curve $C$ does not surround the point $z = 0$, then, as the point $z$ goes around the entire curve $C$, the point $w = \sqrt[n]{z}$ for a chosen value of the root also moves along a closed curve and returns to the initial value of the argument. But if the curve $C$ surrounds the origin, then, as the point $z$ goes around the entire curve $C$ in the positive sense, the argument of $z$ increases by $2\pi$ and the corresponding point $w = \sqrt[n]{z}$ does not return to the initial position. It will return there only after the point $z$ goes $n$ times around the entire curve $C$. If a domain $D$ does not contain a closed curve surrounding the point $z = 0$, then one can singe out $n$ continuous single-valued functions, each of which takes only one of the values $w = \sqrt[n]{z}$; these functions are called the *branches* of the multi-valued function $w = \sqrt[n]{z}$. One cannot single $n$ separate branches of the function $w = \sqrt[n]{z}$ in any neighborhood of the point $z = 0$; accordingly, the point $z = 0$ is called a *branch point* of this function.

2°. The *Zhukovskii function*

$$w = \frac{1}{2}\left(z + \frac{1}{z}\right) \qquad (10.1.2.9)$$

is defined and single-valued for all $z \neq 0$; it is schlicht in any domain that does not simultaneously contain any points $z_1$ and $z_2$ such that $z_1 z_2 = 1$.

3°. The *exponential function* $w = e^z$ is defined by the formula

$$w = e^z = e^{x+iy} = e^x(\cos y + i \sin y). \tag{10.1.2.10}$$

The function $w = e^z$ is analytic everywhere. For the exponential function, the usual differentiation rule is preserved:

$$(e^z)'_z = e^z.$$

The basic property of the exponential function (*addition theorem*) is also preserved:

$$e^{z_1} e^{z_2} = e^{z_1+z_2}.$$

For $x = 0$ and $y = \varphi$, the definition of the exponential function implies the *Euler formula* $e^{i\varphi} = \cos\varphi + i\sin\varphi$, which permits one to write any complex number with modulus $r$ and argument $\varphi$ in the *exponential form*

$$z = r(\cos\varphi + i\sin\varphi) = re^{i\varphi}. \tag{10.1.2.11}$$

The exponential function is $2\pi$-periodic, and the mapping $w = e^z$ is schlicht in the strip $0 \le y < 2\pi$.

4°. The *logarithm* is defined as the inverse of the exponential function: if $e^w = z$, then

$$w = \operatorname{Ln} z. \tag{10.1.2.12}$$

This function is defined for $z \neq 0$. The logarithm satisfies the following relations:

$$\operatorname{Ln} z_1 + \operatorname{Ln} z_2 = \operatorname{Ln}(z_1 z_2), \quad \operatorname{Ln} z_1 - \operatorname{Ln} z_2 = \operatorname{Ln}\frac{z_1}{z_2},$$

$$\operatorname{Ln}(z^n) = n \operatorname{Ln} z, \quad \operatorname{Ln}\sqrt[n]{z} = \frac{1}{n}\operatorname{Ln} z.$$

The exponential form of complex numbers readily shows that the logarithm is infinite-valued:

$$\operatorname{Ln} z = \ln|z| + i \operatorname{Arg} z = \ln|z| + i\arg z + 2\pi k i, \quad k = 0, \pm 1, \pm 2, \dots \tag{10.1.2.13}$$

The value $\ln z = \ln|z| + i\arg z$ is taken to be the principal value of this function. Just as with the function $w = \sqrt[n]{z}$, we see that if the point $z = 0$ is surrounded by a closed curve $C$, then the point $w = \operatorname{Ln} z$ does not return to its initial position after $z$ goes around $C$ in the positive sense, since the argument of $w$ increases by $2\pi i$. Thus if a domain $D$ does not contain a closed curve surrounding the point $z = 0$, then in $D$ one can single out infinitely many continuous and single-valued branches of the multi-valued function $w = \operatorname{Ln} z$; the differences between the values of these branches at each point of the domain have the form $2\pi k i$, where $k$ is an integer. This cannot be done in an arbitrary neighborhood of the point $z = 0$, and this point is called a *branch point* of the logarithm.

5°. *Trigonometric functions* are defined in terms of the exponential function as follows:

$$\cos z = \frac{e^{iz} + e^{-iz}}{2}, \qquad \sin z = \frac{e^{iz} - e^{-iz}}{2i},$$
$$\tan z = \frac{\sin z}{\cos z} = -i\frac{e^{iz} - e^{-iz}}{e^{iz} + e^{-iz}}, \qquad \cot z = \frac{\cos z}{\sin z} = i\frac{e^{iz} + e^{-iz}}{e^{iz} - e^{-iz}}. \tag{10.1.2.14}$$

Properties of the functions $\cos z$ and $\sin z$:
1. They are analytic for any $z$.
2. The usual differentiation rules are valid:
$$(\sin z)'_z = \cos z, \quad (\cos z)'_z = \sin z.$$
3. They are periodic with real period $T = 2\pi$.
4. $\sin z$ is an odd function, and $\cos z$ is an even function.
5. In the complex domain, they are unbounded.
6. The usual trigonometric relations hold:
$$\cos^2 z + \sin^2 z = 1, \quad \cos 2z = \cos^2 z - \sin^2 z, \quad \text{etc.}$$

The function $\tan z$ is analytic everywhere except for the points
$$z_k = \frac{\pi}{2} + k\pi, \quad k = 0, \pm 1, \pm 2, \ldots,$$
and the function $\cot z$ is analytic everywhere except for the points
$$z_k = k\pi, \quad k = 0, \pm 1, \pm 2, \ldots$$

The functions $\tan z$ and $\cot z$ are periodic with real period $T = \pi$.

6°. *Hyperbolic functions* are defined by the formulas

$$\cosh z = \frac{e^z + e^{-z}}{2}, \qquad \sinh z = \frac{e^z - e^{-z}}{2}, \qquad (10.1.2.15)$$
$$\tanh z = \frac{\sinh z}{\cosh z} = \frac{e^z - e^{-z}}{e^z + e^{-z}}, \qquad \coth z = \frac{\cosh z}{\sinh z} = \frac{e^z + e^{-z}}{e^z - e^{-z}}.$$

For real values of the argument, each of these functions coincides with the corresponding real function. Hyperbolic and trigonometric functions are related by the formulas
$$\cosh z = \cos iz, \quad \sinh z = -i \sin iz, \quad \tanh z = -i \tan iz, \quad \coth z = i \cot iz.$$

7°. *Inverse trigonometric and hyperbolic functions* are expressed via the logarithm and hence are infinite-valued:

$$\begin{aligned}
\operatorname{Arccos} z &= -i \operatorname{Ln}(z + \sqrt{z^2 - 1}), & \operatorname{Arcsin} z &= -i \operatorname{Ln}(iz + \sqrt{1 - z^2}), \\
\operatorname{Arctan} z &= -\frac{i}{2} \operatorname{Ln} \frac{1 + iz}{1 - iz}, & \operatorname{Arccot} z &= -\frac{i}{2} \operatorname{Ln} \frac{z + i}{z - i}, \\
\operatorname{arccosh} z &= \operatorname{Ln}(z + \sqrt{z^2 - 1}), & \operatorname{arcsinh} z &= \operatorname{Ln}(z + \sqrt{z^2 - 1}), \\
\operatorname{arctanh} z &= \frac{1}{2} \operatorname{Ln} \frac{1 + z}{1 - z}, & \operatorname{arccoth} z &= \frac{1}{2} \operatorname{Ln} \frac{z + 1}{z - 1}.
\end{aligned} \qquad (10.1.2.16)$$

The principal value of each of these functions is obtained by choosing the principal value of the corresponding logarithmic function.

8°. The *power* $w = z^\gamma$ is defined by the relation
$$z^\gamma = e^{\gamma \operatorname{Ln} z}, \qquad (10.1.2.17)$$
where $\gamma = \alpha + i\beta$ is an arbitrary complex number. Substituting $z = re^{i\varphi}$ into (10.1.2.17), we obtain
$$z^\gamma = e^{\alpha \ln r - \beta(\varphi + 2k\pi)} e^{i\alpha(\varphi + 2k\pi) + i\beta \ln r}, \qquad k = 0, \pm 1, \pm 2, \ldots \qquad (10.1.2.18)$$

It follows from relation (10.1.2.18) that the function $w = z^\gamma$ has infinitely many values for $\beta \neq 0$.

**9°.** The *general exponential function* is defined by the formula

$$w = \gamma^z = e^{z \operatorname{Ln} \gamma} = e^{z \ln |\gamma|} e^{zi \operatorname{Arg} \gamma}, \qquad (10.1.2.19)$$

where $\gamma = \alpha + i\beta$ is an arbitrary nonzero complex number. The function (10.1.2.19) is a set of separate mutually independent single-valued functions that differ from one another by the factors $e^{2k\pi i z}$, $k = 0, \pm 1, \pm 2, \ldots$

**Example 3.** Let us calculate the values of some elementary functions at specific points:
1. $\cos 2i = \frac{1}{2}(e^{2ii} + e^{-2ii}) = \frac{1}{2}(e^2 + e^{-2}) = \cosh 2 \approx 3.7622$.
2. $\ln(-2) = \ln 2 + i\pi$, since $|-2| = 2$ and the principal value of the argument is equal to $\pi$.
3. $\operatorname{Ln}(-2)$ is calculated by formula (10.1.2.13):

$$\operatorname{Ln}(-2) = \ln 2 + i\pi + 2\pi ki = \ln 2 + (1 + 2k)i\pi \quad (k = 0, \pm 1, \pm 2, \ldots).$$

4. $i^i = e^{i \operatorname{Ln} i} = e^{i(i\pi/2 + 2\pi k)} = e^{-\pi/2 - 2\pi k} \quad (k = 0, \pm 1, \pm 2, \ldots)$.

The main elementary functions $w = f(z) = u(x, y) + iv(x, y)$ of the complex variable $z = x + iy$ are given in Table 10.1.

---

**10.1.2-4. Integration of function of complex variable.**

---

Suppose that an oriented curve $C$ connecting points $z = a$ and $z = b$ is given on the complex plane and a function $w = f(z)$ of the complex variable is defined on the curve. We divide the curve $C$ into $n$ parts, $a = z_0, z_1, \ldots, z_{n-1}, z_n = b$, arbitrarily choose $\xi_k \in [z_k, z_{k+1}]$, and compose the integral sum

$$\sum_{k=0}^{n-1} f(\xi_k)(z_{k+1} - z_k).$$

If there exists a limit of this sum as $\max |z_{k+1} - z_k| \to 0$, independent of the construction of the partition and the choice of points $\xi_k$, then this limit is called the *integral* of the function $w = f(z)$ over the curve $C$ and is denoted by

$$\int_C f(z)\, dz. \qquad (10.1.2.20)$$

Properties of the integral of a function of a complex variable:
1. If $\alpha, \beta$ are arbitrary constants, then $\int_C [\alpha f(z) + \beta g(z)]\, dz = \alpha \int_C f(z)\, dz + \beta \int_C g(z)\, dz$.
2. If $\widetilde{C}$ is the same curve as $C$ but with the opposite sense, then $\int_{\widetilde{C}} f(z)\, dz = -\int_C f(z)\, dz$.
3. If $C = C_1 \cup \cdots \cup C_n$, then $\int_C f(z)\, dz = \int_{C_1} f(z)\, dz + \ldots + \int_{C_n} f(z)\, dz$.
4. If $|f(z)| \leq M$ at all points of the curve $C$, then the following *estimate of the absolute value of the integral* holds: $\left|\int_C f(z)\, dz\right| \leq Ml$, where $l$ is the length of the curve $C$.

If $C$ is a piecewise smooth curve and $f(z)$ is bounded and piecewise continuous, then the integral (10.1.2.20) exists. If $z = x + iy$ and $w = u(x, y) + iv(x, y)$, then the computation of the integral (10.1.2.20) is reduced to finding two ordinary curvilinear integrals:

$$\int_C f(z)\, dz = \int_C u(x, y)\, dx - v(x, y)\, dy + i \int_C v(x, y)\, dx + u(x, y)\, dy. \qquad (10.1.2.21)$$

**Remark.** Formula (10.1.2.21) can be rewritten in a form convenient for memorizing:

$$\int_C f(z)\, dz = \int_C (u + iv)(dx + i\, dy).$$

TABLE 10.1

Main elementary functions $w = f(z) = u(x,y) + iv(x,y)$ of complex variable $z = x + iy$

| No. | Complex function $w = f(z)$ | Algebraic form $f(z) = u(x,y) + iv(x,y)$ | Zeros of $n$th order | Singularities |
|---|---|---|---|---|
| 1 | $z$ | $x + iy$ | $z = 0$, $n = 1$ | $z = \infty$ is a first-order pole |
| 2 | $z^2$ | $x^2 - y^2 + i\,2xy$ | $z = 0$, $n = 2$ | $z = \infty$ is a second-order pole |
| 3 | $\dfrac{1}{z - (x_0 + iy_0)}$ ($x_0$, $y_0$ are real numbers) | $\dfrac{x - x_0}{(x-x_0)^2 + (y-y_0)^2} + i\dfrac{-(y-y_0)}{(x-x_0)^2 + (y-y_0)^2}$ | $z = \infty$, $n = 1$ | $z = x_0 + iy_0$ is a first-order pole |
| 4 | $\dfrac{1}{z^2}$ | $\dfrac{x^2 - y^2}{(x^2+y^2)^2} + i\dfrac{-2xy}{(x^2+y^2)^2}$ | $z = \infty$, $n = 2$ | $z = 0$ is a second-order pole |
| 5 | $\sqrt{z}$ | $\pm\left[\left(\dfrac{x+\sqrt{x^2+y^2}}{2}\right)^{1/2} + i\left(\dfrac{-x+\sqrt{x^2+y^2}}{2}\right)^{1/2}\right]$ | $z = 0$ is a branch point | $z = 0$ is a first-order branch point; $z = \infty$ is a first-order branch point |
| 6 | $e^z$ | $e^x \cos y + ie^x \sin y$ | — | $z = \infty$ is an essential singular point |
| 7 | $\operatorname{Ln} z$ | $\ln|z| + i(\arg z + 2k\pi)$, $k = 0, \pm 1, \pm 2, \ldots$ | $z = 1$, $n = 1$ (for the branch corresponding to $k = 0$) | Logarithmic branch points for $z = 0$, $z = \infty$ |
| 8 | $\sin z$ | $\sin x \cosh y + i \cos x \sinh y$ | $z = \pi k$, $n = 1$ ($k = 0, \pm 1, \pm 2, \ldots$) | $z = \infty$ is an essential singular point |
| 9 | $\cos z$ | $\cos x \cosh y + i(-\sin x \sinh y)$ | $z = \tfrac{1}{2}\pi + \pi k$, $n = 1$ ($k = 0, \pm 1, \pm 2, \ldots$) | $z = \infty$ is an essential singular point |
| 10 | $\tan z$ | $\dfrac{\sin 2x}{\cos 2x + \cosh 2y} + i\dfrac{\sinh 2y}{\cos 2x + \cosh 2y}$ | $z = \pi k$, $n = 1$ ($k = 0, \pm 1, \pm 2, \ldots$) | $z = \tfrac{1}{2}\pi + \pi k$ ($k = 0, \pm 1, \pm 2, \ldots$) are first-order poles |

If the curve $C$ is given by the parametric equations $x = x(t)$, $y = y(t)$ ($t_1 \leq t \leq t_2$), then

$$\int_C f(z)\, dz = \int_{t_1}^{t_2} f(z(t)) z'_t(t)\, dt, \qquad (10.1.2.22)$$

where $z = z(t) = x(t) + iy(t)$ is the complex parametric equation of the curve $C$.

If $f(z)$ is an analytic function in a simply connected domain $D$ containing the points $z = a$ and $z = b$, then the *Newton–Leibniz formula* holds:

$$\int_a^b f(z)\, dz = F(b) - F(a), \qquad (10.1.2.23)$$

where $F(z)$ is a primitive of the function $f(z)$, i.e., $F'_z(z) = f(z)$ in the domain $D$.

If $f(z)$ and $g(z)$ are analytic functions in a simply connected domain $D$ and $z = a$ and $z = b$ are arbitrary points of the domain $D$, then the *formula of integration by parts* holds:

$$\int_a^b f(z)\, dg(z) = f(b)g(b) - f(a)g(a) - \int_a^b g(z)\, df(z). \tag{10.1.2.24}$$

If an analytic function $z = g(w)$ determines a single-valued mapping of a curve $\widetilde{C}$ onto a curve $C$, then

$$\int_C f(z)\, dz = \int_{\widetilde{C}} f(g(w)) g'_w(w)\, dw. \tag{10.1.2.25}$$

CAUCHY'S THEOREM FOR A SIMPLY CONNECTED DOMAIN. *If a function $f(z)$ is analytic in a simply connected domain $D$ bounded by a contour $C$ and is continuous in $\overline{D}$, then $\int_C f(z)\, dz = 0$.*

CAUCHY'S THEOREM FOR A MULTIPLY CONNECTED DOMAIN. *If a function $f(z)$ is analytic in a multiply connected domain $D$ bounded by a contour $\Gamma$ consisting of several closed curves and is continuous in $\overline{D}$, then $\int_\Gamma f(z)\, dz = 0$ provided that the sense of all curves forming $\Gamma$ is chosen in such a way that the domain $D$ lies to one side of the contour.*

If a function $f(z)$ is analytic in an $n$-connected domain $D$ and continuous in $\overline{D}$, and $C$ is the boundary of $D$, then for any interior point $z$ of this domain the *Cauchy integral formula* holds:

$$f(z) = \frac{1}{2\pi i} \int_C \frac{f(\xi)}{\xi - z}\, d\xi. \tag{10.1.2.26}$$

(Here integration is in the positive sense of $C$; i.e., the domain $D$ lies to the left of $C$.) Under the same assumptions as above, formula (10.1.2.26) implies expressions for the value of the derivative of arbitrary order of the function $f(z)$ at any interior point $z$ of the domain:

$$f_z^{(n)}(z) = \frac{n!}{2\pi i} \int_C \frac{f(\xi)}{(\xi - z)^{n+1}}\, d\xi \qquad (n = 1, 2, \ldots). \tag{10.1.2.27}$$

For an arbitrary smooth curve $C$, not necessarily closed, and for a function $f(\xi)$ everywhere continuous on $C$, possibly except for finitely many points at which this function has an integrable discontinuity, the right-hand side of formula (10.1.2.26) defines a *Cauchy-type integral*. The function $F(z)$ determined by a Cauchy-type integral is analytic at any point that does not belong to $C$. If $C$ divides the plane into several domains, then the Cauchy-type integral generally determines different analytic functions in these domains.

Formulas (10.1.2.26) and (10.1.2.27) allow one to calculate the integrals

$$\int_C \frac{f(\xi)}{\xi - z}\, d\xi = 2\pi i f(z), \qquad \int_C \frac{f(\xi)}{(\xi - z)^n}\, d\xi = \frac{2\pi i}{n!} f_z^{(n)}(z). \tag{10.1.2.28}$$

**Example 4.** Let us calculate the integral

$$\int_C \operatorname{Im} z\, dz,$$

where $C$ is the semicircle $|z| = 1$, $0 \leq \arg z \leq \pi$ (Fig. 10.4).

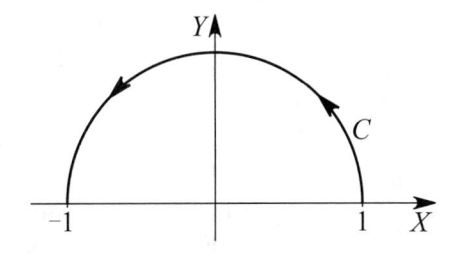

**Figure 10.4.** The semicircle $|z| = 1$, $0 \leq \arg z \leq \pi$.

Using formula (10.1.2.21), we obtain

$$\int_C \operatorname{Im} z\, dz = \int_C y(dx + i\, dy) = \int_C y\, dx + i \int_C y\, dy = \int_1^{-1} \sqrt{1-x^2}\, dx - i \int_1^{-1} x\, dx = -\frac{\pi}{2}.$$

**Example 5.** Let us calculate the integral

$$\int_C \frac{dz}{z - z_0},$$

where $C$ is the circle of radius $R$ centered at a point $z_0$ with anticlockwise sense.

Using the Cauchy integral formula (10.1.2.28), we obtain

$$\int_C \frac{1}{z - z_0}\, dz = 2\pi i.$$

**Example 6.** Let us calculate the integral

$$\int_C \frac{dz}{z^2 + 1},$$

where $C$ is the circle of unit radius centered at the point $i$ with anticlockwise sense.

To apply the Cauchy integral formula (10.1.2.26), we transform the integrand as follows:

$$\frac{1}{1 + z^2} = \frac{1}{(z-i)(z+i)} = \frac{1}{z+i} \frac{1}{z-i} = \frac{f(z)}{z+i}, \quad f(z) = \frac{1}{z+i}.$$

The function $f(z) = 1/(z + i)$ is analytic in the interior of the domain under study and on its boundary; hence the Cauchy integral formula (10.1.2.26) and the first of formulas (10.1.2.28) hold. From the latter formula, we obtain

$$\int_C \frac{dz}{z^2 + 1} = \int_C \frac{f(z)}{z - i}\, dz = 2\pi i f(i) = 2\pi i \frac{1}{2i} = \pi.$$

Formulas (10.1.2.26) and (10.1.2.27) imply the *Cauchy inequalities*

$$|f_z^{(n)}(z)| \leq \frac{n!}{2\pi} \left| \int_C \frac{f(\xi)}{(\xi - z)^{n+1}}\, d\xi \right| \leq \frac{n! M l}{2\pi R^{n+1}}, \qquad (10.1.2.29)$$

where $M = \max_{z \in D} |f(z)|$ is the maximum modulus of the function $f(z)$ in the domain $D$, $R$ is the distance from the point $z$ to the boundary $C$, and $l$ is the length of the boundary $C$.

If, in particular, $f(z)$ is analytic in the disk $D = |z - z_0| < R$, and bounded in $\bar{D}$, then we obtain the inequality

$$|f_z^{(n)}(z_0)| \leq \frac{n! M}{R^n} \qquad (n = 0, 1, 2, \ldots). \qquad (10.1.2.30)$$

MORERA'S THEOREM. *If a function $f(z)$ is continuous in a simply connected domain $D$ and $\int_C f(z)\, dz = 0$ for any closed curve $C$ lying in $D$, then $f(z)$ is analytic in the domain $D$.*

### 10.1.2-5. Taylor and Laurent series.

If a series

$$\sum_{n=0}^{\infty} f_n(z) \qquad (10.1.2.31)$$

of analytic functions in a simply connected domain $D$ converges uniformly in this domain, then its sum is analytic in the domain $D$.

If a series (10.1.2.31) of functions analytic in a domain $D$ and continuous in $\bar{D}$ converges uniformly in $D$, then it can be differentiated termwise any number of times and can be integrated termwise over any piecewise smooth curve $C$ lying in $D$.

## 10.1. BASIC NOTIONS

ABEL'S THEOREM. *If the power series*

$$\sum_{n=0}^{\infty} c_n (z-a)^n \qquad (10.1.2.32)$$

*converges at a point $z_0$, then it also converges at any point $z$ satisfying the condition $|z-a| < |z_0 - a|$. Moreover, the series converges uniformly in any disk $|z-a| < q|z_0 - a|$, where $0 < q < 1$.*

It follows from Abel's theorem that the domain of convergence of a power series is an open disk centered at the point $a$; moreover, this disk can fill the entire plane. The radius of this disk is called the *radius of convergence* of a power series. The sum of the power series inside the disk of convergence is an analytic function.

Remark. The radius of convergence $R$ can be found by the *Cauchy–Hadamard formula*

$$\frac{1}{R} = \overline{\lim_{n \to \infty}} \sqrt[n]{|c_n|},$$

where $\overline{\lim}$ denotes the upper limit.

If a function $f(z)$ is analytic in the open disk $D$ of radius $R$ centered at a point $z = a$, then this function can be represented in this disk by its *Taylor series*

$$f(z) = \sum_{n=0}^{\infty} c_n (z-a)^n,$$

whose coefficients are determined by the formulas

$$c_n = \frac{f_z^{(n)}(a)}{n!} = \frac{1}{2\pi i} \int_C \frac{f(\xi)}{(\xi - z)^{n+1}} \, d\xi \qquad (n = 0, 1, 2, \ldots), \qquad (10.1.2.33)$$

where $C$ is the circle $|z - a| = qR$, $0 < q < 1$. In any closed domain belonging to the disk $D$, the Taylor series converges uniformly. Any power series expansion of an analytic function is its Taylor expansion. The Taylor series expansions of the functions given in Paragraph 10.1.2-3 in powers of $z$ have the form

$$e^z = 1 + z + \frac{z^2}{2!} + \frac{z^3}{3!} + \ldots \qquad (|z| < \infty), \qquad (10.1.2.34)$$

$$\cos z = 1 - \frac{z^2}{2!} + \frac{z^4}{4!} - \ldots, \quad \sin z = z - \frac{z^3}{3!} + \frac{z^5}{5!} - \ldots \quad (|z| < \infty), \qquad (10.1.2.35)$$

$$\cosh z = 1 + \frac{z^2}{2!} + \frac{z^4}{4!} + \ldots, \quad \sinh z = z + \frac{z^3}{3!} + \frac{z^5}{5!} + \ldots \quad (|z| < \infty), \qquad (10.1.2.36)$$

$$\ln(1+z) = z - \frac{z^2}{2!} + \frac{z^3}{3!} - \ldots \qquad (|z| < 1), \qquad (10.1.2.37)$$

$$(1+z)^a = 1 + az + \frac{a(a-1)}{2!} z^2 + \frac{a(a-1)(a-2)}{3!} z^3 + \ldots \qquad (|z| < 1). \qquad (10.1.2.38)$$

The last two expansions are valid for the single-valued branches for which the values of the functions for $z = 0$ are equal to 0 and 1, respectively.

**Remark.** Series expansions (10.1.2.34)–(10.1.2.38) coincide with analogous expansions of the corresponding elementary functions of the real variable (see Paragraph 8.3.2-3).

To obtain the Taylor series for other branches of the multi-valued function Ln(1 + z), one has to add the numbers $2k\pi i$, $k = \pm 1, \pm 2, \ldots$ to the expression in the right-hand side:

$$\operatorname{Ln}(1+z) = z - \frac{z^2}{2!} + \frac{z^3}{3!} - \ldots + 2k\pi.$$

The domain of convergence of the function series $\sum_{n=-\infty}^{\infty} c_n(z-a)^n$ is a circular annulus $K : r < |z - a| < R$, where $0 \le r \le \infty$ and $0 \le R \le \infty$. The sum of the series is an analytic function in the annulus of convergence. Conversely, in any annulus $K$ where the function $f(z)$ is analytic, this function can be represented by the *Laurent series* expansion

$$f(z) = \sum_{n=-\infty}^{\infty} c_n(z-a)^n$$

with coefficients determined by the formulas

$$c_n = \frac{1}{2\pi i} \int_\gamma \frac{f(\xi)}{(\xi - z)^{n+1}} d\xi \quad (n = 0, \pm 1, \pm 2, \ldots), \tag{10.1.2.39}$$

where $\gamma$ is the circle $|z - a| = \rho$, $r < \rho < R$. In any closed domain contained in the annulus $K$, the Laurent series converges uniformly.

The part of the Laurent series with negative numbers,

$$\sum_{n=-\infty}^{-1} c_n(z-a)^n = \sum_{n=1}^{\infty} \frac{c_{-n}}{(z-a)^n}, \tag{10.1.2.40}$$

is called its *principal part*, and the part with nonnegative numbers,

$$\sum_{n=0}^{\infty} c_n(z-a)^n, \tag{10.1.2.41}$$

is called the *regular part*. Any expansion of an analytic function in positive and negative powers of $z - a$ is its Laurent expansion.

**Example 7.** Let us consider Laurent series expansions of the function

$$f(z) = \frac{1}{z(1-z)}$$

in a Laurent series in the domain $0 < |z| < 1$. This function is analytic in the annulus $0 < |z| < 1$ and hence can be expanded in the corresponding Laurent series. We write this function as the sum of elementary fractions:

$$f(z) = \frac{1}{z(1-z)} = \frac{1}{z} + \frac{1}{1-z}.$$

Since $|z| < 1$, we can use formula (10.1.2.39) and obtain the expansion

$$\frac{1}{z(1-z)} = \frac{1}{z} + 1 + z + z^2 + \ldots$$

**Example 8.** Let us consider Laurent series expansions of the function

$$f(z) = e^{1/z}$$

in a Laurent series in a neighborhood of the point $z_0 = 0$. To this end, we use the well-known expansion (10.1.2.34), where we should replace $z$ by $1/z$. Thus we obtain

$$e^{1/z} = 1 + \frac{1}{1!z} + \frac{1}{2!z^2} + \ldots + \frac{1}{n!z^n} + \ldots \quad (z \ne 0).$$

### 10.1.2-6. Zeros and isolated singularities of analytic functions.

A point $z = a$ is called a *zero* of a function $f(z)$ if $f(a) = 0$. If $f(z)$ is analytic at the point $a$ and is not zero identically, then the least order of nonzero coefficients in the Taylor expansion of $f(z)$ centered at $a$, in other words, the number $n$ of the first nonzero derivative $f^{(n)}(a)$, is called the *order of zero* of this function. In a neighborhood of a zero $a$ of order $n$, the Taylor expansion of $f(z)$ has the form

$$f(z) = c_n(z-a)^n + c_{n+1}(z-a)^{n+1} + \ldots \qquad (c_n \neq 0, \ n \geq 1).$$

In this case, $f(z) = c_n(z-a)^n g(z)$, where the function $g(z)$ is analytic at the point $a$ and $g(a) \neq 0$. A first-order zero is said to be *simple*. The point $z = \infty$ is a zero of order $n$ for a function $f(z)$ if $z = 0$ is a zero of order $n$ for $F(z) = f(1/z)$.

If a function $f(z)$ is analytic at a point $a$ and is not identically zero in any neighborhood of $a$, then there exists a neighborhood of $a$ in which $f(z)$ does not have any zeros other than $a$.

UNIQUENESS THEOREM. *If functions $f(z)$ and $g(z)$ are analytic in a domain $D$ and their values coincide on some sequence $a_k$ of points converging to an interior point $a$ of the domain $D$, then $f(z) \equiv g(z)$ everywhere in $D$.*

ROUCHÉ'S THEOREM. *If functions $f(z)$ and $g(z)$ are analytic in a simply connected domain $D$ bounded by a curve $C$, are continuous in $\overline{D}$, and satisfy the inequality $|f(z)| > |g(z)|$ on $C$, then the functions $f(z)$ and $f(z) + g(z)$ have the same number of zeros in $D$.*

A point $a$ is called an *isolated singularity* of a single-valued analytic function $f(z)$ if there exists a neighborhood of this point in which $f(z)$ is analytic everywhere except for the point $a$ itself. The point $a$ is called
1. A *removable singularity* if $\lim_{z \to a} f(z)$ exists and is finite.
2. A *pole* if $\lim_{z \to a} f(z) = \infty$.
3. An *essential singularity* if $\lim_{z \to a} f(z)$ does not exist.

A necessary and sufficient condition for a point $a$ to be a removable singularity of a function $f(z)$ is that the Laurent expansion of $f(z)$ around $a$ does not contain the principal part. If a function $f(z)$ is bounded in a neighborhood of an isolated singularity $a$, then $a$ is a removable singularity of this function.

A necessary and sufficient condition for a point $a$ to be a pole of a function $f(z)$ is that the principal part of the Laurent expansion of $f(z)$ around $a$ contains finitely many terms:

$$f(z) = \frac{c_{-n}}{(z-a)^n} + \ldots + \frac{c_{-1}}{(z-a)} + \sum_{k=0}^{\infty} c_k(z-a)^k. \tag{10.1.2.42}$$

The *order of a pole* $a$ of a function $f(z)$ is defined to be the order of the zero of the function $F(z) = f(1/z)$. If $c_{-n} \neq 0$ in expansion (10.1.2.42), then the order of the pole $a$ of the function $f(z)$ is equal to $n$. For $n = 1$, we have a *simple pole*.

A necessary and sufficient condition for a point $a$ to be an essential singularity of a function $f(z)$ is that the principal part of the Laurent expansion of $f(z)$ around $a$ contains infinitely many nonzero terms.

SOKHOTSKII'S THEOREM. *If $a$ is an essential singularity of a function $f(z)$, then for each complex number $A$ there exists a sequence of points $z_k \to a$ such that $f(z_k) \to A$.*

**Example 9.** Let us consider some functions with different singular points.

$1°$. The function $f(z) = (1 - \cos z)/z^2$ has a removable singularity at the origin, since its Laurent expansion about the origin,

$$\frac{1 - \cos z}{z^2} = \frac{1}{2} - \frac{z^2}{24} + \frac{z^4}{720} - \ldots,$$

does not contain the principal part.

$2°$. The function $f(z) = 1/(1+e^{z^2})$ has infinitely many poles at the points $z = \pm\sqrt{(2k+1)\pi i}$ $(k = 0, \pm 1, \pm 2, \ldots)$. All these poles are simple poles, since the function $1/f(z) = 1 + e^{z^2}$ has simple zeros at these points. (Its derivative is nonzero at these points.)

$3°$. The function $f(z) = \sin(1/z)$ has an essential singularity at the origin, since the principal part of its Laurent expansion

$$\sin\frac{1}{z} = \frac{1}{z} - \frac{1}{z^3 3!} + \ldots$$

contains infinitely many terms.

The following two simplest classes of single-valued analytic functions are distinguished according to the character of singular points.

1. *Entire functions.* A function $f(z)$ is said to be *entire* if it does not have singular points in the finite part of the plane. An entire function can be represented by an everywhere convergent power series

$$f(z) = \sum_{n=0}^{\infty} c_n z^n.$$

An entire function can have only one singular point at $z = \infty$. If this singularity is a pole of order $n$, then $f(z)$ is a polynomial of degree $n$. If $z = \infty$ is an essential singularity, then $f(z)$ is called an *entire transcendental function*. If $z = \infty$ is a regular point (i.e., $f(z)$ is analytic for all $z$), then $f(z)$ is constant (*Liouville's theorem*). All polynomials, the exponential function, $\sin z$, $\cos z$, etc. are examples of entire functions. Sums, differences, and products of entire functions are themselves entire functions.

2. *Meromorphic functions.* A function $f(z)$ is said to be *meromorphic* if it does not have any singularities except for poles. The number of these poles in each finite closed domain $D$ is always finite.

Suppose that a function $f(z)$ is analytic in a neighborhood of the point at infinity. The definition of singular points can be generalized to this function without any changes. But the criteria for the type of a singular point at infinity related to the Laurent expansion are different.

THEOREM. *In the case of a removable singularity at the point at infinity, the Laurent expansion of a function $f(z)$ in a neighborhood of this point does not contain positive powers of $z$. In the case of a pole, it contains finitely many positive powers of $z$. In the case of an essential singularity, it contains infinitely many powers of $z$.*

Let $f(z)$ be a multi-valued function defined in a neighborhood $D$ of a point $z = a$ except possibly for the point $a$ itself, and let $f_1(z), f_2(z) \ldots$ be its branches, which are single-valued continuous functions in the domain where they are defined. The point $a$ is called a *branch point (ramification point)* of the function $f(z)$ if $f(z)$ passes from one branch to another as the point $z$ goes along a closed curve around the point $z$ in a neighborhood of $D$. If the original branch is reached again for the first time after going around this curve $m$ times (in the same sense), then the number $m - 1$ is called the *order* of the branch point, and the point $a$ itself is called a *branch point of order* $m - 1$.

If all branches $f_k(z)$ tend to the same finite or infinite limit as $z \to a$, then the point $a$ is called an *algebraic* branch point. (For example, the point $z = 0$ is an algebraic branch point of the function $f(z) = \sqrt[m]{z}$.) In this case, the single-valued function

$$F(z) = f(z^m + a)$$

has a regular point or a pole for $z = 0$.

If the limit of $f_k(z)$ as $z \to a$ does not exist, then the point $a$ is called a *transcendental* branch point. For example, the point $z = 0$ is a transcendental branch point of the function $f(z) = \exp(\sqrt[m]{1/z})$.

In a neighborhood of a branch point $a$ of finite order, the function $f(z)$ can be expanded in a *fractional power series* (*Puiseux series*)

$$f(z) = \sum_{k=-\infty}^{\infty} c_k (z-a)^{k/m}. \tag{10.1.2.43}$$

If a new branch is obtained each time after going around this curve (in the same sense), then the point $a$ is called a *branch point of infinite order (a logarithmic branch point)*. For example, the points $z = 0$ and $z = \infty$ are logarithmic branch points of the multivalued function $w = \mathrm{Ln}\, z$. A logarithmic branch point is classified as a transcendental branch point.

For $a \neq \infty$, the expansion (10.1.2.43) contains finitely many terms with negative $k$ (infinitely many in the case of a transcendental point).

### 10.1.2-7. Residues.

The *residue* $\mathrm{res}\, f(a)$ of a function $f(z)$ at an isolated singularity $a$ is defined as the number

$$\mathrm{res}\, f(a) = \frac{1}{2\pi i} \oint_C f(z)\, dz, \tag{10.1.2.44}$$

where the integral is taken in the positive sense over a contour $C$ surrounding the point $a$ and containing no other singularities of $f(z)$ in the interior.

**Remark.** Residues are sometimes denoted by $\mathrm{res}[f(z); a]$ or $\mathrm{res}_{z=a} f(z)$.

The residue $\mathrm{res}\, f(a)$ of a function $f(z)$ at a singularity $a$ is equal to the coefficient of $(z-a)^{-1}$ in the Laurent expansion of $f(z)$ in a neighborhood of the point $a$,

$$\mathrm{res}\, f(a) = \frac{1}{2\pi i} \oint_C f(z)\, dz = c_{-1}. \tag{10.1.2.45}$$

Basic rules for finding the residues:
1. The residue of a function at a removable singularity is always zero.
2. If $a$ is a pole of order $n$, then

$$\mathrm{res}\, f(a) = \frac{1}{(n-1)!} \lim_{z \to a} \frac{d^{n-1}}{dz^{n-1}} \left[ f(z)(z-a)^n \right]. \tag{10.1.2.46}$$

3. For a simple pole ($n = 1$),

$$\mathrm{res}\, f(a) = \lim_{z \to a} \left[ f(z)(z-a)^n \right]. \tag{10.1.2.47}$$

4. If $f(z)$ is the quotient of two analytic functions,

$$f(z) = \frac{\varphi(z)}{\psi(z)},$$

in a neighborhood of a point $a$ and $\varphi(a) \neq 0$, $\psi(a) = 0$, but $\psi'_z(a) \neq 0$ (i.e., $a$ is a simple pole of $f(z)$), then

$$\operatorname{res} f(a) = \frac{\varphi(a)}{\psi'_z(a)}. \tag{10.1.2.48}$$

5. If $a$ is an essential singularity of $f(z)$, then to obtain $\operatorname{res} f(a)$, one has to find the coefficient $c_{-1}$ in the Laurent expansion of $f(z)$ in a neighborhood of $a$.

A function $f(z)$ is said to be *continuous on the boundary* $C$ of the domain $D$ if for each boundary point $z_0$ there exists a limit $\lim_{z \to z_0} f(z) = f(z_0)$ as $z \to z_0$, $z \in D$.

CAUCHY'S RESIDUE THEOREM. *Let $f(z)$ be a function continuous on the boundary $C$ of a domain $D$ and analytic in the interior of $D$ everywhere except for finitely many points $a_1, \ldots, a_n$. Then*

$$\int_C f(z)\,dz = 2\pi i \sum_{k=1}^{n} \operatorname{res} f(a_k), \tag{10.1.2.49}$$

*where the integral is taken in the positive sense of $C$.*

The *logarithmic residue* of a function $f(z)$ at a point $a$ is by definition the residue of its logarithmic derivative

$$\left[\ln f(z)\right]'_z = \frac{f'_z(z)}{f(z)}.$$

THEOREM. *The logarithmic derivative $f'_z(z)/f(z)$ has first-order poles at the zeros and poles of $f(z)$. Moreover, the logarithmic residue of $f(z)$ at a zero or a pole of $f(z)$ is equal to the order of the zero or minus the order of the pole, respectively.*

The residue of a function $f(z)$ at infinity is defined as

$$\operatorname{res} f(\infty) = \frac{1}{2\pi i} \oint_\Gamma f(z)\,dz, \tag{10.1.2.50}$$

where $\Gamma$ is a circle of sufficiently large radius $|z| = \rho$ and the integral is taken in the clockwise sense (so that the neighborhood of the point $z = \infty$ remains to the left of the contour, just as in the case of a finite point).

The residue of $f(z)$ at infinity is equal to minus the coefficient of $z^{-1}$ in the Laurent expansion of $f(z)$ in a neighborhood of the point $z = \infty$,

$$\operatorname{res} f(\infty) = -c_{-1}. \tag{10.1.2.51}$$

Note that

$$\operatorname{res} f(\infty) = \lim_{z \to \infty} [-z f(z)], \tag{10.1.2.52}$$

provided that this limit exists.

THEOREM. *If a function $f(z)$ has finitely many singular points $a_1, \ldots, a_n$ in the extended complex plane, then the sum of all its residues, including the residue at infinity, is zero:*

$$\operatorname{res} f(\infty) + \sum_{k=1}^{n} \operatorname{res} f(a_k) = 0. \tag{10.1.2.53}$$

**Example 10.** Let us calculate the integral

$$\oint_C \frac{\ln(z+2)}{z^2}\,dz,$$

where $C$ is the circle $|z| = \frac{1}{2}$.

In the disk $|z| \leq \frac{1}{2}$, there is only one singular point of the integrand, $z = 0$, which is a second-order pole. The residue of $f(z)$ at $z = 0$ is calculated by the formula (10.1.2.46)

$$\operatorname{res} f(0) = \lim_{z \to 0}\left[z^2 \frac{\ln(z+2)}{z^2}\right]'_z = \lim_{z \to 0}[\ln(z+2)]'_z = \lim_{z \to 0}\frac{1}{z+2} = \frac{1}{2}.$$

Using formula (10.1.2.44), we obtain

$$\frac{1}{2} = \frac{1}{2\pi i}\oint_C \frac{\ln(z+2)}{z^2}\,dz, \quad \oint_C \frac{\ln(z+2)}{z^2}\,dz = \pi i.$$

### 10.1.2-8. Calculation of definite integrals.

Suppose that we need to calculate the integral of a real function $f(x)$ over a (finite or infinite) interval $(a, b)$. Let us supplement the interval $(a, b)$ with a curve $\Gamma$ that, together with $(a, b)$, bounds a domain $D$, and then analytically continue the function $f(x)$ into $\overline{D}$. Then the residue theorem can be applied to this analytic continuation of $f(z)$, and by this theorem,

$$\int_a^b f(x)\,dx + \int_\Gamma f(z)\,dz = 2\pi i \Lambda, \qquad (10.1.2.54)$$

where $\Lambda$ is the sum of residues of $f(z)$ in $D$. If $\int_\Gamma f(z)\,dz$ can be calculated or expressed in terms of the desired integral $\int_a^b f(x)\,dx$, then the problem will be solved.

When calculating integrals of the form $\int_{-\infty}^\infty f(x)\,dx$, one should apply (10.1.2.49) to the contour $C$ that consists of the interval $(-R, R)$ of the real axis and the arc $C_R$ of the circle $|z| = R$ in the upper half-plane. Sometimes it is only possible to find the limit as $R \to \infty$ of the integral over the contour $C_R$ rather than to calculate it, and often it turns out that the limit of this integral is equal to zero.

The integral over the curve $\Gamma$ can be estimated using the following lemmas.

JORDAN LEMMA. *If a function $g(z)$ tends to zero uniformly with respect to $\arg z$ along a sequence of circular arcs $C_{R_n} : |z| = R_n$, $\operatorname{Im} z > -a$ (where $R_n \to \infty$ and $a$ is fixed), then*

$$\lim_{n \to \infty}\int_{C_{R_n}} g(z) e^{imz}\,dz = 0 \qquad (10.1.2.55)$$

*for each positive number $m$.*

If a function $f(z)$ is analytic for $|z| > R_0$ and $zf(z) \to 0$ as $|z| \to \infty$ for $y \geq 0$, then

$$\lim_{R \to \infty}\int_{C_R} f(z)\,dz = 0, \qquad (10.1.2.56)$$

where $C_R$ is the arc of the circle $|z| = R$ in the upper half-plane.

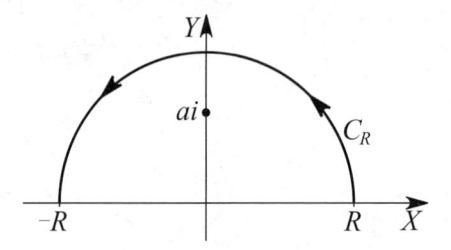

Figure 10.5. The contour to calculate the Laplace integral.

**Example 11 (Laplace integral).** To calculate the integral

$$\int_0^\infty \frac{\cos x}{x^2 + a^2}\, dx,$$

one uses the auxiliary function

$$f(z) = \frac{e^{iz}}{z^2 + a^2} = g(z)e^{iz}, \quad g(z) = \frac{1}{z^2 + a^2}$$

and the contour shown in Fig. 10.5. Since $g(z)$ satisfies the inequality $|g(z)| < (R^2 - a^2)^{-1}$ on $C_R$, it follows that this function uniformly tends to zero as $R \to \infty$, and by the Jordan lemma we obtain

$$\int_{C_R} f(z)\, dz = \int_{C_R} g(z)e^{iz}\, dz \to 0$$

as $R \to \infty$.

By the residue theorem,

$$\int_{-R}^{R} \frac{e^{ix}}{x^2 + a^2}\, dx + \int_{C_R} f(z)\, dz = 2\pi i \frac{e^{-a}}{2ai}$$

for each $R > 0$. (The residue at the singular point $z = ai$ of the function $f(z)$, which is a first-order pole and which is the only singular point of this function lying inside the contour, can be calculated by formula (10.1.2.48).) In the limit as $R \to \infty$, we obtain

$$\int_{-\infty}^{\infty} \frac{e^{ix}}{x^2 + a^2}\, dx = \frac{\pi}{ae^a}.$$

Separating the real part and using the fact that the function is even, we obtain

$$\int_0^\infty \frac{\cos x}{x^2 + a^2}\, dx = \frac{\pi}{2ae^a}.$$

### 10.1.2-9. Analytic continuation.

Let two domains $D_1$ and $D_2$ have a common part $\gamma$ of the boundary, and let single-valued analytic functions $f_1(z)$ and $f_2(z)$, respectively, be given in these domains. The function $f_2(z)$ is called a *direct analytic continuation* of $f_1(z)$ into the domain $D_2$ if there exists a function $f(z)$ analytic in the domain $D_1 \cup \gamma \cup D_2$ and satisfying the condition

$$f(z) = \begin{cases} f_1(z) & \text{for } z \in D_1, \\ f_2(z) & \text{for } z \in D_2. \end{cases} \tag{10.1.2.57}$$

If such a continuation is possible, then the function $f(z)$ is uniquely determined. If the domains are simply connected and the functions $f_1(z)$ and $f_2(z)$ are continuous in $D_1 \cup \gamma$ and $D_2 \cup \gamma$, respectively, and coincide on $\gamma$, then $f_2(z)$ is the direct analytic continuation of $f_1(z)$ into the domain $D_2$. In addition, suppose that the domains $D_1$ and $D_2$ are allowed to have common interior points. A function $f_2(z)$ is called a direct analytic continuation of $f_1(z)$ through $\gamma$ if $f_1(z)$ and $f_2(z)$ are continuous in $D_1 \cup \gamma$ and $D_2 \cup \gamma$, respectively, and their values on $\gamma$ coincide. At the common interior points of $D_1$ and $D_2$, the function determined by relation (10.1.2.57) can be double-valued.

## 10.2. Main Applications

### 10.2.1. Conformal Mappings

**10.2.1-1. Generalities.**

A one-to-one mapping
$$w = f(z) = u(x,y) + iv(x,y) \qquad (10.2.1.1)$$
of a domain $D$ onto a domain $D^*$ is said to be *conformal* if the principal linear part of this mapping at any point of $D$ is an orthogonal orientation-preserving transformation.

Main properties of conformal mappings:

1. *Circular property.* A conformal mapping takes infinitesimal circles to infinitesimal circles (up to higher-order infinitesimals).
2. *Angle preservation property.* A conformal mapping preserves the angles between intersecting curves at points of intersection.

THEOREM. *A function $w = f(z)$ is a conformal mapping of a domain $D$ if and only if it is analytic and schlicht in $D$ and the derivative $f'_z(z)$ vanishes nowhere in $D$.*

The *main problem in the theory of conformal mappings* is as follows: given domains $D$ and $D^*$, construct a function that gives a conformal mapping of one of the domains onto the other.

THE MAIN THEOREM OF THE THEORY OF CONFORMAL MAPPINGS (RIEMANN THEOREM). *For any simply connected domains $D$ and $D^*$ (with boundaries consisting of more than a single point), any points $z_0 \in D$ and $w_0 \in D^*$, and any real number $\alpha_0$, there exists a unique conformal mapping*
$$w = f(z)$$
*of $D$ onto $D^*$ such that*
$$f(z_0) = w_0, \quad \arg f'_z(z_0) = \alpha_0.$$

**10.2.1-2. Boundary correspondence.**

On the boundary $C$ of a domain $D$, let us introduce a real arc length parameter $s$ reckoned from some point of $C$, so that $\zeta = \zeta(s)$ on $C$. If $f(z)$ is a continuous function in the closed domain $\overline{D}$, then on the boundary $C$ one can set
$$f(\zeta) = f[\zeta(s)] = \varphi(s).$$
The function $\varphi(s)$ is called the *boundary function* for $f(z)$.

THEOREM ON THE BOUNDARY CORRESPONDENCE. *Suppose that a function $w = f(z)$ specifies a conformal mapping between domains $D$ and $D^*$. Then the following assertions hold.*

1. *If the boundary of $D^*$ does not have infinite branches, then $f(z)$ is continuous on the boundary of $D$ and the boundary function $w = f(\zeta) = \varphi(s)$ is a continuous one-to-one correspondence between the boundaries of the domains $D$ and $D^*$.*
2. *If the boundaries of $D$ and $D^*$ do not contain infinite branches and have a continuous curvature at each point, then the boundary function $\varphi(s)$ is continuously differentiable.*

BOUNDARY CORRESPONDENCE PRINCIPLE. Let $D$ and $D^*$ be two simply connected domains with boundaries $C$ and $C^*$, and let the domain $D^*$ be bounded. Suppose that a function $w = f(z)$ satisfies the following conditions:

1. It is analytic in $D$ and continuous in $\overline{D}$. If the point at infinity lies in the interior of the domain $D^*$, then the boundary correspondence principle remains valid provided that $w = f(z)$ is continuous in $\overline{D}$ and analytic in $D$ everywhere except for an interior point $z_0$, at which this function has a simple pole.
2. It is a one-to-one sense-preserving mapping of $C$ onto $C^*$.

Then f(z) is a (schlicht) conformal mapping of $D$ onto $D^*$.

**Example 1.** The exponential function $w = e^z$ maps

a) the strip between the straight lines $y = k(x - a_1)$ and $y = k(x - a_2)$ onto the strip lying between the logarithmic spirals (Fig. 10.6). (If $k(a_2 - a_1) = 2\pi$, then the spirals coincide, and we obtain a mapping onto the plane with the spiral cut; for $k(a_2 - a_1) > 2\pi$, the mapping is not schlicht);
b) the strip $0 < \operatorname{Im} z < \pi$ onto the upper half-plane (Fig. 10.7); here the point $\pi i$ is taken to the point $-1$, and the point $0$ is taken to the point $1$;
c) the half-strip $0 < \operatorname{Im} z < \pi$, $\operatorname{Re} z < 0$, onto the half-disk $|w| < 1$, $\operatorname{Im} w > 0$ (Fig. 10.8);
d) a rectangle onto a half-annulus (Fig. 10.9).

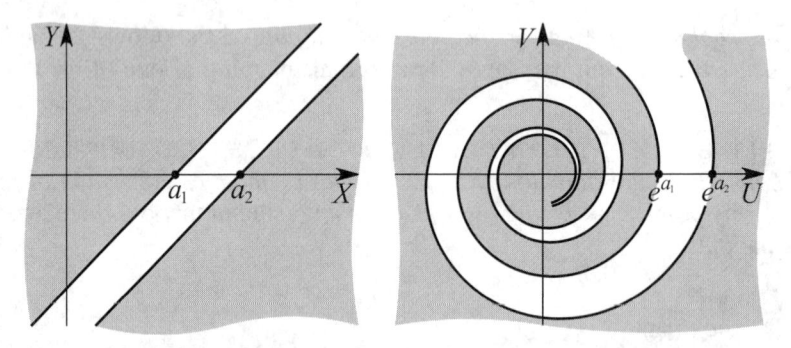

**Figure 10.6.** The exponential function $w = e^z$ maps the strip between the straight lines onto the strip lying between the logarithmic spirals.

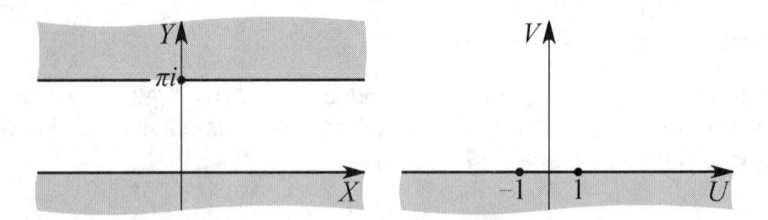

**Figure 10.7.** The exponential function $w = e^z$ maps the strip $0 < \operatorname{Im} z < \pi$ onto the upper half-plane.

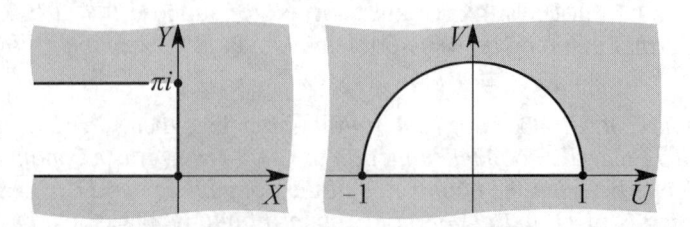

**Figure 10.8.** The exponential function $w = e^z$ maps the half-strip $0 < \operatorname{Im} z < \pi$, $\operatorname{Re} z < 0$, onto the half-disk.

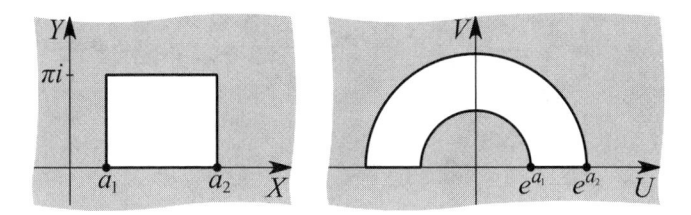

**Figure 10.9.** The exponential function $w = e^z$ maps a rectangle onto a half-annulus.

**Example 2.** The function $w = z^2$ maps the interior of a circle onto the interior of a cardioid (Fig. 10.10). The circle given in polar coordinates by the equation $r = \cos\varphi$ is taken to the cardioid $\rho = \frac{1}{2}(1 + \cos\theta)$, where $\theta = 2\varphi$.

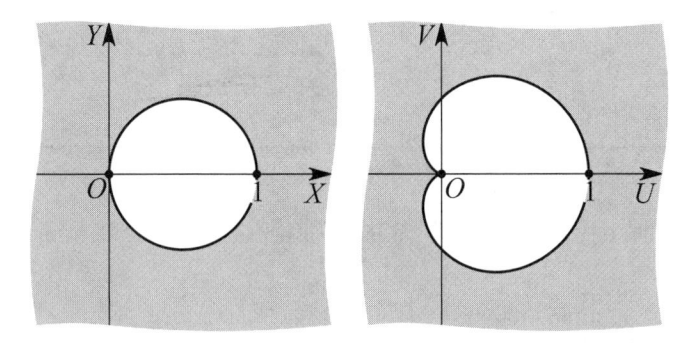

**Figure 10.10.** The function $w = z^2$ maps the interior of a circle onto the interior of a cardioid.

**Example 3.** The function $w = \sqrt{z}$ maps the interior of a circle onto the interior of the right branch of a lemniscate (Fig. 10.11). The circle $r = \cos\varphi$ is taken to the right branch of the lemniscate $\rho = \sqrt{\cos 2\theta}$, where $\theta = \frac{1}{2}\varphi$.

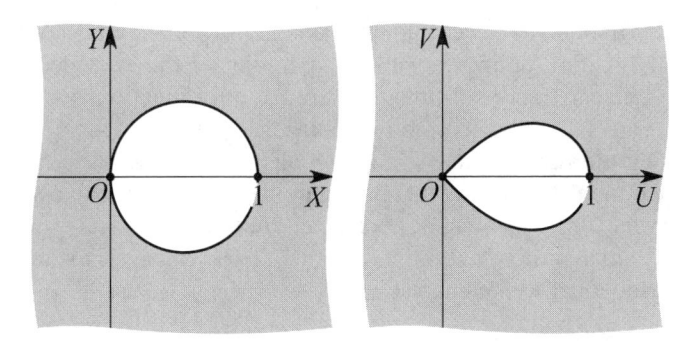

**Figure 10.11.** The function $w = \sqrt{z}$ maps the interior of a circle onto the interior of the right branch of a lemniscate.

**Example 4.** The function $w = -\ln(1-z)$ maps the interior of the unit circle onto the interior of the curve $u = -\ln(2\cos v)$ (Fig. 10.12).

**Example 5.** The function $w = \ln\dfrac{z-1}{1+z}$ maps the upper half-plane onto the strip $0 < \operatorname{Im} z < \pi$ (Fig. 10.13). The function $z = -\coth\frac{1}{2}w$ specifies the inverse mapping of the strip $0 < \operatorname{Im} z < \pi$ onto the upper half-plane.

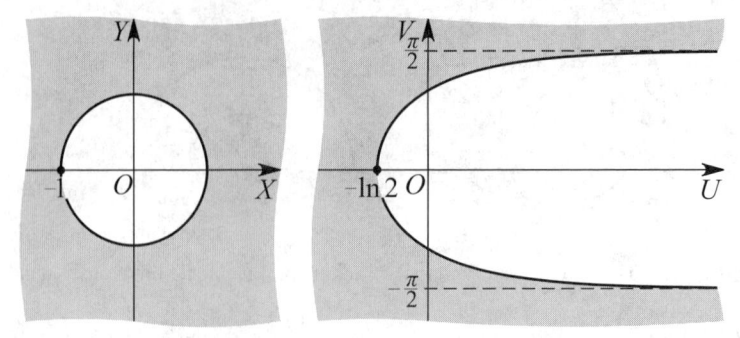

**Figure 10.12.** The function $w = -\ln(1 - z)$ maps the interior of the unit circle onto the interior of the curve $u = -\ln(2 \cos v)$.

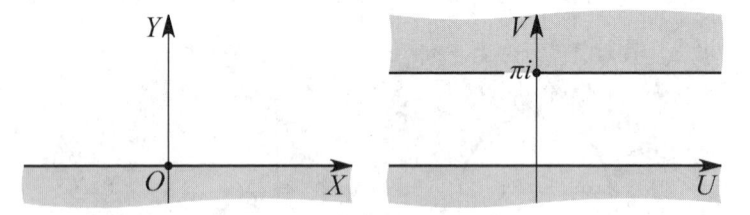

**Figure 10.13.** The function $w = \ln \dfrac{z - 1}{1 + z}$ maps the upper half-plane onto the strip $0 < \operatorname{Im} z < \pi$.

**10.2.1-3. Linear-fractional mappings.**

The mappings given by linear-fractional functions

$$w = \frac{az + b}{cz + d}, \tag{10.2.1.2}$$

where $a$, $b$, $c$, and $d$ are complex constants and $ad - bc \neq 0$, are called *linear-fractional mappings*. The function (10.2.1.2) is defined on the extended complex plane. (Its value at the point $z = -d/c$ is defined to be $\infty$, and the value at the point $z = \infty$ is defined to be $a/c$.) A linear-fractional function defines a schlicht mapping of the extended $z$-plane onto the extended $w$-plane. Linear-fractional functions are the only functions with this property.

Points $z$ and $z^*$ are said to be symmetric about the circle $C_0 : |z - z_0| = R_0$ if they lie on the same ray passing through $z_0$ and $|z - z_0||z^* - z_0| = R_0^2$.

The transformation taking each point $z$ to the point $z^*$ symmetric to $z$ about the circle $C_0$ is called the *symmetry*, or the *inversion*, about the circle.

Points $z$ and $z^*$ are symmetric about a circle $C_0$ if and only if they are the vertices of a pencil of circles orthogonal to the circle $C_0$.

THEOREM. *An arbitrary linear-fractional function*

$$w = \frac{az + b}{cz + d}, \quad ad - bc \neq 0,$$

*defines a schlicht conformal mapping of the extended $z$-plane onto the extended $w$-plane. This mapping transforms any circle on the extended $z$-plane into a circle on the extended $w$-plane (the circular property) and transforms any pair of points symmetric about a circle $C$ into a pair of points symmetric about the image of the circle $C$ (preservation of symmetric points).*

Let us present some formulas that allow one to find the images of straight lines and circles for an arbitrary linear-fractional mapping (10.2.1.2).

1. The straight lines $\operatorname{Re}(\lambda z) = \alpha$ that do not pass through the point $z = -d/c$ ($\alpha \neq -\operatorname{Re}(\lambda d/c)$) are taken to the circles $|w - w_0| = \rho$, where

$$w_0 = \frac{2a\alpha\bar{c} + a\bar{d}\bar{\lambda} + b\bar{c}\lambda}{2\alpha|c|^2 + 2\operatorname{Re}(c\bar{d}\bar{\lambda})}, \quad \rho = \left|\frac{a}{c} - w_0\right|.$$

2. The straight lines $\operatorname{Re}(\lambda z) = -\operatorname{Re}(\lambda d/c)$ passing through the point $z = -d/c$ are taken to the straight lines

$$\operatorname{Re}\left(\frac{ad - bc}{c^2}\lambda\bar{w}\right) = \operatorname{Re}\left(\frac{ad - bc}{c^2}\frac{\lambda\bar{a}}{\bar{c}}\right).$$

3. The circles $|z - z_0| = r$ that do not pass through the point $z = -d/c$ ($r \neq |z_0 + d/c|$) are taken to the circles $|w - w_0| = \rho$, where

$$w_0 = \frac{(az_0 + b)(\bar{c}\bar{z}_0 + \bar{d}) - a\bar{c}r^2}{|cz_0 + d|^2 - |c|^2 r^2}, \quad \rho = \frac{r|ad - bc|}{\left||cz_0 + d|^2 - |c|^2 r^2\right|}.$$

4. The circles $|z - z_0| = |z_0 + d/c|$ are taken to the straight lines

$$\operatorname{Re}\left[\frac{ad - bc}{c(cz_0 + d)}\bar{w}\right] = \frac{|ad - bc|^2 + 2\operatorname{Re}[c(az_0 + b)(\overline{ad} - \overline{bc})]}{2|c(cz_0 + d)|^2}.$$

If a linear-fractional mapping takes four points $z_1$, $z_2$, $z_3$, and $z$ to points $w_1$, $w_2$, $w_3$, and $w$, respectively, then the following relation holds:

$$\frac{w - w_1}{w - w_3}\frac{w_2 - w_3}{w_2 - w_1} = \frac{z - z_1}{z - z_3}\frac{z_2 - z_3}{z_2 - z_1}. \tag{10.2.1.3}$$

THEOREM. *There exists a unique linear-fractional mapping of the extended $z$-plane onto the extended $w$-plane taking three arbitrary distinct points $z_1$, $z_2$, and $z_3$ to three arbitrary distinct points $w_1$, $w_2$, and $w_3$, respectively.*

THEOREM. *Any disk of the extended $z$-plane can be transformed into any disk of the extended $w$-plane by a linear-fractional function.*

**Example 6.** A mapping of the upper half-plane onto the unit disk.

Let $a$ be the point of the upper half-plane which should be taken to the center $w = 0$ of the disk (Fig. 10.14). Then the problem is solved by the linear-fractional function

$$w = e^{i\beta}\frac{z - a}{z - \bar{a}},$$

where $\beta$ is an arbitrary real number. (Changing $\alpha$ means rotating the disk around the center $w = 0$.)

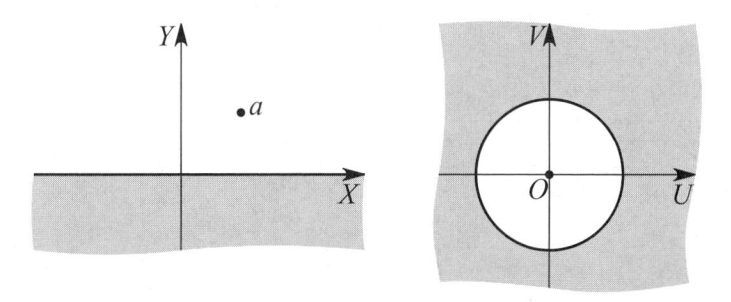

**Figure 10.14.** A mapping of the upper half-plane onto the unit disk.

**Example 7. A mapping of the unit disk onto the upper half-plane.**

Let $a = ih$ be the point of the upper half-plane to which the center $z = 0$ of the disk should be taken (Fig. 10.15). Then the problem is solved by the linear-fractional function

$$w = ih\frac{e^{i\beta} + z}{e^{i\beta} - z},$$

where $\beta$ is an arbitrary real number. (Changing $\beta$ means rotating the disk around the center $w = 0$.)

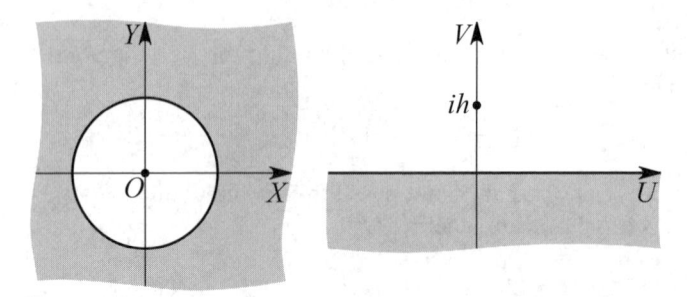

**Figure 10.15.** A mapping of the unit disk onto the upper half-plane.

**Example 8. A mapping of the unit disk onto the unit disk.**

Let $a$ be the point of the disk $|z| < 1$ that should be taken to the center of the disk $|w| < 1$ (Fig. 10.16). Then the problem is solved by the linear-fractional function

$$w = e^{i\beta}\frac{z - a}{1 - z\bar{a}}, \qquad (10.2.1.4)$$

where $\beta$ is an arbitrary real number.

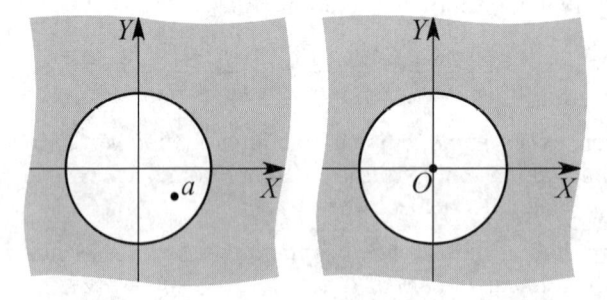

**Figure 10.16.** A mapping of the unit disk onto the unit disk.

Geometrically, $\beta$ is the angle of rotation of the mapping (10.2.1.4) at the point $a$:

$$\beta = \arg\frac{dw}{dz}.$$

If the radius of the disk in the $z$-plane is equal to $R$, then the function $w = f(z)$ mapping this disk onto the disk $|w| < 1$ and satisfying $f(a) = 0$ and $\arg f'_z(a) = \beta$ has the form

$$w = e^{i\beta}\frac{R(z - a)}{R^2 - z\bar{a}}.$$

**10.2.1-4. Mappings determined by the Zhukovskii function.**

The mapping (see also Paragraph 10.1.2-3)

$$w = \frac{1}{2}\left(z + \frac{1}{z}\right) \qquad (10.2.1.5)$$

is conformal except at the points $z = \pm 1$.

The mapping given by the Zhukovskii function is equivalent to the following mappings:

$$z = w + \sqrt{w^2 - 1} \quad \text{or} \quad \frac{w-1}{w+1} = \left(\frac{z-1}{z+1}\right)^2. \tag{10.2.1.6}$$

If we denote $z = re^{i\varphi}$ and $w = u + iv$, then the mapping (10.2.1.5) can be written as

$$u = \frac{1}{2}\left(r + \frac{1}{r}\right)\cos\varphi, \quad v = \frac{1}{2}\left(r - \frac{1}{r}\right)\sin\varphi. \tag{10.2.1.7}$$

The main properties of the mapping (10.2.1.5) are given in Table 10.2.

TABLE 10.2
Properties of mapping $w = \frac{1}{2}\left(z + \frac{1}{z}\right)$, where $z = x + iy = re^{i\varphi}$, $w = u + iv = \rho e^{i\theta}$

| No. | $z$-plane | $w$-plane | Remarks |
|---|---|---|---|
| 1 | Circle $\|z\| = r_0 < 1$ | Ellipse $\dfrac{u^2}{a^2} + \dfrac{v^2}{b^2} = 1$, where $a = \frac{1}{2}(1/r_0 + r_0)$, $b = \frac{1}{2}(1/r_0 - r_0)$ | The ellipse has the negative sense. The foci are at the points $\pm 1$. |
| 2 | Circle $\|z\| = r_0 > 1$ | Ellipse $\dfrac{u^2}{a^2} + \dfrac{v^2}{b^2} = 1$, where $a = \frac{1}{2}(1/r_0 + r_0)$, $b = \frac{1}{2}(1/r_0 - r_0)$ | The ellipse has the positive sense. The foci are at the points $\pm 1$. |
| 3 | Radii $\arg z = \varphi_0 \quad (0 < r < 1)$ | Hyperbolas $\dfrac{u^2}{\cos^2\varphi_0} - \dfrac{v^2}{\sin^2\varphi_0} = 1$ | The foci are at the points $\pm 1$. |
| 4 | Semicircle $\|z\| = 1$, $\operatorname{Im} z \geq 0$ | Segment $\|\operatorname{Re} w\| \leq 1$, $\operatorname{Im} w = 0$ | $u$ decreases with increasing $\varphi$. |
| 5 | Semicircle $\|z\| = 1$, $\operatorname{Im} z \leq 0$ | Segment $\|\operatorname{Re} w\| \leq 1$, $\operatorname{Im} w = 0$ | $u$ increases with increasing $\varphi$. |
| 6 | Segment $0 < \operatorname{Re} z < 1$, $\operatorname{Im} z = 0$ | Half-line $\operatorname{Re} w > 1$, $\operatorname{Im} w = 0$ | — |
| 7 | Segment $-1 < \operatorname{Re} z < 0$, $\operatorname{Im} z = 0$ | Half-line $\operatorname{Re} w < -1$, $\operatorname{Im} w = 0$ | — |
| 8 | Half-line $\operatorname{Re} z > 1$, $\operatorname{Im} z = 0$ | Half-line $\operatorname{Re} w > 1$, $\operatorname{Im} w = 0$ | — |
| 9 | Half-line $\operatorname{Re} z < -1$, $\operatorname{Im} z = 0$ | Half-line $\operatorname{Re} w < -1$, $\operatorname{Im} w = 0$ | — |

**Example 9.** The Zhukovskii function defines the following conformal maps:
1. It maps the interior of the semicircle $|z| < 1$, $\operatorname{Im} z > 0$, onto the lower half-plane (Fig. 10.17). The point $z = 1$ is taken to the point $w = 1$, and the point $z = i$ is taken to the point $w = 0$.
2. It maps the upper half-plane with the disk $|z| < 1$ deleted onto the upper half-plane (Fig. 10.18).
3. It maps the half-annulus $1 < |z| < k$, $\operatorname{Im} z > 0$, onto the half-ellipse given by the equation

$$\left(\frac{2ku}{k^2+1}\right)^2 + \left(\frac{2kv}{k^2-1}\right)^2 = 1$$

on the $w$-plane (Fig. 10.19).

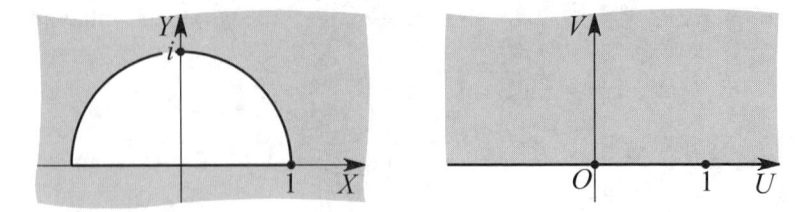

**Figure 10.17.** Zhukovskii function maps the interior of the semicircle $|z| < 1$, $\text{Im } z > 0$, onto the lower half-plane.

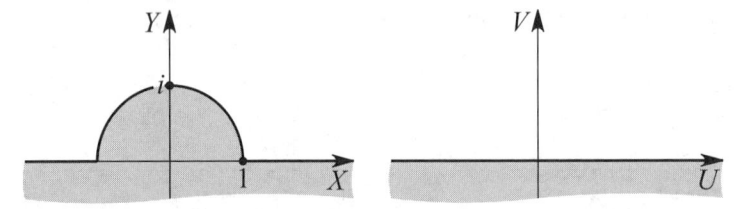

**Figure 10.18.** Zhukovskii function maps the upper half-plane with the disk $|z| < 1$ deleted onto the upper half-plane.

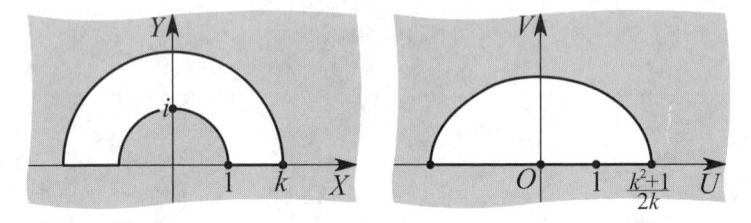

**Figure 10.19.** Zhukovskii function maps the half-annulus $1 < |z| < k$, $\text{Im } z > 0$, onto the half-ellipse.

### 10.2.1-5. Symmetry principle and mapping of polygons.

In a special case, the symmetry principle gives a simple sufficient condition for the existence of an analytic continuation of a function realizing a conformal mapping.

THE RIEMANN–SCHWARZ THEOREM. *Suppose that a function $w = f_1(z)$ realizes a conformal mapping of a domain $D_1$ onto a domain $D_1^*$ and takes a circular arc $C$ of the boundary of $D_1$ to a circular arc $C^*$ of the boundary of $D_1^*$. Then the function $f_1(z)$ admits an analytic continuation $f_2(z)$ through the arc $C$ into a domain $D_2$ symmetric to $D_1$ about $C$, the function $w = f_2(z)$ realizes a conformal mapping of the domain $D_2$ onto the domain $D_2^*$ symmetric to $D_1^*$ about $C^*$, and the function*

$$w = f(z) = \begin{cases} f_1(z) & \text{in } D_1, \\ f_1(z) = f_2(z) & \text{on } C, \\ f_2(z) & \text{in } D_2 \end{cases}$$

*realizes a conformal mapping of the domain $D_1 + C + D_2$ onto the domain $D_1^* + C^* + D_2^*$.*

An arc $C$ is said to be *analytic* if it can be described by parametric equations

$$x = x(t), \quad y = y(t) \quad (\alpha \le t \le \beta)$$

such that $x(t)$ and $y(t)$ are analytic functions of the real variable $t$ on the interval $(\alpha, \beta)$.

SCHWARZ'S ANALYTIC CONTINUATION PRINCIPLE. *Suppose that a function $w = f(z)$ realizes a conformal mapping of a domain $D$ onto a domain $D^*$ and takes an analytic arc $C$ of the boundary of $D$ to an arc $C^*$ of the boundary of $D^*$. Then the function $w = f(z)$ can be analytically continued through the arc $C$.*

**THE SCHWARZ–CHRISTOFFEL THEOREM.** If a function $w = f(z)$ realizes a conformal mapping of the upper half-plane $\operatorname{Im} z > 0$ onto the interior of a bounded polygon $\Delta$ with angles $\pi\alpha_k$ ($0 < \alpha_k \leq 2$, $k = 1, 2, \ldots, n$) at the vertices and if the points $a_k$ of the real axis ($-\infty < a_1 < \ldots < a_n < \infty$) corresponding to the vertices of this polygon are known, then the function $f(z)$ can be represented by the Schwarz–Christoffel integral

$$f(z) = C \int_{z_0}^{z} (z - a_1)^{\alpha_1 - 1}(z - a_2)^{\alpha_2 - 1} \cdots (z - a_n)^{\alpha_n - 1}\, dz + C_1,$$

where $z_0$, $C$, and $C_1$ are some constants.

The Schwarz–Christoffel integral is obtained under the assumption that the points $a_k$ corresponding to the vertices $A_k$ of the polygon are known. In practice, only the vertices of the polygon are given, and the points $a_k$ are unknown. Determining the points $a_k$ is a very difficult task.

Table 10.3 presents some conformal mappings of given domains $D$ onto the unit disk.

TABLE 10.3
Conformal mappings onto the unit disk $|w| \leq 1$, where $z = x + iy = r(\cos\varphi + i\sin\varphi)$

| No. | Domain | Mapping |
|---|---|---|
| 1 | Upper half-plane, $\operatorname{Im} z > 0$ | $w = e^{i\beta}\dfrac{z - a}{z - \bar{a}}$ ($\beta$ is a real number) |
| 2 | Right half-plane, $\operatorname{Re} z > 0$ | $w = e^{i\beta}\dfrac{z - a}{z - \bar{a}}$ ($\beta$ is a real number) |
| 3 | Disk of radius $R$, $|z| < R$ | $w = e^{i\beta}\dfrac{R(z - a)}{R^2 - \bar{a}z}$ ($\beta$ is a real number) |
| 4 | Strip of width $\frac{1}{2}\pi$, $-\frac{1}{4}\pi < \operatorname{Re} z < \frac{1}{4}\pi$ | $w = \tan z$ |
| 5 | Sector of unit disk, $|z| < 1$, $0 < \arg z < \pi\alpha$ | $w = \dfrac{(1 + z^{1/\alpha})^2 - i(1 - z^{1/\alpha})^2}{(1 + z^{1/\alpha})^2 + i(1 - z^{1/\alpha})^2}$ |
| 6 | Plane with cut from $z = 0$ to $z = \infty$ along the positive real axis | $w = \dfrac{\sqrt{z} - i}{\sqrt{z} + i}$ |
| 7 | Exterior of the ellipse, $\dfrac{x^2}{a^2} + \dfrac{y^2}{b^2} = 1$ | $z = R\left(mw + \dfrac{1}{w}\right)$, $R = \dfrac{a+b}{2}$, $m = \dfrac{a-b}{a+b}$ |
| 8 | Exterior of the parabola, $r\cos^2\dfrac{\varphi}{2} = 1$ | $z = \left(\dfrac{2}{w + 1}\right)^2$ |
| 9 | Interior of the parabola, $r\cos^2\dfrac{\varphi}{2} = 1$ | $w = \tan^2\left(\dfrac{\pi}{4}\sqrt{z}\right)$ |
| 10 | Half-disk, $|z| < R$, $\operatorname{Re} z > 0$ | $w = i\dfrac{z^2 + 2Rz - R^2}{z^2 - 2Rz - R^2}$ |

**Remark.** In the items 1, 2, and 3 the point $z = a$ of the domain is taken to the center $w = 0$ of the disk; $\beta$ determines the rotation of the disk about the center $w = 0$.

## 10.2.2. Boundary Value Problems

### 10.2.2-1. Dirichlet problem.

Find a function $u(z)$ harmonic in the domain $D$, continuous in $\overline{D}$, and taking given continuous values $u(\xi)$ on the boundary of $D$.

*Generalized Dirichlet problem.* Given a function $u(\xi)$ defined on the boundary $C$ of a domain $D$ and continuous everywhere except for finitely many points $\xi_1,\ldots,\xi_n$, where it has jump discontinuities, find a function $u(z)$ harmonic and bounded in $D$ and equal to $u(\xi)$ at all points of continuity of this function on $C$.

THEOREM ON THE UNIQUENESS OF A SOLUTION OF THE GENERALIZED DIRICHLET PROBLEM. *In a given domain for a given boundary function $u(\xi)$, there exists at most one solution of the generalized Dirichlet problem.*

THEOREM ON THE EXISTENCE OF A SOLUTION OF THE GENERALIZED DIRICHLET PROBLEM. *For any simply connected domain $D$ and any piecewise continuous boundary function $u(\xi)$ with jump discontinuities, the generalized Dirichlet problem has a solution.*

POISSON'S THEOREM. *The solution of the generalized Dirichlet problem for the unit disk is given by the Poisson integral*

$$u(z) = \frac{1}{2\pi}\int_0^{2\pi} u(e^{it})\frac{1-r^2}{1-2r\cos(t-\varphi)+r^2}\,dt \qquad (z = re^{it}). \tag{10.2.2.1}$$

1°. Let $z_0$ be an arbitrary point of a domain $D$, and let

$$w = f(z;z_0), \qquad f(z_0;z_0) = 0 \tag{10.2.2.2}$$

be a function mapping the domain $D$ onto the unit disk $|w| < 1$. The function

$$g(z;z_0) = \ln\frac{1}{|f(z;z_0)|} \tag{10.2.2.3}$$

is called a *Green's function* of the domain $D$. A Green's function is harmonic everywhere in $D$ except for the point $z_0$ at which it has a pole.

The solution of the generalized Dirichlet problem is given by *Green's formula*

$$u(z) = \frac{1}{2\pi}\int_0^{2\pi} u(\xi)g_{\mathbf{n}}(\xi,z)\,ds, \tag{10.2.2.4}$$

where $g_{\mathbf{n}}$ is the inward normal derivative.

Green's formula expresses the solution of the Dirichlet problem for some domain $D$ in terms of the logarithm of the conformal mapping of the domain $D$ onto the unit disk, i.e., reduces solving the Dirichlet problem to solving the conformal mapping problem. The converse statement is also true. If the solution of the Dirichlet problem is known for some domain $D$, then a conformal mapping of this domain onto the unit disk can be constructed.

2°. Suppose that we need to find a function $f(z)$ that is analytic in the disk $|z| < 1$ and whose real part on the circle takes given values $u(\xi)$ at each point of continuity of the function $u(\xi)$. The solution of this problem is given by the *Schwarz formula*

$$f(z) = \frac{1}{2\pi}\int_0^{2\pi} u(\xi)\frac{\xi+z}{\xi-z}\,dt + iC \qquad (\xi = e^{it}), \tag{10.2.2.5}$$

where $C$ is a real constant. The integral on the right-hand side in (10.2.2.5) is called the *Schwarz integral*.

3°. Suppose that a bounded function $u(t)$ with finitely many points of discontinuity is given on the real axis and there exist finite limits of $u(t)$ as $t \to \pm\infty$. The solution of the Dirichlet problem for the upper half-plane is given by the following *Poisson integral for the half-plane*:

$$u(z) = \frac{1}{\pi} \int_{-\infty}^{+\infty} u(t) \frac{y}{(t-x)^2 + y^2}\, dt. \tag{10.2.2.6}$$

Since

$$\frac{y}{(t-x)^2 + y^2} = \operatorname{Re} \frac{1}{i(t-z)},$$

we can also write out the *Schwarz integral for the half-plane* in the form

$$f(z) = \frac{1}{2\pi} \int_0^{2\pi} u(t) \frac{dt}{t-z} + iC, \tag{10.2.2.7}$$

where $C$ is a real constant.

### 10.2.2-2. Neumann problem.

Find a function $u(z)$ harmonic in the domain $D$ with given normal derivative

$$u_{\mathbf{n}} = u_x \cos\alpha + u_y \sin\alpha = g(\xi) \tag{10.2.2.8}$$

on the boundary $C$ and given value $u(z_0)$ at a point $z_0$ of the domain $\overline{D}$.

It is assumed in (10.2.2.8) that the outward normal is considered and $\alpha$ is the angle between this normal and the axis $OX$. The function $g(\xi)$ is allowed to have only finitely many points of jump discontinuity on $C$; the function $u(z)$ and its first partial derivatives are assumed to be bounded.

*Necessary condition for the solvability of Neumann problem:*

$$\int_C g(\xi)\, ds = 0. \tag{10.2.2.9}$$

If, in addition, we assume that the partial derivatives are continuous in $\overline{D}$, then solving the Neumann problem can be reduced to solving the Dirichlet problem for the conjugate harmonic function. Suppose that $v(z)$ is a harmonic function conjugate to $u(z)$. By the Cauchy–Riemann conditions written for the curve $C$ in the directions of $\mathbf{s}$ and $\mathbf{n}$, we have

$$v_{\mathbf{s}} = u_{\mathbf{n}} = g(\xi).$$

If $v_{\mathbf{s}}$ along the curve $C$ is known, then straightforward integration gives

$$v(\xi) = \int_{\xi_0}^{\xi} v_{\mathbf{s}}\, ds = \int_{\xi_0}^{\xi} g(\xi)\, ds. \tag{10.2.2.10}$$

Now the problem of determining $v(z)$ in the domain $D$ is reduced to the Dirichlet problem. If $v(z)$ is known, then the desired function $u(z)$ can also be obtained by integration.

Now suppose that the domain $D$ is the unit disk. If we set $f(z) = u + iv$, then the function $f(z)$ satisfies the formula

$$f(z) = -\frac{1}{\pi} \int_0^{2\pi} g(e^{it}) \ln(e^{it} - z)\, dt + \text{const.} \tag{10.2.2.11}$$

Separating the real part, we obtain the formula for the desired function:

$$u(z) = -\frac{1}{\pi} \int_0^{2\pi} g(e^{it}) \ln|e^{it} - z|\, dt + \text{const}, \tag{10.2.2.12}$$

which is called the *Dini formula*.

### 10.2.2-3. Cauchy-type integral.

Suppose that $C$ is an arbitrary curve without cusps, not necessarily closed. Let an arbitrary function $f(\xi)$, which is assumed to be finite and integrable, be given on this curve.

The integral

$$F(z) = \frac{1}{2\pi i} \int_C \frac{f(\xi)\, d\xi}{\xi - z} \qquad (10.2.2.13)$$

is called a *Cauchy-type integral*.

The Cauchy-type integral is a function analytic at any point $z$ that does not lie on $C$. If the curve $C$ divides the plane into several domains, then, in general, the Cauchy-type integral defines different analytic functions in these domains.

One says that the function $f(\xi)$ satisfies the *Hölder condition* with exponent $\mu \leq 1$ at a point $\xi = \xi_0$ of the contour $C$ if there exists a constant $M$ such that the inequality

$$|f(\xi) - f(\xi_0)| \leq M|\xi - \xi_0|^\mu \quad (0 < \mu \leq 1) \qquad (10.2.2.14)$$

holds for all points $\xi \in C$ sufficiently close to $\xi_0$. The Hölder condition means that the increment of the function is an infinitesimal of order at least $\mu$ with respect to the increment of the argument.

The *principal value of the integral* is defined as the limit

$$\lim_{r \to 0} \int_{C-c} \frac{f(\xi)\, d\xi}{\xi - \xi_0} = \int_C \frac{f(\xi)\, d\xi}{\xi - \xi_0}, \qquad (10.2.2.15)$$

where $c$ is the segment of the curve $C$ between the points of intersection of $C$ with the circle $|z - \xi_0| = r$.

The *singular integral in the sense of the Cauchy principal value* is defined as the integral given by the formula

$$\int_{C-c} \frac{f(\xi)\, d\xi}{\xi - \xi_0} = \int_C \frac{f(\xi) - f(\xi_0)}{\xi - \xi_0}\, d\xi + f(\xi_0) \ln \frac{b - \xi_0}{a - \xi_0} + i\pi f(\xi_0) + O(r), \qquad (10.2.2.16)$$

where $a$ and $b$ are the endpoints of $C$ and $O(r) \to 0$ as $r \to 0$.

THEOREM. *If the function $f(\xi)$ satisfies the Hölder condition with exponent $\mu \leq 1$ at a point $\xi_0$ which is a regular (nonsingular) point of the contour $C$ and does not coincide with its endpoints, then the Cauchy-type integral exists at this point as a singular integral and its principal value can be expressed in terms of the usual integral by the formula*

$$F(\xi_0) = \frac{1}{2\pi i} \int_C \frac{f(\xi)\, d\xi}{\xi - \xi_0} = \frac{1}{2\pi i} \int_C \frac{f(\xi) - f(\xi_0)}{\xi - \xi_0}\, d\xi + \frac{f(\xi_0)}{2} + \frac{f(\xi_0)}{2\pi i} \ln \frac{b - \xi_0}{a - \xi_0}. \qquad (10.2.2.17)$$

If the curve $C$ is closed, then $a = b$ and formula (10.2.2.17) becomes

$$F(\xi_0) = \frac{1}{2\pi i} \int_C \frac{f(\xi)\, d\xi}{\xi - \xi_0} = \frac{1}{2\pi i} \int_C \frac{f(\xi) - f(\xi_0)}{\xi - \xi_0}\, d\xi + \frac{f(\xi_0)}{2}. \qquad (10.2.2.18)$$

Suppose that the function $f(\xi)$ satisfies the Hölder condition with exponent $\mu \leq 1$ at the point $\xi = \xi_0$ and the point $z$ tends to $\xi_0$ so that the ratio of $h = |z - \xi_0|$ to $d$ ($dh$ is the shortest distance from $z$ to the points of $C$) remains bounded. Then

$$\lim_{z \to \xi_0} \int_C \frac{f(\xi) - f(\xi_0)}{\xi - z}\, d\xi = \int_C \frac{f(\xi) - f(\xi_0)}{\xi - \xi_0}\, d\xi. \qquad (10.2.2.19)$$

SOKHOTSKII'S THEOREM. *Suppose that $\xi_0$ is a regular (nonsingular) point of the contour $C$ and does not coincide with its endpoints, the function $f(\xi)$ satisfies the Hölder condition with exponent $\mu \leq 1$ at this point, and $z \to \xi_0$ so that the ratio $h/d$ remains bounded. Then the Cauchy-type integral has limit values $F^+(\xi_0)$ and $F^-(\xi_0)$ to which this integral tends as $z \to \xi_0$ from the left and, respectively, from the right of $C$, and*

$$F^+(\xi_0) = F(\xi_0) + \frac{1}{2}f(\xi_0), \quad F^-(\xi_0) = F(\xi_0) - \frac{1}{2}f(\xi_0), \qquad (10.2.2.20)$$

where $F(\xi_0)$ is the singular integral (10.2.2.18).

The Cauchy-type integral experiences a jump when passing through the integration contour $C$ at the point $\xi_0$:

$$F^+(\xi_0) - F^-(\xi_0) = f(\xi_0). \qquad (10.2.2.21)$$

The condition

$$F^-(\xi) = 0 \qquad (10.2.2.22)$$

at each point of $C$ is necessary and sufficient for a Cauchy-type integral to be the Cauchy integral.

THEOREM. *If a function $f(\xi)$ satisfies the Hölder condition with exponent $\mu \leq 1$ at each point of a closed contour $C$, then, for its values to be the boundary values of a function analytic in the interior of $C$, it is necessary and sufficient that*

$$\int_C \xi^n f(\xi)\, d\xi = 0 \qquad (n = 0, 1, 2, \ldots). \qquad (10.2.2.23)$$

THEOREM. *If a function $f(\xi)$ satisfies the Hölder condition with exponent $\mu \leq 1$ at each point of a closed contour $C$, then, for the values of $f(\xi)$ to be the boundary values of a function analytic in the interior of $C$, it is necessary and sufficient that*

$$\frac{1}{2\pi i}\int_C \frac{f(\xi)\, d\xi}{\xi - z} = 0 \qquad (10.2.2.24)$$

*for all points $z$ lying in the exterior of $C$.*

THEOREM. *If a function $f(\xi)$ satisfies the Hölder condition with exponent $\mu \leq 1$ at each point of a closed contour $C$, then, for the values of $f(\xi)$ to be the boundary values of a function analytic in the exterior of $C$, it is necessary and sufficient that*

$$\frac{1}{2\pi i}\int_C \frac{f(\xi)\, d\xi}{\xi - z} = f(\infty) \qquad (10.2.2.25)$$

*for all points $z$ lying in the interior of $C$.*

THEOREM. *For the values of a function $f(\xi)$ satisfying the Hölder condition with exponent $\mu \leq 1$ to be the boundary values of a function analytic (a) in the interior of the disk $|z| < 1$ or (b) in the exterior of this disk, it is necessary and sufficient that the following respective conditions hold:*

for all $z$ in the interior of $C$,

$$\frac{1}{2\pi i}\int_C \frac{\overline{f(\xi)}\, d\xi}{\xi - z} = \overline{f(0)}, \qquad (10.2.2.26)$$

for all $z$ in the exterior of $C$

$$\frac{1}{2\pi i}\int_C \frac{\overline{f(\xi)}\, d\xi}{\xi - z} = 0. \qquad (10.2.2.27)$$

**Example (the first main problem of elasticity).**
Let $D$ be the unit disk. Find the elastic equilibrium for given external stresses $F_n = X_n + iY_n$ on the unit circle $C$, where $X_n$ and $Y_n$ are the components of a surface force vector.

The problem is to find functions $\varphi$ and $\psi$ satisfying the boundary condition
$$\varphi(\xi) + \xi\overline{\varphi'_\xi(\xi)} + \overline{\psi(\xi)} = f(\xi),$$
where $f(\xi) = i\int_{\xi_0}^{\xi} F_n\, ds$ is a function given on $C$.

To be definite, we set
$$\psi(0) = \operatorname{Im}\varphi'_\xi(0) = 0.$$

By formula (10.2.2.26), the relation
$$\frac{1}{2\pi i}\int_C \frac{\overline{\psi(\xi)}\, d\xi}{\xi - z} = \frac{1}{2\pi i}\int_C \frac{f(\xi)\, d\xi}{\xi - z} - \frac{1}{2\pi i}\int_C \frac{\varphi(\xi)\, d\xi}{\xi - z} - \frac{1}{2\pi i}\int_C \frac{\xi\overline{\varphi'_\xi(\xi)}}{\xi - z}\, d\xi = 0$$
holds for all $|z| < 1$. Since the function $\varphi(z)$ is analytic in the disk $|z| < 1$, we can use the Cauchy formula and rewrite this relation as
$$\varphi(z) + \frac{1}{2\pi i}\int_C \frac{\xi\overline{\varphi'_\xi(\xi)}}{\xi - z}\, d\xi = \frac{1}{2\pi i}\int_C \frac{f(\xi)\, d\xi}{\xi - z}.$$

Thus we obtain an equation for the function $\varphi(z)$. Omitting the details, we write out the definitive result:
$$\varphi(z) = \frac{1}{2\pi i}\int_C \frac{f(\xi)\, d\xi}{\xi - z} - \frac{z}{4\pi i}\int_C \frac{\overline{f(\xi)}\, d\xi}{\xi^2}.$$

To find the function $\psi(z)$, we pass from the boundary condition $\varphi(\xi) + \xi\overline{\varphi'_\xi(\xi)} + \overline{\psi(\xi)} = f(\xi)$ to the complex conjugate condition and solve it for $\psi(z)$. Thus we obtain
$$\psi(\xi) = \overline{f(\xi)} - \overline{\varphi(\xi)} - \bar\xi\varphi'_\xi(\xi).$$

We calculate the Cauchy-type integral of the expressions in both sides, which is reduced to the Cauchy integral in either case, and obtain
$$\psi(z) = \frac{1}{2\pi i}\int_C \frac{\overline{f(\xi)}\, d\xi}{\xi - z} + \frac{1}{4\pi i z}\int_C \frac{f(\xi)\, d\xi}{\xi^2} - \frac{\varphi'_\xi(\xi)}{z}.$$

### 10.2.2-4. Hilbert–Privalov boundary value problem.

*Privalov boundary value problem.* Given two complex functions $a(\xi) \neq 0$ and $b(\xi)$ satisfying the Hölder condition with exponent $\mu \leq 1$ on a closed curve $C$, find a function $f^-(z)$ analytic in the exterior of $C$ including the point at infinity $z = \infty$ and a function $f^+(z)$ analytic in the interior of $C$ such that the boundary values $f^-(\xi)$ and $f^+(\xi)$ of these functions on $C$ exist and satisfy the relation
$$f^-(\xi) = a(\xi)f^+(\xi) + b(\xi). \qquad (10.2.2.28)$$

If $b(\xi) = 0$, i.e., if the boundary relation has the form
$$f^-(\xi) = a(\xi)f^+(\xi), \qquad (10.2.2.29)$$
then the Privalov boundary value problem is called the *Hilbert boundary value problem*.

The *index (winding number) of a function* $a(\xi)$ is defined to be the integer equal to the net increment of its argument along the closed curve $C$, divided by $2\pi$:
$$\frac{1}{2\pi}\Delta_C \arg a(\xi) = \frac{1}{2\pi i}\int_C d\ln a(\xi). \qquad (10.2.2.30)$$

GAKHOV'S FIRST THEOREM. *The Hilbert problem*
$$f^-(\xi) = a(\xi)f^+(\xi)$$
has a family of solutions depending on $n + 1$ arbitrary constants if the index $n$ of the boundary function $a(\xi)$ is not positive. If the index $n$ is positive, then the problem does not have solutions analytic in the corresponding domains.

The solutions of the Hilbert problem can be written as

$$f^-(\xi) = \left(a_0 + \frac{a_1}{z} + \cdots + \frac{a_n}{z^n}\right) \exp[-F_1^-(z)],$$
$$f^+(\xi) = (a_0 z^n + a_1 z^{n-1} + \cdots + a_n) \exp[-F_1^+(z)],$$
(10.2.2.31)

where

$$F_1(z) = \frac{1}{2\pi i} \int_C \frac{\ln[\xi^n a(\xi)]}{\xi - z} d\xi.$$

The constants $a_0, \ldots, a_n$ in formula (10.2.2.31) are arbitrary, and $a_0$ is determined by the choice of the value $f^-(\infty)$.

GAKHOV'S SECOND THEOREM. *The Privalov problem*

$$f^-(\xi) = a(\xi) f^+(\xi) + b(\xi)$$

*has a family of solutions depending on $n + 1$ arbitrary constants if the index $n$ of the boundary function $a(\xi)$ is not positive. If the index $n$ of the function $a(\xi)$ is positive, then the problem is solvable only if the function $b(\xi)$ satisfies the condition*

$$\int_C \frac{b(\xi) \exp[-F_1^-(\xi)]}{\xi^{k+1}} d\xi = 0 \qquad (k = 1, 2, \ldots, n).$$
(10.2.2.32)

The solutions of the Privalov problem can be written as

$$f^-(\xi) = \left[a_0 + \frac{a_1}{z} + \ldots + \frac{a_n}{z^n} + F_2^-(z)\right] \exp[-F_1^-(z)],$$
$$f^+(\xi) = \left[a_0 z^n + a_1 z^{n-1} + \ldots + a_n + z^n F_2^+(z)\right] \exp[-F_1^+(z)],$$
(10.2.2.33)

where $a_0, \ldots, a_n$ are arbitrary constants and $F_2(z)$ is determined by the formula

$$F_2(z) = -\frac{1}{2\pi i} \int_C \frac{b(\xi) \exp[-F_1(\xi)]}{\xi - z} d\xi.$$

## References for Chapter 10

**Ablowitz, M. J. and Fokas, A. S.,** *Complex Variables: Introduction and Applications* (Cambridge Texts in Applied Mathematics), *2nd Edition*, Cambridge University Press, Cambridge, 2003.

**Berenstein, C. A. and Roger Gay, R.,** *Complex Variables: An Introduction* (Graduate Texts in Mathematics), Springer, New York, 1997.

**Bieberbach, L.,** *Conformal Mapping*, American Mathematical Society, Providence, Rhode Island, 2000.

**Bronshtein, I. N., Semendyayev, K. A., Musiol, G., and Mühlig, H.,** *Handbook of Mathematics, 4th Edition*, Springer, New York, 2004.

**Brown, J. W. and Churchill, R. V.,** *Complex Variables and Applications, 7th Edition*, McGraw-Hill, New York, 2003.

**Caratheodory, C.,** *Conformal Representation*, Dover Publications, New York, 1998.

**Carrier, G. F., Krock, M., and Pearson, C. E.,** *Functions of a Complex Variable: Theory and Technique* (Classics in Applied Mathematics), Society for Industrial & Applied Mathematics, University City Science Center, Philadelphia, 2005.

**Cartan, H.,** *Elementary Theory of Analytic Functions of One or Several Complex Variables*, Dover Publications, New York, 1995.

**Conway, J. B.,** *Functions of One Complex Variable I* (Graduate Texts in Mathematics), *2nd Edition*, Springer, New York, 1995.

**Conway, J. B.,** *Functions of One Complex Variable II* (Graduate Texts in Mathematics), *2nd Edition*, Springer, New York, 1996.

**Dettman, J. W.,** Applied Complex Variables (Mathematics Series), Dover Publications, New York, 1984.

**England, A. H.,** *Complex Variable Methods in Elasticity, Dover Edition*, Dover Publications, New York, 2003.

**Fisher, S. D.,** *Complex Variables* (Dover Books on Mathematics), *2nd Edition*, Dover Publications, New York, 1999.

**Flanigan, F. J.,** *Complex Variables, Dover Ed. Edition*, Dover Publications, New York, 1983.

**Greene, R. E. and Krantz, S. G.,** *Function Theory of One Complex Variable* (Graduate Studies in Mathematics), *Vol. 40, 2nd Edition*, American Mathematical Society, Providence, Rhode Island, 2002.

**Ivanov, V. I. and Trubetskov, M. K.,** *Handbook of Conformal Mapping with Computer-Aided Visualization*, CRC Press, Boca Raton, 1995.

**Korn, G. A and Korn, T. M.,** *Mathematical Handbook for Scientists and Engineers: Definitions, Theorems, and Formulas for Reference and Review*, Dover Edition, Dover Publications, New York, 2000.

**Krantz, S. G.,** *Handbook of Complex Variables*, Birkhäuser, Boston, 1999.

**Lang, S.,** *Complex Analysis* (Graduate Texts in Mathematics), *4th Edition*, Springer, New York, 2003.

**Lavrentiev, M. A. and Shabat, V. B.,** *Methods of the Theory of Functions of a Complex Variable, 5th Edition [in Russian]*, Nauka Publishers, Moscow, 1987.

**LePage, W. R.,** *Complex Variables and the Laplace Transform for Engineers*, Dover Publications, New York, 1980.

**Markushevich, A. I. and Silverman, R. A. (Editor),** *Theory of Functions of a Complex Variable, 2nd Rev. Edition*, American Mathematical Society, Providence, Rhode Island, 2005.

**Narasimhan, R. and Nievergelt, Y.,** *Complex Analysis in One Variable, 2nd Edition*, Birkhäuser, Boston, Basel, Stuttgard, 2000.

**Needham, T.,** *Visual Complex Analysis, Rep. Edition*, Oxford University Press, Oxford, 1999.

**Nehari, Z.,** *Conformal Mapping*, Dover Publications, New York, 1982.

**Paliouras, J. D. and Meadows, D. S.,** *Complex Variables for Scientists and Engineers, Facsimile Edition*, Macmillan Coll. Div., New York, 1990.

**Pierpont, J.,** *Functions of a Complex Variable (Phoenix Edition)*, Dover Publications, New York, 2005.

**Schinzinger, R. and Laura, P. A. A.,** *Conformal Mapping: Methods and Applications*, Dover Publications, New York, 2003.

**Silverman, R. A.,** *Introductory Complex Analysis*, Dover Publications, New York, 1984.

**Spiegel, M. R.,** *Schaum's Outline of Complex Variables*, McGraw-Hill, New York, 1968.

**Sveshnikov, A. G. and Tikhonov, A. N.,** The Theory of Functions of a Complex Variable, Mir Publishers, Moscow, 1982.

**Wunsch, D. A.,** *Complex Variables with Applications, 2nd Edition*, Addison Wesley, Boston, 1993.

# Chapter 11
# Integral Transforms

## 11.1. General Form of Integral Transforms. Some Formulas

### 11.1.1. Integral Transforms and Inversion Formulas

Normally an integral transform has the form

$$\widetilde{f}(\lambda) = \int_a^b \varphi(x,\lambda) f(x)\, dx. \qquad (11.1.1.1)$$

The function $\widetilde{f}(\lambda)$ is called the *transform* of the function $f(x)$ and $\varphi(x,\lambda)$ is called the *kernel* of the integral transform. The function $f(x)$ is called the *inverse transform* of $\widetilde{f}(\lambda)$. The limits of integration $a$ and $b$ are real numbers (usually, $a = 0$, $b = \infty$ or $a = -\infty$, $b = \infty$). For brevity, we rewrite formula (11.1.1.1) as follows: $\widetilde{f}(u) = \mathcal{L}\{f(x)\}$.

General properties of integral transforms (linearity):

$$\mathcal{L}\{kf(x)\} = k\mathcal{L}\{f(x)\},$$
$$\mathcal{L}\{f(x) \pm g(x)\} = \mathcal{L}\{f(x)\} \pm \mathcal{L}\{g(x)\}.$$

Here, $k$ is an arbitrary constant; it is assumed that integral transforms of the functions $f(x)$ and $g(x)$ exist.

In Subsections 11.2–11.6, the most popular (Laplace, Mellin, Fourier, etc.) integral transforms are described. These subsections also describe the corresponding inversion formulas, which normally have the form

$$f(x) = \int_C \psi(x,\lambda) \widetilde{f}(\lambda)\, d\lambda \qquad (11.1.1.2)$$

and make it possible to recover $f(x)$ if $\widetilde{f}(\lambda)$ is given. The integration path $C$ can lie either on the real axis or in the complex plane.

In many cases, to evaluate the integrals in the inversion formula (11.1.1.2)—in particular, to find the inverse Laplace, Mellin, and Fourier transforms—methods of the theory of functions of a complex variable can be applied, including the residue theorem and the Jordan lemma, which are briefly outlined below in Subsection 11.1.2.

### 11.1.2. Residues. Jordan Lemma

11.1.2-1. Residues. Calculation formulas.

The *residue* of a function $f(z)$ holomorphic in a deleted neighborhood of a point $z = a$ (thus, $a$ is an isolated singularity of $f$) of the complex plane $z$ is the number

$$\operatorname*{res}_{z=a} f(z) = \frac{1}{2\pi i} \int_{C_\varepsilon} f(z)\, dz, \quad i^2 = -1,$$

where $C_\varepsilon$ is a circle of sufficiently small radius $\varepsilon$ described by the equation $|z - a| = \varepsilon$.

If the point $z = a$ is a pole of order $n$* of the function $f(z)$, then we have

$$\operatorname*{res}_{z=a} f(z) = \frac{1}{(n-1)!} \lim_{z \to a} \frac{d^{n-1}}{dx^{n-1}} \big[(z-a)^n f(z)\big].$$

For a simple pole, which corresponds to $n = 1$, this implies

$$\operatorname*{res}_{z=a} f(z) = \lim_{z \to a} \big[(z-a)f(z)\big].$$

If $f(z) = \dfrac{\varphi(z)}{\psi(z)}$, where $\varphi(a) \neq 0$ and $\psi(z)$ has a simple zero at the point $z = a$, i.e., $\psi(a) = 0$ and $\psi'_z(a) \neq 0$, then

$$\operatorname*{res}_{z=a} f(z) = \frac{\varphi(a)}{\psi'_z(a)}.$$

### 11.1.2-2. Jordan lemma.

If a function $f(z)$ is continuous in the domain $|z| \geq R_0$, $\operatorname{Im} z \geq \alpha$, where $\alpha$ is a chosen real number, and if $\lim_{z \to \infty} f(z) = 0$, then

$$\lim_{R \to \infty} \int_{C_R} e^{i\lambda z} f(z)\, dz = 0$$

for any $\lambda > 0$, where $C_R$ is the arc of the circle $|z| = R$ that lies in this domain.

▶ For more details about residues and the Jordan lemma, see Paragraphs 10.1.2-7 and 10.1.2-8.

## 11.2. Laplace Transform

### 11.2.1. Laplace Transform and the Inverse Laplace Transform

#### 11.2.1-1. Laplace transform.

The *Laplace transform* of an arbitrary (complex-valued) function $f(x)$ of a real variable $x$ ($x \geq 0$) is defined by

$$\widetilde{f}(p) = \int_0^\infty e^{-px} f(x)\, dx, \qquad (11.2.1.1)$$

where $p = s + i\sigma$ is a complex variable.

The Laplace transform exists for any continuous or piecewise-continuous function satisfying the condition $|f(x)| < M e^{\sigma_0 x}$ with some $M > 0$ and $\sigma_0 \geq 0$. In the following, $\sigma_0$ often means the greatest lower bound of the possible values of $\sigma_0$ in this estimate; this value is called the *growth exponent* of the function $f(x)$.

For any $f(x)$, the transform $\widetilde{f}(p)$ is defined in the half-plane $\operatorname{Re} p > \sigma_0$ and is analytic there.

For brevity, we shall write formula (11.2.1.1) as follows:

$$\widetilde{f}(p) = \mathfrak{L}\{f(x)\}, \qquad \text{or} \qquad \widetilde{f}(p) = \mathfrak{L}\{f(x), p\}.$$

---

* In a neighborhood of this point we have $f(z) \approx \operatorname{const}(z-a)^{-n}$.

## 11.2.1-2. Inverse Laplace transform.

Given the transform $\tilde{f}(p)$, the function $f(x)$ can be found by means of the inverse Laplace transform

$$f(x) = \frac{1}{2\pi i} \int_{c-i\infty}^{c+i\infty} \tilde{f}(p)e^{px}\, dp, \qquad i^2 = -1, \qquad (11.2.1.2)$$

where the integration path is parallel to the imaginary axis and lies to the right of all singularities of $\tilde{f}(p)$, which corresponds to $c > \sigma_0$.

The integral in inversion formula (11.2.1.2) is understood in the sense of the Cauchy principal value:

$$\int_{c-i\infty}^{c+i\infty} \tilde{f}(p)e^{px}\, dp = \lim_{\omega \to \infty} \int_{c-i\omega}^{c+i\omega} \tilde{f}(p)e^{px}\, dp.$$

In the domain $x < 0$, formula (11.2.1.2) gives $f(x) \equiv 0$.

Formula (11.2.1.2) holds for continuous functions. If $f(x)$ has a (finite) jump discontinuity at a point $x = x_0 > 0$, then the left-hand side of (11.2.1.2) is equal to $\frac{1}{2}[f(x_0-0)+f(x_0+0)]$ at this point (for $x_0 = 0$, the first term in the square brackets must be omitted).

For brevity, we write the Laplace inversion formula (11.2.1.2) as follows:

$$f(x) = \mathcal{L}^{-1}\{\tilde{f}(p)\} \qquad \text{or} \qquad f(x) = \mathcal{L}^{-1}\{\tilde{f}(p), x\}.$$

There are tables of direct and inverse Laplace transforms (see Sections T3.1 and T3.2), which are handy in solving linear differential and integral equations.

## 11.2.2. Main Properties of the Laplace Transform. Inversion Formulas for Some Functions

## 11.2.2-1. Convolution theorem. Main properties of the Laplace transform.

$1°$. The *convolution* of two functions $f(x)$ and $g(x)$ is defined as an integral of the form $\int_0^x f(t)g(x-t)\, dt$, and is usually denoted by $f(x) * g(x)$ or

$$f(x) * g(x) = \int_0^x f(t)\, g(x-t)\, dt.$$

By performing substitution $x - t = u$, we see that the convolution is symmetric with respect to the convolved functions: $f(x) * g(x) = g(x) * f(x)$.

The *convolution theorem* states that

$$\mathcal{L}\{f(x) * g(x)\} = \mathcal{L}\{f(x)\}\mathcal{L}\{g(x)\}$$

and is frequently applied to solve Volterra equations with kernels depending on the difference of the arguments.

$2°$. The main properties of the correspondence between functions and their Laplace transforms are gathered in Table 11.1.

$3°$. The Laplace transforms of some functions are listed in Table 11.2; for more detailed tables, see Section T3.1 and the list of references at the end of this chapter.

## TABLE 11.1
### Main properties of the Laplace transform

| No. | Function | Laplace transform | Operation |
|---|---|---|---|
| 1 | $af_1(x) + bf_2(x)$ | $a\widetilde{f}_1(p) + b\widetilde{f}_2(p)$ | Linearity |
| 2 | $f(x/a),\ a > 0$ | $a\widetilde{f}(ap)$ | Scaling |
| 3 | $f(x-a),$ $f(\xi) \equiv 0$ for $\xi < 0$ | $e^{-ap}\widetilde{f}(p)$ | Shift of the argument |
| 4 | $x^n f(x);\ n = 1, 2, \ldots$ | $(-1)^n \widetilde{f}_p^{(n)}(p)$ | Differentiation of the transform |
| 5 | $\dfrac{1}{x} f(x)$ | $\displaystyle\int_p^\infty \widetilde{f}(q)\,dq$ | Integration of the transform |
| 6 | $e^{ax} f(x)$ | $\widetilde{f}(p-a)$ | Shift in the complex plane |
| 7 | $f'_x(x)$ | $p\widetilde{f}(p) - f(+0)$ | Differentiation |
| 8 | $f_x^{(n)}(x)$ | $p^n \widetilde{f}(p) - \displaystyle\sum_{k=1}^n p^{n-k} f_x^{(k-1)}(+0)$ | Differentiation |
| 9 | $x^m f_x^{(n)}(x),\ m = 1, 2, \ldots$ | $(-1)^m \dfrac{d^m}{dp^m}\left[p^n \widetilde{f}(p) - \displaystyle\sum_{k=1}^n p^{n-k} f_x^{(k-1)}(+0)\right]$ | Differentiation |
| 10 | $\dfrac{d^n}{dx^n}\left[x^m f(x)\right],\ m \geq n$ | $(-1)^m p^n \dfrac{d^m}{dp^m} \widetilde{f}(p)$ | Differentiation |
| 11 | $\displaystyle\int_0^x f(t)\,dt$ | $\dfrac{\widetilde{f}(p)}{p}$ | Integration |
| 12 | $\displaystyle\int_0^x f_1(t) f_2(x-t)\,dt$ | $\widetilde{f}_1(p)\widetilde{f}_2(p)$ | Convolution |

## TABLE 11.2
### The Laplace transforms of some functions

| No. | Function, $f(x)$ | Laplace transform, $\widetilde{f}(p)$ | Remarks |
|---|---|---|---|
| 1 | $1$ | $1/p$ | |
| 2 | $x^n$ | $\dfrac{n!}{p^{n+1}}$ | $n = 1, 2, \ldots$ |
| 3 | $x^a$ | $\Gamma(a+1) p^{-a-1}$ | $a > -1$ |
| 4 | $e^{-ax}$ | $(p+a)^{-1}$ | |
| 5 | $x^a e^{-bx}$ | $\Gamma(a+1)(p+b)^{-a-1}$ | $a > -1$ |
| 6 | $\sinh(ax)$ | $\dfrac{a}{p^2 - a^2}$ | |
| 7 | $\cosh(ax)$ | $\dfrac{p}{p^2 - a^2}$ | |
| 8 | $\ln x$ | $-\dfrac{1}{p}(\ln p + \mathcal{C})$ | $\mathcal{C} = 0.5772\ldots$ is the Euler constant |
| 9 | $\sin(ax)$ | $\dfrac{a}{p^2 + a^2}$ | |
| 10 | $\cos(ax)$ | $\dfrac{p}{p^2 + a^2}$ | |
| 11 | $\mathrm{erfc}\left(\dfrac{a}{2\sqrt{x}}\right)$ | $\dfrac{1}{p}\exp(-a\sqrt{p})$ | $a \geq 0$ |
| 12 | $J_0(ax)$ | $\dfrac{1}{\sqrt{p^2 + a^2}}$ | $J_0(x)$ is the Bessel function |

**11.2.2-2. Inverse transforms of rational functions.**

Consider the important case in which the transform is a rational function of the form

$$\widetilde{f}(p) = \frac{R(p)}{Q(p)}, \qquad (11.2.2.1)$$

where $Q(p)$ and $R(p)$ are polynomials in the variable $p$ and the degree of $Q(p)$ exceeds that of $R(p)$.

Assume that the zeros of the denominator are simple, i.e.,

$$Q(p) \equiv \text{const}\,(p - \lambda_1)(p - \lambda_2)\ldots(p - \lambda_n).$$

Then the inverse transform can be determined by the formula

$$f(x) = \sum_{k=1}^{n} \frac{R(\lambda_k)}{Q'(\lambda_k)} \exp(\lambda_k x), \qquad (11.2.2.2)$$

where the primes denote the derivatives.

If $Q(p)$ has multiple zeros, i.e.,

$$Q(p) \equiv \text{const}\,(p - \lambda_1)^{s_1}(p - \lambda_2)^{s_2}\ldots(p - \lambda_m)^{s_m},$$

then

$$f(x) = \sum_{k=1}^{m} \frac{1}{(s_k - 1)!} \lim_{p \to s_k} \frac{d^{s_k-1}}{dp^{s_k-1}} \left[(p - \lambda_k)^{s_k} \widetilde{f}(p) e^{px}\right]. \qquad (11.2.2.3)$$

**Example 1.** The transform

$$\widetilde{f}(p) = \frac{b}{p^2 - a^2} \qquad (a, b \text{ real numbers})$$

can be represented as the fraction (11.2.2.1) with $R(p) = b$ and $Q(p) = (p - a)(p + a)$. The denominator $Q(p)$ has two simple roots, $\lambda_1 = a$ and $\lambda_2 = -a$. Using formula (11.2.2.2) with $n = 2$ and $Q'(p) = 2p$, we obtain the inverse transform in the form

$$f(x) = \frac{b}{2a} e^{ax} - \frac{b}{2a} e^{-ax} = \frac{b}{a} \sinh(ax).$$

**Example 2.** The transform

$$\widetilde{f}(p) = \frac{b}{p^2 + a^2} \qquad (a, b \text{ real numbers})$$

can be written as the fraction (11.2.2.1) with $R(p) = b$ and $Q(p) = (p - ia)(p + ia)$, $i^2 = -1$. The denominator $Q(p)$ has two simple pure imaginary roots, $\lambda_1 = ia$ and $\lambda_2 = -ia$. Using formula (11.2.2.2) with $n = 2$, we find the inverse transform:

$$f(x) = \frac{b}{2ia} e^{iax} - \frac{b}{2ia} e^{-iax} = -\frac{bi}{2a}\bigl[\cos(ax) + i\sin(ax)\bigr] + \frac{bi}{2a}\bigl[\cos(ax) - i\sin(ax)\bigr] = \frac{b}{a}\sin(ax).$$

**Example 3.** The transform

$$\widetilde{f}(p) = ap^{-n},$$

where $n$ is a positive integer, can be written as the fraction (11.2.2.1) with $R(p) = a$ and $Q(p) = p^n$. The denominator $Q(p)$ has one root of multiplicity $n$, $\lambda_1 = 0$. By formula (11.2.2.3) with $m = 1$ and $s_1 = n$, we find the inverse transform:

$$f(x) = \frac{a}{(n-1)!} x^{n-1}.$$

▶ *Fairly detailed tables of inverse Laplace transforms can be found in Section T3.2.*

### 11.2.2-3. Inversion of functions with finitely many singular points.

If the function $\widetilde{f}(p)$ has finitely many singular points, $p_1, p_2, \ldots, p_n$, and tends to zero as $p \to \infty$, then the integral in the Laplace inversion formula (11.2.1.2) may be evaluated using the residue theory by applying the Jordan lemma (see Subsection 11.1.2). In this case

$$f(x) = \sum_{k=1}^{n} \operatorname*{res}_{p=p_k} [\widetilde{f}(p)e^{px}]. \tag{11.2.2.4}$$

Formula (11.2.2.4) can be extended to the case where $\widetilde{f}(p)$ has infinitely many singular points. In this case, $f(x)$ is represented as an infinite series.

## 11.2.3. Limit Theorems. Representation of Inverse Transforms as Convergent Series and Asymptotic Expansions

### 11.2.3-1. Limit theorems.

THEOREM 1. *Let $0 \leq x < \infty$ and $\widetilde{f}(p) = \mathcal{L}\{f(x)\}$ be the Laplace transform of $f(x)$. If a limit of $f(x)$ as $x \to 0$ exists, then*

$$\lim_{x \to 0} f(x) = \lim_{p \to \infty} [p\widetilde{f}(p)].$$

THEOREM 2. *If a limit of $f(x)$ as $x \to \infty$ exists, then*

$$\lim_{x \to \infty} f(x) = \lim_{p \to 0} [p\widetilde{f}(p)].$$

### 11.2.3-2. Representation of inverse transforms as convergent series.

THEOREM 1. *Suppose the transform $\widetilde{f}(p)$ can be expanded into series in negative powers of $p$,*

$$\widetilde{f}(p) = \sum_{n=1}^{\infty} \frac{a_n}{p^n},$$

*convergent for $|p| > R$, where $R$ is an arbitrary positive number; note that the transform tends to zero as $|p| \to \infty$. Then the inverse transform can be obtained by the formula*

$$f(x) = \sum_{n=1}^{\infty} \frac{a_n}{(n-1)!} x^{n-1},$$

*where the series on the right-hand side is convergent for all $x$.*

THEOREM 2. *Suppose the transform $\widetilde{f}(p)$, $|p| > R$, is represented by an absolutely convergent series,*

$$\widetilde{f}(p) = \sum_{n=0}^{\infty} \frac{a_n}{p^{\lambda_n}}, \tag{11.2.3.1}$$

where $\{\lambda_n\}$ is any positive increasing sequence, $0 < \lambda_0 < \lambda_1 < \cdots \to \infty$. Then it is possible to proceed termwise from series (11.2.3.1) to the following inverse transform series:

$$f(x) = \sum_{n=0}^{\infty} \frac{a_n}{\Gamma(\lambda_n)} x^{\lambda_n - 1}, \qquad (11.2.3.2)$$

where $\Gamma(\lambda)$ is the Gamma function. Series (11.2.3.2) is convergent for all real and complex values of $x$ other than zero (if $\lambda_0 \geq 1$, the series is convergent for all $x$).

**11.2.3-3. Representation of inverse transforms as asymptotic expansions as $x \to \infty$.**

1°. Let $p = p_0$ be a singular point of the Laplace transform $\widetilde{f}(p)$ with the greatest real part (it is assumed there is only one such point). If $\widetilde{f}(p)$ can be expanded near $p = p_0$ into an absolutely convergent series,

$$\widetilde{f}(p) = \sum_{n=0}^{\infty} c_n (p - p_0)^{\lambda_n} \qquad (\lambda_0 < \lambda_1 < \cdots \to \infty) \qquad (11.2.3.3)$$

with arbitrary $\lambda_n$, then the inverse transform $f(x)$ can be expressed in the form of the asymptotic expansion

$$f(x) \sim e^{p_0 x} \sum_{n=0}^{\infty} \frac{c_n}{\Gamma(-\lambda_n)} x^{-\lambda_n - 1} \quad \text{as} \quad x \to \infty. \qquad (11.2.3.4)$$

The terms corresponding to nonnegative integer $\lambda_n$ must be omitted from the summation, since $\Gamma(0) = \Gamma(-1) = \Gamma(-2) = \cdots = \infty$.

2°. If the transform $\widetilde{f}(p)$ has several singular points, $p_1, \ldots, p_m$, with the same greatest real part, $\operatorname{Re} p_1 = \cdots = \operatorname{Re} p_m$, then expansions of the form (11.2.3.3) should be obtained for each of these points and the resulting expressions must be added together.

**11.2.3-4. Post–Widder formula.**

In applications, one can find $f(x)$ if the Laplace transform $\widetilde{f}(t)$ on the real semiaxis is known for $t = p \geq 0$. To this end, one uses the Post–Widder formula

$$f(x) = \lim_{n \to \infty} \left[ \frac{(-1)^n}{n!} \left(\frac{n}{x}\right)^{n+1} \widetilde{f}_t^{(n)}\left(\frac{n}{x}\right) \right]. \qquad (11.2.3.5)$$

Approximate inversion formulas are obtained by taking sufficiently large positive integer $n$ in (11.2.3.5) instead of passing to the limit.

## 11.3. Mellin Transform
### 11.3.1. Mellin Transform and the Inversion Formula

**11.3.1-1. Mellin transform.**

Suppose that a function $f(x)$ is defined for positive $x$ and satisfies the conditions

$$\int_0^1 |f(x)| x^{\sigma_1 - 1} \, dx < \infty, \qquad \int_1^{\infty} |f(x)| x^{\sigma_2 - 1} \, dx < \infty$$

for some real numbers $\sigma_1$ and $\sigma_2$, $\sigma_1 < \sigma_2$.

The Mellin transform of $f(x)$ is defined by

$$\hat{f}(s) = \int_0^\infty f(x) x^{s-1}\, dx, \qquad (11.3.1.1)$$

where $s = \sigma + i\tau$ is a complex variable ($\sigma_1 < \sigma < \sigma_2$).

For brevity, we rewrite formula (11.3.1.1) as follows:

$$\hat{f}(s) = \mathfrak{M}\{f(x)\} \qquad \text{or} \qquad \hat{f}(s) = \mathfrak{M}\{f(x), s\}.$$

**11.3.1-2. Inverse Mellin transform.**

Given $\hat{f}(s)$, the function $f(x)$ can be found by means of the *inverse Mellin transform*

$$f(x) = \frac{1}{2\pi i} \int_{\sigma-i\infty}^{\sigma+i\infty} \hat{f}(s) x^{-s}\, ds \qquad (\sigma_1 < \sigma < \sigma_2), \qquad (11.3.1.2)$$

where the integration path is parallel to the imaginary axis of the complex plane $s$ and the integral is understood in the sense of the Cauchy principal value.

Formula (11.3.1.2) holds for continuous functions. If $f(x)$ has a (finite) jump discontinuity at a point $x = x_0 > 0$, then the left-hand side of (11.3.1.2) is equal to $\frac{1}{2}\bigl[f(x_0-0) + f(x_0+0)\bigr]$ at this point (for $x_0 = 0$, the first term in the square brackets must be omitted).

For brevity, we rewrite formula (11.3.1.2) in the form

$$f(x) = \mathfrak{M}^{-1}\{\hat{f}(s)\} \qquad \text{or} \qquad f(x) = \mathfrak{M}^{-1}\{\hat{f}(s), x\}.$$

## 11.3.2. Main Properties of the Mellin Transform. Relation Among the Mellin, Laplace, and Fourier Transforms

**11.3.2-1. Main properties of the Mellin transform.**

1°. The main properties of the correspondence between the functions and their Mellin transforms are gathered in Table 11.3.

2°. The integral relations

$$\int_0^\infty f(x) g(x)\, dx = \mathfrak{M}^{-1}\{\hat{f}(s)\hat{g}(1-s)\},$$

$$\int_0^\infty f(x) g\!\left(\frac{1}{x}\right) dx = \mathfrak{M}^{-1}\{\hat{f}(s)\hat{g}(s)\}$$

hold for fairly general assumptions about the integrability of the functions involved (see Ditkin and Prudnikov, 1965).

**11.3.2-2. Relation among the Mellin, Laplace, and Fourier transforms.**

There are tables of direct and inverse Mellin transforms (see Sections T3.5 and T3.6 and the references listed at the end of the current chapter) that are useful in solving specific integral and differential equations. The Mellin transform is related to the Laplace and Fourier transforms as follows:

$$\mathfrak{M}\{f(x), s\} = \mathfrak{L}\{f(e^x), -s\} + \mathfrak{L}\{f(e^{-x}), s\} = \mathfrak{F}\{f(e^x), is\},$$

which makes it possible to apply much more common tables of direct and inverse Laplace and Fourier transforms.

TABLE 11.3
Main properties of the Mellin transform

| No. | Function | Mellin transform | Operation |
|---|---|---|---|
| 1 | $af_1(x) + bf_2(x)$ | $a\hat{f}_1(s) + b\hat{f}_2(s)$ | Linearity |
| 2 | $f(ax)$, $a > 0$ | $a^{-s}\hat{f}(s)$ | Scaling |
| 3 | $x^a f(x)$ | $\hat{f}(s+a)$ | Shift of the argument of the transform |
| 4 | $f(x^2)$ | $\frac{1}{2}\hat{f}(\frac{1}{2}s)$ | Squared argument |
| 5 | $f(1/x)$ | $\hat{f}(-s)$ | Inversion of the argument of the transform |
| 6 | $x^\lambda f(ax^\beta)$, $a > 0, \beta \neq 0$ | $\frac{1}{\beta} a^{-\frac{s+\lambda}{\beta}} \hat{f}\left(\frac{s+\lambda}{\beta}\right)$ | Power law transform |
| 7 | $f'_x(x)$ | $-(s-1)\hat{f}(s-1)$ | Differentiation |
| 8 | $xf'_x(x)$ | $-s\hat{f}(s)$ | Differentiation |
| 9 | $f_x^{(n)}(x)$ | $(-1)^n \frac{\Gamma(s)}{\Gamma(s-n)} \hat{f}(s-n)$ | Multiple differentiation |
| 10 | $\left(x\frac{d}{dx}\right)^n f(x)$ | $(-1)^n s^n \hat{f}(s)$ | Multiple differentiation |
| 11 | $x^\alpha \int_0^\infty t^\beta f_1(xt) f_2(t)\, dt$ | $\hat{f}_1(s+\alpha)\hat{f}_2(1-s-\alpha+\beta)$ | Complicated integration |
| 12 | $x^\alpha \int_0^\infty t^\beta f_1\left(\frac{x}{t}\right) f_2(t)\, dt$ | $\hat{f}_1(s+\alpha)\hat{f}_2(s+\alpha+\beta+1)$ | Complicated integration |

## 11.4. Various Forms of the Fourier Transform

### 11.4.1. Fourier Transform and the Inverse Fourier Transform

11.4.1-1. Standard form of the Fourier transform.

The *Fourier transform* is defined as follows:

$$\widetilde{f}(u) = \frac{1}{\sqrt{2\pi}} \int_{-\infty}^{\infty} f(x) e^{-iux}\, dx. \qquad (11.4.1.1)$$

For brevity, we rewrite formula (11.4.1.1) as follows:

$$\widetilde{f}(u) = \mathfrak{F}\{f(x)\} \quad \text{or} \quad \widetilde{f}(u) = \mathfrak{F}\{f(x), u\}.$$

Given $\widetilde{f}(u)$, the function $f(x)$ can be found by means of the *inverse Fourier transform*

$$f(x) = \frac{1}{\sqrt{2\pi}} \int_{-\infty}^{\infty} \widetilde{f}(u) e^{iux}\, du. \qquad (11.4.1.2)$$

Formula (11.4.1.2) holds for continuous functions. If $f(x)$ has a (finite) jump discontinuity at a point $x = x_0$, then the left-hand side of (11.4.1.2) is equal to $\frac{1}{2}\left[f(x_0-0) + f(x_0+0)\right]$ at this point.

TABLE 11.4
Main properties of the Fourier transform

| No. | Function | Fourier transform | Operation |
|---|---|---|---|
| 1 | $af_1(x) + bf_2(x)$ | $a\widetilde{f}_1(u) + b\widetilde{f}_2(u)$ | Linearity |
| 2 | $f(x/a)$, $a > 0$ | $a\widetilde{f}(au)$ | Scaling |
| 3 | $x^n f(x)$; $n = 1, 2, \ldots$ | $i^n \widetilde{f}_u^{(n)}(u)$ | Differentiation of the transform |
| 4 | $f''_{xx}(x)$ | $-u^2 \widetilde{f}(u)$ | Differentiation |
| 5 | $f_x^{(n)}(x)$ | $(iu)^n \widetilde{f}(u)$ | Differentiation |
| 6 | $\int_{-\infty}^{\infty} f_1(\xi) f_2(x-\xi)\, d\xi$ | $\widetilde{f}_1(u) \widetilde{f}_2(u)$ | Convolution |

For brevity, we rewrite formula (11.4.1.2) as follows:

$$f(x) = \mathfrak{F}^{-1}\{\widetilde{f}(u)\} \quad \text{or} \quad f(x) = \mathfrak{F}^{-1}\{\widetilde{f}(u), x\}.$$

**11.4.1-2. Asymmetric form of the Fourier transform. Alternative Fourier transform.**

1°. Sometimes it is more convenient to define the Fourier transform by

$$\check{f}(u) = \int_{-\infty}^{\infty} f(x) e^{-iux}\, dx.$$

In this case, the *Fourier inversion formula* reads

$$f(x) = \frac{1}{2\pi} \int_{-\infty}^{\infty} \check{f}(u) e^{iux}\, du.$$

2°. Sometimes the alternative Fourier transform is used (and called merely the *Fourier transform*), which corresponds to the renaming $e^{-iux} \rightleftarrows e^{iux}$ on the right-hand sides of (11.4.1.1) and (11.4.1.2).

**11.4.1-3. Convolution theorem. Main properties of the Fourier transforms.**

1°. The *convolution* of two functions $f(x)$ and $g(x)$ is defined as

$$f(x) * g(x) \equiv \frac{1}{\sqrt{2\pi}} \int_{-\infty}^{\infty} f(x-t) g(t)\, dt.$$

By performing substitution $x - t = u$, we see that the convolution is symmetric with respect to the convolved functions: $f(x) * g(x) = g(x) * f(x)$.

The *convolution theorem* states that

$$\mathfrak{F}\{f(x) * g(x)\} = \mathfrak{F}\{f(x)\} \mathfrak{F}\{g(x)\}.$$

2°. The main properties of the correspondence between functions and their Fourier transforms are gathered in Table 11.4.

## 11.4. Various Forms of the Fourier Transform

**11.4.1-4. $n$-dimensional Fourier transform.**

The Fourier transform admits $n$-dimensional generalization:

$$\widetilde{f}(\mathbf{u}) = \frac{1}{(2\pi)^{n/2}} \int_{\mathbb{R}^n} f(\mathbf{x}) e^{-i(\mathbf{u}\cdot\mathbf{x})} \, d\mathbf{x}, \qquad (\mathbf{u}\cdot\mathbf{x}) = u_1 x_1 + \cdots + u_n x_n, \qquad (11.4.1.3)$$

where $f(\mathbf{x}) = f(x_1, \ldots, x_n)$, $\widetilde{f}(\mathbf{u}) = f(u_1, \ldots, u_n)$, and $d\mathbf{x} = dx_1 \ldots dx_n$.

The corresponding inversion formula is

$$f(\mathbf{x}) = \frac{1}{(2\pi)^{n/2}} \int_{\mathbb{R}^n} \widetilde{f}(\mathbf{u}) e^{i(\mathbf{u}\cdot\mathbf{x})} \, d\mathbf{u}, \qquad d\mathbf{u} = du_1 \ldots du_n.$$

The Fourier transform (11.4.1.3) is frequently used in the theory of linear partial differential equations with constant coefficients ($\mathbf{x} \in \mathbb{R}^n$).

### 11.4.2. Fourier Cosine and Sine Transforms

**11.4.2-1. Fourier cosine transform.**

1°. Let a function $f(x)$ be integrable on the semiaxis $0 \le x < \infty$. The *Fourier cosine transform* is defined by

$$\widetilde{f}_c(u) = \sqrt{\frac{2}{\pi}} \int_0^\infty f(x) \cos(xu) \, dx, \qquad 0 < u < \infty. \qquad (11.4.2.1)$$

For given $\widetilde{f}_c(u)$, the function can be found by means of the *Fourier cosine inversion formula*

$$f(x) = \sqrt{\frac{2}{\pi}} \int_0^\infty \widetilde{f}_c(u) \cos(xu) \, du, \qquad 0 < x < \infty. \qquad (11.4.2.2)$$

The Fourier cosine transform (11.4.2.1) is denoted for brevity by $\widetilde{f}_c(u) = \mathfrak{F}_c\{f(x)\}$.

2°. It follows from formula (11.4.2.2) that the Fourier cosine transform has the property $\mathfrak{F}_c^2 = 1$.

Some other properties of the Fourier cosine transform:

$$\mathfrak{F}_c\{x^{2n} f(x)\} = (-1)^n \frac{d^{2n}}{du^{2n}} \mathfrak{F}_c\{f(x)\}, \qquad n = 1, 2, \ldots;$$

$$\mathfrak{F}_c\{f''(x)\} = -u^2 \mathfrak{F}_c\{f(x)\}.$$

Here, $f(x)$ is assumed to vanish sufficiently rapidly (exponentially) as $x \to \infty$. For the second formula, the condition $f'(0) = 0$ is assumed to hold.

*Parseval's relation for the Fourier cosine transform*:

$$\int_0^\infty \mathfrak{F}_c\{f(x)\} \mathfrak{F}_c\{g(x)\} \, du = \int_0^\infty f(x) g(x) \, dx.$$

There are tables of the Fourier cosine transform (see Section T3.3 and the references listed at the end of the current chapter).

3°. Sometimes the asymmetric form of the Fourier cosine transform is applied, which is given by the pair of formulas

$$\check{f}_c(u) = \int_0^\infty f(x) \cos(xu) \, dx, \qquad f(x) = \frac{2}{\pi} \int_0^\infty \check{f}_c(u) \cos(xu) \, du.$$

### 11.4.2-2. Fourier sine transform.

1°. Let a function $f(x)$ be integrable on the semiaxis $0 \leq x < \infty$. The *Fourier sine transform* is defined by

$$\tilde{f}_s(u) = \sqrt{\frac{2}{\pi}} \int_0^\infty f(x) \sin(xu)\,dx, \qquad 0 < u < \infty. \tag{11.4.2.3}$$

For given $\tilde{f}_s(u)$, the function $f(x)$ can be found by means of the *inverse Fourier sine transform*

$$f(x) = \sqrt{\frac{2}{\pi}} \int_0^\infty \tilde{f}_s(u) \sin(xu)\,du, \qquad 0 < x < \infty. \tag{11.4.2.4}$$

The Fourier sine transform (11.4.2.3) is briefly denoted by $\tilde{f}_s(u) = \mathfrak{F}_s\{f(x)\}$.

2°. It follows from formula (11.4.2.4) that the Fourier sine transform has the property $\mathfrak{F}_s^2 = 1$.

Some other properties of the Fourier sine transform:

$$\mathfrak{F}_s\{x^{2n}f(x)\} = (-1)^n \frac{d^{2n}}{du^{2n}} \mathfrak{F}_s\{f(x)\}, \qquad n = 1, 2, \dots;$$

$$\mathfrak{F}_s\{f''(x)\} = -u^2 \mathfrak{F}_s\{f(x)\}.$$

Here, $f(x)$ is assumed to vanish sufficiently rapidly (exponentially) as $x \to \infty$. For the second formula, the condition $f(0) = 0$ is assumed to hold.

*Parseval's relation for the Fourier sine transform*:

$$\int_0^\infty \mathfrak{F}_s\{f(x)\} \mathfrak{F}_s\{g(x)\}\,du = \int_0^\infty f(x)g(x)\,dx.$$

There are tables of the Fourier cosine transform (see Section T3.4 and the references listed at the end of the current chapter).

3°. Sometimes it is more convenient to apply the asymmetric form of the Fourier sine transform defined by the following two formulas:

$$\check{f}_s(u) = \int_0^\infty f(x)\sin(xu)\,dx, \qquad f(x) = \frac{2}{\pi}\int_0^\infty \check{f}_s(u)\sin(xu)\,du.$$

## 11.5. Other Integral Transforms

### 11.5.1. Integral Transforms Whose Kernels Contain Bessel Functions and Modified Bessel Functions

#### 11.5.1-1. Hankel transform.

1°. The *Hankel transform* is defined as follows:

$$\tilde{f}_\nu(u) = \int_0^\infty x J_\nu(ux) f(x)\,dx, \qquad 0 < u < \infty, \tag{11.5.1.1}$$

where $\nu > -\frac{1}{2}$ and $J_\nu(x)$ is the Bessel function of the first kind of order $\nu$ (see Section SF.6).

For given $\widetilde{f}_\nu(u)$, the function $f(x)$ can be found by means of the *Hankel inversion formula*

$$f(x) = \int_0^\infty u J_\nu(ux) \widetilde{f}_\nu(u)\, du, \qquad 0 < x < \infty. \tag{11.5.1.2}$$

Note that if $f(x) = O(x^\alpha)$ as $x \to 0$, where $\alpha + \nu + 2 > 0$, and $f(x) = O(x^\beta)$ as $x \to \infty$, where $\beta + \frac{3}{2} < 0$, then the integral (11.5.1.1) is convergent.

The inversion formula (11.5.1.2) holds for continuous functions. If $f(x)$ has a (finite) jump discontinuity at a point $x = x_0$, then the left-hand side of (11.5.1.2) is equal to $\frac{1}{2}[f(x_0 - 0) + f(x_0 + 0)]$ at this point.

For brevity, we denote the Hankel transform (11.5.1.1) by $\widetilde{f}_\nu(u) = \mathcal{H}_\nu\{f(x)\}$.

2°. It follows from formula (11.5.1.2) that the Hankel transform has the property $\mathcal{H}_\nu^2 = 1$. Other properties of the Hankel transform:

$$\mathcal{H}_\nu\left\{\frac{1}{x} f(x)\right\} = \frac{u}{2\nu} \mathcal{H}_{\nu-1}\{f(x)\} + \frac{u}{2\nu} \mathcal{H}_{\nu+1}\{f(x)\},$$

$$\mathcal{H}_\nu\{f'(x)\} = \frac{(\nu-1)u}{2\nu} \mathcal{H}_{\nu+1}\{f(x)\} - \frac{(\nu+1)u}{2\nu} \mathcal{H}_{\nu-1}\{f(x)\},$$

$$\mathcal{H}_\nu\left\{f''(x) + \frac{1}{x} f'(x) - \frac{\nu^2}{x^2} f(x)\right\} = -u^2 \mathcal{H}_\nu\{f(x)\}.$$

The conditions

$$\lim_{x \to 0} [x^\nu f(x)] = 0, \quad \lim_{x \to 0} [x^{\nu+1} f'(x)] = 0, \quad \lim_{x \to \infty} [x^{1/2} f(x)] = 0, \quad \lim_{x \to \infty} [x^{1/2} f'(x)] = 0$$

are assumed to hold for the last formula.

*Parseval's relation for the Hankel transform*:

$$\int_0^\infty u \mathcal{H}_\nu\{f(x)\} \mathcal{H}_\nu\{g(x)\}\, du = \int_0^\infty x f(x) g(x)\, dx, \qquad \nu > -\frac{1}{2}.$$

### 11.5.1-2. Meijer transform.

The *Meijer transform* is defined as follows:

$$\hat{f}_\mu(s) = \sqrt{\frac{2}{\pi}} \int_0^\infty \sqrt{sx}\, K_\mu(sx) f(x)\, dx, \qquad 0 < s < \infty,$$

where $K_\mu(x)$ is the modified Bessel function of the second kind (the Macdonald function) of order $\mu$ (see Section SF.7).

For given $\widetilde{f}_\mu(s)$, the function $f(x)$ can be found by means of the *Meijer inversion formula*

$$f(x) = \frac{1}{i\sqrt{2\pi}} \int_{c-i\infty}^{c+i\infty} \sqrt{sx}\, I_\mu(sx) \hat{f}_\mu(s)\, ds, \qquad 0 < x < \infty,$$

where $I_\mu(x)$ is the modified Bessel function of the first kind of order $\mu$ (see Section SF.7). For the Meijer transform, a convolution is defined and an operational calculus is developed.

### 11.5.1-3. Kontorovich–Lebedev transform.

The *Kontorovich–Lebedev transform* is introduced as follows:

$$F(\tau) = \int_0^\infty K_{i\tau}(x) f(x)\, dx, \qquad 0 < \tau < \infty,$$

where $K_\mu(x)$ is the modified Bessel function of the second kind (the Macdonald function) of order $\mu$ (see Section SF.7) and $i = \sqrt{-1}$.

For given $F(\tau)$, the function can be found by means of the *Kontorovich–Lebedev inversion formula*

$$f(x) = \frac{2}{\pi^2 x} \int_0^\infty \tau \sinh(\pi\tau) K_{i\tau}(x) F(\tau)\, d\tau, \qquad 0 < x < \infty.$$

### 11.5.1-4. $Y$-transform.

The $Y$-transform is defined by

$$F_\nu(u) = \int_0^\infty \sqrt{ux}\, Y_\nu(ux) f(x)\, dx,$$

where $Y_\nu(x)$ is the Bessel function of the second kind of order $\nu$.

Given a transform $F_\nu(u)$, the inverse $Y$-transform $f(x)$ is found by the inversion formula

$$f(x) = \int_0^\infty \sqrt{ux}\, \mathbf{H}_\nu(ux) F_\nu(u)\, du,$$

where $\mathbf{H}_\nu(x)$ is the Struve function, which is defined as

$$\mathbf{H}_\nu(x) = \sum_{j=0}^\infty \frac{(-1)^j (x/2)^{\nu+2j+1}}{\Gamma\!\left(j + \frac{3}{2}\right) \Gamma\!\left(\nu + j + \frac{3}{2}\right)}.$$

## 11.5.2. Summary Table of Integral Transforms. Areas of Application of Integral Transforms

### 11.5.2-1. Summary table of integral transforms.

Table 11.5 summarizes the integral transforms considered above and also lists some other integral transforms; for the constraints imposed on the functions and parameters occurring in the integrand, see the references given at the end of this section.

### 11.5.2-2. Areas of application of integral transforms.

Integral transforms are widely used for the evaluation of integrals, summation of series, and solution of various mathematical equations and problems. In particular, the application of an appropriate integral transform to linear ordinary differential, integral, and difference equations reduces the problem to a linear algebraic equation for the transform; and linear partial differential equations are reduced to an ordinary differential equation.

## TABLE 11.5
### Summary table of integral transforms

| Integral transform | Definition | Inversion formula |
|---|---|---|
| Laplace transform | $\widetilde{f}(p) = \int_0^\infty e^{-px} f(x)\, dx$ | $f(x) = \dfrac{1}{2\pi i} \int_{c-i\infty}^{c+i\infty} e^{px} \widetilde{f}(p)\, dp$ |
| Laplace–Carlson transform | $\widetilde{f}(p) = p \int_0^\infty e^{-px} f(x)\, dx$ | $f(x) = \dfrac{1}{2\pi i} \int_{c-i\infty}^{c+i\infty} e^{px} \dfrac{\widetilde{f}(p)}{p}\, dp$ |
| Two-sided Laplace transform | $\widetilde{f}_*(p) = \int_{-\infty}^\infty e^{-px} f(x)\, dx$ | $f(x) = \dfrac{1}{2\pi i} \int_{c-i\infty}^{c+i\infty} e^{px} \widetilde{f}_*(p)\, dp$ |
| Fourier transform | $\widetilde{f}(u) = \dfrac{1}{\sqrt{2\pi}} \int_{-\infty}^\infty e^{-iux} f(x)\, dx$ | $f(x) = \dfrac{1}{\sqrt{2\pi}} \int_{-\infty}^\infty e^{iux} \widetilde{f}(u)\, du$ |
| Fourier sine transform | $\widetilde{f}_s(u) = \sqrt{\dfrac{2}{\pi}} \int_0^\infty \sin(xu) f(x)\, dx$ | $f(x) = \sqrt{\dfrac{2}{\pi}} \int_0^\infty \sin(xu) \widetilde{f}_s(u)\, du$ |
| Fourier cosine transform | $\widetilde{f}_c(u) = \sqrt{\dfrac{2}{\pi}} \int_0^\infty \cos(xu) f(x)\, dx$ | $f(x) = \sqrt{\dfrac{2}{\pi}} \int_0^\infty \cos(xu) \widetilde{f}_c(u)\, du$ |
| Hartley transform | $\widetilde{f}_h(u) = \dfrac{1}{\sqrt{2\pi}} \int_{-\infty}^\infty (\cos xu + \sin xu) f(x)\, dx$ | $f(x) = \dfrac{1}{\sqrt{2\pi}} \int_{-\infty}^\infty (\cos xu + \sin xu) \widetilde{f}_h(u)\, du$ |
| Mellin transform | $\widehat{f}(s) = \int_0^\infty x^{s-1} f(x)\, dx$ | $f(x) = \dfrac{1}{2\pi i} \int_{c-i\infty}^{c+i\infty} x^{-s} \widehat{f}(s)\, ds$ |
| Hankel transform | $\widehat{f}_\nu(w) = \int_0^\infty x J_\nu(xw) f(x)\, dx$ | $f(x) = \int_0^\infty w J_\nu(xw) \widehat{f}_\nu(w)\, dw$ |
| $Y$-transform | $F_\nu(u) = \int_0^\infty \sqrt{ux}\, Y_\nu(ux) f(x)\, dx$ | $f(x) = \int_0^\infty \sqrt{ux}\, \mathbf{H}_\nu(ux) F_\nu(u)\, du$ |
| Meijer transform ($K$-transform) | $\widehat{f}(s) = \sqrt{\dfrac{2}{\pi}} \int_0^\infty \sqrt{sx}\, K_\nu(sx) f(x)\, dx$ | $f(x) = \dfrac{1}{i\sqrt{2\pi}} \int_{c-i\infty}^{c+i\infty} \sqrt{sx}\, I_\nu(sx) \widehat{f}(s)\, ds$ |
| Bochner transform | $\widetilde{f}(r) = \int_0^\infty J_{n/2-1}(2\pi xr) G(x, r) f(x)\, dx,$ $G(x,r) = 2\pi r (x/r)^{n/2}, \quad n = 1, 2, \ldots$ | $f(x) = \int_0^\infty J_{n/2-1}(2\pi rx) G(r, x) \widetilde{f}(r)\, dr$ |
| Weber transform | $F_a(u) = \int_a^\infty W_\nu(xu, au) x f(x)\, dx,$ $W_\nu(\beta, \mu) \equiv J_\nu(\beta) Y_\nu(\mu) - J_\nu(\mu) Y_\nu(\beta)$ | $f(x) = \int_0^\infty \dfrac{W_\nu(xu, au)}{J_\nu^2(au) + Y_\nu^2(au)} u F_a(u)\, du$ |
| Hardy transform | $F(u) = \int_0^\infty C_\nu(xu) x f(x)\, dx,$ $C_\nu(z) \equiv \cos(\pi p) J_\nu(z) + \sin(\pi p) Y_\nu(z)$ | $f(x) = \int_0^\infty \Phi(xu) u F(u)\, du$ $\Phi(z) = \sum_{n=0}^\infty \dfrac{(-1)^n (z/2)^{\nu+2p+2n}}{\Gamma(p+n+1)\Gamma(\nu+p+n+1)}$ |
| Kontorovich–Lebedev transform | $F(\tau) = \int_0^\infty K_{i\tau}(x) f(x)\, dx$ | $f(x) = \dfrac{2}{\pi^2 x} \int_0^\infty \tau \sinh(\pi\tau) K_{i\tau}(x) F(\tau)\, d\tau$ |
| Meler–Fock transform | $\widetilde{F}(\tau) = \int_1^\infty P_{-\frac{1}{2}+i\tau}(x) f(x)\, dx$ | $f(x) = \int_0^\infty \tau \tanh(\pi\tau) P_{-\frac{1}{2}+i\tau}(x) \widetilde{F}(\tau)\, d\tau$ |
| Euler transform of the 1st kind* | $F(x) = \dfrac{1}{\Gamma(\mu)} \int_a^x \dfrac{f(t)\, dt}{(x-t)^{1-\mu}}$ $0 < \mu < 1, \quad x > a$ | $f(x) = \dfrac{1}{\Gamma(1-\mu)} \dfrac{d}{dx} \int_a^x \dfrac{F(t)\, dt}{(x-t)^\mu}$ |

TABLE 11.5 (continued)
Summary table of integral transforms

| Integral transform | Definition | Inversion formula |
|---|---|---|
| Euler transform of the 2nd kind* | $F(x) = \dfrac{1}{\Gamma(\mu)} \displaystyle\int_x^a \dfrac{f(t)\,dt}{(t-x)^{1-\mu}}$ $0 < \mu < 1, \ x < a$ | $f(x) = -\dfrac{1}{\Gamma(1-\mu)} \dfrac{d}{dx} \displaystyle\int_x^a \dfrac{F(t)\,dt}{(t-x)^\mu}$ |
| Gauss transform** | $F(x) = \dfrac{1}{\sqrt{\pi a}} \displaystyle\int_{-\infty}^\infty \exp\left[-\dfrac{(x-t)^2}{a}\right] f(t)\,dt$ | $f(x) = \exp\left(-\dfrac{a}{4}\dfrac{d^2}{dx^2}\right) F(x)$ |
| Hilbert transform*** | $\widehat{F}(s) = \dfrac{1}{\pi} \displaystyle\int_{-\infty}^\infty \dfrac{f(x)}{x-s}\,dx$ | $f(x) = -\dfrac{1}{\pi} \displaystyle\int_{-\infty}^\infty \dfrac{\widehat{F}(s)}{s-x}\,ds$ |

NOTATION: $i = \sqrt{-1}$, $J_\mu(x)$ and $Y_\mu(x)$ are the Bessel functions of the first and the second kind, respectively; $I_\mu(x)$ and $K_\mu(x)$ are the modified Bessel functions of the first and the second kind, respectively; $P_\mu(x)$ is the Legendre spherical function of the second kind; and $\mathbf{H}_\mu(x)$ is the Struve function (see Subsection 11.5.1-4).

REMARKS.

* The Euler transform of the first kind is also known as Riemann–Liouville integral (the left fractional integral of order $\mu$ or, for short, the fractional integral). The Euler transform of the second kind is also called the right fractional integral of order $\mu$.

** If $a = 4$, the Gauss transform is called the Weierstrass transform. In the inversion formula, the exponential is represented by an operator series: $\exp\left(k\dfrac{d^2}{dx^2}\right) \equiv 1 + \sum_{n=1}^\infty \dfrac{k^n}{n!}\dfrac{d^{2n}}{dx^{2n}}$.

*** In the direct and inverse Hilbert transforms, the integrals are understood in the sense of the Cauchy principal value.

Table 11.6 presents various areas of application of integral transforms with literature references.

**Example.**
Consider the Cauchy problem for the integro-differential equation

$$\frac{dy}{dx} + \int_0^x K(x-t)y(t)\,dt = f(x) \qquad (0 \le t < \infty) \tag{11.5.2.1}$$

with the initial condition

$$y = a \quad \text{at} \quad t = 0. \tag{11.5.2.2}$$

Multiply equation (11.5.2.1) by $e^{-px}$ and then integrate with respect to $x$ from zero to infinity. Using properties 7 and 12 of the Laplace transform (Table 11.1) and taking into account the initial condition (11.5.2.2), we obtain a linear algebraic equation for the transform $\widetilde{y}(p)$:

$$p\widetilde{y}(p) - a + \widetilde{K}(p)\widetilde{y}(p) = \widetilde{f}(p).$$

It follows that

$$\widetilde{y}(p) = \frac{\widetilde{f}(p) + a}{p + \widetilde{K}(p)}.$$

By the inversion formula (11.2.1.2), the solution to the original problem (11.5.2.1)–(11.5.2.2) is found in the form

$$y(x) = \frac{1}{2\pi i} \int_{c-i\infty}^{c+i\infty} \frac{\widetilde{f}(p) + a}{p + \widetilde{K}(p)} e^{px}\,dp, \qquad i^2 = -1. \tag{11.5.2.3}$$

Consider the special case of $a = 0$ and $K(x) = \cos(bx)$. From row 10 of Table 11.2 it follows that $\widetilde{K}(p) = \dfrac{p}{p^2 + b^2}$. Rearrange the integrand of (11.5.2.3):

$$\frac{\widetilde{f}(p)}{p + \widetilde{K}(p)} = \frac{p^2 + b^2}{p(p^2 + b^2 + 1)}\widetilde{f}(p) = \left(\frac{1}{p} - \frac{1}{p(p^2 + b^2 + 1)}\right)\widetilde{f}(p).$$

In order to invert this expression, let us use the convolution theorem (see formula 16 of Subsection T3.2.1) as well as formulas 1 and 28 for the inversion of rational functions, Subsection T3.2.1. As a result, we arrive at the solution in the form

$$y(x) = \int_0^x \frac{b^2 + \cos(t\sqrt{b^2+1})}{b^2 + 1} f(x-t)\,dt.$$

TABLE 11.6
Areas of application of integral transforms (first in the last column
come references to appropriate sections of the current book)

| Area of application | Integral transforms | References |
|---|---|---|
| Evaluation of improper integrals | Laplace, Mellin | Paragraph 7.2.8-5; Ditkin & Prudnikov (1965) |
| Summation of series | Laplace (direct and inverse) | Paragraphs 8.1.5-2 and 8.4.4-1 |
| Computation of coefficients of asymptotic expansions | Laplace, Mellin | Paragraph 7.2.9-1 |
| Linear constant- and variable-coefficient ordinary differential equations | Laplace, Mellin, Euler, and others | Paragraphs 12.4.1-3 and 12.4.2-6; Ditkin & Prudnikov (1965); Doetsch (1974); E. Kamke (1977); Sveshnikov & Tikhonov (1970); LePage (1980) |
| Systems of linear constant-coefficient ordinary differential equations | Laplace | Paragraph 12.6.1-4; Doetsch (1974); Ditkin & Prudnikov (1965) |
| Linear equations of mathematical physics | Laplace, Fourier, Fourier sine, Hankel, Kontorovich–Lebedev, and others | Section 14.5; Doetsch (1974); Ditkin & Prudnikov (1965); Antimirov (1993); Sneddon (1995); Zwillinger (1997); Bracewell (1999); Polyanin (2002); Duffy (2004) |
| Linear integral equations | Laplace, Mellin, Fourier, Meler–Fock, Euler, and others | Subsections 16.1.3, 16.2.3, 16.3.2, 16.4.6; Krasnov, Kiselev, & Makarenko (1971); Ditkin & Prudnikov (1965); Samko, Kilbas, & Marichev (1993); Polyanin & Manzhirov (1998) |
| Nonlinear integral equations | Laplace, Mellin, Fourier | Paragraphs 16.5.2-1 and 16.5.3-2; Krasnov, Kiselev, & Makarenko (1971); Polyanin & Manzhirov (1998) |
| Linear difference equations | Laplace | Ditkin & Prudnikov (1965) |
| Linear differential-difference equations | Laplace | Bellman & Roth (1984) |
| Linear integro-differential equations | Laplace, Fourier | Paragraph 11.5.2-2; LePage (1980) |

# References for Chapter 11

**Antimirov, M. Ya.,** *Applied Integral Transforms*, American Mathematical Society, Providence, Rhode Island, 1993.
**Bateman, H. and Erdélyi, A.,** *Tables of Integral Transforms. Vols. 1 and 2*, McGraw-Hill, New York, 1954.
**Beerends, R. J., ter Morschem, H. G., and van den Berg, J. C.,** *Fourier and Laplace Transforms*, Cambridge University Press, Cambridge, 2003.
**Bellman, R. and Roth, R.,** *The Laplace Transform*, World Scientific Publishing Co., Singapore, 1984.
**Bracewell, R.,** *The Fourier Transform and Its Applications, 3rd Edition*, McGraw-Hill, New York, 1999.
**Brychkov, Yu. A. and Prudnikov, A. P.,** *Integral Transforms of Generalized Functions*, Gordon & Breach, New York, 1989.
**Davis, B.,** *Integral Transforms and Their Applications*, Springer-Verlag, New York, 1978.
**Ditkin, V. A. and Prudnikov, A. P.,** *Integral Transforms and Operational Calculus*, Pergamon Press, New York, 1965.
**Doetsch, G.,** *Handbuch der Laplace-Transformation. Theorie der Laplace-Transformation*, Birkhäuser, Basel–Stuttgart, 1950.
**Doetsch, G.,** *Handbuch der Laplace-Transformation. Anwendungen der Laplace-Transformation*, Birkhäuser, Basel–Stuttgart, 1956.
**Doetsch, G.,** *Introduction to the Theory and Application of the Laplace Transformation*, Springer-Verlag, Berlin, 1974.
**Duffy, D. G.,** *Transform Methods for Solving Partial Differential Equations, 2nd Edition*, Chapman & Hall/CRC Press, Boca Raton, 2004.

**Hirschman, I. I. and Widder, D. V.**, *The Convolution Transform*, Princeton University Press, Princeton, New Jersey, 1955.

**Kamke, E.**, *Differentialgleichungen: Lösungsmethoden und Lösungen, I, Gewöhnliche Differentialgleichungen*, B. G. Teubner, Leipzig, 1977.

**Krantz, S. G.**, *Handbook of Complex Variables*, Birkhäuser, Boston, 1999.

**Krasnov, M. L., Kiselev, A. I., and Makarenko, G. I.**, *Problems and Exercises in Integral Equations*, Mir Publishers, Moscow, 1971.

**LePage, W. R.**, *Complex Variables and the Laplace Transform for Engineers*, Dover Publications, New York, 1980.

**Miles, J. W.**, *Integral Transforms in Applied Mathematics*, Cambridge University Press, Cambridge, 1971.

**Oberhettinger, F.**, *Tables of Bessel Transforms*, Springer-Verlag, New York, 1972.

**Oberhettinger, F.**, *Tables of Fourier Transforms and Fourier Transforms of Distributions*, Springer-Verlag, Berlin, 1980.

**Oberhettinger, F.**, *Tables of Mellin Transforms*, Springer-Verlag, New York, 1974.

**Oberhettinger, F. and Badii, L.**, *Tables of Laplace Transforms*, Springer-Verlag, New York, 1973.

**Pinkus, A. and Zafrany, S.**, *Fourier Series and Integral Transforms*, Cambridge University Press, Cambridge, 1997.

**Polyanin, A. D.**, *Handbook of Linear Partial Differential Equations for Engineers and Scientists*, Chapman & Hall/CRC Press, Boca Raton, 2002.

**Polyanin, A. D. and Manzhirov, A. V.**, *Handbook of Integral Equations*, CRC Press, Boca Raton, 1998.

**Poularikas, A. D.**, *The Transforms and Applications Handbook, 2nd Edition*, CRC Press, Boca Raton, 2000.

**Prudnikov, A. P., Brychkov, Yu. A., and Marichev, O. I.**, *Integrals and Series, Vol. 4, Direct Laplace Transform*, Gordon & Breach, New York, 1992.

**Prudnikov, A. P., Brychkov, Yu. A., and Marichev, O. I.**, *Integrals and Series, Vol. 5, Inverse Laplace Transform*, Gordon & Breach, New York, 1992.

**Samko, S. G., Kilbas, A. A., and Marichev, O. I.**, *Fractional Integrals and Derivatives. Theory and Applications*, Gordon & Breach, New York, 1993.

**Sneddon, I.**, *Fourier Transforms*, Dover Publications, New York, 1995.

**Sneddon, I.**, *The Use of Integral Transforms*, McGraw-Hill, New York, 1972.

**Sveshnikov, A. G. and Tikhonov, A. N.**, *Theory of Functions of a Complex Variable* [in Russian], Nauka Publishers, Moscow, 1970.

**Titchmarsh, E. C.**, *Introduction to the Theory of Fourier Integrals, 3rd Edition*, Chelsea Publishing, New York, 1986.

**Zwillinger, D.**, *CRC Standard Mathematical Tables and Formulae, 31st Edition*, CRC Press, Boca Raton, 2002.

**Zwillinger, D.**, *Handbook of Differential Equations, 3rd Edition*, Academic Press, Boston, 1997.

# Chapter 12
# Ordinary Differential Equations

## 12.1. First-Order Differential Equations
### 12.1.1. General Concepts. The Cauchy Problem. Uniqueness and Existence Theorems

**12.1.1-1. Equations solved for the derivative. General solution.**

A *first-order ordinary differential equation*\* solved for the derivative has the form

$$y'_x = f(x, y). \tag{12.1.1.1}$$

Sometimes it is represented in terms of differentials as $dy = f(x, y)\,dx$.

A *solution of a differential equation* is a function $y(x)$ that, when substituted into the equation, turns it into an identity. The *general solution of a differential equation* is the set of all its solutions. In some cases, the general solution can be represented as a function $y = \varphi(x, C)$ that depends on one *arbitrary constant* $C$; specific values of $C$ define specific solutions of the equation (*particular solutions*). In practice, the general solution more frequently appears in implicit form, $\Phi(x, y, C) = 0$, or parametric form, $x = x(t, C)$, $y = y(t, C)$.

Geometrically, the general solution (also called the general integral) of an equation is a family of curves in the $xy$-plane depending on a single parameter $C$; these curves are called *integral curves* of the equation. To each particular solution (particular integral) there corresponds a single curve that passes through a given point in the plane.

For each point $(x, y)$, the equation $y'_x = f(x, y)$ defines a value of $y'_x$, i.e., the slope of the integral curve that passes through this point. In other words, the equation generates a field of directions in the $xy$-plane. From the geometrical point of view, the problem of solving a first-order differential equation involves finding the curves, the slopes of which at each point coincide with the direction of the field at this point.

Figure 12.1 depicts the tangent to an integral curve at a point $(x_0, y_0)$; the slope of the integral curve at this point is determined by the right-hand side of equation (12.1.1.1): $\tan \alpha = f(x_0, y_0)$. The little lines show the field of tangents to the integral curves of the differential equation (12.1.1.1) at other points.

**12.1.1-2. Equations integrable by quadrature.**

To integrate a differential equation in closed form is to represent its solution in the form of formulas written using a predefined bounded set of allowed functions and mathematical operations. A solution is expressed as a quadrature if the set of allowed functions consists of the elementary functions and the functions appearing in the equation and the allowed

---
\* In what follows, we often call an ordinary differential equation a "differential equation" or, even shorter, an "equation."

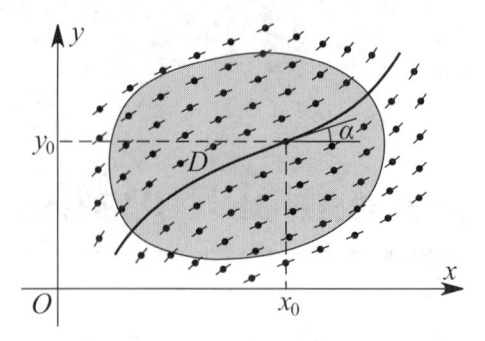

**Figure 12.1.** The direction field of a differential equation and the integral curve passing through a point $(x_0, y_0)$.

mathematical operations are the arithmetic operations, a finite number of function compositions, and the indefinite integral. An equation is said to be integrable by quadrature if its general solution can be expressed in terms of quadratures.

**12.1.1-3. Cauchy problem. The uniqueness and existence theorems.**

The *Cauchy problem*: find a solution of equation (12.1.1.1) that satisfies the *initial condition*

$$y = y_0 \quad \text{at} \quad x = x_0, \tag{12.1.1.2}$$

where $y_0$ and $x_0$ are some numbers.

Geometrical meaning of the Cauchy problem: find an integral curve of equation (12.1.1.1) that passes through the point $(x_0, y_0)$; see Fig. 12.1.

Condition (12.1.1.2) is alternatively written $y(x_0) = y_0$ or $y|_{x=x_0} = y_0$.

THEOREM (EXISTENCE, PEANO). *Let the function $f(x, y)$ be continuous in an open domain $D$ of the $xy$-plane. Then there is at least one integral curve of equation (12.1.1.1) that passes through each point $(x_0, y_0) \in D$; each of these curves can be extended at both ends up to the boundary of any closed domain $D_0 \subset D$ such that $(x_0, y_0)$ belongs to the interior of $D_0$.*

THEOREM (UNIQUENESS). *Let the function $f(x, y)$ be continuous in an open domain $D$ and have in $D$ a bounded partial derivative with respect to $y$ (or the Lipschitz condition holds: $|f(x, y) - f(x, z)| \le M|y - z|$, where $M$ is some positive number). Then there is a unique solution of equation (12.1.1.1) satisfying condition (12.1.1.2).*

**12.1.1-4. Equations not solved for the derivative. The existence theorem.**

A first-order differential equation not solved for the derivative can generally be written as

$$F(x, y, y'_x) = 0. \tag{12.1.1.3}$$

THEOREM (EXISTENCE AND UNIQUENESS). *There exists a unique solution $y = y(x)$ of equation (12.1.1.3) satisfying the conditions $y|_{x=x_0} = y_0$ and $y'_x|_{x=x_0} = t_0$, where $t_0$ is one of the real roots of the equation $F(x_0, y_0, t_0) = 0$ if the following conditions hold in a neighborhood of the point $(x_0, y_0, t_0)$:*

1. *The function $F(x, y, t)$ is continuous in each of the three arguments.*
2. *The partial derivative $F_t$ exists and is nonzero.*
3. *There is a bounded partial derivative with respect to $y$, $|F_y| \le M$.*

*The solution exists for $|x - x_0| \le a$, where $a$ is a (sufficiently small) positive number.*

### 12.1.1-5. Singular solutions.

$1°$. A point $(x, y)$ at which the uniqueness of the solution to equation (12.1.1.3) is violated is called a *singular* point. If conditions 1 and 3 of the existence and uniqueness theorem hold, then
$$F(x, y, t) = 0, \qquad F_t(x, y, t) = 0 \qquad (12.1.1.4)$$
simultaneously at each singular point. Relations (12.1.1.4) define a *t-discriminant curve* in parametric form. In some cases, the parameter $t$ can be eliminated from (12.1.1.4) to give an equation of this curve in implicit form, $\Psi(x, y) = 0$. If a branch $y = \psi(x)$ of the curve $\Psi(x, y) = 0$ consists of singular points and, at the same time, is an integral curve, then this branch is called a *singular integral curve* and the function $y = \psi(x)$ is a *singular solution* of equation (12.1.1.3).

$2°$. The singular solutions can be found by identifying the *envelope of the family of integral curves*, $\Phi(x, y, C) = 0$, of equation (12.1.1.3). The envelope is part of the *C-discriminant curve*, which is defined by the equations
$$\Phi(x, y, C) = 0, \qquad \Phi_C(x, y, C) = 0.$$

The branch of the $C$-discriminant curve at which

(a) there exist bounded partial derivatives, $|\Phi_x| < M_1$ and $|\Phi_y| < M_2$, and
(b) $|\Phi_x| + |\Phi_y| \neq 0$

is the envelope.

### 12.1.1-6. Point transformations.

In the general case, a point transformation is defined by
$$x = F(X, Y), \qquad y = G(X, Y), \qquad (12.1.1.5)$$
where $X$ is the new independent variable, $Y = Y(X)$ is the new dependent variable, and $F$ and $G$ are some (prescribed or unknown) functions.

The derivative $y'_x$ under the point transformation (12.1.1.5) is calculated by
$$y'_x = \frac{G_X + G_Y Y'_X}{F_X + F_Y Y'_X},$$
where the subscripts $X$ and $Y$ denote the corresponding partial derivatives.

Transformation (12.1.1.5) is invertible if $F_X G_Y - F_Y G_X \neq 0$.

Point transformations are used to simplify equations and reduce them to known equations. Sometimes a point transformation allows the reduction of a nonlinear equation to a linear one.

**Example.** The *hodograph transformation* is an important example of a point transformation. It is defined by $x = Y$, $y = X$, which means that $y$ is taken to be the independent variable and $x$ the dependent one. In this case, the derivative is expressed as
$$y'_x = \frac{1}{X'_Y}.$$

Other examples of point transformations can be found in Subsections 12.1.2 and 12.1.4–12.1.6.

## 12.1.2. Equations Solved for the Derivative. Simplest Techniques of Integration

**12.1.2-1. Equations with separated or separable variables.**

1°. An *equation with separated variables* (a *separated equation*) has the form
$$f(y)y'_x = g(x).$$

Equivalently, the equation can be rewritten as $f(y)\,dy = g(x)\,dx$ (the right-hand side depends on $x$ alone and the left-hand side on $y$ alone). The general solution can be obtained by termwise integration:
$$\int f(y)\,dy = \int g(x)\,dx + C,$$
where $C$ is an arbitrary constant.

2°. An *equation with separable variables* (a *separable equation*) is generally represented by
$$f_1(y)g_1(x)y'_x = f_2(y)g_2(x).$$
Dividing the equation by $f_2(y)g_1(x)$, one obtains a separated equation. Integrating yields
$$\int \frac{f_1(y)}{f_2(y)}\,dy = \int \frac{g_2(x)}{g_1(x)}\,dx + C.$$

*Remark.* In termwise division of the equation by $f_2(y)g_1(x)$, solutions corresponding to $f_2(y) = 0$ can be lost.

**12.1.2-2. Equation of the form $y'_x = f(ax + by)$.**

The substitution $z = ax + by$ brings the equation to a separable equation, $z'_x = bf(z) + a$; see Paragraph 12.1.2-1.

**12.1.2-3. Homogeneous equations $y'_x = f(y/x)$.**

1°. A *homogeneous equation* remains the same under simultaneous scaling (dilatation) of the independent and dependent variables in accordance with the rule $x \to \alpha x$, $y \to \alpha y$, where $\alpha$ is an arbitrary constant ($\alpha \neq 0$). Such equations can be represented in the form
$$y'_x = f\left(\frac{y}{x}\right).$$

The substitution $u = y/x$ brings a homogeneous equation to a separable one, $xu'_x = f(u) - u$; see Paragraph 12.1.2-1.

2°. The equations of the form
$$y'_x = f\left(\frac{a_1 x + b_1 y + c_1}{a_2 x + b_2 y + c_2}\right)$$
can be reduced to a homogeneous equation. To this end, for $a_1 x + b_1 y \neq k(a_2 x + b_2 y)$, one should use the change of variables $\xi = x - x_0$, $\eta = y - y_0$, where the constants $x_0$ and $y_0$ are determined by solving the linear algebraic system
$$a_1 x_0 + b_1 y_0 + c_1 = 0,$$
$$a_2 x_0 + b_2 y_0 + c_2 = 0.$$

As a result, one arrives at the following equation for $\eta = \eta(\xi)$:

$$\eta'_\xi = f\left(\frac{a_1\xi + b_1\eta}{a_2\xi + b_2\eta}\right).$$

On dividing the numerator and denominator of the argument of $f$ by $\xi$, one obtains a homogeneous equation whose right-hand side is dependent on the ratio $\eta/\xi$ only:

$$\eta'_\xi = f\left(\frac{a_1 + b_1\eta/\xi}{a_2 + b_2\eta/\xi}\right).$$

For $a_1x + b_1y = k(a_2x + b_2y)$, see the equation of Paragraph 12.1.2-2.

### 12.1.2-4. Generalized homogeneous equations and equations reducible to them.

1°. A *generalized homogeneous equation* (a homogeneous equation in the generalized sense) remains the same under simultaneous scaling of the independent and dependent variables in accordance with the rule $x \to \alpha x$, $y \to \alpha^k y$, where $\alpha \neq 0$ is an arbitrary constant and $k$ is some number. Such equations can be represented in the form

$$y'_x = x^{k-1} f(yx^{-k}).$$

The substitution $u = yx^{-k}$ brings a generalized homogeneous equation to a separable equation, $xu'_x = f(u) - ku$; see Paragraph 12.1.2-1.

**Example.** Consider the equation

$$y'_x = ax^2 y^4 + by^2. \qquad (12.1.2.1)$$

Let us perform the transformation $x = \alpha\bar{x}$, $y = \alpha^k \bar{y}$ and then multiply the resulting equation by $\alpha^{1-k}$ to obtain

$$\bar{y}'_{\bar{x}} = a\alpha^{3(k+1)} \bar{x}^2 \bar{y}^4 + b\alpha^{k+1} \bar{y}^2. \qquad (12.1.2.2)$$

It is apparent that if $k = -1$, the transformed equation (12.1.2.2) is the same as the original one, up to notation. This means that equation (12.1.2.1) is generalized homogeneous of degree $k = -1$. Therefore the substitution $u = xy$ brings it to a separable equation: $xu'_x = au^4 + bu^2 + u$.

2°. The equations of the form

$$y'_x = yf(e^{\lambda x} y)$$

can be reduced to a generalized homogeneous equation. To this end, one should use the change of variable $z = e^x$ and set $\lambda = -k$.

### 12.1.2-5. Linear equation $y'_x + f(x)y = g(x)$.

A first-order *linear equation* is written as

$$y'_x + f(x)y = g(x).$$

The solution is sought in the product form $y = uv$, where $v = v(x)$ is any function that satisfies the "truncated" equation $v'_x + f(x)v = 0$ [as $v(x)$ one takes the particular solution $v = e^{-F}$, where $F = \int f(x)\,dx$]. As a result, one obtains the following separable equation for $u = u(x)$: $v(x)u'_x = g(x)$. Integrating it yields the general solution:

$$y(x) = e^{-F}\left(\int e^F g(x)\,dx + C\right), \qquad F = \int f(x)\,dx,$$

where $C$ is an arbitrary constant.

### 12.1.2-6. Bernoulli equation $y'_x + f(x)y = g(x)y^a$.

A *Bernoulli equation* has the form
$$y'_x + f(x)y = g(x)y^a, \qquad a \neq 0, 1.$$
(For $a = 0$ and $a = 1$, it is a linear equation; see Paragraph 12.1.2-5.) The substitution $z = y^{1-a}$ brings it to a linear equation, $z'_x + (1-a)f(x)z = (1-a)g(x)$, which is discussed in Paragraph 12.1.2-5. With this in view, one can obtain the general integral:
$$y^{1-a} = Ce^{-F} + (1-a)e^{-F}\int e^F g(x)\,dx, \quad \text{where} \quad F = (1-a)\int f(x)\,dx.$$

### 12.1.2-7. Equation of the form $xy'_x = y + f(x)g(y/x)$.

The substitution $u = y/x$ brings the equation to a separable equation, $x^2 u'_x = f(x)g(u)$; see Paragraph 12.1.2-1.

### 12.1.2-8. Darboux equation.

A *Darboux equation* can be represented as
$$\left[f\!\left(\frac{y}{x}\right) + x^a h\!\left(\frac{y}{x}\right)\right] y'_x = g\!\left(\frac{y}{x}\right) + yx^{a-1} h\!\left(\frac{y}{x}\right).$$
Using the substitution $y = xz(x)$ and taking $z$ to be the independent variable, one obtains a Bernoulli equation, which is considered in Paragraph 12.1.2-6:
$$\left[g(z) - zf(z)\right] x'_z = xf(z) + x^{a+1} h(z).$$

▶ Some other first-order equations integrable by quadrature are treated in Section T5.1.

## 12.1.3. Exact Differential Equations. Integrating Factor

### 12.1.3-1. Exact differential equations.

An *exact differential equation* has the form
$$f(x,y)\,dx + g(x,y)\,dy = 0, \quad \text{where} \quad \frac{\partial f}{\partial y} = \frac{\partial g}{\partial x}.$$
The left-hand side of the equation is the total differential of a function of two variables $U(x, y)$.

The general integral, $U(x, y) = C$, where $C$ is an arbitrary constant and the function $U$ is determined from the system:
$$\frac{\partial U}{\partial x} = f, \qquad \frac{\partial U}{\partial y} = g.$$
Integrating the first equation yields $U = \int f(x,y)\,dx + \Psi(y)$ (while integrating, the variable $y$ is treated as a parameter). On substituting this expression into the second equation, one identifies the function $\Psi$ (and hence, $U$). As a result, the general integral of an exact differential equation can be represented in the form
$$\int_{x_0}^{x} f(\xi, y)\,d\xi + \int_{y_0}^{y} g(x_0, \eta)\,d\eta = C,$$
where $x_0$ and $y_0$ are any numbers.

## TABLE 12.1

An integrating factor $\mu = \mu(x, y)$ for some types of ordinary differential equations $f\,dx + g\,dy = 0$, where $f = f(x, y)$ and $g = g(x, y)$. The subscripts $x$ and $y$ indicate the corresponding partial derivatives

| No. | Conditions for $f$ and $g$ | Integrating factor | Remarks |
|---|---|---|---|
| 1 | $f = y\varphi(xy),\ g = x\psi(xy)$ | $\mu = \dfrac{1}{xf - yg}$ | $xf - yg \neq 0$; $\varphi(z)$ and $\psi(z)$ are any functions |
| 2 | $f_x = g_y,\ f_y = -g_x$ | $\mu = \dfrac{1}{f^2 + g^2}$ | $f + ig$ is an analytic function of the complex variable $x + iy$ |
| 3 | $\dfrac{f_y - g_x}{g} = \varphi(x)$ | $\mu = \exp\left[\int \varphi(x)\,dx\right]$ | $\varphi(x)$ is any function |
| 4 | $\dfrac{f_y - g_x}{f} = \varphi(y)$ | $\mu = \exp\left[-\int \varphi(y)\,dy\right]$ | $\varphi(y)$ is any function |
| 5 | $\dfrac{f_y - g_x}{g - f} = \varphi(x + y)$ | $\mu = \exp\left[\int \varphi(z)\,dz\right],\ z = x + y$ | $\varphi(z)$ is any function |
| 6 | $\dfrac{f_y - g_x}{yg - xf} = \varphi(xy)$ | $\mu = \exp\left[\int \varphi(z)\,dz\right],\ z = xy$ | $\varphi(z)$ is any function |
| 7 | $\dfrac{x^2(f_y - g_x)}{yg + xf} = \varphi\left(\dfrac{y}{x}\right)$ | $\mu = \exp\left[-\int \varphi(z)\,dz\right],\ z = \dfrac{y}{x}$ | $\varphi(z)$ is any function |
| 8 | $\dfrac{f_y - g_x}{xg - yf} = \varphi(x^2 + y^2)$ | $\mu = \exp\left[\tfrac{1}{2}\int \varphi(z)\,dz\right],\ z = x^2 + y^2$ | $\varphi(z)$ is any function |
| 9 | $f_y - g_x = \varphi(x)g - \psi(y)f$ | $\mu = \exp\left[\int \varphi(x)\,dx + \int \psi(y)\,dy\right]$ | $\varphi(x)$ and $\psi(y)$ are any functions |
| 10 | $\dfrac{f_y - g_x}{g\omega_x - f\omega_y} = \varphi(\omega)$ | $\mu = \exp\left[\int \varphi(\omega)\,d\omega\right]$ | $\omega = \omega(x, y)$ is any function of two variables |

**Example.** Consider the equation
$$(ay^n + bx)y'_x + by + cx^m = 0, \quad \text{or} \quad (by + cx^m)\,dx + (ay^n + bx)\,dy = 0,$$
defined by the functions $f(x, y) = by + cx^m$ and $g(x, y) = ay^n + bx$. Computing the derivatives, we have
$$\frac{\partial f}{\partial y} = b, \quad \frac{\partial g}{\partial x} = b \quad \Longrightarrow \quad \frac{\partial f}{\partial y} = \frac{\partial g}{\partial x}.$$
Hence the given equation is an exact differential equation. Its solution can be found using the last formula from Paragraph 12.1.3-1 with $x_0 = y_0 = 0$:
$$\frac{a}{n+1}y^{n+1} + bxy + \frac{c}{m+1}x^{m+1} = C.$$

### 12.1.3-2. Integrating factor.

An *integrating factor* for the equation
$$f(x, y)\,dx + g(x, y)\,dy = 0$$
is a function $\mu(x, y) \not\equiv 0$ such that the left-hand side of the equation, when multiplied by $\mu(x, y)$, becomes a total differential, and the equation itself becomes an exact differential equation.

An integrating factor satisfies the first-order partial differential equation,
$$g\frac{\partial \mu}{\partial x} - f\frac{\partial \mu}{\partial y} = \left(\frac{\partial f}{\partial y} - \frac{\partial g}{\partial x}\right)\mu,$$
which is not generally easier to solve than the original equation.

Table 12.1 lists some special cases where an integrating factor can be found in explicit form.

## 12.1.4. Riccati Equation

**12.1.4-1. General Riccati equation. Simplest integrable cases.**

A *Riccati equation* has the general form
$$y'_x = f_2(x)y^2 + f_1(x)y + f_0(x). \qquad (12.1.4.1)$$

If $f_2 \equiv 0$, we have a linear equation (see Paragraph 12.1.2-5), and if $f_0 \equiv 0$, we have a Bernoulli equation (see Paragraph 12.1.2-6 for $a=2$), whose solutions were given previously. For arbitrary $f_2$, $f_1$, and $f_0$, the Riccati equation is not integrable by quadrature.

Listed below are some special cases where the Riccati equation (12.1.4.1) is integrable by quadrature.

1°. The functions $f_2$, $f_1$, and $f_0$ are proportional, i.e.,
$$y'_x = \varphi(x)(ay^2 + by + c),$$
where $a$, $b$, and $c$ are constants. This equation is a separable equation; see Paragraph 12.1.2-1.

2°. The Riccati equation is homogeneous:
$$y'_x = a\frac{y^2}{x^2} + b\frac{y}{x} + c.$$
See Paragraph 12.1.2-3.

3°. The Riccati equation is generalized homogeneous:
$$y'_x = ax^n y^2 + \frac{b}{x}y + cx^{-n-2}.$$
See Paragraph 12.1.2-4 (with $k=-n-1$). The substitution $z = x^{n+1}y$ brings it to a separable equation: $xz'_x = az^2 + (b+n+1)z + c$.

4°. The Riccati equation has the form
$$y'_x = ax^{2n}y^2 + \frac{m-n}{x}y + cx^{2m}.$$
By the substitution $y = x^{m-n}z$, the equation is reduced to a separable equation: $x^{-n-m}z'_x = az^2 + c$.

▶ Some other Riccati equations integrable by quadrature are treated in Section T5.1 (see equations T5.1.6 to T5.1.22).

**12.1.4-2. Polynomial solutions of the Riccati equation.**

Let $f_2 = 1$, $f_1(x)$, and $f_0(x)$ be polynomials. If the degree of the polynomial
$$\Delta = f_1^2 - 2(f_1)'_x - 4f_0$$
is odd, the Riccati equation cannot possess a polynomial solution. If the degree of $\Delta$ is even, the equation involved may possess only the following polynomial solutions:
$$y = -\tfrac{1}{2}\left(f_1 \pm [\sqrt{\Delta}]\right),$$
where $[\sqrt{\Delta}]$ denotes an integer rational part of the expansion of $\sqrt{\Delta}$ in decreasing powers of $x$ (for example, $[\sqrt{x^2 - 2x + 3}] = x - 1$).

### 12.1.4-3. Use of particular solutions to construct the general solution.

$1^\circ$. Given a particular solution $y_0 = y_0(x)$ of the Riccati equation (12.1.4.1), the general solution can be written as

$$y = y_0(x) + \Phi(x)\left[C - \int \Phi(x)f_2(x)\,dx\right]^{-1}, \qquad (12.1.4.2)$$

where $C$ is an arbitrary constant and

$$\Phi(x) = \exp\left\{\int \left[2f_2(x)y_0(x) + f_1(x)\right]dx\right\}. \qquad (12.1.4.3)$$

To the particular solution $y_0(x)$ there corresponds $C = \infty$.

$2^\circ$. Let $y_1 = y_1(x)$ and $y_2 = y_2(x)$ be two different particular solutions of equation (12.1.4.1). Then the general solution can be calculated by

$$y = \frac{Cy_1 + U(x)y_2}{C + U(x)}, \quad \text{where} \quad U(x) = \exp\left[\int f_2(y_1 - y_2)\,dx\right].$$

To the particular solution $y_1(x)$, there corresponds $C = \infty$; and to $y_2(x)$, there corresponds $C = 0$.

$3^\circ$. Let $y_1 = y_1(x)$, $y_2 = y_2(x)$, and $y_3 = y_3(x)$ be three distinct particular solutions of equation (12.1.4.1). Then the general solution can be found without quadrature:

$$\frac{y - y_2}{y - y_1}\frac{y_3 - y_1}{y_3 - y_2} = C.$$

This means that the Riccati equation has a fundamental system of solutions.

### 12.1.4-4. Some transformations.

$1^\circ$. The transformation ($\varphi$, $\psi_1$, $\psi_2$, $\psi_3$, and $\psi_4$ are arbitrary functions)

$$x = \varphi(\xi), \quad y = \frac{\psi_4(\xi)u + \psi_3(\xi)}{\psi_2(\xi)u + \psi_1(\xi)}$$

reduces the Riccati equation (12.1.4.1) to a Riccati equation for $u = u(\xi)$.

$2^\circ$. Let $y_0 = y_0(x)$ be a particular solution of equation (12.1.4.1). Then the substitution $y = y_0 + 1/w$ leads to a linear equation for $w = w(x)$:

$$w'_x + \left[2f_2(x)y_0(x) + f_1(x)\right]w + f_2(x) = 0.$$

For solution of linear equations, see Paragraph 12.1.2-5.

### 12.1.4-5. Reduction of the Riccati equation to a second-order linear equation.

The substitution

$$u(x) = \exp\left(-\int f_2 y\,dx\right)$$

reduces the general Riccati equation (12.1.4.1) to a second-order linear equation:

$$f_2 u''_{xx} - \left[(f_2)'_x + f_1 f_2\right]u'_x + f_0 f_2^2 u = 0,$$

which often may be easier to solve than the original Riccati equation.

#### 12.1.4-6. Reduction of the Riccati equation to the canonical form.

The general Riccati equation (12.1.4.1) can be reduced with the aid of the transformation

$$x = \varphi(\xi), \quad y = \frac{1}{F_2}w - \frac{1}{2}\frac{F_1}{F_2} + \frac{1}{2}\left(\frac{1}{F_2}\right)'_\xi, \quad \text{where} \quad F_i(\xi) = f_i(\varphi)\varphi'_\xi, \qquad (12.1.4.4)$$

to the canonical form

$$w'_\xi = w^2 + \Psi(\xi). \qquad (12.1.4.5)$$

Here, the function $\Psi$ is defined by the formula

$$\Psi(\xi) = F_0 F_2 - \frac{1}{4}F_1^2 + \frac{1}{2}F_1' - \frac{1}{2}F_1\frac{F_2'}{F_2} - \frac{3}{4}\left(\frac{F_2'}{F_2}\right)^2 + \frac{1}{2}\frac{F_2''}{F_2};$$

the prime denotes differentiation with respect to $\xi$.

Transformation (12.1.4.4) depends on a function $\varphi = \varphi(\xi)$ that can be arbitrary. For a specific original Riccati equation, different functions $\varphi$ in (12.1.4.4) will generate different functions $\Psi$ in equation (12.1.4.5). In practice, transformation (12.1.4.4) is most frequently used with $\varphi(\xi) = \xi$.

### 12.1.5. Abel Equations of the First Kind

#### 12.1.5-1. General form of Abel equations of the first kind. Some integrable cases.

An *Abel equation of the first kind* has the general form

$$y'_x = f_3(x)y^3 + f_2(x)y^2 + f_1(x)y + f_0(x), \qquad f_3(x) \not\equiv 0. \qquad (12.1.5.1)$$

In the degenerate case $f_2(x) = f_0(x) = 0$, we have a Bernoulli equation (see Paragraph 12.1.2-6 with $a = 3$). The Abel equation (12.1.5.1) is not integrable in closed form for arbitrary $f_n(x)$.

Listed below are some special cases where the Abel equation of the first kind is integrable by quadrature.

1°. If the functions $f_n(x)$ ($n = 0, 1, 2, 3$) are proportional, i.e., $f_n(x) = a_n g(x)$, then (12.1.5.1) is a separable equation (see Paragraph 12.1.2-1).

2°. The Abel equation is homogeneous:

$$y'_x = a\frac{y^3}{x^3} + b\frac{y^2}{x^2} + c\frac{y}{x} + d.$$

See Paragraph 12.1.2-3.

3°. The Abel equation is generalized homogeneous:

$$y'_x = ax^{2n+1}y^3 + bx^n y^2 + \frac{c}{x}y + dx^{-n-2}.$$

See Paragraph 12.1.2-4 for $k = -n - 1$. The substitution $w = x^{n+1}y$ leads to a separable equation: $xw'_x = aw^3 + bw^2 + (c + n + 1)w + d$.

4°. The Abel equation

$$y'_x = ax^{3n-m}y^3 + bx^{2n}y^2 + \frac{m-n}{x}y + dx^{2m}$$

can be reduced with the substitution $y = x^{m-n}z$ to a separable equation: $x^{-n-m}z'_x = az^3 + bz^2 + c$.

5°. Let $f_0 \equiv 0$, $f_1 \equiv 0$, and $(f_3/f_2)'_x = af_2$ for some constant $a$. Then the substitution $y = f_2 f_3^{-1} u$ leads to a separable equation: $u'_x = f_2^2 f_3^{-1}(u^3 + u^2 + au)$.

6°. If

$$f_0 = \frac{f_1 f_2}{3 f_3} - \frac{2 f_2^3}{27 f_3^2} - \frac{1}{3}\frac{d}{dx}\frac{f_2}{f_3}, \qquad f_n = f_n(x),$$

then the solution of equation (12.1.5.1) is given by

$$y(x) = E\left(C - 2\int f_3 E^2\, dx\right)^{-1/2} - \frac{f_2}{3f_3}, \quad \text{where} \quad E = \exp\left[\int \left(f_1 - \frac{f_2^2}{3f_3}\right) dx\right].$$

For other solvable Abel equations of the first kind, see the books by Kamke (1977) and Polyanin and Zaitsev (2003).

**12.1.5-2. Reduction of the Abel equation of the first kind to the canonical form.**

The transformation

$$y = U(x)\eta(\xi) - \frac{f_2}{3f_3}, \quad \xi = \int f_3 U^2\, dx, \quad \text{where} \quad U(x) = \exp\left[\int\left(f_1 - \frac{f_2^2}{3f_3}\right) dx\right],$$

brings equation (12.1.5.1) to the canonical (normal) form

$$\eta'_\xi = \eta^3 + \Phi(\xi).$$

Here, the function $\Phi(\xi)$ is defined parametrically ($x$ is the parameter) by the relations

$$\Phi = \frac{1}{f_3 U^3}\left(f_0 - \frac{f_1 f_2}{3 f_3} + \frac{2 f_2^3}{27 f_3^2} + \frac{1}{3}\frac{d}{dx}\frac{f_2}{f_3}\right), \quad \xi = \int f_3 U^2\, dx.$$

**12.1.5-3. Reduction to an Abel equation of the second kind.**

Let $y_0 = y_0(x)$ be a particular solution of equation (12.1.5.1). Then the substitution

$$y = y_0 + \frac{E(x)}{z(x)}, \quad \text{where} \quad E(x) = \exp\left[\int (3 f_3 y_0^2 + 2 f_2 y_0 + f_1)\, dx\right],$$

leads to an Abel equation of the second kind:

$$z z'_x = -(3 f_3 y_0 + f_2) E z - f_3 E^2.$$

For equations of this type, see Subsection 12.1.6.

## 12.1.6. Abel Equations of the Second Kind

**12.1.6-1. General form of Abel equations of the second kind. Some integrable cases.**

An *Abel equation of the second kind* has the general form

$$[y + g(x)]y'_x = f_2(x)y^2 + f_1(x)y + f_0(x), \qquad g(x) \not\equiv 0. \tag{12.1.6.1}$$

The Abel equation (12.1.6.1) is not integrable for arbitrary $f_n(x)$ and $g(x)$. Given below are some special cases where the Abel equation of the second kind is integrable by quadrature.

1°. If $g(x) = \text{const}$ and the functions $f_n(x)$ ($n = 0, 1, 2$) are proportional, i.e., $f_n(x) = a_n g(x)$, then (12.1.6.1) is a separable equation (see Paragraph 12.1.2-1).

2°. The Abel equation is homogeneous:

$$(y + sx)y'_x = \frac{a}{x}y^2 + by + cx.$$

See Paragraph 12.1.2-3. The substitution $w = y/x$ leads to a separable equation.

3°. The Abel equation is generalized homogeneous:

$$(y + sx^n)y'_x = \frac{a}{x}y^2 + bx^{n-1}y + cx^{2n-1}.$$

See Paragraph 12.1.2-4 for $k = n$. The substitution $w = yx^{-n}$ leads to a separable equation: $x(w + s)w'_x = (a - n)w^2 + (b - ns)w + c$.

4°. The Abel equation

$$(y + a_2 x + c_2)y'_x = b_1 y + a_1 x + c_1$$

is a special case of the equation treated in Paragraph 12.1.2-3 (see Item 2° with $f(w) = w$ and $b_2 = 1$).

5°. The unnormalized Abel equation

$$[(a_1 x + a_2 x^n)y + b_1 x + b_2 x^n]y'_x = c_2 y^2 + c_1 y + c_0$$

can be reduced to the form (12.1.6.1) by dividing it by $(a_1 x + a_2 x^n)$. Taking $y$ to be the independent variable and $x = x(y)$ to be the dependent one, we obtain the Bernoulli equation

$$(c_2 y^2 + c_1 y + c_0)x'_y = (a_1 y + b_1)x + (a_2 y + b_2)x^n.$$

See Paragraph 12.1.2-6.

6°. The general solution of the Abel equation

$$(y + g)y'_x = f_2 y^2 + f_1 y + f_1 g - f_2 g^2, \qquad f_n = f_n(x), \quad g = g(x),$$

is given by

$$y = -g + CE + E \int (f_1 + g'_x - 2f_2 g)E^{-1}\, dx, \qquad \text{where} \quad E = \exp\left(\int f_2\, dx\right).$$

7°. If $f_1 = 2f_2 g - g'_x$, the general solution of the Abel equation (12.1.6.1) has the form

$$y = -g \pm E\left[2\int (f_0 + gg'_x - f_2 g^2)E^{-2}\, dx + C\right]^{1/2}, \qquad \text{where} \quad E = \exp\left(\int f_2\, dx\right).$$

For other solvable Abel equations of the second kind, see the books by Kamke (1977) and Polyanin and Zaitsev (2003).

### 12.1.6-2. Reduction of the Abel equation of the second kind to the canonical form.

1°. The substitution

$$w = (y + g)E, \quad \text{where} \quad E = \exp\left(-\int f_2\, dx\right), \tag{12.1.6.2}$$

brings equation (12.1.6.1) to the simpler form

$$ww'_x = F_1(x)w + F_0(x), \tag{12.1.6.3}$$

where

$$F_1 = (f_1 - 2f_2 g + g'_x)E, \quad F_0 = (f_0 - f_1 g + f_2 g^2)E^2.$$

2°. In turn, equation (12.1.6.3) can be reduced, by the introduction of the new independent variable

$$z = \int F_1(x)\, dx, \tag{12.1.6.4}$$

to the *canonical form*

$$ww'_z - w = R(z). \tag{12.1.6.5}$$

Here, the function $R(z)$ is defined parametrically ($x$ is the parameter) by the relations

$$R = \frac{F_0(x)}{F_1(x)}, \quad z = \int F_1(x)\, dx.$$

Substitutions (12.1.6.2) and (12.1.6.4), which take the Abel equation to the canonical form, are called *canonical*.

**Remark 1.** The transformation $w = a\hat{w}$, $z = a\hat{z} + b$ brings (12.1.6.5) to a similar equation, $\hat{w}\hat{w}'_{\hat{z}} - \hat{w} = a^{-1}R(a\hat{z} + b)$. Therefore the function $R(z)$ in the right-hand side of the Abel equation (12.1.6.5) can be identified with the two-parameter family of functions $a^{-1}R(az + b)$.

**Remark 2.** Any Abel equations of the second kind related by linear (in $y$) transformations $\tilde{x} = \varphi_1(x)$, $\tilde{y} = \varphi_2(x)y + \varphi_3(x)$ have identical canonical forms (up to the two-parameter family of functions specified in Remark 1).

### 12.1.6-3. Reduction to an Abel equation of the first kind.

The substitution $y + g = 1/u$ leads to an Abel equation of the first kind:

$$u'_x + (f_0 - f_1 g + f_2 g^2)u^3 + (f_1 - 2f_2 g + g'_x)u^2 + f_2 u = 0.$$

For equations of this type, see Subsection 12.1.5.

## 12.1.7. Equations Not Solved for the Derivative

### 12.1.7-1. Method of "integration by differentiation."

In the general case, a first-order equation not solved for the derivative,

$$F(x, y, y'_x) = 0, \tag{12.1.7.1}$$

can be rewritten in the equivalent form

$$F(x, y, t) = 0, \quad t = y'_x. \tag{12.1.7.2}$$

We look for a solution in parametric form: $x = x(t)$, $y = y(t)$. In accordance with the first relation in (12.1.7.2), the differential of $F$ is given by

$$F_x\, dx + F_y\, dy + F_t\, dt = 0. \tag{12.1.7.3}$$

Using the relation $dy = t\, dx$, we eliminate successively $dy$ and $dx$ from (12.1.7.3). As a result, we obtain the system of two first-order ordinary differential equations:

$$\frac{dx}{dt} = -\frac{F_t}{F_x + tF_y}, \qquad \frac{dy}{dt} = -\frac{tF_t}{F_x + tF_y}. \tag{12.1.7.4}$$

By finding a solution of this system, one thereby obtains a solution of the original equation (12.1.7.1) in parametric form, $x = x(t)$, $y = y(t)$.

*Remark 1.* The application of the above method may lead to loss of individual solutions (satisfying the condition $F_x + tF_y = 0$); this issue should be additionally investigated.

*Remark 2.* One of the differential equations of system (12.1.7.4) can be replaced by the algebraic equation $F(x, y, t) = 0$; see equation (12.1.7.2). This technique is used subsequently in Paragraphs 12.1.7-2, 12.1.7-3, and 12.1.7-5.

### 12.1.7-2. Equations of the form $y = f(y'_x)$.

This equation is a special case of equation (12.1.7.1), with $F(x, y, t) = y - f(t)$. The procedure described in Paragraph 12.1.7-1 yields

$$\frac{dx}{dt} = \frac{f'(t)}{t}, \qquad y = f(t). \tag{12.1.7.5}$$

Here, the original equation is used instead of the second equation in system (12.1.7.4); this is valid because the first equation in (12.1.7.4) does not depend on $y$ explicitly.

Integrating the first equation in (12.1.7.5) yields the solution in parametric form,

$$x = \int \frac{f'(t)}{t}\, dt + C, \qquad y = f(t).$$

### 12.1.7-3. Equations of the form $x = f(y'_x)$.

This equation is a special case of equation (12.1.7.1), with $F(x, y, t) = x - f(t)$. The procedure described in Paragraph 12.1.7-1 yields

$$x = f(t), \qquad \frac{dy}{dt} = tf'(t). \tag{12.1.7.6}$$

Here, the original equation is used instead of the first equation in system (12.1.7.4); this is valid because the second equation in (12.1.7.4) does not depend on $x$ explicitly.

Integrating the second equation in (12.1.7.5) yields the solution in parametric form,

$$x = f(t), \qquad y = \int tf'(t)\, dt + C.$$

### 12.1.7-4. Clairaut's equation $y = xy'_x + f(y'_x)$.

*Clairaut's equation* is a special case of equation (12.1.7.1), with $F(x, y, t) = y - xt - f(t)$. It can be rewritten as

$$y = xt + f(t), \qquad t = y'_x. \tag{12.1.7.7}$$

This equation corresponds to the degenerate case $F_x + tF_y \equiv 0$, where system (12.1.7.4) cannot be obtained. One should proceed in the following way: the first relation in (12.1.7.7) gives $dy = x\,dt + t\,dx + f'(t)\,dt$; performing the substitution $dy = t\,dx$, which follows from the second relation in (12.1.7.7), one obtains

$$[x + f'(t)]\,dt = 0.$$

This equation splits into $dt = 0$ and $x + f'(t) = 0$. The solution of the first equation is obvious: $t = C$; it gives the general solution of Clairaut's equation,

$$y = Cx + f(C), \tag{12.1.7.8}$$

which is a family of straight lines. The second equation generates a solution in parametric form,

$$x = -f'(t), \qquad y = -tf'(t) + f(t), \tag{12.1.7.9}$$

which is a singular solution and is the envelope of the family of lines (12.1.7.8).

Remark. There are also "compound" solutions of Clairaut's equation; they consist of part of curve (12.1.7.9) joined with the tangents at finite points; these tangents are defined by formula (12.1.7.8).

### 12.1.7-5. Lagrange's equation $y = xf(y'_x) + g(y'_x)$.

*Lagrange's equation* is a special case of equation (12.1.7.1), with $F(x, y, t) = y - xf(t) - g(t)$. In the special case $f(t) \equiv t$, it coincides with Clairaut's equation; see Paragraph 12.1.7-4.

The procedure described in Paragraph 12.1.7-1 yields

$$\frac{dx}{dt} + \frac{f'(t)}{f(t) - t}x = \frac{g'(t)}{t - f(t)}, \qquad y = xf(t) + g(t). \tag{12.1.7.10}$$

Here, the original equation is used instead of the second equation in system (12.1.7.4); this is valid because the first equation in (12.1.7.4) does not depend on $y$ explicitly.

The first equation of system (12.1.7.10) is linear. Its general solution has the form $x = \varphi(t)C + \psi(t)$; the functions $\varphi$ and $\psi$ are defined in Paragraph 12.1.2-5. Substituting this solution into the second equation in (12.1.7.10), we obtain the general solution of Lagrange's equation in parametric form,

$$x = \varphi(t)C + \psi(t), \qquad y = \bigl[\varphi(t)C + \psi(t)\bigr]f(t) + g(t).$$

Remark. With the above method, solutions of the form $y = t_k x + g(t_k)$, where the $t_k$ are roots of the equation $f(t) - t = 0$, may be lost. These solutions can be particular or singular solutions of Lagrange's equation.

## 12.1.8. Contact Transformations

**12.1.8-1. General form of contact transformations.**

A contact transformation has the form
$$x = F(X, Y, Y'_X),$$
$$y = G(X, Y, Y'_X),$$
(12.1.8.1)

where the functions $F(X, Y, U)$ and $G(X, Y, U)$ are chosen so that the derivative $y'_x$ does not depend on $Y''_{XX}$:
$$y'_x = \frac{y'_X}{x'_X} = \frac{G_X + G_Y Y'_X + G_U Y''_{XX}}{F_X + F_Y Y'_X + F_U Y''_{XX}} = H(X, Y, Y'_X).$$
(12.1.8.2)

The subscripts $X$, $Y$, and $U$ after $F$ and $G$ denote the respective partial derivatives (it is assumed that $F_U \not\equiv 0$ and $G_U \not\equiv 0$).

It follows from (12.1.8.2) that the relation
$$\frac{\partial G}{\partial U}\left(\frac{\partial F}{\partial X} + U\frac{\partial F}{\partial Y}\right) - \frac{\partial F}{\partial U}\left(\frac{\partial G}{\partial X} + U\frac{\partial G}{\partial Y}\right) = 0$$
(12.1.8.3)

holds; the derivative is calculated by
$$y'_x = \frac{G_U}{F_U},$$
(12.1.8.4)

where $G_U/F_U \not\equiv \text{const}$.

The application of contact transformations preserves the order of differential equations. The inverse of a contact transformation can be obtained by solving system (12.1.8.1) and (12.1.8.4) for $X, Y, Y'_X$.

**12.1.8-2. Method for the construction of contact transformations.**

Suppose the function $F = F(X, Y, U)$ in the contact transformation (12.1.8.1) is specified. Then relation (12.1.8.3) can be viewed as a linear partial differential equation for the second function $G$. The corresponding characteristic system of ordinary differential equations (see Subsection 13.1.1),
$$\frac{dX}{1} = \frac{dY}{U} = -\frac{F_U\, dU}{F_X + U F_Y},$$
admits the obvious first integral:
$$F(X, Y, U) = C_1,$$
(12.1.8.5)

where $C_1$ is an arbitrary constant. It follows that, to obtain the general representation of the function $G = G(X, Y, U)$, one has to deal with the ordinary differential equation
$$Y'_X = U,$$
(12.1.8.6)

whose right-hand side is defined in implicit form by (12.1.8.5). Let the first integral of equation (12.1.8.6) have the form
$$\Phi(X, Y, C_1) = C_2.$$

Then the general representation of $G = G(X, Y, U)$ in transformation (12.1.8.1) is given by
$$G = \Psi(F, \widetilde{\Phi}),$$

where $\Psi(F, \widetilde{\Phi})$ is an arbitrary function of two variables, $F = F(X, Y, U)$ and $\widetilde{\Phi} = \Phi(X, Y, F)$.

### 12.1.8-3. Examples of contact transformations.

**Example 1.** *Legendre transformation*:

$$x = Y'_X, \quad y = XY'_X - Y, \quad y'_x = X \quad \text{(direct transformation);}$$
$$X = y'_x, \quad Y = xy'_x - y, \quad Y'_X = x \quad \text{(inverse transformation).}$$

This transformation is used for solving some equations. In particular, the nonlinear equation

$$(xy'_x - y)^a f(y'_x) + yg(y'_x) + xh(y'_x) = 0$$

can be reduced by the Legendre transformation to a Bernoulli equation: $[Xg(X)+h(X)]Y'_X = g(X)Y - f(X)Y^a$ (see Paragraph 12.1.2-6).

**Example 2.** Contact transformation ($a \ne 0$):

$$x = Y'_X + aY, \quad y = be^{aX}Y'_X, \quad y'_x = be^{aX} \quad \text{(direct transformation);}$$
$$X = \frac{1}{a}\ln\frac{y'_x}{b}, \quad Y = \frac{1}{a}\left(x - \frac{y}{y'_x}\right), \quad Y'_X = \frac{y}{y'_x} \quad \text{(inverse transformation).}$$

**Example 3.** Contact transformation ($a \ne 0$):

$$x = Y'_X + aX, \quad y = \tfrac{1}{2}(Y'_X)^2 + aY, \quad y'_x = Y'_X \quad \text{(direct transformation);}$$
$$X = \frac{1}{a}(x - y'_x), \quad Y = \frac{1}{2a}[2y - (y'_x)^2], \quad Y'_X = y'_x \quad \text{(inverse transformation).}$$

## 12.1.9. Approximate Analytic Methods for Solution of Equations

### 12.1.9-1. Method of successive approximations (Picard method).

The method of successive approximations consists of two stages. At the first stage, the Cauchy problem

$$y'_x = f(x, y) \quad \text{(equation)}, \qquad (12.1.9.1)$$
$$y(x_0) = y_0 \quad \text{(initial condition)} \qquad (12.1.9.2)$$

is reduced to the equivalent integral equation:

$$y(x) = y_0 + \int_{x_0}^{x} f(t, y(t))\, dt. \qquad (12.1.9.3)$$

Then a solution of equation (12.1.9.3) is sought using the formula of successive approximations:

$$y_{n+1}(x) = y_0 + \int_{x_0}^{x} f(t, y_n(t))\, dt; \qquad n = 0, 1, 2, \ldots$$

The initial approximation $y_0(x)$ can be chosen arbitrarily; the simplest way is to take $y_0$ to be a number. The iterative process converges as $n \to \infty$, provided the conditions of the theorems in Paragraph 12.1.1-3 are satisfied.

### 12.1.9-2. Method of Taylor series expansion in the independent variable.

A solution of the Cauchy problem (12.1.9.1)–(12.1.9.2) can be sought in the form of the Taylor series in powers of $(x - x_0)$:

$$y(x) = y(x_0) + y'_x(x_0)(x - x_0) + \frac{y''_{xx}(x_0)}{2!}(x - x_0)^2 + \cdots . \qquad (12.1.9.4)$$

The first coefficient $y(x_0)$ in solution (12.1.9.4) is prescribed by the initial condition (12.1.9.2). The values of the derivatives of $y(x)$ at $x = x_0$ are determined from equation (12.1.9.1) and its derivative equations (obtained by successive differentiation), taking into account the initial condition (12.1.9.2). In particular, setting $x = x_0$ in (12.1.9.1) and substituting (12.1.9.2), one obtains the value of the first derivative:

$$y'_x(x_0) = f(x_0, y_0). \qquad (12.1.9.5)$$

Further, differentiating equation (12.1.9.1) yields

$$y''_{xx} = f_x(x, y) + f_y(x, y)y'_x. \qquad (12.1.9.6)$$

On substituting $x = x_0$, as well as the initial condition (12.1.9.2) and the first derivative (12.1.9.5), into the right-hand side of this equation, one calculates the value of the second derivative:

$$y''_{xx}(x_0) = f_x(x_0, y_0) + f(x_0, y_0)f_y(x_0, y_0).$$

Likewise, one can determine the subsequent derivatives of $y$ at $x = x_0$.

Solution (12.1.9.4) obtained by this method can normally be used in only some sufficiently small neighborhood of the point $x = x_0$.

**Example.** Consider the Cauchy problem for the equation

$$y' = e^y + \cos x$$

with the initial condition $y(0) = 0$.

Since $x_0 = 0$, we will be constructing a series in powers of $x$. If follows from the equation that $y'(0) = e^0 + \cos 0 = 2$. Differentiating the original equation yields $y'' = e^y y' - \sin x$. Using the initial condition and the condition $y'(0) = 2$ just obtained, we have $y''(0) = e^0 \times 2 - \sin 0 = 2$. Similarly, we find that $y''' = e^y y'' + e^y (y')^2 - \cos x$, whence $y'''(0) = e^0 \times 2 + e^0 \times 2^2 - \cos 0 = 5$.

Substituting the values of the derivatives at $x = 0$ into series (12.1.9.4), we obtain the desired series representation of the solution: $y = 2x + x^2 + \frac{5}{6}x^3 + \cdots$.

### 12.1.9-3. Method of regular expansion in the small parameter.

Consider a general first-order ordinary differential equation with a small parameter $\varepsilon$:

$$y'_x = f(x, y, \varepsilon). \qquad (12.1.9.7)$$

Suppose the function $f$ is representable as a series in powers of $\varepsilon$:

$$f(x, y, \varepsilon) = \sum_{n=0}^{\infty} \varepsilon^n f_n(x, y). \qquad (12.1.9.8)$$

One looks for a solution of the Cauchy problem for equation (12.1.9.7) with the initial condition (12.1.9.2) as $\varepsilon \to 0$ in the form of a regular expansion in powers of the small parameter:

$$y = \sum_{n=0}^{\infty} \varepsilon^n Y_n(x). \qquad (12.1.9.9)$$

Relation (12.1.9.9) is substituted in equation (12.1.9.7) taking into account (12.1.9.8). Then one expands the functions $f_n$ into a power series in $\varepsilon$ and matches the coefficients of like powers of $\varepsilon$ to obtain a system of equations for $Y_n(x)$:

$$Y_0' = f_0(x, Y_0), \tag{12.1.9.10}$$

$$Y_1' = g(x, Y_0)Y_1 + f_1(x, Y_0), \qquad g(x, y) = \frac{\partial f_0}{\partial y}. \tag{12.1.9.11}$$

Only the first two equations are written out here. The prime denotes differentiation with respect to $x$. The initial conditions for $Y_n$ can be obtained from (12.1.9.2) taking into account (12.1.9.9):

$$Y_0(x_0) = y_0, \qquad Y_1(x_0) = 0.$$

Success in the application of this method is primarily determined by the possibility of constructing a solution of equation (12.1.9.10) for the leading term in the expansion of $Y_0$. It is significant that the remaining terms of the expansion, $Y_n$ with $n \geq 1$, are governed by linear equations with homogeneous initial conditions.

**Remark 1.** Paragraph 12.3.5-2 gives an example of solving a Cauchy problem by the method of regular expansion for a second-order equation and also discusses characteristic features of the method.

**Remark 2.** The methods of scaled coordinates, two-scale expansions, and matched asymptotic expansions are also used to solve problems defined by first-order differential equations with a small parameter. The basic ideas of these methods are given in Subsection 12.3.5.

### 12.1.10. Numerical Integration of Differential Equations

**12.1.10-1. Method of Euler polygonal lines.**

Consider the Cauchy problem for the first-order differential equation

$$y_x' = f(x, y)$$

with the initial condition $y(x_0) = y_0$. Our aim is to construct an approximate solution $y = y(x)$ of this equation on an interval $[x_0, x_*]$.

Let us split the interval $[x_0, x_*]$ into $n$ equal segments of length $\Delta x = \dfrac{x_* - x_0}{n}$. We seek approximate values $y_1, y_2, \ldots, y_n$ of the function $y(x)$ at the partitioning points $x_1, x_2, \ldots, x_n = x_*$.

For a given initial value $y_0 = y(x_0)$ and a sufficiently small $\Delta x$, the values of the unknown function $y_k = y(x_k)$ at the other points $x_k = x_0 + k\Delta x$ are calculated successively by the formula

$$y_{k+1} = y_k + f(x_k, y_k)\Delta x \qquad \text{(Euler polygonal line)},$$

where $k = 0, 1, \ldots, n-1$. The Euler method is a single-step method of the first-order approximation (with respect to the step $\Delta x$).

**12.1.10-2. Single-step methods of the second-order approximation.**

Two single-step methods for solving the Cauchy problem in the second-order approximation are specified by the recurrence formulas

$$y_{k+1} = y_k + f\left(x_k + \tfrac{1}{2}\Delta x,\ y_k + \tfrac{1}{2}f_k\Delta x\right)\Delta x,$$

$$y_{k+1} = y_k + \tfrac{1}{2}\left[f_k + f(x_{k+1},\ y_k + f_k\Delta x)\right]\Delta x,$$

where $f_k = f(x_k, y_k)$; $k = 0, 1, \ldots, n-1$.

**12.1.10-3. Runge–Kutta method of the fourth-order approximation.**

This is one of the widely used methods. The unknown values $y_k$ are successively found by the formulas

$$y_{k+1} = y_k + \tfrac{1}{6}(f_1 + 2f_2 + 2f_3 + f_4)\Delta x,$$

where

$$f_1 = f(x_k, y_k), \qquad f_2 = f(x_k + \tfrac{1}{2}\Delta x, \, y_k + \tfrac{1}{2}f_1\Delta x),$$
$$f_3 = f(x_k + \tfrac{1}{2}\Delta x, \, y_k + \tfrac{1}{2}f_2\Delta x), \qquad f_4 = f(x_k + \Delta x, \, y_k + f_3\Delta x).$$

*Remark 1.* All methods described in Subsection 12.1.10 are special cases of the Runge–Kutta method (a detailed description of this method can be found in the monographs listed at the end of the current chapter).

*Remark 2.* In practice, calculations are performed on the basis of any of the above recurrence formulas with two different steps $\Delta x$, $\tfrac{1}{2}\Delta x$ and an arbitrarily chosen small $\Delta x$. Then one compares the results obtained at common points. If these results coincide within the given order of accuracy, one assumes that the chosen step $\Delta x$ ensures the desired accuracy of calculations. Otherwise, the step is halved and the calculations are performed with the steps $\tfrac{1}{2}\Delta x$ and $\tfrac{1}{4}\Delta x$, after which the results are compared again, etc. (Quite often, one compares the results of calculations with steps varying by ten or more times.)

## 12.2. Second-Order Linear Differential Equations

### 12.2.1. Formulas for the General Solution. Some Transformations

**12.2.1-1. Homogeneous linear equations. Formulas for the general solution.**

$1°$. Consider a second-order homogeneous linear equation in the general form

$$f_2(x)y''_{xx} + f_1(x)y'_x + f_0(x)y = 0. \tag{12.2.1.1}$$

The *trivial solution*, $y = 0$, is a particular solution of the homogeneous linear equation.

Let $y_1(x)$, $y_2(x)$ be a fundamental system of solutions (nontrivial linearly independent particular solutions) of equation (12.2.1.1). Then the general solution is given by

$$y = C_1 y_1(x) + C_2 y_2(x), \tag{12.2.1.2}$$

where $C_1$ and $C_2$ are arbitrary constants.

$2°$. Let $y_1 = y_1(x)$ be any nontrivial particular solution of equation (12.2.1.1). Then its general solution can be represented as

$$y = y_1 \left( C_1 + C_2 \int \frac{e^{-F}}{y_1^2}\, dx \right), \quad \text{where} \quad F = \int \frac{f_1}{f_2}\, dx. \tag{12.2.1.3}$$

$3°$. Consider the equation

$$y''_{xx} + f(x)y = 0,$$

which is written in the canonical form; see Paragraph 12.2.1-3 for the reduction of equations to this form. Let $y_1(x)$ be any nontrivial partial solution of this equation. The general solution can be constructed by formula (12.2.1.3) with $F = 0$ or formula (12.2.1.2) in which

$$y_2(x) = y_1 \int \frac{[f(x) - 1][y_1^2 - (y'_1)^2]}{[y_1^2 + (y'_1)^2]^2}\, dx + \frac{y'_1}{y_1^2 + (y'_1)^2}.$$

Here, $y_1 = y_1(x)$ and the prime denotes differentiation with respect to $x$. The last formula is suitable where $y_1$ vanishes at some points.

4°. The second-order constant coefficient linear equation

$$y''_{xx} + ay'_x + by = 0$$

has the following fundamental system of solutions:

$y_1(x) = \exp(-\tfrac{1}{2}ax) \sinh(\tfrac{1}{2}x\sqrt{a^2 - 4b})$, $\quad y_2(x) = \exp(-\tfrac{1}{2}ax) \cosh(\tfrac{1}{2}x\sqrt{a^2 - 4b})$ $\quad$ if $a^2 > 4b$;

$y_1(x) = \exp(-\tfrac{1}{2}ax) \sin(\tfrac{1}{2}x\sqrt{4b - a^2})$, $\quad y_2(x) = \exp(-\tfrac{1}{2}ax) \cos(\tfrac{1}{2}x\sqrt{4b - a^2})$ $\quad$ if $a^2 < 4b$;

$y_1(x) = \exp(-\tfrac{1}{2}ax)$, $\quad y_2(x) = x \exp(-\tfrac{1}{2}ax)$ $\quad$ if $a^2 = 4b$.

5°. The *Euler equation*

$$x^2 y''_{xx} + axy'_x + by = 0$$

is reduced by the change of variable $x = ke^t$ ($k \neq 0$) to the second-order constant coefficient linear equation $y''_{tt} + (a-1)y'_t + by = 0$, which is treated in Item 4°.

▶ *Solutions to some other second-order linear equations can be found in Section T5.2.*

### 12.2.1-2. Wronskian determinant and Liouville's formula.

The *Wronskian determinant* (or *Wronskian*) is defined by

$$W(x) = \begin{vmatrix} y_1(x) & y_2(x) \\ y'_1(x) & y'_2(x) \end{vmatrix} \equiv y_1(y_2)'_x - y_2(y_1)'_x,$$

where $y_1(x)$, $y_2(x)$ is a fundamental system of solutions of equation (12.2.1.1).

*Liouville's formula*:

$$W(x) = W(x_0) \exp\left[-\int_{x_0}^{x} \frac{f_1(t)}{f_2(t)} dt\right].$$

### 12.2.1-3. Reduction to the canonical form.

1°. The substitution

$$y = u(x) \exp\left(-\frac{1}{2} \int \frac{f_1}{f_2} dx\right) \qquad (12.2.1.4)$$

brings equation (12.2.1.1) to the canonical (or normal) form

$$u''_{xx} + f(x)u = 0, \qquad \text{where} \quad f = \frac{f_0}{f_2} - \frac{1}{4}\left(\frac{f_1}{f_2}\right)^2 - \frac{1}{2}\left(\frac{f_1}{f_2}\right)'_x. \qquad (12.2.1.5)$$

2°. The substitution (12.2.1.4) is a special case of the more general transformation ($\varphi$ is an arbitrary function)

$$x = \varphi(\xi), \quad y = u(\xi)\sqrt{|\varphi'_\xi(\xi)|} \exp\left(-\frac{1}{2}\int \frac{f_1(\varphi)}{f_2(\varphi)} d\varphi\right),$$

which also brings the original equation to the canonical form.

### 12.2.1-4. Reduction to the Riccati equation.

The substitution $u = y'_x/y$ brings the second-order homogeneous linear equation (12.2.1.1) to the Riccati equation:

$$f_2(x)u'_x + f_2(x)u^2 + f_1(x)u + f_0(x) = 0,$$

which is discussed in Subsection 12.1.4.

### 12.2.1-5. Nonhomogeneous linear equations. The existence theorem.

A second-order nonhomogeneous linear equation has the form

$$f_2(x)y''_{xx} + f_1(x)y'_x + f_0(x)y = g(x). \tag{12.2.1.6}$$

THEOREM (EXISTENCE AND UNIQUENESS). *On an open interval $a < x < b$, let the functions $f_2$, $f_1$, $f_0$, and $g$ be continuous and $f_2 \neq 0$. Also let*

$$y(x_0) = A, \quad y'_x(x_0) = B$$

*be arbitrary initial conditions, where $x_0$ is any point such that $a < x_0 < b$, and $A$ and $B$ are arbitrary prescribed numbers. Then a solution of equation (12.2.1.6) exists and is unique. This solutions is defined for all $x \in (a,b)$.*

### 12.2.1-6. Nonhomogeneous linear equations. Formulas for the general solution.

$1°$. The general solution of the nonhomogeneous linear equation (12.2.1.6) is the sum of the general solution of the corresponding homogeneous equation (12.2.1.1) and any particular solution of the nonhomogeneous equation (12.2.1.6).

$2°$. Let $y_1 = y_1(x)$, $y_2 = y_2(x)$ be a fundamental system of solutions of the corresponding homogeneous equation, with $g \equiv 0$. Then the general solution of equation (12.2.1.6) can be represented as

$$y = C_1 y_1 + C_2 y_2 + y_2 \int y_1 \frac{g}{f_2} \frac{dx}{W} - y_1 \int y_2 \frac{g}{f_2} \frac{dx}{W}, \tag{12.2.1.7}$$

where $W = y_1(y_2)'_x - y_2(y_1)'_x$ is the Wronskian determinant.

$3°$. Given a nontrivial particular solution $y_1 = y_1(x)$ of the homogeneous equation (with $g \equiv 0$), a second particular solution $y_2 = y_2(x)$ can be calculated from the formula

$$y_2 = y_1 \int \frac{e^{-F}}{y_1^2} dx, \quad \text{where} \quad F = \int \frac{f_1}{f_2} dx, \quad W = e^{-F}. \tag{12.2.1.8}$$

Then the general solution of equation (12.2.1.6) can be constructed by (12.2.1.7).

$4°$. Let $\bar{y}_1$ and $\bar{y}_2$ be respective solutions of the nonhomogeneous differential equations $L[\bar{y}_1] = g_1(x)$ and $L[\bar{y}_2] = g_2(x)$, which have the same left-hand side but different right-hand sides, where $L[y]$ is the left-hand side of equation (12.2.1.6). Then the function $\bar{y} = \bar{y}_1 + \bar{y}_2$ is a solution of the equation $L[\bar{y}] = g_1(x) + g_2(x)$.

## 12.2.1-7. Reduction to a constant coefficient equation (a special case).

Let $f_2 = 1$, $f_0 \neq 0$, and the condition

$$\frac{1}{|f_0|}\frac{d}{dx}\sqrt{|f_0|} + \frac{f_1}{\sqrt{|f_0|}} = a = \text{const}$$

be satisfied. Then the substitution $\xi = \int \sqrt{|f_0|}\, dx$ leads to a constant coefficient linear equation,

$$y''_{\xi\xi} + a y'_{\xi} + y\, \text{sign}\, f_0 = 0.$$

## 12.2.1-8. Kummer–Liouville transformation.

The transformation

$$x = \alpha(t), \quad y = \beta(t)z + \gamma(t), \tag{12.2.1.9}$$

where $\alpha(t)$, $\beta(t)$, and $\gamma(t)$ are arbitrary sufficiently smooth functions ($\beta \not\equiv 0$), takes any linear differential equation for $y(x)$ to a linear equation for $z = z(t)$. In the special case $\gamma \equiv 0$, a homogeneous equation is transformed to a homogeneous one.

Special cases of transformation (12.2.1.9) are widely used to simplify second- and higher-order linear differential equations.

## 12.2.2. Representation of Solutions as a Series in the Independent Variable

### 12.2.2-1. Equation coefficients are representable in the ordinary power series form.

Let us consider a homogeneous linear differential equation of the general form

$$y''_{xx} + f(x)y'_x + g(x)y = 0. \tag{12.2.2.1}$$

Assume that the functions $f(x)$ and $g(x)$ are representable, in the vicinity of a point $x = x_0$, in the power series form,

$$f(x) = \sum_{n=0}^{\infty} A_n(x-x_0)^n, \quad g(x) = \sum_{n=0}^{\infty} B_n(x-x_0)^n, \tag{12.2.2.2}$$

on the interval $|x - x_0| < R$, where $R$ stands for the minimum radius of convergence of the two series in (12.2.2.2). In this case, the point $x = x_0$ is referred to as an *ordinary point*, and equation (12.2.2.1) possesses two linearly independent solutions of the form

$$y_1(x) = \sum_{n=0}^{\infty} a_n(x-x_0)^n, \quad y_2(x) = \sum_{n=0}^{\infty} b_n(x-x_0)^n. \tag{12.2.2.3}$$

The coefficients $a_n$ and $b_n$ are determined by substituting the series (12.2.2.2) into equation (12.2.2.1) followed by extracting the coefficients of like powers of $(x - x_0)$.*

---
* Prior to that, the terms containing the same powers $(x - x_0)^k$, $k = 0, 1, \ldots$, should be collected.

### 12.2.2-2. Equation coefficients have poles at some point.

Assume that the functions $f(x)$ and $g(x)$ are representable, in the vicinity of a point $x = x_0$, in the form

$$f(x) = \sum_{n=-1}^{\infty} A_n(x-x_0)^n, \qquad g(x) = \sum_{n=-2}^{\infty} B_n(x-x_0)^n, \qquad (12.2.2.4)$$

on the interval $|x-x_0| < R$. In this case, the point $x = x_0$ is referred to as a *regular singular point*. Let $\lambda_1$ and $\lambda_2$ be roots of the quadratic equation

$$\lambda_1^2 + (A_{-1} - 1)\lambda + B_{-2} = 0.$$

There are three cases, depending on the values of the exponents of the singularity.

1. If $\lambda_1 \neq \lambda_2$ and $\lambda_1 - \lambda_2$ is not an integer, equation (12.2.2.1) has two linearly independent solutions of the form

$$y_1(x) = |x-x_0|^{\lambda_1}\left[1 + \sum_{n=1}^{\infty} a_n(x-x_0)^n\right],$$
$$y_2(x) = |x-x_0|^{\lambda_2}\left[1 + \sum_{n=1}^{\infty} b_n(x-x_0)^n\right]. \qquad (12.2.2.5)$$

2. If $\lambda_1 = \lambda_2 = \lambda$, equation (12.2.2.1) possesses two linearly independent solutions:

$$y_1(x) = |x-x_0|^{\lambda}\left[1 + \sum_{n=1}^{\infty} a_n(x-x_0)^n\right],$$
$$y_2(x) = y_1(x) \ln|x-x_0| + |x-x_0|^{\lambda} \sum_{n=0}^{\infty} b_n(x-x_0)^n.$$

3. If $\lambda_1 = \lambda_2 + N$, where $N$ is a positive integer, equation (12.2.2.1) has two linearly independent solutions of the form

$$y_1(x) = |x-x_0|^{\lambda_1}\left[1 + \sum_{n=1}^{\infty} a_n(x-x_0)^n\right],$$
$$y_2(x) = k y_1(x) \ln|x-x_0| + |x-x_0|^{\lambda_2} \sum_{n=0}^{\infty} b_n(x-x_0)^n,$$

where $k$ is a constant to be determined (it may be equal to zero).

To construct the solution in each of the three cases, the following procedure should be performed: substitute the above expressions of $y_1$ and $y_2$ into the original equation (12.2.2.1) and equate the coefficients of $(x-x_0)^n$ and $(x-x_0)^n \ln|x-x_0|$ for different values of $n$ to obtain recurrence relations for the unknown coefficients. From these recurrence relations the solution sought can be found.

## 12.2.3. Asymptotic Solutions

This subsection presents asymptotic solutions, as $\varepsilon \to 0$ ($\varepsilon > 0$), of some second-order linear ordinary differential equations containing arbitrary functions (sufficiently smooth), with the independent variable being real.

12.2.3-1. Equations not containing $y'_x$. Leading asymptotic terms.

1°. Consider the equation
$$\varepsilon^2 y''_{xx} - f(x) y = 0 \qquad (12.2.3.1)$$
on a closed interval $a \leq x \leq b$.

*Case 1.* With the condition $f \neq 0$, the leading terms of the asymptotic expansions of the fundamental system of solutions, as $\varepsilon \to 0$, are given by the formulas

$$y_1 = f^{-1/4} \exp\left(-\frac{1}{\varepsilon} \int \sqrt{f}\, dx\right), \qquad y_2 = f^{-1/4} \exp\left(\frac{1}{\varepsilon} \int \sqrt{f}\, dx\right) \qquad \text{if } f > 0,$$

$$y_1 = (-f)^{-1/4} \cos\left(\frac{1}{\varepsilon} \int \sqrt{-f}\, dx\right), \qquad y_2 = (-f)^{-1/4} \sin\left(\frac{1}{\varepsilon} \int \sqrt{-f}\, dx\right) \qquad \text{if } f < 0.$$

*Case 2.* Discuss the asymptotic solution of equation (12.2.3.1) in the vicinity of the point $x = x_0$, where function $f(x)$ vanishes, $f(x_0) = 0$ (such a point is referred to as a *transition point*). We assume that the function $f$ can be presented in the form

$$f(x) = (x_0 - x)\psi(x), \qquad \text{where} \quad \psi(x) > 0.$$

In this case, the fundamental solutions, as $\varepsilon \to 0$, are described by three different formulas:

$$y_1 = \begin{cases} \dfrac{1}{|f(x)|^{1/4}} \sin\left[\dfrac{1}{\varepsilon}\int_{x_0}^{x} \sqrt{|f(x)|}\, dx + \dfrac{\pi}{4}\right] & \text{if } x - x_0 \geq \delta, \\[4pt] \dfrac{\sqrt{\pi}}{[\varepsilon\psi(x_0)]^{1/6}} \operatorname{Ai}(z) & \text{if } |x - x_0| \leq \delta, \\[4pt] \dfrac{1}{2[f(x)]^{1/4}} \exp\left[-\dfrac{1}{\varepsilon}\int_{x}^{x_0} \sqrt{f(x)}\, dx\right] & \text{if } x_0 - x \geq \delta, \end{cases}$$

$$y_2 = \begin{cases} \dfrac{1}{|f(x)|^{1/4}} \cos\left[\dfrac{1}{\varepsilon}\int_{x_0}^{x} \sqrt{|f(x)|}\, dx + \dfrac{\pi}{4}\right] & \text{if } x - x_0 \geq \delta, \\[4pt] \dfrac{\sqrt{\pi}}{[\varepsilon\psi(x_0)]^{1/6}} \operatorname{Bi}(z) & \text{if } |x - x_0| \leq \delta, \\[4pt] \dfrac{1}{[f(x)]^{1/4}} \exp\left[\dfrac{1}{\varepsilon}\int_{x}^{x_0} \sqrt{f(x)}\, dx\right] & \text{if } x_0 - x \geq \delta, \end{cases}$$

where $\operatorname{Ai}(z)$ and $\operatorname{Bi}(z)$ are the Airy functions of the first and second kind, respectively (see Section 18.8), $z = \varepsilon^{-2/3}[\psi(x_0)]^{1/3}(x_0 - x)$, and $\delta = O(\varepsilon^{2/3})$.

### 12.2.3-2. Equations not containing $y'_x$. Two-term asymptotic expansions.

The two-term asymptotic expansions of the solution of equation (12.2.3.1) with $f > 0$, as $\varepsilon \to 0$, on a closed interval $a \le x \le b$, has the form

$$y_1 = f^{-1/4} \exp\left(-\frac{1}{\varepsilon}\int_{x_0}^{x}\sqrt{f}\,dx\right)\left\{1 - \varepsilon\int_{x_0}^{x}\left[\frac{1}{8}\frac{f''_{xx}}{f^{3/2}} - \frac{5}{32}\frac{(f'_x)^2}{f^{5/2}}\right]dx + O(\varepsilon^2)\right\},$$

$$y_2 = f^{-1/4} \exp\left(\frac{1}{\varepsilon}\int_{x_0}^{x}\sqrt{f}\,dx\right)\left\{1 + \varepsilon\int_{x_0}^{x}\left[\frac{1}{8}\frac{f''_{xx}}{f^{3/2}} - \frac{5}{32}\frac{(f'_x)^2}{f^{5/2}}\right]dx + O(\varepsilon^2)\right\},$$
(12.2.3.2)

where $x_0$ is an arbitrary number satisfying the inequality $a \le x_0 \le b$.

The asymptotic expansions of the fundamental system of solutions of equation (12.2.3.1) with $f < 0$ are derived by separating the real and imaginary parts in either formula (12.2.3.2).

### 12.2.3-3. Equations of special form not containing $y'_x$.

Consider the equation

$$\varepsilon^2 y''_{xx} - x^{m-2} f(x) y = 0 \qquad (12.2.3.3)$$

on a closed interval $a \le x \le b$, where $a < 0$ and $b > 0$, under the conditions that $m$ is a positive integer and $f(x) \ne 0$. In this case, the leading term of the asymptotic solution, as $\varepsilon \to 0$, in the vicinity of the point $x = 0$ is expressed in terms of a simpler model equation, which results from substituting the function $f(x)$ in equation (12.2.3.3) by the constant $f(0)$ (the solution of the model equation is expressed in terms of the Bessel functions of order $1/m$).

We specify below formulas by which the leading terms of the asymptotic expansions of the fundamental system of solutions of equation (12.2.3.3) with $a < x < 0$ and $0 < x < b$ are related (excluding a small vicinity of the point $x = 0$). Three different cases can be extracted.

$1°$. Let $m$ be an even integer and $f(x) > 0$. Then,

$$y_1 = \begin{cases} [f(x)]^{-1/4} \exp\left[\dfrac{1}{\varepsilon}\int_0^x \sqrt{f(x)}\,dx\right] & \text{if } x < 0, \\[2mm] k^{-1}[f(x)]^{-1/4} \exp\left[\dfrac{1}{\varepsilon}\int_0^x \sqrt{f(x)}\,dx\right] & \text{if } x > 0, \end{cases}$$

$$y_2 = \begin{cases} [f(x)]^{-1/4} \exp\left[-\dfrac{1}{\varepsilon}\int_0^x \sqrt{f(x)}\,dx\right] & \text{if } x < 0, \\[2mm] k[f(x)]^{-1/4} \exp\left[-\dfrac{1}{\varepsilon}\int_0^x \sqrt{f(x)}\,dx\right] & \text{if } x > 0, \end{cases}$$

where $f = f(x)$, $k = \sin\left(\dfrac{\pi}{m}\right)$.

$2°$. Let $m$ be an even integer and $f(x) < 0$. Then,

$$y_1 = \begin{cases} |f(x)|^{-1/4} \cos\left[-\dfrac{1}{\varepsilon}\int_0^x \sqrt{|f(x)|}\,dx + \dfrac{\pi}{4}\right] & \text{if } x < 0, \\[2mm] k^{-1}|f(x)|^{-1/4} \cos\left[\dfrac{1}{\varepsilon}\int_0^x \sqrt{|f(x)|}\,dx - \dfrac{\pi}{4}\right] & \text{if } x > 0, \end{cases}$$

$$y_2 = \begin{cases} |f(x)|^{-1/4} \cos\left[-\dfrac{1}{\varepsilon}\int_0^x \sqrt{|f(x)|}\,dx - \dfrac{\pi}{4}\right] & \text{if } x < 0, \\[2mm] k|f(x)|^{-1/4} \cos\left[\dfrac{1}{\varepsilon}\int_0^x \sqrt{|f(x)|}\,dx + \dfrac{\pi}{4}\right] & \text{if } x > 0, \end{cases}$$

where $f = f(x)$, $k = \tan\left(\dfrac{\pi}{2m}\right)$.

$3°$. Let $m$ be an odd integer. Then,

$$y_1 = \begin{cases} |f(x)|^{-1/4} \cos\left[-\dfrac{1}{\varepsilon}\int_0^x \sqrt{|f(x)|}\,dx + \dfrac{\pi}{4}\right] & \text{if } x < 0, \\ \dfrac{1}{2}k^{-1}[f(x)]^{-1/4} \exp\left[\dfrac{1}{\varepsilon}\int_0^x \sqrt{f(x)}\,dx\right] & \text{if } x > 0, \end{cases}$$

$$y_2 = \begin{cases} |f(x)|^{-1/4} \cos\left[-\dfrac{1}{\varepsilon}\int_0^x \sqrt{|f(x)|}\,dx - \dfrac{\pi}{4}\right] & \text{if } x < 0, \\ k[f(x)]^{-1/4} \exp\left[-\dfrac{1}{\varepsilon}\int_0^x \sqrt{f(x)}\,dx\right] & \text{if } x > 0, \end{cases}$$

where $f = f(x)$, $k = \sin\left(\dfrac{\pi}{2m}\right)$.

**12.2.3-4. Equations not containing $y'_x$. Equation coefficients are dependent on $\varepsilon$.**

Consider an equation of the form

$$\varepsilon^2 y''_{xx} - f(x,\varepsilon) y = 0 \qquad (12.2.3.4)$$

on a closed interval $a \le x \le b$ under the condition that $f \ne 0$. Assume that the following asymptotic relation holds:

$$f(x,\varepsilon) = \sum_{k=0}^{\infty} f_k(x) \varepsilon^k, \qquad \varepsilon \to 0.$$

Then the leading terms of the asymptotic expansions of the fundamental system of solutions of equation (12.2.3.4) are given by the formulas

$$y_1 = f_0^{-1/4}(x) \exp\left[-\dfrac{1}{\varepsilon}\int \sqrt{f_0(x)}\,dx + \dfrac{1}{2}\int \dfrac{f_1(x)}{\sqrt{f_0(x)}}\,dx\right][1 + O(\varepsilon)],$$

$$y_2 = f_0^{-1/4}(x) \exp\left[\dfrac{1}{\varepsilon}\int \sqrt{f_0(x)}\,dx + \dfrac{1}{2}\int \dfrac{f_1(x)}{\sqrt{f_0(x)}}\,dx\right][1 + O(\varepsilon)].$$

**12.2.3-5. Equations containing $y'_x$.**

$1°$. Consider an equation of the form

$$\varepsilon y''_{xx} + g(x) y'_x + f(x) y = 0$$

on a closed interval $0 \le x \le 1$. With $g(x) > 0$, the asymptotic solution of this equation, satisfying the boundary conditions $y(0) = C_1$ and $y(1) = C_2$, can be represented in the form

$$y = (C_1 - kC_2) \exp\left[-\varepsilon^{-1} g(0) x\right] + C_2 \exp\left[\int_x^1 \dfrac{f(x)}{g(x)}\,dx\right] + O(\varepsilon),$$

where $k = \exp\left[\int_0^1 \dfrac{f(x)}{g(x)}\,dx\right]$.

2°. Now let us take a look at an equation of the form

$$\varepsilon^2 y''_{xx} + \varepsilon g(x) y'_x + f(x) y = 0 \qquad (12.2.3.5)$$

on a closed interval $a \leq x \leq b$. Assume

$$D(x) \equiv [g(x)]^2 - 4f(x) \neq 0.$$

Then the leading terms of the asymptotic expansions of the fundamental system of solutions of equation (12.2.3.5), as $\varepsilon \to 0$, are expressed by

$$y_1 = |D(x)|^{-1/4} \exp\left[-\frac{1}{2\varepsilon} \int \sqrt{D(x)}\, dx - \frac{1}{2} \int \frac{g'_x(x)}{\sqrt{D(x)}}\, dx\right][1 + O(\varepsilon)],$$

$$y_2 = |D(x)|^{-1/4} \exp\left[\frac{1}{2\varepsilon} \int \sqrt{D(x)}\, dx - \frac{1}{2} \int \frac{g'_x(x)}{\sqrt{D(x)}}\, dx\right][1 + O(\varepsilon)].$$

### 12.2.3-6. Equations of the general form.

The more general equation

$$\varepsilon^2 y''_{xx} + \varepsilon g(x, \varepsilon) y'_x + f(x, \varepsilon) y = 0$$

is reducible, with the aid of the substitution $y = w \exp\left(-\frac{1}{2\varepsilon} \int g\, dx\right)$, to an equation of the form (12.2.3.4),

$$\varepsilon^2 w''_{xx} + (f - \tfrac{1}{4} g^2 - \tfrac{1}{2} \varepsilon g'_x) w = 0,$$

to which the asymptotic formulas given above in Paragraph 12.2.3-4 are applicable.

## 12.2.4. Boundary Value Problems

### 12.2.4-1. First, second, third, and mixed boundary value problems ($x_1 \leq x \leq x_2$).

We consider the second-order nonhomogeneous linear differential equation

$$y''_{xx} + f(x) y'_x + g(x) y = h(x). \qquad (12.2.4.1)$$

1°. *The first boundary value problem*: Find a solution of equation (12.2.4.1) satisfying the boundary conditions

$$y = a_1 \quad \text{at} \quad x = x_1, \qquad y = a_2 \quad \text{at} \quad x = x_2. \qquad (12.2.4.2)$$

(The values of the unknown are prescribed at two distinct points $x_1$ and $x_2$.)

2°. *The second boundary value problem*: Find a solution of equation (12.2.4.1) satisfying the boundary conditions

$$y'_x = a_1 \quad \text{at} \quad x = x_1, \qquad y'_x = a_2 \quad \text{at} \quad x = x_2. \qquad (12.2.4.3)$$

(The values of the derivative of the unknown are prescribed at two distinct points $x_1$ and $x_2$.)

3°. *The third boundary value problem*: Find a solution of equation (12.2.4.1) satisfying the boundary conditions
$$y'_x - k_1 y = a_1 \quad \text{at} \quad x = x_1,$$
$$y'_x + k_2 y = a_2 \quad \text{at} \quad x = x_2. \tag{12.2.4.4}$$

4°. *The third boundary value problem*: Find a solution of equation (12.2.4.1) satisfying the boundary conditions
$$y = a_1 \quad \text{at} \quad x = x_1, \qquad y'_x = a_2 \quad \text{at} \quad x = x_2. \tag{12.2.4.5}$$

(The unknown itself is prescribed at one point, and its derivative at another point.)

Conditions (12.2.4.2), (12.2.4.3), (12.2.4.4), and (12.2.4.5) are called *homogeneous* if $a_1 = a_2 = 0$.

12.2.4-2. Simplification of boundary conditions. The self-adjoint form of equations.

1°. Nonhomogeneous boundary conditions can be reduced to homogeneous ones by the change of variable $z = A_2 x^2 + A_1 x + A_0 + y$ (the constants $A_2$, $A_1$, and $A_0$ are selected using the method of undetermined coefficients). In particular, the nonhomogeneous boundary conditions of the first kind (12.2.4.2) can be reduced to homogeneous boundary conditions by the linear change of variable
$$z = y - \frac{a_2 - a_1}{x_2 - x_1}(x - x_1) - a_1.$$

2°. On multiplying by $p(x) = \exp\left[\int f(x)\,dx\right]$, one reduces equation (12.2.4.1) to the self-adjoint form:
$$[p(x)y'_x]'_x + q(x)y = r(x). \tag{12.2.4.6}$$

Without loss of generality, we can further consider equation (12.2.4.6) instead of (12.2.4.1). We assume that the functions $p$, $p'_x$, $q$, and $r$ are continuous on the interval $x_1 \leq x \leq x_2$, and $p$ is positive.

12.2.4-3. Green's function. Linear problems for nonhomogeneous equations.

The *Green's function* of the first boundary value problem for equation (12.2.4.6) with homogeneous boundary conditions (12.2.4.2) is a function of two variables $G(x, s)$ that satisfies the following conditions:

1°. $G(x, s)$ is continuous in $x$ for fixed $s$, with $x_1 \leq x \leq x_2$ and $x_1 \leq s \leq x_2$.

2°. $G(x, s)$ is a solution of the homogeneous equation (12.2.4.6), with $r = 0$, for all $x_1 < x < x_2$ exclusive of the point $x = s$.

3°. $G(x, s)$ satisfies the homogeneous boundary conditions $G(x_1, s) = G(x_2, s) = 0$.

4°. The derivative $G'_x(x, s)$ has a jump of $1/p(s)$ at the point $x = s$, that is,
$$G'_x(x, s)\big|_{x \to s,\, x > s} - G'_x(x, s)\big|_{x \to s,\, x < s} = \frac{1}{p(s)}.$$

For the second, third, and mixed boundary value problems, the Green's function is defined likewise except that in 3° the homogeneous boundary conditions (12.2.4.3), (12.2.4.4), and (12.2.4.5), with $a_1 = a_2 = 0$, are adopted, respectively.

The solution of the nonhomogeneous equation (12.2.4.6) subject to appropriate homogeneous boundary conditions is expressed in terms of the Green's function as follows:*

$$y(x) = \int_{x_1}^{x_2} G(x,s) r(s)\, ds.$$

### 12.2.4-4. Representation of the Green's function in terms of particular solutions.

We consider the first boundary value problem. Let $y_1 = y_1(x)$ and $y_2 = y_2(x)$ be linearly independent particular solutions of the homogeneous equation (12.2.4.6), with $r = 0$, that satisfy the conditions

$$y_1(x_1) = 0, \quad y_2(x_2) = 0.$$

(Each of the solutions satisfies one of the homogeneous boundary conditions.)

The Green's function is expressed in terms of solutions of the homogeneous equation as follows:

$$G(x,s) = \begin{cases} \dfrac{y_1(x) y_2(s)}{p(s) W(s)} & \text{for } x_1 \le x \le s, \\ \dfrac{y_1(s) y_2(x)}{p(s) W(s)} & \text{for } s \le x \le x_2, \end{cases} \qquad (12.2.4.7)$$

where $W(x) = y_1(x) y_2'(x) - y_1'(x) y_2(x)$ is the Wronskian determinant.

*Remark.* Formula (12.2.4.7) can also be used to construct the Green's functions for the second, third, and mixed boundary value problems. To this end, one should find two linearly independent solutions, $y_1 = y_1(x)$ and $y_2 = y_2(x)$, of the homogeneous equation; the former satisfies the corresponding homogeneous boundary condition at $x = x_1$ and the latter satisfies the one at $x = x_2$.

## 12.2.5. Eigenvalue Problems

### 12.2.5-1. Sturm–Liouville problem.

Consider the second-order homogeneous linear differential equation

$$[p(x) y_x']_x' + [\lambda s(x) - q(x)] y = 0 \qquad (12.2.5.1)$$

subject to linear boundary conditions of the general form

$$\begin{aligned} \alpha_1 y_x' + \beta_1 y = 0 &\quad \text{at} \quad x = x_1, \\ \alpha_2 y_x' + \beta_2 y = 0 &\quad \text{at} \quad x = x_2. \end{aligned} \qquad (12.2.5.2)$$

It is assumed that the functions $p$, $p_x'$, $s$, and $q$ are continuous, and $p$ and $s$ are positive on an interval $x_1 \le x \le x_2$. It is also assumed that $|\alpha_1| + |\beta_1| > 0$ and $|\alpha_2| + |\beta_2| > 0$.

The *Sturm–Liouville problem*: Find the values $\lambda_n$ of the parameter $\lambda$ at which problem (12.2.5.1), (12.2.5.2) has a nontrivial solution. Such $\lambda_n$ are called *eigenvalues* and the corresponding solutions $y_n = y_n(x)$ are called *eigenfunctions* of the Sturm–Liouville problem (12.2.5.1), (12.2.5.2).

---

* The homogeneous boundary value problem — with $r(x) = 0$ and $a_1 = a_2 = 0$ — is assumed to have only the trivial solution.

**12.2.5-2. General properties of the Sturm–Liouville problem (12.2.5.1), (12.2.5.2).**

$1°$. There are infinitely (countably) many eigenvalues. All eigenvalues can be ordered so that $\lambda_1 < \lambda_2 < \lambda_3 < \cdots$. Moreover, $\lambda_n \to \infty$ as $n \to \infty$; hence, there can only be a finite number of negative eigenvalues. Each eigenvalue has multiplicity 1.

$2°$. The eigenfunctions are defined up to a constant factor. Each eigenfunction $y_n(x)$ has precisely $n-1$ zeros on the open interval $(x_1, x_2)$.

$3°$. Any two eigenfunctions $y_n(x)$ and $y_m(x)$, $n \neq m$, are orthogonal with weight $s(x)$ on the interval $x_1 \leq x \leq x_2$:

$$\int_{x_1}^{x_2} s(x) y_n(x) y_m(x)\, dx = 0 \quad \text{if} \quad n \neq m.$$

$4°$. An arbitrary function $F(x)$ that has a continuous derivative and satisfies the boundary conditions of the Sturm–Liouville problem can be decomposed into an absolutely and uniformly convergent series in the eigenfunctions

$$F(x) = \sum_{n=1}^{\infty} F_n y_n(x),$$

where the Fourier coefficients $F_n$ of $F(x)$ are calculated by

$$F_n = \frac{1}{\|y_n\|^2} \int_{x_1}^{x_2} s(x) F(x) y_n(x)\, dx, \quad \|y_n\|^2 = \int_{x_1}^{x_2} s(x) y_n^2(x)\, dx.$$

$5°$. If the conditions
$$q(x) \geq 0, \quad \alpha_1 \beta_1 \leq 0, \quad \alpha_2 \beta_2 \geq 0 \qquad (12.2.5.3)$$
hold true, there are no negative eigenvalues. If $q \equiv 0$ and $\beta_1 = \beta_2 = 0$, the least eigenvalue is $\lambda_1 = 0$, to which there corresponds an eigenfunction $y_1 = \text{const}$. In the other cases where conditions (12.2.5.3) are satisfied, all eigenvalues are positive.

$6°$. The following asymptotic formula is valid for eigenvalues as $n \to \infty$:

$$\lambda_n = \frac{\pi^2 n^2}{\Delta^2} + O(1), \quad \Delta = \int_{x_1}^{x_2} \sqrt{\frac{s(x)}{p(x)}}\, dx. \qquad (12.2.5.4)$$

Paragraphs 12.2.5-3 through 12.2.5-6 will describe special properties of the Sturm–Liouville problem that depend on the specific form of the boundary conditions.

**Remark 1.** Equation (12.2.5.1) can be reduced to the case where $p(x) \equiv 1$ and $s(x) \equiv 1$ by the change of variables

$$\zeta = \int \sqrt{\frac{s(x)}{p(x)}}\, dx, \quad u(\zeta) = \big[p(x) s(x)\big]^{1/4} y(x).$$

In this case, the boundary conditions are transformed to boundary conditions of similar form.

**Remark 2.** The second-order linear equation

$$\varphi_2(x) y''_{xx} + \varphi_1(x) y'_x + [\lambda + \varphi_0(x)] y = 0$$

can be represented in the form of equation (12.2.5.1) where $p(x)$, $s(x)$, and $q(x)$ are given by

$$p(x) = \exp\left[\int \frac{\varphi_1(x)}{\varphi_2(x)}\, dx\right], \quad s(x) = \frac{1}{\varphi_2(x)} \exp\left[\int \frac{\varphi_1(x)}{\varphi_2(x)}\, dx\right], \quad q(x) = -\frac{\varphi_0(x)}{\varphi_2(x)} \exp\left[\int \frac{\varphi_1(x)}{\varphi_2(x)}\, dx\right].$$

## TABLE 12.2
Example estimates of the first eigenvalue $\lambda_1$ in Sturm–Liouville problems with boundary conditions of the first kind $y(0) = y(1) = 0$ obtained using the Rayleigh–Ritz principle [the right-hand side of relation (12.2.5.6)]

| Equation | Test function | $\lambda_1$, approximate | $\lambda_1$, exact |
|---|---|---|---|
| $y''_{xx} + \lambda(1+x^2)^{-2} y = 0$ | $z = \sin \pi x$ | 15.337 | 15.0 |
| $y''_{xx} + \lambda(4-x^2)^{-2} y = 0$ | $z = \sin \pi x$ | 135.317 | 134.837 |
| $[(1+x)^{-1} y'_x]'_x + \lambda y = 0$ | $z = \sin \pi x$ | 7.003 | 6.772 |
| $\left(\sqrt{1+x}\, y'_x\right)'_x + \lambda y = 0$ | $z = \sin \pi x$ | 11.9956 | 11.8985 |
| $y''_{xx} + \lambda(1+\sin \pi x) y = 0$ | $z = \sin \pi x$ <br> $z = x(1-x)$ | $0.54105\,\pi^2$ <br> $0.55204\,\pi^2$ | $0.54032\,\pi^2$ <br> $0.54032\,\pi^2$ |

### 12.2.5-3. Problems with boundary conditions of the first kind.

Let us note some special properties of the Sturm–Liouville problem that is the first boundary value problem for equation (12.2.5.1) with the boundary conditions

$$y = 0 \quad \text{at} \quad x = x_1, \qquad y = 0 \quad \text{at} \quad x = x_2. \tag{12.2.5.5}$$

$1^\circ$. For $n \to \infty$, the asymptotic relation (12.2.5.4) can be used to estimate the eigenvalues $\lambda_n$. In this case, the asymptotic formula

$$\frac{y_n(x)}{\|y_n\|} = \left[\frac{4}{\Delta^2 p(x) s(x)}\right]^{1/4} \sin\left[\frac{\pi n}{\Delta} \int_{x_1}^{x} \sqrt{\frac{s(x)}{p(x)}}\, dx\right] + O\left(\frac{1}{n}\right), \quad \Delta = \int_{x_1}^{x_2} \sqrt{\frac{s(x)}{p(x)}}\, dx$$

holds true for the eigenfunctions $y_n(x)$.

$2^\circ$. If $q \geq 0$, the following upper estimate holds for the least eigenvalue (*Rayleigh–Ritz principle*):

$$\lambda_1 \leq \frac{\int_{x_1}^{x_2} \left[p(x)(z'_x)^2 + q(x) z^2\right] dx}{\int_{x_1}^{x_2} s(x) z^2\, dx}, \tag{12.2.5.6}$$

where $z = z(x)$ is any twice differentiable function that satisfies the conditions $z(x_1) = z(x_2) = 0$. The equality in (12.2.5.6) is attained if $z = y_1(x)$, where $y_1(x)$ is the eigenfunction corresponding to the eigenvalue $\lambda_1$. One can take $z = (x-x_1)(x_2-x)$ or $z = \sin\left[\dfrac{\pi(x-x_1)}{x_2-x_1}\right]$ in (12.2.5.6) to obtain specific estimates.

It is significant to note that the left-hand side of (12.2.5.6) usually gives a fairly precise estimate of the first eigenvalue (see Table 12.2).

$3^\circ$. The extension of the interval $[x_1, x_2]$ leads to decreasing in eigenvalues.

$4^\circ$. Let the inequalities

$$0 < p_{\min} \leq p(x) \leq p_{\max}, \quad 0 < s_{\min} \leq s(x) \leq s_{\max}, \quad 0 < q_{\min} \leq q(x) \leq q_{\max}$$

be satisfied. Then the following bilateral estimates hold:

$$\frac{p_{\min}}{s_{\max}} \frac{\pi^2 n^2}{(x_2 - x_1)^2} + \frac{q_{\min}}{s_{\max}} \leq \lambda_n \leq \frac{p_{\max}}{s_{\min}} \frac{\pi^2 n^2}{(x_2 - x_1)^2} + \frac{q_{\max}}{s_{\min}}.$$

**5°.** In engineering calculations for eigenvalues, the approximate formula

$$\lambda_n = \frac{\pi^2 n^2}{\Delta^2} + \frac{1}{x_2 - x_1} \int_{x_1}^{x_2} \frac{q(x)}{s(x)} \, dx, \qquad \Delta = \int_{x_1}^{x_2} \sqrt{\frac{s(x)}{p(x)}} \, dx \qquad (12.2.5.7)$$

may be quite useful. This formula provides an exact result if $p(x)s(x) = \text{const}$ and $q(x)/s(x) = \text{const}$ (in particular, for constant equation coefficients, $p = p_0$, $q = q_0$, and $s = s_0$) and gives a correct asymptotic behavior of (12.2.5.4) for any $p(x)$, $q(x)$, and $s(x)$. In addition, relation (12.2.5.7) gives two correct leading asymptotic terms as $n \to \infty$ if $p(x) = \text{const}$ and $s(x) = \text{const}$ [and also if $p(x)s(x) = \text{const}$].

**6°.** Suppose $p(x) = s(x) = 1$ and the function $q = q(x)$ has a continuous derivative. The following asymptotic relations hold for eigenvalues $\lambda_n$ and eigenfunctions $y_n(x)$ as $n \to \infty$:

$$\sqrt{\lambda_n} = \frac{\pi n}{x_2 - x_1} + \frac{1}{\pi n} Q(x_1, x_2) + O\left(\frac{1}{n^2}\right),$$

$$y_n(x) = \sin \frac{\pi n(x - x_1)}{x_2 - x_1} - \frac{1}{\pi n}\left[(x_1 - x)Q(x, x_2) + (x_2 - x)Q(x_1, x)\right] \cos \frac{\pi n(x - x_1)}{x_2 - x_1} + O\left(\frac{1}{n^2}\right),$$

where

$$Q(u, v) = \frac{1}{2} \int_u^v q(x) \, dx. \qquad (12.2.5.8)$$

**7°.** Let us consider the eigenvalue problem for the equation with a small parameter

$$y''_{xx} + [\lambda + \varepsilon q(x)]y = 0 \qquad (\varepsilon \to 0)$$

subject to the boundary conditions (12.2.5.5) with $x_1 = 0$ and $x_2 = 1$. We assume that $q(x) = q(-x)$.

This problem has the following eigenvalues and eigenfunctions:

$$\lambda_n = \pi^2 n^2 - \varepsilon A_{nn} + \frac{\varepsilon^2}{\pi^2} \sum_{k \neq n} \frac{A_{nk}^2}{n^2 - k^2} + O(\varepsilon^3), \qquad A_{nk} = 2 \int_0^1 q(x) \sin(\pi n x) \sin(\pi k x) \, dx;$$

$$y_n(x) = \sqrt{2} \sin(\pi n x) - \varepsilon \frac{\sqrt{2}}{\pi^2} \sum_{k \neq n} \frac{A_{nk}}{n^2 - k^2} \sin(\pi k x) + O(\varepsilon^2).$$

Here, the summation is carried out over $k$ from 1 to $\infty$. The next term in the expansion of $y_n$ can be found in Nayfeh (1973).

### 12.2.5-4. Problems with boundary conditions of the second kind.

Let us note some special properties of the Sturm–Liouville problem that is the second boundary value problem for equation (12.2.5.1) with the boundary conditions

$$y'_x = 0 \quad \text{at} \quad x = x_1, \qquad y'_x = 0 \quad \text{at} \quad x = x_2.$$

**1°.** If $q > 0$, the upper estimate (12.2.5.6) is valid for the least eigenvalue, with $z = z(x)$ being any twice-differentiable function that satisfies the conditions $z'_x(x_1) = z'_x(x_2) = 0$. The equality in (12.2.5.6) is attained if $z = y_1(x)$, where $y_1(x)$ is the eigenfunction corresponding to the eigenvalue $\lambda_1$.

**2°.** Suppose $p(x) = s(x) = 1$ and the function $q = q(x)$ has a continuous derivative. The following asymptotic relations hold for eigenvalues $\lambda_n$ and eigenfunctions $y_n(x)$ as $n \to \infty$:

$$\sqrt{\lambda_n} = \frac{\pi(n-1)}{x_2 - x_1} + \frac{1}{\pi(n-1)} Q(x_1, x_2) + O\left(\frac{1}{n^2}\right),$$

$$y_n(x) = \cos\frac{\pi(n-1)(x-x_1)}{x_2 - x_1} + \frac{1}{\pi(n-1)}\Big[(x_1 - x)Q(x, x_2)$$
$$+ (x_2 - x)Q(x_1, x)\Big] \sin\frac{\pi(n-1)(x-x_1)}{x_2 - x_1} + O\left(\frac{1}{n^2}\right),$$

where $Q(u, v)$ is given by (12.2.5.8).

### 12.2.5-5. Problems with boundary conditions of the third kind.

We consider the third boundary value problem for equation (12.2.5.1) subject to condition (12.2.5.2) with $\alpha_1 = \alpha_2 = 1$. We assume that $p(x) = s(x) = 1$ and the function $q = q(x)$ has a continuous derivative.

The following asymptotic formulas hold for eigenvalues $\lambda_n$ and eigenfunctions $y_n(x)$ as $n \to \infty$:

$$\sqrt{\lambda_n} = \frac{\pi(n-1)}{x_2 - x_1} + \frac{1}{\pi(n-1)}\big[Q(x_1, x_2) - \beta_1 + \beta_2\big] + O\left(\frac{1}{n^2}\right),$$

$$y_n(x) = \cos\frac{\pi(n-1)(x-x_1)}{x_2 - x_1} + \frac{1}{\pi(n-1)}\Big\{(x_1 - x)\big[Q(x, x_2) + \beta_2\big]$$
$$+ (x_2 - x)\big[Q(x_1, x) - \beta_1\big]\Big\} \sin\frac{\pi(n-1)(x-x_1)}{x_2 - x_1} + O\left(\frac{1}{n^2}\right),$$

where $Q(u, v)$ is defined by (12.2.5.8).

### 12.2.5-6. Problems with mixed boundary conditions.

Let us note some special properties of the Sturm–Liouville problem that is the mixed boundary value problem for equation (12.2.5.1) with the boundary conditions

$$y'_x = 0 \quad \text{at} \quad x = x_1, \qquad y = 0 \quad \text{at} \quad x = x_2.$$

**1°.** If $q \geq 0$, the upper estimate (12.2.5.6) is valid for the least eigenvalue, with $z = z(x)$ being any twice-differentiable function that satisfies the conditions $z'_x(x_1) = 0$ and $z(x_2) = 0$. The equality in (12.2.5.6) is attained if $z = y_1(x)$, where $y_1(x)$ is the eigenfunction corresponding to the eigenvalue $\lambda_1$.

**2°.** Suppose $p(x) = s(x) = 1$ and the function $q = q(x)$ has a continuous derivative. The following asymptotic relations hold for eigenvalues $\lambda_n$ and eigenfunctions $y_n(x)$ as $n \to \infty$:

$$\sqrt{\lambda_n} = \frac{\pi(2n-1)}{2(x_2 - x_1)} + \frac{2}{\pi(2n-1)} Q(x_1, x_2) + O\left(\frac{1}{n^2}\right),$$

$$y_n(x) = \cos\frac{\pi(2n-1)(x-x_1)}{2(x_2 - x_1)} + \frac{2}{\pi(2n-1)}\Big[(x_1 - x)Q(x, x_2)$$
$$+ (x_2 - x)Q(x_1, x)\Big] \sin\frac{\pi(2n-1)(x-x_1)}{2(x_2 - x_1)} + O\left(\frac{1}{n^2}\right),$$

where $Q(u, v)$ is defined by (12.2.5.8).

## 12.2.6. Theorems on Estimates and Zeros of Solutions

**12.2.6-1. Theorems on estimates of solutions.**

Let $f_n(x)$ and $g_n(x)$ ($n = 1, 2$) be continuous functions on the interval $a \le x \le b$ and let the following inequalities hold:

$$0 \le f_1(x) \le f_2(x), \quad 0 \le g_1(x) \le g_2(x).$$

If $y_n = y_n(x)$ are some solutions to the linear equations

$$y_n'' = f_n(x) y_n + g_n(x) \quad (n = 1, 2)$$

and $y_1(a) \le y_2(a)$ and $y_1'(a) \le y_2'(a)$, then $y_1(x) \le y_2(x)$ and $y_1'(x) \le y_2'(x)$ on each interval $a \le x \le a_1$, where $y_2(x) > 0$.

**12.2.6-2. Sturm comparison theorem on zeros of solutions.**

Consider the equation

$$[f(x) y']' + g(x) y = 0 \quad (a \le x \le b), \tag{12.2.6.1}$$

where the function $f(x)$ is positive and continuously differentiable, and the function $g(x)$ is continuous.

THEOREM (COMPARISON, STURM). Let $y_n = y_n(x)$ be nonzero solutions of the linear equations

$$[f_n(x) y_n']' + g_n(x) y_n = 0 \quad (n = 1, 2)$$

and let the inequalities $f_1(x) \ge f_2(x) > 0$ and $g_1(x) \le g_2(x)$ hold. Then the function $y_2$ has at least one zero lying between any two adjacent zeros, $x_1$ and $x_2$, of the function $y_1$ (it is assumed that the identities $f_1 \equiv f_2$ and $g_1 \equiv g_2$ are not satisfied on any interval simultaneously).

COROLLARY 1. If $g(x) \le 0$ or there exists a constant $k_1$ such that

$$f(x) \ge k_1 > 0, \quad g(x) < k_1 \left( \frac{\pi}{b-a} \right)^2,$$

then every nontrivial solution to equation (12.2.6.1) has no more than one zero on the interval $[a, b]$.

COROLLARY 2. If there exists a constant $k_2$ such that

$$0 < f(x) \le k_2, \quad g(x) > k_2 \left( \frac{\pi m}{b-a} \right)^2, \quad \text{where} \quad m = 1, 2, \dots,$$

then every nontrivial solution to equation (12.2.6.1) has at least $m$ zeros on the interval $[a, b]$.

**12.2.6-3. Qualitative behavior of solutions as $x \to \infty$.**

Consider the equation
$$y'' + f(x)y = 0, \qquad (12.2.6.2)$$
where $f(x)$ is a continuous function for $x \geq a$.

$1°$. For $f(x) \leq 0$, every nonzero solution has no more than one zero, and hence $y \neq 0$ for sufficiently large $x$.

If $f(x) \leq 0$ for all $x$ and $f(x) \not\equiv 0$, then $y \equiv 0$ is the only solution bounded for all $x$.

$2°$. Suppose $f(x) \geq k^2 > 0$. Then every nontrivial solution $y(x)$ and its derivative $y'(x)$ have infinitely many zeros, with the distance between any adjacent zeros remaining finite.

If $f(x) \to k^2 > 0$ for $x \to \infty$ and $f' \geq 0$, then the solutions of the equation for large $x$ behave similarly to those of the equation $y'' + k^2 y = 0$.

$3°$. Let $f(x) \to -\infty$ for $|x| \to \infty$. Then every nonzero solution has only finitely many zeros, and $|y'/y| \to \infty$ as $|x| \to \infty$. There are two linearly independent solutions, $y_1$ and $y_2$, such that $y_1 \to 0$, $y_1' \to 0$, $y_2 \to \infty$, and $y_2' \to -\infty$ as $x \to -\infty$, and there are two linearly independent solutions, $\bar{y}_1$ and $\bar{y}_2$, such that $\bar{y}_1 \to 0$, $\bar{y}_1' \to 0$, $\bar{y}_2 \to \infty$, and $\bar{y}_2' \to \infty$ as $x \to \infty$.

$4°$. If the function $f$ in equation (12.2.6.2) is continuous, monotonic, and positive, then the amplitude of each solution decreases (resp., increases) as $f$ increases (resp., decreases).

## 12.3. Second-Order Nonlinear Differential Equations

### 12.3.1. Form of the General Solution. Cauchy Problem

**12.3.1-1. Equations solved for the derivative. General solution.**

A *second-order ordinary differential equation* solved for the highest derivative has the form
$$y''_{xx} = f(x, y, y'_x). \qquad (12.3.1.1)$$

The general solution of this equation depends on two arbitrary constants, $C_1$ and $C_2$. In some cases, the general solution can be written in explicit form, $y = \varphi(x, C_1, C_2)$, but more often implicit or parametric forms of the general solution are encountered.

**12.3.1-2. Cauchy problem. The existence and uniqueness theorem.**

*Cauchy problem*: Find a solution of equation (12.3.1.1) satisfying the *initial conditions*
$$y(x_0) = y_0, \quad y'_x(x_0) = y_1. \qquad (12.3.1.2)$$

(At a point $x = x_0$, the value of the unknown function, $y_0$, and its derivative, $y_1$, are prescribed.)

EXISTENCE AND UNIQUENESS THEOREM. *Let $f(x, y, z)$ be a continuous function in all its arguments in a neighborhood of a point $(x_0, y_0, y_1)$ and let $f$ have bounded partial derivatives $f_y$ and $f_z$ in this neighborhood, or the Lipschitz condition is satisfied: $|f(x, y, z) - f(x, \bar{y}, \bar{z})| \leq A(|y - \bar{y}| + |z - \bar{z}|)$, where $A$ is some positive number. Then a solution of equation (12.3.1.1) satisfying the initial conditions (12.3.1.2) exists and is unique.*

## 12.3.2. Equations Admitting Reduction of Order

### 12.3.2-1. Equations not containing $y$ explicitly.

In the general case, an equation that does not contain $y$ implicitly has the form

$$F(x, y'_x, y''_{xx}) = 0. \tag{12.3.2.1}$$

Such equations remain unchanged under an arbitrary translation of the dependent variable: $y \to y + \text{const}$. The substitution $y'_x = z(x)$, $y''_{xx} = z'_x(x)$ brings (12.3.2.1) to a first-order equation: $F(x, z, z'_x) = 0$.

### 12.3.2-2. Equations not containing $x$ explicitly (autonomous equations).

In the general case, an equation that does not contain $x$ implicitly has the form

$$F(y, y'_x, y''_{xx}) = 0. \tag{12.3.2.2}$$

Such equations remain unchanged under an arbitrary translation of the independent variable: $x \to x + \text{const}$. Using the substitution $y'_x = w(y)$, where $y$ plays the role of the independent variable, and taking into account the relations $y''_{xx} = w'_x = w'_y y'_x = w'_y w$, one can reduce (12.3.2.2) to a first-order equation: $F(y, w, ww'_y) = 0$.

**Example 1.** Consider the autonomous equation

$$y''_{xx} = f(y),$$

which often arises in the theory of heat and mass transfer and combustion. The change of variable $y'_x = w(y)$ leads to a separable first-order equation: $ww'_y = f(y)$. Integrating yields $w^2 = 2F(w) + C_1$, where $F(w) = \int f(w)\, dw$. Solving for $w$ and returning to the original variable, we obtain the separable equation $y'_x = \pm\sqrt{2F(w) + C_1}$. Its general solution is expressed as

$$\int \frac{dy}{\sqrt{2F(w) + C_1}} = \pm x + C_2, \quad \text{where} \quad F(w) = \int f(w)\, dw.$$

**Remark.** The equation $y''_{xx} = f(y + ax^2 + bx + c)$ is reduced by the change of variable $u = y + ax^2 + bx + c$ to an autonomous equation, $u''_{xx} = f(u) + 2a$.

### 12.3.2-3. Equations of the form $F(ax + by, y'_x, y''_{xx}) = 0$.

Such equations are invariant under simultaneous translations of the independent and dependent variables in accordance with the rule $x \to x + bc$, $y \to y - ac$, where $c$ is an arbitrary constant.

For $b = 0$, see equation (12.3.2.1). For $b \neq 0$, the substitution $bw = ax + by$ leads to equation (12.3.2.2): $F(bw, w'_x - a/b, w''_{xx}) = 0$.

### 12.3.2-4. Equations of the form $F(x, xy'_x - y, y''_{xx}) = 0$.

The substitution $w(x) = xy'_x - y$ leads to a first-order equation: $F(x, w, w'_x/x) = 0$.

### 12.3.2-5. Homogeneous equations.

$1°$. The *equations homogeneous in the independent variable* remain unchanged under scaling of the independent variable, $x \to \alpha x$, where $\alpha$ is an arbitrary nonzero number. In the general case, such equations can be written in the form

$$F(y, xy'_x, x^2 y''_{xx}) = 0. \tag{12.3.2.3}$$

The substitution $z(y) = xy'_x$ leads to a first-order equation: $F(y, z, zz'_y - z) = 0$.

$2°$. The *equations homogeneous in the dependent variable* remain unchanged under scaling of the variable sought, $y \to \alpha y$, where $\alpha$ is an arbitrary nonzero number. In the general case, such equations can be written in the form

$$F(x, y'_x/y, y''_{xx}/y) = 0. \tag{12.3.2.4}$$

The substitution $z(x) = y'_x/y$ leads to a first-order equation: $F(x, z, z'_x + z^2) = 0$.

$3°$. The *equations homogeneous in both variables* are invariant under simultaneous scaling (dilatation) of the independent and dependent variables, $x \to \alpha x$ and $y \to \alpha y$, where $\alpha$ is an arbitrary nonzero number. In the general case, such equations can be written in the form

$$F(y/x, y'_x, xy''_{xx}) = 0. \tag{12.3.2.5}$$

The transformation $t = \ln|x|$, $w = y/x$ leads to an autonomous equation (see Paragraph 12.3.2-2): $F(w, w'_t + w, w''_{tt} + w'_t) = 0$.

**Example 2.** The homogeneous equation

$$xy''_{xx} - y'_x = f(y/x)$$

is reduced by the transformation $t = \ln|x|$, $w = y/x$ to the autonomous form: $w''_{tt} = f(w) + w$. For solution of this equation, see Example 1 in Paragraph 12.3.2-2 (the notation of the right-hand side has to be changed there).

### 12.3.2-6. Generalized homogeneous equations.

$1°$. The *generalized homogeneous equations* remain unchanged under simultaneous scaling of the independent and dependent variables in accordance with the rule $x \to \alpha x$ and $y \to \alpha^k y$, where $\alpha$ is an arbitrary nonzero number and $k$ is some number. Such equations can be written in the form

$$F(x^{-k}y, x^{1-k}y'_x, x^{2-k}y''_{xx}) = 0. \tag{12.3.2.6}$$

The transformation $t = \ln x$, $w = x^{-k}y$ leads to an autonomous equation (see Paragraph 12.3.2-2):

$$F\big(w, w'_t + kw, w''_{tt} + (2k-1)w'_t + k(k-1)w\big) = 0.$$

$2°$. The most general form of representation of generalized homogeneous equations is as follows:

$$\mathcal{F}(x^n y^m, xy'_x/y, x^2 y''_{xx}/y) = 0. \tag{12.3.2.7}$$

The transformation $z = x^n y^m$, $u = xy'_x/y$ brings this equation to the first-order equation

$$\mathcal{F}\big(z, u, z(mu+n)u'_z - u + u^2\big) = 0.$$

**Remark.** For $m \neq 0$, equation (12.3.2.7) is equivalent to equation (12.3.2.6) in which $k = -n/m$. To the particular values $n = 0$ and $m = 0$ there correspond equations (12.3.2.3) and (12.3.2.4) homogeneous in the independent and dependent variables, respectively. For $n = -m \neq 0$, we have an equation homogeneous in both variables, which is equivalent to equation (12.3.2.5).

### 12.3.2-7. Equations invariant under scaling–translation transformations.

1°. The equations of the form

$$F(e^{\lambda x} y, e^{\lambda x} y'_x, e^{\lambda x} y''_{xx}) = 0 \qquad (12.3.2.8)$$

remain unchanged under simultaneous translation and scaling of variables, $x \to x + \alpha$ and $y \to \beta y$, where $\beta = e^{-\alpha \lambda}$ and $\alpha$ is an arbitrary number. The substitution $w = e^{\lambda x} y$ brings (12.3.2.8) to an autonomous equation: $F(w, w'_x - \lambda w, w''_{xx} - 2\lambda w'_x + \lambda^2 w) = 0$ (see Paragraph 12.3.2-2).

2°. The equation

$$F(e^{\lambda x} y^n, y'_x/y, y''_{xx}/y) = 0 \qquad (12.3.2.9)$$

is invariant under the simultaneous translation and scaling of variables, $x \to x + \alpha$ and $y \to \beta y$, where $\beta = e^{-\alpha \lambda/n}$ and $\alpha$ is an arbitrary number. The transformation $z = e^{\lambda x} y^n$, $w = y'_x/y$ brings (12.3.2.9) to a first-order equation: $F\bigl(z, w, z(nw + \lambda)w'_z + w^2\bigr) = 0$.

3°. The equation

$$F(x^n e^{\lambda y}, xy'_x, x^2 y''_{xx}) = 0 \qquad (12.3.2.10)$$

is invariant under the simultaneous scaling and translation of variables, $x \to \alpha x$ and $y \to y + \beta$, where $\alpha = e^{-\beta \lambda/n}$ and $\beta$ is an arbitrary number. The transformation $z = x^n e^{\lambda y}$, $w = xy'_x$ brings (12.3.2.10) to a first-order equation: $F\bigl(z, w, z(\lambda w + n)w'_z - w\bigr) = 0$.

▶ *Some other second-order nonlinear equations are treated in Section T5.3.*

### 12.3.2-8. Exact second-order equations.

The second-order equation

$$F(x, y, y'_x, y''_{xx}) = 0 \qquad (12.3.2.11)$$

is said to be exact if it is the total differential of some function, $F = \varphi'_x$, where $\varphi = \varphi(x, y, y'_x)$. If equation (12.3.2.11) is exact, then we have a first-order equation for $y$:

$$\varphi(x, y, y'_x) = C, \qquad (12.3.2.12)$$

where $C$ is an arbitrary constant.

If equation (12.3.2.11) is exact, then $F(x, y, y'_x, y''_{xx})$ must have the form

$$F(x, y, y'_x, y''_{xx}) = f(x, y, y'_x) y''_{xx} + g(x, y, y'_x). \qquad (12.3.2.13)$$

Here, $f$ and $g$ are expressed in terms of $\varphi$ by the formulas

$$f(x, y, y'_x) = \frac{\partial \varphi}{\partial y'_x}, \qquad g(x, y, y'_x) = \frac{\partial \varphi}{\partial x} + \frac{\partial \varphi}{\partial y} y'_x. \qquad (12.3.2.14)$$

By differentiating (12.3.2.14) with respect to $x$, $y$, and $p = y'_x$, we eliminate the variable $\varphi$ from the two formulas in (12.3.2.14). As a result, we have the following test relations for $f$ and $g$:

$$\begin{aligned} f_{xx} + 2p f_{xy} + p^2 f_{yy} &= g_{xp} + p g_{yp} - g_y, \\ f_{xp} + p f_{yp} + 2 f_y &= g_{pp}. \end{aligned} \qquad (12.3.2.15)$$

Here, the subscripts denote the corresponding partial derivatives.

If conditions (12.3.2.15) hold, then equation (12.3.2.11) with $F$ of (12.3.2.13) is exact. In this case, we can integrate the first equation in (12.3.2.14) with respect to $p = y'_x$ to determine $\varphi = \varphi(x, y, y'_x)$:

$$\varphi = \int f(x, y, p)\, dp + \psi(x, y), \tag{12.3.2.16}$$

where $\psi(x, y)$ is an arbitrary function of integration. This function is determined by substituting (12.3.2.16) into the second equation in (12.3.2.14).

**Example 3.** The left-hand side of the equation

$$yy''_{xx} + (y'_x)^2 + 2axyy'_x + ay^2 = 0 \tag{12.3.2.17}$$

can be represented in the form (12.3.2.13), where $f = y$ and $g = p^2 + 2axyp + ay^2$. It is easy to verify that conditions (12.3.2.15) are satisfied. Hence, equation (12.3.2.17) is exact. Using (12.3.2.16), we obtain

$$\varphi = yp + \psi(x, y). \tag{12.3.2.18}$$

Substituting this expression into the second equation in (12.3.2.14) and taking into account the relation $g = p^2 + 2axyp + ay^2$, we find that $2axyp + ay^2 = \psi_x + p\psi_y$. Since $\psi = \psi(x, y)$, we have $2axy = \psi_y$ and $ay^2 = \psi_x$. Integrating yields $\psi = axy^2 + \text{const}$. Substituting this expression into (12.3.2.18) and taking into account relation (12.3.2.12), we find a first integral of equation (12.3.2.17):

$$yp + axy^2 = C_1, \quad \text{where} \quad p = y'_x.$$

Setting $w = y^2$, we arrive at the first-order linear equation $w'_x + 2axw = 2C_1$, which is easy to integrate. Thus, we find the solution of the original equation in the form:

$$y^2 = 2C_1 \exp(-ax^2) \int \exp(ax^2)\, dx + C_2 \exp(-ax^2).$$

## 12.3.3. Methods of Regular Series Expansions with Respect to the Independent Variable

### 12.3.3-1. Method of expansion in powers of the independent variable.

A solution of the Cauchy problem

$$y''_{xx} = f(x, y, y'_x), \tag{12.3.3.1}$$
$$y(x_0) = y_0, \quad y'_x(x_0) = y_1 \tag{12.3.3.2}$$

can be sought in the form of a Taylor series in powers of the difference $(x - x_0)$, specifically:

$$y(x) = y(x_0) + y'_x(x_0)(x - x_0) + \frac{y''_{xx}(x_0)}{2!}(x - x_0)^2 + \frac{y'''_{xxx}(x_0)}{3!}(x - x_0)^3 + \cdots. \tag{12.3.3.3}$$

The first two coefficients $y(x_0)$ and $y'_x(x_0)$ in solution (12.3.3.3) are defined by the initial conditions (12.3.3.2). The values of the subsequent derivatives of $y$ at the point $x = x_0$ are determined from equation (12.3.3.1) and its derivative equations (obtained by successive differentiation of the equation) taking into account the initial conditions (12.3.3.2). In particular, setting $x = x_0$ in (12.3.3.1) and substituting (12.3.3.2), we obtain the value of the second derivative:

$$y''_{xx}(x_0) = f(x_0, y_0, y_1). \tag{12.3.3.4}$$

Further, differentiating (12.3.3.1) yields

$$y'''_{xxx} = f_x(x, y, y'_x) + f_y(x, y, y'_x)y'_x + f_{y'_x}(x, y, y'_x)y''_{xx}. \tag{12.3.3.5}$$

On substituting $x = x_0$, the initial conditions (12.3.3.2), and the expression of $y''_{xx}(x_0)$ of (12.3.3.4) into the right-hand side of equation (12.3.3.5), we calculate the value of the third derivative:

$$y'''_{xxx}(x_0) = f_x(x_0, y_0, y_1) + f_y(x_0, y_0, y_1)y_1 + f(x_0, y_0, y_1)f_{y'_x}(x_0, y_0, y_1).$$

The subsequent derivatives of the unknown are determined likewise.

The thus obtained solution (12.3.3.3) can only be used in a small neighborhood of the point $x = x_0$.

**Example 1.** Consider the following Cauchy problem for a second-order nonlinear equation:

$$y''_{xx} = yy'_x + y^3; \tag{12.3.3.6}$$
$$y(0) = y'_x(0) = 1. \tag{12.3.3.7}$$

Substituting the initial values of the unknown and its derivative (12.3.3.7) into equation (12.3.3.6) yields the initial value of the second derivative:

$$y''_{xx}(0) = 2. \tag{12.3.3.8}$$

Differentiating equation (12.3.3.6) gives

$$y'''_{xxx} = yy''_{xx} + (y'_x)^2 + 3y^2 y'_x. \tag{12.3.3.9}$$

Substituting here the initial values from (12.3.3.7) and (12.3.3.8), we obtain the initial condition for the third derivative:

$$y'''_{xxx}(0) = 6. \tag{12.3.3.10}$$

Differentiating (12.3.3.9) followed by substituting (12.3.3.7), (12.3.3.8), and (12.3.3.10), we find that

$$y''''_{xxxx}(0) = 24. \tag{12.3.3.11}$$

On substituting the initial data (12.3.3.7), (12.3.3.8), (12.3.3.10), and (12.3.3.11) into (12.3.3.3), we arrive at the Taylor series expansion of the solution about $x = 0$:

$$y = 1 + x + x^2 + x^3 + x^4 + \cdots. \tag{12.3.3.12}$$

This geometric series is convergent only for $|x| < 1$.

### 12.3.3-2. Padé approximants.

Suppose the $k + 1$ leading coefficients in the Taylor series expansion of a solution to a differential equation about the point $x = 0$ are obtained by the method presented in Paragraph 12.3.3-1, so that

$$y_{k+1}(x) = a_0 + a_1 x + \cdots + a_k x^k. \tag{12.3.3.13}$$

The partial sum (12.3.3.13) pretty well approximates the solution at small $x$ but is poor for intermediate and large values of $x$, since the series can be slowly convergent or even divergent. This is also related to the fact that $y_k \to \infty$ as $x \to \infty$, while the exact solution can well be bounded.

In many cases, instead of the expansion (12.3.3.13), it is reasonable to consider a Padé approximant $P_M^N(x)$, which is the ratio of two polynomials of degree $N$ and $M$, specifically,

$$P_M^N(x) = \frac{A_0 + A_1 x + \cdots + A_N x^N}{1 + B_1 x + \cdots + B_M x^M}, \quad \text{where} \quad N + M = k. \tag{12.3.3.14}$$

The coefficients $A_1, \ldots, A_N$ and $B_1, \ldots, B_M$ are selected so that the $k+1$ leading terms in the Taylor series expansion of (12.3.3.14) coincide with the respective terms of the expansion (12.3.3.13). In other words, the expansions (12.3.3.13) and (12.3.3.14) must be asymptotically equivalent as $x \to 0$.

In practice, one usually takes $N = M$ (the diagonal sequence). It often turns out that formula (12.3.3.14) pretty well approximates the exact solution on the entire range of $x$ (for sufficiently large $N$).

**Example 2.** Consider the Cauchy problem (12.3.3.6)–(12.3.3.7) again. The Taylor series expansion of the solution about $x = 0$ has the form (12.3.3.12). This geometric series is convergent only for $|x| < 1$.

The diagonal sequence of Padé approximants corresponding to series (12.3.3.12) is

$$P_1^1(x) = \frac{1}{1-x}, \quad P_2^2(x) = \frac{1}{1-x}, \quad P_3^3(x) = \frac{1}{1-x}. \tag{12.3.3.15}$$

It is not difficult to verify that the function $y(x) = \dfrac{1}{1-x}$ is the exact solution of the Cauchy problem (12.3.3.6)–(12.3.3.7). Hence, in this case, the diagonal sequence of Padé approximants recovers the exact solution from only a few terms in the Taylor series.

**Example 3.** Consider the Cauchy problem for a second-order nonlinear equation:

$$y''_{xx} = 2yy'_x; \quad y(0) = 0, \quad y'_x(0) = 1. \tag{12.3.3.16}$$

Following the method presented in Paragraph 12.3.3-1, we obtain the Taylor series expansion of the solution to problem (12.3.3.16) in the form

$$y(x) = x + \tfrac{1}{3}x^3 + \tfrac{2}{15}x^5 + \tfrac{17}{315}x^7 + \cdots. \tag{12.3.3.17}$$

The exact solution of problem (12.3.3.16) is given by $y(x) = \tan x$. Hence it has singularities at $x = \pm\tfrac{1}{2}(2n+1)\pi$. However, any finite segment of the Taylor series (12.3.3.17) does not have any singularities.

With series (12.3.3.17), we construct the diagonal sequence of Padé approximants:

$$P_2^2(x) = \frac{3x}{3-x^2}, \quad P_3^3(x) = \frac{x(x^2-15)}{3(2x^2-5)}, \quad P_4^4(x) = \frac{5x(21-2x^2)}{x^4-45x^2+105}. \tag{12.3.3.18}$$

These Padé approximants have singularities (at the points where the denominators vanish):

$$x \simeq \pm 1.732 \quad \text{for } P_2^2(x),$$
$$x \simeq \pm 1.581 \quad \text{for } P_3^3(x),$$
$$x \simeq \pm 1.571 \text{ and } x \simeq \pm 6.522 \quad \text{for } P_4^4(x).$$

It is apparent that the Padé approximants are attempting to recover the singularities of the exact solution at $x = \pm\pi/2$ and $x = \pm 3\pi/2$.

In Fig. 12.2, the solid line shows the exact solution of problem (12.3.3.16), the dashed line corresponds to the four-term Taylor series solution (12.3.3.17), and the dot-and-dash line depicts the Padé approximants (12.3.3.18). It is evident that the Padé approximant $P_4^4(x)$ gives an accurate numerical approximation of the exact solution on the interval $|x| \leq 2$; everywhere the error is less than 1%, except for a very small neighborhood of the point $x = \pm\pi/2$ (the error is 1% for $x = \pm 1.535$ and 0.84% for $x = \pm 2$).

## 12.3.4. Movable Singularities of Solutions of Ordinary Differential Equations. Painlevé Transcendents

### 12.3.4-1. Preliminary remarks. Singular points of solutions.

1°. Singular points of solutions to ordinary differential equations can be *fixed* or *movable*. The coordinates of fixed singular points remain the same for different solutions of an equation.* The coordinates of movable singular points vary depending on the particular solution selected (i.e., they depend on the initial conditions).

---

* Solutions of linear ordinary differential equations can only have fixed singular points, and their positions are determined by the singularities of the equation coefficients.

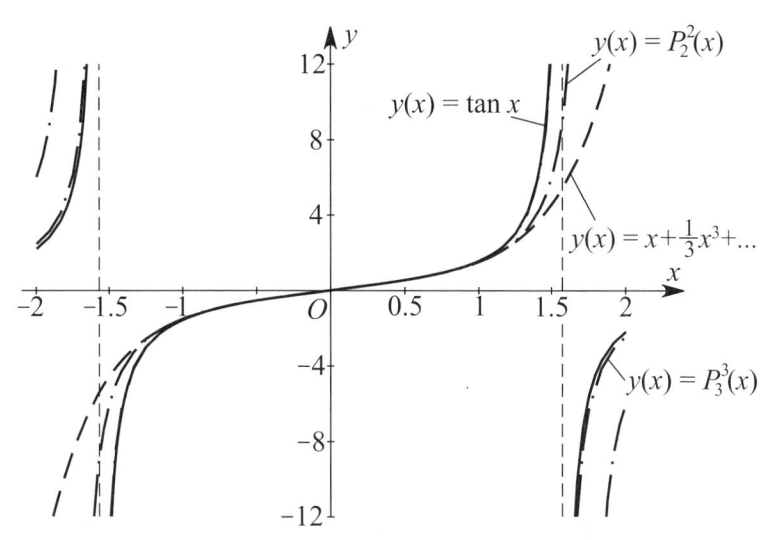

**Figure 12.2.** Comparison of the exact solution to problem (12.3.3.16) with the approximate truncated series solution (12.3.3.17) and associated Padé approximants (12.3.3.18).

Listed below are simple examples of first-order ordinary differential equations and their solutions having movable singularities:

| Equation | Solution | Solution's singularity type |
|---|---|---|
| $y'_z = -y^2$ | $y = 1/(z - z_0)$ | movable pole |
| $y'_z = 1/y$ | $y = 2\sqrt{z - z_0}$ | algebraic branch point |
| $y'_z = e^{-y}$ | $y = \ln(z - z_0)$ | logarithmic branch point |
| $y'_z = -y \ln^2 y$ | $y = \exp[1/(z - z_0)]$ | essential singularity |

Algebraic branch points, logarithmic branch points, and essential singularities are called *movable critical points*.

2°. The Painlevé equations arise from the classification of the following second-order differential equations over the complex plane:

$$y''_{zz} = R(z, y, y'_z),$$

where $R = R(z, y, w)$ is a function rational in $y$ and $w$ and analytic in $z$. It was shown by P. Painlevé (1897–1902) and B. Gambier (1910) that all equations of this type whose solutions do not have movable critical points (but are allowed to have fixed singular points and movable poles) can be reduced to 50 classes of equations. Moreover, 44 classes out of them are integrable by quadrature or admit reduction of order. The remaining 6 equations are irreducible; these are known as the *Painlevé equations* or *Painlevé transcendents*, and their solutions are known as the *Painlevé transcendental functions*.

The canonical forms of the Painlevé transcendents are given below in Paragraphs 12.3.4-2 through 12.3.4-6. Solutions of the first, second, and fourth Painlevé transcendents have movable poles (no fixed singular points). Solutions of the third and fifth Painlevé transcendents have two fixed logarithmic branch points, $z = 0$ and $z = \infty$. Solutions of the sixth Painlevé transcendent have three fixed logarithmic branch points, $z = 0$, $z = 1$, and $z = \infty$.

It is significant that the Painlevé equations often arise in mathematical physics.

### 12.3.4-2. First Painlevé transcendent.

**1°.** The *first Painlevé transcendent* has the form

$$y''_{zz} = 6y^2 + z. \tag{12.3.4.1}$$

The solutions of the first Painlevé transcendent are single-valued functions of $z$.

The solutions of equation (12.3.4.1) can be presented, in the vicinity of movable pole $z_p$, in terms of the series

$$y = \frac{1}{(z-z_p)^2} + \sum_{n=2}^{\infty} a_n(z-z_p)^n,$$

$$a_2 = -\tfrac{1}{10}z_p, \quad a_3 = -\tfrac{1}{6}, \quad a_4 = C, \quad a_5 = 0, \quad a_6 = \tfrac{1}{300}z_p^2,$$

where $z_p$ and $C$ are arbitrary constants; the coefficients $a_j$ ($j \geq 7$) are uniquely defined in terms of $z_p$ and $C$.

**2°.** In a neighborhood of a fixed point $z = z_0$, the solution of the Cauchy problem for the first Painlevé transcendent (12.3.4.1) can be represented by the Taylor series (see Paragraph 12.3.3-1):

$$y = A + B(z-z_0) + \tfrac{1}{2}(6A^2+z_0)(z-z_0)^2 + \tfrac{1}{6}(12AB+1)(z-z_0)^3 + \tfrac{1}{2}(6A^3+B^2+Az_0)(z-z_0)^4 + \cdots,$$

where $A$ and $B$ are initial data of the Cauchy problem, so that $y|_{z=z_0} = A$ and $y'_z|_{z=z_0} = B$.

*Remark.* The solutions of the Cauchy problems for the second and fourth Painlevé transcendents can be expressed likewise (fixed singular points should be excluded from consideration for the remaining Painlevé transcendents).

**3°.** For large values of $|z|$, the following asymptotic formula holds:

$$y \sim z^{1/2} \wp\left(\tfrac{4}{5}z^{5/4} - a;\ 12,\ b\right),$$

where the elliptic Weierstrass function $\wp(\zeta; 12, b)$ is defined implicitly by the integral

$$\zeta = \int \frac{d\wp}{\sqrt{4\wp^3 - 12\wp - b}};$$

$a$ and $b$ are some constants.

### 12.3.4-3. Second Painlevé transcendent.

**1°.** The *second Painlevé transcendent* has the form

$$y''_{zz} = 2y^3 + zy + \alpha. \tag{12.3.4.2}$$

The solutions of the second Painlevé transcendent are single-valued functions of $z$.

The solutions of equation (12.3.4.2) can be represented, in the vicinity of a movable pole $z_p$, in terms of the series

$$y = \frac{m}{z-z_p} + \sum_{n=1}^{\infty} b_n(z-z_p)^n,$$

$$b_1 = -\tfrac{1}{6}mz_p, \quad b_2 = -\tfrac{1}{4}(m+\alpha), \quad b_3 = C, \quad b_4 = \tfrac{1}{72}z_p(m+3\alpha),$$

$$b_5 = \tfrac{1}{3024}\left[(27 + 81\alpha^2 - 2z_p^3)m + 108\alpha - 216Cz_p\right],$$

where $z_p$ and $C$ are arbitrary constants, $m = \pm 1$, and the coefficients $b_n$ ($n \geq 6$) are uniquely defined in terms of $z_p$ and $C$.

**2°.** For fixed $\alpha$, denote the solution by $y(z, \alpha)$. Then the following relation holds:

$$y(z, -\alpha) = -y(z, \alpha), \qquad (12.3.4.3)$$

while the solutions $y(z, \alpha)$ and $y(z, \alpha - 1)$ are related by the Bäcklund transformations:

$$\begin{aligned} y(z, \alpha - 1) &= -y(z, \alpha) + \frac{2\alpha - 1}{2y'_z(z, \alpha) - 2y^2(z, \alpha) - z}, \\ y(z, \alpha) &= -y(z, \alpha - 1) - \frac{2\alpha - 1}{2y'_z(z, \alpha - 1) + 2y^2(z, \alpha - 1) + z}. \end{aligned} \qquad (12.3.4.4)$$

Therefore, in order to study the general solution of equation (12.3.4.2) with arbitrary $\alpha$, it is sufficient to construct the solution for all $\alpha$ out of the band $0 \leq \operatorname{Re} \alpha < \frac{1}{2}$.

Three solutions corresponding to $\alpha$ and $\alpha \pm 1$ are related by the rational formulas

$$y_{\alpha+1} = -\frac{(y_{\alpha-1} + y_\alpha)(4y_\alpha^3 + 2zy_\alpha + 2\alpha + 1) + (2\alpha - 1)y_\alpha}{2(y_{\alpha-1} + y_\alpha)(2y_\alpha^2 + z) + 2\alpha - 1},$$

where $y_\alpha$ stands for $y(z, \alpha)$.

The solutions $y(z, \alpha)$ and $y(z, -\alpha - 1)$ are related by the Bäcklund transformations:

$$\begin{aligned} y(z, -\alpha - 1) &= y(z, \alpha) + \frac{2\alpha + 1}{2y'_z(z, \alpha) + 2y^2(z, \alpha) + z}, \\ y(z, \alpha) &= y(z, -\alpha - 1) - \frac{2\alpha + 1}{2y'_z(z, -\alpha - 1) + 2y^2(z, -\alpha - 1) + z}. \end{aligned}$$

**3°.** For $\alpha = 0$, equation (12.3.4.2) has the trivial solution $y = 0$. Taking into account this fact and relations (12.3.4.3) and (12.3.4.4), we find that the second Painlevé transcendent with $\alpha = \pm 1, \pm 2, \ldots$ has the rational particular solutions

$$y(z, \pm 1) = \mp\frac{1}{z}, \qquad y(z, \pm 2) = \pm\left(\frac{1}{z} - \frac{3z^2}{z^3 + 4}\right), \qquad \ldots$$

For $\alpha = \frac{1}{2}$, equation (12.3.4.2) admits the one-parameter family of solutions:

$$y(z, \tfrac{1}{2}) = -\frac{w'_z}{w}, \quad \text{where} \quad w = \sqrt{z}\left[C_1 J_{1/3}\left(\tfrac{\sqrt{2}}{3} z^{3/2}\right) + C_2 Y_{1/3}\left(\tfrac{\sqrt{2}}{3} z^{3/2}\right)\right]. \qquad (12.3.4.5)$$

(Here, the function $w$ is a solution of the Airy equation, $w''_{zz} + \frac{1}{2}zw = 0$.) It follows from (12.3.4.3)–(12.3.4.5) that the second Painlevé transcendent for all $\alpha = n + \frac{1}{2}$ with $n = 0, \pm 1, \pm 2, \ldots$ has a one-parameter family of solutions that can be expressed in terms of Bessel functions.

### 12.3.4-4. Third Painlevé transcendent.

**1°.** The *third Painlevé transcendent* has the form

$$y''_{zz} = \frac{(y'_z)^2}{y} - \frac{y'_z}{z} + \frac{1}{z}(\alpha y^2 + \beta) + \gamma y^3 + \frac{\delta}{y}. \qquad (12.3.4.6)$$

In terms of the new independent variable $\zeta$ defined by $z = e^\zeta$, the solutions of the transformed equation will be single-valued functions of $\zeta$.

In some special cases, equation (12.3.4.6) can be integrated by quadrature.

2°. Any solution of the Riccati equation

$$y'_z = ky^2 + \frac{\alpha - k}{kz}y + c, \tag{12.3.4.7}$$

where $k^2 = \gamma$, $c^2 = -\delta$, $k\beta + c(\alpha - 2k) = 0$, is a solution of equation (12.3.4.6). Substituting $z = \lambda\tau$, $y = -\dfrac{u'_z}{ku}$, where $\lambda^2 = \dfrac{1}{kc}$, into (12.3.4.7), we obtain a linear equation

$$u''_{\tau\tau} + \frac{k - \alpha}{k\tau}u'_\tau + u = 0,$$

whose general solution is expressed in terms of Bessel functions:

$$u = \tau^{\frac{\alpha}{2k}}\left[C_1 J_{\frac{\alpha}{2k}}(\tau) + C_2 Y_{\frac{\alpha}{2k}}(\tau)\right].$$

### 12.3.4-5. Fourth Painlevé transcendent.

1°. The *fourth Painlevé transcendent* has the form

$$y''_{zz} = \frac{(y'_z)^2}{2y} + \frac{3}{2}y^3 + 4zy^2 + 2(z^2 - \alpha)y + \frac{\beta}{y}. \tag{12.3.4.8}$$

The solutions of the fourth Painlevé transcendent are single-valued functions of $z$.

The Laurent-series expansion of the solution of equation (12.3.4.8) in the vicinity of a movable pole $z_p$ is given by

$$y = \frac{m}{z - z_p} - z_p - \frac{m}{3}(z_p^2 + 2\alpha - 4m)(z - z_p) + C(z - z_p)^2 + \sum_{j=3}^{\infty} a_j(z - z_p)^j,$$

where $m = \pm 1$; $z_p$ and $C$ are arbitrary constants; and the $a_j$ ($j \geq 3$) are uniquely defined in terms of $\alpha$, $\beta$, $z_p$, and $C$.

2°. Two solutions of equation (12.3.4.8) corresponding to different values of the parameters $\alpha$ and $\beta$ are related to each other by the Bäcklund transformations:

$$\widetilde{y} = \frac{1}{2sy}(y'_z - q - 2szy - sy^2), \qquad q^2 = -2\beta,$$

$$y = -\frac{1}{2s\widetilde{y}}(\widetilde{y}'_z - p + 2sz\widetilde{y} + s\widetilde{y}^2), \qquad p^2 = -2\widetilde{\beta},$$

$$2\beta = -(\widetilde{\alpha}s - 1 - \tfrac{1}{2}p)^2, \qquad 4\alpha = -2s - 2\widetilde{\alpha} - 3sp,$$

where $y = y(z, \alpha, \beta)$, $\widetilde{y} = \widetilde{y}(z, \widetilde{\alpha}, \widetilde{\beta})$, and $s$ is an arbitrary parameter.

3°. If the condition $\beta + 2(1 + \alpha m)^2 = 0$, where $m = \pm 1$, is satisfied, then every solution of the Riccati equation

$$y'_z = my^2 + 2mzy - 2(1 + \alpha m)$$

is simultaneously a solution of the fourth Painlevé equation (12.3.4.8).

### 12.3.4-6. Fifth and sixth Painlevé transcendents.

$1°$. The *fifth Painlevé transcendent* has the form

$$y''_{zz} = \frac{3y-1}{2y(y-1)}(y'_z)^2 - \frac{y'_z}{z} + \frac{(y-1)^2}{z^2}\left(\alpha y + \frac{\beta}{y}\right) + \gamma\frac{y}{z} + \frac{\delta y(y+1)}{y-1}.$$

$2°$. The *sixth Painlevé transcendent* has the form

$$y''_{zz} = \frac{1}{2}\left(\frac{1}{y} + \frac{1}{y-1} + \frac{1}{y-z}\right)(y'_z)^2 - \left(\frac{1}{z} + \frac{1}{z-1} + \frac{1}{y-z}\right)y'_z$$
$$+ \frac{y(y-1)(y-z)}{z^2(z-1)^2}\left[\alpha + \beta\frac{z}{y^2} + \gamma\frac{z-1}{(y-1)^2} + \delta\frac{z(z-1)}{(y-z)^2}\right].$$

For details about these equations, see the list of references given at the end of the current chapter.

## 12.3.5. Perturbation Methods of Mechanics and Physics

### 12.3.5-1. Preliminary remarks. A summary table of basic methods.

Perturbation methods are widely used in nonlinear mechanics and theoretical physics for solving problems that are described by differential equations with a small parameter $\varepsilon$. The primary purpose of these methods is to obtain an approximate solution that would be equally suitable at all (small, intermediate, and large) values of the independent variable as $\varepsilon \to 0$.

Equations with a small parameter can be classified according to the following:

(i) the order of the equation remains the same at $\varepsilon = 0$;
(ii) the order of the equation reduces at $\varepsilon = 0$.

For the first type of equations, solutions of related problems* are sufficiently smooth (little varying as $\varepsilon$ decreases). The second type of equation is said to be degenerate at $\varepsilon = 0$, or singularly perturbed. In related problems, thin boundary layers usually arise whose thickness is significantly dependent on $\varepsilon$; such boundary layers are characterized by high gradients of the unknown.

All perturbation methods have a limited domain of applicability; the possibility of using one or another method depends on the type of equations or problems involved. The most commonly used methods are summarized in Table 12.3 (the method of regular series expansions is set out in Paragraph 12.3.5-2). In subsequent paragraphs, additional remarks and specific examples are given for some of the methods. In practice, one usually confines oneself to few leading terms of the asymptotic expansion.

In many problems of nonlinear mechanics and theoretical physics, the independent variable is dimensionless time $t$. Therefore, in this subsection we use the conventional $t$ ($0 \le t < \infty$) instead of $x$.

---

* Further on, we assume that the initial and/or boundary conditions are independent of the parameter $\varepsilon$.

TABLE 12.3
Perturbation methods of nonlinear mechanics and theoretical physics
(the third column gives $n$ leading asymptotic terms with respect to the small parameter $\varepsilon$).

| Method name | Examples of problems solved by the method | Form of the solution sought | Additional conditions and remarks |
|---|---|---|---|
| Method of scaled parameters ($0 \leq t < \infty$) | One looks for periodic solutions of the equation $y''_{tt} + \omega_0^2 y = \varepsilon f(y, y'_t)$; see also Paragraph 12.3.5-3 | $y(t) = \sum_{k=0}^{n-1} \varepsilon^k y_k(z),$ $t = z\left(1 + \sum_{k=1}^{n-1} \varepsilon^k \omega_k\right)$ | Unknowns: $y_k$ and $\omega_k$; $y_{k+1}/y_k = O(1)$; secular terms are eliminated through selection of the constants $\omega_k$ |
| Method of strained coordinates ($0 \leq t < \infty$) | Cauchy problem: $y'_t = f(t, y, \varepsilon);\ y(t_0) = y_0$ ($f$ is of a special form); see also the problem in the method of scaled parameters | $y(t) = \sum_{k=0}^{n-1} \varepsilon^k y_k(z),$ $t = z + \sum_{k=1}^{n-1} \varepsilon^k \varphi_k(z)$ | Unknowns: $y_k$ and $\varphi_k$; $y_{k+1}/y_k = O(1)$, $\varphi_{k+1}/\varphi_k = O(1)$ |
| Averaging method ($0 \leq t < \infty$) | Cauchy problem: $y''_{tt} + \omega_0^2 y = \varepsilon f(y, y'_t),$ $y(0) = y_0,\ y'_t(0) = y_1$; for more general problems, see Paragraph 12.3.5-4, Item 2° | $y = a(t) \cos \varphi(t)$, the amplitude $a$ and phase $\varphi$ are governed by the equations $\frac{da}{dt} = -\frac{\varepsilon}{\omega_0} f_s(a),$ $\frac{d\varphi}{dt} = \omega_0 - \frac{\varepsilon}{a\omega_0} f_c(a)$ | Unknowns: $a$ and $\varphi$; $f_s = \frac{1}{2\pi} \int_0^{2\pi} \sin \varphi F\, d\varphi,$ $f_c = \frac{1}{2\pi} \int_0^{2\pi} \cos \varphi F\, d\varphi,$ $F = f(a \cos \varphi, -a\omega_0 \sin \varphi)$ |
| Krylov–Bogolyubov–Mitropolskii method ($0 \leq t < \infty$) | One looks for periodic solutions of the equation $y''_{tt} + \omega_0^2 y = \varepsilon f(y, y'_t)$; Cauchy problem for this and other equations | $y = a \cos \varphi + \sum_{k=1}^{n-1} \varepsilon^k y_k(a, \varphi),$ $a$ and $\varphi$ are determined by the equations $\frac{da}{dt} = \sum_{k=1}^{n} \varepsilon^k A_k(a),$ $\frac{d\varphi}{dt} = \omega_0 + \sum_{k=1}^{n} \varepsilon^k \Phi_k(a)$ | Unknowns: $y_k$, $A_k$, $\Phi_k$; $y_k$ are $2\pi$-periodic functions of $\varphi$; the $y_k$ are assumed not to contain $\cos \varphi$ |
| Method of two-scale expansions ($0 \leq t < \infty$) | Cauchy problem: $y''_{tt} + \omega_0^2 y = \varepsilon f(y, y'_t),$ $y(0) = y_0,\ y'_t(0) = y_1$; for boundary value problems, see Paragraph 12.3.5-5, Item 2° | $y = \sum_{k=0}^{n-1} \varepsilon^k y_k(\xi, \eta),$ where $\xi = \varepsilon t,\ \eta = t\left(1 + \sum_{k=2}^{n-1} \varepsilon^k \omega_k\right),$ $\frac{d}{dt} = \varepsilon \frac{\partial}{\partial \xi} + (1 + \varepsilon^2 \omega_2 + \cdots)\frac{\partial}{\partial \eta}$ | Unknowns: $y_k$ and $\omega_k$; $y_{k+1}/y_k = O(1)$; secular terms are eliminated through selection of $\omega_k$ |
| Multiple scales method ($0 \leq t < \infty$) | One looks for periodic solutions of the equation $y''_{tt} + \omega_0^2 y = \varepsilon f(y, y'_t)$; Cauchy problem for this and other equations | $y = \sum_{k=0}^{n-1} \varepsilon^k y_k,$ where $y_k = y_k(T_0, T_1, \ldots, T_n),\ T_k = \varepsilon^k t$ $\frac{d}{dt} = \frac{\partial}{\partial T_0} + \varepsilon \frac{\partial}{\partial T_1} + \cdots + \varepsilon^n \frac{\partial}{\partial T_n}$ | Unknowns: $y_k$; $y_{k+1}/y_k = O(1)$; for $n = 1$, this method is equivalent to the averaging method |
| Method of matched asymptotic expansions ($0 \leq x \leq b$) | Boundary value problem: $\varepsilon y''_{xx} + f(x, y) y'_x = g(x, y),$ $y(0) = y_0,\ y(b) = y_b$ ($f$ assumed positive); for other problems, see Paragraph 12.3.5-6, Item 2° | Outer expansion: $y = \sum_{k=0}^{n-1} \sigma_k(\varepsilon) y_k(x),\ O(\varepsilon) \leq x \leq b;$ inner expansion ($z = x/\varepsilon$): $\widetilde{y} = \sum_{k=0}^{n-1} \widetilde{\sigma}_k(\varepsilon) \widetilde{y}_k(z),\ 0 \leq x \leq O(\varepsilon)$ | Unknowns: $y_k, \widetilde{y}_k, \sigma_k, \widetilde{\sigma}_k$; $y_{k+1}/y_k = O(1),$ $\widetilde{y}_{k+1}/\widetilde{y}_k = O(1)$; the procedure of matching expansions is used: $y(x \to 0) = \widetilde{y}(z \to \infty)$ |
| Method of composite expansions ($0 \leq x \leq b$) | Boundary value problem: $\varepsilon y''_{xx} + f(x, y) y'_x = g(x, y),$ $y(0) = y_0,\ y(b) = y_b$ ($f$ assumed positive); boundary value problems for other equations | $y = Y(x, \varepsilon) + \widetilde{Y}(z, \varepsilon),$ $Y = \sum_{k=0}^{n-1} \sigma_k(\varepsilon) Y_k(x),$ $\widetilde{Y} = \sum_{k=0}^{n-1} \widetilde{\sigma}_k(\varepsilon) \widetilde{Y}_k(z),\ z = \frac{x}{\varepsilon};$ here, $\widetilde{Y}_k \to 0$ as $z \to \infty$ | Unknowns: $Y_k, \widetilde{Y}_k, \sigma_k, \widetilde{\sigma}_k$; $Y(b, \varepsilon) = y_b,$ $Y(0, \varepsilon) + \widetilde{Y}(0, \varepsilon) = y_0$; two forms of representation of the equation (in terms of $x$ and $z$) are used to obtain solutions |

**12.3.5-2. Method of regular (direct) expansion in powers of the small parameter.**

We consider an equation of general form with a parameter $\varepsilon$:

$$y''_{tt} + f(t, y, y'_t, \varepsilon) = 0. \tag{12.3.5.1}$$

We assume that the function $f$ can be represented as a series in powers of $\varepsilon$:

$$f(t, y, y'_t, \varepsilon) = \sum_{n=0}^{\infty} \varepsilon^n f_n(t, y, y'_t). \tag{12.3.5.2}$$

Solutions of the Cauchy problem and various boundary value problems for equation (12.3.5.1) with $\varepsilon \to 0$ are sought in the form of a power series expansion:

$$y = \sum_{n=0}^{\infty} \varepsilon^n y_n(t). \tag{12.3.5.3}$$

One should substitute expression (12.3.5.3) into equation (12.3.5.1) taking into account (12.3.5.2). Then the functions $f_n$ are expanded into a power series in the small parameter and the coefficients of like powers of $\varepsilon$ are collected and equated to zero to obtain a system of equations for $y_n$:

$$y''_0 + f_0(t, y_0, y'_0) = 0, \tag{12.3.5.4}$$

$$y''_1 + F(t, y_0, y'_0)y'_1 + G(t, y_0, y'_0)y_1 + f_1(t, y_0, y'_0) = 0, \quad F = \frac{\partial f_0}{\partial y'}, \quad G = \frac{\partial f_0}{\partial y}. \tag{12.3.5.5}$$

Here, only the first two equations are written out. The prime denotes differentiation with respect to $t$. To obtain the initial (or boundary) conditions for $y_n$, the expansion (12.3.5.3) is taken into account.

The success in the application of this method is primarily determined by the possibility of constructing a solution of equation (12.3.5.4) for the leading term $y_0$. It is significant to note that the other terms $y_n$ with $n \geq 1$ are governed by linear equations with homogeneous initial conditions.

**Example 1.** The Duffing equation

$$y''_{tt} + y + \varepsilon y^3 = 0 \tag{12.3.5.6}$$

with initial conditions

$$y(0) = a, \quad y'_t(0) = 0$$

describes the motion of a cubic oscillator, i.e., oscillations of a point mass on a nonlinear spring. Here, $y$ is the deviation of the point mass from the equilibrium and $t$ is dimensionless time.

For $\varepsilon \to 0$, an approximate solution of the problem is sought in the form of the asymptotic expansion (12.3.5.3). We substitute (12.3.5.3) into equation (12.3.5.6) and initial conditions and expand in powers of $\varepsilon$. On equating the coefficients of like powers of the small parameter to zero, we obtain the following problems for $y_0$ and $y_1$:

$$y''_0 + y_0 = 0, \quad y_0 = a, \quad y'_0 = 0;$$
$$y''_1 + y_1 = -y_0^3, \quad y_1 = 0, \quad y'_1 = 0.$$

The solution of the problem for $y_0$ is given by

$$y_0 = a \cos t.$$

Substituting this expression into the equation for $y_1$ and taking into account the identity $\cos^3 t = \frac{1}{4}\cos 3t + \frac{3}{4}\cos t$, we obtain

$$y''_1 + y_1 = -\tfrac{1}{4}a^3(\cos 3t + 3\cos t), \quad y_1 = 0, \quad y'_1 = 0.$$

Integrating yields
$$y_1 = -\tfrac{3}{8}a^3 t \sin t + \tfrac{1}{32}a^3(\cos 3t - 3\cos t).$$

Thus the two-term solution of the original problem is given by
$$y = a\cos t + \varepsilon a^3\left[-\tfrac{3}{8}t\sin t + \tfrac{1}{32}(\cos 3t - 3\cos t)\right] + O(\varepsilon^2).$$

**Remark 1.** The term $t\sin t$ causes $y_1/y_0 \to \infty$ as $t \to \infty$. For this reason, the solution obtained is unsuitable at large times. It can only be used for $\varepsilon t \ll 1$; this results from the condition of applicability of the expansion, $y_0 \gg \varepsilon y_1$.

This circumstance is typical of the method of regular series expansions with respect to the small parameter; in other words, the expansion becomes unsuitable at large values of the independent variable. This method is also inapplicable if the expansion (12.3.5.3) begins with negative powers of $\varepsilon$. Methods that allow avoiding the above difficulties are discussed below in Paragraphs 12.3.5-3 through 12.3.5-5.

**Remark 2.** Growing terms as $t \to \infty$, like $t\sin t$, that narrow down the domain of applicability of asymptotic expansions are called *secular*.

### 12.3.5-3. Method of scaled parameters (Lindstedt–Poincaré method).

We illustrate the characteristic features of the method of scaled parameters with a specific example (the transformation of the independent variable we use here as well as the form of the expansion are specified in the first row of Table 12.3).

**Example 2.** Consider the Duffing equation (12.3.5.6) again. On performing the change of variable
$$t = z(1 + \varepsilon\omega_1 + \cdots),$$
we have
$$y''_{zz} + (1 + \varepsilon\omega_1 + \cdots)^2(y + \varepsilon y^3) = 0. \tag{12.3.5.7}$$

The solution is sought in the series form $y = y_0(z) + \varepsilon y_1(z) + \cdots$. Substituting it into equation (12.3.5.7) and matching the coefficients of like powers of $\varepsilon$, we arrive at the following system of equations for two leading terms of the series:
$$y''_0 + y_0 = 0, \tag{12.3.5.8}$$
$$y''_1 + y_1 = -y_0^3 - 2\omega_1 y_0, \tag{12.3.5.9}$$

where the prime denotes differentiation with respect to $z$.

The general solution of equation (12.3.5.8) is given by
$$y_0 = a\cos(z + b), \tag{12.3.5.10}$$

where $a$ and $b$ are constants of integration. Taking into account (12.3.5.10) and rearranging terms, we reduce equation (12.3.5.9) to
$$y''_1 + y_1 = -\tfrac{1}{4}a^3 \cos[3(z+b)] - 2a\left(\tfrac{3}{8}a^2 + \omega_1\right)\cos(z+b). \tag{12.3.5.11}$$

For $\omega_1 \neq -\tfrac{3}{8}a^2$, the particular solution of equation (12.3.5.11) contains a secular term proportional to $z\cos(z+b)$. In this case, the condition of applicability of the expansion $y_1/y_0 = O(1)$ (see the first row and the last column of Table 12.3) cannot be satisfied at sufficiently large $z$. For this condition to be met, one should set
$$\omega_1 = -\tfrac{3}{8}a^2. \tag{12.3.5.12}$$

In this case, the solution of equation (12.3.5.11) is given by
$$y_1 = \tfrac{1}{32}a^3 \cos[3(z+b)]. \tag{12.3.5.13}$$

Subsequent terms of the expansion can be found likewise.

With (12.3.5.10), (12.3.5.12), and (12.3.5.13), we obtain a solution of the Duffing equation in the form
$$y = a\cos(\omega t + b) + \tfrac{1}{32}\varepsilon a^3 \cos[3(\omega t + b)] + O(\varepsilon^2),$$
$$\omega = \left[1 - \tfrac{3}{8}\varepsilon a^2 + O(\varepsilon^2)\right]^{-1} = 1 + \tfrac{3}{8}\varepsilon a^2 + O(\varepsilon^2).$$

**12.3.5-4. Averaging method (Van der Pol–Krylov–Bogolyubov scheme).**

1°. The averaging method involved two stages. First, the second-order nonlinear equation

$$y''_{tt} + \omega_0^2 y = \varepsilon f(y, y'_t) \qquad (12.3.5.14)$$

is reduced with the transformation

$$y = a \cos\varphi, \quad y'_t = -\omega_0 a \sin\varphi, \quad \text{where} \quad a = a(t), \quad \varphi = \varphi(t),$$

to an equivalent system of two first-order differential equations:

$$\begin{aligned} a'_t &= -\frac{\varepsilon}{\omega_0} f(a\cos\varphi, -\omega_0 a \sin\varphi) \sin\varphi, \\ \varphi'_t &= \omega_0 - \frac{\varepsilon}{\omega_0 a} f(a\cos\varphi, -\omega_0 a \sin\varphi) \cos\varphi. \end{aligned} \qquad (12.3.5.15)$$

The right-hand sides of equations (12.3.5.15) are periodic in $\varphi$, with the amplitude $a$ being a slow function of time $t$. The amplitude and the oscillation character are changing little during the time the phase $\varphi$ changes by $2\pi$.

At the second stage, the right-hand sides of equations (12.3.5.15) are being averaged with respect to $\varphi$. This procedure results in an approximate system of equations:

$$\begin{aligned} a'_t &= -\frac{\varepsilon}{\omega_0} f_s(a), \\ \varphi'_t &= \omega_0 - \frac{\varepsilon}{\omega_0 a} f_c(a), \end{aligned} \qquad (12.3.5.16)$$

where

$$f_s(a) = \frac{1}{2\pi}\int_0^{2\pi} \sin\varphi\, f(a\cos\varphi, -\omega_0 a\sin\varphi)\,d\varphi,$$

$$f_c(a) = \frac{1}{2\pi}\int_0^{2\pi} \cos\varphi\, f(a\cos\varphi, -\omega_0 a\sin\varphi)\,d\varphi.$$

System (12.3.5.16) is substantially simpler than the original system (12.3.5.15) — the first equation in (12.3.5.16), for the oscillation amplitude $a$, is a separable equation and, hence, can readily be integrated; then the second equation in (12.3.5.16), which is linear in $\varphi$, can also be integrated.

Note that the Krylov–Bogolyubov–Mitropolskii method (see the fourth row in Table 12.3) generalizes the above approach and allows obtaining subsequent asymptotic terms as $\varepsilon \to 0$.

2°. Below we outline the general scheme of the averaging method. We consider the second-order nonlinear equation with a small parameter:

$$y''_{tt} + g(t, y, y'_t) = \varepsilon f(t, y, y'_t). \qquad (12.3.5.17)$$

Equation (12.3.5.17) should first be transformed to the equivalent system of equations

$$\begin{aligned} y'_t &= u, \\ u'_t &= -g(t, y, u) + \varepsilon f(t, y, u). \end{aligned} \qquad (12.3.5.18)$$

Suppose the general solution of the "truncated" system (12.3.5.18), with $\varepsilon = 0$, is known:

$$y_0 = \varphi(t, C_1, C_2), \quad u_0 = \psi(t, C_1, C_2), \qquad (12.3.5.19)$$

where $C_1$ and $C_2$ are constants of integration. Taking advantage of the method of variation of constants, we pass from the variables $y$, $u$ in (12.3.5.18) to Lagrange's variables $x_1$, $x_2$ according to the formulas

$$y = \varphi(t, x_1, x_2), \quad u = \psi(t, x_1, x_2), \tag{12.3.5.20}$$

where $\varphi$ and $\psi$ are the same functions that define the general solution of the "truncated" system (12.3.5.19). Transformation (12.3.5.20) allows the reduction of system (12.3.5.18) to the *standard form*

$$\begin{aligned} x_1' &= \varepsilon F_1(t, x_1, x_2), \\ x_2' &= \varepsilon F_2(t, x_1, x_2). \end{aligned} \tag{12.3.5.21}$$

Here, the prime denotes differentiation with respect to $t$ and

$$F_1 = \frac{\varphi_2 f(t, \varphi, \psi)}{\varphi_2 \psi_1 - \varphi_1 \psi_2}, \quad F_2 = -\frac{\varphi_1 f(t, \varphi, \psi)}{\varphi_2 \psi_1 - \varphi_1 \psi_2}; \quad \varphi_k = \frac{\partial \varphi}{\partial x_k}, \quad \psi_k = \frac{\partial \psi}{\partial x_k},$$
$$\varphi = \varphi(t, x_1, x_2), \quad \psi = \psi(t, x_1, x_2).$$

It is significant to note that system (12.3.5.21) is equivalent to the original equation (12.3.5.17). The unknowns $x_1$ and $x_2$ are slow functions of time.

As a result of averaging, system (12.3.5.21) is replaced by a simpler, approximate autonomous system of equations:

$$\begin{aligned} x_1' &= \varepsilon \mathcal{F}_1(x_1, x_2), \\ x_2' &= \varepsilon \mathcal{F}_2(x_1, x_2), \end{aligned} \tag{12.3.5.22}$$

where

$$\mathcal{F}_k(x_1, x_2) = \frac{1}{T} \int_0^T F_k(t, x_1, x_2)\, dt \quad \text{if } F_k \text{ is a } T\text{-periodic function of } t;$$

$$\mathcal{F}_k(x_1, x_2) = \lim_{T \to \infty} \frac{1}{T} \int_0^T F_k(t, x_1, x_2)\, dt \quad \text{if } F_k \text{ is not periodic in } t.$$

**Remark 1.** The averaging method is applicable to equations (12.3.5.14) and (12.3.5.17) with nonsmooth right-hand sides.

**Remark 2.** The averaging method has rigorous mathematical substantiation. There is also a procedure that allows finding subsequent asymptotic terms. For this procedure, e.g., see the books by Bogolyubov and Mitropolskii (1974), Zhuravlev and Klimov (1988), and Arnold, Kozlov, and Neishtadt (1993).

12.3.5-5. Method of two-scale expansions (Cole–Kevorkian scheme).

1°. We illustrate the characteristic features of the method of two-scale expansions with a specific example. Thereafter we outline possible generalizations and modifications of the method.

**Example 3.** Consider the Van der Pol equation

$$y_{tt}'' + y = \varepsilon(1 - y^2)y_t'. \tag{12.3.5.23}$$

The solution is sought in the form (see the fifth row in Table 12.3):

$$y = y_0(\xi, \eta) + \varepsilon y_1(\xi, \eta) + \varepsilon^2 y_2(\xi, \eta) + \cdots,$$
$$\xi = \varepsilon t, \quad \eta = \left(1 + \varepsilon^2 \omega_2 + \cdots\right)t. \tag{12.3.5.24}$$

On substituting (12.3.5.24) into (12.3.5.23) and on matching the coefficients of like powers of $\varepsilon$, we obtain the following system for two leading terms:

$$\frac{\partial^2 y_0}{\partial \eta^2} + y_0 = 0, \tag{12.3.5.25}$$

$$\frac{\partial^2 y_1}{\partial \eta^2} + y_1 = -2\frac{\partial^2 y_0}{\partial \xi \partial \eta} + (1 - y_0^2)\frac{\partial y_0}{\partial \eta}. \tag{12.3.5.26}$$

The general solution of equation (12.3.5.25) is given by

$$y_0 = A(\xi)\cos\eta + B(\xi)\sin\eta. \tag{12.3.5.27}$$

The dependence of $A$ and $B$ on the slow variable $\xi$ is not being established at this stage.

We substitute (12.3.5.27) into the right-hand side of equation (12.3.5.26) and perform elementary manipulations to obtain

$$\frac{\partial^2 y_1}{\partial \eta^2} + y_1 = \left[-2B'_\xi + \tfrac{1}{4}B(4 - A^2 - B^2)\right]\cos\eta + \left[2A'_\xi - \tfrac{1}{4}A(4 - A^2 - B^2)\right]\sin\eta$$
$$+ \tfrac{1}{4}(B^3 - 3A^2 B)\cos 3\eta + \tfrac{1}{4}(A^3 - 3AB^2)\sin 3\eta. \tag{12.3.5.28}$$

The solution of this equation must not contain unbounded terms as $\eta \to \infty$; otherwise the necessary condition $y_1/y_0 = O(1)$ is not satisfied. Therefore the coefficients of $\cos\eta$ and $\sin\eta$ must be set equal to zero:

$$\begin{aligned} -2B'_\xi + \tfrac{1}{4}B(4 - A^2 - B^2) &= 0, \\ 2A'_\xi - \tfrac{1}{4}A(4 - A^2 - B^2) &= 0. \end{aligned} \tag{12.3.5.29}$$

Equations (12.3.5.29) serve to determine $A = A(\xi)$ and $B = B(\xi)$. We multiply the first equation in (12.3.5.29) by $-B$ and the second by $A$ and add them together to obtain

$$r'_\xi - \tfrac{1}{8}r(4 - r^2) = 0, \qquad \text{where} \quad r^2 = A^2 + B^2. \tag{12.3.5.30}$$

The integration by separation of variables yields

$$r^2 = \frac{4r_0^2}{r_0^2 + (4 - r_0^2)e^{-\xi}}, \tag{12.3.5.31}$$

where $r_0$ is the initial oscillation amplitude.

On expressing $A$ and $B$ in terms of the amplitude $r$ and phase $\varphi$, we have $A = r\cos\varphi$ and $B = -r\sin\varphi$. Substituting these expressions into either of the two equations in (12.3.5.29) and using (12.3.5.30), we find that $\varphi'_\xi = 0$ or $\varphi = \varphi_0 = \text{const}$. Therefore the leading asymptotic term can be represented as

$$y_0 = r(\xi)\cos(\eta + \varphi_0),$$

where $\xi = \varepsilon t$ and $\eta = t$, and the function $r(\xi)$ is determined by (12.3.5.31).

2°. The method of two-scale expansions can also be used for solving boundary value problems where the small parameter appears together with the highest derivative as a factor (such problems for $0 \leq x \leq a$ are indicated in the seventh row of Table 12.3 and in Paragraph 12.3.5-6). In the case where a boundary layer arises near the point $x = 0$ (and its thickness has an order of magnitude of $\varepsilon$), the solution is sought in the form

$$y = y_0(\xi, \eta) + \varepsilon y_1(\xi, \eta) + \varepsilon^2 y_2(\xi, \eta) + \cdots,$$
$$\xi = x, \quad \eta = \varepsilon^{-1}\left[g_0(x) + \varepsilon g_1(x) + \varepsilon^2 g_2(x) + \cdots\right],$$

where the functions $y_k = y_k(\xi, \eta)$ and $g_k = g_k(x)$ are to be determined. The derivative with respect to $x$ is calculated in accordance with the rule

$$\frac{d}{dx} = \frac{\partial}{\partial \xi} + \eta'_x \frac{\partial}{\partial \eta} = \frac{\partial}{\partial \xi} + \frac{1}{\varepsilon}\left(g'_0 + \varepsilon g'_1 + \varepsilon^2 g'_2 + \cdots\right)\frac{\partial}{\partial \eta}.$$

Additional conditions are imposed on the asymptotic terms in the domain under consideration; namely, $y_{k+1}/y_k = O(1)$ and $g_{k+1}/g_k = O(1)$ for $k = 0, 1, \ldots$, and $g_0(x) \to x$ as $x \to 0$.

**Remark.** The two-scale method is also used to solve problems that arise in mechanics and physics and are described by partial differential equations.

**12.3.5-6. Method of matched asymptotic expansions.**

1°. We illustrate the characteristic features of the method of matched asymptotic expansions with a specific example (the form of the expansions is specified in the seventh row of Table 12.3). Thereafter we outline possible generalizations and modifications of the method.

**Example 4.** Consider the linear boundary value problem

$$\varepsilon y''_{xx} + y'_x + f(x)y = 0, \qquad (12.3.5.32)$$

$$y(0) = a, \quad y(1) = b, \qquad (12.3.5.33)$$

where $0 < f(0) < \infty$.

At $\varepsilon = 0$ equation (12.3.5.32) degenerates; the solution of the resulting first-order equation

$$y'_x + f(x)y = 0 \qquad (12.3.5.34)$$

cannot meet the two boundary conditions (12.3.5.33) simultaneously. It can be shown that the condition at $x = 0$ has to be omitted in this case (a boundary layer arises near this point).

The leading asymptotic term of the outer expansion, $y = y_0(x) + O(\varepsilon)$, is determined by equation (12.3.5.34). The solution of (12.3.5.34) that satisfies the second boundary condition in (12.3.5.33) is given by

$$y_0(x) = b \exp\left[\int_x^1 f(\xi)\, d\xi\right]. \qquad (12.3.5.35)$$

We seek the leading term of the inner expansion, in the boundary layer adjacent to the left boundary, in the following form (see the seventh row and third column in Table 12.3):

$$\widetilde{y} = \widetilde{y}_0(z) + O(\varepsilon), \quad z = x/\varepsilon, \qquad (12.3.5.36)$$

where $z$ is the extended variable. Substituting (12.3.5.36) into (12.3.5.32) and extracting the coefficient of $\varepsilon^{-1}$, we obtain

$$\widetilde{y}_0'' + \widetilde{y}_0' = 0, \qquad (12.3.5.37)$$

where the prime denotes differentiation with respect to $z$. The solution of equation (12.3.5.37) that satisfies the first boundary condition in (12.3.5.33) is given by

$$\widetilde{y}_0 = a - C + Ce^{-z}. \qquad (12.3.5.38)$$

The constant of integration $C$ is determined from the condition of matching the leading terms of the outer and inner expansions:

$$y_0(x \to 0) = \widetilde{y}_0(z \to \infty). \qquad (12.3.5.39)$$

Substituting (12.3.5.35) and (12.3.5.38) into condition (12.3.5.39) yields

$$C = a - be^{\langle f \rangle}, \qquad \text{where} \quad \langle f \rangle = \int_0^1 f(x)\, dx. \qquad (12.3.5.40)$$

Taking into account relations (12.3.5.35), (12.3.5.36), (12.3.5.38), and (12.3.5.40), we represent the approximate solution in the form

$$y = \begin{cases} be^{\langle f \rangle} + \left(a - be^{\langle f \rangle}\right) e^{-x/\varepsilon} & \text{for } 0 \leq x \leq O(\varepsilon), \\ b \exp\left[\int_x^1 f(\xi)\, d\xi\right] & \text{for } O(\varepsilon) \leq x \leq 1. \end{cases} \qquad (12.3.5.41)$$

It is apparent that inside the thin boundary layer, whose thickness is proportional to $\varepsilon$, the solution rapidly changes by a finite value, $\Delta = be^{\langle f \rangle} - a$.

To determine the function $y$ on the entire interval $x \in [0, 1]$ using formula (12.3.5.41), one has to "switch" at some intermediate point $x = x_0$ from one part of the solution to the other. Such switching is not convenient and, in practice, one often resorts to a *composite solution* instead of using the double formula (12.3.5.41). In similar cases, a composite solution is defined as

$$y = y_0(x) + \widetilde{y}_0(z) - A, \qquad A = \lim_{x \to 0} y_0(x) = \lim_{z \to \infty} \widetilde{y}_0(z).$$

In the problem under consideration, we have $A = be^{\langle f \rangle}$ and hence the composite solution becomes

$$y = \left(a - be^{\langle f \rangle}\right) e^{-x/\varepsilon} + b \exp\left[\int_x^1 f(\xi)\, d\xi\right].$$

For $\varepsilon \ll x \leq 1$, this solution transforms to the outer solution $y_0(x)$ and for $0 \leq x \ll \varepsilon$, to the inner solution, thus providing an approximate representation of the unknown over the entire domain.

$2°$. We now consider an equation of the general form

$$\varepsilon y''_{xx} = F(x, y, y'_x) \qquad (12.3.5.42)$$

subject to boundary conditions (12.3.5.33).

For the leading term of the outer expansion $y = y_0(x) + \cdots$, we have the equation

$$F(x, y_0, y'_0) = 0.$$

In the general case, when using the method of matched asymptotic expansions, the position of the boundary layer and the form of the inner (extended) variable have to be determined in the course of the solution of the problem.

First we assume that the boundary layer is located near the left boundary. In (12.3.5.42), we make a change of variable $z = x/\delta(\varepsilon)$ and rewrite the equation as

$$y''_{zz} = \frac{\delta^2}{\varepsilon} F\left(\delta z, y, \frac{1}{\delta} y'_z\right). \qquad (12.3.5.43)$$

The function $\delta = \delta(\varepsilon)$ is selected so that the right-hand side of equation (12.3.5.43) has a nonzero limit value as $\varepsilon \to 0$, provided that $z$, $y$, and $y'_z$ are of the order of 1.

**Example 5.** For $F(x, y, y'_x) = -kx^\lambda y'_x + y$, where $0 \le \lambda < 1$, the substitution $z = x/\delta(\varepsilon)$ brings equation (12.3.5.42) to

$$y''_{zz} = -\frac{\delta^{1+\lambda}}{\varepsilon} k z^\lambda y'_z + \frac{\delta^2}{\varepsilon} y.$$

In order that the right-hand side of this equation has a nonzero limit value as $\varepsilon \to 0$, one has to set $\delta^{1+\lambda}/\varepsilon = 1$ or $\delta^{1+\lambda}/\varepsilon = \text{const}$, where const is any positive number. It follows that $\delta = \varepsilon^{\frac{1}{1+\lambda}}$.

The leading asymptotic term of the inner expansion in the boundary layer, $y = \widetilde{y}_0(z) + \cdots$, is determined by the equation $\widetilde{y}''_0 + k z^\lambda \widetilde{y}'_0 = 0$, where the prime denotes differentiation with respect to $z$.

If the position of the boundary layer is selected incorrectly, the outer and inner expansions cannot be matched. In this situation, one should consider the case where an arbitrary boundary layer is located on the right (this case is reduced to the previous one with the change of variable $x = 1 - z$). In Example 5 above, the boundary layer is on the left if $k > 0$ and on the right if $k < 0$.

There is a procedure for matching subsequent asymptotic terms of the expansion (see the seventh row and last column in Table 12.3). In its general form, this procedure can be represented as

*inner expansion of the outer expansion ($y$-expansion for $x \to 0$)*

$=$ *outer expansion of the inner expansion ($\widetilde{y}$-expansion for $z \to \infty$).*

**Remark 1.** The method of matched asymptotic expansions can also be applied to construct periodic solutions of singularly perturbed equations (e.g., in the problem of relaxation oscillations of the Van der Pol oscillator).

**Remark 2.** Two boundary layers can arise in some problems (e.g., in cases where the right-hand side of equation (12.3.5.42) does not explicitly depend on $y'_x$).

**Remark 3.** The method of matched asymptotic expansions is also used for solving equations (in semi-infinite domains) that do not degenerate at $\varepsilon = 0$. In such cases, there are no boundary layers; the original variable is used in the inner domain, and an extended coordinate is introduced in the outer domain.

**Remark 4.** The method of matched asymptotic expansions is successfully applied for the solution of various problems in mathematical physics that are described by partial differential equations; in particular, it plays an important role in the theory of heat and mass transfer and in hydrodynamics.

## 12.3.6. Galerkin Method and Its Modifications (Projection Methods)

### 12.3.6-1. General form of an approximate solution.

Consider a boundary value problem for the equation

$$\mathfrak{F}[y] - f(x) = 0 \qquad (12.3.6.1)$$

with linear homogeneous boundary conditions* at the points $x=x_1$ and $x=x_2$ ($x_1 \leq x \leq x_2$). Here, $\mathfrak{F}$ is a linear or nonlinear differential operator of the second order (or a higher order operator); $y = y(x)$ is the unknown function and $f = f(x)$ is a given function. It is assumed that $\mathfrak{F}[0] = 0$.

Let us choose a sequence of linearly independent functions (called *basis functions*)

$$\varphi = \varphi_n(x) \qquad (n = 1, 2, \ldots, N) \qquad (12.3.6.2)$$

satisfying the same boundary conditions as $y = y(x)$. According to all methods that will be considered below, an approximate solution of equation (12.3.6.1) is sought as a linear combination

$$y_N = \sum_{n=1}^{N} A_n \varphi_n(x), \qquad (12.3.6.3)$$

with the unknown coefficients $A_n$ to be found in the process of solving the problem.

The finite sum (12.3.6.3) is called an *approximation function*. The remainder term $R_N$ obtained after the finite sum has been substituted into the left-hand side of equation (12.3.6.1),

$$R_N = \mathfrak{F}[y_N] - f(x). \qquad (12.3.6.4)$$

If the remainder $R_N$ is identically equal to zero, then the function $y_N$ is the exact solution of equation (12.3.6.1). In general, $R_N \not\equiv 0$.

### 12.3.6-2. Galerkin method.

In order to find the coefficients $A_n$ in (12.3.6.3), consider another sequence of linearly independent functions

$$\psi = \psi_k(x) \qquad (k = 1, 2, \ldots, N). \qquad (12.3.6.5)$$

Let us multiply both sides of (12.3.6.4) by $\psi_k$ and integrate the resulting relation over the region $V = \{x_1 \leq x \leq x_2\}$, in which we seek the solution of equation (12.3.6.1). Next, we equate the corresponding integrals to zero (for the exact solutions, these integrals are equal to zero). Thus, we obtain the following system of linear algebraic equations for the unknown coefficients $A_n$:

$$\int_{x_1}^{x_2} \psi_k R_N \, dx = 0 \qquad (k = 1, 2, \ldots, N). \qquad (12.3.6.6)$$

Relations (12.3.6.6) mean that the approximation function (12.3.6.3) satisfies equation (12.3.6.1) "on the average" (i.e., in the integral sense) with weights $\psi_k$. Introducing

---

* Nonhomogeneous boundary conditions can be reduced to homogeneous ones by the change of variable $z = A_2 x^2 + A_1 x + A_0 + y$ (the constants $A_2$, $A_1$, and $A_0$ are selected using the method of undetermined coefficients).

the scalar product $\langle g, h \rangle = \int_{x_1}^{x_2} gh\, dx$ of arbitrary functions $g$ and $h$, we can consider equations (12.3.6.6) as the condition of orthogonality of the remainder $R_N$ to all weight functions $\psi_k$.

The Galerkin method can be applied not only to boundary value problems, but also to eigenvalue problems (in the latter case, one takes $f = \lambda y$ and seeks eigenfunctions $y_n$, together with eigenvalues $\lambda_n$).

Mathematical justification of the Galerkin method for specific boundary value problems can be found in the literature listed at the end of Chapter 12. Below we describe some other methods that are in fact special cases of the Galerkin method.

**Remark.** Most often, one takes suitable sequences of polynomials or trigonometric functions as $\varphi_n(x)$ in the approximation function (12.3.6.3).

### 12.3.6-3. Bubnov–Galerkin method, the moment method, the least squares method.

$1°$. The sequences of functions (12.3.6.2) and (12.3.6.5) in the Galerkin method can be chosen arbitrarily. In the case of equal functions,

$$\varphi_k(x) = \psi_k(x) \qquad (k = 1, 2, \ldots, N), \tag{12.3.6.7}$$

the method is often called the *Bubnov–Galerkin method*.

$2°$. The *moment method* is the Galerkin method with the weight functions (12.3.6.5) being powers of $x$,

$$\psi_k = x^k. \tag{12.3.6.8}$$

$3°$. Sometimes, the functions $\psi_k$ are expressed in terms of $\varphi_k$ by the relations

$$\psi_k = \mathfrak{F}[\varphi_k] \qquad (k = 1, 2, \ldots),$$

where $\mathfrak{F}$ is the differential operator of equation (12.3.6.1). This version of the Galerkin method is called the *least squares method*.

### 12.3.6-4. Collocation method.

In the collocation method, one chooses a sequence of points $x_k$, $k = 1, \ldots, N$, and imposes the condition that the remainder (12.3.6.4) be zero at these points,

$$R_N = 0 \quad \text{at} \quad x = x_k \qquad (k = 1, \ldots, N). \tag{12.3.6.9}$$

When solving a specific problem, the points $x_k$, at which the remainder $R_N$ is set equal to zero, are regarded as most significant. The number of collocation points $N$ is taken equal to the number of the terms of the series (12.3.6.3). This allows one to obtain a complete system of algebraic equations for the unknown coefficients $A_n$ (for linear boundary value problems, this algebraic system is linear).

Note that the collocation method is a special case of the Galerkin method with the sequence (12.3.6.5) consisting of the Dirac delta functions:

$$\psi_k = \delta(x - x_k).$$

In the collocation method, there is no need to calculate integrals, and this essentially simplifies the procedure of solving nonlinear problems (although usually this method yields less accurate results than other modifications of the Galerkin method).

**Example.** Consider the boundary value problem for the linear variable-coefficient second-order ordinary differential equation

$$y''_{xx} + g(x)y - f(x) = 0 \qquad (12.3.6.10)$$

subject to the boundary conditions of the first kind

$$y(-1) = y(1) = 0. \qquad (12.3.6.11)$$

Assume that the coefficients of equation (12.3.6.10) are smooth even functions, so that $f(x) = f(-x)$ and $g(x) = g(-x)$. We use the collocation method for the approximate solution of problem (12.3.6.10)–(12.3.6.11).

1°. Take the polynomials

$$y_n(x) = x^{2n-2}(1 - x^2), \qquad n = 1, 2, \ldots N,$$

as the basis functions; they satisfy the boundary conditions (12.3.6.11), $y_n(\pm 1) = 0$.

Let us consider three collocation points

$$x_1 = -\sigma, \quad x_2 = 0, \quad x_3 = \sigma \qquad (0 < \sigma < 1) \qquad (12.3.6.12)$$

and confine ourselves to two basis functions ($N = 2$), so that the approximation function is taken in the form

$$y(x) = A_1(1 - x^2) + A_2 x^2(1 - x^2). \qquad (12.3.6.13)$$

Substituting (12.3.6.13) in the left-hand side of equation (12.3.6.10) yields the remainder

$$R(x) = A_1\left[-2 + (1 - x^2)g(x)\right] + A_2\left[2 - 12x^2 + x^2(1 - x^2)g(x)\right] - f(x).$$

It must vanish at the collocation points (12.3.6.12). Taking into account the properties $f(\sigma) = f(-\sigma)$ and $g(\sigma) = g(-\sigma)$, we obtain two linear algebraic equations for the coefficients $A_1$ and $A_2$:

$$\begin{aligned} A_1\left[-2 + g(0)\right] + 2A_2 - f(0) &= 0 \qquad \text{(at } x = 0\text{)}, \\ A_1\left[-2 + (1 - \sigma^2)g(\sigma)\right] + A_2\left[2 - 12\sigma^2 + \sigma^2(1 - \sigma^2)g(\sigma)\right] - f(\sigma) &= 0 \qquad \text{(at } x = \pm\sigma\text{)}. \end{aligned} \qquad (12.3.6.14)$$

2°. To be specific, let us take the following functions entering equation (12.3.6.10):

$$f(x) = -1, \quad g(x) = 1 + x^2. \qquad (12.3.6.15)$$

On solving the corresponding system of algebraic equations (12.3.6.14), we find the coefficients

$$A_1 = \frac{\sigma^4 + 11}{\sigma^4 + 2\sigma^2 + 11}, \qquad A_2 = -\frac{\sigma^2}{\sigma^4 + 2\sigma^2 + 11}. \qquad (12.3.6.16)$$

In Fig. 12.3, the solid line depicts the numerical solution to problem (12.3.6.10)–(12.3.6.11), with the functions (12.3.6.15), obtained by the shooting method (see Paragraph 12.3.7-3). The dashed lines 1 and 2 show the approximate solutions obtained by the collocation method using the formulas (12.3.6.13), (12.3.6.16) with $\sigma = \frac{1}{2}$ (equidistant points) and $\sigma = \frac{\sqrt{2}}{2}$ (Chebyshev points, see Subsection 12.4.4), respectively. It is evident that both cases provide good coincidence of the approximate and numerical solutions; the use of Chebyshev points gives a more accurate result.

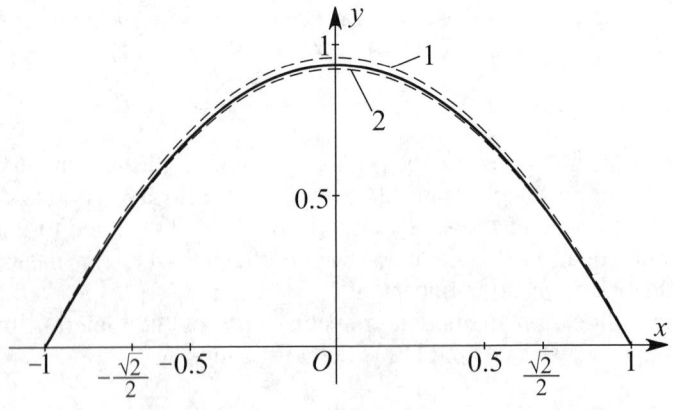

**Figure 12.3.** Comparison of the numerical solution of problem (12.3.6.10), (12.3.6.11), (12.3.6.15) with the approximate analytical solution (12.3.6.13), (12.3.6.16) obtained with the collocation method.

**Remark.** The theorem of convergence of the collocation method for linear boundary value problems is given in Subsection 12.4.4, where $n$th-order differential equations are considered.

### 12.3.6-5. Method of partitioning the domain.

The domain $V = \{x_1 \leq x \leq x_2\}$ is split into $N$ subdomains: $V_k = \{x_{k1} \leq x \leq x_{k2}\}$, $k = 1, \ldots, N$. In this method, the weight functions are chosen as follows:

$$\psi_k(x) = \begin{cases} 1 & \text{for } x \in V_k, \\ 0 & \text{for } x \notin V_k. \end{cases}$$

The subdomains $V_k$ are chosen according to the specific properties of the problem under consideration and can generally be arbitrary (the union of all subdomains $V_k$ may differ from the domain $V$, and some $V_k$ and $V_m$ may overlap).

### 12.3.6-6. Least squared error method.

Sometimes, in order to find the coefficients $A_n$ of the approximation function (12.3.6.3), one uses the least squared error method based on the minimization of the functional:

$$\Phi = \int_{x_1}^{x_2} R_N^2 \, dx \to \min . \tag{12.3.6.17}$$

For given functions $\varphi_n$ in (12.3.6.3), the integral $\Phi$ is a function with respect to the coefficients $A_n$. The corresponding necessary conditions of minimum in (12.3.6.17) have the form

$$\frac{\partial \Phi}{\partial A_n} = 0 \quad (n = 1, \ldots, N).$$

This is a system of algebraic (transcendental) equations for the coefficients $A_n$.

## 12.3.7. Iteration and Numerical Methods

### 12.3.7-1. Method of successive approximations (Cauchy problem).

The method of successive approximations is implemented in two steps. First, the Cauchy problem

$$y''_{xx} = f(x, y, y'_x) \quad \text{(equation)}, \tag{12.3.7.1}$$
$$y(x_0) = y_0, \quad y'_x(x_0) = y'_0 \quad \text{(initial conditions)} \tag{12.3.7.2}$$

is reduced to an equivalent system of integral equations by the introduction of the new variable $u(x) = y'_x$. These integral equations have the form

$$u(x) = y'_0 + \int_{x_0}^{x} f(t, y(t), u(t)) \, dt, \quad y(x) = y_0 + \int_{x_0}^{x} u(t) \, dt. \tag{12.3.7.3}$$

Then the solution of system (12.3.7.3) is sought by means of successive approximations defined by the following recurrence formulas:

$$u_{n+1}(x) = y'_0 + \int_{x_0}^{x} f(t, y_n(t), u_n(t)) \, dt, \quad y_{n+1}(x) = y_0 + \int_{x_0}^{x} u_n(t) \, dt; \quad n = 0, 1, 2, \ldots$$

As the initial approximation, one can take $y_0(x) = y_0$ and $u_0(x) = y'_0$.

### 12.3.7-2. Runge–Kutta method (Cauchy problem).

For the numerical integration of the Cauchy problem (12.3.7.1)–(12.3.7.2), one often uses the Runge–Kutta method.

Let $\Delta x$ be sufficiently small. We introduce the following notation:

$$x_k = x_0 + k\Delta x, \quad y_k = y(x_k), \quad y'_k = y'_x(x_k), \quad f_k = f(x_k, y_k, y'_k); \qquad k = 0, 1, 2, \dots$$

The desired values $y_k$ and $y'_k$ are successively found by the formulas

$$y_{k+1} = y_k + y'_k \Delta x + \tfrac{1}{6}(f_1 + f_2 + f_3)(\Delta x)^2,$$
$$y'_{k+1} = y'_k + \tfrac{1}{6}(f_1 + 2f_2 + 2f_3 + f_4)\Delta x,$$

where

$$f_1 = f(x_k, y_k, y'_k),$$
$$f_2 = f\left(x_k + \tfrac{1}{2}\Delta x, \; y_k + \tfrac{1}{2}y'_k\Delta x, \; y'_k + \tfrac{1}{2}f_1\Delta x\right),$$
$$f_3 = f\left(x_k + \tfrac{1}{2}\Delta x, \; y_k + \tfrac{1}{2}y'_k\Delta x + \tfrac{1}{4}f_1(\Delta x)^2, \; y'_k + \tfrac{1}{2}f_2\Delta x\right),$$
$$f_4 = f\left(x_k + \Delta x, \; y_k + y'_k\Delta x + \tfrac{1}{2}f_2(\Delta x)^2, \; y'_k + f_3\Delta x\right).$$

In practice, the step $\Delta x$ is determined in the same way as for first-order equations (see Remark 2 in Paragraph 12.1.10-3).

### 12.3.7-3. Shooting method (boundary value problems).

In order to solve the boundary value problem for equation (12.3.7.1) with the boundary conditions

$$y(x_1) = y_1, \qquad y(x_2) = y_2, \tag{12.3.7.4}$$

one considers an auxiliary Cauchy problem for equation (12.3.7.1) with the initial conditions

$$y(x_1) = y_1, \qquad y'_x(x_1) = a. \tag{12.3.7.5}$$

(The solution of this Cauchy problem can be obtained by the Runge–Kutta method or some other numerical method.) The parameter $a$ is chosen so that the value of the solution $y = y(x, a)$ at the point $x = x_2$ coincides with the value required by the second boundary condition in (12.3.7.4):

$$y(x_2, a) = y_2.$$

In a similar way one constructs the solution of the boundary value problem with mixed boundary conditions

$$y(x_1) = y_1, \qquad y'_x(x_2) + ky(x_2) = y_2. \tag{12.3.7.6}$$

In this case, one also considers the auxiliary Cauchy problem (12.3.7.1), (12.3.7.5). The parameter $a$ is chosen so that the solution $y = y(x, a)$ satisfies the second boundary condition in (12.3.7.6) at the point $x = x_2$.

**12.3.7-4. Method of accelerated convergence in eigenvalue problems.**

Consider the Sturm–Liouville problem for the second-order nonhomogeneous linear equation

$$[f(x)y'_x]'_x + [\lambda g(x) - h(x)]y = 0 \tag{12.3.7.7}$$

with linear homogeneous boundary conditions of the first kind

$$y(0) = y(1) = 0. \tag{12.3.7.8}$$

It is assumed that the functions $f$, $f'_x$, $g$, $h$ are continuous and $f > 0$, $g > 0$.

First, using the Rayleigh–Ritz principle, one finds an upper estimate for the first eigenvalue $\lambda_1^0$ [this value is determined by the right-hand side of relation (12.2.5.6)]. Then, one solves numerically the Cauchy problem for the auxiliary equation

$$[f(x)y'_x]'_x + [\lambda_1^0 g(x) - h(x)]y = 0 \tag{12.3.7.9}$$

with the boundary conditions

$$y(0) = 0, \quad y'_x(0) = 1. \tag{12.3.7.10}$$

The function $y(x, \lambda_1^0)$ satisfies the condition $y(x_0, \lambda_1^0) = 0$, where $x_0 < 1$. The criterion of closeness of the exact and approximate solutions, $\lambda_1$ and $\lambda_1^0$, has the form of the inequality $|1 - x_0| \leq \delta$, where $\delta$ is a sufficiently small given constant. If this inequality does not hold, one constructs a refinement for the approximate eigenvalue on the basis of the formula

$$\lambda_1^1 = \lambda_1^0 - \varepsilon_0 f(1)\frac{[y'_x(1)]^2}{\|y\|^2}, \quad \varepsilon_0 = 1 - x_0, \tag{12.3.7.11}$$

where $\|y\|^2 = \int_0^1 g(x)y^2(x)\,dx$. Then the value $\lambda_1^1$ is substituted for $\lambda_1^0$ in the Cauchy problem (12.3.7.9)–(12.3.7.10). As a result, a new solution $y$ and a new point $x_1$ are found; and one has to check whether the criterion $|1 - x_1| \leq \delta$ holds. If this inequality is violated, one refines the approximate eigenvalue by means of the formula

$$\lambda_1^2 = \lambda_1^1 - \varepsilon_1 f(1)\frac{[y'_x(1)]^2}{\|y\|^2}, \quad \varepsilon_1 = 1 - x_1, \tag{12.3.7.12}$$

and repeats the above procedure.

Remark 1. Formulas of the type (12.3.7.11) are obtained by a perturbation method based on a transformation of the independent variable $x$ (see Paragraph 12.3.5-2). If $x_n > 1$, the functions $f$, $g$, and $h$ are smoothly extended to the interval $(1, \xi]$, where $\xi \geq x_n$.

Remark 2. The algorithm described above has the property of accelerated convergence $\varepsilon_{n+1} \sim \varepsilon_n^2$, which ensures that the relative error of the approximate solution becomes $10^{-4}$ to $10^{-8}$ after two or three iterations for $\varepsilon_0 \sim 0.1$. This method is quite effective for high-precision calculations, is fail-safe, and guarantees against accumulation of roundoff errors.

Remark 3. In a similar way, one can compute subsequent eigenvalues $\lambda_m$, $m = 2, 3, \ldots$ (to that end, a suitable initial approximation $\lambda_m^0$ should be chosen).

Remark 4. A similar computation scheme can also be used in the case of boundary conditions of the second and the third kinds, periodic boundary conditions, etc. (see the reference below).

**Example 1.** The eigenvalue problem for the equation
$$y''_{xx} + \lambda(1+x^2)^{-2}y = 0$$
with the boundary conditions (12.3.7.8) admits an exact analytic solution and has eigenvalues $\lambda_1 = 15$, $\lambda_2 = 63$, ..., $\lambda_n = 16n^2 - 1$.

According to the Rayleigh–Ritz principle, formula (12.2.5.6) for $z = \sin(\pi x)$ yields the approximate value $\lambda_1^0 = 15.33728$. The solution of the Cauchy problem (12.3.7.9)–(12.3.7.10) with $f(x) = 1$, $g(x) = \lambda(1+x^2)^{-2}$, $h(x) = 0$ yields $x_0 = 0.983848$, $1 - x_0 = 0.016152$, $\|y\|^2 = 0.024585$, $y'_x(x_0) = -0.70622822$.

The first iteration for the first eigenvalue is determined by (12.3.7.11) and results in the value $\lambda_1^1 = 14.99245$ with the relative error $\Delta\lambda/\lambda_1^1 = 5 \times 10^{-4}$.

The second iteration results in $\lambda_1^2 = 14.999986$ with the relative error $\Delta\lambda/\lambda_1^2 < 10^{-6}$.

**Example 2.** Consider the eigenvalue problem for the equation
$$(\sqrt{1+x}\, y'_x)'_x + \lambda y = 0$$
with the boundary conditions (12.3.7.8).

The Rayleigh–Ritz principle yields $\lambda_1^0 = 11.995576$. The next two iterations result in the values $\lambda_1^1 = 11.898578$ and $\lambda_1^2 = 11.898458$. For the relative error we have $\Delta\lambda/\lambda_1^2 < 10^{-5}$.

▶ For more details about finite-difference methods and other numerical methods, see, for instance, the books by Lambert (1973), Keller (1976), Schiesser (1993), and Zwillinger (1997).

## 12.4. Linear Equations of Arbitrary Order

### 12.4.1. Linear Equations with Constant Coefficients

12.4.1-1. Homogeneous linear equations.

An $n$th-order homogeneous linear equation with constant coefficients has the general form
$$y_x^{(n)} + a_{n-1} y_x^{(n-1)} + \cdots + a_1 y'_x + a_0 y = 0. \tag{12.4.1.1}$$

The general solution of this equation is determined by the roots of the characteristic equation
$$P(\lambda) = 0, \quad \text{where} \quad P(\lambda) = \lambda^n + a_{n-1}\lambda^{n-1} + \cdots + a_1\lambda + a_0. \tag{12.4.1.2}$$

The following cases are possible:

1°. All roots $\lambda_1, \lambda_2, \ldots, \lambda_n$ of the characteristic equation (12.4.1.2) are real and distinct. Then the general solution of the homogeneous linear differential equation (12.4.1.1) has the form
$$y = C_1 \exp(\lambda_1 x) + C_2 \exp(\lambda_2 x) + \cdots + C_n \exp(\lambda_n x).$$

2°. There are $m$ equal real roots $\lambda_1 = \lambda_2 = \cdots = \lambda_m$ ($m \leq n$), and the other roots are real and distinct. In this case, the general solution is given by
$$y = \exp(\lambda_1 x)(C_1 + C_2 x + \cdots + C_m x^{m-1})$$
$$+ C_{m+1} \exp(\lambda_{m+1} x) + C_{m+2} \exp(\lambda_{m+2} x) + \cdots + C_n \exp(\lambda_n x).$$

3°. There are $m$ equal complex conjugate roots $\lambda = \alpha \pm i\beta$ ($2m \leq n$), and the other roots are real and distinct. In this case, the general solution is
$$y = \exp(\alpha x)\cos(\beta x)(A_1 + A_2 x + \cdots + A_m x^{m-1})$$
$$+ \exp(\alpha x)\sin(\beta x)(B_1 + B_2 x + \cdots + B_m x^{m-1})$$
$$+ C_{2m+1} \exp(\lambda_{2m+1} x) + C_{2m+2} \exp(\lambda_{2m+2} x) + \cdots + C_n \exp(\lambda_n x),$$
where $A_1, \ldots, A_m, B_1, \ldots, B_m, C_{2m+1}, \ldots, C_n$ are arbitrary constants.

4°. In the general case, where there are $r$ different roots $\lambda_1, \lambda_2, \ldots, \lambda_r$ of multiplicities $m_1, m_2, \ldots, m_r$, respectively, the left-hand side of the characteristic equation (12.4.1.2) can be represented as the product

$$P(\lambda) = (\lambda - \lambda_1)^{m_1}(\lambda - \lambda_2)^{m_2} \ldots (\lambda - \lambda_r)^{m_r},$$

where $m_1 + m_2 + \cdots + m_r = n$. The general solution of the original equation is given by the formula

$$y = \sum_{k=1}^{r} \exp(\lambda_k x)(C_{k,0} + C_{k,1}x + \cdots + C_{k,m_k-1}x^{m_k-1}),$$

where $C_{k,l}$ are arbitrary constants.

If the characteristic equation (12.4.1.2) has complex conjugate roots, then in the above solution, one should extract the real part on the basis of the relation $\exp(\alpha \pm i\beta) = e^{\alpha}(\cos\beta \pm i\sin\beta)$.

**Example 1.** Find the general solution of the linear third-order equation

$$y''' + ay'' - y' - ay = 0.$$

Its characteristic equation is $\lambda^3 + a\lambda^2 - \lambda - a = 0$, or, in factorized form,

$$(\lambda + a)(\lambda - 1)(\lambda + 1) = 0.$$

Depending on the value of the parameter $a$, three cases are possible.

1. Case $a \neq \pm 1$. There are three different roots, $\lambda_1 = -a$, $\lambda_2 = -1$, and $\lambda_3 = 1$. The general solution of the differential equation is expressed as $y = C_1 e^{-ax} + C_2 e^{-x} + C_3 e^{x}$.
2. Case $a = 1$. There is a double root, $\lambda_1 = \lambda_2 = -1$, and a simple root, $\lambda_3 = 1$. The general solution of the differential equation has the form $y = (C_1 + C_2 x)e^{-x} + C_3 e^{x}$.
3. Case $a = -1$. There is a double root, $\lambda_1 = \lambda_2 = 1$, and a simple root, $\lambda_3 = -1$. The general solution of the differential equation is expressed as $y = (C_1 + C_2 x)e^{x} + C_3 e^{-x}$.

**Example 2.** Consider the linear fourth-order equation

$$y''''_{xxxx} - y = 0.$$

Its characteristic equation, $\lambda^4 - 1 = 0$, has four distinct roots, two real and two pure imaginary,

$$\lambda_1 = 1, \quad \lambda_2 = -1, \quad \lambda_3 = i, \quad \lambda_4 = -i.$$

Therefore the general solution of the equation in question has the form (see Item 3°)

$$y = C_1 e^{x} + C_2 e^{-x} + C_3 \sin x + C_4 \cos x.$$

### 12.4.1-2. Nonhomogeneous linear equations. Forms of particular solutions.

1°. An $n$th-order nonhomogeneous linear equation with constant coefficients has the general form

$$y_x^{(n)} + a_{n-1} y_x^{(n-1)} + \cdots + a_1 y'_x + a_0 y = f(x). \quad (12.4.1.3)$$

The general solution of this equation is the sum of the general solution of the corresponding homogeneous equation with $f(x) \equiv 0$ (see Paragraph 12.4.1-1) and any particular solution of the nonhomogeneous equation (12.4.1.3).

If all the roots $\lambda_1, \lambda_2, \ldots, \lambda_n$ of the characteristic equation (12.4.1.2) are different, equation (12.4.1.3) has the general solution:

$$y = \sum_{\nu=1}^{n} C_\nu e^{\lambda_\nu x} + \sum_{\nu=1}^{n} \frac{e^{\lambda_\nu x}}{P'_\lambda(\lambda_\nu)} \int f(x) e^{-\lambda_\nu x}\, dx \quad (12.4.1.4)$$

(for complex roots, the real part should be taken).

In the general case, if the characteristic equation (12.4.1.2) has multiple roots, the solution to equation (12.4.1.3) can be constructed using formula (12.4.2.5).

### TABLE 12.4
Forms of particular solutions of the constant coefficient nonhomogeneous linear equation
$y_x^{(n)} + a_{n-1} y_x^{(n-1)} + \cdots + a_1 y_x' + a_0 y = f(x)$ that correspond to some special forms of the function $f(x)$

| Form of the function $f(x)$ | Roots of the characteristic equation $\lambda^n + a_{n-1}\lambda^{n-1} + \cdots + a_1 \lambda + a_0 = 0$ | Form of a particular solution $y = \widetilde{y}(x)$ |
|---|---|---|
| $P_m(x)$ | Zero is not a root of the characteristic equation (i.e., $a_0 \neq 0$) | $\widetilde{P}_m(x)$ |
|  | Zero is a root of the characteristic equation (multiplicity $r$) | $x^r \widetilde{P}_m(x)$ |
| $P_m(x) e^{\alpha x}$ ($\alpha$ is a real constant) | $\alpha$ is not a root of the characteristic equation | $\widetilde{P}_m(x) e^{\alpha x}$ |
|  | $\alpha$ is a root of the characteristic equation (multiplicity $r$) | $x^r \widetilde{P}_m(x) e^{\alpha x}$ |
| $P_m(x) \cos \beta x + Q_n(x) \sin \beta x$ | $i\beta$ is not a root of the characteristic equation | $\widetilde{P}_\nu(x) \cos \beta x + \widetilde{Q}_\nu(x) \sin \beta x$ |
|  | $i\beta$ is a root of the characteristic equation (multiplicity $r$) | $x^r [\widetilde{P}_\nu(x) \cos \beta x + \widetilde{Q}_\nu(x) \sin \beta x]$ |
| $[P_m(x) \cos \beta x + Q_n(x) \sin \beta x] e^{\alpha x}$ | $\alpha + i\beta$ is not a root of the characteristic equation | $[\widetilde{P}_\nu(x) \cos \beta x + \widetilde{Q}_\nu(x) \sin \beta x] e^{\alpha x}$ |
|  | $\alpha + i\beta$ is a root of the characteristic equation (multiplicity $r$) | $x^r [\widetilde{P}_\nu(x) \cos \beta x + \widetilde{Q}_\nu(x) \sin \beta x] e^{\alpha x}$ |

*Notation*: $P_m$ and $Q_n$ are polynomials of degrees $m$ and $n$ with given coefficients; $\widetilde{P}_m$, $\widetilde{P}_\nu$, and $\widetilde{Q}_\nu$ are polynomials of degrees $m$ and $\nu$ whose coefficients are determined by substituting the particular solution into the basic equation; $\nu = \max(m, n)$; and $\alpha$ and $\beta$ are real numbers, $i^2 = -1$.

2°. Table 12.4 lists the forms of particular solutions corresponding to some special forms of functions on the right-hand side of the linear nonhomogeneous equation.

3°. Consider the Cauchy problem for equation (12.4.1.3) subject to the homogeneous initial conditions
$$y(0) = y_x'(0) = \ldots = y_x^{(n-1)}(0) = 0. \tag{12.4.1.5}$$
Let $y(x)$ be the solution of problem (12.4.1.3), (12.4.1.5) for arbitrary $f(x)$ and let $u(x)$ be the solution of the auxiliary, simpler problem (12.4.1.3), (12.4.1.5) with $f(x) \equiv 1$, so that $u(x) = y(x)|_{f(x) \equiv 1}$. Then the formula
$$y(x) = \int_0^x f(t) u_x'(x - t)\, dt$$
holds. It is called the *Duhamel integral*.

### 12.4.1-3. Solution of the Cauchy problem using the Laplace transform.

Consider the Cauchy problem for equation (12.4.1.3) with arbitrary initial conditions
$$y(0) = y_0, \quad y_x'(0) = y_1, \quad \ldots, \quad y_x^{(n-1)}(0) = y_{n-1}, \tag{12.4.1.6}$$
where $y_0, y_1, \ldots, y_{n-1}$ are given constants.

Problem (12.4.1.3), (12.4.1.6) can be solved using the Laplace transform based on the formulas (for details, see Section 11.2)

$$\widetilde{y}(p) = \mathcal{L}\{y(x)\}, \quad \widetilde{f}(p) = \mathcal{L}\{f(x)\}, \quad \text{where} \quad \mathcal{L}\{f(x)\} \equiv \int_0^\infty e^{-px} f(x)\, dx.$$

To this end, let us multiply equation (12.4.1.3) by $e^{-px}$ and then integrate with respect to $x$ from zero to infinity. Taking into account the differentiation rule

$$\mathcal{L}\{y_x^{(n)}(x)\} = p^n \widetilde{y}(p) - \sum_{k=1}^n p^{n-k} y_x^{(k-1)}(+0)$$

and the initial conditions (12.4.1.6), we arrive at a linear algebraic equation for the transform $\widetilde{y}(p)$:

$$P(p)\widetilde{y}(p) - Q(p) = \widetilde{f}(p), \qquad (12.4.1.7)$$

where

$$P(p) = p^n + a_{n-1} p^{n-1} + \cdots + a_1 p + a_0, \quad Q(p) = b_{n-1} p^{n-1} + \cdots + b_1 p + b_0,$$
$$b_k = y_{n-k-1} + a_{n-1} y_{n-k-2} + \cdots + a_{k+2} y_1 + a_{k+1} y_0, \quad k = 0, 1, \ldots, n-1.$$

The polynomial $P(p)$ coincides with the characteristic polynomial (12.4.1.2) at $\lambda = p$.

The solution of equation (12.4.1.7) is given by the formula

$$\widetilde{y}(p) = \frac{\widetilde{f}(p) + Q(p)}{P(p)}. \qquad (12.4.1.8)$$

On applying the Laplace inversion formula (see in Section 11.2) to (12.4.1.8), we obtain a solution to problem (12.4.1.3), (12.4.1.6) in the form

$$y(x) = \frac{1}{2\pi i} \int_{c-i\infty}^{c+i\infty} \frac{\widetilde{f}(p) + Q(p)}{P(p)} e^{px}\, dp. \qquad (12.4.1.9)$$

Since the transform $\widetilde{y}(p)$ (12.4.1.8) is a rational function, the inverse Laplace transform (12.4.1.9) can be obtained using the formulas from Paragraph 11.2.2-2 or the tables of Section T3.2.

**Remark.** In practice, the solution method for the Cauchy problem based on the Laplace transform leads to the solution faster than the direct application of general formulas like (12.4.1.4), where one has to determine the coefficients $C_1, \ldots, C_n$.

**Example 3.** Consider the following Cauchy problem for a homogeneous fourth-order equation:

$$y''''_{xxxx} + a^4 y = 0; \quad y(0) = y'_x(0) = y'''_{xxx}(0) = 0, \quad y''_{xx}(0) = b.$$

Using the Laplace transform reduces this problem to a linear algebraic equation for $\widetilde{y}(p)$: $(p^4 + a^4)\widetilde{y}(p) - bp = 0$. It follows that

$$\widetilde{y}(p) = \frac{bp}{p^4 + a^4}.$$

In order to invert this expression, let us use the table of inverse Laplace transforms T3.2.2 (see row 52) and take into account that a constant multiplier can be taken outside the transform operator to obtain the solution to the original Cauchy problem in the form

$$y(x) = \frac{b}{a^2} \sin\left(\frac{ax}{\sqrt{2}}\right) \sinh\left(\frac{ax}{\sqrt{2}}\right).$$

## 12.4.2. Linear Equations with Variable Coefficients

**12.4.2-1. Homogeneous linear equations. Structure of the general solution.**

The general solution of the $n$th-order homogeneous linear differential equation

$$f_n(x)y_x^{(n)} + f_{n-1}(x)y_x^{(n-1)} + \cdots + f_1(x)y_x' + f_0(x)y = 0 \qquad (12.4.2.1)$$

has the form
$$y = C_1 y_1(x) + C_2 y_2(x) + \cdots + C_n y_n(x). \qquad (12.4.2.2)$$

Here, $y_1(x), y_2(x), \ldots, y_n(x)$ is a fundamental system of solutions (the $y_k$ are linearly independent particular solutions, $y_k \not\equiv 0$); $C_1, C_2, \ldots, C_n$ are arbitrary constants.

**12.4.2-2. Utilization of particular solutions for reducing the order of the equation.**

1°. Let $y_1 = y_1(x)$ be a nontrivial particular solution of equation (12.4.2.1). The substitution

$$y = y_1(x) \int z(x)\,dx$$

results in a linear equation of order $n-1$ for the function $z(x)$.

2°. Let $y_1 = y_1(x)$ and $y_2 = y_2(x)$ be two nontrivial linearly independent solutions of equation (12.4.2.1). The substitution

$$y = y_1 \int y_2 w\,dx - y_2 \int y_1 w\,dx$$

results in a linear equation of order $n-2$ for $w(x)$.

3°. Suppose that $m$ linearly independent solutions $y_1(x), y_2(x), \ldots, y_m(x)$ of equation (12.4.2.1) are known. Then one can reduce the order of the equation to $n-m$ by successive application of the following procedure. The substitution $y = y_m(x) \int z(x)\,dx$ leads to an equation of order $n-1$ for the function $z(x)$ with known linearly independent solutions:

$$z_1 = \left(\frac{y_1}{y_m}\right)_x', \quad z_2 = \left(\frac{y_2}{y_m}\right)_x', \quad \ldots, \quad z_{m-1} = \left(\frac{y_{m-1}}{y_m}\right)_x'.$$

The substitution $z = z_{m-1}(x) \int w(x)\,dx$ yields an equation of order $n-2$. Repeating this procedure $m$ times, we arrive at a homogeneous linear equation of order $n-m$.

**12.4.2-3. Wronskian determinant and Liouville formula.**

The *Wronskian determinant* (or simply, *Wronskian*) is the function defined as

$$W(x) = \begin{vmatrix} y_1(x) & \cdots & y_n(x) \\ y_1'(x) & \cdots & y_n'(x) \\ \cdots & \cdots & \cdots \\ y_1^{(n-1)}(x) & \cdots & y_n^{(n-1)}(x) \end{vmatrix}, \qquad (12.4.2.3)$$

where $y_1(x), \ldots, y_n(x)$ is a fundamental system of solutions of the homogeneous equation (12.4.2.1); $y_k^{(m)}(x) = \dfrac{d^m y_k}{dx^m}$, $m = 1, \ldots, n-1$; $k = 1, \ldots, n$.

The following *Liouville formula* holds:

$$W(x) = W(x_0) \exp\left[-\int_{x_0}^{x} \frac{f_{n-1}(t)}{f_n(t)}\,dt\right].$$

**12.4.2-4. Nonhomogeneous linear equations. Construction of the general solution.**

1°. The general nonhomogeneous $n$th-order linear differential equation has the form

$$f_n(x)y_x^{(n)} + f_{n-1}(x)y_x^{(n-1)} + \cdots + f_1(x)y_x' + f_0(x)y = g(x). \tag{12.4.2.4}$$

The general solution of the nonhomogeneous equation (12.4.2.4) can be represented as the sum of its particular solution and the general solution of the corresponding homogeneous equation (12.4.2.1).

2°. Let $y_1(x), \ldots, y_n(x)$ be a fundamental system of solutions of the homogeneous equation (12.4.2.1), and let $W(x)$ be the Wronskian determinant (12.4.2.3). Then the general solution of the nonhomogeneous linear equation (12.4.2.4) can be represented as

$$y = \sum_{\nu=1}^{n} C_\nu y_\nu(x) + \sum_{\nu=1}^{n} y_\nu(x) \int \frac{W_\nu(x)\,dx}{f_n(x)W(x)}, \tag{12.4.2.5}$$

where $W_\nu(x)$ is the determinant obtained by replacing the $\nu$th column of the matrix (12.4.2.3) by the column vector with the elements $0, 0, \ldots, 0, g$.

3°. Let $y(x,\sigma)$ be the solution to the Cauchy problem for the homogeneous equation (12.4.2.1) with nonhomogeneous initial conditions at $x = \sigma$:

$$y(\sigma) = y_x'(\sigma) = \cdots = y_x^{(n-2)}(\sigma) = 0, \quad y_x^{(n-1)}(\sigma) = 1,$$

where $\sigma$ is an arbitrary parameter. Then a particular solution of the nonhomogeneous linear equation (12.4.2.4) with homogeneous boundary conditions

$$y(x_0) = y_x'(x_0) = \cdots = y_x^{(n-1)}(x_0) = 0$$

is given by the *Cauchy formula*

$$\bar{y}(x) = \int_{x_0}^{x} y(x,\sigma) \frac{g(\sigma)}{f_n(\sigma)}\,d\sigma.$$

4°. *Superposition principle.* The solution of a nonhomogeneous linear equation

$$\mathbf{L}[y] = \sum_{k=1}^{m} g_k(x), \qquad \mathbf{L}[y] \equiv f_n(x)y_x^{(n)} + f_{n-1}(x)y_x^{(n-1)} + \cdots + f_1(x)y_x' + f_0(x)y$$

is determined by adding together the solutions,

$$y = \sum_{k=1}^{m} y_k,$$

of $m$ (simpler) equations,

$$\mathbf{L}[y_k] = g_k(x), \quad k = 1, 2, \ldots, m,$$

corresponding to respective nonhomogeneous terms in the original equation.

## 12.4.2-5. Euler equation.

1°. The nonhomogeneous Euler equation has the form

$$x^n y_x^{(n)} + a_{n-1} x^{n-1} y_x^{(n-1)} + \cdots + a_1 x y_x' + a_0 y = f(x).$$

The substitution $x = be^t$ ($b \neq 0$) leads to a constant coefficient linear equation of the form (12.4.1.3).

2°. Particular solutions of the homogeneous Euler equation [with $f(x) \equiv 0$] are sought in the form $y = x^k$. If all $k$ are real and distinct, its general solution is expressed as

$$y(x) = C_1 |x|^{k_1} + C_2 |x|^{k_2} + \cdots + C_n |x|^{k_n}.$$

**Remark.** To a pair of complex conjugate values $k = \alpha \pm i\beta$ there corresponds a pair of particular solutions: $y = |x|^\alpha \sin(\beta |x|)$ and $y = |x|^\alpha \cos(\beta |x|)$.

## 12.4.2-6. Solution of equations using the Laplace transform. Laplace equation.

1°. Some classes of equations (12.4.2.1) or (12.4.2.4) with polynomial coefficients

$$f_k(x) = \sum_{m=0}^{s_k} a_{km} x^m$$

may be solved using the Laplace transform (see Paragraph 12.4.1-3 and Section 11.2). To this end, one uses the following formula for the Laplace transform of the product of a power function and a derivative of the unknown function:

$$\mathcal{L}\left\{x^m y_x^{(n)}(x)\right\} = (-1)^m \frac{d^m}{dp^m} \left[p^n \widetilde{y}(p) - \sum_{k=1}^{n} p^{n-k} y_x^{(k-1)}(+0)\right]. \tag{12.4.2.6}$$

The right-hand side contains initial data $y_x^{(m)}(+0)$, $m = 0, 1, \ldots, n-1$ (specified in the Cauchy problem). As a result, one arrives at a linear ordinary differential equation, with respect to $p$, for the transform $\widetilde{y}(p)$; the order of this equation is equal to $\max_{1 \leq k \leq n}\{s_k\}$, the highest degree of the polynomials that determine the equation coefficients. In some cases, the equation for $\widetilde{y}(p)$ turns out to be simpler than the initial equation for $y(x)$ and can be solved in closed form. The desired function $y(x)$ is found by inverting the transform $\widetilde{y}(p)$ using the formulas from Paragraph 11.2.2-2 of the tables from Section T3.2.

2°. Consider the *Laplace equation*

$$(a_n + b_n x) y_x^{(n)} + (a_{n-1} + b_{n-1} x) y_x^{(n-1)} + \cdots + (a_1 + b_1 x) y_x' + (a_0 + b_0 x) y = 0, \tag{12.4.2.7}$$

whose coefficients are linear functions of the independent variable $x$. The application of the Laplace transform, in view of formulas (12.4.2.6), brings it to a linear first-order ordinary differential equation for the transform $\widetilde{y}(p)$.

**Example 1.** Consider a special case of equation (12.4.2.7):
$$xy''_{xx} + y'_x + axy = 0. \qquad (12.4.2.8)$$
Denote $y(0) = y_0$ and $y'_x(0) = y_1$. Let us apply the Laplace transform to this equation using formulas (12.4.2.6). On rearrangement, we obtain a linear first-order equation for $\widetilde{y}(p)$:
$$-(p^2\widetilde{y} - y_0 p - y_1)'_p + (p\widetilde{y} - y_0) - a\widetilde{y}'_p = 0 \quad \Longrightarrow \quad (p^2 + a)\widetilde{y}'_p + p\widetilde{y} = 0.$$
Its general solution is expressed as
$$\widetilde{y} = \frac{C}{\sqrt{p^2 + a}}, \qquad (12.4.2.9)$$
where $C$ is an arbitrary constant. Applying the inverse Laplace transform to (12.4.2.9) and taking into account formulas 19 and 20 from Subsection T3.2.3, we find a solution to the original equation (12.4.2.8):
$$y(x) = \begin{cases} CJ_0(x\sqrt{a}) & \text{if } a > 0, \\ CI_0(x\sqrt{-a}) & \text{if } a < 0, \end{cases} \qquad (12.4.2.10)$$
where $J_0(x)$ is the Bessel function of the first kind and $I_0(x)$ is the modified Bessel function of the first kind.

In this case, only one solution (12.4.2.10) has been obtained. This is due to the fact that the other solution goes to infinity as $x \to 0$, and hence formula (12.4.2.6) cannot be applied to it; this formula is only valid for finite initial values of the function and its derivatives.

### 12.4.2-7. Solution of equations using the Laplace integral.

Solutions to linear differential equations with polynomial coefficients can sometimes be represented as a *Laplace integral* in the form
$$y(x) = \int_{\mathcal{K}} e^{px} u(p) \, dp. \qquad (12.4.2.11)$$
For now, no assumptions are made about the domain of integration $\mathcal{K}$; it could be a segment of the real axis or a curve in the complex plane.

Let us exemplify the usage of the Laplace integral (12.4.2.11) by considering equation (12.4.2.7). It follows from (12.4.2.11) that
$$y_x^{(k)}(x) = \int_{\mathcal{K}} e^{px} p^k u(p) \, dp,$$
$$xy_x^{(k)}(x) = \int_{\mathcal{K}} x e^{px} p^k u(p) \, dp = \left[e^{px} p^k u(p)\right]_{\mathcal{K}} - \int_{\mathcal{K}} e^{px} \frac{d}{dp}\left[p^k u(p)\right] dp.$$
Substituting these expressions into (12.4.2.7) yields
$$\int_{\mathcal{K}} e^{px} \left\{ \sum_{k=0}^{n} a_k p^k u(p) - \sum_{k=0}^{n} b_k \frac{d}{dp}\left[p^k u(p)\right] \right\} dp + \sum_{k=0}^{n} b_k \left[e^{px} p^k u(p)\right]_{\mathcal{K}} = 0. \qquad (12.4.2.12)$$
This equation is satisfied if the expression in braces vanishes, thus resulting in a linear first-order ordinary differential equation for $u(p)$:
$$u(p) \sum_{k=0}^{n} a_k p^k - \frac{d}{dp}\left[u(p) \sum_{k=0}^{n} b_k p^k\right] = 0. \qquad (12.4.2.13)$$
The remaining term in (12.4.2.12) must also vanish:
$$\left[\sum_{k=0}^{n} b_k e^{px} p^k u(p)\right]_{\mathcal{K}} = 0. \qquad (12.4.2.14)$$
This condition can be met by appropriately selecting the path of integration $\mathcal{K}$. Consider the example below to illustrate the aforesaid.

**Example 2.** The linear variable-coefficient second-order equation
$$xy''_{xx} + (x+a+b)y'_x + ay = 0 \quad (a>0, \ b>0) \tag{12.4.2.15}$$
is a special case of equation (12.4.2.7) with $n=2$, $a_2=0$, $a_1=a+b$, $a_0=a$, $b_2=b_1=1$, and $b_0=0$. On substituting these values into (12.4.2.13), we arrive at an equation for $u(p)$:
$$p(p+1)u'_p - [(a+b-2)p+a-1]u = 0.$$
Its solution is given by
$$u(p) = p^{a-1}(p+1)^{b-1}. \tag{12.4.2.16}$$
It follows from condition (12.4.2.14), in view of formula (12.4.2.16), that
$$\left[e^{px}(p+p^2)u(p)\right]_\alpha^\beta = \left[e^{px}p^a(p+1)^b\right]_\alpha^\beta = 0, \tag{12.4.2.17}$$
where a segment of the real axis, $\mathcal{K}=[\alpha,\beta]$, has been chosen to be the path of integration. Condition (12.4.2.17) is satisfied if we set $\alpha=-1$ and $\beta=0$. Consequently, one of the solutions to equation (12.4.2.15) has the form
$$y(x) = \int_{-1}^{0} e^{px} p^{a-1}(p+1)^{b-1}\, dp. \tag{12.4.2.18}$$

**Remark 1.** If $a$ is noninteger, it is necessary to separate the real and imaginary parts in (12.4.2.18) to obtain real solutions.

**Remark 2.** By setting $\alpha=-\infty$ and $\beta=0$ in (12.4.2.17), one can find a second solution to equation (12.4.2.15) (at least for $x>0$).

## 12.4.3. Asymptotic Solutions of Linear Equations

This subsection presents asymptotic solutions, as $\varepsilon \to 0$ ($\varepsilon>0$), of some higher-order linear ordinary differential equations containing arbitrary functions (sufficiently smooth), with the independent variable being real.

### 12.4.3-1. Fourth-order linear equations.

1°. Consider the equation
$$\varepsilon^4 y''''_{xxxx} - f(x)y = 0$$
on a closed interval $a \le x \le b$. With the condition $f>0$, the leading terms of the asymptotic expansions of the fundamental system of solutions, as $\varepsilon \to 0$, are given by the formulas
$$y_1 = [f(x)]^{-3/8} \exp\left\{-\frac{1}{\varepsilon}\int [f(x)]^{1/4}\, dx\right\}, \quad y_2 = [f(x)]^{-3/8} \exp\left\{\frac{1}{\varepsilon}\int [f(x)]^{1/4}\, dx\right\},$$
$$y_3 = [f(x)]^{-3/8} \cos\left\{\frac{1}{\varepsilon}\int [f(x)]^{1/4}\, dx\right\}, \quad y_4 = [f(x)]^{-3/8} \sin\left\{\frac{1}{\varepsilon}\int [f(x)]^{1/4}\, dx\right\}.$$

2°. Now consider the "biquadratic" equation
$$\varepsilon^4 y''''_{xxxx} - 2\varepsilon^2 g(x) y''_{xx} - f(x)y = 0. \tag{12.4.3.1}$$
Introduce the notation
$$D(x) = [g(x)]^2 + f(x).$$
In the range where the conditions $f(x) \ne 0$ and $D(x) \ne 0$ are satisfied, the leading terms of the asymptotic expansions of the fundamental system of solutions of equation (12.4.3.1) are described by the formulas
$$y_k = [\lambda_k(x)]^{-1/2}[D(x)]^{-1/4} \exp\left\{\frac{1}{\varepsilon}\int \lambda_k(x)\, dx - \frac{1}{2}\int \frac{[\lambda_k(x)]'_x}{\sqrt{D(x)}}\, dx\right\}; \quad k=1,2,3,4,$$
where
$$\lambda_1(x) = \sqrt{g(x) + \sqrt{D(x)}}, \quad \lambda_2(x) = -\sqrt{g(x) + \sqrt{D(x)}},$$
$$\lambda_3(x) = \sqrt{g(x) - \sqrt{D(x)}}, \quad \lambda_4(x) = -\sqrt{g(x) - \sqrt{D(x)}}.$$

### 12.4.3-2. Higher-order linear equations.

1°. Consider an equation of the form
$$\varepsilon^n y_x^{(n)} - f(x)y = 0$$
on a closed interval $a \leq x \leq b$. Assume that $f \neq 0$. Then the leading terms of the asymptotic expansions of the fundamental system of solutions, as $\varepsilon \to 0$, are given by
$$y_m = [f(x)]^{-\frac{1}{2} + \frac{1}{2n}} \exp\left\{\frac{\omega_m}{\varepsilon} \int [f(x)]^{\frac{1}{n}} dx\right\}[1 + O(\varepsilon)],$$
where $\omega_1, \omega_2, \ldots, \omega_n$ are roots of the equation $\omega^n = 1$:
$$\omega_m = \cos\left(\frac{2\pi m}{n}\right) + i\sin\left(\frac{2\pi m}{n}\right), \qquad m = 1, 2, \ldots, n.$$

2°. Now consider an equation of the form
$$\varepsilon^n y_x^{(n)} + \varepsilon^{n-1} f_{n-1}(x) y_x^{(n-1)} + \cdots + \varepsilon f_1(x) y_x' + f_0(x) y = 0 \qquad (12.4.3.2)$$
on a closed interval $a \leq x \leq b$. Let $\lambda_m = \lambda_m(x)$ ($m = 1, 2, \ldots, n$) be the roots of the characteristic equation
$$P(x, \lambda) \equiv \lambda^n + f_{n-1}(x)\lambda^{n-1} + \cdots + f_1(x)\lambda + f_0(x) = 0.$$
Let all the roots of the characteristic equation be different on the interval $a \leq x \leq b$, i.e., the conditions $\lambda_m(x) \neq \lambda_k(x)$, $m \neq k$, are satisfied, which is equivalent to the fulfillment of the conditions $P_\lambda(x, \lambda_m) \neq 0$. Then the leading terms of the asymptotic expansions of the fundamental system of solutions of equation (12.4.3.2), as $\varepsilon \to 0$, are given by
$$y_m = \exp\left\{\frac{1}{\varepsilon}\int \lambda_m(x)\,dx - \frac{1}{2}\int [\lambda_m(x)]_x' \frac{P_{\lambda\lambda}(x, \lambda_m(x))}{P_\lambda(x, \lambda_m(x))}\,dx\right\},$$
where
$$P_\lambda(x, \lambda) \equiv \frac{\partial P}{\partial \lambda} = n\lambda^{n-1} + (n-1)f_{n-1}(x)\lambda^{n-2} + \cdots + 2\lambda f_2(x) + f_1(x),$$
$$P_{\lambda\lambda}(x, \lambda) \equiv \frac{\partial^2 P}{\partial \lambda^2} = n(n-1)\lambda^{n-2} + (n-1)(n-2)f_{n-1}(x)\lambda^{n-3} + \cdots + 6\lambda f_3(x) + 2f_2(x).$$

## 12.4.4. Collocation Method and Its Convergence

### 12.4.4-1. Statement of the problem. Approximate solution.

1°. Consider the linear boundary value problem defined by the equation
$$Ly \equiv y_x^{(n)} + f_{n-1}(x) y_x^{(n-1)} + \cdots + f_1(x) y_x' + f_0(x) y = g(x), \qquad -1 < x < 1, \qquad (12.4.4.1)$$
and the boundary conditions
$$\sum_{j=0}^{n-1} [\alpha_{ij} y_x^{(j)}(-1) + \beta_{ij} y_x^{(j)}(1)] = 0, \qquad i = 1, \ldots, n. \qquad (12.4.4.2)$$

2°. We seek an approximate solution to problem (12.4.4.1)–(12.4.4.2) in the form
$$y_m(x) = A_1 \varphi_1(x) + A_2 \varphi_2(x) + \cdots + A_m \varphi_m(x),$$
where $\varphi_k(x)$ is a polynomial of degree $n + k - 1$ that satisfies the boundary conditions (12.4.4.2). The coefficients $A_k$ are determined by the linear system of algebraic equations
$$[Ly_m - g(x)]_{x=x_i} = 0, \qquad i = 1, \ldots, m, \qquad (12.4.4.3)$$
with *Chebyshev nodes* $x_i = \cos\left(\frac{2i-1}{2m}\pi\right)$, $i = 1, \ldots, m$.

#### 12.4.4-2. Convergence theorem for the collocation method.

THEOREM. *Let the functions $f_j(x)$ ($j = 0, \ldots, n-1$) and $g(x)$ be continuous on the interval $[-1, 1]$ and let the boundary value problem (12.4.4.1)–(12.4.4.2) have a unique solution, $y(x)$. Then there exists an $m_0$ such that system (12.4.4.3) is uniquely solvable for $m \geq m_0$; and the limit relations*

$$\max_{-1 \leq x \leq 1} \left| y_m^{(k)}(x) - y^{(k)}(x) \right| \to 0, \qquad k = 0, 1, \ldots, n-1;$$

$$\left\{ \int_{-1}^{1} \frac{\left| y_m^{(n)}(x) - y^{(n)}(x) \right|^2}{\sqrt{1-x^2}} \, dx \right\}^{1/2} \to 0$$

*hold for $m \to \infty$.*

**Remark.** A similar result holds true if the nodes are roots of some orthogonal polynomials with some weight function. If the nodes are equidistant, the method diverges.

## 12.5. Nonlinear Equations of Arbitrary Order

### 12.5.1. Structure of the General Solution. Cauchy Problem

#### 12.5.1-1. Equations solved for the highest derivative. General solution.

An *nth-order differential equation solved for the highest derivative* has the form

$$y_x^{(n)} = f(x, y, y_x', \ldots, y_x^{(n-1)}). \qquad (12.5.1.1)$$

The general solution of this equation depends on $n$ arbitrary constants $C_1, \ldots, C_n$. In some cases, the general solution can be written in explicit form as

$$y = \varphi(x, C_1, \ldots, C_n). \qquad (12.5.1.2)$$

#### 12.5.1-2. Cauchy problem. The existence and uniqueness theorem.

The *Cauchy problem*: find a solution of equation (12.5.1.1) with the *initial conditions*

$$y(x_0) = y_0, \quad y_x'(x_0) = y_0^{(1)}, \quad \ldots, \quad y_x^{(n-1)}(x_0) = y_0^{(n-1)}. \qquad (12.5.1.3)$$

(At a point $x_0$, the values of the unknown function $y(x)$ and all its derivatives of orders $\leq n-1$ are prescribed.)

EXISTENCE AND UNIQUENESS THEOREM. *Suppose the function $f(x, y, z_1, \ldots, z_{n-1})$ is continuous in all its arguments in a neighborhood of the point $(x_0, y_0, y_0^{(1)}, \ldots, y_0^{(n-1)})$ and has bounded derivatives with respect to $y, z_1, \ldots, z_{n-1}$ in this neighborhood. Then a solution of equation (12.5.1.1) satisfying the initial conditions (12.5.1.3) exists and is unique.*

## 12.5.1-3. Construction of a differential equation by a given general solution.

Suppose a general solution (12.5.1.2) of an unknown $n$th-order ordinary differential equation is given. The equation corresponding to the general solution can be obtained by eliminating the arbitrary constants $C_1, \ldots, C_n$ from the identities

$$y = \varphi(x, C_1, \ldots, C_n),$$
$$y'_x = \varphi'_x(x, C_1, \ldots, C_n),$$
$$\ldots\ldots\ldots\ldots\ldots\ldots\ldots\ldots\ldots$$
$$y^{(n)}_x = \varphi^{(n)}_x(x, C_1, \ldots, C_n),$$

obtained by differentiation from formula (12.5.1.2).

## 12.5.1-4. Reduction of an $n$th-order equation to a system of $n$ first-order equation.

The differential equation (12.5.1.1) is equivalent to the following system of $n$ first-order equations:

$$y'_0 = y_1, \quad y'_1 = y_2, \quad \ldots, \quad y'_{n-2} = y_{n-1}, \quad y'_{n-1} = f(x, y_0, y_1, \ldots, y_{n-1}),$$

where the notation $y_0 \equiv y$ is adopted.

## 12.5.2. Equations Admitting Reduction of Order

### 12.5.2-1. Equations not containing $y, y'_x, \ldots, y^{(k)}_x$ explicitly.

An equation that does not explicitly contain the unknown function and its derivatives up to order $k$ inclusive can generally be written as

$$F\big(x, y^{(k+1)}_x, \ldots, y^{(n)}_x\big) = 0 \qquad (1 \leq k+1 < n). \tag{12.5.2.1}$$

Such equations are invariant under arbitrary translations of the unknown function, $y \to y + \text{const}$ (the form of such equations is also preserved under the transformation $u(x) = y + a_k x^k + \cdots + a_1 x + a_0$, where the $a_m$ are arbitrary constants). The substitution $z(x) = y^{(k+1)}_x$ reduces (12.5.2.1) to an equation whose order is by $k+1$ smaller than that of the original equation, $F\big(x, z, z'_x, \ldots, z^{(n-k-1)}_x\big) = 0$.

### 12.5.2-2. Equations not containing $x$ explicitly (autonomous equations).

An equation that does not explicitly contain $x$ has in the general form

$$F\big(y, y'_x, \ldots, y^{(n)}_x\big) = 0. \tag{12.5.2.2}$$

Such equations are invariant under arbitrary translations of the independent variable, $x \to x + \text{const}$. The substitution $y'_x = w(y)$ (where $y$ plays the role of the independent variable) reduces by one the order of an autonomous equation. Higher derivatives can be expressed in terms of $w$ and its derivatives with respect to the new independent variable, $y''_{xx} = ww'_y$, $y'''_{xxx} = w^2 w''_{yy} + w(w'_y)^2, \ldots$

### 12.5.2-3. Equations of the form $F(ax+by, y'_x, \ldots, y^{(n)}_x) = 0$.

Such equations are invariant under simultaneous translations of the independent variable and the unknown function, $x \to x + bc$ and $y \to y - ac$, where $c$ is an arbitrary constant.

For $b = 0$, see equation (12.5.2.1). For $b \neq 0$, the substitution $w(x) = y + (a/b)x$ leads to an autonomous equation of the form (12.5.2.2).

### 12.5.2-4. Equations of the form $F(x, xy'_x - y, y''_{xx}, \ldots, y^{(n)}_x) = 0$.

The substitution $w(x) = xy'_x - y$ reduces the order of this equation by one.

This equation is a special case of the equation

$$F(x, xy'_x - my, y^{(m+1)}_x, \ldots, y^{(n)}_x) = 0, \quad \text{where} \quad m = 1, 2, \ldots, n-1. \quad (12.5.2.3)$$

The substitution $w(x) = xy'_x - my$ reduces by one the order of equation (12.5.2.3).

### 12.5.2-5. Homogeneous equations.

$1°$. *Equations homogeneous in the independent variable* are invariant under scaling of the independent variable, $x \to \alpha x$, where $\alpha$ is an arbitrary constant ($\alpha \neq 0$). In general, such equations can be written in the form

$$F(y, xy'_x, x^2 y''_{xx}, \ldots, x^n y^{(n)}_x) = 0.$$

The substitution $z(y) = xy'_x$ reduces by one the order of this equation.

$2°$. *Equations homogeneous in the unknown function* are invariant under scaling of the unknown function, $y \to \alpha y$, where $\alpha$ is an arbitrary constant ($\alpha \neq 0$). Such equations can be written in the general form

$$F(x, y'_x/y, y''_{xx}/y, \ldots, y^{(n)}_x/y) = 0.$$

The substitution $z(x) = y'_x/y$ reduces by one the order of this equation.

$3°$. *Equations homogeneous in both variables* are invariant under simultaneous scaling (dilatation) of the independent and dependent variables, $x \to \alpha x$ and $y \to \alpha y$, where $\alpha$ is an arbitrary constant ($\alpha \neq 0$). Such equations can be written in the general form

$$F(y/x, y'_x, xy''_{xx}, \ldots, x^{n-1} y^{(n)}_x) = 0.$$

The transformation $t = \ln|x|$, $w = y/x$ leads to an autonomous equation considered in Paragraph 12.5.2-2.

### 12.5.2-6. Generalized homogeneous equations.

$1°$. *Generalized homogeneous equations* (*equations homogeneous in the generalized sense*) are invariant under simultaneous scaling of the independent variable and the unknown function, $x \to \alpha x$ and $y \to \alpha^k y$, where $\alpha \neq 0$ is an arbitrary constant and $k$ is a given number. Such equations can be written in the general form

$$F(x^{-k} y, x^{1-k} y'_x, \ldots, x^{n-k} y^{(n)}_x) = 0.$$

The transformation $t = \ln x$, $w = x^{-k} y$ leads to an autonomous equation considered in Paragraph 12.5.2-2.

2°. The most general form of generalized homogeneous equations is

$$\mathcal{F}(x^n y^m, xy'_x/y, \ldots, x^n y^{(n)}_x/y) = 0.$$

The transformation $z = x^n y^m$, $u = xy'_x/y$ reduces the order of this equation by one.

### 12.5.2-7. Equations of the form $F(e^{\lambda x} y^n, y'_x/y, y''_{xx}/y, \ldots, y^{(n)}_x/y) = 0$.

Such equations are invariant under simultaneous translation and scaling of variables, $x \to x + \alpha$ and $y \to \beta y$, where $\beta = \exp(-\alpha\lambda/n)$ and $\alpha$ is an arbitrary constant. The transformation $z = e^{\lambda x} y^n$, $w = y'_x/y$ leads to an equation of order $n - 1$.

### 12.5.2-8. Equations of the form $F(x^n e^{\lambda y}, xy'_x, x^2 y''_{xx}, \ldots, x^n y^{(n)}_x) = 0$.

Such equations are invariant under simultaneous scaling and translation of variables, $x \to \alpha x$ and $y \to y + \beta$, where $\alpha = \exp(-\beta\lambda/n)$ and $\beta$ is an arbitrary constant. The transformation $z = x^n e^{\lambda y}$, $w = xy'_x$ leads to an equation of order $n - 1$.

### 12.5.2-9. Other equations.

Consider the nonlinear differential equation

$$F(x, L_1[y], \ldots, L_k[y]) = 0, \tag{12.5.2.4}$$

where the $L_s[y]$ are linear homogeneous differential forms,

$$L_s[y] = \sum_{m=0}^{n_s} \varphi^{(s)}_m(x) y^{(m)}_x, \qquad s = 1, \ldots, k.$$

Let $y_0 = y_0(x)$ be a common particular solution of the linear equations

$$L_s[y_0] = 0 \qquad (s = 1, \ldots, k).$$

Then the substitution

$$w = \varphi(x)\big[y_0(x) y'_x - y'_0(x) y\big] \tag{12.5.2.5}$$

with an arbitrary function $\varphi(x)$ reduces by one the order of equation (12.5.2.4).

**Example.** Consider the third-order equation

$$xy'''_{xxx} = f(xy'_x - 2y).$$

It can be represented in the form (12.5.2.4) with

$$k = 2, \quad F(x, u, w) = xu - f(w), \quad L_1[y] = y'''_{xxx}, \quad L_2[y] = xy'_x - 2y.$$

The linear equations $L_k[y] = 0$ are

$$y'''_{xxx} = 0, \quad xy'_x - 2y = 0.$$

These equations have a common particular solution $y_0 = x^2$. Therefore, the substitution $w = xy'_x - 2y$ (see formula (12.5.2.5) with $\varphi(x) = 1/x$) leads to a second-order autonomous equation: $w''_{xx} = f(w)$. For the solution of this equation, see Example 1 in Subsection 12.3.2.

## 12.6. Linear Systems of Ordinary Differential Equations

### 12.6.1. Systems of Linear Constant-Coefficient Equations

**12.6.1-1. Systems of first-order linear homogeneous equations. The general solution.**

1°. In general, a homogeneous linear system of constant-coefficient first-order ordinary differential equations has the form

$$\begin{aligned} y_1' &= a_{11}y_1 + a_{12}y_2 + \cdots + a_{1n}y_n, \\ y_2' &= a_{21}y_1 + a_{22}y_2 + \cdots + a_{2n}y_n, \\ &\cdots\cdots\cdots\cdots\cdots\cdots\cdots\cdots\cdots\cdots\cdots\cdots\cdots \\ y_n' &= a_{n1}y_1 + a_{n2}y_2 + \cdots + a_{nn}y_n, \end{aligned} \tag{12.6.1.1}$$

where a prime stands for the derivative with respect to $x$. In the sequel, all the coefficients $a_{ij}$ of the system are assumed to be real numbers.

The homogeneous system (12.6.1.1) has the trivial particular solution $y_1 = y_2 = \cdots = y_n = 0$.

*Superposition principle for a homogeneous system*: any linear combination of particular solutions of system (12.6.1.1) is also a solution of this system.

The general solution of the system of differential equations (12.6.1.1) is the sum of its $n$ linearly independent (nontrivial) particular solutions multiplied by an arbitrary constant.

System (12.6.1.1) can be reduced to a single homogeneous linear constant-coefficient $n$th-order equation; see Paragraph 12.7.1-3.

2°. For brevity (and clearness), system (12.6.1.1) is conventionally written in vector-matrix form:

$$\mathbf{y}' = \mathbf{a}\mathbf{y}, \tag{12.6.1.2}$$

where $\mathbf{y} = (y_1, y_2, \ldots, y_n)^\mathrm{T}$ is the column vector of the unknowns and $\mathbf{a} = (a_{ij})$ is the matrix of the equation coefficients. The superscript T denotes the transpose of a matrix or a vector. So, for example, a row vector is converted into a column vector:

$$(y_1, y_2)^\mathrm{T} \equiv \begin{pmatrix} y_1 \\ y_2 \end{pmatrix}.$$

The right-hand side of equation (12.6.1.2) is the product of the $n \times n$ square matrix $\mathbf{a}$ by the $n \times 1$ matrix (column vector) $\mathbf{y}$.

Let $\mathbf{y}_k = (y_{k1}, y_{k2}, \ldots, y_{kn})^\mathrm{T}$ be linearly independent particular solutions* of the homogeneous system (12.6.1.1), where $k = 1, 2, \ldots, n$; the first subscript in $y_{km} = y_{km}(x)$ denotes the number of the solution and the second subscript indicates the component of the vector solution. Then the general solution of the homogeneous system (12.6.1.2) is expressed as

$$\mathbf{y} = C_1\mathbf{y}_1 + C_2\mathbf{y}_2 + \cdots + C_n\mathbf{y}_n. \tag{12.6.1.3}$$

A method for the construction of particular solutions that can be used to obtain the general solution by formula (12.6.1.3) is presented below.

---

* This means that the condition $\det |y_{mk}(x)| \neq 0$ holds.

**12.6.1-2. Systems of first-order linear homogeneous equations. Particular solutions.**

Particular solutions to system (12.6.1.1) are determined by the roots of the characteristic equation

$$\Delta(\lambda) = 0, \quad \text{where} \quad \Delta(\lambda) \equiv \begin{vmatrix} a_{11}-\lambda & a_{12} & \dots & a_{1n} \\ a_{21} & a_{22}-\lambda & \dots & a_{2n} \\ \dots & \dots & \dots & \dots \\ a_{n1} & a_{n2} & \dots & a_{nn}-\lambda \end{vmatrix}. \tag{12.6.1.4}$$

The following cases are possible:

$1°$. Let $\lambda = \lambda_k$ be a simple real root of the characteristic equation (12.6.1.4). The corresponding particular solution of the homogeneous linear system of equations (12.6.1.1) has the exponential form

$$y_1 = A_1 e^{\lambda x}, \quad y_2 = A_2 e^{\lambda x}, \quad \dots, \quad y_n = A_n e^{\lambda x}, \tag{12.6.1.5}$$

where the coefficients $A_1, A_2, \dots, A_n$ are determined by solving the associated homogeneous system of algebraic equations obtained by substituting expressions (12.6.1.5) into the differential equation (12.6.1.1) and dividing by $e^{\lambda x}$:

$$\begin{aligned} (a_{11}-\lambda)A_1 + a_{12}A_2 + \cdots + a_{1n}A_n &= 0, \\ a_{21}A_1 + (a_{22}-\lambda)A_2 + \cdots + a_{2n}A_n &= 0, \\ &\dotsb \\ a_{n1}A_1 + a_{n2}A_2 + \cdots + (a_{nn}-\lambda)A_n &= 0. \end{aligned} \tag{12.6.1.6}$$

The solution of this system is unique to within a constant factor.

If all roots of the characteristic equation $\lambda_1, \lambda_2, \dots, \lambda_n$ are real and distinct, then the general solution of system (12.6.1.1) has the form

$$\begin{aligned} y_1 &= C_1 A_{11} e^{\lambda_1 x} + C_2 A_{12} e^{\lambda_2 x} + \cdots + C_n A_{1n} e^{\lambda_n x}, \\ y_2 &= C_1 A_{21} e^{\lambda_1 x} + C_2 A_{22} e^{\lambda_2 x} + \cdots + C_n A_{2n} e^{\lambda_n x}, \\ &\dotsb \\ y_n &= C_1 A_{n1} e^{\lambda_1 x} + C_2 A_{n2} e^{\lambda_2 x} + \cdots + C_n A_{nn} e^{\lambda_n x}, \end{aligned} \tag{12.6.1.7}$$

where $C_1, C_2, \dots, C_n$ are arbitrary constants. The second subscript in $A_{mk}$ indicates a coefficient corresponding to the root $\lambda_k$.

$2°$. For each simple complex root, $\lambda = \alpha \pm i\beta$, of the characteristic equation (12.6.1.4), the corresponding particular solution is obtained in the same way as in the simple real root case; the associated coefficients $A_1, A_2, \dots, A_n$ in (12.6.1.5) will be complex. Separating the real and imaginary parts in (12.6.1.5) results in two real particular solutions to system (12.6.1.1); the same two solutions are obtained if one takes the complex conjugate root, $\bar{\lambda} = \alpha - i\beta$.

$3°$. Let $\lambda$ be a real root of the characteristic equation (12.6.1.4) of multiplicity $m$. The corresponding particular solution of system (12.6.1.1) is sought in the form

$$y_1 = P_1(x)e^{\lambda x}, \quad y_2 = P_2(x)e^{\lambda x}, \quad \dots, \quad y_n = P_n(x)e^{\lambda x}, \tag{12.6.1.8}$$

where the $P_k(x) = \sum_{i=0}^{m-1} B_{ki} x^i$ are polynomials of degree $m-1$. The coefficients of these polynomials result from the substitution of expressions (12.6.1.8) into equations (12.6.1.1); after dividing by $e^{\lambda x}$ and collecting like terms, one obtains $n$ equations, each representing a polynomial equated to zero. By equating the coefficients of all resulting polynomials to zero, one arrives at a linear algebraic system of equations for the coefficients $B_{ki}$; the solution to this system will contain $m$ free parameters.

$4°$. For a multiple complex, $\lambda = \alpha + i\beta$, of multiplicity $m$, the corresponding particular solution is sought, just as in the case of a multiple real root, in the form (12.6.1.8); here the coefficients $B_{ki}$ of the polynomials $P_k(x)$ will be complex. Finally, in order to obtain real solutions of the original system (12.6.1.1), one separates the real and imaginary parts in formulas (12.6.1.8), thus obtaining two particular solutions with $m$ free parameters each. The two solutions correspond to the complex conjugate roots $\lambda = \alpha \pm i\beta$.

$5°$. In the general case, where the characteristic equation (12.6.1.4) has simple and multiple, real and complex roots (see Items $1°$–$4°$), the general solution to system (12.6.1.1) is obtained as the sum of all particular solutions multiplied by arbitrary constants.

**Example 1.** Consider the homogeneous system of two linear differential equations

$$y_1' = y_1 + 4y_2,$$
$$y_2' = y_1 + y_2.$$

The associated characteristic equations,

$$\begin{vmatrix} 1-\lambda & 4 \\ 1 & 1-\lambda \end{vmatrix} = \lambda^2 - 2\lambda - 3 = 0,$$

has distinct real roots:

$$\lambda_1 = 3, \quad \lambda_2 = -1.$$

The system of algebraic equations (12.6.1.6) for solution coefficients becomes

$$(1-\lambda)A_1 + 4A_2 = 0,$$
$$A_1 + (1-\lambda)A_2 = 0. \tag{12.6.1.9}$$

Substituting the first root, $\lambda = 3$, into system (12.6.1.9) yields $A_1 = 2A_2$. We can set $A_1 = 2$ and $A_2 = 1$, since the solution is determined to within a constant factor. Thus the first particular solution of the homogeneous system of linear ordinary differential equations (12.6.1.9) has the form

$$y_1 = 2e^{3x}, \quad y_2 = e^{3x}. \tag{12.6.1.10}$$

The second particular solution, corresponding to $\lambda = -1$, is found in the same way:

$$y_1 = -2e^{-x}, \quad y_2 = e^{-x}. \tag{12.6.1.11}$$

The sum of the two particular solutions (12.6.1.10), (12.6.1.11) multiplied by arbitrary constants, $C_1$ and $C_2$, gives the general solution to the original homogeneous system of linear ordinary differential equations:

$$y_1 = 2C_1 e^{3x} - 2C_2 e^{-x}, \quad y_2 = C_1 e^{3x} + C_2 e^{-x}.$$

**Example 2.** Consider the system of ordinary differential equations

$$y_1' = -y_2,$$
$$y_2' = 2y_1 + 2y_2. \tag{12.6.1.12}$$

The characteristic equation,

$$\begin{vmatrix} -\lambda & -1 \\ 2 & 2-\lambda \end{vmatrix} = \lambda^2 - 2\lambda + 2 = 0$$

has complex conjugate roots:

$$\lambda_1 = 1+i, \quad \lambda_2 = 1-i.$$

The algebraic system (12.6.1.6) for the complex coefficients $A_1$ and $A_2$ becomes

$$-\lambda A_1 - A_2 = 0,$$
$$2A_1 + (2-\lambda)A_2 = 0.$$

With $\lambda = 1+i$, one nonzero solution is given by $A_1 = 1$ and $A_2 = -1-i$. The corresponding complex solution to system (12.6.1.12) has the form

$$y_1 = e^{(1+i)x}, \quad y_2 = (-1-i)e^{(1+i)x}.$$

Separating the real and imaginary parts, taking into account the formulas

$$e^{(1+i)x} = e^x(\cos x + i \sin x) = e^x \cos x + ie^x \sin x,$$
$$(-1-i)e^{(1+i)x} = -(1+i)e^x(\cos x + i \sin x) = e^x(\sin x - \cos x) - ie^x(\sin x + \cos x),$$

and making linear combinations from them, one arrives at the general solution to the original system (12.6.1.12):

$$y_1 = C_1 e^x \cos x + C_2 e^x \sin x,$$
$$y_2 = C_1 e^x (\sin x - \cos x) - C_2 e^x (\sin x + \cos x).$$

**Remark.** Systems of two homogeneous linear constant-coefficient second-order differential equations are treated in detail in Paragraph 12.6.1-7.

### 12.6.1-3. Nonhomogeneous systems of linear first-order equations.

$1°$. In general, a nonhomogeneous linear system of constant-coefficient first-order ordinary differential equations has the form

$$\begin{aligned} y_1' &= a_{11}y_1 + a_{12}y_2 + \cdots + a_{1n}y_n + f_1(x), \\ y_2' &= a_{21}y_1 + a_{22}y_2 + \cdots + a_{2n}y_n + f_2(x), \\ &\quad \cdots\cdots\cdots\cdots\cdots\cdots\cdots\cdots\cdots\cdots\cdots\cdots\cdots\cdots\cdots \\ y_n' &= a_{n1}y_1 + a_{n2}y_2 + \cdots + a_{nn}y_n + f_n(x). \end{aligned} \quad (12.6.1.13)$$

For brevity, the conventional vector notation will also be used:

$$\mathbf{y}' = \mathbf{a}\mathbf{y} + \mathbf{f}(x),$$

where $\mathbf{f}(x) = (f_1(x), f_2(x), \ldots, f_n(x))^{\mathrm{T}}$.

The general solution of this system is the sum of the general solution to the corresponding homogeneous system with $f_k(x) \equiv 0$ [see system (12.6.1.1)] and any particular solution of the nonhomogeneous system (12.6.1.13).

System (12.6.1.13) can also be reduced to a single nonhomogeneous linear constant-coefficient $n$th-order equation; see Paragraph 12.7.1-3.

$2°$. Let $\mathbf{y}_m = (D_{m1}(x), D_{m1}(x), \ldots, D_{mn}(x))^{\mathrm{T}}$ be particular solutions to the homogeneous linear system of constant-coefficient first-order ordinary differential equations (12.6.1.1) that satisfy the special initial conditions

$$y_m(0) = 1, \quad y_k(0) = 0 \quad \text{if} \quad k \neq m; \quad m, k = 1, \ldots, n.$$

Then the general solution to the nonhomogeneous system (12.6.1.13) is expressed as

$$y_k(x) = \sum_{m=1}^{n} \int_0^x f_m(t) D_{mk}(x-t)\, dt + \sum_{m=1}^{n} C_m D_{mk}(x), \quad k = 1, \ldots, n. \quad (12.6.1.14)$$

Alternatively, the general solution to the nonhomogeneous linear system of equations (12.6.1.13) can be obtained using the formulas from Paragraph 12.6.2-2.

The solution of the Cauchy problem for the nonhomogeneous system (12.6.1.13) with arbitrary initial conditions,

$$y_1(0) = y_1^\circ, \quad y_2(0) = y_2^\circ, \quad \ldots, \quad y_n(0) = y_n^\circ, \quad (12.6.1.15)$$

is determined by formulas (12.6.1.14) with $C_m = y_m^\circ$, $m = 1, \ldots, n$.

### 12.6.1-4. Homogeneous linear systems of higher-order differential equations.

An arbitrary system of homogeneous linear systems of constant-coefficient ordinary differential equations consists of $n$ equations, each representing a linear combination of unknowns, $y_k$, and their derivatives, $y'_k, y''_k, \ldots, y_k^{(m_k)}$, $k = 1, 2, \ldots, n$.

The general solution of such systems is a linear combination of particular solutions multiplied by arbitrary constants. In total, such a system has $m_1 + m_2 + \cdots + m_n$ linearly independent particular solutions (the system is assumed to be consistent and nondegenerate, so that the constituent equations are linearly independent).

Particular solutions of the system are sought in the form (12.6.1.5). On substituting these expressions into the differential equations and dividing by $e^{\lambda x}$, one obtains a homogeneous linear algebraic system for coefficients $A_1, A_2, \ldots, A_n$. For this system to have nontrivial solutions, the determinant of the system must vanish. This results in an algebraic equation for the exponent $\lambda$; in physics, this equation is called a *dispersion equation*. To different roots of the dispersion equation there correspond different particular solutions of the original system of equations. For simple real and complex-conjugate roots, the procedure of finding particular solutions is the same as in the case of a linear system of first-order equations (12.6.1.1).

**Example 3.** Consider the linear system of constant-coefficient second-order equations

$$y''_1 + y'_2 + ay_2 = 0,$$
$$y''_2 + y'_1 + ay_1 = 0. \tag{12.6.1.16}$$

Particular solutions are sought in the form

$$y_1 = A_1 e^{\lambda x}, \quad y_2 = A_2 e^{\lambda x}. \tag{12.6.1.17}$$

Substituting (12.6.1.17) into (12.6.1.16) yields a homogeneous linear algebraic system for the coefficients $A_1$ and $A_2$:

$$\lambda^2 A_1 + (\lambda + a) A_2 = 0,$$
$$(\lambda + a) A_1 + \lambda^2 A_2 = 0. \tag{12.6.1.18}$$

For this system to have nontrivial solutions, its determinant must vanish. This results in the dispersion equation

$$\begin{vmatrix} \lambda^2 & \lambda + a \\ \lambda + a & \lambda^2 \end{vmatrix} = \lambda^4 - (\lambda + a)^2 = 0.$$

Its roots are

$$\lambda_{1,2} = \tfrac{1}{2} \pm \sqrt{\tfrac{1}{4} + a}, \quad \lambda_{3,4} = -\tfrac{1}{2} \pm \sqrt{\tfrac{1}{4} - a}. \tag{12.6.1.19}$$

Let us confine ourselves to the simplest case of $-\tfrac{1}{4} < a < \tfrac{1}{4}$, where all roots of the dispersion equation are real and distinct. It follows from the system of algebraic equations (12.6.1.18) that $A_1 = \lambda + a$ and $A_2 = -\lambda^2$, where $\lambda = \lambda_n$. Substituting these values into (12.6.1.17) yields the particular solutions $y_{1n} = (\lambda_n + a) e^{\lambda_n x}$, $y_{2n} = -\lambda_n^2 e^{\lambda_n x}$ ($n = 1, 2, 3, 4$). A linear combination of the particular solutions gives the general solution of system (12.6.1.16):

$$y_1 = C_1(\lambda_1 + a)e^{\lambda_1 x} + C_2(\lambda_2 + a)e^{\lambda_2 x} + C_3(\lambda_3 + a)e^{\lambda_3 x} + C_4(\lambda_4 + a)e^{\lambda_4 x},$$
$$y_2 = -C_1 \lambda_1^2 e^{\lambda_1 x} - C_2 \lambda_2^2 e^{\lambda_2 x} - C_3 \lambda_3^2 e^{\lambda_3 x} - C_4 \lambda_4^2 e^{\lambda_4 x},$$

where $C_1, C_2, C_3$, and $C_4$ are arbitrary constants, and the roots $\lambda_n$ are determined by formulas (12.6.1.19).

**Remark.** Paragraph 12.6.1-6 (see Item 2°) presents a method for the solution of systems of arbitrary homogeneous linear constant-coefficients ordinary differential equations using the Laplace transform.

### 12.6.1-5. Nonhomogeneous higher-order linear systems. D'Alembert's method.

Consider the system of two linear constant-coefficient $m$th-order differential equations

$$y_1^{(m)} = a_{11}y_1 + a_{12}y_2 + f_1(x),$$
$$y_2^{(m)} = a_{21}y_1 + a_{22}y_2 + f_2(x). \qquad (12.6.1.20)$$

Let us multiply the second equation of system (12.6.1.20) by $k$ and add it termwise to the first equation to obtain, after rearrangement,

$$(y_1 + ky_2)^{(m)} = (a_{11} + ka_{21})\left(y_1 + \frac{a_{12} + ka_{22}}{a_{11} + ka_{21}}y_2\right) + f_1(x) + kf_2(x). \qquad (12.6.1.21)$$

Let us take the constant $k$ so that $\dfrac{a_{12} + ka_{22}}{a_{11} + ka_{21}} = k$, which results in a quadratic equation for $k$:

$$a_{21}k^2 + (a_{11} - a_{22})k - a_{12} = 0. \qquad (12.6.1.22)$$

In this case, (12.6.1.21) is a nonhomogeneous linear constant-coefficient equation for $z = y_1 + ky_2$: $z^{(m)} = (a_{11} + ka_{21})z + f_1(x) + kf_2(x)$. Integrating this equation yields

$$y_1 + ky_2 = C_1\varphi_1(x, k) + \cdots + C_m\varphi_m(x, k) + \psi(x, k).$$

It follows that if the roots of the quadratic equation (12.6.1.22) are distinct, we have two relations,

$$y_1 + k_1y_2 = C_1\varphi_1(x, k_1) + \cdots + C_m\varphi_m(x, k_1) + \psi(x, k_1),$$
$$y_1 + k_2y_2 = C_{m+1}\varphi_1(x, k_2) + \cdots + C_{2m}\varphi_m(x, k_2) + \psi(x, k_2),$$

which represent a linear algebraic system of equations for the functions $y_1$ and $y_2$.

*Remark 1.* The above method for the solution of system (12.6.1.20) is known as *D Alembert's method*. The quantity $z = y_1 + ky_2$ in the above reasoning gives an example of an integrable combination (see Paragraph 12.7.2-1).

*Remark 2.* The more complicated system where $y_1$ and $y_2$ on the right-hand side are replaced by the derivatives of the same order, $y_1^{(n)}$ and $y_2^{(n)}$, can be treated likewise.

*Remark 3.* System (12.6.1.20) can be solved using the Laplace transform (see Paragraph 12.6.1-6).

### 12.6.1-6. Usage of the Laplace transform for solving linear systems of equations.

$1°$. To solve the Cauchy problem for the nonhomogeneous linear system of differential equations (12.6.1.13) with the initial conditions (12.6.1.15), one can use the Laplace transform, based on the following formulas (for details, see Section 11.2):

$$\widetilde{y}_k(p) = \mathcal{L}\{y_k(x)\}, \quad \widetilde{f}_k(p) = \mathcal{L}\{f_k(x)\}, \quad \text{where} \quad \mathcal{L}\{f(x)\} \equiv \int_0^\infty e^{-px} f(x)\, dx.$$

To this end, one should multiply each equation in (12.6.1.13) by $e^{-px}$ and then integrate with respect to $x$ from zero to infinity. In view of the differentiation rule $\mathcal{L}\{y_k'(x)\} =$

$p\widetilde{y}_k(p) - y_k(0)$ and the initial conditions (12.6.1.15), one arrives at a nonhomogeneous linear system of algebraic equations for the transforms $\widetilde{y}_k(p)$:

$$(a_{11} - p)\widetilde{y}_1 + a_{12}\widetilde{y}_2 + \cdots + a_{1n}\widetilde{y}_n = -\widetilde{f}_1(p) - y_1^\circ,$$
$$a_{21}\widetilde{y}_1 + (a_{22} - p)\widetilde{y}_2 + \cdots + a_{2n}\widetilde{y}_n = -\widetilde{f}_2(p) - y_2^\circ, \qquad (12.6.1.23)$$
$$\cdots\cdots\cdots\cdots\cdots\cdots\cdots\cdots\cdots\cdots\cdots\cdots\cdots\cdots\cdots\cdots$$
$$a_{n1}\widetilde{y}_1 + a_{n2}\widetilde{y}_2 + \cdots + (a_{nn} - p)\widetilde{y}_n = -\widetilde{f}_n(p) - y_n^\circ.$$

The solution of this system is obtained by Kramer's rule and is given by

$$\widetilde{y}_k = \frac{\Delta_k(p)}{\Delta(p)}; \qquad k = 1, \ldots, n, \qquad (12.6.1.24)$$

where $\Delta(p)$ is the determinant of the basic matrix of system (12.6.1.23), coinciding with the determinant in (12.6.1.4) with $\lambda = p$, and $\Delta_k(p)$ is the determinant of the matrix obtained from the basic matrix by replacing its $k$th column with the column of free terms of system (12.6.1.23).

On applying the Laplace inversion formula (see Section 11.2) to (12.6.1.24), one obtains a solution to the Cauchy problem (12.6.1.13), (12.6.1.15) in the form

$$y_k(x) = \frac{1}{2\pi i}\int_{c-i\infty}^{c+i\infty} \frac{\Delta_k(p)}{\Delta(p)} e^{px}\, dp; \qquad k = 1, \ldots, n.$$

The formulas from Paragraph 11.2.2-2 and tables from Section T3.2 can be used to find the inverse Laplace transform of the function $\Delta_k(p)/\Delta(p)$.

$2^\circ$. The Laplace transform is also suitable for the solution of systems of constant-coefficient second- and higher-order ordinary differential equations.

**Example 4.** Consider the Cauchy problem for the nonhomogeneous linear system of constant-coefficient second-order differential equations

$$\sum_{k=1}^{n}\left(a_{mk}y_k'' + b_{mk}y_k' + c_{mk}y_k\right) = f_m(x), \qquad m = 1, 2, \ldots, n,$$

subject to the initial conditions

$$y_k(0) = \alpha_k, \qquad y_k'(0) = \beta_k, \qquad k = 1, 2, \ldots, n.$$

The Laplace transform reduces this problem to a linear system of algebraic equations for the transform $\widetilde{y}_k(p)$:

$$\sum_{k=1}^{n}\left(a_{mk}p^2 + b_{mk}p + c_{mk}\right)\widetilde{y}_k(p) = \widetilde{f}_m(p) + \sum_{k=1}^{n}\left[(a_{mk}p + b_{mk})\alpha_k + \beta_k\right], \qquad m = 1, 2, \ldots, n.$$

The solution to this system can be obtained using Kramer's rule. By applying then the inverse Laplace transform to the resulting expressions of $\widetilde{y}_k(p)$, one obtains the solution to the Cauchy problem.

---

**12.6.1-7. Classification of equilibrium points of two-dimensional linear systems.**

Let us study the behavior of solutions near the equilibrium (also called stationary, steady-state, or fixed) point $x = y = 0$ for the system of two homogeneous linear constant-coefficient equations

$$x_t' = a_{11}x + a_{12}y,$$
$$y_t' = a_{21}x + a_{22}y. \qquad (12.6.1.25)$$

## 12.6. LINEAR SYSTEMS OF ORDINARY DIFFERENTIAL EQUATIONS

By convention, for clearness and convenience of interpretation of the results, $t$ will be used to designate the independent variable and will be treated as time. A solution $x = x(t)$, $y = y(t)$ of system (12.6.1.25) plotted in the plane $x, y$ (the phase plane) is called a (phase) trajectory of the system.

A solution to system (12.6.1.25) will be sought in the form

$$x = k_1 e^{\lambda t}, \quad y = k_2 e^{\lambda t}. \tag{12.6.1.26}$$

On substituting (12.6.1.26) into (12.6.1.25), one obtains the characteristic equation for the exponent $\lambda$:

$$\begin{vmatrix} a_{11} - \lambda & a_{12} \\ a_{21} & a_{22} - \lambda \end{vmatrix} = 0, \quad \text{or} \quad \lambda^2 - (a_{11} + a_{22})\lambda + a_{11}a_{22} - a_{12}a_{21} = 0. \tag{12.6.1.27}$$

The coefficients $k_1$ and $k_2$ are found as

$$k_1 = Ca_{12}, \quad k_2 = C(\lambda - a_{11}), \tag{12.6.1.28}$$

where $C$ is an arbitrary constant. To two different roots of the quadratic equation (12.6.1.27) there correspond two pairs of coefficients (12.6.1.28).

Denote the discriminant of the quadratic equation (12.6.1.27) by

$$D = (a_{11} - a_{22})^2 + 4a_{12}a_{21}. \tag{12.6.1.29}$$

Three situations are possible.

1°. If $D > 0$, the roots of the characteristic equation (12.6.1.27) are real and distinct ($\lambda_1 \ne \lambda_2$):

$$\lambda_{1,2} = \tfrac{1}{2}(a_{11} + a_{22}) \pm \tfrac{1}{2}\sqrt{D}.$$

The general solution of system (12.6.1.25) is expressed as

$$\begin{aligned} x &= C_1 a_{12} e^{\lambda_1 t} + C_2 a_{12} e^{\lambda_2 t}, \\ y &= C_1 (\lambda_1 - a_{11}) e^{\lambda_1 t} + C_2 (\lambda_2 - a_{11}) e^{\lambda_2 t}, \end{aligned} \tag{12.6.1.30}$$

where $C_1$ and $C_2$ are arbitrary constants. For $C_1 = 0$, $C_2 \ne 0$ and $C_2 = 0$, $C_1 \ne 0$, the trajectories in the phase plane $x, y$ are straight lines. Four cases are possible here.

1.1. Two negative real roots, $\lambda_1 < 0$ and $\lambda_2 < 0$. This corresponds to $a_{11} + a_{22} < 0$ and $a_{11}a_{22} - a_{12}a_{21} > 0$. The equilibrium point is asymptotically stable and all trajectories starting within a small neighborhood of the origin tend to the origin as $t \to \infty$. To $C_1 = 0$, $C_2 \ne 0$ and $C_2 = 0$, $C_1 \ne 0$ there correspond straight lines passing through the origin. Figure 12.4 $a$ depicts the arrangement of the phase trajectories near an equilibrium point called a *stable node* (or a *sink*). The direction of motion along the trajectories with increasing $t$ is shown by arrows.

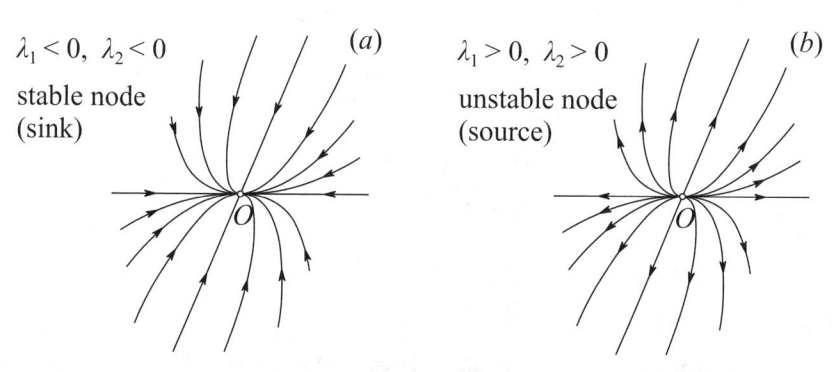

**Figure 12.4.** Phase trajectories of a system of differential equations near an equilibrium point of the node type.

**1.2.** $\lambda_1 > 0$ and $\lambda_2 > 0$. This corresponds to $a_{11} + a_{22} > 0$ and $a_{11}a_{22} - a_{12}a_{21} > 0$. The phase trajectories in the vicinity of the equilibrium point have the same pattern as in the preceding case; however, the trajectories go in the opposite direction, away from the equilibrium point; see Fig. 12.4 *b*. An equilibrium point of this type is called an *unstable node* (or a *source*).

**1.3.** $\lambda_1 > 0$ and $\lambda_2 < 0$ (or $\lambda_1 < 0$ and $\lambda_2 > 0$). This corresponds to $a_{11}a_{22} - a_{12}a_{21} < 0$. In this case, the equilibrium point is also unstable, since the trajectory (12.6.1.30) with $C_2 = 0$ goes beyond a small neighborhood of the origin as $t$ increases. If $C_1 C_2 \neq 0$, then the trajectories leave the neighborhood of the origin as $t \to -\infty$ and $t \to \infty$. An equilibrium point of this type is called a *saddle* (or a *hyperbolic point*); see Fig. 12.5.

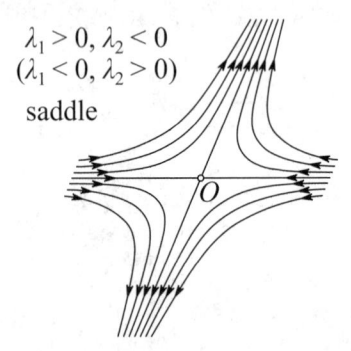

**Figure 12.5.** Phase trajectories of a system of differential equations near an equilibrium point of the saddle type.

**1.4.** $\lambda_1 = 0$ and $\lambda_2 = a_{11} + a_{22} \neq 0$. This corresponds to $a_{11}a_{22} - a_{12}a_{21} = 0$. The general solution of system (12.6.1.25) is expressed as

$$x = C_1 a_{12} + C_2 a_{12} e^{(a_{11}+a_{22})t},$$
$$y = -C_1 a_{11} + C_2 a_{22} e^{(a_{11}+a_{22})t},$$
(12.6.1.31)

where $C_1$ and $C_2$ are arbitrary constants. By eliminating time $t$ from (12.6.1.31), one obtains a family of parallel lines defined by the equation $a_{22}x - a_{12}y = a_{12}(a_{11} + a_{22})C_1$. To $C_2 = 0$ in (12.6.1.31) there corresponds a one-parameter family of equilibrium points that lie on the straight line $a_{11}x + a_{12}y = 0$.

(i) If $\lambda_2 < 0$, then the trajectories approach the equilibrium point lying as $t \to \infty$; see Fig. 12.6. The equilibrium point $x = y = 0$ is stable (or neutrally stable) — there is no asymptotic stability.

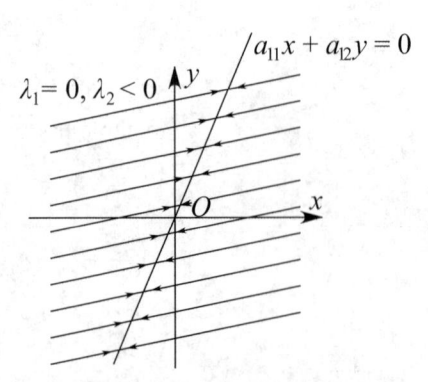

**Figure 12.6.** Phase trajectories of a system of differential equations near a set of equilibrium points located on a straight line.

## 12.6. LINEAR SYSTEMS OF ORDINARY DIFFERENTIAL EQUATIONS

(ii) If $\lambda_2 > 0$, the trajectories have the same pattern as in case (i), but they go, as $t \to \infty$, in the opposite direction, away from the equilibrium point. The point $x = y = 0$ is unstable.

2°. If $D < 0$, the characteristic equation (12.6.1.27) has complex-conjugate roots:

$$\lambda_{1,2} = \alpha \pm i\beta, \qquad \alpha = \tfrac{1}{2}(a_{11} + a_{22}), \qquad \beta = \tfrac{1}{2}\sqrt{|D|}, \qquad i^2 = -1.$$

The general solution of system (12.6.1.25) has the form

$$\begin{aligned} x &= e^{\alpha t}[C_1 \cos(\beta t) + C_2 \sin(\beta t)], \\ y &= e^{\alpha t}[C_1^* \cos(\beta t) + C_2^* \sin(\beta t)], \end{aligned} \qquad (12.6.1.32)$$

where $C_1$ and $C_2$ are arbitrary constants, and $C_1^*$ and $C_2^*$ are defined by linear combinations of $C_1$ and $C_2$. The following cases are possible.

2.1. For $\alpha < 0$, the trajectories in the phase plane are spirals asymptotically approaching the origin of coordinates (the equilibrium point) as $t \to \infty$; see Fig. 12.7 a. Therefore the equilibrium point is asymptotically stable and is called a *stable focus* (also a *stable spiral point* or a *spiral sink*). A focus is characterized by the fact that the tangent to a trajectory changes its direction all the way to the equilibrium point.

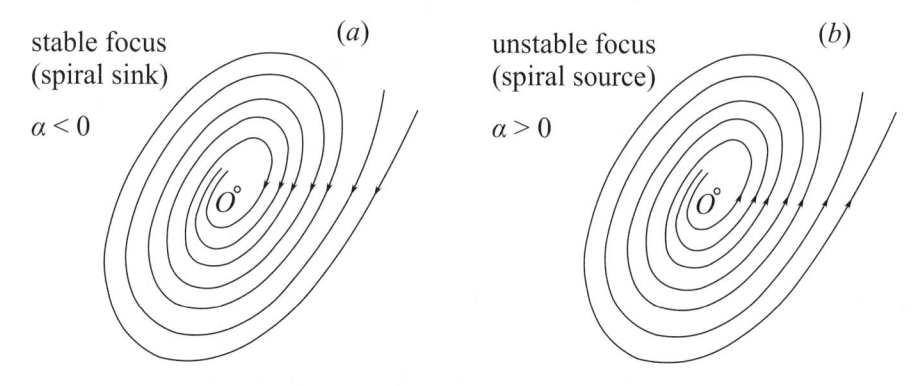

**Figure 12.7.** Phase trajectories of a system of differential equations near an equilibrium point of the focus type.

2.2. For $\alpha > 0$, the phase trajectories are also spirals, but unlike the previous case they spiral away from the origin as $t \to \infty$; see Fig. 12.7 b. Therefore such an equilibrium point is called an *unstable focus* (also an *unstable spiral point* or a *spiral source*).

2.3. At $\alpha = 0$, the phase trajectories are closed curves, containing the equilibrium point inside (see Fig. 12.8). Such an equilibrium point is called a *center*. A center is a stable equilibrium point. Note that there is no asymptotic stability in this case.

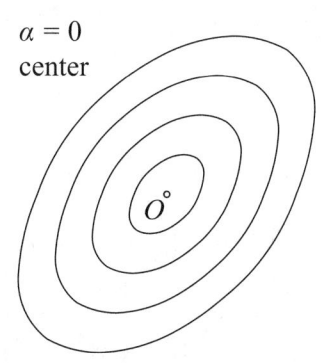

**Figure 12.8.** Phase trajectories of a system of differential equations near an equilibrium point of the center type.

$3°$. If $D = 0$, the characteristic equation (12.6.1.27) has a double real root, $\lambda_1 = \lambda_2 = \frac{1}{2}(a_{11} + a_{22})$. The following cases are possible.

3.1. If $\lambda_1 = \lambda_2 = \lambda < 0$, the general solution of system (12.6.1.25) has the form

$$x = a_{12}(C_1 + C_2 t)e^{\lambda t},$$
$$y = [(\lambda - a_{11})C_1 + C_2 + C_2(\lambda - a_{11})t]e^{\lambda t}, \quad (12.6.1.33)$$

where $C_1$ and $C_2$ are arbitrary constants.

Since there is a rapidly decaying factor, $e^{\lambda t}$, all trajectories tend to the equilibrium point as $t \to \infty$; see Fig. 12.9 $a$. To $C_2 = 0$ there corresponds a straight line in the phase plane $x, y$. The equilibrium point is asymptotically stable and is called a *stable node* (a *sink*). Such a node is in intermediate position between a node from Item 1.1 and a focus from Item 2.1.

(a) $\lambda_1 = \lambda_2 < 0$
stable node
(sink)

(b) $\lambda_1 = \lambda_2 > 0$
unstable node
(source)

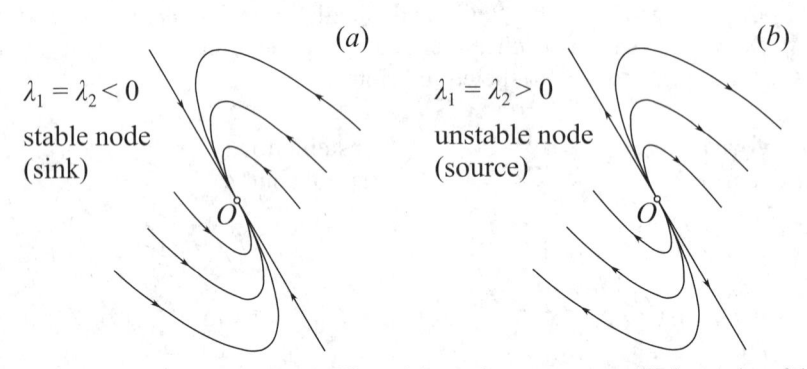

**Figure 12.9.** Phase trajectories of a system of differential equations near an equilibrium point of the node type in the case of a double root, $\lambda_1 = \lambda_2$.

3.2. If $\lambda_1 = \lambda_2 = \lambda > 0$, the general solution of system (12.6.1.25) is determined by formulas (12.6.1.33). The phase trajectories are similar to those from Item 3.1, but they go in the opposite direction, as $t \to \infty$, rapidly away from the equilibrium point. Such an equilibrium point is called an *unstable node* (a *source*); see Fig. 12.9 $b$.

3.3. If $\lambda_1 = \lambda_2 = 0$, which corresponds to

$$a_{11} + a_{22} = 0 \quad \text{and} \quad a_{11}a_{22} - a_{12}a_{21} = 0$$

simultaneously, the general solution of system (12.6.1.25) is obtained by substituting $\lambda = 0$ into (12.6.1.33) and has the form

$$x = a_{12}C_1 + a_{12}C_2 t,$$
$$y = C_2 - a_{11}C_1 - a_{11}C_2 t.$$

For $a_{12} \neq 0$ all trajectories are parallel straight lines. As $t \to \pm\infty$, the trajectories go away from the origin. The equilibrium point is unstable.

For clearness, the classification results for equilibrium points of systems of two linear constant-coefficient differential equations (12.6.1.25) are summarized in Table 12.5.

**Remark.** For general definitions of a stable and an unstable equilibrium point, see Section 7.3.

TABLE 12.5
Classification of equilibrium points for systems of constant-coefficient equations (12.6.1.25);
the symbols ○ and ∗ indicate stable and unstable equilibrium points, respectively, where not clearly stated

| Discriminant $D$, formula (12.6.1.29) | Roots of quadratic equation (12.6.1.27), $\lambda_1$ and $\lambda_2$ | Conditions for coefficients of differential equations (12.6.1.25) | Type of equilibrium points or shape of phase trajectories |
|---|---|---|---|
| $D > 0$ | $\lambda_1 < 0, \lambda_2 < 0, \lambda_1 \neq \lambda_2$<br>$\lambda_1 > 0, \lambda_2 > 0, \lambda_1 \neq \lambda_2$<br>roots have unlike signs<br>$\lambda_1 = 0, \lambda_2 < 0$<br>$\lambda_1 = 0, \lambda_2 > 0$ | $a_{11}+a_{22} < 0,\ a_{11}a_{22}-a_{12}a_{21} > 0$<br>$a_{11}+a_{22} > 0,\ a_{11}a_{22}-a_{12}a_{21} > 0$<br>$a_{11}a_{22}-a_{12}a_{21} < 0$<br>$a_{11}+a_{22} < 0,\ a_{11}a_{22}-a_{12}a_{21} = 0$<br>$a_{11}+a_{22} > 0,\ a_{11}a_{22}-a_{12}a_{21} = 0$ | stable node<br>unstable node<br>saddle∗<br>parallel lines°<br>parallel lines∗ |
| $D < 0$ | $\lambda_{1,2} = \alpha \pm i\beta,\ \alpha > 0$<br>$\lambda_{1,2} = \alpha \pm i\beta,\ \alpha < 0$<br>$\lambda_{1,2} = \pm i\beta$, imaginary roots | $a_{11}+a_{22} < 0,\ (a_{11}-a_{22})^2 + 4a_{12}a_{21} < 0$<br>$a_{11}+a_{22} > 0,\ (a_{11}-a_{22})^2 + 4a_{12}a_{21} < 0$<br>$a_{11}+a_{22} = 0,\ a_{11}a_{22}-a_{12}a_{21} > 0$ | stable focus<br>unstable focus<br>center° |
| $D = 0$ | $\lambda_1 = \lambda_2 < 0$<br>$\lambda_1 = \lambda_2 > 0$<br>$\lambda_1 = \lambda_2 = 0$ | $a_{11}+a_{22} < 0,\ (a_{11}-a_{22})^2 + 4a_{12}a_{21} = 0$<br>$a_{11}+a_{22} > 0,\ (a_{11}-a_{22})^2 + 4a_{12}a_{21} = 0$<br>$a_{11}+a_{22} = 0,\ a_{11}a_{22}-a_{12}a_{21} = 0$ | stable node<br>unstable node<br>saddle∗<br>parallel lines∗ |

## 12.6.2. Systems of Linear Variable-Coefficient Equations

**12.6.2-1. Homogeneous systems of linear first-order equations.**

1°. In general, a homogeneous linear system of variable-coefficient first-order ordinary differential equations has the form

$$\begin{aligned} y_1' &= f_{11}(x)y_1 + f_{12}(x)y_2 + \cdots + f_{1n}(x)y_n, \\ y_2' &= f_{21}(x)y_1 + f_{22}(x)y_2 + \cdots + f_{2n}(x)y_n, \\ &\cdots\cdots\cdots\cdots\cdots\cdots\cdots\cdots\cdots\cdots\cdots\cdots\cdots\cdots\cdots \\ y_n' &= f_{n1}(x)y_1 + f_{n2}(x)y_2 + \cdots + f_{nn}(x)y_n, \end{aligned} \qquad (12.6.2.1)$$

where the prime denotes a derivative with respect to $x$. It is assumed further on that the functions $f_{ij}(x)$ are continuous of an interval $a \leq x \leq b$ (intervals are allowed with $a = -\infty$ or/and $b = +\infty$).

Any homogeneous linear system of the form (12.6.2.1) has the trivial particular solution $y_1 = y_2 = \cdots = y_n = 0$.

*Superposition principle for a homogeneous system*: any linear combination of particular solutions to system (12.6.2.1) is also a solution to this system.

2°. Let

$$\mathbf{y}_k = (y_{k1}, y_{k2}, \ldots, y_{kn})^T, \qquad y_{km} = y_{km}(x); \quad k, m = 1, 2, \ldots, n \qquad (12.6.2.2)$$

be nontrivial particular solutions of the homogeneous system of equations (12.6.2.1). Solutions (12.6.2.2) are linearly independent if the *Wronskian determinant* is nonzero:

$$W(x) \equiv \begin{vmatrix} y_{11}(x) & y_{12}(x) & \cdots & y_{1n}(x) \\ y_{21}(x) & y_{22}(x) & \cdots & y_{2n}(x) \\ \vdots & \vdots & \ddots & \vdots \\ y_{n1}(x) & y_{n2}(x) & \cdots & y_{nn}(x) \end{vmatrix} \neq 0. \qquad (12.6.2.3)$$

If condition (12.6.2.3) is satisfied, the general solution of the homogeneous system (12.6.2.1) is expressed as

$$\mathbf{y} = C_1 \mathbf{y}_1 + C_2 \mathbf{y}_2 + \cdots + C_n \mathbf{y}_n, \qquad (12.6.2.4)$$

where $C_1, C_2, \ldots, C_n$ are arbitrary constants. The vector functions $\mathbf{y}_1, \mathbf{y}_2, \ldots, \mathbf{y}_n$ in (12.6.2.4) are called *fundamental solutions* of system (12.6.2.1).

3°. Suppose condition (12.6.2.3) is met. Then the *Liouville formula*

$$W(x) = W(x_0) \exp\left[\int_{x_0}^{x} \left(\sum_{s=1}^{n} f_{ss}(t)\right) dt\right]$$

holds.

COROLLARY. *Particular solutions (12.6.2.2) are linearly independent on the interval $[a, b]$ if and only if there exists a point $x_0 \in [a, b]$ such that the Wronskian determinant is nonzero at $x_0$: $W(x_0) \neq 0$.*

4°. Suppose a nontrivial particular solution of system (12.6.2.1),

$$\mathbf{y}_1 = (u_1, u_2, \ldots, u_n)^\mathrm{T}, \quad u_m = u_m(x), \quad m = 1, 2, \ldots, n,$$

is known. Then the number of unknowns can be reduced by one. To this end, one considers the auxiliary homogeneous linear system of $n-1$ equations

$$z_k' = \sum_{q=2}^{n} \left[f_{kq}(x) - \frac{u_k(x)}{u_1(x)} f_{1q}(x)\right] z_q, \quad k = 2, \ldots, n. \tag{12.6.2.5}$$

Let

$$\mathbf{z}_p = (z_{p2}, z_{p3}, \ldots, z_{pn})^\mathrm{T}, \quad z_{mk} = z_{mk}(x); \quad p = 2, \ldots, n,$$

be a fundamental system of solutions to system (12.6.2.5) and let

$$\mathbf{Z}_p = (0, z_{p2}, z_{p3}, \ldots, z_{pn})^\mathrm{T}, \quad F_p(x) = \int \left(\frac{1}{u_1(x)} \sum_{s=2}^{n} f_{1s}(x) z_{ps}(x)\right) dx, \quad p = 2, \ldots, n,$$

the vector $\mathbf{Z}_p$ having an additional component compared with $\mathbf{z}_p$. Then the vector functions $\mathbf{y}_p = F_p(x)\mathbf{y}_1 + \mathbf{Z}_p$ ($p = 2, \ldots, n$) together with $\mathbf{y}_1$ form a fundamental system of solutions to the initial homogeneous system of equations (12.6.2.1).

---

**12.6.2-2. Nonhomogeneous systems of linear first-order equations.**

1°. In general, a nonhomogeneous linear system of variable-coefficient first-order differential equations has the form

$$\begin{aligned} y_1' &= f_{11}(x)y_1 + f_{12}(x)y_2 + \cdots + f_{1n}(x)y_n + g_1(x), \\ y_2' &= f_{21}(x)y_1 + f_{22}(x)y_2 + \cdots + f_{2n}(x)y_n + g_2(x), \\ &\quad \ldots\ldots\ldots\ldots\ldots\ldots\ldots\ldots\ldots\ldots\ldots\ldots\ldots\ldots\ldots\ldots\ldots\ldots \\ y_n' &= f_{n1}(x)y_1 + f_{n2}(x)y_2 + \cdots + f_{nn}(x)y_n + g_n(x). \end{aligned} \tag{12.6.2.6}$$

Alternatively, the system can be written in the short vector-matrix notation as

$$\mathbf{y}' = \mathbf{f}(x)\mathbf{y} + \mathbf{g}(x),$$

with $\mathbf{f}(x) = \big(f_{ij}(x)\big)$ being the matrix of equation coefficients and $\mathbf{g}(x) = \big(g_1(x), g_2(x), \ldots, g_n(x)\big)^\mathrm{T}$ being the vector function defining the nonhomogeneous part of the equations.

EXISTENCE AND UNIQUENESS THEOREM. *Let the functions $f_{ij}(x)$ and $g_i(x)$ be continuous on an interval $a < x < b$. Then, for any set of values $x°$, $y_1°, \ldots, y_n°$, where $a < x° < b$, there exists a unique solution $y_1 = y_1(x), \ldots, y_n = y_n(x)$ satisfying the initial conditions*

$$y_1(x°) = y_1°, \quad \ldots, \quad y_n(x°) = y_n°,$$

*and this solution is defined on the whole interval $a < x < b$.*

2°. Let

$$\bar{\mathbf{y}} = (\bar{y}_1, \bar{y}_2, \ldots, \bar{y}_n)^T, \quad \bar{y}_k = \bar{y}_k(x); \quad k = 1, 2, \ldots, n,$$

be a particular solution to the nonhomogeneous system of equations (12.6.2.6). The general solution of this system is the sum of the general solution of the corresponding homogeneous system (12.6.2.1), which corresponds to $g_k(x) \equiv 0$ in (12.6.2.6), and any particular solution of the nonhomogeneous system (12.6.2.6), or

$$\mathbf{y} = C_1 \mathbf{y}_1 + C_2 \mathbf{y}_2 + \cdots + C_n \mathbf{y}_n + \bar{\mathbf{y}}, \tag{12.6.2.7}$$

where $\mathbf{y}_1, \mathbf{y}_2, \ldots, \mathbf{y}_n$ are linearly independent solutions of the homogeneous system (12.6.2.1).

3°. Given a fundamental system of solutions $y_{km}(x)$ (12.6.2.2) of the homogeneous system (12.6.2.1), a particular solution of the nonhomogeneous system (12.6.2.6) is found as

$$\bar{y}_k = \sum_{m=1}^{n} y_{mk}(x) \int \frac{W_m(x)}{W(x)} \, dx, \quad k = 1, 2, \ldots, n,$$

where $W_m(x)$ is the determinant obtained by replacing the $m$th row in the Wronskian determinant (12.6.2.3) by the row of free terms, $g_1(x), g_2(x), \ldots, g_n(x)$, of equation (12.6.2.6). The general solution of the nonhomogeneous system (12.6.2.6) is given by (12.6.2.7).

4°. *Superposition principle for a nonhomogeneous system.* A particular solution, $\mathbf{y} = \bar{\mathbf{y}}$, of the nonhomogeneous system of linear differential equations,

$$\mathbf{y}' = \mathbf{f}(x)\mathbf{y} + \sum_{k=1}^{m} \mathbf{g}_k(x),$$

is given by the sum

$$\mathbf{y} = \sum_{k=1}^{m} \mathbf{y}_k,$$

where the $\mathbf{y}_k$ are particular solutions of $m$ (simpler) systems of equations

$$\mathbf{y}'_k = \mathbf{f}(x)\mathbf{y}_k + \mathbf{g}_k(x), \quad k = 1, 2, \ldots, m,$$

corresponding to individual nonhomogeneous terms of the original system.

**12.6.2-3. Euler system of ordinary differential equations.**

1°. A homogeneous Euler system is a homogeneous linear system of ordinary differential equations composed by linear combinations of the following terms:

$$y_k, \quad xy'_k, \quad x^2 y''_k, \quad \ldots, \quad x^{m_k} y_k^{(m_k)}; \qquad k = 1, 2, \ldots, n.$$

Such a system is invariant under scaling in the independent variable (i.e., it preserves its form under the change of variable $x \to \alpha x$, where $\alpha$ is any nonzero number). A nonhomogeneous Euler system contains additional terms, given functions.

The substitution $x = be^t$ ($b \neq 0$) brings an Euler system, both homogeneous and nonhomogeneous, to a constant-coefficient linear system of equations.

**Example.** In general, a nonhomogeneous Euler system of second-order equations has the form

$$\sum_{k=1}^{n} \left( a_{mk} x^2 \frac{d^2 y_k}{dx^2} + b_{mk} x \frac{dy_k}{dx} + c_{mk} y_k \right) = f_m(x), \qquad m = 1, 2, \ldots, n. \qquad (12.6.2.8)$$

The substitutions $x = \pm e^t$ bring this system to a constant-coefficient linear system,

$$\sum_{k=1}^{n} \left[ a_{mk} \frac{d^2 y_k}{dt^2} + (b_{mk} - a_{mk}) \frac{dy_k}{dt} + c_{mk} y_k \right] = f_m(\pm e^t), \qquad m = 1, 2, \ldots, n,$$

which can be solved using, for example, the Laplace transform (see Example 4 from Paragraph 12.6.1-6).

2°. Particular solutions to a homogeneous Euler system (for system (12.6.2.8), corresponding to $f_m(x) \equiv 0$) are sought in the form of power functions:

$$y_1 = A_1 x^\sigma, \quad y_2 = A_2 x^\sigma, \quad \ldots, \quad y_n = A_n x^\sigma, \qquad (12.6.2.9)$$

where the coefficients $A_1, A_2, \ldots, A_n$ are determined by solving the associated homogeneous system of algebraic equations obtained by substituting expressions (12.6.2.9) into the differential equations of the system in question and dividing by $x^\sigma$. Since the system is homogeneous, for it to have nontrivial solutions, its determinant must vanish. This results in a dispersion equation for the exponent $\sigma$.

## 12.7. Nonlinear Systems of Ordinary Differential Equations

### 12.7.1. Solutions and First Integrals. Uniqueness and Existence Theorems

**12.7.1-1. Systems solved for the derivative. A solution and the general solution.**

We will be dealing with a system of first-order ordinary differential equations solved for the derivatives

$$y'_k = f_k(x, y_1, \ldots, y_n), \qquad k = 1, \ldots, n. \qquad (12.7.1.1)$$

Here and henceforth throughout the current chapter, the prime denotes a derivative with respect to the independent variable $x$.

A set of numbers $x, y_1, \ldots, y_n$ is convenient to treat as a point in the $(n+1)$-dimensional space.

For brevity, system (12.7.1.1) is conventionally written in vector form:

$$\mathbf{y}' = \mathbf{f}(x, \mathbf{y}),$$

where $\mathbf{y}$ and $\mathbf{f}$ are the vectors defined as $\mathbf{y} = (y_1, \ldots, y_n)^T$ and $\mathbf{f} = (f_1, \ldots, f_n)^T$.

A *solution* (also an *integral* or an *integral curve*) of a system of differential equations (12.7.1.1) is a set of functions $y_1 = y_1(x), \ldots, y_n = y_n(x)$ such that, when substituted into all equations (12.7.1.1), they turn them into identities. The *general solution of a system of differential equations* is the set of all its solutions. In the general case, the general solution of system (12.7.1.1) depends on $n$ arbitrary constants.

### 12.7.1-2. Existence and uniqueness theorems.

THEOREM (EXISTENCE, PEANO). *Let the functions $f_k(x, y_1, \ldots, y_n)$ ($k = 1, \ldots, n$) be continuous in a domain $G$ of the $(n+1)$-dimensional space of the variables $x, y_1, \ldots, y_n$. Then there is at least one integral curve passing through every point $M(x°, y_1°, \ldots, y_n°)$ in $G$. Each of such curves can be extended on both ends up to the boundary of any closed domain completely belonging to $G$ and containing the point $M$ inside.*

Remark. If there is more than one integral curve passing through the point $M$, there are infinitely many integral curves passing through $M$.

THEOREM (UNIQUENESS). *There is a unique integral curve passing through the point $M(x°, y_1°, \ldots, y_n°)$ if the functions $f_k$ have partial derivatives with respect to all $y_m$, continuous in $x, y_1, \ldots, y_n$ in the domain $G$, or if each function $f_k$ in $G$ satisfies the Lipschitz condition:*

$$|f_k(x, \bar{y}_1, \ldots, \bar{y}_n) - f_k(x, y_1, \ldots, y_n)| \leq A \sum_{m=1}^{n} |\bar{y}_m - y_m|,$$

*where $A$ is some positive number.*

### 12.7.1-3. Reduction of systems of equations to a single equation.

Suppose the right-hand sides of equations (12.7.1.1) are $n$ times differentiable in all variables. Then system (12.7.1.1) can be reduced to a single $n$th-order equation. Indeed, using the chain rule, let us differentiate the first equation of system (12.7.1.1) with respect to $x$ to get

$$y_1'' = \frac{\partial f_1}{\partial x_1} + \frac{\partial f_1}{\partial y_1} y_1' + \cdots + \frac{\partial f_1}{\partial y_n} y_n'. \tag{12.7.1.2}$$

Then change the first derivatives $y_k'$ in (12.7.1.2) to $f_k(x, y_1, \ldots, y_n)$ [the right-hand sides of equations (12.7.1.1)] to obtain

$$y_1'' = F_2(x, y_1, \ldots, y_n), \tag{12.7.1.3}$$

where $F_2(x, y_1, \ldots, y_n) \equiv \frac{\partial f_1}{\partial x_1} + \frac{\partial f_1}{\partial y_1} f_1 + \cdots + \frac{\partial f_1}{\partial y_n} f_n$. Now differentiate equation (12.7.1.3) with respect to $x$ and replace the first derivatives $y_k'$ on the right-hand side of the resulting equation by $f_k$. As a result, we obtain

$$y_1''' = F_3(x, y_1, \ldots, y_n),$$

where $F_3(x, y_1, \ldots, y_n) \equiv \frac{\partial F_2}{\partial x_1} + \frac{\partial F_2}{\partial y_1} f_1 + \cdots + \frac{\partial F_2}{\partial y_n} f_n$. Repeating this procedure as many times as required, one arrives at the following system of equations:

$$y_1' = F_1(x, y_1, \ldots, y_n),$$
$$y_1'' = F_2(x, y_1, \ldots, y_n),$$
$$\cdots\cdots\cdots\cdots\cdots\cdots\cdots\cdots$$
$$y_1^{(n)} = F_n(x, y_1, \ldots, y_n),$$

where

$$F_1(x, y_1, \ldots, y_n) \equiv f_1(x, y_1, \ldots, y_n), \quad F_{k+1}(x, y_1, \ldots, y_n) \equiv \frac{\partial F_k}{\partial x_1} + \frac{\partial F_k}{\partial y_1} f_1 + \cdots + \frac{\partial F_k}{\partial y_n} f_n.$$

Expressing $y_2, y_3, \ldots, y_n$ from the $n-1$ first equations of this system in terms of $x$, $y_1$, $y_1'$, $\ldots, y_1^{(n-1)}$ and then substituting the resulting expressions into the last equation of system (12.7.1.1), one finally arrives at an $n$th-order equation:

$$y_1^{(n)} = \Phi(x, y_1, y_1', \ldots, y_1^{(n-1)}). \tag{12.7.1.4}$$

**Remark 1.** If (12.7.1.1) is a linear system of first-order differential equations, then (12.7.1.4) is a linear $n$th-order equation.

**Remark 2.** Any equation of the form (12.7.1.4) can be reduced to a system on $n$ first-order equations (see Paragraph 12.5.1-4).

### 12.7.1-4. First integrals. Using them to reduce system dimension.

$1°$. A relation of the form

$$\Psi(x, y_1, \ldots, y_n) = C, \tag{12.7.1.5}$$

where $C$ is an arbitrary constant, is called a *first integral* of system (12.7.1.1) if its left-hand side $\Phi$, generally not identically constant, is turned into a constant by any particular solution, $y_1, \ldots, y_n$, of system (12.7.1.1). In the sequel, we consider only continuously differentiable functions $\Psi(x, y_1, \ldots, y_n)$ in a given domain of variation of its arguments.

THEOREM. *An expression of the form (12.7.1.5) is a first integral of system (12.7.1.1) if and only if the function $\Psi = \Psi(x, y_1, \ldots, y_n)$ satisfies the relation*

$$\frac{\partial \Psi}{\partial x} + \sum_{k=1}^{n} \frac{\partial \Psi}{\partial y_k} f_k(x, y_1, \ldots, y_n) = 0.$$

This relation may be treated as a first-order partial differential equation for $\Psi$.

Different first integrals of system (12.7.1.1) are called *independent* if the Jacobian of their left-hand sides is nonzero.

System (12.7.1.1) admits $n$ independent first integrals if the conditions of the uniqueness theorem from Paragraph 12.7.1-2 are met.

$2°$. Given a first integral (12.7.1.5) of system (12.7.1.1), it may be treated as an implicit specification of one of the unknowns. Solving (12.7.1.5), for example, for $y_n$ yields $y_n = G(x, y_1, \ldots, y_{n-1})$. Substituting this expression into the first $n-1$ equations of system (12.7.1.1), one obtains a system in $n-1$ variables with one arbitrary constant.

Likewise, given $m$ independent first integrals of system (12.7.1.1),

$$\Psi_k(x, y_1, \ldots, y_n) = C_k, \quad k = 1, \ldots, m \quad (m < n),$$

the system may be reduced to a system of $n-m$ first-order equations in $n-m$ unknowns.

## 12.7.2. Integrable Combinations. Autonomous Systems of Equations

**12.7.2-1. Integrable combinations.**

In some cases, first integrals of systems of differential equations may be obtained by finding *integrable combinations*. An integrable combination is a differential equation that is easy to integrate and is a consequence of the equations of the system under consideration. Most commonly, an integrable combination is an equation of the form

$$d\,\Psi(x, y_1, \ldots, y_n) = 0 \qquad (12.7.2.1)$$

or an equation reducible by a change of variables to one of the integrable types of equations in one unknown.

**Example 1.** Consider the nonlinear system

$$ay_1' = (b-c)y_2 y_3, \qquad by_2' = (c-a)y_1 y_3, \qquad cy_3' = (a-b)y_1 y_2, \qquad (12.7.2.2)$$

where $a$, $b$, and $c$ are some constants. Such systems arise in the theory of motion of a rigid body.

Let us multiply the first equation by $y_1$, the second by $y_2$, and the third by $y_3$ and add together to obtain

$$ay_1 y_1' + by_2 y_2' + cy_3 y_3' = 0 \implies d(ay_1^2 + by_2^2 + cy_3^2) = 0.$$

Integrating yields a first integral:

$$ay_1^2 + by_2^2 + cy_3^2 = C_1. \qquad (12.7.2.3)$$

Now multiply the first equation of the system by $ay_1$, the second by $by_2$, and the third by $cy_3$ and add together to obtain

$$a^2 y_1 y_1' + b^2 y_2 y_2' + c^2 y_3 y_3' = 0 \implies d(a^2 y_1^2 + b^2 y_2^2 + c^2 y_3^2) = 0.$$

Integrating yields another first integral:

$$a^2 y_1^2 + b^2 y_2^2 + c^2 y_3^2 = C_2. \qquad (12.7.2.4)$$

If the case $a = b = c$, where system (12.7.2.2) can be integrated directly, does not take place, the above two first integrals (12.7.2.3) and (12.7.2.4) are independent. Hence, using them, one can express $y_2$ and $y_3$ in terms of $y_1$ and then substitute the resulting expressions into the first equation of system (12.7.2.2). As a result, one arrives at a single separable first-order differential equation for $y_1$.

In this example, the integrable combinations have the form (12.7.2.1).

**Example 2.** A specific example of finding an integrable combination reducible with a change of variables to a simpler, integrable equation in one unknown can be found in Paragraph 12.6.1-5.

**12.7.2-2. Autonomous systems and their reduction to systems of lower dimension.**

$1^\circ$. A system of equations is called *autonomous* if the right-hand sides of the equations do not depend explicitly on $x$. In general, such systems have the form

$$y_k' = f_k(y_1, \ldots, y_n), \qquad k = 1, \ldots, n. \qquad (12.7.2.5)$$

If $\mathbf{y}(x)$ is a solution of the autonomous system (12.7.2.5), then the function $\mathbf{y}(x + C)$, where $C$ is an arbitrary constant, is also a solution of this system.

A point $\mathbf{y}^\circ = (y_1^\circ, \ldots, y_n^\circ)$ is called an *equilibrium point* (or a *stationary point*) of the autonomous system (12.7.2.5) if

$$f_k(y_1^\circ, \ldots, y_n^\circ) = 0, \qquad k = 1, \ldots, n.$$

To an equilibrium point there corresponds a special, simplest particular solution when all unknowns are constant:

$$y_1 = y_1^\circ, \quad \ldots, \quad y_n = y_n^\circ, \qquad k = 1, \ldots, n.$$

$2^\circ$. Any $n$-dimensional autonomous system of the form (12.7.2.5) can be reduced to an $(n-1)$-dimensional system of equations independent of $x$. To this end, one should select one of the equations and divide the other $n-1$ equations of the system by it.

**Example 3.** The autonomous system of two first-order equations
$$y'_x = f_1(y, z), \quad z'_x = f_2(y, z) \tag{12.7.2.6}$$
is reduced by dividing the first equation by the second to a single equation for $y = y(z)$:
$$y'_z = \frac{f_1(y, z)}{f_2(y, z)}. \tag{12.7.2.7}$$
If the general solution of equation (12.7.2.7) is obtained in the form
$$y = \varphi(z, C_1), \tag{12.7.2.8}$$
then $z = z(x)$ is found in implicit form from the second equation in (12.7.2.6) by quadrature:
$$\int \frac{dz}{f_2(\varphi(z, C_1), z)} = x + C_2. \tag{12.7.2.9}$$
Formulas (12.7.2.8)–(12.7.2.9) determine the general solution of system (12.7.2.6) with two arbitrary constants, $C_1$ and $C_2$.

**Remark.** The dependent variables $y$ and $z$ in the autonomous system (12.7.2.6) are often called *phase variables*; the plane $y, z$ they form is called a *phase plane*, which serves to display integral curves of equation (12.7.2.7).

## 12.7.3. Elements of Stability Theory

**12.7.3-1. Lyapunov stability. Asymptotic stability.**

1°. In many applications, time $t$ plays the role of the independent variable, and the associated system of differential equations is conventionally written in the following notation:
$$x'_k = f_k(t, x_1, \ldots, x_n), \quad k = 1, \ldots, n. \tag{12.7.3.1}$$
Here the $x_k = x_k(t)$ are unknown functions that may be treated as coordinates of a moving point in an $n$-dimensional space.

Let us supply system (12.7.3.1) with initial conditions
$$x_k = x_k^\circ \quad \text{at} \quad t = t^\circ \quad (k = 1, \ldots, n). \tag{12.7.3.2}$$
Denote by
$$x_k = \varphi_k(t; t^\circ, x_1^\circ, \ldots, x_n^\circ), \quad k = 1, \ldots, n, \tag{12.7.3.3}$$
the solution of system (12.7.3.1) with the initial conditions (12.7.3.2).

A solution (12.7.3.3) of system (12.7.3.1) is called *Lyapunov stable* if for any $\varepsilon > 0$ there exists a $\delta > 0$ such that if
$$|x_k^\circ - \tilde{x}_k^\circ| < \delta, \quad k = 1, \ldots, n, \tag{12.7.3.4}$$
then the following inequalities hold for $t^\circ \le t < \infty$:
$$\left| \varphi_k(t; t^\circ, x_1^\circ, \ldots, x_n^\circ) - \varphi_k(t; t^\circ, \tilde{x}_1^\circ, \ldots, \tilde{x}_n^\circ) \right| < \varepsilon, \quad k = 1, \ldots, n.$$

Any solution which is not stable is called *unstable*. Solution (12.7.3.3) is called unperturbed and the solution $\varphi_k(t; t^\circ, \tilde{x}_1^\circ, \ldots, \tilde{x}_n^\circ)$ is called perturbed. Geometrically, Lyapunov stability means that the trajectory of the perturbed solution stays at all times $t \ge t^\circ$ within a small neighborhood of the associated unperturbed solution.

2°. A solution (12.7.3.3) of system (12.7.3.1) is called *asymptotically stable* if it is Lyapunov stable and, in addition, with inequalities (12.7.3.4) met, satisfies the conditions
$$\lim_{t \to \infty} \left| \varphi_k(t; t^\circ, x_1^\circ, \ldots, x_n^\circ) - \varphi_k(t; t^\circ, \tilde{x}_1^\circ, \ldots, \tilde{x}_n^\circ) \right| = 0, \quad k = 1, \ldots, n. \tag{12.7.3.5}$$

**Remark.** Condition (12.7.3.5) alone does not suffice for the solution to be Lyapunov stable.

3°. In stability analysis, it is normally assumed, without loss of generality, that $t^\circ = x_1^\circ = \cdots = x_n^\circ = 0$ (this can be achieved by shifting each of the variables by a constant value). Further, with the changes of variables
$$z_k = x_k - \varphi_k(t; t^\circ, x_1^\circ, \ldots, x_n^\circ) \quad (k = 1, \ldots, n),$$
the stability analysis of any solution is reduced to that of the zero solution $z_1 = \cdots = z_n = 0$.

## 12.7.3-2. Theorems of stability and instability by first approximation.

In studying stability of the trivial solution $x_1 = \cdots = x_n = 0$ of system (12.7.3.1) the following method is often employed. The right-hand sides of the equations are approximated by the principal (linear) terms of the expansion into Taylor series about the equilibrium point:

$$f_k(t, x_1, \ldots, x_n) \approx a_{k1}(t)x_1 + \cdots + a_{kn}(t)x_n, \quad a_{km}(t) = \frac{\partial f_k}{\partial x_m}\bigg|_{x_1=\cdots=x_n=0}, \quad k = 1, \ldots, n.$$

Then a stability analysis of the resulting simplified, linear system is performed. The question arises: Is it possible to draw correct conclusions about the stability of the original nonlinear system (12.7.3.1) from the analysis of the linearized system? Two theorems stated below give a partial answer to this question.

THEOREM (STABILITY BY FIRST APPROXIMATION). *Suppose in the system*

$$x'_k = a_{k1}x_1 + \cdots + a_{kn}x_n + \psi_k(t, x_1, \ldots, x_n), \quad k = 1, \ldots, n, \tag{12.7.3.6}$$

*the functions $\psi_k$ are defined and continuous in a domain $t \geq 0$, $|x_k| \leq b$ ($k = 1, \ldots, n$) and, in addition, the inequality*

$$\sum_{k=1}^{n} |\psi_k| \leq A \sum_{k=1}^{n} |x_k| \tag{12.7.3.7}$$

*holds for some constant $A$. In particular, this implies that $\psi_k(t, 0, \ldots, 0) = 0$, and therefore*

$$x_1 = \cdots = x_n = 0 \tag{12.7.3.8}$$

*is a solution of system (12.7.3.6). Suppose further that*

$$\frac{\sum_{k=1}^{n} |\psi_k|}{\sum_{k=1}^{n} |x_k|} \to 0 \quad \text{as} \quad \sum_{k=1}^{n} |x_k| \to 0 \quad \text{and} \quad t \to \infty, \tag{12.7.3.9}$$

*and the real parts of all roots of the characteristic equation*

$$\det |a_{ij} - \lambda \delta_{ij}| = 0, \quad \delta_{ij} = \begin{cases} 1 & \text{if } i = j, \\ 0 & \text{if } i \neq j \end{cases} \tag{12.7.3.10}$$

*are negative. Then solution (12.7.3.8) is stable.*

Remark 1. Necessary and sufficient conditions for the real parts of all roots of the characteristic equation (12.7.3.10) to be negative are established by Hurwitz's theorem, which allows avoiding its solution.

Remark 2. In the above system, the $a_{ij}$, $x_k$, and $\psi_k$ may be complex valued.

THEOREM (INSTABILITY BY FIRST APPROXIMATION). *Suppose conditions (12.7.3.7) and (12.7.3.9) are met and the conditions for the functions $\psi_k$ from the previous theorem are also met. If at least one root of the characteristic equation (12.7.3.10) has a positive real part, then the equilibrium point (12.7.3.8) of system (12.7.3.6) is unstable.*

**Example 1.** Consider the following two-dimensional system of the form (12.7.3.6) with real coefficients:

$$\begin{aligned} x'_t &= a_{11}x + a_{12}y + \psi_1(t, x, y), \\ y'_t &= a_{21}x + a_{22}y + \psi_2(t, x, y). \end{aligned} \tag{12.7.3.11}$$

We assume that the functions $\psi_1$ and $\psi_2$ satisfy conditions (12.7.3.7) and (12.7.3.9).

The characteristic equation of the linearized system (obtained by setting $\psi_1 = \psi_2 = 0$) is given by

$$\lambda^2 - b\lambda + c = 0, \quad \text{where} \quad b = a_{11} + a_{22}, \quad c = a_{11}a_{22} - a_{12}a_{21}. \tag{12.7.3.12}$$

1. Using the theorem of stability by first approximation and examining the roots of the quadratic equation (12.7.3.12), we obtain two sufficient stability conditions for system (12.7.3.11):

$$b < 0, \quad 0 < \tfrac{1}{4}b^2 < c \quad \text{(complex roots with negative real part)};$$
$$b < 0, \quad 0 < c < \tfrac{1}{4}b^2 \quad \text{(negative real roots)}.$$

The two conditions can be combined into one:

$$b < 0, \quad c > 0, \quad \text{or} \quad a_{11} + a_{22} < 0, \quad a_{11}a_{22} - a_{12}a_{21} > 0.$$

These inequalities define the second quadrant in the plane $b, c$; see Fig. 12.10.

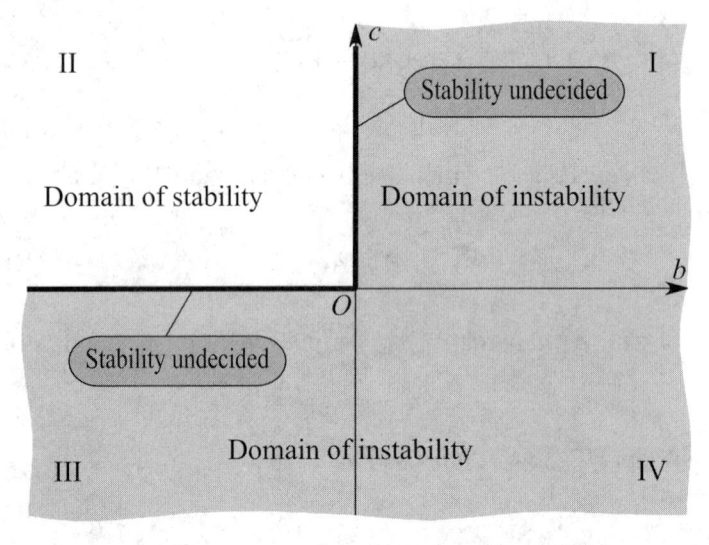

**Figure 12.10.** Domains of stability and instability of the trivial solution of system (12.7.3.11).

2. Using the theorem of instability by first approximation and examining the roots of the quadratic equation (12.7.3.12), we get three sufficient instability conditions for system (12.7.3.11):

$$b > 0, \quad 0 < \tfrac{1}{4}b^2 < c \quad \text{(complex roots with positive real part)};$$
$$b > 0, \quad 0 < c < \tfrac{1}{4}b^2 \quad \text{(positive real roots)};$$
$$c < 0, \quad b \text{ is any} \quad \text{(real roots with different signs)}.$$

The first two conditions can be combined into one:

$$b > 0, \quad c > 0, \quad \text{or} \quad a_{11} + a_{22} > 0, \quad a_{11}a_{22} - a_{12}a_{21} > 0.$$

The domain of instability of system (12.7.3.11) covers the first, third, and fourth quadrants in the plane $b, c$ (shaded in Fig. 12.10).

3. The conditions obtained above in Items 1 and 2 do not cover the whole domain of variation of the parameters $a_{ij}$. Stability or instability is not established for the boundary of the second quarter (shown by thick solid line in Fig. 12.10). This corresponds to the cases

$$b = 0, \quad c \geq 0 \quad \text{(two pure imaginary or two zero roots)};$$
$$c = 0, \quad b \leq 0 \quad \text{(one zero root and one negative real or zero root)}.$$

Specific examples of such systems are considered below in Paragraph 12.7.3-3.

**Remark.** When the conditions of Item 1 or 2 hold, the phase trajectories of the nonlinear system (12.7.3.11) have the same qualitative arrangement in a neighborhood of the equilibrium point $x = y = 0$ as that of the phase trajectories of the linearized system (with $\psi_2 = \psi_1 = 0$). A detailed classification of equilibrium points of linear systems with associated arrangements of the phase trajectories can be found in Paragraph 12.6.1-7.

**12.7.3-3. Lyapunov function. Theorems of stability and instability.**

In the cases where the theorems of stability and instability by first approximation fail to resolve the issue of stability for a specific system of nonlinear differential equations, more subtle methods must be used. Such methods are considered below.

A *Lyapunov function* for system of equations (12.7.3.1) is a differentiable function $V = V(x_1, \ldots, x_n)$ such that

1) $V > 0$ if $\sum_{k=1}^{n} x_k^2 \neq 0$, $V = 0$ if $x_1 = \cdots = x_n = 0$;

2) $\dfrac{dV}{dt} = \sum_{k=1}^{n} f_k(t, x_1, \ldots, x_n) \dfrac{\partial V}{\partial x_k} \leq 0$ for $t \geq 0$.

**Remark.** The derivative with respect to $t$ in the definition of a Lyapunov function is taken along an integral curve of system (12.7.3.1).

THEOREM (STABILITY, LYAPUNOV). *Let system (12.7.3.1) have the trivial solution $x_1 = x_2 = \cdots = x_n = 0$. This solution is stable if there exists a Lyapunov function for the system.*

THEOREM (ASYMPTOTIC STABILITY, LYAPUNOV). *Let system (12.7.3.1) have the trivial solution $x_1 = \cdots = x_n = 0$. This solution is asymptotically stable if there exists a Lyapunov function satisfying the additional condition*

$$\frac{dV}{dt} \leq -\beta < 0 \quad \text{with} \quad \sum_{k=1}^{n} x_k^2 \geq \varepsilon_1 > 0, \quad t \geq \varepsilon_2 \geq 0,$$

*where $\varepsilon_1$ and $\varepsilon_2$ are any positive numbers.*

**Example 2.** Let us perform a stability analysis of the two-dimensional system

$$x_t' = -ay - x\varphi(x, y), \quad y_t' = bx - y\psi(x, y),$$

where $a > 0$, $b > 0$, $\varphi(x, y) \geq 0$, and $\psi(x, y) \geq 0$ ($\varphi$ and $\psi$ are continuous functions).

A Lyapunov function will be sought in the form $V = Ax^2 + By^2$, where $A$ and $B$ are constants to be determined. The first condition characterizing a Lyapunov function will be satisfied automatically if $A > 0$ and $B > 0$ (it will be shown later that these inequalities do hold). To verify the second condition, let us compute the derivative:

$$\frac{dV}{dt} = f_1(x, y) \frac{\partial V}{\partial x} + f_2(x, y) \frac{\partial V}{\partial y} = -2Ax[ay + x\varphi(x, y)] + 2By[bx - y\psi(x, y)]$$

$$= 2(Bb - Aa)xy - 2Ax^2\varphi(x, y) - 2By^2\psi(x, y).$$

Setting here $A = b > 0$ and $B = a > 0$ (thus satisfying the first condition), we obtain the inequality

$$\frac{dV}{dt} = -2bx^2\varphi(x, y) - 2ay^2\psi(x, y) \leq 0.$$

This means that the second condition characterizing a Lyapunov function is also met. Hence, the trivial solution of the system in question is stable.

**Example 3.** Let us perform a stability analysis for the trivial solution of the nonlinear system

$$x_t' = -xy^2, \quad y_t' = yx^4.$$

Let us show that the $V(x, y) = x^4 + y^2$ is a Lyapunov function for the system. Indeed, both conditions are satisfied:

1) $x^4 + y^2 > 0$ if $x^2 + y^2 \neq 0$, $V(0, 0) = 0$ if $x = y = 0$;

2) $\dfrac{dV}{dt} = -4x^4y^2 + 2x^4y^2 = -2x^4y^2 \leq 0.$

Hence the trivial solution of the system is stable.

**Remark.** No stability analysis of the systems considered in Examples 2 and 3 is possible based on the theorem of stability by first approximation.

THEOREM (INSTABILITY, CHETAEV). *Suppose there exists a differentiable function* $W = W(x_1, \ldots, x_n)$ *that possesses the following properties:*

*1. In an arbitrarily small domain* $R$ *containing the origin of coordinates, there exists a subdomain* $R_+ \subset R$ *in which* $W > 0$, *with* $W = 0$ *on part of the boundary of* $R_+$ *in* $R$.

*2. The condition*

$$\frac{dW}{dt} = \sum_{k=1}^{n} f_k(t, x_1, \ldots, x_n) \frac{\partial W}{\partial x_k} > 0$$

*holds in* $R_+$ *and, moreover, in the domain of the variables where* $W \geq \alpha > 0$, *the inequality* $\dfrac{dW}{dt} \geq \beta > 0$ *holds.*

*Then the trivial solution* $x_1 = \cdots = x_n = 0$ *of system (12.7.3.1) is unstable.*

**Example 4.** Perform a stability analysis of the nonlinear system

$$x'_t = y^3 \varphi(x, y, t) + x^5, \qquad y'_t = x^3 \varphi(x, y, t) + y^5,$$

where $\varphi(x, y, t)$ is an arbitrary continuous function.

Let us show that the $W = x^4 - y^4$ satisfies the conditions of the Chetaev theorem. We have:
1. $W > 0$ for $|x| > |y|$, $W = 0$ for $|x| = |y|$.
2. $\dfrac{dW}{dt} = 4x^3[y^3 \varphi(x, y, t) + x^5] - 4y^3[x^3 \varphi(x, y, t) + y^5] = 4(x^8 - y^8) > 0$ for $|x| > |y|$.

Moreover, if $W \geq \alpha > 0$, we have $\dfrac{dW}{dt} = 4\alpha(x^4 + y^4) \geq 4\alpha^2 = \beta > 0$. It follows that the equilibrium point $x = y = 0$ of the system in question is unstable.

# References for Chapter 12

**Akulenko, L. D. and Nesterov, S. V.,** *High Precision Methods in Eigenvalue Problems and Their Applications*, Chapman & Hall/CRC Press, Boca Raton, 2005.

**Arnold, V. I., Kozlov, V. V., and Neishtadt, A. I.,** *Mathematical Aspects of Classical and Celestial Mechanics, Dynamical System III*, Springer-Verlag, Berlin, 1993.

**Bakhvalov, N. S.,** *Numerical Methods: Analysis, Algebra, Ordinary Differential Equations*, Mir Publishers, Moscow, 1977.

**Bogolyubov, N. N. and Mitropol'skii, Yu. A.,** *Asymptotic Methods in the Theory of Nonlinear Oscillations* [in Russian], Nauka Publishers, Moscow, 1974.

**Boyce, W. E. and DiPrima, R. C.,** *Elementary Differential Equations and Boundary Value Problems*, 8th Edition, John Wiley & Sons, New York, 2004.

**Braun, M.,** *Differential Equations and Their Applications*, 4th Edition, Springer-Verlag, New York, 1993.

**Cole, G. D.,** *Perturbation Methods in Applied Mathematics*, Blaisdell Publishing Company, Waltham, MA, 1968.

**Dormand, J. R.,** *Numerical Methods for Differential Equations: A Computational Approach*, CRC Press, Boca Raton, 1996.

**El'sgol'ts, L. E.,** *Differential Equations*, Gordon & Breach, New York, 1961.

**Fedoryuk, M. V.,** *Asymptotic Analysis. Linear Ordinary Differential Equations*, Springer-Verlag, Berlin, 1993.

**Finlayson, B. A.,** *The Method of Weighted Residuals and Variational Principles*, Academic Press, New York, 1972.

**Grimshaw, R.,** *Nonlinear Ordinary Differential Equations*, CRC Press, Boca Raton, 1991.

**Gromak, V. I.,** *Painlevé Differential Equations in the Complex Plane*, Walter de Gruyter, Berlin, 2002.

**Gromak, V. I. and Lukashevich, N. A.,** *Analytical Properties of Solutions of Painlevé Equations* [in Russian], Universitetskoe, Minsk, 1990.

**Ince, E. L.,** *Ordinary Differential Equations*, Dover Publications, New York, 1964.

**Kamke, E.,** *Differentialgleichungen: Lösungsmethoden und Lösungen, I, Gewöhnliche Differentialgleichungen*, B. G. Teubner, Leipzig, 1977.

**Kantorovich, L. V. and Krylov, V. I.,** *Approximate Methods of Higher Analysis* [in Russian], Fizmatgiz, Moscow, 1962.

**Keller, H. B.**, *Numerical Solutions of Two Point Boundary Value Problems*, Society for Industrial & Applied Mathematics, Philadelphia, 1976.

**Kevorkian, J. and Cole, J. D.**, *Multiple Scale and Singular Perturbation Methods*, Springer-Verlag, New York, 1996.

**Korn, G. A. and Korn, T. M.**, *Mathematical Handbook for Scientists and Engineers, 2nd Edition*, Dover Publications, New York, 2000.

**Lambert, J. D.**, *Computational Methods in Ordinary Differential Equations*, Cambridge University Press, New York, 1973.

**Lee, H. J. and Schiesser, W. E.**, *Ordinary and Partial Differential Equation Routines in C, C++, Fortran, Java, Maple, and MATLAB*, Chapman & Hall/CRC Press, Boca Raton, 2004.

**Levitan, B. M. and Sargsjan, I. S.**, *Sturm–Liouville and Dirac Operators*, Kluwer Academic, Dordrecht, 1990.

**Marchenko, V. A.**, *Sturm–Liouville Operators and Applications*, Birkhäuser Verlag, Basel, 1986.

**Murphy, G. M.**, *Ordinary Differential Equations and Their Solutions*, D. Van Nostrand, New York, 1960.

**Nayfeh, A. H.**, *Introduction to Perturbation Techniques*, John Wiley & Sons, New York, 1981.

**Nayfeh, A. H.**, *Perturbation Methods*, Wiley-Interscience, New York, 1973.

**Petrovskii, I. G.**, *Lectures on the Theory of Ordinary Differential Equations* [in Russian], Nauka Publishers, Moscow, 1970.

**Polyanin, A. D. and Zaitsev, V. F.**, *Handbook of Exact Solutions for Ordinary Differential Equations, 2nd Edition*, Chapman & Hall/CRC Press, Boca Raton, 2003.

**Schiesser, W. E.**, *Computational Mathematics in Engineering and Applied Science: ODEs, DAEs, and PDEs*, CRC Press, Boca Raton, 1993.

**Tenenbaum, M. and Pollard, H.**, *Ordinary Differential Equations*, Dover Publications, New York, 1985.

**Wasov, W.**, *Asymptotic Expansions for Ordinary Differential Equations*, John Wiley & Sons, New York, 1965.

**Zhuravlev, V. Ph. and Klimov, D. M.**, *Applied Methods in Oscillation Theory* [in Russian], Nauka Publishers, Moscow, 1988.

**Zwillinger, D.**, *Handbook of Differential Equations, 3rd Edition*, Academic Press, New York, 1997.

# Chapter 13
# First-Order Partial Differential Equations

## 13.1. Linear and Quasilinear Equations
### 13.1.1. Characteristic System. General Solution

**13.1.1-1. Equations with two independent variables. General solution. Examples.**

$1°$. A *first-order quasilinear partial differential equation with two independent variables* has the general form

$$f(x,y,w)\frac{\partial w}{\partial x} + g(x,y,w)\frac{\partial w}{\partial y} = h(x,y,w). \tag{13.1.1.1}$$

Such equations are encountered in various applications (continuum mechanics, gas dynamics, hydrodynamics, heat and mass transfer, wave theory, acoustics, multiphase flows, chemical engineering, etc.).

If two independent integrals,

$$u_1(x,y,w) = C_1, \qquad u_2(x,y,w) = C_2, \tag{13.1.1.2}$$

of the *characteristic system*

$$\frac{dx}{f(x,y,w)} = \frac{dy}{g(x,y,w)} = \frac{dw}{h(x,y,w)} \tag{13.1.1.3}$$

are known, then the *general solution* of equation (13.1.1.1) is given by

$$\Phi(u_1, u_2) = 0, \tag{13.1.1.4}$$

where $\Phi(u,v)$ is an arbitrary function of two variables. With equation (13.1.1.4) solved for $u_1$ or $u_2$, we often specify the general solution in the form

$$u_k = \Psi(u_{3-k}),$$

where $k = 1, 2$ and $\Psi(u)$ is an arbitrary function of one variable.

$2°$. For linear equations (13.1.1.1) with the functions $f$, $g$, and $h$ independent of the unknown $w$, the first integrals (13.1.1.2) of the characteristic system (13.1.1.3) have a simple structure (one integral is independent of $w$ and the other is linear in $w$):

$$U(x,y) = C_1, \qquad w - V(x,y) = C_2.$$

In this case the general solution can be written in explicit form

$$w = V(x,y) + \Psi(U(x,y)),$$

where $\Psi(U)$ is an arbitrary function of one variable.

## TABLE 13.1
General solutions to some special types of linear and quasilinear first-order partial differential equations; $\Psi(u)$ is an arbitrary function. The subscripts $x$ and $y$ indicate the corresponding partial derivatives

| No. | Equations | General solutions | Notation, remarks |
|---|---|---|---|
| 1 | $w_x + [f(x)y + g(x)]w_y = 0$ | $w = \Psi\left(e^{-F}y - \int e^{-F}g(x)\,dx\right)$ | $F = \int f(x)\,dx$ |
| 2 | $w_x + [f(x)y + g(x)y^k]w_y = 0$ | $w = \Psi\left(e^{-F}y^{1-k} - (1-k)\int e^{-F}g(x)\,dx\right)$ | $F = (1-k)\int f(x)\,dx$ |
| 3 | $w_x + [f(x)e^{\lambda y} + g(x)]w_y = 0$ | $w = \Psi\left(e^{-\lambda y}E + \lambda \int f(x)E\,dx\right)$ | $E = \exp\left(\lambda \int g(x)\,dx\right)$ |
| 4 | $f(x)w_x + g(y)w_y = 0$ | $w = \Psi\left(\int \frac{dx}{f(x)} - \int \frac{dy}{g(y)}\right)$ | |
| 5 | $aw_x + bw_y = f(x)g(w)$ | $\int \frac{dw}{g(w)} = \frac{1}{a}\int f(x)\,dx + \Psi(bx - ay)$ | solution in implicit form |
| 6 | $f(x)w_x + g(y)w_y = h(w)$ | $\int \frac{dw}{h(w)} = \int \frac{dx}{f(x)} + \Psi(u)$ | $u = \int \frac{dx}{f(x)} - \int \frac{dy}{g(y)}$ |
| 7 | $w_x + f(w)w_y = 0$ | $y = xf(w) + \Psi(w)$ | solution in implicit form |
| 8 | $w_x + [f(w) + ay]w_y = 0$ | $x = \frac{1}{a}\ln\bigl|ay + f(w)\bigr| + \Psi(w),\ a \neq 0$ | solution in implicit form |
| 9 | $w_x + [f(w) + g(x)]w_y = 0$ | $y = xf(w) + \int g(x)\,dx + \Psi(w)$ | solution in implicit form |

**Example 1.** Consider the linear constant coefficient equation

$$\frac{\partial w}{\partial x} + a\frac{\partial w}{\partial y} = 0.$$

The characteristic system for this equation is

$$\frac{dx}{1} = \frac{dy}{a} = \frac{dw}{0}.$$

It has two independent integrals:

$$y - ax = C_1, \quad w = C_2.$$

Hence, the general solution of the original equation is given by $\Phi(y - ax, w) = 0$. On solving this equation for $w$, one obtains the general solution in explicit form

$$w = \Psi(y - ax).$$

It is the traveling wave solution.

**Example 2.** Consider the quasilinear equation

$$\frac{\partial w}{\partial x} + aw\frac{\partial w}{\partial y} = 1.$$

The characteristic system

$$\frac{dx}{1} = \frac{dy}{aw} = \frac{dw}{1}$$

has two independent integrals:

$$x - w = C_1, \quad 2y - aw^2 = C_2.$$

Hence, the general solution of the original equation is given by

$$\Phi(x - w, 2y - aw^2) = 0.$$

3°. Table 13.1 lists general solutions to some linear and quasilinear first-order partial differential equations in two independent variables.

▶ In Sections T7.1–T7.2, many more first-order linear and quasilinear partial differential equations in two independent variables are considered than in Table 13.1.

### 13.1.1-2. Construction of a quasilinear equations when given its general solution.

Given a set of functions
$$w = F\bigl(x, y, \Psi(G(x, y))\bigr), \tag{13.1.1.5}$$
where $F(x, y, \Psi)$ and $G(x, y)$ are prescribed and $\Psi(G)$ is arbitrary, there exists a quasilinear first-order partial differential equation such that the set of functions (13.1.1.5) is its general solution. To prove this statement, let us differentiate (13.1.1.5) with respect to $x$ and $y$ and then eliminate the partial derivative $\Psi_G$ from the resulting expression to obtain

$$\frac{w_x - F_x}{G_x} = \frac{w_y - F_y}{G_y}. \tag{13.1.1.6}$$

On solving the relation $w = F(x, y, \Psi)$ [see (13.1.1.5)] for $\Psi$ and substituting the resulting expression into (13.1.1.6), one arrives at the desired partial differentiable equation.

**Example 3.** Let us construct a partial differential equation whose general solution is given by
$$w = x^k \Psi(ax^n + by^m), \tag{13.1.1.7}$$
where $\Psi(z)$ is an arbitrary function.

Differentiating (13.1.1.7) with respect to $x$ and $y$ yields the relations $w_x = kx^{k-1}\Psi + anx^{k+n-1}\Psi_z$ and $w_y = bmx^k y^{m-1} \Psi_z$. Eliminating $\Psi_z$ from them gives

$$\frac{w_x - kx^{k-1}\Psi}{anx^{n-1}} = \frac{w_y}{bmy^{m-1}}. \tag{13.1.1.8}$$

Solving the original relation (13.1.1.7) for $\Psi$, we get $\Psi = x^{-k}w$. Substituting this expression into (13.1.1.8) and rearranging, we arrive at the desired equation

$$bmxy^{m-1}\frac{\partial w}{\partial x} - anx^n \frac{\partial w}{\partial y} = bkmy^{m-1},$$

whose general solution is the function (13.1.1.7).

### 13.1.1-3. Equations with $n$ independent variables. General solution.

A first-order quasilinear partial differential equation with $n$ independent variables has the general form

$$f_1(x_1, \ldots, x_n, w)\frac{\partial w}{\partial x_1} + \cdots + f_n(x_1, \ldots, x_n, w)\frac{\partial w}{\partial x_n} = g(x_1, \ldots, x_n, w). \tag{13.1.1.9}$$

Let $n$ independent integrals,
$$u_1(x_1, \ldots, x_n, w) = C_1, \quad \ldots, \quad u_n(x_1, \ldots, x_n, w) = C_n,$$
of the characteristic system
$$\frac{dx_1}{f_1(x_1, \ldots, x_n, w)} = \cdots = \frac{dx_n}{f_n(x_1, \ldots, x_n, w)} = \frac{dw}{g(x_1, \ldots, x_n, w)}$$
be known. Then the general solution of equation (13.1.1.9) is given by
$$\Phi(u_1, \ldots, u_n) = 0,$$
where $\Phi$ is an arbitrary function of $n$ variables.

## 13.1.2. Cauchy Problem. Existence and Uniqueness Theorem

**13.1.2-1. Cauchy problem.**

Consider two formulations of the Cauchy problem.

1°. *Generalized Cauchy problem.* Find a solution $w = w(x, y)$ of equation (13.1.1.1) satisfying the initial conditions

$$x = h_1(\xi), \quad y = h_2(\xi), \quad w = h_3(\xi), \qquad (13.1.2.1)$$

where $\xi$ is a parameter ($\alpha \le \xi \le \beta$) and the $h_k(\xi)$ are given functions.

Geometric interpretation: find an integral surface of equation (13.1.1.1) passing through the line defined parametrically by equation (13.1.2.1).

2°. *Classical Cauchy problem.* Find a solution $w = w(x, y)$ of equation (13.1.1.1) satisfying the initial condition

$$w = \varphi(y) \quad \text{at} \quad x = 0, \qquad (13.1.2.2)$$

where $\varphi(y)$ is a given function.

It is convenient to represent the classical Cauchy problem as a generalized Cauchy problem by rewriting condition (13.1.2.2) in the parametric form

$$x = 0, \quad y = \xi, \quad w = \varphi(\xi). \qquad (13.1.2.3)$$

**13.1.2-2. Procedure of solving the Cauchy problem.**

The procedure of solving the Cauchy problem (13.1.1.1), (13.1.2.1) involves several steps. First, two independent integrals (13.1.1.2) of the characteristic system (13.1.1.3) are determined. Then, to find the constants of integration $C_1$ and $C_2$, the initial data (13.1.2.1) must be substituted into the integrals (13.1.1.2) to obtain

$$u_1\bigl(h_1(\xi), h_2(\xi), h_3(\xi)\bigr) = C_1, \qquad u_2\bigl(h_1(\xi), h_2(\xi), h_3(\xi)\bigr) = C_2. \qquad (13.1.2.4)$$

Eliminating $C_1$ and $C_2$ from (13.1.1.2) and (13.1.2.4) yields

$$\begin{aligned} u_1(x, y, w) &= u_1\bigl(h_1(\xi), h_2(\xi), h_3(\xi)\bigr), \\ u_2(x, y, w) &= u_2\bigl(h_1(\xi), h_2(\xi), h_3(\xi)\bigr). \end{aligned} \qquad (13.1.2.5)$$

Formulas (13.1.2.5) are a parametric form of the solution of the Cauchy problem (13.1.1.1), (13.1.2.1). In some cases, one may succeed in eliminating the parameter $\xi$ from relations (13.1.2.5), thus obtaining the solution in an explicit form.

**Example 1.** Consider the Cauchy problem for linear equation

$$\frac{\partial w}{\partial x} + a\frac{\partial w}{\partial y} = bw \qquad (13.1.2.6)$$

subjected to the initial condition (13.1.2.2).

The corresponding characteristic system for equation (13.1.2.6),

$$\frac{dx}{1} = \frac{dy}{a} = \frac{dw}{bw},$$

has two independent integrals

$$y - ax = C_1, \qquad we^{-bx} = C_2. \qquad (13.1.2.7)$$

Represent the initial condition (13.1.2.2) in parametric form (13.1.2.3) and then substitute the data (13.1.2.3) into the integrals (13.1.2.7). As a result, for the constants of integration we obtain $C_1 = \xi$ and $C_2 = \varphi(\xi)$. Substituting these expressions into (13.1.2.7), we arrive at the solution of the Cauchy problem (13.1.2.6), (13.1.2.2) in parametric form:

$$y - ax = \xi, \qquad we^{-bx} = \varphi(\xi).$$

By eliminating the parameter $\xi$ from these relations, we obtain the solution of the Cauchy problem (13.1.2.6), (13.1.2.2) in explicit form:

$$w = e^{bx}\varphi(y - ax).$$

### 13.1.2-3. Existence and uniqueness theorem.

Let $\mathfrak{G}_0$ be a domain in the $xy$-plane and let $\mathfrak{G}$ be a cylindrical domain of the $xyw$-space obtained from $\mathfrak{G}_0$ by adding the coordinate $w$, with the condition $|w| < A_1$ being satisfied. Let the coefficients $f$, $g$, and $h$ of equation (13.1.1.1) be continuously differentiable functions of $x$, $y$, and $w$ in $\mathfrak{G}$ and let $x = h_1(\xi)$, $y = h_2(\xi)$, and $w = h_3(\xi)$ be continuously differentiable functions of $\xi$ for $|\xi| < A_2$ defining a curve $C$ in $\mathfrak{G}$ with a simple projection $C_0$ onto $\mathfrak{G}_0$. Suppose that $(h'_1)^2 + (h'_2)^2 \neq 0$ (the prime stands for the derivative with respect to $\xi$) and $fh'_2 - gh'_1 \neq 0$ on $C$. Then there exists a subdomain $\overline{\mathfrak{G}}_0 \subset \mathfrak{G}_0$ containing $C_0$ where there exists a continuously differentiable function $w = w(x,y)$ satisfying the differential equation (13.1.1.1) in $\overline{\mathfrak{G}}_0$ and the initial condition (13.1.2.1) on $C_0$. This function is unique.

It is important to note that this theorem has a local character, i.e., the existence of a solution is guaranteed in some "sufficiently narrow," unknown neighborhood of the line $C$ (see the remark at the end of Example 2).

**Example 2.** Consider the Cauchy problem for Hopf's equation

$$\frac{\partial w}{\partial x} + w \frac{\partial w}{\partial y} = 0 \qquad (13.1.2.8)$$

subject to the initial condition (13.1.2.2).

First, we rewrite the initial condition (13.1.2.2) in the parametric form (13.1.2.3). Solving the characteristic system

$$\frac{dx}{1} = \frac{dy}{w} = \frac{dw}{0}, \qquad (13.1.2.9)$$

we find two independent integrals,

$$w = C_1, \quad y - wx = C_2. \qquad (13.1.2.10)$$

Using the initial conditions (13.1.2.3), we find that $C_1 = \varphi(\xi)$ and $C_2 = \xi$. Substituting these expressions into (13.1.2.10) yields the solution of the Cauchy problem (13.1.2.8), (13.1.2.2) in the parametric form

$$w = \varphi(\xi), \qquad (13.1.2.11)$$
$$y = \xi + \varphi(\xi)x. \qquad (13.1.2.12)$$

The characteristics (13.1.2.12) are straight lines in the $xy$-plane with slope $\varphi(\xi)$ that intersect the $y$-axis at the points $\xi$. On each characteristic, the function $w$ has the same value equal to $\varphi(\xi)$ (generally, $w$ takes different values on different characteristics).

For $\varphi'(\xi) > 0$, different characteristics do not intersect and, hence, formulas (13.1.2.11) and (13.1.2.12) define a unique solution. As an example, we consider the initial profile

$$\varphi(\xi) = \begin{cases} w_1 & \text{for } \xi \leq 0, \\ \dfrac{w_2 \xi^2 + \varepsilon w_1}{\xi^2 + \varepsilon} & \text{for } \xi > 0, \end{cases} \qquad (13.1.2.13)$$

where $w_1 < w_2$ and $\varepsilon > 0$. Formulas (13.1.2.11)–(13.1.2.13) give a unique smooth solution in the entire half-plane $x > 0$. In the domain filled by the characteristics $y = \xi + w_1 x$ (for $\xi \leq 0$), the solution is constant, i.e.,

$$w = w_1 \quad \text{for} \quad y/x \leq w_1. \qquad (13.1.2.14)$$

For $\xi > 0$, the solution is determined by relations (13.1.2.11)–(13.1.2.13).

Let us look how this solution is transformed in the limit case $\varepsilon \to 0$, which corresponds to the piecewise-continuous initial profile

$$\varphi(\xi) = \begin{cases} w_1 & \text{for } \xi \leq 0, \\ w_2 & \text{for } \xi > 0, \end{cases} \quad \text{where} \quad w_1 < w_2. \qquad (13.1.2.15)$$

We further assume that $\xi > 0$ [for $\xi \leq 0$, formula (13.1.2.14) is valid]. If $\xi = \text{const} \neq 0$ and $\varepsilon \to 0$, it follows from (13.1.2.13) that $\varphi(\xi) = w_2$. Hence, in the domain filled by the characteristics $y = \xi + w_2 x$ (for $\xi > 0$), the solution is constant, i.e., we have

$$w = w_2 \quad \text{for} \quad y/x \geq w_2 \quad (\text{as } \varepsilon \to 0). \qquad (13.1.2.16)$$

For $\xi \to 0$, the function $\varphi$ can assume any value between $w_1$ and $w_2$ depending on the ratio of the small parameters $\varepsilon$ and $\xi$; the first term on the right-hand side of equation (13.1.2.12) can be neglected. As a result,

we find from equations (13.1.2.11) and (13.1.2.12) that the solution has the following asymptotic behavior in explicit form:
$$w = y/x \quad \text{for} \quad w_1 \le y/x \le w_2 \quad (\text{as } \varepsilon \to 0). \tag{13.1.2.17}$$

By combining relations (13.1.2.14), (13.1.2.16), and (13.1.2.17) together, we obtain the solution of the Cauchy problem for equation (13.1.2.8) subject to the initial conditions (13.1.2.15) in the form

$$w(x, y) = \begin{cases} w_1 & \text{for } y \le w_1 x, \\ y/x & \text{for } w_1 x \le y \le w_2 x, \\ w_2 & \text{for } y \ge w_2 x. \end{cases} \tag{13.1.2.18}$$

Figure 13.1 shows characteristics of equation (13.1.2.8) that satisfy condition (13.1.2.15) with $w_1 = \frac{1}{2}$ and $w_2 = 2$. This figure also depicts the dependence of $w$ on $y$ (for $x = x_0 = 1$). In applications, such a solution is referred to as a centered rarefaction wave (see also Subsection 13.1.3).

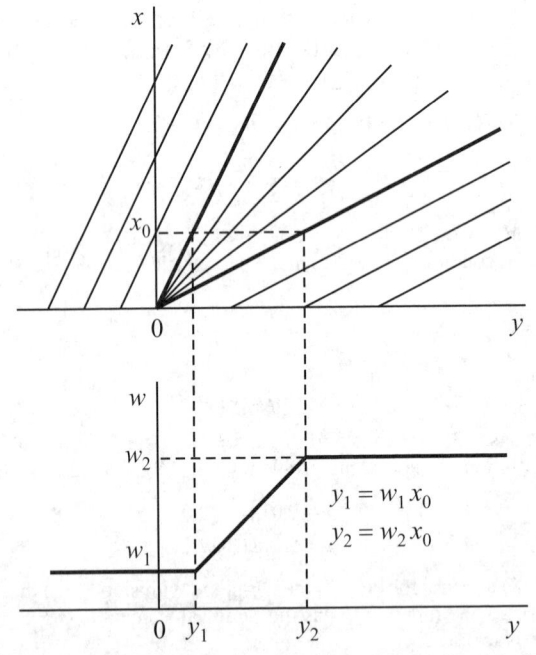

**Figure 13.1.** Characteristics of the Cauchy problem (13.1.2.8), (13.1.2.2) with the initial profile (13.1.2.15) and the dependence of the unknown $w$ on the coordinate $y$ for $w_1 = \frac{1}{2}$, $w_2 = 2$, and $x_0 = 1$.

**Remark.** If there is an interval where $\varphi'(\xi) < 0$, then the characteristics intersect in some domain. According to equation (13.1.2.11), at the point of intersection of two characteristics defined by two distinct values $\xi_1$ and $\xi_2$ of the parameter, the function $w$ takes two distinct values equal to $\varphi(\xi_1)$ and $\varphi(\xi_2)$, respectively. Therefore, the solution is not unique in the domain of intersecting characteristics. This example illustrates the local character of the existence and uniqueness theorem. These issues are discussed in Subsections 13.1.3 and 13.1.4 in more detail.

### 13.1.3. Qualitative Features and Discontinuous Solutions of Quasilinear Equations

#### 13.1.3-1. Model equation of gas dynamics.

Consider a quasilinear equation of the special form*

$$\frac{\partial w}{\partial x} + f(w)\frac{\partial w}{\partial y} = 0, \tag{13.1.3.1}$$

---

* Equations of the general form are discussed in Subsection 13.1.4.

## 13.1. LINEAR AND QUASILINEAR EQUATIONS

which represents a conservation law of mass (or another quantity) and is often encountered in continuum mechanics, gas dynamics, hydrodynamics, wave theory, acoustics, multiphase flows, and chemical engineering. This equation is a model for numerous processes of mass transfer: sorption and chromatography, two-phase flows in porous media, flow of water in river, street traffic development, flow of liquid films along inclined surfaces, etc. The independent variables $x$ and $y$ in equation (13.1.3.1) usually play the role of time and spatial coordinate, respectively, $w$ is the density of the quantity being transferred, and $f(w)$ is the rate of $w$.

**13.1.3-2. Solution of the Cauchy problem. Rarefaction wave. Wave "overturn."**

$1°$. The solution $w = w(x, y)$ of the Cauchy problem for equation (13.1.3.1) subject to the initial condition

$$w = \varphi(y) \quad \text{at} \quad x = 0 \quad (-\infty < y < \infty) \tag{13.1.3.2}$$

can be represented in the parametric form

$$y = \xi + \mathcal{F}(\xi)x, \quad w = \varphi(\xi), \tag{13.1.3.3}$$

where $\mathcal{F}(\xi) = f(\varphi(\xi))$.

Consider the characteristics $y = \xi + \mathcal{F}(\xi)x$ in the $yx$-plane for various values of the parameter $\xi$. These are straight lines with slope $\mathcal{F}(\xi)$. Along each of these lines, the unknown function is constant, $w = \varphi(\xi)$. In the special case $f = a = \text{const}$, the equation in question is linear; solution (13.1.3.3) can be written explicitly as $w = \varphi(y - ax)$, thus representing a traveling wave with a fixed profile. The dependence of $f$ on $w$ leads to a typical nonlinear effect: distortion of the profile of the traveling wave.

We further consider the domain $x \geq 0$ and assume* that $f > 0$ for $w > 0$ and $f'_w > 0$. In this case, the greater values of $w$ propagate faster than the smaller values. If the initial profile satisfies the condition $\varphi'(y) > 0$ for all $y$, then the characteristics in the $yx$-plane that come from the $y$-axis inside the domain $x > 0$ are divergent lines, and hence there exists a unique solution for all $x > 0$. In physics, such solutions are referred to as rarefaction waves.

**Example 1.** Figures 13.2 and 13.3 illustrate characteristics and the evolution of a rarefaction wave for Hopf's equation [for $f(w) = w$ in (13.1.3.1)] with the initial profile

$$\varphi(y) = \frac{4}{\pi} \arctan(y - 2) + 2. \tag{13.1.3.4}$$

It is apparent that the solution is smooth for all $x > 0$.

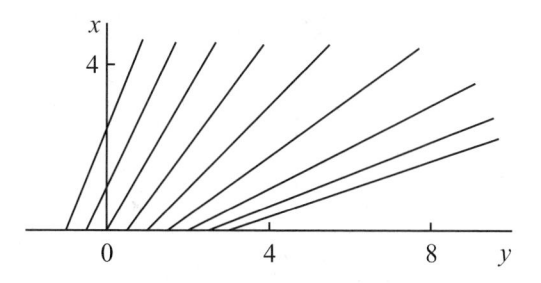

**Figure 13.2.** Characteristics for the Hopf's equation (13.1.2.8) with the initial profile (13.1.3.4).

---

* By the change $x = -\widetilde{x}$ the consideration of the domain $x \leq 0$ can be reduced to that of the domain $\widetilde{x} \geq 0$. The case $f < 0$ can be reduced to the case $f > 0$ by the change $y = -\widetilde{y}$.

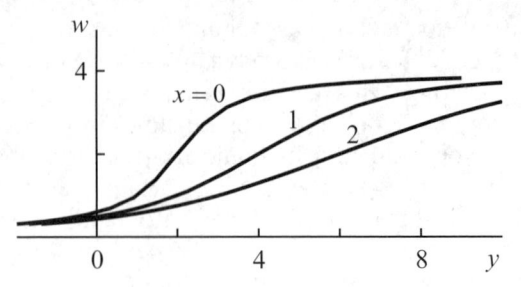

**Figure 13.3.** The evolution of a rarefaction wave for the Hopf's equation (13.1.2.8) with the initial profile (13.1.3.4).

2°. Let us now look at what happens if $\varphi'(y) < 0$ on some interval of the $y$-axis. Let $y_1$ and $y_2$ be points of this interval such that $y_1 < y_2$. Then $f(y_1) > f(y_2)$. It follows from the first relation in (13.1.3.3) that the characteristics issuing from the points $y_1$ and $y_2$ intersect at the "time instant"

$$x_* = \frac{y_2 - y_1}{f(w_1) - f(w_2)}, \qquad \text{where} \quad w_1 = \varphi(y_1), \quad w_2 = \varphi(y_2).$$

Since $w$ has different values on these characteristics, the solution cannot be continuously extended to $x > x_*$. If $\varphi'(y) < 0$ on a bounded interval, then there exists $x_{\min} = \min_{y_1, y_2} x_*$ such that the characteristics intersect in the domain $x > x_{\min}$ (see Fig. 13.4). Therefore, the front part of the wave where its profile is a decreasing function of $y$ will "overturn" with time. The time $x_{\min}$ when the overturning begins is defined by

$$x_{\min} = -\frac{1}{\mathcal{F}'(\xi_0)},$$

where $\xi_0$ is determined by the condition $|\mathcal{F}'(\xi_0)| = \max |\mathcal{F}'(\xi)|$ for $\mathcal{F}'(\xi) < 0$, and the wave is also said to *break*. A formal extension of the solution to the domain $x > x_{\min}$ makes this solution nonunique. The boundary of the uniqueness domain in the $yx$-plane is the envelope of the characteristics. This boundary can be represented in parametric form as

$$y = \xi + \mathcal{F}(\xi)x, \qquad 0 = 1 + \mathcal{F}'(\xi)x.$$

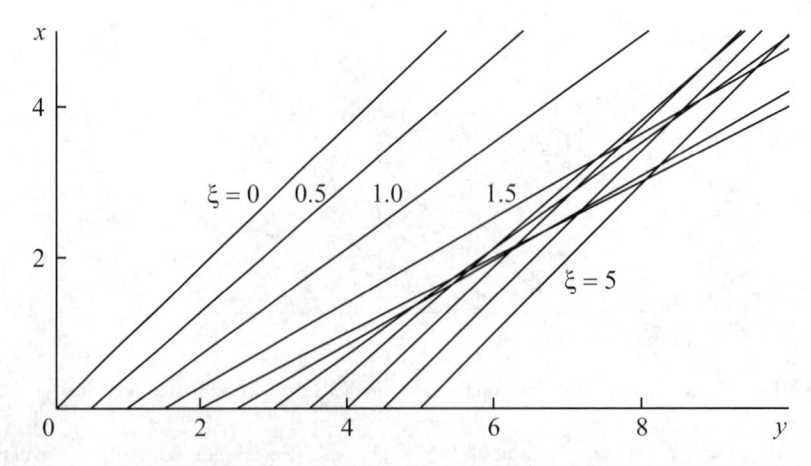

**Figure 13.4.** Characteristics for the Hopf's equation (13.1.2.8) with the initial profile (13.1.3.5).

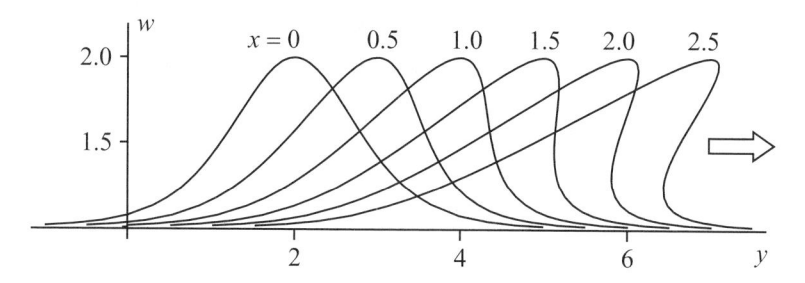

**Figure 13.5.** The evolution of a solitary wave for the Hopf's equation (13.1.2.8) with the initial profile (13.1.3.5).

**Example 2.** Figure 13.5 illustrates the evolution of a solitary wave with the initial profile

$$\varphi(y) = \cosh^{-2}(y-2) + 1 \qquad (13.1.3.5)$$

for equation (13.1.3.1) with $f(w) = w$. It is apparent that for $x > x_{\min}$, where $x_{\min} = \frac{3}{4}\sqrt{3} \approx 1.3$, the wave "overturns" (the wave profile becomes triple-valued).

### 13.1.3-3. Shock waves. Jump conditions.

In most applications where the equation under consideration is encountered, the unknown function $w(x, y)$ is the density of a medium and must be unique for its nature. In these cases, one has to deal with a generalized (nonsmooth) solution describing a step-shaped shock wave rather than a continuous smooth solution. The many-valued part of the wave profile is replaced by an appropriate discontinuity, as shown in Fig. 13.6. It should be emphasized that a discontinuity can occur for arbitrarily smooth functions $f(w)$ and $\varphi(y)$ entering equation (13.1.3.1) and the initial condition (13.1.3.2).

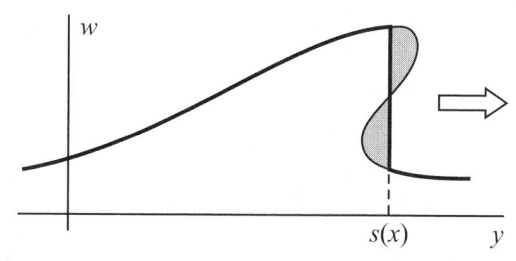

**Figure 13.6.** Replacement of the many-valued part of the wave profile by a discontinuity that cuts off domains with equal areas (shaded) from the profile of a breaking wave.

In what follows, we assume that $w(x, y)$ experiences a jump discontinuity at the line $y = s(x)$ in the $yx$-plane. On both sides of the discontinuity the function $w(x, y)$ is smooth and single-valued; as before, it is described by equations (13.1.3.3). The speed of propagation of the discontinuity, $V$, is expressed as $V = s'$ (the prime stands for the derivative) and must satisfy the condition

$$V = \frac{F(w_2) - F(w_1)}{w_2 - w_1}, \qquad F(w) = \int f(w)\, dw, \qquad (13.1.3.6)$$

where the subscript 1 refers to values before the discontinuity and the subscript 2 to those after the discontinuity. In applications, relation (13.1.3.6), expressing a conservation law at discontinuity, is conventionally referred to as the Rankine–Hugoniot jump condition (this condition is derived below in Paragraph 13.1.3-4).

The continuous wave "overturns" (breaks), thus resulting in a discontinuity if and only if the propagation velocity $f(w)$ decreases as $y$ increases, i.e., the inequalities

$$f(w_2) < V < f(w_1) \tag{13.1.3.7}$$

are satisfied. Conditions (13.1.3.7) have the geometric meaning that the characteristics issuing from the $x$-axis (these characteristics "carry" information about the initial data) must intersect the line of discontinuity (see Fig. 13.7). In this case, the discontinuous solution is stable with respect to small perturbations of the initial profile (i.e., the corresponding solution varies only slightly).

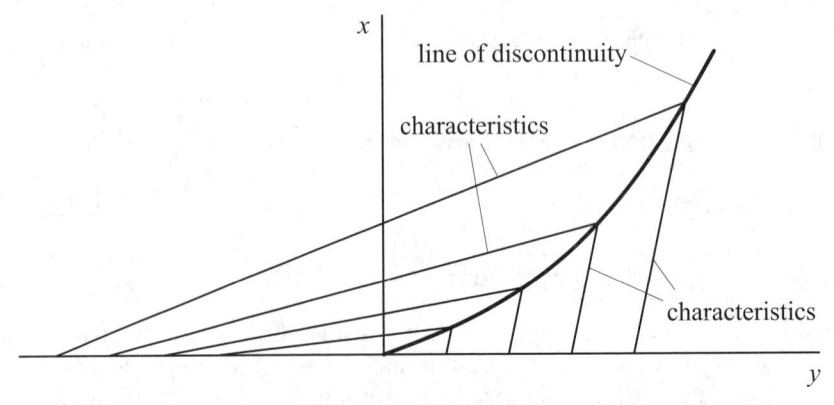

**Figure 13.7.** Mutual arrangement of characteristics and lines of discontinuity in the case of a stable shock wave.

The position of the point of discontinuity in the $yw$-plane may be determined geometrically by following Whitham's rule: the discontinuity must cut off domains with equal areas from the overturning wave profile (these domains are shaded in Fig. 13.6). Mathematically, the position of the point of discontinuity can be determined from the equations

$$\begin{aligned} s(x) &= \xi_1 + \mathcal{F}_1 x, \\ s(x) &= \xi_2 + \mathcal{F}_2 x, \\ w_2 \mathcal{F}_2 - w_1 \mathcal{F}_1 &= F(w_2) - F(w_1) + \frac{\mathcal{F}_2 - \mathcal{F}_1}{\xi_2 - \xi_1} \int_{\xi_1}^{\xi_2} w \, d\xi. \end{aligned} \tag{13.1.3.8}$$

Here, $w$ and $\mathcal{F}$ are defined as functions of $\xi$ by $w = \varphi(\xi)$ and $\mathcal{F} = f(w)$, the function $F(w)$ is introduced in equation (13.1.3.6), and the subscripts 1 and 2 refer to the values of the corresponding quantities at $\xi = \xi_1$ and $\xi_2$. Equations (13.1.3.8) permit one to determine the dependences $s = s(x)$, $\xi_1 = \xi_1(x)$, and $\xi_2 = \xi_2(x)$. It is possible to show that the jump condition (13.1.3.6) follows from the last equation in (13.1.3.8).

**Example 3.** For Hopf's equation, which corresponds to $f(w) = w$ in equation (13.1.3.1), the jump condition (13.1.3.6) can be represented as

$$V = \frac{w_1 + w_2}{2}.$$

Here, we take into account the relation $F(w) = \frac{1}{2} w^2$. System (13.1.3.8), which determines the position of the point of discontinuity, becomes

$$\begin{aligned} s(x) &= \xi_1 + \varphi(\xi_1) x, \\ s(x) &= \xi_2 + \varphi(\xi_2) x, \\ \frac{\varphi(\xi_1) + \varphi(\xi_2)}{2} &= \frac{1}{\xi_2 - \xi_1} \int_{\xi_1}^{\xi_2} \varphi(\xi) \, d\xi, \end{aligned}$$

where the function $\varphi(\xi)$ specifies the initial wave profile.

Figure 13.8 illustrates the formation of the shock wave described by the generalized solution of Hopf's equation with $f(w) = w$ and generated from a solitary wave with the smooth initial profile (13.1.3.5).

The nonsmooth "step" curves depicted in Fig. 13.8 (for $x = 0.15$, $0.20$, and $0.25$) are obtained from the smooth (but many-valued) curves shown in Fig. 13.5 by means of Whitham's rule of equal areas.

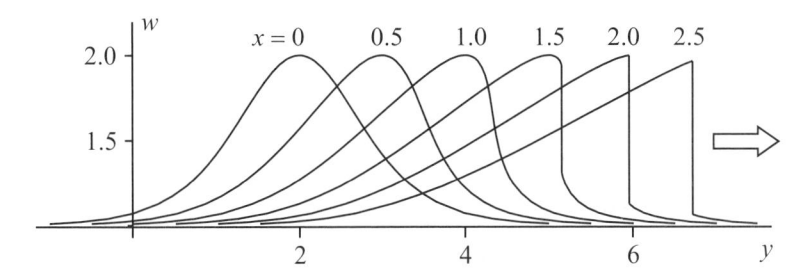

**Figure 13.8.** The formation of a shock wave generated from a solitary wave with the smooth initial profile.

### 13.1.3-4. Utilization of integral relations for determining generalized solutions.

Generalized solutions which are described by piecewise-smooth (piecewise-continuous) functions may formally be introduced by considering the following equation written in an integral form:

$$-\iint_D \left[ w \frac{\partial \psi}{\partial x} + F(w) \frac{\partial \psi}{\partial y} \right] dy\, dx = 0. \tag{13.1.3.9}$$

Here, $D$ is an arbitrary rectangle in the $yx$-plane, $\psi = \psi(x, y)$ is any "test" function with continuous first derivatives in $D$ that is zero at the boundary of $D$, and the function $F(w)$ is defined in equation (13.1.3.6). If $w$ and $F(w)$ are continuously differentiable, then equation (13.1.3.9) is equivalent to the original differential equation (13.1.3.1). Indeed, multiplying equation (13.1.3.1) by $\psi$, integrating over the domain $D$, and then integrating by parts, we obtain equation (13.1.3.9). Conversely, integrating (13.1.3.9) by parts yields

$$\iint_D \left[ \frac{\partial w}{\partial x} + \frac{\partial F(w)}{\partial y} \right] \psi\, dy\, dx = 0.$$

Since this equation must be satisfied for any test function $\psi$ and since $F'(w) = f(w)$, we obtain the original equation (13.1.3.1). However, equation (13.1.3.9) has a wider class of solutions since the admissible functions $w(x, y)$ need not necessarily be differentiable. The functions $w(x, y)$ satisfying the integral relation (13.1.3.9) for all test functions $\psi$ are referred to as generalized (or weak) solutions of equation (13.1.3.1).

The use of generalized solutions is convenient for the description of discontinuities, since it permits one to obtain jump conditions automatically. Consider a solution of equation (13.1.3.9) continuously differentiable in two portions $D_1$ and $D_2$ of the rectangle $D$, which has a jump discontinuity at the interface $\Gamma$ between $D_1$ and $D_2$. Integrating equation (13.1.3.9) by parts in each of the subdomains $D_1$ and $D_2$ yields

$$\iint_{D_1} \left[ \frac{\partial w}{\partial x} + \frac{\partial F(w)}{\partial y} \right] \psi\, dy\, dx + \iint_{D_2} \left[ \frac{\partial w}{\partial x} + \frac{\partial F(w)}{\partial y} \right] \psi\, dy\, dx + \int_\Gamma \{[w]\, dy - [F(w)]\, dx\} \psi = 0,$$

where $[w] = w_2 - w_1$ and $[F(w)] = F(w_2) - F(w_1)$ are jumps of $w$ and $F(w)$ across $\Gamma$. The curvilinear integral over $\Gamma$ is formed by the boundary terms of the integrals over $D_1$ and $D_2$ that result from the integration by parts. Since the relation obtained must be valid for all test functions $\psi$, it follows that equation (13.1.3.1) is valid inside each of the subdomains $D_1$ and $D_2$ and, moreover, the relation

$$[w]\,dy - [F(w)]\,dx = 0 \qquad (\text{on } \Gamma)$$

must hold. Assuming as before that the line of discontinuity is defined by the equation $y = s(x)$, we arrive at the jump condition (13.1.3.6).

It is worth noting that condition (13.1.3.7) does not follow from the integral relation (13.1.3.9) but is deduced from the additional condition of stability of the solution.

### 13.1.3-5. Conservation laws. Viscosity solutions.

Point out also other ways of introducing generalized solutions.

$1°$. Generalized solution may be introduced using the conservation law

$$\frac{d}{dx}\int_{y_1}^{y_2} w\,dy + F(w_2) - F(w_1) = 0, \qquad (13.1.3.10)$$

where $w = w(x, y)$ and $w_n = w(x, y_n)$ ($n = 1, 2$). Just as in equation (13.1.3.6), the function $F(w)$ is defined as $F(w) = \int f(w)\,dw$. Relation (13.1.3.10) is assumed to hold for any $y_1$ and $y_2$. It has a simple physical interpretation: the rate of change of the total value of $w$ distributed over the interval $(y_1, y_2)$ is compensated for by the "flux" of the function $F(w)$ through the endpoints of the interval.

Let $w$ be a continuously differentiable solution of the conservation law. Then, differentiating equation (13.1.3.10) with respect to $y_2$ and setting $y_2 = y$, we arrive at equation (13.1.3.1). The conservation law (13.1.3.10) is convenient for the reason that is admits discontinuous solutions. It is not difficult to show that in this case the jump condition (13.1.3.6) must hold. For this reason, conservation laws like (13.1.3.10) are sometimes used as the basis for determining generalized solutions.

$2°$. An alternative approach to determining generalized solutions involves the consideration of an auxiliary equation of the parabolic type of the form

$$\frac{\partial w}{\partial x} + f(w)\frac{\partial w}{\partial y} = \varepsilon\frac{\partial^2 w}{\partial x^2}, \qquad \varepsilon > 0. \qquad (13.1.3.11)$$

The generalized solution of the Cauchy problem (13.1.3.1), (13.1.3.2) (for a finite initial profile) is defined as the limit of the solution of equation (13.1.3.11) with the same initial condition (13.1.3.2) as $\varepsilon \to 0$. It is shown by Oleinik (1957) and Gelfand (1959) that the above definitions of the generalized solution leads to the same results.

The parameter $\varepsilon$ plays the role of "viscosity" (by analogy with the viscosity of a fluid), which "smooths out" the jump, thus making the profile of the unknown $w$ continuous. Therefore, the above construction, based on proceeding to the limit as $\varepsilon \to 0$, is called the method of vanishing viscosity and the limit function obtained is called the viscosity solution. Equation (13.1.3.11) with small $\varepsilon$ is not infrequently used as a basis for numerical simulation of discontinuous solutions of equation (13.1.3.1); in this case, one need not specially separate in the numerical scheme a domain of discontinuity.

*Remark.* In specific problems, first-order quasilinear equations are often a consequence of integral conservation laws, having clear physical interpretation. In such cases, one should introduce generalized solutions on the basis of these conservation laws; for example, see Whitham (1974) and Rozhdestvenskii and Yanenko (1983). The thus obtained nonsmooth generalized solutions may differ from those described above.

**13.1.3-6. Hopf's formula for the generalized solution.**

Below we give a general formula for a generalized solution of the Cauchy problem (13.1.3.1), (13.1.3.2), describing discontinuous solutions that satisfy the stability condition (13.1.3.7). As above, we assume that $x \geq 0$ and $f > 0$ for $w > 0$; $f'_w > 0$.

Consider the function

$$Z(s) = \min_{w}\{ws - F(w)\}, \quad \text{where} \quad F(w) = \int f(w)\,dw. \tag{13.1.3.12}$$

We set

$$H(x, y, \eta) = \int_0^\eta \varphi(\bar{\eta})\,d\bar{\eta} + x Z\left(\frac{y - \eta}{x}\right). \tag{13.1.3.13}$$

This is a continuous function of $\eta$ for fixed $x$ and $y$. It can be shown that for fixed $x$ and with the exception of a countable set of values of $y$, function (13.1.3.13) has a unique minimum with respect to $\eta$. Denote the position of this minimum by $\eta = \xi$, where $\xi = \xi(x, y)$. The stable generalized solution of equation (13.1.3.1) subject to the initial condition (13.1.3.2) is given by

$$w(x, y) = \mathcal{Z}\left(\frac{y - \xi}{x}\right), \quad \text{where} \quad \mathcal{Z}(s) = \frac{dZ}{ds}. \tag{13.1.3.14}$$

The function $Z = Z(s)$ defined by relation (13.1.3.12) can be represented in the parametric form

$$s = f(w), \quad Z = ws - \int f(w)\,dw. \tag{13.1.3.15}$$

Hence follows the parametric representation for its derivative $\mathcal{Z} = \mathcal{Z}(s)$:

$$s = f(w), \quad \mathcal{Z} = w. \tag{13.1.3.16}$$

The position of the minimum $\eta = \xi(x, y)$ of function (13.1.3.13) is determined by the condition $H_\eta = 0$, which results in the following equation for $\xi$:

$$\varphi(\xi) - \mathcal{Z}\left(\frac{y - \xi}{x}\right) = 0. \tag{13.1.3.17}$$

To illustrate the utilization of the above formulas, we consider two cases.

1°. Let the algebraic (or transcendental) equation (13.1.3.17) have a unique solution $\xi = \xi(x, y)$ in some domain of the $xy$-plane. We set $s = (y-\xi)/x$ in (13.1.3.16) and consider these relations in conjunction with equation (13.1.3.17). Eliminating the functions $f(w)$ and $\mathcal{Z}$ from these equations yields a solution of the problem in the parametric form (13.1.3.3). In this case, we obtain a smooth (classical) solution describing a rarefaction wave.

2°. Let the algebraic (or transcendental) equation (13.1.3.17) have two different solutions, $\xi_1$ and $\xi_2$, that are functions of $x$ and $y$. For both cases, solution (13.1.3.3) is valid, where either $\xi = \xi_1$ or $\xi = \xi_2$. At each point $(x, y)$, we choose that solution $\xi_n$ ($n = 1, 2$) which minimizes the function $H(x, y, \xi_n)$ defined by equation (13.1.3.13). In this case, we obtain a discontinuous (generalized) solution describing a shock wave.

### 13.1.3-7. Problem of propagation of a signal.

In the problem of propagation of a signal and other physical applications, one seeks a solution of equation (13.1.3.1) subject to the conditions

$$\begin{aligned} w &= w_0 & \text{at} \quad x &= 0 & \text{(initial condition)}, \\ w &= g(x) & \text{at} \quad y &= 0 & \text{(boundary condition)}, \end{aligned} \qquad (13.1.3.18)$$

where $w_0$ is some constant and $g(x)$ is a prescribed function. One considers the domain $x > 0$, $y > 0$, where $x$ plays the role of time and $y$ the role of the spatial coordinate. It is assumed that $f(w) > 0$.

The characteristics of this problem issue from the positive semiaxis $y$ and the positive semiaxis $x$ (see Fig. 13.9). We have $w = w_0$ at the characteristics issuing from the $y$-axis. Hence, these characteristics are straight lines defined by $y - a_0 x = \text{const}$, where $a_0 = f(w_0)$. It follows that

$$w = w_0 \quad \text{for} \quad y > a_0 x. \qquad (13.1.3.19)$$

As far as the characteristics issuing from the $x$-axis are concerned, we assume that one of the characteristics starts from a point $x = \tau$. The solution of equation (13.1.3.1) subject to conditions (13.1.3.18) can be represented in the parametric form

$$y = \mathcal{G}(\tau)(x - \tau), \quad w = g(\tau), \qquad (13.1.3.20)$$

where $\mathcal{G}(\tau) = f\big(g(\tau)\big)$. This solution can be related to solution (13.1.3.3) of the Cauchy problem (13.1.3.1), (13.1.3.2) by setting

$$\xi = -\tau \mathcal{G}(\tau), \quad \varphi(\xi) = g(\tau), \quad \mathcal{F}(\xi) = \mathcal{G}(\tau). \qquad (13.1.3.21)$$

This corresponds to the continuation of characteristics through the points $y = 0$, $x = \tau$ to the $y$-axis and to the designation of the points of intersection by $y = \xi$. In this case, the problem of propagation of a signal is formulated as a Cauchy problem.

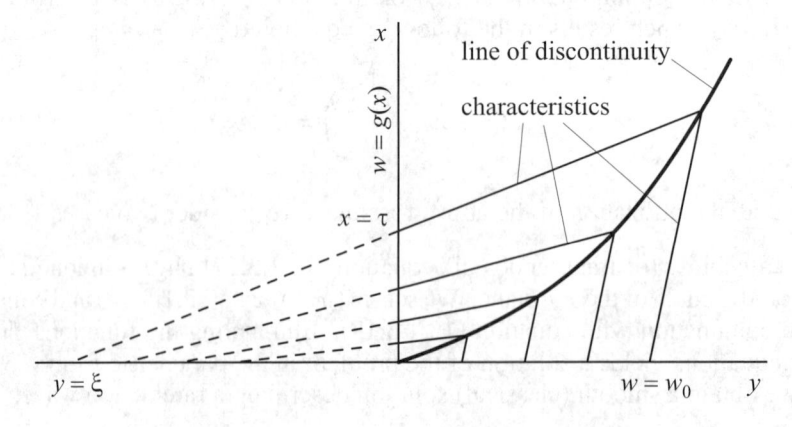

Figure 13.9. Characteristics of the problem of propagation of a signal.

Each domain of nonuniqueness in solution (13.1.3.20) should be replaced by a jump discontinuity. If

$$\mathcal{G}(+0) > a_0, \quad \text{where} \quad a_0 = f(w_0),$$

such a domain arises instantaneously, since the first characteristic $y = \mathcal{G}(+0)x$ is ahead of the last characteristic $y = a_0 x$ of the unperturbed domain. In this case, the discontinuity appears at the origin of coordinates and the relation

$$G - G_0 = (w - w_0)\mathcal{G} - \frac{1}{x - \tau} \int_0^\tau \left[ G(\bar{\tau}) - G_0 \right] d\bar{\tau} \qquad (13.1.3.22)$$

holds. The quantities $w$, $\mathcal{G}$, and $G$ are functions of $\tau$ in the domain behind the discontinuity and are given by

$$w = g(\tau), \quad \mathcal{G} = f\big(g(\tau)\big), \quad G = F\big(g(\tau)\big).$$

The subscript 0 refers to the values of these variables ahead of the discontinuity, $w = w_0$, $\mathcal{G}_0 = f(w_0)$, and $G_0 = F(w_0)$.

Relations (13.1.3.20) describe the solution in the perturbed domain behind the discontinuity. Equation (13.1.3.22) serves to determine $\tau(x)$ at the point of discontinuity; by setting this value into relations (13.1.3.20), we find both the position of the discontinuity and the value of $w$ immediately behind it.

If $g(x)$ remains constant and equal to $w_c$, then for $a_c > a_0$, where $a_c = f(w_c)$, the solution has a jump discontinuity propagating at a constant velocity and separating two homogeneous domains with $w = w_c$ and $w = w_0$.

## 13.1.4. Quasilinear Equations of General Form. Generalized Solution, Jump Condition, and Stability Condition

13.1.4-1. Quasilinear equations in conservative form.

In the general case, the quasilinear equation

$$\frac{\partial w}{\partial x} + f(x, y, w) \frac{\partial w}{\partial y} = g(x, y, w) \qquad (13.1.4.1)$$

can be represented in an equivalent, conservative form as

$$\frac{\partial w}{\partial x} + \frac{\partial}{\partial y} F(x, y, w) = G(x, y, w), \qquad (13.1.4.2)$$

where

$$F(x, y, w) = \int_{w_0}^{w} f(x, y, t)\, dt, \quad G(x, y, w) = g(x, y, w) + \int_{w_0}^{w} \frac{\partial}{\partial y} \left[ f(x, y, t) \right] dt, \quad (13.1.4.3)$$

and $w_0$ is an arbitrary number. In what follows we assume that the functions $f$ and $g$ are continuous and have continuous first derivatives.

As was shown by examples in Subsections 13.1.2 and 13.1.3, characteristics of equation (13.1.4.1) can intersect in some domain, which results in the nonuniqueness of the solution and the absence of a physical interpretation of this solution. For this reason, one has to make use of a generalized solution, described by a discontinuous function instead of a classical smooth solution.

We consider the class of functions $w(x, y) \in \mathcal{K}$ satisfying the following conditions:

1°. In any bounded portion of the half-plane $x \geq 0$, there exists a finite number of lines and points of discontinuity; outside these lines and points, the function $w(x, y)$ is continuous and has continuous first derivatives.

2°. At the lines of discontinuity, $y = y(x)$, the left and right limit values of $w$ exist: $w(x, y-0)$ and $w(x, y + 0)$.

### 13.1.4-2. Generalized solution. Jump condition and stability condition.

A generalized solution may be introduced in the following manner. Let $\psi(x,y) \in C_1$ be a continuous finite function (which vanishes outside a finite portion of the $xy$-plane) having continuous first derivatives.

Multiply equation (13.1.4.1) by $\psi(x,y)$ and integrate the resulting relation over the half-plane $\Omega = \{0 \le x < \infty, -\infty < y < \infty\}$. On integrating by parts, we obtain

$$\iint_\Omega \left[ w \frac{\partial \psi}{\partial x} + F(x,y,w) \frac{\partial \psi}{\partial y} + G(x,y,w)\psi(x,y) \right] dy\, dx + \int_{-\infty}^{\infty} w(0,y)\psi(0,y)\, dy = 0. \tag{13.1.4.4}$$

The function $F(x,y,w)$ is defined in equation (13.1.4.3). The integral relation (13.1.4.4) does not contain derivatives of the unknown function and does not lose its meaning for discontinuous $w(x,y)$. The function $w(x,y) \in \mathcal{K}$ will be called the generalized solution of equation (13.1.4.1) if inequalities (13.1.4.4) hold for any finite $\psi(x,y) \in C_1$.

*Basic properties of the stable generalized solution*:

1°. In the domain where the solution $w$ is continuously differentiable, equations (13.1.4.1) and (13.1.4.4) are equivalent.

2°. Let $y = y(x)$ be the equation of a discontinuity line of $w(x,y)$. Then the Rankine–Hugoniot jump condition must hold. It expresses the speed of motion of the discontinuity line via the solution parameters ahead of and behind the discontinuity as

$$V = \frac{[F(x,y,w)]}{[w]} \equiv \frac{F(x,y(x),w_2(x)) - F(x,y(x),w_1(x))}{w_2(x) - w_1(x)}, \tag{13.1.4.5}$$

where

$$V = y'(x), \quad w_1(x) = w(x,y(x)-0), \quad w_2(x) = w(x,y(x)+0).$$

3°. For $f'_w(x,y,w) \ne 0$, the generalized solution stable with respect to small perturbations of the initial profile (it is stable solutions that are physically realizable) must satisfy the condition

$$f(x,y(x),w_2(x)) \le V \le f(x,y(x),w_1(x)). \tag{13.1.4.6}$$

The stability condition (13.1.4.6) has the geometrical meaning that the characteristics issuing from the $x$-axis (these characteristics "carry" information about the initial data) must intersect the discontinuity line (see Fig. 13.7). This condition is very important since it allows for the existence of a stable generalized solution and provides its uniqueness.

The properties of Items 1° and 2° follow from the integral relation (13.1.4.4), and the condition of Item 3° is additional [it cannot be deduced from the integral relation (13.1.4.4)]. If the stability condition of Item 3° is not imposed, then various generalized solutions satisfying Items 1° and 2° may be constructed.

**Example 4.** Consider the Cauchy problem for equation (13.1.2.8) with the initial condition (13.1.2.15). We set

$$\overline{w}(x,y) = \begin{cases} w_1 & \text{for } y < Vx, \\ w_2 & \text{for } y > Vx, \end{cases} \quad \text{where } V = \frac{w_1 + w_2}{2}. \tag{13.1.4.7}$$

This function is constant from the left and right of the discontinuity line $y = Vx$, where the jump condition (13.1.4.5) is met [since $F(x,y,w) = \frac{1}{2}w^2$], and satisfies the initial condition (13.1.2.15). Hence, $\overline{w}$ is a generalized solution.

Figure 13.10 shows the discontinuity line and characteristics corresponding to solution (13.1.4.7). One can see that the characteristics "issue" from the discontinuity line and do not intersect the $x$-axis. Therefore, solution (13.1.4.7) is unstable, does not satisfy condition (13.1.4.6), and is not physically realizable. A stable solution of this problem was constructed earlier; see relation (13.1.2.18).

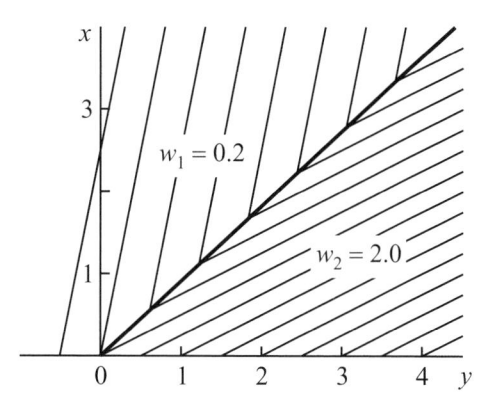

**Figure 13.10.** Characteristics and the discontinuity line for an unstable discontinuous solution (13.1.4.7).

If $f'_w(x, y, w)$ is not a function of fixed sign, the stability condition for the generalized solution becomes more complicated:

$$\frac{F(x, y, w_*) - F(x, y, w_2)}{w_* - w_2} \leq V \leq \frac{F(x, y, w_*) - F(x, y, w_1)}{w_* - w_1},$$
$$y = y(x), \quad w_1 < w_* < w_2,$$

where $w_*$ is any value from the interval $(w_1, w_2)$.

**Remark.** Point out also other ways of defining generalized solutions (using conservation laws and viscosity solutions).

### 13.1.4-3. Method for constructing stable generalized solutions.

Consider the Cauchy problem for the quasilinear equation

$$\frac{\partial w}{\partial x} + \frac{\partial}{\partial y} F(x, y, w) = 0 \tag{13.1.4.8}$$

subject to the initial condition

$$w = \varphi(y) \quad \text{at} \quad x = 0. \tag{13.1.4.9}$$

It is assumed that the function $F(x, y, w)$ is continuously differentiable with respect to all its arguments for $x \geq 0$, $-\infty < y < \infty$ and any bounded $w$. We also assume that the second derivative $F_{ww}$ is positive. Let the functions $\varphi(y)$ and $\varphi'(y)$ be piecewise-continuous for any finite $y$.

The characteristic system for equation (13.1.4.8) has the form

$$y'_x = F_w(x, y, w), \quad w'_x = -F_y(x, y, w), \tag{13.1.4.10}$$

where $F_w$ and $F_y$ are the partial derivatives of the function $F$ with respect to $w$ and $y$.

Suppose the functions

$$y(x) = Y(x, \tau, \xi, \eta), \quad w(x) = W(x, \tau, \xi, \eta) \tag{13.1.4.11}$$

are solutions of system (13.1.4.10) satisfying the boundary conditions

$$y(0) = \eta, \quad y(\tau) = \xi. \tag{13.1.4.12}$$

Here, $\eta$ and $\xi$ are arbitrary numbers and $\tau > 0$. We assume that problem (13.1.4.10), (13.1.4.12) has a unique bounded solution.

The stable generalized solution of the Cauchy problem (13.1.4.8), (13.1.4.9) is given by

$$w(x, y - 0) = W(x, x, y, \xi_-(x, y)),$$
$$w(x, y + 0) = W(x, x, y, \xi_+(x, y)), \qquad (13.1.4.13)$$

where $\xi_-(x, y)$ and $\xi_+(x, y)$ denote, respectively, the greatest lower bound and the least upper bound of the set of values $\{\xi = \xi_n\}$ for which the function

$$I(x, y, \xi) = \int_0^\xi \left[\varphi(\eta) - W(0, x, y, \eta)\right] d\eta \qquad (13.1.4.14)$$

takes the minimum value for fixed $x$ and $y$ ($x > 0$). If function (13.1.4.14) takes the minimum value for a single $\xi = \xi_1$, then $\xi_- = \xi_+$ and relation (13.1.4.14) describes the classical smooth solution.

## 13.2. Nonlinear Equations

### 13.2.1. Solution Methods

13.2.1-1. Complete, general, and singular integrals.

A *nonlinear first-order partial differential equation* with two independent variables has the general form

$$F(x, y, w, p, q) = 0, \qquad \text{where} \quad p = \frac{\partial w}{\partial x}, \quad q = \frac{\partial w}{\partial y}. \qquad (13.2.1.1)$$

Such equations are encountered in analytical mechanics, calculus of variations, optimal control, differential games, dynamic programming, geometric optics, differential geometry, and other fields.

In this subsection, we consider only smooth solutions $w = w(x, y)$ of equation (13.2.1.1), which are continuously differentiable with respect to both arguments (Subsection 13.2.3 deals with nonsmooth solutions).

1°. Let a particular solution of equation (13.2.1.1),

$$w = \Xi(x, y, C_1, C_2), \qquad (13.2.1.2)$$

depending on two parameters $C_1$ and $C_2$, be known. The two-parameter family of solutions (13.2.1.2) is called a *complete integral* of equation (13.2.1.1) if the rank of the matrix

$$M = \begin{pmatrix} \Xi_1 & \Xi_{x1} & \Xi_{y1} \\ \Xi_2 & \Xi_{x2} & \Xi_{y2} \end{pmatrix} \qquad (13.2.1.3)$$

is equal to two in the domain being considered (for example, this is valid if $\Xi_{x1}\Xi_{y2} - \Xi_{x2}\Xi_{y1} \neq 0$). In equation (13.2.1.3), $\Xi_n$ denotes the partial derivative of $\Xi$ with respect to $C_n$ ($n = 1, 2$), $\Xi_{xn}$ is the second partial derivative with respect to $x$ and $C_n$, and $\Xi_{yn}$ is the second partial derivative with respect to $y$ and $C_n$.

In some cases, a complete integral can be found using the method of undetermined coefficients by presetting an appropriate structure of the particular solution sought. (The complete integral is determined by the differential equation nonuniquely.)

**Example 1.** Consider the equation
$$\frac{\partial w}{\partial x} = a\left(\frac{\partial w}{\partial y}\right)^n + b.$$
We seek a particular solution as the sum $w = C_1 y + C_2 + C_3 x$. Substituting this expression into the equation yields the relation $C_3 = aC_1^n + b$ for the coefficients $C_1$ and $C_3$. With this relation, we find a complete integral in the form $w = C_1 y + (aC_1^n + b)x + C_2$.

A complete integral of equation (13.2.1.1) is often written in implicit form:*
$$\Xi(x, y, w, C_1, C_2) = 0. \tag{13.2.1.4}$$

2°. The *general integral* of equation (13.2.1.1) can be represented in parametric form by using the complete integral (13.2.1.2) [or (13.2.1.4)] and the two equations
$$\begin{aligned} C_2 &= f(C_1), \\ \frac{\partial \Xi}{\partial C_1} &+ \frac{\partial \Xi}{\partial C_2} f'(C_1) = 0, \end{aligned} \tag{13.2.1.5}$$
where $f$ is an arbitrary function and the prime stands for the derivative. In a sense, the general integral plays the role of the general solution depending on an arbitrary function (the questions whether it describes all solutions calls for further analysis).

**Example 2.** For the equation considered in the first example, the general integral can be written in parametric form by using the relations
$$w = C_1 y + (aC_1^n + b)x + C_2, \quad C_2 = f(C_1), \quad y + anC_1^{n-1}x + f'(C_1) = 0.$$
Eliminating $C_2$ from these relations and renaming $C_1$ by $C$, one can represent the general integral in a more graphic manner in the form
$$\begin{aligned} w &= Cy + (aC^n + b)x + f(C), \\ y &= -anC^{n-1}x + f'(C). \end{aligned}$$

3°. *Singular integrals* of equation (13.2.1.1) can be found without invoking a complete integral by eliminating $p$ and $q$ from the following system of three algebraic (or transcendental) equations:
$$F = 0, \quad F_p = 0, \quad F_q = 0,$$
where the first equation coincides with equation (13.2.1.1).

---

**13.2.1-2. Method of separation of variables. Equations of special form.**

The method of separation of variables implies searching for a complete integral as the sum or product of functions of various arguments. Such solutions are called *additive separable* and *multiplicative separable*, respectively. Presented below are structures of complete integrals for some classes of nonlinear equations admitting separation of variables.

1°. If the equation does not depend explicitly on $y$ and $w$, i.e.,
$$F(x, w_x, w_y) = 0,$$
then one can seek a complete integral in the form of the sum of two functions with different arguments
$$w = C_1 y + C_2 + u(x).$$
The new unknown function $u$ is determined by solving the following ordinary differential equation:
$$F(x, u'_x, C_1) = 0.$$
Expressing $u'_x$ from this equation in terms of $x$, one arrives at a separable differential equation for $u = u(x)$.

---
\* In equations (13.2.1.2) and (13.2.1.4), the symbol $\Xi$ denotes different functions.

2°. Consider an equation with separated variables
$$F_1(x, w_x) = F_2(y, w_y).$$
Then one can seek a complete integral as the sum of two functions with different arguments,
$$w = u(x) + v(y) + C_1,$$
which are determined by the following two ordinary differential equations:
$$F_1(x, u'_x) = C_2,$$
$$F_2(y, v'_y) = C_2.$$

3°. Let the equation have the form (generalizes the equation of Item 2°)
$$F_1(x, w_x) + F_2(y, w_y) = aw.$$
Then one can seek a complete integral as the sum of two functions with different arguments,
$$w = u(x) + v(y) + C_1,$$
which are determined by the following two ordinary differential equations:
$$F_1(x, u'_x) - au = aC_1 + C_2,$$
$$F_2(y, v'_y) - av = -C_2,$$
where $C_1$ is an arbitrary constant.

4°. Suppose the equation can be rewritten in the form
$$F\big(\varphi(x, w_x),\, y,\, w_y\big) = 0.$$
Then one can seek a complete integral as the sum of two functions with different arguments,
$$w(x, y) = u(x) + v(y) + C_1,$$
which are determined by the following two ordinary differential equations:
$$\varphi(x, u'_x) = C_2,$$
$$F(C_2, y, v'_y) = 0,$$
where $C_2$ is an arbitrary constant.

5°. Let the equation have the form
$$F_1(x, w_x/w) = w^k F_2(y, w_y/w).$$
Then one can seek a complete integral in the form of the product of two functions with different arguments,
$$w = u(x)v(y),$$
which are determined by the following two ordinary differential equations:
$$F_1(x, u'_x/u) = C_1 u^k,$$
$$F_2(y, v'_y/v) = C_1 v^{-k},$$
where $C_1$ is an arbitrary constant.

6°. Table 13.2 lists complete integrals of the above and some other nonlinear equations of general form involving arbitrary functions with several arguments.

▶ Section T7.3 presents complete integrals for many more nonlinear first-order partial differential equations with two independent variables than in Table 13.2.

### TABLE 13.2
Complete integrals for some special types of nonlinear first-order partial differential equations; $C_1$ and $C_2$ are arbitrary constants

| No. | Equations and comments | Complete integrals | Auxiliary equations |
|---|---|---|---|
| 1 | $F(w_x, w_y) = 0$, does not depend on $x$, $y$, and $w$ implicitly | $w = C_1 + C_2 x + C_3 y$ | $F(C_2, C_3) = 0$ |
| 2 | $F(x, w_x, w_y) = 0$, does not depend on $y$ and $w$ implicitly | $w = C_1 y + C_2 + u(x)$ | $F(x, u'_x, C_1) = 0$ |
| 3 | $F(w, w_x, w_y) = 0$, does not depend on $x$ and $y$ implicitly | $w = u(z)$, $z = C_1 x + C_2 y$ | $F(u, C_1 u'_z, C_2 u'_z) = 0$ |
| 4 | $F_1(x, w_x) = F_2(y, w_y)$, separated equation | $w = u(x) + v(y) + C_1$ | $F_1(x, u'_x) = C_2$, $F_2(y, v'_y) = C_2$ |
| 5 | $F_1(x, w_x) + F_2(y, w_y) = aw$, generalizes equation 4 | $w = u(x) + v(y)$ | $F_1(x, u'_x) - au = C_1$, $F_2(y, v'_y) - av = -C_1$ |
| 6 | $F_1(x, w_x) = e^{aw} F_2(y, w_y)$, generalizes equation 4 | $w = u(x) + v(y)$ | $F_1(x, u'_x) = C_1 e^{au}$, $F_2(y, v'_y) = C_1 e^{-av}$ |
| 7 | $F_1(x, w_x/w) = w^k F_2(y, w_y/w)$, can be reduced to equation 6 by the change of variable $w = e^z$ | $w = u(x) v(y)$ | $F_1(x, u'_x/u) = C_1 u^k$, $F_2(y, v'_y/v) = C_1 v^{-k}$ |
| 8 | $w = x w_x + y w_y + F(w_x, w_y)$, Clairaut equation | $w = C_1 x + C_2 y + F(C_1, C_2)$ | — |
| 9 | $F(x, w_x, w_y, w - y w_y) = 0$, generalizes equation 2 | $w = C_1 y + u(x)$ | $F(x, u'_x, C_1, u) = 0$ |
| 10 | $F(w, w_x, w_y, x w_x + y w_y) = 0$, generalizes equations 3 and 8 | $w = u(z)$, $z = C_1 x + C_2 y$ | $F(u, C_1 u'_z, C_2 u'_z, z u'_z) = 0$ |
| 11 | $F\bigl(\varphi(x, w_x), y, w_y\bigr) = 0$, generalizes equation 4 | $w = u(x) + v(y) + C_1$ | $\varphi(x, u'_x) = C_2$, $F(C_2, y, v'_y) = 0$ |

### 13.2.1-3. Lagrange–Charpit method.

Suppose that a first integral,

$$\Phi(x, y, w, p, q) = C_1, \qquad (13.2.1.6)$$

of the characteristic system of ordinary differential equations

$$\frac{dx}{F_p} = \frac{dy}{F_q} = \frac{dw}{p F_p + q F_q} = -\frac{dp}{F_x + p F_w} = -\frac{dq}{F_y + q F_w} \qquad (13.2.1.7)$$

is known. Here,

$$p = \frac{\partial w}{\partial x}, \quad q = \frac{\partial w}{\partial y}, \quad F_x = \frac{\partial F}{\partial x}, \quad F_y = \frac{\partial F}{\partial y}, \quad F_w = \frac{\partial F}{\partial w}, \quad F_p = \frac{\partial F}{\partial p}, \quad F_q = \frac{\partial F}{\partial q}.$$

We assume that solution (13.2.1.6) and equation (13.2.1.1) can be solved for the derivatives $p$ and $q$, i.e.,

$$p = \varphi_1(x, y, w, C_1), \quad q = \varphi_2(x, y, w, C_1). \qquad (13.2.1.8)$$

The first equation of this system can be treated as an ordinary differential equation with independent variable $x$ and parameter $y$. On finding the solution of this equation depending on an arbitrary function $\psi(y)$, one substitutes this solution into the second equation to arrive at an ordinary differential equation for $\psi$. On determining $\psi(y)$ and on substituting it into the general solution of the first equation of (13.2.1.8), one finds a complete integral of equation (13.2.1.1). In a similar manner, one can start solving system (13.2.1.8) with the second equation, treating it as an ordinary differential equation with independent variable $y$ and parameter $x$.

**Example 3.** Consider the equation
$$ywp^2 - q = 0, \qquad \text{where} \quad p = \frac{\partial w}{\partial x}, \quad q = \frac{\partial w}{\partial y}.$$
In this case, the characteristic system (13.2.1.7) has the form
$$\frac{dx}{2ywp} = -\frac{dy}{1} = \frac{dw}{2ywp^2 - q} = -\frac{dp}{yp^3} = -\frac{dq}{wp^2 + yp^2 q}.$$
By making use of the original equation, we simplify the denominator of the third ratio to obtain an integrable combination: $dw/(ywp^2) = -dp/(yp^3)$. This yields the first integral $p = C_1/w$. Solving the original equation for $q$, we obtain the system
$$p = \frac{C_1}{w}, \qquad q = \frac{C_1^2 y}{w}.$$
The general solution of the first equation has the form $w^2 = 2C_1 x + \psi(y)$, where $\psi(y)$ is an arbitrary function. With this solution, it follows from the second equation of the system that $\psi'(y) = 2C_1^2 y$. Thus, $\psi(y) = C_1^2 y^2 + C_2$. Finally, we arrive at a complete integral of the form
$$w^2 = 2C_1 x + C_1^2 y^2 + C_2.$$

Note that the general solution of the completely integrable Pfaff equation (see Subsection 15.14.2)
$$dw = \varphi_1(x, y, w, C_1)\, dx + \varphi_2(x, y, w, C_1)\, dy \tag{13.2.1.9}$$
is a complete integral of equation (13.2.1.1). Here, the functions $\varphi_1$ and $\varphi_2$ are the same as in system (13.2.1.8).

**Remark.** The relation $F(x, y, w, p, q) = C$ is an obvious first integral of the characteristic system (13.2.1.7). Hence, the function $\Phi$ determining the integral (13.2.1.6) must differ from $F$. However, the use of relation (13.2.1.1) makes it possible to reduce the order of system (13.2.1.7) by one.

---

**13.2.1-4. Construction of a complete integral with the aid of two first integrals.**

Suppose two independent first integrals,
$$\Phi(x, y, w, p, q) = C_1, \qquad \Psi(x, y, w, p, q) = C_2, \tag{13.2.1.10}$$
of the characteristic system of ordinary differential equations (13.2.1.7) are known. Assume that the functions $F$, $\Phi$, and $\Psi$ determining equation (13.2.1.1) and the integrals (13.2.1.10) satisfy the two conditions

(a) $J \equiv \dfrac{\partial(F, \Phi, \Psi)}{\partial(w, p, q)} \neq 0,$

(b) $[\Phi, \Psi] \equiv \begin{vmatrix} \Phi_p & \Phi_x + p\Phi_w \\ \Psi_p & \Psi_x + p\Psi_w \end{vmatrix} + \begin{vmatrix} \Phi_q & \Phi_y + q\Phi_w \\ \Psi_q & \Psi_y + q\Psi_w \end{vmatrix} \equiv 0,$ \hfill (13.2.1.11)

where $J$ is the Jacobian of $F$, $\Phi$, and $\Psi$ with respect to $w$, $p$, and $q$, and $[\Phi, \Psi]$ is the *Jacobi–Mayer bracket*. In this case, relations (13.2.1.1) and (13.2.1.10) form a parametric representation of the complete integral of equation (13.2.1.1) ($p$ and $q$ are considered to be parameters). Eliminating $p$ and $q$ from equations (13.2.1.1) and (13.2.1.10) followed by solving the obtained relation for $w$ yields a complete integral in an explicit form $w = w(x, y, C_1, C_2)$.

**Example 4.** Consider the equation
$$pq - aw = 0, \quad \text{where} \quad p = \frac{\partial w}{\partial x}, \quad q = \frac{\partial w}{\partial y}.$$

The characteristic system (13.2.1.7) has the form
$$\frac{dx}{q} = \frac{dy}{p} = \frac{dw}{2pq} = \frac{dp}{ap} = \frac{dq}{aq}.$$

Equating the first ratio with the fifth one and the second ratio with the fourth one, we obtain the first integrals
$$q - ax = C_1, \quad p - ay = C_2.$$

Thus, $F = pq - aw$, $\Phi = q - ax$, and $\Psi = p - ax$. These functions satisfy conditions (13.2.1.11). Solving the equation and the first integrals for $w$ yields a complete integral of the form
$$w = \frac{1}{a}(ax + C_1)(ay + C_2).$$

### 13.2.1-5. Case where the equation does not depend on $w$ explicitly.

Suppose the original equation does not contain the unknown first explicitly, i.e., it has the form
$$F(x, y, p, q) = 0. \tag{13.2.1.12}$$

$1°$. Given a one-parameter family of solutions $w = \overline{\Xi}(x, y, C_1)$ such that $\overline{\Xi}'_1 \neq \text{const}$, a complete integral is given by $w = \overline{\Xi}(x, y, C_1) + C_2$.

$2°$. The first integral may be sought in the form
$$\Phi(x, y, p, q) = C_1$$

similar to that of equation (13.2.1.12). In this case, the characteristic system (13.2.1.7) is represented as
$$\frac{dx}{F_p} = \frac{dy}{F_q} = -\frac{dp}{F_x} = -\frac{dq}{F_y}.$$

The corresponding Pfaff equation (13.2.1.9) becomes
$$dw = \varphi_1(x, y, C_1) \, dx + \varphi_2(x, y, C_1) \, dy.$$

One may integrate this equation in quadrature, thus arriving at the following expression for the complete integral:
$$w = \int_{x_0}^{x} \varphi_1(t, y, C_1) \, dt + \int_{y_0}^{y} \varphi_2(x_0, s, C_1) \, ds + C_2, \tag{13.2.1.13}$$

where $x_0$ and $y_0$ are arbitrary numbers.

$3°$. Suppose that equation (13.2.1.12) can be solved for $p$ or $q$, for example,
$$p = -\mathcal{H}(x, y, q).$$

Then, by differentiating this relation with respect to $y$, we obtain a quasilinear equation for the derivative $q$ in the form
$$\frac{\partial q}{\partial x} + \frac{\partial}{\partial y} \mathcal{H}(x, y, q) = 0, \quad q = \frac{\partial w}{\partial y}.$$

This equation is simpler than the original one; qualitative features of it and solution methods can be found in Section 13.1.1.

### 13.2.1-6. Hamilton–Jacobi equation.

Equation (13.2.1.1) solved for one of the derivatives, e.g.,

$$p + \mathcal{H}(x, y, w, q) = 0, \quad \text{where} \quad p = \frac{\partial w}{\partial x}, \quad q = \frac{\partial w}{\partial y}, \qquad (13.2.1.14)$$

is commonly referred to as the *Hamilton–Jacobi equation** and the function $\mathcal{H}$ as the *Hamiltonian*. Equations of the form (13.2.1.14) are frequently encountered in various fields of mechanics, control theory, and differential games, where the variable $x$ usually plays the role of time and the variable $y$ the role of the spatial coordinate. To the Hamilton–Jacobi equation (13.2.1.14) there corresponds the function $F(x, y, w, p, q) = p + \mathcal{H}(x, y, w, q)$ in equation (13.2.1.1).

The characteristic system (13.2.1.7) for equation (13.2.1.14) can be reduced, by taking into account the relation $p = -\mathcal{H}$, to a simpler system consisting of three differential equations,

$$y'_x = \mathcal{H}_q, \quad w'_x = q\mathcal{H}_q - \mathcal{H}, \quad q'_x = -q\mathcal{H}_w - \mathcal{H}_y, \qquad (13.2.1.15)$$

which are independent of $p$; the left-hand sides of these equations are derivatives with respect to $x$.

## 13.2.2. Cauchy Problem. Existence and Uniqueness Theorem

### 13.2.2-1. Statement of the problem. Solution procedure.

Consider the Cauchy problem for equation (13.2.1.1) subject to the initial conditions

$$x = h_1(\xi), \quad y = h_2(\xi), \quad w = h_3(\xi), \qquad (13.2.2.1)$$

where $\xi$ is a parameter ($\alpha \leq \xi \leq \beta$) and the $h_k(\xi)$ are given functions.

The solution of this problem is carried out in several steps:

$1°$. First, one determines additional initial conditions for the derivatives,

$$p = p_0(\xi), \quad q = q_0(\xi). \qquad (13.2.2.2)$$

To this end, one must solve the algebraic (or transcendental) system of equations

$$F\big(h_1(\xi), h_2(\xi), h_3(\xi), p_0, q_0\big) = 0, \qquad (13.2.2.3)$$

$$p_0 h'_1(\xi) + q_0 h'_2(\xi) - h'_3(\xi) = 0 \qquad (13.2.2.4)$$

for $p_0$ and $q_0$. Equation (13.2.2.3) results from substituting the initial data (13.2.2.1) into the original equation (13.2.1.1). Equation (13.2.2.4) is a consequence of the dependence of $w$ on $x$ and $y$ and the relation $dw = p\,dx + q\,dy$, where $dx$, $dy$, and $dw$ are calculated in accordance with the initial data (13.2.2.1).

$2°$. One solves the autonomous system

$$\frac{dx}{F_p} = \frac{dy}{F_q} = \frac{dw}{pF_p + qF_q} = -\frac{dp}{F_x + pF_w} = -\frac{dq}{F_y + qF_w} = d\tau, \qquad (13.2.2.5)$$

which is obtained from (13.2.1.7) by introducing the additional variable $\tau$ (playing the role of time).

---

* The Hamilton–Jacobi equation often means equation (13.2.1.14) that does not depend on $w$ explicitly, i.e., the equation $p + \mathcal{H}(x, y, q) = 0$.

3°. Finally, one determines the constant of integration from the initial conditions

$$x = h_1(\xi), \quad y = h_2(\xi), \quad w = h_3(\xi), \quad p = p_0(\xi), \quad q = q_0(\xi) \quad \text{at} \quad \tau = 0, \quad (13.2.2.6)$$

which are the combination of conditions (13.2.2.1) and (13.2.2.2). This results in the three functions

$$x = x(\tau, \xi), \quad y = y(\tau, \xi), \quad w = w(\tau, \xi), \quad (13.2.2.7)$$

which give the solution of the stated Cauchy problem in parametric form ($\tau$ and $\xi$ are parameters).

### 13.2.2-2. Existence and uniqueness theorem.

Let the function $F = F(x, y, w, p, q)$, determining equation (13.2.1.1), be twice continuously differentiable with respect to all five arguments (in the domain being considered), with $F_p^2 + F_q^2 \neq 0$. Let the functions $h_1(\xi)$, $h_2(\xi)$, and $h_3(\xi)$, determining the initial conditions (13.2.2.1), be twice differentiable with respect to $\xi$, with $(h_1')^2 + (h_2')^2 \neq 0$. Assume that the function $p_0(\xi)$ and $q_0(\xi)$, determining the additional initial condition (13.2.2.2), satisfy the system (13.2.2.3), (13.2.2.4). Moreover, we adopt the *transversality condition*

$$\Delta \equiv F_p h_2' - F_q h_1' \neq 0,$$

where the functions $h_1$, $h_2$, $p$, and $q$ are defined by equations (13.2.2.1) and (13.2.2.2), and the prime denotes the derivative with respect to $\xi$. Under the adopted assumptions, there exists a unique, twice continuously differentiable solution of equation (13.2.1.1) satisfying the initial conditions (13.2.2.1) and (13.2.2.2).

*Remark 1.* This theorem has a local character, i.e., the existence of a unique solution of the Cauchy problem is merely guaranteed in some neighborhood of the line defined by the initial data (13.2.2.1) together with the additional conditions (13.2.2.2).

*Remark 2.* The algebraic (or transcendental) system (13.2.2.3), (13.2.2.4) can have several solutions (see Example 2 at the end of this subsection) that result in distinct additional conditions (13.2.2.2) for the derivatives $p$ and $q$. Each pair of these additional conditions will generate a solution of the Cauchy problem (13.2.1.1), (13.2.2.1).

*Remark 3.* For nonlinear equations, the global solution of the Cauchy problem (13.2.1.1), (13.2.2.1) may be nonunique for another reason — because the characteristics in the $xy$-plane may intersect (see Example 1 in Subsection 13.2.3). Such a situation is discussed in Subsections 13.1.3 and 13.1.4, where quasilinear equations are considered.

### 13.2.2-3. Cauchy problem for the Hamilton–Jacobi equation.

The initial condition for the Hamilton–Jacobi equation (13.2.1.14) is usually stated in the form

$$w = \varphi(y) \quad \text{at} \quad x = L. \quad (13.2.2.8)$$

In this case, the solution of the Cauchy problem is reduced to the solution of the characteristic system (13.2.1.15) subject to the initial conditions

$$y = \xi, \quad w = \varphi(\xi), \quad q = \varphi'(\xi) \quad \text{at} \quad x = L, \quad (13.2.2.9)$$

where the prime stands for the derivative with respect to the parameter $\xi$.

## 13.2.2-4. Examples of solving the Cauchy problem.

Consider a few specific examples.

**Example 1.** Find a solution of the equation

$$aw = pq, \qquad \text{where} \quad p = \frac{\partial w}{\partial x}, \quad q = \frac{\partial w}{\partial y}, \qquad (13.2.2.10)$$

satisfying the initial condition

$$w = f(y) \quad \text{at} \quad x = 0. \qquad (13.2.2.11)$$

Rewrite the initial condition (13.2.2.11) in parametric form:

$$x = 0, \quad y = \xi, \quad w = f(\xi). \qquad (13.2.2.12)$$

The system of equations (13.2.2.3), (13.2.2.4) for $p_0(\xi)$ and $q_0(\xi)$ is written as

$$af(\xi) = p_0 q_0, \quad q_0 - f'(\xi) = 0.$$

It follows that

$$p_0 = a\frac{f(\xi)}{f'(\xi)}, \quad q_0 = f'(\xi). \qquad (13.2.2.13)$$

If $F = pq - aw$, the characteristic system (13.2.2.5) is written as

$$\frac{dx}{q} = \frac{dy}{p} = \frac{dw}{2pq} = \frac{dp}{ap} = \frac{dq}{aq} = d\tau. \qquad (13.2.2.14)$$

Its solution is given by the relations

$$p = C_1 e^{a\tau}, \quad q = C_2 e^{a\tau}, \quad x = \frac{C_2}{a}e^{a\tau} + C_3, \quad y = \frac{C_1}{a}e^{a\tau} + C_4, \quad w = \frac{C_1 C_2}{a}e^{2a\tau} + C_5; \qquad (13.2.2.15)$$

it is easy to integrate the last two equations first.

By using the initial conditions (13.2.2.12), (13.2.2.13), which must be satisfied at $\tau = 0$, we find the constants of integration in the form

$$C_1 = a\frac{f(\xi)}{f'(\xi)}, \quad C_2 = f'(\xi), \quad C_3 = -\frac{f'(\xi)}{a}, \quad C_4 = \xi - \frac{f(\xi)}{f'(\xi)}, \quad C_5 = 0.$$

Substituting theses values into (13.2.2.15) yields a solution of the Cauchy problem (13.2.2.10), (13.2.2.11) in the parametric form

$$x = \frac{1}{a}f'(\xi)(e^{a\tau} - 1), \quad y = \frac{f(\xi)}{f'(\xi)}(e^{a\tau} - 1) + \xi, \quad w = f(\xi)e^{2a\tau}.$$

**Example 2.** Find a solution of the equation

$$\left(\frac{\partial w}{\partial x}\right)^2 + \left(\frac{\partial w}{\partial y}\right)^2 = a^2, \qquad (13.2.2.16)$$

passing through the circle

$$x^2 + y^2 = b^2, \quad w = 0. \qquad (13.2.2.17)$$

By introducing a parameter $\xi$, we rewrite the equation of the circle in the form

$$x = b\sin\xi, \quad y = b\cos\xi, \quad w = 0. \qquad (13.2.2.18)$$

In this case, the equations (13.2.2.3), (13.2.2.4) for determining additional initial conditions are

$$p_0^2 + q_0^2 = a^2, \quad p_0 \cos\xi - \sin\xi\, q_0 = 0.$$

Whence,

$$p_0 = \varepsilon a \sin\xi, \quad q_0 = \varepsilon a \cos\xi, \quad \text{where} \quad \varepsilon = \pm 1. \qquad (13.2.2.19)$$

For $F = p^2 + q^2 - a^2$, system (13.2.2.5) is represented as

$$\frac{dx}{2p} = \frac{dy}{2q} = \frac{dw}{2(p^2 + q^2)} = -\frac{dp}{0} = -\frac{dq}{0} = d\tau. \qquad (13.2.2.20)$$

The general solution is given by (the last two equations are integrated first)

$$p = C_1, \quad q = C_2, \quad x = 2C_1\tau + C_3, \quad y = 2C_2\tau + C_4, \quad w = 2(C_1^2 + C_2^2)\tau + C_5. \tag{13.2.2.21}$$

Using the initial conditions (13.2.2.18), (13.2.2.19), which must be satisfied at $\tau = 0$, we find the constants of integration

$$C_1 = \varepsilon a \sin \xi, \quad C_2 = \varepsilon a \cos \xi, \quad C_3 = b \sin \xi, \quad C_4 = b \cos \xi, \quad C_5 = 0, \quad \text{where} \quad \varepsilon = \pm 1.$$

Substituting these values into (13.2.2.21) yields a solution of the Cauchy problem (13.2.2.16), (13.2.2.17) in the parametric form

$$x = (2\varepsilon a\tau + b)\sin \xi, \quad y = (2\varepsilon a\tau + b)\cos \xi, \quad w = 2a^2\tau.$$

On eliminating the parameters $\xi$ and $\tau$, we can rewrite this solution in a more graphic form,

$$a^2(x^2 + y^2) = (ab \pm w)^2. \tag{13.2.2.22}$$

Relation (13.2.2.22) describes two circular coaxial cones in the space $(x, y, w)$. The circle (13.2.2.17) is a base of the cones. The common axis coincides with the $w$-axis. The vertices of the cones have the coordinates $w = \pm ab$.

It is significant that solution (13.2.2.22) is a many-valued function.

### 13.2.3. Generalized Viscosity Solutions and Their Applications

13.2.3-1. Preliminary remarks.

Subsections 13.2.1 and 13.2.2 dealt with classical smooth solutions $w = w(x, y)$, having continuous derivatives with respect to both arguments. However, in optimal control, differential games, and some other applications, problems arise whose solutions are continuous but nonsmooth functions. To describe and construct generalized solutions of this sort, approaches other than those outlined above are required. It should be noted that for determining generalized solutions to nonlinear equations of the general forms (13.2.1.1) and (13.2.1.14), it turns out to be ineffective to use graphic constructs such as integral relations and conservation laws, which are frequently encountered in the theory of quasilinear equations (see Subsections 13.1.3 and 13.1.4).

Note that nonsmoothness of a solution can be caused by (i) the intersection of characteristics in the $xy$-plane (see Example 1 below), (ii) nonsmoothness of initial conditions, and/or (iii) nonsmoothness of the functions $F$ and $\mathcal{H}$ determining equations (13.2.1.1) and (13.2.1.14).

13.2.3-2. Viscosity solutions based on the use of a parabolic equation.

The solution of the Cauchy problem for equation (13.2.1.14) subject to the initial condition

$$w = \varphi(y) \quad \text{at} \quad x = 0 \tag{13.2.3.1}$$

may be approximated by the solution of the following second-order partial differential equation of the parabolic type:

$$\frac{\partial u}{\partial x} + \mathcal{H}\left(x, y, u, \frac{\partial u}{\partial y}\right) = \varepsilon \frac{\partial^2 u}{\partial y^2} \quad (\varepsilon > 0). \tag{13.2.3.2}$$

This equation is subject to the same initial condition (13.2.3.1). For a fairly wide class of functions $\mathcal{H}$ and $\varphi$, the Cauchy problem for equation (13.2.3.2) is known to have a unique solution. In the theory of Hamilton–Jacobi equations, this fact is used to determine the solution of the Cauchy problem (13.2.1.14), (13.2.3.1) as a limit of the solution of problem

(13.2.3.2), (13.2.3.1): $w(x,y) = \lim\limits_{\varepsilon \to 0} u(x,y)$. Just as in the theory of quasilinear equations (see Subsection 13.1.3), this structure, based on proceeding to the limit as $\varepsilon \to 0$, is referred to as the method of vanishing viscosity, and the limit function as the viscosity solution of the Hamilton–Jacobi equation.

The vanishing viscosity method can be implemented, for example, by a numerical solution of problem (13.2.3.2), (13.2.3.1) for sufficiently small $\varepsilon$ (in this case, one need not search for singular points, at which the smoothness of the solution is violated). However, this method is difficult to use for constructing analytical solutions, since one has to treat a more complex, second-order partial differential equation.

### 13.2.3-3. Viscosity solutions based on test functions and differential inequalities.

A continuous function $w = w(x,y)$ is called the viscosity solution of the initial value problem (13.2.1.1), (13.2.3.1) in a layer $0 \le x \le L$ if the following two conditions are satisfied:

1°. The function $w = w(x,y)$ satisfies the initial condition (13.2.3.1).

2°. Let $\psi(x,y)$ be any continuously differentiable test function. If $(x^\circ, y^\circ)$ is a local extremum point of the difference

$$w(x,y) - \psi(x,y), \qquad (13.2.3.3)$$

then the following relations hold at this point:

$$\begin{aligned} F(x^\circ, y^\circ, w^\circ, \psi_x^\circ, \psi_y^\circ) \ge 0 &\quad \text{if } (x^\circ, y^\circ) \text{ is a local minimum point,} \\ F(x^\circ, y^\circ, w^\circ, \psi_x^\circ, \psi_y^\circ) \le 0 &\quad \text{if } (x^\circ, y^\circ) \text{ is a local maximum point.} \end{aligned} \qquad (13.2.3.4)$$

Only those local extremum points that lie within the layer in question ($0 < x^\circ < L$) are to be examined.

Note that it is not necessary that a test function $\psi(x,y)$ exists for which the difference (13.2.3.3) has a local extremum. But if such a function does exist, then condition (13.2.3.4) must hold.

If the Cauchy problem has a smooth classical solution, then it coincides with the viscosity generalized solution.

In optimal control and differential games, terminal value problems are encountered apart from initial value problems. In terminal value problems, the solution of equations (13.2.1.1) and (13.2.1.14) is sought in the layer $0 \le x \le L$ and the "initial" condition is set at the right endpoint $x = L$. For these problems, the inequalities in (13.2.3.4) must be changed for the opposites. The terminal value problems can be reduced to initial value problems by introducing a new independent variable $z = L - x$ instead of $x$.

### 13.2.3-4. Local structure of generalized viscosity solutions.

A generalized solution, $w(x,y)$, consists of regular and singular points. In a neighborhood of a regular point, the function $w(x,y)$ is a solution in the classical sense (such twice continuously differentiable solutions are discussed in the existence and uniqueness theorem in Subsection 13.2.2). All points that are not regular are called singular points.

Let $D$ be a sufficiently small neighborhood of a singular point $(x_*, y_*)$. Usually, it happens that the singular points form a smooth curve $\Gamma$, which passes through $(x_*, y_*)$ and

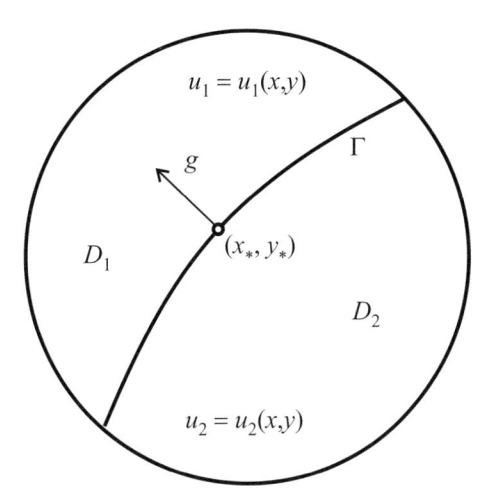

**Figure 13.11.** Generalized viscosity solution undergoes a discontinuity along the singular curve $\Gamma$; the solution is smooth in the subdomains $D_1$ and $D_2$.

divides the domain $D$ into two subdomains $D_1$ and $D_2$ (see Fig. 13.11). On both sides of $\Gamma$, the generalized solution $w$ is defined by different classical solutions $u_1$ and $u_2$, i.e.,

$$w(x,y) = \begin{cases} u_1(x,y) & \text{if } (x,y) \in D_1, \\ u_2(x,y) & \text{if } (x,y) \in D_2. \end{cases} \quad (13.2.3.5)$$

The solutions $u_1$ and $u_2$ join together along the interface $\Gamma$ in a continuous but nonsmooth manner. When crossing $\Gamma$, the derivatives of the generalized solution, $\partial w/\partial x$ and $\partial w/\partial y$, undergo a discontinuity. We assume that the smooth components $u_1$ and $u_2$ of the generalized solution are smoothly expended to the entire domain $D$. Then the equation of the curve $\Gamma$, formed by the singular points, can be represented as

$$g(x,y) = 0, \quad \text{where} \quad g(x,y) = u_2(x,y) - u_1(x,y). \quad (13.2.3.6)$$

The gradient of $g$, directed along the normal to $\Gamma$, is given by

$$\nabla g = (p_2 - p_1)\mathbf{e}_x + (q_2 - q_1)\mathbf{e}_y, \quad p_n = \frac{\partial u_n}{\partial x}, \quad q_n = \frac{\partial u_n}{\partial y},$$

where $\mathbf{e}_x$ and $\mathbf{e}_y$ are the direction cosines along the $x$- and $y$-axes, respectively. Two situations are possible.

$1°$. The vector $\nabla g$ is directed from $D_2$ toward $D_1$. In this case, the following statements are valid:

(A) The generalized solution in $D$ can be represented as $w = \min[u_1, u_2]$; see Fig. 13.12.

(B) There is no smooth test function $\psi(x,y)$ such that a local minimum of the difference (13.2.3.3) is attained at singular points comprising $\Gamma$.

(C) For the one-parameter family of test functions

$$\psi(x,y) = \lambda u_1(x,y) + (1-\lambda)u_2(x,y), \quad 0 \leq \lambda \leq 1, \quad (13.2.3.7)$$

the maximum

$$\max_{(x,y) \in D} \big[w(x,y) - \psi(x,y)\big] \quad (13.2.3.8)$$

is attained at $(x,y) \in \Gamma$.

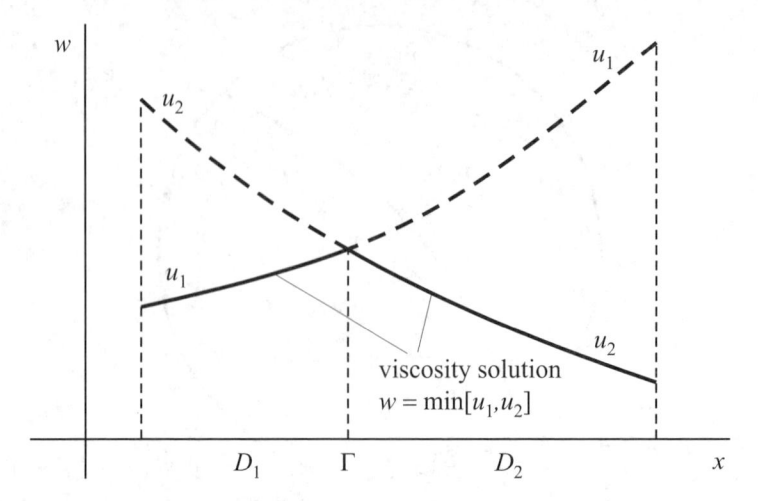

**Figure 13.12.** Graphic construction of a generalized viscosity solution by joining together smooth solutions.

**Remark.** For the generalized solution of the form $w = \min[u_1, u_2]$, one need not check the first inequality of (13.2.3.4); as far as the second inequality of (13.2.3.4) is concerned, it suffices to check this inequality against the one-parameter family of test functions defined by equation (13.2.3.7).

2°. The vector $\nabla g$ is directed from $D_1$ toward $D_2$. In this case, the generalized solution can be represented in the form $w = \max[u_1, u_2]$ and one need to check only the first inequality of (13.2.3.4) against the one-parameter family of test functions of equation (13.2.3.7).

**13.2.3-5. Generalized classical method of characteristics.**

Consider the Cauchy problem for the Hamilton–Jacobi equation of the form

$$\frac{\partial w}{\partial x} + \mathcal{H}\left(x, y, \frac{\partial w}{\partial y}\right) = 0, \qquad (13.2.3.9)$$
$$w = \varphi(y) \quad \text{at} \quad x = L.$$

It is assumed here that (i) the function $\mathcal{H}(x, y, q)$ is convex with respect to $q$ for all $x \in (0, L]$, $y \in \mathbb{R}$; (ii) the function $\mathcal{H}(x, y, q)$ is continuously differentiable with respect to $x$, $y$, and $q$; and (iii) the second derivatives $\mathcal{H}_{xy}$ and $\mathcal{H}_{xq}$ exist.
By

$$y = Y(x, \xi), \quad w = W(x, \xi), \quad q = Q(x, \xi) \qquad (13.2.3.10)$$

we denote a solution of the characteristic system (13.2.1.15) satisfying condition (13.2.2.9).

Let $\{\xi_n = \xi_n(x, y)\}$ be the set of functions obtained by solving the first equation of (13.2.3.10) for the parameter $\xi$. The subscript $n$ indicates the number of such functions.

The classical method of characteristics can be used to construct a generalized viscosity solution with the help of the relation

$$w(x, y) = \max_{\xi \in \{\xi_n\}} W(x, \xi) \qquad (13.2.3.11)$$

for all $x \in (0, L]$, $y \in \mathbb{R}$. To the value $n = 1$ there corresponds the classical smooth solution.

### 13.2.3-6. Examples of viscosity (nonsmooth) solutions.

Below we solve two problems that are encountered in the theory of differential games.

**Example 1.** Consider the terminal value problem for the Hamilton–Jacobi equation

$$\frac{\partial w}{\partial x} + \sqrt{1 + \left(\frac{\partial w}{\partial y}\right)^2} = 0 \tag{13.2.3.12}$$

with the initial condition

$$w = \tfrac{1}{2}y^2 \quad \text{at} \quad x = L. \tag{13.2.3.13}$$

The solution is sought in the domain $0 \le x \le L$.

The characteristic system (13.2.1.15) for equation (13.2.3.12) with the Hamiltonian $\mathcal{H}(x, y, w, q) = \sqrt{1 + q^2}$ has the form

$$y'_x = \frac{q}{\sqrt{1+q^2}}, \quad w'_x = -\frac{1}{\sqrt{1+q^2}}, \quad q'_x = 0. \tag{13.2.3.14}$$

The initial conditions are obtained from (13.2.2.9) for $\varphi(\xi) = \tfrac{1}{2}\xi^2$:

$$y = \xi, \quad w = \tfrac{1}{2}\xi^2, \quad q = \xi \quad \text{at} \quad x = L. \tag{13.2.3.15}$$

Integrating equation (13.2.3.14) and applying conditions (13.2.3.15), we obtain a solution of the Cauchy problem (13.2.3.12), (13.2.3.13) of the form

$$y = \frac{\xi(x-L)}{\sqrt{1+\xi^2}} + \xi, \quad w = \frac{L-x}{\sqrt{1+\xi^2}} + \tfrac{1}{2}\xi^2. \tag{13.2.3.16}$$

Figure 13.13 displays the characteristics $y(x, \xi)$ in the $xy$-plane for $L = 2$ and $\xi = 0, \pm 0.2, \pm 0.4, \ldots, \pm 1.0$. It is apparent that characteristics intersect. In this example, it is possible to construct a local classical solution of problem (13.2.3.12), (13.2.3.13). But this solution cannot be extended to the entire layer $0 \le x \le L$, that is, there is no global classical solution. Pay attention to the fact that the Hamiltonian $\mathcal{H} = \sqrt{1+q^2}$ of equation (13.2.3.12) and the function determining the initial condition (13.2.3.13) are infinitely differentiable functions.

The viscosity solution of the Cauchy problem (13.2.3.12), (13.2.3.13) has the form

$$w(x, y) = \max_{q \in \mathbb{R}} \left[ qy + (L-x)\sqrt{1+q^2} - \tfrac{1}{2}q^2 \right], \tag{13.2.3.17}$$

where $0 \le x \le L$ and $y$ is any number. Level curves of this function are depicted in Fig. 13.14. The heavy line indicates the set of singular points, at which the solution is nondifferentiable.

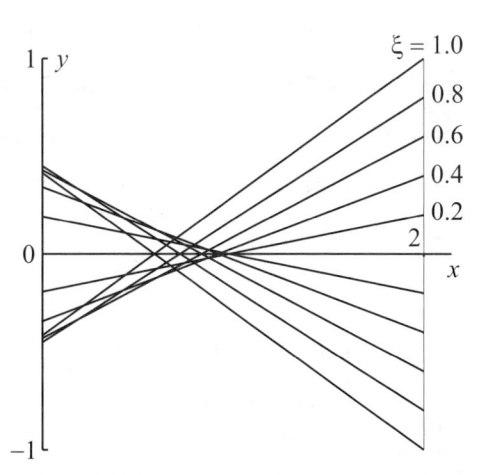

**Figure 13.13.** Characteristics of the terminal value problem (13.2.3.12)–(13.2.3.13) for the Hamilton–Jacobi equation.

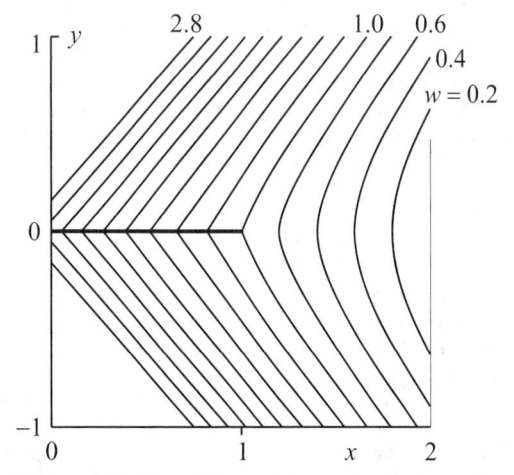

**Figure 13.14.** Level lines of the terminal value problem (13.2.3.12)–(13.2.3.13) for the Hamilton–Jacobi equation.

**Example 2.** Consider the terminal value problems for the more general Hamilton–Jacobi equation

$$\frac{\partial w}{\partial x} + \mathcal{H}\left(\frac{\partial w}{\partial y}\right) = 0 \qquad (0 \le x \le L) \tag{13.2.3.18}$$

with an arbitrary initial condition

$$w = \varphi(y) \quad \text{at} \quad x = L. \tag{13.2.3.19}$$

The following two statements hold:

1°. Let the Hamiltonian satisfy the Lipschitz condition

$$|\mathcal{H}(q_2) - \mathcal{H}(q_1)| \le \beta |q_2 - q_1| \qquad \text{for any} \quad q_1, q_2 \in \mathbb{R}, \tag{13.2.3.20}$$

and let the function $\varphi(y)$ be convex. Then the function

$$w(x,y) = \sup_{q \in \mathbb{R}} \left[qy + (L-x)\mathcal{H}(q) - \varphi^*(q)\right]$$

is the viscosity solution of problem (13.2.3.18), (13.2.3.19). The function $\varphi^*$ is the conjugate of $\varphi$, i.e.,

$$\varphi^*(q) = \sup_{x \in \mathbb{R}} \left[qx - \varphi(x)\right].$$

2°. Let the Hamiltonian $\mathcal{H}$ be convex and satisfy the Lipschitz condition (13.2.3.20). Let the function $\varphi(y)$ be continuous. Then the function

$$w(x,y) = \sup_{t \in \mathbb{R}} \left[\varphi(y + (L-x)t) - (L-x)\mathcal{H}^*(t)\right]$$

is the viscosity solution of problem (13.2.3.18), (13.2.3.19). The function

$$\mathcal{H}^*(t) = \sup_{q \in \mathbb{R}} \left[qt - \mathcal{H}(q)\right]$$

is the conjugate of the Hamiltonian.

# References for Chapter 13

**Bardi, M. and Dolcetta, I. C.,** *Optimal Control and Viscosity Solutions of Hamilton–Jacobi–Bellman Equations*, Birkhäuser Verlag, Boston, 1998.
**Barron, E. N. and Jensen, R.,** Generalized viscosity solutions for Hamilton–Jacobi equations with time-measurable Hamiltonians, *J. Different. Equations*, Vol. 68, No. 1, pp. 10–21, 1987.
**Courant, R. and Hilbert, D.,** *Methods of Mathematical Physics, Vol. 2*, Wiley-Interscience, New York, 1989.
**Crandall, M. G., Evans, L. C., and Lions, P.-L.,** Some properties of viscosity solutions of Hamilton–Jacobi equations, *Trans. Amer. Math. Soc.*, Vol. 283, No. 2, pp. 487–502, 1984.
**Gelfand, I. M.,** Some problems of the theory of quasi-linear equations, *Uspekhi Matem. Nauk*, Vol. 14, No. 2, pp. 87–158, 1959 [*Amer. Math. Soc. Translation, Series 2*, pp. 295–381, 1963].
**Hopf, E.,** Generalized solutions of nonlinear equations of first order, *J. Math. Mech.*, Vol. 14, pp. 951–973, 1965.
**Jeffery, A.,** *Quasilinear Hyperbolic Systems and Waves*, Pitman, London, 1976.
**Kamke, E.,** *Differentialgleichungen: Lösungsmethoden und Lösungen, II, Partielle Differentialgleichungen Erster Ordnung für eine gesuchte Funktion*, Akad. Verlagsgesellschaft Geest & Portig, Leipzig, 1965.
**Kruzhkov, S. N.,** Generalized solutions of nonlinear first order equations with several variables [in Russian], *Mat. Sbornik*, Vol. 70, pp. 394–415, 1966.
**Lax, P. D.,** *Hyperbolic systems of conservation laws and the mathematical theory of shock waves* [reprint from the classical paper of 1957], Society for Industrial & Applied Mathematics, Philadelphia, 1997.
**Melikyan, A. A.,** *Generalized Characteristics of First Order PDEs: Applications in Optimal Control and Differential Games*, Birkhäuser Verlag, Boston, 1998.
**Oleinik, O. A.,** Discontinuous solutions of nonlinear differential equations, *Uspekhi Matem. Nauk*, Vol. 12, No. 3, pp. 3–73, 1957 [*Amer. Math. Soc. Translation, Series 2*, Vol. 26, pp. 95–172, 1963].
**Petrovskii, I. G.,** *Lectures on the Theory of Ordinary Differential Equations* [in Russian], Nauka Publishers, Moscow, 1970.
**Polyanin, A. D., Zaitsev, V. F., and Moussiaux, A.,** *Handbook of First Order Partial Differential Equations*, Taylor & Francis, London, 2002.
**Rhee, H., Aris, R., and Amundson, N. R.,** *First Order Partial Differential Equations, Vols. 1 and 2*, Prentice Hall, Englewood Cliffs, New Jersey, 1986 and 1989.
**Rozhdestvenskii, B. L. and Yanenko, N. N.,** *Systems of Quasilinear Equations and Their Applications to Gas Dynamics*, American Mathematical Society, Providence, Rhode Island, 1983.
**Smoller, J.,** *Shock Waves and Reaction-Diffusion Equations*, Springer-Verlag, New York, 1994.
**Subbotin, A. I.,** *Generalized Solutions of First Order PDEs: the Dynamical Optimization Perspective*, Birkhäuser Verlag, Boston, 1995.
**Whitham, G. B.,** *Linear and Nonlinear Waves*, Wiley, New York, 1974.

# Chapter 14
# Linear Partial Differential Equations

## 14.1. Classification of Second-Order Partial Differential Equations

### 14.1.1. Equations with Two Independent Variables

**14.1.1-1. Examples of equations encountered in applications.**

Three basic types of partial differential equations are distinguished—*parabolic*, *hyperbolic*, and *elliptic*. The solutions of the equations pertaining to each of the types have their own characteristic qualitative differences.

The simplest example of a *parabolic equation* is the *heat equation*

$$\frac{\partial w}{\partial t} - \frac{\partial^2 w}{\partial x^2} = 0, \tag{14.1.1.1}$$

where the variables $t$ and $x$ play the role of time and the spatial coordinate, respectively. Note that equation (14.1.1.1) contains only one highest derivative term. Frequently encountered particular solutions of equation (14.1.1.1) can be found in Paragraph T8.1.1-1.

The simplest example of a *hyperbolic equation* is the *wave equation*

$$\frac{\partial^2 w}{\partial t^2} - \frac{\partial^2 w}{\partial x^2} = 0, \tag{14.1.1.2}$$

where the variables $t$ and $x$ play the role of time and the spatial coordinate, respectively. Note that the highest derivative terms in equation (14.1.1.2) differ in sign.

The simplest example of an *elliptic equation* is the *Laplace equation*

$$\frac{\partial^2 w}{\partial x^2} + \frac{\partial^2 w}{\partial y^2} = 0, \tag{14.1.1.3}$$

where $x$ and $y$ play the role of the spatial coordinates. Note that the highest derivative terms in equation (14.1.1.3) have like signs. Frequently encountered particular solutions of equation (14.1.1.3) can be found in Paragraph T8.3.1-1.

Any linear partial differential equation of the second-order with two independent variables can be reduced, by appropriate manipulations, to a simpler equation that has one of the three highest derivative combinations specified above in examples (14.1.1.1), (14.1.1.2), and (14.1.1.3).

**14.1.1-2. Types of equations. Characteristic equations.**

Consider a second-order partial differential equation with two independent variables that has the general form

$$a(x,y)\frac{\partial^2 w}{\partial x^2} + 2b(x,y)\frac{\partial^2 w}{\partial x \partial y} + c(x,y)\frac{\partial^2 w}{\partial y^2} = F\left(x, y, w, \frac{\partial w}{\partial x}, \frac{\partial w}{\partial y}\right), \tag{14.1.1.4}$$

where $a$, $b$, $c$ are some functions of $x$ and $y$ that have continuous derivatives up to the second-order inclusive.*

Given a point $(x, y)$, equation (14.1.1.4) is said to be

$$\begin{aligned} &\textit{parabolic} & &\text{if } b^2 - ac = 0, \\ &\textit{hyperbolic} & &\text{if } b^2 - ac > 0, \\ &\textit{elliptic} & &\text{if } b^2 - ac < 0 \end{aligned}$$

at this point.

In order to reduce equation (14.1.1.4) to a canonical form, one should first write out the characteristic equation

$$a\,(dy)^2 - 2b\,dx\,dy + c\,(dx)^2 = 0,$$

which splits into two equations

$$a\,dy - \left(b + \sqrt{b^2 - ac}\right)dx = 0 \tag{14.1.1.5}$$

and

$$a\,dy - \left(b - \sqrt{b^2 - ac}\right)dx = 0, \tag{14.1.1.6}$$

and then find their general integrals.

**Remark.** The characteristic equations (14.1.1.5)–(14.1.1.5) may be used if $a \not\equiv 0$. If $a \equiv 0$, the simpler equations

$$dx = 0,$$
$$2b\,dy - c\,dx = 0$$

should be used; the first equation has the obvious general solution $x = C$.

### 14.1.1-3. Canonical form of parabolic equations (case $b^2 - ac = 0$).

In this case, equations (14.1.1.5) and (14.1.1.6) coincide and have a common general integral,

$$\varphi(x, y) = C.$$

By passing from $x$, $y$ to new independent variables $\xi$, $\eta$ in accordance with the relations

$$\xi = \varphi(x, y), \qquad \eta = \eta(x, y),$$

where $\eta = \eta(x, y)$ is any twice differentiable function that satisfies the condition of nondegeneracy of the Jacobian $\frac{D(\xi, \eta)}{D(x, y)}$ in the given domain, we reduce equation (14.1.1.4) to the canonical form

$$\frac{\partial^2 w}{\partial \eta^2} = F_1\left(\xi, \eta, w, \frac{\partial w}{\partial \xi}, \frac{\partial w}{\partial \eta}\right). \tag{14.1.1.7}$$

As $\eta$, one can take $\eta = x$ or $\eta = y$.

It is apparent that the transformed equation (14.1.1.7) has only one highest-derivative term, just as the heat equation (14.1.1.1).

**Remark.** In the degenerate case where the function $F_1$ does not depend on the derivative $\partial_\xi w$, equation (14.1.1.7) is an ordinary differential equation for the variable $\eta$, in which $\xi$ serves as a parameter.

---

* The right-hand side of equation (14.1.1.4) may be nonlinear. The classification and the procedure of reducing such equations to a canonical form are only determined by the left-hand side of the equation.

**14.1.1-4. Canonical forms of hyperbolic equations (case $b^2 - ac > 0$).**

The general integrals
$$\varphi(x, y) = C_1, \qquad \psi(x, y) = C_2$$
of equations (14.1.1.5) and (14.1.1.6) are real and different. These integrals determine two different families of real characteristics.

By passing from $x$, $y$ to new independent variables $\xi$, $\eta$ in accordance with the relations
$$\xi = \varphi(x, y), \qquad \eta = \psi(x, y),$$
we reduce equation (14.1.1.4) to
$$\frac{\partial^2 w}{\partial \xi \partial \eta} = F_2\left(\xi, \eta, w, \frac{\partial w}{\partial \xi}, \frac{\partial w}{\partial \eta}\right).$$

This is the so-called *first canonical form of a hyperbolic equation.*

The transformation
$$\xi = t + z, \qquad \eta = t - z$$
brings the above equation to another canonical form,
$$\frac{\partial^2 w}{\partial t^2} - \frac{\partial^2 w}{\partial z^2} = F_3\left(t, z, w, \frac{\partial w}{\partial t}, \frac{\partial w}{\partial z}\right),$$
where $F_3 = 4F_2$. This is the so-called *second canonical form of a hyperbolic equation.* Apart from notation, the left-hand side of the last equation coincides with that of the wave equation (14.1.1.2).

In some cases, reduction of an equation to a canonical form allows finding its general solution.

**Example.** The equation
$$kx\frac{\partial^2 w}{\partial x^2} + \frac{\partial^2 w}{\partial x \partial y} = 0$$
is a special case of equation (14.1.1.4) with $a = kx$, $b = \frac{1}{2}$, $c = 0$, and $F = 0$. The characteristic equations
$$kx\, dy - dx = 0,$$
$$dy = 0$$
have the general integrals $ky - \ln|x| = C_1$ and $y = C_2$. Switching to the new independent variables
$$\xi = ky - \ln|x|, \qquad \eta = y$$
reduces the original equation to the canonical form
$$\frac{\partial^2 w}{\partial \xi \partial \eta} = k\frac{\partial w}{\partial \xi}.$$
Integrating with respect to $\xi$ yields the linear first-order equation
$$\frac{\partial w}{\partial \eta} = kw + f(\eta)$$
where $f(\eta)$ is an arbitrary function. Its general solution is expressed as
$$w = e^{k\eta} g(\xi) + e^{k\eta} \int e^{-k\eta} f(\eta)\, d\eta,$$
where $g(\xi)$ is an arbitrary function.

### 14.1.1-5. Canonical form of elliptic equations (case $b^2 - ac < 0$).

In this case the general integrals of equations (14.1.1.5) and (14.1.1.6) are complex conjugate; these determine two families of complex characteristics.

Let the general integral of equation (14.1.1.5) have the form

$$\varphi(x,y) + i\psi(x,y) = C, \qquad i^2 = -1,$$

where $\varphi(x,y)$ and $\psi(x,y)$ are real-valued functions.

By passing from $x$, $y$ to new independent variables $\xi$, $\eta$ in accordance with the relations

$$\xi = \varphi(x,y), \qquad \eta = \psi(x,y),$$

we reduce equation (14.1.1.4) to the canonical form

$$\frac{\partial^2 w}{\partial \xi^2} + \frac{\partial^2 w}{\partial \eta^2} = F_4\left(\xi, \eta, w, \frac{\partial w}{\partial \xi}, \frac{\partial w}{\partial \eta}\right).$$

Apart from notation, the left-hand side of the last equation coincides with that of the Laplace equation (14.1.1.3).

### 14.1.1-6. Linear constant-coefficient partial differential equations.

1°. When reduced to a canonical form, linear homogeneous constant-coefficient partial differential equations

$$a\frac{\partial^2 w}{\partial x^2} + 2b\frac{\partial^2 w}{\partial x \partial y} + c\frac{\partial^2 w}{\partial y^2} + p\frac{\partial w}{\partial x} + q\frac{\partial w}{\partial y} + sw = 0 \qquad (14.1.1.8)$$

admit further simplifications. In general, the substitution

$$w(x,y) = \exp(\beta_1 \xi + \beta_2 \eta) u(\xi, \eta) \qquad (14.1.1.9)$$

can be used. Here, $\xi$ and $\eta$ are new variables used to reduce equation (14.1.1.8) to a canonical form (see Paragraphs 14.1.1-3 to 14.1.1-5); the coefficients $\beta_1$ and $\beta_2$ in (14.1.1.9) are chosen so that there is only one first derivative remaining in a parabolic equation or both first derivatives vanish in a hyperbolic or an elliptic equation. For final results, see Table 14.1.

2°. The coefficients $k$ and $k_1$ in the reduced hyperbolic and elliptic equations (see the third, fourth, and fifth rows in Table 14.1) are expressed as

$$k = \frac{2bpq - aq^2 - cp^2}{16a(b^2 - ac)^2} - \frac{s}{4a(b^2 - ac)}, \qquad k_1 = \frac{s}{4b^2} + \frac{cp^2 - 2bpq}{16b^4}. \qquad (14.1.1.10)$$

If the coefficients in equation (14.1.1.8) satisfy the relation

$$2bpq - aq^2 - cp^2 - 4s(b^2 - ac) = 0,$$

then $k = 0$; in this case with $a \neq 0$, the general solution of the corresponding hyperbolic equation has the form

$$w(x,y) = \exp(\beta_1 \xi + \beta_2 \eta)\bigl[f(\xi) + g(\eta)\bigr], \qquad D = b^2 - ac > 0,$$

$$\xi = ay - (b + \sqrt{D})x, \qquad \eta = ay - (b - \sqrt{D})x,$$

$$\beta_1 = \frac{aq - bp}{4aD} + \frac{p}{4a\sqrt{D}}, \qquad \beta_2 = \frac{aq - bp}{4aD} - \frac{p}{4a\sqrt{D}},$$

where $f(\xi)$ and $g(\xi)$ are arbitrary functions.

TABLE 14.1
Reduction of linear homogeneous constant-coefficient partial differential equations (14.1.1.8) using transformation (14.1.1.9); the constants $k$ and $k_1$ are given by formulas (14.1.1.10)

| Type of equation, conditions on coefficients | Variables $\xi$ and $\eta$ in transformation (14.1.1.9) | Coefficients $\beta_1$ and $\beta_2$ in transformation (14.1.1.9) | Reduced equation |
|---|---|---|---|
| Parabolic equation, $a=b=0$, $c\neq 0$, $p\neq 0$ | $\xi = -\dfrac{c}{p}x$, $\eta = y$ | $\beta_1 = \dfrac{4cs-q^2}{4c^2}$, $\beta_2 = -\dfrac{q}{2c}$ | $u_\xi - u_{\eta\eta} = 0$ |
| Parabolic equation, $b^2-ac=0$ ($aq-bp\neq 0$, $\|a\|+\|b\|\neq 0$) | $\xi = \dfrac{a(ay-bx)}{bp-aq}$, $\eta = x$ | $\beta_1 = \dfrac{4as-p^2}{4a^2}$, $\beta_2 = -\dfrac{p}{2a}$ | $u_\xi - u_{\eta\eta} = 0$ |
| Hyperbolic equation, $a\neq 0$, $D = b^2-ac > 0$ | $\xi = ay - (b+\sqrt{D})x$, $\eta = ay - (b-\sqrt{D})x$ | $\beta_{1,2} = \dfrac{aq-bp}{4aD} \pm \dfrac{p}{4a\sqrt{D}}$ | $u_{\xi\eta} + ku = 0$ |
| Hyperbolic equation, $a=0$, $b\neq 0$ | $\xi = x$, $\eta = 2by - cx$ | $\beta_1 = \dfrac{cp-2bq}{4b^2}$, $\beta_2 = -\dfrac{p}{4b^2}$ | $u_{\xi\eta} + k_1 u = 0$ |
| Elliptic equation, $D = b^2-ac < 0$ | $\xi = ay-bx$, $\eta = \sqrt{\|D\|}\,x$ | $\beta_1 = \dfrac{aq-bp}{2aD}$, $\beta_2 = -\dfrac{p}{2a\sqrt{\|D\|}}$ | $u_{\xi\xi} + u_{\eta\eta} + 4ku = 0$ |
| Ordinary differential equation, $b^2-ac=0$, $aq-bp=0$ | $\xi = ay-bx$, $\eta = x$ | $\beta_1 = \beta_2 = 0$ | $aw_{\eta\eta} + pw_\eta + sw = 0$ |

3°. In the degenerate case $b^2 - ac = 0$, $aq - bp = 0$ (where the original equation is reduced to an ordinary differential equation; see the last row in Table 14.1), the general solution of equation (14.1.1.8) is expressed as

$$w = \exp\left(-\frac{px}{2a}\right)\left[f(ay-bx)\exp\left(\frac{x\sqrt{\lambda}}{2a}\right) + g(ay-bx)\exp\left(-\frac{x\sqrt{\lambda}}{2a}\right)\right] \quad \text{if} \quad \lambda = p^2 - 4as > 0,$$

$$w = \exp\left(-\frac{px}{2a}\right)\left[f(ay-bx)\sin\left(\frac{x\sqrt{|\lambda|}}{2a}\right) + g(ay-bx)\cos\left(\frac{x\sqrt{|\lambda|}}{2a}\right)\right] \quad \text{if} \quad \lambda = p^2 - 4as < 0,$$

$$w = \exp\left(-\frac{px}{2a}\right)\left[f(ay-bx) + xg(ay-bx)\right] \quad \text{if} \quad 4as - p^2 = 0,$$

where $f(z)$ and $g(z)$ are arbitrary functions.

### 14.1.2. Equations with Many Independent Variables

Let us consider a second-order partial differential equation with $n$ independent variables $x_1, \ldots, x_n$ that has the form

$$\sum_{i,j=1}^{n} a_{ij}(\mathbf{x}) \frac{\partial^2 w}{\partial x_i \partial x_j} = F\left(\mathbf{x}, w, \frac{\partial w}{\partial x_1}, \ldots, \frac{\partial w}{\partial x_n}\right), \qquad (14.1.2.1)$$

where the $a_{ij}$ are some functions that have continuous derivatives with respect to all variables to the second-order inclusive, and $\mathbf{x} = \{x_1, \ldots, x_n\}$. [The right-hand side of equation (14.1.2.1) may be nonlinear. The left-hand side only is required for the classification of this equation.]

At a point $\mathbf{x} = \mathbf{x}_0$, the following quadratic form is assigned to equation (14.1.2.1):

$$Q = \sum_{i,j=1}^{n} a_{ij}(\mathbf{x}_0) \xi_i \xi_j. \qquad (14.1.2.2)$$

TABLE 14.2
Classification of equations with many independent variables

| Type of equation (14.1.2.1) at a point $\mathbf{x} = \mathbf{x}_0$ | Coefficients of the canonical form (14.1.2.4) |
|---|---|
| Parabolic (in the broad sense) | At least one coefficient of the $c_i$ is zero |
| Hyperbolic (in the broad sense) | All $c_i$ are nonzero and some $c_i$ differ in sign |
| Elliptic | All $c_i$ are nonzero and have like signs |

By an appropriate linear nondegenerate transformation

$$\xi_i = \sum_{k=1}^{n} \beta_{ik}\eta_k \qquad (i = 1, \ldots, n) \qquad (14.1.2.3)$$

the quadratic form (14.1.2.2) can be reduced to the canonical form

$$Q = \sum_{i=1}^{n} c_i \eta_i^2, \qquad (14.1.2.4)$$

where the coefficients $c_i$ assume the values 1, –1, and 0. The number of negative and zero coefficients in (14.1.2.4) does not depend on the way in which the quadratic form is reduced to the canonical form.

Table 14.2 presents the basic criteria according to which the equations with many independent variables are classified.

Suppose all coefficients of the highest derivatives in (14.1.2.1) are constant, $a_{ij}$ = const. By introducing the new independent variables $y_1, \ldots, y_n$ in accordance with the formulas $y_i = \sum_{k=1}^{n} \beta_{ik} x_k$, where the $\beta_{ik}$ are the coefficients of the linear transformation (14.1.2.3), we reduce equation (14.1.2.1) to the canonical form

$$\sum_{i=1}^{n} c_i \frac{\partial^2 w}{\partial y_i^2} = F_1\left(\mathbf{y}, w, \frac{\partial w}{\partial y_1}, \ldots, \frac{\partial w}{\partial y_n}\right). \qquad (14.1.2.5)$$

Here, the coefficients $c_i$ are the same as in the quadratic form (14.1.2.4), and $\mathbf{y} = \{y_1, \ldots, y_n\}$.

Remark 1. Among the parabolic equations, it is conventional to distinguish the parabolic equations in the narrow sense, i.e., the equations for which only one of the coefficients, $c_k$, is zero, while the other $c_i$ is the same, and in this case the right-hand side of equation (14.1.2.5) must contain the first-order partial derivative with respect to $y_k$.

Remark 2. In turn, the hyperbolic equations are divided into normal hyperbolic equations — for which all $c_i$ but one have like signs — and ultrahyperbolic equations — for which there are two or more positive $c_i$ and two or more negative $c_i$.

Specific equations of parabolic, elliptic, and hyperbolic types will be discussed further in Section 14.2.

## 14.2. Basic Problems of Mathematical Physics

### 14.2.1. Initial and Boundary Conditions. Cauchy Problem. Boundary Value Problems

Every equation of mathematical physics governs infinitely many qualitatively similar phenomena or processes. This follows from the fact that differential equations have infinitely many particular solutions. The specific solution that describes the physical phenomenon under study is separated from the set of particular solutions of the given differential equation by means of the initial and boundary conditions.

Throughout this section, we consider linear equations in the $n$-dimensional Euclidean space $\mathbb{R}^n$ or in an open domain $V \in \mathbb{R}^n$ (exclusive of the boundary) with a sufficiently smooth boundary $S = \partial V$.

### 14.2.1-1. Parabolic equations. Initial and boundary conditions.

In general, a linear second-order partial differential equation of the parabolic type with $n$ independent variables can be written as

$$\frac{\partial w}{\partial t} - L_{\mathbf{x},t}[w] = \Phi(\mathbf{x}, t), \qquad (14.2.1.1)$$

where

$$L_{\mathbf{x},t}[w] \equiv \sum_{i,j=1}^{n} a_{ij}(\mathbf{x}, t) \frac{\partial^2 w}{\partial x_i \partial x_j} + \sum_{i=1}^{n} b_i(\mathbf{x}, t) \frac{\partial w}{\partial x_i} + c(\mathbf{x}, t) w, \qquad (14.2.1.2)$$

$$\mathbf{x} = \{x_1, \ldots, x_n\}, \quad \sum_{i,j=1}^{n} a_{ij}(\mathbf{x}, t) \xi_i \xi_j \geq \sigma \sum_{i=1}^{n} \xi_i^2, \quad \sigma > 0.$$

Parabolic equations govern unsteady thermal, diffusion, and other phenomena dependent on time $t$.

Equation (14.2.1.1) is called homogeneous if $\Phi(\mathbf{x}, t) \equiv 0$.

*Cauchy problem* ($t \geq 0$, $\mathbf{x} \in \mathbb{R}^n$). Find a function $w$ that satisfies equation (14.2.1.1) for $t > 0$ and the initial condition

$$w = f(\mathbf{x}) \quad \text{at} \quad t = 0. \qquad (14.2.1.3)$$

*Boundary value problem*\* ($t \geq 0$, $\mathbf{x} \in V$). Find a function $w$ that satisfies equation (14.2.1.1) for $t > 0$, the initial condition (14.2.1.3), and the boundary condition

$$\Gamma_{\mathbf{x},t}[w] = g(\mathbf{x}, t) \quad \text{at} \quad \mathbf{x} \in S \quad (t > 0). \qquad (14.2.1.4)$$

In general, $\Gamma_{\mathbf{x},t}$ is a first-order linear differential operator in the space variables $\mathbf{x}$ with coefficient dependent on $\mathbf{x}$ and $t$. The basic types of boundary conditions are described in Subsection 14.2.2.

The initial condition (14.2.1.3) is called homogeneous if $f(\mathbf{x}) \equiv 0$. The boundary condition (14.2.1.4) is called homogeneous if $g(\mathbf{x}, t) \equiv 0$.

### 14.2.1-2. Hyperbolic equations. Initial and boundary conditions.

Consider a second-order linear partial differential equation of the hyperbolic type with $n$ independent variables of the general form

$$\frac{\partial^2 w}{\partial t^2} + \varphi(\mathbf{x}, t) \frac{\partial w}{\partial t} - L_{\mathbf{x},t}[w] = \Phi(\mathbf{x}, t), \qquad (14.2.1.5)$$

---

\* *Boundary value problems* for parabolic and hyperbolic equations are sometimes called *mixed* or *initial-boundary value problems*.

where the linear differential operator $L_{\mathbf{x},t}$ is defined by (14.2.1.2). Hyperbolic equations govern unsteady wave processes, which depend on time $t$.

Equation (14.2.1.5) is said to be homogeneous if $\Phi(\mathbf{x}, t) \equiv 0$.

*Cauchy problem* ($t \geq 0$, $\mathbf{x} \in \mathbb{R}^n$). Find a function $w$ that satisfies equation (14.2.1.5) for $t > 0$ and the initial conditions

$$\begin{aligned} w &= f_0(\mathbf{x}) \quad \text{at} \quad t = 0, \\ \partial_t w &= f_1(\mathbf{x}) \quad \text{at} \quad t = 0. \end{aligned} \qquad (14.2.1.6)$$

*Boundary value problem* ($t \geq 0$, $\mathbf{x} \in V$). Find a function $w$ that satisfies equation (14.2.1.5) for $t > 0$, the initial conditions (14.2.1.6), and boundary condition (14.2.1.4).

The initial conditions (14.2.1.6) are called homogeneous if $f_0(\mathbf{x}) \equiv 0$ and $f_1(\mathbf{x}) \equiv 0$.

*Generalized Cauchy problem.* In the generalized Cauchy problem for a hyperbolic equation with two independent variables, values of the unknown function and its first derivatives are prescribed on a curve in the $(x, t)$ plane. Alternatively, values of the unknown function and its derivative along the normal to this curve may be prescribed. For more details, see Paragraph 14.8.4-4.

*Goursat problem.* On the characteristics of a hyperbolic equation with two independent variables, the values of the unknown function $w$ are prescribed; for details, see Paragraph 14.8.4-5).

### 14.2.1-3. Elliptic equations. Boundary conditions.

In general, a second-order linear partial differential equation of elliptic type with $n$ independent variables can be written as

$$-L_\mathbf{x}[w] = \Phi(\mathbf{x}), \qquad (14.2.1.7)$$

where

$$L_\mathbf{x}[w] \equiv \sum_{i,j=1}^n a_{ij}(\mathbf{x}) \frac{\partial^2 w}{\partial x_i \partial x_j} + \sum_{i=1}^n b_i(\mathbf{x}) \frac{\partial w}{\partial x_i} + c(\mathbf{x})w, \qquad (14.2.1.8)$$

$$\sum_{i,j=1}^n a_{ij}(\mathbf{x}) \xi_i \xi_j \geq \sigma \sum_{i=1}^n \xi_i^2, \quad \sigma > 0.$$

Elliptic equations govern steady-state thermal, diffusion, and other phenomena independent of time $t$.

Equation (14.2.1.7) is said to be homogeneous if $\Phi(\mathbf{x}) \equiv 0$.

*Boundary value problem.* Find a function $w$ that satisfies equation (14.2.1.7) and the boundary condition

$$\Gamma_\mathbf{x}[w] = g(\mathbf{x}) \quad \text{at} \quad \mathbf{x} \in S. \qquad (14.2.1.9)$$

In general, $\Gamma_\mathbf{x}$ is a first-order linear differential operator in the space variables $\mathbf{x}$. The basic types of boundary conditions are described below in Subsection 14.2.2.

The boundary condition (14.2.1.9) is called homogeneous if $g(\mathbf{x}) \equiv 0$. The boundary value problem (14.2.1.7)–(14.2.1.9) is said to be homogeneous if $\Phi \equiv 0$ and $g \equiv 0$.

TABLE 14.3
Boundary conditions for various boundary value problems specified by
parabolic and hyperbolic equations in two independent variables ($x_1 \leq x \leq x_2$)

| Type of problem | Boundary condition at $x = x_1$ | Boundary condition at $x = x_2$ |
|---|---|---|
| First boundary value problem | $w = g_1(t)$ | $w = g_2(t)$ |
| Second boundary value problem | $\partial_x w = g_1(t)$ | $\partial_x w = g_2(t)$ |
| Third boundary value problem | $\partial_x w + \beta_1 w = g_1(t)$ ($\beta_1 < 0$) | $\partial_x w + \beta_2 w = g_2(t)$ ($\beta_2 > 0$) |
| Mixed boundary value problem | $w = g_1(t)$ | $\partial_x w = g_2(t)$ |
| Mixed boundary value problem | $\partial_x w = g_1(t)$ | $w = g_2(t)$ |

## 14.2.2. First, Second, Third, and Mixed Boundary Value Problems

For any (parabolic, hyperbolic, and elliptic) second-order partial differential equations, it is conventional to distinguish four basic types of boundary value problems, depending on the form of the boundary conditions (14.2.1.4) [see also the analogous condition (14.2.1.9)]. For simplicity, here we confine ourselves to the case where the coefficients $a_{ij}$ of equations (14.2.1.1) and (14.2.1.5) have the special form

$$a_{ij}(\mathbf{x}, t) = a(\mathbf{x}, t)\delta_{ij}, \quad \delta_{ij} = \begin{cases} 1 & \text{if } i = j, \\ 0 & \text{if } i \neq j. \end{cases}$$

This situation is rather frequent in applications; such coefficients are used to describe various phenomena (processes) in isotropic media.

The function $w(\mathbf{x}, t)$ takes prescribed values at the boundary $S$ of the domain:

$$w(\mathbf{x}, t) = g_1(\mathbf{x}, t) \quad \text{for} \quad \mathbf{x} \in S. \tag{14.2.2.1}$$

*Second boundary value problem.* The derivative along the (outward) normal is prescribed at the boundary $S$ of the domain:

$$\frac{\partial w}{\partial N} = g_2(\mathbf{x}, t) \quad \text{for} \quad \mathbf{x} \in S. \tag{14.2.2.2}$$

In heat transfer problems, where $w$ is temperature, the left-hand side of the boundary condition (14.2.2.2) is proportional to the heat flux per unit area of the surface $S$.

*Third boundary value problem.* A linear relationship between the unknown function and its normal derivative is prescribed at the boundary $S$ of the domain:

$$\frac{\partial w}{\partial N} + k(\mathbf{x}, t)w = g_3(\mathbf{x}, t) \quad \text{for} \quad \mathbf{x} \in S. \tag{14.2.2.3}$$

Usually, it is assumed that $k(\mathbf{x}, t) = \text{const}$. In mass transfer problems, where $w$ is concentration, the boundary condition (14.2.2.3) with $g_3 \equiv 0$ describes a surface chemical reaction of the first order.

*Mixed boundary value problems.* Conditions of various types, listed above, are set at different portions of the boundary $S$.

If $g_1 \equiv 0$, $g_2 \equiv 0$, or $g_3 \equiv 0$, the respective boundary conditions (14.2.2.1), (14.2.2.2), (14.2.2.3) are said to be homogeneous.

Boundary conditions for various boundary value problems for parabolic and hyperbolic equations in two independent variables $x$ and $t$ are displayed in Table 14.3. The equation coefficients are assumed to be continuous, with the coefficients of the highest derivatives being nonzero in the range $x_1 \leq x \leq x_2$ considered.

*Remark.* For elliptic equations, the first boundary value problem is often called the *Dirichlet problem*, and the second boundary value problem is called the *Neumann problem*.

## 14.3. Properties and Exact Solutions of Linear Equations

### 14.3.1. Homogeneous Linear Equations and Their Particular Solutions

**14.3.1-1. Preliminary remarks.**

For brevity, in this paragraph a homogeneous linear partial differential equation will be written as

$$\mathcal{L}[w] = 0. \tag{14.3.1.1}$$

For second-order linear parabolic and hyperbolic equations, the linear differential operator $\mathcal{L}[w]$ is defined by the left-hand side of equations (14.2.1.1) and (14.2.1.5), respectively. It is assumed that equation (14.3.1.1) is an arbitrary homogeneous linear partial differential equation of any order in the variables $t, x_1, \ldots, x_n$ with sufficiently smooth coefficients.

A linear operator $\mathcal{L}$ possesses the properties

$$\mathcal{L}[w_1 + w_2] = \mathcal{L}[w_1] + \mathcal{L}[w_2],$$
$$\mathcal{L}[Aw] = A\mathcal{L}[w], \quad A = \text{const.}$$

An arbitrary homogeneous linear equation (14.3.1.1) has a trivial solution, $w \equiv 0$.

A function $w$ is called a *classical solution* of equation (14.3.1.1) if $w$, when substituted into (14.3.1.1), turns the equation into an identity and if all partial derivatives of $w$ that occur in (14.3.1.1) are continuous; the notion of a classical solution is directly linked to the range of the independent variables. In what follows, we usually write "solution" instead of "classical solution" for brevity.

**14.3.1-2. Usage of particular solutions for the construction of other solutions.**

Below are some properties of particular solutions of homogeneous linear equations.

$1^\circ$. Let $w_1 = w_1(\mathbf{x}, t)$, $w_2 = w_2(\mathbf{x}, t)$, ..., $w_k = w_k(\mathbf{x}, t)$ be any particular solutions of the homogeneous equation (14.3.1.1). Then the linear combination

$$w = A_1 w_1 + A_2 w_2 + \cdots + A_k w_k \tag{14.3.1.2}$$

with arbitrary constants $A_1, A_2, \ldots, A_k$ is also a solution of equation (14.3.1.1); in physics, this property is known as the *principle of linear superposition*.

Suppose $\{w_k\}$ is an infinite sequence of solutions of equation (14.3.1.1). Then the series $\sum\limits_{k=1}^{\infty} w_k$, irrespective of its convergence, is called a *formal solution* of (14.3.1.1). If the solutions $w_k$ are classical, the series is uniformly convergent, and the sum of the series has all the necessary particular derivatives, then the sum of the series is a classical solution of equation (14.3.1.1).

$2^\circ$. Let the coefficients of the linear differential operator $\mathcal{L}$ be independent of time $t$. If equation (14.3.1.1) has a particular solution $\widetilde{w} = \widetilde{w}(\mathbf{x}, t)$, then the partial derivatives of $\widetilde{w}$ with respect to time,*

$$\frac{\partial \widetilde{w}}{\partial t}, \quad \frac{\partial^2 \widetilde{w}}{\partial t^2}, \quad \ldots, \quad \frac{\partial^k \widetilde{w}}{\partial t^k}, \quad \ldots,$$

are also solutions of equation (14.3.1.1).

---

\* Here and in what follows, it is assumed that the particular solution $\widetilde{w}$ is differentiable sufficiently many times with respect to $t$ and $x_1, \ldots, x_n$ (or the parameters).

3°. Let the coefficients of the linear differential operator $\mathcal{L}$ be independent of the space variables $x_1, \ldots, x_n$. If equation (14.3.1.1) has a particular solution $\widetilde{w} = \widetilde{w}(\mathbf{x}, t)$, then the partial derivatives of $\widetilde{w}$ with respect to the space coordinates

$$\frac{\partial \widetilde{w}}{\partial x_1}, \quad \frac{\partial \widetilde{w}}{\partial x_2}, \quad \frac{\partial \widetilde{w}}{\partial x_3}, \quad \ldots, \quad \frac{\partial^2 \widetilde{w}}{\partial x_1^2}, \quad \frac{\partial^2 \widetilde{w}}{\partial x_1 \partial x_2}, \quad \ldots, \quad \frac{\partial^{k+m} \widetilde{w}}{\partial x_2^k \partial x_3^m}, \quad \ldots$$

are also solutions of equation (14.3.1.1).

If the coefficients of $\mathcal{L}$ are independent of only one space coordinate, say $x_1$, and equation (14.3.1.1) has a particular solution $\widetilde{w} = \widetilde{w}(\mathbf{x}, t)$, then the partial derivatives

$$\frac{\partial \widetilde{w}}{\partial x_1}, \quad \frac{\partial^2 \widetilde{w}}{\partial x_1^2}, \quad \ldots, \quad \frac{\partial^k \widetilde{w}}{\partial x_1^k}, \quad \ldots$$

are also solutions of equation (14.3.1.1).

4°. Let the coefficients of the linear differential operator $\mathcal{L}$ be constant and let equation (14.3.1.1) have a particular solution $\widetilde{w} = \widetilde{w}(\mathbf{x}, t)$. Then any particular derivatives of $\widetilde{w}$ with respect to time and the space coordinates (inclusive mixed derivatives)

$$\frac{\partial \widetilde{w}}{\partial t}, \quad \frac{\partial \widetilde{w}}{\partial x_1}, \quad \ldots, \quad \frac{\partial^2 \widetilde{w}}{\partial x_2^2}, \quad \frac{\partial^2 \widetilde{w}}{\partial t \partial x_1}, \quad \ldots, \quad \frac{\partial^k \widetilde{w}}{\partial x_3^k}, \quad \ldots$$

are solutions of equation (14.3.1.1).

5°. Suppose equation (14.3.1.1) has a particular solution dependent on a parameter $\mu$, $\widetilde{w} = \widetilde{w}(\mathbf{x}, t; \mu)$, and the coefficients of the linear differential operator $\mathcal{L}$ are independent of $\mu$ (but can depend on time and the space coordinates). Then, by differentiating $\widetilde{w}$ with respect to $\mu$, one obtains other solutions of equation (14.3.1.1),

$$\frac{\partial \widetilde{w}}{\partial \mu}, \quad \frac{\partial^2 \widetilde{w}}{\partial \mu^2}, \quad \ldots, \quad \frac{\partial^k \widetilde{w}}{\partial \mu^k}, \quad \ldots$$

**Example 1.** The linear equation

$$\frac{\partial w}{\partial t} = a \frac{\partial^2 w}{\partial x^2} + bw$$

has a particular solution

$$w(x, t) = \exp[\mu x + (a\mu^2 + b)t],$$

where $\mu$ is an arbitrary constant. Differentiating this equation with respect to $\mu$ yields another solution

$$w(x, t) = (x + 2a\mu t) \exp[\mu x + (a\mu^2 + b)t].$$

Let some constants $\mu_1, \ldots, \mu_k$ belong to the range of the parameter $\mu$. Then the sum

$$w = A_1 \widetilde{w}(\mathbf{x}, t; \mu_1) + \cdots + A_k \widetilde{w}(\mathbf{x}, t; \mu_k), \tag{14.3.1.3}$$

where $A_1, \ldots, A_k$ are arbitrary constants, is also a solution of the homogeneous linear equation (14.3.1.1). The number of terms in sum (14.3.1.3) can be both finite and infinite.

6°. Another effective way of constructing solutions involves the following. The particular solution $\widetilde{w}(\mathbf{x}, t; \mu)$, which depends on the parameter $\mu$ (as before, it is assumed that the coefficients of the linear differential operator $\mathcal{L}$ are independent of $\mu$), is first multiplied by an arbitrary function $\varphi(\mu)$. Then the resulting expression is integrated with respect to $\mu$ over some interval $[\alpha, \beta]$. Thus, one obtains a new function,

$$\int_\alpha^\beta \widetilde{w}(\mathbf{x}, t; \mu) \varphi(\mu) \, d\mu,$$

which is also a solution of the original homogeneous linear equation.

The properties listed in Items 1°–6° enable one to use known particular solutions to construct other particular solutions of homogeneous linear equations of mathematical physics.

TABLE 14.4
Homogeneous linear partial differential equations that admit multiplicative separable solutions

| No. | Form of equation (14.3.1.1) | Form of particular solutions |
|---|---|---|
| 1 | Equation coefficients are constant | $w(\mathbf{x}, t) = A \exp(\lambda t + \beta_1 x_1 + \cdots + \beta_n x_n)$, $\lambda, \beta_1, \ldots, \beta_n$ are related by an algebraic equation |
| 2 | Equation coefficients are independent of time $t$ | $w(\mathbf{x}, t) = e^{\lambda t} \psi(\mathbf{x})$, $\lambda$ is an arbitrary constant, $\mathbf{x} = \{x_1, \ldots, x_n\}$ |
| 3 | Equation coefficients are independent of the coordinates $x_1, \ldots, x_n$ | $w(\mathbf{x}, t) = \exp(\beta_1 x_1 + \cdots + \beta_n x_n)\psi(t)$, $\beta_1, \ldots, \beta_n$ are arbitrary constants |
| 4 | Equation coefficients are independent of the coordinates $x_1, \ldots, x_k$ | $w(\mathbf{x}, t) = \exp(\beta_1 x_1 + \cdots + \beta_k x_k)\psi(t, x_{k+1}, \ldots, x_n)$, $\beta_1, \ldots, \beta_k$ are arbitrary constants |
| 5 | $L_t[w] + L_\mathbf{x}[w] = 0$, operator $L_t$ depends on only $t$, operator $L_\mathbf{x}$ depends on only $\mathbf{x}$ | $w(\mathbf{x}, t) = \varphi(t)\psi(\mathbf{x})$, $\varphi(t)$ satisfies the equation $L_t[\varphi] + \lambda\varphi = 0$, $\psi(\mathbf{x})$ satisfies the equation $L_\mathbf{x}[\psi] - \lambda\psi = 0$ |
| 6 | $L_t[w] + L_1[w] + \cdots + L_n[w] = 0$, operator $L_t$ depends on only $t$, operator $L_k$ depends on only $x_k$ | $w(\mathbf{x}, t) = \varphi(t)\psi_1(x_1)\ldots\psi_n(x_n)$, $\varphi(t)$ satisfies the equation $L_t[\varphi] + \lambda\varphi = 0$, $\psi_k(x_k)$ satisfies the equation $L_k[\psi_k] + \beta_k\psi_k = 0$, $\lambda + \beta_1 + \cdots + \beta_n = 0$ |
| 7 | $f_0(x_1)L_t[w] + \sum_{k=1}^{n} f_k(x_1)L_k[w] = 0$, operator $L_t$ depends on only $t$, operator $L_k$ depends on only $x_k$ | $w(\mathbf{x}, t) = \varphi(t)\psi_1(x_1)\ldots\psi_n(x_n)$, $L_t[\varphi] + \lambda\varphi = 0$, $L_k[\psi_k] + \beta_k\psi_k = 0$, $k = 2, \ldots, n$, $f_1(x_1)L_1[\psi_1] - \left[\lambda f_0(x_1) + \sum_{k=2}^{n} \beta_k f_k(x_1)\right]\psi_1 = 0$ |
| 8 | $\dfrac{\partial w}{\partial t} + L_{1,t}[w] + \cdots + L_{n,t}[w] = 0$, where $L_{k,t}[w] = \sum_{s=0}^{m_k} f_{ks}(x_k, t)\dfrac{\partial^s w}{\partial x_k^s}$ | $w(\mathbf{x}, t) = \psi_1(x_1, t)\psi_2(x_2, t)\ldots\psi_n(x_n, t)$, $\dfrac{\partial \psi_k}{\partial t} + L_{k,t}[\psi_k] = \lambda_k(t)\psi_k$, $k = 1, \ldots, n$, $\lambda_1(t) + \lambda_2(t) + \cdots + \lambda_n(t) = 0$ |

### 14.3.1-3. Multiplicative and additive separable solutions.

1°. Many homogeneous linear partial differential equations have solutions that can be represented as the product of functions depending on different arguments. Such solutions are referred to as *multiplicative separable solutions*; very commonly these solutions are briefly, but less accurately, called just *separable solutions*.

Table 14.4 presents the most commonly encountered types of homogeneous linear differential equations with many independent variables that admit exact separable solutions. Linear combinations of particular solutions that correspond to different values of the separation parameters, $\lambda, \beta_1, \ldots, \beta_n$, are also solutions of the equations in question. For brevity, the word "operator" is used to denote "linear differential operator."

For a constant coefficient equation (see the first row in Table 14.4), the separation parameters must satisfy the algebraic equation

$$D(\lambda, \beta_1, \ldots, \beta_n) = 0, \qquad (14.3.1.4)$$

which results from substituting the solution into equation (14.3.1.1). In physical applications, equation (14.3.1.4) is usually referred to as a *dispersion equation*. Any $n$ of the $n + 1$ separation parameters in (14.3.1.4) can be treated as arbitrary.

**Example 2.** Consider the linear equation
$$\frac{\partial^2 w}{\partial t^2} + k\frac{\partial w}{\partial t} = a^2\frac{\partial^2 w}{\partial x^2} + b\frac{\partial w}{\partial x} + cw.$$
A particular solution is sought in the form
$$w = A\exp(\beta x + \lambda t).$$
This results in the dispersion equation $\lambda^2 + k\lambda = a^2\beta^2 + b\beta + c$, where one of the two parameters $\beta$ or $\lambda$ can be treated as arbitrary.

For more complex multiplicative separable solutions to this equation, see Subsection 14.4.1.

Note that constant coefficient equations also admit more sophisticated solutions; see the second and third rows, the last column.

The eighth row of Table 14.4 presents the case of *incomplete separation of variables* where the solution is separated with respect to the space variables $x_1, \ldots, x_n$, but is not separated with respect to time $t$.

**Remark 1.** For stationary equations that do not depend on $t$, one should set $\lambda = 0$, $L_t[w] \equiv 0$, and $\varphi(t) \equiv 1$ in rows 1, 6, and 7 of Table 14.4.

**Remark 2.** Multiplicative separable solutions play an important role in the theory of linear partial differential equations; they are used for finding solutions to stationary and nonstationary boundary value problems; see Sections 14.4 and 14.7–14.9.

2°. Linear partial differential equations of the form
$$L_t[w] + L_\mathbf{x}[w] = f(\mathbf{x}) + g(t),$$
where $L_t$ is a linear differential operator that depends on only $t$ and $L_\mathbf{x}$ is a linear differential operator that depends on only $\mathbf{x}$, have solutions that can be represented as the sum of functions depending on different arguments
$$w = u(\mathbf{x}) + v(t).$$
Such solutions are referred to as *additive separable solutions*.

**Example 3.** The equation from Example 2 admits an exact additive separable solution $w = u(x) + v(t)$ with $u(x)$ and $v(t)$ described by the linear constant-coefficient ordinary differential equations
$$a^2 u''_{xx} + bu'_x + cu = C,$$
$$v''_{tt} + kv'_t - cv = C,$$
where $C$ is an arbitrary constant, which are easy to integrate. A more general partial differential equation with variable coefficients $a = a(x)$, $b = b(x)$, $k = k(t)$, and $c = $ const also admits an additive separable solution.

---

**14.3.1-4. Solutions in the form of infinite series in $t$.**

1°. The equation
$$\frac{\partial w}{\partial t} = M[w],$$
where $M$ is an arbitrary linear differential operator of the second (or any) order that only depends on the space variables, has the formal series solution
$$w(\mathbf{x},t) = f(\mathbf{x}) + \sum_{k=1}^{\infty} \frac{t^k}{k!} M^k[f(\mathbf{x})], \qquad M^k[f] = M\big[M^{k-1}[f]\big],$$
where $f(\mathbf{x})$ is an arbitrary infinitely differentiable function. This solution satisfies the initial condition $w(\mathbf{x},0) = f(\mathbf{x})$.

**Example 4.** Consider the heat equation
$$\frac{\partial w}{\partial t} = a \frac{\partial^2 w}{\partial x^2}.$$
In this case we have $M = a \frac{\partial^2}{\partial x^2}$. Therefore the formal series solution has the form
$$w(x,t) = f(x) + \sum_{k=1}^{\infty} \frac{(at)^k}{k!} f_x^{(2n)}(x), \qquad f_x^{(m)} = \frac{d^m}{dx^m} f(x).$$
If the function $f(x)$ is taken as a polynomial of degree $n$, the solution will also be a polynomial of degree $n$. For example, setting $f(x) = Ax^2 + Bx + C$, we obtain the particular solution
$$w(x,t) = A(x^2 + 2at) + Bx + C.$$

2°. The equation
$$\frac{\partial^2 w}{\partial t^2} = M[w],$$
where $M$ is a linear differential operator, just as in Item 1°, has a formal solution represented by the sum of two series as
$$w(\mathbf{x}, t) = \sum_{k=0}^{\infty} \frac{t^{2k}}{(2k)!} M^k[f(\mathbf{x})] + \sum_{k=0}^{\infty} \frac{t^{2k+1}}{(2k+1)!} M^k[g(\mathbf{x})],$$
where $f(\mathbf{x})$ and $g(\mathbf{x})$ are arbitrary infinitely differentiable functions. This solution satisfies the initial conditions $w(\mathbf{x}, 0) = f(\mathbf{x})$ and $\partial_t w(\mathbf{x}, 0) = g(\mathbf{x})$.

## 14.3.2. Nonhomogeneous Linear Equations and Their Particular Solutions

14.3.2-1. Simplest properties of nonhomogeneous linear equations.

For brevity, we write a nonhomogeneous linear partial differential equation in the form
$$\mathcal{L}[w] = \Phi(\mathbf{x}, t), \tag{14.3.2.1}$$
where the linear differential operator $\mathcal{L}$ is defined above (see the beginning of Paragraph 14.3.1-1).

Below are the simplest properties of particular solutions of the nonhomogeneous equation (14.3.2.1).

1°. If $\widetilde{w}_\Phi(\mathbf{x}, t)$ is a particular solution of the nonhomogeneous equation (14.3.2.1) and $\widetilde{w}_0(\mathbf{x}, t)$ is a particular solution of the corresponding homogeneous equation (14.3.1.1), then the sum
$$A\widetilde{w}_0(\mathbf{x}, t) + \widetilde{w}_\Phi(\mathbf{x}, t),$$
where $A$ is an arbitrary constant, is also a solution of the nonhomogeneous equation (14.3.2.1). The following, more general statement holds: The general solution of the nonhomogeneous equation (14.3.2.1) is the sum of the general solution of the corresponding homogeneous equation (14.3.1.1) and any particular solution of the nonhomogeneous equation (14.3.2.1).

2°. Suppose $w_1$ and $w_2$ are solutions of nonhomogeneous linear equations with the same left-hand side and different right-hand sides, i.e.,
$$\mathcal{L}[w_1] = \Phi_1(\mathbf{x}, t), \qquad \mathcal{L}[w_2] = \Phi_2(\mathbf{x}, t).$$
Then the function $w = w_1 + w_2$ is a solution of the equation
$$\mathcal{L}[w] = \Phi_1(\mathbf{x}, t) + \Phi_2(\mathbf{x}, t).$$

### 14.3.2-2. Fundamental and particular solutions of stationary equations.

Consider the second-order linear stationary (time-independent) nonhomogeneous equation

$$L_\mathbf{x}[w] = -\Phi(\mathbf{x}). \tag{14.3.2.2}$$

Here, $L_\mathbf{x}$ is a linear differential operator of the second (or any) order of general form whose coefficients are dependent on $\mathbf{x}$, where $\mathbf{x} \in \mathbb{R}^n$.

A distribution $\mathscr{E} = \mathscr{E}(\mathbf{x}, \mathbf{y})$ that satisfies the equation with a special right-hand side

$$L_\mathbf{x}[\mathscr{E}] = -\delta(\mathbf{x} - \mathbf{y}) \tag{14.3.2.3}$$

is called a *fundamental solution* corresponding to the operator $L_\mathbf{x}$. In (14.3.2.3), $\delta(\mathbf{x})$ is an $n$-dimensional Dirac delta function and the vector quantity $\mathbf{y} = \{y_1, \ldots, y_n\}$ appears in equation (14.3.2.3) as an $n$-dimensional free parameter. It is assumed that $\mathbf{y} \in \mathbb{R}^n$.

The $n$-dimensional Dirac delta function possesses the following basic properties:

1. $\delta(\mathbf{x}) = \delta(x_1)\delta(x_2)\ldots\delta(x_n),$

2. $\displaystyle\int_{\mathbb{R}^n} \Phi(\mathbf{y})\delta(\mathbf{x} - \mathbf{y})\, d\mathbf{y} = \Phi(\mathbf{x}),$

where $\delta(x_k)$ is the one-dimensional Dirac delta function, $\Phi(\mathbf{x})$ is an arbitrary continuous function, and $d\mathbf{y} = dy_1 \ldots dy_n$.

For constant coefficient equations, a fundamental solution always exists; it can be found by means of the $n$-dimensional Fourier transform (see Paragraph 11.4.1-4).

The fundamental solution $\mathscr{E} = \mathscr{E}(\mathbf{x}, \mathbf{y})$ can be used to construct a particular solution of the linear stationary nonhomogeneous equation (14.3.2.2) for arbitrary continuous $\Phi(\mathbf{x})$; this particular solution is expressed as follows:

$$w(\mathbf{x}) = \int_{\mathbb{R}^n} \Phi(\mathbf{y})\mathscr{E}(\mathbf{x}, \mathbf{y})\, d\mathbf{y}. \tag{14.3.2.4}$$

*Remark 1.* The fundamental solution $\mathscr{E}$ is not unique; it is defined up to an additive term $w_0 = w_0(\mathbf{x})$, which is an arbitrary solution of the homogeneous equation $L_\mathbf{x}[w_0] = 0$.

*Remark 2.* For constant coefficient differential equations, the fundamental solution possesses the property $\mathscr{E}(\mathbf{x}, \mathbf{y}) = \mathscr{E}(\mathbf{x} - \mathbf{y})$.

*Remark 3.* The right-hand sides of equations (14.3.2.2) and (14.3.2.3) are often prefixed with the plus sign. In this case, formula (14.3.2.4) remains valid.

*Remark 4.* Particular solutions of linear nonstationary nonhomogeneous equations can be expressed in terms of the fundamental solution of the Cauchy problem; see Section 14.6.

**Example 1.** For the two- and three-dimensional Poisson equations, the fundamental solutions have the forms

| Equations | | Fundamental solutions |
|---|---|---|
| $\dfrac{\partial^2 w}{\partial x_1^2} + \dfrac{\partial^2 w}{\partial x_2^2} = -\Phi(x_1, x_2)$ | $\implies$ | $\mathscr{E}(x_1, x_2, y_1, y_2) = \dfrac{1}{2\pi}\ln\dfrac{1}{\rho},$ |
| $\dfrac{\partial^2 w}{\partial x_1^2} + \dfrac{\partial^2 w}{\partial x_2^2} + \dfrac{\partial^2 w}{\partial x_3^2} = -\Phi(x_1, x_2, x_3)$ | $\implies$ | $\mathscr{E}(x_1, x_2, x_3, y_1, y_2, y_3) = \dfrac{1}{4\pi r},$ |

where $\rho = \sqrt{(x_1 - y_1)^2 + (x_2 - y_2)^2}$ and $r = \sqrt{(x_1 - y_1)^2 + (x_2 - y_2)^2 + (x_3 - y_3)^2}$.

**Example 2.** The two-dimensional Helmholtz equation

$$\frac{\partial^2 w}{\partial x_1^2} + \frac{\partial^2 w}{\partial x_2^2} + \lambda w = -\Phi(x_1, x_2)$$

has the following fundamental solutions:

$$\mathscr{E}(x_1, x_2, y_1, y_2) = \frac{1}{2\pi} K_0(k\rho) \quad \text{if} \quad \lambda = -k^2 < 0,$$

$$\mathscr{E}(x_1, x_2, y_1, y_2) = -\frac{i}{4} H_0^{(2)}(k\rho) \quad \text{if} \quad \lambda = k^2 > 0,$$

where $\rho = \sqrt{(x_1 - y_1)^2 + (x_2 - y_2)^2}$, $K_0(z)$ is the modified Bessel function of the second kind, $H_0^{(2)}(z)$ is the Hankel functions of the second kind of order 0, $k > 0$, and $i^2 = -1$.

**Example 3.** The three-dimensional Helmholtz equation

$$\frac{\partial^2 w}{\partial x_1^2} + \frac{\partial^2 w}{\partial x_2^2} + \frac{\partial^2 w}{\partial x_3^2} + \lambda w = -\Phi(x_1, x_2, x_3)$$

has the following fundamental solutions:

$$\mathscr{E}(x_1, x_2, x_3, y_1, y_2, y_3) = \frac{1}{4\pi r} \exp(-kr) \quad \text{if} \quad \lambda = -k^2 < 0,$$

$$\mathscr{E}(x_1, x_2, x_3, y_1, y_2, y_3) = \frac{1}{4\pi r} \exp(-ikr) \quad \text{if} \quad \lambda = k^2 > 0,$$

where $r = \sqrt{(x_1 - y_1)^2 + (x_2 - y_2)^2 + (x_3 - y_3)^2}$, $k > 0$, and $i^2 = -1$.

## 14.3.3. General Solutions of Some Hyperbolic Equations

**14.3.3-1. D'Alembert's solution for the wave equation.**

The wave equation

$$\frac{\partial^2 w}{\partial t^2} - a^2 \frac{\partial^2 w}{\partial x^2} = 0 \tag{14.3.3.1}$$

has the general solution

$$w = \varphi(x + at) + \psi(x - at), \tag{14.3.3.2}$$

where $\varphi(x)$ and $\psi(x)$ are arbitrary twice continuously differentiable functions. Solution (14.3.3.2) has the physical interpretation of two *traveling waves* of arbitrary shape that propagate to the right and to the left along the $x$-axis with a constant speed $a$.

**14.3.3-2. Laplace cascade method for hyperbolic equation in two variables.**

A general linear hyperbolic equation with two independent variables can be reduced to an equation of the form (see Subsection 14.1.1):

$$\frac{\partial^2 w}{\partial x \partial y} + a(x, y) \frac{\partial w}{\partial x} + b(x, y) \frac{\partial w}{\partial y} + c(x, y) w = f(x, y). \tag{14.3.3.3}$$

Sometimes it is possible to obtain formulas determining all solutions to equation (14.3.3.3). First consider two special cases.

1°. *First special case.* Suppose the identity

$$g \equiv \frac{\partial a}{\partial x} + ab - c \equiv 0 \tag{14.3.3.4}$$

is valid; for brevity, the arguments of the functions are omitted. Then equation (14.3.3.3) can be rewritten in the form

$$\frac{\partial u}{\partial x} + bu = f, \tag{14.3.3.5}$$

where
$$u = \frac{\partial w}{\partial y} + aw. \tag{14.3.3.6}$$

Equation (14.3.3.5) is a linear first-order ordinary differential equation in $x$ for $u$ (the variable $y$ appears in the equation as a parameter) and is easy to integrate. Further substituting $u$ into (14.3.3.6) yields a linear first-order ordinary differential equation in $y$ for $w$ (now $x$ appears in the equation as a parameter). On solving this equation, one obtains the general solution of the original equation (14.3.3.3) subject to condition (14.3.3.4):

$$w = \exp\left(-\int a\,dy\right)\left\{\varphi(x) + \int\left[\psi(y) + \int f\exp\left(\int b\,dx\right)dx\right]\exp\left(\int a\,dy - \int b\,dx\right)dy\right\},$$

where $\varphi(x)$ and $\psi(y)$ are arbitrary functions.

2°. *Second special case.* Suppose the identity
$$h \equiv \frac{\partial b}{\partial y} + ab - c \equiv 0 \tag{14.3.3.7}$$

holds true. Proceeding in the same ways as in the first special case, one obtains the general solution to (14.3.3.3):

$$w = \exp\left(-\int b\,dx\right)\left\{\psi(y) + \int\left[\varphi(x) + \int f\exp\left(\int a\,dy\right)dy\right]\exp\left(\int b\,dx - \int a\,dy\right)dx\right\}.$$

3°. *Laplace cascade method.* In the case $g \neq 0$, consider the new equation of the form (14.3.3.3),

$$L_1[w_1] \equiv \frac{\partial^2 w_1}{\partial x \partial y} + a_1(x,y)\frac{\partial w_1}{\partial x} + b_1(x,y)\frac{\partial w_1}{\partial y} + c_1(x,y)w_1 = f_1(x,y), \tag{14.3.3.8}$$

where

$$a_1 = a - \frac{\partial \ln g}{\partial y}, \quad b_1 = b, \quad c_1 = c - \frac{\partial a}{\partial x} + \frac{\partial b}{\partial y} - b\frac{\partial \ln g}{\partial y}, \quad f_1 = \left(a - \frac{\partial \ln g}{\partial y}\right)f.$$

If one manages to find $w_1$, the corresponding solution to the original equation (14.3.3.3) can be obtained by the formula

$$w = \frac{1}{g}\left(\frac{\partial w_1}{\partial x} + bw_1 - f\right).$$

For equation (14.3.3.8), the functions similar to (14.3.3.4) and (14.3.3.7) are expressed as

$$g_1 = 2g - h - \frac{\partial^2 \ln g}{\partial x \partial y}, \quad h_1 = h.$$

If $g_1 \equiv 0$, the function $w_1$ can be found using the technique described above. If $g_1 \not\equiv 0$, one proceeds to the construction, in the same way as above, of the equation $L_2[w_2] = f_2$, and so on. In the case $h \not\equiv 0$, a similar chain of equations may be constructed: $L_1^*[w_1^*] = f_1^*$, $L_2^*[w_2^*] = f_2^*$, etc.

If at some step, $g_k$ or $h_k$ vanishes, it is possible to obtain the general solution of equation (14.3.3.3).

**Example.** Consider the *Euler–Darboux equation*

$$\frac{\partial^2 w}{\partial x \partial y} - \frac{\alpha}{x-y}\frac{\partial w}{\partial x} + \frac{\beta}{x-y}\frac{\partial w}{\partial y} = 0.$$

We will show that its general solution can be obtained if at least one of the numbers $\alpha$ or $\beta$ is integer.

With the notation adopted for equation (14.3.3.3) and function (14.3.3.4), we have

$$a(x,y) = -\frac{\alpha}{x-y}, \quad b(x,y) = \frac{\beta}{x-y}, \quad c(x,y) = f(x,y) = 0, \quad g = \frac{\alpha(1-\beta)}{(x-y)^2},$$

which means that $g \equiv 0$ if $\alpha = 0$ or $\beta = 1$. If $g \not\equiv 0$, we construct equation (14.3.3.8), where

$$a_1 = -\frac{2+\alpha}{x-y}, \quad b_1 = \frac{\beta}{x-y}, \quad c_1 = -\frac{\alpha+\beta}{(x-y)^2}.$$

If follows that

$$g_1 = \frac{(1+\alpha)(2-\beta)}{(x-y)^2},$$

and hence $g_1 \equiv 0$ at $\alpha = -1$ or $\beta = 2$. Similarly, it can be shown that $g_k \equiv 0$ at $\alpha = -k$ or $\beta = k+1$ ($k = 0, 1, 2, \ldots$). If we use the other sequence of auxiliary equations, $L_k^*[w_k^*] = f_k^*$, it can be shown that the above holds for $\alpha = 1, 2, \ldots$ and $\beta = 0, -1, -2, \ldots$

## 14.4. Method of Separation of Variables (Fourier Method)

### 14.4.1. Description of the Method of Separation of Variables. General Stage of Solution

**14.4.1-1. Scheme of solving boundary value problems by separation of variables.**

Many linear problems of mathematical physics can be solved by separation of variables. Figure 14.1 depicts the scheme of application of this method to solve boundary value problems for second-order homogeneous linear equations of the parabolic and hyperbolic type with homogeneous boundary conditions and nonhomogeneous initial conditions. For simplicity, problems with two independent variables $x$ and $t$ are considered, with $x_1 \leq x \leq x_2$ and $t \geq 0$.

The scheme presented in Fig. 14.1 applies to boundary value problems for second-order linear homogeneous partial differential equations of the form

$$\alpha(t)\frac{\partial^2 w}{\partial t^2} + \beta(t)\frac{\partial w}{\partial t} = a(x)\frac{\partial^2 w}{\partial x^2} + b(x)\frac{\partial w}{\partial x} + \big[c(x) + \gamma(t)\big]w \qquad (14.4.1.1)$$

with homogeneous linear boundary conditions,

$$\begin{aligned} s_1 \partial_x w + k_1 w = 0 &\quad \text{at} \quad x = x_1, \\ s_2 \partial_x w + k_2 w = 0 &\quad \text{at} \quad x = x_2, \end{aligned} \qquad (14.4.1.2)$$

and arbitrary initial conditions,

$$\begin{aligned} w = f_0(x) &\quad \text{at} \quad t = 0, &\qquad (14.4.1.3)\\ \partial_t w = f_1(x) &\quad \text{at} \quad t = 0. &\qquad (14.4.1.4) \end{aligned}$$

For parabolic equations, which correspond to $\alpha(t) \equiv 0$ in (14.4.1.1), only the initial condition (14.4.1.3) is set.

## 14.4. METHOD OF SEPARATION OF VARIABLES (FOURIER METHOD)

**Original problem** = $\begin{cases} \text{partial differential equation for } w = w(x, t), \\ \text{initial conditions, and boundary conditions} \end{cases}$

*Search for a particular solution in the form of the product of two functions with different arguments*

**Setting** $w(x, t) = \varphi(x)\, \psi(t)$

*Substitute into the equation and separate the variables (this is assumed to be feasible)*

**Arriving at an equation:** $F_1(x, \varphi, \varphi'_x, \varphi''_{xx}) = F_2(t, \psi, \psi'_t, \psi''_{tt})$

*Both sides of the equation must be equal to the same constant, $-\lambda$*

**Obtaining equations:** $F_1(x, \varphi, \varphi'_x, \varphi''_{xx}) = -\lambda, \quad F_2(t, \psi, \psi'_t, \psi''_{tt}) = -\lambda$

*Seek a solution to the first equation that satisfies the boundary conditions (eigenvalue problem)*

**Finding the eigenvalues** $\lambda_n$
**and eigenfunctions** $\varphi_n(x) = \varphi(x, \lambda_n); \ n = 1, 2, \ldots$

*Solve the ordinary differential equation for $\psi$*

**Finding solutions** $\psi_n(t) = \psi(t, \lambda_n)$
**corresponding to the eigenvalues** $\lambda = \lambda_n$

*Notation: $\psi_n(t) = A_n \psi_{n1}(t) + B_n \psi_{n2}(t)$. The principle of linear superposition is used*

**Searching for a series solution:** $w(x, t) = \varphi_1(x)\, \psi_1(t) + \varphi_2(x)\, \psi_2(t) + \cdots$

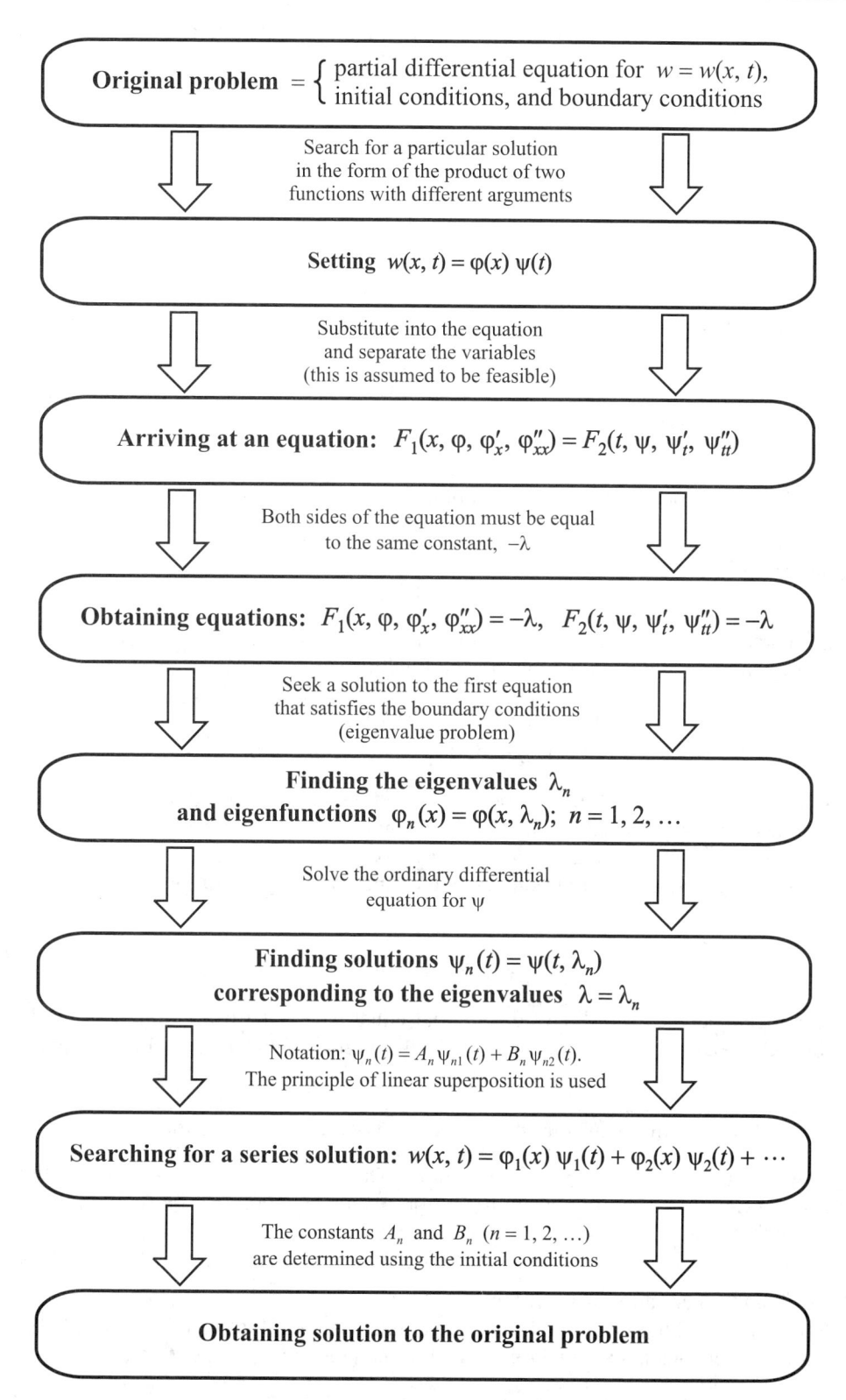

*The constants $A_n$ and $B_n$ $(n = 1, 2, \ldots)$ are determined using the initial conditions*

**Obtaining solution to the original problem**

**Figure 14.1.** Scheme of solving linear boundary value problems by separation of variables (for parabolic equations, the function $F_2$ does not depend on $\psi''_{tt}$, and all $B_n = 0$).

Below we consider the basic stages of the method of separation of variables in more detail. We assume that the coefficients of equation (14.4.1.1) and boundary conditions (14.4.1.2) meet the following requirements:

$$\alpha(t),\ \beta(t),\ \gamma(t),\ a(x),\ b(x),\ c(x) \text{ are continuous functions,}$$
$$\alpha(t) \geq 0,\quad a(x) > 0,\quad |s_1| + |k_1| > 0,\quad |s_2| + |k_2| > 0.$$

**Remark 1.** The method of separation of variables is also used to solve linear boundary value problems for elliptic equations of the form (14.4.1.1), provided that $\alpha(t) < 0$, $a(x) > 0$ and with the boundary conditions (14.4.1.2) in $x$ and similar boundary conditions in $t$. In this case, all results obtained for the general stage of solution, described in Paragraphs 14.4.1-2 and 14.4.1-3, remain valid; for details, see Subsection 14.4.4.

**Remark 2.** In various applications, equations of the form (14.4.1.1) may arise with the coefficient $b(x)$ going to infinity at the boundary, $b(x) \to \infty$ as $x \to x_1$, with the other coefficients being continuous. In this case, the first boundary condition in (14.4.1.2) should be replaced with a condition of boundedness of the solution as $x \to x_1$. This may occur in spatial problems with central or axial symmetry where the solution depends only on the radial coordinate.

### 14.4.1-2. Derivation of equations and boundary conditions for particular solutions.

The approach is based on searching for particular solutions of equation (14.4.1.1) in the product form

$$w(x,t) = \varphi(x)\,\psi(t). \tag{14.4.1.5}$$

After separation of the variables and elementary manipulations, one arrives at the following linear ordinary differential equations for the functions $\varphi = \varphi(x)$ and $\psi = \psi(t)$:

$$a(x)\varphi''_{xx} + b(x)\varphi'_x + [\lambda + c(x)]\varphi = 0, \tag{14.4.1.6}$$
$$\alpha(t)\psi''_{tt} + \beta(t)\psi'_t + [\lambda - \gamma(t)]\psi = 0. \tag{14.4.1.7}$$

These equations contain a free parameter $\lambda$ called the separation constant. With the notation adopted in Fig. 14.1, equations (14.4.1.6) and (14.4.1.7) can be rewritten as follows: $\varphi F_1(x, \varphi, \varphi'_x, \varphi''_{xx}) + \lambda \varphi = 0$ and $\psi F_2(t, \psi, \psi'_t, \psi''_{tt}) + \lambda \psi = 0$.

Substituting (14.4.1.5) into (14.4.1.2) yields the boundary conditions for $\varphi = \varphi(x)$:

$$\begin{aligned} s_1 \varphi'_x + k_1 \varphi = 0 &\quad \text{at} \quad x = x_1, \\ s_2 \varphi'_x + k_2 \varphi = 0 &\quad \text{at} \quad x = x_2. \end{aligned} \tag{14.4.1.8}$$

The homogeneous linear ordinary differential equation (14.4.1.6) in conjunction with the homogeneous linear boundary conditions (14.4.1.8) makes up an eigenvalue problem.

### 14.4.1-3. Solution of eigenvalue problems. Orthogonality of eigenfunctions.

Suppose $\widetilde{\varphi}_1 = \widetilde{\varphi}_1(x, \lambda)$ and $\widetilde{\varphi}_2 = \widetilde{\varphi}_2(x, \lambda)$ are linearly independent particular solutions of equation (14.4.1.6). Then the general solution of this equation can be represented as the linear combination

$$\varphi = C_1 \widetilde{\varphi}_1(x, \lambda) + C_2 \widetilde{\varphi}_2(x, \lambda), \tag{14.4.1.9}$$

where $C_1$ and $C_2$ are arbitrary constants.

Substituting solution (14.4.1.9) into the boundary conditions (14.4.1.8) yields the following homogeneous linear algebraic system of equations for $C_1$ and $C_2$:

$$\begin{aligned} \varepsilon_{11}(\lambda)C_1 + \varepsilon_{12}(\lambda)C_2 = 0, \\ \varepsilon_{21}(\lambda)C_1 + \varepsilon_{22}(\lambda)C_2 = 0, \end{aligned} \tag{14.4.1.10}$$

where $\varepsilon_{ij}(\lambda) = \left[s_i(\widetilde{\varphi}_j)'_x + k_i\widetilde{\varphi}_j\right]_{x=x_i}$. For system (14.4.1.10) to have nontrivial solutions, its determinant must be zero; we have

$$\varepsilon_{11}(\lambda)\varepsilon_{22}(\lambda) - \varepsilon_{12}(\lambda)\varepsilon_{21}(\lambda) = 0. \tag{14.4.1.11}$$

Solving the transcendental equation (14.4.1.11) for $\lambda$, one obtains the *eigenvalues* $\lambda = \lambda_n$, where $n = 1, 2, \ldots$ For these values of $\lambda$, equation (14.4.1.6) has nontrivial solutions,

$$\varphi_n(x) = \varepsilon_{12}(\lambda_n)\widetilde{\varphi}_1(x, \lambda_n) - \varepsilon_{11}(\lambda_n)\widetilde{\varphi}_2(x, \lambda_n), \tag{14.4.1.12}$$

which are called *eigenfunctions* (these functions are defined up to a constant multiplier).

To facilitate the further analysis, we represent equation (14.4.1.6) in the form

$$[p(x)\varphi'_x]'_x + [\lambda\rho(x) - q(x)]\varphi = 0, \tag{14.4.1.13}$$

where

$$p(x) = \exp\left[\int \frac{b(x)}{a(x)} \, dx\right], \quad q(x) = -\frac{c(x)}{a(x)} \exp\left[\int \frac{b(x)}{a(x)} \, dx\right], \quad \rho(x) = \frac{1}{a(x)} \exp\left[\int \frac{b(x)}{a(x)} \, dx\right]. \tag{14.4.1.14}$$

It follows from the adopted assumptions (see the end of Paragraph 14.4.1-1) that $p(x)$, $p'_x(x)$, $q(x)$, and $\rho(x)$ are continuous functions, with $p(x) > 0$ and $\rho(x) > 0$.

The eigenvalue problem (14.4.1.13), (14.4.1.8) is known to possess the following properties:

1. All eigenvalues $\lambda_1, \lambda_2, \ldots$ are real, and $\lambda_n \to \infty$ as $n \to \infty$; consequently, the number of negative eigenvalues is finite.

2. The system of eigenfunctions $\varphi_1(x), \varphi_2(x), \ldots$ is orthogonal on the interval $x_1 \leq x \leq x_2$ with weight $\rho(x)$, i.e.,

$$\int_{x_1}^{x_2} \rho(x)\varphi_n(x)\varphi_m(x) \, dx = 0 \quad \text{for} \quad n \neq m. \tag{14.4.1.15}$$

3. If

$$q(x) \geq 0, \quad s_1 k_1 \leq 0, \quad s_2 k_2 \geq 0, \tag{14.4.1.16}$$

there are no negative eigenvalues. If $q \equiv 0$ and $k_1 = k_2 = 0$, the least eigenvalue is $\lambda_1 = 0$ and the corresponding eigenfunction is $\varphi_1 = \text{const}$. Otherwise, all eigenvalues are positive, provided that conditions (14.4.1.16) are satisfied; the first inequality in (14.4.1.16) is satisfied if $c(x) \leq 0$.

Subsection 12.2.5 presents some estimates for the eigenvalues $\lambda_n$ and eigenfunctions $\varphi_n(x)$.

▶ *The procedure of constructing solutions to nonstationary boundary value problems is further different for parabolic and hyperbolic equations; see Subsections 14.4.2 and 14.4.3 below for results (elliptic equations are treated in Subsection 14.4.4).*

## 14.4.2. Problems for Parabolic Equations: Final Stage of Solution

**14.4.2-1. Series solutions of boundary value problems for parabolic equations.**

Consider the problem for the parabolic equation

$$\frac{\partial w}{\partial t} = a(x)\frac{\partial^2 w}{\partial x^2} + b(x)\frac{\partial w}{\partial x} + \left[c(x) + \gamma(t)\right]w \tag{14.4.2.1}$$

(this equation is obtained from (14.4.1.1) in the case $\alpha(t)\equiv 0$ and $\beta(t)=1$) with homogeneous linear boundary conditions (14.4.1.2) and initial condition (14.4.1.3).

First, one searches for particular solutions to equation (14.4.2.1) in the product form (14.4.1.5), where the function $\varphi(x)$ is obtained by solving an eigenvalue problem for the ordinary differential equation (14.4.1.6) with the boundary conditions (14.4.1.8). The solution of equation (14.4.1.7) with $\alpha(t) \equiv 0$ and $\beta(t) = 1$ corresponding to the eigenvalues $\lambda = \lambda_n$ and satisfying the normalizing conditions $\psi_n(0) = 1$ has the form

$$\psi_n(t) = \exp\left[-\lambda_n t + \int_0^t \gamma(\xi)\,d\xi\right]. \tag{14.4.2.2}$$

Then the solution of the nonstationary boundary value problem (14.4.2.1), (14.4.1.2), (14.4.1.3) is sought in the series form

$$w(x,t) = \sum_{n=1}^{\infty} A_n \varphi_n(x)\,\psi_n(t), \tag{14.4.2.3}$$

where the $A_n$ are arbitrary constants and the functions $w_n(x,t) = \varphi_n(x)\,\psi_n(t)$ are particular solutions (14.4.2.1) satisfying the boundary conditions (14.4.1.2). By the principle of linear superposition, series (14.4.2.3) is also a solution of the original partial differential equation that satisfies the boundary conditions.

To determine the coefficients $A_n$, we substitute series (14.4.2.3) into the initial condition (14.4.1.3), thus obtaining

$$\sum_{n=1}^{\infty} A_n \varphi_n(x) = f_0(x).$$

Multiplying this equation by $\rho(x)\varphi_n(x)$, where the weight function $\rho(x)$ is defined in (14.4.1.14), then integrating the resulting relation with respect to $x$ over the interval $x_1 \le x \le x_2$, and taking into account the properties (14.4.1.15), we find

$$A_n = \frac{1}{\|\varphi_n\|^2} \int_{x_1}^{x_2} \rho(x)\varphi_n(x)f_0(x)\,dx, \quad \|\varphi_n\|^2 = \int_{x_1}^{x_2} \rho(x)\varphi_n^2(x)\,dx. \tag{14.4.2.4}$$

Relations (14.4.2.3), (14.4.2.2), (14.4.2.4), and (14.4.1.12) give a formal solution of the nonstationary boundary value problem (14.4.2.1), (14.4.1.2), (14.4.1.3).

**Example.** Consider the first (Dirichlet) boundary value problem on the interval $0 \le x \le l$ for the heat equation

$$\frac{\partial w}{\partial t} = \frac{\partial^2 w}{\partial x^2} \tag{14.4.2.5}$$

with the general initial condition (14.4.1.3) and the homogeneous boundary conditions

$$w = 0 \quad \text{at} \quad x = 0, \qquad w = 0 \quad \text{at} \quad x = l. \tag{14.4.2.6}$$

The function $\psi(t)$ in the particular solution (14.4.1.5) is found from (14.4.2.2), where $\gamma(t) = 0$:

$$\psi_n(t) = \exp(-\lambda_n t). \tag{14.4.2.7}$$

The functions $\varphi_n(x)$ are determined by solving the eigenvalue problem (14.4.1.6), (14.4.1.8) with $a(x) = 1$, $b(x) = c(x) = 0$, $s_1 = s_2 = 0$, $k_1 = k_2 = 1$, $x_1 = 0$, and $x_2 = l$:

$$\varphi''_{xx} + \lambda\varphi = 0; \qquad \varphi = 0 \quad \text{at} \quad x = 0, \qquad \varphi = 0 \quad \text{at} \quad x = l.$$

So we obtain the eigenfunctions and eigenvalues:

$$\varphi_n(x) = \sin\left(\frac{n\pi x}{l}\right), \quad \lambda_n = \left(\frac{n\pi}{l}\right)^2, \quad n = 1, 2, \ldots \qquad (14.4.2.8)$$

The solution to problem (14.4.2.5)–(14.4.2.6), (14.4.1.3) is given by formulas (14.4.2.3), (14.4.2.4). Taking into account that $\|\varphi_n\|^2 = l/2$, we obtain

$$w(x,t) = \sum_{n=1}^{\infty} A_n \sin\left(\frac{n\pi x}{l}\right) \exp\left(-\frac{n^2\pi^2 t}{l^2}\right), \quad A_n = \frac{2}{l}\int_0^l f_0(\xi)\sin\left(\frac{n\pi\xi}{l}\right)d\xi. \qquad (14.4.2.9)$$

If the function $f_0(x)$ is twice continuously differentiable and the compatibility conditions (see Paragraph 14.4.2-2) are satisfied, then series (14.4.2.9) is convergent and admits termwise differentiation, once with respect to $t$ and twice with respect to $x$. In this case, formula (14.4.2.9) gives the classical smooth solution of problem (14.4.2.5)–(14.4.2.6), (14.4.1.3). [If $f_0(x)$ is not as smooth as indicated or if the compatibility conditions are not met, then series (14.4.2.9) may converge to a discontinuous function, thus giving only a generalized solution.]

**Remark.** For the solution of linear nonhomogeneous parabolic equations with nonhomogeneous boundary conditions, see Section 14.7.

### 14.4.2-2. Conditions of compatibility of initial and boundary conditions.

Suppose the function $w$ has a continuous derivative with respect to $t$ and two continuous derivatives with respect to $x$ and is a solution of problem (14.4.2.1), (14.4.1.2), (14.4.1.3). Then the boundary conditions (14.4.1.2) and the initial condition (14.4.1.3) must be consistent; namely, the following compatibility conditions must hold:

$$[s_1 f_0' + k_1 f_0]_{x=x_1} = 0, \quad [s_2 f_0' + k_2 f_0]_{x=x_2} = 0. \qquad (14.4.2.10)$$

If $s_1 = 0$ or $s_2 = 0$, then the additional compatibility conditions

$$\begin{aligned}{}[a(x) f_0'' + b(x) f_0']_{x=x_1} &= 0 \quad \text{if} \quad s_1 = 0, \\ [a(x) f_0'' + b(x) f_0']_{x=x_2} &= 0 \quad \text{if} \quad s_2 = 0 \end{aligned} \qquad (14.4.2.11)$$

must also hold; the primes denote the derivatives with respect to $x$.

## 14.4.3. Problems for Hyperbolic Equations: Final Stage of Solution

### 14.4.3-1. Series solution of boundary value problems for hyperbolic equations.

For hyperbolic equations, the solution of the boundary value problem (14.4.1.1)–(14.4.1.4) is sought in the series form

$$w(x,t) = \sum_{n=1}^{\infty} \varphi_n(x)\left[A_n \psi_{n1}(t) + B_n \psi_{n2}(t)\right]. \qquad (14.4.3.1)$$

Here, $A_n$ and $B_n$ are arbitrary constants. The functions $\psi_{n1}(t)$ and $\psi_{n2}(t)$ are particular solutions of the linear equation (14.4.1.7) for $\psi$ (with $\lambda = \lambda_n$) that satisfy the conditions

$$\psi_{n1}(0) = 1, \quad \psi_{n1}'(0) = 0; \quad \psi_{n2}(0) = 0, \quad \psi_{n2}'(0) = 1. \qquad (14.4.3.2)$$

The functions $\varphi_n(x)$ and $\lambda_n$ are determined by solving the eigenvalue problem (14.4.1.6), (14.4.1.8).

Substituting solution (14.4.3.1) into the initial conditions (14.4.1.3)–(14.4.1.4) yields

$$\sum_{n=1}^{\infty} A_n \varphi_n(x) = f_0(x), \qquad \sum_{n=1}^{\infty} B_n \varphi_n(x) = f_1(x).$$

Multiplying these equations by $\rho(x)\varphi_n(x)$, where the weight function $\rho(x)$ is defined in (14.4.1.14), then integrating the resulting relations with respect to $x$ on the interval $x_1 \le x \le x_2$, and taking into account the properties (14.4.1.15), we obtain the coefficients of series (14.4.3.1) in the form

$$A_n = \frac{1}{\|\varphi_n\|^2} \int_{x_1}^{x_2} \rho(x)\varphi_n(x) f_0(x)\, dx, \qquad B_n = \frac{1}{\|\varphi_n\|^2} \int_{x_1}^{x_2} \rho(x)\varphi_n(x) f_1(x)\, dx. \qquad (14.4.3.3)$$

The quantity $\|\varphi_n\|$ is defined in (14.4.2.4).

Relations (14.4.3.1), (14.4.1.12), and (14.4.3.3) give a formal solution of the nonstationary boundary value problem (14.4.1.1)–(14.4.1.4) for $\alpha(t) > 0$.

**Example.** Consider a mixed boundary value problem on the interval $0 \le x \le l$ for the wave equation

$$\frac{\partial^2 w}{\partial t^2} = \frac{\partial^2 w}{\partial x^2} \qquad (14.4.3.4)$$

with the general initial conditions (14.4.1.3)–(14.4.1.4) and the homogeneous boundary conditions

$$w = 0 \quad \text{at} \quad x = 0, \qquad \partial_x w = 0 \quad \text{at} \quad x = l. \qquad (14.4.3.5)$$

The functions $\psi_{n1}(t)$ and $\psi_{n2}(t)$ are determined by the linear equation [see (14.4.1.7) with $\alpha(t) = 1$, $\beta(t) = \gamma(t) = 0$, and $\lambda = \lambda_n$]

$$\psi''_{tt} + \lambda \psi = 0$$

with the initial conditions (14.4.3.2). We find

$$\psi_{n1}(t) = \cos\left(\sqrt{\lambda_n}\, t\right), \qquad \psi_{n2}(t) = \frac{1}{\sqrt{\lambda_n}} \sin\left(\sqrt{\lambda_n}\, t\right). \qquad (14.4.3.6)$$

The functions $\varphi_n(x)$ are determined by solving the eigenvalue problem (14.4.1.6), (14.4.1.8) with $a(x) = 1$, $b(x) = c(x) = 0$, $s_1 = k_2 = 0$, $s_2 = k_1 = 1$, $x_1 = 0$, and $x_2 = l$:

$$\varphi''_{xx} + \lambda\varphi = 0; \qquad \varphi = 0 \quad \text{at} \quad x = 0, \qquad \varphi'_x = 0 \quad \text{at} \quad x = l.$$

So we obtain the eigenfunctions and eigenvalues:

$$\varphi_n(x) = \sin(\mu_n x), \qquad \mu_n = \sqrt{\lambda_n} = \frac{\pi(2n-1)}{2l}, \qquad n = 1, 2, \ldots \qquad (14.4.3.7)$$

The solution to problem (14.4.3.4)–(14.4.3.5), (14.4.1.3)–(14.4.1.4) is given by formulas (14.4.3.1) and (14.4.3.3). Taking into account that $\|\varphi_n\|^2 = l/2$, we have

$$w(x,t) = \sum_{n=1}^{\infty} \left[ A_n \cos(\mu_n t) + B_n \sin(\mu_n t) \right] \sin(\mu_n x), \qquad \mu_n = \frac{\pi(2n-1)}{2l}, \qquad (14.4.3.8)$$

$$A_n = \frac{2}{l} \int_0^l f_0(x) \sin(\mu_n x)\, dx, \qquad B_n = \frac{2}{l\mu_n} \int_0^l f_1(x) \sin(\mu_n x)\, dx.$$

If $f_0(x)$ and $f_1(x)$ have three and two continuous derivatives, respectively, and the compatibility conditions are met (see Paragraph 14.4.3-2), then series (14.4.3.8) is convergent and admits double termwise differentiation. In this case, formula (14.4.3.8) gives the classical smooth solution of problem (14.4.3.4)–(14.4.3.5), (14.4.1.3)–(14.4.1.4).

**Remark.** For the solution of linear nonhomogeneous hyperbolic equations with nonhomogeneous boundary conditions, see Section 14.8.

**14.4.3-2. Conditions of compatibility of initial and boundary conditions.**

Suppose $w$ is a twice continuously differentiable solution of problem (14.4.1.1)–(14.4.1.4). Then conditions (14.4.2.10) and (14.4.2.11) must hold. In addition, the following conditions of compatibility of the boundary conditions (14.4.1.2) and initial condition (14.4.1.4) must be satisfied:

$$[s_1 f_1' + k_1 f_1]_{x=x_1} = 0, \qquad [s_2 f_1' + k_2 f_1]_{x=x_2} = 0.$$

## 14.4.4. Solution of Boundary Value Problems for Elliptic Equations

**14.4.4-1. Solution of special problem for elliptic equations.**

Now consider a boundary value problem for the elliptic equation

$$a(x)\frac{\partial^2 w}{\partial x^2} + \alpha(y)\frac{\partial^2 w}{\partial y^2} + b(x)\frac{\partial w}{\partial x} + \beta(y)\frac{\partial w}{\partial y} + [c(x) + \gamma(y)]w = 0 \qquad (14.4.4.1)$$

with homogeneous boundary conditions (14.4.1.2) in $x$ and the following mixed (homogeneous and nonhomogeneous) boundary conditions in $y$:

$$\begin{aligned} \sigma_1 \partial_y w + \nu_1 w &= 0 && \text{at} \quad y = y_1, \\ \sigma_2 \partial_y w + \nu_2 w &= f(x) && \text{at} \quad y = y_2. \end{aligned} \qquad (14.4.4.2)$$

We assume that the coefficients of equation (14.4.4.1) and boundary conditions (14.4.1.2) and (14.4.4.2) meet the following requirements:

$a(x)$, $b(x)$, $c(x)$ $\alpha(y)$, $\beta(y)$, and $\gamma(t)$ are continuous functions,
$a(x) > 0$, $\alpha(y) > 0$, $|s_1| + |k_1| > 0$, $|s_2| + |k_2| > 0$, $|\sigma_1| + |\nu_1| > 0$, $|\sigma_2| + |\nu_2| > 0$.

The approach is based on searching for particular solutions of equation (14.4.4.1) in the product form

$$w(x, y) = \varphi(x)\,\psi(y). \qquad (14.4.4.3)$$

As before, we first arrive at the eigenvalue problem (14.4.1.6), (14.4.1.8) for the function $\varphi = \varphi(x)$; the solution procedure is detailed in Paragraph 14.4.1-3. Further on, we assume the $\lambda_n$ and $\varphi_n(x)$ have been found. The functions $\psi_n = \psi_n(y)$ are determined by solving the linear ordinary differential equation

$$\alpha(y)\psi_{yy}'' + \beta(y)\psi_y' + [\gamma(y) - \lambda_n]\psi = 0 \qquad (14.4.4.4)$$

subject to the homogeneous boundary condition

$$\sigma_1 \partial_y \psi + \nu_1 \psi = 0 \quad \text{at} \quad y = y_1, \qquad (14.4.4.5)$$

which is a consequence of the first condition (14.4.4.2). The functions $\psi_n$ are determined up to a constant factor.

Taking advantage of the principle of linear superposition, we seek the solution to the boundary value problem (14.4.4.1), (14.4.4.2), (14.4.1.2) in the series form

$$w(x, y) = \sum_{n=1}^{\infty} A_n \varphi_n(x)\psi_n(y), \qquad (14.4.4.6)$$

where $A_n$ are arbitrary constants. By construction, series (14.4.4.6) will satisfy equation (14.4.4.1) with the boundary conditions (14.4.1.2) and the first boundary condition (14.4.4.2). In order to find the series coefficients $A_n$, substitute (14.4.4.6) into the second boundary condition (14.4.4.2) to obtain

$$\sum_{n=1}^{\infty} A_n B_n \varphi_n(x) = f(x), \qquad B_n = \sigma_2 \frac{d\psi_n}{dy}\bigg|_{y=y_2} + \nu_2 \psi_n(y_2). \qquad (14.4.4.7)$$

Further, we follow the same procedure as in Paragraph 14.4.2-1. Specifically, multiplying (14.4.4.7) by $\rho(x)\varphi_n(x)$, then integrating the resulting relation with respect to $x$ over the interval $x_1 \le x \le x_2$, and taking into account the properties (14.4.1.15), we obtain

$$A_n = \frac{1}{B_n \|\varphi_n\|^2} \int_{x_1}^{x_2} \rho(x)\varphi_n(x)f(x)\,dx, \qquad \|\varphi_n\|^2 = \int_{x_1}^{x_2} \rho(x)\varphi_n^2(x)\,dx, \qquad (14.4.4.8)$$

where the weight function $\rho(x)$ is defined in (14.4.1.14).

**Example.** Consider the first (Dirichlet) boundary value problem for the Laplace equation

$$\frac{\partial^2 w}{\partial x^2} + \frac{\partial^2 w}{\partial y^2} = 0 \qquad (14.4.4.9)$$

subject to the boundary conditions

$$\begin{array}{llll} w = 0 & \text{at} \ \ x = 0, & w = 0 & \text{at} \ \ x = l_1; \\ w = 0 & \text{at} \ \ y = 0, & w = f(x) & \text{at} \ \ y = l_2 \end{array} \qquad (14.4.4.10)$$

in a rectangular domain $0 \le x \le l_1,\ 0 \le y \le l_2$.

Particular solutions to equation (14.4.4.9) are sought in the form (14.4.4.3). We have the following eigenvalue problem for $\varphi(x)$:

$$\varphi_{xx}'' + \lambda\varphi = 0; \qquad \varphi = 0 \ \ \text{at} \ \ x = 0, \qquad \varphi = 0 \ \ \text{at} \ \ x = l_1.$$

On solving this problem, we find the eigenfunctions with respective eigenvalues

$$\varphi_n(x) = \sin(\mu_n x), \qquad \mu_n = \sqrt{\lambda_n} = \frac{\pi n}{l_1}, \qquad n = 1, 2, \ldots \qquad (14.4.4.11)$$

The functions $\psi_n = \psi_n(y)$ are determined by solving the following problem for a linear ordinary differential equation with homogeneous boundary conditions:

$$\psi_{yy}'' - \lambda_n \psi = 0; \qquad \psi = 0 \ \ \text{at} \ \ y = 0. \qquad (14.4.4.12)$$

It is a special case of problem (14.4.4.4)–(14.4.4.5) with $\alpha(y) = 1$, $\beta(y) = \gamma(y) = 0$, $\sigma_1 = 0$, and $\nu_1 = 1$. The nontrivial solutions of problem (14.4.4.12) are expressed as

$$\psi_n(y) = \sinh(\mu_n y), \qquad \mu_n = \sqrt{\lambda_n} = \frac{\pi n}{l_1}, \qquad n = 1, 2, \ldots \qquad (14.4.4.13)$$

Using formulas (14.4.4.6), (14.4.4.8), (14.4.4.11), (14.4.4.13) and taking into account the relations $B_n = \psi_n(l_2) = \sinh(\mu_n l_2)$, $\rho(x) = 1$, and $\|\varphi_n\|^2 = l/2$, we find the solution of the original problem (14.4.4.9)–(14.4.4.10) in the form

$$w(x,y) = \sum_{n=1}^{\infty} A_n \sin(\mu_n x) \sinh(\mu_n y), \qquad A_n = \frac{2}{l_1 \sinh(\mu_n l_2)} \int_0^{l_1} f(x) \sin(\mu_n x)\,dx, \qquad \mu_n = \frac{\pi n}{l_1}.$$

TABLE 14.5
Description of auxiliary problems for equation (14.4.4.1) and problems for associated functions $\varphi(x)$ and $\psi(y)$ that determine particular solutions of the form (14.4.4.3). The abbreviation HBC below stands for a "homogeneous boundary condition"

| Auxiliary problem | Functions vanishing in the boundary conditions (14.4.4.14) | Eigenvalue problem with homogeneous boundary conditions | Another problem with one homogeneous boundary condition (for $\lambda_n$ found) |
|---|---|---|---|
| Problem 1 | $f_2(y) = f_3(x) = f_4(x) = 0$, function $f_1(y)$ prescribed | functions $\psi_n(y)$ and values $\lambda_n$ to be determined | functions $\varphi_n(x)$ satisfy an HBC at $x = x_2$ |
| Problem 2 | $f_1(y) = f_3(x) = f_4(x) = 0$, function $f_2(y)$ prescribed | functions $\psi_n(y)$ and values $\lambda_n$ to be determined | functions $\varphi(x)$ satisfy an HBC at $x = x_1$ |
| Problem 3 | $f_1(y) = f_2(y) = f_4(x) = 0$, function $f_3(x)$ prescribed | functions $\varphi_n(x)$ and values $\lambda_n$ to be determined | functions $\psi(y)$ satisfy an HBC at $y = y_2$ |
| Problem 4 | $f_1(y) = f_2(y) = f_3(x) = 0$, function $f_4(x)$ prescribed | functions $\varphi_n(x)$ and values $\lambda_n$ to be determined | functions $\psi(y)$ satisfy an HBC at $y = y_1$ |

**14.4.4-2. Generalization to the case of nonhomogeneous boundary conditions.**

Now consider the linear boundary value problem for the elliptic equation (14.4.4.1) with general nonhomogeneous boundary conditions

$$s_1\partial_x w + k_1 w = f_1(y) \quad \text{at} \quad x = x_1, \quad s_2\partial_x w + k_2 w = f_2(y) \quad \text{at} \quad x = x_2,$$
$$\sigma_1\partial_y w + \nu_1 w = f_3(x) \quad \text{at} \quad y = y_1, \quad \sigma_2\partial_y w + \nu_2 w = f_4(x) \quad \text{at} \quad y = y_2. \quad (14.4.4.14)$$

The solution to this problem is the sum of solutions to four simpler auxiliary problems for equation (14.4.4.1), each corresponding to three homogeneous and one nonhomogeneous boundary conditions in (14.4.4.14); see Table 14.5. Each auxiliary problem is solved using the procedure given in Paragraph 14.4.4-1, beginning with the search for solutions in the form of the product of functions with different arguments (14.4.4.3), determined by equations (14.4.1.6) and (14.4.4.4). The separation parameter $\lambda$ is determined by the solution of a eigenvalue problem with homogeneous boundary conditions; see Table 14.5. The solution to each of the auxiliary problems is sought in the series form (14.4.4.6).

*Remark.* For the solution of linear nonhomogeneous elliptic equations subject to nonhomogeneous boundary conditions, see Section 14.9.

## 14.5. Integral Transforms Method

Various integral transforms are widely used to solve linear problems of mathematical physics. The Laplace transform and the Fourier transform are in most common use (its and other integral transforms are considered in Chapter 11 in detail).

### 14.5.1. Laplace Transform and Its Application in Mathematical Physics

**14.5.1-1. Laplace and inverse Laplace transforms. Laplace transforms for derivatives.**

The Laplace transform of an arbitrary (complex-valued) function $f(t)$ of a real variable $t$ ($t \geq 0$) is defined by

$$\widetilde{f}(p) = \mathfrak{L}\{f(t)\}, \quad \text{where} \quad \mathfrak{L}\{f(t)\} \equiv \int_0^\infty e^{-pt} f(t)\,dt, \quad (14.5.1.1)$$

where $p = s + i\sigma$ is a complex variable, $i^2 = -1$.

Given the transform $\tilde{f}(p)$, the function $f(t)$ can be found by means of the inverse Laplace transform

$$f(t) = \mathcal{L}^{-1}\{\tilde{f}(p)\}, \quad \text{where} \quad \mathcal{L}^{-1}\{\tilde{f}(p)\} \equiv \frac{1}{2\pi i}\int_{c-i\infty}^{c+i\infty} \tilde{f}(p)e^{pt}\, dp, \qquad (14.5.1.2)$$

where the integration path is parallel to the imaginary axis and lies to the right of all singularities of $\tilde{f}(p)$, which corresponds to $c > \sigma_0$.

In order to solve nonstationary boundary value problems, the following Laplace transform formulas for derivatives will be required:

$$\begin{aligned}\mathcal{L}\{f'(t)\} &= p\tilde{f}(p) - f(0),\\ \mathcal{L}\{f''(t)\} &= p^2\tilde{f}(p) - pf(0) - f'(0),\end{aligned} \qquad (14.5.1.3)$$

where $f(0)$ and $f'(0)$ are the initial conditions.

More details on the properties of the Laplace transform and the inverse Laplace transform can be found in Section 11.2. The Laplace transforms of some functions are listed in Section T3.1. Tables of inverse Laplace transforms are listed in Section T3.2. Such tables are convenient to use in solving linear problems for partial differential equations.

### 14.5.1-2. Solution procedure for linear problems using the Laplace transform.

Figure 14.2 shows schematically how one can utilize the Laplace transforms to solve boundary value problems for linear parabolic or hyperbolic equations with two independent variables in the case where the equation coefficients are independent of $t$ (the same procedure can be applied for solving linear problems characterized by higher-order equations). Here and henceforth, the short notation $\tilde{w}(x,p) = \mathcal{L}\{w(x,t)\}$ will be used; the arguments of $\tilde{w}$ may be omitted.

It is significant that with the Laplace transform, the original problem for a partial differential equation is reduced to a simpler problem for an ordinary differential equation with parameter $p$; the derivatives with respect to $t$ are replaced by appropriate algebraic expressions taking into account the initial conditions; see formulas (14.5.1.3).

### 14.5.1-3. Solving linear problems for parabolic equations with the Laplace transform.

Consider a linear nonstationary boundary value problem for the parabolic equation

$$\frac{\partial w}{\partial t} = a(x)\frac{\partial^2 w}{\partial x^2} + b(x)\frac{\partial w}{\partial x} + c(x)w + \Phi(x,t) \qquad (14.5.1.4)$$

with the initial condition (14.4.1.3) and the general nonhomogeneous boundary conditions

$$\begin{aligned}s_1 \partial_x w + k_1 w &= g_1(t) \quad \text{at} \quad x = x_1,\\ s_2 \partial_x w + k_2 w &= g_2(t) \quad \text{at} \quad x = x_2.\end{aligned} \qquad (14.5.1.5)$$

The application of the Laplace transform results in the problem defined by the ordinary differential equation for $x$ ($p$ is treated as a parameter)

$$a(x)\frac{\partial^2 \tilde{w}}{\partial x^2} + b(x)\frac{\partial \tilde{w}}{\partial x} + [c(x) - p]\tilde{w} + f_0(x) + \tilde{\Phi}(x,p) = 0 \qquad (14.5.1.6)$$

```
┌─────────────────────────────────────────────────────────────────┐
│  Original problem = { partial differential equation for w = w(x, t), │
│                       initial conditions, and boundary conditions   │
└─────────────────────────────────────────────────────────────────┘
                   ⬇  Application of the Laplace transform (PL.5.1.1)  ⬇
                              with respect to t
┌─────────────────────────────────────────────────────────────────┐
│  Problem for transform = { ordinary differential equation            │
│                            for w̃ = w̃(x, p), boundary conditions    │
└─────────────────────────────────────────────────────────────────┘
                   ⬇   Solution of the ordinary differential equation  ⬇
                              with the boundary conditions
┌─────────────────────────────────────────────────────────────────┐
│            Finding the transform  w̃ = w̃(x, t)                     │
└─────────────────────────────────────────────────────────────────┘
                   ⬇         Application of the inverse               ⬇
                              Laplace transform (PL.5.1.2)
┌─────────────────────────────────────────────────────────────────┐
│   Obtaining solution to the original problem:  w = w(x, t)         │
└─────────────────────────────────────────────────────────────────┘
```

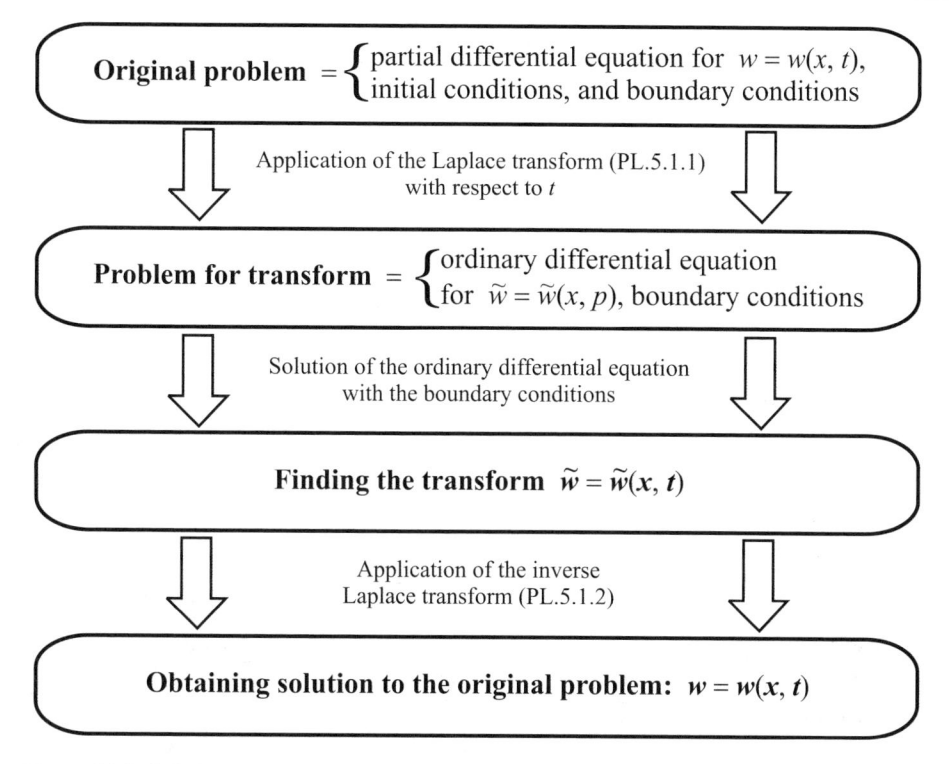

**Figure 14.2.** Solution procedure for linear boundary value problems using the Laplace transform.

with the boundary conditions

$$s_1 \partial_x \widetilde{w} + k_1 \widetilde{w} = \widetilde{g}_1(p) \quad \text{at} \quad x = x_1, \qquad (14.5.1.7)$$
$$s_2 \partial_x \widetilde{w} + k_2 \widetilde{w} = \widetilde{g}_2(p) \quad \text{at} \quad x = x_2.$$

Notation employed: $\widetilde{\Phi}(x,p) = \mathcal{L}\{\Phi(x,t)\}$ and $\widetilde{g}_n(p) = \mathcal{L}\{g_n(t)\}$ ($n = 1, 2$). On solving problem (14.5.1.6)–(14.5.1.7), one should apply to the resulting solution $\widetilde{w} = \widetilde{w}(x,p)$ the inverse Laplace transform (14.5.1.2) to obtain the solution, $w = w(x,t)$, of the original problem.

**Example 1.** Consider the first boundary value problem for the heat equation:

$$\frac{\partial w}{\partial t} = \frac{\partial^2 w}{\partial x^2} \qquad (x > 0, \ t > 0),$$
$$w = 0 \quad \text{at} \quad t = 0 \quad \text{(initial condition)},$$
$$w = w_0 \quad \text{at} \quad x = 0 \quad \text{(boundary condition)},$$
$$w \to 0 \quad \text{at} \quad x \to \infty \quad \text{(boundary condition)}.$$

We apply the Laplace transform with respect to $t$. Let us multiply the equation, the initial condition, and the boundary conditions by $e^{-pt}$ and then integrate with respect to $t$ from zero to infinity. Taking into account the relations

$\mathcal{L}\{\partial_t w\} = p\widetilde{w} - w|_{t=0} = p\widetilde{w}$   (used are first property (14.5.1.3) and the initial condition),

$\mathcal{L}\{w_0\} = w_0 \mathcal{L}\{1\} = w_0/p$   (used are property 1 of Subsection T3.1.1 and the relation $\mathcal{L}\{1\} = 1/p$, see Property 1 of Subsection T3.1.2)

we arrive at the following problem for a second-order linear ordinary differential equation with parameter $p$:

$$\widetilde{w}''_{xx} - p\widetilde{w} = 0,$$
$$\widetilde{w} = w_0/p \quad \text{at} \quad x = 0 \quad \text{(boundary condition)},$$
$$\widetilde{w} \to 0 \quad \text{at} \quad x \to \infty \quad \text{(boundary condition)}.$$

Integrating the equation yields the general solution $\widetilde{w} = A_1(p)e^{-x\sqrt{p}} + A_2(p)e^{x\sqrt{p}}$. Using the boundary conditions, we determine the constants, $A_1(p) = w_0/p$ and $A_2(p) = 0$. Thus, we have

$$\widetilde{w} = \frac{w_0}{p}e^{-x\sqrt{p}}.$$

Let us apply the inverse Laplace transform to both sides of this relation. We refer to the table in Subsection T3.2.5, row 20 (where $x$ must be replaced by $t$ and then $a$ by $x^2$), to find the inverse transform of the right-hand side. Finally, we obtain the solution of the original problem in the form

$$w = w_0 \operatorname{erfc}\left(\frac{x}{2\sqrt{t}}\right).$$

**14.5.1-4. Solving linear problems for hyperbolic equations by the Laplace transform.**

Consider a linear nonstationary boundary value problem defined by the hyperbolic equation

$$\frac{\partial^2 w}{\partial t^2} + \varphi(x)\frac{\partial w}{\partial t} = a(x)\frac{\partial^2 w}{\partial x^2} + b(x)\frac{\partial w}{\partial x} + c(x)w + \Phi(x,t) \qquad (14.5.1.8)$$

with the initial conditions (14.4.1.3), (14.4.1.4) and the general boundary conditions (14.5.1.5). The application of the Laplace transform results in the problem defined by the ordinary differential equation for $x$ ($p$ is treated as a parameter)

$$a(x)\frac{\partial^2 \widetilde{w}}{\partial x^2} + b(x)\frac{\partial \widetilde{w}}{\partial x} + [c(x) - p^2 - p\varphi(x)]\widetilde{w} + f_0(x)[p + \varphi(x)] + f_1(x) + \widetilde{\Phi}(x,p) = 0 \qquad (14.5.1.9)$$

with the boundary conditions (14.5.1.7). On solving this problem, one should apply the inverse Laplace transform to the resulting solution, $\widetilde{w} = \widetilde{w}(x, p)$.

## 14.5.2. Fourier Transform and Its Application in Mathematical Physics

**14.5.2-1. Fourier transform and its properties.**

The Fourier transform is defined as follows:

$$\widetilde{f}(u) = \mathfrak{F}\{f(x)\}, \quad \text{where} \quad \mathfrak{F}\{f(x)\} \equiv \frac{1}{\sqrt{2\pi}}\int_{-\infty}^{\infty} f(x)e^{-iux}\,dx, \qquad i^2 = -1. \quad (14.5.2.1)$$

This relation is meaningful for any function $f(x)$ absolutely integrable on the interval $(-\infty, \infty)$.

Given $\widetilde{f}(u)$, the function $f(x)$ can be found by means of the inverse Fourier transform

$$f(x) = \mathfrak{F}^{-1}\{\widetilde{f}(u)\}, \quad \text{where} \quad \mathfrak{F}^{-1}\{\widetilde{f}(u)\} \equiv \frac{1}{\sqrt{2\pi}}\int_{-\infty}^{\infty} \widetilde{f}(u)e^{iux}\,du, \qquad (14.5.2.2)$$

where the integral is understood in the sense of the Cauchy principal value.

The main properties of the correspondence between functions and their Fourier transforms are gathered in Table 11.4.

**14.5.2-2. Solving linear problems of mathematical physics by the Fourier transform.**

The Fourier transform is usually employed to solve boundary value problems for linear partial differential equations whose coefficients are independent of the space variable $x$, $-\infty < x < \infty$.

The scheme for solving linear boundary value problems with the help of the Fourier transform is similar to that used in solving problems with help of the Laplace transform. With the Fourier transform, the derivatives with respect to $x$ in the equation are replaced by appropriate algebraic expressions; see Property 4 or 5 in Table 11.4. In the case of two independent variables, the problem for a partial differential equation is reduced to a simpler problem for an ordinary differential equation with parameter $u$. On solving the latter problem, one determines the transform. After that, by applying the inverse Fourier transform, one obtains the solution of the original boundary value problem.

**Example 2.** Consider the following Cauchy problem for the heat equation:

$$\frac{\partial w}{\partial t} = \frac{\partial^2 w}{\partial x^2} \quad (-\infty < x < \infty),$$
$$w = f(x) \quad \text{at} \quad t = 0 \quad \text{(initial condition)}.$$

We apply the Fourier transform with respect to the space variable $x$. Setting $\widetilde{w} = \mathfrak{F}\{w(x,t)\}$ and taking into account the relation $\mathfrak{F}\{\partial_{xx}w\} = -u^2\widetilde{w}$ (see Property 4 in Table 11.4), we arrive at the following problem for a linear first-order ordinary differential equation in $t$ with parameter $u$:

$$\widetilde{w}'_t + u^2 \widetilde{w} = 0,$$
$$\widetilde{w} = \widetilde{f}(u) \quad \text{at} \quad t = 0,$$

where $\widetilde{f}(u)$ is defined by (14.5.2.1). On solving this problem for the transform $\widetilde{w}$, we find

$$\widetilde{w} = \widetilde{f}(u) e^{-u^2 t}.$$

Let us apply the inversion formula to both sides of this equation. After some calculations, we obtain the solution of the original problem in the form

$$w = \frac{1}{\sqrt{2\pi}} \int_{-\infty}^{\infty} \widetilde{f}(u) e^{-u^2 t} e^{iux} \, du = \frac{1}{2\pi} \int_{-\infty}^{\infty} \left[ \int_{-\infty}^{\infty} f(\xi) e^{-iu\xi} \, d\xi \right] e^{-u^2 t + iux} \, du$$
$$= \frac{1}{2\pi} \int_{-\infty}^{\infty} f(\xi) \, d\xi \int_{-\infty}^{\infty} e^{-u^2 t + iu(x-\xi)} \, du = \frac{1}{\sqrt{2\pi t}} \int_{-\infty}^{\infty} f(\xi) \exp\left[-\frac{(x-\xi)^2}{4t}\right] d\xi.$$

At the last stage we used the relation $\displaystyle\int_{-\infty}^{\infty} \exp(-a^2 u^2 + bu) \, du = \frac{\sqrt{\pi}}{|a|} \exp\left(\frac{b^2}{4a^2}\right)$.

# 14.6. Representation of the Solution of the Cauchy Problem via the Fundamental Solution

## 14.6.1. Cauchy Problem for Parabolic Equations

**14.6.1-1. General formula for the solution of the Cauchy problem.**

Let $\mathbf{x} = \{x_1, \ldots, x_n\}$ and $\mathbf{y} = \{y_1, \ldots, y_n\}$, where $\mathbf{x} \in \mathbb{R}^n$ and $\mathbf{y} \in \mathbb{R}^n$.

Consider a nonhomogeneous linear equation of the parabolic type with an arbitrary right-hand side,

$$\frac{\partial w}{\partial t} - L_{\mathbf{x},t}[w] = \Phi(\mathbf{x}, t), \tag{14.6.1.1}$$

where the second-order linear differential operator $L_{\mathbf{x},t}$ is defined by relation (14.2.1.2).

The solution of the Cauchy problem for equation (14.6.1.1) with an arbitrary initial condition,
$$w = f(\mathbf{x}) \quad \text{at} \quad t = 0,$$
can be represented as the sum of two integrals,
$$w(\mathbf{x},t) = \int_0^t \int_{\mathbb{R}^n} \Phi(\mathbf{y},\tau)\mathscr{E}(\mathbf{x},\mathbf{y},t,\tau)\,d\mathbf{y}\,d\tau + \int_{\mathbb{R}^n} f(\mathbf{y})\mathscr{E}(\mathbf{x},\mathbf{y},t,0)\,d\mathbf{y}, \quad d\mathbf{y} = dy_1 \ldots dy_n.$$

Here, $\mathscr{E} = \mathscr{E}(\mathbf{x},\mathbf{y},t,\tau)$ is the fundamental solution of the Cauchy problem that satisfies, for $t > \tau \geq 0$, the homogeneous linear equation
$$\frac{\partial \mathscr{E}}{\partial t} - L_{\mathbf{x},t}[\mathscr{E}] = 0 \tag{14.6.1.2}$$
with the nonhomogeneous initial condition of special form
$$\mathscr{E} = \delta(\mathbf{x} - \mathbf{y}) \quad \text{at} \quad t = \tau. \tag{14.6.1.3}$$

The quantities $\tau$ and $\mathbf{y}$ appear in problem (14.6.1.2)–(14.6.1.3) as free parameters, and $\delta(\mathbf{x}) = \delta(x_1)\ldots\delta(x_n)$ is the $n$-dimensional Dirac delta function.

**Remark 1.** If the coefficients of the differential operator $L_{\mathbf{x},t}$ in (14.6.1.2) are independent of time $t$, then the fundamental solution of the Cauchy problem depends on only three arguments, $\mathscr{E}(\mathbf{x},\mathbf{y},t,\tau) = \mathscr{E}(\mathbf{x},\mathbf{y},t-\tau)$.

**Remark 2.** If the differential operator $L_{\mathbf{x},t}$ has constant coefficients, then the fundamental solution of the Cauchy problem depends on only two arguments, $\mathscr{E}(\mathbf{x},\mathbf{y},t,\tau) = \mathscr{E}(\mathbf{x}-\mathbf{y},t-\tau)$.

**14.6.1-2. Fundamental solution allowing incomplete separation of variables.**

Consider the special case where the differential operator $L_{\mathbf{x},t}$ in equation (14.6.1.1) can be represented as the sum
$$L_{\mathbf{x},t}[w] = L_{1,t}[w] + \cdots + L_{n,t}[w], \tag{14.6.1.4}$$
where each term depends on a single space coordinate and time,
$$L_{k,t}[w] \equiv a_k(x_k,t)\frac{\partial^2 w}{\partial x_k^2} + b_k(x_k,t)\frac{\partial w}{\partial x_k} + c_k(x_k,t)w, \quad k = 1, \ldots, n.$$

Equations of this form are often encountered in applications. The fundamental solution of the Cauchy problem for the $n$-dimensional equation (14.6.1.1) with operator (14.6.1.4) can be represented in the product form
$$\mathscr{E}(\mathbf{x},\mathbf{y},t,\tau) = \prod_{k=1}^n \mathscr{E}_k(x_k,y_k,t,\tau), \tag{14.6.1.5}$$
where $\mathscr{E}_k = \mathscr{E}_k(x_k,y_k,t,\tau)$ are the fundamental solutions satisfying the one-dimensional equations
$$\frac{\partial \mathscr{E}_k}{\partial t} - L_{k,t}[\mathscr{E}_k] = 0 \quad (k = 1, \ldots, n)$$
with the initial conditions
$$\mathscr{E}_k = \delta(x_k - y_k) \quad \text{at} \quad t = \tau.$$

In this case, the fundamental solution of the Cauchy problem (14.6.1.5) admits incomplete separation of variables; the fundamental solution is separated in the space variables $x_1, \ldots, x_n$ but not in time $t$.

**Example 1.** Consider the two-dimensional heat equation
$$\frac{\partial w}{\partial t} = \frac{\partial^2 w}{\partial x_1^2} + \frac{\partial^2 w}{\partial x_2^2}.$$
The fundamental solutions of the corresponding one-dimensional heat equations are expressed as

| Equations | Fundamental solutions |
|---|---|
| $\dfrac{\partial w}{\partial t} = \dfrac{\partial^2 w}{\partial x_1^2} \implies$ | $\mathscr{E}_1(x_1, y_1, t, \tau) = \dfrac{1}{2\sqrt{\pi(t-\tau)}} \exp\left[-\dfrac{(x_1-y_1)^2}{4(t-\tau)}\right],$ |
| $\dfrac{\partial w}{\partial t} = \dfrac{\partial^2 w}{\partial x_2^2} \implies$ | $\mathscr{E}_2(x_2, y_2, t, \tau) = \dfrac{1}{2\sqrt{\pi(t-\tau)}} \exp\left[-\dfrac{(x_2-y_2)^2}{4(t-\tau)}\right].$ |

Multiplying $\mathscr{E}_1$ and $\mathscr{E}_2$ together gives the fundamental solution of the two-dimensional heat equation:
$$\mathscr{E}(x_1, x_2, y_1, y_2, t, \tau) = \frac{1}{4\pi(t-\tau)} \exp\left[-\frac{(x_1-y_1)^2 + (x_2-y_2)^2}{4(t-\tau)}\right].$$

**Example 2.** The fundamental solution of the equation
$$\frac{\partial w}{\partial t} = \sum_{k=1}^{n} a_k(t) \frac{\partial^2 w}{\partial x_k^2}, \qquad 0 < a_k(t) < \infty,$$
is given by formula (14.6.1.5) with
$$\mathscr{E}_k(x_k, y_k, t, \tau) = \frac{1}{2\sqrt{\pi T_k}} \exp\left[-\frac{(x_k-y_k)^2}{4T_k}\right], \qquad T_k = \int_\tau^t a_k(\eta)\, d\eta.$$
In the derivation of this formula it was taken into account that the corresponding one-dimensional equations could be reduced to the ordinary constant-coefficient heat equation by passing from $x_k, t$ to the new variables $x_k, T_k$.

### 14.6.2. Cauchy Problem for Hyperbolic Equations

Consider a nonhomogeneous linear equation of the hyperbolic type with an arbitrary right-hand side,
$$\frac{\partial^2 w}{\partial t^2} + \varphi(\mathbf{x}, t)\frac{\partial w}{\partial t} - L_{\mathbf{x},t}[w] = \Phi(\mathbf{x}, t), \tag{14.6.2.1}$$
where the second-order linear differential operator $L_{\mathbf{x},t}$ is defined by relation (14.2.1.2) with $\mathbf{x} \in \mathbb{R}^n$.

The solution of the Cauchy problem for equation (14.6.2.1) with general initial conditions,
$$w = f_0(\mathbf{x}) \quad \text{at} \quad t = 0,$$
$$\partial_t w = f_1(\mathbf{x}) \quad \text{at} \quad t = 0,$$
can be represented as the sum
$$w(\mathbf{x}, t) = \int_0^t \int_{\mathbb{R}^n} \Phi(\mathbf{y}, \tau) \mathscr{E}(\mathbf{x}, \mathbf{y}, t, \tau)\, d\mathbf{y}\, d\tau - \int_{\mathbb{R}^n} f_0(\mathbf{y}) \left[\frac{\partial}{\partial \tau} \mathscr{E}(\mathbf{x}, \mathbf{y}, t, \tau)\right]_{\tau=0} d\mathbf{y}$$
$$+ \int_{\mathbb{R}^n} \big[f_1(\mathbf{y}) + f_0(\mathbf{y})\varphi(\mathbf{y}, 0)\big] \mathscr{E}(\mathbf{x}, \mathbf{y}, t, 0)\, d\mathbf{y}, \qquad d\mathbf{y} = dy_1 \ldots dy_n.$$
Here, $\mathscr{E} = \mathscr{E}(\mathbf{x}, \mathbf{y}, t, \tau)$ is the fundamental solution of the Cauchy problem that satisfies, for $t > \tau \geq 0$, the homogeneous linear equation
$$\frac{\partial^2 \mathscr{E}}{\partial t^2} + \varphi(\mathbf{x}, t)\frac{\partial \mathscr{E}}{\partial t} - L_{\mathbf{x},t}[\mathscr{E}] = 0 \tag{14.6.2.2}$$
with the semihomogeneous initial conditions of special form
$$\mathscr{E} = 0 \qquad \text{at} \quad t = \tau,$$
$$\partial_t \mathscr{E} = \delta(\mathbf{x} - \mathbf{y}) \qquad \text{at} \quad t = \tau. \tag{14.6.2.3}$$
The quantities $\tau$ and $\mathbf{y}$ appear in problem (14.6.2.2)–(14.6.2.3) as free parameters ($\mathbf{y} \in \mathbb{R}^n$).

**Remark 1.** If the coefficients of the differential operator $L_{\mathbf{x},t}$ in (14.6.2.2) are independent of time $t$, then the fundamental solution of the Cauchy problem depends on only three arguments, $\mathscr{E}(\mathbf{x},\mathbf{y},t,\tau) = \mathscr{E}(\mathbf{x},\mathbf{y},t-\tau)$. Here, $\frac{\partial}{\partial \tau} \mathscr{E}(\mathbf{x},\mathbf{y},t,\tau)\big|_{\tau=0} = -\frac{\partial}{\partial t} \mathscr{E}(\mathbf{x},\mathbf{y},t)$.

**Remark 2.** If the differential operator $L_{\mathbf{x},t}$ has constant coefficients, then the fundamental solution of the Cauchy problem depends on only two arguments, $\mathscr{E}(\mathbf{x},\mathbf{y},t,\tau) = \mathscr{E}(\mathbf{x}-\mathbf{y},t-\tau)$.

**Example 1.** For the one-, two-, and three-dimensional wave equations, the fundamental solutions have the forms

*Equations*                                    *Fundamental solutions*

$$\frac{\partial^2 w}{\partial t^2} - \frac{\partial^2 w}{\partial x^2} = 0 \quad \Longrightarrow \quad \mathscr{E}(x,t,y,\tau) = \frac{1}{2}\vartheta(t-\tau-|x-y|), \quad \vartheta(z) = \begin{cases} 1 & \text{if } z \geq 0, \\ 0 & \text{if } z < 0; \end{cases}$$

$$\frac{\partial^2 w}{\partial t^2} - \frac{\partial^2 w}{\partial x_1^2} - \frac{\partial^2 w}{\partial x_2^2} = 0 \quad \Longrightarrow \quad \mathscr{E}(x_1,x_2,t,y_1,y_2,\tau) = \frac{\vartheta(t-\tau-\rho)}{2\pi\sqrt{(t-\tau)^2 - \rho^2}};$$

$$\frac{\partial^2 w}{\partial t^2} - \frac{\partial^2 w}{\partial x_1^2} - \frac{\partial^2 w}{\partial x_2^2} - \frac{\partial^2 w}{\partial x_3^2} = 0 \quad \Longrightarrow \quad \mathscr{E}(x_1,x_2,x_3,t,y_1,y_2,y_3,\tau) = \frac{1}{2\pi}\delta\big((t-\tau)^2 - r^2\big),$$

where $\vartheta(z)$ is the Heaviside unit step function ($\vartheta = 0$ for $z < 0$ and $\vartheta = 1$ for $z \geq 0$), $\rho = \sqrt{(x_1-y_1)^2 + (x_2-y_2)^2}$, $r = \sqrt{(x_1-y_1)^2 + (x_2-y_2)^2 + (x_3-y_3)^2}$, and $\delta(z)$ is the Dirac delta function.

**Example 2.** The one-dimensional Klein–Gordon equation

$$\frac{\partial^2 w}{\partial t^2} = \frac{\partial^2 w}{\partial x^2} - bw$$

has the fundamental solutions

$$\mathscr{E}(x,t,y,\tau) = \tfrac{1}{2}\vartheta\big(t-\tau-|x-y|\big) J_0\big(k\sqrt{(t-\tau)^2 - (x-y)^2}\big) \quad \text{for} \quad b = k^2 > 0,$$

$$\mathscr{E}(x,t,y,\tau) = \tfrac{1}{2}\vartheta\big(t-\tau-|x-y|\big) I_0\big(k\sqrt{(t-\tau)^2 - (x-y)^2}\big) \quad \text{for} \quad b = -k^2 < 0,$$

where $\vartheta(z)$ is the Heaviside unit step function, $J_0(z)$ is the Bessel function, $I_0(z)$ is the modified Bessel function, and $k > 0$.

**Example 3.** The two-dimensional Klein–Gordon equation

$$\frac{\partial^2 w}{\partial t^2} = \frac{\partial^2 w}{\partial x_1^2} + \frac{\partial^2 w}{\partial x_2^2} - bw$$

has the fundamental solutions

$$\mathscr{E}(x_1,x_2,t,y_1,y_2,\tau) = \vartheta(t-\tau-\rho)\frac{\cos\big(k\sqrt{(t-\tau)^2 - \rho^2}\big)}{2\pi\sqrt{(t-\tau)^2 - \rho^2}} \quad \text{for} \quad b = k^2 > 0,$$

$$\mathscr{E}(x_1,x_2,t,y_1,y_2,\tau) = \vartheta(t-\tau-\rho)\frac{\cosh\big(k\sqrt{(t-\tau)^2 - \rho^2}\big)}{2\pi\sqrt{(t-\tau)^2 - \rho^2}} \quad \text{for} \quad b = -k^2 < 0,$$

where $\vartheta(z)$ is the Heaviside unit step function and $\rho = \sqrt{(x_1-y_1)^2 + (x_2-y_2)^2}$.

## 14.7. Boundary Value Problems for Parabolic Equations with One Space Variable. Green's Function

### 14.7.1. Representation of Solutions via the Green's Function

#### 14.7.1-1. Statement of the problem ($t \geq 0$, $x_1 \leq x \leq x_2$).

In general, a nonhomogeneous linear differential equation of the parabolic type with variable coefficients in one dimension can be written as

$$\frac{\partial w}{\partial t} - L_{x,t}[w] = \Phi(x,t), \tag{14.7.1.1}$$

where

$$L_{x,t}[w] \equiv a(x,t)\frac{\partial^2 w}{\partial x^2} + b(x,t)\frac{\partial w}{\partial x} + c(x,t)w, \qquad a(x,t) > 0. \tag{14.7.1.2}$$

Consider the nonstationary boundary value problem for equation (14.7.1.1) with an initial condition of general form

$$w = f(x) \quad \text{at} \quad t = 0, \tag{14.7.1.3}$$

and arbitrary nonhomogeneous linear boundary conditions

$$\alpha_1 \frac{\partial w}{\partial x} + \beta_1 w = g_1(t) \quad \text{at} \quad x = x_1, \tag{14.7.1.4}$$

$$\alpha_2 \frac{\partial w}{\partial x} + \beta_2 w = g_2(t) \quad \text{at} \quad x = x_2. \tag{14.7.1.5}$$

By appropriately choosing the coefficients $\alpha_1, \alpha_2, \beta_1$, and $\beta_2$ in (14.7.1.4) and (14.7.1.5), we obtain the first, second, third, and mixed boundary value problems for equation (14.7.1.1).

**14.7.1-2. Representation of the problem solution in terms of the Green's function.**

The solution of the nonhomogeneous linear boundary value problem (14.7.1.1)–(14.7.1.5) can be represented as

$$w(x,t) = \int_0^t \int_{x_1}^{x_2} \Phi(y,\tau) G(x,y,t,\tau)\, dy\, d\tau + \int_{x_1}^{x_2} f(y) G(x,y,t,0)\, dy$$
$$+ \int_0^t g_1(\tau) a(x_1,\tau) \Lambda_1(x,t,\tau)\, d\tau + \int_0^t g_2(\tau) a(x_2,\tau) \Lambda_2(x,t,\tau)\, d\tau. \tag{14.7.1.6}$$

Here, $G(x,y,t,\tau)$ is the Green's function that satisfies, for $t > \tau \geq 0$, the homogeneous equation

$$\frac{\partial G}{\partial t} - L_{x,t}[G] = 0 \tag{14.7.1.7}$$

with the nonhomogeneous initial condition of special form

$$G = \delta(x - y) \quad \text{at} \quad t = \tau \tag{14.7.1.8}$$

and the homogeneous boundary conditions

$$\alpha_1 \frac{\partial G}{\partial x} + \beta_1 G = 0 \quad \text{at} \quad x = x_1, \tag{14.7.1.9}$$

$$\alpha_2 \frac{\partial G}{\partial x} + \beta_2 G = 0 \quad \text{at} \quad x = x_2. \tag{14.7.1.10}$$

The quantities $y$ and $\tau$ appear in problem (14.7.1.7)–(14.7.1.10) as free parameters, with $x_1 \leq y \leq x_2$, and $\delta(x)$ is the Dirac delta function.

The initial condition (14.7.1.8) implies the limit relation

$$f(x) = \lim_{t \to \tau} \int_{x_1}^{x_2} f(y) G(x,y,t,\tau)\, dy$$

for any continuous function $f = f(x)$.

**TABLE 14.6**
Expressions of the functions $\Lambda_1(x,t,\tau)$ and $\Lambda_2(x,t,\tau)$ involved in the integrands of the last two terms in solution (14.7.1.6)

| Type of problem | Form of boundary conditions | Functions $\Lambda_m(x,t,\tau)$ |
|---|---|---|
| First boundary value problem ($\alpha_1 = \alpha_2 = 0$, $\beta_1 = \beta_2 = 1$) | $w = g_1(t)$ at $x = x_1$ <br> $w = g_2(t)$ at $x = x_2$ | $\Lambda_1(x,t,\tau) = \partial_y G(x,y,t,\tau)\big|_{y=x_1}$ <br> $\Lambda_2(x,t,\tau) = -\partial_y G(x,y,t,\tau)\big|_{y=x_2}$ |
| Second boundary value problem ($\alpha_1 = \alpha_2 = 1$, $\beta_1 = \beta_2 = 0$) | $\partial_x w = g_1(t)$ at $x = x_1$ <br> $\partial_x w = g_2(t)$ at $x = x_2$ | $\Lambda_1(x,t,\tau) = -G(x,x_1,t,\tau)$ <br> $\Lambda_2(x,t,\tau) = G(x,x_2,t,\tau)$ |
| Third boundary value problem ($\alpha_1 = \alpha_2 = 1$, $\beta_1 < 0$, $\beta_2 > 0$) | $\partial_x w + \beta_1 w = g_1(t)$ at $x = x_1$ <br> $\partial_x w + \beta_2 w = g_2(t)$ at $x = x_2$ | $\Lambda_1(x,t,\tau) = -G(x,x_1,t,\tau)$ <br> $\Lambda_2(x,t,\tau) = G(x,x_2,t,\tau)$ |
| Mixed boundary value problem ($\alpha_1 = \beta_2 = 0$, $\alpha_2 = \beta_1 = 1$) | $w = g_1(t)$ at $x = x_1$ <br> $\partial_x w = g_2(t)$ at $x = x_2$ | $\Lambda_1(x,t,\tau) = \partial_y G(x,y,t,\tau)\big|_{y=x_1}$ <br> $\Lambda_2(x,t,\tau) = G(x,x_2,t,\tau)$ |
| Mixed boundary value problem ($\alpha_1 = \beta_2 = 1$, $\alpha_2 = \beta_1 = 0$) | $\partial_x w = g_1(t)$ at $x = x_1$ <br> $w = g_2(t)$ at $x = x_2$ | $\Lambda_1(x,t,\tau) = -G(x,x_1,t,\tau)$ <br> $\Lambda_2(x,t,\tau) = -\partial_y G(x,y,t,\tau)\big|_{y=x_2}$ |

The functions $\Lambda_1(x,t,\tau)$ and $\Lambda_2(x,t,\tau)$ involved in the integrands of the last two terms in solution (14.7.1.6) can be expressed in terms of the Green's function $G(x,y,t,\tau)$. The corresponding formulas for $\Lambda_m(x,t,\tau)$ are given in Table 14.6 for the basic types of boundary value problems.

It is significant that the Green's function $G$ and the functions $\Lambda_1$, $\Lambda_2$ are independent of the functions $\Phi$, $f$, $g_1$, and $g_2$ that characterize various nonhomogeneities of the boundary value problem.

If the coefficients of equation (14.7.1.1)–(14.7.1.2) are independent of time $t$, i.e., the conditions
$$a = a(x), \quad b = b(x), \quad c = c(x) \tag{14.7.1.11}$$
hold, then the Green's function depends on only three arguments,
$$G(x,y,t,\tau) = G(x,y,t-\tau).$$
In this case, the functions $\Lambda_m$ depend on only two arguments, $\Lambda_m = \Lambda_m(x, t-\tau)$, $m = 1, 2$.

Formula (14.7.1.6) also remains valid for the problem with boundary conditions of the third kind if $\beta_1 = \beta_1(t)$ and $\beta_2 = \beta_2(t)$. Here, the relation between $\Lambda_m$ ($m = 1, 2$) and the Green's function $G$ is the same as that in the case of constants $\beta_1$ and $\beta_2$; the Green's function itself is now different.

The condition that the solution must vanish at infinity, $w \to 0$ as $x \to \infty$, is often set for the first, second, and third boundary value problems that are considered on the interval $x_1 \leq x < \infty$. In this case, the solution is calculated by formula (14.7.1.6) with $\Lambda_2 = 0$ and $\Lambda_1$ specified in Table 14.6.

## 14.7.2. Problems for Equation $s(x)\dfrac{\partial w}{\partial t} = \dfrac{\partial}{\partial x}\left[p(x)\dfrac{\partial w}{\partial x}\right] - q(x)w + \Phi(x,t)$

**14.7.2-1. General formulas for solving nonhomogeneous boundary value problems.**

Consider linear equations of the special form
$$s(x)\frac{\partial w}{\partial t} = \frac{\partial}{\partial x}\left[p(x)\frac{\partial w}{\partial x}\right] - q(x)w + \Phi(x,t). \tag{14.7.2.1}$$

TABLE 14.7
Expressions of the functions $\Lambda_1(x,t)$ and $\Lambda_2(x,t)$ involved in the integrands of the last two terms in solutions (14.7.2.2) and (14.8.2.2); the modified Green's function $\mathcal{G}(x,\xi,t)$ for parabolic equations of the form (14.7.2.1) are found by formula (14.7.2.3), and that for hyperbolic equations of the form (14.8.2.1), by formula (14.8.2.3)

| Type of problem | Form of boundary conditions | Functions $\Lambda_m(x,t)$ |
|---|---|---|
| First boundary value problem $(\alpha_1 = \alpha_2 = 0, \ \beta_1 = \beta_2 = 1)$ | $w = g_1(t)$ at $x = x_1$ <br> $w = g_2(t)$ at $x = x_2$ | $\Lambda_1(x,t) = \partial_\xi \mathcal{G}(x,\xi,t)\big|_{\xi=x_1}$ <br> $\Lambda_2(x,t) = -\partial_\xi \mathcal{G}(x,\xi,t)\big|_{\xi=x_2}$ |
| Second boundary value problem $(\alpha_1 = \alpha_2 = 1, \ \beta_1 = \beta_2 = 0)$ | $\partial_x w = g_1(t)$ at $x = x_1$ <br> $\partial_x w = g_2(t)$ at $x = x_2$ | $\Lambda_1(x,t) = -\mathcal{G}(x,x_1,t)$ <br> $\Lambda_2(x,t) = \mathcal{G}(x,x_2,t)$ |
| Third boundary value problem $(\alpha_1 = \alpha_2 = 1, \ \beta_1 < 0, \ \beta_2 > 0)$ | $\partial_x w + \beta_1 w = g_1(t)$ at $x = x_1$ <br> $\partial_x w + \beta_2 w = g_2(t)$ at $x = x_2$ | $\Lambda_1(x,t) = -\mathcal{G}(x,x_1,t)$ <br> $\Lambda_2(x,t) = \mathcal{G}(x,x_2,t)$ |
| Mixed boundary value problem $(\alpha_1 = \beta_2 = 0, \ \alpha_2 = \beta_1 = 1)$ | $w = g_1(t)$ at $x = x_1$ <br> $\partial_x w = g_2(t)$ at $x = x_2$ | $\Lambda_1(x,t) = \partial_\xi \mathcal{G}(x,\xi,t)\big|_{\xi=x_1}$ <br> $\Lambda_2(x,t) = \mathcal{G}(x,x_2,t)$ |
| Mixed boundary value problem $(\alpha_1 = \beta_2 = 1, \ \alpha_2 = \beta_1 = 0)$ | $\partial_x w = g_1(t)$ at $x = x_1$ <br> $w = g_2(t)$ at $x = x_2$ | $\Lambda_1(x,t) = -\mathcal{G}(x,x_1,t)$ <br> $\Lambda_2(x,t) = -\partial_\xi \mathcal{G}(x,\xi,t)\big|_{\xi=x_2}$ |

They are often encountered in heat and mass transfer theory and chemical engineering sciences. Throughout this subsection, we assume that the functions $s$, $p$, $p'_x$, and $q$ are continuous and $s > 0$, $p > 0$, and $x_1 \leq x \leq x_2$.

The solution of equation (14.7.2.1) under the initial condition (14.7.1.3) and the arbitrary linear nonhomogeneous boundary conditions (14.7.1.4)–(14.7.1.5) can be represented as the sum

$$w(x,t) = \int_0^t \int_{x_1}^{x_2} \Phi(\xi,\tau)\mathcal{G}(x,\xi,t-\tau)\,d\xi\,d\tau + \int_{x_1}^{x_2} s(\xi)f(\xi)\mathcal{G}(x,\xi,t)\,d\xi$$
$$+ p(x_1)\int_0^t g_1(\tau)\Lambda_1(x,t-\tau)\,d\tau + p(x_2)\int_0^t g_2(\tau)\Lambda_2(x,t-\tau)\,d\tau. \qquad (14.7.2.2)$$

Here, the modified Green's function is given by

$$\mathcal{G}(x,\xi,t) = \sum_{n=1}^{\infty} \frac{y_n(x)y_n(\xi)}{\|y_n\|^2}\exp(-\lambda_n t), \qquad \|y_n\|^2 = \int_{x_1}^{x_2} s(x)y_n^2(x)\,dx, \qquad (14.7.2.3)$$

where the $\lambda_n$ and $y_n(x)$ are the eigenvalues and corresponding eigenfunctions of the following Sturm–Liouville problem for a second-order linear ordinary differential equation:

$$[p(x)y'_x]'_x + [\lambda s(x) - q(x)]y = 0,$$
$$\alpha_1 y'_x + \beta_1 y = 0 \quad \text{at} \quad x = x_1, \qquad (14.7.2.4)$$
$$\alpha_2 y'_x + \beta_2 y = 0 \quad \text{at} \quad x = x_2.$$

The functions $\Lambda_1(x,t)$ and $\Lambda_2(x,t)$ that occur in the integrands of the last two terms in solution (14.7.2.2) are expressed via the Green's function (14.7.2.3). The corresponding formulas for $\Lambda_m(x,t)$ are given in Table 14.7 for the basic types of boundary value problems.

**14.7.2-2. Properties of Sturm–Liouville problem (14.7.2.4). Heat equation example.**

1°. There are infinitely many eigenvalues. All eigenvalues are real and different and can be ordered so that $\lambda_1 < \lambda_2 < \lambda_3 < \cdots$, with $\lambda_n \to \infty$ as $n \to \infty$ (therefore, there can exist only finitely many negative eigenvalues). Each eigenvalue is of multiplicity 1.

2°. The different eigenfunctions $y_n(x)$ and $y_m(x)$ are orthogonal with weight $s(x)$ on the interval $x_1 \leq x \leq x_2$:

$$\int_{x_1}^{x_2} s(x) y_n(x) y_m(x)\, dx = 0 \quad \text{for} \quad n \neq m.$$

3°. If the conditions

$$q(x) \geq 0, \quad \alpha_1 \beta_1 \leq 0, \quad \alpha_2 \beta_2 \geq 0 \tag{14.7.2.5}$$

are satisfied, there are no negative eigenvalues. If $q \equiv 0$ and $\beta_1 = \beta_2 = 0$, then $\lambda_1 = 0$ is the least eigenvalue, to which there corresponds the eigenfunction $\varphi_1 = \text{const}$. Otherwise, all eigenvalues are positive, provided that conditions (14.7.2.5) are satisfied.

Other general and special properties of the Sturm–Liouville problem (14.7.2.4) are given in Subsection 12.2.5; various asymptotic and approximate formulas for eigenvalues and eigenfunctions can also be found there.

**Example.** Consider the first boundary value problem in the domain $0 \leq x \leq l$ for the heat equation with a source

$$\frac{\partial w}{\partial t} = a \frac{\partial^2 w}{\partial x^2} - bw$$

under the initial condition (14.7.1.3) and boundary conditions

$$\begin{aligned} w &= g_1(t) \quad \text{at} \quad x = 0, \\ w &= g_2(t) \quad \text{at} \quad x = l. \end{aligned} \tag{14.7.2.6}$$

The above equation is a special case of equation (14.7.2.1) with $s(x)=1$, $p(x)=a$, $q(x)=b$, and $\Phi(x,t)=0$. The corresponding Sturm–Liouville problem (14.7.2.4) has the form

$$a y''_{xx} + (\lambda - b) y = 0, \quad y = 0 \quad \text{at} \quad x = 0, \quad y = 0 \quad \text{at} \quad x = l.$$

The eigenfunctions and eigenvalues are found to be

$$y_n(x) = \sin\left(\frac{\pi n x}{l}\right), \quad \lambda_n = b + \frac{a\pi^2 n^2}{l^2}, \quad n = 1, 2, \ldots$$

Using formula (14.7.2.3) and taking into account that $\|y_n\|^2 = l/2$, we obtain the Green's function

$$\mathcal{G}(x, \xi, t) = \frac{2}{l} e^{-bt} \sum_{n=1}^{\infty} \sin\left(\frac{\pi n x}{l}\right) \sin\left(\frac{\pi n \xi}{l}\right) \exp\left(-\frac{a\pi^2 n^2}{l^2} t\right).$$

Substituting this expression into (14.7.2.2) with $p(x_1) = p(x_2) = s(\xi) = 1$, $x_1 = 0$, and $x_2 = l$ and taking into account the formulas

$$\Lambda_1(x,t) = \partial_\xi \mathcal{G}(x, \xi, t)\big|_{\xi = x_1}, \quad \Lambda_2(x,t) = -\partial_\xi \mathcal{G}(x, \xi, t)\big|_{\xi = x_2}$$

(see the first row in Table 14.7), one obtains the solution to the problem in question.

▶ Solutions to various boundary value problems for parabolic equations can be found in Section T8.1.

## 14.8. Boundary Value Problems for Hyperbolic Equations with One Space Variable. Green's Function. Goursat Problem

### 14.8.1. Representation of Solutions via the Green's Function

**14.8.1-1. Statement of the problem ($t \geq 0$, $x_1 \leq x \leq x_2$).**

In general, a one-dimensional nonhomogeneous linear differential equation of hyperbolic type with variable coefficients is written as

$$\frac{\partial^2 w}{\partial t^2} + \varphi(x,t)\frac{\partial w}{\partial t} - L_{x,t}[w] = \Phi(x,t), \tag{14.8.1.1}$$

where the operator $L_{x,t}[w]$ is defined by (14.7.1.2).

Consider the nonstationary boundary value problem for equation (14.8.1.1) with the initial conditions

$$\begin{aligned} w &= f_0(x) \quad \text{at} \quad t = 0, \\ \partial_t w &= f_1(x) \quad \text{at} \quad t = 0 \end{aligned} \tag{14.8.1.2}$$

and arbitrary nonhomogeneous linear boundary conditions

$$\alpha_1 \frac{\partial w}{\partial x} + \beta_1 w = g_1(t) \quad \text{at} \quad x = x_1, \tag{14.8.1.3}$$

$$\alpha_2 \frac{\partial w}{\partial x} + \beta_2 w = g_2(t) \quad \text{at} \quad x = x_2. \tag{14.8.1.4}$$

**14.8.1-2. Representation of the problem solution in terms of the Green's function.**

The solution of problem (14.8.1.1), (14.8.1.2), (14.8.1.3), (14.8.1.4) can be represented as the sum

$$\begin{aligned} w(x,t) &= \int_0^t \int_{x_1}^{x_2} \Phi(y,\tau) G(x,y,t,\tau)\,dy\,d\tau \\ &\quad - \int_{x_1}^{x_2} f_0(y)\left[\frac{\partial}{\partial \tau} G(x,y,t,\tau)\right]_{\tau=0} dy + \int_{x_1}^{x_2} \bigl[f_1(y) + f_0(y)\varphi(y,0)\bigr] G(x,y,t,0)\,dy \\ &\quad + \int_0^t g_1(\tau) a(x_1,\tau)\Lambda_1(x,t,\tau)\,d\tau + \int_0^t g_2(\tau) a(x_2,\tau)\Lambda_2(x,t,\tau)\,d\tau. \end{aligned} \tag{14.8.1.5}$$

Here, the Green's function $G(x,y,t,\tau)$ is determined by solving the homogeneous equation

$$\frac{\partial^2 G}{\partial t^2} + \varphi(x,t)\frac{\partial G}{\partial t} - L_{x,t}[G] = 0 \tag{14.8.1.6}$$

with the semihomogeneous initial conditions

$$\begin{aligned} G &= 0 \quad \text{at} \quad t = \tau, \\ \partial_t G &= \delta(x-y) \quad \text{at} \quad t = \tau, \end{aligned} \tag{14.8.1.7}\\ \tag{14.8.1.8}$$

and the homogeneous boundary conditions

$$\alpha_1 \frac{\partial G}{\partial x} + \beta_1 G = 0 \quad \text{at} \quad x = x_1, \tag{14.8.1.9}$$

$$\alpha_2 \frac{\partial G}{\partial x} + \beta_2 G = 0 \quad \text{at} \quad x = x_2. \tag{14.8.1.10}$$

The quantities $y$ and $\tau$ appear in problem (14.8.1.6)–(14.8.1.8), (14.8.1.9), (14.8.1.10) as free parameters ($x_1 \leq y \leq x_2$), and $\delta(x)$ is the Dirac delta function.

The functions $\Lambda_1(x,t,\tau)$ and $\Lambda_2(x,t,\tau)$ involved in the integrands of the last two terms in solution (14.8.1.5) can be expressed via the Green's function $G(x,y,t,\tau)$. The corresponding formulas for $\Lambda_m(x,t,\tau)$ are given in Table 14.6 for the basic types of boundary value problems.

It is significant that the Green's function $G$ and $\Lambda_1$, $\Lambda_2$ are independent of the functions $\Phi$, $f_0$, $f_1$, $g_1$, and $g_2$ that characterize various nonhomogeneities of the boundary value problem.

If the coefficients of equation (14.8.1.1) are independent of time $t$, then the Green's function depends on only three arguments, $G(x,y,t,\tau) = G(x,y,t-\tau)$. In this case, one can set $\frac{\partial}{\partial \tau}G(x,y,t,\tau)\big|_{\tau=0} = -\frac{\partial}{\partial t}G(x,y,t)$ in solution (14.8.1.5).

## 14.8.2. Problems for Equation
$$s(x)\frac{\partial^2 w}{\partial t^2} = \frac{\partial}{\partial x}\left[p(x)\frac{\partial w}{\partial x}\right] - q(x)w + \Phi(x,t)$$

**14.8.2-1. General relations for solving nonhomogeneous boundary value problems.**

Consider linear equations of the special form

$$s(x)\frac{\partial^2 w}{\partial t^2} = \frac{\partial}{\partial x}\left[p(x)\frac{\partial w}{\partial x}\right] - q(x)w + \Phi(x,t). \tag{14.8.2.1}$$

It is assumed that the functions $s$, $p$, $p'_x$, and $q$ are continuous and the inequalities $s > 0$, $p > 0$ hold for $x_1 \leq x \leq x_2$.

The solution of equation (14.8.2.1) under the general initial conditions (14.8.1.2) and the arbitrary linear nonhomogeneous boundary conditions (14.8.1.3)–(14.8.1.4) can be represented as the sum

$$\begin{aligned}
w(x,t) &= \int_0^t \int_{x_1}^{x_2} \Phi(\xi,\tau)\mathcal{G}(x,\xi,t-\tau)\,d\xi\,d\tau \\
&\quad + \frac{\partial}{\partial t}\int_{x_1}^{x_2} s(\xi)f_0(\xi)\mathcal{G}(x,\xi,t)\,d\xi + \int_{x_1}^{x_2} s(\xi)f_1(\xi)\mathcal{G}(x,\xi,t)\,d\xi \\
&\quad + p(x_1)\int_0^t g_1(\tau)\Lambda_1(x,t-\tau)\,d\tau + p(x_2)\int_0^t g_2(\tau)\Lambda_2(x,t-\tau)\,d\tau. 
\end{aligned} \tag{14.8.2.2}$$

Here, the modified Green's function is determined by

$$\mathcal{G}(x,\xi,t) = \sum_{n=1}^{\infty} \frac{y_n(x)y_n(\xi)\sin\left(t\sqrt{\lambda_n}\right)}{\|y_n\|^2 \sqrt{\lambda_n}}, \qquad \|y_n\|^2 = \int_{x_1}^{x_2} s(x)y_n^2(x)\,dx, \tag{14.8.2.3}$$

where the $\lambda_n$ and $y_n(x)$ are the eigenvalues and corresponding eigenfunctions of the Sturm–Liouville problem for the second-order linear ordinary differential equation

$$[p(x)y'_x]'_x + [\lambda s(x) - q(x)]y = 0,$$
$$\alpha_1 y'_x + \beta_1 y = 0 \quad \text{at} \quad x = x_1, \qquad (14.8.2.4)$$
$$\alpha_2 y'_x + \beta_2 y = 0 \quad \text{at} \quad x = x_2.$$

The functions $\Lambda_1(x, t)$ and $\Lambda_2(x, t)$ that occur in the integrands of the last two terms in solution (14.8.2.2) are expressed in terms of the Green's function of (14.8.2.3). The corresponding formulas for $\Lambda_m(x, t)$ are given in Table 14.7 for the basic types of boundary value problems.

14.8.2-2. Properties of the Sturm–Liouville problem. The Klein–Gordon equation.

The general and special properties of the Sturm–Liouville problem (14.8.2.4) are given in Subsection 12.2.5; various asymptotic and approximate formulas for eigenvalues and eigenfunctions can also be found there.

**Example.** Consider the second boundary value problem in the domain $0 \le x \le l$ for the Klein–Gordon equation

$$\frac{\partial^2 w}{\partial t^2} = a^2 \frac{\partial^2 w}{\partial x^2} - bw,$$

under the initial conditions (14.8.1.2) and boundary conditions

$$\partial_x w = g_1(t) \quad \text{at} \quad x = 0,$$
$$\partial_x w = g_2(t) \quad \text{at} \quad x = l.$$

The Klein–Gordon equation is a special case of equation (14.8.2.1) with $s(x) = 1$, $p(x) = a^2$, $q(x) = b$, and $\Phi(x, t) = 0$. The corresponding Sturm–Liouville problem (14.8.2.4) has the form

$$a^2 y''_{xx} + (\lambda - b)y = 0, \qquad y'_x = 0 \quad \text{at} \quad x = 0, \qquad y'_x = 0 \quad \text{at} \quad x = l.$$

The eigenfunctions and eigenvalues are found to be

$$y_{n+1}(x) = \cos\left(\frac{\pi n x}{l}\right), \qquad \lambda_{n+1} = b + \frac{a\pi^2 n^2}{l^2}, \qquad n = 0, 1, \ldots$$

Using formula (14.8.2.4) and taking into account that $\|y_1\|^2 = l$ and $\|y_n\|^2 = l/2$ ($n = 1, 2, \ldots$), we find the Green's function:

$$\mathcal{G}(x, \xi, t) = \frac{1}{l\sqrt{b}} \sin(t\sqrt{b}) + \frac{2}{l} \sum_{n=1}^{\infty} \cos\left(\frac{\pi n x}{l}\right) \cos\left(\frac{\pi n \xi}{l}\right) \frac{\sin\left(t\sqrt{(a\pi n/l)^2 + b}\right)}{\sqrt{(a\pi n/l)^2 + b}}.$$

Substituting this expression into (14.8.2.3) with $p(x_1) = p(x_2) = s(\xi) = 1$, $x_1 = 0$, and $x_2 = l$ and taking into account the formulas

$$\Lambda_1(x, t) = -\mathcal{G}(x, x_1, t), \qquad \Lambda_2(x, t) = \mathcal{G}(x, x_2, t)$$

(see the second row in Table 14.7), one obtains the solution to the problem in question.

▶ *Solutions to various boundary value problems for hyperbolic equations can be found in Section T8.2.*

## 14.8.3. Problems for Equation $\dfrac{\partial^2 w}{\partial t^2} + a(t)\dfrac{\partial w}{\partial t} = b(t)\left\{\dfrac{\partial}{\partial x}\left[p(x)\dfrac{\partial w}{\partial x}\right] - q(x)w\right\} + \Phi(x,t)$

### 14.8.3-1. General relations to solve nonhomogeneous boundary value problems.

Consider the generalized telegraph equation of the form

$$\frac{\partial^2 w}{\partial t^2} + a(t)\frac{\partial w}{\partial t} = b(t)\left\{\frac{\partial}{\partial x}\left[p(x)\frac{\partial w}{\partial x}\right] - q(x)w\right\} + \Phi(x,t). \qquad (14.8.3.1)$$

It is assumed that the functions $p$, $p'_x$, and $q$ are continuous and $p > 0$ for $x_1 \le x \le x_2$.

The solution of equation (14.8.3.1) under the general initial conditions (14.8.1.2) and the arbitrary linear nonhomogeneous boundary conditions (14.8.1.3)–(14.8.1.4) can be represented as the sum

$$\begin{aligned}w(x,t) &= \int_0^t \int_{x_1}^{x_2} \Phi(\xi,\tau) G(x,\xi,t,\tau)\,d\xi\,d\tau \\ &\quad - \int_{x_1}^{x_2} f_0(\xi)\left[\frac{\partial}{\partial \tau} G(x,\xi,t,\tau)\right]_{\tau=0} d\xi + \int_{x_1}^{x_2}\bigl[f_1(\xi)+a(0)f_0(\xi)\bigr]G(x,\xi,t,0)\,d\xi \\ &\quad + p(x_1)\int_0^t g_1(\tau)b(\tau)\Lambda_1(x,t,\tau)\,d\tau + p(x_2)\int_0^t g_2(\tau)b(\tau)\Lambda_2(x,t,\tau)\,d\tau. \quad (14.8.3.2)\end{aligned}$$

Here, the modified Green's function is determined by

$$G(x,\xi,t,\tau) = \sum_{n=1}^{\infty} \frac{y_n(x) y_n(\xi)}{\|y_n\|^2} U_n(t,\tau), \qquad \|y_n\|^2 = \int_{x_1}^{x_2} y_n^2(x)\,dx, \qquad (14.8.3.3)$$

where the $\lambda_n$ and $y_n(x)$ are the eigenvalues and corresponding eigenfunctions of the Sturm–Liouville problem for the following second-order linear ordinary differential equation with homogeneous boundary conditions:

$$\begin{aligned}&[p(x)y'_x]'_x + [\lambda - q(x)]y = 0, \\ &\alpha_1 y'_x + \beta_1 y = 0 \quad \text{at}\quad x=x_1, \qquad (14.8.3.4)\\ &\alpha_2 y'_x + \beta_2 y = 0 \quad \text{at}\quad x=x_2.\end{aligned}$$

The functions $U_n = U_n(t,\tau)$ are determined by solving the Cauchy problem for the linear ordinary differential equation

$$\begin{aligned}&U''_n + a(t) U'_n + \lambda_n b(t) U_n = 0, \\ &U_n\big|_{t=\tau} = 0, \quad U'_n\big|_{t=\tau} = 1.\end{aligned} \qquad (14.8.3.5)$$

The prime denotes the derivative with respect to $t$, and $\tau$ is a free parameter occurring in the initial conditions.

The functions $\Lambda_1(x,t)$ and $\Lambda_2(x,t)$ that occur in the integrands of the last two terms in solution (14.8.3.2) are expressed in terms of the Green's function of (14.8.3.3). The corresponding formulas will be specified below when studying specific boundary value problems.

The general and special properties of the Sturm–Liouville problem (14.8.3.4) are detailed in Subsection 12.2.5. Asymptotic and approximate formulas for eigenvalues and eigenfunctions are also presented there.

**14.8.3-2. First, second, third, and mixed boundary value problems.**

1°. *First boundary value problem.* The solution of equation (14.8.3.1) with the initial conditions (14.8.1.2) and boundary conditions (14.8.1.3)–(14.8.1.4) for $\alpha_1 = \alpha_2 = 0$ and $\beta_1 = \beta_2 = 1$ is given by relations (14.8.3.2) and (14.8.3.3), where

$$\Lambda_1(x,t,\tau) = \frac{\partial}{\partial \xi} G(x,\xi,t,\tau)\Big|_{\xi=x_1}, \quad \Lambda_2(x,t,\tau) = -\frac{\partial}{\partial \xi} G(x,\xi,t,\tau)\Big|_{\xi=x_2}.$$

2°. *Second boundary value problem.* The solution of equation (14.8.3.1) with the initial conditions (14.8.1.2) and boundary conditions (14.8.1.3)–(14.8.1.4) for $\alpha_1 = \alpha_2 = 1$ and $\beta_1 = \beta_2 = 0$ is given by relations (14.8.3.2) and (14.8.3.3) with

$$\Lambda_1(x,t,\tau) = -G(x,x_1,t,\tau), \quad \Lambda_2(x,t,\tau) = G(x,x_2,t,\tau).$$

3°. *Third boundary value problem.* The solution of equation (14.8.3.1) with the initial conditions (14.8.1.2) and boundary conditions (14.8.1.3)–(14.8.1.4) for $\alpha_1 = \alpha_2 = 1$ and $\beta_1 \beta_2 \neq 0$ is given by relations (14.8.3.2) and (14.8.3.3) in which

$$\Lambda_1(x,t,\tau) = -G(x,x_1,t,\tau), \quad \Lambda_2(x,t,\tau) = G(x,x_2,t,\tau).$$

4°. *Mixed boundary value problem.* The solution of equation (14.8.3.1) with the initial conditions (14.8.1.2) and boundary conditions (14.8.1.3)–(14.8.1.4) for $\alpha_1 = \beta_2 = 0$ and $\alpha_2 = \beta_1 = 1$ is given by relations (14.8.3.2) and (14.8.3.3) with

$$\Lambda_1(x,t,\tau) = \frac{\partial}{\partial \xi} G(x,\xi,t,\tau)\Big|_{\xi=x_1}, \quad \Lambda_2(x,t,\tau) = G(x,x_2,t,\tau).$$

5°. *Mixed boundary value problem.* The solution of equation (14.8.3.1) with the initial conditions (14.8.1.2) and boundary conditions (14.8.1.3)–(14.8.1.4) for $\alpha_1 = \beta_2 = 1$ and $\alpha_2 = \beta_1 = 0$ is given by relations (14.8.3.2) and (14.8.3.3) with

$$\Lambda_1(x,t,\tau) = -G(x,x_1,t,\tau), \quad \Lambda_2(x,t,\tau) = -\frac{\partial}{\partial \xi} G(x,\xi,t,\tau)\Big|_{\xi=x_2}.$$

## 14.8.4. Generalized Cauchy Problem with Initial Conditions Set Along a Curve

**14.8.4-1. Statement of the generalized Cauchy problem. Basic property of a solution.**

Consider the general linear hyperbolic equation in two independent variables which is reduced to the first canonical form (see Paragraph 14.1.1-4):

$$\frac{\partial^2 w}{\partial x \partial y} + a(x,y)\frac{\partial w}{\partial x} + b(x,y)\frac{\partial w}{\partial y} + c(x,y)w = f(x,y), \qquad (14.8.4.1)$$

where $a(x,y)$, $b(x,y)$, $c(x,y)$, and $f(x,y)$ are continuous functions.

Let a segment of a curve in the $xy$-plane be defined by

$$y = \varphi(x) \qquad (\alpha \leq x \leq \beta), \qquad (14.8.4.2)$$

where $\varphi(x)$ is continuously differentiable, with $\varphi'(x) \neq 0$ and $\varphi'(x) \neq \infty$.

The *generalized Cauchy problem* for equation (14.8.4.1) with initial conditions defined along a curve (14.8.4.2) is stated as follows: find a solution to equation (14.8.4.1) that satisfies the conditions

$$w(x,y)|_{y=\varphi(x)} = g(x), \qquad \frac{\partial w}{\partial x}\bigg|_{y=\varphi(x)} = h_1(x), \qquad \frac{\partial w}{\partial y}\bigg|_{y=\varphi(x)} = h_2(x), \qquad (14.8.4.3)$$

where $g(x)$, $h_1(x)$, and $h_2(x)$ are given continuous functions, related by the compatibility condition

$$g'_x(x) = h_1(x) + h_2(x)\varphi'_x(x). \qquad (14.8.4.4)$$

*Basic property of the generalized Cauchy problem*: the value of the solution at any point $M(x_0, y_0)$ depends only on the values of the functions $g(x)$, $h_1(x)$, and $h_2(x)$ on the arc $AB$, cut off on the given curve (14.8.4.2) by the characteristics $x = x_0$ and $y = y_0$, and on the values of $a(x,y)$, $b(x,y)$, $c(x,y)$, and $f(x,y)$ in the curvilinear triangle $AMB$; see Fig. 14.3. The domain of influence on the solution at $M(x_0, y_0)$ is shaded for clarity.

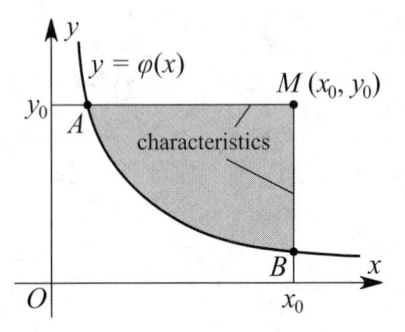

**Figure 14.3.** Domain of influence of the solution to the generalized Cauchy problem at a point $M$.

**Remark 1.** Rather than setting two derivatives in the boundary conditions (14.8.4.3), it suffices to set either of them, with the other being uniquely determined from the compatibility condition (14.8.4.4).

**Remark 2.** Instead of the last two boundary conditions in (14.8.4.3), the value of the derivative along the normal to the curve (14.8.4.2) can be used:

$$\frac{\partial w}{\partial n}\bigg|_{y=\varphi(x)} \equiv \frac{1}{\sqrt{1+[\varphi'_x(x)]^2}} \left[\frac{\partial w}{\partial y} - \varphi'_x(x)\frac{\partial w}{\partial x}\right]_{y=\varphi(x)} = h_3(x). \qquad (14.8.4.5)$$

Denoting $w_x|_{y=\varphi(x)} = h_1(x)$ and $w_y|_{y=\varphi(x)} = h_2(x)$, we have

$$h_2(x) - \varphi'_x(x)h_1(x) = h_3(x)\sqrt{1+[\varphi'_x(x)]^2}. \qquad (14.8.4.6)$$

The functions $h_1(x)$ and $h_2(x)$ can be found from (14.8.4.4) and (14.8.4.6). Further substituting their expressions into (14.8.4.3), one arrives at the standard formulation of the generalized Cauchy problem, where the compatibility condition for initial data (14.8.4.4) will be satisfied automatically.

### 14.8.4-2. Riemann function.

A *Riemann function*, $\mathcal{R} = \mathcal{R}(x, y; x_0, y_0)$, corresponding to equation (14.8.4.1) is defined as a solution to the equation

$$\frac{\partial^2 \mathcal{R}}{\partial x \partial y} - \frac{\partial}{\partial x}\big[a(x,y)\mathcal{R}\big] - \frac{\partial}{\partial y}\big[b(x,y)\mathcal{R}\big] + c(x,y)\mathcal{R} = 0 \qquad (14.8.4.7)$$

that satisfies the conditions

$$\mathcal{R} = \exp\left[\int_{y_0}^{y} a(x_0, \xi)\, d\xi\right] \quad \text{at} \quad x = x_0, \qquad \mathcal{R} = \exp\left[\int_{x_0}^{x} b(\xi, y_0)\, d\xi\right] \quad \text{at} \quad y = y_0 \tag{14.8.4.8}$$

at the characteristics $x = x_0$ and $y = y_0$. Here, $(x_0, y_0)$ is an arbitrary point from the domain of equation (14.8.4.1). The $x_0$ and $y_0$ appear in problem (14.8.4.7)–(14.8.4.8) as parameters in the boundary conditions only.

THEOREM. *If the functions $a$, $b$, $c$ and the partial derivatives $a_x$, $b_y$ are all continuous, then the Riemann function $\mathcal{R}(x, y; x_0, y_0)$ exists. Moreover, the function $\mathcal{R}(x_0, y_0, x, y)$, obtained by swapping the parameters and the arguments, is a solution to the homogeneous equation (14.8.4.1), with $f = 0$.*

Remark. It is significant that the Riemann function depends on neither the shape of the curve (14.8.4.2) nor the initial data set on it (14.8.4.3).

**Example 1.** The Riemann function for the equation $w_{xy} = 0$ is just $\mathcal{R} \equiv 1$.

**Example 2.** The Riemann function for the equation

$$w_{xy} + cw = 0 \qquad (c = \text{const}) \tag{14.8.4.9}$$

is expressed via the Bessel function $J_0(z)$ as

$$\mathcal{R} = J_0\left(\sqrt{4c(x_0 - x)(y_0 - y)}\right).$$

Remark. Any linear constant-coefficient partial differential equation of the parabolic type in two independent variables can be reduced to an equation of the form (14.8.4.9); see Paragraph 14.1.1-6.

**14.8.4-3. Solution of the generalized Cauchy problem via the Riemann function.**

Given a Riemann function, the solution to the generalized Cauchy problem (14.8.4.1)–(14.8.4.3) at any point $(x_0, y_0)$ can written as

$$w(x_0, y_0) = \frac{1}{2}(w\mathcal{R})_A + \frac{1}{2}(w\mathcal{R})_B + \frac{1}{2}\int_{AB}\left(\mathcal{R}\frac{\partial w}{\partial x} - w\frac{\partial \mathcal{R}}{\partial x} + 2bw\mathcal{R}\right) dx$$
$$- \frac{1}{2}\int_{AB}\left(\mathcal{R}\frac{\partial w}{\partial y} - w\frac{\partial \mathcal{R}}{\partial y} + 2aw\mathcal{R}\right) dy + \iint_{\triangle AMB} f\mathcal{R}\, dx\, dy.$$

The first two terms on the right-hand side are evaluated as the points $A$ and $B$. The third and fourth terms are curvilinear integrals over the arc $AB$; the arc is defined by equation (14.8.4.2), and the integrands involve quantities defined by the initial conditions (14.8.4.3). The last integral is taken over the curvilinear triangular domain $AMB$.

## 14.8.5. Goursat Problem (a Problem with Initial Data of Characteristics)

**14.8.5-1. Statement of the Goursat problem. Basic property of the solution.**

The *Goursat problem* for equation (14.8.4.1) is stated as follows: find a solution to equation (14.8.4.1) that satisfies the conditions at characteristics

$$w(x, y)|_{x=x_1} = g(y), \qquad w(x, y)|_{y=y_1} = h(x), \tag{14.8.5.1}$$

where $g(y)$ and $h(x)$ are given continuous functions that match each other at the point of intersection of the characteristics, so that

$$g(y_1) = h(x_1).$$

*Basic properties of the Goursat problem*: the value of the solution at any point $M(x_0, y_0)$ depends only on the values of $g(y)$ at the segment $AN$ (which is part of the characteristic $x = x_1$), the values of $h(x)$ at the segment $BN$ (which is part of the characteristic $y = y_1$), and the values of the functions $a(x, y)$, $b(x, y)$, $c(x, y)$, and $f(x, y)$ in the rectangle $NAMB$; see Fig. 14.4. The domain of influence on the solution at the point $M(x_0, y_0)$ is shaded for clarity.

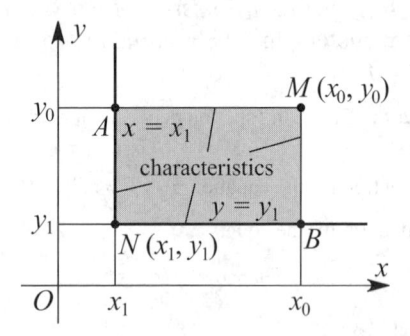

**Figure 14.4.** Domain of influence of the solution to the Goursat problem at a point $M$.

**14.8.5-2. Solution representation for the Goursat problem via the Riemann function.**

Given a Riemann function (see Paragraph 14.8.4-2), the solution to the Goursat problem (14.8.4.1), (14.8.5.1) at any point $(x_0, y_0)$ can be written as

$$w(x_0, y_0) = (w\mathcal{R})_N + \int_N^A \mathcal{R}(g'_y + bg)\, dy + \int_N^B \mathcal{R}(h'_x + ah)\, dx + \iint_{NAMB} f\mathcal{R}\, dx\, dy.$$

The first term on the right-hand side is evaluated at the point of intersection of the characteristics $(x_1, y_1)$. The second and third terms are integrals along the characteristics $y = y_1$ ($x_1 \leq x \leq x_0$) and $x = x_1$ ($y_1 \leq y \leq y_0$); these involve the initial data of (14.8.5.1). The last integral is taken over the rectangular domain $NAMB$ defined by the inequalities $x_1 \leq x \leq x_0$, $y_1 \leq y \leq y_0$.

The Goursat problem for hyperbolic equations reduced to the second canonical form (see Paragraph 14.1.1-4) is treated similarly.

**Example.** Consider the Goursat problem for the wave equation

$$\frac{\partial^2 w}{\partial t^2} - a^2 \frac{\partial^2 w}{\partial x^2} = 0$$

with the boundary conditions prescribed on its characteristics

$$\begin{aligned} w &= f(x) \quad \text{for} \quad x - at = 0 \quad (0 \leq x \leq b), \\ w &= g(x) \quad \text{for} \quad x + at = 0 \quad (0 \leq x \leq c), \end{aligned} \qquad (14.8.5.2)$$

where $f(0) = g(0)$.

Substituting the values set on the characteristics (14.8.5.2) into the general solution of the wave equation, $w = \varphi(x - at) + \psi(x + at)$, we arrive to a system of linear algebraic equations for $\varphi(x)$ and $\psi(x)$. As a result, the solution to the Goursat problem is obtained in the form

$$w(x, t) = f\left(\frac{x + at}{2}\right) + g\left(\frac{x - at}{2}\right) - f(0).$$

The solution propagation domain is the parallelogram bounded by the four lines

$$x - at = 0, \quad x + at = 0, \quad x - at = 2c, \quad x + at = 2b.$$

## 14.9. Boundary Value Problems for Elliptic Equations with Two Space Variables

### 14.9.1. Problems and the Green's Functions for Equation

$$a(x)\frac{\partial^2 w}{\partial x^2} + \frac{\partial^2 w}{\partial y^2} + b(x)\frac{\partial w}{\partial x} + c(x)w = -\Phi(x,y)$$

**14.9.1-1. Statements of boundary value problems.**

Consider two-dimensional boundary value problems for the equation

$$a(x)\frac{\partial^2 w}{\partial x^2} + \frac{\partial^2 w}{\partial y^2} + b(x)\frac{\partial w}{\partial x} + c(x)w = -\Phi(x,y) \tag{14.9.1.1}$$

with general boundary conditions in $x$,

$$\begin{aligned}\alpha_1 \frac{\partial w}{\partial x} - \beta_1 w &= f_1(y) \quad \text{at} \quad x = x_1, \\ \alpha_2 \frac{\partial w}{\partial x} + \beta_2 w &= f_2(y) \quad \text{at} \quad x = x_2,\end{aligned} \tag{14.9.1.2}$$

and different boundary conditions in $y$. It is assumed that the coefficients of equation (14.9.1.1) and the boundary conditions (14.9.1.2) meet the requirements

$a(x)$, $b(x)$, $c(x)$ are continuous ($x_1 \le x \le x_2$); $\quad a > 0, \quad |\alpha_1| + |\beta_1| > 0, \quad |\alpha_2| + |\beta_2| > 0$.

**14.9.1-2. Relations for the Green's function.**

In the general case, the Green's function can be represented as

$$G(x,y,\xi,\eta) = \rho(\xi) \sum_{n=1}^{\infty} \frac{u_n(x)u_n(\xi)}{\|u_n\|^2} \Psi_n(y,\eta;\lambda_n). \tag{14.9.1.3}$$

Here,

$$\rho(x) = \frac{1}{a(x)} \exp\left[\int \frac{b(x)}{a(x)}\,dx\right], \quad \|u_n\|^2 = \int_{x_1}^{x_2} \rho(x) u_n^2(x)\,dx, \tag{14.9.1.4}$$

and the $\lambda_n$ and $u_n(x)$ are the eigenvalues and eigenfunctions of the homogeneous boundary value problem for the ordinary differential equation

$$\begin{aligned}a(x)u''_{xx} + b(x)u'_x + [\lambda + c(x)]u &= 0, &(14.9.1.5)\\ \alpha_1 u'_x - \beta_1 u &= 0 \quad \text{at} \quad x = x_1, &(14.9.1.6)\\ \alpha_2 u'_x + \beta_2 u &= 0 \quad \text{at} \quad x = x_2. &(14.9.1.7)\end{aligned}$$

The functions $\Psi_n$ for various boundary conditions in $y$ are specified in Table 14.8.

Equation (14.9.1.5) can be rewritten in self-adjoint form as

$$[p(x)u'_x]'_x + [\lambda \rho(x) - q(x)]u = 0, \tag{14.9.1.8}$$

TABLE 14.8
The functions $\Psi_n$ in (14.9.1.3) for various boundary conditions.* Notation: $\sigma_n = \sqrt{\lambda_n}$

| Domain | Boundary conditions | Function $\Psi_n(y, \eta; \lambda_n)$ |
|---|---|---|
| $-\infty < y < \infty$ | $\|w\| < \infty$ for $y \to \pm\infty$ | $\dfrac{1}{2\sigma_n} e^{-\sigma_n\|y-\eta\|}$ |
| $0 \leq y < \infty$ | $w = 0$ for $y = 0$ | $\dfrac{1}{\sigma_n} \begin{cases} e^{-\sigma_n y} \sinh(\sigma_n \eta) & \text{for } y > \eta, \\ e^{-\sigma_n \eta} \sinh(\sigma_n y) & \text{for } \eta > y \end{cases}$ |
| $0 \leq y < \infty$ | $\partial_y w = 0$ for $y = 0$ | $\dfrac{1}{\sigma_n} \begin{cases} e^{-\sigma_n y} \cosh(\sigma_n \eta) & \text{for } y > \eta, \\ e^{-\sigma_n \eta} \cosh(\sigma_n y) & \text{for } \eta > y \end{cases}$ |
| $0 \leq y < \infty$ | $\partial_y w - \beta_3 w = 0$ for $y = 0$ | $\dfrac{1}{\sigma_n(\sigma_n + \beta_3)} \begin{cases} e^{-\sigma_n y}[\sigma_n \cosh(\sigma_n \eta) + \beta_3 \sinh(\sigma_n \eta)] & \text{for } y > \eta, \\ e^{-\sigma_n \eta}[\sigma_n \cosh(\sigma_n y) + \beta_3 \sinh(\sigma_n y)] & \text{for } \eta > y \end{cases}$ |
| $0 \leq y \leq h$ | $w = 0$ at $y = 0$, $w = 0$ at $y = h$ | $\dfrac{1}{\sigma_n \sinh(\sigma_n h)} \begin{cases} \sinh(\sigma_n \eta) \sinh[\sigma_n(h-y)] & \text{for } y > \eta, \\ \sinh(\sigma_n y) \sinh[\sigma_n(h-\eta)] & \text{for } \eta > y \end{cases}$ |
| $0 \leq y \leq h$ | $\partial_y w = 0$ at $y = 0$, $\partial_y w = 0$ at $y = h$ | $\dfrac{1}{\sigma_n \sinh(\sigma_n h)} \begin{cases} \cosh(\sigma_n \eta) \cosh[\sigma_n(h-y)] & \text{for } y > \eta, \\ \cosh(\sigma_n y) \cosh[\sigma_n(h-\eta)] & \text{for } \eta > y \end{cases}$ |
| $0 \leq y \leq h$ | $w = 0$ at $y = 0$, $\partial_y w = 0$ at $y = h$ | $\dfrac{1}{\sigma_n \cosh(\sigma_n h)} \begin{cases} \sinh(\sigma_n \eta) \cosh[\sigma_n(h-y)] & \text{for } y > \eta, \\ \sinh(\sigma_n y) \cosh[\sigma_n(h-\eta)] & \text{for } \eta > y \end{cases}$ |

where the functions $p(x)$ and $q(x)$ are given by

$$p(x) = \exp\left[\int \frac{b(x)}{a(x)} dx\right], \quad q(x) = -\frac{c(x)}{a(x)} \exp\left[\int \frac{b(x)}{a(x)} dx\right],$$

and $\rho(x)$ is defined in (14.9.1.4).

The eigenvalue problem (14.9.1.8), (14.9.1.6), (14.9.1.7) possesses the following properties:

1°. All eigenvalues $\lambda_1, \lambda_2, \ldots$ are real and $\lambda_n \to \infty$ as $n \to \infty$.

2°. The system of eigenfunctions $\{u_1(x), u_2(x), \ldots\}$ is orthogonal on the interval $x_1 \leq x \leq x_2$ with weight $\rho(x)$, that is,

$$\int_{x_1}^{x_2} \rho(x) u_n(x) u_m(x) \, dx = 0 \quad \text{for} \quad n \neq m.$$

3°. If the conditions

$$q(x) \geq 0, \quad \alpha_1 \beta_1 \geq 0, \quad \alpha_2 \beta_2 \geq 0 \qquad (14.9.1.9)$$

are satisfied, there are no negative eigenvalues. If $q \equiv 0$ and $\beta_1 = \beta_2 = 0$, then the least eigenvalue is $\lambda_0 = 0$ and the corresponding eigenfunction is $u_0 = \text{const}$; in this case, the summation in (14.9.1.3) must start with $n = 0$. In the other cases, if conditions (14.9.1.9) are satisfied, all eigenvalues are positive; for example, the first inequality in (14.9.1.9) holds if $c(x) \leq 0$.

Subsection 12.2.5 presents some relations for estimating the eigenvalues $\lambda_n$ and eigenfunctions $u_n(x)$.

---

* For unbounded domains, the condition of boundedness of the solution as $y \to \pm\infty$ is set; in Table 14.8, this condition is omitted.

## 14.9. Boundary Value Problems for Elliptic Equations with Two Space Variables

**Example.** Consider a boundary value problem for the Laplace equation

$$\frac{\partial^2 w}{\partial x^2} + \frac{\partial^2 w}{\partial y^2} = 0$$

in a strip $0 \leq x \leq l$, $-\infty < y < \infty$ with mixed boundary conditions

$$w = f_1(y) \quad \text{at} \quad x = 0, \qquad \frac{\partial w}{\partial x} = f_2(y) \quad \text{at} \quad x = l.$$

This equation is a special case of equation (14.9.1.1) with $a(x) = 1$ and $b(x) = c(x) = \Phi(x, t) = 0$. The corresponding Sturm–Liouville problem (14.9.1.5)–(14.9.1.7) is written as

$$u''_{xx} + \lambda y = 0, \qquad u = 0 \quad \text{at} \quad x = 0, \qquad u'_x = 0 \quad \text{at} \quad x = l.$$

The eigenfunctions and eigenvalues are found as

$$u_n(x) = \sin\left[\frac{\pi(2n-1)x}{l}\right], \qquad \lambda_n = \frac{\pi^2(2n-1)^2}{l^2}, \qquad n = 1, 2, \ldots$$

Using formulas (14.9.1.3) and (14.9.1.4) and taking into account the identities $\rho(\xi) = 1$ and $\|y_n\|^2 = l/2$ ($n = 1, 2, \ldots$) and the expression for $\Psi_n$ from the first row in Table 14.8, we obtain the Green's function in the form

$$G(x, y, \xi, \eta) = \frac{1}{l} \sum_{n=1}^{\infty} \frac{1}{\sigma_n} \sin(\sigma_n x) \sin(\sigma_n \xi) e^{-\sigma_n |y-\eta|}, \qquad \sigma_n = \sqrt{\lambda_n} = \frac{\pi(2n-1)}{l}.$$

### 14.9.2. Representation of Solutions to Boundary Value Problems via the Green's Functions

#### 14.9.2-1. First boundary value problem.

The solution of the first boundary value problem for equation (14.9.1.1) with the boundary conditions

$$w = f_1(y) \quad \text{at} \quad x = x_1, \qquad w = f_2(y) \quad \text{at} \quad x = x_2,$$
$$w = f_3(x) \quad \text{at} \quad y = 0, \qquad w = f_4(x) \quad \text{at} \quad y = h$$

is expressed in terms of the Green's function as

$$w(x, y) = a(x_1) \int_0^h f_1(\eta) \left[\frac{\partial}{\partial \xi} G(x, y, \xi, \eta)\right]_{\xi=x_1} d\eta - a(x_2) \int_0^h f_2(\eta) \left[\frac{\partial}{\partial \xi} G(x, y, \xi, \eta)\right]_{\xi=x_2} d\eta$$
$$+ \int_{x_1}^{x_2} f_3(\xi) \left[\frac{\partial}{\partial \eta} G(x, y, \xi, \eta)\right]_{\eta=0} d\xi - \int_{x_1}^{x_2} f_4(\xi) \left[\frac{\partial}{\partial \eta} G(x, y, \xi, \eta)\right]_{\eta=h} d\xi$$
$$+ \int_{x_1}^{x_2} \int_0^h \Phi(\xi, \eta) G(x, y, \xi, \eta) \, d\eta \, d\xi.$$

#### 14.9.2-2. Second boundary value problem.

The solution of the second boundary value problem for equation (14.9.1.1) with boundary conditions

$$\partial_x w = f_1(y) \quad \text{at} \quad x = x_1, \qquad \partial_x w = f_2(y) \quad \text{at} \quad x = x_2,$$
$$\partial_y w = f_3(x) \quad \text{at} \quad y = 0, \qquad \partial_y w = f_4(x) \quad \text{at} \quad y = h$$

is expressed in terms of the Green's function as

$$w(x,y) = -a(x_1) \int_0^h f_1(\eta) G(x,y,x_1,\eta)\, d\eta + a(x_2) \int_0^h f_2(\eta) G(x,y,x_2,\eta)\, d\eta$$
$$- \int_{x_1}^{x_2} f_3(\xi) G(x,y,\xi,0)\, d\xi + \int_{x_1}^{x_2} f_4(\xi) G(x,y,\xi,h)\, d\xi$$
$$+ \int_{x_1}^{x_2} \int_0^h \Phi(\xi,\eta) G(x,y,\xi,\eta)\, d\eta\, d\xi.$$

#### 14.9.2-3. Third boundary value problem.

The solution of the third boundary value problem for equation (14.9.1.1) in terms of the Green's function is represented in the same way as the solution of the second boundary value problem (the Green's function is now different).

▶ Solutions of various boundary value problems for elliptic equations can be found in Section T8.3.

## 14.10. Boundary Value Problems with Many Space Variables. Representation of Solutions via the Green's Function

### 14.10.1. Problems for Parabolic Equations

#### 14.10.1-1. Statement of the problem.

In general, a nonhomogeneous linear differential equation of the parabolic type in $n$ space variables has the form

$$\frac{\partial w}{\partial t} - L_{\mathbf{x},t}[w] = \Phi(\mathbf{x},t), \qquad (14.10.1.1)$$

where

$$L_{\mathbf{x},t}[w] \equiv \sum_{i,j=1}^n a_{ij}(\mathbf{x},t) \frac{\partial^2 w}{\partial x_i \partial x_j} + \sum_{i=1}^n b_i(\mathbf{x},t) \frac{\partial w}{\partial x_i} + c(\mathbf{x},t)w,$$
$$\mathbf{x} = \{x_1,\ldots,x_n\}, \quad \sum_{i,j=1}^n a_{ij}(\mathbf{x},t)\xi_i\xi_j \geq \sigma \sum_{i=1}^n \xi_i^2, \quad \sigma > 0. \qquad (14.10.1.2)$$

Let $V$ be some simply connected domain in $\mathbb{R}^n$ with a sufficiently smooth boundary $S = \partial V$. We consider the nonstationary boundary value problem for equation (14.10.1.1) in the domain $V$ with an arbitrary initial condition,

$$w = f(\mathbf{x}) \quad \text{at} \quad t = 0, \qquad (14.10.1.3)$$

and nonhomogeneous linear boundary conditions,

$$\Gamma_{\mathbf{x},t}[w] = g(\mathbf{x},t) \quad \text{for} \quad \mathbf{x} \in S. \qquad (14.10.1.4)$$

In the general case, $\Gamma_{\mathbf{x},t}$ is a first-order linear differential operator in the space coordinates with coefficients dependent on $\mathbf{x}$ and $t$.

### 14.10.1-2. Representation of the problem solution in terms of the Green's function.

The solution of the nonhomogeneous linear boundary value problem defined by (14.10.1.1)–(14.10.1.4) can be represented as the sum

$$w(\mathbf{x}, t) = \int_0^t \int_V \Phi(\mathbf{y}, \tau) G(\mathbf{x}, \mathbf{y}, t, \tau) \, dV_y \, d\tau + \int_V f(\mathbf{y}) G(\mathbf{x}, \mathbf{y}, t, 0) \, dV_y$$
$$+ \int_0^t \int_S g(\mathbf{y}, \tau) H(\mathbf{x}, \mathbf{y}, t, \tau) \, dS_y \, d\tau, \qquad (14.10.1.5)$$

where $G(\mathbf{x}, \mathbf{y}, t, \tau)$ is the Green's function; for $t > \tau \geq 0$, it satisfies the homogeneous equation

$$\frac{\partial G}{\partial t} - L_{\mathbf{x},t}[G] = 0 \qquad (14.10.1.6)$$

with the nonhomogeneous initial condition of special form

$$G = \delta(\mathbf{x} - \mathbf{y}) \quad \text{at} \quad t = \tau \qquad (14.10.1.7)$$

and the homogeneous boundary condition

$$\Gamma_{\mathbf{x},t}[G] = 0 \quad \text{for} \quad \mathbf{x} \in S. \qquad (14.10.1.8)$$

The vector $\mathbf{y} = \{y_1, \ldots, y_n\}$ appears in problem (14.10.1.6)–(14.10.1.8) as an $n$-dimensional free parameter ($\mathbf{y} \in V$), and $\delta(\mathbf{x} - \mathbf{y}) = \delta(x_1 - y_1) \ldots \delta(x_n - y_n)$ is the $n$-dimensional Dirac delta function. The Green's function $G$ is independent of the functions $\Phi$, $f$, and $g$ that characterize various nonhomogeneities of the boundary value problem. In (14.10.1.5), the integration is performed everywhere with respect to $\mathbf{y}$, with $dV_y = dy_1 \ldots dy_n$.

The function $H(\mathbf{x}, \mathbf{y}, t, \tau)$ involved in the integrand of the last term in solution (14.10.1.5) can be expressed via the Green's function $G(\mathbf{x}, \mathbf{y}, t, \tau)$. The corresponding formulas for $H(\mathbf{x}, \mathbf{y}, t, \tau)$ are given in Table 14.9 for the three basic types of boundary value problems; in the third boundary value problem, the coefficient $k$ can depend on $\mathbf{x}$ and $t$. The boundary conditions of the second and third kind, as well as the solution of the first boundary value problem, involve operators of differentiation along the conormal of operator (14.10.1.2); these operators act as follows:

$$\frac{\partial G}{\partial M_x} \equiv \sum_{i,j=1}^n a_{ij}(\mathbf{x}, t) N_j \frac{\partial G}{\partial x_i}, \qquad \frac{\partial G}{\partial M_y} \equiv \sum_{i,j=1}^n a_{ij}(\mathbf{y}, \tau) N_j \frac{\partial G}{\partial y_i}, \qquad (14.10.1.9)$$

where $\mathbf{N} = \{N_1, \ldots, N_n\}$ is the unit outward normal to the surface $S$. In the special case where $a_{ii}(\mathbf{x}, t) = 1$ and $a_{ij}(\mathbf{x}, t) = 0$ for $i \neq j$, operator (14.10.1.9) coincides with the ordinary operator of differentiation along the outward normal to $S$.

TABLE 14.9
The form of the function $H(\mathbf{x}, \mathbf{y}, t, \tau)$ for the basic types of nonstationary boundary value problems

| Type of problem | Form of boundary condition (14.10.1.4) | Function $H(\mathbf{x}, \mathbf{y}, t, \tau)$ |
| --- | --- | --- |
| First boundary value problem | $w = g(\mathbf{x}, t)$ for $\mathbf{x} \in S$ | $H(\mathbf{x}, \mathbf{y}, t, \tau) = -\dfrac{\partial G}{\partial M_y}(\mathbf{x}, \mathbf{y}, t, \tau)$ |
| Second boundary value problem | $\dfrac{\partial w}{\partial M_x} = g(\mathbf{x}, t)$ for $\mathbf{x} \in S$ | $H(\mathbf{x}, \mathbf{y}, t, \tau) = G(\mathbf{x}, \mathbf{y}, t, \tau)$ |
| Third boundary value problem | $\dfrac{\partial w}{\partial M_x} + kw = g(\mathbf{x}, t)$ for $\mathbf{x} \in S$ | $H(\mathbf{x}, \mathbf{y}, t, \tau) = G(\mathbf{x}, \mathbf{y}, t, \tau)$ |

If the coefficient of equation (14.10.1.6) and the boundary condition (14.10.1.8) are independent of $t$, then the Green's function depends on only three arguments, $G(\mathbf{x}, \mathbf{y}, t, \tau) = G(\mathbf{x}, \mathbf{y}, t - \tau)$.

**Remark.** Let $S_i$ ($i = 1, \ldots, p$) be different portions of the surface $S$ such that $S = \sum_{i=1}^{p} S_i$ and let boundary conditions of various types be set on the $S_i$,

$$\Gamma_{\mathbf{x},t}^{(i)}[w] = g_i(\mathbf{x}, t) \quad \text{for} \quad \mathbf{x} \in S_i, \quad i = 1, \ldots, p. \tag{14.10.1.10}$$

Then formula (14.10.1.5) remains valid but the last term in (14.10.1.5) must be replaced by the sum

$$\sum_{i=1}^{p} \int_0^t \int_{S_i} g_i(\mathbf{y}, \tau) H_i(\mathbf{x}, \mathbf{y}, t, \tau) \, dS_y \, d\tau. \tag{14.10.1.11}$$

## 14.10.2. Problems for Hyperbolic Equations

### 14.10.2-1. Statement of the problem.

The general nonhomogeneous linear differential hyperbolic equation in $n$ space variables can be written as

$$\frac{\partial^2 w}{\partial t^2} + \varphi(\mathbf{x}, t) \frac{\partial w}{\partial t} - L_{\mathbf{x},t}[w] = \Phi(\mathbf{x}, t), \tag{14.10.2.1}$$

where the operator $L_{\mathbf{x},t}[w]$ is explicitly defined in (14.10.1.2).

We consider the nonstationary boundary value problem for equation (14.10.2.1) in the domain $V$ with arbitrary initial conditions,

$$w = f_0(\mathbf{x}) \quad \text{at} \quad t = 0, \tag{14.10.2.2}$$

$$\partial_t w = f_1(\mathbf{x}) \quad \text{at} \quad t = 0, \tag{14.10.2.3}$$

and the nonhomogeneous linear boundary condition (14.10.1.4).

### 14.10.2-2. Representation of the problem solution in terms of the Green's function.

The solution of the nonhomogeneous linear boundary value problem defined by (14.10.2.1)–(14.10.2.3), (14.10.1.4) can be represented as the sum

$$w(\mathbf{x}, t) = \int_0^t \int_V \Phi(\mathbf{y}, \tau) G(\mathbf{x}, \mathbf{y}, t, \tau) \, dV_y \, d\tau - \int_V f_0(\mathbf{y}) \left[ \frac{\partial}{\partial \tau} G(\mathbf{x}, \mathbf{y}, t, \tau) \right]_{\tau=0} dV_y$$

$$+ \int_V \left[ f_1(\mathbf{y}) + f_0(\mathbf{y}) \varphi(\mathbf{y}, 0) \right] G(\mathbf{x}, \mathbf{y}, t, 0) \, dV_y$$

$$+ \int_0^t \int_S g(\mathbf{y}, \tau) H(\mathbf{x}, \mathbf{y}, t, \tau) \, dS_y \, d\tau. \tag{14.10.2.4}$$

Here, $G(\mathbf{x}, \mathbf{y}, t, \tau)$ is the Green's function; for $t > \tau \geq 0$ it satisfies the homogeneous equation

$$\frac{\partial^2 G}{\partial t^2} + \varphi(\mathbf{x}, t) \frac{\partial G}{\partial t} - L_{\mathbf{x},t}[G] = 0 \tag{14.10.2.5}$$

with the semihomogeneous initial conditions

$$G = 0 \quad \text{at} \quad t = \tau,$$
$$\partial_t G = \delta(\mathbf{x} - \mathbf{y}) \quad \text{at} \quad t = \tau,$$

and the homogeneous boundary condition (14.10.1.8).

If the coefficients of equation (14.10.2.5) and the boundary condition (14.10.1.8) are independent of time $t$, then the Green's function depends on only three arguments, $G(\mathbf{x},\mathbf{y},t,\tau) = G(\mathbf{x},\mathbf{y},t-\tau)$. In this case, one can set $\frac{\partial}{\partial \tau}G(\mathbf{x},\mathbf{y},t,\tau)\big|_{\tau=0} = -\frac{\partial}{\partial t}G(\mathbf{x},\mathbf{y},t)$ in solution (14.10.2.4).

The function $H(\mathbf{x},\mathbf{y},t,\tau)$ involved in the integrand of the last term in solution (14.10.2.4) can be expressed via the Green's function $G(\mathbf{x},\mathbf{y},t,\tau)$. The corresponding formulas for $H$ are given in Table 14.9 for the three basic types of boundary value problems; in the third boundary value problem, the coefficient $k$ can depend on $\mathbf{x}$ and $t$.

*Remark.* Let $S_i$ ($i = 1, \ldots, p$) be different portions of the surface $S$ such that $S = \sum_{i=1}^{p} S_i$ and let boundary conditions of various types (14.10.1.10) be set on the $S_i$. Then formula (14.10.2.4) remains valid but the last term in (14.10.2.4) must be replaced by the sum (14.10.1.11).

## 14.10.3. Problems for Elliptic Equations

**14.10.3-1. Statement of the problem.**

In general, a nonhomogeneous linear elliptic equation can be written as

$$-L_{\mathbf{x}}[w] = \Phi(\mathbf{x}), \tag{14.10.3.1}$$

where

$$L_{\mathbf{x}}[w] \equiv \sum_{i,j=1}^{n} a_{ij}(\mathbf{x})\frac{\partial^2 w}{\partial x_i \partial x_j} + \sum_{i=1}^{n} b_i(\mathbf{x})\frac{\partial w}{\partial x_i} + c(\mathbf{x})w. \tag{14.10.3.2}$$

Two-dimensional problems correspond to $n = 2$ and three-dimensional problems to $n = 3$.

We consider equation (14.10.3.1)–(14.10.3.2) in a domain $V$ and assume that the equation is subject to the general linear boundary condition

$$\Gamma_{\mathbf{x}}[w] = g(\mathbf{x}) \quad \text{for} \quad \mathbf{x} \in S. \tag{14.10.3.3}$$

The solution of the stationary problem (14.10.3.1)–(14.10.3.3) can be obtained by passing in (14.10.1.5) to the limit as $t \to \infty$. To this end, one should start with equation (14.10.1.1), whose coefficients are independent of $t$, and take the homogeneous initial condition (14.10.1.3), with $f(\mathbf{x}) = 0$, and the stationary boundary condition (14.10.1.4).

**14.10.3-2. Representation of the problem solution in terms of the Green's function.**

The solution of the linear boundary value problem (14.10.3.1)–(14.10.3.3) can be represented as the sum

$$w(\mathbf{x}) = \int_V \Phi(\mathbf{y})G(\mathbf{x},\mathbf{y})\,dV_y + \int_S g(\mathbf{y})H(\mathbf{x},\mathbf{y})\,dS_y. \tag{14.10.3.4}$$

Here, the Green's function $G(\mathbf{x},\mathbf{y})$ satisfies the nonhomogeneous equation of special form

$$-L_{\mathbf{x}}[G] = \delta(\mathbf{x} - \mathbf{y}) \tag{14.10.3.5}$$

with the homogeneous boundary condition

$$\Gamma_{\mathbf{x}}[G] = 0 \quad \text{for} \quad \mathbf{x} \in S. \tag{14.10.3.6}$$

The vector $\mathbf{y} = \{y_1, \ldots, y_n\}$ appears in problem (14.10.3.5), (14.10.3.6) as an $n$-dimensional free parameter ($\mathbf{y} \in V$). Note that $G$ is independent of the functions $\Phi$ and $g$ characterizing various nonhomogeneities of the original boundary value problem.

The function $H(\mathbf{x}, \mathbf{y})$ involved in the integrand of the second term in solution (14.10.3.4) can be expressed via the Green's function $G(\mathbf{x}, \mathbf{y})$. The corresponding formulas for $H$ are given in Table 14.10 for the three basic types of boundary value problems. The boundary conditions of the second and third kind, as well as the solution of the first boundary value problem, involve operators of differentiation along the conormal of operator (14.10.3.2); these operators are defined by (14.10.1.9); in this case, the coefficients $a_{ij}$ depend on $\mathbf{x}$ only.

TABLE 14.10
The form of the function $H(\mathbf{x}, \mathbf{y})$ involved in the integrand of the last term in solution (14.10.3.4) for the basic types of stationary boundary value problems

| Type of problem | Form of boundary condition (14.10.3.3) | Function $H(\mathbf{x}, \mathbf{y})$ |
| --- | --- | --- |
| First boundary value problem | $w = g(\mathbf{x})$ for $\mathbf{x} \in S$ | $H(\mathbf{x}, \mathbf{y}) = -\dfrac{\partial G}{\partial M_y}(\mathbf{x}, \mathbf{y})$ |
| Second boundary value problem | $\dfrac{\partial w}{\partial M_x} = g(\mathbf{x})$ for $\mathbf{x} \in S$ | $H(\mathbf{x}, \mathbf{y}) = G(\mathbf{x}, \mathbf{y})$ |
| Third boundary value problem | $\dfrac{\partial w}{\partial M_x} + kw = g(\mathbf{x})$ for $\mathbf{x} \in S$ | $H(\mathbf{x}, \mathbf{y}) = G(\mathbf{x}, \mathbf{y})$ |

*Remark.* For the second boundary value problem with $c(\mathbf{x}) \equiv 0$, the thus defined Green's function must not necessarily exist; see Polyanin (2002).

## 14.10.4. Comparison of the Solution Structures for Boundary Value Problems for Equations of Various Types

Table 14.11 lists brief formulations of boundary value problems for second-order equations of elliptic, parabolic, and hyperbolic types. The coefficients of the differential operators $L_\mathbf{x}$ and $\Gamma_\mathbf{x}$ in the space variables $x_1, \ldots, x_n$ are assumed to be independent of time $t$; these operators are the same for the problems under consideration.

TABLE 14.11
Formulations of boundary value problems for equations of various types

| Type of equation | Form of equation | Initial conditions | Boundary conditions |
| --- | --- | --- | --- |
| Elliptic | $-L_\mathbf{x}[w] = \Phi(\mathbf{x})$ | not set | $\Gamma_\mathbf{x}[w] = g(\mathbf{x})$ for $\mathbf{x} \in S$ |
| Parabolic | $\partial_t w - L_\mathbf{x}[w] = \Phi(\mathbf{x}, t)$ | $w = f(\mathbf{x})$ at $t = 0$ | $\Gamma_\mathbf{x}[w] = g(\mathbf{x}, t)$ for $\mathbf{x} \in S$ |
| Hyperbolic | $\partial_{tt} w - L_\mathbf{x}[w] = \Phi(\mathbf{x}, t)$ | $w = f_0(\mathbf{x})$ at $t = 0$, $\partial_t w = f_1(\mathbf{x})$ at $t = 0$ | $\Gamma_\mathbf{x}[w] = g(\mathbf{x}, t)$ for $\mathbf{x} \in S$ |

Below are the respective general formulas defining the solutions of these problems with zero initial conditions ($f = f_0 = f_1 = 0$):

$$w_0(\mathbf{x}) = \int_V \Phi(\mathbf{y}) G_0(\mathbf{x}, \mathbf{y}) \, dV_y \quad + \quad \int_S g(\mathbf{y}) \mathcal{H}\big[G_0(\mathbf{x}, \mathbf{y})\big] \, dS_y,$$

$$w_1(\mathbf{x}, t) = \int_0^t \int_V \Phi(\mathbf{y}, \tau) G_1(\mathbf{x}, \mathbf{y}, t - \tau) \, dV_y \, d\tau + \int_0^t \int_S g(\mathbf{y}, \tau) \mathcal{H}\big[G_1(\mathbf{x}, \mathbf{y}, t - \tau)\big] \, dS_y \, d\tau,$$

$$w_2(\mathbf{x}, t) = \int_0^t \int_V \Phi(\mathbf{y}, \tau) G_2(\mathbf{x}, \mathbf{y}, t - \tau) \, dV_y \, d\tau + \int_0^t \int_S g(\mathbf{y}, \tau) \mathcal{H}\big[G_2(\mathbf{x}, \mathbf{y}, t - \tau)\big] \, dS_y \, d\tau,$$

where the $G_n$ are the Green's functions, and the subscripts 0, 1, and 2 refer to the elliptic, parabolic, and hyperbolic problem, respectively. All solutions involve the same operator $\mathcal{H}[G]$; it is explicitly defined in Subsections 14.10.1–14.10.3 (see also Section 14.7) for different boundary conditions.

It is apparent that the solutions of the parabolic and hyperbolic problems with zero initial conditions have the same structure. The structure of the solution to the problem for a parabolic equation differs from that for an elliptic equation by the additional integration with respect to $t$.

## 14.11. Construction of the Green's Functions. General Formulas and Relations

### 14.11.1. Green's Functions of Boundary Value Problems for Equations of Various Types in Bounded Domains

14.11.1-1. Expressions of the Green's function in terms of infinite series.

Table 14.12 lists the Green's functions of boundary value problems for second-order equations of various types in a bounded domain $V$. It is assumed that $L_\mathbf{x}$ is a second-order linear self-adjoint differential operator (e.g., see Zwillinger, 1997) in the space variables $x_1, \ldots, x_n$, and $\Gamma_\mathbf{x}$ is a zeroth- or first-order linear boundary operator that can define a boundary condition of the first, second, or third kind; the coefficients of the operators $L_\mathbf{x}$ and $\Gamma_\mathbf{x}$ can depend on the space variables but are independent of time $t$. The coefficients $\lambda_k$ and the functions $u_k(\mathbf{x})$ are determined by solving the homogeneous eigenvalue problem

$$L_\mathbf{x}[u] + \lambda u = 0, \qquad (14.11.1.1)$$

$$\Gamma_\mathbf{x}[u] = 0 \quad \text{for} \quad \mathbf{x} \in S. \qquad (14.11.1.2)$$

It is apparent from Table 14.12 that, given the Green's function in the problem for a parabolic (or hyperbolic) equation, one can easily construct the Green's functions of the corresponding problems for elliptic and hyperbolic (or parabolic) equations. In particular, the Green's function of the problem for an elliptic equation can be expressed via the Green's function of the problem for a parabolic equation as follows:

$$G_0(\mathbf{x}, \mathbf{y}) = \int_0^\infty G_1(\mathbf{x}, \mathbf{y}, t) \, dt. \qquad (14.11.1.3)$$

Here, the fact that all $\lambda_k$ are positive is taken into account; for the second boundary value problem, it is assumed that $\lambda = 0$ is not an eigenvalue of problem (14.11.1.1)–(14.11.1.2).

TABLE 14.12
The Green's functions of boundary value problems for equations of various types in bounded domains. In all problems, the operators $L_\mathbf{x}$ and $\Gamma_\mathbf{x}$ are the same; $\mathbf{x} = \{x_1, \ldots, x_n\}$

| Equation | Initial and boundary conditions | Green's function |
|---|---|---|
| Elliptic equation $-L_\mathbf{x}[w] = \Phi(\mathbf{x})$ | $\Gamma_\mathbf{x}[w] = g(\mathbf{x})$ for $\mathbf{x} \in S$ (no initial condition required) | $G(\mathbf{x}, \mathbf{y}) = \sum_{k=1}^{\infty} \dfrac{u_k(\mathbf{x}) u_k(\mathbf{y})}{\|u_k\|^2 \lambda_k}, \quad \lambda_k \neq 0$ |
| Parabolic equation $\partial_t w - L_\mathbf{x}[w] = \Phi(\mathbf{x}, t)$ | $w = f(\mathbf{x})$ at $t = 0$ $\Gamma_\mathbf{x}[w] = g(\mathbf{x}, t)$ for $\mathbf{x} \in S$ | $G(\mathbf{x}, \mathbf{y}, t) = \sum_{k=1}^{\infty} \dfrac{u_k(\mathbf{x}) u_k(\mathbf{y})}{\|u_k\|^2} \exp(-\lambda_k t)$ |
| Hyperbolic equation $\partial_{tt} w - L_\mathbf{x}[w] = \Phi(\mathbf{x}, t)$ | $w = f_0(\mathbf{x})$ at $t = 0$ $w = f_1(\mathbf{x})$ at $t = 0$ $\Gamma_\mathbf{x}[w] = g(\mathbf{x}, t)$ for $\mathbf{x} \in S$ | $G(\mathbf{x}, \mathbf{y}, t) = \sum_{k=1}^{\infty} \dfrac{u_k(\mathbf{x}) u_k(\mathbf{y})}{\|u_k\|^2 \sqrt{\lambda_k}} \sin(t\sqrt{\lambda_k})$ |

14.11.1-2. Some remarks and generalizations.

*Remark 1.* Formula (14.11.1.3) can also be used if the domain $V$ is infinite. In this case, one should make sure that the integral on the right-hand side is convergent.

*Remark 2.* Suppose the equations given in the first column of Table 14.12 contain $-L_\mathbf{x}[w] - \beta w$ instead of $-L_\mathbf{x}[w]$, with $\beta$ being a free parameter. Then the $\lambda_k$ in the expressions of the Green's function in the third column of Table 14.12 must be replaced by $\lambda_k - \beta$; just as previously, the $\lambda_k$ and $u_k(\mathbf{x})$ were determined by solving the eigenvalue problem (14.11.1.1)–(14.11.1.2).

*Remark 3.* The formulas for the Green's functions presented in Table 14.12 will also hold for boundary value problems described by equations of the fourth or higher order in the space variables; provided that the eigenvalue problem for equation (14.11.1.1) subject to appropriate boundary conditions is self-adjoint.

## 14.11.2. Green's Functions Admitting Incomplete Separation of Variables

14.11.2-1. Boundary value problems for rectangular domains.

1°. Consider the parabolic equation

$$\frac{\partial w}{\partial t} = L_{1,t}[w] + \cdots + L_{n,t}[w] + \Phi(\mathbf{x}, t), \qquad (14.11.2.1)$$

where each term $L_{m,t}[w]$ depends on only one space variable, $x_m$, and time $t$:

$$L_{m,t}[w] \equiv a_m(x_m, t) \frac{\partial^2 w}{\partial x_m^2} + b_m(x_m, t) \frac{\partial w}{\partial x_m} + c_m(x_m, t) w, \qquad m = 1, \ldots, n.$$

For equation (14.11.2.1) we set the initial condition of general form

$$w = f(\mathbf{x}) \quad \text{at} \quad t = 0. \qquad (14.11.2.2)$$

Consider the domain $V = \{\alpha_m \leq x_m \leq \beta_m, \ m = 1, \ldots, n\}$, which is an $n$-dimensional parallelepiped. We set the following boundary conditions at the faces of the parallelepiped:

$$s_m^{(1)} \frac{\partial w}{\partial x_m} + k_m^{(1)}(t) w = g_m^{(1)}(\mathbf{x}, t) \quad \text{at} \quad x_m = \alpha_m,$$
$$s_m^{(2)} \frac{\partial w}{\partial x_m} + k_m^{(2)}(t) w = g_m^{(2)}(\mathbf{x}, t) \quad \text{at} \quad x_m = \beta_m. \qquad (14.11.2.3)$$

## 14.11. CONSTRUCTION OF THE GREEN'S FUNCTIONS. GENERAL FORMULAS AND RELATIONS

By appropriately choosing the coefficients $s_m^{(1)}$, $s_m^{(2)}$ and functions $k_m^{(1)} = k_m^{(1)}(t)$, $k_m^{(2)} = k_m^{(2)}(t)$, we can obtain the boundary conditions of the first, second, or third kind. For infinite domains, the boundary conditions corresponding to $\alpha_m = -\infty$ or $\beta_m = \infty$ are omitted.

2°. The Green's function of the nonstationary $n$-dimensional boundary value problem (14.11.2.1)–(14.11.2.3) can be represented in the product form

$$G(\mathbf{x}, \mathbf{y}, t, \tau) = \prod_{m=1}^{n} G_m(x_m, y_m, t, \tau), \qquad (14.11.2.4)$$

where the Green's functions $G_m = G_m(x_m, y_m, t, \tau)$ satisfy the one-dimensional equations

$$\frac{\partial G_m}{\partial t} - L_{m,t}[G_m] = 0 \qquad (m = 1, \ldots, n)$$

with the initial conditions

$$G_m = \delta(x_m - y_m) \quad \text{at} \quad t = \tau$$

and the homogeneous boundary conditions

$$s_m^{(1)} \frac{\partial G_m}{\partial x_m} + k_m^{(1)}(t) G_m = 0 \quad \text{at} \quad x_m = \alpha_m,$$

$$s_m^{(2)} \frac{\partial G_m}{\partial x_m} + k_m^{(2)}(t) G_m = 0 \quad \text{at} \quad x_m = \beta_m.$$

Here, $y_m$ and $\tau$ are free parameters ($\alpha_m \leq y_m \leq \beta_m$ and $t \geq \tau \geq 0$), and $\delta(x)$ is the Dirac delta function.

It can be seen that the Green's function (14.11.2.4) admits incomplete separation of variables; it separates in the space variables $x_1, \ldots, x_n$ but not in time $t$.

**Example.** Consider the boundary value problem for the two-dimensional nonhomogeneous heat equation

$$\frac{\partial w}{\partial t} = \frac{\partial^2 w}{\partial x_1^2} + \frac{\partial^2 w}{\partial x_2^2} + \Phi(x_1, x_2, t)$$

with initial condition (14.11.2.2) and the nonhomogeneous mixed boundary conditions

$$w = g_1(x_2, t) \quad \text{at} \quad x_1 = 0, \qquad w = h_1(x_2, t) \quad \text{at} \quad x_1 = l_1;$$

$$\frac{\partial w}{\partial x_2} = g_2(x_1, t) \quad \text{at} \quad x_2 = 0, \qquad \frac{\partial w}{\partial x_2} = h_2(x_1, t) \quad \text{at} \quad x_2 = l_2.$$

The Green's functions of the corresponding homogeneous one-dimensional heat equations with homogeneous boundary conditions are expressed as

*Equations and boundary conditions*              *Green's functions*

$$\frac{\partial w}{\partial t} = \frac{\partial^2 w}{\partial x_1^2}, \quad w = 0 \text{ at } x_1 = 0, l_1 \implies G_1 = \frac{2}{l_1} \sum_{m=1}^{\infty} \sin(\lambda_m x_1) \sin(\lambda_m y_1) e^{-\lambda_m^2 (t-\tau)}, \quad \lambda_m = \frac{m\pi}{l_1};$$

$$\frac{\partial w}{\partial t} = \frac{\partial^2 w}{\partial x_2^2}, \quad \frac{\partial w}{\partial x_2} = 0 \text{ at } x_2 = 0, l_2 \implies G_2 = \frac{1}{l_2} + \frac{2}{l_2} \sum_{n=1}^{\infty} \sin(\sigma_n x_2) \sin(\sigma_n y_2) e^{-\sigma_n^2 (t-\tau)}, \quad \sigma_n = \frac{n\pi}{l_2}.$$

Multiplying $G_1$ and $G_2$ together gives the Green's function for the original two-dimensional problem:

$$G(x_1, x_2, y_1, y_2, t, \tau) = \frac{4}{l_1 l_2} \left[ \sum_{m=1}^{\infty} \sin(\lambda_m x_1) \sin(\lambda_m y_1) e^{-\lambda_m^2 (t-\tau)} \right] \left[ \frac{1}{2} + \sum_{n=1}^{\infty} \sin(\sigma_n x_2) \sin(\sigma_n y_2) e^{-\sigma_n^2 (t-\tau)} \right].$$

**14.11.2-2. Boundary value problems for a arbitrary cylindrical domain.**

1°. Consider the parabolic equation

$$\frac{\partial w}{\partial t} = L_{\mathbf{x},t}[w] + M_{z,t}[w] + \Phi(\mathbf{x}, z, t), \qquad (14.11.2.5)$$

where $L_{\mathbf{x},t}$ is an arbitrary second-order linear differential operator in $x_1, \ldots, x_n$ with coefficients dependent on $\mathbf{x}$ and $t$, and $M_{z,t}$ is an arbitrary second-order linear differential operator in $z$ with coefficients dependent on $z$ and $t$.

For equation (14.11.2.5) we set the general initial condition (14.11.2.2), where $f(\mathbf{x})$ must be replaced by $f(\mathbf{x}, z)$.

We assume that the space variables belong to a cylindrical domain $V = \{\mathbf{x} \in D, z_1 \leq z \leq z_2\}$ with arbitrary cross-section $D$. We set the boundary conditions*

$$\begin{aligned}
\Gamma_1[w] &= g_1(\mathbf{x}, t) & \text{at} \quad z &= z_1 & (\mathbf{x} \in D), \\
\Gamma_2[w] &= g_2(\mathbf{x}, t) & \text{at} \quad z &= z_2 & (\mathbf{x} \in D), \\
\Gamma_3[w] &= g_3(\mathbf{x}, z, t) & \text{for} \quad \mathbf{x} &\in \partial D & (z_1 \leq z \leq z_2),
\end{aligned} \qquad (14.11.2.6)$$

where the linear boundary operators $\Gamma_k$ ($k = 1, 2, 3$) can define boundary conditions of the first, second, or third kind; in the last case, the coefficients of the differential operators $\Gamma_k$ can be dependent on $t$.

2°. The Green's function of problem (14.11.2.5)–(14.11.2.6), (14.11.2.2) can be represented in the product form

$$G(\mathbf{x}, \mathbf{y}, z, \zeta, t, \tau) = G_L(\mathbf{x}, \mathbf{y}, t, \tau) G_M(z, \zeta, t, \tau), \qquad (14.11.2.7)$$

where $G_L = G_L(\mathbf{x}, \mathbf{y}, t, \tau)$ and $G_M = G_M(z, \zeta, t, \tau)$ are auxiliary Green's functions; these can be determined from the following two simpler problems with fewer independent variables:

*Problem on the cross-section $D$:*
$$\begin{cases}
\dfrac{\partial G_L}{\partial t} = L_{\mathbf{x},t}[G_L] & \text{for} \quad \mathbf{x} \in D, \\
G_L = \delta(\mathbf{x} - \mathbf{y}) & \text{at} \quad t = \tau, \\
\Gamma_3[G_L] = 0 & \text{for} \quad \mathbf{x} \in \partial D,
\end{cases}$$

*Problem on the interval $z_1 \leq z \leq z_2$:*
$$\begin{cases}
\dfrac{\partial G_M}{\partial t} = M_{z,t}[G_M] & \text{for} \quad z_1 < z < z_2, \\
G_M = \delta(z - \zeta) & \text{at} \quad t = \tau, \\
\Gamma_k[G_M] = 0 & \text{at} \quad z = z_k \ (k = 1, 2).
\end{cases}$$

Here, $\mathbf{y}$, $\zeta$, and $\tau$ are free parameters ($\mathbf{y} \in D$, $z_1 \leq \zeta \leq z_2$, $t \geq \tau \geq 0$).

It can be seen that the Green's function (14.11.2.7) admits incomplete separation of variables; it separates in the space variables $\mathbf{x}$ and $z$ but not in time $t$.

## 14.11.3. Construction of Green's Functions via Fundamental Solutions

**14.11.3-1. Elliptic equations. Fundamental solution.**

Consider the elliptic equation

$$L_{\mathbf{x}}[w] + \frac{\partial^2 w}{\partial z^2} = \Phi(\mathbf{x}, z), \qquad (14.11.3.1)$$

---

* If $z_1 = -\infty$ or $z_2 = \infty$, the corresponding boundary condition is to be omitted.

where $\mathbf{x} = \{x_1, \ldots, x_n\} \in \mathbb{R}^n$, $z \in \mathbb{R}^1$, and $L_\mathbf{x}[w]$ is a linear differential operator that depends on $x_1, \ldots, x_n$ but is independent of $z$. For subsequent analysis it is significant that the homogeneous equation (with $\Phi \equiv 0$) does not change under the replacement of $z$ by $-z$ and $z$ by $z + \text{const}$.

Let $\mathscr{E} = \mathscr{E}(\mathbf{x}, \mathbf{y}, z - \zeta)$ be a fundamental solution of equation (14.11.3.1), which means that

$$L_\mathbf{x}[\mathscr{E}] + \frac{\partial^2 \mathscr{E}}{\partial z^2} = \delta(\mathbf{x} - \mathbf{y})\delta(z - \zeta).$$

Here, $\mathbf{y} = \{y_1, \ldots, y_n\} \in \mathbb{R}^n$ and $\zeta \in \mathbb{R}^1$ are free parameters.

The fundamental solution of equation (14.11.3.1) is an even function in the last argument, i.e.,

$$\mathscr{E}(\mathbf{x}, \mathbf{y}, z) = \mathscr{E}(\mathbf{x}, \mathbf{y}, -z).$$

Below, Paragraphs 14.11.3-2 and 14.11.3-3 present relations that permit one to express the Green's functions of some boundary value problems for equation (14.11.3.1) via its fundamental solution.

**14.11.3-2. Domain: $\mathbf{x} \in \mathbb{R}^n$, $0 \le z < \infty$. Problems for elliptic equations.**

$1°$. *First boundary value problem.* The boundary condition:

$$w = f(\mathbf{x}) \quad \text{at} \quad z = 0.$$

Green's function:

$$G(\mathbf{x}, \mathbf{y}, z, \zeta) = \mathscr{E}(\mathbf{x}, \mathbf{y}, z - \zeta) - \mathscr{E}(\mathbf{x}, \mathbf{y}, z + \zeta). \qquad (14.11.3.2)$$

Domain of the free parameters: $\mathbf{y} \in \mathbb{R}^n$ and $0 \le \zeta < \infty$.

**Example 1.** Consider the first boundary value problem in the half-space $-\infty < x_1, x_2 < \infty$, $0 \le x_3 < \infty$ for the three-dimensional Laplace equation

$$\frac{\partial^2 w}{\partial x_1^2} + \frac{\partial^2 w}{\partial x_2^2} + \frac{\partial^2 w}{\partial x_3^2} = 0$$

under boundary condition

$$w = f(x_1, x_2) \quad \text{at} \quad x_3 = 0.$$

The fundamental solution for the Laplace equation has the form

$$\mathscr{E} = \frac{1}{4\pi \sqrt{(x_1 - y_1)^2 + (x_2 - y_2)^2 + (x_3 - y_3)^2}}.$$

In terms of the notation adopted for equation (14.11.3.1) and its fundamental solution, we have $x_3 = z$, $y_3 = \zeta$, and $\mathscr{E} = \mathscr{E}(x_1, y_1, x_2, y_2, z - \zeta)$. Using formula (14.11.3.2), we obtain the Green's function for the first boundary value problem in the half-space:

$$G(x_1, y_1, x_2, y_2, z, \zeta) = \mathscr{E}(x_1, y_1, x_2, y_2, z - \zeta) - \mathscr{E}(x_1, y_1, x_2, y_2, z + \zeta)$$
$$= \frac{1}{4\pi \sqrt{(x_1 - y_1)^2 + (x_2 - y_2)^2 + (x_3 - y_3)^2}} - \frac{1}{4\pi \sqrt{(x_1 - y_1)^2 + (x_2 - y_2)^2 + (x_3 + y_3)^2}}.$$

$2°$. *Second boundary value problem.* The boundary condition:

$$\partial_z w = f(\mathbf{x}) \quad \text{at} \quad z = 0.$$

Green's function:
$$G(\mathbf{x},\mathbf{y},z,\zeta) = \mathscr{E}(\mathbf{x},\mathbf{y},z-\zeta) + \mathscr{E}(\mathbf{x},\mathbf{y},z+\zeta).$$

**Example 2.** The Green's function of the second boundary value problem for the three-dimensional Laplace equation in the half-space $-\infty < x_1, x_2 < \infty, 0 \leq x_3 < \infty$ is expressed as
$$G(x_1,y_1,x_2,y_2,z,\zeta) = \frac{1}{4\pi\sqrt{(x_1-y_1)^2+(x_2-y_2)^2+(x_3-y_3)^2}} + \frac{1}{4\pi\sqrt{(x_1-y_1)^2+(x_2-y_2)^2+(x_3+y_3)^2}}.$$

It is obtained using the same reasoning as in Example 1.

3°. *Third boundary value problem.* The boundary condition:
$$\partial_z w - kw = f(\mathbf{x}) \quad \text{at} \quad z = 0.$$

Green's function:
$$G(\mathbf{x},\mathbf{y},z,\zeta) = \mathscr{E}(\mathbf{x},\mathbf{y},z-\zeta) + \mathscr{E}(\mathbf{x},\mathbf{y},z+\zeta) - 2k\int_0^\infty e^{-ks}\mathscr{E}(\mathbf{x},\mathbf{y},z+\zeta+s)\,ds$$
$$= \mathscr{E}(\mathbf{x},\mathbf{y},z-\zeta) + \mathscr{E}(\mathbf{x},\mathbf{y},z+\zeta) - 2k\int_{z+\zeta}^\infty e^{-k(\sigma-z-\zeta)}\mathscr{E}(\mathbf{x},\mathbf{y},\sigma)\,d\sigma.$$

**14.11.3-3. Domain: $\mathbf{x} \in \mathbb{R}^n$, $0 \leq z \leq l$. Problems for elliptic equations.**

1°. *First boundary value problem.* Boundary conditions:
$$w = f_1(\mathbf{x}) \quad \text{at} \quad z=0, \qquad w = f_2(\mathbf{x}) \quad \text{at} \quad z=l.$$

Green's function:
$$G(\mathbf{x},\mathbf{y},z,\zeta) = \sum_{n=-\infty}^{\infty}\bigl[\mathscr{E}(\mathbf{x},\mathbf{y},z-\zeta+2nl) - \mathscr{E}(\mathbf{x},\mathbf{y},z+\zeta+2nl)\bigr]. \qquad (14.11.3.3)$$

Domain of the free parameters: $\mathbf{y} \in \mathbb{R}^n$ and $0 \leq \zeta \leq l$.

2°. *Second boundary value problem.* Boundary conditions:
$$\partial_z w = f_1(\mathbf{x}) \quad \text{at} \quad z=0, \qquad \partial_z w = f_2(\mathbf{x}) \quad \text{at} \quad z=l.$$

Green's function:
$$G(\mathbf{x},\mathbf{y},z,\zeta) = \sum_{n=-\infty}^{\infty}\bigl[\mathscr{E}(\mathbf{x},\mathbf{y},z-\zeta+2nl) + \mathscr{E}(\mathbf{x},\mathbf{y},z+\zeta+2nl)\bigr]. \qquad (14.11.3.4)$$

3°. *Mixed boundary value problem.* The unknown function and its derivative are prescribed at the left and right end, respectively:
$$w = f_1(\mathbf{x}) \quad \text{at} \quad z=0, \qquad \partial_z w = f_2(\mathbf{x}) \quad \text{at} \quad z=l.$$

Green's function:
$$G(\mathbf{x},\mathbf{y},z,\zeta) = \sum_{n=-\infty}^{\infty}(-1)^n\bigl[\mathscr{E}(\mathbf{x},\mathbf{y},z-\zeta+2nl) - \mathscr{E}(\mathbf{x},\mathbf{y},z+\zeta+2nl)\bigr]. \qquad (14.11.3.5)$$

4°. *Mixed boundary value problem.* The derivative and the unknown function itself are prescribed at the left and right end, respectively:
$$\partial_z w = f_1(\mathbf{x}) \quad \text{at} \quad z=0, \qquad w = f_2(\mathbf{x}) \quad \text{at} \quad z=l.$$

Green's function:
$$G(\mathbf{x},\mathbf{y},z,\zeta) = \sum_{n=-\infty}^{\infty}(-1)^n\bigl[\mathscr{E}(\mathbf{x},\mathbf{y},z-\zeta+2nl) + \mathscr{E}(\mathbf{x},\mathbf{y},z+\zeta+2nl)\bigr]. \qquad (14.11.3.6)$$

**Remark.** One should make sure that series (14.11.3.3)–(14.11.3.6) are convergent; in particular, for the three-dimensional Laplace equation, series (14.11.3.3), (14.11.3.5), and (14.11.3.6) are convergent and series (14.11.3.4) is divergent.

## 14.11.3-4. Boundary value problems for parabolic equations.

Let $\mathbf{x} \in \mathbb{R}^n$, $z \in \mathbb{R}^1$, and $t \geq 0$. Consider the parabolic equation

$$\frac{\partial w}{\partial t} = L_{\mathbf{x},t}[w] + \frac{\partial^2 w}{\partial z^2} + \Phi(\mathbf{x}, z, t), \qquad (14.11.3.7)$$

where $L_{\mathbf{x},t}[w]$ is a linear differential operator that depends on $x_1, \ldots, x_n$ and $t$ but is independent of $z$.

Let $\mathscr{E} = \mathscr{E}(\mathbf{x}, \mathbf{y}, z - \zeta, t, \tau)$ be a fundamental solution of the Cauchy problem for equation (14.11.3.7), i.e.,

$$\frac{\partial \mathscr{E}}{\partial t} = L_{\mathbf{x},t}[\mathscr{E}] + \frac{\partial^2 \mathscr{E}}{\partial z^2} \quad \text{for} \quad t > \tau,$$
$$\mathscr{E} = \delta(\mathbf{x} - \mathbf{y})\,\delta(z - \zeta) \quad \text{at} \quad t = \tau.$$

Here, $\mathbf{y} \in \mathbb{R}^n$, $\zeta \in \mathbb{R}^1$, and $\tau \geq 0$ are free parameters.

The fundamental solution of the Cauchy problem possesses the property

$$\mathscr{E}(\mathbf{x}, \mathbf{y}, z, t, \tau) = \mathscr{E}(\mathbf{x}, \mathbf{y}, -z, t, \tau).$$

Table 14.13 presents formulas that permit one to express the Green's functions of some nonstationary boundary value problems for equation (14.11.3.7) via the fundamental solution of the Cauchy problem.

TABLE 14.13
Representation of the Green's functions of some nonstationary boundary value problems in terms of the fundamental solution of the Cauchy problem

| Boundary value problems | Boundary conditions | Green's functions |
|---|---|---|
| First problem $\mathbf{x} \in \mathbb{R}^n, z \in \mathbb{R}^1$ | $G = 0$ at $z = 0$ | $G(\mathbf{x}, \mathbf{y}, z, \zeta, t, \tau) = \mathscr{E}(\mathbf{x}, \mathbf{y}, z-\zeta, t, \tau) - \mathscr{E}(\mathbf{x}, \mathbf{y}, z+\zeta, t, \tau)$ |
| Second problem $\mathbf{x} \in \mathbb{R}^n, z \in \mathbb{R}^1$ | $\partial_z G = 0$ at $z = 0$ | $G(\mathbf{x}, \mathbf{y}, z, \zeta, t, \tau) = \mathscr{E}(\mathbf{x}, \mathbf{y}, z-\zeta, t, \tau) + \mathscr{E}(\mathbf{x}, \mathbf{y}, z+\zeta, t, \tau)$ |
| Third problem $\mathbf{x} \in \mathbb{R}^n, z \in \mathbb{R}^1$ | $\partial_z G - kG = 0$ at $z = 0$ | $G(\mathbf{x}, \mathbf{y}, z, \zeta, t, \tau) = \mathscr{E}(\mathbf{x}, \mathbf{y}, z-\zeta, t, \tau) + \mathscr{E}(\mathbf{x}, \mathbf{y}, z+\zeta, t, \tau)$ $-2k \int_0^\infty e^{-ks} \mathscr{E}(\mathbf{x}, \mathbf{y}, z+\zeta+s, t, \tau)\,ds$ |
| First problem $\mathbf{x} \in \mathbb{R}^n, 0 \leq z \leq l$ | $G = 0$ at $z = 0$, $G = 0$ at $z = l$ | $G(\mathbf{x}, \mathbf{y}, z, \zeta, t, \tau) = \sum_{n=-\infty}^{\infty} \big[\mathscr{E}(\mathbf{x}, \mathbf{y}, z-\zeta+2nl, t, \tau) - \mathscr{E}(\mathbf{x}, \mathbf{y}, z+\zeta+2nl, t, \tau)\big]$ |
| Second problem $\mathbf{x} \in \mathbb{R}^n, 0 \leq z \leq l$ | $\partial_z G = 0$ at $z = 0$, $\partial_z G = 0$ at $z = l$ | $G(\mathbf{x}, \mathbf{y}, z, \zeta, t, \tau) = \sum_{n=-\infty}^{\infty} \big[\mathscr{E}(\mathbf{x}, \mathbf{y}, z-\zeta+2nl, t, \tau) + \mathscr{E}(\mathbf{x}, \mathbf{y}, z+\zeta+2nl, t, \tau)\big]$ |
| Mixed problem $\mathbf{x} \in \mathbb{R}^n, 0 \leq z \leq l$ | $G = 0$ at $z = 0$, $\partial_z G = 0$ at $z = l$ | $G(\mathbf{x}, \mathbf{y}, z, \zeta, t, \tau) = \sum_{n=-\infty}^{\infty} (-1)^n \big[\mathscr{E}(\mathbf{x}, \mathbf{y}, z-\zeta+2nl, t, \tau) - \mathscr{E}(\mathbf{x}, \mathbf{y}, z+\zeta+2nl, t, \tau)\big]$ |
| Mixed problem $\mathbf{x} \in \mathbb{R}^n, 0 \leq z \leq l$ | $\partial_z G = 0$ at $z = 0$, $G = 0$ at $z = l$ | $G(\mathbf{x}, \mathbf{y}, z, \zeta, t, \tau) = \sum_{n=-\infty}^{\infty} (-1)^n \big[\mathscr{E}(\mathbf{x}, \mathbf{y}, z-\zeta+2nl, t, \tau) + \mathscr{E}(\mathbf{x}, \mathbf{y}, z+\zeta+2nl, t, \tau)\big]$ |

## 14.12. Duhamel's Principles in Nonstationary Problems
### 14.12.1. Problems for Homogeneous Linear Equations

**14.12.1-1. Parabolic equations with two independent variables.**

Consider the problem for the homogeneous linear equation of parabolic type
$$\frac{\partial w}{\partial t} = a(x)\frac{\partial^2 w}{\partial x^2} + b(x)\frac{\partial w}{\partial x} + c(x)w \qquad (14.12.1.1)$$
with the homogeneous initial condition
$$w = 0 \quad \text{at} \quad t = 0 \qquad (14.12.1.2)$$
and the boundary conditions
$$s_1 \partial_x w + k_1 w = g(t) \quad \text{at} \quad x = x_1, \qquad (14.12.1.3)$$
$$s_2 \partial_x w + k_2 w = 0 \quad \text{at} \quad x = x_2. \qquad (14.12.1.4)$$

By appropriately choosing the values of the coefficients $s_1$, $s_2$, $k_1$, and $k_2$ in (14.12.1.3) and (14.12.1.4), one can obtain the first, second, third, and mixed boundary value problems for equation (14.12.1.1).

The solution of problem (14.12.1.1)–(14.12.1.4) with the nonstationary boundary condition (14.12.1.3) at $x = x_1$ can be expressed by the formula (*Duhamel's first principle*)
$$w(x,t) = \frac{\partial}{\partial t}\int_0^t u(x, t-\tau)g(\tau)\,d\tau = \int_0^t \frac{\partial u}{\partial t}(x, t-\tau)g(\tau)\,d\tau \qquad (14.12.1.5)$$
in terms of the solution $u(x,t)$ of the auxiliary problem for equation (14.12.1.1) with the initial and boundary conditions (14.12.1.2) and (14.12.1.4), for $u$ instead of $w$, and the following simpler stationary boundary condition at $x = x_1$:
$$s_1 \partial_x u + k_1 u = 1 \quad \text{at} \quad x = x_1. \qquad (14.12.1.6)$$

*Remark.* A similar formula also holds for the homogeneous boundary condition at $x = x_1$ and a nonhomogeneous nonstationary boundary condition at $x = x_2$.

*Example.* Consider the first boundary value problem for the heat equation
$$\frac{\partial w}{\partial t} = \frac{\partial^2 w}{\partial x^2} \qquad (14.12.1.7)$$
with the homogeneous initial condition (14.12.1.2) and the boundary condition
$$w = g(t) \quad \text{at} \quad x = 0. \qquad (14.12.1.8)$$
(The second boundary condition is not required in this case; $0 \le x < \infty$.)

First consider the following auxiliary problem for the heat equation with the homogeneous initial condition and a simpler boundary condition:
$$\frac{\partial u}{\partial t} = \frac{\partial^2 u}{\partial x^2}, \qquad u = 0 \quad \text{at} \quad t = 0, \qquad u = 1 \quad \text{at} \quad x = 0.$$
This problem has a self-similar solution of the form
$$w = w(z), \qquad z = xt^{-1/2},$$
where the function $w(z)$ is determined by the following ordinary differential equation and boundary conditions:
$$u''_{zz} + \tfrac{1}{2} z u'_z = 0, \qquad u = 1 \quad \text{at} \quad z = 0, \qquad u = 0 \quad \text{at} \quad z = \infty.$$
Its solution is expressed as
$$u(z) = \text{erfc}\left(\frac{z}{2}\right) \quad \Longrightarrow \quad u(x,t) = \text{erfc}\left(\frac{x}{2\sqrt{t}}\right),$$
where $\text{erfc}\, z = \dfrac{2}{\sqrt{\pi}}\displaystyle\int_z^\infty \exp(-\xi^2)\,d\xi$ is the complementary error function. Substituting the obtained expression of $u(x,t)$ into (14.12.1.5), we obtain the solution to the first boundary value problem for the heat equation (14.12.1.7) with the initial condition (14.12.1.2) and an arbitrary boundary condition (14.12.1.8) in the form
$$w(x,t) = \frac{x}{2\sqrt{\pi}}\int_0^t \exp\left[-\frac{x^2}{4(t-\tau)}\right]\frac{g(\tau)\,d\tau}{(t-\tau)^{3/2}}.$$

### 14.12.1-2. Hyperbolic equations with two independent variables.

Consider the problem for the homogeneous linear hyperbolic equation

$$\frac{\partial^2 w}{\partial t^2} + \varphi(x)\frac{\partial w}{\partial t} = a(x)\frac{\partial^2 w}{\partial x^2} + b(x)\frac{\partial w}{\partial x} + c(x)w \tag{14.12.1.9}$$

with the homogeneous initial conditions

$$\begin{aligned} w &= 0 \quad \text{at} \quad t = 0, \\ \partial_t w &= 0 \quad \text{at} \quad t = 0, \end{aligned} \tag{14.12.1.10}$$

and the boundary conditions (14.12.1.3) and (14.12.1.4).

The solution of problem (14.12.1.9), (14.12.1.10), (14.12.1.3), (14.12.1.4) with the nonstationary boundary condition (14.12.1.3) at $x = x_1$ can be expressed by formula (14.12.1.5) in terms of the solution $u(x, t)$ of the auxiliary problem for equation (14.12.1.9) with the initial conditions (14.12.1.10) and boundary condition (14.12.1.4), for $u$ instead of $w$, and the simpler stationary boundary condition (14.12.1.6) at $x = x_1$.

In this case, the remark made in Paragraph 14.12.1-1 remains valid.

### 14.12.1-3. Second-order equations with several independent variables.

Duhamel's first principle can also be used to solve homogeneous linear equations of the parabolic or hyperbolic type with many space variables,

$$\frac{\partial^k w}{\partial t^k} = \sum_{i,j=1}^{n} a_{ij}(\mathbf{x}) \frac{\partial^2 w}{\partial x_i \partial x_j} + \sum_{i=1}^{n} b_i(\mathbf{x}) \frac{\partial w}{\partial x_i} + c(\mathbf{x})w, \tag{14.12.1.11}$$

where $k = 1, 2$ and $\mathbf{x} = \{x_1, \ldots, x_n\}$.

Let $V$ be some bounded domain in $\mathbb{R}^n$ with a sufficiently smooth surface $S = \partial V$. The solution of the boundary value problem for equation (14.12.1.11) in $V$ with the homogeneous initial conditions (14.12.1.2) if $k = 1$ or (14.12.1.10) if $k = 2$, and the nonhomogeneous linear boundary condition

$$\Gamma_{\mathbf{x}}[w] = g(t) \quad \text{for} \quad \mathbf{x} \in S, \tag{14.12.1.12}$$

is given by

$$w(\mathbf{x}, t) = \frac{\partial}{\partial t} \int_0^t u(\mathbf{x}, t - \tau) g(\tau) \, d\tau = \int_0^t \frac{\partial u}{\partial t}(\mathbf{x}, t - \tau) g(\tau) \, d\tau.$$

Here, $u(\mathbf{x}, t)$ is the solution of the auxiliary problem for equation (14.12.1.11) with the same initial conditions, (14.12.1.2) or (14.12.1.10), for $u$ instead of $w$, and the simpler stationary boundary condition

$$\Gamma_{\mathbf{x}}[u] = 1 \quad \text{for} \quad \mathbf{x} \in S.$$

Note that (14.12.1.12) can represent a boundary condition of the first, second, or third kind; the coefficients of the operator $\Gamma_{\mathbf{x}}$ are assumed to be independent of $t$.

## 14.12.2. Problems for Nonhomogeneous Linear Equations

### 14.12.2-1. Parabolic equations.

The solution of the nonhomogeneous linear equation

$$\frac{\partial w}{\partial t} = \sum_{i,j=1}^{n} a_{ij}(\mathbf{x}) \frac{\partial^2 w}{\partial x_i \partial x_j} + \sum_{i=1}^{n} b_i(\mathbf{x}) \frac{\partial w}{\partial x_i} + c(\mathbf{x})w + \Phi(\mathbf{x}, t)$$

with the homogeneous initial condition (14.12.1.2) and the homogeneous boundary condition

$$\Gamma_\mathbf{x}[w] = 0 \quad \text{for} \quad \mathbf{x} \in S \qquad (14.12.2.1)$$

can be represented in the form (*Duhamel's second principle*)

$$w(\mathbf{x}, t) = \int_0^t U(\mathbf{x}, t - \tau, \tau) \, d\tau. \qquad (14.12.2.2)$$

Here, $U(\mathbf{x}, t, \tau)$ is the solution of the auxiliary problem for the homogeneous equation

$$\frac{\partial U}{\partial t} = \sum_{i,j=1}^{n} a_{ij}(\mathbf{x}) \frac{\partial^2 U}{\partial x_i \partial x_j} + \sum_{i=1}^{n} b_i(\mathbf{x}) \frac{\partial U}{\partial x_i} + c(\mathbf{x})U$$

with the boundary condition (14.12.2.1), in which $w$ must be substituted by $U$, and the nonhomogeneous initial condition

$$U = \Phi(\mathbf{x}, \tau) \quad \text{at} \quad t = 0,$$

where $\tau$ is a parameter.

Note that (14.12.2.1) can represent a boundary condition of the first, second, or third kind; the coefficients of the operator $\Gamma_\mathbf{x}$ are assumed to be independent of $t$.

### 14.12.2-2. Hyperbolic equations.

The solution of the nonhomogeneous linear equation

$$\frac{\partial^2 w}{\partial t^2} + \varphi(\mathbf{x}) \frac{\partial w}{\partial t} = \sum_{i,j=1}^{n} a_{ij}(\mathbf{x}) \frac{\partial^2 w}{\partial x_i \partial x_j} + \sum_{i=1}^{n} b_i(\mathbf{x}) \frac{\partial w}{\partial x_i} + c(\mathbf{x})w + \Phi(\mathbf{x}, t)$$

with the homogeneous initial conditions (14.12.1.10) and homogeneous boundary condition (14.12.2.1) can be expressed by formula (14.12.2.2) in terms of the solution $U = U(\mathbf{x}, t, \tau)$ of the auxiliary problem for the homogeneous equation

$$\frac{\partial^2 U}{\partial t^2} + \varphi(\mathbf{x}) \frac{\partial U}{\partial t} = \sum_{i,j=1}^{n} a_{ij}(\mathbf{x}) \frac{\partial^2 U}{\partial x_i \partial x_j} + \sum_{i=1}^{n} b_i(\mathbf{x}) \frac{\partial U}{\partial x_i} + c(\mathbf{x})U$$

with the homogeneous initial and boundary conditions, (14.12.1.2) and (14.12.2.1), where $w$ must be replaced by $U$, and the nonhomogeneous initial condition

$$\partial_t U = \Phi(\mathbf{x}, \tau) \quad \text{at} \quad t = 0,$$

where $\tau$ is a parameter.

Note that (14.12.2.1) can represent a boundary condition of the first, second, or third kind.

## 14.13. Transformations Simplifying Initial and Boundary Conditions

### 14.13.1. Transformations That Lead to Homogeneous Boundary Conditions

A linear problem with arbitrary nonhomogeneous boundary conditions,

$$\Gamma_{\mathbf{x},t}^{(k)}[w] = g_k(\mathbf{x},t) \quad \text{for} \quad \mathbf{x} \in S_k, \tag{14.13.1.1}$$

can be reduced to a linear problem with homogeneous boundary conditions. To this end, one should perform the change of variable

$$w(\mathbf{x},t) = \psi(\mathbf{x},t) + u(\mathbf{x},t), \tag{14.13.1.2}$$

where $u$ is a new unknown function and $\psi$ is any function that satisfies the nonhomogeneous boundary conditions (14.13.1.1),

$$\Gamma_{\mathbf{x},t}^{(k)}[\psi] = g_k(\mathbf{x},t) \quad \text{for} \quad \mathbf{x} \in S_k. \tag{14.13.1.3}$$

Table 14.14 gives examples of such transformations for linear boundary value problems with one space variable for parabolic and hyperbolic equations. In the third boundary value problem, it is assumed that $k_1 < 0$ and $k_2 > 0$.

TABLE 14.14
Simple transformations of the form $w(x,t) = \psi(x,t) + u(x,t)$ that lead to homogeneous boundary conditions in problems with one space variables ($0 \leq x \leq l$)

| No. | Problems | Boundary conditions | Function $\psi(x,t)$ |
|---|---|---|---|
| 1 | First boundary value problem | $w = g_1(t)$ at $x = 0$ <br> $w = g_2(t)$ at $x = l$ | $\psi(x,t) = g_1(t) + \dfrac{x}{l}\big[g_2(t) - g_1(t)\big]$ |
| 2 | Second boundary value problem | $\partial_x w = g_1(t)$ at $x = 0$ <br> $\partial_x w = g_2(t)$ at $x = l$ | $\psi(x,t) = x g_1(t) + \dfrac{x^2}{2l}\big[g_2(t) - g_1(t)\big]$ |
| 3 | Third boundary value problem | $\partial_x w + k_1 w = g_1(t)$ at $x = 0$ <br> $\partial_x w + k_2 w = g_2(t)$ at $x = l$ | $\psi(x,t) = \dfrac{(k_2 x - 1 - k_2 l)g_1(t) + (1 - k_1 x)g_2(t)}{k_2 - k_1 - k_1 k_2 l}$ |
| 4 | Mixed boundary value problem | $w = g_1(t)$ at $x = 0$ <br> $\partial_x w = g_2(t)$ at $x = l$ | $\psi(x,t) = g_1(t) + x g_2(t)$ |
| 5 | Mixed boundary value problem | $\partial_x w = g_1(t)$ at $x = 0$ <br> $w = g_2(t)$ at $x = l$ | $\psi(x,t) = (x - l)g_1(t) + g_2(t)$ |

Note that the selection of the function $\psi$ is of a purely algebraic nature and is not connected with the equation in question; there are infinitely many suitable functions $\psi$ that satisfy condition (14.13.1.3). Transformations of the form (14.13.1.2) can often be used at the first stage of solving boundary value problems.

## 14.13.2. Transformations That Lead to Homogeneous Initial and Boundary Conditions

A linear problem with nonhomogeneous initial and boundary conditions can be reduced to a linear problem with homogeneous initial and boundary conditions. To this end, one should introduce a new dependent variable $u$ by formula (14.13.1.2), where the function $\psi$ must satisfy nonhomogeneous initial and boundary conditions.

Below we specify some simple functions $\psi$ that can be used in transformation (14.13.1.2) to obtain boundary value problems with homogeneous initial and boundary conditions. To be specific, we consider a parabolic equation with one space variable and the general initial condition

$$w = f(x) \quad \text{at} \quad t = 0. \tag{14.13.2.1}$$

*1. First boundary value problem*: the initial condition is (14.13.2.1) and the boundary conditions are given in row 1 of Table 14.14. Suppose that the initial and boundary conditions are compatible, i.e., $f(0) = g_1(0)$ and $f(l) = g_2(0)$. Then, in transformation (14.13.1.2), one can take

$$\psi(x,t) = f(x) + g_1(t) - g_1(0) + \frac{x}{l}\big[g_2(t) - g_1(t) + g_1(0) - g_2(0)\big].$$

*2. Second boundary value problem*: the initial condition is (14.13.2.1) and the boundary conditions are given in row 2 of Table 14.14. Suppose that the initial and boundary conditions are compatible, i.e., $f'(0) = g_1(0)$ and $f'(l) = g_2(0)$. Then, in transformation (14.13.1.2), one can set

$$\psi(x,t) = f(x) + x\big[g_1(t) - g_1(0)\big] + \frac{x^2}{2l}\big[g_2(t) - g_1(t) + g_1(0) - g_2(0)\big].$$

*3. Third boundary value problem*: the initial condition is (14.13.2.1) and the boundary conditions are given in row 3 of Table 14.14. If the initial and boundary conditions are compatible, then, in transformation (14.13.1.2), one can take

$$\psi(x,t) = f(x) + \frac{(k_2 x - 1 - k_2 l)[g_1(t) - g_1(0)] + (1 - k_1 x)[g_2(t) - g_2(0)]}{k_2 - k_1 - k_1 k_2 l} \qquad (k_1 < 0, \ k_2 > 0).$$

*4. Mixed boundary value problem*: the initial condition is (14.13.2.1) and the boundary conditions are given in row 4 of Table 14.14. Suppose that the initial and boundary conditions are compatible, i.e., $f(0) = g_1(0)$ and $f'(l) = g_2(0)$. Then, in transformation (14.13.1.2), one can set

$$\psi(x,t) = f(x) + g_1(t) - g_1(0) + x\big[g_2(t) - g_2(0)\big].$$

*5. Mixed boundary value problem*: the initial condition is (14.13.2.1) and the boundary conditions are given in row 5 of Table 14.14. Suppose that the initial and boundary conditions are compatible, i.e., $f'(0) = g_1(0)$ and $f(l) = g_2(0)$. Then, in transformation (14.13.1.2), one can take

$$\psi(x,t) = f(x) + (x - l)\big[g_1(t) - g_1(0)\big] + g_2(t) - g_2(0).$$

## References for Chapter 14

Akulenko, L. D. and Nesterov, S. V., *High Precision Methods in Eigenvalue Problems and Their Applications*, Chapman & Hall/CRC Press, Boca Raton, 2004.

Butkovskiy, A. G., *Green's Functions and Transfer Functions Handbook*, Halstead Press–John Wiley & Sons, New York, 1982.

**Carslaw, H. S. and Jaeger, J. C.**, *Conduction of Heat in Solids*, Clarendon Press, Oxford, 1984.
**Constanda, C.**, *Solution Techniques for Elementary Partial Differential Equations*, Chapman & Hall/CRC Press, Boca Raton, 2002.
**Courant, R. and Hilbert, D.**, *Methods of Mathematical Physics, Vol. 2*, Wiley-Interscience, New York, 1989.
**Dezin, A. A.**, *Partial Differential Equations. An Introduction to a General Theory of Linear Boundary Value Problems*, Springer-Verlag, Berlin, 1987.
**Ditkin, V. A. and Prudnikov, A. P.**, *Integral Transforms and Operational Calculus*, Pergamon Press, New York, 1965.
**Duffy, D. G.**, *Transform Methods for Solving Partial Differential Equations, 2nd Edition*, Chapman & Hall/CRC Press, Boca Raton, 2004.
**Farlow, S. J.**, *Partial Differential Equations for Scientists and Engineers*, John Wiley & Sons, New York, 1982.
**Guenther, R. B. and Lee, J. W.**, *Partial Differential Equations of Mathematical Physics and Integral Equations*, Dover Publications, New York, 1996.
**Haberman, R.**, *Elementary Applied Partial Differential Equations with Fourier Series and Boundary Value Problems*, Prentice-Hall, Englewood Cliffs, New Jersey, 1987.
**Hanna, J. R. and Rowland, J. H.**, *Fourier Series, Transforms, and Boundary Value Problems*, Wiley-Interscience, New York, 1990.
**Kanwal, R. P.**, *Generalized Functions. Theory and Technique*, Academic Press, Orlando, 1983.
**Leis, R.**, *Initial-Boundary Value Problems in Mathematical Physics*, John Wiley & Sons, Chichester, 1986.
**Mikhlin, S. G. (Editor)**, *Linear Equations of Mathematical Physics*, Holt, Rinehart and Winston, New York, 1967.
**Miller, W., Jr.**, *Symmetry and Separation of Variables*, Addison-Wesley, London, 1977.
**Moon, P. and Spencer, D. E.**, *Field Theory Handbook, Including Coordinate Systems, Differential Equations and Their Solutions, 3rd Edition*, Springer-Verlag, Berlin, 1988.
**Morse, P. M. and Feshbach, H.**, *Methods of Theoretical Physics, Vols. 1 and 2*, McGraw-Hill, New York, 1953.
**Petrovsky, I. G.**, *Lectures on Partial Differential Equations*, Dover Publications, New York, 1991.
**Polyanin, A. D.**, *Handbook of Linear Partial Differential Equations for Engineers and Scientists*, Chapman & Hall/CRC Press, Boca Raton, 2002.
**Sneddon, I. N.**, *Fourier Transformations*, Dover Publications, New York, 1995.
**Stakgold, I.**, *Boundary Value Problems of Mathematical Physics. Vols. 1 and 2*, Society for Industrial & Applied Mathematics, Philadelphia, 2000.
**Strauss, W. A.**, *Partial Differential Equations. An Introduction*, John Wiley & Sons, New York, 1992.
**Tikhonov, A. N. and Samarskii, A. A.**, *Equations of Mathematical Physics*, Dover Publications, New York, 1990.
**Vladimirov, V. S.**, *Equations of Mathematical Physics*, Dekker, New York, 1971.
**Zauderer, E.**, *Partial Differential Equations of Applied Mathematics*, Wiley-Interscience, New York, 1989.
**Zwillinger, D.**, *Handbook of Differential Equations, 3rd Edition*, Academic Press, New York, 1997.

# Chapter 15
# Nonlinear Partial Differential Equations

## 15.1. Classification of Second-Order Nonlinear Equations

### 15.1.1. Classification of Semilinear Equations in Two Independent Variables

A *second-order semilinear partial differential equation* in two independent variables has the form

$$a(x,y)\frac{\partial^2 w}{\partial x^2} + 2b(x,y)\frac{\partial^2 w}{\partial x \partial y} + c(x,y)\frac{\partial^2 w}{\partial y^2} = f\left(x, y, w, \frac{\partial w}{\partial x}, \frac{\partial w}{\partial y}\right). \qquad (15.1.1.1)$$

This equation is classified according to the sign of the discriminant

$$\delta = b^2 - ac, \qquad (15.1.1.2)$$

where the arguments of the equation coefficients are omitted for brevity. Given a point $(x, y)$, equation (15.1.1.1) is

$$\begin{array}{ll} parabolic & \text{if } \delta = 0, \\ hyperbolic & \text{if } \delta > 0, \\ elliptic & \text{if } \delta < 0. \end{array} \qquad (15.1.1.3)$$

The reduction of equation (15.1.1.1) to a canonical form on the basis of the solution of the characteristic equations entirely coincides with that used for linear equations (see Subsection 14.1.1).

The classification of semilinear equations of the form (15.1.1.1) does not depend on their solutions — it is determined solely by the coefficients of the highest derivatives on the left-hand side.

### 15.1.2. Classification of Nonlinear Equations in Two Independent Variables

15.1.2-1. Nonlinear equations of general form.

In general, a *second-order nonlinear partial differential equation* in two independent variables has the form

$$F\left(x, y, w, \frac{\partial w}{\partial x}, \frac{\partial w}{\partial y}, \frac{\partial^2 w}{\partial x^2}, \frac{\partial^2 w}{\partial x \partial y}, \frac{\partial^2 w}{\partial y^2}\right) = 0. \qquad (15.1.2.1)$$

Denote

$$a = \frac{\partial F}{\partial p}, \quad b = \frac{1}{2}\frac{\partial F}{\partial q}, \quad c = \frac{\partial F}{\partial r}, \quad \text{where} \quad p = \frac{\partial^2 w}{\partial x^2}, \quad q = \frac{\partial^2 w}{\partial x \partial y}, \quad r = \frac{\partial^2 w}{\partial y^2}. \tag{15.1.2.2}$$

Let us select a specific solution $w = w(x, y)$ of equation (15.1.2.1) and calculate $a$, $b$, and $c$ by formulas (15.1.2.2) at some point $(x, y)$, and substitute the resulting expressions into (15.1.1.2). Depending on the sign of the discriminant $\delta$, the type of nonlinear equation (15.1.2.1) at the point $(x, y)$ is determined according to (15.1.1.3): if $\delta = 0$, the equation is parabolic, if $\delta > 0$, it is hyperbolic, and if $\delta < 0$, it is elliptic. In general, the coefficients $a$, $b$, and $c$ of the nonlinear equation (15.1.2.1) depend not only on the selection of the point $(x, y)$, but also on the selection of the specific solution. Therefore, it is impossible to determine the sign of $\delta$ without knowing the solution $w(x, y)$. To put it differently, the type of a nonlinear equation can be different for different solutions at the same point $(x, y)$.

A line $\varphi(x, y) = C$ is called a *characteristic* of the nonlinear equation (15.1.2.1) if it is an integral curve of the *characteristic equation*

$$a\,(dy)^2 - 2b\,dx\,dy + c\,(dx)^2 = 0. \tag{15.1.2.3}$$

The form of characteristics depends on the selection of a specific solution.

In individual special cases, the type of a nonlinear equation [other than the semilinear equation (15.1.1.1)] may be independent of the selection of solutions.

**Example.** Consider the nonhomogeneous *Monge–Ampère* equation

$$\left(\frac{\partial^2 w}{\partial x \partial y}\right)^2 - \frac{\partial^2 w}{\partial x^2}\frac{\partial^2 w}{\partial y^2} = f(x, y).$$

It is a special case of equation (15.1.2.1) with

$$F(x, y, p, q, r) \equiv q^2 - pr - f(x, y) = 0, \qquad p = \frac{\partial^2 w}{\partial x^2}, \quad q = \frac{\partial^2 w}{\partial x \partial y}, \quad r = \frac{\partial^2 w}{\partial y^2}. \tag{15.1.2.4}$$

Using formulas (15.1.2.2) and (15.1.2.4), we find the discriminant (15.1.1.2):

$$\delta = q^2 - pr = f(x, y). \tag{15.1.2.5}$$

Here, the relation between the highest derivatives and $f(x, y)$ defined by equation (15.1.2.4) has been taken into account.

From (15.1.2.5) and (15.1.1.3) it follows that the type of the nonhomogeneous equation Monge–Ampère at a point $(x, y)$ depends solely on the sign of $f(x, y)$ and is independent of the selection of a particular solution. At the points where $f(x, y) = 0$, the equation is of parabolic type; at the points where $f(x, y) > 0$, the equation is of hyperbolic type; and at the points where $f(x, y) < 0$, it is elliptic.

### 15.1.2-2. Quasilinear equations.

A *second-order quasilinear partial differential equation* in two independent variables has the form

$$a(x, y, w, \xi, \eta)p + 2b(x, y, w, \xi, \eta)q + c(x, y, w, \xi, \eta)r = f(x, y, w, \xi, \eta), \tag{15.1.2.6}$$

with the short notation

$$\xi = \frac{\partial w}{\partial x}, \quad \eta = \frac{\partial w}{\partial y}, \quad p = \frac{\partial^2 w}{\partial x^2}, \quad q = \frac{\partial^2 w}{\partial x \partial y}, \quad r = \frac{\partial^2 w}{\partial y^2}.$$

Consider a curve $\mathcal{C}_0$ defined in the $x, y$ plane parametrically as

$$x = x(\tau), \quad y = y(\tau). \tag{15.1.2.7}$$

Let us fix a set of boundary conditions on this curve, thus defining the initial values of the unknown function and its first derivatives:

$$w = w(\tau), \quad \xi = \xi(\tau), \quad \eta = \eta(\tau) \quad (w'_\tau = \xi x'_\tau + \eta y'_\tau). \tag{15.1.2.8}$$

The derivative with respect to $\tau$ is obtained by the chain rule, since $w = w(x, y)$. It can be shown that the given set of functions (15.1.2.8) uniquely determines the values of the second derivatives $p$, $q$, and $r$ (and also higher derivatives) at each point of the curve (15.1.2.7), satisfying the condition

$$a(y'_x)^2 - 2by'_x + c \neq 0 \quad (y'_x = y'_\tau/x'_\tau). \tag{15.1.2.9}$$

Here and henceforth, the arguments of the functions $a$, $b$, and $c$ are omitted.

Indeed, bearing in mind that $\xi = \xi(x, y)$ and $\eta = \eta(x, y)$, let us differentiate the second and the third equation in (15.1.2.8) with respect to the parameter $\tau$:

$$\xi'_\tau = px'_\tau + qy'_\tau, \quad \eta'_\tau = qx'_\tau + ry'_\tau. \tag{15.1.2.10}$$

On solving relations (15.1.2.6) and (15.1.2.10) for $p$, $q$, and $r$, we obtain formulas for the second derivatives at the points of the curve (15.1.2.7):

$$
\begin{aligned}
p &= \frac{c(x'_\tau \xi'_\tau - y'_\tau \eta'_\tau) - 2by'_\tau \xi'_\tau + f(y'_\tau)^2}{a(y'_\tau)^2 - 2bx'_\tau y'_\tau + c(x'_\tau)^2}, \\
q &= \frac{ay'_\tau \xi'_\tau + cx'_\tau \eta'_\tau - fx'_\tau y'_\tau}{a(y'_\tau)^2 - 2bx'_\tau y'_\tau + c(x'_\tau)^2}, \\
r &= \frac{a(y'_\tau \eta'_\tau - x'_\tau \xi'_\tau) - 2bx'_\tau \eta'_\tau + f(x'_\tau)^2}{a(y'_\tau)^2 - 2bx'_\tau y'_\tau + c(x'_\tau)^2}.
\end{aligned}
\tag{15.1.2.11}
$$

The third derivatives at the points of the curve (15.1.2.7) can be calculated in a similar way. To this end, one differentiates (15.1.2.6) and (15.1.2.11) with respect to $\tau$ and then expresses the third derivatives from the resulting relations. This procedure can also be extended to higher derivatives. Consequently, the solution to equation (15.1.2.6) can be represented as a Taylor series about the points of the curve (15.1.2.7) that satisfy condition (15.1.2.9).

The singular points at which the denominators in the formulas for the second derivatives (15.1.2.11) vanish satisfy the characteristic equation (15.1.2.3). Conditions of the form (15.1.2.8) cannot be arbitrarily set on the characteristic curves, which are described by equation (15.1.2.3). The additional conditions of vanishing of the numerators in formulas (15.1.2.11) must be used; in this case, the second derivatives will be finite.

## 15.2. Transformations of Equations of Mathematical Physics

### 15.2.1. Point Transformations: Overview and Examples

#### 15.2.1-1. General form of point transformations.

Let $w = w(x, y)$ be a function of independent variables $x$ and $y$. In general, a *point transformation* is defined by the formulas

$$x = X(\xi, \eta, u), \quad y = Y(\xi, \eta, u), \quad w = W(\xi, \eta, u), \tag{15.2.1.1}$$

where $\xi$ and $\eta$ are new independent variables, $u = u(\xi, \eta)$ is a new dependent variable, and the functions $X, Y, W$ may be either given or unknown (have to be found).

A point transformation not only preserves the order or the equation to which it is applied, but also mostly preserves the structure of the equation, since the highest-order derivatives of the new variables are linearly dependent on the highest-order derivatives of the original variables.

Transformation (15.2.1.1) is invertible if

$$\begin{vmatrix} \frac{\partial X}{\partial x} & \frac{\partial X}{\partial y} & \frac{\partial X}{\partial w} \\ \frac{\partial Y}{\partial x} & \frac{\partial Y}{\partial y} & \frac{\partial Y}{\partial w} \\ \frac{\partial W}{\partial x} & \frac{\partial W}{\partial y} & \frac{\partial W}{\partial w} \end{vmatrix} \neq 0.$$

In the general case, a point transformation (15.2.1.1) reduces a second-order equation with two independent variables

$$F\left(x, y, w, \frac{\partial w}{\partial x}, \frac{\partial w}{\partial y}, \frac{\partial^2 w}{\partial x^2}, \frac{\partial^2 w}{\partial x \partial y}, \frac{\partial^2 w}{\partial y^2}\right) = 0 \qquad (15.2.1.2)$$

to an equation

$$G\left(\xi, \eta, u, \frac{\partial u}{\partial \xi}, \frac{\partial u}{\partial \eta}, \frac{\partial^2 u}{\partial \xi^2}, \frac{\partial^2 u}{\partial \xi \partial \eta}, \frac{\partial^2 u}{\partial \eta^2}\right) = 0. \qquad (15.2.1.3)$$

If $u = u(\xi, \eta)$ is a solution of equation (15.2.1.3), then formulas (15.2.1.1) define the corresponding solution of equation (15.2.1.2) in parametric form.

Point transformations are employed to simplify equations and their reduction to known equations.

### 15.2.1-2. Linear transformations.

*Linear point transformations* (or simply *linear transformations*),

$$x = X(\xi, \eta), \quad y = Y(\xi, \eta), \quad w = f(\xi, \eta)u + g(\xi, \eta), \qquad (15.2.1.4)$$

are most commonly used.

The simplest linear transformations of the independent variables are

$x = \xi + x_0,$     $y = \eta + y_0$     (*translation transformation*),
$x = k_1 \xi,$     $y = k_2 \eta$     (*scaling transformation*),
$x = \xi \cos \alpha - \eta \sin \alpha,$     $y = \xi \sin \alpha + \eta \cos \alpha$     (*rotation transformation*).

These transformations correspond to the translation of the origin of coordinates to the point $(x_0, y_0)$, scaling (extension or contraction) along the $x$- and $y$-axes, and the rotation of the coordinate system through the angle $\alpha$, respectively. These transformations do not affect the dependent variable ($w = u$).

Linear transformations (15.2.1.4) are employed to simplify linear and nonlinear equations and to reduce equations to the canonical forms (see Subsections 14.1.1 and 15.1.1).

**Example 1.** The nonlinear equation
$$\frac{\partial w}{\partial t} = a\frac{\partial}{\partial x}\left(w^m \frac{\partial w}{\partial x}\right) + [xf(t) + g(t)]\frac{\partial w}{\partial x} + h(t)w$$
can be simplified to obtain
$$\frac{\partial u}{\partial \tau} = \frac{\partial}{\partial z}\left(u^m \frac{\partial u}{\partial z}\right)$$
with the help of the transformation
$$w(x,t) = u(z,\tau)H(t), \quad z = xF(t) + \int g(t)F(t)\,dt, \quad \tau = \int F^2(t)H^m(t)\,dt,$$
where
$$F(t) = \exp\left[\int f(t)\,dt\right], \quad H(t) = \exp\left[\int h(t)\,dt\right].$$

### 15.2.1-3. Simple nonlinear point transformations.

Point transformations can be used for the reduction of nonlinear equations to linear ones. The simplest nonlinear transformations have the form

$$w = W(u) \tag{15.2.1.5}$$

and do not affect the independent variables ($x = \xi$ and $y = \eta$). Combinations of transformations (15.2.1.4) and (15.2.1.5) are also used quite often.

**Example 2.** The nonlinear equation
$$\frac{\partial w}{\partial t} = \frac{\partial^2 w}{\partial x^2} + a\left(\frac{\partial w}{\partial x}\right)^2 + f(x,t)$$
can be reduced to the linear equation
$$\frac{\partial u}{\partial t} = \frac{\partial^2 u}{\partial x^2} + af(x,t)u$$
for the function $u = u(x,t)$ by means of the transformation $u = \exp(aw)$.

## 15.2.2. Hodograph Transformations (Special Point Transformations)

In some cases, nonlinear equations and systems of partial differential equations can be simplified by means of the *hodograph transformations*, which are special cases of point transformations.

### 15.2.2-1. One of the independent variables is taken to be the dependent one.

For an equation with two independent variables $x, y$ and an unknown function $w = w(x,y)$, the hodograph transformation consists of representing the solution in implicit form

$$x = x(y,w) \tag{15.2.2.1}$$

[or $y = y(x,w)$]. Thus, $y$ and $w$ are treated as independent variables, while $x$ is taken to be the dependent variable. The hodograph transformation (15.2.2.1) does not change the order of the equation and belongs to the class of point transformations (equivalently, it can be represented as $x = \widetilde{w}, y = \widetilde{y}, w = \widetilde{x}$).

**Example 1.** Consider the nonlinear second-order equation

$$\frac{\partial w}{\partial y}\left(\frac{\partial w}{\partial x}\right)^2 = a\frac{\partial^2 w}{\partial x^2}. \tag{15.2.2.2}$$

Let us seek its solution in implicit form. Differentiating relation (15.2.2.1) with respect to both variables as an implicit function and taking into account that $w = w(x, y)$, we get

$$1 = x_w w_x \quad \text{(differentiation in } x\text{)},$$
$$0 = x_w w_y + x_y \quad \text{(differentiation in } y\text{)},$$
$$0 = x_{ww} w_x^2 + x_w w_{xx} \quad \text{(double differentiation in } x\text{)},$$

where the subscripts indicate the corresponding partial derivatives. We solve these relations to express the "old" derivatives through the "new" ones,

$$w_x = \frac{1}{x_w}, \quad w_y = -\frac{x_y}{x_w}, \quad w_{xx} = -\frac{w_x^2 x_{ww}}{x_w} = -\frac{x_{ww}}{x_w^3}.$$

Substituting these expressions into (15.2.2.2), we obtain the linear heat equation:

$$\frac{\partial x}{\partial y} = a\frac{\partial^2 x}{\partial w^2}.$$

### 15.2.2-2. Method of conversion to an equivalent system of equations.

In order to investigate equations with the unknown function $w = w(x, y)$, it may be useful to convert the original equation to an equivalent system of equations for $w(x, y)$ and $v = v(x, y)$ (the elimination of $v$ from the system results in the original equation) and then apply the hodograph transformation

$$x = x(w, v), \quad y = y(w, v), \tag{15.2.2.3}$$

where $w$, $v$ are treated as the independent variables and $x$, $y$ as the dependent variables.

Let us illustrate this by examples of specific equations of mathematical physics.

**Example 2.** Rewrite the *stationary Khokhlov–Zabolotskaya equation* (it arises in acoustics and nonlinear mechanics)

$$\frac{\partial^2 w}{\partial x^2} + a\frac{\partial}{\partial y}\left(w\frac{\partial w}{\partial y}\right) = 0 \tag{15.2.2.4}$$

as the system of equations

$$\frac{\partial w}{\partial x} = \frac{\partial v}{\partial y}, \quad -aw\frac{\partial w}{\partial y} = \frac{\partial v}{\partial x}. \tag{15.2.2.5}$$

We now take advantage of the hodograph transformation (15.2.2.3), which amounts to taking $w$, $v$ as the independent variables and $x$, $y$ as the dependent variables. Differentiating each relation in (15.2.2.3) with respect to $x$ and $y$ (as composite functions) and eliminating the partial derivatives $x_w$, $x_v$, $y_w$, $y_v$ from the resulting relations, we obtain

$$\frac{\partial x}{\partial w} = \frac{1}{J}\frac{\partial v}{\partial y}, \quad \frac{\partial x}{\partial v} = -\frac{1}{J}\frac{\partial w}{\partial y}, \quad \frac{\partial y}{\partial w} = -\frac{1}{J}\frac{\partial v}{\partial x}, \quad \frac{\partial y}{\partial v} = \frac{1}{J}\frac{\partial w}{\partial x}, \quad \text{where} \quad J = \frac{\partial w}{\partial x}\frac{\partial v}{\partial y} - \frac{\partial w}{\partial y}\frac{\partial v}{\partial x}. \tag{15.2.2.6}$$

Using (15.2.2.6) to eliminate the derivatives $w_x$, $w_y$, $v_x$, $v_y$ from (15.2.2.5), we arrive at the system

$$\frac{\partial y}{\partial v} = \frac{\partial x}{\partial w}, \quad -aw\frac{\partial x}{\partial v} = \frac{\partial y}{\partial w}. \tag{15.2.2.7}$$

Let us differentiate the first equation in $w$ and the second in $v$, and then eliminate the mixed derivative $y_{wv}$. As a result, we obtain the following linear equation for the function $x = x(w, v)$:

$$\frac{\partial^2 x}{\partial w^2} + aw\frac{\partial^2 x}{\partial v^2} = 0. \tag{15.2.2.8}$$

Similarly, from system (15.2.2.7), we obtain another linear equation for the function $y = y(w, v)$,

$$\frac{\partial^2 y}{\partial v^2} + \frac{\partial}{\partial w}\left(\frac{1}{aw}\frac{\partial y}{\partial w}\right) = 0. \tag{15.2.2.9}$$

Given a particular solution $x = x(w, v)$ of equation (15.2.2.8), we substitute this solution into system (15.2.2.7) and find $y = y(w, v)$ by straightforward integration. Eliminating $v$ from (15.2.2.3), we obtain an exact solution $w = w(x, y)$ of the nonlinear equation (15.2.2.4).

**Remark.** Equation (15.2.2.8) with an arbitrary $a$ admits a simple particular solution, namely,

$$x = C_1 w v + C_2 w + C_3 v + C_4, \tag{15.2.2.10}$$

where $C_1, \ldots, C_4$ are arbitrary constants. Substituting this solution into system (15.2.2.7), we obtain

$$\frac{\partial y}{\partial v} = C_1 v + C_2, \quad \frac{\partial y}{\partial w} = -a(C_1 w + C_3) w. \tag{15.2.2.11}$$

Integrating the first equation in (15.2.2.11) yields $y = \frac{1}{2} C_1 v^2 + C_2 v + \varphi(w)$. Substituting this solution into the second equation in (15.2.2.11), we find the function $\varphi(w)$, and consequently

$$y = \frac{1}{2} C_1 v^2 + C_2 v - \frac{1}{3} a C_1 w^3 - \frac{1}{2} a C_3 w^2 + C_5. \tag{15.2.2.12}$$

Formulas (15.2.2.10) and (15.2.2.12) define an exact solution of equation (15.2.2.4) in parametric form ($v$ is the parameter).

In a similar way, one can construct more complex solutions of equation (15.2.2.4) in parametric form.

**Example 3.** Consider the *Born–Infeld equation*

$$\left[1 - \left(\frac{\partial w}{\partial t}\right)^2\right] \frac{\partial^2 w}{\partial x^2} + 2 \frac{\partial w}{\partial x} \frac{\partial w}{\partial t} \frac{\partial^2 w}{\partial x \partial t} - \left[1 + \left(\frac{\partial w}{\partial x}\right)^2\right] \frac{\partial^2 w}{\partial t^2} = 0, \tag{15.2.2.13}$$

which is used in nonlinear electrodynamics (field theory).

By introducing the new variables

$$\xi = x - t, \quad \eta = x + t, \quad u = \frac{\partial w}{\partial \xi}, \quad v = \frac{\partial w}{\partial \eta},$$

equation (15.2.2.13) can be rewritten as the equivalent system

$$\frac{\partial u}{\partial \eta} - \frac{\partial v}{\partial \xi} = 0,$$

$$v^2 \frac{\partial u}{\partial \xi} - (1 + 2uv) \frac{\partial u}{\partial \eta} + u^2 \frac{\partial v}{\partial \eta} = 0.$$

The hodograph transformation, where $u, v$ are taken to be the independent variables and $\xi, \eta$ the dependent ones, leads to the linear system

$$\frac{\partial \xi}{\partial v} - \frac{\partial \eta}{\partial u} = 0,$$
$$v^2 \frac{\partial \eta}{\partial v} + (1 + 2uv) \frac{\partial \xi}{\partial v} + u^2 \frac{\partial \xi}{\partial u} = 0. \tag{15.2.2.14}$$

Eliminating $\eta$ yields the linear second-order equation

$$u^2 \frac{\partial^2 \xi}{\partial u^2} + (1 + 2uv) \frac{\partial^2 \xi}{\partial u \partial v} + v^2 \frac{\partial^2 \xi}{\partial v^2} + 2u \frac{\partial \xi}{\partial u} + 2v \frac{\partial \xi}{\partial v} = 0.$$

Assuming that the solution of interest is in the domain of hyperbolicity, we write out the equation of characteristics (see Subsection 14.1.1):

$$u^2 \, dv^2 - (1 + 2uv) \, du \, dv + v^2 \, du^2 = 0.$$

This equation has the integrals $r = C_1$ and $s = C_2$, where

$$r = \frac{\sqrt{1 + 4uv} - 1}{2v}, \quad s = \frac{\sqrt{1 + 4uv} - 1}{2u}. \tag{15.2.2.15}$$

Passing in (15.2.2.14) to the new variables (15.2.2.15), we obtain

$$r^2 \frac{\partial \xi}{\partial r} + \frac{\partial \eta}{\partial r} = 0,$$
$$\frac{\partial \xi}{\partial s} + s^2 \frac{\partial \eta}{\partial s} = 0. \tag{15.2.2.16}$$

Eliminating $\eta$ yields the simple equation

$$\frac{\partial^2 \xi}{\partial r \partial s} = 0,$$

whose general solution is the sum of two arbitrary functions with different arguments: $\xi = f(r) + g(s)$. The function $\eta$ is determined from system (15.2.2.16).

▶ In Paragraph 15.14.4-4, the hodograph transformation is used for the linearization of gas-dynamic systems of equations.

## 15.2.3. Contact Transformations.* Legendre and Euler Transformations

**15.2.3-1. General form of contact transformations.**

Consider functions of two variables $w = w(x, y)$. A common property of contact transformations is the dependence of the original variables on the new variables and their first derivatives:

$$x = X\left(\xi, \eta, u, \frac{\partial u}{\partial \xi}, \frac{\partial u}{\partial \eta}\right), \quad y = Y\left(\xi, \eta, u, \frac{\partial u}{\partial \xi}, \frac{\partial u}{\partial \eta}\right), \quad w = W\left(\xi, \eta, u, \frac{\partial u}{\partial \xi}, \frac{\partial u}{\partial \eta}\right), \tag{15.2.3.1}$$

where $u = u(\xi, \eta)$. The functions $X$, $Y$, and $W$ in (15.2.3.1) cannot be arbitrary and are selected so as to ensure that the first derivatives of the original variables depend only on the transformed variables and, possibly, their first derivatives,

$$\frac{\partial w}{\partial x} = U\left(\xi, \eta, u, \frac{\partial u}{\partial \xi}, \frac{\partial u}{\partial \eta}\right), \quad \frac{\partial w}{\partial y} = V\left(\xi, \eta, u, \frac{\partial u}{\partial \xi}, \frac{\partial u}{\partial \eta}\right). \tag{15.2.3.2}$$

Contact transformations (15.2.3.1)–(15.2.3.2) do not increase the order of the equations to which they are applied.

In general, a contact transformation (15.2.3.1)–(15.2.3.2) reduces a second-order equation in two independent variables

$$F\left(x, y, w, \frac{\partial w}{\partial x}, \frac{\partial w}{\partial y}, \frac{\partial^2 w}{\partial x^2}, \frac{\partial^2 w}{\partial x \partial y}, \frac{\partial^2 w}{\partial y^2}\right) = 0 \tag{15.2.3.3}$$

to an equation of the form

$$G\left(\xi, \eta, u, \frac{\partial u}{\partial \xi}, \frac{\partial u}{\partial \eta}, \frac{\partial^2 u}{\partial \xi^2}, \frac{\partial^2 u}{\partial \xi \partial \eta}, \frac{\partial^2 u}{\partial \eta^2}\right) = 0. \tag{15.2.3.4}$$

In some cases, equation (15.2.3.4) turns out to be more simple than (15.2.3.3). If $u = u(\xi, \eta)$ is a solution of equation (15.2.3.4), then formulas (15.2.3.1) define the corresponding solution of equation (15.2.3.3) in parametric form.

*Remark.* It is significant that the contact transformations are defined regardless of the specific equations.

**15.2.3-2. Legendre transformation.**

An important special case of contact transformations is the *Legendre transformation* defined by the relations

$$x = \frac{\partial u}{\partial \xi}, \quad y = \frac{\partial u}{\partial \eta}, \quad w = x\xi + y\eta - u, \tag{15.2.3.5}$$

where $\xi$ and $\eta$ are the new independent variables, and $u = u(\xi, \eta)$ is the new dependent variable.

Differentiating the last relation in (15.2.3.5) with respect to $x$ and $y$ and taking into account the other two relations, we obtain the first derivatives:

$$\frac{\partial w}{\partial x} = \xi, \quad \frac{\partial w}{\partial y} = \eta. \tag{15.2.3.6}$$

---

* Prior to reading this section, it is recommended that Subsection 12.1.8 be read first.

With (15.2.3.5)–(15.2.3.6), we find the second derivatives

$$\frac{\partial^2 w}{\partial x^2} = J\frac{\partial^2 u}{\partial \eta^2}, \quad \frac{\partial^2 w}{\partial x \partial y} = -J\frac{\partial^2 u}{\partial \xi \partial \eta}, \quad \frac{\partial^2 w}{\partial y^2} = J\frac{\partial^2 u}{\partial \xi^2},$$

where

$$J = \frac{\partial^2 w}{\partial x^2}\frac{\partial^2 w}{\partial y^2} - \left(\frac{\partial^2 w}{\partial x \partial y}\right)^2, \quad \frac{1}{J} = \frac{\partial^2 u}{\partial \xi^2}\frac{\partial^2 u}{\partial \eta^2} - \left(\frac{\partial^2 u}{\partial \xi \partial \eta}\right)^2.$$

The Legendre transformation (15.2.3.5), with $J \neq 0$, allows us to rewrite a general second-order equation with two independent variables

$$F\left(x, y, w, \frac{\partial w}{\partial x}, \frac{\partial w}{\partial y}, \frac{\partial^2 w}{\partial x^2}, \frac{\partial^2 w}{\partial x \partial y}, \frac{\partial^2 w}{\partial y^2}\right) = 0 \qquad (15.2.3.7)$$

in the form

$$F\left(\frac{\partial u}{\partial \xi}, \frac{\partial u}{\partial \eta}, \xi\frac{\partial u}{\partial \xi} + \eta\frac{\partial u}{\partial \eta} - u, \xi, \eta, J\frac{\partial^2 u}{\partial \eta^2}, -J\frac{\partial^2 u}{\partial \xi \partial \eta}, J\frac{\partial^2 u}{\partial \xi^2}\right) = 0. \qquad (15.2.3.8)$$

Sometimes equation (15.2.3.8) may be simpler than (15.2.3.7).

Let $u = u(\xi, \eta)$ be a solution of equation (15.2.3.8). Then the formulas (15.2.3.5) define the corresponding solution of equation (15.2.3.7) in parametric form.

**Remark.** The Legendre transformation may result in the loss of solutions for which $J = 0$.

**Example 1.** The equation of steady-state transonic gas flow

$$a\frac{\partial w}{\partial x}\frac{\partial^2 w}{\partial x^2} + \frac{\partial^2 w}{\partial y^2} = 0$$

is reduced by the Legendre transformation (15.2.3.5) to the linear equation with variable coefficients

$$a\xi\frac{\partial^2 u}{\partial \eta^2} + \frac{\partial^2 u}{\partial \xi^2} = 0.$$

**Example 2.** The Legendre transformation (15.2.3.5) reduces the nonlinear equation

$$f\left(\frac{\partial w}{\partial x}, \frac{\partial w}{\partial y}\right)\frac{\partial^2 w}{\partial x^2} + g\left(\frac{\partial w}{\partial x}, \frac{\partial w}{\partial y}\right)\frac{\partial^2 w}{\partial x \partial y} + h\left(\frac{\partial w}{\partial x}, \frac{\partial w}{\partial y}\right)\frac{\partial^2 w}{\partial y^2} = 0$$

to the following linear equation with variable coefficients:

$$f(\xi, \eta)\frac{\partial^2 u}{\partial \eta^2} - g(\xi, \eta)\frac{\partial^2 u}{\partial \xi \partial \eta} + h(\xi, \eta)\frac{\partial^2 u}{\partial \xi^2} = 0.$$

### 15.2.3-3. Euler transformation.

The *Euler transformation* belongs to the class of contact transformations and is defined by the relations

$$x = \frac{\partial u}{\partial \xi}, \quad y = \eta, \quad w = x\xi - u. \qquad (15.2.3.9)$$

Differentiating the last relation in (15.2.3.9) with respect to $x$ and $y$ and taking into account the other two relations, we find that

$$\frac{\partial w}{\partial x} = \xi, \quad \frac{\partial w}{\partial y} = -\frac{\partial u}{\partial \eta}. \qquad (15.2.3.10)$$

Differentiating these expressions in $x$ and $y$, we find the second derivatives:

$$w_{xx} = \frac{1}{u_{\xi\xi}}, \quad w_{xy} = -\frac{u_{\xi\eta}}{u_{\xi\xi}}, \quad w_{yy} = \frac{u_{\xi\eta}^2 - u_{\xi\xi}u_{\eta\eta}}{u_{\xi\xi}}. \tag{15.2.3.11}$$

The subscripts indicate the corresponding partial derivatives.

The Euler transformation (15.2.3.9) is employed in finding solutions and linearization of certain nonlinear partial differential equations.

The Euler transformation (15.2.3.9) allows us to reduce a general second-order equation with two independent variables

$$F\left(x, y, w, \frac{\partial w}{\partial x}, \frac{\partial w}{\partial y}, \frac{\partial^2 w}{\partial x^2}, \frac{\partial^2 w}{\partial x \partial y}, \frac{\partial^2 w}{\partial y^2}\right) = 0 \tag{15.2.3.12}$$

to the equation

$$F\left(u_\xi, \eta, \xi u_\xi - u, \xi, -u_\eta, \frac{1}{u_{\xi\xi}}, -\frac{u_{\xi\eta}}{u_{\xi\xi}}, \frac{u_{\xi\eta}^2 - u_{\xi\xi}u_{\eta\eta}}{u_{\xi\xi}}\right) = 0. \tag{15.2.3.13}$$

In some cases, equation (15.2.3.13) may become simpler than equation (15.2.3.12).

Let $u = u(\xi, \eta)$ be a solution of equation (15.2.3.13). Then formulas (15.2.3.9) define the corresponding solution of equation (15.2.3.12) in parametric form.

**Remark.** The Euler transformation may result in the loss of solutions for which $w_{xx} = 0$.

**Example 3.** The nonlinear equation

$$\frac{\partial w}{\partial y} \frac{\partial^2 w}{\partial x^2} + a = 0$$

is reduced by the Euler transformation (15.2.3.9)–(15.2.3.11) to the linear heat equation

$$\frac{\partial u}{\partial \eta} = a \frac{\partial^2 u}{\partial \xi^2}.$$

**Example 4.** The nonlinear equation

$$\frac{\partial^2 w}{\partial x \partial y} = a \frac{\partial w}{\partial y} \frac{\partial^2 w}{\partial x^2} \tag{15.2.3.14}$$

can be linearized with the help of the Euler transformation (15.2.3.9)–(15.2.3.11) to obtain

$$\frac{\partial^2 u}{\partial \xi \partial \eta} = a \frac{\partial u}{\partial \eta}.$$

Integrating this equation yields the general solution

$$u = f(\xi) + g(\eta)e^{a\xi}, \tag{15.2.3.15}$$

where $f(\xi)$ and $g(\eta)$ are arbitrary functions.

Using (15.2.3.9) and (15.2.3.15), we obtain the general solution of the original equation (15.2.3.14) in parametric form:

$$w = x\xi - f(\xi) - g(y)e^{a\xi},$$
$$x = f'_\xi(\xi) + ag(y)e^{a\xi}.$$

**Remark.** In the degenerate case $a = 0$, the solution $w = \varphi(y)x + \psi(y)$ is lost, where $\varphi(y)$ and $\psi(y)$ are arbitrary functions; see also the previous remark.

**15.2.3-4. Legendre transformation with many variables.**

For a function of many variables $w = w(x_1, \ldots, x_n)$, the Legendre transformation and its inverse are defined as

*Legendre transformation*
$$x_1 = X_1, \quad \ldots, \quad x_{k-1} = X_{k-1},$$
$$x_k = \frac{\partial W}{\partial X_k}, \quad \ldots, \quad x_n = \frac{\partial W}{\partial X_n},$$
$$w(\mathbf{x}) = \sum_{i=k}^{n} X_i \frac{\partial W}{\partial X_i} - W(\mathbf{X}),$$

*Inverse Legendre transformation*
$$X_1 = x_1, \quad \ldots, \quad X_{k-1} = x_{k-1},$$
$$X_k = \frac{\partial w}{\partial x_k}, \quad \ldots, \quad X_n = \frac{\partial w}{\partial x_n},$$
$$W(\mathbf{X}) = \sum_{i=k}^{n} x_i \frac{\partial w}{\partial x_i} - w(\mathbf{x}),$$

where $\mathbf{x} = \{x_1, \ldots, x_n\}$, $\mathbf{X} = \{X_1, \ldots, X_n\}$, and the derivatives are related by

$$\frac{\partial w}{\partial x_1} = -\frac{\partial W}{\partial X_1}, \quad \ldots, \quad \frac{\partial w}{\partial x_{k-1}} = -\frac{\partial W}{\partial X_{k-1}}.$$

## 15.2.4. Bäcklund Transformations. Differential Substitutions

**15.2.4-1. Bäcklund transformations for second-order equations.**

Let $w = w(x, y)$ be a solution of the equation

$$F_1\left(x, y, w, \frac{\partial w}{\partial x}, \frac{\partial w}{\partial y}, \frac{\partial^2 w}{\partial x^2}, \frac{\partial^2 w}{\partial x \partial y}, \frac{\partial^2 w}{\partial y^2}\right) = 0, \tag{15.2.4.1}$$

and let $u = u(x, y)$ be a solution of another equation

$$F_2\left(x, y, u, \frac{\partial u}{\partial x}, \frac{\partial u}{\partial y}, \frac{\partial^2 u}{\partial x^2}, \frac{\partial^2 u}{\partial x \partial y}, \frac{\partial^2 u}{\partial y^2}\right) = 0. \tag{15.2.4.2}$$

Equations (15.2.4.1) and (15.2.4.2) are said to be related by the *Bäcklund transformation*

$$\begin{aligned}\Phi_1\left(x, y, w, \frac{\partial w}{\partial x}, \frac{\partial w}{\partial y}, u, \frac{\partial u}{\partial x}, \frac{\partial u}{\partial y}\right) &= 0, \\ \Phi_2\left(x, y, w, \frac{\partial w}{\partial x}, \frac{\partial w}{\partial y}, u, \frac{\partial u}{\partial x}, \frac{\partial u}{\partial y}\right) &= 0\end{aligned} \tag{15.2.4.3}$$

if the compatibility of the pair (15.2.4.1), (15.2.4.3) implies equation (15.2.4.2), and the compatibility of the pair (15.2.4.2), (15.2.4.3) implies (15.2.4.1). If, for some specific solution $u = u(x, y)$ of equation (15.2.4.2), one succeeds in solving equations (15.2.4.3) for $w = w(x, y)$, then this function $w = w(x, y)$ will be a solution of equation (15.2.4.1).

Bäcklund transformations may preserve the form of equations* (such transformations are used for obtaining new solutions) or establish relations between solutions of different equations (such transformations are used for obtaining solutions of one equation from solutions of another equation).

---

* In such cases, these are referred to as *auto-Bäcklund transformations*.

**Example 1.** The *Burgers equation*

$$\frac{\partial w}{\partial t} = w \frac{\partial w}{\partial x} + \frac{\partial^2 w}{\partial x^2} \tag{15.2.4.4}$$

is related to the linear heat equation

$$\frac{\partial u}{\partial t} = \frac{\partial^2 u}{\partial x^2} \tag{15.2.4.5}$$

by the Bäcklund transformation

$$\frac{\partial u}{\partial x} - \frac{1}{2} uw = 0,$$
$$\frac{\partial u}{\partial t} - \frac{1}{2} \frac{\partial (uw)}{\partial x} = 0. \tag{15.2.4.6}$$

Eliminating $w$ from (15.2.4.6), we obtain equation (15.2.4.5).

Conversely, let $u(x,t)$ be a nonzero solution of the heat equation (15.2.4.5). Dividing (15.2.4.5) by $u$, differentiating the resulting equation with respect to $x$, and taking into account that $(u_t/u)_x = (u_x/u)_t$, we obtain

$$\left(\frac{u_x}{u}\right)_t = \left(\frac{u_{xx}}{u}\right)_x. \tag{15.2.4.7}$$

From the first equation in (15.2.4.6) we have

$$\frac{u_x}{u} = \frac{w}{2} \quad \Longrightarrow \quad \frac{u_{xx}}{u} - \left(\frac{u_x}{u}\right)^2 = \frac{w_x}{2} \quad \Longrightarrow \quad \frac{u_{xx}}{u} = \frac{w_x}{2} + \frac{1}{4} w^2. \tag{15.2.4.8}$$

Replacing the expressions in parentheses in (15.2.4.7) with the right-hand sides of the first and the last relation (15.2.4.8), we obtain the Burgers equation (15.2.4.4).

**Example 2.** Let us demonstrate that *Liouville's equation*

$$\frac{\partial^2 w}{\partial x \partial y} = e^{\lambda w} \tag{15.2.4.9}$$

is connected with the linear wave equation

$$\frac{\partial^2 u}{\partial x \partial y} = 0 \tag{15.2.4.10}$$

by the Bäcklund transformation

$$\frac{\partial u}{\partial x} - \frac{\partial w}{\partial x} = \frac{2k}{\lambda} \exp\left[\frac{1}{2}\lambda(w+u)\right],$$
$$\frac{\partial u}{\partial y} + \frac{\partial w}{\partial y} = -\frac{1}{k} \exp\left[\frac{1}{2}\lambda(w-u)\right], \tag{15.2.4.11}$$

where $k \neq 0$ is an arbitrary constant.

Indeed, let us differentiate the first relation of (15.2.4.11) with respect to $y$ and the second equation with respect to $x$. Then, taking into account that $u_{yx} = u_{xy}$ and $w_{yx} = w_{xy}$ and eliminating the combinations of the first derivatives using (15.2.4.11), we obtain

$$\frac{\partial^2 u}{\partial x \partial y} - \frac{\partial^2 w}{\partial x \partial y} = k \exp\left[\frac{1}{2}\lambda(w+u)\right]\left(\frac{\partial u}{\partial y} + \frac{\partial w}{\partial y}\right) = -\exp(\lambda w),$$
$$\frac{\partial^2 u}{\partial x \partial y} + \frac{\partial^2 w}{\partial x \partial y} = \frac{\lambda}{2k} \exp\left[\frac{1}{2}\lambda(w-u)\right]\left(\frac{\partial u}{\partial x} - \frac{\partial w}{\partial x}\right) = \exp(\lambda w). \tag{15.2.4.12}$$

Adding relations (15.2.4.12) together, we get the linear equation (15.2.4.10). Subtracting the latter equation from the former gives the nonlinear equation (15.2.4.9).

**Example 3.** The nonlinear heat equation with a exponential source

$$w_{xx} + w_{yy} = ae^{\beta w}$$

is connected with the Laplace equation

$$u_{xx} + u_{yy} = 0$$

by the Bäcklund transformation

$$u_x + \tfrac{1}{2}\beta w_y = \left(\tfrac{1}{2}a\beta\right)^{1/2} \exp\left(\tfrac{1}{2}\beta w\right) \sin u,$$
$$u_y - \tfrac{1}{2}\beta w_x = \left(\tfrac{1}{2}a\beta\right)^{1/2} \exp\left(\tfrac{1}{2}\beta w\right) \cos u.$$

This fact can be proved in a similar way as in Example 2.

Remark 1. It is significant that unlike the contact transformations, the Bäcklund transformations are determined by the specific equations (a Bäcklund transformation that relates given equations does not always exist).

Remark 2. For two $n$th-order evolution equations of the forms

$$\frac{\partial w}{\partial t} = F_1\left(x, w, \frac{\partial w}{\partial x}, \ldots, \frac{\partial^n w}{\partial x^n}\right),$$
$$\frac{\partial u}{\partial t} = F_2\left(x, u, \frac{\partial u}{\partial x}, \ldots, \frac{\partial^n u}{\partial x^n}\right),$$

a Bäcklund transformation is sometimes sought in the form

$$\Phi\left(x, w, \frac{\partial w}{\partial x}, \ldots, \frac{\partial^m w}{\partial x^m}, u, \frac{\partial u}{\partial x}, \ldots, \frac{\partial^k u}{\partial x^k}\right) = 0$$

containing derivatives in only one variable $x$ (the second variable, $t$, is present implicitly through the functions $w, u$). This transformation can be regarded as an ordinary differential equation in one of the dependent variables.

### 15.2.4-2. Nonlocal transformations based on conservation laws.

Consider a differential equation written as a conservation law,

$$\frac{\partial}{\partial x}\left[F\left(w, \frac{\partial w}{\partial x}, \frac{\partial w}{\partial y}, \ldots\right)\right] + \frac{\partial}{\partial y}\left[G\left(w, \frac{\partial w}{\partial x}, \frac{\partial w}{\partial y}, \ldots\right)\right] = 0. \quad (15.2.4.13)$$

The transformation

$$dz = F(w, w_x, w_y, \ldots)\, dy - G(w, w_x, w_y, \ldots)\, dx, \quad d\eta = dy \quad (15.2.4.14)$$
$$\left(dz = \frac{\partial z}{\partial x} dx + \frac{\partial z}{\partial y} dy \;\Longrightarrow\; \frac{\partial z}{\partial x} = -G, \; \frac{\partial z}{\partial y} = F\right)$$

determines the passage from the variables $x$ and $y$ to the new independent variables $z$ and $\eta$ according to the rule

$$\frac{\partial}{\partial x} = -G\frac{\partial}{\partial z}, \quad \frac{\partial}{\partial y} = \frac{\partial}{\partial \eta} + F\frac{\partial}{\partial z}.$$

Here, $F$ and $G$ are the same as in (15.2.4.13). The transformation (15.2.4.14) preserves the order of the equation under consideration.

Remark. Often one may encounter transformations (15.2.4.14) that are supplemented with a transformation of the unknown function in the form $u = \varphi(w)$.

Example 4. Consider the nonlinear heat equation

$$\frac{\partial w}{\partial t} = \frac{\partial}{\partial x}\left[f(w)\frac{\partial w}{\partial x}\right], \quad (15.2.4.15)$$

which represents a special case of equation (15.2.4.13) for $y = t$, $F = f(w)w_x$, and $G = -w$.

In this case, transformation (15.2.4.14) has the form
$$dz = w\,dx + [f(w)w_x]\,dt, \quad d\eta = dt \qquad (15.2.4.16)$$
and determines a transformation from the variables $x$ and $y$ to the new independent variables $z$ and $\eta$ according to the rule
$$\frac{\partial}{\partial x} = w\frac{\partial}{\partial z}, \quad \frac{\partial}{\partial t} = \frac{\partial}{\partial \eta} + [f(w)w_x]\frac{\partial}{\partial z}.$$
Applying transformation (15.2.4.16) to equation (15.2.4.15), we obtain
$$\frac{\partial w}{\partial \eta} = w^2 \frac{\partial}{\partial z}\left[f(w)\frac{\partial w}{\partial z}\right]. \qquad (15.2.4.17)$$
The substitution $w = 1/u$ reduces (15.2.4.17) to an equation of the form (15.2.4.15),
$$\frac{\partial u}{\partial \eta} = \frac{\partial}{\partial z}\left[\frac{1}{u^2}f\!\left(\frac{1}{u}\right)\frac{\partial u}{\partial z}\right].$$

In the special case of $f(w) = aw^{-2}$, the nonlinear equation (15.2.4.15) is reduced to the linear equation $u_\eta = au_{zz}$ by the transformation (15.2.4.16).

## 15.2.5. Differential Substitutions

In mathematical physics, apart from the Bäcklund transformations, one sometimes resorts to the so-called *differential substitutions*. For second-order differential equations, differential substitutions have the form
$$w = \Psi\!\left(x, y, u, \frac{\partial u}{\partial x}, \frac{\partial u}{\partial y}\right).$$

A differential substitution increases the order of an equation (if it is inserted into an equation for $w$) and allows us to obtain solutions of one equation from those of another. The relationship between the solutions of the two equations is generally not invertible and is, in a sense, unilateral. A differential substitution may decrease the order of an equation (if it is inserted into an equation for $u$). A differential substitution may be obtained as a consequence of a Bäcklund transformation (although this is not always the case). A differential substitution may decrease the order of an equation (when the equation for $u$ is regarded as the original one).

In general, differential substitutions are defined by formulas (15.2.3.1), where the function $X, Y$, and $W$ can be defined arbitrarily.

**Example 1.** Consider once again the Burgers equation (15.2.4.4). The first relation in (15.2.4.6) can be rewritten as the differential substitution (the *Hopf–Cole transformation*)
$$w = \frac{2u_x}{u}. \qquad (15.2.5.1)$$
Substituting (15.2.5.1) into (15.2.4.4), we obtain the equation
$$\frac{2u_{tx}}{u} - \frac{2u_t u_x}{u^2} = \frac{2u_{xxx}}{u} - \frac{2u_x u_{xx}}{u^2},$$
which can be converted to
$$\frac{\partial}{\partial x}\left[\frac{1}{u}\left(\frac{\partial u}{\partial t} - \frac{\partial^2 u}{\partial x^2}\right)\right] = 0. \qquad (15.2.5.2)$$
Thus, using formula (15.2.5.1), one can transform each solution of the linear heat equation (15.2.4.5) into a solution of the Burgers equation (15.2.4.4). The converse is not generally true. Indeed, it follows from (15.2.5.2) that a solution of equation (15.2.4.4) generates a solution of the more general equation
$$\frac{\partial u}{\partial t} - \frac{\partial^2 u}{\partial x^2} = f(t)u,$$
where $f(t)$ is a function of $t$.

**Example 2.** The equation of a steady-state laminar hydrodynamic boundary layer at a flat plate has the form (see Schlichting, 1981)

$$\frac{\partial w}{\partial y}\frac{\partial^2 w}{\partial x \partial y} - \frac{\partial w}{\partial x}\frac{\partial^2 w}{\partial y^2} = a\frac{\partial^3 w}{\partial y^3}, \tag{15.2.5.3}$$

where $w$ is the stream function, $x$ and $y$ are the coordinates along and across the flow, and $a$ is the kinematic viscosity of the fluid.

The *von Mises transformation* (a differential substitution)

$$\xi = x, \quad \eta = w, \quad u(\xi, \eta) = \frac{\partial w}{\partial y}, \quad \text{where} \quad w = w(x, y), \tag{15.2.5.4}$$

decreases the order of equation (15.2.5.3) and brings it to the simpler nonlinear heat equation

$$\frac{\partial u}{\partial \xi} = a\frac{\partial}{\partial \eta}\left(u\frac{\partial u}{\partial \eta}\right). \tag{15.2.5.5}$$

When deriving equation (15.2.5.5), the following formulas for the computation of the derivatives have been used:

$$\frac{\partial}{\partial y} = u\frac{\partial}{\partial \eta}, \quad \frac{\partial}{\partial x} = \frac{\partial}{\partial \xi} + \frac{\partial w}{\partial x}\frac{\partial}{\partial \eta}, \quad \frac{\partial w}{\partial y} = u, \quad \frac{\partial^2 w}{\partial y^2} = u\frac{\partial u}{\partial \eta},$$

$$\frac{\partial w}{\partial y}\frac{\partial^2 w}{\partial x \partial y} - \frac{\partial w}{\partial x}\frac{\partial^2 w}{\partial y^2} = u\frac{\partial u}{\partial \xi}, \quad \frac{\partial^3 w}{\partial y^3} = u\frac{\partial}{\partial \eta}\left(u\frac{\partial u}{\partial \eta}\right).$$

## 15.3. Traveling-Wave Solutions, Self-Similar Solutions, and Some Other Simple Solutions. Similarity Method

### 15.3.1. Preliminary Remarks

There are a number of methods for the construction of exact solutions to equations of mathematical physics that are based on the reduction of the original equations to equations in fewer dependent and/or independent variables. The main idea is to find such variables and, by passing to them, to obtain simpler equations. In particular, in this way, finding exact solutions of some partial differential equations in two independent variables may be reduced to finding solutions of appropriate ordinary differential equations (or systems of ordinary differential equations). Naturally, the ordinary differential equations thus obtained do not give all solutions of the original partial differential equation, but provide only a class of solutions with some specific properties.

The simplest classes of exact solutions described by ordinary differential equations involve traveling-wave solutions and self-similar solutions. The existence of such solutions is usually due to the invariance of the equations in question under translations and scaling transformations.

Traveling-wave solutions and self-similar solutions often occur in various applications. Below we consider some characteristic features of such solutions.

It is assumed that the unknown $w$ depends on two variables, $x$ and $t$, where $t$ plays the role of time and $x$ is a spatial coordinate.

### 15.3.2. Traveling-Wave Solutions. Invariance of Equations Under Translations

**15.3.2-1. General form of traveling-wave solutions.**

*Traveling-wave solutions*, by definition, are of the form

$$w(x, t) = W(z), \quad z = kx - \lambda t, \tag{15.3.2.1}$$

where $\lambda/k$ plays the role of the wave propagation velocity (the sign of $\lambda$ can be arbitrary, the value $\lambda = 0$ corresponds to a stationary solution, and the value $k = 0$ corresponds to a space-homogeneous solution). Traveling-wave solutions are characterized by the fact that the profiles of these solutions at different time* instants are obtained from one another by appropriate shifts (translations) along the $x$-axis. Consequently, a Cartesian coordinate system moving with a constant speed can be introduced in which the profile of the desired quantity is stationary. For $k > 0$ and $\lambda > 0$, the wave (15.3.2.1) travels along the $x$-axis to the right (in the direction of increasing $x$).

A traveling-wave solution is found by directly substituting the representation (15.3.2.1) into the original equation and taking into account the relations $w_x = kW'$, $w_t = -\lambda W'$, etc. (the prime denotes a derivative with respect to $z$).

Traveling-wave solutions occur for equations that do not explicitly involve independent variables,

$$F\left(w, \frac{\partial w}{\partial x}, \frac{\partial w}{\partial t}, \frac{\partial^2 w}{\partial x^2}, \frac{\partial^2 w}{\partial x \partial t}, \frac{\partial^2 w}{\partial t^2}, \dots\right) = 0. \tag{15.3.2.2}$$

Substituting (15.3.2.1) into (15.3.2.2), we obtain an autonomous ordinary differential equation for the function $W(z)$:

$$F(W, kW', -\lambda W', k^2 W'', -k\lambda W'', \lambda^2 W'', \dots) = 0,$$

where $k$ and $\lambda$ are arbitrary constants.

**Example 1.** The nonlinear heat equation

$$\frac{\partial w}{\partial t} = \frac{\partial}{\partial x}\left[f(w)\frac{\partial w}{\partial x}\right] \tag{15.3.2.3}$$

admits a traveling-wave solution. Substituting (15.3.2.1) into (15.3.2.3), we arrive at the ordinary differential equation

$$k^2[f(W)W']' + \lambda W' = 0.$$

Integrating this equation twice yields its solution in implicit form:

$$k^2 \int \frac{f(W)\,dW}{\lambda W + C_1} = -z + C_2,$$

where $C_1$ and $C_2$ are arbitrary constants.

**Example 2.** Consider the homogeneous Monge–Ampère equation

$$\left(\frac{\partial^2 w}{\partial x \partial t}\right)^2 - \frac{\partial^2 w}{\partial x^2}\frac{\partial^2 w}{\partial t^2} = 0. \tag{15.3.2.4}$$

Inserting (15.3.2.1) into this equation, we obtain an identity. Therefore, equation (15.3.2.4) admits solutions of the form

$$w = W(kx - \lambda t),$$

where $W(z)$ is an arbitrary function and $k$ and $\lambda$ are arbitrary constants.

### 15.3.2-2. Invariance of solutions and equations under translation transformations.

Traveling-wave solutions (15.3.2.1) are invariant under the translation transformations

$$x = \bar{x} + C\lambda, \quad t = \bar{t} + Ck, \tag{15.3.2.5}$$

where $C$ is an arbitrary constant.

---

* We also use the term *traveling-wave solution* in the cases where the variable $t$ plays the role of a spatial coordinate.

It should be observed that equations of the form (15.3.2.2) are invariant (i.e., preserve their form) under transformation (15.3.2.5); furthermore, these equations are also invariant under general translations in both independent variables:

$$x = \bar{x} + C_1, \quad t = \bar{t} + C_2, \tag{15.3.2.6}$$

where $C_1$ and $C_2$ are arbitrary constants. The property of the invariance of specific equations under translation transformations (15.3.2.5) or (15.3.2.6) is inseparably linked with the existence of traveling-wave solutions to such equations (the former implies the latter).

**Remark 1.** Traveling-wave solutions, which stem from the invariance of equations under translations, are simplest *invariant solutions*.

**Remark 2.** The condition of invariance of equations under translations is not a necessary condition for the existence of traveling-wave solutions. It can be verified directly that the second-order equation

$$F(w, w_x, w_t, xw_x + tw_t, w_{xx}, w_{xt}, w_{tt}) = 0$$

does not admit transformations of the form (15.3.2.5) and (15.3.2.6) but has an exact traveling-wave solution (15.3.2.1) described by the ordinary differential equation

$$F(W, kW', -\lambda W', zW', k^2 W'', -k\lambda W'', \lambda^2 W'') = 0.$$

**15.3.2-3. Functional equation describing traveling-wave solutions.**

Let us demonstrate that traveling-wave solutions can be defined as solutions of the functional equation

$$w(x, t) = w(x + C\lambda, t + Ck), \tag{15.3.2.7}$$

where $k$ and $\lambda$ are some constants and $C$ is an arbitrary constant. Equation (15.3.2.7) states that the unknown function does not change under increasing both arguments by proportional quantities, with $C$ being the coefficient of proportionality.

For $C = 0$, equation (15.3.2.7) turns into an identity. Let us expand (15.3.2.7) into a series in powers of $C$ about $C = 0$, then divide the resulting expression by $C$, and proceed to the limit as $C \to 0$ to obtain the linear first-order partial differential equation

$$\lambda \frac{\partial w}{\partial x} + k \frac{\partial w}{\partial t} = 0.$$

The general solution to this equation is constructed by the method of characteristics (see Paragraph 13.1.1-1) and has the form (15.3.2.1), which was to be proved.

## 15.3.3. Self-Similar Solutions. Invariance of Equations Under Scaling Transformations

**15.3.3-1. General form of self-similar solutions. Similarity method.**

By definition, a *self-similar solution* is a solution of the form

$$w(x, t) = t^\alpha U(\zeta), \quad \zeta = xt^\beta. \tag{15.3.3.1}$$

The profiles of these solutions at different time instants are obtained from one another by a similarity transformation (like scaling).

Self-similar solutions exist if the scaling of the independent and dependent variables,

$$t = C\bar{t}, \quad x = C^k\bar{x}, \quad w = C^m\bar{w}, \quad \text{where } C \neq 0 \text{ is an arbitrary constant,} \quad (15.3.3.2)$$

for some $k$ and $m$ ($|k| + |m| \neq 0$), is equivalent to the identical transformation. This means that the original equation

$$F(x, t, w, w_x, w_t, w_{xx}, w_{xt}, w_{tt}, \ldots) = 0, \quad (15.3.3.3)$$

when subjected to transformation (15.3.3.2), turns into the same equation in the new variables,

$$F(\bar{x}, \bar{t}, \bar{w}, \bar{w}_{\bar{x}}, \bar{w}_{\bar{t}}, \bar{w}_{\bar{x}\bar{x}}, \bar{w}_{\bar{x}\bar{t}}, \bar{w}_{\bar{t}\bar{t}}, \ldots) = 0. \quad (15.3.3.4)$$

Here, the function $F$ is the same as in the original equation (15.3.3.3); it is assumed that equation (15.3.3.3) is independent of the parameter $C$.

Let us find the connection between the parameters $\alpha$, $\beta$ in solution (15.3.3.1) and the parameters $k$, $m$ in the scaling transformation (15.3.3.2). Suppose

$$w = \Phi(x, t) \quad (15.3.3.5)$$

is a solution of equation (15.3.3.3). Then the function

$$\bar{w} = \Phi(\bar{x}, \bar{t}) \quad (15.3.3.6)$$

is a solution of equation (15.3.3.4).

In view of the explicit form of solution (15.3.3.1), if follows from (15.3.3.6) that

$$\bar{w} = \bar{t}^\alpha U(\bar{x}\bar{t}^\beta). \quad (15.3.3.7)$$

Using (15.3.3.2) to return to the new variables in (15.3.3.7), we get

$$w = C^{m-\alpha} t^\alpha U(C^{-k-\beta} x t^\beta). \quad (15.3.3.8)$$

By construction, this function satisfies equation (15.3.3.3) and hence is its solution. Let us require that solution (15.3.3.8) coincide with (15.3.3.1), so that the condition for the uniqueness of the solution holds for any $C \neq 0$. To this end, we must set

$$\alpha = m, \quad \beta = -k. \quad (15.3.3.9)$$

In practice, the above existence criterion is checked: if a pair of $k$ and $m$ in (15.3.3.2) has been found, then a self-similar solution is defined by formulas (15.3.3.1) with parameters (15.3.3.9).

The method for constructing self-similar solutions on the basis of scaling transformations (15.3.3.2) is called the *similarity method*. It is significant that these transformations involve the arbitrary constant $C$ as a parameter.

To make easier to understand, Fig. 15.1 depicts the basic stages for constructing self-similar solutions.

## 15.3. Traveling-Wave, Self-Similar, and Other Simple Solutions. Similarity Method

```
┌─────────────────────────────────────────────────────────────────┐
│ Original equation: F(x, t, w, w_x, w_t, w_xx, w_xt, w_tt, ...) = 0 │
└─────────────────────────────────────────────────────────────────┘
                    ↓  Look for a self-similar solution  ↓
┌─────────────────────────────────────────────────────────────────┐
│ Perform transformation: t = C t̄,  x = C^k x̄,  w = C^m w̄        │
└─────────────────────────────────────────────────────────────────┘
                    ↓  Here C is a free parameter     ↓
                       and k, m are some numbers
┌─────────────────────────────────────────────────────────────────┐
│ Select k and m so that the equation preserves its form          │
└─────────────────────────────────────────────────────────────────┘
                    ↓  Hence, we obtain equation:     ↓
                  F(x̄, t̄, w̄, w̄_x̄, w̄_t̄, w̄_x̄x̄, w̄_x̄t̄, w̄_t̄t̄, ...) = 0
┌─────────────────────────────────────────────────────────────────┐
│ Use self-similar solution: w(x, t) = t^m U(ζ),  ζ = x t^{-k}    │
└─────────────────────────────────────────────────────────────────┘
                    ↓  Substitute into the original equation  ↓
┌─────────────────────────────────────────────────────────────────┐
│ Obtain an ordinary differential equation for U(ζ)               │
└─────────────────────────────────────────────────────────────────┘
```

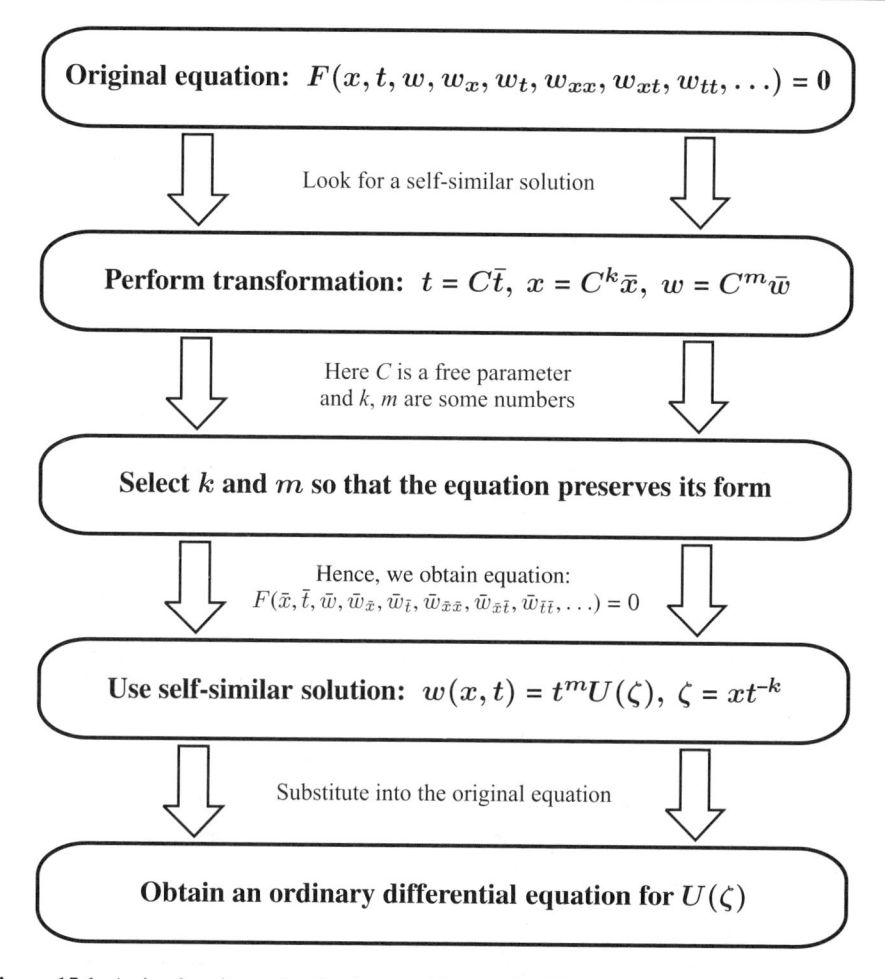

**Figure 15.1.** A simple scheme that is often used in practice for constructing self-similar solutions.

### 15.3.3-2. Examples of self-similar solutions to mathematical physics equations.

**Example 1.** Consider the heat equation with a nonlinear power-law source term

$$\frac{\partial w}{\partial t} = a \frac{\partial^2 w}{\partial x^2} + b w^n. \tag{15.3.3.10}$$

The scaling transformation (15.3.3.2) converts equation (15.3.3.10) into

$$C^{m-1} \frac{\partial \bar{w}}{\partial \bar{t}} = a C^{m-2k} \frac{\partial^2 \bar{w}}{\partial \bar{x}^2} + b C^{mn} \bar{w}^n.$$

Equating the powers of $C$ yields the following system of linear algebraic equations for the constants $k$ and $m$:

$$m - 1 = m - 2k = mn.$$

This system admits a unique solution: $k = \frac{1}{2}$, $m = \frac{1}{1-n}$. Using this solution together with relations (15.3.3.1) and (15.3.3.9), we obtain self-similar variables in the form

$$w = t^{1/(1-n)} U(\zeta), \quad \zeta = x t^{-1/2}.$$

Inserting these into (15.3.3.10), we arrive at the following ordinary differential equation for the function $U(\zeta)$:

$$a U''_{\zeta\zeta} + \frac{1}{2} \zeta U'_\zeta + \frac{1}{n-1} U + b U^n = 0.$$

**Example 2.** Consider the nonlinear equation

$$\frac{\partial^2 w}{\partial t^2} = a\frac{\partial}{\partial x}\left(w^n \frac{\partial w}{\partial x}\right), \tag{15.3.3.11}$$

which occurs in problems of wave and gas dynamics. Inserting (15.3.3.2) into (15.3.3.11) yields

$$C^{m-2}\frac{\partial^2 \bar{w}}{\partial \bar{t}^2} = aC^{mn+m-2k}\frac{\partial}{\partial \bar{x}}\left(\bar{w}^n \frac{\partial \bar{w}}{\partial \bar{x}}\right).$$

Equating the powers of $C$ results in a single linear equation, $m - 2 = mn + m - 2k$. Hence, we obtain $k = \frac{1}{2}mn + 1$, where $m$ is arbitrary. Further, using (15.3.3.1) and (15.3.3.9), we find self-similar variables:

$$w = t^m U(\zeta), \quad \zeta = xt^{-\frac{1}{2}mn-1} \quad (m \text{ is arbitrary}).$$

Substituting these into (15.3.3.11), one obtains an ordinary differential equation for the function $U(\zeta)$.

Table 15.1 gives examples of self-similar solutions to some other nonlinear equations of mathematical physics.

TABLE 15.1
Some nonlinear equations of mathematical physics that admit self-similar solutions

| Equation | Equation name | Form of solutions | Determining equation |
|---|---|---|---|
| $\frac{\partial w}{\partial t} = \frac{\partial}{\partial x}\left[f(w)\frac{\partial w}{\partial x}\right]$ | Unsteady heat equation | $w = w(z),\ z = xt^{-1/2}$ | $[f(w)w']' + \frac{1}{2}zw' = 0$ |
| $\frac{\partial w}{\partial t} = a\frac{\partial}{\partial x}\left(w^n \frac{\partial w}{\partial x}\right) + bw^k$ | Heat equation with source | $w = t^p u(z),\ z = xt^q,$ $p = \frac{1}{1-k},\ q = \frac{k-n-1}{2(1-k)}$ | $a(u^n u')' - qzu'$ $+ bu^k - pu = 0$ |
| $\frac{\partial w}{\partial t} = a\frac{\partial^2 w}{\partial x^2} + bw\frac{\partial w}{\partial x}$ | Burgers equation | $w = t^{-1/2}u(z),\ z = xt^{-1/2}$ | $au'' + buu' + \frac{1}{2}zu' + \frac{1}{2}u = 0$ |
| $\frac{\partial w}{\partial t} = a\frac{\partial^2 w}{\partial x^2} + b\left(\frac{\partial w}{\partial x}\right)^2$ | Potential Burgers equation | $w = w(z),\ z = xt^{-1/2}$ | $aw'' + b(w')^2 + \frac{1}{2}zw' = 0$ |
| $\frac{\partial w}{\partial t} = a\left(\frac{\partial w}{\partial x}\right)^k \frac{\partial^2 w}{\partial x^2}$ | Filtration equation | $w = t^p u(z),\ z = xt^q,$ $p = -\frac{(k+2)q+1}{k},\ q$ is any | $a(u')^k u'' = qzu' + pu$ |
| $\frac{\partial w}{\partial t} = f\left(\frac{\partial w}{\partial x}\right)\frac{\partial^2 w}{\partial x^2}$ | Filtration equation | $w = t^{1/2}u(z),\ z = xt^{-1/2}$ | $2f(u')u'' + zu' - u = 0$ |
| $\frac{\partial^2 w}{\partial t^2} = \frac{\partial}{\partial x}\left[f(w)\frac{\partial w}{\partial x}\right]$ | Wave equation | $w = w(z),\ z = x/t$ | $(z^2 w')' = [f(w)w']'$ |
| $\frac{\partial^2 w}{\partial t^2} = a\frac{\partial}{\partial x}\left(w^n \frac{\partial w}{\partial x}\right)$ | Wave equation | $w = t^{2k}u(z),\ z = xt^{-nk-1},$ $k$ is any | $\frac{2k(2k-1)}{(nk+1)^2}u + \frac{nk-4k+2}{nk+1}zu'$ $+ z^2 u'' = \frac{a}{(nk+1)^2}(u^n u')'$ |
| $\frac{\partial^2 w}{\partial x^2} + \frac{\partial^2 w}{\partial y^2} = aw^n$ | Heat equation with source | $w = x^{\frac{2}{1-n}}u(z),\ z = y/x$ | $(1+z^2)u'' - \frac{2(1+n)}{1-n}zu'$ $+ \frac{2(1+n)}{(1-n)^2}u - au^n = 0$ |
| $\frac{\partial^2 w}{\partial x^2} + a\frac{\partial w}{\partial y}\frac{\partial^2 w}{\partial y^2} = 0$ | Equation of steady transonic gas flow | $w = x^{-3k-2}u(z),\ z = x^k y,$ $k$ is any | $\frac{a}{k+1}u'u'' + \frac{k^2}{k+1}z^2 u''$ $- 5kzu' + 3(3k+2)u = 0$ |
| $\frac{\partial w}{\partial t} = a\frac{\partial^3 w}{\partial x^3} + bw\frac{\partial w}{\partial x}$ | Korteweg–de Vries equation | $w = t^{-2/3}u(z),\ z = xt^{-1/3}$ | $au''' + buu' + \frac{1}{3}zu' + \frac{2}{3}u = 0$ |
| $\frac{\partial w}{\partial y}\frac{\partial^2 w}{\partial x \partial y} - \frac{\partial w}{\partial x}\frac{\partial^2 w}{\partial y^2} = a\frac{\partial^3 w}{\partial y^3}$ | Boundary-layer equation | $w = x^{\lambda+1}u(z),\ z = x^\lambda y,$ $\lambda$ is any | $(2\lambda+1)(u')^2 - (\lambda+1)uu''$ $= au'''$ |

The above method for constructing self-similar solutions is also applicable to systems of partial differential equations. Let us illustrate this by a specific example.

## 15.3. TRAVELING-WAVE, SELF-SIMILAR, AND OTHER SIMPLE SOLUTIONS. SIMILARITY METHOD

**Example 3.** Consider the system of equations of a steady-state laminar boundary hydrodynamic boundary layer at a flat plate (see Schlichting, 1981)

$$u\frac{\partial u}{\partial x} + v\frac{\partial u}{\partial y} = a\frac{\partial^2 u}{\partial y^2},$$
$$\frac{\partial u}{\partial x} + \frac{\partial v}{\partial y} = 0.$$
(15.3.3.12)

Let us scale the independent and dependent variables in (15.3.3.12) according to

$$x = C\bar{x}, \quad y = C^k\bar{y}, \quad u = C^m\bar{u}, \quad v = C^n\bar{v}. \tag{15.3.3.13}$$

Multiplying these relations by appropriate constant factors, we have

$$\bar{u}\frac{\partial \bar{u}}{\partial \bar{x}} + C^{n-m-k+1}\bar{v}\frac{\partial \bar{u}}{\partial \bar{y}} = C^{-m-2k+1}a\frac{\partial^2 \bar{u}}{\partial \bar{y}^2},$$
$$\frac{\partial \bar{u}}{\partial \bar{x}} + C^{n-m-k+1}\frac{\partial \bar{v}}{\partial \bar{y}} = 0.$$
(15.3.3.14)

Let us require that the form of the equations of the transformed system (15.3.3.14) coincide with that of the original system (15.3.3.12). This condition results in two linear algebraic equations, $n - m - k + 1 = 0$ and $-2k - m + 1 = 0$. On solving them for $m$ and $n$, we obtain

$$m = 1 - 2k, \quad n = -k, \tag{15.3.3.15}$$

where the exponent $k$ can be chosen arbitrarily. To find a self-similar solution, let us use the procedure outlined in Fig. 15.1. The following renaming should be done: $x \to y$, $t \to x$, $w \to u$ (for $u$) and $x \to y$, $t \to x$, $w \to v$, $m \to n$ (for $v$). This results in

$$u(x,y) = x^{1-2k}U(\zeta), \quad v(x,y) = x^{-k}V(\zeta), \quad \zeta = yx^{-k}, \tag{15.3.3.16}$$

where $k$ is an arbitrary constant. Inserting (15.3.3.16) into the original system (15.3.3.12), we arrive at a system of ordinary differential equations for $U = U(\zeta)$ and $V = V(\zeta)$:

$$U\big[(1-2k)U - k\zeta U'_\zeta\big] + VU'_\zeta = aU''_{\zeta\zeta},$$
$$(1-2k)U - k\zeta U'_\zeta + V'_\zeta = 0.$$

### 15.3.3-3. More general approach based on solving a functional equation.

The algorithm for the construction of a self-similar solution, presented in Paragraph 15.3.3-1, relies on representing this solution in the form (15.3.3.1) explicitly. However, there is a more general approach that allows the derivation of relation (15.3.3.1) directly from the condition of the invariance of equation (15.3.3.3) under transformations (15.3.3.2).

Indeed, let us assume that transformations (15.3.3.2) convert equation (15.3.3.3) into the same equation (15.3.3.4). Let (15.3.3.5) be a solution of equation (15.3.3.3). Then (15.3.3.6) will be a solution of equation (15.3.3.4). Switching back to the original variables (15.3.3.2) in (15.3.3.6), we obtain

$$w = C^m \Phi\big(C^{-k}x, C^{-1}t\big). \tag{15.3.3.17}$$

By construction, this function satisfies equation (15.3.3.3) and hence is its solution. Let us require that solution (15.3.3.17) coincide with (15.3.3.5), so that the uniqueness condition for the solution is met for any $C \neq 0$. This results in the functional equation

$$\Phi(x,t) = C^m \Phi\big(C^{-k}x, C^{-1}t\big). \tag{15.3.3.18}$$

For $C = 1$, equation (15.3.3.18) is satisfied identically. Let us expand (15.3.3.18) in a power series in $C$ about $C = 1$, then divide the resulting expression by $(C - 1)$, and proceed to the limit as $C \to 1$. This results in a linear first-order partial differential equation for $\Phi$:

$$kx\frac{\partial \Phi}{\partial x} + t\frac{\partial \Phi}{\partial t} - m\Phi = 0. \tag{15.3.3.19}$$

The associated characteristic system of ordinary differential equations (see Paragraph 13.1.1-1) has the form

$$\frac{dx}{kx} = \frac{dt}{t} = \frac{d\Phi}{m\Phi}.$$

Its first integrals are

$$xt^{-k} = A_1, \quad t^{-m}\Phi = A_2,$$

where $A_1$ and $A_2$ are arbitrary constants. The general solution of the partial differential equation (15.3.3.19) is sought in the form $A_2 = U(A_1)$, where $U(A)$ is an arbitrary function (see Paragraph 13.1.1-1). As a result, one obtains a solution of the functional equation (15.3.3.18) in the form

$$\Phi(x,t) = t^m U(\zeta), \quad \zeta = xt^{-k}. \tag{15.3.3.20}$$

Substituting (15.3.3.20) into (15.3.3.5) yields the self-similar solution (15.3.3.1) with parameters (15.3.3.9).

### 15.3.3-4. Some remarks.

**Remark 1.** Self-similar solutions (15.3.3.1) with $\alpha = 0$ arise in problems with simple initial and boundary conditions of the form

$$w = w_1 \quad \text{at} \quad t = 0 \quad (x > 0), \qquad w = w_2 \quad \text{at} \quad x = 0 \quad (t > 0),$$

where $w_1$ and $w_2$ are some constants.

**Remark 2.** Self-similar solutions, which stem from the invariance of equations under scaling transformations, are considered among the simplest *invariant solutions*.

The condition for the existence of a transformation (15.3.3.2) preserving the form of the given equation is sufficient for the existence of a self-similar solution. However, this condition is not necessary: there are equations that do not admit transformations of the form (15.3.3.2) but have self-similar solutions.

For example, the equation

$$a\frac{\partial^2 w}{\partial x^2} + b\frac{\partial^2 w}{\partial t^2} = (bx^2 + at^2)f(w)$$

has a self-similar solution

$$w = w(z), \quad z = xt \quad \Longrightarrow \quad w'' - f(w) = 0,$$

but does not admit transformations of the form (15.3.3.2). In this equation, $a$ and $b$ can be arbitrary functions of the arguments $x, t, w, w_x, w_t, w_{xx}, \ldots$

**Remark 3.** Traveling-wave solutions are closely related to self-similar solutions. Indeed, setting

$$t = \ln \tau, \quad x = \ln y$$

in (15.3.2.1), we obtain a self-similar representation of a traveling wave:

$$w = W\big(k\ln(y\tau^{-\lambda/k})\big) = U(y\tau^{-\lambda/k}),$$

where $U(z) = W(k \ln z)$.

## 15.3.4. Equations Invariant Under Combinations of Translation and Scaling Transformations, and Their Solutions

### 15.3.4-1. Exponential self-similar (limiting self-similar) solutions.

*Exponential self-similar solutions* are solutions of the form

$$w(x,t) = e^{\alpha t} V(\xi), \quad \xi = xe^{\beta t}. \tag{15.3.4.1}$$

Exponential self-similar solutions exist if equation (15.3.3.3) is invariant under transformations of the form

$$t = \bar{t} + \ln C, \quad x = C^k \bar{x}, \quad w = C^m \bar{w}, \tag{15.3.4.2}$$

where $C > 0$ is an arbitrary constant, for some $k$ and $m$. Transformation (15.3.4.2) is a combination of a translation transformation in $t$ and scaling transformations in $x$ and $w$. It should be emphasized that these transformations contain an arbitrary parameter $C$ while the equation concerned is independent of $C$.

Let us find the relation between the parameters $\alpha$, $\beta$ in solution (15.3.4.1) and the parameters $k$, $m$ in the scaling transformation (15.3.4.2). Let $w = \Phi(x, t)$ be a solution of equation (15.3.3.3). Then the function $\bar{w} = \Phi(\bar{x}, \bar{t})$ is a solution of equation (15.3.3.4). In view of the explicit form of solution (15.3.4.1), we have

$$\bar{w} = e^{\alpha \bar{t}} V(\bar{x} e^{\beta \bar{t}}).$$

Going back to the original variables, using (15.3.4.2), we obtain

$$w = C^{m-\alpha} e^{\alpha t} V\left(C^{-k-\beta} x e^{\beta t}\right).$$

Let us require that this solution coincide with (15.3.4.1), which means that the uniqueness condition for the solution must be satisfied for any $C \neq 0$. To this end, we set

$$\alpha = m, \quad \beta = -k. \tag{15.3.4.3}$$

In practice, exponential self-similar solutions are sought using the above existence criterion: if $k$ and $m$ in (15.3.4.2) are known, then the new variables have the form (15.3.4.1) with parameters (15.3.4.3).

**Remark.** Sometimes solutions of the form (15.3.4.1) are also called *limiting self-similar solutions*.

**Example 1.** Let us show that the nonlinear heat equation

$$\frac{\partial w}{\partial t} = a \frac{\partial}{\partial x}\left(w^n \frac{\partial w}{\partial x}\right) \tag{15.3.4.4}$$

admits an exponential self-similar solution. Inserting (15.3.4.2) into (15.3.4.4) yields

$$C^m \frac{\partial \bar{w}}{\partial \bar{t}} = a C^{mn+m-2k} \frac{\partial}{\partial \bar{x}}\left(\bar{w}^n \frac{\partial \bar{w}}{\partial \bar{x}}\right).$$

Equating the powers of $C$ gives one linear equation: $m = mn + m - 2k$. It follows that $k = \frac{1}{2}mn$, where $m$ is any number. Further using formulas (15.3.4.1) and (15.3.4.3) and setting, without loss of generality, $m = 2$ (this is equivalent to scaling in $t$), we find the new variables

$$w = e^{2t} V(\xi), \quad \xi = x e^{-nt}. \tag{15.3.4.5}$$

Substituting these expressions into (15.3.4.4), we arrive at an ordinary differential equation for $V(\xi)$:

$$a(V^n V'_\xi)'_\xi + n\xi V'_\xi - 2V = 0.$$

**Example 2.** With the above approach, it can be shown that the nonlinear wave equation (15.3.3.11) also has an exponential self-similar solution of the form (15.3.4.5).

TABLE 15.2
Invariant solutions that may be obtained using combinations of translation and scaling transformations preserving the form of equations ($C$ is an arbitrary constant, $C > 0$)

| No. | Invariant transformations | Form of invariant solutions | Example of an equation |
|---|---|---|---|
| 1 | $t = \bar{t} + Ck,\ x = \bar{x} + C\lambda$ | $w = U(z),\ z = kx - \lambda t$ | $\frac{\partial w}{\partial t} = \frac{\partial}{\partial x}\left[f(w)\frac{\partial w}{\partial x}\right]$ |
| 2 | $t = C\bar{t},\ x = C^k \bar{x},\ w = C^m \bar{w}$ | $w = t^m U(z),\ z = xt^{-k}$ | see Table 15.1 |
| 3 | $t = \bar{t} + \ln C,\ x = C^k \bar{x},\ w = C^m \bar{w}$ | $w = e^{mt} U(z),\ z = xe^{-kt}$ | equation (15.3.4.4) <br> ($k = \tfrac{1}{2}mn$, $m$ is any) |
| 4 | $t = C\bar{t},\ x = \bar{x} + k\ln C,\ w = C^m \bar{w}$ | $w = t^m U(z),\ z = x - k\ln t$ | equation (15.3.4.4) <br> ($m = -1/n$, $k$ is any) |
| 5 | $t = C\bar{t},\ x = C^\beta \bar{x},\ w = \bar{w} + \alpha \ln C$ | $w = U(z) + \alpha \ln t,\ z = xt^{-\beta}$ | $\frac{\partial w}{\partial t} = \frac{\partial}{\partial x}\left(e^w \frac{\partial w}{\partial x}\right)$ <br> ($\alpha = 2\beta - 1$, $\beta$ is any) |
| 6 | $t = C\bar{t},\ x = \bar{x} + \beta \ln C,\ w = \bar{w} + \alpha \ln C$ | $w = U(z) + \alpha \ln t,\ z = x - \beta \ln t$ | $\left(\frac{\partial^2 w}{\partial x \partial t}\right)^2 - \frac{\partial^2 w}{\partial x^2}\frac{\partial^2 w}{\partial t^2} = 0$ <br> ($\alpha$ and $\beta$ are any) |
| 7 | $t = \bar{t} + C,\ x = \bar{x} + C\lambda,\ w = \bar{w} + Ck$ | $w = U(z) + kt,\ z = x - \lambda t$ | $\frac{\partial w}{\partial t} = f\left(\frac{\partial w}{\partial x}\right)\frac{\partial^2 w}{\partial x^2}$ <br> ($k$ and $\lambda$ are any) |
| 8 | $t = \bar{t} + \ln C,\ x = \bar{x} + k\ln C,\ w = C^m \bar{w}$ | $w = e^{mt} U(z),\ z = x - kt$ | $\left(\frac{\partial^2 w}{\partial x \partial t}\right)^2 - \frac{\partial^2 w}{\partial x^2}\frac{\partial^2 w}{\partial t^2} = 0$ <br> ($k$ and $m$ are any) |

**15.3.4-2. Other solutions obtainable using translation and scaling transformations.**

Table 15.2 lists invariant solutions that may be obtained using combinations of translation and scaling transformations in the independent and dependent variables. The transformations are assumed to preserve the form of equations (the given equation is converted into the same equation). Apart from traveling-wave solutions, self-similar solutions, and exponential self-similar solutions, considered above, another five invariant solutions are presented. The right column of Table 15.2 gives examples of equations that admit the solutions specified.

**Example 3.** Let us show that the nonlinear heat equation (15.3.4.4) admits the solution given in the fourth row of Table 15.2. Perform the transformation
$$t = C\bar{t}, \qquad x = \bar{x} + k\ln C, \qquad w = C^m \bar{w}.$$
This gives
$$C^{m-1}\frac{\partial \bar{w}}{\partial \bar{t}} = aC^{mn+m}\frac{\partial}{\partial \bar{x}}\left(\bar{w}^n \frac{\partial \bar{w}}{\partial \bar{x}}\right).$$
Equating the powers of $C$ results in one linear equation: $m - 1 = mn + m$. It follows that $m = -1/n$, and $k$ is any number. Therefore (see the fourth row in Table 15.2), equation (15.3.4.4) has an invariant solution of the form
$$w = t^{-1/n} U(z), \qquad z = x + k\ln t, \qquad \text{where } k \text{ is any number.} \qquad (15.3.4.6)$$
Substituting (15.3.4.6) into (15.3.4.4) yields the autonomous ordinary differential equation
$$a(U^n U_z')_z' - kU_z' + \frac{1}{n}U = 0.$$
To the special case of $k = 0$ there corresponds a separable equation that results in a solution in the form of the product of functions with different arguments.

The examples considered in Subsections 15.3.2–15.3.4 demonstrate that the existence of exact solutions is due to the fact the partial differential equations concerned are invariant under some transformations (involving one or several parameters) or, what is the same, possess some symmetries. In Section 15.8 below, a general method for the investigation of symmetries of differential equations (the group-theoretic method) will be described that allows finding similar and more complicated invariant solutions on a routine basis.

### 15.3.5. Generalized Self-Similar Solutions

A *generalized self-similar solution* has the form

$$w(x,t) = \varphi(t)u(z), \quad z = \psi(t)x. \tag{15.3.5.1}$$

Formula (15.3.5.1) comprises the above self-similar and exponential self-similar solutions (15.3.3.1) and (15.3.4.1) as special cases.

The procedure of finding generalized self-similar solutions is briefly as follows: after substituting (15.3.5.1) into the given equation, one chooses the functions $\varphi(t)$ and $\psi(t)$ so that $u(z)$ satisfies a single ordinary differential equation.

**Example.** A solution of the nonlinear heat equation (15.3.4.4) will be sought in the form (15.3.5.1). Taking into account that $x = z/\psi(t)$, we find the derivatives

$$w_t = \varphi'_t u + \varphi \psi'_t x u'_z = \varphi'_t u + \frac{\varphi \psi'_t}{\psi} z u'_z, \quad w_x = \varphi \psi u'_z, \quad (w^n w_x)_x = \psi^2 \varphi^{n+1}(u^n u'_z)'_z.$$

Substituting them into (15.3.4.4) and dividing by $\varphi'_t$, we have

$$u + \frac{\varphi \psi'_t}{\varphi'_t \psi} z u'_z = \frac{\psi^2 \varphi^{n+1}}{\varphi'_t}(u^n u'_z)'_z. \tag{15.3.5.2}$$

For this relation to be an ordinary differential equation for $u(z)$, the functional coefficients of $zu'_z$ and $(u^n u'_z)'_z$ must be constant:

$$\frac{\varphi \psi'_t}{\varphi'_t \psi} = a, \quad \frac{\psi^2 \varphi^{n+1}}{\varphi'_t} = b. \tag{15.3.5.3}$$

The function $u(z)$ will satisfy the equation

$$u + azu'_z = b(u^n u'_z)'_z.$$

From the first equation in (15.3.5.3) it follows that

$$\psi = C_1 \varphi^a, \tag{15.3.5.4}$$

where $C_1$ is an arbitrary constant. Substituting the resulting expression into the second equation in (15.3.5.3) and integrating, we obtain

$$\begin{aligned}\frac{C_1^2}{b}t + C_2 &= -\frac{1}{2a+n}\varphi^{-2a-n} \quad \text{for} \quad a \neq -\frac{n}{2}, \\ \frac{C_1^2}{b}t + C_2 &= \ln|\varphi|, \quad \text{for} \quad a = -\frac{n}{2},\end{aligned} \tag{15.3.5.5}$$

where $C_2$ is an arbitrary constant. From (15.3.5.4)–(15.3.5.5) we have, in particular,

$$\varphi(t) = t^{-\frac{1}{2a+n}}, \quad \psi(t) = t^{-\frac{a}{2a+n}} \quad \text{at} \quad C_1 = 1, \quad C_2 = 0, \quad b = -\frac{1}{2a+n};$$

$$\varphi(t) = e^{2t}, \quad \psi(t) = e^{-nt} \quad \text{at} \quad C_1 = 1, \quad C_2 = 0, \quad b = \frac{1}{2}.$$

The first pair of functions $\varphi(t)$ and $\psi(t)$ corresponds to a self-similar solution (with any $a \neq -n/2$), and the second pair, to an exponential self-similar solution.

## 15.4. Exact Solutions with Simple Separation of Variables

### 15.4.1. Multiplicative and Additive Separable Solutions

Separation of variables is the most common approach to solve linear equations of mathematical physics (see Section 14.4). For equations in two independent variables $x$, $t$ and a dependent variable $w$, this approach involves searching for exact solutions in the form of the product of functions depending on different arguments:

$$w(x,t) = \varphi(x)\psi(t). \tag{15.4.1.1}$$

The integration of a few classes of first-order nonlinear partial differential equations is based on searching for exact solutions in the form of the sum of functions depending on different arguments (see Paragraph 13.2.1-2):

$$w(x,t) = \varphi(x) + \psi(t). \tag{15.4.1.2}$$

Some second- and higher-order nonlinear equations of mathematical physics also have exact solutions of the form (15.4.1.1) or (15.4.1.2). Such solutions are called *multiplicative separable* and *additive separable*, respectively.

### 15.4.2. Simple Separation of Variables in Nonlinear Partial Differential Equations

In isolated cases, the separation of variables in nonlinear equations is carried out following the same technique as in linear equations. Specifically, an exact solution is sought in the form of the product or sum of functions depending on different arguments. On substituting it into the equation and performing elementary algebraic manipulations, one obtains an equation with the two sides dependent on different variables (for equations with two variables). Then one concludes that the expressions on each side must be equal to the same constant quantity, called a *separation constant*.

**Example 1.** The heat equation with a power nonlinearity

$$\frac{\partial w}{\partial t} = a\frac{\partial}{\partial x}\left(w^k \frac{\partial w}{\partial x}\right) \tag{15.4.2.1}$$

has a multiplicative separable solution. Substituting (15.4.1.1) into (15.4.2.1) yields

$$\varphi\psi'_t = a\psi^{k+1}(\varphi^k \varphi'_x)'_x.$$

Separating the variables by dividing both sides by $\varphi\psi^{k+1}$, we obtain

$$\frac{\psi'_t}{\psi^{k+1}} = \frac{a(\varphi^k \varphi'_x)'_x}{\varphi}.$$

The left-hand side depends on $t$ alone and the right-hand side on $x$ alone. This is possible only if

$$\frac{\psi'_t}{\psi^{k+1}} = C, \qquad \frac{a(\varphi^k \varphi'_x)'_x}{\varphi} = C, \tag{15.4.2.2}$$

where $C$ is an arbitrary constant (separation constant). On solving the ordinary differential equations (15.4.2.2), we obtain a solution of equation (15.4.2.1) with the form (15.4.1.1).

The procedure for constructing a separable solution (15.4.1.1) of the nonlinear equation (15.4.2.1) is identical to that used in solving linear equations [in particular, equation (15.4.2.1) with $k = 0$]. We refer to similar cases as *simple separation of variables*.

## 15.4. Exact Solutions with Simple Separation of Variables

**Example 2.** The wave equation with an exponential nonlinearity

$$\frac{\partial^2 w}{\partial t^2} = a\frac{\partial}{\partial x}\left(e^{\lambda w}\frac{\partial w}{\partial x}\right) \tag{15.4.2.3}$$

has an additive separable solution. On substituting (15.4.1.2) into (15.4.2.3) and dividing by $e^{\lambda \psi}$, we arrive at the equation

$$e^{-\lambda \psi}\psi''_{tt} = a(e^{\lambda \varphi}\varphi'_x)'_x,$$

whose left-hand side depends on $t$ alone and the right-hand side on $x$ alone. This is possible only if

$$e^{-\lambda \psi}\psi''_{tt} = C, \quad a(e^{\lambda \varphi}\varphi'_x)'_x = C, \tag{15.4.2.4}$$

where $C$ is an arbitrary constant. Solving the ordinary differential equations (15.4.2.4) yields a solution of equation (15.4.2.3) with the form (15.4.1.2).

**Example 3.** The steady-state heat equation in an anisotropic medium with a logarithmic source

$$\frac{\partial}{\partial x}\left[f(x)\frac{\partial w}{\partial x}\right] + \frac{\partial}{\partial y}\left[g(y)\frac{\partial w}{\partial y}\right] = aw\ln w \tag{15.4.2.5}$$

has a multiplicative separable solution

$$w = \varphi(x)\psi(y). \tag{15.4.2.6}$$

On substituting (15.4.2.6) into (15.4.2.5), dividing by $\varphi\psi$, and rearranging individual terms of the resulting equation, we obtain

$$\frac{1}{\varphi}[f(x)\varphi'_x]'_x - a\ln\varphi = -\frac{1}{\psi}[g(y)\psi'_y]'_y + a\ln\psi.$$

The left-hand side of this equation depends only on $x$ and the right-hand only on $y$. By equating both sides to a constant quantity, one obtains ordinary differential equations for $\varphi(x)$ and $\psi(y)$.

Table 15.3 gives other examples of simple, additive or multiplicative, separable solutions for some nonlinear equations.

## 15.4.3. Complex Separation of Variables in Nonlinear Partial Differential Equations

The variables in nonlinear equations often separate more complexly than in linear equations. We exemplify this below.

**Example 1.** Consider the equation with a cubic nonlinearity

$$\frac{\partial w}{\partial t} = a\frac{\partial^2 w}{\partial x^2} + w\left(\frac{\partial w}{\partial x}\right)^2 - bw^3, \tag{15.4.3.1}$$

where $b > 0$. We look for exact solutions in the product form. We substitute (15.4.1.1) into (15.4.3.1) and divide the resulting equation by $\varphi(x)\psi(t)$ to obtain

$$\frac{\psi'_t}{\psi} = a\frac{\varphi''_{xx}}{\varphi} + \psi^2[(\varphi'_x)^2 - b\varphi^2]. \tag{15.4.3.2}$$

In the general case, this expression cannot be represented as the sum of two functions depending on different arguments. This, however, does not mean that equation (15.4.3.1) has no solutions of the form (15.4.1.1).

$1°$. One can make sure by direct check that the functional differential equation (15.4.3.2) has solutions

$$\varphi(x) = Ce^{\pm x\sqrt{b}}, \quad \psi(t) = e^{abt}, \tag{15.4.3.3}$$

where $C$ is an arbitrary constant. Solutions (15.4.3.3) for $\varphi$ make the expression in square brackets in (15.4.3.2) vanish, which allows the separation of variables.

### TABLE 15.3
Some nonlinear equations of mathematical physics that admit additive or multiplicative separable solutions ($C$, $C_1$, and $C_2$ are arbitrary constants)

| Equation | Equation name | Form of solutions | Determining equations |
|---|---|---|---|
| $\frac{\partial w}{\partial t} = a\frac{\partial^2 w}{\partial x^2} + bw\ln w$ | Heat equation with source | $w = \varphi(x)\psi(t)$ | $a\varphi''_{xx}/\varphi - b\ln\varphi =$ $-\psi'_t/\psi + b\ln\psi = C$ |
| $\frac{\partial w}{\partial t} = a\frac{\partial}{\partial x}\left(w^k\frac{\partial w}{\partial x}\right) + bw$ | Heat equation with source | $w = \varphi(x)\psi(t)$ | $(\psi'_t - b\psi)/\psi^{k+1} =$ $a(\varphi^k\varphi'_x)'_x/\varphi = C$ |
| $\frac{\partial w}{\partial t} = a\frac{\partial}{\partial x}\left(w^k\frac{\partial w}{\partial x}\right) + bw^{k+1}$ | Heat equation with source | $w = \varphi(x)\psi(t)$ | $\psi'_t/\psi^{k+1} =$ $a(\varphi^k\varphi'_x)'_x/\varphi + b\varphi^k = C$ |
| $\frac{\partial w}{\partial t} = a\frac{\partial}{\partial x}\left(e^{\lambda w}\frac{\partial w}{\partial x}\right) + b$ | Heat equation with source | $w = \varphi(x) + \psi(t)$ | $e^{-\lambda\psi}(\psi'_t - b) = a(e^{\lambda\varphi}\varphi'_x)'_x = C$ |
| $\frac{\partial w}{\partial t} = a\frac{\partial}{\partial x}\left(e^w\frac{\partial w}{\partial x}\right) + be^w$ | Heat equation with source | $w = \varphi(x) + \psi(t)$ | $e^{-\psi}\psi'_t = a(e^{\varphi}\varphi'_x)'_x + be^{\varphi} = C$ |
| $\frac{\partial w}{\partial t} = a\frac{\partial^2 w}{\partial x^2} + b\left(\frac{\partial w}{\partial x}\right)^2$ | Potential Burgers equation | $w = \varphi(x) + \psi(t)$ | $\psi'_t = a\varphi''_{xx} + b(\varphi'_x)^2 = C$ |
| $\frac{\partial w}{\partial t} = a\left(\frac{\partial w}{\partial x}\right)^k \frac{\partial^2 w}{\partial x^2}$ | Filtration equation | $w = \varphi(x) + \psi(t)$, $w = f(x)g(t)$ | $\psi'_t = a(\varphi'_x)^k \varphi''_{xx} = C_1$, $g'_t/g^{k+1} = a(f'_x)^k f''_{xx}/f = C_2$ |
| $\frac{\partial w}{\partial t} = F\left(\frac{\partial w}{\partial x}\right)\frac{\partial^2 w}{\partial x^2}$ | Filtration equation | $w = \varphi(x) + \psi(t)$ | $\psi'_t = F(\varphi'_x)\varphi''_{xx} = C$ |
| $\frac{\partial^2 w}{\partial t^2} = a\frac{\partial}{\partial x}\left(w^k\frac{\partial w}{\partial x}\right)$ | Wave equation | $w = \varphi(x)\psi(t)$ | $\psi''_{tt}/\psi^{k+1} = a(\varphi^k\varphi'_x)'_x/\varphi = C$ |
| $\frac{\partial^2 w}{\partial t^2} = a\frac{\partial}{\partial x}\left(e^{\lambda w}\frac{\partial w}{\partial x}\right)$ | Wave equation | $w = \varphi(x) + \psi(t)$ | $e^{-\lambda\psi}\psi''_{tt} = a(e^{\lambda\varphi}\varphi'_x)'_x = C$ |
| $\frac{\partial^2 w}{\partial t^2} = a\frac{\partial^2 w}{\partial x^2} + bw\ln w$ | Wave equation with source | $w = \varphi(x)\psi(t)$ | $\psi''_{tt}/\psi - b\ln\psi =$ $a\varphi''_{xx}/\varphi + b\ln\varphi = C$ |
| $\frac{\partial^2 w}{\partial x^2} + a\frac{\partial}{\partial y}\left(w^k\frac{\partial w}{\partial y}\right) = 0$ | Anisotropic steady heat equation | $w = \varphi(x)\psi(y)$ | $\varphi''_{xx}/\varphi^{k+1} = -a(\psi^k\psi'_y)'_y/\psi = C$ |
| $\frac{\partial^2 w}{\partial x^2} + a\frac{\partial w}{\partial y}\frac{\partial^2 w}{\partial y^2} = 0$ | Equation of steady transonic gas flow | $w = \varphi(x) + \psi(y)$, $w = f(x)g(y)$ | $\varphi''_{xx} = -a\psi'_y\psi''_{yy} = C_1$, $f''_{xx}/f = -ag'_y g''_{yy}/g = C_2$ |
| $\left(\frac{\partial^2 w}{\partial x\partial y}\right)^2 = \frac{\partial^2 w}{\partial x^2}\frac{\partial^2 w}{\partial y^2}$ | Monge–Ampère equation | $w = \varphi(x)\psi(y)$ | $\frac{(\varphi'_x)^2}{\varphi\varphi''_{xx}} = \frac{\psi\psi''_{yy}}{(\psi'_y)^2} = C$ |
| $\frac{\partial w}{\partial t} = a\frac{\partial^3 w}{\partial x^3} + b\left(\frac{\partial w}{\partial x}\right)^2$ | Potential Korteweg–de Vries equation | $w = \varphi(x) + \psi(t)$ | $\psi'_t = a\varphi'''_{xxx} + b(\varphi'_x)^2 = C$ |
| $\frac{\partial w}{\partial y}\frac{\partial^2 w}{\partial x\partial y} - \frac{\partial w}{\partial x}\frac{\partial^2 w}{\partial y^2} = a\frac{\partial^3 w}{\partial y^3}$ | Boundary-layer equation | $w = \varphi(x) + \psi(y)$, $w = f(x)g(y)$ | $-\varphi'_x = a\psi'''_{yyy}/\psi''_{yy} = C_1$, $f'_x = ag'''_{yyy}[(g'_y)^2 - gg''_{yy}]^{-1} = C_2$ |

2°. There is a more general solution of the functional differential equation (15.4.3.2):
$$\varphi(x) = C_1 e^{x\sqrt{b}} + C_2 e^{-x\sqrt{b}}, \quad \psi(t) = e^{abt}\left(C_3 + 4C_1 C_2 e^{2abt}\right)^{-1/2},$$
where $C_1$, $C_2$, and $C_3$ are arbitrary constants. The function $\varphi = \varphi(x)$ is such that both $x$-dependent expressions in (15.4.3.2) are constant simultaneously:
$$\varphi''_{xx}/\varphi = \text{const}, \quad (\varphi'_x)^2 - b\varphi^2 = \text{const}.$$
It is this circumstance that makes it possible to separate the variables.

**Example 2.** Consider the second-order equation with a quadratic nonlinearity
$$\frac{\partial w}{\partial y}\frac{\partial^2 w}{\partial x^2} + a\frac{\partial w}{\partial x}\frac{\partial^2 w}{\partial y^2} = b\frac{\partial^3 w}{\partial x^3} + c\frac{\partial^3 w}{\partial y^3}. \tag{15.4.3.4}$$

We look for additive separable solutions
$$w = f(x) + g(y). \tag{15.4.3.5}$$

Substituting (15.4.3.5) into (15.4.3.4) yields
$$g'_y f''_{xx} + a f'_x g''_{yy} = b f'''_{xxx} + c g'''_{yyy}. \tag{15.4.3.6}$$

This expression cannot be rewritten as the equality of two functions depending on different arguments.

It can be shown that equation (15.4.3.4) has a solution of the form (15.4.3.5):
$$w = C_1 e^{-a\lambda x} + \frac{c\lambda}{a} x + C_2 e^{\lambda y} - ab\lambda y + C_3,$$

where $C_1$, $C_2$, $C_3$, and $\lambda$ are arbitrary constants. The mechanism of separation of variables is different here: both nonlinear terms on the left-hand side in (15.4.3.6) contain terms that cannot be rewritten in additive form but are equal in magnitude and have unlike signs. In adding, the two terms cancel out, thus resulting in separation of variables:

$$\begin{array}{rl} g'_y f''_{xx} &= C_1 C_2 a^2 \lambda^3 e^{\lambda y - a\lambda x} - C_1 b(a\lambda)^3 e^{-a\lambda x} \\ + \quad a f'_x g''_{yy} &= -C_1 C_2 a^2 \lambda^3 e^{\lambda y - a\lambda x} + C_2 c \lambda^3 e^{\lambda y} \\ \hline g'_y f''_{xx} + a f'_x g''_{yy} &= -C_1 b(a\lambda)^3 e^{-a\lambda x} + C_2 c \lambda^3 e^{\lambda y} = b f'''_{xxx} + c g'''_{yyy} \end{array}.$$

**Example 3.** Consider the second-order equation with a cubic nonlinearity
$$(1 + w^2)\left(\frac{\partial^2 w}{\partial x^2} + \frac{\partial^2 w}{\partial y^2}\right) - 2w\left(\frac{\partial w}{\partial x}\right)^2 - 2w\left(\frac{\partial w}{\partial y}\right)^2 = aw(1 - w^2). \tag{15.4.3.7}$$

We seek an exact solution of this equation in the product form
$$w = f(x)g(y). \tag{15.4.3.8}$$

Substituting (15.4.3.8) into (15.4.3.7) yields
$$(1 + f^2 g^2)(g f''_{xx} + f g''_{yy}) - 2fg[g^2(f'_x)^2 + f^2(g'_y)^2] = afg(1 - f^2 g^2). \tag{15.4.3.9}$$

This expression cannot be rewritten as the equality of two functions with different arguments. Nevertheless, equation (15.4.3.7) has solutions of the form (15.4.3.8). One can make sure by direct check that the functions $f = f(x)$ and $g = g(y)$ satisfying the nonlinear ordinary differential equations
$$\begin{aligned} (f'_x)^2 &= A f^4 + B f^2 + C, \\ (g'_y)^2 &= C g^4 + (a - B) g^2 + A, \end{aligned} \tag{15.4.3.10}$$

where $A$, $B$, and $C$ are arbitrary constants, reduce equation (15.4.3.9) to an identity; to verify this, one should use the relations $f''_{xx} = 2Af^3 + Bf$ and $g''_{yy} = 2Cg^3 + (a - B)g$ that follow from (15.4.3.10).

**Remark.** By the change of variable $u = 4 \arctan w$ equation (15.4.3.7) can be reduced to a nonlinear heat equation with a sinusoidal source, $\Delta u = a \sin u$.

The examples considered above illustrate some specific features of separable solutions to nonlinear equations. Section 15.5 outlines fairly general methods for constructing similar and more complicated solutions to nonlinear partial differential equations.

# 15.5. Method of Generalized Separation of Variables

## 15.5.1. Structure of Generalized Separable Solutions

**15.5.1-1. General form of solutions. The classes of nonlinear equations considered.**

To simplify the presentation, we confine ourselves to the case of mathematical physics equations in two independent variables $x$, $y$ and a dependent variable $w$ (one of the independent variables can play the role of time).

Linear separable equations of mathematical physics admit exact solutions in the form

$$w(x,y) = \varphi_1(x)\psi_1(y) + \varphi_2(x)\psi_2(y) + \cdots + \varphi_n(x)\psi_n(y), \tag{15.5.1.1}$$

where $w_i = \varphi_i(x)\psi_i(y)$ are particular solutions; the functions $\varphi_i(x)$, as well as the functions $\psi_i(y)$, with different numbers $i$, are not related to one another.

Many nonlinear partial differential equations with quadratic or power nonlinearities,

$$f_1(x)g_1(y)\Pi_1[w] + f_2(x)g_2(y)\Pi_2[w] + \cdots + f_m(x)g_m(y)\Pi_m[w] = 0, \tag{15.5.1.2}$$

also have exact solutions of the form (15.5.1.1). In (15.5.1.2), the $\Pi_i[w]$ are differential forms that are the products of nonnegative integer powers of the function $w$ and its partial derivatives $\partial_x w, \partial_y w, \partial_{xx} w, \partial_{xy} w, \partial_{yy} w, \partial_{xxx} w$, etc. We will refer to solutions (15.5.1.1) of nonlinear equations (15.5.1.2) as *generalized separable solutions*. Unlike linear equations, in nonlinear equations the functions $\varphi_i(x)$ with different subscripts $i$ are usually related to one another [and to functions $\psi_j(y)$]. In general, the functions $\varphi_i(x)$ and $\psi_j(y)$ in (15.5.1.1) are not known in advance and are to be identified. Subsection 15.4.2 gives simple examples of exact solutions of the form (15.5.1.1) with $n = 1$ and $n = 2$ (for $\psi_1 = \varphi_2 = 1$) to some nonlinear equations.

Note that most common of the generalized separable solutions are solutions of the special form

$$w(x,y) = \varphi(x)\psi(y) + \chi(x);$$

the independent variables on the right-hand side can be swapped. In the special case of $\chi(x) = 0$, this is a multiplicative separable solution, and if $\varphi(x) = 1$, this is an additive separable solution.

*Remark.* Expressions of the form (15.5.1.1) are often used in applied and computational mathematics for constructing approximate solutions to differential equations by the Galerkin method (and its modifications).

### 15.5.1-2. General form of functional differential equations.

In general, on substituting expression (15.5.1.1) into the differential equation (15.5.1.2), one arrives at a functional differential equation

$$\Phi_1(X)\Psi_1(Y) + \Phi_2(X)\Psi_2(Y) + \cdots + \Phi_k(X)\Psi_k(Y) = 0 \tag{15.5.1.3}$$

for the $\varphi_i(x)$ and $\psi_i(y)$. The functionals $\Phi_j(X)$ and $\Psi_j(Y)$ depend only on $x$ and $y$, respectively,

$$\begin{aligned}\Phi_j(X) &\equiv \Phi_j\big(x, \varphi_1, \varphi_1', \varphi_1'', \ldots, \varphi_n, \varphi_n', \varphi_n''\big), \\ \Psi_j(Y) &\equiv \Psi_j\big(y, \psi_1, \psi_1', \psi_1'', \ldots, \psi_n, \psi_n', \psi_n''\big).\end{aligned} \tag{15.5.1.4}$$

Here, for simplicity, the formulas are written out for the case of a second-order equation (15.5.1.2); for higher-order equations, the right-hand sides of relations (15.5.1.4) will contain higher-order derivatives of $\varphi_i$ and $\psi_j$.

Subsections 15.5.3 and 15.5.4 outline two different methods for solving functional differential equations of the form (15.5.1.3)–(15.5.1.4).

*Remark.* Unlike ordinary differential equations, equation (15.5.1.3)–(15.5.1.4) involves several functions (and their derivatives) with different arguments.

## 15.5.2. Simplified Scheme for Constructing Solutions Based on Presetting One System of Coordinate Functions

15.5.2-1. Description of the simplified scheme.

To construct exact solutions of equations (15.5.1.2) with quadratic or power nonlinearities that do not depend explicitly on $x$ (all $f_i$ constant), it is reasonable to use the following simplified approach. As before, we seek solutions in the form of finite sums (15.5.1.1). We assume that the system of coordinate functions $\{\varphi_i(x)\}$ is governed by linear differential equations with constant coefficients. The most common solutions of such equations are of the forms

$$\varphi_i(x) = x^i, \quad \varphi_i(x) = e^{\lambda_i x}, \quad \varphi_i(x) = \sin(\alpha_i x), \quad \varphi_i(x) = \cos(\beta_i x). \tag{15.5.2.1}$$

Finite chains of these functions (in various combinations) can be used to search for separable solutions (15.5.1.1), where the quantities $\lambda_i$, $\alpha_i$, and $\beta_i$ are regarded as free parameters. The other system of functions $\{\psi_i(y)\}$ is determined by solving the nonlinear equations resulting from substituting (15.5.1.1) into the equation under consideration [or into equation (15.5.1.3)–(15.5.1.4)].

This simplified approach lacks the generality of the methods outlined in Subsections 15.5.3–15.5.4. However, specifying one of the systems of coordinate functions, $\{\varphi_i(x)\}$, simplifies the procedure of finding exact solutions substantially. The drawback of this approach is that some solutions of the form (15.5.1.1) can be overlooked. It is significant that the overwhelming majority of generalized separable solutions known to date, for partial differential equations with quadratic nonlinearities, are determined by coordinate functions (15.5.2.1) (usually with $n = 2$).

15.5.2-2. Examples of finding exact solutions of second- and third-order equations.

Below we consider specific examples that illustrate the application of the above simplified scheme to the construction of generalized separable solutions of second- and third-order nonlinear equations.

**Example 1.** Consider a nonhomogeneous Monge–Ampère equation of the form

$$\left(\frac{\partial^2 w}{\partial x \partial y}\right)^2 - \frac{\partial^2 w}{\partial x^2}\frac{\partial^2 w}{\partial y^2} = f(x). \tag{15.5.2.2}$$

We look for generalized separable solutions with the form

$$w(x, y) = \varphi(x) y^k + \psi(x), \quad k \neq 0. \tag{15.5.2.3}$$

On substituting (15.5.2.3) into (15.5.2.2) and collecting terms, we obtain

$$[k^2(\varphi'_x)^2 - k(k-1)\varphi \varphi''_{xx}] y^{2k-2} - k(k-1)\varphi \psi''_{xx} y^{k-2} - f(x) = 0. \tag{15.5.2.4}$$

This equation can be satisfied only if $k = 1$ or $k = 2$.

*First case.* If $k = 1$, (15.5.2.4) reduces one equation

$$(\varphi'_x)^2 - f(x) = 0.$$

It has two solutions: $\varphi(x) = \pm \int \sqrt{f(x)}\, dx$. They generate two solutions of equation (15.5.2.2) in the form (15.5.2.3):

$$w(x, y) = \pm y \int \sqrt{f(x)}\, dx + \psi(x),$$

where $\psi(x)$ is an arbitrary function.

*Second case.* If $k = 2$, equating the functional coefficients of the different powers of $y$ to zero, we obtain two equations:
$$2(\varphi_x')^2 - \varphi\varphi_{xx}'' = 0,$$
$$2\varphi\psi_{xx}'' - f(x) = 0.$$

Their general solutions are given by
$$\varphi(x) = \frac{1}{C_1 x + C_2}, \quad \psi(x) = \frac{1}{2}\int_0^x (x-t)(C_1 t + C_2) f(t)\, dt + C_3 x + C_4.$$

Here, $C_1$, $C_2$, $C_3$, and $C_4$ are arbitrary constants.

Table 15.4 lists generalized separable solutions of various nonhomogeneous Monge–Ampère equations of the form

$$\left(\frac{\partial^2 w}{\partial x \partial y}\right)^2 - \frac{\partial^2 w}{\partial x^2}\frac{\partial^2 w}{\partial y^2} = F(x, y). \tag{15.5.2.5}$$

Equations of this form are encountered in differential geometry, gas dynamics, and meteorology.

**Example 2.** Consider the third-order nonlinear equation

$$\frac{\partial^2 w}{\partial x \partial t} + \left(\frac{\partial w}{\partial x}\right)^2 - w\frac{\partial^2 w}{\partial x^2} = a\frac{\partial^3 w}{\partial x^3}, \tag{15.5.2.6}$$

which is encountered in hydrodynamics.

We look for exact solutions of the form

$$w = \varphi(t)e^{\lambda x} + \psi(t), \quad \lambda \neq 0. \tag{15.5.2.7}$$

On substituting (15.5.2.7) into (15.5.2.6), we have

$$\varphi_t' - \lambda\varphi\psi = a\lambda^2\varphi.$$

We now solve this equation for $\psi$ and substitute the resulting expression into (15.5.2.7) to obtain a solution of equation (15.5.2.6) in the form

$$w = \varphi(t)e^{\lambda x} + \frac{1}{\lambda}\frac{\varphi_t'(t)}{\varphi(t)} - a\lambda,$$

where $\varphi(t)$ is an arbitrary function and $\lambda$ is an arbitrary constant.

## 15.5.3. Solution of Functional Differential Equations by Differentiation

### 15.5.3-1. Description of the method.

Below we describe a procedure for constructing solutions to functional differential equations of the form (15.5.1.3)–(15.5.1.4). It involves three successive stages.

$1°$. Assume that $\Psi_k \not\equiv 0$. We divide equation (15.5.1.3) by $\Psi_k$ and differentiate with respect to $y$. This results in a similar equation but with fewer terms:
$$\widetilde{\Phi}_1(X)\widetilde{\Psi}_1(Y) + \widetilde{\Phi}_2(X)\widetilde{\Psi}_2(Y) + \cdots + \widetilde{\Phi}_{k-1}(X)\widetilde{\Psi}_{k-1}(Y) = 0,$$
$$\widetilde{\Phi}_j(X) = \Phi_j(X), \quad \widetilde{\Psi}_j(Y) = [\Psi_j(Y)/\Psi_k(Y)]_y'.$$

We repeat the above procedure $(k-3)$ times more to obtain the separable two-term equation

$$\widehat{\Phi}_1(X)\widehat{\Psi}_1(Y) + \widehat{\Phi}_2(X)\widehat{\Psi}_2(Y) = 0. \tag{15.5.3.1}$$

TABLE 15.4
Exact solutions of a nonhomogeneous Monge–Ampère equation of the form (15.5.2.5); $f(x)$, $g(x)$, and $h(x)$ are arbitrary functions; $C_1$, $C_2$, $C_3$, and $\beta$ are arbitrary constants; $a$, $b$, $k$, and $\lambda$ are some numbers ($k \neq -1, -2$).

| No. | Function $F(x,y)$ | Solution $w(x,y)$ |
|---|---|---|
| 1 | $f(x)$ | $C_1 y^2 + C_2 xy + \dfrac{C_2^2}{4C_1} x^2 - \dfrac{1}{2C_1} \int_a^x (x-t) f(t)\, dt$ |
| 2 | $f(x)$ | $\dfrac{1}{x+C_1}\left(C_2 y^2 + C_3 y + \dfrac{C_3^2}{4C_2}\right) - \dfrac{1}{2C_2} \int_a^x (x-t)(t+C_1) f(t)\, dt$ |
| 3 | $f(x)y$ | $C_1 y^2 + \varphi(x) y + \psi(x)$ ($\varphi$ and $\psi$ are expressed as quadratures) |
| 4 | $f(x)y$ | $\dfrac{1}{x+C_1} y^2 + \varphi(x) y + \psi(x)$ ($\varphi$ and $\psi$ are expressed as quadratures) |
| 5 | $f(x)y$ | $C_1 y^3 - \dfrac{1}{6 C_1} \int_a^x (x-t) f(t)\, dt$ |
| 6 | $f(x)y$ | $\dfrac{y^3}{(x+C_1)^2} - \dfrac{1}{6} \int_a^x (x-t)(t+C_1)^2 f(t)\, dt$ |
| 7 | $f(x)y^2$ | $\varphi(x) y^2 + \psi(x) y + \chi(x)$ ($\varphi$, $\psi$, and $\chi$ are determined by ODEs) |
| 8 | $f(x)y^2$ | $C_1 y^4 - \dfrac{1}{12 C_1} \int_a^x (x-t) f(t)\, dt$ |
| 9 | $f(x)y^2$ | $\dfrac{y^4}{(x+C_1)^3} - \dfrac{1}{12} \int_a^x (x-t)(t+C_1)^3 f(t)\, dt$ |
| 10 | $f(x)y^2 + g(x)y + h(x)$ | $\varphi(x) y^2 + \psi(x) y + \chi(x)$ ($\varphi$, $\psi$, and $\chi$ are determined by ODEs) |
| 11 | $f(x)y^k$ | $\dfrac{C_1 y^{k+2}}{(k+1)(k+2)} - \dfrac{1}{C_1} \int_a^x (x-t) f(t)\, dt$ |
| 12 | $f(x)y^k$ | $\dfrac{y^{k+2}}{(x+C_1)^{k+1}} - \dfrac{1}{(k+1)(k+2)} \int_a^x (x-t)(t+C_1)^{k+1} f(t)\, dt$ |
| 13 | $f(x)y^{2k+2} + g(x)y^k$ | $\varphi(x) y^{k+2} - \dfrac{1}{(k+1)(k+2)} \int_a^x (x-t) \dfrac{g(t)}{\varphi(t)}\, dt$ ($\varphi$ is determined by an ODE) |
| 14 | $f(x)e^{\lambda y}$ | $C_1 e^{\beta x + \lambda y} - \dfrac{1}{C_1 \lambda^2} \int_a^x (x-t) e^{-\beta t} f(t)\, dt$ |
| 15 | $f(x)e^{2\lambda y} + g(x)e^{\lambda y}$ | $\varphi(x) e^{\lambda y} - \dfrac{1}{\lambda^2} \int_a^x (x-t) \dfrac{g(t)}{\varphi(t)}\, dt$ ($\varphi$ is determined by an ODE) |
| 16 | $f(x)g(y) + \lambda^2$ | $C_1 \int_a^x (x-t) f(t)\, dt - \dfrac{1}{C_1} \int_b^y (y-\xi) g(\xi)\, d\xi \pm \lambda xy$ |

Three cases must be considered.

*Nondegenerate case*: $|\widehat{\Phi}_1(X)| + |\widehat{\Phi}_2(X)| \neq 0$ and $|\widehat{\Psi}_1(Y)| + |\widehat{\Psi}_2(Y)| \neq 0$. Then the solutions of equation (15.5.3.1) are determined by the ordinary differential equations

$$\widehat{\Phi}_1(X) + C\widehat{\Phi}_2(X) = 0, \quad C\widehat{\Psi}_1(Y) - \widehat{\Psi}_2(Y) = 0,$$

where $C$ is an arbitrary constant. The equations $\widehat{\Phi}_2 = 0$ and $\widehat{\Psi}_1 = 0$ correspond to the limit case $C = \infty$.

*Two degenerate cases*:

$$\widehat{\Phi}_1(X) \equiv 0, \quad \widehat{\Phi}_2(X) \equiv 0 \implies \widehat{\Psi}_{1,2}(Y) \text{ are any functions};$$
$$\widehat{\Psi}_1(Y) \equiv 0, \quad \widehat{\Psi}_2(Y) \equiv 0 \implies \widehat{\Phi}_{1,2}(X) \text{ are any functions.}$$

2°. The solutions of the two-term equation (15.5.3.1) should be substituted into the original functional differential equation (15.5.1.3) to "remove" redundant constants of integration [these arise because equation (15.5.3.1) is obtained from (15.5.1.3) by differentiation].

3°. The case $\Psi_k \equiv 0$ should be treated separately (since we divided the equation by $\Psi_k$ at the first stage). Likewise, we have to study all other cases where the functionals by which the intermediate functional differential equations were divided vanish.

**Remark 1.** The functional differential equation (15.5.1.3) happens to have no solutions.

**Remark 2.** At each subsequent stage, the number of terms in the functional differential equation can be reduced by differentiation with respect to either $y$ or $x$. For example, we can assume at the first stage that $\Phi_k \not\equiv 0$. On dividing equation (15.5.1.3) by $\Phi_k$ and differentiating with respect to $x$, we again obtain a similar equation that has fewer terms.

### 15.5.3-2. Examples of constructing exact generalized separable solutions.

Below we consider specific examples illustrating the application of the above method of constructing exact generalized separable solutions of nonlinear equations.

**Example 1.** The equations of a laminar boundary layer on a flat plate are reduced to a single third-order nonlinear equation for the stream function (see Schlichting, 1981, and Loitsyanskiy, 1996):

$$\frac{\partial w}{\partial y}\frac{\partial^2 w}{\partial x \partial y} - \frac{\partial w}{\partial x}\frac{\partial^2 w}{\partial y^2} = a\frac{\partial^3 w}{\partial y^3}. \tag{15.5.3.2}$$

We look for generalized separable solutions to equation (15.5.3.2) in the form

$$w(x, y) = \varphi(x)\psi(y) + \chi(x). \tag{15.5.3.3}$$

On substituting (15.5.3.3) into (15.5.3.2) and canceling by $\varphi$, we arrive at the functional differential equation

$$\varphi'_x[(\psi'_y)^2 - \psi\psi''_{yy}] - \chi'_x\psi''_{yy} = a\psi'''_{yyy}. \tag{15.5.3.4}$$

We differentiate (15.5.3.4) with respect to $x$ to obtain

$$\varphi''_{xx}[(\psi'_y)^2 - \psi\psi''_{yy}] = \chi''_{xx}\psi''_{yy}. \tag{15.5.3.5}$$

*Nondegenerate case.* On separating the variables in (15.5.3.5), we get

$$\chi''_{xx} = C_1\varphi''_{xx},$$
$$(\psi'_y)^2 - \psi\psi''_{yy} - C_1\psi''_{yy} = 0.$$

Integrating yields

$$\psi(y) = C_4 e^{\lambda y} - C_1, \quad \varphi(x) \text{ is any function}, \quad \chi(x) = C_1\varphi(x) + C_2 x + C_3, \tag{15.5.3.6}$$

where $C_1, \ldots, C_4$, and $\lambda$ are constants of integration. On substituting (15.5.3.6) into (15.5.3.4), we establish the relationship between constants to obtain $C_2 = -a\lambda$. Ultimately, taking into account the aforesaid and formulas (15.5.3.3) and (15.5.3.6), we arrive at a solution of equation (15.5.3.2) of the form (15.5.3.3):

$$w(x, y) = \varphi(x)e^{\lambda y} - a\lambda x + C,$$

where $\varphi(x)$ is an arbitrary function and $C$ and $\lambda$ are arbitrary constants ($C = C_3$, $C_4 = 1$).

*Degenerate case.* It follows from (15.5.3.5) that

$$\varphi''_{xx} = 0, \quad \chi''_{xx} = 0, \quad \psi(y) \text{ is any function}. \tag{15.5.3.7}$$

Integrating the first two equations in (15.5.3.7) twice yields

$$\varphi(x) = C_1 x + C_2, \quad \chi(x) = C_3 x + C_4, \tag{15.5.3.8}$$

where $C_1, \ldots, C_4$ are arbitrary constants.

Substituting (15.5.3.8) into (15.5.3.4), we arrive at an ordinary differential equation for $\psi = \psi(y)$:

$$C_1(\psi_y')^2 - (C_1\psi + C_3)\psi_{yy}'' = a\psi_{yyy}'''. \tag{15.5.3.9}$$

Formulas (15.5.3.3) and (15.5.3.8) together with equation (15.5.3.9) determine an exact solution of equation (15.5.3.2).

**Example 2.** The two-dimensional stationary equations of motion of a viscous incompressible fluid are reduced to a single fourth-order nonlinear equation for the stream function (see Loitsyanskiy, 1996):

$$\frac{\partial w}{\partial y}\frac{\partial}{\partial x}(\Delta w) - \frac{\partial w}{\partial x}\frac{\partial}{\partial y}(\Delta w) = a\Delta\Delta w, \quad \Delta w = \frac{\partial^2 w}{\partial x^2} + \frac{\partial^2 w}{\partial y^2}. \tag{15.5.3.10}$$

Here, $a$ is the kinematic viscosity of the fluid and $x$, $y$ are Cartesian coordinates.

We seek exact separable solutions of equation (15.5.3.10) in the form

$$w = f(x) + g(y). \tag{15.5.3.11}$$

Substituting (15.5.3.11) into (15.5.3.10) yields

$$g_y' f_{xxx}''' - f_x' g_{yyy}''' = a f_{xxxx}'''' + a g_{yyyy}''''. \tag{15.5.3.12}$$

Differentiating (15.5.3.12) with respect to $x$ and $y$, we obtain

$$g_{yy}'' f_{xxxx}'''' - f_{xx}'' g_{yyyy}'''' = 0. \tag{15.5.3.13}$$

*Nondegenerate case.* If $f_{xx}'' \not\equiv 0$ and $g_{yy}'' \not\equiv 0$, we separate the variables in (15.5.3.13) to obtain the ordinary differential equations

$$f_{xxxx}'''' = C f_{xx}'', \tag{15.5.3.14}$$
$$g_{yyyy}'''' = C g_{yy}'', \tag{15.5.3.15}$$

which have different solutions depending on the value of the integration constant $C$.

$1°$. Solutions of equations (15.5.3.14) and (15.5.3.15) for $C = 0$:

$$\begin{aligned} f(x) &= A_1 + A_2 x + A_3 x^2 + A_4 x^3, \\ g(y) &= B_1 + B_2 y + B_3 y^2 + B_4 y^3, \end{aligned} \tag{15.5.3.16}$$

where $A_k$ and $B_k$ are arbitrary constants ($k = 1, 2, 3, 4$). On substituting (15.5.3.16) into (15.5.3.12), we evaluate the integration constants. Three cases are possible:

$$\begin{aligned} A_4 = B_4 = 0, & \quad A_n, B_n \text{ are any numbers} & (n = 1, 2, 3); \\ A_k = 0, & \quad B_k \text{ are any numbers} & (k = 1, 2, 3, 4); \\ B_k = 0, & \quad A_k \text{ are any numbers} & (k = 1, 2, 3, 4). \end{aligned}$$

The first two sets of constants determine two simple solutions (15.5.3.11) of equation (15.5.3.10):

$$\begin{aligned} w &= C_1 x^2 + C_2 x + C_3 y^2 + C_4 y + C_5, \\ w &= C_1 y^3 + C_2 y^2 + C_3 y + C_4, \end{aligned}$$

where $C_1, \ldots, C_5$ are arbitrary constants.

$2°$. Solutions of equations (15.5.3.14) and (15.5.3.15) for $C = \lambda^2 > 0$:

$$\begin{aligned} f(x) &= A_1 + A_2 x + A_3 e^{\lambda x} + A_4 e^{-\lambda x}, \\ g(y) &= B_1 + B_2 y + B_3 e^{\lambda y} + B_4 e^{-\lambda y}. \end{aligned} \tag{15.5.3.17}$$

Substituting (15.5.3.17) into (15.5.3.12), dividing by $\lambda^3$, and collecting terms, we obtain

$$A_3(a\lambda - B_2)e^{\lambda x} + A_4(a\lambda + B_2)e^{-\lambda x} + B_3(a\lambda + A_2)e^{\lambda y} + B_4(a\lambda - A_2)e^{-\lambda y} = 0.$$

Equating the coefficients of the exponentials to zero, we find

$$A_3 = A_4 = B_3 = 0, \quad A_2 = a\lambda \qquad \text{(case 1)},$$
$$A_3 = B_3 = 0, \quad A_2 = a\lambda, \quad B_2 = -a\lambda \qquad \text{(case 2)},$$
$$A_3 = B_4 = 0, \quad A_2 = -a\lambda, \quad B_2 = -a\lambda \qquad \text{(case 3)}.$$

(The other constants are arbitrary.) These sets of constants determine three solutions of the form (15.5.3.11) for equation (15.5.3.10):

$$w = C_1 e^{-\lambda y} + C_2 y + C_3 + a\lambda x,$$
$$w = C_1 e^{-\lambda x} + a\lambda x + C_2 e^{-\lambda y} - a\lambda y + C_3,$$
$$w = C_1 e^{-\lambda x} - a\lambda x + C_2 e^{\lambda y} - a\lambda y + C_3,$$

where $C_1$, $C_2$, $C_3$, and $\lambda$ are arbitrary constants.

3°. Solution of equations (15.5.3.14) and (15.5.3.15) for $C = -\lambda^2 < 0$:

$$f(x) = A_1 + A_2 x + A_3 \cos(\lambda x) + A_4 \sin(\lambda x),$$
$$g(y) = B_1 + B_2 y + B_3 \cos(\lambda y) + B_4 \sin(\lambda y). \tag{15.5.3.18}$$

Substituting (15.5.3.18) into (15.5.3.12) does not yield new real solutions.

*Degenerate cases.* If $f''_{xx} \equiv 0$ or $g''_{yy} \equiv 0$, equation (15.5.3.13) becomes an identity for any $g = g(y)$ or $f = f(x)$, respectively. These cases should be treated separately from the nondegenerate case. For example, if $f''_{xx} \equiv 0$, we have $f(x) = Ax + B$, where $A$ and $B$ are arbitrary numbers. Substituting this $f$ into (15.5.3.12), we arrive at the equation $-Ag'''_{yyy} = ag''''_{yyyy}$. Its general solution is given by $g(y) = C_1 \exp(-Ay/a) + C_2 y^2 + C_3 y + C_4$. Thus, we obtain another solution of the form (15.5.3.11) for equation (15.5.3.10):

$$w = C_1 e^{-\lambda y} + C_2 y^2 + C_3 y + C_4 + a\lambda x \qquad (A = a\lambda, \; B = 0).$$

## 15.5.4. Solution of Functional-Differential Equations by Splitting

15.5.4-1. Preliminary remarks. Description of the method.

As one reduces the number of terms in the functional differential equation (15.5.1.3)–(15.5.1.4) by differentiation, redundant constants of integration arise. These constants must be "removed" at the final stage. Furthermore, the resulting equation can be of a higher order than the original equation. To avoid these difficulties, it is convenient to reduce the solution of the functional differential equation to the solution of a bilinear functional equation of a standard form and solution of a system of ordinary differential equations. Thus, the original problem splits into two simpler problems. Below we outline the basic stages of the splitting method.

1°. At the first stage, we treat equation (15.5.1.3) as a purely functional equation that depends on two variables $X$ and $Y$, where $\Phi_n = \Phi_n(X)$ and $\Psi_n = \Psi_n(Y)$ are unknown quantities ($n = 1, \ldots, k$).

It can be shown* that the bilinear functional equation (15.5.1.3) has $k - 1$ different solutions:

$$\Phi_i(X) = C_{i,1}\Phi_{m+1}(X) + C_{i,2}\Phi_{m+2}(X) + \cdots + C_{i,k-m}\Phi_k(X), \quad i = 1, \ldots, m;$$
$$\Psi_{m+j}(Y) = -C_{1,j}\Psi_1(Y) - C_{2,j}\Psi_2(Y) - \cdots - C_{m,j}\Psi_m(Y), \quad j = 1, \ldots, k - m;$$
$$m = 1, 2, \ldots, k - 1;$$
$$\tag{15.5.4.1}$$

where $C_{i,j}$ are arbitrary constants. The functions $\Phi_{m+1}(X), \ldots, \Phi_k(X), \Psi_1(Y), \ldots, \Psi_m(Y)$ on the right-hand sides of formulas (15.5.4.1) are defined arbitrarily. It is apparent that for fixed $m$, solution (15.5.4.1) contains $m(k - m)$ arbitrary constants.

---

* These solutions can be obtained by differentiation following the procedure outlined in Subsection 15.5.3, and by induction. Another simple method for finding solutions is described in Paragraph 15.5.4-2, Item 3°.

2°. At the second stage, we successively substitute the $\Phi_i(X)$ and $\Psi_j(Y)$ of (15.5.1.4) into all solutions (15.5.4.1) to obtain systems of ordinary differential equations* for the unknown functions $\varphi_p(x)$ and $\psi_q(y)$. Solving these systems, we get generalized separable solutions of the form (15.5.1.1).

**Remark 1.** It is important that, for fixed $k$, the bilinear functional equation (15.5.1.3) used in the splitting method is the same for different classes of original nonlinear mathematical physics equations.

**Remark 2.** For fixed $m$, solution (15.5.4.1) contains $m(k-m)$ arbitrary constants $C_{i,j}$. Given $k$, the solutions having the maximum number of arbitrary constants are defined by

| Solution number | Number of arbitrary constants | Conditions on $k$ |
|---|---|---|
| $m = \frac{1}{2}k$ | $\frac{1}{4}k^2$ | if $k$ is even, |
| $m = \frac{1}{2}(k \pm 1)$ | $\frac{1}{4}(k^2 - 1)$ | if $k$ is odd. |

It is these solutions of the bilinear functional equation that most frequently result in nontrivial generalized separable solution in nonlinear partial differential equations.

**Remark 3.** The bilinear functional equation (15.5.1.3) and its solutions (15.5.4.1) play an important role in the method of functional separation of variables.

For visualization, the main stages of constructing generalized separable solutions by the splitting method are displayed in Fig. 15.2.

### 15.5.4-2. Solutions of simple functional equations and their application.

Below we give solutions to two simple bilinear functional equations of the form (15.5.1.3) that will be used subsequently for solving specific nonlinear partial differential equations.

1°. The functional equation
$$\Phi_1 \Psi_1 + \Phi_2 \Psi_2 + \Phi_3 \Psi_3 = 0, \qquad (15.5.4.2)$$

where $\Phi_i$ are all functions of the same argument and $\Psi_i$ are all functions of another argument, has two solutions:
$$\begin{aligned}\Phi_1 &= A_1 \Phi_3, & \Phi_2 &= A_2 \Phi_3, & \Psi_3 &= -A_1 \Psi_1 - A_2 \Psi_2; \\ \Psi_1 &= A_1 \Psi_3, & \Psi_2 &= A_2 \Psi_3, & \Phi_3 &= -A_1 \Phi_1 - A_2 \Phi_2.\end{aligned} \qquad (15.5.4.3)$$

The arbitrary constants are renamed as follows: $A_1 = C_{1,1}$ and $A_2 = C_{2,1}$ in the first solution, and $A_1 = -1/C_{1,2}$ and $A_2 = C_{1,1}/C_{1,2}$ in the second solution. The functions on the right-hand sides of the formulas in (15.5.4.3) are assumed to be arbitrary.

2°. The functional equation
$$\Phi_1 \Psi_1 + \Phi_2 \Psi_2 + \Phi_3 \Psi_3 + \Phi_4 \Psi_4 = 0, \qquad (15.5.4.4)$$

where $\Phi_i$ are all functions of the same argument and $\Psi_i$ are all functions of another argument, has a solution
$$\begin{aligned}\Phi_1 &= A_1 \Phi_3 + A_2 \Phi_4, & \Phi_2 &= A_3 \Phi_3 + A_4 \Phi_4, \\ \Psi_3 &= -A_1 \Psi_1 - A_3 \Psi_2, & \Psi_4 &= -A_2 \Psi_1 - A_4 \Psi_2\end{aligned} \qquad (15.5.4.5)$$

dependent on four arbitrary constants $A_1, \ldots, A_4$; see solution (15.5.4.1) with $k = 4$, $m = 2$, $C_{1,1} = A_1$, $C_{1,2} = A_2$, $C_{2,1} = A_3$, and $C_{2,2} = A_4$. The functions on the right-hand sides of the solutions in (15.5.4.3) are assumed to be arbitrary.

---

* Such systems are usually overdetermined.

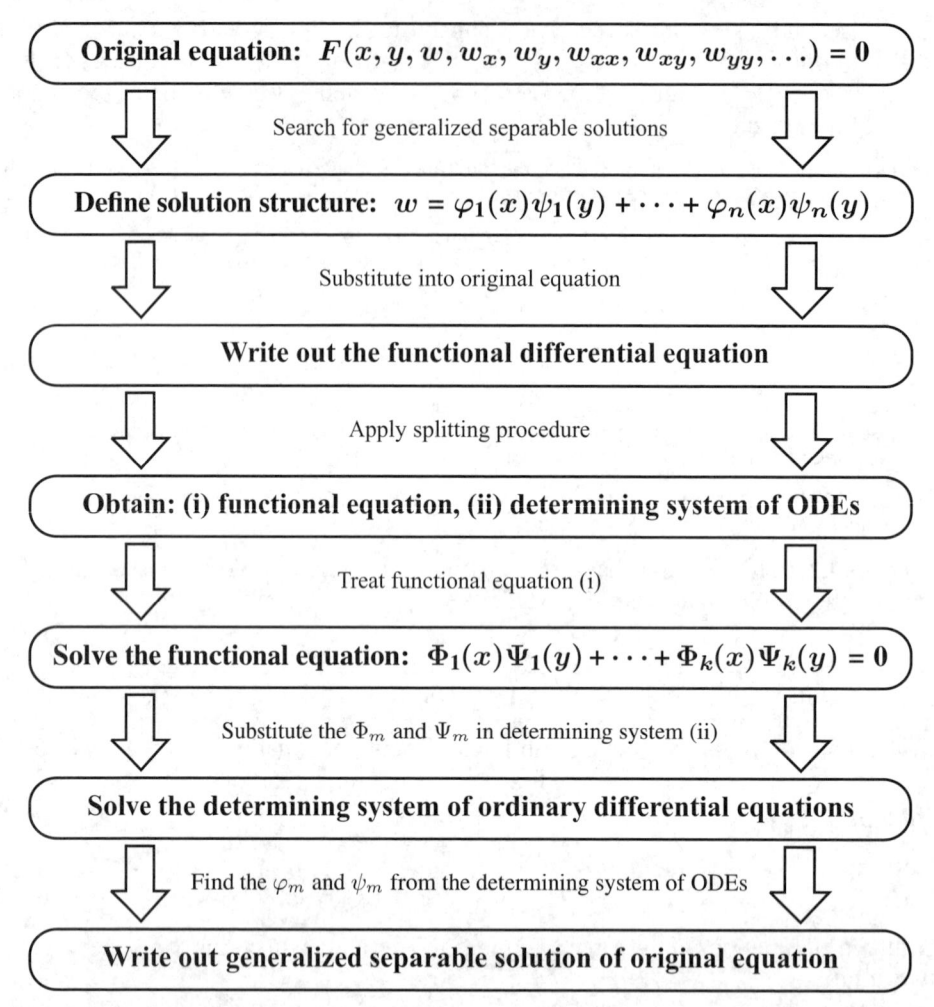

```
┌─────────────────────────────────────────────────────────────────────┐
│ Original equation: $F(x, y, w, w_x, w_y, w_{xx}, w_{xy}, w_{yy}, \ldots) = 0$ │
└─────────────────────────────────────────────────────────────────────┘
                    ⇩  Search for generalized separable solutions  ⇩
┌─────────────────────────────────────────────────────────────────────┐
│ Define solution structure: $w = \varphi_1(x)\psi_1(y) + \cdots + \varphi_n(x)\psi_n(y)$ │
└─────────────────────────────────────────────────────────────────────┘
                    ⇩  Substitute into original equation  ⇩
┌─────────────────────────────────────────────────────────────────────┐
│               Write out the functional differential equation        │
└─────────────────────────────────────────────────────────────────────┘
                    ⇩  Apply splitting procedure  ⇩
┌─────────────────────────────────────────────────────────────────────┐
│       Obtain: (i) functional equation, (ii) determining system of ODEs │
└─────────────────────────────────────────────────────────────────────┘
                    ⇩  Treat functional equation (i)  ⇩
┌─────────────────────────────────────────────────────────────────────┐
│ Solve the functional equation: $\Phi_1(x)\Psi_1(y) + \cdots + \Phi_k(x)\Psi_k(y) = 0$ │
└─────────────────────────────────────────────────────────────────────┘
                    ⇩  Substitute the $\Phi_m$ and $\Psi_m$ in determining system (ii)  ⇩
┌─────────────────────────────────────────────────────────────────────┐
│        Solve the determining system of ordinary differential equations │
└─────────────────────────────────────────────────────────────────────┘
                    ⇩  Find the $\varphi_m$ and $\psi_m$ from the determining system of ODEs  ⇩
┌─────────────────────────────────────────────────────────────────────┐
│          Write out generalized separable solution of original equation │
└─────────────────────────────────────────────────────────────────────┘
```

**Figure 15.2.** General scheme for constructing generalized separable solutions by the splitting method. Abbreviation: ODE stands for ordinary differential equation.

Equation (15.5.4.4) also has two other solutions:

$$\begin{aligned} &\Phi_1 = A_1\Phi_4, \quad \Phi_2 = A_2\Phi_4, \quad \Phi_3 = A_3\Phi_4, \quad \Psi_4 = -A_1\Psi_1 - A_2\Psi_2 - A_3\Psi_3; \\ &\Psi_1 = A_1\Psi_4, \quad \Psi_2 = A_2\Psi_4, \quad \Psi_3 = A_3\Psi_4, \quad \Phi_4 = -A_1\Phi_1 - A_2\Phi_2 - A_3\Phi_3 \end{aligned} \quad (15.5.4.6)$$

involving three arbitrary constants. In the first solution, $A_1 = C_{1,1}$, $A_2 = C_{2,1}$, and $A_3 = C_{3,1}$, and in the second solution, $A_1 = -1/C_{1,3}$, $A_2 = C_{1,1}/C_{1,3}$, and $A_3 = C_{1,2}/C_{1,3}$.

Solutions (15.5.4.6) will sometimes be called *degenerate*, to emphasize the fact that they contain fewer arbitrary constants than solution (15.5.4.5).

3°. Solutions of the functional equation

$$\Phi_1\Psi_1 + \Phi_2\Psi_2 + \Phi_3\Psi_3 + \Phi_4\Psi_4 + \Phi_5\Psi_5 = 0 \qquad (15.5.4.7)$$

can be found by formulas (15.5.4.1) with $k = 5$. Below is a simple technique for finding solutions, which is quite useful in practice, based on equation (15.5.4.7) itself. Let us assume that $\Phi_1$, $\Phi_2$, and $\Phi_3$ are linear combinations of $\Phi_4$ and $\Phi_5$:

$$\Phi_1 = A_1\Phi_4 + B_1\Phi_5, \quad \Phi_2 = A_2\Phi_4 + B_2\Phi_5, \quad \Phi_3 = A_3\Phi_4 + B_3\Phi_5, \qquad (15.5.4.8)$$

where $A_n$, $B_n$ are arbitrary constants. Let us substitute (15.5.4.8) into (15.5.4.7) and collect the terms proportional to $\Phi_4$ and $\Phi_5$ to obtain

$$(A_1\Psi_1 + A_2\Psi_2 + A_3\Psi_3 + \Psi_4)\Phi_4 + (B_1\Psi_1 + B_2\Psi_2 + B_3\Psi_3 + \Psi_5)\Phi_5 = 0.$$

Equating the expressions in parentheses to zero, we have

$$\begin{aligned}\Psi_4 &= -A_1\Psi_1 - A_2\Psi_2 - A_3\Psi_3,\\ \Psi_5 &= -B_1\Psi_1 - B_2\Psi_2 - B_3\Psi_3.\end{aligned} \quad (15.5.4.9)$$

Formulas (15.5.4.8) and (15.5.4.9) give solutions to equation (15.5.4.7). Other solutions are found likewise.

**Example 1.** Consider the nonlinear hyperbolic equation

$$\frac{\partial^2 w}{\partial t^2} = a\frac{\partial}{\partial x}\left(w\frac{\partial w}{\partial x}\right) + f(t)w + g(t), \quad (15.5.4.10)$$

where $f(t)$ and $g(t)$ are arbitrary functions. We look for generalized separable solutions of the form

$$w(x,t) = \varphi(x)\psi(t) + \chi(t). \quad (15.5.4.11)$$

Substituting (15.5.4.11) into (15.5.4.10) and collecting terms yield

$$a\psi^2(\varphi\varphi_x')_x' + a\psi\chi\varphi_{xx}'' + (f\psi - \psi_{tt}'')\varphi + f\chi + g - \chi_{tt}'' = 0.$$

This equation can be represented as a functional equation (15.5.4.4) in which

$$\begin{aligned}&\Phi_1 = (\varphi\varphi_x')_x', \quad \Phi_2 = \varphi_{xx}'', \quad \Phi_3 = \varphi, \quad \Phi_4 = 1,\\ &\Psi_1 = a\psi^2, \quad \Psi_2 = a\psi\chi, \quad \Psi_3 = f\psi - \psi_{tt}'', \quad \Psi_4 = f\chi + g - \chi_{tt}''.\end{aligned} \quad (15.5.4.12)$$

On substituting (15.5.4.12) into (15.5.4.5), we obtain the following overdetermined system of ordinary differential equations for the functions $\varphi = \varphi(x)$, $\psi = \psi(t)$, and $\chi = \chi(t)$:

$$\begin{aligned}(\varphi\varphi_x')_x' &= A_1\varphi + A_2, & \varphi_{xx}'' &= A_3\varphi + A_4,\\ f\psi - \psi_{tt}'' &= -A_1 a\psi^2 - A_3 a\psi\chi, & f\chi + g - \chi_{tt}'' &= -A_2 a\psi^2 - A_4 a\psi\chi.\end{aligned} \quad (15.5.4.13)$$

The first two equations in (15.5.4.13) are compatible only if

$$A_1 = 6B_2, \quad A_2 = B_1^2 - 4B_0 B_2, \quad A_3 = 0, \quad A_4 = 2B_2, \quad (15.5.4.14)$$

where $B_0$, $B_1$, and $B_2$ are arbitrary constants, and the solution is given by

$$\varphi(x) = B_2 x^2 + B_1 x + B_0. \quad (15.5.4.15)$$

On substituting the expressions (15.5.4.14) into the last two equations in (15.5.4.13), we obtain the following system of equations for $\psi(t)$ and $\chi(t)$:

$$\begin{aligned}\psi_{tt}'' &= 6aB_2\psi^2 + f(t)\psi,\\ \chi_{tt}'' &= [2aB_2\psi + f(t)]\chi + a(B_1^2 - 4B_0 B_2)\psi^2 + g(t).\end{aligned} \quad (15.5.4.16)$$

Relations (15.5.4.11), (15.5.4.15) and system (15.5.4.16) determine a generalized separable solution of equation (15.5.4.10). The first equation in (15.5.4.16) can be solved independently; it is linear if $B_2 = 0$ and is integrable by quadrature for $f(t) = $ const. The second equation in (15.5.4.16) is linear in $\chi$ (for $\psi$ known).

Equation (15.5.4.10) does not have other solutions with the form (15.5.4.11) if $f$ and $g$ are arbitrary functions and $\varphi \not\equiv 0$, $\psi \not\equiv 0$, and $\chi \not\equiv 0$.

**Remark.** It can be shown that equation (15.5.4.10) has a more general solution with the form

$$w(x, y) = \varphi_1(x)\psi_1(t) + \varphi_2(x)\psi_2(t) + \psi_3(t), \qquad \varphi_1(x) = x^2, \quad \varphi_2(x) = x, \qquad (15.5.4.17)$$

where the functions $\psi_i = \psi_i(t)$ are determined by the ordinary differential equations

$$\begin{aligned} \psi_1'' &= 6a\psi_1^2 + f(t)\psi_1, \\ \psi_2'' &= [6a\psi_1 + f(t)]\psi_2, \\ \psi_3'' &= [2a\psi_1 + f(t)]\psi_3 + a\psi_2^2 + g(t). \end{aligned} \qquad (15.5.4.18)$$

(The prime denotes a derivative with respect to $t$.) The second equation in (15.5.4.18) has a particular solution $\psi_2 = \psi_1$. Hence, its general solution can be represented as (see Polyanin and Zaitsev, 2003)

$$\psi_2 = C_1\psi_1 + C_2\psi_1 \int \frac{dt}{\psi_1^2}.$$

The solution obtained in Example 1 corresponds to the special case $C_2 = 0$.

**Example 2.** Consider the nonlinear equation

$$\frac{\partial^2 w}{\partial x \partial t} + \left(\frac{\partial w}{\partial x}\right)^2 - w\frac{\partial^2 w}{\partial x^2} = a\frac{\partial^3 w}{\partial x^3}, \qquad (15.5.4.19)$$

which arises in hydrodynamics (see Polyanin and Zaitsev, 2004).

We look for exact solutions of the form

$$w = \varphi(t)\theta(x) + \psi(t). \qquad (15.5.4.20)$$

Substituting (15.5.4.20) into (15.5.4.19) yields

$$\varphi_t'\theta_x' - \varphi\psi\theta_{xx}'' + \varphi^2\left[(\theta_x')^2 - \theta\theta_{xx}''\right] - a\varphi\theta_{xxx}''' = 0.$$

This functional differential equation can be reduced to the functional equation (15.5.4.4) by setting

$$\begin{aligned} \Phi_1 &= \varphi_t', & \Phi_2 &= \varphi\psi, & \Phi_3 &= \varphi^2, & \Phi_4 &= a\varphi, \\ \Psi_1 &= \theta_x', & \Psi_2 &= -\theta_{xx}'', & \Psi_3 &= (\theta_x')^2 - \theta\theta_{xx}'', & \Psi_4 &= -\theta_{xxx}'''. \end{aligned} \qquad (15.5.4.21)$$

On substituting these expressions into (15.5.4.5), we obtain the system of ordinary differential equations

$$\begin{aligned} \varphi_t' &= A_1\varphi^2 + A_2a\varphi, & \varphi\psi &= A_3\varphi^2 + A_4a\varphi, \\ (\theta_x')^2 - \theta\theta_{xx}'' &= -A_1\theta_x' + A_3\theta_{xx}'', & \theta_{xxx}''' &= A_2\theta_x' - A_4\theta_{xx}''. \end{aligned} \qquad (15.5.4.22)$$

It can be shown that the last two equations in (15.5.4.22) are compatible only if the function $\theta$ and its derivative are linearly dependent,

$$\theta_x' = B_1\theta + B_2. \qquad (15.5.4.23)$$

The six constants $B_1$, $B_2$, $A_1$, $A_2$, $A_3$, and $A_4$ must satisfy the three conditions

$$\begin{aligned} B_1(A_1 + B_2 - A_3B_1) &= 0, \\ B_2(A_1 + B_2 - A_3B_1) &= 0, \\ B_1^2 + A_4B_1 - A_2 &= 0. \end{aligned} \qquad (15.5.4.24)$$

Integrating (15.5.4.23) yields

$$\theta = \begin{cases} B_3\exp(B_1x) - \dfrac{B_2}{B_1} & \text{if } B_1 \neq 0, \\ B_2x + B_3 & \text{if } B_1 = 0, \end{cases} \qquad (15.5.4.25)$$

where $B_3$ is an arbitrary constant.

The first two equations in (15.5.4.22) lead to the following expressions for $\varphi$ and $\psi$:

$$\varphi = \begin{cases} \dfrac{A_2a}{C\exp(-A_2at) - A_1} & \text{if } A_2 \neq 0, \\ -\dfrac{1}{A_1t + C} & \text{if } A_2 = 0, \end{cases} \qquad \psi = A_3\varphi + A_4a, \qquad (15.5.4.26)$$

where $C$ is an arbitrary constant.

Formulas (15.5.4.25), (15.5.4.26) and relations (15.5.4.24) allow us to find the following solutions of equation (15.5.4.19) with the form (15.5.4.20):

$$w = \frac{x + C_1}{t + C_2} + C_3 \qquad \text{if} \quad A_2 = B_1 = 0, \ B_2 = -A_1;$$

$$w = \frac{C_1 e^{-\lambda x} + 1}{\lambda t + C_2} + a\lambda \qquad \text{if} \quad A_2 = 0, \ B_1 = -A_4, \ B_2 = -A_1 - A_3 A_4;$$

$$w = C_1 e^{-\lambda(x + a\beta t)} + a(\lambda + \beta) \qquad \text{if} \quad A_1 = A_3 = B_2 = 0, \ A_2 = B_1^2 + A_4 B_1;$$

$$w = \frac{a\beta + C_1 e^{-\lambda x}}{1 + C_2 e^{-a\lambda\beta t}} + a(\lambda - \beta) \qquad \text{if} \quad A_1 = A_3 B_1 - B_2, \ A_2 = B_1^2 + A_4 B_1,$$

where $C_1, C_2, C_3, \beta$, and $\lambda$ are arbitrary constants (these can be expressed in terms of the $A_k$ and $B_k$).

The analysis of the second solution (15.5.4.6) of the functional equation (15.5.4.4) in view of (15.5.4.21) leads to the following two more general solutions of the differential equation (15.5.4.19):

$$w = \frac{x}{t + C_1} + \psi(t),$$

$$w = \varphi(t) e^{-\lambda x} - \frac{\varphi'_t(t)}{\lambda \varphi(t)} + a\lambda,$$

where $\varphi(t)$ and $\psi(t)$ are arbitrary functions, and $C_1$ and $\lambda$ are arbitrary constants.

### 15.5.5. Titov–Galaktionov Method

**15.5.5-1. Method description. Linear subspaces invariant under a nonlinear operator.**

Consider the nonlinear evolution equation

$$\frac{\partial w}{\partial t} = F[w], \tag{15.5.5.1}$$

where $F[w]$ is a nonlinear differential operator with respect to $x$,

$$F[w] \equiv F\left(x, w, \frac{\partial w}{\partial x}, \ldots, \frac{\partial^m w}{\partial x^m}\right). \tag{15.5.5.2}$$

*Definition.* A finite-dimensional linear subspace

$$\mathscr{L}_n = \{\varphi_1(x), \ldots, \varphi_n(x)\} \tag{15.5.5.3}$$

formed by linear combinations of linearly independent functions $\varphi_1(x), \ldots, \varphi_n(x)$ is called invariant under the operator $F$ if $F[\mathscr{L}_n] \subseteq \mathscr{L}_n$. This means that there exist functions $f_1, \ldots, f_n$ such that

$$F\left[\sum_{i=1}^{n} C_i \varphi_i(x)\right] = \sum_{i=1}^{n} f_i(C_1, \ldots, C_n) \varphi_i(x) \tag{15.5.5.4}$$

for arbitrary constants $C_1, \ldots, C_n$.

Let the linear subspace (15.5.5.3) be invariant under the operator $F$. Then equation (15.5.5.1) possesses generalized separable solutions of the form

$$w(x, t) = \sum_{i=1}^{n} \psi_i(t) \varphi_i(x). \tag{15.5.5.5}$$

Here, the functions $\psi_1(t), \ldots, \psi_n(t)$ are described by the autonomous system of ordinary differential equations

$$\psi'_i = f_i(\psi_1, \ldots, \psi_n), \qquad i = 1, \ldots, n, \tag{15.5.5.6}$$

where the prime denotes a derivative with respect to $t$.

The following example illustrates the scheme for constructing generalized separable solutions.

**Example 1.** Consider the nonlinear second-order parabolic equation

$$\frac{\partial w}{\partial t} = a\frac{\partial^2 w}{\partial x^2} + \left(\frac{\partial w}{\partial x}\right)^2 + kw^2 + bw + c. \qquad (15.5.5.7)$$

Obviously, the nonlinear differential operator

$$F[w] = aw_{xx} + (w_x)^2 + kw^2 + bw + c \qquad (15.5.5.8)$$

for $k > 0$ has a two-dimensional invariant subspace $\mathscr{L}_2 = \{1, \cos(x\sqrt{k}\,)\}$. Indeed, for arbitrary $C_1$ and $C_2$ we have

$$F\bigl[C_1 + C_2\cos(x\sqrt{k}\,)\bigr] = k(C_1^2 + C_2^2) + bC_1 + c + C_2(2kC_1 - ak + b)\cos(x\sqrt{k}\,).$$

Therefore, there is a generalized separable solution of the form

$$w(x,t) = \psi_1(t) + \psi_2(t)\cos(x\sqrt{k}\,), \qquad (15.5.5.9)$$

where the functions $\psi_1(t)$ and $\psi_2(t)$ are determined by the autonomous system of ordinary differential equations

$$\begin{aligned}\psi_1' &= k(\psi_1^2 + \psi_2^2) + b\psi_1 + c, \\ \psi_2' &= \psi_2(2k\psi_1 - ak + b).\end{aligned} \qquad (15.5.5.10)$$

**Remark 1.** Example 3 below shows how one can find all two-dimensional linear subspaces invariant under the nonlinear differential operator (15.5.5.8).

**Remark 2.** For $k > 0$, the nonlinear differential operator (15.5.5.8) has a three-dimensional invariant subspace $\mathscr{L}_3 = \{1, \sin(x\sqrt{k}\,), \cos(x\sqrt{k}\,)\}$; see Example 3.

**Remark 3.** For $k < 0$, the nonlinear differential operator (15.5.5.8) has a three-dimensional invariant subspace $\mathscr{L}_3 = \{1, \sinh(x\sqrt{-k}\,), \cosh(x\sqrt{-k}\,)\}$; see Example 3.

**Remark 4.** A more general equation (15.5.5.7), with $a = a(t)$, $b = b(t)$, and $c = c(t)$ being arbitrary functions, and $k = \text{const} < 0$, also admits a generalized separable solution of the form (15.5.5.9), where the functions $\psi_1(t)$ and $\psi_2(t)$ are determined by the system of ordinary differential equations (15.5.5.10).

### 15.5.5-2. Some generalizations.

Likewise, one can consider a more general equation of the form

$$L_1[w] = L_2[U], \qquad U = F[w], \qquad (15.5.5.11)$$

where $L_1[w]$ and $L_2[U]$ are linear differential operators with respect to $t$,

$$L_1[w] \equiv \sum_{i=0}^{s_1} a_i(t)\frac{\partial^i w}{\partial t^i}, \qquad L_2[U] \equiv \sum_{j=0}^{s_2} b_j(t)\frac{\partial^j U}{\partial t^j}, \qquad (15.5.5.12)$$

and $F[w]$ is a nonlinear differential operator with respect to $x$,

$$F[w] \equiv F\left(t, x, w, \frac{\partial w}{\partial x}, \ldots, \frac{\partial^m w}{\partial x^m}\right), \qquad (15.5.5.13)$$

and may depend on $t$ as a parameter.

Let the linear subspace (15.5.5.3) be invariant under the operator $F$, i.e., for arbitrary constants $C_1, \ldots, C_n$ the following relation holds:

$$F\left[\sum_{i=1}^{n} C_i \varphi_i(x)\right] = \sum_{i=1}^{n} f_i(t, C_1, \ldots, C_n) \varphi_i(x). \quad (15.5.5.14)$$

Then equation (15.5.5.11) possesses generalized separable solutions of the form (15.5.5.5), where the functions $\psi_1(t), \ldots, \psi_n(t)$ are described by the system of ordinary differential equations

$$L_1[\psi_i(t)] = L_2[f_i(t, \psi_1, \ldots, \psi_n)], \quad i = 1, \ldots, n. \quad (15.5.5.15)$$

**Example 2.** Consider the equation

$$a_2(t)\frac{\partial^2 w}{\partial t^2} + a_1(t)\frac{\partial w}{\partial t} = \frac{\partial w}{\partial x}\frac{\partial^2 w}{\partial x^2}, \quad (15.5.5.16)$$

which, in the special case of $a_2(t) = k_2$ and $a_1(t) = k_1/t$, is used for describing transonic gas flows (where $t$ plays the role of a spatial variable).

Equation (15.5.5.16) is a special case of equation (15.5.5.11), where $L_1[w] = a_2(t)w_{tt} + a_1(t)w_t$, $L_2[U] = U$, and $F[w] = w_x w_{xx}$. It can be shown that the nonlinear differential operator $F[w] = w_x w_{xx}$ admits the three-dimensional invariant subspace $\mathscr{L}_3 = \{1, x^{3/2}, x^3\}$. Therefore, equation (15.5.5.16) possesses generalized separable solutions of the form

$$w(x,t) = \psi_1(t) + \psi_2(t)x^{3/2} + \psi_3(t)x^3,$$

where the functions $\psi_1(t)$, $\psi_2(t)$, and $\psi_3(t)$ are described by the system of ordinary differential equations

$$a_2(t)\psi_1'' + a_1(t)\psi_1' = \tfrac{9}{8}\psi_2^2,$$
$$a_2(t)\psi_2'' + a_1(t)\psi_2' = \tfrac{45}{4}\psi_2\psi_3,$$
$$a_2(t)\psi_3'' + a_1(t)\psi_3' = 18\psi_3^2.$$

**Remark.** The operator $F[w] = w_x w_{xx}$ also has a four-dimensional invariant subspace $\mathscr{L}_4 = \{1, x, x^2, x^3\}$. Therefore, equation (15.5.5.16) has a generalized separable solution of the form

$$w(x,t) = \psi_1(t) + \psi_2(t)x + \psi_3(t)x^2 + \psi_4(t)x^3.$$

### 15.5.5-3. How to find linear subspaces invariant under a given nonlinear operator.

The most difficult part in using the Titov–Galaktionov method for the construction of exact solutions to specific equations is to find linear subspaces invariant under a given nonlinear operator.

In order to determine basis functions $\varphi_i = \varphi_i(x)$, let us substitute the linear combination $\sum_{i=1}^{n} C_i \varphi_i(x)$ into the differential operator (15.5.5.2). This is assumed to result in an expression like

$$F\left[\sum_{i=1}^{n} C_i \varphi_i(x)\right] = A_1(C)\Phi_1(X) + A_2(C)\Phi_2(X) + \cdots + A_k(C)\Phi_k(X)$$
$$+ B_1(C)\varphi_1(x) + B_2(C)\varphi_2(x) + \cdots + B_n(C)\varphi_n(x), \quad (15.5.5.17)$$

where $A_j(C)$ and $B_i(C)$ depend on $C_1, \ldots, C_n$ only, and the functionals $\Phi_j(X)$ depend on $x$ and are independent of $C_1, \ldots, C_n$:

$$A_j(C) \equiv A_j(C_1, \ldots, C_n), \quad j = 1, \ldots, k,$$
$$B_i(C) \equiv B_i(C_1, \ldots, C_n), \quad i = 1, \ldots, n, \quad (15.5.5.18)$$
$$\Phi_j(X) \equiv \Phi_j\bigl(x, \varphi_1, \varphi_1', \varphi_1'', \ldots, \varphi_n, \varphi_n', \varphi_n''\bigr).$$

Here, for simplicity, the formulas are written out for the case of a second-order differential operator. For higher-order operators, the right-hand sides of relations (15.5.5.18) will contain higher-order derivatives of $\varphi_i$. The functionals and functions $\Phi_1(X), \ldots, \Phi_k(X)$, $\varphi_1(x), \ldots, \varphi_n(x)$ together are assumed to be linearly independent, and the $A_j(C)$ are linearly independent functions of $C_1, \ldots, C_n$.

The basis functions are determined by solving the (usually overdetermined) system of ordinary differential equations

$$\Phi_j\big(x, \varphi_1, \varphi'_1, \varphi''_1, \ldots, \varphi_n, \varphi'_n, \varphi''_n\big) = p_{j,1}\varphi_1 + p_{j,2}\varphi_2 + \cdots + p_{j,n}\varphi_n, \quad j = 1, \ldots, k, \tag{15.5.5.19}$$

where $p_{j,i}$ are some constants independent of the parameters $C_1, \ldots, C_n$. If for some collection of the constants $p_{i,j}$, system (15.5.5.19) is solvable (in practice, it suffices to find a particular solution), then the functions $\varphi_i = \varphi_i(x)$ define a linear subspace invariant under the nonlinear differential operator (15.5.5.2). In this case, the functions appearing on the right-hand side of (15.5.5.4) are given by

$$\begin{aligned} f_i(C_1, \ldots, C_n) &= p_{1,i} A_1(C_1, \ldots, C_n) + p_{2,i} A_2(C_1, \ldots, C_n) + \cdots \\ &\quad + p_{k,i} A_k(C_1, \ldots, C_n) + B_i(C_1, \ldots, C_n). \end{aligned}$$

*Remark.* The analysis of nonlinear differential operators is useful to begin with looking for two-dimensional invariant subspaces of the form $\mathscr{L}_2 = \{1, \varphi(x)\}$.

*Proposition 1.* Let a nonlinear differential operator $F[w]$ admit a two-dimensional invariant subspace of the form $\mathscr{L}_2 = \{1, \varphi(x)\}$, where $\varphi(x) = p\varphi_1(x) + q\varphi_2(x)$, $p$ and $q$ are arbitrary constants, and the functions $1, \varphi_1(x), \varphi_2(x)$ are linearly independent. Then the operator $F[w]$ also admits a three-dimensional invariant subspace $\mathscr{L}_2 = \{1, \varphi_1(x), \varphi_2(x)\}$.

*Proposition 2.* Let two nonlinear differential operators $F_1[w]$ and $F_2[w]$ admit one and the same invariant subspace $\mathscr{L}_n = \{\varphi_1(x), \ldots, \varphi_n(x)\}$. Then the nonlinear operator $pF_1[w] + qF_2[w]$, where $p$ and $q$ are arbitrary constants, also admits the same invariant subspace.

*Example 3.* Consider the nonlinear differential operator (15.5.5.8). We look for its invariant subspaces of the form $\mathscr{L}_2 = \{1, \varphi(x)\}$. We have

$$F[C_1 + C_2\varphi(x)] = C_2^2[(\varphi'_x)^2 + k\varphi^2] + C_2 a \varphi''_{xx} + kC_1^2 + bC_1 + c + (bC_2 + 2kC_1C_2)\varphi.$$

Here, $\Phi_1(X) = (\varphi'_x)^2 + k\varphi^2$ and $\Phi_2(X) = a\varphi''_{xx}$. Hence, the basis function $\varphi(x)$ is determined by the overdetermined system of ordinary differential equations

$$\begin{aligned} (\varphi'_x)^2 + k\varphi^2 &= p_1 + p_2 \varphi, \\ \varphi''_{xx} &= p_3 + p_4 \varphi, \end{aligned} \tag{15.5.5.20}$$

where $p_1 = p_{1,1}$, $p_2 = p_{1,2}$, $p_3 = p_{2,1}/a$, and $p_4 = p_{2,2}/a$. Let us investigate system (15.5.5.20) for consistency. To this end, we differentiate the first equation with respect to $x$ and then divide by $\varphi'_x$ to obtain $\varphi''_{xx} = -k\varphi + p_2/2$. Using this relation to eliminate the second derivative from the second equation in (15.5.5.20), we get $(p_4 + k)\varphi + p_3 - \frac{1}{2}p_2 = 0$. For this equation to be satisfied, the following identities must hold:

$$p_4 = -k, \quad p_3 = \tfrac{1}{2}p_2. \tag{15.5.5.21}$$

The simultaneous solution of system (15.5.5.20) under condition (15.5.5.21) is given by

$$\begin{aligned} \varphi(x) &= px^2 + qx & &\text{if } k = 0 & &(p_1 = q^2,\ p_2 = 4p), \\ \varphi(x) &= p\sin(x\sqrt{k}) + q\cos(x\sqrt{k}) & &\text{if } k > 0 & &(p_1 = kp^2 + kq^2,\ p_2 = 0), \\ \varphi(x) &= p\sinh(x\sqrt{-k}) + q\cosh(x\sqrt{-k}) & &\text{if } k < 0 & &(p_1 = -kp^2 + kq^2,\ p_2 = 0), \end{aligned} \tag{15.5.5.22}$$

where $p$ and $q$ are arbitrary constants.

Since formulas (15.5.5.22) involve two arbitrary parameters $p$ and $q$, it follows from Proposition 1 that the nonlinear differential operator (15.5.5.8) admits the following invariant subspaces:

$$\begin{aligned} \mathscr{L}_3 &= \{1, x, x^2\} & &\text{if } k = 0, \\ \mathscr{L}_3 &= \{1, \sin(x\sqrt{k}), \cos(x\sqrt{k})\} & &\text{if } k > 0, \\ \mathscr{L}_3 &= \{1, \sinh(x\sqrt{-k}), \cosh(x\sqrt{-k})\} & &\text{if } k < 0. \end{aligned}$$

## 15.6. Method of Functional Separation of Variables

### 15.6.1. Structure of Functional Separable Solutions. Solution by Reduction to Equations with Quadratic Nonlinearities

15.6.1-1. Structure of functional separable solutions.

Suppose a nonlinear equation for $w = w(x, y)$ is obtained from a linear mathematical physics equation for $z = z(x, y)$ by a nonlinear change of variable $w = F(z)$. Then, if the linear equation for $z$ admits separable solutions, the transformed nonlinear equation for $w$ will have exact solutions of the form

$$w(x, y) = F(z), \quad \text{where} \quad z = \sum_{m=1}^{n} \varphi_m(x) \psi_m(y). \tag{15.6.1.1}$$

It is noteworthy that many nonlinear partial differential equations that are not reducible to linear equations have exact solutions of the form (15.6.1.1) as well. We will call such solutions *functional separable solutions*. In general, the functions $\varphi_m(x)$, $\psi_m(y)$, and $F(z)$ in (15.6.1.1) are not known in advance and are to be identified.

*Main idea*: The functional differential equation resulting from the substitution of (15.6.1.1) in the original partial differential equation should be reduced to the standard bilinear functional equation (15.5.1.3) or to a functional differential equation of the form (15.5.1.3)–(15.5.1.4), and then the results of Subsections 15.5.3–15.5.5 should be used.

Remark. The function $F(z)$ can be determined by a single ordinary differential equation or by an overdetermined system of equations; both possibilities must be taken into account.

15.6.1-2. Solution by reduction to equations with quadratic (or power) nonlinearities.

In some cases, solutions of the form (15.6.1.1) can be searched for in two stages. First, one looks for a transformation that would reduce the original equation to an equation with a quadratic (or power) nonlinearity. Then the methods outlined in Subsections 15.5.3–15.5.5 are used for finding solutions of the resulting equation.

Table 15.5 gives examples of nonlinear heat equations with power, exponential, and logarithmic nonlinearities reducible, by simple substitutions of the form $w = F(z)$, to quadratically nonlinear equations. For these equations, it can be assumed that the form of the function $F(z)$ in solution (15.6.1.1) is given a priori.

### 15.6.2. Special Functional Separable Solutions. Generalized Traveling-Wave Solutions

15.6.2-1. Special functional separable and generalized traveling-wave solutions.

To simplify the analysis, some of the functions in (15.6.1.1) can be specified a priori and the other functions will be defined in the analysis. We call such solutions *special functional separable solutions*.

A generalized separable solution (see Section 15.5) is a functional separable solution of the special form corresponding to $F(z) = z$.

Below we consider two simplest functional separable solutions of special forms:

$$\begin{aligned} w &= F(z), \quad z = \varphi_1(x)y + \varphi_2(x); \\ w &= F(z), \quad z = \varphi(x) + \psi(y). \end{aligned} \tag{15.6.2.1}$$

TABLE 15.5
Some nonlinear heat equations reducible to quadratically nonlinear equations by a transformation of the form $w = F(z)$; the constant $\sigma$ is expressed in terms of the coefficients of the transformed equation

| Original equation | Transformation | Transformed equation | Form of solutions |
|---|---|---|---|
| $\frac{\partial w}{\partial t} = a\frac{\partial}{\partial x}\left(w^n \frac{\partial w}{\partial x}\right) + bw + cw^{1-n}$ | $w = z^{1/n}$ | $\frac{\partial z}{\partial t} = az\frac{\partial^2 z}{\partial x^2} + \frac{a}{n}\left(\frac{\partial z}{\partial x}\right)^2 + bnz + cn$ | $z = \varphi(t)x^2 + \psi(t)x + \chi(t)$ |
| $\frac{\partial w}{\partial t} = a\frac{\partial}{\partial x}\left(w^n \frac{\partial w}{\partial x}\right) + bw^{n+1} + cw$ | $w = z^{1/n}$ | $\frac{\partial z}{\partial t} = az\frac{\partial^2 z}{\partial x^2} + \frac{a}{n}\left(\frac{\partial z}{\partial x}\right)^2 + bnz^2 + cnz$ | $z = \varphi(t)e^{\sigma x} + \psi(t)e^{-\sigma x} + \chi(t)$ $z = \varphi(t)\sin(\sigma x) + \psi(t)\cos(\sigma x) + \chi(t)$ |
| $\frac{\partial w}{\partial t} = a\frac{\partial}{\partial x}\left(e^{\lambda w} \frac{\partial w}{\partial x}\right) + b + ce^{-\lambda w}$ | $w = \frac{1}{\lambda}\ln z$ | $\frac{\partial z}{\partial t} = az\frac{\partial^2 z}{\partial x^2} + b\lambda z + c\lambda$ | $z = \varphi(t)x^2 + \psi(t)x + \chi(t)$ |
| $\frac{\partial w}{\partial t} = a\frac{\partial}{\partial x}\left(e^{\lambda w} \frac{\partial w}{\partial x}\right) + be^{\lambda w} + c$ | $w = \frac{1}{\lambda}\ln z$ | $\frac{\partial z}{\partial t} = az\frac{\partial^2 z}{\partial x^2} + bz^2 + c\lambda z$ | $z = \varphi(t)e^{\sigma x} + \psi(t)e^{-\sigma x} + \chi(t)$ $z = \varphi(t)\sin(\sigma x) + \psi(t)\cos(\sigma x) + \chi(t)$ |
| $\frac{\partial w}{\partial t} = a\frac{\partial^2 w}{\partial x^2} + bw\ln w + cw$ | $w = e^z$ | $\frac{\partial z}{\partial t} = a\frac{\partial^2 z}{\partial x^2} + a\left(\frac{\partial z}{\partial x}\right)^2 + bz + c$ | $z = \varphi(t)x^2 + \psi(t)x + \chi(t)$ |
| $\frac{\partial w}{\partial t} = a\frac{\partial^2 w}{\partial x^2} + bw\ln^2 w + cw$ | $w = e^z$ | $\frac{\partial z}{\partial t} = a\frac{\partial^2 z}{\partial x^2} + a\left(\frac{\partial z}{\partial x}\right)^2 + bz^2 + c$ | $z = \varphi(t)e^{\sigma x} + \psi(t)e^{-\sigma x} + \chi(t)$ $z = \varphi(t)\sin(\sigma x) + \psi(t)\cos(\sigma x) + \chi(t)$ |

The first solution (15.6.2.1) will be called a *generalized traveling-wave solution* ($x$ and $y$ can be swapped). After substituting this solution into the original equation, one should eliminate $y$ with the help of the expression for $z$. This will result in a functional differential equation with two arguments, $x$ and $z$. Its solution may be obtained with the methods outlined in Subsections 15.5.3–15.5.5.

Remark 1. In functional separation of variables, searching for solutions in the forms $w = F(\varphi(x) + \psi(y))$ [it is the second solution in (15.6.2.1)] and $w = F(\varphi(x)\psi(y))$ leads to equivalent results because the two forms are functionally equivalent. Indeed, we have $F(\varphi(x)\psi(y)) = F_1(\varphi_1(x) + \psi_1(y))$, where $F_1(z) = F(e^z)$, $\varphi_1(x) = \ln \varphi(x)$, and $\psi_1(y) = \ln \psi(y)$.

Remark 2. In constructing functional separable solutions with the form $w = F(\varphi(x) + \psi(y))$ [it is the second solution in (15.6.2.1)], it is assumed that $\varphi \not\equiv \text{const}$ and $\psi \not\equiv \text{const}$.

**Example 1.** Consider the third-order nonlinear equation

$$\frac{\partial w}{\partial y}\frac{\partial^2 w}{\partial x \partial y} - \frac{\partial w}{\partial x}\frac{\partial^2 w}{\partial y^2} = a\left(\frac{\partial^2 w}{\partial y^2}\right)^{n-1}\frac{\partial^3 w}{\partial y^3},$$

which describes a boundary layer of a power-law fluid on a flat plate; $w$ is the stream function, $x$ and $y$ are coordinates along and normal to the plate, and $n$ is a rheological parameter (the value $n = 1$ corresponds to a Newtonian fluid). Searching for solutions in the form

$$w = w(z), \qquad z = \varphi(x)y + \psi(x)$$

leads to the equation $\varphi'_x(w'_z)^2 = a\varphi^{2n}(w''_{zz})^{n-1}w'''_{zzz}$, which is independent of $\psi$. Separating the variables and integrating yields

$$\varphi(x) = (ax + C)^{1/(1-2n)}, \qquad \psi(x) \text{ is arbitrary,}$$

and $w = w(z)$ is determined by solving the ordinary differential equation $(w'_z)^2 = (1 - 2n)(w''_{zz})^{n-1}w'''_{zzz}$.

### 15.6.2-2. General scheme for constructing generalized traveling-wave solutions.

For visualization, the general scheme for constructing generalized traveling-wave solutions for evolution equations is displayed in Fig. 15.3.

```
┌─────────────────────────────────────────────────────────────────┐
│  Original equation:  $w_t = H(t, w, w_x, w_{xx}, \ldots, w_x^{(n)})$  │
└─────────────────────────────────────────────────────────────────┘
                    ⇓  Search for generalized traveling-wave solutions  ⇓
┌─────────────────────────────────────────────────────────────────┐
│  Define solution structure:  $w = F(z)$, where $z = \varphi(t)x + \psi(x)$  │
└─────────────────────────────────────────────────────────────────┘
                    ⇓  Substitute into original equation and replace $x$ by $(z-\psi)/\varphi$  ⇓
┌─────────────────────────────────────────────────────────────────┐
│  Write out the functional differential equation in two arguments  │
└─────────────────────────────────────────────────────────────────┘
                    ⇓  Apply splitting procedure  ⇓
┌─────────────────────────────────────────────────────────────────┐
│  Obtain (i) functional equation, (ii) determining system of ODEs │
└─────────────────────────────────────────────────────────────────┘
                    ⇓  Treat functional equation (i)  ⇓
┌─────────────────────────────────────────────────────────────────┐
│  Solve the functional equation:  $\Phi_1(z)\Psi_1(t) + \cdots + \Phi_k(z)\Psi_k(t) = 0$  │
└─────────────────────────────────────────────────────────────────┘
                    ⇓  Substitute the $\Phi_m$ and $\Psi_m$ in determining system (ii)  ⇓
┌─────────────────────────────────────────────────────────────────┐
│  Solve the determining system of ordinary differential equations │
└─────────────────────────────────────────────────────────────────┘
                    ⇓  Find the functions $\varphi$, $\psi$, and $F$  ⇓
┌─────────────────────────────────────────────────────────────────┐
│  Write out generalized traveling-wave solution of original equation │
└─────────────────────────────────────────────────────────────────┘
```

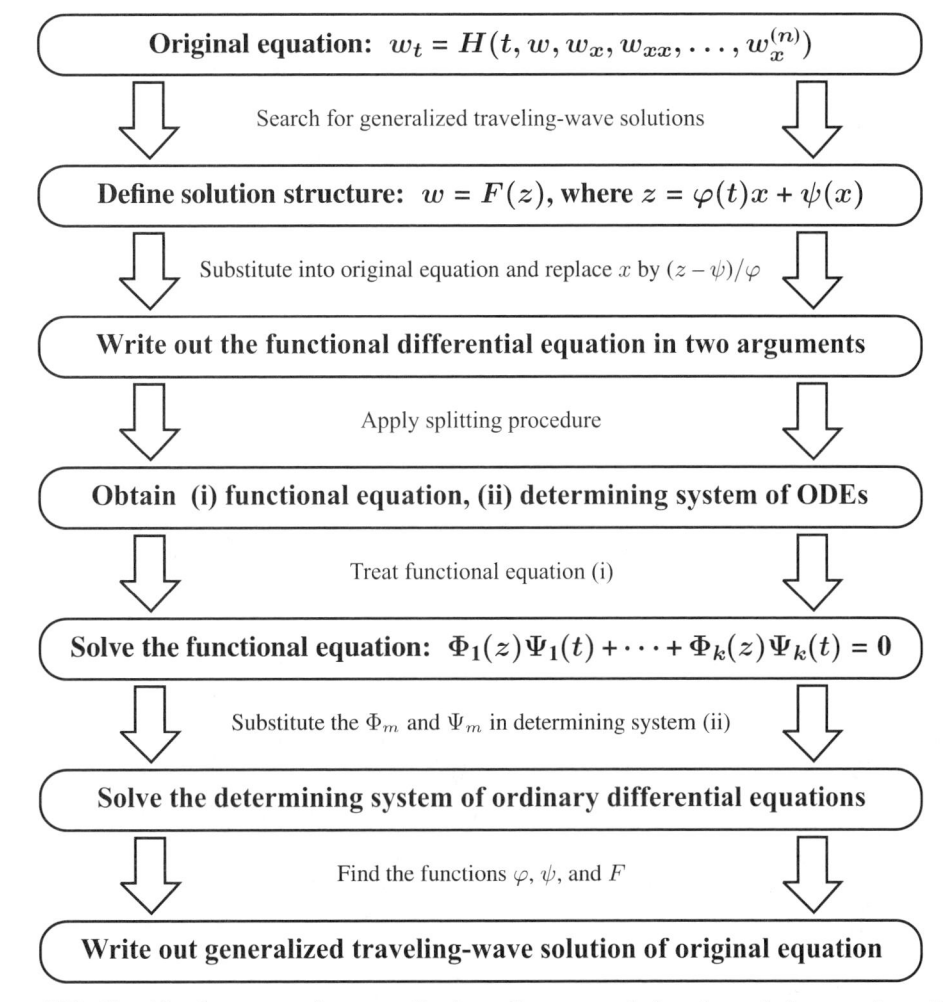

**Figure 15.3.** Algorithm for constructing generalized traveling-wave solutions for evolution equations. Abbreviation: ODE stands for ordinary differential equation.

**Example 2.** Consider the nonstationary heat equation with a nonlinear source

$$\frac{\partial w}{\partial t} = \frac{\partial^2 w}{\partial x^2} + \mathcal{F}(w). \tag{15.6.2.2}$$

We look for functional separable solutions of the special form

$$w = w(z), \quad z = \varphi(t)x + \psi(t). \tag{15.6.2.3}$$

The functions $w(z)$, $\varphi(t)$, $\psi(t)$, and $\mathcal{F}(w)$ are to be determined.

On substituting (15.6.2.3) into (15.6.2.2) and on dividing by $w'_z$, we have

$$\varphi'_t x + \psi'_t = \varphi^2 \frac{w''_{zz}}{w'_z} + \frac{\mathcal{F}(w)}{w'_z}. \tag{15.6.2.4}$$

We express $x$ from (15.6.2.3) in terms of $z$ and substitute into (15.6.2.4) to obtain a functional differential equation with two variables, $t$ and $z$:

$$-\psi'_t + \frac{\psi}{\varphi}\varphi'_t - \frac{\varphi'_t}{\varphi}z + \varphi^2 \frac{w''_{zz}}{w'_z} + \frac{\mathcal{F}(w)}{w'_z} = 0,$$

which can be treated as the functional equation (15.5.4.4), where

$$\Phi_1 = -\psi'_t + \frac{\psi}{\varphi}\varphi'_t, \quad \Phi_2 = -\frac{\varphi'_t}{\varphi}, \quad \Phi_3 = \varphi^2, \quad \Phi_4 = 1,$$

$$\Psi_1 = 1, \quad \Psi_2 = z, \quad \Psi_3 = \frac{w''_{zz}}{w'_z}, \quad \Psi_4 = \frac{\mathcal{F}(w)}{w'_z}.$$

Substituting these expressions into relations (15.5.4.5) yields the system of ordinary differential equations

$$-\psi'_t + \frac{\psi}{\varphi}\varphi'_t = A_1\varphi^2 + A_2, \quad -\frac{\varphi'_t}{\varphi} = A_3\varphi^2 + A_4,$$
$$\frac{w''_{zz}}{w'_z} = -A_1 - A_3 z, \quad \frac{\mathcal{F}(w)}{w'_z} = -A_2 - A_4 z,$$
(15.6.2.5)

where $A_1, \ldots, A_4$ are arbitrary constants.

*Case 1.* For $A_4 \neq 0$, the solution of system (15.6.2.5) is given by

$$\varphi(t) = \pm\left(C_1 e^{2A_4 t} - \frac{A_3}{A_4}\right)^{-1/2},$$
$$\psi(t) = -\varphi(t)\left[A_1 \int \varphi(t)\,dt + A_2 \int \frac{dt}{\varphi(t)} + C_2\right],$$
(15.6.2.6)
$$w(z) = C_3 \int \exp\!\left(-\tfrac{1}{2}A_3 z^2 - A_1 z\right) dz + C_4,$$
$$\mathcal{F}(w) = -C_3(A_4 z + A_2)\exp\!\left(-\tfrac{1}{2}A_3 z^2 - A_1 z\right),$$

where $C_1, \ldots, C_4$ are arbitrary constants. The dependence $\mathcal{F} = \mathcal{F}(w)$ is defined by the last two relations in parametric form ($z$ is considered the parameter). If $A_3 \neq 0$ in (15.6.2.6), the source function is expressed in terms of elementary functions and the inverse of the error function.

In the special case $A_3 = C_4 = 0$, $A_1 = -1$, and $C_3 = 1$, the source function can be represented in explicit form as

$$\mathcal{F}(w) = -w(A_4 \ln w + A_2).$$

*Case 2.* For $A_4 = 0$, the solution to the first two equations in (15.6.2.5) is given by

$$\varphi(t) = \pm\frac{1}{\sqrt{2A_3 t + C_1}}, \quad \psi(t) = \frac{C_2}{\sqrt{2A_3 t + C_1}} - \frac{A_1}{A_3} - \frac{A_2}{3A_3}(2A_3 t + C_1),$$

and the solutions to the other equations are determined by the last two formulas in (15.6.2.6) where $A_4 = 0$.

*Remark.* The algorithm presented in Fig. 15.3 can also be used for finding exact solutions of the more general form $w = \sigma(t)F(z) + \varphi_1(t)x + \psi_2(t)$, where $z = \varphi_1(t)x + \psi_2(t)$. For an example of this sort of solution, see Subsection 15.7.2 (Example 1).

## 15.6.3. Differentiation Method

### 15.6.3-1. Basic ideas of the method. Reduction to a standard equation.

In general, the substitution of expression (15.6.1.1) into the nonlinear partial differential equation under study leads to a functional differential equation with three arguments — two arguments are usual, $x$ and $y$, and the third is composite, $z$. In some cases, the resulting equation can be reduced by differentiation to a standard functional differential equation with two arguments (either $x$ or $y$ is eliminated). To solve the two-argument equation, one can use the methods outlined in Subsections 15.5.3–15.5.5.

### 15.6.3-2. Examples of constructing functional separable solutions.

Below we consider specific examples illustrating the application of the differentiation method for constructing functional separable solutions of nonlinear equations.

**Example 1.** Consider the nonlinear heat equation

$$\frac{\partial w}{\partial t} = \frac{\partial}{\partial x}\left[\mathcal{F}(w)\frac{\partial w}{\partial x}\right].$$
(15.6.3.1)

We look for exact solutions with the form

$$w = w(z), \quad z = \varphi(x) + \psi(t).$$
(15.6.3.2)

On substituting (15.6.3.2) into (15.6.3.1) and dividing by $w'_z$, we obtain the functional differential equation with three variables

$$\psi'_t = \varphi''_{xx}\mathcal{F}(w) + (\varphi'_x)^2 H(z), \tag{15.6.3.3}$$

where

$$H(z) = \mathcal{F}(w)\frac{w''_{zz}}{w'_z} + \mathcal{F}'_z(w), \qquad w = w(z). \tag{15.6.3.4}$$

Differentiating (15.6.3.3) with respect to $x$ yields

$$\varphi'''_{xxx}\mathcal{F}(w) + \varphi'_x\varphi''_{xx}[\mathcal{F}'_z(w) + 2H(z)] + (\varphi'_x)^3 H'_z = 0. \tag{15.6.3.5}$$

This functional differential equation with two variables can be treated as the functional equation (15.5.4.2). This three-term functional equation has two different solutions. Accordingly, we consider two cases.

*Case 1.* The solutions of the functional differential equation (15.6.3.5) are determined from the system of ordinary differential equations

$$\begin{aligned}\mathcal{F}'_z + 2H &= 2A_1\mathcal{F}, \qquad H'_z = A_2\mathcal{F}, \\ \varphi'''_{xxx} + 2A_1\varphi'_x\varphi''_{xx} + A_2(\varphi'_x)^3 &= 0,\end{aligned} \tag{15.6.3.6}$$

where $A_1$ and $A_2$ are arbitrary constants.

The first two equations (15.6.3.6) are linear and independent of the third equation. Their general solution is given by

$$\mathcal{F} = \begin{cases} e^{A_1 z}(B_1 e^{kz} + B_2 e^{-kz}) & \text{if } A_1^2 > 2A_2, \\ e^{A_1 z}(B_1 + B_2 z) & \text{if } A_1^2 = 2A_2, \\ e^{A_1 z}[B_1 \sin(kz) + B_2 \cos(kz)] & \text{if } A_1^2 < 2A_2, \end{cases} \quad H = A_1\mathcal{F} - \tfrac{1}{2}\mathcal{F}'_z, \quad k = \sqrt{|A_1^2 - 2A_2|}. \tag{15.6.3.7}$$

Substituting $H$ of (15.6.3.7) into (15.6.3.4) yields an ordinary differential equation for $w = w(z)$. On integrating this equation, we obtain

$$w = C_1 \int e^{A_1 z}|\mathcal{F}(z)|^{-3/2} dz + C_2, \tag{15.6.3.8}$$

where $C_1$ and $C_2$ are arbitrary constants. The expression of $\mathcal{F}$ in (15.6.3.7) together with expression (15.6.3.8) defines the function $\mathcal{F} = \mathcal{F}(w)$ in parametric form.

Without full analysis, we will study the case $A_2 = 0$ ($k = A_1$) and $A_1 \neq 0$ in more detail. It follows from (15.6.3.7) and (15.6.3.8) that

$$\mathcal{F}(z) = B_1 e^{2A_1 z} + B_2, \quad H = A_1 B_2, \quad w(z) = C_3(B_1 + B_2 e^{-2A_1 z})^{-1/2} + C_2 \quad (C_1 = A_1 B_2 C_3). \tag{15.6.3.9}$$

Eliminating $z$ yields

$$\mathcal{F}(w) = \frac{B_2 C_3^2}{C_3^2 - B_1 w^2}. \tag{15.6.3.10}$$

The last equation in (15.6.3.6) with $A_2 = 0$ has the first integral $\varphi''_{xx} + A_1(\varphi'_x)^2 = \text{const}$. The corresponding general solution is given by

$$\begin{aligned}\varphi(x) &= -\frac{1}{2A_1}\ln\left[\frac{D_2}{D_1}\frac{1}{\sinh^2\left(A_1\sqrt{D_2}\,x + D_3\right)}\right] && \text{for } D_1 > 0 \text{ and } D_2 > 0, \\ \varphi(x) &= -\frac{1}{2A_1}\ln\left[-\frac{D_2}{D_1}\frac{1}{\cos^2\left(A_1\sqrt{-D_2}\,x + D_3\right)}\right] && \text{for } D_1 > 0 \text{ and } D_2 < 0, \\ \varphi(x) &= -\frac{1}{2A_1}\ln\left[-\frac{D_2}{D_1}\frac{1}{\cosh^2\left(A_1\sqrt{D_2}\,x + D_3\right)}\right] && \text{for } D_1 < 0 \text{ and } D_2 > 0,\end{aligned} \tag{15.6.3.11}$$

where $D_1$, $D_2$, and $D_3$ are constants of integration. In all three cases, the following relations hold:

$$(\varphi'_x)^2 = D_1 e^{-2A_1\varphi} + D_2, \qquad \varphi''_{xx} = -A_1 D_1 e^{-2A_1\varphi}. \tag{15.6.3.12}$$

We substitute (15.6.3.9) and (15.6.3.12) into the original functional differential equation (15.6.3.3). With reference to the expression of $z$ in (15.6.3.2), we obtain the following equation for $\psi = \psi(t)$:

$$\psi'_t = -A_1 B_1 D_1 e^{2A_1\psi} + A_1 B_2 D_2.$$

Its general solution is given by

$$\psi(t) = \frac{1}{2A_1}\ln\frac{B_2 D_2}{D_4 \exp(-2A_1^2 B_2 D_2 t) + B_1 D_1}, \tag{15.6.3.13}$$

where $D_4$ is an arbitrary constant.

Formulas (15.6.3.2), (15.6.3.9) for $w$, (15.6.3.11), and (15.6.3.13) define three solutions of the nonlinear equation (15.6.3.1) with $\mathcal{F}(w)$ of the form (15.6.3.10) [recall that these solutions correspond to the special case $A_2 = 0$ in (15.6.3.7) and (15.6.3.8)].

*Case 2*. The solutions of the functional differential equation (15.6.3.5) are determined from the system of ordinary differential equations

$$\varphi'''_{xxx} = A_1(\varphi'_x)^3, \quad \varphi'_x \varphi''_{xx} = A_2(\varphi'_x)^3,$$
$$A_1\mathcal{F} + A_2(\mathcal{F}'_z + 2H) + H'_z = 0. \tag{15.6.3.14}$$

The first two equations in (15.6.3.14) are compatible in the two cases

$$A_1 = A_2 = 0 \implies \varphi(x) = B_1 x + B_2,$$
$$A_1 = 2A_2^2 \implies \varphi(x) = -\frac{1}{A_2} \ln|B_1 x + B_2|. \tag{15.6.3.15}$$

The first solution in (15.6.3.15) eventually leads to the traveling-wave solution $w = w(B_1 x + B_2 t)$ of equation (15.6.3.1) and the second solution to the self-similar solution of the form $w = \widetilde{w}(x^2/t)$. In both cases, the function $\mathcal{F}(w)$ in (15.6.3.1) is arbitrary.

**Example 2.** Consider the *nonlinear Klein–Gordon equation*

$$\frac{\partial^2 w}{\partial t^2} - \frac{\partial^2 w}{\partial x^2} = \mathcal{F}(w). \tag{15.6.3.16}$$

We look for functional separable solutions in additive form:

$$w = w(z), \qquad z = \varphi(x) + \psi(t). \tag{15.6.3.17}$$

Substituting (15.6.3.17) into (15.6.3.16) yields

$$\psi''_{tt} - \varphi''_{xx} + \left[(\psi'_t)^2 - (\varphi'_x)^2\right] g(z) = h(z), \tag{15.6.3.18}$$

where

$$g(z) = w''_{zz}/w'_z, \quad h(z) = \mathcal{F}(w(z))/w'_z. \tag{15.6.3.19}$$

On differentiating (15.6.3.18) first with respect to $t$ and then with respect to $x$ and on dividing by $\psi'_t \varphi'_x$, we have

$$2(\psi''_{tt} - \varphi''_{xx}) g'_z + \left[(\psi'_t)^2 - (\varphi'_x)^2\right] g''_{zz} = h''_{zz}.$$

Eliminating $\psi''_{tt} - \varphi''_{xx}$ from this equation with the aid of (15.6.3.18), we obtain

$$\left[(\psi'_t)^2 - (\varphi'_x)^2\right](g''_{zz} - 2gg'_z) = h''_{zz} - 2g'_z h. \tag{15.6.3.20}$$

This relation holds in the following cases:

$$\begin{aligned} g''_{zz} - 2gg'_z = 0, \quad h''_{zz} - 2g'_z h = 0 & \quad \text{(case 1)}, \\ (\psi'_t)^2 = A\psi + B, \quad (\varphi'_x)^2 = -A\varphi + B - C, \quad h''_{zz} - 2g'_z h = (Az + C)(g''_{zz} - 2gg'_z) & \quad \text{(case 2)}, \end{aligned} \tag{15.6.3.21}$$

where $A$, $B$, and $C$ are arbitrary constants. We consider both cases.

*Case 1*. The first two equations in (15.6.3.21) enable one to determine $g(z)$ and $h(z)$. Integrating the first equation once yields $g'_z = g^2 + \text{const}$. Further, the following cases are possible:

$$g = k, \tag{15.6.3.22a}$$
$$g = -1/(z + C_1), \tag{15.6.3.22b}$$
$$g = -k \tanh(kz + C_1), \tag{15.6.3.22c}$$
$$g = -k \coth(kz + C_1), \tag{15.6.3.22d}$$
$$g = k \tan(kz + C_1), \tag{15.6.3.22e}$$

where $C_1$ and $k$ are arbitrary constants.

The second equation in (15.6.3.21) has a particular solution $h = g(z)$. Hence, its general solution is expressed by [e.g., see Polyanin and Zaitsev (2003)]:

$$h = C_2 g(z) + C_3 g(z) \int \frac{dz}{g^2(z)}, \tag{15.6.3.23}$$

where $C_2$ and $C_3$ are arbitrary constants.

The functions $w(z)$ and $\mathcal{F}(w)$ are found from (15.6.3.19) as

$$w(z) = B_1 \int G(z)\, dz + B_2, \quad \mathcal{F}(w) = B_1 h(z) G(z), \quad \text{where} \quad G(z) = \exp\left[\int g(z)\, dz\right], \tag{15.6.3.24}$$

and $B_1$ and $B_2$ are arbitrary constants ($\mathcal{F}$ is defined parametrically).

Let us dwell on the case (15.6.3.22b). According to (15.6.3.23),

$$h = A_1(z + C_1)^2 + \frac{A_2}{z + C_1}, \qquad (15.6.3.25)$$

where $A_1 = -C_3/3$ and $A_2 = -C_2$ are any numbers. Substituting (15.6.3.22b) and (15.6.3.25) into (15.6.3.24) yields

$$w = B_1 \ln|z + C_1| + B_2, \quad \mathcal{F} = A_1 B_1 (z + C_1) + \frac{A_2 B_1}{(z + C_1)^2}.$$

Eliminating $z$, we arrive at the explicit form of the right-hand side of equation (15.6.3.16):

$$\mathcal{F}(w) = A_1 B_1 e^u + A_2 B_1 e^{-2u}, \quad \text{where} \quad u = \frac{w - B_2}{B_1}. \qquad (15.6.3.26)$$

For simplicity, we set $C_1 = 0$, $B_1 = 1$, and $B_2 = 0$ and denote $A_1 = a$ and $A_2 = b$. Thus, we have

$$w(z) = \ln|z|, \quad \mathcal{F}(w) = ae^w + be^{-2w}, \quad g(z) = -1/z, \quad h(z) = az^2 + b/z. \qquad (15.6.3.27)$$

It remains to determine $\psi(t)$ and $\varphi(x)$. We substitute (15.6.3.27) into the functional differential equation (15.6.3.18). Taking into account (15.6.3.17), we find

$$[\psi''_{tt}\psi - (\psi'_t)^2 - a\psi^3 - b] - [\varphi''_{xx}\varphi - (\varphi'_x)^2 + a\varphi^3] + (\psi''_{tt} - 3a\psi^2)\varphi - \psi(\varphi''_{xx} + 3a\varphi^2) = 0. \qquad (15.6.3.28)$$

Differentiating (15.6.3.28) with respect to $t$ and $x$ yields the separable equation*

$$(\psi'''_{ttt} - 6a\psi\psi'_t)\varphi'_x - (\varphi'''_{xxx} + 6a\varphi\varphi'_x)\psi'_t = 0,$$

whose solution is determined by the ordinary differential equations

$$\psi'''_{ttt} - 6a\psi\psi'_t = A\psi'_t,$$
$$\varphi'''_{xxx} + 6a\varphi\varphi'_x = A\varphi'_x,$$

where $A$ is the separation constant. Each equation can be integrated twice, thus resulting in

$$(\psi'_t)^2 = 2a\psi^3 + A\psi^2 + C_1\psi + C_2,$$
$$(\varphi'_x)^2 = -2a\varphi^3 + A\varphi^2 + C_3\varphi + C_4, \qquad (15.6.3.29)$$

where $C_1, \ldots, C_4$ are arbitrary constants. Eliminating the derivatives from (15.6.3.28) using (15.6.3.29), we find that the arbitrary constants are related by $C_3 = -C_1$ and $C_4 = C_2 + b$. So, the functions $\psi(t)$ and $\varphi(x)$ are determined by the first-order nonlinear autonomous equations

$$(\psi'_t)^2 = 2a\psi^3 + A\psi^2 + C_1\psi + C_2,$$
$$(\varphi'_x)^2 = -2a\varphi^3 + A\varphi^2 - C_1\varphi + C_2 + b.$$

The solutions of these equations are expressed in terms of elliptic functions.

For the other cases in (15.6.3.22), the analysis is performed in a similar way. Table 15.6 presents the final results for the cases (15.6.3.22a)–(15.6.3.22e).

*Case 2.* Integrating the third and fourth equations in (15.6.3.21) yields

$$\psi = \pm\sqrt{B}\,t + D_1, \qquad \varphi = \pm\sqrt{B - C}\,t + D_2 \qquad \text{if} \quad A = 0;$$
$$\psi = \frac{1}{4A}(At + D_1)^2 - \frac{B}{A}, \quad \varphi = -\frac{1}{4A}(Ax + D_2)^2 + \frac{B - C}{A} \quad \text{if} \quad A \neq 0; \qquad (15.6.3.30)$$

where $D_1$ and $D_2$ are arbitrary constants. In both cases, the function $\mathcal{F}(w)$ in equation (15.6.3.16) is arbitrary. The first row in (15.6.3.30) corresponds to the traveling-wave solution $w = w(kx + \lambda t)$. The second row leads to a solution of the form $w = w(x^2 - t^2)$.

---

* To solve equation (15.6.3.28), one can use the solution of functional equation (15.5.4.4) [see (15.5.4.5)].

TABLE 15.6
Nonlinear Klein–Gordon equations $\partial_{tt}w - \partial_{xx}w = \mathcal{F}(w)$ admitting functional separable solutions of the form $w = w(z)$, $z = \varphi(x) + \psi(t)$. Notation: $A$, $C_1$, and $C_2$ are arbitrary constants; $\sigma = 1$ for $z > 0$ and $\sigma = -1$ for $z < 0$.

| No. | Right-hand side $\mathcal{F}(w)$ | Solution $w(z)$ | Equations for $\psi(t)$ and $\varphi(x)$ |
|---|---|---|---|
| 1 | $aw \ln w + bw$ | $e^z$ | $(\psi'_t)^2 = C_1 e^{-2\psi} + a\psi - \frac{1}{2}a + b + A$, <br> $(\varphi'_x)^2 = C_2 e^{-2\varphi} - a\varphi + \frac{1}{2}a + A$ |
| 2 | $ae^w + be^{-2w}$ | $\ln|z|$ | $(\psi'_t)^2 = 2a\psi^3 + A\psi^2 + C_1\psi + C_2$, <br> $(\varphi'_x)^2 = -2a\varphi^3 + A\varphi^2 - C_1\varphi + C_2 + b$ |
| 3 | $a \sin w + b\left(\sin w \ln \tan \frac{w}{4} + 2 \sin \frac{w}{4}\right)$ | $4 \arctan e^z$ | $(\psi'_t)^2 = C_1 e^{2\psi} + C_2 e^{-2\psi} + b\psi + a + A$, <br> $(\varphi'_x)^2 = -C_2 e^{2\varphi} - C_1 e^{-2\varphi} - b\varphi + A$ |
| 4 | $a \sinh w + b\left(\sinh w \ln \tanh \frac{w}{4} + 2 \sinh \frac{w}{2}\right)$ | $2 \ln\left|\coth \frac{z}{2}\right|$ | $(\psi'_t)^2 = C_1 e^{2\psi} + C_2 e^{-2\psi} - \sigma b\psi + a + A$, <br> $(\varphi'_x)^2 = C_2 e^{2\varphi} + C_1 e^{-2\varphi} + \sigma b\varphi + A$ |
| 5 | $a \sinh w + 2b\left(\sinh w \arctan e^{w/2} + \cosh \frac{w}{2}\right)$ | $2 \ln\left|\tan \frac{z}{2}\right|$ | $(\psi'_t)^2 = C_1 \sin 2\psi + C_2 \cos 2\psi + \sigma b\psi + a + A$, <br> $(\varphi'_x)^2 = -C_1 \sin 2\varphi + C_2 \cos 2\varphi - \sigma b\varphi + A$ |

## 15.6.4. Splitting Method. Solutions of Some Nonlinear Functional Equations and Their Applications

15.6.4-1. Three-argument functional equations of special form. Splitting method.

The substitution of the expression

$$w = F(z), \quad z = \varphi(x) + \psi(t) \qquad (15.6.4.1)$$

into a nonlinear partial differential equation sometimes leads to functional differential equations of the form

$$f(t) + \Phi_1(x)\Psi_1(z) + \cdots + \Phi_k(x)\Psi_k(z) = 0, \qquad (15.6.4.2)$$

where $\Phi_j(x)$ and $\Psi_j(z)$ are functionals dependent on the variables $x$ and $z$, respectively,

$$\begin{aligned}\Phi_j(x) &\equiv \Phi_j\big(x, \varphi, \varphi'_x, \ldots, \varphi_x^{(n)}\big), \\ \Psi_j(z) &\equiv \Psi_j\big(F, F'_z, \ldots, F_z^{(n)}\big).\end{aligned} \qquad (15.6.4.3)$$

It is reasonable to solve equation (15.6.4.2) by the splitting method. At the first stage, we treat (15.6.4.2) as a purely functional equation, thus disregarding (15.6.4.3). Differentiating (15.6.4.2) with respect to $x$ yields the standard bilinear functional differential equation in two independent variables $x$ and $z$:

$$\Phi'_1(x)\Psi_1(z) + \cdots + \Phi'_k(x)\Psi_k(z) + \Phi_1(x)\varphi'(x)\Psi'_1(z) + \cdots + \varphi'(x)\Phi_k(x)\Psi'_k(z) = 0, \quad (15.6.4.4)$$

which can be solved using the results of Subsections 15.5.3–15.5.5. Then, substituting the solutions $\Phi_m(x)$ and $\Psi_m(z)$ into (15.6.4.2) and taking into account the second relation in (15.6.4.1), we find the function $f(t)$. Further, substituting the functionals (15.6.4.3) into the solutions of the functional equation (15.6.4.2), we obtain determining systems of ordinary differential equations for $F(z)$, $\varphi(x)$, and $\psi(t)$.

Below, we discuss several types of three-argument functional equations of the form (15.6.4.2) that arise most frequently in the functional separation of variables in nonlinear equations of mathematical physics. The results are used for constructing exact solutions for some classes of nonlinear heat and wave equations.

**15.6.4-2. Functional equation $f(t) + g(x) = Q(z)$, with $z = \varphi(x) + \psi(t)$.**

Here, one of the two functions $f(t)$ and $\psi(t)$ is prescribed and the other is assumed unknown, also one of the functions $g(x)$ and $\psi(x)$ is prescribed and the other is unknown, and the function $Q(z)$ is assumed unknown.*

Differentiating the equation with respect to $x$ and $t$ yields $Q''_{zz} = 0$. Consequently, the solution is given by

$$f(t) = A\psi(x) + B, \quad g(x) = A\varphi(x) - B + C, \quad Q(z) = Az + C, \tag{15.6.4.5}$$

where $A$, $B$, and $C$ are arbitrary constants.

**15.6.4-3. Functional equation $f(t) + g(x) + h(x)Q(z) + R(z) = 0$, with $z = \varphi(x) + \psi(t)$.**

Differentiating the equation with respect to $x$ yields the two-argument equation

$$g'_x + h'_x Q + h\varphi'_x Q'_z + \varphi'_x R'_z = 0. \tag{15.6.4.6}$$

Such equations were discussed in Subsections 15.5.3 and 15.5.4. Hence, the following relations hold [see formulas (15.5.4.4) and (15.5.4.5)]:

$$\begin{aligned} g'_x &= A_1 h \varphi'_x + A_2 \varphi'_x, \\ h'_x &= A_3 h \varphi'_x + A_4 \varphi'_x, \\ Q'_z &= -A_1 - A_3 Q, \\ R'_z &= -A_2 - A_4 Q, \end{aligned} \tag{15.6.4.7}$$

where $A_1, \ldots, A_4$ are arbitrary constants. By integrating system (15.6.4.7) and substituting the resulting solutions into the original functional equation, one obtains the results given below.

*Case 1*. If $A_3 = 0$ in (15.6.4.7), the corresponding solution of the functional equation is given by

$$\begin{aligned} f &= -\tfrac{1}{2} A_1 A_4 \psi^2 + (A_1 B_1 + A_2 + A_4 B_3)\psi - B_2 - B_1 B_3 - B_4, \\ g &= \tfrac{1}{2} A_1 A_4 \varphi^2 + (A_1 B_1 + A_2)\varphi + B_2, \\ h &= A_4 \varphi + B_1, \\ Q &= -A_1 z + B_3, \\ R &= \tfrac{1}{2} A_1 A_4 z^2 - (A_2 + A_4 B_3)z + B_4, \end{aligned} \tag{15.6.4.8}$$

where $A_k$ and $B_k$ are arbitrary constants and $\varphi = \varphi(x)$ and $\psi = \psi(t)$ are arbitrary functions.

---

* In similar equations with a composite argument, it is assumed that $\varphi(x) \neq \text{const}$ and $\psi(t) \neq \text{const}$.

*Case 2.* If $A_3 \neq 0$ in (15.6.4.7), the corresponding solution of the functional equation is

$$f = -B_1 B_3 e^{-A_3\psi} + \left(A_2 - \frac{A_1 A_4}{A_3}\right)\psi - B_2 - B_4 - \frac{A_1 A_4}{A_3^2},$$

$$g = \frac{A_1 B_1}{A_3} e^{A_3\varphi} + \left(A_2 - \frac{A_1 A_4}{A_3}\right)\varphi + B_2,$$

$$h = B_1 e^{A_3\varphi} - \frac{A_4}{A_3}, \qquad (15.6.4.9)$$

$$Q = B_3 e^{-A_3 z} - \frac{A_1}{A_3},$$

$$R = \frac{A_4 B_3}{A_3} e^{-A_3 z} + \left(\frac{A_1 A_4}{A_3} - A_2\right)z + B_4,$$

where $A_k$ and $B_k$ are arbitrary constants and $\varphi = \varphi(x)$ and $\psi = \psi(t)$ are arbitrary functions.

*Case 3.* In addition, the functional equation has two degenerate solutions:

$$f = A_1\psi + B_1, \quad g = A_1\varphi + B_2, \quad h = A_2, \quad R = -A_1 z - A_2 Q - B_1 - B_2, \quad (15.6.4.10a)$$

where $\varphi = \varphi(x)$, $\psi = \psi(t)$, and $Q = Q(z)$ are arbitrary functions; $A_1$, $A_2$, $B_1$, and $B_2$ are arbitrary constants; and

$$f = A_1\psi + B_1, \quad g = A_1\varphi + A_2 h + B_2, \quad Q = -A_2, \quad R = -A_1 z - B_1 - B_2, \quad (15.6.4.10b)$$

where $\varphi = \varphi(x)$, $\psi = \psi(t)$, and $h = h(x)$ are arbitrary functions; and $A_1$, $A_2$, $B_1$, and $B_2$ are arbitrary constants. The degenerate solutions (15.6.4.10a) and (15.6.4.10b) can be obtained directly from the original equation or its consequence (15.6.4.6) using formulas (15.5.4.6).

**Example 1.** Consider the nonstationary heat equation with a nonlinear source

$$\frac{\partial w}{\partial t} = \frac{\partial^2 w}{\partial x^2} + \mathcal{F}(w). \qquad (15.6.4.11)$$

We look for exact solutions of the form

$$w = w(z), \quad z = \varphi(x) + \psi(t). \qquad (15.6.4.12)$$

Substituting (15.6.4.12) into (15.6.4.11) and dividing by $w'_z$ yields the functional differential equation

$$\psi'_t = \varphi''_{xx} + (\varphi'_x)^2 \frac{w''_{zz}}{w'_z} + \frac{\mathcal{F}(w(z))}{w'_z}.$$

Let us solve it by the splitting method. To this end, we represent this equation as the functional equation $f(t) + g(x) + h(x)Q(z) + R(z) = 0$, where

$$f(t) = -\psi'_t, \quad g(x) = \varphi''_{xx}, \quad h(x) = (\varphi'_x)^2, \quad Q(z) = w''_{zz}/w'_z, \quad R(z) = f(w(z))/w'_z. \qquad (15.6.4.13)$$

On substituting the expressions of $g$ and $h$ of (15.6.4.13) into (15.6.4.8)–(15.6.4.10), we arrive at overdetermined systems of equations for $\varphi = \varphi(x)$.

*Case 1.* The system

$$\varphi''_{xx} = \tfrac{1}{2} A_1 A_4 \varphi^2 + (A_1 B_1 + A_2)\varphi + B_2,$$

$$(\varphi'_x)^2 = A_4\varphi + B_1$$

following from (15.6.4.8) and corresponding to $A_3 = 0$ in (15.6.4.7) is consistent in the cases

$$\begin{array}{ll} \varphi = C_1 x + C_2 & \text{for} \quad A_2 = -A_1 C_1^2, \ A_4 = B_2 = 0, \ B_1 = C_1^2, \\ \varphi = \tfrac{1}{4} A_4 x^2 + C_1 x + C_2 & \text{for} \quad A_1 = A_2 = 0, \ B_1 = C_1^2 - A_4 C_2, \ B_2 = \tfrac{1}{2} A_4, \end{array} \qquad (15.6.4.14)$$

where $C_1$ and $C_2$ are arbitrary constants.

The first solution in (15.6.4.14) with $A_1 \neq 0$ leads to a right-hand side of equation (15.6.4.11) containing the inverse of the error function [the form of the right-hand side is identified from the last two relations in (15.6.4.8) and (15.6.4.13)]. The second solution in (15.6.4.14) corresponds to the right-hand side of $\mathcal{F}(w) = k_1 w \ln w + k_2 w$ in (15.6.4.11). In both cases, the first relation in (15.6.4.8) is, taking into account that $f = -\psi'_t$, a first-order linear solution with constant coefficients, whose solution is an exponential plus a constant.

*Case 2.* The system

$$\varphi''_{xx} = \frac{A_1 B_1}{A_3} e^{A_3 \varphi} + \left(A_2 - \frac{A_1 A_4}{A_3}\right) \varphi + B_2,$$

$$(\varphi'_x)^2 = B_1 e^{A_3 \varphi} - \frac{A_4}{A_3},$$

following from (15.6.4.9) and corresponding to $A_3 \neq 0$ in (15.6.4.7), is consistent in the following cases:

$\varphi = \pm\sqrt{-A_4/A_3}\, x + C_1$ for $A_2 = A_1 A_4/A_3$, $B_1 = B_2 = 0$,

$\varphi = -\dfrac{2}{A_3} \ln|x| + C_1$ for $A_1 = \tfrac{1}{2}A_3^2$, $A_2 = A_4 = B_2 = 0$, $B_1 = 4A_3^{-2} e^{-A_3 C_1}$,

$\varphi = -\dfrac{2}{A_3} \ln\left|\cos\left(\tfrac{1}{2}\sqrt{A_3 A_4}\, x + C_1\right)\right| + C_2$ for $A_1 = \tfrac{1}{2}A_3^2$, $A_2 = \tfrac{1}{2}A_3 A_4$, $B_2 = 0$, $A_3 A_4 > 0$,

$\varphi = -\dfrac{2}{A_3} \ln\left|\sinh\left(\tfrac{1}{2}\sqrt{-A_3 A_4}\, x + C_1\right)\right| + C_2$ for $A_1 = \tfrac{1}{2}A_3^2$, $A_2 = \tfrac{1}{2}A_3 A_4$, $B_2 = 0$, $A_3 A_4 < 0$,

$\varphi = -\dfrac{2}{A_3} \ln\left|\cosh\left(\tfrac{1}{2}\sqrt{-A_3 A_4}\, x + C_1\right)\right| + C_2$ for $A_1 = \tfrac{1}{2}A_3^2$, $A_2 = \tfrac{1}{2}A_3 A_4$, $B_2 = 0$, $A_3 A_4 < 0$,

where $C_1$ and $C_2$ are arbitrary constants. The right-hand sides of equation (15.6.4.11) corresponding to these solutions are represented in parametric form.

*Case 3.* Traveling-wave solutions of the nonlinear heat equation (15.6.4.11) and solutions of the linear equation (15.6.4.11) with $\mathcal{F}'_w = \text{const}$ correspond to the degenerate solutions of the functional equation (15.6.4.10).

**15.6.4-4. Functional equation $f(t) + g(x)Q(z) + h(x)R(z) = 0$, with $z = \varphi(x) + \psi(t)$.**

Differentiating with respect to $x$ yields the two-argument functional differential equation

$$g'_x Q + g\varphi'_x Q'_z + h'_x R + h\varphi'_x R'_z = 0, \qquad (15.6.4.15)$$

which coincides with equation (15.5.4.4), up to notation.

*Nondegenerate case.* Equation (15.6.4.15) can be solved using formulas (15.5.4.5). In this way, we arrive at the system of ordinary differential equations

$$\begin{aligned} g'_x &= (A_1 g + A_2 h)\varphi'_x, \\ h'_x &= (A_3 g + A_4 h)\varphi'_x, \\ Q'_z &= -A_1 Q - A_3 R, \\ R'_z &= -A_2 Q - A_4 R, \end{aligned} \qquad (15.6.4.16)$$

where $A_1, \ldots, A_4$ are arbitrary constants.

The solution of equation (15.6.4.16) is given by

$$\begin{aligned} g(x) &= A_2 B_1 e^{k_1 \varphi} + A_2 B_2 e^{k_2 \varphi}, \\ h(x) &= (k_1 - A_1) B_1 e^{k_1 \varphi} + (k_2 - A_1) B_2 e^{k_2 \varphi}, \\ Q(z) &= A_3 B_3 e^{-k_1 z} + A_3 B_4 e^{-k_2 z}, \\ R(z) &= (k_1 - A_1) B_3 e^{-k_1 z} + (k_2 - A_1) B_4 e^{-k_2 z}, \end{aligned} \qquad (15.6.4.17)$$

where $B_1, \ldots, B_4$ are arbitrary constants and $k_1$ and $k_2$ are roots of the quadratic equation

$$(k - A_1)(k - A_4) - A_2 A_3 = 0. \qquad (15.6.4.18)$$

In the degenerate case $k_1 = k_2$, the terms $e^{k_2\varphi}$ and $e^{-k_2 z}$ in (15.6.4.17) must be replaced by $\varphi e^{k_1\varphi}$ and $ze^{-k_1 z}$, respectively. In the case of purely imaginary or complex roots, one should extract the real (or imaginary) part of the roots in solution (15.6.4.17).

On substituting (15.6.4.17) into the original functional equation, one obtains conditions that must be met by the free coefficients and identifies the function $f(t)$, specifically,

$$
\begin{aligned}
B_2 = B_4 = 0 &\implies f(t) = [A_2 A_3 + (k_1 - A_1)^2] B_1 B_3 e^{-k_1 \psi}, \\
B_1 = B_3 = 0 &\implies f(t) = [A_2 A_3 + (k_2 - A_1)^2] B_2 B_4 e^{-k_2 \psi}, \\
A_1 = 0 &\implies f(t) = (A_2 A_3 + k_1^2) B_1 B_3 e^{-k_1 \psi} + (A_2 A_3 + k_2^2) B_2 B_4 e^{-k_2 \psi}.
\end{aligned}
\qquad (15.6.4.19)
$$

Solution (15.6.4.17), (15.6.4.19) involves arbitrary functions $\varphi = \varphi(x)$ and $\psi = \psi(t)$.

*Degenerate case.* In addition, the functional equation has two degenerate solutions,

$$
f = B_1 B_2 e^{A_1 \psi}, \quad g = A_2 B_1 e^{-A_1 \varphi}, \quad h = B_1 e^{-A_1 \varphi}, \quad R = -B_2 e^{A_1 z} - A_2 Q,
$$

where $\varphi = \varphi(x)$, $\psi = \psi(t)$, and $Q = Q(z)$ are arbitrary functions; $A_1$, $A_2$, $B_1$, and $B_2$ are arbitrary constants; and

$$
f = B_1 B_2 e^{A_1 \psi}, \quad h = -B_1 e^{-A_1 \varphi} - A_2 g, \quad Q = A_2 B_2 e^{A_1 z}, \quad R = B_2 e^{A_1 z},
$$

where $\varphi = \varphi(x)$, $\psi = \psi(t)$, and $g = g(x)$ are arbitrary functions; and $A_1$, $A_2$, $B_1$, and $B_2$ are arbitrary constants. The degenerate solutions can be obtained immediately from the original equation or its consequence (15.6.4.15) using formulas (15.5.4.6).

**Example 2.** For the nonlinear heat equation (15.6.3.1), searching for exact solutions in the form $w = w(z)$, with $z = \varphi(x) + \psi(t)$, leads to the functional equation (15.6.3.3), which coincides with the equation $f(t) + g(x)Q(z) + h(x)R(z) = 0$ if

$$
f(t) = -\psi'_t, \quad g(x) = \varphi''_{xx}, \quad h(x) = (\varphi'_x)^2, \quad Q(z) = \mathcal{F}(w), \quad R(z) = \frac{[\mathcal{F}(w) w'_z]'_z}{w'_z}, \quad w = w(z).
$$

## 15.7. Direct Method of Symmetry Reductions of Nonlinear Equations

### 15.7.1. Clarkson–Kruskal Direct Method

**15.7.1-1. Simplified scheme. Examples of constructing exact solutions.**

The basic idea of the simplified scheme is as follows: for an equation with the unknown function $w = w(x,t)$, an exact solution is sought in the form

$$
w = f(t)u(z) + g(x,t), \quad z = \varphi(t)x + \psi(t). \qquad (15.7.1.1)
$$

The functions $f(t)$, $g(x,t)$, $\varphi(t)$, and $\psi(t)$ are found in the subsequent analysis and are chosen in such a way that, ultimately, the function $u(z)$ would satisfy a *single ordinary differential equation*.

Below we consider some cases in which it is possible to construct exact solutions of nonlinear equations of the form (15.7.1.1).

**Example 1.** Consider the *generalized Burgers–Korteweg–de Vries equation*

$$\frac{\partial w}{\partial t} = a\frac{\partial^n w}{\partial x^n} + bw\frac{\partial w}{\partial x}. \tag{15.7.1.2}$$

We seek its exact solution in the form (15.7.1.1). Inserting (15.7.1.1) into (15.7.1.2), we obtain

$$af\varphi^n u_z^{(n)} + bf^2 \varphi u u_z' + f(bg\varphi - \varphi_t' x - \psi_t')u_z' + (bfg_x - f_t')u + ag_x^{(n)} + bgg_x - g_t = 0. \tag{15.7.1.3}$$

Equating the functional coefficients of $u_z^{(n)}$ and $uu_z'$ in (15.7.1.3), we get

$$f = \varphi^{n-1}. \tag{15.7.1.4}$$

Further, equating the coefficient of $u_z'$ to zero, we obtain

$$g = \frac{1}{b\varphi}(\varphi_t' x + \psi_t'). \tag{15.7.1.5}$$

Inserting the expressions (15.7.1.4) and (15.7.1.5) into (15.7.1.3), we arrive at the relation

$$\varphi^{2n-1}(au_z^{(n)} + buu_z') + (2-n)\varphi^{n-2}\varphi_t' u + \frac{1}{b\varphi^2}\left[(2\varphi_t^2 - \varphi\varphi_{tt})x + 2\varphi_t\psi_t - \varphi\psi_{tt}\right] = 0.$$

Dividing each term by $\varphi^{2n-1}$ and then eliminating $x$ with the help of the relation $x = (z - \psi)/\varphi$, we obtain

$$au_z^{(n)} + buu_z' + (2-n)\varphi^{-n-1}\varphi_t' u + \frac{1}{b}\varphi^{-2n-2}(2\varphi_t^2 - \varphi\varphi_{tt})z + \frac{1}{b}\varphi^{-2n-2}(\varphi\psi\varphi_{tt} - \varphi^2\psi_{tt} + 2\varphi\varphi_t\psi_t - 2\psi\varphi_t^2) = 0. \tag{15.7.1.6}$$

Let us require that the functional coefficient of $u$ and the last term be constant,

$$\varphi^{-n-1}\varphi_t' = -A, \qquad \varphi^{-2n-2}(\varphi\psi\varphi_{tt} - \varphi^2\psi_{tt} + 2\varphi\varphi_t\psi_t - 2\psi\varphi_t^2) = B,$$

where $A$ and $B$ are arbitrary. As a result, we arrive at the following system of ordinary differential equations for $\varphi$ and $\psi$:

$$\varphi_t = -A\varphi^{n+1},$$
$$\psi_{tt} + 2A\varphi^n\psi_t + A^2(1-n)\varphi^{2n}\psi = -B\varphi^{2n}. \tag{15.7.1.7}$$

Using (15.7.1.6) and (15.7.1.7), we obtain an equation for $u(z)$,

$$au_z^{(n)} + buu_z' + A(n-2)u + \frac{A^2}{b}(1-n)z + \frac{B}{b} = 0. \tag{15.7.1.8}$$

For $A \neq 0$, the general solution of equations (15.7.1.7) has the form

$$\varphi(t) = (Ant + C_1)^{-\frac{1}{n}},$$
$$\psi(t) = C_2(Ant + C_1)^{\frac{n-1}{n}} + C_3(Ant + C_1)^{-\frac{1}{n}} + \frac{B}{A^2(n-1)}, \tag{15.7.1.9}$$

where $C_1$, $C_2$, and $C_3$ are arbitrary constants.

Formulas (15.7.1.1), (15.7.1.4), (15.7.1.5), and (15.7.1.9), together with equation (15.7.1.8), describe an exact solution of the generalized Burgers–Korteweg–de Vries equation (15.7.1.2).

**Example 2.** Consider the *Boussinesq equation*

$$\frac{\partial^2 w}{\partial t^2} + \frac{\partial}{\partial x}\left(w\frac{\partial w}{\partial x}\right) + a\frac{\partial^4 w}{\partial x^4} = 0. \tag{15.7.1.10}$$

Just as in Example 1, we seek its solutions in the form (15.7.1.1), where the functions $f(t)$, $g(x,t)$, $\varphi(t)$, and $\psi(t)$ are found in the subsequent analysis. Substituting (15.7.1.1) into (15.7.1.10) yields

$$af\varphi^4 u'''' + f^2\varphi^2 uu'' + f(z_t^2 + g\varphi^2)u'' + f^2\varphi^2(u')^2 + (fz_{tt} + 2fg_x\varphi + 2f_t z_t)u'$$
$$+ (fg_{xx} + f_{tt})u + g_{tt} + gg_{xx} + g_x^2 + ag_x^{(4)} = 0. \tag{15.7.1.11}$$

Equating the functional coefficients of $u''''$ and $uu''$, we get

$$f = \varphi^2. \tag{15.7.1.12}$$

Equating the functional coefficient of $u''$ to zero and taking into account (15.7.1.12), we obtain

$$g = -\frac{1}{\varphi^2}(\varphi'_t x + \psi'_t)^2. \qquad (15.7.1.13)$$

Substituting the expressions (15.7.1.12) and (15.7.1.13) into (15.7.1.11), we arrive at the relation

$$\varphi^6(au'''' + uu'' + u'^2) + \varphi^2(x\varphi_{tt} + \psi_{tt})u' + 2\varphi\varphi_{tt}u - \left[\varphi^{-2}(\varphi_t x + \psi_t)^2\right]_{tt} + 6\varphi^{-4}\varphi_t^2(\varphi_t x + \psi_t)^2 = 0.$$

Let us perform the double differentiation of the expression in square brackets and then divide all terms by $\varphi^6$. Excluding $x$ with the help of the relation $x = (z - \psi)/\varphi$, we get

$$au'''' + uu'' + (u')^2 + \varphi^{-5}(\varphi_{tt}z + \varphi\psi_{tt} - \psi\varphi_{tt})u' + 2\varphi^{-5}\varphi_{tt}u + \cdots = 0. \qquad (15.7.1.14)$$

Let us require that the functional coefficient of $u'$ be a function of only one variable, $z$, i.e.,

$$\varphi^{-5}(\varphi_{tt}z + \varphi\psi_{tt} - \psi\varphi_{tt}) = \varphi^{-5}\varphi_{tt}z + \varphi^{-5}(\varphi\psi_{tt} - \psi\varphi_{tt}) \equiv Az + B,$$

where $A$ and $B$ are arbitrary constants. Hence, we obtain the following system of ordinary differential equations for the functions $\varphi$ and $\psi$:

$$\begin{aligned}\varphi_{tt} &= A\varphi^5, \\ \psi_{tt} &= (A\psi + B)\varphi^4.\end{aligned} \qquad (15.7.1.15)$$

Let us eliminate the second and the third derivatives of the functions $\varphi$ and $\psi$ from (15.7.1.14). As a result, we arrive at the following ordinary differential equation for the function $u(z)$:

$$au'''' + uu'' + (u')^2 + (Az + B)u' + 2Au - 2(Az + B)^2 = 0. \qquad (15.7.1.16)$$

Formulas (15.7.1.1), (15.7.1.12), and (15.7.1.13), together with equations (15.7.1.15)–(15.7.1.16), describe an exact solution of the Boussinesq equation (15.7.1.10).

### 15.7.1-2. Description of the Clarkson–Kruskal direct method.

1°. The basic idea of the method is the following: for an equation with the unknown function $w = w(x, t)$, an exact solution is sought in the form

$$w(x, t) = f(x, t)u(z) + g(x, t), \qquad z = z(x, t). \qquad (15.7.1.17)$$

The functions $f(x, t)$, $g(x, t)$, and $z(x, t)$ are determined in the subsequent analysis, so that ultimately one obtains a *single ordinary differential equation* for the function $u(z)$.

2°. Inserting (15.7.1.17) into a nonlinear partial differential equation with a quadratic or a power nonlinearity, we obtain

$$\Phi_1(x, t)\Pi_1[u] + \Phi_2(x, t)\Pi_2[u] + \cdots + \Phi_m(x, t)\Pi_m[u] = 0. \qquad (15.7.1.18)$$

Here, $\Pi_k[u]$ are differential forms that are the products of nonnegative integer powers of the function $u$ and its derivatives $u'_z$, $u''_{zz}$, etc., and $\Phi_k(x, t)$ depend on the functions $f(x, t)$, $g(x, t)$, and $z(x, t)$ and their partial derivatives with respect to $x$ and $t$. Suppose that the differential form $\Pi_1[u]$ contains the highest-order derivative with respect to $z$. Then the function $\Phi_1(x, t)$ is used as a normalizing factor. This means that the following relations should hold:

$$\Phi_k(x, t) = \Gamma_k(z)\Phi_1(x, t), \qquad k = 1, \ldots, m, \qquad (15.7.1.19)$$

where $\Gamma_k(z)$ are functions to be determined.

3°. The representation of a solution in the form (15.7.1.17) has "redundant" generality and the functions $f$, $g$, $u$, and $z$ are ambiguously determined. In order to remove the ambiguity, we use the following three degrees of freedom in the determination of the above functions:

(a) if $f = f(x,t)$ has the form $f = f_0(x,t)\Omega(z)$, then we can take $\Omega \equiv 1$, which corresponds to the replacement $u(z) \to u(z)/\Omega(z)$;

(b) if $g = g(x,t)$ has the form $g = g_0(x,t) + f(x,t)\Omega(z)$, then we can take $\Omega \equiv 0$, which corresponds to the replacement $u(z) \to u(z) - \Omega(z)$;

(c) if $z = z(x,t)$ is determined by an equation of the form $\Omega(z) = h(x,y)$, where $\Omega(z)$ is any invertible function, then we can take $\Omega(z) = z$, which corresponds to the replacement $z \to \Omega^{-1}(z)$.

4°. Having determined the functions $\Gamma_k(z)$, we substitute (15.7.1.19) into (15.7.1.18) to obtain an ordinary differential equation for $u(z)$,

$$\Pi_1[u] + \Gamma_2(z)\Pi_2[u] + \cdots + \Gamma_m(z)\Pi_m[u] = 0. \tag{15.7.1.20}$$

Below we illustrate the main points of the Clarkson–Kruskal direct method by an example.

**Example 3.** We seek a solution of the Boussinesq equation (15.7.1.10) in the form (15.7.1.17). We have

$$afz_x^4 u'''' + a(6fz_x^2 z_{xx} + 4f_x z_x^3)u''' + f^2 z_x^2 uu'' + \cdots = 0. \tag{15.7.1.21}$$

Here, we have written out only the first three terms and have omitted the arguments of the functions $f$ and $z$. The functional coefficients of $u''''$ and $uu''$ should satisfy the condition [see (15.7.1.19)]:

$$f^2 z_x^2 = afz_x^4 \Gamma_3(z),$$

where $\Gamma_3(z)$ is a function to be determined. Hence, using the degree of freedom mentioned in Item 3°(a), we choose

$$f = z_x^2, \quad \Gamma_3(z) = 1/a. \tag{15.7.1.22}$$

Similarly, the functional coefficients of $u''''$ and $u'''$ must satisfy the condition

$$6fz_x^2 z_{xx} + 4f_x z_x^3 = fz_x^4 \Gamma_2(z), \tag{15.7.1.23}$$

where $\Gamma_2(z)$ is another function to be determined. Hence, with (15.7.1.22), we find

$$14\, z_{xx}/z_x = \Gamma_2(z) z_x.$$

Integrating with respect to $x$ yields

$$\ln z_x = I(z) + \ln \widetilde{\varphi}(t), \quad I(z) = \frac{1}{14}\int \Gamma_2(z)\,dz,$$

where $\widetilde{\varphi}(t)$ is an arbitrary function. Integrate again to obtain

$$\int e^{-I(z)}\,dz = \widetilde{\varphi}(t)x + \widetilde{\psi}(t),$$

where $\widetilde{\psi}(t)$ is another arbitrary function. We have a function of $z$ on the left and, therefore, using the degree of freedom mentioned in Item 3°(c), we obtain

$$z = x\varphi(t) + \psi(t), \tag{15.7.1.24}$$

where $\varphi(t)$ and $\psi(t)$ are to be determined.

From formulas (15.7.1.22)–(15.7.1.24) it follows that

$$f = \varphi^2(t), \quad \Gamma_2(z) = 0. \tag{15.7.1.25}$$

Substituting (15.7.1.24) and (15.7.1.25) into (15.7.1.17), we obtain a solution of the form (15.7.1.1) with the function $f$ defined by (15.7.1.12). Thus, the general approach based on the representation of a solution in the form (15.7.1.17) ultimately leads us to the same result as the approach based on the more simple formula (15.7.1.1).

**Remark 1.** In a similar way, it can be shown that formulas (15.7.1.1) and (15.7.1.17) used for the construction of an exact solution of the generalized Burgers–Korteweg–de Vries equation (15.7.1.2) lead us to the same result.

**Remark 2.** The above examples clearly show that it is more reasonable to perform the initial analysis of specific equations on the basis of the simpler formula (15.7.1.1) rather than the general formula (15.7.1.17).

**Remark 3.** A more general scheme of the Clarkson–Kruskal direct method is as follows: for an equation with the unknown function $w = w(x,t)$, an exact solution is sought in the form

$$w(x,t) = F(x,t,u(z)), \quad z = z(x,t). \tag{15.7.1.26}$$

The functions $F(x,t,u)$ and $z(x,t)$ should be chosen so as to obtain ultimately a single ordinary differential equation for $u(z)$. Unlike formulas (15.7.1.1) and (15.7.1.17), the relationship between the functions $w$ and $u$ in (15.7.1.26) can be nonlinear.

## 15.7.2. Some Modifications and Generalizations

### 15.7.2-1. Symmetry reductions based on the generalized separation of variables.

$1°$. The Clarkson–Kruskal direct method based on the representation of solutions in the forms (15.7.1.17) and (15.7.1.26) attaches particular significance to the function $u = u(z)$, because the choice of the other functions is meant to ensure a *single ordinary differential equation* for $u(z)$. However, in some cases it is reasonable to combine these methods with the ideas of the generalized and functional separation of variables, with all determining functions being regarded as equally important. Then the function $u(z)$ is described by an *overdetermined system of equations*.

$2°$. Exact solutions of nonlinear partial differential equations with quadratic or power nonlinearities may be sought in the form (15.7.1.1) with $g(x,t) = g_1(t)x + g_0(t)$. Substituting (15.7.1.1) into an equation under consideration, we replace $x$ by $x = [z - \psi(t)]/\varphi(t)$. As a result, we obtain a functional differential equation with two arguments, $t$ and $z$. Its solution can sometimes be obtained by the differentiation and splitting methods outlined in Subsections 15.5.3 and 15.5.4.

**Example 1.** Consider the *equation of an axisymmetric steady hydrodynamic boundary layer*

$$\frac{\partial w}{\partial y}\frac{\partial^2 w}{\partial x \partial y} - \frac{\partial w}{\partial x}\frac{\partial^2 w}{\partial y^2} = \frac{\partial}{\partial y}\left(y \frac{\partial^2 w}{\partial y^2}\right) + \mathcal{F}(x), \tag{15.7.2.1}$$

where $w$ is the stream function, $y = \frac{1}{4}r^2$, and $x$ and $r$ are axial and radial coordinates.

Its solution is sought in the form

$$w(x,y) = f(x)u(z) + g(x), \quad z = \varphi(x)y + \psi(x). \tag{15.7.2.2}$$

Let us substitute this expression into equation (15.7.2.1) and eliminate $y$ using the relation $y = [z - \psi(x)]/\varphi(x)$. After the division by $\varphi^2 f$, we arrive at the functional differential equation

$$(zu''_{zz})'_z - \psi u'''_{zzz} + f'_x u u''_{zz} + g'_x u''_{zz} - \frac{(f\varphi)'_x}{\varphi}(u'_z)^2 + \frac{\mathcal{F}}{f\varphi^2} = 0. \tag{15.7.2.3}$$

General methods for solving such equations are outlined in Section 15.5. Here we use a simplified scheme for the construction of exact solutions. Assume that the functional coefficients of $uu''_{zz}$, $u''_{zz}$, $(u'_z)^2$, and 1 are linear combinations of the coefficients 1 and $\psi$ of the highest-order terms $(zu''_{zz})'_z$ and $u'''_{zzz}$, respectively. We have

$$f'_x = A_1 + B_1\psi,$$
$$g'_x = A_2 + B_2\psi,$$
$$-(f\varphi)'_x/\varphi = A_3 + B_3\psi,$$
$$\mathcal{F}/(f\varphi^2) = A_4 + B_4\psi,$$

$$\tag{15.7.2.4}$$

where $A_k$ and $B_k$ are arbitrary constants. Let us substitute the expressions of (15.7.2.4) into (15.7.2.3) and sum up the terms proportional to $\psi$ (it is assumed that $\psi \neq \text{const}$). Equating the functional coefficient of $\psi$ to zero, we obtain the following overdetermined system:

$$(zu''_{zz})'_z + A_1 u u''_{zz} + A_2 u''_{zz} + A_3 (u'_z)^2 + A_4 = 0, \qquad (15.7.2.5)$$

$$-u'''_{zzz} + B_1 u u''_{zz} + B_2 u''_{zz} + B_3 (u'_z)^2 + B_4 = 0. \qquad (15.7.2.6)$$

*Case 1.* Let

$$A_1 = A_3 = A_4 = 0, \quad A_2 = -n. \qquad (15.7.2.7)$$

Then the solution of equation (15.7.2.5) has the form

$$u(z) = \frac{C_1}{n(n+1)} z^{n+1} + C_2 z + C_3, \qquad (15.7.2.8)$$

where $C_1, C_2$, and $C_3$ are integration constants. The solution (15.7.2.8) of equation (15.7.2.5) can be a solution of equation (15.7.2.6) only if the following conditions are satisfied:

$$n = -2, \quad B_1 = B_3, \quad C_1 = -4/B_1, \quad C_2^2 = -B_4/B_1, \quad C_3 = -B_2/B_1. \qquad (15.7.2.9)$$

Let us insert the coefficients (15.7.2.7), (15.7.2.9) into system (15.7.2.4). Integrating yields

$$g(x) = 2x - C_3 f, \quad \varphi = \frac{C_4}{f^2}, \quad \psi = -\frac{C_1}{4} f'_x, \quad \mathcal{F} = -(C_2 C_4)^2 \frac{f'_x}{f^3}, \qquad (15.7.2.10)$$

where $f = f(x)$ is an arbitrary function.

Formulas (15.7.2.2), (15.7.2.8), (15.7.2.10) define an exact solution of the axisymmetric boundary layer equation (15.7.2.1).

*Case 2.* For

$$B_1 = B_3 = B_4 = 0, \quad B_2 = -\lambda, \quad A_2 = 0, \quad A_3 = -A_1, \quad A_4 = \lambda^2/A_1 \qquad (15.7.2.11)$$

a common solution of system (15.7.2.5), (15.7.2.6) can be written in the form

$$u(z) = \frac{1}{A_1}(C_1 e^{-\lambda z} + \lambda z - 3). \qquad (15.7.2.12)$$

A solution of system (15.7.2.4) with coefficients (15.7.2.11) is described by the formulas

$$f = A_1 x + C_2, \quad \varphi = C_3, \quad \psi = -\frac{1}{\lambda} g'_x, \quad \mathcal{F} = \frac{(C_3 \lambda)^2}{A_1}(A_1 x + C_2), \qquad (15.7.2.13)$$

where $C_1, C_2$, and $C_3$ are arbitrary constants and $g = g(x)$ is an arbitrary function.

Formulas (15.7.2.2), (15.7.2.12), (15.7.2.13) define an exact solution of the axisymmetric boundary layer equation (15.7.2.1).

*Case 3.* System (15.7.2.5)–(15.7.2.6) also admits solutions of the form

$$u(z) = C_1 z^2 + C_2 z + C_3,$$

with constants $C_1, C_2$, and $C_3$ related to $A_n$ and $B_n$. The corresponding solution is easier to obtain directly from the original equation (15.7.2.1) by substituting $w = \varphi_2(x) y^2 + \varphi_1(x) y + \varphi_0(x)$ into it, which corresponds to the method of generalized separation of variables. This results in the solution

$$w(x, y) = C_1 y^2 + \varphi(x) y + \frac{1}{4C_1} \varphi^2(x) - \frac{1}{2C_1} \int \mathcal{F}(x)\, dx - x + C_3,$$

where $\varphi(x)$ is an arbitrary function and $C_1$ and $C_3$ are arbitrary constants.

**Example 2.** Consider the equation with a cubic nonlinearity

$$\frac{\partial w}{\partial t} + \sigma w \frac{\partial w}{\partial x} = a \frac{\partial^2 w}{\partial x^2} + b_3 w^3 + b_2 w^2 + b_1 w + b_0. \qquad (15.7.2.14)$$

Let us seek its solution in the form

$$w(x, t) = f(x, t) u(z) + \lambda, \quad z = z(x, t), \qquad (15.7.2.15)$$

where the functions $f = f(x,t)$, $z = z(x,t)$, and $u = u(z)$, as well as the constant $\lambda$, are to be determined. Substituting (15.7.2.15) into equation (15.7.2.14), we obtain

$$afz_x^2 u'' - \sigma f^2 z_x uu' + (afz_{xx} + 2af_x z_x - \sigma\lambda f z_x - fz_t)u' + b_3 f^3 u^3$$
$$+ (3b_3\lambda f^2 + b_2 f^2 - \sigma f f_x)u^2 + (3b_3\lambda^2 f + 2b_2\lambda f + b_1 f + af_{xx} - \sigma\lambda f_x - f_t)u \qquad (15.7.2.16)$$
$$+ b_3\lambda^3 + b_2\lambda^2 + b_1\lambda + b_0 = 0.$$

From the overdetermined system of ordinary differential equations resulting from the condition of proportionality of the three functions $u''$, $uu'$, and $u^3$ and that of the two functions $u'$ and $u^2$, it follows that

$$u(z) = 1/z, \qquad (15.7.2.17)$$

where the constant factor is taken equal to unity [this factor can be included in $f$, since formula (15.7.2.15) contains the product of $u$ and $f$]. Let us substitute (15.7.2.17) into (15.7.2.16) and represent the resulting expression as a finite expansion in negative powers of $z$. Equating the functional coefficient of $z^{-3}$ to zero, we obtain

$$f = \beta z_x, \qquad (15.7.2.18)$$

where $\beta$ is a root of the quadratic equation

$$b_3\beta^2 + \sigma\beta + 2a = 0. \qquad (15.7.2.19)$$

Equating the functional coefficients of the other powers of $z$ to zero and taking into account (15.7.2.18), we find that

$$z_t - (3a + \beta\sigma)z_{xx} + (\sigma\lambda + \beta b_2 + 3\beta b_3 \lambda)z_x = 0 \qquad \text{(coefficient of } z^{-2}\text{)},$$
$$z_{xt} - az_{xxx} + \sigma\lambda z_{xx} - (b_1 + 2\lambda b_2 + 3b_3\lambda^2)z_x = 0 \qquad \text{(coefficient of } z^{-1}\text{)}, \qquad (15.7.2.20)$$
$$b_3\lambda^3 + b_2\lambda^2 + b_1\lambda + b_0 = 0 \qquad \text{(coefficient of } z^0\text{)}.$$

Here, the first two linear partial differential equations form an overdetermined system for the function $z(x,t)$, while the last cubic equation serves for the determination of the constant $\lambda$.

Using (15.7.2.15), (15.7.2.17), and (15.7.2.18), we can write out a solution of equation (15.7.2.14) in the form

$$w(x,t) = \frac{\beta}{z}\frac{\partial z}{\partial x} + \lambda. \qquad (15.7.2.21)$$

Let $\beta$ be a root of the quadratic equation (15.7.2.19), and $\lambda$ be a root of the last (cubic) equation in (15.7.2.20). According to the value of the constant $b_3$, one should consider two cases.

$1^\circ$. *Case $b_3 \neq 0$.* From the first two equations in (15.7.2.20), one obtains

$$z_t + p_1 z_{xx} + p_2 z_x = 0,$$
$$z_{xxx} + q_1 z_{xx} + q_2 z_x = 0,$$

where

$$p_1 = -\beta\sigma - 3a, \quad p_2 = \lambda\sigma + \beta b_2 + 3\beta\lambda b_3, \quad q_1 = -\frac{\beta b_2 + 3\beta\lambda b_3}{\beta\sigma + 2a}, \quad q_2 = -\frac{3b_3\lambda^2 + 2b_2\lambda + b_1}{\beta\sigma + 2a}.$$

Four situations are possible.

1.1. For $q_2 \neq 0$ and $q_1^2 \neq 4q_2$, we have

$$z(x,t) = C_1 \exp(k_1 x + s_1 t) + C_2 \exp(k_2 x + s_2 t) + C_3,$$
$$k_n = -\tfrac{1}{2}q_1 \pm \tfrac{1}{2}\sqrt{q_1^2 - 4q_2}, \quad s_n = -k_n^2 p_1 - k_n p_2,$$

where $C_1$, $C_2$, and $C_3$ are arbitrary constants; $n = 1, 2$.

1.2. For $q_2 \neq 0$ and $q_1^2 = 4q_2$,

$$z(x,t) = C_1 \exp(kx + s_1 t) + C_2(kx + s_2 t)\exp(kx + s_1 t) + C_3,$$
$$k = -\tfrac{1}{2}q_1, \quad s_1 = -\tfrac{1}{4}p_1 q_1^2 + \tfrac{1}{2}p_2 q_1, \quad s_2 = -\tfrac{1}{2}p_1 q_1^2 + \tfrac{1}{2}p_2 q_1.$$

1.3. For $q_2 = 0$ and $q_1 \neq 0$,

$$z(x,t) = C_1(x - p_2 t) + C_2 \exp[-q_1 x + q_1(p_2 - p_1 q_1)t] + C_3.$$

1.4. For $q_2 = q_1 = 0$,

$$z(x,t) = C_1(x - p_2 t)^2 + C_2(x - p_2 t) - 2C_1 p_1 t + C_3.$$

$2^\circ$. *Case $b_3 = 0$, $b_2 \neq 0$.* The solutions are determined by (15.7.2.21), where

$$\beta = -\frac{2a}{\sigma}, \quad z(x,t) = C_1 + C_2 \exp\left[Ax + A\left(\frac{b_1\sigma}{2b_2} + \frac{2ab_2}{\sigma}\right)t\right], \quad A = \frac{\sigma(b_1 + 2b_2\lambda)}{2ab_2},$$

and $\lambda = \lambda_{1,2}$ are roots of the quadratic equation $b_2\lambda^2 + b_1\lambda + b_0 = 0$.

## 15.7.2-2. Similarity reductions in equations with three or more independent variables.

The procedure of the construction of exact solutions to nonlinear equations with three or more independent variables sometimes involves (at intermediate stages) the solution of functional differential equations considered in Subsections 15.5.3 and 15.5.4.

**Example 3.** Consider the nonlinear nonstationary wave equation anisotropic in one of the directions

$$\frac{\partial^2 w}{\partial t^2} = a\frac{\partial^2 w}{\partial x^2} + \frac{\partial}{\partial y}\left[(bw+c)\frac{\partial w}{\partial y}\right]. \tag{15.7.2.22}$$

Let us seek its solution in the form

$$w = U(z) + f(x,t), \quad z = y + g(x,t). \tag{15.7.2.23}$$

Substituting (15.7.2.23) into equation (15.7.2.22), we get

$$[(bU + ag_x^2 - g_t^2 + bf + c)U_z']_z' + (ag_{xx} - g_{tt})U_z' + af_{xx} - f_{tt} = 0.$$

Suppose that the functions $f$ and $g$ satisfy the following overdetermined system of equations:

$$af_{xx} - f_{tt} = C_1, \tag{15.7.2.24}$$

$$ag_{xx} - g_{tt} = C_2, \tag{15.7.2.25}$$

$$ag_x^2 - g_t^2 + bf = C_3, \tag{15.7.2.26}$$

where $C_1$, $C_2$, and $C_3$ are arbitrary constants. Then the function $U(z)$ is determined by the autonomous ordinary differential equation

$$[(bU + c + C_3)U_z']_z' + C_2 U_z' + C_1 = 0. \tag{15.7.2.27}$$

The general solutions of equations (15.7.2.24)–(15.7.2.25) are expressed as

$$f = \varphi_1(\xi) + \psi_1(\eta) - \tfrac{1}{2}C_1 t^2,$$

$$g = \varphi_2(\xi) + \psi_2(\eta) - \tfrac{1}{2}C_2 t^2,$$

$$\xi = x + t\sqrt{a}, \quad \eta = x - t\sqrt{a}.$$

Let us insert these expressions into equation (15.7.2.26) and then eliminate $t$ with the help of the formula $t = \dfrac{\xi - \eta}{2\sqrt{a}}$. After simple transformations, we obtain a functional differential equation with two arguments,

$$b\varphi_1(\xi) + C_2\xi\varphi_2'(\xi) - k\xi^2 - C_3 + b\psi_1(\eta) + C_2\eta\psi_2'(\eta) - k\eta^2 + \psi_2'(\eta)[4a\varphi_2'(\xi) - C_2\xi] + \eta[2k\xi - C_2\varphi_2'(\xi)] = 0, \tag{15.7.2.28}$$

where

$$k = \frac{1}{8a}(bC_1 + 2C_2^2).$$

Equation (15.7.2.28) can be solved by the splitting method described in Section 15.5. According to the simplified scheme, set

$$b\varphi_1(\xi) + C_2\xi\varphi_2'(\xi) - k\xi^2 - C_3 = A_1,$$

$$4a\varphi_2'(\xi) - C_2\xi = A_2, \tag{15.7.2.29}$$

$$2k\xi - C_2\varphi_2'(\xi) = A_3,$$

where $A_1$, $A_2$, and $A_3$ are constants. The common solution of system (15.7.2.29) has the form

$$\varphi_1(\xi) = -\frac{C_2^2}{8ab}\xi^2 - \frac{BC_2}{b}\xi + \frac{A_1+C_3}{b}, \quad \varphi_2(\xi) = \frac{C_2}{8a}\xi^2 + B\xi \tag{15.7.2.30}$$

and corresponds to the following values of the constants:

$A_1$ is arbitrary, $A_2 = 4aB$, $A_3 = -BC_2$, $B$ is arbitrary, $C_1 = -\dfrac{C_2^2}{b}$, $C_2$ and $C_3$ are arbitrary, $k = \dfrac{C_2^2}{8a}$. (15.7.2.31)

From (15.7.2.28) and (15.7.2.29) we obtain an equation that establishes a relation between the functions $\psi_1$ and $\psi_2$,

$$A_1 + b\psi_1(\eta) + C_2\eta\psi_2'(\eta) - k\eta^2 + A_2\psi_2'(\eta) + A_3\eta = 0. \tag{15.7.2.32}$$

Hence, taking into account (15.7.2.31), we get

$$\psi_1(\eta) = -\frac{1}{b}(C_2\eta + 4aB)\psi_2'(\eta) + \frac{1}{b}\left(\frac{C_2^2}{8a}\eta^2 + BC_2\eta - A_1\right), \quad \psi_2(\eta) \text{ is an arbitrary function.}$$

Ultimately, we find the functions that determine solution (15.7.2.23):

$$f(x,t) = -\frac{C_2^2}{2\sqrt{a}\,b}xt + \frac{C_2^2}{2b}t^2 - \frac{2\sqrt{a}\,BC_2}{b}t + \frac{C_3}{b} - \frac{1}{b}(C_2\eta + 4aB)\psi_2'(\eta),$$

$$g(x,t) = \frac{C_2}{8a}\left(x^2 + 2\sqrt{a}\,xt - 3at^2\right) + B(x + \sqrt{a}\,t) + \psi_2(\eta), \quad \eta = x - t\sqrt{a}.$$

**Remark.** In the special case of $a = 1$, $b < 0$, and $c > 0$, equation (15.7.2.22) describes spatial transonic flows of an ideal polytropic gas.

## 15.8. Classical Method of Studying Symmetries of Differential Equations

**Preliminary remarks.** The classical method of studying symmetries of differential equations* presents a routine procedure that allows to obtain the following:

(i) transformations under which equations are invariant (such transformations bring the given equation to itself);

(ii) new variables (dependent and independent) in which equations become considerably simpler.

The transformations of (i) convert a solution of an equation to the same or another solution of this equation. In the former case, we have an *invariant solution*, which can be found by symmetry reduction, rewriting the equation in new, fewer variables. In the latter case, we have noninvariant solutions, which may be "multiplied" to a family of solutions.

**Remark 1.** In a sense, the classical method of symmetry analysis of differential equations may be treated as significant extension of the similarity method, outlined in Subsection 15.3.3.

**Remark 2.** Subsections 15.8.1–15.8.3 give a description of the classical method in a nontraditional way, with minimal use of the special (group) terminology, for the reader's easier understanding. Subsection 15.8.4 will explain the origin of the term "Lie group analysis."

### 15.8.1. One-Parameter Transformations and Their Local Properties

15.8.1-1. One-parameter transformations. Infinitesimal operator.

We will consider invertible transformations of the form

$$\begin{aligned}
\bar{x} &= \varphi_1(x, y, w, \varepsilon), & \bar{x}|_{\varepsilon=0} &= x, \\
\bar{y} &= \varphi_2(x, y, w, \varepsilon), & \bar{y}|_{\varepsilon=0} &= y, \\
\bar{w} &= \psi(x, y, w, \varepsilon), & \bar{w}|_{\varepsilon=0} &= w,
\end{aligned} \qquad (15.8.1.1)$$

where $\varphi_1$, $\varphi_2$, and $\psi$ are sufficiently smooth functions of their arguments, and $\varepsilon$ is a real parameter. It is assumes that the successive application (composition) of two transformations of the form (15.8.1.1) with parameters $\varepsilon$ and $\bar{\varepsilon}$ is equivalent to a single transformation of the form with parameter $\varepsilon + \bar{\varepsilon}$ (this means that such transformations have the group property).

---

* It is also known as the *Lie group analysis of differential equations* or the *classical method of symmetry reductions*.

**Remark.** In the special case of transformations in the plane, the functions $\varphi_1$ and $\varphi_2$ in (15.8.1.1) are independent of $w$, and $\psi = w$ (i.e., $\bar{w} = w$).

The expansion of (15.8.1.1) into truncated Taylor series in $\varepsilon$ about $\varepsilon = 0$ to linear terms gives

$$\bar{x} \simeq x + \varepsilon\xi(x, y, w), \quad \bar{y} \simeq y + \varepsilon\eta(x, y, w), \quad \bar{w} \simeq w + \varepsilon\zeta(x, y, w), \qquad (15.8.1.2)$$

where

$$\xi(x,y,w) = \frac{\partial\varphi_1}{\partial\varepsilon}\bigg|_{\varepsilon=0}, \quad \eta(x,y,w) = \frac{\partial\varphi_2}{\partial\varepsilon}\bigg|_{\varepsilon=0}, \quad \zeta(x,y,w) = \frac{\partial\psi}{\partial\varepsilon}\bigg|_{\varepsilon=0}.$$

The linear first-order differential operator

$$X = \xi(x,y,w)\frac{\partial}{\partial x} + \eta(x,y,w)\frac{\partial}{\partial y} + \zeta(x,y,w)\frac{\partial}{\partial w}, \qquad (15.8.1.3)$$

which corresponds to the infinitesimal transformation (15.8.1.2), is called an *infinitesimal operator*.*

THEOREM (LIE). *Suppose the coordinates $\xi(x,y,w)$, $\eta(x,y,w)$, $\zeta(x,y,z)$ of the infinitesimal operator (15.8.1.3) are known. Then the transformation (15.8.1.1), having the group property, can be completely recovered by solving the Lie equations*

$$\frac{d\varphi_1}{d\varepsilon} = \xi(\varphi_1, \varphi_2, \psi), \quad \frac{d\varphi_2}{d\varepsilon} = \eta(\varphi_1, \varphi_2, \psi), \quad \frac{d\psi}{d\varepsilon} = \zeta(\varphi_1, \varphi_2, \psi)$$

*with the initial conditions*

$$\varphi_1|_{\varepsilon=0} = x, \quad \varphi_2|_{\varepsilon=0} = y, \quad \psi|_{\varepsilon=0} = w.$$

**15.8.1-2. Invariant of an infinitesimal operator. Transformations in the plane.**

An *invariant of the operator* (15.8.1.3) is a function $I(x,y,w)$ that satisfies the condition

$$I(\bar{x}, \bar{y}, \bar{w}) = I(x, y, w).$$

Let us expand this equation in Taylor series in the small parameter $\varepsilon$, divide the resulting relation by $\varepsilon$, and the proceed to the limit as $\varepsilon \to 0$ to obtain a linear partial differential equation for $I$:

$$XI = \xi(x,y,w)\frac{\partial I}{\partial x} + \eta(x,y,w)\frac{\partial I}{\partial y} + \zeta(x,y,w)\frac{\partial I}{\partial w} = 0. \qquad (15.8.1.4)$$

Let the associated characteristic system of ordinary differential equations (see Paragraph 13.1.1-3)

$$\frac{dx}{\xi(x,y,w)} = \frac{dy}{\eta(x,y,w)} = \frac{dw}{\zeta(x,y,w)} \qquad (15.8.1.5)$$

---

* In the literature, it is also known as an *infinitesimal generator* or a *group generator*. In this book, we call an operator (15.8.1.3) an infinitesimal operator, a group generator, or, for short, an operator.

TABLE 15.7
One-parameter transformations in the plane

| Name | Transformation | Transformation | Invariant |
|---|---|---|---|
| Translation in the $x$-axis | $\bar{x} = x + \varepsilon,\ \bar{y} = y$ | $X = \frac{\partial}{\partial x}$ | $I_1 = y$ |
| Translation along the straight line $ax + by = 0$ | $\bar{x} = x + b\varepsilon,\ \bar{y} = y - a\varepsilon$ | $X = b\frac{\partial}{\partial x} - a\frac{\partial}{\partial y}$ | $I_1 = ax + by$ |
| Rotation | $\bar{x} = x\cos\varepsilon + y\sin\varepsilon,$ $\bar{y} = y\cos\varepsilon - x\sin\varepsilon$ | $X = y\frac{\partial}{\partial x} - x\frac{\partial}{\partial y}$ | $I_1 = x^2 + y^2$ |
| Galileo transformation | $\bar{x} = x + \varepsilon y,\ \bar{y} = y$ | $X = y\frac{\partial}{\partial x}$ | $I_1 = y$ |
| Lorentz transformation | $\bar{x} = x\cosh\varepsilon + y\sinh\varepsilon,$ $\bar{y} = y\cosh\varepsilon + x\sinh\varepsilon$ | $X = y\frac{\partial}{\partial x} + x\frac{\partial}{\partial y}$ | $I_1 = y^2 - x^2$ |
| Uniform extension | $\bar{x} = xe^\varepsilon,\ \bar{y} = ye^\varepsilon$ | $X = x\frac{\partial}{\partial x} + y\frac{\partial}{\partial y}$ | $I_1 = y/x$ |
| Nonuniform extension | $\bar{x} = xe^{a\varepsilon},\ \bar{y} = ye^{b\varepsilon}$ | $X = ax\frac{\partial}{\partial x} + by\frac{\partial}{\partial y}$ | $I_1 = \|y\|^a \|x\|^{-b}$ |

have the functionally independent integrals

$$I_1(x, y, w) = C_1, \quad I_2(x, y, w) = C_2, \qquad (15.8.1.6)$$

where $C_1$ and $C_2$ are arbitrary constants. Then the general solution of equation (15.8.1.4) has the from

$$I = \Psi(I_1, I_2), \qquad (15.8.1.7)$$

where $\Psi(I_1, I_2)$ is an arbitrary function of two arguments, $I_1 = I_1(x, y, w)$ and $I_2 = I_2(x, y, w)$.

This means that the operator (15.8.1.3) has two functionally independent invariants, $I_1$ and $I_2$, and any function $\Phi(x, y, w)$ invariant under the operator (15.8.1.3) can be represented as a function of the two invariants.

Table 15.7 lists the most common transformations in the plane and the corresponding operator (15.8.1.3) and invariants; only one invariant is specified, with the other being the same: $I_2 = w$.

### 15.8.1-3. Formulas for derivatives. Coordinates of the first and second prolongations.

In the new variables (15.8.1.1), the first derivatives become

$$\frac{\partial \bar{w}}{\partial \bar{x}} \simeq \frac{\partial w}{\partial x} + \varepsilon \zeta_1, \quad \frac{\partial \bar{w}}{\partial \bar{y}} \simeq \frac{\partial w}{\partial y} + \varepsilon \zeta_2. \qquad (15.8.1.8)$$

Here, $\zeta_1$ and $\zeta_2$ are the coordinates of the first prolongation, which are found as

$$\zeta_1 = D_x(\zeta) - w_x D_x(\xi) - w_y D_x(\eta) = \zeta_x + (\zeta_w - \xi_x)w_x - \eta_x w_y - \xi_w w_x^2 - \eta_w w_x w_y,$$
$$\zeta_2 = D_y(\zeta) - w_x D_y(\xi) - w_y D_y(\eta) = \zeta_y - \xi_y w_x + (\zeta_w - \eta_y)w_y - \xi_w w_x w_y - \eta_w w_y^2,$$
$$(15.8.1.9)$$

where $D_x$ and $D_y$ are the total differential operators with respect to $x$ and $y$:

$$D_x = \frac{\partial}{\partial x} + w_x\frac{\partial}{\partial w} + w_{xx}\frac{\partial}{\partial w_x} + w_{xy}\frac{\partial}{\partial w_y} + \cdots,$$
$$D_y = \frac{\partial}{\partial y} + w_y\frac{\partial}{\partial w} + w_{xy}\frac{\partial}{\partial w_x} + w_{yy}\frac{\partial}{\partial w_y} + \cdots.$$
$$(15.8.1.10)$$

Let us verify that the first formula in (15.8.1.8) holds. Obviously,
$$\bar{w}_x = \bar{w}_{\bar{x}}\bar{x}_x + \bar{w}_{\bar{y}}\bar{y}_x, \quad \bar{w}_y = \bar{w}_{\bar{x}}\bar{x}_y + \bar{w}_{\bar{y}}\bar{y}_y. \tag{15.8.1.11}$$

Differentiating (15.8.1.2) with respect to $x$ and $y$ and retaining terms to the first order of $\varepsilon$, we have
$$\begin{aligned} \bar{x}_x &= 1 + \varepsilon D_x \xi, & \bar{x}_y &= \varepsilon D_y \xi, \\ \bar{y}_x &= \varepsilon D_x \eta, & \bar{y}_y &= 1 + \varepsilon D_y \eta, \\ \bar{w}_x &= w_x + \varepsilon D_x \zeta, & \bar{w}_y &= w_y + \varepsilon D_y \zeta. \end{aligned} \tag{15.8.1.12}$$

In order to calculate $\bar{w}_{\bar{x}}$, let us eliminate $\bar{w}_{\bar{y}}$ from (15.8.1.11) and then substitute the derivatives $\bar{x}_x, \bar{x}_y, \bar{y}_x, \bar{y}_y, \bar{w}_x, \bar{w}_y$ for their respective expression from (15.8.1.12) to obtain
$$\bar{w}_{\bar{x}} = \frac{w_x + \varepsilon(D_x\zeta + w_x D_y \eta - w_y D_x \eta) + \varepsilon^2(D_x\zeta D_y\eta - D_x\eta D_y\zeta)}{1 + \varepsilon(D_x\xi + D_y\eta) + \varepsilon^2(D_x\xi D_y\eta - D_x\eta D_y\xi)}.$$

Expanding into a series in $\varepsilon$, we have
$$\bar{w}_{\bar{x}} \simeq w_x + \varepsilon\zeta_1, \quad \zeta_1 = D_x\zeta - w_x D_x\xi - w_y D_x\eta,$$

which was to be proved. The coordinate $\zeta_2$ is calculated likewise.

The second derivatives in the new variables (15.8.1.1) are calculated as
$$\frac{\partial^2 \bar{w}}{\partial \bar{x}^2} \simeq \frac{\partial^2 w}{\partial x^2} + \varepsilon\zeta_{11}, \quad \frac{\partial^2 \bar{w}}{\partial \bar{x} \partial \bar{y}} \simeq \frac{\partial^2 w}{\partial x \partial y} + \varepsilon\zeta_{12}, \quad \frac{\partial^2 \bar{w}}{\partial \bar{y}^2} \simeq \frac{\partial^2 w}{\partial y^2} + \varepsilon\zeta_{22}. \tag{15.8.1.13}$$

Here, the $\zeta_{ij}$ are the coordinates of the second prolongation and are found as
$$\begin{aligned} \zeta_{11} &= D_x(\zeta_1) - w_{xx}D_x(\xi) - w_{xy}D_x(\eta), \\ \zeta_{12} &= D_y(\zeta_1) - w_{xx}D_y(\xi) - w_{xy}D_y(\eta), \\ \zeta_{22} &= D_y(\zeta_2) - w_{xy}D_y(\xi) - w_{yy}D_y(\eta), \end{aligned}$$

or, in detailed form,
$$\begin{aligned}
\zeta_{11} =\ & \zeta_{xx} + (2\zeta_{wx} - \xi_{xx})w_x - \eta_{xx}w_y + (\zeta_{ww} - 2\xi_{wx})w_x^2 - 2\eta_{wx}w_x w_y - \\
& - \xi_{ww}w_x^3 - \eta_{ww}w_x^2 w_y + (\zeta_w - 2\xi_x - 3\xi_w w_x - \eta_w w_y)w_{xx} - 2(\eta_x + \eta_w w_x)w_{xy}, \\
\zeta_{12} =\ & \zeta_{xy} + (\zeta_{wy} - \xi_{xy})w_x + (\zeta_{wx} - \eta_{xy})w_y - \xi_{wy}w_x^2 - \\
& - (\zeta_{ww} - \xi_{wx} - \eta_{wy})w_x w_y - \eta_{wx}w_y^2 - \xi_{ww}w_x^2 w_y - \eta_{ww}w_x w_y^2 - \\
& - (\xi_y + \xi_w w_y)w_{xx} + (\zeta_w - \xi_x - \eta_y - 2\xi_w w_x - 2\eta_w w_y)w_{xy} - (\eta_x + \eta_w w_x)w_{yy}, \\
\zeta_{22} =\ & \zeta_{yy} - \xi_{yy}w_x + (2\zeta_{wy} - \eta_{yy})w_y - 2\xi_{wy}w_x w_y + (\zeta_{ww} - 2\eta_{wy})w_y^2 - \\
& - \xi_{ww}w_x w_y^2 - \eta_{ww}w_y^3 - 2(\xi_y + \xi_w w_y)w_{xy} + (\zeta_w - 2\eta_y - \xi_w w_x - 3\eta_w w_y)w_{yy}.
\end{aligned} \tag{15.8.1.14}$$

The above formulas for the coordinates of the first and second prolongation, (15.8.1.9) and (15.8.1.14), will be required later for the analysis of differential equations.

## 15.8.2. Symmetries of Nonlinear Second-Order Equations. Invariance Condition

**15.8.2-1. Invariance condition. Splitting in derivatives.**

We will consider second-order partial differential equations in two independent variables
$$F\left(x, y, w, \frac{\partial w}{\partial x}, \frac{\partial w}{\partial y}, \frac{\partial^2 w}{\partial x^2}, \frac{\partial^2 w}{\partial x \partial y}, \frac{\partial^2 w}{\partial y^2}\right) = 0. \tag{15.8.2.1}$$

The procedure for finding symmetries* of equation (15.8.2.1) consists of several stages. At the first stage, one requires that equation (15.8.2.1) must be invariant (preserve its form) under transformations (15.8.1.1), so that

$$F\left(\bar{x}, \bar{y}, \bar{w}, \frac{\partial \bar{w}}{\partial \bar{x}}, \frac{\partial \bar{w}}{\partial \bar{y}}, \frac{\partial^2 \bar{w}}{\partial \bar{x}^2}, \frac{\partial^2 \bar{w}}{\partial \bar{x} \partial \bar{y}}, \frac{\partial^2 \bar{w}}{\partial \bar{y}^2}\right) = 0. \tag{15.8.2.2}$$

Let us expand this equation into a series in $\varepsilon$ about $\varepsilon = 0$, taking into account that the leading term vanishes, according to (15.8.2.1). Using formulas (15.8.1.2), (15.8.1.8), (15.8.1.13) and retaining the terms to the first-order of $\varepsilon$, we obtain

$$\underset{2}{X}F\left(x, y, w, \frac{\partial w}{\partial x}, \frac{\partial w}{\partial y}, \frac{\partial^2 w}{\partial x^2}, \frac{\partial^2 w}{\partial x \partial y}, \frac{\partial^2 w}{\partial y^2}\right)\bigg|_{F=0} = 0, \tag{15.8.2.3}$$

where

$$\underset{2}{X}F = \xi \frac{\partial F}{\partial x} + \eta \frac{\partial F}{\partial y} + \zeta \frac{\partial F}{\partial w} + \zeta_1 \frac{\partial F}{\partial w_x} + \zeta_2 \frac{\partial F}{\partial w_y} + \zeta_{11} \frac{\partial F}{\partial w_{xx}} + \zeta_{12} \frac{\partial F}{\partial w_{xy}} + \zeta_{22} \frac{\partial F}{\partial w_{yy}}. \tag{15.8.2.4}$$

The coordinates of the first, $\zeta_i$, and the second, $\zeta_{ij}$, prolongation are defined by formulas (15.8.1.9) and (15.8.1.14). Relation (15.8.2.3) is called the *invariance condition*, and the operator $\underset{2}{X}$ is called the *second prolongation of the operator* (15.8.1.3).

At the second stage, either the derivative $\frac{\partial^2 w}{\partial y^2}$ or $\frac{\partial^2 w}{\partial x^2}$ is eliminated from (15.8.2.3) using equation (15.8.2.1). The resulting relation is then written as a polynomial in the "independent variables," the various combinations of the products of the derivatives (involving various powers of $w_x$, $w_y$, $w_{xx}$, and $w_{xy}$):

$$\sum A_{k_1 k_2 k_3 k_4}(w_x)^{k_1}(w_y)^{k_2}(w_{xx})^{k_3}(w_{xy})^{k_4} = 0, \tag{15.8.2.5}$$

where the functional coefficients $A_{k_1 k_2 k_3 k_4}$ are dependent on $x$, $y$, $w$, $\xi$, $\eta$, $\zeta$ and the derivatives of $\xi$, $\eta$, $\zeta$ only and are independent of the derivatives of $w$. Equation (15.8.2.5) is satisfied if all $A_{k_1 k_2 k_3 k_4}$ are zero. Thus, the invariance condition is split to an overdetermined determining system, resulting from equating all functional coefficients of the various products of the remaining derivatives to zero (recall that the unknowns $\xi$, $\eta$, and $\zeta$ are independent of $w_x$, $w_y$, $w_{xx}$, and $w_{xy}$).

At the third stage, one solves the determining system and finds admissible coordinates $\xi$, $\eta$, and $\zeta$ of the infinitesimal operator (15.8.1.3).

**Remark 1.** It should be noted that the functional coefficients $A_{k_1 k_2 k_3 k_4}$ and the determining system are linear in the unknowns $\xi$, $\eta$, and $\zeta$.

**Remark 2.** An invariant $I$ that is a solution of equation (15.8.1.4) is also a solution of the equation $\underset{2}{X}I = 0$.

The procedure for finding symmetries of differential equations is illustrated below by specific examples.

---

**15.8.2-2. Examples of finding symmetries of nonlinear equations.**

**Example 1.** Consider the two-dimensional stationary heat equation with a nonlinear source

$$\frac{\partial^2 w}{\partial x^2} + \frac{\partial^2 w}{\partial y^2} = f(w). \tag{15.8.2.6}$$

The corresponding left-hand side of equation (15.8.2.1) is $F = w_{xx} + w_{yy} - f(w)$.

---

\* A symmetry of an equation is a transformation that preserves its form.

## 15.8. Classical Method of Studying Symmetries of Differential Equations

Infinitesimal operators X admitted by the equations are sought in the form (15.8.1.4), where the coordinates $\xi = \xi(x, y, w)$, $\eta = \eta(x, y, w)$, $\zeta = \zeta(x, y, w)$ are yet unknown and are to be determined in the subsequent analysis. In view of identity $F = w_{xx} + w_{yy} - f(w)$, the invariance condition (15.8.2.3)–(15.8.2.4) is written as

$$\zeta_{22} + \zeta_{11} - \zeta f'(w) = 0.$$

Substituting here the expressions of the coordinates of the second prolongation (15.8.1.14) and then replacing $w_{yy}$ by $f(w) - w_{xx}$, which follows from equation (15.8.2.6), we obtain

$$-2\xi_w w_x w_{xx} + 2\eta_w w_y w_{xx} - 2\eta_w w_x w_{xy} - 2\xi_w w_y w_{xy} - 2(\xi_x - \eta_y)w_{xx} - 2(\xi_y + \eta_x)w_{xy}-$$
$$-\xi_{ww} w_x^3 - \eta_{ww} w_x^2 w_y - \xi_{ww} w_x w_y^2 - \eta_{ww} w_y^3 + (\zeta_{ww} - 2\xi_{xw})w_x^2 - 2(\xi_{yw} + \eta_{xw})w_x w_y+$$
$$+ (\zeta_{ww} - 2\eta_{yw})w_y^2 + (2\zeta_{xw} - \xi_{xx} - \xi_{yy} - f\xi_w)w_x + (2\zeta_{yw} - \eta_{xx} - \eta_{yy} - 3f\eta_w)w_y+$$
$$+ \zeta_{xx} + \zeta_{yy} + f(\zeta_w - 2\eta_y) - \zeta f' = 0,$$

where $f = f(w)$ and $f' = df/dw$. Equating the coefficients of all combinations of the derivatives to zero, we have the system

$$\begin{aligned}
w_x w_{xx}: &\quad \xi_w = 0, \\
w_y w_{xx}: &\quad \eta_w = 0, \\
w_{xx}: &\quad \xi_x - \eta_y = 0, \\
w_{xy}: &\quad \xi_y + \eta_x = 0, \\
w_x^2: &\quad \zeta_{ww} - 2\xi_{wx} = 0, \\
w_x w_y: &\quad \eta_{wx} + \xi_{wy} = 0, \\
w_x: &\quad 2\zeta_{wx} - \xi_{xx} - \xi_{yy} - \xi_w f(w) = 0, \\
w_y^2: &\quad \zeta_{ww} - 2\eta_{wy} = 0, \\
w_y: &\quad 2\zeta_{wy} - \eta_{xx} - \eta_{yy} - 3\eta_w f(w) = 0, \\
1: &\quad \zeta_{xx} + \zeta_{yy} - f'(w)\zeta + f(w)(\zeta_w - 2\eta_y) = 0.
\end{aligned} \qquad (15.8.2.7)$$

Here, the left column indicates a combination of the derivatives and the right column gives the associated coefficient. The coefficients of $w_y w_{xy}$, $w_x w_{xy}$, $w_x^3$, $w_x^2 w_y$, $w_x w_y^2$, $w_y^3$ either are among those already listed or are their differential consequences, and therefore they are omitted. It follows from the first, second, and fifth equations and their consequences that

$$\xi = \xi(x, y), \quad \eta = \eta(x, y), \quad \zeta = a(x, y)w + b(x, y). \qquad (15.8.2.8)$$

The third and fourth equations of system (15.8.2.7) give

$$\xi_{xx} + \xi_{yy} = 0, \quad \eta_{xx} + \eta_{yy} = 0. \qquad (15.8.2.9)$$

Substituting (15.8.2.8) into the seventh and ninth equations of (15.8.2.7) and using (15.8.2.9), we find that $a_x = a_y = 0$, whence

$$a(x, y) = a = \text{const}. \qquad (15.8.2.10)$$

In view of (15.8.2.8) and (15.8.2.10), system (15.8.2.7) becomes

$$\begin{aligned}
&\xi_x - \eta_y = 0, \\
&\xi_y + \eta_x = 0, \\
&b_{xx} + b_{yy} - awf'(w) - bf'(w) + f(w)(a - 2\eta_y) = 0.
\end{aligned} \qquad (15.8.2.11)$$

For arbitrary $f$, it follows that $a = b = \eta_y = 0$, and then $\xi = C_1 y + C_2$, $\eta = -C_1 x + C_3$, and $\zeta = 0$. By setting one of the constants to unity and the others to zero, we establish that the original equation admits three different operators:

$$\begin{aligned}
X_1 &= \partial_x & (C_2 = 1, \ C_1 = C_3 = 0); \\
X_2 &= \partial_y & (C_3 = 1, \ C_1 = C_2 = 0); \\
X_3 &= y\partial_x - x\partial_y & (C_1 = 1, \ C_2 = C_3 = 0).
\end{aligned} \qquad (15.8.2.12)$$

The first two operators define a translation along the $x$- and $y$-axis. The third operator represents a rotational symmetry.

Let us dwell on the third equation of system (15.8.2.11). If the relation

$$(aw + b)f'(w) - f(w)(a - 2\eta_y) = 0 \qquad (15.8.2.13)$$

holds, there may exist other solutions of system (15.8.2.11) that will result in operators other than (15.8.2.12). Let us study two cases: $a \neq 0$ and $a = 0$.

*Case 1.* If $a \neq 0$, the solution of equation (15.8.2.13) gives

$$f(w) = C(aw+b)^{1-\frac{2\gamma}{a}},$$

where $\gamma = \eta_y = \text{const}$ and $b = \text{const}$. Therefore, for $f(w) = w^k$, equation (15.8.2.6) admits another operator,

$$X_4 = x\partial_x + y\partial_y + \frac{2}{1-k}w\partial_w,$$

that defines a nonuniform extension.

*Case 2.* If $a = 0$, we have

$$f(w) = Ce^{\lambda w},$$

where $\lambda = \text{const}$. Then $b = -2\eta_y/\lambda$, and the functions $\xi$ and $\eta$ satisfy the first two equations (15.8.2.11), which coincide with the Cauchy–Riemann conditions for analytic functions. The real and the imaginary part of any analytic function, $\Phi(z) = \xi(x,y) + i\eta(x,y)$, of the complex variable $z = x + iy$ satisfies the Cauchy–Riemann conditions. In particular, if $b = \text{const}$ and $f(w) = e^w$, the equation admits another operator,

$$X_4 = x\partial_x + y\partial_y - 2\partial_w,$$

which corresponds to extension in $x$ and $y$ with simultaneous translation in $w$.

**Example 2.** Consider the nonlinear nonstationary heat equation

$$\frac{\partial w}{\partial t} = \frac{\partial}{\partial x}\left[f(w)\frac{\partial w}{\partial x}\right]. \tag{15.8.2.14}$$

In the invariance condition (15.8.2.3)–(15.8.2.4), one should set

$$y = t, \quad F = w_t - f(w)w_{xx} - f'(w)w_x^2, \quad \zeta_{12} = \zeta_{22} = 0.$$

The coordinates of the first and second prolongations, $\zeta_1$, $\zeta_2$, and $\zeta_{11}$, are expressed by (15.8.1.9) and (15.8.1.14) with $y = t$. In the resulting equation, one should replace $w_t$ with the right-hand side of equation (15.8.2.14) and equate the coefficients of the various combinations of the remaining derivatives to zero, thus arriving at the system of equations

$$
\begin{aligned}
w_x w_{xx}: &\quad 2f(w)[\eta_{wx}f(w) + \xi_w] + f'(w)\eta_x = 0, \\
w_{xx}: &\quad \zeta f'(w) - f^2(w)\eta_{xx} - f(w)(2\xi_x - \eta_t) = 0, \\
w_x w_{xt}: &\quad f(w)\eta_w = 0, \\
w_{xt}: &\quad f(w)\eta_x = 0, \\
w_x^4: &\quad f'(w)\eta_w + f(w)\eta_{ww} = 0, \\
w_x^3: &\quad 2[f'(w)]^2\eta_x + f(w)\xi_{ww} + f'(w)\xi_w + 2f(w)f'(w)\eta_{wx} = 0, \\
w_x^2: &\quad f(w)\zeta_{ww} + f''(w)\zeta - 2f(w)\xi_{wx} - f'(w)(2\xi_x - \eta_t) + f'(w)\zeta_w - f(w)f'(w)\eta_{ww} = 0, \\
w_x: &\quad 2f(w)\zeta_{wx} + 2f'(w)\zeta_x - f(w)\xi_{xx} + \xi_t = 0, \\
1: &\quad \zeta_t - f(w)\zeta_{xx} = 0.
\end{aligned}
$$

Here, the left column indicates a combination of the derivatives, and the right column gives the associated equation (to a constant factor); identities and differential consequences are omitted. Since $f(w) \neq 0$, it follows from the third and fourth equations of the system that $\eta = \eta(t)$. Then from the first and second equations, we have

$$\xi = \xi(x,t), \quad \zeta = \frac{f(w)(2\xi_x - \eta_t)}{f'(w)}.$$

With these relations, the remaining equations of the system become

$$[ff'f''' - f(f'')^2 + (f')^2f''](2\xi_x - \eta_t) = 0,$$
$$f[4ff'' - 7(f')^2]\xi_{xx} - (f')^2\xi_t = 0,$$
$$2f\xi_{xxx} - 2\xi_{xt} + \eta_{tt} = 0;$$

the equations have been canceled by nonzero factors. In the general case, for arbitrary function $f$, from the first equation it follows that $2\xi_x - \eta_t = 0$ and from the second it follows that $\xi_t = 0$. The third equation gives $\xi = C_1 + C_2 x$, and then $\eta = 2C_2 t + C_3$. Therefore, for arbitrary $f$, equation (15.8.2.14) admits three operators

$$X_1 = \partial_x \qquad (C_1 = 1, \ C_2 = C_3 = 0);$$
$$X_2 = \partial_t \qquad (C_3 = 1, \ C_1 = C_2 = 0);$$
$$X_3 = 2t\partial_t + x\partial_x \qquad (C_2 = 1, \ C_1 = C_3 = 0).$$

Similarly, it can be shown that the following special forms of $f$ result in additional operators:

1. $f = e^w$: $\qquad X_4 = x\partial_x + 2\partial_w$;
2. $f = w^k$, $k \neq 0, -4/3$: $\qquad X_4 = kx\partial_x + 2w\partial_w$;
3. $f = w^{-4/3}$: $\qquad X_4 = 2x\partial_x - 3w\partial_w, \quad X_5 = x^2\partial_x - 3xw\partial_w$.

**Example 3.** Consider now the *nonlinear wave equation*

$$\frac{\partial^2 w}{\partial t^2} = \frac{\partial}{\partial x}\left[f(w)\frac{\partial w}{\partial x}\right]. \tag{15.8.2.15}$$

In the invariance condition (15.8.2.3)–(15.8.2.4), one should set

$$y = t, \quad F = w_{tt} - f(w)w_{xx} - f'(w)w_x^2, \quad \zeta_2 = \zeta_{12} = 0,$$

and use the coordinates of the first and second prolongations, $\zeta_1$, $\zeta_{11}$, and $\zeta_{22}$, expressed by (15.8.1.9) and (15.8.1.14) with $y = t$. In the resulting equation, one should replace $w_{tt}$ with the right-hand side of equation (15.8.2.15) and equate the coefficients of the various combinations of the remaining derivatives to zero, thus arriving at the system of equations

$$\begin{aligned}
w_x w_{xx}: &\quad f(w)\xi_w = 0, \\
w_t w_{xx}: &\quad f(w)\eta_w = 0, \\
w_{xx}: &\quad f'(w)\zeta + 2f(w)(\eta_t - \xi_x) = 0, \\
w_{xt}: &\quad f(w)\eta_x - \xi_t = 0, \\
w_x^3: &\quad f'(w)\xi_w + f(w)\xi_{ww} = 0, \\
w_x^2 w_t: &\quad f(w)\eta_{ww} - f'(w)\eta_w = 0, \\
w_x^2: &\quad f(w)\zeta_{ww} + f'(w)\zeta_w + f''(w)\zeta - 2f(w)\xi_{wx} - 2f'(w)(\xi_x - \eta_t) = 0, \\
w_x w_t: &\quad 2f'(w)\eta_x + 2f(w)\eta_{wx} - 2\xi_{wt} = 0, \\
w_x: &\quad 2f'(w)\zeta_x - f(w)\xi_{xx} + 2f(w)\zeta_{wx} + \xi_{tt} = 0, \\
w_t^2: &\quad \zeta_{ww} - 2\eta_{wt} = 0, \\
w_t: &\quad f(w)\eta_{xx} + 2\zeta_{wt} - \eta_{tt} = 0, \\
1: &\quad \zeta_{tt} - f(w)\zeta_{xx} = 0.
\end{aligned}$$

Identities and differential consequences have been omitted. Since $f(w) \neq \text{const}$, it follows from the first two equations that $\xi = \xi(x,t)$ and $\eta = \eta(x,t)$. The tenth equation of the system becomes $\zeta_{ww} = 0$, thus giving $\zeta = a(x,t)w + b(x,t)$. As a result, the system becomes

$$\begin{aligned}
&wf'(w)a(x,y) + f'(w)b(x,y) + 2f(w)(\eta_t - \xi_x) = 0, \\
&f'(w)a(x,y) + wf''(w)a(x,y) + f''(w)b(x,y) - 2f'(w)(\xi_x - \eta_t) = 0, \\
&2f'(w)(a_x w + b_x) - f(w)\xi_{xx} + 2f(w)a_x = 0, \\
&2a_t - \eta_{tt} = 0, \\
&a_{tt}w + b_{tt} - f(w)(a_{xx}w + b_{xx}) = 0.
\end{aligned}$$

For arbitrary function $f(w)$, we have $a = b = 0$, $\xi_{xx} = 0$, $\eta_{tt} = 0$, and $\xi_x - \eta_t = 0$. Integrating yields three operators:

$$X_1 = \partial_x, \quad X_2 = \partial_t, \quad X_3 = x\partial_x + t\partial_t.$$

Similarly, it can be established that the following special forms of $f$ result in additional operators:

1. $f = e^w$:  $\quad X_4 = x\partial_x + 2\partial_w$;
2. $f = w^k$, $k \neq 0, -4/3, -4$:  $\quad X_4 = kx\partial_x + 2w\partial_w$;
3. $f = w^{-4/3}$:  $\quad X_4 = 2x\partial_x - 3w\partial_w$, $\quad X_5 = x^2\partial_x - 3xw\partial_w$;
4. $f = w^{-4}$:  $\quad X_4 = 2x\partial_x - w\partial_w$, $\quad X_5 = t^2\partial_t + tw\partial_w$.

The symmetries obtained with the procedure presented can be used to find exact solutions of the differential equations considered (see below).

## 15.8.3. Using Symmetries of Equations for Finding Exact Solutions. Invariant Solutions

### 15.8.3-1. Using symmetries of equations for constructing one-parameter solutions.

Suppose a particular solution,
$$w = g(x, y), \tag{15.8.3.1}$$
of a given equation is known. Let us show that any symmetry of the equation defined by a transformation of the form (15.8.1.1) generates a one-parameter family of solutions (except for the cases where the solution is not mapped into itself by the transformations; see Paragraph 15.8.3-2).

Indeed, since equation (15.8.2.1) converted to the new variables (15.8.1.1) acquires the same form (15.8.2.2), then the transformed equation (15.8.2.2) has a solution
$$\bar{w} = g(\bar{x}, \bar{y}). \tag{15.8.3.2}$$

In (15.8.3.2), going back to the old variables by formulas (15.8.1.1), we obtain a one-parameter solution of the original equation (15.8.2.1).

**Example 1.** The two-dimensional heat equation with an exponential source
$$\frac{\partial^2 w}{\partial x^2} + \frac{\partial^2 w}{\partial y^2} = e^w \tag{15.8.3.3}$$
has a one-dimensional solution
$$w = \ln \frac{2}{x^2}. \tag{15.8.3.4}$$

Equation (15.8.3.3) admits the operator $X_3 = y\partial_x - x\partial_y$ (see Example 1 in Subsection 15.8.2), which defines rotation in the plane. The corresponding transformation is given in Table 15.7. Replacing $x$ in (15.8.3.4) by $\bar{x}$ (from Table 15.7), we obtain a one-parameter solution of equation (15.8.3.3):
$$w = \ln \frac{2}{(x\cos\varepsilon + y\sin\varepsilon)^2},$$
where $\varepsilon$ is a free parameter.

### 15.8.3-2. Procedure for constructing invariant solutions.

Solution (15.8.3.1) of equation (15.8.2.1) is called *invariant* under transformations (15.8.1.1) if it coincides with solution (15.8.3.2), which must be rewritten in terms of the old variables using formulas (15.8.1.1). This means that an invariant solution is converted to itself under the given transformation. The basic stages of constructing invariant solutions are outlined below.

Invariant solutions of equation (15.8.2.1) are sought in the implicit form

$$I(x, y, w) = 0.$$

Then $I(\bar{x}, \bar{y}, \bar{w}) = 0$. Let us find a one-parameter transformation with operator (15.8.1.3) whose coordinates are determined from the invariance condition (15.8.2.3) following the procedure described in Subsection 15.8.2. Find two functionally independent integrals (15.8.1.6) of the characteristic system of ordinary differential equations (15.8.1.5). The general solution of the partial differential equation (15.8.1.4) is determined by formula (15.8.1.7). Setting in this formula $I = 0$ and solving for the invariant $I_2$, we obtain

$$I_2 = \Phi(I_1), \tag{15.8.3.5}$$

where the functions $I_1 = I_1(x, y, w)$ and $I_2 = I_2(x, y, w)$ are known,* and the function $\Phi$ is to be determined. Relation (15.8.3.5) is the basis for the construction of invariant solutions: solving (15.8.3.5) for $w$ and substituting the resulting expression into (15.8.2.1), we arrive at an ordinary differential equation for $\Phi$.

**Example 2.** A well-known and very important special case of invariant solutions is the self-similar solutions (see Subsection 15.3.3); they are based on invariants of scaling groups. The corresponding infinitesimal operator and its invariants are

$$X = ax\frac{\partial}{\partial x} + by\frac{\partial}{\partial y} + cw\frac{\partial}{\partial w}; \qquad I_1 = |y|^a|x|^{-b}, \quad I_2 = |w|^a|x|^{-c}.$$

Substituting the invariants into (15.8.3.5) gives $|w|^a|x|^{-c} = \Phi(|y|^a|x|^{-b})$. On solving this equation for $w$, we obtain the form of the desired solution, $w = |x|^{c/a}\Psi(y|x|^{-b/a})$, where $\Psi(z)$ is an unknown function.

To make it clearer, the general scheme for constructing invariant solutions for evolution second-order equations is depicted in Fig. 15.4. The first-order partial differential equation (15.8.1.4) for finding group invariants is omitted, since the corresponding characteristic system of ordinary differential equations (15.8.1.5) can be immediately used.

---

**15.8.3-3. Examples of constructing invariant solutions to nonlinear equations.**

**Example 3.** Consider once again the stationary heat equation with nonlinear source

$$\frac{\partial^2 w}{\partial x^2} + \frac{\partial^2 w}{\partial y^2} = f(w).$$

1°. Let us dwell on the case $f(w) = w^k$, where the equation admits an additional operator (see Example 1 from Subsection 15.8.2):

$$X_4 = x\partial_x + y\partial_y + \frac{2}{1-k}w\partial_w.$$

In order to find invariants of this operator, we have to consider the linear first-order partial differential equation $X_4 I = 0$, or, in detailed form,

$$x\frac{\partial I}{\partial x} + y\frac{\partial I}{\partial y} + \frac{2}{1-k}w\frac{\partial I}{\partial w} = 0.$$

The corresponding characteristic system of ordinary differential equations,

$$\frac{dx}{x} = \frac{dy}{y} = \frac{1-k}{2}\frac{dw}{w},$$

has the first integrals

$$y/x = C_1, \qquad x^{2/(k-1)}w = C_2,$$

---

* Usually, the invariant that is independent of $w$ is taken to be $I_1$.

```
┌─────────────────────────────────────────────────────────────────┐
│   Original equation:  w_t = H(x, t, w, w_x, w_xx)              │
└─────────────────────────────────────────────────────────────────┘
           ⬇         Calculate the coordinates of the prolonged operator         ⬇
┌─────────────────────────────────────────────────────────────────┐
│  Write out invariance condition: ζ₂ = ξH_x+ηH_t+ζH_w+ζ₁H_{w_x}+ζ₁₁H_{w_xx} │
└─────────────────────────────────────────────────────────────────┘
           ⬇                  Replace w_t by H                    ⬇
┌─────────────────────────────────────────────────────────────────┐
│  Split with respect to powers of the remaining derivatives w_x, w_xx, w_xt │
└─────────────────────────────────────────────────────────────────┘
           ⬇           Derive the determining system of PDEs      ⬇
┌─────────────────────────────────────────────────────────────────┐
│  Solve the (overdetermined) determining system of PDEs for ξ, η, ζ │
└─────────────────────────────────────────────────────────────────┘
           ⬇               Find the functions ξ, η, ζ             ⬇
┌─────────────────────────────────────────────────────────────────┐
│  Write out the characteristic system of ODEs:  dx/ξ = dt/η = dw/ζ │
└─────────────────────────────────────────────────────────────────┘
           ⬇              Solve the characteristic system         ⬇
┌─────────────────────────────────────────────────────────────────┐
│  Find the first integrals:  I₁(x, t, w) = C₁ and I₂(x, t, w) = C₂ │
└─────────────────────────────────────────────────────────────────┘
           ⬇       Look for an invariant solution in the form I₂ = Φ(I₁)       ⬇
┌─────────────────────────────────────────────────────────────────┐
│        Obtain an ODE for Φ = Φ(I₁) from the original equation   │
└─────────────────────────────────────────────────────────────────┘
```

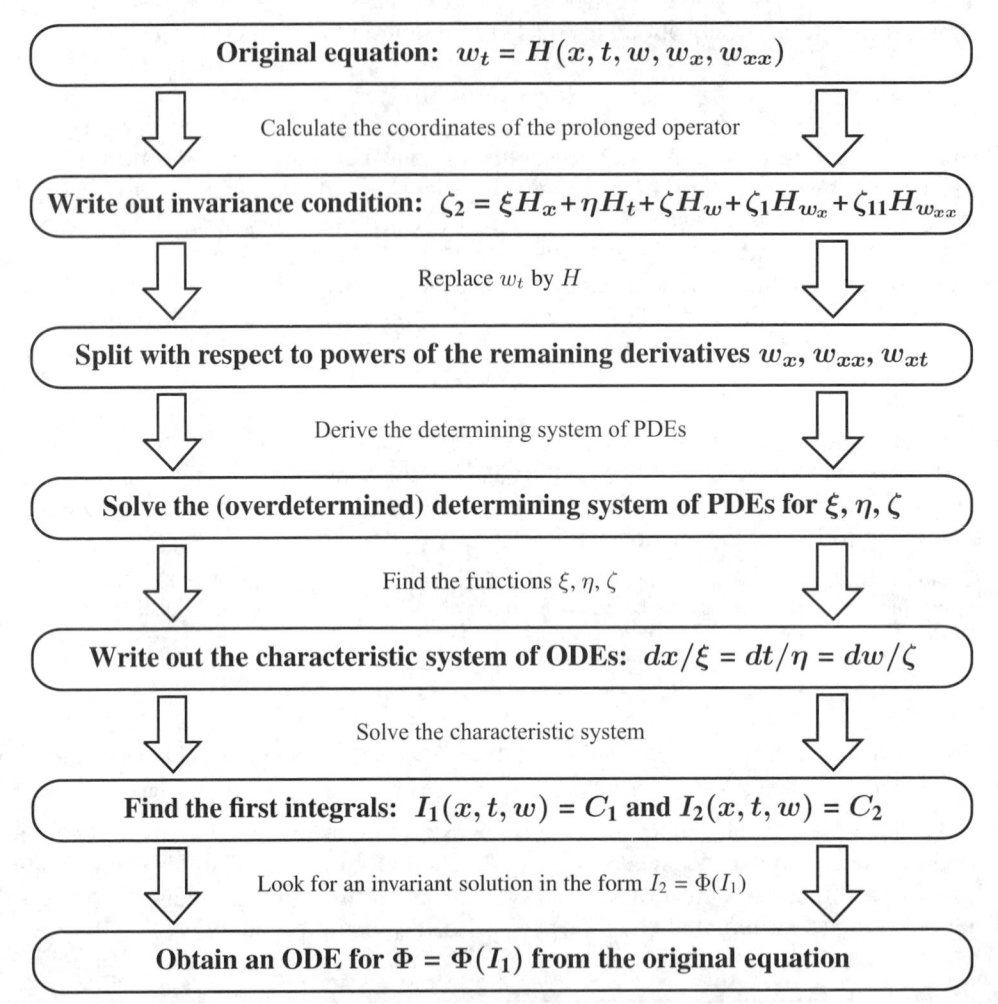

**Figure 15.4.** An algorithm for constructing invariant solutions for evolution second-order equations. Notation: ODE stands for ordinary differential equation and PDE stands for partial differential equation; $\xi = \xi(x, t, w)$, $\eta = \eta(x, t, w)$, $\zeta = \zeta(x, t, w)$; $\zeta_1$, $\zeta_2$, and $\zeta_{11}$ are the coordinates of the prolonged operator, which are defined by formulas (15.8.1.9) and (15.8.1.14) with $y = t$.

where $C_1$, $C_2$ are arbitrary constants. Therefore, the functions $I_1 = y/x$ and $I_2 = x^{2/(k-1)}w$ are invariants of the operator $X_4$.

Assuming that $I_2 = \Phi(I_1)$ and expressing $w$, we find the form of the invariant (self-similar) solution:

$$w = x^{-2/(k-1)}\Phi(y/x). \tag{15.8.3.6}$$

Substituting (15.8.3.6) into the original equation (15.8.2.6) yields a second-order ordinary differential equation for $\Phi(z)$:

$$(k-1)^2(z^2+1)\Phi''_{zz} + 2(k^2-1)z\Phi'_z + 2(k+1)\Phi - (k-1)^2\Phi^k = 0,$$

where $z = y/x$.

2°. The functions $u = x^2 + y^2$ and $w$ are invariants of the operator $X_3$ for the nonlinear heat equation concerned. The substitutions $w = w(u)$ and $u = x^2 + y^2$ lead to an ordinary differential equation describing solutions of the original equation which are invariant under rotation:

$$uw''_{uu} + w'_u = \tfrac{1}{4}f(w).$$

**Remark.** In applications, the polar radius $r = \sqrt{x^2 + y^2}$ is normally used as an invariant instead of $u = x^2 + y^2$.

## 15.8. CLASSICAL METHOD OF STUDYING SYMMETRIES OF DIFFERENTIAL EQUATIONS

**Example 4.** Consider the nonlinear nonstationary heat equation (15.8.2.14).

$1°$. For arbitrary $f(w)$, the equation admits the operator (see Example 2 from Subsection 15.8.2)

$$X_3 = 2t\partial_t + x\partial_x.$$

Invariants of $X_3$ are found for the linear first-order partial differential equation $X_3 I = 0$, or

$$2t\frac{\partial I}{\partial t} + x\frac{\partial I}{\partial x} + 0\frac{\partial I}{\partial w} = 0.$$

The associated characteristic system of ordinary differential equations,

$$\frac{dx}{x} = \frac{dt}{2t} = \frac{dw}{0},$$

has the first integrals

$$xt^{-1/2} = C_1, \quad w = C_2,$$

where $C_1$ and $C_2$ are arbitrary constants. Therefore, the functions $I_1 = xt^{-1/2}$ and $I_2 = w$ are invariants of the operator $X_3$.

Assuming $I_2 = \Phi(I_1)$, we get

$$w = \Phi(z), \quad z = xt^{-1/2}, \tag{15.8.3.7}$$

where $\Phi(z)$ is to be determined in the subsequent analysis. Substituting (15.8.3.7) in the original equation (15.8.2.14) yields the second-order ordinary differential equation

$$2[f(\Phi)\Phi'_z]'_z + z\Phi'_z = 0,$$

which describes an invariant (self-similar) solution.

$2°$. Let us dwell on the case $f(w) = w^k$, where the equation admits the operator

$$X_4 = kx\partial_x + 2w\partial_w.$$

The invariants are described by the first-order partial differential equation $X_4 I = 0$, or

$$0\frac{\partial I}{\partial t} + kx\frac{\partial I}{\partial x} + 2w\frac{\partial I}{\partial w} = 0.$$

The associated characteristic system of ordinary differential equations,

$$\frac{dt}{0} = \frac{dx}{kx} = \frac{dw}{2w},$$

has the first integrals

$$t = C_1, \quad x^{-2/k}w = C_2,$$

where $C_1, C_2$ are arbitrary constants. Therefore, $I_1 = t$ and $I_2 = x^{-2/k}w$ are invariants of the operator $X_4$.

Assuming $I_2 = \theta(I_1)$ and expressing $w$, we get

$$w = x^{2/k}\theta(t), \tag{15.8.3.8}$$

where $\theta(t)$ is to be determined in the subsequent analysis. Substituting (15.8.3.8) in the original equation (15.8.2.14) with $f(w) = w^k$ gives the first-order ordinary differential equation

$$2k\theta'_t = 2(k+2)\theta^{k+1}.$$

Integrating yields

$$\theta(t) = \left[A - \frac{2(k+2)}{k}t\right]^{-1/k},$$

where $A$ is an arbitrary constant. Hence, the solution of equation (15.8.2.14) with $f(w) = w^k$, which is invariant under scaling, has the from

$$w(x,t) = x^{2/k}\left[A - \frac{2(k+2)}{k}t\right]^{-1/k}.$$

TABLE 15.8
Operators, invariants, and solution structures admitted by the nonlinear nonstationary heat equation (15.8.2.14)

| Function $f(w)$ | Operators | Invariants | Solution structure |
|---|---|---|---|
| Arbitrary | $X_1 = \partial_x$, <br> $X_2 = \partial_t$, <br> $X_3 = 2t\partial_t + x\partial_x$ | $I_1 = t, I_2 = w$, <br> $I_1 = x, I_2 = w$, <br> $I_1 = x^2/t, I_2 = w$ | $w = w(t) =$ const, <br> $w = w(x)$, <br> $w = w(z), z = x^2/t$ |
| $e^w$ | $X_4 = x\partial_x + 2\partial_w$ | $I_1 = t, I_2 = w - 2\ln|x|$ | $w = 2\ln|x| + \theta(t)$ |
| $w^k$ ($k \neq 0, -\frac{4}{3}$) | $X_4 = kx\partial_x + 2w\partial_w$ | $I_1 = t, I_2 = w|x|^{-k/2}$ | $w = |x|^{k/2}\theta(t)$ |
| $w^{-4/3}$ | $X_4 = 2x\partial_x - 3w\partial_w$, <br> $X_5 = x^2\partial_x - 3xw\partial_w$ | $I_1 = t, I_2 = wx^{2/3}$, <br> $I_1 = t, I_2 = wx^3$ | $w = x^{-2/3}\theta(t)$, <br> $w = x^{-3}\theta(t)$ |

Table 15.8 summarizes the symmetries of equation (15.8.2.14) (see Example 2 from Subsection 15.8.2 and Example 4 from Subsection 15.8.3).

**Example 5.** Consider the nonlinear wave equation (15.8.2.15). For arbitrary $f(w)$, this equation admits the following operator (see Example 3 from Subsection 15.8.2):

$$X_3 = t\partial_t + x\partial_x.$$

The invariants are found from the linear first-order partial differential equation $X_3 I_1 = 0$, or

$$t\frac{\partial I}{\partial t} + x\frac{\partial I}{\partial x} + 0\frac{\partial I}{\partial w} = 0.$$

The associated characteristic system of ordinary differential equations

$$\frac{dx}{x} = \frac{dt}{t} = \frac{dw}{0}$$

admits the first integrals

$$xt^{-1} = C_1, \quad w = C_2,$$

where $C_1, C_2$ are arbitrary constants. Therefore, $I_1 = xt^{-1}$ and $I_2 = w$ are invariants of the operator $X_3$.

Taking $I_2 = \Phi(I_1)$, we get

$$w = \Phi(y), \quad y = xt^{-1}. \tag{15.8.3.9}$$

The function $\Phi(y)$ is found by substituting (15.8.3.9) in the original equation (15.8.2.15). This results in the ordinary differential equation

$$[f(\Phi)\Phi'_y]'_y = (y^2\Phi'_y)'_y,$$

which defines an invariant (self-similar) solution. This equation has the obvious first integral $f(\Phi)\Phi'_y = y^2\Phi'_y + C$.

Table 15.9 summarizes the symmetries of equation (15.8.2.15) (see Example 3 from Subsection 15.8.2 and Example 5 from Subsection 15.8.3).

### 15.8.3-4. Solutions induced by linear combinations of admissible operators.

If a given equation admits $N$ operators, then we have $N$ associated different invariant solutions. However, when dealing with operators individually, one may overlook solutions that are invariant under a linear superposition of the operators; such solutions may have a significantly different form. In order to find all types of invariant solutions, one should study all possible linear combinations of the admissible operators.

**Example 6.** Consider once again the nonlinear nonstationary heat equation (15.8.2.14).

### TABLE 15.9
Operators, invariants, and solution structures admitted by the nonlinear wave equation (15.8.2.15)

| Functions $f(w)$ | Operators | Invariants | Solution structure |
|---|---|---|---|
| Arbitrary | $X_1 = \partial_x$, <br> $X_2 = \partial_t$, <br> $X_3 = t\partial_t + x\partial_x$ | $I_1 = t$, $I_2 = w$, <br> $I_1 = x$, $I_2 = w$, <br> $I_1 = x/t$, $I_2 = w$ | $w = w(t)$, <br> $w = w(x)$, <br> $w = w(z)$, $z = x/t$ |
| $e^w$ | $X_4 = x\partial_x + 2\partial_w$ | $I_1 = t$, $I_2 = w - 2\ln|x|$ | $w = 2\ln|x| + \theta(t)$ |
| $w^k$ ($k \neq 0, -\frac{4}{3}, -4$) | $X_4 = kx\partial_x + 2w\partial_w$ | $I_1 = t$, $I_2 = w|x|^{-k/2}$ | $w = |x|^{k/2}\theta(t)$ |
| $w^{-4/3}$ | $X_4 = 2x\partial_x - 3w\partial_w$, <br> $X_5 = x^2\partial_x - 3xw\partial_w$ | $I_1 = t$, $I_2 = wx^{2/3}$, <br> $I_1 = t$, $I_2 = wx^3$ | $w = x^{-2/3}\theta(t)$, <br> $w = x^{-3}\theta(t)$ |
| $w^{-4}$ | $X_4 = 2x\partial_x - w\partial_w$, <br> $X_5 = t^2\partial_t + tw\partial_w$ | $I_1 = t$, $I_2 = w|x|^{1/2}$, <br> $I_1 = x$, $I_2 = w/t$ | $w = |x|^{-1/2}\theta(t)$, <br> $w = t\theta(x)$ |

$1°$. For arbitrary $f(w)$, this equation admits three operators (see Table 15.8):

$$X_1 = \partial_t, \quad X_2 = \partial_x, \quad X_3 = 2t\partial_t + x\partial_x.$$

The respective invariant solutions are

$$w = \Phi(x), \quad w = \Phi(t), \quad w = \Phi(x^2/t).$$

However, various linear combinations give another operator,

$$X_{1,2} = X_1 + aX_2 = \partial_t + a\partial_x,$$

where $a \neq 0$ is an arbitrary constant. The solution invariant under this operator is written as

$$w = \Phi(x - at).$$

It is apparent that solutions of this type (traveling waves) are not contained in the invariant solutions associated with the individual operators $X_1$, $X_2$, and $X_3$.

$2°$. If $f(w) = e^w$, apart from the above three operators, there is another one, $X_4 = x\partial_x + 2\partial_w$ (see Table 15.8). In this case, the linear combination

$$X_{3,4} = X_3 + aX_4 = 2t\partial_t + (a+1)x\partial_x + 2a\partial_w$$

gives another invariant solution,

$$w = \Phi(\xi) + a\ln t, \quad \xi = xt^{-\frac{a+1}{2}},$$

where the function $\Phi = \Phi(\xi)$ satisfies the ordinary differential equation

$$(e^\Phi \Phi'_\xi)'_\xi + \tfrac{1}{2}(a+1)\xi\Phi'_\xi = a.$$

$3°$. If $f(w) = w^k$ ($k \neq 0, -4/3$), apart from the three operators from $1°$, there is another one $X_4 = kx\partial_x + 2w\partial_w$. The linear combination

$$X_{3,4} = X_3 + aX_4 = 2t\partial_t + (ak+1)x\partial_x + 2aw\partial_w$$

generates the invariant (self-similar) solution

$$w = t^a \Phi(\zeta), \quad \zeta = xt^{-\frac{ak+1}{2}},$$

where the function $\Phi = \Phi(\zeta)$ satisfies the ordinary differential equation

$$(\Phi^k \Phi'_\zeta)'_\zeta + \tfrac{1}{2}(ak+1)\zeta\Phi'_\zeta = a\Phi.$$

The invariant solutions presented in Items $1°$–$3°$ are not listed in Table 15.8. It is clearly important to consider solutions induced by linear combinations of admissible operators.

## 15.8.4. Some Generalizations. Higher-Order Equations

### 15.8.4-1. One-parameter Lie groups of point transformations. Group generator.

Here we will be considering functions dependent on $n + 1$ variables, $x_1, \ldots, x_n, w$. The brief notation $x = (x_1, \ldots, x_n)$ will be used.

The set of invertible transformations of the form

$$T_\varepsilon = \begin{cases} \bar{x}_i = \varphi_i(x, w, \varepsilon), & \bar{x}_i|_{\varepsilon=0} = x_i, \\ \bar{w} = \psi(x, w, \varepsilon), & \bar{w}|_{\varepsilon=0} = w, \end{cases} \qquad (15.8.4.1)$$

where $\varphi_i$ and $\psi$ are sufficiently smooth functions of their arguments ($i = 1, \ldots, n$) and $\varepsilon$ is a real parameter, is called a *one-parameter continuous point group* of transformations G if for any $\varepsilon_1$ and $\varepsilon_2$ the relation $T_{\varepsilon_1} \circ T_{\varepsilon_2} = T_{\varepsilon_1+\varepsilon_2}$ holds, that is, the successive application (composition) of two transformations of the form (15.8.4.1) with parameters $\varepsilon_1$ and $\varepsilon_2$ is equivalent to a single transformation of the same form with parameter $\varepsilon_1 + \varepsilon_2$.

Further on, we consider local one-parameter continuous Lie groups of point transformations (or, for short, point groups), corresponding to the infinitesimal transformation (15.8.4.1) as $\varepsilon \to 0$. The expansion of (15.8.4.1) into Taylor series in the parameter $\varepsilon$ about $\varepsilon = 0$ to the first order gives

$$\bar{x}_i \simeq x_i + \varepsilon \xi_i(x, w), \quad \bar{w} \simeq w + \varepsilon \zeta(x, w), \qquad (15.8.4.2)$$

where

$$\xi_i(x, w) = \frac{\partial \varphi_i(x, w, \varepsilon)}{\partial \varepsilon}\bigg|_{\varepsilon=0}, \quad \zeta(x, w) = \frac{\partial \psi(x, w, \varepsilon)}{\partial \varepsilon}\bigg|_{\varepsilon=0}.$$

The linear first-order differential operator

$$X = \xi_i(x, w) \frac{\partial}{\partial x_i} + \zeta(x, w) \frac{\partial}{\partial w} \qquad (15.8.4.3)$$

corresponding to the infinitesimal transformation (15.8.4.2) is called a *group generator* (or an *infinitesimal operator*). In formula (15.8.4.3), summation is assumed over the repeated index $i$.

THEOREM (LIE). *Suppose the coordinates $\xi_i(x, w)$ and $\zeta(x, w)$ of the group generator (15.8.4.3) are known. Then the one-parameter group of transformations (15.8.4.1) can be completely recovered by solving the Lie equations*

$$\frac{d\varphi_i}{d\varepsilon} = \xi_i(\varphi, \psi), \quad \frac{d\psi}{d\varepsilon} = \zeta(\varphi, \psi) \quad (i = 1, \ldots, n),$$

with the initial conditions

$$\varphi_i|_{\varepsilon=0} = x_i, \quad \psi|_{\varepsilon=0} = w.$$

Here, the short notation $\varphi = (\varphi_1, \ldots, \varphi_n)$ has been used.

*Remark.* The widely known terms "Lie group analysis of differential equations," "group-theoretic methods," and others are due to the prevailing concept of a local one-parameter Lie group of point transformations. However, in this book, we prefer to use the term "method of symmetry analysis of differential equations."

### 15.8.4-2. Group invariants. Local transformations of derivatives.

A *universal invariant* (or, for short, an *invariant*) of a group (15.8.4.1) and a operator (15.8.4.3) is a function $I(x, w)$ that satisfies the condition $I(\bar{x}, \bar{w}) = I(x, w)$. The expansion in a series in powers of the small parameter $\varepsilon$ gives rise to the linear partial differential equation for $I$:

$$XI = \xi_i(x, w)\frac{\partial I}{\partial x_i} + \zeta(x, w)\frac{\partial I}{\partial w} = 0. \tag{15.8.4.4}$$

From the theory of first-order partial differential equations it follows that the group (15.8.4.1) and the operator (15.8.4.3) have $n$ functionally independent universal invariants. On the other side, this means that any function $F(x, w)$ that is invariant under the group (15.8.4.1) can be written as a function of $n$ invariants.

In the new variable (15.8.4.1), the derivatives are transformed as follows:

$$\frac{\partial \bar{w}}{\partial \bar{x}_i} \simeq \frac{\partial w}{\partial x_i} + \varepsilon \zeta_i, \quad \frac{\partial^2 \bar{w}}{\partial \bar{x}_i \partial \bar{x}_j} \simeq \frac{\partial^2 w}{\partial x_i \partial x_j} + \varepsilon \zeta_{ij}, \quad \frac{\partial^3 \bar{w}}{\partial \bar{x}_i \partial \bar{x}_j \partial \bar{x}_k} \simeq \frac{\partial^3 w}{\partial x_i \partial x_j \partial x_k} + \varepsilon \zeta_{ijk}, \ldots \tag{15.8.4.5}$$

The coordinates $\zeta_i$, $\zeta_{ij}$, and $\zeta_{ijk}$ of the first three prolongations are expressed as

$$\begin{aligned} \zeta_i &= D_i(\zeta) - p_s D_i(\xi_s), \\ \zeta_{ij} &= D_j(\zeta_i) - q_{is} D_j(\xi_s), \\ \zeta_{ijk} &= D_k(\zeta_{ij}) - r_{ijs} D_k(\xi_s), \end{aligned} \tag{15.8.4.6}$$

where summation is assumed over the repeated index $s$ and the following short notation partial derivatives are used:

$$p_i = \frac{\partial w}{\partial x_i}, \quad q_{ij} = \frac{\partial^2 w}{\partial x_i \partial x_j}, \quad r_{ijk} = \frac{\partial^3 w}{\partial x_i \partial x_j \partial x_k},$$

$$D_i = \frac{\partial}{\partial x_i} + p_i \frac{\partial}{\partial w} + q_{ij} \frac{\partial}{\partial p_j} + r_{ijk} \frac{\partial}{\partial q_{jk}} + \cdots,$$

with $D_i$ being the total differential operator with respect to $x_i$.

### 15.8.4-3. Invariant condition. Splitting procedure. Invariant solutions.

We will consider partial differential equations of order $m$ in $n$ independent variables

$$F\left(x, w, \frac{\partial w}{\partial x_i}, \frac{\partial^2 w}{\partial x_i \partial x_j}, \frac{\partial^3 w}{\partial x_i \partial x_j \partial x_k}, \ldots\right) = 0, \tag{15.8.4.7}$$

where $i, j, k = 1, \ldots, n$.

The group analysis of equation (15.8.4.7) consists of several stages. At the first stage, let us require that equation (15.8.4.7) be invariant (must preserve its form) under transformations (15.8.4.1), so that

$$F\left(\bar{x}, \bar{w}, \frac{\partial \bar{w}}{\partial \bar{x}_i}, \frac{\partial^2 \bar{w}}{\partial \bar{x}_i \partial \bar{x}_j}, \frac{\partial^3 \bar{w}}{\partial \bar{x}_i \partial \bar{x}_j \partial \bar{x}_k}, \ldots\right) = 0. \tag{15.8.4.8}$$

Let us expand this expression into a series in powers of $\varepsilon$ about $\varepsilon = 0$ taking into account that the leading term (15.8.4.7) vanishes. Using formulas (15.8.4.1) and (15.8.4.5) and retaining terms to the first order of $\varepsilon$, we get

$$\underset{m}{X} F\left(x, w, \frac{\partial w}{\partial x_i}, \frac{\partial^2 w}{\partial x_i \partial x_j}, \frac{\partial^3 w}{\partial x_i \partial x_j \partial x_k}, \dots\right)\bigg|_{F=0} = 0, \qquad (15.8.4.9)$$

where

$$\underset{m}{X} F = \xi_i \frac{\partial F}{\partial x_i} + \zeta \frac{\partial F}{\partial w} + \zeta_i \frac{\partial F}{\partial w_{x_i}} + \zeta_{ij} \frac{\partial F}{\partial w_{x_i x_j}} + \zeta_{ijk} \frac{\partial F}{\partial w_{x_i x_j x_k}} + \cdots \qquad (15.8.4.10)$$

The coordinates $\zeta_i$, $\zeta_{ij}$, and $\zeta_{ijk}$ of the first three prolongations are defined by formulas (15.8.4.5)–(15.8.4.6). Summation is assumed over repeated indices. Relation (15.8.4.9) is called the *invariance condition* and the operator $\underset{m}{X}$ is called the $m$th prolongation of the group generator; the partial derivatives of $F$ with respect to all $m$ derivatives of $w$ appear last in (15.8.4.10).

At the second stage, one of the highest $m$th-order derivatives is eliminated from (15.8.4.9) using equation (15.8.4.7). The resulting relation is then represented as a polynomial in "independent variables," the various combinations of the remaining derivatives, which are the products of different powers of $w_x$, $w_y$, $w_{xx}$, $w_{xy}$, ... All the coefficients of this polynomial — they depend on $x$, $w$, $\xi_i$, and $\zeta$ only and are independent of the derivatives of $w$ — are further equated to zero. As a result, the invariance condition is split into an overdetermined linear determining system.

The third stage involves solving the determining system and finding admissible coordinates $\xi_i$ and $\zeta$ of the group generator (15.8.4.3).

For $m$th-order equations in two independent variables, the invariant solutions are defined in a similar way as for second-order equations. In this case, the procedure of constructing invariant solutions (for known coordinates of the group generator) is identical to that described in Subsection 15.8.3.

## 15.9. Nonclassical Method of Symmetry Reductions*
### 15.9.1. Description of the Method. Invariant Surface Condition

Consider a second-order equation in two independent variables of the form

$$F\left(x, y, w, \frac{\partial w}{\partial x}, \frac{\partial w}{\partial y}, \frac{\partial^2 w}{\partial x^2}, \frac{\partial^2 w}{\partial x \partial y}, \frac{\partial^2 w}{\partial y^2}\right) = 0. \qquad (15.9.1.1)$$

The results of the classical group analysis (see Section 15.8) can be substantially extended if, instead of finding invariants of an admissible infinitesimal operator X by means of solving the characteristic system of equations

$$\frac{dx}{\xi(x, y, w)} = \frac{dy}{\eta(x, y, w)} = \frac{dw}{\zeta(x, y, w)},$$

one imposes the corresponding *invariant surface condition* (Bluman and Cole, 1969)

$$\xi(x, y, w)\frac{\partial w}{\partial x} + \eta(x, y, w)\frac{\partial w}{\partial y} = \zeta(x, y, w). \qquad (15.9.1.2)$$

---
* Prior to reading this section, the reader may find it useful to get acquainted with Section 15.8.

Equation (15.9.1.1) and condition (15.9.1.2) are supplemented by the invariance condition

$$\underset{2}{\mathrm{X}} F\left(x, y, w, \frac{\partial w}{\partial x}, \frac{\partial w}{\partial y}, \frac{\partial^2 w}{\partial x^2}, \frac{\partial^2 w}{\partial x \partial y}, \frac{\partial^2 w}{\partial y^2}\right)\bigg|_{F=0} = 0, \qquad (15.9.1.3)$$

which coincides with equation (15.8.2.3).

All three equations (15.9.1.1)–(15.9.1.3) are used for the construction of exact solutions of the original equation (15.9.1.1). It should be observed that in this case, the determining equations obtained for the unknown functions $\xi(x, y, w)$, $\eta(x, y, w)$, and $\zeta(x, y, w)$ by the splitting procedure are nonlinear. The symmetries determined by the invariant surface (15.9.1.2) are called *nonclassical*.

Figure 15.5 is intended to clarify the general scheme for constructing of exact solutions of a second-order evolution equation by the nonclassical method on the basis of the invariant surface condition (15.9.1.2).

**Remark.** Apart from the algorithm shown in Fig. 15.5, its modification can also be used. Instead of solving the characteristic system of ordinary differential equations, the derivative $w_t$ is eliminated from (15.9.1.1)–(15.9.1.2) after finding the coordinates $\xi$, $\eta$, and $\zeta$. Then the resulting equation is solved, which can be treated as an ordinary differential equation for $x$ with parameter $t$.

## 15.9.2. Examples: The Newell–Whitehead Equation and a Nonlinear Wave Equation

**Example 1.** Consider the *Newell–Whitehead equation*

$$w_t = w_{xx} + w - w^3, \qquad (15.9.2.1)$$

which corresponds to the left-hand side $F = -w_t + w_{xx} + w - w^3$ of equation (15.9.1.1) with $y = t$.

Without loss of generality, we set $\eta = 1$ in the invariant surface condition (15.9.1.2) with $y = t$, thus assuming that $\eta \neq 0$. We have

$$\frac{\partial w}{\partial t} + \xi(x, t, w)\frac{\partial w}{\partial x} = \zeta(x, t, w). \qquad (15.9.2.2)$$

The invariance condition is obtained by a procedure similar to the classical algorithm (see Subsection 15.8.2). Namely, we apply the operator

$$\underset{2}{\mathrm{X}} = \xi\partial_x + \eta\partial_t + \zeta\partial_w + \zeta_1\partial_{w_x} + \zeta_2\partial_{w_t} + \zeta_{11}\partial_{w_{xx}} + \zeta_{12}\partial_{w_{xt}} + \zeta_{22}\partial_{w_{tt}} \qquad (15.9.2.3)$$

to equation (15.9.2.1). Taking into account that $\partial_x = \partial_t = \partial_{w_x} = \partial_{w_{xt}} = \partial_{w_{tt}} = 0$, since the equation is explicitly independent of $x$, $t$, $w_x$, $w_{xt}$, $w_{tt}$, we get the invariance condition in the form

$$\zeta_2 = \zeta(3w^2 - 1) + \zeta_{11}.$$

Substituting here the expressions (15.8.1.9) and (15.8.1.14) for the coordinates $\zeta_2$ and $\zeta_{11}$ of the first and second prolongations, with $y = t$ and $\eta = 1$, we obtain

$$\zeta_t - \xi_t w_x + \zeta_w w_t - \xi_w w_x w_t = \zeta(-3w^2 + 1) + \zeta_{xx}$$
$$+ (2\zeta_{wx} - \xi_{xx})w_x + (\zeta_{ww} - 2\xi_{wx})w_x^2 - \xi_{ww}w_x^3 + (\zeta_w - 2\xi_x - 3\xi_w w_x)w_{xx}. \qquad (15.9.2.4)$$

Let us express the derivatives $w_t$ and $w_{xx}$ with the help of (15.9.2.1)–(15.9.2.2) via the other quantities:

$$w_t = \zeta - \xi w_x, \qquad w_{xx} = \zeta - \xi w_x + w(w^2 - 1). \qquad (15.9.2.5)$$

Inserting these expressions into the invariance condition (15.9.2.4), we arrive at cubic polynomial in the remaining "independent" derivative $w_x$. Equating all functional coefficients of the various powers of this polynomial to zero, we get the determining system

$$\begin{aligned}
w_x^3: &\quad \xi_{ww} = 0, \\
w_x^2: &\quad \zeta_{ww} - 2(\xi_{wx} - \xi\xi_w) = 0, \\
w_x: &\quad 2\zeta_{wx} - 2\xi_w\zeta - 3w(w^2 - 1)\xi_w - \xi_{xx} + 2\xi\xi_x + \xi_t = 0, \\
1: &\quad \zeta_t - \zeta_{xx} + 2\xi_x\zeta + (2\xi_x - \zeta_w)w(w^2 - 1) + (3w^2 - 1)\zeta = 0,
\end{aligned} \qquad (15.9.2.6)$$

```
┌─────────────────────────────────────────────────────────────────┐
│  Original equation:  $w_t = H(x, t, w, w_x, w_{xx})$      (1)   │
└─────────────────────────────────────────────────────────────────┘
                 ⇓  Impose the invariant surface condition  ⇓
┌─────────────────────────────────────────────────────────────────┐
│  Write out associated 1st-order quasilinear PDE: $\xi w_x + \eta w_y = \zeta$  (2) │
└─────────────────────────────────────────────────────────────────┘
                 ⇓  Calculate the coordinates of the prolonged operator  ⇓
┌─────────────────────────────────────────────────────────────────┐
│  Invariance condition: $\zeta_2 = \xi H_x + \eta H_t + \zeta H_w + \zeta_1 H_{w_x} + \zeta_{11} H_{w_{xx}}$  (3) │
└─────────────────────────────────────────────────────────────────┘
                 ⇓  Eliminate $w_t$ and $w_{xx}$ from (1)–(3)  ⇓
┌─────────────────────────────────────────────────────────────────┐
│  Split the resulting expression in powers of the remaining derivative $w_x$ │
└─────────────────────────────────────────────────────────────────┘
                 ⇓  Derive the determining system of PDEs  ⇓
┌─────────────────────────────────────────────────────────────────┐
│  Solve the determining system of PDEs for $\xi, \eta, \zeta$ (one usually sets $\eta = 1$) │
└─────────────────────────────────────────────────────────────────┘
                 ⇓  Find the functions $\xi, \eta, \zeta$  ⇓
┌─────────────────────────────────────────────────────────────────┐
│  Characteristic system of ODEs corresponding to (2): $dx/\xi = dt/\eta = dw/\zeta$ │
└─────────────────────────────────────────────────────────────────┘
                 ⇓  Solve the characteristic system  ⇓
┌─────────────────────────────────────────────────────────────────┐
│  Find the first integrals: $I_1(x, t, w) = C_1$ and $I_2(x, t, w) = C_2$ │
└─────────────────────────────────────────────────────────────────┘
                 ⇓  Look for a solution in the form $I_2 = \Phi(I_1)$  ⇓
┌─────────────────────────────────────────────────────────────────┐
│  Obtain an ODE for $\Phi = \Phi(I_1)$ from the original equation │
└─────────────────────────────────────────────────────────────────┘
```

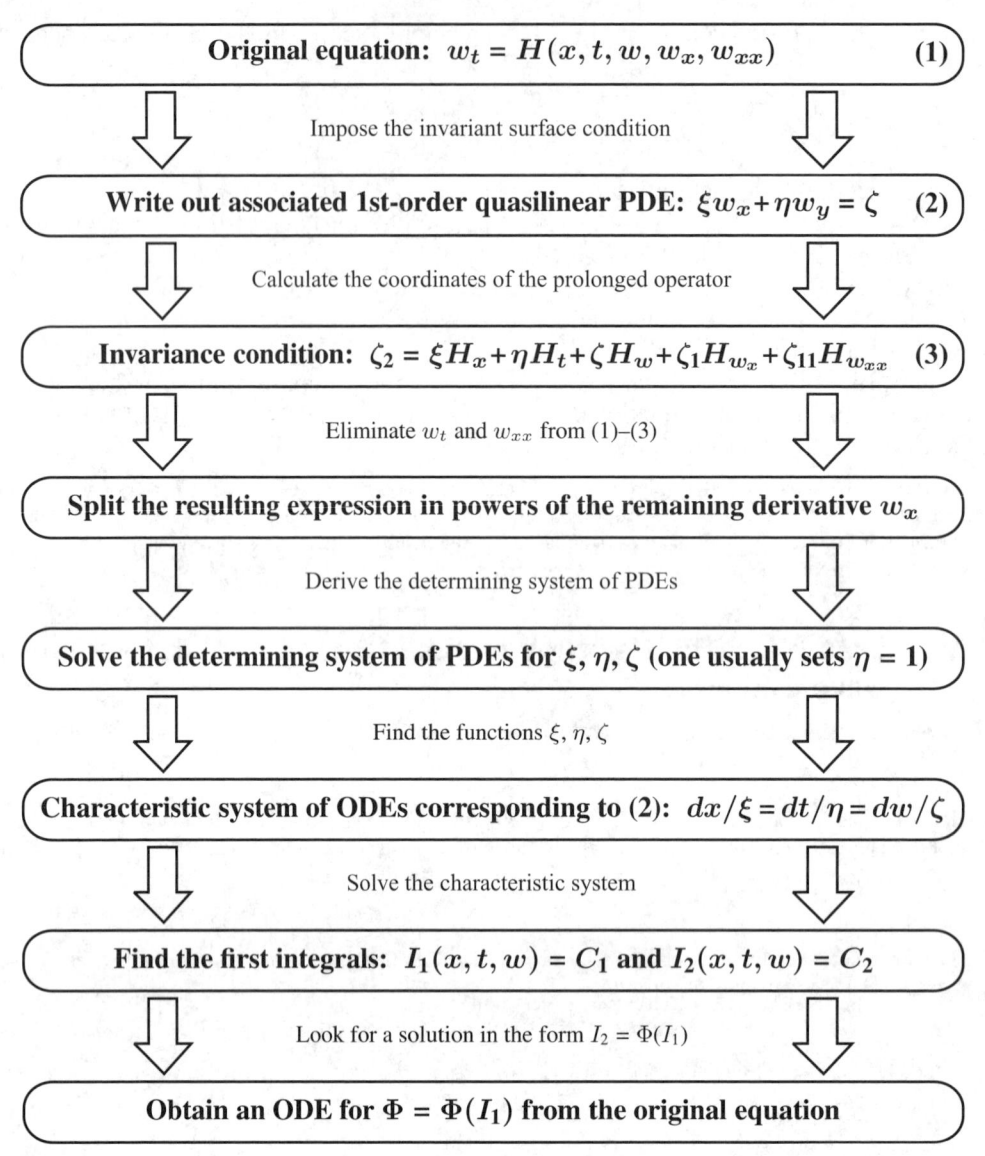

**Figure 15.5.** Algorithm for the construction of exact solutions by a nonclassical method for second-order evolution equations. Here, ODE stands for ordinary differential equation and PDE for partial differential equation.

which consists of only four equations.

The analysis of the system (15.9.2.6) allows us to conclude that

$$\xi = \xi(x, t), \quad \zeta = -\xi_x w,$$

where the function $\xi(x, t)$ satisfies the system

$$\xi_t - 3\xi_{xx} + 2\xi\xi_x = 0,$$
$$\xi_{xt} - \xi_{xxx} + 2\xi_x^2 + 2\xi\xi_x = 0. \qquad (15.9.2.7)$$

The associated invariant surface condition has the form (15.9.2.2):

$$w_t + \xi w_x = -\xi_x w. \qquad (15.9.2.8)$$

We look for a stationary particular solution to equation (15.9.2.7) in the form $\xi = \xi(x)$. Let us differentiate the first equation of (15.9.2.7) with respect to $x$ and then eliminate the third derivative, using the second equation of (15.9.2.7), from the resulting expression. This will give a second-order equation. Eliminating from it the second derivative, using the first equation of (15.9.2.7), after simple rearrangements we obtain

$$(6\xi'_x - 2\xi^2 + 9)\xi'_x = 0. \tag{15.9.2.9}$$

Equating the expression in parentheses in (15.9.2.9) to zero, we get a separable first-order equation. Its general solution can be written as

$$\xi = -\frac{3}{\sqrt{2}} \frac{A \exp(\sqrt{2}\, x) + B}{A \exp(\sqrt{2}\, x) - B} = -\frac{3}{\sqrt{2}} \frac{A \exp(\frac{\sqrt{2}}{2} x) + B \exp(-\frac{\sqrt{2}}{2} x)}{A \exp(\frac{\sqrt{2}}{2} x) - B \exp(-\frac{\sqrt{2}}{2} x)}, \tag{15.9.2.10}$$

where $A$ and $B$ are arbitrary constants.

The characteristic system corresponding to (15.9.2.8),

$$\frac{dt}{1} = \frac{dx}{\xi} = -\frac{dw}{\xi'_x w}, \tag{15.9.2.11}$$

admits the first integrals

$$t + \tfrac{2}{3} \ln \left| A \exp\left(\tfrac{\sqrt{2}}{2} x\right) + B \exp\left(-\tfrac{\sqrt{2}}{2} x\right) \right| = C_1, \quad \xi w = C_2, \tag{15.9.2.12}$$

where $C_1$ and $C_2$ are arbitrary constants. For convenience, we introduce the new constant $\overline{C}_1 = \exp(\tfrac{3}{2} C_1)$ and look for a solution in the form $C_2 = -\tfrac{3}{\sqrt{2}} \overline{C}_1 F(\overline{C}_1)$. Inserting (15.9.2.12) here and taking into account (15.9.2.10), we obtain the solution structure

$$w(x,t) = \left\{ A \exp\left[\tfrac{1}{2}(\sqrt{2}x + 3t)\right] - B \exp\left[\tfrac{1}{2}(-\sqrt{2}x + 3t)\right] \right\} F(z),$$
$$z = A \exp\left[\tfrac{1}{2}(\sqrt{2}x + 3t)\right] + B \exp\left[\tfrac{1}{2}(-\sqrt{2}x + 3t)\right]. \tag{15.9.2.13}$$

Substituting (15.9.2.13) in the original equation, we find the equation for $F = F(z)$:

$$F''_{zz} = 2F^3. \tag{15.9.2.14}$$

The general solution of equation (15.9.2.14) is expressed in terms of elliptic functions. This equation admits the following particular solutions: $F = \pm \dfrac{1}{z + C}$, where $C$ is an arbitrary constant.

**Remark 1.** To the degenerate case $\xi'_x = 0$ in (15.9.2.9) there corresponds a traveling-wave solution.

**Remark 2.** It is apparent from this example that using the invariant surface condition (15.9.1.2) gives much more freedom in determining the coordinates $\xi, \eta, \zeta$ as compared with the classical scheme presented in Section 15.8. This stems from the fact that in the classical scheme, the invariance condition is split with respect to two derivatives, $w_x$ and $w_{xx}$, which are considered to be independent (see Example 2 in Subsection 15.8.2). In the nonclassical scheme, the derivatives $w_x$ and $w_{xx}$ are related by the second equation of (15.9.2.5) and the invariance conditions are split in only one derivative, $w_x$. This is why in the classical scheme the determining system consists of a larger number of equations, which impose additional constraints on the unknown quantities, as compared with the nonclassical scheme. In particular, the classical scheme fails to find exact solutions to equation Newell–Whitehead (15.9.2.1) in the form (15.9.2.13).

**Example 2.** Consider the nonlinear wave equation

$$\frac{\partial^2 w}{\partial t^2} = w \frac{\partial^2 w}{\partial x^2}, \tag{15.9.2.15}$$

which corresponds to the left-hand side $F = w_{tt} - ww_{xx}$ of equation (15.9.1.1) with $y = t$.

Let us add the invariant surface condition (15.9.2.2). Applying the prolonged operator (15.9.2.3) to equation (15.9.2.15) gives the invariance condition. Taking into account that $\partial_x = \partial_t = \partial_{w_x} = \partial_{w_t} = \partial_{w_{xt}} = 0$ (since the equation is independent of $x, t, w_x, w_t, w_{xt}$ explicitly) and $\eta = 1$, we get the invariance condition

$$\zeta_{22} = \zeta w_{xx} + w \zeta_{11}.$$

Substituting here the expressions (15.8.1.14) for the coordinates $\zeta_{11}$ and $\zeta_{22}$ of the second prolongation with $y = t$ and $\eta = 1$, we have

$$\zeta_{tt} - \xi_{tt} w_x + 2\zeta_{wt} w_t - 2\xi_{wt} w_x w_t + \zeta_{ww} w_t^2 - \xi_{ww} w_x w_t^2 - 2(\xi_t + \xi_w w_t) w_{xt} + (\zeta_w - \xi_w w_x) w_{tt}$$
$$= \zeta w_{xx} + w \left[ \zeta_{xx} + (2\zeta_{wx} - \xi_{xx}) w_x + (\zeta_{ww} - 2\xi_{wx}) w_x^2 - \xi_{ww} w_x^3 + (\zeta_w - 2\xi_x - 3\xi_w w_x) w_{xx} \right]. \tag{15.9.2.16}$$

From the invariant surface condition (15.9.2.2) and equation (15.9.2.15) it follows that

$$w_t = \zeta - \xi w_x, \quad w_{tt} = ww_{xx}, \quad w_{xt} = \zeta_x - \xi_x w_x - \xi w_{xx}. \tag{15.9.2.17}$$

The last formula has been obtained by differentiating the first one with respect to $x$. Substituting $w_t$, $w_{tt}$, $w_{xt}$ from (15.9.2.17) into (15.9.2.16) yields a polynomial in two "independent" derivatives, $w_x$ and $w_{xx}$. Equating the functional coefficients of this polynomial to zero, we arrive at the determining system:

$$w_x w_{xx}: \quad (\xi^2 - w)\xi_w = 0,$$
$$w_{xx}: \quad 2\xi\xi_t + 2w\xi_x + 2\xi\xi_w\zeta - \zeta = 0,$$
$$w_x^3: \quad (\xi^2 - w)\xi_{ww} = 0,$$
$$w_x^2: \quad (\xi^2 - w)\zeta_{ww} + 2\xi\xi_{wt} + 2\xi\xi_{ww}\zeta - 2\xi\xi_x\xi_w + 2w\xi_{wx} = 0,$$
$$w_x: \quad \xi_{tt} + 2\xi\zeta_{wt} + 2\xi_{wt}\zeta + 2\xi\zeta\zeta_{ww} + \xi_{ww}\zeta^2 - 2\xi_t\xi_x - 2\xi_x\xi_w\zeta - 2\xi\xi_w\zeta_x + 2w\zeta_{wx} - w\xi_{xx} = 0,$$
$$1: \quad \zeta_{tt} + 2\zeta\zeta_{wt} + \zeta^2\zeta_{ww} - 2\xi_t\zeta_x - 2\xi_w\zeta\zeta_x - w\zeta_{xx} = 0.$$

From the first equation it follows that

$$1) \quad \xi = \xi(x,t); \tag{15.9.2.18}$$
$$2) \quad \xi = \sqrt{w}. \tag{15.9.2.19}$$

Consider both cases.

$1°$. For $\xi = \xi(x,t)$, the third equation of the determining system (15.9.2.18) is satisfied identically. From the second equation it follows that

$$\zeta = 2w\xi_x + 2\xi\xi_t. \tag{15.9.2.20}$$

The fourth equation is also satisfied identically in view of (15.9.2.18) and (15.9.2.20). Substituting (15.9.2.18) and (15.9.2.20) into the fifth and sixth equations of the determining system yields two solutions:

$$\begin{aligned}\xi = \alpha t + \beta, &\quad \zeta = 2\alpha(\alpha t + \beta) \quad \text{(first solution)};\\ \xi = \alpha x + \beta, &\quad \zeta = 2\alpha w \quad \text{(second solution)};\end{aligned} \tag{15.9.2.21}$$

where $\alpha$ and $\beta$ are arbitrary constants.

*First solution.* The characteristic system of ordinary differential equations associated with the first solution (15.9.2.21), with $\alpha = 2$ and $\beta = 0$, has the form

$$\frac{dt}{1} = \frac{dx}{2t} = \frac{dw}{8t}.$$

Find the first integrals: $C_1 = x - t^2$, $C_2 = w - 4t^2$. Following the scheme shown in Fig. 15.5, we look for a solution in the form $w - 4t^2 = \Phi(x - t^2)$. Substituting

$$w = \Phi(z) + 4t^2, \quad z = x - t^2 \tag{15.9.2.22}$$

into (15.9.2.15) yields the following autonomous ordinary differential equation for $\Phi = \Phi(z)$:

$$\Phi\Phi''_{zz} + 2\Phi'_z = 8.$$

It is easy to integrate. Reducing its order gives a separable equation. As a result, one can obtain an exact solution to equation (15.9.2.15) of the form (15.9.2.22).

*Second solution.* The characteristic system of ordinary differential equations associated with the second solution of (15.9.2.21), with $\alpha = 1$ and $\beta = 0$, has the form

$$\frac{dt}{1} = \frac{dx}{x} = \frac{dw}{2w}.$$

Find the first integrals: $C_1 = \ln|x| - t$, $C_2 = w/x^2$. Following the scheme shown in Fig. 15.5, we look for a solution in the form $w/x^2 = \Phi(\ln|x| - t)$. Substituting

$$w = x^2\Phi(z), \quad z = \ln|x| - t$$

into (15.9.2.15) yields the autonomous ordinary differential equation

$$(\Phi - 1)\Phi''_{zz} + 3\Phi\Phi'_z + 2\Phi^2 = 0,$$

which admits order reduction via the substitution $U(\Phi) = \Phi'_z$.

$2°$. The second case of (15.9.2.19) gives rise to the trivial solution $\zeta = 0$ (as follows from the fourth equation of the determining system), which generates the obvious solution $w = \text{const}$.

## 15.10. Differential Constraints Method
### 15.10.1. Description of the Method

**15.10.1-1. Preliminary remarks. A simple example.**

***Basic idea:*** Try to find an exact solution to a complex equation by analyzing it in conjunction with a simpler, auxiliary equation, called a *differential constraint*.

In Subsections 15.4.1 and 15.4.3, we have considered examples of additive separable solutions of nonlinear equations in the form

$$w(x, y) = \varphi(x) + \psi(y). \quad (15.10.1.1)$$

At the initial stage, the functions $\varphi(x)$ and $\psi(y)$ are assumed arbitrary and are to be determined in the subsequent analysis.

Differentiating the expression (15.10.1.1) with respect to $y$, we obtain

$$\frac{\partial w}{\partial y} = f(y) \quad (f = \psi'_y). \quad (15.10.1.2)$$

Conversely, relation (15.10.1.2) implies a representation of the solution in the form (15.10.1.1).

Further, differentiating (15.10.1.2) with respect to $x$ gives

$$\frac{\partial^2 w}{\partial x \partial y} = 0. \quad (15.10.1.3)$$

Conversely, from (15.10.1.3) we obtain a representation of the solution in the form (15.10.1.1).

Thus, the problem of finding exact solutions of the form (15.10.1.1) for a specific partial differential equation may be replaced by an equivalent problem of finding exact solutions of the given equation supplemented with the condition (15.10.1.2) or (15.10.1.3). Such supplementary conditions in the form of one or several differential equations will be called *differential constraints*.

Prior to giving a general description of the differential constraints method, we demonstrate its features by a simple example.

**Example.** Consider the third-order nonlinear equation

$$\frac{\partial w}{\partial y} \frac{\partial^2 w}{\partial x \partial y} - \frac{\partial w}{\partial x} \frac{\partial^2 w}{\partial y^2} = a \frac{\partial^3 w}{\partial y^3}, \quad (15.10.1.4)$$

which occurs in the theory of the hydrodynamic boundary layer. Let us seek a solution of equation (15.10.1.4) satisfying the linear first-order differential constraint

$$\frac{\partial w}{\partial x} = \varphi(y). \quad (15.10.1.5)$$

Here, the function $\varphi(y)$ cannot be arbitrary, in general, but must satisfy the condition of compatibility of equations (15.10.1.4) and (15.10.1.5). The compatibility condition is a differential equation for $\varphi(y)$ and is a consequence of equations (15.10.1.4), (15.10.1.5), and those obtained by their differentiation.

Successively differentiating (15.10.1.5) with respect to different variables, we calculate the derivatives

$$w_{xx} = 0, \quad w_{xy} = \varphi'_y, \quad w_{xxy} = 0, \quad w_{xyy} = \varphi''_{yy}, \quad w_{xyyy} = \varphi'''_{yyy}. \quad (15.10.1.6)$$

Differentiating (15.10.1.4) with respect to $x$ yields

$$w_{xy}^2 + w_y w_{xxy} - w_{xx} w_{yy} - w_x w_{xyy} = a w_{xyyy}. \quad (15.10.1.7)$$

Substituting the derivatives of the function $w$ from (15.10.1.5) and (15.10.1.6) into (15.10.1.7), we obtain the following third-order ordinary differential equation for $\varphi$:

$$(\varphi'_y)^2 - \varphi \varphi''_{yy} = a \varphi'''_{yyy}, \tag{15.10.1.8}$$

which represents the compatibility condition for equations (15.10.1.4) and (15.10.1.5).

In order to construct an exact solution, we integrate equation (15.10.1.5) to obtain

$$w = \varphi(y)x + \psi(y). \tag{15.10.1.9}$$

The function $\psi(y)$ is found by substituting (15.10.1.9) into (15.10.1.4) and taking into account the condition (15.10.1.8). As a result, we arrive at the ordinary differential equation

$$\varphi'_y \psi'_y - \varphi \psi''_{yy} = a \psi'''_{yyy}. \tag{15.10.1.10}$$

Finally, we obtain an exact solution of the form (15.10.1.9), with the functions $\varphi$ and $\psi$ described by equations (15.10.1.8) and (15.10.1.10).

**Remark.** It is easier to obtain the above solution by directly substituting expression (15.10.1.9) into the original equation (15.10.1.4).

---

**15.10.1-2. General description of the differential constraints method.**

The procedure of the construction of exact solutions to nonlinear equations of mathematical physics by the differential constraints method consists of several steps described below.

1°. In the general case, the identification of particular solutions of the equation

$$F\left(x, y, w, \frac{\partial w}{\partial x}, \frac{\partial w}{\partial y}, \frac{\partial^2 w}{\partial x^2}, \frac{\partial^2 w}{\partial x \partial y}, \frac{\partial^2 w}{\partial y^2}, \ldots\right) = 0 \tag{15.10.1.11}$$

is performed by supplementing this equation with an additional differential constraint

$$G\left(x, y, w, \frac{\partial w}{\partial x}, \frac{\partial w}{\partial y}, \frac{\partial^2 w}{\partial x^2}, \frac{\partial^2 w}{\partial x \partial y}, \frac{\partial^2 w}{\partial y^2}, \ldots\right) = 0. \tag{15.10.1.12}$$

The form of the differential constraint (15.10.1.12) may be prescribed on the basis of (i) a priori considerations (for instance, it may be required that the constraint should represent a solvable equation); (ii) certain properties of the equation under consideration (for instance, it may be required that the constraint should follow from symmetries of the equation or the corresponding conservation laws).

2°. In general, the thus obtained overdetermined system (15.10.1.11)–(15.10.1.12) requires a compatibility analysis. If the differential constraint (15.10.1.12) is specified on the basis of a priori considerations, it should allow for sufficient freedom in choosing functions (i.e., involve arbitrary determining functions). The compatibility analysis of system (15.10.1.11)–(15.10.1.12) should provide conditions that specify the structure of the determining functions. These conditions (*compatibility conditions*) are written as a system of ordinary differential equations (or a system of partial differential equations).

Usually, the compatibility analysis is performed by means of differentiating (possibly, several times) equations (15.10.1.11) and (15.10.1.12) with respect to $x$ and $y$ and eliminating the highest-order derivatives from the resulting differential relations and equations (15.10.1.11) and (15.10.1.12). As a result, one arrives at an equation involving powers of lower-order derivatives. Equating the coefficients of all powers of the derivatives to zero, one obtains compatibility conditions connecting the functional coefficients of equations (15.10.1.11) and (15.10.1.12).

3°. One solves the system of differential equations obtained in Item 2° for the determining functions. Then these functions are substituted into the differential constraint (15.10.1.12) to obtain an equation of the form

$$g\left(x, y, w, \frac{\partial w}{\partial x}, \frac{\partial w}{\partial y}, \frac{\partial^2 w}{\partial x^2}, \frac{\partial^2 w}{\partial x \partial y}, \frac{\partial^2 w}{\partial y^2}, \dots\right) = 0. \qquad (15.10.1.13)$$

A differential constraint (15.10.1.13) that is compatible with equation (15.10.1.11) under consideration is called an *invariant manifold* of equation (15.10.1.11).

4°. One should find the general solution of (i) equation (15.10.1.13) or (ii) some equation that follows from equations (15.10.1.11) and (15.10.1.13). The solution thus obtained will involve some arbitrary functions $\{\varphi_m\}$ (these may depend on $x$ and $y$, as well as $w$). Note that in some cases, one can use, instead of the general solution, some particular solutions of equation (15.10.1.13) or equations that follow from (15.10.1.13).

5°. The solution obtained in Item 4° should be substituted into the original equation (15.10.1.11). As a result, one arrives at a functional differential equation from which the functions $\{\varphi_m\}$ should be found. Having found the $\{\varphi_m\}$, one should insert these functions into the solution from Item 4°. Thus, one obtains an exact solution of the original equation (15.10.1.11).

Remark 1. Should the choice of a differential constraint be inadequate, equations (15.10.1.11) and (15.10.1.12) may happen to be incompatible (having no common solutions).

Remark 2. There may be several differential constraints of the form (15.10.1.12).

Remark 3. At the last three steps of the differential constraints method, one has to solve various equations (systems of equations). If no solution can be constructed at one of those steps, one fails to construct an exact solution of the original equation.

For the sake of clarity, the general scheme of the differential constraints method is represented in Figure 15.6.

## 15.10.2. First-Order Differential Constraints

### 15.10.2-1. Second-order evolution equations.

Consider a general second-order evolution equation solved for the highest-order derivative:

$$\frac{\partial^2 w}{\partial x^2} = \mathcal{F}\left(x, t, w, \frac{\partial w}{\partial x}, \frac{\partial w}{\partial t}\right). \qquad (15.10.2.1)$$

Let us supplement this equation with a first-order differential constraint

$$\frac{\partial w}{\partial t} = \mathcal{G}\left(x, t, w, \frac{\partial w}{\partial x}\right). \qquad (15.10.2.2)$$

The condition of compatibility of these equations is obtained by differentiating (15.10.2.1) with respect to $t$ once and differentiating (15.10.2.2) with respect to $x$ twice, and then equating the two resulting expressions for the third derivatives $w_{xxt}$ and $w_{txx}$:

$$D_t \mathcal{F} = D_x^2 \mathcal{G}. \qquad (15.10.2.3)$$

```
┌─────────────────────────────────────────────────────────────────────┐
│  Original equation:  $F(x, y, w, w_x, w_y, w_{xx}, w_{xy}, w_{yy}, \ldots) = 0$  │
└─────────────────────────────────────────────────────────────────────┘
            ⇓  Introduce a supplementary equation  ⇓
┌─────────────────────────────────────────────────────────────────────┐
│  Differential constraint:  $G(x, y, w, w_x, w_y, w_{xx}, w_{xy}, w_{yy}, \ldots) = 0$  │
└─────────────────────────────────────────────────────────────────────┘
            ⇓  Perform compatibility analysis of the two equations  ⇓
┌─────────────────────────────────────────────────────────────────────┐
│  Find compatibility conditions for the equations $F = 0$ and $G = 0$  │
└─────────────────────────────────────────────────────────────────────┘
            ⇓  Obtain equations for the determining functions  ⇓
┌─────────────────────────────────────────────────────────────────────┐
│              Solve the equations for the determining functions       │
└─────────────────────────────────────────────────────────────────────┘
            ⇓  Insert the solution into the differential constraint  ⇓
┌─────────────────────────────────────────────────────────────────────┐
│  Find an invariant manifold:  $g(x, y, w, w_x, w_y, w_{xx}, w_{xy}, w_{yy}, \ldots) = 0$  │
└─────────────────────────────────────────────────────────────────────┘
            ⇓  Solve the equation $g = 0$ for $w$  ⇓
┌─────────────────────────────────────────────────────────────────────┐
│      Insert resulting solution (with arbitrariness) into the original equation  │
└─────────────────────────────────────────────────────────────────────┘
            ⇓  Determine the unknown functions and constants  ⇓
┌─────────────────────────────────────────────────────────────────────┐
│              Obtain an exact solution of the original equation       │
└─────────────────────────────────────────────────────────────────────┘
```

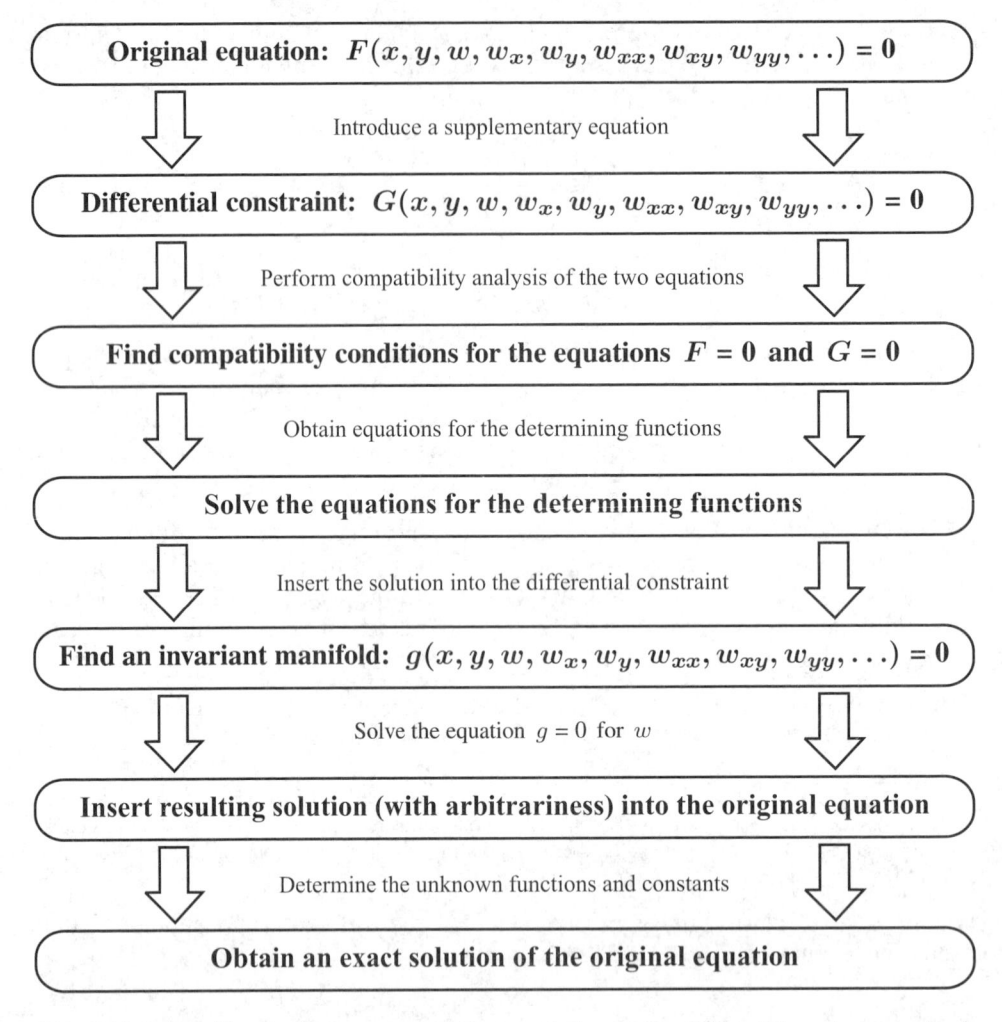

**Figure 15.6.** Algorithm for the construction of exact solutions by the differential constraints method.

Here, $D_t$ and $D_x$ are the total differentiation operators with respect to $t$ and $x$:

$$D_t = \frac{\partial}{\partial t} + w_t \frac{\partial}{\partial w} + w_{xt} \frac{\partial}{\partial w_x} + w_{tt} \frac{\partial}{\partial w_t}, \quad D_x = \frac{\partial}{\partial x} + w_x \frac{\partial}{\partial w} + w_{xx} \frac{\partial}{\partial w_x} + w_{xt} \frac{\partial}{\partial w_t}. \tag{15.10.2.4}$$

The partial derivatives $w_t$, $w_{xx}$, $w_{xt}$, and $w_{tt}$ in (15.10.2.4) should be expressed in terms of $x$, $t$, $w$, and $w_x$ by means of the relations (15.10.2.1), (15.10.2.2) and those obtained by differentiation of (15.10.2.1), (15.10.2.2). As a result, we get

$$w_t = \mathcal{G}, \quad w_{xx} = \mathcal{F}, \quad w_{xt} = D_x \mathcal{G} = \frac{\partial \mathcal{G}}{\partial x} + w_x \frac{\partial \mathcal{G}}{\partial w} + \mathcal{F} \frac{\partial \mathcal{G}}{\partial w_x},$$

$$w_{tt} = D_t \mathcal{G} = \frac{\partial \mathcal{G}}{\partial t} + \mathcal{G} \frac{\partial \mathcal{G}}{\partial w} + w_{xt} \frac{\partial \mathcal{G}}{\partial w_x} = \frac{\partial \mathcal{G}}{\partial t} + \mathcal{G} \frac{\partial \mathcal{G}}{\partial w} + \left( \frac{\partial \mathcal{G}}{\partial x} + w_x \frac{\partial \mathcal{G}}{\partial w} + \mathcal{F} \frac{\partial \mathcal{G}}{\partial w_x} \right) \frac{\partial \mathcal{G}}{\partial w_x}. \tag{15.10.2.5}$$

In the expression for $\mathcal{F}$, the derivative $w_t$ should be replaced by $\mathcal{G}$ by virtue of (15.10.2.2).

**Example 1.** From the class of nonlinear heat equations with a source

$$\frac{\partial w}{\partial t} = \frac{\partial}{\partial x} \left[ f(w) \frac{\partial w}{\partial x} \right] + g(w), \tag{15.10.2.6}$$

let us single out equations possessing invariant manifolds of the simplest form
$$\frac{\partial w}{\partial t} = \varphi(w). \tag{15.10.2.7}$$
Equations (15.10.2.6) and (15.10.2.7) are special cases of (15.10.2.1) and (15.10.2.2) with
$$\mathcal{F} = \frac{w_t - f'(w)w_x^2 - g(w)}{f(w)} = \frac{\varphi(w) - g(w) - f'(w)w_x^2}{f(w)}, \quad \mathcal{G} = \varphi(w).$$
The functions $f(w)$, $g(w)$, and $\varphi(w)$ are unknown in advance and are to be determined in the subsequent analysis.

Using (15.10.2.5) and (15.10.2.4), we find partial derivatives and the total differentiation operators:
$$w_t = \varphi, \quad w_{xx} = \mathcal{F}, \quad w_{xt} = \varphi'w_x, \quad w_{tt} = \varphi\varphi',$$
$$D_t = \frac{\partial}{\partial t} + \varphi\frac{\partial}{\partial w} + \varphi'w_x\frac{\partial}{\partial w_x} + \varphi\varphi'\frac{\partial}{\partial w_t}, \quad D_x = \frac{\partial}{\partial x} + w_x\frac{\partial}{\partial w} + \mathcal{F}\frac{\partial}{\partial w_x} + \varphi'w_x\frac{\partial}{\partial w_t}.$$
We insert the expressions of $D_x$ and $D_t$ into the compatibility conditions (15.10.2.3) and rearrange terms to obtain
$$\left[\frac{(f\varphi)'}{f}\right]' w_x^2 + \frac{\varphi - g}{f}\varphi' - \varphi\left(\frac{\varphi - g}{f}\right)' = 0.$$
In order to ensure that this equality holds true for any $w_x$, one should take
$$\left[\frac{(f\varphi)'}{f}\right]' = 0, \quad \frac{\varphi - g}{f}\varphi' - \varphi\left(\frac{\varphi - g}{f}\right)' = 0. \tag{15.10.2.8}$$

*Nondegenerate case.* Assuming that the function $f = f(w)$ is given, we obtain a three-parameter solution of equations (15.10.2.8) for the functions $g(w)$ and $\varphi(w)$:
$$g(w) = \frac{a + cf}{f}\left(\int f\,dw + b\right), \quad \varphi(w) = \frac{a}{f}\left(\int f\,dw + b\right), \tag{15.10.2.9}$$
where $a$, $b$, and $c$ are arbitrary constants.

We substitute $\varphi(w)$ of (15.10.2.9) into equation (15.10.2.7) and integrate to obtain
$$\int f\,dw = \theta(x)e^{at} - b. \tag{15.10.2.10}$$
Differentiating (15.10.2.10) with respect to $x$ and $t$, we get $w_t = ae^{at}\theta/f$ and $w_x = e^{at}\theta'_x/f$. Substituting these expressions into (15.10.2.6) and taking into account (15.10.2.9), we arrive at the equation $\theta''_{xx} + c\theta = 0$, whose general solution is given by
$$\theta = \begin{cases} C_1\sin(x\sqrt{c}) + C_2\cos(x\sqrt{c}) & \text{if } c > 0, \\ C_1\sinh(x\sqrt{-c}) + C_2\cosh(x\sqrt{-c}) & \text{if } c < 0, \\ C_1x + C_2 & \text{if } c = 0, \end{cases} \tag{15.10.2.11}$$
where $C_1$ and $C_2$ are arbitrary constants. Formulas (15.10.2.10)–(15.10.2.11) describe exact solutions (in implicit form) of equation (15.10.2.6) with $f(w)$ arbitrary and $g(w)$ given by (15.10.2.9).

*Degenerate case.* There also exists a two-parameter solution of equations (15.10.2.8) for the functions $g(w)$ and $\varphi(w)$ (as above, $f$ is assumed arbitrary):
$$g(w) = \frac{b}{f} + c, \quad \varphi(w) = \frac{b}{f},$$
where $b$ and $c$ are arbitrary constants. This solution can be obtained from (15.10.2.9) by renaming variables, $b \to b/a$ and $c \to ac/b$, and letting $a \to 0$. After simple calculations, we obtain the corresponding solution of equation (15.10.2.6) in implicit form:
$$\int f\,dw = bt - \frac{1}{2}cx^2 + C_1x + C_2.$$

The example given below shows that calculations may be performed without the use of the general formulas (15.10.2.3)–(15.10.2.5).

**Remark 1.** In the general case, for a given function $\mathcal{F}$, the compatibility condition (15.10.2.3) is a nonlinear partial differential equation for the function $\mathcal{G}$. This equation admits infinitely many solutions (by the theorem about the local existence of solutions). Therefore, the second-order partial differential equation (15.10.2.1) admits infinitely many compatible first-order differential constraints (15.10.2.2).

**Remark 2.** In the general case, the solution of the first-order partial differential equation (15.10.2.2) reduces to the solution of a system of ordinary differential equations; see Chapter 13.

### 15.10.2-2. Second-order hyperbolic equations.

In a similar way, one can consider second-order hyperbolic equations of the form

$$\frac{\partial^2 w}{\partial x \partial t} = \mathcal{F}\left(x, t, w, \frac{\partial w}{\partial x}, \frac{\partial w}{\partial t}\right), \tag{15.10.2.12}$$

supplemented by a first-order differential constraint (15.10.2.2). Assume that $\mathcal{G}_{w_x} \neq 0$.

A compatibility condition for these equations is obtained by differentiating (15.10.2.12) with respect to $t$ and (15.10.2.2) with respect to $t$ and $x$, and then equating the resulting expressions of the third derivatives $w_{xtt}$ and $w_{ttx}$ to one another:

$$D_t \mathcal{F} = D_x[D_t \mathcal{G}]. \tag{15.10.2.13}$$

Here, $D_t$ and $D_x$ are the total differential operators of (15.10.2.4) in which the partial derivatives $w_t$, $w_{xx}$, $w_{xt}$, and $w_{tt}$ must be expressed in terms of $x$, $t$, $w$, and $w_x$ with the help of relations (15.10.2.12) and (15.10.2.2) and those obtained by differentiating (15.10.2.12) and (15.10.2.2).

Let us show how the second derivatives can be calculated. We differentiate (15.10.2.2) with respect to $x$ and replace the mixed derivative by the right-hand side of (15.10.2.12) to obtain the following expression for the second derivative with respect to $x$:

$$\frac{\partial \mathcal{G}}{\partial x} + w_x \frac{\partial \mathcal{G}}{\partial w} + w_{xx} \frac{\partial \mathcal{G}}{\partial w_x} = \mathcal{F}\left(x, t, w, \frac{\partial w}{\partial x}, \frac{\partial w}{\partial t}\right) \implies \frac{\partial^2 w}{\partial x^2} = \mathcal{H}_1\left(x, t, w, \frac{\partial w}{\partial x}\right). \tag{15.10.2.14}$$

Here and in what follows, we have taken into account that (15.10.2.2) allows us to express the derivative with respect to $t$ through the derivative with respect to $x$. Further, differentiating (15.10.2.2) with respect to $t$ yields

$$\frac{\partial^2 w}{\partial t^2} = \frac{\partial \mathcal{G}}{\partial t} + w_t \frac{\partial \mathcal{G}}{\partial w} + w_{xt} \frac{\partial \mathcal{G}}{\partial w_x} = \frac{\partial \mathcal{G}}{\partial t} + \mathcal{G} \frac{\partial \mathcal{G}}{\partial w} + \mathcal{F} \frac{\partial \mathcal{G}}{\partial w_x} \implies \frac{\partial^2 w}{\partial t^2} = \mathcal{H}_2\left(x, t, w, \frac{\partial w}{\partial x}\right). \tag{15.10.2.15}$$

Replacing the derivatives $w_t$, $w_{xt}$, $w_{xx}$, and $w_{tt}$ in (15.10.2.4) by their expressions from (15.10.2.2), (15.10.2.12), (15.10.2.14), and (15.10.2.15), we find the total differential operators $D_t$ and $D_x$, which are required for the compatibility condition (15.10.2.13).

**Example 2.** Consider the nonlinear equation

$$\frac{\partial^2 w}{\partial x \partial t} = f(w). \tag{15.10.2.16}$$

Let us supplement (15.10.2.16) with quasilinear differential constraint of the form

$$\frac{\partial w}{\partial x} = \varphi(t) g(w). \tag{15.10.2.17}$$

Differentiating (15.10.2.16) with respect to $x$ and then replacing the first derivative with respect to $x$ by the right-hand side of (15.10.2.17), we get

$$w_{xxt} = \varphi g f'_w. \tag{15.10.2.18}$$

Differentiating further (15.10.2.17) with respect to $x$ and $t$, we obtain two relations

$$w_{xx} = \varphi g'_w w_x = \varphi^2 g g'_w, \tag{15.10.2.19}$$
$$w_{xt} = \varphi'_t g + \varphi g'_w w_t. \tag{15.10.2.20}$$

Eliminating the mixed derivative from (15.10.2.20) using equation (15.10.2.16), we find the first derivative with respect to $t$:

$$w_t = \frac{f - \varphi'_t g}{\varphi g'_w}. \tag{15.10.2.21}$$

Differentiating (15.10.2.19) with respect to $t$ and replacing $w_t$ by the right-hand side of (15.10.2.21), we have

$$w_{xxt} = 2\varphi\varphi'_t gg'_w + \varphi^2(gg'_w)'_w w_t = 2\varphi\varphi'_t gg'_w + \varphi(gg'_w)'_w \frac{f - \varphi'_t g}{g'_w}. \tag{15.10.2.22}$$

Equating now the third derivatives (15.10.2.18) and (15.10.2.22), canceling them by $\varphi$, and performing simple rearrangements, we get the determining equation

$$\varphi'_t g[(g'_w)^2 - gg''_{ww}] = gg'_w f'_w - f(gg'_w)'_w, \tag{15.10.2.23}$$

which has two different solutions.

*Solution 1.* Equation (15.10.2.23) is satisfied identically for any $\varphi = \varphi(t)$ if

$$(g'_w)^2 - gg''_{ww} = 0,$$
$$gg'_w f'_w - f(gg'_w)'_w = 0.$$

The general solution of this system has the form

$$f(w) = ae^{\lambda w}, \quad g(w) = be^{\lambda w/2}, \tag{15.10.2.24}$$

where $a$, $b$, and $\lambda$ are arbitrary constants. For simplicity, we set

$$a = b = 1, \quad \lambda = -2. \tag{15.10.2.25}$$

Substitute $g(w)$ defined by (15.10.2.24)–(15.10.2.25) into the differential constraint (15.10.2.17) and integrate the resulting equation to obtain

$$w = \ln[\varphi(t)x + \psi(t)], \tag{15.10.2.26}$$

where $\psi(t)$ is an arbitrary function. Substituting (15.10.2.26) into the original equation (15.10.2.16) with the right-hand side (15.10.2.24)–(15.10.2.25), we arrive at a linear ordinary differential equation for $\psi(t)$:

$$\psi\varphi'_t - \varphi\psi'_t = 1.$$

The general solution of this equation is expressed as

$$\psi(t) = C\varphi(t) - \varphi(t)\int \frac{dt}{\varphi^2(t)}, \tag{15.10.2.27}$$

where $C$ is an arbitrary constant.

Thus, formulas (15.10.2.26)–(15.10.2.27), where $\varphi(t)$ is an arbitrary function, define an exact solution to the nonlinear equation $w_{xt} = e^{-2w}$.

*Solution 2.* The second solution is determined by the linear relation

$$\varphi(t) = at + b, \tag{15.10.2.28}$$

where $a$ and $b$ are arbitrary constants. In this case, the functions $f(w)$ and $g(w)$ are related by (15.10.2.23), with $\varphi'_t = a$. Integrating (15.10.2.17) with constraint (15.10.2.28) yields the solution structure

$$w = w(z), \quad z = (at + b)x + \psi(t), \tag{15.10.2.29}$$

where $\psi(t)$ is an arbitrary function. Inserting it into the original equation (15.10.2.16) and changing $x$ to $z$ with the help of (15.10.2.29), we obtain

$$[az + (at + b)\psi'_t - a\psi]w''_{zz} + aw'_z = f(w). \tag{15.10.2.30}$$

In order for this relation to be an ordinary differential equation for $w = w(z)$, one should set

$$(at + b)\psi'_t - a\psi = \text{const}.$$

Integrating yields $\psi(t)$ in the form

$$\psi(t) = ct + d, \tag{15.10.2.31}$$

where $c$ and $d$ are arbitrary constants.

Formulas (15.10.2.29) and (15.10.2.31) define a solution to equation (15.10.2.16) for arbitrary $f(w)$. The function $w(z)$ is described by equation (15.10.2.30) with constraint (15.10.2.31). To the special case $a = d = 0$ there corresponds a traveling-wave solution and to the case $b = c = d = 0$, a self-similar solution.

### 15.10.2-3. Second-order equations of general form.

Consider a second-order equation of the general form

$$\mathcal{F}(x, t, w, w_x, w_t, w_{xx}, w_{xt}, w_{tt}) = 0 \tag{15.10.2.32}$$

with a first-order differential constraint

$$\mathcal{G}(x, t, w, w_x, w_t) = 0. \tag{15.10.2.33}$$

Let us successively differentiate equations (15.10.2.32) and (15.10.2.33) with respect to both variables so as to obtain differential relations involving second and third derivatives. We get

$$D_x \mathcal{F} = 0, \quad D_t \mathcal{F} = 0, \quad D_x \mathcal{G} = 0, \quad D_t \mathcal{G} = 0, \quad D_x[D_x \mathcal{G}] = 0, \quad D_x[D_t \mathcal{G}] = 0, \quad D_t[D_t \mathcal{G}] = 0. \tag{15.10.2.34}$$

The compatibility condition for (15.10.2.32) and (15.10.2.33) can be found by eliminating the derivatives $w_t$, $w_{xx}$, $w_{xt}$, $w_{tt}$, $w_{xxx}$, $w_{xxt}$, $w_{xtt}$, and $w_{ttt}$ from the nine equations of (15.10.2.32)–(15.10.2.34). In doing so, we obtain an expression of the form

$$\mathcal{H}(x, t, w, w_x) = 0. \tag{15.10.2.35}$$

If the left-hand side of (15.10.2.35) is a polynomial in $w_x$, then the compatibility conditions result from equating the functional coefficients of the polynomial to zero.

## 15.10.3. Second- and Higher-Order Differential Constraints

Constructing exact solutions of nonlinear partial differential equations with the help of second- and higher-order differential constraints requires finding exact solutions of these differential constraints. The latter is generally rather difficult or even impossible. For this reason, one employs some special differential constraints that involve derivatives with respect to only one variable. In practice, one considers second-order ordinary differential equations in, say, $x$ and the other variable, $t$, is involved implicitly or is regarded as a parameter, so that integration constants depend on $t$.

The problem of compatibility of a second-order evolution equation

$$\frac{\partial w}{\partial t} = \mathcal{F}_1 \left( x, t, w, \frac{\partial w}{\partial x}, \frac{\partial^2 w}{\partial x^2} \right)$$

with a similar differential constraint

$$\frac{\partial w}{\partial t} = \mathcal{F}_2 \left( x, t, w, \frac{\partial w}{\partial x}, \frac{\partial^2 w}{\partial x^2} \right)$$

may be reduced to a problem with the first-order differential constraint considered in Paragraph 15.10.2-1. To that end, one should first eliminate the second derivative $w_{xx}$ from the equations. Then the resulting first-order equation is examined together with the original equation (or the original differential constraint).

## 15.10. Differential Constraints Method

**Example.** From the class of nonlinear heat equations with a source

$$\frac{\partial w}{\partial t} = \frac{\partial}{\partial x}\left[f_1(w)\frac{\partial w}{\partial x}\right] + f_2(w) \tag{15.10.3.1}$$

one singles out equations that admit invariant manifolds of the form

$$\frac{\partial^2 w}{\partial x^2} = g_1(w)\left(\frac{\partial w}{\partial x}\right)^2 + g_2(w). \tag{15.10.3.2}$$

The functions $f_2(w)$, $f_1(w)$, $g_2(w)$, and $g_1(w)$ are to be determined in the further analysis.

Eliminating the second derivative from (15.10.3.1) and (15.10.3.2), we obtain

$$\frac{\partial w}{\partial t} = \varphi(w)\left(\frac{\partial w}{\partial x}\right)^2 + \psi(w), \tag{15.10.3.3}$$

where

$$\varphi(w) = f_1(w)g_1(w) + f_1'(w), \quad \psi(w) = f_1(w)g_2(w) + f_2(w). \tag{15.10.3.4}$$

The condition of invariance of the manifold (15.10.3.2) under equation (15.10.3.1) is obtained by differentiating (15.10.3.2) with respect to $t$:

$$w_{xxt} = 2g_1 w_x w_{xt} + g_1' w_x^2 w_t + g_2' w_t.$$

The derivatives $w_{xxt}$, $w_{xt}$, and $w_t$ should be eliminated from this relation with the help of equations (15.10.3.2) and (15.10.3.3) and those obtained by their differentiation. As a result, we get

$$(2\varphi g_1^2 + 3\varphi' g_1 + \varphi g_1' + \varphi'')w_x^4 + (4\varphi g_1 g_2 + 5\varphi' g_2 + \varphi g_2' - g_1\psi' - \psi g_1' + \psi'')w_x^2 + 2\varphi g_2^2 + \psi' g_2 - \psi g_2' = 0.$$

Equating the coefficients of like powers of $w_x$ to zero, one obtains three equations, which, for convenience, may be written in the form

$$(\varphi' + \varphi g_1)' + 2g_1(\varphi' + \varphi g_1) = 0,$$
$$4g_2(\varphi' + \varphi g_1) + (\varphi g_2 - \psi g_1)' + \psi'' = 0, \tag{15.10.3.5}$$
$$\varphi = -\tfrac{1}{2}(\psi/g_2)'.$$

The first equation can be satisfied by taking $\varphi' + \varphi g_1 = 0$. The corresponding particular solution of system (15.10.3.5) has the form

$$\varphi = -\frac{1}{2}\mu', \quad \psi = \mu g_2, \quad g_1 = -\frac{\mu''}{\mu'}, \quad g_2 = \left(2C_1 + \frac{C_2}{\sqrt{|\mu|}}\right)\frac{1}{\mu'}, \tag{15.10.3.6}$$

where $\mu = \mu(w)$ is an arbitrary function.

Taking into account (15.10.3.4), we find the functional coefficients of the original equation (15.10.3.1) and the invariant set (15.10.3.2):

$$f_1 = \left(C_3 - \frac{1}{2}w\right)\mu', \quad f_2 = (\mu - f_1)g_2, \quad g_1 = -\frac{\mu''}{\mu'}, \quad g_2 = \left(2C_1 + \frac{C_2}{\sqrt{|\mu|}}\right)\frac{1}{\mu'}. \tag{15.10.3.7}$$

Equation (15.10.3.2), together with (15.10.3.7), admits the first integral

$$w_x^2 = \left[4C_1\mu + 4C_2\sqrt{|\mu|} + 2\sigma_t'(t)\right]\frac{1}{(\mu')^2}, \tag{15.10.3.8}$$

where $\sigma(t)$ is an arbitrary function. Let us eliminate $w_x^2$ from (15.10.3.3) by means of (15.10.3.8) and substitute the functions $\varphi$ and $\psi$ from (15.10.3.6) to obtain the equation

$$\mu' w_t = -C_2\sqrt{|\mu|} - \sigma_t'(t). \tag{15.10.3.9}$$

Let us dwell on the special case $C_2 = C_3 = 0$. Integrating equation (15.10.3.9) and taking into account that $\mu_t = \mu' w_t$ yield

$$\mu = -\sigma(t) + \theta(x), \tag{15.10.3.10}$$

where $\theta(x)$ is an arbitrary function. Substituting (15.10.3.10) into (15.10.3.8) and taking into account the relation $\mu_x = \mu' w_x$, we obtain

$$\theta_x^2 - 4C_1\theta = 2\sigma_t - 4C_1\sigma.$$

Equating both sides of this equation to zero and integrating the resulting ordinary differential equations, we find the functions on the right-hand side of (15.10.3.10):

$$\sigma(t) = A\exp(2C_1 t), \quad \theta(x) = C_1(x+B)^2, \tag{15.10.3.11}$$

where $A$ and $B$ are arbitrary constants. Thus, an exact solution of equation (15.10.3.1) with the functions $f_1$ and $f_2$ from (15.10.3.7) can be represented in implicit form for $C_2 = C_3 = 0$ as follows:

$$\mu(w) = -A\exp(2C_1 t) + C_1(x+B)^2.$$

In the solution and the determining relations (15.10.3.7), the function $\mu(w)$ can be chosen arbitrarily.

TABLE 15.10
Second-order differential constraints corresponding to some
classes of exact solutions representable in explicit form

| No. | Type of solution | Structure of solution | Differential constraints |
|---|---|---|---|
| 1 | Additive separable solution | $w = \varphi(x) + \psi(y)$ | $w_{xy} = 0$ |
| 2 | Multiplicative separable solution | $w = \varphi(x)\psi(y)$ | $ww_{xy} - w_x w_y = 0$ |
| 3 | Generalized separable solution | $w = \varphi(x)y^2 + \psi(x)y + \chi(x)$ | $w_{yy} - f(x) = 0$ |
| 4 | Generalized separable solution | $w = \varphi(x)\psi(y) + \chi(x)$ | $w_{yy} - f(y)w_y = 0$ <br> $w_{xy} - g(x)w_y = 0$ |
| 5 | Functional separable solution | $w = f(z),\ z = \varphi(x)y + \psi(x)$ | $w_{yy} - g(w)w_y^2 = 0$ |
| 6 | Functional separable solution | $w = f(z),\ z = \varphi(x) + \psi(y)$ | $ww_{xy} - g(w)w_x w_y = 0$ |

## 15.10.4. Connection Between the Differential Constraints Method and Other Methods

The differential constraints method is one of the most general methods for the construction of exact solutions to nonlinear partial differential equations. Many other methods can be treated as its particular cases.*

### 15.10.4-1. Generalized/functional separation of variables vs. differential constraints.

Table 15.10 lists examples of second-order differential constraints that are essentially equivalent to most common forms of separable solutions. For functional separable solutions (rows 5 and 6), the function $g$ can be expressed through $f$.

Searching for a generalized separable solution of the form $w(x,y) = \varphi_1(x)\psi_1(y) + \cdots + \varphi_n(x)\psi_n(y)$, with $2n$ unknown functions, is equivalent to prescribing a differential constraint of order $2n$; in general, the number of unknown functions $\varphi_i(x)$, $\psi_i(y)$ corresponds to the order of the differential equation representing the differential constraint.

For the types of solutions listed in Table 15.10, it is preferable to use the methods of generalized and functional separation of variables, since these methods require less steps where it is necessary to solve intermediate differential equations. Furthermore, the method of differential constraints is ill-suited for the construction of exact solutions of third- and higher-order equations since they lead to cumbersome computations and rather complex equations (often, the original equations are simpler).

*Remark.* The direct specification of a solution structure, on which the methods of generalized and functional separation of variables are based, may be treated as the use of a zeroth-order differential constraint.

### 15.10.4-2. Generalized similarity reductions and differential constraints.

Consider a generalized similarity reduction based on a prescribed form of the desired solution,

$$w(x,t) = F(x,t,u(z)), \quad z = z(x,t), \qquad (15.10.4.1)$$

---

* The basic difficulty in applying the differential constraints method is due to the great generality of its statements and the necessity of selecting differential constraints suitable for specific classes of equations. This is why for the construction of exact solutions of nonlinear equations, it is often preferable to use simpler (but less general) methods.

where $F(x, t, u)$ and $z(x, t)$ should be selected so as to obtain ultimately a single ordinary differential equation for $u(z)$; see Remark 3 in Subsection 15.7.1-2.

Let us show that employing the solution structure (15.10.4.1) is equivalent to searching for a solution with the help of a first-order quasilinear differential constraint

$$\xi(x,t)\frac{\partial w}{\partial t} + \eta(x,t)\frac{\partial w}{\partial x} = \zeta(x,t,w). \tag{15.10.4.2}$$

Indeed, first integrals of the characteristic system of ordinary differential equations

$$\frac{dt}{\xi(x,t)} = \frac{dx}{\eta(x,t)} = \frac{dw}{\zeta(x,t,w)}$$

have the form

$$z(x,t) = C_1, \qquad \varphi(x,t,w) = C_2, \tag{15.10.4.3}$$

where $C_1$ and $C_2$ are arbitrary constants. Therefore, the general solution of equation (15.10.4.2) can be written as follows:

$$\varphi(x,t,w) = u(z(x,t)), \tag{15.10.4.4}$$

where $u(z)$ is an arbitrary function. On solving (15.10.4.4) for $w$, we obtain a representation of the solution in the form (15.10.4.1).

**15.10.4-3. Nonclassical method of symmetry reductions and differential constraints.**

The nonclassical method of symmetry reductions can be restated in terms of the differential constraints method. This can be demonstrated by the following example with a general second-order equation

$$F\left(x, y, w, \frac{\partial w}{\partial x}, \frac{\partial w}{\partial y}, \frac{\partial^2 w}{\partial x^2}, \frac{\partial^2 w}{\partial x \partial y}, \frac{\partial^2 w}{\partial y^2}\right) = 0. \tag{15.10.4.5}$$

Let us supplement equation (15.10.4.5) with two differential constraints

$$\xi\frac{\partial w}{\partial x} + \eta\frac{\partial w}{\partial y} = \zeta, \tag{15.10.4.6}$$

$$\xi\frac{\partial F}{\partial x} + \eta\frac{\partial F}{\partial y} + \zeta\frac{\partial F}{\partial w} + \zeta_1\frac{\partial F}{\partial w_x} + \zeta_2\frac{\partial F}{\partial w_y} + \zeta_{11}\frac{\partial F}{\partial w_{xx}} + \zeta_{12}\frac{\partial F}{\partial w_{xy}} + \zeta_{22}\frac{\partial F}{\partial w_{yy}} = 0, \tag{15.10.4.7}$$

where $\xi = \xi(x, y, w)$, $\eta = \eta(x, y, w)$, and $\zeta = \zeta(x, y, w)$ are unknown functions, and the coordinates of the first and the second prolongations $\zeta_i$ and $\zeta_{ij}$ are defined by formulas (15.8.1.9) and (15.8.1.14). The differential constraint (15.10.4.7) coincides with the invariance condition for equation (15.10.4.5); see (15.8.2.3)–(15.8.2.4).

The method for the construction of exact solutions to equation (15.10.4.5) based on using the first-order partial differential equation (15.10.4.6) and the invariance condition (15.10.4.7) corresponds to the nonclassical method of symmetry reductions (see Subsection 15.9.1).

## 15.11. Painlevé Test for Nonlinear Equations of Mathematical Physics

### 15.11.1. Solutions of Partial Differential Equations with a Movable Pole. Method Description

***Basic idea:*** Solutions of partial differential equations are sought in the form of series expansions containing a movable pole singularity. The position of the pole is defined by an arbitrary function.

**15.11.1-1. Simple scheme for studying nonlinear partial differential equations.**

For a clearer exposition, we will be considering equations of mathematical physics in two independent variables, $x$ and $t$, and one dependent variable, $w$, explicitly independent of $x$ and $t$.

A solution sought in a small neighborhood of a manifold $x - x_0(t) = 0$ is the form of the following series expansion (Jimbo, Kruskal, and Miwa, 1982):

$$w(x,t) = \frac{1}{\xi^p} \sum_{m=0}^{\infty} w_m(t) \xi^m, \quad \xi = x - x_0(t). \tag{15.11.1.1}$$

Here, the exponent $p$ is a positive integer, so that the movable singularity is of the pole type. The function $x_0(t)$ is assumed arbitrary, and the $w_m$ are assumed to depend on derivatives of $x_0(t)$.

The representation (15.11.1.1) is substituted into the given equation. The exponent $p$ and the leading term $u_0(t)$ are first determined from the balance of powers in the expansion. Terms with like powers of $\xi$ are further collected. In the resulting polynomial, the coefficients of the different powers of $\xi$ are all equated to zero to obtain a system of ordinary differential equations for the functions $w_m(t)$.

The solution obtained is general if the expansion (15.11.1.1) involves arbitrary functions, with the number of them equal to the order of the equation in question.

**15.11.1-2. General scheme for analysis of nonlinear partial differential equations.**

A solution of a partial differential equation is sought in a neighborhood of a singular manifold $\xi(x, t) = 0$ in the form of a generalized series expansion symmetric in the independent variables (Weiss, Tabor, and Carnevalle, 1983):

$$w(x,t) = \frac{1}{\xi^p} \sum_{m=0}^{\infty} w_m(x,t) \xi^m, \quad \xi = \xi(x,t), \tag{15.11.1.2}$$

where $\xi_t \xi_x \neq 0$. Here and henceforth, the subscripts $x$ and $t$ denote partial derivatives; the function $\xi(x, t)$ is assumed arbitrary and the $w_m$ are assumed to be dependent on derivatives of $\xi(x, t)$.

Expansion (15.11.1.1) is a special case of expansion (15.11.1.2), when the equation of the singular manifold has been resolved for $x$.

The requirement that there are no movable singularities implies that $p$ is a positive integer. The solution will be general if the number of arbitrary functions appearing in the coefficients $w_m(x,t)$ and the expansion variable $\xi(x,t)$ coincides with the order of the equation.

Substituting (15.11.1.2) into the equation, collect terms with like powers of $\xi$, and equating the coefficients to zero, one arrives at the following recurrence relations for the expansion coefficients:

$$k_m w_m = \Phi_m(w_0, w_1, \ldots, w_{m-1}, \xi_t, \xi_x, \ldots). \tag{15.11.1.3}$$

Here, $k_m$ are polynomials of degree $n$ with integer argument $m$ of the form

$$k_m = (m+1)(m-m_1)(m-m_2)\ldots(m-m_{n-1}), \tag{15.11.1.4}$$

where $n$ is the order of the equation concerned.

If the roots of the polynomial, $m_1, m_2, \ldots, m_{n-1}$, called *Fuchs indices* (*resonances*), are all nonnegative integers and the consistency conditions

$$\Phi_{m=m_j} = 0 \qquad (j = 1, 2, \ldots, n-1) \tag{15.11.1.5}$$

hold, the equation is said to pass the *Painlevé test*. Equations that pass the Painlevé test are often classified as integrable, which is supported by the fact that such equations are reducible to linear equations in many known cases.

15.11.1-3. Basic steps of the Painlevé test for nonlinear equations.

For nonlinear equations of mathematical physics, the Painlevé test is convenient to carry out in several steps. At the first and second steps, one determines the leading term in the expansion (15.11.1.1) and the Fuchs indices; this allows to verify the necessary conditions for the Painlevé test without making full computations. For the sake of clarity, the basic steps in performing the Painlevé test for nonlinear equations, using the expansion (15.11.1.1), are shown in Figure 15.7.

**Remark.** An equation fails to pass the Painlevé test if any of the following conditions holds: $p < 0$, $p$ is noninteger, $p$ is complex, $m_j < 0$, $m_j$ is noninteger, or $m_j$ is complex (at least for one $j$, where $j = 1, \ldots, n-1$).

15.11.1-4. Some remarks. Truncated expansions.

The numerous researchers have established that many known integrable nonlinear equations of mathematical physics pass the Painlevé test. New equations possessing this property have also been found.

As a simple check whether a specific equation passes the Painlevé test, one may use the simple scheme based on the expansion (15.11.1.1). The associated important technical simplifications as compared with the expansion (15.11.1.2) are due to the fact that $(w_m)_x = 0$ and $\xi_x = 1$.

The general expansion (15.11.1.2), involving more cumbersome but yet more informative computations, can prove useful after the Painlevé property has been established at the simple check. It may help reveal many important properties of equations and their solutions.

In some cases, a truncated expansion,

$$w = \frac{w_0}{\xi^p} + \frac{w_1}{\xi^{p-1}} + \cdots + w_p, \tag{15.11.1.6}$$

can be useful for constructing exact solutions and finding a Bäcklund transformation linearizing the original equation. This expansion corresponds to zero values of the expansion coefficients $w_m$ with $m > p$ in (15.11.1.2); see Examples 1 and 2 in Subsection 15.11.2.

The Painlevé test for a nonlinear partial differential equation can be performed for special classes of its exact solutions, usually traveling-wave solutions and self-similar solutions, which are determined by ordinary differential equations. If the ordinary differential equation obtained fails the Painlevé test, then the original partial differential equation also fails the test. If the ordinary differential equation passes the Painlevé test, then the original partial differential equation normally also passes the test.

```
┌─────────────────────────────────────────────────────────────────────┐
│  Original equation:  $F(x, t, w, w_x, w_t, w_{xx}, \ldots) = 0$     │
└─────────────────────────────────────────────────────────────────────┘
          ⇓    First step: look for the leading term of expansion (15.11.1.2)    ⇓
┌─────────────────────────────────────────────────────────────────────┐
│  Substitute into equation: $w = w_0 \xi^{-p}$, $w_0 = w_0(x,t)$, $\xi = \xi(x,t)$ │
└─────────────────────────────────────────────────────────────────────┘
          ⇓    Find the constant $p$ and the function $w_0(x,t)$    ⇓
┌─────────────────────────────────────────────────────────────────────┐
│  If $p$ is a positive integer, proceed to the second step           │
└─────────────────────────────────────────────────────────────────────┘
          ⇓    Look for Fuchs indices (resonances)    ⇓
┌─────────────────────────────────────────────────────────────────────┐
│  Substitute $w = w_0 \xi^{-p} + w_m \xi^{m-p}$ in the leading terms of the equation │
└─────────────────────────────────────────────────────────────────────┘
          ⇓    Collect terms proportional to $w_m$    ⇓
┌─────────────────────────────────────────────────────────────────────┐
│  This results in the expression: $k_m w_m \xi^q + \cdots$           │
└─────────────────────────────────────────────────────────────────────┘
          ⇓    $k_m$ is a polynomial with respect to $m$    ⇓
┌─────────────────────────────────────────────────────────────────────┐
│  Factorize the polynomial: $k_m = (m+1)(m-m_1)\ldots(m-m_{n-1})$    │
└─────────────────────────────────────────────────────────────────────┘
          ⇓    $n$ is the equation order and the $m_j$ are resonances    ⇓
┌─────────────────────────────────────────────────────────────────────┐
│  If $m_1, \ldots, m_{n-1}$ are all nonnegative integers, proceed to the third step │
└─────────────────────────────────────────────────────────────────────┘
          ⇓    Substitute expansion (15.11.1.2) into the equation    ⇓
┌─────────────────────────────────────────────────────────────────────┐
│  Obtain recurrence relations (15.11.1.3) for expansion coefficients $w_m$ │
└─────────────────────────────────────────────────────────────────────┘
          ⇓    Check the consistency conditions (15.11.1.5)    ⇓
┌─────────────────────────────────────────────────────────────────────┐
│  If these conditions are satisfied, the equation passes the Painlevé test │
└─────────────────────────────────────────────────────────────────────┘
```

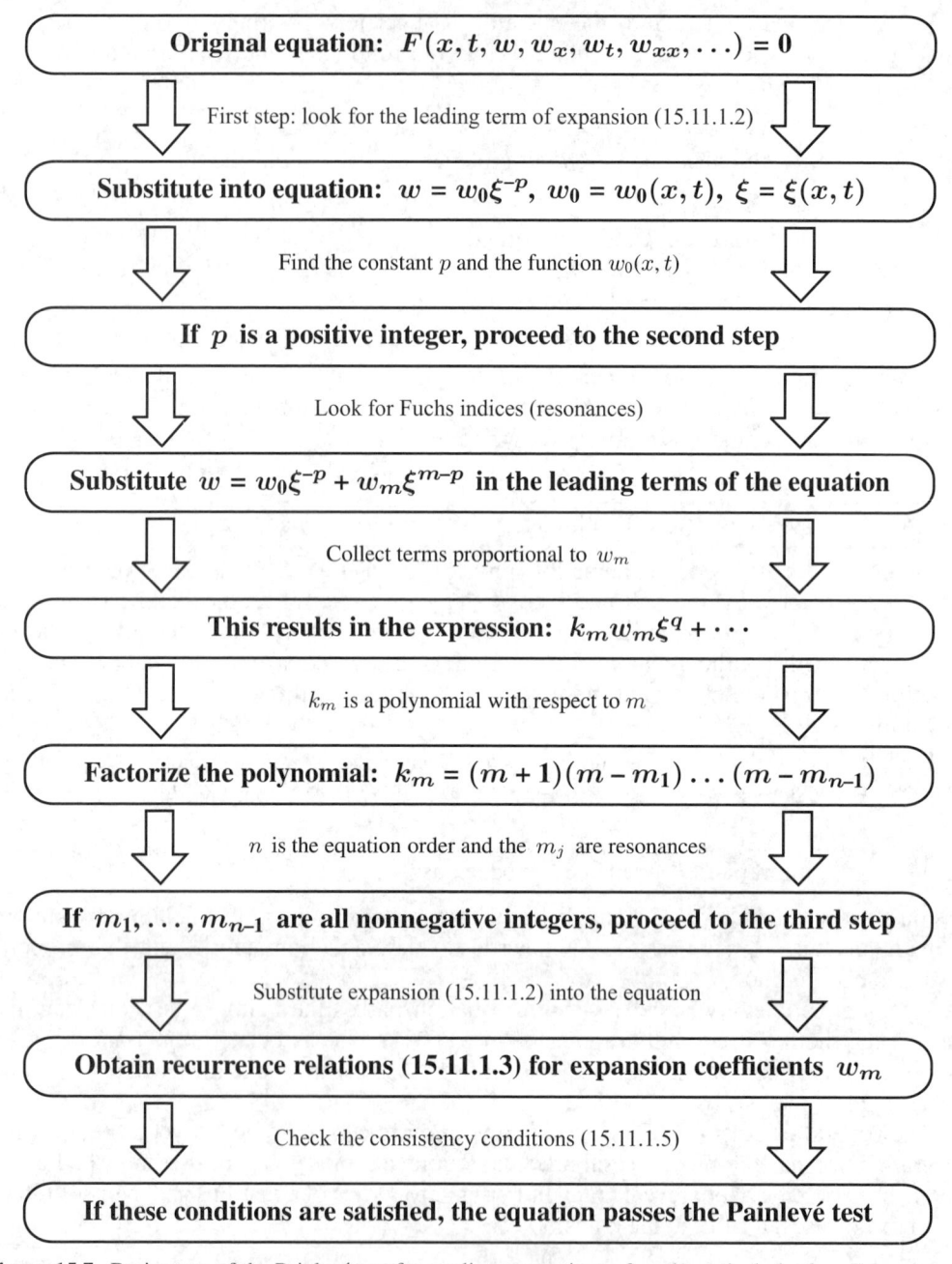

**Figure 15.7.** Basic steps of the Painlevé test for nonlinear equations of mathematical physics. It is assumed that $\xi = x - x_0(t)$ for the simple scheme and $\xi = \xi(x,t)$ for the general scheme.

### 15.11.2. Examples of Performing the Painlevé Test and Truncated Expansions for Studying Nonlinear Equations

This Subsection treats some common nonlinear equations of mathematical physics. For their analysis, the simple scheme of the Painlevé test will be used first; this scheme is based on the expansion (15.11.1.1) [see also the scheme in Fig. 15.7 with $\xi = x - x_0(t)$]. The truncated expansion (15.11.1.6) will then be used for constructing Bäcklund transformations.

**Example 1.** Consider the Burgers equation

$$\frac{\partial w}{\partial t} + w\frac{\partial w}{\partial x} = \nu \frac{\partial^2 w}{\partial x^2}. \tag{15.11.2.1}$$

*First step.* Substitute the leading term of expansion (15.11.1.1) in equation (15.11.2.1) and multiply the resulting relation by $\xi^{p+2}$ (the product $\xi^{p+2} w_{xx}$ is equal to unity by order of magnitude). This results in

$$w_0' \xi^2 + p w_0 x_0' \xi - p w_0^2 \xi^{1-p} = \nu p(p+1) w_0,$$

where $\xi = x - x_0$, $x_0 = x_0(t)$, $w_0 = w_0(t)$, and the prime denotes a derivative with respect to $t$. We find from the balance of the leading terms, which corresponds to dropping two leftmost terms, that

$$p = 1, \quad w_0 = -2\nu \quad (m = 0). \tag{15.11.2.2}$$

Since $p$ is a positive integer, the first necessary condition of the Painlevé test is satisfied.

*Second step.* In order to find the Fuchs indices (resonances), we substitute the binomial

$$w = -2\nu \xi^{-1} + w_m \xi^{m-1}$$

into the leading terms $ww_x$ and $\nu w_{xx}$ of equation (15.11.2.1). Collecting coefficients of like powers of $w_m$, we get

$$\nu(m+1)(m-2) w_m \xi^{m-3} + \cdots = 0.$$

Equating $(m+1)(m-2)$ to zero yields the Fuchs index

$$m_1 = 2.$$

Since it is a positive integer, the second necessary condition of the Painlevé test is satisfied.

*Third step.* We substitute the expansion (15.11.1.1) (according to the second step, we have to consider terms up to number $m = 2$ inclusive),

$$w = -2\nu \xi^{-1} + w_1 + w_2 \xi + \cdots,$$

into the Burgers equation (15.11.2.1), collect terms of like powers of $\xi = x - x_0(t)$, and then equate the coefficients of the different powers to zero to obtain a system of equations for the $w_m$:

$$\begin{aligned} \xi^{-2}: & \quad 2\nu(w_1 - x_0') = 0, \\ \xi^{-1}: & \quad 0 \times w_2 = 0. \end{aligned} \tag{15.11.2.3}$$

From the second relation in (15.11.2.3) it follows that the function $w_2 = w_2(t)$ can be chosen arbitrarily. Therefore the Burgers equation (15.11.2.1) passes the Painlevé test and its solution has two arbitrary functions, $x_0 = x_0(t)$ and $w_2 = w_2(t)$, as required.

It follows from the first relation in (15.11.2.3) that $w_1 = x_0'(t)$. The solution to equation (15.11.2.1) can be written as

$$w(x,t) = -\frac{2\nu}{x - x_0(t)} + x_0'(t) + w_2(t)[x - x_0(t)]^2 + \cdots,$$

where $x_0(t)$ and $w_2(t)$ are arbitrary functions.

*Cole–Hopf transformation.* For further analysis of the Burgers equation (15.11.2.1), we use a truncated expansion of the general form (15.11.1.6) with $p = 1$:

$$w = \frac{w_0}{\xi} + w_1, \tag{15.11.2.4}$$

where $w_0 = w_0(x,t)$, $w_1 = w_1(x,t)$, and $\xi = \xi(x,t)$. Substitute (15.11.2.4) in (15.11.2.1) and collect the terms of equal powers in $\xi$ to obtain

$$\xi^{-3} w_0 \xi_x (w_0 + 2\nu \xi_x) + \xi^{-2} \left[ w_0 \xi_t - w_0(w_0)_x + w_0 w_1 \xi_x - 2\nu(w_0)_x \xi_x - \nu w_0 \xi_{xx} \right]$$
$$+ \xi^{-1} \left[ -(w_0)_t - w_0(w_1)_x - w_1(w_0)_x + \nu(w_0)_{xx} \right] - (w_1)_t - w_1(w_1)_x + \nu(w_1)_{xx} = 0,$$

where the subscripts $x$ and $t$ denote partial derivatives. Equating the coefficients of like powers of $\xi$ to zero, we get the system of equations

$$\begin{aligned} w_0 + 2\nu \xi_x &= 0, \\ w_0 \xi_t - w_0(w_0)_x + w_0 w_1 \xi_x - 2\nu(w_0)_x \xi_x - \nu w_0 \xi_{xx} &= 0, \\ (w_0)_t + w_0(w_1)_x + w_1(w_0)_x - \nu(w_0)_{xx} &= 0, \\ (w_1)_t + w_1(w_1)_x - \nu(w_1)_{xx} &= 0, \end{aligned} \tag{15.11.2.5}$$

where it has been taken into account that $\xi_x \neq 0$.

If follows from the first equation of (15.11.2.5) that

$$w_0 = -2\nu \xi_x. \tag{15.11.2.6}$$

Substituting this into the second and third equations of (15.11.2.5), after some rearrangements we obtain

$$\begin{aligned}\xi_t + w_1 \xi_x - \nu \xi_{xx} &= 0, \\ (\xi_t + w_1 \xi_x - \nu \xi_{xx})_x &= 0.\end{aligned} \tag{15.11.2.7}$$

It is obvious that if the first equation of (15.11.2.7) is valid, then the second is satisfied identically. The last equation in (15.11.2.5) is the Burgers equation.

Hence, formula (15.11.2.4) in view of (15.11.2.6) can be rewritten as

$$w = -2\nu \frac{\partial}{\partial x} \ln \xi + w_1, \tag{15.11.2.8}$$

where the functions $w$ and $w_1$ satisfy the Burgers equation and the function $\xi$ is described by the first equation of (15.11.2.7). Given a solution $w_1$ of the Burgers equation, formula (15.11.2.8) allows obtaining other solutions of it by solving the first equation in (15.11.2.7), which is linear in $\xi$.

Taking into account that $w_1 = 0$ is a particular solution of the Burgers equation, let us substitute it into (15.11.2.7) and (15.11.2.8). This results in the Cole–Hopf transformation

$$w = -2\nu \frac{\xi_x}{\xi}.$$

This transformation allows constructing solutions of the nonlinear Burgers equation (15.11.2.1) via solutions of the linear heat equation

$$\xi_t = \nu \xi_{xx}.$$

**Example 2.** Consider the Korteweg–de Vries equation

$$\frac{\partial w}{\partial t} + w \frac{\partial w}{\partial x} + \frac{\partial^3 w}{\partial x^3} = 0. \tag{15.11.2.9}$$

*First step.* Let us substitute the leading term of the expansion (15.11.1.1) into equation (15.11.2.9) and then multiply the resulting relation by $\xi^{p+3}$ (the product $\xi^{p+3} w_{xxx}$ gives a zeroth order quantity) to obtain

$$w_0' \xi^3 + p w_0 x_0' \xi^2 - p w_0^2 \xi^{2-p} - p(p+1)(p+2) w_0 = 0,$$

where $\xi = x - x_0$, $x_0 = x_0(t)$, and $w_0 = w_0(t)$. From the balance of the highest-order terms (only the last two terms are taken into account) it follows that

$$p = 2, \quad w_0 = -12 \quad (m = 0).$$

Since $p$ is a positive integer, the first necessary condition of the Painlevé test is satisfied.

*Second step.* To fine the Fuchs indices (resonances), we substitute the binomial

$$w = -12 \xi^{-2} + w_m \xi^{m-2}$$

into the leading terms of equation (15.11.2.9), where the second and the third term are taken into account. Isolating the term proportional to $w_m$, we have

$$(m+1)(m-4)(m-6) w_m \xi^{m-5} + \cdots = 0.$$

Equating $(m+1)(m-4)(m-6)$ to zero gives the Fuchs indices

$$m_1 = 4, \quad m_2 = 6.$$

Since they are both positive integers, the second necessary condition of the Painlevé test is satisfied.

*Third step.* We substitute the expansion (15.11.1.1), while considering, according to the second step, the terms up to number $m = 6$ inclusive,

$$w = -12 \xi^{-2} + w_1 \xi^{-1} + w_2 + w_3 \xi + w_4 \xi^2 + w_5 \xi^3 + w_6 \xi^4 + \cdots, \quad \xi = x - x_0(t) \tag{15.11.2.10}$$

into equation (15.11.2.9). Then we collect the terms of equal powers in $\xi$ and equate the coefficients of the different powers of $\xi$ to zero to arrive at a system of equations for the $w_m$:

$$\begin{aligned}
\xi^{-4}:\quad & 2w_1 = 0, \\
\xi^{-3}:\quad & 24w_2 - 24x_0' - w_1^2 = 0, \\
\xi^{-2}:\quad & 12w_3 + w_1 x_0' - w_1 w_2 = 0, \\
\xi^{-1}:\quad & 0 \times w_4 + w_1' = 0, \\
1:\quad & -6w_5 + w_2' - w_3 x_0' + w_1 w_4 + w_2 w_3 = 0, \\
\xi:\quad & 0 \times w_6 + w_3' - 2w_4 x_0' + w_3^2 + 2w_1 w_5 + 2w_2 w_4 = 0.
\end{aligned} \qquad (15.11.2.11)$$

Simple computations show that the equations with resonances corresponding to the powers $\xi^{-1}$ and $\xi$ are satisfied identically. Hence, the Korteweg–de Vries equation (15.11.2.9) passes the Painlevé test.

The solution of equation (15.11.2.11) results in the following expansion coefficients in (15.11.2.10):

$$w_1 = 0, \quad w_2 = x_0'(t), \quad w_3 = 0, \quad w_4 = w_4(t), \quad w_5 = \tfrac{1}{6}x_0''(t), \quad w_6 = w_6(t),$$

where $x_0(t)$, $w_4(t)$, and $w_6(t)$ are arbitrary functions.

*Truncated series expansion and the Bäcklund transformation.* For further analysis, let us use a truncated expansion of the general form (15.11.1.6) with $p = 2$:

$$w = \frac{w_0}{\xi^2} + \frac{w_1}{\xi} + w_2. \qquad (15.11.2.12)$$

Substituting (15.11.2.12) into (15.11.2.9) and equating the functional coefficients of the different powers of $\xi$ to zero, in the same way as in Example 2, we arrive at the Bäcklund transformation

$$\begin{aligned}
& w = 12(\ln \xi)_{xx} + w_2, \\
& \xi_t \xi_x + w_2 \xi_x^2 + 4\xi_x \xi_{xxx} - 3\xi_{xx}^2 = 0, \\
& \xi_{xt} + w_2 \xi_{xx} + \xi_{xxx} = 0, \\
& (w_2)_t + w_2 (w_2)_x + (w_2)_{xxx} = 0.
\end{aligned} \qquad (15.11.2.13)$$

It relates the solutions $w$ and $w_2$ of the Korteweg–de Vries equations. Eliminating $w_2$ from the second and third equations in (15.11.2.13), one can derive an equation for $\xi$, which can further be reduced, via suitable transformations, to a system of linear equations.

### 15.11.3. Construction of Solutions of Nonlinear Equations That Fail the Painlevé Test, Using Truncated Expansions

In some cases truncated expansions of the form (15.11.1.6) may be effective in finding exact solutions to nonlinear equations of mathematical physics that fail the Painlevé test. In such cases, the expansion parameter $p$ must be a positive integer; it is determined in the same way as at the first step of performing the Painlevé test. We illustrate this by a specific example below.

**Example.** Consider the nonlinear diffusion equation with a cubic source

$$w_t = w_{xx} - 2w^3. \qquad (15.11.3.1)$$

*First step.* Let us substitute the leading term of the expansion (15.11.1.1) into equation (15.11.3.1) and then multiply the resulting relation by $\xi^{p+2}$ to obtain

$$w_0' \xi^2 + p w_0 x_0' \xi = p(p+1) w_0 - 2 w_0^3 \xi^{2-2p},$$

where $\xi = x - x_0$, $x_0 = x_0(t)$, and $w_0 = w_0(t)$. From the balance of the highest-order terms (both terms on the right-hand side are taken into account) it follows that

$$p = 1, \quad w_0 = \pm 1 \quad (m = 0). \qquad (15.11.3.2)$$

Since $p$ is a positive integer, the first necessary condition of the Painlevé test is satisfied.

*Second step.* The equation is invariant under the substitution of $-w$ for $w$. Hence, it suffices to consider only the positive value of $w_0$ in (15.11.3.2). Therefore, in order to find resonances, we substitute the binomial

$$w = \xi^{-1} + w_m \xi^{m-1}$$

in the leading terms $w_{xx}$ and $bw^3$ of equation (15.11.3.1). Collecting the terms proportional to $w_m$, we get

$$(m+1)(m-4)w_m \xi^{m-3} + \cdots = 0.$$

Equating $(m+1)(m-4)$ to zero gives the Fuchs index $m_1 = 4$. Since it is integer and positive, the second necessary condition of the Painlevé test is satisfied.

*Third step.* We substitute the expansion (15.11.1.1) into equation (15.11.3.1); according to the second step, the terms up to number $m = 4$ inclusive must be taken into account. It can be shown that the consistency condition (15.11.1.5) is not satisfied, and therefore the equation in question fails the Painlevé test.

*Using a truncated expansion for finding exact solutions.* For further analysis, we use a truncated expansion of the general form (15.11.1.6) with $p = 1$, which from the first step. This results in formula (15.11.2.4). Substituting (15.11.2.4) into the diffusion equation (15.11.3.1) and collecting the terms of equal powers in $\xi$, we obtain

$$\xi^{-3}\left(2w_0 \xi_x^2 - 2w_0^3\right) + \xi^{-2}\left[w_0 \xi_t - 2(w_0)_x \xi_x - w_0 \xi_{xx} - 6w_0^2 w_1\right]$$
$$+ \xi^{-1}\left[-(w_0)_t + (w_0)_{xx} - 6w_0 w_1^2\right] - (w_1)_t + (w_1)_{xx} - 2w_1^3 = 0.$$

Equating the coefficients of like powers of $\xi$ to zero, we arrive at the system of equations

$$w_0(\xi_x^2 - w_0^2) = 0,$$
$$w_0 \xi_t - 2(w_0)_x \xi_x - w_0 \xi_{xx} - 6w_0^2 w_1 = 0,$$
$$-(w_0)_t + (w_0)_{xx} - 6w_0 w_1^2 = 0, \qquad (15.11.3.3)$$
$$(w_1)_t - (w_1)_{xx} + 2w_1^3 = 0.$$

From the first equation in (15.11.3.3) we have

$$w_0 = \xi_x. \qquad (15.11.3.4)$$

The other solution differs in sign only and gives rise to the same result, and therefore is not considered. Substituting (15.11.3.4) into the second and third equations of (15.11.3.3) and canceling by nonzero factors, we obtain

$$\xi_t - 3\xi_{xx} - 6 w_1 \xi_x = 0,$$
$$-\xi_{xt} + \xi_{xxx} - 6w_1^2 \xi_x = 0. \qquad (15.11.3.5)$$

The latter equation in (15.11.3.3), which coincides with the original equation (15.11.3.1), is satisfied if

$$w_1 = 0. \qquad (15.11.3.6)$$

On inserting (15.11.3.4) and (15.11.3.6) in (15.11.2.4), we get the following representation for solutions:

$$w = \frac{\xi_x}{\xi}, \qquad (15.11.3.7)$$

where the function $\xi$ is determined by an overdetermined linear system of equations resulting from the substitution of (15.11.3.6) into (15.11.3.5):

$$\xi_t - 3\xi_{xx} = 0,$$
$$-\xi_{xt} + \xi_{xxx} = 0. \qquad (15.11.3.8)$$

Differentiate the first equation with respect to $x$ and then eliminate the mixed derivative $w_{xt}$ using the second equation to obtain $\xi_{xxx} = 0$. It follows that

$$\xi = \varphi_2(t)x^2 + \varphi_1(t)x + \varphi_0(t). \qquad (15.11.3.9)$$

In order to determine the functions $\varphi_k(t)$, let us substitute (15.11.3.9) into equations (15.11.3.8) to obtain

$$\varphi_2' x^2 + \varphi_1' x + \varphi_0' - 6\varphi_2 = 0,$$
$$-\varphi_2' x - \varphi_1' = 0.$$

Equating the functional coefficients of the different powers of $x$ to zero and integrating the resulting equations, we get

$$\varphi_2 = C_2, \quad \varphi_1 = C_1, \quad \varphi_0 = 6C_2 t + C_0, \qquad (15.11.3.10)$$

where $C_0$, $C_1$, and $C_2$ are arbitrary constants. Substituting (15.11.3.9) into (15.11.3.7) and taking into account (15.11.3.10), we find an exact solution of equation (15.11.3.1) in the form

$$w = \frac{2C_2 x + C_1}{C_2 x^2 + C_1 x + 6C_2 t + C_0}.$$

## 15.12. Methods of the Inverse Scattering Problem (Soliton Theory)

**Preliminary remarks.** The methods of the inverse scattering problem rely on "implicit" linearization of equations. Main idea: Instead of the original nonlinear equation in the unknown $w$, one considers an auxiliary overdetermined linear system of equation for a (vector) function $\varphi$, with the coefficients of this system generally dependent on $w$ and the derivatives of $w$ with respect to the independent variables. The linear system for $\varphi$ is chosen so that the compatibility condition for its equations gives rise to the original nonlinear equation for $w$.

### 15.12.1. Method Based on Using Lax Pairs

15.12.1-1. Method description. Consistency condition. Lax pairs.

We will be studying a nonlinear evolution equation of the form

$$\frac{\partial w}{\partial t} = F(w), \tag{15.12.1.1}$$

where the right-hand side $F(w)$ depends on $w$ and its derivatives with respect to $x$.

Consider two auxiliary linear differential equations, one corresponding to an eigenvalue problem and involving derivatives with respect to the space variable $x$ only,

$$L\varphi = \lambda\varphi, \tag{15.12.1.2}$$

and the other describing the evolution of the eigenfunction in time,

$$\frac{\partial \varphi}{\partial t} = -M\varphi. \tag{15.12.1.3}$$

The coefficients of the linear differential operators L and M in equations (15.12.1.2) and (15.12.1.3) depend on $w$ and its derivatives with respect to $x$.

Since system (15.12.1.2)–(15.12.1.3) is overdetermined (there are two equations for $\varphi$), the operators L and M cannot be arbitrary — they must satisfy a compatibility condition. In order to find this condition, let us first differentiate (15.12.1.2) with respect to $t$. Assuming that the eigenvalues $\lambda$ are independent of time $t$, we have

$$L_t\varphi + L\varphi_t = \lambda\varphi_t.$$

Replacing $\varphi_t$ here by the right-hand side (15.12.1.3), we get

$$L_t\varphi - LM\varphi = -\lambda M\varphi.$$

Taking into account the relations $\lambda M\varphi = M(\lambda\varphi)$ and $\lambda\varphi = L\varphi$, we arrive at the compatibility condition

$$L_t\varphi = LM\varphi - ML\varphi,$$

which can be rewritten in the form of an operator equation:

$$\frac{\partial L}{\partial t} = LM - ML. \tag{15.12.1.4}$$

The linear operators L and M [or the linear equations (15.12.1.2) and (15.12.1.3)] are said to form a *Lax pair* for the nonlinear equation (15.12.1.1) if the compatibility condition (15.12.1.4) coincides with equation (15.12.1.1). The right-hand side of equation (15.12.1.4) represents the *commutator of the operators* L and M, which is denoted by [L, M] = LM−ML for short.

Thus, if a suitable Lax pair is found, the analysis of the nonlinear equation (15.12.1.1) can be reduced to that of two simpler, linear equations, (15.12.1.2) and (15.12.1.3).

**Remark.** The operator M in equations (15.12.1.3) and (15.12.1.3) is defined to an additive function of time; it can be changed according to the rule

$$M \implies M + p(t),$$

where $p(t)$ is an arbitrary function. This function is found in solving a Cauchy problem for equation (15.12.1.1); see Paragraph 15.12.3-2.

### 15.12.1-2. Examples of Lax pairs for nonlinear equations of mathematical physics.

**Example 1.** Let us show that a Lax pair for the Korteweg–de Vries equation

$$\frac{\partial w}{\partial t} + \frac{\partial^3 w}{\partial x^3} - 6w \frac{\partial w}{\partial x} = 0 \tag{15.12.1.5}$$

is formed by the operators

$$L = w - \frac{\partial^2}{\partial x^2}, \quad M = 4\frac{\partial^3}{\partial x^3} - 6w\frac{\partial}{\partial x} - 3\frac{\partial w}{\partial x} + p(t), \tag{15.12.1.6}$$

which generate the linear equations

$$\begin{aligned}\varphi_{xx} + (\lambda - w)\varphi &= 0, \\ \varphi_t + 4\varphi_{xxx} - 6w\varphi_x - 3w_x\varphi + p(t)\varphi &= 0.\end{aligned} \tag{15.12.1.7}$$

Here, $p(t)$ is an arbitrary function.

It is not difficult to verify that the following formulas hold:

$$\begin{aligned}LM(\varphi) &= -4\varphi_{xxxxx} + 10w\varphi_{xxx} + [15w_x - p(t)]\varphi_{xx} + (12w_{xx} - 6w^2)\varphi_x \\ &\quad + [3w_{xxx} - 3ww_x + wp(t)]\varphi, \\ ML(\varphi) &= -4\varphi_{xxxxx} + 10w\varphi_{xxx} + [15w_x - p(t)]\varphi_{xx} + (12w_{xx} - 6w^2)\varphi_x \\ &\quad + [4w_{xxx} - 9ww_x + wp(t)]\varphi,\end{aligned}$$

where $\varphi(x,t)$ is an arbitrary function. It follows that

$$LM(\varphi) - ML(\varphi) = (-w_{xxx} + 6ww_x)\varphi. \tag{15.12.1.8}$$

From (15.12.1.6) and (15.12.1.8) we obtain

$$L_t = w_t, \quad LM - ML = -w_{xxx} + 6ww_x.$$

On inserting these expressions into (15.12.1.4), we arrive at the Korteweg–de Vries equation (15.12.1.5).

**Remark.** A procedure for solving the Cauchy problem for equation (15.12.1.5) is outlined in Subsection 15.12.3.

The linear equations (15.12.1.2) and (15.12.1.3) for the auxiliary function $\varphi$, which form a Lax pair, can be vector; in this case, the linear operators L and M are represented by matrices. In other words, the individual equations (15.12.1.2) and (15.12.1.3) may be replaced by appropriate systems of linear equations.

**Example 2.** The sinh-*Gordon equation*

$$w_{xt} = 4a \sinh w$$

can be represented as a vector Lax pair

$$\begin{aligned}L\varphi &= \lambda\varphi, \\ \varphi_t &= -M\varphi,\end{aligned}$$

where

$$\varphi = \begin{pmatrix} \varphi_1 \\ \varphi_2 \end{pmatrix}, \quad L = \begin{pmatrix} 0 & \partial_x + \tfrac{1}{2}w_x \\ \partial_x - \tfrac{1}{2}w_x & 0 \end{pmatrix}, \quad M = \frac{a}{\lambda}\begin{pmatrix} 0 & e^w \\ e^{-w} & 0 \end{pmatrix}.$$

The determination of a Lax pair for a given nonlinear equation is a very complex problem that is basically solvable for isolated equations only. Therefore, the "implicit" linearization of equations is usually realized using a simpler method, described in Subsection 15.12.2.

## 15.12.2. Method Based on a Compatibility Condition for Systems of Linear Equations

**15.12.2-1. General scheme. Compatibility condition. Systems of two equations.**

Consider two systems of linear equations
$$\varphi_x = A\varphi, \qquad (15.12.2.1)$$
$$\varphi_t = B\varphi, \qquad (15.12.2.2)$$
where $\varphi$ is an $n$-dimensional vector and A and B are $n \times n$ matrices. The right-hand sides of systems (15.12.2.1) and (15.12.2.2) cannot be arbitrary — they must satisfy a compatibility condition. To find this condition, let us differentiate systems (15.12.2.1) and (15.12.2.2) with respect to $t$ and $x$, respectively, and eliminate the mixed derivative $\varphi_{xt}$ from the resulting equations. Then replacing the derivatives $\varphi_x$ and $\varphi_t$ by the right-hand sides of (15.12.2.1) and (15.12.2.2), we obtain
$$A_t - B_x + [A, B] = 0, \qquad (15.12.2.3)$$
where $[A, B] = AB - BA$. It turns out that, given a matrix A, there is a simple deductive procedure for finding B as a result of which the compatibility condition (15.12.2.3) becomes a nonlinear evolution equation.

Let us dwell on the special case where the vector function $\varphi$ two components, $\varphi = \begin{pmatrix} \varphi_1 \\ \varphi_2 \end{pmatrix}$. We choose a linear system of equations (15.12.2.1) in the form
$$\begin{aligned} (\varphi_1)_x &= -i\lambda \varphi_1 + f\varphi_2, \\ (\varphi_2)_x &= g\varphi_1 + i\lambda \varphi_2, \end{aligned} \qquad (15.12.2.4)$$
where $\lambda$ is the spectral parameter, $f$ and $g$ are some (generally complex valued) functions of two variables $x$ and $t$, and $i^2 = -1$. As system (15.12.2.2) we take the most general linear system involving the derivatives with respect to $t$:
$$\begin{aligned} (\varphi_1)_t &= A\varphi_1 + B\varphi_2, \\ (\varphi_2)_t &= C\varphi_1 + D\varphi_2, \end{aligned} \qquad (15.12.2.5)$$
where $A$, $B$, $C$, and $D$ are some functions (dependent on the variables $x$, $t$ and the parameter $\lambda$) to be determined in the subsequent analysis. Differentiating equations (15.12.2.4) with respect to $t$ and equations (15.12.2.5) with respect to $x$ and assuming that $(\varphi_{1,2})_{xt} = (\varphi_{1,2})_{tx}$, we obtain compatibility conditions in the form
$$\begin{aligned} A_x &= Cf - Bg, \\ B_x + 2i\lambda B &= f_t - (A - D)f, \\ C_x - 2i\lambda C &= g_t + (A - D)g, \\ -D_x &= Cf - Bg. \end{aligned} \qquad (15.12.2.6)$$

For simplicity, we set
$$D = -A.$$
In this case, the first and last equations in (15.12.2.6) coincide, so that (15.12.2.6) turns into a system of three determining equations:
$$\begin{aligned} A_x &= Cf - Bg, \\ B_x + 2i\lambda B &= f_t - 2Af, \\ C_x - 2i\lambda C &= g_t + 2Ag. \end{aligned} \qquad (15.12.2.7)$$

The functions $A$, $B$, and $C$ must be expressed in terms of $f$ and $g$. The general solution of system (15.12.2.7) for arbitrary functions $f$ and $g$ cannot be found. So let us look for particular solutions in the form finite expansions in positive and negative powers of the parameter $\lambda$.

15.12.2-2. **Solution of the determining equations in the form of polynomials in $\lambda$.**

The simplest polynomial representations of the unknown functions that give rise to nontrivial results are quadratic in the spectral parameter $\lambda$:

$$\begin{aligned} A &= A_2\lambda^2 + A_1\lambda + A_0, \\ B &= B_2\lambda^2 + B_1\lambda + B_0, \\ C &= C_2\lambda^2 + C_1\lambda + C_0. \end{aligned} \tag{15.12.2.8}$$

Let us substitute (15.12.2.8) into (15.12.2.7) and collect the terms with equal powers in $\lambda$ to obtain

$$\lambda^2(A_{2x} - C_2 f + B_2 g) + \lambda(A_{1x} - C_1 f + B_1 g) + A_{0x} - C_0 f + B_0 g = 0, \tag{15.12.2.9}$$

$$2i\lambda^3 B_2 + \lambda^2(B_{2x} + 2iB_1 + 2A_2 f) \\ + \lambda(B_{1x} + 2iB_0 + 2A_1 f) + B_{0x} + 2A_0 f - f_t = 0, \tag{15.12.2.10}$$

$$-2i\lambda^3 C_2 + \lambda^2(C_{2x} - 2iC_1 - 2A_2 g) \\ + \lambda(C_{1x} - 2iC_0 - 2A_1 g) + C_{0x} - 2A_0 g - g_t = 0. \tag{15.12.2.11}$$

Let us equate the coefficients of like powers of $\lambda$ to zero starting from the highest power. Setting the coefficients of $\lambda^3$ to zero gives

$$B_2 = C_2 = 0. \tag{15.12.2.12}$$

Equating the coefficients of $\lambda^2$ to zero and taking into account (15.12.2.12), we find that

$$A_2 = a = \text{const}, \quad B_1 = iaf, \quad C_1 = iag. \tag{15.12.2.13}$$

Setting the coefficient of $\lambda$ in (15.12.2.9) to zero and then replacing $B_1$ and $C_1$ in accordance with (15.12.2.13), we have $A_{1x} = 0$, whence $A_1 = a_1 = \text{const}$. For simplicity, we dwell on the special case $a_1 = 0$ (arbitrary $a_1$ gives rise to more general results), so that

$$A_1 = 0. \tag{15.12.2.14}$$

By equating to zero the coefficients of $\lambda$ in the equations obtained from (15.12.2.10) and (15.12.2.11) and taking into account (15.12.2.13) and (15.12.2.14), we get

$$B_0 = -\tfrac{1}{2}af_x, \quad C_0 = \tfrac{1}{2}ag_x. \tag{15.12.2.15}$$

Setting the coefficient of $\lambda^0$ in (15.12.2.9) to zero and then integrating, we find that $A_0 = \tfrac{1}{2}afg + a_0$, where $a_0$ is an arbitrary constant. As before, we set $a_0 = 0$ for simplicity, which results in

$$A_0 = \tfrac{1}{2}afg. \tag{15.12.2.16}$$

Then equations (15.12.2.10) and (15.12.2.11) in view of (15.12.2.15) and (15.12.2.16) become

$$\begin{aligned} f_t &= -\tfrac{1}{2}af_{xx} + af^2 g, \\ g_t &= \tfrac{1}{2}ag_{xx} - afg^2. \end{aligned} \tag{15.12.2.17}$$

Substituting (15.12.2.8) and (15.12.2.12)–(15.12.2.17) into (15.12.2.5) yields

$$\begin{aligned} (\varphi_1)_t &= a(\lambda^2 + \tfrac{1}{2}fg)\varphi_1 + a(i\lambda f - \tfrac{1}{2}f_x)\varphi_2, \\ (\varphi_2)_t &= a(i\lambda g + \tfrac{1}{2}g_x)\varphi_1 - a(\lambda^2 + \tfrac{1}{2}fg)\varphi_2. \end{aligned} \tag{15.12.2.18}$$

Thus, two linear systems (15.12.2.4) and (15.12.2.18) are compatible if the functions $f$ and $g$ satisfy the cubically nonlinear system (15.12.2.17). For

$$g = -k\bar{f}, \quad a = 2i, \tag{15.12.2.19}$$

where $k$ is a real constant and the bar over a symbol stands for its complex conjugate, both equations (15.12.2.18) turn into one and the same nonlinear Schrödinger equation

$$if_t = f_{xx} + 2k|f|^2 f \quad (f\bar{f} = |f|^2). \tag{15.12.2.20}$$

Likewise, one can use other polynomials in $\lambda$ and determine the associated linear systems generating nonlinear evolution equations.

**Example 1.** Searching for solutions of the determining system (15.12.2.7) in the form third-order polynomials in $\lambda$ results in

$$\begin{aligned}
A &= a_3\lambda^3 + a_2\lambda^2 + \tfrac{1}{2}(a_3 fg + a_1)\lambda + \tfrac{1}{2}a_2 fg - \tfrac{1}{4}ia_3(fg_x - gf_x) + a_0, \\
B &= ia_3 f\lambda^2 + (ia_2 f - \tfrac{1}{2}a_3 f_x)\lambda + ia_1 f + \tfrac{1}{2}ia_3 f^2 g - \tfrac{1}{2}a_2 f_x - \tfrac{1}{4}a_3 f_{xx}, \\
C &= ia_3 g\lambda^2 + (ia_2 g + \tfrac{1}{2}a_3 g_x)\lambda + ia_1 g + \tfrac{1}{2}ia_3 fg^2 + \tfrac{1}{2}a_2 g_x - \tfrac{1}{4}ia_3 g_{xx},
\end{aligned} \tag{15.12.2.21}$$

where $a_0$, $a_1$, $a_2$, and $a_3$ are arbitrary constants. The evolution equations for $f$ and $g$ corresponding to (15.12.2.21) are

$$\begin{aligned}
f_t + \tfrac{1}{4}ia_3(f_{xxx} - 6fgf_x) + \tfrac{1}{2}a_2(f_{xx} - 2f^2 g) - ia_1 f_x - 2a_0 f &= 0, \\
g_t + \tfrac{1}{4}ia_3(g_{xxx} - 6fgg_x) - \tfrac{1}{2}a_2(g_{xx} - 2fg^2) - ia_1 g_x + 2a_0 g &= 0.
\end{aligned} \tag{15.12.2.22}$$

Consider two important special cases leading to interesting nonlinear equations of mathematical physics.

1°. If

$$a_0 = a_1 = a_2 = 0, \quad a_3 = -4i, \quad g = 1$$

the second equation of (15.12.2.22) is satisfied identically, and the first equation of (15.12.2.22) translates into the Korteweg–de Vries equation

$$f_t + f_{xxx} - 6ff_x = 0.$$

2°. If

$$a_0 = a_1 = a_2 = 0, \quad a_3 = -4i, \quad g = \pm f,$$

both equations (15.12.2.22) translate into one and the same modified Korteweg–de Vries equation

$$f_t + f_{xxx} \mp 6f^2 f_x = 0.$$

**Example 2.** Now we will look for a solution of the determining system (15.12.2.7) in the form of the simple one-term expansion in negative powers of $\lambda$:

$$A = a(x,t)\lambda^{-1}, \quad B = b(x,t)\lambda^{-1}, \quad C = c(x,t)\lambda^{-1}. \tag{15.12.2.23}$$

This results in the relations

$$a_x = \tfrac{1}{2}i(fg)_t, \quad b = -\tfrac{1}{2}if_t, \quad c = \tfrac{1}{2}ig_t. \tag{15.12.2.24}$$

The evolution equations for functions $f$ and $g$ corresponding to (15.12.2.24) are written as

$$\begin{aligned}
f_{xt} &= -4iaf, \\
g_{xt} &= -4iag.
\end{aligned} \tag{15.12.2.25}$$

If we set

$$a = \tfrac{1}{4}i\cos w, \quad b = c = \tfrac{1}{4}i\sin w, \quad f = -g = -\tfrac{1}{2}w_x, \tag{15.12.2.26}$$

then the three relations (15.12.2.24) are reduced to one and the same equation, the sine-Gordon equation

$$w_{xt} = \sin w, \tag{15.12.2.27}$$

and the two equations (15.12.2.25) coincide and give a differential consequence of equation (15.12.2.27):

$$w_{xxt} = w_x \cos w.$$

Thus, the linear system of equations (15.12.2.4) and (15.12.2.5), whose coefficients are defined by formulas (15.12.2.23) and (15.12.2.26) with $D = -A$, is compatible if the function $w$ satisfies the sine-Gordon equation (15.12.2.27).

**Remark.** Sometimes the determining equations (15.12.2.1)–(15.12.2.2) are also called a Lax pair, by analogy with (15.12.1.2)–(15.12.1.3).

## 15.12.3. Solution of the Cauchy Problem by the Inverse Scattering Problem Method

**15.12.3-1. Preliminary remarks. The direct and inverse scattering problems.**

The solution of the Cauchy problem for nonlinear equations admitting a Lax pair or another "implicit" linearization (see Subsection 15.12.2) falls into several successive steps. Two of them involve the solution of the direct and the inverse scattering problem for auxiliary linear equations. Summarized below are relevant results for the linear stationary Schrödinger equation

$$\varphi''_{xx} + [\lambda - f(x)]\varphi = 0 \qquad (-\infty < x < \infty). \tag{15.12.3.1}$$

It is assumed that the function $f(x)$, called the *potential*, vanishes as $x \to \pm\infty$ and the condition $\int_{-\infty}^{\infty}(1+|x|)|f(x)|\,dx < \infty$ holds.

*Direct scattering problem.* Consider the linear eigenvalue problem for the ordinary differential equation (15.12.3.1).

Eigenvalues can be of two types:

$$\begin{aligned} \lambda_n &= -\varkappa_n^2, \quad n = 1, 2, \ldots, N \quad \text{(discrete spectrum)}, \\ \lambda &= k^2 \quad -\infty < k < \infty \quad \text{(continuous spectrum)}. \end{aligned} \tag{15.12.3.2}$$

It is known that if $\min f(x) < \lambda < 0$, equation (15.12.3.1) has a discrete spectrum of eigenvalues, and if $f(x) < 0$ and $\lambda > 0$, it has a continuous spectrum.

Let $\lambda_n = -\varkappa_n^2$ be discrete eigenvalues numbered so that

$$\lambda_1 < \lambda_2 < \cdots < \lambda_N < 0,$$

and let $\varphi_n = \varphi_n(x)$ be the associated eigenfunctions, which vanish as $x \to \pm\infty$ and are square summable. Eigenfunctions are defined up to a constant factor. If an eigenfunction is fixed by its asymptotic behavior for negative $x$,

$$\varphi_n \to \exp(\varkappa_n x) \quad \text{as} \quad x \to -\infty,$$

then the leading asymptotic term in the expansion of $\varphi_n$ for large positive $x$ is expressed as

$$\varphi_n \to c_n \exp(-\varkappa_n x) \quad \text{as} \quad x \to \infty. \tag{15.12.3.3}$$

The eigenfunction $\varphi_n$ has $n-1$ zero and moreover

$$c_n = (-1)^{n-1}|c_n|.$$

For continuous spectrum, $\lambda = k^2$, the behavior of the wave function $\varphi$ at infinity is specified by a linear combination of the exponentials $\exp(\pm ikx)$, since $f \to 0$ as $|x| \to \infty$. The conditions at infinity are

$$\begin{aligned} \varphi &\to e^{-ikx} & \text{as} \quad x &\to -\infty, \\ \varphi &\to a(k)e^{-ikx} + b(k)e^{ikx} & \text{as} \quad x &\to \infty, \end{aligned} \tag{15.12.3.4}$$

and equation (15.12.3.1) uniquely determines the functions $a(k)$ and $b(k)$; note that they are related by the simple constraint $|a|^2 - |b|^2 = 1$. The first term in the second condition of (15.12.3.4) corresponds to a refracted wave, and the second term to a reflected wave. Therefore, the quotient

$$r(k) = \frac{b(k)}{a(k)} \tag{15.12.3.5}$$

is called the *coefficient of reflection* (also *reflection factor* or *reflectance*).

The set of quantities

$$S = \{\varkappa_n, c_n, r(k);\ n = 1, \ldots, N\} \qquad (15.12.3.6)$$

appearing in relations (15.12.3.3)–(15.12.3.5) is called the *scattering data*. It is the determination of these data for a given potential $f(x)$ that is the purpose of the direct scattering problem. The scattering data completely determine the eigenvalue spectrum of the stationary Schrödinger equation (15.12.3.1).

Note the useful formulas

$$|a(k)| = \left(1 - |r(k)|^2\right)^{-1},$$
$$\arg a(k) = \frac{1}{i}\sum_{n=1}^{N}\frac{k - i\varkappa_n}{k + i\varkappa_n} - \frac{1}{\pi}\int_{-\infty}^{\infty}\frac{\ln|a(s)|}{s - k}\,ds, \qquad (15.12.3.7)$$

which can be used to restore the function $a(k)$ for a given reflection factor $r(k)$. The integral in the second formula of (15.12.3.7) is understood as a Cauchy principal value integral, and $i\varkappa_n$ are zeros of the function $a(k)$ analytic in the upper half-plane [the $\varkappa_n$ appear in the asymptotic formulas (15.12.3.3)].

***Inverse scattering problem.*** The mapping of the potentials of equation (15.12.3.1) into the scattering data (15.12.3.6) is unique and invertible. The procedure of restoring $f(x)$ for given $S$ is the subject matter of the inverse scattering problem. Summarized below are the most important results of studying this problem.

To determine the potential, one has to solve first the Gel'fand–Levitan–Marchenko integral equation

$$K(x, y) + \Phi(x, y) + \int_x^{\infty} K(x, z)\Phi(z, y)\,dz = 0, \qquad (15.12.3.8)$$

where the function $\Phi(x, y) = \overline{\Phi}(x + y)$ is determined via the scattering data (15.12.3.6) as follows:

$$\Phi(x, y) = \sum_{n=1}^{N}\frac{c_n}{ia'(i\varkappa_n)}e^{-\varkappa_n(x+y)} + \frac{1}{2\pi}\int_{-\infty}^{\infty}r(k)e^{ik(x+y)}\,dk,\quad a'(k) = \frac{da}{dk}. \qquad (15.12.3.9)$$

The potential is expressed in terms of the solution to the linear integral equation (15.12.3.8) as

$$f(x) = -2\frac{d}{dx}K(x, x). \qquad (15.12.3.10)$$

Note that the direct and inverse scattering problems are usually taught in courses on quantum mechanics. For more details about these problems, see, for example, the book by Novikov, Manakov, Pitaevskii, and Zakharov (1984).

---

**15.12.3-2. Solution of the Cauchy problem by the inverse scattering problem method.**

Figure 15.8 depicts the general scheme of solving the Cauchy problem for nonlinear equations using the method of the inverse scattering problem. It is assumed that either a Lax pair (15.12.1.2)–(15.12.1.3) for the given equation has been obtained at an earlier stage or a

```
┌─────────────────────────────────────────────────────────────┐
│  Cauchy problem = nonlinear equation + initial condition    │
└─────────────────────────────────────────────────────────────┘
           ⇓           First step: solve the direct scattering problem           ⇓
┌─────────────────────────────────────────────────────────────┐
│          Use the stationary equation from the Lax pair      │
└─────────────────────────────────────────────────────────────┘
           ⇓         Substitute in it the function w from the initial condition         ⇓
┌─────────────────────────────────────────────────────────────┐
│  Determine the stationary scattering data: $S = \{\varkappa_n,\, c_n,\, r(k)\}$  │
└─────────────────────────────────────────────────────────────┘
           ⇓           Second step: look for the dependence
                       of the scattering data on time                            ⇓
┌─────────────────────────────────────────────────────────────┐
│        Treat the nonstationary equation from the Lax pair   │
└─────────────────────────────────────────────────────────────┘
           ⇓      Determine the asymptotic behavior of the solutions for large $x$      ⇓
┌─────────────────────────────────────────────────────────────┐
│  Find the nonstationary scattering data: $S(t) = \{\varkappa_n,\, c_n(t),\, r(k,t)\}$  │
└─────────────────────────────────────────────────────────────┘
           ⇓           Third step: solve the inverse scattering problem           ⇓
┌─────────────────────────────────────────────────────────────┐
│     Write out the Gel'fand–Levitan–Marchenko integral equation    │
└─────────────────────────────────────────────────────────────┘
           ⇓                  Solve the integral equation                         ⇓
┌─────────────────────────────────────────────────────────────┐
│  With the solution obtained, find the solution of the Cauchy problem  │
└─────────────────────────────────────────────────────────────┘
```

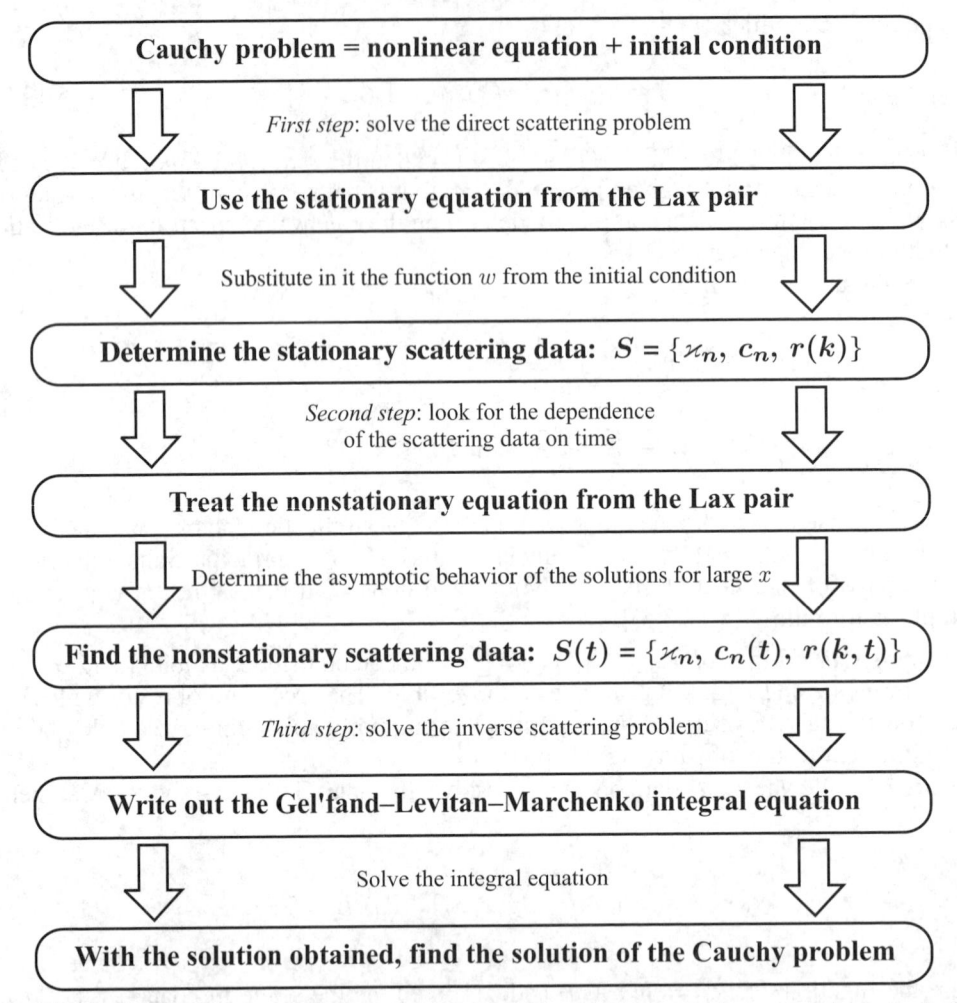

**Figure 15.8.** Main stages of solving the Cauchy problem for nonlinear equations using the method of the inverse scattering problem.

representation of the equation as the compatibility condition for the overdetermined system of linear equations (15.12.2.1)–(15.12.2.2) is known.

Let us employ this scheme for the solution of the Cauchy problem for the Korteweg–de Vries equation

$$\frac{\partial w}{\partial t} + \frac{\partial^3 w}{\partial x^3} - 6w\frac{\partial w}{\partial x} = 0 \qquad (15.12.3.11)$$

with the initial condition

$$w = f(x) \quad \text{at} \quad t = 0 \quad (-\infty < x < \infty). \qquad (15.12.3.12)$$

The function $f(x) < 0$ is assumed to satisfy the same conditions as the potential in equation (15.12.3.1).

The solution of the Cauchy problem (15.12.3.11), (15.12.3.12) falls into several successive steps (see Figure 15.8). The Korteweg–de Vries equation is represented in the form of a Lax pair (15.12.1.7) and the results for the solution of the direct and inverse scattering problems are used.

*First step.* One considers the linear eigenvalue problem — the direct scattering problem — for the auxiliary ordinary differential equation (15.12.3.1) that results from the substitution of the initial profile (15.12.3.12) into the first equation of the Lax pair (15.12.1.7) for the Korteweg–de Vries equation (15.12.3.11). On solving this problem, one determines the scattering data (15.12.3.6).

*Second step.* For $t > 0$, the function $w = w(x,t)$ must appear in the first equation (15.12.1.7) instead of the initial profile $f(x)$. In the associated nonstationary problem, the eigenvalues (15.12.3.2) are preserved, since they are independent of $t$ (see Paragraph 15.12.1-1), but the eigenfunctions $\varphi = \varphi(x,t)$ are changed.

For continuous eigenvalues $\lambda > 0$, the asymptotic behavior of the eigenfunctions, according to (15.12.3.4), is as follows:

$$\begin{aligned}\varphi &\to e^{-ikx} & \text{as} \quad x \to -\infty, \\ \varphi &\to a(k,t)e^{-ikx} + b(k,t)e^{ikx} & \text{as} \quad x \to \infty.\end{aligned} \quad (15.12.3.13)$$

Substitute the first asymptotic relation of (15.12.3.13) into the second equation of the Lax pair (15.12.1.7) and take into account that $w \to 0$ as $x \to -\infty$ to obtain the function $p(t)$:

$$p(t) = -4ik^3 = \text{const}. \quad (15.12.3.14)$$

In order to determine the time dependences, $a = a(k,t)$ and $b = b(k,t)$, let us consider the second equation of the Lax pair (15.12.1.7), where $p(t)$ is given by (15.12.3.14). Taking into account that $w \to 0$ as $x \to \infty$, we get the linearized equation

$$\varphi_t + 4\varphi_{xxx} - 4ik^3\varphi = 0. \quad (15.12.3.15)$$

Substituting the second asymptotic relation of (15.12.3.13) into (15.12.3.15) and equating the coefficients of the exponentials $e^{-ikx}$ and $e^{ikx}$, we arrive at the linear ordinary differential equations

$$a'_t = 0, \quad b'_t - 8ik^3 b = 0.$$

Integrating them gives the coefficient of reflection

$$r(k,t) = \frac{b(t,k)}{a(t,k)} = r(k,0)e^{8ik^3 t}. \quad (15.12.3.16)$$

The asymptotic behavior the eigenfunctions for discrete eigenvalues, with $k = i\varkappa_n$, is determined in a similar way using the linearized equation (15.12.3.15).

Consequently, we obtain the time dependences of the scattering data:

$$S(t) = \{\varkappa_n,\ c_n(t) = c_n(0)e^{8\varkappa_n^3 t},\ r(k,t) = r(k,0)e^{8ik^3 t};\ n = 1,\ldots,N\}. \quad (15.12.3.17)$$

*Third step.* Using the scattering data (15.12.3.17), we introduce the following function by analogy with (15.12.3.9):

$$\Phi(x,y;t) = \frac{1}{2\pi}\int_{-\infty}^{\infty} r(k,0)e^{i[8k^3 t + k(x+y)]}\,dk + \sum_{n=1}^{N} \frac{c_n(0)}{ia'(i\varkappa_n)} e^{8\varkappa_n^3 t - \varkappa_n(x+y)}. \quad (15.12.3.18)$$

In this case, $\Phi$ depends on time parametrically. This function plays the same role for the first equation of the Lax pair (15.12.1.7) for $t > 0$ as the function (15.12.3.9) for equation (15.12.3.1); recall that equation (15.12.3.1) coincides with the first equation of the Lax pair (15.12.1.7), where $w$ is substituted for by its value in the initial condition (15.12.3.12).

In order to recover the function $w(x,t)$ from the scattering data (15.12.3.17), one has first to solve the linear integral equation

$$K(x,y;t) + \Phi(x,y;t) + \int_x^\infty K(x,z;t)\Phi(z,y;t)\,dz = 0, \qquad (15.12.3.19)$$

obtained from (15.12.3.8) by simple renaming. This equation contains the function $\Phi$ of (15.12.3.18). Then, one should use the formula

$$w(x,t) = -2\frac{d}{dx}K(x,x;t), \qquad (15.12.3.20)$$

obtained by renaming from (15.12.3.10), to determine the solution of the Cauchy problem for the Korteweg–de Vries equation (15.12.3.11)–(15.12.3.12).

**15.12.3-3. $N$-soliton solution of the Korteweg–de Vries equation.**

Let us find an exact solution of the Korteweg–de Vries equation (15.12.3.11) for the case of nonreflecting potentials, with zero coefficient of reflection (15.12.3.5). We proceed from the linear integral equation (15.12.3.19).

Setting $r(k,0) = 0$ in (15.12.3.18), we have

$$\Phi(x,y;t) = \sum_{n=1}^N \gamma_n e^{-\varkappa_n(x+y)+8\varkappa_n^3 t}, \qquad \gamma_n = \frac{c_n(0)}{ia'(i\varkappa_n)} > 0.$$

Substituting this function into (15.12.3.19), after simple rearrangements we obtain the equation

$$K(x,y;t) + \sum_{n=1}^N \Gamma_n(t)e^{-\varkappa_n(x+y)} + \sum_{n=1}^N \Gamma_n(t)e^{-\varkappa_n y}\int_x^\infty e^{-\varkappa_n z}K(x,z;t)\,dz = 0, \quad (15.12.3.21)$$

where

$$\Gamma_n(t) = \gamma_n e^{8\varkappa_n^3 t}. \qquad (15.12.3.22)$$

The solution of the integral equation (15.12.3.21) is sought in the form of a finite sum:

$$K(x,y;t) = \sum_{n=1}^N e^{-\varkappa_n y}K_n(x,t). \qquad (15.12.3.23)$$

Inserting (15.12.3.23) in (15.12.3.21) and integrating, we have

$$\sum_{n=1}^N e^{-\varkappa_n y}K_n(x,t) + \sum_{n=1}^N \Gamma_n(t)e^{-\varkappa_n(x+y)} + \sum_{n=1}^N \Gamma_n(t)e^{-\varkappa_n y}\sum_{m=1}^N K_m(x,t)\frac{e^{-(\varkappa_n+\varkappa_m)x}}{\varkappa_n+\varkappa_m} = 0.$$

Rewriting this equation in the form $\sum_{n=1}^N \psi_n(x,t)e^{-\varkappa_n y} = 0$ and then setting $\psi_n(x,t) = 0$, we arrive at a nonhomogeneous system of linear algebraic equations for $K_n(x,t)$:

$$K_n(x,t) + \Gamma_n(t)\sum_{m=1}^N \frac{1}{\varkappa_n+\varkappa_m}e^{-(\varkappa_n+\varkappa_m)x}K_m(x,t) = -\Gamma_n(t)e^{-\varkappa_n x}, \quad n = 1,\ldots,N.$$

$$(15.12.3.24)$$

Using Cramer's rule, we rewrite the solution to system (15.12.3.24) as the ratio of determinants

$$K_n(x,t) = \frac{\det \mathbb{A}^{(n)}(x,t)}{\det \mathbb{A}(x,t)}, \tag{15.12.3.25}$$

where $\mathbb{A}$ is the matrix with entries

$$\begin{aligned}
A_{n,m}(x,t) &= \delta_{nm} + \frac{\Gamma_n(t)}{\varkappa_n + \varkappa_m} e^{-(\varkappa_n + \varkappa_m)x} \\
&= \delta_{nm} + \frac{\gamma_n}{\varkappa_n + \varkappa_m} e^{-(\varkappa_n + \varkappa_m)x + 8\varkappa_n^3 t},
\end{aligned} \tag{15.12.3.26}$$

$$\delta_{nm} = \begin{cases} 1 & \text{if } n = m, \\ 0 & \text{if } n \neq m, \end{cases}$$

and $\mathbb{A}^{(n)}(x,t)$ is the matrix obtained from $\mathbb{A}$ by substituting the right-hand sides of equations (15.12.3.24) for the $n$th column. Substituting (15.12.3.25) in (15.12.3.23) yields

$$K(x,y;t) = \sum_{n=1}^{N} e^{-\varkappa_n y} \frac{\det \mathbb{A}^{(n)}(x,t)}{\det \mathbb{A}(x,t)}.$$

Further, by setting $y = x$, we get

$$K(x,x;t) = \frac{1}{\det \mathbb{A}(x,t)} \sum_{n=1}^{N} e^{-\varkappa_n x} \det \mathbb{A}^{(n)}(x,t) = \frac{\partial}{\partial x} \ln \det \mathbb{A}(x,t).$$

In view of (15.12.3.20), we now find a solution to the Korteweg–de Vries equation (15.12.3.11) for the case of nonreflecting potentials in the form

$$w(x,t) = -2 \frac{\partial^2}{\partial x^2} \ln \det \mathbb{A}(x,t). \tag{15.12.3.27}$$

This solution contains $2N$ free parameters $\varkappa_n, \gamma_n$ ($n = 1, \ldots, N$) and is called an $N$-*soliton solution*.

**Example.** In the special case $N = 1$, the matrix $\mathbb{A}(x,t)$ is characterized by a single element, defined by $n = m = 1$ in (15.12.3.26):

$$A_{1,1}(x,t) = 1 + \frac{\gamma}{2\varkappa} e^{-2\varkappa x + 8\varkappa^3 t}.$$

Substituting this expression into (15.12.3.27), we get the *one-soliton solution* of the Korteweg–de Vries equation

$$w(x,t) = -\frac{2\varkappa^2}{\cosh^2[\varkappa(x - 4\varkappa^2 t) + \xi_0]}, \quad \xi_0 = \frac{1}{2\varkappa} \ln \frac{\gamma}{2\varkappa}. \tag{15.12.3.28}$$

This solution represents a solitary wave traveling to the right with a constant $v = 4\varkappa^2$ and rapidly decaying at infinity.

For the $N$-soliton solution (15.12.3.27), the following asymptotic relations hold:

$$w(x,t) \approx -2 \sum_{n=1}^{N} \frac{\varkappa_n^2}{\cosh^2\left[\varkappa_n(x - v_n t) + \xi_n^{\pm}\right]} \quad \text{as} \quad t \to \pm\infty, \tag{15.12.3.29}$$

where $v_n = 4\varkappa_n^2$ is the speed of the $n$th component. From the comparison of formulas (15.12.3.28) and (15.12.3.29) it follows that for large times, the $N$-soliton solution gets broken into $N$ independent one-soliton solutions, and solitons with a higher amplitude travel with a higher speed.

## 15.13. Conservation Laws and Integrals of Motion

### 15.13.1. Basic Definitions and Examples

**15.13.1-1. General form of conservation laws.**

Consider a partial differential equation with two independent variables

$$F\left(x, t, w, \frac{\partial w}{\partial x}, \frac{\partial w}{\partial t}, \frac{\partial^2 w}{\partial x^2}, \frac{\partial^2 w}{\partial x \partial t}, \frac{\partial^2 w}{\partial x^2}, \dots\right) = 0. \qquad (15.13.1.1)$$

A *conservation law* for this equation has the form

$$\frac{\partial T}{\partial t} + \frac{\partial X}{\partial x} = 0, \qquad (15.13.1.2)$$

where

$$T = T\left(x, t, w, \frac{\partial w}{\partial x}, \frac{\partial w}{\partial t}, \dots\right), \quad X = X\left(x, t, w, \frac{\partial w}{\partial x}, \frac{\partial w}{\partial t}, \dots\right). \qquad (15.13.1.3)$$

The left-hand side of the conservation law (15.13.1.2) must vanish for all (sufficiently smooth) solutions of equation (15.13.1.1). In simplest cases, the substitution of relations (15.13.1.3) into the conservation law (15.13.1.2) followed by differentiation and elementary transformations leads to a relation that coincides with (15.13.1.1) up to a functional factor. The quantities $T$ and $X$ in (15.13.1.2) are called a *density* and a *flow*, respectively.

The density and the flow in the conservation law (15.13.1.2) are not uniquely determined; they can be changed according to the rules $T \Longrightarrow T + \varphi(x)$, $X \Longrightarrow X + \psi(t)$, where $\varphi(x)$ and $\psi(t)$ are arbitrary functions, or, in general, by the rules

$$T \Longrightarrow aT + \frac{\partial \Phi}{\partial x}, \quad X \Longrightarrow aX - \frac{\partial \Phi}{\partial t},$$

where $\Phi = \Phi(x, t)$ is an arbitrary function of two variables and $a \neq 0$ is any number.

For nonstationary equations with $n$ spatial variables $x_1, \dots, x_n$, conservation laws have the form

$$\frac{\partial T}{\partial t} + \sum_{k=1}^{n} \frac{\partial X_k}{\partial x_k} = 0.$$

Partial differential equations can have several (sometimes infinitely many) conservation laws or none at all.

**15.13.1-2. Integrals of motion.**

If the total variation of the quantity $X$ on the interval $a \leq x \leq b$ is equal to zero, which means that the relation

$$X(a) = X(b)$$

holds, then the following "integral of motion" takes place:

$$\int_a^b T \, dx = \text{const} \qquad \text{(for all } t\text{)}. \qquad (15.13.1.4)$$

For many specific equations, relations of the form (15.13.1.4) have a clear physical meaning and are used for approximate analytical solution of the corresponding problems, as well as for the verification of results obtained by numerical methods.

## 15.13.1-3. Conservation laws for some nonlinear equations of mathematical physics.

**Example 1.** The Korteweg–de Vries equation

$$\frac{\partial w}{\partial t} + \frac{\partial^3 w}{\partial x^3} - 6w\frac{\partial w}{\partial x} = 0$$

admits infinitely many conservation laws of the form (15.13.1.2). The first three are determined by

$$T_1 = w, \qquad X_1 = w_{xx} - 3w^2;$$
$$T_2 = w^2, \qquad X_2 = 2ww_{xx} - w_x^2 - 4w^3;$$
$$T_3 = w_x^2 + 2w^3, \qquad X_3 = 2w_x w_{xxx} - w_{xx}^2 + 6w^2 w_{xx} - 12ww_x^2 - 9w^4,$$

where the subscripts denote partial derivatives with respect to $x$.

**Example 2.** The sine-Gordon equation

$$\frac{\partial^2 w}{\partial x \partial t} - \sin w = 0$$

also has infinitely many conservation laws. The first three are described by the formulas

$$T_1 = w_x^2, \qquad X_1 = 2\cos w;$$
$$T_2 = w_x^4 - 4w_{xx}^2, \qquad X_2 = 4w_x^2 \cos w;$$
$$T_3 = 3w_x^6 - 12w_x^2 w_{xx}^2 + 16w_x^3 w_{xxx} + 24w_{xxx}^2, \qquad X_3 = (2w_x^4 - 24w_{xx}^2)\cos w.$$

**Example 3.** The nonhomogeneous Monge–Ampère equation

$$\left(\frac{\partial^2 w}{\partial x \partial y}\right)^2 - \frac{\partial^2 w}{\partial x^2}\frac{\partial^2 w}{\partial y^2} = F(x, y),$$

where $F(x, y)$ is an arbitrary function, admits the conservation law

$$\frac{\partial}{\partial x}\left(\frac{\partial w}{\partial x}\frac{\partial^2 w}{\partial y^2}\right) + \frac{\partial}{\partial y}\left(-\frac{\partial w}{\partial x}\frac{\partial^2 w}{\partial x \partial y} + \int_a^y F(x, z)\,dz\right) = 0.$$

## 15.13.2. Equations Admitting Variational Formulation. Noetherian Symmetries

### 15.13.2-1. Lagrangian. Euler–Lagrange equation. Noetherian symmetries.

Here we consider second-order equations in two independent variables

$$F\left(x, y, w, \frac{\partial w}{\partial x}, \frac{\partial w}{\partial y}, \frac{\partial^2 w}{\partial x^2}, \frac{\partial^2 w}{\partial x \partial y}, \frac{\partial^2 w}{\partial x^2}\right) = 0 \qquad (15.13.2.1)$$

that admit the variational formulation of minimizing a functional of the form

$$Z[w] = \int_S L(x, y, w, w_x, w_y)\,dx\,dy. \qquad (15.13.2.2)$$

The function $L = L(x, y, w, w_x, w_y)$ is called a *Lagrangian*.

It is well known that a minimum of the functional (15.13.2.2) corresponds to the *Euler–Lagrange equation*

$$\frac{\partial L}{\partial w} - D_x\left(\frac{\partial L}{\partial w_x}\right) - D_y\left(\frac{\partial L}{\partial w_y}\right) = 0, \qquad (15.13.2.3)$$

where $D_x$ and $D_y$ are the total differential operators in $x$ and $y$:

$$D_x = \frac{\partial}{\partial x} + w_x \frac{\partial}{\partial w} + w_{xx} \frac{\partial}{\partial w_x} + w_{xy} \frac{\partial}{\partial w_y},$$

$$D_y = \frac{\partial}{\partial y} + w_y \frac{\partial}{\partial w} + w_{xy} \frac{\partial}{\partial w_x} + w_{yy} \frac{\partial}{\partial w_y}.$$

The original equation (15.13.2.1) must be a consequence of equation (15.13.2.3).

A symmetry that preserves the differential form

$$\Omega = L(x, y, w, w_x, w_y) \, dx \, dy$$

is called a *Noetherian symmetry of the Lagrangian L*. In order to obtain Noetherian symmetries, one should find point transformations

$$\bar{x} = f_1(x, y, w, \varepsilon), \quad \bar{y} = f_2(x, y, w, \varepsilon), \quad \bar{w} = g(x, y, w, \varepsilon) \qquad (15.13.2.4)$$

such that preserve the differential form, $\bar{\Omega} = \Omega$, i.e.,

$$\bar{L} \, d\bar{x} \, d\bar{y} = L \, dx \, dy. \qquad (15.13.2.5)$$

Calculating the differentials $d\bar{x}, d\bar{y}$ and taking into account (15.13.2.4), we obtain

$$d\bar{x} = D_x f_1 \, dx, \quad d\bar{y} = D_y f_2 \, dy,$$

and therefore, relation (15.13.2.5) can be rewritten as

$$(L - \bar{L} D_x f_1 D_y f_2) \, dx \, dy = 0,$$

which is equivalent to

$$L - \bar{L} D_x f_1 D_y f_2 = 0. \qquad (15.13.2.6)$$

Let us associate the point transformation (15.13.2.4) with the prolongation operator

$$X = \xi \partial_x + \eta \partial_y + \zeta \partial_w + \zeta_1 \partial_{w_x} + \zeta_2 \partial_{w_y}, \qquad (15.13.2.7)$$

where the coordinates of the first prolongation, $\zeta_1$ and $\zeta_2$, are defined by formulas (15.8.1.9). Then, by the usual procedure, from (15.13.2.6) one obtains the invariance condition in the form

$$X(L) + L(D_x \xi + D_y \eta) = 0. \qquad (15.13.2.8)$$

Noetherian symmetries are determined by (15.13.2.8).

Each Noetherian symmetry operator $X$ generates a conservation law,

$$D_x \left( \xi L + (\zeta - \xi w_x - \eta w_y) \frac{\partial L}{\partial w_x} \right) + D_y \left( \eta L + (\zeta - \xi w_x - \eta w_y) \frac{\partial L}{\partial w_y} \right) = 0. \qquad (15.13.2.9)$$

## 15.13.2-2. Examples of constructing conservation laws using Noetherian symmetries.

**Example 1.** Consider the stationary heat equation with nonlinear source

$$\frac{\partial^2 w}{\partial x^2} + \frac{\partial^2 w}{\partial y^2} - f(w) = 0. \tag{15.13.2.10}$$

This equation admits the variational formulation of minimizing the functional (15.13.2.2) with the Lagrangian

$$L = w_x^2 + w_y^2 + 2F(w), \qquad F(w) = \int f(w)\,dw. \tag{15.13.2.11}$$

This can be verified by substituting (15.13.2.11) into the Euler–Lagrange equation (15.13.2.3). In (15.13.2.11), it is assumed that $F(w) \geq 0$.

Substituting (15.13.2.11) into the invariance condition (15.13.2.8) and taking into account (15.13.2.7) and (15.8.1.9), after some rearrangements we obtain a polynomial in the derivatives $w_x$ and $w_y$:

$$-\xi_w w_x^3 - \eta_w w_x^2 w_y - \xi_w w_x w_y^2 - \eta_w w_y^3 + (2\zeta_w - \xi_x + \eta_y)w_x^2 - 2(\eta_x + \xi_y)w_x w_y$$
$$+ (2\zeta_w - \eta_y + \xi_x)w_y^2 + 2(\zeta_x + F\xi_w)w_x + 2(\zeta_y + F\eta_w)w_y + 2f\zeta + 2F(\xi_x + \eta_y) = 0. \tag{15.13.2.12}$$

Equating the functional coefficients of $w_x^3$, $w_x^2 w_y$, $w_x$, and $w_y$ to zero, we find that $\xi_w = 0$, $\eta_w = 0$, $\zeta_x = 0$, and $\zeta_y = 0$. Consequently,

$$\xi = \xi(x,y), \qquad \eta = \eta(x,y), \qquad \zeta = \zeta(w). \tag{15.13.2.13}$$

In (15.13.2.12), equating the coefficients of the remaining powers of the derivatives to zero, we obtain

$$\begin{aligned} w_x^2: & \quad 2\zeta_w - \xi_x + \eta_y = 0, \\ w_y^2: & \quad 2\zeta_w - \eta_y + \xi_x = 0, \\ w_x w_y: & \quad \eta_x + \xi_y = 0, \\ 1: & \quad f\zeta + F(\xi_x + \eta_y) = 0. \end{aligned} \tag{15.13.2.14}$$

For arbitrary $f = f(w)$, it follows from the last equation in (15.13.2.14) that

$$\zeta = 0, \qquad \xi_x + \eta_y = 0. \tag{15.13.2.15}$$

From the first equation in (15.13.2.14) and the second equation in (15.13.2.15), with $\zeta = 0$, it follows that $\xi_x = 0$, $\eta_y = 0$ or $\xi = \xi(y)$, $\eta = \eta(x)$. Substituting these expressions into the third equation of (15.13.2.14), we get

$$\xi = C_1 y + C_2, \qquad \eta = -C_1 x + C_3,$$

where $C_1$, $C_2$, and $C_3$ are arbitrary constants. Therefore, for arbitrary $f(w)$, Noetherian symmetries of the Lagrangian (15.13.2.11) are defined by the three operators

$$\begin{aligned} X_1 &= \partial_x & (C_2 = 1,\ C_1 = C_3 = 0); \\ X_2 &= \partial_y & (C_3 = 1,\ C_1 = C_2 = 0); \\ X_3 &= y\partial_x - x\partial_y & (C_1 = 1,\ C_2 = C_3 = 0). \end{aligned} \tag{15.13.2.16}$$

In accordance with (15.13.2.9), these operators determine three conservation laws:

$$\begin{aligned} & D_x(-w_x^2 + w_y^2 + 2F) + D_y(-2w_x w_y) = 0 & (\xi = 1,\ \eta = \zeta = 0); \\ & D_x(-2w_x w_y) + D_y(w_x^2 - w_y^2 + 2F) = 0 & (\eta = 1,\ \xi = \zeta = 0); \\ & D_x(-yw_x^2 + yw_y^2 + 2xw_x w_y + 2yF) \\ & \quad + D_y(-xw_x^2 + xw_y^2 - 2yw_x w_y - 2xF) = 0 & (\xi = y,\ \eta = -x,\ \zeta = 0), \end{aligned} \tag{15.13.2.17}$$

with the function $F = F(w)$ defined by (15.13.2.11).

**Remark 1.** The operators (15.13.2.16) could be found by a symmetry analysis of the original differential equation (15.13.2.10), as in Example 1 in Subsection 15.8.2.

**Remark 2.** In the variational formulation of equation (15.13.2.11), it is assumed that $F(w) \geq 0$. However, the conservation laws (15.13.2.17) are valid for any $F(w)$. This situation is also typical of other equations; to obtain conservation laws, it usually suffices that the equation in question is representable as the Euler–Lagrange equation (15.13.2.3) (the variational formulation may be unnecessary).

**Example 2.** The *equation of minimal surfaces*
$$(1 + w_y^2)w_{xx} - 2w_x w_y w_{xy} + (1 + w_x^2)w_{yy} = 0$$
corresponds to the functional (15.13.2.2) with Lagrangian
$$L = \sqrt{1 + w_x^2 + w_y^2}.$$
The admissible point operators
$$X_1 = \partial_x, \quad X_2 = \partial_y, \quad X_3 = x\partial_x + y\partial_y + w\partial_w, \quad X_4 = y\partial_x - x\partial_y, \quad X_5 = \partial_w$$
are found from the invariance condition (15.13.2.8), as was the case in Example 1 (the procedure is also described in Subsection 15.13.2). These operators determine Noetherian symmetries and correspond to conservation laws:

$$X_1: \quad D_x\left(L - w_x \frac{\partial L}{\partial w_x}\right) + D_y\left(-w_x \frac{\partial L}{\partial w_y}\right) = 0,$$

$$X_2: \quad D_x\left(-w_y \frac{\partial L}{\partial w_x}\right) + D_y\left(L - w_y \frac{\partial L}{\partial w_y}\right) = 0,$$

$$X_3: \quad D_x\left(Lx + (w - xw_x - yw_y)\frac{\partial L}{\partial w_x}\right) + D_y\left(Ly + (w - xw_x - yw_y)\frac{\partial L}{\partial w_y}\right) = 0,$$

$$X_4: \quad D_x\left(Ly + (yw_x - xw_y)\frac{\partial L}{\partial w_x}\right) + D_x\left(-Ly + (yw_x - xw_y)\frac{\partial L}{\partial w_y}\right) = 0,$$

$$X_5: \quad D_x\left(\frac{w_x}{\sqrt{1 + w_x^2 + w_y^2}}\right) + D_y\left(\frac{w_y}{\sqrt{1 + w_x^2 + w_y^2}}\right) = 0.$$

## 15.14. Nonlinear Systems of Partial Differential Equations

### 15.14.1. Overdetermined Systems of Two Equations

#### 15.14.1-1. Overdetermined systems of first-order equations in one unknown.

Consider an overdetermined system of two quasilinear first-order equations
$$\begin{aligned} z_x &= F(x, y, z), \\ z_y &= G(x, y, z), \end{aligned} \tag{15.14.1.1}$$
in one unknown function $z = z(x, y)$.

In the general case, the system is unsolvable. To derive a necessary consistency condition for system (15.14.1.1), let us differentiate the first equation with respect to $y$ and the second with respect to $x$. Eliminate the first derivatives from the resulting relations using the initial equations of the system to obtain
$$\begin{aligned} z_{xy} &= F_y + F_z z_y = F_y + GF_z, \\ z_{yx} &= G_x + G_z z_x = G_x + FG_z. \end{aligned}$$

Equating the expressions of the second derivatives $z_{xy}$ and $z_{yx}$ to each other, we obtain a necessary condition for consistency of system (15.14.1.1):
$$F_y + GF_z = G_x + FG_z. \tag{15.14.1.2}$$

Consider two possible situations.

*First case.* Suppose the substitution of the right-hand sides of equations (15.14.1.1) into the necessary condition (15.14.1.2) results in an equation with $x$, $y$, and $z$. Treating it as an algebraic (transcendental) equation for $z$, we find $z = z(x, y)$. The direct substitution of the expression $z = z(x, y)$ into both equations (15.14.1.1) gives an answer to the question whether it is a solution of the system in question or not.

**Example 1.** Consider the quadratically nonlinear system

$$z_x = yz,$$
$$z_y = z^2 + axz. \qquad (15.14.1.3)$$

In this case, the consistency condition (15.14.1.2) can be written, after simple rearrangements, in the form

$$z(a - 1 + yz) = 0.$$

This equations is not satisfied identically and gives rise to two possible expressions for $z$:

$$z = 0 \quad \text{or} \quad z = \frac{1-a}{y}.$$

Substituting them into (15.14.1.3), we see that the former is a solution of the system and the latter is not.

*Second case.* Suppose the substitution of the right-hand sides of equations (15.14.1.1) into the necessary condition (15.14.1.2) turns it into an identity. In this case, the system has a family of solutions that depends at least on one arbitrary constant. If the partial derivatives $F_y$, $F_z$, $G_x$, $G_z$ are continuous and condition (15.14.1.2) is satisfied identically, then system (15.14.1.1) has a unique solution $z = z(x, y)$ that takes a given value, $z = z_0$, at $x = x_0$ and $y = y_0$.

A simple way to find solutions in this case is as follows. The first equation of system (15.14.1.1) is treated as an ordinary differential equation in $x$ with parameter $y$. One finds its solution, where the role of the arbitrary constant is played by an arbitrary function $\varphi(y)$. Substituting this solution into the second equation of system (15.14.1.1), one determines the function $\varphi(y)$.

**Example 2.** Consider the nonlinear system

$$z_x = ae^{y-z},$$
$$z_y = be^{y-z} + 1. \qquad (15.14.1.4)$$

In this case, the consistency condition (15.14.1.2) is satisfied identically.

To solve the first equation, let us make the change of variable $w = e^z$ to obtain the linear equation $w_x = ae^y$. Its general solution has the form $w = ae^y x + \varphi(y)$, where $\varphi(y)$ is an arbitrary function. Going back to the original variable, we find the general solution of the first equation of system (15.14.1.4):

$$z = \ln[ae^y x + \varphi(y)]. \qquad (15.14.1.5)$$

Substituting this solution into the second equation of system (15.14.1.4), we obtain a linear first-order equation for $\varphi = \varphi(y)$:

$$\varphi'_y = \varphi + be^y.$$

Its general solution is $\varphi = (by + C)e^y$, where $C$ is an arbitrary constant. Substituting this solution into (15.14.1.5), we obtain the following solution of system (15.14.1.4):

$$z = \ln[ae^y x + (by + C)e^y] = y + \ln(ax + by + C).$$

### 15.14.1-2. Other overdetermined systems of equations in one unknown.

Section 15.10 describes a consistency analysis for overdetermined systems of partial differential equations in the context of the differential constraints method; one of the equations there is called a differential constraint. Detailed consideration is given to systems consisting of one second-order equation and one first-order equation (see Subsection 15.10.2), and also to some systems consisting of two second-order equations (see Subsection 15.10.3). Examples of solving such systems are also given.

Table 15.11 presents some overdetermined systems of equations with consistency conditions; the derivation of the consistency conditions for these systems can be found in Section 15.10.

TABLE 15.11
Some overdetermined systems of equations

| No. | First equation of system | Second equation of system | Consistency conditions |
|---|---|---|---|
| 1 | $\frac{\partial w}{\partial t} = \frac{\partial}{\partial x}\left[f(w)\frac{\partial w}{\partial x}\right] + g(w)$ | $\frac{\partial w}{\partial t} = \varphi(w)$ | (15.10.2.8) |
| 2 | $\frac{\partial^2 w}{\partial x \partial t} = f(w)$ | $\frac{\partial w}{\partial x} = \varphi(t)g(w)$ | (15.10.2.23) |
| 3 | $\frac{\partial w}{\partial t} = \frac{\partial}{\partial x}\left[f_1(w)\frac{\partial w}{\partial x}\right] + f_2(w)$ | $\frac{\partial^2 w}{\partial x^2} = g_1(w)\left(\frac{\partial w}{\partial x}\right)^2 + g_2(w)$ | (15.10.3.5) |
| 4 | $\frac{\partial w}{\partial y}\frac{\partial^2 w}{\partial x \partial y} - \frac{\partial w}{\partial x}\frac{\partial^2 w}{\partial y^2} = a\frac{\partial^3 w}{\partial y^3}$ | $\frac{\partial w}{\partial x} = \varphi(y)$ | (15.10.1.8) |

## 15.14.2. Pfaffian Equations and Their Solutions. Connection with Overdetermined Systems

**15.14.2-1. Pfaffian equations.**

A *Pfaffian equation* is an equation of the form

$$P(x,y,z)\,dx + Q(x,y,z)\,dy + R(x,y,z)\,dz = 0. \tag{15.14.2.1}$$

Equation (15.14.2.1) always has a solution $x = x_0$, $y = y_0$, $z = z_0$, where $x_0$, $y_0$, and $z_0$ are arbitrary constants. Such simple solutions are not considered below.

We will distinguish between the following two cases:

1. Find a two-dimensional solution to the Pfaffian equation, when the three variables $x$, $y$, $z$ are connected by a single relation (a certain condition must hold for such a solution to exist).

2. Find a one-dimensional solution to the Pfaffian equation, when the three variables $x$, $y$, $z$ are connected by two relations.

**15.14.2-2. Condition for integrability of the Pfaffian equation by a single relation.**

Let a solution of the Pfaffian equation be representable in the form $z = z(x,y)$, where $z$ is the unknown function and $x$, $y$ are independent variables. From equation (15.14.2.1) we find the expression for the differential:

$$dz = -\frac{P}{R}\,dx - \frac{Q}{R}\,dy. \tag{15.14.2.2}$$

On the other hand, since $z = z(x,y)$, we have

$$dz = \frac{\partial z}{\partial x}\,dx + \frac{\partial z}{\partial y}\,dy. \tag{15.14.2.3}$$

Equating the right-hand sides of (15.14.2.2) and (15.14.2.3) to each other and taking into account the independence of the differentials $dx$ and $dy$, we obtain an overdetermined system of equations of the form (15.14.1.1):

$$z_x = -P/R, \quad z_y = -Q/R. \tag{15.14.2.4}$$

A consistency condition for the system can be found by setting $F = -P/R$ and $G = -Q/R$ in (15.14.1.2). After some rearrangements we obtain the equation

$$R\left(\frac{\partial P}{\partial y} - \frac{\partial Q}{\partial x}\right) + P\left(\frac{\partial Q}{\partial z} - \frac{\partial R}{\partial y}\right) + Q\left(\frac{\partial R}{\partial x} - \frac{\partial P}{\partial z}\right) = 0. \tag{15.14.2.5}$$

If condition (15.14.2.5) is satisfied identically, the Pfaffian equation (15.14.2.1) is integrable by one relation of the form

$$U(x, y, z) = C, \tag{15.14.2.6}$$

where $C$ is an arbitrary constant. In this case, the Pfaffian equation is said to be *completely integrable*. The left-hand side of a completely integrable Pfaffian equation (15.14.2.1) can be represented in the form

$$P(x, y, z)\,dx + Q(x, y, z)\,dy + R(x, y, z)\,dz \equiv \mu(x, y, z)\,dU(x, y, z),$$

where $\mu(x, y, z)$ is an *integrating factor*.

A solution of a completely integrable Pfaffian equation can be found by solving the overdetermined system (15.14.2.4) employing the method presented in Subsection 15.14.1. Alternatively, the following equivalent technique can be used: it is first assumed that $x = \text{const}$ in equation (15.14.2.1), which corresponds to $dx = 0$. Then the resulting ordinary differential equation is solved for $z = z(y)$, where $x$ is treated as a parameter, and the constant of integration is regarded as an arbitrary function of $x$: $C = \varphi(x)$. Finally, by substituting the resulting solution into the original equation (15.14.2.1), one finds the function $\varphi(x)$.

**Example 1.** Consider the Pfaffian equation

$$y(xz + a)\,dx + x(y + b)\,dy + x^2 y\,dz = 0. \tag{15.14.2.7}$$

The integrability condition (15.14.2.5) is satisfied identically. Let us set $dx = 0$ in equation (15.14.2.7) to obtain the ordinary differential equation

$$(y + b)\,dy + xy\,dz = 0. \tag{15.14.2.8}$$

Treating $x$ as a parameter, we find the general solution of the separable equation (15.14.2.8):

$$z = -\frac{1}{x}(y + b\ln|y|) + \varphi(x), \tag{15.14.2.9}$$

where $\varphi(x)$ is the constant of integration, dependent on $x$. On substituting (15.14.2.9) into the original equation (15.14.2.7), we arrive at a linear ordinary differential equation for $\varphi(x)$:

$$x^2 \varphi'_x + x\varphi + a = 0.$$

Its general solution is expressed as

$$\varphi(x) = -\frac{a}{x}\ln|x| + \frac{C}{x},$$

where $C$ is an arbitrary constant. Substituting this $\varphi(x)$ into (15.14.2.9) yields the solution of the Pfaffian equation (15.14.2.7):

$$z = -\frac{1}{x}(y + a\ln|x| + b\ln|y| - C).$$

This solutions can be equivalently represented in the form of an integral (15.14.2.6) as $xz+y+a\ln|x|+b\ln|y|=C$.

*Geometric interpretation.* Let us introduce the vector $\mathbf{F} = \{P, Q, R\}$. Then the condition of complete integrability (15.14.2.5) of the Pfaffian equation (15.14.2.1) can be written in the dot product form: $\mathbf{F} \cdot \operatorname{curl} \mathbf{F} = 0$. Solution (15.14.2.6) represents a one-parameter family of surfaces orthogonal to $\mathbf{F}$.

### 15.14.2-3. Pfaffian equations not satisfying the integrability condition.

Consider now Pfaffian equations that do not satisfy the condition of integrability by one relation. In this case, relation (15.14.2.5) is not satisfied identically and there are two different methods for the investigation of such equations.

*First method.* Relation (15.14.2.5) is treated as an algebraic (transcendental) equation for one of the variables. For example, solving it for $z$, we get a relation $z = z(x, y)$.* The direct substitution of this solution into (15.14.2.1), with $x$ and $y$ regarded as independent variables, answers the question whether its a solution of the Pfaffian equation. The thus obtained solutions (if any) do not contain a free parameter.

*Second method.* One-dimensional solutions are sought in the form of two relations. One relation is prescribed, for example, in the form

$$z = f(x), \tag{15.14.2.10}$$

where $f(x)$ is an arbitrary function. Using it, one eliminates $z$ from the Pfaffian equation (15.14.2.1) to obtain an ordinary differential equation for $y = y(x)$:

$$Q(x, y, f(x))y'_x + P(x, y, f(x)) + R(x, y, f(x))f'_x(x) = 0.$$

Suppose the general solution of this equation has the form

$$\Phi(x, y, C) = 0, \tag{15.14.2.11}$$

where $C$ is an arbitrary constant. Then formulas (15.14.2.10), (15.14.2.11) define a one-dimensional solution of the Pfaffian equation in the form of two relations involving one arbitrary function and one free parameter.

**Example 2.** Consider the Pfaffian equation

$$y\,dx + dy + x\,dz = 0. \tag{15.14.2.12}$$

Substituting $P = y$, $Q = 1$, and $R = x$ into the left-hand side of condition (15.14.2.5), we find that $x + 1 \not\equiv 0$. Therefore, equation (15.14.2.12) is not integrable by one relation.

Let us look for one-dimensional solutions by choosing one relation in the form (15.14.2.10). Consequently, we arrive at the ordinary differential equation

$$y'_x + y + xf'_x(x) = 0.$$

Its general solution has the form

$$y = Ce^{-x} - e^{-x}\int e^x x f'_x(x)\,dx, \tag{15.14.2.13}$$

where $C$ is an arbitrary constant. Formulas (15.14.2.10) and (15.14.2.13) represent a one-dimensional solution of the Pfaffian equation (15.14.2.12) involving and arbitrary function $f(x)$ and an arbitrary constant $C$.

**Remark.** For equations that are not completely integrable, the first relation can be chosen in a more general form than (15.14.2.10):
$$z = f(x, y),$$
where $f(x, y)$ is an arbitrary function of two arguments. Using it to eliminate $z$ from the Pfaffian equation, we obtain an ordinary differential equation for $y = y(x)$. However, it is impossible to find the general solution of this equation in closed form even for very simple equations, including equation (15.14.2.12).

---

\* There may be several such relations, and even infinitely many.

## 15.14. Nonlinear Systems of Partial Differential Equations

### TABLE 15.12
Systems equations of the form (15.14.3.1) admitting invariant solutions that are found using combinations of translation and scaling transformations ($C$ is an arbitrary constant)

| Systems equations | Invariant transformations | Form of invariant solutions |
|---|---|---|
| $\frac{\partial u}{\partial x} = e^{au} f(au-bw)$, $\frac{\partial w}{\partial t} = e^{bw} g(au-bw)$ | $t = C^{ab}\bar{t}$, $x = C^{ab}\bar{x}$, $u = \bar{u} - b\ln C$, $w = \bar{w} - a\ln C$ | $u = y(\xi) - \frac{1}{a}\ln t$, $w = z(\xi) - \frac{1}{b}\ln t$, $\xi = \frac{x}{t}$ |
| $\frac{\partial u}{\partial x} = u^k f(u^n w^m)$, $\frac{\partial w}{\partial t} = w^s g(u^n w^m)$ | $t = C^{n(s-1)}\bar{t}$, $x = C^{m(1-k)}\bar{x}$, $u = C^m \bar{u}$, $w = C^{-n} \bar{w}$ | $u = t^{\frac{m}{n(s-1)}} y(\xi)$, $w = t^{\frac{1}{1-s}} z(\xi)$, $\xi = xt^{\frac{m(k-1)}{n(s-1)}}$ |
| $\frac{\partial u}{\partial x} = u^k f(u^n w^m)$, $\frac{\partial w}{\partial t} = w g(u^n w^m)$ | $t = \bar{t} + \ln C$, $x = C^{m(1-k)}\bar{x}$, $u = C^m \bar{u}$, $w = C^{-n} \bar{w}$ | $u = e^{mt} y(\xi)$, $w = e^{-nt} z(\xi)$, $\xi = e^{m(k-1)t} x$ |

### 15.14.3. Systems of First-Order Equations Describing Convective Mass Transfer with Volume Reaction*

**15.14.3-1. Traveling-wave solutions and some other invariant solutions.**

This subsection presents exact solutions to some classes of nonlinear systems of first-order equations of the form

$$\frac{\partial u}{\partial x} = F_1(u, w), \qquad \frac{\partial w}{\partial t} = F_2(u, w). \qquad (15.14.3.1)$$

**Remark.** Equations of convective mass transfer in two-component systems with volume chemical reaction without diffusion,

$$\frac{\partial u}{\partial \tau} + a_1 \frac{\partial u}{\partial \xi} = F_1(u, w), \qquad \frac{\partial w}{\partial \tau} + a_2 \frac{\partial w}{\partial \xi} = F_2(u, w),$$

are reduced to such systems of equations. This is achieved by using the characteristic variables

$$x = \frac{\xi - a_2 \tau}{a_1 - a_2}, \qquad t = \frac{\xi - a_1 \tau}{a_2 - a_1} \qquad (a_1 \neq a_2).$$

It is apparent that system (15.14.3.1) admits exact traveling-wave solutions

$$u = u(z), \qquad w = w(z), \qquad z = kx - \lambda t,$$

where $k$ and $\lambda$ are arbitrary constants, and the functions $u(z)$ and $w(z)$ are determined by the autonomous system of ordinary differential equations

$$ku'_z - F_1(u, w) = 0, \qquad \lambda w'_z + F_2(u, w) = 0.$$

Further on, $f(\dots)$ and $g(\dots)$ are arbitrary functions of their arguments.

Table 15.12 lists some systems of the form (15.14.3.1) admitting other invariant solutions that are found using combinations of translation and scaling transformations in the independent and dependent variables; these transformations preserve the form of the equations (i.e., a given system of equations is converted into an identical one). The procedure for constructing such solutions for a single equation is detailed in Section 15.3.

---

* Paragraphs 15.14.3-3 and 15.14.3-4 were written by P. G. Bedrikovetsky.

**15.14.3-2. Systems reducible to an ordinary differential equation.**

Consider the system of equations of the special form

$$\frac{\partial u}{\partial x} = f(w)u, \qquad \frac{\partial w}{\partial t} = g(w)u^k. \qquad (15.14.3.2)$$

In two special cases, $k = 0$ and $f(w) = \text{const}$, one of the two equations can be solved independently from the other, and consequently system (15.14.3.2) is easy to integrate. Further on, it is assumed that $k \neq 0$ and $f(w) \neq \text{const}$.

Let us first simplify system (15.14.3.2) via the change of variables

$$U = u^k, \qquad W = \int \frac{dw}{g(w)}. \qquad (15.14.3.3)$$

We obtain

$$\frac{\partial U}{\partial x} = \Phi(W)U, \qquad \frac{\partial W}{\partial t} = U, \qquad (15.14.3.4)$$

where the function $\Phi(W)$ is defined parametrically by the formulas

$$\Phi = kf(w), \qquad W = \int \frac{dw}{g(w)}, \qquad (15.14.3.5)$$

with $w$ treated as a parameter. Substituting $U$ in the first equation of system (15.14.3.4) by the left-hand side of the second equation of that system, we arrive at a second-order equation for $W$:

$$\frac{\partial^2 W}{\partial x \partial t} = \Phi(W) \frac{\partial W}{\partial t}. \qquad (15.14.3.6)$$

Integrating with respect to $t$ gives

$$\frac{\partial W}{\partial x} = \int \Phi(W) \, dW + \theta(x), \qquad (15.14.3.7)$$

where $\theta(x)$ is an arbitrary function.

Going back in (15.14.3.7) to the original variable $w$ by formulas (15.14.3.3), we get

$$\frac{\partial w}{\partial x} = kg(w) \int \frac{f(w)}{g(w)} \, dw + \theta(x)g(w). \qquad (15.14.3.8)$$

Equation (15.14.3.8) may be treated as a first-order ordinary differential equation in $x$. On finding its general solution, one should replace the constant of integration $C$ in it by an arbitrary function of time $\psi(t)$, since $w$ is dependent on $x$ and $t$.

**Example 1.** Consider the quadratically nonlinear system

$$\frac{\partial u}{\partial x} = awu, \qquad \frac{\partial w}{\partial t} = bwu, \qquad (15.14.3.9)$$

which is a special case of system (15.14.3.2) with $k = 1$, $f(w) = aw$, and $g(w) = bw$. Substituting these functions into (15.14.3.8), we obtain the Bernoulli equation

$$\frac{\partial w}{\partial x} = aw^2 + b\theta(x)w.$$

Its general solution is given by

$$w = \frac{\xi}{\psi(t) - a \int \xi \, dx}, \quad \xi = \exp\left(b \int \theta(x) \, dx\right), \quad (15.14.3.10)$$

where $\psi(t)$ is an arbitrary function; it has been taken into account that $w$ is dependent on $x$ and $t$. Since formula (15.14.3.10) involves an arbitrary function $\theta(x)$, it is convenient to introduce the notation $\varphi(x) = -\int \xi \, dx$. Consequently, we have the general solution of system (15.14.3.9) in the form

$$u = -\frac{\psi'_t(t)}{a\varphi(x) + b\psi(t)}, \quad w = -\frac{\varphi'_x(x)}{a\varphi(x) + b\psi(t)},$$

where $\varphi(x)$ and $\psi(t)$ are arbitrary functions; here, $\psi$ has been renamed $b\psi$ as compared with (15.14.3.10).

**Example 2.** To the special case $\theta(x) = \text{const}$ in (15.14.3.8) there correspond special solutions of system (15.14.3.2) of the form

$$w = w(z), \quad u = [\psi'(t)]^{1/k} v(z), \quad z = x + \psi(t),$$

involving an arbitrary function $\psi(t)$; the prime denotes a derivative. The functions $w(z)$ and $v(z)$ are determined by the autonomous system of ordinary differential equations

$$v'_z = f(w)v, \quad w'_z = g(w)v^k.$$

Its general solution can be written out in implicit form:

$$\int \frac{dw}{g(w)[kF(w) + C_1]} = z + C_2, \quad v = [kF(w) + C_1]^{1/k}, \quad F(w) = \int \frac{f(w)}{g(w)} \, dw.$$

▶ Exact solutions to a number of other first-order nonlinear systems of the form (15.14.3.1) can be found in Section T10.1.

---

**15.14.3-3. Some nonlinear problems of suspension transport in porous media.**

1°. Deep bed filtration of particle suspensions in porous media occurs during sea/produced water injection in oil reservoirs, drilling fluid invasion into reservoir productive zones, sand filtration in gravel packs, fines migration in oil fields, bacteria, virus or contaminant transport in groundwater, industrial filtering, etc. The basic features of the process are the particle capture by the porous medium and the consequent permeability reduction. The particles are captured due to size exclusion, surface sorption, electrical forces, sedimentation, diffusion, etc.

For a single component system, the governing equations consist of the mass balance for retained and suspended particles, and of capture kinetics:

$$\frac{\partial}{\partial t}(u + w) + \frac{\partial u}{\partial x} = 0,$$
$$\frac{\partial w}{\partial t} = f(w)u, \quad (15.14.3.11)$$

where $u$ is the suspended concentration and $w$ is a retained (deposited) concentration, and $f(w)$ is the filtration coefficient.

Let us transform the first equation of system (15.14.3.11) in the following way. Replace $w_t$ in the first equation by the right-hand side of the second equation and then pass from $x, t$ to the new characteristic variables $\xi = -x$, $\eta = x - t$. Consequently, we obtain a system of equations of the form (15.14.3.2):

$$\frac{\partial u}{\partial \xi} = f(w)u, \quad \frac{\partial w}{\partial \eta} = -f(w)u.$$

Its solution is reduced to integrating an ordinary differential equation.

2°. The problem of particle suspension injection into a particle-free porous reservoir is described by system (15.14.3.11) with the following initial and boundary conditions:

$$u = w = 0 \quad \text{at} \quad t = 0, \qquad u = 1 \quad \text{at} \quad x = 0. \tag{15.14.3.12}$$

The problem (15.14.3.11)–(15.14.3.12) has exact analytical solution. The introduction of the potential

$$\Phi(w) = \int_0^w \frac{dz}{f(z)} \tag{15.14.3.13}$$

allows transforming the second equation (15.14.3.11) to the form

$$u = \frac{\partial \Phi}{\partial t}. \tag{15.14.3.14}$$

The substitution of (15.14.3.14) into the first equation (15.14.3.11) and its integration with respect to $t$ from 0 to $t$ taking into account the initial conditions (15.14.3.12) results in the following first-order quasilinear partial differential equation:

$$\frac{\partial w}{\partial t} + \frac{\partial w}{\partial x} = -f(w)w. \tag{15.14.3.15}$$

Fixing $u = 1$ in the second equation of (15.14.3.11) for the inlet $x = 0$ provides a boundary condition for equation (15.14.3.15):

$$\Phi(w) = t \quad \text{at} \quad x = 0. \tag{15.14.3.16}$$

The solution of problem (15.14.3.15)–(15.14.3.16) is obtained by the method of characteristics (see Subsections 13.1.1–13.1.2) and can be represented in the implicit form

$$\int_w^{\Phi^{-1}(t-x)} \frac{dz}{z f(z)} = x. \tag{15.14.3.17}$$

Here, $\Phi^{-1}(w)$ is the inverse of the integral (15.14.3.13).

Now let us obtain the expression for the suspended concentration $u(x,t)$. Taking the time derivative of both sides of (15.14.3.17), we obtain

$$\left( \frac{1}{w f(w)} \frac{\partial w}{\partial t} \right)_{x,t} - \left( \frac{1}{w f(w)} \frac{\partial w}{\partial t} \right)_{0, t-x} = 0.$$

Replacing here the partial derivative by the left-hand side of the second equation in (15.14.3.11), we have

$$\frac{u(x,t)}{w(x,t)} = \frac{u(0, t-x)}{w(0, t-x)}, \tag{15.14.3.18}$$

so the ratio $u/w$ does not change along the characteristics. Taking into account the boundary conditions (15.14.3.12) and (15.14.3.16) in (15.14.3.18), we obtain the expression for the suspended concentration

$$u(x,t) = \frac{w(x,t)}{\Phi^{-1}(t-x)}. \tag{15.14.3.19}$$

Finally, formulas (15.14.3.17), (15.14.3.19), and (15.14.3.13) form an exact solution for problem (15.14.3.11)–(15.14.3.12) for $x < t$. For $x > t$, we have $u = w = 0$.

**Example 3.** Consider system (15.14.3.11) with the linear filtration coefficient
$$f(w) = 1 - w. \tag{15.14.3.20}$$
In this case the potential (15.14.3.13) is
$$\Phi(w) = -\ln(1-w).$$
From (15.14.3.16) it is possible to calculate the value on the boundary
$$w(0,t) = \Phi^{-1}(t) = 1 - e^{-t}$$
and the left-hand side of (15.14.3.17) can be determined explicitly. Finally, in view of (15.14.3.19), we find the solution to problem (15.14.3.11)–(15.14.3.12) with condition (15.14.3.20) in the form
$$u = \frac{e^{t-x}}{e^x + e^{t-x} - 1}, \qquad w = \frac{e^{t-x} - 1}{e^x + e^{t-x} - 1}.$$

### 15.14.3-4. Some generalizations.

$1°$. Consider a more general system than (15.14.3.11),
$$\frac{\partial}{\partial t}g(u,w) + a\frac{\partial u}{\partial x} = f_1(w)u,$$
$$\frac{\partial w}{\partial t} = f_2(w)u.$$

On eliminating $u$ from the first equation using the second equation and on integrating with respect to $t$, we obtain a nonlinear first-order partial differential equation for $w(x,t)$:
$$g\left(\frac{1}{f_2(w)}\frac{\partial w}{\partial t}, w\right) + \frac{a}{f_2(w)}\frac{\partial w}{\partial x} = \int \frac{f_1(w)}{f_2(w)}\,dw + \theta(x), \tag{15.14.3.21}$$

where $\theta(x)$ is an arbitrary function. If $\theta = \text{const}$, a complete integral of equation (15.14.3.21) is sought in the form $w = w(C_1 x + C_2 t + C_3)$. In this case, one can obtain the general integral (containing an arbitrary function) of equation (15.14.3.21) in parametric form by the method described in Subsection 13.2.1, after solving the associated ordinary differential equation.

$2°$. System (15.14.3.11) admits the following generalization:
$$\frac{\partial}{\partial t}\left[u + G(w_1, \ldots, w_n)\right] + \frac{\partial u}{\partial x} = 0, \tag{15.14.3.22}$$
$$\frac{\partial w_k}{\partial t} = F_k(w_1, \ldots, w_n)u, \quad k = 1, \ldots, n. \tag{15.14.3.23}$$

From equations (15.14.3.23) we have
$$\frac{1}{F_1(w_1, \ldots, w_n)}\frac{\partial w_1}{\partial t} = \cdots = \frac{1}{F_n(w_1, \ldots, w_n)}\frac{\partial w_n}{\partial t} = u. \tag{15.14.3.24}$$

We look for solutions of system (15.14.3.22)–(15.14.3.23) of the special form
$$w_1 = w_1(w_n), \quad \ldots, \quad w_{n-1} = w_{n-1}(w_n). \tag{15.14.3.25}$$

The functions $w_1, \ldots, w_{n-1}$ are thus assumed to be expressible in terms of $w_n$. From (15.14.3.24), in view of (15.14.3.25), we find the following system of $n - 1$ ordinary differential equations:

$$\frac{dw_k}{dw_n} = \frac{F_k(w_1, \ldots, w_n)}{F_n(w_1, \ldots, w_n)}, \quad k = 1, \ldots, n-1. \qquad (15.14.3.26)$$

Further, assuming that a solution of system (15.14.3.26) has been obtained and the functions (15.14.3.25) are known, we substitute them into (15.14.3.22)–(15.14.3.23) to arrive at the system of two equations

$$\begin{aligned}\frac{\partial}{\partial t}\bigl[u + g(w_n)\bigr] + \frac{\partial u}{\partial x} &= 0, \\ \frac{\partial w_n}{\partial t} &= f_n(w_n)u,\end{aligned} \qquad (15.14.3.27)$$

where

$$g(w_n) = G\bigl(w_1(w_n), \ldots, w_{n-1}(w_n), w_n\bigr), \quad f_n(w_n) = F_n\bigl(w_1(w_n), \ldots, w_{n-1}(w_n), w_n\bigr).$$

Introducing the new dependent variable

$$w = g(w_n) \equiv G\bigl(w_1(w_n), \ldots, w_{n-1}(w_n), w_n\bigr)$$

in (15.14.3.27), we obtain system (15.14.3.11) in which the function $f = f(w)$ is defined parametrically,

$$f = g'(w_n) f_n(w_n), \quad w = g(w_n),$$

with $w_n$ being the parameter.

**Remark 1.** Solutions of the above special form arise, for example, in problems with initial and boundary conditions,

$$w_1 = \cdots = w_n = u = 0 \quad \text{at} \quad t = 0,$$
$$u = 1 \quad \text{at} \quad x = 0.$$

(The initial conditions correspond to the absence of particle in the reservoir at the initial time and the boundary condition corresponds to a given suspended concentration in the injected fluid.) In this case, the system ordinary differential equations (15.14.3.26) is solved under the following initial conditions:

$$w_1 = \cdots = w_{n-1} = 0 \quad \text{at} \quad w_n = 0.$$

**Remark 2.** In problems on flows of $n$-component fluids through porous media, the mass balance for retained and suspended particles is governed by equation (15.14.3.22) in which the function $G$ is the sum of individual components:

$$G(w_1, \ldots, w_n) = \sum_{k=1}^{n} w_k;$$

equations (15.14.3.23) define the capture kinetics.

### 15.14.4. First-Order Hyperbolic Systems of Quasilinear Equations. Systems of Conservation Laws of Gas Dynamic Type*

15.14.4-1. Systems of two equations. Systems in the form of conservation laws.

1°. Consider the system of two quasilinear equations of the form

$$\begin{aligned}f_1(u,w)\frac{\partial u}{\partial t} + g_1(u,w)\frac{\partial w}{\partial t} + h_1(u,w)\frac{\partial u}{\partial x} + k_1(u,w)\frac{\partial w}{\partial x} &= 0, \\ f_2(u,w)\frac{\partial u}{\partial t} + g_2(u,w)\frac{\partial w}{\partial t} + h_2(u,w)\frac{\partial u}{\partial x} + k_2(u,w)\frac{\partial w}{\partial x} &= 0,\end{aligned} \qquad (15.14.4.1)$$

---

* Subsection 15.14.4 was written by A. P. Pires and P. G. Bedrikovetsky.

where $x$ and $t$ are independent variables, $u = u(x,t)$ and $w = w(x,t)$ are two unknown functions, and $f_k(u,w)$ and $g_k(u,w)$ are prescribed functions ($k = 1, 2, 3, 4$).

For any $f_k(u,w)$ and $g_k(u,w)$, system (15.14.4.1) admits the following simplest solutions:
$$u = C_1, \quad w = C_2, \tag{15.14.4.2}$$
where $C_1$ and $C_2$ are arbitrary constants.

2°. The main mathematical models in continuum mechanics and theoretical physics have the form of systems of conservation laws. Usually mass, momentum, and energy for phases and/or components are conserved.

A quasilinear system of two conservation laws in two independent variables has the form
$$\frac{\partial F_1(u,w)}{\partial t} + \frac{\partial G_1(u,w)}{\partial x} = 0,$$
$$\frac{\partial F_2(u,w)}{\partial t} + \frac{\partial G_2(u,w)}{\partial x} = 0. \tag{15.14.4.3}$$

It is a special case of system (15.14.4.1). It admits simple solutions of the form (15.14.4.2).

### 15.14.4-2. Self-similar continuous solutions. Hyperbolic systems.

System (15.14.4.1) is invariant under the transformation $(x,t) \to (ax, at)$, where $a$ is any number ($a \neq 0$). Therefore, it admits a self-similar solution of the form
$$u = u(\xi), \quad w = w(\xi), \quad \xi = \frac{x}{t}. \tag{15.14.4.4}$$

The substitution of (15.14.4.4) into (15.14.4.1) results in the following system of ordinary differential equations:
$$(h_1 - \xi f_1)u'_\xi + (k_1 - \xi g_1)w'_\xi = 0,$$
$$(h_2 - \xi f_2)u'_\xi + (k_2 - \xi g_2)w'_\xi = 0, \tag{15.14.4.5}$$

where the arguments of the functions $f_m$, $g_m$, $h_m$, and $k_m$ ($m = 1, 2$) are omitted for brevity.

System (15.14.4.5) may be treated as an algebraic system for the derivatives $u'_\xi$ and $w'_\xi$; for it to have a nonzero solution, its determinant must be equal to zero:
$$(f_1 g_2 - f_2 g_1)\xi^2 - (f_1 k_2 + g_2 h_1 - f_2 k_1 - g_1 h_2)\xi + h_1 k_2 - h_2 k_1 = 0. \tag{15.14.4.6}$$

System (15.14.4.1) is called *strictly hyperbolic* (or, for short, *hyperbolic*) if the discriminant of the quadratic equation (15.14.4.6), with respect to $\xi$, is positive:
$$(f_1 k_2 + g_2 h_1 - f_2 k_1 - g_1 h_2)^2 - 4(f_1 g_2 - f_2 g_1)(h_1 k_2 - h_2 k_1) > 0. \tag{15.14.4.7}$$

In this case, equation (15.14.4.6) has two different roots:
$$\xi_{1,2} = \xi_{1,2}(u,w). \tag{15.14.4.8}$$

Substituting either root (15.14.4.8) into (15.14.4.5) and taking $u$ to be a parameter along the integral curve, we obtain
$$(k_1 - \xi_{1,2} g_1)\frac{dw}{du} + h_1 - \xi_{1,2} f_1 = 0. \tag{15.14.4.9}$$

The ordinary differential equation (15.14.4.9) provides two sets of continuous solutions for system (15.14.4.1). The procedure for constructing these solutions consists of two steps: (i) the dependence $w = w(u)$, different for each set, is found from (15.14.4.9); and (ii) this dependence is then substituted into formula (15.14.4.8), in the left-hand side of which $\xi_{1,2}$ is replaced by $x/t$. This results in a solution defined in implicit form:

$$\xi_{1,2}(u, w) = x/t, \quad w = w(u).$$

**Example 1.** The system of equations describing one-dimensional longitudinal oscillations of an elastic bar consists of the equations of balance of mass and momentum:

$$\frac{\partial u}{\partial t} - \frac{\partial w}{\partial x} = 0,$$
$$\frac{\partial w}{\partial t} - \frac{\partial \sigma(u)}{\partial x} = 0. \tag{15.14.4.10}$$

Here, $u$ is the deformation gradient (strain), $v$ is the strain rate, and $\sigma(u)$ is the stress.

Self-similar solutions are sought in the form (15.14.4.4). This results in the system of ordinary differential equations

$$\xi u'_\xi + w'_\xi = 0,$$
$$\sigma'_u(u) u'_\xi + \xi w'_\xi = 0. \tag{15.14.4.11}$$

Equating the determinant of this system to zero, we obtain the quadratic equation $\xi^2 = \sigma'_u(u)$, whose roots are

$$\xi_{1,2} = \pm\sqrt{\sigma'_u(u)}. \tag{15.14.4.12}$$

The system is assumed to be hyperbolic, which implies that $\sigma'_u(u) > 0$. Substituting (15.14.4.12) into (15.14.4.11) yields a separable first-order ordinary differential equation:

$$w'_u \pm \sqrt{\sigma'_u(u)} = 0.$$

Integrating gives its general solution

$$w = C \mp \int \sqrt{\sigma'_u(u)}\, du, \tag{15.14.4.13}$$

where $C$ is an arbitrary constant (two different constants, corresponding to the plus and minus sign). Formula (15.14.4.13) together with

$$\pm\sqrt{\sigma'_u(u)} = x/t \tag{15.14.4.14}$$

defines two one-parameter self-similar solutions of system (15.14.4.10) of the form (15.14.4.4) in implicit form.

### 15.14.4-3. Simple Riemann waves.

In the case of self-similar solutions of the form (15.14.4.4), one of the unknowns can be expressed in terms of the other by eliminating the independent variable $\xi$ to obtain, for example,

$$w = w(u). \tag{15.14.4.15}$$

Assuming initially that the unknowns are functionally related via (15.14.4.15), one can obtain a wider class of exact solutions to system (15.14.4.1) than the class of self-similar solutions. Indeed, substituting (15.14.4.15) into (15.14.4.1) gives two equations for one function $u = u(x, t)$:

$$(f_1 + g_1 w'_u)\frac{\partial u}{\partial t} + (h_1 + k_1 w'_u)\frac{\partial u}{\partial x} = 0,$$
$$(f_2 + g_2 w'_u)\frac{\partial u}{\partial t} + (h_2 + k_2 w'_u)\frac{\partial u}{\partial x} = 0. \tag{15.14.4.16}$$

The dependence (15.14.4.15) is chosen so that the two equations of (15.14.4.16) coincide. This condition results in the following first-order ordinary differential equation for $w = w(u)$ quadratically nonlinear in the derivative:

$$(g_1 k_2 - g_2 k_1)(w'_u)^2 + (f_1 k_2 + g_1 h_2 - f_2 k_1 - g_2 h_1) w'_u + f_1 h_2 - f_2 h_1 = 0. \qquad (15.14.4.17)$$

Treating (15.14.4.17) as a quadratic equation in $w'_u$, we assume its discriminant to be positive:

$$(f_1 k_2 + g_1 h_2 - f_2 k_1 - g_2 h_1)^2 - 4(f_1 h_2 - f_2 h_1)(g_1 k_2 - g_2 k_1) > 0.$$

This condition is equivalent to condition (15.14.4.7) and implies hat the system in question, (15.14.4.1), is strictly hyperbolic. In this case, equation (15.14.4.17) has two different real roots and is reducible to first-order ordinary differential equations of the standard form

$$w'_u = \Lambda_m(u, w), \quad m = 1, 2. \qquad (15.14.4.18)$$

Having found a solution $w = w(u)$ of this equation for any $m$, we substitute it into the first (or the second) equation (15.14.4.16). As a result, we obtain a quasilinear first-order partial differential equation for $u(x, t)$:

$$(f_1 + g_1 \Lambda_m)\frac{\partial u}{\partial t} + (h_1 + k_1 \Lambda_m)\frac{\partial u}{\partial x} = 0, \quad w = w(u). \qquad (15.14.4.19)$$

The general solution of this equation can be obtained by the method of characteristics (see Subsection 13.1.1). This solution depends on one arbitrary function and is called a *simple Riemann wave*. To each of the two equations (15.14.4.18) there corresponds an equation (15.14.4.19) and a Riemann wave.

Suppose equations (15.14.4.18) have the integrals

$$\mathcal{R}_m(u, w) = C_m, \quad m = 1, 2,$$

where $C_1$ and $C_2$ are arbitrary constants. The functions $\mathcal{R}_m(u, w)$ are called *Riemann invariants*. Another definition of Riemann functions is given in Paragraph 15.14.4-6.

**Example 2.** We look for exact solutions of system (15.14.4.10) of the special form (15.14.4.15), implying that one of the unknown functions can be expressed via the other. Substituting (15.14.4.15) into (15.14.4.10) gives

$$\frac{\partial u}{\partial t} - w'_u \frac{\partial u}{\partial x} = 0, \quad w'_u \frac{\partial u}{\partial t} - \sigma'_u(u) \frac{\partial u}{\partial x} = 0. \qquad (15.14.4.20)$$

Requiring that the two equations coincide, we get the condition

$$(w'_u)^2 = \sigma'_u(u).$$

It follows that there are two possible relations between the unknowns:

$$w \mp \int \sqrt{\sigma'_u(u)}\, du = C_m, \quad m = 1, 2. \qquad (15.14.4.21)$$

The left-hand sides of these relations represent Riemann functions. On eliminating $w$ from (15.14.4.20), using (15.14.4.21), we arrive at the quasilinear first-order partial differential equation

$$\frac{\partial u}{\partial t} \mp \sqrt{\sigma'_u(u)} \frac{\partial u}{\partial x} = 0. \qquad (15.14.4.22)$$

A detailed analysis of this equation is given in Subsection 13.1.3; the general solution is found by the method of characteristics and can be represented in implicit form:

$$x \pm t\sqrt{\sigma'_u(u)} = \Phi(u), \qquad (15.14.4.23)$$

where $\Phi(u)$ is an arbitrary function.

Formulas (15.14.4.21) and (15.14.4.23) define two exact solutions of system (15.14.4.10), each containing one arbitrary function and one arbitrary constant. In the special case $\Phi(u) \equiv 0$, this solution turns into the self-similar solution considered above. In the case of a linear system, with $\sigma(u) = a^2 u$, formulas (15.14.4.23) give two traveling-wave solutions

$$u = \Psi_{1,2}(x \pm at),$$

which represent waves propagating in opposite directions and having an arbitrary undistorted profile.

The evolution of a Riemann wave with the initial profile

$$u = \varphi(x) \quad \text{at} \quad t = 0 \quad (-\infty < x < \infty) \tag{15.14.4.24}$$

for the lower sign in equation (15.14.4.22) is described parametrically by the formulas

$$x = \xi + \mathcal{F}(\xi)t, \quad u = \varphi(\xi), \tag{15.14.4.25}$$

where $\mathcal{F}(\xi) = \sqrt{\sigma'_u(u)}\big|_{u=\varphi(\xi)}$. The second unknown, $w$, is expressed in terms of $u$ by formula (15.14.4.21).

**Example 3.** A one-dimensional ideal adiabatic (isentropic) gas flow is governed by the system of two equations

$$\frac{\partial \rho}{\partial t} + \frac{\partial (\rho v)}{\partial x} = 0,$$
$$\frac{\partial (\rho v)}{\partial t} + \frac{\partial [\rho v^2 + p(\rho)]}{\partial x} = 0. \tag{15.14.4.26}$$

Here, $\rho = \rho(x,t)$ is the density, $v = v(x,t)$ is the velocity, and $p$ is the pressure. The first equation (15.14.4.26) represents the law of conservation of mass in fluid mechanics and is referred to as a continuity equation. The second equation (15.14.4.26) represents the law of conservation of momentum. The equation of state is given in the form $p = p(\rho)$. For an ideal polytropic gas, $p = A\rho^\gamma$, where the constant $\gamma$ is the adiabatic exponent.

Exact solutions of system (15.14.4.26) of the form $v = v(\rho)$ are given by the formulas

$$v = \pm \int \sqrt{p'_\rho(\rho)}\, \frac{d\rho}{\rho} + C_m,$$
$$x \mp t \left[ \int \sqrt{p'_\rho(\rho)}\, \frac{d\rho}{\rho} + \sqrt{p'_\rho(\rho)} + C_m \right] = \Phi(\rho),$$

where $\Phi(\rho)$ is an arbitrary function and the $C_m$ are arbitrary constants (different solutions have the different constants, $m = 1, 2$). The Riemann functions are obtained from the first relation by expressing $C_m$ via $\rho$ and $v$.

For an ideal polytropic gas, which corresponds to $p = A\rho^\gamma$, the exact solution of system (15.14.4.26) with $\gamma \neq 1$ becomes

$$v = \pm \frac{2}{\gamma - 1} \sqrt{A\gamma}\, \rho^{\frac{\gamma-1}{2}} + C_m,$$
$$x \mp t \left( \frac{\gamma + 1}{\gamma - 1} \sqrt{A\gamma}\, \rho^{\frac{\gamma-1}{2}} + C_m \right) = \Phi(\rho).$$

**Remark.** System (15.14.4.26) with $\rho = h$ and $p(\rho) = \frac{1}{2}gh^2$, where $v$ is the horizontal velocity averaged over the height $h$ of the water level and $g$ is the acceleration due to gravity, governs the dynamics of shallow water.

---

### 15.14.4-4. Linearization of gas dynamic systems by the hodograph transformation.

The hodograph transformation is used in gas dynamics and the theory of jets for the linearization of equations and finding solutions of certain boundary value problems.

For a quasilinear system of two equations (15.14.4.1), the hodograph transformation has the form

$$x = x(u, w), \quad t = t(u, w); \tag{15.14.4.27}$$

this means that $u$, $w$ are now treated as the independent variables and $x$, $t$ as the dependent variables.

Differentiating each relation in (15.14.4.27) with respect to $x$ and $t$ (as composite functions) and eliminating the partial derivatives $u_t$, $w_t$, $u_x$, $w_x$ from the resulting relations, we obtain

$$\frac{\partial u}{\partial t} = -J\frac{\partial x}{\partial w}, \quad \frac{\partial w}{\partial t} = J\frac{\partial x}{\partial u}, \quad \frac{\partial u}{\partial x} = J\frac{\partial t}{\partial w}, \quad \frac{\partial w}{\partial x} = -J\frac{\partial t}{\partial u}, \quad (15.14.4.28)$$

where $J = \dfrac{\partial u}{\partial x}\dfrac{\partial w}{\partial t} - \dfrac{\partial u}{\partial t}\dfrac{\partial w}{\partial x}$ is the Jacobian of the functions $u = u(x,t)$ and $w = w(x,t)$.
Replacing the derivatives in (15.14.4.1) with the help of relations (15.14.4.28) and dividing by $J$, we arrive at the linear system

$$\begin{aligned} g_1(u,w)\frac{\partial x}{\partial u} - k_1(u,w)\frac{\partial t}{\partial u} - f_1(u,w)\frac{\partial x}{\partial w} + h_1(u,w)\frac{\partial t}{\partial w} &= 0, \\ g_2(u,w)\frac{\partial x}{\partial u} - k_2(u,w)\frac{\partial t}{\partial u} - f_2(u,w)\frac{\partial x}{\partial w} + h_2(u,w)\frac{\partial t}{\partial w} &= 0. \end{aligned} \quad (15.14.4.29)$$

**Remark.** The hodograph transformation (15.14.4.27) is unusable if $J \equiv 0$. In this degenerate case, the quantities $u$ and $w$ are functionally related, and hence they cannot be taken as independent variables. In this case, relation (15.14.4.15) holds; it determines simple Riemann waves. This is why in using the hodograph transformation (15.14.4.27), one loses solutions corresponding to simple Riemann waves.

### 15.14.4-5. Cauchy and Riemann problems. Qualitative features of solutions.

*Cauchy problem* ($t \geq 0$, $-\infty < x < \infty$). Find functions $u = u(x,t)$, $w = w(x,t)$ that solve system (15.14.4.1) for $t > 0$ and satisfy the initial conditions

$$u(x,0) = \varphi_1(x), \quad w(x,0) = \varphi_2(x), \quad (15.14.4.30)$$

where $\varphi_1(x)$ and $\varphi_2(x)$ are prescribed functions. The Cauchy problem is also often referred to as an initial value problem.

*Riemann problem* ($t \geq 0$, $-\infty < x < \infty$). Find functions $u = u(x,t)$, $w = w(x,t)$ that solve system (15.14.4.1) for $t > 0$ and satisfy piecewise-smooth initial conditions of the special form

$$u(x,0) = \begin{cases} u_L & \text{if } x < 0, \\ u_R & \text{if } x > 0, \end{cases} \quad w(x,0) = \begin{cases} w_L & \text{if } x < 0, \\ w_R & \text{if } x > 0. \end{cases} \quad (15.14.4.31)$$

Here, $u_L$, $u_R$, $w_L$, and $w_R$ are prescribed constant quantities.

Since equations (15.14.4.1) and the boundary conditions (15.14.4.31) do not change under the transformation $(x,t) \to (ax, at)$, where $a$ is any positive number, the solution of the Riemann problem (15.14.4.1), (15.14.4.31) is self-similar. The unknown functions are sought in the form (15.14.4.4) and satisfy the ordinary differential equations (15.14.4.5). When passing to the self-similar variable $\xi = x/t$, the initial conditions (15.14.4.31) are transformed into boundary conditions of the form

$$u \to u_L, \quad w \to w_L \quad \text{as} \quad \xi \to -\infty; \quad u \to u_R, \quad w \to w_R \quad \text{as} \quad \xi \to \infty.$$

Solutions of the Riemann problem (15.14.4.1), (15.14.4.31) can be either continuous (such solutions are called *rarefaction waves*) or discontinuous (in this case, they describe *shock waves*). In both cases, the solution for each of the unknowns (e.g., for $u$) consists of two types of segments: those where the solution is constant, $u = \text{const}$, and those where it smoothly changes according to the law $u = u(x/t)$. The endpoints of these segments lie on straight lines $x = a_n t$ in the plane $(x, t)$. At these points, individual segments of solutions either join together (for continuous solutions) or agree with one another based on conservation laws (for discontinuous solutions); for details see Paragraphs 15.14.4-8 and 15.14.4-10.

Let us explain why shock waves, corresponding to discontinuous solutions, arise by an example. Consider the system of equations for one-dimensional longitudinal oscillations of an elastic bar (15.14.4.10). It was shown above that this system admits simple Riemann waves, which are described by the quasilinear first-order partial differential equation (15.14.4.22). This equation coincides, up to notation, with the model equation of gas dynamics (13.1.3.1), which was studied in detail in Subsection 13.1.3. If the function $\mathcal{F}(\xi)$ in solution (15.14.4.25) is nonmonotonic (this function is determined by the initial profile of the wave), then the characteristic lines defined by the first formula in (15.14.4.25) will intersect in the $(x, t)$ plane for various $\xi$. This results in nonuniqueness in determining the function $u = u(x, t)$, which contradicts physical laws. Therefore, in order to avoid nonuniqueness of physical quantities, discontinuous solutions have to be considered.

---

**15.14.4-6. Reduction of systems to the canonical form. Riemann invariants.**

1°. Eliminating the derivative $w_t$ and then the derivative $u_t$ from (15.14.4.1), we arrive after simple rearrangements to the *canonical form* of a gas dynamics system:

$$\begin{aligned}\frac{\partial u}{\partial t} + p_1(u, w)\frac{\partial u}{\partial x} + q_1(u, w)\frac{\partial w}{\partial x} &= 0, \\ \frac{\partial w}{\partial t} + p_2(u, w)\frac{\partial u}{\partial x} + q_2(u, w)\frac{\partial w}{\partial x} &= 0,\end{aligned} \qquad (15.14.4.32)$$

where

$$p_1 = \frac{h_1 g_2 - g_1 h_2}{f_1 g_2 - g_1 f_2}, \quad q_1 = \frac{k_1 g_2 - g_1 k_2}{f_1 g_2 - g_1 f_2}, \quad p_2 = \frac{f_1 h_2 - h_1 f_2}{f_1 g_2 - g_1 f_2}, \quad q_2 = \frac{f_1 k_2 - k_1 f_2}{f_1 g_2 - g_1 f_2}.$$

The functions $f_m$, $g_m$, $h_m$, $k_m$, $p_m$, and $q_m$ ($m = 1, 2$) depend on $u$ and $w$; it is assumed that $f_1 g_2 - g_1 f_2 \not\equiv 0$.

2°. For further simplifications of system (15.14.4.32), we proceed as follows. Multiply the first equation by $b_1 = b_1(u, w)$ and the second by $b_2 = b_2(u, w)$ and add up to obtain

$$b_1 \frac{\partial u}{\partial t} + b_2 \frac{\partial w}{\partial t} + (b_1 p_1 + b_2 p_2)\frac{\partial u}{\partial x} + (b_1 q_1 + b_2 q_2)\frac{\partial w}{\partial x} = 0. \qquad (15.14.4.33)$$

Impose the following constraints of the functions $b_1$ and $b_2$:

$$\begin{aligned} b_1 p_1 + b_2 p_2 &= \lambda b_1, \\ b_1 q_1 + b_2 q_2 &= \lambda b_2. \end{aligned} \qquad (15.14.4.34)$$

These constraints represent a linear homogeneous algebraic system of equations for $b_m$. For this system to have nontrivial solutions, its determinant must be zero. This results in a quadratic equation for the eigenvalues $\lambda = \lambda(u, w)$:

$$\lambda^2 - (p_1 + q_2)\lambda + p_1 q_2 - p_2 q_1 = 0. \qquad (15.14.4.35)$$

For hyperbolic systems (15.14.4.32), equation (15.14.4.35) has two different real roots:

$$\lambda_{1,2} = \tfrac{1}{2}(p_1 + q_2) \pm \tfrac{1}{2}\sqrt{(p_1 - q_2)^2 + 4p_2 q_1}. \qquad (15.14.4.36)$$

For each eigenvalue $\lambda_m$ of (15.14.4.36), from (15.14.4.34) we find the associated eigenfunction

$$b_1^{(m)} = \varphi_m(\lambda_m - q_2), \quad b_2^{(m)} = \varphi_m q_1, \qquad (15.14.4.37)$$

where $\varphi_m = \varphi_m(u, w)$ is an arbitrary function (it will be defined later); $m = 1, 2$.

From (15.14.4.33), in view of (15.14.4.34), we obtain two equations for the two roots (15.14.4.36):

$$b_1^{(m)}\left(\frac{\partial u}{\partial t} + \lambda_m \frac{\partial u}{\partial x}\right) + b_2^{(m)}\left(\frac{\partial w}{\partial t} + \lambda_m \frac{\partial w}{\partial x}\right) = 0, \quad m = 1, 2. \qquad (15.14.4.38)$$

The function $\varphi_m$ in (15.14.4.37) can be determined from the conditions

$$b_1^{(m)} = \frac{\partial \mathcal{R}_m}{\partial u}, \quad b_2^{(m)} = \frac{\partial \mathcal{R}_m}{\partial w}. \qquad (15.14.4.39)$$

On differentiating the first relation (15.14.4.39) with respect to $w$ and the second with respect to $u$, we equate the mixed derivatives $(\mathcal{R}_m)_{uw}$ and $(\mathcal{R}_m)_{wu}$. In view of (15.14.4.37), we obtain the following linear first-order partial differential equation for $\varphi_m$:

$$\frac{\partial}{\partial w}[\varphi_m(\lambda_m - q_2)] = \frac{\partial}{\partial u}(\varphi_m q_1). \qquad (15.14.4.40)$$

This equation can be solved by the method of characteristics; see Subsection 13.1.1. Assuming that a solution of equation (15.14.4.40) has been obtained (any nontrivial solution can be taken) and taking into account formulas (15.14.4.37), we find the functions $\mathcal{R}_m = \mathcal{R}_m(u, w)$ from system (15.14.4.39). Replacing $b_1^{(m)}$ and $b_2^{(m)}$ in (15.14.4.38) by the right-hand sides of (15.14.4.39), we get two equations:

$$\begin{aligned}\frac{\partial \mathcal{R}_1}{\partial t} + \tilde{\lambda}_1(\mathcal{R}_1, \mathcal{R}_2)\frac{\partial \mathcal{R}_1}{\partial x} &= 0, \\ \frac{\partial \mathcal{R}_2}{\partial t} + \tilde{\lambda}_2(\mathcal{R}_1, \mathcal{R}_2)\frac{\partial \mathcal{R}_2}{\partial x} &= 0,\end{aligned} \qquad (15.14.4.41)$$

where $\tilde{\lambda}_m(\mathcal{R}_1, \mathcal{R}_2) = \lambda_m(u, w)$, $m = 1, 2$. The functions $\mathcal{R}_1$ and $\mathcal{R}_2$ appearing in system (15.14.4.41) are called *Riemann invariants*.

System (15.14.4.41) admits two exact solutions:

$$\begin{aligned}\mathcal{R}_1 &= C_1, \quad x - \tilde{\lambda}_2(C_1, \mathcal{R}_2)t = \Phi_1(\mathcal{R}_2), \\ \mathcal{R}_2 &= C_2, \quad x - \tilde{\lambda}_1(\mathcal{R}_1, C_2)t = \Phi_2(\mathcal{R}_1),\end{aligned}$$

where $C_m$ are arbitrary constants and $\Phi_m(\mathcal{R}_{3-m})$ are arbitrary functions ($m = 1, 2$).

If the function $\tilde{\lambda}_1$ in (15.14.4.41) is independent of $\mathcal{R}_2$, then the solution of system (15.14.4.41) is reduced to successive integration of two quasilinear first-order partial differential equations.

**Remark 1.** Sometimes it is more convenient to use the formulas
$$b_1^{(m)} = \varphi_m p_2, \qquad b_2^{(m)} = \varphi_m(\lambda_m - p_1) \tag{15.14.4.37a}$$
rather than (15.14.4.37). In this case, equation (15.14.4.40) is replaced by
$$\frac{\partial}{\partial w}(\varphi_m p_2) = \frac{\partial}{\partial u}[\varphi_m(\lambda_m - p_1)]. \tag{15.14.4.40a}$$

**Remark 2.** The Riemann functions are not uniquely defined. Transformations of the dependent variables
$$\mathcal{R}_1 = \Psi_1(z_1), \qquad \mathcal{R}_2 = \Psi_2(z_2),$$
where $\Psi_1(z_1)$ and $\Psi_2(z_2)$ are arbitrary functions, preserve the general form of system (15.14.4.41); what changes are the functions $\widetilde{\lambda}_m$ only. This is due to the functional arbitrariness in determining the functions $\varphi_m$ from equations (15.14.4.40) or (15.14.4.40a). This arbitrariness can be used for the selection of most simple Riemann functions.

**Example 4.** Consider the system of equation of longitudinal oscillations of an elastic bar (15.14.4.10), which is a special case of system (15.14.4.32) with
$$p_1 = 0, \qquad q_1 = -1, \qquad p_2 = -\sigma'(u), \qquad q_2 = 0. \tag{15.14.4.42}$$

To determine the eigenvalues and the corresponding eigenfunctions, we use formulas (15.14.4.36), (15.14.4.37), (15.14.4.42) and equation (15.14.4.40). As a result, we obtain
$$\lambda_{1,2} = \pm\sqrt{\sigma'(u)}; \quad b_1^{(m)} = \sqrt{\sigma'(u)}, \quad b_2^{(m)} = \mp 1, \quad \varphi_m = \pm 1 \quad (m = 1, 2).$$
The Riemann functions are found from equations (15.14.4.39):
$$\mathcal{R}_1 = -w + \int \sqrt{\sigma'(u)}\, du, \qquad \mathcal{R}_2 = w + \int \sqrt{\sigma'(u)}\, du.$$

Note that the Riemann functions coincide, up to notation, with the integrals (15.14.4.21) obtained earlier from other considerations.

**Example 5.** Consider the system of equations (15.14.4.26) describing one-dimensional flow of an ideal adiabatic gas. First, reduce the system to the canonical form
$$\frac{\partial \rho}{\partial t} + v\frac{\partial \rho}{\partial x} + \rho\frac{\partial v}{\partial x} = 0,$$
$$\frac{\partial v}{\partial t} + \frac{p'(\rho)}{\rho}\frac{\partial \rho}{\partial x} + v\frac{\partial v}{\partial x} = 0.$$
This is a special case of system (15.14.4.32) with
$$u = \rho, \quad w = v, \quad p_1 = v, \quad q_1 = \rho, \quad p_2 = p'(\rho)/\rho, \quad q_2 = v. \tag{15.14.4.43}$$

To determine the eigenvalues and the corresponding eigenfunctions, we use formulas (15.14.4.36), (15.14.4.37), (15.14.4.43) and equation (15.14.4.40). As a result, we obtain
$$\lambda_{1,2} = v \pm \sqrt{p'(\rho)}; \quad b_1^{(m)} = \frac{\sqrt{p'(\rho)}}{\rho}, \quad b_2^{(m)} = \pm 1, \quad \varphi_m = \pm\frac{1}{\rho} \quad (m = 1, 2).$$
The Riemann functions are found from equations (15.14.4.39):
$$\mathcal{R}_1 = v + \int \frac{\sqrt{p'(\rho)}}{\rho}\, d\rho, \qquad \mathcal{R}_2 = -v + \int \frac{\sqrt{p'(\rho)}}{\rho}\, d\rho.$$

3°. Let us show that the Riemann invariants are constant along the rarefaction waves for hyperbolic systems. The substitution of the self-similar solution forms $\mathcal{R}_m = \mathcal{R}_m(\xi)$, where $\xi = x/t$, into system (15.14.4.41) results in the following system of two ordinary differential equations:
$$[\xi - \lambda_m(\mathcal{R}_1, \mathcal{R}_2)]\frac{d\mathcal{R}_m}{d\xi} = 0 \quad (m = 1, 2). \tag{15.14.4.44}$$

The equality $\xi = \lambda_1(\mathcal{R}_1, \mathcal{R}_2)$ takes place along the first rarefaction wave. Hence, the first factor in the second equation of (15.14.4.44) is nonzero. Therefore, the second factor in the second equation of (15.14.4.44) is zero. It follows that $\mathcal{R}_2 = $ const along the first rarefaction wave. Along the second rarefaction wave, $\mathcal{R}_1$ is constant.

### 15.14.4-7. Hyperbolic $n \times n$ systems of conservation laws. Exact solutions.

Here we consider systems of conservation laws of the form

$$\frac{\partial \mathbf{F}(\mathbf{u})}{\partial t} + \frac{\partial \mathbf{G}(\mathbf{u})}{\partial x} = 0, \qquad (15.14.4.45)$$

where $\mathbf{u} = \mathbf{u}(x, t)$ is a vector function of two scalar variables, and $\mathbf{F} = \mathbf{F}(\mathbf{u})$ and $\mathbf{G} = \mathbf{G}(\mathbf{u})$ are vector functions,

$$\mathbf{u} = (u_1, \ldots, u_n)^T, \quad u_i = u_i(x, t);$$
$$\mathbf{F} = (F_1, \ldots, F_n)^T, \quad F_i = F_i(\mathbf{u});$$
$$\mathbf{G} = (G_1, \ldots, G_n)^T, \quad G_i = G_i(\mathbf{u}).$$

Here and henceforth, $(u_1, \ldots, u_n)^T$ stands for a column vector with components $u_1, \ldots, u_n$.

Note three important types of exact solutions to system (15.14.4.45):

1. For any $\mathbf{F}$ and $\mathbf{G}$, system (15.14.4.45) admits the following simplest solutions:

$$\mathbf{u} = \mathbf{C}, \qquad (15.14.4.46)$$

where $\mathbf{C}$ is an arbitrary constant vector.

2. System (15.14.4.45) admits self-similar solutions of the form

$$\mathbf{u} = \mathbf{u}(\xi), \quad \xi = x/t. \qquad (15.14.4.47)$$

The procedure for finding them is analogous to that used in Paragraph 15.14.4-2 (see also Paragraph 15.14.4-10).

3. System (15.14.4.45) also admits more general exact solutions of the form

$$u_1 = u_1(u_n), \quad \ldots, \quad u_{n-1} = u_{n-1}(u_n), \qquad (15.14.4.48)$$

where all the components are functionally related and can be expressed in terms of one of them. After substituting (15.14.4.48) into the original system (15.14.4.45), we require that all the resulting equations coincide. As a result, we obtain a system of $(n-1)$ ordinary differential equations for determining the dependences (15.14.4.48) and one first-order partial differential equation for $u_n = u_n(x, t)$. The mentioned procedure is described in detail in Paragraph 15.14.4-3 for the case of a two-equation system.

Let us show that some systems of conservation laws can be represented as systems of ordinary differential equations along curves $x = x(t)$ called characteristic curves.

Differentiating both sides of system (15.14.4.45) yields

$$\frac{\partial \mathbf{u}}{\partial t} + \mathbf{A}\frac{\partial \mathbf{u}}{\partial x} = 0, \qquad (15.14.4.49)$$

where $\mathbf{A} = \widetilde{\mathbf{F}}^{-1}(\mathbf{u})\widetilde{\mathbf{G}}(\mathbf{u})$, $\widetilde{\mathbf{F}}(\mathbf{u})$ is the matrix with entries $\frac{\partial F_i}{\partial u_j}$, $\widetilde{\mathbf{G}}(\mathbf{u})$ is the matrix with entries $\frac{\partial G_i}{\partial u_j}$, and $\widetilde{\mathbf{F}}^{-1}$ is the inverse of the matrix $\widetilde{\mathbf{F}}$.

Let us multiply each scalar equation in (15.14.4.49) by $b_i = b_i(\mathbf{u})$ and take the sum. On rearranging terms under the summation sign, we obtain

$$\sum_{i=1}^{n} b_i \frac{\partial u_i}{\partial t} + \sum_{i,j=1}^{n} b_j a_{ji} \frac{\partial u_i}{\partial x} = 0, \qquad (15.14.4.50)$$

where $a_{ij} = a_{ij}(\mathbf{u})$ are the entries of the matrix $\mathbf{A}$.

If $\mathbf{b} = (b_1, \ldots, b_n)$ is a left eigenvector of the matrix $\mathbf{A}(\mathbf{u})$ that corresponds to an eigenvalue $\lambda = \lambda(\mathbf{u})$, so that

$$\sum_{j=1}^{n} b_j a_{ji} = \lambda b_i,$$

then equation (15.14.4.50) can be rewritten in the form

$$\sum_{i=1}^{n} b_i \left( \frac{\partial u_i}{\partial t} + \lambda \frac{\partial u_i}{\partial x} \right) = 0. \qquad (15.14.4.51)$$

Thus, system (15.14.4.49) is transformed to a linear combination of total derivatives of the unknowns $u_i$ with respect to $t$ along the direction $(\lambda, 1)$ on the plane $(x, t)$, i.e., the total time derivatives are taken along the trajectories having the velocity $\lambda$:

$$\sum_{i=1}^{n} b_i \frac{du_i}{dt} = 0, \quad \frac{dx}{dt} = \lambda, \qquad (15.14.4.52)$$

where

$$b_i = b_i(\mathbf{u}), \quad \lambda = \lambda(\mathbf{u}), \quad x = x(t), \quad \frac{du_i}{dt} = \frac{\partial u_i}{\partial t} + \frac{dx}{dt} \frac{\partial u_i}{\partial x}.$$

Equations (15.14.4.52) are called *differential relations on characteristics*. The second equation in (15.14.4.52) explains why an eigenvalue $\lambda$ is called a *characteristic velocity*.

System (15.14.4.49) is called *hyperbolic* if the matrix $\mathbf{A}$ has $n$ real eigenvalues and $n$ linearly independent left eigenvectors. If all eigenvalues are distinct for any $\mathbf{u}$, then system (15.14.4.49) is said to be *strictly hyperbolic*.

**Remark 1.** If the hyperbolic system (15.14.4.49) is linear and the coefficients of the matrix $\mathbf{A}$ are constant, then the eigenvalues $\lambda_k$ are constant and the characteristic lines in the $(x, t)$ plane become straight lines:

$$x = \lambda_k t + \text{const}.$$

Since all eigenvalues $\lambda_k$ are different, the general solution of system (15.14.4.49) can be represented as the sum of particular solutions as follows:

$$\mathbf{u} = \phi_1(x - \lambda_1 t)\mathbf{r}^1 + \cdots + \phi_n(x - \lambda_n t)\mathbf{r}^n,$$

where the $\phi_k(\xi_k)$ are arbitrary functions, $\xi_k = x - \lambda_k t$, and $\mathbf{r}^k$ is the right eigenvector of $\mathbf{A}$ corresponding to the eigenvalue $\lambda_k$, $k = 1, \ldots, n$. The particular solutions $\mathbf{u}_k = \phi_k(x - \lambda_k t)\mathbf{r}^k$ are called traveling-wave solutions. Each of these solutions represents a wave that travels in the $\mathbf{r}^k$-direction with velocity $\lambda_k$.

**Remark 2.** The characteristic form (15.14.4.51) of the hyperbolic system (15.14.4.49) forms the basis for the numerical characteristics method which allows the solution of system (15.14.4.49) in its domain of continuity. Suppose that we already have a solution $\mathbf{u}(x, t)$ for all values of $x$ and a fixed time $t$. To construct a solution at a point $(x, t + \Delta t)$, we find the points $(x - \lambda_k \Delta t, t)$ from which the characteristics arrive at the point $(x, t + \Delta t)$. Since the $\mathbf{u}(x - \lambda_k \Delta t, t)$ are known, relations (15.14.4.52) can be regarded as a system of $n$ linear equations in the $n$ unknowns $\mathbf{u}(x, t + \Delta t)$. Thus, a solution for the time $t + \Delta t$ can be found.

### 15.14.4-8. Shock waves. Rankine–Hugoniot jump conditions.

Let us consider a discontinuity along a trajectory $x_f(t)$ and obtain the mass balance condition along a discontinuity (shock wave). The region $x > x_f(t)$ is conventionally assumed to lie ahead of the shock, and the region $x < x_f(t)$ is assumed to lie behind the shock. The shock speed $D$ is determined by the relation

$$D = \frac{dx_f}{dt}. \qquad (15.14.4.53)$$

## 15.14. NONLINEAR SYSTEMS OF PARTIAL DIFFERENTIAL EQUATIONS

To refer to the value of a quantity, $A$, behind the shock, the minus superscript will be used, $A^-$, since this value corresponds to negative $x$ in the initial value formulation. Likewise, the value of $A$ ahead of the shock will be denoted $A^+$.

For an arbitrary quasilinear system in the form of conservation laws (15.14.4.45), the balance equations for a shock can be represented as

$$[F_i(\mathbf{u})]D = [G_i(\mathbf{u})], \quad i = 1, \ldots, n, \qquad (15.14.4.54)$$

where $[A] = A^+ - A^-$ stands for the jump of a quantity $A$ at the shock. Formulas (15.14.4.54) are called the *Rankine–Hugoniot jump conditions*; they are derived from integral consequences of the equations in question in much the same ways as for a single equation (see Paragraph 13.1.3-4).

**Example 6.** The Rankine–Hugoniot conditions for an isentropic gas flow (15.14.4.26) follow from (15.14.4.54). We have

$$\begin{aligned}[\rho]D &= [\rho v], \\ [\rho v]D &= [\rho v^2 + p(\rho)].\end{aligned} \qquad (15.14.4.55)$$

Here, the density and the velocity ahead of the shock are denoted $\rho^+$ and $v^+$, while those behind the shock are $\rho^-$ and $v^-$.

Eliminating the shock speed $D$ from (15.14.4.55), we obtain the relation

$$v^+ - v^- = \pm \sqrt{\frac{(\rho^+ - \rho^-)\big[p(\rho^+) - p(\rho^-)\big]}{\rho^- \rho^+}}. \qquad (15.14.4.56)$$

Each of the signs before the square root in (15.14.4.56) corresponds to a branch of the locus of points that can be connected with a given point $(v^-, \rho^-)$ by a shock (see Fig. 15.9a). For an ideal polytropic gas, one should set $p = A\rho^\gamma$ in formulas (15.14.4.55) and (15.14.4.56).

Let us determine the set of states $(v^+, \rho^+)$ reachable by a shock from a given point $(v^-, \rho^-)$. Express the point $(v^+, \rho^+)$ ahead of the shock via the solution of the transcendental system (15.14.4.55) to obtain

$$v^+ = v^+(v^-, \rho^-, D), \qquad \rho^+ = \rho^+(v^-, \rho^-, D).$$

The graphs of the solution determined by (15.14.4.56) are shown in Figs. 15.9a and 15.9b. The solid lines correspond to the minus sign before the radical and represent stable (evolutionary) shocks, while the dashed lines correspond to the plus sign and represent unstable (nonevolutionary) shocks; see below.

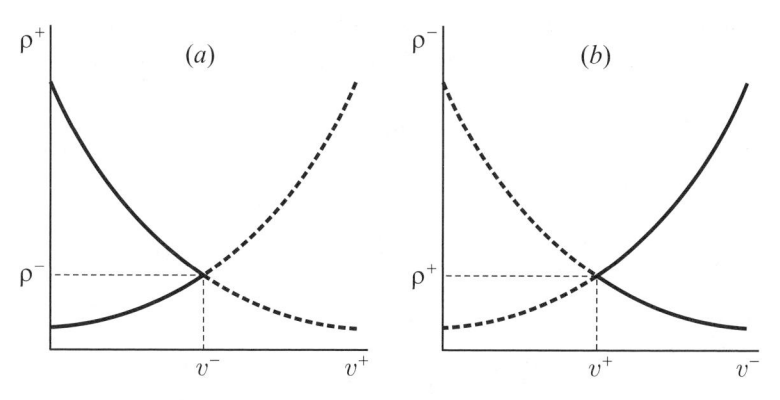

**Figure 15.9.** Loci of points that can be connected by a shock wave: (*a*) with a given state $(v^-, \rho^-)$ and (*b*) with a given state $(v^+, \rho^+)$. The solid lines correspond, respectively, to the minus and plus signs in formula (15.14.4.56) (for $p = A\rho^\gamma$) before the radical for cases (*a*) and (*b*).

### 15.14.4-9. Shock waves. Evolutionary conditions. Lax condition.

In general, discontinuities of solutions are surfaces where conditions are imposed that relate the quantities on both sides of the surfaces. For hyperbolic systems in the conservation-law form (15.14.4.45), these relations have the form (15.14.4.54) and involve the discontinuity velocity $D$.

The evolutionary conditions are necessary conditions for unique solvability of the problem of the discontinuity interaction with small perturbations depending on the $x$-coordinate normal to the discontinuity surface. For hyperbolic systems, a one-dimensional small perturbation can be represented as a superposition of $n$ waves, each being a traveling wave propagating at a characteristic velocity $\lambda_i^{\mp}$. This allows us to classify all these waves into incoming and outgoing ones, depending on the sign of the difference $\lambda_i^{\mp} - D$. Incoming waves are fully determined by the initial conditions, while outgoing ones must be determined from the linearized boundary conditions at the shock.

We consider below the stability of a shock with respect to a small perturbation. This kind of stability is determined by incoming waves. For this reason, we focus below on incoming waves.

Let $m^+$ and $m^-$ be the numbers of incoming waves from the right and left of the shock, respectively. It can be shown that if the relation

$$m^+ + m^- - 1 = n \tag{15.14.4.57}$$

holds, the problem of the discontinuity interaction with small perturbations is uniquely solvable. Relation (15.14.4.57) is called the *Lax condition*. If (15.14.4.57) holds, the corresponding discontinuity is called *evolutionary*; otherwise, it is called *nonevolutionary*. For evolutionary discontinuities, small incoming perturbations generate small outgoing perturbations and small changes in the discontinuity velocity.

If

$$m^+ + m^- - 1 > n,$$

then either such discontinuities do not exist or the perturbed quantities cannot be uniquely determined (the given conditions are underdetermined).

If

$$m^+ + m^- - 1 < n,$$

then the problem of the discontinuity interaction with small perturbations has no solution in the linear approximation. Previous studies of various physical problems have shown that the interaction of nonevolutionary discontinuities with small perturbations results in their disintegration into two or more evolutionary discontinuities.

The evolutionary condition (15.14.4.57) can be rewritten in the form of inequalities relating the shock speed $D$ and the velocities $\lambda_i^{\mp}$ of small disturbances. Let us enumerate the characteristic velocities on both sides of the discontinuity so that

$$\lambda_1(\mathbf{u}) \leq \lambda_2(\mathbf{u}) \leq \cdots \leq \lambda_n(\mathbf{u}).$$

A shock is called a $k$-shock if both $k$th characteristics are incoming:

$$\begin{aligned} &\text{if } i > k, \text{ then } D < \lambda_i^{\mp}; \\ &\text{if } i < k, \text{ then } D > \lambda_i^{\mp}; \\ &\text{if } i = k, \text{ then } \lambda_i^+ < D < \lambda_i^-. \end{aligned} \tag{15.14.4.58}$$

Below is another, equivalent statement of the Lax condition: $n+1$ inequalities out of the $2n$ inequalities

$$\lambda_k^+ \leq D \leq \lambda_k^- \qquad (k = 1, \ldots, n) \tag{15.14.4.59}$$

must hold.

For two-equation systems, the shock evolutionarity requires that three out of four inequalities

$$\lambda_1^+ \leq D \leq \lambda_1^-, \quad \lambda_2^+ \leq D \leq \lambda_2^-, \tag{15.14.4.60}$$

must hold.

**Example 7.** The Lax condition for the adiabatic gas flow equations (15.14.4.26) are obtained by substituting the eigenvalue expressions $\lambda_{1,2} = v \pm \sqrt{p'(\rho)}$ (see Example 5) into inequalities (15.14.4.60). We have

$$v^+ \pm \sqrt{p'(\rho^+)} < D < v^- \pm \sqrt{p'(\rho^-)}. \tag{15.14.4.61}$$

The shock evolutionarity requires that three of the four inequalities in (15.14.4.61) hold. Substituting the equation of state for a polytropic ideal gas, $p = A r^\gamma$, into (15.14.4.61), we obtain the following evolutionarity criterion:

$$v^+ \pm \sqrt{A\gamma(\rho^+)^{\gamma-1}} < D < v^- \pm \sqrt{A\gamma(\rho^-)^{\gamma-1}}. \tag{15.14.4.62}$$

For the adiabatic gas flow system (15.14.4.26), the solution vector is $\mathbf{u} = (\rho, v)^T$. Figure 15.9a shows the locus of points $\mathbf{u}^+$ that can be connected by a shock to the point $\mathbf{u}^-$; it is divided into the evolutionary part (solid line) and nonevolutionary part (dashed line). It can be shown that a shock issuing from the point $(v^-, \rho^-)$ and passing through any point $(v^+, \rho^+)$ of the solid part of the locus of first-family shocks obeys the Lax conditions (15.14.4.62). The shock speed $D$ of the first family decreases from $\lambda_1(\mathbf{u}^-)$, for points $\mathbf{u}^+$ tending to point $\mathbf{u}^-$, to $v^-$ as $\rho^+ \to 0$ and $v^+ \to -\infty$. Along the locus of the second-family shocks, the speed decreases from $\lambda_2(\mathbf{u}^-)$, for points $\mathbf{u}^+$ tending to $\mathbf{u}^-$, to $-\infty$ as $\rho^+ \to \infty$ and $v^+ \to -\infty$.

Figure 15.9b depicts the locus of points $\mathbf{u}^-$ that can be connected by a shock to the point $\mathbf{u}^+$. The evolutionary part of the locus is shown by a solid line; the dashed line shows the nonevolutionary part. The shock speed $D$ of the first family increases from $\lambda_1(\mathbf{u}^+)$, for points $\mathbf{u}^-$ tending to $\mathbf{u}^+$, to $\infty$ as $\rho^- \to \infty$ and $v^- \to \infty$. Along the locus of the second family shocks, the speed increases from $\lambda_2(\mathbf{u}^+)$, for points $\mathbf{u}^-$ tending to $\mathbf{u}^+$, to $v^+$ as $\rho^- \to 0$ and $v^- \to \infty$.

---

**15.14.4-10. Solutions for the Riemann problem. Solutions describing shock waves.**

1°. Self-similar solutions (15.14.4.47) together with the simplest solutions (15.14.4.46) enable the solution of the Riemann problem. This problem is formulated as follows: find a function $\mathbf{u} = \mathbf{u}(x,t)$ that solves system (15.14.4.45) for $t > 0$ and satisfies the following initial condition of a special form:

$$\mathbf{u} = \begin{cases} \mathbf{u}_L & \text{if } x < 0 \\ \mathbf{u}_R & \text{if } x > 0 \end{cases} \quad \text{at} \quad t = 0. \tag{15.14.4.63}$$

Here, $\mathbf{u}_L$ and $\mathbf{u}_R$ are two prescribed constant vectors.

With the self-similar variable $\xi = x/t$, the initial condition (15.14.4.63) is transformed into the boundary conditions

$$\mathbf{u} \to \mathbf{u}_L \quad \text{as} \quad \xi \to -\infty, \quad \mathbf{u} \to \mathbf{u}_R \quad \text{as} \quad \xi \to \infty. \tag{15.14.4.64}$$

2°. Without loss of generality, we will be considering the case $\mathbf{F}(\mathbf{u}) = \mathbf{u}$. The substitution of the self-similar form $\mathbf{u}(x,t) = \mathbf{u}(\xi)$ into (15.14.4.45) with $\mathbf{F}(\mathbf{u}) = \mathbf{u}$ yields

$$\left(\widetilde{\mathbf{G}} - \xi \mathbf{I}\right) \mathbf{u}'_\xi = 0, \tag{15.14.4.65}$$

where $\widetilde{\mathbf{G}} = \widetilde{\mathbf{G}}(\mathbf{u})$ is the matrix with entries $G_{ij} = \frac{\partial G_i}{\partial u_j}$ and $\mathbf{I}$ is the identity matrix.

Hence, the velocity vector for the continuous solution $\mathbf{u}(\xi)$ is a right eigenvector of the matrix $\widetilde{\mathbf{G}}$ for any point $\mathbf{u}$, and the corresponding eigenvalue equals the self-similar coordinate:

$$\xi = \lambda_k, \quad \mathbf{u}'_\xi = \alpha \mathbf{r}^k. \tag{15.14.4.66}$$

Here, $\lambda_k = \lambda_k(\mathbf{u})$ is a root of the algebraic equation $\det |\widetilde{\mathbf{G}} - \lambda \mathbf{I}| = 0$, $\mathbf{r}^k = \mathbf{r}^k(\mathbf{u})$ is a solution of the corresponding degenerate linear system of equations $(\widetilde{\mathbf{G}} - \lambda \mathbf{I})\mathbf{r} = 0$, and $\alpha = \alpha(\mathbf{u})$ is a positive function that will be defined below.

Differentiating both sides of the first equation (15.14.4.66) with respect to $\xi$ yields

$$\alpha = \frac{1}{\langle \nabla \lambda_k, \mathbf{r}^k \rangle}, \quad \langle \nabla \lambda_k, \mathbf{r}^k \rangle = \frac{\partial \lambda_k}{\partial u_1} r_1^k + \cdots + \frac{\partial \lambda_k}{\partial u_n} r_n^k.$$

Any $n \times n$ hyperbolic system allows for $n$ continuous solutions [of system (15.14.4.65)] corresponding to $n$ characteristic velocities $\lambda = \lambda_k$. The continuous solutions are determined by $n$ systems of ordinary differential equations. Each system is represented by a phase portrait in the $n$-dimensional $\mathbf{u}$-space. A solution/trajectory that corresponds to a characteristic velocity $\lambda_k$ is called a $k$th rarefaction wave.

3°. A trajectory of solution (15.14.4.47) in the space $\mathbf{u} = (u_1, \ldots, u_n)^T$ is called a *solution path*. The path is parametrized by the self-similar coordinate $\xi$. The path connects the point $\mathbf{u} = \mathbf{u}_L$ with the point $\mathbf{u} = \mathbf{u}_R$. The self-similar coordinate $\xi$ monotonically increases along the path varying from $-\infty$ at $\mathbf{u} = \mathbf{u}_L$ to $+\infty$ at $\mathbf{u} = \mathbf{u}_R$. The path consists of continuous segments representing solutions of the ordinary differential equations (15.14.4.65) (rarefaction waves), line segments that connect two points $\mathbf{u}^-$ and $\mathbf{u}^+$ satisfying the Rankine–Hugoniot conditions (15.14.4.54) and evolutionary conditions (15.14.4.59), and rest points $\mathbf{u}(\xi) = \text{const}$.

**Example 8.** Consider a solution consisting of two shocks and one rarefaction. The structural formula* for the solution path is $\mathbf{u}_L \to 1 - 2 \to \mathbf{u}_R$; specifically,

$$\mathbf{u}(x,t) = \begin{cases} \mathbf{u}_L & \text{if } -\infty < x/t < D_1, \\ \mathbf{u}_1 & \text{if } D_1 < x/t < \lambda_2(\mathbf{u}_1), \\ \mathbf{u}^{(2)}(\xi) & \text{if } \lambda_2(\mathbf{u}_1) < x/t < \lambda_2(\mathbf{u}_2), \\ \mathbf{u}_2 & \text{if } \lambda_2(\mathbf{u}_2) < x/t < D_2, \\ \mathbf{u}_R & \text{if } D_2 < x/t < \infty. \end{cases}$$

The shock speed $D_1$ (resp., $D_2$) can be found from the Hugoniot condition by setting $\mathbf{u}^- = \mathbf{u}_L$ and $\mathbf{u}^+ = \mathbf{u}_1$ (resp., $\mathbf{u}^- = \mathbf{u}_2$ and $\mathbf{u}^+ = \mathbf{u}_R$). Points 1 and 2 are located on the same rarefaction curve. The vector $\mathbf{u}^{(2)}(\xi)$ is a second-family rarefaction wave that is described by the system of ordinary differential equations (15.14.4.65) with $\xi = \lambda_2(\mathbf{u})$.

Figure 15.10$a$ depicts a sequence of rarefactions and shocks in the plane $(x, t)$. Figure 15.10$b$ shows the profile of the solution component $u_i$ along the $x$-axis. The self-similar curves $\mathbf{u} = \mathbf{u}(\xi)$ coincide with the profiles $\mathbf{u}(x, t = 1)$. For $t > 1$, the graphs of $\mathbf{u}(x, t)$ are obtained from the self-similar curves by extending them along the axis $x$ by a factor of $t$.

**Example 9.** Let us discuss an adiabatic gas flow in a tube in front of an impermeable piston moving with a velocity $v_L$ (see Fig. 15.11). The initial state is defined by prescribing initial values of the velocity and density:

$$v = 0, \quad \rho = \rho_R \quad \text{at} \quad \xi = \infty \quad (\xi = x/t). \tag{15.14.4.67}$$

The piston is impermeable; therefore, the gas velocity in front of the shock is equal to the piston velocity (Fig. 15.12$a$):

$$v = v_L \quad \text{at} \quad \xi = v_L. \tag{15.14.4.68}$$

The gas density in front of the piston is unknown in this problem.

---

* In structural formulas like $\mathbf{u}_L \to 1 - 2 \to \mathbf{u}_R$, the symbol "$\to$" stands for a shock wave and "$-$" stands for a rarefaction.

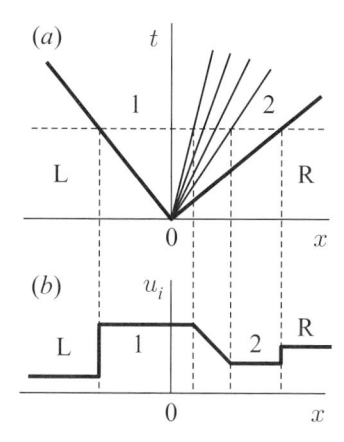

## 15.14. NONLINEAR SYSTEMS OF PARTIAL DIFFERENTIAL EQUATIONS

**Figure 15.10.** Solution for the Riemann problem: (*a*) centered waves in the $(x, t)$ plane; (*b*) the $u_i$ profile.

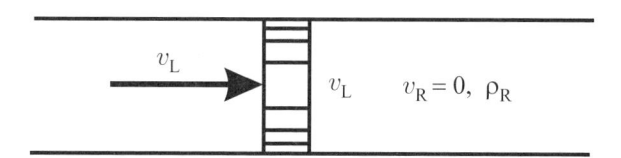

**Figure 15.11.** Constant velocity piston motion in a tube.

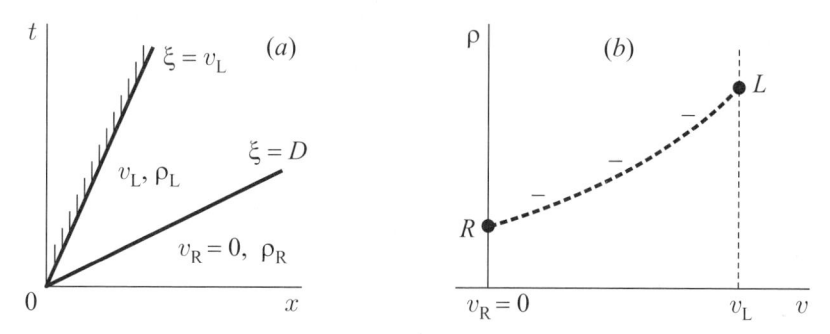

**Figure 15.12.** Solution of the piston problem: (*a*) shock wave in the $(x, t)$ plane; (*b*) shock locus in the phase plane.

Figure 15.12*b* shows the locus of points that can be connected by a shock to the point $(v_R = 0, \rho_R)$. This locus is a first-family shock. The intersection of the locus with the line $v = v_L$ defines the value $\rho_L$. Hence, $\rho_L$ can be found from the equation

$$v_L^2 = \frac{[p(\rho_L) - p(\rho_R)](\rho_L - \rho_R)}{\rho_L \rho_R}, \qquad (15.14.4.69)$$

which has been obtained by taking the square of equation (15.14.4.56). There exists a root $\rho_L$ of (15.14.4.69) such that $\rho_L > \rho_R$. Hence, the gas is compressed ahead of the piston (Fig. 15.12*a*). The expression for the shock speed can be found from the first Hugoniot condition (15.14.4.55):

$$D = \frac{\rho_L v_L}{\rho_L - \rho_R} > v_L. \qquad (15.14.4.70)$$

The shock speed exceeds the piston velocity of (15.14.4.70) for $\rho_L > \rho_R$. Both characteristics of the first family as well as the characteristic ahead of the shock from the second family arrive at the shock, so that the Lax condition is satisfied. It can be proved that there are no other configurations that satisfy the initial-boundary conditions (15.14.4.67), (15.14.4.68).

## 15.14.5. Systems of Second-Order Equations of Reaction-Diffusion Type

**15.14.5-1. Traveling-wave solutions and some other invariant solutions.**

In this subsection, we consider systems of nonlinear second-order equations

$$\frac{\partial u}{\partial t} = a_1 \frac{\partial^2 u}{\partial x^2} + F_1(u, w),$$
$$\frac{\partial w}{\partial t} = a_2 \frac{\partial^2 w}{\partial x^2} + F_2(u, w), \qquad (15.14.5.1)$$

which often arise in the theory of mass exchange of reactive media, combustion theory, mathematical biology, and biophysics.

It is apparent that systems (15.14.5.1) admit exact traveling-wave solutions

$$u = u(z), \quad w = w(z), \quad z = kx - \lambda t,$$

where $k$ and $\lambda$ are arbitrary constants, and the functions $u(z)$ and $w(z)$ are determined by the autonomous system of ordinary differential equations

$$a_1 k^2 u''_{zz} + \lambda u'_z + F_1(u, w) = 0,$$
$$a_2 k^2 w''_{zz} + \lambda w'_z + F_2(u, w) = 0.$$

Table 15.13 lists some systems of the form (15.14.5.1) admitting other invariant solutions that are found using combinations of translation and scaling transformations in the independent and dependent variables; these transformations preserve the form of the equations (i.e., a given system of equations is converted into an identical one). The procedure for constructing such solutions for a single equation is detailed in Section 15.3.

TABLE 15.13
Systems of equations of the form (15.14.5.1) admitting invariant solutions that are found using combinations of translation and scaling transformations ($C$ is an arbitrary constant)

| Systems of equations | Invariant transformations | Form of invariant solutions |
|---|---|---|
| $\frac{\partial u}{\partial t} = a\frac{\partial^2 u}{\partial x^2} + e^{\lambda u} f(\lambda u - \sigma w),$ $\frac{\partial w}{\partial t} = b\frac{\partial^2 w}{\partial x^2} + e^{\sigma w} g(\lambda u - \sigma w)$ | $t = C^{2\lambda\sigma}\bar{t},\ x = C^{\lambda\sigma}\bar{x},$ $u = \bar{u} - 2\sigma \ln C,$ $w = \bar{w} - 2\lambda \ln C$ | $u = y(\xi) - \frac{1}{\lambda}\ln t,\ w = z(\xi) - \frac{1}{\sigma}\ln t,\ \xi = \frac{x}{\sqrt{t}}$ |
| $\frac{\partial u}{\partial t} = a\frac{\partial^2 u}{\partial x^2} + u^{1+kn} f(u^n w^m),$ $\frac{\partial w}{\partial t} = b\frac{\partial^2 w}{\partial x^2} + w^{1-km} g(u^n w^m)$ | $t = C^{-kmn}\bar{t},$ $x = C^{-\frac{1}{2}kmn}\bar{x},$ $u = C^m \bar{u},\ w = C^{-n}\bar{w}$ | $u = t^{-\frac{1}{kn}} y(\xi),\ w = t^{\frac{1}{km}} z(\xi),\ \xi = \frac{x}{\sqrt{t}}$ |
| $\frac{\partial u}{\partial t} = a\frac{\partial^2 u}{\partial x^2} + uf(u^n w^m),$ $\frac{\partial w}{\partial t} = b\frac{\partial^2 w}{\partial x^2} + wg(u^n w^m)$ | $t = \bar{t} + \ln C,$ $x = \bar{x} + \lambda \ln C,$ $u = C^m \bar{u},\ w = C^{-n}\bar{w}$ | $u = e^{mt} y(\xi),\ w = e^{-nt} z(\xi),\ \xi = x - \lambda t$ |

**15.14.5-2. Generalized separable solutions.**

In some cases, nonlinear systems admit generalized separable solutions of the form

$$u = \varphi_1(t)\theta(x, t) + \psi_1(t),$$
$$w = \varphi_2(t)\theta(x, t) + \psi_2(t), \qquad (15.14.5.2)$$

where the functions $\varphi_1(t)$, $\varphi_2(t)$, $\psi_1(t)$, and $\psi_2(t)$ are selected so that both equations of system (15.14.5.1) are reduced to one and the same equation for $\theta(x, t)$.

## 15.14. Nonlinear Systems of Partial Differential Equations

**Example 1.** Consider the system

$$\frac{\partial u}{\partial t} = a\frac{\partial^2 u}{\partial x^2} + uf(bu - cw) + g_1(bu - cw),$$
$$\frac{\partial w}{\partial t} = a\frac{\partial^2 w}{\partial x^2} + wf(bu - cw) + g_2(bu - cw),$$
(15.14.5.3)

where $f(z)$, $g(z)$, and $h(z)$ are arbitrary functions.

Exact solutions are sought in the form (15.14.5.2). Let us require that the arguments of the functions appearing in the system must depend on time $t$ only, so that

$$\frac{\partial}{\partial x}(bu - cw) = [b\varphi_1(t) - c\varphi_2(t)]\frac{\partial}{\partial x}\theta(x, t) = 0.$$

It follows that

$$\varphi_1(t) = c\varphi(t), \qquad \varphi_2(t) = b\varphi(t). \tag{15.14.5.4}$$

Substituting (15.14.5.2), in view of (15.14.5.4), into (15.14.5.3) after elementary rearrangements, we obtain

$$\frac{\partial\theta}{\partial t} = a\frac{\partial^2\theta}{\partial x^2} + \left[f(b\psi_1 - c\psi_2) - \frac{\varphi'}{\varphi}\right]\theta + \frac{1}{c\varphi}\left[\psi_1 f(b\psi_1 - c\psi_2) + g_1(b\psi_1 - c\psi_2) - \psi_1'\right],$$
$$\frac{\partial\theta}{\partial t} = a\frac{\partial^2\theta}{\partial x^2} + \left[f(b\psi_1 - c\psi_2) - \frac{\varphi'}{\varphi}\right]\theta + \frac{1}{b\varphi}\left[\psi_2 f(b\psi_1 - c\psi_2) + g_2(b\psi_1 - c\psi_2) - \psi_2'\right].$$
(15.14.5.5)

For the equations of (15.14.5.5) to coincide, let us equate the expressions in square brackets to zero. As a result, we obtain an autonomous system of ordinary differential equations for $\varphi = \varphi(t)$, $\psi_1 = \psi_1(t)$, and $\psi_2 = \psi_2(t)$:

$$\varphi' = \varphi f(b\psi_1 - c\psi_2),$$
$$\psi_1' = \psi_1 f(b\psi_1 - c\psi_2) + g_1(b\psi_1 - c\psi_2),$$
$$\psi_2' = \psi_2 f(b\psi_1 - c\psi_2) + g_2(b\psi_1 - c\psi_2).$$
(15.14.5.6)

From (15.14.5.5)–(15.14.5.6) it follows that $\theta = \theta(x, t)$ satisfies the linear heat equation

$$\frac{\partial\theta}{\partial t} = a\frac{\partial^2\theta}{\partial x^2}. \tag{15.14.5.7}$$

From the first equation in (15.14.5.6) we can express the function $\varphi$ via the other two:

$$\varphi = C_1 \exp\left(\int f(b\psi_1 - c\psi_2)\,dt\right). \tag{15.14.5.8}$$

(The constant of integration $C_1$ can be set equal to 1, since $\varphi$ and $\theta$ appear in the solution in the form of a product.) As a result, we obtain an exact solution of system (15.14.5.2) in the form

$$u = \psi_1(t) + c\exp\left[\int f(b\psi_1 - c\psi_2)\,dt\right]\theta(x, t),$$
$$w = \psi_2(t) + b\exp\left[\int f(b\psi_1 - c\psi_2)\,dt\right]\theta(x, t),$$
(15.14.5.9)

where the functions $\psi_1 = \psi_1(t)$ and $\psi_2 = \psi_2(t)$ are described by the last two equations of system (15.14.5.6), and the function $\theta(x, t)$ is a solution of the linear heat equation (15.14.5.7).

**Remark.** Now we show how the functions $\varphi(t)$ and $\psi(t)$, appearing in (15.14.5.9), can be found. Multiply the second equation (15.14.5.6) by $b$ and the third by $-c$ and add up to obtain a separable first-order ordinary differential equation for the linear combination of the unknowns:

$$z_t' = zf(z) + bg_1(z) - cg_2(z), \qquad z = b\psi_1 - c\psi_2. \tag{15.14.5.10}$$

Its general solution is expressed in implicit form as

$$\int \frac{dz}{zf(z) + bg_1(z) - cg_2(z)} = t + C_2. \tag{15.14.5.11}$$

For given functions $f(z)$, $g(z)$, and $h(z)$, from (15.14.5.11) we find $z = z(t)$. Substituting it into the second equation (15.14.5.6), we get a linear equation for $\psi_1$. Solving this equation and taking into account the relation for $z$, $\psi_1$, and $\psi_2$ in (15.14.5.10), we obtain

$$\psi_1 = C_3 F(t) + F(t)\int \frac{g_1(z)}{F(t)}\,dt, \qquad \psi_2 = \frac{b}{c}\psi_1 - \frac{1}{c}z, \qquad F(t) = \exp\left[\int f(z)\,dt\right], \qquad z = z(t).$$

**Example 2.** Consider now the system

$$\frac{\partial u}{\partial t} = a\frac{\partial^2 u}{\partial x^2} + uf\left(\frac{u}{w}\right),$$

$$\frac{\partial w}{\partial t} = a\frac{\partial^2 w}{\partial x^2} + wg\left(\frac{u}{w}\right).$$

An exact solution is sought in the form (15.14.5.2) with $\varphi_2(t) = 1$ and $\psi_1(t) = \psi_2(t) = 0$. After simple rearrangements we get

$$u = \varphi(t)\exp\left[\int g(\varphi(t))\,dt\right]\theta(x,t), \quad w = \exp\left[\int g(\varphi(t))\,dt\right]\theta(x,t),$$

where the function $\varphi = \varphi(t)$ is described by the separable first-order nonlinear ordinary differential equation

$$\varphi'_t = [f(\varphi) - g(\varphi)]\varphi, \qquad (15.14.5.12)$$

and the function $\theta = \theta(x,t)$ satisfies the linear heat equation (15.14.5.7).

The general solution of equation (15.14.5.12) is written out in implicit form as

$$\int \frac{d\varphi}{[f(\varphi) - g(\varphi)]\varphi} = t + C.$$

▶ Exact solutions to some other second-order nonlinear systems of the form (15.14.5.1) can be found in Subsection T10.3.1.

# References for Chapter 15

**Ablowitz, M. J. and Clarkson, P. A.**, *Solitons, Non-Linear Evolution Equations and Inverse Scattering*, Cambridge University Press, Cambridge, 1991.

**Ablowitz, M. J. and Segur, H.**, *Solitons and the Inverse Scattering Transform*, Society for Industrial and Applied Mathematics, Philadelphia, 1981.

**Ames, W. F.**, *Nonlinear Partial Differential Equations in Engineering*, Vols. 1 and 2, Academic Press, New York, 1967 and 1972.

**Andreev, V. K., Kaptsov, O. V., Pukhnachov, V. V., and Rodionov, A. A.**, *Applications of Group-Theoretical Methods in Hydrodynamics*, Kluwer Academic, Dordrecht, 1998.

**Barenblatt, G. I.**, *Dimensional Analysis*, Gordon & Breach, New York, 1989.

**Bluman, G. W. and Cole, J. D.**, *Similarity Methods for Differential Equations*, Springer-Verlag, New York, 1974.

**Bluman, G. W. and Cole, J. D.**, The general similarity solution of the heat equation, *J. Math. Mech.*, Vol. 18, pp. 1025–1042, 1969.

**Bluman, G. W. and Kumei, S.**, *Symmetries and Differential Equations*, Springer-Verlag, New York, 1989.

**Bullough, R. K. and Caudrey, P. J. (Editors)**, *Solitons*, Springer-Verlag, Berlin, 1980.

**Burde, G. I.**, The construction of special explicit solutions of the boundary-layer equations. Steady flows, *Quart. J. Mech. Appl. Math.*, Vol. 47, No. 2, pp. 247–260, 1994.

**Calogero, F. and Degasperis, A.**, *Spectral Transform and Solitons: Tolls to Solve and Investigate Nonlinear Evolution Equations*, North Holland, Amsterdam, 1982.

**Clarkson, P. A. and Kruskal, M. D.**, New similarity reductions of the Boussinesq equation, *J. Math. Phys.*, Vol. 30, No. 10, pp. 2201–2213, 1989.

**Conte, R. (Editor)**, *The Painlevé Property. One Century Later*, Springer-Verlag, New York, 1999.

**Courant, R. and Hilbert, D.**, *Methods of Mathematical Physics*, Vol. 2, Wiley-Interscience, New York, 1989.

**Faddeev, L. D. and Takhtajan, L. A.**, *Hamiltonian Methods in the Theory of Solitons*, Springer-Verlag, Berlin, 1987.

**Fushchich, W. I., Shtelen, W. M., and Serov, N. I.**, *Symmetry Analysis and Exact Solutions of the Equations of Mathematical Physics*, Kluwer Academic, Dordrecht, 1993.

**Galaktionov, V. A.**, Invariant subspaces and new explicit solutions to evolution equations with quadratic nonlinearities, *Proc. Roy. Soc. Edinburgh*, Vol. 125A, No. 2, pp. 225–448, 1995.

**Galaktionov, V. A.**, On new exact blow-up solutions for nonlinear heat conduction equations with source and applications, *Differential and Integral Equations*, Vol. 3, No. 5, pp. 863–874, 1990.

**Gardner, C. S., Greene, J. M., Kruskal, M. D., and Miura, R. M.**, Method for solving the Korteweg–de Vries equation, *Phys. Rev. Lett.*, Vol. 19, No. 19, pp. 1095–1097, 1967.

**Ibragimov, N. H. (Editor),** *CRC Handbook of Lie Group Analysis of Differential Equations, Vols 1–3*, CRC Press, Boca Raton, 1994–1996.

**Jimbo, M., Kruskal, M. D., and Miwa, T.,** Painlevé test for the self-dual Yang–Mills equation, *Phys. Lett.,* Ser. A, Vol. 92, No. 2, pp. 59–60, 1982.

**Kudryashov, N. A.,** *Analytical Theory of Nonlinear Differential Equations* [in Russian], Institut kompjuternyh issledovanii, Moscow–Izhevsk, 2004.

**Loitsyanskiy, L. G.,** *Mechanics of Liquids and Gases*, Begell House, New York, 1996.

**Novikov, S. P., Manakov, S. V., Pitaevskii, L. B., and Zakharov, V. E.,** *Theory of Solitons. The Inverse Scattering Method*, Plenum Press, New York, 1984.

**Olver, P. J.,** *Application of Lie Groups to Differential Equations, 2nd Edition*, Springer-Verlag, New York, 1993.

**Ovsiannikov, L. V.,** *Group Analysis of Differential Equations*, Academic Press, New York, 1982.

**Polyanin, A. D. and Zaitsev, V. F.,** *Handbook of Exact Solutions for Ordinary Differential Equations, 2nd Edition*, Chapman & Hall/CRC Press, Boca Raton, 2003.

**Polyanin, A. D. and Zaitsev, V. F.,** *Handbook of Nonlinear Partial Differential Equations*, Chapman & Hall/CRC Press, Boca Raton, 2004.

**Polyanin, A. D., Zaitsev, V. F., and Zhurov, A. I.,** *Methods for the Solution of Nonlinear Equations of Mathematical Physics and Mechanics* [in Russian], Fizmatlit, Moscow, 2005.

**Rogers, C. and Shadwick, W. F.,** *Bäcklund Transformations and Their Applications*, Academic Press, New York, 1982.

**Rozhdestvenskii, B. L. and Yanenko, N. N.,** *Systems of Quasilinear Equations and Their Applications to Gas Dynamics*, American Mathematical Society, Providence, Rhode Island, 1983.

**Schlichting, H.,** *Boundary Layer Theory*, McGraw-Hill, New York, 1981.

**Sedov, L. I.,** *Similarity and Dimensional Methods in Mechanics*, CRC Press, Boca Raton, 1993.

**Sidorov, A. F., Shapeev, V. P., and Yanenko, N. N.,** *Method of Differential Constraints and its Applications in Gas Dynamics* [in Russian], Nauka Publishers, Novosibirsk, 1984.

**Tabor, M.,** *Integrability in Nonlinear Dynamics: An Introduction*, Wiley-Interscience, New York, 1989.

**Vinogradov, A. M.,** Local symmetries and conservation laws, *Acta Appl. Math.,* Vol. 2, No. 7, pp. 21–78, 1984.

**Weiss, J.,** The Painlevé property for partial differential equations. II: Bäcklund transformation, Lax pairs, and the Schwarzian derivative, *J. Math. Phys.,* Vol. 24, No. 6, pp. 1405–1413, 1983.

**Weiss, J., Tabor, M., and Carnevalle, G.,** The Painlevé property for partial differential equations, *J. Math. Phys.,* Vol. 24, No. 3, pp. 522–526, 1983.

**Whitham, G. B.,** *Linear and Nonlinear Waves*, Wiley, New York, 1974.

**Zwillinger, D.,** *Handbook of Differential Equations, 3rd Edition*, Academic Press, San Diego, 1998.

# Chapter 16
# Integral Equations

## 16.1. Linear Integral Equations of the First Kind with Variable Integration Limit

### 16.1.1. Volterra Equations of the First Kind

**16.1.1-1. Some definitions. Function and kernel classes.**

A Volterra linear integral equation of the first kind has the general form

$$\int_a^x K(x,t)y(t)\,dt = f(x), \qquad (16.1.1.1)$$

where $y(x)$ is the unknown function ($a \leq x \leq b$), $K(x,t)$ is the kernel of the integral equation, and $f(x)$ is a given function, the *right-hand side* of equation (16.1.1.1). The functions $y(x)$ and $f(x)$ are usually assumed to be continuous or square integrable on $[a,b]$. The kernel $K(x,t)$ is usually assumed either to be continuous on the square $S = \{a \leq x \leq b, a \leq t \leq b\}$ or to satisfy the condition

$$\int_a^b \int_a^b K^2(x,t)\,dx\,dt = B^2 < \infty, \qquad (16.1.1.2)$$

where $B$ is a constant, that is, to be square integrable on this square. It is assumed in (16.1.1.2) that $K(x,t) \equiv 0$ for $t > x$.

The kernel $K(x,t)$ is said to be *degenerate* if it can be represented in the form $K(x,t) = g_1(x)h_1(t) + \cdots + g_n(x)h_n(t)$. The kernel $K(x,t)$ of an integral equation is called *difference kernel* if it depends only on the difference of the arguments, $K(x,t) = K(x-t)$.

Polar kernels

$$K(x,t) = L(x,t)(x-t)^{-\beta} + M(x,t), \qquad 0 < \beta < 1, \qquad (16.1.1.3)$$

and logarithmic kernels (kernels with logarithmic singularity)

$$K(x,t) = L(x,t)\ln(x-t) + M(x,t), \qquad (16.1.1.4)$$

where the functions $L(x,t)$ and $M(x,t)$ are continuous on $S$ and $L(x,x) \not\equiv 0$, are often considered as well.

Polar and logarithmic kernels form a class of kernels with weak singularity. Equations containing such kernels are called *equations with weak singularity*.

In case the functions $K(x,t)$ and $f(x)$ are continuous, the right-hand side of equation (16.1.1.1) must satisfy the following conditions:

1°. If $K(a,a) \neq 0$, then $f(x)$ must be constrained by $f(a) = 0$.

2°. If $K(a,a) = K'_x(a,a) = \cdots = K_x^{(n-1)}(a,a) = 0$, $0 < |K_x^{(n)}(a,a)| < \infty$, then the right-hand side of the equation must satisfy the conditions $f(a) = f'_x(a) = \cdots = f_x^{(n)}(a) = 0$.

3°. If $K(a,a) = K'_x(a,a) = \cdots = K_x^{(n-1)}(a,a) = 0$, $K_x^{(n)}(a,a) = \infty$, then the right-hand side of the equation must satisfy the conditions $f(a) = f'_x(a) = \cdots = f_x^{(n-1)}(a) = 0$.

For polar kernels of the form (16.1.1.4) and continuous $f(x)$, no additional conditions are imposed on the right-hand side of the integral equation.

16.1.1-2. Existence and uniqueness of a solution.

Assume that in equation (16.1.1.1) the functions $f(x)$ and $K(x,t)$ are continuous together with their first derivatives on $[a,b]$ and on $S$, respectively. If $K(x,x) \neq 0$ ($x \in [a,b]$) and $f(a) = 0$, then there exists a unique continuous solution $y(x)$ of equation (16.1.1.1).

Remark. A Volterra equation of the first kind can be treated as a Fredholm equation of the first kind whose kernel $K(x,t)$ vanishes for $t > x$ (see Section 16.3).

## 16.1.2. Equations with Degenerate Kernel: $K(x,t) = g_1(x)h_1(t) + \cdots + g_n(x)h_n(t)$

16.1.2-1. Equations with kernel of the form $K(x,t) = g_1(x)h_1(t) + g_2(x)h_2(t)$.

Any equation of this type can be rewritten in the form

$$g_1(x) \int_a^x h_1(t)y(t)\,dt + g_2(x) \int_a^x h_2(t)y(t)\,dt = f(x). \tag{16.1.2.1}$$

It is assumed that $g_1(x)/g_2(x) \neq \text{const}$, $h_1(t)/h_2(t) \neq \text{const}$, $0 < g_1^2(a) + g_2^2(a) < \infty$, and $f(a) = 0$.

The change of variables

$$u(x) = \int_a^x h_1(t)y(t)\,dt, \tag{16.1.2.2}$$

followed by the integration by parts in the second integral in (16.1.2.1) with regard to the relation $u(a) = 0$, yields the following Volterra equation of the second kind:

$$[g_1(x)h_1(x) + g_2(x)h_2(x)]u(x) - g_2(x)h_1(x)\int_a^x \left[\frac{h_2(t)}{h_1(t)}\right]'_t u(t)\,dt = h_1(x)f(x). \tag{16.1.2.3}$$

The substitution

$$w(x) = \int_a^x \left[\frac{h_2(t)}{h_1(t)}\right]'_t u(t)\,dt \tag{16.1.2.4}$$

reduces equation (16.1.2.3) to the first-order linear ordinary differential equation

$$[g_1(x)h_1(x) + g_2(x)h_2(x)]w'_x - g_2(x)h_1(x)\left[\frac{h_2(x)}{h_1(x)}\right]'_x w = f(x)h_1(x)\left[\frac{h_2(x)}{h_1(x)}\right]'_x. \tag{16.1.2.5}$$

1°. In the case $g_1(x)h_1(x) + g_2(x)h_2(x) \not\equiv 0$, the solution of equation (16.1.2.5) satisfying the condition $w(a) = 0$ [this condition is a consequence of the substitution (16.1.2.4)] has the form

$$w(x) = \Phi(x) \int_a^x \left[\frac{h_2(t)}{h_1(t)}\right]'_t \frac{f(t)h_1(t)\,dt}{\Phi(t)[g_1(t)h_1(t) + g_2(t)h_2(t)]}, \tag{16.1.2.6}$$

$$\Phi(x) = \exp\left\{\int_a^x \left[\frac{h_2(t)}{h_1(t)}\right]'_t \frac{g_2(t)h_1(t)\,dt}{g_1(t)h_1(t) + g_2(t)h_2(t)}\right\}. \tag{16.1.2.7}$$

Let us differentiate relation (16.1.2.4) and substitute the function (16.1.2.6) into the resulting expression. After integrating by parts with regard to the relations $f(a) = 0$ and $w(a) = 0$, for $f \not\equiv \text{const } g_2$ we obtain

$$u(x) = \frac{g_2(x)h_1(x)\Phi(x)}{g_1(x)h_1(x) + g_2(x)h_2(x)} \int_a^x \left[\frac{f(t)}{g_2(t)}\right]'_t \frac{dt}{\Phi(t)}.$$

Using formula (16.1.2.2), we find a solution of the original equation in the form

$$y(x) = \frac{1}{h_1(x)} \frac{d}{dx}\left\{\frac{g_2(x)h_1(x)\Phi(x)}{g_1(x)h_1(x) + g_2(x)h_2(x)} \int_a^x \left[\frac{f(t)}{g_2(t)}\right]'_t \frac{dt}{\Phi(t)}\right\}, \tag{16.1.2.8}$$

where the function $\Phi(x)$ is given by (16.1.2.7).

If $f(x) \equiv \text{const } g_2(x)$, the solution is given by formulas (16.1.2.8) and (16.1.2.7), in which the subscript 1 must be changed by 2 and vice versa.

$2^\circ$. In the case $g_1(x)h_1(x) + g_2(x)h_2(x) \equiv 0$, the solution has the form

$$y(x) = \frac{1}{h_1}\frac{d}{dx}\left[\frac{(f/g_2)'_x}{(g_1/g_2)'_x}\right] = -\frac{1}{h_1}\frac{d}{dx}\left[\frac{(f/g_2)'_x}{(h_2/h_1)'_x}\right].$$

**16.1.2-2. Equations with general degenerate kernel.**

A Volterra equation of the first kind with general degenerate kernel has the form

$$\sum_{m=1}^n g_m(x) \int_a^x h_m(t)y(t)\,dt = f(x). \tag{16.1.2.9}$$

Using the notation

$$w_m(x) = \int_a^x h_m(t)y(t)\,dt, \qquad m = 1,\ldots,n, \tag{16.1.2.10}$$

we can rewrite equation (16.1.2.9) as follows:

$$\sum_{m=1}^n g_m(x)w_m(x) = f(x). \tag{16.1.2.11}$$

On differentiating formulas (16.1.2.10) and eliminating $y(x)$ from the resulting equations, we arrive at the following linear differential equations for the functions $w_m = w_m(x)$:

$$h_1(x)w'_m = h_m(x)w'_1, \qquad m = 2,\ldots,n \tag{16.1.2.12}$$

(the prime stands for the derivative with respect to $x$) with the initial conditions
$$w_m(a) = 0, \qquad m = 1, \ldots, n.$$
Any solution of system (16.1.2.11), (16.1.2.12) determines a solution of the original integral equation (16.1.2.9) by each of the expressions
$$y(x) = \frac{w'_m(x)}{h_m(x)}, \qquad m = 1, \ldots, n,$$
which can be obtained by differentiating formula (16.1.2.10).

System (16.1.2.11), (16.1.2.12) can be reduced to a linear $(n-1)$st-order differential equation for any function $w_m(x)$ $(m = 1, \ldots, n)$ by multiple differentiation of equation (16.1.2.11) with regard to (16.1.2.12).

### 16.1.3. Equations with Difference Kernel: $K(x,t) = K(x-t)$

#### 16.1.3-1. Solution method based on the Laplace transform.

Volterra equations of the first kind with kernel depending on the difference of the arguments have the form
$$\int_0^x K(x-t)y(t)\,dt = f(x). \qquad (16.1.3.1)$$

To solve these equations, the Laplace transform can be used (see Section 11.2). In what follows we need the transforms of the kernel and the right-hand side; they are given by the formulas
$$\widetilde{K}(p) = \int_0^\infty K(x)e^{-px}\,dx, \quad \widetilde{f}(p) = \int_0^\infty f(x)e^{-px}\,dx. \qquad (16.1.3.2)$$

Applying the Laplace transform $\mathfrak{L}$ to equation (16.1.3.1) and taking into account the fact that an integral with kernel depending on the difference of the arguments is transformed to the product by the rule (see Paragraph 11.2.2-1)
$$\mathfrak{L}\left\{\int_0^x K(x-t)y(t)\,dt\right\} = \widetilde{K}(p)\widetilde{y}(p),$$
we obtain the following equation for the transform $\widetilde{y}(p)$:
$$\widetilde{K}(p)\widetilde{y}(p) = \widetilde{f}(p). \qquad (16.1.3.3)$$

The solution of equation (16.1.3.3) is given by the formula
$$\widetilde{y}(p) = \frac{\widetilde{f}(p)}{\widetilde{K}(p)}. \qquad (16.1.3.4)$$

On applying the Laplace inversion formula (if it is applicable) to (16.1.3.4), we obtain a solution of equation (16.1.3.1) in the form
$$y(x) = \frac{1}{2\pi i}\int_{c-i\infty}^{c+i\infty} \frac{\widetilde{f}(p)}{\widetilde{K}(p)}e^{px}\,dp. \qquad (16.1.3.5)$$

To evaluate the corresponding integrals, tables of direct and inverse Laplace transforms can be applied (see Sections T3.1 and T3.2), and, in many cases, to find the inverse transform, methods of the theory of functions of a complex variable are applied, including formulas for the calculation of residues and the Jordan lemma (see Subsection 11.1.2).

**16.1.3-2. Case in which the transform of the solution is a rational function.**

Consider the important special case in which the transform (16.1.3.4) of the solution is a rational function of the form

$$\widetilde{y}(p) = \frac{\widetilde{f}(p)}{\widetilde{K}(p)} \equiv \frac{R(p)}{Q(p)},$$

where $Q(p)$ and $R(p)$ are polynomials in the variable $p$ and the degree of $Q(p)$ exceeds that of $R(p)$.

If the zeros of the denominator $Q(p)$ are simple, i.e.,

$$Q(p) \equiv \text{const}\,(p - \lambda_1)(p - \lambda_2)\dots(p - \lambda_n),$$

and $\lambda_i \ne \lambda_j$ for $i \ne j$, then the solution has the form

$$y(x) = \sum_{k=1}^{n} \frac{R(\lambda_k)}{Q'(\lambda_k)} \exp(\lambda_k x),$$

where the prime stands for the derivatives.

**Example 1.** Consider the Volterra integral equation of the first kind

$$\int_0^x e^{-a(x-t)} y(t)\, dt = A \sinh(bx).$$

We apply the Laplace transform to this equation and obtain (see Subsections T3.1.1 and T3.1.4)

$$\frac{1}{p+a}\widetilde{y}(p) = \frac{Ab}{p^2 - b^2}.$$

This implies

$$\widetilde{y}(p) = \frac{Ab(p+a)}{p^2 - b^2} = \frac{Ab(p+a)}{(p-b)(p+b)}.$$

We have $Q(p) = (p-b)(p+b)$, $R(p) = Ab(p+a)$, $\lambda_1 = b$, and $\lambda_2 = -b$. Therefore, the solution of the integral equation has the form

$$y(x) = \tfrac{1}{2} A(b+a)e^{bx} + \tfrac{1}{2} A(b-a)e^{-bx} = Aa\sinh(bx) + Ab\cosh(bx).$$

**16.1.3-3. Convolution representation of a solution.**

In solving Volterra integral equations of the first kind with difference kernel $K(x-t)$ by means of the Laplace transform, it is sometimes useful to apply the following approach.

Let us represent the transform (16.1.3.4) of a solution in the form

$$\widetilde{y}(p) = \widetilde{N}(p)\widetilde{M}(p)\widetilde{f}(p), \qquad \widetilde{N}(p) \equiv \frac{1}{\widetilde{K}(p)\widetilde{M}(p)}. \tag{16.1.3.6}$$

If we can find a function $\widetilde{M}(p)$ for which the inverse transforms

$$\mathcal{L}^{-1}\{\widetilde{M}(p)\} = M(x), \qquad \mathcal{L}^{-1}\{\widetilde{N}(p)\} = N(x) \tag{16.1.3.7}$$

exist and can be found in a closed form, then the solution can be written as the convolution

$$y(x) = \int_0^x N(x-t) F(t)\, dt, \qquad F(t) = \int_0^t M(t-s) f(s)\, ds. \tag{16.1.3.8}$$

**Example 2.** Consider the equation

$$\int_0^x \sin(k\sqrt{x-t}\,) y(t)\,dt = f(x), \qquad f(0) = 0. \tag{16.1.3.9}$$

Applying the Laplace transform, we obtain (see Subsections T3.1.1 and T3.1.6)

$$\tilde{y}(p) = \frac{2}{\sqrt{\pi}\,k} p^{3/2} \exp(\alpha/p) \tilde{f}(p), \qquad \alpha = \tfrac{1}{4}k^2. \tag{16.1.3.10}$$

Let us rewrite the right-hand side of (16.1.3.10) in the equivalent form

$$\tilde{y}(p) = \frac{2}{\sqrt{\pi}\,k} p^2 \left[ p^{-1/2} \exp(\alpha/p) \right] \tilde{f}(p), \qquad \alpha = \tfrac{1}{4}k^2, \tag{16.1.3.11}$$

where the factor in the square brackets corresponds to $\widetilde{M}(p)$ in formula (16.1.3.6) and $\widetilde{N}(p) = \text{const}\,p^2$.

By applying the Laplace inversion formula according to the above scheme to formula (16.1.3.11) with regard to the relation (see Subsections T3.2.1 and T3.2.5)

$$\mathcal{L}^{-1}\{p^2\tilde{\varphi}(p)\} = \frac{d^2}{dx^2}\varphi(x), \qquad \mathcal{L}^{-1}\{p^{-1/2}\exp(\alpha/p)\} = \frac{1}{\sqrt{\pi x}}\cosh(k\sqrt{x}\,),$$

we find the solution

$$y(x) = \frac{2}{\pi k}\frac{d^2}{dx^2}\int_0^x \frac{\cosh(k\sqrt{x-t}\,)}{\sqrt{x-t}} f(t)\,dt.$$

### 16.1.3-4. Application of an auxiliary equation.

Consider the equation

$$\int_a^x K(x-t) y(t)\,dt = f(x), \tag{16.1.3.12}$$

where the kernel $K(x)$ has an integrable singularity at $x = 0$.

Let $w = w(x)$ be the solution of the simpler auxiliary equation with $f(x) \equiv 1$ and $a = 0$,

$$\int_0^x K(x-t) w(t)\,dt = 1. \tag{16.1.3.13}$$

Then the solution of the original equation (16.1.3.12) with arbitrary right-hand side can be expressed as follows via the solution of the auxiliary equation (16.1.3.13):

$$y(x) = \frac{d}{dx}\int_a^x w(x-t) f(t)\,dt = f(a) w(x-a) + \int_a^x w(x-t) f'_t(t)\,dt. \tag{16.1.3.14}$$

**Example 3.** Consider the generalized Abel equation

$$\int_a^x \frac{y(t)\,dt}{(x-t)^\mu} = f(x), \qquad 0 < \mu < 1. \tag{16.1.3.15}$$

We seek a solution of the corresponding auxiliary equation

$$\int_0^x \frac{w(t)\,dt}{(x-t)^\mu} = 1, \qquad 0 < \mu < 1, \tag{16.1.3.16}$$

by the method of indeterminate coefficients in the form

$$w(x) = Ax^\beta. \tag{16.1.3.17}$$

Let us substitute (16.1.3.17) into (16.1.3.16) and then perform the change of variable $t = x\xi$ in the integral. Taking into account the relationship

$$B(p,q) = \int_0^1 \xi^{p-1}(1-\xi)^{1-q}\,d\xi = \frac{\Gamma(p)\Gamma(q)}{\Gamma(p+q)}$$

between the beta and gamma functions, we obtain
$$A \frac{\Gamma(\beta+1)\Gamma(1-\mu)}{\Gamma(2+\beta-\mu)} x^{\beta+1-\mu} = 1.$$
From this relation we find the coefficients $A$ and $\beta$:
$$\beta = \mu - 1, \quad A = \frac{1}{\Gamma(\mu)\Gamma(1-\mu)} = \frac{\sin(\pi\mu)}{\pi}. \tag{16.1.3.18}$$
Formulas (16.1.3.17) and (16.1.3.18) define the solution of the auxiliary equation (16.1.3.16) and make it possible to find the solution of the generalized Abel equation (16.1.3.15) by means of formula (16.1.3.14) as follows:
$$y(x) = \frac{\sin(\pi\mu)}{\pi} \frac{d}{dx} \int_a^x \frac{f(t)\,dt}{(x-t)^{1-\mu}} = \frac{\sin(\pi\mu)}{\pi} \left[ \frac{f(a)}{(x-a)^{1-\mu}} + \int_a^x \frac{f'_t(t)\,dt}{(x-t)^{1-\mu}} \right].$$

## 16.1.4. Reduction of Volterra Equations of the First Kind to Volterra Equations of the Second Kind

**16.1.4-1. First method.**

Suppose that the kernel and the right-hand side of the equation
$$\int_a^x K(x,t) y(t)\,dt = f(x) \tag{16.1.4.1}$$
have continuous derivatives with respect to $x$ and that the condition $K(x,x) \ne 0$ holds. In this case, after differentiating relation (16.1.4.1) and dividing the resulting expression by $K(x,x)$, we arrive at the following Volterra equation of the second kind:
$$y(x) + \int_a^x \frac{K'_x(x,t)}{K(x,x)} y(t)\,dt = \frac{f'_x(x)}{K(x,x)}. \tag{16.1.4.2}$$
Equations of this type are considered in Section 16.2.

If $K(x,x) \equiv 0$, then, on differentiating equation (16.1.4.1) with respect to $x$ twice and assuming that $K'_x(x,t)|_{t=x} \ne 0$, we obtain the Volterra equation of the second kind
$$y(x) + \int_a^x \frac{K''_{xx}(x,t)}{K'_x(x,t)|_{t=x}} y(t)\,dt = \frac{f''_{xx}(x)}{K'_x(x,t)|_{t=x}}.$$
If $K'_x(x,x) \equiv 0$, we can again apply differentiation, and so on.

**16.1.4-2. Second method.**

Let us introduce the new variable
$$Y(x) = \int_a^x y(t)\,dt$$
and integrate the right-hand side of equation (16.1.4.1) by parts taking into account the relation $f(a) = 0$. After dividing the resulting expression by $K(x,x)$, we arrive at the Volterra equation of the second kind
$$Y(x) - \int_a^x \frac{K'_t(x,t)}{K(x,x)} Y(t)\,dt = \frac{f(x)}{K(x,x)},$$
for which the condition $K(x,x) \ne 0$ must hold.

## 16.1.5. Method of Quadratures

**16.1.5-1. Quadrature formulas.**

The *method of quadratures* is a method for constructing an approximate solution of an integral equation based on the replacement of integrals by finite sums according to some formula. Such formulas are called *quadrature formulas* and, in general, have the form

$$\int_a^b \psi(x)\,dx = \sum_{i=1}^n A_i \psi(x_i) + \varepsilon_n[\psi], \tag{16.1.5.1}$$

where $x_i$ ($i = 1, \ldots, n$) are the abscissas of the partition points of the integration interval $[a, b]$, or *quadrature (interpolation) nodes*, $A_i$ ($i = 1, \ldots, n$) are numerical coefficients independent of the choice of the function $\psi(x)$, and $\varepsilon_n[\psi]$ is the remainder (the truncation error) of formula (16.1.5.1). As a rule, $A_i \geq 0$ and $\sum_{i=1}^n A_i = b - a$.

There are quite a few quadrature formulas of the form (16.1.5.1). The following formulas are the simplest and most frequently used in practice.

*Rectangle rule*:

$$A_1 = A_2 = \cdots = A_{n-1} = h, \quad A_n = 0,$$
$$h = \frac{b-a}{n-1}, \quad x_i = a + h(i-1) \quad (i = 1, 2, \ldots, n). \tag{16.1.5.2}$$

*Trapezoidal rule*:

$$A_1 = A_n = \tfrac{1}{2}h, \quad A_2 = A_3 = \cdots = A_{n-1} = h,$$
$$h = \frac{b-a}{n-1}, \quad x_i = a + h(i-1) \quad (i = 1, 2, \ldots, n). \tag{16.1.5.3}$$

*Simpson's rule* (or *prismoidal formula*):

$$A_1 = A_{2m+1} = \tfrac{1}{3}h, \quad A_2 = \cdots = A_{2m} = \tfrac{4}{3}h, \quad A_3 = \cdots = A_{2m-1} = \tfrac{2}{3}h,$$
$$h = \frac{b-a}{n-1}, \quad x_i = a + h(i-1) \quad (n = 2m+1, \ i = 1, \ldots, n), \tag{16.1.5.4}$$

where $m$ is a positive integer.

In formulas (16.1.5.2)–(16.1.5.4), $h$ is a constant integration step.

The quadrature formulas due to Chebyshev and Gauss with various numbers of interpolation nodes are also widely applied. Let us illustrate these formulas by an example.

**Example.** For the interval $[-1, 1]$, the parameters in formula (16.1.5.1) acquire the following values:

*Chebyshev's formula* ($n = 6$):

$$A_1 = A_2 = \cdots = \frac{2}{n} = \frac{1}{3}, \quad x_1 = -x_6 = -0.8662468181,$$
$$x_2 = -x_5 = -0.4225186538, \quad x_3 = -x_4 = -0.2666354015. \tag{16.1.5.5}$$

*Gauss' formula* ($n = 7$):

$$A_1 = A_7 = 0.1294849662, \quad A_2 = A_6 = 0.2797053915,$$
$$A_3 = A_5 = 0.3818300505, \quad A_4 = 0.4179591837,$$
$$x_1 = -x_7 = -0.9491079123, \quad x_2 = -x_6 = -0.7415311856,$$
$$x_3 = -x_5 = -0.4058451514, \quad x_4 = 0. \tag{16.1.5.6}$$

Note that a vast literature is devoted to quadrature formulas, and the reader can find books of interest [e.g., see Bakhvalov (1973), Korn and Korn (2000)].

**16.1.5-2. General scheme of the method.**

Let us solve the Volterra integral equation of the first kind

$$\int_a^x K(x,t)y(t)\,dt = f(x), \qquad f(a) = 0, \tag{16.1.5.7}$$

on an interval $a \leq x \leq b$ by the method of quadratures. The procedure of constructing the solution involves two stages:

1°. First, we determine the initial value $y(a)$. To this end, we differentiate equation (16.1.5.7) with respect to $x$, thus obtaining

$$K(x,x)y(x) + \int_a^x K'_x(x,t)y(t)\,dt = f'_x(x).$$

By setting $x = a$, we find that

$$y_1 = y(a) = \frac{f'_x(a)}{K(a,a)} = \frac{f'_x(a)}{K_{11}}.$$

2°. Let us choose a constant integration step $h$ and consider the discrete set of points $x_i = a + h(i-1)$, $i = 1, \ldots, n$. For $x = x_i$, equation (16.1.5.7) acquires the form

$$\int_a^{x_i} K(x_i, t)y(t)\,dt = f(x_i), \qquad i = 2, \ldots, n. \tag{16.1.5.8}$$

Applying the quadrature formula (16.1.5.1) to the integral in (16.1.5.8) and choosing $x_j$ ($j = 1, \ldots, i$) to be the nodes in $t$, we arrive at the system of equations

$$\sum_{j=1}^{i} A_{ij} K(x_i, x_j) y(x_j) = f(x_i) + \varepsilon_i[y], \qquad i = 2, \ldots, n, \tag{16.1.5.9}$$

where $A_{ij}$ are the coefficients of the quadrature formula on the interval $[a, x_i]$ and $\varepsilon_i[y]$ is the truncation error. Assume that $\varepsilon_i[y]$ are small and neglect them; then we obtain a system of linear algebraic equations in the form

$$\sum_{j=1}^{i} A_{ij} K_{ij} y_j = f_i, \qquad i = 2, \ldots, n, \tag{16.1.5.10}$$

where $K_{ij} = K(x_i, x_j)$ ($j = 1, \ldots, i$), $f_i = f(x_i)$, and $y_j$ are approximate values of the unknown function at the nodes $x_i$.

Now system (16.1.5.10) permits one, provided that $A_{ii} K_{ii} \neq 0$ ($i = 2, \ldots, n$), to successively find the desired approximate values by the formulas

$$y_1 = \frac{f'_x(a)}{K_{11}}, \quad y_2 = \frac{f_2 - A_{21}K_{21}y_1}{A_{22}K_{22}}, \quad \ldots, \quad y_n = \frac{f_n - \sum_{j=1}^{n-1} A_{nj}K_{nj}y_j}{A_{nn}K_{nn}},$$

whose specific form depends on the choice of the quadrature formula.

### 16.1.5-3. Algorithm based on the trapezoidal rule.

According to the trapezoidal rule (16.1.5.3), we have

$$A_{i1} = A_{ii} = \tfrac{1}{2}h, \quad A_{i2} = \cdots = A_{i,i-1} = h, \quad i = 2,\ldots,n.$$

The application of the trapezoidal rule in the general scheme leads to the following step algorithm:

$$y_1 = \frac{f'_x(a)}{K_{11}}, \quad f'_x(a) = \frac{-3f_1 + 4f_2 - f_3}{2h},$$

$$y_i = \frac{2}{K_{ii}}\left(\frac{f_i}{h} - \sum_{j=1}^{i-1}\beta_j K_{ij} y_j\right), \quad \beta_j = \begin{cases} \tfrac{1}{2} & \text{for } j = 1, \\ 1 & \text{for } j > 1, \end{cases} \quad i = 2,\ldots,n,$$

where the notation coincides with that introduced in Paragraph 16.1.5-2. The trapezoidal rule is quite simple and effective and frequently used in practice for solving integral equations with variable limit of integration.

## 16.2. Linear Integral Equations of the Second Kind with Variable Integration Limit

### 16.2.1. Volterra Equations of the Second Kind

#### 16.2.1-1. Some definitions. Equations for the resolvent.

A Volterra linear integral equation of the second kind has the general form

$$y(x) - \int_a^x K(x,t) y(t)\,dt = f(x), \tag{16.2.1.1}$$

where $y(x)$ is the unknown function ($a \le x \le b$), $K(x,t)$ is the kernel of the integral equation, and $f(x)$ is the *right-hand side* of the integral equation. The function classes to which $y(x)$, $f(x)$, and $K(x,t)$ can belong are defined in Paragraph 16.1.1-1. In these function classes, there exists a unique solution of the Volterra integral equation of the second kind.

Equation (16.2.1.1) is said to be *homogeneous* if $f(x) \equiv 0$ and *nonhomogeneous* otherwise.

The kernel $K(x,t)$ is said to be *degenerate* if it can be represented in the form $K(x,t) = g_1(x)h_1(t) + \cdots + g_n(x)h_n(t)$. The kernel $K(x,t)$ of an integral equation is called *difference kernel* if it depends only on the difference of the arguments, $K(x,t) = K(x-t)$.

**Remark 1.** A homogeneous Volterra integral equation of the second kind has only the trivial solution.

**Remark 2.** A Volterra equation of the second kind can be regarded as a Fredholm equation of the second kind whose kernel $K(x,t)$ vanishes for $t > x$ (see Section 16.4).

#### 16.2.1-2. Structure of the solution. The resolvent.

The solution of equation (16.2.1.1) can be presented in the form

$$y(x) = f(x) + \int_a^x R(x,t) f(t)\,dt, \tag{16.2.1.2}$$

where the resolvent $R(x,t)$ is independent of $f(x)$ and the lower limit of integration $a$; it is determined by the kernel of the integral equation alone.

#### 16.2.1-3. Relationship between solutions of some integral equations.

Let us present two useful formulas that express the solution of one integral equation via the solutions of other integral equations.

1°. Assume that the Volterra equation of the second kind with kernel $K(x,t)$ has a resolvent $R(x,t)$. Then the Volterra equation of the second kind with kernel $K^*(x,t) = -K(t,x)$ has the resolvent $R^*(x,t) = -R(t,x)$.

2°. Assume that two Volterra equations of the second kind with kernels $K_1(x,t)$ and $K_2(x,t)$ are given and that resolvents $R_1(x,t)$ and $R_2(x,t)$ correspond to these equations. In this case the Volterra equation with kernel

$$K(x,t) = K_1(x,t) + K_2(x,t) - \int_t^x K_1(x,s) K_2(s,t)\, ds \tag{16.2.1.3}$$

has the resolvent

$$R(x,t) = R_1(x,t) + R_2(x,t) + \int_t^x R_1(s,t) R_2(x,s)\, ds. \tag{16.2.1.4}$$

Note that in formulas (16.2.1.3) and (16.2.1.4), the integration is performed with respect to different pairs of variables.

### 16.2.2. Equations with Degenerate Kernel: $K(x,t) = g_1(x)h_1(t) + \cdots + g_n(x)h_n(t)$

#### 16.2.2-1. Equations with kernel of the form $K(x,t) = \varphi(x) + \psi(x)(x-t)$.

1°. The solution of a Volterra equation (see Paragraph 16.2.1-1) with kernel of this type can be expressed by the formula

$$y = w''_{xx},$$

where $w = w(x)$ is the solution of the second-order linear nonhomogeneous ordinary differential equation

$$w''_{xx} - \varphi(x) w'_x - \psi(x) w = f(x),$$

with the initial conditions $w(a) = w'_x(a) = 0$.

2°. For a degenerate kernel of the above form, the resolvent can be defined by the formula

$$R(x,t) = u''_{xx}, \tag{16.2.2.1}$$

where the auxiliary function $u$ is the solution of the homogeneous linear second-order ordinary differential equation

$$u''_{xx} - \varphi(x) u'_x - \psi(x) u = 0 \tag{16.2.2.2}$$

with the following initial conditions at $x = t$:

$$u\big|_{x=t} = 0, \quad u'_x\big|_{x=t} = 1. \tag{16.2.2.3}$$

The parameter $t$ occurs only in the initial conditions (16.2.2.3), and equation (16.2.2.2) itself is independent of $t$.

**16.2.2-2. Equations with kernel of the form $K(x,t) = \varphi(t) + \psi(t)(t-x)$.**

For a degenerate kernel of the above form, the resolvent is determined by the expression

$$R(x,t) = -v''_{tt}, \tag{16.2.2.4}$$

where the auxiliary function $v$ is the solution of the homogeneous linear second-order ordinary differential equation

$$v''_{tt} + \varphi(t)v'_t + \psi(t)v = 0 \tag{16.2.2.5}$$

with the following initial conditions at $t = x$:

$$v\big|_{t=x} = 0, \quad v'_t\big|_{t=x} = 1. \tag{16.2.2.6}$$

The point $x$ occurs only in the initial data (16.2.2.6) as a parameter, and equation (16.2.2.5) itself is independent of $x$.

**16.2.2-3. Equations with degenerate kernel of the general form.**

In this case, the Volterra equation of the second kind can be represented in the form

$$y(x) - \sum_{m=1}^{n} g_m(x) \int_a^x h_m(t)y(t)\,dt = f(x). \tag{16.2.2.7}$$

Let us introduce the notation

$$w_j(x) = \int_a^x h_j(t)y(t)\,dt, \quad j = 1,\ldots,n, \tag{16.2.2.8}$$

and rewrite equation (16.2.2.7) as follows:

$$y(x) = \sum_{m=1}^{n} g_m(x)w_m(x) + f(x). \tag{16.2.2.9}$$

On differentiating the expressions (16.2.2.8) with regard to formula (16.2.2.9), we arrive at the following system of linear differential equations for the functions $w_j = w_j(x)$:

$$w'_j = h_j(x)\Big[\sum_{m=1}^{n} g_m(x)w_m + f(x)\Big], \quad j = 1,\ldots,n,$$

with the initial conditions

$$w_j(a) = 0, \quad j = 1,\ldots,n.$$

Once the solution of this system is found, the solution of the original integral equation (16.2.2.7) is defined by formula (16.2.2.9) or any of the expressions

$$y(x) = \frac{w'_j(x)}{h_j(x)}, \quad j = 1,\ldots,n,$$

which can be obtained from formula (16.2.2.8) by differentiation.

## 16.2.3. Equations with Difference Kernel: $K(x,t) = K(x-t)$

16.2.3-1. Solution method based on the Laplace transform.

Volterra equations of the second kind with kernel depending on the difference of the arguments have the form

$$y(x) - \int_0^x K(x-t)y(t)\,dt = f(x). \tag{16.2.3.1}$$

Applying the Laplace transform $\mathfrak{L}$ to equation (16.2.3.1) and taking into account the fact that by the convolution theorem (see Subsection 11.2.2) the integral with kernel depending on the difference of the arguments is transformed into the product $\widetilde{K}(p)\widetilde{y}(p)$, we arrive at the following equation for the transform of the unknown function:

$$\widetilde{y}(p) - \widetilde{K}(p)\widetilde{y}(p) = \widetilde{f}(p). \tag{16.2.3.2}$$

The solution of equation (16.2.3.2) is given by the formula

$$\widetilde{y}(p) = \frac{\widetilde{f}(p)}{1 - \widetilde{K}(p)}, \tag{16.2.3.3}$$

which can be written equivalently in the form

$$\widetilde{y}(p) = \widetilde{f}(p) + \widetilde{R}(p)\widetilde{f}(p), \qquad \widetilde{R}(p) = \frac{\widetilde{K}(p)}{1 - \widetilde{K}(p)}. \tag{16.2.3.4}$$

On applying the Laplace inversion formula to (16.2.3.4), we obtain the solution of equation (16.2.3.1) in the form

$$y(x) = f(x) + \int_0^x R(x-t)f(t)\,dt,$$
$$R(x) = \frac{1}{2\pi i}\int_{c-i\infty}^{c+i\infty} \widetilde{R}(p)e^{px}\,dp. \tag{16.2.3.5}$$

To calculate the corresponding integrals, tables of direct and inverse Laplace transforms can be applied (see Sections T3.1 and T3.2), and, in many cases, to find the inverse transform, methods of the theory of functions of a complex variable are applied, including formulas for the calculation of residues and the Jordan lemma (see Subsection 11.1.2).

Remark. If the lower limit of the integral in the Volterra equation with kernel depending on the difference of the arguments is equal to $a$, then this equation can be reduced to equation (16.2.3.1) by the change of variables $x = \bar{x} - a$, $t = \bar{t} - a$.

Figure 16.1 depicts the principal scheme of solving Volterra integral equations of the second kind with difference kernel by means of the Laplace integral transform.

**Example.** Consider the equation

$$y(x) + A\int_0^x \sin[\lambda(x-t)]y(t)\,dt = f(x), \tag{16.2.3.6}$$

which is a special case of equation (16.2.3.1) for $K(x) = -A\sin(\lambda x)$.

We first apply the table of Laplace transforms (see Subsection T3.1.6) and obtain the transform of the kernel of the integral equation in the form

$$\widetilde{K}(p) = -\frac{A\lambda}{p^2 + \lambda^2}.$$

```
┌─────────────────────────────────────────────────────────────────┐
│  Original integral equation:  $y(x) - \int_0^x K(x-t)\,y(t)\,dt = f(x)$  │
└─────────────────────────────────────────────────────────────────┘
```

⬇ Application of the Laplace transform
$$\widetilde{f}(p) = \int_0^\infty e^{-px} f(x)\,dx$$ ⬇

```
┌─────────────────────────────────────────────────────────────────┐
│  Algebraic equation for the transform:  $\widetilde{y}(p) - \widetilde{K}(p)\,\widetilde{y}(p) = \widetilde{f}(p)$  │
└─────────────────────────────────────────────────────────────────┘
```

⬇ Solution of the equation for the transform ⬇

```
┌─────────────────────────────────────────────────────────────────┐
│  $\widetilde{y}(p) = \dfrac{\widetilde{f}(p)}{1 - \widetilde{K}(p)} \equiv \widetilde{f}(p) + \dfrac{\widetilde{K}(p)}{1 - \widetilde{K}(p)}\widetilde{f}(p)$  │
└─────────────────────────────────────────────────────────────────┘
```

⬇ Application of the inverse Laplace transform
$$f(x) = \frac{1}{2\pi i}\int_{c-i\infty}^{c+i\infty} e^{px}\widetilde{f}(p)\,dp$$ ⬇

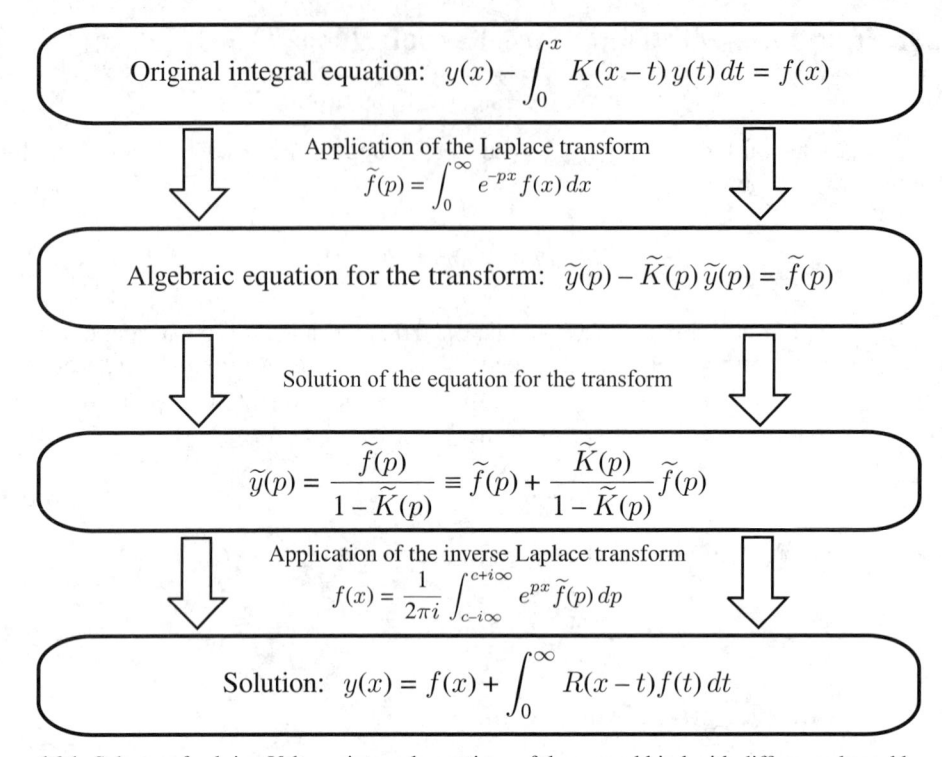

```
┌─────────────────────────────────────────────────────────────────┐
│  Solution:  $y(x) = f(x) + \int_0^\infty R(x-t)f(t)\,dt$  │
└─────────────────────────────────────────────────────────────────┘
```

**Figure 16.1.** Scheme of solving Volterra integral equations of the second kind with difference kernel by means of the Laplace integral transform; $R(x)$ is the inverse transform of the function $\widetilde{R}(p) = \widetilde{K}(p)/[1 - \widetilde{K}(p)]$.

Next, by formula (16.2.3.4) we find the transform of the resolvent:

$$\widetilde{R}(p) = -\frac{A\lambda}{p^2 + \lambda(A+\lambda)}.$$

Furthermore, applying the table of inverse Laplace transforms (see Subsection T3.2.2) we obtain the resolvent

$$R(x) = \begin{cases} -\dfrac{A\lambda}{k}\sin(kx) & \text{for } \lambda(A+\lambda) > 0, \\ -\dfrac{A\lambda}{k}\sinh(kx) & \text{for } \lambda(A+\lambda) < 0, \end{cases} \quad \text{where} \quad k = |\lambda(A+\lambda)|^{1/2}.$$

Moreover, in the special case $\lambda = -A$, we have $R(x) = A^2 x$. On substituting the expressions for the resolvent into formula (16.2.3.5), we find the solution of the integral equation (16.2.3.6). In particular, for $\lambda(A+\lambda) > 0$, this solution has the form

$$y(x) = f(x) - \frac{A\lambda}{k}\int_0^x \sin\bigl[k(x-t)\bigr] f(t)\,dt, \qquad k = \sqrt{\lambda(A+\lambda)}. \tag{16.2.3.7}$$

**Remark.** The Laplace transform can be applied to solve systems of Volterra integral equations of the form

$$y_m(x) - \sum_{k=1}^n \int_0^x K_{mk}(x-t)y_k(t)\,dt = f_m(x), \qquad m = 1,\ldots,n.$$

### 16.2.3-2. Method based on the solution of an auxiliary equation.

Consider the integral equation

$$Ay(x) + B\int_a^x K(x-t)y(t)\,dt = f(x). \tag{16.2.3.8}$$

Let $w = w(x)$ be a solution of the simpler auxiliary equation with $f(x) \equiv 1$ and $a = 0$,

$$Aw(x) + B \int_0^x K(x-t)w(t)\,dt = 1. \tag{16.2.3.9}$$

In this case, the solution of the original equation (16.2.3.8) with an arbitrary right-hand side can be expressed via the solution of the auxiliary equation (16.2.3.9) by the formula

$$y(x) = \frac{d}{dx} \int_a^x w(x-t)f(t)\,dt = f(a)w(x-a) + \int_a^x w(x-t)f'_t(t)\,dt.$$

## 16.2.4. Construction of Solutions of Integral Equations with Special Right-Hand Side

In this section we describe some approaches to the construction of solutions of integral equations with special right-hand sides. These approaches are based on the application of auxiliary solutions that depend on a free parameter.

### 16.2.4-1. General scheme.

Consider a linear equation, which we shall write in the following brief form:

$$\mathbf{L}[y] = f_g(x, \lambda), \tag{16.2.4.1}$$

where $\mathbf{L}$ is a linear operator (integral, differential, etc.) that acts with respect to the variable $x$ and is independent of the parameter $\lambda$, and $f_g(x, \lambda)$ is a given function that depends on the variable $x$ and the parameter $\lambda$.

Suppose that the solution of equation (16.2.4.1) is known:

$$y = y(x, \lambda). \tag{16.2.4.2}$$

Let $\mathbf{M}$ be a linear operator (integral, differential, etc.) that acts with respect to the parameter $\lambda$ and is independent of the variable $x$. Consider the (usual) case in which $\mathbf{M}$ commutes with $\mathbf{L}$. We apply the operator $\mathbf{M}$ to equation (16.2.4.1) and find that the equation

$$\mathbf{L}[w] = f_M(x), \qquad f_M(x) = \mathbf{M}\left[f_g(x, \lambda)\right] \tag{16.2.4.3}$$

has the solution

$$w = \mathbf{M}\left[y(x, \lambda)\right]. \tag{16.2.4.4}$$

By choosing the operator $\mathbf{M}$ in a different way, we can obtain solutions for other right-hand sides of equation (16.2.4.1). The original function $f_g(x, \lambda)$ is called the *generating function* for the operator $\mathbf{L}$.

### 16.2.4-2. Generating function of exponential form.

Consider a linear equation with exponential right-hand side

$$\mathbf{L}[y] = e^{\lambda x}. \tag{16.2.4.5}$$

Suppose that the solution is known and is given by formula (16.2.4.2). In Table 16.1 we present solutions of the equation $\mathbf{L}[y] = f(x)$ with various right-hand sides; these solutions are expressed via the solution of equation (16.2.4.5).

**Remark 1.** When applying the formulas indicated in the table, we need not know the left-hand side of the linear equation (16.2.4.5) (the equation can be integral, differential, etc.) provided that a particular solution of this equation for the exponential right-hand side is known. It is only of importance that the left-hand side of the equation is independent of the parameter $\lambda$.

**Remark 2.** When applying formulas indicated in the table, the convergence of the integrals occurring in the resulting solution must be verified.

**Example 1.** We seek a solution of the equation with exponential right-hand side

$$y(x) + \int_x^\infty K(x-t)y(t)\,dt = e^{\lambda x} \tag{16.2.4.6}$$

in the form $y(x,\lambda) = k e^{\lambda x}$ by the method of indeterminate coefficients. Then we obtain

$$y(x,\lambda) = \frac{1}{B(\lambda)} e^{\lambda x}, \qquad B(\lambda) = 1 + \int_0^\infty K(-z) e^{\lambda z}\,dz. \tag{16.2.4.7}$$

It follows from row 3 of Table 16.1 that the solution of the equation

$$y(x) + \int_x^\infty K(x-t)y(t)\,dt = Ax \tag{16.2.4.8}$$

has the form

$$y(x) = \frac{A}{D}x - \frac{AC}{D^2}, \quad \text{where} \quad D = 1 + \int_0^\infty K(-z)\,dz, \quad C = \int_0^\infty zK(-z)\,dz.$$

For such a solution to exist, it is necessary that the improper integrals of the functions $K(-z)$ and $zK(-z)$ exist. This holds if the function $K(-z)$ decreases more rapidly than $z^{-2}$ as $z \to \infty$. Otherwise a solution can be nonexistent. It is of interest that for functions $K(-z)$ with power-law growth as $z \to \infty$ in the case $\lambda < 0$, the solution of equation (16.2.4.6) exists and is given by formula (16.2.4.7), whereas equation (16.2.4.8) does not have a solution. Therefore, we must be careful when using formulas from Table 16.1 and verify the convergence of the integrals occurring in the solution.

It follows from row 15 of Table 16.1 that the solution of the equation

$$y(x) + \int_x^\infty K(x-t)y(t)\,dt = A\sin(\lambda x) \tag{16.2.4.9}$$

is given by the formula

$$y(x) = \frac{A}{B_c^2 + B_s^2}\left[B_c \sin(\lambda x) - B_s \cos(\lambda x)\right],$$

where

$$B_c = 1 + \int_0^\infty K(-z)\cos(\lambda z)\,dz, \qquad B_s = \int_0^\infty K(-z)\sin(\lambda z)\,dz.$$

### 16.2.4-3. Power-law generating function.

Consider the linear equation with power-law right-hand side

$$\mathbf{L}\,[y] = x^\lambda. \tag{16.2.4.10}$$

Suppose that the solution is known and is given by formula (16.2.4.2). In Table 16.2, solutions of the equation $\mathbf{L}\,[y] = f(x)$ with various right-hand sides are presented, which can be expressed via the solution of equation (16.2.4.10).

## 16.2. Linear Integral Equations of the Second Kind with Variable Integration Limit

### TABLE 16.1
Solutions of the equation $\mathbf{L}[y] = f(x)$ with generating function of the exponential form

| No. | Right-hand side $f(x)$ | Solution $y$ | Solution method |
|---|---|---|---|
| 1 | $e^{\lambda x}$ | $y(x, \lambda)$ | Original equation |
| 2 | $A_1 e^{\lambda_1 x} + \cdots + A_n e^{\lambda_n x}$ | $A_1 y(x, \lambda_1) + \cdots + A_n y(x, \lambda_n)$ | Follows from linearity |
| 3 | $Ax + B$ | $A \dfrac{\partial}{\partial \lambda}\left[y(x,\lambda)\right]_{\lambda=0} + B y(x, 0)$ | Follows from linearity and the results of row 4 |
| 4 | $Ax^n$, $n = 0, 1, 2, \ldots$ | $A\left\{\dfrac{\partial^n}{\partial \lambda^n}\left[y(x,\lambda)\right]\right\}_{\lambda=0}$ | Follows from the results of row 6 for $\lambda = 0$ |
| 5 | $\dfrac{A}{x+a}$, $a > 0$ | $A \displaystyle\int_0^\infty e^{-a\lambda} y(x, -\lambda)\, d\lambda$ | Integration with respect to the parameter $\lambda$ |
| 6 | $Ax^n e^{\lambda x}$, $n = 0, 1, 2, \ldots$ | $A \dfrac{\partial^n}{\partial \lambda^n}\left[y(x,\lambda)\right]$ | Differentiation with respect to the parameter $\lambda$ |
| 7 | $a^x$ | $y(x, \ln a)$ | Follows from row 1 |
| 8 | $A \cosh(\lambda x)$ | $\tfrac{1}{2} A [y(x, \lambda) + y(x, -\lambda)]$ | Linearity and relations to the exponential |
| 9 | $A \sinh(\lambda x)$ | $\tfrac{1}{2} A [y(x, \lambda) - y(x, -\lambda)]$ | Linearity and relations to the exponential |
| 10 | $Ax^m \cosh(\lambda x)$, $m = 1, 3, 5, \ldots$ | $\tfrac{1}{2} A \dfrac{\partial^m}{\partial \lambda^m}[y(x, \lambda) - y(x, -\lambda)]$ | Differentiation with respect to $\lambda$ and relation to the exponential |
| 11 | $Ax^m \cosh(\lambda x)$, $m = 2, 4, 6, \ldots$ | $\tfrac{1}{2} A \dfrac{\partial^m}{\partial \lambda^m}[y(x, \lambda) + y(x, -\lambda)]$ | Differentiation with respect to $\lambda$ and relation to the exponential |
| 12 | $Ax^m \sinh(\lambda x)$, $m = 1, 3, 5, \ldots$ | $\tfrac{1}{2} A \dfrac{\partial^m}{\partial \lambda^m}[y(x, \lambda) + y(x, -\lambda)]$ | Differentiation with respect to $\lambda$ and relation to the exponential |
| 13 | $Ax^m \sinh(\lambda x)$, $m = 2, 4, 6, \ldots$ | $\tfrac{1}{2} A \dfrac{\partial^m}{\partial \lambda^m}[y(x, \lambda) - y(x, -\lambda)]$ | Differentiation with respect to $\lambda$ and relation to the exponential |
| 14 | $A \cos(\beta x)$ | $A \operatorname{Re}[y(x, i\beta)]$ | Selection of the real part for $\lambda = i\beta$ |
| 15 | $A \sin(\beta x)$ | $A \operatorname{Im}[y(x, i\beta)]$ | Selection of the imaginary part for $\lambda = i\beta$ |
| 16 | $Ax^n \cos(\beta x)$, $n = 1, 2, 3, \ldots$ | $A \operatorname{Re}\left\{\dfrac{\partial^n}{\partial \lambda^n}[y(x, \lambda)]\right\}_{\lambda = i\beta}$ | Differentiation with respect to $\lambda$ and selection of the real part for $\lambda = i\beta$ |
| 17 | $Ax^n \sin(\beta x)$, $n = 1, 2, 3, \ldots$ | $A \operatorname{Im}\left\{\dfrac{\partial^n}{\partial \lambda^n}[y(x, \lambda)]\right\}_{\lambda = i\beta}$ | Differentiation with respect to $\lambda$ and selection of the imaginary part for $\lambda = i\beta$ |
| 18 | $A e^{\mu x} \cos(\beta x)$ | $A \operatorname{Re}[y(x, \mu + i\beta)]$ | Selection of the real part for $\lambda = \mu + i\beta$ |
| 19 | $A e^{\mu x} \sin(\beta x)$ | $A \operatorname{Im}[y(x, \mu + i\beta)]$ | Selection of the imaginary part for $\lambda = \mu + i\beta$ |
| 20 | $Ax^n e^{\mu x} \cos(\beta x)$, $n = 1, 2, 3, \ldots$ | $A \operatorname{Re}\left\{\dfrac{\partial^n}{\partial \lambda^n}[y(x, \lambda)]\right\}_{\lambda = \mu + i\beta}$ | Differentiation with respect to $\lambda$ and selection of the real part for $\lambda = \mu + i\beta$ |
| 21 | $Ax^n e^{\mu x} \sin(\beta x)$, $n = 1, 2, 3, \ldots$ | $A \operatorname{Im}\left\{\dfrac{\partial^n}{\partial \lambda^n}[y(x, \lambda)]\right\}_{\lambda = \mu + i\beta}$ | Differentiation with respect to $\lambda$ and selection of the imaginary part for $\lambda = \mu + i\beta$ |

TABLE 16.2
Solutions of the equation $\mathbf{L}[y] = f(x)$ with generating function of power-law form

| No. | Right-hand side $f(x)$ | Solution $y$ | Solution method |
|---|---|---|---|
| 1 | $x^\lambda$ | $y(x,\lambda)$ | Original equation |
| 2 | $\sum_{k=0}^{n} A_k x^k$ | $\sum_{k=0}^{n} A_k y(x,k)$ | Follows from linearity |
| 3 | $A \ln x + B$ | $A \dfrac{\partial}{\partial \lambda}\bigl[y(x,\lambda)\bigr]_{\lambda=0} + By(x,0)$ | Follows from linearity and from the results of row 4 |
| 4 | $A \ln^n x$, $n = 0, 1, 2, \ldots$ | $A\left\{\dfrac{\partial^n}{\partial\lambda^n}\bigl[y(x,\lambda)\bigr]\right\}_{\lambda=0}$ | Follows from the results of row 5 for $\lambda = 0$ |
| 5 | $Ax^\lambda \ln^n x$, $n = 0, 1, 2, \ldots$ | $A \dfrac{\partial^n}{\partial\lambda^n}\bigl[y(x,\lambda)\bigr]$ | Differentiation with respect to the parameter $\lambda$ |
| 6 | $A \cos(\beta \ln x)$ | $A \operatorname{Re}\bigl[y(x, i\beta)\bigr]$ | Selection of the real part for $\lambda = i\beta$ |
| 7 | $A \sin(\beta \ln x)$ | $A \operatorname{Im}\bigl[y(x, i\beta)\bigr]$ | Selection of the imaginary part for $\lambda = i\beta$ |
| 8 | $Ax^\mu \cos(\beta \ln x)$ | $A \operatorname{Re}\bigl[y(x, \mu + i\beta)\bigr]$ | Selection of the real part for $\lambda = \mu + i\beta$ |
| 9 | $Ax^\mu \sin(\beta \ln x)$ | $A \operatorname{Im}\bigl[y(x, \mu + i\beta)\bigr]$ | Selection of the imaginary part for $\lambda = \mu + i\beta$ |

**Example 2.** We seek a solution of the equation with power-law right-hand side

$$y(x) + \int_0^x \frac{1}{x} K\!\left(\frac{t}{x}\right) y(t)\, dt = x^\lambda$$

in the form $y(x,\lambda) = kx^\lambda$ by the method of indeterminate coefficients. We finally obtain

$$y(x,\lambda) = \frac{1}{1 + B(\lambda)} x^\lambda, \qquad B(\lambda) = \int_0^1 K(t) t^\lambda\, dt.$$

It follows from row 3 of Table 16.2 that the solution of the equation with logarithmic right-hand side

$$y(x) + \int_0^x \frac{1}{x} K\!\left(\frac{t}{x}\right) y(t)\, dt = A \ln x$$

has the form

$$y(x) = \frac{A}{1 + I_0} \ln x - \frac{A I_1}{(1 + I_0)^2}, \quad \text{where} \quad I_0 = \int_0^1 K(t)\, dt, \quad I_1 = \int_0^1 K(t) \ln t\, dt.$$

**Remark.** The cases where the generating function is defined sine or cosine are treated likewise.

### 16.2.5. Method of Model Solutions

16.2.5-1. Preliminary remarks.*

Consider a linear equation that we briefly write out in the form

$$\mathbf{L}[y(x)] = f(x), \qquad (16.2.5.1)$$

where $\mathbf{L}$ is a linear (integral) operator, $y(x)$ is an unknown function, and $f(x)$ is a known function.

---

* Before reading this section, it is useful to look over Subsection 16.2.4.

We first define arbitrarily a test solution

$$y_0 = y_0(x, \lambda), \qquad (16.2.5.2)$$

which depends on an auxiliary parameter $\lambda$ (it is assumed that the operator $\mathbf{L}$ is independent of $\lambda$ and $y_0 \neq \text{const}$). By means of equation (16.2.5.1) we define the right-hand side that corresponds to the test solution (16.2.5.2):

$$f_0(x, \lambda) = \mathbf{L}\,[y_0(x, \lambda)].$$

Let us multiply equation (16.2.5.1), for $y = y_0$ and $f = f_0$, by some function $\varphi(\lambda)$ and integrate the resulting relation with respect to $\lambda$ over an interval $[a, b]$. We finally obtain

$$\mathbf{L}\,[y_\varphi(x)] = f_\varphi(x), \qquad (16.2.5.3)$$

where

$$y_\varphi(x) = \int_a^b y_0(x, \lambda)\varphi(\lambda)\,d\lambda, \qquad f_\varphi(x) = \int_a^b f_0(x, \lambda)\varphi(\lambda)\,d\lambda. \qquad (16.2.5.4)$$

It follows from formulas (16.2.5.3) and (16.2.5.4) that, for the right-hand side $f = f_\varphi(x)$, the function $y = y_\varphi(x)$ is a solution of the original equation (16.2.5.1). Since the choice of the function $\varphi(\lambda)$ (as well as of the integration interval) is arbitrary, the function $f_\varphi(x)$ can be arbitrary in principle. Here the main problem is how to choose a function $\varphi(\lambda)$ to obtain a given function $f_\varphi(x)$. This problem can be solved if we can find a test solution such that the right-hand side of equation (16.2.5.1) is the kernel of a known inverse integral transform (we denote such a test solution by $Y(x, \lambda)$ and call it a *model solution*).

**16.2.5-2. Description of the method.**

Indeed, let $\mathfrak{P}$ be an invertible integral transform that takes each function $f(x)$ to the corresponding transform $F(\lambda)$ by the rule

$$F(\lambda) = \mathfrak{P}\{f(x)\}. \qquad (16.2.5.5)$$

Assume that the inverse transform $\mathfrak{P}^{-1}$ has the kernel $\psi(x, \lambda)$ and acts as follows:

$$\mathfrak{P}^{-1}\{F(\lambda)\} = f(x), \qquad \mathfrak{P}^{-1}\{F(\lambda)\} \equiv \int_a^b F(\lambda)\psi(x, \lambda)\,d\lambda. \qquad (16.2.5.6)$$

The limits of integration $a$ and $b$ and the integration path in (16.2.5.6) may well lie in the complex plane.

Suppose that we succeeded in finding a model solution $Y(x, \lambda)$ of the auxiliary problem for equation (16.2.5.1) whose right-hand side is the kernel of the inverse transform $\mathfrak{P}^{-1}$:

$$\mathbf{L}\,[Y(x, \lambda)] = \psi(x, \lambda). \qquad (16.2.5.7)$$

Let us multiply equation (16.2.5.7) by $F(\lambda)$ and integrate with respect to $\lambda$ within the same limits that stand in the inverse transform (16.2.5.6). Taking into account the fact that the operator $\mathbf{L}$ is independent of $\lambda$ and applying the relation $\mathfrak{P}^{-1}\{F(\lambda)\} = f(x)$, we obtain

$$\mathbf{L}\left[\int_a^b Y(x, \lambda)F(\lambda)\,d\lambda\right] = f(x).$$

Therefore, the solution of equation (16.2.5.1) for an arbitrary function $f(x)$ on the right-hand side is expressed via a solution of the simpler auxiliary equation (16.2.5.7) by the formula

$$y(x) = \int_a^b Y(x,\lambda) F(\lambda)\, d\lambda, \qquad (16.2.5.8)$$

where $F(\lambda)$ is the transform (16.2.5.5) of the function $f(x)$.

For the right-hand side of the auxiliary equation (16.2.5.7) we can take, for instance, exponential, power-law, and trigonometric functions, which are the kernels of the Laplace, Mellin, and sine and cosine Fourier transforms (up to a constant factor). Sometimes it is rather easy to find a model solution by means of the method of indeterminate coefficients (by prescribing its structure). Afterward, to construct a solution of the equation with an arbitrary right-hand side, we can apply formulas written out below in Paragraphs 16.2.5-3–16.2.5-6.

16.2.5-3. Model solution in the case of an exponential right-hand side.

Assume that we have found a model solution $Y = Y(x,\lambda)$ that corresponds to the exponential right-hand side:

$$\mathbf{L}[Y(x,\lambda)] = e^{\lambda x}. \qquad (16.2.5.9)$$

Consider two cases:

1°. *Equations on the semiaxis*, $0 \le x < \infty$. Let $\widetilde{f}(p)$ be the Laplace transform of the function $f(x)$:

$$\widetilde{f}(p) = \mathcal{L}\{f(x)\}, \qquad \mathcal{L}\{f(x)\} \equiv \int_0^\infty f(x) e^{-px}\, dx. \qquad (16.2.5.10)$$

The solution of equation (16.2.5.1) for an arbitrary right-hand side $f(x)$ can be expressed via the solution of the simpler auxiliary equation with exponential right-hand side (16.2.5.9) for $\lambda = p$ by the formula

$$y(x) = \frac{1}{2\pi i} \int_{c-i\infty}^{c+i\infty} Y(x,p)\widetilde{f}(p)\, dp. \qquad (16.2.5.11)$$

2°. *Equations on the entire axis*, $-\infty < x < \infty$. Let $\widetilde{f}(u)$ the Fourier transform of the function $f(x)$:

$$\widetilde{f}(u) = \mathfrak{F}\{f(x)\}, \qquad \mathfrak{F}\{f(x)\} \equiv \frac{1}{\sqrt{2\pi}} \int_{-\infty}^\infty f(x) e^{-iux}\, dx. \qquad (16.2.5.12)$$

The solution of equation (16.2.5.1) for an arbitrary right-hand side $f(x)$ can be expressed via the solution of the simpler auxiliary equation with exponential right-hand side (16.2.5.9) for $\lambda = iu$ by the formula

$$y(x) = \frac{1}{\sqrt{2\pi}} \int_{-\infty}^\infty Y(x,iu)\widetilde{f}(u)\, du. \qquad (16.2.5.13)$$

In the calculation of the integrals on the right-hand sides in (16.2.5.11) and (16.2.5.13), methods of the theory of functions of a complex variable are applied, including formulas for the calculation of residues and the Jordan lemma (see Subsection 11.1.2).

**Remark 1.** The structure of a model solution $Y(x,\lambda)$ can differ from that of the kernel of the Laplace or Fourier inversion formula.

**Remark 2.** When applying the method under consideration, the left-hand side of equation (16.2.5.1) need not be known (the equation can be integral, differential, functional, etc.) if a particular solution of this equation is known for the exponential right-hand side. Here only the most general information is important, namely, that the equation is linear, and its left-hand side is independent of the parameter $\lambda$.

**Example 1.** Consider the following Volterra equation of the second kind with difference kernel:

$$y(x) + \int_x^\infty K(x-t)y(t)\,dt = f(x). \tag{16.2.5.14}$$

This equation cannot be solved by direct application of the Laplace transform because the convolution theorem cannot be used here.

In accordance with the method of model solutions, we consider the auxiliary equation with exponential right-hand side

$$y(x) + \int_x^\infty K(x-t)y(t)\,dt = e^{px}. \tag{16.2.5.15}$$

Its solution has the form (see Example 1 of Section 16.2.4)

$$Y(x,p) = \frac{1}{1+\widetilde{K}(-p)}e^{px}, \qquad \widetilde{K}(-p) = \int_0^\infty K(-z)e^{pz}\,dz. \tag{16.2.5.16}$$

This, by means of formula (16.2.5.11), yields a solution of equation (16.2.5.14) for an arbitrary right-hand side,

$$y(x) = \frac{1}{2\pi i}\int_{c-i\infty}^{c+i\infty}\frac{\widetilde{f}(p)}{1+\widetilde{K}(-p)}e^{px}\,dp, \tag{16.2.5.17}$$

where $\widetilde{f}(p)$ is the Laplace transform (16.2.5.10) of the function $f(x)$.

### 16.2.5-4. Model solution in the case of a power-law right-hand side.

Suppose that we have succeeded in finding a model solution $Y = Y(x,s)$ that corresponds to a power-law right-hand side of the equation:

$$\mathbf{L}\,[Y(x,s)] = x^{-s}, \qquad \lambda = -s. \tag{16.2.5.18}$$

Let $\hat{f}(s)$ be the Mellin transform of the function $f(x)$:

$$\hat{f}(s) = \mathfrak{M}\{f(x)\}, \qquad \mathfrak{M}\{f(x)\} \equiv \int_0^\infty f(x)x^{s-1}\,dx. \tag{16.2.5.19}$$

The solution of equation (16.2.5.1) for an arbitrary right-hand side $f(x)$ can be expressed via the solution of the simpler auxiliary equation with power-law right-hand side (16.2.5.18) by the formula

$$y(x) = \frac{1}{2\pi i}\int_{c-i\infty}^{c+i\infty} Y(x,s)\hat{f}(s)\,ds. \tag{16.2.5.20}$$

In the calculation of the corresponding integrals on the right-hand side of formula (16.2.5.20), one can use tables of inverse Mellin transforms (e.g., see Section T3.6, as well as methods of the theory of functions of a complex variable, including formulas for the calculation of residues and the Jordan lemma (see Subsection 11.1.2).

**Example 2.** Consider the equation

$$y(x) + \int_0^x \frac{1}{x} K\left(\frac{t}{x}\right) y(t)\, dt = f(x). \tag{16.2.5.21}$$

In accordance with the method of model solutions, we consider the following auxiliary equation with power-law right-hand side:

$$y(x) + \int_0^x \frac{1}{x} K\left(\frac{t}{x}\right) y(t)\, dt = x^{-s}.$$

Its solution has the form (see Example 2 for $\lambda = -s$ in Section 16.2.4)

$$Y(x, s) = \frac{1}{1 + B(s)} x^{-s}, \qquad B(s) = \int_0^1 K(t) t^{-s}\, dt.$$

This, by means of formula (16.2.5.20), yields the solution of equation (16.2.5.21) for an arbitrary right-hand side:

$$y(x) = \frac{1}{2\pi i} \int_{c-i\infty}^{c+i\infty} \frac{\hat{f}(s)}{1 + B(s)} x^{-s}\, ds,$$

where $\hat{f}(s)$ is the Mellin transform (16.2.5.19) of the function $f(x)$.

## 16.2.6. Successive Approximation Method

### 16.2.6-1. General scheme.

$1°$. Consider a Volterra integral equation of the second kind

$$y(x) - \int_a^x K(x, t) y(t)\, dt = f(x). \tag{16.2.6.1}$$

Assume that $f(x)$ is continuous on the interval $[a, b]$ and the kernel $K(x, t)$ is continuous for $a \le x \le b$ and $a \le t \le x$.

Let us seek the solution by the successive approximation method. To this end, we set

$$y(x) = f(x) + \sum_{n=1}^{\infty} \varphi_n(x), \tag{16.2.6.2}$$

where the $\varphi_n(x)$ are determined by the formulas

$$\varphi_1(x) = \int_a^x K(x, t) f(t)\, dt,$$

$$\varphi_2(x) = \int_a^x K(x, t) \varphi_1(t)\, dt = \int_a^x K_2(x, t) f(t)\, dt,$$

$$\varphi_3(x) = \int_a^x K(x, t) \varphi_2(t)\, dt = \int_a^x K_3(x, t) f(t)\, dt, \quad \text{etc.}$$

Here

$$K_n(x, t) = \int_a^x K(x, z) K_{n-1}(z, t)\, dz, \tag{16.2.6.3}$$

where $n = 2, 3, \ldots$, and we have the relations $K_1(x, t) \equiv K(x, t)$ and $K_n(x, t) = 0$ for $t > x$. The functions $K_n(x, t)$ given by formulas (16.2.6.3) are called *iterated kernels*. These kernels satisfy the relation

$$K_n(x, t) = \int_a^x K_m(x, s) K_{n-m}(s, t)\, ds, \tag{16.2.6.4}$$

where $m$ is an arbitrary positive integer less than $n$.

**2°**. The successive approximations can be implemented in a more general scheme:

$$y_n(x) = f(x) + \int_a^x K(x,t) y_{n-1}(t)\,dt, \qquad n = 1, 2, \ldots, \tag{16.2.6.5}$$

where the function $y_0(x)$ is continuous on the interval $[a, b]$. The functions $y_1(x), y_2(x), \ldots$ which are obtained from (16.2.6.5) are also continuous on $[a, b]$.

Under the assumptions adopted in Item 1° for $f(x)$ and $K(x,t)$, the sequence $\{y_n(x)\}$ converges, as $n \to \infty$, to the continuous solution $y(x)$ of the integral equation. A successful choice of the "zeroth" approximation $y_0(x)$ can result in a rapid convergence of the procedure.

**Remark 1.** In the special case $y_0(x) = f(x)$, this method becomes that described in Item 1°.

**Remark 2.** If the kernel $K(x,t)$ is square integrable on the square $S = \{a \le x \le b,\ a \le t \le b\}$ and $f(x) \in L_2(a, b)$, then the successive approximations are mean-square convergent to the solution $y(x) \in L_2(a, b)$ of the integral equation (16.2.6.1) for any initial approximation $y_0(x) \in L_2(a, b)$.

### 16.2.6-2. Formula for the resolvent.

The resolvent of the integral equation (16.2.6.1) is determined via the iterated kernels by the formula

$$R(x, t) = \sum_{n=1}^{\infty} K_n(x, t),$$

where the convergent series on the right-hand side is called the *Neumann series* of the kernel $K(x, t)$. Now the solution of the Volterra equation of the second kind (16.2.6.1) can be rewritten in the traditional form

$$y(x) = f(x) + \int_a^x R(x, t) f(t)\,dt.$$

**Remark.** In the case of a kernel with weak singularity (see Paragraph 16.1.1-1), the solution of equation (16.2.6.1) can be obtained by the successive approximation method. In this case the kernels $K_n(x, t)$ are continuous starting from some $n$. For $\beta < \frac{1}{2}$, even the kernel $K_2(x, t)$ is continuous.

## 16.2.7. Method of Quadratures

### 16.2.7-1. General scheme of the method.

Let us consider the linear Volterra integral equation of the second kind

$$y(x) - \int_a^x K(x,t) y(t)\,dt = f(x), \tag{16.2.7.1}$$

on an interval $a \le x \le b$. Assume that the kernel and the right-hand side of the equation are continuous functions.

From equation (16.2.7.1) we find that $y(a) = f(a)$. Let us choose a constant integration step $h$ and consider the discrete set of points $x_i = a + h(i - 1)$, $i = 1, \ldots, n$. For $x = x_i$, equation (16.2.7.1) acquires the form

$$y(x_i) - \int_a^{x_i} K(x_i, t) y(t)\,dt = f(x_i), \qquad i = 1, \ldots, n. \tag{16.2.7.2}$$

Applying the quadrature formula (see Subsection 16.1.5) to the integral in (16.2.7.2) and choosing $x_j$ ($j = 1, \ldots, i$) to be the nodes in $t$, we arrive at the system of equations

$$y(x_i) - \sum_{j=1}^{i} A_{ij} K(x_i, x_j) y(x_j) = f(x_i) + \varepsilon_i[y], \qquad i = 2, \ldots, n, \tag{16.2.7.3}$$

where $\varepsilon_i[y]$ is the truncation error and $A_{ij}$ are the coefficients of the quadrature formula on the interval $[a, x_i]$ (see Subsection 16.1.5). Suppose that $\varepsilon_i[y]$ are small and neglect them; then we obtain a system of linear algebraic equations in the form

$$y_1 = f_1, \qquad y_i - \sum_{j=1}^{i} A_{ij} K_{ij} y_j = f_i, \qquad i = 2, \ldots, n, \tag{16.2.7.4}$$

where $K_{ij} = K(x_i, x_j)$, $f_i = f(x_i)$, and $y_i$ are approximate values of the unknown function $y(x)$ at the nodes $x_i$.

From (16.2.7.4) we obtain the recurrent formula

$$y_1 = f_1, \qquad y_i = \frac{f_i + \sum_{j=1}^{i-1} A_{ij} K_{ij} y_j}{1 - A_{ii} K_{ii}}, \qquad i = 2, \ldots, n, \tag{16.2.7.5}$$

valid under the condition $1 - A_{ii} K_{ii} \neq 0$, which can always be ensured by an appropriate choice of the nodes and by guaranteeing that the coefficients $A_{ii}$ are sufficiently small.

**16.2.7-2. Application of the trapezoidal rule.**

According to the trapezoidal rule (see Section 16.1.5), we have

$$A_{i1} = A_{ii} = \tfrac{1}{2} h, \qquad A_{i2} = \cdots = A_{i,i-1} = h, \qquad i = 2, \ldots, n.$$

The application of the trapezoidal rule in the general scheme leads to the following step algorithm:

$$y_1 = f_1, \qquad y_i = \frac{f_i + h \sum_{j=1}^{i-1} \beta_j K_{ij} y_j}{1 - \tfrac{1}{2} h K_{ii}}, \qquad i = 2, \ldots, n,$$

$$x_i = a + (i-1)h, \qquad n = \frac{b-a}{h} + 1, \qquad \beta_j = \begin{cases} \tfrac{1}{2} & \text{for } j = 1, \\ 1 & \text{for } j > 1, \end{cases}$$

where the notation coincides with that introduced in Paragraph 16.2.7-1. The trapezoidal rule is quite simple and effective, and frequently used in practice.

## 16.3. Linear Integral Equations of the First Kind with Constant Limits of Integration

### 16.3.1. Fredholm Integral Equations of the First Kind

**16.3.1-1. Some definitions. Function and kernel classes.**

A linear integral equation of the first kind with constant limits of integration have the general form

$$\int_a^b K(x, t) y(t) \, dt = f(x), \tag{16.3.1.1}$$

where $y(x)$ is the unknown function ($a \le x \le b$), $K(x,t)$ is the *kernel* of the integral equation, and $f(x)$ is a given function, which is called the *right-hand side* of equation (16.3.1.1). The functions $y(x)$ and $f(x)$ are usually assumed to be continuous or square integrable on $[a,b]$. If the kernel of the integral equation (16.3.1.1) is continuous on the square $S = \{a \le x \le b, a \le t \le b\}$ or at least square integrable on this square, i.e.,

$$\int_a^b \int_a^b K^2(x,t)\,dx\,dt = B^2 < \infty, \tag{16.3.1.2}$$

where $B$ is a constant, then this kernel is called a *Fredholm kernel*. Equations of the form (16.3.1.1) with constant integration limits and Fredholm kernel are called *Fredholm equations of the first kind*.

The kernel $K(x,t)$ of an integral equation is said to be *degenerate* if it can be represented in the form $K(x,t) = g_1(x)h_1(t) + \cdots + g_n(x)h_n(t)$. The kernel $K(x,t)$ of an integral equation is called a *difference kernel* if it depends only on the difference of the arguments: $K(x,t) = K(x-t)$. The kernel $K(x,t)$ of an integral equation is said to be *symmetric* if it satisfies the condition $K(x,t) = K(t,x)$.

The integral equation obtained from (16.3.1.1) by replacing the kernel $K(x,t)$ with $K(t,x)$ is said to be *transposed* to (16.3.1.1).

The integral equation of the first kind with difference kernel on the entire axis ($a = -\infty$, $b = \infty$) and semiaxis ($a = 0$, $b = \infty$) is referred to as an *equation of convolution type of the first kind* and a *Wiener–Hopf integral equation of the first kind*, respectively.

16.3.1-2. Integral equations of the first kind with weak singularity.

If the kernel of the integral equation (16.3.1.1) is polar, i.e., if

$$K(x,t) = L(x,t)|x-t|^{-\beta} + M(x,t), \qquad 0 < \beta < 1, \tag{16.3.1.3}$$

or logarithmic, i.e.,

$$K(x,t) = L(x,t)\ln|x-t| + M(x,t), \tag{16.3.1.4}$$

where the functions $L(x,t)$ and $M(x,t)$ are continuous on $S$ and $L(x,x) \not\equiv 0$, then $K(x,t)$ is called a *kernel with weak singularity*, and the equation itself is called an *equation with weak singularity*.

Kernels with logarithmic singularity and polar kernels with $0 < \alpha < \frac{1}{2}$ are Fredholm kernels.

## 16.3.2. Method of Integral Transforms

The method of integral transforms enables one to reduce some integral equations on the entire axis and on the semiaxis to algebraic equations for transforms. These algebraic equations can readily be solved for the transform of the desired function. The solution of the original integral equation is then obtained by applying the inverse integral transform.

16.3.2-1. Equation with difference kernel on the entire axis.

Consider an integral equation of the form

$$\int_{-\infty}^{\infty} K(x-t)y(t)\,dt = f(x), \qquad -\infty < x < \infty. \tag{16.3.2.1}$$

Let us apply the Fourier transform (see Subsection 11.4.1) to equation (16.3.2.1). In this case, taking into account the convolution theorem (see Paragraph 11.4.1-3), we obtain

$$\sqrt{2\pi}\,\tilde{K}(u)\tilde{y}(u) = \tilde{f}(u). \tag{16.3.2.2}$$

Thus, by means of the Fourier transform we have reduced the solution of the original integral equation (16.3.2.1) to the solution of the algebraic equation (16.3.2.2) for the Fourier transform of the desired solution. The solution of the latter equation has the form

$$\tilde{y}(u) = \frac{1}{\sqrt{2\pi}}\frac{\tilde{f}(u)}{\tilde{K}(u)}, \tag{16.3.2.3}$$

where the function $\tilde{f}(u)/\tilde{K}(u)$ must belong to the space $L_2(-\infty,\infty)$.

Thus, the Fourier transform of the solution of the original integral equation is expressed via the Fourier transforms of known functions, namely, the kernel and the right-hand side of the equation. The solution itself can be expressed via its Fourier transform by means of the Fourier inversion formula:

$$y(x) = \frac{1}{\sqrt{2\pi}}\int_{-\infty}^{\infty}\tilde{y}(u)e^{iux}\,du = \frac{1}{2\pi}\int_{-\infty}^{\infty}\frac{\tilde{f}(u)}{\tilde{K}(u)}e^{iux}\,du. \tag{16.3.2.4}$$

**16.3.2-2. Equations with kernel $K(x,t) = K(x/t)$ on the semiaxis.**

The integral equation of the first kind

$$\int_0^{\infty} K(x/t)y(t)\,dt = f(x), \quad 0 \le x < \infty, \tag{16.3.2.5}$$

can be reduced to the form (16.3.2.1) by the change of variables $x = e^\xi$, $t = e^\tau$, $w(\tau) = ty(t)$. The solution to this equation can also be obtained by straightforward application of the Mellin transform, and this method is applied in a similar situation in the next section.

**16.3.2-3. Equations with kernel $K(x,t) = K(xt)$ on the semiaxis.**

Consider the equation

$$\int_0^{\infty} K(xt)y(t)\,dt = f(x), \quad 0 \le x < \infty. \tag{16.3.2.6}$$

By changing variables $x = e^\xi$ and $t = e^{-\tau}$ this equation can be reduced to the form (16.3.2.1), but it is more convenient here to apply the Mellin transform (see Section 11.3). On multiplying equation (16.3.2.6) by $x^{s-1}$ and integrating with respect to $x$ from 0 to $\infty$, we obtain

$$\int_0^{\infty} y(t)\,dt\int_0^{\infty} K(xt)x^{s-1}\,dx = \int_0^{\infty} f(x)x^{s-1}\,dx.$$

We make the change of variables $z = xt$ in the inner integral of the double integral. This implies the relation

$$\hat{K}(s)\int_0^{\infty} y(t)t^{-s}\,dt = \hat{f}(s). \tag{16.3.2.7}$$

Taking into account the formula

$$\int_0^\infty y(t)t^{-s}\,dt = \hat{y}(1-s),$$

we can rewrite equation (16.3.2.7) in the form

$$\hat{K}(s)\hat{y}(1-s) = \hat{f}(s). \tag{16.3.2.8}$$

Replacing $1-s$ by $s$ in (16.3.2.8) and solving the resulting relation for $\hat{y}(s)$, we obtain the transform

$$\hat{y}(s) = \frac{\hat{f}(1-s)}{\hat{K}(1-s)}$$

of the desired solution.

Applying the Mellin inversion formula (if it is applicable), we obtain the solution of the integral equation (16.3.2.6) in the form

$$y(x) = \frac{1}{2\pi i}\int_{c-i\infty}^{c+i\infty}\frac{\hat{f}(1-s)}{\hat{K}(1-s)}x^{-s}\,ds.$$

### 16.3.3. Regularization Methods

#### 16.3.3-1. Lavrentiev regularization method.

Consider the Fredholm equation of the first kind

$$\int_a^b K(x,t)y(t)\,dt = f(x), \qquad a \le x \le b, \tag{16.3.3.1}$$

where $f(x) \in L_2(a,b)$ and $y(x) \in L_2(a,b)$. The kernel $K(x,t)$ is square integrable, symmetric, and positive definite, that is, for all $\varphi(x) \in L_2(a,b)$, we have

$$\int_a^b \int_a^b K(x,t)\varphi(x)\varphi(t)\,dx\,dt \ge 0,$$

where the equality is attained only for $\varphi(x) \equiv 0$.

In the above classes of functions and kernels, the problem of finding a solution of equation (16.3.3.1) is ill-posed, i.e., unstable with respect to small variations in the right-hand side of the integral equation.

Following the Lavrentiev regularization method, along with equation (16.3.3.1) we consider the regularized equation

$$\varepsilon y_\varepsilon(x) + \int_a^b K(x,t)y_\varepsilon(t)\,dt = f(x), \qquad a \le x \le b, \tag{16.3.3.2}$$

where $\varepsilon > 0$ is the regularization parameter. This equation is a Fredholm equation of the second kind, so it can be solved by the methods presented in Section 16.4, whence the solution exists and is unique.

On taking a sufficiently small $\varepsilon$ in equation (16.3.3.2), we find a solution $y_\varepsilon(x)$ of the equation and substitute this solution into equation (16.3.3.1), thus obtaining

$$\int_a^b K(x,t) y_\varepsilon(t)\,dt = f_\varepsilon(x), \qquad a \le x \le b. \tag{16.3.3.3}$$

If the function $f_\varepsilon(x)$ thus obtained differs only slightly from $f(x)$, that is,

$$\|f(x) - f_\varepsilon(x)\| \le \delta, \tag{16.3.3.4}$$

where $\delta$ is a prescribed small positive number, then the solution $y_\varepsilon(x)$ is regarded as a sufficiently good approximate solution of equation (16.3.3.1).

The parameter $\delta$ usually defines the error of the initial data provided that the right-hand side of equation (16.3.3.1) is defined or determined by an experiment with some accuracy.

For the case in which, for a given $\varepsilon$, condition (16.3.3.4) fails, we must choose another value of the regularization parameter and repeat the above procedure.

### 16.3.3-2. Tikhonov regularization method.

Consider the Fredholm integral equation of the first kind

$$\int_a^b K(x,t) y(t)\,dt = f(x), \qquad c \le x \le d. \tag{16.3.3.5}$$

Assume that $K(x,t)$ is any function square-integrable in the domain $\{a \le t \le b,\ c \le x \le d\}$, $f(x) \in L_2(c,d)$, and $y(x) \in L_2(a,b)$. The problem of finding the solution of equation (16.3.3.5) is also ill-posed in the above sense.

Following the Tikhonov (zero-order) regularization method, along with (16.3.3.5) we consider the following Fredholm integral equation of the second kind (see Section 16.4):

$$\varepsilon y_\varepsilon(x) + \int_a^b K^*(x,t) y_\varepsilon(t)\,dt = f^*(x), \qquad a \le x \le b, \tag{16.3.3.6}$$

where

$$K^*(x,t) = K^*(t,x) = \int_c^d K(s,x) K(s,t)\,ds, \qquad f^*(x) = \int_c^d K(s,x) f(s)\,ds,$$

and the positive number $\varepsilon$ is the regularization parameter. Equation (16.3.3.6) is said to be a *regularized integral equation*, and its solution exists and is unique.

Taking a sufficiently small $\varepsilon$ in equation (16.3.3.6), we find a solution $y_\varepsilon(x)$ of the equation and substitute this solution into equation (16.3.3.5), thus obtaining

$$\int_a^b K(x,t) y_\varepsilon(t)\,dt = f_\varepsilon(x), \qquad c \le x \le d.$$

By comparing the right-hand side with the given $f(x)$ using formula (16.3.3.4), we either regard $f_\varepsilon(x)$ as a satisfactory approximate solution obtained in accordance with the above simple algorithm, or continue the procedure for a new value of the regularization parameter.

## 16.4. Linear Integral Equations of the Second Kind with Constant Limits of Integration

### 16.4.1. Fredholm Integral Equations of the Second Kind. Resolvent

**16.4.1-1. Some definitions. The eigenfunctions of a Fredholm integral equation.**

Linear integral equations of the second kind with constant limits of integration have the general form

$$y(x) - \lambda \int_a^b K(x,t) y(t)\, dt = f(x), \qquad (16.4.1.1)$$

where $y(x)$ is the unknown function ($a \leq x \leq b$), $K(x,t)$ is the *kernel* of the integral equation, and $f(x)$ is a given function, which is called the *right-hand side* of equation (16.4.1.1). For convenience of analysis, a number $\lambda$ is traditionally singled out in equation (16.4.1.1), which is called the *parameter of integral equation*. The classes of functions and kernels under consideration were defined above in Paragraphs 16.3.1-1 and 16.3.1-2. Note that equations of the form (16.4.1.1) with constant limits of integration and with Fredholm kernels or kernels with weak singularity are called *Fredholm equations of the second kind* and *equations with weak singularity of the second kind*, respectively.

Equation (16.4.1.1) is said to be *homogeneous* if $f(x) \equiv 0$ and *nonhomogeneous* otherwise.

A number $\lambda$ is called a *characteristic value* of the integral equation (16.4.1.1) if there exist nontrivial solutions of the corresponding homogeneous equation. The nontrivial solutions themselves are called the *eigenfunctions* of the integral equation corresponding to the characteristic value $\lambda$. If $\lambda$ is a characteristic value, the number $1/\lambda$ is called an *eigenvalue* of the integral equation (16.4.1.1). A value of the parameter $\lambda$ is said to be *regular* if for this value the above homogeneous equation has only the trivial solution. Sometimes the characteristic values and the eigenfunctions of a Fredholm integral equation are called the *characteristic values* and the *eigenfunctions of the kernel* $K(x,t)$.

The kernel $K(x,t)$ of the integral equation (16.4.1.1) is called a *degenerate kernel* if it has the form $K(x,t) = g_1(x)h_1(t) + \cdots + g_n(x)h_n(t)$, a *difference kernel* if it depends on the difference of the arguments ($K(x,t) = K(x-t)$), and a *symmetric kernel* if it satisfies the condition $K(x,t) = K(t,x)$.

The *transposed* integral equation is obtained from (16.4.1.1) by replacing the kernel $K(x,t)$ by $K(t,x)$.

The integral equation of the second kind with difference kernel on the entire axis ($a = -\infty$, $b = \infty$) and semiaxis ($a = 0$, $b = \infty$) are referred to as an *equation of convolution type of the second kind* and a *Wiener–Hopf integral equation of the second kind*, respectively.

**16.4.1-2. Structure of the solution. The resolvent.**

The solution of equation (16.4.1.1) can be presented in the form

$$y(x) = f(x) + \lambda \int_a^b R(x,t;\lambda) f(t)\, dt,$$

where the resolvent $R(x,t;\lambda)$ is independent of $f(x)$ and is determined by the kernel of the integral equation.

## 16.4.2. Fredholm Equations of the Second Kind with Degenerate Kernel

**16.4.2-1. Simplest degenerate kernel.**

Consider Fredholm integral equations of the second kind with the simplest degenerate kernel:

$$y(x) - \lambda \int_a^b g(x)h(t)y(t)\,dt = f(x), \qquad a \le x \le b. \tag{16.4.2.1}$$

We seek a solution of equation (16.4.2.1) in the form

$$y(x) = f(x) + \lambda A g(x). \tag{16.4.2.2}$$

On substituting the expressions (16.4.2.2) into equation (16.4.2.1), after simple algebraic manipulations we obtain

$$A\left[1 - \lambda \int_a^b h(t)g(t)\,dt\right] = \int_a^b f(t)h(t)\,dt. \tag{16.4.2.3}$$

Both integrals occurring in equation (16.4.2.3) are supposed to exist. On the basis of (16.4.2.1) and (16.4.2.3) and taking into account the fact that the unique characteristic value $\lambda_1$ of equation (16.4.2.1) is given by the expression

$$\lambda_1 = \left[\int_a^b h(t)g(t)\,dt\right]^{-1}, \tag{16.4.2.4}$$

we obtain the following results:

1°. If $\lambda \ne \lambda_1$, then for an arbitrary right-hand side there exists a unique solution of equation (16.4.2.1), which can be written in the form

$$y(x) = f(x) + \frac{\lambda \lambda_1 f_1}{\lambda_1 - \lambda} g(x), \qquad f_1 = \int_a^b f(t)h(t)\,dt. \tag{16.4.2.5}$$

2°. If $\lambda = \lambda_1$ and $f_1 = 0$, then any solution of equation (16.4.2.1) can be represented in the form

$$y = f(x) + C y_1(x), \qquad y_1(x) = g(x), \tag{16.4.2.6}$$

where $C$ is an arbitrary constant and $y_1(x)$ is an eigenfunction that corresponds to the characteristic value $\lambda_1$.

3°. If $\lambda = \lambda_1$ and $f_1 \ne 0$, then there are no solutions.

**16.4.2-2. Degenerate kernel in the general case.**

In the general case, a Fredholm integral equation of the second kind with degenerate kernel has the form

$$y(x) - \lambda \int_a^b \left[\sum_{k=1}^n g_k(x)h_k(t)\right] y(t)\,dt = f(x), \qquad n = 2, 3, \ldots \tag{16.4.2.7}$$

Let us rewrite equation (16.4.2.7) in the form

$$y(x) = f(x) + \lambda \sum_{k=1}^{n} g_k(x) \int_a^b h_k(t) y(t)\, dt, \qquad n = 2, 3, \ldots \qquad (16.4.2.8)$$

We assume that equation (16.4.2.8) has a solution and introduce the notation

$$A_k = \int_a^b h_k(t) y(t)\, dt. \qquad (16.4.2.9)$$

In this case we have

$$y(x) = f(x) + \lambda \sum_{k=1}^{n} A_k g_k(x), \qquad (16.4.2.10)$$

and hence the solution of the integral equation with degenerate kernel is reduced to the definition of the constants $A_k$.

Let us multiply equation (16.4.2.10) by $h_m(x)$ and integrate with respect to $x$ from $a$ to $b$. We obtain the following system of linear algebraic equations for the coefficients $A_k$:

$$A_m - \lambda \sum_{k=1}^{n} s_{mk} A_k = f_m, \qquad m = 1, \ldots, n, \qquad (16.4.2.11)$$

where

$$s_{mk} = \int_a^b h_m(x) g_k(x)\, dx, \quad f_m = \int_a^b f(x) h_m(x)\, dx; \quad m, k = 1, \ldots, n. \qquad (16.4.2.12)$$

Once we construct a solution of system (16.4.2.11), we obtain a solution of the integral equation with degenerate kernel (16.4.2.7) as well. The values of the parameter $\lambda$ at which the determinant of system (16.4.2.11) vanishes are characteristic values of the integral equation (16.4.2.7), and it is clear that there are just $n$ such values counted according to their multiplicities.

**Example.** Let us solve the integral equation

$$y(x) - \lambda \int_{-\pi}^{\pi} (x \cos t + t^2 \sin x + \cos x \sin t) y(t)\, dt = x, \qquad -\pi \le x \le \pi. \qquad (16.4.2.13)$$

Let us denote

$$A_1 = \int_{-\pi}^{\pi} y(t) \cos t\, dt, \quad A_2 = \int_{-\pi}^{\pi} t^2 y(t)\, dt, \quad A_3 = \int_{-\pi}^{\pi} y(t) \sin t\, dt, \qquad (16.4.2.14)$$

where $A_1$, $A_2$, and $A_3$ are unknown constants. Then equation (16.4.2.13) can be rewritten in the form

$$y(x) = A_1 \lambda x + A_2 \lambda \sin x + A_3 \lambda \cos x + x. \qquad (16.4.2.15)$$

On substituting the expression (16.4.2.15) into relations (16.4.2.14), we obtain

$$A_1 = \int_{-\pi}^{\pi} (A_1 \lambda t + A_2 \lambda \sin t + A_3 \lambda \cos t + t) \cos t\, dt,$$

$$A_2 = \int_{-\pi}^{\pi} (A_1 \lambda t + A_2 \lambda \sin t + A_3 \lambda \cos t + t) t^2\, dt,$$

$$A_3 = \int_{-\pi}^{\pi} (A_1 \lambda t + A_2 \lambda \sin t + A_3 \lambda \cos t + t) \sin t\, dt.$$

On calculating the integrals occurring in these equations, we obtain the following system of algebraic equations for the unknowns $A_1$, $A_2$, and $A_3$:
$$A_1 - \lambda\pi A_3 = 0,$$
$$A_2 + 4\lambda\pi A_3 = 0, \qquad (16.4.2.16)$$
$$-2\lambda\pi A_1 - \lambda\pi A_2 + A_3 = 2\pi.$$

System (16.4.2.16) has the unique solution
$$A_1 = \frac{2\lambda\pi^2}{1+2\lambda^2\pi^2}, \quad A_2 = -\frac{8\lambda\pi^2}{1+2\lambda^2\pi^2}, \quad A_3 = \frac{2\pi}{1+2\lambda^2\pi^2}.$$

On substituting the above values of $A_1$, $A_2$, and $A_3$ into (16.4.2.15), we obtain the solution of the original integral equation:
$$y(x) = \frac{2\lambda\pi}{1+2\lambda^2\pi^2}(\lambda\pi x - 4\lambda\pi\sin x + \cos x) + x.$$

## 16.4.3. Solution as a Power Series in the Parameter. Method of Successive Approximations

### 16.4.3-1. Iterated kernels.

Consider the Fredholm integral equation of the second kind:
$$y(x) - \lambda \int_a^b K(x,t)y(t)\,dt = f(x), \qquad a \le x \le b. \qquad (16.4.3.1)$$

We seek the solution in the form of a series in powers of the parameter $\lambda$:
$$y(x) = f(x) + \sum_{n=1}^{\infty} \lambda^n \psi_n(x). \qquad (16.4.3.2)$$

Substitute series (16.4.3.2) into equation (16.4.3.1). On matching the coefficients of like powers of $\lambda$, we obtain a recurrent system of equations for the functions $\psi_n(x)$. The solution of this system yields
$$\psi_1(x) = \int_a^b K(x,t)f(t)\,dt,$$
$$\psi_2(x) = \int_a^b K(x,t)\psi_1(t)\,dt = \int_a^b K_2(x,t)f(t)\,dt,$$
$$\psi_3(x) = \int_a^b K(x,t)\psi_2(t)\,dt = \int_a^b K_3(x,t)f(t)\,dt, \quad \text{etc.}$$

Here
$$K_n(x,t) = \int_a^b K(x,z)K_{n-1}(z,t)\,dz, \qquad (16.4.3.3)$$

where $n = 2, 3, \ldots$, and we have $K_1(x,t) \equiv K(x,t)$. The functions $K_n(x,t)$ defined by formulas (16.4.3.3) are called *iterated kernels*. These kernels satisfy the relation
$$K_n(x,t) = \int_a^b K_m(x,s)K_{n-m}(s,t)\,ds, \qquad (16.4.3.4)$$

where $m$ is an arbitrary positive integer less than $n$.

## 16.4. LINEAR INTEGRAL EQUATIONS OF THE SECOND KIND WITH CONSTANT LIMITS OF INTEGRATION

The iterated kernels $K_n(x,t)$ can be directly expressed via $K(x,t)$ by the formula

$$K_n(x,t) = \underbrace{\int_a^b \int_a^b \cdots \int_a^b}_{n-1} K(x,s_1)K(s_1,s_2)\ldots K(s_{n-1},t)\, ds_1\, ds_2 \ldots ds_{n-1}.$$

All iterated kernels $K_n(x,t)$, beginning with $K_2(x,t)$, are continuous functions on the square $S = \{a \leq x \leq b, a \leq t \leq b\}$ if the original kernel $K(x,t)$ is square integrable on $S$. If $K(x,t)$ is symmetric, then all iterated kernels $K_n(x,t)$ are also symmetric.

### 16.4.3-2. Method of successive approximations.

The results of Subsection 16.4.3-1 can also be obtained by means of the method of successive approximations. To this end, one should use the recurrent formula

$$y_n(x) = f(x) + \lambda \int_a^b K(x,t)y_{n-1}(t)\, dt, \qquad n = 1, 2, \ldots,$$

with the zeroth approximation $y_0(x) = f(x)$.

### 16.4.3-3. Construction of the resolvent.

The resolvent of the integral equation (16.4.3.1) is defined via the iterated kernels by the formula

$$R(x,t;\lambda) = \sum_{n=1}^{\infty} \lambda^{n-1} K_n(x,t), \qquad (16.4.3.5)$$

where the series on the right-hand side is called the *Neumann series of the kernel* $K(x,t)$. It converges to a unique square integrable solution of equation (16.4.3.1) provided that

$$|\lambda| < \frac{1}{B}, \qquad B = \sqrt{\int_a^b \int_a^b K^2(x,t)\, dx\, dt}. \qquad (16.4.3.6)$$

If, in addition, we have

$$\int_a^b K^2(x,t)\, dt \leq A, \qquad a \leq x \leq b,$$

where $A$ is a constant, then the Neumann series converges absolutely and uniformly on $[a,b]$.

A solution of a Fredholm equation of the second kind of the form (16.4.3.1) is expressed by the formula

$$y(x) = f(x) + \lambda \int_a^b R(x,t;\lambda)f(t)\, dt, \qquad a \leq x \leq b. \qquad (16.4.3.7)$$

Inequality (16.4.3.6) is essential for the convergence of the series (16.4.3.5). However, a solution of equation (16.4.3.1) can exist for values $|\lambda| > 1/B$ as well.

**Example.** Let us solve the integral equation

$$y(x) - \lambda \int_0^1 xt y(t)\, dt = f(x), \qquad 0 \le x \le 1,$$

by the method of successive approximations. Here we have $K(x,t) = xt$, $a = 0$, and $b = 1$. We successively define

$$K_1(x,t) = xt, \quad K_2(x,t) = \int_0^1 (xz)(zt)\, dz = \frac{xt}{3}, \quad K_3(x,t) = \frac{1}{3}\int_0^1 (xz)(zt)\, dz = \frac{xt}{3^2}, \quad \ldots, \quad K_n(x,t) = \frac{xt}{3^{n-1}}.$$

According to formula (16.4.3.5) for the resolvent, we obtain

$$R(x,t;\lambda) = \sum_{n=1}^{\infty} \lambda^{n-1} K_n(x,t) = xt \sum_{n=1}^{\infty} \left(\frac{\lambda}{3}\right)^{n-1} = \frac{3xt}{3-\lambda},$$

where $|\lambda| < 3$, and it follows from formula (16.4.3.7) that the solution of the integral equation can be rewritten in the form

$$y(x) = f(x) + \lambda \int_0^1 \frac{3xt}{3-\lambda} f(t)\, dt, \qquad 0 \le x \le 1, \quad \lambda \ne 3.$$

### 16.4.4. Fredholm Theorems and the Fredholm Alternative

#### 16.4.4-1. Fredholm theorems.

THEOREM 1. *If $\lambda$ is a regular value, then both the Fredholm integral equation of the second kind and the transposed equation are solvable for any right-hand side, and both the equations have unique solutions. The corresponding homogeneous equations have only the trivial solutions.*

THEOREM 2. *For the nonhomogeneous integral equation to be solvable, it is necessary and sufficient that the right-hand side $f(x)$ satisfies the conditions*

$$\int_a^b f(x)\psi_k(x)\, dx = 0, \qquad k = 1, \ldots, n,$$

*where $\psi_k(x)$ is a complete set of linearly independent solutions of the corresponding transposed homogeneous equation.*

THEOREM 3. *If $\lambda$ is a characteristic value, then both the homogeneous integral equation and the transposed homogeneous equation have nontrivial solutions. The number of linearly independent solutions of the homogeneous integral equation is finite and is equal to the number of linearly independent solutions of the transposed homogeneous equation.*

THEOREM 4. *A Fredholm equation of the second kind has at most countably many characteristic values, whose only possible accumulation point is the point at infinity.*

#### 16.4.4-2. Fredholm alternative.

The Fredholm theorems imply the so-called Fredholm alternative, which is most frequently used in the investigation of integral equations.

THE FREDHOLM ALTERNATIVE. *Either the nonhomogeneous equation is solvable for any right-hand side or the corresponding homogeneous equation has nontrivial solutions.*

The first part of the alternative holds if the given value of the parameter is regular and the second if it is characteristic.

**Remark.** The Fredholm theory is also valid for integral equations of the second kind with weak singularity.

## 16.4.5. Fredholm Integral Equations of the Second Kind with Symmetric Kernel

**16.4.5-1. Characteristic values and eigenfunctions.**

Integral equations whose kernels are *symmetric*, that is, satisfy the condition $K(x,t) = K(t,x)$, are called *symmetric integral equations*.

Each symmetric kernel that is not identically zero has at least one characteristic value.

For any $n$, the set of characteristic values of the $n$th iterated kernel coincides with the set of $n$th powers of the characteristic values of the first kernel.

The eigenfunctions of a symmetric kernel corresponding to distinct characteristic values are orthogonal, i.e., if

$$\varphi_1(x) = \lambda_1 \int_a^b K(x,t)\varphi_1(t)\,dt, \quad \varphi_2(x) = \lambda_2 \int_a^b K(x,t)\varphi_2(t)\,dt, \quad \lambda_1 \neq \lambda_2,$$

then

$$(\varphi_1, \varphi_2) = 0, \quad (\varphi, \psi) \equiv \int_a^b \varphi(x)\psi(x)\,dx.$$

The characteristic values of a symmetric kernel are real.

The eigenfunctions can be normalized; namely, we can divide each characteristic function by its norm. If several linearly independent eigenfunctions correspond to the same characteristic value, say, $\varphi_1(x), \ldots, \varphi_n(x)$, then each linear combination of these functions is an eigenfunction as well, and these linear combinations can be chosen so that the corresponding eigenfunctions are orthonormal.

Indeed, the function

$$\psi_1(x) = \frac{\varphi_1(x)}{\|\varphi_1\|}, \quad \|\varphi_1\| = \sqrt{(\varphi_1, \varphi_1)},$$

has the norm equal to one, i.e., $\|\psi_1\| = 1$. Let us form a linear combination $\alpha\psi_1 + \varphi_2$ and choose $\alpha$ so that

$$(\alpha\psi_1 + \varphi_2, \psi_1) = 0,$$

i.e.,

$$\alpha = -\frac{(\varphi_2, \psi_1)}{(\psi_1, \psi_1)} = -(\varphi_2, \psi_1).$$

The function

$$\psi_2(x) = \frac{\alpha\psi_1 + \varphi_2}{\|\alpha\psi_1 + \varphi_2\|}$$

is orthogonal to $\psi_1(x)$ and has the unit norm. Next, we choose a linear combination $\alpha\psi_1 + \beta\psi_2 + \varphi_3$, where the constants $\alpha$ and $\beta$ can be found from the orthogonality relations

$$(\alpha\psi_1 + \beta\psi_2 + \varphi_3, \psi_1) = 0, \quad (\alpha\psi_1 + \beta\psi_2 + \varphi_3, \psi_2) = 0.$$

For the coefficients $\alpha$ and $\beta$ thus defined, the function

$$\psi_3 = \frac{\alpha\psi_1 + \beta\psi_2 + \varphi_2}{\|\alpha\psi_1 + \beta\psi_2 + \varphi_3\|}$$

is orthogonal to $\psi_1$ and $\psi_2$ and has the unit norm, and so on.

As was noted above, the eigenfunctions corresponding to distinct characteristic values are orthogonal. Hence, the sequence of eigenfunctions of a symmetric kernel can be made orthonormal.

In what follows we assume that the sequence of eigenfunctions of a symmetric kernel is orthonormal.

We also assume that the characteristic values are always numbered in the increasing order of their absolute values. Thus, if

$$\lambda_1, \lambda_2, \ldots, \lambda_n, \ldots \tag{16.4.5.1}$$

is the sequence of characteristic values of a symmetric kernel, and if a sequence of eigenfunctions

$$\varphi_1, \varphi_2, \ldots, \varphi_n, \ldots \tag{16.4.5.2}$$

corresponds to the sequence (16.4.5.1) so that

$$\varphi_n(x) - \lambda_n \int_a^b K(x,t)\varphi_n(t)\,dt = 0, \tag{16.4.5.3}$$

then

$$\int_a^b \varphi_i(x)\varphi_j(x)\,dx = \begin{cases} 1 & \text{for } i = j, \\ 0 & \text{for } i \neq j, \end{cases} \tag{16.4.5.4}$$

and

$$|\lambda_1| \leq |\lambda_2| \leq \cdots \leq |\lambda_n| \leq \cdots . \tag{16.4.5.5}$$

If there are infinitely many characteristic values, then it follows from the fourth Fredholm theorem that their only accumulation point is the point at infinity, and hence $\lambda_n \to \infty$ as $n \to \infty$.

The set of all characteristic values and the corresponding normalized eigenfunctions of a symmetric kernel is called the *system of characteristic values and eigenfunctions* of the kernel. The system of eigenfunctions is said to be *incomplete* if there exists a nonzero square integrable function that is orthogonal to all functions of the system. Otherwise, the system of eigenfunctions is said to be *complete*.

### 16.4.5-2. Bilinear series.

Assume that a kernel $K(x,t)$ admits an expansion in a uniformly convergent series with respect to the orthonormal system of its eigenfunctions:

$$K(x,t) = \sum_{k=1}^{\infty} a_k(x)\varphi_k(t) \tag{16.4.5.6}$$

for all $x$ in the case of a continuous kernel or for almost all $x$ in the case of a square integrable kernel.

We have

$$a_k(x) = \int_a^b K(x,t)\varphi_k(t)\,dt = \frac{\varphi_k(x)}{\lambda_k}, \tag{16.4.5.7}$$

and hence

$$K(x,t) = \sum_{k=1}^{\infty} \frac{\varphi_k(x)\varphi_k(t)}{\lambda_k}. \tag{16.4.5.8}$$

Conversely, if the series

$$\sum_{k=1}^{\infty} \frac{\varphi_k(x)\varphi_k(t)}{\lambda_k} \tag{16.4.5.9}$$

is uniformly convergent, then formula (16.4.5.8) holds.

The following assertion holds: the bilinear series (16.4.5.9) converges in mean-square to the kernel $K(x,t)$.

If a symmetric kernel $K(x,t)$ has finitely many characteristic values $\lambda_1, \ldots, \lambda_n$, then it is degenerate, because in this case, there are only $n$ terms remaining in the sum (16.4.5.8).

A kernel $K(x,t)$ is said to be *positive definite* if for all functions $\varphi(x)$ that are not identically zero we have

$$\int_a^b \int_a^b K(x,t)\varphi(x)\varphi(t)\,dx\,dt > 0,$$

and the above quadratic functional vanishes for $\varphi(x) = 0$ only. Such a kernel has positive characteristic values only.

Each symmetric positive definite continuous kernel can be decomposed in a bilinear series in eigenfunctions that is absolutely and uniformly convergent with respect to the variables $x,t$. The assertion remains valid if we assume that the kernel has finitely many negative characteristic values.

If a kernel $K(x,t)$ is symmetric, continuous on the square $S = \{a \leq x \leq b,\ a \leq t \leq b\}$, and has uniformly bounded partial derivatives on this square, then this kernel can be expanded in a uniformly convergent bilinear series in eigenfunctions.

### 16.4.5-3. Hilbert–Schmidt theorem.

If a function $f(x)$ can be represented in the form

$$f(x) = \int_a^b K(x,t)g(t)\,dt, \tag{16.4.5.10}$$

where the symmetric kernel $K(x,t)$ is square integrable and $g(t)$ is a square integrable function, then $f(x)$ can be represented by its *Fourier series* with respect to the orthonormal system of eigenfunctions of the kernel $K(x,t)$:

$$f(x) = \sum_{k=1}^{\infty} a_k \varphi_k(x), \tag{16.4.5.11}$$

where

$$a_k = \int_a^b f(x)\varphi_k(x)\,dx, \qquad k = 1, 2, \ldots$$

Moreover, if

$$\int_a^b K^2(x,t)\,dt \leq A < \infty, \tag{16.4.5.12}$$

then the series (16.4.5.11) is absolutely and uniformly convergent for any function $f(x)$ of the form (16.4.5.10).

*Remark.* In the Hilbert–Schmidt theorem, the completeness of the system of eigenfunctions is not assumed.

### 16.4.5-4. Bilinear series of iterated kernels.

By the definition of the iterated kernels, we have

$$K_m(x,t) = \int_a^b K(x,z)K_{m-1}(z,t)\,dz, \qquad m = 2, 3, \ldots \qquad (16.4.5.13)$$

The Fourier coefficients $a_k(t)$ of the kernel $K_m(x,t)$, regarded as a function of the variable $x$, with respect to the orthonormal system of eigenfunctions of the kernel $K(x,t)$ are equal to

$$a_k(t) = \int_a^b K_m(x,t)\varphi_k(x)\,dx = \frac{\varphi_k(t)}{\lambda_k^m}. \qquad (16.4.5.14)$$

On applying the Hilbert–Schmidt theorem to (16.4.5.13), we obtain

$$K_m(x,t) = \sum_{k=1}^{\infty} \frac{\varphi_k(x)\varphi_k(t)}{\lambda_k^m}, \qquad m = 2, 3, \ldots \qquad (16.4.5.15)$$

In formula (16.4.5.15), the sum of the series is understood as the limit in mean-square. If, in addition to the above assumptions, inequality (16.4.5.12) is satisfied, then the series in (16.4.5.15) is uniformly convergent.

### 16.4.5-5. Solution of the nonhomogeneous equation.

Let us represent an integral equation

$$y(x) - \lambda \int_a^b K(x,t)y(t)\,dt = f(x), \qquad a \le x \le b, \qquad (16.4.5.16)$$

where the parameter $\lambda$ is not a characteristic value, in the form

$$y(x) - f(x) = \lambda \int_a^b K(x,t)y(t)\,dt \qquad (16.4.5.17)$$

and apply the Hilbert–Schmidt theorem to the function $y(x) - f(x)$:

$$y(x) - f(x) = \sum_{k=1}^{\infty} A_k \varphi_k(x),$$

$$A_k = \int_a^b [y(x) - f(x)]\varphi_k(x)\,dx = \int_a^b y(x)\varphi_k(x)\,dx - \int_a^b f(x)\varphi_k(x)\,dx = y_k - f_k.$$

Taking into account the expansion (16.4.5.8), we obtain

$$\lambda \int_a^b K(x,t)y(t)\,dt = \lambda \sum_{k=1}^{\infty} \frac{y_k}{\lambda_k} \varphi_k(x),$$

and thus

$$\lambda \frac{y_k}{\lambda_k} = y_k - f_k, \qquad y_k = \frac{\lambda_k f_k}{\lambda_k - \lambda}, \qquad A_k = \frac{\lambda f_k}{\lambda_k - \lambda}. \qquad (16.4.5.18)$$

Hence,
$$y(x) = f(x) + \lambda \sum_{k=1}^{\infty} \frac{f_k}{\lambda_k - \lambda} \varphi_k(x). \qquad (16.4.5.19)$$

However, if $\lambda$ is a characteristic value, i.e.,
$$\lambda = \lambda_p = \lambda_{p+1} = \cdots = \lambda_q, \qquad (16.4.5.20)$$

then for $k \neq p, p+1, \ldots, q$, the terms (16.4.5.19) preserve their form. For $k = p, p+1, \ldots, q$, formula (16.4.5.18) implies the relation $f_k = A_k(\lambda - \lambda_k)/\lambda$, and by (16.4.5.20) we obtain $f_p = f_{p+1} = \cdots = f_q = 0$. The last relation means that
$$\int_a^b f(x)\varphi_k(x)\,dx = 0$$

for $k = p, p+1, \ldots, q$, i.e., the right-hand side of the equation must be orthogonal to the eigenfunctions that correspond to the characteristic value $\lambda$.

In this case, the solutions of equations (16.4.5.16) have the form
$$y(x) = f(x) + \lambda \sum_{k=1}^{\infty} \frac{f_k}{\lambda_k - \lambda} \varphi_k(x) + \sum_{k=p}^{q} C_k \varphi_k(x), \qquad (16.4.5.21)$$

where the terms in the first of the sums (16.4.5.21) with indices $k = p, p+1, \ldots, q$ must be omitted (for these indices, $f_k$ and $\lambda - \lambda_k$ vanish in this sum simultaneously). The coefficients $C_k$ in the second sum are arbitrary constants.

### 16.4.5-6. Fredholm alternative for symmetric equations.

The above results can be unified in the following alternative form.

A symmetric integral equation
$$y(x) - \lambda \int_a^b K(x,t)y(t)\,dt = f(x), \qquad a \leq x \leq b, \qquad (16.4.5.22)$$

for a given $\lambda$, either has a unique square integrable solution for an arbitrarily given function $f(x) \in L_2(a,b)$, in particular, $y = 0$ for $f = 0$, or the corresponding homogeneous equation has finitely many linearly independent solutions $Y_1(x), \ldots, Y_r(x), r > 0$.

For the second case, the nonhomogeneous equation has a solution if and only if the right-hand side $f(x)$ is orthogonal to all the functions $Y_1(x), \ldots, Y_r(x)$ on the interval $[a,b]$. Here the solution is defined only up to an arbitrary additive linear combination $A_1 Y_1(x) + \cdots + A_r Y_r(x)$.

### 16.4.5-7. Resolvent of a symmetric kernel.

The solution of a Fredholm equation of the second kind (16.4.5.22) can be written in the form
$$y(x) = f(x) + \lambda \int_a^b R(x,t;\lambda)f(t)\,dt, \qquad (16.4.5.23)$$

where the resolvent $R(x,t;\lambda)$ is given by the series

$$R(x,t;\lambda) = \sum_{k=1}^{\infty} \frac{\varphi_k(x)\varphi_k(t)}{\lambda_k - \lambda}. \qquad (16.4.5.24)$$

Here the collections $\varphi_k(x)$ and $\lambda_k$ form the system of eigenfunctions and characteristic values of (16.4.5.22). It follows from formula (16.4.5.24) that the resolvent of a symmetric kernel has only simple poles.

### 16.4.5-8. Extremal properties of characteristic values and eigenfunctions.

Let us introduce the notation

$$(u,w) = \int_a^b u(x)w(x)\,dx, \quad \|u\|^2 = (u,u),$$

$$(Ku,u) = \int_a^b \int_a^b K(x,t)u(x)u(t)\,dx\,dt,$$

where $(u,w)$ is the *inner product* of functions $u(x)$ and $w(x)$, $\|u\|$ is the *norm* of a function $u(x)$, and $(Ku,u)$ is the *quadratic form* generated by the kernel $K(x,t)$.

Let $\lambda_1$ be the characteristic value of the symmetric kernel $K(x,t)$ with minimum absolute value and let $y_1(x)$ be the eigenfunction corresponding to this value. Then

$$\frac{1}{|\lambda_1|} = \max_{y \neq 0} \frac{|(Ky,y)|}{\|y\|^2}; \qquad (16.4.5.25)$$

in particular, the maximum is attained, and $y = y_1$ is a maximum point.

Let $\lambda_1, \dots, \lambda_n$ be the first $n$ characteristic values of a symmetric kernel $K(x,t)$ (in the ascending order of their absolute values) and let $y_1(x), \dots, y_n(x)$ be orthonormal eigenfunctions corresponding to $\lambda_1, \dots, \lambda_n$, respectively. Then the formula

$$\frac{1}{|\lambda_{n+1}|} = \max \frac{|(Ky,y)|}{\|y\|^2} \qquad (16.4.5.26)$$

is valid for the characteristic value $\lambda_{n+1}$ following $\lambda_n$. The maximum is taken over the set of functions $y$ which are orthogonal to all $y_1, \dots, y_n$ and are not identically zero, that is, $y \neq 0$,

$$(y,y_j) = 0, \quad j = 1, \dots, n; \qquad (16.4.5.27)$$

in particular, the maximum in (16.4.5.26) is attained, and $y = y_{n+1}$ is a maximum point, where $y_{n+1}$ is any eigenfunction corresponding to the characteristic value $\lambda_{n+1}$ which is orthogonal to $y_1, \dots, y_n$.

**Remark.** For a positive definite kernel $K(x,t)$, the symbol of modulus on the right-hand sides of (16.4.5.26) and (16.4.5.27) can be omitted.

### 16.4.5-9. Skew-symmetric integral equations.

By a *skew-symmetric integral equation* we mean an equation whose kernel is skew-symmetric, i.e., an equation of the form

$$y(x) - \lambda \int_a^b K(x,t)y(t)\,dt = f(x) \qquad (16.4.5.28)$$

whose kernel $K(x,t)$ has the property $K(t,x) = -K(x,t)$.

Equation (16.4.5.28) with the skew-symmetric kernel has at least one characteristic value, and all its characteristic values are purely imaginary.

## 16.4.6. Methods of Integral Transforms

**16.4.6-1. Equation with difference kernel on the entire axis.**

Consider an integral equation of convolution type of the second kind with one kernel

$$y(x) + \frac{1}{\sqrt{2\pi}} \int_{-\infty}^{\infty} K(x-t)y(t)\, dt = f(x), \qquad -\infty < x < \infty. \tag{16.4.6.1}$$

Let us apply the (alternative) Fourier transform to equation (16.4.6.1). In this case, taking into account the convolution theorem (see Paragraph 11.4.1-3), we obtain

$$\mathcal{Y}(u)[1 + \mathcal{K}(u)] = \mathcal{F}(u). \tag{16.4.6.2}$$

Thus, on applying the Fourier transform we reduce the solution of the original integral equation (16.4.6.1) to the solution of the algebraic equation (16.4.6.2) for the transform of the unknown function. The solution of equation (16.4.6.2) has the form

$$\mathcal{Y}(u) = \frac{\mathcal{F}(u)}{1 + \mathcal{K}(u)}. \tag{16.4.6.3}$$

Formula (16.4.6.3) gives the transform of the solution of the original integral equation in terms of the transforms of the known functions, namely, the kernel and the right-hand side of the equation. The solution itself can be obtained by applying the Fourier inversion formula:

$$y(x) = \frac{1}{\sqrt{2\pi}} \int_{-\infty}^{\infty} \mathcal{Y}(u) e^{-iux}\, du = \frac{1}{\sqrt{2\pi}} \int_{-\infty}^{\infty} \frac{\mathcal{F}(u)}{1 + \mathcal{K}(u)} e^{-iux}\, du. \tag{16.4.6.4}$$

In fact, formula (16.4.6.4) solves the problem; however, sometimes it is not convenient because it requires the calculation of the transform $\mathcal{F}(u)$ for each right-hand side $f(x)$. In many cases, the representation of the solution of the nonhomogeneous integral equation via the resolvent of the original equation is more convenient. To obtain the desired representation, we note that formula (16.4.6.3) can be transformed to the expression

$$\mathcal{Y}(u) = [1 - \mathcal{R}(u)]\mathcal{F}(u), \qquad \mathcal{R}(u) = \frac{\mathcal{K}(u)}{1 + \mathcal{K}(u)}. \tag{16.4.6.5}$$

On the basis of (16.4.6.5), by applying the Fourier inversion formula and the convolution theorem (for transforms) we obtain

$$y(x) = f(x) - \frac{1}{\sqrt{2\pi}} \int_{-\infty}^{\infty} R(x-t)f(t)\, dt, \tag{16.4.6.6}$$

where the resolvent $R(x-t)$ of the integral equation (16.4.6.1) is given by the relation

$$R(x) = \frac{1}{\sqrt{2\pi}} \int_{-\infty}^{\infty} \frac{\mathcal{K}(u)}{1 + \mathcal{K}(u)} e^{-iux}\, du. \tag{16.4.6.7}$$

Thus, to determine the solution of the original integral equation (16.4.6.1), it suffices to find the function $R(x)$ by formula (16.4.6.7). To calculate direct and inverse Fourier transforms, one can use the corresponding tables from Sections T3.3 and T3.4, and the books by Bateman and Erdélyi (1954) and by Ditkin and Prudnikov (1965).

**Example.** Let us solve the integral equation

$$y(x) - \lambda \int_{-\infty}^{\infty} \exp(\alpha|x-t|) y(t)\, dt = f(x), \qquad -\infty < x < \infty, \tag{16.4.6.8}$$

which is a special case of equation (16.4.6.1) with kernel $K(x-t)$ given by the expression

$$K(x) = -\sqrt{2\pi}\, \lambda e^{-\alpha|x|}, \qquad \alpha > 0. \tag{16.4.6.9}$$

Let us find the function $R(x)$. To this end, we calculate the integral

$$\mathcal{K}(u) = -\int_{-\infty}^{\infty} \lambda e^{-\alpha|x|} e^{iux}\, dx = -\frac{2\alpha\lambda}{u^2 + \alpha^2}. \tag{16.4.6.10}$$

In this case, formula (16.4.6.5) implies

$$\mathcal{R}(u) = \frac{\mathcal{K}(u)}{1 + \mathcal{K}(u)} = -\frac{2\alpha\lambda}{u^2 + \alpha^2 - 2\alpha\lambda}, \tag{16.4.6.11}$$

and hence

$$R(x) = \frac{1}{\sqrt{2\pi}} \int_{-\infty}^{\infty} \mathcal{R}(u) e^{-iux}\, du = -\sqrt{\frac{2}{\pi}} \int_{-\infty}^{\infty} \frac{\alpha\lambda}{u^2 + \alpha^2 - 2\alpha\lambda} e^{-iux}\, du. \tag{16.4.6.12}$$

Assume that $\lambda < \tfrac{1}{2}\alpha$. In this case the integral (16.4.6.12) makes sense and can be calculated by means of the theory of residues on applying the Jordan lemma (see Subsection 11.1.2). After some algebraic manipulations, we obtain

$$R(x) = -\sqrt{2\pi}\, \frac{\alpha\lambda}{\sqrt{\alpha^2 - 2\alpha\lambda}} \exp\left(-|x|\sqrt{\alpha^2 - 2\alpha\lambda}\right) \tag{16.4.6.13}$$

and finally, in accordance with (16.4.6.6), we obtain

$$y(x) = f(x) + \frac{\alpha\lambda}{\sqrt{\alpha^2 - 2\alpha\lambda}} \int_{-\infty}^{\infty} \exp\left(-|x-t|\sqrt{\alpha^2 - 2\alpha\lambda}\right) f(t)\, dt, \qquad -\infty < x < \infty. \tag{16.4.6.14}$$

### 16.4.6-2. Equation with the kernel $K(x, t) = t^{-1} Q(x/t)$ on the semiaxis.

Here we consider the following equation on the semiaxis:

$$y(x) - \int_0^\infty \frac{1}{t} Q\left(\frac{x}{t}\right) y(t)\, dt = f(x). \tag{16.4.6.15}$$

To solve this equation we apply the Mellin transform which is defined as follows (see also Section 11.3):

$$\hat{f}(s) = \mathfrak{M}\{f(x), s\} \equiv \int_0^\infty f(x) x^{s-1}\, dx, \tag{16.4.6.16}$$

where $s = \sigma + i\tau$ is a complex variable ($\sigma_1 < \sigma < \sigma_2$) and $\hat{f}(s)$ is the transform of the function $f(x)$. In what follows, we briefly denote the Mellin transform by $\mathfrak{M}\{f(x)\} \equiv \mathfrak{M}\{f(x), s\}$.

For known $\hat{f}(s)$, the original function can be found by means of the Mellin inversion formula

$$f(x) = \mathfrak{M}^{-1}\{\hat{f}(s)\} \equiv \frac{1}{2\pi i} \int_{c-i\infty}^{c+i\infty} \hat{f}(s) x^{-s}\, ds, \qquad \sigma_1 < c < \sigma_2, \tag{16.4.6.17}$$

where the integration path is parallel to the imaginary axis of the complex plane $s$ and the integral is understood in the sense of the Cauchy principal value.

On applying the Mellin transform to equation (16.4.6.15) and taking into account the fact that the integral with such a kernel is transformed into the product by the rule (see Subsection 11.3.2)

$$\mathfrak{M}\left\{\int_0^\infty \frac{1}{t} Q\!\left(\frac{x}{t}\right) y(t)\, dt\right\} = \hat{Q}(s)\hat{y}(s),$$

we obtain the following equation for the transform $\hat{y}(s)$:

$$\hat{y}(s) - \hat{Q}(s)\hat{y}(s) = \hat{f}(s).$$

The solution of this equation is given by the formula

$$\hat{y}(s) = \frac{\hat{f}(s)}{1 - \hat{Q}(s)}. \tag{16.4.6.18}$$

On applying the Mellin inversion formula to equation (16.4.6.18) we obtain the solution of the original integral equation

$$y(x) = \frac{1}{2\pi i} \int_{c-i\infty}^{c+i\infty} \frac{\hat{f}(s)}{1 - \hat{Q}(s)} x^{-s}\, ds. \tag{16.4.6.19}$$

This solution can also be represented via the resolvent in the form

$$y(x) = f(x) + \int_0^\infty \frac{1}{t} N\!\left(\frac{x}{t}\right) f(t)\, dt, \tag{16.4.6.20}$$

where we have used the notation

$$N(x) = \mathfrak{M}^{-1}\{\hat{N}(s)\}, \qquad \hat{N}(s) = \frac{\hat{Q}(s)}{1 - \hat{Q}(s)}. \tag{16.4.6.21}$$

Under the application of this analytical method of solution, the following technical difficulties can occur: (a) in the calculation of the transform for a given kernel $K(x)$ and (b) in the calculation of the solution for the known transform $\hat{y}(s)$. To find the corresponding integrals, tables of direct and inverse Mellin transforms are applied (e.g., see Sections T3.5 and T3.6). In many cases, the relationship between the Mellin transform and the Fourier and Laplace transforms is first used:

$$\mathfrak{M}\{f(x), s\} = \mathfrak{F}\{f(e^x), is\} = \mathfrak{L}\{f(e^x), -s\} + \mathfrak{L}\{f(e^{-x}), s\}, \tag{16.4.6.22}$$

and then tables of direct and inverse Fourier transforms and Laplace transforms are applied (see Sections T3.1–T3.4).

---

**16.4.6-3. Equation with the kernel $K(x,t) = t^\beta Q(xt)$ on the semiaxis.**

Consider the following equation on the semiaxis:

$$y(x) - \int_0^\infty t^\beta Q(xt) y(t)\, dt = f(x). \tag{16.4.6.23}$$

To solve this equation, we apply the Mellin transform. On multiplying equation (16.4.6.23) by $x^{s-1}$ and integrating with respect to $x$ from zero to infinity, we obtain

$$\int_0^\infty y(x)x^{s-1}\,dx - \int_0^\infty y(t)t^\beta\,dt \int_0^\infty Q(xt)x^{s-1}\,dx = \int_0^\infty f(x)x^{s-1}\,dx. \qquad (16.4.6.24)$$

Let us make the change of variables $z = xt$. We finally obtain

$$\hat{y}(s) - \hat{Q}(s)\int_0^\infty y(t)t^{\beta-s}\,dt = \hat{f}(s). \qquad (16.4.6.25)$$

Taking into account the relation

$$\int_0^\infty y(t)t^{\beta-s}\,dt = \hat{y}(1+\beta-s),$$

we rewrite equation (16.4.6.25) in the form

$$\hat{y}(s) - \hat{Q}(s)\hat{y}(1+\beta-s) = \hat{f}(s). \qquad (16.4.6.26)$$

On replacing $s$ by $1+\beta-s$ in equation (16.4.6.26), we obtain

$$\hat{y}(1+\beta-s) - \hat{Q}(1+\beta-s)\hat{y}(s) = \hat{f}(1+\beta-s). \qquad (16.4.6.27)$$

Let us eliminate $\hat{y}(1+\beta-s)$ from (16.4.6.26) and (16.4.6.27), and then solve the resulting equation for $\hat{y}(s)$. We thus find the transform of the solution:

$$\hat{y}(s) = \frac{\hat{f}(s) + \hat{Q}(s)\hat{f}(1+\beta-s)}{1 - \hat{Q}(s)\hat{Q}(1+\beta-s)}.$$

On applying the Mellin inversion formula, we obtain the solution of the integral equation (16.4.6.23) in the form

$$y(x) = \frac{1}{2\pi i}\int_{c-i\infty}^{c+i\infty} \frac{\hat{f}(s) + \hat{Q}(s)\hat{f}(1+\beta-s)}{1 - \hat{Q}(s)\hat{Q}(1+\beta-s)} x^{-s}\,ds.$$

## 16.4.7. Method of Approximating a Kernel by a Degenerate One

### 16.4.7-1. Approximation of the kernel.

For the approximate solution of the Fredholm integral equation of the second kind

$$y(x) - \int_a^b K(x,t)y(t)\,dt = f(x), \qquad a \le x \le b, \qquad (16.4.7.1)$$

where, for simplicity, the functions $f(x)$ and $K(x,t)$ are assumed to be continuous, it is useful to replace the kernel $K(x,t)$ by a close degenerate kernel

$$K_{(n)}(x,t) = \sum_{k=0}^n g_k(x)h_k(t). \qquad (16.4.7.2)$$

Let us indicate several ways to perform such a change. If the kernel $K(x, t)$ is differentiable with respect to $x$ on $[a, b]$ sufficiently many times, then, for a degenerate kernel $K_{(n)}(x, t)$, we can take a finite segment of the Taylor series:

$$K_{(n)}(x, t) = \sum_{m=0}^{n} \frac{(x - x_0)^m}{m!} K_x^{(m)}(x_0, t), \qquad (16.4.7.3)$$

where $x_0 \in [a, b]$. A similar trick can be applied for the case in which $K(x, t)$ is differentiable with respect to $t$ on $[a, b]$ sufficiently many times.

To construct a degenerate kernel, a finite segment of the double Fourier series can be used:

$$K_{(n)}(x, t) = \sum_{p=0}^{n} \sum_{q=0}^{n} a_{pq} (x - x_0)^p (t - t_0)^q, \qquad (16.4.7.4)$$

where

$$a_{pq} = \frac{1}{p!\, q!} \frac{\partial^{p+q}}{\partial x^p \partial t^q} K(x, t) \bigg|_{\substack{x=x_0 \\ t=t_0}}, \qquad a \le x_0 \le b, \quad a \le t_0 \le b.$$

A continuous kernel $K(x, t)$ admits an approximation by a trigonometric polynomial of period $2l$, where $l = b - a$.

For instance, we can set

$$K_{(n)}(x, t) = \frac{1}{2} a_0(t) + \sum_{k=1}^{n} a_k(t) \cos\left(\frac{k\pi x}{l}\right), \qquad (16.4.7.5)$$

where the $a_k(t)$ ($k = 0, 1, 2, \dots$) are the Fourier coefficients

$$a_k(t) = \frac{2}{l} \int_a^b K(x, t) \cos\left(\frac{p\pi x}{l}\right) dx. \qquad (16.4.7.6)$$

A similar decomposition can be obtained by interchanging the roles of the variables $x$ and $t$. A finite segment of the double Fourier series can also be applied by setting, for instance,

$$a_k(t) \approx \frac{1}{2} a_{k0} + \sum_{m=1}^{n} a_{km} \cos\left(\frac{m\pi t}{l}\right), \qquad k = 0, 1, \dots, n, \qquad (16.4.7.7)$$

and it follows from formulas (16.4.7.5)–(16.4.7.7) that

$$K_{(n)}(x, t) = \frac{1}{4} a_{00} + \frac{1}{2} \sum_{k=1}^{n} a_{k0} \cos\left(\frac{k\pi x}{l}\right) + \frac{1}{2} \sum_{m=1}^{n} a_{0m} \cos\left(\frac{m\pi t}{l}\right)$$

$$+ \sum_{k=1}^{n} \sum_{m=1}^{n} a_{km} \cos\left(\frac{k\pi x}{l}\right) \cos\left(\frac{m\pi t}{l}\right),$$

where

$$a_{km} = \frac{4}{l^2} \int_a^b \int_a^b K(x, t) \cos\left(\frac{k\pi x}{l}\right) \cos\left(\frac{m\pi t}{l}\right) dx\, dt. \qquad (16.4.7.8)$$

One can also use other methods of interpolating and approximating the kernel $K(x, t)$.

### 16.4.7-2. Approximate solution.

If $K_{(n)}(x,t)$ is an approximate degenerate kernel for a given exact kernel $K(x,t)$ and if a function $f_n(x)$ is close to $f(x)$, then the solution $y_n(x)$ of the integral equation

$$y_n(x) - \int_a^b K_{(n)}(x,t) y_n(t)\, dt = f_n(x) \tag{16.4.7.9}$$

can be regarded as an approximation to the solution $y(x)$ of equation (16.4.7.1).

Assume that the following error estimates hold:

$$\int_a^b |K(x,t) - K_{(n)}(x,t)|\, dt \le \varepsilon, \qquad |f(x) - f_n(x)| \le \delta.$$

Next, let the resolvent $R_n(x,t)$ of equation (16.4.7.9) satisfy the relation

$$\int_a^b |R_n(x,t)|\, dt \le M_n$$

for $a \le x \le b$. Finally, assume that the following inequality holds:

$$q = \varepsilon(1 + M_n) < 1.$$

In this case, equation (16.4.7.1) has a unique solution $y(x)$ and

$$|y(x) - y_n(x)| \le \varepsilon \frac{N(1+M_n)^2}{1-q} + \delta, \qquad N = \max_{a \le x \le b} |f(x)|. \tag{16.4.7.10}$$

**Example.** Let us find an approximate solution of the equation

$$y(x) - \int_0^{1/2} e^{-x^2 t^2} y(t)\, dt = 1. \tag{16.4.7.11}$$

Applying the expansion in a double Taylor series, we replace the kernel $K(x,t) = e^{-x^2 t^2}$ with the degenerate kernel

$$K_{(2)}(x,t) = 1 - x^2 t^2 + \tfrac{1}{2} x^4 t^4.$$

Hence, instead of equation (16.4.7.11) we obtain

$$y_2(x) = 1 + \int_0^{1/2} \left(1 - x^2 t^2 + \tfrac{1}{2} x^4 t^4\right) y_2(t)\, dt. \tag{16.4.7.12}$$

Therefore,

$$y_2(x) = 1 + A_1 + A_2 x^2 + A_3 x^4, \tag{16.4.7.13}$$

where

$$A_1 = \int_0^{1/2} y_2(x)\, dx, \qquad A_2 = -\int_0^{1/2} x^2 y_2(x)\, dx, \qquad A_3 = \frac{1}{2} \int_0^{1/2} x^4 y_2(x)\, dx. \tag{16.4.7.14}$$

From (16.4.7.13) and (16.4.7.14) we obtain a system of three equations with three unknowns; to the fourth decimal place, the solution is

$$A_1 = 0.9930, \qquad A_2 = -0.0833, \qquad A_3 = 0.0007.$$

Hence,

$$y(x) \approx y_2(x) = 1.9930 - 0.0833\, x^2 + 0.0007\, x^4, \qquad 0 \le x \le \tfrac{1}{2}. \tag{16.4.7.15}$$

An error estimate for the approximate solution (16.4.7.15) can be performed by formula (16.4.7.10).

## 16.4.8. Collocation Method

### 16.4.8-1. General remarks.

Let us rewrite the Fredholm integral equation of the second kind in the form

$$\varepsilon[y(x)] \equiv y(x) - \lambda \int_a^b K(x,t)y(t)\,dt - f(x) = 0. \tag{16.4.8.1}$$

Let us seek an approximate solution of equation (16.4.8.1) in the special form

$$Y_n(x) = \Phi(x, A_1, \ldots, A_n) \tag{16.4.8.2}$$

with free parameters $A_1, \ldots, A_n$ (undetermined coefficients). On substituting the expression (16.4.8.2) into equation (16.4.8.1), we obtain the residual

$$\varepsilon[Y_n(x)] = Y_n(x) - \lambda \int_a^b K(x,t)Y_n(t)\,dt - f(x). \tag{16.4.8.3}$$

If $y(x)$ is an exact solution, then, clearly, the residual $\varepsilon[y(x)]$ is zero. Therefore, one tries to choose the parameters $A_1, \ldots, A_n$ so that, in a sense, the residual $\varepsilon[Y_n(x)]$ is as small as possible. The residual $\varepsilon[Y_n(x)]$ can be minimized in several ways. Usually, to simplify the calculations, a function $Y_n(x)$ linearly depending on the parameters $A_1, \ldots, A_n$ is taken. On finding the parameters $A_1, \ldots, A_n$, we obtain an approximate solution (16.4.8.2). If

$$\lim_{n\to\infty} Y_n(x) = y(x), \tag{16.4.8.4}$$

then, by taking a sufficiently large number of parameters $A_1, \ldots, A_n$, we find that the solution $y(x)$ can be found with an arbitrary prescribed precision.

Now let us go to the description of a concrete method of construction of an approximate solution $Y_n(x)$.

### 16.4.8-2. Approximate solution.

We set

$$Y_n(x) = \varphi_0(x) + \sum_{i=1}^n A_i \varphi_i(x), \tag{16.4.8.5}$$

where $\varphi_0(x), \varphi_1(x), \ldots, \varphi_n(x)$ are given functions (*coordinate functions*) and $A_1, \ldots, A_n$ are indeterminate coefficients, and assume that the functions $\varphi_i(x)$ ($i = 1, \ldots, n$) are linearly independent. Note that, in particular, we can take $\varphi_0(x) = f(x)$ or $\varphi_0(x) \equiv 0$. On substituting the expression (16.4.8.5) into the left-hand side of equation (16.4.8.1), we obtain the residual

$$\varepsilon[Y_n(x)] = \varphi_0(x) + \sum_{i=1}^n A_i \varphi_i(x) - f(x) - \lambda \int_a^b K(x,t)\left[\varphi_0(t) + \sum_{i=1}^n A_i \varphi_i(t)\right]dt,$$

or

$$\varepsilon[Y_n(x)] = \psi_0(x,\lambda) + \sum_{i=1}^n A_i \psi_i(x,\lambda), \tag{16.4.8.6}$$

where

$$\psi_0(x, \lambda) = \varphi_0(x) - f(x) - \lambda \int_a^b K(x, t)\varphi_0(t)\, dt,$$
$$\psi_i(x, \lambda) = \varphi_i(x) - \lambda \int_a^b K(x, t)\varphi_i(t)\, dt, \qquad i = 1, \ldots, n.$$
(16.4.8.7)

According to the collocation method, we require that the residual $\varepsilon[Y_n(x)]$ be zero at the given system of *the collocation points* $x_1, \ldots, x_n$ on the interval $[a, b]$, i.e., we set

$$\varepsilon[Y_n(x_j)] = 0, \qquad j = 1, \ldots, n,$$

where
$$a \le x_1 < x_2 < \cdots < x_{n-1} < x_n \le b.$$

It is common practice to set $x_1 = a$ and $x_n = b$.

This, together with formula (16.4.8.6), implies the linear algebraic system

$$\sum_{i=1}^n A_i \psi_i(x_j, \lambda) = -\psi_0(x_j, \lambda), \qquad j = 1, \ldots, n,$$
(16.4.8.8)

for the coefficients $A_1, \ldots, A_n$. If the determinant of system (16.4.8.8) is nonzero, $\det[\psi_i(x_j, \lambda)] \ne 0$, then system (16.4.8.8) uniquely determines the numbers $A_1, \ldots, A_n$, and hence makes it possible to find the approximate solution $Y_n(x)$ by formula (16.4.8.5).

### 16.4.8-3. Eigenfunctions of the equation.

On equating the determinant with zero, we obtain the relation

$$\det[\psi_i(x_j, \lambda)] = 0,$$

which enables us to find approximate values $\widetilde{\lambda}_k$ ($k = 1, \ldots, n$) for the characteristic values of the kernel $K(x, t)$.

If we set
$$f(x) \equiv 0, \qquad \varphi_0(x) \equiv 0, \qquad \lambda = \widetilde{\lambda}_k,$$

then, instead of system (16.4.8.8), we obtain the homogeneous system

$$\sum_{i=1}^n \widetilde{A}_i^{(k)} \psi_i(x_j, \widetilde{\lambda}_k) = 0, \qquad j = 1, \ldots, n.$$
(16.4.8.9)

On finding nonzero solutions $\widetilde{A}_i^{(k)}$ ($i = 1, \ldots, n$) of system (16.4.8.9), we obtain approximate eigenfunctions for the kernel $K(x, t)$:

$$\widetilde{Y}_n^{(k)}(x) = \sum_{i=1}^n \widetilde{A}_i^{(k)} \varphi_i(x),$$

that correspond to its characteristic value $\lambda_k \approx \widetilde{\lambda}_k$.

**Example.** Let us solve the equation

$$y(x) - \int_0^1 \frac{t^2 y(t)}{x^2 + t^2} \, dt = x \arctan \frac{1}{x} \qquad (16.4.8.10)$$

by the collocation method.

We set
$$Y_2(x) = A_1 + A_2 x.$$

On substituting this expression into equation (16.4.8.10), we obtain the residual

$$\varepsilon[Y_2(x)] = -A_1 x \arctan \frac{1}{x} + A_2 \left[ x - \frac{1}{2} + \frac{x^2}{2} \ln\left(1 + \frac{1}{x^2}\right) \right] - x \arctan \frac{1}{x}.$$

On choosing the collocation points $x_1 = 0$ and $x_2 = 1$ and taking into account the relations

$$\lim_{x \to 0} x \arctan \frac{1}{x} = 0, \quad \lim_{x \to 0} x^2 \ln\left(1 + \frac{1}{x^2}\right) = 0,$$

we obtain the following system for the coefficients $A_1$ and $A_2$:

$$0 \times A_1 - \tfrac{1}{2} A_2 = 0,$$
$$-\tfrac{\pi}{4} A_1 + \tfrac{1}{2}(1 + \ln 2) A_2 = \tfrac{\pi}{4}.$$

This implies $A_2 = 0$ and $A_1 = -1$. Thus,

$$Y_2(x) = -1. \qquad (16.4.8.11)$$

We can readily verify that the approximate solution (16.4.8.11) thus obtained is exact.

## 16.4.9. Method of Least Squares

### 16.4.9-1. Description of the method.

By analogy with the collocation method, for the equation

$$\varepsilon[y(x)] \equiv y(x) - \lambda \int_a^b K(x,t) y(t) \, dt - f(x) = 0 \qquad (16.4.9.1)$$

we set

$$Y_n(x) = \varphi_0(x) + \sum_{i=1}^n A_i \varphi_i(x), \qquad (16.4.9.2)$$

where $\varphi_0(x), \varphi_1(x), \ldots, \varphi_n(x)$ are given functions, $A_1, \ldots, A_n$ are indeterminate coefficients, and $\varphi_i(x)$ $(i = 1, \ldots, n)$ are linearly independent.

On substituting (16.4.9.2) into the left-hand side of equation (16.4.9.1), we obtain the residual

$$\varepsilon[Y_n(x)] = \psi_0(x, \lambda) + \sum_{i=1}^n A_i \psi_i(x, \lambda), \qquad (16.4.9.3)$$

where $\psi_0(x, \lambda)$ and $\psi_i(x, \lambda)$ $(i = 1, \ldots, n)$ are defined by formulas (16.4.8.7).

According to the method of least squares, the coefficients $A_i$ $(i = 1, \ldots, n)$ can be found from the condition for the minimum of the integral

$$I = \int_a^b \{\varepsilon[Y_n(x)]\}^2 \, dx = \int_a^b \left[ \psi_0(x, \lambda) + \sum_{i=1}^n A_i \psi_i(x, \lambda) \right]^2 dx. \qquad (16.4.9.4)$$

This requirement leads to the algebraic system of equations

$$\frac{\partial I}{\partial A_j} = 0, \qquad j = 1, \ldots, n, \tag{16.4.9.5}$$

and hence, on the basis of (16.4.9.4), by differentiating with respect to the parameters $A_1, \ldots, A_n$ under the integral sign, we obtain

$$\frac{1}{2}\frac{\partial I}{\partial A_j} = \int_a^b \psi_j(x,\lambda)\left[\psi_0(x,\lambda) + \sum_{i=1}^n A_i\psi_i(x,\lambda)\right]dx = 0, \qquad j = 1, \ldots, n. \tag{16.4.9.6}$$

Using the notation

$$c_{ij}(\lambda) = \int_a^b \psi_i(x,\lambda)\psi_j(x,\lambda)\,dx, \tag{16.4.9.7}$$

we can rewrite system (16.4.9.6) in the form of the *normal system of the method of least squares*:

$$\begin{aligned} c_{11}(\lambda)A_1 + c_{12}(\lambda)A_2 + \cdots + c_{1n}(\lambda)A_n &= -c_{10}(\lambda), \\ c_{21}(\lambda)A_1 + c_{22}(\lambda)A_2 + \cdots + c_{2n}(\lambda)A_n &= -c_{20}(\lambda), \\ &\cdots\cdots\cdots\cdots\cdots\cdots\cdots\cdots\cdots\cdots\cdots \\ c_{n1}(\lambda)A_1 + c_{n2}(\lambda)A_2 + \cdots + c_{nn}(\lambda)A_n &= -c_{n0}(\lambda). \end{aligned} \tag{16.4.9.8}$$

Note that if $\varphi_0(x) \equiv 0$, then $\psi_0(x) = -f(x)$. Moreover, since $c_{ij}(\lambda) = c_{ji}(\lambda)$, the matrix of system (16.4.9.8) is symmetric.

**16.4.9-2. Construction of eigenfunctions.**

The method of least squares can also be applied for the approximate construction of characteristic values and eigenfunctions of the kernel $K(x,s)$, similarly to the way in which it can be done in the collocation method. Namely, by setting $f(x) \equiv 0$ and $\varphi_0(x) \equiv 0$, which implies $\psi_0(x) \equiv 0$, we determine approximate values of the characteristic values from the algebraic equation

$$\det[c_{ij}(\lambda)] = 0.$$

After this, approximate eigenfunctions can be found from the homogeneous system of the form (16.4.9.8), where, instead of $\lambda$, the corresponding approximate value is substituted.

## 16.4.10. Bubnov–Galerkin Method

**16.4.10-1. Description of the method.**

Let

$$\varepsilon[y(x)] \equiv y(x) - \lambda \int_a^b K(x,t)y(t)\,dt - f(x) = 0. \tag{16.4.10.1}$$

Similarly to the above reasoning, we seek an approximate solution of equation (16.4.10.1) in the form of a finite sum

$$Y_n(x) = f(x) + \sum_{i=1}^n A_i\varphi_i(x), \qquad i = 1, \ldots, n, \tag{16.4.10.2}$$

## 16.4. Linear Integral Equations of the Second Kind with Constant Limits of Integration

where the $\varphi_i(x)$ ($i = 1, \ldots, n$) are some given linearly independent functions (*coordinate functions*) and $A_1, \ldots, A_n$ are indeterminate coefficients. On substituting the expression (16.4.10.2) into the left-hand side of equation (16.4.10.1), we obtain the residual

$$\varepsilon[Y_n(x)] = \sum_{j=1}^{n} A_j \left[ \varphi_j(x) - \lambda \int_a^b K(x,t)\varphi_j(t)\,dt \right] - \lambda \int_a^b K(x,t)f(t)\,dt. \quad (16.4.10.3)$$

According to the Bubnov–Galerkin method, the coefficients $A_i$ ($i = 1, \ldots, n$) are defined from the condition that the residual is orthogonal to all coordinate functions $\varphi_1(x), \ldots, \varphi_n(x)$. This gives the system of equations

$$\int_a^b \varepsilon[Y_n(x)]\varphi_i(x)\,dx = 0, \qquad i = 1, \ldots, n,$$

or, by virtue of (16.4.10.3),

$$\sum_{j=1}^{n} (\alpha_{ij} - \lambda\beta_{ij})A_j = \lambda\gamma_i, \qquad i = 1, \ldots, n, \quad (16.4.10.4)$$

where

$$\alpha_{ij} = \int_a^b \varphi_i(x)\varphi_j(x)\,dx, \qquad \beta_{ij} = \int_a^b \int_a^b K(x,t)\varphi_i(x)\varphi_j(t)\,dt\,dx,$$

$$\gamma_i = \int_a^b \int_a^b K(x,t)\varphi_i(x)f(t)\,dt\,dx, \qquad i,j = 1, \ldots, n.$$

If the determinant of system (16.4.10.4)

$$D(\lambda) = \det[\alpha_{ij} - \lambda\beta_{ij}]$$

is nonzero, then this system uniquely determines the coefficients $A_1, \ldots, A_n$. In this case, formula (16.4.10.2) gives an approximate solution of the integral equation (16.4.10.1).

### 16.4.10-2. Characteristic values.

The equation $D(\lambda) = 0$ gives approximate characteristic values $\widetilde{\lambda}_1, \ldots, \widetilde{\lambda}_n$ of the integral equation. On finding nonzero solutions of the homogeneous linear system

$$\sum_{j=1}^{n} (\alpha_{ij} - \widetilde{\lambda}_k \beta_{ij})\widetilde{A}_j^{(k)} = 0, \qquad i = 1, \ldots, n,$$

we can construct approximate eigenfunctions $\widetilde{Y}_n^{(k)}(x)$ corresponding to characteristic values $\widetilde{\lambda}_k$:

$$\widetilde{Y}_n^{(k)}(x) = \sum_{i=1}^{n} \widetilde{A}_i^{(k)} \varphi(x).$$

It can be shown that the Bubnov–Galerkin method is equivalent to the replacement of the kernel $K(x,t)$ by some degenerate kernel $K_{(n)}(x,t)$. Therefore, for the approximate solution $Y_n(x)$ we have an error estimate similar to that presented in Subsection 16.4.7-2.

**Example.** Let us find the first two characteristic values of the integral equation

$$\varepsilon[y(x)] \equiv y(x) - \lambda \int_0^1 K(x,t) y(t) \, dt = 0,$$

where

$$K(x,t) = \begin{cases} t & \text{for } t \le x, \\ x & \text{for } t > x. \end{cases} \tag{16.4.10.5}$$

On the basis of (16.4.10.5), we have

$$\varepsilon[y(x)] = y(x) - \lambda \left\{ \int_0^x t y(t) \, dt + \int_x^1 x y(t) \, dt \right\}.$$

We set

$$Y_2(x) = A_1 x + A_2 x^2.$$

In this case

$$\varepsilon[Y_2(x)] = A_1 x + A_2 x^2 - \lambda \left[ \tfrac{1}{3} A_1 x^3 + \tfrac{1}{4} A_2 x^4 + x \left( \tfrac{1}{2} A_1 + \tfrac{1}{3} A_2 \right) - \left( \tfrac{1}{2} A_1 x^3 + \tfrac{1}{3} A_2 x^4 \right) \right] =$$
$$= A_1 \left[ \left(1 - \tfrac{1}{2} \lambda \right) x + \tfrac{1}{6} \lambda x^3 \right] + A_2 \left( -\tfrac{1}{3} \lambda x + x^2 + \tfrac{1}{12} \lambda x^4 \right).$$

On orthogonalizing the residual $\varepsilon[Y_2(x)]$, we obtain the system

$$\int_0^1 \varepsilon[Y_2(x)] x \, dx = 0,$$

$$\int_0^1 \varepsilon[Y_2(x)] x^2 \, dx = 0,$$

or the following homogeneous system of two algebraic equations with two unknowns:

$$A_1(120 - 48\lambda) + A_2(90 - 35\lambda) = 0,$$
$$A_1(630 - 245\lambda) + A_2(504 - 180\lambda) = 0. \tag{16.4.10.6}$$

On equating the determinant of system (16.4.10.6) with zero, we obtain the following equation for the characteristic values:

$$D(\lambda) \equiv \begin{vmatrix} 120 - 48\lambda & 90 - 35\lambda \\ 630 - 245\lambda & 504 - 180\lambda \end{vmatrix} = 0.$$

Hence,

$$\lambda^2 - 26.03\lambda + 58.15 = 0. \tag{16.4.10.7}$$

Equations (16.4.10.7) imply

$$\widetilde{\lambda}_1 = 2.462\ldots \quad \text{and} \quad \widetilde{\lambda}_2 = 23.568\ldots$$

For comparison we present the exact characteristic values:

$$\lambda_1 = \tfrac{1}{4} \pi^2 = 2.467\ldots \quad \text{and} \quad \lambda_2 = \tfrac{9}{4} \pi^2 = 22.206\ldots,$$

which can be obtained from the solution of the following boundary value problem equivalent to the original equation:

$$y''_{xx}(x) + \lambda y(x) = 0; \quad y(0) = 0, \quad y'_x(1) = 0.$$

Thus, the error of $\widetilde{\lambda}_1$ is approximately equal to 0.2% and that of $\widetilde{\lambda}_2$, to 6%.

### 16.4.11. Quadrature Method

#### 16.4.11-1. General scheme for Fredholm equations of the second kind.

In the solution of an integral equation, the reduction to the solution of systems of algebraic equations obtained by replacing the integrals with finite sums is one of the most effective tools. The method of quadratures is related to the approximation methods. It is widespread in practice because it is rather universal with respect to the principle of constructing algorithms for solving both linear and nonlinear equations.

## 16.4. Linear Integral Equations of the Second Kind with Constant Limits of Integration

Just as in the case of Volterra equations, the method is based on a quadrature formula (see Subsection 16.1.5):

$$\int_a^b \varphi(x)\,dx = \sum_{j=1}^n A_j \varphi(x_j) + \varepsilon_n[\varphi], \qquad (16.4.11.1)$$

where the $x_j$ are the nodes of the quadrature formula, $A_j$ are given coefficients that do not depend on the function $\varphi(x)$, and $\varepsilon_n[\varphi]$ is the error of replacement of the integral by the sum (the truncation error).

If in the Fredholm integral equation of the second kind

$$y(x) - \lambda \int_a^b K(x,t)y(t)\,dt = f(x), \qquad a \le x \le b, \qquad (16.4.11.2)$$

we set $x = x_i$ ($i = 1, \ldots, n$), then we obtain the following relation that is the basic formula for the method under consideration:

$$y(x_i) - \lambda \int_a^b K(x_i,t)y(t)\,dt = f(x_i), \qquad i = 1, \ldots, n. \qquad (16.4.11.3)$$

Applying the quadrature formula (16.4.11.1) to the integral in (16.4.11.3), we arrive at the following system of equations:

$$y(x_i) - \lambda \sum_{j=1}^n A_j K(x_i, x_j) y(x_j) = f(x_i) + \lambda \varepsilon_n[y]. \qquad (16.4.11.4)$$

By neglecting the small term $\lambda \varepsilon_n[y]$ in this formula, we obtain the system of linear algebraic equations for approximate values $y_i$ of the solution $y(x)$ at the nodes $x_1, \ldots, x_n$:

$$y_i - \lambda \sum_{j=1}^n A_j K_{ij} y_j = f_i, \qquad i = 1, \ldots, n, \qquad (16.4.11.5)$$

where $K_{ij} = K(x_i, x_j)$, $f_i = f(x_i)$.

The solution of system (16.4.11.5) gives the values $y_1, \ldots, y_n$, which determine an approximate solution of the integral equation (16.4.11.2) on the entire interval $[a,b]$ by interpolation. Here for the approximate solution we can take the function obtained by linear interpolation, i.e., the function that coincides with $y_i$ at the points $x_i$ and is linear on each of the intervals $[x_i, x_{i+1}]$. Moreover, for an analytic expression of the approximate solution to the equation, a function

$$\widetilde{y}(x) = f(x) + \lambda \sum_{j=1}^n A_j K(x, x_j) y_j \qquad (16.4.11.6)$$

can be chosen, which also takes the values $y_1, \ldots, y_n$ at the points $x_1, \ldots, x_n$.

**Example.** Consider the equation

$$y(x) - \tfrac{1}{2} \int_0^1 xt\, y(t)\,dt = \tfrac{5}{6} x.$$

Let us choose the nodes $x_1 = 0$, $x_2 = \frac{1}{2}$, $x_3 = 1$ and calculate the values of the right-hand side $f(x) = \frac{5}{6}x$ and of the kernel $K(x,t) = xt$ at these nodes:
$$f(0) = 0, \quad f\left(\tfrac{1}{2}\right) = \tfrac{5}{12}, \quad f(1) = \tfrac{5}{6},$$
$$K(0,0) = 0, \quad K\left(0,\tfrac{1}{2}\right) = 0, \quad K(0,1) = 0, \quad K\left(\tfrac{1}{2},0\right) = 0, \quad K\left(\tfrac{1}{2},\tfrac{1}{2}\right) = \tfrac{1}{4},$$
$$K\left(\tfrac{1}{2},1\right) = \tfrac{1}{2}, \quad K(1,0) = 0, \quad K\left(1,\tfrac{1}{2}\right) = \tfrac{1}{2}, \quad K(1,1) = 1.$$

On applying Simpson's rule (see Subsection 16.1.5)
$$\int_0^1 F(x)\,dx \approx \tfrac{1}{6}\left[F(0) + 4F\left(\tfrac{1}{2}\right) + F(1)\right]$$
to determine the approximate values $y_i$ ($i = 1, 2, 3$) of the solution $y(x)$ at the nodes $x_i$ we obtain the system
$$y_1 = 0,$$
$$\tfrac{11}{12}y_2 - \tfrac{1}{24}y_3 = \tfrac{5}{12},$$
$$-\tfrac{2}{12}y_2 + \tfrac{11}{12}y_3 = \tfrac{5}{6},$$
whose solution is $y_1 = 0$, $y_2 = \tfrac{1}{2}$, $y_3 = 1$. In accordance with the expression (16.4.11.6), the approximate solution can be presented in the form
$$\widetilde{y}(x) = \tfrac{5}{6}x + \tfrac{1}{2} \times \tfrac{1}{6}\left(0 + 4 \times \tfrac{1}{2} \times \tfrac{1}{2}x + 1 \times 1 \times x\right) = x.$$
We can readily verify that it coincides with the exact solution.

### 16.4.11-2. Construction of the eigenfunctions.

The method of quadratures can also be applied for solutions of homogeneous Fredholm equations of the second kind. In this case, system (16.4.11.5) becomes homogeneous ($f_i = 0$) and has a nontrivial solution only if its determinant $D(\lambda)$ is equal to zero. The algebraic equation $D(\lambda) = 0$ of degree $n$ for $\lambda$ makes it possible to find the roots $\widetilde{\lambda}_1, \ldots, \widetilde{\lambda}_n$, which are approximate values of $n$ characteristic values of the equation. The substitution of each value $\widetilde{\lambda}_k$ ($k = 1, \ldots, n$) into (16.4.11.5) for $f_i \equiv 0$ leads to the system of equations
$$y_i^{(k)} - \widetilde{\lambda}_k \sum_{j=1}^n A_j K_{ij} y_j^{(k)} = 0, \qquad i = 1, \ldots, n,$$
whose nonzero solutions $y_i^{(k)}$ make it possible to obtain approximate expressions for the eigenfunctions of the integral equation:
$$\widetilde{y}_k(x) = \widetilde{\lambda}_k \sum_{j=1}^n A_j K(x, x_j) y_j^{(k)}.$$

If $\lambda$ differs from each of the roots $\widetilde{\lambda}_k$, then the nonhomogeneous system of linear algebraic equations (16.4.11.5) has a unique solution. In the same case, the homogeneous system of equations (16.4.11.5) has only the trivial solution.

## 16.4.12. Systems of Fredholm Integral Equations of the Second Kind

### 16.4.12-1. Some remarks.

A system of Fredholm integral equations of the second kind has the form
$$y_i(x) - \lambda \sum_{j=1}^n \int_a^b K_{ij}(x,t) y_j(t)\,dt = f_i(x), \qquad a \le x \le b, \quad i = 1, \ldots, n. \qquad (16.4.12.1)$$

Assume that the kernels $K_{ij}(x,t)$ are continuous or square integrable on the square $S = \{a \leq x \leq b,\ a \leq t \leq b\}$ and the right-hand sides $f_i(x)$ are continuous or square integrable on $[a,b]$. We also assume that the functions $y_i(x)$ to be defined are continuous or square integrable on $[a,b]$ as well. The theory developed above for Fredholm equations of the second kind can be completely extended to such systems. In particular, it can be shown that for systems (16.4.12.1), the successive approximations converge in mean-square to the solution of the system if $\lambda$ satisfies the inequality

$$|\lambda| < \frac{1}{B_*}, \qquad (16.4.12.2)$$

where

$$\sum_{i=1}^{n}\sum_{j=1}^{n}\int_a^b\int_a^b |K_{ij}(x,t)|^2\,dx\,dt = B_*^2 < \infty. \qquad (16.4.12.3)$$

If the kernel $K_{ij}(x,t)$ satisfies the additional condition

$$\int_a^b K_{ij}^2(x,t)\,dt \leq A_{ij}, \qquad a \leq x \leq b, \qquad (16.4.12.4)$$

where $A_{ij}$ are some constants, then the successive approximations converge absolutely and uniformly.

If all kernels $K_{ij}(x,t)$ are degenerate, then system (16.4.12.1) can be reduced to a linear algebraic system. It can be established that for a system of Fredholm integral equations, all Fredholm theorems are satisfied.

**16.4.12-2. Method of reducing a system of equations to a single equation.**

System (16.4.12.1) can be transformed into a single Fredholm integral equation of the second kind. Indeed, let us introduce the functions $Y(x)$ and $F(x)$ on $[a,\ nb-(n-1)a]$ by setting

$$Y(x) = y_i\big(x-(i-1)(b-a)\big), \qquad F(x) = f_i\big(x-(i-1)(b-a)\big),$$

for

$$(i-1)b-(i-2)a \leq x \leq ib-(i-1)a.$$

Let us define a kernel $K(x,t)$ on the square $\{a \leq x \leq nb-(n-1)a,\ a \leq t \leq nb-(n-1)a\}$ as follows:

$$K(x,t) = K_{ij}\big(x-(i-1)(b-a),\ t-(j-1)(b-a)\big)$$

for

$$(i-1)b-(i-2)a \leq x \leq ib-(i-1)a, \qquad (j-1)b-(j-2)a \leq t \leq jb-(j-1)a.$$

Now system (16.4.12.1) can be rewritten as the single Fredholm equation

$$Y(x) - \lambda \int_a^{nb-(n-1)a} K(x,t)Y(t)\,dt = F(x), \qquad a \leq x \leq nb-(n-1)a.$$

If the kernels $K_{ij}(x,t)$ are square integrable on the square $S = \{a \leq x \leq b,\ a \leq t \leq b\}$ and the right-hand sides $f_i(x)$ are square integrable on $[a,b]$, then the kernel $K(x,t)$ is square integrable on the new square

$$S_n = \{a < x < nb-(n-1)a,\ a < t < nb-(n-1)a\},$$

and the right-hand side $F(x)$ is square integrable on $[a,\ nb-(n-1)a]$.

If condition (16.4.12.4) is satisfied, then the kernel $K(x,t)$ satisfies the inequality

$$\int_a^b K^2(x,t)\, dt \leq A_*, \qquad a < x < nb - (n-1)a,$$

where $A_*$ is a constant.

## 16.5. Nonlinear Integral Equations

### 16.5.1. Nonlinear Volterra and Urysohn Integral Equations

**16.5.1-1. Nonlinear integral equations with variable integration limit.**

*Nonlinear Volterra integral equations* have the form

$$\int_a^x K(x,t,y(t))\, dt = F(x,y(x)), \tag{16.5.1.1}$$

where $K(x,t,y(t))$ is the kernel of the integral equation and $y(x)$ is the unknown function ($a \leq x \leq b$). All functions in (16.5.1.1) are usually assumed to be continuous.

**16.5.1-2. Nonlinear integral equations with constant integration limits.**

*Nonlinear Urysohn integral equations* have the form

$$\int_a^b K(x,t,y(t))\, dt = F(x,y(x)), \qquad \alpha \leq x \leq \beta, \tag{16.5.1.2}$$

where $K(x,t,y(t))$ is the kernel of the integral equation and $y(x)$ is the unknown function. Usually, all functions in (16.5.1.2) are assumed to be continuous and the case of $\alpha = a$ and $\beta = b$ is considered.

Conditions for existence and uniqueness of the solution of an Urysohn equation are discussed below in Paragraphs 16.5.3-4 and 16.5.3-5.

**Remark.** A feature of nonlinear equations is that it frequently has several solutions.

### 16.5.2. Nonlinear Volterra Integral Equations

**16.5.2-1. Method of integral transforms.**

Consider a Volterra integral equation with quadratic nonlinearity

$$\mu y(x) - \lambda \int_0^x y(x-t) y(t)\, dt = f(x). \tag{16.5.2.1}$$

To solve this equation, the Laplace transform can be applied, which, with regard to the convolution theorem (see Section 11.2), leads to a quadratic equation for the transform $\widetilde{y}(p) = \mathcal{L}\{y(x)\}$:

$$\mu \widetilde{y}(p) - \lambda \widetilde{y}^2(p) = \widetilde{f}(p).$$

This implies

$$\widetilde{y}(p) = \frac{\mu \pm \sqrt{\mu^2 - 4\lambda \widetilde{f}(p)}}{2\lambda}. \tag{16.5.2.2}$$

The inverse Laplace transform $y(x) = \mathcal{L}^{-1}\{\widetilde{y}(p)\}$ (if it exists) is a solution to equation (16.5.2.1). Note that for the two different signs in formula (16.5.2.2), there are two corresponding solutions of the original equation.

**Example 1.** Consider the integral equation

$$\int_0^x y(x-t)y(t)\,dt = Ax^m, \qquad m > -1.$$

Applying the Laplace transform to the equation under consideration with regard to the relation $\mathcal{L}\{x^m\} = \Gamma(m+1)p^{-m-1}$, we obtain

$$\tilde{y}^2(p) = A\Gamma(m+1)p^{-m-1},$$

where $\Gamma(m)$ is the gamma function. On extracting the square root of both sides of the equation, we obtain

$$\tilde{y}(p) = \pm\sqrt{A\Gamma(m+1)}\,p^{-\frac{m+1}{2}}.$$

Applying the Laplace inversion formula, we obtain two solutions to the original integral equation:

$$y_1(x) = -\frac{\sqrt{A\Gamma(m+1)}}{\Gamma\left(\dfrac{m+1}{2}\right)}\,x^{\frac{m-1}{2}}, \qquad y_2(x) = \frac{\sqrt{A\Gamma(m+1)}}{\Gamma\left(\dfrac{m+1}{2}\right)}\,x^{\frac{m-1}{2}}.$$

### 16.5.2-2. Method of differentiation for integral equations.

Sometimes, differentiation (possibly multiple) of a nonlinear integral equation with subsequent elimination of the integral terms by means of the original equation makes it possible to reduce this equation to a nonlinear ordinary differential equation. Below we briefly list some equations of this type.

1°. The equation

$$y(x) + \int_a^x f\bigl(t, y(t)\bigr)\,dt = g(x) \tag{16.5.2.3}$$

can be reduced by differentiation to the nonlinear first-order equation

$$y'_x + f(x, y) - g'_x(x) = 0$$

with the initial condition $y(a) = g(a)$.

2°. The equation

$$y(x) + \int_a^x (x-t) f\bigl(t, y(t)\bigr)\,dt = g(x) \tag{16.5.2.4}$$

can be reduced by double differentiation (with the subsequent elimination of the integral term by using the original equation) to the nonlinear second-order equation:

$$y''_{xx} + f(x, y) - g''_{xx}(x) = 0. \tag{16.5.2.5}$$

The initial conditions for the function $y = y(x)$ have the form

$$y(a) = g(a), \qquad y'_x(a) = g'_x(a). \tag{16.5.2.6}$$

3°. The equation

$$y(x) + \int_a^x e^{\lambda(x-t)} f\bigl(t, y(t)\bigr)\,dt = g(x) \tag{16.5.2.7}$$

can be reduced by differentiation to the nonlinear first-order equation

$$y'_x + f(x, y) - \lambda y + \lambda g(x) - g'_x(x) = 0. \tag{16.5.2.8}$$

The desired function $y = y(x)$ must satisfy the initial condition $y(a) = g(a)$.

4°. Equations of the form

$$y(x) + \int_a^x \cosh[\lambda(x-t)] f(t, y(t))\, dt = g(x), \qquad (16.5.2.9)$$

$$y(x) + \int_a^x \sinh[\lambda(x-t)] f(t, y(t))\, dt = g(x), \qquad (16.5.2.10)$$

$$y(x) + \int_a^x \cos[\lambda(x-t)] f(t, y(t))\, dt = g(x), \qquad (16.5.2.11)$$

$$y(x) + \int_a^x \sin[\lambda(x-t)] f(t, y(t))\, dt = g(x) \qquad (16.5.2.12)$$

can also be reduced to second-order ordinary differential equations by double differentiation. For these equations, see the book by Polyanin and Manzhirov (1998).

### 16.5.2-3. Successive approximation method.

In many cases, the successive approximation method can be applied successfully to solve various types of integral equations. The principles of constructing the iteration process are the same as in the case of linear equations. For Volterra equations of the form

$$y(x) - \int_a^x K(x, t, y(t))\, dt = f(x), \qquad a \le x \le b, \qquad (16.5.2.13)$$

the corresponding recurrent expression has the form

$$y_{k+1}(x) = f(x) + \int_a^x K(x, t, y_k(t))\, dt, \qquad k = 0, 1, 2, \dots \qquad (16.5.2.14)$$

It is customary to take the initial approximation either in the form $y_0(x) \equiv 0$ or in the form $y_0(x) = f(x)$.

In contrast to the case of linear equations, the successive approximation method has a smaller domain of convergence. Let us present the convergence conditions for the iteration process (16.5.2.14) that are simultaneously the existence conditions for a solution of equation (16.5.2.13). To be definite, we assume that $y_0(x) = f(x)$.

If for any $z_1$ and $z_2$ we have the relation

$$|K(x, t, z_1) - K(x, t, z_2)| \le \varphi(x, t)|z_1 - z_2|$$

and the relation

$$\left| \int_a^x K(x, t, f(t))\, dt \right| \le \psi(x)$$

holds, where

$$\int_a^x \psi^2(t)\, dt \le N^2, \qquad \int_a^b \int_a^x \varphi^2(x, t)\, dt\, dx \le M^2,$$

for some constants $N$ and $M$, then the successive approximations converge to a unique solution of equation (16.5.2.13) almost everywhere absolutely and uniformly.

**Example 2.** Let us apply the successive approximation method to solve the equation
$$y(x) = \int_0^x \frac{1 + y^2(t)}{1 + t^2} \, dt.$$

If $y_0(x) \equiv 0$, then

$$y_1(x) = \int_0^x \frac{dt}{1 + t^2} = \arctan x,$$

$$y_2(x) = \int_0^x \frac{1 + \arctan^2 t}{1 + t^2} \, dt = \arctan x + \tfrac{1}{3} \arctan^3 x,$$

$$y_3(x) = \int_0^x \frac{1 + \arctan t + \tfrac{1}{3} \arctan^3 t}{1 + t^2} \, dt = \arctan x + \tfrac{1}{3} \arctan^3 x + \tfrac{2}{3 \cdot 5} \arctan^5 x + \tfrac{1}{7 \cdot 9} \arctan^7 x.$$

On continuing this process, we can observe that $y_k(x) \to \tan(\arctan x) = x$ as $k \to \infty$, i.e., $y(x) = x$. The substitution of this result into the original equation shows the validity of the result.

### 16.5.2-4. Newton–Kantorovich method.

A merit of the iteration methods when applied to Volterra linear equations of the second kind is their unconditional convergence under weak restrictions on the kernel and the right-hand side. When solving nonlinear equations, the applicability domain of the method of simple iterations is smaller, and if the process is still convergent, then, in many cases, the rate of convergence can be very low. An effective method that makes it possible to overcome the indicated complications is the Newton–Kantorovich method.

Let us apply the Newton–Kantorovich method to solve a nonlinear Volterra equation of the form

$$y(x) = f(x) + \int_a^x K(x, t, y(t)) \, dt. \tag{16.5.2.15}$$

We obtain the following iteration process:

$$y_k(x) = y_{k-1}(x) + \varphi_{k-1}(x), \qquad k = 1, 2, \ldots, \tag{16.5.2.16}$$

$$\varphi_{k-1}(x) = \varepsilon_{k-1}(x) + \int_a^x K'_y(x, t, y_{k-1}(t)) \varphi_{k-1}(t) \, dt, \tag{16.5.2.17}$$

$$\varepsilon_{k-1}(x) = f(x) + \int_a^x K(x, t, y_{k-1}(t)) \, dt - y_{k-1}(x). \tag{16.5.2.18}$$

The algorithm is based on the solution of the linear integral equation (16.5.2.17) for the correction $\varphi_{k-1}(x)$ with the kernel and right-hand side that vary from step to step. This process has a high rate of convergence, but it is rather complicated because we must solve a new equation at each step of iteration. To simplify the problem, we can replace equation (16.5.2.17) with the equation

$$\varphi_{k-1}(x) = \varepsilon_{k-1}(x) + \int_a^x K'_y(x, t, y_0(t)) \varphi_{k-1}(t) \, dt \tag{16.5.2.19}$$

or with the equation

$$\varphi_{k-1}(x) = \varepsilon_{k-1}(x) + \int_a^x K'_y(x, t, y_m(t)) \varphi_{k-1}(t) \, dt, \tag{16.5.2.20}$$

whose kernels do not vary. In equation (16.5.2.20), $m$ is fixed and satisfies the condition $m < k - 1$.

It is reasonable to apply equation (16.5.2.19) with an appropriately chosen initial approximation. Otherwise we can stop at some $m$th approximation and, beginning with this approximation, apply the simplified equation (16.5.2.20). The iteration process thus obtained is the modified Newton–Kantorovich method. In principle, it converges somewhat slower than the original process (16.5.2.16)–(16.5.2.18); however, it is not so cumbersome in the calculations.

**Example 3.** Let us apply the Newton–Kantorovich method to solve the equation
$$y(x) = \int_0^x [ty^2(t) - 1]\, dt.$$

The derivative of the integrand with respect to $y$ has the form
$$K'_y(t, y(t)) = 2ty(t).$$

For the zero approximation we take $y_0(x) \equiv 0$. According to (16.5.2.17) and (16.5.2.18) we obtain $\varphi_0(x) = -x$ and $y_1(x) = -x$. Furthermore, $y_2(x) = y_1(x) + \varphi_1(x)$. By (16.5.2.18) we have
$$\varepsilon_1(x) = \int_0^x [t(-t)^2 - 1]\, dt + x = \tfrac{1}{4}x^4.$$

The equation for the correction has the form
$$\varphi_1(x) = -2\int_0^x t^2 \varphi_1(t)\, dt + \tfrac{1}{4}x^4$$

and can be solved by any of the known methods for Volterra linear equations of the second kind. In the case under consideration, we apply the successive approximation method, which leads to the following results (the number of the step is indicated in the superscript):

$$\varphi_1^{(0)} = \tfrac{1}{4}x^4,$$

$$\varphi_1^{(1)} = \tfrac{1}{4}x^4 - 2\int_0^x \tfrac{1}{4}t^6\, dt = \tfrac{1}{4}x^4 - \tfrac{1}{14}x^7,$$

$$\varphi_1^{(2)} = \tfrac{1}{4}x^4 - 2\int_0^x t^2\left(\tfrac{1}{4}t^4 - \tfrac{1}{14}t^7\right) dt = \tfrac{1}{4}x^4 - \tfrac{1}{14}x^7 + \tfrac{1}{70}x^{10}.$$

We restrict ourselves to the second approximation and obtain
$$y_2(x) = -x + \tfrac{1}{4}x^4 - \tfrac{1}{14}x^7 + \tfrac{1}{70}x^{10}$$

and then pass to the third iteration step of the Newton–Kantorovich method:
$$y_3(x) = y_2(x) + \varphi_2(x),$$
$$\varepsilon_2(x) = \tfrac{1}{160}x^{10} - \tfrac{1}{1820}x^{13} - \tfrac{1}{7840}x^{16} + \tfrac{1}{9340}x^{19} + \tfrac{1}{107800}x^{22},$$
$$\varphi_2(x) = \varepsilon_2(x) + 2\int_0^x t\left(-t + \tfrac{1}{4}t^4 - \tfrac{1}{14}t^7 + \tfrac{1}{70}t^{10}\right)\varphi_2(t)\, dt.$$

When solving the last equation, we restrict ourselves to the zero approximation and obtain
$$y_3(x) = -x + \tfrac{1}{4}x^4 - \tfrac{1}{14}x^7 + \tfrac{23}{112}x^{10} - \tfrac{1}{1820}x^{13} - \tfrac{1}{7840}x^{16} + \tfrac{1}{9340}x^{19} + \tfrac{1}{107800}x^{22}.$$

The application of the successive approximation method to the original equation leads to the same result at the fourth step.

As usual, in the numerical solution the integral is replaced by a quadrature formula. The main difficulty of the implementation of the method in this case is in evaluating the derivative of the kernel. The problem can be simplified if the kernel is given as an analytic expression that can be differentiated in the analytic form. However, if the kernel is given by a table, then the evaluation must be performed numerically.

### 16.5.2-5. Collocation method.

When applied to the solution of a nonlinear Volterra equation

$$\int_a^x K(x,t,y(t))\,dt = f(x), \qquad a \le x \le b, \tag{16.5.2.21}$$

the *collocation method* is as follows. The interval $[a,b]$ is divided into $N$ parts on each of which the desired solution can be presented by a function of a certain form

$$\widetilde{y}(x) = \Phi(x, A_1, \ldots, A_m), \tag{16.5.2.22}$$

involving free parameters $A_i$, $i = 1, \ldots, m$.

On the $(k+1)$st part $x_k \le x \le x_{k+1}$, where $k = 0, 1, \ldots, N-1$, the solution can be written in the form

$$\int_{x_k}^x K(x,t,\widetilde{y}(t))\,dt = f(x) - \Psi_k(x), \tag{16.5.2.23}$$

where the integral

$$\Psi_k(x) = \int_a^{x_k} K(x,t,\widetilde{y}(t))\,dt \tag{16.5.2.24}$$

can always be calculated for the approximate solution $\widetilde{y}(x)$, which is known on the interval $a \le x \le x_k$ and was previously obtained for $k-1$ parts. The initial value $y(a)$ of the desired solution can be found by an auxiliary method or is assumed to be given.

To solve equation (16.5.2.23), representation (16.5.2.22) is applied, and the free parameters $A_i$ ($i = 1, \ldots, m$) can be defined from the condition that the residuals vanish:

$$\varepsilon(A_i, x_{k,j}) = \int_{x_k}^{x_{k,j}} K(x_{k,j}, t, \Phi(t, A_1, \ldots, A_m))\,dt - f(x_{k,j}) - \Psi_k(x_{k,j}), \tag{16.5.2.25}$$

where $x_{k,j}$ ($j = 1, \ldots, m$) are the nodes that correspond to the partition of the interval $[x_k, x_{k+1}]$ into $m$ parts (subintervals). System (16.5.2.25) is a system of $m$ equations for $A_1, \ldots, A_m$.

For convenience of the calculations, it is reasonable to present the desired solution on any part as a polynomial

$$\widetilde{y}(x) = \sum_{i=1}^m A_i \varphi_i(x), \tag{16.5.2.26}$$

where $\varphi_i(x)$ are linearly independent coordinate functions. For the functions $\varphi_i(x)$, power and trigonometric polynomials are frequently used; for instance, $\varphi_i(x) = x^{i-1}$.

In applications, the concrete form of the functions $\varphi_i(x)$ in formula (16.5.2.26), as well as the form of the functions $\Phi$ in (16.5.2.20), can sometimes be given on the basis of physical reasoning or defined by the structure of the solution of a simpler model equation.

### 16.5.2-6. Quadrature method.

To solve a nonlinear Volterra equation, we can apply the method based on the use of quadrature formulas. The procedure of constructing the approximate system of equations is the same as in the linear case (see Subsection 16.2.7).

**1°.** We consider the nonlinear Volterra equation of the form

$$y(x) - \int_a^x K(x, t, y(t))\, dt = f(x) \qquad (16.5.2.27)$$

on an interval $a \leq x \leq b$. Assume that $K(x, t, y(t))$ and $f(x)$ are continuous functions.

From equation (16.5.2.27) we find that $y(a) = f(a)$. Let us choose a constant integration step $h$ and consider the discrete set of points $x_i = a + h(i-1)$, where $i = 1, \ldots, n$. For $x = x_i$, equation (16.5.2.27) becomes

$$y(x_i) - \int_a^{x_i} K(x_i, t, y(t))\, dt = f(x_i). \qquad (16.5.2.28)$$

Applying the quadrature formula (see Subsection 16.1.5) to the integral in (16.5.2.28), choosing $x_j$ ($j = 1, \ldots, i$) to be the nodes in $t$, and neglecting the truncation error, we arrive at the following system of nonlinear algebraic (or transcendental) equations:

$$y_1 = f_1, \quad y_i - \sum_{j=1}^{i} A_{ij} K_{ij}(y_j) = f_i, \quad i = 2, \ldots, n, \qquad (16.5.2.29)$$

where $A_{ij}$ are the coefficients of the quadrature formula on the interval $[a, x_i]$; $y_i$ are the approximate values of the solution $y(x)$ at the nodes $x_i$; $f_i = f(x_i)$; and $K_{ij}(y_j) = K(x_i, t_j, y_j)$.

Relations (16.5.2.29) can be rewritten as a sequence of recurrent nonlinear equations,

$$y_1 = f_1, \quad y_i - A_{ii} K_{ii}(y_i) = f_i + \sum_{j=1}^{i-1} A_{ij} K_{ij}(y_j), \quad i = 2, \ldots, n, \qquad (16.5.2.30)$$

for the approximate values of the desired solution at the nodes.

**2°.** When applied to the Volterra equation of the second kind in the Hammerstein form

$$y(x) - \int_a^x Q(x, t)\Phi(t, y(t))\, dt = f(x), \qquad (16.5.2.31)$$

the main relations of the quadrature method have the form ($x_1 = a$)

$$y_1 = f_1, \quad y_i - \sum_{j=1}^{i} A_{ij} Q_{ij} \Phi_j(y_j) = f_i, \quad i = 2, \ldots, n, \qquad (16.5.2.32)$$

where $Q_{ij} = Q(x_i, t_j)$ and $\Phi_j(y_j) = \Phi(t_j, y_j)$. These relations lead to the sequence of nonlinear recurrent equations

$$y_1 = f_1, \quad y_i - A_{ii} Q_{ii} \Phi_i(y_i) = f_i + \sum_{j=1}^{i-1} A_{ij} Q_{ij} \Phi_j(y_j), \quad i = 2, \ldots, n, \qquad (16.5.2.33)$$

whose solutions give approximate values of the desired function.

**Example 4.** In the solution of the equation

$$y(x) - \int_0^x e^{-(x-t)} y^2(t)\, dt = e^{-x}, \qquad 0 \le x \le 0.1,$$

where $Q(x,t) = e^{-(x-t)}$, $\Phi(t, y(t)) = y^2(t)$, and $f(x) = e^{-x}$, the approximate expression has the form

$$y(x_i) - \int_0^{x_i} e^{-(x_i-t)} y^2(t)\, dt = e^{-x_i}.$$

On applying the trapezoidal rule to evaluate the integral (with step $h = 0.02$) and finding the solution at the nodes $x_i = 0, 0.02, 0.04, 0.06, 0.08, 0.1$, we obtain, according to (16.5.2.33), the following system of computational relations:

$$y_1 = f_1, \quad y_i - 0.01\, Q_{ii} y_i^2 = f_i + \sum_{j=1}^{i-1} 0.02\, Q_{ij} y_j^2, \quad i = 2, \ldots, 6.$$

Thus, to find an approximate solution, we must solve a quadratic equation for each value $y_i$, which makes it possible to write out the answer

$$y_i = 50 \pm 50 \left[ 1 - 0.04 \left( f_i + \sum_{j=1}^{i-1} 0.02\, Q_{ij} y_j^2 \right) \right]^{1/2}, \quad i = 2, \ldots, 6.$$

## 16.5.3. Equations with Constant Integration Limits

**16.5.3-1. Nonlinear equations with degenerate kernels.**

1°. Consider a Urysohn equation of the form

$$y(x) = \int_a^b Q(x,t) \Phi(t, y(t))\, dt, \tag{16.5.3.1}$$

where $Q(x,t)$ and $\Phi(t,y)$ are given functions and $y(x)$ is the unknown function.

Let the kernel $Q(x,t)$ be degenerate, i.e.,

$$Q(x,t) = \sum_{k=1}^m g_k(x) h_k(t). \tag{16.5.3.2}$$

In this case equation (16.5.3.1) becomes

$$y(x) = \sum_{k=1}^m g_k(x) \int_a^b h_k(t) \Phi(t, y(t))\, dt. \tag{16.5.3.3}$$

We write

$$A_k = \int_a^b h_k(t) \Phi(t, y(t))\, dt, \qquad k = 1, \ldots, m, \tag{16.5.3.4}$$

where the constants $A_k$ are yet unknown. Then it follows from (16.5.3.3) that

$$y(x) = \sum_{k=1}^m A_k g_k(x). \tag{16.5.3.5}$$

On substituting the expression (16.5.3.5) for $y(x)$ into relations (16.5.3.4), we obtain (in the general case) $m$ transcendental equations of the form

$$A_k = \Psi_k(A_1, \ldots, A_m), \qquad k = 1, \ldots, m, \tag{16.5.3.6}$$

which contain $m$ unknown numbers $A_1, \ldots, A_m$.

For the case in which $\Phi(t, y)$ is a polynomial in $y$, i.e.,

$$\Phi(t, y) = p_0(t) + p_1(t)y + \cdots + p_n(t)y^n, \tag{16.5.3.7}$$

where $p_0(t), \ldots, p_n(t)$ are, for instance, continuous functions of $t$ on the interval $[a, b]$, system (16.5.3.6) becomes a system of nonlinear algebraic equations for $A_1, \ldots, A_m$.

The number of solutions of the integral equation (16.5.3.3) is equal to the number of solutions of system (16.5.3.6). Each solution of system (16.5.3.6) generates a solution (16.5.3.5) of the integral equation.

2°. Consider the Urysohn equation with a degenerate kernel of the special form

$$y(x) + \int_a^b \left\{ \sum_{k=1}^n g_k(x) f_k(t, y(t)) \right\} dt = h(x). \tag{16.5.3.8}$$

Its solution has the form

$$y(x) = h(x) + \sum_{k=1}^n \lambda_k g_k(x), \tag{16.5.3.9}$$

where the constants $\lambda_k$ can be defined by solving the algebraic (or transcendental) system of equations

$$\lambda_m + \int_a^b f_m\left(t, h(t) + \sum_{k=1}^n \lambda_k g_k(t)\right) dt = 0, \quad m = 1, \ldots, n. \tag{16.5.3.10}$$

To different roots of this system, there are different corresponding solutions of the nonlinear integral equation. It may happen that (real) solutions are absent.

A solution of the Urysohn equation with a degenerate kernel in the general form

$$f(x, y(x)) + \int_a^b \left\{ \sum_{k=1}^n g_k(x, y(x)) h_k(t, y(t)) \right\} dt = 0 \tag{16.5.3.11}$$

can be represented in the implicit form

$$f(x, y(x)) + \sum_{k=1}^n \lambda_k g_k(x, y(x)) = 0, \tag{16.5.3.12}$$

where the parameters $\lambda_k$ are determined from the system of algebraic (or transcendental) equations:

$$\lambda_k - H_k(\vec{\lambda}) = 0, \quad k = 1, \ldots, n,$$
$$H_k(\vec{\lambda}) = \int_a^b h_k(t, y(t)) \, dt, \quad \vec{\lambda} = \{\lambda_1, \ldots, \lambda_n\}. \tag{16.5.3.13}$$

Into system (16.5.3.13), we must substitute the function $y(x) = y(x, \vec{\lambda})$, which can be obtained by solving equation (16.5.3.12).

The number of solutions of the integral equation is defined by the number of solutions obtained from (16.5.3.12) and (16.5.3.13). It can happen that there is no solution.

**Example 1.** Let us solve the integral equation

$$y(x) = \lambda \int_0^1 xty^3(t)\,dt \qquad (16.5.3.14)$$

with parameter $\lambda$. We write

$$A = \int_0^1 ty^3(t)\,dt. \qquad (16.5.3.15)$$

In this case, it follows from (16.5.3.14) that

$$y(x) = \lambda A x. \qquad (16.5.3.16)$$

On substituting $y(x)$ in the form (16.5.3.16) into relation (16.5.3.15), we obtain

$$A = \int_0^1 t\lambda^3 A^3 t^3\,dt.$$

Hence,

$$A = \tfrac{1}{5}\lambda^3 A^3. \qquad (16.5.3.17)$$

For $\lambda > 0$, equation (16.5.3.17) has three solutions:

$$A_1 = 0, \quad A_2 = \left(\frac{5}{\lambda^3}\right)^{1/2}, \quad A_3 = -\left(\frac{5}{\lambda^3}\right)^{1/2}.$$

Hence, the integral equation (16.5.3.14) also has three solutions for any $\lambda > 0$:

$$y_1(x) \equiv 0, \quad y_2(x) = \left(\frac{5}{\lambda^3}\right)^{1/2} x, \quad y_3(x) = -\left(\frac{5}{\lambda^3}\right)^{1/2} x.$$

For $\lambda \leq 0$, equation (16.5.3.17) has only the trivial solution $y(x) \equiv 0$.

### 16.5.3-2. Method of integral transforms.

1°. Consider the following nonlinear integral equation with quadratic nonlinearity on a semiaxis:

$$\mu y(x) - \lambda \int_0^\infty \frac{1}{t} y\!\left(\frac{x}{t}\right) y(t)\,dt = f(x). \qquad (16.5.3.18)$$

To solve this equation, the Mellin transform can be applied, which, with regard to the convolution theorem (see Section 11.3), leads to a quadratic equation for the transform $\hat{y}(s) = \mathfrak{M}\{y(x)\}$:

$$\mu \hat{y}(s) - \lambda \hat{y}^2(s) = \hat{f}(s).$$

This implies

$$\hat{y}(s) = \frac{\mu \pm \sqrt{\mu^2 - 4\lambda \hat{f}(s)}}{2\lambda}. \qquad (16.5.3.19)$$

The inverse transform $y(x) = \mathfrak{M}^{-1}\{\hat{y}(s)\}$ obtained by means of the Mellin inversion formula (if it exists) is a solution of equation (16.5.3.18). To different signs in the formula for the images (16.5.3.19), there are two corresponding solutions of the original equation.

2°. By applying the Mellin transform, one can solve nonlinear integral equations of the form

$$y(x) - \lambda \int_0^\infty t^\beta y(xt) y(t)\,dt = f(x). \qquad (16.5.3.20)$$

The Mellin transform (see Table 11.3 in Section 11.3) reduces (16.5.3.20) to the following functional equation for the transform $\hat{y}(s) = \mathfrak{M}\{y(x)\}$:

$$\hat{y}(s) - \lambda \hat{y}(s)\hat{y}(1-s+\beta) = \hat{f}(s). \qquad (16.5.3.21)$$

On replacing $s$ by $1-s+\beta$ in (16.5.3.21), we obtain the relationship
$$\hat{y}(1-s+\beta) - \lambda\hat{y}(s)\hat{y}(1-s+\beta) = \hat{f}(1-s+\beta). \tag{16.5.3.22}$$
On eliminating the quadratic term from (16.5.3.21) and (16.5.3.22), we obtain
$$\hat{y}(s) - \hat{f}(s) = \hat{y}(1-s+\beta) - \hat{f}(1-s+\beta). \tag{16.5.3.23}$$
We express $\hat{y}(1-s+\beta)$ from this relation and substitute it into (16.5.3.21). We arrive at the quadratic equation
$$\lambda\hat{y}^2(s) - \left[1 + \hat{f}(s) - \hat{f}(1-s+\beta)\right]\hat{y}(s) + \hat{f}(s) = 0. \tag{16.5.3.24}$$
On solving (16.5.3.24) for $\hat{y}(s)$, by means of the Mellin inversion formula we can find a solution of the original integral equation (16.5.3.20).

### 16.5.3-3. Method of differentiating for integral equations.

1°. Consider the equation
$$y(x) + \int_a^b |x-t| f(t, y(t))\, dt = g(x). \tag{16.5.3.25}$$
Let us remove the modulus in the integrand:
$$y(x) + \int_a^x (x-t) f(t, y(t))\, dt + \int_x^b (t-x) f(t, y(t))\, dt = g(x). \tag{16.5.3.26}$$
Differentiating (16.5.3.26) with respect to $x$ yields
$$y'_x(x) + \int_a^x f(t, y(t))\, dt - \int_x^b f(t, y(t))\, dt = g'_x(x). \tag{16.5.3.27}$$
Differentiating (16.5.3.27), we arrive at a second-order ordinary differential equation for $y = y(x)$:
$$y''_{xx} + 2f(x,y) = g''_{xx}(x). \tag{16.5.3.28}$$

Let us derive the boundary conditions for equation (16.5.3.28). We assume that $-\infty < a < b < \infty$. By setting $x = a$ and $x = b$ in (16.5.3.26), we obtain the relations
$$\begin{aligned} y(a) + \int_a^b (t-a) f(t, y(t))\, dt &= g(a), \\ y(b) + \int_a^b (b-t) f(t, y(t))\, dt &= g(b). \end{aligned} \tag{16.5.3.29}$$

Let us solve equation (16.5.3.28) for $f(x,y)$ and substitute the result into (16.5.3.29). Integrating by parts yields the desired boundary conditions for $y(x)$:
$$\begin{aligned} y(a) + y(b) + (b-a)\left[g'_x(b) - y'_x(b)\right] &= g(a) + g(b), \\ y(a) + y(b) + (a-b)\left[g'_x(a) - y'_x(a)\right] &= g(a) + g(b). \end{aligned} \tag{16.5.3.30}$$

Let us point out a useful consequence of (16.5.3.30):
$$y'_x(a) + y'_x(b) = g'_x(a) + g'_x(b),$$
which can be used together with one of conditions (16.5.3.30).

Equation (16.5.3.28) under the boundary conditions (16.5.3.30) determines the solution of the original integral equation (there may be several solutions).

2°. The equations

$$y(x) + \int_a^b e^{\lambda|x-t|} f(t, y(t))\, dt = g(x),$$

$$y(x) + \int_a^b \sinh(\lambda|x-t|) f(t, y(t))\, dt = g(x),$$

$$y(x) + \int_a^b \sin(\lambda|x-t|) f(t, y(t))\, dt = g(x),$$

can also be reduced to second-order ordinary differential equations by means of the differentiation. For these equations, see the book by Polyanin and Manzhirov (1998).

### 16.5.3-4. Successive approximation method.

Consider the nonlinear Urysohn integral equation in the canonical form

$$y(x) = \int_a^b \mathcal{K}(x, t, y(t))\, dt, \qquad a \le x \le b. \tag{16.5.3.31}$$

The iteration process for this equation is constructed by the formula

$$y_k(x) = \int_a^b \mathcal{K}(x, t, y_{k-1}(t))\, dt, \qquad k = 1, 2, \ldots \tag{16.5.3.32}$$

If the function $\mathcal{K}(x, t, y)$ is jointly continuous together with the derivative $\mathcal{K}'_y(x, t, y)$ (with respect to the variables $x$, $t$, and $\rho$, $a \le x \le b$, $a \le t \le b$, and $|y| \le \rho$) and if

$$\int_a^b \sup_y |\mathcal{K}(x, t, y)|\, dt \le \rho, \qquad \int_a^b \sup_y |\mathcal{K}'_y(x, t, y)|\, dt \le \beta < 1, \tag{16.5.3.33}$$

then for any continuous function $y_0(x)$ of the initial approximation from the domain $\{|y| \le \rho,\ a \le x \le b\}$, the successive approximations (16.5.3.32) converge to a continuous solution $y^*(x)$, which lies in the same domain and is unique in this domain. The rate of convergence is defined by the inequality

$$|y^*(x) - y_k(x)| \le \frac{\beta^k}{1-\beta} \sup_x |y_1(x) - y_0(x)|, \qquad a \le x \le b, \tag{16.5.3.34}$$

which gives an *a priori* estimate for the error of the $k$th approximation. The *a posteriori* estimate (which is, in general, more precise) has the form

$$|y^*(x) - y_k(x)| \le \frac{\beta}{1-\beta} \sup_x |y_k(x) - y_{k-1}(x)|, \qquad a \le x \le b. \tag{16.5.3.35}$$

A solution of an equation of the form (16.5.3.31) with an additional term $f(x)$ on the right-hand side can be constructed in a similar manner.

**Example 2.** Let us apply the successive approximation method to solve the equation

$$y(x) = \int_0^1 xt y^2(t)\, dt - \tfrac{5}{12} x + 1.$$

The recurrent formula has the form

$$y_k(x) = \int_0^1 xt y_{k-1}^2(t)\, dt - \tfrac{5}{12} x + 1, \qquad k = 1, 2, \ldots$$

For the initial approximation we take $y_0(x) = 1$. The calculation yields

$$y_1(x) = 1 + 0.083\, x, \quad y_2(x) = 1 + 0.14\, x, \quad y_3(x) = 1 + 0.18\, x, \quad \ldots$$
$$y_8(x) = 1 + 0.27\, x, \quad y_9(x) = 1 + 0.26\, x, \quad y_{10}(x) = 1 + 0.29\, x, \quad \ldots$$
$$y_{16}(x) = 1 + 0.318\, x, \quad y_{17}(x) = 1 + 0.321\, x, \quad y_{18}(x) = 1 + 0.323\, x, \quad \ldots$$

Thus, the approximations tend to the exact solution $y(x) = 1 + \tfrac{1}{3} x$. We see that the rate of convergence of the iteration process is fairly small. Note that in Paragraph 16.5.3-5, the equation in question is solved by a more efficient method.

### 16.5.3-5. Newton–Kantorovich method.

We consider the Newton–Kantorovich method in connection with the Urysohn equation in the canonical form (16.5.3.31). The iteration process is constructed as follows:

$$y_k(x) = y_{k-1}(x) + \varphi_{k-1}(x), \qquad k = 1, 2, \ldots, \tag{16.5.3.36}$$

$$\varphi_{k-1}(x) = \varepsilon_{k-1}(x) + \int_a^b \mathcal{K}'_y\bigl(x, t, y_{k-1}(t)\bigr) \varphi_{k-1}(t)\, dt, \tag{16.5.3.37}$$

$$\varepsilon_{k-1}(x) = \int_a^b \mathcal{K}\bigl(x, t, y_{k-1}(t)\bigr)\, dt - y_{k-1}(x). \tag{16.5.3.38}$$

At each step of the algorithm, a linear integral equation for the correction $\varphi_{k-1}(x)$ is solved. Under some conditions, the process (16.5.3.36) has a high rate of convergence; however, it is rather complicated because at each iteration we must obtain the new kernel $\mathcal{K}'_y\bigl(x, t, y_{k-1}(t)\bigr)$ for equations (16.5.3.37).

The algorithm can be simplified by using the equation

$$\varphi_{k-1}(x) = \varepsilon_{k-1}(x) + \int_a^b \mathcal{K}'_y\bigl(x, t, y_0(t)\bigr) \varphi_{k-1}(t)\, dt \tag{16.5.3.39}$$

instead of (16.5.3.37). If the initial approximation is chosen successfully, then the difference between the integral operators in (16.5.3.37) and (16.5.3.39) is small, and the kernel in (16.5.3.39) remains the same in the course of the solution.

The successive approximation method that consists of the application of formulas (16.5.3.36), (16.5.3.38), and (16.5.3.39) is called the *modified Newton–Kantorovich method*. In principle, its rate of convergence is less than that of the original (unmodified) method; however, this version of the method is less complicated in calculations, and therefore it is frequently preferable.

Let the function $\mathcal{K}(x, t, y)$ be jointly continuous together with the derivatives $\mathcal{K}'_y(x, t, y)$ and $\mathcal{K}''_{yy}(x, t, y)$ with respect to the variables $x, t, y$, where $a \leq x \leq b$ and $a \leq t \leq b$, and let the following conditions hold:

$1^\circ$. For the initial approximation $y_0(x)$, the resolvent $\mathcal{R}(x,t)$ of the linear integral equation (16.5.3.37) with the kernel $\mathcal{K}'_y(x,t,y_0(t))$ satisfies the condition

$$\int_a^b |\mathcal{R}(x,t)|\,dt \le A < \infty, \qquad a \le x \le b.$$

$2^\circ$. The residual $\varepsilon_0(x)$ of equation (16.5.3.38) for the approximation $y_0(x)$ satisfies the inequality

$$|\varepsilon_0(x)| = \left| \int_a^b \mathcal{K}(x,t,y_0(t))\,dt - y_0(x) \right| \le B < \infty.$$

$3^\circ$. In the domain $|y(x) - y_0(x)| \le 2(1+A)B$, the following relation holds:

$$\int_a^b \sup_y |\mathcal{K}''_{yy}(x,t,y)|\,dt \le D < \infty.$$

$4^\circ$. The constants $A$, $B$, and $D$ satisfy the condition

$$H = (1+A)^2 BD \le \tfrac{1}{2}.$$

In this case, under assumptions $1^\circ$–$4^\circ$, the process (16.5.3.36) converges to a solution $y^*(x)$ of equation (16.5.3.31) in the domain

$$|y(x) - y_0(x)| \le (1 - \sqrt{1-2H})H^{-1}(1-A)B, \qquad a \le x \le b.$$

This solution is unique in the domain

$$|y(x) - y_0(x)| \le 2(1+A)B, \qquad a \le x \le b.$$

The rate of convergence is determined by the estimate

$$|y^*(x) - y_k(x)| \le 2^{1-k}(2H)^{2^k-1}(1-A)B, \qquad a \le x \le b.$$

Thus, the above conditions establish the convergence of the algorithm and the existence, the position, and the uniqueness domain of a solution of the nonlinear equation (16.5.3.31). These conditions impose certain restrictions on the initial approximation $y_0(x)$ whose choice is an important independent problem that has no unified approach. As usual, the initial approximation is determined either by more detailed *a priori* analysis of the equation under consideration or by physical reasoning implied by the essence of the problem described by this equation. Under a successful choice of the initial approximation, the Newton–Kantorovich method provides a high rate of convergence of the iteration process to obtain an approximate solution with given accuracy.

**Remark.** Let the right-hand side of equation (16.5.3.31) contain an additional term $f(x)$. Then such an equation can be represented in the form (16.5.3.31), where the integrand is $K(x,t,y(t)) + (b-a)^{-1}f(x)$.

**Example 3.** Let us apply the Newton–Kantorovich method to solve the equation

$$y(x) = \int_0^1 xty^2(t)\,dt - \tfrac{5}{12}x + 1. \qquad (16.5.3.40)$$

For the initial approximation we take $y_0(x) = 1$. According to (16.5.3.38), we find the residual

$$\varepsilon_0(x) = \int_0^1 xty_0^2(t)\,dt - \tfrac{5}{12}x + 1 - y_0(x) = x\int_0^1 t\,dt - \tfrac{5}{12}x + 1 - 1 = \tfrac{1}{12}x.$$

The $y$-derivative of the kernel $\mathcal{K}(x,t,y) = xty^2(t)$, which is needed in the calculations, has the form $\mathcal{K}'_y(x,t,y) = 2xty(t)$. According to (16.5.3.37), we form the following equation for $\varphi_0(x)$:

$$\varphi_0(x) = \tfrac{1}{12}x + 2x\int_0^1 ty_0(t)\varphi_0(t)\,dt,$$

where the kernel turns out to be degenerate, which makes it possible to obtain the solution $\varphi_0(x) = \tfrac{1}{4}x$ directly.

Now we define the first approximation to the desired function:

$$y_1(x) = y_0(x) + \varphi_0(x) = 1 + \tfrac{1}{4}x.$$

We continue the iteration process and obtain

$$\varepsilon_1(x) = \int_0^1 xt\left(1+\tfrac{1}{4}t\right)dt + \left(1-\tfrac{5}{12}x\right) - \left(1+\tfrac{1}{4}x\right) = \tfrac{1}{64}x.$$

The equation for $\varphi_1(x)$ has the form

$$\varphi_1(x) = \tfrac{1}{64}x + 2x\int_0^1 t\left(1+\tfrac{1}{4}t\right)dt + \left(1-\tfrac{5}{12}x\right) - \left(1+\tfrac{1}{4}x\right),$$

and the solution is $\varphi_1(x) = \tfrac{3}{40}x$. Hence, $y_2(x) = 1 + \tfrac{1}{4}x + \tfrac{3}{40}x = 1 + 0.325\,x$. The maximal difference between the exact solution $y(x) = 1 + \tfrac{1}{3}x$ and the approximate solution $y_2(x)$ is observed at $x=1$ and is less than 0.5%.

This solution is not unique. The other solution can be obtained by taking the function $y_0(x) = 1 + 0.8\,x$ for the initial approximation. In this case we can repeat the above sequence of approximations and obtain the following results (the numerical coefficient of $x$ is rounded):

$$y_1(x) = 1 + 0.82\,x, \quad y_2(x) = 1 + 1.13\,x, \quad y_3(x) = 1 + 0.98\,x, \quad \ldots,$$

and the subsequent approximations tend to the exact solution $y(x) = 1 + x$.

We see that the rate of convergence of the iteration process performed by the Newton–Kantorovich method is significantly higher than that performed by the method of successive approximations (see Example 2 in Paragraph 16.5.3-4).

To estimate the rate of convergence of the performed iteration process, we can compare the above results with the realization of the modified Newton–Kantorovich method. In connection with the latter, for the above versions of the approximations we can obtain

$y_n(x) = 1 + k_n x$;

| $k_0$ | $k_1$ | $k_2$ | $k_3$ | $k_4$ | $k_5$ | $k_6$ | $k_7$ | $k_8$ | ... |
|---|---|---|---|---|---|---|---|---|---|
| 0 | 0.25 | 0.69 | 0.60 | 0.51 | 0.44 | 0.38 | 0.36 | 0.345 | ... |

The iteration process converges to the exact solution $y(x) = 1 + \tfrac{1}{3}x$.

We see that the modified Newton–Kantorovich method is less efficient than the Newton–Kantorovich method, but more efficient than the method of successive approximations (see Example 2).

### 16.5.3-6. Quadrature method.

To solve an arbitrary nonlinear equation, we can apply the method based on the application of quadrature formulas. The procedure of composing the approximating system of equations is the same as in the linear case (see Subsection 16.4.11). We consider this procedure for an example of the Urysohn equation

$$y(x) - \int_a^b K(x,t,y(t))\,dt = f(x), \qquad a \le x \le b, \qquad (16.5.3.41)$$

We set $x = x_i$ ($i = 1,\ldots,n$). Then we obtain

$$y(x_i) - \int_a^b K(x_i,t,y(t))\,dt = f(x_i). \qquad i = 1,\ldots,n. \qquad (16.5.3.42)$$

On applying the quadrature formula from Subsection 16.4.11 and neglecting the approximation error, we transform relations (16.5.3.42) into the nonlinear system of algebraic (or transcendental) equations

$$y_i - \sum_{j=1}^{n} A_j K_{ij}(y_j) = f_i, \qquad i = 1, \ldots, n, \tag{16.5.3.43}$$

for the approximate values $y_i$ of the solution $y(x)$ at the nodes $x_1, \ldots, x_n$, where $f_i = f(x_i)$ and $K_{ij}(y_j) = K(x_i, t_j, y_j)$, and $A_j$ are the coefficients of the quadrature formula.

The solution of the nonlinear system (16.5.3.43) gives values $y_1, \ldots, y_n$ for which by interpolation we find an approximate solution of the integral equation (16.5.3.41) on the entire interval $[a, b]$. For the analytic expression of an approximate solution, we can take the function

$$\widetilde{y}(x) = f(x) + \sum_{j=1}^{n} A_j K(x, x_j, y_j).$$

## References for Chapter 16

**Atkinson, K. E.**, *Numerical Solution of Integral Equations of the Second Kind*, Cambridge University Press, Cambridge, 1997.
**Bakhvalov, N. S.**, *Numerical Methods* [in Russian], Nauka Publishers, Moscow, 1973.
**Bateman, H. and Erdélyi, A.**, *Tables of Integral Transforms. Vols. 1 and 2*, McGraw-Hill, New York, 1954.
**Bitsadze, A. V.**, *Integral Equation of the First Kind*, World Scientific Publishing Co., Singapore, 1995.
**Brunner, H.**, *Collocation Methods for Volterra Integral and Related Functional Differential Equations*, Cambridge University Press, Cambridge, 2004.
**Cochran, J. A.**, *The Analysis of Linear Integral Equations*, McGraw-Hill, New York, 1972.
**Corduneanu, C.**, *Integral Equations and Applications*, Cambridge University Press, Cambridge, 1991.
**Courant, R. and Hilbert, D.**, *Methods of Mathematical Physics. Vol. 1*, Interscience, New York, 1953.
**Delves, L. M. and Mohamed, J. L.**, *Computational Methods for Integral Equations*, Cambridge University Press, Cambridge, 1985.
**Demidovich, B. P., Maron, I. A., and Shuvalova, E. Z.**, *Numerical Methods. Approximation of Functions and Differential and Integral Equations* [in Russian], Fizmatgiz, Moscow, 1963.
**Ditkin, V. A. and Prudnikov, A. P.**, *Integral Transforms and Operational Calculus*, Pergamon Press, New York, 1965.
**Dzhuraev, A.**, *Methods of Singular Integral Equations*, Wiley, New York, 1992.
**Gakhov, F. D. and Cherskii, Yu. I.**, *Equations of Convolution Type* [in Russian], Nauka Publishers, Moscow, 1978.
**Gohberg, I. C. and Krein, M. G.**, *The Theory of Volterra Operators in a Hilbert Space and Its Applications* [in Russian], Nauka Publishers, Moscow, 1967.
**Golberg, A. (Editor)**, *Numerical Solution of Integral Equations*, Plenum Press, New York, 1990.
**Gorenflo, R. and Vessella, S.**, *Abel Integral Equations: Analysis and Applications*, Springer-Verlag, Berlin, 1991.
**Goursat, E.**, *Cours d'Analyse Mathématique, III*, 3$^{me}$ éd., Gauthier–Villars, Paris, 1923.
**Gripenberg, G., Londen, S.-O., and Staffans, O.**, *Volterra Integral and Functional Equations*, Cambridge University Press, Cambridge, 1990.
**Hackbusch, W.**, *Integral Equations: Theory and Numerical Treatment*, Birkhäuser Verlag, Boston, 1995.
**Jerry, A. J.**, *Introduction to Integral Equations with Applications*, Marcel Dekker, New York, 1985.
**Kantorovich, L. V. and Akilov, G. P.**, *Functional Analysis in Normed Spaces*, Macmillan, New York, 1964.
**Kantorovich, L. V. and Krylov, V. I.**, *Approximate Methods of Higher Analysis*, Interscience, New York, 1958.
**Kanwal, R. P.**, *Linear Integral Equations*, Birkhäuser Verlag, Boston, 1997.
**Kolmogorov, A. N. and Fomin, S. V.**, *Introductory Real Analysis*, Prentice-Hall, Englewood Cliffs, New Jersey, 1970.
**Kondo, J.**, *Integral Equations*, Clarendon Press, Oxford, 1991.
**Korn, G. A. and Korn, T. M.**, *Mathematical Handbook for Scientists and Engineers*, Dover Publications, New York, 2000.

Krasnosel'skii, M. A., *Topological Methods in the Theory of Nonlinear Integral Equations*, Macmillan, New York, 1964.

Krasnov, M. L., Kiselev, A. I., and Makarenko, G. I., *Problems and Exercises in Integral Equations*, Mir Publishers, Moscow, 1971.

Krein, M. G., Integral equations on a half-line with kernels depending upon the difference of the arguments [in Russian], *Uspekhi Mat. Nauk*, Vol. 13, No. 5 (83), pp. 3–120, 1958.

Krylov, V. I., Bobkov, V. V., and Monastyrnyi, P. I., *Introduction to the Theory of Numerical Methods. Integral Equations, Problems, and Improvement of Convergence* [in Russian], Nauka i Tekhnika, Minsk, 1984.

Kythe, P. K. and Puri, P., *Computational Methods for Linear Integral Equations*, Birkhäuser Verlag, Boston, 2002.

Ladopoulos, E. G., *Singular Integral Equations: Linear and Non-Linear Theory and Its Applications in Science and Engineering*, Springer-Verlag, Berlin, 2000.

Lavrentiev, M. M., *Some Improperly Posed Problems of Mathematical Physics*, Springer-Verlag, New York, 1967.

Lovitt, W. V., *Linear Integral Equations*, Dover Publications, New York, 1950.

Mikhlin, S. G., *Linear Integral Equations*, Hindustan Publishing, Delhi, 1960.

Mikhlin, S. G. and Prössdorf, S., *Singular Integral Operators*, Springer-Verlag, Berlin, 1986.

Mikhlin, S. G. and Smolitskiy, K. L., *Approximate Methods for Solution of Differential and Integral Equations*, American Elsevier, New York, 1967.

Muskhelishvili N. I., *Singular Integral Equations: Boundary Problems of Function Theory and Their Applications to Mathematical Physics*, Dover Publications, New York, 1992.

Petrovskii, I. G., *Lectures on the Theory of Integral Equations*, Graylock Press, Rochester, 1957.

Pipkin, A. C., *A Course on Integral Equations*, Springer-Verlag, New York, 1991.

Polyanin, A. D. and Manzhirov, A. V., *A Handbook of Integral Equations*, CRC Press, Boca Raton, 1998.

Porter, D. and Stirling, D. S. G., *Integral Equations: A Practical Treatment, from Spectral Theory to Applications*, Cambridge University Press, Cambridge, 1990.

Precup, R., *Methods in Nonlinear Integral Equations*, Kluwer Academic, Dordrecht, 2002.

Prössdorf, S. and Silbermann, B., *Numerical Analysis for Integral and Related Operator Equations*, Birkhäuser Verlag, Basel, 1991.

Sakhnovich, L. A., *Integral Equations with Difference Kernels on Finite Intervals*, Birkhäuser Verlag, Basel, 1996.

Samko, S. G., Kilbas, A. A., and Marichev, O. I., *Fractional Integrals and Derivatives. Theory and Applications*, Gordon & Breach, New York, 1993.

Tricomi, F. G., *Integral Equations*, Dover Publications, New York, 1985.

Tslaf, L. Ya., *Variational Calculus and Integral Equations* [in Russian], Nauka Publishers, Moscow, 1970.

Verlan', A. F. and Sizikov, V. S., *Integral Equations: Methods, Algorithms, and Programs* [in Russian], Naukova Dumka, Kiev, 1986.

Zabreyko, P. P., Koshelev, A. I., et al., *Integral Equations: A Reference Text*, Noordhoff International Publishing, Leyden, 1975.

# Chapter 17

# Difference Equations and Other Functional Equations

## 17.1. Difference Equations of Integer Argument

### 17.1.1. First-Order Linear Difference Equations of Integer Argument

**17.1.1-1. First-order homogeneous linear difference equations. General solution.**

Let $y_n = y(n)$ be a function of integer argument $n = 0, 1, 2, \ldots$ A *first-order homogeneous linear difference equation* has the form

$$y_{n+1} + a_n y_n = 0. \tag{17.1.1.1}$$

Its general solution can be written in the form

$$y_n = C u_n, \quad u_n = (-1)^n a_0 a_1 \ldots a_{n-1}, \quad n = 1, 2, \ldots, \tag{17.1.1.2}$$

where $C = y_0$ is an arbitrary constant and $u_n$ is a particular solution.

**17.1.1-2. First-order nonhomogeneous linear difference equations. General solution.**

A *first-order nonhomogeneous linear difference equation* has the form

$$y_{n+1} + a_n y_n = f_n. \tag{17.1.1.3}$$

The general solution of the nonhomogeneous linear equation (17.1.1.3) can be represented as the sum of the general solution (17.1.1.2) of the corresponding homogeneous equation (17.1.1.1) and a particular solution $\widetilde{y}_n$ of the nonhomogeneous equation (17.1.1.3):

$$y_n = C u_n + \widetilde{y}_n, \quad n = 1, 2, \ldots, \tag{17.1.1.4}$$

where $C = y_0$ is an arbitrary constant, $u_n$ is defined by (17.1.1.2), and

$$\widetilde{y}_n = \sum_{j=0}^{n-1} \frac{u_n}{u_{j+1}} f_j = f_{n-1} - a_{n-1} f_{n-2} + a_{n-2} a_{n-1} f_{n-3} - \cdots + (-1)^{n-1} a_1 a_2 \ldots a_{n-1} f_0. \tag{17.1.1.5}$$

**17.1.1-3. First-order linear difference equations with constant coefficients.**

A *first-order linear difference equation with constant coefficients* has the form

$$y_{n+1} - a y_n = f_n.$$

Using (17.1.1.2), (17.1.1.4), and (17.1.1.5) for $a_n = -a$, we obtain its general solution

$$y_n = C a^n + \sum_{j=0}^{n-1} a^{n-j-1} f_j.$$

## 17.1.2. First-Order Nonlinear Difference Equations of Integer Argument

**17.1.2-1. First-order nonlinear equations. General and particular solutions.**

Let $y_n = y(n)$ be a function of integer argument $n = 0, \pm 1, \pm 2, \ldots$ A *first-order nonlinear difference equation*, in the general case, has the form

$$F(n, y_n, y_{n+1}) = 0. \qquad (17.1.2.1)$$

A *solution of the difference equation* (17.1.2.1) is defined as a discrete function $y_n$ that, being substituted into the equation, turns it into identity. The *general solution of a difference equation* is the set of all its solutions. The general solution of equation (17.1.2.1) depends on an arbitrary constant $C$. The general solution can be written either in explicit form

$$y_n = \varphi(n, C) \qquad (17.1.2.2)$$

or in implicit form $\Phi(n, y_n, C) = 0$. Specific values of $C$ define specific solutions of the equation (*particular solutions*).

Any constant solution $y_n = \xi$ of equation (17.1.2.1), with $\xi$ independent of $n$, is called an *equilibrium solution*.

**17.1.2-2. Cauchy's problem and its solution.**

A difference equation resolved with respect to the leading term $y_{n+1}$ has the form

$$y_{n+1} = f(n, y_n). \qquad (17.1.2.3)$$

The Cauchy problem consists of finding a solution of this equation with a given initial value of $y_0$.

The next value $y_1$ is calculated by substituting the initial value into the right-hand side of equation (17.1.2.3) for $n = 0$:

$$y_1 = f(0, y_0). \qquad (17.1.2.4)$$

Then, taking $n = 1$ in (17.1.2.3), we get

$$y_2 = f(1, y_1). \qquad (17.1.2.5)$$

Substituting the previous value (17.1.2.4) into this relation, we find $y_2 = f\bigl(1, f(0, y_0)\bigr)$. Taking $n = 2$ in (17.1.2.3) and using the calculated value $y_2$, we find $y_3$, etc. In a similar way, one finds subsequent values $y_4, y_5, \ldots$

**Example.** Consider the Cauchy problem for the nonlinear difference equation

$$y_{n+1} = a y_n^\beta; \qquad y_0 = 1.$$

Consecutive calculations yield

$$y_1 = a, \quad y_2 = a^{\beta+1}, \quad y_3 = a^{\beta^2+\beta+1}, \quad \ldots, \quad y_n = a^{\beta^{n-1}+\beta^{n-2}+\cdots+\beta+1} = a^{\frac{\beta^n-1}{\beta-1}}.$$

**Remark.** As a rule, solutions of nonlinear difference equations cannot be found in closed form (i.e., in terms of a single, not a recurrent, formula).

### 17.1.2-3. Riccati difference equation.

The *Riccati difference equation* has the general form

$$y_n y_{n+1} = a_n y_{n+1} + b_n y_n + c_n, \qquad n = 0, 1, \ldots, \qquad (17.1.2.6)$$

with the constants $a_n$, $b_n$, $c_n$ satisfying the condition $a_n b_n + c_n \neq 0$.

1°. The substitution

$$y_n = \frac{u_{n+1}}{u_n} + a_n, \qquad u_0 = 1,$$

leads us to the linear second-order difference equation

$$u_{n+2} + (a_{n+1} - b_n) u_{n+1} - (a_n b_n + c_n) u_n = 0$$

with the initial conditions

$$u_0 = 1, \qquad u_1 = y_0 - a_0.$$

2°. Let $y_n^*$ be a particular solution of equation (17.1.2.6). Then the substitution

$$z_n = \frac{1}{y_n - y_n^*}, \qquad n = 0, 1, \ldots,$$

reduces equation (17.1.2.6) to the first-order linear nonhomogeneous difference equation

$$z_{n+1} + \frac{(y_n^* - a_n)^2}{a_n b_n + c_n} z_n + \frac{y_n^* - a_n}{a_n b_n + c_n} = 0.$$

With regard to the solution of this equation see Paragraph 17.1.1-2.

3°. Let $y_n^{(1)}$ and $y_n^{(2)}$ be two particular solutions of equation (17.1.2.6) with $y_n^{(1)} \neq y_n^{(2)}$. Then the substitution

$$w_n = \frac{1}{y_n - y_n^{(1)}} + \frac{1}{y_n^{(1)} - y_n^{(2)}}, \qquad n = 0, 1, \ldots,$$

reduces equation (17.1.2.6) to the first-order linear homogeneous difference equation

$$w_{n+1} + \frac{(y_n^{(1)} - a_n)^2}{a_n b_n + c_n} w_n = 0, \qquad n = 0, 1, \ldots$$

With regard to the solution of this equation see Paragraph 17.1.1-1.

### 17.1.2-4. Logistic difference equation.

Consider the initial-value problem for the *logistic difference equation*

$$y_{n+1} = a y_n \left(1 - \frac{y_n}{b}\right), \qquad n = 0, 1, \ldots,$$

$$y_0 = \lambda, \qquad (17.1.2.7)$$

where $0 < a \leq 4$, $b > 0$, and $0 \leq \lambda \leq b$.

**1°.** Let
$$a = b = 4, \quad \lambda = 4\sin^2\theta \quad (0 \le \theta \le \tfrac{\pi}{2}).$$

Then problem (17.1.2.7) has the closed-form solution
$$y_n = 4\sin^2(2^n\theta), \quad n = 0, 1, \ldots$$

**2°.** Let
$$a = 4, \quad b = 1, \quad \lambda = \sin^2\theta \quad (0 \le \theta \le \tfrac{\pi}{2}).$$

Then problem (17.1.2.7) has the closed-form solution
$$y_n = \sin^2(2^n\theta), \quad n = 0, 1, \ldots$$

**3°.** Let $0 \le a \le 4$ and $b = 1$. In this case, the solutions of the logistic equation have the following properties:
  (a) There are equilibrium solutions $y_n = 0$ and $y_n = (a-1)/a$.
  (b) If $0 \le y_0 \le 1$, then $0 \le y_n \le 1$.
  (c) If $a = 0$, then $y_n = 0$.
  (d) If $0 < a \le 1$, then $y_n \to 0$ as $n \to \infty$.
  (e) If $1 < a \le 3$, then $y_n \to (a-1)/a$ as $n \to \infty$.
  (f) If $3 < a < 3.449\ldots$, then $y_n$ oscillates between the two points:
$$y_\pm = \frac{1}{2a}(a + 1 \pm \sqrt{a^2 - 2a - 3}).$$

**17.1.2-5. Graphical construction of solutions to nonlinear difference equations.**

Consider nonlinear difference equations of special form
$$y_{n+1} = f(y_n), \quad n = 0, 1, \ldots \tag{17.1.2.8}$$

The points $y_0, y_1, y_2, \ldots$ are constructed on the plane $(y, z)$ on the basis of the graph $z = f(y)$ and the straight line $z = y$, called the *iteration axis*.

Figure 17.1 shows the result of constructing the points $P_0, P_1, P_2, \ldots$ on the graph of the function $z = f(y)$ with the abscissas $y_0, y_1, y_2, \ldots$ determined by equation (17.1.2.8).

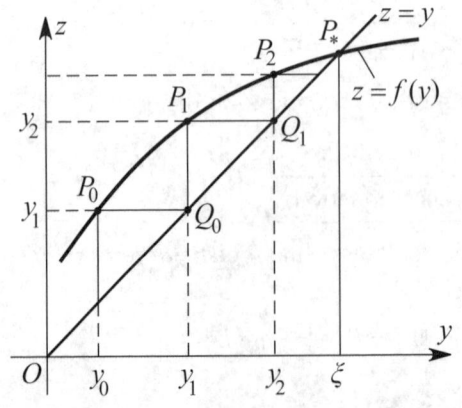

**Figure 17.1.** Construction, using the graph of the function $z = f(y)$, of the points with abscissas $y_0, y_1, y_2, \ldots$ that satisfy the difference equations (17.1.2.8).

This construction consists of the following steps:

1. Through the point $P_0 = (y_0, y_1)$ with $y_1 = f(y_0)$, we draw a horizontal line. This line crosses the iteration axis at the point $Q_0 = (y_1, y_1)$.
2. Through the point $Q_0$, we draw a vertical line. This line crosses the graph of the function $f(y)$ at the point $P_1 = (y_1, y_2)$ with $y_2 = f(y_1)$.
3. Repeating the operations of steps 1 and 2, we obtain the following sequence on the graph of $f(y)$:

$$P_0 = (y_0, f(y_0)), \quad P_1 = (y_1, f(y_1)), \quad P_2 = (y_2, f(y_2)), \quad \ldots$$

In the case under consideration, for $n \to \infty$, the points $y_n$ converge to a fixed $\xi$, which determines an equilibrium solution satisfying the algebraic (transcendental) equation $\xi = f(\xi)$.

**17.1.2-6. Convergence to a fixed point. Qualitative behavior of solutions.**

A fixed point of a mapping $f$ of a set $I$ is a point $\xi \in I$ such that $f(\xi) = \xi$.

BRAUER FIXED POINT THEOREM. *If $f(y)$ is a continuous function on the interval $I = \{a \leq y \leq b\}$ and $f(I) \subset I$, then $f(y)$ has a fixed point in $I$.*

A set $E$ is called the *domain of attraction of a fixed point* $\xi$ of a function $f(y)$ if the sequence $y_{n+1} = f(y_n)$ converges to $\xi$ for any $y_0 \in E$.

If $\xi = f(\xi)$ and $|f'(\xi)| < 1$, then $\xi$ is an attracting fixed point: there is a neighborhood of $\xi$ belonging to its domain of attraction.

Figure 17.2 illustrates the qualitative behavior of sequences (17.1.2.8) starting from points sufficiently close to a fixed point $\xi$ such that $f'(\xi) \neq 0$ and $|f'(\xi)| \neq 1$.

According to the behavior of the iteration process in a neighborhood of the fixed point, the cases represented in Fig. 17.2 may be called one-dimensional analogues of a "stable node" (for $0 < f'(\xi) < 1$; see Fig. 17.2 a), "stable focus" (for $-1 < f'(\xi) < 0$; see Fig. 17.2 b), "unstable node" (for $1 < f'(\xi)$; see Fig. 17.2 c), or "unstable focus" (for $f'(\xi) < -1$; see Fig. 17.2 d).

## 17.1.3. Second-Order Linear Difference Equations with Constant Coefficients

**17.1.3-1. Homogeneous linear equations.**

A second-order homogeneous linear difference equation with constant coefficients has the form

$$ay_{n+2} + by_{n+1} + cy_n = 0. \tag{17.1.3.1}$$

The general solution of this equation is determined by the roots of the quadratic equation

$$a\lambda^2 + b\lambda + c = 0. \tag{17.1.3.2}$$

1°. Let $b^2 - 4ac > 0$. Then the quadratic equation (17.1.3.2) has two different real roots

$$\lambda_1 = \frac{-b + \sqrt{b^2 - 4ac}}{2a}, \quad \lambda_2 = \frac{-b - \sqrt{b^2 - 4ac}}{2a},$$

and the general solution of the difference equation (17.1.3.1) is given by the formula

$$y_n = C_1 \frac{\lambda_1 \lambda_2^n - \lambda_1^n \lambda_2}{\lambda_1 - \lambda_2} + C_2 \frac{\lambda_1^n - \lambda_2^n}{\lambda_1 - \lambda_2}, \tag{17.1.3.3}$$

where $C_1$ and $C_2$ are arbitrary constants. Solution (17.1.3.3) satisfies the initial conditions $y_0 = C_1$, $y_1 = C_2$.

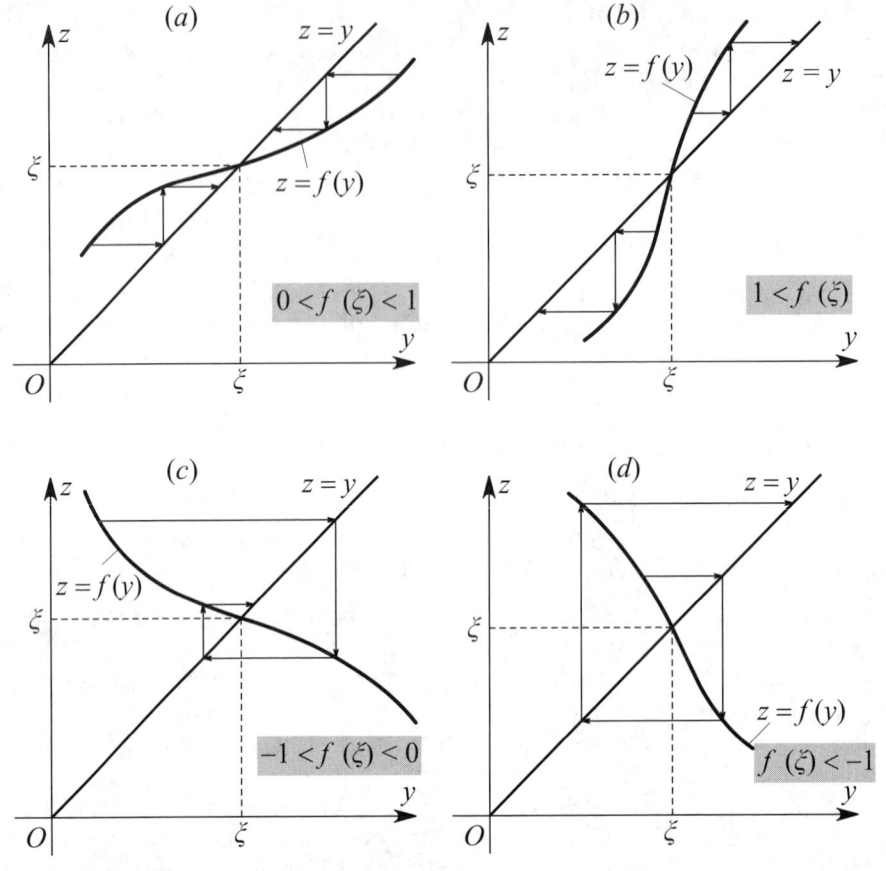

**Figure 17.2.** Iterative sequences $y_{n+1} = f(y_n)$ in a neighborhood of a fixed point $\xi = f(\xi)$. Qualitative analysis of different cases using the graphs of the function $z = f(y)$.

$2°$. Let $b^2 - 4ac = 0$. Then the quadratic equation (17.1.3.2) has one double real root

$$\lambda = -\frac{b}{2a},$$

and the general solution of the difference equation (17.1.3.1) has the form

$$y_n = C_1(1-n)\lambda^n + C_2 n \lambda^{n-1}.$$

This formula can be obtained from (17.1.3.3) by taking $\lambda_1 = \lambda$, $\lambda_2 = \lambda(1-\varepsilon)$ and passing to the limit as $\varepsilon \to 0$.

$3°$. Let $b^2 - 4ac < 0$. Then the quadratic equation (17.1.3.2) has two complex conjugate roots

$$\lambda_1 = \rho(\cos\varphi + i\sin\varphi), \quad \lambda_2 = \rho(\cos\varphi - i\sin\varphi),$$

$$\rho = \sqrt{\frac{c}{a}}, \quad \tan\varphi = -\frac{1}{b}\sqrt{4ac - b^2},$$

and the general solution of the difference equation (17.1.3.1) has the form

$$y_n = C_1 \rho^n \frac{\sin[(n-1)\varphi]}{\sin\varphi} + C_2 \rho^{n-1} \frac{\sin(n\varphi)}{\sin\varphi}.$$

This formula can also be obtained from (17.1.3.3) by expressing $\lambda_1$ and $\lambda_2$ in terms of $\rho$ and $\varphi$.

## 17.1.3-2. Nonhomogeneous linear equations.

A second-order nonhomogeneous linear difference equation with constant coefficients has the form
$$ay_{n+2} + by_{n+1} + cy_n = f_n. \tag{17.1.3.4}$$

The general solution of this equation is given by
$$y_n = C_1 \frac{\lambda_1 \lambda_2^n - \lambda_1^n \lambda_2}{\lambda_1 - \lambda_2} + C_2 \frac{\lambda_1^n - \lambda_2^n}{\lambda_1 - \lambda_2} + \widetilde{y}_n,$$

where
$$\widetilde{y}_0 = \widetilde{y}_1 = 0, \quad \widetilde{y}_n = \frac{1}{a} \sum_{k=0}^{n-2} \frac{\lambda_1^{n-k-1} - \lambda_2^{n-k-1}}{\lambda_1 - \lambda_2} f_k,$$

$C_1$ and $C_2$ are arbitrary constants, and $\lambda_1, \lambda_2$ are the roots of the quadratic equation (17.1.3.2).

## 17.1.3-3. Boundary value problem.

The solution of the boundary value problem for equation*
$$ay_{n+1} + by_n + cy_{n-1} = f_n, \quad n = 1, 2, \ldots, N-1,$$

with the boundary conditions
$$y_0 = \mu_1, \quad y_{N-1} = \mu_2$$

is given by the formula
$$y_n = \frac{(\lambda_1 \lambda_2)^n (\lambda_1^{N-n} - \lambda_2^{N-n})}{\lambda_1^N - \lambda_2^N} \mu_1 + \frac{\lambda_1^n - \lambda_2^n}{\lambda_1^N - \lambda_2^N} \mu_2$$
$$- \frac{1}{a} \sum_{k=1}^{n-1} \frac{(\lambda_1 \lambda_2)^{n-k} (\lambda_1^{N-n} - \lambda_2^{N-n})(\lambda_1^k - \lambda_2^k)}{(\lambda_1 - \lambda_2)(\lambda_1^N - \lambda_2^N)} f_k - \frac{1}{a} \sum_{k=n}^{N-1} \frac{(\lambda_1^{N-k} - \lambda_2^{N-k})(\lambda_1^n - \lambda_2^n)}{(\lambda_1 - \lambda_2)(\lambda_1^N - \lambda_2^N)} f_k.$$

## 17.1.4. Second-Order Linear Difference Equations with Variable Coefficients

### 17.1.4-1. Second-order homogeneous linear difference equations. General solution.

1°. A second-order homogeneous linear difference equation with variable coefficients has the form
$$a_n y_{n+2} + b_n y_{n+1} + c_n y_n = 0. \tag{17.1.4.1}$$

The *trivial solution* $y_n = 0$ is a particular solution of the homogeneous linear equation. Let $y_n^{(1)}$, $y_n^{(2)}$ be particular solutions of equation (17.1.4.1) satisfying the condition
$$y_0^{(1)} y_1^{(2)} - y_0^{(2)} y_1^{(1)} \neq 0. \tag{17.1.4.2}$$

Then the general solution of equation (17.1.4.1) is given by
$$y_n = C_1 y_n^{(1)} + C_2 y_n^{(2)}, \tag{17.1.4.3}$$

where $C_1$ and $C_2$ are arbitrary constants.

---
* This equation is obtained from (17.1.3.4) by shifting the subscript of the sought function by unity.

**Remark.** If condition (17.1.4.2) holds, then the solutions $y_n^{(1)}$, $y_n^{(2)}$ are linearly independent and for all $n$ the following inequality holds: $\Delta_n = y_n^{(1)} y_{n+1}^{(2)} - y_n^{(2)} y_{n+1}^{(1)} \ne 0$. It is convenient to single out two linearly independent solutions $y_n^{(1)}$, $y_n^{(2)}$ with the help of the initial conditions

$$y_0^{(1)} = 1, \quad y_1^{(1)} = 0; \qquad y_0^{(2)} = 0, \quad y_1^{(2)} = 1.$$

2°. Let $y_n^*$ be a nontrivial particular solution of equation (17.1.4.1). Then the replacement

$$y_n = y_n^* u_n \qquad (17.1.4.4)$$

yields the equation

$$a_n y_{n+2}^* u_{n+2} + b_n y_{n+1}^* u_{n+1} + c_n y_n^* u_n = 0. \qquad (17.1.4.5)$$

Taking into account that equation (17.1.4.1) holds for $y_n^*$, and substituting

$$b_n y_{n+1}^* = -a_n y_{n+2}^* - c_n y_n^*$$

into (17.1.4.5), we find, after simple transformations, that

$$a_n y_{n+2}^* (u_{n+2} - u_{n+1}) - c_n y_n^* (u_{n+1} - u_n) = 0.$$

Introducing a new variable by

$$w_n = u_{n+1} - u_n, \qquad (17.1.4.6)$$

we come to the homogeneous first-order difference equation

$$a_n y_{n+2}^* w_{n+1} - c_n y_n^* w_n = 0.$$

Solving this equation (see Paragraph 17.1.1-1), one finds a solution of the nonhomogeneous first-order equation with constant coefficients (17.1.4.6) (see Paragraph 17.1.1-3), and then, using (17.1.4.4), one finds a solution of the original equation.

### 17.1.4-2. Second-order nonhomogeneous linear equations. General solution.

1°. A second-order nonhomogeneous linear difference equation with variable coefficients has the form

$$a_n y_{n+2} + b_n y_{n+1} + c_n y_n = f_n. \qquad (17.1.4.7)$$

The general solution of the nonhomogeneous linear equation (17.1.4.7) can be represented as a sum of the general solution (17.1.4.3) of the corresponding homogeneous equation (17.1.4.1) and a particular solution $\tilde{y}_n$ of the nonhomogeneous equation (17.1.4.7):

$$y_n = C_1 y_n^{(1)} + C_2 y_n^{(2)} + \tilde{y}_n,$$

where

$$\tilde{y}_0 = \tilde{y}_1 = 0, \quad \tilde{y}_n = \sum_{j=0}^{n-2} \frac{y_{j+1}^{(1)} y_n^{(2)} - y_n^{(1)} y_{j+1}^{(2)}}{y_{j+1}^{(1)} y_{j+2}^{(2)} - y_{j+2}^{(1)} y_{j+1}^{(2)}} \frac{f_j}{a_j}, \quad n = 2, 3, \dots$$

2°. Let $y_n^*$ be a nontrivial particular solution of the homogeneous equation (17.1.4.1). Then the substitutions (17.1.4.4) and (17.1.4.6) yield a nonhomogeneous first-order difference equation

$$a_n y_{n+2}^* w_{n+1} - c_n y_n^* w_n = f_n.$$

With regard to the solution of this equation see Paragraph 17.1.1-2.

3°. *Superposition principle.* Let $y_n^{(1)}$ and $y_n^{(2)}$ be solutions of two nonhomogeneous linear difference equations with the same left-hand sides and different right-hand sides:

$$a_n y_{n+2} + b_n y_{n+1} + c_n y_n = f_n,$$
$$a_n y_{n+2} + b_n y_{n+1} + c_n y_n = g_n.$$

Then $\alpha y_n^{(1)} + \beta y_n^{(2)}$ is a solution of the equation

$$a_n y_{n+2} + b_n y_{n+1} + c_n y_n = \alpha f_n + \beta g_n,$$

where $\alpha$ and $\beta$ are arbitrary constants.

## 17.1.5. Linear Difference Equations of Arbitrary Order with Constant Coefficients

17.1.5-1. Homogeneous linear equations.

An $m$th-order homogeneous linear difference equation with constant coefficients has the general form
$$a_m y_{n+m} + a_{m-1} y_{n+m-1} + \cdots + a_1 y_{n+1} + a_0 y_n = 0. \qquad (17.1.5.1)$$

The general solution of this equation is determined by the roots of the characteristic equation
$$a_m \lambda^m + a_{m-1} \lambda^{m-1} + \cdots + a_1 \lambda + a_0 = 0. \qquad (17.1.5.2)$$

The following cases are possible:

1°. All roots $\lambda_1, \lambda_2, \ldots, \lambda_m$ of the characteristic equation (17.1.5.2) are real and distinct. Then the general solution of the homogeneous linear differential equation (17.1.5.1) has the form
$$y_n = C_1 \lambda_1^n + C_2 \lambda_2^n + \cdots + C_m \lambda_m^n,$$
where $C_1, C_2, \ldots, C_m$ are arbitrary constants.

2°. There are $k$ equal real roots $\lambda_1 = \lambda_2 = \cdots = \lambda_k$ ($k \leq m$), and the other roots are real and distinct. In this case, the general solution is given by
$$y_n = (C_1 + C_2 n + \cdots + C_k n^{k-1}) \lambda_1^n + C_{k+1} \lambda_{k+1}^n + \cdots + C_m \lambda_m^n.$$

3°. There are $k$ pairs of distinct complex conjugate roots $\lambda_j = \rho_j (\cos \varphi_j \pm i \sin \varphi_j)$ ($j = 1, \ldots, k$; $2k \leq m$), and the other roots are real and distinct. In this case, the general solution is
$$y_n = \rho_1^n [A_1 \cos(n\varphi_1) + B_1 \sin(n\varphi_1)] + \cdots + \rho_k^n [A_k \cos(n\varphi_k) + B_k \sin(n\varphi_k)] \\ + C_{2k+1} \lambda_{2k+1}^n + \cdots + C_m \lambda_m^n,$$
where $A_1, \ldots, A_k, B_1, \ldots, B_k, C_{2k+1}, \ldots, C_m$ are arbitrary constants.

4°. In the general case, if there are $r$ different roots $\lambda_1, \lambda_2, \ldots, \lambda_r$ of multiplicities $k_1, k_2, \ldots, k_r$, respectively, the left-hand side of the characteristic equation (17.1.5.2) can be represented as the product
$$P(\lambda) = a_m (\lambda - \lambda_1)^{k_1} (\lambda - \lambda_2)^{k_2} \ldots (\lambda - \lambda_r)^{k_r},$$
where $k_1 + k_2 + \cdots + k_r = m$. The general solution of the original equation is given by the formula
$$y_n = \sum_{s=1}^{r} (C_{s,1} + C_{s,2} n + \cdots + C_{s,k_s} n^{k_s - 1}) \lambda_s^n,$$
where $C_{s,k}$ are arbitrary constants.

If the characteristic equation (17.1.5.2) has complex conjugate roots of the form $\lambda = \rho e^{\pm i \varphi} = \rho (\cos \varphi \pm i \sin \varphi)$, then in the above solution, one should extract the real part on the basis of the relation $\lambda^n = \rho^n e^{\pm i n \varphi} = \rho^n [\cos(n\varphi) \pm i \sin(n\varphi)]$.

### 17.1.5-2. Nonhomogeneous linear equations.

$1°$. An $m$th-order nonhomogeneous linear difference equation with constant coefficients has the general form

$$a_m y_{n+m} + a_{m-1} y_{n+m-1} + \cdots + a_1 y_{n+1} + a_0 y_n = f_n. \qquad (17.1.5.3)$$

The general solution of this equation can be represented as the sum $y_n = Y_n + \widetilde{y}_n$, where $Y_n$ is the general solution of the corresponding homogeneous equation (for $f_n \equiv 0$) and $\widetilde{y}_n$ is any particular solution of the nonhomogeneous equation (17.1.5.3). The general solution of the corresponding homogeneous equation is constructed with the help of the formulas from Paragraph 17.1.5-1, and a particular solution of the nonhomogeneous equation for an arbitrary function $f_n$ is constructed with the help of the formulas from Paragraph 17.1.6-2.

$2°$. If the roots $\lambda_1, \lambda_2, \ldots, \lambda_m$ of the characteristic equation (17.1.5.2) are mutually distinct, the particular solution of the nonhomogeneous difference equation (17.1.5.3) has the form

$$\widetilde{y}_n = \sum_{\nu=m}^{n} f_{n-\nu} \sum_{k=1}^{m} \frac{\lambda_k^{\nu-1}}{P'(\lambda_k)}, \qquad (17.1.5.4)$$

where $P(\lambda)$ is the characteristic polynomial [coinciding with the left-hand side of equation (17.1.5.2)], and the prime indicates its derivative

$$P'(\lambda) \equiv a_m m \lambda^{m-1} + a_{m-1}(m-1)\lambda^{m-2} + \cdots + 2a_2 \lambda + a_1.$$

In the case of complex conjugate roots, solution (17.1.5.4) should be split into the real and the imaginary parts.

## 17.1.6. Linear Difference Equations of Arbitrary Order with Variable Coefficients

### 17.1.6-1. Homogeneous linear difference equations.

An $m$th-order homogeneous linear difference equation with variable coefficients has the form

$$a_m(n) y_{n+m} + a_{m-1}(n) y_{n+m-1} + \cdots + a_1(n) y_{n+1} + a_0(n) y_n = 0. \qquad (17.1.6.1)$$

Functions $u_n^{(1)}, u_n^{(2)}, \ldots, u_n^{(m)}$ are called *linearly independent solutions* of equation (17.1.6.1) if
1) they take finite values and satisfy equation (17.1.6.1),
2) the relation

$$C_1 u_n^{(1)} + C_2 u_n^{(2)} + \cdots + C_m u_n^{(m)} = 0, \quad \text{for all} \quad n = 1, 2, \ldots,$$

with constants $C_1, C_2, \ldots, C_m$ implies that $C_1 = \cdots = C_m = 0$.

If $u_n^{(1)}, u_n^{(2)}, \ldots, u_n^{(m)}$ are linearly independent solutions of equation (17.1.6.1), then the determinant

$$\Delta_n = \begin{vmatrix} u_n^{(1)} & u_{n+1}^{(1)} & \cdots & u_{n+m-1}^{(1)} \\ u_n^{(2)} & u_{n+1}^{(2)} & \cdots & u_{n+m-1}^{(2)} \\ \cdots & \cdots & & \cdots \\ u_n^{(m)} & u_{n+1}^{(m)} & \cdots & u_{n+m-1}^{(m)} \end{vmatrix} \qquad (17.1.6.2)$$

differs from zero for all admissible $n$. Conversely, if the determinant (17.1.6.2) for solutions $u_n^{(1)}, u_n^{(2)}, \ldots, u_n^{(m)}$ of equation (17.1.6.1) differs from zero for some $n$, then the solutions are linearly independent.

If $u_n^{(1)}, u_n^{(2)}, \ldots, u_n^{(m)}$ are linearly independent solutions of equation (17.1.6.1), then the general solution of this equation has the form

$$y_n = C_1 u_n^{(1)} + C_2 u_n^{(2)} + \cdots + C_m u_n^{(m)}, \tag{17.1.6.3}$$

where $C_1, C_2, \ldots, C_m$ are arbitrary constants.

A solution is determined by prescribing the initial values of the sought function at $m$ points.

### 17.1.6-2. Nonhomogeneous linear difference equations.

$1°$. An $m$th-order nonhomogeneous linear difference equation with variable coefficients has the form

$$a_m(n)y_{n+m} + a_{m-1}(n)y_{n+m-1} + \cdots + a_1(n)y_{n+1} + a_0(n)y_n = f_n. \tag{17.1.6.4}$$

The general solution of this equation can be represented as a sum of the general solution of the corresponding homogeneous equation (17.1.6.1) and a particular solution $\widetilde{y}_n$ of the nonhomogeneous equation (17.1.6.4):

$$y_n = C_1 u_n^{(1)} + C_2 u_n^{(2)} + \cdots + C_m u_n^{(m)} + \widetilde{y}_n.$$

A particular solution can be determined by the formula

$$\widetilde{y}_0 = \widetilde{y}_1 = \cdots = \widetilde{y}_{m-1} = 0, \quad \widetilde{y}_n = \sum_{j=0}^{n-m} \frac{A_{m,n,j}}{B_{m,j}} \frac{f_j}{a_m(j)}, \quad n = m, m+1, \ldots,$$

where

$$A_{m,n,j} = \begin{vmatrix} u_{j+1}^{(1)} & u_{j+1}^{(2)} & \cdots & u_{j+1}^{(m)} \\ \cdots & \cdots & \cdots & \cdots \\ u_{j+m-1}^{(1)} & u_{j+m-1}^{(2)} & \cdots & u_{j+m-1}^{(m)} \\ u_n^{(1)} & u_n^{(2)} & \cdots & u_n^{(m)} \end{vmatrix}, \quad B_{m,j} = \begin{vmatrix} u_{j+1}^{(1)} & u_{j+2}^{(1)} & \cdots & u_{j+m}^{(1)} \\ \cdots & \cdots & \cdots & \cdots \\ u_{j+1}^{(m-1)} & u_{j+2}^{(m-1)} & \cdots & u_{j+m}^{(m-1)} \\ u_{j+1}^{(m)} & u_{j+2}^{(m)} & \cdots & u_{j+m}^{(m)} \end{vmatrix}.$$

$2°$. *Superposition principle*. Let $y_n^{(1)}$ and $y_n^{(2)}$ be solutions of two nonhomogeneous linear difference equations with the same left-hand side and different right-hand sides:

$$a_m(n)y_{n+m} + a_{m-1}(n)y_{n+m-1} + \cdots + a_1(n)y_{n+1} + a_0(n)y_n = f_n,$$
$$a_m(n)y_{n+m} + a_{m-1}(n)y_{n+m-1} + \cdots + a_1(n)y_{n+1} + a_0(n)y_n = g_n.$$

Then $\alpha y_n^{(1)} + \beta y_n^{(2)}$ is a solution of the equation

$$a_m(n)y_{n+m} + a_{m-1}(n)y_{n+m-1} + \cdots + a_1(n)y_{n+1} + a_0(n)y_n = \alpha f_n + \beta g_n,$$

where $\alpha$ and $\beta$ are arbitrary constants.

## 17.1.7. Nonlinear Difference Equations of Arbitrary Order

**17.1.7-1. Difference equations of $m$th order. General solution.**

Let $y_n = y(n)$ be a function of integer argument $n = 0, \pm 1, \pm 2, \ldots$ An *$m$th-order difference equation*, in the general case, has the form

$$F(n, y_n, y_{n+1}, \ldots, y_{n+m}) = 0. \tag{17.1.7.1}$$

A *solution of the difference equation* (17.1.7.1) is a discrete function $y_n$ that, being substituted into the equation, turns it into identity. The *general solution of a difference equation* is the set of all its solutions. The general solution of equation (17.1.7.1) depends on $m$ arbitrary constants $C_1, \ldots, C_m$. The general solution can be written in explicit form as

$$y_n = \varphi(n, C_1, \ldots, C_m), \tag{17.1.7.2}$$

or in implicit form as $\Phi(n, y_n, C_1, \ldots, C_m) = 0$. Specific values of $C_1, \ldots, C_m$ define specific solutions of the equation (*particular solutions*).

Any constant solution $y_n = \xi$ of equation (17.1.7.1), where $\xi$ is independent of $n$, is called an *equilibrium solution*.

*Remark.* The term *difference equation* was introduced in numerical mathematics in connection with the investigation of equations of the form

$$G(n, y_n, \Delta y_n, \ldots, \Delta^m y_n) = 0 \tag{17.1.7.3}$$

with finite differences*

$$\Delta y_n = y_{n+1} - y_n, \quad \Delta^2 y_n = y_{n+2} - 2y_{n+1} + y_n, \quad \Delta^m y_n = \Delta^{m-1} \Delta y_n. \tag{17.1.7.4}$$

The replacement of the finite differences in (17.1.7.3), by their explicit expressions in terms of the values of the sought function according to (17.1.7.4), brings us to equations of the form (17.1.7.1).

**17.1.7-2. Construction of a difference equation by a given general solution.**

Suppose that the general solution of a difference equation is given in the form (17.1.7.2). Then the corresponding $m$th-order difference equation with this solution can be constructed by eliminating the arbitrary constants $C_1, \ldots, C_m$ from the relations

$$y_n = \varphi(n, C_1, \ldots, C_m),$$
$$y_{n+1} = \varphi(n+1, C_1, \ldots, C_m),$$
$$\ldots\ldots\ldots\ldots\ldots\ldots\ldots\ldots\ldots\ldots$$
$$y_{n+m} = \varphi(n+m, C_1, \ldots, C_m).$$

**17.1.7-3. Cauchy's problem and its solution. The step method.**

A difference equation resolved with respect to the leading term $y_{n+m}$ has the form

$$y_{n+m} = f(n, y_n, y_{n+1}, \ldots, y_{n+m-1}). \tag{17.1.7.5}$$

The Cauchy problem consists of finding a solution of this equation with given initial values of $y_0, y_1, \ldots, y_{m-1}$.

---

* Finite differences are used for the approximation of derivatives in differential equations.

The next value, $y_m$, is calculated by substituting the initial values into the right-hand side of equation (17.1.7.5) with $n = 0$. We have $y_m = f(0, y_0, y_1, \ldots, y_{m-1})$. Then one takes $n = 1$ in (17.1.7.5). We get

$$y_{m+1} = f(1, y_1, y_2, \ldots, y_m). \tag{17.1.7.6}$$

Substituting the initial values $y_1, \ldots, y_{m-1}$ and the calculated value $y_m$ into this relation, we find $y_{m+1}$. Further, taking $n = 2$ in (17.1.7.6) and using the initial values $y_2, \ldots, y_{m-1}$ and the calculated values $y_m$, $y_{m+1}$, we find $y_{m+2}$. In a similar way, we consecutively find all subsequent values $y_{m+3}, y_{m+4}, \ldots$

The above method of solving difference equations is called the *step method*.

## 17.2. Linear Difference Equations with a Single Continuous Variable

### 17.2.1. First-Order Linear Difference Equations

17.2.1-1. Homogeneous linear difference equations. General properties of solutions.

1°. A *first-order homogeneous linear difference equation* has the form

$$y(x+1) - f(x)y(x) = 0, \tag{17.2.1.1}$$

where $f(x)$ is a given continuous real-valued function of real argument and $y(x)$ is an unknown real-valued function, $0 \le x < \infty$.

Let $y_1 = y_1(x)$ be a nontrivial particular solution of equation (17.2.1.1), $y_1(x) \not\equiv 0$. Then the general solution of equation (17.2.1.1) is given by

$$y(x) = \Theta(x)y_1(x), \tag{17.2.1.2}$$

where $\Theta(x) = \Theta(x+1)$ is an arbitrary 1-periodic function.

**Example 1.** The equation with constant coefficients

$$y(x+1) - ay(x) = 0, \tag{17.2.1.3}$$

with $a > 0$, admits a particular solution $y_0(x) = a^x$. Therefore, the general solution of equation (17.2.1.3) has the form

$$y(x) = \Theta(x)a^x, \tag{17.2.1.4}$$

where $\Theta(x)$ is an arbitrary periodic function with unit period. In the special case of $a = 1$, the general solution of equation (17.2.1.3) is an arbitrary 1-periodic function.

**Remark 1.** When using formula (17.2.1.2) for obtaining continuous particular solutions from a given continuous particular solution, one should consider not only continuous but also discontinuous or unbounded periodic functions $\Theta(x)$. For instance, the equation

$$y(x+1) + y(x) = 0, \tag{17.2.1.5}$$

which determines *antiperiodic functions* of unit period, has continuous particular solutions

$$y_1(x) = \cos(\pi x), \quad y_2(x) = \sin(\pi x).$$

In order to pass from the first of these particular solutions to the second with the help of (17.2.1.2), one should take the unbounded periodic function $\Theta(x) = \tan(\pi x)$, which is undefined at the points $x = \frac{1}{2} + n$, $n = 0, \pm 1, \pm 2, \ldots$

See also Remark 2.

**Remark 2.** The general solution of a difference equation may be represented by quite different formulas. Thus, the general solution of equation (17.2.1.5) can be represented by the following formulas:

$$y(x) = (-1)^{[x]}\Theta(x), \qquad (17.2.1.6a)$$

$$y(x) = \Theta_1(x)\cos(\pi x) + \Theta_2(x)\sin(\pi x), \qquad (17.2.1.6b)$$

$$y(x) = \Theta_3\left(\frac{x+1}{2}\right) - \Theta_3\left(\frac{x}{2}\right), \qquad (17.2.1.6c)$$

where $\Theta(x)$, $\Theta_1(x)$, $\Theta_2(x)$, and $\Theta_3(x)$ are arbitrary periodic functions of period 1, $[x]$ is the integer part of $x$.

Let us show that formulas (17.2.1.6a), (17.2.1.6b), and (17.2.1.6c) are equivalent.

First, we note that (17.2.1.6b) contains one redundant arbitrary function and can be written as

$$y(x) = \overline{\Theta}_1(x)\cos(\pi x) = \overline{\Theta}_2(x)\sin(\pi x),$$
$$\overline{\Theta}_1(x) = \Theta_1(x) + \tan(\pi x)\Theta_2(x), \qquad \overline{\Theta}_2(x) = \cot(\pi x)\Theta_1(x) + \Theta_2(x).$$

Here, $\overline{\Theta}_1(x)$ and $\overline{\Theta}_2(x)$ are arbitrary 1-periodic functions, since $\tan(\pi x)$ and $\cot(\pi x)$ have period 1. Therefore, without the loss of generality, we can take $\Theta_1(x) \equiv 0$ in (17.2.1.6b) (or $\Theta_2(x) \equiv 0$). And we should consider bounded, as well as unbounded, periodic functions $\Theta_2(x)$ (or $\Theta_1(x)$).

Formula (17.2.1.6a) with a continuous $\Theta(x)$ yields discontinuous solutions, in general (for $\Theta(x)$ different from zero at integer points). This formula can be represented in the form (17.2.1.6b):

$$y(x) = \sin(\pi x)\Theta_2(x), \qquad \Theta_2(x) = \frac{(-1)^{[x]}}{\sin(\pi x)}\Theta(x).$$

Here, $\Theta_2(x)$ is an arbitrary periodic function of period 1, since the function $(-1)^{[x]}/\sin(\pi x)$ has period 1.

Formula (17.2.1.6c) can be written in the form (17.2.1.6a):

$$y(x) = (-1)^{[x]}\Theta(x), \qquad \Theta(x) = (-1)^{[x]}\left[\Theta_3\left(\frac{x+1}{2}\right) - \Theta_3\left(\frac{x}{2}\right)\right].$$

For the investigation of continuous (smooth) solutions it is more convenient to use formulas (17.2.1.6b) and (17.2.1.6c), where $\Theta_1(x)$, $\Theta_2(x)$, and $\Theta_3(x)$ are arbitrary continuous (smooth) 1-periodic functions.

2°. Consider the equation

$$y(x+1) - af(x)y(x) = 0 \qquad (17.2.1.7)$$

with a parameter $a$. Let $y_1 = y_1(x)$ be a nontrivial particular solution of equation (17.2.1.7) with $a = 1$. Then the following results hold:

For $a > 0$, the general solution of equation (17.2.1.7) has the form

$$y(x) = a^x \Theta(x) y_1(x), \qquad (17.2.1.8)$$

where $\Theta(x)$ is an arbitrary periodic function with period 1.

For $a < 0$, the general solution of equation (17.2.1.7) can be represented by any of the formulas

$$y(x) = (-1)^{[x]}|a|^x \Theta(x) y_1(x),$$
$$y(x) = |a|^x \left[\Theta_1(x)\cos(\pi x) + \Theta_2(x)\sin(\pi x)\right] y_1(x), \qquad (17.2.1.9)$$
$$y(x) = |a|^x \left[\Theta_3\left(\frac{x+1}{2}\right) - \Theta_3\left(\frac{x}{2}\right)\right] y_1(x),$$

which generalize formulas (17.2.1.6).

3°. Let $y_1 = y_1(x)$ be a solution of equation (17.2.1.1). Then the equation

$$y(x+1) - f(x+a)y(x) = 0$$

admits the solution

$$y(x) = y_1(x+a).$$

**4°.** Let $y_1 = y_1(x)$ be a positive solution of equation (17.2.1.1). Then the equation

$$y(x+1) - [f(x)]^k y(x) = 0,$$

for any $k$, admits the solution

$$y(x) = [y_1(x)]^k.$$

**5°.** Let $y_1 = y_1(x)$ be a solution of equation (17.2.1.1). Then the equation

$$y(x+1) - \varphi(x)f(x)y(x) = 0,$$

where $\varphi(x) = \varphi(x+1)$ is an arbitrary positive 1-periodic function, admits the solution

$$y(x) = y_1(x)[\varphi(x)]^{x+1}.$$

**6°.** A solution of the linear homogeneous difference equation

$$y(x+1) + f_1(x)f_2(x)\ldots f_n(x)y(x) = 0$$

can be represented as the product

$$y(x) = y_1(x)y_2(x)\ldots y_n(x),$$

where $y_k(x)$ are solutions of the linear homogeneous difference equations

$$y_k(x+1) + f_k(x)y_k(x) = 0, \qquad k = 1, \ldots, n.$$

**7°.** The equation

$$y(x+a) - f(x)y(x) = 0$$

can be reduced to an equation of the form (17.2.1.1) with the help of the transformation $z = x/a$, $y(x) = w(z)$. And we obtain

$$w(z+1) - f(az)w(z) = 0.$$

### 17.2.1-2. Linear difference equations with rational and exponential functions.

Below we give some particular solutions of some homogeneous linear difference equations with rational and exponential functions. Their general solutions can be obtained as a product of a particular solution and an arbitrary 1-periodic function; see (17.2.1.2).

**1°.** The general solution of the first-order homogeneous linear difference equation with constant coefficients (17.2.1.3) is determined by (17.2.1.8) and (17.2.1.9) with $y_1(x) \equiv 1$.

**2°.** The equation

$$y(x+1) - xy(x) = 0$$

admits a particular solution

$$y(x) = \Gamma(x), \qquad \Gamma(x) = \int_0^\infty t^{x-1}e^{-t}\,dt,$$

where $\Gamma(x)$ is the gamma-function.

3°. Consider the first-order equation with rational coefficients
$$P_n(x)y(x+1) - Q_m(x)y(x) = 0,$$
where $P_n(x)$ and $Q_m(x)$ are given polynomials of degrees $n$ and $m$, respectively. Suppose that these polynomials are represented in the form
$$P_n(x) = a(x - \nu_1)(x - \nu_2)\ldots(x - \nu_n),$$
$$Q_m(x) = b(x - \mu_1)(x - \mu_2)\ldots(x - \mu_m), \quad ab > 0.$$
Direct verification shows that the function
$$y(x) = \left(\frac{b}{a}\right)^x \frac{\Gamma(x-\mu_1)\Gamma(x-\mu_2)\ldots\Gamma(x-\mu_m)}{\Gamma(x-\nu_1)\Gamma(x-\nu_2)\ldots\Gamma(x-\nu_n)}$$
is a particular solution of the equation under consideration, where $\Gamma(x)$ is the gamma-function. This solution can have polar singularities at the points $x = \mu_k - s$ ($k = 1, 2, \ldots$; $s = 0, 1, \ldots$).

4°. The equation
$$y(x+1) - e^{\lambda x} y(x) = 0$$
with the parameter $\lambda$ admits the solution
$$y(x) = \exp\bigl(\tfrac{1}{2}\lambda x^2 - \tfrac{1}{2}\lambda x\bigr).$$

5°. The equation
$$y(x+1) - e^{\mu x^2 + \lambda x} y(x) = 0$$
with the parameters $\mu$ and $\lambda$ admits the solution
$$y(x) = \exp\bigl[\tfrac{1}{3}\mu x^3 + \tfrac{1}{2}(\lambda - \mu)x^2 + \tfrac{1}{6}(\mu - 3\lambda)x\bigr].$$

6°. The equation
$$y(x+1) - \exp[P_n(x)]y(x) = 0, \qquad P_n(x) = \sum_{k=1}^{n} b_k x^k,$$
has a particular solution of the form
$$y(x) = \exp[Q_{n+1}(x)], \qquad Q_{n+1}(x) = \sum_{k=1}^{n+1} c_k x^k,$$
where $c_k$ can be found by the method of indefinite coefficients.

7°. The particular solutions from Items 1°–6° allow us to obtain solutions of more intricate linear difference equations with the help of formulas from Paragraph 17.2.1-1.

**Example 2.** Consider the equation
$$y(x+1) - (x+a)^k y(x) = 0$$
with the parameters $a$ and $k$. As a starting point, we take the solution from Item 1° corresponding to the special case of the equation with $a = 0$, $k = 1$. Consecutive utilization of the formulas from Items 3° and 4° of Paragraph 17.2.1-1 yields the following solution of the equation under consideration:
$$y(x) = [\Gamma(x+a)]^k,$$
where $\Gamma(x)$ is the gamma-function.

8°. The general solution of equation (17.2.1.1) with arbitrary $f(x)$ can be constructed with the help of formulas from Paragraph 17.2.1-3.

**17.2.1-3. Homogeneous linear difference equations. Cauchy's problem.**

1°. *Cauchy's problem*: Find a solution of equation (17.2.1.1) with the initial condition

$$y(x) = \varphi(x) \quad \text{for} \quad 0 \leq x < 1, \tag{17.2.1.10}$$

where $\varphi(x)$ is a given continuous function defined on the interval $0 \leq x \leq 1$.

A solution of problem (17.2.1.1), (17.2.1.10) is obtained by the step method: on the interval $1 \leq x < 2$, the solution is constructed from equation (17.2.1.1) with the initial condition (17.2.1.10) taken into account; on the interval $2 \leq x < 3$, one utilizes equation (17.2.1.1) and the solution obtained for $1 \leq x < 2$; on the interval $3 \leq x < 4$, one uses equation (17.2.1.1) and the solution obtained for $2 \leq x < 3$; etc. As a result, we get

$$\begin{aligned} y(x) &= \varphi(x) & \text{for} \quad & 0 \leq x < 1, \\ y(x) &= f(x-1)\varphi(x-1) & \text{for} \quad & 1 \leq x < 2, \\ y(x) &= f(x-1)f(x-2)\varphi(x-2) & \text{for} \quad & 2 \leq x < 3, \\ y(x) &= f(x-1)f(x-2)\ldots f(x-n)\varphi(x-n) & \text{for} \quad & n \leq x < n+1, \end{aligned} \tag{17.2.1.11}$$

where $n = 3, 4, \ldots$

The sequence of formulas (17.2.1.11) that determine a solution of the Cauchy problem (17.2.1.1), (17.2.1.10) can be written as a single formula

$$y(x) = \varphi(\{x\}) \prod_{k=1}^{[x]} f(x-k), \tag{17.2.1.12}$$

where $[x]$ and $\{x\}$ denote, respectively, the integer and the fractional parts of $x$ ($x = [x] + \{x\}$), and the product over the empty set of indexes (for $[x] = 0$) is assumed equal to unity.

Solution (17.2.1.12) is continuous if it is continuous at the integer points $x = 1, 2, \ldots,$ and this brings us to the condition

$$\varphi(1) = f(0)\varphi(0). \tag{17.2.1.13}$$

**Example 3.** Consider the Cauchy problem for equation (17.2.1.1), where $f(x) = f(x+1)$ is an arbitrary (nonnegative) 1-periodic function. In the initial condition (17.2.1.10), take

$$\varphi(x) = \Theta(x)[f(x)]^{x+1},$$

where $\Theta(x) = \Theta(x+1)$ is an arbitrary 1-periodic function. It is easy to check that the continuity condition (17.2.1.13) is satisfied. Using (17.2.1.11), we obtain the solution of the problem in closed form

$$y(x) = \Theta(x)[f(x)]^{x+1} \quad (0 \leq x < \infty). \tag{17.2.1.14}$$

**Remark.** The general solution (17.2.1.4) of the equation with constant coefficients (17.2.1.3) can be obtained by substituting $f(x) = a > 0$ into (17.2.1.14) and making the transformation $a\Theta(x) \to \Theta(x)$.

2°. The general solution of the homogeneous linear difference equation (17.2.1.1) is obtained by replacing $\varphi(\{x\})$ with $\Theta(x)$ in (17.2.1.12), where $\Theta(x)$ is an arbitrary 1-periodic function.

### 17.2.1-4. Nonhomogeneous linear difference equations. General solution.

**1°.** Consider a *first-order nonhomogeneous linear difference equation*

$$y(x+1) - f(x)y(x) = g(x), \qquad (17.2.1.15)$$

where $f(x)$ and $g(x)$ are given continuous functions, $y(x)$ is the sought function, $0 \le x < \infty$.

The general solution of the nonhomogeneous equation (17.2.1.15) can be represented as the sum

$$y(x) = u(x) + \tilde{y}(x),$$

where the first term $u(x)$ is the general solution of the corresponding homogeneous equation (with $g \equiv 0$), and the second term $\tilde{y}(x)$ is a particular solution of equation (17.2.1.15).

A formula for the general solution of equation (17.2.1.15) is given in Paragraph 17.2.1-7, Item 2°.

**2°.** Let $g(x) = g(x+1)$ be a 1-periodic function and let $y_1(x)$ be a solution of equation (17.2.1.15) in the special case of $g(x) \equiv 1$. Then the function

$$y(x) = g(x) y_1(x) \qquad (17.2.1.16)$$

is a solution of equation (17.2.1.15).

**Example 4.** Consider the difference equation, which is a special case of equation (17.2.1.15) with $f(x) \equiv a > 0$:

$$y(x+1) - ay(x) = g(x), \qquad (17.2.1.17)$$

where $g(x) = g(x+1)$ is a given 1-periodic function. Equation (17.2.1.17) with $g(x) \equiv 1$ admits the particular solution

$$y_1(x) = \begin{cases} x & \text{if } a = 1, \\ \dfrac{1}{1-a} & \text{if } a \ne 1. \end{cases}$$

The corresponding particular solution of equation (17.2.1.17) is found with the help of (17.2.1.16), and the general solution has the form

$$y(x) = \begin{cases} \Theta(x) + x g(x) & \text{if } a = 1, \\ \Theta(x) a^x + \dfrac{1}{1-a} g(x) & \text{if } a \ne 1, \end{cases}$$

where $\Theta(x)$ is an arbitrary 1-periodic function.

**3°.** Consider the difference equation which is a special case of (17.2.1.15) with $f(x) \equiv 1$:

$$y(x+1) - y(x) = g(x). \qquad (17.2.1.18)$$

Let $x \in (a, \infty)$ with an arbitrary $a$. Suppose that the function $g(x)$ is monotone, strictly convex (or strictly concave), and satisfies the condition

$$\lim_{x \to \infty} [g(x+1) - g(x)] = 0,$$

and let $x_0 \in (a, \infty)$ be an arbitrary fixed point. Then for every $y_0$, there exists exactly one function $y(x)$ (monotone and strictly convex/concave) satisfying equation (17.2.1.18), together with the condition

$$y(x_0) = y_0.$$

This solution is given by the formulas

$$y(x) = y_0 + (x - x_0) g(x_0) - \sum_{n=0}^{\infty} \Big\{ g(x+n) - g(x_0+n) - (x-x_0)\big[g(x_0+n+1) - g(x_0+n)\big] \Big\}.$$

4°. The functional equation
$$y(x+1) + y(x) = g(x),$$
after the transformation
$$y(x) = u\left(\frac{x+1}{2}\right) - u\left(\frac{x}{2}\right), \quad \xi = \frac{x}{2},$$
is reduced to an equation of the form (17.2.1.18):
$$u(\xi+1) - u(\xi) = g(2\xi).$$

**17.2.1-5. Nonhomogeneous linear equations with right-hand sides of special form.**

1°. The equation
$$y(x+1) - y(x) = \sum_{k=0}^{n} a_k x^k$$
with a polynomial right-hand side admits the particular solution
$$\widetilde{y}(x) = \sum_{k=0}^{n} \frac{a_k}{k+1} B_{k+1}(x),$$
where $B_k(x)$ are *Bernoulli polynomials*.

The Bernoulli polynomials are defined with the help of the generating function
$$\frac{te^{xt}}{e^t - 1} \equiv \sum_{n=0}^{\infty} B_n(x) \frac{t^n}{n!} \qquad (|t| < 2\pi).$$

The first six Bernoulli polynomials have the form
$$B_0(x) = 1, \quad B_1(x) = x - \tfrac{1}{2}, \quad B_2(x) = x^2 - x + \tfrac{1}{6}, \quad B_3(x) = x^3 - \tfrac{3}{2}x^2 + \tfrac{1}{2}x,$$
$$B_4(x) = x^4 - 2x^3 + x^2 - \tfrac{1}{30}, \quad B_5(x) = x^5 - \tfrac{5}{2}x^4 + \tfrac{5}{3}x^3 - \tfrac{1}{6}x.$$

See also Subsection 18.18.1.

2°. For the equation with polynomial right-hand side
$$y(x+1) - ay(x) = \sum_{k=0}^{n} b_k x^k, \qquad a \neq 1, \tag{17.2.1.19}$$

a particular solution is sought by the method of indefinite coefficients in the form of a polynomial of degree $n$.

A particular solution of equation (17.2.1.19) may also be defined by the formula
$$\widetilde{y}(x) = \sum_{k=0}^{n} b_k \left[\frac{d^k}{d\lambda^k}\left(\frac{e^{\lambda x}}{e^\lambda - a}\right)\right]_{\lambda=0}.$$

3°. For equations with rational right-hand side
$$y(x+1) - ay(x) = \frac{b}{x+c},$$
where $c \geq 0$ and $a \notin [0, 1]$, a particular solution is written in the form of the integral
$$\widetilde{y}(x) = b \int_0^1 \frac{\lambda^{x+c-1}}{\lambda - a} \, d\lambda.$$

4°. For the equation with exponential right-hand side
$$y(x+1) - ay(x) = \sum_{k=0}^{n} b_k e^{\lambda_k x},$$
there is a particular solution of the form
$$\widetilde{y}(x) = \begin{cases} \displaystyle\sum_{k=0}^{n} \frac{b_k}{e^{\lambda_k} - a} e^{\lambda_k x} & \text{if } a \neq e^{\lambda_m}, \\[2ex] b_m x e^{\lambda_m(x-1)} + \displaystyle\sum_{k=0,\, k \neq m}^{n} \frac{b_k}{e^{\lambda_k} - a} e^{\lambda_k x} & \text{if } a = e^{\lambda_m}, \end{cases}$$
where $m = 0, 1, \ldots, n$.

5°. For the equation with sinusoidal right-hand side
$$y(x+1) - ay(x) = b\sin(\beta x),$$
there is a particular solution of the form
$$y(x) = \frac{b}{a^2 + 1 - 2a\cos\beta} \big[(\cos\beta - a)\sin(\beta x) - \sin\beta \cos(\beta x)\big].$$

6°. For the equation with cosine in the right-hand side
$$y(x+1) - ay(x) = b\cos(\beta x),$$
there is a particular solution of the form
$$\widetilde{y}(x) = \frac{b}{a^2 + 1 - 2a\cos\beta} \big[(\cos\beta - a)\cos(\beta x) + \sin\beta \sin(\beta x)\big].$$

**17.2.1-6. Nonhomogeneous linear equations with the right-hand side of general form.**

1°. Consider equation (17.2.1.18) whose right-hand side is an analytic function that admits the expansion in power series:
$$g(x) = \sum_{k=0}^{\infty} a_k x^k, \qquad (17.2.1.20)$$
which converges on the entire complex plane (it follows that $\lim_{k \to \infty} |a_k|^{1/k} = 0$). If
$$\varlimsup_{k \to \infty} \big(k!\,|a_k|\big)^{1/k} < 2\pi, \qquad (17.2.1.21)$$
then the solution of equation (17.2.1.18) can be represented as a convergent series
$$\widetilde{y}(x) = \sum_{k=0}^{\infty} \frac{a_k}{k+1} B_{k+1}(x),$$
where $B_k(x)$ are Bernoulli polynomials (see Paragraph 17.2.1-5, Item 1°).

Below we state a more general result that requires no conditions of the type (17.2.1.21).

2°. Suppose that the right-hand side of equation (17.2.1.18) is an analytic function admitting expansion by power series (17.2.1.20) convergent on the entire complex plane (such functions are called entire functions). In this case, the solution of equation (17.2.1.18) can be represented in the form

$$\widetilde{y}(x) = \sum_{k=0}^{\infty} a_k \beta_k(x),$$

where $\beta_k(x)$ *are Hurwitz functions*, defined by

$$\beta_k(x) = \frac{k!}{2\pi i} \int_{|t|=1} \frac{(e^{tx}-1)\,dt}{(e^t-1)t^{k+1}} + k! \sum_{s=-k,\,s\neq 0}^{k} \frac{e^{2is\pi x}-1}{(2is\pi)^{s+1}} = P_{k+1}(x) + \Theta_k(x),$$

$P_{n+1}(x)$ is a polynomial of degree $n+1$, $\Theta_k(x)$ is an entire 1-periodic function, $|t|=1$ is the unit circle, and $i^2 = -1$.

### 17.2.1-7. Nonhomogeneous linear difference equations. Cauchy's problem.

1°. Consider the Cauchy problem for a nonhomogeneous linear difference equation of special form (17.2.1.18) with the initial condition (17.2.1.10), where $g(x)$ is a continuous function defined for $x \geq 0$, and $\varphi(x)$ is a given continuous function defined on the segment $0 \leq x \leq 1$.

In order to find a solution of this Cauchy problem, the step method can be used: on the interval $1 \leq x < 2$ one constructs a solution from equation (17.2.1.18) and the boundary condition (17.2.1.10); on the interval $2 \leq x < 3$, one uses equation (17.2.1.18) and the solution obtained for $1 \leq x < 2$; on the interval $3 \leq x < 4$, one uses equation (17.2.1.18) and the solution obtained for $2 \leq x < 3$, etc. As a result, we have

$$y(x) = \varphi(x-n) + g(x-n) + g(x-n+1) + \cdots + g(x-1), \qquad n \leq x < n+1, \quad (17.2.1.22)$$

where $n = 1, 2, \ldots$

Solution (17.2.1.22) is continuous if it is continuous at integer points $x = 1, 2, \ldots$, and this leads us to the condition

$$\varphi(1) = \varphi(0) + g(0). \qquad (17.2.1.23)$$

The solution is continuously differentiable if the functions $g(x)$ and $\varphi(x)$ are continuously differentiable and, together with (17.2.1.23), the additional condition

$$\varphi'(1) = \varphi'(0) + g'(0)$$

for the corresponding one-sided derivatives is satisfied.

2°. In a similar way, one considers the Cauchy problem for a nonhomogeneous linear difference equation of general form (17.2.1.15) with the initial condition (17.2.1.10). As a result, we obtain a solution of the form

$$y(x) = \varphi(\{x\}) \prod_{j=1}^{[x]} f(x-j) + \sum_{i=1}^{[x]} g(x-i) \prod_{j=1}^{i-1} f(x-j),$$

where $[x]$ and $\{x\}$ denote, respectively, the integer and the fractional parts of $x$ ($x = [x] + \{x\}$); the product and the sum over the empty index set (for $[x] = 0$) are assumed equal to 1 and 0, respectively.

If a nontrivial particular solution $y_1(x)$ of the homogeneous equation (17.2.1.1) is known, then the solution of the Cauchy problem for the nonhomogeneous equation (17.2.1.15) with the initial condition (17.2.1.10) is given by

$$y(x) = y_1(x) \left[ \frac{\varphi(\{x\})}{y_1(0)} + \sum_{i=1}^{[x]} \frac{g(x-i)}{y_1(x-i+1)} \right].$$

## 17.2.2. Second-Order Linear Difference Equations with Integer Differences

**17.2.2-1. Linear homogeneous difference equations with constant coefficients.**

*A second-order homogeneous linear integer-difference equation with constant coefficients* has the form

$$ay(x+2) + by(x+1) + cy(x) = 0, \qquad ac \neq 0. \qquad (17.2.2.1)$$

This equation has the trivial solution $y(x) \equiv 0$.

The general solution of the difference equation (17.2.2.1) is determined by the roots of the characteristic equation

$$a\lambda^2 + b\lambda + c = 0. \qquad (17.2.2.2)$$

$1°$. For $b^2 - 4ac > 0$, the quadratic equation (17.2.2.2) has two distinct roots:

$$\lambda_1 = \frac{-b + \sqrt{b^2 - 4ac}}{2a}, \qquad \lambda_2 = \frac{-b - \sqrt{b^2 - 4ac}}{2a}.$$

The general solution of the difference equation (17.2.2.1) is given by

$$\begin{aligned} y(x) &= \Theta_1(x)\lambda_1^x + \Theta_2(x)\lambda_2^x & \text{if} \quad ab < 0, \ ac > 0; \\ y(x) &= \Theta_1(x)\lambda_1^x + \Theta_2(x)|\lambda_2|^x \cos(\pi x) & \text{if} \quad ac < 0; \\ y(x) &= \Theta_1(x)|\lambda_1|^x \cos(\pi x) + \Theta_2(x)|\lambda_2|^x \cos(\pi x) & \text{if} \quad ab > 0, \ ac > 0, \end{aligned} \qquad (17.2.2.3)$$

where $\Theta_1(x)$ and $\Theta_2(x)$ are arbitrary 1-periodic functions, $\Theta_k(x) = \Theta_k(x+1)$, $k = 1, 2$.

$2°$. For $b^2 - 4ac = 0$, the quadratic equation (17.2.2.2) has one double real root

$$\lambda = -\frac{b}{2a},$$

and the general solution of the difference equation (17.2.2.1) is given by

$$\begin{aligned} y &= \big[\Theta_1(x) + x\Theta_2(x)\big] \lambda^x & \text{if} \quad ab < 0, \\ y &= \big[\Theta_1(x) + x\Theta_2(x)\big] |\lambda|^x \cos(\pi x) & \text{if} \quad ab > 0. \end{aligned} \qquad (17.2.2.4)$$

$3°$. For $b^2 - 4ac < 0$, the quadratic equation (17.2.2.2) has two complex conjugate roots

$$\lambda_{1,2} = \rho(\cos\beta \pm i\sin\beta), \qquad \rho = \frac{c}{a}, \qquad \beta = \arccos\left(-\frac{b}{2\sqrt{ac}}\right),$$

and the general solution of the difference equation (17.2.2.1) has the form

$$y = \Theta_1(x)\rho^x \cos(\beta x) + \Theta_2(x)\rho^x \sin(\beta x), \qquad (17.2.2.5)$$

where $\Theta_1(x)$ and $\Theta_2(x)$ are arbitrary 1-periodic functions.

### 17.2.2-2. Linear nonhomogeneous difference equations with constant coefficients.

**1°.** *A second-order nonhomogeneous linear integer-difference equation with constant coefficients* has the form

$$ay(x+2) + by(x+1) + cy(x) = f(x), \qquad ac \neq 0. \tag{17.2.2.6}$$

Let $y_1(x)$ and $y_2(x)$ be two particular solutions of the homogeneous equation (17.2.2.1). According to the roots of the characteristic equation (17.2.2.2), these solutions are defined by (17.2.2.3)–(17.2.2.5), respectively, with $\Theta_1(x) \equiv 1$, $\Theta_2(x) \equiv 0$ and $\Theta_1(x) \equiv 0$, $\Theta_2(x) \equiv 1$.

A particular solution of the nonhomogeneous equation (17.2.2.6) satisfying zero initial conditions

$$y(x) = 0, \quad y(x+1) = 0 \quad \text{for} \quad 0 \leq x < 1$$

has the form

$$\widetilde{y}(x) = \frac{1}{a} \sum_{j=1}^{[x]-1} \frac{\Delta(x-j, x)}{\Delta(x-j, x-j+1)} f(x-j), \qquad \Delta(x, z) = \begin{vmatrix} y_1(x) & y_2(x) \\ y_1(z) & y_2(z) \end{vmatrix}. \tag{17.2.2.7}$$

**2°.** The solution of the Cauchy problem for the nonhomogeneous equation (17.2.2.6) with arbitrary initial conditions

$$y(x) = \varphi(x), \quad y(x+1) = \psi(x) \quad \text{for} \quad 0 \leq x < 1 \tag{17.2.2.8}$$

is the sum of the particular solution (17.2.2.7) and the function

$$u(x) = -\frac{1}{\Delta(\{x\}, \{x\}+1)} \begin{vmatrix} 0 & y_1(x) & y_2(x) \\ \varphi(\{x\}) & y_1(\{x\}) & y_2(\{x\}) \\ \psi(\{x\}) & y_1(\{x\}+1) & y_2(\{x\}+1) \end{vmatrix},$$

which is a solution of the homogeneous equation (17.2.2.1) with the initial conditions (17.2.2.8).

### 17.2.2-3. Linear homogeneous difference equations with variable coefficients.

**1°.** *A second-order linear homogeneous integer-difference equation* has the form

$$a(x)y(x+2) + b(x)y(x+1) + c(x)y(x) = 0, \qquad a(x)c(x) \neq 0. \tag{17.2.2.9}$$

The *trivial solution*, $y(x) = 0$, is a particular solution of the homogeneous linear equation. Let $y_1(x)$, $y_2(x)$ be two particular solutions of equation (17.2.2.9) with the condition*

$$D(x) \equiv y_1(x)y_2(x+1) - y_2(x)y_1(x+1) \neq 0. \tag{17.2.2.10}$$

Then the general solution of equation (17.2.2.9) is given by

$$y(x) = \Theta_1(x)y_1(x) + \Theta_2(x)y_2(x), \tag{17.2.2.11}$$

where $\Theta_1(x)$ and $\Theta_2(x)$ are arbitrary 1-periodic functions, $\Theta_{1,2}(x) = \Theta_{1,2}(x+1)$.

---

* Condition (17.2.2.10) may be violated at singular points of equation (17.2.2.9); for details see Paragraph 17.2.3-3.

**2°.** Let $y_0(x)$ be a nontrivial particular solution of equation (17.2.2.9). Then the substitution

$$y(x) = y_0(x)u(x) \qquad (17.2.2.12)$$

results in the equation

$$a(x)y_0(x+2)u(x+2) + b(x)y_0(x+1)u(x+1) + c(x)y_0(x)u(x) = 0. \qquad (17.2.2.13)$$

Taking into account that $y_0(x)$ satisfies equation (17.2.2.9), let us substitute the expression

$$b(x)y_0(x+1) = -a(x)y_0(x+2) - c(x)y_0(x)$$

into (17.2.2.13). Then, after simple transformations, we obtain

$$a(x)y_0(x+2)[u(x+2) - u(x+1)] - c(x)y_0(x)[u(x+1) - u(x)] = 0.$$

Introducing a new variable by

$$w(x) = u(x+1) - u(x) \qquad (17.2.2.14)$$

we come to the first-order difference equation

$$a(x)y_0(x+2)w(x+1) - c(x)y_0(x)w(x) = 0.$$

After solving this equation, one solves the nonhomogeneous first-order equation with constant coefficients (17.2.2.14), and then, using (17.2.2.12), one finds a solution of the original equation.

---

**17.2.2-4. Linear nonhomogeneous difference equations with variable coefficients.**

**1°.** A *second-order linear nonhomogeneous difference equation* with integer differences has the form

$$a(x)y(x+2) + b(x)y(x+1) + c(x)y(x) = f(x), \qquad a(x)c(x) \not\equiv 0. \qquad (17.2.2.15)$$

The general solution of the nonhomogeneous equation (17.2.2.15) is given by the sum

$$y(x) = u(x) + \widetilde{y}(x),$$

where $u(x)$ is the general solution of the corresponding homogeneous equation (with $f \equiv 0$), and $\widetilde{y}(x)$ is a particular solution of equation (17.2.2.15). The general solution of the homogeneous equation is defined by the right-hand side of (17.2.2.11).

Every solution of equation (17.2.2.15) is uniquely determined by given values of the sought function on the interval $[0, 2)$.

**2°.** A particular solution $\widetilde{y}(x)$ of the linear nonhomogeneous difference equation

$$a(x)y(x+2) + b(x)y(x+1) + c(x)y(x) = \sum_{k=1}^{n} f_k(x)$$

can be represented by the sum

$$\widetilde{y}(x) = \sum_{k=1}^{n} \widetilde{y}_k(x),$$

where $\widetilde{y}_k(x)$ are particular solutions of the linear nonhomogeneous difference equations

$$a(x)y_k(x+2) + b(x)y_k(x+1) + c(x)y_k(x) = f_k(x).$$

3°. Let $y_1(x)$ and $y_2(x)$ be particular solutions of the corresponding linear homogeneous equation (17.2.2.9) that satisfy the condition (17.2.2.10). A particular solution of the linear nonhomogeneous equation (17.2.2.15) can be sought in the form

$$\widetilde{y}(x) = \varphi_1(x)y_1(x) + \varphi_2(x)y_2(x), \qquad (17.2.2.16)$$

where $\varphi_1(x)$ and $\varphi_2(x)$ are functions to be determined.

For what follows, we need the identity

$$\Delta[\varphi(x)\psi(x)] \equiv \psi(x+1)\Delta\varphi(x) + \varphi(x)\Delta\psi(x), \qquad (17.2.2.17)$$

where $\Delta\varphi(x)$ is the standard notation for the difference,

$$\Delta\varphi(x) \equiv \varphi(x+1) - \varphi(x).$$

From (17.2.2.16), using the identity (17.2.2.17), we obtain

$$\Delta\widetilde{y}(x) = y_1(x+1)\Delta\varphi_1(x) + y_2(x+1)\Delta\varphi_2(x) + \varphi_1(x)\Delta y_1(x) + \varphi_2(x)\Delta y_2(x). \qquad (17.2.2.18)$$

As one of the equations to determine the functions $\varphi_1(x)$ and $\varphi_2(x)$ we take

$$y_1(x+1)\Delta\varphi_1(x) + y_2(x+1)\Delta\varphi_2(x) = 0. \qquad (17.2.2.19)$$

In view of the above considerations, from (17.2.2.16) and (17.2.2.18) we find that

$$\begin{aligned}\widetilde{y}(x+1) &= \varphi_1(x)y_1(x+1) + \varphi_2(x)y_2(x+1), \\ \widetilde{y}(x+2) &= \varphi_1(x+1)y_1(x+2) + \varphi_2(x+1)y_2(x+2) \\ &= \varphi_1(x)y_1(x+2) + \varphi_2(x)y_2(x+2) + y_1(x+2)\Delta\varphi_1(x) + y_2(x+2)\Delta\varphi_2(x),\end{aligned} \qquad (17.2.2.20)$$

where the second relation is obtained from the first one by replacing $x$ with $x+1$.

Substituting (17.2.2.16) and (17.2.2.20) into equation (17.2.2.15) and taking into account that $y_1(x)$ and $y_2(x)$ are particular solutions of the linear homogeneous equation (17.2.2.9), we obtain

$$y_1(x+2)\Delta\varphi_1(x) + y_2(x+2)\Delta\varphi_2(x) = f(x)/a(x). \qquad (17.2.2.21)$$

Relations (17.2.2.19) and (17.2.2.21) form a system of linear algebraic equations for the differences $\Delta\varphi_1(x)$ and $\Delta\varphi_2(x)$. The solution of this system has the form

$$\Delta\varphi_1(x) = -\frac{f(x)y_2(x+1)}{a(x)D(x+1)}, \quad \Delta\varphi_2(x) = -\frac{f(x)y_1(x+1)}{a(x)D(x+1)}, \qquad (17.2.2.22)$$

where $D(x)$ is the determinant introduced in (17.2.2.10).

Thus, the construction of a particular solution of the second-order nonhomogeneous equation (17.2.2.15) amounts to finding a solution of two independent nonhomogeneous first-order equations (17.2.2.22) considered in detail in Paragraphs 17.2.1-4–17.2.1-7.

4°. The structure of particular solutions of second-order difference equations with specific (polynomial, exponential, sinusoidal, etc.) right-hand sides is described in Paragraph 17.2.3-2.

## 17.2.3. Linear $m$th-Order Difference Equations with Integer Differences

**17.2.3-1. Linear homogeneous difference equations with constant coefficients.**

An *$m$th-order linear integer-difference homogeneous equation with constant coefficients* has the form

$$a_m y(x+m) + a_{m-1} y(x+m-1) + \cdots + a_1 y(x+1) + a_0 y(x) = 0, \qquad (17.2.3.1)$$

where $a_0 a_m \neq 0$.

Let us write out its characteristic equation,

$$a_m \lambda^m + a_{m-1} \lambda^{m-1} + \cdots + a_1 \lambda + a_0 = 0. \qquad (17.2.3.2)$$

Consider the following cases:

1°. All roots $\lambda_1, \lambda_2, \ldots, \lambda_n$ of equation (17.2.3.2) are real and mutually distinct. Then the general solution of the original equation (17.2.3.1) has the form

$$y(x) = \Theta_1(x) \lambda_1^x + \Theta_2(x) \lambda_2^x + \cdots + \Theta_n(x) \lambda_n^x, \qquad (17.2.3.3)$$

where $\Theta_1(x), \Theta_2(x), \ldots, \Theta_n(x)$ are arbitrary 1-periodic functions, $\Theta_k(x) = \Theta_k(x+1)$, $k = 1, 2, \ldots, n$.

For $\Theta_k(x) \equiv C_k$, formula (17.2.3.3) yields the particular solution

$$y(x) = C_1 \lambda_1^x + C_2 \lambda_2^x + \cdots + C_n \lambda_n^x,$$

where $C_1, C_2, \ldots, C_n$ are arbitrary constants.

2°. There are $k$ equal real roots: $\lambda_1 = \lambda_2 = \cdots = \lambda_k$ ($k \leq m$), and the other roots are real and mutually distinct. In this case, the solution of the difference equation (17.2.3.1) is defined by

$$\begin{aligned} y &= \left[ \Theta_1(x) + x \Theta_2(x) + \cdots + x^{k-1} \Theta_k(x) \right] \lambda_1^x \\ &\quad + \Theta_{k+1}(x) \lambda_{k+1}^x + \Theta_{k+2}(x) \lambda_{k+2}^x + \cdots + \Theta_m(x) \lambda_m^x. \end{aligned} \qquad (17.2.3.4)$$

3°. There are $k$ equal complex conjugate roots: $\lambda = \rho(\cos\beta \pm i \sin\beta)$ ($2k \leq m$), and the other roots are real and mutually distinct. In this case, for $\Theta_n(x) \equiv \text{const}_n$ the solution of the functional equation has the form

$$\begin{aligned} y &= \rho^x \cos(\beta x)(A_1 + A_2 x + \cdots + A_k x^{k-1}) + \\ &\quad + \rho^x \sin(\beta x)(B_1 + B_2 x + \cdots + B_k x^{k-1}) + \\ &\quad + C_{2k+1} \lambda_{2k+1}^x + C_{2k+2} \lambda_{2k+2}^x + \cdots + C_m \lambda_m^x, \end{aligned}$$

where $A_1, \ldots, A_k, B_1, \ldots, B_k, C_{2k+1}, \ldots, C_m$ are arbitrary constants. In the general case, the arbitrary constants involved in this solution should be replaced by arbitrary 1-periodic functions.

4°. In a similar way, one can consider the general situation.

**17.2.3-2. Linear nonhomogeneous difference equations with constant coefficients.**

An *mth-order linear nonhomogeneous integer-difference equation* with constant coefficients has the form

$$a_m y(x+m) + a_{m-1} y(x+m-1) + \cdots + a_1 y(x+1) + a_0 y(x) = f(x), \qquad (17.2.3.5)$$

where $a_0 a_m \neq 0$.

The general solution of the nonhomogeneous equation (17.2.3.5) is given by the sum

$$y(x) = u(x) + \widetilde{y}(x),$$

where $u(x)$ is the general solution of the corresponding homogeneous equation (with $f \equiv 0$), and $\widetilde{y}(x)$ is a particular solution of the nonhomogeneous equation (17.2.3.5). Regarding the solution of the homogeneous equation see Paragraph 17.2.3-1.

Below, we consider some methods for the construction of a particular solution of the nonhomogeneous equation (17.2.3.5).

1°. Table 17.1 lists the forms of particular solutions corresponding to some special cases of the function on the right-hand side of the linear nonhomogeneous difference equation (17.2.3.5).

2°. Let

$$P(\lambda) = a_m \lambda^m + a_{m-1} \lambda^{m-1} + \cdots + a_1 \lambda + a_0$$

be the characteristic polynomial of equation (17.2.3.5). Consider the function

$$g(\lambda) = \frac{1}{P(\lambda)} = \sum_{k=0}^{\infty} g_k \lambda^k, \qquad |\lambda| < |\lambda_1|, \qquad (17.2.3.6)$$

where $\lambda_1$ is the root of the equation $P(\lambda) = 0$ with the smallest absolute value. Then $\overline{\lim}_{k \to \infty} |g_k|^{1/k} = |\lambda_1|^{-1}$.

From (17.2.3.6), it follows that

$$a_0 g_0 = 1, \quad a_1 g_0 + a_0 g_1 = 0, \quad \ldots, \quad \sum_{k=0}^{s-1} a_{s-k-1} g_k = 0, \quad s \geq 2, \quad a_\nu = 0 \text{ if } \nu > m.$$

If

$$\overline{\lim}_{k \to \infty} |f(x+k)|^{1/k} = \sigma_f < |\lambda_1|,$$

then the series

$$y(x) = g_0 f(x) + g_1 f(x+1) + \cdots + g_k f(x+k) + \cdots$$

is convergent and its sum gives a solution of equation (17.2.3.5).

3°. Let the right-hand side of equation (17.2.3.5) can be represented by the integral

$$f(x) = \int_L F(\lambda) \lambda^{x-1} \, d\lambda, \qquad (17.2.3.7)$$

where the line of integration $L$ does not cross the roots of the characteristic polynomial $P(\lambda)$. Direct verification shows that the integral

$$y(x) = \int_L \frac{F(\lambda)}{P(\lambda)} \lambda^{x-1} \, d\lambda \qquad (17.2.3.8)$$

represents a particular solution of equation (17.2.3.5).

## TABLE 17.1
Forms of particular solutions of the linear nonhomogeneous difference equation with constant coefficients $a_m y(x+m) + a_{m-1} y(x+m-1) + \cdots + a_1 y(x+1) + a_0 y(x) = f(x)$ in some special cases of the function $f(x)$; $a_0 a_m \neq 0$

| Form of the function $f(x)$ | Roots of the characteristic equation $a_m \lambda^m + a_{m-1} \lambda^{m-1} + \cdots + a_1 \lambda + a_0 = 0$ | Form of a particular solution $y = \widetilde{y}(x)$ |
|---|---|---|
| $x^n$ | $\lambda = 1$ is not a root of the characteristic equation (i.e., $a_m + a_{m-1} + \cdots + a_1 + a_0 \neq 0$) | $\sum_{k=0}^{n} b_k x^k$ |
|  | $\lambda = 1$ is a root of the characteristic equation (multiplicity $r$) | $\sum_{k=0}^{n+r} b_k x^k$ |
| $e^{\beta x}$ ($\beta$ is a real constant) | $\lambda = e^\beta$ is not a root of the characteristic equation | $b e^{\beta x}$ |
|  | $\lambda = e^\beta$ is a root of the characteristic equation (multiplicity $r$) | $e^{\beta x} \sum_{k=0}^{r} b_k x^k$ |
| $x^n e^{\beta x}$ ($\beta$ is a real constant) | $\beta$ is not a root of the characteristic equation | $e^{\beta x} \sum_{k=0}^{n} b_k x^k$ |
|  | $\beta$ is a root of the characteristic equation (multiplicity $r$) | $e^{\beta x} \sum_{k=0}^{n+r} b_k x^k$ |
| $P_m(x) \cos \beta x + Q_n(x) \sin \beta x$ | $i\beta$ is not a root of the characteristic equation | $\widetilde{P}_\nu(x) \cos \beta x + \widetilde{Q}_\nu(x) \sin \beta x$ |
|  | $i\beta$ is a root of the characteristic equation (multiplicity $r$) | $\widetilde{P}_{\nu+r}(x) \cos \beta x + \widetilde{Q}_{\nu+r}(x) \sin \beta x$ |
| $[P_m(x) \cos \beta x + Q_n(x) \sin \beta x] e^{\alpha x}$ | $\alpha + i\beta$ is not a root of the characteristic equation | $[\widetilde{P}_\nu(x) \cos \beta x + \widetilde{Q}_\nu(x) \sin \beta x] e^{\alpha x}$ |
|  | $\alpha + i\beta$ is a root of the characteristic equation (multiplicity $r$) | $[\widetilde{P}_{\nu+r}(x) \cos \beta x + \widetilde{Q}_{\nu+r}(x) \sin \beta x] e^{\alpha x}$ |

**Notation**: $P_m(x)$ and $Q_n(x)$ are polynomials of degrees $m$ and $n$ with given coefficients; $\widetilde{P}_m(x)$, $\widetilde{P}_\nu(x)$, and $\widetilde{Q}_\nu(x)$ are polynomials of degrees $m$ and $\nu$ whose coefficients are determined by substituting the particular solution into the basic equation; $\nu = \max(m, n)$; and $\alpha$ and $\beta$ are real numbers, $i^2 = -1$.

**Example 1.** Taking $F(\lambda) = a\lambda^b$, $b \geq 0$, and $L = \{0 \leq x \leq 1\}$ in (17.2.3.7), we have

$$f(x) = \int_0^1 a\lambda^b \lambda^{x-1} \, d\lambda = \frac{a}{x+b}.$$

Therefore, if the characteristic polynomial $P(\lambda)$ has no roots on the segment $[0, 1]$, then a particular solution of equation (17.2.3.5) with the right-hand side

$$f(x) = \frac{a}{x+b} \tag{17.2.3.9}$$

can be obtained in the form

$$y(x) = a \int_0^1 \frac{\lambda^b}{P(\lambda)} \lambda^{x-1} \, d\lambda. \tag{17.2.3.10}$$

**Remark.** For $L = \{0 \leq x < \infty\}$, the representation (17.2.3.7) may be regarded as the Mellin transformation that maps $F(\lambda)$ into $f(x)$.

4°. Let $y(x)$ be a solution of equation (17.2.3.5). Then $u(x) = \beta y'_x(x)$ is a solution of the nonhomogeneous equation

$$a_m u(x+m) + a_{m-1} u(x+m-1) + \cdots + a_1 u(x+1) + a_0 u(x) = \beta f'_x(x).$$

**Example 2.** In order to find a solution of equation (17.2.3.5) with the right-hand side

$$f(x) = \frac{a}{(x+b)^2} = -\frac{d}{dx}\frac{a}{x+b}, \qquad (17.2.3.11)$$

let us multiply (17.2.3.9)–(17.2.3.10) by $\beta = -1$ and then differentiate the resulting expressions. Thus, we obtain the following particular solution of the nonhomogeneous equation (17.2.3.5) with the right-hand side (17.2.3.11):

$$y(x) = -a \int_0^1 \frac{\ln \lambda}{P(\lambda)} \lambda^{x+b-1} \, d\lambda. \qquad (17.2.3.12)$$

Consecutively differentiating expressions (17.2.3.11) and (17.2.3.12), we obtain a solution of equation (17.2.3.5) with the right-hand side

$$f(x) = \frac{a}{(x+b)^n} = \frac{(-1)^{n-1}}{(n-1)!} \frac{d^{n-1}}{dx^{n-1}}\left(\frac{a}{x+b}\right).$$

This solution has the form

$$y(x) = a\frac{(-1)^{n-1}}{(n-1)!} \int_0^1 \frac{(\ln \lambda)^{n-1}}{P(\lambda)} \lambda^{x+b-1} \, d\lambda.$$

5°. In Paragraph 17.2.3-4, Item 3°, there is a formula that allows us to obtain a particular solution of the nonhomogeneous equation (17.2.3.5) with an arbitrary right-hand side.

### 17.2.3-3. Linear homogeneous difference equations with variable coefficients.

1°. An *mth-order linear homogeneous integer-difference equation* with variable coefficients has the form

$$a_m(x)y(x+m) + a_{m-1}(x)y(x+m-1) + \cdots + a_1(x)y(x+1) + a_0(x)y(x) = 0, \quad (17.2.3.13)$$

where $a_0(x)a_m(x) \not\equiv 0$.

This equation admits the trivial solution $y(x) \equiv 0$.

The set $E$ of all singular points of equation (17.2.3.13) consists of points of three classes:
1) zeroes of the function $a_0(x)$, denoted by $\mu_1, \mu_2, \ldots$ ;
2) zeroes of the function $a_m(x-m)$, denoted by $\nu_1, \nu_2, \ldots$ ;
3) singular points of the coefficients of the equation, denoted by $\eta_1, \eta_2, \ldots$

The points of the set

$$S(E) = \{\mu_s - n, \ \nu_s + n, \ \eta_s - n, \ \eta_s + m + n; \ n = 0, 1, 2, \ldots, \ s = 1, 2, 3, \ldots\}$$

are called *comparable* with singular points of equation (17.2.3.13).

Let

$$y_1 = y_1(x), \quad y_2 = y_2(x), \quad \ldots, \quad y_m = y_m(x) \qquad (17.2.3.14)$$

be particular solutions of equation (17.2.3.13). Then the function

$$y = \Theta_1(x)y_1(x) + \Theta_2(x)y_2(x) + \cdots + \Theta_m(x)y_m(x) \qquad (17.2.3.15)$$

with arbitrary 1-periodic functions $\Theta_1(x), \Theta_2(x), \ldots, \Theta_m(x)$ is also a solution of equation (17.2.3.13).

The *Casoratti determinant* is the function defined as

$$D(x) = \begin{vmatrix} y_1(x) & y_2(x) & \cdots & y_m(x) \\ y_1(x+1) & y_2(x+1) & \cdots & y_m(x+1) \\ \cdots & \cdots & \cdots & \cdots \\ y_1(x+m-1) & y_2(x+m-1) & \cdots & y_m(x+m-1) \end{vmatrix}. \qquad (17.2.3.16)$$

THEOREM (CASORATTI). *Formula (17.2.3.15) gives the general solution of the linear homogeneous difference equation (17.2.3.13) if and only if for any point $x \notin S(E)$ such that $x + k \notin S(E)$ for $k = 0, 1, \ldots, m$, the condition $D(x) \neq 0$ holds.*

The Casoratti determinant (17.2.3.16) satisfies the first-order difference equation

$$D(x+1) = (-1)^m \frac{a_0(x)}{a_m(x)} D(x). \qquad (17.2.3.17)$$

$2°$. Let $y_0 = y_0(x)$ be a nontrivial particular solution of equation (17.2.3.13). Then the order of equation (17.2.3.13) can be reduced by unity. Indeed, making the replacement

$$y(x) = y_0(x) u(x) \qquad (17.2.3.18)$$

in equation (17.2.3.13), we get

$$\sum_{k=0}^{m} a_k(x) y_0(x+k) u(x+k) = 0. \qquad (17.2.3.19)$$

Let us transform this relation with the help of the Abel identity

$$\sum_{k=0}^{m} F_k G_k = -\sum_{k=0}^{m-1} (G_{k+1} - G_k) \sum_{s=0}^{k} F_s + G_m \sum_{k=0}^{m} F_k,$$

in which we take $F_k = a_k(x) y_0(x+k)$ and $G_k = u(x+k)$. As a result, we get

$$-\sum_{k=0}^{m-1} r_k(x)[u(x+k+1) - u(x+k)] + u(x+m) \sum_{k=0}^{m} a_k(x) y_0(x+k) = 0, \qquad (17.2.3.20)$$

where

$$r_k(x) = \sum_{s=0}^{k} a_s(x) y_0(x+s), \qquad k = 0, 1, \ldots, m-1.$$

Taking into account that the second sum in (17.2.3.20) is equal to zero (since $y_0$ is a particular solution of the equation under consideration) and setting

$$w(x) = u(x+1) - u(x) \qquad (17.2.3.21)$$

in (17.2.3.20), we come to an $(m-1)$th-order difference equation

$$\sum_{k=0}^{m-1} r_k(x) w(x+k) = 0.$$

### 17.2.3-4. Linear nonhomogeneous difference equations with variable coefficients.

$1°$. An *mth-order linear nonhomogeneous integer-difference equation* with variable coefficients has the form

$$a_m(x) y(x+m) + a_{m-1}(x) y(x+m-1) + \cdots + a_1(x) y(x+1) + a_0(x) y(x) = f(x), \qquad (17.2.3.22)$$

where $a_0(x) a_m(x) \not\equiv 0$.

The general solution of the nonhomogeneous equation (17.2.3.22) is given by the sum

$$y(x) = u(x) + \widetilde{y}(x), \qquad (17.2.3.23)$$

where $u(x)$ is the general solution of the corresponding homogeneous equation (with $f \equiv 0$) and $\widetilde{y}(x)$ is a particular solution of the nonhomogeneous equation (17.2.3.22). The general solution of the homogeneous equation is defined by the right-hand side of (17.2.3.15).

Every solution of equation (17.2.3.22) is uniquely determined by prescribing the values of the sought function on the interval $[0, m)$.

**2°.** A particular solution $\widetilde{y}(x)$ of the linear nonhomogeneous difference equation

$$a_m(x)y(x+m) + a_{m-1}(x)y(x+m-1) + \cdots + a_1(x)y(x+1) + a_0(x)y(x) = \sum_{k=1}^{n} f_k(x)$$

can be represented by the sum

$$\widetilde{y}(x) = \sum_{k=1}^{n} \widetilde{y}_k(x),$$

where $\widetilde{y}_k(x)$ are particular solutions of the linear nonhomogeneous difference equations

$$a_m(x)y(x+m) + a_{m-1}(x)y(x+m-1) + \cdots + a_1(x)y(x+1) + a_0(x)y(x) = f_k(x).$$

**3°.** The solution of the Cauchy problem for the nonhomogeneous equation (17.2.3.22) with arbitrary initial conditions

$$y(x+j) = \varphi_j(x) \quad \text{for} \quad 0 \le x < 1, \quad j = 0, 1, \ldots, m-1, \tag{17.2.3.24}$$

is given by the sum (17.2.3.23), where

$$u(x) = -\frac{1}{D(\{x\})} \begin{vmatrix} 0 & y_1(x) & \cdots & y_m(x) \\ \varphi_0(\{x\}) & y_1(\{x\}) & \cdots & y_m(\{x\}) \\ \cdots & \cdots & \cdots & \cdots \\ \varphi_{m-1}(\{x\}) & y_1(\{x\}+m-1) & \cdots & y_m(\{x\}+m-1) \end{vmatrix}$$

is a solution of the homogeneous equation (17.2.3.13) with the boundary conditions (17.2.3.24), and

$$\widetilde{y}(x) = \sum_{j=m}^{[x]} \frac{D_*(x-j+1)}{D(x-j+1)} \frac{f(x-j)}{a_m(x-j)} \tag{17.2.3.25}$$

is a particular evolution of the nonhomogeneous equation (17.2.3.22) with zero initial conditions $\widetilde{y}(x) = 0$ for $0 \le x < m$. Formula (17.2.3.25) contains the determinant

$$D_*(t+1) = \begin{vmatrix} y_1(t+1) & \cdots & y_m(t+1) \\ \cdots & \cdots & \cdots \\ y_1(t+m-1) & \cdots & y_m(t+m-1) \\ y_1(x) & \cdots & y_m(x) \end{vmatrix},$$

which is obtained from the determinant $D(t+1)$ by replacing the last row $[y_1(t+m), \ldots, y_m(t+m)]$ with the row $[y_1(x), \ldots, y_m(x)]$. Note that $D_*(t+1) = 0$ for $x-m+1 \le t \le x-1$.

---

**17.2.3-5. Equations reducible to equations with constant coefficients.**

**1°.** The difference equation with variable coefficients

$$a_m f(x+m) y(x+m) + a_{m-1} f(x+m-1) y(x+m-1) + \cdots$$
$$+ a_1 f(x+1) y(x+1) + a_0 f(x) y(x) = g(x)$$

can be reduced, with the help of the replacement

$$y(x) = f(x) u(x),$$

to the equation with constant coefficients
$$a_m u(x+m) + a_{m-1} u(x+m-1) + \cdots + a_1 u(x+1) + a_0 u(x) = g(x).$$

2°. The difference equation with variable coefficients
$$a_m y(x+m) + a_{m-1} f(x) y(x+m-1) + a_{m-2} f(x) f(x-1) y(x+m-2) + \cdots$$
$$+ a_0 f(x) f(x-1) \ldots f(x-m+1) y(x) = g(x)$$
can be reduced to a nonhomogeneous equation with constant coefficients with the help of the replacement
$$y(x) = u(x) \exp[\varphi(x-m)],$$
with $\varphi(x)$ satisfying the auxiliary first-order difference equation
$$\varphi(x+1) - \varphi(x) = \ln f(x).$$
The resulting equation with constant coefficients has the form
$$a_m u(x+m) + a_{m-1} u(x-m-1) + \cdots + a_1 u(x+1) + a_0 u(x) = g(x) \exp[-\varphi(x)].$$

3°. The difference equation with variable coefficients
$$a_m f(x) f(x+1) \ldots f(x+m-1) y(x+m) + a_{m-1} f(x) f(x+1) \ldots f(x+m-2) y(x+m-1) + \cdots$$
$$+ a_1 f(x) y(x+1) + a_0 y(x) = g(x)$$
can be reduced to a nonhomogeneous equation with constant coefficients with the help of the replacement
$$y(x) = u(x) \exp[-\psi(x)],$$
where $\varphi(x)$ is a function satisfying the auxiliary first-order difference equation
$$\psi(x+1) - \psi(x) = \ln f(x).$$
The resulting equation with constant coefficients has the form
$$a_m u(x+m) + a_{m-1} u(x-m-1) + \cdots + a_1 u(x+1) + a_0 u(x) = g(x) \exp[\psi(x)].$$

## 17.2.4. Linear $m$th-Order Difference Equations with Arbitrary Differences

### 17.2.4-1. Linear homogeneous difference equations.

1°. A *linear homogeneous difference equation with constant coefficients*, in the case of arbitrary differences, has the form
$$a_m y(x+h_m) + a_{m-1} y(x+h_{m-1}) + \cdots + a_1 y(x+h_1) + a_0 y(x+h_0) = 0, \qquad (17.2.4.1)$$
where $a_0 a_m \ne 0$, $m \ge 1$, $h_i \ne h_j$ for $i \ne j$; the coefficients $a_k$ and the differences $h_k$ are complex numbers and $x$ is a complex variable.

Equation (17.2.4.1) can be reduced to an equation with integer differences if the quantities $h_k - h_0$ are commensurable in the sense that there is a common constant $q$ such that $h_k - h_0 = q N_k$ with integer $N_k$. Indeed, we have
$$y(x+h_k) = y(x+h_0+qN_k) = y\left(q\left(\frac{x+h_0}{q}+N_k\right)\right) = w(z+N_k),$$
where the new variables have the form $z = (x+h_0)/q$ and $w(z) = y(qz)$. As a result, we obtain an equation with integer differences $N_k$ for the function $w(z)$. In particular, this situation takes place if $h_k - h_0$ are rational numbers: $h_k - h_0 = p_k/r_k$, where $p_k$ and $r_k$ are positive integers ($k = 1, 2, \ldots, m$). In this case, one can take $q = 1/r$, where $r$ is the common denominator of the fractions $p_k/r_k$.

**2°.** In what follows, we assume that the numbers $h_k - h_0$ are not commensurable ($k = 1, \ldots, m$). We seek particular solutions of equation (17.2.4.1) in the form $y = e^{tx}$. Substituting this expression into (17.2.4.1) and dividing the result by $e^{tx}$, we obtain the transcendental equation

$$A(t) \equiv a_m e^{h_m t} + a_{m-1} e^{h_{m-1} t} + \cdots + a_1 e^{h_1 t} + a_0 e^{h_0 t} = 0, \qquad (17.2.4.2)$$

where $A(t)$ is the characteristic function.

It is known that equation (17.2.4.2) has infinitely many roots.

**3°.** Let $h_k$ be real ordered numbers, $h_0 < h_1 < \cdots < h_m$, and $t = t_1 + it_2$, where $t_1 = \operatorname{Re} t$ and $t_2 = \operatorname{Im} t$. Then the following statements hold:

(a) There exist constants $\gamma_1$ and $\gamma_2$ such that the following estimates are valid:

$$|A(t)| > \tfrac{1}{2}|a_0| e^{h_0 t_1} \quad \text{if} \quad t_1 \leq \gamma_1,$$
$$|A(t)| > \tfrac{1}{2}|a_m| e^{h_m t_1} \quad \text{if} \quad t_1 \geq \gamma_2.$$

(b) All roots of equation (17.2.4.2) belong to the vertical strip $\gamma_1 < t_1 < \gamma_2$.

(c) Let $\beta_1, \beta_2, \ldots, \beta_n, \ldots$ be roots of equation (17.2.4.2), and $|\beta_1| \leq |\beta_2| \leq \cdots \leq |\beta_n| \leq \cdots$ Then the following limit relation holds:

$$\lim_{n \to \infty} \frac{\beta_n}{n} = \frac{2\pi}{h_m - h_0}.$$

**4°.** A root $\beta_k$ of multiplicity $n_k$ of the characteristic equation (17.2.4.2) corresponds to exactly $n_k$ linearly independent solutions of equation (17.2.4.1):

$$e^{\beta_k x}, \quad x e^{\beta_k x}, \quad \ldots, \quad x^{n_k - 1} e^{\beta_k x} \qquad (k = 1, 2, \ldots). \qquad (17.2.4.3)$$

Equation (17.2.4.1) admits infinitely many solutions of the form (17.2.4.3), since the characteristic function (17.2.4.2) has infinitely many roots. In order to single out different classes of solutions, it is convenient to use a condition that characterizes the order of their growth for large values of the argument.

The class of *functions of exponential growth* of finite degree $\sigma$ is denoted by $[1, \sigma]$ and is defined as the set of all entire functions* $f(x)$ satisfying the condition

$$e^{(\sigma - \varepsilon)|x|} < |f(x)| < e^{(\sigma + \varepsilon)|x|} \qquad (17.2.4.4)$$

for any $\varepsilon > 0$, where the right inequality in (17.2.4.4) should hold for all sufficiently large $x$: $|x| > R(\varepsilon)$, while the left inequality in (17.2.4.4) should hold for some sequence $x = x_n = x_n(\varepsilon) \to \infty$.

The parameter $\sigma$ can be found from the relations

$$\sigma = \varlimsup_{r \to \infty} \frac{M(r)}{r} = \varlimsup_{k \to \infty} \left( k! |b_k| \right)^{1/k}, \quad M(r) = \max_{|x| = r} |f(x)| \qquad (0 \leq \sigma < \infty),$$

where $b_k$ are the coefficients in the power series expansion of the function $f(x)$ (see the footnote).

---

* An entire function is a function that is analytic on the entire complex plane (except, possibly, the infinite point). Any such function can be expanded in power series, $f(x) = \sum_{k=0}^{\infty} b_k x^k$, convergent on the entire complex plane, i.e., $\lim_{k \to \infty} |b_k|^{1/k} = 0$.

THEOREM. *Any solution of equation (17.2.4.1) in the class of exponentially growing functions of a finite degree $\sigma$ can be represented in the form*

$$y(x) = \sum_{|\beta_k| \leq \sigma} \sum_{s=0}^{h_k-1} C_{ks} x^s e^{\beta_k x} \equiv \sum_{|\beta_k| \leq \sigma} P_k(x) e^{\beta_k x}, \quad (17.2.4.5)$$

*where the sum is over all zeroes of the characteristic function (17.2.4.2) in the circle $|t| \leq \sigma$; $C_{ks}$ are arbitrary constants, and $P_k(x)$ are arbitrary polynomials of degrees $\leq n_k - 1$.*

*Corollary 1.* A necessary and sufficient condition for the existence of polynomial solutions of equation (17.2.4.1) is that the characteristic function (17.2.4.2) has zero root $\beta_1 = 0$. In this case, the coefficients of equation (17.2.4.1) should satisfy the condition $a_m + a_{m-1} + \cdots + a_1 + a_0 = 0$.

*Corollary 2.* There is only one solution $y(x) \equiv 0$ in the class $[1, \sigma]$ only if $\sigma < \min\{|\beta_1|, |\beta_2|, \ldots\}$.

### 17.2.4-2. Linear nonhomogeneous difference equations.

1°. A *linear nonhomogeneous difference equation with constant coefficients*, in the case of arbitrary differences, has the form

$$a_m y(x + h_m) + a_{m-1} y(x + h_{m-1}) + \cdots + a_1 y(x + h_1) + a_0 y(x + h_0) = f(x), \quad (17.2.4.6)$$

where $a_0 a_m \neq 0$, $m \geq 1$, and $h_i \neq h_j$ for $i \neq j$.

Let $f(x)$ be a function of exponential growth of degree $\sigma$. Then equation (17.2.4.6) always has a solution $\tilde{y}(x)$ in the class $[1, \sigma]$. The general solution of equation (17.2.4.6) in the class $[1, \sigma]$ can be represented as the sum of the general solution (17.2.4.5) of the homogeneous equation (17.2.4.1) and a particular solution $\tilde{y}(x)$ of the nonhomogeneous equation (17.2.4.6).

2°. Suppose that the right-hand side of the equation is the polynomial

$$f(x) = \sum_{s=0}^{n} b_s x^s, \quad b_n \neq 0, \quad n \geq 0. \quad (17.2.4.7)$$

Then equation (17.2.4.6) has a particular solution of the form

$$\tilde{y}(x) = x^\mu \sum_{s=0}^{n} c_s x^s, \quad c_n \neq 0, \quad (17.2.4.8)$$

where $t = 0$ is a root of the characteristic function (17.2.4.2), the multiplicity of this root being equal to $\mu$. The coefficients $c_n$ in (17.2.4.8) can be found by the method of indefinite coefficients.

3°. Suppose that the right-hand side of the equation has the form

$$f(x) = e^{px} \sum_{s=0}^{n} b_s x^s, \quad b_n \neq 0, \quad n \geq 0. \quad (17.2.4.9)$$

Then equation (17.2.4.6) has a particular solution of the form

$$\tilde{y}(x) = e^{px} x^\mu \sum_{s=0}^{n} c_s x^s, \quad c_n \neq 0, \quad (17.2.4.10)$$

where $t = p$ is a root of the characteristic function (17.2.4.2), its multiplicity being equal to $\mu$. The coefficients $c_n$ in (17.2.4.10) can be found by the method of indefinite coefficients.

**17.2.4-3. Equations reducible to equations with constant coefficients.**

$1°$. The difference equation with variable coefficients

$$a_m f(x+h_m) y(x+h_m) + a_{m-1} f(x+h_{m-1}) y(x+h_{m-1}) + \cdots$$
$$+ a_1 f(x+h_1) y(x+h_1) + a_0 f(x+h_0) y(x+h_0) = g(x)$$

can be reduced, with the help of the replacement

$$y(x) = f(x) u(x),$$

to a difference equation with constant coefficients

$$a_m u(x+h_m) + a_{m-1} u(x+h_{m-1}) + \cdots + a_1 u(x+h_1) + a_0 u(x+h_0) = g(x).$$

$2°$. Two other difference equations with variable coefficients can be obtained from equations considered in Paragraph 17.2.3-5 (Items $2°$ and $3°$), where the quantities $x+m$, $x+m-1$, ..., $x+1$, $x$ should be replaced, respectively, by $x+h_m$, $x+h_{m-1}$, ..., $x+h_1$, $x+h_0$.

## 17.3. Linear Functional Equations

### 17.3.1. Iterations of Functions and Their Properties

**17.3.1-1. Definition of iterations.**

Consider a function $f(x)$ defined on a set $I$ and suppose that

$$f(I) \subset I. \qquad (17.3.1.1)$$

A set $I$ for which (17.3.1.1) holds is called a *submodulus set* for the function $f(x)$. If we have $f(I) = I$, then $I$ is called a *modulus set* for the function $f(x)$.

For a function $f(x)$ defined on a set $I$ and satisfying the condition (17.3.1.1), by $f^{[n]}(x)$ we denote the $n$th iteration defined by the relations

$$f^{[0]}(x) = x, \quad f^{[n+1]}(x) = f(f^{[n]}(x)), \qquad x \in I, \quad n = 0, 1, 2, \ldots \qquad (17.3.1.2)$$

For an invertible function $f(x)$, one can define its iterations also for negative values of the iteration index $n$:

$$f^{[n-1]}(x) = f^{-1}(f^{[n]}(x)), \qquad x \in I, \quad n = 0, -1, -2, \ldots, \qquad (17.3.1.3)$$

where $f^{-1}$ denotes the function inverse to $f$. In view of the relations (17.3.1.2) and (17.3.1.3), we also have

$$f^{[1]}(x) = f(x), \quad f^{[n]}(f^{[m]}(x)) = f^{[n+m]}(x). \qquad (17.3.1.4)$$

### 17.3.1-2. Fixed points of a function. Some classes of functions.

A point $\xi$ is called a *fixed point* of the function $f(x)$ if $f(\xi) = \xi$. A point $\xi$ is called an *attractive fixed point* of the function $f(x)$ if there exists a neighborhood $U$ of $\xi$ such that $\lim\limits_{n\to\infty} f^{[n]}(x) = \xi$ for any $x \in U$. If, in addition, we have

$$|f(x) - \xi| \leq \varepsilon |x - \xi|, \quad 0 < \varepsilon < 1,$$

for $x \in U$, then $\xi$ is called a *strongly attractive fixed point*.

If $f(x)$ is differentiable at a fixed point $x = \xi$ and

$$|f'(\xi)| < 1,$$

then $\xi$ is a strongly attractive fixed point.

For $a < b$, let $I$ be any of the sets

$$a < x < b, \quad a \leq x < b, \quad a < x \leq b, \quad a \leq x \leq b.$$

One or both endpoints $a$ and $b$ may be infinite. The closure of $I$ is denoted by $\overline{I}$.

Denote by $S_\xi^m[I]$ (briefly $S_\xi^m$) the class of functions $f(x)$ satisfying the following conditions:

1°. $f(x)$ has continuous derivatives up to the order $m$ in $I$.

2°. $f(x)$ satisfies the inequalities

$$(f(x) - x)(\xi - x) > 0 \quad \text{for} \quad x \in I, \; x \neq \xi; \tag{17.3.1.5}$$
$$(f(x) - \xi)(\xi - x) < 0 \quad \text{for} \quad x \in I, \; x \neq \xi, \tag{17.3.1.6}$$

where $\xi \in \overline{I}$.

Denote by $R_\xi^m[I]$ (briefly $R_\xi^m$) the class of functions $f(x)$ belonging to $S_\xi^m$ and strictly increasing on $I$. Fig. 17.3 represents a function in $S_\xi^m$ and Fig. 17.4 represents a function in $R_\xi^m$.

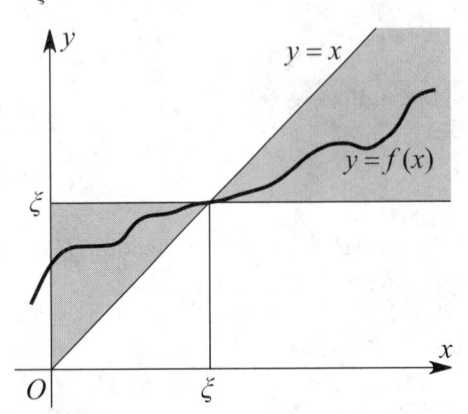

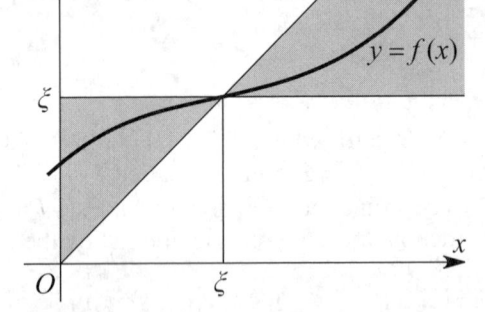

**Figure 17.3.** A function belonging to $S_\xi^m$.

**Figure 17.4.** A function belonging to $R_\xi^m$.

**Remark.** If $\xi = +\infty$, then in the definition of the classes $S_\infty^m$ and $R_\infty^m$, the condition (17.3.1.6) is superfluous and (17.3.1.5) is replaced by $f(x) > x$ for $x \in I$. Analogously, if $\xi = -\infty$, then (17.3.1.5) is replaced by $f(x) < x$ for $x \in I$.

The following statements hold:

(i) If $f(x) \in S_\xi^m$, then $\xi$ is a fixed point of $f(x)$.

(ii) If $f(x) \in S_\xi^m$, then for any $x_0 \in I$ the sequence $f^{[n]}(x_0)$ is monotonic (and strictly monotonic whenever $x_0 \neq \xi$) and

$$\lim_{n\to\infty} f^{[n]}(x_0) = \xi.$$

**17.3.1-3. Asymptotic properties of iterations in a neighborhood of a fixed point.**

**1°.** Let $f(0) = 0$, $0 < f(x) < x$ for $0 < x < x_0$, and suppose that in a neighborhood of the fixed point the function $f(x)$ can be represented in the form

$$f(x) = x - ax^k + bx^m + o(x^m),$$

where $1 < k < m$ and $a, b > 0$. Then the following limit relation holds:

$$\lim_{n \to \infty} n^{\frac{1}{k-1}} f^{[n]}(x) = [a(k-1)]^{-\frac{1}{k-1}}, \qquad 0 < x < x_0. \tag{17.3.1.7}$$

**Example.** Consider the function $f(x) = \sin x$. We have $0 < \sin x < x$ for $0 < x < \infty$ and $\sin x = x - \frac{1}{6}x^3 + \frac{1}{120}x^5 + o(x^5)$, which corresponds to the values $a = \frac{1}{6}$ and $k = 3$. Substituting these values into (17.3.1.7), we obtain

$$\lim_{n \to \infty} \sqrt{n} \sin^{[n]} x = \sqrt{3}, \qquad 0 < x < \infty.$$

**2°.** Let $f(0) = 0$, $0 < f(x) < x$ for $0 < x < x_0$, and suppose that in a neighborhood of the fixed point the function $f(x)$ can be represented in the form

$$f(x) = \lambda x + ax^k + bx^m + o(x^m),$$

where $0 < \lambda < 1$ and $1 < k < m$. Then the following limit relation holds:

$$\lim_{n \to \infty} \frac{f^{[n]}(x) - \lambda^n x}{\lambda^n} = \frac{ax^k}{\lambda - \lambda^k}, \qquad 0 < x < x_0.$$

**17.3.1-4. Representation of iterations by power series.**

Let $f(x)$ be a function with a fixed point $\xi = f(\xi)$ and suppose that in a neighborhood of that point $f(x)$ can be represented by the series

$$f(x) = \xi + \sum_{j=1}^{\infty} a_j (x - \xi)^j \tag{17.3.1.8}$$

with a nonzero radius of convergence. For any integer $N > 0$, there is a neighborhood $U$ of the point $\xi$ in which all iterations $f^{[n]}(x)$ for integer $0 \le n \le N$ are defined and also admit the representation

$$f^{[n]}(x) = \xi + \sum_{j=1}^{\infty} A_{nj} (x - \xi)^j. \tag{17.3.1.9}$$

The coefficients $A_{nj}$ can be uniquely expressed through the coefficients $a_j$ with the help of formal power series and the relations $f(f^{[n]}(x)) = f^{[n]}(f(x))$.

For $a_1 \ne 0$ and $|a_1| \ne 1$, the first three coefficients of the series (17.3.1.9) have the form

$$A_{n1} = a_1^n, \qquad A_{n2} = \frac{a_1^{2n} - a_1^n}{a_1^2 - a_1} a_2,$$

$$A_{n3} = \frac{a_1^{3n} - a_1^n}{a_1^3 - a_1} a_3 + 2\frac{(a_1^{2n} - a_1^n)(a_1^n - a_1)}{(a_1^3 - a_1)(a_1^2 - a_1)} a_2, \qquad n = 2, 3, \ldots$$

For $a_1 = 1$, we have

$$A_{n1} = 1, \quad A_{n2} = na_2, \quad A_{n3} = na_3 + n(n-1)a_2^2, \quad n = 2, 3, \ldots$$

The series (17.3.1.8) has a nonzero radius of convergence if and only if there exist constants $A > 0$ and $B > 0$ such that $|a_j| \le AB^{j-1}$, $j = 1, 2, \ldots$. Under these conditions, the series (17.3.1.8) is convergent for $|x - \xi| < 1/B$, and the series (17.3.1.9) is convergent for $|x - \xi| < R_n$, where

$$R_n = \begin{cases} \dfrac{A-1}{B(A^n - 1)} & \text{if } A \ne 1, \\ \dfrac{1}{nB} & \text{if } A = 1. \end{cases}$$

### 17.3.2. Linear Homogeneous Functional Equations

**17.3.2-1. Equations of general form. Possible cases.**

A linear homogeneous functional equation has the form

$$y(f(x)) = g(x)y(x), \qquad (17.3.2.1)$$

where $f(x)$ and $g(x) \ne 0$ are known functions and the function $y(x)$ is to be found.

Let $f(x) \in R_\xi^0$, where $x \in I$ and $\xi \in I$, and let $g(x)$ be continuous on $I$, $g(x) \ne 0$ for $x \in I$, $x \ne \xi$.

Consider the sequence of functions

$$G_n(x) = \prod_{k=0}^{n-1} g(f^{[k]}(x)), \qquad n = 1, 2, \ldots \qquad (17.3.2.2)$$

Three cases may occur.
(i) The limit

$$G(x) = \lim_{n \to \infty} G_n(x) \qquad (17.3.2.3)$$

exists on $I$. Moreover, $G(x)$ is continuous and $G(x) \ne 0$.
(ii) There exists an interval $I_0 \subset I$ such that

$$\lim_{n \to \infty} G_n(x) = 0$$

uniformly on $I_0$.
(iii) Neither of the cases (i) or (ii) occurs.

THEOREM. *In case* (i), *the homogeneous functional equation* (17.3.2.1) *has a one-parameter family of continuous solutions on* $I$. *For any* $a$, *there exists exactly one continuous function* $y(x)$ *satisfying equation* (17.3.2.1) *and the condition*

$$y(\xi) = a.$$

*This solution has the form*

$$y(x) = \frac{a}{G(x)}. \qquad (17.3.2.4)$$

*In case (ii), equation (17.3.2.1) has a continuous solution depending on an arbitrary function, and every continuous solution $y(x)$ on $I$ satisfies the condition $\lim_{x \to \xi} y(x) = 0$.*

*In case (iii), equation (17.3.2.1) has a single continuous solution, $y(x) \equiv 0$.*

In order to decide which of the cases (i), (ii), or (iii) takes place, the following simple criteria can be used.

(i) This case takes place if $\xi$ is a strongly attractive fixed point of $f(x)$ and there exist positive constants $\delta$, $\mu$, and $M$ such that

$$|g(x) - 1| \leq M|x - \xi|^\mu \quad \text{for} \quad x \in I \cap (\xi - \delta, \xi + \delta).$$

(ii) This case takes place if $|g(\xi)| < 1$.
(iii) This case takes place if $|g(\xi)| > 1$.

**Remark.** For $g(\xi) = 1$, any of the three cases (i), (ii), or (iii) may take place.

### 17.3.2-2. Schröder–Koenigs functional equation.

Consider the *Schröder–Koenigs equation*

$$y(f(x)) = sy(x), \quad s \neq 0, \tag{17.3.2.5}$$

which is a special case of the linear homogeneous functional equation (17.3.2.1) for $g(x) = s = \text{const}$.

$1°$. Let $s > 0$, $s \neq 1$ and let $\bar{y}(x)$ be a solution of equation (17.3.2.5) satisfying the condition $\bar{y}(x) \neq 0$. Then the function

$$y(x) = \bar{y}(x) \Theta\left(\frac{\ln |\bar{y}(x)|}{\ln s}\right),$$

where $\Theta(z)$ is an arbitrary 1-periodic function, is also a solution of equation (17.3.2.5).

$2°$. Let $f(x)$ be defined on a submodulus set $I$ and let $h(x)$ be a one-to-one mapping of $I$ onto a set $I_1$. Let

$$p(x) = h(f(h^{-1}(x))),$$

where $h^{-1}$ is the inverse function of $h$. If $\psi(x)$ is a solution of the equation

$$\psi(p(x)) = s\psi(x)$$

on $I_1$, then the function

$$y(x) = \psi(h(x))$$

satisfies equation (17.3.2.5) on $I$.

On the basis of this statement, a fixed point $\xi$ can be moved to the origin. Indeed, if $\xi$ is finite, we can take $h(x) = x - \xi$. If $\xi = \infty$, we take $h(x) = 1/x$. Thus, we can assume that $0$ is a fixed point of $f(x)$.

3°. Suppose that $f(x) \in R_0^2[I]$, $0 \in I$, and

$$f'(0) = s, \quad 0 < s < 1.$$

Then for each $\sigma \in (-\infty, \infty)$, there exists a unique continuously differentiable solution of equation (17.3.2.5) on $I$ satisfying the condition

$$y'(0) = \sigma. \tag{17.3.2.6}$$

This solution is given by the formula

$$y(x) = \sigma \lim_{n \to \infty} s^{-n} f^{[n]}(x), \tag{17.3.2.7}$$

and for $\sigma \neq 0$ it is strictly monotone on $I$ (it is an increasing function for $\sigma > 0$, and it is a decreasing function for $\sigma < 0$).

4°. For an invertible function $f$, (17.3.2.5) can be reduced to a similar equation with the help of the transformation $z = f(x)$:

$$y(f^{-1}(z)) = s^{-1} y(z).$$

5°. For $s > 0$, $s \neq 0$, the replacement $y(x) = s^{u(x)/c}$ reduces the Schröder–Koenigs equation (17.3.2.5) to the Abel equation for the function $u(x)$ (see equation (17.3.3.5), in which $y$ should be replaced by $u$). In Subsection 17.3.3, Item 4°, there is a description of the method for constructing continuous monotone solutions of the Abel equation to within certain arbitrary functions.

### 17.3.2-3. Automorphic functions.

Consider the special case of equation (17.3.2.5) for $s = 1$:

$$y(f(x)) = y(x). \tag{17.3.2.8}$$

Solutions of this equation are called *automorphic functions*. If $f(x)$ is invertible on a modulus set $I$, then the general solution of equation (17.3.2.8) may be written in the form

$$y(x) = \sum_{n=-\infty}^{\infty} \varphi(f^{[n]}(x)), \tag{17.3.2.9}$$

where $\varphi(x)$ is an arbitrary function on $I$ such that the series (17.3.2.9) is convergent.

## 17.3.3. Linear Nonhomogeneous Functional Equations

### 17.3.3-1. Equations of general form. Possible cases.

1°. Consider a linear nonhomogeneous functional equation of the general form

$$y(f(x)) = g(x)y(x) + F(x). \tag{17.3.3.1}$$

Let $x \in I$ and $\xi \in I$, $f(x) \in R_\xi^0$. Suppose that $g(x)$ and $F(x)$ are continuous functions on $I$ and $g(x) \neq 0$ for $x \in I$, $x \neq \xi$. In accordance with the investigation of the corresponding homogeneous equation with $F(x) \equiv 0$, the following three cases are possible (the notation used here is in agreement with that of Paragraph 17.3.2-1):

(i) Equation (17.3.3.1) on $I$ either has a one-parameter family of continuous solutions or no solutions at all. If (17.3.3.1) has a continuous solution $y_0(x)$ on $I$, then the general continuous solution on $I$ is given by

$$y(x) = y_0(x) + \frac{a}{G(x)},$$

where $a$ is an arbitrary constant and $G(x)$ is given by (17.3.2.3).

(ii) Equation (17.3.3.1) on $I$ has a continuous solution depending on an arbitrary function, or has no continuous solutions on $I$.

(iii) Equation (17.3.3.1) on $I$ either has a continuous solution or no solutions at all.

$2°$. Let $x \in I, \xi \in I$, and $f(x) \in R_\xi^0[I]$. Suppose that $g(x)$ and $F(x)$ are continuous functions, $g(x) \neq 0$ for $x \neq \xi$, and $g(\xi) > 1$. Then equation (17.3.3.1) has a unique continuous solution that can be represented by the series

$$y(x) = -\sum_{n=0}^{\infty} \frac{F(f^{[n]}(x))}{G_{n+1}(x)}, \tag{17.3.3.2}$$

where the function $G_n(x)$ is defined by (17.3.2.2).

**17.3.3-2. Equations of special form with $g(x) = $ const.**

$1°$. Consider the equation

$$y(f(x)) + y(x) = F(x), \tag{17.3.3.3}$$

which is a special case of equation (17.3.3.1) with $g(x) = -1$.

(A) Let $\xi \in I$, $f(x) \in R_\xi^0[I]$, and suppose that $F(x)$ is a continuous function. If there exists a continuous solution of equation (17.3.3.3), it can be represented by the power series

$$y(x) = \frac{1}{2}F(\xi) + \sum_{n=0}^{\infty}(-1)^n \left[F(f^{[n]}(x)) - F(\xi)\right].$$

(B) Suppose that the assumptions of Item (A) hold and, moreover, there exist positive constants $\delta$, $\kappa$, and $C$ such that

$$|F(x) - F(\xi)| \leq C|x - \xi|^\kappa, \quad x \in (\xi - \delta, \xi + \delta) \cap I,$$

and $\xi$ is a strongly attractive fixed point of $f(x)$. Then equation (17.3.3.3) has a continuous solution on $I$.

$2°$. Consider the functional equation

$$y(f(x)) - \lambda y(x) = F(x), \quad \lambda \neq 0. \tag{17.3.3.4}$$

Let $x \in I = (a, b]$. Assume that on the interval $I$ the function $f(x)$ is continuous, strongly increasing, and satisfies the condition $a < f(x) < x$, and $F(x)$ is a function of bounded variation.

Depending on the value of the parameter $\lambda$, the following cases are possible.

*Case* $|\lambda| > 1$. There is a unique solution

$$y(x) = -\sum_{n=0}^{\infty} \frac{F(f^{[n]}(x))}{\lambda^{n+1}},$$

which coincides with (17.3.3.2) for $g(x) = \lambda$ and $G_{n+1}(x) = \lambda^n$.

*Case* $0 < |\lambda| < 1$. For any $x_0 \in I$ and any test function $\varphi(z)$ that has bounded variation on $[f(x_0), x_0]$ and satisfies the condition

$$\varphi(f(x_0)) - \lambda\varphi(x_0) = F(x_0),$$

there is a single function of bounded variation for which equation (17.3.3.4) holds. This solution has the same variation on the interval $[f(x_0), x_0]$ as the test function $\varphi(x)$.

*Case* $\lambda = 1$. If $\lim_{x \to a+0} F(x) = 0$ and there is a point $x_0 \in I$ such that the series $\sum_{n=0}^{\infty} F(f^{[n]}(x_0))$ is convergent, then equation (17.3.3.4) has a one-parameter family of solutions

$$y(x) = C - \sum_{n=0}^{\infty} F(f^{[n]}(x))$$

(here, $C$ is an arbitrary constant), for which there exists a finite limit $\lim_{x \to a+0} y(x)$.

*Case* $\lambda = -1$. See Item 1°.

### 17.3.3-3. Abel functional equation.

*The Abel functional equation* has the form

$$y(f(x)) = y(x) + c, \quad c \neq 0. \tag{17.3.3.5}$$

**Remark.** Without the loss of generality, we can take $c = 1$ (this can be achieved by passing to the normalized unknown function $\bar{y} = y/c$).

1°. Let $\bar{y}(x)$ be a solution of equation (17.3.3.5). Then the function

$$y(x) = \bar{y}(x) + \Theta\left(\frac{\bar{y}(x)}{c}\right),$$

where $\Theta(z)$ is an arbitrary 1-periodic function, is also a solution of equation (17.3.3.5).

2°. Suppose that the following conditions hold:
 (i) $f(x)$ is strictly increasing and continuous for $0 \leq x \leq a$;
 (ii) $f(0) = 0$ and $f(x) < x$ for $0 < x < a$;
 (iii) the derivative $f'(x)$ exists, has bounded variation on the interval $0 < x < a$, and $\lim_{x \to +0} f'(x) = 1$. Then, for any $x, x_0 \in (0, a]$, there exists the limit

$$y(x) = \lim_{n \to \infty} \frac{f^{[n]}(x) - f^{[n]}(x_0)}{f^{[n-1]}(x_0) - f^{[n]}(x_0)}, \tag{17.3.3.6}$$

which defines a monotonically increasing function satisfying the Abel equation (17.3.3.5) with $c = -1$ (this is the *Lévy solution*).

$3°$. Let $f(x)$ be defined on a submodulus interval $I$ and suppose that there is a sequence $d_n$ for which
$$\lim_{n\to\infty} \frac{f^{[n+1]}(x) - f^{[n]}(x)}{d_n} = c, \qquad x \in I.$$
If for all $x \in I$ there exists the limit
$$y(x) = \lim_{n\to\infty} \frac{f^{[n]}(x) - f^{[n]}(x_0)}{d_n}, \qquad (17.3.3.7)$$
where $x_0$ is an arbitrary fixed point from the interval $I$, then the function $y(x)$ satisfies equation (17.3.3.5).

$4°$. Let us describe a procedure for the construction of monotone solutions of the Abel equation (17.3.3.5). Assume that the function $f(x)$ is continuous, strictly increasing on the segment $[0, a]$, and the following conditions hold: $f(0) = 0$ and $f(x) < x \le a$ for $x \ne 0$.

We define a function $\varphi(y)$ on the semiaxis $[0, \infty)$ by
$$\varphi(y) = f^{[n]}\big(\varphi_0(\{y\})\big), \qquad n = [y], \qquad (17.3.3.8)$$
where $[y]$ and $\{y\}$ are, respectively, the integer and the fractional parts of $y$, and $\varphi_0(y)$ is any continuous strictly decreasing function on $[0, 1)$ such that $\varphi_0(0) = a$, $\varphi_0(1 - 0) = f(a)$.

On the intervals $[n, n+1)$ ($n = 0, 1, \ldots$), the function (17.3.3.8) takes the values
$$\varphi(y) = f^{[n]}\big(\varphi_0(y - n)\big), \qquad n \le y < n + 1, \qquad (17.3.3.9)$$
and therefore is continuous and strictly decreasing, being a superposition of continuous strictly increasing functions and a continuous strictly decreasing function. At integer points, too, the function $\varphi(y)$ is continuous, since (17.3.3.8) and (17.3.3.9) imply that
$$\varphi(n - 0) = f^{[n-1]}\big(\varphi_0(1 - 0)\big) = f^{[n-1]}\big(f(a)\big) = f^{[n]}(a),$$
$$\varphi(n) = f^{[n]}\big(\varphi_0(0)\big) = f^{[n]}(a).$$

Therefore, the function $\varphi(y)$ is also continuous and strictly decreasing on the entire semiaxis $[0, \infty)$. By continuity, we can define the value $\varphi(\infty) = 0$.

It follows that there exists the inverse function $\varphi^{-1}$ defined by the relations
$$x = \varphi(y), \qquad y = \varphi^{-1}(x) \qquad (0 \le x \le a,\ 0 \le y \le \infty), \qquad (17.3.3.10)$$
and $\varphi^{-1}(0) = \infty$, $\varphi^{-1}(a) = 0$.

From (17.3.3.8), it follows that the function $\varphi(y)$ satisfies the functional equation
$$\varphi(y + 1) = f\big(\varphi(y)\big), \qquad 0 \le y < \infty. \qquad (17.3.3.11)$$

Applying $\varphi^{-1}$ to both sides of (17.3.3.11) and then passing from $y$ to a new variable $x$ with the help of (17.3.3.10), we come to the Abel equation (17.3.3.5) with $c = 1$, where $y(x) = \varphi^{-1}(x)$. Thus, finding a solution of the Abel equation is reduced to the inversion of the known function (17.3.3.8).

$5°$. The Abel equation (17.3.3.5) can be reduced to the Schrödinger–Koenig equation
$$u(f(x)) = su(x),$$
with the help of the replacement $u(x) = s^{y(x)/c}$; see equation (17.3.2.5).

## 17.3.4. Linear Functional Equations Reducible to Linear Difference Equations with Constant Coefficients

**17.3.4-1. Functional equations with arguments proportional to $x$.**

Consider the linear equation

$$a_m y(b_m x) + a_{m-1} y(b_{m-1} x) + \cdots + a_1 y(b_1 x) + a_0 y(b_0 x) = f(x). \qquad (17.3.4.1)$$

The transformation

$$y(x) = w(z), \quad z = \ln x$$

reduces (17.3.4.1) to a linear difference equation with constant coefficients

$$a_m w(z+h_m) + a_{m-1} w(z+h_{m-1}) + \cdots + a_1 w(z+h_1) + a_0 w(z+h_0) = f(e^z), \qquad h_k = \ln b_k.$$

Note that the homogeneous functional equation (17.3.4.1) with $f(x) \equiv 0$ admits particular solutions of power type, $y(x) = Cx^\beta$, where $C$ is an arbitrary constant and $\beta$ is a root of the transcendental equation

$$a_m b_m^\beta + a_{m-1} b_{m-1}^\beta + \cdots + a_1 b_1^\beta + a_0 b_0^\beta = 0.$$

**17.3.4-2. Functional equations with powers of $x$ as arguments.**

Consider the linear equation

$$a_m y(x^{n_m}) + a_{m-1} y(x^{n_{m-1}}) + \cdots + a_1 y(x^{n_1}) + a_0 y(x^{n_0}) = f(x). \qquad (17.3.4.2)$$

The transformation

$$y(x) = w(z), \quad z = \ln \ln x$$

reduces (17.3.4.2) to a linear difference equation with constant coefficients

$$a_m w(z+h_m) + a_{m-1} w(z+h_{m-1}) + \cdots + a_1 w(z+h_1) + a_0 w(z+h_0) = f(e^{e^z}), \qquad h_k = \ln \ln n_k.$$

Note that the homogeneous functional equation (17.3.4.2) admits particular solutions of the form $y(x) = C|\ln x|^p$, where $C$ is an arbitrary constant and $p$ is a root of the transcendental equation

$$a_m |n_m|^p + a_{m-1} |n_{m-1}|^p + \cdots + a_1 |n_1|^p + a_0 |n_0|^p = 0.$$

**17.3.4-3. Functional equations with exponential functions of $x$ as arguments.**

Consider the linear equation

$$a_m y(e^{\lambda_m x}) + a_{m-1} y(e^{\lambda_{m-1} x}) + \cdots + a_1 y(e^{\lambda_1 x}) + a_0 y(e^{\lambda_0 x}) = f(x).$$

The transformation

$$y(x) = w(\ln z), \quad z = \ln x$$

reduces this equation to a linear difference equation with constant coefficients

$$a_m w(z+h_m) + a_{m-1} w(z+h_{m-1}) + \cdots + a_1 w(z+h_1) + a_0 w(z+h_0) = f(e^z), \qquad h_k = \ln \lambda_k.$$

**17.3.4-4. Equations containing iterations of the unknown function.**

Consider the linear homogeneous equation

$$a_m y^{[m]}(x) + a_{m-1} y^{[m-1]}(x) + \cdots + a_1 y(x) + a_0 x = 0$$

with successive iterations of the unknown function, $y^{[2]}(x) = y(y(x))$, ..., $y^{[n]}(x) = y(y^{[n-1]}(x))$.

A solution of this equation is sought in the parametric form

$$x = w(t), \quad y = w(t+1). \tag{17.3.4.3}$$

Then the original equation is reduced to the following $m$th-order linear equation with constant coefficients (equation of the type (17.2.3.1) with integer differences):

$$a_m w(t+m) + a_{m-1} w(t+m-1) + \cdots + a_1 w(t+1) + a_0 w(t) = 0.$$

**Example.** Consider the functional equation

$$y(y(x)) + ay(x) + bx = 0. \tag{17.3.4.4}$$

The transformation (17.3.4.3) reduces this equation to a difference equation of the form (17.2.2.1). Let $\lambda_1$ and $\lambda_2$ be distinct real roots of the characteristic equation

$$\lambda^2 + a\lambda + b = 0, \tag{17.3.4.5}$$

$\lambda_1 \neq \lambda_2$. Then the general solution of the functional equation (17.3.4.4) in parametric form can be written as

$$\begin{aligned} x &= \Theta_1(t)\lambda_1^t + \Theta_2(t)\lambda_2^t, \\ y &= \Theta_1(t)\lambda_1^{t+1} + \Theta_2(t)\lambda_2^{t+1}, \end{aligned} \tag{17.3.4.6}$$

where $\Theta_1(t)$ and $\Theta_2(t)$ are arbitrary 1-periodic functions, $\Theta_k(t) = \Theta_k(t+1)$.

Taking $\Theta_1(t) = C_1$ and $\Theta_2(t) = C_2$ in (17.3.4.6), where $C_1$ and $C_2$ are arbitrary constants, and eliminating the parameter $t$, we obtain a particular solution of equation (17.3.4.4) in implicit form:

$$\frac{\lambda_2 x - y}{\lambda_2 - \lambda_1} = C_1 \left[ \frac{\lambda_1 x - y}{C_2(\lambda_1 - \lambda_2)} \right]^\gamma, \quad \gamma = \frac{\ln \lambda_1}{\ln \lambda_2}.$$

**17.3.4-5. Babbage equation and involutory functions.**

1°. Functions satisfying the *Babbage equation*

$$y(y(x)) = x \tag{17.3.4.7}$$

are called *involutory functions*.

Equation (17.3.4.7) is a special case of (17.3.4.4) with $a = 0$, $b = -1$. The corresponding characteristic equation (17.3.3.5) has the roots $\lambda_{1,2} = \pm 1$. The parametric representation of solution (17.3.4.6) contains the complex quantity $(-1)^t = e^{i\pi t} = \cos(\pi t) + i\sin(\pi t)$ and is, therefore, not very convenient.

2°. The following statements hold:

(i) On the interval $x \in (a,b)$, there is a continuous decreasing solution of equation (17.3.4.7) depending on an arbitrary function. This solution has the form

$$y(x) = \begin{cases} \varphi(x) & \text{for } x \in (a,c], \\ \varphi^{-1}(x) & \text{for } x \in (c,b), \end{cases}$$

where $c \in (a,b)$ is an arbitrary point, and $\varphi(x)$ is an arbitrary continuous decreasing function on $(a,c]$ such that

$$\lim_{x \to a+0} \varphi(x) = b, \quad \varphi(c) = c.$$

(ii) The only increasing solution of equation (17.3.4.7) has the form $y(x) \equiv x$.

3°. Solution in parametric form:
$$x = \Theta\left(\frac{t}{2}\right), \quad y = \Theta\left(\frac{t+1}{2}\right),$$
where $\Theta(t) = \Theta(t+1)$ is an arbitrary periodic function with period 1.

4°. Solution in parametric form (an alternative representation):
$$x = \Theta_1(t) + \Theta_2(t)\sin(\pi t),$$
$$y = \Theta_1(t) - \Theta_2(t)\sin(\pi t),$$
where $\Theta_1(t)$ and $\Theta_2(t)$ are arbitrary 1-periodic functions, $\Theta_k(t) = \Theta_k(t+1)$, $k = 1, 2$. In this solution, the functions $\sin(\pi t)$ can be replaced by $\cos(\pi t)$.

## 17.4. Nonlinear Difference and Functional Equations with a Single Variable

### 17.4.1. Nonlinear Difference Equations with a Single Variable

**17.4.1-1. Riccati difference equation.**

The *Riccati difference equation* has the general form
$$y(x)y(x+1) = a(x)y(x+1) + b(x)y(x) + c(x), \qquad (17.4.1.1)$$
where the functions $a(x)$, $b(x)$, $c(x)$ satisfy the condition $a(x)b(x) + c(x) \not\equiv 0$.

1°. The substitution
$$y(x) = \frac{u(x+1)}{u(x)} + a(x)$$
reduces the Riccati equation to the linear second-order difference equation
$$u(x+2) + [a(x+1) - b(x)]u(x+1) - [a(x)b(x) + c(x)]u(x) = 0.$$

2°. Let $y_0(x)$ be a particular solution of equation (17.4.1.1). Then the substitution
$$z(x) = \frac{1}{y(x) - y_0(x)}$$
reduces equation (17.4.1.1) to the first-order linear nonhomogeneous difference equation
$$z(x+1) + \frac{[y_0(x) - a(x)]^2}{a(x)b(x) + c(x)}z(x) + \frac{y_0(x) - a(x)}{a(x)b(x) + c(x)} = 0.$$

**17.4.1-2. First-order nonlinear difference equations with no explicit dependence on $x$.**

Consider the nonlinear functional equation
$$y(x+1) = f(y(x)), \qquad 0 \le x < \infty, \qquad (17.4.1.2)$$
where $f(y)$ is a continuous function.

Let $y(x)$ be a function defined on the semiaxis $[0, \infty)$ by
$$y(x) = f^{[n]}(\varphi(\{x\})), \quad n = [x], \qquad (17.4.1.3)$$
where $[x]$ and $\{x\}$ are, respectively, the integer and the fractional parts of $x$, and $\varphi(x)$ is any continuous function on $[0, 1)$ such that $\varphi(0) = a$, $\varphi(1-0) = f(a)$, $a$ is an arbitrary constant.

Direct verification shows that the function (17.4.1.3) satisfies equation (17.4.1.2) and is continuous on the semiaxis $[0, \infty)$.

**17.4.1-3. Nonlinear $m$th-order difference equations.**

$1°$. In the general case, a *nonlinear difference equation* has the form

$$F\big(x,\, y(x+h_0),\, y(x+h_1),\, \ldots,\, y(x+h_m)\big) = 0. \qquad (17.4.1.4)$$

This equation contains the unknown function with different values of its argument differing from one another by constants. The constants $h_0, h_1, \ldots, h_m$ are called differences or deviations of the argument. If all $h_k$ are integers, then (17.4.1.4) is called an equation with integer differences. If $h_k = k$ ($k = 0, 1, \ldots, m$) and the equation explicitly involves $y(x)$ and $y(x+m)$, then it is called a difference equation of order $m$.

$2°$. An $m$th-order difference equation resolved with respect to the leading term $y(x+m)$ has the form

$$y(x+m) = f\big(x,\, y(x),\, y(x+1),\, \ldots,\, y(x+m-1)\big). \qquad (17.4.1.5)$$

The Cauchy problem for this equation consists of finding its solution with a given initial distribution of the unknown function on the interval $0 \leq x \leq m$:

$$y = \varphi(x) \quad \text{at} \quad 0 \leq x \leq m. \qquad (17.4.1.6)$$

The values of $y(x)$ for $m \leq x \leq m+1$ are calculated by substituting the initial values (17.4.1.6) into the right-hand side of equation (17.4.1.5) for $0 \leq x \leq 1$. We have

$$y(x+m) = f\big(x,\, \varphi(x),\, \varphi(x+1),\, \ldots,\, \varphi(x+m-1)\big). \qquad (17.4.1.7)$$

Then, replacing $x$ by $x+1$ in equation (17.4.1.5), we obtain

$$y(x+m+1) = f\big(x+1,\, y(x+1),\, y(x+2),\, \ldots,\, y(x+m)\big). \qquad (17.4.1.8)$$

The values of $y(x)$ for $m+1 \leq x \leq m+2$ are determined by the right-hand side of (17.4.1.8) for $0 \leq x \leq 1$, the quantities $y(x+1), \ldots, y(x+m-1)$ are determined by the initial condition (17.4.1.6), and $y(x+m)$ is taken from (17.4.1.7).

In order to find $y(x)$ for $m+2 \leq x \leq m+3$, we replace $x$ by $x+1$ in equation (17.4.1.8) and consider this equation on the interval $0 \leq x \leq 1$. Using the initial conditions for $y(x+2)$, $\ldots, y(x+m-1)$ and the quantities $y(x+m),\, y(x+m+1)$ found on the previous steps, we find $y(x+m+2)$.

In a similar way, one can find $y(x)$ for $m+3 \leq x \leq m+4$, etc. This method for solving difference equations is called the *step method*.

## 17.4.2. Reciprocal (Cyclic) Functional Equations

**17.4.2-1. Reciprocal functional equations depending on $y(x)$ and $y(a-x)$.**

$1°$. Consider a *reciprocal equation* of the form

$$F\big(x,\, y(x),\, y(a-x)\big) = 0.$$

Replacing $x$ by $a-x$, we obtain a similar equation containing the unknown function with the same arguments:

$$F\big(a-x,\, y(a-x),\, y(x)\big) = 0.$$

Eliminating $y(a-x)$ from this equation and the original equation, we come to the usual algebraic (or transcendental) equation of the form $\Psi\big(x, y(x)\big) = 0$.

In other words, solutions of the original functional equation $y = y(x)$ are defined in a parametric manner by means of two algebraic (or transcendental) equations:

$$F(x, y, t) = 0, \qquad F(a-x, t, y) = 0, \qquad (17.4.2.1)$$

where $t$ is a parameter.

**Remark.** System (17.4.2.1), and therefore the original functional equation, may have several (even infinitely many) solutions or no solutions at all.

**Example.** Consider the nonlinear equation
$$y^2(x) = f(x)y(a-x). \qquad (17.4.2.2)$$

Replacing $x$ by $a - x$ in (17.4.2.2), we get
$$y^2(a-x) = f(a-x)y(x). \qquad (17.4.2.3)$$

Eliminating $y(a-x)$ from (17.4.2.2)–(17.4.2.3), we obtain two solutions of the original equation:
$$y(x) = [f^2(x)f(a-x)]^{1/3} \quad \text{and} \quad y(x) \equiv 0.$$

2°. Consider a reciprocal equation of the form
$$F(x, y(x), y(a/x)) = 0.$$

Replacing $x$ by $a/x$, we obtain a similar equation with the unknown function having the same arguments:
$$F(a/x, y(a/x), y(x)) = 0.$$

Eliminating $y(a/x)$ from this and the original equation, we come to the usual algebraic (or transcendental) equation of the form $\Psi(x, y(x)) = 0$.

In other words, solutions of the original functional equation $y = y(x)$ are defined in a parametric manner by means of a system of two algebraic (or transcendental) equations:
$$F(x, y, t) = 0, \qquad F(a/x, t, y) = 0,$$

where $t$ is a parameter.

### 17.4.2-2. Reciprocal (cyclic) functional equations of general form.

Reciprocal functional equations have the form
$$F\big(x, y(x), y(\varphi(x)), y(\varphi^{[2]}(x)), \ldots, y(\varphi^{[n-1]}(x))\big) = 0. \qquad (17.4.2.4)$$

Here we use the notation $\varphi^{[n]}(x) = \varphi(\varphi^{[n-1]}(x))$ and $\varphi(x)$ is a cyclic function satisfying the condition
$$\varphi^{[n]}(x) = x. \qquad (17.4.2.5)$$

The value $n$ is called the order of a cyclic (reciprocal) equation.

Successively replacing $n$ times the argument $x$ by $\varphi(x)$ in the functional equation (17.4.2.4), we obtain the following system (the original equation coincides with the first equation of this system):
$$F(x, y_0, y_1, \ldots, y_{n-1}) = 0,$$
$$F(\varphi(x), y_1, y_2, \ldots, y_0) = 0, \qquad (17.4.2.6)$$
$$\ldots\ldots\ldots\ldots\ldots\ldots\ldots\ldots\ldots,$$
$$F(\varphi^{[n-1]}(x), y_{n-1}, y_0, \ldots, y_{n-2}) = 0,$$

where we have set $y_0 = y(x)$, $y_1 = y(\varphi(x))$, ..., $y_{n-1} = y(\varphi^{[n-1]}(x))$; condition (17.4.2.5) implies that $y_n = y_0$.

Eliminating $y_1, y_2, \ldots, y_{n-1}$ from the system of nonlinear algebraic (or transcendental) equations (17.4.2.6), we obtain a solution of the functional equation (17.4.2.4) in implicit form $\Psi(x, y_0) = 0$, where $y_0 = y(x)$.

## 17.4.3. Nonlinear Functional Equations Reducible to Difference Equations

**17.4.3-1. Functional equations with arguments proportional to $x$.**

Consider the equation
$$F\big(x,\, y(a_1 x),\, y(a_2 x),\, \ldots,\, y(a_n x)\big) = 0. \tag{17.4.3.1}$$

The transformation
$$y(x) = w(z), \qquad z = \ln x$$
reduces (17.4.3.1) to the difference equation
$$F\big(e^z,\, w(z+h_1),\, w(z+h_2),\, \ldots,\, w(z+h_n)\big) = 0, \qquad h_k = \ln a_k.$$

**17.4.3-2. Functional equations with powers of $x$ as arguments.**

Consider the equation
$$F\big(x,\, y(x^{n_1}),\, y(x^{n_2}),\, \ldots,\, y(x^{n_m})\big) = 0. \tag{17.4.3.2}$$

The transformation
$$y(x) = w(z), \qquad z = \ln \ln x$$
reduces (17.4.3.2) to the difference equation
$$F\big(e^{e^z},\, w(z+h_1),\, w(z+h_2),\, \ldots,\, w(z+h_n)\big) = 0, \qquad h_k = \ln \ln n_k.$$

**17.4.3-3. Functional equations with exponential functions of $x$ as arguments.**

Consider the equation
$$F\big(x,\, y(e^{\lambda_1 x}),\, y(e^{\lambda_2 x}),\, \ldots,\, y(e^{\lambda_n x})\big) = 0.$$

The transformation
$$y(x) = w(\ln z), \qquad z = \ln x$$
reduces this equation to the difference equation
$$F\big(e^z,\, w(z+h_1),\, w(z+h_2),\, \ldots,\, w(z+h_n)\big) = 0, \qquad h_k = \ln \lambda_k.$$

**17.4.3-4. Equations containing iterations of the unknown function.**

Consider the equation
$$F\big(x,\, y(x),\, y^{[2]}(x),\, \ldots,\, y^{[n]}(x)\big) = 0,$$
where $y^{[2]}(x) = y\big(y(x)\big), \ldots, y^{[n]}(x) = y\big(y^{[n-1]}(x)\big)$.

We seek its solution in parametric form
$$x = w(t), \qquad y = w(t+1).$$

Then the original equation is reduced to the following $n$th-order difference equation:
$$F\big(w(t),\, w(t+1),\, w(t+2),\, \ldots,\, w(t+n)\big) = 0.$$

## 17.4.4. Power Series Solution of Nonlinear Functional Equations

In some situations, a solution of a functional equation can be found in the form of power series expansion.

Consider the nonlinear functional equation

$$y(x) - F\big(x, y(\varphi(x))\big) = 0. \tag{17.4.4.1}$$

Suppose that the following conditions hold:

1) there exist $a$ and $b$ such that $\varphi(a) = a$, $F(a, b) = b$;
2) the function $\varphi(x)$ is analytic in a neighborhood of $a$ and $|\varphi'(a)| < 1$;
3) the function $F$ is analytic in a neighborhood of the point $(a, b)$ and $|\frac{\partial F}{\partial b}(a, b)| < 1$.

Then the formal power series solution of equation (17.4.4.1),

$$y(x) = b + \sum_{k=1}^{\infty} c_k (x - a)^k, \tag{17.4.4.2}$$

has a positive radius of convergence.

The formal solution is obtained by substituting the expansions

$$\varphi(x) = a + \sum_{k=1}^{\infty} a_k (x - a)^k,$$

$$F(x, y) = b + \sum_{i,j=1}^{\infty} b_{ij} (x - a)^i (y - b)^j,$$

and (17.4.4.2) into equation (17.4.4.1). Gathering the terms with the same powers of the difference $\xi = (x - a)$ and then equating to zero the coefficients of different powers of $\xi$, we obtain a triangular system of algebraic equations for the coefficients $c_k$.

▶ See also Section T12.2, which gives exact solutions of some nonlinear difference and functional equations with one independent variable.

## 17.5. Functional Equations with Several Variables

### 17.5.1. Method of Differentiation in a Parameter

17.5.1-1. Classes of equations. Description of the method.

Consider linear functional equations of the form

$$w(x, t) = \theta(x, t, a) \, w\big(\varphi(x, t, a), \psi(x, t, a)\big), \tag{17.5.1.1}$$

where $x$ and $t$ are independent variables, $w = w(x, t)$ is the function to be found, $\theta = \theta(x, t, a)$, $\varphi = \varphi(x, t, a)$, $\psi = \psi(x, t, a)$ are given functions, and $a$ is a free parameter that can take any value (on some interval). Assume that for a particular $a = a_0$, we have

$$\theta(x, t, a_0) = 1, \quad \varphi(x, t, a_0) = x, \quad \psi(x, t, a_0) = t, \tag{17.5.1.2}$$

i.e., for $a = a_0$ the functional equation (17.5.1.1) turns into identity.

Let us expand (17.5.1.1) in powers of the small parameter $a$ in a neighborhood of $a_0$, taking into account (17.5.1.2), and then divide the resulting equation by $a - a_0$ and pass to the limit for $a \to a_0$. As a result, we obtain a first-order linear partial differential equation for the function $w$:

$$\varphi_a^\circ(x,t)\frac{\partial w}{\partial x} + \psi_a^\circ(x,t)\frac{\partial w}{\partial t} + \theta_a^\circ(x,t)w = 0, \qquad (17.5.1.3)$$

where we have used the following notation:

$$\varphi_a^\circ(x,t) = \frac{\partial \varphi}{\partial a}\bigg|_{a=a_0}, \quad \psi_a^\circ(x,t) = \frac{\partial \psi}{\partial a}\bigg|_{a=a_0}, \quad \theta_a^\circ(x,t) = \frac{\partial \theta}{\partial a}\bigg|_{a=a_0}.$$

In order to solve equation (17.5.1.3), one should consider the corresponding system of equations for characteristics (see Subsection 13.1.1):

$$\frac{dx}{\varphi_a^\circ(x,t)} = \frac{dt}{\psi_a^\circ(x,t)} = -\frac{dw}{\theta_a^\circ(x,t)w}. \qquad (17.5.1.4)$$

Let

$$u_1(x,t) = C_1, \quad u_2(x,t,w) = C_2 \qquad (17.5.1.5)$$

be independent integrals of the characteristic system (17.5.1.4). Then the general solution of equation (17.5.1.3) has the form

$$u_2(x,t,w) = F\big(u_1(x,t)\big), \qquad (17.5.1.6)$$

where $F(z)$ is an arbitrary function. The function $w$ should be expressed from (17.5.1.6) and substituted into the original equation (17.5.1.1) for verification [there is a possibility of redundant solutions; it is also possible that a solution of the partial differential equation (17.5.1.3) is not a solution of the functional equation (17.5.1.1); see Example 3 below].

Remark 1. Equation (17.5.1.3) can be obtained from (17.5.1.1) by differentiation in the parameter $a$, after which one should take $a = a_0$.

Remark 2. It is convenient to choose the second integral in (17.5.1.5) to be a linear function of $w$, i.e., $u_2(x,t,w) = \xi(x,t)w$, and rewrite (17.5.1.6) as a relation resolved with respect to $w$.

### 17.5.1-2. Examples of solutions of some specific functional equations.

**Example 1.** Self-similar solutions, which often occur in mathematical physics, may be defined as solutions invariant with respect to the scaling transformation, i.e., solutions satisfying the functional equation

$$w(x,t) = a^k w(a^m x, a^n t), \qquad (17.5.1.7)$$

where $k, m, n$ are given constants, and $a > 0$ is an arbitrary constant.

Equation (17.5.1.7) turns into identity for $a = 1$. Differentiating (17.5.1.7) in $a$ and taking $a = 1$, we come to the first-order partial differential equation

$$mx\frac{\partial w}{\partial x} + nt\frac{\partial w}{\partial t} + kw = 0. \qquad (17.5.1.8)$$

The first integrals of the corresponding characteristic system of ordinary differential equations

$$\frac{dx}{mx} = \frac{dt}{nt} = -\frac{dw}{kw}$$

can be written in the form

$$xt^{-m/n} = C_1, \quad t^{k/n}w = C_2 \quad (n \neq 0).$$

Therefore, the general solution of the partial differential equation (17.5.1.8) has the form
$$w(x,t) = t^{-k/n}F(z), \quad z = xt^{-m/n}, \tag{17.5.1.9}$$
where $F(z)$ is an arbitrary function. Direct verification shows that expression (17.5.1.9) is a solution of the functional equation (17.5.1.7).

**Example 2.** Consider the functional equation
$$w(x,t) = a^k w(a^m x, t + \beta \ln a), \tag{17.5.1.10}$$
where $k, m, \beta$ are given constants, $a > 0$ is an arbitrary constant.

Equation (17.5.1.10) turns into identity for $a = 1$. Differentiating (17.5.1.10) in $a$ and taking $a = 1$, we come to the first-order partial differential equation
$$mx\frac{\partial w}{\partial x} + \beta\frac{\partial w}{\partial t} + kw = 0. \tag{17.5.1.11}$$

The corresponding characteristic system of ordinary differential equations
$$\frac{dx}{mx} = \frac{dt}{\beta} = -\frac{dw}{kw}$$
admits the first integrals
$$x\exp(-mt/\beta) = C_1, \quad w\exp(kt/\beta) = C_2.$$
Therefore, the general solution of the partial differential equation (17.5.1.11) has the form
$$w(x,t) = \exp(-kt/\beta)F(z), \quad z = x\exp(-mt/\beta), \tag{17.5.1.12}$$
where $F(z)$ is an arbitrary function. Direct verification shows that (17.5.1.12) is a solution of the functional equation (17.5.1.10).

**Example 3.** Now consider the functional equation
$$w(x,t) = a^k w(x + (1-a)t, a^n t), \tag{17.5.1.13}$$
where $a > 0$ is arbitrary and $n$ is a constant.

Equation (17.5.1.13) turns into identity for $a = 1$. Differentiating (17.5.1.13) in $a$ and taking $a = 1$, we come to the first-order partial differential equation
$$-t\frac{\partial w}{\partial x} + nt\frac{\partial w}{\partial t} + kw = 0. \tag{17.5.1.14}$$

The corresponding characteristic system
$$-\frac{dx}{t} = \frac{dt}{nt} = -\frac{dw}{kw}$$
has the first integrals
$$t + nx = C_1, \quad wt^{k/n} = C_2.$$
Therefore, the general solution of the partial differential equation (17.5.1.14) has the form
$$w(x,t) = t^{-k/n}F(nx + t), \tag{17.5.1.15}$$
where $F(z)$ is an arbitrary function.

Substituting (17.5.1.15) into the original equation (17.5.1.13) and dividing the result by $t^{-k/n}$, we obtain
$$F(nx + t) = F(nx + \sigma t), \quad \sigma = (1-a)n + a^n. \tag{17.5.1.16}$$
Hence, for $F(z) \neq \text{const}$ we have $\sigma = 1$ or
$$(1-a)n + a^n = 1. \tag{17.5.1.17}$$
Since (17.5.1.16) must hold for all $a > 0$, it follows that (17.5.1.17), too, must hold for all $a > 0$. This can take place only if
$$n = 1.$$
In this case, the solution of equation (17.5.1.13) is given by the following formula [see (17.5.1.15) for $n = 1$]:
$$w(x,t) = t^{-k}F(x + t), \tag{17.5.1.18}$$
where $F(z)$ is an arbitrary function.

If $n \neq 1$, then equation (17.5.1.13) admits only a degenerate solution $w(x,t) = Ct^{-k/n}$, where $C$ is an arbitrary constant [the degenerate solution corresponds to $F = \text{const}$ in (17.5.1.16)].

## 17.5.2. Method of Differentiation in Independent Variables

### 17.5.2-1. Preliminary remarks.

1°. In some situations, differentiation in independent variables can be used to eliminate some arguments of the functional equation under consideration and reduce it to an ordinary differential equation (see Example 1 below). The solution obtained in this way should be then inserted into the original equation in order to get rid of redundant integration constants, which may appear due to the differentiation.

2°. In some situations, differentiation in independent variables should be combined with the multiplication (division) of the equation and the results of its differentiation by suitable functions. Sometimes it is useful to take logarithm of the equation or the results of its transformation (see Example 2 below).

3°. In some situations, differentiation of a functional equation in independent variables allows us to eliminate some arguments and reduce the equation to a simpler functional equation whose solution is known (see Subsection 17.5.5).

### 17.5.2-2. Examples of solutions of some specific functional equations.

**Example 1.** Consider the *Pexider equation*
$$f(x) + g(y) = h(x+y), \tag{17.5.2.1}$$
where $f(x)$, $g(y)$, $h(z)$ are the functions to be found.

Differentiating the functional equation (17.5.2.1) in $x$ and $y$, we come to the ordinary differential equation $h''_{zz}(z) = 0$, where $z = x + y$. Its solution is the linear function
$$h(z) = az + b. \tag{17.5.2.2}$$
Substituting this expression into (17.5.2.1), we obtain
$$f(x) + g(y) = ax + ay + b.$$
Separating the variables, we find the functions $f$ and $g$:
$$f(x) = ax + b + c,$$
$$g(y) = ay - c. \tag{17.5.2.3}$$
Thus, the solution of the Pexider equation (17.5.2.1) is given by the formulas (17.5.2.2), (17.5.2.3), where $a$, $b$, $c$ are arbitrary constants.

**Example 2.** Consider the nonlinear functional equation
$$f(x+y) = f(x) + f(y) + af(x)f(y), \quad a \neq 0, \tag{17.5.2.4}$$
which occurs in the theory of probability with $a = -1$.

Differentiating both sides of this equation in $x$ and $y$, we get
$$f''_{zz}(z) = af'_x(x)f'_y(y), \tag{17.5.2.5}$$
where $z = x + y$. Taking the logarithm of both sides of equation (17.5.2.5) and differentiating the resulting relation in $x$ and $y$, we come to the ordinary differential equation
$$[\ln f''_{zz}(z)]''_{zz} = 0. \tag{17.5.2.6}$$
Integrating (17.5.2.6) in $z$ twice, we get
$$f''_{zz}(z) = C_1 \exp(C_2 z), \tag{17.5.2.7}$$
where $C_1$ and $C_2$ are arbitrary constants. Substituting (17.5.2.7) into (17.5.2.5), we obtain the equation
$$C_1 \exp[C_2(x+y)] = af'_x(x)f'_y(y),$$
which admits separation of variables. Integration yields
$$f(x) = A\exp(C_2 x) + B, \quad A = \pm\frac{1}{C_2}\sqrt{\frac{C_1}{a}}. \tag{17.5.2.8}$$
Substituting (17.5.2.8) into the original equation (17.5.2.4), we find the values of the constants: $A = -B = 1/a$ and $C_2 = \beta$ is an arbitrary constant. As a result, we obtain the desired solution
$$f(x) = \frac{1}{a}(e^{\beta x} - 1).$$

### 17.5.3. Method of Substituting Particular Values of Independent Arguments

In order to solve functional equations with several independent variables, one often uses the method of substituting particular values of independent arguments. This procedure is aimed at reducing the problem for a functional equation with several independent variables to a problem for a system of algebraic (transcendental) equations with a single independent variable.

Basic ideas of this method can be demonstrated by the example of a fairly general functional equation of the form

$$\Phi\big(f(x), f(x+y), f(x-y), x, y\big) = 0. \qquad (17.5.3.1)$$

$1°$. Taking $y = 0$ in this equation, we get

$$\Phi\big(f(x), f(x), f(x), x, 0\big) = 0. \qquad (17.5.3.2)$$

If the left-hand side of this relation does not vanish identically for all $f(x)$, then this equation can be resolved with respect to $f(x)$. Then the function $f(x)$ obtained in this way should be inserted into the original equation (17.5.3.1) and one should find conditions under which this function is its solution.

Now, assume that the left-hand side of (17.5.3.2) is identically equal to zero for all $f(x)$.

$2°$. In equation (17.5.3.1), we consecutively take

$$x = 0, \quad y = t; \qquad x = t, \quad y = 2t; \qquad x = t, \quad y = -2t.$$

We obtain a system of algebraic (transcendental) equations

$$\begin{aligned}
\Phi\big(a, f(t), f(-t), 0, t\big) &= 0, \\
\Phi\big(f(t), f(3t), f(-t), t, 2t\big) &= 0, \\
\Phi\big(f(t), f(-t), f(3t), t, -2t\big) &= 0
\end{aligned} \qquad (17.5.3.3)$$

for the unknown quantities $f(t)$, $f(-t)$, $f(3t)$, where $a = f(0)$. Resolving system (17.5.3.3) with respect to $f(t)$ [or with respect to $f(-t)$ or $f(3x)$], we obtain an admissible solution that should be inserted into the original equation for verification.

$3°$. In some cases the following trick can be used. In equation (17.5.3.1), one consecutively takes

$$x = 0, \quad y = t; \qquad x = t + a, \quad y = a; \qquad x = a, \quad y = t + a, \qquad (17.5.3.4)$$

where $a$ is a free parameter. We obtain the system of equations

$$\begin{aligned}
\Phi\big(f(0), f(t), f(-t), 0, t\big) &= 0, \\
\Phi\big(f(t+a), f(t+2a), f(t), t+a, a\big) &= 0, \\
\Phi\big(f(a), f(t+2a), f(-t), a, t+a\big) &= 0.
\end{aligned} \qquad (17.5.3.5)$$

Letting $f(0) = C_1$ and $f(a) = C_2$, where $C_1$ and $C_2$ are arbitrary constants, we eliminate $f(t)$ and $f(t+2a)$ from system (17.5.3.5) (it is assumed that this is possible). As a result, we come to the reciprocal equation

$$\Psi\big(f(t+a), f(-t), t, a, C_1, C_2\big) = 0. \qquad (17.5.3.6)$$

In order to find a solution of equation (17.5.3.6), we replace $t$ with $-t - a$. We get

$$\Psi\big(f(-t), f(t+a), -t-a, a, C_1, C_2\big) = 0. \qquad (17.5.3.7)$$

Further, eliminating $f(t+a)$ from equations (17.5.3.6)–(17.5.3.7), we come to the algebraic (transcendental) equation for $f(-t)$.

Remark. For the sake of analysis, it is sometimes convenient to choose suitable values of the parameter $a$ in order to simplify system (17.5.3.5).

**Example.** Consider the equation

$$f(x+y) + f(x-y) = 2f(x)\cos y. \qquad (17.5.3.8)$$

This equation holds as identity for $y = 0$ and any $f(x)$.

Substituting (17.5.3.4) into (17.5.3.8), we get

$$\begin{aligned} f(t) + f(-t) &= 2C_1 \cos t, \\ f(t+2a) + f(t) &= 2f(t+a)\cos a, \\ f(t+2a) + f(-t) &= 2C_2 \cos(t+a), \end{aligned} \qquad (17.5.3.9)$$

where $C_1 = f(0)$, $C_2 = f(a)$.

System (17.5.3.9) becomes much simpler for $a = \pi/2$. In this case, $\cos a = 0$ and the function $f(t+a)$ is "dropped" from the equations; summing up the first two equations term by term and subtracting the third equation from the resulting relation, we immediately find a solution of the functional equation (17.5.3.8) in the form

$$f(t) = C_1 \cos t + C_2 \sin t. \qquad (17.5.3.10)$$

Verification shows that the function (17.5.3.10) is indeed a solution of the functional equation (17.5.3.8).

$4°$. Now consider a functional equation more general than (17.5.3.1),

$$\Phi\big(f(x), f(y), f(x+y), f(x-y), x, y\big) = 0. \qquad (17.5.3.11)$$

Letting $y = 0$, we get

$$\Phi\big(f(x), a, f(x), f(x), x, 0\big) = 0, \qquad (17.5.3.12)$$

where $a = f(0)$. If the left-hand side of this relation does not vanish identically for all $f(x)$, then it can be resolved with respect to $f(x)$. Then, inserting an admissible solution $f(x)$ into the original equation (17.5.3.11), we find possible values of the parameter $a$ (there are cases in which the equation has no solutions).

$5°$. If the left-hand side of (17.5.3.11) identically vanishes for all $f(x)$ and $a$, the following approach can be used. In (17.5.3.11), we consecutively take

$$x = 0, \quad y = t; \qquad x = t, \quad y = 2t; \qquad x = 2t, \quad y = t; \qquad x = t, \quad y = t.$$

We get

$$\begin{aligned} \Phi\big(a, f(t), f(t), f(-t), 0, t\big) &= 0, \\ \Phi\big(f(t), f(2t), f(3t), f(-t), t, 2t\big) &= 0, \\ \Phi\big(f(2t), f(t), f(3t), f(t), 2t, t\big) &= 0, \\ \Phi\big(f(t), f(t), f(2t), a, t, t\big) &= 0, \end{aligned} \qquad (17.5.3.13)$$

where $a = f(0)$. Eliminating $f(-t)$, $f(2t)$, and $f(3t)$ from system (17.5.3.13), we come to an equation for $f(t)$. The solution obtained in this manner should be inserted into the original equation (17.5.3.11).

## 17.5.4. Method of Argument Elimination by Test Functions

**17.5.4-1. Classes of equations. Description of the method.**

Consider linear functional equations of the form
$$w(x,t) = \theta(x,t,a)\, w\big(\varphi(x,t,a),\, \psi(x,t,a)\big), \qquad (17.5.4.1)$$
where $x$ and $t$ are independent variables, $w = w(x,t)$ is the function to be found, $\theta = \theta(x,t,a)$, $\varphi = \varphi(x,t,a)$, $\psi = \psi(x,t,a)$ are given functions, and $a$ is a free parameter, which can take any value (on some interval).

Instead of equation (17.5.4.1), consider an auxiliary more general functional equation
$$w(x,t) = \theta(x,t,\xi)\, w\big(\varphi(x,t,\xi),\, \psi(x,t,\xi)\big), \qquad (17.5.4.2)$$
where $\xi = \xi(x,t)$ is an arbitrary function.

*Basic idea*: If an exact solution of equation (17.5.4.2) can somehow be obtained, this function will also be a solution of the original equation (17.5.4.1) [since (17.5.4.1) is a special case of equation (17.5.4.2) with $\xi = a$].

In view that the function $\xi = \xi(x,t)$ can be arbitrary, let us first take a test function so that it satisfies the condition
$$\psi(x,t,\xi) = b, \qquad (17.5.4.3)$$
where $b$ is a constant (usually, it is convenient to take $b = 1$ or $b = 0$). Resolving (17.5.4.3) with respect to $\xi$ and substituting the test function $\xi = \xi(x,t)$ thus obtained into (17.5.4.2), we have
$$w(x,t) = \theta(x,t,\xi(x,t))\, \Phi\big(\varphi(x,t,\xi(x,t))\big), \qquad (17.5.4.4)$$
where $\Phi(\varphi) \equiv w(\varphi, b)$.

Expression (17.5.4.4) is crucial for the construction of an exact solution of the original functional equation: this expression should be substituted into (17.5.4.2) and one should find out for which functions $\Phi(\varphi)$ it is indeed a solution of the equation for arbitrary $\xi = \xi(x,t)$ (in this connection, some constraints on the structure of the determining functions $\theta$, $\varphi$, $\psi$ may appear).

**Remark 1.** Expression (17.5.4.4) may be substituted directly into the original equation (17.5.4.1).

**Remark 2.** Condition (17.5.4.3) corresponds to the elimination of the second argument (it is replaced by a constant) in the right-hand side of equation (17.5.4.2).

**Remark 3.** Instead of (17.5.4.3), a similar condition $\varphi(x,t,\xi) = b$ can be used for choosing the test function $\xi = \xi(x,t)$.

**17.5.4-2. Examples of solutions of specific functional equations.**

**Example 1.** Consider the functional equation
$$w(x,t) = a^k w(a^m x, a^n t), \qquad (17.5.4.5)$$
($k$, $m$, $n$ are given constants, $a > 0$ is an arbitrary constant), which is a special case of equation (17.5.4.1) for $\theta(x,t,a) = a^k$, $\varphi(x,t,a) = a^m x$, $\psi(x,t,a) = a^n t$.

Following the scheme described in Paragraph 17.5.4-1, let us use the auxiliary equation
$$w(x,t) = \xi^k w(\xi^m x, \xi^n t) \qquad (17.5.4.6)$$
and the test function $\xi$ defined, according to (17.5.4.3), from the condition
$$\xi^n t = 1 \qquad (b = 1). \qquad (17.5.4.7)$$
Hence, we find that $\xi = t^{-1/n}$. Substituting this expression into (17.5.4.6), we get
$$w(x,t) = t^{-k/n} \Phi(t^{-m/n} x), \qquad (17.5.4.8)$$
where $\Phi(\varphi) \equiv w(\varphi, 1)$.

It is easy to show by direct verification that (17.5.4.8) is a solution of the functional equation (17.5.4.5) for an arbitrary function $\Phi$ and coincides (to within notation) with solution (17.5.1.9) obtained by the method of differentiation in a parameter.

**Remark.** Instead of 1 on the right-hand side of (17.5.4.7) we can take any constant $b \neq 0$ and obtain the same result (to within notation of the arbitrary function $\Phi$).

**Example 2.** Consider the functional equation

$$w(x,t) = a^k w(a^m x, t + \beta \ln a), \tag{17.5.4.9}$$

($k$, $m$, $\beta$ are given constants, $a > 0$ is an arbitrary constant), which is a special case of equation (17.5.4.1) for $\theta(x,t,a) = a^k$, $\varphi(x,t,a) = a^m x$, $\psi(x,t,a) = t + \beta \ln a$.

Following the above scheme, consider a more general auxiliary equation

$$w(x,t) = \xi^k w(\xi^m x, t + \beta \ln \xi). \tag{17.5.4.10}$$

The test function $\xi$ is found from the condition

$$t + \beta \ln \xi = 0 \qquad (b = 0).$$

We have $\xi = \exp(-t/\beta)$. Substituting this expression into (17.5.4.10), we get

$$w(x,t) = e^{-kt/\beta} \Phi(xe^{-mt/\beta}), \tag{17.5.4.11}$$

where $\Phi(\varphi) \equiv w(\varphi, 0)$. Direct verification shows that (17.5.4.11) is a solution of the functional equation (17.5.4.9) for an arbitrary function $\Phi$ and coincides with solution (17.5.1.12) obtained by the method of differentiation in a parameter.

**Example 3.** Now consider the functional equation

$$w(x,t) = a^k w\big(x + (1-a)t,\ a^n t\big), \tag{17.5.4.12}$$

($a > 0$ is arbitrary, $n$ is a constant), which is a special case of equation (17.5.4.1) for $\theta(x,t,a) = a^k$, $\varphi(x,t,a) = x + (1-a)t$, $\psi(x,t,a) = a^n t$.

Following the scheme described above, consider the auxiliary equation

$$w(x,t) = \xi^k w\big(x + (1-\xi)t,\ \xi^n t\big) \tag{17.5.4.13}$$

and define the test function $\xi$ from the condition (17.5.4.7), according to (17.5.4.3). We have $\xi = t^{-1/n}$. Substituting this expression into (17.5.4.13), we get

$$w(x,t) = t^{-k/n} \Phi(z), \qquad z = x + t - t^{(n-1)/n}, \tag{17.5.4.14}$$

where $\Phi(\varphi) \equiv w(\varphi, 1)$.

Substituting (17.5.4.14) into the original equation (17.5.4.12) and dividing the result by $t^{-k/n}$, we find that

$$\Phi\big(x + t - t^{(n-1)/n}\big) = \Phi\big(x + (1 - a + a^n)t - a^{n-1} t^{(n-1)/n}\big). \tag{17.5.4.15}$$

Since this relation must hold for all $a > 0$, there are two possibilities:

$$\begin{aligned} &1) \quad n \text{ is arbitrary}, \quad \Phi = C = \text{const}; \\ &2) \quad n = 1, \quad \Phi \text{ is arbitrary}. \end{aligned} \tag{17.5.4.16}$$

In the second case, which corresponds to $n = 1$ in the functional equation (17.5.4.12), its solution can be written in the form

$$w(x,t) = t^{-k} F(x + t), \tag{17.5.4.17}$$

where $F(z)$ is an arbitrary function, $F(z) = \Phi(z-1)$. We see that expression (17.5.4.17) coincides with solution (17.5.1.18) obtained by the method of differentiation in a parameter.

**Remark 1.** The results of solving specific functional equations obtained in Subsection 17.5.4 by the elimination of an argument coincide with those obtained for the same equations in Subsection 17.5.1 by the method of differentiation in a parameter. However, it should be observed that the intermediate results, when solving equation (17.5.4.12) by these methods, may not coincide [cf. (17.5.4.15) and (17.5.1.16)].

**Remark 2.** The method of elimination of an argument is much simpler than that of differentiation in a parameter, since the former only requires to solve algebraic (transcendental) equations of the form (17.5.4.3) with respect to $\xi$ and does not require solutions of the corresponding first-order partial differential equations (see Subsection 17.5.1).

## 17.5.5. Bilinear Functional Equations and Nonlinear Functional Equations Reducible to Bilinear Equations

**17.5.5-1. Bilinear functional equations.**

1°. A binomial bilinear functional equation has the form

$$f_1(x)g_1(y) + f_2(x)g_2(y) = 0, \qquad (17.5.5.1)$$

where $f_n = f_n(x)$ and $g_n = g_n(y)$ ($n = 1, 2$) are unknown functions of different arguments. In this section, it is assumed that $f_n \not\equiv 0$, $g_n \not\equiv 0$.

Separating the variables in (17.5.5.1), we find the solution:

$$f_1 = Af_2, \quad g_2 = -Ag_1, \qquad (17.5.5.2)$$

where $A$ is an arbitrary constant. The functions on the right-hand sides in (17.5.5.2) are assumed arbitrary.

2°. The trinomial bilinear functional equation

$$f_1(x)g_1(y) + f_2(x)g_2(y) + f_3(x)g_3(y) = 0, \qquad (17.5.5.3)$$

where $f_n = f_n(x)$ and $g_n = g_n(y)$ ($n = 1, 2, 3$) are unknown functions, has two solutions:

$$\begin{aligned} f_1 = A_1 f_3, \quad f_2 = A_2 f_3, \quad g_3 = -A_1 g_1 - A_2 g_2; \\ g_1 = A_1 g_3, \quad g_2 = A_2 g_3, \quad f_3 = -A_1 f_1 - A_2 f_2, \end{aligned} \qquad (17.5.5.4)$$

where $A_1$, and $A_2$ are arbitrary constants. The functions on the right-hand sides of the equations in (17.5.5.4) are assumed arbitrary.

3°. The quadrinomial functional equation

$$f_1(x)g_1(y) + f_2(x)g_2(y) + f_3(x)g_3(y) + f_4(x)g_4(y) = 0, \qquad (17.5.5.5)$$

where all $f_i$ are functions of the same argument and all $g_i$ are functions of another argument, has a solution

$$\begin{aligned} f_1 = A_1 f_3 + A_2 f_4, \quad f_2 = A_3 f_3 + A_4 f_4, \\ g_3 = -A_1 g_1 - A_3 g_2, \quad g_4 = -A_2 g_1 - A_4 g_2 \end{aligned} \qquad (17.5.5.6)$$

depending on four arbitrary constants $A_1, \ldots, A_4$. The functions on the right-hand sides of the solutions in (17.5.5.6) are assumed arbitrary.

Equation (17.5.5.5) has two other solutions:

$$\begin{aligned} f_1 = A_1 f_4, \quad f_2 = A_2 f_4, \quad f_3 = A_3 f_4, \quad g_4 = -A_1 g_1 - A_2 g_2 - A_3 g_3; \\ g_1 = A_1 g_4, \quad g_2 = A_2 g_4, \quad g_3 = A_3 g_4, \quad f_4 = -A_1 f_1 - A_2 f_2 - A_3 f_3 \end{aligned} \qquad (17.5.5.7)$$

involving three arbitrary constants.

4°. Consider a bilinear functional equation of the general form

$$f_1(x)g_1(y) + f_2(x)g_2(y) + \cdots + f_k(x)g_k(y) = 0, \qquad (17.5.5.8)$$

where $f_n = f_n(x)$ and $g_n = g_n(y)$ are unknown functions ($n = 1, \ldots, k$).

It can be shown that the bilinear functional equation (17.5.5.8) has $k - 1$ different solutions:

$$f_i(x) = A_{i,1}f_{m+1}(x) + A_{i,2}f_{m+2}(x) + \cdots + A_{i,k-m}f_k(x), \quad i = 1, \ldots, m;$$
$$g_{m+j}(y) = -A_{1,j}g_1(y) - A_{2,j}g_2(y) - \cdots - A_{m,j}g_m(y), \quad j = 1, \ldots, k-m;$$
$$m = 1, 2, \ldots, k-1,$$
(17.5.5.9)

where the $A_{i,j}$ are arbitrary constants. The functions $f_{m+1}(x), \ldots, f_k(x), g_1(y), \ldots, g_m(y)$ on the right-hand sides of solutions (17.5.5.9) can be chosen arbitrarily. It is obvious that for a fixed $m$, solution (17.5.5.9) contains $m(k - m)$ arbitrary constants.

For a fixed $m$, solution (17.5.5.9) contains $m(k - m)$ arbitrary constants $A_{i,j}$. Given $k$, the solutions having the maximum number of arbitrary constants are defined by

| Solution number | Number of arbitrary constants | Conditions on $k$ |
|---|---|---|
| $m = \frac{1}{2}k$ | $\frac{1}{4}k^2$ | if $k$ is even, |
| $m = \frac{1}{2}(k \pm 1)$ | $\frac{1}{4}(k^2 - 1)$ | if $k$ is odd. |

**Remark 1.** Formulas (17.5.5.9) imply that equation (17.5.5.8) may hold only if the functions $f_n$ (and $g_n$) are linearly dependent.

**Remark 2.** The bilinear functional equation (17.5.5.8) and its solutions (17.5.5.9) play an important role in the methods of generalized and functional separation of variables for nonlinear PDEs (see Section 15.5).

### 17.5.5-2. Functional-differential equations reducible to a bilinear equation.

Consider a nonlinear functional-differential equation of the form

$$f_1(x)g_1(y) + f_2(x)g_2(y) + \cdots + f_k(x)g_k(y) = 0, \tag{17.5.5.10}$$

where $f_i(x)$ and $g_i(x)$ are given function of the form

$$\begin{aligned} f_i(x) &\equiv F_i\big(x, \varphi_1, \varphi_1', \varphi_1'', \ldots, \varphi_n, \varphi_n', \varphi_n''\big), & \varphi_p &= \varphi_p(x); \\ g_i(y) &\equiv G_i\big(y, \psi_1, \psi_1', \psi_1'', \ldots, \psi_m, \psi_m', \psi_m''\big), & \psi_q &= \psi_q(y). \end{aligned} \tag{17.5.5.11}$$

The problem is to find the functions $\varphi_p = \varphi_p(x)$ and $\psi_q = \psi_q(y)$ depending on different variables. Here, for simplicity, we consider an equation that contains only second-order derivatives; in the general case, the right-hand sides of (17.5.5.11) may contain higher-order derivatives of $\varphi_p = \varphi_p(x)$ and $\psi_q = \psi_q(y)$.

The functional-differential equation (17.5.5.10)–(17.5.5.11) is solved by the method of splitting. On the first stage, we treat (17.5.5.10) as a purely functional equation that depends on two variables $x$ and $y$, where $f_i = f_i(x)$ and $g_i = g_i(y)$ are unknown quantities. The solutions of this equation are described by (17.5.5.9). On the second stage, we successively substitute the functions $f_i(x)$ and $g_i(y)$ from (17.5.5.11) into all solutions (17.5.5.9) to obtain systems of ordinary differential equations for the unknown functions $\varphi_p(x)$ and $\psi_q(y)$. Solving these systems, we get solutions of the functional-differential equation (17.5.5.10)–(17.5.5.11).

**Remark.** The method of splitting will be used in Paragraph 17.5.5-3 for the construction of solutions of some nonlinear functional equations.

**17.5.5-3. Nonlinear equations containing the complex argument $z = \varphi(x) + \psi(t)$.**

Here, we discuss some nonlinear functional equations with two variables. Such equations often arise when the method of functional separation of variables is used for finding solutions of nonlinear equations of mathematical physics.

$1°$. Consider a functional equation of the form

$$f(t) + g(x) + h(x)Q(z) + R(z) = 0, \qquad \text{where} \quad z = \varphi(x) + \psi(t). \qquad (17.5.5.12)$$

Here, one of the two functions $f(t)$ and $\psi(t)$ is prescribed and the other is assumed unknown; also one of the functions $g(x)$ and $\varphi(x)$ is prescribed and the other is unknown, and the functions $h(x)$, $Q(z)$, and $R(z)$ are assumed unknown.*

Differentiating equation (17.5.5.12) with respect to $x$, we obtain the two-argument equation

$$g'_x + h'_x Q + h\varphi'_x Q'_z + \varphi'_x R'_z = 0. \qquad (17.5.5.13)$$

Such equations were discussed in Paragraph 17.5.5-2; their solutions are found with the help of (17.5.5.6) and (17.5.5.7). Hence, we obtain the following system of ordinary differential equations [see formulas (17.5.5.6)]:

$$\begin{aligned} g'_x &= A_1 h \varphi'_x + A_2 \varphi'_x, \\ h'_x &= A_3 h \varphi'_x + A_4 \varphi'_x, \\ Q'_z &= -A_1 - A_3 Q, \\ R'_z &= -A_2 - A_4 Q, \end{aligned} \qquad (17.5.5.14)$$

where $A_1, \ldots, A_4$ are arbitrary constants. Integrating the system of ODEs (17.5.5.14) and substituting the resulting solutions into the original functional equation, one obtains the following results.

*Case 1*. If $A_3 = 0$ in (17.5.5.14), then the corresponding solution of the functional equation is given by

$$\begin{aligned} f &= -\tfrac{1}{2} A_1 A_4 \psi^2 + (A_1 B_1 + A_2 + A_4 B_3)\psi - B_2 - B_1 B_3 - B_4, \\ g &= \tfrac{1}{2} A_1 A_4 \varphi^2 + (A_1 B_1 + A_2)\varphi + B_2, \\ h &= A_4 \varphi + B_1, \\ Q &= -A_1 z + B_3, \\ R &= \tfrac{1}{2} A_1 A_4 z^2 - (A_2 + A_4 B_3) z + B_4, \end{aligned} \qquad (17.5.5.15)$$

where $A_k$ and $B_k$ are arbitrary constants and $\varphi = \varphi(x)$ and $\psi = \psi(t)$ are arbitrary functions.

*Case 2*. If $A_3 \ne 0$ in (17.5.5.14), then the corresponding solution of the functional

---

* In similar equations with a composite argument, it is assumed that $\varphi(x) \ne \text{const}$ and $\psi(y) \ne \text{const}$.

equation is

$$f = -B_1 B_3 e^{-A_3 \psi} + \left(A_2 - \frac{A_1 A_4}{A_3}\right)\psi - B_2 - B_4 - \frac{A_1 A_4}{A_3^2},$$

$$g = \frac{A_1 B_1}{A_3} e^{A_3 \varphi} + \left(A_2 - \frac{A_1 A_4}{A_3}\right)\varphi + B_2,$$

$$h = B_1 e^{A_3 \varphi} - \frac{A_4}{A_3}, \qquad (17.5.5.16)$$

$$Q = B_3 e^{-A_3 z} - \frac{A_1}{A_3},$$

$$R = \frac{A_4 B_3}{A_3} e^{-A_3 z} + \left(\frac{A_1 A_4}{A_3} - A_2\right)z + B_4,$$

where $A_k$ and $B_k$ are arbitrary constants and $\varphi = \varphi(x)$ and $\psi = \psi(t)$ are arbitrary functions.

*Case 3.* In addition, the functional equation has two degenerate solutions [formulas (17.5.5.7) are used]:

$$f = A_1 \psi + B_1, \quad g = A_1 \varphi + B_2, \quad h = A_2, \quad R = -A_1 z - A_2 Q - B_1 - B_2, \quad (17.5.5.17)$$

where $\varphi = \varphi(x)$, $\psi = \psi(t)$, and $Q = Q(z)$ are arbitrary functions, $A_1$, $A_2$, $B_1$, and $B_2$ are arbitrary constants, and

$$f = A_1 \psi + B_1, \quad g = A_1 \varphi + A_2 h + B_2, \quad Q = -A_2, \quad R = -A_1 z - B_1 - B_2, \quad (17.5.5.18)$$

where $\varphi = \varphi(x)$, $\psi = \psi(t)$, and $h = h(x)$ are arbitrary functions, $A_1$, $A_2$, $B_1$, and $B_2$ are arbitrary constants.

2°. Consider a functional equation of the form

$$f(t) + g(x)Q(z) + h(x)R(z) = 0, \qquad \text{where} \quad z = \varphi(x) + \psi(t). \qquad (17.5.5.19)$$

Differentiating (17.5.5.19) in $x$, we get the two-argument functional-differential equation

$$g'_x Q + g\varphi'_x Q'_z + h'_x R + h\varphi'_x R'_z = 0, \qquad (17.5.5.20)$$

which coincides with equation (17.5.5.5), up to notation.

*Nondegenerate case.* Equation (17.5.5.20) can be solved with the help of formulas (17.5.5.6)–(17.5.5.7). In this way, we arrive at the system of ordinary differential equations

$$\begin{aligned} g'_x &= (A_1 g + A_2 h)\varphi'_x, \\ h'_x &= (A_3 g + A_4 h)\varphi'_x, \\ Q'_z &= -A_1 Q - A_3 R, \\ R'_z &= -A_2 Q - A_4 R, \end{aligned} \qquad (17.5.5.21)$$

where $A_1, \ldots, A_4$ are arbitrary constants.

The solution of equation (17.5.5.21) is given by

$$\begin{aligned} g(x) &= A_2 B_1 e^{k_1 \varphi} + A_2 B_2 e^{k_2 \varphi}, \\ h(x) &= (k_1 - A_1) B_1 e^{k_1 \varphi} + (k_2 - A_1) B_2 e^{k_2 \varphi}, \\ Q(z) &= A_3 B_3 e^{-k_1 z} + A_3 B_4 e^{-k_2 z}, \\ R(z) &= (k_1 - A_1) B_3 e^{-k_1 z} + (k_2 - A_1) B_4 e^{-k_2 z}, \end{aligned} \qquad (17.5.5.22)$$

where $B_1, \ldots, B_4$ are arbitrary constants and $k_1$ and $k_2$ are the roots of the quadratic equation
$$(k - A_1)(k - A_4) - A_2 A_3 = 0. \tag{17.5.5.23}$$
In the degenerate case $k_1 = k_2$, the terms $e^{k_2 \varphi}$ and $e^{-k_2 z}$ in (17.5.5.22) should be replaced with $\varphi e^{k_1 \varphi}$ and $z e^{-k_1 z}$, respectively. In the case of purely imaginary or complex roots, one should separate the real (or imaginary) part of the roots in solution (17.5.5.22).

On substituting (17.5.5.22) into the original functional equation, one obtains conditions for the free coefficients and identifies the function $f(t)$, namely,

$$\begin{aligned} B_2 = B_4 = 0 &\implies f(t) = [A_2 A_3 + (k_1 - A_1)^2] B_1 B_3 e^{-k_1 \psi}, \\ B_1 = B_3 = 0 &\implies f(t) = [A_2 A_3 + (k_2 - A_1)^2] B_2 B_4 e^{-k_2 \psi}, \\ A_1 = 0 &\implies f(t) = (A_2 A_3 + k_1^2) B_1 B_3 e^{-k_1 \psi} + (A_2 A_3 + k_2^2) B_2 B_4 e^{-k_2 \psi}. \end{aligned} \tag{17.5.5.24}$$

Solution (17.5.5.22), (17.5.5.24) involves arbitrary functions $\varphi = \varphi(x)$ and $\psi = \psi(t)$.

*Degenerate case.* In addition, the functional equation has two degenerate solutions [formulas (17.5.5.7) are used],
$$f = B_1 B_2 e^{A_1 \psi}, \quad g = A_2 B_1 e^{-A_1 \varphi}, \quad h = B_1 e^{-A_1 \varphi}, \quad R = -B_2 e^{A_1 z} - A_2 Q,$$
where $\varphi = \varphi(x)$, $\psi = \psi(t)$, and $Q = Q(z)$ are arbitrary functions; $A_1$, $A_2$, $B_1$, and $B_2$ are arbitrary constants; and
$$f = B_1 B_2 e^{A_1 \psi}, \quad h = -B_1 e^{-A_1 \varphi} - A_2 g, \quad Q = A_2 B_2 e^{A_1 z}, \quad R = B_2 e^{A_1 z},$$
where $\varphi = \varphi(x)$, $\psi = \psi(t)$, and $g = g(x)$ are arbitrary functions; and $A_1$, $A_2$, $B_1$, and $B_2$ are arbitrary constants.

3°. Consider a more general functional equation of the form
$$f(t) + g_1(x) Q_1(z) + \cdots + g_n(x) Q_n(z) = 0, \quad \text{where} \quad z = \varphi(x) + \psi(t). \tag{17.5.5.25}$$

By differentiation in $x$, this equation can be reduced to a functional differential equation, which may be regarded as a bilinear functional equation of the form (17.5.5.8). Using formulas (17.5.5.9) for the construction of its solution, one can first obtain a system of ODEs and then find solutions of the original equation (17.5.5.25).

4°. Consider a functional equation of the form
$$f_1(t) g_1(x) + \cdots + f_m(t) g_m(x) + h_1(x) Q_1(z) + \cdots + h_n(x) Q_n(z) = 0, \quad z = \varphi(x) + \psi(t). \tag{17.5.5.26}$$

Assume that $g_m(x) \not\equiv 0$. Dividing equation (17.5.5.26) by $g_m(x)$ and differentiating the result in $x$, we come to an equation of the form
$$f_1(t) \bar{g}_1(x) + \cdots + f_{m-1}(t) \bar{g}_{m-1}(x) + \sum_{i=1}^{2n} s_i(x) R_i(z) = 0$$

with a smaller number of functions $f_i(t)$. Proceeding in this way, we can eliminate all functions $f_i(t)$ and obtain a functional-differential equation with two variables of the form (17.5.5.10)–(17.5.5.11), which can be reduced to the standard bilinear functional equation by the method of splitting.

**17.5.5-4. Nonlinear equations containing the complex argument $z = \varphi(t)\theta(x) + \psi(t)$.**

Consider a functional equation of the form

$$[\alpha_1(t)\theta(x) + \beta_1(t)]R_1(z) + \cdots + [\alpha_n(t)\theta(x) + \beta_n(t)]R_n(z) = 0, \quad z = \varphi(t)\theta(x) + \psi(t). \tag{17.5.5.27}$$

Passing in (17.5.5.27) from the variables $x$ and $t$ to new variables $z$ and $t$ [the function $\theta$ is replaced by $(z - \psi)/\varphi$], we come to the bilinear equation of the form (17.5.5.8):

$$\sum_{i=1}^{n} \alpha_i(t) z R_i(z) + \sum_{i=1}^{n} [\varphi(t)\beta_i(t) - \psi(t)\alpha_i(t)] R_i(z) = 0.$$

*Remark.* Instead of the expressions $\alpha_i(t)\theta(x) + \beta_i(t)$ in (17.5.5.27) linearly depending on the function $\theta(x)$, one can consider polynomials of $\theta(x)$ with coefficients depending on $t$.

▶ See also Section T12.3 for exact solutions of some linear and nonlinear difference and functional equations with several independent variables.

## References for Chapter 17

**Aczél, J.,** *Functional Equations: History, Applications and Theory*, Kluwer Academic, Dordrecht, 2002.
**Aczél, J.,** *Lectures on Functional Equations and Their Applications*, Dover Publications, New York, 2006.
**Aczél, J.,** Some general methods in the theory of functional equations with a single variable. New applications of functional equations [in Russian], *Uspekhi Mat. Nauk*, Vol. 11, No. 3 (69), pp. 3–68, 1956.
**Aczél, J. and Dhombres, J.,** *Functional Equations in Several Variables*, Cambridge University Press, Cambridge, 1989.
**Agarwal, R. P.,** *Difference Equations and Inequalities, 2nd Edition*, Marcel Dekker, New York, 2000.
**Balasubrahmanyan, R. and Lau, K.,** *Functional Equations in Probability Theory*, Academic Press, San Diego, 1991.
**Belitskii, G. R. and Tkachenko, V.,** *One-Dimensional Functional Equations*, Birkhäuser Verlag, Boston, 2003.
**Castillo, E. and Ruiz-Cobo, M. R.,** *Functional Equations and Modelling in Science and Engineering*, Marcel Dekker, New York, 1992.
**Castillo, E. and Ruiz-Cobo, R.,** *Functional Equations in Applied Sciences*, Elsevier, New York, 2005.
**Czerwik, S.,** *Functional Equations and Inequalities in Several Variables*, World Scientific Publishing Co., Singapore, 2002.
**Daroczy, Z. and Pales, Z. (Editors),** *Functional Equations – Results and Advances*, Kluwer Academic, Dordrecht, 2002.
**Eichhorn, W.,** *Functional Equations in Economics*, Addison Wesley, Reading, Massachusetts, 1978.
**Elaydi, S.,** *An Introduction to Difference Equations, 3rd Edition*, Springer-Verlag, New York, 2005.
**Goldberg, S.,** *Introduction to Difference Equations*, Dover Publications, New York, 1986.
**Jarai, A.,** *Regularity Properties of Functional Equations in Several Variables*, Springer-Verlag, New York, 2005.
**Kelley, W. and Peterson, A.,** *Difference Equations. An Introduction with Applications, 2nd Edition*, Academic Press, New York, 2000.
**Kuczma, M.,** *An Introduction to the Theory of Functional Equations and Inequalities*, Polish Scientific Publishers, Warsaw, 1985.
**Kuczma, M.,** *Functional Equations in a Single Variable*, Polish Scientific Publishers, Warsaw, 1968.
**Kuczma, M., Choczewski, B., and Ger, R.,** *Iterative Functional Equations*, Cambridge University Press, Cambridge, 1990.
**Mickens, R.,** *Difference Equations, 2nd Edition*, CRC Press, Boca Raton, 1991.
**Mirolyubov, A. A. and Soldatov, M. A.,** *Linear Homogeneous Difference Equations* [in Russian], Nauka Publishers, Moscow, 1981.
**Mirolyubov, A. A. and Soldatov, M. A.,** *Linear Nonhomogeneous Difference Equations* [in Russian], Nauka Publishers, Moscow, 1986.
**Nechepurenko, M. I.,** *Iterations of Real Functions and Functional Equations* [in Russian], Institute of Computational Mathematics and Mathematical Geophysics, Novosibirsk, 1997.

**Pelyukh, G. P. and Sharkovskii, O. M.,** *Introduction to the Theory of Functional Equations* [in Russian], Naukova Dumka, Kiev, 1974.

**Polyanin, A. D.,** *Functional Equations*, From Website *EqWorld — The World of Mathematical Equations*, http://eqworld.ipmnet.ru/en/solutions/fe.htm.

**Polyanin, A. D. and Manzhirov, A. V.,** *Handbook of Integral Equations: Exact Solutions (Supplement. Some Functional Equations)* [in Russian], Faktorial, Moscow, 1998.

**Polyanin, A. D. and Zaitsev, V. F.,** *Handbook of Nonlinear Partial Differential Equations* (Sections S.4 and S.5), Chapman & Hall/CRC Press, Boca Raton, 2004.

**Polyanin, A. D., Zaitsev, V. F., and Zhurov, A. I.,** *Methods for the Solution of Nonlinear Equations of Mathematical Physics and Mechanics* (Chapter 16. Solution of Some Functional Equations) [in Russian], Fizmatlit, Moscow, 2005.

**Rassias, T. M.,** *Functional Equations and Inequalities*, Kluwer Academic, Dordrecht, 2000.

**Samarskii, A. A. and Karamzin, Yu. N.,** *Difference Equations* [in Russian], Nauka Publishers, Moscow, 1978.

**Samarskii, A. A. and Nikolaev, E. S.,** *Methods for the Solution of Grid Equations* [in Russian], Nauka Publishers, Moscow, 1972.

**Sedaghat, H.,** *Nonlinear Difference Equations Theory with Applications to Social Science Models*, Kluwer Academic, Dordrecht, 2003.

**Sharkovsky, A. N., Maistrenko, Yu. L., and Romanenko, E. Yu.,** *Difference Equations and Their Applications*, Kluwer Academic, Dordrecht, 1993.

**Small, C. G.,** *Functional Equations and How to Solve Them*, Springer-Verlag, Berlin, 2007.

**Smital, J. and Dravecky, J.,** *On Functions and Functional Equations*, Adam Hilger, Bristol-Philadelphia, 1988.

**Zwillinger, D.,** *CRC Standard Mathematical Tables and Formulae, 31st Edition*, Chapman & Hall/CRC Press, Boca Raton, 2003.

# Chapter 18

# Special Functions and Their Properties

▶ Throughout Chapter 18 it is assumed that $n$ is a positive integer unless otherwise specified.

## 18.1. Some Coefficients, Symbols, and Numbers

### 18.1.1. Binomial Coefficients

#### 18.1.1-1. Definitions.

$$C_n^k = \binom{n}{k} = \frac{n!}{k!\,(n-k)!}, \quad \text{where} \quad k = 1,\ldots,n;$$

$$C_a^0 = 1, \quad C_a^k = \binom{a}{k} = (-1)^k \frac{(-a)_k}{k!} = \frac{a(a-1)\ldots(a-k+1)}{k!}, \quad \text{where} \quad k = 1, 2, \ldots$$

Here $a$ is an arbitrary real number.

#### 18.1.1-2. Generalization. Some properties.

General case:

$$C_a^b = \frac{\Gamma(a+1)}{\Gamma(b+1)\Gamma(a-b+1)}, \quad \text{where} \quad \Gamma(x) \text{ is the gamma function.}$$

Properties:

$$C_a^0 = 1, \quad C_n^k = 0 \quad \text{for} \quad k = -1, -2, \ldots \text{ or } k > n,$$

$$C_a^{b+1} = \frac{a}{b+1} C_{a-1}^b = \frac{a-b}{b+1} C_a^b, \quad C_a^b + C_a^{b+1} = C_{a+1}^{b+1},$$

$$C_{-1/2}^n = \frac{(-1)^n}{2^{2n}} C_{2n}^n = (-1)^n \frac{(2n-1)!!}{(2n)!!},$$

$$C_{1/2}^n = \frac{(-1)^{n-1}}{n 2^{2n-1}} C_{2n-2}^{n-1} = \frac{(-1)^{n-1}}{n} \frac{(2n-3)!!}{(2n-2)!!},$$

$$C_{n+1/2}^{2n+1} = (-1)^n 2^{-4n-1} C_{2n}^n, \quad C_{2n+1/2}^n = 2^{-2n} C_{4n+1}^{2n},$$

$$C_n^{1/2} = \frac{2^{2n+1}}{\pi C_{2n}^n}, \quad C_n^{n/2} = \frac{2^{2n}}{\pi} C_n^{(n-1)/2}.$$

## 18.1.2. Pochhammer Symbol

**18.1.2-1. Definition.**

$$(a)_n = a(a+1)\ldots(a+n-1) = \frac{\Gamma(a+n)}{\Gamma(a)} = (-1)^n \frac{\Gamma(1-a)}{\Gamma(1-a-n)}.$$

**18.1.2-2. Some properties ($k = 1, 2, \ldots$).**

$$(a)_0 = 1, \quad (a)_{n+k} = (a)_n(a+n)_k, \quad (n)_k = \frac{(n+k-1)!}{(n-1)!},$$

$$(a)_{-n} = \frac{\Gamma(a-n)}{\Gamma(a)} = \frac{(-1)^n}{(1-a)_n}, \quad \text{where } a \neq 1, \ldots, n;$$

$$(1)_n = n!, \quad (1/2)_n = 2^{-2n}\frac{(2n)!}{n!}, \quad (3/2)_n = 2^{-2n}\frac{(2n+1)!}{n!},$$

$$(a+mk)_{nk} = \frac{(a)_{mk+nk}}{(a)_{mk}}, \quad (a+n)_n = \frac{(a)_{2n}}{(a)_n}, \quad (a+n)_k = \frac{(a)_k(a+k)_n}{(a)_n}.$$

## 18.1.3. Bernoulli Numbers

**18.1.3-1. Definition.**

The *Bernoulli numbers* are defined by the recurrence relation

$$B_0 = 1, \quad \sum_{k=0}^{n-1} C_n^k B_k = 0, \quad n = 2, 3, \ldots$$

Numerical values:

$B_0 = 1, \quad B_1 = -\frac{1}{2}, \quad B_2 = \frac{1}{6}, \quad B_4 = -\frac{1}{30}, \quad B_6 = \frac{1}{42}, \quad B_8 = -\frac{1}{30}, \quad B_{10} = \frac{5}{66}, \quad \ldots,$
$B_{2m+1} = 0$ for $m = 1, 2, \ldots$

All odd-numbered Bernoulli numbers but $B_1$ are zero; all even-numbered Bernoulli numbers have alternating signs.

The Bernoulli numbers are the values of Bernoulli polynomials at $x = 0$: $B_n = B_n(0)$.

**18.1.3-2. Generating function.**

Generating function:

$$\frac{x}{e^x - 1} = \sum_{n=0}^{\infty} B_n \frac{x^n}{n!}, \quad |x| < 2\pi.$$

This relation may be regarded as a definition of the Bernoulli numbers.

The following expansions may be used to calculate the Bernoulli numbers:

$$\tan x = \sum_{n=1}^{\infty} |B_{2n}| \frac{2^{2n}(2^{2n}-1)}{(2n)!} x^{2n}, \quad |x| < \frac{\pi}{2};$$

$$\cot x = \sum_{n=0}^{\infty} (-1)^n B_{2n} \frac{2^{2n}}{(2n)!} x^{2n-1}, \quad |x| < \pi.$$

## 18.1.4. Euler Numbers

**18.1.4-1. Definition.**

The *Euler numbers* $E_n$ are defined by the recurrence relation

$$\sum_{k=0}^{n} C_{2n}^{2k} E_{2k} = 0 \quad \text{(even numbered)},$$

$$E_{2n+1} = 0 \quad \text{(odd numbered)},$$

where $n = 0, 1, \ldots$
Numerical values:

$E_0 = 1$, $\quad E_2 = -1$, $\quad E_4 = 5$, $\quad E_6 = -61$, $\quad E_8 = 1385$, $\quad E_{10} = -50251$, $\quad \ldots,$
$E_{2n+1} = 0 \quad$ for $\quad n = 0, 1, \ldots$

All Euler numbers are integer, the odd-numbered Euler numbers are zero, and the even-numbered Euler numbers have alternating signs.

The Euler numbers are expressed via the values of Euler polynomials at $x = 1/2$: $E_n = 2^n E_n(1/2)$, where $n = 0, 1, \ldots$

**18.1.4-2. Generating function. Integral representation.**

Generating function:

$$\frac{e^x}{e^{2x} + 1} = \sum_{n=0}^{\infty} E_n \frac{x^n}{n!}, \quad |x| < 2\pi.$$

This relation may be regarded as a definition of the Euler numbers.
Representation via a definite integral:

$$E_{2n} = (-1)^n 2^{2n+1} \int_0^\infty \frac{t^{2n} dt}{\cosh(\pi t)}.$$

# 18.2. Error Functions. Exponential and Logarithmic Integrals

## 18.2.1. Error Function and Complementary Error Function

**18.2.1-1. Integral representations.**

Definitions:

$$\operatorname{erf} x = \frac{2}{\sqrt{\pi}} \int_0^x \exp(-t^2)\, dt \quad \text{(error function, also called probability integral)},$$

$$\operatorname{erfc} x = 1 - \operatorname{erf} x = \frac{2}{\sqrt{\pi}} \int_x^\infty \exp(-t^2)\, dt \quad \text{(complementary error function)}.$$

Properties:

$\operatorname{erf}(-x) = -\operatorname{erf} x; \quad \operatorname{erf}(0) = 0, \quad \operatorname{erf}(\infty) = 1; \quad \operatorname{erfc}(0) = 1, \quad \operatorname{erfc}(\infty) = 0.$

**18.2.1-2. Expansions as $x \to 0$ and $x \to \infty$. Definite integral.**

Expansion of erf $x$ into series in powers of $x$ as $x \to 0$:

$$\operatorname{erf} x = \frac{2}{\sqrt{\pi}} \sum_{k=0}^{\infty} (-1)^k \frac{x^{2k+1}}{k!(2k+1)} = \frac{2}{\sqrt{\pi}} \exp(-x^2) \sum_{k=0}^{\infty} \frac{2^k x^{2k+1}}{(2k+1)!!}.$$

Asymptotic expansion of erfc $x$ as $x \to \infty$:

$$\operatorname{erfc} x = \frac{1}{\sqrt{\pi}} \exp(-x^2) \left[ \sum_{m=0}^{M-1} (-1)^m \frac{\left(\frac{1}{2}\right)_m}{x^{2m+1}} + O(|x|^{-2M-1}) \right], \qquad M = 1, 2, \ldots$$

Integral:

$$\int_0^x \operatorname{erf} t \, dt = x \operatorname{erf} x - \frac{1}{2} + \frac{1}{2} \exp(-x^2).$$

## 18.2.2. Exponential Integral

**18.2.2-1. Integral representations.**

Definition:

$$\operatorname{Ei}(x) = \int_{-\infty}^x \frac{e^t}{t} \, dt = - \int_{-x}^{\infty} \frac{e^{-t}}{t} \, dt \qquad \text{for} \quad x < 0,$$

$$\operatorname{Ei}(x) = \lim_{\varepsilon \to +0} \left( \int_{-\infty}^{-\varepsilon} \frac{e^t}{t} \, dt + \int_{\varepsilon}^{x} \frac{e^t}{t} \, dt \right) \qquad \text{for} \quad x > 0.$$

Other integral representations:

$$\operatorname{Ei}(-x) = -e^{-x} \int_0^{\infty} \frac{x \sin t + t \cos t}{x^2 + t^2} \, dt \qquad \text{for} \quad x > 0,$$

$$\operatorname{Ei}(-x) = e^{-x} \int_0^{\infty} \frac{x \sin t - t \cos t}{x^2 + t^2} \, dt \qquad \text{for} \quad x < 0,$$

$$\operatorname{Ei}(-x) = -x \int_1^{\infty} e^{-xt} \ln t \, dt \qquad \text{for} \quad x > 0,$$

$$\operatorname{Ei}(x) = C + \ln x + \int_0^x \frac{e^t - 1}{t} \, dt \qquad \text{for} \quad x > 0,$$

where $C = 0.5772\ldots$ is the Euler constant.

**18.2.2-2. Expansions as $x \to 0$ and $x \to \infty$.**

Expansion into series in powers of $x$ as $x \to 0$:

$$\operatorname{Ei}(x) = \begin{cases} C + \ln(-x) + \sum_{k=1}^{\infty} \dfrac{x^k}{k!\,k} & \text{if } x < 0, \\[2mm] C + \ln x + \sum_{k=1}^{\infty} \dfrac{x^k}{k!\,k} & \text{if } x > 0. \end{cases}$$

Asymptotic expansion as $x \to \infty$:

$$\operatorname{Ei}(-x) = e^{-x} \sum_{k=1}^{n} (-1)^k \frac{(k-1)!}{x^k} + R_n, \qquad R_n < \frac{n!}{x^n}.$$

## 18.2.3. Logarithmic Integral

**18.2.3-1. Integral representations.**

Definition:

$$\operatorname{li}(x) = \begin{cases} \displaystyle\int_0^x \frac{dt}{\ln t} & \text{if } 0 < x < 1, \\ \displaystyle\lim_{\varepsilon \to +0}\left(\int_0^{1-\varepsilon} \frac{dt}{\ln t} + \int_{1+\varepsilon}^x \frac{dt}{\ln t}\right) & \text{if } x > 1. \end{cases}$$

**18.2.3-2. Limiting properties. Relation to the exponential integral.**

For small $x$,

$$\operatorname{li}(x) \approx \frac{x}{\ln(1/x)}.$$

For large $x$,

$$\operatorname{li}(x) \approx \frac{x}{\ln x}.$$

Asymptotic expansion as $x \to 1$:

$$\operatorname{li}(x) = C + \ln|\ln x| + \sum_{k=1}^{\infty} \frac{\ln^k x}{k!\, k}.$$

Relation to the exponential integral:

$$\operatorname{li} x = \operatorname{Ei}(\ln x), \quad x < 1;$$
$$\operatorname{li}(e^x) = \operatorname{Ei}(x), \quad x < 0.$$

## 18.3. Sine Integral and Cosine Integral. Fresnel Integrals

### 18.3.1. Sine Integral

**18.3.1-1. Integral representations. Properties.**

Definition:

$$\operatorname{Si}(x) = \int_0^x \frac{\sin t}{t}\, dt, \qquad \operatorname{si}(x) = -\int_x^\infty \frac{\sin t}{t}\, dt = \operatorname{Si}(x) - \frac{\pi}{2}.$$

Specific values:

$$\operatorname{Si}(0) = 0, \quad \operatorname{Si}(\infty) = \frac{\pi}{2}, \quad \operatorname{si}(\infty) = 0.$$

Properties:

$$\operatorname{Si}(-x) = -\operatorname{Si}(x), \quad \operatorname{si}(x) + \operatorname{si}(-x) = -\pi, \quad \lim_{x \to -\infty} \operatorname{si}(x) = -\pi.$$

**18.3.1-2. Expansions as $x \to 0$ and $x \to \infty$.**

Expansion into series in powers of $x$ as $x \to 0$:

$$\operatorname{Si}(x) = \sum_{k=1}^{\infty} \frac{(-1)^{k+1} x^{2k-1}}{(2k-1)(2k-1)!}.$$

Asymptotic expansion as $x \to \infty$:

$$\operatorname{si}(x) = -\cos x \left[ \sum_{m=0}^{M-1} \frac{(-1)^m (2m)!}{x^{2m+1}} + O(|x|^{-2M-1}) \right] + \sin x \left[ \sum_{m=1}^{N-1} \frac{(-1)^m (2m-1)!}{x^{2m}} + O(|x|^{-2N}) \right],$$

where $M, N = 1, 2, \ldots$

## 18.3.2. Cosine Integral

**18.3.2-1. Integral representations.**

Definition:

$$\operatorname{Ci}(x) = -\int_x^{\infty} \frac{\cos t}{t}\, dt = \mathcal{C} + \ln x + \int_0^x \frac{\cos t - 1}{t}\, dt,$$

where $\mathcal{C} = 0.5772\ldots$ is the Euler constant.

**18.3.2-2. Expansions as $x \to 0$ and $x \to \infty$.**

Expansion into series in powers of $x$ as $x \to 0$:

$$\operatorname{Ci}(x) = \mathcal{C} + \ln x + \sum_{k=1}^{\infty} \frac{(-1)^k x^{2k}}{2k\,(2k)!}.$$

Asymptotic expansion as $x \to \infty$:

$$\operatorname{Ci}(x) = \cos x \left[ \sum_{m=1}^{M-1} \frac{(-1)^m (2m-1)!}{x^{2m}} + O(|x|^{-2M}) \right] + \sin x \left[ \sum_{m=0}^{N-1} \frac{(-1)^m (2m)!}{x^{2m+1}} + O(|x|^{-2N-1}) \right],$$

where $M, N = 1, 2, \ldots$

## 18.3.3. Fresnel Integrals

**18.3.3-1. Integral representations.**

Definitions:

$$S(x) = \frac{1}{\sqrt{2\pi}} \int_0^x \frac{\sin t}{\sqrt{t}}\, dt = \sqrt{\frac{2}{\pi}} \int_0^{\sqrt{x}} \sin t^2\, dt,$$

$$C(x) = \frac{1}{\sqrt{2\pi}} \int_0^x \frac{\cos t}{\sqrt{t}}\, dt = \sqrt{\frac{2}{\pi}} \int_0^{\sqrt{x}} \cos t^2\, dt.$$

### 18.3.3-2. Expansions as $x \to 0$ and $x \to \infty$.

Expansion into series in powers of $x$ as $x \to 0$:

$$S(x) = \sqrt{\frac{2}{\pi}} \, x \sum_{k=0}^{\infty} \frac{(-1)^k x^{2k+1}}{(4k+3)(2k+1)!},$$

$$C(x) = \sqrt{\frac{2}{\pi}} \, x \sum_{k=0}^{\infty} \frac{(-1)^k x^{2k}}{(4k+1)(2k)!}.$$

Asymptotic expansion as $x \to \infty$:

$$S(x) = \frac{1}{2} - \frac{\cos x}{\sqrt{2\pi x}} P(x) - \frac{\sin x}{\sqrt{2\pi x}} Q(x),$$

$$C(x) = \frac{1}{2} + \frac{\sin x}{\sqrt{2\pi x}} P(x) - \frac{\cos x}{\sqrt{2\pi x}} Q(x),$$

$$P(x) = 1 - \frac{1 \times 3}{(2x)^2} + \frac{1 \times 3 \times 5 \times 7}{(2x)^4} - \cdots, \quad Q(x) = \frac{1}{2x} - \frac{1 \times 3 \times 5}{(2x)^3} + \cdots.$$

## 18.4. Gamma Function, Psi Function, and Beta Function

### 18.4.1. Gamma Function

#### 18.4.1-1. Integral representations. Simplest properties.

The gamma function, $\Gamma(z)$, is an analytic function of the complex argument $z$ everywhere except for the points $z = 0, -1, -2, \ldots$

For $\operatorname{Re} z > 0$,

$$\Gamma(z) = \int_0^{\infty} t^{z-1} e^{-t} \, dt.$$

For $-(n+1) < \operatorname{Re} z < -n$, where $n = 0, 1, 2, \ldots,$

$$\Gamma(z) = \int_0^{\infty} \left[ e^{-t} - \sum_{m=0}^{n} \frac{(-1)^m}{m!} \right] t^{z-1} \, dt.$$

Simplest properties:

$$\Gamma(z+1) = z\Gamma(z), \quad \Gamma(n+1) = n!, \quad \Gamma(1) = \Gamma(2) = 1.$$

Fractional values of the argument:

$$\Gamma\!\left(\tfrac{1}{2}\right) = \sqrt{\pi}, \quad \Gamma\!\left(n + \tfrac{1}{2}\right) = \frac{\sqrt{\pi}}{2^n}(2n-1)!!,$$

$$\Gamma\!\left(-\tfrac{1}{2}\right) = -2\sqrt{\pi}, \quad \Gamma\!\left(\tfrac{1}{2} - n\right) = (-1)^n \frac{2^n \sqrt{\pi}}{(2n-1)!!}.$$

## 18.4.1-2. Euler, Stirling, and other formulas.

Euler formula

$$\Gamma(z) = \lim_{n \to \infty} \frac{n!\, n^z}{z(z+1)\dots(z+n)} \qquad (z \neq 0, -1, -2, \dots).$$

Symmetry formulas:

$$\Gamma(z)\Gamma(-z) = -\frac{\pi}{z\sin(\pi z)}, \qquad \Gamma(z)\Gamma(1-z) = \frac{\pi}{\sin(\pi z)},$$

$$\Gamma\left(\frac{1}{2}+z\right)\Gamma\left(\frac{1}{2}-z\right) = \frac{\pi}{\cos(\pi z)}.$$

Multiple argument formulas:

$$\Gamma(2z) = \frac{2^{2z-1}}{\sqrt{\pi}} \Gamma(z)\Gamma\left(z+\frac{1}{2}\right),$$

$$\Gamma(3z) = \frac{3^{3z-1/2}}{2\pi} \Gamma(z)\Gamma\left(z+\frac{1}{3}\right)\Gamma\left(z+\frac{2}{3}\right),$$

$$\Gamma(nz) = (2\pi)^{(1-n)/2} n^{nz-1/2} \prod_{k=0}^{n-1} \Gamma\left(z+\frac{k}{n}\right).$$

Asymptotic expansion (*Stirling formula*):

$$\Gamma(z) = \sqrt{2\pi}\, e^{-z} z^{z-1/2} \left[1 + \tfrac{1}{12} z^{-1} + \tfrac{1}{288} z^{-2} + O(z^{-3})\right] \qquad (|\arg z| < \pi).$$

## 18.4.2. Psi Function (Digamma Function)

### 18.4.2-1. Definition. Integral representations.

Definition:

$$\psi(z) = \frac{d \ln \Gamma(z)}{dz} = \frac{\Gamma'_z(z)}{\Gamma(z)}.$$

The psi function is the logarithmic derivative of the gamma function and is also called the *digamma function*.

Integral representations (Re $z > 0$):

$$\psi(z) = \int_0^\infty \left[e^{-t} - (1+t)^{-z}\right] t^{-1}\, dt,$$

$$\psi(z) = \ln z + \int_0^\infty \left[t^{-1} - (1 - e^{-t})^{-1}\right] e^{-tz}\, dt,$$

$$\psi(z) = -\mathcal{C} + \int_0^1 \frac{1 - t^{z-1}}{1 - t}\, dt,$$

where $\mathcal{C} = -\psi(1) = 0.5772\dots$ is the Euler constant.

Values for integer argument:

$$\psi(1) = -\mathcal{C}, \qquad \psi(n) = -\mathcal{C} + \sum_{k=1}^{n-1} k^{-1} \qquad (n = 2, 3, \dots).$$

### 18.4.2-2. Properties. Asymptotic expansion as $z \to \infty$.

Functional relations:

$$\psi(z) - \psi(1+z) = -\frac{1}{z},$$
$$\psi(z) - \psi(1-z) = -\pi \cot(\pi z),$$
$$\psi(z) - \psi(-z) = -\pi \cot(\pi z) - \frac{1}{z},$$
$$\psi\left(\tfrac{1}{2}+z\right) - \psi\left(\tfrac{1}{2}-z\right) = \pi \tan(\pi z),$$
$$\psi(mz) = \ln m + \frac{1}{m}\sum_{k=0}^{m-1}\psi\left(z+\frac{k}{m}\right).$$

Asymptotic expansion as $z \to \infty$ ($|\arg z| < \pi$):

$$\psi(z) = \ln z - \frac{1}{2z} - \frac{1}{12z^2} + \frac{1}{120z^4} - \frac{1}{252z^6} + \cdots = \ln z - \frac{1}{2z} - \sum_{n=1}^{\infty}\frac{B_{2n}}{2nz^{2n}},$$

where the $B_{2n}$ are Bernoulli numbers.

## 18.4.3. Beta Function

### 18.4.3-1. Integral representation. Relationship with the gamma function.

Definition:

$$B(x,y) = \int_0^1 t^{x-1}(1-t)^{y-1}\,dt,$$

where $\operatorname{Re} x > 0$ and $\operatorname{Re} y > 0$.

Relationship with the gamma function:

$$B(x,y) = \frac{\Gamma(x)\Gamma(y)}{\Gamma(x+y)}.$$

### 18.4.3-2. Some properties.

$$B(x,y) = B(y,x);$$
$$B(x,y+1) = \frac{y}{x}B(x+1,y) = \frac{y}{x+y}B(x,y);$$
$$B(x,1-x) = \frac{\pi}{\sin(\pi x)}, \quad 0 < x < 1;$$
$$\frac{1}{B(n,m)} = mC_{n+m-1}^{n-1} = nC_{n+m-1}^{m-1},$$

where $n$ and $m$ are positive integers.

## 18.5. Incomplete Gamma and Beta Functions

### 18.5.1. Incomplete Gamma Function

18.5.1-1. Integral representations. Recurrence formulas.

Definitions:
$$\gamma(\alpha, x) = \int_0^x e^{-t} t^{\alpha-1} \, dt, \qquad \operatorname{Re} \alpha > 0,$$
$$\Gamma(\alpha, x) = \int_x^\infty e^{-t} t^{\alpha-1} \, dt = \Gamma(\alpha) - \gamma(\alpha, x).$$

Recurrence formulas:
$$\gamma(\alpha + 1, x) = \alpha \gamma(\alpha, x) - x^\alpha e^{-x},$$
$$\gamma(\alpha + 1, x) = (x + \alpha)\gamma(\alpha, x) + (1 - \alpha)x\gamma(\alpha - 1, x),$$
$$\Gamma(\alpha + 1, x) = \alpha \Gamma(\alpha, x) + x^\alpha e^{-x}.$$

Special cases:
$$\gamma(n + 1, x) = n! \left[1 - e^{-x} \left(\sum_{k=0}^n \frac{x^k}{k!}\right)\right], \qquad n = 0, 1, \ldots;$$
$$\Gamma(n + 1, x) = n! \, e^{-x} \sum_{k=0}^n \frac{x^k}{k!}, \qquad n = 0, 1, \ldots;$$
$$\Gamma(-n, x) = \frac{(-1)^n}{n!} \left[\Gamma(0, x) - e^{-x} \sum_{k=0}^{n-1} (-1)^k \frac{k!}{x^{k+1}}\right], \qquad n = 1, 2, \ldots$$

18.5.1-2. Expansions as $x \to 0$ and $x \to \infty$. Relation to other functions.

Asymptotic expansions as $x \to 0$:
$$\gamma(\alpha, x) = \sum_{n=0}^\infty \frac{(-1)^n x^{\alpha+n}}{n!(\alpha + n)},$$
$$\Gamma(\alpha, x) = \Gamma(\alpha) - \sum_{n=0}^\infty \frac{(-1)^n x^{\alpha+n}}{n!(\alpha + n)}.$$

Asymptotic expansions as $x \to \infty$:
$$\gamma(\alpha, x) = \Gamma(\alpha) - x^{\alpha-1} e^{-x} \left[\sum_{m=0}^{M-1} \frac{(1-\alpha)_m}{(-x)^m} + O\big(|x|^{-M}\big)\right],$$
$$\Gamma(\alpha, x) = x^{\alpha-1} e^{-x} \left[\sum_{m=0}^{M-1} \frac{(1-\alpha)_m}{(-x)^m} + O\big(|x|^{-M}\big)\right] \qquad \left(-\tfrac{3}{2}\pi < \arg x < \tfrac{3}{2}\pi\right).$$

Asymptotic formulas as $\alpha \to \infty$:

$$\gamma(x,\alpha) = \Gamma(\alpha)\left[\Phi\left(2\sqrt{x} - \sqrt{\alpha-1}\right) + O\left(\frac{1}{\sqrt{\alpha}}\right)\right], \quad \Phi(x) = \frac{1}{\sqrt{2\pi}}\int_{-\infty}^{x} \exp\left(-\frac{1}{2}t^2\right) dt;$$

$$\gamma(x,\alpha) = \Gamma(\alpha)\left[\Phi\left(3\sqrt{\alpha}\,z\right) + O\left(\frac{1}{\alpha}\right)\right], \quad z = \left(\frac{x}{\alpha}\right)^{1/3} - 1 + \frac{1}{9\alpha}.$$

Representation of the error function, complementary error function, and exponential integral in terms of the gamma functions:

$$\operatorname{erf} x = \frac{1}{\sqrt{\pi}}\gamma\left(\frac{1}{2}, x^2\right), \quad \operatorname{erfc} x = \frac{1}{\sqrt{\pi}}\Gamma\left(\frac{1}{2}, x^2\right), \quad \operatorname{Ei}(-x) = -\Gamma(0, x).$$

### 18.5.2. Incomplete Beta Function

**18.5.2-1. Integral representation.**

Definitions:

$$B_x(a,b) = \int_0^x t^{a-1}(1-t)^{b-1}\, dt, \quad I_x(a,b) = \frac{B_x(a,b)}{B(a,b)},$$

where $\operatorname{Re} a > 0$ and $\operatorname{Re} b > 0$, and $B(a,b) = B_1(a,b)$ is the beta function.

**18.5.2-2. Some properties.**

Symmetry:

$$I_x(a,b) + I_{1-x}(b,a) = 1.$$

recurrence formulas:

$$I_x(a,b) = xI_x(a-1,b) + (1-x)I_x(a,b-1),$$
$$(a+b)I_x(a,b) = aI_x(a+1,b) + bI_x(a,b+1),$$
$$(a+b-ax)I_x(a,b) = a(1-x)I_x(a+1,b-1) + bI_x(a,b+1).$$

## 18.6. Bessel Functions (Cylindrical Functions)
### 18.6.1. Definitions and Basic Formulas

**18.6.1-1. Bessel functions of the first and the second kind.**

The *Bessel function of the first kind*, $J_\nu(x)$, and the *Bessel function of the second kind*, $Y_\nu(x)$ (also called the *Neumann function*), are solutions of the Bessel equation

$$x^2 y''_{xx} + x y'_x + (x^2 - \nu^2) y = 0$$

and are defined by the formulas

$$J_\nu(x) = \sum_{k=0}^{\infty} \frac{(-1)^k (x/2)^{\nu+2k}}{k!\,\Gamma(\nu+k+1)}, \quad Y_\nu(x) = \frac{J_\nu(x)\cos\pi\nu - J_{-\nu}(x)}{\sin\pi\nu}. \qquad (18.6.1.1)$$

The formula for $Y_\nu(x)$ is valid for $\nu \neq 0, \pm 1, \pm 2, \ldots$ (the cases $\nu \neq 0, \pm 1, \pm 2, \ldots$ are discussed in what follows).

The general solution of the Bessel equation has the form $Z_\nu(x) = C_1 J_\nu(x) + C_2 Y_\nu(x)$ and is called the *cylinder function*.

**18.6.1-2. Some formulas.**

$$2\nu Z_\nu(x) = x[Z_{\nu-1}(x) + Z_{\nu+1}(x)],$$

$$\frac{d}{dx} Z_\nu(x) = \frac{1}{2}[Z_{\nu-1}(x) - Z_{\nu+1}(x)] = \pm\left[\frac{\nu}{x} Z_\nu(x) - Z_{\nu\pm1}(x)\right],$$

$$\frac{d}{dx}[x^\nu Z_\nu(x)] = x^\nu Z_{\nu-1}(x), \qquad \frac{d}{dx}[x^{-\nu} Z_\nu(x)] = -x^{-\nu} Z_{\nu+1}(x),$$

$$\left(\frac{1}{x}\frac{d}{dx}\right)^n [x^\nu J_\nu(x)] = x^{\nu-n} J_{\nu-n}(x), \qquad \left(\frac{1}{x}\frac{d}{dx}\right)^n [x^{-\nu} J_\nu(x)] = (-1)^n x^{-\nu-n} J_{\nu+n}(x),$$

$$J_{-n}(x) = (-1)^n J_n(x), \qquad Y_{-n}(x) = (-1)^n Y_n(x), \qquad n = 0, 1, 2, \ldots$$

**18.6.1-3. Bessel functions for $\nu = \pm n \pm \frac{1}{2}$, where $n = 0, 1, 2, \ldots$**

$$J_{1/2}(x) = \sqrt{\frac{2}{\pi x}} \sin x, \qquad J_{-1/2}(x) = \sqrt{\frac{2}{\pi x}} \cos x,$$

$$J_{3/2}(x) = \sqrt{\frac{2}{\pi x}} \left(\frac{1}{x} \sin x - \cos x\right), \qquad J_{-3/2}(x) = \sqrt{\frac{2}{\pi x}} \left(-\frac{1}{x} \cos x - \sin x\right),$$

$$J_{n+1/2}(x) = \sqrt{\frac{2}{\pi x}} \left[\sin\left(x - \frac{n\pi}{2}\right) \sum_{k=0}^{[n/2]} \frac{(-1)^k (n+2k)!}{(2k)!(n-2k)!(2x)^{2k}} \right.$$

$$\left. + \cos\left(x - \frac{n\pi}{2}\right) \sum_{k=0}^{[(n-1)/2]} \frac{(-1)^k (n+2k+1)!}{(2k+1)!(n-2k-1)!(2x)^{2k+1}}\right],$$

$$J_{-n-1/2}(x) = \sqrt{\frac{2}{\pi x}} \left[\cos\left(x + \frac{n\pi}{2}\right) \sum_{k=0}^{[n/2]} \frac{(-1)^k (n+2k)!}{(2k)!(n-2k)!(2x)^{2k}} \right.$$

$$\left. - \sin\left(x + \frac{n\pi}{2}\right) \sum_{k=0}^{[(n-1)/2]} \frac{(-1)^k (n+2k+1)!}{(2k+1)!(n-2k-1)!(2x)^{2k+1}}\right],$$

$$Y_{1/2}(x) = -\sqrt{\frac{2}{\pi x}} \cos x, \qquad Y_{-1/2}(x) = \sqrt{\frac{2}{\pi x}} \sin x,$$

$$Y_{n+1/2}(x) = (-1)^{n+1} J_{-n-1/2}(x), \qquad Y_{-n-1/2}(x) = (-1)^n J_{n+1/2}(x),$$

where $[A]$ is the integer part of the number $A$.

**18.6.1-4. Bessel functions for $\nu = \pm n$, where $n = 0, 1, 2, \ldots$**

Let $\nu = n$ be an arbitrary integer. The relations

$$J_{-n}(x) = (-1)^n J_n(x), \qquad Y_{-n}(x) = (-1)^n Y_n(x)$$

are valid. The function $J_n(x)$ is given by the first formula in (18.6.1.1) with $\nu = n$, and $Y_n(x)$ can be obtained from the second formula in (18.6.1.1) by proceeding to the limit

$\nu \to n$. For nonnegative $n$, $Y_n(x)$ can be represented in the form

$$Y_n(x) = \frac{2}{\pi} J_n(x) \ln \frac{x}{2} - \frac{1}{\pi} \sum_{k=0}^{n-1} \frac{(n-k-1)!}{k!} \left(\frac{2}{x}\right)^{n-2k} - \frac{1}{\pi} \sum_{k=0}^{\infty} (-1)^k \left(\frac{x}{2}\right)^{n+2k} \frac{\psi(k+1) + \psi(n+k+1)}{k!(n+k)!},$$

where $\psi(1) = -\mathcal{C}$, $\psi(n) = -\mathcal{C} + \sum_{k=1}^{n-1} k^{-1}$, $\mathcal{C} = 0.5772\ldots$ is the Euler constant, and $\psi(x) = [\ln \Gamma(x)]'_x$ is the logarithmic derivative of the gamma function, also known as the digamma function.

### 18.6.1-5. Wronskians and similar formulas.

$$W(J_\nu, J_{-\nu}) = -\frac{2}{\pi x} \sin(\pi\nu), \quad W(J_\nu, Y_\nu) = \frac{2}{\pi x},$$

$$J_\nu(x) J_{-\nu+1}(x) + J_{-\nu}(x) J_{\nu-1}(x) = \frac{2 \sin(\pi\nu)}{\pi x}, \quad J_\nu(x) Y_{\nu+1}(x) - J_{\nu+1}(x) Y_\nu(x) = -\frac{2}{\pi x}.$$

Here, the notation $W(f, g) = f g'_x - f'_x g$ is used.

## 18.6.2. Integral Representations and Asymptotic Expansions

### 18.6.2-1. Integral representations.

The functions $J_\nu(x)$ and $Y_\nu(x)$ can be represented in the form of definite integrals (for $x > 0$):

$$\pi J_\nu(x) = \int_0^\pi \cos(x \sin \theta - \nu \theta) \, d\theta - \sin \pi\nu \int_0^\infty \exp(-x \sinh t - \nu t) \, dt,$$

$$\pi Y_\nu(x) = \int_0^\pi \sin(x \sin \theta - \nu \theta) \, d\theta - \int_0^\infty (e^{\nu t} + e^{-\nu t} \cos \pi\nu) e^{-x \sinh t} \, dt.$$

For $|\nu| < \frac{1}{2}$, $x > 0$,

$$J_\nu(x) = \frac{2^{1+\nu} x^{-\nu}}{\pi^{1/2} \Gamma(\frac{1}{2} - \nu)} \int_1^\infty \frac{\sin(xt) \, dt}{(t^2 - 1)^{\nu+1/2}},$$

$$Y_\nu(x) = -\frac{2^{1+\nu} x^{-\nu}}{\pi^{1/2} \Gamma(\frac{1}{2} - \nu)} \int_1^\infty \frac{\cos(xt) \, dt}{(t^2 - 1)^{\nu+1/2}}.$$

For $\nu > -\frac{1}{2}$,

$$J_\nu(x) = \frac{2(x/2)^\nu}{\pi^{1/2} \Gamma(\frac{1}{2} + \nu)} \int_0^{\pi/2} \cos(x \cos t) \sin^{2\nu} t \, dt \quad \text{(Poisson's formula)}.$$

For $\nu = 0$, $x > 0$,

$$J_0(x) = \frac{2}{\pi} \int_0^\infty \sin(x \cosh t) \, dt, \qquad Y_0(x) = -\frac{2}{\pi} \int_0^\infty \cos(x \cosh t) \, dt.$$

For integer $\nu = n = 0, 1, 2, \ldots$,

$$J_n(x) = \frac{1}{\pi} \int_0^\pi \cos(nt - x \sin t)\, dt \quad \text{(Bessel's formula)},$$

$$J_{2n}(x) = \frac{2}{\pi} \int_0^{\pi/2} \cos(x \sin t) \cos(2nt)\, dt,$$

$$J_{2n+1}(x) = \frac{2}{\pi} \int_0^{\pi/2} \sin(x \sin t) \sin[(2n+1)t]\, dt.$$

**18.6.2-2. Asymptotic expansions as $|x| \to \infty$.**

$$J_\nu(x) = \sqrt{\frac{2}{\pi x}} \left\{ \cos\left(\frac{4x - 2\nu\pi - \pi}{4}\right) \left[ \sum_{m=0}^{M-1} (-1)^m (\nu, 2m)(2x)^{-2m} + O(|x|^{-2M}) \right] \right.$$
$$\left. - \sin\left(\frac{4x - 2\nu\pi - \pi}{4}\right) \left[ \sum_{m=0}^{M-1} (-1)^m (\nu, 2m+1)(2x)^{-2m-1} + O(|x|^{-2M-1}) \right] \right\},$$

$$Y_\nu(x) = \sqrt{\frac{2}{\pi x}} \left\{ \sin\left(\frac{4x - 2\nu\pi - \pi}{4}\right) \left[ \sum_{m=0}^{M-1} (-1)^m (\nu, 2m)(2x)^{-2m} + O(|x|^{-2M}) \right] \right.$$
$$\left. + \cos\left(\frac{4x - 2\nu\pi - \pi}{4}\right) \left[ \sum_{m=0}^{M-1} (-1)^m (\nu, 2m+1)(2x)^{-2m-1} + O(|x|^{-2M-1}) \right] \right\},$$

where $(\nu, m) = \dfrac{1}{2^{2m} m!} (4\nu^2 - 1)(4\nu^2 - 3^2) \ldots [4\nu^2 - (2m-1)^2] = \dfrac{\Gamma(\frac{1}{2} + \nu + m)}{m!\, \Gamma(\frac{1}{2} + \nu - m)}$.

For nonnegative integer $n$ and large $x$,

$$\sqrt{\pi x}\, J_{2n}(x) = (-1)^n (\cos x + \sin x) + O(x^{-2}),$$
$$\sqrt{\pi x}\, J_{2n+1}(x) = (-1)^{n+1} (\cos x - \sin x) + O(x^{-2}).$$

**18.6.2-3. Asymptotic for large $\nu$ ($\nu \to \infty$).**

$$J_\nu(x) \simeq \frac{1}{\sqrt{2\pi\nu}} \left(\frac{ex}{2\nu}\right)^\nu, \quad Y_\nu(x) \simeq -\sqrt{\frac{2}{\pi\nu}} \left(\frac{ex}{2\nu}\right)^{-\nu},$$

where $x$ is fixed,

$$J_\nu(\nu) \simeq \frac{2^{1/3}}{3^{2/3}\Gamma(2/3)} \frac{1}{\nu^{1/3}}, \quad Y_\nu(\nu) \simeq -\frac{2^{1/3}}{3^{1/6}\Gamma(2/3)} \frac{1}{\nu^{1/3}}.$$

**18.6.2-4. Integrals with Bessel functions.**

$$\int_0^x x^\lambda J_\nu(x)\, dx = \frac{x^{\lambda+\nu+1}}{2^\nu (\lambda+\nu+1)\Gamma(\nu+1)} F\left(\frac{\lambda+\nu+1}{2}, \frac{\lambda+\nu+3}{2}; \nu+1; -\frac{x^2}{4}\right), \quad \operatorname{Re}(\lambda+\nu) > -1,$$

where $F(a,b,c;x)$ is the hypergeometric series (see Section 18.10.1),

$$\int_0^x x^\lambda Y_\nu(x)\,dx = -\frac{\cos(\nu\pi)\Gamma(-\nu)}{2^\nu \pi(\lambda+\nu+1)} x^{\lambda+\nu+1} F\left(\frac{\lambda+\nu+1}{2}, \nu+1, \frac{\lambda+\nu+3}{2}; -\frac{x^2}{4}\right)$$
$$-\frac{2^\nu \Gamma(\nu)}{\lambda-\nu+1} x^{\lambda-\nu+1} F\left(\frac{\lambda-\nu+1}{2}, 1-\nu, \frac{\lambda-\nu+3}{2}; -\frac{x^2}{4}\right), \qquad \text{Re}\,\lambda > |\text{Re}\,\nu| - 1.$$

## 18.6.3. Zeros and Orthogonality Properties of Bessel Functions

### 18.6.3-1. Zeros of Bessel functions.

Each of the functions $J_\nu(x)$ and $Y_\nu(x)$ has infinitely many real zeros (for real $\nu$). All zeros are simple, except possibly for the point $x = 0$.

The zeros $\gamma_m$ of $J_0(x)$, i.e., the roots of the equation $J_0(\gamma_m) = 0$, are approximately given by

$$\gamma_m = 2.4 + 3.13\,(m-1) \qquad (m = 1, 2, \ldots),$$

with a maximum error of 0.2%.

### 18.6.3-2. Orthogonality properties of Bessel functions.

1°. Let $\mu = \mu_m$ be positive roots of the Bessel function $J_\nu(\mu)$, where $\nu > -1$ and $m = 1, 2, 3, \ldots$ Then the set of functions $J_\nu(\mu_m r/a)$ is orthogonal on the interval $0 \le r \le a$ with weight $r$:

$$\int_0^a J_\nu\!\left(\frac{\mu_m r}{a}\right) J_\nu\!\left(\frac{\mu_k r}{a}\right) r\,dr = \begin{cases} 0 & \text{if } m \ne k, \\ \tfrac{1}{2} a^2 \left[J_\nu'(\mu_m)\right]^2 = \tfrac{1}{2} a^2 J_{\nu+1}^2(\mu_m) & \text{if } m = k. \end{cases}$$

2°. Let $\mu = \mu_m$ be positive zeros of the Bessel function derivative $J_\nu'(\mu)$, where $\nu > -1$ and $m = 1, 2, 3, \ldots$ Then the set of functions $J_\nu(\mu_m r/a)$ is orthogonal on the interval $0 \le r \le a$ with weight $r$:

$$\int_0^a J_\nu\!\left(\frac{\mu_m r}{a}\right) J_\nu\!\left(\frac{\mu_k r}{a}\right) r\,dr = \begin{cases} 0 & \text{if } m \ne k, \\ \dfrac{1}{2} a^2 \left(1 - \dfrac{\nu^2}{\mu_m^2}\right) J_\nu^2(\mu_m) & \text{if } m = k. \end{cases}$$

3°. Let $\mu = \mu_m$ be positive roots of the transcendental equation $\mu J_\nu'(\mu) + s J_\nu(\mu) = 0$, where $\nu > -1$ and $m = 1, 2, 3, \ldots$ Then the set of functions $J_\nu(\mu_m r/a)$ is orthogonal on the interval $0 \le r \le a$ with weight $r$:

$$\int_0^a J_\nu\!\left(\frac{\mu_m r}{a}\right) J_\nu\!\left(\frac{\mu_k r}{a}\right) r\,dr = \begin{cases} 0 & \text{if } m \ne k, \\ \dfrac{1}{2} a^2 \left(1 + \dfrac{s^2 - \nu^2}{\mu_m^2}\right) J_\nu^2(\mu_m) & \text{if } m = k. \end{cases}$$

4°. Let $\mu = \mu_m$ be positive roots of the transcendental equation

$$J_\nu(\lambda_m b) Y_\nu(\lambda_m a) - J_\nu(\lambda_m a) Y_\nu(\lambda_m b) = 0 \qquad (\nu > -1,\; m = 1, 2, 3, \ldots).$$

Then the set of functions

$$Z_\nu(\lambda_m r) = J_\nu(\lambda_m r) Y_\nu(\lambda_m a) - J_\nu(\lambda_m a) Y_\nu(\lambda_m r), \qquad m = 1, 2, 3, \ldots;$$

satisfying the conditions $Z_\nu(\lambda_m a) = Z_\nu(\lambda_m b) = 0$ is orthogonal on the interval $a \leq r \leq b$ with weight $r$:

$$\int_a^b Z_\nu(\lambda_m r) Z_\nu(\lambda_k r) r\, dr = \begin{cases} 0 & \text{if } m \neq k, \\ \dfrac{2}{\pi^2 \lambda_m^2} \dfrac{J_\nu^2(\lambda_m a) - J_\nu^2(\lambda_m b)}{J_\nu^2(\lambda_m b)} & \text{if } m = k. \end{cases}$$

5°. Let $\mu = \mu_m$ be positive roots of the transcendental equation

$$J_\nu'(\lambda_m b) Y_\nu'(\lambda_m a) - J_\nu'(\lambda_m a) Y_\nu'(\lambda_m b) = 0 \qquad (\nu > -1,\ m = 1, 2, 3, \ldots).$$

Then the set of functions

$$Z_\nu(\lambda_m r) = J_\nu(\lambda_m r) Y_\nu'(\lambda_m a) - J_\nu'(\lambda_m a) Y_\nu(\lambda_m r), \qquad m = 1, 2, 3, \ldots;$$

satisfying the conditions $Z_\nu'(\lambda_m a) = Z_\nu'(\lambda_m b) = 0$ is orthogonal on the interval $a \leq r \leq b$ with weight $r$:

$$\int_a^b Z_\nu(\lambda_m r) Z_\nu(\lambda_k r) r\, dr = \begin{cases} 0 & \text{if } m \neq k, \\ \dfrac{2}{\pi^2 \lambda_m^2} \left[ \left(1 - \dfrac{\nu^2}{b^2 \lambda_m^2}\right) \dfrac{[J_\nu'(\lambda_m a)]^2}{[J_\nu'(\lambda_m b)]^2} - \left(1 - \dfrac{\nu^2}{a^2 \lambda_m^2}\right) \right] & \text{if } m = k. \end{cases}$$

## 18.6.4. Hankel Functions (Bessel Functions of the Third Kind)

### 18.6.4-1. Definition.

The *Hankel functions of the first kind and the second kind* are related to Bessel functions by

$$H_\nu^{(1)}(z) = J_\nu(z) + i Y_\nu(z),$$
$$H_\nu^{(2)}(z) = J_\nu(z) - i Y_\nu(z),$$

where $i^2 = -1$.

### 18.6.4-2. Expansions as $z \to 0$ and $z \to \infty$.

Asymptotics for $z \to 0$:

$$H_0^{(1)}(z) \simeq \dfrac{2i}{\pi} \ln z, \qquad H_\nu^{(1)}(z) \simeq -\dfrac{i}{\pi} \dfrac{\Gamma(\nu)}{(z/2)^\nu} \qquad (\operatorname{Re} \nu > 0),$$

$$H_0^{(2)}(z) \simeq -\dfrac{2i}{\pi} \ln z, \qquad H_\nu^{(2)}(z) \simeq \dfrac{i}{\pi} \dfrac{\Gamma(\nu)}{(z/2)^\nu} \qquad (\operatorname{Re} \nu > 0).$$

Asymptotics for $|z| \to \infty$:

$$H_\nu^{(1)}(z) \simeq \sqrt{\dfrac{2}{\pi z}} \exp\left[i\left(z - \tfrac{1}{2}\pi\nu - \tfrac{1}{4}\pi\right)\right] \qquad (-\pi < \arg z < 2\pi),$$

$$H_\nu^{(2)}(z) \simeq \sqrt{\dfrac{2}{\pi z}} \exp\left[-i\left(z - \tfrac{1}{2}\pi\nu - \tfrac{1}{4}\pi\right)\right] \qquad (-2\pi < \arg z < \pi).$$

## 18.7. Modified Bessel Functions

### 18.7.1. Definitions. Basic Formulas

#### 18.7.1-1. Modified Bessel functions of the first and the second kind.

The *modified Bessel functions of the first kind*, $I_\nu(x)$, and the *modified Bessel functions of the second kind*, $K_\nu(x)$ (also called the *Macdonald function*), of order $\nu$ are solutions of the modified Bessel equation

$$x^2 y''_{xx} + x y'_x - (x^2 + \nu^2) y = 0$$

and are defined by the formulas

$$I_\nu(x) = \sum_{k=0}^{\infty} \frac{(x/2)^{2k+\nu}}{k!\,\Gamma(\nu+k+1)}, \qquad K_\nu(x) = \frac{\pi}{2} \frac{I_{-\nu}(x) - I_\nu(x)}{\sin(\pi\nu)},$$

(see below for $K_\nu(x)$ with $\nu = 0, 1, 2, \dots$).

#### 18.7.1-2. Some formulas.

The modified Bessel functions possess the properties

$$K_{-\nu}(x) = K_\nu(x); \qquad I_{-n}(x) = (-1)^n I_n(x), \quad n = 0, 1, 2, \dots$$
$$2\nu I_\nu(x) = x[I_{\nu-1}(x) - I_{\nu+1}(x)], \qquad 2\nu K_\nu(x) = -x[K_{\nu-1}(x) - K_{\nu+1}(x)],$$
$$\frac{d}{dx} I_\nu(x) = \frac{1}{2}[I_{\nu-1}(x) + I_{\nu+1}(x)], \qquad \frac{d}{dx} K_\nu(x) = -\frac{1}{2}[K_{\nu-1}(x) + K_{\nu+1}(x)].$$

#### 18.7.1-3. Modified Bessel functions for $\nu = \pm n \pm \frac{1}{2}$, where $n = 0, 1, 2, \dots$

$$I_{1/2}(x) = \sqrt{\frac{2}{\pi x}} \sinh x, \qquad I_{-1/2}(x) = \sqrt{\frac{2}{\pi x}} \cosh x,$$
$$I_{3/2}(x) = \sqrt{\frac{2}{\pi x}} \left(-\frac{1}{x} \sinh x + \cosh x\right), \qquad I_{-3/2}(x) = \sqrt{\frac{2}{\pi x}} \left(-\frac{1}{x} \cosh x + \sinh x\right),$$
$$I_{n+1/2}(x) = \frac{1}{\sqrt{2\pi x}} \left[ e^x \sum_{k=0}^{n} \frac{(-1)^k (n+k)!}{k!\,(n-k)!\,(2x)^k} - (-1)^n e^{-x} \sum_{k=0}^{n} \frac{(n+k)!}{k!\,(n-k)!\,(2x)^k} \right],$$
$$I_{-n-1/2}(x) = \frac{1}{\sqrt{2\pi x}} \left[ e^x \sum_{k=0}^{n} \frac{(-1)^k (n+k)!}{k!\,(n-k)!\,(2x)^k} + (-1)^n e^{-x} \sum_{k=0}^{n} \frac{(n+k)!}{k!\,(n-k)!\,(2x)^k} \right],$$
$$K_{\pm 1/2}(x) = \sqrt{\frac{\pi}{2x}} e^{-x}, \qquad K_{\pm 3/2}(x) = \sqrt{\frac{\pi}{2x}} \left(1 + \frac{1}{x}\right) e^{-x},$$
$$K_{n+1/2}(x) = K_{-n-1/2}(x) = \sqrt{\frac{\pi}{2x}} e^{-x} \sum_{k=0}^{n} \frac{(n+k)!}{k!\,(n-k)!\,(2x)^k}.$$

### 18.7.1-4. Modified Bessel functions for $\nu = n$, where $n = 0, 1, 2, \ldots$

If $\nu = n$ is a nonnegative integer, then

$$K_n(x) = (-1)^{n+1} I_n(x) \ln \frac{x}{2} + \frac{1}{2} \sum_{m=0}^{n-1} (-1)^m \left(\frac{x}{2}\right)^{2m-n} \frac{(n-m-1)!}{m!}$$

$$+ \frac{1}{2}(-1)^n \sum_{m=0}^{\infty} \left(\frac{x}{2}\right)^{n+2m} \frac{\psi(n+m+1) + \psi(m+1)}{m!(n+m)!}; \quad n = 0, 1, 2, \ldots,$$

where $\psi(z)$ is the logarithmic derivative of the gamma function; for $n = 0$, the first sum is dropped.

### 18.7.1-5. Wronskians and similar formulas.

$$W(I_\nu, I_{-\nu}) = -\frac{2}{\pi x} \sin(\pi\nu), \quad W(I_\nu, K_\nu) = -\frac{1}{x},$$

$$I_\nu(x) I_{-\nu+1}(x) - I_{-\nu}(x) I_{\nu-1}(x) = -\frac{2 \sin(\pi\nu)}{\pi x}, \quad I_\nu(x) K_{\nu+1}(x) + I_{\nu+1}(x) K_\nu(x) = \frac{1}{x},$$

where $W(f, g) = f g'_x - f'_x g$.

## 18.7.2. Integral Representations and Asymptotic Expansions

### 18.7.2-1. Integral representations.

The functions $I_\nu(x)$ and $K_\nu(x)$ can be represented in terms of definite integrals:

$$I_\nu(x) = \frac{x^\nu}{\pi^{1/2} 2^\nu \Gamma(\nu + \frac{1}{2})} \int_{-1}^{1} \exp(-xt)(1-t^2)^{\nu-1/2} \, dt \quad (x > 0, \ \nu > -\tfrac{1}{2}),$$

$$K_\nu(x) = \int_0^\infty \exp(-x \cosh t) \cosh(\nu t) \, dt \quad (x > 0),$$

$$K_\nu(x) = \frac{1}{\cos\left(\frac{1}{2}\pi\nu\right)} \int_0^\infty \cos(x \sinh t) \cosh(\nu t) \, dt \quad (x > 0, \ -1 < \nu < 1),$$

$$K_\nu(x) = \frac{1}{\sin\left(\frac{1}{2}\pi\nu\right)} \int_0^\infty \sin(x \sinh t) \sinh(\nu t) \, dt \quad (x > 0, \ -1 < \nu < 1).$$

For integer $\nu = n$,

$$I_n(x) = \frac{1}{\pi} \int_0^\pi \exp(x \cos t) \cos(nt) \, dt \quad (n = 0, 1, 2, \ldots),$$

$$K_0(x) = \int_0^\infty \cos(x \sinh t) \, dt = \int_0^\infty \frac{\cos(xt)}{\sqrt{t^2 + 1}} \, dt \quad (x > 0).$$

#### 18.7.2-2. Asymptotic expansions as $x \to \infty$.

$$I_\nu(x) = \frac{e^x}{\sqrt{2\pi x}}\left\{1 + \sum_{m=1}^{M}(-1)^m \frac{(4\nu^2-1)(4\nu^2-3^2)\ldots[4\nu^2-(2m-1)^2]}{m!\,(8x)^m}\right\},$$

$$K_\nu(x) = \sqrt{\frac{\pi}{2x}}\,e^{-x}\left\{1 + \sum_{m=1}^{M} \frac{(4\nu^2-1)(4\nu^2-3^2)\ldots[4\nu^2-(2m-1)^2]}{m!\,(8x)^m}\right\}.$$

The terms of the order of $O(x^{-M-1})$ are omitted in the braces.

#### 18.7.2-3. Integrals with modified Bessel functions.

$$\int_0^x x^\lambda I_\nu(x)\,dx = \frac{x^{\lambda+\nu+1}}{2^\nu(\lambda+\nu+1)\Gamma(\nu+1)}\,F\!\left(\frac{\lambda+\nu+1}{2},\,\frac{\lambda+\nu+3}{2},\,\nu+1;\,\frac{x^2}{4}\right), \qquad \operatorname{Re}(\lambda+\nu) > -1,$$

where $F(a,b,c;x)$ is the hypergeometric series (see Subsection 18.10.1),

$$\int_0^x x^\lambda K_\nu(x)\,dx = \frac{2^{\nu-1}\Gamma(\nu)}{\lambda-\nu+1}x^{\lambda-\nu+1}F\!\left(\frac{\lambda-\nu+1}{2},\,1-\nu,\,\frac{\lambda-\nu+3}{2};\,\frac{x^2}{4}\right)$$
$$+ \frac{2^{-\nu-1}\Gamma(-\nu)}{\lambda+\nu+1}x^{\lambda+\nu+1}F\!\left(\frac{\lambda+\nu+1}{2},\,1+\nu,\,\frac{\lambda+\nu+3}{2};\,\frac{x^2}{4}\right), \qquad \operatorname{Re}\lambda > |\operatorname{Re}\nu|-1.$$

## 18.8. Airy Functions
### 18.8.1. Definition and Basic Formulas
#### 18.8.1-1. Airy functions of the first and the second kinds.

The *Airy function of the first kind*, $\operatorname{Ai}(x)$, and the *Airy function of the second kind*, $\operatorname{Bi}(x)$, are solutions of the Airy equation

$$y''_{xx} - xy = 0$$

and are defined by the formulas

$$\operatorname{Ai}(x) = \frac{1}{\pi}\int_0^\infty \cos\!\left(\tfrac{1}{3}t^3 + xt\right)dt,$$
$$\operatorname{Bi}(x) = \frac{1}{\pi}\int_0^\infty \left[\exp\!\left(-\tfrac{1}{3}t^3 + xt\right) + \sin\!\left(\tfrac{1}{3}t^3 + xt\right)\right]dt.$$

Wronskian: $W\{\operatorname{Ai}(x), \operatorname{Bi}(x)\} = 1/\pi$.

#### 18.8.1-2. Relation to the Bessel functions and the modified Bessel functions.

$$\operatorname{Ai}(x) = \tfrac{1}{3}\sqrt{x}\left[I_{-1/3}(z) - I_{1/3}(z)\right] = \pi^{-1}\sqrt{\tfrac{1}{3}x}\,K_{1/3}(z), \qquad z = \tfrac{2}{3}x^{3/2},$$
$$\operatorname{Ai}(-x) = \tfrac{1}{3}\sqrt{x}\left[J_{-1/3}(z) + J_{1/3}(z)\right],$$
$$\operatorname{Bi}(x) = \sqrt{\tfrac{1}{3}x}\left[I_{-1/3}(z) + I_{1/3}(z)\right],$$
$$\operatorname{Bi}(-x) = \sqrt{\tfrac{1}{3}x}\left[J_{-1/3}(z) - J_{1/3}(z)\right].$$

## 18.8.2. Power Series and Asymptotic Expansions

**18.8.2-1. Power series expansions as $x \to 0$.**

$$\text{Ai}(x) = c_1 f(x) - c_2 g(x),$$
$$\text{Bi}(x) = \sqrt{3}\,[c_1 f(x) + c_2 g(x)],$$
$$f(x) = 1 + \frac{1}{3!}x^3 + \frac{1 \times 4}{6!}x^6 + \frac{1 \times 4 \times 7}{9!}x^9 + \cdots = \sum_{k=0}^{\infty} 3^k \left(\tfrac{1}{3}\right)_k \frac{x^{3k}}{(3k)!},$$
$$g(x) = x + \frac{2}{4!}x^4 + \frac{2 \times 5}{7!}x^7 + \frac{2 \times 5 \times 8}{10!}x^{10} + \cdots = \sum_{k=0}^{\infty} 3^k \left(\tfrac{2}{3}\right)_k \frac{x^{3k+1}}{(3k+1)!},$$

where $c_1 = 3^{-2/3}/\Gamma(2/3) \approx 0.3550$ and $c_2 = 3^{-1/3}/\Gamma(1/3) \approx 0.2588$.

**18.8.2-2. Asymptotic expansions as $x \to \infty$.**

For large values of $x$, the leading terms of asymptotic expansions of the Airy functions are

$$\text{Ai}(x) \simeq \tfrac{1}{2}\pi^{-1/2} x^{-1/4} \exp(-z), \quad z = \tfrac{2}{3}x^{3/2},$$
$$\text{Ai}(-x) \simeq \pi^{-1/2} x^{-1/4} \sin\!\left(z + \tfrac{\pi}{4}\right),$$
$$\text{Bi}(x) \simeq \pi^{-1/2} x^{-1/4} \exp(z),$$
$$\text{Bi}(-x) \simeq \pi^{-1/2} x^{-1/4} \cos\!\left(z + \tfrac{\pi}{4}\right).$$

# 18.9. Degenerate Hypergeometric Functions (Kummer Functions)

## 18.9.1. Definitions and Basic Formulas

**18.9.1-1. Degenerate hypergeometric functions $\Phi(a, b; x)$ and $\Psi(a, b; x)$.**

The *degenerate hypergeometric functions (Kummer functions)* $\Phi(a, b; x)$ and $\Psi(a, b; x)$ are solutions of the degenerate hypergeometric equation

$$xy''_{xx} + (b - x)y'_x - ay = 0.$$

In the case $b \ne 0, -1, -2, -3, \ldots$, the function $\Phi(a, b; x)$ can be represented as Kummer's series:

$$\Phi(a, b; x) = 1 + \sum_{k=1}^{\infty} \frac{(a)_k}{(b)_k} \frac{x^k}{k!},$$

where $(a)_k = a(a+1)\ldots(a+k-1)$, $(a)_0 = 1$.

Table 18.1 presents some special cases where $\Phi$ can be expressed in terms of simpler functions.

The function $\Psi(a, b; x)$ is defined as follows:

$$\Psi(a, b; x) = \frac{\Gamma(1-b)}{\Gamma(a-b+1)} \Phi(a, b; x) + \frac{\Gamma(b-1)}{\Gamma(a)} x^{1-b} \Phi(a-b+1, 2-b;\, x).$$

Table 18.2 presents some special cases where $\Psi$ can be expressed in terms of simpler functions.

TABLE 18.1
Special cases of the Kummer function $\Phi(a,b;z)$

| $a$ | $b$ | $z$ | $\Phi$ | Conventional notation |
|---|---|---|---|---|
| $a$ | $a$ | $x$ | $e^x$ | |
| $1$ | $2$ | $2x$ | $\dfrac{1}{x}e^x \sinh x$ | |
| $a$ | $a+1$ | $-x$ | $ax^{-a}\gamma(a,x)$ | Incomplete gamma function $\gamma(a,x) = \int_0^x e^{-t} t^{a-1}\, dt$ |
| $\dfrac{1}{2}$ | $\dfrac{3}{2}$ | $-x^2$ | $\dfrac{\sqrt{\pi}}{2}\operatorname{erf} x$ | Error function $\operatorname{erf} x = \dfrac{2}{\sqrt{\pi}} \int_0^x \exp(-t^2)\, dt$ |
| $-n$ | $\dfrac{1}{2}$ | $\dfrac{x^2}{2}$ | $\dfrac{n!}{(2n)!}\left(-\dfrac{1}{2}\right)^{-n} H_{2n}(x)$ | Hermite polynomials $H_n(x) = (-1)^n e^{x^2} \dfrac{d^n}{dx^n}\left(e^{-x^2}\right),$ $n = 0, 1, 2, \ldots$ |
| $-n$ | $\dfrac{3}{2}$ | $\dfrac{x^2}{2}$ | $\dfrac{n!}{(2n+1)!}\left(-\dfrac{1}{2}\right)^{-n} H_{2n+1}(x)$ | |
| $-n$ | $b$ | $x$ | $\dfrac{n!}{(b)_n} L_n^{(b-1)}(x)$ | Laguerre polynomials $L_n^{(\alpha)}(x) = \dfrac{e^x x^{-\alpha}}{n!} \dfrac{d^n}{dx^n}\left(e^{-x} x^{n+\alpha}\right),$ $\alpha = b-1,$ $(b)_n = b(b+1)\ldots(b+n-1)$ |
| $\nu + \dfrac{1}{2}$ | $2\nu+1$ | $2x$ | $\Gamma(1+\nu) e^x \left(\dfrac{x}{2}\right)^{-\nu} I_\nu(x)$ | Modified Bessel functions $I_\nu(x)$ |
| $n+1$ | $2n+2$ | $2x$ | $\Gamma\left(n+\dfrac{3}{2}\right) e^x \left(\dfrac{x}{2}\right)^{-n-\frac{1}{2}} I_{n+\frac{1}{2}}(x)$ | |

### 18.9.1-2. Kummer transformation and linear relations.

Kummer transformation:
$$\Phi(a,b;x) = e^x \Phi(b-a,b;-x), \qquad \Psi(a,b;x) = x^{1-b}\Psi(1+a-b, 2-b; x).$$

Linear relations for $\Phi$:
$$(b-a)\Phi(a-1,b;x) + (2a-b+x)\Phi(a,b;x) - a\Phi(a+1,b;x) = 0,$$
$$b(b-1)\Phi(a,b-1;x) - b(b-1+x)\Phi(a,b;x) + (b-a)x\Phi(a,b+1;x) = 0,$$
$$(a-b+1)\Phi(a,b;x) - a\Phi(a+1,b;x) + (b-1)\Phi(a,b-1;x) = 0,$$
$$b\Phi(a,b;x) - b\Phi(a-1,b;x) - x\Phi(a,b+1;x) = 0,$$
$$b(a+x)\Phi(a,b;x) - (b-a)x\Phi(a,b+1;x) - ab\Phi(a+1,b;x) = 0,$$
$$(a-1+x)\Phi(a,b;x) + (b-a)\Phi(a-1,b;x) - (b-1)\Phi(a,b-1;x) = 0.$$

Linear relations for $\Psi$:
$$\Psi(a-1,b;x) - (2a-b+x)\Psi(a,b;x) + a(a-b+1)\Psi(a+1,b;x) = 0,$$
$$(b-a-1)\Psi(a,b-1;x) - (b-1+x)\Psi(a,b;x) + x\Psi(a,b+1;x) = 0,$$
$$\Psi(a,b;x) - a\Psi(a+1,b;x) - \Psi(a,b-1;x) = 0,$$
$$(b-a)\Psi(a,b;x) - x\Psi(a,b+1;x) + \Psi(a-1,b;x) = 0,$$
$$(a+x)\Psi(a,b;x) + a(b-a-1)\Psi(a+1,b;x) - x\Psi(a,b+1;x) = 0,$$
$$(a-1+x)\Psi(a,b;x) - \Psi(a-1,b;x) + (a-c+1)\Psi(a,b-1;x) = 0.$$

## TABLE 18.2
Special cases of the Kummer function $\Psi(a, b; z)$

| $a$ | $b$ | $z$ | $\Psi$ | Conventional notation |
|---|---|---|---|---|
| $1-a$ | $1-a$ | $x$ | $e^x \Gamma(a, x)$ | Incomplete gamma function $\Gamma(a, x) = \int_x^\infty e^{-t} t^{a-1}\, dt$ |
| $\frac{1}{2}$ | $\frac{1}{2}$ | $x^2$ | $\sqrt{\pi} \exp(x^2)\, \mathrm{erfc}\, x$ | Complementary error function $\mathrm{erfc}\, x = \frac{2}{\sqrt{\pi}} \int_x^\infty \exp(-t^2)\, dt$ |
| $1$ | $1$ | $-x$ | $-e^{-x}\, \mathrm{Ei}(x)$ | Exponential integral $\mathrm{Ei}(x) = \int_{-\infty}^x \frac{e^t}{t}\, dt$ |
| $1$ | $1$ | $-\ln x$ | $-x^{-1}\, \mathrm{li}\, x$ | Logarithmic integral $\mathrm{li}\, x = \int_0^x \frac{dt}{t}$ |
| $\frac{1}{2} - \frac{n}{2}$ | $\frac{3}{2}$ | $x^2$ | $2^{-n} x^{-1} H_n(x)$ | Hermite polynomials $H_n(x) = (-1)^n e^{x^2} \frac{d^n}{dx^n}\left(e^{-x^2}\right)$, $n = 0, 1, 2, \ldots$ |
| $\nu + \frac{1}{2}$ | $2\nu + 1$ | $2x$ | $\pi^{-1/2}(2x)^{-\nu} e^x K_\nu(x)$ | Modified Bessel functions $K_\nu(x)$ |

### 18.9.1-3. Differentiation formulas and Wronskian.

Differentiation formulas:

$$\frac{d}{dx}\Phi(a, b; x) = \frac{a}{b}\Phi(a+1, b+1; x), \qquad \frac{d^n}{dx^n}\Phi(a, b; x) = \frac{(a)_n}{(b)_n}\Phi(a+n, b+n; x),$$

$$\frac{d}{dx}\Psi(a, b; x) = -a\Psi(a+1, b+1; x), \qquad \frac{d^n}{dx^n}\Psi(a, b; x) = (-1)^n (a)_n \Psi(a+n, b+n; x).$$

Wronskian:

$$W(\Phi, \Psi) = \Phi \Psi'_x - \Phi'_x \Psi = -\frac{\Gamma(b)}{\Gamma(a)} x^{-b} e^x.$$

### 18.9.1-4. Degenerate hypergeometric functions for $n = 0, 1, 2, \ldots$

$$\Psi(a, n+1; x) = \frac{(-1)^{n-1}}{n!\, \Gamma(a-n)} \left\{ \Phi(a, n+1; x) \ln x \right.$$
$$\left. + \sum_{r=0}^\infty \frac{(a)_r}{(n+1)_r} [\psi(a+r) - \psi(1+r) - \psi(1+n+r)] \frac{x^r}{r!} \right\} + \frac{(n-1)!}{\Gamma(a)} \sum_{r=0}^{n-1} \frac{(a-n)_r}{(1-n)_r} \frac{x^{r-n}}{r!},$$

where $n = 0, 1, 2, \ldots$ (the last sum is dropped for $n = 0$), $\psi(z) = [\ln \Gamma(z)]'_z$ is the logarithmic derivative of the gamma function,

$$\psi(1) = -\mathcal{C}, \qquad \psi(n) = -\mathcal{C} + \sum_{k=1}^{n-1} k^{-1},$$

where $\mathcal{C} = 0.5772\ldots$ is the Euler constant.

If $b < 0$, then the formula
$$\Psi(a, b; x) = x^{1-b} \Psi(a - b + 1, 2 - b; x)$$
is valid for any $x$.

For $b \neq 0, -1, -2, -3, \ldots$, the general solution of the degenerate hypergeometric equation can be represented in the form
$$y = C_1 \Phi(a, b; x) + C_2 \Psi(a, b; x),$$
and for $b = 0, -1, -2, -3, \ldots$, in the form
$$y = x^{1-b} [C_1 \Phi(a - b + 1, 2 - b; x) + C_2 \Psi(a - b + 1, 2 - b; x)].$$

## 18.9.2. Integral Representations and Asymptotic Expansions

**18.9.2-1. Integral representations.**

$$\Phi(a, b; x) = \frac{\Gamma(b)}{\Gamma(a)\Gamma(b-a)} \int_0^1 e^{xt} t^{a-1} (1-t)^{b-a-1} \, dt \quad \text{(for } b > a > 0\text{)},$$

$$\Psi(a, b; x) = \frac{1}{\Gamma(a)} \int_0^\infty e^{-xt} t^{a-1} (1+t)^{b-a-1} \, dt \quad \text{(for } a > 0, \, x > 0\text{)},$$

where $\Gamma(a)$ is the gamma function.

**18.9.2-2. Asymptotic expansion as $|x| \to \infty$.**

$$\Phi(a, b; x) = \frac{\Gamma(b)}{\Gamma(a)} e^x x^{a-b} \left[ \sum_{n=0}^N \frac{(b-a)_n (1-a)_n}{n!} x^{-n} + \varepsilon \right], \quad x > 0,$$

$$\Phi(a, b; x) = \frac{\Gamma(b)}{\Gamma(b-a)} (-x)^{-a} \left[ \sum_{n=0}^N \frac{(a)_n (a-b+1)_n}{n!} (-x)^{-n} + \varepsilon \right], \quad x < 0,$$

$$\Psi(a, b; x) = x^{-a} \left[ \sum_{n=0}^N (-1)^n \frac{(a)_n (a-b+1)_n}{n!} x^{-n} + \varepsilon \right], \quad -\infty < x < \infty,$$

where $\varepsilon = O(x^{-N-1})$.

**18.9.2-3. Integrals with degenerate hypergeometric functions.**

$$\int \Phi(a, b; x) \, dx = \frac{b-1}{a-1} \Psi(a - 1, b - 1; x) + C,$$

$$\int \Psi(a, b; x) \, dx = \frac{1}{1-a} \Psi(a - 1, b - 1; x) + C,$$

$$\int x^n \Phi(a, b; x) \, dx = n! \sum_{k=1}^{n+1} \frac{(-1)^{k+1} (1-b)_k x^{n-k+1}}{(1-a)_k (n-k+1)!} \Phi(a - k, b - k; x) + C,$$

$$\int x^n \Psi(a, b; x) \, dx = n! \sum_{k=1}^{n+1} \frac{(-1)^{k+1} x^{n-k+1}}{(1-a)_k (n-k+1)!} \Psi(a - k, b - k; x) + C.$$

### 18.9.3. Whittaker Functions

The *Whittaker functions* $M_{k,\mu}(x)$ and $W_{k,\mu}(x)$ are linearly independent solutions of the Whittaker equation:
$$y''_{xx} + \left[-\tfrac{1}{4} + \tfrac{1}{2}k + \left(\tfrac{1}{4} - \mu^2\right)x^{-2}\right]y = 0.$$
The Whittaker functions are expressed in terms of degenerate hypergeometric functions as
$$M_{k,\mu}(x) = x^{\mu+1/2}e^{-x/2}\Phi\left(\tfrac{1}{2} + \mu - k, \, 1 + 2\mu; \, x\right),$$
$$W_{k,\mu}(x) = x^{\mu+1/2}e^{-x/2}\Psi\left(\tfrac{1}{2} + \mu - k, \, 1 + 2\mu; \, x\right).$$

## 18.10. Hypergeometric Functions

### 18.10.1. Various Representations of the Hypergeometric Function

#### 18.10.1-1. Representations of the hypergeometric function via hypergeometric series.

The *hypergeometric function* $F(\alpha,\beta,\gamma;x)$ is a solution of the Gaussian hypergeometric equation
$$x(x-1)y''_{xx} + [(\alpha+\beta+1)x - \gamma]y'_x + \alpha\beta y = 0.$$

For $\gamma \neq 0, -1, -2, -3, \ldots$, the function $F(\alpha,\beta,\gamma;x)$ can be expressed in terms of the hypergeometric series:
$$F(\alpha,\beta,\gamma;x) = 1 + \sum_{k=1}^{\infty} \frac{(\alpha)_k(\beta)_k}{(\gamma)_k} \frac{x^k}{k!}, \quad (\alpha)_k = \alpha(\alpha+1)\ldots(\alpha+k-1),$$
which certainly converges for $|x| < 1$.

If $\gamma$ is not an integer, then the general solution of the hypergeometric equation can be written in the form
$$y = C_1 F(\alpha,\beta,\gamma;x) + C_2 x^{1-\gamma} F(\alpha-\gamma+1, \, \beta-\gamma+1, \, 2-\gamma; \, x).$$

Table 18.3 shows some special cases where $F$ can be expressed in term of elementary functions.

#### 18.10.1-2. Integral representation.

For $\gamma > \beta > 0$, the hypergeometric function can be expressed in terms of a definite integral:
$$F(\alpha,\beta,\gamma;x) = \frac{\Gamma(\gamma)}{\Gamma(\beta)\Gamma(\gamma-\beta)} \int_0^1 t^{\beta-1}(1-t)^{\gamma-\beta-1}(1-tx)^{-\alpha}\, dt,$$
where $\Gamma(\beta)$ is the gamma function.

### 18.10.2. Basic Properties

#### 18.10.2-1. Linear transformation formulas.

$$F(\alpha,\beta,\gamma;x) = F(\beta,\alpha,\gamma;x),$$
$$F(\alpha,\beta,\gamma;x) = (1-x)^{\gamma-\alpha-\beta} F(\gamma-\alpha, \, \gamma-\beta, \, \gamma; \, x),$$
$$F(\alpha,\beta,\gamma;x) = (1-x)^{-\alpha} F\left(\alpha, \, \gamma-\beta, \, \gamma; \, \frac{x}{x-1}\right),$$
$$F(\alpha,\beta,\gamma;x) = (1-x)^{-\beta} F\left(\beta, \, \gamma-\alpha, \, \gamma; \, \frac{x}{x-1}\right).$$

TABLE 18.3
Some special cases where the hypergeometric function $F(\alpha,\beta,\gamma;z)$
can be expressed in terms of elementary functions.

| $\alpha$ | $\beta$ | $\gamma$ | $z$ | $F$ |
|---|---|---|---|---|
| $-n$ | $\beta$ | $\gamma$ | $x$ | $\sum_{k=0}^{n} \frac{(-n)_k (\beta)_k}{(\gamma)_k} \frac{x^k}{k!}$, where $n=1,2,\ldots$ |
| $-n$ | $\beta$ | $-n-m$ | $x$ | $\sum_{k=0}^{n} \frac{(-n)_k (\beta)_k}{(-n-m)_k} \frac{x^k}{k!}$, where $n=1,2,\ldots$ |
| $\alpha$ | $\beta$ | $\beta$ | $x$ | $(1-x)^{-\alpha}$ |
| $\alpha$ | $\alpha+1$ | $\frac{1}{2}\alpha$ | $x$ | $(1+x)(1-x)^{-\alpha-1}$ |
| $\alpha$ | $\alpha+\frac{1}{2}$ | $2\alpha+1$ | $x$ | $\left(\frac{1+\sqrt{1-x}}{2}\right)^{-2\alpha}$ |
| $\alpha$ | $\alpha+\frac{1}{2}$ | $2\alpha$ | $x$ | $\frac{1}{\sqrt{1-x}}\left(\frac{1+\sqrt{1-x}}{2}\right)^{1-2\alpha}$ |
| $\alpha$ | $\alpha+\frac{1}{2}$ | $\frac{3}{2}$ | $x^2$ | $\frac{(1+x)^{1-2\alpha}-(1-x)^{1-2\alpha}}{2x(1-2\alpha)}$ |
| $\alpha$ | $\alpha+\frac{1}{2}$ | $\frac{1}{2}$ | $-\tan^2 x$ | $\cos^{2\alpha} x \cos(2\alpha x)$ |
| $\alpha$ | $\alpha+\frac{1}{2}$ | $\frac{1}{2}$ | $x^2$ | $\frac{1}{2}\left[(1+x)^{-2\alpha}+(1-x)^{-2\alpha}\right]$ |
| $\alpha$ | $\alpha-\frac{1}{2}$ | $2\alpha-1$ | $x$ | $2^{2\alpha-2}\left(1+\sqrt{1-x}\right)^{2-2\alpha}$ |
| $\alpha$ | $2-\alpha$ | $\frac{3}{2}$ | $\sin^2 x$ | $\frac{\sin[(2\alpha-2)x]}{(\alpha-1)\sin(2x)}$ |
| $\alpha$ | $1-\alpha$ | $\frac{1}{2}$ | $-x^2$ | $\frac{\left(\sqrt{1+x^2}+x\right)^{2\alpha-1}+\left(\sqrt{1+x^2}-x\right)^{2\alpha-1}}{2\sqrt{1+x^2}}$ |
| $\alpha$ | $1-\alpha$ | $\frac{3}{2}$ | $\sin^2 x$ | $\frac{\sin[(2\alpha-1)x]}{(\alpha-1)\sin(2x)}$ |
| $\alpha$ | $1-\alpha$ | $\frac{1}{2}$ | $\sin^2 x$ | $\frac{\cos[(2\alpha-1)x]}{\cos x}$ |
| $\alpha$ | $-\alpha$ | $\frac{1}{2}$ | $-x^2$ | $\frac{1}{2}\left[\left(\sqrt{1+x^2}+x\right)^{2\alpha}+\left(\sqrt{1+x^2}-x\right)^{2\alpha}\right]$ |
| $\alpha$ | $-\alpha$ | $\frac{1}{2}$ | $\sin^2 x$ | $\cos(2\alpha x)$ |
| $1$ | $1$ | $2$ | $-x$ | $\frac{1}{x}\ln(x+1)$ |
| $\frac{1}{2}$ | $1$ | $\frac{3}{2}$ | $x^2$ | $\frac{1}{2x}\ln\frac{1+x}{1-x}$ |
| $\frac{1}{2}$ | $1$ | $\frac{3}{2}$ | $-x^2$ | $\frac{1}{x}\arctan x$ |
| $\frac{1}{2}$ | $\frac{1}{2}$ | $\frac{3}{2}$ | $x^2$ | $\frac{1}{x}\arcsin x$ |
| $\frac{1}{2}$ | $\frac{1}{2}$ | $\frac{3}{2}$ | $-x^2$ | $\frac{1}{x}\operatorname{arcsinh} x$ |
| $n+1$ | $n+m+1$ | $n+m+l+2$ | $x$ | $\frac{(-1)^m (n+m+l+1)!}{n!\,l!\,(n+m)!\,(m+l)!}\frac{d^{n+m}}{dx^{n+m}}\left\{(1-x)^{m+l}\frac{d^l F}{dx^l}\right\}$, $F=-\frac{\ln(1-x)}{x}$, $n,m,l=0,1,2,\ldots$ |

### 18.10.2-2. Gauss's linear relations for contiguous functions.

$$(\beta - \alpha)F(\alpha, \beta, \gamma; x) + \alpha F(\alpha + 1, \beta, \gamma; x) - \beta F(\alpha, \beta + 1, \gamma; x) = 0,$$
$$(\gamma - \alpha - 1)F(\alpha, \beta, \gamma; x) + \alpha F(\alpha + 1, \beta, \gamma; x) - (\gamma - 1)F(\alpha, \beta, \gamma - 1; x) = 0,$$
$$(\gamma - \beta - 1)F(\alpha, \beta, \gamma; x) + \beta F(\alpha, \beta + 1, \gamma; x) - (\gamma - 1)F(\alpha, \beta, \gamma - 1; x) = 0,$$
$$(\gamma - \alpha - \beta)F(\alpha, \beta, \gamma; x) + \alpha(1 - x)F(\alpha + 1, \beta, \gamma; x) - (\gamma - \beta)F(\alpha, \beta - 1, \gamma; x) = 0,$$
$$(\gamma - \alpha - \beta)F(\alpha, \beta, \gamma; x) - (\gamma - \alpha)F(\alpha - 1, \beta, \gamma; x) + \beta(1 - x)F(\alpha, \beta + 1, \gamma; x) = 0.$$

### 18.10.2-3. Differentiation formulas.

$$\frac{d}{dx}F(\alpha, \beta, \gamma; x) = \frac{\alpha\beta}{\gamma}F(\alpha + 1, \beta + 1, \gamma + 1; x),$$
$$\frac{d^n}{dx^n}F(\alpha, \beta, \gamma; x) = \frac{(\alpha)_n (\beta)_n}{(\gamma)_n}F(\alpha + n, \beta + n, \gamma + n; x),$$
$$\frac{d^n}{dx^n}\left[x^{\gamma-1}F(\alpha, \beta, \gamma; x)\right] = (\gamma - n)_n x^{\gamma-n-1}F(\alpha, \beta, \gamma - n; x),$$
$$\frac{d^n}{dx^n}\left[x^{\alpha+n-1}F(\alpha, \beta, \gamma; x)\right] = (\alpha)_n x^{\alpha-1}F(\alpha + n, \beta, \gamma; x),$$

where $(\alpha)_n = \alpha(\alpha + 1)\ldots(\alpha + n - 1)$.

See Abramowitz and Stegun (1964) and Bateman and Erdélyi (1953, Vol. 1) for more detailed information about hypergeometric functions.

## 18.11. Legendre Polynomials, Legendre Functions, and Associated Legendre Functions

### 18.11.1. Legendre Polynomials and Legendre Functions

#### 18.11.1-1. Implicit and recurrence formulas for Legendre polynomials and functions.

The *Legendre polynomials* $P_n(x)$ and the *Legendre functions* $Q_n(x)$ are solutions of the second-order linear ordinary differential equation

$$(1 - x^2)y''_{xx} - 2xy'_x + n(n + 1)y = 0.$$

The *Legendre polynomials* $P_n(x)$ and the Legendre functions $Q_n(x)$ are defined by the formulas

$$P_n(x) = \frac{1}{n!\, 2^n} \frac{d^n}{dx^n}(x^2 - 1)^n,$$
$$Q_n(x) = \frac{1}{2}P_n(x) \ln\frac{1+x}{1-x} - \sum_{m=1}^{n} \frac{1}{m} P_{m-1}(x) P_{n-m}(x).$$

The polynomials $P_n = P_n(x)$ can be calculated using the formulas

$$P_0(x) = 1, \quad P_1(x) = x, \quad P_2(x) = \frac{1}{2}(3x^2 - 1),$$
$$P_3(x) = \frac{1}{2}(5x^3 - 3x), \quad P_4(x) = \frac{1}{8}(35x^4 - 30x^2 + 3),$$
$$P_{n+1}(x) = \frac{2n+1}{n+1} x P_n(x) - \frac{n}{n+1} P_{n-1}(x).$$

The first five functions $Q_n = Q_n(x)$ have the form

$$Q_0(x) = \frac{1}{2}\ln\frac{1+x}{1-x}, \quad Q_1(x) = \frac{x}{2}\ln\frac{1+x}{1-x} - 1,$$

$$Q_2(x) = \frac{1}{4}(3x^2 - 1)\ln\frac{1+x}{1-x} - \frac{3}{2}x, \quad Q_3(x) = \frac{1}{4}(5x^3 - 3x)\ln\frac{1+x}{1-x} - \frac{5}{2}x^2 + \frac{2}{3},$$

$$Q_4(x) = \frac{1}{16}(35x^4 - 30x^2 + 3)\ln\frac{1+x}{1-x} - \frac{35}{8}x^3 + \frac{55}{24}x.$$

The polynomials $P_n(x)$ have the explicit representation

$$P_n(x) = 2^{-n} \sum_{m=0}^{[n/2]} (-1)^m C_n^m C_{2n-2m}^n x^{n-2m},$$

where $[A]$ stands for the integer part of a number $A$.

18.11.1-2. Integral representation. Useful formulas.

Integral representation of the Legendre polynomials (*Laplace integral*):

$$P_n(x) = \frac{1}{\pi}\int_0^\pi (x \pm \sqrt{x^2-1}\cos t)^n \, dt, \qquad x > 1.$$

Integral representation of the Legendre polynomials (*Dirichlet–Mehler integral*):

$$P_n(\cos\theta) = \frac{\sqrt{2}}{\pi}\int_0^\theta \frac{\cos\left[(n+\tfrac{1}{2})\psi\right] d\psi}{\sqrt{\cos\psi - \cos\theta}}, \qquad 0 < \theta < \pi, \quad n = 0, 1, \ldots$$

Integral representation of the Legendre functions:

$$Q_n(x) = 2^n \int_x^\infty \frac{(t-x)^n}{(t^2-1)^{n+1}} \, dt, \qquad x > 1.$$

Properties:
$$P_n(-x) = (-1)^n P_n(x), \quad Q_n(-x) = (-1)^{n+1} Q_n(x).$$

Recurrence relations:
$$(n+1)P_{n+1}(x) - (2n+1)xP_n(x) + nP_{n-1}(x) = 0,$$
$$(x^2-1)\frac{d}{dx}P_n(x) = n[xP_n(x) - P_{n-1}(x)] = \frac{n(n+1)}{2n+1}[P_{n+1}(x) - P_{n-1}(x)].$$

Values of the Legendre polynomials and their derivatives at $x = 0$:

$$P_{2m}(0) = (-1)^m \frac{(2m-1)!!}{2^m m!}, \quad P_{2m+1}(0) = 0, \quad P'_{2m}(0) = 0, \quad P'_{2m+1}(0) = (-1)^m \frac{(2m+1)!!}{2^m m!}.$$

Asymptotic formula as $n \to \infty$:

$$P_n(\cos\theta) \approx \left(\frac{2}{\pi n \sin\theta}\right)^{1/2} \sin\left[\left(n+\frac{1}{2}\right)\theta + \frac{\pi}{4}\right], \qquad 0 < \theta < \pi.$$

### 18.11.1-3. Zeros and orthogonality of the Legendre polynomials.

The polynomials $P_n(x)$ (with natural $n$) have exactly $n$ real distinct zeros; all zeros lie on the interval $-1 < x < 1$. The zeros of $P_n(x)$ and $P_{n+1}(x)$ alternate with each other. The function $Q_n(x)$ has exactly $n+1$ zeros, which lie on the interval $-1 < x < 1$.

The functions $P_n(x)$ form an orthogonal system on the interval $-1 \leq x \leq 1$, with

$$\int_{-1}^{1} P_n(x) P_m(x)\, dx = \begin{cases} 0 & \text{if } n \neq m, \\ \dfrac{2}{2n+1} & \text{if } n = m. \end{cases}$$

### 18.11.1-4. Generating functions.

The generating function for Legendre polynomials is

$$\frac{1}{\sqrt{1-2sx+s^2}} = \sum_{n=0}^{\infty} P_n(x) s^n \qquad (|s| < 1).$$

The generating function for Legendre functions is

$$\frac{1}{\sqrt{1-2sx+s^2}} \ln\left[\frac{x-s+\sqrt{1-2sx+s^2}}{\sqrt{1-x^2}}\right] = \sum_{n=0}^{\infty} Q_n(x) s^n \qquad (|s| < 1,\ x > 1).$$

## 18.11.2. Associated Legendre Functions with Integer Indices and Real Argument

### 18.11.2-1. Formulas for associated Legendre functions. Differential equation.

The *associated Legendre functions* $P_n^m(x)$ of order $m$ are defined by the formulas

$$P_n^m(x) = (1-x^2)^{m/2} \frac{d^m}{dx^m} P_n(x), \qquad n = 1, 2, 3, \ldots, \quad m = 0, 1, 2, \ldots$$

It is assumed by definition that $P_n^0(x) = P_n(x)$.

Properties:

$$P_n^m(x) = 0 \quad \text{if} \quad m > n, \qquad P_n^m(-x) = (-1)^{n-m} P_n^m(x).$$

The associated Legendre functions $P_n^m(x)$ have exactly $n-m$ real zeros, which and lie on the interval $-1 < x < 1$.

The associated Legendre functions $P_n^m(x)$ with low indices:

$$P_1^1(x) = (1-x^2)^{1/2}, \quad P_2^1(x) = 3x(1-x^2)^{1/2}, \quad P_2^2(x) = 3(1-x^2),$$
$$P_3^1(x) = \tfrac{3}{2}(5x^2-1)(1-x^2)^{1/2}, \quad P_3^2(x) = 15x(1-x^2), \quad P_3^3(x) = 15(1-x^2)^{3/2}.$$

The associated Legendre functions $P_n^m(x)$ with $n > m$ are solutions of the linear ordinary differential equation

$$(1-x^2) y''_{xx} - 2x y'_x + \left[n(n+1) - \frac{m^2}{1-x^2}\right] y = 0.$$

**18.11.2-2. Orthogonality of the associated Legendre functions.**

The functions $P_n^m(x)$ form an orthogonal system on the interval $-1 \leq x \leq 1$, with

$$\int_{-1}^{1} P_n^m(x) P_k^m(x)\, dx = \begin{cases} 0 & \text{if } n \neq k, \\ \dfrac{2}{2n+1} \dfrac{(n+m)!}{(n-m)!} & \text{if } n = k. \end{cases}$$

The functions $P_n^m(x)$ (with $m \neq 0$) are orthogonal on the interval $-1 \leq x \leq 1$ with weight $(1-x^2)^{-1}$, that is,

$$\int_{-1}^{1} \frac{P_n^m(x) P_n^k(x)}{1-x^2}\, dx = \begin{cases} 0 & \text{if } m \neq k, \\ \dfrac{(n+m)!}{m(n-m)!} & \text{if } m = k. \end{cases}$$

### 18.11.3. Associated Legendre Functions. General Case

**18.11.3-1. Definitions. Basic formulas.**

In the general case, the associated Legendre functions of the first and the second kind, $P_\nu^\mu(z)$ and $Q_\nu^\mu(z)$, are linearly independent solutions of the Legendre equation

$$(1-z^2) y''_{zz} - 2z y'_z + \left[\nu(\nu+1) - \frac{\mu^2}{1-z^2}\right] y = 0,$$

where the parameters $\nu$ and $\mu$ and the variable $z$ can assume arbitrary real or complex values.

For $|1-z| < 2$, the formulas

$$P_\nu^\mu(z) = \frac{1}{\Gamma(1-\mu)} \left(\frac{z+1}{z-1}\right)^{\mu/2} F\left(-\nu, 1+\nu, 1-\mu; \frac{1-z}{2}\right),$$

$$Q_\nu^\mu(z) = A \left(\frac{z-1}{z+1}\right)^{\frac{\mu}{2}} F\left(-\nu, 1+\nu, 1+\mu; \frac{1-z}{2}\right) + B \left(\frac{z+1}{z-1}\right)^{\frac{\mu}{2}} F\left(-\nu, 1+\nu, 1-\mu; \frac{1-z}{2}\right),$$

$$A = e^{i\mu\pi} \frac{\Gamma(-\mu)\Gamma(1+\nu+\mu)}{2\,\Gamma(1+\nu-\mu)}, \quad B = e^{i\mu\pi} \frac{\Gamma(\mu)}{2}, \quad i^2 = -1,$$

are valid, where $F(a, b, c; z)$ is the hypergeometric series (see Section 18.10).

For $|z| > 1$,

$$P_\nu^\mu(z) = \frac{2^{-\nu-1} \Gamma(-\frac{1}{2} - \nu)}{\sqrt{\pi}\, \Gamma(-\nu-\mu)} z^{-\nu+\mu-1} (z^2-1)^{-\mu/2} F\left(\frac{1+\nu-\mu}{2}, \frac{2+\nu-\mu}{2}, \frac{2\nu+3}{2}; \frac{1}{z^2}\right)$$
$$+ \frac{2^\nu \Gamma(\frac{1}{2}+\nu)}{\Gamma(1+\nu-\mu)} z^{\nu+\mu} (z^2-1)^{-\mu/2} F\left(-\frac{\nu+\mu}{2}, \frac{1-\nu-\mu}{2}, \frac{1-2\nu}{2}; \frac{1}{z^2}\right),$$

$$Q_\nu^\mu(z) = e^{i\pi\mu} \frac{\sqrt{\pi}\, \Gamma(\nu+\mu+1)}{2^{\nu+1} \Gamma(\nu+\frac{3}{2})} z^{-\nu-\mu-1} (z^2-1)^{\mu/2} F\left(\frac{2+\nu+\mu}{2}, \frac{1+\nu+\mu}{2}, \frac{2\nu+3}{2}; \frac{1}{z^2}\right).$$

The functions $P_\nu(z) \equiv P_\nu^0(z)$ and $Q_\nu(z) \equiv Q_\nu^0(z)$ are called the *Legendre functions*. For $n = 1, 2, \ldots$,

$$P_\nu^n(z) = (z^2-1)^{n/2} \frac{d^n}{dz^n} P_\nu(z), \quad Q_\nu^n(z) = (z^2-1)^{n/2} \frac{d^n}{dz^n} Q_\nu(z).$$

### 18.11.3-2. Relations between associated Legendre functions.

$$P_\nu^\mu(z) = P_{-\nu-1}^\mu(z), \qquad P_\nu^n(z) = \frac{\Gamma(\nu+n+1)}{\Gamma(\nu-n+1)} P_\nu^{-n}(z), \qquad n = 0, 1, 2, \ldots,$$

$$P_{\nu+1}^\mu(z) = \frac{2\nu+1}{\nu-\mu+1} z P_\nu^\mu(z) - \frac{\nu+\mu}{\nu-\mu+1} P_{\nu-1}^\mu(z),$$

$$P_{\nu+1}^\mu(z) = P_{\nu-1}^\mu(z) + (2\nu+1)(z^2-1)^{1/2} P_\nu^{\mu-1}(z),$$

$$(z^2-1)\frac{d}{dz} P_\nu^\mu(z) = \nu z P_\nu^\mu(z) - (\nu+m) P_{\nu-1}^\mu(z),$$

$$Q_\nu^\mu(z) = \frac{\pi}{2\sin(\mu\pi)} e^{i\pi\mu} \left[ P_\nu^\mu(z) - \frac{\Gamma(1+\nu+\mu)}{\Gamma(1+\nu-\mu)} P_\nu^{-\mu}(z) \right],$$

$$Q_\nu^\mu(z) = e^{i\pi\mu} \left(\frac{\pi}{2}\right)^{1/2} \Gamma(\nu+\mu+1)(z^2-1)^{-1/4} P_{-\mu-1/2}^{-\nu-1/2}\left(\frac{z}{\sqrt{z^2-1}}\right), \qquad \operatorname{Re} z > 0.$$

### 18.11.3-3. Integral representations.

For $\operatorname{Re}(-\mu) > \operatorname{Re}\nu > -1$,

$$P_\nu^\mu(z) = \frac{2^{-\nu}(z^2-1)^{-\mu/2}}{\Gamma(\nu+1)\Gamma(-\mu-\nu)} \int_0^\infty (z+\cosh t)^{\mu-\nu-1}(\sinh t)^{2\nu+1}\, dt,$$

where $z$ does not lie on the real axis between $-1$ and $\infty$.

For $\mu < 1/2$,

$$P_\nu^\mu(z) = \frac{2^\mu (z^2-1)^{-\mu/2}}{\sqrt{\pi}\,\Gamma(\frac{1}{2}-\mu)} \int_0^\pi \left(z+\sqrt{z^2-1}\cos t\right)^{\nu+\mu}(\sin t)^{-2\mu}\, dt,$$

where $z$ does not lie on the real axis between $-1$ and $1$.

For $\operatorname{Re}\nu > -1$ and $\operatorname{Re}(\nu+\mu+1) > 0$,

$$Q_\nu^\mu(z) = e^{\pi\mu i} \frac{\Gamma(\nu+\mu+1)(z^2-1)^{-\mu/2}}{2^{\nu+1}\Gamma(\nu+1)} \int_0^\pi (z+\cos t)^{\mu-\nu-1}(\sin t)^{2\nu+1}\, dt,$$

where $z$ does not lie on the real axis between $-1$ and $1$.

For $n = 0, 1, 2, \ldots$,

$$P_\nu^n(z) = \frac{\Gamma(\nu+n+1)}{\pi\Gamma(\nu+1)} \int_0^\pi \left(z+\sqrt{z^2-1}\cos t\right)^\nu \cos(nt)\, dt, \qquad \operatorname{Re} z > 0;$$

$$Q_\nu^n(z) = (-1)^n \frac{\Gamma(\nu+n+1)}{2^{\nu+1}\Gamma(\nu+1)} (z^2-1)^{-n/2} \int_0^\pi (z+\cos t)^{n-\nu-1}(\sin t)^{2\nu+1}\, dt, \qquad \operatorname{Re}\nu > -1.$$

Note that $z \neq x$, $-1 < x < 1$, in the latter formula.

### 18.11.3-4. Modified associated Legendre functions.

The *modified associated Legendre functions*, on the cut $z = x$, $-1 < x < 1$, of the real axis are defined by the formulas

$$\mathsf{P}_\nu^\mu(x) = \tfrac{1}{2}\left[e^{\frac{1}{2}i\mu\pi} P_\nu^\mu(x+i0) + e^{-\frac{1}{2}i\mu\pi} P_\nu^\mu(x-i0)\right],$$

$$\mathsf{Q}_\nu^\mu(x) = \tfrac{1}{2} e^{-i\mu\pi}\left[e^{-\frac{1}{2}i\mu\pi} Q_\nu^\mu(x+i0) + e^{\frac{1}{2}i\mu\pi} Q_\nu^\mu(x-i0)\right].$$

Notation:
$$\mathsf{P}_\nu(x) = \mathsf{P}_\nu^0(x), \qquad \mathsf{Q}_\nu(x) = \mathsf{Q}_\nu^0(x).$$

**18.11.3-5. Trigonometric expansions.**

For $-1 < x < 1$, the modified associated Legendre functions can be represented in the form of the trigonometric series:

$$\mathsf{P}_\nu^\mu(\cos\theta) = \frac{2^{\mu+1}}{\sqrt{\pi}} \frac{\Gamma(\nu+\mu+1)}{\Gamma(\nu+\frac{3}{2})} (\sin\theta)^\mu \sum_{k=0}^{\infty} \frac{(\frac{1}{2}+\mu)_k (1+\nu+\mu)_k}{k!\,(\nu+\frac{3}{2})_k} \sin[(2k+\nu+\mu+1)\theta],$$

$$\mathsf{Q}_\nu^\mu(\cos\theta) = \sqrt{\pi}\,2^\mu \frac{\Gamma(\nu+\mu+1)}{\Gamma(\nu+\frac{3}{2})} (\sin\theta)^\mu \sum_{k=0}^{\infty} \frac{(\frac{1}{2}+\mu)_k (1+\nu+\mu)_k}{k!\,(\nu+\frac{3}{2})_k} \cos[(2k+\nu+\mu+1)\theta],$$

where $0 < \theta < \pi$.

**18.11.3-6. Some relations for the modified associated Legendre functions.**

For $0 < x < 1$,

$$\mathsf{P}_\nu^\mu(-x) = \mathsf{P}_\nu^\mu(x)\cos[\pi(\nu+\mu)] - 2\pi^{-1}\mathsf{Q}_\nu^\mu(x)\sin[\pi(\nu+\mu)],$$

$$\mathsf{Q}_\nu^\mu(-x) = -\mathsf{Q}_\nu^\mu(x)\cos[\pi(\nu+\mu)] - \tfrac{1}{2}\pi\,\mathsf{P}_\nu^\mu(x)\sin[\pi(\nu+\mu)].$$

For $-1 < x < 1$,

$$\mathsf{P}_{\nu+1}^\mu(x) = \frac{2\nu+1}{\nu-\mu+1}\,x\,\mathsf{P}_\nu^\mu(x) - \frac{\nu+\mu}{\nu-\mu+1}\,\mathsf{P}_{\nu-1}^\mu(x),$$

$$\mathsf{P}_{\nu+1}^\mu(x) = \mathsf{P}_{\nu-1}^\mu(x) - (2\nu+1)(1-x^2)^{1/2}\,\mathsf{P}_\nu^{\mu-1}(x),$$

$$\mathsf{P}_{\nu+1}^\mu(x) = x\,\mathsf{P}_\nu^\mu(x) - (\nu+\mu)(1-x^2)^{1/2}\,\mathsf{P}_\nu^{\mu-1}(x),$$

$$\frac{d}{dx}\mathsf{P}_\nu^\mu(x) = \frac{\nu x}{x^2-1}\mathsf{P}_\nu^\mu(x) - \frac{\nu+\mu}{x^2-1}\mathsf{P}_{\nu-1}^\mu(x).$$

Wronskian:

$$\mathsf{P}_\nu^\mu(x)\frac{d}{dx}\mathsf{Q}_\nu^\mu(x) - \mathsf{Q}_\nu^\mu(x)\frac{d}{dx}\mathsf{P}_\nu^\mu(x) = \frac{k}{1-x^2}, \qquad k = 2^{2\mu}\frac{\Gamma\!\left(\frac{\nu+\mu+1}{2}\right)\Gamma\!\left(\frac{\nu+\mu+2}{2}\right)}{\Gamma\!\left(\frac{\nu-\mu+1}{2}\right)\Gamma\!\left(\frac{\nu-\mu+2}{2}\right)}.$$

For $n = 1, 2, \ldots$,

$$\mathsf{P}_\nu^n(x) = (-1)^n (1-x^2)^{n/2}\frac{d^n}{dx^n}\mathsf{P}_\nu(x), \qquad \mathsf{Q}_\nu^n(x) = (-1)^n (1-x^2)^{n/2}\frac{d^n}{dx^n}\mathsf{Q}_\nu(x).$$

## 18.12. Parabolic Cylinder Functions
### 18.12.1. Definitions. Basic Formulas

**18.12.1-1. Differential equation. Formulas for the parabolic cylinder functions.**

The *Weber parabolic cylinder function* $D_\nu(z)$ is a solution of the linear ordinary differential equation:

$$y''_{zz} + \left(-\tfrac{1}{4}z^2 + \nu + \tfrac{1}{2}\right)y = 0,$$

where the parameter $\nu$ and the variable $z$ can assume arbitrary real or complex values. Another linearly independent solution of this equation is the function $D_{-\nu-1}(iz)$; if $\nu$ is noninteger, then $D_\nu(-z)$ can also be taken as a linearly independent solution.

The parabolic cylinder functions can be expressed in terms of degenerate hypergeometric functions as

$$D_\nu(z) = 2^{1/2}\exp\!\left(-\tfrac{1}{4}z^2\right)\left[\frac{\Gamma\!\left(\frac{1}{2}\right)}{\Gamma\!\left(\frac{1}{2}-\frac{\nu}{2}\right)}\Phi\!\left(-\tfrac{\nu}{2},\tfrac{1}{2};\tfrac{1}{2}z^2\right) + 2^{-1/2}\frac{\Gamma\!\left(-\frac{1}{2}\right)}{\Gamma\!\left(-\frac{\nu}{2}\right)}z\,\Phi\!\left(\tfrac{1}{2}-\tfrac{\nu}{2},\tfrac{3}{2};\tfrac{1}{2}z^2\right)\right].$$

### 18.12.1-2. Special cases.

For nonnegative integer $\nu = n$, we have

$$D_n(z) = \frac{1}{2^{n/2}} \exp\left(-\frac{z^2}{4}\right) H_n\left(\frac{z}{\sqrt{2}}\right), \quad n = 0, 1, 2, \ldots;$$

$$H_n(z) = (-1)^n \exp(z^2) \frac{d^n}{dz^n} \exp(-z^2),$$

where $H_n(z)$ is the Hermitian polynomial of order $n$.

Connection with the error function:

$$D_{-1}(z) = \sqrt{\frac{\pi}{2}} \exp\left(\frac{z^2}{4}\right) \operatorname{erfc}\left(\frac{z}{\sqrt{2}}\right),$$

$$D_{-2}(z) = \sqrt{\frac{\pi}{2}} z \exp\left(\frac{z^2}{4}\right) \operatorname{erfc}\left(\frac{z}{\sqrt{2}}\right) - \exp\left(-\frac{z^2}{4}\right).$$

## 18.12.2. Integral Representations, Asymptotic Expansions, and Linear Relations

### 18.12.2-1. Integral representations and the asymptotic expansion.

Integral representations:

$$D_\nu(z) = \sqrt{2/\pi}\, \exp\left(\tfrac{1}{4}z^2\right) \int_0^\infty t^\nu \exp\left(-\tfrac{1}{2}t^2\right) \cos\left(zt - \tfrac{1}{2}\pi\nu\right) dt \quad \text{for} \quad \operatorname{Re}\nu > -1,$$

$$D_\nu(z) = \frac{1}{\Gamma(-\nu)} \exp\left(-\tfrac{1}{4}z^2\right) \int_0^\infty t^{-\nu-1} \exp\left(-zt - \tfrac{1}{2}t^2\right) dt \quad \text{for} \quad \operatorname{Re}\nu < 0.$$

Asymptotic expansion as $|z| \to \infty$:

$$D_\nu(z) = z^\nu \exp\left(-\tfrac{1}{4}z^2\right) \left[\sum_{n=0}^N \frac{(-2)^n \left(-\tfrac{\nu}{2}\right)_n \left(\tfrac{1}{2} - \tfrac{\nu}{2}\right)_n}{n!} \frac{1}{z^{2n}} + O\left(|z|^{-2N-2}\right)\right] \quad \text{for} \quad |\arg z| < \frac{3\pi}{4},$$

where $(a)_0 = 1$, $(a)_n = a(a+1)\ldots(a+n-1)$ for $n = 1, 2, 3, \ldots$

### 18.12.2-2. Recurrence relations.

$$D_{\nu+1}(z) - zD_\nu(z) + \nu D_{\nu-1}(z) = 0,$$

$$\frac{d}{dz}D_\nu(z) + \frac{1}{2}zD_\nu(z) - \nu D_{\nu-1}(z) = 0,$$

$$\frac{d}{dz}D_\nu(z) - \frac{1}{2}zD_\nu(z) + D_{\nu+1}(z) = 0.$$

## 18.13. Elliptic Integrals

### 18.13.1. Complete Elliptic Integrals

**18.13.1-1. Definitions. Properties. Conversion formulas.**

*Complete elliptic integral of the first kind*:

$$\mathsf{K}(k) = \int_0^{\pi/2} \frac{d\alpha}{\sqrt{1 - k^2 \sin^2 \alpha}} = \int_0^1 \frac{dx}{\sqrt{(1-x^2)(1-k^2 x^2)}}.$$

*Complete elliptic integral of the second kind*:

$$\mathsf{E}(k) = \int_0^{\pi/2} \sqrt{1 - k^2 \sin^2 \alpha}\, d\alpha = \int_0^1 \frac{\sqrt{1 - k^2 x^2}}{\sqrt{1 - x^2}}\, dx.$$

The argument $k$ is called the *elliptic modulus* ($k^2 < 1$).
Notation:
$$k' = \sqrt{1 - k^2}, \quad \mathsf{K}'(k) = \mathsf{K}(k'), \quad \mathsf{E}'(k) = \mathsf{E}(k'),$$

where $k'$ is the *complementary modulus*.
Properties:
$$\mathsf{K}(-k) = \mathsf{K}(k), \quad \mathsf{E}(-k) = \mathsf{E}(k);$$
$$\mathsf{K}(k) = \mathsf{K}'(k'), \quad \mathsf{E}(k) = \mathsf{E}'(k');$$
$$\mathsf{E}(k)\mathsf{K}'(k) + \mathsf{E}'(k)\mathsf{K}(k) - \mathsf{K}(k)\mathsf{K}'(k) = \frac{\pi}{2}.$$

Conversion formulas for complete elliptic integrals:

$$\mathsf{K}\left(\frac{1-k'}{1+k'}\right) = \frac{1+k'}{2}\mathsf{K}(k),$$

$$\mathsf{E}\left(\frac{1-k'}{1+k'}\right) = \frac{1}{1+k'}\big[\mathsf{E}(k) + k'\mathsf{K}(k)\big],$$

$$\mathsf{K}\left(\frac{2\sqrt{k}}{1+k}\right) = (1+k)\mathsf{K}(k),$$

$$\mathsf{E}\left(\frac{2\sqrt{k}}{1+k}\right) = \frac{1}{1+k}\big[2\mathsf{E}(k) - (k')^2\mathsf{K}(k)\big].$$

**18.13.1-2. Representation of complete elliptic integrals in series form.**

Representation of complete elliptic integrals in the form of series in powers of the modulus $k$:

$$\mathsf{K}(k) = \frac{\pi}{2}\left\{1 + \left(\frac{1}{2}\right)^2 k^2 + \left(\frac{1\times 3}{2\times 4}\right)^2 k^4 + \cdots + \left[\frac{(2n-1)!!}{(2n)!!}\right]^2 k^{2n} + \cdots\right\},$$

$$\mathsf{E}(k) = \frac{\pi}{2}\left\{1 - \left(\frac{1}{2}\right)^2 \frac{k^2}{1} - \left(\frac{1\times 3}{2\times 4}\right)^2 \frac{k^4}{3} - \cdots - \left[\frac{(2n-1)!!}{(2n)!!}\right]^2 \frac{k^{2n}}{2n-1} - \cdots\right\}.$$

Representation of complete elliptic integrals in the form of series in powers of the complementary modulus $k' = \sqrt{1-k^2}$:

$$K(k) = \frac{\pi}{1+k'}\left\{1 + \left(\frac{1}{2}\right)^2\left(\frac{1-k'}{1+k'}\right)^2 + \left(\frac{1\times 3}{2\times 4}\right)^2\left(\frac{1-k'}{1+k'}\right)^4 + \cdots + \left[\frac{(2n-1)!!}{(2n)!!}\right]^2\left(\frac{1-k'}{1+k'}\right)^{2n} + \cdots\right\},$$

$$K(k) = \ln\frac{4}{k'} + \left(\frac{1}{2}\right)^2\left(\ln\frac{4}{k'} - \frac{2}{1\times 2}\right)(k')^2 + \left(\frac{1\times 3}{2\times 4}\right)^2\left(\ln\frac{4}{k'} - \frac{2}{1\times 2} - \frac{2}{3\times 4}\right)(k')^4$$

$$+ \left(\frac{1\times 3\times 5}{2\times 4\times 6}\right)^2\left(\ln\frac{4}{k'} - \frac{2}{1\times 2} - \frac{2}{3\times 4} - \frac{2}{5\times 6}\right)(k')^6 + \cdots;$$

$$E(k) = \frac{\pi(1+k')}{4}\left\{1 + \frac{1}{2^2}\left(\frac{1-k'}{1+k'}\right)^2 + \frac{1^2}{(2\times 4)^2}\left(\frac{1-k'}{1+k'}\right)^4 + \cdots + \left[\frac{(2n-3)!!}{(2n)!!}\right]^2\left(\frac{1-k'}{1+k'}\right)^{2n} + \cdots\right\},$$

$$E(k) = 1 + \frac{1}{2}\left(\ln\frac{4}{k'} - \frac{1}{1\times 2}\right)(k')^2 + \frac{1^2\times 3}{2^2\times 4}\left(\ln\frac{4}{k'} - \frac{2}{1\times 2} - \frac{1}{3\times 4}\right)(k')^4$$

$$+ \frac{1^2\times 3^2\times 5}{2^2\times 4^2\times 6}\left(\ln\frac{4}{k'} - \frac{2}{1\times 2} - \frac{2}{3\times 4} - \frac{1}{5\times 6}\right)(k')^6 + \cdots.$$

### 18.13.1-3. Differentiation formulas. Differential equations.

Differentiation formulas:

$$\frac{d\,K(k)}{dk} = \frac{E(k)}{k(k')^2} - \frac{K(k)}{k}, \qquad \frac{d\,E(k)}{dk} = \frac{E(k) - K(k)}{k}.$$

The functions $K(k)$ and $K'(k)$ satisfy the second-order linear ordinary differential equation

$$\frac{d}{dk}\left[k(1-k^2)\frac{d\,K}{dk}\right] - k\,K = 0.$$

The functions $E(k)$ and $E'(k) - K'(k)$ satisfy the second-order linear ordinary differential equation

$$(1-k^2)\frac{d}{dk}\left(k\frac{d\,E}{dk}\right) + k\,E = 0.$$

## 18.13.2. Incomplete Elliptic Integrals (Elliptic Integrals)

### 18.13.2-1. Definitions. Properties.

*Elliptic integral of the first kind*:

$$F(\varphi, k) = \int_0^\varphi \frac{d\alpha}{\sqrt{1 - k^2\sin^2\alpha}} = \int_0^{\sin\varphi} \frac{dx}{\sqrt{(1-x^2)(1-k^2x^2)}}.$$

*Elliptic integral of the second kind*:

$$E(\varphi, k) = \int_0^\varphi \sqrt{1 - k^2\sin^2\alpha}\,d\alpha = \int_0^{\sin\varphi} \frac{\sqrt{1 - k^2x^2}}{\sqrt{1-x^2}}\,dx.$$

*Elliptic integral of the third kind*:

$$\Pi(\varphi, n, k) = \int_0^\varphi \frac{d\alpha}{(1 - n\sin^2 \alpha)\sqrt{1 - k^2 \sin^2 \alpha}} = \int_0^{\sin\varphi} \frac{dx}{(1 - nx^2)\sqrt{(1 - x^2)(1 - k^2 x^2)}}.$$

The quantity $k$ is called the *elliptic modulus* ($k^2 < 1$), $k' = \sqrt{1 - k^2}$ is the *complementary modulus*, and $n$ is the *characteristic parameter*.

Complete elliptic integrals:

$$\mathsf{K}(k) = F\left(\frac{\pi}{2}, k\right), \qquad \mathsf{E}(k) = E\left(\frac{\pi}{2}, k\right),$$
$$\mathsf{K}'(k) = F\left(\frac{\pi}{2}, k'\right), \qquad \mathsf{E}'(k) = E\left(\frac{\pi}{2}, k'\right).$$

Properties of elliptic integrals:

$$F(-\varphi, k) = -F(\varphi, k), \qquad F(n\pi \pm \varphi, k) = 2n\,\mathsf{K}(k) \pm F(\varphi, k);$$
$$E(-\varphi, k) = -E(\varphi, k), \qquad E(n\pi \pm \varphi, k) = 2n\,\mathsf{E}(k) \pm E(\varphi, k).$$

**18.13.2-2. Conversion formulas.**

Conversion formulas for elliptic integrals (first set):

$$F\left(\psi, \frac{1}{k}\right) = kF(\varphi, k),$$
$$E\left(\psi, \frac{1}{k}\right) = \frac{1}{k}\left[E(\varphi, k) - (k')^2 F(\varphi, k)\right],$$

where the angles $\varphi$ and $\psi$ are related by $\sin\psi = k\sin\varphi$, $\cos\psi = \sqrt{1 - k^2 \sin^2 \varphi}$.

Conversion formulas for elliptic integrals (second set):

$$F\left(\psi, \frac{1-k'}{1+k'}\right) = (1 + k')F(\varphi, k),$$
$$E\left(\psi, \frac{1-k'}{1+k'}\right) = \frac{2}{1+k'}\left[E(\varphi, k) + k' F(\varphi, k)\right] - \frac{1-k'}{1+k'}\sin\psi,$$

where the angles $\varphi$ and $\psi$ are related by $\tan(\psi - \varphi) = k'\tan\varphi$.

Transformation formulas for elliptic integrals (third set):

$$F\left(\psi, \frac{2\sqrt{k}}{1+k}\right) = (1 + k)F(\varphi, k),$$
$$E\left(\psi, \frac{2\sqrt{k}}{1+k}\right) = \frac{1}{1+k}\left[2E(\varphi, k) - (k')^2 F(\varphi, k) + 2k\frac{\sin\varphi\cos\varphi}{1 + k\sin^2\varphi}\sqrt{1 - k^2 \sin^2 \varphi}\right],$$

where the angles $\varphi$ and $\psi$ are related by $\sin\psi = \dfrac{(1+k)\sin\varphi}{1 + k\sin^2\varphi}$.

### 18.13.2-3. Trigonometric expansions.

Trigonometric expansions for small $k$ and $\varphi$:

$$F(\varphi, k) = \frac{2}{\pi} \mathsf{K}(k)\varphi - \sin\varphi \cos\varphi \left(a_0 + \frac{2}{3}a_1 \sin^2\varphi + \frac{2\times 4}{3\times 5}a_2 \sin^4\varphi + \cdots\right),$$

$$a_0 = \frac{2}{\pi}\mathsf{K}(k) - 1, \quad a_n = a_{n-1} - \left[\frac{(2n-1)!!}{(2n)!!}\right]^2 k^{2n};$$

$$E(\varphi, k) = \frac{2}{\pi}\mathsf{E}(k)\varphi - \sin\varphi \cos\varphi \left(b_0 + \frac{2}{3}b_1 \sin^2\varphi + \frac{2\times 4}{3\times 5}b_2 \sin^4\varphi + \cdots\right),$$

$$b_0 = 1 - \frac{2}{\pi}\mathsf{E}(k), \quad b_n = b_{n-1} - \left[\frac{(2n-1)!!}{(2n)!!}\right]^2 \frac{k^{2n}}{2n-1}.$$

Trigonometric expansions for $k \to 1$:

$$F(\varphi, k) = \frac{2}{\pi}\mathsf{K}'(k) \ln \tan\left(\frac{\varphi}{2} + \frac{\pi}{4}\right) - \frac{\tan\varphi}{\cos\varphi}\left(a'_0 - \frac{2}{3}a'_1 \tan^2\varphi + \frac{2\times 4}{3\times 5}a'_2 \tan^4\varphi - \cdots\right),$$

$$a'_0 = \frac{2}{\pi}\mathsf{K}'(k) - 1, \quad a'_n = a'_{n-1} - \left[\frac{(2n-1)!!}{(2n)!!}\right]^2 (k')^{2n};$$

$$E(\varphi, k) = \frac{2}{\pi}\mathsf{E}'(k) \ln \tan\left(\frac{\varphi}{2} + \frac{\pi}{4}\right) + \frac{\tan\varphi}{\cos\varphi}\left(b'_0 - \frac{2}{3}b'_1 \tan^2\varphi + \frac{2\times 4}{3\times 5}b'_2 \tan^4\varphi - \cdots\right),$$

$$b'_0 = \frac{2}{\pi}\mathsf{E}'(k) - 1, \quad b'_n = b'_{n-1} - \left[\frac{(2n-1)!!}{(2n)!!}\right]^2 \frac{(k')^{2n}}{2n-1}.$$

## 18.14. Elliptic Functions

An *elliptic function* is a function that is the inverse of an elliptic integral. An elliptic function is a doubly periodic meromorphic function of a complex variable. All its periods can be written in the form $2m\omega_1 + 2n\omega_2$ with integer $m$ and $n$, where $\omega_1$ and $\omega_2$ are a pair of (primitive) half-periods. The ratio $\tau = \omega_2/\omega_1$ is a complex quantity that may be considered to have a positive imaginary part, $\mathrm{Im}\,\tau > 0$.

Throughout the rest of this section, the following brief notation will be used: $\mathsf{K} = \mathsf{K}(k)$ and $\mathsf{K}' = \mathsf{K}(k')$ are complete elliptic integrals with $k' = \sqrt{1-k^2}$.

### 18.14.1. Jacobi Elliptic Functions

#### 18.14.1-1. Definitions. Simple properties. Special cases.

When the upper limit $\varphi$ of the incomplete elliptic integral of the first kind

$$u = \int_0^\varphi \frac{d\alpha}{\sqrt{1-k^2\sin^2\alpha}} = F(\varphi, k)$$

is treated as a function of $u$, the following notation is used:

$$u = \mathrm{am}\,\varphi.$$

## 18.14. ELLIPTIC FUNCTIONS

Naming: $\varphi$ is the *amplitude* and $u$ is the *argument*.
*Jacobi elliptic functions*:

$$\operatorname{sn} u = \sin \varphi = \sin \operatorname{am} u \quad \textit{(sine amplitude)},$$

$$\operatorname{cn} u = \cos \varphi = \cos \operatorname{am} u \quad \textit{(cosine amplitude)},$$

$$\operatorname{dn} u = \sqrt{1 - k^2 \sin^2 \varphi} = \frac{d\varphi}{du} \quad \textit{(delta amlplitude)}.$$

Along with the brief notations $\operatorname{sn} u$, $\operatorname{cn} u$, $\operatorname{dn} u$, the respective full notations are also used: $\operatorname{sn}(u, k)$, $\operatorname{cn}(u, k)$, $\operatorname{dn}(u, k)$.

Simple properties:

$$\operatorname{sn}(-u) = -\operatorname{sn} u, \quad \operatorname{cn}(-u) = \operatorname{cn} u, \quad \operatorname{dn}(-u) = \operatorname{dn} u;$$

$$\operatorname{sn}^2 u + \operatorname{cn}^2 u = 1, \quad k^2 \operatorname{sn}^2 u + \operatorname{dn}^2 u = 1, \quad \operatorname{dn}^2 u - k^2 \operatorname{cn}^2 u = 1 - k^2,$$

where $i^2 = -1$.

Jacobi functions for special values of the modulus ($k = 0$ and $k = 1$):

$$\operatorname{sn}(u, 0) = \sin u, \quad \operatorname{cn}(u, 0) = \cos u, \quad \operatorname{dn}(u, 0) = 1;$$

$$\operatorname{sn}(u, 1) = \tanh u, \quad \operatorname{cn}(u, 1) = \frac{1}{\cosh u}, \quad \operatorname{dn}(u, 1) = \frac{1}{\cosh u}.$$

Jacobi functions for special values of the argument:

$$\operatorname{sn}(\tfrac{1}{2}\mathsf{K}, k) = \frac{1}{\sqrt{1+k'}}, \quad \operatorname{cn}(\tfrac{1}{2}\mathsf{K}, k) = \sqrt{\frac{k'}{1+k'}}, \quad \operatorname{dn}(\tfrac{1}{2}\mathsf{K}, k) = \sqrt{k'};$$

$$\operatorname{sn}(\mathsf{K}, k) = 1, \quad \operatorname{cn}(\mathsf{K}, k) = 0, \quad \operatorname{dn}(\mathsf{K}, k) = k'.$$

### 18.14.1-2. Reduction formulas.

$$\operatorname{sn}(u \pm \mathsf{K}) = \pm \frac{\operatorname{cn} u}{\operatorname{dn} u}, \quad \operatorname{cn}(u \pm \mathsf{K}) = \mp k' \frac{\operatorname{sn} u}{\operatorname{dn} u}, \quad \operatorname{dn}(u \pm \mathsf{K}) = \frac{k'}{\operatorname{dn} u};$$

$$\operatorname{sn}(u \pm 2\mathsf{K}) = -\operatorname{sn} u, \quad \operatorname{cn}(u \pm 2\mathsf{K}) = -\operatorname{cn} u, \quad \operatorname{dn}(u \pm 2\mathsf{K}) = \operatorname{dn} u;$$

$$\operatorname{sn}(u + i\mathsf{K}') = \frac{1}{k \operatorname{sn} u}, \quad \operatorname{cn}(u + i\mathsf{K}') = -\frac{i \operatorname{dn} u}{k \operatorname{sn} u}, \quad \operatorname{dn}(u + i\mathsf{K}') = -i \frac{\operatorname{cn} u}{\operatorname{sn} u};$$

$$\operatorname{sn}(u + 2i\mathsf{K}') = \operatorname{sn} u, \quad \operatorname{cn}(u + 2i\mathsf{K}') = -\operatorname{cn} u, \quad \operatorname{dn}(u + 2i\mathsf{K}') = -\operatorname{dn} u;$$

$$\operatorname{sn}(u + \mathsf{K} + i\mathsf{K}') = \frac{\operatorname{dn} u}{k \operatorname{cn} u}, \quad \operatorname{cn}(u + \mathsf{K} + i\mathsf{K}') = -i \frac{k'}{k \operatorname{cn} u}, \quad \operatorname{dn}(u + \mathsf{K} + i\mathsf{K}') = ik' \frac{\operatorname{sn} u}{\operatorname{cn} u};$$

$$\operatorname{sn}(u + 2\mathsf{K} + 2i\mathsf{K}') = -\operatorname{sn} u, \quad \operatorname{cn}(u + 2\mathsf{K} + 2i\mathsf{K}') = \operatorname{cn} u, \quad \operatorname{dn}(u + 2\mathsf{K} + 2i\mathsf{K}') = -\operatorname{dn} u.$$

### 18.14.1-3. Periods, zeros, poses, and residues.

TABLE 18.4
Periods, zeros, poles, and residues of the Jacobian elliptic functions
($m, n = 0, \pm 1, \pm 2, \ldots;\ i^2 = -1$)

| Functions | Periods | Zeros | Poles | Residues |
|---|---|---|---|---|
| $\operatorname{sn} u$ | $4m\mathsf{K} + 2n\mathsf{K}' i$ | $2m\mathsf{K} + 2n\mathsf{K}' i$ | $2m\mathsf{K} + (2n+1)\mathsf{K}' i$ | $(-1)^m \frac{1}{k}$ |
| $\operatorname{cn} u$ | $(4m+2)\mathsf{K} + 2n\mathsf{K}' i$ | $(2m+1)\mathsf{K} + 2n\mathsf{K}' i$ | $2m\mathsf{K} + (2n+1)\mathsf{K}' i$ | $(-1)^{m-1} \frac{i}{k}$ |
| $\operatorname{dn} u$ | $2m\mathsf{K} + 4n\mathsf{K}' i$ | $(2m+1)\mathsf{K} + (2n+1)\mathsf{K}' i$ | $2m\mathsf{K} + (2n+1)\mathsf{K}' i$ | $(-1)^{n-1} i$ |

### 18.14.1-4. Double-argument formulas.

$$\operatorname{sn}(2u) = \frac{2\operatorname{sn} u \operatorname{cn} u \operatorname{dn} u}{1 - k^2 \operatorname{sn}^4 u} = \frac{2\operatorname{sn} u \operatorname{cn} u \operatorname{dn} u}{\operatorname{cn}^2 u + \operatorname{sn}^2 u \operatorname{dn}^2 u},$$

$$\operatorname{cn}(2u) = \frac{\operatorname{cn}^2 u - \operatorname{sn}^2 u \operatorname{dn}^2 u}{1 - k^2 \operatorname{sn}^4 u} = \frac{\operatorname{cn}^2 u - \operatorname{sn}^2 u \operatorname{dn}^2 u}{\operatorname{cn}^2 u + \operatorname{sn}^2 u \operatorname{dn}^2 u},$$

$$\operatorname{dn}(2u) = \frac{\operatorname{dn}^2 u - k^2 \operatorname{sn}^2 u \operatorname{cn}^2 u}{1 - k^2 \operatorname{sn}^4 u} = \frac{\operatorname{dn}^2 u + \operatorname{cn}^2 u (\operatorname{dn}^2 u - 1)}{\operatorname{dn}^2 u - \operatorname{cn}^2 u (\operatorname{dn}^2 u - 1)}.$$

### 18.14.1-5. Half-argument formulas.

$$\operatorname{sn}^2 \frac{u}{2} = \frac{1}{k^2} \frac{1 - \operatorname{dn} u}{1 + \operatorname{cn} u} = \frac{1 - \operatorname{cn} u}{1 + \operatorname{dn} u},$$

$$\operatorname{cn}^2 \frac{u}{2} = \frac{\operatorname{cn} u + \operatorname{dn} u}{1 + \operatorname{dn} u} = \frac{1 - k^2}{k^2} \frac{1 - \operatorname{dn} u}{\operatorname{dn} u - \operatorname{cn} u},$$

$$\operatorname{dn}^2 \frac{u}{2} = \frac{\operatorname{cn} u + \operatorname{dn} u}{1 + \operatorname{cn} u} = (1 - k^2) \frac{1 - \operatorname{cn} u}{\operatorname{dn} u - \operatorname{cn} u}.$$

### 18.14.1-6. Argument addition formulas.

$$\operatorname{sn}(u \pm v) = \frac{\operatorname{sn} u \operatorname{cn} v \operatorname{dn} v \pm \operatorname{sn} v \operatorname{cn} u \operatorname{dn} u}{1 - k^2 \operatorname{sn}^2 u \operatorname{sn}^2 v},$$

$$\operatorname{cn}(u \pm v) = \frac{\operatorname{cn} u \operatorname{cn} v \mp \operatorname{sn} u \operatorname{sn} v \operatorname{dn} u \operatorname{dn} v}{1 - k^2 \operatorname{sn}^2 u \operatorname{sn}^2 v},$$

$$\operatorname{dn}(u \pm v) = \frac{\operatorname{dn} u \operatorname{dn} v \mp k^2 \operatorname{sn} u \operatorname{sn} v \operatorname{cn} u \operatorname{cn} v}{1 - k^2 \operatorname{sn}^2 u \operatorname{sn}^2 v}.$$

### 18.14.1-7. Conversion formulas.

Table 18.5 presents conversion formulas for Jacobi elliptic functions. If $k > 1$, then $k_1 = 1/k < 1$. Elliptic functions with real modulus can be reduced, using the first set of conversion formulas, to elliptic functions with a modulus lying between 0 and 1.

### 18.14.1-8. Descending Landen transformation (Gauss's transformation).

Notation:

$$\mu = \left| \frac{1 - k'}{1 + k'} \right|, \quad v = \frac{u}{1 + \mu}.$$

Descending transformations:

$$\operatorname{sn}(u, k) = \frac{(1 + \mu) \operatorname{sn}(v, \mu^2)}{1 + \mu \operatorname{sn}^2(v, \mu^2)}, \quad \operatorname{cn}(u, k) = \frac{\operatorname{cn}(v, \mu^2) \operatorname{dn}(v, \mu^2)}{1 + \mu \operatorname{sn}^2(v, \mu^2)}, \quad \operatorname{dn}(u, k) = \frac{\operatorname{dn}^2(v, \mu^2) + \mu - 1}{1 + \mu - \operatorname{dn}^2(v, \mu^2)}.$$

### TABLE 18.5
Conversion formulas for Jacobi elliptic functions. Full notation is used: sn$(u,k)$, cn$(u,k)$, dn$(u,k)$

| $u_1$ | $k_1$ | sn$(u_1,k_1)$ | cn$(u_1,k_1)$ | dn$(u_1,k_1)$ |
|---|---|---|---|---|
| $ku$ | $\dfrac{1}{k}$ | $k\,\text{sn}(u,k)$ | $\text{dn}(u,k)$ | $\text{cn}(u,k)$ |
| $iu$ | $k'$ | $i\dfrac{\text{sn}(u,k)}{\text{cn}(u,k)}$ | $\dfrac{1}{\text{cn}(u,k)}$ | $\dfrac{\text{dn}(u,k)}{\text{cn}(u,k)}$ |
| $k'u$ | $i\dfrac{k}{k'}$ | $k'\dfrac{\text{sn}(u,k)}{\text{dn}(u,k)}$ | $\dfrac{\text{cn}(u,k)}{\text{dn}(u,k)}$ | $\dfrac{1}{\text{dn}(u,k)}$ |
| $iku$ | $i\dfrac{k'}{k}$ | $ik\dfrac{\text{sn}(u,k)}{\text{dn}(u,k)}$ | $\dfrac{1}{\text{dn}(u,k)}$ | $\dfrac{\text{cn}(u,k)}{\text{dn}(u,k)}$ |
| $ik'u$ | $\dfrac{1}{k'}$ | $ik'\dfrac{\text{sn}(u,k)}{\text{cn}(u,k)}$ | $\dfrac{\text{dn}(u,k)}{\text{cn}(u,k)}$ | $\dfrac{1}{\text{cn}(u,k)}$ |
| $(1+k)u$ | $\dfrac{2\sqrt{k}}{1+k}$ | $\dfrac{(1+k)\,\text{sn}(u,k)}{1+k\,\text{sn}^2(u,k)}$ | $\dfrac{\text{cn}(u,k)\,\text{dn}(u,k)}{1+k\,\text{sn}^2(u,k)}$ | $\dfrac{1-k\,\text{sn}^2(u,k)}{1+k\,\text{sn}^2(u,k)}$ |
| $(1+k')u$ | $\dfrac{1-k'}{1+k'}$ | $\dfrac{(1+k')\,\text{sn}(u,k)\,\text{cn}(u,k)}{\text{dn}(u,k)}$ | $\dfrac{1-(1+k')\,\text{sn}^2(u,k)}{\text{dn}(u,k)}$ | $\dfrac{1-(1-k')\,\text{sn}^2(u,k)}{\text{dn}(u,k)}$ |

#### 18.14.1-9. Ascending Landen transformation.

Notation:
$$\mu = \frac{4k}{(1+k)^2}, \qquad \sigma = \left|\frac{1-k}{1+k}\right|, \qquad v = \frac{u}{1+\sigma}.$$

Ascending transformations:
$$\text{sn}(u,k) = (1+\sigma)\frac{\text{sn}(v,\mu)\,\text{cn}(v,\mu)}{\text{dn}(v,\mu)}, \quad \text{cn}(u,k) = \frac{1+\sigma}{\mu}\frac{\text{dn}^2(v,\mu)-\sigma}{\text{dn}(v,\mu)}, \quad \text{dn}(u,k) = \frac{1-\sigma}{\mu}\frac{\text{dn}^2(v,\mu)+\sigma}{\text{dn}(v,\mu)}.$$

#### 18.14.1-10. Series representation.

Representation Jacobi functions in the form of power series in $u$:

$$\text{sn}\,u = u - \frac{1}{3!}(1+k^2)u^3 + \frac{1}{5!}(1+14k^2+k^4)u^5 - \frac{1}{7!}(1+135k^2+135k^4+k^6)u^7 + \cdots,$$
$$\text{cn}\,u = 1 - \frac{1}{2!}u^2 + \frac{1}{4!}(1+4k^2)u^4 - \frac{1}{6!}(1+44k^2+16k^4)u^6 + \cdots,$$
$$\text{dn}\,u = 1 - \frac{1}{2!}k^2 u^2 + \frac{1}{4!}k^2(4+k^2)u^4 - \frac{1}{6!}k^2(16+44k^2+k^4)u^6 + \cdots,$$
$$\text{am}\,u = u - \frac{1}{3!}k^2 u^3 + \frac{1}{5!}k^2(4+k^2)u^5 - \frac{1}{7!}k^2(16+44k^2+k^4)u^7 + \cdots.$$

These functions converge for $|u| < |\mathsf{K}(k')|$.

Representation Jacobi functions in the form of trigonometric series:

$$\text{sn}\,u = \frac{2\pi}{k\,\mathsf{K}\sqrt{q}}\sum_{n=1}^{\infty}\frac{q^n}{1-q^{2n-1}}\sin\left[(2n-1)\frac{\pi u}{2\,\mathsf{K}}\right],$$

$$\operatorname{cn} u = \frac{2\pi}{k\mathsf{K}\sqrt{q}} \sum_{n=1}^{\infty} \frac{q^n}{1+q^{2n-1}} \cos\left[(2n-1)\frac{\pi u}{2\mathsf{K}}\right],$$

$$\operatorname{dn} u = \frac{\pi}{2\mathsf{K}} + \frac{2\pi}{\mathsf{K}} \sum_{n=1}^{\infty} \frac{q^n}{1+q^{2n}} \cos\left(\frac{n\pi u}{\mathsf{K}}\right),$$

$$\operatorname{am} u = \frac{\pi u}{2\mathsf{K}} + 2 \sum_{n=1}^{\infty} \frac{1}{n} \frac{q^n}{1+q^{2n}} \sin\left(\frac{n\pi u}{\mathsf{K}}\right),$$

where $q = \exp(-\pi \mathsf{K}'/\mathsf{K})$, $\mathsf{K} = \mathsf{K}(k)$, $\mathsf{K}' = \mathsf{K}(k')$, and $k' = \sqrt{1-k^2}$.

#### 18.14.1-11. Derivatives and integrals.

Derivatives:

$$\frac{d}{du} \operatorname{sn} u = \operatorname{cn} u \operatorname{dn} u, \qquad \frac{d}{du} \operatorname{cn} u = -\operatorname{sn} u \operatorname{dn} u, \qquad \frac{d}{du} \operatorname{dn} u = -k^2 \operatorname{sn} u \operatorname{cn} u.$$

Integrals:

$$\int \operatorname{sn} u \, du = \frac{1}{k} \ln(\operatorname{dn} u - k \operatorname{cn} u) = -\frac{1}{k} \ln(\operatorname{dn} u + k \operatorname{cn} u),$$

$$\int \operatorname{cn} u \, du = \frac{1}{k} \arccos(\operatorname{dn} u) = \frac{1}{k} \arcsin(k \operatorname{sn} u),$$

$$\int \operatorname{dn} u \, du = \arcsin(\operatorname{sn} u) = \operatorname{am} u.$$

The arbitrary additive constant $C$ in the integrals is omitted.

### 18.14.2. Weierstrass Elliptic Function

#### 18.14.2-1. Infinite series representation. Some properties.

The Weierstrass elliptic function (or Weierstrass $\wp$-function) is defined as

$$\wp(z) = \wp(z|\omega_1, \omega_2) = \frac{1}{z^2} + \sum_{m,n} \left[ \frac{1}{(z - 2m\omega_1 - 2n\omega_2)^2} - \frac{1}{(2m\omega_1 + 2n\omega_2)^2} \right],$$

where the summation is assumed over all integer $m$ and $n$, except for $m = n = 0$. This function is a complex, double periodic function of a complex variable $z$ with periods $2\omega_1$ and $2\omega_1$:

$$\wp(-z) = \wp(z),$$
$$\wp(z + 2m\omega_1 + 2n\omega_2) = \wp(z),$$

where $m, n = 0, \pm 1, \pm 2, \ldots$ and $\operatorname{Im}(\omega_2/\omega_1) \neq 0$. The series defining the Weierstrass $\wp$-function converges everywhere except for second-order poles located at $z_{mn} = 2m\omega_1 + 2n\omega_2$.

Argument addition formula:

$$\wp(z_1 + z_2) = -\wp(z_1) - \wp(z_2) + \frac{1}{4} \left[ \frac{\wp'(z_1) - \wp'(z_2)}{\wp(z_1) - \wp(z_2)} \right]^2.$$

### 18.14.2-2. Representation in the form of a definite integral.

The Weierstrass function $\wp = \wp(z, g_2, g_3) = \wp(z|\omega_1, \omega_2)$ is defined implicitly by the elliptic integral:

$$z = \int_\wp^\infty \frac{dt}{\sqrt{4t^3 - g_2 t - g_3}} = \int_\wp^\infty \frac{dt}{2\sqrt{(t-e_1)(t-e_2)(t-e_3)}}.$$

The parameters $g_2$ and $g_3$ are known as the *invariants*.

The parameters $e_1, e_2, e_3$, which are the roots of the cubic equation $4z^3 - g_2 z - g_3 = 0$, are related to the half-periods $\omega_1, \omega_2$ and invariants $g_2, g_3$ by

$$e_1 = \wp(\omega_1), \quad e_2 = \wp(\omega_1 + \omega_2), \quad e_1 = \wp(\omega_2),$$
$$e_1 + e_2 + e_3 = 0, \quad e_1 e_2 + e_1 e_3 + e_2 e_3 = -\tfrac{1}{4}g_2, \quad e_1 e_2 e_3 = \tfrac{1}{4}g_3.$$

Homogeneity property:

$$\wp(z, g_2, g_3) = \lambda^2 \wp(\lambda z, \lambda^{-4} g_2, \lambda^{-6} g_3).$$

### 18.14.2-3. Representation as a Laurent series. Differential equations.

The Weierstrass $\wp$-function can be expanded into a Laurent series:

$$\wp(z) = \frac{1}{z^2} + \frac{g_2}{20} z^2 + \frac{g_3}{28} z^4 + \frac{g_2^2}{1200} z^6 + \frac{3 g_2 g_3}{6160} z^8 + \cdots = \frac{1}{z^2} + \sum_{k=2}^\infty a_k z^{2k-2},$$

$$a_k = \frac{3}{(k-3)(2k+1)} \sum_{m=2}^{k-2} a_m a_{k-m} \quad \text{for} \quad k \geq 4, \quad 0 < |z| < \min(|\omega_1|, |\omega_2|).$$

The Weierstrass $\wp$-function satisfies the first-order and second-order nonlinear differential equations:

$$(\wp'_z)^2 = 4\wp^3 - g_2 \wp - g_3,$$
$$\wp''_{zz} = 6\wp^2 - \tfrac{1}{2} g_2.$$

### 18.14.2-4. Connection with Jacobi elliptic functions.

Direct and inverse representations of the Weierstrass elliptic function via Jacobi elliptic functions:

$$\wp(z) = e_1 + (e_1 - e_3) \frac{\operatorname{cn}^2 w}{\operatorname{sn}^2 w} = e_2 + (e_1 - e_3) \frac{\operatorname{dn}^2 w}{\operatorname{sn}^2 w} = e_3 + \frac{e_1 - e_3}{\operatorname{sn}^2 w};$$

$$\operatorname{sn} w = \sqrt{\frac{e_1 - e_3}{\wp(z) - e_3}}, \quad \operatorname{cn} w = \sqrt{\frac{\wp(z) - e_1}{\wp(z) - e_3}}, \quad \operatorname{dn} w = \sqrt{\frac{\wp(z) - e_2}{\wp(z) - e_3}};$$

$$w = z\sqrt{e_1 - e_3} = \mathsf{K} z / \omega_1.$$

The parameters are related by

$$k = \sqrt{\frac{e_2 - e_3}{e_1 - e_3}}, \quad k' = \sqrt{\frac{e_1 - e_2}{e_1 - e_3}}, \quad \mathsf{K} = \omega_1 \sqrt{e_1 - e_3}, \quad i\mathsf{K}' = \omega_2 \sqrt{e_1 - e_3}.$$

## 18.15. Jacobi Theta Functions

### 18.15.1. Series Representation of the Jacobi Theta Functions. Simplest Properties

**18.15.1-1. Definition of the Jacobi theta functions.**

The *Jacobi theta functions* are defined by the following series:

$$\vartheta_1(v) = \vartheta_1(v,q) = \vartheta_1(v|\tau) = 2\sum_{n=0}^{\infty}(-1)^n q^{(n+1/2)^2} \sin[(2n+1)\pi v] = i\sum_{n=-\infty}^{\infty}(-1)^n q^{(n-1/2)^2} e^{i\pi(2n-1)v},$$

$$\vartheta_2(v) = \vartheta_2(v,q) = \vartheta_2(v|\tau) = 2\sum_{n=0}^{\infty} q^{(n+1/2)^2} \cos[(2n+1)\pi v] = \sum_{n=-\infty}^{\infty} q^{(n-1/2)^2} e^{i\pi(2n-1)v},$$

$$\vartheta_3(v) = \vartheta_3(v,q) = \vartheta_3(v|\tau) = 1 + 2\sum_{n=0}^{\infty} q^{n^2} \cos(2n\pi v) = \sum_{n=-\infty}^{\infty} q^{n^2} e^{2i\pi nv},$$

$$\vartheta_4(v) = \vartheta_4(v,q) = \vartheta_4(v|\tau) = 1 + 2\sum_{n=0}^{\infty}(-1)^n q^{n^2} \cos(2n\pi v) = \sum_{n=-\infty}^{\infty}(-1)^n q^{n^2} e^{2i\pi nv},$$

where $v$ is a complex variable and $q = e^{i\pi\tau}$ is a complex parameter ($\tau$ has a positive imaginary part).

**18.15.1-2. Simplest properties.**

The Jacobi theta functions are periodic entire functions that possess the following properties:

$\vartheta_1(v)$ odd, has period 2, vanishes at $v = m + n\tau$;

$\vartheta_2(v)$ even, has period 2, vanishes at $v = m + n\tau + \frac{1}{2}$;

$\vartheta_3(v)$ even, has period 1, vanishes at $v = m + (n + \frac{1}{2})\tau + \frac{1}{2}$;

$\vartheta_4(v)$ even, has period 1, vanishes at $v = m + (n + \frac{1}{2})\tau$.

Here, $m, n = 0, \pm 1, \pm 2, \ldots$

*Remark.* The theta functions are not elliptic functions. The very good convergence of their series allows the computation of various elliptic integrals and elliptic functions using the relations given above in Paragraph 18.15.1-1.

### 18.15.2. Various Relations and Formulas. Connection with Jacobi Elliptic Functions

**18.15.2-1. Linear and quadratic relations.**

Linear relations (first set):

$$\vartheta_1\left(v + \frac{1}{2}\right) = \vartheta_2(v), \quad \vartheta_2\left(v + \frac{1}{2}\right) = -\vartheta_1(v),$$

$$\vartheta_3\left(v + \frac{1}{2}\right) = \vartheta_4(v), \quad \vartheta_4\left(v + \frac{1}{2}\right) = \vartheta_3(v),$$

$$\vartheta_1\left(v + \frac{\tau}{2}\right) = ie^{-i\pi\left(v+\frac{\tau}{4}\right)}\vartheta_4(v), \quad \vartheta_2\left(v + \frac{\tau}{2}\right) = e^{-i\pi\left(v+\frac{\tau}{4}\right)}\vartheta_3(v),$$

$$\vartheta_3\left(v + \frac{\tau}{2}\right) = e^{-i\pi\left(v+\frac{\tau}{4}\right)}\vartheta_2(v), \quad \vartheta_4\left(v + \frac{\tau}{2}\right) = ie^{-i\pi\left(v+\frac{\tau}{4}\right)}\vartheta_1(v).$$

Linear relations (second set):

$$\vartheta_1(v|\tau+1) = e^{i\pi/4}\vartheta_1(v|\tau), \qquad \vartheta_2(v|\tau+1) = e^{i\pi/4}\vartheta_2(v|\tau),$$
$$\vartheta_3(v|\tau+1) = \vartheta_4(v|\tau), \qquad \vartheta_4(v|\tau+1) = \vartheta_3(v|\tau),$$
$$\vartheta_1\left(\frac{v}{\tau}\Big|-\frac{1}{\tau}\right) = \frac{1}{i}\sqrt{\frac{\tau}{i}}\,e^{i\pi v^2/\tau}\vartheta_1(v|\tau), \qquad \vartheta_2\left(\frac{v}{\tau}\Big|-\frac{1}{\tau}\right) = \sqrt{\frac{\tau}{i}}\,e^{i\pi v^2/\tau}\vartheta_4(v|\tau),$$
$$\vartheta_3\left(\frac{v}{\tau}\Big|-\frac{1}{\tau}\right) = \sqrt{\frac{\tau}{i}}\,e^{i\pi v^2/\tau}\vartheta_3(v|\tau), \qquad \vartheta_4\left(\frac{v}{\tau}\Big|-\frac{1}{\tau}\right) = \sqrt{\frac{\tau}{i}}\,e^{i\pi v^2/\tau}\vartheta_2(v|\tau).$$

Quadratic relations:

$$\vartheta_1^2(v)\vartheta_2^2(0) = \vartheta_4^2(v)\vartheta_3^2(0) - \vartheta_3^2(v)\vartheta_4^2(0),$$
$$\vartheta_1^2(v)\vartheta_3^2(0) = \vartheta_4^2(v)\vartheta_2^2(0) - \vartheta_2^2(v)\vartheta_4^2(0),$$
$$\vartheta_1^2(v)\vartheta_4^2(0) = \vartheta_3^2(v)\vartheta_2^2(0) - \vartheta_2^2(v)\vartheta_3^2(0),$$
$$\vartheta_4^2(v)\vartheta_4^2(0) = \vartheta_3^2(v)\vartheta_3^2(0) - \vartheta_2^2(v)\vartheta_2^2(0).$$

**18.15.2-2. Representation of the theta functions in the form of infinite products.**

$$\vartheta_1(v) = 2q_0 q^{1/4} \sin(\pi v) \prod_{n=1}^{\infty} \left[1 - 2q^{2n}\cos(2\pi v) + q^{4n}\right],$$
$$\vartheta_2(v) = 2q_0 q^{1/4} \cos(\pi v) \prod_{n=1}^{\infty} \left[1 + 2q^{2n}\cos(2\pi v) + q^{4n}\right],$$
$$\vartheta_3(v) = q_0 \prod_{n=1}^{\infty} \left[1 + 2q^{2n-1}\cos(2\pi v) + q^{4n-2}\right],$$
$$\vartheta_4(v) = q_0 \prod_{n=1}^{\infty} \left[1 - 2q^{2n-1}\cos(2\pi v) + q^{4n-2}\right],$$

where $q_0 = \prod_{n=1}^{\infty}(1 - q^{2n})$.

**18.15.2-3. Connection with Jacobi elliptic functions.**

Representations of Jacobi elliptic functions in terms of the theta functions:

$$\mathrm{sn}\,w = \frac{\vartheta_3(0)}{\vartheta_2(0)}\frac{\vartheta_1(v)}{\vartheta_4(v)}, \qquad \mathrm{cn}\,w = \frac{\vartheta_4(0)}{\vartheta_2(0)}\frac{\vartheta_2(v)}{\vartheta_4(v)}, \qquad \mathrm{dn}\,w = \frac{\vartheta_4(0)}{\vartheta_3(0)}\frac{\vartheta_3(v)}{\vartheta_4(v)}, \qquad w = 2\mathsf{K}\,v.$$

The parameters are related by

$$k = \frac{\vartheta_2^2(0)}{\vartheta_3^2(0)}, \qquad k' = \frac{\vartheta_4^2(0)}{\vartheta_3^2(0)}, \qquad \mathsf{K} = \frac{\pi}{2}\vartheta_3^2(0), \qquad \mathsf{K}' = -i\tau\,\mathsf{K}.$$

TABLE 18.6
The Mathieu functions $ce_n = ce_n(x,q)$ and $se_n = se_n(x,q)$ (for odd $n$, functions $ce_n$ and $se_n$ are $2\pi$-periodic, and for even $n$, they are $\pi$-periodic); definite eigenvalues $a = a_n(q)$ and $a = b_n(q)$ correspond to each value of parameter $q$

| Mathieu functions | Recurrence relations for coefficients | Normalization conditions |
|---|---|---|
| $ce_{2n} = \sum_{m=0}^{\infty} A_{2m}^{2n} \cos 2mx$ | $qA_2^{2n} = a_{2n}A_0^{2n};$ $qA_4^{2n} = (a_{2n}-4)A_2^{2n} - 2qA_0^{2n};$ $qA_{2m+2}^{2n} = (a_{2n}-4m^2)A_{2m}^{2n}$ $-qA_{2m-2}^{2n}, \quad m \geq 2$ | $(A_0^{2n})^2 + \sum_{m=0}^{\infty}(A_{2m}^{2n})^2$ $= \begin{cases} 2 & \text{if } n=0 \\ 1 & \text{if } n \geq 1 \end{cases}$ |
| $ce_{2n+1} = \sum_{m=0}^{\infty} A_{2m+1}^{2n+1} \cos(2m+1)x$ | $qA_3^{2n+1} = (a_{2n+1}-1-q)A_1^{2n+1};$ $qA_{2m+3}^{2n+1} = [a_{2n+1}-(2m+1)^2]A_{2m+1}^{2n+1}$ $-qA_{2m-1}^{2n+1}, \quad m \geq 1$ | $\sum_{m=0}^{\infty}(A_{2m+1}^{2n+1})^2 = 1$ |
| $se_{2n} = \sum_{m=0}^{\infty} B_{2m}^{2n} \sin 2mx,$ $se_0 = 0$ | $qB_4^{2n} = (b_{2n}-4)B_2^{2n};$ $qB_{2m+2}^{2n} = (b_{2n}-4m^2)B_{2m}^{2n}$ $-qB_{2m-2}^{2n}, \quad m \geq 2$ | $\sum_{m=0}^{\infty}(B_{2m}^{2n})^2 = 1$ |
| $se_{2n+1} = \sum_{m=0}^{\infty} B_{2m+1}^{2n+1} \sin(2m+1)x$ | $qB_3^{2n+1} = (b_{2n+1}-1-q)B_1^{2n+1};$ $qB_{2m+3}^{2n+1} = [b_{2n+1}-(2m+1)^2]B_{2m+1}^{2n+1}$ $-qB_{2m-1}^{2n+1}, \quad m \geq 1$ | $\sum_{m=0}^{\infty}(B_{2m+1}^{2n+1})^2 = 1$ |

## 18.16. Mathieu Functions and Modified Mathieu Functions

### 18.16.1. Mathieu Functions

#### 18.16.1-1. Mathieu equation and Mathieu functions.

The Mathieu functions $ce_n(x,q)$ and $se_n(x,q)$ are periodical solutions of the Mathieu equation
$$y''_{xx} + (a - 2q\cos 2x)y = 0.$$

Such solutions exist for definite values of parameters $a$ and $q$ (those values of $a$ are referred to as eigenvalues). The Mathieu functions are listed in Table 18.6.

#### 18.16.1-2. Properties of the Mathieu functions.

The Mathieu functions possess the following properties:

$$ce_{2n}(x,-q) = (-1)^n ce_{2n}\left(\frac{\pi}{2}-x, q\right), \quad ce_{2n+1}(x,-q) = (-1)^n se_{2n+1}\left(\frac{\pi}{2}-x, q\right),$$
$$se_{2n}(x,-q) = (-1)^{n-1} se_{2n}\left(\frac{\pi}{2}-x, q\right), \quad se_{2n+1}(x,-q) = (-1)^n ce_{2n+1}\left(\frac{\pi}{2}-x, q\right).$$

Selecting sufficiently large number $m$ and omitting the term with the maximum number in the recurrence relations (indicated in Table 18.6), we can obtain approximate relations for eigenvalues $a_n$ (or $b_n$) with respect to parameter $q$. Then, equating the determinant of the corresponding homogeneous linear system of equations for coefficients $A_m^n$ (or $B_m^n$) to zero, we obtain an algebraic equation for finding $a_n(q)$ (or $b_n(q)$).

For fixed real $q \neq 0$, eigenvalues $a_n$ and $b_n$ are all real and different, while

$$\text{if } q > 0 \text{ then } a_0 < b_1 < a_1 < b_2 < a_2 < \cdots ;$$
$$\text{if } q < 0 \text{ then } a_0 < a_1 < b_1 < b_2 < a_2 < a_3 < b_3 < b_4 < \cdots .$$

The eigenvalues possess the properties

$$a_{2n}(-q) = a_{2n}(q), \quad b_{2n}(-q) = b_{2n}(q), \quad a_{2n+1}(-q) = b_{2n+1}(q).$$

Tables of the eigenvalues $a_n = a_n(q)$ and $b_n = b_n(q)$ can be found in Abramowitz and Stegun (1964, chap. 20).

The solution of the Mathieu equation corresponding to eigenvalue $a_n$ (or $b_n$) has $n$ zeros on the interval $0 \leq x < \pi$ ($q$ is a real number).

### 18.16.1-3. Asymptotic expansions as $q \to 0$ and $q \to \infty$.

Listed below are two leading terms of asymptotic expansions of the Mathieu functions $ce_n(x, q)$ and $se_n(x, q)$, as well as of the corresponding eigenvalues $a_n(q)$ and $b_n(q)$, as $q \to 0$:

$$ce_0(x, q) = \frac{1}{\sqrt{2}}\left(1 - \frac{q}{2}\cos 2x\right), \quad a_0(q) = -\frac{q^2}{2} + \frac{7q^4}{128};$$

$$ce_1(x, q) = \cos x - \frac{q}{8}\cos 3x, \quad a_1(q) = 1 + q;$$

$$ce_2(x, q) = \cos 2x + \frac{q}{4}\left(1 - \frac{\cos 4x}{3}\right), \quad a_2(q) = 4 + \frac{5q^2}{12};$$

$$ce_n(x, q) = \cos nx + \frac{q}{4}\left[\frac{\cos(n+2)x}{n+1} - \frac{\cos(n-2)x}{n-1}\right], \quad a_n(q) = n^2 + \frac{q^2}{2(n^2-1)} \quad (n \geq 3);$$

$$se_1(x, q) = \sin x - \frac{q}{8}\sin 3x, \quad b_1(q) = 1 - q;$$

$$se_2(x, q) = \sin 2x - q\frac{\sin 4x}{12}, \quad b_2(q) = 4 - \frac{q^2}{12};$$

$$se_n(x, q) = \sin nx - \frac{q}{4}\left[\frac{\sin(n+2)x}{n+1} - \frac{\sin(n-2)x}{n-1}\right], \quad b_n(q) = n^2 + \frac{q^2}{2(n^2-1)} \quad (n \geq 3).$$

Asymptotic results as $q \to \infty$ ($-\pi/2 < x < \pi/2$):

$$a_n(q) \approx -2q + 2(2n+1)\sqrt{q} + \tfrac{1}{4}(2n^2 + 2n + 1),$$
$$b_{n+1}(q) \approx -2q + 2(2n+1)\sqrt{q} + \tfrac{1}{4}(2n^2 + 2n + 1),$$
$$ce_n(x, q) \approx \lambda_n q^{-1/4}\cos^{-n-1} x \left[\cos^{2n+1}\xi \exp(2\sqrt{q}\sin x) + \sin^{2n+1}\xi \exp(-2\sqrt{q}\sin x)\right],$$
$$se_{n+1}(x, q) \approx \mu_{n+1} q^{-1/4}\cos^{-n-1} x \left[\cos^{2n+1}\xi \exp(2\sqrt{q}\sin x) - \sin^{2n+1}\xi \exp(-2\sqrt{q}\sin x)\right],$$

where $\lambda_n$ and $\mu_n$ are some constants independent of the parameter $q$, and $\xi = \tfrac{1}{2}x + \tfrac{\pi}{4}$.

## 18.16.2. Modified Mathieu Functions

The modified Mathieu functions $\mathrm{Ce}_n(x,q)$ and $\mathrm{Se}_n(x,q)$ are solutions of the modified Mathieu equation

$$y''_{xx} - (a - 2q\cosh 2x)y = 0,$$

with $a = a_n(q)$ and $a = b_n(q)$ being the eigenvalues of the Mathieu equation (see Subsection 18.16.1).

The modified Mathieu functions are defined as

$$\mathrm{Ce}_{2n+p}(x,q) = \mathrm{ce}_{2n+p}(ix,q) = \sum_{k=0}^{\infty} A_{2k+p}^{2n+p} \cosh[(2k+p)x],$$

$$\mathrm{Se}_{2n+p}(x,q) = -i\,\mathrm{se}_{2n+p}(ix,q) = \sum_{k=0}^{\infty} B_{2k+p}^{2n+p} \sinh[(2k+p)x],$$

where $p$ may be equal to 0 and 1, and coefficients $A_{2k+p}^{2n+p}$ and $B_{2k+p}^{2n+p}$ are indicated in Subsection 18.16.1.

## 18.17. Orthogonal Polynomials

All zeros of each of the orthogonal polynomials $\mathcal{P}_n(x)$ considered in this section are real and simple. The zeros of the polynomials $\mathcal{P}_n(x)$ and $\mathcal{P}_{n+1}(x)$ are alternating.

For Legendre polynomials see Subsection 18.11.1.

### 18.17.1. Laguerre Polynomials and Generalized Laguerre Polynomials

#### 18.17.1-1. Laguerre polynomials.

The Laguerre polynomials $L_n = L_n(x)$ satisfy the second-order linear ordinary differential equation

$$xy''_{xx} + (1-x)y'_x + ny = 0$$

and are defined by the formulas

$$L_n(x) = \frac{1}{n!} e^x \frac{d^n}{dx^n}\left(x^n e^{-x}\right) = \frac{(-1)^n}{n!}\left[x^n - n^2 x^{n-1} + \frac{n^2(n-1)^2}{2!}x^{n-2} + \cdots\right].$$

The first four polynomials have the form

$$L_0(x) = 1, \quad L_1(x) = -x+1, \quad L_2(x) = \tfrac{1}{2}(x^2 - 4x + 2), \quad L_3(x) = \tfrac{1}{6}(-x^3 + 9x^2 - 18x + 6).$$

To calculate $L_n(x)$ for $n \geq 2$, one can use the recurrence formulas

$$L_{n+1}(x) = \frac{1}{n+1}\left[(2n+1-x)L_n(x) - nL_{n-1}(x)\right].$$

The functions $L_n(x)$ form an orthonormal system on the interval $0 < x < \infty$ with weight $e^{-x}$:

$$\int_0^{\infty} e^{-x} L_n(x) L_m(x)\,dx = \begin{cases} 0 & \text{if } n \neq m, \\ 1 & \text{if } n = m. \end{cases}$$

The generating function is

$$\frac{1}{1-s}\exp\left(-\frac{sx}{1-s}\right) = \sum_{n=0}^{\infty} L_n(x) s^n, \qquad |s| < 1.$$

### 18.17.1-2. Generalized Laguerre polynomials.

The generalized Laguerre polynomials $L_n^\alpha = L_n^\alpha(x)$ ($\alpha > -1$) satisfy the equation

$$xy''_{xx} + (\alpha + 1 - x)y'_x + ny = 0$$

and are defined by the formulas

$$L_n^\alpha(x) = \frac{1}{n!} x^{-\alpha} e^x \frac{d^n}{dx^n}\left(x^{n+\alpha} e^{-x}\right) = \sum_{m=0}^{n} C_{n+\alpha}^{n-m} \frac{(-x)^m}{m!} = \sum_{m=0}^{n} \frac{\Gamma(n+\alpha+1)}{\Gamma(m+\alpha+1)} \frac{(-x)^m}{m!(n-m)!}.$$

Notation: $L_n^0(x) = L_n(x)$.
Special cases:

$$L_0^\alpha(x) = 1, \qquad L_1^\alpha(x) = \alpha + 1 - x, \qquad L_n^{-n}(x) = (-1)^n \frac{x^n}{n!}.$$

To calculate $L_n^\alpha(x)$ for $n \geq 2$, one can use the recurrence formulas

$$L_{n+1}^\alpha(x) = \frac{1}{n+1}\left[(2n + \alpha + 1 - x)L_n^\alpha(x) - (n+\alpha)L_{n-1}^\alpha(x)\right].$$

Other recurrence formulas:

$$L_n^\alpha(x) = L_{n-1}^\alpha(x) + L_n^{\alpha-1}(x), \qquad \frac{d}{dx}L_n^\alpha(x) = -L_{n-1}^{\alpha+1}(x), \qquad x\frac{d}{dx}L_n^\alpha(x) = nL_n^\alpha(x) - (n+\alpha)L_{n-1}^\alpha(x).$$

The functions $L_n^\alpha(x)$ form an orthogonal system on the interval $0 < x < \infty$ with weight $x^\alpha e^{-x}$:

$$\int_0^\infty x^\alpha e^{-x} L_n^\alpha(x) L_m^\alpha(x)\, dx = \begin{cases} 0 & \text{if } n \neq m, \\ \frac{\Gamma(\alpha+n+1)}{n!} & \text{if } n = m. \end{cases}$$

The generating function is

$$(1-s)^{-\alpha-1} \exp\left(-\frac{sx}{1-s}\right) = \sum_{n=0}^{\infty} L_n^\alpha(x) s^n, \qquad |s| < 1.$$

## 18.17.2. Chebyshev Polynomials and Functions

### 18.17.2-1. Chebyshev polynomials of the first kind.

The *Chebyshev polynomials of the first kind* $T_n = T_n(x)$ satisfy the second-order linear ordinary differential equation

$$(1-x^2)y''_{xx} - xy'_x + n^2 y = 0 \qquad (18.17.2.1)$$

and are defined by the formulas

$$T_n(x) = \cos(n \arccos x) = \frac{(-2)^n n!}{(2n)!} \sqrt{1-x^2} \frac{d^n}{dx^n}\left[(1-x^2)^{n-\frac{1}{2}}\right]$$

$$= \frac{n}{2} \sum_{m=0}^{[n/2]} (-1)^m \frac{(n-m-1)!}{m!(n-2m)!} (2x)^{n-2m} \qquad (n = 0, 1, 2, \dots),$$

where $[A]$ stands for the integer part of a number $A$.

An alternative representation of the Chebyshev polynomials:
$$T_n(x) = \frac{(-1)^n}{(2n-1)!!}(1-x^2)^{1/2}\frac{d^n}{dx^n}(1-x^2)^{n-1/2}.$$

The first five Chebyshev polynomials of the first kind are

$T_0(x) = 1$, $\quad T_1(x) = x$, $\quad T_2(x) = 2x^2 - 1$, $\quad T_3(x) = 4x^3 - 3x$, $\quad T_4(x) = 8x^4 - 8x^2 + 1$.

The recurrence formulas:
$$T_{n+1}(x) = 2xT_n(x) - T_{n-1}(x), \qquad n \geq 2.$$

The functions $T_n(x)$ form an orthogonal system on the interval $-1 < x < 1$, with
$$\int_{-1}^{1}\frac{T_n(x)T_m(x)}{\sqrt{1-x^2}}\,dx = \begin{cases} 0 & \text{if } n \neq m, \\ \frac{1}{2}\pi & \text{if } n = m \neq 0, \\ \pi & \text{if } n = m = 0. \end{cases}$$

The generating function is
$$\frac{1-sx}{1-2sx+s^2} = \sum_{n=0}^{\infty}T_n(x)s^n \qquad (|s| < 1).$$

The functions $T_n(x)$ have only real simple zeros, all lying on the interval $-1 < x < 1$.

The normalized Chebyshev polynomials of the first kind, $2^{1-n}T_n(x)$, deviate from zero least of all. This means that among all polynomials of degree $n$ with the leading coefficient 1, it is the maximum of the modulus $\max\limits_{-1\leq x\leq 1}|2^{1-n}T_n(x)|$ that has the least value, the maximum being equal to $2^{1-n}$.

### 18.17.2-2. Chebyshev polynomials of the second kind.

The *Chebyshev polynomials of the second kind* $U_n = U_n(x)$ satisfy the second-order linear ordinary differential equation
$$(1-x^2)y''_{xx} - 3xy'_x + n(n+2)y = 0$$
and are defined by the formulas
$$U_n(x) = \frac{\sin[(n+1)\arccos x]}{\sqrt{1-x^2}} = \frac{2^n(n+1)!}{(2n+1)!}\frac{1}{\sqrt{1-x^2}}\frac{d^n}{dx^n}(1-x^2)^{n+1/2}$$
$$= \sum_{m=0}^{[n/2]}(-1)^m\frac{(n-m)!}{m!(n-2m)!}(2x)^{n-2m} \qquad (n = 0, 1, 2, \ldots).$$

The first five Chebyshev polynomials of the second kind are

$U_0(x) = 1$, $\quad U_1(x) = 2x$, $\quad U_2(x) = 4x^2 - 1$, $\quad U_3(x) = 8x^3 - 4x$, $\quad U_4(x) = 16x^4 - 12x^2 + 1$.

The recurrence formulas:
$$U_{n+1}(x) = 2xU_n(x) - U_{n-1}(x), \qquad n \geq 2.$$

The generating function is
$$\frac{1}{1-2sx+s^2} = \sum_{n=0}^{\infty}U_n(x)s^n \qquad (|s| < 1).$$

The Chebyshev polynomials of the first and second kind are related by
$$U_n(x) = \frac{1}{n+1}\frac{d}{dx}T_{n+1}(x).$$

## 18.17.2-3. Chebyshev functions of the second kind.

The *Chebyshev functions of the second kind*,

$$U_0(x) = \arcsin x,$$

$$U_n(x) = \sin(n \arccos x) = \frac{\sqrt{1-x^2}}{n} \frac{dT_n(x)}{dx} \quad (n = 1, 2, \dots),$$

just as the Chebyshev polynomials, also satisfy the differential equation (18.17.2.1).
The first five the Chebyshev functions are

$$U_0(x) = 0, \quad U_1(x) = \sqrt{1-x^2}, \quad U_2(x) = 2x\sqrt{1-x^2},$$
$$U_3(x) = (4x^2-1)\sqrt{1-x^2}, \quad U_5(x) = (8x^3-4x)\sqrt{1-x^2}.$$

The recurrence formulas:

$$U_{n+1}(x) = 2x\, U_n(x) - U_{n-1}(x), \quad n \geq 2.$$

The functions $U_n(x)$ form an orthogonal system on the interval $-1 < x < 1$, with

$$\int_{-1}^{1} \frac{U_n(x)\, U_m(x)}{\sqrt{1-x^2}}\, dx = \begin{cases} 0 & \text{if } n \neq m \text{ or } n = m = 0, \\ \frac{1}{2}\pi & \text{if } n = m \neq 0. \end{cases}$$

The generating function is

$$\frac{\sqrt{1-x^2}}{1-2sx+s^2} = \sum_{n=0}^{\infty} U_{n+1}(x) s^n \quad (|s| < 1).$$

## 18.17.3. Hermite Polynomials

### 18.17.3-1. Various representations of the Hermite polynomials.

The *Hermite polynomials* $H_n = H_n(x)$ satisfy the second-order linear ordinary differential equation

$$y''_{xx} - 2x y'_x + 2n y = 0$$

and is defined by the formulas

$$H_n(x) = (-1)^n \exp(x^2) \frac{d^n}{dx^n} \exp(-x^2) = \sum_{m=0}^{[n/2]} (-1)^m \frac{n!}{m!(n-2m)!} (2x)^{n-2m}.$$

The first five polynomials are

$$H_0(x)=1, \quad H_1(x)=2x, \quad H_2(x)=4x^2-2, \quad H_3(x)=8x^3-12x, \quad H_4(x)=16x^4-48x^2+12.$$

Recurrence formulas:

$$H_{n+1}(x) = 2x H_n(x) - 2n H_{n-1}(x), \quad n \geq 2;$$

$$\frac{d}{dx} H_n(x) = 2n H_{n-1}(x).$$

Integral representation:

$$H_{2n}(x) = \frac{(-1)^n 2^{2n+1}}{\sqrt{\pi}} \exp(x^2) \int_0^\infty \exp(-t^2) t^{2n} \cos(2xt)\, dt,$$

$$H_{2n+1}(x) = \frac{(-1)^n 2^{2n+2}}{\sqrt{\pi}} \exp(x^2) \int_0^\infty \exp(-t^2) t^{2n+1} \sin(2xt)\, dt,$$

where $n = 0, 1, 2, \dots$

### 18.17.3-2. Orthogonality. The generating function. An asymptotic formula.

The functions $H_n(x)$ form an orthogonal system on the interval $-\infty < x < \infty$ with weight $e^{-x^2}$:

$$\int_{-\infty}^{\infty} \exp(-x^2) H_n(x) H_m(x) \, dx = \begin{cases} 0 & \text{if } n \neq m, \\ \sqrt{\pi} \, 2^n n! & \text{if } n = m. \end{cases}$$

Generating function:

$$\exp(-s^2 + 2sx) = \sum_{n=0}^{\infty} H_n(x) \frac{s^n}{n!}.$$

Asymptotic formula as $n \to \infty$:

$$H_n(x) \approx 2^{\frac{n+1}{2}} n^{\frac{n}{2}} e^{-\frac{n}{2}} \exp(x^2) \cos\left(\sqrt{2n+1}\, x - \tfrac{1}{2}\pi n\right).$$

### 18.17.3-3. Hermite functions.

The *Hermite functions* $h_n(x)$ are introduced by the formula

$$h_n(x) = \exp\left(-\tfrac{1}{2}x^2\right) H_n(x) = (-1)^n \exp\left(\tfrac{1}{2}x^2\right) \frac{d^n}{dx^n} \exp(-x^2), \qquad n = 0, 1, 2, \dots$$

The Hermite functions satisfy the second-order linear ordinary differential equation

$$h''_{xx} + (2n + 1 - x^2)h = 0.$$

The functions $h_n(x)$ form an orthogonal system on the interval $-\infty < x < \infty$ with weight 1:

$$\int_{-\infty}^{\infty} h_n(x) h_m(x) \, dx = \begin{cases} 0 & \text{if } n \neq m, \\ \sqrt{\pi} \, 2^n n! & \text{if } n = m. \end{cases}$$

## 18.17.4. Jacobi Polynomials and Gegenbauer Polynomials

### 18.17.4-1. Jacobi polynomials.

The *Jacobi polynomials*, $P_n^{\alpha,\beta}(x)$, are solutions of the second-order linear ordinary differential equation

$$(1-x^2)y''_{xx} + \left[\beta - \alpha - (\alpha + \beta + 2)x\right]y'_x + n(n + \alpha + \beta + 1)y = 0$$

and are defined by the formulas

$$P_n^{\alpha,\beta}(x) = \frac{(-1)^n}{2^n n!}(1-x)^{-\alpha}(1+x)^{-\beta} \frac{d^n}{dx^n}\left[(1-x)^{\alpha+n}(1+x)^{\beta+n}\right]$$

$$= 2^{-n} \sum_{m=0}^{n} C_{n+\alpha}^{m} C_{n+\beta}^{n-m} (x-1)^{n-m}(x+1)^{m},$$

where the $C_b^a$ are binomial coefficients.

The generating function:

$$2^{\alpha+\beta} R^{-1}(1-s+R)^{-\alpha}(1+s+R)^{-\beta} = \sum_{n=0}^{\infty} P_n^{\alpha,\beta}(x) s^n, \quad R = \sqrt{1-2xs+s^2}, \quad |s| < 1.$$

The Jacobi polynomials are orthogonal on the interval $-1 \leq x \leq 1$ with weight $(1-x)^\alpha (1+x)^\beta$:

$$\int_{-1}^{1} (1-x)^\alpha (1+x)^\beta P_n^{\alpha,\beta}(x) P_m^{\alpha,\beta}(x)\, dx = \begin{cases} 0 & \text{if } n \neq m, \\ \dfrac{2^{\alpha+\beta+1}}{\alpha+\beta+2n+1} \dfrac{\Gamma(\alpha+n+1)\Gamma(\beta+n+1)}{n!\,\Gamma(\alpha+\beta+n+1)} & \text{if } n = m. \end{cases}$$

For $\alpha > -1$ and $\beta > -1$, all zeros of the polynomial $P_n^{\alpha,\beta}(x)$ are simple and lie on the interval $-1 < x < 1$.

### 18.17.4-2. Gegenbauer polynomials.

The *Gegenbauer polynomials* (also called *ultraspherical polynomials*), $C_n^{(\lambda)}(x)$, are solutions of the second-order linear ordinary differential equation

$$(1-x^2) y''_{xx} - (2\lambda+1) x y'_x + n(n+2\lambda) y = 0$$

and are defined by the formulas

$$C_n^{(\lambda)}(x) = \frac{(-2)^n}{n!} \frac{\Gamma(n+\lambda)\Gamma(n+2\lambda)}{\Gamma(\lambda)\Gamma(2n+2\lambda)} (1-x^2)^{-\lambda+1/2} \frac{d^n}{dx^n} (1-x^2)^{n+\lambda-1/2}$$

$$= \sum_{m=0}^{[n/2]} (-1)^m \frac{\Gamma(n-m+\lambda)}{\Gamma(\lambda)\, m!\, (n-2m)!} (2x)^{n-2m}.$$

Recurrence formulas:

$$C_{n+1}^{(\lambda)}(x) = \frac{2(n+\lambda)}{n+1} x C_n^{(\lambda)}(x) - \frac{n+2\lambda-1}{n+1} C_{n-1}^{(\lambda)}(x);$$

$$C_n^{(\lambda)}(-x) = (-1)^n C_n^{(\lambda)}(x), \quad \frac{d}{dx} C_n^{(\lambda)}(x) = 2\lambda C_{n-1}^{(\lambda+1)}(x).$$

The generating function:

$$\frac{1}{(1-2xs+s^2)^\lambda} = \sum_{n=0}^{\infty} C_n^{(\lambda)}(x) s^n.$$

The Gegenbauer polynomials are orthogonal on the interval $-1 \leq x \leq 1$ with weight $(1-x^2)^{\lambda-1/2}$:

$$\int_{-1}^{1} (1-x^2)^{\lambda-1/2} C_n^{(\lambda)}(x) C_m^{(\lambda)}(x)\, dx = \begin{cases} 0 & \text{if } n \neq m, \\ \dfrac{\pi \Gamma(2\lambda+n)}{2^{2\lambda-1}(\lambda+n) n!\, \Gamma^2(\lambda)} & \text{if } n = m. \end{cases}$$

## 18.18. Nonorthogonal Polynomials

### 18.18.1. Bernoulli Polynomials

**18.18.1-1. Definition. Basic properties.**

The *Bernoulli polynomials* $B_n(x)$ are introduced by the formula

$$B_n(x) = \sum_{k=0}^{n} C_n^k B_k x^{n-k} \quad (n = 0, 1, 2, \ldots),$$

where $C_n^k$ are the binomial coefficients and $B_n$ are Bernoulli numbers (see Subsection 18.1.3).

The Bernoulli polynomials can be defined using the recurrence relation

$$B_0(x) = 1, \quad \sum_{k=0}^{n-1} C_n^k B_k(x) = nx^{n-1}, \quad n = 2, 3, \ldots$$

The first six Bernoulli polynomials are given by

$$B_0(x) = 1, \quad B_1(x) = x - \tfrac{1}{2}, \quad B_2(x) = x^2 - x + \tfrac{1}{6}, \quad B_3(x) = x^3 - \tfrac{3}{2}x^2 + \tfrac{1}{2}x,$$

$$B_4(x) = x^4 - 2x^3 + x^2 - \tfrac{1}{30}, \quad B_5(x) = x^5 - \tfrac{5}{2}x^4 + \tfrac{5}{3}x^3 - \tfrac{1}{6}x.$$

Basic properties:

$$B_n(x+1) - B_n(x) = nx^{n-1}, \quad B'_{n+1}(x) = (n+1)B_n(x),$$
$$B_n(1-x) = (-1)^n B_n(x), \quad (-1)^n E_n(-x) = E_n(x) + nx^{n-1},$$

where the prime denotes a derivative with respect to $x$, and $n = 0, 1, \ldots$

Multiplication and addition formulas:

$$B_n(mx) = m^{n-1} \sum_{k=0}^{m-1} B_n\left(x + \frac{k}{m}\right),$$

$$B_n(x+y) = \sum_{k=0}^{n} C_n^k B_k(x) y^{n-k},$$

where $n = 0, 1, \ldots$ and $m = 1, 2, \ldots$

**18.18.1-2. Generating function. Fourier series expansions. Integrals.**

The generating function is expressed as

$$\frac{te^{xt}}{e^t - 1} \equiv \sum_{n=0}^{\infty} B_n(x) \frac{t^n}{n!} \quad (|t| < 2\pi).$$

This relation may be used as a definition of the Bernoulli polynomials.

Fourier series expansions:

$$B_n(x) = -2\frac{n!}{(2\pi)^n} \sum_{k=1}^{\infty} \frac{\cos(2\pi kx - \tfrac{1}{2}\pi n)}{k^n}, \qquad (n=1,\ 0<x<1;\ n>1,\ 0\le x\le 1);$$

$$B_{2n-1}(x) = 2(-1)^n \frac{(2n-1)!}{(2\pi)^{2n-1}} \sum_{k=1}^{\infty} \frac{\sin(2k\pi x)}{k^{2n-1}} \qquad (n=1,\ 0<x<1;\ n>1,\ 0\le x\le 1);$$

$$B_{2n}(x) = 2(-1)^n \frac{(2n)!}{(2\pi)^{2n}} \sum_{k=1}^{\infty} \frac{\cos(2k\pi x)}{k^{2n}} \qquad (n=1,2,\ldots,\ 0\le x\le 1).$$

Integrals:

$$\int_a^x B_n(t)\,dt = \frac{B_{n+1}(x) - B_{n+1}(a)}{n+1},$$

$$\int_0^1 B_m(t) B_n(t)\,dt = (-1)^{n-1} \frac{m!\,n!}{(m+n)!} B_{m+n},$$

where $m$ and $n$ are positive integers and $B_n$ are Bernoulli numbers.

### 18.18.2. Euler Polynomials

**18.18.2-1. Definition. Basic properties.**

Definition:

$$E_n(x) = \sum_{k=0}^{n} C_n^k \frac{E_k}{2^n} \left(x - \frac{1}{2}\right)^{n-k} \qquad (n = 0, 1, 2, \ldots),$$

where $C_n^k$ are the binomial coefficients and $E_n$ are Euler numbers.

The first six Euler polynomials are given by

$$E_0(x) = 1, \quad E_1(x) = x - \tfrac{1}{2}, \quad E_2(x) = x^2 - x, \quad E_3(x) = x^3 - \tfrac{3}{2}x^2 + \tfrac{1}{4},$$

$$E_4(x) = x^4 - 2x^3 + x, \quad E_5(x) = x^5 - \tfrac{5}{2}x^4 + \tfrac{5}{2}x^2 - \tfrac{1}{2}.$$

Basic properties:

$$E_n(x+1) + E_n(x) = 2x^n, \quad E'_{n+1} = (n+1)E_n(x),$$
$$E_n(1-x) = (-1)^n E_n(x), \quad (-1)^{n+1} E_n(-x) = E_n(x) - 2x^n,$$

where the prime denotes a derivative with respect to $x$, and $n = 0, 1, \ldots$

Multiplication and addition formulas:

$$E_n(mx) = m^n \sum_{k=0}^{m-1} (-1)^k E_n\!\left(x + \frac{k}{m}\right), \quad n = 0, 1, \ldots,\ m = 1, 3, \ldots;$$

$$E_n(mx) = -\frac{2}{n+1} m^n \sum_{k=0}^{m-1} (-1)^k E_{n+1}\!\left(x + \frac{k}{m}\right), \quad n = 0, 1, \ldots,\ m = 2, 4, \ldots;$$

$$E_n(x+y) = \sum_{k=0}^{n} C_n^k E_k(x) y^{n-k}, \quad n = 0, 1, \ldots$$

**18.18.2-2. Generating function. Fourier series expansions. Integrals.**

The generating function is expressed as

$$\frac{2e^{xt}}{e^t+1} \equiv \sum_{n=0}^{\infty} E_n(x)\frac{t^n}{n!} \qquad (|t| < \pi).$$

This relation may be used as a definition of the Euler polynomials.

Fourier series expansions:

$$E_n(x) = 4\frac{n!}{\pi^{n+1}} \sum_{k=0}^{\infty} \frac{\sin\big((2k+1)\pi x - \tfrac{1}{2}\pi n\big)}{(2k+1)^{n+1}} \qquad (n=0,\ 0<x<1;\ n>0,\ 0\le x\le 1);$$

$$E_{2n}(x) = 4(-1)^n \frac{(2n)!}{\pi^{2n+1}} \sum_{k=0}^{\infty} \frac{\sin\big((2k+1)\pi x\big)}{(2k+1)^{2n+1}} \qquad (n=0,\ 0<x<1;\ n>0,\ 0\le x\le 1);$$

$$E_{2n-1}(x) = 4(-1)^n \frac{(2n-1)!}{\pi^{2n}} \sum_{k=0}^{\infty} \frac{\cos\big((2k+1)\pi x\big)}{(2k+1)^{2n}} \qquad (n=1,2,\ldots,\ 0\le x\le 1).$$

Integrals:

$$\int_a^x E_n(t)\,dt = \frac{E_{n+1}(x) - E_{n+1}(a)}{n+1},$$

$$\int_0^1 E_m(t)E_n(t)\,dt = 4(-1)^n(2^{m+n+2} - 1)\frac{m!\,n!}{(m+n+2)!}B_{m+n+2},$$

where $m, n = 0, 1, \ldots$ and $B_n$ are Bernoulli numbers. The Euler polynomials are orthogonal for even $n+m$.

Connection with the Bernoulli polynomials:

$$E_{n-1}(x) = \frac{2^n}{n}\left[B_n\!\left(\frac{x+1}{2}\right) - B_n\!\left(\frac{x}{2}\right)\right] = \frac{2}{n}\left[B_n(x) - 2^n B_n\!\left(\frac{x}{2}\right)\right],$$

where $n = 1, 2, \ldots$

# References for Chapter 18

**Abramowitz, M. and Stegun, I. A. (Editors)**, *Handbook of Mathematical Functions with Formulas, Graphs and Mathematical Tables*, National Bureau of Standards Applied Mathematics, Washington, D.C., 1964.

**Bateman, H. and Erdélyi, A.**, *Higher Transcendental Functions, Vol. 1 and Vol. 2*, McGraw-Hill, New York, 1953.

**Bateman, H. and Erdélyi, A.**, *Higher Transcendental Functions, Vol. 3*, McGraw-Hill, New York, 1955.

**Gradshteyn, I. S. and Ryzhik, I. M.**, *Tables of Integrals, Series, and Products*, Academic Press, New York, 1980.

**Magnus, W., Oberhettinger, F., and Soni, R. P.**, *Formulas and Theorems for the Special Functions of Mathematical Physics, 3rd Edition*, Springer-Verlag, Berlin, 1966.

**McLachlan, N. W.**, *Bessel Functions for Engineers*, Clarendon Press, Oxford, 1955.

**Polyanin, A. D. and Zaitsev, V. F.**, *Handbook of Exact Solutions for Ordinary Differential Equations, 2nd Edition*, Chapman & Hall/CRC Press, Boca Raton, 2003.

**Slavyanov, S. Yu. and Lay, W.**, *Special Functions: A Unified Theory Based on Singularities*, Oxford University Press, Oxford, 2000.

**Weisstein, E. W.**, *CRC Concise Encyclopedia of Mathematics, 2nd Edition*, CRC Press, Boca Raton, 2003.

**Zwillinger, D.**, *CRC Standard Mathematical Tables and Formulae, 31st Edition*, CRC Press, Boca Raton, 2002.

# Chapter 19

# Calculus of Variations and Optimization

## 19.1. Calculus of Variations and Optimal Control

### 19.1.1. Some Definitions and Formulas

#### 19.1.1-1. Notion of functional.

Let a class $M$ of functions $x(t)$ be given. If for each function $x(t) \in M$ there is a certain number $\mathcal{J}$ assigned to $x(t)$ according to some law, then one says that a *functional* $\mathcal{J} = \mathcal{J}[x]$ is defined on $M$.

**Example 1.** Let $M = C^1[t_0, t_1]$ be the class of functions $x(t)$ defined on the interval $[t_0, t_1]$ and continuously differentiable on this interval. Then

$$\mathcal{J}[x] = \int_{t_0}^{t_1} \sqrt{1 + [x'_t(t)]^2}\, dt$$

is a functional defined on this class of functions. Geometrically, this functional expresses the length of the curve $x = x(t)$ with endpoints $A(t_0, x(t_0))$ and $B(t_1, x(t_1))$.

Calculus of variations established conditions under which functionals attain their extrema.

Suppose that a functional $\mathcal{J} = \mathcal{J}[x]$ attains its minimum or maximum at a function $\hat{x}$. A *strong (zero-order) neighborhood of* $\hat{x}$ is the set of continuous comparison functions (or trial functions) $x$ such that

$$|x(t) - \hat{x}(t)| < \varepsilon \quad (t_1 \le t \le t_2)$$

for a given $\varepsilon > 0$. A *weak (first-order) neighborhood of* $\hat{x}$ is the set of piecewise continuous comparison functions $x$ such that

$$|x(t) - \hat{x}(t)| + |x'_t(t) - \hat{x}'_t(t)| < \varepsilon \quad (t_1 \le t \le t_2)$$

for a given $\varepsilon > 0$. If $\hat{x}(t)$ minimizes the functional $\mathcal{J}[x]$ in a strong (resp., weak) neighborhood of itself, then it is called a point of *strong* (resp., *weak*) *minimum of the functional* $\mathcal{J}[x]$. If $\hat{x}$ maximizes the functional $\mathcal{J}[x]$ in a strong (resp., weak) neighborhood of itself, then it is called a point of *strong* (resp., *weak*) *maximum of the functional* $\mathcal{J}[x]$. Any strong extremum is also a weak extremum. Strong and weak extrema are *relative extrema*. The extremum of the functional $\mathcal{J}[x]$ over the entire domain where it is defined is called an *absolute extremum*. An absolute extremum is also a relative extremum.

*The function classes conventionally used in calculus of variations:*

1. The class $C[t_0, t_1]$ of continuous functions on the interval $[t_0, t_1]$ with the norm $\|x(t)\|_0 = \max\limits_{t_1 \le t \le t_2} |x(t)|$.
2. The class $C^1[t_0, t_1]$ of functions with continuous first derivative on the interval $[t_0, t_1]$ with the norm $\|x(t)\|_1 = \max\limits_{t_1 \le t \le t_2} |x(t)| + \max\limits_{t_1 \le t \le t_2} |x'_t(t)|$.

3. The class $C^n[t_0, t_1]$ of functions with continuous $n$th derivative on the interval $[t_0, t_1]$ with the norm $\|x(t)\|_n = \sum_{k=1}^{n} \max_{t_1 \leq t \leq t_2} |x_t^{(k)}(t)|$.

When stating variational problems, one should indicate which kind of extremum is to be found and specify the function class in which it is sought.

**Remark 1.** The class $PC^1[t_0, t_1]$ of continuous functions with piecewise continuous derivative on the interval $[t_0, t_1]$ is equipped with the norm indicated in item 1. The class $PC^n[t_0, t_1]$ of continuous functions with piecewise continuous $n$th derivative on the interval $[t_0, t_1]$ is equipped with the norm indicated in item 2.

**Remark 2.** We do not specify function classes whenever the results are valid for an arbitrary normed space.

### 19.1.1-2. First variation of functional.

The difference
$$\delta x = x - x_0 \tag{19.1.1.1}$$
of two functions $x(t)$ and $x_0(t)$ in a given function class $M$ is called the *variation* (or *increment*) of the argument $x(t)$ of the functional $\mathcal{J}[x]$.

The difference
$$\Delta \mathcal{J} \equiv \Delta \mathcal{J}[x] = \mathcal{J}[x + \delta x] - \mathcal{J}[x] \tag{19.1.1.2}$$
is called the *increment of the functional* $\mathcal{J}[x]$ corresponding to the increment $\delta x$ of the argument.

*First definition of the variation of a functional.* If the increment (19.1.1.2) can be represented as
$$\Delta \mathcal{J} = \mathcal{L}[x, \delta x] + \beta[x, \delta x]\|\delta x\|, \tag{19.1.1.3}$$
where $\mathcal{L}[x, \delta x]$ is a functional linear in $\delta x$ and $\beta[x, \delta x] \to 0$ as $\|\delta x\| \to 0$, then the linear part $\mathcal{L}[x, \delta x]$ of the increment of the functional is called the *variation of the functional* and is denoted by $\delta \mathcal{J}$. In this case, the functional $\Delta \mathcal{J}[x]$ is said to be *differentiable* at the point $x(t)$.

*Second definition of the variation of a functional.* Consider the functional $\mathcal{F}(\alpha) = \mathcal{J}[x + \alpha \delta x]$. The value
$$\delta \mathcal{J} = \mathcal{F}'_\alpha(0) \tag{19.1.1.4}$$
of its derivative with respect to the parameter $\alpha$ at $\alpha = 0$ is called the *variation of the functional* $\mathcal{J}[x]$ at the point $x(t)$.

If the variation of a functional exists as the principal linear part of its increment, i.e., in the sense of the first definition, then the variation understood as the value of the derivative with respect to the parameter $\alpha$ at $\alpha = 0$ also exists and these definitions coincide.

**Remark.** The second definition of the variation of a functional is somewhat wider than the first: there exist functionals whose increments do not have principal linear parts, but their variations in the sense of the second definition still exist.

### 19.1.1-3. Second variation of functional.

A functional $\mathcal{J}[x, y]$ depending on two arguments (that belong to some linear space) is said to be *bilinear* if it is a linear functional of each of the arguments. If we set $y \equiv x$ in a bilinear functional, then the resulting functional $\mathcal{J}[x, x]$ is said to be *quadratic*. A bilinear functional in a finite-dimensional space is called a *bilinear form*.

A quadratic functional $\mathcal{J}[x,x]$ is said to be *positive definite* if $\mathcal{J}[x,x] > 0$ for all $x \neq 0$.

A quadratic functional $\mathcal{J}[x,x]$ is said to be *strongly positive* if there is a constant $C > 0$ such that $\mathcal{J}[x,x] \geq C\|x\|^2$ for all $x$.

**Example 2.** The integral
$$\int_{t_0}^{t_1}\int_{t_0}^{t_1} K(s,t)x(s)y(t)\,ds\,dt,$$
where $K(s,t)$ is a fixed function of two variables, is a bilinear functional on $C[t_0, t_1]$.

*First definition of the second variation of a functional.* If the increment (19.1.1.2) of a functional can be represented as
$$\Delta \mathcal{J} = \mathcal{L}_1[\delta x] + \frac{1}{2}\mathcal{L}_2[\delta x] + \beta\|\delta x\|^2, \tag{19.1.1.5}$$

where $\mathcal{L}_1[\delta x]$ is a linear functional, $\mathcal{L}_2[\delta x]$ is a quadratic functional, and $\beta \to 0$ as $\|\delta x\| \to 0$, then the quadratic functional $\mathcal{L}_2[\delta x]$ is called the *second variation* (second differential) of the functional $\mathcal{J}[x]$ and is denoted by $\delta^2 \mathcal{J}$.

*Second definition of the second variation of a functional.* Consider the functional $\mathcal{F}(\alpha) = \mathcal{J}[x + \alpha \delta x]$. The value
$$\delta^2 \mathcal{J} = \mathcal{F}''_{\alpha\alpha}(0) \tag{19.1.1.6}$$
of its second derivative with respect to the parameter $\alpha$ at $\alpha = 0$ is called the *second-order variation of the functional* $\mathcal{J}[x]$ at the point $x(t)$.

The first and second variations of a functional permit stating necessary conditions for the minimum or maximum of a functional.

*Necessary conditions for the minimum or maximum of a functional:*
1. The first variation must be zero, $\delta \mathcal{J} = 0$.
2. The second-order variation must be nonnegative, $\delta^2 \mathcal{J} \geq 0$, in the case of minimum; and nonpositive, $\delta^2 \mathcal{J} \leq 0$, in the case of maximum.

## 19.1.2. Simplest Problem of Calculus of Variations

### 19.1.2-1. Statement of problem.

The extremal problem
$$\mathcal{J}[x] \equiv \int_{t_0}^{t_1} L(t, x, x'_t)\,dt \to \text{extremum}, \qquad x = x(t), \quad x(t_0) = x_0, \quad x(t_1) = x_1 \tag{19.1.2.1}$$

is called the *simplest problem of calculus of variations*. The function $L(t, x, x'_t)$ is called the *Lagrangian*, and the functional $\mathcal{J}$ is referred to as a *classical integral functional*. One usually assumes that the Lagrangian is jointly continuous in the arguments and has continuous partial derivatives of order $\leq 3$. The extremum in the problem is sought in the set of continuously differentiable functions $x(t) \in C^1[t_0, t_1]$ on the interval $[t_0, t_1]$ satisfying the *boundary* (or *endpoint*) conditions $x(t_0) = x_0$ and $x(t_1) = x_1$. Such functions are said to be *admissible*.

An admissible function $\hat{x}(t)$ provides a *weak local minimum* (resp., *maximum*) in problem (19.1.2.1) if it provides a local minimum (resp., maximum) in the space $C^1[t_0, t_1]$, i.e., if there exists a $\delta > 0$ such that the inequality
$$\mathcal{J}[x] \geq \mathcal{J}[\hat{x}] \quad (\text{resp.,} \quad \mathcal{J}[x] \leq \mathcal{J}[\hat{x}]) \tag{19.1.2.2}$$
holds for any admissible function $x(t) \in C^1[t_0, t_1]$ such that $\|x(t) - \hat{x}(t)\|_1 < \delta$.

The classical calculus of variations deals not only with weak extrema, but also with strong extrema. In the latter case, one considers functions $x(t)$ in the class $PC^1[t_0, t_1]$; i.e., the extremum is sought in the class of piecewise continuously differentiable functions satisfying the endpoint conditions.

An admissible function $\hat{x}(t) \in PC^1[t_0, t_1]$ provides a *strong local minimum* (resp., *maximum*) in problem (19.1.2.1) if there exists a $\delta > 0$ such that the inequality

$$\mathcal{J}[x] \geq \mathcal{J}[\hat{x}] \quad (\text{resp.,} \quad \mathcal{J}[x] \leq \mathcal{J}[\hat{x}]) \tag{19.1.2.3}$$

holds for any admissible function $x(t) \in PC^1[t_0, t_1]$ such that $\|x - \hat{x}\|_0 < \delta$.

Since the set of admissible functions is wider for a strong extremum than for a weak extremum, the following assertion is true: if $\hat{x}(t) \in C^1[t_0, t_1]$ provides a strong extremum, then it also provides a weak extremum. It follows that, for such functions, a necessary condition for weak extremum is a necessary condition for strong extremum, and a sufficient condition for strong extremum is a sufficient condition for weak extremum.

---

**19.1.2-2. First-order necessary condition for extremum. Euler equation.**

For the simplest problem of calculus of variations, the variation $\delta \mathcal{J}$ of the classical functional $\mathcal{J}[x]$ has the form

$$\delta \mathcal{J}[x, h] = \int_{t_0}^{t_1} \left( \frac{\partial L}{\partial x} - \frac{d}{dt} \frac{\partial L}{\partial x'_t} \right) h \, dt, \tag{19.1.2.4}$$

where $h(t)$ is an arbitrary smooth function satisfying the conditions $h(t_0) = h(t_1) = 0$ and

$$\frac{d}{dt} = \frac{\partial}{\partial t} + x'_t \frac{\partial}{\partial x} + x''_{tt} \frac{\partial}{\partial x'_t}.$$

A necessary condition for admissible function $x(t)$ to provide a weak extremum in problem (19.1.2.1) is that $\delta \mathcal{J} = 0$, i.e., that the function $x(t)$ satisfies the *Euler equation*

$$\frac{\partial L}{\partial x} - \frac{d}{dt} \frac{\partial L}{\partial x'_t} = 0. \tag{19.1.2.5}$$

Here we assume that the functions $L$, $\partial L/\partial x$, $\partial L/\partial x'_t$ are continuous as functions of three variables $(t, x, \text{and } x'_t)$, and $\partial L/\partial x'_t \in C^1[t_0, t_1]$. The solutions of the Euler equation (19.1.2.5) are called *extremals*. Admissible functions satisfying the Euler equation are called *admissible extremals*.

The Euler equation (19.1.2.5) in expanded form reads

$$L_x - L_{tx'_t} - x'_t L_{xx'_t} - x''_{tt} L_{x'_t x'_t} = 0. \tag{19.1.2.6}$$

From now on, the subscripts $t$, $x$, and $x'_t$ indicate the corresponding partial derivatives. If $L_{x'_t x'_t} \neq 0$, then equation (19.1.2.6) is a second-order differential equation, so that its general solution depends on two arbitrary constants. The values of these constants are in general determined by the boundary conditions $x(t_0) = x_0$ and $x(t_1) = x_1$. The boundary value problem

$$L_x - \frac{d}{dt} L_{x'_t} = 0, \quad x = x(t), \quad x(t_0) = x_0, \quad x(t_1) = x_1$$

does not always have a solution; if a solution exists, it may be nonunique.

The Euler equation (19.1.2.5) for a classical functional is a second-order differential equation, and so every solution $x(t)$ of this equation must have the second derivative $x''_{tt}(t)$. But sometimes a function on which a classical functional $\mathcal{J}[x]$ attains an extremum is not twice differentiable.

An extremal also satisfies the equation

$$\frac{d}{dt}\left(L - x'_t L_{x'_t}\right) - L_x = 0. \tag{19.1.2.7}$$

The points of an extremal $x(t)$ at which $L_{x'_t x'_t} \neq 0$ are said to be *regular*. If all points of an extremal are regular, then the extremal itself is said to be *regular (or nonsingular)*. For regular extremals, the Euler equation can be reduced to the form

$$x''_{tt} = f(t, x, x'_t).$$

**Remark 1.** An extremal $x(t)$ can have a break point only if $L_{x'_t x'_t} = 0$.

**Remark 2.** Extremals, i.e., functions "suspected for extremum," should not be confused with functions that actually provide an extremum.

### 19.1.2-3. Integration of Euler equation.

$1°$. The Lagrangian $L$ is independent of $x'_t$; i.e., $L(t, x, x'_t) \equiv L(t, x)$.
The Euler equation (19.1.2.5) becomes

$$L_x = 0. \tag{19.1.2.8}$$

In general, the solution of this finite equation does not satisfy the boundary conditions $x(t_0) = x_0$ and $x(t_1) = x_1$. There exists an extremal only in the exceptional cases in which the curve (19.1.2.8) passes through the boundary points $(t_0, x_0)$ and $(t_1, x_1)$.

$2°$. The Lagrangian $L$ depends on $x'_t$ linearly; i.e., $L(t, x, x'_t) \equiv M(t, x) + N(t, x)x'_t$.
The Euler equation (19.1.2.5) becomes

$$M_x - N_t = 0, \tag{19.1.2.9}$$

where derivative $M_x$ and function $N$ are evaluated at $x = x(t)$. In general, the curve determined by equation (19.1.2.9) does not satisfy the boundary conditions, and hence, as a rule, the variational problem does not have a solution in the class of continuous functions.

If equation (19.1.2.9) is satisfied identically in a domain $D$ on the plane $OXY$, then $L(t, x, x'_t) dt = M(t, x) dt + N(t, x) dx$ is an exact differential; consequently, $\mathcal{J}[x]$ is independent of the integration path and has a constant value for all $x$. In this case, the variational problem becomes meaningless.

$3°$. The Lagrangian $L$ is independent of $x$; i.e., $L(t, x, x'_t) \equiv L(t, x'_t)$.
The Euler equation (19.1.2.5) becomes

$$\frac{d}{dt} L_{x'_t} = 0, \tag{19.1.2.10}$$

whence it follows that

$$L_{x'_t} = \text{const}. \tag{19.1.2.11}$$

Equation (19.1.2.11) is a first-order differential equation. By integrating this equation, we obtain the extremals of the problem.

**Remark.** Relation (19.1.2.11) is called the *momentum conservation law*.

4°. The Lagrangian $L$ is independent of $t$; i.e., $L(t, x, x'_t) \equiv L(x, x'_t)$.
The Euler equation (19.1.2.6) becomes

$$L_x - x'_t L_{xx'_t} - x''_{tt} L_{x'_t x'_t} = 0, \tag{19.1.2.12}$$

whence we readily obtain a first integral of the Euler equation:

$$L - x'_t L_{x'_t} = \text{const.} \tag{19.1.2.13}$$

**Remark.** Relation (19.1.2.13) is called the *energy conservation law*.

5°. The Lagrangian $L$ depends only on $x'_t$; i.e., $L(t, x, x'_t) \equiv L(x'_t)$.
The Euler equation (19.1.2.6) becomes

$$x''_{tt} L_{x'_t x'_t} = 0. \tag{19.1.2.14}$$

In this case, the extremals are given by the equations

$$x(t) = C_1 t + C_2, \tag{19.1.2.15}$$

where $C_1$ and $C_2$ are arbitrary constants.

**Example 1.** Let us give an example in which there exists a unique admissible extremal that provides a global extremum.
Let

$$\mathcal{J}[x] = \int_0^1 (x'_t)^2 \, dt \to \min, \qquad x = x(t), \quad x(0) = 0, \quad x(1) = 1.$$

The Euler equation becomes $2x''_{tt} = 0$. The extremals are given by the equation $x(t) = C_1 t + C_2$. The unique admissible extremal is the function $\hat{x}(t) = t$, which provides the global minimum. Suppose that $x'_t(t) \in C^1[0, 1]$, $x(0) = 0$, $x(1) = 1$. Then

$$\mathcal{J}[x] = \mathcal{J}[\hat{x} + h] = \int_0^1 (1 + h'_t)^2 \, dt = \mathcal{J}[\hat{x}] + \int_0^1 (h'_t)^2 \, dt \geq \mathcal{J}[\hat{x}]$$

for an arbitrary function $h(t)$ such that $h(0) = 0$ and $h(1) = 0$.

**Example 2.** Let us give an example in which there exists a unique admissible extremal that provides a weak extremum but does not provide a strong extremum.
Let

$$\mathcal{J}[x] = \int_0^1 (x'_t)^3 \, dt \to \min; \qquad x = x(t), \quad x(0) = 0, \quad x(1) = 1.$$

The Euler equation becomes $6x'_t x''_{tt} = 0$. The extremals are given by the equation $x(t) = C_1 t + C_2$. The unique admissible extremal is the function $\hat{x}(t) = t$, which provides a weak local minimum. Indeed,

$$\mathcal{J}[x] = \mathcal{J}[\hat{x} + h] = \int_0^1 (1 + h'_t)^3 \, dt = \mathcal{J}[\hat{x}] + \int_0^1 (h'_t)^2 (3 + h'_t) \, dt$$

for an arbitrary function $h(t)$ such that $h(0) = h(1) = 0$. Thus, if $\|h\|_1 \leq 3$, then $3 + h'_t > 0$ and hence $\mathcal{J}[x] \geq \mathcal{J}[\hat{x}]$.
Consider the sequence of functions

$$g_n(t) = \begin{cases} -\sqrt{n}, & t \in [0, 1/n), \\ 0, & t \in [1/n, 1/2], \\ 2/\sqrt{n}, & t \in (1/2, 1]. \end{cases} \qquad h_n(t) = \int_0^t g_n(\tau) \, d\tau \quad (n = 2, 3, \ldots).$$

Obviously, $h_n(0) = h_n(1) = 0$ and $\|h(t)\|_0 \to 0$ as $n \to \infty$. We set $x_n(t) = h'_t(t) + h_n(t)$ and obtain a sequence of functions $x_n(t)$ such that $x_n(0) = 0$, $x_n(1) = 1$, $x_n(t) \to \hat{x}(t)$ in $C[0, 1]$, and

$$\mathcal{J}[x_n] = \mathcal{J}[\hat{x} + h_n] = \int_0^1 (1 + h'_t)^3 \, dt = 1 + 3\int_0^1 g_n^2 \, dt + \int_0^1 g_n^3 \, dt$$

$$= 1 + \int_0^{1/n} (3n - n^{3/2}) \, dt + \int_{1/2}^1 \left(\frac{12}{n} - \frac{8}{n\sqrt{n}}\right) dt = -\sqrt{n} + O(1).$$

In this case $\mathcal{J}[x_n] \to -\infty$ as $n \to \infty$.

**Example 3 (Hilbert's example).** In this example, a unique admissible extremal exists that provides a global extremum but is not a differentiable function.

Let
$$\mathcal{J}[x] = \int_0^1 t^{2/3}(x'_t)^2 \, dt \to \min; \qquad x = x(t), \quad x(0) = 0, \quad x(1) = 1.$$

The Euler equation becomes $\frac{d}{dt}(2t^{2/3}x'_t) = 0$. The extremals are given by the equation $x(t) = C_1 t^{1/3} + C_2$. The unique extremal satisfying the endpoint conditions is the function $\hat{x}(t) = t^{1/3}$, which does not belong to the class $C^1[0,1]$. Nevertheless, it provides the global minimum in this problem on the set of all absolutely continuous functions $x(t)$ that satisfy the boundary conditions and for which the integral $\mathcal{J}$ is finite. Indeed,

$$\mathcal{J}[x] = \mathcal{J}[\hat{x} + h] = \int_0^1 t^{2/3}\left(\frac{1}{3t^{2/3}} + h'_t\right)^2 dt = \mathcal{J}[\hat{x}] + \frac{2}{3}\int_0^1 h'_t \, dt + \int_0^1 t^{2/3}(h'_t)^2 \, dt \geq \mathcal{J}[\hat{x}].$$

**Example 4 (Weierstrass's example).** In this example, the problem has no solutions and no admissible extremals even in the class of absolutely continuous functions.

Let
$$\mathcal{J}[x] = \int_0^1 t^2 (x'_t)^2 \, dt \to \min; \qquad x = x(t), \quad x(0) = 0, \quad x(1) = 1.$$

The Euler equation becomes $\frac{d}{dt}(2t^2 x'_t) = 0$. The extremals are given by the equation $x(t) = C_1 t^{-1} + C_2$; none of them satisfies the boundary condition $x(0) = 0$. Furthermore, $\mathcal{J}[x] \geq 0$, and $\mathcal{J}[x] > 0$ for any absolutely continuous function $x(t)$. Obviously, the lower bound in this problem is zero. To prove this, it suffices to consider the sequence of admissible functions $x_n(t) = \arctan(nt)/\arctan n$. One has

$$\mathcal{J}[x] = \int_0^1 t^2 \frac{n^2}{(1+n^2t^2)^2 \arctan^2 n} \, dt \leq \int_0^{1/n} \frac{dt}{\arctan^2 n} + \int_{1/n}^1 \frac{dt}{n^2 t^2 \arctan^2 n} \to 0.$$

**Example 5.** In this example, there exists a unique admissible extremal that does not provide an extremum.

Let
$$\mathcal{J}[x] = \int_0^{3\pi/2} \left[(x'_t)^2 - x^2\right] dt \to \min; \qquad x = x(t), \quad x(0) = x(3\pi/2) = 0.$$

The Euler equation becomes $x''_{tt} + x = 0$. The extremals are given by the equation $x(t) = C_1 \sin t + C_2 \cos t$. The only extremal satisfying the endpoint conditions is $\hat{x}(t) = 0$.

The admissible extremal $\hat{x}(t) = 0$ does not provide a local minimum. Consider the sequence of admissible functions $x_n(t) = n^{-1} \sin(2t/3)$. Obviously, $x_n(t) \to 0$ in $C^1[0, 3\pi/2]$, but

$$\mathcal{J}[x_n] = -\frac{5\pi}{12n^2} < 0 = \mathcal{J}[\hat{x}].$$

This example shows that the Euler equation is a necessary but not sufficient condition for extremum.

### 19.1.2-4. Broken extremals. Weierstrass–Erdmann conditions.

If the Euler equation has a piecewise smooth solution, i.e., if $x(t)$ has corner points (the function $x'_t(t)$ has a discontinuity), then the *Weierstrass–Erdmann conditions*

$$L_{x'_t}[c, x(c), x'_t(c-0)] = L_{x'_t}[c, x(c), x'_t(c+0)], \tag{19.1.2.16}$$

$$L[c, x(c), x'_t(c-0)] - x'_t(c-0) L_{x'_t}[c, x(c), x'_t(c-0)]$$
$$= L[c, x(c), x'_t(c+0)] - x'_t(c+0) L_{x'_t}[c, x(c), x'_t(c+0)] \tag{19.1.2.17}$$

hold at each point $c$ that is the abscissa of a corner point.

If $x'_t L_{x'_t} \neq 0$ ($t_0 \leq t \leq t_1$), then the Euler equation has only smooth solutions.

Curves consisting of pieces of extremals and satisfying the Weierstrass–Erdmann conditions at the corner points are called *broken extremals*.

### 19.1.2-5. Second-order necessary conditions.

*Legendre condition:* If an extremal provides a minimum (resp., maximum) of the functional, then the following inequality holds:

$$L_{x'_t x'_t} \geq 0 \quad (\text{resp.,} \ L_{x'_t x'_t} \leq 0) \qquad (t_0 \leq t \leq t_1). \tag{19.1.2.18}$$

*Weierstrass condition:* If a curve $x(t)$ provides a strong minimum (resp., maximum) of the classical functional, then the *Weierstrass function*

$$E(t, x, x'_t, k) \equiv L(t, x, k) - L(t, x, x'_t) - (k - x'_t) L_{x'_t}(t, x, x'_t) \tag{19.1.2.19}$$

is nonnegative (resp., nonpositive) for arbitrary finite values $k$ at all points $(t, x)$ of the extremal.

The equation

$$L_{xx} h + L_{xx'_t} h'_t - \frac{d}{dt}\left(L_{xx'_t} h + L_{x'_t x'_t} h'_t\right) = 0 \tag{19.1.2.20}$$

is called the *Jacobi equation*. Here $h(t)$ is an arbitrary smooth function satisfying the conditions $h(t_0) = h(t_1) = 0$.

If the Legendre condition $L_{x'_t x'_t} \neq 0$ ($t_0 \leq t \leq t_1$) is satisfied, then the Jacobi equation is a second-order linear equation that can be solved for the second derivative. It follows from the conditions $h(t_0) = h(t_1) = 0$ that $h(t) \equiv 0$. A point $\tau$ is said to be *conjugate* to the point $t_0$ if there exists a nontrivial solution of the Jacobi equation such that $h(t_0) = h(\tau) = 0$.

*Jacobi condition:* If the extremal $x(t)$ ($t_0 \leq t \leq t_1$) provides an extremum of the functional (19.1.2.1), then it does not contain points conjugate to $t_0$.

### 19.1.2-6. Multidimensional case.

The vector problem is posed and necessary conditions for an extremum are stated by analogy with the one-dimensional simplest problem of calculus of variations. Let $\mathbf{x}(t) = (x_1(t), \ldots, x_n(t))$ be an $n$-dimensional vector function, and let the Lagrangian $L$ be a function of $2n + 1$ variables: $L = L(t, \mathbf{x}, \mathbf{x}'_t)$.

The vector problem has the form

$$\mathcal{J}[\mathbf{x}] \equiv \int_{t_0}^{t_1} L(t, \mathbf{x}, \mathbf{x}'_t) \, dt \to \text{extremum}; \quad \mathbf{x} = \mathbf{x}(t), \ x_i(t_j) = x_{ij} \ (i = 1, 2, \ldots, n, \ j = 0, 1).$$
$$\tag{19.1.2.21}$$

A necessary condition for admissible vector function $\mathbf{x}(t)$ to provide weak extremum in problem (19.1.2.21) is that the function $\mathbf{x}(t)$ satisfies the system of $n$ *Euler differential equations*

$$L_{x_i} - \frac{d}{dt} L_{(x_i)'_t} = 0 \quad (i = 1, 2, \ldots, n), \tag{19.1.2.22}$$

where we assume that the functions $L$, $L_x$, $L_{x'_t}$ are continuous as functions of $2n+1$ variables $(t, x_i, \text{ and } (x_i)'_t, i = 1, 2, \ldots, n)$, and $L_{(x_i)'_t} \in C^1[t_0, t_1]$. The solutions of the system of Euler differential equations (19.1.2.22) are called *extremals*. Admissible functions satisfying the system of Euler differential equations are called *admissible extremals*.

The derivation of conditions (19.1.2.5) and (19.1.2.22) from the relation $\delta \mathcal{J} = 0$ is based on the fundamental lemmas of calculus of variations.

LEMMA 1 (FUNDAMENTAL LEMMA OR LAGRANGE'S LEMMA). Let $f(t) \in C[t_0, t_1]$ and assume that
$$\int_{t_0}^{t_1} f(t)g(t)\,dt = 0$$
for each function $g(t) \in C^1[t_0, t_1]$ such that $g(t_0) = g(t_1) = 0$. Then $f(t) \equiv 0$ on $[t_0, t_1]$.

LEMMA 2. Let $f(t) \in C[t_0, t_1]$ and assume that
$$\int_{t_0}^{t_1} f(t)g(t)\,dt = 0$$
for each function $g(t) \in C^1[t_0, t_1]$ such that $\int_{t_0}^{t_1} g(t)\,dt = 0$. Then $f(t) = \text{const}$ on $[t_0, t_1]$.

### 19.1.2-7. Weierstrass–Erdmann conditions in multidimensional case.

If the system of Euler differential equations has a piecewise smooth solution, i.e., if $\mathbf{x}(t)$ has corner points (the function $\mathbf{x}'_t(t)$ has discontinuities), then the *Weierstrass–Erdmann conditions*

$$L_{(x_i)'_t}[c, \mathbf{x}(c), \mathbf{x}'_t(c-0)] = L_{(x_i)'_t}[c, \mathbf{x}(c), \mathbf{x}'_t(c+0)] \qquad (i = 1, 2, \ldots, n),$$

$$L[c, \mathbf{x}(c), \mathbf{x}'_t(c-0)] - \sum_{i=1}^{n}(x_i)'_t\Big|_{t=c-0} L_{(x_i)'_t}[c, \mathbf{x}(c), \mathbf{x}'_t(c-0)]$$
$$= L[c, \mathbf{x}(c), \mathbf{x}'_t(c+0)] - \sum_{i=1}^{n}(x_i)'_t\Big|_{t=c+0} L_{(x_i)'_t}[c, \mathbf{x}(c), \mathbf{x}'_t(c+0)]$$

hold at each point $c$ that is the abscissa of a corner point.

If $(x_i)'_t L_{(x_i)'_t} \neq 0$ ($t_0 \leq t \leq t_1$, $i = 1, 2, \ldots, n$), then the system of Euler differential equations has only smooth solutions.

Curves consisting of pieces of extremals and satisfying the Weierstrass–Erdmann conditions at the corner points are called *broken extremals*.

### 19.1.2-8. Higher-order necessary conditions in multidimensional case.

We consider the simplest problem

$$\mathcal{J}[x] \equiv \int_{t_0}^{t_1} L(t, \mathbf{x}, \mathbf{x}'_t)\,dt \to \min \text{ (or max)}, \qquad \mathbf{x} = \mathbf{x}(t), \quad \mathbf{x}(t_0) = \mathbf{x}_0, \quad \mathbf{x}(t_1) = \mathbf{x}_1.$$
(19.1.2.23)

*Legendre condition:* If an extremal provides a minimum (resp., maximum) of the functional, then the following inequality holds:

$$L_{\mathbf{x}'_t \mathbf{x}'_t} \equiv \left[L_{(x_i)'_t (x_j)'_t}\right] \geq 0 \quad (\text{resp., } L_{\mathbf{x}'_t \mathbf{x}'_t} \leq 0) \qquad (t_0 \leq t \leq t_1;\ i, j = 1, 2, \ldots, n). \quad (19.1.2.24)$$

*Strengthened Legendre condition:* If an extremal provides a minimum (resp., maximum) of the functional, then the following inequality holds:

$$L_{\mathbf{x}'_t \mathbf{x}'_t} > 0 \quad (\text{resp., } L_{\mathbf{x}'_t \mathbf{x}'_t} < 0) \qquad (t_0 \leq t \leq t_1;\ i, j = 1, 2, \ldots, n). \quad (19.1.2.25)$$

**Remark.** In the vector case, $L_{\mathbf{x}'_t \mathbf{x}'_t}$ is an $n \times n$ matrix. The condition $L_{\mathbf{x}'_t \mathbf{x}'_t} \geq 0$ means that the matrix is positive semidefinite, and the condition $L_{\mathbf{x}'_t \mathbf{x}'_t} > 0$ means that the matrix is positive definite.

The Jacobi equation for the vector problem has a form similar to that in the one-dimensional case, i.e.,

$$L_{\mathbf{xx}}\mathbf{h} + L_{\mathbf{xx}'_t}\mathbf{h}'_t - (L_{\mathbf{xx}'_t}\mathbf{h} + L_{\mathbf{x}'_t\mathbf{x}'_t}\mathbf{h}'_t)'_t = 0, \qquad (19.1.2.26)$$

where $L_{\mathbf{xx}} \equiv [L_{(x_i)(x_j)}]$, $L_{\mathbf{xx}'_t} \equiv [L_{(x_i)(x_j)'_t}]$, and $\mathbf{h} = \mathbf{h}(t)$ is an column-vector with components $h_i(t)$, which are arbitrary smooth functions satisfying the conditions $h_i(t_0) = h_i(t_1) = 0$ ($i = 1, 2, \ldots, n$).

Suppose that the strengthened Legendre condition is satisfied on an extremal $\mathbf{x}(t)$. A point $\tau$ is said to be *conjugate to the point* $t_0$ if there exists a nontrivial solution $\mathbf{h}(t)$ of the Jacobi equation such that $\mathbf{h}(t_0) = \mathbf{h}(\tau) = 0$. The *Jacobi condition* (resp., *the strengthened Jacobi condition*) is said to be satisfied on an extremal $\mathbf{x}(t)$ if the interval $(t_0, t_1)$ (resp., the half-open interval $(t_0, t_1]$) does not contain points conjugate to $t_0$.

We find the fundamental system of solutions of the Jacobi equation for the functions $\mathbf{x}(t)$ in the matrix form

$$H(t, t_0) \equiv (\mathbf{h}^1(t), \ldots, \mathbf{h}^n(t)), \qquad \mathbf{h}^i(t) \equiv \begin{pmatrix} h^i_1(t) \\ \vdots \\ h^i_n(t) \end{pmatrix},$$

where $H(t_0, t_0) = 0$ and $H'_t(t_0, t_0) \neq 0$ or $H'_t(t_0, t_0) = I$. Column-vectors $\mathbf{h}^i(t)$ are the solutions of the Jacobi system of equations. A point $\tau$ is conjugate to the point $t_0$ if and only if the matrix $H(\tau, t_0)$ is degenerate.

*Necessary conditions for weak minimum (resp., maximum):*

1. If $\mathbf{x}(t)$ provides a weak minimum (resp., maximum), then the function $\mathbf{x}(t)$ is an admissible extremal on which the Legendre and Jacobi conditions are satisfied.
2. If $\mathbf{x}(t)$ provides a strong local minimum (resp., maximum), then the *Weierstrass function*

$$E(t, \mathbf{x}, \mathbf{x}'_t, \mathbf{k}) = L(t, \mathbf{x}, \mathbf{k}) - L(t, \mathbf{x}, \mathbf{x}'_t) - \sum_{i=1}^{n}[k_i - (x_i)'_t]L_{(x_i)'_t}(t, \mathbf{x}, \mathbf{x}'_t) \qquad (19.1.2.27)$$

is nonnegative (resp., nonpositive) for arbitrary finite values $\mathbf{k} = (k_1, \ldots, k_n)$ at all points $(t, \mathbf{x})$ of the extremal.

*Sufficient condition for the weak minimum or maximum:* If the strengthened Legendre and Jacobi conditions are satisfied on an admissible extremal, then this extremal provides a weak local minimum (or maximum).

**Example 6.**

$$\mathcal{J}[\mathbf{x}] = \int_0^1 \left\{ [(x_1)'_t]^2 + [(x_2)'_t]^2 + 2x_1 x_2 \right\} dt \to \min;$$

$$x_1(0) = x_2(0) = 0, \qquad x_1(1) = \sin 1, \qquad x_2(1) = -\sin 1 \quad (i = 1, 2).$$

The Lagrangian is $L = [(x_1)'_t]^2 + [(x_2)'_t]^2 + 2x_1 x_2$.

A necessary condition is given by the system (19.1.2.22) of Euler equations

$$-(x_1)''_{tt} + x_2 = 0, \quad -(x_2)''_{tt} + x_1 = 0 \quad \Longrightarrow \quad (x_1)''''_{tttt} = x_1, \quad (x_2)''''_{tttt} = x_2.$$

The general solution of the Euler equations is

$$x_1(t) = C_1 \sinh t + C_2 \cosh t + C_3 \sin t + C_4 \cos t,$$
$$x_2(t) = C_1 \sinh t + C_2 \cosh t - C_3 \sin t - C_4 \cos t.$$

It follows from the initial conditions that $C_1 = C_2 = C_4 = 0$ and $C_3 = 1$. The only admissible extremal is $\hat{x}_1(t) = \sin t$, $\hat{x}_2(t) = -\sin t$.

Let us apply sufficient conditions. The strengthened Legendre condition (19.1.2.25) is satisfied. The system of Jacobi equations (19.1.2.26) coincides with the system of Euler equations, i.e.,

$$-(h_1)''_{tt} + h_2 = 0, \quad -(h_2)''_{tt} + h_1 = 0.$$

For $H(t,0)$, such that $H(0,0) = 0$ and $H'_t(0,0) = I$, we take the matrix

$$\begin{pmatrix} \frac{1}{2}(\sinh t + \sin t) & \frac{1}{2}(\sinh t - \sin t) \\ \frac{1}{2}(\sinh t - \sin t) & \frac{1}{2}(\sinh t + \sin t) \end{pmatrix}.$$

Then

$$\det H(t,0) = \sinh t \sin t.$$

The conjugate points are $\tau = k\pi$, $k = 1, 2, 3, \ldots$

The half-open interval $(0, 1]$ does not contain conjugate points, and so the strengthened Jacobi condition is satisfied. Thus the vector $\hat{\mathbf{x}}(t) = (\hat{x}_1(t), \hat{x}_2(t))$ provides a local minimum of the functional $\mathcal{J}$.

### 19.1.2-9. Bolza problem.

$1°$. The *Bolza problem* is the following extremal problem without constraints in the space $C^1[t_0, t_1]$:

$$\mathcal{B}[x] \equiv \int_{t_0}^{t_1} L(t, x, x'_t) \, dt + l(x(t_0), x(t_1)) \to \text{extremum}. \tag{19.1.2.28}$$

The function $L(t, x, x'_t)$ is called the *Lagrangian*, the function $l = l(x(t_0), x(t_1))$ is called the *terminal cost function*, and the functional $\mathcal{B}$ is called the *Bolza functional*. The Bolza problem is an elementary problem of classical calculus of variations. Any functions of class $C^1[t_0, t_1]$ are admissible.

An admissible function $\hat{x}(t) \in C^1[t_0, t_1]$ provides a *weak local minimum (maximum)* in problem (19.1.2.28) if there exists a $\delta > 0$ such that

$$\mathcal{B}[x] \geq \mathcal{B}[\hat{x}] \quad (\mathcal{B}[x] \leq \mathcal{B}[\hat{x}]) \tag{19.1.2.29}$$

for any admissible function $x(t)$ satisfying $\|x - \hat{x}\|_1 < \delta$ (see Paragraph 19.1.2-1).

A necessary condition for a weak extremum of the Bolza functional has the same character as the necessary condition in the classical problem; i.e., the function providing an extremum of the Bolza functional must be a solution of a boundary value problem for the Euler equation (19.1.2.5). The boundary conditions are given by the relations

$$L_{x'_t}(t_0, x(t_0), x'_t(t_0)) = \frac{\partial l}{\partial [x(t_0)]}, \quad L_{x'_t}(t_1, x(t_1), x'_t(t_1)) = -\frac{\partial l}{\partial [x(t_1)]} \tag{19.1.2.30}$$

and are called the *transversality conditions*. If the function part of the Bolza functional is lacking, i.e., $l \equiv 0$, then conditions (19.1.2.30) acquire the form $L_{x'_t}(t_0, x(t_0), x'_t(t_0)) = L_{x'_t}(t_1, x(t_1), x'_t(t_1)) = 0$, which means in many applied problems that the extremal trajectories are orthogonal to the boundary vertical lines $t = t_0$ and $t = t_1$. If a fixing condition is given at one endpoint of the interval, then the fixing condition at this endpoint and the transversality condition at the other endpoint serve as the boundary conditions for the Euler equation.

2°. The $n$-dimensional problem is posed and necessary conditions for extremum are stated by analogy with the one-dimensional Bolza problem. Let $\mathbf{x}(t) = (x_1(t), \ldots, x_n(t))$ be an $n$-dimensional vector function, and let the Lagrangian $L$ be a function of $2n + 1$ variables: $L = L(t, \mathbf{x}, \mathbf{x}'_t)$.

Then the vector problem reads

$$\int_{t_0}^{t_1} L(t, \mathbf{x}, \mathbf{x}'_t)\, dt + l(\mathbf{x}(t_0), \mathbf{x}(t_1)) \to \text{extremum}. \tag{19.1.2.31}$$

**Example 7.** Consider the problem

$$\mathcal{B}[x] = \int_0^1 \left[(x'_t)^2 - x\right] dt + x^2(1) \to \min.$$

The Euler equation (19.1.2.5) has the form

$$-1 - 2x''_{tt} = 0,$$

and the transversality conditions (19.1.2.30) are as follows:

$$2x'_t(0) = 0, \quad 2x'_t(1) = -2x(1).$$

The extremals are given by the equation $x(t) = -\frac{1}{4}t^2 + C_1 t + C_2$. The unique extremal satisfying the transversality conditions is $\hat{x}(t) = \frac{1}{4}(3-t^2)$. This admissible extremal provides the absolute minimum in the problem. Indeed,

$$\mathcal{B}[\hat{x}+h] - \mathcal{B}[\hat{x}] = \int_0^1 2\hat{x}'_t h'_t\, dt + \int_0^1 (h'_t)^2\, dt - \int_0^1 h\, dt + 2\hat{x}(1)h(1) + [h'_t(1)]^2 = \int_0^1 (h'_t)^2\, dt + [h'_t(1)]^2 \geq 0$$

for an arbitrary function $h(t) \in C^1[t_0, t_1]$. Thus $\hat{x}(t) = \frac{1}{4}(3-t^2)$ provides the absolute minimum in the problem. Furthermore, $\mathcal{B}[\hat{x}] = -\frac{1}{3}$.

### 19.1.3. Isoperimetric Problem

**19.1.3-1. Statement of problem. Necessary condition for extremum.**

The *isoperimetric problem* (with fixed endpoints) in calculus of variations is the following extremal problem in the space $C^1[t_0, t_1]$ (or $PC^1[t_0, t_1]$):

$$\mathcal{J}_0[x] \equiv \int_{t_0}^{t_1} f_0(t, x, x'_t)\, dt \to \text{extremum}; \tag{19.1.3.1}$$

$$\mathcal{J}_i[x] \equiv \int_{t_0}^{t_1} f_i(t, x, x'_t)\, dt = \alpha_i \quad (i = 1, 2, \ldots, m); \tag{19.1.3.2}$$

$$x(t_0) = x_0, \quad x(t_1) = x_1, \tag{19.1.3.3}$$

where $\alpha_1, \ldots, \alpha_m \in \mathbb{R}$ are given numbers. Constraints of the form (19.1.3.2) are said to be *isoperimetric*. The functions $f_i(t, x, x'_t)$ $(i = 1, 2, \ldots, m)$ are called the *Lagrangians* of the problem. Functions $x(t) \in C^1[t_0, t_1]$ satisfying the isoperimetric conditions (19.1.3.2) and conditions (19.1.3.3) at the endpoints are said to be *admissible*.

An admissible function $\hat{x}(t)$ provides a *weak local minimum* (resp., *maximum*) in problem (19.1.3.1) if there exists a $\delta > 0$ such that the inequality

$$\mathcal{J}_0[x] \geq \mathcal{J}_0[\hat{x}] \quad (\text{resp.,} \ \mathcal{J}_0[x] \leq \mathcal{J}_0[\hat{x}])$$

holds for any admissible function $x(t) \in C^1[t_0, t_1]$ satisfying $\|x - \hat{x}\|_1 < \delta$.

*Necessary condition for extremum (the Lagrange multiplier rule):*

Let $f_i(t, x, x'_t)$ ($i = 1, 2, \ldots, m$) be functions continuous together with their partial derivatives $(f_i)_x$ and $(f_i)_{x'_t}$, and let $x(t) \in C^1[t_0, t_1]$. If $x(t)$ provides a weak local extremum in the isometric problem (19.1.3.1), then there exist Lagrange multipliers $\lambda_0, \lambda_1, \ldots, \lambda_m$, not all zero simultaneously, such that for the Lagrangian

$$L(t, x, x'_t) = \sum_{i=0}^{m} \lambda_i f_i(t, x, x'_t) \tag{19.1.3.4}$$

the Euler equation

$$L_x - \frac{d}{dt} L_{x'_t} = 0 \tag{19.1.3.5}$$

is satisfied. If the regularity condition is satisfied [the functions $-\frac{d}{dt}(f_i)_{x'_t} + (f_i)_x$ ($i = 1, 2, \ldots, m$), are linearly independent], then $\lambda_0 \neq 0$.

One of the best-known isoperimetric problems, after which the entire class of problems was named, is the *Dido problem*.

**Example 1.** Find a smooth curve of given length fixed at two points of a straight line and, together with the segment of the straight line, bounding the largest area.

The formalized problem is

$$\int_{-T_0}^{T_0} x \, dt \to \max; \qquad \int_{-T_0}^{T_0} \sqrt{1 + (x'_t)^2} \, dt = l, \qquad x = x(t), \quad x(-T_0) = x(T_0) = 0,$$

where $T_0$ is given.

The Lagrangian (19.1.3.4) has the form $L = \lambda_0 x + \lambda \sqrt{1 + (x'_t)^2}$.

A necessary condition is given by the Euler equation (19.1.3.5)

$$\lambda_0 - \frac{d}{dt} \frac{\lambda x'_t}{\sqrt{1 + (x'_t)^2}} = 0.$$

Elementary calculations result in the first-order equation

$$(C + x)\sqrt{1 + (x'_t)^2} = -\lambda.$$

We integrate this equation and obtain

$$(C + x)^2 + (C_1 + t)^2 = \lambda^2,$$

which is the equation of a circle. It follows from the endpoint conditions $x(-T_0) = x(T_0)$ that $C_1 = 0$, i.e., $(C + x)^2 + t^2 = \lambda^2$.

The unknown constants $C$ and $\lambda$ are uniquely determined by the condition $x(T_0) = 0$ and the isoperimetric condition $\int_{-T_0}^{T_0} \sqrt{1 + (x'_t)^2} \, dt = l$.

For $2T_0 < l < \pi T_0$, there is a unique (up to the sign) extremal that is an arc of length $l$ of the circle passing through the points $(\pm T_0, 0)$ and centered on the $OX$-axis. Since this is a maximization problem, we must take the extremal lying in the upper half-plane. For $l < 2T_0$, there are no admissible functions in the problem, and for $l > \pi T_0$, there are no extremals.

### 19.1.3-2. Higher-order necessary conditions. Sufficient conditions.

Consider the isoperimetric problem

$$\begin{aligned} \mathcal{J}_0[x] &\to \min \ (\text{or} \ \max); \\ \mathcal{J}_i[x] &= \alpha_i \qquad (i = 1, 2, \ldots, m); \\ x(t_0) &= x_0, \quad x(t_1) = x_1. \end{aligned} \tag{19.1.3.6}$$

Let $x(t) \in C^2[t_0, t_1]$ be an extremal of problem (19.1.3.6) with $\lambda_0 = 1$; i.e., the Euler equation (19.1.3.4) is satisfied on this extremal for the Lagrangian

$$L(t, x, x'_t) = f_0(t, x, x'_t) + \sum_{i=1}^{m} \lambda_i f_i(t, x, x'_t)$$

with some Lagrange multipliers $\lambda_i$.

*Legendre condition:* If an extremal provides a minimum (resp., maximum) of the functional, then the following inequality holds:

$$L_{x'_t x'_t} \geq 0 \quad (\text{resp.,} \quad L_{x'_t x'_t} \leq 0) \quad (t_0 \leq t \leq t_1). \quad (19.1.3.7)$$

*Strengthened Legendre condition:* If an extremal provides a minimum (resp., maximum) of the functional, then the following inequality holds:

$$L_{x'_t x'_t} > 0 \quad (\text{resp.,} \quad L_{x'_t x'_t} < 0) \quad (t_0 \leq t \leq t_1). \quad (19.1.3.8)$$

The equation

$$x L_{xx} + x'_t L_{x'_t x} - \frac{d}{dt}\left(x L_{x'_t x} + x'_t L_{x'_t x'_t}\right) + \sum_{i=1}^{m} \mu_i g_i = 0, \quad g_i = -\frac{d}{dt}(f_i)_{x'_t} + (f_i)_x \quad (19.1.3.9)$$

is called the (inhomogeneous) *Jacobi equation* for isoperimetric problem (19.1.3.6) on the extremal $x(t)$; $\mu_i$ are Lagrange multipliers ($i = 1, 2, \ldots, n$).

Suppose that the strengthened Legendre condition (19.1.3.8) is satisfied on the extremal $x(t)$. A point $\tau$ is said to be *conjugate to the point* $t_0$ if there exists a nontrivial solution of the Jacobi equation such that

$$\int_0^{\tau} g_i(t) h(t)\, dt = 0 \quad (i = 1, 2, \ldots, m),$$

where $h(t)$ is an arbitrary smooth function satisfying the conditions $h(0) = h(\tau) = 0$.

We say that the *Jacobi condition* (resp., *strengthened Jacobi condition*) is satisfied on the extremal $x(t)$ if the interval $(t_0, t_1)$ (resp., the half-interval $(t_0, t_1]$) does not contain points conjugate to $t_0$.

A point $\tau$ is conjugate to $t_0$ if and only if the matrix

$$H(\tau) = \begin{pmatrix} h_0(\tau) & \cdots & h_m(\tau) \\ \int_{t_0}^{\tau} g_1(t) h_0(t)\, dt & \cdots & \int_{t_0}^{\tau} g_1(t) h_m(t)\, dt \\ \vdots & \ddots & \vdots \\ \int_{t_0}^{\tau} g_m(t) h_0(t)\, dt & \cdots & \int_{t_0}^{\tau} g_m(t) h_m(t)\, dt \end{pmatrix}$$

is degenerate.

*Necessary conditions for weak minimum (maximum):*

Suppose that the Lagrangians $f_i(t, x, x'_t)$ ($i = 0, 1, \ldots, m$) in problem (19.1.3.6) are sufficiently smooth. If $x(t) \in C^2[t_0, t_1]$ provides a weak minimum (resp., maximum) in problem (19.1.3.6) and the regularity condition is satisfied (i.e., the functions $g_i(t)$ are linearly independent on any of the intervals $[t_0, \tau]$ and $[\tau, t_1]$ for any $\tau$), then $x(t)$ is an extremal of problem (19.1.3.6) and the Legendre and Jacobi conditions are satisfied on $x(t)$.

*Sufficient conditions for a strong minimum (resp., maximum):*

Suppose that the Lagrangian

$$L = f_0 + \sum_{i=1}^{m} \mu_i f_i$$

is sufficiently smooth and the strengthened Legendre and Jacobi conditions, as well as the regularity condition, are satisfied on an admissible extremal $x(t)$. Then $x(t)$ provides a strong minimum (resp., maximum).

THEOREM. *Suppose that the functional $\mathcal{J}_0$ in problem (19.1.3.6) is quadratic, i.e.,*

$$\mathcal{J}_0[x] = \int_{t_0}^{t_1} \left[A_0(x'_t)^2 + B_0 x^2\right] dt,$$

*and the functionals $\mathcal{J}_i$ are linear,*

$$\mathcal{J}_i[x] = \int_{t_0}^{t_1} \left[a_i(x'_t)^2 + b_i x^2\right] dt \qquad (i = 1, 2, \ldots, m).$$

*Moreover, assume that the functions $A_0, a_1, \ldots, a_m$ are continuously differentiable, the functions $B_0, b_1, \ldots, b_m$ are continuous, and the strengthened Legendre condition and the regularity condition are satisfied. If the Jacobi condition does not hold, then the lower bound in the problem is $-\infty$ (resp., the upper bound is $+\infty$). If the Jacobi condition holds, then there exists a unique admissible extremal that provides the absolute minimum (resp., maximum).*

**Example 2.** Consider the problem

$$\mathcal{J} = \int_0^{2\pi} \left[(x'_t)^2 - x^2\right] dt \to \min; \qquad \int_0^{2\pi} x \, dt, \qquad x(0) = x(2\pi) = 0.$$

A necessary condition is given by the Lagrange multiplier rule (19.1.3.5): $x''_{tt} + x - \lambda = 0$. The general solution of the resulting equation with the condition $x(0) = 0$ taken into account is $x(t) = A \sin t + B(\cos t - 1)$. The set of admissible extremals always contains the admissible extremal $\hat{x}(t) \equiv 0$.

The Legendre condition (19.1.3.8) is satisfied: $L_{x'_t x'_t}(t, \hat{x}, \hat{x}'_t) = 2 > 0$. The Jacobi equation (19.1.3.9) coincides with the Euler equation (19.1.3.5). The solution $h_0(t)$ of the homogeneous equation with the conditions $h_0(0) = 0$ and $(h_0)'_t(0) = 1$ is the function $\sin t$. The solution $h_1(t)$ of the homogeneous equation $x''_{tt} + x + 1 = 0$ with the conditions $h_1(0) = 0$ and $(h_1)'_t(0) = 0$ is the function $\cos t - 1$. The matrix $H(\tau)$ acquires the form

$$H(\tau) = \begin{pmatrix} h_0(\tau) & h_0(\tau) \\ \int_0^\tau h_0 g_1 \, dt & \int_0^\tau h_m g_1 \, dt \end{pmatrix} = \begin{pmatrix} \sin \tau & \cos \tau - 1 \\ 1 - \cos \tau & \sin \tau - \tau \end{pmatrix}.$$

Thus the conjugate points are the solutions of the equation

$$\det H(\tau) = 2 - 2\cos \tau - \tau \sin \tau = 0 \quad \Leftrightarrow \quad \sin \frac{\tau}{2} = 0, \quad \frac{\tau}{2} = \tan \frac{\tau}{2}.$$

The conjugate point nearest to zero is $\tau = 2\pi$.

Thus the admissible extremals have the form $\hat{x}(t) = C \sin t$ and provide the absolute minimum $\mathcal{J}[\hat{x}] = 0$.

## 19.1.4. Problems with Higher Derivatives

**19.1.4-1. Statement of problem. Necessary condition for extremum.**

A *problem with higher derivatives* (with fixed endpoints) in classical calculus of variations is the following extremal problem in the space $C^n[t_0, t_1]$:

$$\mathcal{J}[x] = \int_{t_0}^{t_1} f_0(t, x, x'_t, \ldots, x_t^{(n)}) \, dt \to \text{extremum}; \tag{19.1.4.1}$$

$$x_t^{(k)}(t_j) = x_j^k \qquad (k = 0, 1, \ldots, n-1, \ j = 0, 1). \tag{19.1.4.2}$$

Here $L$ is a function of $n + 2$ variables, which is called the *Lagrangian*. Functions $x(t) \in C^n[t_0, t_1]$ satisfying conditions (19.1.4.2) at the endpoints are said to be *admissible*.

An admissible function $\hat{x}(t)$ is said to provide a *weak local minimum* (or *maximum*) in problem (19.1.4.1) if there exists a $\delta > 0$ such that the inequality

$$\mathcal{J}[x] \ge \mathcal{J}[\hat{x}] \quad (\text{resp.}, \ \mathcal{J}[x] \le \mathcal{J}[\hat{x}])$$

holds for any admissible function $x(t) \in C^n[t_0, t_1]$ satisfying $\|x - \hat{x}\|_n < \delta$.

An admissible function $\hat{x}(t) \in PC^n[t_0, t_1]$ is said to provide a *strong minimum* (resp., *maximum*) in problem (19.1.4.1) if there exists an $\varepsilon > 0$ such that the inequality

$$\mathcal{J}[x] \ge \mathcal{J}[\hat{x}] \quad (\text{resp.}, \ \mathcal{J}[x] \le \mathcal{J}[\hat{x}])$$

holds for any admissible function $x(t) \in PC^n[t_0, t_1]$ satisfying $\|x(t) - \hat{x}(t)\|_{n-1} < \varepsilon$.

*Necessary condition for extremum:*

Suppose that the Lagrangian $L$ is continuous together with its derivatives with respect to $x, x'_t, \ldots, x_t^{(n)}$ (the smoothness condition) for all $t \in [t_0, t_1]$. If the function $x(t)$ provides a local extremum in problem (19.1.4.1), then $L_{x_t^{(k)}} \in C^k[t_0, t_1]$ $(k = 1, 2, \ldots, n)$ and the *Euler–Poisson equation* holds:

$$\sum_{k=0}^{n} (-1)^k \frac{d^k}{dt^k} L_{x_t^{(k)}} = 0 \qquad (t_0 \le t \le t_1). \tag{19.1.4.3}$$

For $n = 1$, the Euler–Poisson equation coincides with the Euler equation (19.1.2.5). For $n = 2$, the Euler–Poisson equation has the form

$$\frac{d^2}{dt^2} L_{x''_{tt}} - \frac{d}{dt} L_{x'_t} + L_x = 0.$$

The general solution of equation (19.1.4.3) contains $2n$ arbitrary constants. These constants can be determined from the boundary conditions (19.1.4.2).

**19.1.4-2. Higher-order necessary and sufficient conditions.**

Consider the problem

$$\mathcal{J}[x] = \int_{t_0}^{t_1} f_0(t, x, x'_t, \ldots, x_t^{(n)}) \, dt \to \min \ (\text{or } \max); \tag{19.1.4.4}$$

$$x_t^{(k)}(t_j) = x_j^k \quad (k = 0, 1, \ldots, n-1, \ j = 0, 1)$$

with higher derivatives, where $L$ is the function of $n+1$ variables. Suppose that $x(t) \in C^{2n}[t_0, t_1]$ is an extremal of problem (19.1.4.4), i.e., the Euler–Poisson equation is satisfied on this extremal.

*Legendre condition:* If an extremal provides a minimum (resp., maximum) of the functional, then the following inequality holds:

$$L_{x_t^{(n)} x_t^{(n)}} \geq 0 \quad (\text{resp.,} \ L_{x_t^{(n)} x_t^{(n)}} \leq 0) \qquad (t_0 \leq t \leq t_1). \tag{19.1.4.5}$$

*Strengthened Legendre condition:* If an extremal provides a minimum (resp., maximum) of the functional, then the following inequality holds:

$$L_{x_t^{(n)} x_t^{(n)}} > 0 \quad (\text{resp.,} \ L_{x_t^{(n)} x_t^{(n)}} < 0) \qquad (t_0 \leq t \leq t_1). \tag{19.1.4.6}$$

The functional $\mathcal{J}$ has the second derivative at the point $x(t)$: $\mathcal{J}''_{tt}[x, x] = \mathcal{K}[x]$, where

$$\mathcal{K}[x] = \int_{t_0}^{t_1} \sum_{i,j=0}^{n} L_{x_t^{(i)} x_t^{(j)}}(t, x, x'_t, \dots, x_t^{(n)}) x_t^{(i)} x_t^{(j)} \, dt. \tag{19.1.4.7}$$

The Euler–Poisson equation (19.1.4.3) for the functional $\mathcal{K}$ is called the *Jacobi equation* for problem (19.1.4.4) on the extremal $\hat{x}(t)$.

For a quadratic functional $\mathcal{K}$ of the form

$$\mathcal{K}[x] = \int_{t_0}^{t_1} \sum_{i=0}^{n} L_{x_t^{(i)} x_t^{(i)}}(t, x, x'_t, \dots, x_t^{(n)}) \left(x_t^{(i)}\right)^2 dt, \tag{19.1.4.8}$$

the Jacobi equation reads

$$\sum_{i=0}^{n} (-1)^i \frac{d^i}{dt^i} \left[L_{x_t^{(i)} x_t^{(i)}} x_t^{(i)}\right] = 0.$$

Suppose that the strengthened Legendre condition (19.1.4.6) is satisfied on an extremal $x(t)$. A point $\tau$ is said to be *conjugate to the point* $t_0$ if there exists a nontrivial solution $h(t)$ of the Jacobi equation such that $h_t^{(i)}(t_0) = h_t^{(i)}(\tau) = 0$ ($i = 0, 1, \dots, n-1$). One says that the *Jacobi condition* (resp., *the strengthened Jacobi condition*) is satisfied on the extremal $x(t)$ if the interval $(t_0, t_1)$ (resp., the half-interval $(t_0, t_1]$) does not contain points conjugate to $t_0$.

The Jacobi equation is a $2n$th-order linear equation that can be solved for the higher derivative. Suppose that $h_1(t), \dots, h_n(t)$ are solutions of the Jacobi equation such that $H(t_0) = 0$ and $H_t^{(n)}(t_0)$ is a nondegenerate matrix, where

$$H(\tau) = \begin{pmatrix} h_1(\tau) & \cdots & h_n(\tau) \\ \vdots & \ddots & \vdots \\ [h_1(\tau)]_t^{(n-1)} & \cdots & [h_n(\tau)]_t^{(n-1)} \end{pmatrix}, \quad H_t^{(n)}(\tau) = \begin{pmatrix} [h_1(\tau)]_t^{(n)} & \cdots & [h_n(\tau)]_t^{(n)} \\ \vdots & \ddots & \vdots \\ [h_1(\tau)]_t^{(2n-1)} & \cdots & [h_n(\tau)]_t^{(2n-1)} \end{pmatrix}.$$

A point $\tau$ is conjugate to $t_0$ if and only if the matrix $H(\tau)$ is degenerate.

*Necessary conditions for weak minimum (resp., maximum):*

Suppose that the Lagrangian $L$ of problem (19.1.4.4) satisfies the smoothness condition. If a function $x(t) \in C^{2n}[t_0, t_1]$ provides a weak minimum (resp., maximum), then $x(t)$ is an extremal and the Legendre and Jacobi conditions hold on $x(t)$.

*Sufficient conditions for strong minimum (resp., maximum):*

Suppose that the Lagrangian $L$ is sufficiently smooth. If $x(t) \in C^{2n}[t_0, t_1]$ is an admissible extremal and the strengthened Legendre condition and the strengthened Jacobi condition are satisfied on $x(t)$, then $x(t)$ provides a strong minimum (resp., maximum) in problem (19.1.4.4).

For quadratic functionals of the form (19.1.4.8), the problem can be examined completely.

THEOREM. *Suppose that the functional has the form (19.1.4.8), $L_{x_t^{(i)} x_t^{(i)}} \in C^i[t_0, t_1]$, and the strengthened Legendre condition is satisfied. If the Jacobi condition does not hold, then the lower bound in the problem is $-\infty$ (the upper bound is $+\infty$). If the Jacobi condition is satisfied, then there exists a unique admissible extremal that provides the absolute minimum (maximum).*

**Example 3.** Consider the problem

$$\mathcal{J}[x] = \int_0^{2\pi} \left[(x_{tt}'')^2 - (x_t')^2\right] dt \to \text{extremum}; \quad x = x(t), \quad x(0) = x(2\pi) = x_t'(0) = x_t'(2\pi) = 0.$$

A necessary condition is given by the Euler–Poisson equation (19.1.4.3): $x_{tttt}'''' + x_{tt}'' = 0$. The general solution of this equation is $x(t) = C_1 \sin t + C_2 \cos t + C_3 t + C_4$. The set of admissible extremals always contains the admissible extremal $\hat{x}(t) \equiv 0$.

The Legendre condition $L_{x_t' x_t'}''(t, \hat{x}, \hat{x}_t', \hat{x}_{tt}'') = 2 > 0$ is satisfied. The Jacobi equation coincides with the Euler-Poisson equation. If we set $h_1(t) = 1 - \cos t$ and $h_2(t) = \sin t - t$, then the matrix $H(t)$ acquires the form

$$H(t) = \begin{pmatrix} h_1(t) & h_2(t) \\ [h_1(t)]_t' & [h_2(t)]_t' \end{pmatrix} = \begin{pmatrix} 1 - \cos t & \sin t - t \\ \sin t & \cos t - 1 \end{pmatrix}.$$

Then $H(0) = 0$ and

$$\det H_{tt}''(0) = \begin{pmatrix} [h_1(t)]_{tt}'' & [h_2(t)]_{tt}'' \\ [h_1(t)]_{ttt}''' & [h_2(t)]_{ttt}''' \end{pmatrix} = \begin{pmatrix} 1 & 0 \\ 0 & -1 \end{pmatrix} \neq 0.$$

Thus the conjugate points are the solutions of the equation

$$\det H(t) = 2(\cos t - 1) - t \sin t = 0 \iff \sin \frac{t}{2} = 0, \quad \frac{t}{2} = \tan \frac{t}{2}.$$

The conjugate point nearest to zero is $t_1 = 2\pi$.

Thus the admissible extremals have the form $\hat{x}(t) = C(1 - \cos t)$ and provide the absolute minimum $\mathcal{J}[\hat{x}] = 0$.

## 19.1.5. Lagrange Problem

### 19.1.5-1. Lagrange principle.

The *Lagrange problem* is the following problem:

$$\mathcal{B}_0(\gamma) \to \min;$$

$$\mathcal{B}_i(\gamma) \leq 0 \quad (i = 1, 2, \ldots, m'), \quad (19.1.5.1)$$

$$\mathcal{B}_i(\gamma) = 0 \quad (i = m' + 1, m' + 2, \ldots, m),$$

$$(\mathbf{x}_\alpha)_t' - \varphi(t, \mathbf{x}) = 0 \quad \text{for all } t \in T, \quad (19.1.5.2)$$

where $\mathbf{x} \equiv \mathbf{x(t)} \equiv (\mathbf{x}_\alpha, \mathbf{x}_\beta) \equiv (x_1(t), \ldots, x_n(t)) \in PC^1(\Gamma, \mathbb{R}^n)$, $\mathbf{x}_\alpha \equiv (x_1(t), \ldots, x_k(t)) \in PC^1(\Gamma, \mathbb{R}^k)$, $\mathbf{x}_\beta \equiv (x_{k+1}(t), \ldots, x_n(t)) \in PC^1(\Gamma, \mathbb{R}^{n-k})$, $\gamma = (\mathbf{x}, t_0, t_1)$, $\varphi \in PC(\Gamma, \mathbb{R}^n)$, $t_0, t_1 \in \Gamma$, $t_0 < t_1$, $\Gamma$ is a given finite interval, and

$$\mathcal{B}_i(\mathbf{x}, t_0, t_1) = \int_{t_0}^{t_1} f_i(t, \mathbf{x}, (\mathbf{x}_\beta)'_t) \, dt + \psi_i(t_0, \mathbf{x}(t_0), t_1, \mathbf{x}(t_1)) \quad (i = 0, 1, \ldots, m).$$

Here $PC(\Gamma, \mathbb{R}^n)$ is the space of piecewise continuous vector functions on the closed interval $\Gamma$, and $PC^1(\Gamma, \mathbb{R}^n)$ is the space of continuous vector functions with piecewise continuous derivative on $\Gamma$.

The constraint (19.1.5.2) is the differential equation that is called the *differential constraint*. The differential constraint can be imposed on all coordinates $\mathbf{x}$ (i.e., $k = n$ in (19.1.5.2)) or be lacking altogether ($k = 0$). The element $\gamma$ is called an *admissible element*.

An admissible element $\hat{\gamma} = (\hat{\mathbf{x}}, \hat{t}_0, \hat{t}_1)$ provides a *weak local minimum* in the Lagrange problem if there exists a $\delta > 0$ such that the inequality $\mathcal{B}_0(\gamma) \geq \mathcal{B}_0(\hat{\gamma})$ holds for any admissible element $\gamma$ satisfying the condition $\|\gamma - \hat{\gamma}\|_{C^1} < \delta$, $|t - \hat{t}_0| < \delta$, and $|t - \hat{t}_1| < \delta$, where $\|\mathbf{x}\|_{C^1} = \max_{t \in T} |\mathbf{x}| + \max_{t \in T} |\mathbf{x}'_t|$.

### 19.1.5-2. Necessary conditions for extremum. Euler–Lagrange theorem.

Suppose that $\hat{\gamma}$ provides a weak local minimum in the Lagrange problem (19.1.5.1), and, moreover, the functions $\varphi = (\varphi_1, \ldots, \varphi_n)$ and $f_i$ ($i = 0, 1, \ldots, m$) and their partial derivatives are continuous in $x$ in a neighborhood of $\{(t, \hat{\mathbf{x}} | t \in \Gamma\}$ and the functions $\psi_i$ ($i = 0, 1, \ldots, m$) are continuously differentiable in a neighborhood of the point $(\hat{t}_0, \hat{\mathbf{x}}(\hat{t}_0), \hat{t}_1, \hat{\mathbf{x}}(\hat{t}_1))$ (the smoothness condition).

Then there exist Lagrange multipliers $\lambda_i$ ($i = 0, \ldots, m$) and $p_j \equiv p_j(t) \in PC^1(T)$ ($j = 1, \ldots, k$) that are not zero simultaneously, such that the Lagrange function

$$\Lambda = \int_{t_0}^{t_1} \left\{ \sum_{i=0}^{m} \lambda_i f_i(t, \mathbf{x}, (\mathbf{x}_\beta)'_t) + \sum_{i=1}^{k} p_i \left[ (x_i)'_t - \varphi_i(t, \mathbf{x}) \right] \right\} dt + \sum_{i=0}^{m} \lambda_i \psi_i(t_0, \mathbf{x}(t_0), t_1, \mathbf{x}(t_1))$$

satisfies the following conditions:
1. The conditions of stationarity with respect to $\mathbf{x}$, i.e., the Euler equations

$$\frac{dp_i}{dt} + \sum_{j=1}^{k} p_j \frac{\partial \varphi_j}{\partial x_i} = \sum_{j=0}^{m} \lambda_j \frac{\partial f_j}{\partial x_i} \quad (i = 1, 2, \ldots, k) \quad \text{for all } t \in T,$$

where all derivatives with respect to $x_k$ are evaluated at $(t, \hat{\mathbf{x}})$.
2. The conditions of transversality with respect to $\mathbf{x}$,

$$p_i(\hat{t}_j) = (-1)^j \sum_{j=0}^{k} \lambda_j \frac{\partial \psi_j}{\partial x_i(t_j)} \quad (j = 0, 1; \ i = 1, 2, \ldots, k),$$

where all derivatives with respect to $x_i(t_k)$ ($k = 0, 1$) are evaluated at $(\hat{t}_0, \hat{\mathbf{x}}(\hat{t}_0), \hat{t}_1, \hat{\mathbf{x}}(\hat{t}_1))$.
3. The conditions of stationarity with respect to $t_k$ (only for movable endpoints of the integration interval),

$$\Lambda_{t_k}(\hat{t}_k) = 0 \quad (k = 0, 1).$$

4. The complementary slackness conditions

$$\lambda_i B_i(\hat{\gamma}) = 0 \quad (i = 1, 2, \ldots, m').$$

5. The nonnegativity conditions

$$\lambda_i \geq 0 \quad (i = 1, 2, \ldots, m').$$

### 19.1.6. Pontryagin Maximum Principle

**19.1.6-1. Statement of problem.**

The optimal control problem (in Pontryagin's form) is the problem

$$\mathcal{B}_0(\omega) \to \min;$$
$$\mathcal{B}_i(\omega) \leq 0 \quad (i = 1, 2, \ldots, m'), \tag{19.1.6.1}$$
$$\mathcal{B}_i(\omega) = 0 \quad (i = m' + 1, m' + 2, \ldots, m),$$
$$\mathbf{x}'_t - \varphi(t, \mathbf{x}, \mathbf{u}) = 0 \quad \text{for all } t \in T, \tag{19.1.6.2}$$
$$\mathbf{u} \in U \quad \text{for all } t \in \Gamma, \tag{19.1.6.3}$$

where $\mathbf{x} \equiv \mathbf{x(t)} \in PC^1(\Gamma, \mathbb{R}^n)$, $\mathbf{u} \equiv \mathbf{u}(t) \in PC(\Gamma, \mathbb{R}^r)$, $\omega = (\mathbf{x}, \mathbf{u}, t_0, t_1)$, $\varphi \in PC(\Gamma, \mathbb{R}^n)$, $t_0, t_1 \in \Gamma$, $t_0 < t_1$, $\Gamma$ is a given finite interval, $U \subset \mathbb{R}^r$ is an arbitrary set, $T \subset \Gamma$ is the set of continuity points of $\mathbf{u}$, and

$$\mathcal{B}_i(\mathbf{x}, \mathbf{u}, t_0, t_1) = \int_{t_0}^{t_1} f_i(t, \mathbf{x}, \mathbf{u})\, dt + \psi_i(t_0, \mathbf{x}(t_0), t_1, \mathbf{x}(t_1)) \quad (i = 0, 1, \ldots, m).$$

Here $PC(\Gamma, \mathbb{R}^n)$ is the space of piecewise continuous vector functions on the closed interval $\Gamma$, and $PC^1(\Gamma, \mathbb{R}^n)$ is the space of continuous vector functions with piecewise continuous derivative on $\Gamma$.

The vector function $\mathbf{x} = (x_1(t), \ldots, x_n(t))$ is called the *phase variable*, and the vector function $\mathbf{u} = (u_1(t), \ldots, u_r(t))$ is called the *control*. The constraint (19.1.6.2) is a differential equation that is called a *differential constraint*. In contrast with the Lagrange problem, this problem contains the inclusion-type constraint (19.1.6.3), which should be satisfied at all points $t \in \Gamma$, and, moreover, the phase variable $\mathbf{x}$ can be less smooth.

An element $\omega = (\mathbf{x}, \mathbf{u}, t_0, t_1)$ for which all conditions and constraints of the problem are satisfied is called an *admissible controlled process*. An admissible controlled process $\hat{\omega} = (\hat{\mathbf{x}}, \hat{\mathbf{u}}, \hat{t}_0, \hat{t}_1)$ is called a (locally) *optimal process (or a process optimal in the strong sense)* if there exists a $\delta > 0$ such that $\mathcal{B}_0(\omega) \geq \mathcal{B}_0(\hat{\omega})$ for any admissible controlled process $\omega = (\mathbf{x}, \mathbf{u}, t_0, t_1)$ such that

$$\|\omega - \hat{\omega}\|_C < \delta, \quad |t - \hat{t}_0| < \delta, \quad |t - \hat{t}_1| < \delta,$$

where $\|\mathbf{x}\|_C = \max\limits_{t \in \Gamma} |\mathbf{x}|$.

**19.1.6-2. Necessary conditions for extremum. Pontryagin maximum principle.**

Suppose that $\hat{\omega} = (\hat{\mathbf{x}}, \hat{\mathbf{u}}, \hat{t}_0, \hat{t}_1)$ is an optimal (in the strong sense) process in the optimal control problem (19.1.6.1) and, moreover, the functions $\varphi = (\varphi_1, \ldots, \varphi_n)$ and $f_i$ ($i = 0, 1, \ldots, m$) and their partial derivatives are continuous in $x$ in a neighborhood of the Cartesian product of $\{(t, \hat{\mathbf{x}} | t \in \Gamma\}$ by $U$ and the functions $\psi_i$ ($i = 0, 1, \ldots, m$) are continuously differentiable in a neighborhood of the point $(\hat{t}_0, \hat{\mathbf{x}}(\hat{t}_0), \hat{t}_1, \hat{\mathbf{x}}(\hat{t}_1))$ (the smoothness condition).

Then there exist Lagrange multipliers $\lambda_i$ ($i = 0, \ldots, m$) and $p_k \equiv p_k(t) \in PC^1(\Gamma)$ ($k = 1, \ldots, n$), which are not zero simultaneously, such that the Lagrange function

$$\Lambda = \int_{t_0}^{t_1} \left\{ \sum_{i=0}^{m} \lambda_i f_i(t, \mathbf{x}, \mathbf{u}) + \sum_{i=0}^{n} p_i \left[(x_i)'_t - \varphi_i(t, \mathbf{x}, \mathbf{u})\right] \right\} dt + \sum_{i=0}^{m} \lambda_i \psi_i(t_0, \mathbf{x}(t_0), t_1, \mathbf{x}(t_1))$$

satisfies the following conditions:
1. The conditions of stationarity with respect to $\mathbf{x}$, i.e., the Euler equations

$$\frac{dp_i}{dt} + \sum_{k=1}^{n} p_k \frac{\partial \varphi_k}{\partial x_i} = \sum_{k=0}^{m} \lambda_k \frac{\partial f_k}{\partial x_i} \quad (i = 1, 2, \ldots, n) \quad \text{for all } t \in T,$$

where all derivatives with respect to $x_k$ are evaluated at $(t, \hat{\mathbf{x}}, \hat{\mathbf{u}})$.

2. The conditions of transversality with respect to $\mathbf{x}$,

$$p_i(\hat{t}_k) = (-1)^k \sum_{k=0}^{m} \lambda_k \frac{\partial \psi_k}{\partial x_i(t_k)} \quad (k = 0, 1;\ i = 1, 2, \ldots, n),$$

where all derivatives with respect to $x_i(t_k)$ ($k = 0, 1$) are evaluated at $(\hat{t}_0, \hat{\mathbf{x}}(\hat{t}_0), \hat{t}_1, \hat{\mathbf{x}}(\hat{t}_1))$.

3. The condition of optimality with respect to $\mathbf{u}$,

$$\max_{\mathbf{u} \in U} H(t, \hat{\mathbf{x}}, \mathbf{u}, p) = H(t, \hat{\mathbf{x}}, \hat{\mathbf{u}}, p),$$

where $H(t, \mathbf{x}, \mathbf{u}, p) = \sum_{i=1}^{n} p_i \varphi_i(t, \mathbf{x}, \mathbf{u}) - \sum_{i=0}^{m} \lambda_i f_i(t, \mathbf{x}, \mathbf{u})$ is the *Pontryagin function*.

4. The conditions of stationarity with respect to $t_k$ (only for movable endpoints of the integration interval),

$$\Lambda_{t_k}(\hat{t}_k) = 0 \quad (k = 0, 1).$$

5. The complementary slackness conditions

$$\lambda_i B_i(\hat{\omega}) = 0 \quad (i = 1, 2, \ldots, m').$$

6. The nonnegativity conditions

$$\lambda_i \geq 0 \quad (i = 1, 2, \ldots, m').$$

The Pontryagin maximum principle permits one to obtain all necessary conditions in calculus of variations: the Euler equations, the Legendre condition, the Weierstrass condition, and the Lagrange multiplier rule.

## 19.2. Mathematical Programming

### 19.2.1. Linear Programming

**19.2.1-1. Statement of linear programming problem.**

The *general mathematical programming problem* is the problem of finding the values of $n$ variables $\mathbf{x} = (x_1, \ldots, x_n)$ that provide an extremum of the objective function

$$Z(\mathbf{x}) \to \text{extremum} \qquad (19.2.1.1)$$

and satisfy the system of constraints

$$\begin{aligned}
\varphi_i(\mathbf{x}) &= 0 \quad \text{for} \quad i = 1, 2, \ldots, k, \\
\varphi_i(\mathbf{x}) &\leq 0 \quad \text{for} \quad i = k+1, k+2, \ldots, l, \\
\varphi_i(\mathbf{x}) &\geq 0 \quad \text{for} \quad i = l+1, l+2, \ldots, m.
\end{aligned} \qquad (19.2.1.2)$$

Any $n$-dimensional vector $\mathbf{x}$ satisfying the system of constraints is called a *feasible solution* of the mathematical programming problem. The set of feasible solutions of the problem is called the *feasible region*. A feasible solution minimizing (maximizing) the objective function is called an *optimal solution*.

If the objective function (19.2.1.1) and the constraints (19.2.1.2) are linear, then the mathematical programming problem is called a *linear programming problem*.

**Example 1 (the product mix problem).**

Suppose that $n$ types of products are manufactured from $m$ types of resources, whose respective supplies are $b_1, \ldots, b_m$. Let $a_{ij}$ ($i = 1, 2, \ldots, m;\ j = 1, 2, \ldots, n$) be the withdrawal of resource $i$ per unit output of product $j$, and let $c_j$ ($j = 1, 2, \ldots, n$) be the profit per unit output of product $j$. The objective is to find a product mix maximizing the profit.

If the vector of the variables is denoted by $\mathbf{x} = (x_1, \ldots, x_n)$, where $x_j$ ($j = 1, 2, \ldots, n$) is the output of product $j$, then the mathematical model of the problem has the form

$$Z(\mathbf{x}) = \sum_{j=1}^{n} c_j x_j \to \max$$

under the constraints

$$\begin{aligned}
a_{11} x_1 + a_{12} x_2 + \cdots + a_{1n} x_n &\leq b_1, \\
a_{21} x_1 + a_{22} x_2 + \cdots + a_{2n} x_n &\leq b_2, \\
&\cdots\cdots\cdots\cdots\cdots\cdots\cdots \\
a_{m1} x_1 + a_{m2} x_2 + \cdots + a_{mn} x_n &\leq b_m, \\
x_j &\geq 0 \quad (j = 1, 2, \cdots, n).
\end{aligned}$$

**Example 2 (the diet problem).**

The animal diet includes feeds of $n$ types. Animals must receive nutrients of $m$ types daily; the amount of nutrient $j$ must be at least $b_i$ ($i = 1, 2, \ldots, m$). Let $a_{ij}$ ($i = 1, 2, \ldots, m;\ j = 1, 2, \ldots, n$) be the content of nutrient $i$ per unit of feed $j$, and let $c_j$ ($j = 1, 2, \ldots, n$) be the cost per unit of feed $j$. The objective is to find a diet minimizing the daily cost.

If the vector of variables is denoted by $\mathbf{x} = (x_1, \ldots, x_n)$, where $x_j$ ($j = 1, 2, \ldots, n$) is the amount of feed $j$ in the daily diet, then the mathematical model of the problem has the form

$$Z(\mathbf{x}) = \sum_{j=1}^{n} c_j x_j \to \min$$

under the constraints

$$\begin{aligned}
a_{11} x_1 + a_{12} x_2 + \cdots + a_{1n} x_n &\geq b_1, \\
a_{21} x_1 + a_{22} x_2 + \cdots + a_{2n} x_n &\geq b_2, \\
&\cdots\cdots\cdots\cdots\cdots\cdots\cdots \\
a_{m1} x_1 + a_{m2} x_2 + \cdots + a_{mn} x_n &\geq b_m, \\
x_j &\geq 0 \quad (j = 1, 2, \ldots, n).
\end{aligned}$$

To develop a general method for solving linear programming problems, one reduces various forms of linear programming problems to the *standard form*

$$Z(\mathbf{x}) = \sum_{j=1}^{n} c_j x_j \to \text{extremum},$$

$$a_{11}x_1 + a_{12}x_2 + \cdots + a_{1n}x_n = b_1,$$
$$a_{21}x_1 + a_{22}x_2 + \cdots + a_{2n}x_n = b_2, \tag{19.2.1.3}$$
$$\cdots\cdots\cdots\cdots\cdots\cdots\cdots\cdots\cdots\cdots\cdots\cdots\cdots$$
$$a_{m1}x_1 + a_{m2}x_2 + \cdots + a_{mn}x_n = b_m,$$
$$x_j \geq 0 \quad (j = 1, 2, \ldots, n);$$
$$b_i \geq 0 \quad (i = 1, 2, \ldots, m).$$

This form of the linear programming problem differs from other forms in that the system of constraints is a system of equations and all variables are nonnegative.

Any linear programming problem can be reduced to the standard problem by the following rules:

1. By introducing additional variables, one can reduce inequality constraints to equality constraints. The inequality $a_{i1}x_1 + a_{i2}x_2 + \cdots + a_{in}x_n \geq b_i$ is replaced by the equation $a_{i1}x_1 + a_{i2}x_2 + \cdots + a_{in}x_n - x_{n+1} = b_i$ and the nonnegativity condition for the additional *surplus variable* $x_{n+1} \geq 0$. The inequality $a_{i1}x_1 + a_{i2}x_2 + \cdots + a_{in}x_n \leq b_i$ is replaced by the equation $a_{i1}x_1 + a_{i2}x_2 + \cdots + a_{in}x_n + x_{n+1} = b_i$ and the nonnegativity condition for the additional *slack variable* $x_{n+1} \geq 0$. The additional variables occur with zero coefficients in the objective function.

2. If some variable $x_j$ can take arbitrary values, it can be represented as the difference $x_j = \tilde{x}_j - \tilde{\tilde{x}}_j$ of nonnegative variables $\tilde{x}_j \geq 0$ and $\tilde{\tilde{x}}_j \geq 0$.

**Remark.** In the standard linear programming problem, the objective function can be either minimized or maximized. To pass from the minimization problem to the maximization problem or vice versa, it suffices to change the signs of the coefficients of the objective function, i.e., pass from the function $Z(\mathbf{x})$ to the function $-Z(\mathbf{x})$. The resulting problem and the original problem have the same optimal solution, and the values of the objective functions on this solution differ only in sign.

In some cases, it may be necessary to pass from the standard problem to the *symmetric problem*, which has no equality constraints and in matrix form reads

$$Z(\mathbf{x}) = CX \to \max(\min),$$
$$AX \leq B \quad (AX \geq B), \quad X \geq O, \tag{19.2.1.4}$$

where the following notation is used:

$$C = (c_1, \ldots, c_n), \quad A = \begin{pmatrix} a_{11} & a_{12} & \cdots & a_{1n} \\ a_{21} & a_{22} & \cdots & a_{2n} \\ \vdots & \vdots & \ddots & \vdots \\ a_{m1} & a_{m2} & \cdots & a_{mn} \end{pmatrix}, \quad X = \begin{pmatrix} x_1 \\ x_2 \\ \vdots \\ x_n \end{pmatrix}, \quad B = \begin{pmatrix} b_1 \\ b_2 \\ \vdots \\ b_m \end{pmatrix},$$

and $O$ is the zero column matrix.

### 19.2.1-2. Graphical method for solving linear programming problems.

Consider the linear programming problem

$$Z(\mathbf{x}) = \sum_{j=1}^{n} c_j x_j \to \text{extremum},$$

$$a_{i1}x_1 + a_{i2}x_2 + \cdots + a_{in}x_n \leq b_i \quad \text{for} \quad i = 1, 2, \ldots, k,$$
$$a_{i1}x_1 + a_{i2}x_2 + \cdots + a_{in}x_n \geq b_i \quad \text{for} \quad i = k+1, k+2, \ldots, m, \quad (19.2.1.5)$$
$$x_j \geq 0 \quad (j = 1, 2, \ldots, n)$$
$$b_i \geq 0 \quad (i = 1, 2, \ldots, m).$$

It is expedient to use the graphical method when solving problems with two variables of the form

$$Z(\mathbf{x}) = c_1 x_1 + c_2 x_2 \to \text{extremum},$$
$$a_{i1}x_1 + a_{i2}x_2 \leq b_i \quad \text{for} \quad i = 1, 2, \ldots, k,$$
$$a_{i1}x_1 + a_{i2}x_2 \geq b_i \quad \text{for} \quad i = k+1, k+2, \ldots, m, \quad (19.2.1.6)$$
$$x_j \geq 0 \quad (j = 1, 2),$$
$$b_i \geq 0 \quad (i = 1, 2, \ldots, m).$$

This method is based on the possibility of graphically representing the feasible region of problem (19.2.1.6) and finding the optimal solution in the feasible region. Geometrically, each inequality in the system of constraints determines a half-plane bounded by the line $a_{i1}x_1 + a_{i2}x_2 = b_i$ ($i = 1, 2, \ldots, m$). The nonnegativity conditions determine half-planes with boundary lines $x_1 = 0$ and $x_2 = 0$.

The feasible region of the problem is the common part of the half-planes of solutions to all inequalities in the system of constraints. If the system of constraints is consistent, then the feasible region is a convex polygon.

A line on which the objective function of the problem takes a constant value is called a *level line*. In the general case, a level line is determined by the equation $c_1 x_1 + c_2 x_2 = l$, where $l = \text{const}$. All level lines are parallel to one another. Their common normal is $\mathbf{N} = \text{grad } Z(x_1, x_2) = (c_1, c_2)$. The values of the objective function on the level lines increase in the direction of the normal $\mathbf{N}$ and decrease in the opposite direction.

A level line that meets the feasible region is called a *support line* if the feasible region lies in one of the half-planes into which the line divides the plane.

If $n = 3$ in the system of constraints (19.2.1.5), then each inequality geometrically determines a half-space with boundary plane $a_{i1}x_1 + a_{i2}x_2 + a_{i3}x_3 = b_i$ ($i = 1, 2, \ldots, m$) in three-dimensional space. The nonnegativity conditions determine half-spaces with boundary planes $x_1 = 0$, $x_2 = 0$, and $x_3 = 0$. If the system of constraints is consistent, then the feasible region is a convex polyhedron.

If $n > 3$ in the system of constraints (19.2.1.5), then each inequality geometrically determines a half-space with boundary hyperplane $a_{i1}x_1 + a_{i2}x_2 + \cdots + a_{in}x_n = b_i$ ($i = 1, 2, \ldots, m$) in the $n$-dimensional space. The nonnegativity conditions determine half-spaces with boundary hyperplanes $x_1 = 0, \ldots, x_n = 0$.

### 19.2.1-3. Simplex method.

*Foundations of the simplex method:*
1. The feasible region in a linear programming problem is a convex set with finitely many corners (vertices).
2. One of the vertices of the feasible region is an optimal solution of the linear programming problem.
3. Algebraically, the vertices of the feasible region represent several basic solutions of the system of constraints.

The simplex method solves the linear programming problem by purposefully progressing from one feasible solution to another.

*Main stages of the simplex method:*
1. Search of the initial basic solution.
2. Verification of the optimality criterion for the basic solution or verification of the condition that the function is unbounded.
3. Passage from one basic solution to another at which the value of the objective function is closer to the optimal value.

Suppose that a linear programming problem (19.2.1.3) is given and the matrix of the system of equations contains an identity $m \times m$ submatrix. The variables corresponding to the columns of the identity submatrix are usually called the *basic variables*, and the other variables are said to be *nonbasic*. The quantity

$$\Delta_j = \sum_{i=1}^{m} a_{ij} c_i - c_j$$

is called the *reduced cost* of the variable $x_j$.

THEOREM 1. *If the condition*

$$\Delta_j < 0 \quad (\text{resp.,} \quad \Delta_j > 0) \quad (j = 1, 2, \ldots, n) \quad (19.2.1.7)$$

*holds for some nonbasic variable, then one can obtain a new feasible solution for which the value of the objective function is greater (resp., less) than the original value. Here the following two cases are possible:*
1. *If all entries in the column corresponding to the variable to be entered in the basic set are nonpositive, then the linear programming problem does not have any solutions.*
2. *If the column corresponding to the variable to be entered in the basic set contains at least one positive element, then a new feasible solution can be obtained.*

THEOREM 2. *If the condition*

$$\Delta_j \geq 0 \quad (\text{resp.,} \quad \Delta_j \leq 0) \quad (j = 1, 2, \ldots, n) \quad (19.2.1.8)$$

*is satisfied, then the current feasible solution is optimal.*

Remark 1. The reduced costs of basic variables are always zero.

Remark 2. A linear programming problem has infinitely many optimal solutions if the reduced cost of at least one of the nonbasic variables is zero for an optimal solution.

The variable $x_k$ to be entered into the basic set is chosen, on the basis of the optimality criterion, from the condition
1. $\min_j \{\Delta_j\}$ in the maximization problem;
2. $\max_j \{\Delta_j\}$ in the minimization problem.

The variable $x_r$ to be deleted from the basic set is chosen from the condition

$$\theta = \min_i \frac{b_i}{a_{ik}} = \frac{b_r}{a_{rk}}; \quad a_{ik} > 0 \quad (i = 1, 2, \ldots, m). \tag{19.2.1.9}$$

The entry $a_{rk}$ is called the *pivot entry*, the row containing the pivot entry is called the *pivot row*, and the column containing the pivot entry is called the *pivot column*.

To pass to a new feasible solution, one should perform the following actions:

1. The entries of the row to be entered, corresponding to the pivot row, are computed by the formulas

$$\widetilde{a}_{rj} = \frac{a_{rj}}{a_{rk}} \quad (j = 1, 2, \ldots, n).$$

2. The elements of any other $i$th row are recomputed by the formulas

$$\widetilde{a}_{ij} = \frac{a_{ij}a_{rk} - a_{rj}a_{ik}}{a_{rk}} \quad (i = 1, 2, \ldots, m;\ j = 1, 2, \ldots, n).$$

3. The values of the basic variables of the new feasible solution are determined by the formulas

$$\widetilde{b}_i = \begin{cases} b_i/a_{ik} & \text{for } i = r, \\ (b_i a_{rk} - b_r a_{ik})/a_{rk} & \text{for } i \ne r \end{cases} \quad (i = 1, 2, \ldots, m).$$

**Example 3.** Let us solve the linear programming problem

$$Z(\mathbf{x}) = 5x_1 + 2x_2 + x_3 \to \max,$$
$$x_1 + 0x_2 + 0x_3 - x_4 + 0x_5 = 1,$$
$$0x_1 + x_2 + x_3 + x_4 + 0x_5 = 2,$$
$$0x_1 + 4x_2 + 0x_3 + 5x_4 + x_5 = 13,$$
$$x_j \ge 0 \quad (j = 1, 2, \ldots, 5).$$

Since the system of constraints in this problem contains an identity submatrix in the first, third, and fifth columns, it follows that the variables $x_1, x_3, x_5$ are basic and the variables $x_2$ and $x_4$ are nonbasic. The reduced costs of the nonbasic variables are $x_2 = -1 < 0$ and $x_4 = -4 < 0$; hence the variable $x_4$ should be entered into the basic set. Since $\theta = \min\{2/1, 15/5\} = 2$, which corresponds to the variable $x_3$, this variable should be deleted from the basic set. The new basic set consists of the variables $x_1, x_4$, and $x_5$, and the system of constraints becomes

$$\begin{cases} x_1 + x_2 + x_3 + 0x_4 + 0x_5 = 3, \\ 0x_1 + x_2 + x_3 + x_4 + 0x_5 = 2, \\ 0x_1 - x_2 - 5x_3 + 0x_4 + x_5 = 3. \end{cases}$$

The reduced costs of the nonbasic variables are $x_2 = 3 > 0$ and $x_3 = 4 > 0$; hence the solution $\mathbf{x} = (3, 0, 0, 2, 3)$ is optimal, and $Z_{\min} = 15$ in this case.

If it is difficult to find a feasible initial solution of problem (19.2.1.3), then it is expedient to use the method of *artificial basic variables*. For the original problem, one constructs an extended problem by introducing *artificial variables*. These are nonnegative variables introduced in the equality constraints so as to obtain a feasible initial solution. Each artificial variable is introduced in the left-hand side of one of the equations in the system of constraints with coefficient +1 and in the objective function with coefficient $-M$ (in a maximization problem) or $+M$ (in a minimization problem), where $M \gg 1$. In the general case, the extended minimization problem has the form

$$Z(\mathbf{x}, \mathbf{y}) = \sum_{j=1}^{n} c_j x_j + M \sum_{k=1}^{m} y_k \to \min,$$

$$a_{11}x_1 + a_{12}x_2 + \cdots + a_{1n}x_n + y_1 = b_1,$$
$$a_{21}x_1 + a_{22}x_2 + \cdots + a_{2n}x_n + y_2 = b_2, \qquad (19.2.1.10)$$
$$\dotsb\dotsb\dotsb\dotsb\dotsb\dotsb\dotsb\dotsb\dotsb\dotsb\dotsb$$
$$a_{m1}x_1 + a_{m2}x_2 + \cdots + a_{mn}x_n + y_m = b_m,$$
$$x_j \geq 0 \quad (j = 1, 2, \ldots, n);$$
$$y_i \geq 0, \quad b_i \geq 0 \quad (i = 1, 2, \ldots, m).$$

THEOREM 1 (A SOLUTION OPTIMALITY TEST). *If the extended linear programming problem has an optimal solution* $(x_1^*, \ldots, x_n^*, 0, \ldots, 0)$ *in which all artificial variables are zero, then the original problem has the optimal solution* $(x_1^*, \ldots, x_n^*)$ *obtained from the optimal solution of the extended problem by omitting the artificial variables.*

THEOREM 2 (A TEST FOR THE INFEASIBILITY OF THE PROBLEM). *If the extended linear programming problem has an optimal solution in which at least one artificial variable is nonzero, then the original problem is infeasible (i.e., has no solutions, since the system of constraints is inconsistent).*

THEOREM 3 (A TEST FOR THE UNBOUNDEDNESS OF THE PROBLEM). *If the extended linear programming problem is unbounded (i.e., has no solutions, since the objective function is unbounded), then so is the original problem.*

### 19.2.1-4. Duality in linear programming.

To each linear programming problem, called the *primal*, one can assign another problem, called the *dual*. The primal and dual problems are related in such a way that the solution of either of them can be directly obtained from the solution of the other.

Suppose that the primal has the form

$$Z(\mathbf{x}) = \sum_{j=1}^{n} c_j x_j \to \max, \qquad (19.2.1.11)$$

$$\sum_{j=1}^{n} a_{ij} x_j \leq b_i \quad (i = 1, 2, \ldots, m), \qquad (19.2.1.12)$$

$$x_j \geq 0 \quad (j = 1, 2, \ldots, n). \qquad (19.2.1.13)$$

Then the dual has the form

$$W(\mathbf{y}) = \sum_{i=1}^{m} b_i y_i \to \min, \qquad (19.2.1.14)$$

$$\sum_{i=1}^{m} a_{ij} y_i \geq c_j \quad (j = 1, 2, \ldots, n), \qquad (19.2.1.15)$$

$$y_i \geq 0 \quad (j = 1, 2, \ldots, m). \qquad (19.2.1.16)$$

The rules for constructing the dual problem are as follows:
1. In all constraints of the primal problem, the constant terms must be on the right-hand side, and the terms with the unknowns must be on the left-hand side.
2. All inequality constraints must have the same inequality signs.
3. The matrix
$$A = \begin{pmatrix} a_{11} & a_{12} & \cdots & a_{1n} \\ a_{21} & a_{22} & \cdots & a_{2n} \\ \vdots & \vdots & \ddots & \vdots \\ a_{m1} & a_{m2} & \cdots & a_{mn} \end{pmatrix}$$
of coefficients of the unknowns in the system of constraints (19.2.1.12) of the primal problem and the similar matrix
$$A^T = \begin{pmatrix} a_{11} & a_{21} & \cdots & a_{m1} \\ a_{12} & a_{22} & \cdots & a_{m2} \\ \vdots & \vdots & \ddots & \vdots \\ a_{1n} & a_{2n} & \cdots & a_{mn} \end{pmatrix}$$
for the dual problem are obtained from each other by transposition.
4. If the inequality constraints of the primal problem have the $\leq$ signs, then the objective function $Z(\mathbf{x})$ must be maximized; if the signs are $\geq$, then the objective function $Z(\mathbf{x})$ must be minimized.
5. To each constraint in the primal problem, there corresponds an unknown in the dual problem. The unknown corresponding to an inequality constraint must be nonnegative, and the unknown corresponding to an equality constraint may have any sign.
6. The number of variables in the dual problem is equal to the number of functional constraints (19.2.1.12) in the primal problem, and the number of constraints in system (19.2.1.15) of the dual problem is equal to the number of variables in the primal problem.
7. The coefficients of the unknowns in the objective function (19.2.1.14) of the dual problem are equal to the constant terms in system (19.2.1.12) of constraints in the primal problem. The right-hand sides of constraints (19.2.1.15) in the dual problem are equal to the coefficients of the unknowns in the objective function (19.2.1.11) of the primal problem.
8. The objective function $W(\mathbf{y})$ of the dual problem is to be minimized if $Z(\mathbf{x})$ is to be maximized (i.e., if $Z(\mathbf{x}) \to \max$, then $W(\mathbf{y}) \to \min$), and vice versa.

Duality theorems establish a relationship between the optimal solutions of a pair of dual problems.

FIRST DUALITY THEOREM. *For a pair of mutually dual linear programming problems, one of the following alternative cases holds:*
1. *If one of the problems has an optimal solution, then so does the other, and the objective values on the optimal solutions are the same for both problems.*
2. *If one of the problems is unbounded, then the other is infeasible.*

SECOND DUALITY THEOREM. *Let $\mathbf{x} = (x_1, \ldots, x_n)$ be a feasible solution of the primal problem (19.2.1.11)–(19.2.1.13), and let $\mathbf{y} = (y_1, \ldots, y_m)$ be a feasible solution of the dual problem (19.2.1.14)–(19.2.1.16). These solutions are optimal in the respective problems if and only if*

$$x_j \left( \sum_{i=1}^{m} a_{ij} y_i - c_j \right) = 0 \qquad (j = 1, 2, \ldots, n); \qquad (19.2.1.17)$$

$$y_i\left(\sum_{j=1}^{n} a_{ij}x_j - b_i\right) = 0 \qquad (i = 1, 2, \ldots, m). \tag{19.2.1.18}$$

In other words, if the $i$th constraint in the primal problem is inactive (holds with strict inequality) for an optimal solution of the primal problem, then the $i$th coordinate of the optimal solution of the dual problem is zero, and, conversely, if the $i$th coordinate of the optimal solution of the dual problem is nonzero, then the $i$th constraint in the primal problem is active (holds with equality) for the optimal solution of the primal problem.

**Example 4.** Suppose that the primal problem has the form

$$Z(\mathbf{x}) = -2x_1 + 4x_2 + 14x_3 + 2x_4 \to \min,$$
$$-2x_1 - x_2 + x_3 + 2x_4 = 6,$$
$$-x_1 + 2x_2 + 4x_3 - 5x_4 = 30,$$
$$x_j \geq 0 \quad (j = 1, 2, 3, 4).$$

Then the dual problem is

$$W(\mathbf{y}) = 6y_1 + 30y_2 \to \max,$$
$$-2y_1 - y_2 \leq -2,$$
$$-y_1 + 2y_2 \leq 4,$$
$$y_1 + 4y_2 \leq 14,$$
$$2y_1 - 5y_2 \leq 2.$$

The optimal solution of the dual problem is $\mathbf{y}_{\text{opt}} = (2, 3)$. After the substitution of the optimal solution into the system of constraints in the dual problem, we see that the first and last constraints are satisfied with strict inequality,

$$-2 \times 2 - 3 < -2 \quad \Rightarrow x_1^* = 0,$$
$$-2 + 2 \times 3 = 4 \quad \Rightarrow x_2^* > 0,$$
$$2 + 4 \times 3 = 14 \quad \Rightarrow x_3^* > 0,$$
$$2 \times 2 - 5 \times 3 < 2 \quad \Rightarrow x_4^* = 0.$$

By the second duality theorem, the corresponding coordinates of the optimal solution of the primal problem are zero, $x_1^* = x_4^* = 0$. The system of constraints of the primal problem acquires the form

$$-x_2 + x_3 = 6,$$
$$2x_2 + 4x_3 = 30,$$

which implies that $x_2^* = 1$, $x_3^* = 7$.

Thus $\mathbf{x}_{\text{opt}} = (0, 1, 7, 0)$ is the optimal solution of the primal problem.

### 19.2.1-5. Transportation problem. General statement of problem.

Suppose that $m$ supply sources $A_1, \ldots, A_m$ have amounts $a_1, \ldots, a_m$ of identical goods that must be shipped to $n$ consumers $B_1, \ldots, B_n$ with respective demands $b_1, \ldots, b_n$ for these goods. Let $c_{ij}$ ($i = 1, 2, \ldots, m$; $j = 1, 2, \ldots, n$) be the unit transportation cost from supply source $i$ to consumer destination $j$. The problem is to find a flow of least cost that ships from supply sources to consumer destinations.

The input data of the transportation problem are usually presented in the form of the table shown in Fig. 19.1.

Let $x_{ij}$ ($i = 1, 2, \ldots, m$; $j = 1, \ldots, n$) be the flow from supply source $i$ to consumer destination $j$. The *mathematical model of the transportation problem* in the general case has the form

$$Z(\mathbf{x}) = \sum_{i=1}^{m} \sum_{j=1}^{n} c_{ij} x_{ij} \to \min, \tag{19.2.1.19}$$

|  | Consumers | | | |
|---|---|---|---|---|
| Sources | $B_1$ $b_1$ | $B_2$ $b_2$ | $\cdots$ | $B_n$ $b_n$ |
| $A_1$  $a_1$ | $c_{11}$ | $c_{12}$ | $\cdots$ | $c_{1n}$ |
| $A_2$  $a_2$ | $c_{21}$ | $c_{22}$ | $\cdots$ | $c_{2n}$ |
| $\vdots$ | $\vdots$ | $\vdots$ | $\ddots$ | $\vdots$ |
| $A_m$  $a_m$ | $c_{m1}$ | $c_{m2}$ | $\cdots$ | $c_{mn}$ |

**Figure 19.1.** The input data of the transportation problem.

$$\sum_{j=1}^{n} x_{ij} = a_i \quad (i = 1, 2, \ldots, m), \tag{19.2.1.20}$$

$$\sum_{i=1}^{m} x_{ij} = b_j \quad (j = 1, 2, \ldots, n), \tag{19.2.1.21}$$

$$x_{ij} > 0, \quad a_i \geq 0, \quad b_j \geq 0 \quad (i = 1, 2, \ldots, m; \; j = 1, 2, \ldots, n). \tag{19.2.1.22}$$

The objective function (19.2.1.19) is the total transportation cost to be minimized. Equations (19.2.1.20) mean that each supply source must ship all goods present at that source. Equations (19.2.1.21) mean that the demands of all consumers must be completely satisfied. Inequalities (19.2.1.22) are the conditions that all variables in the problem are nonnegative.

A necessary and sufficient condition for the solvability of (19.2.1.19)–(19.2.1.22) is the *balance condition*

$$\sum_{i=1}^{m} a_i = \sum_{j=1}^{n} b_j. \tag{19.2.1.23}$$

A transportation problem satisfying relation (19.2.1.23) is said to be *balanced*, and the corresponding model is said to be *closed*. If this relation does not hold, then the problem is said to be *unbalanced* and the corresponding model is said to be *open*.

Under the balance condition (19.2.1.23), the rank of the system of equations (19.2.1.20), (19.2.1.20) is equal to $n + m - 1$; therefore, $n + m - 1$ out of the $mn$ unknown variables must be basic variables. It follows that for any feasible basic flow the number of occupied cells in the transportation tableau shown in Fig. 19. 1 is equal to $n + m - 1$; these cells are usually said to be *basic* and the other cells are said to be *nonbasic*.

There are various methods for obtaining a feasible initial solution (a feasible shipment) in the transportation problem, including the cross-out method, the northwest corner rule, the minimal cost method, Vogel's approximation method, etc. Of these methods, the minimal cost method is simplest and most convenient.

*Minimal cost method.* The method consists of several steps of the same type. At each step, one fills only one cell in the tableau corresponding to the minimal cost $\min_{i,j}\{c_{ij}\}$ and excludes only one row (supply source) or column (consumer destination) from subsequent considerations. A supply source is excluded if its supply of goods is completely exhausted. A consumer is eliminated if his demand is completely satisfied. At each step, either a supply source or a consumer is excluded. If a supply source is yet to be excluded but its supply of goods is already zero, then, at the step where it should (but cannot) supply goods, a basic zero is written in the corresponding cell and only after this the supply source is excluded. Consumers are treated similarly.

*Reduction of an unbalanced transportation problem to a balanced transportation problem.*

1. If the total supply exceeds the total demand, i.e.,

$$\sum_{i=1}^{m} a_i > \sum_{j=1}^{n} b_j,$$

then one introduces fictitious consumer $n+1$ with demand

$$b_{n+1} = \sum_{i=1}^{m} a_i - \sum_{j=1}^{n} b_j$$

equal to the difference between the total supply and total demand and with zero unit transportation costs $c_{i(n+1)} = 0$ $(i = 1, 2, \ldots, m)$.

2. If the total demand exceeds the total supply, i.e.,

$$\sum_{i=1}^{m} a_i < \sum_{j=1}^{n} b_j,$$

then one introduces fictitious supply source $m+1$ with supply

$$a_{m+1} = \sum_{j=1}^{n} b_j - \sum_{i=1}^{m} a_i$$

equal to the difference between the total demand and the total supply and with zero unit transportation costs $c_{(m+1)j} = 0$ $(j = 1, 2, \ldots, n)$.

**Remark.** When constructing the initial basic solution, the supply of the fictitious source and the demands of the fictitious consumer are the last to be assigned, even though they correspond to minimum (zero) unit transportation costs.

### 19.2.1-6. Method of potentials.

A balanced transportation problem can be solved as a linear programming problem. But there are less cumbersome methods for solving the transportation problem. The most widely used method is the method of potentials.

If a feasible solution $X \equiv [x_{ij}]$ $(i = 1, 2, \ldots, m;\ j = 1, 2, \ldots, n)$ of the transportation problem is optimal, then there exist *supplier potentials* $u_i$ $(i = 1, 2, \ldots, m)$ and consumer potentials $v_j$ $(j = 1, 2, \ldots, n)$ satisfying the following conditions:

$$u_i + v_j = c_{ij} \quad \text{for basic cells,} \qquad (19.2.1.24)$$
$$u_i + v_j \leq c_{ij} \quad \text{for nonbasic cells.} \qquad (19.2.1.25)$$

Relations (19.2.1.24) are used as a system of equations for the potentials. This system has $n + m$ unknowns and $n + m - 1$ equations. Since the number of unknowns exceeds the number of equations by one, one of the variables can be taken arbitrarily, and the others are then found from the system.

Inequalities (19.2.1.25) are used to verify the optimality of a basic solution. To this end, the *reduced costs*

$$\Delta_{ij} = u_i + v_j - c_{ij} \qquad (19.2.1.26)$$

of nonbasic cells are used.

**Remark 1.** For basic cells, the quantities $\Delta_{ij}$ are zero, $\Delta_{ij} = 0$.

**Remark 2.** The variables $u_i$ ($i = 1, 2, \ldots, m$) are dual to the respective constraints in (19.2.1.20), and the variables $v_j$ ($j = 1, 2, \ldots, n$) are dual to the respective constraints in (19.2.1.21). The dual problem has the form

$$\sum_{i=1}^{m} a_i u_i + \sum_{j=1}^{n} b_j v_j \to \min,$$

$$u_i + v_j \leq c_{ij} \quad (i = 1, 2, \ldots, m, \ j = 1, 2, \ldots, n).$$

THEOREM (AN OPTIMALITY CRITERION). *An admissible solution is optimal if the reduced costs are nonpositive for all cells in the tableau.*

A *cycle* is a sequence $(i_1, j_1), (i_1, j_2), (i_2, j_2), \ldots, (i_k, j_1)$ of cells in the transportation problem tableau in which exactly two neighboring cells lie in the same row or column and, moreover, the first and last cells also lie in the same row or column.

If the current basic solution is not optimal, then one needs to pass to a new solution with smaller value of the objective function. To this end, in the tableau one takes the cell with the largest positive reduced cost

$$\max\{\Delta_{ij}\} = \Delta_{lk}.$$

Next, one constructs a cycle including this cell and some basic cells. The cells in the cycle are alternately marked by "+" and "−" signs starting from the cell with the largest positive reduced cost. For the "−" cells, one finds the quantity $\theta = \min\{x_{ij}\}$. Next, one performs a shift (redistribution of goods) over the cycle by $\theta$. The "−" cell at which $\min\{x_{ij}\}$ is attained becomes empty. If the minimum is attained at several cells, then one of them becomes empty and the other cells are filled with basic zeros, so that the number of occupied cells remains equal to $n + m - 1$.

**Example 5.** Let us solve the transportation problem with the following input data (see Fig. 19.2):

| Sources | Consumers | | | |
|---|---|---|---|---|
| | $B_1$ 200 | $B_2$ 200 | $B_3$ 300 | $B_4$ 400 |
| $A_1$ 200 | 4 | 3 | 2 | 1 |
| $A_2$ 300 | 2 | 3 | 5 | 6 |
| $A_3$ 500 | 6 | 7 | 9 | 12 |

**Figure 19.2.** Input data for Example 5.

The total demand $\sum_{j=1}^{4} b_j = 200+200+300+400 = 1100$ exceeds the total supply $\sum_{i=1}^{3} a_i = 200+300+500 = 1000$ by 100 units; hence the transportation problem is unbalanced. One should introduce the fictitious fourth supply source with supply $a_4 = 100$ and with zero unit transportation costs (see Fig. 19.3).

The initial basic solution $\mathbf{x}_1$ is found by the minimal cost method (Fig. 19.3). On this solution, the objective function takes the value

$$Z(\mathbf{x}_1) = 1 \times 200 + 2 \times 200 + 3 \times 100 + 7 \times 100 + 9 \times 300 + 12 \times 100 + 0 \times 100 = 5300.$$

## 19.2. MATHEMATICAL PROGRAMMING

| Sources | Consumers | | | |
|---|---|---|---|---|
| | $B_1$ 200 | $B_2$ 200 | $B_3$ 300 | $B_4$ 400 |
| $A_1$ 200 | 4 | 3 | 2 | 1 <br> 200 |
| $A_2$ 300 | 2 <br> 200 | 3 <br> 100 | 5 | 6 |
| $A_3$ 500 | 6 | 7 <br> 100 | 9 <br> 300 | 12 <br> 100 |
| $A_4$ 100 | 0 | 0 | 0 | 0 <br> 100 |

**Figure 19.3.** Introduction of the fictitious fourth supply source. The initial basis solution.

To verify whether the basic solution is optimal, we find the potentials. To this end, we use system (19.2.1.24), which acquires the form

$$u_1 + v_4 = 1,$$
$$u_2 + v_1 = 2,$$
$$u_2 + v_2 = 3,$$
$$u_3 + v_2 = 7,$$
$$u_3 + v_3 = 9,$$
$$u_3 + v_4 = 12,$$
$$u_4 + v_4 = 0.$$

Let $u_3 = 0$. Then the other potentials are determined uniquely: $v_2 = 7$, $v_3 = 9$, $v_4 = 12$, $u_1 = -11$, $u_4 = -12$, $u_2 = -4$, and $v_1 = 6$. The values of the potentials are written down in the tableau for the solution of the transportation problem (see Fig. 19.4).

| Sources | Consumers | | | | $u_i$ |
|---|---|---|---|---|---|
| | $B_1$ 200 | $B_2$ 200 | $B_3$ 300 | $B_4$ 400 | |
| $A_1$ 200 | 4 | 3 | 2 | 1 <br> 200 | $-11$ |
| $A_2$ 300 | 2 <br> 200 | 3 <br> $-$ 100 | 5 | 6 <br> $+$ | $-4$ |
| $A_3$ 500 | 6 | 7 <br> $+$ 100 | 9 <br> 300 | 12 <br> 100 $-$ | 0 |
| $A_4$ 100 | 0 | 0 | 0 | 0 <br> 100 | $-12$ |
| $v_j$ | 6 | 7 | 9 | 12 | |

**Figure 19.4.** Determination of potentials. Construction of the cycle.

Next, we verify whether the solution $\mathbf{x}_1$ is optimal. To this end, using (19.2.1.26), we find the reduced costs $\Delta_{ij}$ for all empty (nonbasic) cells of the tableau. The solution $\mathbf{x}_1$ is not optimal since there is a positive reduced cost $\Delta_{24} = u_2 + v_4 - c_{24} = -4 + 12 - 6 = 2$.

For cell $(2, 4)$ with positive reduced cost, we construct the cycle $(2, 4), (3, 4), (3, 2), (2, 2), (2, 4)$. The cycle is shown in Fig. 19.4. At the corners of the cycle, we alternately write the "+" and "−" signs, starting from cell $(2, 4)$. For the cells with the "−" sign, we have $\theta = \min\{100, 100\}$. Then we perform a shift (redistribution of

| Sources | Consumers | | | | $u_i$ |
|---|---|---|---|---|---|
| | $B_1$ 200 | $B_2$ 200 | $B_3$ 300 | $B_4$ 400 | |
| $A_1$ 200 | 4 | 3 | 2 | 1 <br> 200 | $-5$ |
| $A_2$ 300 | 2 <br> 200 | 3 | 5 | 6 <br> 100 | 0 |
| $A_3$ 500 | 6 | 7 <br> 200 | 9 <br> 300 | 12 | 4 |
| $A_4$ 100 | 0 | 0 | 0 | 0 <br> 100 | $-6$ |
| $v_j$ | 2 | 3 | 5 | 6 | |

**Figure 19.5.** The second basis solution.

goods) over the cycle by $\theta = 100$ and obtain the second basic solution $\mathbf{x}_2$ (see Fig. 19.5). Once the system of potentials has been found (Fig. 19.5), we see that the solution $\mathbf{x}_2$ is optimal. The objective function takes the value
$$Z(\mathbf{x}_2) = 1 \times 200 + 2 \times 200 + 6 \times 100 + 7 \times 200 + 9 \times 300 + 0 \times 100 = 5200$$
on this solution. Thus, $Z_{\min}(X) = 5200$ for
$$X^* = \begin{pmatrix} 0 & 0 & 0 & 200 \\ 200 & 0 & 0 & 100 \\ 0 & 200 & 300 & 0 \end{pmatrix}.$$

### 19.2.1-7. Game theory.

Mathematical models of conflict situations are called *games*, and their participants are called *players*. Mathematical models of conflict situations and various methods for solving problems that arise in these situations are constructed in *game theory*.

According to the number of players, the games are divided into *two-person* and *n-person games*. In $n$-person games, the players' interests may coincide. In this case, they can cooperate and form coalitions. Such games are called *coalition games*.

A player's *strategy* is a set of rules uniquely determining the player's behavior in each specific situation arising in the game. A strategy ensuring the maximum possible mean payoff for a player in repeated games is said to be *optimal*. The number of possible strategies of each player can be either finite or infinite. Depending on this, the games are divided into *finite* and *infinite games*.

Games in which one of the players is indifferent to the results are usually called "*games with nature*."

A two-person game in which the payoff of one player is equal to the loss of the other player is called an *antagonistic two-person zero-sum game*. Suppose that two players $A$ and $B$ have finitely many pure strategies: player $A$ can choose any of $m$ strategies $A_1, \ldots, A_m$, and player $B$ can choose any of $n$ strategies $B_1, \ldots, B_n$. These strategies determine the *payoff matrix*

$$\begin{pmatrix} a_{11} & a_{12} & \cdots & a_{1n} \\ a_{21} & a_{22} & \cdots & a_{2n} \\ \vdots & \vdots & \ddots & \vdots \\ a_{m1} & a_{m2} & \cdots & a_{mn} \end{pmatrix}, \qquad (19.2.1.27)$$

where $a_{ij}$ is the payoff (positive or negative) of player $A$ against player $B$ if player $A$ uses the pure strategy $A_i$ and player $B$ used the pure strategy $B_j$.

**Remark.** The sum of payoffs of both players is zero for each move. (That is why the game is called a *zero-sum game*.)

Let $\alpha_i = \min_j\{a_{ij}\}$ be the minimum possible payoff of player $A$ if he uses the pure strategy $A_i$. If player $A$ acts reasonably, he must choose a strategy $A_i$ for which $\alpha_i$ is maximal,

$$\alpha = \max_i\{\alpha_i\} = \max_i \min_j\{a_{ij}\}. \tag{19.2.1.28}$$

The number $\alpha$ is called the *lower price of the game*. Let $\beta_j = \max_i\{a_{ij}\}$ be the maximum possible loss of player $B$ if he uses the pure strategy $B_j$. If player $B$ acts reasonably, he must choose a strategy $B_j$ for which $\beta_j$ is minimal,

$$\beta = \min_j \beta_j = \min_j \max_i\{a_{ij}\}. \tag{19.2.1.29}$$

The number $\beta$ is called the *upper price of the game*.

**Remark.** The principle for constructing the strategies of player $A$ (the first player) based on the maximization of minimal payoffs is called the *maximin principle*. The principle for constructing the strategies of player $B$ (the second player) based on the minimization of maximal losses is called the *minimax principle*.

The lower price of the game is the guaranteed minimal payoff of player $A$ if he follows the maximin principle. The upper price of the game is the guaranteed maximal loss of player $B$ if he follows the minimax principle.

THEOREM. *In a two-person zero-sum game, the lower price $\alpha$ and the upper price $\beta$ satisfy the inequality*

$$\alpha \leq \beta. \tag{19.2.1.30}$$

If $\alpha = \beta$, then such a game is called the *game with a saddle point*, and a pair $(A_{i,\text{opt}}, B_{j,\text{opt}})$ of optimal strategies is called a *saddle point of the payoff matrix*. The entry $v = a_{ij}$ corresponding to a saddle point $(A_{i,\text{opt}}, B_{j,\text{opt}})$ is called the *game value*. If a game has a saddle point, then one says that the game can be solved in pure strategies.

**Remark.** There can be several saddle points, but they all have the same value.

If the payoff matrix has no saddle points, i.e., the strict inequality $\alpha < \beta$ holds, then the search of a solution of the game leads to a complex strategy in which a player randomly uses two or more strategies with certain frequencies. Such complex strategies are said to be *mixed*.

The strategies of player $A$ are determined by the set $\mathbf{x} = (x_1, \ldots, x_m)$ of probabilities that the player uses the respective pure strategies $A_1, \ldots, A_m$. For player $B$, the strategies are determined by the set $\mathbf{y} = (y_1, \ldots, y_n)$ of probabilities that the player uses the respective pure strategies $B_1, \ldots, B_n$. These sets of probabilities must satisfy the identity

$$\sum_{i=1}^{m} x_i = \sum_{j=1}^{n} y_j = 1.$$

The expectation of the payoff of player $A$ is given by the function

$$H(\mathbf{x}, \mathbf{y}) = \sum_{i=1}^{m} \sum_{j=1}^{n} a_{ij} x_i y_j. \tag{19.2.1.31}$$

THE VON NEUMANN MINIMAX THEOREM. *There exist optimal mixed strategies* $\mathbf{x}^*$ *and* $\mathbf{y}^*$, *i.e., strategies such that*

$$H(\mathbf{x}, \mathbf{y}^*) \leq H(\mathbf{x}^*, \mathbf{y}^*) \leq H(\mathbf{x}^*, \mathbf{y}) \qquad (19.2.1.32)$$

for any probabilities $\mathbf{x}$ and $\mathbf{y}$.

The number $v = H(\mathbf{x}^*, \mathbf{y}^*)$ is called the *game price in mixed strategies*.

MINIMAX THEOREM FOR ANTAGONISTIC TWO-PERSON ZERO-SUM GAMES. *For any payoff matrix (19.2.1.27)*,

$$v = \max_{x_1,\ldots,x_m} \left( \min_{y_1,\ldots,y_n} \sum_{i=1}^{m} \sum_{j=1}^{n} a_{ij} x_i y_j \right) = \min_{y_1,\ldots,y_n} \left( \max_{x_1,\ldots,x_m} \sum_{i=1}^{m} \sum_{j=1}^{n} a_{ij} x_i y_j \right). \qquad (19.2.1.33)$$

### 19.2.1-8. Relationship between game theory and linear programming.

Without loss of generality, we can assume that $v > 0$. This can be ensured if we add the same positive constant $a > 0$ to all entries $a_{ij}$ of the payoff matrix (19.2.1.27); in this case, only the game price varies (increases by $a > 0$), while the optimal solution remains the same.

An antagonistic two-person zero-sum game can be reduced to a linear programming problem by the change of variables

$$\begin{aligned} v &= \frac{1}{Z_{\min}} = \frac{1}{W_{\max}}, \\ x_i &= v X_i \quad (i = 1, 2, \ldots, m); \\ y_j &= v Y_j \quad (j = 1, 2, \ldots, n). \end{aligned} \qquad (19.2.1.34)$$

The quantities $Z_{\min}$, $W_{\max}$, $X_i$, and $Y_j$ form a solution of the following pair of dual problems:

$$\begin{aligned} Z &= X_1 + X_2 + \cdots + X_m \to \min, \\ a_{11} X_1 &+ a_{21} X_2 + \cdots + a_{m1} X_m \geq 1, \\ a_{12} X_1 &+ a_{22} X_2 + \cdots + a_{m2} X_m \geq 1, \\ &\qquad \cdots\cdots\cdots\cdots\cdots\cdots\cdots\cdots\cdots \\ a_{1n} X_1 &+ a_{2n} X_2 + \cdots + a_{mn} X_m \geq 1, \\ X_i &\geq 0 \quad (i = 1, 2, \ldots, m); \end{aligned} \qquad (19.2.1.35)$$

$$\begin{aligned} W &= Y_1 + Y_2 + \cdots + Y_n \to \max, \\ a_{11} Y_1 &+ a_{12} Y_2 + \cdots + a_{1n} Y_n \leq 1, \\ a_{21} Y_1 &+ a_{22} Y_2 + \cdots + a_{2n} Y_n \leq 1, \\ &\qquad \cdots\cdots\cdots\cdots\cdots\cdots\cdots\cdots\cdots \\ a_{m1} Y_1 &+ a_{m2} Y_2 + \cdots + a_{mn} Y_n \leq 1, \\ Y_j &\geq 0 \quad (j = 1, 2, \ldots, n). \end{aligned} \qquad (19.2.1.36)$$

## 19.2.2. Nonlinear Programming

**19.2.2-1. General statement of nonlinear programming problem.**

The *nonlinear programming problem* is the problem of finding $n$ variables $\mathbf{x} = (x_1, \ldots, x_n)$ that provide an extremum of the objective function

$$Z(\mathbf{x}) = f(\mathbf{x}) \to \text{extremum} \tag{19.2.2.1}$$

and satisfy the system of constraints

$$\begin{aligned}
\varphi_i(\mathbf{x}) &= 0 \quad \text{for} \quad i = 1, 2, \ldots, k, \\
\varphi_i(\mathbf{x}) &\leq 0 \quad \text{for} \quad i = k+1, k+2, \ldots, l, \\
\varphi_i(\mathbf{x}) &\geq 0 \quad \text{for} \quad i = l+1, l+2, \ldots, m.
\end{aligned} \tag{19.2.2.2}$$

Here the objective function (19.2.2.1) and/or at least one of the functions $\varphi_i(\mathbf{x})$ ($i = 1, 2, \ldots, m$) is nonlinear.

Depending on the properties of the functions $f(\mathbf{x})$ and $\varphi_i(\mathbf{x})$, the following types of problems are distinguished:
1. *Convex programming.*
2. *Quadratic programming.*
3. *Geometric programming.*

A necessary condition for the maximum of the function

$$Z(\mathbf{x}) = f(\mathbf{x}) \tag{19.2.2.3}$$

under the inequality constraints

$$\varphi_i(\mathbf{x}) \leq 0 \quad (i = 1, 2, \ldots, m)$$

is that there exist $m+1$ nonnegative Lagrange multipliers $\lambda_0, \lambda_1, \ldots, \lambda_m$ that are not simultaneously zero and satisfy the conditions

$$\begin{aligned}
\lambda_i &\geq 0 \quad (i = 0, 1, \ldots, m), \\
\lambda_i \varphi_i(\mathbf{x}) &= 0 \quad (i = 1, 2, \ldots, m), \\
\lambda_0 f_{x_j} &+ \sum_{i=1}^{m} \lambda_i (\varphi_i)_{x_j} = 0,
\end{aligned} \tag{19.2.2.4}$$

where derivatives $f_{x_j}$ and $(\varphi_i)_{x_j}$ are evaluated at $\mathbf{x}$.

One of the most widely used methods of nonlinear programming is the *penalty function method*. This method approximates a problem with constraints by a problem without constraints and with objective function that penalizes infeasibility. The higher the penalties, the closer the problem of maximizing the penalty function is to the original problem.

**19.2.2-2. Dynamic programming.**

*Dynamic programming* is the branch of mathematical programming dealing with multistage optimal decision-making problems.

The general outline of a multistage optimal decision-making process is as follows. Consider a controlled system $S$ taken by the control from an initial state $s_0$ to a state $\widetilde{s}$. Let

$x_k$ ($k = 1, 2, \ldots, n$) be the control at the $k$th stage, let $\mathbf{x} = (x_1, \ldots, x_n)$ be the control taking the system $S$ from the state $s_0$ to the state $\widetilde{s}$, and let $s_k$ be the state of the system after the $k$th control step. The efficiency of the control is characterized by an objective function that depends on the initial state and the control,

$$Z = F(s_0, \mathbf{x}). \tag{19.2.2.5}$$

We assume that
1. The state $s_k$ depends only on the preceding state $s_{k-1}$ and the control $x_k$ at the $k$th step,

$$s_k = \varphi_k(s_{k-1}, x_k) \quad (k = 1, 2, \ldots, n). \tag{19.2.2.6}$$

2. The objective function (19.2.2.5) is an additive function of the performance factor at each step.
If the performance factor at the $k$th step is

$$Z_k = f_k(s_{k-1}, x_k) \quad (k = 1, 2, \ldots, n), \tag{19.2.2.7}$$

then the objective function (19.2.2.5) can be written as

$$Z = \sum_{k=1}^{n} f_k(s_{k-1}, x_k). \tag{19.2.2.8}$$

*The dynamic programming problem.* Find an admissible control $\mathbf{x}$ taking the system $S$ from the state $s_0$ to the state $\widetilde{s}$ and maximizing (or minimizing) the objective function (19.2.2.8).

THEOREM (BELLMAN'S OPTIMALITY PRINCIPLE). *For any state $s$ of the system after any number of steps, one should choose the control at the current step so as to ensure that this control, together with the optimal control at all subsequent steps, leads to the optimal payoff at all remaining steps, including the current step.*

Let $Z_k^*(s_{k-1})$ be the conditional maximum of the objective function obtained under the optimal control at $n-k-1$ steps starting from the $k$th step until the end under the assumption that the system was in the state $s_{k-1}$ by the beginning of the $k$th step. The equations

$$Z_n^*(s_{n-1}) = \max_{x_n}\{f_n(s_{n-1}, x_n)\},$$

$$Z_k^*(s_{k-1}) = \max_{x_k}\{f_k(s_{k-1}, x_k) + Z_{k+1}^*(s_k)\} \quad (k = n-1, n-2, \ldots, 1)$$

are called the *Bellman equations*. The Bellman equations for the dynamic programming problem and for any $n$ and $s_0$ permit finding a solution, which is given by

$$Z_{\max} = Z_1^*(s_0).$$

## References for Chapter 19

Akhiezer, N. I., *Calculus of Variations*, Taylor & Francis, London, New York, 1988.
Avriel, M., *Nonlinear Programming: Analysis and Methods*, Dover Publications, New York, 2003.
Bazaraa, M. S., Sherali, H. D., and Shetty, C. M., *Nonlinear Programming: Theory and Algorithms*, Wiley, New York, 2006.
Belegundu, A. D. and Chandrupatla, T. R., *Optimization Concepts and Applications in Engineering*, Bk&CD ROM Edition, Prentice Hall, Englewood Cliffs, New Jersey, 1999.

Bellman, R., *Adaptive Control Processes: A Guided Tour*, Princeton University Press, Princeton, New Jersey, 1961.

Bellman, R., *Dynamic Programming, Dover Edition*, Dover Publications, New York, 2003.

Bellman, R. and Dreyfus, S. E., *Applied Dynamic Programming*, Princeton University Press, Princeton, New Jersey, 1962.

Bertsekas, D. P., *Nonlinear Programming, 2nd Edition*, Athena Scientific, Belmont, Massachusetts, 1999.

Bertsimas, D. and Tsitsiklis, J. N., *Introduction to Linear Optimization* (Athena Scientific Series in Optimization and Neural Computation, Vol. 6), Athena Scientific, Belmont, Massachusetts, 1997.

Bolza, O., *Lectures on the Calculus of Variations, 3rd Edition*, American Mathematical Society, Providence, Rhode Island, 2000.

Bonnans, J. F., Gilbert, J. C., Lemarechal, C., and Sagastizabal, C. A., *Numerical Optimization*, Springer, New York, 2003.

Boyd, S. and Vandenberghe, L., *Convex Optimization*, Cambridge University Press, Cambridge, 2004.

Brechteken-Mandersch, U., *Introduction to the Calculus of Variations*, Chapman & Hall/CRC Press, Boca Raton, 1991.

Brinkhuis, J. and Tikhomirov, V., *Optimization: Insights and Applications*, Princeton University Press, Princeton, New Jersey, 2005.

Bronshtein, I. N., Semendyayev, K. A., Musiol, G., and Mühlig, H., *Handbook of Mathematics, 4th Edition*, Springer, New York, 2004.

Bronson, R. and Naadimuthu, G., *Schaum's Outline of Operations Research, 2nd Edition*, McGraw-Hill, New York, 1997.

van Brunt, B., *The Calculus of Variations*, Springer, New York, 2003.

Calvert, J. E., *Linear Programming*, Brooks Cole, Stamford, 1989.

Chong, E. K. P. and Žak, S. H., *An Introduction to Optimization, 2nd Edition*, Wiley, New York, 2001.

Chvatal, V., *Linear Programming* (Series of Books in the Mathematical Sciences), W. H. Freeman, New York, 1983.

Cooper, L., *Applied Nonlinear Programming for Engineers and Scientists*, Aloray, Goshen, 1974.

Dacorogna, B., *Introduction to the Calculus of Variations*, Imperial College Press, London, 2004.

Darst, R. B., *Introduction to Linear Programming* (Pure and Applied Mathematics (Marcel Dekker)), CRC Press, Boca Raton, 1990.

Denardo, E. V., *Dynamic Programming: Models and Applications*, Dover Publications, New York, 2003.

Dreyfus, S. E., *Dynamic Programming and the Calculus of Variations*, Academic Press, New York, 1965.

Elsgolts, L., *Differential Equations and the Calculus of Variations*, University Press of the Pacific, Honolulu, Hawaii, 2003.

Ewing, G. M., *Calculus of Variations with Applications* (Mathematics Series), Dover Publications, New York, 1985.

Fletcher, R., *Practical Methods of Optimization, 2nd Edition*, Wiley, New York, 2000.

Fomin, S. V. and Gelfand, I. M., *Calculus of Variations*, Dover Publications, New York, 2000.

Fox, C., *An Introduction to the Calculus of Variations*, Dover Publications, New York, 1987.

Galeev, E. M. and Tikhomirov, V. M., *Optimization: Theory, Examples, and Problems* [in Russian], Editorial URSS, Moscow, 2000.

Gass, S. I., *An Illustrated Guide to Linear Programming, Rep. Edition*, Dover Publications, New York, 1990.

Gass, S. I., *Linear Programming: Methods and Applications, 5th Edition*, Dover Publications, New York, 2003.

Gill, Ph. E., Murray, W., and Wright, M. H., *Practical Optimization, Rep. Edition*, Academic Press, New York, 1982.

Giusti, E., *Direct Methods in the Calculus of Variations*, World Scientific Publishing Co., Hackensack, New Jersey, 2003.

Glicksman, A. M., *Introduction to Linear Programming and the Theory of Games*, Dover Publications, New York, 2001.

Hillier, F. S. and Lieberman, G. J., *Introduction to Operations Research*, McGraw-Hill, New York, 2002.

Horst, R. and Pardalos, P. M. (Editors), *Handbook of Global Optimization*, Kluwer Academic, Dordrecht, 1995.

Jensen, P. A. and Bard, J. F., *Operations Research Models and Methods, Bk&CD-Rom Edition*, Wiley, New York, 2002.

Jost, J. and Li-Jost, X., *Calculus of Variations* (Cambridge Studies in Advanced Mathematics), Cambridge University Press, Cambridge, 1999.

Kolman, B. and Beck, R. E., *Elementary Linear Programming with Applications, 2nd Edition* (Computer Science and Scientific Computing), Academic Press, New York, 1995.

Korn, G. A. and Korn, T. M., *Mathematical Handbook for Scientists and Engineers: Definitions, Theorems, and Formulas for Reference and Review, 2nd Rev. Edition*, Dover Publications, New York, 2000.

**Krasnov, M. L., G. K., Makarenko, G. K.,, and Kiselev, A. I.,** *Problems and Exercises in the Calculus of Variations,* Imported Publications, Inc., New York, 1985.

**Lebedev, L. P. and Cloud, M. J.,** *The Calculus of Variations and Functional Analysis with Optimal Control and Applications in Mechanics* (Series on Stability, Vibration and Control of Systems, Series A, Vol. 12), World Scientific Publishing Co., Hackensack, New Jersey, 2003.

**Liberti, L. and Maculan, N. (Editors),** *Global Optimization: From Theory to Implementation (Nonconvex Optimization and Its Applications),* Springer, New York, 2006.

**Luenberger, D. G.,** *Linear and Nonlinear Programming, 2nd Edition,* Springer, New York, 2003.

**MacCluer, C. R.,** *Calculus of Variations: Mechanics, Control, and Other Applications,* Prentice Hall, Englewood Cliffs, New Jersey, 2004.

**Mangasarian, O. L.,** *Nonlinear Programming* (Classics in Applied Mathematics, Vol. 10), Society for Industrial & Applied Mathematics, University City Science Center, Philadelphia, 1994.

**Marlow, W. H.,** *Mathematics for Operations Research,* Dover Publications, New York, 1993.

**Morse, Ph. M. and Kimball, G. E.,** *Methods of Operations Research,* Dover Publications, New York, 2003.

**Moser, J.,** *Selected Chapters in the Calculus of Variations: Lecture Notes by Oliver Knill* (Lectures in Mathematics. ETH Zurich), Birkhäuser Verlag, Basel, Stuttgart, 2003.

**Murty, K. G.,** *Linear Programming, Rev. Edition,* Wiley, New York, 1983.

**Nash, S. G. and Sofer, A.,** *Linear and Nonlinear Programming,* McGraw-Hill, New York, 1995.

**Nocedal, J. and Wright, S.,** *Numerical Optimization,* Springer, New York, 2000.

**Padberg, M.,** *Linear Optimization and Extensions (Algorithms and Combinatorics), 2nd Edition,* Springer, New York, 1999.

**Pannell, D. J.,** *Introduction to Practical Linear Programming, Bk&Disk Edition,* Wiley, New York, 1996.

**Pardalos, P. M. and Resende, M. G. C. (Editors),** *Handbook of Applied Optimization,* Oxford University Press, Oxford, 2002.

**Pardalos, P. M. and Romeijn, H. E. (Editors),** *Handbook of Global Optimization, Vol. 2 (Nonconvex Optimization and Its Applications),* Springer, New York, 2002.

**Pierre, D. A.,** *Optimization Theory with Applications,* Dover Publications, New York, 1987.

**Rao, S. S.,** *Engineering Optimization: Theory and Practice, 3rd Edition,* Wiley, New York, 1996.

**Rardin, R. L.,** *Optimization in Operations Research,* Prentice Hall, Englewood Cliffs, New Jersey, 1997.

**Ross, S. M.,** *Applied Probability Models with Optimization Applications, Rep. Edition* (Dover Books on Mathematics), Dover Publications, New York, 1992.

**Ruszczynski, A.,** *Nonlinear Optimization,* Princeton University Press, Princeton, New Jersey, 2006.

**Sagan, H.,** *Introduction to the Calculus of Variations, Rep. Edition,* Dover Publications, New York, 1992.

**Shenoy, G. V.,** *Linear Programming: Methods and Applications,* Halsted Press, New York, 1989.

**Simon, C. P. and Blume, L.,** *Mathematics for Economists,* W. W. Norton & Company, New York, 1994.

**Smith, D. R.,** *Variational Methods in Optimization* (Dover Books on Mathematics), Dover Publications, New York, 1998.

**Strayer, J. K.,** *Linear Programming and Its Applications* (Undergraduate Texts in Mathematics), Springer-Verlag, Berlin, 1989.

**Sundaram, R. K.,** *A First Course in Optimization Theory,* Cambridge University Press, Cambridge, 1996.

**Taha, H. A.,** *Operations Research: An Introduction, 7th Edition,* Prentice Hall, Englewood Cliffs, New Jersey, 2002.

**Tslaf, L. Ya.,** *Calculus of Variations and Integral Equations, 3rd Edition* [in Russian], Lan, Moscow, 2005.

**Tuckey, C.,** *Nonstandard Methods in the Calculus of Variations,* Chapman & Hall/CRC Press, Boca Raton, 1993.

**Vasilyev, F. P. and Ivanitskiy, A. Y.,** *In-Depth Analysis of Linear Programming,* Springer, New York, 2001.

**Venkataraman, P.,** *Applied Optimization with MATLAB Programming,* Wiley, New York, 2001.

**Wan, F.,** *Introduction to the Calculus of Variations and Its Applications, 2nd Edition,* Chapman & Hall/CRC Press, Boca Raton, 1995.

**Weinstock, R.,** *Calculus of Variations,* Dover Publications, New York, 1974.

**Winston, W. L.,** *Operations Research: Applications and Algorithms (with CD-ROM and InfoTrac), 4th Edition,* Duxbury Press, Boston, 2003.

**Young, L. C.,** *Lecture on the Calculus of Variations and Optimal Control Theory,* American Mathematical Society, Providence, Rhode Island, 2000.

# Chapter 20
# Probability Theory

## 20.1. Simplest Probabilistic Models
### 20.1.1. Probabilities of Random Events

**20.1.1-1. Random events. Basic definitions.**

The simplest indivisible mutually exclusive outcomes of an experiment are called *elementary events* $\omega$. The set of all elementary outcomes, which we denote by the symbol $\Omega$, is called the *space of elementary events* or the *sample space*. Any subset of $\Omega$ is called a *random event* $A$ (or simply an *event* $A$). Elementary events that belong to $A$ are said to *favor* $A$. In any probabilistic model, a certain condition set $\Sigma$ is assumed to be fixed.

An event $A$ *implies* an event $B$ ($A \subseteq B$) if $B$ occurs in each realization of $\Sigma$ for which $A$ occurs. Events $A$ and $B$ are said to be *equivalent* ($A = B$) if $A$ implies $B$ and $B$ implies $A$, i.e., if, for each realization of $\Sigma$, both events $A$ and $B$ occur or do not occur simultaneously.

The *intersection* $C = A \cap B = AB$ of events $A$ and $B$ is the event that both $A$ and $B$ occur. The elementary outcomes of the intersection $AB$ are the elementary outcomes that simultaneously belong to $A$ and $B$.

The *union* $C = A \cup B = A + B$ of events $A$ and $B$ is the event that at least one of the events $A$ or $B$ occurs. The elementary outcomes of the union $A + B$ are the elementary outcomes that belong to at least one of the events $A$ and $B$.

The *difference* $C = A \setminus B = A - B$ of events $A$ and $B$ is the event that $A$ occurs and $B$ does not occur. The elementary outcomes of the difference $A \setminus B$ are the elementary outcomes of $A$ that do not belong to $B$.

The event that $A$ does not occur is called the *complement* of $A$, or the *complementary event*, and is denoted by $\overline{A}$. The elementary outcomes of $\overline{A}$ are the elementary outcomes that do not belong to the event $A$.

An event is said to be *sure* if it necessarily occurs for each realization of the condition set $\Sigma$. Obviously, the sure event is equivalent to the space of elementary events, and hence the sure event should be denoted by the symbol $\Omega$.

An event is said to be *impossible* if it cannot occur for any realization of the condition set $\Sigma$. Obviously, the impossible event does not contain any elementary outcome and hence should be denoted by the symbol $\varnothing$.

Two events $A$ and $\overline{A}$ are said to be *opposite* if they simultaneously satisfy the following two conditions:

$$A \cup \overline{A} = \Omega, \quad A \cap \overline{A} = \varnothing.$$

Events $A$ and $B$ are said to be *incompatible*, or *mutually exclusive*, if their simultaneous realization is impossible, i.e., if $A \cap B = \varnothing$.

Events $H_1, \ldots, H_n$ are said to form a *complete group of events*, or to be *collectively exhaustive*, if at least one of them necessarily occurs for each realization of the condition set $\Sigma$, i.e., if

$$H_1 \cup \cdots \cup H_n = \Omega.$$

Events $H_1, \ldots, H_n$ form a *complete group of pairwise incompatible events* (are *hypotheses*) if exactly one of the events necessarily occurs for each realization of the condition set $\Sigma$, i.e., if
$$H_1 \cup \cdots \cup H_n = \Omega \quad \text{and} \quad H_i \cap H_j = \varnothing \quad (i \neq j).$$

Main properties of random events:
1. $A \cup B = B \cup A$ and $A \cap B = B \cap A$ (commutativity).
2. $(A \cup B) \cap C = (A \cap C) \cup (B \cap C)$ and $(A \cap B) \cup C = (A \cup C) \cap (B \cup C)$ (distributivity).
3. $(A \cup B) \cup C = A \cup (B \cup C)$ and $(A \cap B) \cap C = A \cap (B \cap C)$ (associativity).
4. $A \cup A = A$ and $A \cap A = A$.
5. $A \cup \Omega = \Omega$ and $A \cap \Omega = A$.
6. $A \cup \overline{A} = \Omega$ and $A \cap \overline{A} = \varnothing$.
7. $\overline{\varnothing} = \Omega, \overline{\Omega} = \varnothing$, and $\overline{\overline{A}} = A$;
8. $A \backslash B = A \cap \overline{B}$.
9. $\overline{A \cup B} = \overline{A} \cap \overline{B}$ and $\overline{A \cap B} = \overline{A} \cup \overline{B}$ (*de Morgan's laws*).

### 20.1.1-2. Axiomatic definition of probability.

In the case of an uncountable sample space $\Omega$, not all subsets of $\Omega$ but only subsets belonging to certain classes, called algebras of sets and $\sigma$-algebras, are viewed as events.

A class $\mathcal{F}$ of subsets of the space $\Omega$ is called an *algebra of sets (events)* if $\varnothing \in \mathcal{F}, \Omega \in \mathcal{F}$, and the following conditions are satisfied:
1. If $A \in \mathcal{F}$, then $\overline{A} \in \mathcal{F}$.
2. If $A \in \mathcal{F}$ and $B \in \mathcal{F}$, then $A \cup B \in \mathcal{F}$ and $A \cap B \in \mathcal{F}$.

The simplest example of an algebra of events is the system $\mathcal{F} = \{\varnothing, \Omega\}$. Indeed, applying any of the operations listed above to any elements of the class $\mathcal{F}$, we obtain an element of this class: $\varnothing \cup \Omega = \Omega, \varnothing \cap \Omega = \varnothing, \overline{\varnothing} = \Omega$, and $\varnothing = \overline{\Omega}$.

An algebra of sets $\mathcal{F}$ of subsets of the space $\Omega$ is called a *$\sigma$-algebra* if the following condition is satisfied: if $A_n \in \mathcal{F}, n = 1, 2, \ldots$, then $\bigcup_n A_n \in \mathcal{F}$ and $\bigcap_n A_n \in \mathcal{F}$.

The elements of $\mathcal{F}$ are called *random events*.

The *probability of an event* is defined to be a single-valued real function $P(A)$ defined on the $\sigma$-algebra of events $\mathcal{F}$ and satisfying the following three axioms:
1. *Nonnegativity*: $P(A) \geq 0$ for any $A \in \mathcal{F}$.
2. *Normalization*: $P(\Omega) = 1$.
3. *Additivity*: $P\left(\bigcup_n A_n\right) = \sum_n P(A_n)$, provided that $A_i \cap A_j = \varnothing$ whenever $i \neq j$.

A *probability space* is a triple $(\Omega, \mathcal{F}, P)$, where $\Omega = \Omega(\omega)$ is a space of elementary events, $\mathcal{F}$ is a $\sigma$-algebra of subsets of $\Omega$, called random events, and $P(A)$ is a probability defined on the $\sigma$-algebra $\mathcal{F}$.

Properties of probability:
1. The probability of an impossible event is zero; i.e., $P(\varnothing) = 0$.
2. The probability of the event $\overline{A}$ opposite to an event $A$ is equal to $P(\overline{A}) = 1 - P(A)$.
3. Probability is a bounded function; i.e., $0 \leq P(A) \leq 1$.
4. If an event $A$ implies an event $B$ ($A \subseteq B$), then $P(A) \leq P(B)$.
5. If events $H_1, \ldots, H_n$ form a complete group of pairwise incompatible events, then
$$\sum_{i=1}^n P(H_i) = 1.$$

### 20.1.1-3. Discrete probability space. Classical definition of probability.

Suppose that $\Omega = \{\omega_1, \ldots, \omega_n\}$ is a finite space. The $\sigma$-algebra $\mathcal{F}$ of events includes all $2^n$ subsets of $\Omega$. To each elementary event $\omega_i \in \Omega$ ($i = 1, 2, \ldots, n$) there corresponds a number $p(\omega_i)$, called the probability of the elementary event $\omega_i$. Thus a real function satisfying the following two conditions is defined on the set $\Omega$:

1. *Nonnegativity condition*: $p(\omega_i) \geq 0$ for any $\omega_i \in \Omega$.
2. *Normalization condition*: $\sum_{i=1}^{n} p(\omega_i) = 1$.

The *probability* $P(A)$ of an event $A$ for any subset $A \subset \Omega$ is defined to be the sum of probabilities of the elementary events that form $A$; i.e.,

$$P(A) = \sum_{\omega_i \in A} p(\omega_i). \tag{20.1.1.1}$$

The triple $(\Omega, \mathcal{F}, P)$ thus defined is a finite *discrete probability space*.

A special case of the definition of probability (20.1.1.1) is the *classical definition of probability*, in which all elementary events are equiprobable: $p(\omega_1) = \cdots = p(\omega_n) = 1/n$. Then the probability $P(A)$ of an event $A = \{\omega_{i_1}, \ldots, \omega_{i_m}\}$ is equal to the ratio of the number of elementary events $\omega_i$ contained in $A$ (the number of outcomes that favor $A$) to the total number of elementary events in $\Omega$ (the number of all possible outcomes),

$$P(A) = \frac{|A|}{|\Omega|} = \frac{m}{n}. \tag{20.1.1.2}$$

**Example 1.** Let two dice be thrown. Under the assumption that the elementary events are equiprobable, find the probability of the event $A$ that the sum of numbers shown is greater than 10. Obviously, the sample space can be represented as $\Omega = \{(i, j) : i, j = 1, 2, 3, 4, 5, 6\}$, where $i$ is the number shown by the first die and $j$ is the number shown by the second die. The total number of elementary events is $|\Omega| = 36$. The event $A$ corresponds to the subset $A = \{(5, 6), (6, 5), (6, 6)\}$ of $\Omega$. Since $|A| = 3$, formula (20.1.1.2) gives $P(A) = |A|/|\Omega| = 3/36 = 1/12$.

### 20.1.1-4. Sampling without replacement.

Let $\Delta = \{1, \ldots, n\}$ be the set of $n$ numbers, and let $\omega = (i_1, \ldots, i_m)$ be an ordered sequence of $m$ elements of $\Delta$. *Random sampling without replacement* is the sampling scheme in which

$$\Omega = \{\omega = (i_1, \ldots, i_m) : i_k \in \Delta, \ k = 1, \ldots, m, \text{ and all } i_1, \ldots, i_m \text{ are distinct}\} \tag{20.1.1.3}$$

and all elementary events $\omega$ are equiprobable.

When calculating probabilities by formula (20.1.1.2), the following combinatorial formulas are often useful. Suppose that a set $\Delta = \{a_1, \ldots, a_n\}$ of $n$ elements is given. Subsets of $\Delta$ are called *combinations*. The number of distinct combinations of $n$ elements of $\Delta$ taken $m$ at a time is denoted by $C_n^m$ or $\binom{n}{m}$. The following formula holds:

$$C_n^m \equiv \binom{n}{m} = \frac{n!}{(n-m)!\, m!}. \tag{20.1.1.4}$$

Ordered sequences $a_{i_1}, \ldots, a_{i_m}$ of distinct elements of $\Delta$ are called *arrangements*, or *permutations* of $n$ elements taken $m$ at a time. The number of arrangements of $m$ out of $n$

elements, i.e., the number of ordered sequences of $m$ distinct elements selected from $n$ elements, is denoted by $A_n^m$. One has

$$A_n^m = n(n-1)\ldots(n-m+1) = \frac{n!}{(n-m)!}. \qquad (20.1.1.5)$$

Arrangements with $m = n$ are called *permutations*. The number of distinct permutations $P_n$ on $n$ elements is given by the formula

$$P_n = n!. \qquad (20.1.1.6)$$

### 20.1.1-5. Sampling with replacement.

The sampling scheme in which

$$\Omega = \{\omega = (i_1,\ldots,i_m) : i_k \in \Delta,\ k = 1, 2, \ldots, n\} \qquad (20.1.1.7)$$

and all elementary events $\omega$ are equiprobable is called *random sampling with replacement*.

If, for random sampling of $m$ out of $n$ elements with replacement, no subsequent ordering is performed (i.e., each of the $n$ elements can occur $0, 1, \ldots,$ or $m$ times in any combination), then one speaks of *combinations with repetitions*. The number $\overline{C}_n^m$ of all distinct combinations with repetitions of $n$ elements taken $m$ at a time is given by the formula

$$\overline{C}_n^m = C_{n+m-1}^m. \qquad (20.1.1.8)$$

If, for random sampling of $m$ out of $n$ elements with replacement, the chosen elements are ordered in some way, then one speaks of *arrangements with repetitions*. The number $\overline{A}_n^m$ of distinct arrangements with repetitions of $n$ elements taken $m$ at a time is given by the formula

$$\overline{A}_n^m = n^m. \qquad (20.1.1.9)$$

Suppose that a set of $n$ elements contains $k$ distinct elements, of which the first occurs $n_1$ times, the second occurs $n_2$ times, ..., and the $k$th occurs $n_k$ times, $n_1 + \cdots + n_k = n$. Permutations of $n$ elements of this set are called *permutations with repetitions on $n$ elements*. The number $P_n(n_1,\ldots,n_k)$ of permutations with repetitions on $n$ elements is given by the formula

$$P_n(n_1,\ldots,n_k) = \frac{n!}{n_1!\ldots n_k!}. \qquad (20.1.1.10)$$

**Example 2.** Consider the set $\{1, 2, 3\}$ of $n = 3$ elements. The elements of this set give $P_3 = 3! = 6$ permutations: $(1, 2, 3), (1, 3, 2), (2, 1, 3), (2, 3, 1), (3, 1, 2), (3, 2, 1)$. For $m = 2$, there are

1. $C_3^2 \equiv \binom{3}{2} = 3$ combinations without repetitions $[(1, 2), (1, 3), (2, 3)]$.
2. $\overline{C}_3^2 = C_{3+2-1}^2 = C_4^2 = 6$ combinations with repetitions $[(1, 2), (1, 3), (2, 3), (1, 1), (2, 2), (3, 3)]$.
3. $A_3^2 = 3$ arrangements without repetitions $[(1, 2), (1, 3), (2, 3), (2, 1), (3, 2), (3, 1)]$.
4. $\overline{A}_3^2 = 3^2 = 9$ arrangements with repetitions $[(1, 2), (1, 3), (2, 3), (2, 1), (3, 2), (3, 1), (1, 1), (2, 2), (3, 3)]$.

### 20.1.1-6. Geometric definition of probability.

Let $\Omega$ be a set of positive finite measure $\mu(\Omega)$ in the $n$-dimensional Euclidean space, and let the $\sigma$-algebra $\mathcal{F}$ consist of all measurable (i.e., having a measure) subsets $A \subseteq \Omega$. The *geometric probability* of an event $A$ is defined to be the ratio of the measure of $A$ to that of $\Omega$,

$$P(A) = \frac{\mu(A)}{\mu(\Omega)} \qquad (A \in \mathcal{F}). \qquad (20.1.1.11)$$

The notion of geometric probability is not invariant under transformations of the domain $\Omega$ and depends on how the measure $\mu(A)$ is introduced.

**Example 3.** A point is randomly thrown into a disk of radius $R = 1$. Find the probability of the event that the point lands in the disk of radius $r = \frac{1}{2}$ centered at the same point.

*First method.* Let $A$ be the event that the point lands in the smaller disk. We find the probability $P(A)$ as the ratio of the area of the smaller disk to that of the larger disk:

$$P(A) = \frac{\pi r^2}{\pi R^2} = \frac{1}{4}.$$

*Second method.* Consider the polar coordinate system in which the position of a point is determined by the angle $\varphi$ between the position vector of the point and the axis $OX$ and by the distance $\rho$ from the point to the origin. Since all points equidistant from the center either belong or do not belong to the smaller disk simultaneously, it follows that the probability of landing in this disk is equal to the ratio of the radii:

$$P(A) = \frac{r}{R} = \frac{1}{2}.$$

Thus we have obtained two different answers in the same problem. The cause is that the notion of geometric probability is not invariant under transformations of the domain $\Omega$ and depends on how the measure $\mu(A)$ is introduced.

## 20.1.2. Conditional Probability and Simplest Formulas

**20.1.2-1. Probability of the union of events.**

The probability of realization of at least one of two events $H_1$ and $H_2$ is given by the formula

$$P(H_1 \cup H_2) = P(H_1) + P(H_2) - P(H_1 \cap H_2). \tag{20.1.2.1}$$

In particular, for $H_1 \cap H_2 = \varnothing$, we have

$$P(H_1 \cup H_2) = P(H_1) + P(H_2). \tag{20.1.2.2}$$

The probability of realization of at least one of $n$ events is given by the formula

$$P(H_1 \cup \cdots \cup H_n) = \sum_{k=1}^{n} P(H_k) - \sum_{1 \le k_1 < k_2 \le n} P(H_{k_1} \cap H_{k_2})$$
$$+ \sum_{1 \le k_1 < k_2 < k_3 \le n} P(H_{k_1} \cap H_{k_2} \cap H_{k_3}) - \cdots + (-1)^{n-1} P(H_1 \cap \cdots \cap H_n). \tag{20.1.2.3}$$

**20.1.2-2. Conditional probability.**

The *conditional probability* $P(A|H)$, or $P_H(A)$, of an event $A$ given the occurrence of some other event $H$ is defined by the formula

$$P(A|H) = \frac{P(A \cap H)}{P(H)}, \quad P(H) > 0. \tag{20.1.2.4}$$

The conditional probability $P(A|H)$ can be treated as the probability of the event $A$ under the condition that the event $H$ occurs.

Relation (20.1.2.4) can be written as the "probability multiplication theorem"

$$P(A \cap H) = P(H) P(A|H). \tag{20.1.2.5}$$

The formula

$$P(A_1 \cap \cdots \cap A_n) = P(A_1) P(A_2|A_1) P(A_3|A_1 \cap A_2) \ldots P(A_n|A_1 \cap \cdots \cap A_{n-1})$$

is a generalization of (20.1.2.5).

### 20.1.2-3. Independence of events.

Two random events $A$ and $B$ are said to be *statistically independent* if the conditional probability of $A$, given $B$, coincides with the unconditional probability of $A$,

$$P(A|B) = P(A). \qquad (20.1.2.6)$$

Random events $A_1, \ldots, A_n$ are jointly statistically independent if the relation

$$P\left(\bigcap_{k=1}^{m} A_{i_k}\right) = \prod_{k=1}^{m} P(A_{i_k}) \qquad (20.1.2.7)$$

holds whenever $1 \leq i_1 < \cdots < i_m \leq n$ and $m \leq n$.

The pairwise independence of the events $A_i$ and $A_j$ for all $i \neq j$ ($i, j = 1, 2, \ldots, n$) does not imply that the events $A_1, \ldots, A_n$ are jointly independent.

**Example 1.** Suppose that the experiment is to draw one of four balls. Let three of them be labeled by the numbers 1, 2, and 3, and let the fourth ball bear all these numbers. By $A_i$ ($i = 1, 2, 3$) we denote the event that the chosen ball bears the number $i$. Are the events $A_1$, $A_2$, and $A_3$ dependent?

Since each number is encountered twice, $P(A_1) = P(A_2) = P(A_3) = 1/2$. Since any two distinct numbers are present only on one of the balls, we have $P(A_1 A_2) = P(A_2 A_3) = P(A_1 A_3) = 1/4$, and hence the events $A_1$, $A_2$, and $A_3$ are pairwise independent. All three distinct numbers are present only on one of the balls, and $P(A_1 A_2 A_3) = 1/4 \neq P(A_1) P(A_2) P(A_3) = 1/8$.

Thus we see that the events $A_1$, $A_2$, and $A_3$ are jointly dependent, even though they are pairwise independent.

### 20.1.2-4. Total probability formula. Bayes' formula.

Suppose that a complete group of pairwise incompatible events $H_1, \ldots, H_n$ is given and the unconditional probabilities $P(H_1), \ldots, P(H_n)$, as well as the conditional probabilities $P(A|H_1), \ldots, P(A|H_n)$ of an event $A$, are known. Then the probability of $A$ can be determined by the *total probability formula*

$$P(A) = \sum_{k=1}^{n} P(H_k) P(A|H_k). \qquad (20.1.2.8)$$

**Example 2.** There are two urns; the first urn contains $a$ white and $b$ black balls, and the second urn contains $c$ white and $d$ black balls. We take one ball from the first urn and put it into the second urn. After this, we draw one ball from the second urn. Find the probability of the event that this ball is white.

Let $A$ be the event of drawing a white ball. Consider the following complete group of events:

$H_1$, a white ball is taken from the first urn and put into the second urn.

$H_2$, a black ball is taken from the first urn and put into the second urn. Obviously,

$$P(H_1) = \frac{a}{a+b}, \quad P(H_2) = \frac{b}{a+b}; \quad P(A|H_1) = \frac{c+1}{c+d+1}, \quad P(A|H_2) = \frac{c}{c+d+1}.$$

Now by the total probability formula (20.1.2.8) we obtain

$$P(A) = P(H_1) P(A|H_1) + P(H_2) P(A|H_2) = \frac{a}{a+b} \frac{c+1}{c+d+1} + \frac{b}{a+b} \frac{c}{c+d+1}.$$

If it is known that the event $A$ has occurred but it is unknown which of the events $H_1, \ldots, H_n$ has occurred, then *Bayes' formula* is used:

$$P(H_k|A) = \frac{P(H_k) P(A|H_k)}{P(A)} \qquad (k = 1, 2, \ldots, n). \qquad (20.1.2.9)$$

**Example 3.** The urn contains one ball. It is known that this ball is either white or black. We put a black ball into the urn, then thoroughly shuffle the balls and draw one of the balls, which turns out to be black. Find the probability of the event that the ball remaining in the urn is black.

We consider the following random events that form a complete group of pairwise incompatible events:

$H_1$, the urn initially contained a white ball.

$H_2$, the urn initially contained a black ball.

We have $P(H_1) = P(H_2) = 1/2$. Suppose that the event $A$ is that a black ball is drawn. Then $P(A|H_1) = 1/2$ and $P(A|H_2) = 1$. The desired probability of the event $H_2|A$ is determined by Bayes' formula:

$$P(H_2|A) = \frac{P(H_2)P(A|H_2)}{P(A)} = \frac{1 \times \frac{1}{2}}{1 \times \frac{1}{2} + \frac{1}{2} \times \frac{1}{2}} = \frac{2}{3}.$$

### 20.1.3. Sequences of Trials

**20.1.3-1. Mathematical model of a sequence of $n$ independent trials.**

Trials in which events occurring in distinct trials are independent are said to be *independent*. Each trial $S_k$ can be viewed as a probability space $(\Omega_k, \mathcal{F}_k, P_k)$. Independent trials are described by the probability space $(\Omega, \mathcal{F}, P)$ that is the direct product of the probability spaces $(\Omega_k, \mathcal{F}_k, P_k)$. Here the probability of each event $A$ of the form $A = A_1 \times \cdots \times A_n$ is defined as $P(A) = P(A_1) \ldots P(A_n)$.

A sequence of $n$ independent trials is also called a *Bernoulli scheme*. Let $p_k = P(A_k)$. Then the probability of the event that $n_1$ events $A_1$, $n_2$ events $A_2$, ..., and $n_k$ events $A_k$ occur in $n$ independent trials is equal to

$$P_n(n_1, \ldots, n_k) = \frac{n!}{n_1! \ldots n_k!} p_1^{n_1} \ldots p_k^{n_k}. \tag{20.1.3.1}$$

The probability (20.1.3.1) is the coefficient of $x_1^{n_1} \ldots x_k^{n_k}$ in the expansion of the polynomial $(p_1 x_1 + \cdots + p_k x_k)^n$ in powers of $x_1, \ldots, x_k$.

The probability $P_n(n_1, \ldots, n_k)$ can be found by the technique of generating functions, since this probability is the coefficient of $z_1^{n_1} \ldots z_k^{n_k}$ in the generating function

$$\varphi_X(z_1, \ldots, z_k) = (p_1 z_1 + \cdots + p_k z_k)^n.$$

**20.1.3-2. Bernoulli process.**

The special case of the Bernoulli scheme with $N = 2$ is called the *Bernoulli process*. In this case, some event $A$ occurs with probability $p = P(A)$ (the probability of "success") and does not occur with probability $q = P(\overline{A}) = 1 - P(A) = 1 - p$ (the probability of "failure") in each trial. If $\mu_n$ is the number of occurrences of the event $A$ (the number of "successes") in $n$ independent Bernoulli trials, then the probability that $A$ occurs exactly $k$ times is given by the formula

$$P(\mu_n = k) = p_n(k) = C_n^k p^k (1-p)^{n-k} \qquad (k = 0, 1, \ldots, n). \tag{20.1.3.2}$$

Relation (20.1.3.2) is called the *Bernoulli formula (binomial distribution)*.

The probability that the event occurs at least $m$ times in $n$ independent trials is calculated by the formula

$$p_n(k \geq m) = \sum_{k=m}^{n} p_n(k) = 1 - \sum_{k=0}^{m-1} p_n(k).$$

The probability that the event occurs at least once in $n$ independent trials is calculated by the formula
$$p_n(k \geq 1) = 1 - (1-p)^n.$$
The number $n$ of independent trials necessary for the event to occur at least once with probability at least $P$ is given by the formula
$$n \geq \frac{\ln(1-P)}{\ln(1-p)}.$$

**Example (Banach's problem).** A smoker mathematician has two matchboxes on him, each of which initially contains exactly $n$ matches. Each time he needs to light a cigarette, he selects a matchbox at random. Find the probability of the event that as the mathematician takes out an empty box for the first time, precisely $k$ matches will be left in the other box ($k \leq n$).

The mathematician has taken matches $2n - k$ times, $n$ out of them from the box that is eventually empty. This scheme corresponds to the scheme of $2n - k$ independent Bernoulli trials with $n$ "successes." The probability of a "success" in a single trial is equal to 0.5. The desired probability can be found by the formula
$$P(\mu_{2n-k} = n) = p_{2n-k}(n) = C_n^{2n-k} p^n (1-p)^{n-k} = C_n^{2n-k} \left(\frac{1}{2}\right)^{2n-k}.$$

### 20.1.3-3. Limit theorems for Bernoulli process.

It is very difficult to use Bernoulli's formula (20.1.3.2) for large $n$ and $m$. In this case, one has to use approximate formulas for calculating $p_n(k)$ with desired accuracy.

*Poisson formula.* If the number of independent trials increases unboundedly ($n \to \infty$) and the probability $p$ simultaneously decays ($p \to 0$) so that their product $np$ is a constant ($np = \lambda = \text{const}$), then the probability $P(\mu_n = k) = p_n(k)$ satisfies the limit relation
$$\lim_{n \to \infty} P(\mu_n = k) = \frac{\lambda^k}{k!} e^{-\lambda}. \tag{20.1.3.3}$$

*Local theorem of de Moivre–Laplace.* Suppose that $n \to \infty$, $p = \text{const}$, $0 < p < 1$, and $0 < c_1 \leq x_{n,k} = (k - np)[np(1-p)]^{-1/2} \leq c_2 < \infty$; then
$$P(\mu_n = k) = \frac{1}{\sqrt{2\pi np(1-p)}} e^{-x_{n,k}^2/2}[1 + O(1/\sqrt{n})] \tag{20.1.3.4}$$
uniformly with respect to $x_{n,k} \in [c_1, c_2]$.

*Integral theorem of de Moivre–Laplace.* Suppose that $n \to \infty$ and $p = \text{const}$, $0 < p < 1$; then
$$P\left[x_1 < \frac{\mu_n - np}{\sqrt{np(1-p)}} < x_2\right] \to \frac{1}{\sqrt{2\pi}} \int_{x_1}^{x_2} e^{-t^2/2}\, dt \tag{20.1.3.5}$$
uniformly with respect to $x_1$ and $x_2$.

The approximate formula (20.1.3.3) is normally used for $n \geq 50$ and $np \leq 10$. The approximate formulas (20.1.3.4) and (20.1.3.5) are used for $np(1-p) > 9$. The error in formulas (20.1.3.4) and (20.1.3.5) can increase owing to the fact that the prelimit distribution is discrete. This error is of the order of $O([np(1-p)]^{-1/2})$.

The limit expression in (20.1.3.4) can readily be expressed via the probability density
$$\varphi(x) = \frac{1}{\sqrt{2\pi}} e^{-x^2/2}$$
of the standard normal distribution, and the limit expression in (20.1.3.5) can be expressed in terms of the cumulative distribution function
$$\Phi(x) = \frac{1}{\sqrt{2\pi}} \int_{-\infty}^{x} e^{-t^2/2}\, dt$$
of the standard normal distribution.

## 20.2. Random Variables and Their Characteristics

### 20.2.1. One-Dimensional Random Variables

**20.2.1-1. Notion of random variable. Distribution function of random variable.**

A *random variable* $X$ is a real function $X = X(\omega)$, $\omega \in \Omega$, on a probability space $\Omega$ such that the set $\{\omega : X(\omega) \leq x\}$ belongs to the $\sigma$-algebra $\mathcal{F}$ of events for each real $x$.

Any rule (table, function, graph, or otherwise) that permits one to find the probabilities of events $A \subseteq \mathcal{F}$ is usually called the distribution law of a random variable. In general, random variables can be *discrete* or *continuous*.

The *cumulative distribution function* of a random variable $X$ is the function $F_X(x)$ whose value at each point $x$ is equal to the probability of the event $\{X < x\}$:

$$F_X(x) = F(x) = P(X < x). \tag{20.2.1.1}$$

Properties of the cumulative distribution function:
1. $F(x)$ is bounded, i.e., $0 \leq F(x) \leq 1$.
2. $F(x)$ is a nondecreasing function for $x \in (-\infty, \infty)$; i.e., if $x_2 > x_1$, then $F(x_2) \geq F(x_1)$.
3. $\lim_{x \to -\infty} F(x) = F(-\infty) = 0$.
4. $\lim_{x \to +\infty} F(x) = F(+\infty) = 1$.
5. The probability that a random variable $X$ lies in the interval $[x_1, x_2)$ is equal to the increment of its cumulative distribution function on this interval; i.e.,

$$P(x_1 \leq X < x_2) = F(x_2) - F(x_1).$$

6. $F(x)$ is left continuous; i.e., $\lim_{x \to x_0 - 0} F(x) = F(x_0)$.

**20.2.1-2. Discrete random variables.**

A random variable $X$ is said to be *discrete* if the set of its possible values (the spectrum of the discrete random variable) is at most countable. A discrete distribution is determined by a finite or countable set of probabilities $P(X = x_i)$ such that

$$\sum_i P(X = x_i) = 1.$$

To define a discrete random variable, it is necessary to specify the values $x_1, x_2, \ldots$ and the corresponding probabilities $p_1, p_2, \ldots$, where $p_i = P(X = x_i)$.

*Remark.* In what follows, we assume that the values of a discrete random variable $X$ are arranged in ascending order.

In this case, the cumulative distribution function of a discrete random variable $X$ is the step function defined as the sum

$$F(x) = \sum_{x_i < x} P(X = x_i). \tag{20.2.1.2}$$

It is often convenient to write out the cumulative distribution function using the function $\theta(x)$ such that $\theta(x) = 1$ for $x > 0$ and $\theta(x) = 0$ for $x \leq 0$:

$$F(x) = \sum_i P(X = x_i)\theta(x - x_i).$$

For discrete random variables, one can introduce the notion of probability density function $p(x)$ by setting

$$p(x) = \sum_i P(X = x_i)\delta(x - x_i),$$

where $\delta(x)$ is the delta function.

### 20.2.1-3. Continuous random variables. Probability density function.

A random variable $X$ is said to be *continuous* if its cumulative distribution function $F_X(x)$ can be represented in the form

$$F(x) = \int_{-\infty}^{x} p(y)\, dy. \qquad (20.2.1.3)$$

The function $p(x)$ is called the *probability density function* of the random variable $X$. Obviously, relation (20.2.1.3) is equivalent to the relation

$$p(x) = \lim_{\Delta x \to 0} \frac{P(x \leq X \leq x + \Delta x)}{\Delta x} = \frac{dF(x)}{dx}. \qquad (20.2.1.4)$$

The differential $dF(x) = p(x)\, dx \approx P(x \leq X < x + dx)$ is called a *probability element*.

Properties of the probability density function:
1. $p(x) \geq 0$.
2. $P(x_1 \leq X < x_2) = \int_{x_1}^{x_2} p(y)\, dy$.
3. $\int_{-\infty}^{+\infty} p(x)\, dx = 1$.
4. $P(x \leq \xi < x + \Delta x) \approx p(x)\Delta x$.
5. For continuous random variables, one always has $P(X = x) = 0$, but the event $\{X = x\}$ is not necessarily impossible.
6. For continuous random variables,

$$P(x_1 \leq X < x_2) = P(x_1 < X < x_2) = P(x_1 < X \leq x_2) = P(x_1 \leq X \leq x_2).$$

### 20.2.1-4. Unified description of probability distribution.

Discrete and continuous probability distributions can be studied simultaneously if the probability of each event $\{a \leq X < b\}$ is represented in terms of the integral

$$P(a \leq X < b) = \int_a^b dF(x), \qquad (20.2.1.5)$$

where $F(x) = P(X < x)$ is the cumulative distribution function of the random variable $X$. For a continuous distribution, the integral (20.2.1.5) becomes the Riemann integral. For a discrete distribution, the integral can be reduced to the form

$$P(a \leq X < b) = \sum_{a \leq x_i < b} P(X = x_i).$$

In particular, the integral can also be used in the case of mixed distributions, i.e., distributions that are partially continuous and partially discrete.

### 20.2.1-5. Symmetric random variables.

A random variable $X$ is said to be *symmetric* if the condition

$$P(X < -x) = P(X > x) \qquad (20.2.1.6)$$

holds for all $x$.

Properties of symmetric random variables:
1. $P(|X| < x) = F(x) - F(-x) = 2F(x) - 1$.
2. $F(0) = 0.5$.
3. If a moment $\alpha_{2k+1}$ of odd order about the origin (see Paragraph 20.2.2-3) exists, then it is zero.
4. If $t_\gamma$ is the quantile of level $\gamma$ (see Paragraph 20.2.2-5), then $t_\gamma = -t_{1-\gamma}$.

A random variable $Y$ is said to be *symmetric about its expected value* if the random variable $X = Y - E\{Y\}$ is symmetric, where $E\{Y\}$ is the expected value of a random variable $Y$ (see Paragraph 20.2.2-1).

Properties of random variables symmetric about the expected value:
1. $P(|Y - E\{Y\}| < x) = 2F_Y(x + E\{Y\}) - 1$.
2. $F_Y(E\{Y\}) = 0.5$.
3. If a central moment $\mu_{2k+1}$ of odd order (see Paragraph 20.2.2-3) exists, then it is equal to zero.

### 20.2.1-6. Functions of random variables.

Suppose that a random variable $Y$ is related to a random variable $X$ by a functional dependence $Y = f(X)$. If $X$ is discrete, then, obviously, $Y$ is also discrete. To find the distribution law of the random variable $Y$, it suffices to calculate the values $f(x_i)$. If there are repeated values among $y_i = f(x_i)$, then these repeated values are taken into account only once, the corresponding probabilities being added.

If $X$ is a continuous random variable with probability density function $p_X(x)$, then, in general, the random variable $Y$ is also continuous. The cumulative distribution function of $Y$ is given by the formula

$$F_Y(y) = P(\eta < y) = P[f(x) < y] = \int_{f(x)<y} p_X(x)\,dx. \qquad (20.2.1.7)$$

If the function $y = f(x)$ is differentiable and monotone on the entire range of the argument $x$, then the probability density function $p_Y(y)$ of the random variable $Y$ is given by the formula

$$p_Y(y) = p_X[\psi(y)]|\psi'_y(y)|, \qquad (20.2.1.8)$$

where $\psi$ is the inverse function of $f(x)$.

If $f(x)$ is a nonmonotonic function, then the inverse function is nonunique and the probability density function of the random variable $y$ is the sum of as many terms as there are values (for a given $y$) of the inverse function:

$$p_Y(y) = \sum_{i=1}^{k} p_X[\psi_i(y)]\big|[\psi_i(y)]'_y\big|, \qquad (20.2.1.9)$$

where $\psi_1(y), \ldots, \psi_k(y)$ are the values of the inverse function for a given $y$.

**Example 1.** Suppose that a random variable $X$ has the probability density

$$p_X(x) = \frac{1}{\sqrt{2\pi}} e^{-x^2/2}.$$

Find the distribution of the random variable $Y = X^2$.

In this case, $y = f(x) = x^2$. According to (20.2.1.7), we obtain

$$F_Y(y) = \int_{x^2 < y} \frac{1}{\sqrt{2\pi}} e^{-x^2/2}\, dx = \frac{1}{\sqrt{2\pi}} \int_{-\sqrt{y}}^{\sqrt{y}} e^{-x^2/2}\, dx = \frac{2}{\sqrt{2\pi}} \int_{0}^{\sqrt{y}} e^{-x^2/2}\, dx = \frac{1}{\sqrt{2\pi}} \int_{0}^{y} \frac{e^{-t/2}}{\sqrt{t}}\, dt.$$

**Example 2.** Suppose that a random variable $X$ has the probability density

$$p_X(x) = \frac{1}{\sqrt{2\pi}\sigma} \exp\left[-\frac{(x-a)^2}{2\sigma^2}\right].$$

Find the probability density of the random variable $Y = e^X$.

For $y > 0$, the cumulative distribution function of the random variable $Y = e^X$ is determined by the relations

$$F_Y(y) = P(Y < y) = P(e^X < y) = P(X < \ln y) = F_X(\ln y).$$

We differentiate this relation and obtain

$$p_Y(y) = \frac{dF_Y(y)}{dy} = \frac{dF_X(\ln y)}{dy} = p_X(\ln y)\frac{1}{y} = \frac{1}{\sqrt{2\pi}\sigma y} \exp\left[-\frac{(\ln y - a)^2}{2\sigma^2}\right] \text{ for } y > 0.$$

The distribution of $Y$ is called the *log-normal distribution*.

**Example 3.** Suppose that a random variable $X$ has the probability density $p_X(x)$ for $x \in (-\infty, \infty)$. Then the probability density of the random variable $Y = |X|$ is given by the formula $p_Y(y) = p_X(x) + p_X(-x)$ $(y \geq 0)$. In particular, if $X$ is symmetric, then $p_Y(y) = 2p_X(y)$ $(y \geq 0)$.

### 20.2.2. Characteristics of One-Dimensional Random Variables

#### 20.2.2-1. Expectation.

The *expectation (expected value)* $E\{X\}$ *of a discrete or continuous random variable* $X$ is the expression given by the formula

$$E\{X\} = \int_{-\infty}^{+\infty} x\, dF(x) = \begin{cases} \displaystyle\sum_i x_i p_i & \text{in the discrete case,} \\ \displaystyle\int_{-\infty}^{+\infty} x p(x)\, dx & \text{in the continuous case.} \end{cases} \qquad (20.2.2.1)$$

For the existence of the expectation (20.2.2.1), it is necessary that the corresponding series or integral converge absolutely.

The expectation is the main characteristic defining the "position" of a random variable, i.e., the number near which its possible values are concentrated.

We note that the expectation is not a function of the variable $x$ but a functional describing the properties of the distribution of the random variable $X$. There are distributions for which the expectation does not exist.

**Example 1.** For the Cauchy distribution given by the probability density function

$$p(x) = \frac{1}{\pi(1+x^2)}, \quad x \in (-\infty, +\infty),$$

the expectation does not exist because the integral $\int_{-\infty}^{+\infty} |x|/\pi(1+x^2)\, dx$ diverges.

### 20.2.2-2. Expectation of function of random variable.

If a random variable $Y$ is related to a random variable $X$ by a functional dependence $Y = f(X)$, then the expectation of the random variable $Y = f(X)$ can be determined by two methods. The first method is to construct the distribution of the random variable $Y$ and then use already known formulas to find $E\{Y\}$. The second method is to use the formulas

$$E\{Y\} = E\{f(X)\} = \begin{cases} \sum_i f(x_i) p_i & \text{in the discrete case,} \\ \int_{-\infty}^{+\infty} f(x) p(x)\, dx & \text{in the continuous case} \end{cases} \quad (20.2.2.2)$$

if these expressions exist in the sense of absolute convergence.

**Example 2.** Suppose that a random variable $X$ is uniformly distributed in the interval $(-\pi/2, \pi/2)$, i.e., $p(x) = 1/\pi$ for $x \in (-\pi/2, \pi/2)$. Then the expectation of the random variable $Y = \sin(X)$ is equal to

$$E\{Y\} = \int_{-\infty}^{+\infty} f(x) p(x)\, dx = \int_{-\pi/2}^{\pi/2} \frac{1}{\pi} \sin x\, dx = 0.$$

Properties of the expectation:
1. $E\{C\} = C$ for any real $C$.
2. $E\{\alpha X + \beta Y\} = \alpha E\{X\} + \beta E\{Y\}$ for any real $\alpha$ and $\beta$.
3. $E\{X\} \le E\{Y\}$ if $X(\omega) \le Y(\omega)$, $\omega \in \Omega$.
4. $E\left\{\sum_{k=1}^{\infty} X_k\right\} = \sum_{k=1}^{\infty} E\{X_k\}$ if the series $\sum_{k=1}^{\infty} E\{|X_k|\}$ converges.
5. $g(E\{X\}) \le E\{g(X)\}$ for convex functions $g(X)$.
6. Any bounded random variable has a finite expectation.
7. $|E\{X\}| \le E\{|X|\}$.
8. The *Cauchy–Schwarz inequality* $(E\{|XY|\})^2 \le (E\{X\})^2 (E\{Y\})^2$ holds.
9. $E\left\{\prod_{k=1}^{n} X_k\right\} = \prod_{k=1}^{n} E\{X_k\}$ for mutually independent random variables $X_1, \dots, X_n$.

### 20.2.2-3. Moments.

The expectation $E\{(X-a)^k\}$ is called the *k*th *moment* of a random variable $X$ about $a$. The moments about zero are usually referred to simply as the moments of a random variable. (Sometimes they are called *initial moments*.) The $k$th moment satisfies the relation

$$\alpha_k = E\{X^k\} = \int_{-\infty}^{+\infty} x^k\, dF(x) = \begin{cases} \sum_i x_i^k p_i & \text{in the discrete case,} \\ \int_{-\infty}^{+\infty} x^k p(x)\, dx & \text{in the continuous case.} \end{cases} \quad (20.2.2.3)$$

If $a = E\{X\}$, then the $k$th moment of the random variable $X$ about $a$ is called the *k*th *central moment*. The $k$th central moment satisfies the relation

$$\mu_k = E\{(X - E\{X\})^k\} = \begin{cases} \sum_i (x_i - E\{X\})^k p_i & \text{in the discrete case,} \\ \int_{-\infty}^{+\infty} (x - E\{X\})^k p(x)\, dx & \text{in the continuous case.} \end{cases} \quad (20.2.2.4)$$

In particular, $\mu_0 = 1$ for any random variable.

The number $m_k = E\{|X - a|^k\}$ is called the *kth absolute moment* of $X$ about $a$.

The existence of a $k$th moment $\alpha_k$ or $\mu_k$ implies the existence of the moments $\alpha_m$ and $\mu_m$ of orders $m \leq k$; if the integral (or series) for $\alpha_k$ or $\mu_k$ diverges, then all integrals (series) for $\alpha_m$ and $\mu_m$ of orders $m \geq k$ also diverge.

There is a simple relationship between the central and initial moments:

$$\mu_k = \sum_{m=0}^{k} C_k^m \alpha_m (\alpha_1)^{k-m}, \quad \alpha_0 = 1; \quad \alpha_k = \sum_{m=0}^{k} C_k^m \mu_m (\alpha_1)^{k-m}. \qquad (20.2.2.5)$$

Relations (20.2.2.5) can be represented in the following easy-to-memorize symbolic form: $\mu_k = (\alpha - \alpha_1)^k$, $\alpha_k = (\mu + \alpha_1)^k$, where it is assumed that after the right-hand sides have been multiplied out according to the binomial formula, the expressions $\alpha^m$ and $\mu^m$ are replaced by $\alpha_m$ and $\mu_m$, respectively.

If the probability distribution is symmetric about its expectation, then all existing central moments $\mu_k$ of even order $k$ are zero.

The probability distribution is uniquely determined by the moments $\alpha_0, \alpha_1, \ldots$ provided that they all exist and the series $\sum_{m=0}^{\infty} |\alpha_m| t^m / m!$ converges for some $t > 0$.

### 20.2.2-4. Variance.

The *variance* of a random variable is the measure $\mathrm{Var}\{X\}$ of the deviation of a random variable $X$ from its expectation $E\{X\}$, determined by the relation

$$\mathrm{Var}\{X\} = E\{(X - E\{X\})^2\}. \qquad (20.2.2.6)$$

The variance $\mathrm{Var}\{X\}$ is the second central moment of the random variable $X$. The variance can be determined by the formulas

$$\mathrm{Var}\{X\} = \int_{-\infty}^{+\infty} (x - E\{X\})^2 \, dF(x)$$

$$= \begin{cases} \sum_i (x_i - E\{X\})^2 p_i & \text{in the discrete case,} \\ \int_{-\infty}^{+\infty} (x - E\{X\})^2 p(x) \, dx & \text{in the continuous case.} \end{cases} \qquad (20.2.2.7)$$

The variance characterizes the spread in values of the random variable $X$ about its expectation.

Properties of the variance:
1. $\mathrm{Var}\{C\} = 0$ for any real $C$.
2. The variance is nonnegative: $\mathrm{Var}\{X\} \geq 0$.
3. $\mathrm{Var}\{\alpha X + \beta\} = \alpha^2 \mathrm{Var}\{X\}$ for any real numbers $\alpha$ and $\beta$.
4. $\mathrm{Var}\{X\} = E\{X^2\} - (E\{X\})^2$.
5. $\min_m E\{(X - m)^2\} = \mathrm{Var}\{X\}$ and is attained for $m = E\{X\}$.
6. $\mathrm{Var}\{X_1 + \cdots + X_n\} = \mathrm{Var}\{X_1\} + \cdots + \mathrm{Var}\{X_n\}$ for pairwise independent random variables $X_1, \ldots, X_n$.
7. If $X$ and $Y$ are independent random variables, then

$$\mathrm{Var}\{XY\} = \mathrm{Var}\{X\} \mathrm{Var}\{Y\} + \mathrm{Var}\{X\}(E\{Y\})^2 + \mathrm{Var}\{Y\}(E\{X\})^2.$$

**20.2.2-5. Numerical characteristics of random variables.**

A *quantile of level* $\gamma$ of a one-dimensional distribution is a number $t_\gamma$ for which the value of the corresponding distribution function is equal to $\gamma$; i.e.,
$$P(X < t_\gamma) = F(t_\gamma) = \gamma \quad (0 < \gamma < 1). \tag{20.2.2.8}$$

Quantiles exist for each probability distribution, but they are not necessarily uniquely determined. Quantiles are widely used in statistics. The quantile $t_{1/2}$ is called the *median* Med$\{X\}$. For $n = 4$, the quantiles $t_{m/n}$ are called *quartiles*, for $n = 10$, they are called *deciles*, and for $n = 100$, they are called *percentiles*.

A *mode* Mode$\{X\}$ *of a continuous probability distribution* is a point of maximum of the probability density function $p(x)$. A *mode of a discrete probability distribution* is a value Mode$\{X\}$ preceded and followed by values associated with probabilities smaller than $p(\text{Mode}\{X\})$.

Distributions with one, two, or more modes are said to be *unimodal*, *bimodal*, or *multimodal*, respectively.

The *standard deviation (root-mean-square deviation)* of a random variable $X$ is the square root of its variance,
$$\sigma = \sqrt{\text{Var}\{X\}}.$$
The standard deviation has the same dimension as the random variable itself.

The *coefficient of variation* is the ratio of the standard deviation to the expected value,
$$v = \frac{\sigma}{E\{X\}}.$$

The *asymmetry coefficient*, or *skewness*, is defined by the formula
$$\gamma_1 = \frac{\mu_3}{(\mu_2)^{3/2}}. \tag{20.2.2.9}$$

If $\gamma_1 > 0$, then the distribution curve is more flattened to the right of the mode Mode$\{X\}$; if $\gamma_1 < 0$, then the distribution curve is more flattened to the left of the mode Mode$\{X\}$ (see Fig. 20.1). (As a rule, this applies to continuous random variables.)

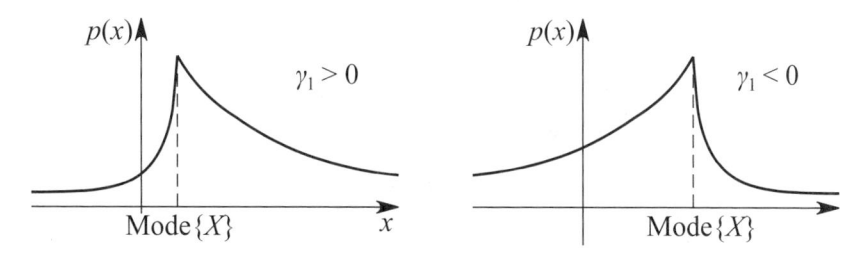

**Figure 20.1.** Relationship of the distribution curve and the asymmetry coefficient.

The *excess coefficient*, or *excess*, or *kurtosis*, is defined by the formula
$$\gamma_2 = \frac{\mu_4}{\mu_2^2} - 3. \tag{20.2.2.10}$$

One says that for $\gamma_2 = 0$ the distribution has a normal excess, for $\gamma_2 > 0$ the distribution has a positive excess, and for $\gamma_2 < 0$ the distribution has a negative excess.

Remark. The coefficients $\gamma_1^2$ and $\gamma_2 + 3$ or $(\gamma_2 + 3)/2$ are often used instead of $\gamma_1$ and $\gamma_2$.

*Pearson's first skewness coefficient* for a unimodal distribution is defined by the formula
$$s = \frac{E\{X\} - \text{Mode}\{X\}}{\sigma}.$$

### 20.2.2-6. Characteristic functions. Semi-invariants.

1°. The *characteristic function* of a random variable $X$ is the expectation of the random variable $e^{itX}$, i.e.,

$$f(t) = E\{e^{itX}\} = \int_{-\infty}^{+\infty} e^{itx}\, dF(x)$$

$$= \begin{cases} \sum_j e^{itx_j} p_j & \text{in the discrete case,} \\ \int_{-\infty}^{+\infty} e^{itx} p(x)\, dx & \text{in the continuous case,} \end{cases} \qquad (20.2.2.11)$$

where $t$ is a real variable ranging from $-\infty$ to $+\infty$ and $i$ is the imaginary unit, $i^2 = -1$.

Properties of characteristic functions:
1. The cumulative distribution function is uniquely determined by the characteristic function.
2. The characteristic function is uniformly continuous on the entire real line.
3. $|f(t)| \le f(0) = 1$.
4. $\overline{f(-t)} = f(t)$.
5. $f(t)$ is a real function if and only if the random variable $X$ is symmetric.
6. The characteristic function of the sum of two independent random variables is equal to the product of their characteristic functions.
7. If a random variable $X$ has a $k$th absolute moment, then the characteristic function of $X$ is $k$ times differentiable and the relation $f^{(m)}(0) = i^m E\{X^m\}$ holds for $m \le k$.
8. If $x_1$ and $x_2$ are points of continuity of the cumulative distribution function $F(x)$, then

$$F(x_2) - F(x_1) = \frac{1}{2\pi} \lim_{T \to \infty} \int_{-T}^{T} \frac{e^{-itx_1} - e^{-itx_2}}{it} f(t)\, dt. \qquad (20.2.2.12)$$

9. If $\int_{-\infty}^{+\infty} |f(t)|\, dt < \infty$, then the cumulative distribution function $F(x)$ has a probability density function $p(x)$, which is given by the formula

$$p(x) = \frac{1}{2\pi} \int_{-\infty}^{+\infty} e^{-itx} f(t)\, dt. \qquad (20.2.2.13)$$

If the probability distribution has a $k$th moment $\alpha_k$, then there exist *semi-invariants* (*cumulants*) $\tau_1, \ldots, \tau_k$ determined by the relation

$$\ln f(t) = \sum_{l=1}^{k} \tau_l \frac{(it)^l}{l!} + o(t^k). \qquad (20.2.2.14)$$

The semi-invariants $\tau_1, \ldots, \tau_k$ can be calculated by the formulas

$$\tau_l = i^{-l} \left. \frac{\partial^l \ln f(t)}{\partial t^l} \right|_{t=0}.$$

### 20.2.2-7. Generating functions.

The *generating function of a numerical sequence* $a_0, a_1, \ldots$ is defined as the power series

$$\varphi_X(z) = \sum_{n=0}^{\infty} a_n z^n, \tag{20.2.2.15}$$

where $z$ is either a formal variable or a complex or real number. If $X$ is a random variable whose absolute moments of any order are finite, then the series

$$\sum_{n=0}^{\infty} E\{X^n\} \frac{z^n}{n!} \tag{20.2.2.16}$$

is called the *moment-generating function* of the random variable $X$.

If $X$ is a nonnegative random variable taking integer values, then the formulas

$$\varphi_X(z) = E\{z^X\} = \sum_{n=0}^{\infty} P(X = n) z^n \tag{20.2.2.17}$$

define the *probability-generating function*, or simply the *generating function of the random variable* $X$. The generating function of a random variable $X$ is related to its characteristic function $f(t)$ by the formula

$$f(t) = \varphi_X(e^{it}). \tag{20.2.2.18}$$

## 20.2.3. Main Discrete Distributions

### 20.2.3-1. Binomial distribution.

A random variable $X$ has the *binomial distribution* with parameters $(n, p)$ (see Fig. 20.2) if

$$P(X = k) = C_n^k p^k (1-p)^{n-k}, \quad k = 0, 1, \ldots, n, \tag{20.2.3.1}$$

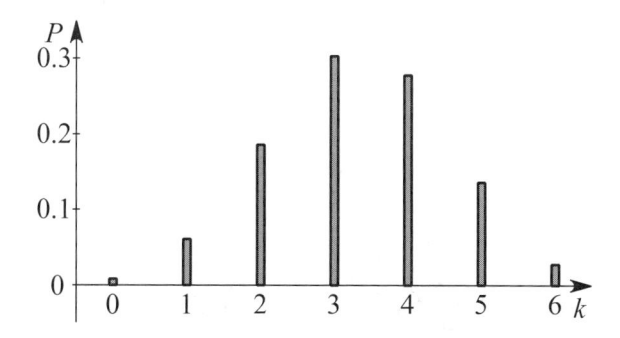

where $0 < p < 1$, $n \geq 1$.

**Figure 20.2.** Binomial distribution for $p = 0.55$, $n = 6$.

The cumulative distribution function, the probability-generating function, and the characteristic function have the form

$$F(x) = \begin{cases} 1 & \text{for } x > n, \\ \sum_{k=1}^{m} C_n^k p^k (1-p)^{n-k} & \text{for } m \leq x < m+1 \ (m = 1, 2, \ldots, n-1), \\ 0 & \text{for } x < 0, \end{cases} \quad (20.2.3.2)$$

$$\varphi_X(z) = (1 - p + pz)^n,$$
$$f(t) = (1 - p + pe^{it})^n,$$

and the numerical characteristics are given by the formulas

$$E\{X\} = np, \quad \text{Var}\{X\} = np(1-p), \quad \gamma_1 = \frac{1-2p}{\sqrt{np(1-p)}}, \quad \gamma_2 = \frac{1-6p(1-p)}{np(1-p)}.$$

The binomial distribution is a model of random experiments consisting of $n$ independent identical Bernoulli trials. If $X_1, \ldots, X_n$ are independent random variables, each of which can take only two values 1 or 0 with probabilities $p$ and $q = 1-p$, respectively, then the random variable $X = \sum_{k=1}^{n} X_k$ has the binomial distribution with parameters $(n, p)$.

The binomial distribution is asymptotically normal with parameters $(np, np(1-p))$ as $n \to \infty$ (the *de Moivre–Laplace limit theorem*, which is a special case of the central limit theorem, see Paragraph 20.3.2-2); specifically,

$$P(X = k) = C_n^k p^k (1-p)^{n-k} \approx \frac{1}{\sqrt{np(1-p)}} \varphi\left[\frac{k-np}{\sqrt{np(1-p)}}\right] \quad \text{as} \quad \frac{(k-np)^3}{[np(1-p)]^4} \to 0,$$

$$P(k_1 \leq X \leq k_2) \approx \Phi\left[\frac{k_2 - np}{\sqrt{np(1-p)}}\right] - \Phi\left[\frac{k_1 - np}{\sqrt{np(1-p)}}\right] \quad \text{as} \quad \frac{(k_{1,2} - np)^3}{[np(1-p)]^4} \to 0,$$

where $\varphi(x)$ and $\Phi(x)$ are the probability density function and the cumulative distribution function of the standard normal distribution (see Paragraph 20.2.4-3).

### 20.2.3-2. Geometric distribution.

A random variable $X$ has a *geometric distribution* with parameter $p$ $(0 < p < 1)$ (see Fig. 20.3) if

$$P(X = k) = p(1-p)^k, \quad k = 0, 1, 2, \ldots \quad (20.2.3.3)$$

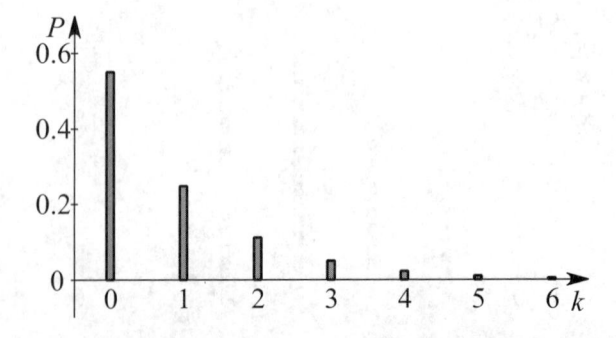

**Figure 20.3.** Geometric distribution for $p = 0.55$.

The probability-generating function and the characteristic function have the form
$$\varphi_X(z) = p[1-(1-p)z]^{-1},$$
$$f(t) = p[1-(1-p)e^{it}]^{-1},$$
and the numerical characteristics can be calculated by the formulas
$$E\{X\} = \frac{1-p}{p}, \quad \alpha_2 = \frac{(1-p)(2-p)}{p^2}, \quad \text{Var}\{X\} = \frac{1-p}{p^2}, \quad \gamma_1 = \frac{2-p}{\sqrt{1-p}}, \quad \gamma_2 = 6 + \frac{p^2}{1-p}.$$

The geometric distribution describes a random variable $X$ equal to the number of failures before the first success in a sequence of Bernoulli trials with probability $p$ of success in each trial.

The geometric distribution is the only discrete distribution that is *memoryless*, i.e., satisfies the relation
$$P(X > s+t | X > t) = P(X > s)$$
for all $s, t > 0$. This property permits one to view the geometric distribution as the discrete analog of the exponential distribution.

### 20.2.3-3. Hypergeometric distribution.

A random variable $X$ has the *hypergeometric distribution* with parameters $(N, p, n)$ (see Fig. 20.4) if
$$P(X = k) = \frac{C_{Np}^k C_{N(1-p)}^{n-k}}{C_N^n}, \quad k = 0, 1, \ldots, n, \tag{20.2.3.4}$$
where $0 < p < 1$, $0 \le n \le N$, $N > 0$.

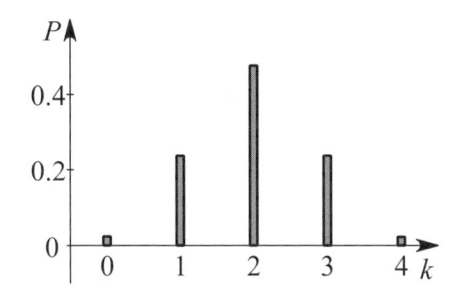

**Figure 20.4.** Hypergeometric distribution for $p = 0.5$, $N = 10$, $n = 4$.

The numerical characteristics are given by the formulas
$$E\{X\} = np, \quad \text{Var}\{X\} = \frac{N-n}{N-1} np(1-p).$$

A typical scheme in which the hypergeometric distribution arises is as follows: $n$ elements are randomly drawn without replacement from a population of $N$ elements containing exactly $Np$ elements of type $I$ and $N(1-p)$ elements of type $II$. The number of elements of type $I$ in the sample is described by the hypergeometric distribution.

If $n \ll N$ (in practice, $n < 0.1N$), then
$$\frac{C_{Np}^k C_{N(1-p)}^{n-k}}{C_N^n} \approx C_n^k p^k (1-p)^{n-k};$$
i.e., the hypergeometric distribution tends to the binomial distribution.

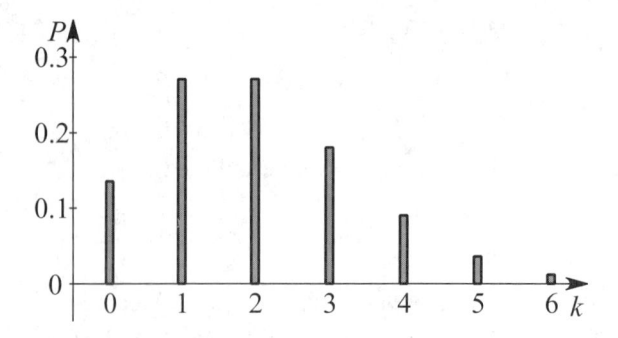

**Figure 20.5.** Poisson distribution for $\lambda = 2$.

### 20.2.3-4. Poisson distribution.

A random variable $X$ has the *Poisson distribution* with parameter $\lambda$ ($\lambda > 0$) (see Fig. 20.5) if

$$P(X = k) = \frac{\lambda^k}{k!} e^{-\lambda}, \quad k = 0, 1, 2, \ldots \qquad (20.2.3.5)$$

The cumulative distribution function of the Poisson distribution at the points $k = 0, 1, 2, \ldots$ is given by the formula

$$F(k) = \frac{1}{k!} \int_\lambda^\infty y^k e^{-y}\, dy = 1 - S_{k+1}(\lambda),$$

where $S_{k+1}(\lambda)$ is the value at the point $\lambda$ of the cumulative distribution function of the gamma distribution with parameter $k + 1$. In particular, $P(X = k) = S_k(\lambda) - S_{k+1}(\lambda)$. The sum of independent random variables $X_1, \ldots, X_n$, obeying the Poisson distributions with parameters $\lambda_1, \ldots, \lambda_n$, respectively, has the Poisson distribution with parameter $\lambda_1 + \cdots + \lambda_n$.

The probability-generating function and the characteristic function have the form

$$\varphi_X(z) = e^{\lambda(z-1)},$$
$$f(t) = e^{\lambda(e^{it}-1)},$$

and the numerical characteristics are given by the expressions

$$E\{X\} = \lambda, \quad \mathrm{Var}\{X\} = \lambda, \quad \alpha_2 = \lambda^2 + \lambda, \quad \alpha_3 = \lambda(\lambda^2 + 3\lambda + 1),$$
$$\alpha_4 = \lambda(\lambda^3 + 6\lambda^2 + 7\lambda + 1), \quad \mu_3 = \lambda, \quad \mu_4 = 3\lambda^2 + \lambda, \quad \gamma_1 = \lambda^{-1/2}, \quad \gamma_2 = \lambda^{-1}.$$

The Poisson distribution is the limit distribution for many discrete distributions such as the hypergeometric distribution, the binomial distribution, the negative binomial distribution, distributions arising in problems of arrangement of particles in cells, etc. The Poisson distribution is an acceptable model for describing the random number of occurrences of certain events on a given time interval in a given domain in space.

### 20.2.3-5. Negative binomial distribution.

A random variable $X$ has the *negative binomial distribution* with parameters $(r, p)$ (see Fig. 20.6) if

$$P(X = k) = C_{r+k-1}^{r-1} p^r (1-p)^k, \quad k = 0, 1, \ldots, r, \qquad (20.2.3.6)$$

where $0 < p < 1$, $r > 0$.

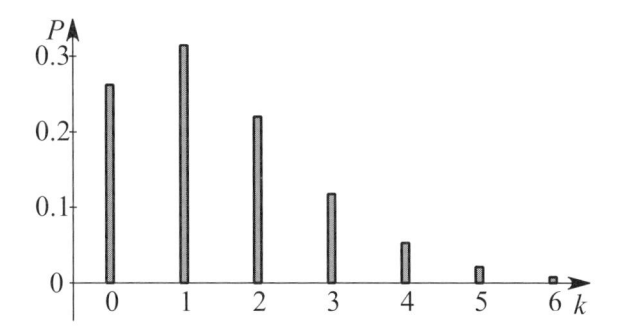

**Figure 20.6.** Negative binomial distribution for $p = 0.8$, $n = 6$.

The probability-generating function and the characteristic function have the form

$$\varphi_X(z) = \left[\frac{p}{1-(1-p)z}\right]^r,$$

$$f(t) = \left[\frac{p}{1-(1-p)e^{it}}\right]^r,$$

and the numerical characteristics can be calculated by the formulas

$$E\{X\} = \frac{r(1-p)}{p}, \quad \text{Var}\{X\} = \frac{r(1-p)}{p^2}, \quad \gamma_1 = \frac{2-p}{\sqrt{r(1-p)}}, \quad \gamma_2 = \frac{6}{r} + \frac{p^2}{r(1-p)}.$$

The negative binomial distribution describes the number $X$ of failures before the $r$th success in a Bernoulli process with probability $p$ of success on each trial. For $r = 1$, the negative binomial distribution coincides with the geometric distribution.

### 20.2.4. Continuous Distributions

**20.2.4-1 Uniform distribution.**

A random variable $X$ is *uniformly* distributed on the interval $[a, b]$ (Fig. 20.7a) if

$$p(x) = \frac{1}{b-a} \quad \text{for } x \in [a, b]. \tag{20.2.4.1}$$

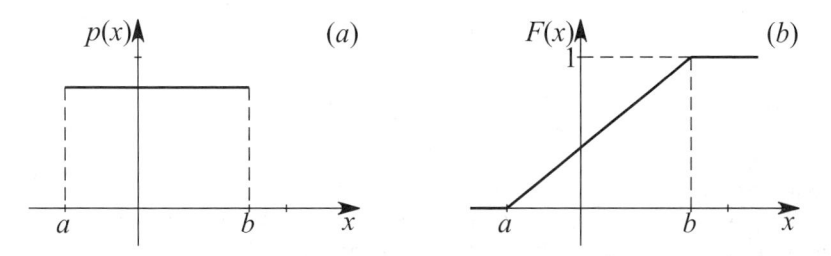

**Figure 20.7.** Probability density (*a*) and cumulate distribution (*b*) functions of uniform distribution.

The cumulative distribution function (see Fig. 20.7b) and the characteristic function have the form

$$F(x) = \begin{cases} 0 & \text{for } x \leq a, \\ \dfrac{x-a}{b-a} & \text{for } a < x \leq b, \\ 1 & \text{for } x > b, \end{cases} \quad (20.2.4.2)$$

$$f(t) = \frac{1}{t(b-a)}(e^{itb} - e^{ita}),$$

and the numerical characteristics are given by the expressions

$$E\{X\} = \frac{a+b}{2}, \quad \text{Var}\{X\} = \frac{(b-a)^2}{12}, \quad \gamma_1 = 0, \quad \gamma_2 = -1.2, \quad \text{Med}\{X\} = \frac{a+b}{2}(a+b).$$

The uniform distribution does not have a mode.

### 20.2.4-2. Exponential distribution.

A random variable $X$ has the *exponential distribution* with parameter $\lambda > 0$ (Fig. 20.8a) if

$$p(x) = \lambda e^{-\lambda x}, \quad x > 0. \quad (20.2.4.3)$$

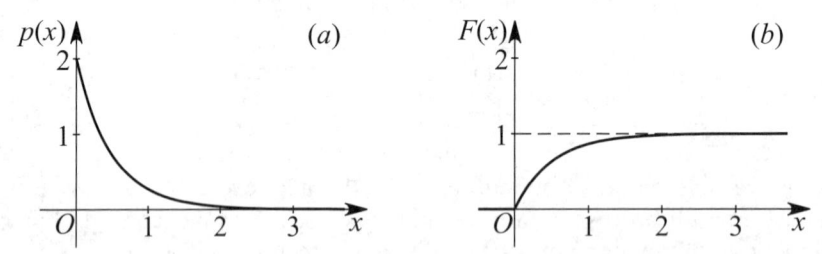

**Figure 20.8.** Probability density (a) and cumulate distribution (b) functions of exponential distribution for $\lambda = 2$.

The cumulative distribution function (see Fig. 20.8b) and the characteristic function have the form

$$F(x) = \begin{cases} 1 - e^{-\lambda x} & \text{for } x > 0, \\ 0 & \text{for } x \leq 0, \end{cases} \quad (20.2.4.4)$$

$$f(t) = \left(1 - \frac{it}{\lambda}\right)^{-1},$$

and the numerical characteristics are given by the formulas

$$E\{X\} = \frac{1}{\lambda}, \quad \alpha_2 = \frac{2}{\lambda^2}, \quad \text{Med}\{X\} = \frac{\ln 2}{\lambda}, \quad \text{Var}\{X\} = \frac{1}{\lambda^2}, \quad \gamma_1 = 2, \quad \gamma_2 = 6.$$

The exponential distribution is the continuous analog of the geometric distribution and is memoryless:

$$P(X > t + s | X > s) = P(X > s).$$

The exponential distribution is closely related to Poisson processes: if a flow of events is described by a Poisson process, then the time intervals between successive events are independent random variables obeying the exponential distribution. The exponential distribution is used in queuing theory and theory of reliability.

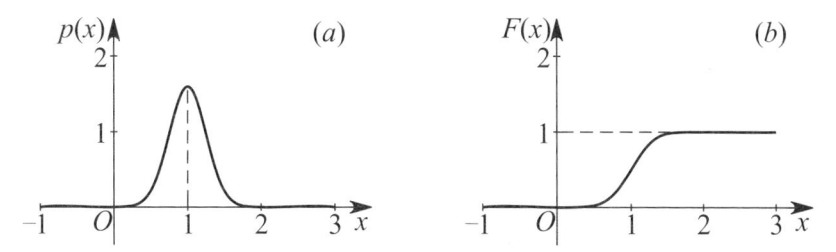

**Figure 20.9.** Probability density (*a*) and cumulate distribution (*b*) functions of normal distribution for $a = 1$, $\sigma = 1/4$.

### 20.2.4-3. Normal distribution.

A random variable $X$ has the *normal distribution* with parameters $(a, \sigma^2)$ (see Fig. 20.9*a*) if its probability density function has the form

$$p(x) = \frac{1}{\sqrt{2\pi}\sigma} \exp\left[-\frac{(x-a)^2}{2\sigma^2}\right], \quad x \in (-\infty, \infty). \tag{20.2.4.5}$$

The cumulative distribution function (see Fig. 20.9*b*) and the characteristic function have the form

$$F(x) = \frac{1}{\sqrt{2\pi}\sigma} \int_{-\infty}^{x} \exp\left[-\frac{(t-a)^2}{2\sigma^2}\right] dt,$$

$$f(t) = \exp\left[iat - \frac{\sigma^2 t^2}{2}\right], \tag{20.2.4.6}$$

and the numerical characteristics are given by the formulas

$$E\{X\} = a, \quad \text{Var}\{X\} = \sigma^2, \quad \text{Mode}\{X\} = \text{Med}\{X\} = a, \quad \gamma_1 = 0, \quad \gamma_2 = 0,$$

$$\mu_k = \begin{cases} 0, & k = 2m-1, \, m = 1, 2, \ldots \\ (2k-1)!! \, \sigma^{2k}, & k = 2m, \, m = 1, 2, \ldots \end{cases}$$

The linear transformation $Y = \frac{X-a}{\sigma}$ reduces the normal distribution with parameters $(a, \sigma^2)$ and cumulative distribution function $F(x)$ to the standard normal distribution with parameters $(0, 1)$ and cumulative distribution function

$$\Phi(x) = \frac{1}{\sqrt{2\pi}} \int_{-\infty}^{x} e^{-t^2/2} \, dt; \tag{20.2.4.7}$$

moreover, $\Phi(-x) = 1 - \Phi(x)$.

**Remark 1.** The values of the cumulative distribution function $\Phi(x)$ of the standard normal distribution are computed by the function NORMSDIST(z) in EXCEL software; for example, for $\Phi(2)$, the function call NORMSDIST(2) returns the value 0.9972.

**Remark 2.** The values of the cumulative distribution function $F(x)$ of the normal distribution are computed by the function pnorm(x,mu,sigma) in MATHCAD software; to compute $\Phi(x)$, one should use pnorm(x,0,1). For example, $\Phi(2) = \text{pnorm}(2,0,1) = 0.9972$.

The probability that a random variable $X$ normally distributed with parameters $(m, \sigma^2)$ lies in the interval $(a, b)$ is given by the formula

$$P(a < \xi < b) = P\left(\frac{a-m}{\sigma} < \frac{\xi-m}{\sigma} < \frac{b-m}{\sigma}\right) = \Phi\left(\frac{b-m}{\sigma}\right) - \Phi\left(\frac{a-m}{\sigma}\right). \tag{20.2.4.8}$$

A normally distributed random variable takes values close to its expectation with large probability; this is expressed by the *sigma rule*

$$P(|X - m| \geq k\sigma) = 2[1 - \Phi(k)] = \begin{cases} 0.3173 & \text{for } k = 1, \\ 0.0456 & \text{for } k = 2, \\ 0.0027 & \text{for } k = 3. \end{cases}$$

The three-sigma rule is most frequently used.

The fundamental role of the normal distribution is due to the fact that, under mild assumptions, the distribution of a sum of random variables is asymptotically normal as the number of terms increases. The corresponding conditions are given in the central limit theorem.

### 20.2.4-4. Cauchy distribution.

A random variable $X$ obeys the *Cauchy distribution* with parameters $(a, \lambda)$ ($\lambda > 0$) (see Fig. 20.10a) if

$$p(x) = \frac{\lambda}{\pi[\lambda^2 + (x - a)^2]}, \quad x \in (-\infty, \infty). \tag{20.2.4.9}$$

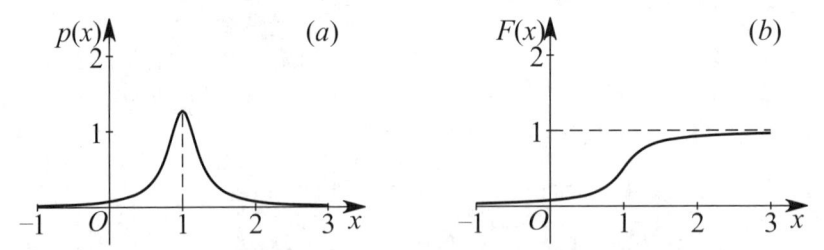

**Figure 20.10.** Probability density (a) and cumulate distribution (b) functions of Cauchy distribution for $a = 1$, $\lambda = 4$.

The cumulative distribution function has the form (see Fig. 20.10b)

$$F(x) = \frac{1}{\pi} \arctan \frac{x - a}{\lambda} + \frac{1}{2}. \tag{20.2.4.10}$$

The numerical characteristics of a random variable that has a Cauchy distribution do not exist in the usual sense. The expectation exists only in the sense of the Cauchy principal value (see Paragraph 10.2.2-3) and is given by the formula

$$E\{X\} = \lim_{T \to \infty} \frac{\lambda}{\pi} \int_{-T}^{T} \frac{x\, dx}{\lambda^2 + (x - a)^2} = a.$$

### 20.2.4-5. Chi-square distribution.

A random variable $X = \chi^2(n)$ has the *chi-square distribution* with $n$ degrees of freedom if its probability density function has the form (see Fig. 20.11a)

$$p(x) = \begin{cases} \dfrac{1}{2^{n/2}\Gamma(\alpha/2)} x^{n/2-1} e^{-x/2} & \text{for } x > 0, \\ 0 & \text{for } x \leq 0. \end{cases} \tag{20.2.4.11}$$

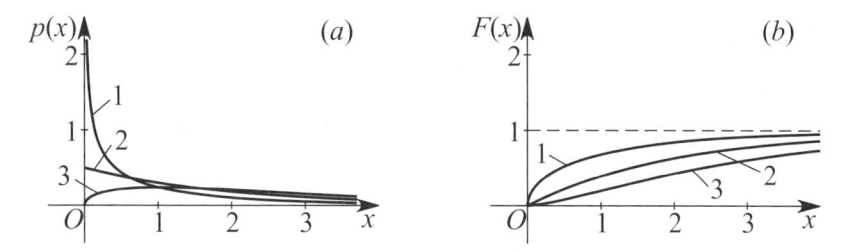

**Figure 20.11.** Probability density (*a*) and cumulate distribution (*b*) functions of chi-square distribution for $n = 1$ (curve 1), $n = 2$ (curve 2), and $n = 3$ (curve 3).

The cumulative distribution function can be written as (see Fig. 20.11*b*)

$$F(x) = \frac{1}{2^{n/2}\Gamma(\alpha/2)} \int_0^x \xi^{n/2-1} e^{-\xi/2}\, d\xi, \qquad (20.2.4.12)$$

where $\Gamma(x)$ is a Gamma function.

**Remark.** The values $\chi^2(x, n)$ of the cumulative distribution function of the chi-square distribution with $n$ degrees of freedom can be obtained using the expression 1 − CHIDIST(x; deg_freedom) in EXCEL software. For example, for the chi-square distribution with 10 degrees of freedom at the point $x = 2$, one gets $\chi^2(2, 10) =$ 1 − CHIDIST(2; 10) = 0.0037. A similar result is obtained if we use the function pchisq(x,n) in MATHCAD software: $\chi^2(2, 10) =$ pchisq(2, 10) = 0.0037.

*Main property of the chi-square distribution.* For an arbitrary $n$, the sum

$$X = \sum_{k=1}^{n} X_k^2,$$

of squares of independent random variables obeying the standard normal distribution has the chi-square distribution with $n$ degrees of freedom.

THEOREM ON DECOMPOSITION. *Suppose that the sum $\sum_{k=1}^{n} X_k^2$ of squares of independent standard normally distributed random variables is expressed as the sum of $L$ quadratic forms $y_j(X_1, \ldots, X_n)$ of ranks $n_j$, respectively. The variables $y_1, \ldots, y_L$ are independent and obey the chi-square distributions with $n_1, \ldots, n_L$ degrees of freedom if and only if $n_1 + \cdots + n_L = n$.*

THEOREM ON ADDITION, OR STABILITY PROPERTY. *The sum of $L$ independent random variables $y_1, \ldots, y_L$ obeying the chi-square distributions with $n_1, \ldots, n_L$ degrees of freedom, respectively, has the chi-square distribution with $n = n_1 + \cdots + n_L$ degrees of freedom.*

The characteristic function has the form

$$f(t) = (1 - 2it)^{-n/2},$$

and the numerical characteristics are given by the formulas

$$E\{\chi^2(n)\} = n, \quad \operatorname{Var}\{\chi^2(n)\} = 2n, \quad \alpha_k = n(n+2)\cdot\ldots\cdot[n+2(k-1)],$$

$$\gamma_1 = 2\sqrt{\frac{2}{n}}, \quad \gamma_2 = \frac{12}{n}, \quad \operatorname{Mode}\{\chi^2(n)\} = n - 2 \quad (n \geq 2).$$

Relationship with other distributions:

1. For $n = 1$, formula (20.2.4.11) gives the probability density function of the square $X^2$ of a random variable with the standard normal distribution.
2. For $n = 2$, formula (20.2.4.11) gives the exponential distribution with parameter $\lambda = \frac{1}{2}$.
3. As $n \to \infty$, the random variable $X = \chi^2(n)$ has an asymptotically normal distribution with parameters $(n, 2n)$.
4. As $n \to \infty$, the random variable $\sqrt{2\chi^2(n)}$ has an asymptotically normal distribution with parameters $(\sqrt{2n-1}, 1)$.

For the quantiles (denoted by $\chi^2_\gamma$ or $\chi^2_\gamma(n)$), one has the approximation formula

$$\chi^2_\gamma(n) \approx \frac{1}{2}(\sqrt{2n-1} + t_\gamma)^2 \quad (n \geq 30),$$

where $t_\gamma$ is the quantile of the standard normal distribution.

For $\gamma$ close to 0 or 1, it is more expedient to use the approximation given by the formula

$$\chi^2_\gamma(n) \approx n\left(1 - \frac{2}{9n} + t_\gamma\sqrt{\frac{2}{9n}}\right)^3.$$

The quantiles $\chi^2_\gamma(n)$ are tabulated; they can also be computed in EXCEL, MATHCAD, and other software.

**Remark.** Tables often list $\chi^2_{1-\gamma}(n)$ rather than $\chi^2_\gamma(n)$.

### 20.2.4-6. Student's $t$-distribution.

A random variable $X = t(n)$ has *Student's distribution (t-distribution) with $n$ degrees of freedom* $(n > 0)$ if its probability density function has the form (see Fig. 20.12a)

$$p(x) = \frac{\Gamma(\frac{n+1}{2})}{\sqrt{n\pi}\,\Gamma(\frac{n}{2})}\left(1 + \frac{x^2}{n}\right)^{-\frac{n+1}{2}}, \quad x \in (-\infty, \infty). \tag{20.2.4.13}$$

where $\Gamma(x)$ is Gamma function.

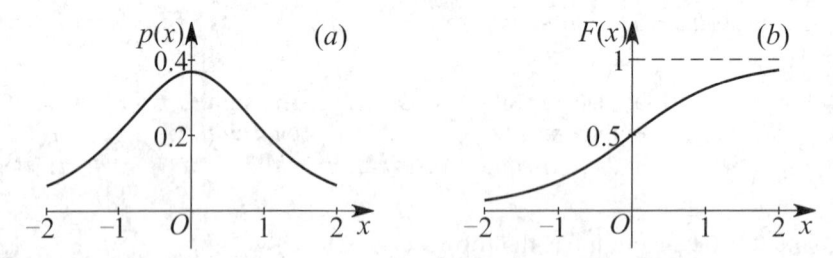

**Figure 20.12.** Probability density (*a*) and cumulate distribution (*b*) functions of Student's $t$-distribution for $n = 3$.

The cumulative distribution function has the form (see Fig. 20.12b)

$$F(x) = \frac{\Gamma(\frac{n+1}{2})}{\sqrt{n\pi}\,\Gamma(\frac{n}{2})}\int_{-\infty}^{x}\left(1 + \frac{\xi^2}{n}\right)^{-\frac{n+1}{2}} d\xi. \tag{20.2.4.14}$$

**Remark.** The values of Student's distribution function $t(n)$ with $n$ degrees of freedom can be computed, for example, by using the function pt(x,n) in MATHCAD software.

*Main property of Student's distribution.* If $\eta$ and $\chi^2(n)$ are independent random variables and $\eta$ has the standard normal distribution, then the random variable

$$t(n) = \eta\sqrt{\frac{n}{\chi^2(n)}}$$

has Student's distribution with $n$ degrees of freedom.

The numerical characteristics are given by the formulas

$$E\{t(n)\} = 0 \quad (n > 1), \quad \text{Var}\{t(n)\} = \begin{cases} \dfrac{n}{n-2} & \text{for } n > 2, \\ 0 & \text{for } n \leq 2, \end{cases}$$

$$\text{Mode}\{t(n)\} = \text{Med}\{t(n)\} = 0 \quad (n > 1),$$

$$\alpha_{2k-1} = 0, \quad \alpha_{2k} = \frac{n^k \Gamma(n/2 - k)\Gamma(k + 1/2)}{\sqrt{\pi}\,\Gamma(n/2)} \quad (2k < n),$$

$$\gamma_1 = 0, \quad \gamma_2 = \frac{3(n-2)}{n-4} \quad (n > 4).$$

Relationship with other distributions:
1. For $n = 1$, Student's distribution coincides with the Cauchy distribution.
2. As $n \to \infty$, Student's distribution is asymptotically normal with parameters $(0, 1)$.

The quantiles of Student's distribution are denoted by $t_\gamma(n)$ and satisfy

$$t_\gamma(n) = -t_{1-\gamma}(n), \quad |t_{1-\gamma}(n)| = t_{1-\gamma/2}(n).$$

The quantiles $t_\gamma(n)$ are tabulated; they can also be computed in EXCEL, MATHCAD, and other software.

Student's distribution is used when testing the hypothesis about the mean of a normally distributed population with unknown variance.

### 20.2.5. Multivariate Random Variables

**20.2.5-1. Distribution of bivariate random variable.**

Suppose that random variables $X_1, \ldots, X_n$ are defined on a probability space $(\Omega, \mathcal{F}, P)$; then one says that an *$n$-dimensional random vector* $\mathbf{X} = (X_1, \ldots, X_n)$ or a *system of random variables* is given. The random variables $X_1, \ldots, X_n$ can be viewed as the coordinates of points in an $n$-dimensional space.

The *distribution function* $F(x_1, x_2) = F_{X_1, X_2}(x_1, x_2)$ *of a two-dimensional random vector* $(X_1, X_2)$, or the *joint distribution function of the random variables* $X_1$ and $X_2$, is defined as the probability of the simultaneous occurrence (intersection) of the events $(X_1 < x_1)$ and $(X_2 < x_2)$; i.e.,

$$F(x_1, x_2) = F_{X_1, X_2}(x_1, x_2) = P(X_1 < x_1, X_2 < x_2). \tag{20.2.5.1}$$

Geometrically, $F(x_1, x_2)$ can be interpreted as the probability that the random point $(X_1, X_2)$ lies in the lower left infinite quadrant with vertex $(x_1, x_2)$ (see Fig. 20.13).

Given the joint distribution of random variables $X_1$ and $X_2$, one can find the distributions of each of the random variables $X_1$ and $X_2$, known as the *marginal distributions*:

$$\begin{aligned} F_{X_1}(x_1) &= P(X_1 < x_1) = P(X_1 < x_1, X_2 < +\infty) = F_{X_1, X_2}(x_1, +\infty), \\ F_{X_2}(x_2) &= P(X_2 < x_2) = P(X_1 < +\infty, X_2 < x_2) = F_{X_1, X_2}(+\infty, x_2). \end{aligned} \tag{20.2.5.2}$$

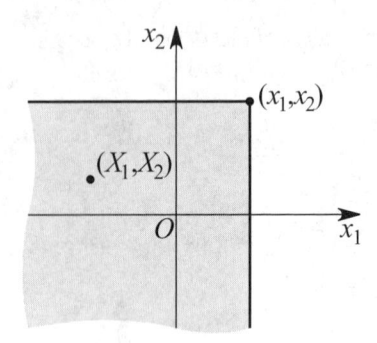

**Figure 20.13.** Geometrically interpretation of the distribution function $F_{X_1,X_2}(x_1,x_2)$.

The marginal distributions do not completely characterize the two-dimensional random variable $(X_1, X_2)$; i.e., the joint distribution of the random variables $X_1$ and $X_2$ cannot in general be reconstructed from the marginal distributions.

Properties of the joint distribution function of random variables $X_1$ and $X_2$:
1. The function $F(x_1, x_2)$ is a nondecreasing function of each of the arguments.
2. $F(x_1, -\infty) = F(-\infty, x_2) = F(-\infty, -\infty) = 0$.
3. $F(+\infty, +\infty) = 1$.
4. The probability that the random vector lies in a rectangle with sides parallel to the coordinate axes is

$$P(a_1 \leq X_1 < b_1, a_2 \leq X_2 < b_2) = F(b_1, b_2) - F(b_1, a_2) - F(a_1, b_2) + F(a_1, a_2).$$

5. The function $F(x_1, x_2)$ is left continuous in each of the arguments.

### 20.2.5-2. Discrete bivariate random variables.

A bivariate random variable $(X_1, X_2)$ is said to be *discrete* if each of the random variables $X_1$ and $X_2$ is discrete.

If the random variable $X_1$ takes the values $x_{11}, \ldots, x_{1m}$ and the random variable $X_2$ takes the values $x_{21}, \ldots, x_{2n}$, then the random vector $(X_1, X_2)$ can take only the pairs of values $(x_{1i}, x_{2j})$ ($i = 1, \ldots, m, j = 1, \ldots, n$). It is convenient to describe the distribution of a bivariate discrete random variable using the distribution matrix shown in Fig. 20.14.

| $X_1 \backslash X_2$ | $x_{21}$ | $\ldots$ | $x_{2n}$ | $P_{X_1}$ |
|---|---|---|---|---|
| $x_{11}$ | $p_{11}$ | $\ldots$ | $p_{1n}$ | $P_{X_1,1}$ |
| $\ldots$ | $\ldots$ | $\ldots$ | $\ldots$ | $\ldots$ |
| $x_{1m}$ | $p_{m1}$ | $\ldots$ | $p_{mn}$ | $P_{X_1,m}$ |
| $P_{X_2}$ | $P_{X_2,1}$ | $\ldots$ | $P_{X_2,n}$ | |

**Figure 20.14.** Distribution matrix.

The entries $p_{ij} = P(X_1 = x_{1i}, X_2 = x_{2j})$ of the distribution matrix are the probabilities of the simultaneous occurrence of the events $(X_1 = x_{1i})$ and $(X_2 = x_{2j})$; $P_{X_1,i} = p_{i1} + \cdots + p_{in}$ is the probability that the random variable $X_1$ takes the value $x_{1i}$; $P_{X_2,j} = p_{1j} + \cdots + p_{mj}$

## 20.2. RANDOM VARIABLES AND THEIR CHARACTERISTICS

is the probability that the random variable $X_2$ takes the value $x_{2j}$; the last column (resp., row) shows the distribution of the random variable $X_1$ (resp., $X_2$).

The distribution function of a discrete bivariate random variable can be determined by the formula

$$F(x_1, x_2) = \sum_{\substack{x_{1i} < x_1 \\ x_{2j} < x_2}} p_{ij}. \qquad (20.2.5.3)$$

### 20.2.5-3. Continuous bivariate random variables.

A bivariate random variable $(X_1, X_2)$ is said to be *continuous* if its joint distribution function $F(x_1, x_2)$ can be represented as

$$F(x_1, x_2) = \int_{-\infty}^{x_2} \int_{-\infty}^{x_1} p(y_1, y_2) \, dy_1 \, dy_2, \qquad (20.2.5.4)$$

where the *joint probability function* $p(x_1, x_2) = p_{X_1, X_2}(x_1, x_2)$ is piecewise continuous.

The joint probability function can be expressed via the joint distribution function as follows:

$$p_{X_1, X_2}(x_1, x_2) = p(x_1, x_2) = F_{x_1 x_2}(x_1, x_2). \qquad (20.2.5.5)$$

Formulas (20.2.5.4) and (20.2.5.5) establish a one-to-one correspondence (up to sets of probability zero) between the joint probability functions and the joint distribution functions of continuous bivariate random variables. The differential $p(x_1, x_2) \, dx_1 \, dx_2$ is called a *probability element*. Up to higher-order infinitesimals, the probability element is equal to the probability for the random variable $(X_1, X_2)$ to lie in the infinitesimal rectangle $(x_1, x_1 + \Delta x_1) \times (x_2, x_2 + \Delta x_2)$.

The probability density function of the two-dimensional random variable $(X_1, X_2)$, which is also called the *joint probability function* of the random variables $X_1$ and $X_2$, determines the probability density functions of the random variables $X_1$ and $X_2$, which are called the marginal probability functions of the two-dimensional random variable $(X_1, X_2)$, by the formulas

$$p_{X_1}(x_1) = \int_{-\infty}^{+\infty} p_{X_1, X_2}(x_1, x_2) \, dx_2, \quad p_{X_2}(x_2) = \int_{-\infty}^{+\infty} p_{X_1, X_2}(x_1, x_2) \, dx_1. \quad (20.2.5.6)$$

In the general case, the joint probability function cannot be reconstructed from the marginal probability functions, and hence the latter do not completely characterize the bivariate random variable $(X_1, X_2)$.

Properties of the joint probability function of random variables $X_1$ and $X_2$:

1. The function $p(x_1, x_2)$ is nonnegative; i.e., $p(x_1, x_2) \geq 0$.
2. $\int_{-\infty}^{+\infty} \int_{-\infty}^{+\infty} p(x_1, x_2) \, dx_1 \, dx_2 = 1$.
3. $P(a_1 < X_1 < b_1, a_2 < X_2 < b_2) = \int_{a_1}^{b_1} dx_1 \int_{a_2}^{b_2} p(x_1, x_2) \, dx_2 = \int_{a_2}^{b_2} dx_2 \int_{a_1}^{b_1} p(x_1, x_2) \, dx_1$.
4. The probability for a two-dimensional random variable $(X_1, X_2)$ to lie in a domain $D \subset R^2$ is numerically equal to the volume of the curvilinear cylinder with base $D$ bounded above by the surface of the joint probability function:

$$P[(X_1, X_2) \in D] = \iint_{(x_1, x_2) \in D} p_{X_1, X_2}(x_1, x_2) \, dx_1 \, dx_2.$$

Random variables $X_1$ and $X_2$ are said to be *independent* if the relation

$$P(X_1 \in S_1, X_2 \in S_2) = P(X_1 \in S_1) P(X_2 \in S_2) \qquad (20.2.5.7)$$

holds for any measurable sets $S_1$ and $S_2$.

THEOREM 1. *Random variables $X_1$ and $X_2$ are independent if and only if*

$$F_{X_1,X_2}(x_1, x_2) = F_{X_1}(x_1) F_{X_2}(x_2).$$

THEOREM 2. *Random variables $X_1$ and $X_2$ are independent if and only if the characteristic function of the bivariate random variable $(X_1, X_2)$ is equal to the product of the characteristic functions of $X_1$ and $X_2$,*

$$f_{X_1,X_2}(x_1, x_2) = f_{X_1}(x_1) f_{X_2}(x_2).$$

20.2.5-4. **Numerical characteristics of bivariate random variables.**

The *expectation* of a function $g(X_1, X_2)$ of a bivariate random variable $(X_1, X_2)$ is defined as the expression computed by the formula

$$E\{g(X_1, X_2)\} = \begin{cases} \sum_i \sum_j g(x_{1i}, x_{2j}) p_{ij} & \text{in the discrete case,} \\ \int_{-\infty}^{+\infty} \int_{-\infty}^{+\infty} g(x_1, x_2) p(x_1, x_2) \, dx_1 \, dx_2 & \text{in the continuous case,} \end{cases} \qquad (20.2.5.8)$$

if these expressions exist in the sense of absolute convergence; otherwise, one says that $E\{g(X_1, X_2)\}$ does not exist.

The *moment of order* $r_1 + r_2$ of a two-dimensional random variable $(X_1, X_2)$ about a point $(a_1, a_2)$ is defined as the expectation $E\{(X_1 - a_1)^{r_1} (X_2 - a_2)^{r_2}\}$.

If $a_1 = a_2 = 0$, then the moment of order $r_1 + r_2$ of a two-dimensional random variable $(X_1, X_2)$ is called simply the moment, or the *initial moment*. The initial moment of order $r_1 + r_2$ is usually denoted by $\alpha_{r_1, r_2}$; i.e., $\alpha_{r_1, r_2} = E\{X_1^{r_1} X_2^{r_2}\}$.

The first initial moments are the expectations of the random variables $X_1$ and $X_2$; i.e., $\alpha_{1,0} = E\{X_1^1 X_2^0\} = E\{X_1\}$ and $\alpha_{0,1} = E\{X_1^0 X_2^1\} = E\{X_2\}$. The point $(E\{X_1\}, E\{X_2\})$ on the $OXY$-plane characterizes the position of the random point $(X_1, X_2)$; this position spreads about the point $(E\{X_1\}, E\{X_2\})$. Obviously, the first central moments are zero.

The second initial moments are given by the formulas

$$\alpha_{2,0} = \alpha_2(X_1), \quad \alpha_{0,2} = \alpha_2(X_2), \quad \alpha_{1,1} = E\{X_1 X_2\}.$$

If $a_1 = E\{X_1\}$ and $a_2 = E\{X_2\}$, then the moment of order $r_1 + r_2$ of the bivariate random variable $(X_1, X_2)$ is called the *central moment*. The central moment of order $r_1 + r_2$ is usually denoted by $\mu_{r_1, r_2}$; i.e., $\mu_{r_1, r_2} = E\{(X_1 - E\{X_1\})^{r_1} (X_2 - E\{X_2\})^{r_2}\}$.

The second central moments are of special interest and have special names and notation:

$$\lambda_{11} = \mu_{2,0} = \text{Var}\{X_1\}, \quad \lambda_{22} = \mu_{0,2} = \text{Var}\{X_2\},$$

$$\lambda_{12} = \lambda_{21} = \mu_{1,1} = E\{(X_1 - E\{X_1\})(X_2 - E\{X_2\})\}.$$

The first two of these moments are the variances of the respective random variables, and the third moment is called the *covariance* and will be considered below.

### 20.2.5-5. Covariance and correlation of two random variables.

The *covariance* (correlation moment, or mixed second moment) $\mathrm{Cov}(X_1, X_2)$ of random variables $X_1$ and $X_2$ is defined as the central moment of order $(1 + 1)$:

$$\mathrm{Cov}(X_1, X_2) = \alpha_{1,1} = E\{(X_1 - E\{X_1\})(X_2 - E\{X_2\})\}. \tag{20.2.5.9}$$

Properties of the covariance:

1. $\mathrm{Cov}(X_1, X_2) = \mathrm{Cov}(X_2, X_1)$.
2. $\mathrm{Cov}(X, X) = \mathrm{Var}\{X\}$.
3. If the random variables $X_1$ and $X_2$ are independent, then $\mathrm{Cov}(X_1, X_2) = 0$. If $\mathrm{Cov}(X_1, X_2) \neq 0$, then the random variables $X_1$ and $X_2$ are dependent.
4. If $Y_1 = a_1 X_1 + b_1$ and $Y_2 = a_2 X_2 + b_2$, then $\mathrm{Cov}(Y_1, Y_2) = a_1 a_2 \mathrm{Cov}(X_1, X_2)$.
5. $\mathrm{Cov}(X_1, X_2) = E\{X_1 X_2\} - E\{X_1\} E\{X_2\}$.
6. $|\mathrm{Cov}(X_1, X_2)| \leq \sqrt{\mathrm{Var}\{X_1\} \mathrm{Var}\{X_2\}}$. Moreover, $\mathrm{Cov}(X_1, X_2) = \pm\sqrt{\mathrm{Var}\{X_1\} \mathrm{Var}\{X_2\}}$ if and only if the random variables $X_1$ and $X_2$ are linearly dependent.
7. $\mathrm{Var}\{X_1 + X_2\} = \mathrm{Var}\{X_1\} + \mathrm{Var}\{X_2\} + 2\mathrm{Cov}(X_1, X_2)$.

If $\mathrm{Cov}(X_1, X_2) = 0$, then the random variables $X_1$ and $X_2$ are said to be *uncorrelated*; if $\mathrm{Cov}(X_1, X_2) \neq 0$, then they are *correlated*. Independent random variables are always uncorrelated, but correlated random variables are not necessarily independent in general.

**Example 1.** Suppose that we throw two dice. Let $X_1$ be the number of spots on top of the first die, and let $X_2$ be the number of spots on top of the second die. We consider the random variables $Y_1 = X_1 + X_2$ and $Y_2 = X_1 - X_2$ (the sum and difference of points obtained). Then

$$\mathrm{Cov}(Y_1, Y_2) = E\{(X_1 + X_2 - E\{X_1 + X_2\})(X_1 - X_2 - E\{X_1 - X_2\})\} =$$
$$= E\{(X_1 - E\{X_1\})^2 - (X_2 - E\{X_2\})^2\} = \mathrm{Var}\{X_1\} - \mathrm{Var}\{X_2\} = 0,$$

since $X_1$ and $X_2$ are identically distributed and hence $\mathrm{Var}\{X_1\} = \mathrm{Var}\{X_2\}$. But $Y_1$ and $Y_2$ are obviously dependent; for example, if $Y_1 = 2$ then one necessarily has $Y_2 = 0$.

The covariance of random variables $X_1$ and $X_2$ characterizes both the degree of their dependence on each other and their spread around the point $(E\{X_1\}, E\{X_2\})$. The covariance of $X_1$ and $X_2$ has the dimension equal to the product of dimensions of $X_1$ and $X_2$. Along with the covariance of random variables $X_1$ and $X_2$, one often uses the correlation $\rho(X_1, X_2)$, which is a dimensionless normalized variable. The *correlation* (or *correlation coefficient*) of random variables $X_1$ and $X_2$ is the ratio of the covariance of $X_1$ and $X_2$ to the product of their standard deviations,

$$\rho(X_1, X_2) = \frac{\mathrm{Cov}(X_1, X_2)}{\sigma_{X_1} \sigma_{X_2}}. \tag{20.2.5.10}$$

The correlation of random variables $X_1$ and $X_2$ indicates the degree of linear dependence between the variables. If $\rho(X_1, X_2) = 0$, then there is no linear relation between the random variables, but there may well be some different relation between them.

Properties of the correlation:

1. $\rho(X_1, X_2) = \rho(X_2, X_1)$.
2. $\rho(X, X) = 1$.
3. If random variables $X_1$ and $X_2$ are independent, then $\rho(X_1, X_2) = 0$. If $\rho(X_1, X_2) \neq 0$, then the random variables $X_1$ and $X_2$ are dependent.
4. If $Y_1 = a_1 X_1 + b_1$ and $Y_2 = a_2 X_2 + b_2$, then $\rho(X_1, X_2) = \pm\rho(X_1, X_2)$.
5. $|\rho(X_1, X_2)| \leq 1$. Moreover, $\rho(X_1, X_2) = \pm 1$ if and only if the random variables $X_1$ and $X_2$ are linearly dependent.

### 20.2.5-6. Conditional distributions.

The joint distribution of random variables $X_1$ and $X_2$ determines the *conditional distribution* of one of the random variables given that the other random variable takes a certain value (or lies in a certain interval). If the joint distribution is discrete, then the conditional distributions of $X_1$ and $X_2$ are also discrete. The conditional distributions are described by the formulas

$$P_{1|2}(x_{1i}|x_{2j}) = P(X_1 = x_{1i}|X_2 = x_{2j}) = \frac{P(X_1 = x_{1i}, X_2 = x_{2j})}{P(X_2 = x_{2j})} = \frac{p_{ij}}{p_{X_2,j}},$$

$$P_{2|1}(x_{2j}|x_{1i}) = P(X_2 = x_{2j}|X_1 = x_{1i}) = \frac{P(X_1 = x_{1i}, X_2 = x_{2j})}{P(X_1 = x_{1i})} = \frac{p_{ij}}{p_{X_1,i}}, \quad (20.2.5.11)$$

$$i = 1, \ldots, m; \ j = 1, \ldots, n.$$

The probabilities $P_{1|2}(x_{1i}|x_{2j})$, $j = 1, \ldots, n$, define the conditional probability mass function of the random variable $X_2$ given $X_1 = x_{1i}$; and the probabilities $P_{2|1}(x_{2j}|x_{1i})$, $i = 1, \ldots, m$, define the conditional probability mass function of the random variable $X_1$ given $X_2 = x_{2j}$. These conditional probability mass functions have the properties of usual probability mass functions; for example, the sum of probabilities in each of them is equal to 1,

$$\sum_i P_{1|2}(x_{1i}|x_{2j}) = \sum_j P_{2|1}(x_{2j}|x_{1i}) = 1.$$

If the joint distribution is continuous, then the conditional distributions of the random variables $X_1$ and $X_2$ are also continuous and are described by the conditional probability density functions

$$p_{1|2}(x_1|x_2) = \frac{p_{X_1,X_2}(x_1,x_2)}{p_{X_2}(x_2)}, \quad p_{2|1}(x_2|x_1) = \frac{p_{X_1,X_2}(x_1,x_2)}{p_{X_1}(x_1)}. \quad (20.2.5.12)$$

The conditional distributions of the random variables $X_1$ and $X_2$ can also be described by the *conditional cumulative distribution functions*

$$F_{X_2}(x_2|X_1 = x_1) = P(X_2 < x_2|X_1 = x_1),$$
$$F_{X_1}(x_1|X_2 = x_2) = P(X_1 < x_1|X_2 = x_2). \quad (20.2.5.13)$$

The total probability formulas for the cumulative distribution functions of continuous random variables have the form

$$F_{X_2}(x_2) = \int_{-\infty}^{+\infty} F_{X_2}(x_2|X_1 = x_1) p_{X_1}(x_1) dx_1,$$
$$F_{X_1}(x_1) = \int_{-\infty}^{+\infty} F_{X_1}(x_1|X_2 = x_2) p_{X_2}(x_2) dx_2. \quad (20.2.5.14)$$

THEOREM ON MULTIPLICATION OF DENSITIES. *The joint probability function for two random variables is equal to the product of the probability density function of one random variable by the conditional probability density function of the other random variable, given the value of the first random variable:*

$$p_{X_1,X_2}(x_1,x_2) = p_{X_2}(x_2)p_{1|2}(x_1 x_2) = p_{X_1}(x_1)p_{2|1}(x_2|x_1). \quad (20.2.5.15)$$

*Bayes' formulas:*

$$P_{1|2}(x_{1i}|x_{2j}) = \frac{p_{X_1}(x_{1i})p_{2|1}(x_{2j}|x_{1i})}{\sum_i P(X_1 = x_{1i}) P_{2|1}(x_{2j}x_{1i})},$$

$$P_{2|1}(x_{2j}|x_{1i}) = \frac{p_{X_2}(x_{2j})p_{1|2}(x_{1i}|x_{2j})}{\sum_j P(X_2 = x_{2j})p_{1|2}(x_{1i}|x_{2j})};$$

(20.2.5.16)

$$p_{1|2}(x_1|x_2) = \frac{p_{X_1}(x_1)p_{2|1}(x_2|x_1)}{\int_{-\infty}^{+\infty} p_{X_1}(x_1)p_{2|1}(x_2|x_1)\,dx_1},$$

$$p_{2|1}(x_2|x_1) = \frac{p_{X_2}(x_2)p_{1|2}(x_1|x_2)}{\int_{-\infty}^{+\infty} p_{X_2}(x_2)p_{1|2}(x_1|x_2)\,dx_2}.$$

(20.2.5.17)

**20.2.5-7. Conditional expectation. Regression.**

The *conditional expectation* of a discrete random variable $X_2$, given $X_1 = x_1$ (where $x_1$ is a possible value of the random variable $X_1$), is defined to be the sum of products of possible values of $X_2$ by their conditional probabilities,

$$E\{X_2|X_1 = x_1\} = \sum_j x_{2j}p_{2|1}(x_{2j}|x_1). \quad (20.2.5.18)$$

For continuous random variables,

$$E\{(X_2|X_1 = x_1\} = \int_{-\infty}^{+\infty} x_2 p_{2|1}(x_2|x_1)\,dx_2. \quad (20.2.5.19)$$

Properties of the conditional expectation:
1. If random variables $X$ and $Y$ are independent, then their conditional expectations coincide with the unconditional expectations; i.e., $E\{Y|X = x\} = E\{Y\}$ and $E\{X|Y = y\} = E\{X\}$.
2. $E\{f(X)h(Y)|X = x\} = f(x)E\{h(Y)|X = x\}$.
3. *Additivity of the conditional expectation*:

$$E\{Y_1 + Y_2|X\} = E\{Y_1|X = x\} + E\{Y_2|X = x\}.$$

A function $g_2(X_1)$ is called the *best mean-square approximation* to a random variable $X_2$ if the expectation $E\{[X_2 - g_2(X_1)]^2\}$ takes the least possible value; the function $g_2(x_1)$ is called the *mean-square regression* of $X_2$ on $X_1$.

The conditional expectation $E\{X_2|X_1\}$ is a function of $X_1$,

$$E\{X_2|X_1\} = g_2(X_1). \quad (20.2.5.20)$$

It is called the *regression function* of $X_2$ on $X_1$ and is the mean-square regression of $X_2$ on $X_1$.

In a majority of cases, it suffices to approximate the regression (20.2.5.20) by the linear function

$$\tilde{g}_2(X_1) = \alpha + \beta_{21}X_1 = E\{X_2\} + \beta_{21}(X_1 - E\{X_1\}).$$

Here the coefficient $\beta_{21} = \rho_{12}\sigma_{X_2}/\sigma_{X_1}$ is called the *regression coefficient* of $X_2$ on $X_1$ ($\rho_{12} = \rho(X_1, X_2)$). The number $\sigma_{X_2}^2(1-\rho_{12}^2)$ is called the *residual standard deviation* of the random variable $X_2$ with respect to the random variable $X_1$; this number characterizes the error arising if $X_2$ is replaced by the linear function $g_2(X_1) = \alpha + \beta_{21}X_1$.

**Remark 1.** The regression (20.2.5.20) can be approximated more precisely by a polynomial of degree $k > 1$ (parabolic regression of order $k$) or some other nonlinear functions (exponential regression, logarithmic regression, etc.).

**Remark 2.** If $X_2$ is taken for the independent variable, then we obtain the mean-square regression

$$E\{X_1|X_2\} = g_1(X_2)$$

of $X_1$ on $X_2$ and the linear regression

$$\tilde{g}_1(X_2) = E\{X_1\} + \beta_{12}(X_2 - E\{X_2\}), \quad \beta_{12} = \rho_{12}\frac{\sigma_{X_1}}{\sigma_{X_2}}$$

of $X_1$ on $X_2$.

**Remark 3.** All regression lines pass through the point $(E\{X_1\}, E\{X_2\})$.

### 20.2.5-8. Distribution function of multivariate random variable.

The probability $P(X_1 < x_1, \ldots, X_n < x_n)$ treated as a function of a point $\mathbf{x} = (x_1, \ldots, x_n)$ of the $n$-dimensional space and denoted by

$$F_{\mathbf{X}}(\mathbf{x}) = F(\mathbf{x}) = P(X_1 < x_1, \ldots, X_n < x_n) \tag{20.2.5.21}$$

is called the *multiple (or joint) distribution function* of the $n$-dimensional random vector $\mathbf{X} = (X_1, \ldots, X_n)$.

Properties of the joint distribution function of a random vector $\mathbf{X}$:

1. $F(\mathbf{x})$ is a nondecreasing function in each of the arguments.
2. If at least one of the arguments $x_1, \ldots, x_n$ is equal to $-\infty$, then the joint distribution function is equal to zero.
3. The $m$-dimensional distribution function of the subsystem of $m < n$ random variables $X_1, \ldots, X_m$ can be determined if the arguments corresponding to the remaining random variables $X_{m+1}, \ldots, X_n$ are set to $+\infty$,

$$F_{X_1, \ldots, X_m}(x_1, \ldots, x_m) = F_{\mathbf{X}}(x_1, \ldots, x_m, -\infty, \ldots, +\infty).$$

(The $m$-dimensional distribution function $F_{X_1, \ldots, X_m}(x_1, \ldots, x_m)$ is usually called the *marginal distribution function*.)

4. The function $F_{\mathbf{X}}(\mathbf{x})$ is left continuous in each of the arguments.

An $n$-dimensional random variable $\mathbf{X}$ is said to be *discrete* if each of the random variables $X_1, X_2, \ldots, X_n$ is discrete. The distribution of a subsystem $X_1, \ldots, X_m$ of random variables and the conditional distributions are defined as in Paragraphs 20.2.5-6 and 20.2.5-7.

An $n$-dimensional random variable $\mathbf{X}$ is said to be *continuous* if its distribution function $F(\mathbf{x})$ can be written in the form

$$F(\mathbf{x}) = \int\limits_{-\infty < y_1 < x_1} \cdots \int\limits_{-\infty < y_n < x_n} p(\mathbf{y})\, d\mathbf{y}, \tag{20.2.5.22}$$

where $d\mathbf{y} = dy_1 \ldots dy_n$ and the function $p(\mathbf{x})$, called the *multiple (or joint) probability function* of the random variables $X_1, \ldots, X_n$, is piecewise continuous. The joint probability function can be expressed via the joint distribution function by the formula

$$p(\mathbf{x}) = \frac{\partial^n F_{\mathbf{X}}(\mathbf{x})}{\partial x_1 \ldots \partial x_n}; \tag{20.2.5.23}$$

i.e., the joint probability function is the $n$th mixed partial derivative (one differentiation in each of the arguments) of the joint distribution function.

Formulas (20.2.5.22) and (20.2.5.23) establish a one-to-one correspondence (up to sets of probability zero) between the joint probability functions and the joint distribution functions of continuous multivariate random variables. The differential $p(\mathbf{x})\, d\mathbf{x}$ is called a *probability element*. The joint probability function of $n$ random variables $X_1, X_2, \ldots, X_n$ has the same properties as the joint probability function of two random variables $X_1$ and $X_2$ (see Paragraph 20.2.1-4.). The marginal probability functions and conditional probability functions obtained from a continuous $n$-dimensional probability distribution are defined precisely as in Paragraphs 20.2.1-4 and 20.2.1-8.

**Remark 1.** The distribution of a system of two or more multivariate random variables $\mathbf{X}_1 = (X_{11}, X_{12}, \ldots)$ and $\mathbf{X}_2 = (X_{21}, X_{22}, \ldots)$ is the joint distribution of all variables $X_{11}, X_{12}, \ldots ; X_{21}, X_{22}, \ldots ; \ldots$

**Remark 2.** A joint distribution can be discrete in some random variables and continuous in the other random variables.

### 20.2.5-9. Numerical characteristics of multivariate random variables.

The *expectation* of a function $g(\mathbf{X})$ of a multivariate random variable $\mathbf{X}$ is defined by the formula

$$E\{g(\mathbf{X})\} = \begin{cases} \displaystyle\sum_{i_1} \cdots \sum_{i_n} g(x_{1i_1}, \ldots, x_{2i_n}) p_{i_1 i_2 \ldots i_n} & \text{in the discrete case,} \\ \displaystyle\int_{-\infty}^{+\infty} \cdots \int_{-\infty}^{+\infty} g(\mathbf{x}) p(\mathbf{x})\, d\mathbf{x} & \text{in the continuous case} \end{cases} \quad (20.2.5.24)$$

if these expressions exist in the sense of absolute convergence; otherwise, one says that $E\{g(\mathbf{X})\}$ does not exist.

The *moment of order* $r_1 + \cdots + r_n$ of a random variable $\mathbf{X}$ about a point $(a_1, \ldots, a_n)$ is defined as the expectation $E\{(X_1 - a_1)^{r_1} \ldots (X_n - a_n)^{r_n}\}$.

For $a_1 = \cdots = a_n = 0$, the moment of order $r_1 + \cdots + r_n$ of an $n$-dimensional random variable $\mathbf{X}$ is called the *initial moment* and is denoted by

$$\alpha_{r_1 \ldots r_n} = E\{X_1^{r_1} \ldots X_n^{r_n}\}.$$

The first initial moments are the expectations of the coordinates $X_1, \ldots, X_n$. The point $(E\{X_1\}, \ldots, E\{X_n\})$ in the space $\mathbb{R}^n$ characterizes the position of the random point $(X_1, \ldots, X_n)$, which spreads about the point $(E\{X_1\}, \ldots, E\{X_n\})$. The first central moments are naturally zero.

If $a_1 = E\{X_1\}, \ldots, a_n = E\{X_n\}$, then the moment of order $r_1 + \cdots + r_n$ of the $n$-dimensional random variable $\mathbf{X}$ is called the *central moment* and is denoted by

$$\mu_{r_1 \ldots r_n} = E\{(X_1 - E\{X_1\})^{r_1} \ldots (X_n - E\{X_n\})^{r_n}\}.$$

The second central moments have the following notation:

$$\lambda_{ij} = \lambda_{ji} = E\{(X_i - E\{X_i\})(X_j - E\{X_j\})\} = \begin{cases} \operatorname{Var}\{X_i\} = \sigma_i^2 & \text{for } i = j, \\ \operatorname{Cov}(X_i, X_j) & \text{for } i \neq j. \end{cases} \quad (20.2.5.25)$$

The moments $\lambda_{ij}$ given by relation (20.2.5.25) determine the *covariance matrix* (*matrix of moments*) $[\lambda_{ij}]$. Obviously, the covariance matrix is real and symmetric; its determinant $\det[\lambda_{ij}]$ is called the *generalized variance of the $n$-dimensional distribution*. The correlations

$$\rho_{ij} = \rho(X_i, X_j) = \frac{\operatorname{Cov}(X_i, X_j)}{\sigma_{X_i} \sigma_{X_j}} = \frac{\lambda_{ij}}{\sqrt{\lambda_{ii} \lambda_{jj}}} \quad (i, j = 1, 2, \ldots, n) \quad (20.2.5.26)$$

determine the *correlation matrix* $[\rho_{ij}]$ of the $n$-dimensional distribution provided that all variances $\text{Var}\{X_i\}$ are nonzero. Obviously, the correlation matrix is real and symmetric. The quantity $\sqrt{\det[\rho_{ij}]}$ is called the spread coefficient.

### 20.2.5-10. Regression.

A function $g_1(X_2, \ldots, X_n)$ is called the *best mean-square approximation* to a random variable $X_1$ if the expectation $E\{[X_1 - g_1(X_2, \ldots, X_n)]^2\}$ takes the least possible value. The function $g_1(x_2, \ldots, x_n)$ is called the *mean-square regression* of $X_1$ on $X_2, \ldots, X_n$.

The *conditional expectation* $E\{X_1 | X_2, \ldots, X_n\}$ is a function of $X_2, \ldots, X_n$,

$$E\{X_1 | X_2, \ldots, X_n\} = g_1(X_2, \ldots, X_n). \tag{20.2.5.27}$$

It is called the *regression function* of $X_1$ on $X_2, \ldots, X_n$ and is the mean-square regression of $X_1$ on $X_2, \ldots, X_n$.

In a majority of cases, it suffices to approximate the regression (20.2.5.27) by the linear function

$$\widetilde{g}_i = E\{X_i\} + \sum_{j \neq i} \beta_{ij}(X_j - E\{X_j\}). \tag{20.2.5.28}$$

Relation (20.2.5.28) determines the linear regression of $X_i$ on the other $n-1$ variables. The regression coefficients $\beta_{ij}$ are determined by the relation

$$\beta_{ij} = -\frac{\Lambda_{ij}}{\Lambda_{ii}},$$

where $\Lambda_{ij}$ are the entries of the inverse of the covariance matrix. The measure of correlation between $X_i$ and the other $n-1$ variables is the *multiple correlation coefficient*

$$\rho(X_i, \widetilde{g}_i) = \sqrt{1 - \frac{1}{\lambda_{ii}\Lambda_{ii}}}.$$

The *residual* of $X_i$ with respect to the other $n-1$ variables is defined as the random variable $\Delta_i = X_i - \widetilde{g}_i$. It satisfies the relations

$$\text{Cov}(\Delta_i, X_j) = \begin{cases} 0 & \text{for } i \neq j, \\ \text{Var}\{\Delta_i\} & \text{for } i = j \text{ (residual variance)}. \end{cases}$$

### 20.2.5-11. Characteristic functions.

The *characteristic function* of a random variable $\mathbf{X}$ is defined as the expectation of the random variable $\exp\left(i \sum_{j=1}^{n} t_j X_j\right)$:

$$f_{\mathbf{X}}(\mathbf{t}) = f(\mathbf{t}) = E\left\{\exp\left(i \sum_{j=1}^{n} t_j X_j\right)\right\}, \tag{20.2.5.29}$$

where $\mathbf{t} = (t_1, \ldots, t_n)$, $i$ is the imaginary unit, $i^2 = -1$.

For continuous random variables,

$$f(\mathbf{t}) = \int_{-\infty}^{+\infty} \cdots \int_{-\infty}^{+\infty} \exp\left(i \sum_{j=1}^{n} t_j x_j\right) p(\mathbf{x}) \, d\mathbf{x}.$$

The inversion formula for a continuous distribution has the form

$$p(\mathbf{x}) = \frac{1}{(2\pi)^n} \int_{-\infty}^{+\infty} \cdots \int_{-\infty}^{+\infty} \exp\left(-i \sum_{j=1}^{n} t_j x_j\right) f(\mathbf{t}) \, d\mathbf{t},$$

where $d\mathbf{t} = dt_1 \ldots dt_n$.

If the initial moments of a random variable $\mathbf{X}$ exist, then

$$E\{X_1^{r_1} \ldots X_n^{r_n}\} = i^{-\sum_{j=1}^{n} r_j} \left. \frac{\partial^{r_1 + \cdots + r_n} f(\mathbf{t})}{\partial t_1^{r_1} \ldots \partial t_n^{r_n}} \right|_{t_1 = \cdots = t_n = 0}.$$

The characteristic function corresponding to the $m$-dimensional marginal distribution of $m$ out of $n$ variables $X_1, \ldots, X_n$ can be obtained from the characteristic function (20.2.5.29) if the variables $t_j$ corresponding to the random variables that are not contained in the $m$-dimensional marginal distribution are replaced by zeros.

CONTINUITY THEOREM FOR CHARACTERISTIC FUNCTIONS. *The weak convergence $F_{X_n} \to F$ of a sequence of distribution functions $F_1(x), F_2(x), \ldots$ is equivalent to the uniform convergence $f_n(t) \to f(t)$ of characteristic functions on each finite interval.*

### 20.2.5-12. Independency of random variables.

Random variables $X_1, \ldots, X_n$ are said to be *independent* if the events $\{X_1 \in S_1\}, \ldots, \{X_n \in S_n\}$ are independent for any measurable sets $S_1, \ldots, S_n$. For this, it is necessary and sufficient that

$$P(X_1 \in S_1, \ldots, X_n \in S_n) = \prod_{k=1}^{n} P(X_k \in S_k). \tag{20.2.5.30}$$

Relation (20.2.5.30) is equivalent to one of the following three:

1. In the general case: for any $\mathbf{x} \in \mathbb{R}^n$,

$$F_{\mathbf{X}}(\mathbf{x}) = \prod_{k=1}^{n} F_{X_k}(x_k).$$

2. For absolutely continuous distributions: for any $\mathbf{x} \in \mathbb{R}^n$ (except possibly for a set of measure zero),

$$p_{\mathbf{X}}(\mathbf{x}) = \prod_{k=1}^{n} p_{X_k}(x_k).$$

3. For discrete distributions: for any $\mathbf{x} \in \mathbb{R}^n$,

$$P(X_1 = x_1, \ldots, X_n = x_n) = \prod_{k=1}^{n} P(X_k = x_k).$$

The joint distribution of independent random variables is uniquely determined by their individual distributions. Independent random variables are uncorrelated, but the converse is not true in general.

Random variables $X_1, \ldots, X_n$ are independent if and only if the characteristic function of the multivariate random variable $\mathbf{X}$ is equal to the product of the characteristic functions of the random variables $X_1, \ldots, X_n$.

## 20.3. Limit Theorems

### 20.3.1. Convergence of Random Variables

**20.3.1-1. Convergence in probability.**

A sequence of random variables $X_1, X_2, \ldots$ is said to *converge in probability* to a random variable $X$ ($X_n \xrightarrow{P} X$) if

$$\lim_{n \to \infty} P(|X_n - X| \geq \varepsilon) = 0 \qquad (20.3.1.1)$$

for each $\varepsilon > 0$, i.e., if for any $\varepsilon > 0$ and $\delta > 0$ there exists a number $N$, depending on $\varepsilon$ and $\delta$, such that the inequality

$$P(|X_n - X| > \varepsilon) < \delta$$

holds for $n > N$. A sequence of $k$-dimensional random variables $\mathbf{X}_n$ is said to converge in probability to a random variable $\mathbf{X}$ if each coordinate of the random variable $\mathbf{X}_n$ converges in probability to the respective coordinate of the random variable $\mathbf{X}$.

**20.3.1-2. Almost sure convergence (convergence with probability 1).**

A sequence of random variables $X_1, X_2, \ldots$ is said to *converge almost surely* (or *with probability* 1) to a random variable $X$ ($X_n \xrightarrow{a.s.} X$) if

$$P[\omega \in \Omega : \lim_{n \to \infty} X_n(\omega) = X(\omega)] = 1. \qquad (20.3.1.2)$$

A sequence $X_n \to X$ converges almost surely if and only if

$$P\Big(\bigcup_{m=1}^{\infty} \{|X_{n+m} - X| \geq \varepsilon\}\Big) \xrightarrow[n \to \infty]{} 0$$

for each $\varepsilon > 0$.

Convergence almost surely implies convergence in probability. The converse statement is not true in general.

**20.3.1-3. Convergence in mean.**

A sequence of random variables $X_1, X_2, \ldots$ with finite $p$th initial moments ($p = 1, 2, \ldots$) is said to *converge in $p$th mean* to a random variable $X$ ($E\{X^p\} < \infty$) if

$$\lim_{n \to \infty} E\{|X_n - X|^p\} = 0. \qquad (20.3.1.3)$$

Convergence in $p$th mean, for $p = 2$ is called convergence in mean square. If $X_n \to X$ in $p$th mean then $X_n \to X$ in $p_1$th mean for all $p_1 \leq p$.

Convergence in $p$th mean implies convergence in probability. The converse statement is not true in general.

## 20.3.1-4. Convergence in distribution.

Suppose that a sequence $F_1(x), F_2(x), \ldots$ of cumulative distribution functions converges to a distribution function $F(x)$,

$$\lim_{n \to \infty} F_n(x) = F(x), \tag{20.3.1.4}$$

for every point $x$ at which $F(x)$. In this case, we say that the sequence $X_1, X_2, \ldots$ of the corresponding random variables converges to the random variable $X$ *in distribution*. The random variables $X_1, X_2, \ldots$ can be defined on different probability spaces.

A sequence $F_1(x), F_2(x), \ldots$ of distribution functions *weakly converges* to a distribution function $F(x)$ ($F_n \to F$) if

$$\lim_{n \to \infty} E\{h(X_n)\} = E\{h(X)\} \tag{20.3.1.5}$$

for any bounded continuous function $h$ as $n \to \infty$.

Convergence in distribution and weak convergence of distribution functions are equivalent.

The weak convergence $F_{X_n} \to F$ for random variables having a probability density function means the convergence

$$\int_{-\infty}^{+\infty} g(x) p_{X_n}(x)\, dx \to \int_{-\infty}^{+\infty} g(x) p(x)\, dx \tag{20.3.1.6}$$

for any bounded continuous function $g(x)$.

## 20.3.2. Limit Theorems

### 20.3.2-1. Law of large numbers.

The law of large numbers consists of several theorems establishing the stability of average results and revealing conditions for this stability to occur.

The notion of convergence in probability is most often used for the case in which the limit random variable $X$ has the degenerate distribution concentrated at a point $a$ ($P(\xi = a) = 1$) and

$$X_n = \frac{1}{n} \sum_{k=1}^{n} Y_k, \tag{20.3.2.1}$$

where $Y_1, Y_2, \ldots$ are arbitrary random variables.

A sequence $Y_1, Y_2, \ldots$ satisfies the *weak law of large numbers* if the limit relation

$$\lim_{n \to \infty} P\left(\left|\frac{1}{n}\sum_{k=1}^{n} Y_k - a\right| \geq \varepsilon\right) \equiv \lim_{n \to \infty} P(|X_n - a| \geq \varepsilon) = 0 \tag{20.3.2.2}$$

holds for any $\varepsilon > 0$.

If the relation

$$P\left(\omega \in \Omega : \lim_{n \to \infty} \frac{1}{n} \sum_{k=1}^{n} Y_k = a\right) \equiv P\left(\omega \in \Omega : \lim_{n \to \infty} X_n = a\right) = 1 \tag{20.3.2.3}$$

is satisfied instead of (20.3.2.2), i.e., the sequence $X_n$ converges to the number $a$ with probability 1, then the sequence $Y_1, Y_2, \ldots$ satisfies the *strong law of large numbers*.

*Markov inequality.* For any nonnegative random variable $X$ that has an expectation $E\{X\}$, the inequality

$$P(X \geq \varepsilon) \leq \frac{E\{X\}}{\varepsilon^2} \qquad (20.3.2.4)$$

holds for each $\varepsilon > 0$.

*Chebyshev inequality.* For any random variable $X$ with finite variance, the inequality

$$P(|X - E\{X\}| \geq \varepsilon) \leq \frac{\text{Var}\{X\}}{\varepsilon^2} \qquad (20.3.2.5)$$

holds for each $\varepsilon > 0$.

CHEBYSHEV THEOREM. *If $X_1, X_2, \ldots$ is a sequence of independent random variables with uniformly bounded finite variances, $\text{Var}\{X_1\} \leq C$, $\text{Var}\{X_2\} \leq C, \ldots$, then the limit relation*

$$\lim_{n \to \infty} P\left(\left|\frac{1}{n}\sum_{k=1}^{n} X_k - \frac{1}{n}\sum_{k=1}^{n} E\{X_k\}\right| < \varepsilon\right) = 1 \qquad (20.3.2.6)$$

*holds for each $\varepsilon > 0$.*

BERNOULLI THEOREM. *Let $\mu_n$ be the number of occurrences of an event $A$ (the number of successes) in $n$ independent trials, and let $p = P(A)$ be the probability of the occurrence of the event $A$ (the probability of success) in each of the trials. Then the sequence of relative frequencies $\mu_n/n$ of the occurrence of the event $A$ in $n$ independent trials converges in probability to $p = P(A)$ as $n \to \infty$; i.e., the limit relation*

$$\lim_{n \to \infty} P\left(\left|\frac{\mu_n}{n} - p\right| < \varepsilon\right) = 1 \qquad (20.3.2.7)$$

*holds for each $\varepsilon > 0$.*

POISSON THEOREM. *If in a sequence of independent trials the probability that an event $A$ occurs in the $k$th trial is equal to $p_k$, then*

$$\lim_{n \to \infty} P\left(\left|\frac{\mu_n}{n} - \frac{p_1 + \cdots + p_n}{n}\right| < \varepsilon\right) = 1. \qquad (20.3.2.8)$$

KOLMOGOROV THEOREM. *If a sequence of independent random variables $X_1, X_2, \ldots$ satisfies the condition*

$$\sum_{k=1}^{\infty} \frac{\text{Var}\{X_k\}}{k^2} < +\infty, \qquad (20.3.2.9)$$

*then it obeys the strong law of large numbers.*

The existence of the expectation is a necessary and sufficient condition for the strong law of large numbers to apply to a sequence of independent identically distributed random variables.

### 20.3.2-2. Central limit theorem.

A random variable $X_n$ with distribution function $F_{X_n}$ is *asymptotically normally distributed* if there exists a sequence of pairs of real numbers $m_n$, $\sigma_n^2$ such that the random variables $(X_n - m_n/\sigma_n)$ converge in probability to a standard normal variable. This occurs if and only if the limit relation

$$\lim_{n \to \infty} P(X_n + a\sigma_n < X_n < X_n + b\sigma_n) = \Phi(b) - \Phi(a), \qquad (20.3.2.10)$$

where $\Phi(x)$ is the distribution function of the standard normal law, holds for any $a$ and $b$ ($b > a$).

LYAPUNOV CENTRAL LIMIT THEOREM. *If $X_1, \ldots, X_n$ is a sequence of independent random variables satisfying the Lyapunov condition*

$$\lim_{n \to \infty} \frac{\sum_{k=1}^{n} \alpha_3(X_k)}{\sqrt{\sum_{k=1}^{n} \mathrm{Var}\{X_k\}}} = 0,$$

*where $\alpha_3(X_k)$ is the third initial moment of the random variable $X_k$, then the sequence of random variables*

$$Y_n = \frac{\sum_{k=1}^{n}(X_k - E\{X_k\})}{\sqrt{\sum_{k=1}^{n} \mathrm{Var}\{X_k\}}}$$

*converges in distribution to the normal law, i.e., the following limit exists:*

$$\lim_{n \to \infty} P\left( \frac{\sum_{k=1}^{n}(X_k - E\{X_k\})}{\sqrt{\sum_{k=1}^{n} \mathrm{Var}\{X_k\}}} < t \right) = \frac{1}{\sqrt{2\pi}} \int_{-\infty}^{t} e^{-t^2/2}\, dt = \Phi(t). \qquad (20.3.2.11)$$

LINDEBERG CENTRAL LIMIT THEOREM. *Let $X_1, X_2, \ldots$ be a sequence of independent identically distributed random variables with finite expectation $E\{X_k\} = m$ and finite variance $\sigma^2$. Then, as $n \to \infty$, the random variable $\frac{1}{n}\sum_{k=1}^{n} X_k$ has an asymptotically normal probability distribution with parameters $(m, \sigma^2/n)$.*

Let $\mu_n$ be the number of occurrences of an event $A$ (the number of successes) in $n$ independent trials, and let $p = P(A)$ be the probability of the occurrence of the event $A$ (the probability of success) in each of the trials. Then the sequence of relative frequencies $\mu_n/n$ has an asymptotically normal probability distribution with parameters $(p, p(1-p)/n)$.

## 20.4. Stochastic Processes

### 20.4.1. Theory of Stochastic Processes

#### 20.4.1-1. Notion of stochastic process.

Let a family

$$\xi(t) = \xi(\omega, t), \quad \omega \in \Omega, \qquad (20.4.1.1)$$

of random variables depending on a parameter $t \in T$ be given on a probability space $(\Omega, \mathcal{F}, P)$. The variable $\xi(t)$, $t \in T$, can be treated as a random function of the variable $t \in T$. The values of this function are the values of the random variable (20.4.1.1). The random function $\xi(t)$ of the independent variable $t$ is called a *stochastic process*. If a random outcome $\omega \in \Omega$ occurs, then the actual process is described by the corresponding trajectory, which is called a *realization* of the process, or a *sample function*.

A stochastic process can be simply a numerical function $\xi(t)$ of time admitting different realizations $\xi(\omega, t)$ (*one-dimensional stochastic process*) or a vector function $\Xi(t)$ (*multi-dimensional, or vector, stochastic process*). The study of a multidimensional stochastic process can be reduced to the study of one-dimensional stochastic processes by a transformation taking $\Xi(t) = (\xi_1(t), \ldots, \xi_n(t))$ to the auxiliary process

$$\xi_{\mathbf{a}}(t) = \Xi(t) \cdot \mathbf{a} = \sum_{i=1}^{n} a_i \xi_i(t), \qquad (20.4.1.2)$$

where $\mathbf{a} = (a_1, \ldots, a_n)$ is an arbitrary $k$-dimensional vector. Therefore, the study of one-dimensional stochastic processes $\xi(t)$ is the main point in the theory of stochastic processes. If the parameter $t$ ranges in some interval of the real line $R$, then the stochastic process is called a *stochastic process with continuous time*, and if the parameter $t$ takes integer values, then the process is called a *stochastic process with discrete time* (a *random sequence*).

To describe a stochastic process, one should specify an infinite set of compatible finite-dimensional probability distributions of the random vectors $\Xi(t)$ corresponding to all possible finite subsets $\mathbf{t} = (t_1, \ldots, t_n)$ of values of the argument.

**Remark.** Specifying compatible finite-dimensional probability distributions of random vectors may be insufficient for specifying the probabilities of events depending on the values of $\xi(t)$ on an infinite set of values of the parameter $t$; i.e., this does not uniquely determine the stochastic process $\xi(t)$.

**Example.** Suppose that $\xi(t) = \cos(\omega t + \Phi)$, $0 \leq t \leq 1$, is a harmonic oscillation with random phase $\Phi$, $Z$ is a random variable uniformly distributed on the interval $[0, 1]$, and the stochastic process $\zeta(t)$, $0 \leq t \leq 1$, is given by the relation

$$\zeta(t) = \begin{cases} \xi(t) & \text{for } t \neq Z, \\ \xi(t) + 3 & \text{for } t = Z. \end{cases}$$

Since $P[(Z = t_1) \cup \cdots \cup (Z = t_n)] = 0$ for any finite set $\mathbf{t} = (t_1, \ldots, t_n)$, we see that all finite-dimensional distributions of the stochastic processes $\xi(t)$ and $\zeta(t)$ are the same. At the same time, these processes differ from each other.

Specifying the set of finite-dimensional probability distributions often permits one to clarify whether there exists at least one stochastic process $\xi(t)$ with finite-dimensional distributions whose realizations satisfy a certain property (for example, are continuous or differentiable).

### 20.4.1-2. Correlation function.

Let $\xi(t)$ and $\zeta(t)$ be real stochastic processes.

The *autocorrelation function* of a stochastic process is defined as the function

$$B_{\xi\xi}(t, s) = E\big\{ [\xi(t) - E\{\xi(t)\}] [\xi(s) - E\{\xi(s)\}] \big\}, \qquad (20.4.1.3)$$

which is the second central moment function.

**Remark.** The autocorrelation function of a stochastic process is also called the *covariance function*.

The mixed second moment, i.e., the function

$$B_{\xi\zeta}(t, s) = E\{\xi(t)\zeta(s)\}, \qquad (20.4.1.4)$$

of values of $\xi(t)$ and $\zeta(t)$ at two points is called the *cross-correlation function* (*cross-covariance function*).

The mixed central moment (covariance), i.e., the function

$$b_{\xi\zeta}(t,s) = E\{[\xi(t) - E\{\xi(t)\}][\zeta(s) - E\{\zeta(s)\}]\}, \qquad (20.4.1.5)$$

is called the *central cross-correlation function*.

The correlation coefficient, i.e., the function

$$\rho_{\xi\zeta}(t,s) = \frac{b_{\xi\zeta}(t,s)}{\sqrt{\text{Var}\{\xi(t)\}\text{Var}\{\zeta(s)\}}}, \qquad (20.4.1.6)$$

is called the *normalized cross-correlation function*.

The following relations hold:

$$E\{\xi^2(t)\} = B_{\xi\xi}(t,t) \geq 0,$$
$$|B_{\xi\xi}(t,s)|^2 \leq B_{\xi\xi}(t,t)B_{\xi\xi}(s,s),$$
$$\text{Var}\{\xi(t)\} = B_{\xi\xi}(t,t) - [E\{\xi(t)\}]^2,$$
$$\text{Cov}[\xi(t),\zeta(s)] = B_{\xi\zeta}(t,s) - \xi(t)\zeta(s).$$

Suppose that $\xi(t)$ and $\zeta(t)$ are *complex* stochastic processes (essentially, two-dimensional stochastic processes). The autocorrelation function and the cross-correlation function are determined by the relations

$$B_{\xi\xi}(t,s) = E\{\overline{\xi(t)}\xi(s)\} = \overline{B_{\xi\xi}(s,t)}, \qquad B_{\xi\zeta}(t,s) = E\{\overline{\xi(t)}\zeta(s)\} = \overline{B_{\zeta\xi}(s,t)}, \quad (20.4.1.7)$$

where $\overline{\xi(t)}$, $\overline{B_{\xi\xi}(s,t)}$, and $\overline{B_{\zeta\xi}(s,t)}$ are the function conjugate to a function $\xi(t)$, $B_{\xi\xi}(s,t)$, and $B_{\zeta\xi}(s,t)$, respectively.

20.4.1-3. Differentiation and integration of stochastic process.

To *differentiate a stochastic process* is to calculate the limit

$$\lim_{h \to 0} \frac{\xi(t+h) - \xi(t)}{h} = \xi'_t(t). \qquad (20.4.1.8)$$

If the limit is understood in the sense of convergence in mean square (resp., with probability 1), then the differentiation is also understood in mean square (resp., with probability 1).

The following formulas hold:

$$E\{\xi'_t(t)\} = \frac{dE\{\xi(t)\}}{dt}, \qquad B_{\xi'_t\xi'_t}(t,s) = \frac{\partial^2 B_{\xi\xi}(t,s)}{\partial t \partial s}, \qquad B_{\xi'_t\xi}(t,s) = \frac{\partial B_{\xi\xi}(t,s)}{\partial t}. \quad (20.4.1.9)$$

The integral

$$\int_a^b \xi(t)\,dt \qquad (20.4.1.10)$$

of a stochastic process $\xi(t)$ defined on an interval $[a,b]$ with autocorrelation function $B_{\xi\xi}(t,s)$ is the limit in mean square of the integral sums

$$\sum_{k=1}^n \xi(s_k)(t_k - t_{k-1})$$

as the diameter of the partition $a = t_0 < t_1 < \cdots < t_n = b$, where $s_k \in [t_k, t_{k-1}]$ tends to zero.

For the integral (20.4.1.10) to exist, it is necessary and sufficient that the following limit exist:

$$\lim_{\lambda \to 0} \sum_{k=1}^{n} \sum_{l=1}^{n} B_{\xi\xi}(s_k, s_l)(t_k - t_{k-1})(t_l - t_{l-1}),$$

where $\lambda = \max_{k}(t_k - t_{k-1})$.

In particular, the integral (20.4.1.10) exists if the following repeated integral exists:

$$\int_{a}^{b} \int_{a}^{b} B_{\xi\xi}(s, t) \, dt \, ds.$$

## 20.4.2. Models of Stochastic Processes

### 20.4.2-1. Stationary stochastic process.

A stochastic process $\xi(t)$ is said to be *stationary* if its probability characteristics remain the same in the course of time, i.e., are invariant under time shifts $t \to t + a$, $\xi(t) \to \xi(t + a)$ for any given $a$ (real or integer for a stochastic process with continuous or discrete time, respectively).

For a stationary process, the mean value (the expectation)

$$E\{\xi(t)\} = E\{\xi(0)\} = m$$

is a constant, and the correlation function is determined by the relation

$$E\{\overline{\xi(t)}\xi(t + \tau)\} = B_{\xi\xi}(\tau), \tag{20.4.2.1}$$

where $\overline{\xi(t)}$ is the function conjugate to a function $\xi(t)$. The correlation function is positive definite:

$$\sum_{k=1}^{n} \sum_{j=1}^{n} c_k \bar{c}_j B_{\xi\xi}(t_k - t_j) = E\left\{\left|\sum_{k=1}^{n} c_k \xi(t_k)\right|\right\} \geq 0.$$

In this case, the following relations hold:

$$\begin{aligned} B_{\xi\xi}(\tau) &= \overline{B_{\xi\xi}(-\tau)}, \quad B_{\xi\zeta}(\tau) = \overline{B_{\zeta\xi}(-\tau)} \\ |B_{\xi\xi}(\tau)| &\leq B_{\xi\xi}(0), \quad |B_{\xi\zeta}(\tau)|^2 \leq B_{\xi\xi}(0) B_{\zeta\zeta}(0), \end{aligned} \tag{20.4.2.2}$$

where $\overline{B_{\xi\xi}(s,t)}$ and $\overline{B_{\zeta\zeta}(s,t)}$ are the function conjugate to a function $B_{\xi\xi}(s,t)$ and $B_{\zeta\zeta}(s,t)$, respectively.

Stochastic processes for which $E\{\xi(t)\}$ and $E\{\overline{\xi(t)}\xi(t+\tau)\}$ are independent of $t$ are called *stationary stochastic processes in the wide sense*. Stochastic processes, all of whose characteristics remain the same in the course of time, are called *stationary stochastic processes in the narrow sense*.

KHINCHIN'S THEOREM. The correlation function $B_{\xi\xi}(\tau)$ of a stationary stochastic process with continuous time can always be represented in the form

$$B_{\xi\xi}(\tau) = \int_{-\infty}^{\infty} e^{i\tau\lambda} \, dF(\lambda), \tag{20.4.2.3}$$

where $F(\lambda)$ is a monotone nondecreasing function, $i$ is the imaginary unit, and $i^2 = -1$.

If $B_{\xi\xi}(\tau)$ decreases sufficiently rapidly as $|\tau| \to \infty$ (as happens most often in applications provided that $\xi(t)$ is understood as the difference $\xi(t) - E\{\xi(t)\}$, i.e., it is assumed that $E\{\xi(t)\} = 0$), then the integral on the right-hand side in (20.4.2.3) becomes the Fourier integral

$$B_{\xi\xi}(\tau) = \int_{-\infty}^{\infty} e^{i\tau\lambda} f(\lambda)\, d\lambda, \tag{20.4.2.4}$$

where $f(\lambda) = F'_\lambda(\lambda)$ is a monotone nondecreasing function. The function $F(\lambda)$ is called the *spectral function* of the stationary stochastic process, and the function $f(\lambda)$ is called its *spectral density*. The process $\xi(t)$ itself admits the *spectral resolution*

$$\xi(t) = \int_{-\infty}^{\infty} e^{it\lambda} dZ(\lambda), \tag{20.4.2.5}$$

where $Z(\lambda)$ is a random function with uncorrelated increments (i.e., a function such that $E\{dZ(\lambda_1)\overline{dZ(\lambda_2)}\} = 0$ for $\lambda_1 \neq \lambda_2$) satisfying the condition $|E\{dZ(\lambda)\}|^2 = dF(\lambda)$ and the integral is understood as the mean-square limit of the corresponding sequence of integral sums.

### 20.4.2-2. Markov processes.

A stochastic process $\xi(t)$ is called a *Markov process* if for two arbitrary times $t_0$ and $t_1$, $t_0 < t_1$, the conditional distribution of $\xi(t_1)$ given all values of $\xi(t)$ for $t \leq t_0$ depends only on $\xi(t_0)$. This property is called the *Markov property* or the *absence of aftereffect*.

The probability of a transition from state $i$ to state $j$ in time $t$ is called the *transition probability* $p_{ij}(t)$ ($t \geq 0$). The transition probability satisfies the relation

$$p_{ij}(t) = P[\xi(t) = j | \xi(0) = i]. \tag{20.4.2.6}$$

Suppose that the initial probability distribution

$$p_i^0 = P[\xi(0) = i], \qquad i = 0, \pm 1, \pm 2, \ldots$$

is given. In this case, the joint probability distribution of the random variables $\xi(t_1), \ldots, \xi(t_n)$ for any $0 = t_0 < t_1 < \cdots < t_n$ is given by

$$P[\xi(t_1) = j_1, \ldots, \xi(t_n) = j_n] = \sum_i p_i^0 p_{ij_1}(t_1 - t_0) p_{ij_2}(t_2 - t_1) \ldots p_{j_{n-1}j_n}(t_n - t_{n-1}); \tag{20.4.2.7}$$

and, in particular, the probability that the system at time $t > 0$ is in state $j$ is

$$p_j(t) = \sum_i p_i^0 p_{ij}(t), \qquad j = 0, \pm 1, \pm 2, \ldots$$

The dependence of the transition probabilities $p_{ij}(t)$ on time $t \geq 0$ is given by the formula

$$p_{ij}(s+t) = \sum_k p_{ik}(s) p_{kj}(t), \qquad i, j = 0, \pm 1, \pm 2, \ldots \tag{20.4.2.8}$$

Suppose that $\lambda_{ij} = [p_{ij}(t)]'_t\big|_{t=0}$, $j = 0, \pm 1, \pm 2, \ldots$ The parameters $\lambda_{ij}$ satisfy the condition

$$\lambda_{ii} = \lim_{h \to 0} \frac{p_{ii}(h) - 1}{h} = -\sum_{i \neq j} \lambda_{ij}, \qquad \lambda_{ij} = \lim_{h \to 0} \frac{p_{ij}(h)}{h} \geq 0 \quad (i \neq j). \tag{20.4.2.9}$$

THEOREM. *Under condition (20.4.2.9), the transition probabilities satisfy the system of differential equations*

$$[p_{ij}(t)]'_t = \sum_k \lambda_{ik} p_{kj}(t), \qquad i,j = 0, \pm 1, \pm 2, \ldots \qquad (20.4.2.10)$$

The system of differential equations (20.4.2.10) is called the system of *backward Kolmogorov equations*.

THEOREM. *The transition probabilities $p_{ij}(t)$ satisfy the system of differential equations*

$$[p_{ij}(t)]'_t = \sum_k \lambda_{kj} p_{ik}(t), \qquad i,j = 0, \pm 1, \pm 2, \ldots \qquad (20.4.2.11)$$

The system of differential equations (20.4.2.11) is called the system of *forward Kolmogorov equations*.

### 20.4.2-3. Poisson processes.

For a flow of events, let $\Lambda(t)$ be the expectation of the number of events on the interval $[0, t)$. The number of events in the half-open interval $[a, b)$ is a Poisson random variable with parameter

$$\Lambda(b) - \Lambda(a).$$

The probability structure of a Poisson process is completely determined by the function $\Lambda(t)$.

The *Poisson process* is a stochastic process $\xi(t)$, $t \geq 0$, with independent increments having the Poisson distribution; i.e.,

$$P[\xi(t) - \xi(s) = k] = \frac{[\Lambda(t) - \Lambda(s)]^k}{k!} e^{\Lambda(t) - \Lambda(s)}$$

for all $0 \leq s \leq t$, $k = 0, 1, 2, \ldots$, and $t \geq 0$.

A *Poisson point process* is a stochastic process for which the numbers of points (counting multiplicities) in any disjoint measurable sets of the phase space are independent random variables with the Poisson distribution.

In queueing theory, it is often assumed that the incoming traffic is a Poisson point process. The *simplest point process* is defined as the Poisson point process characterized by the following three properties:

1. Stationarity.
2. Memorylessness.
3. Orderliness.

*Stationarity* means that, for any finite group of disjoint time intervals, the probability that a given number of events occurs on each of these time intervals depends only on these numbers and on the duration of the time intervals, but is independent of any shift of all time intervals by the same value. In particular, the probability that $k$ event occurs on the time interval from $\tau$ to $\tau + t$ is independent of $\tau$ and is a function only of the variables $k$ and $t$.

*Memorylessness* means that the probability of the occurrence of $k$ events on the time interval from $\tau$ to $\tau + t$ is independent of how many times and how the events occurred earlier. This means that the conditional probability of the occurrence of events on the time interval from $\tau$ to $\tau + t$ under any possible assumptions concerning the occurrence of the events before time $\tau$ coincides with the unconditional probability. In particular, memorylessness means that the occurrences of any number of events on disjoint time intervals are independent.

*Orderliness* expresses the requirement that the occurrence of two or more events on a small time interval is practically impossible.

### 20.4.2-4. Birth–death processes.

Suppose that a system can be in one of the states

$$E_0, E_1, E_2, \ldots,$$

and the set of these states is finite or countable. In the course of time, the states of the system vary; on a time interval of length $h$, the system passes from the state $E_n$ to the state $E_{n+1}$ with probability

$$\lambda_n h + o(h)$$

and to the state $E_{n-1}$ with probability

$$\upsilon_n h + o(h).$$

The probability to stay at the same state $E_n$ on a time interval of length $h$ is equal to

$$1 - \lambda_n h - \upsilon_n h + o(h).$$

It is assumed that the constants $\lambda_n$ and $\upsilon_n$ depend only on $n$ and are independent of $t$ and of how the system arrived at this state.

The stochastic process described above is called a *birth–death process*. If the relation

$$\upsilon_n = 0$$

holds for any $n \geq 1$, then the process is called a *pure birth process*. If the relation

$$\lambda_n = 0$$

holds for any $n \geq 1$, then the process is called the *death process*.

Let $p_k(t)$ be the probability that the system is in the state $E_k$ at time $t$. Then the birth–death process is described by the system of differential equations

$$\begin{aligned} [p_0(t)]'_t &= -\lambda_0 p_0(t) + \upsilon_1 p_1(t), \\ [p_k(t)]'_t &= -(\lambda_k + \upsilon_k) p_k(t) + \lambda_{k-1} p_{k-1}(t) + \upsilon_{k+1} p_{k+1}(t), \qquad k \geq 1. \end{aligned} \qquad (20.4.2.12)$$

**Example 1.** Consider the system consisting of the states $E_0$ and $E_1$. The system of differential equations for the probabilities $p_0(t)$ and $p_1(t)$ has the form

$$\begin{aligned} [p_0(t)]'_t &= -\lambda p_0(t) + \upsilon p_1(t), \\ [p_1(t)]'_t &= \lambda p_0(t) - \upsilon p_1(t). \end{aligned}$$

The solution of the system of equations with the initial conditions $p_0(0) = 1$, $p_1(0) = 0$ has the form

$$\begin{aligned} [p_0(t)]'_t &= \frac{\upsilon}{\upsilon + \lambda} \left[ 1 + \frac{\upsilon}{\lambda} e^{-(\upsilon + \lambda)t} \right], \\ [p_1(t)]'_t &= \frac{\lambda}{\upsilon + \lambda} \left[ 1 - \frac{\upsilon}{\lambda} e^{-(\upsilon + \lambda)t} \right]. \end{aligned}$$

FELLER THEOREM. *For the solution $p_k(t)$ of the pure birth equations to satisfy the relation*

$$\sum_{k=0}^{\infty} p_k(t) = 1, \qquad (20.4.2.13)$$

*for all $t$, it is necessary and sufficient that the following series diverge:*

$$\sum_{k=0}^{\infty} \frac{1}{\lambda_k}. \qquad (20.4.2.14)$$

In the case of a pure birth process, the system of equations (20.4.2.11) can be solved by simple successive integration, because the differential equations have the form of simple recursive relations. In the general case, it is already impossible to find the function $p_k(t)$ successively.

The relation

$$\sum_{k=0}^{\infty} p_k(t) = 1$$

holds for all $t$ if the series

$$\sum_{k=1}^{\infty} \prod_{i=1}^{k} \frac{v_i}{\lambda_i} \qquad (20.4.2.15)$$

diverges. If, in addition, the series

$$\sum_{k=1}^{\infty} \prod_{i=1}^{k} \frac{\lambda_i - 1}{v_i} \qquad (20.4.2.16)$$

converges, then there exist limits

$$p_k = \lim_{t \to \infty} p_k(t) \qquad (20.4.2.17)$$

for all $t$.

If relation (20.4.2.17) holds, then system (20.4.2.12) becomes

$$\begin{aligned} &-\lambda_0 p_0 + v_1 p_1 = 0, \\ &-(\lambda_k + v_k) p_k + \lambda_{k-1} p_{k-1} + v_{k+1} p_{k+1} = 0, \qquad k \geq 1. \end{aligned} \qquad (20.4.2.18)$$

The solutions of system (20.4.2.18) have the form

$$p_k = \frac{\lambda_{k-1}}{v_k} p_{k-1} = \prod_{i=1}^{k} \frac{\lambda_i - 1}{v_i} p_0. \qquad (20.4.2.19)$$

The constant $p_0$ is determined by the normalization condition $\sum_{k=0}^{\infty} p_k(t) = 1$:

$$p_0 = \left(1 + \sum_{k=1}^{\infty} \prod_{i=1}^{k} \frac{\lambda_i - 1}{v_i}\right). \qquad (20.4.2.20)$$

**Example 2.** Servicing with queue.

A Poisson flow of jobs with parameter $\lambda$ arrives at $n$ identical servers. A server serves a job in random time with the probability distribution

$$H(x) = 1 - e^{-vx}.$$

If there is at least one free server when a job arrives, then servicing starts immediately. But if all servers are occupied, then the new jobs enter a queue. The conditions of the problem satisfy the assumptions of the theory of birth–death processes. In this problem, $\lambda_k = \lambda$ for any $k$, $v_k = kv$ for $k \leq n$, and $v_k = nv$ for $k \geq n$.

By formulas (20.4.2.19) and (20.4.2.20), we have

$$p_k = \begin{cases} \dfrac{\rho^k}{k!} p_0 & \text{for } k \leq n, \\ \dfrac{\rho^k}{n! \, n^{k-n}} p_0 & \text{for } k \geq n, \end{cases}$$

where $\rho = \lambda/v$.

The constant $p_0$ is determined by the relation

$$p_0 = \left[ \sum_{k=0}^{n} \frac{\rho^k}{k!} + \frac{\rho^n}{n!} \sum_{k=n+1}^{\infty} \left( \frac{\rho}{n} \right)^{k-n} \right]^{-1}.$$

If $\rho < n$, then

$$p_0 = \left[ 1 + \sum_{k=1}^{n} \frac{\rho^k}{k!} + \frac{\rho^{n+1}}{n!(n-\rho)!} \right]^{-1}.$$

But if $\rho \geq n$, the series in the parentheses is divergent and $p_k = 0$ for all $k$; i.e., in this case the queue to be served increases in time without bound.

**Example 3.** Maintenance of machines by a team of workers.

A team of $l$ workers maintains $n$ identical machines. The machines fail independently; the probability of a failure in the time interval $(t, t+h)$ is equal to $\lambda h + o(h)$. The probability that a machine will be repaired on the interval $(t, t+h)$ is equal to $vh + o(h)$. Each worker can repair only one machine; each machine can be repaired only by one worker. Find the probability of the event that a given number of machines is out of operation at a given time.

Let $E_k$ be the event that exactly $k$ machines are out of operation at a given time. Obviously, the system can be only in the states $E_0, \ldots, E_n$. We deal with a birth–death process such that

$$\lambda_k = \begin{cases} (n-k)\lambda & \text{for } 0 \leq k < n, \\ 0 & \text{for } k = n, \end{cases} \qquad v_k = \begin{cases} kv & \text{for } 1 \leq k < l, \\ lv & \text{for } l \leq k \leq n. \end{cases}$$

By formulas (20.4.2.19) and (20.4.2.20), we have

$$p_k = \begin{cases} \dfrac{n!}{k!(n-k)!} \rho^k p_0 & \text{for } 1 \leq k \leq l, \\[2mm] \dfrac{n!}{l^{n-k} l!(n-k)!} \rho^k p_0 & \text{for } l \leq k \leq n, \end{cases}$$

where $\rho = \lambda/v$. The constant $p_0$ is determined by the relation

$$p_0 = \left[ \sum_{k=0}^{l} \frac{n!}{k!(n-k)!} \rho^k + \sum_{k=l+1}^{n} \frac{n!}{l^{n-k} l!(n-k)!} \rho^k \right]^{-1}.$$

# References for Chapter 20

**Bain, L. J. and Engelhardt, M.,** *Introduction to Probability and Mathematical Statistics, 2nd Edition* (Duxbury Classic), Duxbury Press, Boston, 2000.

**Bean, M. A.,** *Probability: The Science of Uncertainty with Applications to Investments, Insurance, and Engineering*, Brooks Cole, Stamford, 2000.

**Bertsekas, D. P. and Tsitsiklis, J. N.,** *Introduction to Probability*, Athena Scientific, Belmont, Massachusetts, 2002.

**Beyer, W. H. (Editor),** *CRC Standard Probability and Statistics Tables and Formulae*, CRC Press, Boca Raton, 1990.

**Burlington, R. S. and May, D.,** *Handbook of Probability and Statistics With Tables, 2nd Edition*, McGraw-Hill, New York, 1970.

**Chung, K. L.,** *A Course in Probability Theory Revised, 2nd Edition*, Academic Press, Boston, 2000.

**DeGroot, M. H. and Schervish, M. J.,** *Probability and Statistics, 3rd Edition*, Addison Wesley, Boston, 2001.

**Devore, J. L.,** *Probability and Statistics for Engineering and the Sciences (with CD-ROM and InfoTrac), 6th Edition*, Duxbury Press, Boston, 2003.

**Freund, J. E.,** *Introduction to Probability, Rep. Edition* (Dover Books on Advanced Mathematics), Dover Publications, New York, 1993.

**Garcia, L.,** *Probability and Random Processes for Electrical Engineering: Student Solutions Manual, 2nd Edition*, Addison Wesley, Boston, 1993.

**Gnedenko, B. V. and Khinchin, A. Ya.,** *An Elementary Introduction to the Theory of Probability, 5th Edition*, Dover Publications, New York, 1962.

**Ghahramani, S.,** *Fundamentals of Probability, with Stochastic Processes, 3rd Edition*, Prentice Hall, Englewood Cliffs, New Jersey, 2004.

**Goldberg, S.,** *Probability: An Introduction*, Dover Publications, New York, 1987.

**Grimmett, G. R. and Stirzaker, D. R.**, *Probability and Random Processes, 3rd Edition*, Oxford University Press, Oxford, 2001.
**Hines, W. W., Montgomery, D. C., Goldsman, D. M., and Borror, C. M.**, *Probability and Statistics in Engineering, 4th Edition*, Wiley, New York, 2003.
**Hsu, H.**, *Schaum's Outline of Probability, Random Variables, and Random Processes*, McGraw-Hill, New York, 1996.
**Kokoska, S. and Zwillinger, D. (Editors)**, *CRC Standard Probability and Statistics Tables and Formulae, Student Edition*, Chapman & Hall/CRC, Boca Raton, 2000.
**Lange, K.**, *Applied Probability*, Springer, New York, 2004.
**Ledermann, W.**, *Probability* (Handbook of Applicable Mathematics), Wiley, New York, 1981.
**Lipschutz, S.**, *Schaum's Outline of Probability, 2nd Edition*, McGraw-Hill, New York, 2000.
**Lipschutz, S. and Schiller, J.**, *Schaum's Outline of Introduction to Probability and Statistics*, McGraw-Hill, New York, 1998.
**Mendenhall, W., Beaver, R. J., and Beaver, B. M.**, *Introduction to Probability and Statistics (with CD-ROM), 12th Edition*, Duxbury Press, Boston, 2005.
**Milton, J. S. and Arnold, J. C.**, *Introduction to Probability and Statistics: Principles and Applications for Engineering and the Computing Sciences, 2nd Edition*, McGraw-Hill, New York, 2002.
**Montgomery, D. C. and Runger, G. C.**, *Applied Statistics and Probability for Engineers, Student Solutions Manual, 4th Edition*, Wiley, New York, 2006.
**Pfeiffer, P. E.**, *Concepts of Probability Theory, 2nd Rev. Edition*, Dover Publications, New York, 1978.
**Pitman, J.**, *Probability*, Springer, New York, 1993.
**Ross, S. M.**, *Applied Probability Models with Optimization Applications* (Dover Books on Mathematics), *Rep. Edition*, Dover Publications, New York, 1992.
**Ross, S. M.**, *Introduction to Probability Models, 8th Edition*, Academic Press, Boston, 2002.
**Ross, S. M.**, *A First Course in Probability, 7th Edition*, Prentice Hall, Englewood Cliffs, New Jersey, 2005.
**Rozanov, Y. A.**, *Probability Theory: A Concise Course*, Dover Publications, New York, 1977.
**Scheaffer, R. L. and McClave, J. T.**, *Probability and Statistics for Engineers, 4th Edition* (Statistics), Duxbury Press, Boston, 1994.
**Seely, J. A.**, *Probability and Statistics for Engineering and Science, 6th Edition*, Brooks Cole, Stamford, 2003.
**Shiryaev, A. N.**, *Probability, 2nd Edition* (Graduate Texts in Mathematics), Springer, New York, 1996.
**Ventsel, H.**, *Théorie des Probabilités*, Mir Publishers, Moscow, 1987.
**Ventsel, H. and Ovtcharov, L.**, *Problémes Appliqués de la Théorie des Probabilités*, Mir Publishers, Moscow, 1988.
**Walpole, R. E., Myers, R. H., Myers, S. L., and Ye, K.**, *Probability & Statistics for Engineers & Scientists, 8th Edition*, Prentice Hall, Englewood Cliffs, New Jersey, 2006.
**Yates, R. D. and Goodman, D.J.**, *Probability and Stochastic Processes: A Friendly Introduction for Electrical and Computer Engineers, 2nd Edition*, Wiley, New York, 2004.
**Zwillinger, D. (Editor)**, *CRC Standard Mathematical Tables and Formulae, 31st Edition*, Chapman & Hall/CRC, Boca Raton, 2001.

# Chapter 21

# Mathematical Statistics

## 21.1. Introduction to Mathematical Statistics

### 21.1.1. Basic Notions and Problems of Mathematical Statistics

**21.1.1-1. Basic problems of mathematical statistics.**

The term "statistics" derives from the Latin word "status," meaning "state." Statistics comprises three major divisions: collection of statistical data, their statistical analysis, and development of mathematical methods for processing and using the statistical data to draw scientific and practical conclusions. It is the last division that is commonly known as *mathematical statistics*.

The original material for a statistical study is a set of results specially gathered for this study or a set of results of specially performed experiments. The following problems arise in this connection.

1. *Estimating the unknown probability of a random event*
2. *Finding the unknown theoretical distribution function*

   The problem is stated as follows. Given the values $x_1, \ldots, x_n$ of a random variable $X$ obtained in $n$ independent trials, find, at least approximately, the unknown distribution function $F(x)$ of the random variable $X$.

3. *Determining the unknown parameters of the theoretical distribution function*

   The problem is stated as follows. A random variable $X$ has the distribution function $F(x; \theta_1, \ldots, \theta_k)$ depending on $k$ parameters $\theta_1, \ldots, \theta_k$, whose values are unknown. The main goal is to estimate the unknown parameters $\theta_1, \ldots, \theta_k$ using only the results $X_1, \ldots, X_n$ of observations of the random variable $X$.

   Instead of seeking approximate values of the unknown parameters $\theta_1, \ldots, \theta_k$ in the form of functions $\theta_1^*, \ldots, \theta_k^*$, in a number of problems it is preferable to seek functions $\theta_{i,L}^*$ and $\theta_{i,R}^*$ ($i = 1, 2, \ldots, k$) depending on the results of observations and known variables and such that with sufficient reliability one can claim that $\theta_{i,L}^* < \theta_i^* < \theta_{i,R}^*$ ($i = 1, 2, \ldots, k$). The functions $\theta_{i,L}^*$ and $\theta_{i,R}^*$ ($i = 1, 2, \ldots, k$) are called the *confidence boundaries* for $\theta_1^*, \ldots, \theta_k^*$.

4. *Testing statistical hypotheses*

   The problem is stated as follows. Some reasoning suggests that the distribution function of a random variable $X$ is $F(x)$; the question is whether the observed values are compatible with the hypothesis to be tested that the random variable $X$ has the distribution $F(x)$.

5. *Estimation of dependence*

   A sequence of observations is performed simultaneously for two random variables $X$ and $Y$. The results of observations are given by pairs of values $x_1, y_1, x_2, y_2, \ldots, x_n, y_n$. It is required to find a functional or correlation relationship between $X$ and $Y$.

**21.1.1-2. Population and sample.**

The set of all possible results of observations that can be made under a given set of conditions is called the *population*. In some problems, the population is treated as a random variable $X$.

An example of population is the entire population of a country. In this population, we are, for example, interested in the age of people. Another example of population is the set of parts produced by a given machine. These parts can be either accepted or rejected.

The number of entities in a population is called its size and is usually denoted by the symbol $N$.

A set of entities randomly selected from a population is called a *sample*. A sample must be representative of the population; i.e., it must show the right proportions characteristic of the population. This is achieved by the randomness of the selection when all entities in the population can be selected with the same probability.

The number of elements in a sample is called its *size* and is usually denoted by the symbol $n$. The elements of a sample will be denoted by $X_1, \ldots, X_n$.

Note that sampling itself can be performed by various methods. Having selected an element and measured its value, one can delete this element from the population so that it cannot be selected in subsequent trials (sampling without replacement). Alternatively, after measuring the value of an element, one can return it to the population (samples with replacement). Obviously, for a sufficiently large population size the difference between sampling with and without replacement disappears.

21.1.1-3. Theoretical distribution function.

Each element $X_i$ in a sample has the distribution function $F(x)$, and the elements $X_1, \ldots, X_n$ are assumed to be independent for sampling with replacement or as $n \to \infty$. A sample $X_1, \ldots, X_n$ is interpreted as a set of $n$ independent identically distributed random variables with distribution function $F(x)$ or as $n$ independent realizations of an observable random variable $X$ with distribution function $F(x)$. The distribution function $F(x)$ is called the *theoretical distribution function*.

The joint distribution function $F_{X_1,\ldots,X_n}(x_1, \ldots, x_n)$ of the sample $X_1, \ldots, X_n$ is given by the formula

$$F_{X_1,\ldots,X_n}(x_1, \ldots, x_n) = P(X_1 < x_1, \ldots, X_n < x_n) = F(x_1)F(x_2)\ldots F(x_n). \qquad (21.1.1.1)$$

## 21.1.2. Simplest Statistical Transformations

21.1.2-1. Series of order statistics.

By arranging the elements of a sample $X_1, \ldots, X_n$ in ascending order, $X_{(1)} \leq \cdots \leq X_{(n)}$, we obtain the *series of order statistics* $X_{(1)}, \ldots, X_{(n)}$. Obviously, this transformation does not lead to a loss of information about the theoretical distribution function. The variables $X_{(1)}$ and $X_{(n)}$ are called the *extreme order statistics*.

The difference

$$R = X_{(n)}^* - X_{(1)}^* \qquad (21.1.2.1)$$

of the extreme order statistics is called the *range statistic*, or the *sample range* $R$.

The series of order statistics is used to construct the empirical distribution function (see Paragraph 21.1.2-6).

21.1.2-2. Statistical series.

If a sample $X_1, \ldots, X_n$ contains coinciding elements, which may happen in observations of a discrete random variable, then it is expedient to group the elements. For a common value of several variates in the sample $X_1, \dots, X_n$, the size of the corresponding group of coinciding elements is called the *frequency* or the *weight* of this variate value. By $n_i$ we denote the number of occurrences of the $i$th variate value.

The set $Z_1, \dots, Z_L$ of distinct variate values arranged in ascending order with the corresponding frequencies $n_1, \dots, n_L$ represents the sample $X_1, \dots, X_n$ and is called a *statistical series* (see Example 1 in Paragraph 21.1.2-7).

### 21.1.2-3. Interval series.

Interval series are used in observations of continuous random variables. In this case, the entire sample range is divided into finitely many *bins*, or *class intervals*, and then the number of variates in each bin is calculated.

The ordered sequence of class intervals with the corresponding frequencies or relative frequencies of occurrences of variates in each of these intervals is called an *interval series*. It is convenient to represent an interval series as a table with two rows (e.g., see Example 2 in Paragraph 21.1.2-7). The first row of the table contains the class intervals $[x_0, x_1)$, $[x_1, x_2)$, $\dots$, $[x_{L-1}, x_L)$, which are usually chosen to have the same length. The interval length $h$ is usually determined by the *Sturges formula*

$$h = \frac{X^*_{(n)} - X^*_{(1)}}{1 + \log_2 n}, \tag{21.1.2.2}$$

where $1 + \log_2 n = L$ is the number of intervals ($\log_2 n \approx 3.322 \lg n$). The second row of the interval series contains the frequencies or relative frequencies of occurrences of the sample elements in each of these intervals.

**Remark.** It is recommended to take $X_{(1)} - \frac{1}{2}h$ for the left boundary of the first interval.

### 21.1.2-4. Relative frequencies.

Let $H$ be the event that the value of a random variable $X$ belongs to a set $S_H$. Suppose also that a random sample $X_1, \dots, X_n$ is given. The number $n_H$ of sample elements lying in $S_H$ is called the *frequency of the event* $H$. The ratio of the frequency $n_H$ to the sample size is called the *relative frequency* and is denoted by

$$p^*_H = \frac{n_H}{n}. \tag{21.1.2.3}$$

Since a random sample can be treated as the result of a sequence of $n$ Bernoulli trials (Paragraph 20.1.3-2), it follows that the random variable $n_H$ has the binomial distribution with parameter $p = P(H)$, where $P(H)$ is the probability of the event $H$. One has

$$E\{p^*_H\} = P(H), \quad \mathrm{Var}\{p^*_H\} = \frac{P(H)[1 - P(H)]}{n}. \tag{21.1.2.4}$$

The relative frequency $p^*_H$ is an unbiased consistent estimator for the corresponding probability $P(H)$. As $n \to \infty$, the estimator $p^*_H$ is asymptotically normal with the parameters (21.1.2.4).

Let $H_i$ ($i = 1, 2, \dots, L$) be the random events that the random variable takes the value $Z_i$ (in the discrete case) or lies in the $i$th interval of the interval series (in the continuous case), and let $n_i$ and $p_i^*$ be their frequencies and relative frequencies, respectively. The *cumulative frequencies* $N_l$ are determined by the formula

$$N_l = \sum_{i=1}^{l} n_i. \tag{21.1.2.5}$$

The *cumulative relative frequencies* $W_l$ are given by the expression

$$W_l = \sum_{i=1}^{l} p_i^* = \frac{N_l}{n}. \tag{21.1.2.6}$$

### 21.1.2-5. Notion of statistic.

To make justified statistical conclusions, one needs a sample of sufficiently large size $n$. Obviously, it is rather difficult to use and store such samples. The notion of statistic allows one to avoid these problems.

A *statistic* $S = (S_1, \ldots, S_k)$ is an arbitrary $k$-dimensional function of the sample $X_1, \ldots, X_n$:

$$S_i = S_i(X_1, \ldots, X_n) \qquad (i = 1, 2, \ldots, k). \tag{21.1.2.7}$$

Being a function of the random vector $(X_1, \ldots, X_n)$, the statistic $S = (S_1, \ldots, S_k)$ is also a random vector, and its distribution function

$$F_{S_1,\ldots,S_k}(x_1, \ldots, x_n) = P(S_1 < x_1, \ldots, S_k < x_k)$$

is given by the formula

$$F_{S_1,\ldots,S_k}(x_1, \ldots, x_n) = \sum P(y_1) \ldots P(y_n)$$

for a discrete random variable $X$ and by the formula

$$F_{S_1,\ldots,S_k}(x_1, \ldots, x_n) = \int \ldots \int p(y_1) \ldots p(y_n) \, dy_1 \ldots dy_n$$

for a continuous random variable, where the summation or integration extends over all possible values $y_1, \ldots, y_n$ (in the discrete case, each $y_i$ belongs to the set $Z_1, \ldots, Z_L$) satisfying the inequalities

$$S_1(y_1, \ldots, y_n) < x_1, \quad S_2(y_1, \ldots, y_n) < x_2, \quad \ldots, \quad S_k(y_1, \ldots, y_n) < x_k.$$

### 21.1.2-6. Empirical distribution function.

The *empirical (sample) distribution function* corresponding to a random sample $X_1, \ldots, X_n$ is defined for each real $x$ by the formula

$$F_n^*(x) = \frac{\mu_n(X_1, \ldots, X_n; x)}{n}, \tag{21.1.2.8}$$

where $\mu_n(X_1, \ldots, X_n; x)$ is the number of sample elements whose values are less than $x$. It is a nondecreasing step function such that $F_n^*(-\infty) = 0$ and $F_n^*(+\infty) = 1$. Since each $X_i$ is less than $x$ with probability $p_x = F_n^*(x)$, while $X_i$ themselves are independent, it follows that $\mu_n(X_1, \ldots, X_n; x)$ is an integer random variable distributed according to the binomial law

$$P(\mu_n(X_1, \ldots, X_n; x) = k) = C_n^k [F(x)]^k [1 - F(x)]^{n-k}$$

with $E\{F_n^*(x)\} = F(x)$ and $\text{Var}\{F_n^*(x)\} = F(x)[1 - F(x)]$. By the *Glivenko–Cantelli theorem*,

$$D_n = \sup_x \left| F_n^*(x) - F(x) \right| \xrightarrow{\text{a.s.}} 0 \qquad (21.1.2.9)$$

as $n \to \infty$; i.e., the variable $D_n$ converges to 0 with probability 1 or almost surely (see Paragraph 20.3.1-2). The random variable $D_n$ measures how close $F_n^*(x)$ and $F(x)$ are. The empirical distribution function $F_n^*(x)$ is an unbiased consistent estimator of the theoretical distribution function.

If a sample is given by a statistical series, then the following formula can be used:

$$F^*(x) = \sum_{Z_i < x} p_i^*. \qquad (21.1.2.10)$$

It is convenient to construct the empirical distribution function $F_n^*(x)$ using the series of order statistics $X_{(1)} \leq \ldots \leq X_{(n)}$. In this case,

$$F_n^*(x) = \begin{cases} 0 & \text{if } x \leq X_{(1)}, \\ k/n & \text{if } X_{(k)} < x \leq X_{(k+1)}, \\ 1 & \text{if } x > X_{(n)}; \end{cases} \qquad (21.1.2.11)$$

i.e., the function $F_n^*(x)$ is constant on each interval $(X_{(k)}, X_{(k+1)}]$ and increases by $1/n$ at the point $X_{(k)}$.

### 21.1.2-7. Graphical representation of statistical distribution.

$1°$. A broken line passing through the points with coordinates $(Z_i, n_i)$ ($i = 1, 2, \ldots, L$), where $Z_i$ are the variate values in a statistical series and $n_i$ are the corresponding frequencies, is called the *frequency polygon* or a *distribution polygon*.

If the relative frequencies $p_1^* = n_1/n, \ldots, p_L^* = n_L/n$ are used instead of the frequencies $n_i$ ($n_1 + \cdots + n_L = n$), then the polygon is called the *relative frequency polygon*.

**Example 1.** For the statistical series

| $Z_j$ | 0 | 1 | 2 | 3 | 4 | 5 |
|---|---|---|---|---|---|---|
| $p_j^*$ | 0.1 | 0.15 | 0.3 | 0.25 | 0.15 | 0.05 |

the relative frequency polygon has the form shown in Fig. 21.1.

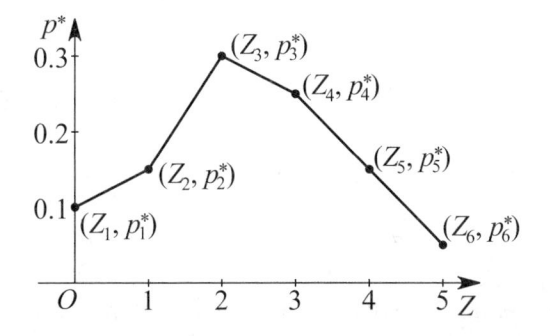

**Figure 21.1.** Example of a relative frequency polygon.

2°. The bar graph consisting of rectangles whose bases are class intervals of length $\Delta_i = x_{i+1} - x_i$ and whose heights are equal to the frequency densities $n_i/\Delta_i$ is called the *frequency histogram*. The area of a frequency histogram is equal to the size of the corresponding random sample.

The bar graph consisting of rectangles whose bases are class intervals of length $\Delta_i = x_{i+1} - x_i$ and whose heights are equal to the relative frequency densities $p_i^*(x)/\Delta_i = n_i/(n\Delta_i)$ is called the *relative frequency histogram*. The area of the relative frequency histogram is equal to 1. The relative frequency histogram is an estimator of the probability density.

**Example 2.** For the interval series

| $[x_i, x_{i+1})$ | $[0, 5)$ | $[5, 10)$ | $[10, 15)$ | $[15, 20)$ | $[20, 25)$ |
|---|---|---|---|---|---|
| $n_i$ | 4 | 6 | 12 | 10 | 8 |

the relative frequency histogram has the form shown in Fig. 21.2.

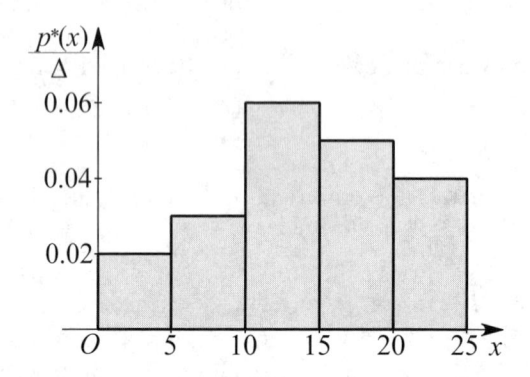

**Fig. 21.2.** Example of a relative frequency histogram.

### 21.1.2-8. Main distributions of mathematical statistics.

The normal distribution, the chi-square distribution, and the Student distribution were considered in Paragraphs 20.2.4-3, 20.2.4-5, and 20.2.4-6, respectively.

1°. A random variable $\Psi$ has a *Fisher–Snedecor distribution*, or an *F-distribution*, with $n_1$ and $n_2$ degrees of freedom if

$$\Psi = \frac{n_2 \chi_1^2}{n_1 \chi_2^2}, \qquad (21.1.2.12)$$

where $\chi_1^2$ and $\chi_2^2$ are independent random variables obeying the chi-square distribution with $n_1$ and $n_2$ degrees of freedom. The $F$-distribution is characterized by the probability density function

$$\Psi(x) = \frac{\Gamma\left(\frac{n_1+n_2}{2}\right)}{\Gamma\left(\frac{n_1}{2}\right)\Gamma\left(\frac{n_2}{2}\right)} n_1^{\frac{n_1}{2}} n_2^{\frac{n_2}{2}} x^{\frac{n_1}{2}-1} (n_2 + n_1 x)^{-\frac{n_1+n_2}{2}} \qquad (x > 0). \qquad (21.1.2.13)$$

where $\Gamma(x)$ is Gamma function. The quantiles of the $F$-distribution are usually denoted by $\phi_\alpha$.

2°. The *Kolmogorov distribution function* has the form

$$K(x) = \sum_{k=-\infty}^{\infty} (-1)^k e^{-2k^2 x^2} \qquad (x > 0). \qquad (21.1.2.14)$$

The Kolmogorov distribution is the distribution of the random variable $\eta = \max_{0 \le t \le 1} |\xi(t)|$, where $\xi(t)$ is a *Wiener process* on the interval $0 \le t \le 1$ with fixed endpoints $\xi(0) = 0$ and $\xi(1) = 0$.

## 21.1.3. Numerical Characteristics of Statistical Distribution

### 21.1.3-1. Sample moments.

The *kth sample moment* of a random sample $X_1, \ldots, X_n$ is defined as

$$\alpha_k^* = \frac{1}{n} \sum_{i=1}^{n} X_i^k. \qquad (21.1.3.1)$$

The *kth sample central moment* of a random sample $X_1, \ldots, X_n$ is defined as

$$\mu_k^* = \frac{1}{n} \sum_{i=1}^{n} (X_i - \alpha_1^*)^k. \qquad (21.1.3.2)$$

The sample moments satisfy the following formulas:

$$E\{\alpha_k^*\} = \alpha_k, \quad \mathrm{Var}\{\alpha_k^*\} = \frac{\alpha_{2k} - \alpha_k^2}{n}, \qquad (21.1.3.3)$$

$$E\{\mu_k^*\} = \mu_k + O(1/n),$$
$$\mathrm{Var}\{\mu_k^*\} = \frac{\mu_{2k} - 2k\mu_{k-1}\mu_{k+1} - \mu_k^2 + k^2\mu_2\mu_{k-1}^2}{n} + O(1/n^2). \qquad (21.1.3.4)$$

The sample moment $\alpha_k^*$ is an unbiased consistent estimator of the corresponding population moment $\alpha_k$. The sample central moment $\mu_k^*$ is a biased consistent estimator of the corresponding population central moment $\mu_k$.

If there exists a moment $\mu_{2k}$, then the sample moment $\mu_k^*$ is asymptotically normally distributed with parameters $(\alpha_k, (\alpha_{2k} - \alpha_k^2)/n)$ as $n \to \infty$.

Unbiased consistent estimators for $\mu_3$ and $\mu_4$ are given by

$$\mu_3^* = \frac{n^2 \alpha_3^*}{(n-1)(n-2)}, \quad \mu_4^* = \frac{n(n^2 - 2n + 3)\alpha_4^* - 3n(2n-3)(\alpha_4^*)^2}{(n-1)(n-2)(n-3)}. \qquad (21.1.3.5)$$

### 21.1.3-2. Sample mean.

The *sample mean* of a random sample $X_1, \ldots, X_n$ is defined as the first-order sample moment, i.e.,

$$m^* = \alpha_1^* = \frac{1}{n} \sum_{i=1}^{n} X_i. \qquad (21.1.3.6)$$

The sample mean of a random sample $X_1, \ldots, X_n$ is also denoted by $\overline{X}$. It satisfies the following formulas:

$$E\{m^*\} = m \quad (m = \alpha_1), \quad \mathrm{Var}\{m^*\} = \frac{\sigma^2}{n}, \qquad (21.1.3.7)$$

$$E\{(m^* - m)^3\} = \frac{\mu_3}{n^2}, \quad E\{(m^* - m)^4\} = \frac{3(n-1)\sigma^2 + \mu_4}{n^3}. \qquad (21.1.3.8)$$

The sample mean $m^*$ is an unbiased consistent estimator of the population expectation $E\{X\} = m$. If the population variance $\sigma^2$ exists, then the sample mean $m^*$ is asymptotically normally distributed with parameters $(m, \sigma^2/n)$.

The sample mean for the function $Y = f(X)$ of a random variable $X$ is

$$\overline{Y} = \frac{1}{n} \sum_{i=1}^{n} f(X_i).$$

### 21.1.3-3. Sample variances.

The statistic

$$\sigma^{2*} = \alpha_2^* - (m^*)^2 = \frac{1}{n}\sum_{i=1}^{n}(X_i - m^*)^2 \qquad (21.1.3.9)$$

is called the *sample variance (variance of the empirical distribution)* of the sample $X_1, \ldots, X_n$. Suppose that $\alpha_k = E\{X_1^k\}$ and $\mu_k = E\{(X_1 - \alpha_1)^k\}$. If $\alpha_4 < \infty$, then the sample variance (21.1.3.9) has the properties of asymptotic unbiasness and consistency.

The statistic

$$s^{2*} = \frac{n}{n-1}\left[\alpha_2^* - (m^*)^2\right] = \frac{1}{n-1}\sum_{i=1}^{n}(X_i - m^*)^2 = (s^*)^2 \qquad (21.1.3.10)$$

is called the *adjusted sample variance* and $s^*$ is called the *sample mean-square deviation* of the sample $X_1, \ldots, X_n$. They satisfy the formulas

$$E\{\sigma^{2*}\} = \frac{n-1}{n}\sigma^2, \quad E\{s^{2*}\} = \sigma^2, \quad \mathrm{Var}\{s^{2*}\} = \frac{1}{n}\left(\mu_4 - \frac{n-3}{n-4}\sigma^4\right). \qquad (21.1.3.11)$$

The statistic $s^{2*}$ is an unbiased estimator of the variance $\mu_2$.

### 21.1.3-4. Characteristics of asymmetry and excessî

The number

$$\gamma_1^* = \frac{\mu_3^*}{(\mu_2^*)^{3/2}} = \frac{1}{n\sigma^{3*}}\sum_{i=1}^{n}(X_i - m^*)^3 \qquad (21.1.3.12)$$

is called the *sample asymmetry coefficient* of a random sample $X_1, \ldots, X_n$.

The number

$$\gamma_2^* = \frac{\mu_4^*}{\mu_2^{*2}} - 3 = \frac{1}{n\sigma^{4*}}\sum_{i=1}^{n}(X_i - m^*)^4 - 3 \qquad (21.1.3.13)$$

is called the *sample excess coefficient* of a random sample $X_1, \ldots, X_n$.

The sample excess coefficient is used in criteria for testing the hypothesis $H_0 : \gamma_2 \neq 0$, which implies that the distribution of the random variable $X_i$ differs from the normal distribution.

The statistics $\gamma_1^*$ and $\gamma_2^*$ are consistent estimators of asymmetry and excess.

## 21.2. Statistical Estimation
### 21.2.1. Estimators and Their Properties

#### 21.2.1-1. Notion of estimator.

A *statistical estimator* $\theta^*$ (or simply an *estimator*) of an unknown parameter $\theta$ for a sample $X_1, \ldots, X_n$ is a function $\theta^* = \theta^*(X_1, \ldots, X_n)$ depending only on the sample $X_1, \ldots, X_n$.

An estimator is a random variable and varies depending on the sample. Just as any random variable, it has a distribution function $F_{\theta^*}(x)$. The distribution law of the statistic $\theta^* = \theta^*(X_1, \ldots, X_n)$ can be found by well-known methods of probability theory.

Estimators producing separate points in the space of parameters or parametric functions to be estimated are called *point estimators*.

Estimators producing sets of points in the space of parameters or parametric functions to be estimated are called *interval estimators*.

## 21.2.1-2. Unbiased estimators.

The difference $E\{\theta^*\} - \theta$ is called the *bias of the estimator* $\theta^*$ of the parameter $\theta$. The estimator $\theta^*$ is said to be *unbiased* if its expectation is equal to the parameter to be estimated, i.e., if $E\{\theta^*\} = \theta$; otherwise, the estimator $\theta^*$ is said to be *biased*.

**Example 1.** The sample moments $\alpha_k^*$ are unbiased estimators of the moments $\alpha_k$.

**Example 2.** The sample variance $\sigma^{2*}$ is a biased estimator of the variance $\sigma^2$, because
$$E\{\sigma^{2*}\} = \frac{n-1}{n}\sigma^2 \neq \sigma^2.$$

If an estimator is not unbiased, then it either overestimates or underestimates $\theta$. In both cases, this results in systematic errors of the same sign in the estimate of the parameter $\theta$.

If $E\{\theta_n^*\} \to \theta$ as $n \to \infty$, then the estimator $\theta^*$ is said to be *asymptotically unbiased*.

## 21.2.1-3. Efficiency of estimators.

An unbiased estimator $\theta_n^*$ is said to be *efficient* if it has the least variance among all possible unbiased estimators of the parameter $\theta$ for random samples of the same size.

*Cramér–Rao inequality.* Let $\theta^*$ be an unbiased estimator of a parameter $\theta$. Then (under additional regularity conditions imposed on the family $F(x;\theta)$) the variance of $\theta^*$ satisfies the inequality
$$\text{Var}\{\theta^*\} \geq \frac{1}{nI(\theta)}, \qquad (21.2.1.1)$$
where $I(\theta)$ is the *Fisher information*, determined in the continuous case by the formula
$$I(\theta) = E\left\{\left[\frac{\partial}{\partial \theta}\ln p(x;\theta)\right]^2\right\}$$
and in the discrete case by the formula
$$I(\theta) = E\left\{\left[\frac{\partial}{\partial \theta}\ln P(X;\theta)\right]^2\right\}.$$

The Cramér–Rao inequality determines a lower bound for the variance of an unbiased estimator. The variable
$$\varepsilon(\theta) = \frac{1}{nI(\theta)\text{Var}\{\theta^*\}} \qquad (21.2.1.2)$$
is called the *efficiency* $\varepsilon(\theta)$ of an unbiased estimator $\theta^*$. An unbiased estimator $\theta^*$ is said to be *(Cramér–Rao) efficient* if $\varepsilon(\theta) = 1$ for any $\theta$.

**Remark.** The Cramér–Rao inequality for biased estimators has the form
$$\text{Var}\{\theta_n^*\} \geq \frac{1}{nI(\theta)}\left[1 + \frac{\partial}{\partial \theta}(E\{\theta_n^*\} - \theta)\right]. \qquad (21.2.1.3)$$

## 21.2.1-4. Consistency of estimators.

An estimator $\theta^*$ is said to be *consistent* if it converges to the estimated parameter as the sample size increases. The convergence in question can be of different types: in probability, with probability 1, in mean-square, etc. As a rule, convergence in probability is used; i.e., an estimator $\theta^*$ is said to be consistent if for each $\varepsilon > 0$ and for all possible values of the unknown parameter $\theta$ this estimator satisfies the relation
$$P\{|\theta^* - \theta| > \varepsilon\} \to 0 \quad \text{as} \quad n \to \infty. \qquad (21.2.1.4)$$

The consistency of an estimator justifies an increase in the size of a random sample, since in this case the probability of a large error in the estimate of the parameter $\theta$ decreases.

### 21.2.1-5. Sufficient estimators.

1°. A statistic $S = (S_1(X_1, \ldots, X_n), \ldots, S_k(X_1, \ldots, X_n))$ is said to be *sufficient* for $\theta$ if the conditional distribution $F_{X_1, \ldots, X_n}(x_1, \ldots, x_n | S = s)$ is independent of the parameter $\theta$.

NEYMAN–FISHER THEOREM. *A statistic $S$ is sufficient for a parameter $\theta$ if and only if the likelihood function (see Paragraph 21.2.2-2) has the form*

$$L(X_1, \ldots, X_n; \theta) = A(X_1, \ldots, X_n) B(S; \theta), \qquad (21.2.1.5)$$

*where $A(X_1, \ldots, X_n)$ depends only on the sample $X_1, \ldots, X_n$ and the function $B(S, \theta)$ depends only on $S$ and $\theta$.*

THEOREM. *If $S$ is a sufficient statistic and $\theta^*$ is an estimator of a parameter $\theta$, then the conditional expectation $\theta_S^* = E\{\theta^* | S\}$ is an unbiased estimator of the parameter $\theta$, depends only on the sufficient statistic $S$, and satisfies the inequality*

$$\operatorname{Var}\{\theta_S^*\} \leq \operatorname{Var}\{\theta^*\} \qquad (21.2.1.6)$$

*for all $\theta$.*

2°. A statistic $S = S(X_1, \ldots, X_n)$ is said to be *complete* for a family of distributions $F(x; \theta)$ if the fact that the relation

$$E\{h(s)\} = E\{h(S(X_1, \ldots, X_n))\}$$
$$= \int_{-\infty}^{+\infty} \cdots \int_{-\infty}^{+\infty} h(S(x_1, \ldots, x_n)) p(x_1; \theta) \ldots p(x_n; \theta) \, dx_1 \ldots dx_n \qquad (21.2.1.7)$$

is satisfied for all $\theta$ implies that $h(s) \equiv 0$.

THEOREM (MINIMALITY OF VARIANCE OF AN ESTIMATOR DEPENDING ON A COMPLETE SUFFICIENT STATISTIC). *Let $S$ be a complete sufficient statistic, and let $\theta^*$ be an unbiased estimator of an unknown parameter $\theta$. Then*

$$\theta_S^* = E\{\theta^* | S\} \qquad (21.2.1.8)$$

*is the unique unbiased estimator with minimal variance.*

### 21.2.1-6. Main statistical estimators.

Let $X_1, \ldots, X_n$ be a random sample of a normal population with parameters $(a, \sigma^2)$. Then

a) the statistic $\dfrac{X_i - a}{\sigma}$ has the standard normal distribution;

b) the statistic $\dfrac{(m^* - a)\sqrt{n}}{\sigma}$ has the standard normal distribution;

c) the statistic $\dfrac{(m^* - a)\sqrt{n}}{\sqrt{s^{2*}}} = \dfrac{(m^* - a)\sqrt{n-1}}{\sqrt{\sigma^{2*}}}$, which is called *Student's ratio*, has the $t$-distribution with $n - 1$ degrees of freedom;

d) the statistic $\dfrac{X_i - a}{\sqrt{s^{2*}}} = \dfrac{X_i - a}{\sqrt{\sigma^{2*}}} \sqrt{\dfrac{n-1}{n}}$ has the $t$-distribution with $n-1$ degrees of freedom;

e) the statistic $\dfrac{(n-1)s^{2*}}{\sigma^2} = \dfrac{n\sigma^{2*}}{\sigma^2} = \dfrac{1}{\sigma^2}\sum_{k=1}^{n}(X_k - m^*)^2$ has the chi-square distribution with $n - 1$ degrees of freedom;

f) the statistic $\dfrac{1}{\sigma^2}\sum_{k=1}^{n}(X_k - a)^2$ has the chi-square distribution with $n$ degrees of freedom.

**Remark.** The quantiles $u_\alpha$, $\chi^2_\alpha$, $t_\alpha$, and $\phi_\alpha$ of the normal distribution, the chi-square distribution, the $t$-distribution, and the $F$-distribution can be found in the corresponding tables or calculated in EXCEL, MATHCAD, and other software.

**Example 3.** For a normal distribution, the $\gamma$-quantile $u_\gamma$ can be determined using NORMINV($\gamma$;m;$\sigma$) in EXCEL software, where, in the examples illustrating the use of software, $\gamma$ is the confidence level, $\alpha = 1 - \gamma$ is the significance level, $m$ is the expectation, and $\sigma$ is the standard deviation. The function qnorm($\gamma, m, \sigma$) in MATHCAD software can also be used.

**Example 4.** For the chi-square distribution with $n$ degrees of freedom, the $\gamma$-quantile $\chi^2_\gamma(n)$ can be found using the function CHIINV($\alpha, n$) in EXCEL software. The function qchisq($\gamma, n$) in MATHCAD software can also be used.

**Example 5.** For the $t$-distribution with $n$ degrees of freedom, the $\gamma$-quantile $t_\gamma$ can be found using the function TINV($2\alpha, n$) in EXCEL software. The function qt($\gamma, n$) in MATHCAD software can also be used.

**Example 6.** For the $F$-distribution with $n_1$ and $n_2$ degrees of freedom, the $\gamma$-quantile $\phi_\gamma$ can be found using the function FINV($\alpha, n_1, n_2$) in EXCEL software. The function qF($\gamma, n_1, n_2$) in MATHCAD software can also be used.

## 21.2.2. Estimation Methods for Unknown Parameters

### 21.2.2-1. Method of moments.

If the theoretical distribution function $F(x)$ of a population belongs to a $k$-parameter family $F(x; \theta_1, \ldots, \theta_k)$ with unknown parameters $\theta_1, \ldots, \theta_k$, then any numerical characteristic is a function of the parameters $\theta_1, \ldots, \theta_k$. For the known distribution function, one can find the first $k$ theoretical moments

$$\alpha_i = \alpha_i(\theta_1, \ldots, \theta_k) \qquad (i = 1, 2, \ldots, k) \tag{21.2.2.1}$$

if these moments exist. The *method of moments* works as follows: for a large sample size $k$, the theoretical moments $\alpha_1, \ldots, \alpha_k$ in system (21.2.2.1) are replaced by the sample moments $\alpha_1^*, \ldots, \alpha_k^*$; then this system is solved for $\theta_1, \ldots, \theta_k$ and the estimates of the unknown parameters are obtained. Thus the estimators $\theta_1^*, \ldots, \theta_k^*$ of the unknown parameters $\theta_1, \ldots, \theta_k$ in the method of moments are obtained from the system of equations

$$\alpha_i^* = \alpha_1(\theta_1^*, \ldots, \theta_k^*) \qquad (i = 1, 2, \ldots, k). \tag{21.2.2.2}$$

The estimators obtained by the method of moments are, as a rule, consistent.

**Example 1.** A sample $X_1, \ldots, X_n$ is selected from a population with theoretical distribution function having the exponential density $p(x) = p(x, \theta) = \theta e^{-\theta x}$ ($x \geq 0$). Since $\alpha_1 = 1/\theta$ for the exponential law, we see that the method of moments gives $\alpha_1^* = 1/\theta^*$, which implies that $\theta^* = 1/\alpha_1^* = 1/\overline{X}$.

It should be noted that the efficiency of estimators obtained by the method of moments is, as a rule, less than 1, and these estimators are even biased. Since the estimators obtained by the method of moments are rather simple, they are often used as initial approximations for finding more efficient estimators.

### 21.2.2-2. Maximum likelihood estimation.

Maximum likelihood estimation is the most popular estimation method. It is based on conditions for the extremum of a function of one or several random variables. The likelihood function is usually taken for such a function.

The *likelihood function* is defined as the function
$$L(X_1,\ldots,X_n) = L(X_1,\ldots,X_n;\theta)$$
$$= \begin{cases} P(X_1;\theta)P(X_2;\theta)\ldots P(X_n;\theta) & \text{in the discrete case,} \\ p(x_1;\theta)p(x_2;\theta)\ldots p(x_n;\theta) & \text{in the continuous case.} \end{cases} \quad (21.2.2.3)$$

In the likelihood function $L(X_1,\ldots,X_n;\theta)$, the sample elements $X_1, \ldots, X_n$ are fixed parameters and $\theta$ is an argument.

The *maximum likelihood estimator* is a value $\theta^*$ such that
$$L(X_1,\ldots,X_n;\theta^*) = \max_\theta L(X_1,\ldots,X_n;\theta). \quad (21.2.2.4)$$

Since $L$ and $\ln L$ attain maximum values for the same values of the argument $\theta$, it is convenient to use the logarithm of the likelihood function rather than the function itself in practical implementations of the maximum likelihood method.

The equation
$$\frac{\partial}{\partial \theta} \ln[L(X_1,\ldots,X_n;\theta)] = 0 \quad (21.2.2.5)$$
is called the *likelihood equation*.

If the theoretical distribution function $F(X_1,\ldots,X_n;\theta_1,\ldots,\theta_k)$ depends on several parameters $\theta_1,\ldots,\theta_k$, then equation (21.2.2.5) should be replaced in the maximum likelihood method by the system of likelihood equations

$$\frac{\partial}{\partial \theta_1} \ln[L(X_1,\ldots,X_n;\theta_1,\ldots,\theta_k)] = 0,$$
$$\frac{\partial}{\partial \theta_2} \ln[L(X_1,\ldots,X_n;\theta_1,\ldots,\theta_k)] = 0, \quad (21.2.2.6)$$
$$\ldots\ldots\ldots\ldots\ldots\ldots\ldots\ldots\ldots\ldots\ldots\ldots\ldots$$
$$\frac{\partial}{\partial \theta_k} \ln[L(X_1,\ldots,X_n;\theta_1,\ldots,\theta_k)] = 0.$$

Properties of estimators obtained by the maximum likelihood method:

1. *Coincidence of the efficient estimator with the estimator obtained by the maximum likelihood method.* If there exists an efficient estimator (or a set of jointly efficient estimators), then it is the unique solution of the likelihood equation (or system of equations).
2. *Asymptotic efficiency of the maximum likelihood estimator.* Under certain conditions on the family $F(x;\theta)$ that guarantee the possibility of differentiating in the integrand and expanding $\partial/\partial\theta \ln p(x,\theta)$ in a Taylor series up to the first term, the likelihood equation has a solution that is asymptotically normal with parameters $(\theta, 1/(nI))$ as $n \to \infty$, where $I$ is the Fisher information.

### 21.2.2-3. Least-squares method.

The *least-squares method* for obtaining an estimator $\theta^*$ of the parameter $\theta$ is based on the minimization of the sum of squares of deviations of the sample data from the desired estimator of $\theta$; i.e., it is required to find a value $\theta^*$ minimizing the sum
$$S(\theta) = \sum_{i=1}^{n}(X_i - \theta)^2 \to \min. \quad (21.2.2.7)$$

The least-squares method is the simplest method for finding estimators of the parameter $\theta$.

**Example 2.** Let us estimate the parameter $\theta$ of the Poisson distribution

$$P(X = k) = \frac{\lambda^k}{k!} e^{-\lambda}.$$

The function $S(\theta) = \sum_{i=1}^{n}(X_i - \theta)^2$ has a minimum at the point $\widehat{\theta} = \frac{1}{n}\sum_{i=1}^{n} X_i$. Thus, then least-squares estimator of the parameter $\theta$ in the Poisson distribution is $\theta^* = \frac{1}{n}\sum_{i=1}^{n} X_i$.

**Remark.** Estimators of unknown parameters can also be found by the following methods: the minimal distance method, the method of nomograms, the minimal risk method, the maximum a posteriori (conditional) probability method, etc.

### 21.2.3. Interval Estimators (Confidence Intervals)

**21.2.3-1. Confidence interval.**

An estimator $\theta^*$ of a parameter $\theta$ of the distribution of an observable random variable $X$ (the estimator is a statistic, i.e., function of a random sample) is itself a random variable with its own distribution law and numerical characteristics (parameters) of the distribution. For a small number of observations, the following problems may arise:

1. What error can arise from the replacement of the parameter $\theta$ by its point estimator $\theta^*$?
2. What is the probability that the errors obtained are not beyond any prescribed limits?

An *interval estimator (confidence interval)* is an interval $(\theta_L^*, \theta_R^*)$, determined by the sample, such that with some probability close to unity this interval contains the value of the population parameter to be estimated; i.e.,

$$P(\theta_L^* \le \theta \le \theta_R^*) = \gamma, \tag{21.2.3.1}$$

where $\theta_L^*$ and $\theta_R^*$ are the lower and upper (left and right) boundaries of the confidence interval of the parameter $\theta$ and $\gamma$ is the *confidence level*.

The confidence level and the significance level satisfy the relation

$$\gamma + \alpha = 1. \tag{21.2.3.2}$$

The confidence intervals can be constructed from a given estimator $\theta^*$ and a given confidence level $\gamma$ by various methods. In practice, the following two types of confidence intervals are used: two-sided and one-sided.

**21.2.3-2. Confidence interval in the case of normal samples.**

$1°$. *Confidence interval for estimating the expectation given variance.* Suppose that a random sample $X_1, \ldots, X_n$ is selected from a population $X$ with a normal distribution law with unknown expectation $a$ and known variance $\sigma^2$. To estimate the expectation, we use the statistic $m^*$ (the sample mean), which has the normal distribution with parameters $(a, \sigma^2/n)$. Then the statistic $(m^* - a)\sqrt{n}/\sigma$ has the normal distribution with parameters $(0, 1)$.

The confidence interval for the expectation $m$ given variance $\sigma^2$ has the form

$$\left( m^* - u_{\frac{1+\gamma}{2}} \frac{\sigma}{\sqrt{n}};\ m^* + u_{\frac{1+\gamma}{2}} \frac{\sigma}{\sqrt{n}} \right), \tag{21.2.3.3}$$

where $u_{\frac{1+\gamma}{2}}$ is the $\frac{1+\gamma}{2}$-quantile of the normal distribution with parameters $(0, 1)$.

**2°. Confidence interval for the expectation if the variance is unknown.** Suppose that a random sample $X_1, \ldots, X_n$ is selected from a population $X$ normally distributed with unknown variance $\sigma^2$ and unknown expectation $a$. To estimate the expectation, we use the statistic

$$T = \sqrt{\frac{n}{s^{2*}}}\,(m^* - a)$$

having the $t$-distribution (Student's distribution) with $L = n - 1$ degrees of freedom.

The confidence interval for the expectation for the case in which the variance is unknown has the form

$$\left(m^* - \sqrt{\frac{s^{2*}}{n}}\,t_{\frac{1+\gamma}{2}};\; m^* + \sqrt{\frac{s^{2*}}{n}}\,t_{\frac{1+\gamma}{2}}\right). \tag{21.2.3.4}$$

**Remark.** For a sufficiently large sample size $n$, the difference between the confidence intervals obtained by (21.2.3.3) and (21.2.3.4) is small, since Student's distribution tends to the normal distribution as $n \to \infty$.

**3°. Confidence interval for the variance of a population with a normal distribution and known expectation.**

Suppose that a sample $X_1, \ldots, X_n$ is selected from a normal population with unknown variance and known expectation equal to $a$. For the estimator of the unknown variance $\sigma^2$ the statistic $\sigma_0^2 = \frac{1}{n}\sum_{k=1}^{n}(X_k - a)^2$ is used. In this case, the statistic $\chi^2 = n\sigma_0^2/\sigma^2$ has the chi-square distribution with $n$ degrees of freedom.

The confidence interval for the variance, given expectation, has the form

$$\left(\frac{n\sigma_0^2}{\chi^2_{\frac{1+\gamma}{2}}(n)};\; \frac{n\sigma_0^2}{\chi^2_{\frac{1-\gamma}{2}}(n)}\right). \tag{21.2.3.5}$$

**4°. Confidence interval for the variance of a population with normal distribution and unknown expectation.** For the estimator of the unknown variance $\sigma^2$ the sample variance $s^{2*}$ is used. In this case, the statistic $\chi^2 = (n-1)s^{2*}/\sigma^2$ has the chi-square distribution with $n-1$ degrees of freedom. The confidence interval for the variance given that the expectation is unknown has the form

$$\left(\frac{(n-1)s^{2*}}{\chi^2_{\frac{1+\gamma}{2}}(n-1)};\; \frac{(n-1)s^{2*}}{\chi^2_{\frac{1-\gamma}{2}}(n-1)}\right). \tag{21.2.3.6}$$

## 21.3. Statistical Hypothesis Testing

### 21.3.1. Statistical Hypothesis. Test

#### 21.3.1-1. Statistical hypothesis.

Any assumption concerning the form of the population distribution law or numerical values of the parameters of the distribution law is called a *statistical hypothesis*. Any statistical hypothesis uniquely determining the distribution law is said to be *simple*; otherwise, it is said to be *composite*.

A statistical hypothesis is said to be *parametric* if it contains an assumption concerning the range of the unknown parameters. If a hypothesis does not contain any assumption concerning the range of the unknown parameters, then such a hypothesis is said to be *nonparametric*.

The hypothesis $H_0$ to be tested is called the *null hypothesis*, and the opposite hypothesis $H_1$ is called the *alternative hypothesis*.

**Example 1.** The hypothesis that the theoretical distribution function is normal with zero expectation is a parametric hypothesis.

**Example 2.** The hypothesis that the theoretical distribution function is normal is a nonparametric hypothesis.

**Example 3.** The hypothesis $H_0$ that the variance of a random variable $X$ is equal to $\sigma_0^2$, i.e., $H_0 : \text{Var}\{X\} = \sigma_0^2$, is simple. For an alternative hypothesis one can take one of the following hypotheses: $H_1 : \text{Var}\{X\} > \sigma_0^2$ (composite hypothesis), $H_1 : \text{Var}\{X\} < \sigma_0^2$ (composite hypothesis), $H_1 : \text{Var}\{X\} \neq \sigma_0^2$ (composite hypothesis), or $H_1 : \text{Var}\{X\} = \sigma_1^2$ (simple hypothesis).

### 21.3.1-2. Statistical test. Type I and Type II errors.

$1°$. A *statistical test* (or simply a *test*) is a rule that permits one, on the basis of a sample $X_1, \ldots, X_n$ alone, to accept or reject the null hypothesis $H_0$ (respectively, reject or accept the alternative hypothesis $H_1$). Any test is characterized by two disjoint regions:

1. The *critical region* $W$ is the region in the $n$-dimensional space $\mathbb{R}^n$ such that if the sample $X_1, \ldots, X_n$ lies in this region, then the null hypothesis $H_0$ is rejected (and the alternative hypothesis $H_1$ is accepted).
2. The *acceptance region* $\overline{W}$ ($\overline{W} = \mathbb{R}^n \setminus W$) is the region in the $n$-dimensional space $\mathbb{R}^n$ such that if the sample $X_1, \ldots, X_n$ lies in this region, then the null hypothesis $H_0$ is accepted (and the alternative hypothesis $H_1$ is rejected).

$2°$. Suppose that there are two hypotheses $H_0$ and $H_1$, i.e., two disjoint subsets $\Gamma_0$ and $\Gamma_1$ are singled out from the set of all distribution functions. We consider the null hypothesis $H_0$ that the sample $X_1, \ldots, X_n$ is drawn from a population with theoretical distribution function $F(x)$ belonging to the subset $\Gamma_0$ and the alternative hypothesis that the sample is drawn from a population with theoretical distribution function $F(x)$ belonging to the subset $\Gamma_1$. Suppose, also, that a test for verifying these hypotheses is given; i.e., the critical region $W$ and the admissible region $\overline{W}$ are given. Since the sample is random, there may be errors of two types:

i) *Type I error* is the error of accepting the hypothesis $H_1$ (the hypothesis $H_0$ is rejected), while the null hypothesis $H_0$ is true.

ii) *Type II error* is the error of accepting the hypothesis $H_0$ (the hypothesis $H_1$ is rejected), while the alternative hypothesis $H_1$ is true.

The probability $\alpha$ of Type I error is called the *false positive rate*, or *size* of the test, and is determined by the formula

$$\alpha = P[(X_1, \ldots, X_n) \in W]$$
$$= \begin{cases} \sum P(X_1)P(X_2)\ldots P(X_n) & \text{in the discrete case,} \\ \int \ldots \int p(X_1)p(X_2)\ldots p(X_n)\, dx_1 \ldots dx_n & \text{in the continuous case;} \end{cases}$$

here $P(x)$ or $p(x)$ is the distribution series or the distribution density of the random variable $X$ under the assumption that the null hypothesis $H_0$ is true, and the summation or integration is performed over all points $(x_1, \ldots, x_n) \in W$. The number $1 - \alpha$ is called the *specificity* of the test. If the hypothesis $H_0$ is composite, then the size $\alpha = \alpha[F(x)]$ depends on the actual theoretical distribution function $F(x) \in \Gamma_0$. If, moreover, $H_0$ is a parametric hypothesis, i.e., $\Gamma_0$ is a parametric family of distribution functions $F(x; \theta)$ depending on the parameter $\theta$ with range $\Theta_0 \in \Theta$, where $\Theta$ is the region of all possible values $\theta$, then, instead of notation $\alpha[F(x)]$, the notation $\alpha(\theta)$ is used under the assumption that $\theta \in \Theta_0$.

The probability $\widetilde{\beta}$ of Type II error is called the *false negative rate*. The *power* $\beta = 1 - \widetilde{\beta}$ of the test is the probability that Type II error does not occur, i.e., the probability of rejecting the false hypothesis $H_0$ and accepting the hypothesis $H_1$. The test power is determined by the same formula as the test specificity, but in this case, the distribution series $P(x)$ or the density function $p(x)$ are taken under the assumption that the alternative hypothesis $H_1$ is true. If the hypothesis $H_1$ is composite, then the power $\beta = \beta[F(x)]$ depends on the actual theoretical distribution function $F(x) \in \Gamma_1$. If, moreover, $H_1$ is a parametric hypothesis, then, instead of the notation $\beta[F(x)]$, the notation $\beta(\theta)$ is used under the assumption $\theta \in \Theta_1$, where $\Theta_1$ is the range of the unknown parameter $\theta$ under the assumption that the hypothesis $H_1$ is true.

The difference between the test specificity and the test power is that the specificity $1 - \alpha[F(x)]$ is determined for the theoretical distribution functions $F(x) \in \Gamma_0$, and the power $\beta(\theta)$ is determined for the theoretical distribution functions $F(x) \in \Gamma_1$.

3°. Depending on the form of the alternative hypothesis $H_1$, the critical regions are classified as one-sided (right-sided and left-sided) and two-sided:

1. The right-sided critical region (Fig. 21.3a) consisting of the interval $(t_{\text{cr}}^R; \infty)$, where the boundary $t_{\text{cr}}^R$ is determined by the condition

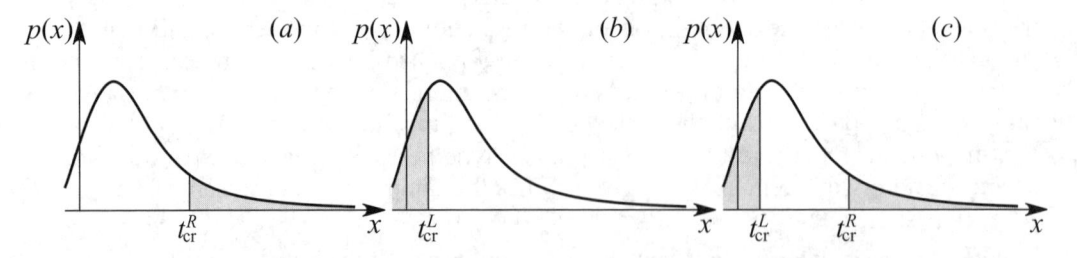

$$P[S(X_1, \ldots, X_n) > t_{\text{cr}}^R] = \alpha; \qquad (21.3.1.1)$$

**Figure 21.3.** Right-sided (*a*), left-sided (*b*), and two-sided (*c*) critical region.

2. The left-sided critical region (Fig. 21.3*b*) consisting of the interval $(-\infty; t_{\text{cr}}^L)$, where the boundary $t_{\text{cr}}^L$ is determined by the condition

$$P[S(X_1, \ldots, X_n) < t_{\text{cr}}^L] = \alpha; \qquad (21.3.1.2)$$

3. The two-sided critical region (Fig. 21.3*c*) consisting of the intervals $(-\infty; t_{\text{cr}}^L)$ and $(t_{\text{cr}}^R; \infty)$, where the points $t_{\text{cr}}^L$ and $t_{\text{cr}}^R$ are determined by the conditions

$$P[S(X_1, \ldots, X_n) < t_{\text{cr}}^L] = \frac{\alpha}{2} \quad \text{and} \quad P[S(X_1, \ldots, X_n) > t_{\text{cr}}^R] = \frac{\alpha}{2}. \qquad (21.3.1.3)$$

21.3.1-3. Simple hypotheses.

Suppose that a sample $X_1, \ldots, X_n$ is selected from a population with theoretical distribution function $F(x)$ about which there are two simple hypotheses, the null hypothesis $H_0 : F(x) = F_0(x)$ and the alternative hypothesis $H_1 : F(x) = F_1(x)$, where $F_0(x)$ and $F_1(x)$ are known distribution functions. In this case, there is a test that is most powerful for a given size $\alpha$;

this is called the *likelihood ratio test*. The likelihood ratio test is based on the statistic called the *likelihood ratio*,

$$\Lambda = \Lambda(X_1, \ldots, X_n) = \frac{L_1(X_1, \ldots, X_n)}{L_0(X_1, \ldots, X_n)}, \qquad (21.3.1.4)$$

where $L_0(X_1, \ldots, X_n)$ is the likelihood function under the assumption that the null hypothesis $H_0$ is true, and $L_1(X_1, \ldots, X_n)$ is the likelihood function under the assumption that the alternative hypothesis $H_1$ is true.

The critical region $W$ of the likelihood ratio test consists of all points $(x_1, \ldots, x_n)$ for which $\Lambda(X_1, \ldots, X_n)$ is larger than a critical value $C$.

NEYMAN–PEARSON LEMMA. *Of all tests of given size $\alpha$ testing two simple hypotheses $H_0$ and $H_1$, the likelihood ratio test is most powerful.*

### 21.3.1-4. Sequential analysis. Wald test.

*Sequential analysis* is the method of statistical analysis in which the sample size is not fixed in advance but is determined in the course of experiment. The ideas of sequential analysis are most often used for testing statistical hypotheses. Suppose that observations $X_1, X_2, \ldots$ are performed successively; after each trial, one can stop the trials and accept one of the hypotheses $H_0$ and $H_1$. The hypothesis $H_0$ is that the random variables $X_i$ have the probability distribution with density $p_0(x)$ in the continuous case or the probability distribution determined by probabilities $P_0(X_i)$ in the discrete case. The hypothesis $H_1$ is that the random variables $X_i$ have the probability distribution with density $p_1(x)$ in the continuous case or the probability distribution determined by probabilities $P_1(X_i)$ in the discrete case.

WALD TEST. *Of all tests with given size $\alpha$, power $\beta$, finite mean number $N_0$ of observations under the assumption that the hypotheses $H_0$ is true, and finite mean number $N_1$ of observations under the assumption that the hypothesis $H_1$ is true, the sequential likelihood ratio test minimizes both $N_0$ and $N_1$.*

The decision in the Wald test is made as follows. One specifies critical values $A$ and $B$, $0 < A < B$. The result $X_1$ of the first observation determines the logarithm of the likelihood ratio

$$\lambda(X_1) = \begin{cases} \ln \dfrac{P_1(X_1)}{P_0(X_1)} & \text{in the discrete case,} \\ \ln \dfrac{p_1(X_1)}{p_0(X_1)} & \text{in the continuous case.} \end{cases}$$

If $\lambda(X_1) \geq B$, then the hypothesis $H_1$ is accepted; if $\lambda(X_1) \leq A$, then the hypothesis $H_1$ is accepted; and if $A < \lambda(X_1) < B$, then the second trial is performed. The logarithm of the likelihood ratio

$$\lambda(X_1, X_2) = \lambda(X_1) + \lambda(X_2)$$

is again determined. If $\lambda(X_1, X_2) \geq B$, then the hypothesis $H_1$ is accepted; if $\lambda(X_1, X_2) \leq A$, then the hypothesis $H_1$ is accepted; and if $A < \lambda(X_1, X_2) < B$, then the third trial is performed. The logarithm of the likelihood ratio

$$\lambda(X_1, X_2, X_3) = \lambda(X_1) + \lambda(X_2) + \lambda(X_3)$$

is again determined, and so on. The graphical scheme of trials is shown in Fig. 21.4.

For the size $\alpha$ and the power $\beta$ of the Wald test, the following approximate estimates hold:

$$\alpha \approx \frac{1 - e^A}{e^B - e^A}, \quad \beta \approx \frac{e^B(1 - e^A)}{e^B - e^A}.$$

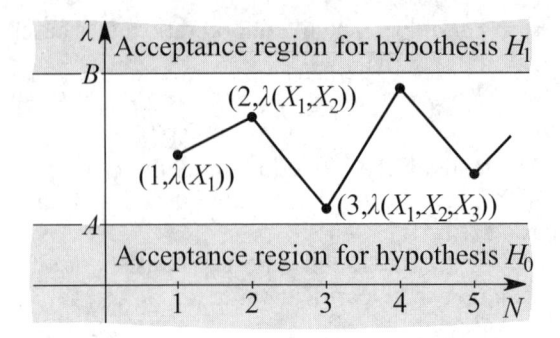

Figure 21.4. The graphical scheme of the Wald test.

For given $\alpha$ and $\beta$, these estimates result in the following approximate expressions for the critical values $A$ and $B$:
$$A \approx \ln \frac{\beta}{1-\alpha}, \quad B \approx \ln \frac{\beta}{\alpha}.$$

For the mean numbers $N_0$ and $N_1$ of observations, the following approximate estimates hold under the assumptions that the hypothesis $H_0$ or $H_1$ is true:
$$N_0 \approx \frac{\alpha B + (1-\alpha)A}{\mathrm{M}[\lambda(X)|H_0]}, \quad N_1 \approx \frac{\beta B + (1-\beta)A}{\mathrm{M}[\lambda(X)|H_1]},$$

where
$$E\{\lambda(X)|H_0\} = \sum_{i=1}^{L} \ln \frac{P_1(b_i)}{P_0(b_i)} P_0(b_i), \quad E\{\lambda(X)|H_1\} = \sum_{i=1}^{L} \ln \frac{P_1(b_i)}{P_0(b_i)} P_1(b_i)$$

in the discrete case and
$$E\{\lambda(X)|H_0\} = \int_{-\infty}^{\infty} \ln \frac{p_1(x)}{p_0(x)} p_0(x)\, dx, \quad E\{\lambda(X)|H_1\} = \int_{-\infty}^{\infty} \ln \frac{p_1(x)}{p_0(x)} p_1(x)\, dx$$

in the continuous case.

### 21.3.2. Goodness-of-Fit Tests

**21.3.2-1. Statement of problem.**

Suppose that there is a random sample $X_1, \ldots, X_n$ drawn from a population $X$ with unknown theoretical distribution function. It is required to test the simple nonparametric hypothesis $H_0 : F(x) = F_0(x)$ against the composite alternative hypothesis $H_1 : F(x) \ne F_0(x)$, where $F_0(x)$ is a given theoretical distribution function. There are several methods for solving this problem that differ in the form of the measure of discrepancy between the empirical and hypothetical distribution laws. For example, in the Kolmogorov test (see Paragraph 21.3.2-2) and the Smirnov test (see Paragraph 21.3.2-3), this measure is a function of the difference between the empirical distribution function $F^*(x)$ and the theoretical distribution function $F(x)$, i.e.,
$$\rho = \rho[F^*(x) - F(x)];$$

and in the $\chi^2$-test, this measure is a function of the difference between the theoretical probabilities $p_i^T = P(H_i)$ of the random events $H_1, \ldots, H_L$ and their relative frequencies $p_i^* = n_i/n$, i.e.,
$$\rho = \rho(p_i^T - p_i^*).$$

### 21.3.2-2. Kolmogorov test.

To test a hypothesis concerning the distribution law, the statistic

$$\rho = \rho(X_1, \ldots, X_n) = \sqrt{n} \sup_{-\infty < x < \infty} |F^*(x) - F(x)| \qquad (21.3.2.1)$$

is used to measure the compatibility (goodness of fit) of the hypothesis in the *Kolmogorov test*. A right-sided region is chosen to be the critical region in the Kolmogorov test. For a given size $\alpha$, the boundary $t_{\text{cr}}^R$ of the right-sided critical region can be found from the relation

$$t_{\text{cr}}^R = F^{-1}(1 - \alpha).$$

Table 21.1 presents values depending on the size and calculated by formula (21.3.2.1).

TABLE 21.1
Boundary of right-sided critical region

| $\alpha$ | 0.5 | 0.1 | 0.05 | 0.01 | 0.001 |
|---|---|---|---|---|---|
| $t_{\text{cr}}^R$ | 0.828 | 1.224 | 1.385 | 1.627 | 1.950 |

As $n \to \infty$, the distribution of the statistic $\rho$ converges to the Kolmogorov distribution and the boundary $t_{\text{cr}}^R$ of the right-sided critical region coincides with the $(1 - \alpha)$-quantile $k_{1-\alpha}$ of the Kolmogorov distribution.

The advantages of the Kolmogorov test are its simplicity and the absence of complicated calculations. But this test has several essential drawbacks:

1. The use of the test requires considerable a priori information about the theoretical law of distribution; i.e., in addition to the form of the distribution law, one must know the values of all parameters of the distribution.
2. The test deals only with the maximal deviation of the empirical distribution function from the theoretical one and does not take into account the variations of this deviation on the entire range of the random sample.

### 21.3.2-3. Smirnov test ($\omega^2$-test).

In contrast to the Kolmogorov test, the *Smirnov test* takes the mean value of a function of the difference between the empirical and theoretical distribution functions on the entire domain of the distribution function to be the measure of discrepancy between the empirical distribution function and the theoretical one; this eliminates the drawback of the Kolmogorov test.

In the general case, the statistic

$$\omega^2 = \omega^2(X_1, \ldots, X_n) = \int_{-\infty}^{\infty} [F^*(x) - F(x)]^2 \, dF(x) \qquad (21.3.2.2)$$

is used. Using the series $X_1^*, \ldots, X_n^*$ of order statistics, one can rewrite the statistic $\omega^2$ in the form

$$\omega^2 = \frac{1}{n} \sum_{i=1}^{n} \left[ F^*(X_i^*) - \frac{2i-1}{2n} \right]^2 + \frac{1}{12n^2}. \qquad (21.3.2.3)$$

A right-sided region is chosen to be the critical region in the Smirnov test. For a given size $\alpha$, the boundary $t_{\text{cr}}^R$ of the right-sided critical region can be found from the relation

$$t_{\text{cr}}^R = F^{-1}(1-\alpha). \tag{21.3.2.4}$$

Table 21.2 presents the values of $t_{\text{cr}}^R$ depending on the size and calculated by formula (21.3.2.4).

TABLE 21.2
Boundary of right-sided critical region

| $\alpha$ | 0.5 | 0.1 | 0.05 | 0.01 | 0.001 |
|---|---|---|---|---|---|
| $t_{\text{cr}}^R$ | 0.118 | 0.347 | 0.461 | 0.620 | 0.744 |

As $n \to \infty$, the distribution of the statistic $\omega^2$ converges to the $\omega^2$-distribution and the boundary $t_{\text{cr}}^R$ of the right-sided critical region coincides with the $(1-\alpha)$-quantile of an $\omega^2$-distribution.

### 21.3.2-4. Pearson test ($\chi^2$-test).

1°. The $\chi^2$-*test* is used to measure the compatibility (goodness of fit) of the theoretical probabilities $p_k = P(H_k)$ of random events $H_1, \ldots, H_L$ with their relative frequencies $p_k^* = n_k/n$ in a sample of $n$ independent observations. The $\chi^2$-test permits comparing the theoretical distribution of the population with its empirical distribution.

The goodness of fit is measured by the statistic

$$\chi^2 = \sum_{k=1}^{L} \frac{(n_k - np_k)^2}{np_k} = \sum_{k=1}^{L} \frac{n_k^2}{np_k} - n, \tag{21.3.2.5}$$

whose distribution as $n \to \infty$ tends to the chi-square distribution with $v = L - 1$ degrees of freedom. According to the $\chi^2$-test, there are no grounds to reject the theoretical probabilities for a given confidence level $\gamma$ if the inequality $\chi^2 < \chi_\gamma^2(v)$ holds, where $\chi_\gamma^2(v)$ is the $\gamma$-quantile of a $\chi^2$-distribution with $v$ degrees of freedom. For $v > 30$, instead of the chi-square distribution, one can use the normal distribution of the random variable $\sqrt{2\chi^2}$ with expectation $\sqrt{2v-1}$ and variance 1.

Remark. The condition $n_k > 5$ is a necessary condition for the $\chi^2$-test to be used.

2°. $\chi^2$-*test with estimated parameters.*
Suppose that $X_1, \ldots, X_n$ is a sample drawn from a population $X$ with unknown distribution function $F(x)$. We test the null hypothesis $H_0$ stating that the population is distributed according to the law with the distribution function $F(x)$ equal to the function $F_0(x)$, i.e., the null hypothesis $H_0 : F(x) = F_0(x)$ is tested. Then the alternative hypothesis is $H_1 : F(x) \neq F_0(x)$.

In this case, the statistic (21.3.2.5) as $n \to \infty$ tends to the chi-square distribution with $v = L - q - 1$ degrees of freedom, where $q$ is the number of estimated parameters. Thus, for example, $q = 2$ for the normal distribution and $q = 1$ for the Poisson distribution. The null hypothesis $H_0$ for a given confidence level $\alpha$ is accepted if $\chi^2 < \chi_\alpha^2(L-q-1)$.

## 21.3.3. Problems Related to Normal Samples

**21.3.3-1. Testing hypotheses about numerical values of parameters of normal distribution.**

Suppose that a random sample $X_1, \ldots, X_n$ is drawn from a population $X$ with normal distribution. Table 21.3 presents several tests for hypotheses about numerical values of the parameters of the normal distribution.

TABLE 21.3
Several tests related to normal populations with parameters $(a, \sigma^2)$

| No. | Hypothesis to be tested | Test statistic | Statistic distribution | Critical region for a given size |
|---|---|---|---|---|
| 1 | $H_0: a = a_0$, $H_1: a \neq a_0$ | $U = \dfrac{m^* - a}{\sigma}\sqrt{n}$ | standard normal | $|U| > u_{1-\alpha/2}$ |
| 2 | $H_0: a \leq a_0$, $H_1: a > a_0$ | $U = \dfrac{m^* - a}{\sigma}\sqrt{n}$ | standard normal | $U > u_{1-\alpha}$ |
| 3 | $H_0: a \geq a_0$, $H_1: a < a_0$ | $U = \dfrac{m^* - a}{\sigma}\sqrt{n}$ | standard normal | $U > -u_{1-\alpha}$ |
| 4 | $H_0: a = a_0$, $H_1: a \neq a_0$ | $T = \sqrt{\dfrac{n}{s^{2*}}}(m^* - a)$ | $t$-distribution with $n-1$ degrees of freedom | $|T| > t_{1-\alpha/2}$ |
| 5 | $H_0: a \leq a_0$, $H_1: a > a_0$ | $T = \sqrt{\dfrac{n}{s^{2*}}}(m^* - a)$ | $t$-distribution with $n-1$ degrees of freedom | $T > t_{1-\alpha}$ |
| 6 | $H_0: a \geq a_0$, $H_1: a < a_0$ | $T = \sqrt{\dfrac{n}{s^{2*}}}(m^* - a)$ | $t$-distribution with $n-1$ degrees of freedom | $T > -t_{1-\alpha}$ |
| 7 | $H_0: \sigma^2 = \sigma_0^2$, $H_1: \sigma^2 \neq \sigma_0^2$ | $\chi^2 = \dfrac{s^{2*}}{\sigma_0^2}(n-1)$ | $\chi^2$-distribution with $n-1$ degrees of freedom | $\chi^2_{\alpha/2} > \chi^2_{1-\alpha/2}$, $\chi^2 > \chi^2_{1-\alpha/2}$ |
| 8 | $H_0: \sigma^2 \leq \sigma_0^2$, $H_1: \sigma^2 > \sigma_0^2$ | $\chi^2 = \dfrac{s^{2*}}{\sigma_0^2}(n-1)$ | $\chi^2$-distribution with $n-1$ degrees of freedom | $\chi^2 > \chi^2_{1-\alpha}$ |
| 9 | $H_0: \sigma^2 \geq \sigma_0^2$, $H_1: \sigma^2 < \sigma_0^2$ | $\chi^2 = \dfrac{s^{2*}}{\sigma_0^2}(n-1)$ | $\chi^2$-distribution with $n-1$ degrees of freedom | $\chi^2 < \chi^2_{\alpha}$ |

Remark 1. In items 1–6 $\sigma^2$ is known.
Remark 2. In items 1–3 $u_\alpha$ is $\alpha$-quantile of standard normal distribution.

**21.3.3-2. Goodness-of-fit tests.**

Suppose that a sample $X_1, \ldots, X_n$ is drawn from a population $X$ with theoretical distribution function $F(x)$. It is required to test the composite null hypothesis, $H_0: F(x)$ is normal with unknown parameters $(a, \sigma^2)$, against the composite alternative hypothesis, $H_1: F(x)$ is not normal. Since the parameters $a$ and $\sigma^2$ are decisive for the normal law, the sample mean $m^*$ (or $\overline{X}$) and the adjusted sample variance $s^{2*}$ are used to estimate these parameters.

1°. *Romanovskii test.* To test the null hypothesis, the following *statistic (Romanovskii ratio)* is used:

$$\rho_{\text{rom}} = \rho_{\text{rom}}(X_1, \ldots, X_n) = \frac{\chi^2(m) - m}{\sqrt{2m}}, \qquad (21.3.3.1)$$

where $m$ is the number of degrees of freedom. If $|\rho_{\text{rom}}| \le 3$, then the null hypothesis should be accepted; if $|\rho_{\text{rom}}| > 3$, then the null hypothesis should be rejected. This test has a fixed size $\alpha = 0.0027$.

2°. *Test for excluding outliers.* To test the null hypothesis, the following statistic is used:

$$x^* = \frac{1}{s^*} \max_{1 \le i \le n} |X_i - m^*|, \qquad (21.3.3.2)$$

where $s^*$ is the sample mean-square deviation. The null hypothesis $H_0$ for a given size $\alpha$ is accepted if $x^* < x_{1-\alpha}$, where $x_\alpha$ is the quantile of the statistic $x^*$ under the assumption that the sample is normal. The values of $x_\alpha$ for various $n$ and $\alpha$ can be found in statistical tables.

3°. *Test based on the sample first absolute central moment.* The test is based on the statistic

$$\mu^* = \mu^*(X_1, \ldots, X_n) = \frac{1}{ns^*} \sum_{i=1}^{n} |X_i - m^*|. \qquad (21.3.3.3)$$

Under the assumption that the null hypothesis $H_0$ is true, the distribution of the statistic $\mu^*$ depends only on the sample size $n$ and is independent of the parameters $a$ and $\sigma^2$. The null hypothesis $H_0$ for a given size $\alpha$ is accepted if $\mu_{\alpha/2} < \mu^* < \mu_{1-\alpha/2}$, where $\mu_\alpha$ is the $\alpha$-quantile of the statistic $\mu^*$. The values of $\mu_\alpha$ for various $n$ and $\alpha$ can be found in statistical tables.

4°. *Test based on the sample asymmetry coefficient.* The test is based on the statistic

$$\gamma_1^* = \gamma_1^*(X_1, \ldots, X_n) = \frac{1}{n(s^*)^3} \sum_{i=1}^{n} (X_i - m^*)^3. \qquad (21.3.3.4)$$

The null hypothesis $H_0$ for a given size $\alpha$ is accepted if $|\gamma_1^*| < \gamma_{1,1-\alpha/2}$, where $\gamma_{1,\alpha}$ is the $\alpha$-quantile of the statistic $\gamma_1^*$. The values of $\gamma_{1,\alpha}$ for various $n$ and $\alpha$ can be found in statistical tables.

5°. *Test based on sample excess coefficient.* The test verifies the closeness of the sample excess (the test statistic)

$$\gamma_2^* = \gamma_2^*(X_1, \ldots, X_n) = \frac{1}{n(s^*)^4} \sum_{i=1}^{n} (X_i - m^*)^4 \qquad (21.3.3.5)$$

and the theoretical excess $\gamma_2 + 3 = E\{(X - E\{X\})^4\}/(\text{Var}\{X\})^2$, equal to 3 for the normal law. The null hypothesis $H_0$ for a given size $\alpha$ is accepted if the inequality $\gamma_{2,\alpha/2} < \gamma_2^* < \gamma_{2,1-\alpha/2}$ holds, where $\gamma_{2,\alpha}$ is the $\alpha$-quantile of the statistic $\gamma_2^*$. The values of $\gamma_{2,\alpha}$ for various $n$ and $\alpha$ can be found in statistical tables.

---

**21.3.3-3. Comparison of expectations of two normal populations.**

Suppose that $X$ and $Y$ are two populations with known variances $\sigma_1^2$ and $\sigma_2^2$ and unknown expectations $a_1$ and $a_2$. Two independent samples $X_1, \ldots, X_n$ and $Y_1, \ldots, Y_k$ are drawn from the populations and the sample expectations (means) $m_1^*$ and $m_2^*$ are calculated. The hypothesis that the expectations are equal to each other is tested using the statistic

$$U = \frac{m_1^* - m_2^*}{\sqrt{\sigma_1^2/n + \sigma_2^2/k}}, \tag{21.3.3.6}$$

which has a normal distribution with parameters $(0, 1)$ under the assumption that the null hypothesis $H_0 : a_X = a_Y$ is true.

If the variances of the populations are unknown, then either the sample size should be sufficiently large for obtaining a reliable accurate estimator or the variances should coincide; otherwise, the known tests are inefficient. If the variances of the populations are equal to each other, $\sigma_1^2 = \sigma_2^2$, then one can test the null hypothesis $H_0 : a_X = a_Y$ using the statistic

$$T = (m_1^* - m_2^*)\left[\left(\frac{1}{n} + \frac{1}{k}\right)\frac{s_1^{2*}(n-1) + s_2^{2*}(k-1)}{n+k-2}\right]^{-\frac{1}{2}}, \tag{21.3.3.7}$$

which has the $t$-distribution (Student's distribution) with $v = n + k - 2$ degrees of freedom.

The choice of the critical region depends on the form of the alternative hypothesis:

1. For the alternative hypothesis $H_1 : a_1 > a_2$, one should choose a right-sided critical region.
2. For the alternative hypothesis $H_1 : a_1 < a_2$, one should choose a left-sided critical region.
3. For the alternative hypothesis $H_1 : a_1 \neq a_2$, one should choose a two-sided critical region.

21.3.3-4. **Tests for variances to be equal.**

Suppose that there are $L$ independent samples

$$X_{11}, \ldots, X_{1n_1}; \; X_{21}, \ldots, X_{2n_2}; \; \ldots; \; X_{L1}, \ldots, X_{Ln_L}$$

of sizes $n_1, \ldots, n_L$ drawn from distinct normal populations with unknown expectation $a_1, \ldots, a_L$ and unknown variances $\sigma_1^2, \ldots, \sigma_L^2$. It is required to test the simple hypothesis $H_0 : \sigma_1^2 = \cdots = \sigma_L^2$ (the variances of all populations are the same) against the alternative hypothesis $H_1$ that some variances are different.

$1°$. *Bartlett's test.* The statistic in this test has the form

$$b = N \ln\left[\frac{1}{N}\sum_{i=1}^{L}(n_i - 1)s_i^{2*}\right] - \sum_{i=1}^{L}(n_i - 1)\ln s_i^{2*}, \tag{21.3.3.8}$$

where

$$N = \sum_{i=1}^{L}(n_i - 1), \quad s_j^{2*} = \frac{1}{n_j - 1}\sum_{j=1}^{n_j}(X_{ij} - L_j^*)^2, \quad L_j^* = \frac{1}{n_j}\sum_{j=1}^{n_j}X_{ij}.$$

The statistic $b$ permits reducing the problem of testing the hypothesis that the variances of normal samples are equal to each other to the problem of testing the hypothesis that the expectations of approximately normal samples are equal to each other. If the null hypothesis $H_0$ is true and all $n > 5$, then the ratio

$$B = b\left[1 + \frac{1}{3(L-1)} + \left(\sum_{i=1}^{L}\frac{1}{n_i - 1} - \frac{1}{N}\right)\right]^{-1}$$

is distributed approximately according to the chi-square law with $L - 1$ degrees of freedom. The null hypothesis $H_0$ for a given size $\alpha$ is accepted if $B < \chi_{1-\alpha}^2(L-1)$, where $\chi_\alpha^2$ is the $\alpha$-quantile of the chi-square distribution with $L - 1$ degrees of freedom.

2°. *Cochran's test.* If all samples have the same volume ($n_1 = \cdots = n_L = n$), then the null hypothesis $H_0$ is tested against the alternative hypothesis $H_1$ using the *Cochran statistic*

$$G = \frac{s_{\max}^{2*}}{s_1^{2*} + \cdots + s_L^{2*}}, \qquad (21.3.3.9)$$

where $s_{\max}^{2*} = \max_{1 \leq i \leq L} s_i^{2*}$.

Cochran's test is, in general, less powerful than the Bartlett test, but it is simpler. The null hypothesis $H_0$ for a given size $\alpha$ is accepted if $G < G_\alpha$, where $G_\alpha$ is the $\alpha$-quantile of the statistic $G$. The values of $G_\alpha$ for various $\alpha$, $L$, and $v = n - 1$ can be found in statistical tables.

3°. *Fisher's test.* For $L = 2$, to test the null hypothesis $H_0$ that the variances of two samples coincide, it is most expedient to use Fisher's test based on the statistic

$$\Psi = \frac{s_2^{2*}}{s_1^{2*}}, \qquad (21.3.3.10)$$

where $s_1^{2*}$ and $s_2^{2*}$ are the adjusted sample variances of the two samples. The statistic $\Psi$ has the $F$-distribution (Fisher–Snedecor distribution) with $n_2 - 1$ and $n_1 - 1$ degrees of freedom.

The one-sided Fisher test verifies the null hypothesis $H_0 : \sigma_1^2 = \sigma_2^2$ against the alternative hypothesis $H_1 : \sigma_1^2 < \sigma_2^2$; the critical region of the one-sided Fisher test for a given size $\alpha$ is determined by the inequality $\Psi > \Psi_{1-\alpha}(n_2 - 1, n_1 - 1)$.

The two-sided Fisher test verifies the null hypothesis $H_0 : \sigma_1^2 = \sigma_2^2$ against the alternative hypothesis $H_1 : \sigma_1^2 \neq \sigma_2^2$; the critical region of the two-sided Fisher test for a given size $\alpha$ is determined by the two inequalities $\Psi_{\alpha/2} < \Psi < \Psi_{1-\alpha/2}$, where $\Psi_\alpha$ is the $\alpha$-quantile of the $F$-distribution with parameters $n_2 - 1$ and $n_1 - 1$.

### 21.3.3-5. Sample correlation.

Suppose that a sample $\mathbf{X}_1, \ldots, \mathbf{X}_n$ is two-dimensional and its elements $\mathbf{X}_i = (X_{i1}, X_{i2})$ are two-dimensional random variables with joint normal distribution with means $a_1$ and $a_2$, variances $\sigma_1^2$ and $\sigma_2^2$, and correlation $r$. It is required to test the hypothesis that the components $X_{(1)}$ and $X_{(2)}$ of the vector $\mathbf{X}$ are independent, i.e., test the hypothesis that the correlation is zero.

Estimation of the correlation $r$ is based on the *sample correlation*

$$r^* = \frac{\sum_{i=1}^{n}(X_{i1} - m_1^*)(X_{i2} - m_2^*)}{\sqrt{\sum_{i=1}^{n}(X_{i1} - m_1^*)^2 \sum_{j=1}^{n}(X_{i2} - m_2^*)^2}}, \qquad (21.3.3.11)$$

where

$$m_1^* = \frac{1}{n}\sum_{i=1}^{n} X_{i1}, \quad m_2^* = \frac{1}{n}\sum_{i=1}^{n} X_{i2}.$$

The distribution of the statistic $r^*$ depends only on the sample size $n$, and the statistic $r^*$ itself is a consistent asymptotically efficient estimator of the correlation $r$.

The null hypothesis $H_0 : r = 0$ that $X_{(1)}$ and $X_{(2)}$ are independent against the alternative hypothesis $H_1 : r \neq 0$ ($X_{(1)}$ and $X_{(2)}$ are dependent) is accepted if the inequality $r_{\alpha/2} < r^* < r_{1-\alpha/2}$ is satisfied. Here $r_\alpha$ is the $\alpha$-quantile of the sample correlation under the assumption that the null hypothesis $H_0$ is true; the relation $r_\alpha = -r_{1-\alpha}$ holds because of symmetry.

To construct the confidence intervals, one should use the *Fisher transformation*

$$y = \operatorname{arctanh} r^* = \frac{1}{2} \ln \frac{1+r^*}{1-r^*}, \tag{21.3.3.12}$$

which, for $n > 10$, is approximately normal with parameters

$$E\{y\} \approx \frac{1}{2} \ln \frac{1+r}{1-r} + \frac{r}{2(n-3)}, \quad \operatorname{Var}\{y\} \approx \frac{1}{n-3}.$$

**21.3.3-6. Regression analysis.**

Let $X_1, \ldots, X_n$ be the results of $n$ independent observations

$$X_i = \sum_{j=1}^{L} \theta_j f_j(t_i) + \varepsilon_i,$$

where $f_1(t), \ldots, f_L(t)$ are known functions, $\theta_1, \ldots, \theta_L$ are unknown parameters, and $\varepsilon_1, \ldots, \varepsilon_n$ are random errors known to be independent and normally distributed with zero mean and with the same unknown variance $\sigma^2$.

The regression parameters are subject to the following constraints:
1. The number of observations $n$ is greater than the number $L$ of unknown parameters.
2. The vectors

$$\mathbf{f}_i = (f_i(t_1), \ldots, f_i(t_n)) \qquad (i = 1, 2, \ldots, L) \tag{21.3.3.13}$$

must be linearly independent.

$1°$. *Estimation of unknown parameters $\theta_1, \ldots, \theta_L$ and construction of (one-dimensional) confidence intervals for them.*

To solve this problem, we consider the sum of squares

$$S^2 = S^2(\theta_1, \ldots, \theta_L) = \sum_{i=1}^{n} [X_i - \theta_1 f_1(t_i) - \cdots - \theta_L f_L(t_i)]^2. \tag{21.3.3.14}$$

The estimators $\theta_1^*, \ldots, \theta_L^*$ form a solution of the system of equations

$$\theta_1^* \sum_{j=1}^{n} f_i(t_j) + \cdots + \theta_L^* \sum_{j=1}^{n} f_i(t_j) = \sum_{j=1}^{n} X_j f_i(t_j). \tag{21.3.3.15}$$

The estimators $\theta_1^*, \ldots, \theta_L^*$ are linear and efficient; in particular, they are unbiased and have the minimal variance among all possible estimators.

**Remark.** If we omit the requirement that the errors $\varepsilon_1, \ldots, \varepsilon_n$ are normally distributed and only assume that they are uncorrelated and have zero expectation and the same variance $\sigma^2$, then the estimators $\theta_1^*, \ldots, \theta_L^*$ are linear, unbiased, and have the minimal variance in the class of all linear estimators.

The confidence intervals for a given confidence level $\gamma$ for the unknown parameters $\theta_1, \ldots, \theta_L$ have the form

$$|\theta_i - \theta_i^*| < t_{(1+\gamma)/2} \sqrt{c_i^2 s_0^{2*}}, \tag{21.3.3.16}$$

where $t_\gamma$ is the $\gamma$-quantile of the $t$-distribution with $n - L$ degrees of freedom,

$$s_0^{2*} = \frac{1}{n-L} \min_{\theta_1, \ldots, \theta_L} S^2(\theta_1, \ldots, \theta_L) = \frac{S^2(\theta_1^*, \ldots, \theta_L^*)}{n-L}, \quad c_i^2 = \sum_{i=1}^{n} c_{ij}^2,$$

and $c_{ij}$ are the coefficients in the representation $\theta_i^* = \sum_{j=1}^{n} c_{ij} X_j$.

System (21.3.3.15) can be solved in the simplest way if the vectors (21.3.3.13) are orthogonal. In this case, system (21.3.3.15) splits into separate equations

$$\theta_i^* \sum_{j=1}^n f_i^2(t_j) = \sum_{j=1}^n X_j f_i(t_j).$$

Then the estimators $\theta_1^*, \ldots, \theta_L^*$ are independent, linear, and efficient.

**Remark.** If we omit the requirement that the errors $\varepsilon_1, \ldots, \varepsilon_n$ are normally distributed, then the estimators $\theta_1^*, \ldots, \theta_L^*$ are uncorrelated, linear, and unbiased and have the minimal variance in the class of all linear estimators.

2°. *Testing the hypothesis that some $\theta_i$ are zero.* Suppose that it is required to test the null hypothesis $H_0 : \theta_{k+1} = \cdots = \theta_L = 0$ ($0 \leq k < L$). This problem can be solved using the statistic

$$\Psi = \frac{s_1^{2*}}{s_0^{2*}},$$

where

$$s_0^{2*} = \frac{1}{n-L} \min_{\theta_1,\ldots,\theta_L} S^2(\theta_1,\ldots,\theta_L), \quad s_1^{2*} = \frac{S_2^2 - (n-L)s_0^{2*}}{L-k},$$

$$S_2^2 = \min_{\theta_1,\ldots,\theta_k} S^2(\theta_1,\ldots,\theta_k, 0, \ldots, 0).$$

The hypothesis $H_0$ for a given size $\gamma$ is accepted if $\Psi < \Psi_{1-\gamma}$, where $\Psi_\gamma$ is the $\gamma$-quantile of the $F$-distribution with parameters $L-k$ and $n-L$.

3°. *Finding the estimator $x^*(t)$ of the regression $x(t) = \sum\limits_{i=1}^L \theta_i f_i(t)$ at an arbitrary time and the construction of confidence intervals.*

The estimator $x^*(t)$ of the regression $x(t)$ is obtained if $\theta_i$ in $x(t)$ are replaced by their estimators:

$$x^*(t) = \sum_{i=1}^L \theta_i^* f_i(t).$$

The estimator $x^*(t)$ is a linear, efficient, normally distributed, and unbiased estimator of the regression $x(t)$.

The confidence interval of confidence level $\gamma$ is given by the inequality

$$|x(t) - x^*(t)| < t_{(1+\gamma)/2} \sqrt{c(t) s_0^{2*}}, \qquad (21.3.3.17)$$

where $c(t) = \sum\limits_{i=1}^L c_{ij}(t) f_i(t)$ and $t_\gamma$ is the $\alpha$-quantile of the $t$-distribution with $n-L$ degrees of freedom.

**Example 1.** Consider a linear regression $x(t) = \theta_1 + \theta_2 t$.

1°. The estimators $\theta_1^*$ and $\theta_2^*$ of the unknown parameters $\theta_1$ and $\theta_2$ are given by the formulas

$$\theta_1^* = \sum_{j=1}^n c_{1j} X_j, \quad \theta_2^* = \sum_{j=1}^n c_{2j} X_j,$$

where

$$c_{1j} = \frac{\sum\limits_{k=1}^n t_k^2 - t_j \sum\limits_{k=1}^n t_k}{n \sum\limits_{k=1}^n t_k^2 - \left(\sum\limits_{k=1}^n t_k\right)^2}, \quad c_{2j} = \frac{n t_j - \sum\limits_{k=1}^n t_k}{n \sum\limits_{k=1}^n t_k^2 - \left(\sum\limits_{k=1}^n t_k\right)^2}.$$

The statistic $s_0^{2*}$, up to the factor $(n-2)/\sigma^2$, has the $\chi^2$-distribution with $n-2$ degrees of freedom and is determined by the formula

$$s_0^{2*} = \frac{1}{n-2} \sum_{j=1}^{n} (X_j - \theta_1^* - \theta_2^* t_j)^2.$$

The confidence intervals of confidence level $\gamma$ for the parameters $\theta_i$ are given by the formula

$$|\theta_i - \theta_i^*| < t_{(1+\gamma)/2} \sqrt{c_i^2 s_0^{2*}}, \quad i = 1, 2,$$

where $t_\gamma$ is the $\gamma$-quantile of the $t$-distribution with $n-2$ degrees of freedom.

2°. We test the null hypothesis $H_0 : \theta_2 = 0$, i.e., the hypothesis that $x(t)$ is independent of time. The value of $S_2^2$ is given by the formula

$$S_2^2 = \sum_{i=1}^{n} (X_i - m^*)^2, \quad m^* = \frac{1}{n} \sum_{i=1}^{n} X_i,$$

and the value of $s_1^{2*}$ is given by the formula

$$s_1^{2*} = S_2^2 - (n-2) s_0^{2*}.$$

Thus, the hypothesis $H_0$ for a given confidence level $\gamma$ is accepted if $\phi < \phi_\gamma$, where $\phi_\gamma$ is the $\gamma$-quantile of an $F$-distribution with parameters 1 and $n-2$.

3°. The estimator $x^*(t)$ of the regression $x(t)$ has the form

$$x^*(t) = \theta_1^* + \theta_2^* t.$$

The coefficient $c(t)$ is determined by the formula

$$c(t) = \sum_{j=1}^{n} (c_{1j} + c_{2j} t)^2 = \sum_{j=1}^{n} c_{1j}^2 + 2t \sum_{j=1}^{n} c_{1j} c_{2j} + t^2 \sum_{j=2}^{n} c_{1j}^2 = b_0 + b_1 t + b_2 t^2.$$

Thus the boundaries of the confidence interval for a given confidence level $\gamma$ are given by the formulas

$$x_L^*(t) = \theta_1^* + \theta_2^* t - t_{(1+\gamma)/2} \sqrt{s_0^{2*}(b_0 + b_1 t + b_2 t^2)},$$
$$x_R^*(t) = \theta_1^* + \theta_2^* t + t_{(1+\gamma)/2} \sqrt{s_0^{2*}(b_0 + b_1 t + b_2 t^2)},$$

where $t_\gamma$ is the $\gamma$-quantile of the $t$-distribution with $n-2$ degrees of freedom.

### 21.3.3-7. Analysis of variance.

*Analysis of variance* is a statistical method for clarifying the influence of several factors on experimental results and for planning subsequent experiments.

1°. *The simplest problem of analysis of variance.* Suppose that there are $L$ independent samples

$$X_{11}, \ldots, X_{1n_1}; \; X_{21}, \ldots, X_{2n_1}; \; \ldots; \; X_{L1}, X_{L2}, \ldots, X_{Ln_1}, \quad (21.3.3.18)$$

drawn from normal populations with unknown expectations $a_1, \ldots, a_L$ and unknown but equal variances $\sigma^2$. It is necessary to test the null hypothesis $H_0 : a_1 = \cdots = a_L$ that all theoretical expectations $a_i$ are the same against the alternative hypothesis $H_1$ that some theoretical expectations are different.

The intragroup variances are determined by the formulas

$$s_{(i)}^{2*} = \frac{1}{n_i - 1} \sum_{j=1}^{n_j} (X_{ij} - m_i^*)^2 \quad (i = 1, 2, \ldots, n), \quad (21.3.3.19)$$

where $m_i^* = \frac{1}{n_j} \sum_{j=1}^{n_j} X_{ij}$ is the sample mean of the corresponding sample. The random variable $(n_i - 1)s_{(i)}^{2*}/\sigma^2$ has the chi-square distribution with $n_i - 1$ degrees of freedom. An unbiased estimator of the unknown variance $\sigma^2$ is given by the statistic

$$s_0^{2*} = \frac{\sum_{i=1}^{L}(n_i - 1)s_{(i)}^{2*}}{\sum_{i=1}^{L}(n_i - 1)}, \qquad (21.3.3.20)$$

called the *residual variance*.

The *intergroup sample variance* is defined to be the statistic

$$s_1^{2*} = \frac{1}{L - 1}\sum_{i=1}^{L}(m_i^* - m^*)^2 n_i, \qquad (21.3.3.21)$$

where $m^*$ is the common sample mean of the generalized sample. The statistic $s_1^{2*}$ is independent of $s_0^{2*}$ and is an unbiased estimator of the unknown variance $\sigma^2$. The random variable $s_0^{2*}\sum_{i=1}^{L}(n_i - 1)/\sigma^2$ is distributed by the chi-square law with $\sum_{i=1}^{L}(n_i - 1)$ degrees of freedom, and the random variable $s_1^{2*}(L - 1)/\sigma^2$ is distributed by the chi-square law with $L - 1$ degrees of freedom.

According to the one-sided Fisher test, the null hypothesis $H_0$ must be accepted for a given confidence level $\gamma$ if $\Psi = s_1^{2*}/s_0^{2*} < \Psi_\gamma$, where $\Psi_\gamma$ is the $\gamma$-quantile of the $F$-distribution with parameters $L - 1$ and $\sum_{i=1}^{L}(n_i - 1)$.

2°. *Multifactor analysis of variance.* We consider two-factor analysis of variance. Suppose that the first factor acts at $L_1$ levels and the second factor acts at $L_2$ levels (the two-factor $(L_1, L_2)$-level model of analysis of variance). Suppose that we have $n_{ij}$ observations in which the first factor acted at the $i$th level and the second factor acted at the $j$th level. The observation results $X_{ijk}$ are independent normally distributed random variables with the same (unknown) variance $\sigma^2$ and unknown expectations $a_{ij}$. It is required to test the null hypothesis $H_0$ that the first and second factors do not affect the results of observations, i.e., all $a_{ij}$ are the same.

The action of two factors at levels $L_1$ and $L_2$ is identified with the action of a single factor at $L_1 L_2$ levels; then to test the hypothesis $H_0$, it is expedient to use the one-factor $L_1 L_2$-level model. The statistics $s_0^{2*}$ and $s_1^{2*}$ are determined by the formulas

$$s_0^{2*} = \frac{\sum_{i=1}^{L_1}\sum_{j=1}^{L_2}\sum_{k=1}^{n_{ij}}(X_{ijk} - m_{ij}^*)^2}{\sum_{i=1}^{L_1}\sum_{j=1}^{L_2}(n_{ij} - 1)}, \qquad (21.3.3.22)$$

$$s_1^{2*} = \frac{1}{L_1 L_2 - 1}\sum_{i=1}^{L_1}\sum_{j=1}^{L_2}n_{ij}(m_{ij}^* - m^*)^2,$$

where $m_{ij}^* = \frac{1}{n_{ij}} \sum_{k=1}^{n_{ij}} X_{ijk}$ is the mean over the observations in which the first and second factors acted at levels $i$ and $j$ and $m^*$ is the common mean of all observations

$$m^* = \sum_{i=1}^{L_1} \sum_{j=1}^{L_2} \sum_{k=1}^{n_{ij}} X_{ijk} \bigg/ \sum_{i=1}^{L_1} \sum_{j=1}^{L_2} n_{ij}.$$

The hypothesis $H_0$ that there is no influence of both factors at all levels must be accepted for a given confidence level $\gamma$ if $\phi = s_1^{2*}/s_0^{2*} < \phi_\gamma$, where $\phi_\gamma$ is the $\gamma$-quantile of the $F$-distribution with parameters $\sum_{i=1}^{L_1} \sum_{j=1}^{L_2} (n_{ij} - 1)$ and $L_1 L_2 - 1$.

Analysis of variance also permits testing the hypothesis that the theoretical expectation can be represented in the form $a_{ij} = a_i^{(1)} + a_j^{(2)}$, where $a_i^{(1)}$ and $a_j^{(2)}$ are unknown expectations introduced by the first and second factors under the action at levels $i$ and $j$.

## References for Chapter 21

**Arora, P. N. and Anand, S. K.,** *Mathematical and Statistical Tables and Formulae*, Anmol Publications Pvt Ltd, Delhi, 2002.

**Bain, L. J. and Engelhardt, M.,** *Introduction to Probability and Mathematical Statistics, 2nd Edition* (Duxbury Classic), Duxbury Press, Boston, 2000.

**Bean, M. A.,** *Probability: The Science of Uncertainty with Applications to Investments, Insurance, and Engineering*, Brooks Cole, Stamford, 2000.

**Beyer, W. H. (Editor),** *CRC Standard Probability and Statistics Tables and Formulae*, CRC Press, Boca Raton, 1990.

**Bruning, L. and Kintz, B. L.,** *Computational Handbook of Statistics, 4th Edition*, Allyn & Bacon, Boston, 1997.

**Bulmer, M. G.,** *Principles of Statistics*, Dover Publishers, New York, 1979.

**Burlington, R. S. and May, D.,** *Handbook of Probability and Statistics With Tables, 2nd Edition*, McGraw-Hill, New York, 1970.

**Craig, R. V. and Hogg, A. T.,** *Introduction to Mathematical Statistics*, Macmillan Coll. Div., New York, 1970.

**DeGroot, M. H. and Schervish, M. J.,** *Probability and Statistics, 3rd Edition*, Addison Wesley, Boston, 2001.

**Devore, J. L.,** *Probability and Statistics for Engineering and the Sciences (with CD-ROM and InfoTrac), 6th Edition*, Duxbury Press, Boston, 2003.

**Fisher, R. A.,** *Statistical Tables for Biological, Agricultural and Medical Research*, Longman Group United Kingdom, Harlow, 1995.

**Freedman, D., Pisani, R., and Purves, R.,** *Statistics, 3rd Edition*, W. W. Norton & Company, New York, 1997.

**Graham, A.,** *Teach Yourself Statistics, 2nd Edition*, McGraw-Hill, New York, 2003.

**Hines, W. W., Montgomery, D. C., Goldsman, D. M., and Borror, C. M.,** *Probability and Statistics in Engineering, 4th Edition*, Wiley, New York, 2003.

**Hogg, R. V., Craig, A., and McKean, J. W.,** *Introduction to Mathematical Statistics, 6th Edition*, Prentice Hall, Englewood Cliffs, New Jersey, 2004.

**Kapadia, A. S., Chan, W., and Moye, L. A.,** *Mathematical Statistics with Applications* (Statistics: Textbooks and Monographs), Chapman & Hall/CRC, Boca Raton, 2005.

**Kokoska, S. and Zwillinger, D. (Editors),** *CRC Standard Probability and Statistics Tables and Formulae, Student Edition*, Chapman & Hall/CRC, Boca Raton, 2000.

**Langley, R.,** *Practical Statistics Simply Explained, Rev. Edition* (Dover Books Explaining Science), Dover Publishers, New York, 1971.

**Larsen, R. J. and Marx, M. L.,** *An Introduction to Mathematical Statistics and Its Applications, 4th Edition*, Prentice Hall, Englewood Cliffs, New Jersey, 2005.

**Laurencelle, L. and Dupuis, F.,** *Statistical Tables: Exlained and Applied*, World Scientific Publishing Co., Hackensack, New Jersey, 2002.

**Lindley, D. V. and Scott, W. F.,** *New Cambridge Statistical Tables, 2nd Edition*, Cambridge University Press, Cambridge, 1995.

**Lipschutz, S. and Schiller, J.,** *Schaum's Outline of Introduction to Probability and Statistics*, McGraw-Hill, New York, 1998.

**Martinez, W. L. and Martinez, A. R.,** *Computational Statistics Handbook with MATLAB*, Chapman & Hall/CRC, Boca Raton, 2001.
**McClave, J.T. and Sincich, T.,** *Statistics, 10th Edition*, Prentice Hall, Englewood Cliffs, New Jersey, 2005.
**Mendenhall, W., Beaver, R. J., and Beaver, B. M.,** *Introduction to Probability and Statistics (with CD-ROM), 12th Edition*, Duxbury Press, Boston, 2005.
**Mendenhall, W. and Sincich, T. L.,** *Statistics for Engineering and the Sciences, 4th Edition*, Prentice Hall, Englewood Cliffs, New Jersey, 1995.
**Milton, J. S. and Arnold, J. C.,** *Introduction to Probability and Statistics: Principles and Applications for Engineering and the Computing Sciences, 2nd Edition*, McGraw-Hill, New York, 2002.
**Miller, I. and Miller, M.,** *John E. Freund's Mathematical Statistics with Applications, 7th Edition*, Prentice Hall, Englewood Cliffs, New Jersey, 2003.
**Montgomery, D. C. and Runger, G. C.,** *Applied Statistics and Probability for Engineers, Student Solutions Manual, 4th Edition*, Wiley, New York, 2006.
**Montgomery, D. C., Runger, G. C., and Hubele, N. F.,** *Engineering Statistics, 3th Edition*, Wiley, New York, 2003.
**Murdoch, J. and Barnes, J. A.,** *Statistical Tables for Science, Engineering, Management and Business Studies*, Palgrave Macmillan, New York, 1986.
**Neave, H. R.,** *Statistics Tables: For Mathematicians, Engineers, Economists and the Behavioural Management Sciences* (Textbook Binding), Routledge, London, 1998.
**Owen, D.B.,** *Handbook of Statistical Tables* (Addison-Wesley Series in Statistics), Addison Wesley, Boston, 1962.
**Pestman, W. R. and Alberink, I. B.,** *Mathematical Statistics: Problems and Detailed Solutions* (De Gruyter Textbook), Walter de Gruyter, Berlin, New York, 1998.
**Pham, H. (Editor),** *Springer Handbook of Engineering Statistics*, Springer, New York, 2006.
**Rice, J. A.,** *Mathematical Statistics and Data Analysis, 2nd Edition*, Duxbury Press, Boston, 1994.
**Rohlf, F. J. and Sokal, R. R.,** *Statistical Tables, 3rd Edition*, W. H. Freeman, New York, 1994.
**Rose, C. and Smith, M. D.,** *Mathematical Statistics with MATHEMATICA*, Springer, New York, 2002.
**Scheaffer, R. L. and McClave, J. T.,** *Probability and Statistics for Engineers, 4th Edition* (Statistics), Duxbury Press, Boston, 1994.
**Seely, J. A.,** *Probability and Statistics for Engineering and Science, 6th Edition*, Brooks Cole, Stamford, 2003.
**Shao, J.,** *Mathematical Statistics, 2nd Edition*, Springer, New York, 2003.
**Shao, J.,** *Mathematical Statistics: Exercises and Solutions*, Springer, New York, 2005.
**Stigler, S. M.,** *Statistics on the Table: The History of Statistical Concepts and Methods, Reprint edition*, Harvard University Press, Cambridge, Massachusetts, 2002.
**Terrell, G. R.,** *Mathematical Statistics: A Unified Introduction* (Springer Texts in Statistics), Springer, New York, 1999.
**Ventsel, H.,** *Théorie des Probabilités*, Mir, Moscou, 1987.
**Ventsel, H. and Ovtcharov, L.,** *Problémes appliqués de la théorie des probabilités*, Mir, Moscou, 1988.
**Wackerly, D., Mendenhall, W., and Scheaffer, R. L.,** *Mathematical Statistics with Applications, 6th Edition*, Duxbury Press, Boston, 2001.
**Walpole, R. E., Myers, R. H., Myers, S. L., and Ye, K.,** *Probability & Statistics for Engineers & Scientists, 8th Edition*, Prentice Hall, Englewood Cliffs, New Jersey, 2006.
**Witte, R. S. and Witte, J. S.,** *Statistics, 7th Edition*, Wiley, New York, 2003.
**Wonnacott, T. H. and Wonnacott, R. J.,** *Introductory Statistics, 5th Edition*, Wiley, New York, 1990.
**Zwillinger, D. (Editor),** *CRC Standard Mathematical Tables and Formulae, 31st Edition*, Chapman & Hall/CRC, Boca Raton, 2001.

# Part II
# Mathematical Tables

# Chapter T1
# Finite Sums and Infinite Series

## T1.1. Finite Sums
### T1.1.1. Numerical Sum

**T1.1.1-1. Progressions.**

Arithmetic progression:

1. $\displaystyle\sum_{k=0}^{n-1}(a+bk) = an + \frac{bn(n-1)}{2}.$

Geometric progression:

2. $\displaystyle\sum_{k=1}^{n} aq^{k-1} = a\frac{q^n-1}{q-1}.$

Arithmetic-geometric progression:

3. $\displaystyle\sum_{k=0}^{n-1}(a+bk)q^k = \frac{a(1-q^n) - b(n-1)q^n}{1-q} + \frac{bq(1-q^{n-1})}{(1-q)^2}.$

**T1.1.1-2. Sums of powers of natural numbers having the form $\sum k^m$.**

1. $\displaystyle\sum_{k=1}^{n} k = \frac{n(n+1)}{2}.$

2. $\displaystyle\sum_{k=1}^{n} k^2 = \frac{1}{6}n(n+1)(2n+1).$

3. $\displaystyle\sum_{k=1}^{n} k^3 = \frac{1}{4}n^2(n+1)^2.$

4. $\displaystyle\sum_{k=1}^{n} k^4 = \frac{1}{30}n(n+1)(2n+1)(3n^2+3n-1).$

5. $\displaystyle\sum_{k=1}^{n} k^5 = \frac{1}{12}n^2(n+1)^2(2n^2+2n-1).$

6. $\displaystyle\sum_{k=1}^{n} k^m = \frac{n^{m+1}}{m+1} + \frac{n^m}{2} + \frac{1}{2}C_m^1 B_2 n^{m-1} + \frac{1}{4}C_m^3 B_4 n^{m-3} + \frac{1}{6}C_m^5 B_6 n^{m-5} + \cdots.$

Here the $C_m^k$ are binomial coefficients and the $B_{2k}$ are Bernoulli numbers; the last term in the sum contains $n$ or $n^2$.

T1.1.1-3. **Alternating sums of powers of natural numbers, $\sum(-1)^k k^m$.**

1. $\sum_{k=1}^{n}(-1)^k k = (-1)^n \left[\frac{n-1}{2}\right]$;   $[m]$ stands for the integer part of $m$.

2. $\sum_{k=1}^{n}(-1)^k k^2 = (-1)^n \frac{n(n+1)}{2}$.

3. $\sum_{k=1}^{n}(-1)^k k^3 = \frac{1}{8}\left[1 + (-1)^n(4n^3 + 6n^2 - 1)\right]$.

4. $\sum_{k=1}^{n}(-1)^k k^4 = (-1)^n \frac{1}{2}(n^4 + 2n^3 - n)$.

5. $\sum_{k=1}^{n}(-1)^k k^5 = \frac{1}{4}\left[-1 + (-1)^n(2n^5 + 5n^4 - 5n^2 + 1)\right]$.

T1.1.1-4. **Other sums containing integers.**

1. $\sum_{k=0}^{n}(2k+1) = (n+1)^2$.

2. $\sum_{k=0}^{n}(2k+1)^2 = \frac{1}{3}(n+1)(2n+1)(2n+3)$.

3. $\sum_{k=1}^{n} k(k+1) = \frac{1}{3}n(n+1)(n+2)$.

4. $\sum_{k=1}^{n}(k+a)(k+b) = \frac{1}{6}n(n+1)(2n+1+3a+3b) + nab$.

5. $\sum_{k=1}^{n} k\, k! = (n+1)! - 1$.

6. $\sum_{k=0}^{n}(-1)^k(2k+1) = (-1)^n(n+1)$.

7. $\sum_{k=0}^{n}(-1)^k(2k+1)^2 = 2(-1)^n(n+1)^2 - \frac{1}{2}\left[1 + (-1)^n\right]$.

T1.1.1-5. **Sums containing binomial coefficients.**

▶ *Throughout Paragraph T1.1.1-5, it is assumed that $m = 1, 2, 3, \ldots$*

1. $\sum_{k=0}^{n} C_n^k = 2^n$.

2. $\displaystyle\sum_{k=0}^{n} C_{m+k}^{m} = C_{n+m+1}^{m+1}.$

3. $\displaystyle\sum_{k=0}^{n} (-1)^k C_m^k = (-1)^n C_{m-1}^n.$

4. $\displaystyle\sum_{k=0}^{n} (k+1) C_n^k = 2^{n-1}(n+2).$

5. $\displaystyle\sum_{k=1}^{n} (-1)^{k+1} k C_n^k = 0.$

6. $\displaystyle\sum_{k=1}^{n} \frac{(-1)^{k+1}}{k} C_n^k = \sum_{m=1}^{n} \frac{1}{m}.$

7. $\displaystyle\sum_{k=1}^{n} \frac{(-1)^{k+1}}{k+1} C_n^k = \frac{n}{n+1}.$

8. $\displaystyle\sum_{k=0}^{n} \frac{1}{k+1} C_n^k = \frac{2^{n+1}-1}{n+1}.$

9. $\displaystyle\sum_{k=0}^{n} \frac{a^{k+1}}{k+1} C_n^k = \frac{(a+1)^{n+1}-1}{n+1}.$

10. $\displaystyle\sum_{k=0}^{p} C_n^k C_m^{p-k} = C_{n+m}^p;\quad$ $m$ and $p$ are natural numbers.

11. $\displaystyle\sum_{k=0}^{n-p} C_n^k C_n^{p+k} = \frac{(2n)!}{(n-p)!\,(n+p)!}.$

12. $\displaystyle\sum_{k=0}^{n} (C_n^k)^2 = C_{2n}^n.$

13. $\displaystyle\sum_{k=0}^{2n} (-1)^k (C_{2n}^k)^2 = (-1)^n C_{2n}^n.$

14. $\displaystyle\sum_{k=0}^{2n+1} (-1)^k (C_{2n+1}^k)^2 = 0.$

15. $\displaystyle\sum_{k=1}^{n} k (C_n^k)^2 = \frac{(2n-1)!}{[(n-1)!]^2}.$

---

T1.1.1-6. Other numerical sums.

1. $\displaystyle\sum_{k=1}^{n-1} \sin\frac{\pi k}{n} = \cot\frac{\pi}{2n}.$

2. $\displaystyle\sum_{k=1}^{n} \sin^{2m}\frac{\pi k}{2n} = \frac{n}{2^{2m}}C_{2m}^{m} + \frac{1}{2}$, $\quad m < 2n$.

3. $\displaystyle\sum_{k=0}^{n-1} (-1)^k \cos^m \frac{\pi k}{n} = \frac{1}{2}\left[1-(-1)^{m+n}\right]$, $\quad m = 0, 1, \ldots, n-1$.

4. $\displaystyle\sum_{k=0}^{n-1} (-1)^k \cos^n \frac{\pi k}{n} = \frac{n}{2^{n-1}}$.

## T1.1.2. Functional Sums

T1.1.2-1. Sums involving hyperbolic functions.

1. $\displaystyle\sum_{k=0}^{n-1} \sinh(kx+a) = \sinh\left(\frac{n-1}{2}x+a\right)\frac{\sinh(nx/2)}{\sinh(x/2)}$.

2. $\displaystyle\sum_{k=0}^{n-1} \cosh(kx+a) = \cosh\left(\frac{n-1}{2}x+a\right)\frac{\sinh(nx/2)}{\sinh(x/2)}$.

3. $\displaystyle\sum_{k=0}^{n-1} (-1)^k \sinh(kx+a) = \frac{1}{2\cosh(x/2)}\left[\sinh\left(a-\frac{x}{2}\right) + (-1)^n \sinh\left(\frac{2n-1}{2}x+a\right)\right]$.

4. $\displaystyle\sum_{k=0}^{n-1} (-1)^k \cosh(kx+a) = \frac{1}{2\cosh(x/2)}\left[\cosh\left(a-\frac{x}{2}\right) + (-1)^n \cosh\left(\frac{2n-1}{2}x+a\right)\right]$.

5. $\displaystyle\sum_{k=1}^{n-1} k\sinh(kx+a) = -\frac{1}{\sinh^2(x/2)}\left\{n\sinh[(n-1)x+a] - (n-1)\sinh(nx+a) - \sinh a\right\}$.

6. $\displaystyle\sum_{k=1}^{n-1} k\cosh(kx+a) = -\frac{1}{\sinh^2(x/2)}\left\{n\cosh[(n-1)x+a] - (n-1)\cosh(nx+a) - \cosh a\right\}$.

7. $\displaystyle\sum_{k=1}^{n-1} (-1)^k k \sinh(kx+a) = \frac{1}{\cosh^2(x/2)}\left\{(-1)^{n-1} n \sinh[(n-1)x+a]\right.$
$\left. + (-1)^{n-1}(n-1)\sinh(nx+a) - \sinh a\right\}$.

8. $\displaystyle\sum_{k=1}^{n-1} (-1)^k k \cosh(kx+a) = \frac{1}{\cosh^2(x/2)}\left\{(-1)^{n-1} n \cosh[(n-1)x+a]\right.$
$\left. + (-1)^{n-1}(n-1)\cosh(nx+a) - \cosh a\right\}$.

9. $\displaystyle\sum_{k=0}^{n} C_n^k \sinh(kx+a) = 2^n \cosh^n \frac{x}{2}\sinh\left(\frac{nx}{2}+a\right)$.

10. $\displaystyle\sum_{k=0}^{n} C_n^k \cosh(kx+a) = 2^n \cosh^n \frac{x}{2}\cosh\left(\frac{nx}{2}+a\right)$.

11. $\displaystyle\sum_{k=1}^{n-1} a^k \sinh(kx) = \frac{a \sinh x - a^n \sinh(nx) + a^{n+1} \sinh[(n-1)x]}{1 - 2a \cosh x + a^2}$.

12. $\displaystyle\sum_{k=0}^{n-1} a^k \cosh(kx) = \frac{1 - a \cosh x - a^n \cosh(nx) + a^{n+1} \cosh[(n-1)x]}{1 - 2a \cosh x + a^2}$.

13. $\displaystyle\sum_{k=1}^{n} \frac{1}{2^k} \tanh \frac{x}{2^k} = \coth x - \frac{1}{2^n} \coth \frac{x}{2^n}$.

14. $\displaystyle\sum_{k=0}^{n-1} 2^k \tanh(2^k x) = 2^n \coth(2^n x) - \coth x$.

T1.1.2-2. Sums involving trigonometric functions.

1. $\displaystyle\sum_{k=1}^{n} \sin(2kx) = \sin[(n+1)x] \sin(nx) \operatorname{cosec} x$.

2. $\displaystyle\sum_{k=0}^{n} \cos(2kx) = \sin[(n+1)x] \cos(nx) \operatorname{cosec} x$.

3. $\displaystyle\sum_{k=1}^{n} \sin[(2k-1)x] = \sin^2(nx) \operatorname{cosec} x$.

4. $\displaystyle\sum_{k=1}^{n} \cos[(2k-1)x] = \sin(nx) \cos(nx) \operatorname{cosec} x$.

5. $\displaystyle\sum_{k=0}^{n-1} \sin(kx + a) = \sin\left(\frac{n-1}{2}x + a\right) \sin \frac{nx}{2} \operatorname{cosec} \frac{x}{2}$.

6. $\displaystyle\sum_{k=0}^{n-1} \cos(kx + a) = \cos\left(\frac{n-1}{2}x + a\right) \sin \frac{nx}{2} \operatorname{cosec} \frac{x}{2}$.

7. $\displaystyle\sum_{k=0}^{2n-1} (-1)^k \cos(kx + a) = \sin\left(\frac{2n-1}{2}x + a\right) \sin(nx) \sec \frac{x}{2}$.

8. $\displaystyle\sum_{k=1}^{n} (-1)^{k+1} \sin[(2k-1)x] = (-1)^{n+1} \frac{\sin(2nx)}{2 \cos x}$.

9. $\displaystyle\sum_{k=1}^{n} (-1)^k \cos(2kx) = -\frac{1}{2} + (-1)^n \frac{\cos[(2n+1)x]}{2 \cos x}$.

10. $\displaystyle\sum_{k=1}^{n} \sin^2(kx) = \frac{n}{2} - \frac{\cos[(n+1)x] \sin(nx)}{2 \sin x}$.

11. $\displaystyle\sum_{k=1}^{n} \cos^2(kx) = \frac{n}{2} + \frac{\cos[(n+1)x]\sin(nx)}{2\sin x}$.

12. $\displaystyle\sum_{k=1}^{n-1} k\sin(2kx) = \frac{\sin(2nx)}{4\sin^2 x} - \frac{n\cos[(2n-1)x]}{2\sin x}$.

13. $\displaystyle\sum_{k=1}^{n-1} k\cos(2kx) = \frac{n\sin[(2n-1)x]}{2\sin x} - \frac{1-\cos(2nx)}{4\sin^2 x}$.

14. $\displaystyle\sum_{k=1}^{n-1} a^k \sin(kx) = \frac{a\sin x - a^n \sin(nx) + a^{n+1}\sin[(n-1)x]}{1 - 2a\cos x + a^2}$.

15. $\displaystyle\sum_{k=0}^{n-1} a^k \cos(kx) = \frac{1 - a\cos x - a^n \cos(nx) + a^{n+1}\cos[(n-1)x]}{1 - 2a\cos x + a^2}$.

16. $\displaystyle\sum_{k=0}^{n} C_n^k \sin(kx+a) = 2^n \cos^n \frac{x}{2} \sin\left(\frac{nx}{2} + a\right)$.

17. $\displaystyle\sum_{k=0}^{n} C_n^k \cos(kx+a) = 2^n \cos^n \frac{x}{2} \cos\left(\frac{nx}{2} + a\right)$.

18. $\displaystyle\sum_{k=0}^{n} (-1)^k C_n^k \sin(kx+a) = (-2)^n \sin^n \frac{x}{2} \sin\left(\frac{nx}{2} + \frac{\pi n}{2} + a\right)$.

19. $\displaystyle\sum_{k=0}^{n} (-1)^k C_n^k \cos(kx+a) = (-2)^n \sin^n \frac{x}{2} \cos\left(\frac{nx}{2} + \frac{\pi n}{2} + a\right)$.

20. $\displaystyle\sum_{k=1}^{n} \left(2^k \sin^2 \frac{x}{2^k}\right)^2 = \left(2^n \sin^2 \frac{x}{2^n}\right)^2 - \sin^2 x$.

21. $\displaystyle\sum_{k=0}^{n} \frac{1}{2^k} \tan \frac{x}{2^k} = \frac{1}{2^n} \cot \frac{x}{2^n} - 2\cot(2x)$.

## T1.2. Infinite Series

### T1.2.1. Numerical Series

T1.2.1-1. Progressions.

1. $\displaystyle\sum_{k=0}^{\infty} aq^k = \frac{a}{1-q}$, $|q| < 1$.

2. $\displaystyle\sum_{k=0}^{\infty} (a+bk)q^k = \frac{a}{1-q} + \frac{bq}{(1-q)^2}$, $|q| < 1$.

**T1.2.1-2. Other numerical series.**

1. $\sum_{n=0}^{\infty} \dfrac{(-1)^n}{n+1} = \ln 2.$

2. $\sum_{n=0}^{\infty} \dfrac{(-1)^n}{2n+1} = \dfrac{\pi}{4}.$

3. $\sum_{n=1}^{\infty} \dfrac{1}{n(n+1)} = 1.$

4. $\sum_{n=1}^{\infty} \dfrac{(-1)^n}{n(n+1)} = 1 - 2\ln 2.$

5. $\sum_{n=1}^{\infty} \dfrac{1}{n(n+2)} = \dfrac{3}{4}.$

6. $\sum_{n=1}^{\infty} \dfrac{(-1)^n}{n(n+2)} = -\dfrac{1}{4}.$

7. $\sum_{n=1}^{\infty} \dfrac{1}{(2n-1)(2n+1)} = \dfrac{1}{2}.$

8. $\sum_{n=1}^{\infty} \dfrac{1}{n^2} = \dfrac{\pi^2}{6}.$

9. $\sum_{n=1}^{\infty} \dfrac{(-1)^{n+1}}{n^2} = \dfrac{\pi^2}{12}.$

10. $\sum_{n=1}^{\infty} \dfrac{1}{(2n-1)^2} = \dfrac{\pi^2}{8}.$

11. $\sum_{n=1}^{\infty} \dfrac{1}{n^2+a^2} = \dfrac{\pi}{2a}\coth(\pi a) - \dfrac{1}{2a^2}.$

12. $\sum_{n=1}^{\infty} \dfrac{1}{n^2-a^2} = -\dfrac{\pi}{2a}\cot(\pi a) + \dfrac{1}{2a^2}.$

13. $\sum_{k=1}^{\infty} \dfrac{1}{k^{2n}} = \dfrac{2^{2n-1}\pi^{2n}}{(2n)!}|B_{2n}|;$ the $B_{2n}$ are Bernoulli numbers.

14. $\sum_{k=1}^{\infty} \dfrac{(-1)^{k+1}}{k^{2n}} = \dfrac{(2^{2n-1}-1)\pi^{2n}}{(2n)!}|B_{2n}|;$ the $B_{2n}$ are Bernoulli numbers.

15. $\sum_{k=1}^{\infty} \dfrac{1}{(2k-1)^{2n}} = \dfrac{(2^{2n-1}-1)\pi^{2n}}{2(2n)!}|B_{2n}|;$ the $B_{2n}$ are Bernoulli numbers.

16. $\displaystyle\sum_{k=1}^{\infty} \frac{1}{k2^k} = \ln 2.$

17. $\displaystyle\sum_{k=0}^{\infty} \frac{(-1)^k}{n^{2k}} = \frac{n^2}{n^2 + 1}.$

18. $\displaystyle\sum_{k=0}^{\infty} \frac{1}{k!} = e = 2.71828\ldots$

19. $\displaystyle\sum_{k=0}^{\infty} \frac{(-1)^k}{k!} = \frac{1}{e} = 0.36787\ldots$

20. $\displaystyle\sum_{k=1}^{\infty} \frac{k}{(k+1)!} = 1.$

## T1.2.2. Functional Series

T1.2.2-1. Power series.

1. $\displaystyle\sum_{k=0}^{\infty} x^k = \frac{1}{1-x}, \quad |x| < 1.$

2. $\displaystyle\sum_{k=1}^{\infty} k x^k = \frac{x}{(1-x)^2}, \quad |x| < 1.$

3. $\displaystyle\sum_{k=1}^{\infty} k^2 x^k = \frac{x(x+1)}{(1-x)^3}, \quad |x| < 1.$

4. $\displaystyle\sum_{k=1}^{\infty} k^3 x^k = \frac{x(1 + 4x + x^2)}{(1-x)^4}, \quad |x| < 1.$

5. $\displaystyle\sum_{k=0}^{\infty} (\pm 1)^k k^n x^k = \left( x \frac{d}{dx} \right)^n \frac{1}{1 \mp x}, \quad |x| < 1.$

6. $\displaystyle\sum_{k=1}^{\infty} \frac{x^k}{k} = -\ln(1-x), \quad -1 \le x < 1.$

7. $\displaystyle\sum_{k=1}^{\infty} (-1)^{k-1} \frac{x^k}{k} = \ln(1+x), \quad |x| < 1.$

8. $\displaystyle\sum_{k=1}^{\infty} \frac{x^{2k-1}}{2k-1} = \frac{1}{2} \ln \frac{1+x}{1-x}, \quad |x| < 1.$

9. $\displaystyle\sum_{k=1}^{\infty} (-1)^{k-1} \frac{x^{2k-1}}{2k-1} = \arctan x, \quad |x| \le 1.$

10. $\displaystyle\sum_{k=1}^{\infty} \frac{x^k}{k^2} = -\int_0^x \frac{\ln(1-t)}{t}\,dt, \quad |x| \leq 1.$

11. $\displaystyle\sum_{k=1}^{\infty} \frac{x^{k+1}}{k(k+1)} = x + (1-x)\ln(1-x), \quad |x| \leq 1.$

12. $\displaystyle\sum_{k=1}^{\infty} \frac{x^{k+2}}{k(k+2)} = \frac{x}{2} + \frac{x^2}{4} + \frac{1}{2}(1-x^2)\ln(1-x), \quad |x| \leq 1.$

13. $\displaystyle\sum_{k=0}^{\infty} \frac{x^k}{k!} = e^x, \quad x \text{ is any number.}$

14. $\displaystyle\sum_{k=0}^{\infty} \frac{x^{2k}}{(2k)!} = \cosh x, \quad x \text{ is any number.}$

15. $\displaystyle\sum_{k=0}^{\infty} (-1)^k \frac{x^{2k}}{(2k)!} = \cos x, \quad x \text{ is any number.}$

16. $\displaystyle\sum_{k=0}^{\infty} \frac{x^{2k+1}}{(2k+1)!} = \sinh x, \quad x \text{ is any number.}$

17. $\displaystyle\sum_{k=0}^{\infty} (-1)^k \frac{x^{2k+1}}{(2k+1)!} = \sin x, \quad x \text{ is any number.}$

18. $\displaystyle\sum_{k=0}^{\infty} \frac{x^{k+1}}{k!(k+1)} = e^x - 1, \quad x \text{ is any number.}$

19. $\displaystyle\sum_{k=0}^{\infty} \frac{x^{k+2}}{k!(k+2)} = (x-1)e^x + 1, \quad x \text{ is any number.}$

20. $\displaystyle\sum_{k=0}^{\infty} (-1)^k \frac{x^{2k+1}}{k!(2k+1)} = \frac{\sqrt{\pi}}{2}\,\mathrm{erf}\,x, \quad x \text{ is any number.}$

21. $\displaystyle\sum_{k=0}^{\infty} \frac{(k+a)^n}{k!} x^k = \left[\frac{d^n}{dt^n}\exp(at + xe^t)\right]_{t=0}, \quad x \text{ is any number.}$

22. $\displaystyle\sum_{k=1}^{\infty} \frac{2^{2k}(2^{2k}-1)|B_{2k}|}{(2k)!} x^{2k-1} = \tan x; \quad \text{the } B_{2k} \text{ are Bernoulli numbers, } |x| < \pi/2.$

23. $\displaystyle\sum_{k=1}^{\infty} (-1)^{k-1}\frac{2^{2k}(2^{2k}-1)|B_{2k}|}{(2k)!} x^{2k-1} = \tanh x; \quad \text{the } B_{2k} \text{ are Bernoulli numbers, } |x| < \pi/2.$

24. $\displaystyle\sum_{k=1}^{\infty} \frac{2^{2k}|B_{2k}|}{(2k)!} x^{2k-1} = \frac{1}{x} - \cot x; \quad \text{the } B_{2k} \text{ are Bernoulli numbers, } 0 < |x| < \pi.$

25. $\displaystyle\sum_{k=1}^{\infty} (-1)^{k-1}\frac{2^{2k}|B_{2k}|}{(2k)!} x^{2k-1} = \coth x - \frac{1}{x}; \quad \text{the } B_{2k} \text{ are Bernoulli numbers, } |x| < \pi.$

T1.2.2-2. **Trigonometric series in one variable involving sine.**

1. $\sum_{k=1}^{\infty} \frac{1}{k} \sin(kx) = \frac{1}{2}(\pi - x), \quad 0 < x < 2\pi.$

2. $\sum_{k=1}^{\infty} \frac{(-1)^{k-1}}{k} \sin(kx) = \frac{1}{2}x, \quad -\pi < x < \pi.$

3. $\sum_{k=1}^{\infty} \frac{a^k}{k} \sin(kx) = \arctan \frac{a \sin x}{1 - a \cos x}, \quad 0 < x < 2\pi, \ |a| \le 1.$

4. $\sum_{k=0}^{\infty} \frac{1}{2k+1} \sin(kx) = \frac{\pi}{4} \cos \frac{x}{2} - \sin \frac{x}{2} \ln\left(\cot^2 \frac{x}{4}\right), \quad 0 < x < 2\pi.$

5. $\sum_{k=0}^{\infty} \frac{(-1)^k}{2k+1} \sin(kx) = -\frac{1}{4} \cos \frac{x}{2} \ln\left(\cot^2 \frac{x+\pi}{4}\right) - \frac{\pi}{4} \sin \frac{x}{2}, \quad -\pi < x < \pi.$

6. $\sum_{k=1}^{\infty} \frac{1}{k^2} \sin(kx) = -\int_0^x \ln\left(2 \sin \frac{t}{2}\right) dt, \quad 0 \le x < \pi.$

7. $\sum_{k=1}^{\infty} \frac{(-1)^k}{k^2} \sin(kx) = -\int_0^x \ln\left(2 \cos \frac{t}{2}\right) dt, \quad -\pi < x < \pi.$

8. $\sum_{k=1}^{\infty} \frac{1}{k(k+1)} \sin(kx) = (\pi - x) \sin^2 \frac{x}{2} + \sin x \ln\left(2 \sin \frac{x}{2}\right), \quad 0 \le x \le 2\pi.$

9. $\sum_{k=1}^{\infty} \frac{(-1)^k}{k(k+1)} \sin(kx) = -x \cos^2 \frac{x}{2} + \sin x \ln\left(2 \cos \frac{x}{2}\right), \quad -\pi \le x \le \pi.$

10. $\sum_{k=1}^{\infty} \frac{k}{k^2 + a^2} \sin(kx) = \frac{\pi}{2 \sinh(\pi a)} \sinh[a(\pi - x)], \quad 0 < x < 2\pi.$

11. $\sum_{k=1}^{\infty} (-1)^{k+1} \frac{k}{k^2 + a^2} \sin(kx) = \frac{\pi}{2 \sinh(\pi a)} \sinh(ax), \quad -\pi < x < \pi.$

12. $\sum_{k=1}^{\infty} \frac{k}{k^2 - a^2} \sin(kx) = \frac{\pi}{2 \sin(\pi a)} \sin[a(\pi - x)], \quad 0 < x < 2\pi.$

13. $\sum_{k=1}^{\infty} (-1)^{k+1} \frac{k}{k^2 - a^2} \sin(kx) = \frac{\pi}{2 \sin(\pi a)} \sin(ax), \quad -\pi < x < \pi.$

14. $\sum_{k=2}^{\infty} (-1)^k \frac{k}{k^2 - 1} \sin(kx) = \frac{1}{4} \sin x + \frac{1}{2} x \cos x, \quad -\pi < x < \pi.$

15. $\sum_{k=1}^{\infty} \frac{1}{k^{2n+1}} \sin(kx) = \frac{(-1)^{n-1}(2\pi)^{2n+1}}{2(2n+1)!} B_{2n+1}\left(\frac{x}{2\pi}\right),$

where $0 \le x \le 2\pi$ for $n = 1, 2, \ldots$; $0 < x < 2\pi$ for $n = 0$; and the $B_n(x)$ are Bernoulli polynomials.

16. $\sum_{k=1}^{\infty} \frac{(-1)^k}{k^{2n+1}} \sin(kx) = \frac{(-1)^{n-1}(2\pi)^{2n+1}}{2(2n+1)!} B_{2n+1}\left(\frac{x+\pi}{2\pi}\right),$

where $-\pi < x \leq \pi$ for $n = 0, 1, \ldots$; the $B_n(x)$ are Bernoulli polynomials.

17. $\sum_{k=1}^{\infty} \frac{1}{k!} \sin(kx) = \exp(\cos x) \sin(\sin x), \quad x$ is any number.

18. $\sum_{k=1}^{\infty} \frac{(-1)^k}{k!} \sin(kx) = -\exp(-\cos x) \sin(\sin x), \quad x$ is any number.

19. $\sum_{k=0}^{\infty} \frac{1}{(2k)!} \sin(kx) = \sin\left(\sin \frac{x}{2}\right) \sinh\left(\cos \frac{x}{2}\right), \quad x$ is any number.

20. $\sum_{k=0}^{\infty} \frac{(-1)^k}{(2k)!} \sin(kx) = -\sin\left(\cos \frac{x}{2}\right) \sinh\left(\sin \frac{x}{2}\right), \quad x$ is any number.

21. $\sum_{k=0}^{\infty} \frac{a^k}{k!} \sin(kx) = \exp(k \cos x) \sin(k \sin x), \quad |a| \leq 1, \; x$ is any number.

22. $\sum_{k=0}^{\infty} a^k \sin(kx) = \frac{a \sin x}{1 - 2a \cos x + a^2}, \quad |a| < 1, \; x$ is any number.

23. $\sum_{k=1}^{\infty} k a^k \sin(kx) = \frac{a(1 - a^2) \sin x}{(1 - 2a \cos x + a^2)^2}, \quad |a| < 1, \; x$ is any number.

24. $\sum_{k=1}^{\infty} \frac{1}{k} \sin(kx + a) = \frac{1}{2}(\pi - x) \cos a - \ln\left(2 \sin \frac{x}{2}\right) \sin a, \quad 0 < x < 2\pi.$

25. $\sum_{k=1}^{\infty} \frac{(-1)^{k-1}}{k} \sin(kx + a) = \frac{1}{2} x \cos a + \ln\left(2 \cos \frac{x}{2}\right) \sin a, \quad -\pi < x < \pi.$

26. $\sum_{k=1}^{\infty} \frac{\sin[(2k-1)x]}{2k-1} = \frac{\pi}{4}, \quad 0 < x < \pi.$

27. $\sum_{k=1}^{\infty} (-1)^{k-1} \frac{\sin[(2k-1)x]}{2k-1} = \frac{1}{2} \ln \tan\left(\frac{x}{2} + \frac{\pi}{4}\right), \quad -\frac{\pi}{2} < x < \frac{\pi}{2}.$

28. $\sum_{k=1}^{\infty} a^{2k-1} \frac{\sin[(2k-1)x]}{2k-1} = \frac{1}{2} \arctan \frac{2a \sin x}{1 - a^2}, \quad 0 < x < 2\pi, \; |a| \leq 1.$

29. $\sum_{k=1}^{\infty} (-1)^{k-1} a^{2k-1} \frac{\sin[(2k-1)x]}{2k-1} = \frac{1}{4} \ln \frac{1 + 2a \sin x + a^2}{1 - 2a \sin x + a^2}, \quad 0 < x < \pi, \; |a| \leq 1.$

30. $\sum_{k=1}^{\infty} (-1)^k \frac{\sin[(k+1)x]}{k(k+1)} = \sin x - \frac{1}{2} x (1 + \cos x) - \sin x \ln \left|2 \cos \frac{x}{2}\right|.$

31. $\displaystyle\sum_{k=0}^{\infty} a^{2k+1} \sin[(2k+1)x] = \frac{a(1+a^2)\sin x}{(1+a^2)^2 - 4a^2 \cos^2 x}$, $\quad |a| < 1$, $x$ is any number.

32. $\displaystyle\sum_{k=0}^{\infty} (-1)^k a^{2k+1} \sin[(2k+1)x] = \frac{a(1-a^2)\sin x}{(1+a^2)^2 - 4a^2 \sin^2 x}$, $\quad |a| < 1$, $x$ is any number.

33. $\displaystyle\sum_{k=1}^{\infty} \frac{\sin[2(k+1)x]}{k(k+1)} = \sin(2x) - (\pi - 2x)\sin^2 x - \sin x \cos x \ln(4\sin^2 x)$, $\quad 0 \le x \le \pi$.

34. $\displaystyle\sum_{k=1}^{\infty} (-1)^k \frac{\sin[(2k+1)x]}{(2k+1)^2} = \begin{cases} \frac{1}{4}\pi x & \text{if } -\frac{1}{2}\pi \le x \le \frac{1}{2}\pi, \\ \frac{1}{4}\pi(\pi - x) & \text{if } \frac{1}{2}\pi \le x \le \frac{3}{2}\pi. \end{cases}$

| T1.2.2-3. Trigonometric series in one variable involving cosine. |
|---|

1. $\displaystyle\sum_{k=1}^{\infty} \frac{1}{k} \cos(kx) = -\ln\left(2\sin\frac{x}{2}\right)$, $\quad 0 < x < 2\pi$.

2. $\displaystyle\sum_{k=1}^{\infty} \frac{(-1)^{k-1}}{k} \cos(kx) = \ln\left(2\cos\frac{x}{2}\right)$, $\quad -\pi < x < \pi$.

3. $\displaystyle\sum_{k=1}^{\infty} \frac{a^k}{k} \cos(kx) = \ln\frac{1}{\sqrt{1-2a\cos x + a^2}}$, $\quad 0 < x < 2\pi$, $|a| \le 1$.

4. $\displaystyle\sum_{k=0}^{\infty} \frac{1}{2k+1} \cos(kx) = \frac{\pi}{4}\sin\frac{x}{2} + \cos\frac{x}{2}\ln\left(\cot^2\frac{x}{4}\right)$, $\quad 0 < x < 2\pi$.

5. $\displaystyle\sum_{k=0}^{\infty} \frac{(-1)^k}{2k+1} \cos(kx) = -\frac{1}{4}\sin\frac{x}{2}\ln\left(\cot^2\frac{x+\pi}{4}\right) + \frac{\pi}{4}\cos\frac{x}{2}$, $\quad -\pi < x < \pi$.

6. $\displaystyle\sum_{k=1}^{\infty} \frac{1}{k^2} \cos(kx) = \frac{1}{12}(3x^2 - 6\pi x + 2\pi^2)$, $\quad 0 \le x \le 2\pi$.

7. $\displaystyle\sum_{k=1}^{\infty} \frac{(-1)^k}{k^2} \cos(kx) = \frac{1}{12}(3x^2 - \pi^2)$, $\quad -\pi \le x \le \pi$.

8. $\displaystyle\sum_{k=1}^{\infty} \frac{1}{k(k+1)} \cos(kx) = \frac{1}{2}(x-\pi)\sin x - 2\sin^2\frac{x}{2}\ln\left(2\sin\frac{x}{2}\right) + 1$, $\quad 0 \le x \le 2\pi$.

9. $\displaystyle\sum_{k=1}^{\infty} \frac{(-1)^k}{k(k+1)} \cos(kx) = -\frac{1}{2}x\sin x - 2\cos^2\frac{x}{2}\ln\left(2\cos\frac{x}{2}\right) + 1$, $\quad -\pi \le x \le \pi$.

10. $\displaystyle\sum_{k=1}^{\infty} \frac{1}{k^2 + a^2} \cos(kx) = \frac{\pi}{2a\sinh(\pi a)}\cosh[a(\pi - x)] - \frac{1}{2a^2}$, $\quad 0 \le x \le 2\pi$.

11. $\displaystyle\sum_{k=1}^{\infty} \frac{1}{k^2 - a^2} \cos(kx) = -\frac{\pi}{2a\sin(\pi a)}\cos[a(\pi - x)] + \frac{1}{2a^2}$, $\quad 0 \le x \le 2\pi$.

12. $\displaystyle\sum_{k=2}^{\infty} \frac{(-1)^k}{k^2-1} \cos(kx) = \frac{1}{2} - \frac{1}{4}\cos x - \frac{1}{2}x \sin x, \quad -\pi \le x \le \pi.$

13. $\displaystyle\sum_{k=2}^{\infty} \frac{k}{k^2-1} \cos(kx) = -\frac{1}{2} - \frac{1}{4}\cos x - \cos x \ln\left(2 \sin \frac{x}{2}\right), \quad 0 < x < 2\pi.$

14. $\displaystyle\sum_{k=1}^{\infty} \frac{1}{k^{2n}} \cos(kx) = \frac{(-1)^{n-1}(2\pi)^{2n}}{2(2n)!} B_{2n}\left(\frac{x}{2\pi}\right),$

    where $0 \le x \le 2\pi$ for $n = 1, 2, \ldots$; the $B_n(x)$ are Bernoulli polynomials.

15. $\displaystyle\sum_{k=1}^{\infty} \frac{(-1)^k}{k^{2n}} \cos(kx) = \frac{(-1)^{n-1}(2\pi)^{2n}}{2(2n)!} B_{2n}\left(\frac{x+\pi}{2\pi}\right),$

    where $-\pi \le x \le \pi$ for $n = 1, 2, \ldots$; the $B_n(x)$ are Bernoulli polynomials.

16. $\displaystyle\sum_{k=0}^{\infty} \frac{1}{k!} \cos(kx) = \exp(\cos x) \cos(\sin x), \quad x$ is any number.

17. $\displaystyle\sum_{k=0}^{\infty} \frac{(-1)^k}{k!} \cos(kx) = \exp(-\cos x) \cos(\sin x), \quad x$ is any number.

18. $\displaystyle\sum_{k=0}^{\infty} \frac{1}{(2k)!} \cos(kx) = \cos\left(\sin \frac{x}{2}\right) \cosh\left(\cos \frac{x}{2}\right), \quad x$ is any number.

19. $\displaystyle\sum_{k=0}^{\infty} \frac{(-1)^k}{(2k)!} \cos(kx) = \cos\left(\cos \frac{x}{2}\right) \cosh\left(\sin \frac{x}{2}\right), \quad x$ is any number.

20. $\displaystyle\sum_{k=0}^{\infty} \frac{a^k}{k!} \cos(kx) = \exp(a \cos x) \cos(a \sin x), \quad |a| \le 1, \; x$ is any number.

21. $\displaystyle\sum_{k=0}^{\infty} a^k \cos(kx) = \frac{1 - a \cos x}{1 - 2a \cos x + a^2}, \quad |a| < 1, \; x$ is any number.

22. $\displaystyle\sum_{k=1}^{\infty} k a^k \cos(kx) = \frac{a(1+a^2)\cos x - 2a^2}{(1 - 2a \cos x + a^2)^2}, \quad |a| < 1, \; x$ is any number.

23. $\displaystyle\sum_{k=1}^{\infty} \frac{1}{k} \cos(kx + a) = \frac{1}{2}(x - \pi) \sin a - \ln\left(2 \sin \frac{x}{2}\right) \cos a, \quad 0 < x < 2\pi.$

24. $\displaystyle\sum_{k=1}^{\infty} \frac{(-1)^{k-1}}{k} \cos(kx + a) = -\frac{1}{2} x \sin a + \ln\left(2 \cos \frac{x}{2}\right) \cos a, \quad -\pi < x < \pi.$

25. $\displaystyle\sum_{k=1}^{\infty} \frac{\cos[(2k-1)x]}{2k-1} = \frac{1}{2} \ln \cot \frac{x}{2}, \quad 0 < x < \pi.$

26. $\displaystyle\sum_{k=1}^{\infty} (-1)^{k-1} \frac{\cos[(2k-1)x]}{2k-1} = \frac{\pi}{4}, \quad 0 < x < \pi.$

27. $\displaystyle\sum_{k=1}^{\infty} a^{2k-1} \frac{\cos[(2k-1)x]}{2k-1} = \frac{1}{4} \ln \frac{1 + 2a \cos x + a^2}{1 - 2a \cos x + a^2}$, $\quad 0 < x < 2\pi$, $|a| \le 1$.

28. $\displaystyle\sum_{k=1}^{\infty} (-1)^{k-1} a^{2k-1} \frac{\cos[(2k-1)x]}{2k-1} = \frac{1}{2} \arctan \frac{2a \cos x}{1 - a^2}$, $\quad 0 < x < \pi$, $|a| \le 1$.

29. $\displaystyle\sum_{k=1}^{\infty} \frac{\cos[(2k-1)x]}{(2k-1)^2} = \frac{\pi}{4}\left(\frac{\pi}{2} - |x|\right)$, $\quad -\pi \le x \le \pi$.

30. $\displaystyle\sum_{k=1}^{\infty} (-1)^k \frac{\cos[(k+1)x]}{k(k+1)} = \cos x - \frac{1}{2} x \sin x - (1 + \cos x) \ln\left|2 \cos \frac{x}{2}\right|$.

31. $\displaystyle\sum_{k=0}^{\infty} a^{2k+1} \cos[(2k+1)x] = \frac{a(1-a^2)\cos x}{(1+a^2)^2 - 4a^2 \cos^2 x}$, $\quad |a| < 1$, $x$ is any number.

32. $\displaystyle\sum_{k=0}^{\infty} (-1)^k a^{2k+1} \cos[(2k+1)x] = \frac{a(1+a^2)\cos x}{(1+a^2)^2 - 4a^2 \sin^2 x}$, $\quad |a| < 1$, $x$ is any number.

33. $\displaystyle\sum_{k=1}^{\infty} \frac{\cos[2(k+1)x]}{k(k+1)} = \cos(2x) - \left(\frac{\pi}{2} - x\right) \sin(2x) + \sin^2 x \ln(4 \sin^2 x)$, $\quad 0 \le x \le \pi$.

**T1.2.2-4. Trigonometric series in two variables.**

1. $\displaystyle\sum_{k=1}^{\infty} \frac{1}{k} \sin(kx) \sin(ky) = \frac{1}{2} \ln\left|\sin \frac{x+y}{2} \operatorname{cosec} \frac{x-y}{2}\right|$, $\quad x \pm y \ne 0, 2\pi, 4\pi, \ldots$

2. $\displaystyle\sum_{k=1}^{\infty} \frac{(-1)^k}{k} \sin(kx) \sin(ky) = \frac{1}{2} \ln\left|\cos \frac{x+y}{2} \sec \frac{x-y}{2}\right|$, $\quad x \pm y \ne \pi, 3\pi, 5\pi, \ldots$

3. $\displaystyle\sum_{k=1}^{\infty} \frac{1}{k^2} \sin(kx) \sin(ky) = \begin{cases} \frac{1}{2} x(\pi - y) & \text{if } -y \le x \le y, \\ \frac{1}{2} y(\pi - x) & \text{if } y \le x \le 2\pi - y. \end{cases}$ Here, $0 < y < \pi$.

4. $\displaystyle\sum_{k=1}^{\infty} \frac{(-1)^{k+1}}{k^2} \sin(kx) \sin(ky) = \frac{1}{2} xy$, $\quad |x \pm y| \le \pi$.

5. $\displaystyle\sum_{k=1}^{\infty} \frac{a^k}{k} \sin(kx) \sin(ky) = \frac{1}{4} \ln \frac{4a \sin^2[(x+y)/2] + (a-1)^2}{4a \sin^2[(x-y)/2] + (a-1)^2}$, $\quad 0 < a < 1$.

6. $\displaystyle\sum_{k=1}^{\infty} \frac{1}{k^2} \sin^2(kx) \sin^2(ky) = \frac{1}{2} \pi x$, $\quad 0 \le x \le y \le \frac{\pi}{2}$.

7. $\displaystyle\sum_{k=1}^{\infty} \frac{1}{k} \cos(kx) \cos(ky) = -\frac{1}{2} \ln|2(\cos x - \cos y)|$, $\quad x \pm y \ne 0, 2\pi, 4\pi, \ldots$

8. $\displaystyle\sum_{k=1}^{\infty} \frac{(-1)^k}{k} \cos(kx) \cos(ky) = -\frac{1}{2} \ln|2(\cos x + \cos y)|$, $\quad x \pm y \ne \pi, 3\pi, 5\pi, \ldots$

9. $\displaystyle\sum_{k=1}^{\infty} \frac{1}{k} \sin(kx)\cos(ky) = \begin{cases} -\frac{1}{2} & \text{if } 0 < x < y, \\ \frac{1}{4}(\pi - 2y) & \text{if } x = y, \\ \frac{1}{2}(\pi - x) & \text{if } y < x < \pi. \end{cases}$ Here, $0 < y < \pi$.

10. $\displaystyle\sum_{k=1}^{\infty} \frac{1}{k^2} \cos(kx)\cos(ky) = \begin{cases} \frac{1}{12}\left[3x^2 + 3(y-\pi)^2 - \pi^2\right] & \text{if } 0 \leq x \leq y, \\ \frac{1}{12}\left[3y^2 + 3(x-\pi)^2 - \pi^2\right] & \text{if } y \leq x \leq \pi. \end{cases}$

Here, $0 < y < \pi$.

11. $\displaystyle\sum_{k=1}^{\infty} \frac{(-1)^k}{k^2} \cos(kx)\cos(ky) = \begin{cases} \frac{1}{12}\left[3(x^2+y^2) - \pi^2\right] & \text{if } -(\pi-y) \leq x \leq \pi-y, \\ \frac{1}{12}\left[3(x-\pi)^2 + 3(y-\pi)^2 - \pi^2\right] & \text{if } \pi-y \leq x \leq \pi+y. \end{cases}$

Here, $0 < y < \pi$.

## References for Chapter T1

**Dwight, H. B.**, *Tables of Integrals and Other Mathematical Data*, Macmillan, New York, 1961.
**Gradshteyn, I. S. and Ryzhik, I. M.**, *Tables of Integrals, Series, and Products*, 6th Edition, Academic Press, New York, 2000.
**Hansen, E. R.**, *A Table of Series and Products*, Printice Hall, Englewood Cliffs, London, 1975.
**Mangulis, V.**, *Handbook of Series for Scientists and Engineers*, Academic Press, New York, 1965.
**Prudnikov, A. P., Brychkov, Yu. A., and Marichev, O. I.**, *Integrals and Series, Vol. 1, Elementary Functions*, Gordon & Breach, New York, 1986.
**Zwillinger, D.**, *CRC Standard Mathematical Tables and Formulae, 31st Edition*, CRC Press, Boca Raton, 2002.

# Chapter T2
# Integrals

## T2.1. Indefinite Integrals

▶ *Throughout Section T2.1, the integration constant $C$ is omitted for brevity.*

### T2.1.1. Integrals Involving Rational Functions

**T2.1.1-1. Integrals involving $a + bx$.**

1. $\displaystyle\int \frac{dx}{a + bx} = \frac{1}{b} \ln|a + bx|$.

2. $\displaystyle\int (a + bx)^n \, dx = \frac{(a + bx)^{n+1}}{b(n + 1)}, \quad n \neq -1$.

3. $\displaystyle\int \frac{x \, dx}{a + bx} = \frac{1}{b^2}\big(a + bx - a \ln|a + bx|\big)$.

4. $\displaystyle\int \frac{x^2 \, dx}{a + bx} = \frac{1}{b^3}\left[\frac{1}{2}(a + bx)^2 - 2a(a + bx) + a^2 \ln|a + bx|\right]$.

5. $\displaystyle\int \frac{dx}{x(a + bx)} = -\frac{1}{a} \ln\left|\frac{a + bx}{x}\right|$.

6. $\displaystyle\int \frac{dx}{x^2(a + bx)} = -\frac{1}{ax} + \frac{b}{a^2} \ln\left|\frac{a + bx}{x}\right|$.

7. $\displaystyle\int \frac{x \, dx}{(a + bx)^2} = \frac{1}{b^2}\left(\ln|a + bx| + \frac{a}{a + bx}\right)$.

8. $\displaystyle\int \frac{x^2 \, dx}{(a + bx)^2} = \frac{1}{b^3}\left(a + bx - 2a \ln|a + bx| - \frac{a^2}{a + bx}\right)$.

9. $\displaystyle\int \frac{dx}{x(a + bx)^2} = \frac{1}{a(a + bx)} - \frac{1}{a^2} \ln\left|\frac{a + bx}{x}\right|$.

10. $\displaystyle\int \frac{x \, dx}{(a + bx)^3} = \frac{1}{b^2}\left[-\frac{1}{a + bx} + \frac{a}{2(a + bx)^2}\right]$.

**T2.1.1-2. Integrals involving $a + x$ and $b + x$.**

1. $\displaystyle\int \frac{a + x}{b + x} \, dx = x + (a - b) \ln|b + x|$.

2. $\int \dfrac{dx}{(a+x)(b+x)} = \dfrac{1}{a-b} \ln\left|\dfrac{b+x}{a+x}\right|$, $a \neq b$. For $a = b$, see Integral 2 with $n = -2$ in Paragraph T2.1.1-1.

3. $\int \dfrac{x\,dx}{(a+x)(b+x)} = \dfrac{1}{a-b}\bigl(a\ln|a+x| - b\ln|b+x|\bigr)$.

4. $\int \dfrac{dx}{(a+x)(b+x)^2} = \dfrac{1}{(b-a)(b+x)} + \dfrac{1}{(a-b)^2}\ln\left|\dfrac{a+x}{b+x}\right|$.

5. $\int \dfrac{x\,dx}{(a+x)(b+x)^2} = \dfrac{b}{(a-b)(b+x)} - \dfrac{a}{(a-b)^2}\ln\left|\dfrac{a+x}{b+x}\right|$.

6. $\int \dfrac{x^2\,dx}{(a+x)(b+x)^2} = \dfrac{b^2}{(b-a)(b+x)} + \dfrac{a^2}{(a-b)^2}\ln|a+x| + \dfrac{b^2-2ab}{(b-a)^2}\ln|b+x|$.

7. $\int \dfrac{dx}{(a+x)^2(b+x)^2} = -\dfrac{1}{(a-b)^2}\left(\dfrac{1}{a+x} + \dfrac{1}{b+x}\right) + \dfrac{2}{(a-b)^3}\ln\left|\dfrac{a+x}{b+x}\right|$.

8. $\int \dfrac{x\,dx}{(a+x)^2(b+x)^2} = \dfrac{1}{(a-b)^2}\left(\dfrac{a}{a+x} + \dfrac{b}{b+x}\right) + \dfrac{a+b}{(a-b)^3}\ln\left|\dfrac{a+x}{b+x}\right|$.

9. $\int \dfrac{x^2\,dx}{(a+x)^2(b+x)^2} = -\dfrac{1}{(a-b)^2}\left(\dfrac{a^2}{a+x} + \dfrac{b^2}{b+x}\right) + \dfrac{2ab}{(a-b)^3}\ln\left|\dfrac{a+x}{b+x}\right|$.

T2.1.1-3. Integrals involving $a^2 + x^2$.

1. $\int \dfrac{dx}{a^2+x^2} = \dfrac{1}{a}\arctan\dfrac{x}{a}$.

2. $\int \dfrac{dx}{(a^2+x^2)^2} = \dfrac{x}{2a^2(a^2+x^2)} + \dfrac{1}{2a^3}\arctan\dfrac{x}{a}$.

3. $\int \dfrac{dx}{(a^2+x^2)^3} = \dfrac{x}{4a^2(a^2+x^2)^2} + \dfrac{3x}{8a^4(a^2+x^2)} + \dfrac{3}{8a^5}\arctan\dfrac{x}{a}$.

4. $\int \dfrac{dx}{(a^2+x^2)^{n+1}} = \dfrac{x}{2na^2(a^2+x^2)^n} + \dfrac{2n-1}{2na^2}\int \dfrac{dx}{(a^2+x^2)^n}$; $n = 1, 2, \ldots$

5. $\int \dfrac{x\,dx}{a^2+x^2} = \dfrac{1}{2}\ln(a^2+x^2)$.

6. $\int \dfrac{x\,dx}{(a^2+x^2)^2} = -\dfrac{1}{2(a^2+x^2)}$.

7. $\int \dfrac{x\,dx}{(a^2+x^2)^3} = -\dfrac{1}{4(a^2+x^2)^2}$.

8. $\int \dfrac{x\,dx}{(a^2+x^2)^{n+1}} = -\dfrac{1}{2n(a^2+x^2)^n}$; $n = 1, 2, \ldots$

9. $\int \dfrac{x^2\,dx}{a^2+x^2} = x - a\arctan\dfrac{x}{a}$.

10. $\int \dfrac{x^2\,dx}{(a^2+x^2)^2} = -\dfrac{x}{2(a^2+x^2)} + \dfrac{1}{2a}\arctan\dfrac{x}{a}$.

11. $\int \dfrac{x^2\,dx}{(a^2+x^2)^3} = -\dfrac{x}{4(a^2+x^2)^2} + \dfrac{x}{8a^2(a^2+x^2)} + \dfrac{1}{8a^3}\arctan\dfrac{x}{a}$.

12. $\int \dfrac{x^2\,dx}{(a^2+x^2)^{n+1}} = -\dfrac{x}{2n(a^2+x^2)^n} + \dfrac{1}{2n}\int \dfrac{dx}{(a^2+x^2)^n};\quad n=1,2,\ldots$

13. $\int \dfrac{x^3\,dx}{a^2+x^2} = \dfrac{x^2}{2} - \dfrac{a^2}{2}\ln(a^2+x^2)$.

14. $\int \dfrac{x^3\,dx}{(a^2+x^2)^2} = \dfrac{a^2}{2(a^2+x^2)} + \dfrac{1}{2}\ln(a^2+x^2)$.

15. $\int \dfrac{x^3\,dx}{(a^2+x^2)^{n+1}} = -\dfrac{1}{2(n-1)(a^2+x^2)^{n-1}} + \dfrac{a^2}{2n(a^2+x^2)^n};\quad n=2,3,\ldots$

16. $\int \dfrac{dx}{x(a^2+x^2)} = \dfrac{1}{2a^2}\ln\dfrac{x^2}{a^2+x^2}$.

17. $\int \dfrac{dx}{x(a^2+x^2)^2} = \dfrac{1}{2a^2(a^2+x^2)} + \dfrac{1}{2a^4}\ln\dfrac{x^2}{a^2+x^2}$.

18. $\int \dfrac{dx}{x(a^2+x^2)^3} = \dfrac{1}{4a^2(a^2+x^2)^2} + \dfrac{1}{2a^4(a^2+x^2)} + \dfrac{1}{2a^6}\ln\dfrac{x^2}{a^2+x^2}$.

19. $\int \dfrac{dx}{x^2(a^2+x^2)} = -\dfrac{1}{a^2 x} - \dfrac{1}{a^3}\arctan\dfrac{x}{a}$.

20. $\int \dfrac{dx}{x^2(a^2+x^2)^2} = -\dfrac{1}{a^4 x} - \dfrac{x}{2a^4(a^2+x^2)} - \dfrac{3}{2a^5}\arctan\dfrac{x}{a}$.

21. $\int \dfrac{dx}{x^3(a^2+x^2)^2} = -\dfrac{1}{2a^4 x^2} - \dfrac{1}{2a^4(a^2+x^2)} - \dfrac{1}{a^6}\ln\dfrac{x^2}{a^2+x^2}$.

22. $\int \dfrac{dx}{x^2(a^2+x^2)^3} = -\dfrac{1}{a^6 x} - \dfrac{x}{4a^4(a^2+x^2)^2} - \dfrac{7x}{8a^6(a^2+x^2)} - \dfrac{15}{8a^7}\arctan\dfrac{x}{a}$.

23. $\int \dfrac{dx}{x^3(a^2+x^2)^3} = -\dfrac{1}{2a^6 x^2} - \dfrac{1}{a^6(a^2+x^2)} - \dfrac{1}{4a^4(a^2+x^2)^2} - \dfrac{3}{2a^8}\ln\dfrac{x^2}{a^2+x^2}$.

**T2.1.1-4. Integrals involving $a^2 - x^2$.**

1. $\int \dfrac{dx}{a^2-x^2} = \dfrac{1}{2a}\ln\left|\dfrac{a+x}{a-x}\right|$.

2. $\int \dfrac{dx}{(a^2-x^2)^2} = \dfrac{x}{2a^2(a^2-x^2)} + \dfrac{1}{4a^3}\ln\left|\dfrac{a+x}{a-x}\right|$.

3. $\int \dfrac{dx}{(a^2-x^2)^3} = \dfrac{x}{4a^2(a^2-x^2)^2} + \dfrac{3x}{8a^4(a^2-x^2)} + \dfrac{3}{16a^5}\ln\left|\dfrac{a+x}{a-x}\right|$.

4. $\int \dfrac{dx}{(a^2-x^2)^{n+1}} = \dfrac{x}{2na^2(a^2-x^2)^n} + \dfrac{2n-1}{2na^2}\int \dfrac{dx}{(a^2-x^2)^n};\quad n=1,2,\ldots$

5. $\int \dfrac{x\,dx}{a^2-x^2} = -\dfrac{1}{2}\ln|a^2-x^2|$.

6. $\int \dfrac{x\,dx}{(a^2-x^2)^2} = \dfrac{1}{2(a^2-x^2)}.$

7. $\int \dfrac{x\,dx}{(a^2-x^2)^3} = \dfrac{1}{4(a^2-x^2)^2}.$

8. $\int \dfrac{x\,dx}{(a^2-x^2)^{n+1}} = \dfrac{1}{2n(a^2-x^2)^n}; \quad n=1,2,\ldots$

9. $\int \dfrac{x^2\,dx}{a^2-x^2} = -x + \dfrac{a}{2}\ln\left|\dfrac{a+x}{a-x}\right|.$

10. $\int \dfrac{x^2\,dx}{(a^2-x^2)^2} = \dfrac{x}{2(a^2-x^2)} - \dfrac{1}{4a}\ln\left|\dfrac{a+x}{a-x}\right|.$

11. $\int \dfrac{x^2\,dx}{(a^2-x^2)^3} = \dfrac{x}{4(a^2-x^2)^2} - \dfrac{x}{8a^2(a^2-x^2)} - \dfrac{1}{16a^3}\ln\left|\dfrac{a+x}{a-x}\right|.$

12. $\int \dfrac{x^2\,dx}{(a^2-x^2)^{n+1}} = \dfrac{x}{2n(a^2-x^2)^n} - \dfrac{1}{2n}\int \dfrac{dx}{(a^2-x^2)^n}; \quad n=1,2,\ldots$

13. $\int \dfrac{x^3\,dx}{a^2-x^2} = -\dfrac{x^2}{2} - \dfrac{a^2}{2}\ln|a^2-x^2|.$

14. $\int \dfrac{x^3\,dx}{(a^2-x^2)^2} = \dfrac{a^2}{2(a^2-x^2)} + \dfrac{1}{2}\ln|a^2-x^2|.$

15. $\int \dfrac{x^3\,dx}{(a^2-x^2)^{n+1}} = -\dfrac{1}{2(n-1)(a^2-x^2)^{n-1}} + \dfrac{a^2}{2n(a^2-x^2)^n}; \quad n=2,3,\ldots$

16. $\int \dfrac{dx}{x(a^2-x^2)} = \dfrac{1}{2a^2}\ln\left|\dfrac{x^2}{a^2-x^2}\right|.$

17. $\int \dfrac{dx}{x(a^2-x^2)^2} = \dfrac{1}{2a^2(a^2-x^2)} + \dfrac{1}{2a^4}\ln\left|\dfrac{x^2}{a^2-x^2}\right|.$

18. $\int \dfrac{dx}{x(a^2-x^2)^3} = \dfrac{1}{4a^2(a^2-x^2)^2} + \dfrac{1}{2a^4(a^2-x^2)} + \dfrac{1}{2a^6}\ln\left|\dfrac{x^2}{a^2-x^2}\right|.$

T2.1.1-5. Integrals involving $a^3+x^3$.

1. $\int \dfrac{dx}{a^3+x^3} = \dfrac{1}{6a^2}\ln\dfrac{(a+x)^2}{a^2-ax+x^2} + \dfrac{1}{a^2\sqrt{3}}\arctan\dfrac{2x-a}{a\sqrt{3}}.$

2. $\int \dfrac{dx}{(a^3+x^3)^2} = \dfrac{x}{3a^3(a^3+x^3)} + \dfrac{2}{3a^3}\int \dfrac{dx}{a^3+x^3}.$

3. $\int \dfrac{x\,dx}{a^3+x^3} = \dfrac{1}{6a}\ln\dfrac{a^2-ax+x^2}{(a+x)^2} + \dfrac{1}{a\sqrt{3}}\arctan\dfrac{2x-a}{a\sqrt{3}}.$

4. $\int \dfrac{x\,dx}{(a^3+x^3)^2} = \dfrac{x^2}{3a^3(a^3+x^3)} + \dfrac{1}{3a^3}\int \dfrac{x\,dx}{a^3+x^3}.$

5. $\int \dfrac{x^2\,dx}{a^3+x^3} = \dfrac{1}{3}\ln|a^3+x^3|.$

6. $\int \dfrac{dx}{x(a^3+x^3)} = \dfrac{1}{3a^3} \ln\left|\dfrac{x^3}{a^3+x^3}\right|.$

7. $\int \dfrac{dx}{x(a^3+x^3)^2} = \dfrac{1}{3a^3(a^3+x^3)} + \dfrac{1}{3a^6} \ln\left|\dfrac{x^3}{a^3+x^3}\right|.$

8. $\int \dfrac{dx}{x^2(a^3+x^3)} = -\dfrac{1}{a^3 x} - \dfrac{1}{a^3} \int \dfrac{x\,dx}{a^3+x^3}.$

9. $\int \dfrac{dx}{x^2(a^3+x^3)^2} = -\dfrac{1}{a^6 x} - \dfrac{x^2}{3a^6(a^3+x^3)} - \dfrac{4}{3a^6} \int \dfrac{x\,dx}{a^3+x^3}.$

T2.1.1-6. Integrals involving $a^3 - x^3$.

1. $\int \dfrac{dx}{a^3-x^3} = \dfrac{1}{6a^2} \ln \dfrac{a^2+ax+x^2}{(a-x)^2} + \dfrac{1}{a^2\sqrt{3}} \arctan \dfrac{2x+a}{a\sqrt{3}}.$

2. $\int \dfrac{dx}{(a^3-x^3)^2} = \dfrac{x}{3a^3(a^3-x^3)} + \dfrac{2}{3a^3} \int \dfrac{dx}{a^3-x^3}.$

3. $\int \dfrac{x\,dx}{a^3-x^3} = \dfrac{1}{6a} \ln \dfrac{a^2+ax+x^2}{(a-x)^2} - \dfrac{1}{a\sqrt{3}} \arctan \dfrac{2x+a}{a\sqrt{3}}.$

4. $\int \dfrac{x\,dx}{(a^3-x^3)^2} = \dfrac{x^2}{3a^3(a^3-x^3)} + \dfrac{1}{3a^3} \int \dfrac{x\,dx}{a^3-x^3}.$

5. $\int \dfrac{x^2\,dx}{a^3-x^3} = -\dfrac{1}{3} \ln|a^3-x^3|.$

6. $\int \dfrac{dx}{x(a^3-x^3)} = \dfrac{1}{3a^3} \ln\left|\dfrac{x^3}{a^3-x^3}\right|.$

7. $\int \dfrac{dx}{x(a^3-x^3)^2} = \dfrac{1}{3a^3(a^3-x^3)} + \dfrac{1}{3a^6} \ln\left|\dfrac{x^3}{a^3-x^3}\right|.$

8. $\int \dfrac{dx}{x^2(a^3-x^3)} = -\dfrac{1}{a^3 x} + \dfrac{1}{a^3} \int \dfrac{x\,dx}{a^3-x^3}.$

9. $\int \dfrac{dx}{x^2(a^3-x^3)^2} = -\dfrac{1}{a^6 x} - \dfrac{x^2}{3a^6(a^3-x^3)} + \dfrac{4}{3a^6} \int \dfrac{x\,dx}{a^3-x^3}.$

T2.1.1-7. Integrals involving $a^4 \pm x^4$.

1. $\int \dfrac{dx}{a^4+x^4} = \dfrac{1}{4a^3\sqrt{2}} \ln \dfrac{a^2+ax\sqrt{2}+x^2}{a^2-ax\sqrt{2}+x^2} + \dfrac{1}{2a^3\sqrt{2}} \arctan \dfrac{ax\sqrt{2}}{a^2-x^2}.$

2. $\int \dfrac{x\,dx}{a^4+x^4} = \dfrac{1}{2a^2} \arctan \dfrac{x^2}{a^2}.$

3. $\int \dfrac{x^2\,dx}{a^4+x^4} = -\dfrac{1}{4a\sqrt{2}} \ln \dfrac{a^2+ax\sqrt{2}+x^2}{a^2-ax\sqrt{2}+x^2} + \dfrac{1}{2a\sqrt{2}} \arctan \dfrac{ax\sqrt{2}}{a^2-x^2}.$

4. $\int \dfrac{dx}{a^4 - x^4} = \dfrac{1}{4a^3} \ln \left| \dfrac{a+x}{a-x} \right| + \dfrac{1}{2a^3} \arctan \dfrac{x}{a}$.

5. $\int \dfrac{x\,dx}{a^4 - x^4} = \dfrac{1}{4a^2} \ln \left| \dfrac{a^2 + x^2}{a^2 - x^2} \right|$.

6. $\int \dfrac{x^2\,dx}{a^4 - x^4} = \dfrac{1}{4a} \ln \left| \dfrac{a+x}{a-x} \right| - \dfrac{1}{2a} \arctan \dfrac{x}{a}$.

## T2.1.2. Integrals Involving Irrational Functions

**T2.1.2-1. Integrals involving $x^{1/2}$.**

1. $\int \dfrac{x^{1/2}\,dx}{a^2 + b^2 x} = \dfrac{2}{b^2} x^{1/2} - \dfrac{2a}{b^3} \arctan \dfrac{bx^{1/2}}{a}$.

2. $\int \dfrac{x^{3/2}\,dx}{a^2 + b^2 x} = \dfrac{2x^{3/2}}{3b^2} - \dfrac{2a^2 x^{1/2}}{b^4} + \dfrac{2a^3}{b^5} \arctan \dfrac{bx^{1/2}}{a}$.

3. $\int \dfrac{x^{1/2}\,dx}{(a^2 + b^2 x)^2} = -\dfrac{x^{1/2}}{b^2(a^2 + b^2 x)} + \dfrac{1}{ab^3} \arctan \dfrac{bx^{1/2}}{a}$.

4. $\int \dfrac{x^{3/2}\,dx}{(a^2 + b^2 x)^2} = \dfrac{2x^{3/2}}{b^2(a^2 + b^2 x)} + \dfrac{3a^2 x^{1/2}}{b^4(a^2 + b^2 x)} - \dfrac{3a}{b^5} \arctan \dfrac{bx^{1/2}}{a}$.

5. $\int \dfrac{dx}{(a^2 + b^2 x)x^{1/2}} = \dfrac{2}{ab} \arctan \dfrac{bx^{1/2}}{a}$.

6. $\int \dfrac{dx}{(a^2 + b^2 x)x^{3/2}} = -\dfrac{2}{a^2 x^{1/2}} - \dfrac{2b}{a^3} \arctan \dfrac{bx^{1/2}}{a}$.

7. $\int \dfrac{dx}{(a^2 + b^2 x)^2 x^{1/2}} = \dfrac{x^{1/2}}{a^2(a^2 + b^2 x)} + \dfrac{1}{a^3 b} \arctan \dfrac{bx^{1/2}}{a}$.

8. $\int \dfrac{x^{1/2}\,dx}{a^2 - b^2 x} = -\dfrac{2}{b^2} x^{1/2} + \dfrac{2a}{b^3} \ln \left| \dfrac{a + bx^{1/2}}{a - bx^{1/2}} \right|$.

9. $\int \dfrac{x^{3/2}\,dx}{a^2 - b^2 x} = -\dfrac{2x^{3/2}}{3b^2} - \dfrac{2a^2 x^{1/2}}{b^4} + \dfrac{a^3}{b^5} \ln \left| \dfrac{a + bx^{1/2}}{a - bx^{1/2}} \right|$.

10. $\int \dfrac{x^{1/2}\,dx}{(a^2 - b^2 x)^2} = \dfrac{x^{1/2}}{b^2(a^2 - b^2 x)} - \dfrac{1}{2ab^3} \ln \left| \dfrac{a + bx^{1/2}}{a - bx^{1/2}} \right|$.

11. $\int \dfrac{x^{3/2}\,dx}{(a^2 - b^2 x)^2} = \dfrac{3a^2 x^{1/2} - 2b^2 x^{3/2}}{b^4(a^2 - b^2 x)} - \dfrac{3a}{2b^5} \ln \left| \dfrac{a + bx^{1/2}}{a - bx^{1/2}} \right|$.

12. $\int \dfrac{dx}{(a^2 - b^2 x)x^{1/2}} = \dfrac{1}{ab} \ln \left| \dfrac{a + bx^{1/2}}{a - bx^{1/2}} \right|$.

13. $\int \dfrac{dx}{(a^2 - b^2 x)x^{3/2}} = -\dfrac{2}{a^2 x^{1/2}} + \dfrac{b}{a^3} \ln \left| \dfrac{a + bx^{1/2}}{a - bx^{1/2}} \right|$.

14. $\int \dfrac{dx}{(a^2 - b^2 x)^2 x^{1/2}} = \dfrac{x^{1/2}}{a^2(a^2 - b^2 x)} + \dfrac{1}{2a^3 b} \ln \left| \dfrac{a + bx^{1/2}}{a - bx^{1/2}} \right|$.

### T2.1.2-2. Integrals involving $(a + bx)^{p/2}$.

1. $\int (a + bx)^{p/2}\, dx = \dfrac{2}{b(p + 2)} (a + bx)^{(p+2)/2}.$

2. $\int x(a + bx)^{p/2}\, dx = \dfrac{2}{b^2}\left[ \dfrac{(a + bx)^{(p+4)/2}}{p + 4} - \dfrac{a(a + bx)^{(p+2)/2}}{p + 2}\right].$

3. $\int x^2(a + bx)^{p/2}\, dx = \dfrac{2}{b^3}\left[ \dfrac{(a + bx)^{(p+6)/2}}{p + 6} - \dfrac{2a(a + bx)^{(p+4)/2}}{p + 4} + \dfrac{a^2(a + bx)^{(p+2)/2}}{p + 2}\right].$

### T2.1.2-3. Integrals involving $(x^2 + a^2)^{1/2}$.

1. $\int (x^2 + a^2)^{1/2}\, dx = \dfrac{1}{2} x(a^2 + x^2)^{1/2} + \dfrac{a^2}{2} \ln\left[ x + (x^2 + a^2)^{1/2}\right].$

2. $\int x(x^2 + a^2)^{1/2}\, dx = \dfrac{1}{3}(a^2 + x^2)^{3/2}.$

3. $\int (x^2 + a^2)^{3/2}\, dx = \dfrac{1}{4} x(a^2 + x^2)^{3/2} + \dfrac{3}{8} a^2 x(a^2 + x^2)^{1/2} + \dfrac{3}{8} a^4 \ln\left| x + (x^2 + a^2)^{1/2}\right|.$

4. $\int \dfrac{1}{x}(x^2 + a^2)^{1/2}\, dx = (a^2 + x^2)^{1/2} - a \ln\left| \dfrac{a + (x^2 + a^2)^{1/2}}{x}\right|.$

5. $\int \dfrac{dx}{\sqrt{x^2 + a^2}} = \ln\left[ x + (x^2 + a^2)^{1/2}\right].$

6. $\int \dfrac{x\, dx}{\sqrt{x^2 + a^2}} = (x^2 + a^2)^{1/2}.$

7. $\int (x^2 + a^2)^{-3/2}\, dx = a^{-2} x(x^2 + a^2)^{-1/2}.$

### T2.1.2-4. Integrals involving $(x^2 - a^2)^{1/2}$.

1. $\int (x^2 - a^2)^{1/2}\, dx = \dfrac{1}{2} x(x^2 - a^2)^{1/2} - \dfrac{a^2}{2} \ln\left| x + (x^2 - a^2)^{1/2}\right|.$

2. $\int x(x^2 - a^2)^{1/2}\, dx = \dfrac{1}{3}(x^2 - a^2)^{3/2}.$

3. $\int (x^2 - a^2)^{3/2}\, dx = \dfrac{1}{4} x(x^2 - a^2)^{3/2} - \dfrac{3}{8} a^2 x(x^2 - a^2)^{1/2} + \dfrac{3}{8} a^4 \ln\left| x + (x^2 - a^2)^{1/2}\right|.$

4. $\int \dfrac{1}{x}(x^2 - a^2)^{1/2}\, dx = (x^2 - a^2)^{1/2} - a \arccos\left| \dfrac{a}{x}\right|.$

5. $\int \dfrac{dx}{\sqrt{x^2 - a^2}} = \ln\left| x + (x^2 - a^2)^{1/2}\right|.$

6. $\int \dfrac{x\, dx}{\sqrt{x^2 - a^2}} = (x^2 - a^2)^{1/2}.$

7. $\int (x^2 - a^2)^{-3/2}\, dx = -a^{-2} x(x^2 - a^2)^{-1/2}.$

**T2.1.2-5. Integrals involving $(a^2 - x^2)^{1/2}$.**

1. $\displaystyle\int (a^2 - x^2)^{1/2}\, dx = \frac{1}{2}x(a^2 - x^2)^{1/2} + \frac{a^2}{2} \arcsin \frac{x}{a}.$

2. $\displaystyle\int x(a^2 - x^2)^{1/2}\, dx = -\frac{1}{3}(a^2 - x^2)^{3/2}.$

3. $\displaystyle\int (a^2 - x^2)^{3/2}\, dx = \frac{1}{4}x(a^2 - x^2)^{3/2} + \frac{3}{8}a^2 x(a^2 - x^2)^{1/2} + \frac{3}{8}a^4 \arcsin \frac{x}{a}.$

4. $\displaystyle\int \frac{1}{x}(a^2 - x^2)^{1/2}\, dx = (a^2 - x^2)^{1/2} - a \ln\left|\frac{a + (a^2 - x^2)^{1/2}}{x}\right|.$

5. $\displaystyle\int \frac{dx}{\sqrt{a^2 - x^2}} = \arcsin \frac{x}{a}.$

6. $\displaystyle\int \frac{x\, dx}{\sqrt{a^2 - x^2}} = -(a^2 - x^2)^{1/2}.$

7. $\displaystyle\int (a^2 - x^2)^{-3/2}\, dx = a^{-2} x(a^2 - x^2)^{-1/2}.$

**T2.1.2-6. Integrals involving arbitrary powers. Reduction formulas.**

1. $\displaystyle\int \frac{dx}{x(ax^n + b)} = \frac{1}{bn} \ln\left|\frac{x^n}{ax^n + b}\right|.$

2. $\displaystyle\int \frac{dx}{x\sqrt{x^n + a^2}} = \frac{2}{an} \ln\left|\frac{x^{n/2}}{\sqrt{x^n + a^2} + a}\right|.$

3. $\displaystyle\int \frac{dx}{x\sqrt{x^n - a^2}} = \frac{2}{an} \arccos\left|\frac{a}{x^{n/2}}\right|.$

4. $\displaystyle\int \frac{dx}{x\sqrt{ax^{2n} + bx^n}} = -\frac{2\sqrt{ax^{2n} + bx^n}}{bnx^n}.$

▶ *The parameters $a$, $b$, $p$, $m$, and $n$ below in Integrals 5–8 can assume arbitrary values, except for those at which denominators vanish in successive applications of a formula. Notation: $w = ax^n + b$.*

5. $\displaystyle\int x^m(ax^n + b)^p\, dx = \frac{1}{m + np + 1}\left(x^{m+1} w^p + npb \int x^m w^{p-1}\, dx\right).$

6. $\displaystyle\int x^m(ax^n + b)^p\, dx = \frac{1}{bn(p+1)}\left[-x^{m+1} w^{p+1} + (m + n + np + 1)\int x^m w^{p+1}\, dx\right].$

7. $\displaystyle\int x^m(ax^n + b)^p\, dx = \frac{1}{b(m+1)}\left[x^{m+1} w^{p+1} - a(m + n + np + 1)\int x^{m+n} w^p\, dx\right].$

8. $\displaystyle\int x^m(ax^n + b)^p\, dx = \frac{1}{a(m + np + 1)}\left[x^{m-n+1} w^{p+1} - b(m - n + 1)\int x^{m-n} w^p\, dx\right].$

## T2.1.3. Integrals Involving Exponential Functions

1. $\int e^{ax}\,dx = \dfrac{1}{a}e^{ax}$.

2. $\int a^x\,dx = \dfrac{a^x}{\ln a}$.

3. $\int xe^{ax}\,dx = e^{ax}\left(\dfrac{x}{a} - \dfrac{1}{a^2}\right)$.

4. $\int x^2 e^{ax}\,dx = e^{ax}\left(\dfrac{x^2}{a} - \dfrac{2x}{a^2} + \dfrac{2}{a^3}\right)$.

5. $\int x^n e^{ax}\,dx = e^{ax}\left[\dfrac{1}{a}x^n - \dfrac{n}{a^2}x^{n-1} + \dfrac{n(n-1)}{a^3}x^{n-2} - \cdots + (-1)^{n-1}\dfrac{n!}{a^n}x + (-1)^n\dfrac{n!}{a^{n+1}}\right]$, $n = 1, 2, \ldots$

6. $\int P_n(x)e^{ax}\,dx = e^{ax}\displaystyle\sum_{k=0}^{n}\dfrac{(-1)^k}{a^{k+1}}\dfrac{d^k}{dx^k}P_n(x)$, where $P_n(x)$ is an arbitrary polynomial of degree $n$.

7. $\int \dfrac{dx}{a + be^{px}} = \dfrac{x}{a} - \dfrac{1}{ap}\ln|a + be^{px}|$.

8. $\int \dfrac{dx}{ae^{px} + be^{-px}} = \begin{cases} \dfrac{1}{p\sqrt{ab}}\arctan\left(e^{px}\sqrt{\dfrac{a}{b}}\right) & \text{if } ab > 0, \\ \dfrac{1}{2p\sqrt{-ab}}\ln\left(\dfrac{b + e^{px}\sqrt{-ab}}{b - e^{px}\sqrt{-ab}}\right) & \text{if } ab < 0. \end{cases}$

9. $\int \dfrac{dx}{\sqrt{a + be^{px}}} = \begin{cases} \dfrac{1}{p\sqrt{a}}\ln\dfrac{\sqrt{a + be^{px}} - \sqrt{a}}{\sqrt{a + be^{px}} + \sqrt{a}} & \text{if } a > 0, \\ \dfrac{2}{p\sqrt{-a}}\arctan\dfrac{\sqrt{a + be^{px}}}{\sqrt{-a}} & \text{if } a < 0. \end{cases}$

## T2.1.4. Integrals Involving Hyperbolic Functions

T2.1.4-1. Integrals involving $\cosh x$.

1. $\int \cosh(a + bx)\,dx = \dfrac{1}{b}\sinh(a + bx)$.

2. $\int x \cosh x\,dx = x \sinh x - \cosh x$.

3. $\int x^2 \cosh x\,dx = (x^2 + 2)\sinh x - 2x \cosh x$.

4. $\int x^{2n} \cosh x\,dx = (2n)!\displaystyle\sum_{k=1}^{n}\left[\dfrac{x^{2k}}{(2k)!}\sinh x - \dfrac{x^{2k-1}}{(2k-1)!}\cosh x\right]$.

5. $\int x^{2n+1} \cosh x \, dx = (2n+1)! \sum_{k=0}^{n} \left[ \dfrac{x^{2k+1}}{(2k+1)!} \sinh x - \dfrac{x^{2k}}{(2k)!} \cosh x \right].$

6. $\int x^p \cosh x \, dx = x^p \sinh x - p x^{p-1} \cosh x + p(p-1) \int x^{p-2} \cosh x \, dx.$

7. $\int \cosh^2 x \, dx = \tfrac{1}{2} x + \tfrac{1}{4} \sinh 2x.$

8. $\int \cosh^3 x \, dx = \sinh x + \tfrac{1}{3} \sinh^3 x.$

9. $\int \cosh^{2n} x \, dx = C_{2n}^{n} \dfrac{x}{2^{2n}} + \dfrac{1}{2^{2n-1}} \sum_{k=0}^{n-1} C_{2n}^{k} \dfrac{\sinh[2(n-k)x]}{2(n-k)}, \quad n = 1, 2, \ldots$

10. $\int \cosh^{2n+1} x \, dx = \dfrac{1}{2^{2n}} \sum_{k=0}^{n} C_{2n+1}^{k} \dfrac{\sinh[(2n-2k+1)x]}{2n-2k+1} = \sum_{k=0}^{n} C_{n}^{k} \dfrac{\sinh^{2k+1} x}{2k+1},$
    $n = 1, 2, \ldots$

11. $\int \cosh^p x \, dx = \dfrac{1}{p} \sinh x \cosh^{p-1} x + \dfrac{p-1}{p} \int \cosh^{p-2} x \, dx.$

12. $\int \cosh ax \cosh bx \, dx = \dfrac{1}{a^2 - b^2} (a \cosh bx \sinh ax - b \cosh ax \sinh bx).$

13. $\int \dfrac{dx}{\cosh ax} = \dfrac{2}{a} \arctan\left(e^{ax}\right).$

14. $\int \dfrac{dx}{\cosh^{2n} x} = \dfrac{\sinh x}{2n-1} \left[ \dfrac{1}{\cosh^{2n-1} x} \right.$
    $\left. + \sum_{k=1}^{n-1} \dfrac{2^k (n-1)(n-2) \ldots (n-k)}{(2n-3)(2n-5) \ldots (2n-2k-1)} \dfrac{1}{\cosh^{2n-2k-1} x} \right], \quad n = 1, 2, \ldots$

15. $\int \dfrac{dx}{\cosh^{2n+1} x} = \dfrac{\sinh x}{2n} \left[ \dfrac{1}{\cosh^{2n} x} \right.$
    $\left. + \sum_{k=1}^{n-1} \dfrac{(2n-1)(2n-3) \ldots (2n-2k+1)}{2^k (n-1)(n-2) \ldots (n-k)} \dfrac{1}{\cosh^{2n-2k} x} \right] + \dfrac{(2n-1)!!}{(2n)!!} \operatorname{arctan} \sinh x,$
    $n = 1, 2, \ldots$

16. $\int \dfrac{dx}{a + b \cosh x} = \begin{cases} -\dfrac{\operatorname{sign} x}{\sqrt{b^2 - a^2}} \arcsin \dfrac{b + a \cosh x}{a + b \cosh x} & \text{if } a^2 < b^2, \\ \dfrac{1}{\sqrt{a^2 - b^2}} \ln \dfrac{a + b + \sqrt{a^2 - b^2} \tanh(x/2)}{a + b - \sqrt{a^2 - b^2} \tanh(x/2)} & \text{if } a^2 > b^2. \end{cases}$

T2.1.4-2. Integrals involving $\sinh x$.

1. $\int \sinh(a + bx) \, dx = \dfrac{1}{b} \cosh(a + bx).$

2. $\int x \sinh x \, dx = x \cosh x - \sinh x.$

3. $\int x^2 \sinh x \, dx = (x^2 + 2) \cosh x - 2x \sinh x.$

4. $\int x^{2n} \sinh x \, dx = (2n)! \left[ \sum_{k=0}^{n} \frac{x^{2k}}{(2k)!} \cosh x - \sum_{k=1}^{n} \frac{x^{2k-1}}{(2k-1)!} \sinh x \right].$

5. $\int x^{2n+1} \sinh x \, dx = (2n+1)! \sum_{k=0}^{n} \left[ \frac{x^{2k+1}}{(2k+1)!} \cosh x - \frac{x^{2k}}{(2k)!} \sinh x \right].$

6. $\int x^p \sinh x \, dx = x^p \cosh x - p x^{p-1} \sinh x + p(p-1) \int x^{p-2} \sinh x \, dx.$

7. $\int \sinh^2 x \, dx = -\tfrac{1}{2} x + \tfrac{1}{4} \sinh 2x.$

8. $\int \sinh^3 x \, dx = -\cosh x + \tfrac{1}{3} \cosh^3 x.$

9. $\int \sinh^{2n} x \, dx = (-1)^n C_{2n}^n \frac{x}{2^{2n}} + \frac{1}{2^{2n-1}} \sum_{k=0}^{n-1} (-1)^k C_{2n}^k \frac{\sinh[2(n-k)x]}{2(n-k)}, \quad n = 1, 2, \ldots$

10. $\int \sinh^{2n+1} x \, dx = \frac{1}{2^{2n}} \sum_{k=0}^{n} (-1)^k C_{2n+1}^k \frac{\cosh[(2n-2k+1)x]}{2n-2k+1}$
    $= \sum_{k=0}^{n} (-1)^{n+k} C_n^k \frac{\cosh^{2k+1} x}{2k+1}, \quad n = 1, 2, \ldots$

11. $\int \sinh^p x \, dx = \frac{1}{p} \sinh^{p-1} x \cosh x - \frac{p-1}{p} \int \sinh^{p-2} x \, dx.$

12. $\int \sinh ax \sinh bx \, dx = \frac{1}{a^2 - b^2} \left( a \cosh ax \sinh bx - b \cosh bx \sinh ax \right).$

13. $\int \frac{dx}{\sinh ax} = \frac{1}{a} \ln \left| \tanh \frac{ax}{2} \right|.$

14. $\int \frac{dx}{\sinh^{2n} x} = \frac{\cosh x}{2n-1} \left[ -\frac{1}{\sinh^{2n-1} x} \right.$
    $\left. + \sum_{k=1}^{n-1} (-1)^{k-1} \frac{2^k (n-1)(n-2) \ldots (n-k)}{(2n-3)(2n-5) \ldots (2n-2k-1)} \frac{1}{\sinh^{2n-2k-1} x} \right], \quad n = 1, 2, \ldots$

15. $\int \frac{dx}{\sinh^{2n+1} x} = \frac{\cosh x}{2n} \left[ -\frac{1}{\sinh^{2n} x} \right.$
    $\left. + \sum_{k=1}^{n-1} (-1)^{k-1} \frac{(2n-1)(2n-3) \ldots (2n-2k+1)}{2^k (n-1)(n-2) \ldots (n-k)} \frac{1}{\sinh^{2n-2k} x} \right] + (-1)^n \frac{(2n-1)!!}{(2n)!!} \ln \tanh \frac{x}{2},$
    $n = 1, 2, \ldots$

16. $\int \frac{dx}{a + b \sinh x} = \frac{1}{\sqrt{a^2 + b^2}} \ln \frac{a \tanh(x/2) - b + \sqrt{a^2 + b^2}}{a \tanh(x/2) - b - \sqrt{a^2 + b^2}}.$

17. $\int \frac{Ax + B \sinh x}{a + b \sinh x} dx = \frac{B}{b} x + \frac{Ab - Ba}{b \sqrt{a^2 + b^2}} \ln \frac{a \tanh(x/2) - b + \sqrt{a^2 + b^2}}{a \tanh(x/2) - b - \sqrt{a^2 + b^2}}.$

T2.1.4-3. Integrals involving $\tanh x$ or $\coth x$.

1. $\int \tanh x \, dx = \ln \cosh x.$

2. $\int \tanh^2 x \, dx = x - \tanh x.$

3. $\int \tanh^3 x \, dx = -\frac{1}{2} \tanh^2 x + \ln \cosh x.$

4. $\int \tanh^{2n} x \, dx = x - \sum_{k=1}^{n} \frac{\tanh^{2n-2k+1} x}{2n - 2k + 1}, \quad n = 1, 2, \ldots$

5. $\int \tanh^{2n+1} x \, dx = \ln \cosh x - \sum_{k=1}^{n} \frac{(-1)^k C_n^k}{2k \cosh^{2k} x} = \ln \cosh x - \sum_{k=1}^{n} \frac{\tanh^{2n-2k+2} x}{2n - 2k + 2},$
   $n = 1, 2, \ldots$

6. $\int \tanh^p x \, dx = -\frac{1}{p-1} \tanh^{p-1} x + \int \tanh^{p-2} x \, dx.$

7. $\int \coth x \, dx = \ln |\sinh x|.$

8. $\int \coth^2 x \, dx = x - \coth x.$

9. $\int \coth^3 x \, dx = -\frac{1}{2} \coth^2 x + \ln |\sinh x|.$

10. $\int \coth^{2n} x \, dx = x - \sum_{k=1}^{n} \frac{\coth^{2n-2k+1} x}{2n - 2k + 1}, \quad n = 1, 2, \ldots$

11. $\int \coth^{2n+1} x \, dx = \ln |\sinh x| - \sum_{k=1}^{n} \frac{C_n^k}{2k \sinh^{2k} x} = \ln |\sinh x| - \sum_{k=1}^{n} \frac{\coth^{2n-2k+2} x}{2n - 2k + 2},$
    $n = 1, 2, \ldots$

12. $\int \coth^p x \, dx = -\frac{1}{p-1} \coth^{p-1} x + \int \coth^{p-2} x \, dx.$

## T2.1.5. Integrals Involving Logarithmic Functions

1. $\int \ln ax \, dx = x \ln ax - x.$

2. $\int x \ln x \, dx = \frac{1}{2} x^2 \ln x - \frac{1}{4} x^2.$

3. $\int x^p \ln ax \, dx = \begin{cases} \dfrac{1}{p+1} x^{p+1} \ln ax - \dfrac{1}{(p+1)^2} x^{p+1} & \text{if } p \neq -1, \\ \frac{1}{2} \ln^2 ax & \text{if } p = -1. \end{cases}$

4. $\int (\ln x)^2 \, dx = x(\ln x)^2 - 2x \ln x + 2x.$

5. $\int x(\ln x)^2 \, dx = \frac{1}{2}x^2(\ln x)^2 - \frac{1}{2}x^2 \ln x + \frac{1}{4}x^2.$

6. $\int x^p (\ln x)^2 \, dx = \begin{cases} \dfrac{x^{p+1}}{p+1}(\ln x)^2 - \dfrac{2x^{p+1}}{(p+1)^2} \ln x + \dfrac{2x^{p+1}}{(p+1)^3} & \text{if } p \neq -1, \\ \dfrac{1}{3} \ln^3 x & \text{if } p = -1. \end{cases}$

7. $\int (\ln x)^n \, dx = \dfrac{x}{n+1} \sum_{k=0}^{n} (-1)^k (n+1)n \ldots (n-k+1)(\ln x)^{n-k}, \quad n = 1, 2, \ldots$

8. $\int (\ln x)^q \, dx = x(\ln x)^q - q \int (\ln x)^{q-1} \, dx, \quad q \neq -1.$

9. $\int x^n (\ln x)^m \, dx = \dfrac{x^{n+1}}{m+1} \sum_{k=0}^{m} \dfrac{(-1)^k}{(n+1)^{k+1}} (m+1)m \ldots (m-k+1)(\ln x)^{m-k},$
$n, m = 1, 2, \ldots$

10. $\int x^p (\ln x)^q \, dx = \dfrac{1}{p+1} x^{p+1}(\ln x)^q - \dfrac{q}{p+1} \int x^p (\ln x)^{q-1} \, dx, \quad p, q \neq -1.$

11. $\int \ln(a + bx) \, dx = \dfrac{1}{b}(ax + b)\ln(ax + b) - x.$

12. $\int x \ln(a + bx) \, dx = \dfrac{1}{2}\left(x^2 - \dfrac{a^2}{b^2}\right)\ln(a + bx) - \dfrac{1}{2}\left(\dfrac{x^2}{2} - \dfrac{a}{b}x\right).$

13. $\int x^2 \ln(a + bx) \, dx = \dfrac{1}{3}\left(x^3 - \dfrac{a^3}{b^3}\right)\ln(a + bx) - \dfrac{1}{3}\left(\dfrac{x^3}{3} - \dfrac{ax^2}{2b} + \dfrac{a^2 x}{b^2}\right).$

14. $\int \dfrac{\ln x \, dx}{(a + bx)^2} = -\dfrac{\ln x}{b(a + bx)} + \dfrac{1}{ab} \ln \dfrac{x}{a + bx}.$

15. $\int \dfrac{\ln x \, dx}{(a + bx)^3} = -\dfrac{\ln x}{2b(a + bx)^2} + \dfrac{1}{2ab(a + bx)} + \dfrac{1}{2a^2 b} \ln \dfrac{x}{a + bx}.$

16. $\int \dfrac{\ln x \, dx}{\sqrt{a + bx}} = \begin{cases} \dfrac{2}{b}\left[(\ln x - 2)\sqrt{a + bx} + \sqrt{a} \ln \dfrac{\sqrt{a + bx} + \sqrt{a}}{\sqrt{a + bx} - \sqrt{a}}\right] & \text{if } a > 0, \\ \dfrac{2}{b}\left[(\ln x - 2)\sqrt{a + bx} + 2\sqrt{-a} \arctan \dfrac{\sqrt{a + bx}}{\sqrt{-a}}\right] & \text{if } a < 0. \end{cases}$

17. $\int \ln(x^2 + a^2) \, dx = x \ln(x^2 + a^2) - 2x + 2a \arctan(x/a).$

18. $\int x \ln(x^2 + a^2) \, dx = \dfrac{1}{2}\left[(x^2 + a^2)\ln(x^2 + a^2) - x^2\right].$

19. $\int x^2 \ln(x^2 + a^2) \, dx = \dfrac{1}{3}\left[x^3 \ln(x^2 + a^2) - \dfrac{2}{3}x^3 + 2a^2 x - 2a^3 \arctan(x/a)\right].$

## T2.1.6. Integrals Involving Trigonometric Functions

**T2.1.6-1. Integrals involving $\cos x$ ($n = 1, 2, \ldots$).**

1. $\displaystyle\int \cos(a + bx)\, dx = \frac{1}{b} \sin(a + bx).$

2. $\displaystyle\int x \cos x\, dx = \cos x + x \sin x.$

3. $\displaystyle\int x^2 \cos x\, dx = 2x \cos x + (x^2 - 2) \sin x.$

4. $\displaystyle\int x^{2n} \cos x\, dx = (2n)! \left[ \sum_{k=0}^{n} (-1)^k \frac{x^{2n-2k}}{(2n-2k)!} \sin x + \sum_{k=0}^{n-1} (-1)^k \frac{x^{2n-2k-1}}{(2n-2k-1)!} \cos x \right].$

5. $\displaystyle\int x^{2n+1} \cos x\, dx = (2n+1)! \sum_{k=0}^{n} \left[ (-1)^k \frac{x^{2n-2k+1}}{(2n-2k+1)!} \sin x + \frac{x^{2n-2k}}{(2n-2k)!} \cos x \right].$

6. $\displaystyle\int x^p \cos x\, dx = x^p \sin x + p x^{p-1} \cos x - p(p-1) \int x^{p-2} \cos x\, dx.$

7. $\displaystyle\int \cos^2 x\, dx = \tfrac{1}{2} x + \tfrac{1}{4} \sin 2x.$

8. $\displaystyle\int \cos^3 x\, dx = \sin x - \tfrac{1}{3} \sin^3 x.$

9. $\displaystyle\int \cos^{2n} x\, dx = \frac{1}{2^{2n}} C_{2n}^{n} x + \frac{1}{2^{2n-1}} \sum_{k=0}^{n-1} C_{2n}^{k} \frac{\sin[(2n-2k)x]}{2n-2k}.$

10. $\displaystyle\int \cos^{2n+1} x\, dx = \frac{1}{2^{2n}} \sum_{k=0}^{n} C_{2n+1}^{k} \frac{\sin[(2n-2k+1)x]}{2n-2k+1}.$

11. $\displaystyle\int \frac{dx}{\cos x} = \ln \left| \tan \left( \frac{x}{2} + \frac{\pi}{4} \right) \right|.$

12. $\displaystyle\int \frac{dx}{\cos^2 x} = \tan x.$

13. $\displaystyle\int \frac{dx}{\cos^3 x} = \frac{\sin x}{2 \cos^2 x} + \frac{1}{2} \ln \left| \tan \left( \frac{x}{2} + \frac{\pi}{4} \right) \right|.$

14. $\displaystyle\int \frac{dx}{\cos^n x} = \frac{\sin x}{(n-1) \cos^{n-1} x} + \frac{n-2}{n-1} \int \frac{dx}{\cos^{n-2} x}, \quad n > 1.$

15. $\displaystyle\int \frac{x\, dx}{\cos^{2n} x} = \sum_{k=0}^{n-1} \frac{(2n-2)(2n-4)\ldots(2n-2k+2)}{(2n-1)(2n-3)\ldots(2n-2k+3)} \frac{(2n-2k)x \sin x - \cos x}{(2n-2k+1)(2n-2k) \cos^{2n-2k+1} x}$
$\qquad + \dfrac{2^{n-1}(n-1)!}{(2n-1)!!} \left( x \tan x + \ln |\cos x| \right).$

16. $\displaystyle\int \cos ax \cos bx\, dx = \frac{\sin[(b-a)x]}{2(b-a)} + \frac{\sin[(b+a)x]}{2(b+a)}, \quad a \neq \pm b.$

17. $\displaystyle\int \frac{dx}{a+b\cos x} = \begin{cases} \dfrac{2}{\sqrt{a^2-b^2}} \arctan \dfrac{(a-b)\tan(x/2)}{\sqrt{a^2-b^2}} & \text{if } a^2 > b^2, \\ \dfrac{1}{\sqrt{b^2-a^2}} \ln \left|\dfrac{\sqrt{b^2-a^2}+(b-a)\tan(x/2)}{\sqrt{b^2-a^2}-(b-a)\tan(x/2)}\right| & \text{if } b^2 > a^2. \end{cases}$

18. $\displaystyle\int \frac{dx}{(a+b\cos x)^2} = \frac{b\sin x}{(b^2-a^2)(a+b\cos x)} - \frac{a}{b^2-a^2}\int \frac{dx}{a+b\cos x}.$

19. $\displaystyle\int \frac{dx}{a^2+b^2\cos^2 x} = \frac{1}{a\sqrt{a^2+b^2}} \arctan \frac{a\tan x}{\sqrt{a^2+b^2}}.$

20. $\displaystyle\int \frac{dx}{a^2-b^2\cos^2 x} = \begin{cases} \dfrac{1}{a\sqrt{a^2-b^2}} \arctan \dfrac{a\tan x}{\sqrt{a^2-b^2}} & \text{if } a^2 > b^2, \\ \dfrac{1}{2a\sqrt{b^2-a^2}} \ln\left|\dfrac{\sqrt{b^2-a^2}-a\tan x}{\sqrt{b^2-a^2}+a\tan x}\right| & \text{if } b^2 > a^2. \end{cases}$

21. $\displaystyle\int e^{ax}\cos bx\, dx = e^{ax}\left(\frac{b}{a^2+b^2}\sin bx + \frac{a}{a^2+b^2}\cos bx\right).$

22. $\displaystyle\int e^{ax}\cos^2 x\, dx = \frac{e^{ax}}{a^2+4}\left(a\cos^2 x + 2\sin x\cos x + \frac{2}{a}\right).$

23. $\displaystyle\int e^{ax}\cos^n x\, dx = \frac{e^{ax}\cos^{n-1} x}{a^2+n^2}(a\cos x + n\sin x) + \frac{n(n-1)}{a^2+n^2}\int e^{ax}\cos^{n-2} x\, dx.$

**T2.1.6-2. Integrals involving $\sin x$ ($n = 1, 2, \ldots$).**

1. $\displaystyle\int \sin(a+bx)\, dx = -\frac{1}{b}\cos(a+bx).$

2. $\displaystyle\int x\sin x\, dx = \sin x - x\cos x.$

3. $\displaystyle\int x^2 \sin x\, dx = 2x\sin x - (x^2 - 2)\cos x.$

4. $\displaystyle\int x^3 \sin x\, dx = (3x^2 - 6)\sin x - (x^3 - 6x)\cos x.$

5. $\displaystyle\int x^{2n} \sin x\, dx = (2n)!\left[\sum_{k=0}^{n}(-1)^{k+1}\frac{x^{2n-2k}}{(2n-2k)!}\cos x + \sum_{k=0}^{n-1}(-1)^k \frac{x^{2n-2k-1}}{(2n-2k-1)!}\sin x\right].$

6. $\displaystyle\int x^{2n+1} \sin x\, dx = (2n+1)! \sum_{k=0}^{n}\left[(-1)^{k+1}\frac{x^{2n-2k+1}}{(2n-2k+1)!}\cos x + (-1)^k \frac{x^{2n-2k}}{(2n-2k)!}\sin x\right].$

7. $\displaystyle\int x^p \sin x\, dx = -x^p \cos x + px^{p-1}\sin x - p(p-1)\int x^{p-2}\sin x\, dx.$

8. $\displaystyle\int \sin^2 x\, dx = \tfrac{1}{2}x - \tfrac{1}{4}\sin 2x.$

9. $\int x \sin^2 x \, dx = \frac{1}{4}x^2 - \frac{1}{4}x \sin 2x - \frac{1}{8} \cos 2x.$

10. $\int \sin^3 x \, dx = -\cos x + \frac{1}{3} \cos^3 x.$

11. $\int \sin^{2n} x \, dx = \frac{1}{2^{2n}} C_{2n}^n x + \frac{(-1)^n}{2^{2n-1}} \sum_{k=0}^{n-1} (-1)^k C_{2n}^k \frac{\sin[(2n-2k)x]}{2n-2k},$

   where $C_m^k = \dfrac{m!}{k!(m-k)!}$ are binomial coefficients $(0! = 1)$.

12. $\int \sin^{2n+1} x \, dx = \frac{1}{2^{2n}} \sum_{k=0}^{n} (-1)^{n+k+1} C_{2n+1}^k \frac{\cos[(2n-2k+1)x]}{2n-2k+1}.$

13. $\int \dfrac{dx}{\sin x} = \ln \left| \tan \dfrac{x}{2} \right|.$

14. $\int \dfrac{dx}{\sin^2 x} = -\cot x.$

15. $\int \dfrac{dx}{\sin^3 x} = -\dfrac{\cos x}{2 \sin^2 x} + \dfrac{1}{2} \ln \left| \tan \dfrac{x}{2} \right|.$

16. $\int \dfrac{dx}{\sin^n x} = -\dfrac{\cos x}{(n-1) \sin^{n-1} x} + \dfrac{n-2}{n-1} \int \dfrac{dx}{\sin^{n-2} x}, \quad n > 1.$

17. $\int \dfrac{x \, dx}{\sin^{2n} x} = -\sum_{k=0}^{n-1} \dfrac{(2n-2)(2n-4)\ldots(2n-2k+2)}{(2n-1)(2n-3)\ldots(2n-2k+3)} \dfrac{\sin x + (2n-2k)x \cos x}{(2n-2k+1)(2n-2k) \sin^{2n-2k+1} x}$
   $+ \dfrac{2^{n-1}(n-1)!}{(2n-1)!!} \left( \ln |\sin x| - x \cot x \right).$

18. $\int \sin ax \sin bx \, dx = \dfrac{\sin[(b-a)x]}{2(b-a)} - \dfrac{\sin[(b+a)x]}{2(b+a)}, \quad a \neq \pm b.$

19. $\int \dfrac{dx}{a + b \sin x} = \begin{cases} \dfrac{2}{\sqrt{a^2 - b^2}} \arctan \dfrac{b + a \tan x/2}{\sqrt{a^2 - b^2}} & \text{if } a^2 > b^2, \\ \dfrac{1}{\sqrt{b^2 - a^2}} \ln \left| \dfrac{b - \sqrt{b^2 - a^2} + a \tan x/2}{b + \sqrt{b^2 - a^2} + a \tan x/2} \right| & \text{if } b^2 > a^2. \end{cases}$

20. $\int \dfrac{dx}{(a + b \sin x)^2} = \dfrac{b \cos x}{(a^2 - b^2)(a + b \sin x)} + \dfrac{a}{a^2 - b^2} \int \dfrac{dx}{a + b \sin x}.$

21. $\int \dfrac{dx}{a^2 + b^2 \sin^2 x} = \dfrac{1}{a\sqrt{a^2 + b^2}} \arctan \dfrac{\sqrt{a^2 + b^2} \tan x}{a}.$

22. $\int \dfrac{dx}{a^2 - b^2 \sin^2 x} = \begin{cases} \dfrac{1}{a\sqrt{a^2 - b^2}} \arctan \dfrac{\sqrt{a^2 - b^2} \tan x}{a} & \text{if } a^2 > b^2, \\ \dfrac{1}{2a\sqrt{b^2 - a^2}} \ln \left| \dfrac{\sqrt{b^2 - a^2} \tan x + a}{\sqrt{b^2 - a^2} \tan x - a} \right| & \text{if } b^2 > a^2. \end{cases}$

23. $\int \dfrac{\sin x \, dx}{\sqrt{1 + k^2 \sin^2 x}} = -\dfrac{1}{k} \arcsin \dfrac{k \cos x}{\sqrt{1 + k^2}}.$

24. $\displaystyle\int \frac{\sin x\, dx}{\sqrt{1-k^2\sin^2 x}} = -\frac{1}{k}\ln\left|k\cos x + \sqrt{1-k^2\sin^2 x}\,\right|.$

25. $\displaystyle\int \sin x\sqrt{1+k^2\sin^2 x}\, dx = -\frac{\cos x}{2}\sqrt{1+k^2\sin^2 x} - \frac{1+k^2}{2k}\arcsin\frac{k\cos x}{\sqrt{1+k^2}}.$

26. $\displaystyle\int \sin x\sqrt{1-k^2\sin^2 x}\, dx$
$$= -\frac{\cos x}{2}\sqrt{1-k^2\sin^2 x} - \frac{1-k^2}{2k}\ln\left|k\cos x + \sqrt{1-k^2\sin^2 x}\,\right|.$$

27. $\displaystyle\int e^{ax}\sin bx\, dx = e^{ax}\left(\frac{a}{a^2+b^2}\sin bx - \frac{b}{a^2+b^2}\cos bx\right).$

28. $\displaystyle\int e^{ax}\sin^2 x\, dx = \frac{e^{ax}}{a^2+4}\left(a\sin^2 x - 2\sin x\cos x + \frac{2}{a}\right).$

29. $\displaystyle\int e^{ax}\sin^n x\, dx = \frac{e^{ax}\sin^{n-1} x}{a^2+n^2}(a\sin x - n\cos x) + \frac{n(n-1)}{a^2+n^2}\int e^{ax}\sin^{n-2} x\, dx.$

**T2.1.6-3. Integrals involving $\sin x$ and $\cos x$.**

1. $\displaystyle\int \sin ax\cos bx\, dx = -\frac{\cos[(a+b)x]}{2(a+b)} - \frac{\cos[(a-b)x]}{2(a-b)},\quad a\neq\pm b.$

2. $\displaystyle\int \frac{dx}{b^2\cos^2 ax + c^2\sin^2 ax} = \frac{1}{abc}\arctan\left(\frac{c}{b}\tan ax\right).$

3. $\displaystyle\int \frac{dx}{b^2\cos^2 ax - c^2\sin^2 ax} = \frac{1}{2abc}\ln\left|\frac{c\tan ax + b}{c\tan ax - b}\right|.$

4. $\displaystyle\int \frac{dx}{\cos^{2n} x\sin^{2m} x} = \sum_{k=0}^{n+m-1} C_{n+m-1}^{k}\frac{\tan^{2k-2m+1} x}{2k-2m+1},\quad n,m=1,2,\ldots$

5. $\displaystyle\int \frac{dx}{\cos^{2n+1} x\sin^{2m+1} x} = C_{n+m}^{m}\ln|\tan x| + \sum_{k=0}^{n+m} C_{n+m}^{k}\frac{\tan^{2k-2m} x}{2k-2m},\quad n,m=1,2,\ldots$

**T2.1.6-4. Reduction formulas.**

▶ The parameters $p$ and $q$ below can assume any values, except for those at which the denominators on the right-hand side vanish.

1. $\displaystyle\int \sin^p x\cos^q x\, dx = -\frac{\sin^{p-1} x\cos^{q+1} x}{p+q} + \frac{p-1}{p+q}\int \sin^{p-2} x\cos^q x\, dx.$

2. $\displaystyle\int \sin^p x\cos^q x\, dx = \frac{\sin^{p+1} x\cos^{q-1} x}{p+q} + \frac{q-1}{p+q}\int \sin^p x\cos^{q-2} x\, dx.$

3. $\displaystyle\int \sin^p x\cos^q x\, dx = \frac{\sin^{p-1} x\cos^{q-1} x}{p+q}\left(\sin^2 x - \frac{q-1}{p+q-2}\right)$
$$+ \frac{(p-1)(q-1)}{(p+q)(p+q-2)}\int \sin^{p-2} x\cos^{q-2} x\, dx.$$

4. $\displaystyle\int \sin^p x \cos^q x\, dx = \frac{\sin^{p+1} x \cos^{q+1} x}{p+1} + \frac{p+q+2}{p+1} \int \sin^{p+2} x \cos^q x\, dx.$

5. $\displaystyle\int \sin^p x \cos^q x\, dx = -\frac{\sin^{p+1} x \cos^{q+1} x}{q+1} + \frac{p+q+2}{q+1} \int \sin^p x \cos^{q+2} x\, dx.$

6. $\displaystyle\int \sin^p x \cos^q x\, dx = -\frac{\sin^{p-1} x \cos^{q+1} x}{q+1} + \frac{p-1}{q+1} \int \sin^{p-2} x \cos^{q+2} x\, dx.$

7. $\displaystyle\int \sin^p x \cos^q x\, dx = \frac{\sin^{p+1} x \cos^{q-1} x}{p+1} + \frac{q-1}{p+1} \int \sin^{p+2} x \cos^{q-2} x\, dx.$

T2.1.6-5. Integrals involving $\tan x$ and $\cot x$.

1. $\displaystyle\int \tan x\, dx = -\ln|\cos x|.$

2. $\displaystyle\int \tan^2 x\, dx = \tan x - x.$

3. $\displaystyle\int \tan^3 x\, dx = \tfrac{1}{2} \tan^2 x + \ln|\cos x|.$

4. $\displaystyle\int \tan^{2n} x\, dx = (-1)^n x - \sum_{k=1}^{n} \frac{(-1)^k (\tan x)^{2n-2k+1}}{2n-2k+1}, \quad n = 1, 2, \ldots$

5. $\displaystyle\int \tan^{2n+1} x\, dx = (-1)^{n+1} \ln|\cos x| - \sum_{k=1}^{n} \frac{(-1)^k (\tan x)^{2n-2k+2}}{2n-2k+2}, \quad n = 1, 2, \ldots$

6. $\displaystyle\int \frac{dx}{a + b\tan x} = \frac{1}{a^2+b^2}\big(ax + b\ln|a\cos x + b\sin x|\big).$

7. $\displaystyle\int \frac{\tan x\, dx}{\sqrt{a + b\tan^2 x}} = \frac{1}{\sqrt{b-a}} \arccos\left(\sqrt{1 - \frac{a}{b}} \cos x\right), \quad b > a,\ b > 0.$

8. $\displaystyle\int \cot x\, dx = \ln|\sin x|.$

9. $\displaystyle\int \cot^2 x\, dx = -\cot x - x.$

10. $\displaystyle\int \cot^3 x\, dx = -\tfrac{1}{2} \cot^2 x - \ln|\sin x|.$

11. $\displaystyle\int \cot^{2n} x\, dx = (-1)^n x + \sum_{k=1}^{n} \frac{(-1)^k (\cot x)^{2n-2k+1}}{2n-2k+1}, \quad n = 1, 2, \ldots$

12. $\displaystyle\int \cot^{2n+1} x\, dx = (-1)^n \ln|\sin x| + \sum_{k=1}^{n} \frac{(-1)^k (\cot x)^{2n-2k+2}}{2n-2k+2}, \quad n = 1, 2, \ldots$

13. $\displaystyle\int \frac{dx}{a + b\cot x} = \frac{1}{a^2+b^2}\big(ax - b\ln|a\sin x + b\cos x|\big).$

## T2.1.7. Integrals Involving Inverse Trigonometric Functions

1. $\int \arcsin \dfrac{x}{a}\, dx = x \arcsin \dfrac{x}{a} + \sqrt{a^2 - x^2}$.

2. $\int \left(\arcsin \dfrac{x}{a}\right)^2 dx = x \left(\arcsin \dfrac{x}{a}\right)^2 - 2x + 2\sqrt{a^2 - x^2}\, \arcsin \dfrac{x}{a}$.

3. $\int x \arcsin \dfrac{x}{a}\, dx = \dfrac{1}{4}(2x^2 - a^2) \arcsin \dfrac{x}{a} + \dfrac{x}{4}\sqrt{a^2 - x^2}$.

4. $\int x^2 \arcsin \dfrac{x}{a}\, dx = \dfrac{x^3}{3} \arcsin \dfrac{x}{a} + \dfrac{1}{9}(x^2 + 2a^2)\sqrt{a^2 - x^2}$.

5. $\int \arccos \dfrac{x}{a}\, dx = x \arccos \dfrac{x}{a} - \sqrt{a^2 - x^2}$.

6. $\int \left(\arccos \dfrac{x}{a}\right)^2 dx = x \left(\arccos \dfrac{x}{a}\right)^2 - 2x - 2\sqrt{a^2 - x^2}\, \arccos \dfrac{x}{a}$.

7. $\int x \arccos \dfrac{x}{a}\, dx = \dfrac{1}{4}(2x^2 - a^2) \arccos \dfrac{x}{a} - \dfrac{x}{4}\sqrt{a^2 - x^2}$.

8. $\int x^2 \arccos \dfrac{x}{a}\, dx = \dfrac{x^3}{3} \arccos \dfrac{x}{a} - \dfrac{1}{9}(x^2 + 2a^2)\sqrt{a^2 - x^2}$.

9. $\int \arctan \dfrac{x}{a}\, dx = x \arctan \dfrac{x}{a} - \dfrac{a}{2} \ln(a^2 + x^2)$.

10. $\int x \arctan \dfrac{x}{a}\, dx = \dfrac{1}{2}(x^2 + a^2) \arctan \dfrac{x}{a} - \dfrac{ax}{2}$.

11. $\int x^2 \arctan \dfrac{x}{a}\, dx = \dfrac{x^3}{3} \arctan \dfrac{x}{a} - \dfrac{ax^2}{6} + \dfrac{a^3}{6} \ln(a^2 + x^2)$.

12. $\int \text{arccot}\, \dfrac{x}{a}\, dx = x\, \text{arccot}\, \dfrac{x}{a} + \dfrac{a}{2} \ln(a^2 + x^2)$.

13. $\int x\, \text{arccot}\, \dfrac{x}{a}\, dx = \dfrac{1}{2}(x^2 + a^2)\, \text{arccot}\, \dfrac{x}{a} + \dfrac{ax}{2}$.

14. $\int x^2\, \text{arccot}\, \dfrac{x}{a}\, dx = \dfrac{x^3}{3}\, \text{arccot}\, \dfrac{x}{a} + \dfrac{ax^2}{6} - \dfrac{a^3}{6} \ln(a^2 + x^2)$.

# T2.2. Tables of Definite Integrals

▶ *Throughout Section T2.2 it is assumed that $n$ is a positive integer, unless otherwise specified.*

## T2.2.1. Integrals Involving Power-Law Functions

**T2.2.1-1. Integrals over a finite interval.**

1. $\displaystyle\int_0^1 \dfrac{x^n\, dx}{x + 1} = (-1)^n \left[ \ln 2 + \sum_{k=1}^{n} \dfrac{(-1)^k}{k} \right]$.

2. $\int_0^1 \dfrac{dx}{x^2 + 2x\cos\beta + 1} = \dfrac{\beta}{2\sin\beta}.$

3. $\int_0^1 \dfrac{(x^a + x^{-a})\,dx}{x^2 + 2x\cos\beta + 1} = \dfrac{\pi\sin(a\beta)}{\sin(\pi a)\sin\beta}, \quad |a| < 1,\ \beta \neq (2n+1)\pi.$

4. $\int_0^1 x^a(1-x)^{1-a}\,dx = \dfrac{\pi a(1-a)}{2\sin(\pi a)}, \quad -1 < a < 1.$

5. $\int_0^1 \dfrac{dx}{x^a(1-x)^{1-a}} = \dfrac{\pi}{\sin(\pi a)}, \quad 0 < a < 1.$

6. $\int_0^1 \dfrac{x^a\,dx}{(1-x)^a} = \dfrac{\pi a}{\sin(\pi a)}, \quad -1 < a < 1.$

7. $\int_0^1 x^{p-1}(1-x)^{q-1}\,dx \equiv B(p,q) = \dfrac{\Gamma(p)\Gamma(q)}{\Gamma(p+q)}, \quad p,q > 0.$

8. $\int_0^1 x^{p-1}(1-x^q)^{-p/q}\,dx = \dfrac{\pi}{q\sin(\pi p/q)}, \quad q > p > 0.$

9. $\int_0^1 x^{p+q-1}(1-x^q)^{-p/q}\,dx = \dfrac{\pi p}{q^2 \sin(\pi p/q)}, \quad q > p.$

10. $\int_0^1 x^{q/p-1}(1-x^q)^{-1/p}\,dx = \dfrac{\pi}{q\sin(\pi/p)}, \quad p > 1,\ q > 0.$

11. $\int_0^1 \dfrac{x^{p-1} - x^{-p}}{1 - x}\,dx = \pi\cot(\pi p), \quad |p| < 1.$

12. $\int_0^1 \dfrac{x^{p-1} - x^{-p}}{1 + x}\,dx = \dfrac{\pi}{\sin(\pi p)}, \quad |p| < 1.$

13. $\int_0^1 \dfrac{x^p - x^{-p}}{x - 1}\,dx = \dfrac{1}{p} - \pi\cot(\pi p), \quad |p| < 1.$

14. $\int_0^1 \dfrac{x^p - x^{-p}}{1 + x}\,dx = \dfrac{1}{p} - \dfrac{\pi}{\sin(\pi p)}, \quad |p| < 1.$

15. $\int_0^1 \dfrac{x^{1+p} - x^{1-p}}{1 - x^2}\,dx = \dfrac{\pi}{2}\cot\left(\dfrac{\pi p}{2}\right) - \dfrac{1}{p}, \quad |p| < 1.$

16. $\int_0^1 \dfrac{x^{1+p} - x^{1-p}}{1 + x^2}\,dx = \dfrac{1}{p} - \dfrac{\pi}{2\sin(\pi p/2)}, \quad |p| < 1.$

17. $\int_0^1 \dfrac{dx}{\sqrt{(1 + a^2 x)(1 - x)}} = \dfrac{2}{a}\arctan a.$

18. $\int_0^1 \dfrac{dx}{\sqrt{(1 - a^2 x)(1 - x)}} = \dfrac{1}{a}\ln\dfrac{1 + a}{1 - a}.$

19. $\int_{-1}^1 \dfrac{dx}{(a - x)\sqrt{1 - x^2}} = \dfrac{\pi}{\sqrt{a^2 - 1}}, \quad 1 < a.$

20. $\int_0^1 \dfrac{x^n\,dx}{\sqrt{1-x}} = \dfrac{2(2n)!!}{(2n+1)!!}, \quad n = 1, 2, \ldots$

21. $\int_0^1 \dfrac{x^{n-1/2}\,dx}{\sqrt{1-x}} = \dfrac{\pi(2n-1)!!}{(2n)!!}, \quad n = 1, 2, \ldots$

22. $\int_0^1 \dfrac{x^{2n}\,dx}{\sqrt{1-x^2}} = \dfrac{\pi}{2}\,\dfrac{1\times 3\times\ldots\times(2n-1)}{2\times 4\times\ldots\times(2n)}, \quad n = 1, 2, \ldots$

23. $\int_0^1 \dfrac{x^{2n+1}\,dx}{\sqrt{1-x^2}} = \dfrac{2\times 4\times\ldots\times(2n)}{1\times 3\times\ldots\times(2n+1)}, \quad n = 1, 2, \ldots$

24. $\int_0^1 \dfrac{x^{\lambda-1}\,dx}{(1+ax)(1-x)^\lambda} = \dfrac{\pi}{(1+a)^\lambda \sin(\pi\lambda)}, \quad 0 < \lambda < 1, \quad a > -1.$

25. $\int_0^1 \dfrac{x^{\lambda-1/2}\,dx}{(1+ax)^\lambda(1-x)^\lambda} = 2\pi^{-1/2}\Gamma\!\left(\lambda+\tfrac{1}{2}\right)\Gamma(1-\lambda)\cos^{2\lambda} k\,\dfrac{\sin[(2\lambda-1)k]}{(2\lambda-1)\sin k},$
$k = \arctan\sqrt{a}, \quad -\tfrac{1}{2} < \lambda < 1, \quad a > 0.$

**T2.2.1-2. Integrals over an infinite interval.**

1. $\int_0^\infty \dfrac{dx}{ax^2 + b} = \dfrac{\pi}{2\sqrt{ab}}.$

2. $\int_0^\infty \dfrac{dx}{x^4 + 1} = \dfrac{\pi\sqrt{2}}{4}.$

3. $\int_0^\infty \dfrac{x^{a-1}\,dx}{x+1} = \dfrac{\pi}{\sin(\pi a)}, \quad 0 < a < 1.$

4. $\int_0^\infty \dfrac{x^{\lambda-1}\,dx}{(1+ax)^2} = \dfrac{\pi(1-\lambda)}{a^\lambda \sin(\pi\lambda)}, \quad 0 < \lambda < 2.$

5. $\int_0^\infty \dfrac{x^{\lambda-1}\,dx}{(x+a)(x+b)} = \dfrac{\pi(a^{\lambda-1} - b^{\lambda-1})}{(b-a)\sin(\pi\lambda)}, \quad 0 < \lambda < 2.$

6. $\int_0^\infty \dfrac{x^{\lambda-1}(x+c)\,dx}{(x+a)(x+b)} = \dfrac{\pi}{\sin(\pi\lambda)}\left(\dfrac{a-c}{a-b}a^{\lambda-1} + \dfrac{b-c}{b-a}b^{\lambda-1}\right), \quad 0 < \lambda < 1.$

7. $\int_0^\infty \dfrac{x^\lambda\,dx}{(x+1)^3} = \dfrac{\pi\lambda(1-\lambda)}{2\sin(\pi\lambda)}, \quad -1 < \lambda < 2.$

8. $\int_0^\infty \dfrac{x^{\lambda-1}\,dx}{(x^2+a^2)(x^2+b^2)} = \dfrac{\pi\left(b^{\lambda-2} - a^{\lambda-2}\right)}{2(a^2-b^2)\sin(\pi\lambda/2)}, \quad 0 < \lambda < 4.$

9. $\int_0^\infty \dfrac{x^{p-1} - x^{q-1}}{1-x}\,dx = \pi[\cot(\pi p) - \cot(\pi q)], \quad p, q > 0.$

10. $\int_0^\infty \dfrac{x^{\lambda-1}\,dx}{(1+ax)^{n+1}} = (-1)^n \dfrac{\pi C_{\lambda-1}^n}{a^\lambda \sin(\pi\lambda)}, \quad 0 < \lambda < n+1, \quad C_{\lambda-1}^n = \dfrac{(\lambda-1)(\lambda-2)\ldots(\lambda-n)}{n!}.$

11. $\int_0^\infty \frac{x^m\,dx}{(a+bx)^{n+1/2}} = 2^{m+1} m! \frac{(2n-2m-3)!!}{(2n-1)!!} \frac{a^{m-n+1/2}}{b^{m+1}},$
    $a, b > 0, \quad n, m = 1, 2, \ldots, \quad m < b - \frac{1}{2}.$

12. $\int_0^\infty \frac{dx}{(x^2+a^2)^n} = \frac{\pi}{2} \frac{(2n-3)!!}{(2n-2)!!} \frac{1}{a^{2n-1}}, \quad n = 1, 2, \ldots$

13. $\int_0^\infty \frac{(x+1)^{\lambda-1}}{(x+a)^{\lambda+1}}\,dx = \frac{1-a^{-\lambda}}{\lambda(a-1)}, \quad a > 0.$

14. $\int_0^\infty \frac{x^{a-1}\,dx}{x^b+1} = \frac{\pi}{b\sin(\pi a/b)}, \quad 0 < a \le b.$

15. $\int_0^\infty \frac{x^{a-1}\,dx}{(x^b+1)^2} = \frac{\pi(a-b)}{b^2 \sin[\pi(a-b)/b]}, \quad a < 2b.$

16. $\int_0^\infty \frac{x^{\lambda-1/2}\,dx}{(x+a)^\lambda(x+b)^\lambda} = \sqrt{\pi}(\sqrt{a}+\sqrt{b})^{1-2\lambda} \frac{\Gamma(\lambda-1/2)}{\Gamma(\lambda)}, \quad \lambda > 0.$

17. $\int_0^\infty \frac{1-x^a}{1-x^b} x^{c-1}\,dx = \frac{\pi \sin A}{b \sin C \sin(A+C)}, \quad A = \frac{\pi a}{b}, \quad C = \frac{\pi c}{b}; \quad a+c<b, \quad c>0.$

18. $\int_0^\infty \frac{x^{a-1}\,dx}{(1+x^2)^{1-b}} = \tfrac{1}{2} B\!\left(\tfrac{1}{2}a, 1-b-\tfrac{1}{2}a\right), \quad \tfrac{1}{2}a + b < 1, \quad a > 0.$

19. $\int_0^\infty \frac{x^{2m}\,dx}{(ax^2+b)^n} = \frac{\pi(2m-1)!!(2n-2m-3)!!}{2(2n-2)!!\,a^m b^{n-m-1}\sqrt{ab}}, \quad a, b > 0, \quad n > m+1.$

20. $\int_0^\infty \frac{x^{2m+1}\,dx}{(ax^2+b)^n} = \frac{m!(n-m-2)!}{2(n-1)!a^{m+1}b^{n-m-1}}, \quad ab > 0, \quad n > m+1 \ge 1.$

21. $\int_0^\infty \frac{x^{\mu-1}\,dx}{(1+ax^p)^\nu} = \frac{1}{pa^{\mu/p}} B\!\left(\frac{\mu}{p}, \nu - \frac{\mu}{p}\right), \quad p > 0, \quad 0 < \mu < p\nu.$

22. $\int_0^\infty \left(\sqrt{x^2+a^2}-x\right)^n dx = \frac{na^{n+1}}{n^2-1}, \quad n = 2, 3, \ldots$

23. $\int_0^\infty \frac{dx}{\left(x+\sqrt{x^2+a^2}\right)^n} = \frac{n}{a^{n-1}(n^2-1)}, \quad n = 2, 3, \ldots$

24. $\int_0^\infty x^m \left(\sqrt{x^2+a^2}-x\right)^n dx = \frac{m!\,na^{n+m+1}}{(n-m-1)(n-m+1)\ldots(n+m+1)},$
    $n, m = 1, 2, \ldots, \quad 0 \le m \le n-2.$

25. $\int_0^\infty \frac{x^m\,dx}{\left(x+\sqrt{x^2+a^2}\right)^n} = \frac{m!\,n}{(n-m-1)(n-m+1)\ldots(n+m+1)a^{n-m-1}}, \quad n=2, 3, \ldots$

## T2.2.2. Integrals Involving Exponential Functions

1. $\int_0^\infty e^{-ax}\,dx = \frac{1}{a}, \quad a > 0.$

2. $\int_0^1 x^n e^{-ax}\, dx = \dfrac{n!}{a^{n+1}} - e^{-a} \sum_{k=0}^{n} \dfrac{n!}{k!} \dfrac{1}{a^{n-k+1}}, \qquad a > 0, \ n = 1, 2, \ldots$

3. $\int_0^\infty x^n e^{-ax}\, dx = \dfrac{n!}{a^{n+1}}, \qquad a > 0, \ n = 1, 2, \ldots$

4. $\int_0^\infty \dfrac{e^{-ax}}{\sqrt{x}}\, dx = \sqrt{\dfrac{\pi}{a}}, \qquad a > 0.$

5. $\int_0^\infty x^{\nu-1} e^{-\mu x}\, dx = \dfrac{\Gamma(\nu)}{\mu^\nu}, \qquad \mu, \nu > 0.$

6. $\int_0^\infty \dfrac{dx}{1 + e^{ax}} = \dfrac{\ln 2}{a}.$

7. $\int_0^\infty \dfrac{x^{2n-1}\, dx}{e^{px} - 1} = (-1)^{n-1} \left(\dfrac{2\pi}{p}\right)^{2n} \dfrac{B_{2n}}{4n}, \qquad n = 1, 2, \ldots;$
the $B_m$ are Bernoulli numbers.

8. $\int_0^\infty \dfrac{x^{2n-1}\, dx}{e^{px} + 1} = (1 - 2^{1-2n}) \left(\dfrac{2\pi}{p}\right)^{2n} \dfrac{|B_{2n}|}{4n}, \qquad n = 1, 2, \ldots$

9. $\int_{-\infty}^\infty \dfrac{e^{-px}\, dx}{1 + e^{-qx}} = \dfrac{\pi}{q \sin(\pi p/q)}, \qquad q > p > 0 \text{ or } 0 > p > q.$

10. $\int_0^\infty \dfrac{e^{ax} + e^{-ax}}{e^{bx} + e^{-bx}}\, dx = \dfrac{\pi}{2b \cos\left(\dfrac{\pi a}{2b}\right)}, \qquad b > a.$

11. $\int_0^\infty \dfrac{e^{-px} - e^{-qx}}{1 - e^{-(p+q)x}}\, dx = \dfrac{\pi}{p+q} \cot \dfrac{\pi p}{p+q}, \qquad p, q > 0.$

12. $\int_0^\infty \left(1 - e^{-\beta x}\right)^\nu e^{-\mu x}\, dx = \dfrac{1}{\beta} B\left(\dfrac{\mu}{\beta}, \nu + 1\right).$

13. $\int_0^\infty \exp(-ax^2)\, dx = \dfrac{1}{2}\sqrt{\dfrac{\pi}{a}}, \qquad a > 0.$

14. $\int_0^\infty x^{2n+1} \exp(-ax^2)\, dx = \dfrac{n!}{2a^{n+1}}, \qquad a > 0, \ n = 1, 2, \ldots$

15. $\int_0^\infty x^{2n} \exp(-ax^2)\, dx = \dfrac{1 \times 3 \times \ldots \times (2n-1)\sqrt{\pi}}{2^{n+1} a^{n+1/2}}, \qquad a > 0, \ n = 1, 2, \ldots$

16. $\int_{-\infty}^\infty \exp(-a^2 x^2 \pm bx)\, dx = \dfrac{\sqrt{\pi}}{|a|} \exp\left(\dfrac{b^2}{4a^2}\right).$

17. $\int_0^\infty \exp\left(-ax^2 - \dfrac{b}{x^2}\right) dx = \dfrac{1}{2}\sqrt{\dfrac{\pi}{a}} \exp(-2\sqrt{ab}), \qquad a, b > 0.$

18. $\int_0^\infty \exp(-x^a)\, dx = \dfrac{1}{a} \Gamma\left(\dfrac{1}{a}\right), \qquad a > 0.$

### T2.2.3. Integrals Involving Hyperbolic Functions

1. $\int_0^\infty \dfrac{dx}{\cosh ax} = \dfrac{\pi}{2|a|}$.

2. $\int_0^\infty \dfrac{dx}{a + b\cosh x} = \begin{cases} \dfrac{2}{\sqrt{b^2 - a^2}} \arctan \dfrac{\sqrt{b^2 - a^2}}{a + b} & \text{if } |b| > |a|, \\ \dfrac{1}{\sqrt{a^2 - b^2}} \ln \dfrac{a + b + \sqrt{a^2 - b^2}}{a + b - \sqrt{a^2 + b^2}} & \text{if } |b| < |a|. \end{cases}$

3. $\int_0^\infty \dfrac{x^{2n}\, dx}{\cosh ax} = \left(\dfrac{\pi}{2a}\right)^{2n+1} |E_{2n}|, \quad a > 0;$ the $B_m$ are Bernoulli numbers.

4. $\int_0^\infty \dfrac{x^{2n}}{\cosh^2 ax}\, dx = \dfrac{\pi^{2n}(2^{2n} - 2)}{a(2a)^{2n}} |B_{2n}|, \quad a > 0$.

5. $\int_0^\infty \dfrac{\cosh ax}{\cosh bx}\, dx = \dfrac{\pi}{2b \cos\left(\dfrac{\pi a}{2b}\right)}, \quad b > |a|$.

6. $\int_0^\infty x^{2n} \dfrac{\cosh ax}{\cosh bx}\, dx = \dfrac{\pi}{2b} \dfrac{d^{2n}}{da^{2n}} \dfrac{1}{\cos\left(\tfrac{1}{2}\pi a/b\right)}, \quad b > |a|, \quad n = 1, 2, \ldots$

7. $\int_0^\infty \dfrac{\cosh ax \cosh bx}{\cosh(cx)}\, dx = \dfrac{\pi}{c} \cdot \dfrac{\cos\left(\dfrac{\pi a}{2c}\right) \cos\left(\dfrac{\pi b}{2c}\right)}{\cos\left(\dfrac{\pi a}{c}\right) + \cos\left(\dfrac{\pi b}{c}\right)}, \quad c > |a| + |b|$.

8. $\int_0^\infty \dfrac{x\, dx}{\sinh ax} = \dfrac{\pi^2}{2a^2}, \quad a > 0$.

9. $\int_0^\infty \dfrac{dx}{a + b\sinh x} = \dfrac{1}{\sqrt{a^2 + b^2}} \ln \dfrac{a + b + \sqrt{a^2 + b^2}}{a + b - \sqrt{a^2 + b^2}}, \quad ab \neq 0$.

10. $\int_0^\infty \dfrac{\sinh ax}{\sinh bx}\, dx = \dfrac{\pi}{2b} \tan\left(\dfrac{\pi a}{2b}\right), \quad b > |a|$.

11. $\int_0^\infty x^{2n} \dfrac{\sinh ax}{\sinh bx}\, dx = \dfrac{\pi}{2b} \dfrac{d^{2n}}{dx^{2n}} \tan\left(\dfrac{\pi a}{2b}\right), \quad b > |a|, \quad n = 1, 2, \ldots$

12. $\int_0^\infty \dfrac{x^{2n}}{\sinh^2 ax}\, dx = \dfrac{\pi^{2n}}{a^{2n+1}} |B_{2n}|, \quad a > 0$.

### T2.2.4. Integrals Involving Logarithmic Functions

1. $\int_0^1 x^{a-1} \ln^n x\, dx = (-1)^n n!\, a^{-n-1}, \quad a > 0, \quad n = 1, 2, \ldots$

2. $\int_0^1 \dfrac{\ln x}{x + 1}\, dx = -\dfrac{\pi^2}{12}$.

3. $\int_0^1 \dfrac{x^n \ln x}{x + 1}\, dx = (-1)^{n+1} \left[\dfrac{\pi^2}{12} + \sum_{k=1}^n \dfrac{(-1)^k}{k^2}\right], \quad n = 1, 2, \ldots$

4. $\int_0^1 \dfrac{x^{\mu-1} \ln x}{x+a}\,dx = \dfrac{\pi a^{\mu-1}}{\sin(\pi\mu)}\bigl[\ln a - \pi \cot(\pi\mu)\bigr], \quad 0 < \mu < 1.$

5. $\int_0^1 |\ln x|^{\mu}\,dx = \Gamma(\mu+1), \quad \mu > -1.$

6. $\int_0^\infty x^{\mu-1} \ln(1+ax)\,dx = \dfrac{\pi}{\mu a^{\mu} \sin(\pi\mu)}, \quad -1 < \mu < 0.$

7. $\int_0^1 x^{2n-1} \ln(1+x)\,dx = \dfrac{1}{2n}\sum_{k=1}^{2n} \dfrac{(-1)^{k-1}}{k}, \quad n = 1, 2, \ldots$

8. $\int_0^1 x^{2n} \ln(1+x)\,dx = \dfrac{1}{2n+1}\left[\ln 4 + \sum_{k=1}^{2n+1} \dfrac{(-1)^k}{k}\right], \quad n = 0, 1, \ldots$

9. $\int_0^1 x^{n-1/2} \ln(1+x)\,dx = \dfrac{2\ln 2}{2n+1} + \dfrac{4(-1)^n}{2n+1}\left[\pi - \sum_{k=0}^{n} \dfrac{(-1)^k}{2k+1}\right], \quad n = 1, 2, \ldots$

10. $\int_0^\infty \ln \dfrac{a^2+x^2}{b^2+x^2}\,dx = \pi(a-b), \quad a, b > 0.$

11. $\int_0^\infty \dfrac{x^{p-1} \ln x}{1+x^q}\,dx = -\dfrac{\pi^2 \cos(\pi p/q)}{q^2 \sin^2(\pi p/q)}, \quad 0 < p < q.$

12. $\int_0^\infty e^{-\mu x} \ln x\,dx = -\dfrac{1}{\mu}(\mathcal{C} + \ln \mu), \quad \mu > 0, \quad \mathcal{C} = 0.5772\ldots$

## T2.2.5. Integrals Involving Trigonometric Functions

**T2.2.5-1. Integrals over a finite interval.**

1. $\int_0^{\pi/2} \cos^{2n} x\,dx = \dfrac{\pi}{2} \dfrac{1 \times 3 \times \ldots \times (2n-1)}{2 \times 4 \times \ldots \times (2n)}, \quad n = 1, 2, \ldots$

2. $\int_0^{\pi/2} \cos^{2n+1} x\,dx = \dfrac{2 \times 4 \times \ldots \times (2n)}{1 \times 3 \times \ldots \times (2n+1)}, \quad n = 1, 2, \ldots$

3. $\int_0^{\pi/2} x \cos^n x\,dx = -\sum_{k=0}^{m-1} \dfrac{(n-2k+1)(n-2k+3)\ldots(n-1)}{(n-2k)(n-2k+2)\ldots n} \dfrac{1}{n-2k}$
$+ \begin{cases} \dfrac{\pi}{2} \dfrac{(2m-2)!!}{(2m-1)!!} & \text{if } n = 2m-1, \\ \dfrac{\pi^2}{8} \dfrac{(2m-1)!!}{(2m)!!} & \text{if } n = 2m, \end{cases} \quad m = 1, 2, \ldots$

4. $\int_0^{\pi} \dfrac{dx}{(a+b\cos x)^{n+1}} = \dfrac{\pi}{2^n(a+b)^n \sqrt{a^2-b^2}} \sum_{k=0}^{n} \dfrac{(2n-2k-1)!!\,(2k-1)!!}{(n-k)!\,k!} \left(\dfrac{a+b}{a-b}\right)^k, \quad a > |b|.$

5. $\int_0^{\pi/2} \sin^{2n} x\,dx = \dfrac{\pi}{2} \dfrac{1 \times 3 \times \ldots \times (2n-1)}{2 \times 4 \times \ldots \times (2n)}, \quad n = 1, 2, \ldots$

6. $\int_0^{\pi/2} \sin^{2n+1} x\, dx = \dfrac{2 \times 4 \times \ldots \times (2n)}{1 \times 3 \times \ldots \times (2n+1)}$, $\quad n = 1, 2, \ldots$

7. $\int_0^\pi x \sin^\mu x\, dx = \dfrac{\pi^2}{2^{\mu+1}} \dfrac{\Gamma(\mu+1)}{\left[\Gamma\left(\mu+\frac{1}{2}\right)\right]^2}$, $\quad \mu > -1$.

8. $\int_0^{\pi/2} \dfrac{\sin x\, dx}{\sqrt{1-k^2 \sin^2 x}} = \dfrac{1}{2k} \ln \dfrac{1+k}{1-k}$.

9. $\int_0^{\pi/2} \sin^{2n+1} x \cos^{2m+1} x\, dx = \dfrac{n!\, m!}{2(n+m+1)!}$, $\quad n, m = 1, 2, \ldots$

10. $\int_0^{\pi/2} \sin^{p-1} x \cos^{q-1} x\, dx = \tfrac{1}{2} B\left(\tfrac{1}{2}p, \tfrac{1}{2}q\right)$.

11. $\int_0^{2\pi} (a \sin x + b \cos x)^{2n}\, dx = 2\pi \dfrac{(2n-1)!!}{(2n)!!} \left(a^2 + b^2\right)^n$, $\quad n = 1, 2, \ldots$

12. $\int_0^\pi \dfrac{\sin x\, dx}{\sqrt{a^2 + 1 - 2a \cos x}} = \begin{cases} 2 & \text{if } 0 \leq a \leq 1, \\ 2/a & \text{if } 1 < a. \end{cases}$

13. $\int_0^{\pi/2} (\tan x)^{\pm \lambda}\, dx = \dfrac{\pi}{2 \cos\left(\frac{1}{2}\pi \lambda\right)}$, $\quad |\lambda| < 1$.

T2.2.5-2. Integrals over an infinite interval.

1. $\int_0^\infty \dfrac{\cos ax}{\sqrt{x}}\, dx = \sqrt{\dfrac{\pi}{2a}}$, $\quad a > 0$.

2. $\int_0^\infty \dfrac{\cos ax - \cos bx}{x}\, dx = \ln\left|\dfrac{b}{a}\right|$, $\quad ab \neq 0$.

3. $\int_0^\infty \dfrac{\cos ax - \cos bx}{x^2}\, dx = \tfrac{1}{2}\pi(b-a)$, $\quad a, b \geq 0$.

4. $\int_0^\infty x^{\mu-1} \cos ax\, dx = a^{-\mu} \Gamma(\mu) \cos\left(\tfrac{1}{2}\pi\mu\right)$, $\quad a > 0, \; 0 < \mu < 1$.

5. $\int_0^\infty \dfrac{\cos ax}{b^2 + x^2}\, dx = \dfrac{\pi}{2b} e^{-ab}$, $\quad a, b > 0$.

6. $\int_0^\infty \dfrac{\cos ax}{b^4 + x^4}\, dx = \dfrac{\pi\sqrt{2}}{4b^3} \exp\left(-\dfrac{ab}{\sqrt{2}}\right) \left[\cos\left(\dfrac{ab}{\sqrt{2}}\right) + \sin\left(\dfrac{ab}{\sqrt{2}}\right)\right]$, $\quad a, b > 0$.

7. $\int_0^\infty \dfrac{\cos ax}{(b^2 + x^2)^2}\, dx = \dfrac{\pi}{4b^3}(1+ab)e^{-ab}$, $\quad a, b > 0$.

8. $\int_0^\infty \dfrac{\cos ax\, dx}{(b^2 + x^2)(c^2 + x^2)} = \dfrac{\pi\left(be^{-ac} - ce^{-ab}\right)}{2bc\left(b^2 - c^2\right)}$, $\quad a, b, c > 0$.

9. $\int_0^\infty \cos(ax^2)\, dx = \dfrac{1}{2}\sqrt{\dfrac{\pi}{2a}}$, $\quad a > 0$.

10. $\int_0^\infty \cos(ax^p)\, dx = \frac{\Gamma(1/p)}{pa^{1/p}} \cos \frac{\pi}{2p}, \quad a > 0, \quad p > 1.$

11. $\int_0^\infty \frac{\sin ax}{x}\, dx = \frac{\pi}{2} \operatorname{sign} a.$

12. $\int_0^\infty \frac{\sin^2 ax}{x^2}\, dx = \frac{\pi}{2}|a|.$

13. $\int_0^\infty \frac{\sin ax}{\sqrt{x}}\, dx = \sqrt{\frac{\pi}{2a}}, \quad a > 0.$

14. $\int_0^\infty x^{\mu-1} \sin ax\, dx = a^{-\mu}\Gamma(\mu) \sin\left(\tfrac{1}{2}\pi\mu\right), \quad a > 0, \quad 0 < \mu < 1.$

15. $\int_0^\infty \sin(ax^2)\, dx = \frac{1}{2}\sqrt{\frac{\pi}{2a}}, \quad a > 0.$

16. $\int_0^\infty \sin(ax^p)\, dx = \frac{\Gamma(1/p)}{pa^{1/p}} \sin \frac{\pi}{2p}, \quad a > 0, \quad p > 1.$

17. $\int_0^\infty \frac{\sin x \cos ax}{x}\, dx = \begin{cases} \frac{\pi}{2} & \text{if } |a| < 1, \\ \frac{\pi}{4} & \text{if } |a| = 1, \\ 0 & \text{if } 1 < |a|. \end{cases}$

18. $\int_0^\infty \frac{\tan ax}{x}\, dx = \frac{\pi}{2} \operatorname{sign} a.$

19. $\int_0^\infty e^{-ax} \sin bx\, dx = \frac{b}{a^2 + b^2}, \quad a > 0.$

20. $\int_0^\infty e^{-ax} \cos bx\, dx = \frac{a}{a^2 + b^2}, \quad a > 0.$

21. $\int_0^\infty \exp(-ax^2) \cos bx\, dx = \frac{1}{2}\sqrt{\frac{\pi}{a}} \exp\left(-\frac{b^2}{4a}\right).$

22. $\int_0^\infty \cos(ax^2) \cos bx\, dx = \sqrt{\frac{\pi}{8a}} \left[\cos\left(\frac{b^2}{4a}\right) + \sin\left(\frac{b^2}{4a}\right)\right], \quad a, b > 0.$

23. $\int_0^\infty (\cos ax + \sin ax) \cos(b^2 x^2)\, dx = \frac{1}{b}\sqrt{\frac{\pi}{8}} \exp\left(-\frac{a^2}{2b}\right), \quad a, b > 0.$

24. $\int_0^\infty [\cos ax + \sin ax] \sin(b^2 x^2)\, dx = \frac{1}{b}\sqrt{\frac{\pi}{8}} \exp\left(-\frac{a^2}{2b}\right), \quad a, b > 0.$

## References for Chapter T2

**Bronshtein, I. N. and Semendyayev, K. A.**, *Handbook of Mathematics, 4th Edition*, Springer-Verlag, Berlin, 2004.

**Dwight, H. B.**, *Tables of Integrals and Other Mathematical Data*, Macmillan, New York, 1961.

**Gradshteyn, I. S. and Ryzhik, I. M.**, *Tables of Integrals, Series, and Products, 6th Edition*, Academic Press, New York, 2000.

**Prudnikov, A. P., Brychkov, Yu. A., and Marichev, O. I.**, *Integrals and Series, Vol. 1, Elementary Functions*, Gordon & Breach, New York, 1986.

**Prudnikov, A. P., Brychkov, Yu. A., and Marichev, O. I.**, *Integrals and Series, Vol. 2, Special Functions*, Gordon & Breach, New York, 1986.

**Prudnikov, A. P., Brychkov, Yu. A., and Marichev, O. I.**, *Integrals and Series, Vol. 3, More Special Functions*, Gordon & Breach, New York, 1988.

**Zwillinger, D.**, *CRC Standard Mathematical Tables and Formulae, 31st Edition*, CRC Press, Boca Raton, 2002.

# Chapter T3
# Integral Transforms

## T3.1. Tables of Laplace Transforms
### T3.1.1. General Formulas

| No. | Original function, $f(x)$ | Laplace transform, $\widetilde{f}(p) = \int_0^\infty e^{-px} f(x)\,dx$ |
|---|---|---|
| 1 | $af_1(x) + bf_2(x)$ | $a\widetilde{f}_1(p) + b\widetilde{f}_2(p)$ |
| 2 | $f(x/a)$, $a>0$ | $a\widetilde{f}(ap)$ |
| 3 | $\begin{cases} 0 & \text{if } 0 < x < a, \\ f(x-a) & \text{if } a < x \end{cases}$ | $e^{-ap}\widetilde{f}(p)$ |
| 4 | $x^n f(x);\ n = 1, 2, \ldots$ | $(-1)^n \dfrac{d^n}{dp^n}\widetilde{f}(p)$ |
| 5 | $\dfrac{1}{x}f(x)$ | $\int_p^\infty \widetilde{f}(q)\,dq$ |
| 6 | $e^{ax}f(x)$ | $\widetilde{f}(p-a)$ |
| 7 | $\sinh(ax)f(x)$ | $\tfrac{1}{2}\left[\widetilde{f}(p-a) - \widetilde{f}(p+a)\right]$ |
| 8 | $\cosh(ax)f(x)$ | $\tfrac{1}{2}\left[\widetilde{f}(p-a) + \widetilde{f}(p+a)\right]$ |
| 9 | $\sin(\omega x)f(x)$ | $-\tfrac{i}{2}\left[\widetilde{f}(p-i\omega) - \widetilde{f}(p+i\omega)\right]$, $i^2 = -1$ |
| 10 | $\cos(\omega x)f(x)$ | $\tfrac{1}{2}\left[\widetilde{f}(p-i\omega) + \widetilde{f}(p+i\omega)\right]$, $i^2 = -1$ |
| 11 | $f(x^2)$ | $\dfrac{1}{\sqrt{\pi}}\int_0^\infty \exp\!\left(-\dfrac{p^2}{4t^2}\right)\widetilde{f}(t^2)\,dt$ |
| 12 | $x^{a-1}f\!\left(\dfrac{1}{x}\right)$, $a>-1$ | $\int_0^\infty (t/p)^{a/2} J_a(2\sqrt{pt})\,\widetilde{f}(t)\,dt$ |
| 13 | $f(a\sinh x)$, $a>0$ | $\int_0^\infty J_p(at)\widetilde{f}(t)\,dt$ |
| 14 | $f(x+a) = f(x)$ (periodic function) | $\dfrac{1}{1-e^{ap}}\int_0^a f(x)e^{-px}\,dx$ |
| 15 | $f(x+a) = -f(x)$ (antiperiodic function) | $\dfrac{1}{1+e^{-ap}}\int_0^a f(x)e^{-px}\,dx$ |
| 16 | $f'_x(x)$ | $p\widetilde{f}(p) - f(+0)$ |
| 17 | $f^{(n)}_x(x)$ | $p^n \widetilde{f}(p) - \sum_{k=1}^n p^{n-k} f^{(k-1)}_x(+0)$ |

| No. | Original function, $f(x)$ | Laplace transform, $\widetilde{f}(p) = \int_0^\infty e^{-px} f(x)\, dx$ |
|---|---|---|
| 18 | $x^m f_x^{(n)}(x), \quad m \geq n$ | $\left(-\dfrac{d}{dp}\right)^m \left[p^n \widetilde{f}(p)\right]$ |
| 19 | $\dfrac{d^n}{dx^n}\left[x^m f(x)\right], \quad m \geq n$ | $(-1)^m p^n \dfrac{d^m}{dp^m} \widetilde{f}(p)$ |
| 20 | $\displaystyle\int_0^x f(t)\, dt$ | $\dfrac{\widetilde{f}(p)}{p}$ |
| 21 | $\displaystyle\int_0^x (x-t) f(t)\, dt$ | $\dfrac{1}{p^2} \widetilde{f}(p)$ |
| 22 | $\displaystyle\int_0^x (x-t)^\nu f(t)\, dt, \quad \nu > -1$ | $\Gamma(\nu+1) p^{-\nu-1} \widetilde{f}(p)$ |
| 23 | $\displaystyle\int_0^x e^{-a(x-t)} f(t)\, dt$ | $\dfrac{1}{p+a} \widetilde{f}(p)$ |
| 24 | $\displaystyle\int_0^x \sinh\left[a(x-t)\right] f(t)\, dt$ | $\dfrac{a \widetilde{f}(p)}{p^2 - a^2}$ |
| 25 | $\displaystyle\int_0^x \sin\left[a(x-t)\right] f(t)\, dt$ | $\dfrac{a \widetilde{f}(p)}{p^2 + a^2}$ |
| 26 | $\displaystyle\int_0^x f_1(t) f_2(x-t)\, dt$ | $\widetilde{f}_1(p) \widetilde{f}_2(p)$ |
| 27 | $\displaystyle\int_0^x \dfrac{1}{t} f(t)\, dt$ | $\dfrac{1}{p} \displaystyle\int_p^\infty \widetilde{f}(q)\, dq$ |
| 28 | $\displaystyle\int_x^\infty \dfrac{1}{t} f(t)\, dt$ | $\dfrac{1}{p} \displaystyle\int_0^p \widetilde{f}(q)\, dq$ |
| 29 | $\displaystyle\int_0^\infty \dfrac{1}{\sqrt{t}} \sin\left(2\sqrt{xt}\right) f(t)\, dt$ | $\dfrac{\sqrt{\pi}}{p\sqrt{p}} \widetilde{f}\!\left(\dfrac{1}{p}\right)$ |
| 30 | $\dfrac{1}{\sqrt{x}} \displaystyle\int_0^\infty \cos\left(2\sqrt{xt}\right) f(t)\, dt$ | $\dfrac{\sqrt{\pi}}{\sqrt{p}} \widetilde{f}\!\left(\dfrac{1}{p}\right)$ |
| 31 | $\displaystyle\int_0^\infty \dfrac{1}{\sqrt{\pi x}} \exp\!\left(-\dfrac{t^2}{4x}\right) f(t)\, dt$ | $\dfrac{1}{\sqrt{p}} \widetilde{f}(\sqrt{p})$ |
| 32 | $\displaystyle\int_0^\infty \dfrac{t}{2\sqrt{\pi x^3}} \exp\!\left(-\dfrac{t^2}{4x}\right) f(t)\, dt$ | $\widetilde{f}(\sqrt{p})$ |
| 33 | $f(x) - a \displaystyle\int_0^x f\!\left(\sqrt{x^2 - t^2}\right) J_1(at)\, dt$ | $\widetilde{f}\!\left(\sqrt{p^2 + a^2}\right)$ |
| 34 | $f(x) + a \displaystyle\int_0^x f\!\left(\sqrt{x^2 - t^2}\right) I_1(at)\, dt$ | $\widetilde{f}\!\left(\sqrt{p^2 - a^2}\right)$ |

## T3.1.2. Expressions with Power-Law Functions

| No. | Original function, $f(x)$ | Laplace transform, $\widetilde{f}(p) = \int_0^\infty e^{-px} f(x)\, dx$ |
|---|---|---|
| 1 | $1$ | $\dfrac{1}{p}$ |
| 2 | $\begin{cases} 0 & \text{if } 0 < x < a, \\ 1 & \text{if } a < x < b, \\ 0 & \text{if } b < x \end{cases}$ | $\dfrac{1}{p}\left(e^{-ap} - e^{-bp}\right)$ |
| 3 | $x$ | $\dfrac{1}{p^2}$ |
| 4 | $\dfrac{1}{x+a}$ | $-e^{ap}\operatorname{Ei}(-ap)$ |
| 5 | $x^n, \quad n = 1, 2, \ldots$ | $\dfrac{n!}{p^{n+1}}$ |
| 6 | $x^{n-1/2}, \quad n = 1, 2, \ldots$ | $\dfrac{1 \times 3 \times \ldots \times (2n-1)\sqrt{\pi}}{2^n p^{n+1/2}}$ |
| 7 | $\dfrac{1}{\sqrt{x+a}}$ | $\sqrt{\dfrac{\pi}{p}}\, e^{ap}\operatorname{erfc}\left(\sqrt{ap}\right)$ |
| 8 | $\dfrac{\sqrt{x}}{x+a}$ | $\sqrt{\dfrac{\pi}{p}} - \pi\sqrt{a}\, e^{ap}\operatorname{erfc}\left(\sqrt{ap}\right)$ |
| 9 | $(x+a)^{-3/2}$ | $2a^{-1/2} - 2(\pi p)^{1/2} e^{ap}\operatorname{erfc}\left(\sqrt{ap}\right)$ |
| 10 | $x^{1/2}(x+a)^{-1}$ | $(\pi/p)^{1/2} - \pi a^{1/2} e^{ap}\operatorname{erfc}\left(\sqrt{ap}\right)$ |
| 11 | $x^{-1/2}(x+a)^{-1}$ | $\pi a^{-1/2} e^{ap}\operatorname{erfc}\left(\sqrt{ap}\right)$ |
| 12 | $x^\nu, \quad \nu > -1$ | $\Gamma(\nu+1) p^{-\nu-1}$ |
| 13 | $(x+a)^\nu, \quad \nu > -1$ | $p^{-\nu-1} e^{-ap} \Gamma(\nu+1, ap)$ |
| 14 | $x^\nu (x+a)^{-1}, \quad \nu > -1$ | $k e^{ap} \Gamma(-\nu, ap), \qquad k = a^\nu \Gamma(\nu+1)$ |
| 15 | $(x^2 + 2ax)^{-1/2}(x+a)$ | $a e^{ap} K_1(ap)$ |

## T3.1.3. Expressions with Exponential Functions

| No. | Original function, $f(x)$ | Laplace transform, $\widetilde{f}(p) = \int_0^\infty e^{-px} f(x)\, dx$ |
|---|---|---|
| 1 | $e^{-ax}$ | $(p+a)^{-1}$ |
| 2 | $xe^{-ax}$ | $(p+a)^{-2}$ |
| 3 | $x^{\nu-1} e^{-ax}, \quad \nu > 0$ | $\Gamma(\nu)(p+a)^{-\nu}$ |
| 4 | $\dfrac{1}{x}\left(e^{-ax} - e^{-bx}\right)$ | $\ln(p+b) - \ln(p+a)$ |
| 5 | $\dfrac{1}{x^2}\left(1 - e^{-ax}\right)^2$ | $(p+2a)\ln(p+2a) + p\ln p - 2(p+a)\ln(p+a)$ |
| 6 | $\exp(-ax^2), \quad a > 0$ | $(\pi b)^{1/2} \exp(bp^2) \operatorname{erfc}(p\sqrt{b}), \qquad a = \dfrac{1}{4b}$ |
| 7 | $x\exp(-ax^2)$ | $2b - 2\pi^{1/2} b^{3/2} p \operatorname{erfc}(p\sqrt{b}), \qquad a = \dfrac{1}{4b}$ |

| No. | Original function, $f(x)$ | Laplace transform, $\widetilde{f}(p) = \int_0^\infty e^{-px} f(x)\,dx$ |
|---|---|---|
| 8 | $\exp(-a/x)$, $\quad a \geq 0$ | $2\sqrt{a/p}\, K_1(2\sqrt{ap})$ |
| 9 | $\sqrt{x}\exp(-a/x)$, $\quad a \geq 0$ | $\frac{1}{2}\sqrt{\pi/p^3}\,(1 + 2\sqrt{ap})\exp(-2\sqrt{ap})$ |
| 10 | $\dfrac{1}{\sqrt{x}}\exp(-a/x)$, $\quad a \geq 0$ | $\sqrt{\pi/p}\exp(-2\sqrt{ap})$ |
| 11 | $\dfrac{1}{x\sqrt{x}}\exp(-a/x)$, $\quad a > 0$ | $\sqrt{\pi/a}\exp(-2\sqrt{ap})$ |
| 12 | $x^{\nu-1}\exp(-a/x)$, $\quad a > 0$ | $2(a/p)^{\nu/2} K_\nu(2\sqrt{ap})$ |
| 13 | $\exp(-2\sqrt{ax})$ | $p^{-1} - (\pi a)^{1/2} p^{-3/2} e^{a/p}\operatorname{erfc}(\sqrt{a/p})$ |
| 14 | $\dfrac{1}{\sqrt{x}}\exp(-2\sqrt{ax})$ | $(\pi/p)^{1/2} e^{a/p}\operatorname{erfc}(\sqrt{a/p})$ |

## T3.1.4. Expressions with Hyperbolic Functions

| No. | Original function, $f(x)$ | Laplace transform, $\widetilde{f}(p) = \int_0^\infty e^{-px} f(x)\,dx$ |
|---|---|---|
| 1 | $\sinh(ax)$ | $\dfrac{a}{p^2 - a^2}$ |
| 2 | $\sinh^2(ax)$ | $\dfrac{2a^2}{p^3 - 4a^2 p}$ |
| 3 | $\dfrac{1}{x}\sinh(ax)$ | $\dfrac{1}{2}\ln\dfrac{p+a}{p-a}$ |
| 4 | $x^{\nu-1}\sinh(ax)$, $\quad \nu > -1$ | $\frac{1}{2}\Gamma(\nu)\left[(p-a)^{-\nu} - (p+a)^{-\nu}\right]$ |
| 5 | $\sinh(2\sqrt{ax})$ | $\dfrac{\sqrt{\pi a}}{p\sqrt{p}} e^{a/p}$ |
| 6 | $\sqrt{x}\sinh(2\sqrt{ax})$ | $\pi^{1/2} p^{-5/2}\left(\frac{1}{2}p + a\right) e^{a/p}\operatorname{erf}(\sqrt{a/p}) - a^{1/2} p^{-2}$ |
| 7 | $\dfrac{1}{\sqrt{x}}\sinh(2\sqrt{ax})$ | $\pi^{1/2} p^{-1/2} e^{a/p}\operatorname{erf}(\sqrt{a/p})$ |
| 8 | $\dfrac{1}{\sqrt{x}}\sinh^2(\sqrt{ax})$ | $\frac{1}{2}\pi^{1/2} p^{-1/2}\left(e^{a/p} - 1\right)$ |
| 9 | $\cosh(ax)$ | $\dfrac{p}{p^2 - a^2}$ |
| 10 | $\cosh^2(ax)$ | $\dfrac{p^2 - 2a^2}{p^3 - 4a^2 p}$ |
| 11 | $x^{\nu-1}\cosh(ax)$, $\quad \nu > 0$ | $\frac{1}{2}\Gamma(\nu)\left[(p-a)^{-\nu} + (p+a)^{-\nu}\right]$ |
| 12 | $\cosh(2\sqrt{ax})$ | $\dfrac{1}{p} + \dfrac{\sqrt{\pi a}}{p\sqrt{p}} e^{a/p}\operatorname{erf}(\sqrt{a/p})$ |
| 13 | $\sqrt{x}\cosh(2\sqrt{ax})$ | $\pi^{1/2} p^{-5/2}\left(\frac{1}{2}p + a\right) e^{a/p}$ |
| 14 | $\dfrac{1}{\sqrt{x}}\cosh(2\sqrt{ax})$ | $\pi^{1/2} p^{-1/2} e^{a/p}$ |
| 15 | $\dfrac{1}{\sqrt{x}}\cosh^2(\sqrt{ax})$ | $\frac{1}{2}\pi^{1/2} p^{-1/2}\left(e^{a/p} + 1\right)$ |

## T3.1.5. Expressions with Logarithmic Functions

| No. | Original function, $f(x)$ | Laplace transform, $\widetilde{f}(p) = \int_0^\infty e^{-px} f(x)\, dx$ |
|---|---|---|
| 1 | $\ln x$ | $-\dfrac{1}{p}(\ln p + \mathcal{C})$, $\mathcal{C} = 0.5772\ldots$ is the Euler constant |
| 2 | $\ln(1 + ax)$ | $-\dfrac{1}{p} e^{p/a} \operatorname{Ei}(-p/a)$ |
| 3 | $\ln(x + a)$ | $\dfrac{1}{p}\left[\ln a - e^{ap} \operatorname{Ei}(-ap)\right]$ |
| 4 | $x^n \ln x, \quad n = 1, 2, \ldots$ | $\dfrac{n!}{p^{n+1}}\left(1 + \tfrac{1}{2} + \tfrac{1}{3} + \cdots + \tfrac{1}{n} - \ln p - \mathcal{C}\right)$, $\mathcal{C} = 0.5772\ldots$ is the Euler constant |
| 5 | $\dfrac{1}{\sqrt{x}} \ln x$ | $-\sqrt{\pi/p}\left[\ln(4p) + \mathcal{C}\right]$ |
| 6 | $x^{n-1/2} \ln x, \quad n = 1, 2, \ldots$ | $\dfrac{k_n}{p^{n+1/2}}\left[2 + \tfrac{2}{3} + \tfrac{2}{5} + \cdots + \tfrac{2}{2n-1} - \ln(4p) - \mathcal{C}\right]$, $k_n = 1 \times 3 \times 5 \times \ldots \times (2n-1) \dfrac{\sqrt{\pi}}{2^n}$, $\mathcal{C} = 0.5772\ldots$ |
| 7 | $x^{\nu-1} \ln x, \quad \nu > 0$ | $\Gamma(\nu) p^{-\nu}\left[\psi(\nu) - \ln p\right]$, $\psi(\nu)$ is the logarithmic derivative of the gamma function |
| 8 | $(\ln x)^2$ | $\dfrac{1}{p}\left[(\ln x + \mathcal{C})^2 + \tfrac{1}{6}\pi^2\right]$, $\mathcal{C} = 0.5772\ldots$ |
| 9 | $e^{-ax} \ln x$ | $-\dfrac{\ln(p+a) + \mathcal{C}}{p + a}$, $\mathcal{C} = 0.5772\ldots$ |

## T3.1.6. Expressions with Trigonometric Functions

| No. | Original function, $f(x)$ | Laplace transform, $\widetilde{f}(p) = \int_0^\infty e^{-px} f(x)\, dx$ |
|---|---|---|
| 1 | $\sin(ax)$ | $\dfrac{a}{p^2 + a^2}$ |
| 2 | $|\sin(ax)|, \quad a > 0$ | $\dfrac{a}{p^2 + a^2} \coth\left(\dfrac{\pi p}{2a}\right)$ |
| 3 | $\sin^{2n}(ax), \quad n = 1, 2, \ldots$ | $\dfrac{a^{2n}(2n)!}{p\left[p^2 + (2a)^2\right]\left[p^2 + (4a)^2\right] \ldots \left[p^2 + (2na)^2\right]}$ |
| 4 | $\sin^{2n+1}(ax), \quad n = 1, 2, \ldots$ | $\dfrac{a^{2n+1}(2n+1)!}{\left[p^2 + a^2\right]\left[p^2 + 3^2 a^2\right] \ldots \left[p^2 + (2n+1)^2 a^2\right]}$ |
| 5 | $x^n \sin(ax), \quad n = 1, 2, \ldots$ | $\dfrac{n!\, p^{n+1}}{(p^2 + a^2)^{n+1}} \displaystyle\sum_{0 \le 2k \le n} (-1)^k C_{n+1}^{2k+1}\left(\dfrac{a}{p}\right)^{2k+1}$ |
| 6 | $\dfrac{1}{x} \sin(ax)$ | $\arctan\left(\dfrac{a}{p}\right)$ |
| 7 | $\dfrac{1}{x} \sin^2(ax)$ | $\tfrac{1}{4} \ln\left(1 + 4a^2 p^{-2}\right)$ |

| No. | Original function, $f(x)$ | Laplace transform, $\widetilde{f}(p) = \int_0^\infty e^{-px} f(x)\, dx$ |
|---|---|---|
| 8 | $\dfrac{1}{x^2}\sin^2(ax)$ | $a\arctan(2a/p) - \tfrac{1}{4}p\ln(1+4a^2p^{-2})$ |
| 9 | $\sin(2\sqrt{ax})$ | $\dfrac{\sqrt{\pi a}}{p\sqrt{p}}e^{-a/p}$ |
| 10 | $\dfrac{1}{x}\sin(2\sqrt{ax})$ | $\pi\operatorname{erf}(\sqrt{a/p})$ |
| 11 | $\cos(ax)$ | $\dfrac{p}{p^2+a^2}$ |
| 12 | $\cos^2(ax)$ | $\dfrac{p^2+2a^2}{p(p^2+4a^2)}$ |
| 13 | $x^n\cos(ax),\quad n=1,2,\ldots$ | $\dfrac{n!\,p^{n+1}}{(p^2+a^2)^{n+1}}\displaystyle\sum_{0\le 2k\le n+1}(-1)^k C_{n+1}^{2k}\left(\dfrac{a}{p}\right)^{2k}$ |
| 14 | $\dfrac{1}{x}[1-\cos(ax)]$ | $\tfrac{1}{2}\ln(1+a^2p^{-2})$ |
| 15 | $\dfrac{1}{x}[\cos(ax)-\cos(bx)]$ | $\dfrac{1}{2}\ln\dfrac{p^2+b^2}{p^2+a^2}$ |
| 16 | $\sqrt{x}\cos(2\sqrt{ax})$ | $\tfrac{1}{2}\pi^{1/2}p^{-5/2}(p-2a)e^{-a/p}$ |
| 17 | $\dfrac{1}{\sqrt{x}}\cos(2\sqrt{ax})$ | $\sqrt{\pi/p}\,e^{-a/p}$ |
| 18 | $\sin(ax)\sin(bx)$ | $\dfrac{2abp}{[p^2+(a+b)^2][p^2+(a-b)^2]}$ |
| 19 | $\cos(ax)\sin(bx)$ | $\dfrac{b(p^2-a^2+b^2)}{[p^2+(a+b)^2][p^2+(a-b)^2]}$ |
| 20 | $\cos(ax)\cos(bx)$ | $\dfrac{p(p^2+a^2+b^2)}{[p^2+(a+b)^2][p^2+(a-b)^2]}$ |
| 21 | $\dfrac{ax\cos(ax)-\sin(ax)}{x^2}$ | $p\arctan\dfrac{a}{x}-a$ |
| 22 | $e^{bx}\sin(ax)$ | $\dfrac{a}{(p-b)^2+a^2}$ |
| 23 | $e^{bx}\cos(ax)$ | $\dfrac{p-b}{(p-b)^2+a^2}$ |
| 24 | $\sin(ax)\sinh(ax)$ | $\dfrac{2a^2p}{p^4+4a^4}$ |
| 25 | $\sin(ax)\cosh(ax)$ | $\dfrac{a(p^2+2a^2)}{p^4+4a^4}$ |
| 26 | $\cos(ax)\sinh(ax)$ | $\dfrac{a(p^2-2a^2)}{p^4+4a^4}$ |
| 27 | $\cos(ax)\cosh(ax)$ | $\dfrac{p^3}{p^4+4a^4}$ |

## T3.1.7. Expressions with Special Functions

| No. | Original function, $f(x)$ | Laplace transform, $\widetilde{f}(p) = \int_0^\infty e^{-px} f(x)\,dx$ |
|---|---|---|
| 1 | $\operatorname{erf}(ax)$ | $\dfrac{1}{p}\exp(b^2 p^2)\operatorname{erfc}(bp), \qquad b=\dfrac{1}{2a}$ |
| 2 | $\operatorname{erf}(\sqrt{ax})$ | $\dfrac{\sqrt{a}}{p\sqrt{p+a}}$ |
| 3 | $e^{ax}\operatorname{erf}(\sqrt{ax})$ | $\dfrac{\sqrt{a}}{\sqrt{p}(p-a)}$ |
| 4 | $\operatorname{erf}(\tfrac{1}{2}\sqrt{a/x})$ | $\dfrac{1}{p}\left[1-\exp(-\sqrt{ap})\right]$ |
| 5 | $\operatorname{erfc}(\sqrt{ax})$ | $\dfrac{\sqrt{p+a}-\sqrt{a}}{p\sqrt{p+a}}$ |
| 6 | $e^{ax}\operatorname{erfc}(\sqrt{ax})$ | $\dfrac{1}{p+\sqrt{ap}}$ |
| 7 | $\operatorname{erfc}(\tfrac{1}{2}\sqrt{a/x})$ | $\dfrac{1}{p}\exp(-\sqrt{ap})$ |
| 8 | $\operatorname{Ci}(x)$ | $\dfrac{1}{2p}\ln(p^2+1)$ |
| 9 | $\operatorname{Si}(x)$ | $\dfrac{1}{p}\operatorname{arccot} p$ |
| 10 | $\operatorname{Ei}(-x)$ | $-\dfrac{1}{p}\ln(p+1)$ |
| 11 | $J_0(ax)$ | $\dfrac{1}{\sqrt{p^2+a^2}}$ |
| 12 | $J_\nu(ax), \qquad \nu>-1$ | $\dfrac{a^\nu}{\sqrt{p^2+a^2}\left(p+\sqrt{p^2+a^2}\right)^\nu}$ |
| 13 | $x^n J_n(ax), \qquad n=1,2,\ldots$ | $1\times 3\times 5\times\ldots\times(2n-1)a^n(p^2+a^2)^{-n-1/2}$ |
| 14 | $x^\nu J_\nu(ax), \qquad \nu>-\tfrac{1}{2}$ | $2^\nu \pi^{-1/2}\Gamma(\nu+\tfrac{1}{2})a^\nu(p^2+a^2)^{-\nu-1/2}$ |
| 15 | $x^{\nu+1} J_\nu(ax), \qquad \nu>-1$ | $2^{\nu+1}\pi^{-1/2}\Gamma(\nu+\tfrac{3}{2})a^\nu p(p^2+a^2)^{-\nu-3/2}$ |
| 16 | $J_0(2\sqrt{ax})$ | $\dfrac{1}{p}e^{-a/p}$ |
| 17 | $\sqrt{x}J_1(2\sqrt{ax})$ | $\dfrac{\sqrt{a}}{p^2}e^{-a/p}$ |
| 18 | $x^{\nu/2}J_\nu(2\sqrt{ax}), \qquad \nu>-1$ | $a^{\nu/2}p^{-\nu-1}e^{-a/p}$ |
| 19 | $J_0(a\sqrt{x^2+bx})$ | $\dfrac{1}{\sqrt{p^2+a^2}}\exp\left(bp-b\sqrt{p^2+a^2}\right)$ |
| 20 | $I_0(ax)$ | $\dfrac{1}{\sqrt{p^2-a^2}}$ |
| 21 | $I_\nu(ax), \qquad \nu>-1$ | $\dfrac{a^\nu}{\sqrt{p^2-a^2}\left(p+\sqrt{p^2-a^2}\right)^\nu}$ |

| No. | Original function, $f(x)$ | Laplace transform, $\widetilde{f}(p) = \int_0^\infty e^{-px} f(x)\, dx$ |
|---|---|---|
| 22 | $x^\nu I_\nu(ax)$, $\quad \nu > -\frac{1}{2}$ | $2^\nu \pi^{-1/2} \Gamma\!\left(\nu + \frac{1}{2}\right) a^\nu \left(p^2 - a^2\right)^{-\nu - 1/2}$ |
| 23 | $x^{\nu+1} I_\nu(ax)$, $\quad \nu > -1$ | $2^{\nu+1} \pi^{-1/2} \Gamma\!\left(\nu + \frac{3}{2}\right) a^\nu p \left(p^2 - a^2\right)^{-\nu - 3/2}$ |
| 24 | $I_0(2\sqrt{ax})$ | $\dfrac{1}{p} e^{a/p}$ |
| 25 | $\dfrac{1}{\sqrt{x}} I_1(2\sqrt{ax})$ | $\dfrac{1}{\sqrt{a}}\left(e^{a/p} - 1\right)$ |
| 26 | $x^{\nu/2} I_\nu(2\sqrt{ax})$, $\quad \nu > -1$ | $a^{\nu/2} p^{-\nu - 1} e^{a/p}$ |
| 27 | $Y_0(ax)$ | $-\dfrac{2}{\pi} \dfrac{\operatorname{arcsinh}(p/a)}{\sqrt{p^2 + a^2}}$ |
| 28 | $K_0(ax)$ | $\dfrac{\ln\!\left(p + \sqrt{p^2 - a^2}\right) - \ln a}{\sqrt{p^2 - a^2}}$ |

## T3.2. Tables of Inverse Laplace Transforms

### T3.2.1. General Formulas

| No. | Laplace transform, $\widetilde{f}(p)$ | Inverse transform, $f(x) = \dfrac{1}{2\pi i} \int_{c-i\infty}^{c+i\infty} e^{px} \widetilde{f}(p)\, dp$ |
|---|---|---|
| 1 | $\widetilde{f}(p + a)$ | $e^{-ax} f(x)$ |
| 2 | $\widetilde{f}(ap)$, $\quad a > 0$ | $\dfrac{1}{a} f\!\left(\dfrac{x}{a}\right)$ |
| 3 | $\widetilde{f}(ap + b)$, $\quad a > 0$ | $\dfrac{1}{a} \exp\!\left(-\dfrac{b}{a} x\right) f\!\left(\dfrac{x}{a}\right)$ |
| 4 | $\widetilde{f}(p - a) + \widetilde{f}(p + a)$ | $2 f(x) \cosh(ax)$ |
| 5 | $\widetilde{f}(p - a) - \widetilde{f}(p + a)$ | $2 f(x) \sinh(ax)$ |
| 6 | $e^{-ap} \widetilde{f}(p)$, $\quad a \geq 0$ | $\begin{cases} 0 & \text{if } 0 \leq x < a, \\ f(x - a) & \text{if } a < x \end{cases}$ |
| 7 | $p \widetilde{f}(p)$ | $\dfrac{df(x)}{dx}$, $\quad$ if $f(+0) = 0$ |
| 8 | $\dfrac{1}{p} \widetilde{f}(p)$ | $\displaystyle\int_0^x f(t)\, dt$ |
| 9 | $\dfrac{1}{p + a} \widetilde{f}(p)$ | $e^{-ax} \displaystyle\int_0^x e^{at} f(t)\, dt$ |
| 10 | $\dfrac{1}{p^2} \widetilde{f}(p)$ | $\displaystyle\int_0^x (x - t) f(t)\, dt$ |
| 11 | $\dfrac{\widetilde{f}(p)}{p(p + a)}$ | $\dfrac{1}{a} \displaystyle\int_0^x \left[1 - e^{a(x-t)}\right] f(t)\, dt$ |
| 12 | $\dfrac{\widetilde{f}(p)}{(p + a)^2}$ | $\displaystyle\int_0^x (x - t) e^{-a(x-t)} f(t)\, dt$ |
| 13 | $\dfrac{\widetilde{f}(p)}{(p + a)(p + b)}$ | $\dfrac{1}{b - a} \displaystyle\int_0^x \left[e^{-a(x-t)} - e^{-b(x-t)}\right] f(t)\, dt$ |

## T3.2. TABLES OF INVERSE LAPLACE TRANSFORMS

| No. | Laplace transform, $\widetilde{f}(p)$ | Inverse transform, $f(x) = \dfrac{1}{2\pi i}\int_{c-i\infty}^{c+i\infty} e^{px}\widetilde{f}(p)\,dp$ |
|---|---|---|
| 14 | $\dfrac{\widetilde{f}(p)}{(p+a)^2+b^2}$ | $\dfrac{1}{b}\int_0^x e^{-a(x-t)}\sin[b(x-t)]f(t)\,dt$ |
| 15 | $\dfrac{1}{p^n}\widetilde{f}(p),\quad n=1,2,\ldots$ | $\dfrac{1}{(n-1)!}\int_0^x (x-t)^{n-1} f(t)\,dt$ |
| 16 | $\widetilde{f}_1(p)\widetilde{f}_2(p)$ | $\int_0^x f_1(t)f_2(x-t)\,dt$ |
| 17 | $\dfrac{1}{\sqrt{p}}\widetilde{f}\!\left(\dfrac{1}{p}\right)$ | $\int_0^\infty \dfrac{\cos(2\sqrt{xt}\,)}{\sqrt{\pi x}} f(t)\,dt$ |
| 18 | $\dfrac{1}{p\sqrt{p}}\widetilde{f}\!\left(\dfrac{1}{p}\right)$ | $\int_0^\infty \dfrac{\sin(2\sqrt{xt}\,)}{\sqrt{\pi t}} f(t)\,dt$ |
| 19 | $\dfrac{1}{p^{2\nu+1}}\widetilde{f}\!\left(\dfrac{1}{p}\right)$ | $\int_0^\infty (x/t)^\nu J_{2\nu}\!\left(2\sqrt{xt}\,\right) f(t)\,dt$ |
| 20 | $\dfrac{1}{p}\widetilde{f}\!\left(\dfrac{1}{p}\right)$ | $\int_0^\infty J_0\!\left(2\sqrt{xt}\,\right) f(t)\,dt$ |
| 21 | $\dfrac{1}{p}\widetilde{f}\!\left(p+\dfrac{1}{p}\right)$ | $\int_0^x J_0\!\left(2\sqrt{xt-t^2}\,\right) f(t)\,dt$ |
| 22 | $\dfrac{1}{p^{2\nu+1}}\widetilde{f}\!\left(p+\dfrac{a}{p}\right),\quad -\dfrac{1}{2}<\nu\le 0$ | $\int_0^x \left(\dfrac{x-t}{at}\right)^\nu J_{2\nu}\!\left(2\sqrt{axt-at^2}\,\right) f(t)\,dt$ |
| 23 | $\widetilde{f}(\sqrt{p}\,)$ | $\int_0^\infty \dfrac{t}{2\sqrt{\pi x^3}}\exp\!\left(-\dfrac{t^2}{4x}\right) f(t)\,dt$ |
| 24 | $\dfrac{1}{\sqrt{p}}\widetilde{f}(\sqrt{p}\,)$ | $\dfrac{1}{\sqrt{\pi x}}\int_0^\infty \exp\!\left(-\dfrac{t^2}{4x}\right) f(t)\,dt$ |
| 25 | $\widetilde{f}(p+\sqrt{p}\,)$ | $\dfrac{1}{2\sqrt{\pi}}\int_0^x \dfrac{t}{(x-t)^{3/2}}\exp\!\left[-\dfrac{t^2}{4(x-t)}\right] f(t)\,dt$ |
| 26 | $\widetilde{f}(\sqrt{p^2+a^2}\,)$ | $f(x) - a\int_0^x f\!\left(\sqrt{x^2-t^2}\,\right) J_1(at)\,dt$ |
| 27 | $\widetilde{f}(\sqrt{p^2-a^2}\,)$ | $f(x) + a\int_0^x f\!\left(\sqrt{x^2-t^2}\,\right) I_1(at)\,dt$ |
| 28 | $\dfrac{\widetilde{f}(\sqrt{p^2+a^2}\,)}{\sqrt{p^2+a^2}}$ | $\int_0^x J_0\!\left(a\sqrt{x^2-t^2}\,\right) f(t)\,dt$ |
| 29 | $\dfrac{\widetilde{f}(\sqrt{p^2-a^2}\,)}{\sqrt{p^2-a^2}}$ | $\int_0^x I_0\!\left(a\sqrt{x^2-t^2}\,\right) f(t)\,dt$ |
| 30 | $\widetilde{f}(\sqrt{(p+a)^2-b^2}\,)$ | $e^{-ax}f(x) + be^{-ax}\int_0^x f\!\left(\sqrt{x^2-t^2}\,\right) I_1(bt)\,dt$ |
| 31 | $\widetilde{f}(\ln p)$ | $\int_0^\infty \dfrac{x^{t-1}}{\Gamma(t)} f(t)\,dt$ |
| 32 | $\dfrac{1}{p}\widetilde{f}(\ln p)$ | $\int_0^\infty \dfrac{x^t}{\Gamma(t+1)} f(t)\,dt$ |
| 33 | $\widetilde{f}(p-ia)+\widetilde{f}(p+ia),\ i^2=-1$ | $2f(x)\cos(ax)$ |
| 34 | $i\!\left[\widetilde{f}(p-ia)-\widetilde{f}(p+ia)\right],\ i^2=-1$ | $2f(x)\sin(ax)$ |

| No. | Laplace transform, $\tilde{f}(p)$ | Inverse transform, $f(x) = \dfrac{1}{2\pi i}\int_{c-i\infty}^{c+i\infty} e^{px}\tilde{f}(p)\,dp$ |
|---|---|---|
| 35 | $\dfrac{d\tilde{f}(p)}{dp}$ | $-xf(x)$ |
| 36 | $\dfrac{d^n \tilde{f}(p)}{dp^n}$ | $(-x)^n f(x)$ |
| 37 | $p^n \dfrac{d^m \tilde{f}(p)}{dp^m}, \quad m \geq n$ | $(-1)^m \dfrac{d^n}{dx^n}\left[x^m f(x)\right]$ |
| 38 | $\displaystyle\int_p^\infty \tilde{f}(q)\,dq$ | $\dfrac{1}{x} f(x)$ |
| 39 | $\dfrac{1}{p}\displaystyle\int_0^p \tilde{f}(q)\,dq$ | $\displaystyle\int_x^\infty \dfrac{f(t)}{t}\,dt$ |
| 40 | $\dfrac{1}{p}\displaystyle\int_p^\infty \tilde{f}(q)\,dq$ | $\displaystyle\int_0^x \dfrac{f(t)}{t}\,dt$ |

## T3.2.2. Expressions with Rational Functions

| No. | Laplace transform, $\tilde{f}(p)$ | Inverse transform, $f(x) = \dfrac{1}{2\pi i}\int_{c-i\infty}^{c+i\infty} e^{px}\tilde{f}(p)\,dp$ |
|---|---|---|
| 1 | $\dfrac{1}{p}$ | $1$ |
| 2 | $\dfrac{1}{p+a}$ | $e^{-ax}$ |
| 3 | $\dfrac{1}{p^2}$ | $x$ |
| 4 | $\dfrac{1}{p(p+a)}$ | $\dfrac{1}{a}\left(1-e^{-ax}\right)$ |
| 5 | $\dfrac{1}{(p+a)^2}$ | $xe^{-ax}$ |
| 6 | $\dfrac{p}{(p+a)^2}$ | $(1-ax)e^{-ax}$ |
| 7 | $\dfrac{1}{p^2-a^2}$ | $\dfrac{1}{a}\sinh(ax)$ |
| 8 | $\dfrac{p}{p^2-a^2}$ | $\cosh(ax)$ |
| 9 | $\dfrac{1}{(p+a)(p+b)}$ | $\dfrac{1}{a-b}\left(e^{-bx}-e^{-ax}\right)$ |
| 10 | $\dfrac{p}{(p+a)(p+b)}$ | $\dfrac{1}{a-b}\left(ae^{-ax}-be^{-bx}\right)$ |
| 11 | $\dfrac{1}{p^2+a^2}$ | $\dfrac{1}{a}\sin(ax)$ |
| 12 | $\dfrac{p}{p^2+a^2}$ | $\cos(ax)$ |
| 13 | $\dfrac{1}{(p+b)^2+a^2}$ | $\dfrac{1}{a}e^{-bx}\sin(ax)$ |
| 14 | $\dfrac{p}{(p+b)^2+a^2}$ | $e^{-bx}\left[\cos(ax)-\dfrac{b}{a}\sin(ax)\right]$ |

## T3.2. Tables of Inverse Laplace Transforms

| No. | Laplace transform, $\widetilde{f}(p)$ | Inverse transform, $f(x) = \dfrac{1}{2\pi i}\displaystyle\int_{c-i\infty}^{c+i\infty} e^{px}\widetilde{f}(p)\,dp$ |
|---|---|---|
| 15 | $\dfrac{1}{p^3}$ | $\tfrac{1}{2}x^2$ |
| 16 | $\dfrac{1}{p^2(p+a)}$ | $\dfrac{1}{a^2}\left(e^{-ax}+ax-1\right)$ |
| 17 | $\dfrac{1}{p(p+a)(p+b)}$ | $\dfrac{1}{ab(a-b)}\left(a-b+be^{-ax}-ae^{-bx}\right)$ |
| 18 | $\dfrac{1}{p(p+a)^2}$ | $\dfrac{1}{a^2}\left(1-e^{-ax}-axe^{-ax}\right)$ |
| 19 | $\dfrac{1}{(p+a)(p+b)(p+c)}$ | $\dfrac{(c-b)e^{-ax}+(a-c)e^{-bx}+(b-a)e^{-cx}}{(a-b)(b-c)(c-a)}$ |
| 20 | $\dfrac{p}{(p+a)(p+b)(p+c)}$ | $\dfrac{a(b-c)e^{-ax}+b(c-a)e^{-bx}+c(a-b)e^{-cx}}{(a-b)(b-c)(c-a)}$ |
| 21 | $\dfrac{p^2}{(p+a)(p+b)(p+c)}$ | $\dfrac{a^2(c-b)e^{-ax}+b^2(a-c)e^{-bx}+c^2(b-a)e^{-cx}}{(a-b)(b-c)(c-a)}$ |
| 22 | $\dfrac{1}{(p+a)(p+b)^2}$ | $\dfrac{1}{(a-b)^2}\left[e^{-ax}-e^{-bx}+(a-b)xe^{-bx}\right]$ |
| 23 | $\dfrac{p}{(p+a)(p+b)^2}$ | $\dfrac{1}{(a-b)^2}\left\{-ae^{-ax}+\left[a+b(b-a)x\right]e^{-bx}\right\}$ |
| 24 | $\dfrac{p^2}{(p+a)(p+b)^2}$ | $\dfrac{1}{(a-b)^2}\left[a^2 e^{-ax}+b(b-2a-b^2 x+abx)e^{-bx}\right]$ |
| 25 | $\dfrac{1}{(p+a)^3}$ | $\tfrac{1}{2}x^2 e^{-ax}$ |
| 26 | $\dfrac{p}{(p+a)^3}$ | $x\left(1-\tfrac{1}{2}ax\right)e^{-ax}$ |
| 27 | $\dfrac{p^2}{(p+a)^3}$ | $\left(1-2ax+\tfrac{1}{2}a^2 x^2\right)e^{-ax}$ |
| 28 | $\dfrac{1}{p(p^2+a^2)}$ | $\dfrac{1}{a^2}\left[1-\cos(ax)\right]$ |
| 29 | $\dfrac{1}{p\left[(p+b)^2+a^2\right]}$ | $\dfrac{1}{a^2+b^2}\left\{1-e^{-bx}\left[\cos(ax)+\dfrac{b}{a}\sin(ax)\right]\right\}$ |
| 30 | $\dfrac{1}{(p+a)(p^2+b^2)}$ | $\dfrac{1}{a^2+b^2}\left[e^{-ax}+\dfrac{a}{b}\sin(bx)-\cos(bx)\right]$ |
| 31 | $\dfrac{p}{(p+a)(p^2+b^2)}$ | $\dfrac{1}{a^2+b^2}\left[-ae^{-ax}+a\cos(bx)+b\sin(bx)\right]$ |
| 32 | $\dfrac{p^2}{(p+a)(p^2+b^2)}$ | $\dfrac{1}{a^2+b^2}\left[a^2 e^{-ax}-ab\sin(bx)+b^2\cos(bx)\right]$ |
| 33 | $\dfrac{1}{p^3+a^3}$ | $\dfrac{1}{3a^2}e^{-ax}-\dfrac{1}{3a^2}e^{ax/2}\left[\cos(kx)-\sqrt{3}\sin(kx)\right],$ $k=\tfrac{1}{2}a\sqrt{3}$ |
| 34 | $\dfrac{p}{p^3+a^3}$ | $-\dfrac{1}{3a}e^{-ax}+\dfrac{1}{3a}e^{ax/2}\left[\cos(kx)+\sqrt{3}\sin(kx)\right],$ $k=\tfrac{1}{2}a\sqrt{3}$ |

| No. | Laplace transform, $\widetilde{f}(p)$ | Inverse transform, $f(x) = \dfrac{1}{2\pi i}\displaystyle\int_{c-i\infty}^{c+i\infty} e^{px}\widetilde{f}(p)\,dp$ |
|---|---|---|
| 35 | $\dfrac{p^2}{p^3+a^3}$ | $\tfrac{1}{3}e^{-ax} + \tfrac{2}{3}e^{ax/2}\cos(kx), \quad k=\tfrac{1}{2}a\sqrt{3}$ |
| 36 | $\dfrac{1}{(p+a)[(p+b)^2+c^2]}$ | $\dfrac{e^{-ax} - e^{-bx}\cos(cx) + ke^{-bx}\sin(cx)}{(a-b)^2+c^2}, \quad k=\dfrac{a-b}{c}$ |
| 37 | $\dfrac{p}{(p+a)[(p+b)^2+c^2]}$ | $\dfrac{-ae^{-ax} + ae^{-bx}\cos(cx) + ke^{-bx}\sin(cx)}{(a-b)^2+c^2}$, $k=\dfrac{b^2+c^2-ab}{c}$ |
| 38 | $\dfrac{p^2}{(p+a)[(p+b)^2+c^2]}$ | $\dfrac{a^2 e^{-ax} + (b^2+c^2-2ab)e^{-bx}\cos(cx) + ke^{-bx}\sin(cx)}{(a-b)^2+c^2}$, $k=-ac-bc+\dfrac{ab^2-b^3}{c}$ |
| 39 | $\dfrac{1}{p^4}$ | $\tfrac{1}{6}x^3$ |
| 40 | $\dfrac{1}{p^3(p+a)}$ | $\dfrac{1}{a^3} - \dfrac{1}{a^2}x + \dfrac{1}{2a}x^2 - \dfrac{1}{a^3}e^{-ax}$ |
| 41 | $\dfrac{1}{p^2(p+a)^2}$ | $\dfrac{1}{a^2}x(1+e^{-ax}) + \dfrac{2}{a^3}(e^{-ax}-1)$ |
| 42 | $\dfrac{1}{p^2(p+a)(p+b)}$ | $-\dfrac{a+b}{a^2 b^2} + \dfrac{1}{ab}x + \dfrac{1}{a^2(b-a)}e^{-ax} + \dfrac{1}{b^2(a-b)}e^{-bx}$ |
| 43 | $\dfrac{1}{(p+a)^2(p+b)^2}$ | $\dfrac{1}{(a-b)^2}\left[e^{-ax}\left(x+\dfrac{2}{a-b}\right) + e^{-bx}\left(x-\dfrac{2}{a-b}\right)\right]$ |
| 44 | $\dfrac{1}{(p+a)^4}$ | $\tfrac{1}{6}x^3 e^{-ax}$ |
| 45 | $\dfrac{p}{(p+a)^4}$ | $\tfrac{1}{2}x^2 e^{-ax} - \tfrac{1}{6}ax^3 e^{-ax}$ |
| 46 | $\dfrac{1}{p^2(p^2+a^2)}$ | $\dfrac{1}{a^3}[ax - \sin(ax)]$ |
| 47 | $\dfrac{1}{p^4-a^4}$ | $\dfrac{1}{2a^3}[\sinh(ax) - \sin(ax)]$ |
| 48 | $\dfrac{p}{p^4-a^4}$ | $\dfrac{1}{2a^2}[\cosh(ax) - \cos(ax)]$ |
| 49 | $\dfrac{p^2}{p^4-a^4}$ | $\dfrac{1}{2a}[\sinh(ax) + \sin(ax)]$ |
| 50 | $\dfrac{p^3}{p^4-a^4}$ | $\dfrac{1}{2}[\cosh(ax) + \cos(ax)]$ |
| 51 | $\dfrac{1}{p^4+a^4}$ | $\dfrac{1}{a^3\sqrt{2}}(\cosh\xi\sin\xi - \sinh\xi\cos\xi), \quad \xi=\dfrac{ax}{\sqrt{2}}$ |
| 52 | $\dfrac{p}{p^4+a^4}$ | $\dfrac{1}{a^2}\sin\left(\dfrac{ax}{\sqrt{2}}\right)\sinh\left(\dfrac{ax}{\sqrt{2}}\right)$ |
| 53 | $\dfrac{p^2}{p^4+a^4}$ | $\dfrac{1}{a\sqrt{2}}(\cos\xi\sinh\xi + \sin\xi\cosh\xi), \quad \xi=\dfrac{ax}{\sqrt{2}}$ |

| No. | Laplace transform, $\widetilde{f}(p)$ | Inverse transform, $f(x) = \dfrac{1}{2\pi i} \int_{c-i\infty}^{c+i\infty} e^{px} \widetilde{f}(p)\, dp$ |
|---|---|---|
| 54 | $\dfrac{1}{(p^2+a^2)^2}$ | $\dfrac{1}{2a^3}\big[\sin(ax) - ax\cos(ax)\big]$ |
| 55 | $\dfrac{p}{(p^2+a^2)^2}$ | $\dfrac{1}{2a}\, x\sin(ax)$ |
| 56 | $\dfrac{p^2}{(p^2+a^2)^2}$ | $\dfrac{1}{2a}\big[\sin(ax) + ax\cos(ax)\big]$ |
| 57 | $\dfrac{p^3}{(p^2+a^2)^2}$ | $\cos(ax) - \tfrac{1}{2}ax\sin(ax)$ |
| 58 | $\dfrac{1}{\big[(p+b)^2+a^2\big]^2}$ | $\dfrac{1}{2a^3} e^{-bx}\big[\sin(ax) - ax\cos(ax)\big]$ |
| 59 | $\dfrac{1}{(p^2-a^2)(p^2-b^2)}$ | $\dfrac{1}{a^2-b^2}\left[\dfrac{1}{a}\sinh(ax) - \dfrac{1}{b}\sinh(bx)\right]$ |
| 60 | $\dfrac{p}{(p^2-a^2)(p^2-b^2)}$ | $\dfrac{\cosh(ax) - \cosh(bx)}{a^2-b^2}$ |
| 61 | $\dfrac{p^2}{(p^2-a^2)(p^2-b^2)}$ | $\dfrac{a\sinh(ax) - b\sinh(bx)}{a^2-b^2}$ |
| 62 | $\dfrac{p^3}{(p^2-a^2)(p^2-b^2)}$ | $\dfrac{a^2\cosh(ax) - b^2\cosh(bx)}{a^2-b^2}$ |
| 63 | $\dfrac{1}{(p^2+a^2)(p^2+b^2)}$ | $\dfrac{1}{b^2-a^2}\left[\dfrac{1}{a}\sin(ax) - \dfrac{1}{b}\sin(bx)\right]$ |
| 64 | $\dfrac{p}{(p^2+a^2)(p^2+b^2)}$ | $\dfrac{\cos(ax) - \cos(bx)}{b^2-a^2}$ |
| 65 | $\dfrac{p^2}{(p^2+a^2)(p^2+b^2)}$ | $\dfrac{-a\sin(ax) + b\sin(bx)}{b^2-a^2}$ |
| 66 | $\dfrac{p^3}{(p^2+a^2)(p^2+b^2)}$ | $\dfrac{-a^2\cos(ax) + b^2\cos(bx)}{b^2-a^2}$ |
| 67 | $\dfrac{1}{p^n}, \quad n = 1, 2, \ldots$ | $\dfrac{1}{(n-1)!}\, x^{n-1}$ |
| 68 | $\dfrac{1}{(p+a)^n}, \quad n = 1, 2, \ldots$ | $\dfrac{1}{(n-1)!}\, x^{n-1} e^{-ax}$ |
| 69 | $\dfrac{1}{p(p+a)^n}, \quad n = 1, 2, \ldots$ | $a^{-n}\big[1 - e^{-ax} e_n(ax)\big], \quad e_n(z) = 1 + \dfrac{z}{1!} + \cdots + \dfrac{z^n}{n!}$ |
| 70 | $\dfrac{1}{p^{2n}+a^{2n}}, \quad n = 1, 2, \ldots$ | $-\dfrac{1}{na^{2n}} \sum_{k=1}^{n} \exp(a_k x)\big[a_k \cos(b_k x) - b_k \sin(b_k x)\big],$ $a_k = a\cos\varphi_k,\ b_k = a\sin\varphi_k,\ \varphi_k = \dfrac{\pi(2k-1)}{2n}$ |
| 71 | $\dfrac{1}{p^{2n}-a^{2n}}, \quad n = 1, 2, \ldots$ | $\dfrac{1}{na^{2n-1}}\sinh(ax) + \dfrac{1}{na^{2n}}\sum_{k=2}^{n}\exp(a_k x)$ $\times \big[a_k\cos(b_k x) - b_k\sin(b_k x)\big],$ $a_k = a\cos\varphi_k,\ b_k = a\sin\varphi_k,\ \varphi_k = \dfrac{\pi(k-1)}{n}$ |

| No. | Laplace transform, $\widetilde{f}(p)$ | Inverse transform, $f(x) = \dfrac{1}{2\pi i}\displaystyle\int_{c-i\infty}^{c+i\infty} e^{px}\widetilde{f}(p)\,dp$ |
|---|---|---|
| 72 | $\dfrac{1}{p^{2n+1}+a^{2n+1}}$, $\quad n=0,1,\ldots$ | $\dfrac{e^{-ax}}{(2n+1)a^{2n}} - \dfrac{2}{(2n+1)a^{2n+1}}\displaystyle\sum_{k=1}^{n}\exp(a_k x)$ $\times [a_k\cos(b_k x) - b_k\sin(b_k x)]$, $a_k = a\cos\varphi_k$, $b_k = a\sin\varphi_k$, $\varphi_k = \dfrac{\pi(2k-1)}{2n+1}$ |
| 73 | $\dfrac{1}{p^{2n+1}-a^{2n+1}}$, $\quad n=0,1,\ldots$ | $\dfrac{e^{ax}}{(2n+1)a^{2n}} + \dfrac{2}{(2n+1)a^{2n+1}}\displaystyle\sum_{k=1}^{n}\exp(a_k x)$ $\times [a_k\cos(b_k x) - b_k\sin(b_k x)]$, $a_k = a\cos\varphi_k$, $b_k = a\sin\varphi_k$, $\varphi_k = \dfrac{2\pi k}{2n+1}$ |
| 74 | $\dfrac{Q(p)}{P(p)}$, $P(p)=(p-a_1)\ldots(p-a_n)$; $Q(p)$ is a polynomial of degree $\leq n-1$; $a_i \neq a_j$ if $i \neq j$ | $\displaystyle\sum_{k=1}^{n}\dfrac{Q(a_k)}{P'(a_k)}\exp(a_k x)$, (the prime stand for the differentiation) |
| 75 | $\dfrac{Q(p)}{P(p)}$, $P(p)=(p-a_1)^{m_1}\ldots(p-a_n)^{m_n}$; $Q(p)$ is a polynomial of degree $< m_1+m_2+\cdots+m_n-1$; $a_i \neq a_j$ if $i \neq j$ | $\displaystyle\sum_{k=1}^{n}\sum_{l=1}^{m_k}\dfrac{\Phi_{kl}(a_k)}{(m_k-l)!\,(l-1)!}x^{m_k-l}\exp(a_k x)$, $\Phi_{kl}(p) = \dfrac{d^{l-1}}{dp^{l-1}}\left[\dfrac{Q(p)}{P_k(p)}\right]$, $P_k(p) = \dfrac{P(p)}{(p-a_k)^{m_k}}$ |
| 76 | $\dfrac{Q(p)+pR(p)}{P(p)}$, $P(p)=(p^2+a_1^2)\ldots(p^2+a_n^2)$; $Q(p)$ and $R(p)$ are polynomials of degree $\leq 2n-2$; $a_l \neq a_j$, $l \neq j$ | $\displaystyle\sum_{k=1}^{n}\dfrac{Q(ia_k)\sin(a_k x) + a_k R(ia_k)\cos(a_k x)}{a_k P_k(ia_k)}$, $P_m(p) = \dfrac{P(p)}{p^2+a_m^2}$, $i^2=-1$ |

## T3.2.3. Expressions with Square Roots

| No. | Laplace transform, $\widetilde{f}(p)$ | Inverse transform, $f(x) = \dfrac{1}{2\pi i}\displaystyle\int_{c-i\infty}^{c+i\infty} e^{px}\widetilde{f}(p)\,dp$ |
|---|---|---|
| 1 | $\dfrac{1}{\sqrt{p}}$ | $\dfrac{1}{\sqrt{\pi x}}$ |
| 2 | $\sqrt{p-a}-\sqrt{p-b}$ | $\dfrac{e^{bx}-e^{ax}}{2\sqrt{\pi x^3}}$ |
| 3 | $\dfrac{1}{\sqrt{p+a}}$ | $\dfrac{1}{\sqrt{\pi x}}e^{-ax}$ |
| 4 | $\sqrt{\dfrac{p+a}{p}}-1$ | $\tfrac{1}{2}ae^{-ax/2}\left[I_1\!\left(\tfrac{1}{2}ax\right) + I_0\!\left(\tfrac{1}{2}ax\right)\right]$ |
| 5 | $\dfrac{\sqrt{p+a}}{p+b}$ | $\dfrac{e^{-ax}}{\sqrt{\pi x}} + (a-b)^{1/2}e^{-bx}\operatorname{erf}\!\left[(a-b)^{1/2}x^{1/2}\right]$ |
| 6 | $\dfrac{1}{p\sqrt{p}}$ | $2\sqrt{\dfrac{x}{\pi}}$ |
| 7 | $\dfrac{1}{(p+a)\sqrt{p+b}}$ | $(b-a)^{-1/2}e^{-ax}\operatorname{erf}\!\left[(b-a)^{1/2}x^{1/2}\right]$ |

| No. | Laplace transform, $\widetilde{f}(p)$ | Inverse transform, $f(x) = \dfrac{1}{2\pi i} \int_{c-i\infty}^{c+i\infty} e^{px} \widetilde{f}(p)\,dp$ |
|---|---|---|
| 8 | $\dfrac{1}{\sqrt{p}\,(p-a)}$ | $\dfrac{1}{\sqrt{a}} e^{ax} \operatorname{erf}\!\left(\sqrt{ax}\right)$ |
| 9 | $\dfrac{1}{p^{3/2}(p-a)}$ | $a^{-3/2} e^{ax} \operatorname{erf}\!\left(\sqrt{ax}\right) - 2a^{-1}\pi^{-1/2} x^{1/2}$ |
| 10 | $\dfrac{1}{\sqrt{p}+a}$ | $\pi^{-1/2} x^{-1/2} - a e^{a^2 x} \operatorname{erfc}\!\left(a\sqrt{x}\right)$ |
| 11 | $\dfrac{a}{p(\sqrt{p}+a)}$ | $1 - e^{a^2 x} \operatorname{erfc}\!\left(a\sqrt{x}\right)$ |
| 12 | $\dfrac{1}{p+a\sqrt{p}}$ | $e^{a^2 x} \operatorname{erfc}\!\left(a\sqrt{x}\right)$ |
| 13 | $\dfrac{1}{(\sqrt{p}+\sqrt{a})^2}$ | $1 - \dfrac{2}{\sqrt{\pi}} (ax)^{1/2} + (1-2ax) e^{ax}\!\left[\operatorname{erf}\!\left(\sqrt{ax}\right) - 1\right]$ |
| 14 | $\dfrac{1}{p(\sqrt{p}+\sqrt{a})^2}$ | $\dfrac{1}{a} + \left(2x - \dfrac{1}{a}\right) e^{ax} \operatorname{erfc}\!\left(\sqrt{ax}\right) - \dfrac{2}{\sqrt{\pi a}} \sqrt{x}$ |
| 15 | $\dfrac{1}{\sqrt{p}\,(\sqrt{p}+a)^2}$ | $2\pi^{-1/2} x^{1/2} - 2ax e^{a^2 x} \operatorname{erfc}\!\left(a\sqrt{x}\right)$ |
| 16 | $\dfrac{1}{(\sqrt{p}+a)^3}$ | $\dfrac{2}{\sqrt{\pi}}(a^2 x + 1)\sqrt{x} - ax(2a^2 x + 3) e^{a^2 x} \operatorname{erfc}\!\left(a\sqrt{x}\right)$ |
| 17 | $p^{-n-1/2}, \quad n=1,2,\ldots$ | $\dfrac{2^n}{1\times 3\times \ldots \times (2n-1)\sqrt{\pi}} x^{n-1/2}$ |
| 18 | $(p+a)^{-n-1/2}$ | $\dfrac{2^n}{1\times 3\times \ldots \times (2n-1)\sqrt{\pi}} x^{n-1/2} e^{-ax}$ |
| 19 | $\dfrac{1}{\sqrt{p^2+a^2}}$ | $J_0(ax)$ |
| 20 | $\dfrac{1}{\sqrt{p^2-a^2}}$ | $I_0(ax)$ |
| 21 | $\dfrac{1}{\sqrt{p^2+ap+b}}$ | $\exp\!\left(-\tfrac{1}{2}ax\right) J_0\!\left[\left(b - \tfrac{1}{4}a^2\right)^{1/2} x\right]$ |
| 22 | $\left(\sqrt{p^2+a^2}-p\right)^{1/2}$ | $\dfrac{1}{\sqrt{2\pi x^3}} \sin(ax)$ |
| 23 | $\dfrac{1}{\sqrt{p^2+a^2}}\left(\sqrt{p^2+a^2}+p\right)^{1/2}$ | $\dfrac{\sqrt{2}}{\sqrt{\pi x}} \cos(ax)$ |
| 24 | $\dfrac{1}{\sqrt{p^2-a^2}}\left(\sqrt{p^2-a^2}+p\right)^{1/2}$ | $\dfrac{\sqrt{2}}{\sqrt{\pi x}} \cosh(ax)$ |
| 25 | $\left(\sqrt{p^2+a^2}+p\right)^{-n}$ | $n a^{-n} x^{-1} J_n(ax)$ |
| 26 | $\left(\sqrt{p^2-a^2}+p\right)^{-n}$ | $n a^{-n} x^{-1} I_n(ax)$ |
| 27 | $(p^2+a^2)^{-n-1/2}$ | $\dfrac{(x/a)^n J_n(ax)}{1\times 3\times 5\times \ldots \times (2n-1)}$ |
| 28 | $(p^2-a^2)^{-n-1/2}$ | $\dfrac{(x/a)^n I_n(ax)}{1\times 3\times 5\times \ldots \times (2n-1)}$ |

## T3.2.4. Expressions with Arbitrary Powers

| No. | Laplace transform, $\widetilde{f}(p)$ | Inverse transform, $f(x) = \dfrac{1}{2\pi i}\displaystyle\int_{c-i\infty}^{c+i\infty} e^{px}\widetilde{f}(p)\,dp$ |
|---|---|---|
| 1 | $(p+a)^{-\nu}, \quad \nu > 0$ | $\dfrac{1}{\Gamma(\nu)}x^{\nu-1}e^{-ax}$ |
| 2 | $\left[(p+a)^{1/2}+(p+b)^{1/2}\right]^{-2\nu}, \quad \nu > 0$ | $\dfrac{\nu}{(a-b)^{\nu}}x^{-1}\exp\left[-\tfrac{1}{2}(a+b)x\right]I_{\nu}\left[\tfrac{1}{2}(a-b)x\right]$ |
| 3 | $\left[(p+a)(p+b)\right]^{-\nu}, \quad \nu > 0$ | $\dfrac{\sqrt{\pi}}{\Gamma(\nu)}\left(\dfrac{x}{a-b}\right)^{\nu-1/2}\exp\left(-\dfrac{a+b}{2}x\right)I_{\nu-1/2}\left(\dfrac{a-b}{2}x\right)$ |
| 4 | $(p^2+a^2)^{-\nu-1/2}, \quad \nu > -\tfrac{1}{2}$ | $\dfrac{\sqrt{\pi}}{(2a)^{\nu}\Gamma(\nu+\tfrac{1}{2})}x^{\nu}J_{\nu}(ax)$ |
| 5 | $(p^2-a^2)^{-\nu-1/2}, \quad \nu > -\tfrac{1}{2}$ | $\dfrac{\sqrt{\pi}}{(2a)^{\nu}\Gamma(\nu+\tfrac{1}{2})}x^{\nu}I_{\nu}(ax)$ |
| 6 | $p(p^2+a^2)^{-\nu-1/2}, \quad \nu > 0$ | $\dfrac{a\sqrt{\pi}}{(2a)^{\nu}\Gamma(\nu+\tfrac{1}{2})}x^{\nu}J_{\nu-1}(ax)$ |
| 7 | $p(p^2-a^2)^{-\nu-1/2}, \quad \nu > 0$ | $\dfrac{a\sqrt{\pi}}{(2a)^{\nu}\Gamma(\nu+\tfrac{1}{2})}x^{\nu}I_{\nu-1}(ax)$ |
| 8 | $\left[(p^2+a^2)^{1/2}+p\right]^{-\nu} = a^{-2\nu}\left[(p^2+a^2)^{1/2}-p\right]^{\nu}, \quad \nu > 0$ | $\nu a^{-\nu}x^{-1}J_{\nu}(ax)$ |
| 9 | $\left[(p^2-a^2)^{1/2}+p\right]^{-\nu} = a^{-2\nu}\left[p-(p^2-a^2)^{1/2}\right]^{\nu}, \quad \nu > 0$ | $\nu a^{-\nu}x^{-1}I_{\nu}(ax)$ |
| 10 | $p\left[(p^2+a^2)^{1/2}+p\right]^{-\nu}, \quad \nu > 1$ | $\nu a^{1-\nu}x^{-1}J_{\nu-1}(ax) - \nu(\nu+1)a^{-\nu}x^{-2}J_{\nu}(ax)$ |
| 11 | $p\left[(p^2-a^2)^{1/2}+p\right]^{-\nu}, \quad \nu > 1$ | $\nu a^{1-\nu}x^{-1}I_{\nu-1}(ax) - \nu(\nu+1)a^{-\nu}x^{-2}I_{\nu}(ax)$ |
| 12 | $\dfrac{\left(\sqrt{p^2+a^2}+p\right)^{-\nu}}{\sqrt{p^2+a^2}}, \quad \nu > -1$ | $a^{-\nu}J_{\nu}(ax)$ |
| 13 | $\dfrac{\left(\sqrt{p^2-a^2}+p\right)^{-\nu}}{\sqrt{p^2-a^2}}, \quad \nu > -1$ | $a^{-\nu}I_{\nu}(ax)$ |

## T3.2.5. Expressions with Exponential Functions

| No. | Laplace transform, $\widetilde{f}(p)$ | Inverse transform, $f(x) = \dfrac{1}{2\pi i}\displaystyle\int_{c-i\infty}^{c+i\infty} e^{px}\widetilde{f}(p)\,dp$ |
|---|---|---|
| 1 | $p^{-1}e^{-ap}, \quad a > 0$ | $\begin{cases} 0 & \text{if } 0 < x < a, \\ 1 & \text{if } a < x \end{cases}$ |
| 2 | $p^{-1}(1-e^{-ap}), \quad a > 0$ | $\begin{cases} 1 & \text{if } 0 < x < a, \\ 0 & \text{if } a < x \end{cases}$ |
| 3 | $p^{-1}(e^{-ap}-e^{-bp}), \quad 0 \le a < b$ | $\begin{cases} 0 & \text{if } 0 < x < a, \\ 1 & \text{if } a < x < b, \\ 0 & \text{if } b < x \end{cases}$ |
| 4 | $p^{-2}(e^{-ap}-e^{-bp}), \quad 0 \le a < b$ | $\begin{cases} 0 & \text{if } 0 < x < a, \\ x-a & \text{if } a < x < b, \\ b-a & \text{if } b < x \end{cases}$ |
| 5 | $(p+b)^{-1}e^{-ap}, \quad a > 0$ | $\begin{cases} 0 & \text{if } 0 < x < a, \\ e^{-b(x-a)} & \text{if } a < x \end{cases}$ |

| No. | Laplace transform, $\widetilde{f}(p)$ | Inverse transform, $f(x) = \dfrac{1}{2\pi i} \int_{c-i\infty}^{c+i\infty} e^{px} \widetilde{f}(p)\, dp$ |
|---|---|---|
| 6 | $p^{-\nu} e^{-ap}$, $\nu > 0$ | $\begin{cases} 0 & \text{if } 0 < x < a, \\ \dfrac{(x-a)^{\nu-1}}{\Gamma(\nu)} & \text{if } a < x \end{cases}$ |
| 7 | $p^{-1}(e^{ap} - 1)^{-1}$, $a > 0$ | $f(x) = n$ if $na < x < (n+1)a$; $n = 0, 1, 2, \ldots$ |
| 8 | $e^{a/p} - 1$ | $\sqrt{\dfrac{a}{x}}\, I_1(2\sqrt{ax})$ |
| 9 | $p^{-1/2} e^{a/p}$ | $\dfrac{1}{\sqrt{\pi x}} \cosh(2\sqrt{ax})$ |
| 10 | $p^{-3/2} e^{a/p}$ | $\dfrac{1}{\sqrt{\pi a}} \sinh(2\sqrt{ax})$ |
| 11 | $p^{-5/2} e^{a/p}$ | $\sqrt{\dfrac{x}{\pi a}} \cosh(2\sqrt{ax}) - \dfrac{1}{2\sqrt{\pi a^3}} \sinh(2\sqrt{ax})$ |
| 12 | $p^{-\nu-1} e^{a/p}$, $\nu > -1$ | $(x/a)^{\nu/2} I_\nu(2\sqrt{ax})$ |
| 13 | $1 - e^{-a/p}$ | $\sqrt{\dfrac{a}{x}}\, J_1(2\sqrt{ax})$ |
| 14 | $p^{-1/2} e^{-a/p}$ | $\dfrac{1}{\sqrt{\pi x}} \cos(2\sqrt{ax})$ |
| 15 | $p^{-3/2} e^{-a/p}$ | $\dfrac{1}{\sqrt{\pi a}} \sin(2\sqrt{ax})$ |
| 16 | $p^{-5/2} e^{-a/p}$ | $\dfrac{1}{2\sqrt{\pi a^3}} \sin(2\sqrt{ax}) - \sqrt{\dfrac{x}{\pi a}} \cos(2\sqrt{ax})$ |
| 17 | $p^{-\nu-1} e^{-a/p}$, $\nu > -1$ | $(x/a)^{\nu/2} J_\nu(2\sqrt{ax})$ |
| 18 | $\exp(-\sqrt{ap})$, $a > 0$ | $\dfrac{\sqrt{a}}{2\sqrt{\pi}} x^{-3/2} \exp\!\left(-\dfrac{a}{4x}\right)$ |
| 19 | $p \exp(-\sqrt{ap})$, $a > 0$ | $\dfrac{\sqrt{a}}{8\sqrt{\pi}} (a - 6x) x^{-7/2} \exp\!\left(-\dfrac{a}{4x}\right)$ |
| 20 | $\dfrac{1}{p} \exp(-\sqrt{ap})$, $a \geq 0$ | $\operatorname{erfc}\!\left(\dfrac{\sqrt{a}}{2\sqrt{x}}\right)$ |
| 21 | $\sqrt{p}\, \exp(-\sqrt{ap})$, $a > 0$ | $\dfrac{1}{4\sqrt{\pi}} (a - 2x) x^{-5/2} \exp\!\left(-\dfrac{a}{4x}\right)$ |
| 22 | $\dfrac{1}{\sqrt{p}} \exp(-\sqrt{ap})$, $a \geq 0$ | $\dfrac{1}{\sqrt{\pi x}} \exp\!\left(-\dfrac{a}{4x}\right)$ |
| 23 | $\dfrac{1}{p\sqrt{p}} \exp(-\sqrt{ap})$, $a \geq 0$ | $\dfrac{2\sqrt{x}}{\sqrt{\pi}} \exp\!\left(-\dfrac{a}{4x}\right) - \sqrt{a}\, \operatorname{erfc}\!\left(\dfrac{\sqrt{a}}{2\sqrt{x}}\right)$ |
| 24 | $\dfrac{\exp(-k\sqrt{p^2 + a^2})}{\sqrt{p^2 + a^2}}$, $k > 0$ | $\begin{cases} 0 & \text{if } 0 < x < k, \\ J_0(a\sqrt{x^2 - k^2}) & \text{if } k < x \end{cases}$ |
| 25 | $\dfrac{\exp(-k\sqrt{p^2 - a^2})}{\sqrt{p^2 - a^2}}$, $k > 0$ | $\begin{cases} 0 & \text{if } 0 < x < k, \\ I_0(a\sqrt{x^2 - k^2}) & \text{if } k < x \end{cases}$ |

## T3.2.6. Expressions with Hyperbolic Functions

| No. | Laplace transform, $\tilde{f}(p)$ | Inverse transform, $f(x) = \dfrac{1}{2\pi i}\int_{c-i\infty}^{c+i\infty} e^{px}\tilde{f}(p)\,dp$ |
|---|---|---|
| 1 | $\dfrac{1}{p\sinh(ap)}$, $\ a > 0$ | $f(x) = 2n$ if $a(2n-1) < x < a(2n+1)$; $n = 0, 1, 2, \ldots$ $(x > 0)$ |
| 2 | $\dfrac{1}{p^2 \sinh(ap)}$, $\ a > 0$ | $f(x) = 2n(x - an)$ if $a(2n-1) < x < a(2n+1)$; $n = 0, 1, 2, \ldots$ $(x > 0)$ |
| 3 | $\dfrac{\sinh(a/p)}{\sqrt{p}}$ | $\dfrac{1}{2\sqrt{\pi x}}\left[\cosh(2\sqrt{ax}) - \cos(2\sqrt{ax})\right]$ |
| 4 | $\dfrac{\sinh(a/p)}{p\sqrt{p}}$ | $\dfrac{1}{2\sqrt{\pi a}}\left[\sinh(2\sqrt{ax}) - \sin(2\sqrt{ax})\right]$ |
| 5 | $p^{-\nu-1}\sinh(a/p)$, $\ \nu > -2$ | $\tfrac{1}{2}(x/a)^{\nu/2}\left[I_\nu(2\sqrt{ax}) - J_\nu(2\sqrt{ax})\right]$ |
| 6 | $\dfrac{1}{p\cosh(ap)}$, $\ a > 0$ | $f(x) = \begin{cases} 0 & \text{if } a(4n-1) < x < a(4n+1), \\ 2 & \text{if } a(4n+1) < x < a(4n+3), \end{cases}$ $n = 0, 1, 2, \ldots$ $(x > 0)$ |
| 7 | $\dfrac{1}{p^2 \cosh(ap)}$, $\ a > 0$ | $x - (-1)^n(x - 2an)$ if $2n - 1 < x/a < 2n + 1$; $n = 0, 1, 2, \ldots$ $(x > 0)$ |
| 8 | $\dfrac{\cosh(a/p)}{\sqrt{p}}$ | $\dfrac{1}{2\sqrt{\pi x}}\left[\cosh(2\sqrt{ax}) + \cos(2\sqrt{ax})\right]$ |
| 9 | $\dfrac{\cosh(a/p)}{p\sqrt{p}}$ | $\dfrac{1}{2\sqrt{\pi a}}\left[\sinh(2\sqrt{ax}) + \sin(2\sqrt{ax})\right]$ |
| 10 | $p^{-\nu-1}\cosh(a/p)$, $\ \nu > -1$ | $\tfrac{1}{2}(x/a)^{\nu/2}\left[I_\nu(2\sqrt{ax}) + J_\nu(2\sqrt{ax})\right]$ |
| 11 | $\dfrac{1}{p}\tanh(ap)$, $\ a > 0$ | $f(x) = (-1)^{n-1}$ if $2a(n-1) < x < 2an$; $n = 1, 2, \ldots$ |
| 12 | $\dfrac{1}{p}\coth(ap)$, $\ a > 0$ | $f(x) = (2n - 1)$ if $2a(n-1) < x < 2an$; $n = 1, 2, \ldots$ |
| 13 | $\operatorname{arccoth}(p/a)$ | $\dfrac{1}{x}\sinh(ax)$ |

## T3.2.7. Expressions with Logarithmic Functions

| No. | Laplace transform, $\tilde{f}(p)$ | Inverse transform, $f(x) = \dfrac{1}{2\pi i}\int_{c-i\infty}^{c+i\infty} e^{px}\tilde{f}(p)\,dp$ |
|---|---|---|
| 1 | $\dfrac{1}{p}\ln p$ | $-\ln x - \mathcal{C}$, $\mathcal{C} = 0.5772\ldots$ is the Euler constant |
| 2 | $p^{-n-1}\ln p$ | $\left(1 + \tfrac{1}{2} + \tfrac{1}{3} + \cdots + \tfrac{1}{n} - \ln x - \mathcal{C}\right)\dfrac{x^n}{n!}$, $\mathcal{C} = 0.5772\ldots$ is the Euler constant |
| 3 | $p^{-n-1/2}\ln p$ | $k_n\left[2 + \tfrac{2}{3} + \tfrac{2}{5} + \cdots + \tfrac{2}{2n-1} - \ln(4x) - \mathcal{C}\right]x^{n-1/2}$, $k_n = \dfrac{2^n}{1 \times 3 \times 5 \times \ldots \times (2n-1)\sqrt{\pi}}$, $\mathcal{C} = 0.5772\ldots$ |
| 4 | $p^{-\nu}\ln p$, $\ \nu > 0$ | $\dfrac{1}{\Gamma(\nu)}x^{\nu-1}\left[\psi(\nu) - \ln x\right]$, $\psi(\nu)$ is the logarithmic derivative of the gamma function |
| 5 | $\dfrac{1}{p}(\ln p)^2$ | $(\ln x + \mathcal{C})^2 - \tfrac{1}{6}\pi^2$, $\mathcal{C} = 0.5772\ldots$ |

| No. | Laplace transform, $\widetilde{f}(p)$ | Inverse transform, $f(x) = \dfrac{1}{2\pi i}\displaystyle\int_{c-i\infty}^{c+i\infty} e^{px}\widetilde{f}(p)\,dp$ |
|---|---|---|
| 6 | $\dfrac{1}{p^2}(\ln p)^2$ | $x\left[(\ln x + C - 1)^2 + 1 - \tfrac{1}{6}\pi^2\right]$ |
| 7 | $\dfrac{\ln(p+b)}{p+a}$ | $e^{-ax}\left\{\ln(b-a) - \operatorname{Ei}\bigl[(a-b)x\bigr]\right\}$ |
| 8 | $\dfrac{\ln p}{p^2+a^2}$ | $\dfrac{1}{a}\cos(ax)\operatorname{Si}(ax) + \dfrac{1}{a}\sin(ax)\bigl[\ln a - \operatorname{Ci}(ax)\bigr]$ |
| 9 | $\dfrac{p\ln p}{p^2+a^2}$ | $\cos(ax)\bigl[\ln a - \operatorname{Ci}(ax)\bigr] - \sin(ax)\operatorname{Si}(ax)$ |
| 10 | $\ln\dfrac{p+b}{p+a}$ | $\dfrac{1}{x}\left(e^{-ax} - e^{-bx}\right)$ |
| 11 | $\ln\dfrac{p^2+b^2}{p^2+a^2}$ | $\dfrac{2}{x}\bigl[\cos(ax) - \cos(bx)\bigr]$ |
| 12 | $p\ln\dfrac{p^2+b^2}{p^2+a^2}$ | $\dfrac{2}{x}\bigl[\cos(bx) + bx\sin(bx) - \cos(ax) - ax\sin(ax)\bigr]$ |
| 13 | $\ln\dfrac{(p+a)^2+k^2}{(p+b)^2+k^2}$ | $\dfrac{2}{x}\cos(kx)(e^{-bx} - e^{-ax})$ |
| 14 | $p\ln\!\left(\dfrac{1}{p}\sqrt{p^2+a^2}\right)$ | $\dfrac{1}{x^2}\bigl[\cos(ax) - 1\bigr] + \dfrac{a}{x}\sin(ax)$ |
| 15 | $p\ln\!\left(\dfrac{1}{p}\sqrt{p^2-a^2}\right)$ | $\dfrac{1}{x^2}\bigl[\cosh(ax) - 1\bigr] - \dfrac{a}{x}\sinh(ax)$ |

## T3.2.8. Expressions with Trigonometric Functions

| No. | Laplace transform, $\widetilde{f}(p)$ | Inverse transform, $f(x) = \dfrac{1}{2\pi i}\displaystyle\int_{c-i\infty}^{c+i\infty} e^{px}\widetilde{f}(p)\,dp$ |
|---|---|---|
| 1 | $\dfrac{\sin(a/p)}{\sqrt{p}}$ | $\dfrac{1}{\sqrt{\pi x}}\sinh\!\bigl(\sqrt{2ax}\bigr)\sin\!\bigl(\sqrt{2ax}\bigr)$ |
| 2 | $\dfrac{\sin(a/p)}{p\sqrt{p}}$ | $\dfrac{1}{\sqrt{\pi a}}\cosh\!\bigl(\sqrt{2ax}\bigr)\sin\!\bigl(\sqrt{2ax}\bigr)$ |
| 3 | $\dfrac{\cos(a/p)}{\sqrt{p}}$ | $\dfrac{1}{\sqrt{\pi x}}\cosh\!\bigl(\sqrt{2ax}\bigr)\cos\!\bigl(\sqrt{2ax}\bigr)$ |
| 4 | $\dfrac{\cos(a/p)}{p\sqrt{p}}$ | $\dfrac{1}{\sqrt{\pi a}}\sinh\!\bigl(\sqrt{2ax}\bigr)\cos\!\bigl(\sqrt{2ax}\bigr)$ |
| 5 | $\dfrac{1}{\sqrt{p}}\exp(-\sqrt{ap})\sin(\sqrt{ap})$ | $\dfrac{1}{\sqrt{\pi x}}\sin\!\left(\dfrac{a}{2x}\right)$ |
| 6 | $\dfrac{1}{\sqrt{p}}\exp(-\sqrt{ap})\cos(\sqrt{ap})$ | $\dfrac{1}{\sqrt{\pi x}}\cos\!\left(\dfrac{a}{2x}\right)$ |
| 7 | $\arctan\dfrac{a}{p}$ | $\dfrac{1}{x}\sin(ax)$ |
| 8 | $\dfrac{1}{p}\arctan\dfrac{a}{p}$ | $\operatorname{Si}(ax)$ |
| 9 | $p\arctan\dfrac{a}{p} - a$ | $\dfrac{1}{x^2}\bigl[ax\cos(ax) - \sin(ax)\bigr]$ |
| 10 | $\arctan\dfrac{2ap}{p^2+b^2}$ | $\dfrac{2}{x}\sin(ax)\cos\!\bigl(x\sqrt{a^2+b^2}\bigr)$ |

## T3.2.9. Expressions with Special Functions

| No. | Laplace transform, $\widetilde{f}(p)$ | Inverse transform, $f(x) = \dfrac{1}{2\pi i}\int_{c-i\infty}^{c+i\infty} e^{px}\widetilde{f}(p)\,dp$ |
|---|---|---|
| 1 | $\exp(ap^2)\operatorname{erfc}(p\sqrt{a})$ | $\dfrac{1}{\sqrt{\pi a}}\exp\!\left(-\dfrac{x^2}{4a}\right)$ |
| 2 | $\dfrac{1}{p}\exp(ap^2)\operatorname{erfc}(p\sqrt{a})$ | $\operatorname{erf}\!\left(\dfrac{x}{2\sqrt{a}}\right)$ |
| 3 | $\operatorname{erfc}(\sqrt{ap}),\quad a>0$ | $\begin{cases} 0 & \text{if } 0<x<a, \\ \dfrac{\sqrt{a}}{\pi x\sqrt{x-a}} & \text{if } a<x \end{cases}$ |
| 4 | $\dfrac{1}{\sqrt{p}}\operatorname{erfc}(\sqrt{ap}),\quad a>0$ | $\begin{cases} 0 & \text{if } 0<x<a, \\ \dfrac{1}{\sqrt{\pi x}} & \text{if } x>a \end{cases}$ |
| 5 | $e^{ap}\operatorname{erfc}(\sqrt{ap})$ | $\dfrac{\sqrt{a}}{\pi\sqrt{x}\,(x+a)}$ |
| 6 | $\dfrac{1}{\sqrt{p}}e^{ap}\operatorname{erfc}(\sqrt{ap})$ | $\dfrac{1}{\sqrt{\pi(x+a)}}$ |
| 7 | $\dfrac{1}{\sqrt{p}}\operatorname{erf}(\sqrt{ap}),\quad a>0$ | $\begin{cases} \dfrac{1}{\sqrt{\pi x}} & \text{if } 0<x<a, \\ 0 & \text{if } x>a \end{cases}$ |
| 8 | $\operatorname{erf}(\sqrt{a/p})$ | $\dfrac{1}{\pi x}\sin(2\sqrt{ax})$ |
| 9 | $\dfrac{1}{\sqrt{p}}\exp(a/p)\operatorname{erf}(\sqrt{a/p})$ | $\dfrac{1}{\sqrt{\pi x}}\sinh(2\sqrt{ax})$ |
| 10 | $\dfrac{1}{\sqrt{p}}\exp(a/p)\operatorname{erfc}(\sqrt{a/p})$ | $\dfrac{1}{\sqrt{\pi x}}\exp(-2\sqrt{ax})$ |
| 11 | $p^{-a}\gamma(a,bp),\quad a,b>0$ | $\begin{cases} x^{a-1} & \text{if } 0<x<b, \\ 0 & \text{if } b<x \end{cases}$ |
| 12 | $\gamma(a,b/p),\quad a>0$ | $b^{a/2}x^{a/2-1}J_a(2\sqrt{bx})$ |
| 13 | $a^{-p}\gamma(p,a)$ | $\exp(-ae^{-x})$ |
| 14 | $K_0(ap),\quad a>0$ | $\begin{cases} 0 & \text{if } 0<x<a, \\ (x^2-a^2)^{-1/2} & \text{if } a<x \end{cases}$ |
| 15 | $K_\nu(ap),\quad a>0$ | $\begin{cases} 0 & \text{if } 0<x<a, \\ \dfrac{\cosh[\nu\operatorname{arccosh}(x/a)]}{\sqrt{x^2-a^2}} & \text{if } a<x \end{cases}$ |
| 16 | $K_0(a\sqrt{p})$ | $\dfrac{1}{2x}\exp\!\left(-\dfrac{a^2}{4x}\right)$ |
| 17 | $\dfrac{1}{\sqrt{p}}K_1(a\sqrt{p})$ | $\dfrac{1}{a}\exp\!\left(-\dfrac{a^2}{4x}\right)$ |

# T3.3. Tables of Fourier Cosine Transforms
## T3.3.1. General Formulas

| No. | Original function, $f(x)$ | Cosine transform, $\check{f}_c(u) = \int_0^\infty f(x)\cos(ux)\,dx$ |
|---|---|---|
| 1 | $af_1(x) + bf_2(x)$ | $a\check{f}_{1c}(u) + b\check{f}_{2c}(u)$ |
| 2 | $f(ax)$, $a > 0$ | $\dfrac{1}{a}\check{f}_c\!\left(\dfrac{u}{a}\right)$ |
| 3 | $x^{2n}f(x)$, $n = 1, 2, \ldots$ | $(-1)^n \dfrac{d^{2n}}{du^{2n}}\check{f}_c(u)$ |
| 4 | $x^{2n+1}f(ax)$, $n = 0, 1, \ldots$ | $(-1)^n \dfrac{d^{2n+1}}{du^{2n+1}}\check{f}_s(u)$,  $\check{f}_s(u) = \int_0^\infty f(x)\sin(xu)\,dx$ |
| 5 | $f(ax)\cos(bx)$, $a, b > 0$ | $\dfrac{1}{2a}\left[\check{f}_c\!\left(\dfrac{u+b}{a}\right) + \check{f}_c\!\left(\dfrac{u-b}{a}\right)\right]$ |

## T3.3.2. Expressions with Power-Law Functions

| No. | Original function, $f(x)$ | Cosine transform, $\check{f}_c(u) = \int_0^\infty f(x)\cos(ux)\,dx$ |
|---|---|---|
| 1 | $\begin{cases} 1 & \text{if } 0 < x < a, \\ 0 & \text{if } a < x \end{cases}$ | $\dfrac{1}{u}\sin(au)$ |
| 2 | $\begin{cases} x & \text{if } 0 < x < 1, \\ 2 - x & \text{if } 1 < x < 2, \\ 0 & \text{if } 2 < x \end{cases}$ | $\dfrac{4}{u^2}\cos u \, \sin^2\dfrac{u}{2}$ |
| 3 | $\dfrac{1}{a+x}$, $a > 0$ | $-\sin(au)\operatorname{si}(au) - \cos(au)\operatorname{Ci}(au)$ |
| 4 | $\dfrac{1}{a^2 + x^2}$, $a > 0$ | $\dfrac{\pi}{2a}e^{-au}$ |
| 5 | $\dfrac{1}{a^2 - x^2}$, $a > 0$ | $\dfrac{\pi \sin(au)}{2u}$ (the integral is understood in the sense of Cauchy principal value) |
| 6 | $\dfrac{a}{a^2 + (b+x)^2} + \dfrac{a}{a^2 + (b-x)^2}$ | $\pi e^{-au}\cos(bu)$ |
| 7 | $\dfrac{b+x}{a^2 + (b+x)^2} + \dfrac{b-x}{a^2 + (b-x)^2}$ | $\pi e^{-au}\sin(bu)$ |
| 8 | $\dfrac{1}{a^4 + x^4}$, $a > 0$ | $\tfrac{1}{2}\pi a^{-3}\exp\!\left(-\dfrac{au}{\sqrt{2}}\right)\sin\!\left(\dfrac{\pi}{4} + \dfrac{au}{\sqrt{2}}\right)$ |
| 9 | $\dfrac{1}{(a^2+x^2)(b^2+x^2)}$, $a, b > 0$ | $\dfrac{\pi}{2}\dfrac{ae^{-bu} - be^{-au}}{ab(a^2 - b^2)}$ |
| 10 | $\dfrac{x^{2m}}{(x^2+a)^{n+1}}$, $n, m = 1, 2, \ldots$; $n+1 > m \geq 0$ | $(-1)^{n+m}\dfrac{\pi}{2n!}\dfrac{\partial^n}{\partial a^n}\!\left(a^{1/\sqrt{m}}e^{-u\sqrt{a}}\right)$ |
| 11 | $\dfrac{1}{\sqrt{x}}$ | $\sqrt{\dfrac{\pi}{2u}}$ |
| 12 | $\begin{cases} \dfrac{1}{\sqrt{x}} & \text{if } 0 < x < a, \\ 0 & \text{if } a < x \end{cases}$ | $2\sqrt{\dfrac{\pi}{2u}}\,C(au)$,  $C(u)$ is the Fresnel integral |

| No. | Original function, $f(x)$ | Cosine transform, $\check{f}_c(u) = \int_0^\infty f(x)\cos(ux)\,dx$ |
|---|---|---|
| 13 | $\begin{cases} 0 & \text{if } 0 < x < a, \\ \dfrac{1}{\sqrt{x}} & \text{if } a < x \end{cases}$ | $\sqrt{\dfrac{\pi}{2u}}\,[1 - 2C(au)]$, $C(u)$ is the Fresnel integral |
| 14 | $\begin{cases} 0 & \text{if } 0 < x < a, \\ \dfrac{1}{\sqrt{x-a}} & \text{if } a < x \end{cases}$ | $\sqrt{\dfrac{\pi}{2u}}\,[\cos(au) - \sin(au)]$ |
| 15 | $\dfrac{1}{\sqrt{a^2 + x^2}}$ | $K_0(au)$ |
| 16 | $\begin{cases} \dfrac{1}{\sqrt{a^2 - x^2}} & \text{if } 0 < x < a, \\ 0 & \text{if } a < x \end{cases}$ | $\dfrac{\pi}{2}\,J_0(au)$ |
| 17 | $(a^2 + x^2)^{-1/2}\left[(a^2 + x^2)^{1/2} + a\right]^{1/2}$ | $(2u/\pi)^{-1/2} e^{-au}$, $a > 0$ |
| 18 | $x^{-\nu}$, $0 < \nu < 1$ | $\sin\!\left(\tfrac{1}{2}\pi\nu\right)\Gamma(1-\nu)u^{\nu-1}$ |

## T3.3.3. Expressions with Exponential Functions

| No. | Original function, $f(x)$ | Cosine transform, $\check{f}_c(u) = \int_0^\infty f(x)\cos(ux)\,dx$ |
|---|---|---|
| 1 | $e^{-ax}$ | $\dfrac{a}{a^2 + u^2}$ |
| 2 | $\dfrac{1}{x}\left(e^{-ax} - e^{-bx}\right)$ | $\dfrac{1}{2}\ln\dfrac{b^2 + u^2}{a^2 + u^2}$ |
| 3 | $\sqrt{x}\,e^{-ax}$ | $\tfrac{1}{2}\sqrt{\pi}\,(a^2 + u^2)^{-3/4}\cos\!\left(\tfrac{3}{2}\arctan\dfrac{u}{a}\right)$ |
| 4 | $\dfrac{1}{\sqrt{x}}\,e^{-ax}$ | $\sqrt{\dfrac{\pi}{2}}\left[\dfrac{a + (a^2 + u^2)^{1/2}}{a^2 + u^2}\right]^{1/2}$ |
| 5 | $x^n e^{-ax}$, $n = 1, 2, \ldots$ | $\dfrac{a^{n+1} n!}{(a^2 + u^2)^{n+1}} \sum_{0 \le 2k \le n+1} (-1)^k C_{n+1}^{2k}\left(\dfrac{u}{a}\right)^{2k}$ |
| 6 | $x^{n-1/2} e^{-ax}$, $n = 1, 2, \ldots$ | $k_n u \dfrac{\partial^n}{\partial a^n}\dfrac{1}{r\sqrt{r-a}}$, where $r = \sqrt{a^2 + u^2}$, $k_n = (-1)^n \sqrt{\pi/2}$ |
| 7 | $x^{\nu-1} e^{-ax}$ | $\Gamma(\nu)(a^2 + u^2)^{-\nu/2}\cos\!\left(\nu\arctan\dfrac{u}{a}\right)$ |
| 8 | $\dfrac{x}{e^{ax} - 1}$ | $\dfrac{1}{2u^2} - \dfrac{\pi^2}{2a^2 \sinh^2(\pi a^{-1} u)}$ |
| 9 | $\dfrac{1}{x}\left(\dfrac{1}{2} - \dfrac{1}{x} + \dfrac{1}{e^x - 1}\right)$ | $-\dfrac{1}{2}\ln\!\left(1 - e^{-2\pi u}\right)$ |
| 10 | $\exp(-ax^2)$ | $\dfrac{1}{2}\sqrt{\dfrac{\pi}{a}}\exp\!\left(-\dfrac{u^2}{4a}\right)$ |
| 11 | $\dfrac{1}{\sqrt{x}}\exp\!\left(-\dfrac{a}{x}\right)$ | $\sqrt{\dfrac{\pi}{2u}}\,e^{-\sqrt{2au}}\left[\cos(\sqrt{2au}) - \sin(\sqrt{2au})\right]$ |
| 12 | $\dfrac{1}{x\sqrt{x}}\exp\!\left(-\dfrac{a}{x}\right)$ | $\sqrt{\dfrac{\pi}{a}}\,e^{-\sqrt{2au}}\cos(\sqrt{2au})$ |

## T3.3.4. Expressions with Hyperbolic Functions

| No. | Original function, $f(x)$ | Cosine transform, $\check{f}_c(u) = \int_0^\infty f(x)\cos(ux)\,dx$ |
|---|---|---|
| 1 | $\dfrac{1}{\cosh(ax)}, \quad a>0$ | $\dfrac{\pi}{2a\cosh\left(\frac{1}{2}\pi a^{-1}u\right)}$ |
| 2 | $\dfrac{1}{\cosh^2(ax)}, \quad a>0$ | $\dfrac{\pi u}{2a^2\sinh\left(\frac{1}{2}\pi a^{-1}u\right)}$ |
| 3 | $\dfrac{\cosh(ax)}{\cosh(bx)}, \quad |a|<b$ | $\dfrac{\pi}{b}\left[\dfrac{\cos\left(\frac{1}{2}\pi ab^{-1}\right)\cosh\left(\frac{1}{2}\pi b^{-1}u\right)}{\cos\left(\pi ab^{-1}\right)+\cosh\left(\pi b^{-1}u\right)}\right]$ |
| 4 | $\dfrac{1}{\cosh(ax)+\cos b}$ | $\dfrac{\pi\sinh\left(a^{-1}bu\right)}{a\sin b\sinh\left(\pi a^{-1}u\right)}$ |
| 5 | $\exp(-ax^2)\cosh(bx), \quad a>0$ | $\dfrac{1}{2}\sqrt{\dfrac{\pi}{a}}\exp\left(\dfrac{b^2-u^2}{4a}\right)\cos\left(\dfrac{abu}{2a}\right)$ |
| 6 | $\dfrac{x}{\sinh(ax)}$ | $\dfrac{\pi^2}{4a^2\cosh^2\left(\frac{1}{2}\pi a^{-1}u\right)}$ |
| 7 | $\dfrac{\sinh(ax)}{\sinh(bx)}, \quad |a|<b$ | $\dfrac{\pi}{2b}\dfrac{\sin\left(\pi ab^{-1}\right)}{\cos\left(\pi ab^{-1}\right)+\cosh\left(\pi b^{-1}u\right)}$ |
| 8 | $\dfrac{1}{x}\tanh(ax), \quad a>0$ | $\ln\left[\coth\left(\tfrac{1}{4}\pi a^{-1}u\right)\right]$ |

## T3.3.5. Expressions with Logarithmic Functions

| No. | Original function, $f(x)$ | Cosine transform, $\check{f}_c(u) = \int_0^\infty f(x)\cos(ux)\,dx$ |
|---|---|---|
| 1 | $\begin{cases}\ln x & \text{if } 0<x<1,\\ 0 & \text{if } 1<x\end{cases}$ | $-\dfrac{1}{u}\operatorname{Si}(u)$ |
| 2 | $\dfrac{\ln x}{\sqrt{x}}$ | $-\sqrt{\dfrac{\pi}{2u}}\left[\ln(4u)+\mathcal{C}+\dfrac{\pi}{2}\right],$<br>$\mathcal{C}=0.5772\ldots$ is the Euler constant |
| 3 | $x^{\nu-1}\ln x, \quad 0<\nu<1$ | $\Gamma(\nu)\cos\left(\dfrac{\pi\nu}{2}\right)u^{-\nu}\left[\psi(\nu)-\dfrac{\pi}{2}\tan\left(\dfrac{\pi\nu}{2}\right)-\ln u\right]$ |
| 4 | $\ln\left|\dfrac{a+x}{a-x}\right|, \quad a>0$ | $\dfrac{2}{u}\left[\cos(au)\operatorname{Si}(au)-\sin(au)\operatorname{Ci}(au)\right]$ |
| 5 | $\ln\left(1+a^2/x^2\right), \quad a>0$ | $\dfrac{\pi}{u}\left(1-e^{-au}\right)$ |
| 6 | $\ln\dfrac{a^2+x^2}{b^2+x^2}, \quad a,b>0$ | $\dfrac{\pi}{u}\left(e^{-bu}-e^{-au}\right)$ |
| 7 | $e^{-ax}\ln x, \quad a>0$ | $-\dfrac{a\mathcal{C}+\frac{1}{2}a\ln(u^2+a^2)+u\arctan(u/a)}{u^2+a^2}$ |
| 8 | $\ln\left(1+e^{-ax}\right), \quad a>0$ | $\dfrac{a}{2u^2}-\dfrac{\pi}{2u\sinh\left(\pi a^{-1}u\right)}$ |
| 9 | $\ln\left(1-e^{-ax}\right), \quad a>0$ | $\dfrac{a}{2u^2}-\dfrac{\pi}{2u}\coth\left(\pi a^{-1}u\right)$ |

## T3.3.6. Expressions with Trigonometric Functions

| No. | Original function, $f(x)$ | Cosine transform, $\check{f}_c(u) = \int_0^\infty f(x)\cos(ux)\,dx$ |
|---|---|---|
| 1 | $\dfrac{\sin(ax)}{x}, \quad a > 0$ | $\begin{cases} \frac{1}{2}\pi & \text{if } u < a, \\ \frac{1}{4}\pi & \text{if } u = a, \\ 0 & \text{if } u > a \end{cases}$ |
| 2 | $x^{\nu-1}\sin(ax), \quad a > 0,\ \|\nu\| < 1$ | $\pi\,\dfrac{(u+a)^{-\nu} - \|u+a\|^{-\nu}\operatorname{sign}(u-a)}{4\Gamma(1-\nu)\cos\left(\frac{1}{2}\pi\nu\right)}$ |
| 3 | $\dfrac{x\sin(ax)}{x^2+b^2}, \quad a,b > 0$ | $\begin{cases} \frac{1}{2}\pi e^{-ab}\cosh(bu) & \text{if } u < a, \\ -\frac{1}{2}\pi e^{-bu}\sinh(ab) & \text{if } u > a \end{cases}$ |
| 4 | $\dfrac{\sin(ax)}{x(x^2+b^2)}, \quad a,b > 0$ | $\begin{cases} \frac{1}{2}\pi b^{-2}\left[1 - e^{-ab}\cosh(bu)\right] & \text{if } u < a, \\ \frac{1}{2}\pi b^{-2} e^{-bu}\sinh(ab) & \text{if } u > a \end{cases}$ |
| 5 | $e^{-bx}\sin(ax), \quad a,b > 0$ | $\dfrac{1}{2}\left[\dfrac{a+u}{(a+u)^2+b^2} + \dfrac{a-u}{(a-u)^2+b^2}\right]$ |
| 6 | $\dfrac{1}{x}\sin^2(ax), \quad a > 0$ | $\dfrac{1}{4}\ln\left\|1 - 4\dfrac{a^2}{u^2}\right\|$ |
| 7 | $\dfrac{1}{x^2}\sin^2(ax), \quad a > 0$ | $\begin{cases} \frac{1}{4}\pi(2a-u) & \text{if } u < 2a, \\ 0 & \text{if } u > 2a \end{cases}$ |
| 8 | $\dfrac{1}{x}\sin\left(\dfrac{a}{x}\right), \quad a > 0$ | $\dfrac{\pi}{2} J_0(2\sqrt{au})$ |
| 9 | $\dfrac{1}{\sqrt{x}}\sin(a\sqrt{x})\sin(b\sqrt{x}), \quad a,b > 0$ | $\sqrt{\dfrac{\pi}{u}}\,\sin\left(\dfrac{ab}{2u}\right)\sin\left(\dfrac{a^2+b^2}{4u} - \dfrac{\pi}{4}\right)$ |
| 10 | $\sin(ax^2), \quad a > 0$ | $\sqrt{\dfrac{\pi}{8a}}\left[\cos\left(\dfrac{u^2}{4a}\right) - \sin\left(\dfrac{u^2}{4a}\right)\right]$ |
| 11 | $\exp(-ax^2)\sin(bx^2), \quad a > 0$ | $\dfrac{\sqrt{\pi}}{(A^2+B^2)^{1/4}}\exp\left(-\dfrac{Au^2}{A^2+B^2}\right)\sin\left(\varphi - \dfrac{Bu^2}{A^2+B^2}\right),$ $A = 4a,\ B = 4b,\ \varphi = \tfrac{1}{2}\arctan(b/a)$ |
| 12 | $\dfrac{1-\cos(ax)}{x}, \quad a > 0$ | $\dfrac{1}{2}\ln\left\|1 - \dfrac{a^2}{u^2}\right\|$ |
| 13 | $\dfrac{1-\cos(ax)}{x^2}, \quad a > 0$ | $\begin{cases} \frac{1}{2}\pi(a-u) & \text{if } u < a, \\ 0 & \text{if } u > a \end{cases}$ |
| 14 | $x^{\nu-1}\cos(ax), \quad a > 0,\ 0 < \nu < 1$ | $\frac{1}{2}\Gamma(\nu)\cos\left(\tfrac{1}{2}\pi\nu\right)\left[\,\|u-a\|^{-\nu} + (u+a)^{-\nu}\right]$ |
| 15 | $\dfrac{\cos(ax)}{x^2+b^2}, \quad a,b > 0$ | $\begin{cases} \frac{1}{2}\pi b^{-1}e^{-ab}\cosh(bu) & \text{if } u < a, \\ \frac{1}{2}\pi b^{-1}e^{-bu}\cosh(ab) & \text{if } u > a \end{cases}$ |
| 16 | $e^{-bx}\cos(ax), \quad a,b > 0$ | $\dfrac{b}{2}\left[\dfrac{1}{(a+u)^2+b^2} + \dfrac{1}{(a-u)^2+b^2}\right]$ |
| 17 | $\dfrac{1}{\sqrt{x}}\cos(a\sqrt{x})$ | $\sqrt{\dfrac{\pi}{u}}\,\sin\left(\dfrac{a^2}{4u} + \dfrac{\pi}{4}\right)$ |
| 18 | $\dfrac{1}{\sqrt{x}}\cos(a\sqrt{x})\cos(b\sqrt{x})$ | $\sqrt{\dfrac{\pi}{u}}\,\cos\left(\dfrac{ab}{2u}\right)\sin\left(\dfrac{a^2+b^2}{4u} + \dfrac{\pi}{4}\right)$ |
| 19 | $\exp(-bx^2)\cos(ax), \quad b > 0$ | $\dfrac{1}{2}\sqrt{\dfrac{\pi}{b}}\exp\left(-\dfrac{a^2+u^2}{4b}\right)\cosh\left(\dfrac{au}{2b}\right)$ |
| 20 | $\cos(ax^2), \quad a > 0$ | $\sqrt{\dfrac{\pi}{8a}}\left[\cos\left(\tfrac{1}{4}a^{-1}u^2\right) + \sin\left(\tfrac{1}{4}a^{-1}u^2\right)\right]$ |
| 21 | $\exp(-ax^2)\cos(bx^2), \quad a > 0$ | $\dfrac{\sqrt{\pi}}{(A^2+B^2)^{1/4}}\exp\left(-\dfrac{Au^2}{A^2+B^2}\right)\cos\left(\varphi - \dfrac{Bu^2}{A^2+B^2}\right),$ $A = 4a,\ B = 4b,\ \varphi = \tfrac{1}{2}\arctan(b/a)$ |

## T3.3.7. Expressions with Special Functions

| No. | Original function, $f(x)$ | Cosine transform, $\check{f}_c(u) = \int_0^\infty f(x)\cos(ux)\,dx$ |
|---|---|---|
| 1 | $\mathrm{Ei}(-ax)$ | $-\dfrac{1}{u}\arctan\left(\dfrac{u}{a}\right)$ |
| 2 | $\mathrm{Ci}(ax)$ | $\begin{cases} 0 & \text{if } 0<u<a, \\ -\dfrac{\pi}{2u} & \text{if } a<u \end{cases}$ |
| 3 | $\mathrm{si}(ax)$ | $-\dfrac{1}{2u}\ln\left\|\dfrac{u+a}{u-a}\right\|, \quad u\neq a$ |
| 4 | $J_0(ax), \quad a>0$ | $\begin{cases} (a^2-u^2)^{-1/2} & \text{if } 0<u<a, \\ 0 & \text{if } a<u \end{cases}$ |
| 5 | $J_\nu(ax), \quad a>0,\ \nu>-1$ | $\begin{cases} \dfrac{\cos[\nu\arcsin(u/a)]}{\sqrt{a^2-u^2}} & \text{if } 0<u<a, \\ -\dfrac{a^\nu\sin(\pi\nu/2)}{\xi(u+\xi)^\nu} & \text{if } a<u, \end{cases}$ <br> where $\xi=\sqrt{u^2-a^2}$ |
| 6 | $\dfrac{1}{x}J_\nu(ax), \quad a>0,\ \nu>0$ | $\begin{cases} \nu^{-1}\cos[\nu\arcsin(u/a)] & \text{if } 0<u<a, \\ \dfrac{a^\nu\cos(\pi\nu/2)}{\nu(u+\sqrt{u^2-a^2})^\nu} & \text{if } a<u \end{cases}$ |
| 7 | $x^{-\nu}J_\nu(ax), \quad a>0,\ \nu>-\tfrac{1}{2}$ | $\begin{cases} \dfrac{\sqrt{\pi}\,(a^2-u^2)^{\nu-1/2}}{(2a)^\nu\,\Gamma(\nu+\tfrac{1}{2})} & \text{if } 0<u<a, \\ 0 & \text{if } a<u \end{cases}$ |
| 8 | $x^{\nu+1}J_\nu(ax),$ <br> $a>0,\ -1<\nu<-\tfrac{1}{2}$ | $\begin{cases} 0 & \text{if } 0<u<a, \\ \dfrac{2^{\nu+1}\sqrt{\pi}\,a^\nu u}{\Gamma(-\nu-\tfrac{1}{2})(u^2-a^2)^{\nu+3/2}} & \text{if } a<u \end{cases}$ |
| 9 | $J_0(a\sqrt{x}), \quad a>0$ | $\dfrac{1}{u}\sin\left(\dfrac{a^2}{4u}\right)$ |
| 10 | $\dfrac{1}{\sqrt{x}}J_1(a\sqrt{x}), \quad a>0$ | $\dfrac{4}{a}\sin^2\left(\dfrac{a^2}{8u}\right)$ |
| 11 | $x^{\nu/2}J_\nu(a\sqrt{x}), \quad a>0,\ -1<\nu<\tfrac{1}{2}$ | $\left(\dfrac{a}{2}\right)^\nu u^{-\nu-1}\sin\left(\dfrac{a^2}{4u}-\dfrac{\pi\nu}{2}\right)$ |
| 12 | $J_0(a\sqrt{x^2+b^2})$ | $\begin{cases} \dfrac{\cos(b\sqrt{a^2-u^2})}{\sqrt{a^2-u^2}} & \text{if } 0<u<a, \\ 0 & \text{if } a<u \end{cases}$ |
| 13 | $Y_0(ax), \quad a>0$ | $\begin{cases} 0 & \text{if } 0<u<a, \\ -(u^2-a^2)^{-1/2} & \text{if } a<u \end{cases}$ |
| 14 | $x^\nu Y_\nu(ax), \quad a>0,\ \|\nu\|<\tfrac{1}{2}$ | $\begin{cases} 0 & \text{if } 0<u<a, \\ -\dfrac{(2a)^\nu\sqrt{\pi}}{\Gamma(\tfrac{1}{2}-\nu)(u^2-a^2)^{\nu+1/2}} & \text{if } a<u \end{cases}$ |
| 15 | $K_0(a\sqrt{x^2+b^2}), \quad a,b>0$ | $\dfrac{\pi}{2\sqrt{u^2+a^2}}\exp(-b\sqrt{u^2+a^2})$ |

# T3.4. Tables of Fourier Sine Transforms

## T3.4.1. General Formulas

| No. | Original function, $f(x)$ | Sine transform, $\check{f}_s(u) = \int_0^\infty f(x)\sin(ux)\,dx$ |
|---|---|---|
| 1 | $af_1(x) + bf_2(x)$ | $a\check{f}_{1s}(u) + b\check{f}_{2s}(u)$ |
| 2 | $f(ax)$, $a > 0$ | $\dfrac{1}{a}\check{f}_s\left(\dfrac{u}{a}\right)$ |
| 3 | $x^{2n}f(x)$, $n = 1, 2, \ldots$ | $(-1)^n \dfrac{d^{2n}}{du^{2n}}\check{f}_s(u)$ |
| 4 | $x^{2n+1}f(ax)$, $n = 0, 1, \ldots$ | $(-1)^{n+1} \dfrac{d^{2n+1}}{du^{2n+1}}\check{f}_c(u)$, $\check{f}_c(u) = \int_0^\infty f(x)\cos(xu)\,dx$ |
| 5 | $f(ax)\cos(bx)$, $a, b > 0$ | $\dfrac{1}{2a}\left[\check{f}_s\left(\dfrac{u+b}{a}\right) + \check{f}_s\left(\dfrac{u-b}{a}\right)\right]$ |

## T3.4.2. Expressions with Power-Law Functions

| No. | Original function, $f(x)$ | Sine transform, $\check{f}_s(u) = \int_0^\infty f(x)\sin(ux)\,dx$ |
|---|---|---|
| 1 | $\begin{cases} 1 & \text{if } 0 < x < a, \\ 0 & \text{if } a < x \end{cases}$ | $\dfrac{1}{u}[1 - \cos(au)]$ |
| 2 | $\begin{cases} x & \text{if } 0 < x < 1, \\ 2-x & \text{if } 1 < x < 2, \\ 0 & \text{if } 2 < x \end{cases}$ | $\dfrac{4}{u^2}\sin u \sin^2\dfrac{u}{2}$ |
| 3 | $\dfrac{1}{x}$ | $\dfrac{\pi}{2}$ |
| 4 | $\dfrac{1}{a+x}$, $a > 0$ | $\sin(au)\operatorname{Ci}(au) - \cos(au)\operatorname{si}(au)$ |
| 5 | $\dfrac{x}{a^2+x^2}$, $a > 0$ | $\dfrac{\pi}{2}e^{-au}$ |
| 6 | $\dfrac{1}{x(a^2+x^2)}$, $a > 0$ | $\dfrac{\pi}{2a^2}\left(1 - e^{-au}\right)$ |
| 7 | $\dfrac{a}{a^2+(x-b)^2} - \dfrac{a}{a^2+(x+b)^2}$ | $\pi e^{-au}\sin(bu)$ |
| 8 | $\dfrac{x+b}{a^2+(x+b)^2} - \dfrac{x-b}{a^2+(x-b)^2}$ | $\pi e^{-au}\cos(bu)$ |
| 9 | $\dfrac{x}{(x^2+a^2)^n}$, $a > 0$, $n = 1, 2, \ldots$ | $\dfrac{\pi u e^{-au}}{2^{2n-2}(n-1)!\,a^{2n-3}}\displaystyle\sum_{k=0}^{n-2}\dfrac{(2n-k-4)!}{k!(n-k-2)!}(2au)^k$ |
| 10 | $\dfrac{x^{2m+1}}{(x^2+a)^{n+1}}$, $n, m = 0, 1, \ldots;\ 0 \le m \le n$ | $(-1)^{n+m}\dfrac{\pi}{2n!}\dfrac{\partial^n}{\partial a^n}\left(a^m e^{-u\sqrt{a}}\right)$ |
| 11 | $\dfrac{1}{\sqrt{x}}$ | $\sqrt{\dfrac{\pi}{2u}}$ |
| 12 | $\dfrac{1}{x\sqrt{x}}$ | $\sqrt{2\pi u}$ |

| No. | Original function, $f(x)$ | Sine transform, $\check{f}_s(u) = \int_0^\infty f(x)\sin(ux)\,dx$ |
|---|---|---|
| 13 | $x(a^2+x^2)^{-3/2}$ | $uK_0(au)$ |
| 14 | $\dfrac{\left(\sqrt{a^2+x^2}-a\right)^{1/2}}{\sqrt{a^2+x^2}}$ | $\sqrt{\dfrac{\pi}{2u}}\,e^{-au}$ |
| 15 | $x^{-\nu},\quad 0<\nu<2$ | $\cos\left(\tfrac{1}{2}\pi\nu\right)\Gamma(1-\nu)u^{\nu-1}$ |

## T3.4.3. Expressions with Exponential Functions

| No. | Original function, $f(x)$ | Sine transform, $\check{f}_s(u) = \int_0^\infty f(x)\sin(ux)\,dx$ |
|---|---|---|
| 1 | $e^{-ax},\quad a>0$ | $\dfrac{u}{a^2+u^2}$ |
| 2 | $x^n e^{-ax},\quad a>0,\; n=1,2,\ldots$ | $n!\left(\dfrac{a}{a^2+u^2}\right)^{n+1}\sum_{k=0}^{[n/2]}(-1)^k C_{n+1}^{2k+1}\left(\dfrac{u}{a}\right)^{2k+1}$ |
| 3 | $\dfrac{1}{x}e^{-ax},\quad a>0$ | $\arctan\dfrac{u}{a}$ |
| 4 | $\sqrt{x}\,e^{-ax},\quad a>0$ | $\dfrac{\sqrt{\pi}}{2}(a^2+u^2)^{-3/4}\sin\left(\dfrac{3}{2}\arctan\dfrac{u}{a}\right)$ |
| 5 | $\dfrac{1}{\sqrt{x}}e^{-ax},\quad a>0$ | $\sqrt{\dfrac{\pi}{2}}\,\dfrac{\left(\sqrt{a^2+u^2}-a\right)^{1/2}}{\sqrt{a^2+u^2}}$ |
| 6 | $\dfrac{1}{x\sqrt{x}}e^{-ax},\quad a>0$ | $\sqrt{2\pi}\left(\sqrt{a^2+u^2}-a\right)^{1/2}$ |
| 7 | $x^{n-1/2}e^{-ax},\quad a>0,\; n=1,2,\ldots$ | $(-1)^n\sqrt{\dfrac{\pi}{2}}\,\dfrac{\partial^n}{\partial a^n}\left[\dfrac{\left(\sqrt{a^2+u^2}-a\right)^{1/2}}{\sqrt{a^2+u^2}}\right]$ |
| 8 | $x^{\nu-1}e^{-ax},\quad a>0,\;\nu>-1$ | $\Gamma(\nu)(a^2+u^2)^{-\nu/2}\sin\left(\nu\arctan\dfrac{u}{a}\right)$ |
| 9 | $x^{-2}\left(e^{-ax}-e^{-bx}\right),\quad a,b>0$ | $\dfrac{u}{2}\ln\left(\dfrac{u^2+b^2}{u^2+a^2}\right)+b\arctan\left(\dfrac{u}{b}\right)-a\arctan\left(\dfrac{u}{a}\right)$ |
| 10 | $\dfrac{1}{e^{ax}+1},\quad a>0$ | $\dfrac{1}{2u}-\dfrac{\pi}{2a\sinh(\pi u/a)}$ |
| 11 | $\dfrac{1}{e^{ax}-1},\quad a>0$ | $\dfrac{\pi}{2a}\coth\left(\dfrac{\pi u}{a}\right)-\dfrac{1}{2u}$ |
| 12 | $\dfrac{e^{x/2}}{e^x-1}$ | $-\tfrac{1}{2}\tanh(\pi u)$ |
| 13 | $x\exp(-ax^2)$ | $\dfrac{\sqrt{\pi}}{4a^{3/2}}\,u\exp\left(-\dfrac{u^2}{4a}\right)$ |
| 14 | $\dfrac{1}{x}\exp(-ax^2)$ | $\dfrac{\pi}{2}\,\mathrm{erf}\left(\dfrac{u}{2\sqrt{a}}\right)$ |
| 15 | $\dfrac{1}{\sqrt{x}}\exp\left(-\dfrac{a}{x}\right)$ | $\sqrt{\dfrac{\pi}{2u}}\,e^{-\sqrt{2au}}\left[\cos\left(\sqrt{2au}\right)+\sin\left(\sqrt{2au}\right)\right]$ |
| 16 | $\dfrac{1}{x\sqrt{x}}\exp\left(-\dfrac{a}{x}\right)$ | $\sqrt{\dfrac{\pi}{a}}\,e^{-\sqrt{2au}}\sin\left(\sqrt{2au}\right)$ |

## T3.4.4. Expressions with Hyperbolic Functions

| No. | Original function, $f(x)$ | Sine transform, $\check{f}_s(u) = \int_0^\infty f(x)\sin(ux)\,dx$ |
|---|---|---|
| 1 | $\dfrac{1}{\sinh(ax)}$, $\quad a > 0$ | $\dfrac{\pi}{2a}\tanh\!\left(\tfrac{1}{2}\pi a^{-1}u\right)$ |
| 2 | $\dfrac{x}{\sinh(ax)}$, $\quad a > 0$ | $\dfrac{\pi^2 \sinh\!\left(\tfrac{1}{2}\pi a^{-1}u\right)}{4a^2 \cosh^2\!\left(\tfrac{1}{2}\pi a^{-1}u\right)}$ |
| 3 | $\dfrac{1}{x}e^{-bx}\sinh(ax)$, $\quad b > \lvert a\rvert$ | $\tfrac{1}{2}\arctan\!\left(\dfrac{2au}{u^2+b^2-a^2}\right)$ |
| 4 | $\dfrac{1}{x\cosh(ax)}$, $\quad a > 0$ | $\arctan\!\left[\sinh\!\left(\tfrac{1}{2}\pi a^{-1}u\right)\right]$ |
| 5 | $1 - \tanh\!\left(\tfrac{1}{2}ax\right)$, $\quad a > 0$ | $\dfrac{1}{u} - \dfrac{\pi}{a\sinh(\pi a^{-1}u)}$ |
| 6 | $\coth\!\left(\tfrac{1}{2}ax\right) - 1$, $\quad a > 0$ | $\dfrac{\pi}{a}\coth(\pi a^{-1}u) - \dfrac{1}{u}$ |
| 7 | $\dfrac{\cosh(ax)}{\sinh(bx)}$, $\quad \lvert a\rvert < b$ | $\dfrac{\pi}{2b}\dfrac{\sinh(\pi b^{-1}u)}{\cos(\pi ab^{-1}) + \cosh(\pi b^{-1}u)}$ |
| 8 | $\dfrac{\sinh(ax)}{\cosh(bx)}$, $\quad \lvert a\rvert < b$ | $\dfrac{\pi}{b}\dfrac{\sin\!\left(\tfrac{1}{2}\pi ab^{-1}\right)\sinh\!\left(\tfrac{1}{2}\pi b^{-1}u\right)}{\cos(\pi ab^{-1}) + \cosh(\pi b^{-1}u)}$ |

## T3.4.5. Expressions with Logarithmic Functions

| No. | Original function, $f(x)$ | Sine transform, $\check{f}_s(u) = \int_0^\infty f(x)\sin(ux)\,dx$ |
|---|---|---|
| 1 | $\begin{cases}\ln x & \text{if } 0 < x < 1,\\ 0 & \text{if } 1 < x\end{cases}$ | $\dfrac{1}{u}[\operatorname{Ci}(u) - \ln u - \mathcal{C}]$, $\mathcal{C} = 0.5772\ldots$ is the Euler constant |
| 2 | $\dfrac{\ln x}{x}$ | $-\tfrac{1}{2}\pi(\ln u + \mathcal{C})$ |
| 3 | $\dfrac{\ln x}{\sqrt{x}}$ | $-\sqrt{\dfrac{\pi}{2u}}\left[\ln(4u) + \mathcal{C} - \dfrac{\pi}{2}\right]$ |
| 4 | $x^{\nu-1}\ln x$, $\quad \lvert\nu\rvert < 1$ | $\dfrac{\pi u^{-\nu}\left[\psi(\nu) + \tfrac{\pi}{2}\cot\!\left(\tfrac{\pi\nu}{2}\right) - \ln u\right]}{2\Gamma(1-\nu)\cos\!\left(\tfrac{\pi\nu}{2}\right)}$ |
| 5 | $\ln\left\lvert\dfrac{a+x}{a-x}\right\rvert$, $\quad a > 0$ | $\dfrac{\pi}{u}\sin(au)$ |
| 6 | $\ln\dfrac{(x+b)^2 + a^2}{(x-b)^2 + a^2}$, $\quad a,b > 0$ | $\dfrac{2\pi}{u}e^{-au}\sin(bu)$ |
| 7 | $e^{-ax}\ln x$, $\quad a > 0$ | $\dfrac{a\arctan(u/a) - \tfrac{1}{2}u\ln(u^2+a^2) - e^{\mathcal{C}}u}{u^2 + a^2}$ |
| 8 | $\dfrac{1}{x}\ln(1 + a^2 x^2)$, $\quad a > 0$ | $-\pi\operatorname{Ei}\!\left(-\dfrac{u}{a}\right)$ |

## T3.4.6. Expressions with Trigonometric Functions

| No. | Original function, $f(x)$ | Sine transform, $\check{f}_s(u) = \int_0^\infty f(x)\sin(ux)\,dx$ |
|---|---|---|
| 1 | $\dfrac{\sin(ax)}{x}$, $a > 0$ | $\dfrac{1}{2}\ln\left\|\dfrac{u+a}{u-a}\right\|$ |
| 2 | $\dfrac{\sin(ax)}{x^2}$, $a > 0$ | $\begin{cases}\frac{1}{2}\pi u & \text{if } 0 < u < a,\\ \frac{1}{2}\pi a & \text{if } u > a\end{cases}$ |
| 3 | $x^{\nu-1}\sin(ax)$, $a > 0$, $-2 < \nu < 1$ | $\pi\,\dfrac{\|u-a\|^{-\nu} - \|u+a\|^{-\nu}}{4\Gamma(1-\nu)\sin\left(\frac{1}{2}\pi\nu\right)}$, $\nu \neq 0$ |
| 4 | $\dfrac{\sin(ax)}{x^2 + b^2}$, $a, b > 0$ | $\begin{cases}\frac{1}{2}\pi b^{-1} e^{-ab}\sinh(bu) & \text{if } 0 < u < a,\\ \frac{1}{2}\pi b^{-1} e^{-bu}\sinh(ab) & \text{if } u > a\end{cases}$ |
| 5 | $\dfrac{\sin(\pi x)}{1 - x^2}$ | $\begin{cases}\sin u & \text{if } 0 < u < \pi,\\ 0 & \text{if } u > \pi\end{cases}$ |
| 6 | $e^{-ax}\sin(bx)$, $a > 0$ | $\dfrac{a}{2}\left[\dfrac{1}{a^2 + (b-u)^2} - \dfrac{1}{a^2 + (b+u)^2}\right]$ |
| 7 | $x^{-1}e^{-ax}\sin(bx)$, $a > 0$ | $\dfrac{1}{4}\ln\dfrac{(u+b)^2 + a^2}{(u-b)^2 + a^2}$ |
| 8 | $\dfrac{1}{x}\sin^2(ax)$, $a > 0$ | $\begin{cases}\frac{1}{4}\pi & \text{if } 0 < u < 2a,\\ \frac{1}{8}\pi & \text{if } u = 2a,\\ 0 & \text{if } u > 2a\end{cases}$ |
| 9 | $\dfrac{1}{x^2}\sin^2(ax)$, $a > 0$ | $\frac{1}{4}(u + 2a)\ln\|u + 2a\| + \frac{1}{4}(u - 2a)\ln\|u - 2a\| - \frac{1}{2}u\ln u$ |
| 10 | $\exp(-ax^2)\sin(bx)$, $a > 0$ | $\dfrac{1}{2}\sqrt{\dfrac{\pi}{a}}\exp\left(-\dfrac{u^2 + b^2}{4a}\right)\sinh\left(\dfrac{bu}{2a}\right)$ |
| 11 | $\dfrac{1}{x}\sin(ax)\sin(bx)$, $a \geq b > 0$ | $\begin{cases}0 & \text{if } 0 < u < a-b,\\ \frac{\pi}{4} & \text{if } a-b < u < a+b,\\ 0 & \text{if } a+b < u\end{cases}$ |
| 12 | $\sin\left(\dfrac{a}{x}\right)$, $a > 0$ | $\dfrac{\pi\sqrt{a}}{2\sqrt{u}}J_1(2\sqrt{au})$ |
| 13 | $\dfrac{1}{\sqrt{x}}\sin\left(\dfrac{a}{x}\right)$, $a > 0$ | $\sqrt{\dfrac{\pi}{8u}}\left[\sin(2\sqrt{au}) - \cos(2\sqrt{au}) + \exp(-2\sqrt{au})\right]$ |
| 14 | $\exp(-a\sqrt{x})\sin(a\sqrt{x})$, $a > 0$ | $a\sqrt{\dfrac{\pi}{8}}\,u^{-3/2}\exp\left(-\dfrac{a^2}{2u}\right)$ |
| 15 | $\dfrac{\cos(ax)}{x}$, $a > 0$ | $\begin{cases}0 & \text{if } 0 < u < a,\\ \frac{1}{4}\pi & \text{if } u = a,\\ \frac{1}{2}\pi & \text{if } a < u\end{cases}$ |
| 16 | $x^{\nu-1}\cos(ax)$, $a > 0$, $\|\nu\| < 1$ | $\dfrac{\pi(u+a)^{-\nu} - \text{sign}(u-a)\|u-a\|^{-\nu}}{4\Gamma(1-\nu)\cos\left(\frac{1}{2}\pi\nu\right)}$ |
| 17 | $\dfrac{x\cos(ax)}{x^2 + b^2}$, $a, b > 0$ | $\begin{cases}-\frac{1}{2}\pi e^{-ab}\sinh(bu) & \text{if } u < a,\\ \frac{1}{2}\pi e^{-bu}\cosh(ab) & \text{if } u > a\end{cases}$ |
| 18 | $\dfrac{1 - \cos(ax)}{x^2}$, $a > 0$ | $\dfrac{u}{2}\ln\left\|\dfrac{u^2 - a^2}{u^2}\right\| + \dfrac{a}{2}\ln\left\|\dfrac{u+a}{u-a}\right\|$ |
| 19 | $\dfrac{1}{\sqrt{x}}\cos(a\sqrt{x})$ | $\sqrt{\dfrac{\pi}{u}}\cos\left(\dfrac{a^2}{4u} + \dfrac{\pi}{4}\right)$ |
| 20 | $\dfrac{1}{\sqrt{x}}\cos(a\sqrt{x})\cos(b\sqrt{x})$, $a, b > 0$ | $\sqrt{\dfrac{\pi}{u}}\cos\left(\dfrac{ab}{2u}\right)\cos\left(\dfrac{a^2 + b^2}{4u} + \dfrac{\pi}{4}\right)$ |

## T3.4.7. Expressions with Special Functions

| No. | Original function, $f(x)$ | Sine transform, $\check{f}_s(u) = \int_0^\infty f(x)\sin(ux)\,dx$ |
|---|---|---|
| 1 | $\operatorname{erfc}(ax)$, $a>0$ | $\dfrac{1}{u}\left[1-\exp\left(-\dfrac{u^2}{4a^2}\right)\right]$ |
| 2 | $\operatorname{ci}(ax)$, $a>0$ | $-\dfrac{1}{2u}\ln\left\|1-\dfrac{u^2}{a^2}\right\|$ |
| 3 | $\operatorname{si}(ax)$, $a>0$ | $\begin{cases} 0 & \text{if } 0<u<a, \\ -\tfrac{1}{2}\pi u^{-1} & \text{if } a<u \end{cases}$ |
| 4 | $J_0(ax)$, $a>0$ | $\begin{cases} 0 & \text{if } 0<u<a, \\ \dfrac{1}{\sqrt{u^2-a^2}} & \text{if } a<u \end{cases}$ |
| 5 | $J_\nu(ax)$, $a>0$, $\nu>-2$ | $\begin{cases} \dfrac{\sin[\nu\arcsin(u/a)]}{\sqrt{a^2-u^2}} & \text{if } 0<u<a, \\ \dfrac{a^\nu\cos(\pi\nu/2)}{\xi(u+\xi)^\nu} & \text{if } a<u, \end{cases}$ where $\xi=\sqrt{u^2-a^2}$ |
| 6 | $\dfrac{1}{x}J_0(ax)$, $a>0$, $\nu>0$ | $\begin{cases} \arcsin(u/a) & \text{if } 0<u<a, \\ \pi/2 & \text{if } a<u \end{cases}$ |
| 7 | $\dfrac{1}{x}J_\nu(ax)$, $a>0$, $\nu>-1$ | $\begin{cases} \nu^{-1}\sin[\nu\arcsin(u/a)] & \text{if } 0<u<a, \\ \dfrac{a^\nu\sin(\pi\nu/2)}{\nu(u+\sqrt{u^2-a^2})^\nu} & \text{if } a<u \end{cases}$ |
| 8 | $x^\nu J_\nu(ax)$, $a>0$, $-1<\nu<\tfrac{1}{2}$ | $\begin{cases} 0 & \text{if } 0<u<a, \\ \dfrac{\sqrt{\pi}(2a)^\nu}{\Gamma\!\left(\tfrac{1}{2}-\nu\right)(u^2-a^2)^{\nu+1/2}} & \text{if } a<u \end{cases}$ |
| 9 | $x^{-1}e^{-ax}J_0(bx)$, $a>0$ | $\arcsin\left(\dfrac{2u}{\sqrt{(u+b)^2+a^2}+\sqrt{(u-b)^2+a^2}}\right)$ |
| 10 | $\dfrac{J_0(ax)}{x^2+b^2}$, $a,b>0$ | $\begin{cases} b^{-1}\sinh(bu)K_0(ab) & \text{if } 0<u<a, \\ 0 & \text{if } a<u \end{cases}$ |
| 11 | $\dfrac{xJ_0(ax)}{x^2+b^2}$, $a,b>0$ | $\begin{cases} 0 & \text{if } 0<u<a, \\ \tfrac{1}{2}\pi e^{-bu}I_0(ab) & \text{if } a<u \end{cases}$ |
| 12 | $\dfrac{\sqrt{x}J_{2n+1/2}(ax)}{x^2+b^2}$, $a,b>0$, $n=0,1,2,\ldots$ | $\begin{cases} (-1)^n\sinh(bu)K_{2n+1/2}(ab) & \text{if } 0<u<a, \\ 0 & \text{if } a<u \end{cases}$ |
| 13 | $\dfrac{x^\nu J_\nu(ax)}{x^2+b^2}$, $a,b>0$, $-1<\nu<\tfrac{5}{2}$ | $\begin{cases} b^{\nu-1}\sinh(bu)K_\nu(ab) & \text{if } 0<u<a, \\ 0 & \text{if } a<u \end{cases}$ |
| 14 | $\dfrac{x^{1-\nu}J_\nu(ax)}{x^2+b^2}$, $a,b>0$, $\nu>-\tfrac{3}{2}$ | $\begin{cases} 0 & \text{if } 0<u<a, \\ \tfrac{1}{2}\pi b^{-\nu}e^{-bu}I_\nu(ab) & \text{if } a<u \end{cases}$ |
| 15 | $J_0(a\sqrt{x})$, $a>0$ | $\dfrac{1}{u}\cos\!\left(\dfrac{a^2}{4u}\right)$ |
| 16 | $\dfrac{1}{\sqrt{x}}J_1(a\sqrt{x})$, $a>0$ | $\dfrac{2}{a}\sin\!\left(\dfrac{a^2}{4u}\right)$ |
| 17 | $x^{\nu/2}J_\nu(a\sqrt{x})$, $a>0$, $-2<\nu<\tfrac{1}{2}$ | $\dfrac{a^\nu}{2^\nu u^{\nu+1}}\cos\!\left(\dfrac{a^2}{4u}-\dfrac{\pi\nu}{2}\right)$ |

| No. | Original function, $f(x)$ | Sine transform, $\check{f}_s(u) = \int_0^\infty f(x)\sin(ux)\,dx$ |
|---|---|---|
| 18 | $Y_0(ax)$, $a>0$ | $\begin{cases} \dfrac{2\arcsin(u/a)}{\pi\sqrt{a^2-u^2}} & \text{if } 0<u<a, \\ \dfrac{2\left[\ln\left(u-\sqrt{u^2-a^2}\right)-\ln a\right]}{\pi\sqrt{u^2-a^2}} & \text{if } a<u \end{cases}$ |
| 19 | $Y_1(ax)$, $a>0$ | $\begin{cases} 0 & \text{if } 0<u<a, \\ -\dfrac{u}{a\sqrt{u^2-a^2}} & \text{if } a<u \end{cases}$ |
| 20 | $K_0(ax)$, $a>0$ | $\dfrac{\ln\left(u+\sqrt{u^2+a^2}\right)-\ln a}{\sqrt{u^2+a^2}}$ |
| 21 | $xK_0(ax)$, $a>0$ | $\dfrac{\pi u}{2(u^2+a^2)^{3/2}}$ |
| 22 | $x^{\nu+1}K_\nu(ax)$, $a>0$, $\nu>-\tfrac{3}{2}$ | $\sqrt{\pi}\,(2a)^\nu \Gamma\!\left(\nu+\tfrac{3}{2}\right) u(u^2+a^2)^{-\nu-3/2}$ |

## T3.5. Tables of Mellin Transforms

### T3.5.1. General Formulas

| No. | Original function, $f(x)$ | Mellin transform, $\hat{f}(s) = \int_0^\infty f(x)x^{s-1}\,dx$ |
|---|---|---|
| 1 | $af_1(x)+bf_2(x)$ | $a\hat{f}_1(s)+b\hat{f}_2(s)$ |
| 2 | $f(ax)$, $a>0$ | $a^{-s}\hat{f}(s)$ |
| 3 | $x^a f(x)$ | $\hat{f}(s+a)$ |
| 4 | $f(1/x)$ | $\hat{f}(-s)$ |
| 5 | $f(x^\beta)$, $\beta>0$ | $\dfrac{1}{\beta}\hat{f}\!\left(\dfrac{s}{\beta}\right)$ |
| 6 | $f(x^{-\beta})$, $\beta>0$ | $\dfrac{1}{\beta}\hat{f}\!\left(-\dfrac{s}{\beta}\right)$ |
| 7 | $x^\lambda f(ax^\beta)$, $a,\beta>0$ | $\dfrac{1}{\beta}a^{-\frac{s+\lambda}{\beta}}\hat{f}\!\left(\dfrac{s+\lambda}{\beta}\right)$ |
| 8 | $x^\lambda f(ax^{-\beta})$, $a,\beta>0$ | $\dfrac{1}{\beta}a^{\frac{s+\lambda}{\beta}}\hat{f}\!\left(-\dfrac{s+\lambda}{\beta}\right)$ |
| 9 | $f'_x(x)$ | $-(s-1)\hat{f}(s-1)$ |
| 10 | $xf'_x(x)$ | $-s\hat{f}(s)$ |
| 11 | $f^{(n)}_x(x)$ | $(-1)^n \dfrac{\Gamma(s)}{\Gamma(s-n)}\hat{f}(s-n)$ |
| 12 | $\left(x\dfrac{d}{dx}\right)^n f(x)$ | $(-1)^n s^n \hat{f}(s)$ |
| 13 | $\left(\dfrac{d}{dx}x\right)^n f(x)$ | $(-1)^n (s-1)^n \hat{f}(s)$ |
| 14 | $x^\alpha \int_0^\infty t^\beta f_1(xt)f_2(t)\,dt$ | $\hat{f}_1(s+\alpha)\hat{f}_2(1-s-\alpha+\beta)$ |
| 15 | $x^\alpha \int_0^\infty t^\beta f_1\!\left(\dfrac{x}{t}\right)f_2(t)\,dt$ | $\hat{f}_1(s+\alpha)\hat{f}_2(s+\alpha+\beta+1)$ |

## T3.5.2. Expressions with Power-Law Functions

| No. | Original function, $f(x)$ | Mellin transform, $\hat{f}(s) = \int_0^\infty f(x) x^{s-1}\, dx$ |
|---|---|---|
| 1 | $\begin{cases} x & \text{if } 0 < x < 1, \\ 2-x & \text{if } 1 < x < 2, \\ 0 & \text{if } 2 < x \end{cases}$ | $\begin{cases} \dfrac{2(2^s - 1)}{s(s+1)} & \text{if } s \neq 0, \\ 2\ln 2 & \text{if } s = 0, \end{cases}$ $\quad \operatorname{Re} s > -1$ |
| 2 | $\dfrac{1}{x+a},\quad a > 0$ | $\dfrac{\pi a^{s-1}}{\sin(\pi s)},\quad 0 < \operatorname{Re} s < 1$ |
| 3 | $\dfrac{1}{(x+a)(x+b)},\quad a,b > 0$ | $\dfrac{\pi(a^{s-1} - b^{s-1})}{(b-a)\sin(\pi s)},\quad 0 < \operatorname{Re} s < 2$ |
| 4 | $\dfrac{x+a}{(x+b)(x+c)},\quad b,c > 0$ | $\dfrac{\pi}{\sin(\pi s)}\left[\left(\dfrac{b-a}{b-c}\right)b^{s-1} + \left(\dfrac{c-a}{c-b}\right)c^{s-1}\right],$ $0 < \operatorname{Re} s < 1$ |
| 5 | $\dfrac{1}{x^2 + a^2},\quad a > 0$ | $\dfrac{\pi a^{s-2}}{2\sin\left(\tfrac{1}{2}\pi s\right)},\quad 0 < \operatorname{Re} s < 2$ |
| 6 | $\dfrac{1}{x^2 + 2ax\cos\beta + a^2},\quad a > 0,\ |\beta| < \pi$ | $-\dfrac{\pi a^{s-2}\sin[\beta(s-1)]}{\sin\beta\,\sin(\pi s)},\quad 0 < \operatorname{Re} s < 2$ |
| 7 | $\dfrac{1}{(x^2+a^2)(x^2+b^2)},\quad a,b > 0$ | $\dfrac{\pi(a^{s-2} - b^{s-2})}{2(b^2 - a^2)\sin\left(\tfrac{1}{2}\pi s\right)},\quad 0 < \operatorname{Re} s < 4$ |
| 8 | $\dfrac{1}{(1+ax)^{n+1}},\quad a > 0,\ n = 1, 2, \ldots$ | $\dfrac{(-1)^n \pi}{a^s \sin(\pi s)} C_{s-1}^n,\quad 0 < \operatorname{Re} s < n+1$ |
| 9 | $\dfrac{1}{x^n + a^n},\quad a > 0,\ n = 1, 2, \ldots$ | $\dfrac{\pi a^{s-n}}{n \sin(\pi s/n)},\quad 0 < \operatorname{Re} s < n$ |
| 10 | $\dfrac{1-x}{1-x^n},\quad n = 2, 3, \ldots$ | $\dfrac{\pi \sin(\pi/n)}{n \sin(\pi s/n) \sin[\pi(s+1)/n]},\quad 0 < \operatorname{Re} s < n-1$ |
| 11 | $\begin{cases} x^\nu & \text{if } 0 < x < 1, \\ 0 & \text{if } 1 < x \end{cases}$ | $\dfrac{1}{s+\nu},\quad \operatorname{Re} s > -\nu$ |
| 12 | $\dfrac{1-x^\nu}{1-x^{n\nu}},\quad n = 2, 3, \ldots$ | $\dfrac{\pi \sin(\pi/n)}{n\nu \sin\left(\tfrac{\pi s}{n\nu}\right)\sin\left[\tfrac{\pi(s+\nu)}{n\nu}\right]},\quad 0 < \operatorname{Re} s < (n-1)\nu$ |

## T3.5.3. Expressions with Exponential Functions

| No. | Original function, $f(x)$ | Mellin transform, $\hat{f}(s) = \int_0^\infty f(x) x^{s-1}\, dx$ |
|---|---|---|
| 1 | $e^{-ax},\quad a > 0$ | $a^{-s}\Gamma(s),\quad \operatorname{Re} s > 0$ |
| 2 | $\begin{cases} e^{-bx} & \text{if } 0 < x < a, \\ 0 & \text{if } a < x, \end{cases}\quad b > 0$ | $b^{-s}\gamma(s, ab),\quad \operatorname{Re} s > 0$ |
| 3 | $\begin{cases} 0 & \text{if } 0 < x < a, \\ e^{-bx} & \text{if } a < x, \end{cases}\quad b > 0$ | $b^{-s}\Gamma(s, ab)$ |
| 4 | $\dfrac{e^{-ax}}{x+b},\quad a,b > 0$ | $e^{ab} b^{s-1} \Gamma(s)\Gamma(1-s, ab),\quad \operatorname{Re} s > 0$ |
| 5 | $\exp(-ax^\beta),\quad a, \beta > 0$ | $\beta^{-1} a^{-s/\beta} \Gamma(s/\beta),\quad \operatorname{Re} s > 0$ |
| 6 | $\exp(-ax^{-\beta}),\quad a, \beta > 0$ | $\beta^{-1} a^{s/\beta} \Gamma(-s/\beta),\quad \operatorname{Re} s < 0$ |
| 7 | $1 - \exp(-ax^\beta),\quad a, \beta > 0$ | $-\beta^{-1} a^{-s/\beta} \Gamma(s/\beta),\quad -\beta < \operatorname{Re} s < 0$ |
| 8 | $1 - \exp(-ax^{-\beta}),\quad a, \beta > 0$ | $-\beta^{-1} a^{s/\beta} \Gamma(-s/\beta),\quad 0 < \operatorname{Re} s < \beta$ |

## T3.5.4. Expressions with Logarithmic Functions

| No. | Original function, $f(x)$ | Mellin transform, $\hat{f}(s) = \int_0^\infty f(x) x^{s-1}\, dx$ |
|---|---|---|
| 1 | $\begin{cases} \ln x & \text{if } 0 < x < a, \\ 0 & \text{if } a < x \end{cases}$ | $\dfrac{s \ln a - 1}{s^2 a^s}, \quad \operatorname{Re} s > 0$ |
| 2 | $\ln(1 + ax), \quad a > 0$ | $\dfrac{\pi}{s a^s \sin(\pi s)}, \quad -1 < \operatorname{Re} s < 0$ |
| 3 | $\ln\lvert 1 - x\rvert$ | $\dfrac{\pi}{s} \cot(\pi s), \quad -1 < \operatorname{Re} s < 0$ |
| 4 | $\dfrac{\ln x}{x + a}, \quad a > 0$ | $\dfrac{\pi a^{s-1}\left[\ln a - \pi \cot(\pi s)\right]}{\sin(\pi s)}, \quad 0 < \operatorname{Re} s < 1$ |
| 5 | $\dfrac{\ln x}{(x+a)(x+b)}, \quad a, b > 0$ | $\dfrac{\pi\left[a^{s-1}\ln a - b^{s-1}\ln b - \pi \cot(\pi s)(a^{s-1} - b^{s-1})\right]}{(b-a)\sin(\pi s)},$ $0 < \operatorname{Re} s < 1$ |
| 6 | $\begin{cases} x^\nu \ln x & \text{if } 0 < x < 1, \\ 0 & \text{if } 1 < x \end{cases}$ | $-\dfrac{1}{(s+\nu)^2}, \quad \operatorname{Re} s > -\nu$ |
| 7 | $\dfrac{\ln^2 x}{x+1}$ | $\dfrac{\pi^3 \left[2 - \sin^2(\pi s)\right]}{\sin^3(\pi s)}, \quad 0 < \operatorname{Re} s < 1$ |
| 8 | $\begin{cases} \ln^{\nu-1} x & \text{if } 0 < x < 1, \\ 0 & \text{if } 1 < x \end{cases}$ | $\Gamma(\nu)(-s)^{-\nu}, \quad \operatorname{Re} s < 0,\ \nu > 0$ |
| 9 | $\ln(x^2 + 2x\cos\beta + 1), \quad \lvert\beta\rvert < \pi$ | $\dfrac{2\pi \cos(\beta s)}{s \sin(\pi s)}, \quad -1 < \operatorname{Re} s < 0$ |
| 10 | $\ln\left\lvert\dfrac{1+x}{1-x}\right\rvert$ | $\dfrac{\pi}{s} \tan\left(\tfrac{1}{2}\pi s\right), \quad -1 < \operatorname{Re} s < 1$ |
| 11 | $e^{-x} \ln^n x, \quad n = 1, 2, \ldots$ | $\dfrac{d^n}{ds^n} \Gamma(s), \quad \operatorname{Re} s > 0$ |

## T3.5.5. Expressions with Trigonometric Functions

| No. | Original function, $f(x)$ | Mellin transform, $\hat{f}(s) = \int_0^\infty f(x) x^{s-1}\, dx$ |
|---|---|---|
| 1 | $\sin(ax), \quad a > 0$ | $a^{-s} \Gamma(s) \sin\left(\tfrac{1}{2}\pi s\right), \quad -1 < \operatorname{Re} s < 1$ |
| 2 | $\sin^2(ax), \quad a > 0$ | $-2^{-s-1} a^{-s} \Gamma(s) \cos\left(\tfrac{1}{2}\pi s\right), \quad -2 < \operatorname{Re} s < 0$ |
| 3 | $\sin(ax)\sin(bx), \quad a, b > 0,\ a \ne b$ | $\tfrac{1}{2}\Gamma(s) \cos\left(\tfrac{1}{2}\pi s\right)\left[\lvert b-a\rvert^{-s} - (b+a)^{-s}\right],$ $-2 < \operatorname{Re} s < 1$ |
| 4 | $\cos(ax), \quad a > 0$ | $a^{-s} \Gamma(s) \cos\left(\tfrac{1}{2}\pi s\right), \quad 0 < \operatorname{Re} s < 1$ |
| 5 | $\sin(ax)\cos(bx), \quad a, b > 0$ | $\dfrac{\Gamma(s)}{2} \sin\left(\dfrac{\pi s}{2}\right)\left[(a+b)^{-s} + \lvert a-b\rvert^{-s} \operatorname{sign}(a-b)\right],$ $-1 < \operatorname{Re} s < 1$ |
| 6 | $e^{-ax} \sin(bx), \quad a > 0$ | $\dfrac{\Gamma(s)\sin\left[s\arctan(b/a)\right]}{(a^2 + b^2)^{s/2}}, \quad -1 < \operatorname{Re} s$ |
| 7 | $e^{-ax} \cos(bx), \quad a > 0$ | $\dfrac{\Gamma(s)\cos\left[s\arctan(b/a)\right]}{(a^2 + b^2)^{s/2}}, \quad 0 < \operatorname{Re} s$ |
| 8 | $\begin{cases} \sin(a \ln x) & \text{if } 0 < x < 1, \\ 0 & \text{if } 1 < x \end{cases}$ | $-\dfrac{a}{s^2 + a^2}, \quad \operatorname{Re} s > 0$ |
| 9 | $\begin{cases} \cos(a \ln x) & \text{if } 0 < x < 1, \\ 0 & \text{if } 1 < x \end{cases}$ | $\dfrac{s}{s^2 + a^2}, \quad \operatorname{Re} s > 0$ |

| No. | Original function, $f(x)$ | Mellin transform, $\hat{f}(s) = \int_0^\infty f(x)x^{s-1}\,dx$ |
|---|---|---|
| 10 | arctan $x$ | $-\dfrac{\pi}{2s\cos\left(\frac{1}{2}\pi s\right)}$, $\quad -1 < \mathrm{Re}\,s < 0$ |
| 11 | arccot $x$ | $\dfrac{\pi}{2s\cos\left(\frac{1}{2}\pi s\right)}$, $\quad 0 < \mathrm{Re}\,s < 1$ |

## T3.5.6. Expressions with Special Functions

| No. | Original function, $f(x)$ | Mellin transform, $\hat{f}(s) = \int_0^\infty f(x)x^{s-1}\,dx$ |
|---|---|---|
| 1 | erfc $x$ | $\dfrac{\Gamma\left(\frac{1}{2}s + \frac{1}{2}\right)}{\sqrt{\pi}\,s}$, $\quad \mathrm{Re}\,s > 0$ |
| 2 | Ei$(-x)$ | $-s^{-1}\Gamma(s)$, $\quad \mathrm{Re}\,s > 0$ |
| 3 | Si$(x)$ | $-s^{-1}\sin\left(\frac{1}{2}\pi s\right)\Gamma(s)$, $\quad -1 < \mathrm{Re}\,s < 0$ |
| 4 | si$(x)$ | $-4s^{-1}\sin\left(\frac{1}{2}\pi s\right)\Gamma(s)$, $\quad -1 < \mathrm{Re}\,s < 0$ |
| 5 | Ci$(x)$ | $-s^{-1}\cos\left(\frac{1}{2}\pi s\right)\Gamma(s)$, $\quad 0 < \mathrm{Re}\,s < 1$ |
| 6 | $J_\nu(ax)$, $\quad a > 0$ | $\dfrac{2^{s-1}\Gamma\left(\frac{1}{2}\nu + \frac{1}{2}s\right)}{a^s\Gamma\left(\frac{1}{2}\nu - \frac{1}{2}s + 1\right)}$, $\quad -\nu < \mathrm{Re}\,s < \frac{3}{2}$ |
| 7 | $Y_\nu(ax)$, $\quad a > 0$ | $-\dfrac{2^{s-1}}{\pi a^s}\Gamma\left(\dfrac{s}{2} + \dfrac{\nu}{2}\right)\Gamma\left(\dfrac{s}{2} - \dfrac{\nu}{2}\right)\cos\left[\dfrac{\pi(s-\nu)}{2}\right]$, $\quad |\nu| < \mathrm{Re}\,s < \frac{3}{2}$ |
| 8 | $e^{-ax}I_\nu(ax)$, $\quad a > 0$ | $\dfrac{\Gamma(1/2 - s)\Gamma(s + \nu)}{\sqrt{\pi}\,(2a)^s\Gamma(1 + \nu - s)}$, $\quad -\nu < \mathrm{Re}\,s < \frac{1}{2}$ |
| 9 | $K_\nu(ax)$, $\quad a > 0$ | $\dfrac{2^{s-2}}{a^s}\Gamma\left(\dfrac{s}{2} + \dfrac{\nu}{2}\right)\Gamma\left(\dfrac{s}{2} - \dfrac{\nu}{2}\right)$, $\quad |\nu| < \mathrm{Re}\,s$ |
| 10 | $e^{-ax}K_\nu(ax)$, $\quad a > 0$ | $\dfrac{\sqrt{\pi}\,\Gamma(s-\nu)\Gamma(s+\nu)}{(2a)^s\Gamma(s + 1/2)}$, $\quad |\nu| < \mathrm{Re}\,s$ |

# T3.6. Tables of Inverse Mellin Transforms

▶ *See Section T3.5.1 for general formulas.*

## T3.6.1. Expressions with Power-Law Functions

| No. | Direct transform, $\hat{f}(s)$ | Inverse transform, $f(x) = \dfrac{1}{2\pi i}\displaystyle\int_{\sigma-i\infty}^{\sigma+i\infty} \hat{f}(s)x^{-s}\,ds$ |
|---|---|---|
| 1 | $\dfrac{1}{s}$, $\quad \mathrm{Re}\,s > 0$ | $\begin{cases} 1 & \text{if } 0 < x < 1, \\ 0 & \text{if } 1 < x \end{cases}$ |
| 2 | $\dfrac{1}{s}$, $\quad \mathrm{Re}\,s < 0$ | $\begin{cases} 0 & \text{if } 0 < x < 1, \\ -1 & \text{if } 1 < x \end{cases}$ |
| 3 | $\dfrac{1}{s+a}$, $\quad \mathrm{Re}\,s > -a$ | $\begin{cases} x^a & \text{if } 0 < x < 1, \\ 0 & \text{if } 1 < x \end{cases}$ |
| 4 | $\dfrac{1}{s+a}$, $\quad \mathrm{Re}\,s < -a$ | $\begin{cases} 0 & \text{if } 0 < x < 1, \\ -x^a & \text{if } 1 < x \end{cases}$ |
| 5 | $\dfrac{1}{(s+a)^2}$, $\quad \mathrm{Re}\,s > -a$ | $\begin{cases} -x^a \ln x & \text{if } 0 < x < 1, \\ 0 & \text{if } 1 < x \end{cases}$ |

| No. | Direct transform, $\hat{f}(s)$ | Inverse transform, $f(x) = \dfrac{1}{2\pi i}\int_{\sigma-i\infty}^{\sigma+i\infty} \hat{f}(s) x^{-s}\, ds$ |
|---|---|---|
| 6 | $\dfrac{1}{(s+a)^2}$, $\quad \operatorname{Re} s < -a$ | $\begin{cases} 0 & \text{if } 0 < x < 1, \\ x^a \ln x & \text{if } 1 < x \end{cases}$ |
| 7 | $\dfrac{1}{(s+a)(s+b)}$, $\quad \operatorname{Re} s > -a, -b$ | $\begin{cases} \dfrac{x^a - x^b}{b-a} & \text{if } 0 < x < 1, \\ 0 & \text{if } 1 < x \end{cases}$ |
| 8 | $\dfrac{1}{(s+a)(s+b)}$, $\quad -a < \operatorname{Re} s < -b$ | $\begin{cases} \dfrac{x^a}{b-a} & \text{if } 0 < x < 1, \\ \dfrac{x^b}{b-a} & \text{if } 1 < x \end{cases}$ |
| 9 | $\dfrac{1}{(s+a)(s+b)}$, $\quad \operatorname{Re} s < -a, -b$ | $\begin{cases} 0 & \text{if } 0 < x < 1, \\ \dfrac{x^b - x^a}{b-a} & \text{if } 1 < x \end{cases}$ |
| 10 | $\dfrac{1}{(s+a)^2 + b^2}$, $\quad \operatorname{Re} s > -a$ | $\begin{cases} \dfrac{1}{b} x^a \sin\!\left(b \ln \dfrac{1}{x}\right) & \text{if } 0 < x < 1, \\ 0 & \text{if } 1 < x \end{cases}$ |
| 11 | $\dfrac{s+a}{(s+a)^2 + b^2}$, $\quad \operatorname{Re} s > -a$ | $\begin{cases} x^a \cos(b \ln x) & \text{if } 0 < x < 1, \\ 0 & \text{if } 1 < x \end{cases}$ |
| 12 | $\sqrt{s^2 - a^2} - s$, $\quad \operatorname{Re} s > \lvert a\rvert$ | $\begin{cases} -\dfrac{a}{\ln x} I_1(-a \ln x) & \text{if } 0 < x < 1, \\ 0 & \text{if } 1 < x \end{cases}$ |
| 13 | $\sqrt{\dfrac{s+a}{s-a}} - 1$, $\quad \operatorname{Re} s > \lvert a\rvert$ | $\begin{cases} a I_0(-a \ln x) + a I_1(-a \ln x) & \text{if } 0 < x < 1, \\ 0 & \text{if } 1 < x \end{cases}$ |
| 14 | $(s+a)^{-\nu}$, $\quad \operatorname{Re} s > -a,\ \nu > 0$ | $\begin{cases} \dfrac{1}{\Gamma(\nu)} x^a (-\ln x)^{\nu-1} & \text{if } 0 < x < 1, \\ 0 & \text{if } 1 < x \end{cases}$ |
| 15 | $s^{-1}(s+a)^{-\nu}$, $\operatorname{Re} s > 0,\ \operatorname{Re} s > -a,\ \nu > 0$ | $\begin{cases} a^{-\nu}[\Gamma(\nu)]^{-1} \gamma(\nu, -a \ln x) & \text{if } 0 < x < 1, \\ 0 & \text{if } 1 < x \end{cases}$ |
| 16 | $s^{-1}(s+a)^{-\nu}$, $-a < \operatorname{Re} s < 0,\ \nu > 0$ | $\begin{cases} -a^{-\nu}[\Gamma(\nu)]^{-1} \Gamma(\nu, -a \ln x) & \text{if } 0 < x < 1, \\ -a^{-\nu} & \text{if } 1 < x \end{cases}$ |
| 17 | $(s^2 - a^2)^{-\nu}$, $\quad \operatorname{Re} s > \lvert a\rvert,\ \nu > 0$ | $\begin{cases} \dfrac{\sqrt{\pi}\,(-\ln x)^{\nu-1/2} I_{\nu-1/2}(-a \ln x)}{\Gamma(\nu)(2a)^{\nu-1/2}} & \text{if } 0 < x < 1, \\ 0 & \text{if } 1 < x \end{cases}$ |
| 18 | $(a^2 - s^2)^{-\nu}$, $\quad \operatorname{Re} s < \lvert a\rvert,\ \nu > 0$ | $\begin{cases} \dfrac{(-\ln x)^{\nu-1/2} K_{\nu-1/2}(-a \ln x)}{\sqrt{\pi}\,\Gamma(\nu)(2a)^{\nu-1/2}} & \text{if } 0 < x < 1, \\ \dfrac{(\ln x)^{\nu-1/2} K_{\nu-1/2}(a \ln x)}{\sqrt{\pi}\,\Gamma(\nu)(2a)^{\nu-1/2}} & \text{if } 1 < x \end{cases}$ |

## T3.6.2. Expressions with Exponential and Logarithmic Functions

| No. | Direct transform, $\hat{f}(s)$ | Inverse transform, $f(x) = \dfrac{1}{2\pi i}\int_{\sigma-i\infty}^{\sigma+i\infty} \hat{f}(s) x^{-s}\, ds$ |
|---|---|---|
| 1 | $\exp(as^2)$, $\quad a > 0$ | $\dfrac{1}{2\sqrt{\pi a}} \exp\!\left(-\dfrac{\ln^2 x}{4a}\right)$ |
| 2 | $s^{-\nu} e^{-a/s}$, $\quad \operatorname{Re} s > 0;\ a, \nu > 0$ | $\begin{cases} \left\lvert \dfrac{a}{\ln x} \right\rvert^{\frac{1-\nu}{2}} J_{\nu-1}\!\left(2\sqrt{a\lvert\ln x\rvert}\right) & \text{if } 0 < x < 1, \\ 0 & \text{if } 1 < x \end{cases}$ |
| 3 | $\exp(-\sqrt{as})$, $\quad \operatorname{Re} s > 0,\ a > 0$ | $\begin{cases} \dfrac{(a/\pi)^{1/2}}{2\lvert\ln x\rvert^{3/2}} \exp\!\left(-\dfrac{a}{4\lvert\ln x\rvert}\right) & \text{if } 0 < x < 1, \\ 0 & \text{if } 1 < x \end{cases}$ |

| No. | Direct transform, $\hat{f}(s)$ | Inverse transform, $f(x) = \dfrac{1}{2\pi i}\int_{\sigma-i\infty}^{\sigma+i\infty}\hat{f}(s)x^{-s}\,ds$ |
|---|---|---|
| 4 | $\dfrac{1}{s}\exp(-a\sqrt{s})$, $\operatorname{Re} s > 0$ | $\begin{cases}\operatorname{erfc}\left(\dfrac{a}{2\sqrt{|\ln x|}}\right) & \text{if } 0 < x < 1,\\ 0 & \text{if } 1 < x\end{cases}$ |
| 5 | $\dfrac{1}{s}[\exp(-a\sqrt{s})-1]$, $\operatorname{Re} s > 0$ | $\begin{cases}-\operatorname{erf}\left(\dfrac{a}{2\sqrt{|\ln x|}}\right) & \text{if } 0 < x < 1,\\ 0 & \text{if } 1 < x\end{cases}$ |
| 6 | $\sqrt{s}\exp(-\sqrt{as})$, $\operatorname{Re} s > 0$ | $\begin{cases}\dfrac{a-2|\ln x|}{4\sqrt{\pi|\ln x|^5}}\exp\left(-\dfrac{a}{4|\ln x|}\right) & \text{if } 0 < x < 1,\\ 0 & \text{if } 1 < x\end{cases}$ |
| 7 | $\dfrac{1}{\sqrt{s}}\exp(-\sqrt{as})$, $\operatorname{Re} s > 0$ | $\begin{cases}\dfrac{1}{\sqrt{\pi|\ln x|}}\exp\left(-\dfrac{a}{4|\ln x|}\right) & \text{if } 0 < x < 1,\\ 0 & \text{if } 1 < x\end{cases}$ |
| 8 | $\ln\dfrac{s+a}{s+b}$, $\operatorname{Re} s > -a,-b$ | $\begin{cases}\dfrac{x^a - x^b}{\ln x} & \text{if } 0 < x < 1,\\ 0 & \text{if } 1 < x\end{cases}$ |
| 9 | $s^{-\nu}\ln s$, $\operatorname{Re} s > 0$, $\nu > 0$ | $\begin{cases}|\ln x|^{\nu-1}\dfrac{\psi(\nu) - \ln|\ln x|}{\Gamma(\nu)} & \text{if } 0 < x < 1,\\ 0 & \text{if } 1 < x\end{cases}$ |

## T3.6.3. Expressions with Trigonometric Functions

| No. | Direct transform, $\hat{f}(s)$ | Inverse transform, $f(x) = \dfrac{1}{2\pi i}\int_{\sigma-i\infty}^{\sigma+i\infty}\hat{f}(s)x^{-s}\,ds$ |
|---|---|---|
| 1 | $\dfrac{\pi}{\sin(\pi s)}$, $0 < \operatorname{Re} s < 1$ | $\dfrac{1}{x+1}$ |
| 2 | $\dfrac{\pi}{\sin(\pi s)}$, $-n < \operatorname{Re} s < 1-n$, $n = \ldots, -1, 0, 1, 2, \ldots$ | $(-1)^n \dfrac{x^n}{x+1}$ |
| 3 | $\dfrac{\pi^2}{\sin^2(\pi s)}$, $0 < \operatorname{Re} s < 1$ | $\dfrac{\ln x}{x-1}$ |
| 4 | $\dfrac{\pi^2}{\sin^2(\pi s)}$, $n < \operatorname{Re} s < n+1$, $n = \ldots, -1, 0, 1, 2, \ldots$ | $\dfrac{\ln x}{x^n(x-1)}$ |
| 5 | $\dfrac{2\pi^3}{\sin^3(\pi s)}$, $0 < \operatorname{Re} s < 1$ | $\dfrac{\pi^2 + \ln^2 x}{x+1}$ |
| 6 | $\dfrac{2\pi^3}{\sin^3(\pi s)}$, $n < \operatorname{Re} s < n+1$, $n = \ldots, -1, 0, 1, 2, \ldots$ | $\dfrac{\pi^2 + \ln^2 x}{(-x)^n(x+1)}$ |
| 7 | $\sin(s^2/a)$, $a > 0$ | $\dfrac{1}{2}\sqrt{\dfrac{a}{\pi}}\sin\left(\tfrac{1}{4}a|\ln x|^2 - \tfrac{1}{4}\pi\right)$ |
| 8 | $\dfrac{\pi}{\cos(\pi s)}$, $-\tfrac{1}{2} < \operatorname{Re} s < \tfrac{1}{2}$ | $\dfrac{\sqrt{x}}{x+1}$ |
| 9 | $\dfrac{\pi}{\cos(\pi s)}$, $n - \tfrac{1}{2} < \operatorname{Re} s < n + \tfrac{1}{2}$, $n = \ldots, -1, 0, 1, 2, \ldots$ | $(-1)^n \dfrac{x^{1/2-n}}{x+1}$ |
| 10 | $\dfrac{\cos(\beta s)}{s\cos(\pi s)}$, $-1 < \operatorname{Re} s < 0$, $|\beta| < \pi$ | $\dfrac{1}{2\pi}\ln(x^2 + 2x\cos\beta + 1)$ |

| No. | Direct transform, $\hat{f}(s)$ | Inverse transform, $f(x) = \dfrac{1}{2\pi i}\int_{\sigma-i\infty}^{\sigma+i\infty}\hat{f}(s)x^{-s}\,ds$ |
|---|---|---|
| 11 | $\cos(s^2/a)$, $a>0$ | $\dfrac{1}{2}\sqrt{\dfrac{a}{\pi}}\cos\left(\tfrac{1}{4}a\lvert\ln x\rvert^2-\tfrac{1}{4}\pi\right)$ |
| 12 | $\arctan\left(\dfrac{a}{s+b}\right)$, $\operatorname{Re} s>-b$ | $\begin{cases}\dfrac{x^b}{\lvert\ln x\rvert}\sin(a\lvert\ln x\rvert) & \text{if } 0<x<1,\\ 0 & \text{if } 1<x\end{cases}$ |

## T3.6.4. Expressions with Special Functions

| No. | Direct transform, $\hat{f}(s)$ | Inverse transform, $f(x) = \dfrac{1}{2\pi i}\int_{\sigma-i\infty}^{\sigma+i\infty}\hat{f}(s)x^{-s}\,ds$ |
|---|---|---|
| 1 | $\Gamma(s)$, $\operatorname{Re} s>0$ | $e^{-x}$ |
| 2 | $\Gamma(s)$, $-1<\operatorname{Re} s<0$ | $e^{-x}-1$ |
| 3 | $\sin\left(\tfrac{1}{2}\pi s\right)\Gamma(s)$, $-1<\operatorname{Re} s<1$ | $\sin x$ |
| 4 | $\sin(as)\Gamma(s)$, $\operatorname{Re} s>-1$, $\lvert a\rvert<\dfrac{\pi}{2}$ | $\exp(-x\cos a)\sin(x\sin a)$ |
| 5 | $\cos\left(\tfrac{1}{2}\pi s\right)\Gamma(s)$, $0<\operatorname{Re} s<1$ | $\cos x$ |
| 6 | $\cos\left(\tfrac{1}{2}\pi s\right)\Gamma(s)$, $-2<\operatorname{Re} s<0$ | $-2\sin^2(x/2)$ |
| 7 | $\cos(as)\Gamma(s)$, $\operatorname{Re} s>0$, $\lvert a\rvert<\dfrac{\pi}{2}$ | $\exp(-x\cos a)\cos(x\sin a)$ |
| 8 | $\dfrac{\Gamma(s)}{\cos(\pi s)}$, $0<\operatorname{Re} s<\dfrac{1}{2}$ | $e^x\operatorname{erfc}\left(\sqrt{x}\right)$ |
| 9 | $\Gamma(a+s)\Gamma(b-s)$, $-a<\operatorname{Re} s<b$, $a+b>0$ | $\Gamma(a+b)x^a(x+1)^{-a-b}$ |
| 10 | $\Gamma(a+s)\Gamma(b+s)$, $\operatorname{Re} s>-a,-b$ | $2x^{(a+b)/2}K_{a-b}\left(2\sqrt{x}\right)$ |
| 11 | $\dfrac{\Gamma(s)}{\Gamma(s+\nu)}$, $\operatorname{Re} s>0$, $\nu>0$ | $\begin{cases}\dfrac{(1-x)^{\nu-1}}{\Gamma(\nu)} & \text{if } 0<x<1,\\ 0 & \text{if } 1<x\end{cases}$ |
| 12 | $\dfrac{\Gamma(1-\nu-s)}{\Gamma(1-s)}$, $\operatorname{Re} s<1-\nu$, $\nu>0$ | $\begin{cases}0 & \text{if } 0<x<1,\\ \dfrac{(x-1)^{\nu-1}}{\Gamma(\nu)} & \text{if } 1<x\end{cases}$ |
| 13 | $\dfrac{\Gamma(s)}{\Gamma(\nu-s+1)}$, $0<\operatorname{Re} s<\dfrac{\nu}{2}+\dfrac{3}{4}$ | $x^{-\nu/2}J_\nu\left(2\sqrt{x}\right)$ |
| 14 | $\dfrac{\Gamma(s+\nu)\Gamma(s-\nu)}{\Gamma(s+1/2)}$, $\operatorname{Re} s>\lvert\nu\rvert$ | $\pi^{-1/2}e^{-x/2}K_\nu(x/2)$ |
| 15 | $\dfrac{\Gamma(s+\nu)\Gamma(1/2-s)}{\Gamma(1+\nu-s)}$, $-\nu<\operatorname{Re} s<\tfrac{1}{2}$ | $\pi^{1/2}e^{-x/2}I_\nu(x/2)$ |
| 16 | $\psi(s+a)-\psi(s+b)$, $\operatorname{Re} s>-a,-b$ | $\begin{cases}\dfrac{x^b-x^a}{1-x} & \text{if } 0<x<1,\\ 0 & \text{if } 1<x\end{cases}$ |
| 17 | $\Gamma(s)\psi(s)$, $\operatorname{Re} s>0$ | $e^{-x}\ln x$ |
| 18 | $\Gamma(s,a)$, $a>0$ | $\begin{cases}0 & \text{if } 0<x<a,\\ e^{-x} & \text{if } a<x\end{cases}$ |
| 19 | $\Gamma(s)\Gamma(1-s,a)$, $\operatorname{Re} s>0$, $a>0$ | $(x+1)^{-1}e^{-a(x+1)}$ |

| No. | Direct transform, $\hat{f}(s)$ | Inverse transform, $f(x) = \dfrac{1}{2\pi i} \int_{\sigma-i\infty}^{\sigma+i\infty} \hat{f}(s) x^{-s}\, ds$ |
|---|---|---|
| 20 | $\gamma(s,a)$,  $\operatorname{Re} s > 0$, $a > 0$ | $\begin{cases} e^{-x} & \text{if } 0 < x < a, \\ 0 & \text{if } a < x \end{cases}$ |
| 21 | $J_0\!\left(a\sqrt{b^2 - s^2}\right)$,  $a > 0$ | $\begin{cases} 0 & \text{if } 0 < x < e^{-a}, \\ \dfrac{\cos\!\left(b\sqrt{a^2 - \ln^2 x}\right)}{\pi\sqrt{a^2 - \ln^2 x}} & \text{if } e^{-a} < x < e^a, \\ 0 & \text{if } e^a < x \end{cases}$ |
| 22 | $s^{-1} I_0(s)$,  $\operatorname{Re} s > 0$ | $\begin{cases} 1 & \text{if } 0 < x < e^{-1}, \\ \pi^{-1} \arccos(\ln x) & \text{if } e^{-1} < x < e, \\ 0 & \text{if } e < x \end{cases}$ |
| 23 | $I_\nu(s)$,  $\operatorname{Re} s > 0$ | $\begin{cases} -\dfrac{2^\nu \sin(\pi\nu)}{\pi F(x)\sqrt{\ln^2 x - 1}} & \text{if } 0 < x < e^{-1}, \\ \dfrac{\cos[\nu \arccos(\ln x)]}{\pi\sqrt{1 - \ln^2 x}} & \text{if } e^{-1} < x < e, \\ 0 & \text{if } e < x, \end{cases}$<br>where $F(x) = \left(\sqrt{-1 - \ln x} + \sqrt{1 - \ln x}\right)^{2\nu}$ |
| 24 | $s^{-1} I_\nu(s)$,  $\operatorname{Re} s > 0$ | $\begin{cases} \dfrac{2^\nu \sin(\pi\nu)}{\pi\nu F(x)} & \text{if } 0 < x < e^{-1}, \\ \dfrac{\sin[\nu \arccos(\ln x)]}{\pi\nu} & \text{if } e^{-1} < x < e, \\ 0 & \text{if } e < x, \end{cases}$<br>where $F(x) = \left(\sqrt{-1 - \ln x} + \sqrt{1 - \ln x}\right)^{2\nu}$ |
| 25 | $s^{-\nu} I_\nu(s)$,  $\operatorname{Re} s > -\tfrac{1}{2}$ | $\begin{cases} 0 & \text{if } 0 < x < e^{-1}, \\ \dfrac{(1 - \ln^2 x)^{\nu - 1/2}}{\sqrt{\pi}\, 2^\nu \Gamma(\nu + 1/2)} & \text{if } e^{-1} < x < e, \\ 0 & \text{if } e < x \end{cases}$ |
| 26 | $s^{-1} K_0(s)$,  $\operatorname{Re} s > 0$ | $\begin{cases} \operatorname{arccosh}(-\ln x) & \text{if } 0 < x < e^{-1}, \\ 0 & \text{if } e^{-1} < x \end{cases}$ |
| 27 | $s^{-1} K_1(s)$,  $\operatorname{Re} s > 0$ | $\begin{cases} \sqrt{\ln^2 x - 1} & \text{if } 0 < x < e^{-1}, \\ 0 & \text{if } e^{-1} < x \end{cases}$ |
| 28 | $K_\nu(s)$,  $\operatorname{Re} s > 0$ | $\begin{cases} \dfrac{\cosh[\nu \operatorname{arccosh}(-\ln x)]}{\sqrt{\ln^2 x - 1}} & \text{if } 0 < x < e^{-1}, \\ 0 & \text{if } e^{-1} < x \end{cases}$ |
| 29 | $s^{-1} K_\nu(s)$,  $\operatorname{Re} s > 0$ | $\begin{cases} \dfrac{1}{\nu} \sinh[\nu \operatorname{arccosh}(-\ln x)] & \text{if } 0 < x < e^{-1}, \\ 0 & \text{if } e^{-1} < x \end{cases}$ |
| 30 | $s^{-\nu} K_\nu(s)$,  $\operatorname{Re} s > 0$, $\nu > -\tfrac{1}{2}$ | $\begin{cases} \dfrac{\sqrt{\pi}\, (\ln^2 x - 1)^{\nu - 1/2}}{2^\nu \Gamma(\nu + 1/2)} & \text{if } 0 < x < e^{-1}, \\ 0 & \text{if } e^{-1} < x \end{cases}$ |

# References for Chapter T3

**Bateman, H. and Erdélyi, A.,** *Tables of Integral Transforms. Vol. 1*, McGraw-Hill, New York, 1954.

**Bateman, H. and Erdélyi, A.,** *Tables of Integral Transforms. Vol. 2*, McGraw-Hill, New York, 1954.

**Ditkin, V. A. and Prudnikov, A. P.,** *Integral Transforms and Operational Calculus*, Pergamon Press, New York, 1965.

**Oberhettinger, F.,** *Tables of Fourier Transforms and Fourier Transforms of Distributions*, Springer-Verlag, Berlin, 1980.

**Oberhettinger, F.,** *Tables of Mellin Transforms*, Springer-Verlag, New York, 1974.

**Oberhettinger, F. and Badii, L.,** *Tables of Laplace Transforms*, Springer-Verlag, New York, 1973.

**Prudnikov, A. P., Brychkov, Yu. A., and Marichev, O. I.,** *Integrals and Series, Vol. 4, Direct Laplace Transforms*, Gordon & Breach, New York, 1992.

**Prudnikov, A. P., Brychkov, Yu. A., and Marichev, O. I.,** *Integrals and Series, Vol. 5, Inverse Laplace Transforms*, Gordon & Breach, New York, 1992.

# Chapter T4
# Orthogonal Curvilinear Systems of Coordinate

## T4.1. Arbitrary Curvilinear Coordinate Systems
### T4.1.1. General Nonorthogonal Curvilinear Coordinates

T4.1.1-1. Metric tensor. Arc length and volume elements in curvilinear coordinates.

The curvilinear coordinates $x^1$, $x^2$, $x^3$ are defined as functions of the rectangular Cartesian coordinates $x$, $y$, $z$:

$$x^1 = x^1(x,y,z), \quad x^2 = x^2(x,y,z), \quad x^3 = x^3(x,y,z).$$

Using these formulas, one can express $x$, $y$, $z$ in terms of the curvilinear coordinates $x^1$, $x^2$, $x^3$ as follows:

$$x = x(x^1,x^2,x^3), \quad y = y(x^1,x^2,x^3), \quad z = z(x^1,x^2,x^3).$$

The *metric tensor components* $g_{ij}$ are determined by the formulas

$$g_{ij}(x^1,x^2,x^3) = \frac{\partial x}{\partial x^i}\frac{\partial x}{\partial x^j} + \frac{\partial y}{\partial x^i}\frac{\partial y}{\partial x^j} + \frac{\partial z}{\partial x^i}\frac{\partial z}{\partial x^j};$$
$$g_{ij}(x^1,x^2,x^3) = g_{ji}(x^1,x^2,x^3); \quad i,j = 1, 2, 3.$$

The arc length $dl$ between close points $(x,y,z) \equiv (x^1,x^2,x^3)$ and $(x+dx, y+dy, z+dz) \equiv (x^1+dx^1, x^2+dx^2, x^3+dx^3)$ is expressed as

$$(dl)^2 = (dx)^2 + (dy)^2 + (dz)^2 = \sum_{i=1}^{3}\sum_{j=1}^{3} g_{ij}(x^1,x^2,x^3)\,dx^i\,dx^j.$$

The volume of the elementary parallelepiped with vertices at the eight points $(x^1,x^2,x^3)$, $(x^1+dx^1, x^2, x^3)$, $(x^1, x^2+dx^2, x^3)$, $(x^1, x^2, x^3+dx^3)$, $(x^1+dx^1, x^2+dx^2, x^3)$, $(x^1+dx^1, x^2, x^3+dx^3)$, $(x^1, x^2+dx^2, x^3+dx^3)$, $(x^1+dx^1, x^2+dx^2, x^3+dx^3)$ is given by

$$dV = \frac{\partial(x,y,z)}{\partial(x^1,x^2,x^3)}\,dx^1\,dx^2\,dx^3 = \pm\sqrt{\det|g_{ij}|}\,dx^1\,dx^2\,dx^3.$$

Here, the plus sign corresponds to the standard situation where the tangent vectors to the coordinate lines $x^1, x^2, x^3$, pointing toward the direction of growth of the respective coordinate, form a right-handed triple, just as unit vectors $\vec{i}, \vec{j}, \vec{k}$ of a right-handed rectangular Cartesian coordinate system.

### T4.1.1-2. Vector components in Cartesian and curvilinear coordinate systems.

The unit vectors $\vec{i}, \vec{j}, \vec{k}$ of a rectangular Cartesian coordinate system* $x, y, z$ and the unit vectors $\vec{i}_1, \vec{i}_2, \vec{i}_3$ of a curvilinear coordinate system $x^1, x^2, x^3$ are connected by the linear relations

$$\vec{i}_n = \frac{1}{\sqrt{g_{nn}}}\left(\frac{\partial x}{\partial x^n}\vec{i} + \frac{\partial y}{\partial x^n}\vec{j} + \frac{\partial z}{\partial x^n}\vec{k}\right), \quad n = 1, 2, 3;$$

$$\vec{i} = \sqrt{g_{11}}\frac{\partial x^1}{\partial x}\vec{i}_1 + \sqrt{g_{22}}\frac{\partial x^2}{\partial x}\vec{i}_2 + \sqrt{g_{33}}\frac{\partial x^3}{\partial x}\vec{i}_3;$$

$$\vec{j} = \sqrt{g_{11}}\frac{\partial x^1}{\partial y}\vec{i}_1 + \sqrt{g_{22}}\frac{\partial x^2}{\partial y}\vec{i}_2 + \sqrt{g_{33}}\frac{\partial x^3}{\partial y}\vec{i}_3;$$

$$\vec{k} = \sqrt{g_{11}}\frac{\partial x^1}{\partial z}\vec{i}_1 + \sqrt{g_{22}}\frac{\partial x^2}{\partial z}\vec{i}_2 + \sqrt{g_{33}}\frac{\partial x^3}{\partial z}\vec{i}_3.$$

In the general case, the vectors $\vec{i}_1, \vec{i}_2, \vec{i}_3$ are not orthogonal and change their direction from point to point.

The components $v_x, v_y, v_z$ of a vector $\vec{v}$ in a rectangular Cartesian coordinate system $x, y, z$ and the components $v_1, v_2, v_3$ of the same vector in a curvilinear coordinate system $x^1, x^2, x^3$ are related by

$$\vec{v} = v_x\vec{i} + v_y\vec{j} + v_z\vec{k} = v_1\vec{i}_1 + v_2\vec{i}_2 + v_3\vec{i}_3,$$

$$v_n = \sqrt{g_{nn}}\left(\frac{\partial x^n}{\partial x}v_x + \frac{\partial x^n}{\partial y}v_y + \frac{\partial x^n}{\partial z}v_z\right), \quad n = 1, 2, 3;$$

$$v_x = \frac{\partial x}{\partial x^1}\frac{v_1}{\sqrt{g_{11}}} + \frac{\partial x}{\partial x^2}\frac{v_2}{\sqrt{g_{22}}} + \frac{\partial x}{\partial x^3}\frac{v_3}{\sqrt{g_{33}}};$$

$$v_y = \frac{\partial y}{\partial x^1}\frac{v_1}{\sqrt{g_{11}}} + \frac{\partial y}{\partial x^2}\frac{v_2}{\sqrt{g_{22}}} + \frac{\partial y}{\partial x^3}\frac{v_3}{\sqrt{g_{33}}};$$

$$v_z = \frac{\partial z}{\partial x^1}\frac{v_1}{\sqrt{g_{11}}} + \frac{\partial z}{\partial x^2}\frac{v_2}{\sqrt{g_{22}}} + \frac{\partial z}{\partial x^3}\frac{v_3}{\sqrt{g_{33}}}.$$

## T4.1.2. General Orthogonal Curvilinear Coordinates

### T4.1.2-1. Orthogonal coordinates. Length, area, and volume elements.

A system of coordinates is orthogonal if

$$g_{ij}(x^1, x^2, x^3) = 0 \quad \text{for} \quad i \neq j.$$

In this case the third invariant of the metric tensor is given by

$$g = \det|g_{ij}| = g_{11}g_{22}g_{33}.$$

The *Lamé coefficients* $L_k$ of orthogonal curvilinear coordinates are expressed in terms of the components of the metric tensor as

$$L_i = \sqrt{g_{ii}} = \sqrt{\left(\frac{\partial x}{\partial x^i}\right)^2 + \left(\frac{\partial y}{\partial x^i}\right)^2 + \left(\frac{\partial z}{\partial x^i}\right)^2}, \quad i = 1, 2, 3.$$

---

* Here and henceforth the coordinate axes and the respective coordinates of points in space are denoted by the same letters.

Arc length element:
$$dl = \sqrt{(L_1\,dx^1)^2 + (L_2\,dx^2)^2 + (L_3\,dx^3)^2} = \sqrt{g_{11}(dx^1)^2 + g_{22}(dx^2)^2 + g_{33}(dx^3)^2}.$$

The area elements $ds_i$ of the respective coordinate surfaces $x^i$ = const are given by
$$ds_1 = dl_2\,dl_3 = L_2 L_3\,dx^2\,dx^3 = \sqrt{g_{22}g_{33}}\,dx^2\,dx^3,$$
$$ds_2 = dl_1\,dl_3 = L_1 L_3\,dx^1\,dx^3 = \sqrt{g_{11}g_{33}}\,dx^1\,dx^3,$$
$$ds_3 = dl_1\,dl_2 = L_1 L_2\,dx^1\,dx^2 = \sqrt{g_{11}g_{22}}\,dx^1\,dx^2.$$

Volume element:
$$dV = L_1 L_2 L_3\,dx^1\,dx^2\,dx^3 = \sqrt{g_{11}g_{22}g_{33}}\,dx^1\,dx^2\,dx^3.$$

**T4.1.2-2. Basic differential relations in orthogonal curvilinear coordinates.**

In what follows, we present the basic differential operators in the orthogonal curvilinear coordinates $x^1$, $x^2$, $x^3$. The corresponding unit vectors are denoted by $\vec{i}_1, \vec{i}_2, \vec{i}_3$.

The gradient of a scalar $f$ is expressed as
$$\operatorname{grad} f \equiv \nabla f = \frac{1}{\sqrt{g_{11}}}\frac{\partial f}{\partial x^1}\vec{i}_1 + \frac{1}{\sqrt{g_{22}}}\frac{\partial f}{\partial x^2}\vec{i}_2 + \frac{1}{\sqrt{g_{33}}}\frac{\partial f}{\partial x^3}\vec{i}_3.$$

Divergence of a vector $\vec{v} = \vec{i}_1 v_1 + \vec{i}_2 v_2 + \vec{i}_3 v_3$:
$$\operatorname{div}\vec{v} \equiv \nabla\cdot\vec{v} = \frac{1}{\sqrt{g}}\left[\frac{\partial}{\partial x^1}\left(v_1\sqrt{\frac{g}{g_{11}}}\right) + \frac{\partial}{\partial x^2}\left(v_2\sqrt{\frac{g}{g_{22}}}\right) + \frac{\partial}{\partial x^3}\left(v_3\sqrt{\frac{g}{g_{33}}}\right)\right].$$

Gradient of a scalar $f$ along a vector $\vec{v}$:
$$(\vec{v}\cdot\nabla)f = \frac{v_1}{\sqrt{g_{11}}}\frac{\partial f}{\partial x^1} + \frac{v_2}{\sqrt{g_{22}}}\frac{\partial f}{\partial x^2} + \frac{v_3}{\sqrt{g_{33}}}\frac{\partial f}{\partial x^3}.$$

Gradient of a vector $\vec{w}$ along a vector $\vec{v}$:
$$(\vec{v}\cdot\nabla)\vec{w} = \vec{i}_1(\vec{v}\cdot\nabla)w_1 + \vec{i}_2(\vec{v}\cdot\nabla)w_2 + \vec{i}_3(\vec{v}\cdot\nabla)w_3.$$

Curl of a vector $\vec{v}$:
$$\operatorname{curl}\vec{v} \equiv \nabla\times\vec{v} = \vec{i}_1\frac{\sqrt{g_{11}}}{\sqrt{g}}\left[\frac{\partial}{\partial x^2}\left(v_3\sqrt{g_{33}}\right) - \frac{\partial}{\partial x^3}\left(v_2\sqrt{g_{22}}\right)\right]$$
$$+ \vec{i}_2\frac{\sqrt{g_{22}}}{\sqrt{g}}\left[\frac{\partial}{\partial x^3}\left(v_1\sqrt{g_{11}}\right) - \frac{\partial}{\partial x^1}\left(v_3\sqrt{g_{33}}\right)\right]$$
$$+ \vec{i}_3\frac{\sqrt{g_{33}}}{\sqrt{g}}\left[\frac{\partial}{\partial x^1}\left(v_2\sqrt{g_{22}}\right) - \frac{\partial}{\partial x^2}\left(v_1\sqrt{g_{11}}\right)\right].$$

**Remark.** Sometimes curl $\vec{v}$ is denoted by rot $\vec{v}$.

Laplace operator of a scalar $f$:
$$\Delta f \equiv \nabla^2 f = \frac{1}{\sqrt{g}}\left[\frac{\partial}{\partial x^1}\left(\frac{\sqrt{g}}{g_{11}}\frac{\partial f}{\partial x^1}\right) + \frac{\partial}{\partial x^2}\left(\frac{\sqrt{g}}{g_{22}}\frac{\partial f}{\partial x^2}\right) + \frac{\partial}{\partial x^3}\left(\frac{\sqrt{g}}{g_{33}}\frac{\partial f}{\partial x^3}\right)\right].$$

## T4.2. Special Curvilinear Coordinate Systems
### T4.2.1. Cylindrical Coordinates

**T4.2.1-1. Transformations of coordinates and vectors. The metric tensor components.**

The Cartesian coordinates are expressed in terms of the cylindrical ones as
$$x = \rho \cos \varphi, \quad y = \rho \sin \varphi, \quad z = z$$
$$(0 \leq \rho < \infty, \ 0 \leq \varphi < 2\pi, \ -\infty < z < \infty).$$

The cylindrical coordinates are expressed in terms of the cylindrical ones as
$$\rho = \sqrt{x^2 + y^2}, \quad \tan \varphi = y/x, \quad z = z \quad (\sin \varphi = y/\rho).$$

Coordinate surfaces:

$x^2 + y^2 = \rho^2$ (right circular cylinders with their axis coincident with the $z$-axis),
$y = x \tan \varphi$ (half-planes through the $z$-axis),
$z = z$ (planes perpendicular to the $z$-axis).

Direct and inverse transformations of the components of a vector $\vec{v} = v_x \vec{i} + v_y \vec{j} + v_z \vec{k} = v_\rho \vec{i}_\rho + v_\varphi \vec{i}_\varphi + v_z \vec{i}_z$:

$$v_\rho = v_x \cos \varphi + v_y \sin \varphi, \qquad v_x = v_\rho \cos \varphi - v_\varphi \sin \varphi,$$
$$v_\varphi = -v_x \sin \varphi + v_y \cos \varphi, \qquad v_y = v_\rho \sin \varphi + v_\varphi \cos \varphi,$$
$$v_z = v_z; \qquad v_z = v_z.$$

Metric tensor components:
$$g_{\rho\rho} = 1, \quad g_{\varphi\varphi} = \rho^2, \quad g_{zz} = 1, \quad \sqrt{g} = \rho.$$

**T4.2.1-2. Basic differential relations.**

Gradient of a scalar $f$:
$$\nabla f = \frac{\partial f}{\partial \rho} \vec{i}_\rho + \frac{1}{\rho} \frac{\partial f}{\partial \varphi} \vec{i}_\varphi + \frac{\partial f}{\partial z} \vec{i}_z.$$

Divergence of a vector $\vec{v}$:
$$\nabla \cdot \vec{v} = \frac{1}{\rho} \frac{\partial(\rho v_\rho)}{\partial \rho} + \frac{1}{\rho} \frac{\partial v_\varphi}{\partial \varphi} + \frac{\partial v_z}{\partial z}.$$

Gradient of a scalar $f$ along a vector $\vec{v}$:
$$(\vec{v} \cdot \nabla) f = v_\rho \frac{\partial f}{\partial \rho} + \frac{v_\varphi}{\rho} \frac{\partial f}{\partial \varphi} + v_z \frac{\partial f}{\partial z}.$$

Gradient of a vector $\vec{w}$ along a vector $\vec{v}$:
$$(\vec{v} \cdot \nabla) \vec{w} = (\vec{v} \cdot \nabla) w_\rho \vec{i}_\rho + (\vec{v} \cdot \nabla) w_\varphi \vec{i}_\varphi + (\vec{v} \cdot \nabla) w_z \vec{i}_z.$$

Curl of a vector $\vec{v}$:
$$\nabla \times \vec{v} = \left( \frac{1}{\rho} \frac{\partial v_z}{\partial \varphi} - \frac{\partial v_\varphi}{\partial z} \right) \vec{i}_\rho + \left( \frac{\partial v_\rho}{\partial z} - \frac{\partial v_z}{\partial \rho} \right) \vec{i}_\varphi + \frac{1}{\rho} \left[ \frac{\partial(\rho v_\varphi)}{\partial \rho} - \frac{\partial v_\rho}{\partial \varphi} \right] \vec{i}_z.$$

Laplacian of a scalar $f$:
$$\Delta f = \frac{1}{\rho} \frac{\partial}{\partial \rho} \left( \rho \frac{\partial f}{\partial \rho} \right) + \frac{1}{\rho^2} \frac{\partial^2 f}{\partial \varphi^2} + \frac{\partial^2 f}{\partial z^2}.$$

Remark. The cylindrical coordinates $\rho, \varphi$ are also used as *polar coordinates* on the plane $xy$.

## T4.2.2. Spherical Coordinates

**T4.2.2-1. Transformations of coordinates and vectors. The metric tensor components.**

Cartesian coordinates via spherical ones:
$$x = r \sin\theta \cos\varphi, \quad y = r \sin\theta \sin\varphi, \quad z = r \cos\theta$$
$$(0 \le r < \infty, \ 0 \le \theta \le \pi, \ 0 \le \varphi < 2\pi).$$

Spherical coordinates via Cartesian ones:
$$r = \sqrt{x^2 + y^2 + z^2}, \quad \theta = \arccos\frac{z}{r}, \quad \tan\varphi = \frac{y}{x} \quad \left(\sin\varphi = \frac{y}{\sqrt{x^2+y^2}}\right).$$

Coordinate surfaces:
$$\begin{aligned} x^2 + y^2 + z^2 &= r^2 & &\text{(spheres)}, \\ x^2 + y^2 - z^2 \tan^2\theta &= 0 & &\text{(circular cones)}, \\ y &= x \tan\varphi & &\text{(half-planes trough the } z\text{-axis)}. \end{aligned}$$

Direct and inverse transformations of the components of a vector $\vec{v} = v_x\vec{i} + v_y\vec{j} + v_z\vec{k} = v_r\vec{i}_r + v_\theta\vec{i}_\theta + v_\varphi\vec{i}_\varphi$:

$$\begin{aligned} v_r &= v_x \sin\theta \cos\varphi + v_y \sin\theta \sin\varphi + v_z \cos\theta, & v_x &= v_r \sin\theta \cos\varphi + v_\theta \cos\theta \cos\varphi - v_\varphi \sin\varphi, \\ v_\theta &= v_x \cos\theta \cos\varphi + v_y \cos\theta \sin\varphi - v_z \sin\theta, & v_y &= v_r \sin\theta \sin\varphi + v_\theta \cos\theta \sin\varphi + v_\varphi \cos\varphi, \\ v_\varphi &= -v_x \sin\varphi + v_y \cos\varphi; & v_z &= v_r \cos\theta - v_\theta \sin\theta. \end{aligned}$$

The metric tensor components are
$$g_{rr} = 1, \quad g_{\theta\theta} = r^2, \quad g_{\varphi\varphi} = r^2 \sin^2\theta, \quad \sqrt{g} = r^2 \sin\theta.$$

**T4.2.2-2. Basic differential relations.**

Gradient of a scalar $f$:
$$\nabla f = \frac{\partial f}{\partial r}\vec{i}_r + \frac{1}{r}\frac{\partial f}{\partial \vartheta}\vec{i}_\theta + \frac{1}{r \sin\theta}\frac{\partial f}{\partial \varphi}\vec{i}_\varphi.$$

Divergence of a vector $\vec{v}$:
$$\nabla \cdot \vec{v} = \frac{1}{r^2}\frac{\partial}{\partial r}\left(r^2 v_r\right) + \frac{1}{r \sin\theta}\frac{\partial}{\partial \theta}\left(\sin\theta \, v_\theta\right) + \frac{1}{r \sin\varphi}\frac{\partial v_\varphi}{\partial \varphi}.$$

Gradient of a scalar $f$ along a vector $\vec{v}$:
$$(\vec{v} \cdot \nabla)f = v_r \frac{\partial f}{\partial r} + \frac{v_\theta}{r}\frac{\partial f}{\partial \theta} + \frac{v_\varphi}{r \sin\theta}\frac{\partial f}{\partial \varphi}.$$

Gradient of a vector $\vec{w}$ along a vector $\vec{v}$:
$$(\vec{v} \cdot \nabla)\vec{w} = (\vec{v} \cdot \nabla)w_r \vec{i}_r + (\vec{v} \cdot \nabla)w_\theta \vec{i}_\theta + (\vec{v} \cdot \nabla)w_\varphi \vec{i}_\varphi.$$

Curl of a vector $\vec{v}$:
$$\nabla \times \vec{v} = \frac{1}{r \sin\theta}\left[\frac{\partial(\sin\theta \, v_\varphi)}{\partial \theta} - \frac{\partial v_\theta}{\partial \varphi}\right]\vec{i}_r + \frac{1}{r}\left[\frac{1}{\sin\theta}\frac{\partial v_r}{\partial \varphi} - \frac{\partial(rv_\varphi)}{\partial r}\right]\vec{i}_\theta + \frac{1}{r}\left[\frac{\partial(rv_\theta)}{\partial r} - \frac{\partial v_r}{\partial \theta}\right]\vec{i}_\varphi.$$

Laplacian of a scalar $f$:
$$\Delta f = \frac{1}{r^2}\frac{\partial}{\partial r}\left(r^2 \frac{\partial f}{\partial r}\right) + \frac{1}{r^2 \sin\theta}\frac{\partial}{\partial \theta}\left(\sin\theta \frac{\partial f}{\partial \theta}\right) + \frac{1}{r^2 \sin^2\theta}\frac{\partial^2 f}{\partial \varphi^2}.$$

## T4.2.3. Coordinates of a Prolate Ellipsoid of Revolution

**T4.2.3-1. Transformations of coordinates. The metric tensor components.**

1°. Transformations of coordinates:
$$x^2 = a^2(\sigma^2 - 1)(1 - \tau^2)\cos^2\varphi, \quad y^2 = a^2(\sigma^2 - 1)(1 - \tau^2)\sin^2\varphi, \quad z = a\sigma\tau$$
$$(\sigma \geq 1 \geq \tau \geq -1, \ 0 \leq \varphi < 2\pi).$$

Coordinate surfaces (the $z$-axis is the axis revolution):

$$\frac{x^2 + y^2}{a^2(\sigma^2 - 1)} + \frac{z^2}{a^2\sigma^2} = 1 \quad \text{(prolate ellipsoids of revolution)},$$

$$\frac{x^2 + y^2}{a^2(\tau^2 - 1)} + \frac{z^2}{a^2\tau^2} = 1 \quad \text{(hyperboloid of revolution of two sheets)},$$

$$y = x \tan\varphi \quad \text{(half-planes through the } z\text{-axis)}.$$

Metric tensor components:
$$g_{\sigma\sigma} = a^2 \frac{\sigma^2 - \tau^2}{\sigma^2 - 1}, \quad g_{\tau\tau} = a^2 \frac{\sigma^2 - \tau^2}{1 - \tau^2}, \quad g_{\varphi\varphi} = a^2(\sigma^2 - 1)(1 - \tau^2), \quad \sqrt{g} = a^3(\sigma^2 - \tau^2).$$

2°. Special coordinate system $u, v, \varphi$:
$$\sigma = \cosh u, \quad \tau = \cos v, \quad \varphi = \varphi \quad (0 \leq u < \infty, \ 0 \leq v \leq \pi, \ 0 \leq \varphi < 2\pi);$$
$$x = a\sinh u \sin v \cos\varphi, \quad y = a\sinh u \sin v \sin\varphi, \quad z = a\cosh u \cos v.$$

Metric tensor components:
$$g_{uu} = g_{vv} = a^2(\sinh^2 u + \sin^2 v), \quad g_{\varphi\varphi} = a^2 \sinh^2 u \sin^2 v.$$

**T4.2.3-2. Basic differential relations.**

Gradient of a scalar $f$:
$$\nabla f = \frac{1}{a}\sqrt{\frac{\sigma^2 - 1}{\sigma^2 - \tau^2}}\frac{\partial f}{\partial \sigma}\vec{i}_\sigma + \frac{1}{a}\sqrt{\frac{1 - \tau^2}{\sigma^2 - \tau^2}}\frac{\partial f}{\partial \tau}\vec{i}_\tau + \frac{1}{a\sqrt{(1 - \tau^2)(\sigma^2 - 1)}}\frac{\partial f}{\partial \varphi}\vec{i}_\varphi.$$

Divergence of a vector $\vec{v}$:
$$\nabla \cdot \vec{v} = \frac{1}{a(\sigma^2 - \tau^2)}\left\{\frac{\partial}{\partial \sigma}\left[v_\sigma \sqrt{(\sigma^2 - \tau^2)(\sigma^2 - 1)}\right]\right.$$
$$\left. + \frac{\partial}{\partial \tau}\left[v_\tau \sqrt{(\sigma^2 - \tau^2)(1 - \tau^2)}\right] + \frac{\partial}{\partial \varphi}\left[v_\varphi \frac{\sigma^2 - \tau^2}{\sqrt{(\sigma^2 - 1)(1 - \tau^2)}}\right]\right\}.$$

Gradient of a scalar $f$ along a vector $\vec{v}$:
$$(\vec{v} \cdot \nabla)f = \frac{v_\sigma}{a}\sqrt{\frac{\sigma^2 - 1}{\sigma^2 - \tau^2}}\frac{\partial f}{\partial \sigma} + \frac{v_\tau}{a}\sqrt{\frac{1 - \tau^2}{\sigma^2 - \tau^2}}\frac{\partial f}{\partial \tau} + \frac{v_\varphi}{a\sqrt{(\sigma^2 - 1)(1 - \tau^2)}}\frac{\partial f}{\partial \varphi}.$$

Gradient of a vector $\vec{w}$ along a vector $\vec{v}$:
$$(\vec{v} \cdot \nabla)\vec{w} = (\vec{v} \cdot \nabla)w_\sigma \vec{i}_\sigma + (\vec{v} \cdot \nabla)w_\tau \vec{i}_\tau + (\vec{v} \cdot \nabla)w_\varphi \vec{i}_\varphi.$$

Laplacian of a scalar $f$:
$$\Delta f = \frac{1}{a^2(\sigma^2 - \tau^2)}\left\{\frac{\partial}{\partial \sigma}\left[(\sigma^2 - 1)\frac{\partial f}{\partial \sigma}\right] + \frac{\partial}{\partial \tau}\left[(1 - \tau^2)\frac{\partial f}{\partial \tau}\right] + \frac{\sigma^2 - \tau^2}{(\sigma^2 - 1)(1 - \tau^2)}\frac{\partial^2 f}{\partial \varphi^2}\right\}.$$

## T4.2.4. Coordinates of an Oblate Ellipsoid of Revolution

**T4.2.4-1. Transformations of coordinates. The metric tensor components.**

1°. Transformations of coordinates:
$$x^2 = a^2(1+\sigma^2)(1-\tau^2)\cos^2\varphi, \quad y^2 = a^2(1+\sigma^2)(1-\tau^2)\sin^2\varphi, \quad z = a\sigma\tau$$
$$(\sigma \geq 0, \; -1 \leq \tau \leq 1, \; 0 \leq \varphi < 2\pi).$$

Coordinate surfaces (the $z$-axis is the axis of revolution):

$$\frac{x^2+y^2}{a^2(1-\sigma^2)} + \frac{z^2}{a^2\sigma^2} = 1 \quad \text{(oblate ellipsoids of revolution)},$$

$$\frac{x^2+y^2}{a^2(1-\tau^2)} - \frac{z^2}{a^2\tau^2} = 1 \quad \text{(hyperboloid of revolution of one sheet)},$$

$$y = x\tan\varphi \quad \text{(half-planes through the } z\text{-axis)}.$$

Components of the metric tensor:
$$g_{\sigma\sigma} = a^2\frac{\sigma^2+\tau^2}{1+\sigma^2}, \quad g_{\tau\tau} = a^2\frac{\sigma^2+\tau^2}{1-\tau^2}, \quad g_{\varphi\varphi} = a^2(1+\sigma^2)(1-\tau^2), \quad \sqrt{g} = a^3(\sigma^2+\tau^2).$$

2°. Special coordinates system $u, v, \varphi$:
$$\sigma = \sinh u, \quad \tau = \cos v, \quad \varphi = \varphi \quad (0 \leq u < \infty, \; 0 \leq v \leq \pi, \; 0 \leq \varphi < 2\pi),$$
$$x = a\cosh u\sin v\cos\varphi, \quad y = a\cosh u\sin v\sin\varphi, \quad z = a\sinh u\cos v.$$

Components of the metric tensor:
$$g_{uu} = g_{vv} = a^2(\sinh^2 u + \cos^2 v), \quad g_{\varphi\varphi} = a^2\cosh^2 u\sin^2 v.$$

**T4.2.4-2. Basic differential relations.**

Gradient of a scalar $f$:
$$\nabla f = \frac{1}{a}\sqrt{\frac{\sigma^2+1}{\sigma^2+\tau^2}}\frac{\partial f}{\partial \sigma}\vec{i}_\sigma + \frac{1}{a}\sqrt{\frac{1-\tau^2}{\sigma^2+\tau^2}}\frac{\partial f}{\partial \tau}\vec{i}_\tau + \frac{1}{a\sqrt{(1-\tau^2)(\sigma^2+1)}}\frac{\partial f}{\partial \varphi}\vec{i}_\varphi.$$

Divergence of a vector $\vec{v}$:
$$\nabla \cdot \vec{v} = \frac{1}{a(\sigma^2+\tau^2)}\left\{\frac{\partial}{\partial \sigma}\left[v_\sigma\sqrt{(\sigma^2+\tau^2)(\sigma^2+1)}\right]\right.$$
$$\left. + \frac{\partial}{\partial \tau}\left[v_\tau\sqrt{(\sigma^2+\tau^2)(1-\tau^2)}\right] + \frac{\partial}{\partial \varphi}\left[v_\varphi\frac{\sigma^2+\tau^2}{\sqrt{(\sigma^2+1)(1-\tau^2)}}\right]\right\}.$$

Gradient of a scalar $f$ along a vector $\vec{v}$:
$$(\vec{v}\cdot\nabla)f = \frac{v_\sigma}{a}\sqrt{\frac{\sigma^2+1}{\sigma^2+\tau^2}}\frac{\partial f}{\partial \sigma} + \frac{v_\tau}{a}\sqrt{\frac{1-\tau^2}{\sigma^2+\tau^2}}\frac{\partial f}{\partial \tau} + \frac{v_\varphi}{a\sqrt{(\sigma^2+1)(1-\tau^2)}}\frac{\partial f}{\partial \varphi}.$$

Gradient of a vector $\vec{w}$ along a vector $\vec{v}$:
$$(\vec{v}\cdot\nabla)\vec{w} = (\vec{v}\cdot\nabla)w_\sigma\vec{i}_\sigma + (\vec{v}\cdot\nabla)w_\tau\vec{i}_\tau + (\vec{v}\cdot\nabla)w_\varphi\vec{i}_\varphi.$$

Laplacian of a scalar $f$:
$$\Delta f = \frac{1}{a^2(\sigma^2+\tau^2)}\left\{\frac{\partial}{\partial \sigma}\left[(1+\sigma^2)\frac{\partial f}{\partial \sigma}\right] + \frac{\partial}{\partial \tau}\left[(1-\tau^2)\frac{\partial f}{\partial \tau}\right] + \frac{\sigma^2+\tau^2}{(1+\sigma^2)(1-\tau^2)}\frac{\partial^2 f}{\partial \varphi^2}\right\}.$$

## T4.2.5. Coordinates of an Elliptic Cylinder

**1°. Transformations of coordinates:**
$$x = a\sigma\tau, \quad y^2 = a^2(\sigma^2 - 1)(1 - \tau^2), \quad z = z$$
$$(\sigma \geq 1, \quad -1 \leq \tau \leq 1, \quad -\infty < z < \infty).$$

Coordinate surfaces:
$$\frac{x^2}{a^2\sigma^2} + \frac{y^2}{a^2(\sigma^2 - 1)} = 1 \quad \text{(elliptic cylinders)},$$
$$\frac{x^2}{a^2\tau^2} + \frac{y^2}{a^2(\tau^2 - 1)} = 1 \quad \text{(hyperbolic cylinders)},$$
$$z = z \quad \text{(planes parallel to the } xy\text{-plane)}.$$

Components of the metric tensor:
$$g_{\sigma\sigma} = a^2 \frac{\sigma^2 - \tau^2}{\sigma^2 - 1}, \quad g_{\tau\tau} = a^2 \frac{\sigma^2 - \tau^2}{1 - \tau^2}, \quad g_{zz} = 1.$$

**2°. Special coordinate system $u, v, z$:**
$$\sigma = \cosh u, \quad \tau = \cos v, \quad z = z;$$
$$x = a \cosh u \cos v, \quad y = a \sinh u \sin v, \quad z = z$$
$$(0 \leq u < \infty, \quad 0 \leq v \leq \pi, \quad -\infty < z < \infty).$$

Components of the metric tensor:
$$g_{uu} = g_{vv} = a^2(\sinh^2 u + \sin^2 v), \quad g_{zz} = 1.$$

**3°. Laplacian:**
$$\Delta f = \frac{1}{a^2(\sinh^2 u + \sin^2 v)} \left( \frac{\partial^2 f}{\partial u^2} + \frac{\partial^2 f}{\partial v^2} \right) + \frac{\partial^2 f}{\partial z^2}$$
$$= \frac{\sqrt{\sigma^2 - 1}}{a^2(\sigma^2 - \tau^2)} \frac{\partial}{\partial \sigma} \left( \sqrt{\sigma^2 - 1} \frac{\partial f}{\partial \sigma} \right) + \frac{\sqrt{1 - \tau^2}}{a^2(\sigma^2 - \tau^2)} \frac{\partial}{\partial \tau} \left( \sqrt{1 - \tau^2} \frac{\partial f}{\partial \tau} \right) + \frac{\partial^2 f}{\partial z^2}.$$

*Remark.* The elliptic cylinder coordinates $\sigma, \tau$ are also used as *elliptic coordinates* on the plane $xy$.

## T4.2.6. Conical Coordinates

Transformations of coordinates:
$$x = \pm\frac{uvw}{ab}, \quad y^2 = \frac{u^2(v^2 - a^2)(w^2 - a^2)}{a^2(a^2 - b^2)}, \quad z^2 = \frac{u^2(v^2 - b^2)(w^2 - b^2)}{b^2(b^2 - a^2)}$$
$$(b^2 > v^2 > a^2 > w^2).$$

Coordinate surfaces:
$$x^2 + y^2 + z^2 = u^2 \quad \text{(spheres)},$$
$$\frac{x^2}{v^2} + \frac{y^2}{v^2 - a^2} - \frac{z^2}{b^2 - v^2} = 0 \quad \text{(cones with their axes coincident with the } z\text{-axis)},$$
$$\frac{x^2}{w^2} - \frac{y^2}{a^2 - w^2} - \frac{z^2}{b^2 - w^2} = 0 \quad \text{(cones with their axes coincident with the } x\text{-axis)}.$$

Components of the metric tensor:
$$g_{uu} = 1, \quad g_{vv} = \frac{u^2(v^2 - w^2)}{(v^2 - a^2)(b^2 - v^2)}, \quad g_{ww} = \frac{u^2(v^2 - w^2)}{(a^2 - w^2)(b^2 - w^2)}.$$

## T4.2.7. Parabolic Cylinder Coordinates

Transformations of coordinates:

$$x = \sigma\tau, \quad y = \frac{1}{2}(\tau^2 - \sigma^2), \quad z = z.$$

Coordinate surfaces:

$\dfrac{x^2}{\sigma^2} = 2y + \sigma^2$ (right parabolic cylinders with element parallel to the $z$-axis),

$\dfrac{x^2}{\tau^2} = -2y + \tau^2$ (right parabolic cylinders with element parallel to the $z$-axis),

$z = z$ (planes parallel to the $xy$-plane).

Components of the metric tensor:

$$g_{\sigma\sigma} = g_{\tau\tau} = \sigma^2 + \tau^2, \quad g_{zz} = 1.$$

Laplacian of a scalar $f$:

$$\Delta f = \frac{1}{(\sigma^2 + \tau^2)}\left(\frac{\partial^2 f}{\partial \sigma^2} + \frac{\partial^2 f}{\partial \tau^2}\right) + \frac{\partial^2 f}{\partial z^2}.$$

Remark. The parabolic cylinder coordinates $\sigma$, $\tau$ are also used as *parabolic coordinates* on the plane $xy$.

## T4.2.8. Parabolic Coordinates

Transformations of coordinates:

$$x = \sigma\tau\cos\varphi, \quad y = \sigma\tau\sin\varphi, \quad z = \frac{1}{2}(\tau^2 - \sigma^2).$$

Coordinate surfaces (the $z$-axis is the axis of revolution):

$\dfrac{x^2 + y^2}{\sigma^2} = 2z + \sigma^2$ (paraboloids of revolution),

$\dfrac{x^2 + y^2}{\tau^2} = -2z + \tau^2$ (paraboloids of revolution),

$y = x\tan\varphi$ (half-planes through the $z$-axis).

Components of the metric tensor:

$$g_{\sigma\sigma} = g_{\tau\tau} = \sigma^2 + \tau^2, \quad g_{\varphi\varphi} = \sigma^2\tau^2.$$

Laplacian of a scalar $f$:

$$\Delta f = \frac{1}{(\sigma^2 + \tau^2)}\left[\frac{1}{\sigma}\frac{\partial}{\partial \sigma}\left(\sigma\frac{\partial f}{\partial \sigma}\right) + \frac{1}{\tau}\frac{\partial}{\partial \tau}\left(\tau\frac{\partial f}{\partial \tau}\right) + \left(\frac{1}{\sigma^2} + \frac{1}{\tau^2}\right)\frac{\partial^2 f}{\partial \varphi^2}\right].$$

## T4.2.9. Bicylindrical Coordinates

Transformations of coordinates:

$$x = \frac{a \sinh \tau}{\cosh \tau - \cos \sigma}, \quad y = \frac{a \sin \sigma}{\cosh \tau - \cos \sigma}, \quad z = z.$$

Coordinate surfaces:

$x^2 + (y - a \cot \sigma)^2 = a^2(\cot^2 \sigma + 1)$ (right circular cylinders with element parallel to the $z$-axis),
$(x - a \coth \tau)^2 + y^2 = a^2(\coth^2 \tau - 1)$ (right circular cylinders with element parallel to the $z$-axis),
$z = z$ (planes parallel to the $xy$-plane).

Components of the metric tensor:

$$g_{\sigma\sigma} = g_{\tau\tau} = \frac{a^2}{(\cosh \tau - \cos \sigma)^2}, \quad g_{zz} = 1.$$

Laplacian of a scalar $f$:

$$\Delta f = \frac{1}{a^2}(\cosh \tau - \cos \sigma)^2 \left( \frac{\partial^2 f}{\partial \sigma^2} + \frac{\partial^2 f}{\partial \tau^2} \right) + \frac{\partial^2 f}{\partial z^2}.$$

Remark. The bicylindrical coordinates $\sigma$, $\tau$ are also used as *bipolar coordinates* on the plane $xy$.

## T4.2.10. Bipolar Coordinates (in Space)

Transformations of coordinates:

$$x = \frac{a \sin \sigma \cos \varphi}{\cosh \tau - \cos \sigma}, \quad y = \frac{a \sin \sigma \sin \varphi}{\cosh \tau - \cos \sigma}, \quad z = \frac{a \sinh \tau}{\cosh \tau - \cos \sigma}$$
$(-\infty < \tau < \infty, \ 0 \leq \sigma < \pi, \ 0 \leq \varphi < 2\pi).$

Coordinate surfaces (the $z$-axis is the axis of revolution):

$x^2 + y^2 + (z - a \coth \tau)^2 = \dfrac{a^2}{\sinh^2 \tau}$ (spheres with centers on the $z$-axis),

$(\sqrt{x^2 + y^2} - a \cot \sigma)^2 + z^2 = \dfrac{a^2}{\sin^2 \sigma}$ (surfaces obtained by revolution of circular arches $(y - a \cot \sigma)^2 + z^2 = a^2/\sin^2 \sigma$ about the $z$-axis),

$y = x \tan \varphi$ (half-planes through the $z$-axis).

Components of the metric tensor:

$$g_{\sigma\sigma} = g_{\tau\tau} = \frac{a^2}{(\cosh \tau - \cos \sigma)^2}, \quad g_{\varphi\varphi} = \frac{a^2 \sin^2 \sigma}{(\cosh \tau - \cos \sigma)^2}.$$

Laplacian of a scalar $f$:

$$\Delta f = \frac{(\cosh \tau - \cos \sigma)^2}{a^2 \sin \sigma} \left[ \frac{\partial}{\partial \tau} \left( \frac{\sin \sigma}{\cosh \tau - \cos \sigma} \frac{\partial f}{\partial \tau} \right) \right.$$
$$\left. + \frac{\partial}{\partial \sigma} \left( \frac{\sin \sigma}{\cosh \tau - \cos \sigma} \frac{\partial f}{\partial \sigma} \right) + \frac{1}{\sin \sigma (\cosh \tau - \cos \sigma)} \frac{\partial^2 f}{\partial \varphi^2} \right].$$

## T4.2.11. Toroidal Coordinates

Transformations of coordinates:

$$x = \frac{a \sinh \tau \cos \varphi}{\cosh \tau - \cos \sigma}, \quad y = \frac{a \sinh \tau \sin \varphi}{\cosh \tau - \cos \sigma}, \quad z = \frac{a \sin \sigma}{\cosh \tau - \cos \sigma}$$
$$(-\pi \leq \sigma \leq \pi, \ 0 \leq \tau < \infty, \ 0 \leq \varphi < 2\pi).$$

Coordinate surfaces (the $z$-axis is the axis of revolution):

$$x^2 + y^2 + (z - a \cot \sigma)^2 = \frac{a^2}{\sin^2 \sigma} \quad \text{(spheres with centers on the } z\text{-axis),}$$

$$(\sqrt{x^2 + y^2} - a \coth \tau)^2 + z^2 = \frac{a^2}{\sinh^2 \tau} \quad \text{(tori with centers on the } z\text{-axis),}$$

$$y = x \tan \varphi \quad \text{(half-planes through the } z\text{-axis).}$$

Components of the metric tensor:

$$g_{\sigma\sigma} = g_{\tau\tau} = \frac{a^2}{(\cosh \tau - \cos \sigma)^2}, \quad g_{\varphi\varphi} = \frac{a^2 \sinh^2 \tau}{(\cosh \tau - \cos \sigma)^2}.$$

Laplacian of a scalar $f$:

$$\Delta f = \frac{(\cosh \tau - \cos \sigma)^2}{a^2 \sinh \tau} \left[ \frac{\partial}{\partial \sigma}\left(\frac{\sinh \tau}{\cosh \tau - \cos \sigma} \frac{\partial f}{\partial \sigma}\right) \right.$$
$$\left. + \frac{\partial}{\partial \tau}\left(\frac{\sinh \tau}{\cosh \tau - \cos \sigma} \frac{\partial f}{\partial \tau}\right) + \frac{1}{\sinh \tau (\cosh \tau - \cos \sigma)} \frac{\partial^2 f}{\partial \varphi^2} \right].$$

# References for Chapter T4

**Arfken, G. B. and Weber, H. J.**, *Mathematical Methods for Physicists, 4th Edition*, Academic Press, New York, 1995.

**Korn, G. A. and Korn, T. M.**, *Mathematical Handbook for Scientists and Engineers, 2nd Edition*, Dover Publications, New York, 2000.

**Menzel, D. H.**, *Mathematical Physics*, Dover Publications, New York, 1961.

**Moon, P. and Spencer, D. E.**, *Field Theory Handbook, Including Coordinate Systems, Differential Equations and Their Solutions, 3rd Edition*, Springer-Verlag, Berlin, 1988.

**Morse, P. M. and Feshbach, H.**, *Methods of Theoretical Physics, Part I*, McGraw-Hill, New York, 1953.

**Weisstein, E. W.**, *CRC Concise Encyclopedia of Mathematics, 2nd Edition*, CRC Press, Boca Raton, 2003.

**Zwillinger, D.**, *CRC Standard Mathematical Tables and Formulae, 31st Edition*, CRC Press, Boca Raton, 2002.

# Chapter T5
# Ordinary Differential Equations

## T5.1. First-Order Equations

▶ *In this Section we shall often use the term "solution" to mean "general solution."*

**1.** $y'_x = f(y)$.

*Autonomous equation.*

Solution: $x = \int \dfrac{dy}{f(y)} + C$.

Particular solutions: $y = A_k$, where the $A_k$ are roots of the algebraic (transcendental) equation $f(A_k) = 0$.

**2.** $y'_x = f(x)g(y)$.

*Separable equation.*

Solution: $\int \dfrac{dy}{g(y)} = \int f(x)\,dx + C$.

Particular solutions: $y = A_k$, where the $A_k$ are roots of the algebraic (transcendental) equation $g(A_k) = 0$.

**3.** $g(x)y'_x = f_1(x)y + f_0(x)$.

*Linear equation.*
Solution:

$$y = Ce^F + e^F \int e^{-F} \dfrac{f_0(x)}{g(x)}\,dx, \quad \text{where} \quad F(x) = \int \dfrac{f_1(x)}{g(x)}\,dx.$$

**4.** $g(x)y'_x = f_1(x)y + f_0(x)y^k$.

*Bernoulli equation.* Here, $k$ is an arbitrary number. For $k \neq 1$, the substitution $w(x) = y^{1-k}$ leads to a linear equation: $g(x)w'_x = (1-k)f_1(x)w + (1-k)f_0(x)$.
Solution:

$$y^{1-k} = Ce^F + (1-k)e^F \int e^{-F} \dfrac{f_0(x)}{g(x)}\,dx, \quad \text{where} \quad F(x) = (1-k)\int \dfrac{f_1(x)}{g(x)}\,dx.$$

**5.** $y'_x = f(y/x)$.

*Homogeneous equation.* The substitution $u(x) = y/x$ leads to a separable equation: $xu'_x = f(u) - u$.

**6.** $y'_x = ay^2 + bx^n$.

*Special Riccati equation,* $n$ is an arbitrary number.

1°. Solution for $n \neq -2$:

$$y = -\frac{1}{a}\frac{w'_x}{w}, \quad w(x) = \sqrt{x}\left[C_1 J_{\frac{1}{2k}}\left(\frac{1}{k}\sqrt{ab}\,x^k\right) + C_2 Y_{\frac{1}{2k}}\left(\frac{1}{k}\sqrt{ab}\,x^k\right)\right],$$

where $k = \frac{1}{2}(n+2)$; $J_m(z)$ and $Y_m(z)$ are Bessel functions (see Subsection 18.6).

2°. Solution for $n = -2$:

$$y = \frac{\lambda}{x} - x^{2a\lambda}\left(\frac{ax}{2a\lambda + 1}x^{2a\lambda} + C\right)^{-1},$$

where $\lambda$ is a root of the quadratic equation $a\lambda^2 + \lambda + b = 0$.

**7.** $y'_x = y^2 + f(x)y - a^2 - af(x)$.

Particular solution: $y_0 = a$. The general solution can be obtained by formulas given in Item 1° of equation T5.1.23.

**8.** $y'_x = f(x)y^2 + ay - ab - b^2 f(x)$.

Particular solution: $y_0 = b$. The general solution can be obtained by formulas given in Item 1° of equation T5.1.23.

**9.** $y'_x = y^2 + xf(x)y + f(x)$.

Particular solution: $y_0 = -1/x$. The general solution can be obtained by formulas given in Item 1° of equation T5.1.23.

**10.** $y'_x = f(x)y^2 - ax^n f(x)y + anx^{n-1}$.

Particular solution: $y_0 = ax^n$. The general solution can be obtained by formulas given in Item 1° of equation T5.1.23.

**11.** $y'_x = f(x)y^2 + anx^{n-1} - a^2 x^{2n} f(x)$.

Particular solution: $y_0 = ax^n$. The general solution can be obtained by formulas given in Item 1° of equation T5.1.23.

**12.** $y'_x = -(n+1)x^n y^2 + x^{n+1} f(x)y - f(x)$.

Particular solution: $y_0 = x^{-n-1}$. The general solution can be obtained by formulas given in Item 1° of equation T5.1.23.

**13.** $xy'_x = f(x)y^2 + ny + ax^{2n} f(x)$.

Solution: $y = \begin{cases} \sqrt{a}\,x^n \tan\left[\sqrt{a}\int x^{n-1} f(x)\,dx + C\right] & \text{if } a > 0, \\ \sqrt{|a|}\,x^n \tanh\left[-\sqrt{|a|}\int x^{n-1} f(x)\,dx + C\right] & \text{if } a < 0. \end{cases}$

**14.** $xy'_x = x^{2n} f(x) y^2 + [ax^n f(x) - n] y + b f(x)$.

The substitution $z = x^n y$ leads to a separable equation: $z'_x = x^{n-1} f(x)(z^2 + az + b)$.

**15.** $y'_x = f(x) y^2 + g(x) y - a^2 f(x) - a g(x)$.

Particular solution: $y_0 = a$. The general solution can be obtained by formulas given in Item 1° of equation T5.1.23.

**16.** $y'_x = f(x) y^2 + g(x) y + a n x^{n-1} - a^2 x^{2n} f(x) - a x^n g(x)$.

Particular solution: $y_0 = ax^n$. The general solution can be obtained by formulas given in Item 1° of equation T5.1.23.

**17.** $y'_x = a e^{\lambda x} y^2 + a e^{\lambda x} f(x) y + \lambda f(x)$.

Particular solution: $y_0 = -\dfrac{\lambda}{a} e^{-\lambda x}$. The general solution can be obtained by formulas given in Item 1° of equation T5.1.23.

**18.** $y'_x = f(x) y^2 - a e^{\lambda x} f(x) y + a \lambda e^{\lambda x}$.

Particular solution: $y_0 = a e^{\lambda x}$. The general solution can be obtained by formulas given in Item 1° of equation T5.1.23.

**19.** $y'_x = f(x) y^2 + a \lambda e^{\lambda x} - a^2 e^{2\lambda x} f(x)$.

Particular solution: $y_0 = a e^{\lambda x}$. The general solution can be obtained by formulas given in Item 1° of equation T5.1.23.

**20.** $y'_x = f(x) y^2 + \lambda y + a e^{2\lambda x} f(x)$.

Solution: $y = \begin{cases} \sqrt{a}\, e^{\lambda x} \tan\!\left[\sqrt{a} \int e^{\lambda x} f(x)\, dx + C\right] & \text{if } a > 0, \\ \sqrt{|a|}\, e^{\lambda x} \tanh\!\left[-\sqrt{|a|} \int e^{\lambda x} f(x)\, dx + C\right] & \text{if } a < 0. \end{cases}$

**21.** $y'_x = y^2 - f^2(x) + f'_x(x)$.

Particular solution: $y_0 = f(x)$. The general solution can be obtained by formulas given in Item 1° of equation T5.1.23.

**22.** $y'_x = f(x) y^2 - f(x) g(x) y + g'_x(x)$.

Particular solution: $y_0 = g(x)$. The general solution can be obtained by formulas given in Item 1° of equation T5.1.23.

**23.** $y'_x = f(x) y^2 + g(x) y + h(x)$.

*General Riccati equation.*

1°. Given a particular solution $y_0 = y_0(x)$ of the Riccati equation, the general solution can be written as:
$$y = y_0(x) + \Phi(x)\left[C - \int \Phi(x)f_2(x)\,dx\right]^{-1},$$
where
$$\Phi(x) = \exp\left\{\int \left[2f_2(x)y_0(x) + f_1(x)\right]dx\right\}.$$
To the particular solution $y_0(x)$ there corresponds $C = \infty$.

2°. The substitution
$$u(x) = \exp\left(-\int f_2 y\,dx\right)$$
reduces the general Riccati equation to a second-order linear equation:
$$f_2 u''_{xx} - \left[(f_2)'_x + f_1 f_2\right]u'_x + f_0 f_2^2 u = 0,$$
which often may be easier to solve than the original Riccati equation.

3°. For more details about the Riccati equation, see Subsection 12.1.4. Many solvable equations of this form can be found in the books by Kamke (1977) and Polyanin and Zaitsev (2003).

**24.** $yy'_x = y + f(x)$.

*Abel equation of the second kind in the canonical form.* Many solvable equations of this form can be found in the books by Zaitsev and Polyanin (1994) and Polyanin and Zaitsev (2003).

**25.** $yy'_x = f(x)y + g(x)$.

*Abel equation of the second kind.* Many solvable equations of this form can be found in the books by Zaitsev and Polyanin (1994) and Polyanin and Zaitsev (2003).

**26.** $yy'_x = f(x)y^2 + g(x)y + h(x)$.

*Abel equation of the second kind.* Many solvable equations of this form can be found in the books by Zaitsev and Polyanin (1994) and Polyanin and Zaitsev (2003).

▶ In equations T5.1.27–T5.1.48, the functions $f$, $g$, and $h$ are arbitrary composite functions whose arguments can depend on both $x$ and $y$.

**27.** $y'_x = f(ax + by + c)$.

If $b \neq 0$, the substitution $u(x) = ax + by + c$ leads to a separable equation: $u'_x = bf(u) + a$.

**28.** $y'_x = f(y + ax^n + b) - anx^{n-1}$.

The substitution $u = y + ax^n + b$ leads to a separable equation: $u'_x = f(u)$.

**29.** $y'_x = \dfrac{y}{x}f(x^n y^m)$.

*Generalized homogeneous equation.* The substitution $z = x^n y^m$ leads to a separable equation: $xz'_x = nz + mzf(z)$.

**30.** $y'_x = -\dfrac{n}{m}\dfrac{y}{x} + y^k f(x) g(x^n y^m).$

The substitution $z = x^n y^m$ leads to a separable equation: $z'_x = mx^{\frac{n-nk}{m}} f(x) z^{\frac{k+m-1}{m}} g(z).$

**31.** $y'_x = f\!\left(\dfrac{ax + by + c}{\alpha x + \beta y + \gamma}\right).$

See Paragraph 12.1.2-3, Item 2°.

**32.** $y'_x = x^{n-1} y^{1-m} f(ax^n + by^m).$

The substitution $w = ax^n + by^m$ leads to a separable equation: $w'_x = x^{n-1}[an + bmf(w)].$

**33.** $[x^n f(y) + xg(y)] y'_x = h(y).$

This is a Bernoulli equation with respect to $x = x(y)$ (see equation T5.1.4).

**34.** $x[f(x^n y^m) + mx^k g(x^n y^m)] y'_x = y[h(x^n y^m) - nx^k g(x^n y^m)].$

The transformation $t = x^n y^m$, $z = x^{-k}$ leads to a linear equation with respect to $z = z(t)$:
$t[nf(t) + mh(t)] z'_t = -kf(t)z - kmg(t).$

**35.** $x[f(x^n y^m) + my^k g(x^n y^m)] y'_x = y[h(x^n y^m) - ny^k g(x^n y^m)].$

The transformation $t = x^n y^m$, $z = y^{-k}$ leads to a linear equation with respect to $z = z(t)$:
$t[nf(t) + mh(t)] z'_t = -kh(t)z + kng(t).$

**36.** $x[sf(x^n y^m) - mg(x^k y^s)] y'_x = y[ng(x^k y^s) - kf(x^n y^m)].$

The transformation $t = x^n y^m$, $w = x^k y^s$ leads to a separable equation: $tf(t) w'_t = wg(w).$

**37.** $[f(y) + amx^n y^{m-1}] y'_x + g(x) + anx^{n-1} y^m = 0.$

Solution: $\displaystyle\int f(y)\,dy + \int g(x)\,dx + ax^n y^m = C.$

**38.** $y'_x = e^{-\lambda x} f(e^{\lambda x} y).$

The substitution $u = e^{\lambda x} y$ leads to a separable equation: $u'_x = f(u) + \lambda u.$

**39.** $y'_x = e^{\lambda y} f(e^{\lambda y} x).$

The substitution $u = e^{\lambda y} x$ leads to a separable equation: $xu'_x = \lambda u^2 f(u) + u.$

**40.** $y'_x = yf(e^{\alpha x} y^m).$

The substitution $z = e^{\alpha x} y^m$ leads to a separable equation: $z'_x = \alpha z + mzf(z).$

**41.** $y'_x = \dfrac{1}{x} f(x^n e^{\alpha y}).$

The substitution $z = x^n e^{\alpha y}$ leads to a separable equation: $xz'_x = nz + \alpha z f(z).$

**42.** $y'_x = f(x)e^{\lambda y} + g(x)$.

The substitution $u = e^{-\lambda y}$ leads to a linear equation: $u'_x = -\lambda g(x)u - \lambda f(x)$.

**43.** $y'_x = -\dfrac{n}{x} + f(x)g(x^n e^y)$.

The substitution $z = x^n e^y$ leads to a separable equation: $z'_x = f(x)zg(z)$.

**44.** $y'_x = -\dfrac{\alpha}{m}y + y^k f(x)g(e^{\alpha x}y^m)$.

The substitution $z = e^{\alpha x}y^m$ leads to a separable equation:
$$z'_x = m\exp\left[\frac{\alpha}{m}(1-k)x\right]f(x)z^{\frac{k+m-1}{m}}g(z).$$

**45.** $y'_x = e^{\alpha x - \beta y}f(ae^{\alpha x} + be^{\beta y})$.

The substitution $w = ae^{\alpha x} + be^{\beta y}$ leads to a separable equation: $w'_x = e^{\alpha x}[a\alpha + b\beta f(w)]$.

**46.** $[e^{\alpha x}f(y) + a\beta]y'_x + e^{\beta y}g(x) + a\alpha = 0$.

Solution: $\displaystyle\int e^{-\beta y}f(y)\,dy + \int e^{-\alpha x}g(x)\,dx - ae^{-\alpha x - \beta y} = C$.

**47.** $x[f(x^n e^{\alpha y}) + \alpha yg(x^n e^{\alpha y})]y'_x = h(x^n e^{\alpha y}) - nyg(x^n e^{\alpha y})$.

The substitution $t = x^n e^{\alpha y}$ leads to a linear equation with respect to $y = y(t)$: $t[nf(t) + \alpha h(t)]y'_t = -ng(t)y + h(t)$.

**48.** $[f(e^{\alpha x}y^m) + mxg(e^{\alpha x}y^m)]y'_x = y[h(e^{\alpha x}y^m) - \alpha xg(e^{\alpha x}y^m)]$.

The substitution $t = e^{\alpha x}y^m$ leads to a linear equation with respect to $x = x(t)$: $t[\alpha f(t) + mh(t)]x'_t = mg(t)x + f(t)$.

## T5.2. Second-Order Linear Equations

**Preliminary remarks.** A homogeneous linear equation of the second order has the general form
$$f_2(x)y''_{xx} + f_1(x)y'_x + f_0(x)y = 0. \tag{1}$$

Let $y_0 = y_0(x)$ be a nontrivial particular solution ($y_0 \not\equiv 0$) of this equation. Then the general solution of equation (1) can be found from the formula
$$y = y_0\left(C_1 + C_2\int \frac{e^{-F}}{y_0^2}\,dx\right), \quad \text{where} \quad F = \int \frac{f_1}{f_2}\,dx. \tag{2}$$

For specific equations described below, often only particular solutions are given, while the general solutions can be obtained with formula (2).

## T5.2.1. Equations Involving Power Functions

**1.** $y''_{xx} + ay = 0$.

*Equation of free oscillations.*

Solution: $y = \begin{cases} C_1 \sinh(x\sqrt{|a|}) + C_2 \cosh(x\sqrt{|a|}) & \text{if } a < 0, \\ C_1 + C_2 x & \text{if } a = 0, \\ C_1 \sin(x\sqrt{a}) + C_2 \cos(x\sqrt{a}) & \text{if } a > 0. \end{cases}$

**2.** $y''_{xx} - ax^n y = 0$.

$1°$. For $n = -2$, this is the Euler equation T5.2.1.12 (the solution is expressed in terms of elementary function).

$2°$. Assume $2/(n+2) = 2m + 1$, where $m$ is an integer. Then the solution is

$$y = \begin{cases} x(x^{1-2q}D)^{m+1}\left[C_1 \exp\left(\frac{\sqrt{a}}{q}x^q\right) + C_2 \exp\left(-\frac{\sqrt{a}}{q}x^q\right)\right] & \text{if } m \geq 0, \\ (x^{1-2q}D)^{-m}\left[C_1 \exp\left(\frac{\sqrt{a}}{q}x^q\right) + C_2 \exp\left(-\frac{\sqrt{a}}{q}x^q\right)\right] & \text{if } m < 0, \end{cases}$$

where $D = \dfrac{d}{dx}$, $q = \dfrac{n+2}{2} = \dfrac{1}{2m+1}$.

$3°$. For any $n$, the solution is expressed in terms of Bessel functions and modified Bessel functions:

$$y = \begin{cases} C_1 \sqrt{x}\, J_{\frac{1}{2q}}\left(\dfrac{\sqrt{-a}}{q}x^q\right) + C_2 \sqrt{x}\, Y_{\frac{1}{2q}}\left(\dfrac{\sqrt{-a}}{q}x^q\right) & \text{if } a < 0, \\ C_1 \sqrt{x}\, I_{\frac{1}{2q}}\left(\dfrac{\sqrt{a}}{q}x^q\right) + C_2 \sqrt{x}\, K_{\frac{1}{2q}}\left(\dfrac{\sqrt{a}}{q}x^q\right) & \text{if } a > 0, \end{cases}$$

where $q = \frac{1}{2}(n+2)$. The functions $J_\nu(z)$, $Y_\nu(z)$ and $I_\nu(z)$, $K_\nu(z)$ are described in Sections 18.6 and 18.7 in detail; see also equations T5.2.1.13 and T5.2.1.14.

**3.** $y''_{xx} + ay'_x + by = 0$.

*Second-order constant coefficient linear equation.* In physics this equation is called an *equation of damped vibrations.*

Solution: $y = \begin{cases} \exp\left(-\frac{1}{2}ax\right)\left[C_1 \exp\left(\frac{1}{2}\lambda x\right) + C_2 \exp\left(-\frac{1}{2}\lambda x\right)\right] & \text{if } \lambda^2 = a^2 - 4b > 0, \\ \exp\left(-\frac{1}{2}ax\right)\left[C_1 \sin\left(\frac{1}{2}\lambda x\right) + C_2 \cos\left(\frac{1}{2}\lambda x\right)\right] & \text{if } \lambda^2 = 4b - a^2 > 0, \\ \exp\left(-\frac{1}{2}ax\right)(C_1 x + C_2) & \text{if } a^2 = 4b. \end{cases}$

**4.** $y''_{xx} + ay'_x + (bx + c)y = 0$.

$1°$. Solution with $b \neq 0$:

$$y = \exp\left(-\tfrac{1}{2}ax\right)\sqrt{\xi}\left[C_1 J_{1/3}\left(\tfrac{2}{3}\sqrt{b}\,\xi^{3/2}\right) + C_2 Y_{1/3}\left(\tfrac{2}{3}\sqrt{b}\,\xi^{3/2}\right)\right], \quad \xi = x + \frac{4c - a^2}{4b},$$

where $J_{1/3}(z)$ and $Y_{1/3}(z)$ are Bessel functions.

$2°$. For $b = 0$, see equation T5.2.1.3.

**5.** $y''_{xx} + (ax+b)y'_x + (\alpha x^2 + \beta x + \gamma)y = 0.$

The substitution $y = u\exp(sx^2)$, where $s$ is a root of the quadratic equation $4s^2 + 2as + \alpha = 0$, leads to an equation of the form T5.2.1.11: $u''_{xx} + [(a+4s)x+b]u'_x + [(\beta+2bs)x+\gamma+2s]u = 0.$

**6.** $xy''_{xx} + ay'_x + by = 0.$

1°. The solution is expressed in terms of Bessel functions and modified Bessel functions:

$$y = \begin{cases} x^{\frac{1-a}{2}}\left[C_1 J_\nu(2\sqrt{bx}) + C_2 Y_\nu(2\sqrt{bx})\right] & \text{if } bx > 0, \\ x^{\frac{1-a}{2}}\left[C_1 I_\nu(2\sqrt{|bx|}) + C_2 K_\nu(2\sqrt{|bx|})\right] & \text{if } bx < 0, \end{cases}$$

where $\nu = |1-a|$.

2°. For $a = \frac{1}{2}(2n+1)$, where $n = 0, 1, \ldots$, the solution is

$$y = \begin{cases} C_1 \dfrac{d^n}{dx^n}\cos\sqrt{4bx} + C_2 \dfrac{d^n}{dx^n}\sin\sqrt{4bx} & \text{if } bx > 0, \\ C_1 \dfrac{d^n}{dx^n}\cosh\sqrt{4|bx|} + C_2 \dfrac{d^n}{dx^n}\sinh\sqrt{4|bx|} & \text{if } bx < 0. \end{cases}$$

**7.** $xy''_{xx} + ay'_x + bxy = 0.$

1°. The solution is expressed in terms of Bessel functions and modified Bessel functions:

$$y = \begin{cases} x^{\frac{1-a}{2}}\left[C_1 J_\nu(\sqrt{b}\,x) + C_2 Y_\nu(\sqrt{b}\,x)\right] & \text{if } b > 0, \\ x^{\frac{1-a}{2}}\left[C_1 I_\nu(\sqrt{|b|}\,x) + C_2 K_\nu(\sqrt{|b|}\,x)\right] & \text{if } b < 0, \end{cases}$$

where $\nu = \frac{1}{2}|1-a|$.

2°. For $a = 2n$, where $n = 1, 2, \ldots$, the solution is

$$y = \begin{cases} C_1 \left(\dfrac{1}{x}\dfrac{d}{dx}\right)^n \cos(x\sqrt{b}) + C_2 \left(\dfrac{1}{x}\dfrac{d}{dx}\right)^n \sin(x\sqrt{b}) & \text{if } b > 0, \\ C_1 \left(\dfrac{1}{x}\dfrac{d}{dx}\right)^n \cosh(x\sqrt{-b}) + C_2 \left(\dfrac{1}{x}\dfrac{d}{dx}\right)^n \sinh(x\sqrt{-b}) & \text{if } b < 0. \end{cases}$$

**8.** $xy''_{xx} + ny'_x + bx^{1-2n}y = 0.$

For $n = 1$, this is the Euler equation T5.2.1.12. For $n \neq 1$, the solution is

$$y = \begin{cases} C_1 \sin\left(\dfrac{\sqrt{b}}{n-1}x^{1-n}\right) + C_2 \cos\left(\dfrac{\sqrt{b}}{n-1}x^{1-n}\right) & \text{if } b > 0, \\ C_1 \exp\left(\dfrac{\sqrt{-b}}{n-1}x^{1-n}\right) + C_2 \exp\left(\dfrac{-\sqrt{-b}}{n-1}x^{1-n}\right) & \text{if } b < 0. \end{cases}$$

**9.** $xy''_{xx} + ay'_x + bx^n y = 0.$

If $n = -1$ and $b = 0$, we have the Euler equation T5.2.1.12. If $n \neq -1$ and $b \neq 0$, the solution is expressed in terms of Bessel functions:

$$y = x^{\frac{1-a}{2}}\left[C_1 J_\nu\left(\dfrac{2\sqrt{b}}{n+1}x^{\frac{n+1}{2}}\right) + C_2 Y_\nu\left(\dfrac{2\sqrt{b}}{n+1}x^{\frac{n+1}{2}}\right)\right], \qquad \text{where } \nu = \dfrac{|1-a|}{n+1}.$$

**10.** $xy''_{xx} + (b-x)y'_x - ay = 0.$

*Degenerate hypergeometric equation.*

$1°$. If $b \neq 0, -1, -2, -3, \ldots$, Kummer's series is a particular solution:

$$\Phi(a,b;x) = 1 + \sum_{k=1}^{\infty} \frac{(a)_k}{(b)_k} \frac{x^k}{k!},$$

where $(a)_k = a(a+1)\ldots(a+k-1)$, $(a)_0 = 1$. If $b > a > 0$, this solution can be written in terms of a definite integral:

$$\Phi(a,b;x) = \frac{\Gamma(b)}{\Gamma(a)\Gamma(b-a)} \int_0^1 e^{xt} t^{a-1}(1-t)^{b-a-1}\,dt,$$

where $\Gamma(z) = \int_0^\infty e^{-t} t^{z-1}\,dt$ is the gamma function.

If $b$ is not an integer, then the general solution has the form:

$$y = C_1 \Phi(a,b;x) + C_2 x^{1-b} \Phi(a-b+1, 2-b; x).$$

$2°$. For $b \neq 0, -1, -2, -3, \ldots$, the general solution of the degenerate hypergeometric equation can be written in the form

$$y = C_1 \Phi(a,b;x) + C_2 \Psi(a,b;x),$$

while for $b = 0, -1, -2, -3, \ldots$, it can be represented as

$$y = x^{1-b}\left[C_1 \Phi(a-b+1, 2-b; x) + C_2 \Psi(a-b+1, 2-b; x)\right].$$

The functions $\Phi(a,b;x)$ and $\Psi(a,b;x)$ are described in Subsection 18.9 in detail.

**11.** $(a_2 x + b_2)y''_{xx} + (a_1 x + b_1)y'_x + (a_0 x + b_0)y = 0.$

Let the function $\mathcal{J}(a,b;x)$ be an arbitrary solution of the degenerate hypergeometric equation $xy''_{xx} + (b-x)y'_x - ay = 0$ (see T5.2.1.10), and let the function $Z_\nu(x)$ be an arbitrary solution of the Bessel equation $x^2 y''_{xx} + xy'_x + (x^2 - \nu^2)y = 0$ (see T5.2.1.13). The results of solving the original equation are presented in Table T5.1.

TABLE T5.1
Solutions of equation T5.2.1.11 for different values of the determining parameters

| Solution: $y = e^{kx} w(z)$, where $z = \dfrac{x-\mu}{\lambda}$ | | | | | |
|---|---|---|---|---|---|
| Constraints | $k$ | $\lambda$ | $\mu$ | $w$ | Parameters |
| $a_2 \neq 0$, $a_1^2 \neq 4a_0 a_2$ | $\dfrac{\sqrt{D}-a_1}{2a_2}$ | $-\dfrac{a_2}{2a_2 k + a_1}$ | $-\dfrac{b_2}{a_2}$ | $\mathcal{J}(a,b;z)$ | $a = B(k)/(2a_2 k + a_1)$, $b = (a_2 b_1 - a_1 b_2)a_2^{-2}$ |
| $a_2 = 0$, $a_1 \neq 0$ | $-\dfrac{a_0}{a_1}$ | $1$ | $-\dfrac{2b_2 k + b_1}{a_1}$ | $\mathcal{J}(a, \tfrac{1}{2}; \beta z^2)$ | $a = B(k)/(2a_1)$, $\beta = -a_1/(2b_2)$ |
| $a_2 \neq 0$, $a_1^2 = 4a_0 a_2$ | $-\dfrac{a_1}{2a_2}$ | $a_2$ | $-\dfrac{b_2}{a_2}$ | $z^{\nu/2} Z_\nu(\beta\sqrt{z})$ | $\nu = 1 - (2b_2 k + b_1)a_2^{-1}$, $\beta = 2\sqrt{B(k)}$ |
| $a_2 = a_1 = 0$, $a_0 \neq 0$ | $-\dfrac{b_1}{2b_2}$ | $1$ | $\dfrac{b_1^2 - 4b_0 b_2}{4a_0 b_2}$ | $z^{1/2} Z_{1/3}(\beta z^{3/2})$; see also 2.1.2.12 | $\beta = \dfrac{2}{3}\left(\dfrac{a_0}{b_2}\right)^{1/2}$ |
| Notation: $D = a_1^2 - 4a_0 a_2$, $B(k) = b_2 k^2 + b_1 k + b_0$ | | | | | |

**12.** $x^2 y''_{xx} + a x y'_x + b y = 0.$

*Euler equation.* Solution:

$$y = \begin{cases} |x|^{\frac{1-a}{2}} \left( C_1 |x|^\mu + C_2 |x|^{-\mu} \right) & \text{if } (1-a)^2 > 4b, \\ |x|^{\frac{1-a}{2}} \left( C_1 + C_2 \ln |x| \right) & \text{if } (1-a)^2 = 4b, \\ |x|^{\frac{1-a}{2}} \left[ C_1 \sin(\mu \ln |x|) + C_2 \cos(\mu \ln |x|) \right] & \text{if } (1-a)^2 < 4b, \end{cases}$$

where $\mu = \frac{1}{2} |(1-a)^2 - 4b|^{1/2}$.

**13.** $x^2 y''_{xx} + x y'_x + (x^2 - \nu^2) y = 0.$

*Bessel equation.*

1°. Let $\nu$ be an arbitrary noninteger. Then the general solution is given by

$$y = C_1 J_\nu(x) + C_2 Y_\nu(x), \tag{1}$$

where $J_\nu(x)$ and $Y_\nu(x)$ are the Bessel functions of the first and second kind:

$$J_\nu(x) = \sum_{k=0}^\infty \frac{(-1)^k (x/2)^{\nu+2k}}{k!\, \Gamma(\nu+k+1)}, \qquad Y_\nu(x) = \frac{J_\nu(x) \cos \pi\nu - J_{-\nu}(x)}{\sin \pi \nu}. \tag{2}$$

Solution (1) is denoted by $y = Z_\nu(x)$, which is referred to as the cylindrical function.

The functions $J_\nu(x)$ and $Y_\nu(x)$ can be expressed in terms of definite integrals (with $x > 0$):

$$\pi J_\nu(x) = \int_0^\pi \cos(x \sin \theta - \nu \theta)\, d\theta - \sin \pi\nu \int_0^\infty \exp(-x \sinh t - \nu t)\, dt,$$

$$\pi Y_\nu(x) = \int_0^\pi \sin(x \sin \theta - \nu \theta)\, d\theta - \int_0^\infty (e^{\nu t} + e^{-\nu t} \cos \pi\nu) e^{-x \sinh t}\, dt.$$

2°. In the case $\nu = n + \frac{1}{2}$, where $n = 0, 1, 2, \ldots$, the Bessel functions are expressed in terms of elementary functions:

$$J_{n+\frac{1}{2}}(x) = \sqrt{\frac{2}{\pi}} x^{n+\frac{1}{2}} \left( -\frac{1}{x} \frac{d}{dx} \right)^n \frac{\sin x}{x}, \qquad J_{-n-\frac{1}{2}}(x) = \sqrt{\frac{2}{\pi}} x^{n+\frac{1}{2}} \left( \frac{1}{x} \frac{d}{dx} \right)^n \frac{\cos x}{x},$$

$$Y_{n+\frac{1}{2}}(x) = (-1)^{n+1} J_{-n-\frac{1}{2}}(x).$$

The Bessel functions are described in Section 18.6 in detail.

**14.** $x^2 y''_{xx} + x y'_x - (x^2 + \nu^2) y = 0.$

*Modified Bessel equation.* It can be reduced to equation T5.2.1.13 by means of the substitution $x = i\bar{x}$ ($i^2 = -1$).

Solution:

$$y = C_1 I_\nu(x) + C_2 K_\nu(x),$$

where $I_\nu(x)$ and $K_\nu(x)$ are modified Bessel functions of the first and second kind:

$$I_\nu(x) = \sum_{k=0}^\infty \frac{(x/2)^{2k+\nu}}{k!\, \Gamma(\nu+k+1)}, \qquad K_\nu(x) = \frac{\pi}{2} \frac{I_{-\nu}(x) - I_\nu(x)}{\sin \pi \nu}.$$

The modified Bessel functions are described in Subsection 18.7 in detail.

15. $x^2 y''_{xx} + axy'_x + (bx^n + c)y = 0, \quad n \neq 0.$

The case $b = 0$ corresponds to the Euler equation T5.2.1.12.

For $b \neq 0$, the solution is

$$y = x^{\frac{1-a}{2}} \left[ C_1 J_\nu \left( \frac{2}{n} \sqrt{b}\, x^{\frac{n}{2}} \right) + C_2 Y_\nu \left( \frac{2}{n} \sqrt{b}\, x^{\frac{n}{2}} \right) \right],$$

where $\nu = \frac{1}{n} \sqrt{(1-a)^2 - 4c}$; $J_\nu(z)$ and $Y_\nu(z)$ are the Bessel functions of the first and second kind.

16. $x^2 y''_{xx} + axy'_x + x^n(bx^n + c)y = 0.$

The substitution $\xi = x^n$ leads to an equation of the form T5.2.1.11:

$$n^2 \xi y''_{\xi\xi} + n(n - 1 + a) y'_\xi + (b\xi + c) y = 0.$$

17. $x^2 y''_{xx} + (ax + b) y'_x + cy = 0.$

The transformation $x = z^{-1}$, $y = z^k e^z w$, where $k$ is a root of the quadratic equation $k^2 + (1 - a)k + c = 0$, leads to an equation of the form T5.2.1.11:

$$z w''_{zz} + [(2 - b)z + 2k + 2 - a] w'_z + [(1 - b)z + 2k + 2 - a - bk] w = 0.$$

18. $(1 - x^2) y''_{xx} - 2xy'_x + n(n+1) y = 0, \quad n = 0, 1, 2, \ldots$

*Legendre equation.*

The solution is given by

$$y = C_1 P_n(x) + C_2 Q_n(x),$$

where the Legendre polynomials $P_n(x)$ and the Legendre functions of the second kind $Q_n(x)$ are given by the formulas

$$P_n(x) = \frac{1}{n!\, 2^n} \frac{d^n}{dx^n} (x^2 - 1)^n, \quad Q_n(x) = \frac{1}{2} P_n(x) \ln \frac{1+x}{1-x} - \sum_{m=1}^{n} \frac{1}{m} P_{m-1}(x) P_{n-m}(x).$$

The functions $P_n = P_n(x)$ can be conveniently calculated using the recurrence relations

$$P_0(x) = 1, \quad P_1(x) = x, \quad P_2(x) = \frac{1}{2}(3x^2 - 1), \quad \ldots, \quad P_{n+1}(x) = \frac{2n+1}{n+1} x P_n(x) - \frac{n}{n+1} P_{n-1}(x).$$

Three leading functions $Q_n = Q_n(x)$ are

$$Q_0(x) = \frac{1}{2} \ln \frac{1+x}{1-x}, \quad Q_1(x) = \frac{x}{2} \ln \frac{1+x}{1-x} - 1, \quad Q_2(x) = \frac{3x^2 - 1}{4} \ln \frac{1+x}{1-x} - \frac{3}{2} x.$$

The Legendre polynomials and the Legendre functions are described in Subsection 18.11.1 in more detail.

**19.** $(1-x^2)y''_{xx} - 2xy'_x + \nu(\nu+1)y = 0.$

*Legendre equation;* $\nu$ is an arbitrary number. The case $\nu = n$ where $n$ is a nonnegative integer is considered in T5.2.1.18.

The substitution $z = x^2$ leads to the hypergeometric equation. Therefore, with $|x| < 1$ the solution can be written as

$$y = C_1 F\left(-\frac{\nu}{2}, \frac{1+\nu}{2}, \frac{1}{2}; x\right) + C_2 x F\left(\frac{1-\nu}{2}, 1+\frac{\nu}{2}, \frac{3}{2}; x\right),$$

where $F(\alpha, \beta, \gamma; x)$ is the hypergeometric series (see T5.2.1.22).

The Legendre equation is discussed in Subsection 18.11.3 in more detail.

**20.** $(ax^2 + b)y''_{xx} + axy'_x + cy = 0.$

The substitution $z = \int \dfrac{dx}{\sqrt{ax^2+b}}$ leads to a constant coefficient linear equation: $y''_{zz} + cy = 0$.

**21.** $(1-x^2)y''_{xx} + (ax+b)y'_x + cy = 0.$

1°. The substitution $2z = 1 + x$ leads to the hypergeometric equation T5.2.1.22:

$$z(1-z)y''_{zz} + [az + \tfrac{1}{2}(b-a)]y'_z + cy = 0.$$

2°. For $a = -2m-3$, $b = 0$, and $c = \lambda$, the Gegenbauer functions are solutions of the equation.

**22.** $x(x-1)y''_{xx} + [(\alpha + \beta + 1)x - \gamma]y'_x + \alpha\beta y = 0.$

*Gaussian hypergeometric equation.* For $\gamma \neq 0, -1, -2, -3, \ldots$, a solution can be expressed in terms of the hypergeometric series:

$$F(\alpha, \beta, \gamma; x) = 1 + \sum_{k=1}^{\infty} \frac{(\alpha)_k (\beta)_k}{(\gamma)_k} \frac{x^k}{k!}, \quad (\alpha)_k = \alpha(\alpha+1)\ldots(\alpha+k-1),$$

which, *a fortiori*, is convergent for $|x| < 1$.

For $\gamma > \beta > 0$, this solution can be expressed in terms of a definite integral:

$$F(\alpha, \beta, \gamma; x) = \frac{\Gamma(\gamma)}{\Gamma(\beta)\Gamma(\gamma-\beta)} \int_0^1 t^{\beta-1}(1-t)^{\gamma-\beta-1}(1-tx)^{-\alpha}\, dt,$$

where $\Gamma(\beta)$ is the gamma function.

If $\gamma$ is not an integer, the general solution of the hypergeometric equation has the form:

$$y = C_1 F(\alpha, \beta, \gamma; x) + C_2 x^{1-\gamma} F(\alpha - \gamma + 1, \beta - \gamma + 1, 2 - \gamma; x).$$

In the degenerate cases $\gamma = 0, -1, -2, -3, \ldots$, a particular solution of the hypergeometric equation corresponds to $C_1 = 0$ and $C_2 = 1$. If $\gamma$ is a positive integer, another particular solution corresponds to $C_1 = 1$ and $C_2 = 0$. In both these cases, the general solution can be constructed by means of formula (2) given in the preliminary remarks at the beginning of Section T5.2.

Table T5.2 gives the general solutions of the hypergeometric equation for some values of the determining parameters.

The hypergeometric functions $F(\alpha, \beta, \gamma; x)$ are discussed in Section 18.10 in detail.

TABLE T5.2
General solutions of the hypergeometric equation for some values of the determining parameters

| $\alpha$ | $\beta$ | $\gamma$ | Solution: $y = y(x)$ |
|---|---|---|---|
| 0 | $\beta$ | $\gamma$ | $C_1 + C_2 \int |x|^{-\gamma} |x-1|^{\gamma-\beta-1} \, dx$ |
| $\alpha$ | $\alpha + \frac{1}{2}$ | $2\alpha + 1$ | $C_1 (1+\sqrt{1-x})^{-2\alpha} + C_2 x^{-2\alpha} (1+\sqrt{1-x})^{2\alpha}$ |
| $\alpha$ | $\alpha - \frac{1}{2}$ | $\frac{1}{2}$ | $C_1 (1+\sqrt{x})^{1-2\alpha} + C_2 (1-\sqrt{x})^{1-2\alpha}$ |
| $\alpha$ | $\alpha + \frac{1}{2}$ | $\frac{3}{2}$ | $\frac{1}{\sqrt{x}} \left[ C_1 (1+\sqrt{x})^{1-2\alpha} + C_2 (1-\sqrt{x})^{1-2\alpha} \right]$ |
| 1 | $\beta$ | $\gamma$ | $|x|^{1-\gamma} |x-1|^{\gamma-\beta-1} \left( C_1 + C_2 \int |x|^{\gamma-2} |x-1|^{\beta-\gamma} \, dx \right)$ |
| $\alpha$ | $\beta$ | $\alpha$ | $|x-1|^{-\beta} \left( C_1 + C_2 \int |x|^{-\alpha} |x-1|^{\beta-1} \, dx \right)$ |
| $\alpha$ | $\beta$ | $\alpha + 1$ | $|x|^{-\alpha} \left( C_1 + C_2 \int |x|^{\alpha-1} |x-1|^{-\beta} \, dx \right)$ |

**23.** $(1-x^2)^2 y''_{xx} - 2x(1-x^2) y'_x + [\nu(\nu+1)(1-x^2) - \mu^2] y = 0.$

*Legendre equation*, $\nu$ and $\mu$ are arbitrary parameters.

The transformation $x = 1 - 2\xi$, $y = |x^2-1|^{\mu/2} w$ leads to the hypergeometric equation T5.2.1.22:

$$\xi(\xi-1) w''_{\xi\xi} + (\mu+1)(1-2\xi) w'_\xi + (\nu-\mu)(\nu+\mu+1) w = 0$$

with parameters $\alpha = \mu - \nu$, $\beta = \mu + \nu + 1$, $\gamma = \mu + 1$.

In particular, the original equation is integrable by quadrature if $\nu = \mu$ or $\nu = -\mu - 1$.

**24.** $(x-a)^2 (x-b)^2 y''_{xx} - cy = 0, \qquad a \neq b.$

The transformation $\xi = \ln \left| \dfrac{x-a}{x-b} \right|$, $y = (x-b)\eta$ leads to a constant coefficient linear equation: $(a-b)^2 (\eta''_{\xi\xi} - \eta'_\xi) - c\eta = 0$. Therefore, the solution is as follows:

$$y = C_1 |x-a|^{(1+\lambda)/2} |x-b|^{(1-\lambda)/2} + C_2 |x-a|^{(1-\lambda)/2} |x-b|^{(1+\lambda)/2},$$

where $\lambda^2 = 4c(a-b)^{-2} + 1 \neq 0$.

**25.** $(ax^2 + bx + c)^2 y''_{xx} + Ay = 0.$

The transformation $\xi = \displaystyle\int \dfrac{dx}{ax^2 + bx + c}$, $w = \dfrac{y}{\sqrt{|ax^2 + bx + c|}}$ leads to a constant coefficient linear equation of the form T5.2.1.1: $w''_{\xi\xi} + (A + ac - \tfrac{1}{4} b^2) w = 0.$

**26.** $x^2 (ax^n - 1) y''_{xx} + x(apx^n + q) y'_x + (arx^n + s) y = 0.$

Find the roots $A_1, A_2$ and $B_1, B_2$ of the quadratic equations

$$A^2 - (q+1)A - s = 0, \quad B^2 - (p-1)B + r = 0$$

and define parameters $c$, $\alpha$, $\beta$, and $\gamma$ by the relations
$$c = A_1, \quad \alpha = (A_1 + B_1)n^{-1}, \quad \beta = (A_1 + B_2)n^{-1}, \quad \gamma = 1 + (A_1 - A_2)n^{-1}.$$
Then the solution of the original equation has the form $y = x^c u(ax^n)$, where $u = u(z)$ is the general solution of the hypergeometric equation T5.2.1.22: $z(z-1)u''_{zz} + [(\alpha+\beta+1)z - \gamma]u'_z + \alpha\beta u = 0$.

## T5.2.2. Equations Involving Exponential and Other Functions

1. $y''_{xx} + ae^{\lambda x}y = 0, \quad \lambda \neq 0$.

Solution: $y = C_1 J_0(z) + C_2 Y_0(z)$, where $z = 2\lambda^{-1}\sqrt{a}\,e^{\lambda x/2}$; $J_0(z)$ and $Y_0(z)$ are Bessel functions.

2. $y''_{xx} + (ae^x - b)y = 0$.

Solution: $y = C_1 J_{2\sqrt{b}}(2\sqrt{a}\,e^{x/2}) + C_2 Y_{2\sqrt{b}}(2\sqrt{a}\,e^{x/2})$, where $J_\nu(z)$ and $Y_\nu(z)$ are Bessel functions.

3. $y''_{xx} - (ae^{2\lambda x} + be^{\lambda x} + c)y = 0$.

The transformation $z = e^{\lambda x}$, $w = z^{-k}y$, where $k = \sqrt{c}/\lambda$, leads to an equation of the form T5.2.1.11: $\lambda^2 z w''_{zz} + \lambda^2(2k+1)w'_z - (az+b)w = 0$.

4. $y''_{xx} + ay'_x + be^{2ax}y = 0$.

The transformation $\xi = e^{ax}$, $u = ye^{ax}$ leads to a constant coefficient linear equation of the form T5.2.1.1: $u''_{\xi\xi} + ba^{-2}u = 0$.

5. $y''_{xx} - ay'_x + be^{2ax}y = 0$.

The substitution $\xi = e^{ax}$ leads to a constant coefficient linear equation of the form T5.2.1.1: $y''_{\xi\xi} + ba^{-2}y = 0$.

6. $y''_{xx} + ay'_x + (be^{\lambda x} + c)y = 0$.

Solution: $y = e^{-ax/2}\left[C_1 J_\nu(2\lambda^{-1}\sqrt{b}\,e^{\lambda x/2}) + C_2 Y_\nu(2\lambda^{-1}\sqrt{b}\,e^{\lambda x/2})\right]$, where $\nu = \frac{1}{\lambda}\sqrt{a^2 - 4c}$; $J_\nu(z)$ and $Y_\nu(z)$ are Bessel functions.

7. $y''_{xx} - (a - 2q\cosh 2x)y = 0$.

*Modified Mathieu equation.* The substitution $x = i\xi$ leads to the Mathieu equation T5.2.2.8:
$$y''_{\xi\xi} + (a - 2q\cos 2\xi)y = 0.$$
For eigenvalues $a = a_n(q)$ and $a = b_n(q)$, the corresponding solutions of the modified Mathieu equation are
$$\text{Ce}_{2n+p}(x, q) = \text{ce}_{2n+p}(ix, q) = \sum_{k=0}^{\infty} A_{2k+p}^{2n+p} \cosh[(2k+p)x],$$
$$\text{Se}_{2n+p}(x, q) = -i\,\text{se}_{2n+p}(ix, q) = \sum_{k=0}^{\infty} B_{2k+p}^{2n+p} \sinh[(2k+p)x],$$
where $p$ can be either 0 or 1, and the coefficients $A_{2k+p}^{2n+p}$ and $B_{2k+p}^{2n+p}$ are specified in T5.2.2.8.

The functions $\text{Ce}_{2n+p}(x, q)$ and $\text{Se}_{2n+p}(x, q)$ are discussed in Section 18.16 in more detail.

## 8. $y''_{xx} + (a - 2q\cos 2x)y = 0$.

*Mathieu equation.*

$1°$. Given numbers $a$ and $q$, there exists a general solution $y(x)$ and a characteristic index $\mu$ such that

$$y(x + \pi) = e^{2\pi\mu}y(x).$$

For small values of $q$, an approximate value of $\mu$ can be found from the equation

$$\cosh(\pi\mu) = 1 + 2\sin^2\left(\tfrac{1}{2}\pi\sqrt{a}\right) + \frac{\pi q^2}{(1-a)\sqrt{a}}\sin\left(\pi\sqrt{a}\right) + O(q^4).$$

If $y_1(x)$ is the solution of the Mathieu equation satisfying the initial conditions $y_1(0) = 1$ and $y'_1(0) = 0$, the characteristic index can be determined from the relation

$$\cosh(2\pi\mu) = y_1(\pi).$$

The solution $y_1(x)$, and hence $\mu$, can be determined with any degree of accuracy by means of numerical or approximate methods.

The general solution differs depending on the value of $y_1(\pi)$ and can be expressed in terms of two auxiliary periodical functions $\varphi_1(x)$ and $\varphi_2(x)$ (see Table T5.3).

TABLE T5.3
The general solution of the Mathieu equation T5.2.2.8 expressed in terms of auxiliary periodical functions $\varphi_1(x)$ and $\varphi_2(x)$

| Constraint | General solution $y = y(x)$ | Period of $\varphi_1$ and $\varphi_2$ | Index |
|---|---|---|---|
| $y_1(\pi) > 1$ | $C_1 e^{2\mu x}\varphi_1(x) + C_2 e^{-2\mu x}\varphi_2(x)$ | $\pi$ | $\mu$ is a real number |
| $y_1(\pi) < -1$ | $C_1 e^{2\rho x}\varphi_1(x) + C_2 e^{-2\rho x}\varphi_2(x)$ | $2\pi$ | $\mu = \rho + \tfrac{1}{2}i$, $i^2 = -1$, $\rho$ is the real part of $\mu$ |
| $|y_1(\pi)| < 1$ | $(C_1\cos\nu x + C_2\sin\nu x)\varphi_1(x)$ $+(C_1\cos\nu x - C_2\sin\nu x)\varphi_2(x)$ | $\pi$ | $\mu = i\nu$ is a pure imaginary number, $\cos(2\pi\nu) = y_1(\pi)$ |
| $y_1(\pi) = \pm 1$ | $C_1\varphi_1(x) + C_2 x\varphi_2(x)$ | $\pi$ | $\mu = 0$ |

$2°$. In applications, of major interest are periodical solutions of the Mathieu equation that exist for certain values of the parameters $a$ and $q$ (those values of $a$ are referred to as eigenvalues). The most important periodical solutions of the Mathieu equation have the form

$$\operatorname{ce}_{2n}(x,q) = \sum_{m=0}^{\infty} A_{2m}^{2n}\cos(2mx), \quad \operatorname{ce}_{2n+1}(x,q) = \sum_{m=0}^{\infty} A_{2m+1}^{2n+1}\cos\left[(2m+1)x\right];$$

$$\operatorname{se}_{2n}(x,q) = \sum_{m=0}^{\infty} B_{2m}^{2n}\sin(2mx), \quad \operatorname{se}_{2n+1}(x,q) = \sum_{m=0}^{\infty} B_{2m+1}^{2n+1}\sin\left[(2m+1)x\right];$$

where $A_j^i$ and $B_j^i$ are constants determined by recurrence relations.

The Mathieu functions $\operatorname{ce}_{2n}(x,q)$ and $\operatorname{se}_{2n}(x,q)$ are discussed in Section 18.16 in more detail.

**9.** $y''_{xx} + a \tan x \, y'_x + by = 0.$

$1°$. The substitution $\xi = \sin x$ leads to a linear equation of the form T5.2.1.21: $(\xi^2-1)y''_{\xi\xi} + (1-a)\xi y'_\xi - by = 0.$

$2°$. Solution for $a = -2$:

$$y \cos x = \begin{cases} C_1 \sin(kx) + C_2 \cos(kx) & \text{if } b+1 = k^2 > 0, \\ C_1 \sinh(kx) + C_2 \cosh(kx) & \text{if } b+1 = -k^2 < 0. \end{cases}$$

$3°$. Solution for $a = 2$ and $b = 3$: $y = C_1 \cos^3 x + C_2 \sin x \, (1 + 2\cos^2 x).$

## T5.2.3. Equations Involving Arbitrary Functions

▶ Notation: $f = f(x)$ and $g = g(x)$ are arbitrary functions; $a$, $b$, and $\lambda$ are arbitrary parameters.

**1.** $y''_{xx} + f y'_x + a(f-a)y = 0.$

Particular solution: $y_0 = e^{-ax}.$

**2.** $y''_{xx} + xf y'_x - fy = 0.$

Particular solution: $y_0 = x.$

**3.** $xy''_{xx} + (xf+a)y'_x + (a-1)fy = 0.$

Particular solution: $y_0 = x^{1-a}.$

**4.** $xy''_{xx} + [(ax+1)f + ax - 1]y'_x + a^2 x f y = 0.$

Particular solution: $y_0 = (ax+1)e^{-ax}.$

**5.** $xy''_{xx} + [(ax^2 + bx)f + 2]y'_x + bfy = 0.$

Particular solution: $y_0 = a + b/x.$

**6.** $x^2 y''_{xx} + xf y'_x + a(f-a-1)y = 0.$

Particular solution: $y_0 = x^{-a}.$

**7.** $y''_{xx} + (f + ae^{\lambda x})y'_x + ae^{\lambda x}(f + \lambda)y = 0.$

Particular solution: $y_0 = \exp\left(-\dfrac{a}{\lambda}e^{\lambda x}\right).$

**8.** $y''_{xx} - (f^2 + f'_x)y = 0.$

Particular solution: $y_0 = \exp\left(\int f \, dx\right).$

**9.** $y''_{xx} + 2f y'_x + (f^2 + f'_x)y = 0.$

Solution: $y = (C_2 x + C_1)\exp\left(-\int f \, dx\right).$

**10.** $y''_{xx} + (1-a)fy'_x - a(f^2 + f'_x)y = 0.$

Particular solution: $y_0 = \exp\left(a \int f\,dx\right)$.

**11.** $y''_{xx} + fy'_x + (fg - g^2 + g'_x)y = 0.$

Particular solution: $y_0 = \exp\left(-\int g\,dx\right)$.

**12.** $fy''_{xx} - af'_x y'_x - bf^{2a+1}y = 0.$

Solution: $y = C_1 e^u + C_2 e^{-u}$, where $u = \sqrt{b}\int f^a\,dx$.

**13.** $f^2 y''_{xx} + f(f'_x + a)y'_x + by = 0.$

The substitution $\xi = \int f^{-1}dx$ leads to a constant coefficient linear equation: $y''_{\xi\xi} + ay'_\xi + by = 0$.

**14.** $y''_{xx} - f'_x y'_x + a^2 e^{2f} y = 0.$

Solution: $y = C_1 \sin\left(a\int e^f\,dx\right) + C_2 \cos\left(a\int e^f\,dx\right)$.

**15.** $y''_{xx} - f'_x y'_x - a^2 e^{2f} y = 0.$

Solution: $y = C_1 \exp\left(a\int e^f\,dx\right) + C_2 \exp\left(-a\int e^f\,dx\right)$.

## T5.3. Second-Order Nonlinear Equations

### T5.3.1. Equations of the Form $y''_{xx} = f(x, y)$

**1.** $y''_{xx} = f(y).$

*Autonomous equation.*

Solution: $\int\left[C_1 + 2\int f(y)\,dy\right]^{-1/2} dy = C_2 \pm x.$

Particular solutions: $y = A_k$, where $A_k$ are roots of the algebraic (transcendental) equation $f(A_k) = 0$.

**2.** $y''_{xx} = Ax^n y^m.$

*Emden–Fowler equation.*

$1°$. With $m \neq 1$, the Emden–Fowler equation has a particular solution:

$$y = \lambda x^{\frac{n+2}{1-m}}, \quad \text{where } \lambda = \left[\frac{(n+2)(n+m+1)}{A(m-1)^2}\right]^{\frac{1}{m-1}}.$$

$2°$. The transformation $z = x^{n+2}y^{m-1}$, $w = xy'_x/y$ leads to a first-order (Abel) equation: $z[(m-1)w + n + 2]w'_z = -w^2 + w + Az$.

$3°$. The transformation $y = w/t$, $x = 1/t$ leads to the Emden–Fowler equation with the independent variable raised to a different power: $w''_{tt} = At^{-n-m-3}w^m$.

$4°$. The books by Zaitsev and Polyanin (1994) and Polyanin and Zaitsev (2003) present 28 solvable cases of the Emden–Fowler equation (corresponding to some pairs of $n$ and $m$).

**3.** $y''_{xx} + f(x)y = ay^{-3}$.

*Ermakov's equation.* Let $w = w(x)$ be a nontrivial solution of the second-order linear equation $w''_{xx} + f(x)w = 0$. The transformation $\xi = \int \frac{dx}{w^2}$, $z = \frac{y}{w}$ leads to an autonomous equation of the form T5.3.1: $z''_{\xi\xi} = az^{-3}$.

Solution: $C_1 y^2 = aw^2 + w^2 \left( C_2 + C_1 \int \frac{dx}{w^2} \right)^2$.

▶ Further on, $f$, $g$, $h$, and $\psi$ are arbitrary composite functions of their arguments indicated in parentheses after the function name (the arguments can depend on $x$, $y$, $y'_x$).

**4.** $y''_{xx} = f(ay + bx + c)$.

The substitution $w = ay + bx + c$ leads to an equation of the form T5.3.1.1: $w''_{xx} = af(w)$.

**5.** $y''_{xx} = f(y + ax^2 + bx + c)$.

The substitution $w = y + ax^2 + bx + c$ leads to an equation of the form T5.3.1.1: $w''_{xx} = f(w) + 2a$.

**6.** $y''_{xx} = x^{-1} f(yx^{-1})$.

*Homogeneous equation.* The transformation $t = -\ln|x|$, $z = y/x$ leads to an autonomous equation: $z''_{tt} - z'_t = f(z)$.

**7.** $y''_{xx} = x^{-3} f(yx^{-1})$.

The transformation $\xi = 1/x$, $w = y/x$ leads to the equation of the form T5.3.1.1: $w''_{\xi\xi} = f(w)$.

**8.** $y''_{xx} = x^{-3/2} f(yx^{-1/2})$.

Having set $w = yx^{-1/2}$, we obtain $\frac{d}{dx}(xw'_x)^2 = \frac{1}{2}ww'_x + 2f(w)w'_x$. Integrating the latter equation, we arrive at a separable equation.

Solution: $\int \left[ C_1 + \frac{1}{4}w^2 + 2 \int f(w)\,dw \right]^{-1/2} dw = C_2 \pm \ln x$.

**9.** $y''_{xx} = x^{k-2} f(x^{-k} y)$.

*Generalized homogeneous equation.* The transformation $z = x^{-k}y$, $w = xy'_x/y$ leads to a first-order equation: $z(w-k)w'_z = z^{-1} f(z) + w - w^2$.

**10.** $y''_{xx} = yx^{-2} f(x^n y^m)$.

*Generalized homogeneous equation.* The transformation $z = x^n y^m$, $w = xy'_x/y$ leads to a first-order equation: $z(mw + n)w'_z = f(z) + w - w^2$.

**11.** $y''_{xx} = y^{-3} f\left( \dfrac{y}{\sqrt{ax^2 + bx + c}} \right)$.

Setting $u(x) = y(ax^2 + bx + c)^{-1/2}$ and integrating the equation, we obtain a first-order separable equation:

$$(ax^2 + bx + c)^2 (u'_x)^2 = (\tfrac{1}{4}b^2 - ac)u^2 + 2\int u^{-3} f(u)\,du + C_1.$$

**12.** $y''_{xx} = e^{-ax}f(e^{ax}y)$.

The transformation $z = e^{ax}y$, $w = y'_x/y$ leads to a first-order equation: $z(w+a)w'_z = z^{-1}f(z) - w^2$.

**13.** $y''_{xx} = yf(e^{ax}y^m)$.

The transformation $z = e^{ax}y^m$, $w = y'_x/y$ leads to a first-order equation: $z(mw+a)w'_z = f(z) - w^2$.

**14.** $y''_{xx} = x^{-2}f(x^n e^{ay})$.

The transformation $z = x^n e^{ay}$, $w = xy'_x$ leads to a first-order equation: $z(aw+n)w'_z = f(z) + w$.

**15.** $y''_{xx} = \dfrac{\psi''_{xx}}{\psi}y + \psi^{-3}f\left(\dfrac{y}{\psi}\right)$, $\psi = \psi(x)$.

The transformation $\xi = \int \dfrac{dx}{\psi^2}$, $w = \dfrac{y}{\psi}$ leads to an equation of the form T5.3.1.1: $w''_{\xi\xi} = f(w)$.

Solution: $\int \left[C_1 + 2\int f(w)\,dw\right]^{-1/2} dw = C_2 \pm \int \dfrac{dx}{\psi^2(x)}$.

## T5.3.2. Equations of the Form $f(x,y)y''_{xx} = g(x,y,y'_x)$

**1.** $y''_{xx} - y'_x = f(y)$.

*Autonomous equation.* The substitution $w(y) = y'_x$ leads to a first-order equation. For solvable equations of the form in question, see the book by Polyanin and Zaitsev (2003).

**2.** $y''_{xx} + f(y)y'_x + g(y) = 0$.

*Lienard equation.* The substitution $w(y) = y'_x$ leads to a first-order equation. For solvable equations of the form in question, see the book by Polyanin and Zaitsev (2003).

**3.** $y''_{xx} + [ay + f(x)]y'_x + f'_x(x)y = 0$.

Integrating yields a Riccati equation: $y'_x + f(x)y + \tfrac{1}{2}ay^2 = C$.

**4.** $y''_{xx} + [2ay + f(x)]y'_x + af(x)y^2 = g(x)$.

On setting $u = y'_x + ay^2$, we obtain a first-order linear equation: $u'_x + f(x)u = g(x)$.

**5.** $y''_{xx} = ay'_x + e^{2ax}f(y)$.

Solution: $\int \left[C_1 + 2\int f(y)\,dy\right]^{-1/2} dy = C_2 \pm \dfrac{1}{a}e^{ax}$.

**6.** $y''_{xx} = f(y)y'_x$.

Solution: $\int \dfrac{dy}{F(y) + C_1} = C_2 + x$, where $F(y) = \int f(y)\,dy$.

7. $y''_{xx} = [e^{\alpha x} f(y) + \alpha] y'_x$.

The substitution $w(y) = e^{-\alpha x} y'_x$ leads to a first-order separable equation: $w'_y = f(y)$.

Solution: $\displaystyle\int \frac{dy}{F(y) + C_1} = C_2 + \frac{1}{\alpha} e^{\alpha x}$, where $F(y) = \int f(y)\, dy$.

8. $xy''_{xx} = ny'_x + x^{2n+1} f(y)$.

1°. Solution for $n \neq -1$:
$$\int \left[ C_1 + 2\int f(y)\, dy \right]^{-1/2} dy = \pm \frac{x^{n+1}}{n+1} + C_2.$$

2°. Solution for $n = -1$:
$$\int \left[ C_1 + 2\int f(y)\, dy \right]^{-1/2} dy = \pm \ln|x| + C_2.$$

9. $xy''_{xx} = f(y) y'_x$.

The substitution $w(y) = xy'_x/y$ leads to a first-order linear equation: $yw'_y = -w + 1 + f(y)$.

10. $xy''_{xx} = [x^k f(y) + k - 1] y'_x$.

Solution: $\displaystyle\int \frac{dy}{F(y) + C_1} = C_2 + \frac{1}{k} x^k$, where $F(y) = \int f(y)\, dy$.

11. $x^2 y''_{xx} + xy'_x = f(y)$.

The substitution $x = \pm e^t$ leads to an autonomous equation of the form T5.3.1.1: $y''_{tt} = f(y)$.

12. $(ax^2 + b) y''_{xx} + axy'_x + f(y) = 0$.

The substitution $\xi = \displaystyle\int \frac{dx}{\sqrt{ax^2 + b}}$ leads to an autonomous equation of the form T5.3.1.1: $y''_{\xi\xi} + f(y) = 0$.

13. $y''_{xx} = f(y) y'_x + g(x)$.

Integrating yields a first-order equation: $y'_x = \int f(y)\, dy + \int g(x)\, dx + C$.

14. $xy''_{xx} + (n+1) y'_x = x^{n-1} f(yx^n)$.

The transformation $\xi = x^n$, $w = yx^n$ leads to an autonomous equation of the form T5.3.1.1: $n^2 w''_{\xi\xi} = f(w)$.

15. $gy''_{xx} + \tfrac{1}{2} g'_x y'_x = f(y)$, $\quad g = g(x)$.

Integrating yields a first-order separable equation: $g(x)(y'_x)^2 = 2\int f(y)\, dy + C_1$.

Solution for $g(x) \geq 0$:
$$\int \left[ C_1 + 2\int f(y)\, dy \right]^{-1/2} dy = C_2 \pm \int \frac{dx}{\sqrt{g(x)}}.$$

**16.** $y''_{xx} = -ay'_x + e^{ax}f(ye^{ax})$.

The transformation $\xi = e^{ax}$, $w = ye^{ax}$ leads to the equation $w''_{\xi\xi} = a^{-2}f(w)$, which is of the form of T5.3.1.1.

**17.** $xy''_{xx} = f(x^n e^{ay})y'_x$.

The transformation $z = x^n e^{ay}$, $w = xy'_x$ leads to the following first-order separable equation: $z(aw + n)w'_z = [f(z) + 1]w$.

**18.** $x^2 y''_{xx} + xy'_x = f(x^n e^{ay})$.

The transformation $z = x^n e^{ay}$, $w = xy'_x$ leads to the following first-order separable equation: $z(aw + n)w'_z = f(z)$.

**19.** $yy''_{xx} + (y'_x)^2 + f(x)yy'_x + g(x) = 0$.

The substitution $u = y^2$ leads to a linear equation, $u''_{xx} + f(x)u'_x + 2g(x) = 0$, which can be reduced by the change of variable $w(x) = u'_x$ to a first-order linear equation.

**20.** $yy''_{xx} - (y'_x)^2 + f(x)yy'_x + g(x)y^2 = 0$.

The substitution $u = y'_x/y$ leads to a first-order linear equation: $u'_x + f(x)u + g(x) = 0$.

**21.** $yy''_{xx} - n(y'_x)^2 + f(x)y^2 + ay^{4n-2} = 0$.

$1°$. For $n = 1$, this is an equation of the form T5.3.2.22.

$2°$. For $n \neq 1$, the substitution $w = y^{1-n}$ leads to Ermakov's equation T5.3.1.5: $w''_{xx} + (1-n)f(x)w + a(1-n)w^{-3} = 0$.

**22.** $yy''_{xx} - n(y'_x)^2 + f(x)y^2 + g(x)y^{n+1} = 0$.

The substitution $w = y^{1-n}$ leads to a nonhomogeneous linear equation: $w''_{xx} + (1-n)f(x)w + (1-n)g(x) = 0$.

**23.** $yy''_{xx} + a(y'_x)^2 + f(x)yy'_x + g(x)y^2 = 0$.

The substitution $w = y^{a+1}$ leads to a linear equation: $w''_{xx} + f(x)w'_x + (a+1)g(x)w = 0$.

**24.** $yy''_{xx} = f(x)(y'_x)^2$.

The substitution $w(x) = xy'_x/y$ leads to a Bernoulli equation T5.1.4: $xw'_x = w + [f(x)-1]w^2$.

**25.** $y''_{xx} - a(y'_x)^2 + f(x)e^{ay} + g(x) = 0$.

The substitution $w = e^{-ay}$ leads to a nonhomogeneous linear equation: $w''_{xx} - ag(x)w = af(x)$.

**26.** $y''_{xx} - a(y'_x)^2 + be^{4ay} + f(x) = 0$.

The substitution $w = e^{-ay}$ leads to Ermakov's equation T5.3.1.5: $w''_{xx} - af(x)w = abw^{-3}$.

**27.** $y''_{xx} + a(y'_x)^2 - \frac{1}{2}y'_x = e^x f(y)$.

The substitution $w(y) = e^{-x}(y'_x)^2$ leads to a first-order linear equation: $w'_y + 2aw = 2f(y)$.

28. $y''_{xx} + \alpha(y'_x)^2 = [e^{\beta x} f(y) + \beta] y'_x$.

Solution:
$$\int \frac{e^{\alpha y} dy}{F(y) + C_1} = C_2 + \frac{1}{\beta} e^{\beta x}, \quad \text{where} \quad F(y) = \int e^{\alpha y} f(y) dy.$$

29. $y''_{xx} + f(y)(y'_x)^2 + g(y) = 0$.

The substitution $w(y) = (y'_x)^2$ leads to a first-order linear equation: $w'_y + 2f(y)w + 2g(y) = 0$.

30. $y''_{xx} + f(y)(y'_x)^2 - \frac{1}{2} y'_x = e^x g(y)$.

The substitution $w(y) = e^{-x}(y'_x)^2$ leads to a first-order linear equation: $w'_y + 2f(y)w = 2g(y)$.

31. $y''_{xx} = x f(y)(y'_x)^3$.

Taking $y$ to be the independent variable, we obtain a linear equation with respect to $x = x(y)$: $x''_{yy} = -f(y)x$.

32. $y''_{xx} = f(y)(y'_x)^2 + g(x) y'_x$.

Dividing by $y'_x$, we obtain an exact differential equation. Its solution follows from the equation:
$$\ln |y'_x| = \int f(y) dy + \int g(x) dx + C.$$

Solving the latter for $y'_x$, we arrive at a separable equation. In addition, $y = C_1$ is a singular solution, with $C_1$ being an arbitrary constant.

33. $y''_{xx} = f(x) g(x y'_x - y)$.

The substitution $w = x y'_x - y$ leads to a first-order separable equation: $w'_x = x f(x) g(w)$.

34. $y''_{xx} = \frac{y}{x^2} f\left(\frac{x y'_x}{y}\right)$.

The substitution $w(x) = x y'_x / y$ leads to a first-order separable equation: $x w'_x = f(w) + w - w^2$.

35. $g y''_{xx} + \frac{1}{2} g'_x y'_x = f(y) h(y'_x \sqrt{g})$, $\quad g = g(x)$.

The substitution $w(y) = y'_x \sqrt{g}$ leads to a first-order separable equation: $w w'_y = f(y) h(w)$.

36. $y''_{xx} = f(y'^2_x + ay)$.

The substitution $w(y) = (y'_x)^2 + ay$ leads to a first-order separable equation: $w'_y = 2f(w) + a$.

## References for Chapter T5

**Kamke, E.**, *Differentialgleichungen: Lösungsmethoden und Lösungen, I, Gewöhnliche Differentialgleichungen*, B. G. Teubner, Leipzig, 1977.

**Murphy, G. M.**, *Ordinary Differential Equations and Their Solutions*, D. Van Nostrand, New York, 1960.

**Polyanin, A. D. and Zaitsev, V. F.**, *Handbook of Exact Solutions for Ordinary Differential Equations, 2nd Edition*, Chapman & Hall/CRC Press, Boca Raton, 2003.

**Zaitsev, V. F. and Polyanin, A. D.**, *Discrete-Group Methods for Integrating Equations of Nonlinear Mechanics*, CRC Press, Boca Raton, 1994.

# Chapter T6

# Systems of Ordinary Differential Equations

## T6.1. Linear Systems of Two Equations
### T6.1.1. Systems of First-Order Equations

**1.** $x'_t = ax + by, \quad y'_t = cx + dy.$

*System of two constant-coefficient first-order linear homogeneous differential equations.*

Let us write out the characteristic equation
$$\lambda^2 - (a+d)\lambda + ad - bc = 0 \tag{1}$$

and find its discriminant
$$D = (a-d)^2 + 4bc. \tag{2}$$

$1°$. Case $ad - bc \neq 0$. The origin of coordinates $x = y = 0$ is the only one stationary point; it is

$\qquad$ a node if $D = 0$;

$\qquad$ a node if $D > 0$ and $ad - bc > 0$;

$\qquad$ a saddle if $D > 0$ and $ad - bc < 0$;

$\qquad$ a focus if $D < 0$ and $a + d \neq 0$;

$\qquad$ a center if $D < 0$ and $a + d = 0$.

1.1. Suppose $D > 0$. The characteristic equation (1) has two distinct real roots, $\lambda_1$ and $\lambda_2$. The general solution of the original system of differential equations is expressed as
$$x = C_1 b e^{\lambda_1 t} + C_2 b e^{\lambda_2 t},$$
$$y = C_1(\lambda_1 - a) e^{\lambda_1 t} + C_2(\lambda_2 - a) e^{\lambda_2 t},$$

where $C_1$ and $C_2$ are arbitrary constants.

1.2. Suppose $D < 0$. The characteristic equation (1) has two complex conjugate roots, $\lambda_{1,2} = \sigma \pm i\beta$. The general solution of the original system of differential equations is given by
$$x = b e^{\sigma t} \big[C_1 \sin(\beta t) + C_2 \cos(\beta t)\big],$$
$$y = e^{\sigma t} \big\{[(\sigma - a)C_1 - \beta C_2] \sin(\beta t) + [\beta C_1 + (\sigma - a)C_2] \cos(\beta t)\big\},$$

where $C_1$ and $C_2$ are arbitrary constants.

1.3. Suppose $D = 0$ and $a \neq d$. The characteristic equation (1) has two equal real roots, $\lambda_1 = \lambda_2$. The general solution of the original system of differential equations is
$$x = 2b\left(C_1 + \frac{C_2}{a-d} + C_2 t\right) \exp\left(\frac{a+d}{2}t\right),$$
$$y = [(d-a)C_1 + C_2 + (d-a)C_2 t] \exp\left(\frac{a+d}{2}t\right),$$

where $C_1$ and $C_2$ are arbitrary constants.

1.4. Suppose $a = d \neq 0$ and $b = 0$. Solution:
$$x = C_1 e^{at}, \qquad y = (cC_1 t + C_2)e^{at}.$$

1.5. Suppose $a = d \neq 0$ and $c = 0$. Solution:
$$x = (bC_1 t + C_2)e^{at}, \qquad y = C_1 e^{at}.$$

2°. Case $ad - bc = 0$ and $a^2 + b^2 > 0$. The whole of the line $ax + by = 0$ consists of singular points. The system in question may be rewritten in the form
$$x'_t = ax + by, \qquad y'_t = k(ax + by).$$

2.1. Suppose $a + bk \neq 0$. Solution:
$$x = bC_1 + C_2 e^{(a+bk)t}, \qquad y = -aC_1 + kC_2 e^{(a+bk)t}.$$

2.2. Suppose $a + bk = 0$. Solution:
$$x = C_1(bkt - 1) + bC_2 t, \qquad y = k^2 bC_1 t + (bk^2 t + 1)C_2.$$

**2.** $x'_t = a_1 x + b_1 y + c_1, \qquad y'_t = a_2 x + b_2 y + c_2.$

The general solution of this system is given by the sum of its any particular solution and the general solution of the corresponding homogeneous system (see system T6.1.1.1).

1°. Suppose $a_1 b_2 - a_2 b_1 \neq 0$. A particular solution:
$$x = x_0, \qquad y = y_0,$$
where the constants $x_0$ and $y_0$ are determined by solving the linear algebraic system of equations
$$a_1 x_0 + b_1 y_0 + c_1 = 0, \qquad a_2 x_0 + b_2 y_0 + c_2 = 0.$$

2°. Suppose $a_1 b_2 - a_2 b_1 = 0$ and $a_1^2 + b_1^2 > 0$. Then the original system can be rewritten as
$$x'_t = ax + by + c_1, \qquad y'_t = k(ax + by) + c_2.$$

2.1. If $\sigma = a + bk \neq 0$, the original system has a particular solution of the form
$$x = b\sigma^{-1}(c_1 k - c_2)t - \sigma^{-2}(ac_1 + bc_2), \qquad y = kx + (c_2 - c_1 k)t.$$

2.2. If $\sigma = a + bk = 0$, the original system has a particular solution of the form
$$x = \tfrac{1}{2}b(c_2 - c_1 k)t^2 + c_1 t, \qquad y = kx + (c_2 - c_1 k)t.$$

**3.** $x'_t = f(t)x + g(t)y, \qquad y'_t = g(t)x + f(t)y.$

Solution:
$$x = e^F(C_1 e^G + C_2 e^{-G}), \qquad y = e^F(C_1 e^G - C_2 e^{-G}),$$
where $C_1$ and $C_2$ are arbitrary constants, and
$$F = \int f(t)\, dt, \qquad G = \int g(t)\, dt.$$

4. $x'_t = f(t)x + g(t)y$, $\quad y'_t = -g(t)x + f(t)y$.

Solution:
$$x = F(C_1 \cos G + C_2 \sin G), \quad y = F(-C_1 \sin G + C_2 \cos G),$$

where $C_1$ and $C_2$ are arbitrary constants, and

$$F = \exp\left[\int f(t)\,dt\right], \quad G = \int g(t)\,dt.$$

5. $x'_t = f(t)x + g(t)y$, $\quad y'_t = ag(t)x + [f(t) + bg(t)]y$.

The transformation

$$x = \exp\left[\int f(t)\,dt\right]u, \quad y = \exp\left[\int f(t)\,dt\right]v, \quad \tau = \int g(t)\,dt$$

leads to a system of constant coefficient linear differential equations of the form T6.1.1.1:

$$u'_\tau = v, \quad v'_\tau = au + bv.$$

6. $x'_t = f(t)x + g(t)y$, $\quad y'_t = a[f(t) + ah(t)]x + a[g(t) - h(t)]y$.

Let us multiply the first equation by $-a$ and add it to the second equation to obtain

$$y'_t - ax'_t = -ah(t)(y - ax).$$

By setting $U = y - ax$ and then integrating, one obtains

$$y - ax = C_1 \exp\left[-a\int h(t)\,dt\right], \qquad (*)$$

where $C_1$ is an arbitrary constant. On solving $(*)$ for $y$ and on substituting the resulting expression into the first equation of the system, one arrives at a first-order linear differential equation for $x$.

7. $x'_t = f(t)x + g(t)y$, $\quad y'_t = h(t)x + p(t)y$.

$1°$. Let us express $y$ from the first equation and substitute into the second one to obtain a second-order linear equation:

$$gx''_{tt} - (fg + gp + g'_t)x'_t + (fgp - g^2h + fg'_t - f'_t g)x = 0. \qquad (1)$$

This equation is easy to integrate if, for example, the following conditions are met:

1) $fgp - g^2h + fg'_t - f'_t g = 0$;
2) $fgp - g^2h + fg'_t - f'_t g = ag, \quad fg + gp + g'_t = bg$.

In the first case, equation (1) has a particular solution $u = C = $ const. In the second case, it is a constant-coefficient equation.

A considerable number of other solvable cases of equation (1) can be found in the handbooks by Kamke (1977) and Polyanin and Zaitsev (2003).

2°. Suppose a particular solution of the system in question is known,
$$x = x_0(t), \quad y = y_0(t).$$
Then the general solution can be written out in the form
$$x(t) = C_1 x_0(t) + C_2 x_0(t) \int \frac{g(t) F(t) P(t)}{x_0^2(t)} \, dt,$$
$$y(t) = C_1 y_0(t) + C_2 \left[ \frac{F(t) P(t)}{x_0(t)} + y_0(t) \int \frac{g(t) F(t) P(t)}{x_0^2(t)} \, dt \right],$$
where $C_1$ and $C_2$ are arbitrary constants, and
$$F(t) = \exp\left[\int f(t)\, dt\right], \quad P(t) = \exp\left[\int p(t)\, dt\right].$$

## T6.1.2. Systems of Second-Order Equations

**1.** $x''_{tt} = ax + by, \quad y''_{tt} = cx + dy.$

*System of two constant-coefficient second-order linear homogeneous differential equations.*
The characteristic equation has the form
$$\lambda^4 - (a+d)\lambda^2 + ad - bc = 0.$$

1°. Case $ad - bc \neq 0$.

1.1. Suppose $(a-d)^2 + 4bc \neq 0$. The characteristic equation has four distinct roots $\lambda_1, \ldots, \lambda_4$. The general solution of the system in question is written as
$$x = C_1 b e^{\lambda_1 t} + C_2 b e^{\lambda_2 t} + C_3 b e^{\lambda_3 t} + C_4 b e^{\lambda_4 t},$$
$$y = C_1(\lambda_1^2 - a) e^{\lambda_1 t} + C_2(\lambda_2^2 - a) e^{\lambda_2 t} + C_3(\lambda_3^2 - a) e^{\lambda_3 t} + C_4(\lambda_4^2 - a) e^{\lambda_4 t},$$
where $C_1, \ldots, C_4$ are arbitrary constants.

1.2. Solution with $(a-d)^2 + 4bc = 0$ and $a \neq d$:
$$x = 2C_1\left(bt + \frac{2bk}{a-d}\right) e^{kt/2} + 2C_2\left(bt - \frac{2bk}{a-d}\right) e^{-kt/2} + 2bC_3 t e^{kt/2} + 2bC_4 t e^{-kt/2},$$
$$y = C_1(d-a)t e^{kt/2} + C_2(d-a)t e^{-kt/2} + C_3[(d-a)t + 2k] e^{kt/2} + C_4[(d-a)t - 2k] e^{-kt/2},$$
where $C_1, \ldots, C_4$ are arbitrary constants and $k = \sqrt{2(a+d)}$.

1.3. Solution with $a = d \neq 0$ and $b = 0$:
$$x = 2\sqrt{a}\, C_1 e^{\sqrt{a}\, t} + 2\sqrt{a}\, C_2 e^{-\sqrt{a}\, t},$$
$$y = cC_1 t e^{\sqrt{a}\, t} - cC_2 t e^{-\sqrt{a}\, t} + C_3 e^{\sqrt{a}\, t} + C_4 e^{-\sqrt{a}\, t}.$$

1.4. Solution with $a = d \neq 0$ and $c = 0$:
$$x = bC_1 t e^{\sqrt{a}\, t} - bC_2 t e^{-\sqrt{a}\, t} + C_3 e^{\sqrt{a}\, t} + C_4 e^{-\sqrt{a}\, t},$$
$$y = 2\sqrt{a}\, C_1 e^{\sqrt{a}\, t} + 2\sqrt{a}\, C_2 e^{-\sqrt{a}\, t}.$$

$2^\circ$. Case $ad - bc = 0$ and $a^2 + b^2 > 0$. The original system can be rewritten in the form

$$x''_{tt} = ax + by, \quad y''_{tt} = k(ax + by).$$

2.1. Solution with $a + bk \neq 0$:

$$x = C_1 \exp(t\sqrt{a+bk}) + C_2 \exp(-t\sqrt{a+bk}) + C_3 bt + C_4 b,$$
$$y = C_1 k \exp(t\sqrt{a+bk}) + C_2 k \exp(-t\sqrt{a+bk}) - C_3 at - C_4 a.$$

2.2. Solution with $a + bk = 0$:

$$x = C_1 bt^3 + C_2 bt^2 + C_3 t + C_4,$$
$$y = kx + 6C_1 t + 2C_2.$$

2. $x''_{tt} = a_1 x + b_1 y + c_1, \quad y''_{tt} = a_2 x + b_2 y + c_2.$

The general solution of this system is expressed as the sum of its any particular solution and the general solution of the corresponding homogeneous system (see system T6.1.2.1).

$1^\circ$. Suppose $a_1 b_2 - a_2 b_1 \neq 0$. A particular solution:

$$x = x_0, \quad y = y_0,$$

where the constants $x_0$ and $y_0$ are determined by solving the linear algebraic system of equations

$$a_1 x_0 + b_1 y_0 + c_1 = 0, \quad a_2 x_0 + b_2 y_0 + c_2 = 0.$$

$2^\circ$. Suppose $a_1 b_2 - a_2 b_1 = 0$ and $a_1^2 + b_1^2 > 0$. Then the system can be rewritten as

$$x''_{tt} = ax + by + c_1, \quad y''_{tt} = k(ax + by) + c_2.$$

2.1. If $\sigma = a + bk \neq 0$, the original system has a particular solution

$$x = \tfrac{1}{2} b\sigma^{-1}(c_1 k - c_2) t^2 - \sigma^{-2}(ac_1 + bc_2), \quad y = kx + \tfrac{1}{2}(c_2 - c_1 k) t^2.$$

2.2. If $\sigma = a + bk = 0$, the system has a particular solution

$$x = \tfrac{1}{24} b(c_2 - c_1 k) t^4 + \tfrac{1}{2} c_1 t^2, \quad y = kx + \tfrac{1}{2}(c_2 - c_1 k) t^2.$$

3. $x''_{tt} - ay'_t + bx = 0, \quad y''_{tt} + ax'_t + by = 0.$

This system is used to describe the horizontal motion of a pendulum taking into account the rotation of the earth.

Solution with $a^2 + 4b > 0$:

$$x = C_1 \cos(\alpha t) + C_2 \sin(\alpha t) + C_3 \cos(\beta t) + C_4 \sin(\beta t),$$
$$y = -C_1 \sin(\alpha t) + C_2 \cos(\alpha t) - C_3 \sin(\beta t) + C_4 \cos(\beta t),$$

where $C_1, \ldots, C_4$ are arbitrary constants and

$$\alpha = \tfrac{1}{2} a + \tfrac{1}{2}\sqrt{a^2 + 4b}, \quad \beta = \tfrac{1}{2} a - \tfrac{1}{2}\sqrt{a^2 + 4b}.$$

**4.** $x_{tt}'' + a_1 x_t' + b_1 y_t' + c_1 x + d_1 y = k_1 e^{i\omega t}, \qquad y_{tt}'' + a_2 x_t' + b_2 y_t' + c_2 x + d_2 y = k_2 e^{i\omega t}.$

Systems of this type often arise in oscillation theory (e.g., oscillations of a ship and a ship gyroscope). The general solution of this constant-coefficient linear nonhomogeneous system of differential equations is expressed as the sum of its any particular solution and the general solution of the corresponding homogeneous system (with $k_1 = k_2 = 0$).

1°. A particular solution is sought by the method of undetermined coefficients in the form

$$x = A_* e^{i\omega t}, \quad y = B_* e^{i\omega t}.$$

On substituting these expressions into the system of differential equations in question, one arrives at a linear nonhomogeneous system of algebraic equations for the coefficients $A_*$ and $B_*$.

2°. The general solution of a homogeneous system of differential equations is determined by a linear combination of its linearly independent particular solutions, which are sought using the method of undetermined coefficients in the form of exponential functions,

$$x = A e^{\lambda t}, \quad y = B e^{\lambda t}.$$

On substituting these expressions into the system and on collecting the coefficients of the unknowns $A$ and $B$, one obtains

$$(\lambda^2 + a_1 \lambda + c_1)A + (b_1 \lambda + d_1)B = 0,$$
$$(a_2 \lambda + c_2)A + (\lambda^2 + b_2 \lambda + d_2)B = 0.$$

For a nontrivial solution to exist, the determinant of this system must vanish. This requirement results in the characteristic equation

$$(\lambda^2 + a_1 \lambda + c_1)(\lambda^2 + b_2 \lambda + d_2) - (b_1 \lambda + d_1)(a_2 \lambda + c_2) = 0,$$

which is used to determine $\lambda$. If the roots of this equation, $k_1, \dots, k_4$, are all distinct, then the general solution of the original system of differential equations has the form

$$x = -C_1(b_1\lambda_1 + d_1)e^{\lambda_1 t} - C_2(b_1\lambda_2 + d_1)e^{\lambda_2 t} - C_3(b_1\lambda_1 + d_1)e^{\lambda_3 t} - C_4(b_1\lambda_4 + d_1)e^{\lambda_4 t},$$
$$y = C_1(\lambda_1^2 + a_1\lambda_1 + c_1)e^{\lambda_1 t} + C_2(\lambda_2^2 + a_1\lambda_2 + c_1)e^{\lambda_2 t}$$
$$+ C_3(\lambda_3^2 + a_1\lambda_3 + c_1)e^{\lambda_3 t} + C_4(\lambda_4^2 + a_1\lambda_4 + c_1)e^{\lambda_4 t},$$

where $C_1, \dots, C_4$ are arbitrary constants.

**5.** $x_{tt}'' = a(t y_t' - y), \qquad y_{tt}'' = b(t x_t' - x).$

The transformation

$$u = t x_t - x, \quad v = t y_t' - y \qquad (1)$$

leads to a first-order system:

$$u_t' = a t v, \quad v_t' = b t u.$$

The general solution of this system is expressed as

with $ab > 0$: $\begin{cases} u(t) = C_1 a \exp\left(\tfrac{1}{2}\sqrt{ab}\, t^2\right) + C_2 a \exp\left(-\tfrac{1}{2}\sqrt{ab}\, t^2\right), \\ v(t) = C_1 \sqrt{ab}\, \exp\left(\tfrac{1}{2}\sqrt{ab}\, t^2\right) - C_2 \sqrt{ab}\, \exp\left(-\tfrac{1}{2}\sqrt{ab}\, t^2\right); \end{cases}$

with $ab < 0$: $\begin{cases} u(t) = C_1 a \cos\left(\tfrac{1}{2}\sqrt{|ab|}\, t^2\right) + C_2 a \sin\left(\tfrac{1}{2}\sqrt{|ab|}\, t^2\right), \\ v(t) = -C_1 \sqrt{|ab|}\, \sin\left(\tfrac{1}{2}\sqrt{|ab|}\, t^2\right) + C_2 \sqrt{|ab|}\, \cos\left(\tfrac{1}{2}\sqrt{|ab|}\, t^2\right), \end{cases}$ (2)

where $C_1$ and $C_2$ are arbitrary constants. On substituting (2) into (1) and integrating, one arrives at the general solution of the original system in the form

$$x = C_3 t + t \int \frac{u(t)}{t^2}\, dt, \qquad y = C_4 t + t \int \frac{v(t)}{t^2}\, dt,$$

where $C_3$ and $C_4$ are arbitrary constants.

6. $x''_{tt} = f(t)(a_1 x + b_1 y), \qquad y''_{tt} = f(t)(a_2 x + b_2 y).$

Let $k_1$ and $k_2$ be roots of the quadratic equation

$$k^2 - (a_1 + b_2)k + a_1 b_2 - a_2 b_1 = 0.$$

Then, on multiplying the equations of the system by appropriate constants and on adding them together, one can rewrite the system in the form of two independent equations:

$$\begin{aligned} z''_1 &= k_1 f(t) z_1, & z_1 &= a_2 x + (k_1 - a_1) y; \\ z''_2 &= k_2 f(t) z_2, & z_2 &= a_2 x + (k_2 - a_1) y. \end{aligned}$$

Here, a prime stands for a derivative with respect to $t$.

7. $x''_{tt} = f(t)(a_1 x'_t + b_1 y'_t), \qquad y''_{tt} = f(t)(a_2 x'_t + b_2 y'_t).$

Let $k_1$ and $k_2$ be roots of the quadratic equation

$$k^2 - (a_1 + b_2)k + a_1 b_2 - a_2 b_1 = 0.$$

Then, on multiplying the equations of the system by appropriate constants and on adding them together, one can reduce the system to two independent equations:

$$\begin{aligned} z''_1 &= k_1 f(t) z'_1, & z_1 &= a_2 x + (k_1 - a_1) y; \\ z''_2 &= k_2 f(t) z'_2, & z_2 &= a_2 x + (k_2 - a_1) y. \end{aligned}$$

Integrating these equations and returning to the original variables, one arrives at a linear algebraic system for the unknowns $x$ and $y$:

$$a_2 x + (k_1 - a_1) y = C_1 \int \exp\bigl[k_1 F(t)\bigr]\, dt + C_2,$$

$$a_2 x + (k_2 - a_1) y = C_3 \int \exp\bigl[k_2 F(t)\bigr]\, dt + C_4,$$

where $C_1, \dots, C_4$ are arbitrary constants and $F(t) = \int f(t)\, dt$.

8. $x''_{tt} = a f(t)(t y'_t - y), \qquad y''_{tt} = b f(t)(t x'_t - x).$

The transformation

$$u = t x_t - x, \qquad v = t y'_t - y \qquad (1)$$

leads to a system of first-order equations:

$$u'_t = a t f(t) v, \qquad v'_t = b t f(t) u.$$

The general solution of this system is expressed as

if $ab > 0$,
$$\begin{cases} u(t) = C_1 a \exp\left(\sqrt{ab}\int tf(t)\,dt\right) + C_2 a \exp\left(-\sqrt{ab}\int tf(t)\,dt\right), \\ v(t) = C_1 \sqrt{ab}\exp\left(\sqrt{ab}\int tf(t)\,dt\right) - C_2\sqrt{ab}\exp\left(-\sqrt{ab}\int tf(t)\,dt\right); \end{cases}$$

if $ab < 0$,
$$\begin{cases} u(t) = C_1 a \cos\left(\sqrt{|ab|}\int tf(t)\,dt\right) + C_2 a \sin\left(\sqrt{|ab|}\int tf(t)\,dt\right), \\ v(t) = -C_1\sqrt{|ab|}\sin\left(\sqrt{|ab|}\int tf(t)\,dt\right) + C_2\sqrt{|ab|}\cos\left(\sqrt{|ab|}\int tf(t)\,dt\right), \end{cases}$$
(2)

where $C_1$ and $C_2$ are arbitrary constants. On substituting (2) into (1) and integrating, one obtains the general solution of the original system

$$x = C_3 t + t\int \frac{u(t)}{t^2}\,dt, \qquad y = C_4 t + t\int \frac{v(t)}{t^2}\,dt,$$

where $C_3$ and $C_4$ are arbitrary constants.

**9.** $t^2 x''_{tt} + a_1 t x'_t + b_1 t y'_t + c_1 x + d_1 y = 0, \qquad t^2 y''_{tt} + a_2 t x'_t + b_2 t y'_t + c_2 x + d_2 y = 0.$

*Linear system homogeneous in the independent variable (an Euler type system).*

$1°$. The general solution is determined by a linear combination of linearly independent particular solutions that are sought by the method of undetermined coefficients in the form of power-law functions
$$x = A|t|^k, \qquad y = B|t|^k.$$

On substituting these expressions into the system and on collecting the coefficients of the unknowns $A$ and $B$, one obtains

$$[k^2 + (a_1 - 1)k + c_1]A + (b_1 k + d_1)B = 0,$$
$$(a_2 k + c_2)A + [k^2 + (b_2 - 1)k + d_2]B = 0.$$

For a nontrivial solution to exist, the determinant of this system must vanish. This requirement results in the characteristic equation

$$[k^2 + (a_1 - 1)k + c_1][k^2 + (b_2 - 1)k + d_2] - (b_1 k + d_1)(a_2 k + c_2) = 0,$$

which is used to determine $k$. If the roots of this equation, $k_1, \ldots, k_4$, are all distinct, then the general solution of the system of differential equations in question has the form

$$x = -C_1(b_1 k_1 + d_1)|t|^{k_1} - C_2(b_1 k_2 + d_1)|t|^{k_2} - C_3(b_1 k_1 + d_1)|t|^{k_3} - C_4(b_1 k_4 + d_1)|t|^{k_4},$$
$$y = C_1[k_1^2 + (a_1 - 1)k_1 + c_1]|t|^{k_1} + C_2[k_2^2 + (a_1 - 1)k_2 + c_1]|t|^{k_2}$$
$$+ C_3[k_3^2 + (a_1 - 1)k_3 + c_1]|t|^{k_3} + C_4[k_4^2 + (a_1 - 1)k_4 + c_1]|t|^{k_4},$$

where $C_1, \ldots, C_4$ are arbitrary constants.

$2°$. The substitution $t = \sigma e^\tau$ ($\sigma \neq 0$) leads to a system of constant-coefficient linear differential equations:

$$x''_{\tau\tau} + (a_1 - 1)x'_\tau + b_1 y'_\tau + c_1 x + d_1 y = 0,$$
$$y''_{\tau\tau} + a_2 x'_\tau + (b_2 - 1)y'_\tau + c_2 x + d_2 y = 0.$$

**10.** $(\alpha t^2 + \beta t + \gamma)^2 x_{tt}'' = ax + by, \quad (\alpha t^2 + \beta t + \gamma)^2 y_{tt}'' = cx + dy.$

The transformation
$$\tau = \int \frac{dt}{\alpha t^2 + \beta t + \gamma}, \quad u = \frac{x}{\sqrt{|\alpha t^2 + \beta t + \gamma|}}, \quad v = \frac{y}{\sqrt{|\alpha t^2 + \beta t + \gamma|}}$$

leads to a constant-coefficient linear system of equations of the form T6.1.2.1:
$$u_{\tau\tau}'' = (a - \alpha\gamma + \tfrac{1}{4}\beta^2)u + bv,$$
$$v_{\tau\tau}'' = cu + (d - \alpha\gamma + \tfrac{1}{4}\beta^2)v.$$

**11.** $x_{tt}'' = f(t)(tx_t' - x) + g(t)(ty_t' - y), \quad y_{tt}'' = h(t)(tx_t' - x) + p(t)(ty_t' - y).$

The transformation
$$u = tx_t - x, \quad v = ty_t' - y \tag{1}$$

leads to a linear system of first-order equations
$$u_t' = tf(t)u + tg(t)v, \quad v_t' = th(t)u + tp(t)v. \tag{2}$$

In order to find the general solution of this system, it suffices to know its any particular solution (see system T6.1.1.7).

For solutions of some systems of the form (2), see systems T6.1.1.3–T6.1.1.6.

If all functions in (2) are proportional, that is,
$$f(t) = a\varphi(t), \quad g(t) = b\varphi(t), \quad h(t) = c\varphi(t), \quad p(t) = d\varphi(t),$$

then the introduction of the new independent variable $\tau = \int t\varphi(t)\,dt$ leads to a constant-coefficient system of the form T6.1.1.1.

$2°$. Suppose a solution of system (2) has been found in the form
$$u = u(t, C_1, C_2), \quad v = v(t, C_1, C_2), \tag{3}$$

where $C_1$ and $C_2$ are arbitrary constants. Then, on substituting (3) into (1) and integrating, one obtains a solution of the original system:
$$x = C_3 t + t \int \frac{u(t, C_1, C_2)}{t^2}\,dt, \quad y = C_4 t + t \int \frac{v(t, C_1, C_2)}{t^2}\,dt,$$

where $C_3$ and $C_4$ are arbitrary constants.

## T6.2. Linear Systems of Three and More Equations

**1.** $x_t' = ax, \quad y_t' = bx + cy, \quad z_t' = dx + ky + pz.$

Solution:
$$x = C_1 e^{at},$$
$$y = \frac{bC_1}{a-c} e^{at} + C_2 e^{ct},$$
$$z = \frac{C_1}{a-p}\left(d + \frac{bk}{a-c}\right)e^{at} + \frac{kC_2}{c-p}e^{ct} + C_3 e^{pt},$$

where $C_1$, $C_2$, and $C_3$ are arbitrary constants.

**2.** $x'_t = cy - bz, \quad y'_t = az - cx, \quad z'_t = bx - ay.$

**1°.** First integrals:
$$ax + by + cz = A, \tag{1}$$
$$x^2 + y^2 + z^2 = B^2, \tag{2}$$

where $A$ and $B$ are arbitrary constants. It follows that the integral curves are circles formed by the intersection of planes (1) and spheres (2).

**2°.** Solution:
$$x = aC_0 + kC_1 \cos(kt) + (cC_2 - bC_3) \sin(kt),$$
$$y = bC_0 + kC_2 \cos(kt) + (aC_3 - cC_1) \sin(kt),$$
$$z = cC_0 + kC_3 \cos(kt) + (bC_1 - aC_2) \sin(kt),$$

where $k = \sqrt{a^2 + b^2 + c^2}$ and the three of four constants of integration $C_0, \ldots, C_3$ are related by the constraint
$$aC_1 + bC_2 + cC_3 = 0.$$

**3.** $ax'_t = bc(y - z), \quad by'_t = ac(z - x), \quad cz'_t = ab(x - y).$

**1°.** First integral:
$$a^2 x + b^2 y + c^2 z = A,$$

where $A$ is an arbitrary constant. It follows that the integral curves are plane ones.

**2°.** Solution:
$$x = C_0 + kC_1 \cos(kt) + a^{-1}bc(C_2 - C_3) \sin(kt),$$
$$y = C_0 + kC_2 \cos(kt) + ab^{-1}c(C_3 - C_1) \sin(kt),$$
$$z = C_0 + kC_3 \cos(kt) + abc^{-1}(C_1 - C_2) \sin(kt),$$

where $k = \sqrt{a^2 + b^2 + c^2}$ and the three of four constants of integration $C_0, \ldots, C_3$ are related by the constraint
$$a^2 C_1 + b^2 C_2 + c^2 C_3 = 0.$$

**4.** $x'_t = (a_1 f + g)x + a_2 f y + a_3 f z,$
$$y'_t = b_1 f x + (b_2 f + g)y + b_3 f z, \quad z'_t = c_1 f x + c_2 f y + (c_3 f + g)z.$$

Here, $f = f(t)$ and $g = g(t)$.

The transformation
$$x = \exp\left[\int g(t)\,dt\right] u, \quad y = \exp\left[\int g(t)\,dt\right] v, \quad z = \exp\left[\int g(t)\,dt\right] w, \quad \tau = \int f(t)\,dt$$

leads to the system of constant coefficient linear differential equations
$$u'_\tau = a_1 u + a_2 v + a_3 w, \quad v'_\tau = b_1 u + b_2 v + b_3 w, \quad w'_\tau = c_1 u + c_2 v + c_3 w.$$

**5.** $x'_t = h(t)y - g(t)z, \quad y'_t = f(t)z - h(t)x, \quad z'_t = g(t)x - f(t)y.$

**1°.** First integral:
$$x^2 + y^2 + z^2 = C^2,$$

where $C$ is an arbitrary constant.

**2°.** The system concerned can be reduced to a Riccati equation (see Kamke, 1977).

6. $x'_k = a_{k1}x_1 + a_{k2}x_2 + \cdots + a_{kn}x_n; \qquad k = 1, 2, \ldots, n.$

*System of $n$ constant-coefficient first-order linear homogeneous differential equations.*

The general solution of a homogeneous system of differential equations is determined by a linear combination of linearly independent particular solutions, which are sought by the method of undetermined coefficients in the form of exponential functions,

$$x_k = A_k e^{\lambda t}; \qquad k = 1, 2, \ldots, n.$$

On substituting these expressions into the system and on collecting the coefficients of the unknowns $A_k$, one obtains a linear homogeneous system of algebraic equations:

$$a_{k1}A_1 + a_{k2}A_2 + \cdots + (a_{kk} - \lambda)A_k + \cdots + a_{kn}A_n = 0; \qquad k = 1, 2, \ldots, n.$$

For a nontrivial solution to exist, the determinant of this system must vanish. This requirement results in a characteristic equation that serves to determine $\lambda$.

## T6.3. Nonlinear Systems of Two Equations

### T6.3.1. Systems of First-Order Equations

1. $x'_t = x^n F(x, y), \quad y'_t = g(y) F(x, y).$

Solution:
$$x = \varphi(y), \qquad \int \frac{dy}{g(y) F(\varphi(y), y)} = t + C_2,$$

where

$$\varphi(y) = \begin{cases} \left[ C_1 + (1-n) \int \frac{dy}{g(y)} \right]^{\frac{1}{1-n}} & \text{if } n \neq 1, \\ C_1 \exp\left[ \int \frac{dy}{g(y)} \right] & \text{if } n = 1, \end{cases}$$

$C_1$ and $C_2$ are arbitrary constants.

2. $x'_t = e^{\lambda x} F(x, y), \quad y'_t = g(y) F(x, y).$

Solution:
$$x = \varphi(y), \qquad \int \frac{dy}{g(y) F(\varphi(y), y)} = t + C_2,$$

where

$$\varphi(y) = \begin{cases} -\frac{1}{\lambda} \ln\left[ C_1 - \lambda \int \frac{dy}{g(y)} \right] & \text{if } \lambda \neq 0, \\ C_1 + \int \frac{dy}{g(y)} & \text{if } \lambda = 0, \end{cases}$$

$C_1$ and $C_2$ are arbitrary constants.

**3.** $x'_t = F(x, y), \quad y'_t = G(x, y).$

*Autonomous system of general form.*
  Suppose
$$y = y(x, C_1),$$
where $C_1$ is an arbitrary constant, is the general solution of the first-order equation
$$F(x, y)y'_x = G(x, y).$$
Then the general solution of the system in question results in the following dependence for the variable $x$:
$$\int \frac{dx}{F(x, y(x, C_1))} = t + C_2.$$

**4.** $x'_t = f_1(x)g_1(y)\Phi(x, y, t), \quad y'_t = f_2(x)g_2(y)\Phi(x, y, t).$

First integral:
$$\int \frac{f_2(x)}{f_1(x)} dx - \int \frac{g_1(y)}{g_2(y)} dy = C, \qquad (*)$$
where $C$ is an arbitrary constant.
  On solving $(*)$ for $x$ (or $y$) and on substituting the resulting expression into one of the equations of the system concerned, one arrives at a first-order equation for $y$ (or $x$).

**5.** $x = tx'_t + F(x'_t, y'_t), \quad y = ty'_t + G(x'_t, y'_t).$

*Clairaut system.*
  The following are solutions of the system:
  (i) straight lines
$$x = C_1 t + F(C_1, C_2), \quad y = C_2 t + G(C_1, C_2),$$
where $C_1$ and $C_2$ are arbitrary constants;
  (ii) envelopes of these lines;
  (iii) continuously differentiable curves that are formed by segments of curves (i) and (ii).

## T6.3.2. Systems of Second-Order Equations

**1.** $x''_{tt} = xf(ax - by) + g(ax - by), \quad y''_{tt} = yf(ax - by) + h(ax - by).$

Let us multiply the first equation by $a$ and the second one by $-b$ and add them together to obtain the autonomous equation
$$z''_{tt} = zf(z) + ag(z) - bh(z), \quad z = ax - by. \qquad (1)$$

We will consider this equation in conjunction with the first equation of the system,
$$x''_{tt} = xf(z) + g(z). \qquad (2)$$

Equation (1) can be treated separately; its general solution can be written out in implicit form (see Polyanin and Zaitsev, 2003). The function $x = x(t)$ can be determined by solving the linear equation (2), and the function $y = y(t)$ is found as $y = (ax - z)/b$.

**2.** $x''_{tt} = xf(y/x), \quad y''_{tt} = yg(y/x).$

A periodic particular solution:
$$x = C_1 \sin(kt) + C_2 \cos(kt), \quad k = \sqrt{-f(\lambda)},$$
$$y = \lambda[C_1 \sin(kt) + C_2 \cos(kt)],$$

where $C_1$ and $C_2$ are arbitrary constants and $\lambda$ is a root of the transcendental (algebraic) equation
$$f(\lambda) = g(\lambda). \tag{1}$$

2°. Particular solution:
$$x = C_1 \exp(kt) + C_2 \exp(-kt), \quad k = \sqrt{f(\lambda)},$$
$$y = \lambda[C_1 \exp(kt) + C_2 \exp(-kt)],$$

where $C_1$ and $C_2$ are arbitrary constants and $\lambda$ is a root of the transcendental (algebraic) equation (1).

**3.** $x''_{tt} = kxr^{-3}, \quad y''_{tt} = kyr^{-3}, \quad \text{where} \quad r = \sqrt{x^2 + y^2}.$

*Equation of motion of a point mass in the xy-plane under gravity.*

Passing to polar coordinates by the formulas
$$x = r \cos \varphi, \quad y = r \sin \varphi, \quad r = r(t), \quad \varphi = \varphi(t),$$

one may obtain the first integrals
$$r^2 \varphi'_t = C_1, \quad (r'_t)^2 + r^2(\varphi'_t)^2 = -2kr^{-1} + C_2, \tag{1}$$

where $C_1$ and $C_2$ are arbitrary constants. Assuming that $C_1 \neq 0$ and integrating further, one finds that
$$r[C \cos(\varphi - \varphi_0) - k] = C_1^2, \quad C^2 = C_1^2 C_2 + k^2.$$

This is an equation of a conic section. The dependence $\varphi(t)$ may be found from the first equation in (1).

**4.** $x''_{tt} = xf(r), \quad y''_{tt} = yf(r), \quad \text{where} \quad r = \sqrt{x^2 + y^2}.$

*Equation of motion of a point mass in the xy-plane under a central force.*

Passing to polar coordinates by the formulas
$$x = r \cos \varphi, \quad y = r \sin \varphi, \quad r = r(t), \quad \varphi = \varphi(t),$$

one may obtain the first integrals
$$r^2 \varphi'_t = C_1, \quad (r'_t)^2 + r^2(\varphi'_t)^2 = 2 \int r f(r)\, dr + C_2,$$

where $C_1$ and $C_2$ are arbitrary constants. Integrating further, one finds that
$$t + C_3 = \pm \int \frac{r\, dr}{\sqrt{2r^2 F(r) + r^2 C_2 - C_1^2}}, \quad \varphi = C_1 \int \frac{dt}{r} + C_4, \tag{$*$}$$

where $C_3$ and $C_4$ are arbitrary constants and
$$F(r) = \int r f(r)\, dr.$$

It is assumed in the second relation in $(*)$ that the dependence $r = r(t)$ is obtained by solving the first equation in $(*)$ for $r(t)$.

**5.** $x''_{tt} + a(t)x = x^{-3}f(y/x)$, $\quad y''_{tt} + a(t)y = y^{-3}g(y/x)$.

*Generalized Ermakov system.*

1°. First integral:
$$\frac{1}{2}(xy'_t - yx'_t)^2 + \int^{y/x}\left[uf(u) - u^{-3}g(u)\right]du = C,$$

where $C$ is an arbitrary constant.

2°. Suppose $\varphi = \varphi(t)$ is a nontrivial solution of the second-order linear differential equation
$$\varphi''_{tt} + a(t)\varphi = 0. \tag{1}$$

Then the transformation
$$\tau = \int \frac{dt}{\varphi^2(t)}, \quad u = \frac{x}{\varphi(t)}, \quad v = \frac{y}{\varphi(t)} \tag{2}$$

leads to the autonomous system of equations
$$u''_{\tau\tau} = u^{-3}f(v/u), \quad v''_{\tau\tau} = v^{-3}g(v/u). \tag{3}$$

3°. Particular solution of system (3) is
$$u = A\sqrt{C_2\tau^2 + C_1\tau + C_0}, \quad v = Ak\sqrt{C_2\tau^2 + C_1\tau + C_0}, \quad A = \left[\frac{f(k)}{C_0C_2 - \frac{1}{4}C_1^2}\right]^{1/4},$$

where $C_0$, $C_1$, and $C_2$ are arbitrary constants, and $k$ is a root of the algebraic (transcendental) equation
$$k^4 f(k) = g(k).$$

**6.** $x''_{tt} = f(y'_t/x'_t)$, $\quad y''_{tt} = g(y'_t/x'_t)$.

1°. The transformation
$$u = x'_t, \quad w = y'_t \tag{1}$$

leads to a system of the first-order equations
$$u'_t = f(w/u), \quad w'_t = g(w/u). \tag{2}$$

Eliminating $t$ yields a homogeneous first-order equation, whose solution is given by
$$\int \frac{f(\xi)\,d\xi}{g(\xi) - \xi f(\xi)} = \ln|u| + C, \quad \xi = \frac{w}{u}, \tag{3}$$

where $C$ is an arbitrary constant. On solving (3) for $w$, one obtains $w = w(u, C)$. On substituting this expression into the first equation of (2), one can find $u = u(t)$ and then $w = w(t)$. Finally, one can determine $x = x(t)$ and $y = y(t)$ from (1) by simple integration.

**2°.** *The Suslov problem.* The problem of a point particle sliding down an inclined rough plane is described by the equations

$$x''_{tt} = 1 - \frac{kx'_t}{\sqrt{(x'_t)^2 + (y'_t)^2}}, \qquad y''_{tt} = -\frac{ky'_t}{\sqrt{(x'_t)^2 + (y'_t)^2}},$$

which correspond to a special case of the system in question with

$$f(z) = 1 - \frac{k}{\sqrt{1+z^2}}, \qquad g(z) = -\frac{kz}{\sqrt{1+z^2}}.$$

The solution of the corresponding Cauchy problem under the initial conditions

$$x(0) = y(0) = x'_t(0) = 0, \qquad y'_t(0) = 1$$

leads, for the case $k = 1$, to the following dependences $x(t)$ and $y(t)$ written in parametric form:

$$x = -\tfrac{1}{16} + \tfrac{1}{16}\xi^4 - \tfrac{1}{4}\ln\xi, \quad y = \tfrac{2}{3} - \tfrac{1}{2}\xi - \tfrac{1}{6}\xi^3, \quad t = \tfrac{1}{4} - \tfrac{1}{4}\xi^2 - \tfrac{1}{2}\ln\xi \qquad (0 \le \xi \le 1).$$

**7.** $x''_{tt} = x\Phi(x, y, t, x'_t, y'_t), \quad y''_{tt} = y\Phi(x, y, t, x'_t, y'_t).$

**1°.** First integral:

$$xy'_t - yx'_t = C,$$

where $C$ is an arbitrary constant.

*Remark.* The function $\Phi$ can also be dependent on the second and higher derivatives with respect to $t$.

**2°.** Particular solution: $y = C_1 x$, where $C_1$ is an arbitrary constant and the function $x = x(t)$ is determined by the ordinary differential equation

$$x''_{tt} = x\Phi(x, C_1 x, t, x'_t, C_1 x'_t).$$

**8.** $x''_{tt} + x^{-3} f(y/x) = x\Phi(x, y, t, x'_t, y'_t), \quad y''_{tt} + y^{-3} g(y/x) = y\Phi(x, y, t, x'_t, y'_t).$

First integral:

$$\tfrac{1}{2}(xy'_t - yx'_t)^2 + \int^{y/x} \left[ u^{-3} g(u) - u f(u) \right] du = C,$$

where $C$ is an arbitrary constant.

*Remark.* The function $\Phi$ can also be dependent on the second and higher derivatives with respect to $t$.

**9.** $x''_{tt} = F(t, tx'_t - x, ty'_t - y), \quad y''_{tt} = G(t, tx'_t - x, ty'_t - y).$

**1°.** The transformation

$$u = tx_t - x, \qquad v = ty'_t - y \tag{1}$$

leads to a system of first-order equations

$$u'_t = tF(t, u, v), \qquad v'_t = tG(t, u, v). \tag{2}$$

**2°.** Suppose a solution of system (2) has been found in the form

$$u = u(t, C_1, C_2), \qquad v = v(t, C_1, C_2), \tag{3}$$

where $C_1$ and $C_2$ are arbitrary constants. Then, substituting (3) into (1) and integrating, one obtains a solution of the original system,

$$x = C_3 t + t \int \frac{u(t, C_1, C_2)}{t^2}\, dt, \qquad y = C_4 t + t \int \frac{v(t, C_1, C_2)}{t^2}\, dt.$$

**3°.** If the functions $F$ and $G$ are independent of $t$, then, on eliminating $t$ from system (2), one arrives at a first-order equation

$$g(u, v) u'_v = F(u, v).$$

## T6.4. Nonlinear Systems of Three or More Equations

**1.** $ax'_t = (b-c)yz, \quad by'_t = (c-a)zx, \quad cz'_t = (a-b)xy.$

First integrals:
$$ax^2 + by^2 + cz^2 = C_1,$$
$$a^2x^2 + b^2y^2 + c^2z^2 = C_2,$$

where $C_1$ and $C_2$ are arbitrary constants. On solving the first integrals for $y$ and $z$ and on substituting the resulting expressions into the first equation of the system, one arrives at a separable first-order equation.

**2.** $ax'_t = (b-c)yzF(x,y,z,t),$
$$by'_t = (c-a)zxF(x,y,z,t), \quad cz'_t = (a-b)xyF(x,y,z,t).$$

First integrals:
$$ax^2 + by^2 + cz^2 = C_1,$$
$$a^2x^2 + b^2y^2 + c^2z^2 = C_2,$$

where $C_1$ and $C_2$ are arbitrary constants. On solving the first integrals for $y$ and $z$ and on substituting the resulting expressions into the first equation of the system, one arrives at a separable first-order equation; if $F$ is independent of $t$, this equation will be separable.

**3.** $x'_t = cF_2 - bF_3, \quad y'_t = aF_3 - cF_1, \quad z'_t = bF_1 - aF_2, \quad$ where $\quad F_n = F_n(x,y,z).$

First integral:
$$ax + by + cz = C_1,$$

where $C_1$ is an arbitrary constant. On eliminating $t$ and $z$ from the first two equations of the system (using the above first integral), one arrives at the first-order equation

$$\frac{dy}{dx} = \frac{aF_3(x,y,z) - cF_1(x,y,z)}{cF_2(x,y,z) - bF_3(x,y,z)}, \quad \text{where} \quad z = \frac{1}{c}(C_1 - ax - by).$$

**4.** $x'_t = czF_2 - byF_3, \quad y'_t = axF_3 - czF_1, \quad z'_t = byF_1 - axF_2.$

Here, $F_n = F_n(x,y,z)$ are arbitrary functions ($n = 1, 2, 3$).

First integral:
$$ax^2 + by^2 + cz^2 = C_1,$$

where $C_1$ is an arbitrary constant. On eliminating $t$ and $z$ from the first two equations of the system (using the above first integral), one arrives at the first-order equation

$$\frac{dy}{dx} = \frac{axF_3(x,y,z) - czF_1(x,y,z)}{czF_2(x,y,z) - byF_3(x,y,z)}, \quad \text{where} \quad z = \pm\sqrt{\frac{1}{c}(C_1 - ax^2 - by^2)}.$$

**5.** $x'_t = x(cF_2 - bF_3), \quad y'_t = y(aF_3 - cF_1), \quad z'_t = z(bF_1 - aF_2).$

Here, $F_n = F_n(x,y,z)$ are arbitrary functions ($n = 1, 2, 3$).

First integral:
$$|x|^a |y|^b |z|^c = C_1,$$

where $C_1$ is an arbitrary constant. On eliminating $t$ and $z$ from the first two equations of the system (using the above first integral), one may obtain a first-order equation.

**6.** $x'_t = h(z)F_2 - g(y)F_3$,  $y'_t = f(x)F_3 - h(z)F_1$,  $z'_t = g(y)F_1 - f(x)F_2$.

Here, $F_n = F_n(x, y, z)$ are arbitrary functions ($n = 1, 2, 3$).
First integral:
$$\int f(x)\,dx + \int g(y)\,dy + \int h(z)\,dz = C_1,$$
where $C_1$ is an arbitrary constant. On eliminating $t$ and $z$ from the first two equations of the system (using the above first integral), one may obtain a first-order equation.

**7.** $x''_{tt} = \dfrac{\partial F}{\partial x}$,  $y''_{tt} = \dfrac{\partial F}{\partial y}$,  $z''_{tt} = \dfrac{\partial F}{\partial z}$,  where  $F = F(r)$, $r = \sqrt{x^2 + y^2 + z^2}$.

*Equations of motion of a point particle under gravity.*
The system can be rewritten as a single vector equation:
$$\mathbf{r}''_{tt} = \operatorname{grad} F \quad \text{or} \quad \mathbf{r}''_{tt} = \frac{F'(r)}{r}\mathbf{r},$$
where $\mathbf{r} = (x, y, z)$.

1°. First integrals:
$$(\mathbf{r}'_t)^2 = 2F(r) + C_1 \quad \text{(law of conservation of energy)},$$
$$[\mathbf{r} \times \mathbf{r}'_t] = \mathbf{C} \quad \text{(law of conservation of areas)},$$
$$(\mathbf{r} \cdot \mathbf{C}) = 0 \quad \text{(all trajectories are plane curves)}.$$

2°. Solution:
$$\mathbf{r} = \mathbf{a}\, r\cos\varphi + \mathbf{b}\, r\sin\varphi.$$

Here, the constant vectors $\mathbf{a}$ and $\mathbf{b}$ must satisfy the conditions
$$|\mathbf{a}| = |\mathbf{b}| = 1, \quad (\mathbf{a} \cdot \mathbf{b}) = 0,$$
and the functions $r = r(t)$ and $\varphi = \varphi(t)$ are given by
$$t = \int \frac{r\,dr}{\sqrt{2r^2 F(r) + C_1 r^2 - C_3^2}} + C_2, \quad \varphi = C_3 \int \frac{dr}{r\sqrt{2r^2 F(r) + C_1 r^2 - C_3^2}}, \quad C_3 = |\mathbf{C}|.$$

**8.** $x''_{tt} = xF$,  $y''_{tt} = yF$,  $z''_{tt} = zF$,  where  $F = F(x, y, z, t, x'_t, y'_t, z'_t)$.

First integrals (laws of conservation of areas):
$$zy'_t - yz'_t = C_1,$$
$$xz'_t - zx'_t = C_2,$$
$$yx'_t - xy'_t = C_3,$$
where $C_1$, $C_2$, and $C_3$ are arbitrary constants.
Corollary of the conservation laws:
$$C_1 x + C_2 y + C_3 z = 0.$$

This implies that all integral curves are plane ones.

*Remark.* The function $\Phi$ can also be dependent on the second and higher derivatives with respect to $t$.

9. $x''_{tt} = F_1$, $y''_{tt} = F_2$, $z''_{tt} = F_3$, where $F_n = F_n(t, tx'_t - x, ty'_t - y, tz'_t - z)$.

1°. The transformation

$$u = tx_t - x, \quad v = ty'_t - y, \quad w = tz'_t - z \tag{1}$$

leads to the system of first-order equations

$$u'_t = tF_1(t, u, v, w), \quad v'_t = tF_2(t, u, v, w), \quad w'_t = tF_3(t, u, v, w). \tag{2}$$

2°. Suppose a solution of system (2) has been found in the form

$$u(t) = u(t, C_1, C_2, C_3), \quad v(t) = v(t, C_1, C_2, C_3), \quad w(t) = w(t, C_1, C_2, C_3), \tag{3}$$

where $C_1$, $C_2$, and $C_3$ are arbitrary constants. Then, substituting (3) into (1) and integrating, one obtains a solution of the original system:

$$x = C_4 t + t \int \frac{u(t)}{t^2} dt, \quad y = C_5 t + t \int \frac{v(t)}{t^2} dt, \quad z = C_6 t + t \int \frac{w(t)}{t^2} dt,$$

where $C_4$, $C_5$, and $C_6$ are arbitrary constants.

## References for Chapter T6

**Kamke, E.,** *Differentialgleichungen: Lösungsmethoden und Lösungen, I, Gewöhnliche Differentialgleichungen,* B. G. Teubner, Leipzig, 1977.

**Polyanin, A. D.,** *Systems of Ordinary Differential Equations,* From Website *EqWorld — The World of Mathematical Equations,* http://eqworld.ipmnet.ru/en/solutions/sysode.htm.

**Polyanin, A. D. and Zaitsev, V. F.,** *Handbook of Exact Solutions for Ordinary Differential Equations,* 2nd Edition, Chapman & Hall/CRC Press, Boca Raton, 2003.

# Chapter T7
# First-Order Partial Differential Equations

## T7.1. Linear Equations

### T7.1.1. Equations of the Form $f(x,y)\dfrac{\partial w}{\partial x} + g(x,y)\dfrac{\partial w}{\partial y} = 0$

▶ In equations T7.1.1.1–T7.1.1.11, the general solution is expressed in terms of the principal integral $\Xi$ as $w = \Phi(\Xi)$, where $\Phi(\Xi)$ is an arbitrary function.

1. $\dfrac{\partial w}{\partial x} + [f(x)y + g(x)]\dfrac{\partial w}{\partial y} = 0.$

Principal integral: $\Xi = e^{-F}y - \int e^{-F}g(x)\,dx$, where $F = \int f(x)\,dx$.

2. $\dfrac{\partial w}{\partial x} + [f(x)y + g(x)y^k]\dfrac{\partial w}{\partial y} = 0.$

Principal integral: $\Xi = e^{-F}y^{1-k} - (1-k)\int e^{-F}g(x)\,dx$, where $F = (1-k)\int f(x)\,dx$.

3. $\dfrac{\partial w}{\partial x} + [f(x)e^{\lambda y} + g(x)]\dfrac{\partial w}{\partial y} = 0.$

Principal integral: $\Xi = e^{-\lambda y}E + \lambda \int f(x)E\,dx$, where $E = \exp\left[\lambda \int g(x)\,dx\right]$.

4. $f(x)\dfrac{\partial w}{\partial x} + g(y)\dfrac{\partial w}{\partial y} = 0.$

Principal integral: $\Xi = \int \dfrac{dx}{f(x)} - \int \dfrac{dy}{g(y)}$.

5. $[f(y) + amx^ny^{m-1}]\dfrac{\partial w}{\partial x} - [g(x) + anx^{n-1}y^m]\dfrac{\partial w}{\partial y} = 0.$

Principal integral: $\Xi = \int f(y)\,dy + \int g(x)\,dx + ax^ny^m$.

6. $[e^{\alpha x}f(y) + c\beta]\dfrac{\partial w}{\partial x} - [e^{\beta y}g(x) + c\alpha]\dfrac{\partial w}{\partial y} = 0.$

Principal integral: $\Xi = \int e^{-\beta y}f(y)\,dy + \int e^{-\alpha x}g(x)\,dx - ce^{-\alpha x - \beta y}$.

**7.** $\dfrac{\partial w}{\partial x} + f(ax+by+c)\dfrac{\partial w}{\partial y} = 0, \qquad b \neq 0.$

Principal integral: $\Xi = \displaystyle\int \dfrac{dv}{a+bf(v)} - x$, where $v = ax+by+c$.

**8.** $\dfrac{\partial w}{\partial x} + f\!\left(\dfrac{y}{x}\right)\dfrac{\partial w}{\partial y} = 0.$

Principal integral: $\Xi = \displaystyle\int \dfrac{dv}{f(v)-v} - \ln|x|$, where $v = \dfrac{y}{x}$.

**9.** $x\dfrac{\partial w}{\partial x} + yf(x^n y^m)\dfrac{\partial w}{\partial y} = 0.$

Principal integral: $\Xi = \displaystyle\int \dfrac{dv}{v[mf(v)+n]} - \ln|x|$, where $v = x^n y^m$.

**10.** $\dfrac{\partial w}{\partial x} + yf(e^{\alpha x} y^m)\dfrac{\partial w}{\partial y} = 0.$

Principal integral: $\Xi = \displaystyle\int \dfrac{dv}{v[\alpha + mf(v)]} - x$, where $v = e^{\alpha x} y^m$.

**11.** $x\dfrac{\partial w}{\partial x} + f(x^n e^{\alpha y})\dfrac{\partial w}{\partial y} = 0.$

Principal integral: $\Xi = \displaystyle\int \dfrac{dv}{v[n+\alpha f(v)]} - \ln|x|$, where $v = x^n e^{\alpha y}$.

### T7.1.2. Equations of the Form $f(x,y)\dfrac{\partial w}{\partial x} + g(x,y)\dfrac{\partial w}{\partial y} = h(x,y)$

▶ In the solutions of equations T7.1.2.1–T7.1.2.12, $\Phi(z)$ is an arbitrary composite function whose argument $z$ can depend on both $x$ and $y$.

**1.** $a\dfrac{\partial w}{\partial x} + b\dfrac{\partial w}{\partial y} = f(x).$

General solution: $w = \dfrac{1}{a}\displaystyle\int f(x)\,dx + \Phi(bx - ay)$.

**2.** $\dfrac{\partial w}{\partial x} + a\dfrac{\partial w}{\partial y} = f(x)y^k.$

General solution: $w = \displaystyle\int_{x_0}^{x}(y-ax+at)^k f(t)\,dt + \Phi(y-ax)$, where $x_0$ can be taken arbitrarily.

**3.** $\dfrac{\partial w}{\partial x} + a\dfrac{\partial w}{\partial y} = f(x)e^{\lambda y}.$

General solution: $w = e^{\lambda(y-ax)}\displaystyle\int f(x)e^{a\lambda x}\,dx + \Phi(y-ax)$.

**4.** $a\dfrac{\partial w}{\partial x} + b\dfrac{\partial w}{\partial y} = f(x) + g(y)$.

General solution: $w = \dfrac{1}{a}\displaystyle\int f(x)\,dx + \dfrac{1}{b}\displaystyle\int g(y)\,dy + \Phi(bx - ay)$.

**5.** $\dfrac{\partial w}{\partial x} + a\dfrac{\partial w}{\partial y} = f(x)g(y)$.

General solution: $w = \displaystyle\int_{x_0}^{x} f(t)g(y - ax + at)\,dt + \Phi(y - ax)$, where $x_0$ can be taken arbitrarily.

**6.** $\dfrac{\partial w}{\partial x} + a\dfrac{\partial w}{\partial y} = f(x, y)$.

General solution: $w = \displaystyle\int_{x_0}^{x} f(t, y - ax + at)\,dt + \Phi(y - ax)$, where $x_0$ can be taken arbitrarily.

**7.** $\dfrac{\partial w}{\partial x} + [ay + f(x)]\dfrac{\partial w}{\partial y} = g(x)$.

General solution: $w = \displaystyle\int g(x)\,dx + \Phi(u)$, where $u = e^{-ax}y - \displaystyle\int f(x)e^{-ax}\,dx$.

**8.** $\dfrac{\partial w}{\partial x} + [ay + f(x)]\dfrac{\partial w}{\partial y} = g(x)h(y)$.

General solution:
$$w = \int g(x)\, h\!\left(e^{ax}u + e^{ax}\int f(x)e^{-ax}\,dx\right) dx + \Phi(u), \quad \text{where } u = e^{-ax}y - \int f(x)e^{-ax}\,dx.$$

In the integration, $u$ is treated as a parameter.

**9.** $\dfrac{\partial w}{\partial x} + [f(x)y + g(x)y^k]\dfrac{\partial w}{\partial y} = h(x)$.

General solution: $w = \displaystyle\int h(x)\,dx + \Phi(u)$, where
$$u = e^{-F}y^{1-k} - (1-k)\int e^{-F}g(x)\,dx, \quad F = (1-k)\int f(x)\,dx.$$

**10.** $\dfrac{\partial w}{\partial x} + [f(x) + g(x)e^{\lambda y}]\dfrac{\partial w}{\partial y} = h(x)$.

General solution: $w = \displaystyle\int h(x)\,dx + \Phi(u)$, where
$$u = e^{-\lambda y}F(x) + \lambda\int g(x)F(x)\,dx, \quad F(x) = \exp\!\left[\lambda\int f(x)\,dx\right].$$

11. $ax\dfrac{\partial w}{\partial x} + by\dfrac{\partial w}{\partial y} = f(x, y)$.

General solution:
$$w = \dfrac{1}{a}\int \dfrac{1}{x} f\!\left(x,\, u^{1/a} x^{b/a}\right) dx + \Phi(u), \text{ where } u = y^a x^{-b}.$$

In the integration, $u$ is treated as a parameter.

12. $f(x)\dfrac{\partial w}{\partial x} + g(y)\dfrac{\partial w}{\partial y} = h_1(x) + h_2(y)$.

General solution: $w = \displaystyle\int \dfrac{h_1(x)}{f(x)}\,dx + \int \dfrac{h_2(y)}{g(y)}\,dy + \Phi\!\left(\int \dfrac{dx}{f(x)} - \int \dfrac{dy}{g(y)}\right)$.

13. $f(x)\dfrac{\partial w}{\partial x} + g(y)\dfrac{\partial w}{\partial y} = h(x, y)$.

The transformation $\xi = \displaystyle\int \dfrac{dx}{f(x)},\ \eta = \int \dfrac{dy}{g(y)}$ leads to an equation of the form T7.1.2.6 for $w = w(\xi, \eta)$.

14. $f(y)\dfrac{\partial w}{\partial x} + g(x)\dfrac{\partial w}{\partial y} = h(x, y)$.

The transformation $\xi = \displaystyle\int g(x)\,dx,\ \eta = \int f(y)\,dy$ leads to an equation of the form T7.1.2.6 for $w = w(\xi, \eta)$.

### T7.1.3. Equations of the Form
$$f(x, y)\dfrac{\partial w}{\partial x} + g(x, y)\dfrac{\partial w}{\partial y} = h(x, y)w + r(x, y)$$

▶ In the solutions of equations T7.1.3.1–T7.1.3.10, $\Phi(z)$ is an arbitrary composite function whose argument $z$ can depend on both $x$ and $y$.

1. $a\dfrac{\partial w}{\partial x} + b\dfrac{\partial w}{\partial y} = f(x)w$.

General solution: $w = \exp\!\left[\dfrac{1}{a}\int f(x)\,dx\right]\Phi(bx - ay)$.

2. $a\dfrac{\partial w}{\partial x} + b\dfrac{\partial w}{\partial y} = f(x)w + g(x)$.

General solution: $w = \exp\!\left[\dfrac{1}{a}\int f(x)\,dx\right]\!\left\{\Phi(bx-ay) + \dfrac{1}{a}\int g(x)\exp\!\left[-\dfrac{1}{a}\int f(x)\,dx\right]dx\right\}$.

3. $a\dfrac{\partial w}{\partial x} + b\dfrac{\partial w}{\partial y} = [f(x) + g(y)]w$.

General solution: $w = \exp\!\left[\dfrac{1}{a}\int f(x)\,dx + \dfrac{1}{b}\int g(y)\,dy\right]\Phi(bx - ay)$.

**4.** $\dfrac{\partial w}{\partial x} + a \dfrac{\partial w}{\partial y} = f(x,y) w.$

General solution: $w = \exp\left[\displaystyle\int_{x_0}^{x} f(t, y - ax + at)\, dt\right] \Phi(y - ax)$, where $x_0$ can be taken arbitrarily.

**5.** $\dfrac{\partial w}{\partial x} + a \dfrac{\partial w}{\partial y} = f(x,y) w + g(x,y).$

General solution:
$$w = F(x,u)\left[\Phi(u) + \int \frac{g(x, u + ax)}{F(x,u)}\, dx\right], \quad F(x,u) = \exp\left[\int f(x, u + ax)\, dx\right],$$
where $u = y - ax$. In the integration, $u$ is treated as a parameter.

**6.** $ax\dfrac{\partial w}{\partial x} + by \dfrac{\partial w}{\partial y} = f(x) w + g(x).$

General solution: $w = \exp\left[\dfrac{1}{a}\displaystyle\int \dfrac{f(x)\,dx}{x}\right]\left\{\Phi(x^{-b/a} y) + \dfrac{1}{a}\displaystyle\int \dfrac{g(x)}{x} \exp\left[-\dfrac{1}{a}\int \dfrac{f(x)\,dx}{x}\right] dx\right\}.$

**7.** $ax\dfrac{\partial w}{\partial x} + by\dfrac{\partial w}{\partial y} = f(x,y) w.$

General solution:
$$w = \exp\left[\frac{1}{a}\int \frac{1}{x} f\!\left(x, u^{1/a} x^{b/a}\right) dx\right] \Phi(u), \quad \text{where} \quad u = y^a x^{-b}.$$

In the integration, $u$ is treated as a parameter.

**8.** $x\dfrac{\partial w}{\partial x} + ay\dfrac{\partial w}{\partial y} = f(x,y) w + g(x,y).$

General solution:
$$w = F(x,u)\left[\Phi(u) + \int \frac{g(x, ux^a)}{xF(x,u)}\, dx\right], \quad F(x,u) = \exp\left[\int \frac{1}{x} f(x, ux^a)\, dx\right],$$
where $u = yx^{-a}$. In the integration, $u$ is treated as a parameter.

**9.** $f(x)\dfrac{\partial w}{\partial x} + g(y)\dfrac{\partial w}{\partial y} = \bigl[h_1(x) + h_2(y)\bigr] w.$

General solution: $w = \exp\left[\displaystyle\int \dfrac{h_1(x)}{f(x)}\, dx + \int \dfrac{h_2(y)}{g(y)}\, dy\right] \Phi\!\left(\int \dfrac{dx}{f(x)} - \int \dfrac{dy}{g(y)}\right).$

10. $f_1(x)\dfrac{\partial w}{\partial x} + f_2(y)\dfrac{\partial w}{\partial y} = aw + g_1(x) + g_2(y).$

General solution:
$$w = E_1(x)\Phi(u) + E_1(x)\int \dfrac{g_1(x)\,dx}{f_1(x)E_1(x)} + E_2(y)\int \dfrac{g_2(y)\,dy}{f_2(y)E_2(y)},$$

where
$$E_1(x) = \exp\left[a\int \dfrac{dx}{f_1(x)}\right], \quad E_2(y) = \exp\left[a\int \dfrac{dy}{f_2(y)}\right], \quad u = \int \dfrac{dx}{f_1(x)} - \int \dfrac{dy}{f_2(y)}.$$

11. $f(x)\dfrac{\partial w}{\partial x} + g(y)\dfrac{\partial w}{\partial y} = h(x,y)w + r(x,y).$

The transformation $\xi = \displaystyle\int \dfrac{dx}{f(x)}$, $\eta = \displaystyle\int \dfrac{dy}{g(y)}$ leads to an equation of the form T7.1.3.5 for $w = w(\xi, \eta)$.

12. $f(y)\dfrac{\partial w}{\partial x} + g(x)\dfrac{\partial w}{\partial y} = h(x,y)w + r(x,y).$

The transformation $\xi = \displaystyle\int g(x)\,dx$, $\eta = \displaystyle\int f(y)\,dy$ leads to an equation of the form T7.1.3.5 for $w = w(\xi, \eta)$.

## T7.2. Quasilinear Equations

### T7.2.1. Equations of the Form $f(x,y)\dfrac{\partial w}{\partial x} + g(x,y)\dfrac{\partial w}{\partial y} = h(x,y,w)$

▶ In the solutions of equations T7.2.1.1–T7.2.1.12, $\Phi(z)$ is an arbitrary composite function whose argument $z$ can depend on both $x$ and $y$.

1. $\dfrac{\partial w}{\partial x} + a\dfrac{\partial w}{\partial y} = f(x)w + g(x)w^k.$

General solution:
$$w^{1-k} = F(x)\Phi(y - ax) + (1-k)F(x)\int \dfrac{g(x)}{F(x)}\,dx, \quad \text{where } F(x) = \exp\left[(1-k)\int f(x)\,dx\right].$$

2. $\dfrac{\partial w}{\partial x} + a\dfrac{\partial w}{\partial y} = f(x) + g(x)e^{\lambda w}.$

General solution:
$$e^{-\lambda w} = F(x)\Phi(y - ax) - \lambda F(x)\int \dfrac{g(x)}{F(x)}\,dx, \quad \text{where } F(x) = \exp\left[-\lambda \int f(x)\,dx\right].$$

**3.** $a\dfrac{\partial w}{\partial x} + b\dfrac{\partial w}{\partial y} = f(w).$

General solution: $\displaystyle\int \dfrac{dw}{f(w)} = \dfrac{x}{a} + \Phi(bx - ay).$

**4.** $a\dfrac{\partial w}{\partial x} + b\dfrac{\partial w}{\partial y} = f(x)g(w).$

General solution: $\displaystyle\int \dfrac{dw}{g(w)} = \dfrac{1}{a}\int f(x)\,dx + \Phi(bx - ay).$

**5.** $\dfrac{\partial w}{\partial x} + a\dfrac{\partial w}{\partial y} = f(x)g(y)h(w).$

General solution: $\displaystyle\int \dfrac{dw}{h(w)} = \int_{x_0}^{x} f(t)g(y - ax + at)\,dt + \Phi(y - ax),$ where $x_0$ can be taken arbitrarily.

**6.** $ax\dfrac{\partial w}{\partial x} + by\dfrac{\partial w}{\partial y} = f(w).$

General solution: $\displaystyle\int \dfrac{dw}{f(w)} = \dfrac{1}{a}\ln|x| + \Phi\big(|x|^b |y|^{-a}\big).$

**7.** $ay\dfrac{\partial w}{\partial x} + bx\dfrac{\partial w}{\partial y} = f(w).$

General solution: $\displaystyle\int \dfrac{dw}{f(w)} = \dfrac{1}{\sqrt{ab}}\ln\big|\sqrt{ab}\,x + ay\big| + \Phi\big(ay^2 - bx^2\big), \qquad ab > 0.$

**8.** $ax^n\dfrac{\partial w}{\partial x} + by^k\dfrac{\partial w}{\partial y} = f(w).$

General solution: $\displaystyle\int \dfrac{dw}{f(w)} = \dfrac{1}{a(1-n)}x^{1-n} + \Phi(u),$ where $u = \dfrac{1}{a(1-n)}x^{1-n} - \dfrac{1}{b(1-k)}y^{1-k}.$

**9.** $ay^n\dfrac{\partial w}{\partial x} + bx^k\dfrac{\partial w}{\partial y} = f(w).$

General solution:

$$a\int \dfrac{dw}{f(w)} = \int \left(\dfrac{b}{a}\dfrac{n+1}{k+1}x^{k+1} - u\right)^{-\tfrac{n}{n+1}} dx, \quad \text{where } u = \dfrac{b}{a}\dfrac{n+1}{k+1}x^{k+1} - y^{n+1}.$$

In the integration, $u$ is treated as a parameter.

**10.** $ae^{\lambda x}\dfrac{\partial w}{\partial x} + be^{\beta y}\dfrac{\partial w}{\partial y} = f(w).$

General solution: $\displaystyle\int \dfrac{dw}{f(w)} = -\dfrac{1}{a\lambda}e^{-\lambda x} + \Phi(u),$ where $u = a\lambda e^{-\beta y} - b\beta e^{-\lambda x}.$

**11.** $ae^{\lambda y}\dfrac{\partial w}{\partial x} + be^{\beta x}\dfrac{\partial w}{\partial y} = f(w)$.

General solution: $\displaystyle\int \dfrac{dw}{f(w)} = \dfrac{c(\beta x - \lambda y)}{u} + \Phi(u)$, where $u = a\beta e^{\lambda y} - b\lambda e^{\beta x}$.

**12.** $f(x)\dfrac{\partial w}{\partial x} + g(y)\dfrac{\partial w}{\partial y} = h(w)$.

General solution: $\displaystyle\int \dfrac{dw}{h(w)} = \int \dfrac{dx}{f(x)} + \Phi(u)$, where $u = \displaystyle\int \dfrac{dx}{f(x)} - \int \dfrac{dy}{g(y)}$.

**13.** $f(y)\dfrac{\partial w}{\partial x} + g(x)\dfrac{\partial w}{\partial y} = h(w)$.

The transformation $\xi = \displaystyle\int g(x)\,dx$, $\eta = \displaystyle\int f(y)\,dy$ leads to an equation of the form T7.2.1.5:

$$\dfrac{\partial w}{\partial \xi} + \dfrac{\partial w}{\partial \eta} = F(\xi)G(\eta)h(w), \quad \text{where} \quad F(\xi) = \dfrac{1}{g(x)}, \ G(\eta) = \dfrac{1}{f(y)}.$$

## T7.2.2. Equations of the Form $\dfrac{\partial w}{\partial x} + f(x, y, w)\dfrac{\partial w}{\partial y} = 0$

▶ In the solutions of equations T7.2.2.1–T7.2.2.10, $\Phi(w)$ is an arbitrary function.

**1.** $\dfrac{\partial w}{\partial x} + [aw + yf(x)]\dfrac{\partial w}{\partial y} = 0$.

General solution: $yF(x) - aw \displaystyle\int F(x)\,dx = \Phi(w)$, where $F(x) = \exp\left[-\displaystyle\int f(x)\,dx\right]$.

**2.** $\dfrac{\partial w}{\partial x} + [aw + f(y)]\dfrac{\partial w}{\partial y} = 0$.

General solution: $x = \displaystyle\int_{y_0}^{y} \dfrac{dt}{f(t) + aw} + \Phi(w)$.

**3.** $\dfrac{\partial w}{\partial x} + f(w)\dfrac{\partial w}{\partial y} = 0$.

*A model equation of gas dynamics.* This equation is also encountered in hydrodynamics, multiphase flows, wave theory, acoustics, chemical engineering, and other applications.

1°. General solution:
$$y = xf(w) + \Phi(w),$$
where $\Phi$ is an arbitrary function.

2°. The solution of the Cauchy problem with the initial condition
$$w = \varphi(y) \quad \text{at} \quad x = 0$$
can be represented in the parametric form
$$y = \xi + \mathcal{F}(\xi)x, \quad w = \varphi(\xi),$$
where $\mathcal{F}(\xi) = f\bigl(\varphi(\xi)\bigr)$.

3°. Consider the Cauchy problem with the discontinuous initial condition

$$w(0, y) = \begin{cases} w_1 & \text{for } y < 0, \\ w_2 & \text{for } y > 0. \end{cases}$$

It is assumed that $x \geq 0$, $f > 0$, and $f' > 0$ for $w > 0$, $w_1 > 0$, and $w_2 > 0$.

Generalized solution for $w_1 < w_2$:

$$w(x, y) = \begin{cases} w_1 & \text{for } y/x < V_1, \\ f^{-1}(y/x) & \text{for } V_1 \leq y/x \leq V_2, \\ w_2 & \text{for } y/x > V_2, \end{cases} \quad \text{where} \quad V_1 = f(w_1), \ V_2 = f(w_2).$$

Here $f^{-1}$ is the inverse of the function $f$, i.e., $f^{-1}(f(w)) \equiv w$. This solution is continuous in the half-plane $x > 0$ and describes a "rarefaction wave."

Generalized solution for $w_1 > w_2$:

$$w(x, y) = \begin{cases} w_1 & \text{for } y/x < V, \\ w_2 & \text{for } y/x > V, \end{cases} \quad \text{where} \quad V = \frac{1}{w_2 - w_1} \int_{w_1}^{w_2} f(w)\, dw.$$

This solution undergoes a discontinuity along the line $y = Vx$ and describes a "shock wave."

4°. In Subsection 13.1.3, qualitative features of solutions to this equation are considered, including the wave-breaking effect and shock waves. This subsection also presents general formulas that permit one to construct generalized (discontinuous) solutions for arbitrary initial conditions.

4. $\dfrac{\partial w}{\partial x} + [f(w) + ax]\dfrac{\partial w}{\partial y} = 0.$

General solution: $y = xf(w) + \frac{1}{2}ax^2 + \Phi(w)$.

5. $\dfrac{\partial w}{\partial x} + [f(w) + ay]\dfrac{\partial w}{\partial y} = 0.$

General solution: $x = \dfrac{1}{a} \ln|ay + f(w)| + \Phi(w)$.

6. $\dfrac{\partial w}{\partial x} + [f(w) + g(x)]\dfrac{\partial w}{\partial y} = 0.$

General solution: $y = xf(w) + \displaystyle\int g(x)\, dx + \Phi(w)$.

7. $\dfrac{\partial w}{\partial x} + [f(w) + g(y)]\dfrac{\partial w}{\partial y} = 0.$

General solution: $x = \displaystyle\int_{y_0}^{y} \dfrac{dt}{g(t) + f(w)} + \Phi(w)$.

8. $\dfrac{\partial w}{\partial x} + [yf(w) + g(x)]\dfrac{\partial w}{\partial y} = 0.$

General solution: $y \exp[-xf(w)] - \displaystyle\int_{x_0}^{x} g(t)\exp[-tf(w)]\, dt = \Phi(w)$, where $x_0$ can be taken arbitrarily.

9. $\dfrac{\partial w}{\partial x} + \bigl[xf(w) + yg(w) + h(w)\bigr]\dfrac{\partial w}{\partial y} = 0.$

General solution: $y + \dfrac{xf(w) + h(w)}{g(w)} + \dfrac{f(w)}{g^2(w)} = \exp\bigl[g(w)x\bigr]\Phi(w).$

10. $\dfrac{\partial w}{\partial x} + f(x)g(y)h(w)\dfrac{\partial w}{\partial y} = 0.$

General solution: $\displaystyle\int \dfrac{dy}{g(y)} - h(w)\int f(x)\,dx = \Phi(w).$

## T7.2.3. Equations of the Form $\dfrac{\partial w}{\partial x} + f(x, y, w)\dfrac{\partial w}{\partial y} = g(x, y, w)$

▶ In the solutions of equations T7.2.3.1–T7.2.3.11, $\Phi(z)$ is an arbitrary composite function whose argument $z$ can depend on $x$, $y$, and $w$.

1. $\dfrac{\partial w}{\partial x} + aw\dfrac{\partial w}{\partial y} = f(x).$

General solution:

$$y = ax\bigl[w - F(x)\bigr] + a\int F(x)\,dx + \Phi\bigl(w - F(x)\bigr), \quad \text{where } F(x) = \int f(x)\,dx.$$

2. $\dfrac{\partial w}{\partial x} + aw\dfrac{\partial w}{\partial y} = f(y).$

General solution:

$$x = \pm\int_{y_0}^{y} \dfrac{dz}{\sqrt{2aF(z) - 2au}} + \Phi(u), \quad \text{where } F(y) = \int f(y)\,dy, \quad u = F(y) - \tfrac{1}{2}aw^2.$$

3. $\dfrac{\partial w}{\partial x} + \bigl[aw + f(x)\bigr]\dfrac{\partial w}{\partial y} = g(x).$

General solution:

$$y = ax\bigl[w - G(x)\bigr] + a\int G(x)\,dx + F(x) + \Phi\bigl(w - G(x)\bigr),$$

where

$$F(x) = \int f(x)\,dx, \quad G(x) = \int g(x)\,dx.$$

4. $\dfrac{\partial w}{\partial x} + f(w)\dfrac{\partial w}{\partial y} = g(x).$

General solution: $y = \displaystyle\int_{x_0}^{x} f\bigl(G(t) - G(x) + w\bigr)\,dt + \Phi\bigl(w - G(x)\bigr),$ where $G(x) = \displaystyle\int g(x)\,dx.$

5. $\dfrac{\partial w}{\partial x} + f(w)\dfrac{\partial w}{\partial y} = g(y).$

General solution:
$$x = \int_{y_0}^{y} \psi\bigl(G(t) - G(y) + F(w)\bigr)\, dt + \Phi\bigl(F(w) - G(y)\bigr),$$

where $G(y) = \int g(y)\, dy$ and $F(w) = \int f(w)\, dw$. The function $\psi = \psi(z)$ is defined parametrically by $\psi = \dfrac{1}{f(w)}$, $z = F(w)$.

6. $\dfrac{\partial w}{\partial x} + f(w)\dfrac{\partial w}{\partial y} = g(w).$

General solution: $y = \displaystyle\int \dfrac{f(w)}{g(w)}\, dw + \Phi\!\left(x - \int \dfrac{dw}{g(w)}\right).$

7. $\dfrac{\partial w}{\partial x} + \bigl[f(w) + g(x)\bigr]\dfrac{\partial w}{\partial y} = h(x).$

General solution:
$$y = \int_{x_0}^{x} f\bigl(H(t) - H(x) + w\bigr)\, dt + G(x) + \Phi\bigl(w - H(x)\bigr),$$

where
$$G(x) = \int g(x)\, dx, \quad H(x) = \int h(x)\, dx.$$

8. $\dfrac{\partial w}{\partial x} + \bigl[f(w) + g(x)\bigr]\dfrac{\partial w}{\partial y} = h(w).$

General solution:
$$y = \int \dfrac{f(w)}{h(w)}\, dw + \int_{w_0}^{w} \dfrac{g\bigl(H(t) - H(w) + x\bigr)}{h(t)}\, dt + \Phi\bigl(x - H(w)\bigr), \quad \text{where } H(x) = \int \dfrac{dw}{h(w)}.$$

9. $\dfrac{\partial w}{\partial x} + \bigl[f(w) + y g(x)\bigr]\dfrac{\partial w}{\partial y} = h(x).$

General solution:
$$y G(x) - \int G(x) f\bigl(H(t) - H(x) + w\bigr)\, dx = \Phi\bigl(w - H(x)\bigr),$$

where $G(x) = \exp\!\left[-\int g(x)\, dx\right]$ and $H(x) = \int h(x)\, dx$.

10. $\dfrac{\partial w}{\partial x} + f(x, w)\dfrac{\partial w}{\partial y} = g(x).$

General solution: $y = \displaystyle\int_{x_0}^{x} f\bigl(t,\, G(t) - G(x) + w\bigr)\, dt + \Phi\bigl(w - G(x)\bigr),$ where $G(x) = \int g(x)\, dx.$

11. $\dfrac{\partial w}{\partial x} + f(x,w)\dfrac{\partial w}{\partial y} = g(w).$

General solution: $y = \displaystyle\int_{w_0}^{w} \dfrac{f(G(t)-G(w)+x,\, t)}{g(t)}\, dt + \Phi(x - G(w))$, where $G(w) = \displaystyle\int \dfrac{dw}{g(w)}$.

## T7.3. Nonlinear Equations

### T7.3.1. Equations Quadratic in One Derivative

▶ In this subsection, only complete integrals are presented. In order to construct the corresponding general solution, one should use the formulas of Subsection 13.2.1.

1. $\dfrac{\partial w}{\partial x} + a\left(\dfrac{\partial w}{\partial y}\right)^2 = by.$

This equation governs the free vertical drop of a point body near the Earth's surface ($y$ is the vertical coordinate measured downward, $x$ time, $m = \tfrac{1}{2a}$ the mass of the body, and $g = 2ab$ the gravitational acceleration).

Complete integral: $w = -C_1 x \pm \dfrac{2a}{3b}\left(\dfrac{by + C_1}{a}\right)^{3/2} + C_2.$

2. $\dfrac{\partial w}{\partial x} + a\left(\dfrac{\partial w}{\partial y}\right)^2 + by^2 = 0.$

This equation governs free oscillations of a point body of mass $m = 1/(2a)$ in an elastic field with elastic coefficient $k = 2b$ ($x$ is time and $y$ is the displacement from the equilibrium).

Complete integral: $w = -C_1 x + C_2 \pm \displaystyle\int \sqrt{\dfrac{C_1 - by^2}{a}}\, dx + C_2.$

3. $\dfrac{\partial w}{\partial x} + a\left(\dfrac{\partial w}{\partial y}\right)^2 = f(x) + g(y).$

Complete integral: $w = -C_1 x + \displaystyle\int f(x)\, dx + \displaystyle\int \sqrt{\dfrac{g(y) + C_1}{a}}\, dy + C_2.$

4. $\dfrac{\partial w}{\partial x} + a\left(\dfrac{\partial w}{\partial y}\right)^2 = f(x) y + g(x).$

Complete integral:

$w = \varphi(x) y + \displaystyle\int \left[g(x) - a\varphi^2(x)\right] dx + C_1,$ where $\varphi(x) = \displaystyle\int f(x)\, dx + C_2.$

5. $\dfrac{\partial w}{\partial x} + a\left(\dfrac{\partial w}{\partial y}\right)^2 = f(x) w + g(x).$

Complete integral:

$w = F(x)(C_1 + C_2 y) + F(x)\displaystyle\int \left[g(x) - aC_2^2 F^2(x)\right] \dfrac{dx}{F(x)},$ where $F(x) = \exp\left[\displaystyle\int f(x)\, dx\right].$

**6.** $\dfrac{\partial w}{\partial x} - f(w)\left(\dfrac{\partial w}{\partial y}\right)^2 = 0.$

Complete integral in implicit form: $\int f(w)\,dw = C_1^2 x + C_1 y + C_2.$

**7.** $f_1(x)\dfrac{\partial w}{\partial x} + f_2(y)\left(\dfrac{\partial w}{\partial y}\right)^2 = g_1(x) + g_2(y).$

Complete integral: $w = \int \dfrac{g_1(x) - C_1}{f_1(x)}\,dx + \int \sqrt{\dfrac{g_2(y) + C_1}{f_2(y)}}\,dy + C_2.$

**8.** $\dfrac{\partial w}{\partial x} + a\left(\dfrac{\partial w}{\partial y}\right)^2 + b\dfrac{\partial w}{\partial y} = f(x) + g(y).$

Complete integral: $w = -C_1 x + C_2 + \int f(x)\,dx - \dfrac{b}{2a}y \pm \dfrac{1}{2a}\int \sqrt{4ag(y) + b^2 + 4aC_1}\,dy.$

**9.** $\dfrac{\partial w}{\partial x} + a\left(\dfrac{\partial w}{\partial y}\right)^2 + b\dfrac{\partial w}{\partial y} = f(x)y + g(x).$

Complete integral:

$$w = \varphi(x)y + \int \left[g(x) - a\varphi^2(x) - b\varphi(x)\right]dx + C_1, \quad \text{where } \varphi(x) = \int f(x)\,dx + C_2.$$

**10.** $\dfrac{\partial w}{\partial x} + a\left(\dfrac{\partial w}{\partial y}\right)^2 + b\dfrac{\partial w}{\partial y} = f(x)w + g(x).$

Complete integral:

$$w = (C_1 y + C_2)F(x) + F(x)\int \left[g(x) - aC_1^2 F^2(x) - bC_1 F(x)\right]\dfrac{dx}{F(x)}, \quad F(x) = \exp\left[\int f(x)\,dx\right].$$

### T7.3.2. Equations Quadratic in Two Derivatives

**1.** $a\left(\dfrac{\partial w}{\partial x}\right)^2 + b\left(\dfrac{\partial w}{\partial y}\right)^2 = c.$

For $a = b$, this is a *differential equation of light rays*.
Complete integral: $w = C_1 x + C_2 y + C_3$, where $aC_1^2 + bC_2^2 = c.$
An alternative form of the complete integral: $\dfrac{w^2}{c} = \dfrac{(x - C_1)^2}{a} + \dfrac{(y - C_2)^2}{b}.$

**2.** $\left(\dfrac{\partial w}{\partial x}\right)^2 + \left(\dfrac{\partial w}{\partial y}\right)^2 = a - 2by.$

This equation governs parabolic motion of a point mass in vacuum (the coordinate $x$ is measured along the Earth's surface, the coordinate $y$ is measured vertically upward from the Earth's surface, and $a$ is the gravitational acceleration).

Complete integral: $w = C_1 x \pm \tfrac{1}{3b}(a - C_1^2 - 2by)^{3/2} + C_2.$

3. $\left(\dfrac{\partial w}{\partial x}\right)^2 + \left(\dfrac{\partial w}{\partial y}\right)^2 = \dfrac{a}{\sqrt{x^2+y^2}} + b.$

This equation arises from the solution of the two-body problem in celestial mechanics.
   Complete integral:

$$w = \pm \int \sqrt{b + \dfrac{a}{r} - \dfrac{C_1^2}{r^2}}\, dr + C_1 \arctan \dfrac{y}{x} + C_2, \quad \text{where } r = \sqrt{x^2+y^2}.$$

4. $\left(\dfrac{\partial w}{\partial x}\right)^2 + \left(\dfrac{\partial w}{\partial y}\right)^2 = f(x).$

Complete integral: $w = C_1 y + C_2 \pm \displaystyle\int \sqrt{f(x) - C_1^2}\, dx.$

5. $\left(\dfrac{\partial w}{\partial x}\right)^2 + \left(\dfrac{\partial w}{\partial y}\right)^2 = f(x) + g(y).$

Complete integral: $w = \pm \displaystyle\int \sqrt{f(x) + C_1}\, dx \pm \int \sqrt{g_2(y) - C_1}\, dy + C_2.$ The signs before each of the integrals can be chosen independently of each other.

6. $\left(\dfrac{\partial w}{\partial x}\right)^2 + \left(\dfrac{\partial w}{\partial y}\right)^2 = f(x^2 + y^2).$

Hamilton's equation for the plane motion of a point mass under the action of a central force.
   Complete integral:

$$w = C_1 \arctan \dfrac{x}{y} + C_2 \pm \dfrac{1}{2}\int \sqrt{zf(z) - C_1^2}\,\dfrac{dz}{z}, \quad z = x^2 + y^2.$$

7. $\left(\dfrac{\partial w}{\partial x}\right)^2 + \left(\dfrac{\partial w}{\partial y}\right)^2 = f(w).$

Complete integral in implicit form: $\displaystyle\int \dfrac{dw}{\sqrt{f(w)}} = \pm\sqrt{(x+C_1)^2 + (y+C_2)^2}.$

8. $\left(\dfrac{\partial w}{\partial x}\right)^2 + \dfrac{1}{x^2}\left(\dfrac{\partial w}{\partial y}\right)^2 = f(x).$

This equation governs the plane motion of a point mass in a central force field, with $x$ and $y$ being polar coordinates.

Complete integral: $w = C_1 y \pm \displaystyle\int \sqrt{f(x) - \dfrac{C_1^2}{x^2}}\, dx + C_2.$

9. $\left(\dfrac{\partial w}{\partial x}\right)^2 + f(x)\left(\dfrac{\partial w}{\partial y}\right)^2 = g(x).$

Complete integral: $w = C_1 y + C_2 + \displaystyle\int \sqrt{g(x) - C_1^2 f(x)}\, dx.$

**10.** $\left(\dfrac{\partial w}{\partial x}\right)^2 + f(y)\left(\dfrac{\partial w}{\partial y}\right)^2 = g(y).$

Complete integral: $w = C_1 x + C_2 + \displaystyle\int \sqrt{\dfrac{g(y) - C_1^2}{f(y)}}\, dy.$

**11.** $\left(\dfrac{\partial w}{\partial x}\right)^2 + f(w)\left(\dfrac{\partial w}{\partial y}\right)^2 = g(w).$

Complete integral in implicit form: $\displaystyle\int \sqrt{\dfrac{C_1^2 + C_2^2 f(w)}{g(w)}}\, dw = C_1 x + C_2 y + C_3.$
One of the constants $C_1$ or $C_2$ can be set equal to $\pm 1$.

**12.** $f_1(x)\left(\dfrac{\partial w}{\partial x}\right)^2 + f_2(y)\left(\dfrac{\partial w}{\partial y}\right)^2 = g_1(x) + g_2(y).$

*A separable equation.* This equation is encountered in differential geometry in studying geodesic lines of Liouville surfaces. Complete integral:

$$w = \pm \int \sqrt{\dfrac{g_1(x) + C_1}{f_1(x)}}\, dx \pm \int \sqrt{\dfrac{g_2(y) - C_1}{f_2(y)}}\, dy + C_2.$$

The signs before each of the integrals can be chosen independently of each other.

### T7.3.3. Equations with Arbitrary Nonlinearities in Derivatives

**1.** $\dfrac{\partial w}{\partial x} + f\!\left(\dfrac{\partial w}{\partial y}\right) = 0.$

This equation is encountered in optimal control and differential games.

1°. Complete integral: $w = C_1 y - f(C_1) x + C_2.$

2°. On differentiating the equation with respect to $y$, we arrive at a quasilinear equation of the form T7.2.2.3:
$$\dfrac{\partial u}{\partial x} + f'(u)\dfrac{\partial u}{\partial y} = 0, \qquad u = \dfrac{\partial w}{\partial y},$$
which is discussed in detail in Subsection 13.1.3.

3°. The solution of the Cauchy problem with the initial condition $w(0, y) = \varphi(y)$ can be written in parametric form as
$$y = f'(\zeta) x + \xi, \quad w = \big[\zeta f'(\zeta) - f(\zeta)\big] x + \varphi(\xi), \qquad \text{where} \quad \zeta = \varphi'(\xi).$$

See also Examples 1 and 2 in Subsection 13.2.3.

**2.** $\dfrac{\partial w}{\partial x} + f\!\left(\dfrac{\partial w}{\partial y}\right) = g(x).$

Complete integral: $w = C_1 y - f(C_1) x + \displaystyle\int g(x)\, dx + C_2.$

3. $\dfrac{\partial w}{\partial x} + f\left(\dfrac{\partial w}{\partial y}\right) = g(x)y + h(x).$

Complete integral:
$$w = \varphi(x)y + \int \left[h(x) - f(\varphi(x))\right] dx + C_1, \quad \text{where } \varphi(x) = \int g(x)\,dx + C_2.$$

4. $\dfrac{\partial w}{\partial x} + f\left(\dfrac{\partial w}{\partial y}\right) = g(x)w + h(x).$

Complete integral:
$$w = (C_1 y + C_2)\varphi(x) + \varphi(x)\int \left[h(x) - f(C_1 \varphi(x))\right] \dfrac{dx}{\varphi(x)}, \quad \text{where } \varphi(x) = \exp\left[\int g(x)\,dx\right].$$

5. $\dfrac{\partial w}{\partial x} - F\left(x, \dfrac{\partial w}{\partial y}\right) = 0.$

Complete integral: $w = \displaystyle\int F(x, C_1)\,dx + C_1 y + C_2.$

6. $\dfrac{\partial w}{\partial x} + F\left(x, \dfrac{\partial w}{\partial y}\right) = aw.$

Complete integral: $w = e^{ax}(C_1 y + C_2) - e^{ax}\displaystyle\int e^{-ax} F(x, C_1 e^{ax})\,dx.$

7. $\dfrac{\partial w}{\partial x} + F\left(x, \dfrac{\partial w}{\partial y}\right) = g(x)w.$

Complete integral:
$$w = \varphi(x)(C_1 y + C_2) - \varphi(x)\int F(x, C_1 \varphi(x)) \dfrac{dx}{\varphi(x)}, \quad \text{where } \varphi(x) = \exp\left[\int g(x)\,dx\right].$$

8. $F\left(\dfrac{\partial w}{\partial x}, \dfrac{\partial w}{\partial y}\right) = 0.$

Complete integral:
$$w = C_1 x + C_2 y + C_3,$$
where $C_1$ and $C_3$ are arbitrary constants and the constant $C_2$ is related to $C_1$ by $F(C_1, C_2) = 0.$

9. $w = x\dfrac{\partial w}{\partial x} + y\dfrac{\partial w}{\partial y} + F\left(\dfrac{\partial w}{\partial x}, \dfrac{\partial w}{\partial y}\right).$

*Clairaut's equation.* Complete integral: $w = C_1 x + C_2 y + F(C_1, C_2).$

10. $F_1\left(x, \dfrac{\partial w}{\partial x}\right) = F_2\left(y, \dfrac{\partial w}{\partial y}\right).$

A *separable equation*. Complete integral:
$$w = \varphi(x) + \psi(y) + C_1,$$

where the functions $\varphi = \varphi(x)$ and $\psi = \psi(y)$ are determined from the ordinary differential equations
$$F_1(x, \varphi'_x) = C_2, \qquad F_2(y, \psi'_y) = C_2.$$

11. $F_1\left(x, \dfrac{\partial w}{\partial x}\right) + F_2\left(y, \dfrac{\partial w}{\partial y}\right) + aw = 0.$

A *separable equation*. Complete integral:
$$w = \varphi(x) + \psi(y),$$

where the functions $\varphi = \varphi(x)$ and $\psi = \psi(y)$ are determined from the ordinary differential equations
$$F_1(x, \varphi'_x) + a\varphi = C_1, \qquad F_2(y, \psi'_y) + a\psi = -C_1,$$

where $C_1$ is an arbitrary constant. If $a \neq 0$, one can set $C_1 = 0$ in these equations.

12. $F_1\left(x, \dfrac{1}{w}\dfrac{\partial w}{\partial x}\right) + w^k F_2\left(y, \dfrac{1}{w}\dfrac{\partial w}{\partial y}\right) = 0.$

A *separable equation*. Complete integral:
$$w(x, y) = \varphi(x)\psi(y).$$

The functions $\varphi = \varphi(x)$ and $\psi = \psi(y)$ are determined by solving the ordinary differential equations
$$\varphi^{-k} F_1(x, \varphi'_x/\varphi) = C, \qquad \psi^k F_2(y, \psi'_y/\psi) = -C,$$

where $C$ is an arbitrary constant.

13. $F_1\left(x, \dfrac{\partial w}{\partial x}\right) + e^{\lambda w} F_2\left(y, \dfrac{\partial w}{\partial y}\right) = 0.$

A *separable equation*. Complete integral:
$$w(x, y) = \varphi(x) + \psi(y).$$

The functions $\varphi = \varphi(x)$ and $\psi = \psi(y)$ are determined by solving the ordinary differential equations
$$e^{-\lambda \varphi} F_1(x, \varphi'_x) = C, \qquad e^{\lambda \psi} F_2(y, \psi'_y) = -C,$$

where $C$ is an arbitrary constant.

14. $F_1\left(x, \dfrac{1}{w}\dfrac{\partial w}{\partial x}\right) + F_2\left(y, \dfrac{1}{w}\dfrac{\partial w}{\partial y}\right) = k\ln w.$

*A separable equation.* Complete integral:
$$w(x,y) = \varphi(x)\psi(y).$$

The functions $\varphi = \varphi(x)$ and $\psi = \psi(y)$ are determined by solving the ordinary differential equations
$$F_1(x, \varphi'_x/\varphi) - k\ln\varphi = C, \qquad F_2(y, \psi'_y/\psi) - k\ln\psi = -C,$$
where $C$ is an arbitrary constant.

15. $\dfrac{\partial w}{\partial x} + yF_1\left(x, \dfrac{\partial w}{\partial y}\right) + F_2\left(x, \dfrac{\partial w}{\partial y}\right) = 0.$

Complete integral:
$$w = \varphi(x)y - \int F_2(x, \varphi(x))\, dx + C_1,$$
where the function $\varphi(x)$ is determined by solving the ordinary differential equation $\varphi'_x + F_1(x, \varphi) = 0$.

16. $F\left(\dfrac{\partial w}{\partial x} + ay, \dfrac{\partial w}{\partial y} + ax\right) = 0.$

Complete integral: $w = -axy + C_1x + C_2y + C_3$, where $F(C_1, C_2) = 0$.

17. $\left(\dfrac{\partial w}{\partial x}\right)^2 + \left(\dfrac{\partial w}{\partial y}\right)^2 = F\left(x^2 + y^2,\ y\dfrac{\partial w}{\partial x} - x\dfrac{\partial w}{\partial y}\right).$

Complete integral: $w = -C_1 \arctan\dfrac{y}{x} + \dfrac{1}{2}\int\sqrt{\xi F(\xi, C_1) - C_1^2}\,\dfrac{d\xi}{\xi} + C_2$, where $\xi = x^2 + y^2$.

18. $F\left(x, \dfrac{\partial w}{\partial x}, \dfrac{\partial w}{\partial y}\right) = 0.$

Complete integral: $w = C_1 y + \varphi(x, C_1) + C_2$, where the function $\varphi = \varphi(x, C_1)$ is determined from the ordinary differential equation $F(x, \varphi'_x, C_1) = 0$.

19. $F\left(ax + by, \dfrac{\partial w}{\partial x}, \dfrac{\partial w}{\partial y}\right) = 0.$

For $b = 0$, see equation T7.3.3.18. Complete integral for $b \neq 0$:
$$w = C_1 x + \varphi(z, C_1) + C_2, \qquad z = ax + by,$$
where the function $\varphi = \varphi(z)$ is determined from the nonlinear ordinary differential equation $F(z, a\varphi'_z + C_1, b\varphi'_z) = 0$.

20. $F\left(w, \dfrac{\partial w}{\partial x}, \dfrac{\partial w}{\partial y}\right) = 0.$

Complete integral:
$$w = w(z), \qquad z = C_1 x + C_2 y,$$
where $C_1$ and $C_2$ are arbitrary constants and $w = w(z)$ is determined by the autonomous ordinary differential equation $F(w, C_1 w'_z, C_2 w'_z) = 0$.

**21.** $F\left(ax + by + cw, \dfrac{\partial w}{\partial x}, \dfrac{\partial w}{\partial y}\right) = 0.$

For $c = 0$, see equation T7.3.3.19. If $c \neq 0$, then the substitution $cu = ax + by + cw$ leads to an equation of the form T7.3.3.20: $F\left(cu, \dfrac{\partial u}{\partial x} - \dfrac{a}{c}, \dfrac{\partial u}{\partial y} - \dfrac{b}{c}\right) = 0.$

**22.** $F\left(x, \dfrac{\partial w}{\partial x}, \dfrac{\partial w}{\partial y}, w - y\dfrac{\partial w}{\partial y}\right) = 0.$

Complete integral: $w = C_1 y + \varphi(x)$, where the function $\varphi(x)$ is determined from the ordinary differential equation $F(x, \varphi'_x, C_1, \varphi) = 0$.

**23.** $F\left(w, \dfrac{\partial w}{\partial x}, \dfrac{\partial w}{\partial y}, x\dfrac{\partial w}{\partial x} + y\dfrac{\partial w}{\partial y}\right) = 0.$

Complete integral:
$$w = \varphi(\xi), \quad \xi = C_1 x + C_2 y,$$
where the function $\varphi(\xi)$ is determined by solving the nonlinear ordinary differential equation $F(\varphi, C_1 \varphi'_\xi, C_2 \varphi'_\xi, \xi \varphi'_\xi) = 0$.

**24.** $F\left(ax + by, \dfrac{\partial w}{\partial x}, \dfrac{\partial w}{\partial y}, w - x\dfrac{\partial w}{\partial x} - y\dfrac{\partial w}{\partial y}\right) = 0.$

Complete integral:
$$w = C_1 x + C_2 y + \varphi(\xi), \quad \xi = ax + by,$$
where the function $\varphi(\xi)$ is determined by solving the nonlinear ordinary differential equation $F(\xi, a\varphi'_\xi + C_1, b\varphi'_\xi + C_2, \varphi - \xi \varphi'_\xi) = 0$.

**25.** $F\left(x, \dfrac{\partial w}{\partial x}, G\left(y, \dfrac{\partial w}{\partial y}\right)\right) = 0.$

Complete integral:
$$w = \varphi(x, C_1) + \psi(y, C_1) + C_2,$$
where the functions $\varphi$ and $\psi$ are determined by the ordinary differential equations
$$F(x, \varphi'_x, C_1) = 0, \quad G(y, \psi'_y) = C_1.$$

On solving these equations for the derivatives, we obtain linear separable equations, which are easy to integrate.

# References for Chapter T7

**Kamke, E.,** *Differentialgleichungen: Lösungsmethoden und Lösungen, II, Partielle Differentialgleichungen Erster Ordnung für eine gesuchte Funktion*, Akad. Verlagsgesellschaft Geest & Portig, Leipzig, 1965.

**Lopez, G.,** *Partial Differential Equations of First Order and Their Applications to Physics*, World Scientific Publishing Co., Singapore, 2000.

**Polyanin, A. D., Zaitsev, V. F., and Moussiaux, A.,** *Handbook of First Order Partial Differential Equations*, Taylor & Francis, London, 2002.

**Rhee, H., Aris, R., and Amundson, N. R.,** *First Order Partial Differential Equations, Vols. 1 and 2*, Prentice Hall, Englewood Cliffs, New Jersey, 1986 and 1989.

**Tran, D. V., Tsuji, M., and Nguyen, D. T. S.,** *The Characteristic Method and Its Generalizations for First-Order Nonlinear Partial Differential Equations*, Chapman & Hall, London, 1999.

# Chapter T8
# Linear Equations and Problems of Mathematical Physics

## T8.1. Parabolic Equations

### T8.1.1. Heat Equation $\dfrac{\partial w}{\partial t} = a \dfrac{\partial^2 w}{\partial x^2}$

**T8.1.1-1. Particular solutions.**

$$w(x) = Ax + B,$$
$$w(x,t) = A(x^2 + 2at) + B,$$
$$w(x,t) = A(x^3 + 6atx) + B,$$
$$w(x,t) = A(x^4 + 12atx^2 + 12a^2t^2) + B,$$
$$w(x,t) = x^{2n} + \sum_{k=1}^{n} \frac{(2n)(2n-1)\dots(2n-2k+1)}{k!}(at)^k x^{2n-2k},$$
$$w(x,t) = x^{2n+1} + \sum_{k=1}^{n} \frac{(2n+1)(2n)\dots(2n-2k+2)}{k!}(at)^k x^{2n-2k+1},$$
$$w(x,t) = A\exp(a\mu^2 t \pm \mu x) + B,$$
$$w(x,t) = A\frac{1}{\sqrt{t}}\exp\left(-\frac{x^2}{4at}\right) + B,$$
$$w(x,t) = A\frac{x}{t^{3/2}}\exp\left(-\frac{x^2}{4at}\right) + B,$$
$$w(x,t) = A\exp(-a\mu^2 t)\cos(\mu x + B) + C,$$
$$w(x,t) = A\exp(-\mu x)\cos(\mu x - 2a\mu^2 t + B) + C,$$
$$w(x,t) = A\,\mathrm{erf}\!\left(\frac{x}{2\sqrt{at}}\right) + B,$$

where $A$, $B$, $C$, and $\mu$ are arbitrary constants, $n$ is a positive integer, and $\mathrm{erf}\,z \equiv \dfrac{2}{\sqrt{\pi}} \int_0^z \exp(-\xi^2)\,d\xi$ is the error function (probability integral).

**T8.1.1-2. Formulas allowing the construction of particular solutions.**

Suppose $w = w(x,t)$ is a solution of the heat equation. Then the functions

$$w_1 = Aw(\pm\lambda x + C_1,\ \lambda^2 t + C_2) + B,$$
$$w_2 = A\exp(\lambda x + a\lambda^2 t)w(x + 2a\lambda t + C_1,\ t + C_2),$$
$$w_3 = \frac{A}{\sqrt{|\delta + \beta t|}}\exp\!\left[-\frac{\beta x^2}{4a(\delta + \beta t)}\right] w\!\left(\pm\frac{x}{\delta + \beta t},\ \frac{\gamma + \lambda t}{\delta + \beta t}\right), \qquad \lambda\delta - \beta\gamma = 1,$$

where $A$, $B$, $C_1$, $C_2$, $\beta$, $\delta$, and $\lambda$ are arbitrary constants, are also solutions of this equation. The last formula with $\beta = 1$, $\gamma = -1$, $\delta = \lambda = 0$ was obtained with the Appell transformation.

**T8.1.1-3. Cauchy problem and boundary value problems.**

For solutions of the Cauchy problem and various boundary value problems, see Subsection T8.1.2 with $\Phi(x, t) \equiv 0$.

## T8.1.2. Nonhomogeneous Heat Equation $\dfrac{\partial w}{\partial t} = a\dfrac{\partial^2 w}{\partial x^2} + \Phi(x, t)$

**T8.1.2-1. Domain: $-\infty < x < \infty$. Cauchy problem.**

An initial condition is prescribed:
$$w = f(x) \quad \text{at} \quad t = 0.$$

Solution:
$$w(x,t) = \int_{-\infty}^{\infty} f(\xi)G(x,\xi,t)\,d\xi + \int_0^t \int_{-\infty}^{\infty} \Phi(\xi,\tau)G(x,\xi,t-\tau)\,d\xi\,d\tau,$$

where
$$G(x,\xi,t) = \dfrac{1}{2\sqrt{\pi a t}} \exp\left[-\dfrac{(x-\xi)^2}{4at}\right].$$

**T8.1.2-2. Solutions of boundary value problems in terms of the Green's function.**

We consider boundary value problems on an interval $0 \le x \le l$ with the general initial condition
$$w = f(x) \quad \text{at} \quad t = 0$$
and various homogeneous boundary conditions. The solution can be represented in terms of the Green's function as
$$w(x,t) = \int_0^l f(\xi)G(x,\xi,t)\,d\xi + \int_0^t \int_0^l \Phi(\xi,\tau)G(x,\xi,t-\tau)\,d\xi\,d\tau.$$

Here, the upper limit $l$ can be finite or infinite; if $l = \infty$, there is no boundary condition corresponding to it.

Paragraphs T8.1.2-3 through T8.1.2-8 present the Green's functions for various types of homogeneous boundary conditions.

*Remark.* Formulas from Section 14.7 should be used to obtain solutions to corresponding nonhomogeneous boundary value problems.

**T8.1.2-3. Domain: $0 \le x < \infty$. First boundary value problem.**

A boundary condition is prescribed:
$$w = 0 \quad \text{at} \quad x = 0.$$

Green's function:
$$G(x,\xi,t) = \dfrac{1}{2\sqrt{\pi a t}}\left\{\exp\left[-\dfrac{(x-\xi)^2}{4at}\right] - \exp\left[-\dfrac{(x+\xi)^2}{4at}\right]\right\}.$$

### T8.1.2-4. Domain: $0 \leq x < \infty$. Second boundary value problem.

A boundary condition is prescribed:
$$\partial_x w = 0 \quad \text{at} \quad x = 0.$$

Green's function:
$$G(x, \xi, t) = \frac{1}{2\sqrt{\pi a t}} \left\{ \exp\left[-\frac{(x-\xi)^2}{4at}\right] + \exp\left[-\frac{(x+\xi)^2}{4at}\right] \right\}.$$

### T8.1.2-5. Domain: $0 \leq x < \infty$. Third boundary value problem.

A boundary condition is prescribed:
$$\partial_x w - kw = 0 \quad \text{at} \quad x = 0.$$

Green's function:
$$G(x, \xi, t) = \frac{1}{2\sqrt{\pi a t}} \left\{ \exp\left[-\frac{(x-\xi)^2}{4at}\right] + \exp\left[-\frac{(x+\xi)^2}{4at}\right] - 2k \int_0^\infty \exp\left[-\frac{(x+\xi+\eta)^2}{4at} - k\eta\right] d\eta \right\}.$$

### T8.1.2-6. Domain: $0 \leq x \leq l$. First boundary value problem.

Boundary conditions are prescribed:
$$w = 0 \quad \text{at} \quad x = 0, \qquad w = 0 \quad \text{at} \quad x = l.$$

Two forms of representation of the Green's function:
$$G(x, \xi, t) = \frac{2}{l} \sum_{n=1}^\infty \sin\left(\frac{n\pi x}{l}\right) \sin\left(\frac{n\pi \xi}{l}\right) \exp\left(-\frac{an^2\pi^2 t}{l^2}\right)$$
$$= \frac{1}{2\sqrt{\pi a t}} \sum_{n=-\infty}^\infty \left\{ \exp\left[-\frac{(x-\xi+2nl)^2}{4at}\right] - \exp\left[-\frac{(x+\xi+2nl)^2}{4at}\right] \right\}.$$

The first series converges rapidly at large $t$ and the second series at small $t$.

### T8.1.2-7. Domain: $0 \leq x \leq l$. Second boundary value problem.

Boundary conditions are prescribed:
$$\partial_x w = 0 \quad \text{at} \quad x = 0, \qquad \partial_x w = 0 \quad \text{at} \quad x = l.$$

Two forms of representation of the Green's function:
$$G(x, \xi, t) = \frac{1}{l} + \frac{2}{l} \sum_{n=1}^\infty \cos\left(\frac{n\pi x}{l}\right) \cos\left(\frac{n\pi \xi}{l}\right) \exp\left(-\frac{an^2\pi^2 t}{l^2}\right)$$
$$= \frac{1}{2\sqrt{\pi a t}} \sum_{n=-\infty}^\infty \left\{ \exp\left[-\frac{(x-\xi+2nl)^2}{4at}\right] + \exp\left[-\frac{(x+\xi+2nl)^2}{4at}\right] \right\}.$$

The first series converges rapidly at large $t$ and the second series at small $t$.

**T8.1.2-8. Domain: $0 \leq x \leq l$. Third boundary value problem ($k_1 > 0$, $k_2 > 0$).**

Boundary conditions are prescribed:

$$\partial_x w - k_1 w = 0 \quad \text{at} \quad x = 0, \qquad \partial_x w + k_2 w = 0 \quad \text{at} \quad x = l.$$

Green's function:

$$G(x, \xi, t) = \sum_{n=1}^{\infty} \frac{1}{\|y_n\|^2} y_n(x) y_n(\xi) \exp(-a\mu_n^2 t),$$

$$y_n(x) = \cos(\mu_n x) + \frac{k_1}{\mu_n} \sin(\mu_n x), \quad \|y_n\|^2 = \frac{k_2}{2\mu_n^2} \frac{\mu_n^2 + k_1^2}{\mu_n^2 + k_2^2} + \frac{k_1}{2\mu_n^2} + \frac{l}{2}\left(1 + \frac{k_1^2}{\mu_n^2}\right),$$

where $\mu_n$ are positive roots of the transcendental equation $\dfrac{\tan(\mu l)}{\mu} = \dfrac{k_1 + k_2}{\mu^2 - k_1 k_2}$.

## T8.1.3. Equation of the Form $\dfrac{\partial w}{\partial t} = a\dfrac{\partial^2 w}{\partial x^2} + b\dfrac{\partial w}{\partial x} + cw + \Phi(x, t)$

The substitution

$$w(x, t) = \exp(\beta t + \mu x) u(x, t), \qquad \beta = c - \frac{b^2}{4a}, \qquad \mu = -\frac{b}{2a}$$

leads to the nonhomogeneous heat equation

$$\frac{\partial u}{\partial t} = a \frac{\partial^2 u}{\partial x^2} + \exp(-\beta t - \mu x) \Phi(x, t),$$

which is considered in Subsections T8.1.1 and T8.1.2.

## T8.1.4. Heat Equation with Axial Symmetry $\dfrac{\partial w}{\partial t} = a\left(\dfrac{\partial^2 w}{\partial r^2} + \dfrac{1}{r}\dfrac{\partial w}{\partial r}\right)$

This is a heat (diffusion) equation with axial symmetry, where $r = \sqrt{x^2 + y^2}$ is the radial coordinate.

**T8.1.4-1. Particular solutions.**

$$w(r) = A + B \ln r,$$
$$w(r, t) = A + B(r^2 + 4at),$$
$$w(r, t) = A + B(r^4 + 16atr^2 + 32a^2 t^2),$$
$$w(r, t) = A + B\left(r^{2n} + \sum_{k=1}^{n} \frac{4^k [n(n-1)\ldots(n-k+1)]^2}{k!} (at)^k r^{2n-2k}\right),$$
$$w(r, t) = A + B(4at \ln r + r^2 \ln r - r^2),$$
$$w(r, t) = A + \frac{B}{t} \exp\left(-\frac{r^2}{4at}\right),$$

$$w(r,t) = A + B\exp(-a\mu^2 t)J_0(\mu r),$$
$$w(r,t) = A + B\exp(-a\mu^2 t)Y_0(\mu r),$$
$$w(r,t) = A + \frac{B}{t}\exp\left(-\frac{r^2+\mu^2}{4t}\right)I_0\left(\frac{\mu r}{2t}\right),$$
$$w(r,t) = A + \frac{B}{t}\exp\left(-\frac{r^2+\mu^2}{4t}\right)K_0\left(\frac{\mu r}{2t}\right),$$

where $A$, $B$, and $\mu$ are arbitrary constants, $n$ is an arbitrary positive integer, $J_0(z)$ and $Y_0(z)$ are Bessel functions, and $I_0(z)$ and $K_0(z)$ are modified Bessel functions.

**T8.1.4-2. Formulas allowing the construction of particular solutions.**

Suppose $w = w(r,t)$ is a solution of the original equation. Then the functions

$$w_1 = Aw(\pm\lambda r, \lambda^2 t + C) + B,$$
$$w_2 = \frac{A}{\delta+\beta t}\exp\left[-\frac{\beta r^2}{4a(\delta+\beta t)}\right]w\left(\pm\frac{r}{\delta+\beta t}, \frac{\gamma+\lambda t}{\delta+\beta t}\right), \qquad \lambda\delta - \beta\gamma = 1,$$

where $A$, $B$, $C$, $\beta$, $\delta$, and $\lambda$ are arbitrary constants, are also solutions of this equation. The second formula usually may be encountered with $\beta = 1$, $\gamma = -1$, and $\delta = \lambda = 0$.

**T8.1.4-3. Boundary value problems.**

For solutions of various boundary value problems, see Subsection T8.1.5 with $\Phi(r,t) \equiv 0$.

## T8.1.5. Equation of the Form $\frac{\partial w}{\partial t} = a\left(\frac{\partial^2 w}{\partial r^2} + \frac{1}{r}\frac{\partial w}{\partial r}\right) + \Phi(r,t)$

**T8.1.5-1. Solutions of boundary value problems in terms of the Green's function.**

We consider boundary value problems in domain $0 \leq r \leq R$ with the general initial condition

$$w = f(r) \quad \text{at} \quad t = 0 \tag{T8.1.5.1}$$

and various homogeneous boundary conditions (the solutions bounded at $r = 0$ are sought). The solution can be represented in terms of the Green's function as

$$w(x,t) = \int_0^R f(\xi)G(r,\xi,t)\,d\xi + \int_0^t\int_0^R \Phi(\xi,\tau)G(r,\xi,t-\tau)\,d\xi\,d\tau. \tag{T8.1.5.2}$$

**T8.1.5-2. Domain: $0 \leq r \leq R$. First boundary value problem.**

A boundary condition is prescribed:

$$w = 0 \quad \text{at} \quad r = R.$$

Green's function:
$$G(r,\xi,t) = \sum_{n=1}^{\infty} \frac{2\xi}{R^2 J_1^2(\mu_n)} J_0\left(\mu_n \frac{r}{R}\right) J_0\left(\mu_n \frac{\xi}{R}\right) \exp\left(-\frac{a\mu_n^2 t}{R^2}\right),$$

where $\mu_n$ are positive zeros of the Bessel function, $J_0(\mu) = 0$. Below are the numerical values of the first ten roots:

$\mu_1 = 2.4048$, $\quad \mu_2 = 5.5201$, $\quad \mu_3 = 8.6537$, $\quad \mu_4 = 11.7915$, $\quad \mu_5 = 14.9309$,
$\mu_6 = 18.0711$, $\quad \mu_7 = 21.2116$, $\quad \mu_8 = 24.3525$, $\quad \mu_9 = 27.4935$, $\quad \mu_{10} = 30.6346$.

The zeros of the Bessel function $J_0(\mu)$ may be approximated by the formula

$$\mu_n = 2.4 + 3.13(n-1) \qquad (n = 1, 2, 3, \ldots),$$

which is accurate within 0.3%. As $n \to \infty$, we have $\mu_{n+1} - \mu_n \to \pi$.

**T8.1.5-3. Domain: $0 \le r \le R$. Second boundary value problem.**

A boundary condition is prescribed:

$$\partial_r w = 0 \quad \text{at} \quad r = R.$$

Green's function:

$$G(r,\xi,t) = \frac{2}{R^2}\xi + \frac{2}{R^2}\sum_{n=1}^{\infty} \frac{\xi}{J_0^2(\mu_n)} J_0\left(\frac{\mu_n r}{R}\right) J_0\left(\frac{\mu_n \xi}{R}\right) \exp\left(-\frac{a\mu_n^2 t}{R^2}\right),$$

where $\mu_n$ are positive zeros of the first-order Bessel function, $J_1(\mu) = 0$. Below are the numerical values of the first ten roots:

$\mu_1 = 3.8317$, $\quad \mu_2 = 7.0156$, $\quad \mu_3 = 10.1735$, $\quad \mu_4 = 13.3237$, $\quad \mu_5 = 16.4706$,
$\mu_6 = 19.6159$, $\quad \mu_7 = 22.7601$, $\quad \mu_8 = 25.9037$, $\quad \mu_9 = 29.0468$, $\quad \mu_{10} = 32.1897$.

As $n \to \infty$, we have $\mu_{n+1} - \mu_n \to \pi$.

## T8.1.6. Heat Equation with Central Symmetry

$$\frac{\partial w}{\partial t} = a\left(\frac{\partial^2 w}{\partial r^2} + \frac{2}{r}\frac{\partial w}{\partial r}\right)$$

This is the heat (diffusion) equation with central symmetry; $r = \sqrt{x^2 + y^2 + z^2}$ is the radial coordinate.

**T8.1.6-1. Particular solutions.**

$$w(r) = A + Br^{-1},$$
$$w(r,t) = A + B(r^2 + 6at),$$
$$w(r,t) = A + B(r^4 + 20atr^2 + 60a^2t^2),$$

$$w(r,t) = A + B\left[r^{2n} + \sum_{k=1}^{n} \frac{(2n+1)(2n)\ldots(2n-2k+2)}{k!}(at)^k r^{2n-2k}\right],$$

$$w(r,t) = A + 2aBtr^{-1} + Br,$$

$$w(r,t) = Ar^{-1}\exp(a\mu^2 t \pm \mu r) + B,$$

$$w(r,t) = A + \frac{B}{t^{3/2}}\exp\left(-\frac{r^2}{4at}\right),$$

$$w(r,t) = A + \frac{B}{r\sqrt{t}}\exp\left(-\frac{r^2}{4at}\right),$$

$$w(r,t) = Ar^{-1}\exp(-a\mu^2 t)\cos(\mu r + B) + C,$$

$$w(r,t) = Ar^{-1}\exp(-\mu r)\cos(\mu r - 2a\mu^2 t + B) + C,$$

$$w(r,t) = \frac{A}{r}\operatorname{erf}\left(\frac{r}{2\sqrt{at}}\right) + B,$$

where $A$, $B$, $C$, and $\mu$ are arbitrary constants, and $n$ is an arbitrary positive integer.

### T8.1.6-2. Reduction to a constant coefficient equation. Some formulas.

$1°$. The substitution $u(r,t) = rw(r,t)$ brings the original equation with variable coefficients to the constant coefficient equation $\partial_t u = a\partial_{ww}$, which is discussed in Subsection T8.1.1.

$2°$. Suppose $w = w(r,t)$ is a solution of the original equation. Then the functions

$$w_1 = Aw(\pm\lambda r, \lambda^2 t + C) + B,$$

$$w_2 = \frac{A}{|\delta + \beta t|^{3/2}}\exp\left[-\frac{\beta r^2}{4a(\delta + \beta t)}\right] w\left(\pm\frac{r}{\delta + \beta t}, \frac{\gamma + \lambda t}{\delta + \beta t}\right), \qquad \lambda\delta - \beta\gamma = 1,$$

where $A$, $B$, $C$, $\beta$, $\delta$, and $\lambda$ are arbitrary constants, are also solutions of this equation. The second formula may usually be encountered with $\beta = 1$, $\gamma = -1$, and $\delta = \lambda = 0$.

### T8.1.6-3. Boundary value problems.

For solutions of various boundary value problems, see Subsection T8.1.7 with $\Phi(r,t) \equiv 0$.

## T8.1.7. Equation of the Form $\dfrac{\partial w}{\partial t} = a\left(\dfrac{\partial^2 w}{\partial r^2} + \dfrac{2}{r}\dfrac{\partial w}{\partial r}\right) + \Phi(r,t)$

### T8.1.7-1. Solutions of boundary value problems in terms of the Green's function.

Solutions to boundary value problems on an interval $0 \leq x \leq R$ with the general initial condition (T8.1.5.1) and various homogeneous boundary conditions are expressed via the Green's function by formula (T8.1.5.2).

### T8.1.7-2. Domain: $0 \leq r \leq R$. First boundary value problem.

A boundary condition is prescribed:

$$w = 0 \quad \text{at} \quad r = R.$$

Green's function:
$$G(r,\xi,t) = \frac{2\xi}{Rr}\sum_{n=1}^{\infty} \sin\left(\frac{n\pi r}{R}\right)\sin\left(\frac{n\pi\xi}{R}\right)\exp\left(-\frac{an^2\pi^2 t}{R^2}\right).$$

**T8.1.7-3. Domain: $0 \leq r \leq R$. Second boundary value problem.**

A boundary condition is prescribed:
$$\partial_r w = 0 \quad \text{at} \quad r = R.$$

Green's function:
$$G(r,\xi,t) = \frac{3\xi^2}{R^3} + \frac{2\xi}{Rr}\sum_{n=1}^{\infty} \frac{\mu_n^2+1}{\mu_n^2}\sin\left(\frac{\mu_n r}{R}\right)\sin\left(\frac{\mu_n \xi}{R}\right)\exp\left(-\frac{a\mu_n^2 t}{R^2}\right),$$

where $\mu_n$ are positive roots of the transcendental equation $\tan\mu - \mu = 0$. The first five roots are

$$\mu_1 = 4.4934, \quad \mu_2 = 7.7253, \quad \mu_3 = 10.9041, \quad \mu_4 = 14.0662, \quad \mu_5 = 17.2208.$$

## T8.1.8. Equation of the Form $\dfrac{\partial w}{\partial t} = \dfrac{\partial^2 w}{\partial x^2} + \dfrac{1-2\beta}{x}\dfrac{\partial w}{\partial x}$

This dimensionless equation is encountered in problems of the diffusion boundary layer. For $\beta = 0$, $\beta = \frac{1}{2}$, or $\beta = -\frac{1}{2}$, see the equations in Subsections T8.1.4, T8.1.1, or T8.1.6, respectively.

**T8.1.8-1. Particular solutions.**

$$w(x) = A + Bx^{2\beta},$$
$$w(x,t) = A + 4(1-\beta)Bt + Bx^2,$$
$$w(x,t) = A + 16(2-\beta)(1-\beta)Bt^2 + 8(2-\beta)Btx^2 + Bx^4,$$
$$w(x,t) = x^{2n} + \sum_{p=1}^{n} \frac{4^p}{p!} s_{n,p} s_{n-\beta,p} t^p x^{2(n-p)}, \quad s_{q,p} = q(q-1)\ldots(q-p+1),$$
$$w(x,t) = A + 4(1+\beta)Btx^{2\beta} + Bx^{2\beta+2},$$
$$w(x,t) = A + Bt^{\beta-1}\exp\left(-\frac{x^2}{4t}\right),$$
$$w(x,t) = A + B\frac{x^{2\beta}}{t^{\beta+1}}\exp\left(-\frac{x^2}{4t}\right),$$
$$w(x,t) = A + B\gamma\left(\beta, \frac{x^2}{4t}\right),$$
$$w(x,t) = A + B\exp(-\mu^2 t)x^{\beta}J_{\beta}(\mu x),$$
$$w(x,t) = A + B\exp(-\mu^2 t)x^{\beta}Y_{\beta}(\mu x),$$
$$w(x,t) = A + B\frac{x^{\beta}}{t}\exp\left(-\frac{x^2+\mu^2}{4t}\right)I_{\pm\beta}\left(\frac{\mu x}{2t}\right),$$
$$w(x,t) = A + B\frac{x^{\beta}}{t}\exp\left(-\frac{x^2+\mu^2}{4t}\right)K_{\beta}\left(\frac{\mu x}{2t}\right),$$

where $A$, $B$, and $\mu$ are arbitrary constants, $n$ is an arbitrary positive number, $\gamma(\beta, z)$ is the incomplete gamma function, $J_\beta(z)$ and $Y_\beta(z)$ are Bessel functions, and $I_\beta(z)$ and $K_\beta(z)$ are modified Bessel functions.

**T8.1.8-2. Formulas and transformations for constructing particular solutions.**

1°. Suppose $w = w(x,t)$ is a solution of the original equation. Then the functions

$$w_1 = Aw(\pm \lambda x, \lambda^2 t + C),$$

$$w_2 = A|a + bt|^{\beta-1} \exp\left[-\frac{bx^2}{4(a+bt)}\right] w\left(\pm \frac{x}{a+bt}, \frac{c+kt}{a+bt}\right), \qquad ak - bc = 1,$$

where $A$, $C$, $a$, $b$, and $c$ are arbitrary constants, are also solutions of this equation. The second formula usually may be encountered with $a = k = 0$, $b = 1$, and $c = -1$.

2°. The substitution $w = x^{2\beta} u(x,t)$ brings the equation with parameter $\beta$ to an equation of the same type with parameter $-\beta$:

$$\frac{\partial u}{\partial t} = \frac{\partial^2 u}{\partial x^2} + \frac{1 + 2\beta}{x} \frac{\partial u}{\partial x}.$$

**T8.1.8-3. Domain: $0 \leq x < \infty$. First boundary value problem.**

The following initial and boundary conditions are prescribed:

$$w = f(x) \quad \text{at} \quad t = 0, \qquad w = g(t) \quad \text{at} \quad x = 0.$$

Solution for $0 < \beta < 1$:

$$w(x,t) = \frac{x^\beta}{2t} \int_0^\infty f(\xi) \xi^{1-\beta} \exp\left(-\frac{x^2 + \xi^2}{4t}\right) I_\beta\left(\frac{\xi x}{2t}\right) d\xi$$

$$+ \frac{x^{2\beta}}{2^{2\beta+1}\Gamma(\beta+1)} \int_0^t g(\tau) \exp\left[-\frac{x^2}{4(t-\tau)}\right] \frac{d\tau}{(t-\tau)^{1+\beta}}.$$

**T8.1.8-4. Domain: $0 \leq x < \infty$. Second boundary value problem.**

The following initial and boundary conditions are prescribed:

$$w = f(x) \quad \text{at} \quad t = 0, \qquad (x^{1-2\beta} \partial_x w) = g(t) \quad \text{at} \quad x = 0.$$

Solution for $0 < \beta < 1$:

$$w(x,t) = \frac{x^\beta}{2t} \int_0^\infty f(\xi) \xi^{1-\beta} \exp\left(-\frac{x^2 + \xi^2}{4t}\right) I_{-\beta}\left(\frac{\xi x}{2t}\right) d\xi$$

$$- \frac{2^{2\beta-1}}{\Gamma(1-\beta)} \int_0^t g(\tau) \exp\left[-\frac{x^2}{4(t-\tau)}\right] \frac{d\tau}{(t-\tau)^{1-\beta}}.$$

## T8.1.9. Equations of the Diffusion (Thermal) Boundary Layer

**1.** $f(x)\dfrac{\partial w}{\partial x} + g(x)y\dfrac{\partial w}{\partial y} = \dfrac{\partial^2 w}{\partial y^2}.$

This equation is encountered in diffusion boundary layer problems (mass exchange of drops and bubbles with flow).

The transformation ($A$ and $B$ are any numbers)

$$t = \int \frac{h^2(x)}{f(x)}\,dx + A, \quad z = yh(x), \quad \text{where} \quad h(x) = B\exp\left[-\int \frac{g(x)}{f(x)}\,dx\right],$$

leads to a constant coefficient equation, $\partial_t w = \partial_{zz} w$, which is considered in Subsection T8.1.1.

**2.** $f(x)y^{n-1}\dfrac{\partial w}{\partial x} + g(x)y^n\dfrac{\partial w}{\partial y} = \dfrac{\partial^2 w}{\partial y^2}.$

This equation is encountered in diffusion boundary layer problems (mass exchange of solid particles, drops, and bubbles with flow).

The transformation ($A$ and $B$ are any numbers)

$$t = \frac{1}{4}(n+1)^2 \int \frac{h^2(x)}{f(x)}\,dx + A, \quad z = h(x)y^{\frac{n+1}{2}}, \quad \text{where} \quad h(x) = B\exp\left[-\frac{n+1}{2}\int \frac{g(x)}{f(x)}\,dx\right],$$

leads to the simpler equation

$$\frac{\partial w}{\partial t} = \frac{\partial^2 w}{\partial z^2} + \frac{1-2k}{z}\frac{\partial w}{\partial z}, \qquad k = \frac{1}{n+1},$$

which is considered in Subsection T8.1.8.

## T8.1.10. Schrödinger Equation $i\hbar\dfrac{\partial w}{\partial t} = -\dfrac{\hbar^2}{2m}\dfrac{\partial^2 w}{\partial x^2} + U(x)w$

### T8.1.10-1. Eigenvalue problem. Cauchy problem.

Schrödinger's equation is the basic equation of quantum mechanics; $w$ is the wave function, $i^2 = -1$, $\hbar$ is Planck's constant, $m$ is the mass of the particle, and $U(x)$ is the potential energy of the particle in the force field.

$1°$. In discrete spectrum problems, the particular solutions are sought in the form

$$w(x,t) = \exp\left(-\frac{iE_n}{\hbar}t\right)\psi_n(x),$$

where the eigenfunctions $\psi_n$ and the respective energies $E_n$ have to be determined by solving the eigenvalue problem

$$\frac{d^2\psi_n}{dx^2} + \frac{2m}{\hbar^2}[E_n - U(x)]\psi_n = 0, \tag{T8.1.10.1}$$

$$\psi_n \to 0 \text{ at } x \to \pm\infty, \quad \int_{-\infty}^{\infty} |\psi_n|^2\,dx = 1.$$

The last relation is the normalizing condition for $\psi_n$.

2°. In the cases where the eigenfunctions $\psi_n(x)$ form an orthonormal basis in $L_2(\mathbb{R})$, the solution of the Cauchy problem for Schrödinger's equation with the initial condition

$$w = f(x) \quad \text{at} \quad t = 0 \tag{T8.1.10.2}$$

is given by

$$w(x,t) = \int_{-\infty}^{\infty} G(x,\xi,t) f(\xi)\, d\xi, \qquad G(x,\xi,t) = \sum_{n=0}^{\infty} \psi_n(x)\psi_n(\xi) \exp\left(-\frac{iE_n}{\hbar} t\right).$$

Various potentials $U(x)$ are considered below and particular solutions of the boundary value problem (T8.1.10.1) or the Cauchy problem for Schrödinger's equation are presented.

### T8.1.10-2. Free particle: $U(x) = 0$.

The solution of the Cauchy problem with the initial condition (T8.1.10.2) is given by

$$w(x,t) = \frac{1}{2\sqrt{i\pi\tau}} \int_{-\infty}^{\infty} \exp\left[-\frac{(x-\xi)^2}{4i\tau}\right] f(\xi)\, d\xi, \qquad \tau = \frac{\hbar t}{2m}, \qquad \sqrt{ia} = \begin{cases} e^{\pi i/4}\sqrt{|a|} & \text{if } a > 0, \\ e^{-\pi i/4}\sqrt{|a|} & \text{if } a < 0. \end{cases}$$

### T8.1.10-3. Linear potential (motion in a uniform external field): $U(x) = ax$.

Solution of the Cauchy problem with the initial condition (T8.1.10.2):

$$w(x,t) = \frac{1}{2\sqrt{i\pi\tau}} \exp\left(-ib\tau x - \tfrac{1}{3} i b^2 \tau^3\right) \int_{-\infty}^{\infty} \exp\left[-\frac{(x+b\tau^2-\xi)^2}{4i\tau}\right] f(\xi)\, d\xi, \qquad \tau = \frac{\hbar t}{2m}, \qquad b = \frac{2am}{\hbar^2}.$$

### T8.1.10-4. Linear harmonic oscillator: $U(x) = \tfrac{1}{2} m\omega^2 x^2$.

Eigenvalues:
$$E_n = \hbar\omega\left(n + \tfrac{1}{2}\right), \qquad n = 0, 1, \ldots$$

Normalized eigenfunctions:

$$\psi_n(x) = \frac{1}{\pi^{1/4}\sqrt{2^n n!\, x_0}} \exp\left(-\tfrac{1}{2}\xi^2\right) H_n(\xi), \qquad \xi = \frac{x}{x_0}, \qquad x_0 = \sqrt{\frac{\hbar}{m\omega}},$$

where $H_n(\xi)$ are the Hermite polynomials. The functions $\psi_n(x)$ form an orthonormal basis in $L_2(\mathbb{R})$.

### T8.1.10-5. Isotropic free particle: $U(x) = a/x^2$.

Here, the variable $x \geq 0$ plays the role of the radial coordinate, and $a > 0$. The equation with $U(x) = a/x^2$ results from Schrödinger's equation for a free particle with $n$ space coordinates if one passes to spherical (cylindrical) coordinates and separates the angular variables.

The solution of Schrödinger's equation satisfying the initial condition (T8.1.10.2) has the form

$$w(x,t) = \frac{\exp\left[-\tfrac{1}{2} i\pi(\mu+1)\operatorname{sign} t\right]}{2|\tau|} \int_0^{\infty} \sqrt{xy} \exp\left(i\frac{x^2+y^2}{4\tau}\right) J_\mu\left(\frac{xy}{2|\tau|}\right) f(y)\, dy,$$

$$\tau = \frac{\hbar t}{2m}, \qquad \mu = \sqrt{\frac{2am}{\hbar^2} + \frac{1}{4}} \geq 1,$$

where $J_\mu(\xi)$ is the Bessel function.

**T8.1.10-6. Morse potential:** $U(x) = U_0(e^{-2x/a} - 2e^{-x/a})$.

Eigenvalues:
$$E_n = -U_0\left[1 - \frac{1}{\beta}(n+\tfrac{1}{2})\right]^2, \quad \beta = \frac{a\sqrt{2mU_0}}{\hbar}, \quad 0 \le n < \beta - 2.$$

Eigenfunctions:
$$\psi_n(x) = \xi^s e^{-\xi/2}\Phi(-n, 2s+1, \xi), \quad \xi = 2\beta e^{-x/a}, \quad s = \frac{a\sqrt{-2mE_n}}{\hbar},$$

where $\Phi(a,b,\xi)$ is the degenerate hypergeometric function.

In this case the number of eigenvalues (energy levels) $E_n$ and eigenfunctions $\psi_n$ is finite: $n = 0, 1, \ldots, n_{\max}$.

## T8.2. Hyperbolic Equations

### T8.2.1. Wave Equation $\dfrac{\partial^2 w}{\partial t^2} = a^2 \dfrac{\partial^2 w}{\partial x^2}$

This equation is also known as the *equation of vibration of a string*. It is often encountered in elasticity, aerodynamics, acoustics, and electrodynamics.

**T8.2.1-1. General solution. Some formulas.**

1°. General solution:
$$w(x,t) = \varphi(x+at) + \psi(x-at),$$
where $\varphi(x)$ and $\psi(x)$ are arbitrary functions.

2°. If $w(x,t)$ is a solution of the wave equation, then the functions
$$w_1 = Aw(\pm\lambda x + C_1, \pm\lambda t + C_2) + B,$$
$$w_2 = Aw\left(\frac{x-vt}{\sqrt{1-(v/a)^2}}, \frac{t-va^{-2}x}{\sqrt{1-(v/a)^2}}\right),$$
$$w_3 = Aw\left(\frac{x}{x^2 - a^2t^2}, \frac{t}{x^2 - a^2t^2}\right)$$

are also solutions of the equation everywhere these functions are defined ($A$, $B$, $C_1$, $C_2$, $v$, and $\lambda$ are arbitrary constants). The signs at $\lambda$'s in the formula for $w_1$ are taken arbitrarily. The function $w_2$ results from the invariance of the wave equation under the *Lorentz transformations*.

**T8.2.1-2. Domain: $-\infty < x < \infty$. Cauchy problem.**

Initial conditions are prescribed:
$$w = f(x) \quad \text{at} \quad t = 0, \qquad \partial_t w = g(x) \quad \text{at} \quad t = 0.$$

Solution (*D'Alembert's formula*):
$$w(x,t) = \frac{1}{2}[f(x+at) + f(x-at)] + \frac{1}{2a}\int_{x-at}^{x+at} g(\xi)\,d\xi.$$

### T8.2.1-3. Domain: $0 \leq x < \infty$. First boundary value problem.

The following two initial and one boundary conditions are prescribed:

$$w = f(x) \quad \text{at} \quad t = 0, \qquad \partial_t w = g(x) \quad \text{at} \quad t = 0, \qquad w = h(t) \quad \text{at} \quad x = 0.$$

Solution:

$$w(x,t) = \begin{cases} \dfrac{1}{2}[f(x+at) + f(x-at)] + \dfrac{1}{2a}\displaystyle\int_{x-at}^{x+at} g(\xi)\,d\xi & \text{for } t < \dfrac{x}{a}, \\ \dfrac{1}{2}[f(x+at) - f(at-x)] + \dfrac{1}{2a}\displaystyle\int_{at-x}^{x+at} g(\xi)\,d\xi + h\!\left(t - \dfrac{x}{a}\right) & \text{for } t > \dfrac{x}{a}. \end{cases}$$

In the domain $t < x/a$ the boundary conditions have no effect on the solution and the expression of $w(x,t)$ coincides with D'Alembert's solution for an infinite line (see Paragraph T8.2.1-2).

### T8.2.1-4. Domain: $0 \leq x < \infty$. Second boundary value problem.

The following two initial and one boundary conditions are prescribed:

$$w = f(x) \quad \text{at} \quad t = 0, \qquad \partial_t w = g(x) \quad \text{at} \quad t = 0, \qquad \partial_x w = h(t) \quad \text{at} \quad x = 0.$$

Solution:

$$w(x,t) = \begin{cases} \dfrac{1}{2}[f(x+at) + f(x-at)] + \dfrac{1}{2a}[G(x+at) - G(x-at)] & \text{for } t < \dfrac{x}{a}, \\ \dfrac{1}{2}[f(x+at) + f(at-x)] + \dfrac{1}{2a}[G(x+at) + G(at-x)] - aH\!\left(t - \dfrac{x}{a}\right) & \text{for } t > \dfrac{x}{a}, \end{cases}$$

where $G(z) = \displaystyle\int_0^z g(\xi)\,d\xi$ and $H(z) = \displaystyle\int_0^z h(\xi)\,d\xi$.

### T8.2.1-5. Domain: $0 \leq x \leq l$. Boundary value problems.

For solutions of various boundary value problems, see Subsection T8.2.2 for $\Phi(x,t) \equiv 0$.

## T8.2.2. Equation of the Form $\dfrac{\partial^2 w}{\partial t^2} = a^2 \dfrac{\partial^2 w}{\partial x^2} + \Phi(x,t)$

### T8.2.2-1. Solutions of boundary value problems in terms of the Green's function.

We consider boundary value problems on an interval $0 \leq x \leq l$ with the general initial conditions

$$w = f(x) \quad \text{at} \quad t = 0, \qquad \partial_t w = g(x) \quad \text{at} \quad t = 0 \tag{T8.2.2.1}$$

and various homogeneous boundary conditions. The solution can be represented in terms of the Green's function as

$$w(x,t) = \frac{\partial}{\partial t}\int_0^l f(\xi) G(x,\xi,t)\,d\xi + \int_0^l g(\xi) G(x,\xi,t)\,d\xi + \int_0^t\!\!\int_0^l \Phi(\xi,\tau) G(x,\xi,t-\tau)\,d\xi\,d\tau. \tag{T8.2.2.2}$$

Here, the upper limit $l$ can assume any finite values.

Paragraphs T8.2.2-2 through T8.2.2-4 present the Green's functions for various types of homogeneous boundary conditions.

*Remark.* Formulas from Subsections 14.8.1–14.8.2 should be used to obtain solutions to corresponding nonhomogeneous boundary value problems.

**T8.2.2-2. Domain: $0 \leq x \leq l$. First boundary value problem.**

Boundary conditions are prescribed:
$$w = 0 \quad \text{at} \quad x = 0, \qquad w = 0 \quad \text{at} \quad x = l.$$

Green's function:
$$G(x, \xi, t) = \frac{2}{a\pi} \sum_{n=1}^{\infty} \frac{1}{n} \sin\left(\frac{n\pi x}{l}\right) \sin\left(\frac{n\pi \xi}{l}\right) \sin\left(\frac{n\pi a t}{l}\right).$$

**T8.2.2-3. Domain: $0 \leq x \leq l$. Second boundary value problem.**

Boundary conditions are prescribed:
$$\partial_x w = 0 \quad \text{at} \quad x = 0, \qquad \partial_x w = 0 \quad \text{at} \quad x = l.$$

Green's function:
$$G(x, \xi, t) = \frac{t}{l} + \frac{2}{a\pi} \sum_{n=1}^{\infty} \frac{1}{n} \cos\left(\frac{n\pi x}{l}\right) \cos\left(\frac{n\pi \xi}{l}\right) \sin\left(\frac{n\pi a t}{l}\right).$$

**T8.2.2-4. Domain: $0 \leq x \leq l$. Third boundary value problem ($k_1 > 0$, $k_2 > 0$).**

Boundary conditions are prescribed:
$$\partial_x w - k_1 w = 0 \quad \text{at} \quad x = 0, \qquad \partial_x w + k_2 w = 0 \quad \text{at} \quad x = l.$$

Green's function:
$$G(x, \xi, t) = \frac{1}{a} \sum_{n=1}^{\infty} \frac{1}{\lambda_n \|u_n\|^2} \sin(\lambda_n x + \varphi_n) \sin(\lambda_n \xi + \varphi_n) \sin(\lambda_n a t),$$

$$\varphi_n = \arctan \frac{\lambda_n}{k_1}, \qquad \|u_n\|^2 = \frac{l}{2} + \frac{(\lambda_n^2 + k_1 k_2)(k_1 + k_2)}{2(\lambda_n^2 + k_1^2)(\lambda_n^2 + k_2^2)};$$

the $\lambda_n$ are positive roots of the transcendental equation $\cot(\lambda l) = \dfrac{\lambda^2 - k_1 k_2}{\lambda(k_1 + k_2)}$.

## T8.2.3. Klein–Gordon Equation $\dfrac{\partial^2 w}{\partial t^2} = a^2 \dfrac{\partial^2 w}{\partial x^2} - bw$

This equation is encountered in quantum field theory and a number of applications.

### T8.2.3-1. Particular solutions.

$$w(x,t) = \cos(\lambda x)[A\cos(\mu t) + B\sin(\mu t)], \quad b = -a^2\lambda^2 + \mu^2,$$
$$w(x,t) = \sin(\lambda x)[A\cos(\mu t) + B\sin(\mu t)], \quad b = -a^2\lambda^2 + \mu^2,$$
$$w(x,t) = \exp(\pm\mu t)[A\cos(\lambda x) + B\sin(\lambda x)], \quad b = -a^2\lambda^2 - \mu^2,$$
$$w(x,t) = \exp(\pm\lambda x)[A\cos(\mu t) + B\sin(\mu t)], \quad b = a^2\lambda^2 + \mu^2,$$
$$w(x,t) = \exp(\pm\lambda x)[A\exp(\mu t) + B\exp(-\mu t)], \quad b = a^2\lambda^2 - \mu^2,$$
$$w(x,t) = AJ_0(\xi) + BY_0(\xi), \quad \xi = \frac{\sqrt{b}}{a}\sqrt{a^2(t+C_1)^2 - (x+C_2)^2}, \quad b > 0,$$
$$w(x,t) = AI_0(\xi) + BK_0(\xi), \quad \xi = \frac{\sqrt{-b}}{a}\sqrt{a^2(t+C_1)^2 - (x+C_2)^2}, \quad b < 0,$$

where $A$, $B$, $C_1$, and $C_2$ are arbitrary constants, $J_0(\xi)$ and $Y_0(\xi)$ are Bessel functions, and $I_0(\xi)$ and $K_0(\xi)$ are modified Bessel functions.

### T8.2.3-2. Formulas allowing the construction of particular solutions.

Suppose $w = w(x,t)$ is a solution of the Klein–Gordon equation. Then the functions

$$w_1 = Aw(\pm x + C_1, \pm t + C_2) + B,$$
$$w_2 = Aw\left(\frac{x - vt}{\sqrt{1 - (v/a)^2}}, \frac{t - va^{-2}x}{\sqrt{1 - (v/a)^2}}\right),$$

where $A$, $B$, $C_1$, $C_2$, and $v$ are arbitrary constants, are also solutions of this equation. The signs in the formula for $w_1$ are taken arbitrarily.

### T8.2.3-3. Domain: $0 \leq x \leq l$. Boundary value problems.

For solutions of the first and second boundary value problems, see Subsection T8.2.4 for $\Phi(x,t) \equiv 0$.

## T8.2.4. Equation of the Form $\dfrac{\partial^2 w}{\partial t^2} = a^2 \dfrac{\partial^2 w}{\partial x^2} - bw + \Phi(x,t)$

### T8.2.4-1. Solutions of boundary value problems in terms of the Green's function.

Solutions to boundary value problems on an interval $0 \leq x \leq l$ with the general initial conditions (T8.2.2.1) and various homogeneous boundary conditions are expressed via the Green's function by formula (T8.2.2.2).

### T8.2.4-2. Domain: $0 \leq x \leq l$. First boundary value problem.

Boundary conditions are prescribed:

$$w = 0 \quad \text{at} \quad x = 0, \quad w = 0 \quad \text{at} \quad x = l.$$

Green's function for $b > 0$:

$$G(x, \xi, t) = \frac{2}{l}\sum_{n=1}^{\infty} \sin(\lambda_n x)\sin(\lambda_n \xi)\frac{\sin\left(t\sqrt{a^2\lambda_n^2 + b}\right)}{\sqrt{a^2\lambda_n^2 + b}}, \quad \lambda_n = \frac{\pi n}{l}.$$

**T8.2.4-3. Domain: $0 \leq x \leq l$. Second boundary value problem.**

Boundary conditions are prescribed:
$$\partial_x w = 0 \quad \text{at} \quad x = 0, \qquad \partial_x w = 0 \quad \text{at} \quad x = l.$$

Green's function for $b > 0$:
$$G(x, \xi, t) = \frac{1}{l\sqrt{b}} \sin(t\sqrt{b}) + \frac{2}{l} \sum_{n=1}^{\infty} \cos(\lambda_n x) \cos(\lambda_n \xi) \frac{\sin(t\sqrt{a^2 \lambda_n^2 + b})}{\sqrt{a^2 \lambda_n^2 + b}}, \qquad \lambda_n = \frac{\pi n}{l}.$$

## T8.2.5. Equation of the Form $\dfrac{\partial^2 w}{\partial t^2} = a^2 \left( \dfrac{\partial^2 w}{\partial r^2} + \dfrac{1}{r} \dfrac{\partial w}{\partial r} \right) + \Phi(r, t)$

This is the wave equation with axial symmetry, where $r = \sqrt{x^2 + y^2}$ is the radial coordinate.

**T8.2.5-1. Solutions of boundary value problems in terms of the Green's function.**

We consider boundary value problems in domain $0 \leq r \leq R$ with the general initial conditions
$$w = f(r) \quad \text{at} \quad t = 0, \qquad \partial_t w = g(r) \quad \text{at} \quad t = 0, \tag{T8.2.5.1}$$
and various homogeneous boundary conditions at $r = R$ (the solutions bounded at $r = 0$ are sought). The solution can be represented in terms of the Green's function as
$$w(r, t) = \frac{\partial}{\partial t} \int_0^R f(\xi) G(r, \xi, t) \, d\xi$$
$$+ \int_0^R g(\xi) G(r, \xi, t) \, d\xi + \int_0^t \int_0^R \Phi(\xi, \tau) G(r, \xi, t - \tau) \, d\xi \, d\tau. \tag{T8.2.5.2}$$

**T8.2.5-2. Domain: $0 \leq r \leq R$. First boundary value problem.**

A boundary condition is prescribed:
$$w = 0 \quad \text{at} \quad r = R.$$

Green's function:
$$G(r, \xi, t) = \frac{2\xi}{aR} \sum_{n=1}^{\infty} \frac{1}{\lambda_n J_1^2(\lambda_n)} J_0\left(\frac{\lambda_n r}{R}\right) J_0\left(\frac{\lambda_n \xi}{R}\right) \sin\left(\frac{\lambda_n a t}{R}\right),$$
where $\lambda_n$ are positive zeros of the Bessel function, $J_0(\lambda) = 0$. The numerical values of the first ten $\lambda_n$ are specified in Paragraph T8.1.5-2.

**T8.2.5-3. Domain: $0 \leq r \leq R$. Second boundary value problem.**

A boundary condition is prescribed:
$$\partial_r w = 0 \quad \text{at} \quad r = R.$$

Green's function:
$$G(r, \xi, t) = \frac{2 t \xi}{R^2} + \frac{2\xi}{aR} \sum_{n=1}^{\infty} \frac{1}{\lambda_n J_0^2(\lambda_n)} J_0\left(\frac{\lambda_n r}{R}\right) J_0\left(\frac{\lambda_n \xi}{R}\right) \sin\left(\frac{\lambda_n a t}{R}\right),$$
where $\lambda_n$ are positive zeros of the first-order Bessel function, $J_1(\lambda) = 0$. The numerical values of the first ten roots $\lambda_n$ are specified in Paragraph T8.1.5-3.

## T8.2.6. Equation of the Form $\dfrac{\partial^2 w}{\partial t^2} = a^2 \left( \dfrac{\partial^2 w}{\partial r^2} + \dfrac{2}{r} \dfrac{\partial w}{\partial r} \right) + \Phi(r, t)$

This is the equation of vibration of a gas with central symmetry, where $r = \sqrt{x^2 + y^2 + z^2}$ is the radial coordinate.

### T8.2.6-1. General solution for $\Phi(r, t) \equiv 0$.

$$w(t, r) = \frac{\varphi(r + at) + \psi(r - at)}{r},$$

where $\varphi(r_1)$ and $\psi(r_2)$ are arbitrary functions.

### T8.2.6-2. Reduction to a constant coefficient equation.

The substitution $u(r, t) = rw(r, t)$ leads to the nonhomogeneous constant coefficient equation

$$\frac{\partial^2 u}{\partial t^2} = a^2 \frac{\partial^2 u}{\partial r^2} + r\Phi(r, t),$$

which is discussed in Subsection T8.2.1.

### T8.2.6-3. Solutions of boundary value problems in terms of the Green's function.

Solutions to boundary value problems on an interval $0 \leq x \leq R$ with the general initial conditions (T8.2.5.1) and various homogeneous boundary conditions are expressed via the Green's function by formula (T8.2.5.2).

### T8.2.6-4. Domain: $0 \leq r \leq R$. First boundary value problem.

A boundary condition is prescribed:

$$w = 0 \quad \text{at} \quad r = R.$$

Green's function:

$$G(r, \xi, t) = \frac{2\xi}{\pi a r} \sum_{n=1}^{\infty} \frac{1}{n} \sin\left(\frac{n\pi r}{R}\right) \sin\left(\frac{n\pi \xi}{R}\right) \sin\left(\frac{a n \pi t}{R}\right).$$

### T8.2.6-5. Domain: $0 \leq r \leq R$. Second boundary value problem.

A boundary condition is prescribed:

$$\partial_r w = 0 \quad \text{at} \quad r = R.$$

Green's function:

$$G(r, \xi, t) = \frac{3t\xi^2}{R^3} + \frac{2\xi}{ar} \sum_{n=1}^{\infty} \frac{\mu_n^2 + 1}{\mu_n^3} \sin\left(\frac{\mu_n r}{R}\right) \sin\left(\frac{\mu_n \xi}{R}\right) \sin\left(\frac{\mu_n a t}{R}\right),$$

where $\mu_n$ are positive roots of the transcendental equation $\tan \mu - \mu = 0$. The numerical values of the first five roots $\mu_n$ are specified in Paragraph T8.1.7-3.

## T8.2.7. Equations of the Form

$$\frac{\partial^2 w}{\partial t^2} + k\frac{\partial w}{\partial t} = a^2\frac{\partial^2 w}{\partial x^2} + b\frac{\partial w}{\partial x} + cw + \Phi(x,t)$$

1. $\dfrac{\partial^2 w}{\partial t^2} + k\dfrac{\partial w}{\partial t} = a^2\dfrac{\partial^2 w}{\partial x^2} + bw.$

For $k > 0$ and $b < 0$, it is the *telegraph equation*.

The substitution $w(x,t) = \exp(-\tfrac{1}{2}kt)u(x,t)$ leads to the equation

$$\frac{\partial^2 u}{\partial t^2} = a^2\frac{\partial^2 u}{\partial x^2} + \left(b + \tfrac{1}{4}k^2\right)u,$$

which is discussed in Subsection T8.2.4.

2. $\dfrac{\partial^2 w}{\partial t^2} + k\dfrac{\partial w}{\partial t} = a^2\dfrac{\partial^2 w}{\partial x^2} + b\dfrac{\partial w}{\partial x} + cw + \Phi(x,t).$

The substitution $w(x,t) = \exp\!\left(-\tfrac{1}{2}a^{-2}bx - \tfrac{1}{2}kt\right)u(x,t)$ leads to the equation

$$\frac{\partial^2 u}{\partial t^2} = a^2\frac{\partial^2 u}{\partial x^2} + \left(c + \tfrac{1}{4}k^2 - \tfrac{1}{4}a^{-2}b^2\right)u + \exp\!\left(\tfrac{1}{2}a^{-2}bx + \tfrac{1}{2}kt\right)\Phi(x,t),$$

which is discussed in Subsection T8.2.5.

## T8.3. Elliptic Equations

### T8.3.1. Laplace Equation $\Delta w = 0$

The Laplace equation is often encountered in heat and mass transfer theory, fluid mechanics, elasticity, electrostatics, and other areas of mechanics and physics.

The two-dimensional Laplace equation has the following form:

$$\frac{\partial^2 w}{\partial x^2} + \frac{\partial^2 w}{\partial y^2} = 0 \quad \text{in the Cartesian coordinate system,}$$

$$\frac{1}{r}\frac{\partial}{\partial r}\!\left(r\frac{\partial w}{\partial r}\right) + \frac{1}{r^2}\frac{\partial^2 w}{\partial \varphi^2} = 0 \quad \text{in the polar coordinate system,}$$

where $x = r\cos\varphi$, $y = r\sin\varphi$, and $r = \sqrt{x^2 + y^2}$.

---

T8.3.1-1. Particular solutions and methods for their construction.

1°. Particular solutions in the Cartesian coordinate system:

$$w(x,y) = Ax + By + C,$$
$$w(x,y) = A(x^2 - y^2) + Bxy,$$
$$w(x,y) = A(x^3 - 3xy^2) + B(3x^2y - y^3),$$
$$w(x,y) = \frac{Ax + By}{x^2 + y^2} + C,$$
$$w(x,y) = \exp(\pm\mu x)(A\cos\mu y + B\sin\mu y),$$
$$w(x,y) = (A\cos\mu x + B\sin\mu x)\exp(\pm\mu y),$$
$$w(x,y) = (A\sinh\mu x + B\cosh\mu x)(C\cos\mu y + D\sin\mu y),$$
$$w(x,y) = (A\cos\mu x + B\sin\mu x)(C\sinh\mu y + D\cosh\mu y),$$

where $A, B, C, D,$ and $\mu$ are arbitrary constants.

2°. Particular solutions in the polar coordinate system:

$$w(r) = A \ln r + B,$$

$$w(r, \varphi) = \left( A r^m + \frac{B}{r^m} \right)(C \cos m\varphi + D \sin m\varphi),$$

where $A$, $B$, $C$, and $D$ are arbitrary constants, and $m = 1, 2, \ldots$

3°. If $w(x, y)$ is a solution of the Laplace equation, then the functions

$$w_1 = Aw(\pm \lambda x + C_1, \pm \lambda y + C_2) + B,$$
$$w_2 = Aw(x \cos \beta + y \sin \beta, -x \sin \beta + y \cos \beta),$$
$$w_3 = Aw\left( \frac{x}{x^2 + y^2}, \frac{y}{x^2 + y^2} \right)$$

are also solutions everywhere they are defined; $A$, $B$, $C_1$, $C_2$, $\beta$, and $\lambda$ are arbitrary constants. The signs at $\lambda$'s in the formula for $w_1$ are taken arbitrarily.

4°. A fairly general method for constructing particular solutions involves the following. Let $f(z) = u(x, y) + iv(x, y)$ be any analytic function of the complex variable $z = x + iy$ ($u$ and $v$ are real functions of the real variables $x$ and $y$; $i^2 = -1$). Then the real and imaginary parts of $f$ both satisfy the two-dimensional Laplace equation,

$$\Delta_2 u = 0, \qquad \Delta_2 v = 0.$$

Thus, by specifying analytic functions $f(z)$ and taking their real and imaginary parts, one obtains various solutions of the two-dimensional Laplace equation.

---

**T8.3.1-2. Domain: $-\infty < x < \infty$, $0 \le y < \infty$. First boundary value problem.***

A half-plane is considered. A boundary condition is prescribed:

$$w = f(x) \quad \text{at} \quad y = 0.$$

Solution:

$$w(x, y) = \frac{1}{\pi} \int_{-\infty}^{\infty} \frac{y f(\xi) \, d\xi}{(x - \xi)^2 + y^2} = \frac{1}{\pi} \int_{-\pi/2}^{\pi/2} f(x + y \tan \theta) \, d\theta.$$

---

**T8.3.1-3. Domain: $-\infty < x < \infty$, $0 \le y < \infty$. Second boundary value problem.***

A half-plane is considered. A boundary condition is prescribed:

$$\partial_y w = f(x) \quad \text{at} \quad y = 0.$$

Solution:

$$w(x, y) = \frac{1}{\pi} \int_{-\infty}^{\infty} f(\xi) \ln \sqrt{(x - \xi)^2 + y^2} \, d\xi + C,$$

where $C$ is an arbitrary constant.

---

* For the Laplace equation and other elliptic equations, the first boundary value problem is often called the *Dirichlet problem*, and the second boundary value problem is called the *Neumann problem*.

**T8.3.1-4. Domain: $0 \leq x < \infty$, $0 \leq y < \infty$. First boundary value problem.**

A quadrant of the plane is considered. Boundary conditions are prescribed:
$$w = f_1(y) \quad \text{at} \quad x = 0, \qquad w = f_2(x) \quad \text{at} \quad y = 0.$$

Solution:
$$w(x,y) = \frac{4}{\pi} xy \int_0^\infty \frac{f_1(\eta)\eta \, d\eta}{[x^2 + (y-\eta)^2][x^2 + (y+\eta)^2]} + \frac{4}{\pi} xy \int_0^\infty \frac{f_2(\xi)\xi \, d\xi}{[(x-\xi)^2 + y^2][(x+\xi)^2 + y^2]}.$$

**T8.3.1-5. Domain: $-\infty < x < \infty$, $0 \leq y \leq a$. First boundary value problem.**

An infinite strip is considered. Boundary conditions are prescribed:
$$w = f_1(x) \quad \text{at} \quad y = 0, \qquad w = f_2(x) \quad \text{at} \quad y = a.$$

Solution:
$$w(x,y) = \frac{1}{2a} \sin\left(\frac{\pi y}{a}\right) \int_{-\infty}^\infty \frac{f_1(\xi) \, d\xi}{\cosh[\pi(x-\xi)/a] - \cos(\pi y/a)}$$
$$+ \frac{1}{2a} \sin\left(\frac{\pi y}{a}\right) \int_{-\infty}^\infty \frac{f_2(\xi) \, d\xi}{\cosh[\pi(x-\xi)/a] + \cos(\pi y/a)}.$$

**T8.3.1-6. Domain: $-\infty < x < \infty$, $0 \leq y \leq a$. Second boundary value problem.**

An infinite strip is considered. Boundary conditions are prescribed:
$$\partial_y w = f_1(x) \quad \text{at} \quad y = 0, \qquad \partial_y w = f_2(x) \quad \text{at} \quad y = a.$$

Solution:
$$w(x,y) = \frac{1}{2\pi} \int_{-\infty}^\infty f_1(\xi) \ln\{\cosh[\pi(x-\xi)/a] - \cos(\pi y/a)\} \, d\xi$$
$$- \frac{1}{2\pi} \int_{-\infty}^\infty f_2(\xi) \ln\{\cosh[\pi(x-\xi)/a] + \cos(\pi y/a)\} \, d\xi + C,$$

where $C$ is an arbitrary constant.

**T8.3.1-7. Domain: $0 \leq x \leq a$, $0 \leq y \leq b$. First boundary value problem.**

A rectangle is considered. Boundary conditions are prescribed:
$$w = f_1(y) \quad \text{at} \quad x = 0, \qquad w = f_2(y) \quad \text{at} \quad x = a,$$
$$w = f_3(x) \quad \text{at} \quad y = 0, \qquad w = f_4(x) \quad \text{at} \quad y = b.$$

Solution:
$$w(x,y) = \sum_{n=1}^\infty A_n \sinh\left[\frac{n\pi}{b}(a-x)\right] \sin\left(\frac{n\pi}{b}y\right) + \sum_{n=1}^\infty B_n \sinh\left(\frac{n\pi}{b}x\right) \sin\left(\frac{n\pi}{b}y\right)$$
$$+ \sum_{n=1}^\infty C_n \sin\left(\frac{n\pi}{a}x\right) \sinh\left[\frac{n\pi}{a}(b-y)\right] + \sum_{n=1}^\infty D_n \sin\left(\frac{n\pi}{a}x\right) \sinh\left(\frac{n\pi}{a}y\right),$$

where the coefficients $A_n$, $B_n$, $C_n$, and $D_n$ are expressed as

$$A_n = \frac{2}{\lambda_n} \int_0^b f_1(\xi) \sin\left(\frac{n\pi\xi}{b}\right) d\xi, \quad B_n = \frac{2}{\lambda_n} \int_0^b f_2(\xi) \sin\left(\frac{n\pi\xi}{b}\right) d\xi, \quad \lambda_n = b \sinh\left(\frac{n\pi a}{b}\right),$$

$$C_n = \frac{2}{\mu_n} \int_0^a f_3(\xi) \sin\left(\frac{n\pi\xi}{a}\right) d\xi, \quad D_n = \frac{2}{\mu_n} \int_0^a f_4(\xi) \sin\left(\frac{n\pi\xi}{a}\right) d\xi, \quad \mu_n = a \sinh\left(\frac{n\pi b}{a}\right).$$

### T8.3.1-8. Domain: $0 \le r \le R$. First boundary value problem.

A circle is considered. A boundary condition is prescribed:

$$w = f(\varphi) \quad \text{at} \quad r = R.$$

Solution in the polar coordinates:

$$w(r, \varphi) = \frac{1}{2\pi} \int_0^{2\pi} f(\psi) \frac{R^2 - r^2}{r^2 - 2Rr \cos(\varphi - \psi) + R^2} \, d\psi.$$

This formula is conventionally referred to as the *Poisson integral*.

### T8.3.1-9. Domain: $0 \le r \le R$. Second boundary value problem.

A circle is considered. A boundary condition is prescribed:

$$\partial_r w = f(\varphi) \quad \text{at} \quad r = R.$$

Solution in the polar coordinates:

$$w(r, \varphi) = \frac{R}{2\pi} \int_0^{2\pi} f(\psi) \ln \frac{r^2 - 2Rr \cos(\varphi - \psi) + R^2}{R^2} \, d\psi + C,$$

where $C$ is an arbitrary constant; this formula is known as the *Dini integral*.

**Remark.** The function $f(\varphi)$ must satisfy the solvability condition $\int_0^{2\pi} f(\varphi) \, d\varphi = 0$.

## T8.3.2. Poisson Equation $\Delta w + \Phi(\mathbf{x}) = 0$

The two-dimensional Poisson equation has the following form:

$$\frac{\partial^2 w}{\partial x^2} + \frac{\partial^2 w}{\partial y^2} + \Phi(x, y) = 0 \quad \text{in the Cartesian coordinate system,}$$

$$\frac{1}{r} \frac{\partial}{\partial r}\left(r \frac{\partial w}{\partial r}\right) + \frac{1}{r^2} \frac{\partial^2 w}{\partial \varphi^2} + \Phi(r, \varphi) = 0 \quad \text{in the polar coordinate system.}$$

### T8.3.2-1. Domain: $-\infty < x < \infty$, $-\infty < y < \infty$.

Solution:

$$w(x, y) = \frac{1}{2\pi} \int_{-\infty}^{\infty} \int_{-\infty}^{\infty} \Phi(\xi, \eta) \ln \frac{1}{\sqrt{(x - \xi)^2 + (y - \eta)^2}} \, d\xi \, d\eta.$$

**T8.3.2-2. Domain: $-\infty < x < \infty$, $0 \leq y < \infty$. First boundary value problem.**

A half-plane is considered. A boundary condition is prescribed:
$$w = f(x) \quad \text{at} \quad y = 0.$$

Solution:
$$w(x,y) = \frac{1}{\pi} \int_{-\infty}^{\infty} \frac{y f(\xi)\,d\xi}{(x-\xi)^2 + y^2} + \frac{1}{2\pi} \int_0^{\infty}\int_{-\infty}^{\infty} \Phi(\xi,\eta) \ln \frac{\sqrt{(x-\xi)^2+(y+\eta)^2}}{\sqrt{(x-\xi)^2+(y-\eta)^2}}\, d\xi\, d\eta.$$

**T8.3.2-3. Domain: $0 \leq x < \infty$, $0 \leq y < \infty$. First boundary value problem.**

A quadrant of the plane is considered. Boundary conditions are prescribed:
$$w = f_1(y) \quad \text{at} \quad x = 0, \qquad w = f_2(x) \quad \text{at} \quad y = 0.$$

Solution:
$$w(x,y) = \frac{4}{\pi} xy \int_0^{\infty} \frac{f_1(\eta)\eta\,d\eta}{[x^2+(y-\eta)^2][x^2+(y+\eta)^2]} + \frac{4}{\pi} xy \int_0^{\infty} \frac{f_2(\xi)\xi\,d\xi}{[(x-\xi)^2+y^2][(x+\xi)^2+y^2]}$$
$$+ \frac{1}{2\pi} \int_0^{\infty}\int_0^{\infty} \Phi(\xi,\eta) \ln \frac{\sqrt{(x-\xi)^2+(y+\eta)^2}\sqrt{(x+\xi)^2+(y-\eta)^2}}{\sqrt{(x-\xi)^2+(y-\eta)^2}\sqrt{(x+\xi)^2+(y+\eta)^2}}\, d\xi\, d\eta.$$

**T8.3.2-4. Domain: $0 \leq x \leq a$, $0 \leq y \leq b$. First boundary value problem.**

A rectangle is considered. Boundary conditions are prescribed:
$$w = f_1(y) \quad \text{at} \quad x = 0, \qquad w = f_2(y) \quad \text{at} \quad x = a,$$
$$w = f_3(x) \quad \text{at} \quad y = 0, \qquad w = f_4(x) \quad \text{at} \quad y = b.$$

Solution:
$$w(x,y) = \int_0^a \int_0^b \Phi(\xi,\eta) G(x,y,\xi,\eta)\,d\eta\,d\xi$$
$$+ \int_0^b f_1(\eta)\left[\frac{\partial}{\partial \xi} G(x,y,\xi,\eta)\right]_{\xi=0} d\eta - \int_0^b f_2(\eta)\left[\frac{\partial}{\partial \xi} G(x,y,\xi,\eta)\right]_{\xi=a} d\eta$$
$$+ \int_0^a f_3(\xi)\left[\frac{\partial}{\partial \eta} G(x,y,\xi,\eta)\right]_{\eta=0} d\xi - \int_0^a f_4(\xi)\left[\frac{\partial}{\partial \eta} G(x,y,\xi,\eta)\right]_{\eta=b} d\xi.$$

Two forms of representation of the Green's function:
$$G(x,y,\xi,\eta) = \frac{2}{a}\sum_{n=1}^{\infty} \frac{\sin(p_n x)\sin(p_n \xi)}{p_n \sinh(p_n b)} H_n(y,\eta) = \frac{2}{b}\sum_{m=1}^{\infty} \frac{\sin(q_m y)\sin(q_m \eta)}{q_m \sinh(q_m a)} Q_m(x,\xi),$$

where
$$p_n = \frac{\pi n}{a}, \quad H_n(y,\eta) = \begin{cases} \sinh(p_n \eta)\sinh[p_n(b-y)] & \text{for } b \geq y > \eta \geq 0, \\ \sinh(p_n y)\sinh[p_n(b-\eta)] & \text{for } b \geq \eta > y \geq 0; \end{cases}$$
$$q_m = \frac{\pi m}{b}, \quad Q_m(x,\xi) = \begin{cases} \sinh(q_m \xi)\sinh[q_m(a-x)] & \text{for } a \geq x > \xi \geq 0, \\ \sinh(q_m x)\sinh[q_m(a-\xi)] & \text{for } a \geq \xi > x \geq 0. \end{cases}$$

T8.3.2-5. Domain: $0 \leq r \leq R$. First boundary value problem.

A circle is considered. A boundary condition is prescribed:
$$w = f(\varphi) \quad \text{at} \quad r = R.$$

Solution in the polar coordinates:
$$w(r,\varphi) = \frac{1}{2\pi}\int_0^{2\pi} f(\eta)\frac{R^2 - r^2}{r^2 - 2Rr\cos(\varphi-\eta) + R^2}\,d\eta + \int_0^{2\pi}\int_0^R \Phi(\xi,\eta)G(r,\varphi,\xi,\eta)\xi\,d\xi\,d\eta,$$
where
$$G(r,\varphi,\xi,\eta) = \frac{1}{4\pi}\ln\frac{r^2\xi^2 - 2R^2 r\xi\cos(\varphi-\eta) + R^4}{R^2[r^2 - 2r\xi\cos(\varphi-\eta) + \xi^2]}.$$

## T8.3.3. Helmholtz Equation $\Delta w + \lambda w = -\Phi(\mathrm{x})$

Many problems related to steady-state oscillations (mechanical, acoustical, thermal, electromagnetic) lead to the two-dimensional Helmholtz equation. For $\lambda < 0$, this equation describes mass transfer processes with volume chemical reactions of the first order.

The two-dimensional Helmholtz equation has the following form:
$$\frac{\partial^2 w}{\partial x^2} + \frac{\partial^2 w}{\partial y^2} + \lambda w = -\Phi(x,y) \quad \text{in the Cartesian coordinate system,}$$
$$\frac{1}{r}\frac{\partial}{\partial r}\left(r\frac{\partial w}{\partial r}\right) + \frac{1}{r^2}\frac{\partial^2 w}{\partial \varphi^2} + \lambda w = -\Phi(r,\varphi) \quad \text{in the polar coordinate system.}$$

T8.3.3-1. Particular solutions of the homogeneous Helmholtz equation with $\Phi \equiv 0$.

1°. Particular solutions of the homogeneous equation in the Cartesian coordinate system:
$$w = (Ax + B)(C\cos\mu y + D\sin\mu y), \quad \lambda = \mu^2,$$
$$w = (Ax + B)(C\cosh\mu y + D\sinh\mu y), \quad \lambda = -\mu^2,$$
$$w = (A\cos\mu x + B\sin\mu x)(Cy + D), \quad \lambda = \mu^2,$$
$$w = (A\cosh\mu x + B\sinh\mu x)(Cy + D), \quad \lambda = -\mu^2,$$
$$w = (A\cos\mu_1 x + B\sin\mu_1 x)(C\cos\mu_2 y + D\sin\mu_2 y), \quad \lambda = \mu_1^2 + \mu_2^2,$$
$$w = (A\cos\mu_1 x + B\sin\mu_1 x)(C\cosh\mu_2 y + D\sinh\mu_2 y), \quad \lambda = \mu_1^2 - \mu_2^2,$$
$$w = (A\cosh\mu_1 x + B\sinh\mu_1 x)(C\cos\mu_2 y + D\sin\mu_2 y), \quad \lambda = -\mu_1^2 + \mu_2^2,$$
$$w = (A\cosh\mu_1 x + B\sinh\mu_1 x)(C\cosh\mu_2 y + D\sinh\mu_2 y), \quad \lambda = -\mu_1^2 - \mu_2^2,$$
where $A$, $B$, $C$, and $D$ are arbitrary constants.

2°. Particular solutions of the homogeneous equation in the polar coordinate system:
$$w = [AJ_0(\mu r) + BY_0(\mu r)](C\varphi + D), \quad \lambda = \mu^2,$$
$$w = [AI_0(\mu r) + BK_0(\mu r)](C\varphi + D), \quad \lambda = -\mu^2,$$
$$w = [AJ_m(\mu r) + BY_m(\mu r)](C\cos m\varphi + D\sin m\varphi), \quad \lambda = \mu^2,$$
$$w = [AI_m(\mu r) + BK_m(\mu r)](C\cos m\varphi + D\sin m\varphi), \quad \lambda = -\mu^2,$$
where $m = 1, 2, \ldots$; $A$, $B$, $C$, $D$ are arbitrary constants; $J_m(\mu)$ and $Y_m(\mu)$ are Bessel functions; and $I_m(\mu)$ and $K_m(\mu)$ are modified Bessel functions.

3°. Suppose $w = w(x, y)$ is a solution of the homogeneous Helmholtz equation. Then the functions
$$w_1 = w(\pm x + C_1, \pm y + C_2),$$
$$w_2 = w(x\cos\theta + y\sin\theta + C_1, -x\sin\theta + y\cos\theta + C_2),$$

where $C_1$, $C_2$, and $\theta$ are arbitrary constants, are also solutions of the equation. The signs in the formula for $w_1$ are taken arbitrarily.

**T8.3.3-2. Domain: $-\infty < x < \infty$, $-\infty < y < \infty$.**

1°. Solution for $\lambda = -s^2 < 0$:
$$w(x, y) = \frac{1}{2\pi} \int_{-\infty}^{\infty}\int_{-\infty}^{\infty} \Phi(\xi, \eta) K_0(s\varrho)\, d\xi\, d\eta, \qquad \varrho = \sqrt{(x-\xi)^2 + (y-\eta)^2}.$$

2°. Solution for $\lambda = k^2 > 0$:
$$w(x, y) = -\frac{i}{4} \int_{-\infty}^{\infty}\int_{-\infty}^{\infty} \Phi(\xi, \eta) H_0^{(2)}(k\varrho)\, d\xi\, d\eta, \qquad \varrho = \sqrt{(x-\xi)^2 + (y-\eta)^2}.$$

*Remark.* The radiation Sommerfeld conditions at infinity were used to obtain the solution with $\lambda > 0$; see Tikhonov and Samarskii (1990) and Polyanin (2002).

**T8.3.3-3. Domain: $-\infty < x < \infty$, $0 \le y < \infty$. First boundary value problem.**

A half-plane is considered. A boundary condition is prescribed:
$$w = f(x) \quad \text{at} \quad y = 0.$$

Solution:
$$w(x, y) = \int_{-\infty}^{\infty} f(\xi)\left[\frac{\partial}{\partial \eta} G(x, y, \xi, \eta)\right]_{\eta=0} d\xi + \int_0^{\infty}\int_{-\infty}^{\infty} \Phi(\xi, \eta) G(x, y, \xi, \eta)\, d\xi\, d\eta.$$

1°. The Green's function for $\lambda = -s^2 < 0$:
$$G(x, y, \xi, \eta) = \frac{1}{2\pi}\left[K_0(s\varrho_1) - K_0(s\varrho_2)\right],$$
$$\varrho_1 = \sqrt{(x-\xi)^2 + (y-\eta)^2}, \qquad \varrho_2 = \sqrt{(x-\xi)^2 + (y+\eta)^2}.$$

2°. The Green's function for $\lambda = k^2 > 0$:
$$G(x, y, \xi, \eta) = -\frac{i}{4}\left[H_0^{(2)}(k\varrho_1) - H_0^{(2)}(k\varrho_2)\right].$$

*Remark.* The radiation Sommerfeld conditions at infinity were used to obtain the solution with $\lambda > 0$; see Tikhonov and Samarskii (1990) and Polyanin (2002).

**T8.3.3-4. Domain:** $-\infty < x < \infty$, $0 \leq y < \infty$. **Second boundary value problem.**

A half-plane is considered. A boundary condition is prescribed:
$$\partial_y w = f(x) \quad \text{at} \quad y = 0.$$

Solution:
$$w(x,y) = -\int_{-\infty}^{\infty} f(\xi) G(x,y,\xi,0)\, d\xi + \int_0^{\infty}\int_{-\infty}^{\infty} \Phi(\xi,\eta) G(x,y,\xi,\eta)\, d\xi\, d\eta.$$

$1^\circ$. The Green's function for $\lambda = -s^2 < 0$:
$$G(x,y,\xi,\eta) = \frac{1}{2\pi}\big[K_0(s\varrho_1) + K_0(s\varrho_2)\big],$$
$$\varrho_1 = \sqrt{(x-\xi)^2 + (y-\eta)^2}, \quad \varrho_2 = \sqrt{(x-\xi)^2 + (y+\eta)^2}.$$

$2^\circ$. The Green's function for $\lambda = k^2 > 0$:
$$G(x,y,\xi,\eta) = -\frac{i}{4}\big[H_0^{(2)}(k\varrho_1) + H_0^{(2)}(k\varrho_2)\big].$$

**Remark.** The radiation Sommerfeld conditions at infinity were used to obtain the solution with $\lambda > 0$; see Tikhonov and Samarskii (1990) and Polyanin (2002).

**T8.3.3-5. Domain:** $0 \leq x < \infty$, $0 \leq y < \infty$. **First boundary value problem.**

A quadrant of the plane is considered. Boundary conditions are prescribed:
$$w = f_1(y) \quad \text{at} \quad x = 0, \qquad w = f_2(x) \quad \text{at} \quad y = 0.$$

Solution:
$$w(x,y) = \int_0^{\infty} f_1(\eta)\left[\frac{\partial}{\partial \xi} G(x,y,\xi,\eta)\right]_{\xi=0} d\eta + \int_0^{\infty} f_2(\xi)\left[\frac{\partial}{\partial \eta} G(x,y,\xi,\eta)\right]_{\eta=0} d\xi$$
$$+ \int_0^{\infty}\int_0^{\infty} \Phi(\xi,\eta) G(x,y,\xi,\eta)\, d\xi\, d\eta.$$

$1^\circ$. The Green's function for $\lambda = -s^2 < 0$:
$$G(x,y,\xi,\eta) = \frac{1}{2\pi}\big[K_0(s\varrho_1) - K_0(s\varrho_2) - K_0(s\varrho_3) + K_0(s\varrho_4)\big],$$
$$\varrho_1 = \sqrt{(x-\xi)^2 + (y-\eta)^2}, \quad \varrho_2 = \sqrt{(x-\xi)^2 + (y+\eta)^2},$$
$$\varrho_3 = \sqrt{(x+\xi)^2 + (y-\eta)^2}, \quad \varrho_4 = \sqrt{(x+\xi)^2 + (y+\eta)^2}.$$

$2^\circ$. The Green's function for $\lambda = k^2 > 0$:
$$G(x,y,\xi,\eta) = -\frac{i}{4}\big[H_0^{(2)}(k\varrho_1) - H_0^{(2)}(k\varrho_2) - H_0^{(2)}(k\varrho_3) + H_0^{(2)}(k\varrho_4)\big].$$

**T8.3.3-6. Domain: $0 \leq x < \infty$, $0 \leq y < \infty$. Second boundary value problem.**

A quadrant of the plane is considered. Boundary conditions are prescribed:

$$\partial_x w = f_1(y) \quad \text{at} \quad x = 0, \qquad \partial_y w = f_2(x) \quad \text{at} \quad y = 0.$$

Solution:

$$w(x,y) = -\int_0^\infty f_1(\eta) G(x,y,0,\eta)\, d\eta - \int_0^\infty f_2(\xi) G(x,y,\xi,0)\, d\xi$$
$$+ \int_0^\infty \int_0^\infty \Phi(\xi,\eta) G(x,y,\xi,\eta)\, d\xi\, d\eta.$$

1°. The Green's function for $\lambda = -s^2 < 0$:

$$G(x,y,\xi,\eta) = \frac{1}{2\pi}\bigl[K_0(s\varrho_1) + K_0(s\varrho_2) + K_0(s\varrho_3) + K_0(s\varrho_4)\bigr],$$

$$\varrho_1 = \sqrt{(x-\xi)^2 + (y-\eta)^2}, \qquad \varrho_2 = \sqrt{(x-\xi)^2 + (y+\eta)^2},$$
$$\varrho_3 = \sqrt{(x+\xi)^2 + (y-\eta)^2}, \qquad \varrho_4 = \sqrt{(x+\xi)^2 + (y+\eta)^2}.$$

2°. The Green's function for $\lambda = k^2 > 0$:

$$G(x,y,\xi,\eta) = -\frac{i}{4}\bigl[H_0^{(2)}(k\varrho_1) + H_0^{(2)}(k\varrho_2) + H_0^{(2)}(k\varrho_3) + H_0^{(2)}(k\varrho_4)\bigr].$$

**T8.3.3-7. Domain: $0 \leq x \leq a$, $0 \leq y \leq b$. First boundary value problem.**

A rectangle is considered. Boundary conditions are prescribed:

$$w = f_1(y) \quad \text{at} \quad x = 0, \qquad w = f_2(y) \quad \text{at} \quad x = a,$$
$$w = f_3(x) \quad \text{at} \quad y = 0, \qquad w = f_4(x) \quad \text{at} \quad y = b.$$

1°. Eigenvalues of the homogeneous problem with $\Phi \equiv 0$ (it is convenient to label them with a double subscript):

$$\lambda_{nm} = \pi^2\left(\frac{n^2}{a^2} + \frac{m^2}{b^2}\right); \qquad n = 1, 2, \ldots; \quad m = 1, 2, \ldots$$

Eigenfunctions and the norm squared:

$$w_{nm} = \sin\left(\frac{n\pi x}{a}\right)\sin\left(\frac{m\pi y}{b}\right), \qquad \|w_{nm}\|^2 = \frac{ab}{4}.$$

2°. Solution for $\lambda \neq \lambda_{nm}$:

$$w(x,y) = \int_0^a \int_0^b \Phi(\xi,\eta) G(x,y,\xi,\eta)\, d\eta\, d\xi$$
$$+ \int_0^b f_1(\eta)\left[\frac{\partial}{\partial \xi} G(x,y,\xi,\eta)\right]_{\xi=0} d\eta - \int_0^b f_2(\eta)\left[\frac{\partial}{\partial \xi} G(x,y,\xi,\eta)\right]_{\xi=a} d\eta$$
$$+ \int_0^a f_3(\xi)\left[\frac{\partial}{\partial \eta} G(x,y,\xi,\eta)\right]_{\eta=0} d\xi - \int_0^a f_4(\xi)\left[\frac{\partial}{\partial \eta} G(x,y,\xi,\eta)\right]_{\eta=b} d\xi.$$

Two forms of representation of the Green's function:

$$G(x, y, \xi, \eta) = \frac{2}{a} \sum_{n=1}^{\infty} \frac{\sin(p_n x) \sin(p_n \xi)}{\beta_n \sinh(\beta_n b)} H_n(y, \eta) = \frac{2}{b} \sum_{m=1}^{\infty} \frac{\sin(q_m y) \sin(q_m \eta)}{\mu_m \sinh(\mu_m a)} Q_m(x, \xi),$$

where

$$p_n = \frac{\pi n}{a}, \quad \beta_n = \sqrt{p_n^2 - \lambda}, \quad H_n(y, \eta) = \begin{cases} \sinh(\beta_n \eta) \sinh[\beta_n(b-y)] & \text{for } b \geq y > \eta \geq 0, \\ \sinh(\beta_n y) \sinh[\beta_n(b-\eta)] & \text{for } b \geq \eta > y \geq 0; \end{cases}$$

$$q_m = \frac{\pi m}{b}, \quad \mu_m = \sqrt{q_m^2 - \lambda}, \quad Q_m(x, \xi) = \begin{cases} \sinh(\mu_m \xi) \sinh[\mu_m(a-x)] & \text{for } a \geq x > \xi \geq 0, \\ \sinh(\mu_m x) \sinh[\mu_m(a-\xi)] & \text{for } a \geq \xi > x \geq 0. \end{cases}$$

**T8.3.3-8. Domain: $0 \leq r \leq R$. First boundary value problem.**

A circle is considered. A boundary condition is prescribed:

$$w = 0 \quad \text{at} \quad r = R.$$

Eigenvalues of the homogeneous boundary value problem with $\Phi \equiv 0$:

$$\lambda_{nm} = \frac{\mu_{nm}^2}{R^2}; \quad n = 0, 1, 2, \ldots; \quad m = 1, 2, 3, \ldots$$

Here, $\mu_{nm}$ are positive zeros of the Bessel functions, $J_n(\mu) = 0$.
Eigenfunctions:

$$w_{nm}^{(1)} = J_n(r\sqrt{\lambda_{nm}}) \cos n\varphi, \quad w_{nm}^{(2)} = J_n(r\sqrt{\lambda_{nm}}) \sin n\varphi.$$

Eigenfunctions possessing the axial symmetry property: $w_{0m}^{(1)} = J_0(r\sqrt{\lambda_{0m}})$.

**T8.3.3-9. Domain: $0 \leq r \leq R$. Second boundary value problem.**

A circle is considered. A boundary condition is prescribed:

$$\partial_r w = 0 \quad \text{at} \quad r = R.$$

Eigenvalues of the homogeneous boundary value problem with $\Phi \equiv 0$:

$$\lambda_{nm} = \frac{\mu_{nm}^2}{R^2},$$

where $\mu_{nm}$ are roots of the quadratic equation $J_n'(\mu) = 0$.
Eigenfunctions:

$$w_{nm}^{(1)} = J_n(r\sqrt{\lambda_{nm}}) \cos n\varphi, \quad w_{nm}^{(2)} = J_n(r\sqrt{\lambda_{nm}}) \sin n\varphi.$$

Here, $n = 0, 1, 2, \ldots$; for $n \neq 0$, the parameter $m$ assumes the values $m = 1, 2, 3, \ldots$; for $n = 0$, a root $\mu_{00} = 0$ (the corresponding eigenfunction is $w_{00} = 1$).
Eigenfunctions possessing the axial symmetry property: $w_{0m}^{(1)} = J_0(r\sqrt{\lambda_{0m}})$.

## T8.4. Fourth-Order Linear Equations

### T8.4.1. Equation of the Form $\dfrac{\partial^2 w}{\partial t^2} + a^2 \dfrac{\partial^4 w}{\partial x^4} = 0$

This equation is encountered in studying transverse vibration of elastic rods.

**T8.4.1-1. Particular solutions.**

$$w(x,t) = (Ax^3 + Bx^2 + Cx + D)t + A_1 x^3 + B_1 x^2 + C_1 x + D_1,$$
$$w(x,t) = \left[ A \sin(\lambda x) + B \cos(\lambda x) + C \sinh(\lambda x) + D \cos(\lambda x) \right] \sin(\lambda^2 a t),$$
$$w(x,t) = \left[ A \sin(\lambda x) + B \cos(\lambda x) + C \sinh(\lambda x) + D \cos(\lambda x) \right] \cos(\lambda^2 a t),$$

where $A$, $B$, $C$, $D$, $A_1$, $B_1$, $C_1$, $D_1$, and $\lambda$ are arbitrary constants.

**T8.4.1-2. Domain: $-\infty < x < \infty$. Cauchy problem.**

Initial conditions are prescribed:

$$w = f(x) \quad \text{at} \quad t = 0, \qquad \partial_t w = a g''(x) \quad \text{at} \quad t = 0.$$

Boussinesq solution:

$$w(x,t) = \frac{1}{\sqrt{2\pi}} \int_{-\infty}^{\infty} f(x - 2\xi\sqrt{at})(\cos \xi^2 + \sin \xi^2)\, d\xi$$
$$+ \frac{1}{a\sqrt{2\pi}} \int_{-\infty}^{\infty} g(x - 2\xi\sqrt{at})(\cos \xi^2 - \sin \xi^2)\, d\xi.$$

**T8.4.1-3. Domain: $0 \le x < \infty$. Free vibration of a semi-infinite rod.**

The following conditions are prescribed:

$$w = 0 \quad \text{at} \quad t = 0, \qquad \partial_t w = 0 \quad \text{at} \quad t = 0 \qquad \text{(initial conditions)},$$
$$w = f(t) \quad \text{at} \quad x = 0, \qquad \partial_{xx} w = 0 \quad \text{at} \quad x = 0 \qquad \text{(boundary conditions)}.$$

Boussinesq solution:

$$w(x,t) = \frac{1}{\sqrt{\pi}} \int_{x/\sqrt{2at}}^{\infty} f\!\left(t - \frac{x^2}{2a\xi^2}\right)\!\left(\sin \frac{\xi^2}{2} + \cos \frac{\xi^2}{2}\right) d\xi.$$

**T8.4.1-4. Domain: $0 \le x \le l$. Boundary value problems.**

For solutions of various boundary value problems, see Subsection T8.4.2 for $\Phi \equiv 0$.

## T8.4.2. Equation of the Form $\dfrac{\partial^2 w}{\partial t^2} + a^2 \dfrac{\partial^4 w}{\partial x^4} = \Phi(x,t)$

**T8.4.2-1. Domain: $0 \leq x \leq l$. Solution in terms of the Green's function.**

We consider boundary value problems on an interval $0 \leq x \leq l$ with the general initial condition

$$w = f(x) \quad \text{at} \quad t = 0, \qquad \partial_t w = g(x) \quad \text{at} \quad t = 0$$

and various homogeneous boundary conditions. The solution can be represented in terms of the Green's function as

$$w(x,t) = \frac{\partial}{\partial t} \int_0^l f(\xi) G(x,\xi,t)\,d\xi + \int_0^l g(\xi) G(x,\xi,t)\,d\xi + \int_0^t \int_0^l \Phi(\xi,\tau) G(x,\xi,t-\tau)\,d\xi\,d\tau.$$

Paragraphs T8.4.2-2 through T8.4.2-6 present the Green's functions for various types of homogeneous boundary conditions.

**T8.4.2-2. Both ends of the rod are clamped.**

Boundary conditions are prescribed:

$$w = \partial_x w = 0 \quad \text{at} \quad x = 0, \qquad w = \partial_x w = 0 \quad \text{at} \quad x = l.$$

Green's function:

$$G(x,\xi,t) = \frac{4}{al} \sum_{n=1}^{\infty} \frac{\lambda_n^2}{[\varphi_n''(l)]^2} \varphi_n(x) \varphi_n(\xi) \sin(\lambda_n^2 a t),$$

where

$$\varphi_n(x) = \big[\sinh(\lambda_n l) - \sin(\lambda_n l)\big]\big[\cosh(\lambda_n x) - \cos(\lambda_n x)\big]$$
$$- \big[\cosh(\lambda_n l) - \cos(\lambda_n l)\big]\big[\sinh(\lambda_n x) - \sin(\lambda_n x)\big];$$

$\lambda_n$ are positive roots of the transcendental equation $\cosh(\lambda l)\cos(\lambda l) = 1$. The numerical values of the roots can be calculated from the formulas

$$\lambda_n = \frac{\mu_n}{l}, \quad \text{where} \quad \mu_1 = 1.875, \quad \mu_2 = 4.694, \quad \mu_n = \frac{\pi}{2}(2n-1) \quad \text{for} \quad n \geq 3.$$

**T8.4.2-3. Both ends of the rod are hinged.**

Boundary conditions are prescribed:

$$w = \partial_{xx} w = 0 \quad \text{at} \quad x = 0, \qquad w = \partial_{xx} w = 0 \quad \text{at} \quad x = l.$$

Green's function:

$$G(x,\xi,t) = \frac{2l}{a\pi^2} \sum_{n=1}^{\infty} \frac{1}{n^2} \sin(\lambda_n x) \sin(\lambda_n \xi) \sin(\lambda_n^2 a t), \qquad \lambda_n = \frac{\pi n}{l}.$$

**T8.4.2-4. One end of the rod is clamped and the other is hinged.**

Boundary conditions are prescribed:
$$w = \partial_x w = 0 \quad \text{at} \quad x = 0, \qquad w = \partial_{xx} w = 0 \quad \text{at} \quad x = l.$$

Green's function:
$$G(x, \xi, t) = \frac{2}{al} \sum_{n=1}^{\infty} \lambda_n^2 \frac{\varphi_n(x)\varphi_n(\xi)}{|\varphi_n'(l)\varphi_n'''(l)|} \sin(\lambda_n^2 a t),$$

where
$$\varphi_n(x) = \big[\sinh(\lambda_n l) - \sin(\lambda_n l)\big]\big[\cosh(\lambda_n x) - \cos(\lambda_n x)\big]$$
$$- \big[\cosh(\lambda_n l) - \cos(\lambda_n l)\big]\big[\sinh(\lambda_n x) - \sin(\lambda_n x)\big];$$

$\lambda_n$ are positive roots of the transcendental equation $\tan(\lambda l) - \tanh(\lambda l) = 0$.

**T8.4.2-5. One end of the rod is clamped and the other is free.**

Boundary conditions are prescribed:
$$w = \partial_x w = 0 \quad \text{at} \quad x = 0, \qquad \partial_{xx} w = \partial_{xxx} w = 0 \quad \text{at} \quad x = l.$$

Green's function:
$$G(x, \xi, t) = \frac{4}{al} \sum_{n=1}^{\infty} \frac{\varphi_n(x)\varphi_n(\xi)}{\lambda_n^2 \varphi_n^2(l)} \sin(\lambda_n^2 a t),$$

where
$$\varphi_n(x) = \big[\sinh(\lambda_n l) + \sin(\lambda_n l)\big]\big[\cosh(\lambda_n x) - \cos(\lambda_n x)\big]$$
$$- \big[\cosh(\lambda_n l) + \cos(\lambda_n l)\big]\big[\sinh(\lambda_n x) - \sin(\lambda_n x)\big];$$

$\lambda_n$ are positive roots of the transcendental equation $\cosh(\lambda l)\cos(\lambda l) = -1$.

**T8.4.2-6. One end of the rod is hinged and the other is free.**

Boundary conditions are prescribed:
$$w = \partial_{xx} w = 0 \quad \text{at} \quad x = 0, \qquad \partial_{xx} w = \partial_{xxx} w = 0 \quad \text{at} \quad x = l.$$

Green's function:
$$G(x, \xi, t) = \frac{4}{al} \sum_{n=1}^{\infty} \frac{\varphi_n(x)\varphi_n(\xi)}{\lambda_n^2 \varphi_n^2(l)} \sin(\lambda_n^2 a t),$$

where
$$\varphi_n(x) = \sin(\lambda_n l)\sinh(\lambda_n x) + \sinh(\lambda_n l)\sin(\lambda_n x);$$

$\lambda_n$ are positive roots of the transcendental equation $\tan(\lambda l) - \tanh(\lambda l) = 0$.

## T8.4.3. Biharmonic Equation $\Delta\Delta w = 0$

The biharmonic equation is encountered in plane problems of elasticity ($w$ is the Airy stress function). It is also used to describe slow flows of viscous incompressible fluids ($w$ is the stream function).

In the rectangular Cartesian system of coordinates, the biharmonic operator has the form

$$\Delta\Delta \equiv \Delta^2 = \frac{\partial^4}{\partial x^4} + 2\frac{\partial^4}{\partial x^2 \partial y^2} + \frac{\partial^4}{\partial y^4}.$$

### T8.4.3-1. Particular solutions.

$$w(x,y) = (A\cosh\beta x + B\sinh\beta x + Cx\cosh\beta x + Dx\sinh\beta x)(a\cos\beta y + b\sin\beta y),$$
$$w(x,y) = (A\cos\beta x + B\sin\beta x + Cx\cos\beta x + Dx\sin\beta x)(a\cosh\beta y + b\sinh\beta y),$$
$$w(x,y) = Ar^2\ln r + Br^2 + C\ln r + D, \quad r = \sqrt{(x-a)^2 + (y-b)^2},$$

where $A, B, C, D, E, a, b, c$, and $\beta$ are arbitrary constants.

### T8.4.3-2. Various representations of the general solution.

1°. Various representations of the general solution in terms of harmonic functions:

$$w(x,y) = xu_1(x,y) + u_2(x,y),$$
$$w(x,y) = yu_1(x,y) + u_2(x,y),$$
$$w(x,y) = (x^2 + y^2)u_1(x,y) + u_2(x,y),$$

where $u_1$ and $u_2$ are arbitrary functions satisfying the Laplace equation $\Delta u_k = 0$ ($k = 1, 2$).

2°. Complex form of representation of the general solution:

$$w(x,y) = \mathrm{Re}\big[\bar{z}f(z) + g(z)\big],$$

where $f(z)$ and $g(z)$ are arbitrary analytic functions of the complex variable $z = x + iy$; $\bar{z} = x - iy$, $i^2 = -1$. The symbol Re[$A$] stands for the real part of the complex quantity $A$.

### T8.4.3-3. Boundary value problems for the upper half-plane.

1°. Domain: $-\infty < x < \infty$, $0 \leq y < \infty$. The desired function and its derivative along the normal are prescribed at the boundary:

$$w = 0 \quad \text{at} \quad y = 0, \qquad \partial_y w = f(x) \quad \text{at} \quad y = 0.$$

Solution:

$$w(x,y) = \int_{-\infty}^{\infty} f(\xi) G(x-\xi, y)\,d\xi, \qquad G(x,y) = \frac{1}{\pi}\frac{y^2}{x^2 + y^2}.$$

2°. Domain: $-\infty < x < \infty$, $0 \leq y < \infty$. The derivatives of the desired function are prescribed at the boundary:

$$\partial_x w = f(x) \quad \text{at} \quad y = 0, \qquad \partial_y w = g(x) \quad \text{at} \quad y = 0.$$

Solution:

$$w(x,y) = \frac{1}{\pi}\int_{-\infty}^{\infty} f(\xi)\left[\arctan\left(\frac{x-\xi}{y}\right) + \frac{y(x-\xi)}{(x-\xi)^2 + y^2}\right] d\xi + \frac{y^2}{\pi}\int_{-\infty}^{\infty}\frac{g(\xi)\,d\xi}{(x-\xi)^2 + y^2} + C,$$

where $C$ is an arbitrary constant.

### T8.4.3-4. Boundary value problem for a circle.

Domain: $0 \leq r \leq a$. Boundary conditions in the polar coordinate system:

$$w = f(\varphi) \quad \text{at} \quad r = a, \qquad \partial_r w = g(\varphi) \quad \text{at} \quad r = a.$$

Solution:

$$w(r,\varphi) = \frac{1}{2\pi a}(r^2 - a^2)^2 \left[ \int_0^{2\pi} \frac{[a - r\cos(\eta-\varphi)]f(\eta)\,d\eta}{[r^2 + a^2 - 2ar\cos(\eta-\varphi)]^2} - \frac{1}{2}\int_0^{2\pi} \frac{g(\eta)\,d\eta}{r^2 + a^2 - 2ar\cos(\eta-\varphi)} \right].$$

## T8.4.4. Nonhomogeneous Biharmonic Equation $\Delta\Delta w = \Phi(x,y)$

### T8.4.4-1. Domain: $-\infty < x < \infty$, $-\infty < y < \infty$.

Solution:

$$w(x,y) = \int_{-\infty}^{\infty}\int_{-\infty}^{\infty} \Phi(\xi,\eta)\mathscr{E}(x-\xi, y-\eta)\,d\xi\,d\eta, \qquad \mathscr{E}(x,y) = \frac{1}{8\pi}(x^2+y^2)\ln\sqrt{x^2+y^2}.$$

### T8.4.4-2. Domain: $-\infty < x < \infty$, $0 \leq y < \infty$. Boundary value problem.

The upper half-plane is considered. The derivatives are prescribed at the boundary:

$$\partial_x w = f(x) \quad \text{at} \quad y = 0, \qquad \partial_y w = g(x) \quad \text{at} \quad y = 0.$$

Solution:

$$w(x,y) = \frac{1}{\pi}\int_{-\infty}^{\infty} f(\xi)\left[\arctan\left(\frac{x-\xi}{y}\right) + \frac{y(x-\xi)}{(x-\xi)^2 + y^2}\right] d\xi + \frac{y^2}{\pi}\int_{-\infty}^{\infty} \frac{g(\xi)\,d\xi}{(x-\xi)^2 + y^2}$$
$$+ \frac{1}{8\pi}\int_{-\infty}^{\infty} d\xi \int_0^{\infty} \left[\frac{1}{2}(R_+^2 - R_-^2) - R_-^2 \ln\frac{R_+}{R_-}\right]\Phi(\xi,\eta)\,d\eta + C,$$

where $C$ is an arbitrary constant,

$$R_+^2 = (x-\xi)^2 + (y+\eta)^2, \qquad R_-^2 = (x-\xi)^2 + (y-\eta)^2.$$

### T8.4.4-3. Domain: $0 \leq x \leq l_1$, $0 \leq y \leq l_2$. The sides of the plate are hinged.

A rectangle is considered. Boundary conditions are prescribed:

$$w = \partial_{xx}w = 0 \quad \text{at} \quad x = 0, \qquad w = \partial_{xx}w = 0 \quad \text{at} \quad x = l_1,$$
$$w = \partial_{yy}w = 0 \quad \text{at} \quad y = 0, \qquad w = \partial_{yy}w = 0 \quad \text{at} \quad y = l_2.$$

Solution:

$$w(x,y) = \int_0^{l_1}\int_0^{l_2} \Phi(\xi,\eta) G(x,y,\xi,\eta)\,d\eta\,d\xi,$$

where

$$G(x,y,\xi,\eta) = \frac{4}{l_1 l_2}\sum_{n=1}^{\infty}\sum_{m=1}^{\infty}\frac{1}{(p_n^2 + q_m^2)^2}\sin(p_n x)\sin(q_m y)\sin(p_n \xi)\sin(q_m \eta), \quad p_n = \frac{\pi n}{l_1}, \quad q_m = \frac{\pi m}{l_2}.$$

# References for Chapter T8

**Butkovskiy, A. G.,** *Green's Functions and Transfer Functions Handbook*, Halstead Press–John Wiley & Sons, New York, 1982.

**Carslaw, H. S. and Jaeger, J. C.,** *Conduction of Heat in Solids*, Clarendon Press, Oxford, 1984.

**Miller, W., Jr.,** *Symmetry and Separation of Variables*, Addison-Wesley, London, 1977.

**Polyanin, A. D.,** *Handbook of Linear Partial Differential Equations for Engineers and Scientists*, Chapman & Hall/CRC Press, Boca Raton, 2002.

**Sutton, W. G. L.,** On the equation of diffusion in a turbulent medium, *Proc. Roy. Soc., Ser. A*, Vol. 138, No. 988, pp. 48–75, 1943.

**Tikhonov, A. N. and Samarskii, A. A.,** *Equations of Mathematical Physics*, Dover Publications, New York, 1990.

# Chapter T9
# Nonlinear Mathematical Physics Equations

## T9.1. Parabolic Equations

### T9.1.1. Nonlinear Heat Equations of the Form $\dfrac{\partial w}{\partial t} = \dfrac{\partial^2 w}{\partial x^2} + f(w)$

▶ *Equations of this form admit traveling-wave solutions $w = w(z)$, $z = kx + \lambda t$, where $k$ and $\lambda$ are arbitrary constants, and the function $w(z)$ is determined by the second-order autonomous ordinary differential equation $ak^2 w''_{zz} - \lambda w'_z + f(w) = 0$.*

1. $\dfrac{\partial w}{\partial t} = \dfrac{\partial^2 w}{\partial x^2} + aw(1 - w).$

*Fisher's equation.* This equation arises in heat and mass transfer, biology, and ecology.

Traveling-wave solutions ($C$ is an arbitrary constant):

$$w(x, t) = \left[1 + C \exp\left(-\tfrac{5}{6}at \pm \tfrac{1}{6}\sqrt{6a}\, x\right)\right]^{-2},$$

$$w(x, t) = \dfrac{1 + 2C \exp\left(-\tfrac{5}{6}at \pm \tfrac{1}{6}\sqrt{-6a}\, x\right)}{\left[1 + C \exp\left(-\tfrac{5}{6}at \pm \tfrac{1}{6}\sqrt{-6a}\, x\right)\right]^{2}}.$$

2. $\dfrac{\partial w}{\partial t} = \dfrac{\partial^2 w}{\partial x^2} + aw - bw^3.$

$1^\circ$. Solutions with $a > 0$ and $b > 0$:

$$w(x,t) = \pm \sqrt{\dfrac{a}{b}} \, \dfrac{C_1 \exp\!\left(\tfrac{1}{2}\sqrt{2a}\, x\right) - C_2 \exp\!\left(-\tfrac{1}{2}\sqrt{2a}\, x\right)}{C_1 \exp\!\left(\tfrac{1}{2}\sqrt{2a}\, x\right) + C_2 \exp\!\left(-\tfrac{1}{2}\sqrt{2a}\, x\right) + C_3 \exp\!\left(-\tfrac{3}{2}at\right)},$$

$$w(x,t) = \pm \sqrt{\dfrac{a}{b}} \left[\dfrac{2C_1 \exp\!\left(\sqrt{2a}\, x\right) + C_2 \exp\!\left(\tfrac{1}{2}\sqrt{2a}\, x - \tfrac{3}{2}at\right)}{C_1 \exp\!\left(\sqrt{2a}\, x\right) + C_2 \exp\!\left(\tfrac{1}{2}\sqrt{2a}\, x - \tfrac{3}{2}at\right) + C_3} - 1\right],$$

where $C_1$, $C_2$, and $C_3$ are arbitrary constants.

$2^\circ$. Solutions with $a < 0$ and $b > 0$:

$$w(x,t) = \pm \sqrt{\dfrac{|a|}{b}} \, \dfrac{\sin\!\left(\tfrac{1}{2}\sqrt{2|a|}\, x + C_1\right)}{\cos\!\left(\tfrac{1}{2}\sqrt{2|a|}\, x + C_1\right) + C_2 \exp\!\left(-\tfrac{3}{2}at\right)}.$$

$3^\circ$. Solutions with $a = 0$ and $b > 0$:

$$w(x,t) = \pm \sqrt{\dfrac{2}{b}} \, \dfrac{2C_1 x + C_2}{C_1 x^2 + C_2 x + 6C_1 t + C_3}.$$

4°. Solution with $a = 0$ (generalizes the solution of Item 3°):

$$w(x, y) = xu(z), \quad z = t + \tfrac{1}{6}x^2,$$

where the function $u(z)$ is determined by the autonomous ordinary differential equation $u''_{zz} - 9bu^3 = 0$.

5°. For $a = 0$, there is a self-similar solution of the form

$$w(x, t) = t^{-1/2} f(\xi), \quad \xi = xt^{-1/2},$$

where the function $f(\xi)$ is determined by the ordinary differential equation $f''_{\xi\xi} + \tfrac{1}{2}\xi f'_\xi + \tfrac{1}{2} f - b f^3 = 0$.

3. $\dfrac{\partial w}{\partial t} = \dfrac{\partial^2 w}{\partial x^2} - w(1-w)(a-w).$

*Fitzhugh–Nagumo equation.* This equation arises in genetics, biology, and heat and mass transfer.

Solutions:

$$w(x, t) = \frac{A \exp(z_1) + aB \exp(z_2)}{A \exp(z_1) + B \exp(z_2) + C},$$

$$z_1 = \pm\tfrac{\sqrt{2}}{2} x + \left(\tfrac{1}{2} - a\right)t, \quad z_2 = \pm\tfrac{\sqrt{2}}{2} ax + a\left(\tfrac{1}{2}a - 1\right)t,$$

where $A$, $B$, and $C$ are arbitrary constants.

4. $\dfrac{\partial w}{\partial t} = \dfrac{\partial^2 w}{\partial x^2} + aw + bw^m.$

1°. Traveling-wave solutions (the signs are chosen arbitrarily):

$$w(x, t) = \left[\pm\beta + C \exp(\lambda t \pm \mu x)\right]^{\frac{2}{1-m}},$$

where $C$ is an arbitrary constant and $\beta = \sqrt{-\dfrac{b}{a}}$, $\lambda = \dfrac{a(1-m)(m+3)}{2(m+1)}$, $\mu = \sqrt{\dfrac{a(1-m)^2}{2(m+1)}}$.

2°. For $a = 0$, there is a self-similar solution of the form $w(x, t) = t^{1/(1-m)} U(z)$, where $z = xt^{-1/2}$.

5. $\dfrac{\partial w}{\partial t} = \dfrac{\partial^2 w}{\partial x^2} + a + be^{\lambda w}.$

Traveling-wave solutions (the signs are chosen arbitrarily):

$$w(x, t) = -\frac{2}{\lambda} \ln\left[\pm\beta + C \exp\left(\pm\mu x - \tfrac{1}{2}a\lambda t\right)\right], \quad \beta = \sqrt{-\dfrac{b}{a}}, \quad \mu = \sqrt{\dfrac{a\lambda}{2}},$$

where $C$ is an arbitrary constant.

**6.** $\dfrac{\partial w}{\partial t} = \dfrac{\partial^2 w}{\partial x^2} + aw \ln w.$

Functional separable solutions:

$$w(x,t) = \exp\left(Ae^{at}x + \dfrac{A^2}{a}e^{2at} + Be^{at}\right),$$

$$w(x,t) = \exp\left[\tfrac{1}{2} - \tfrac{1}{4}a(x+A)^2 + Be^{at}\right],$$

$$w(x,t) = \exp\left[-\dfrac{a(x+A)^2}{4(1+Be^{-at})} + \dfrac{1}{2B}e^{at}\ln(1+Be^{-at}) + Ce^{at}\right],$$

where $A$, $B$, and $C$ are arbitrary constants.

## T9.1.2. Equations of the Form $\dfrac{\partial w}{\partial t} = \dfrac{\partial}{\partial x}\left[f(w)\dfrac{\partial w}{\partial x}\right] + g(w)$

▶ *Equations of this form admit traveling-wave solutions* $w = w(z)$, $z = kx + \lambda t$, *where $k$ and $\lambda$ are arbitrary constants and the function $w(z)$ is determined by the second-order autonomous ordinary differential equation* $k^2[f(w)w'_z]'_z - \lambda w'_z + f(w) = 0.$

**1.** $\dfrac{\partial w}{\partial t} = a\dfrac{\partial}{\partial x}\left(w^m \dfrac{\partial w}{\partial x}\right).$

This equation occurs in nonlinear problems of heat and mass transfer and flows in porous media.

1°. Solutions:

$$w(x,t) = (\pm kx + k\lambda t + A)^{1/m}, \quad k = \lambda m/a,$$

$$w(x,t) = \left[\dfrac{m(x-A)^2}{2a(m+2)(B-t)}\right]^{\frac{1}{m}},$$

$$w(x,t) = \left[A|t+B|^{-\frac{m}{m+2}} - \dfrac{m}{2a(m+2)}\dfrac{(x+C)^2}{t+B}\right]^{\frac{1}{m}},$$

$$w(x,t) = \left[\dfrac{m(x+A)^2}{\varphi(t)} + B|x+A|^{\frac{m}{m+1}}|\varphi(t)|^{-\frac{m(2m+3)}{2(m+1)^2}}\right]^{\frac{1}{m}}, \quad \varphi(t) = C - 2a(m+2)t,$$

where $A$, $B$, $C$, and $\lambda$ are arbitrary constants. The second solution for $B > 0$ corresponds to *blow-up regime* (the solution increases without bound on a finite time interval).

2°. There are solutions of the following forms:

$w(x,t) = (t+C)^{-1/m}F(x)$     (multiplicative separable solution);

$w(x,t) = t^\lambda G(\xi), \quad \xi = xt^{-\frac{m\lambda+1}{2}}$     (self-similar solution);

$w(x,t) = e^{-2\lambda t}H(\eta), \quad \eta = xe^{\lambda m t}$     (generalized self-similar solution);

$w(x,t) = (t+C)^{-1/m}U(\zeta), \quad \zeta = x + \lambda \ln(t+C),$

where $C$ and $\lambda$ are arbitrary constants.

2. $\dfrac{\partial w}{\partial t} = a\dfrac{\partial}{\partial x}\left(w^m \dfrac{\partial w}{\partial x}\right) + bw.$

By the transformation $w(x,t) = e^{bt}v(x,\tau)$, $\tau = \dfrac{1}{bm}e^{bmt} + C$ the original equation can be reduced to an equation of the form T9.1.2.1:

$$\dfrac{\partial v}{\partial \tau} = a\dfrac{\partial}{\partial x}\left(v^m \dfrac{\partial v}{\partial x}\right).$$

3. $\dfrac{\partial w}{\partial t} = a\dfrac{\partial}{\partial x}\left(w^m \dfrac{\partial w}{\partial x}\right) + bw^{m+1}.$

$1^\circ$. Multiplicative separable solution ($a = b = 1$, $m > 0$):

$$w(x,t) = \begin{cases} \left[\dfrac{2(m+1)}{m(m+2)}\dfrac{\cos^2(\pi x/L)}{(t_0 - t)}\right]^{1/m} & \text{for } |x| \le \dfrac{L}{2}, \\ 0 & \text{for } |x| > \dfrac{L}{2}, \end{cases}$$

where $L = 2\pi(m+1)^{1/2}/m$. This solution describes a blow-up regime that exists on a limited time interval $t \in [0, t_0)$. The solution is localized in the interval $|x| < L/2$.

$2^\circ$. Multiplicative separable solution:

$$w(x,t) = \left(\dfrac{Ae^{\mu x} + Be^{-\mu x} + D}{m\lambda t + C}\right)^{1/m},$$

$$B = \dfrac{\lambda^2(m+1)^2}{4b^2 A(m+2)^2}, \quad D = -\dfrac{\lambda(m+1)}{b(m+2)}, \quad \mu = m\sqrt{-\dfrac{b}{a(m+1)}},$$

where $A$, $C$, and $\lambda$ are arbitrary constants, $ab(m+1) < 0$.

$3^\circ$. Functional separable solutions [it is assumed that $ab(m+1) < 0$]:

$$w(x,t) = \left[F(t) + C_2|F(t)|^{\frac{m+2}{m+1}}e^{\lambda x}\right]^{1/m}, \quad F(t) = \dfrac{1}{C_1 - bmt}, \quad \lambda = \pm m\sqrt{\dfrac{-b}{a(m+1)}},$$

where $C_1$ and $C_2$ are arbitrary constants.

$4^\circ$. There are functional separable solutions of the following forms:

$$w(x,t) = \left[f(t) + g(t)(Ae^{\lambda x} + Be^{-\lambda x})\right]^{1/m}, \quad \lambda = m\sqrt{\dfrac{-b}{a(m+1)}};$$

$$w(x,t) = \left[f(t) + g(t)\cos(\lambda x + C)\right]^{1/m}, \quad \lambda = m\sqrt{\dfrac{b}{a(m+1)}},$$

where $A$, $B$, and $C$ are arbitrary constants.

**4.** $\dfrac{\partial w}{\partial t} = a\dfrac{\partial}{\partial x}\left(w^m \dfrac{\partial w}{\partial x}\right) + bw^{1-m}.$

Functional separable solution:
$$w(x,t) = \left[\dfrac{(x+A)^2}{F(t)} + B|F(t)|^{-\frac{m}{m+2}} - \dfrac{bm^2}{4a(m+1)}F(t)\right]^{1/m}, \quad F(t) = C - \dfrac{2a(m+2)}{m}t,$$
where $A$, $B$, and $C$ are arbitrary constants.

**5.** $\dfrac{\partial w}{\partial t} = a\dfrac{\partial}{\partial x}\left(w^{2n} \dfrac{\partial w}{\partial x}\right) + bw^{1-n}.$

Generalized traveling-wave solution:
$$w(x,t) = \left[\pm\dfrac{x+C_1}{\sqrt{C_2 - kt}} - \dfrac{bn^2}{3a(n+1)}(C_2 - kt)\right]^{1/n}, \quad k = \dfrac{2a(n+1)}{n},$$
where $C_1$ and $C_2$ are arbitrary constants.

**6.** $\dfrac{\partial w}{\partial t} = a\dfrac{\partial}{\partial x}\left(e^{\lambda w} \dfrac{\partial w}{\partial x}\right).$

$1°$. Solutions:
$$w(x,t) = \dfrac{2}{\lambda}\ln\left(\dfrac{\pm x + A}{\sqrt{B - 2at}}\right),$$
$$w(x,t) = \dfrac{1}{\lambda}\ln\dfrac{A + Bx - Cx^2}{D + 2aCt},$$
where $A$, $B$, $C$, and $D$ are arbitrary constants.

$2°$. There are solutions of the following forms:
$$w(x,t) = F(z), \quad z = kx + \beta t \qquad \text{(traveling-wave solution);}$$
$$w(x,t) = G(\xi), \quad \xi = xt^{-1/2} \qquad \text{(self-similar solution);}$$
$$w(x,t) = H(\eta) + 2kt, \quad \eta = xe^{-k\lambda t};$$
$$w(x,t) = U(\zeta) - \lambda^{-1}\ln t, \quad \zeta = x + k\ln t,$$
where $k$ and $\beta$ are arbitrary constants.

**7.** $\dfrac{\partial w}{\partial t} = \dfrac{\partial}{\partial x}\left[f(w)\dfrac{\partial w}{\partial x}\right].$

This equation occurs in nonlinear problems of heat and mass transfer and flows in porous media.

$1°$. Traveling-wave solution in implicit form:
$$k^2 \int \dfrac{f(w)\,dw}{\lambda w + C_1} = kx + \lambda t + C_2,$$
where $C_1$, $C_2$, $k$, and $\lambda$ are arbitrary constants. To $\lambda = 0$ there corresponds a stationary solution.

$2°$. Self-similar solution:
$$w = w(z), \quad z = xt^{-1/2},$$
where the function $w(z)$ is determined by the ordinary differential equation $[f(w)w'_z]'_z + \tfrac{1}{2}zw'_z = 0$.

8. $\dfrac{\partial w}{\partial t} = \dfrac{\partial}{\partial x}\left[f(w)\dfrac{\partial w}{\partial x}\right] + g(w).$

This equation occurs in nonlinear problems of heat and mass transfer with volume reaction.

$1°$. Traveling-wave solutions:

$$w = w(z), \quad z = \pm x + \lambda t,$$

where the function $w(z)$ is determined by the autonomous ordinary differential equation $[f(w)w'_z]'_z - \lambda w'_z + g(w) = 0$.

$2°$. Let the function $f = f(w)$ be arbitrary and let $g = g(w)$ be defined by

$$g(w) = \dfrac{A}{f(w)} + B,$$

where $A$ and $B$ are some numbers. In this case, there is a functional separable solution that is defined implicitly by

$$\int f(w)\,dw = At - \dfrac{1}{2}Bx^2 + C_1 x + C_2,$$

where $C_1$ and $C_2$ are arbitrary constants.

$3°$. Now let $g = g(w)$ be arbitrary and let $f = f(w)$ be defined by

$$f(w) = \dfrac{A_1 A_2 w + B}{g(w)} + \dfrac{A_2 A_3}{g(w)}\int Z\,dw, \qquad (1)$$

$$Z = -A_2 \int \dfrac{dw}{g(w)}, \qquad (2)$$

where $A_1$, $A_2$, and $A_3$ are some numbers. Then there are generalized traveling-wave solutions of the form

$$w = w(Z), \quad Z = \dfrac{\pm x + C_2}{\sqrt{2A_3 t + C_1}} - \dfrac{A_1}{A_3} - \dfrac{A_2}{3A_3}(2A_3 t + C_1),$$

where the function $w(Z)$ is determined by the inversion of (2), and $C_1$ and $C_2$ are arbitrary constants.

$4°$. Let $g = g(w)$ be arbitrary and let $f = f(w)$ be defined by

$$f(w) = \dfrac{1}{g(w)}\left(A_1 w + A_3 \int Z\,dw\right)\exp\left[-A_4 \int \dfrac{dw}{g(w)}\right], \qquad (3)$$

$$Z = \dfrac{1}{A_4}\exp\left[-A_4 \int \dfrac{dw}{g(w)}\right] - \dfrac{A_2}{A_4}, \qquad (4)$$

where $A_1$, $A_2$, $A_3$, and $A_4$ are some numbers ($A_4 \ne 0$). In this case, there are generalized traveling-wave solutions of the form

$$w = w(Z), \quad Z = \varphi(t)x + \psi(t),$$

where the function $w(Z)$ is determined by the inversion of (4),

$$\varphi(t) = \pm\left(C_1 e^{2A_4 t} - \dfrac{A_3}{A_4}\right)^{-1/2}, \quad \psi(t) = -\varphi(t)\left[A_1 \int \varphi(t)\,dt + A_2 \int \dfrac{dt}{\varphi(t)} + C_2\right],$$

and $C_1$ and $C_2$ are arbitrary constants.

5°. Let the functions $f(w)$ and $g(w)$ be as follows:
$$f(w) = \varphi'(w), \qquad g(w) = \frac{a\varphi(w)+b}{\varphi'(w)} + c[a\varphi(w)+b],$$

where $\varphi(w)$ is an arbitrary function and $a$, $b$, and $c$ are any numbers (the prime denotes a derivative with respect to $w$). Then there are functional separable solutions defined implicitly by

$$\varphi(w) = e^{at}\big[C_1 \cos(x\sqrt{ac}) + C_2 \sin(x\sqrt{ac})\big] - \frac{b}{a} \qquad \text{if } ac > 0,$$
$$\varphi(w) = e^{at}\big[C_1 \cosh(x\sqrt{-ac}) + C_2 \sinh(x\sqrt{-ac})\big] - \frac{b}{a} \qquad \text{if } ac < 0.$$

6°. Let $f(w)$ and $g(w)$ be as follows:
$$f(w) = w\varphi'_w(w), \qquad g(w) = a\left[w + 2\frac{\varphi(w)}{\varphi'_w(w)}\right],$$

where $\varphi(w)$ is an arbitrary function and $a$ is any number. Then there are functional separable solutions defined implicitly by

$$\varphi(w) = C_1 e^{2at} - \tfrac{1}{2}a(x+C_2)^2.$$

7°. Let $f(w)$ and $g(w)$ be as follows:
$$f(w) = A\frac{V(z)}{V'_z(z)}, \qquad g(w) = B\big[2z^{-1/2}V'_z(z) + z^{-3/2}V(z)\big],$$

where $V(z)$ is an arbitrary function, $A$ and $B$ are arbitrary constants ($AB \neq 0$), and the function $z = z(w)$ is defined implicitly by

$$w = \int z^{-1/2} V'_z(z)\, dz + C_1; \qquad (5)$$

$C_1$ is an arbitrary constant. Then there are functional separable solutions of the form (5), where
$$z = -\frac{(x+C_3)^2}{4At+C_2} + 2Bt + \frac{BC_2}{2A},$$

and $C_2$ and $C_3$ are arbitrary constants.

### T9.1.3. Burgers Equation and Nonlinear Heat Equation in Radial Symmetric Cases

1. $\dfrac{\partial w}{\partial t} = \dfrac{\partial^2 w}{\partial x^2} + w\dfrac{\partial w}{\partial x}.$

*Burgers equation.* It is used for describing wave processes in acoustics and hydrodynamics.

1°. Solutions:

$$w(x,t) = \lambda + \frac{2}{x + \lambda t + A},$$

$$w(x,t) = \frac{4x + 2A}{x^2 + Ax + 2t + B},$$

$$w(x,t) = \frac{6(x^2 + 2t + A)}{x^3 + 6xt + 3Ax + B},$$

$$w(x,t) = \frac{2\lambda}{1 + A\exp(-\lambda^2 t - \lambda x)},$$

$$w(x,t) = -\lambda + A\frac{\exp[A(x - \lambda t)] - B}{\exp[A(x - \lambda t)] + B},$$

where $A$, $B$, and $\lambda$ are arbitrary constants.

2°. Other solutions can be obtained using the following formula (*Hopf–Cole transformation*):

$$w(x,t) = \frac{2}{u}\frac{\partial u}{\partial x},$$

where $u = u(x,t)$ is a solution of the linear heat equation, $u_t = u_{xx}$ (see Subsection T8.1.1).

3°. The Cauchy problem with the initial condition:

$$w = f(x) \quad \text{at} \quad t = 0 \quad (-\infty < x < \infty).$$

Solution:

$$w(x,t) = 2\frac{\partial}{\partial x}\ln F(x,t), \quad F(x,t) = \frac{1}{\sqrt{4\pi t}}\int_{-\infty}^{\infty}\exp\left[-\frac{(x-\xi)^2}{4t} - \frac{1}{2}\int_0^\xi f(\xi')\,d\xi'\right]d\xi.$$

2. $$\frac{\partial w}{\partial t} = \frac{1}{x^n}\frac{\partial}{\partial x}\left[x^n f(w)\frac{\partial w}{\partial x}\right] + g(w).$$

This is a nonlinear equation of heat and mass transfer in the radial symmetric case ($n = 1$ corresponds to a plane problem and $n = 2$ to a spatial one).

1°. Let the function $f = f(w)$ be arbitrary and let $g = g(w)$ be defined by

$$g(w) = \left(\frac{a}{f(w)} + b\right)\left(\int f(w)\,dw + c\right),$$

where $a$, $b$, and $c$ are some numbers. In this case, there is a functional separable solution that is defined implicitly by

$$\int f(w)\,dw + c = e^{at}z(x),$$

where the function $z = z(x)$ is determined by the linear ordinary differential equation

$$z''_{xx} + \frac{n}{x}z'_x + bz = 0.$$

Its general solution can be expressed in terms of Bessel functions of modified Bessel functions.

2°. Let $f(w)$ and $g(w)$ be defined by

$$f(w) = w\varphi'_w(w), \quad g(w) = a(n+1)w + 2a\frac{\varphi(w)}{\varphi'_w(w)},$$

where $\varphi(w)$ is an arbitrary function (the prime denotes a derivative with respect to $w$). In this case, there is a functional separable solution defined implicitly by

$$\varphi(w) = Ce^{2at} - \tfrac{1}{2}ax^2,$$

where $C$ is an arbitrary constant.

3°. Let $f(w)$ and $g(w)$ be defined as follows:

$$f(w) = a\varphi^{-\frac{n+1}{2}}\varphi'\int \varphi^{\frac{n+1}{2}}\,dw, \quad g(w) = b\frac{\varphi}{\varphi'},$$

where $\varphi = \varphi(w)$ is an arbitrary function (the prime denotes a derivative with respect to $w$). In this case, there is a functional separable solution defined implicitly by

$$\varphi(w) = \frac{bx^2}{Ce^{-bt} - 4a}.$$

4°. Let $f(w)$ and $g(w)$ be as follows:

$$f(w) = A\frac{V(z)}{V'_z(z)}, \quad g(w) = B\left[2z^{-\frac{n+1}{2}}V'_z(z) + (n+1)z^{-\frac{n+3}{2}}V(z)\right],$$

where $V(z)$ is an arbitrary function, $A$ and $B$ are arbitrary constants ($AB \neq 0$), and the function $z = z(w)$ is defined implicitly by

$$w = \int z^{-\frac{n+1}{2}} V'_z(z)\,dz + C_1; \qquad (*)$$

$C_1$ is an arbitrary constant. Then there are functional separable solutions of the form $(*)$ where

$$z = -\frac{x^2}{4At + C_2} + 2Bt + \frac{BC_2}{2A},$$

and $C_2$ is an arbitrary constant.

## T9.1.4. Nonlinear Schrödinger Equations

▶ *In equations T9.1.4.1–T9.1.4.3, $w$ is a complex function of real variables $x$ and $t$; $i^2 = -1$.*

1.  $i\dfrac{\partial w}{\partial t} + \dfrac{\partial^2 w}{\partial x^2} + k|w|^2 w = 0.$

*Schrödinger equation with a cubic nonlinearity.* Here, $k$ is a real number. This equation occurs in various chapters of physics, including nonlinear optics, superconductivity, and plasma physics.

1°. Solutions:
$$w(x,t) = C_1 \exp\{i\,[C_2 x + (kC_1^2 - C_2^2)t + C_3]\},$$
$$w(x,t) = \pm C_1 \sqrt{\frac{2}{k}}\,\frac{\exp[i(C_1^2 t + C_2)]}{\cosh(C_1 x + C_3)},$$
$$w(x,t) = \pm A \sqrt{\frac{2}{k}}\,\frac{\exp[iBx + i(A^2 - B^2)t + iC_1]}{\cosh(Ax - 2ABt + C_2)},$$
$$w(x,t) = \frac{C_1}{\sqrt{t}} \exp\left[i\frac{(x+C_2)^2}{4t} + i(kC_1^2 \ln t + C_3)\right],$$

where $A$, $B$, $C_1$, $C_2$, and $C_3$ are arbitrary real constants. The second and third solutions are valid for $k > 0$. The third solution describes the motion of a soliton in a rapidly decaying case.

2°. $N$-soliton solutions for $k > 0$:
$$w(x,t) = \sqrt{\frac{2}{k}}\,\frac{\det \mathbf{R}(x,t)}{\det \mathbf{M}(x,t)}.$$

Here, $\mathbf{M}(x,t)$ is an $N \times N$ matrix with entries
$$M_{n,k}(x,t) = \frac{1 + \overline{g}_n(x,t)g_n(x,t)}{\overline{\lambda}_n - \lambda_k}, \qquad g_n(x,t) = \gamma_n e^{i(\lambda_n x - \lambda_n^2 t)}, \qquad n,k = 1, \ldots, N,$$

where $\lambda_n$ and $\gamma_n$ are arbitrary complex numbers that satisfy the constraints $\operatorname{Im}\lambda_n > 0$ ($\lambda_n \neq \lambda_k$ if $n \neq k$) and $\gamma_n \neq 0$; the bar over a symbol denotes the complex conjugate. The square matrix $\mathbf{R}(x,t)$ is of order $N + 1$; it is obtained by augmenting $\mathbf{M}(x,t)$ with a column on the right and a row at the bottom. The entries of $\mathbf{R}$ are defined as

$$\begin{aligned}
R_{n,k}(x,t) &= M_{n,k}(x,t) &&\text{for}\quad n,k = 1, \ldots, N &&\text{(bulk of the matrix)},\\
R_{n,N+1}(x,t) &= g_n(x,t) &&\text{for}\quad n = 1, \ldots, N &&\text{(rightmost column)},\\
R_{N+1,n}(x,t) &= 1 &&\text{for}\quad n = 1, \ldots, N &&\text{(bottom row)},\\
R_{N+1,N+1}(x,t) &= 0 &&&&\text{(lower right diagonal entry)}.
\end{aligned}$$

The above solution can be represented, for $t \to \pm\infty$, as the sum of $N$ single-soliton solutions.

3°. For other exact solutions, see equation T9.1.4.18 with $n = 1$ and equation T9.1.4.19 with $f(u) = ku^2$.

4°. The Schrödinger equation with a cubic nonlinearity is integrable by the inverse scattering method; see Ablowitz and Segur (1981) and Novikov et al. (1984).

**2.** $i\dfrac{\partial w}{\partial t} + \dfrac{\partial^2 w}{\partial x^2} + A|w|^{2n}w = 0.$

*Schrödinger equation with a power-law nonlinearity.* The numbers $A$ and $n$ are assumed real.

1°. Solutions:

$$w(x,t) = C_1 \exp\{i\,[C_2 x + (A|C_1|^{2n} - C_2^2)t + C_3]\},$$

$$w(x,t) = \pm\left[\frac{(n+1)C_1^2}{A\cosh^2(C_1 n x + C_2)}\right]^{\frac{1}{2n}} \exp[i(C_1^2 t + C_3)],$$

$$w(x,t) = \frac{C_1}{\sqrt{t}}\exp\left[i\frac{(x+C_2)^2}{4t} + i\left(\frac{AC_1^{2n}}{1-n}t^{1-n} + C_3\right)\right],$$

where $C_1$, $C_2$, and $C_3$ are arbitrary real constants.

2°. There is a self-similar solution of the form $w = t^{-1/(2n)}u(\xi)$, where $\xi = xt^{-1/2}$.

3°. For other exact solutions, see equation T9.1.4.19 with $f(u) = Au^{2n}$.

**3.** $i\dfrac{\partial w}{\partial t} + \dfrac{\partial^2 w}{\partial x^2} + f(|w|)w = 0.$

*Schrödinger equation of general form; $f(u)$ is a real function of a real variable.*

1°. Suppose $w(x,t)$ is a solution of the Schrödinger equation in question. Then the function

$$w_1 = e^{-i(\lambda x + \lambda^2 t + C_1)} w(x + 2\lambda t + C_2, t + C_3),$$

where $C_1$, $C_2$, $C_3$, and $\lambda$ are arbitrary real constants, is also a solution of the equation.

2°. Traveling-wave solution:

$$w(x,t) = C_1 \exp[i\varphi(x,t)], \quad \varphi(x,t) = C_2 x - C_2^2 t + f(|C_1|)t + C_3.$$

3°. Multiplicative separable solution:

$$w(x,t) = u(x)e^{i(C_1 t + C_2)},$$

where the function $u = u(x)$ is defined implicitly by

$$\int \frac{du}{\sqrt{C_1 u^2 - 2F(u) + C_3}} = C_4 \pm x, \quad F(u) = \int u f(|u|)\,du.$$

Here, $C_1, \ldots, C_4$ are arbitrary real constants.

4°. Solution:

$$w(x,t) = U(\xi)e^{i(Ax+Bt+C)}, \quad \xi = x - 2At, \tag{1}$$

where the function $U = U(\xi)$ is determined by the autonomous ordinary differential equation $U''_{\xi\xi} + f(|U|)U - (A^2 + B)U = 0$. Integrating yields the general solution in implicit form:

$$\int \frac{dU}{\sqrt{(A^2+B)U^2 - 2F(U) + C_1}} = C_2 \pm \xi, \quad F(U) = \int U f(|U|)\,dU. \tag{2}$$

Relations (1) and (2) involve arbitrary real constants $A$, $B$, $C$, $C_1$, and $C_2$.

5°. Solution ($A$, $B$, and $C$ are arbitrary constants):

$$w(x,t) = \psi(z)\exp\left[i(Axt - \tfrac{2}{3}A^2 t^3 + Bt + C)\right], \quad z = x - At^2,$$

where the function $\psi = \psi(z)$ is determined by the ordinary differential equation $\psi''_{zz} + f(|\psi|)\psi - (Az+B)\psi = 0$.

6°. Solutions:

$$w(x,t) = \pm\frac{1}{\sqrt{C_1 t}}\exp\left[i\varphi(x,t)\right], \quad \varphi(x,t) = \frac{(x+C_2)^2}{4t} + \int f\left(|C_1 t|^{-1/2}\right)dt + C_3,$$

where $C_1$, $C_2$, and $C_3$ are arbitrary real constants.

7°. Solution:

$$w(x,t) = u(x)\exp\left[i\varphi(x,t)\right], \quad \varphi(x,t) = C_1 t + C_2\int\frac{dx}{u^2(x)} + C_3,$$

where $C_1$, $C_2$, and $C_3$ are arbitrary real constants, and the function $u = u(x)$ is determined by the autonomous ordinary differential equation $u''_{xx} - C_1 u - C_2^2 u^{-3} + f(|u|)u = 0$.

8°. There is an exact solution of the form

$$w(x,t) = u(z)\exp\left[iAt + i\varphi(z)\right], \quad z = kx + \lambda t,$$

where $A$, $k$, and $\lambda$ are arbitrary real constants.

## T9.2. Hyperbolic Equations

### T9.2.1. Nonlinear Wave Equations of the Form $\dfrac{\partial^2 w}{\partial t^2} = a\dfrac{\partial^2 w}{\partial x^2} + f(w)$

1. $\dfrac{\partial^2 w}{\partial t^2} = \dfrac{\partial^2 w}{\partial x^2} + aw + bw^n$.

1°. Traveling-wave solutions for $a > 0$:

$$w(x,t) = \left[\frac{2b\sinh^2 z}{a(n+1)}\right]^{\frac{1}{1-n}}, \quad z = \tfrac{1}{2}\sqrt{a}\,(1-n)(x\sinh C_1 \pm t\cosh C_1) + C_2 \quad \text{if } b(n+1) > 0,$$

$$w(x,t) = \left[-\frac{2b\cosh^2 z}{a(n+1)}\right]^{\frac{1}{1-n}}, \quad z = \tfrac{1}{2}\sqrt{a}\,(1-n)(x\sinh C_1 \pm t\cosh C_1) + C_2 \quad \text{if } b(n+1) < 0,$$

where $C_1$ and $C_2$ are arbitrary constants.

2°. Traveling-wave solutions for $a < 0$ and $b(n+1) > 0$:

$$w(x,t) = \left[-\frac{2b\cos^2 z}{a(n+1)}\right]^{\frac{1}{1-n}}, \quad z = \tfrac{1}{2}\sqrt{|a|}\,(1-n)(x\sinh C_1 \pm t\cosh C_1) + C_2.$$

3°. For $a = 0$, there is a self-similar solution of the form $w = t^{\frac{2}{1-n}} F(z)$, where $z = x/t$.

4°. For other exact solutions of this equation, see equation T9.2.1.7 with $f(w) = aw + bw^n$.

**2.** $\dfrac{\partial^2 w}{\partial t^2} = \dfrac{\partial^2 w}{\partial x^2} + aw^n + bw^{2n-1}$.

Solutions:
$$w(x,t) = \left[\frac{a(1-n)^2}{2(n+1)}(x\sinh C_1 \pm t\cosh C_1 + C_2)^2 - \frac{b(n+1)}{2an}\right]^{\frac{1}{1-n}},$$

$$w(x,t) = \left\{\frac{1}{4}a(1-n)^2\left[(t+C_1)^2 - (x+C_2)^2\right] - \frac{b}{an}\right\}^{\frac{1}{1-n}},$$

where $C_1$ and $C_2$ are arbitrary constants.

**3.** $\dfrac{\partial^2 w}{\partial t^2} = a^2 \dfrac{\partial^2 w}{\partial x^2} + be^{\beta w}$.

1°. Traveling-wave solutions:
$$w(x,t) = \frac{1}{\beta}\ln\left[\frac{2(B^2 - a^2 A^2)}{b\beta(Ax + Bt + C)^2}\right],$$

$$w(x,t) = \frac{1}{\beta}\ln\left[\frac{2(a^2 A^2 - B^2)}{b\beta \cosh^2(Ax + Bt + C)}\right],$$

$$w(x,t) = \frac{1}{\beta}\ln\left[\frac{2(B^2 - a^2 A^2)}{b\beta \sinh^2(Ax + Bt + C)}\right],$$

$$w(x,t) = \frac{1}{\beta}\ln\left[\frac{2(B^2 - a^2 A^2)}{b\beta \cos^2(Ax + Bt + C)}\right],$$

where $A$, $B$, and $C$ are arbitrary constants.

2°. Functional separable solutions:
$$w(x,t) = \frac{1}{\beta}\ln\left(\frac{8a^2 C}{b\beta}\right) - \frac{2}{\beta}\ln\left|(x+A)^2 - a^2(t+B)^2 + C\right|,$$

$$w(x,t) = -\frac{2}{\beta}\ln\left[C_1 e^{\lambda x} \pm \frac{\sqrt{2b\beta}}{2a\lambda}\sinh(a\lambda t + C_2)\right],$$

$$w(x,t) = -\frac{2}{\beta}\ln\left[C_1 e^{\lambda x} \pm \frac{\sqrt{-2b\beta}}{2a\lambda}\cosh(a\lambda t + C_2)\right],$$

$$w(x,t) = -\frac{2}{\beta}\ln\left[C_1 e^{a\lambda t} \pm \frac{\sqrt{-2b\beta}}{2a\lambda}\sinh(\lambda x + C_2)\right],$$

$$w(x,t) = -\frac{2}{\beta}\ln\left[C_1 e^{a\lambda t} \pm \frac{\sqrt{2b\beta}}{2a\lambda}\cosh(\lambda x + C_2)\right],$$

where $A$, $B$, $C$, $C_1$, $C_2$, and $\lambda$ are arbitrary constants.

3°. General solution:
$$w(x,t) = \frac{1}{\beta}\left[f(z) + g(y)\right] - \frac{2}{\beta}\ln\left|k\int \exp[f(z)]\,dz - \frac{b\beta}{8a^2 k}\int \exp[g(y)]\,dy\right|,$$
$$z = x - at, \qquad y = x + at,$$

where $f = f(z)$ and $g = g(y)$ are arbitrary functions and $k$ is an arbitrary constant.

4. $\dfrac{\partial^2 w}{\partial t^2} = \dfrac{\partial^2 w}{\partial x^2} + a e^{\beta w} + b e^{2\beta w}.$

1°. Traveling-wave solutions:

$$w(x,t) = -\frac{1}{\beta} \ln\left[\frac{a\beta}{C_1^2 - C_2^2} + C_3 \exp(C_1 x + C_2 t) + \frac{a^2\beta^2 + b\beta(C_1^2 - C_2^2)}{4C_3(C_1^2 - C_2^2)^2} \exp(-C_1 x - C_2 t)\right],$$

$$w(x,t) = -\frac{1}{\beta} \ln\left[\frac{a\beta}{C_2^2 - C_1^2} + \frac{\sqrt{a^2\beta^2 + b\beta(C_2^2 - C_1^2)}}{C_2^2 - C_1^2} \sin(C_1 x + C_2 t + C_3)\right],$$

where $C_1$, $C_2$, and $C_3$ are arbitrary constants.

2°. For other exact solutions of this equation, see equation T9.2.1.7 with $f(w) = ae^{\beta w} + be^{2\beta w}$.

5. $\dfrac{\partial^2 w}{\partial t^2} = a \dfrac{\partial^2 w}{\partial x^2} + b \sinh(\lambda w).$

*Sinh-Gordon equation.* It arises in some areas of physics.

1°. Traveling-wave solutions:

$$w(x,t) = \pm \frac{2}{\lambda} \ln\left[\tan \frac{b\lambda(kx + \mu t + \theta_0)}{2\sqrt{b\lambda(\mu^2 - ak^2)}}\right],$$

$$w(x,t) = \pm \frac{4}{\lambda} \operatorname{arctanh}\left[\exp \frac{b\lambda(kx + \mu t + \theta_0)}{\sqrt{b\lambda(\mu^2 - ak^2)}}\right],$$

where $k$, $\mu$, and $\theta_0$ are arbitrary constants. It is assumed that $b\lambda(\mu^2 - ak^2) > 0$ in both formulas.

2°. Functional separable solution:

$$w(x,t) = \frac{4}{\lambda} \operatorname{arctanh}\left[f(t)g(x)\right], \qquad \operatorname{arctanh} z = \frac{1}{2} \ln \frac{1+z}{1-z},$$

where the functions $f = f(t)$ and $g = g(x)$ are determined by the first-order autonomous ordinary differential equations

$$(f'_t)^2 = Af^4 + Bf^2 + C, \qquad a(g'_x)^2 = Cg^4 + (B - b\lambda)g^2 + A,$$

where $A$, $B$, and $C$ are arbitrary constants.

3°. For other exact solutions of this equation, see equation T9.2.1.7 with $f(w) = b \sinh(\lambda w)$.

6. $\dfrac{\partial^2 w}{\partial t^2} = a \dfrac{\partial^2 w}{\partial x^2} + b \sin(\lambda w).$

*Sine-Gordon equation.* It arises in differential geometry and various areas of physics.

1°. Traveling-wave solutions:

$$w(x,t) = \frac{4}{\lambda} \arctan\left\{\exp\left[\pm \frac{b\lambda(kx + \mu t + \theta_0)}{\sqrt{b\lambda(\mu^2 - ak^2)}}\right]\right\} \quad \text{if} \quad b\lambda(\mu^2 - ak^2) > 0,$$

$$w(x,t) = -\frac{\pi}{\lambda} + \frac{4}{\lambda} \arctan\left\{\exp\left[\pm \frac{b\lambda(kx + \mu t + \theta_0)}{\sqrt{b\lambda(ak^2 - \mu^2)}}\right]\right\} \quad \text{if} \quad b\lambda(\mu^2 - ak^2) < 0,$$

where $k$, $\mu$, and $\theta_0$ are arbitrary constants. The first expression corresponds to a single-soliton solution.

2°. Functional separable solutions:

$$w(x,t) = \frac{4}{\lambda} \arctan\left[\frac{\mu \sinh(kx+A)}{k\sqrt{a}\cosh(\mu t+B)}\right], \quad \mu^2 = ak^2 + b\lambda > 0;$$

$$w(x,t) = \frac{4}{\lambda} \arctan\left[\frac{\mu \sin(kx+A)}{k\sqrt{a}\cosh(\mu t+B)}\right], \quad \mu^2 = b\lambda - ak^2 > 0;$$

$$w(x,t) = \frac{4}{\lambda} \arctan\left[\frac{\gamma}{\mu}\frac{e^{\mu(t+A)}+ak^2 e^{-\mu(t+A)}}{e^{k\gamma(x+B)}+e^{-k\gamma(x+B)}}\right], \quad \mu^2 = ak^2\gamma^2 + b\lambda > 0,$$

where $A$, $B$, $k$, and $\gamma$ are arbitrary constants.

3°. An $N$-soliton solution is given by ($a=1$, $b=-1$, and $\lambda=1$)

$$w(x,t) = \arccos\left[1 - 2\left(\frac{\partial^2}{\partial x^2} - \frac{\partial^2}{\partial t^2}\right)(\ln F)\right],$$

$$F = \det[M_{ij}], \quad M_{ij} = \frac{2}{a_i+a_j}\cosh\left(\frac{z_i+z_j}{2}\right), \quad z_i = \pm\frac{x - \mu_i t + C_i}{\sqrt{1-\mu_i^2}}, \quad a_i = \pm\sqrt{\frac{1-\mu_i}{1+\mu_i}},$$

where $\mu_i$ and $C_i$ are arbitrary constants.

4°. For other exact solutions of the original equation, see equation T9.2.1.7 with $f(w) = b\sin(\lambda w)$.

5°. The sine-Gordon equation is integrated by the inverse scattering method; see the book by Novikov et al. (1984).

7. $\dfrac{\partial^2 w}{\partial t^2} = \dfrac{\partial^2 w}{\partial x^2} + f(w).$

*Nonlinear Klein–Gordon equation.*

1°. Suppose $w = w(x,t)$ is a solution of the equation in question. Then the functions

$$w_1 = w(\pm x + C_1, \pm t + C_2),$$
$$w_2 = w(x\cosh\beta + t\sinh\beta, t\cosh\beta + x\sinh\beta),$$

where $C_1$, $C_2$, and $\beta$ are arbitrary constants, are also solutions of the equation (the plus or minus signs in $w_1$ are chosen arbitrarily).

2°. Traveling-wave solution in implicit form:

$$\int\left[C_1 + \frac{2}{\lambda^2-k^2}\int f(w)\,dw\right]^{-1/2} dw = kx + \lambda t + C_2,$$

where $C_1$, $C_2$, $k$, and $\lambda$ are arbitrary constants.

3°. Functional separable solution:

$$w = w(\xi), \quad \xi = \tfrac{1}{4}(t+C_1)^2 - \tfrac{1}{4}(x+C_2)^2,$$

where $C_1$ and $C_2$ are arbitrary constants, and the function $w = w(\xi)$ is determined by the ordinary differential equation $\xi w''_{\xi\xi} + w'_\xi - f(w) = 0$.

## T9.2.2. Other Nonlinear Wave Equations

**1.** $\dfrac{\partial^2 w}{\partial t^2} = a\dfrac{\partial}{\partial x}\left(w\dfrac{\partial w}{\partial x}\right).$

1°. Solutions:
$$w(x,t) = \tfrac{1}{2}aA^2 t^2 + Bt + Ax + C,$$
$$w(x,t) = \tfrac{1}{12}aA^{-2}(At+B)^4 + Ct + D + x(At+B),$$
$$w(x,t) = \frac{1}{a}\left(\frac{x+A}{t+B}\right)^2,$$
$$w(x,t) = (At+B)\sqrt{Cx+D},$$
$$w(x,t) = \pm\sqrt{A(x+a\lambda t) + B} + a\lambda^2,$$

where $A, B, C, D$, and $\lambda$ are arbitrary constants.

2°. Generalized separable solution quadratic in $x$:
$$w(x,t) = \frac{1}{at^2}x^2 + \left(\frac{C_1}{t^2} + C_2 t^3\right)x + \frac{aC_1^2}{4t^2} + \frac{C_3}{t} + C_4 t^2 + \tfrac{1}{2}aC_1 C_2 t^3 + \tfrac{1}{54}aC_2^2 t^8,$$

where $C_1, \ldots, C_4$ are arbitrary constants.

3°. Solution:
$$w = U(z) + 4aC_1^2 t^2 + 4aC_1 C_2 t, \qquad z = x + aC_1 t^2 + aC_2 t,$$

where $C_1$ and $C_2$ are arbitrary constants and the function $U(z)$ is determined by the first-order ordinary differential equation $(U - aC_2^2)U'_z - 2C_1 U = 8C_1^2 z + C_3$.

4°. See also equation T9.2.2.5 with $f(w) = aw$.

**2.** $\dfrac{\partial^2 w}{\partial t^2} = a\dfrac{\partial}{\partial x}\left(w^n \dfrac{\partial w}{\partial x}\right) + bw^k.$

There are solutions of the following forms:
$$w(x,t) = U(z), \quad z = \lambda x + \beta t \qquad \text{(traveling-wave solution)};$$
$$w(x,t) = t^{\frac{2}{1-k}} V(\xi), \quad \xi = xt^{\frac{k-n-1}{1-k}} \qquad \text{(self-similar solution)}.$$

**3.** $\dfrac{\partial^2 w}{\partial t^2} = \dfrac{\partial}{\partial x}\left(ae^{\lambda w}\dfrac{\partial w}{\partial x}\right), \qquad a > 0.$

1°. Additive separable solutions:
$$w(x,t) = \frac{1}{\lambda}\ln|Ax+B| + Ct + D,$$
$$w(x,t) = \frac{2}{\lambda}\ln|Ax+B| - \frac{2}{\lambda}\ln|\pm A\sqrt{a}\,t + C|,$$
$$w(x,t) = \frac{1}{\lambda}\ln(aA^2 x^2 + Bx + C) - \frac{2}{\lambda}\ln(aAt+D),$$

$$w(x,t) = \frac{1}{\lambda}\ln(Ax^2+Bx+C) + \frac{1}{\lambda}\ln\left[\frac{p^2}{aA\cos^2(pt+q)}\right],$$

$$w(x,t) = \frac{1}{\lambda}\ln(Ax^2+Bx+C) + \frac{1}{\lambda}\ln\left[\frac{p^2}{aA\sinh^2(pt+q)}\right],$$

$$w(x,t) = \frac{1}{\lambda}\ln(Ax^2+Bx+C) + \frac{1}{\lambda}\ln\left[\frac{-p^2}{aA\cosh^2(pt+q)}\right],$$

where $A$, $B$, $C$, $D$, $p$, and $q$ are arbitrary constants.

2°. There are solutions of the following forms:

$$w(x,t) = F(z), \quad z = kx + \beta t \qquad \text{(traveling-wave solution);}$$
$$w(x,t) = G(\xi), \quad \xi = x/t \qquad \text{(self-similar solution);}$$
$$w(x,t) = H(\eta) + 2(k-1)\lambda^{-1}\ln t, \quad \eta = xt^{-k};$$
$$w(x,t) = U(\zeta) - 2\lambda^{-1}\ln|t|, \quad \zeta = x + k\ln|t|;$$
$$w(x,t) = V(\zeta) - 2\lambda^{-1}t, \quad \eta = xe^t,$$

where $k$ and $\beta$ are arbitrary constants.

**4.** $\dfrac{\partial^2 w}{\partial t^2} = \dfrac{a}{x^n}\dfrac{\partial}{\partial x}\left(x^n \dfrac{\partial w}{\partial x}\right) + f(w), \qquad a > 0.$

To $n = 1$ and $n = 2$ there correspond nonlinear waves with axial and central symmetry, respectively.

Functional separable solution:

$$w = w(\xi), \quad \xi = \sqrt{ak(t+C)^2 - kx^2},$$

where $w(\xi)$ is determined by the ordinary differential equation $w''_{\xi\xi} + (1+n)\xi^{-1}w'_\xi = (ak)^{-1}f(w)$.

**5.** $\dfrac{\partial^2 w}{\partial t^2} = \dfrac{\partial}{\partial x}\left[f(w)\dfrac{\partial w}{\partial x}\right].$

This equation is encountered in wave and gas dynamics.

1°. Traveling-wave solution in implicit form:

$$\lambda^2 w - \int f(w)\,dw = A(x + \lambda t) + B,$$

where $A$, $B$, and $\lambda$ are arbitrary constants.

2°. Self-similar solution:

$$w = w(\xi), \quad \xi = \frac{x+A}{t+B},$$

where the function $w(\xi)$ is determined by the ordinary differential equation $(\xi^2 w'_\xi)'_\xi = [f(w)w'_\xi]'_\xi$, which admits the first integral

$$[\xi^2 - f(w)]w'_\xi = C.$$

To the special case $C = 0$ there corresponds the solution in implicit form: $\xi^2 = f(w)$.

3°. The equation concerned can be reduced to a linear equation; see Item 3° of equation T9.3.3.3, where one should set $g(w) = -1$ and $y = t$.

## T9.3. Elliptic Equations

### T9.3.1. Nonlinear Heat Equations of the Form $\dfrac{\partial^2 w}{\partial x^2} + \dfrac{\partial^2 w}{\partial y^2} = f(w)$

**1.** $\dfrac{\partial^2 w}{\partial x^2} + \dfrac{\partial^2 w}{\partial y^2} = aw + bw^n.$

$1^\circ$. Traveling-wave solutions for $a > 0$:

$$w(x,y) = \left[\frac{2b\sinh^2 z}{a(n+1)}\right]^{\frac{1}{1-n}}, \quad z = \tfrac{1}{2}\sqrt{a}\,(1-n)(x\sin C_1 + y\cos C_1) + C_2 \quad \text{if} \quad b(n+1) > 0,$$

$$w(x,y) = \left[-\frac{2b\cosh^2 z}{a(n+1)}\right]^{\frac{1}{1-n}}, \quad z = \tfrac{1}{2}\sqrt{a}\,(1-n)(x\sin C_1 + y\cos C_1) + C_2 \quad \text{if} \quad b(n+1) < 0,$$

where $C_1$ and $C_2$ are arbitrary constants.

$2^\circ$. Traveling-wave solutions for $a < 0$ and $b(n+1) > 0$:

$$w(x,y) = \left[-\frac{2b\cos^2 z}{a(n+1)}\right]^{\frac{1}{1-n}}, \quad z = \tfrac{1}{2}\sqrt{|a|}\,(1-n)(x\sin C_1 + y\cos C_1) + C_2.$$

$3^\circ$. For $a = 0$, there is a self-similar solution of the form $w = x^{\frac{2}{1-n}} F(z)$, where $z = y/x$.

$4^\circ$. For other exact solutions of this equation, see equation T9.3.1.7 with $f(w) = aw + bw^n$.

**2.** $\dfrac{\partial^2 w}{\partial x^2} + \dfrac{\partial^2 w}{\partial y^2} = aw^n + bw^{2n-1}.$

Solutions:

$$w(x,y) = \left[\frac{a(1-n)^2}{2(n+1)}(x\sin C_1 + y\cos C_1 + C_2)^2 - \frac{b(n+1)}{2an}\right]^{\frac{1}{1-n}},$$

$$w(x,y) = \left\{\frac{1}{4}a(1-n)^2\left[(x+C_1)^2 + (y+C_2)^2\right] - \frac{b}{an}\right\}^{\frac{1}{1-n}},$$

where $C_1$ and $C_2$ are arbitrary constants.

**3.** $\dfrac{\partial^2 w}{\partial x^2} + \dfrac{\partial^2 w}{\partial y^2} = ae^{\beta w}.$

This equation occurs in combustion theory and is a special case of equation T9.3.1.7 with $f(w) = ae^{\beta w}$.

1°. Solutions:

$$w(x, y) = \frac{1}{\beta} \ln\left[\frac{2(A^2 + B^2)}{a\beta(Ax + By + C)^2}\right] \quad \text{if } a\beta > 0,$$

$$w(x, y) = \frac{1}{\beta} \ln\left[\frac{2(A^2 + B^2)}{a\beta \sinh^2(Ax + By + C)}\right] \quad \text{if } a\beta > 0,$$

$$w(x, y) = \frac{1}{\beta} \ln\left[\frac{-2(A^2 + B^2)}{a\beta \cosh^2(Ax + By + C)}\right] \quad \text{if } a\beta < 0,$$

$$w(x, y) = \frac{1}{\beta} \ln\left[\frac{2(A^2 + B^2)}{a\beta \cos^2(Ax + By + C)}\right] \quad \text{if } a\beta > 0,$$

$$w(x, y) = \frac{1}{\beta} \ln\left(\frac{8C}{a\beta}\right) - \frac{2}{\beta} \ln\big|(x + A)^2 + (y + B)^2 - C\big|,$$

where $A$, $B$, and $C$ are arbitrary constants. The first four solutions are of traveling-wave type and the last one is a radial symmetric solution with center at the point $(-A, -B)$.

2°. Functional separable solutions:

$$w(x, y) = -\frac{2}{\beta} \ln\left[C_1 e^{ky} \pm \frac{\sqrt{2a\beta}}{2k} \cos(kx + C_2)\right],$$

$$w(x, y) = \frac{1}{\beta} \ln \frac{2k^2(B^2 - A^2)}{a\beta[A\cosh(kx + C_1) + B\sin(ky + C_2)]^2},$$

$$w(x, y) = \frac{1}{\beta} \ln \frac{2k^2(A^2 + B^2)}{a\beta[A\sinh(kx + C_1) + B\cos(ky + C_2)]^2},$$

where $A$, $B$, $C_1$, $C_2$, and $k$ are arbitrary constants ($x$ and $y$ can be swapped to give another three solutions).

3°. General solution:

$$w(x, y) = -\frac{2}{\beta} \ln \frac{\big|1 - 2a\beta\Phi(z)\overline{\Phi(z)}\big|}{4|\Phi'_z(z)|},$$

where $\Phi = \Phi(z)$ is an arbitrary analytic (holomorphic) function of the complex variable $z = x + iy$ with nonzero derivative, and the bar over a symbol denotes the complex conjugate.

4. $\dfrac{\partial^2 w}{\partial x^2} + \dfrac{\partial^2 w}{\partial y^2} = ae^{\beta w} + be^{2\beta w}$.

1°. Traveling-wave solutions:

$$w(x, y) = -\frac{1}{\beta} \ln\left[-\frac{a\beta}{C_1^2 + C_2^2} + C_3 \exp(C_1 x + C_2 y) + \frac{a^2\beta^2 - b\beta(C_1^2 + C_2^2)}{4C_3(C_1^2 + C_2^2)^2} \exp(-C_1 x - C_2 y)\right],$$

$$w(x, y) = -\frac{1}{\beta} \ln\left[\frac{a\beta}{C_1^2 + C_2^2} + \frac{\sqrt{a^2\beta^2 + b\beta(C_1^2 + C_2^2)}}{C_1^2 + C_2^2} \sin(C_1 x + C_2 y + C_3)\right],$$

where $C_1$, $C_2$, and $C_3$ are arbitrary constants.

2°. For other exact solutions of this equation, see equation T9.3.1.7 with $f(w) = ae^{\beta w} + be^{2\beta w}$.

5. $\dfrac{\partial^2 w}{\partial x^2} + \dfrac{\partial^2 w}{\partial y^2} = \alpha w \ln(\beta w)$.

1°. Solutions:
$$w(x,y) = \frac{1}{\beta} \exp\left[\tfrac{1}{4}\alpha(x+A)^2 + \tfrac{1}{4}\alpha(y+B)^2 + 1\right],$$
$$w(x,y) = \frac{1}{\beta} \exp\left[A(x+B)^2 \pm \sqrt{A\alpha - 4A^2}\,(x+B)(y+C) + (\tfrac{1}{4}\alpha - A)(y+C)^2 + \tfrac{1}{2}\right],$$

where $A$, $B$, and $C$ are arbitrary constants.

2°. There are exact solutions of the following forms:
$$w(x,y) = F(z), \quad z = Ax + By,$$
$$w(x,y) = G(r), \quad r = \sqrt{(x+C_1)^2 + (y+C_2)^2},$$
$$w(x,y) = f(x)g(y).$$

6. $\dfrac{\partial^2 w}{\partial x^2} + \dfrac{\partial^2 w}{\partial y^2} = \alpha \sin(\beta w)$.

1°. Functional separable solution for $\alpha = \beta = 1$:
$$w(x,y) = 4\arctan\left(\cot A \frac{\cosh F}{\cosh G}\right), \quad F = \frac{\cos A}{\sqrt{1+B^2}}(x - By), \quad G = \frac{\sin A}{\sqrt{1+B^2}}(y + Bx),$$

where $A$ and $B$ are arbitrary constants.

2°. Functional separable solution (generalizes the solution of Item 1°):
$$w(x,y) = \frac{4}{\beta} \arctan\left[f(x)g(y)\right],$$

where the functions $f = f(x)$ and $g = g(y)$ are determined by the first-order autonomous ordinary differential equations
$$(f'_x)^2 = Af^4 + Bf^2 + C, \quad (g'_y)^2 = Cg^4 + (\alpha\beta - B)g^2 + A,$$

and $A$, $B$, and $C$ are arbitrary constants.

3°. For other exact solutions of this equation, see equation T9.3.1.7 with $f(w) = \alpha \sin(\beta w)$.

7. $\dfrac{\partial^2 w}{\partial x^2} + \dfrac{\partial^2 w}{\partial y^2} = f(w)$.

This is a *stationary heat equation with a nonlinear source*.

1°. Suppose $w = w(x,y)$ is a solution of the equation in question. Then the functions
$$w_1 = w(\pm x + C_1, \pm y + C_2),$$
$$w_2 = w(x \cos\beta - y \sin\beta,\; x \sin\beta + y \cos\beta),$$

where $C_1$, $C_2$, and $\beta$ are arbitrary constants, are also solutions of the equation (the plus or minus signs in $w_1$ are chosen arbitrarily).

2°. Traveling-wave solution in implicit form:

$$\int \left[ C + \frac{2}{A^2 + B^2} F(w) \right]^{-1/2} dw = Ax + By + D, \qquad F(w) = \int f(w)\,dw,$$

where $A$, $B$, $C$, and $D$ are arbitrary constants.

3°. Solution with central symmetry about the point $(-C_1, -C_2)$:

$$w = w(\zeta), \qquad \zeta = \sqrt{(x + C_1)^2 + (y + C_2)^2},$$

where $C_1$ and $C_2$ are arbitrary constants and the function $w = w(\zeta)$ is determined by the ordinary differential equation $w''_{\zeta\zeta} + \zeta^{-1} w'_\zeta = f(w)$.

## T9.3.2. Equations of the Form $\dfrac{\partial}{\partial x}\left[f(x)\dfrac{\partial w}{\partial x}\right] + \dfrac{\partial}{\partial y}\left[g(y)\dfrac{\partial w}{\partial y}\right] = f(w)$

1. $\dfrac{\partial}{\partial x}\left(ax^n \dfrac{\partial w}{\partial x}\right) + \dfrac{\partial}{\partial y}\left(by^m \dfrac{\partial w}{\partial y}\right) = f(w).$

Functional separable solution for $n \neq 2$ and $m \neq 2$:

$$w = w(r), \qquad r = \left[b(2-m)^2 x^{2-n} + a(2-n)^2 y^{2-m}\right]^{1/2}.$$

Here, the function $w(r)$ is determined by the ordinary differential equation

$$w''_{rr} + A r^{-1} w'_r = B f(w),$$

where $A = \dfrac{4 - nm}{(2-n)(2-m)}$, $B = \dfrac{4}{ab(2-n)^2(2-m)^2}$.

2. $a\dfrac{\partial^2 w}{\partial x^2} + \dfrac{\partial}{\partial y}\left(be^{\mu y}\dfrac{\partial w}{\partial y}\right) = f(w), \qquad ab > 0.$

Functional separable solution for $\mu \neq 0$:

$$w = w(\xi), \qquad \xi = \left[b\mu^2 (x + C_1)^2 + 4ae^{-\mu y}\right]^{1/2},$$

where $C_1$ is an arbitrary constant and the function $w(\xi)$ is defined implicitly by

$$\int \left[C_2 + \frac{2}{ab\mu^2} F(w)\right]^{-1/2} dw = C_3 \pm \xi, \qquad F(w) = \int f(w)\,dw,$$

with $C_2$ and $C_3$ being arbitrary constants.

3. $\dfrac{\partial}{\partial x}\left(ae^{\beta x}\dfrac{\partial w}{\partial x}\right) + \dfrac{\partial}{\partial y}\left(be^{\mu y}\dfrac{\partial w}{\partial y}\right) = f(w), \qquad ab > 0.$

Functional separable solution for $\beta\mu \neq 0$:

$$w = w(\xi), \qquad \xi = \left(b\mu^2 e^{-\beta x} + a\beta^2 e^{-\mu y}\right)^{1/2},$$

where the function $w(\xi)$ is determined by the ordinary differential equation

$$w''_{\xi\xi} - \xi^{-1} w'_\xi = A f(w), \qquad A = 4/(ab\beta^2 \mu^2).$$

4. $\dfrac{\partial}{\partial x}\left[f(x)\dfrac{\partial w}{\partial x}\right] + \dfrac{\partial}{\partial y}\left[g(y)\dfrac{\partial w}{\partial y}\right] = kw\ln w.$

Multiplicative separable solution:

$$w(x,y) = \varphi(x)\psi(y),$$

where the functions $\varphi(x)$ and $\psi(y)$ are determined by the ordinary differential equations

$$[f(x)\varphi'_x]'_x = k\varphi\ln\varphi + C\varphi, \quad [g(y)\psi'_y]'_y = k\psi\ln\psi - C\psi,$$

and $C$ is an arbitrary constant.

### T9.3.3. Equations of the Form $\dfrac{\partial}{\partial x}\left[f(w)\dfrac{\partial w}{\partial x}\right] + \dfrac{\partial}{\partial y}\left[g(w)\dfrac{\partial w}{\partial y}\right] = h(w)$

1. $\dfrac{\partial^2 w}{\partial x^2} + \dfrac{\partial}{\partial y}\left[(\alpha w + \beta)\dfrac{\partial w}{\partial y}\right] = 0.$

*Stationary Khokhlov–Zabolotskaya equation.* It arises in acoustics and nonlinear mechanics.

1°. Solutions:

$$w(x,y) = Ay - \tfrac{1}{2}A^2\alpha x^2 + C_1 x + C_2,$$
$$w(x,y) = (Ax+B)y - \dfrac{\alpha}{12A^2}(Ax+B)^4 + C_1 x + C_2,$$
$$w(x,y) = -\dfrac{1}{\alpha}\left(\dfrac{y+A}{x+B}\right)^2 + \dfrac{C_1}{x+B} + C_2(x+B)^2 - \dfrac{\beta}{\alpha},$$
$$w(x,y) = -\dfrac{1}{\alpha}\left[\beta + \lambda^2 \pm \sqrt{A(y+\lambda x)+B}\,\right],$$
$$w(x,y) = (Ax+B)\sqrt{C_1 y + C_2} - \dfrac{\beta}{\alpha},$$

where $A$, $B$, $C_1$, $C_2$, and $\lambda$ are arbitrary constants.

2°. Generalized separable solution quadratic in $y$ (generalizes the third solution of Item 1°):

$$w(x,y) = -\dfrac{1}{\alpha(x+A)^2}y^2 + \left[\dfrac{B_1}{(x+A)^2} + B_2(x+A)^3\right]y$$
$$+ \dfrac{C_1}{x+A} + C_2(x+A)^2 - \dfrac{\beta}{\alpha} - \dfrac{\alpha B_1^2}{4(x+A)^2} - \tfrac{1}{2}\alpha B_1 B_2(x+A)^3 - \dfrac{1}{54}\alpha B_2^2(x+A)^8,$$

where $A$, $B_1$, $B_2$, $C_1$, and $C_2$ are arbitrary constants.

3°. See also equation T9.3.3.3 with $f(w) = 1$ and $g(w) = \alpha w + \beta$.

**2.** $\dfrac{\partial^2 w}{\partial x^2} + \dfrac{\partial}{\partial y}\left(ae^{\beta w}\dfrac{\partial w}{\partial y}\right) = 0, \quad a > 0.$

1°. Additive separable solutions:

$$w(x,y) = \frac{1}{\beta}\ln(Ay + B) + Cx + D,$$

$$w(x,y) = \frac{1}{\beta}\ln(-aA^2y^2 + By + C) - \frac{2}{\beta}\ln(-aAx + D),$$

$$w(x,y) = \frac{1}{\beta}\ln(Ay^2 + By + C) + \frac{1}{\beta}\ln\left[\frac{p^2}{aA\cosh^2(px + q)}\right],$$

$$w(x,y) = \frac{1}{\beta}\ln(Ay^2 + By + C) + \frac{1}{\beta}\ln\left[\frac{p^2}{-aA\cos^2(px + q)}\right],$$

$$w(x,y) = \frac{1}{\beta}\ln(Ay^2 + By + C) + \frac{1}{\beta}\ln\left[\frac{p^2}{-aA\sinh^2(px + q)}\right],$$

where $A$, $B$, $C$, $D$, $p$, and $q$ are arbitrary constants.

2°. There are exact solutions of the following forms:

$$\begin{aligned}
w(x,y) &= F(r), \quad r = k_1 x + k_2 y;\\
w(x,y) &= G(z), \quad z = y/x;\\
w(x,y) &= H(\xi) - 2(k+1)\beta^{-1}\ln|x|, \quad \xi = y|x|^k;\\
w(x,y) &= U(\eta) - 2\beta^{-1}\ln|x|, \quad \eta = y + k\ln|x|;\\
w(x,y) &= V(\zeta) - 2\beta^{-1}x, \quad \zeta = ye^x,
\end{aligned}$$

where $k$, $k_1$, and $k_2$ are arbitrary constants.

3°. For other solutions, see equation T9.3.3.3 with $f(w) = 1$ and $g(w) = ae^{\beta w}$.

**3.** $\dfrac{\partial}{\partial x}\left[f(w)\dfrac{\partial w}{\partial x}\right] + \dfrac{\partial}{\partial y}\left[g(w)\dfrac{\partial w}{\partial y}\right] = 0.$

This is a *stationary anisotropic heat (diffusion) equation*.

1°. Traveling-wave solution in implicit form:

$$\int\left[A^2 f(w) + B^2 g(w)\right] dw = C_1(Ax + By) + C_2,$$

where $A$, $B$, $C_1$, and $C_2$ are arbitrary constants.

2°. Self-similar solution:

$$w = w(\zeta), \quad \zeta = \frac{x + A}{y + B},$$

where the function $w(\zeta)$ is determined by the ordinary differential equation

$$[f(w)w'_\zeta]'_\zeta + [\zeta^2 g(w)w'_\zeta]'_\zeta = 0. \tag{1}$$

Integrating (1) and taking $w$ to be the independent variable, one obtains the Riccati equation $C\zeta'_w = g(w)\zeta^2 + f(w)$, where $C$ is an arbitrary constant.

3°. The original equation can be represented as the system of the equations

$$f(w)\frac{\partial w}{\partial x} = \frac{\partial v}{\partial y}, \quad -g(w)\frac{\partial w}{\partial y} = \frac{\partial v}{\partial x}. \tag{2}$$

The hodograph transformation

$$x = x(w,v), \quad y = y(w,v),$$

where $w$, $v$ are treated as the independent variables and $x$, $y$ as the dependent ones, brings (2) to the linear system

$$f(w)\frac{\partial y}{\partial v} = \frac{\partial x}{\partial w}, \quad -g(w)\frac{\partial x}{\partial v} = \frac{\partial y}{\partial w}. \tag{3}$$

Eliminating $y$ yields the following linear equation for $x = x(w,v)$:

$$\frac{\partial}{\partial w}\left[\frac{1}{f(w)}\frac{\partial x}{\partial w}\right] + g(w)\frac{\partial^2 x}{\partial v^2} = 0.$$

Likewise, we can obtain another linear equation for $y = y(w,v)$ from system (3).

## T9.4. Other Second-Order Equations
### T9.4.1. Equations of Transonic Gas Flow

1. $a\dfrac{\partial w}{\partial x}\dfrac{\partial^2 w}{\partial x^2} + \dfrac{\partial^2 w}{\partial y^2} = 0.$

This is an *equation of steady transonic gas flow*.

1°. Suppose $w(x,y)$ is a solution of the equation in question. Then the function

$$w_1 = C_1^{-3}C_2^2 w(C_1 x + C_3, C_2 y + C_4) + C_5 y + C_6,$$

where $C_1, \ldots, C_6$ are arbitrary constants, is also a solution of the equation.

2°. Solutions:

$$w(x,y) = C_1 xy + C_2 x + C_3 y + C_4,$$

$$w(x,y) = -\frac{(x+C_1)^3}{3a(y+C_2)^2} + C_3 y + C_4,$$

$$w(x,y) = \frac{a^2 C_1^3}{39}(y+A)^{13} + \frac{2}{3}aC_1^2(y+A)^8(x+B) + 3C_1(y+A)^3(x+B)^2 - \frac{(x+B)^3}{3a(y+A)^2},$$

$$w(x,y) = -aC_1 y^2 + C_2 y + C_3 \pm \frac{4}{3C_1}(C_1 x + C_4)^{3/2},$$

$$w(x,y) = -aA^3 y^2 - \frac{B^2}{aA^2}x + C_1 y + C_2 \pm \frac{4}{3}(Ax + By + C_3)^{3/2},$$

$$w(x,y) = \frac{1}{3}(Ay+B)(2C_1 x + C_2)^{3/2} - \frac{aC_1^3}{12A^2}(Ay+B)^4 + C_3 y + C_4,$$

$$w(x,y) = -\frac{9aA^2}{y+C_1} + 4A\left(\frac{x+C_2}{y+C_1}\right)^{3/2} - \frac{(x+C_2)^3}{3a(y+C_1)^2} + C_3 y + C_4,$$

$$w(x,y) = -\frac{3}{7}aA^2(y+C_1)^7 + 4A(x+C_2)^{3/2}(y+C_1)^{5/2} - \frac{(x+C_2)^3}{3a(y+C_1)^2} + C_3 y + C_4,$$

where $A, B, C_1, \ldots, C_4$ are arbitrary constants (the first solution is degenerate).

3°. There are solutions of the following forms:

$$w(x,y) = y^{-3k-2}U(z), \quad z = xy^k \quad \text{(self-similar solution, } k \text{ is any number);}$$
$$w(x,y) = \varphi_1(y) + \varphi_2(y)x^{3/2} + \varphi_3(y)x^3 \quad \text{(generalized separable solution);}$$
$$w(x,y) = \psi_1(y) + \psi_2(y)x + \psi_3(y)x^2 + \psi_4(y)x^3 \quad \text{(generalized separable solution);}$$
$$w(x,y) = \psi_1(y)\varphi(x) + \psi_2(y) \quad \text{(generalized separable solution).}$$

**2.** $\dfrac{\partial^2 w}{\partial y^2} + \dfrac{a}{y}\dfrac{\partial w}{\partial y} + b\dfrac{\partial w}{\partial x}\dfrac{\partial^2 w}{\partial x^2} = 0.$

1°. Suppose $w(x, y)$ is a solution of the equation in question. Then the function

$$w_1 = C_1^{-3}C_2^2 w(C_1 x + C_3, C_2 y) + C_4 y^{1-a} + C_5,$$

where $C_1, \ldots, C_5$ are arbitrary constants, is also a solution of the equation.

2°. Additive separable solution:

$$w(x,y) = -\frac{bC_1}{4(a+1)}y^2 + C_2 y^{1-a} + C_3 \pm \frac{2}{3C_1}(C_1 x + C_4)^{3/2},$$

where $C_1, \ldots, C_4$ are arbitrary constants.

3°. Generalized separable solutions:

$$w(x,y) = -\frac{9A^2 b}{16(n+1)(2n+1+a)}y^{2n+2} + Ay^n(x+C)^{3/2} + \frac{a-3}{9b}\frac{(x+C)^3}{y^2},$$

where $A$ and $C$ are arbitrary constants, and $n = n_{1,2}$ are roots of the quadratic equation $n^2 + (a-1)n + \tfrac{5}{4}(a-3) = 0$.

4°. Generalized separable solution:

$$w(x,y) = (Ay^{1-a} + B)(2C_1 x + C_2)^{3/2} + 9bC_1^3 \theta(y),$$
$$\theta(y) = -\frac{B^2}{2(a+1)}y^2 - \frac{AB}{3-a}y^{3-a} - \frac{A^2}{2(2-a)(3-a)}y^{4-2a} + C_3 y^{1-a} + C_4,$$

where $A$, $B$, $C_1$, $C_2$, $C_3$, and $C_4$ are arbitrary constants.

5°. There are solutions of the following forms:

$$w(x,y) = y^{-3k-2}U(z), \quad z = xy^k \quad \text{(self-similar solution, } k \text{ is any number);}$$
$$w(x,y) = \varphi_1(y) + \varphi_2(y)x^{3/2} + \varphi_3(y)x^3 \quad \text{(generalized separable solution);}$$
$$w(x,y) = \psi_1(y) + \psi_2(y)x + \psi_3(y)x^2 + \psi_4(y)x^3 \quad \text{(generalized separable solution);}$$
$$w(x,y) = \psi_1(y)\varphi(x) + \psi_2(y) \quad \text{(generalized separable solution).}$$

## T9.4.2. Monge–Ampère Equations

**1.** $\left(\dfrac{\partial^2 w}{\partial x \partial y}\right)^2 - \dfrac{\partial^2 w}{\partial x^2}\dfrac{\partial^2 w}{\partial y^2} = 0.$

*Homogeneous Monge–Ampère equation.*

$1°$. General solution in parametric form:

$$w = tx + \varphi(t)y + \psi(t),$$
$$x + \varphi'(t)y + \psi'(t) = 0,$$

where $t$ is the parameter, and $\varphi = \varphi(t)$ and $\psi = \psi(t)$ are arbitrary functions.

$2°$. Solutions involving one arbitrary function:

$$w(x,y) = \varphi(C_1 x + C_2 y) + C_3 x + C_4 y + C_5,$$
$$w(x,y) = (C_1 x + C_2 y)\varphi\left(\dfrac{y}{x}\right) + C_3 x + C_4 y + C_5,$$
$$w(x,y) = (C_1 x + C_2 y + C_3)\varphi\left(\dfrac{C_4 x + C_5 y + C_6}{C_1 x + C_2 y + C_3}\right) + C_7 x + C_8 y + C_9,$$

where $C_1, \ldots, C_9$ are arbitrary constants and $\varphi = \varphi(z)$ is an arbitrary function.

**2.** $\left(\dfrac{\partial^2 w}{\partial x \partial y}\right)^2 - \dfrac{\partial^2 w}{\partial x^2}\dfrac{\partial^2 w}{\partial y^2} = A.$

*Nonhomogeneous Monge–Ampère equation.*

$1°$. General solution in parametric form for $A = a^2 > 0$:

$$x = \dfrac{\beta - \lambda}{2a}, \quad y = \dfrac{\psi'(\lambda) - \varphi'(\beta)}{2a}, \quad w = \dfrac{(\beta + \lambda)[\psi'(\lambda) - \varphi'(\beta)] + 2\varphi(\beta) - 2\psi(\lambda)}{4a},$$

where $\beta$ and $\lambda$ are the parameters, $\varphi = \varphi(\beta)$ and $\psi = \psi(\lambda)$ are arbitrary functions.

$2°$. Solutions:

$$w(x,y) = \pm\dfrac{\sqrt{A}}{C_2}x(C_1 x + C_2 y) + \varphi(C_1 x + C_2 y) + C_3 x + C_4 y,$$
$$w(x,y) = C_1 y^2 + C_2 xy + \dfrac{1}{4C_1}(C_2^2 - A)x^2 + C_3 y + C_4 x + C_5,$$
$$w(x,y) = \dfrac{1}{x + C_1}\left(C_2 y^2 + C_3 y + \dfrac{C_3^2}{4C_2}\right) - \dfrac{A}{12C_2}(x^3 + 3C_1 x^2) + C_4 y + C_5 x + C_6,$$
$$w(x,y) = \pm\dfrac{2\sqrt{A}}{3C_1 C_2}(C_1 x - C_2^2 y^2 + C_3)^{3/2} + C_4 x + C_5 y + C_6,$$

where $C_1, \ldots, C_6$ are arbitrary constants and $\varphi = \varphi(z)$ is an arbitrary function.

## T9.5. Higher-Order Equations

### T9.5.1. Third-Order Equations

1. $\dfrac{\partial w}{\partial t} + \dfrac{\partial^3 w}{\partial x^3} - 6w\dfrac{\partial w}{\partial x} = 0.$

*Korteweg–de Vries equation.* It is used in many sections of nonlinear mechanics and physics.

1°. Suppose $w(x,t)$ is a solution of the Korteweg–de Vries equation. Then the function

$$w_1 = C_1^2 w(C_1 x + 6C_1 C_2 t + C_3, C_1^3 t + C_4) + C_2,$$

where $C_1, \ldots, C_4$ are arbitrary constants, is also a solution of the equation.

2°. One-soliton solution:

$$w(x,t) = -\dfrac{a}{2\cosh^2\left[\tfrac{1}{2}\sqrt{a}\,(x - at - b)\right]},$$

where $a$ and $b$ are arbitrary constants.

3°. Two-soliton solution:

$$w(x,t) = -2\dfrac{\partial^2}{\partial x^2} \ln\!\left(1 + B_1 e^{\theta_1} + B_2 e^{\theta_2} + AB_1 B_2 e^{\theta_1+\theta_2}\right),$$

$$\theta_1 = a_1 x - a_1^3 t, \quad \theta_2 = a_2 x - a_2^3 t, \quad A = \left(\dfrac{a_1 - a_2}{a_1 + a_2}\right)^2,$$

where $B_1, B_2, a_1$, and $a_2$ are arbitrary constants.

4°. $N$-soliton solution:

$$w(x,t) = -2\dfrac{\partial^2}{\partial x^2}\left\{\ln\det\left[\mathbf{I} + \mathbf{C}(\mathbf{x},\mathbf{t})\right]\right\}.$$

Here, $\mathbf{I}$ is the $N \times N$ identity matrix and $\mathbf{C}(x,t)$ the $N \times N$ symmetric matrix with entries

$$C_{mn}(x,t) = \dfrac{\sqrt{\rho_m(t)\rho_n(t)}}{p_m + p_n}\exp\!\left[-(p_m + p_n)x\right],$$

where the normalizing factors $\rho_n(t)$ are given by

$$\rho_n(t) = \rho_n(0)\exp\!\left(8p_n^3 t\right), \quad n = 1, 2, \ldots, N.$$

The solution involves $2N$ arbitrary constants $p_n$ and $\rho_n(0)$.

The above solution can be represented, for $t \to \pm\infty$, as the sum of $N$ single-soliton solutions.

5°. "One soliton + one pole" solution:

$$w(x,t) = -2p^2\left[\cosh^{-2}(pz) - (1+px)^{-2}\tanh^2(pz)\right]\left[1 - (1+px)^{-1}\tanh(pz)\right]^{-2}, \quad z = x - 4p^2 t - c,$$

where $p$ and $c$ are arbitrary constants.

6°. Rational solutions (algebraic solitons):
$$w(x,t) = \frac{6x(x^3 - 24t)}{(x^3 + 12t)^2},$$
$$w(x,t) = -2\frac{\partial^2}{\partial x^2} \ln(x^6 + 60x^3 t - 720t^2).$$

7°. There is a self-similar solution of the form $w = t^{-2/3} U(z)$, where $z = t^{-1/3} x$.

8°. Solution:
$$w(x,t) = 2\varphi(z) + 2C_1 t, \qquad z = x + 6C_1 t^2 + C_2 t,$$

where $C_1$ and $C_2$ are arbitrary constants, and the function $\varphi(z)$ is determined by the second-order ordinary differential equation $\varphi''_{zz} = 6\varphi^2 - C_2 \varphi - C_1 z + C_3$.

9°. The Korteweg–de Vries equation is solved by the inverse scattering method. Any rapidly decreasing function $F = F(x, y; t)$ as $x \to +\infty$ that simultaneously satisfies the two linear equations
$$\frac{\partial^2 F}{\partial x^2} - \frac{\partial^2 F}{\partial y^2} = 0, \qquad \frac{\partial F}{\partial t} + \left(\frac{\partial}{\partial x} + \frac{\partial}{\partial y}\right)^3 F = 0$$

generates a solution of the Korteweg–de Vries equation in the form
$$w = -2\frac{d}{dx} K(x, x; t),$$

where $K(x, y; t)$ is a solution of the linear Gel'fand–Levitan–Marchenko integral equation
$$K(x, y; t) + F(x, y; t) + \int_x^\infty K(x, z; t) F(z, y; t)\, dz = 0.$$

Time $t$ appears in this equation as a parameter.

2. $\quad \dfrac{\partial w}{\partial t} + \dfrac{\partial^3 w}{\partial x^3} - 6w\dfrac{\partial w}{\partial x} + \dfrac{1}{2t} w = 0.$

Cylindrical Korteweg–de Vries equation.

The transformation
$$w(x,t) = -\frac{x}{12t} - \frac{1}{2t} u(z, \tau), \qquad x = \frac{z}{\tau}, \qquad t = -\frac{1}{2\tau^2}$$

leads to the Korteweg–de Vries equation T9.5.1.1:
$$\frac{\partial u}{\partial \tau} + \frac{\partial^3 u}{\partial z^3} - 6u \frac{\partial u}{\partial z} = 0.$$

3. $\quad \dfrac{\partial w}{\partial t} + \dfrac{\partial^3 w}{\partial x^3} + 6\sigma w^2 \dfrac{\partial w}{\partial x} = 0.$

Modified Korteweg–de Vries equation.

1°. One-soliton solution for $\sigma = 1$:
$$w(x,t) = a + \frac{k^2}{\sqrt{4a^2 + k^2}\cosh z + 2a}, \qquad z = kx - (6a^2 k + k^3)t + b,$$
where $a$, $b$, and $k$ are arbitrary constants.

2°. Two-soliton solution for $\sigma = 1$:
$$w(x,t) = 2\frac{a_1 e^{\theta_1} + a_2 e^{\theta_2} + Aa_2 e^{2\theta_1 + \theta_2} + Aa_1 e^{\theta_1 + 2\theta_2}}{1 + e^{2\theta_1} + e^{2\theta_2} + 2(1-A)e^{\theta_1 + \theta_2} + A^2 e^{2(\theta_1 + \theta_2)}},$$
$$\theta_1 = a_1 x - a_1^3 t + b_1, \quad \theta_2 = a_2 x - a_2^3 t + b_2, \quad A = \left(\frac{a_1 - a_2}{a_1 + a_2}\right)^2,$$
where $a_1$, $a_2$, $b_1$, and $b_2$ are arbitrary constants.

3°. Rational solutions (algebraic solitons) for $\sigma = 1$:
$$w(x,t) = a - \frac{4a}{4a^2 z^2 + 1}, \qquad z = x - 6a^2 t,$$
$$w(x,t) = a - \frac{12a\left(z^4 + \frac{3}{2}a^{-2}z^2 - \frac{3}{16}a^{-4} - 24tz\right)}{4a^2\left(z^3 + 12t - \frac{3}{4}a^{-2}z\right)^2 + 3\left(z^2 + \frac{1}{4}a^{-2}\right)^2},$$
where $a$ is an arbitrary constant.

4°. There is a self-similar solution of the form $w = t^{-1/3}U(z)$, where $z = t^{-1/3}x$.

5°. The modified Korteweg–de Vries equation is solved by the inverse scattering method.

**4.** $\dfrac{\partial w}{\partial y}\dfrac{\partial^2 w}{\partial x \partial y} - \dfrac{\partial w}{\partial x}\dfrac{\partial^2 w}{\partial y^2} = \nu \dfrac{\partial^3 w}{\partial y^3}.$

This is an equation of a steady laminar boundary layer on a flat plate ($w$ is the stream function).

1°. Suppose $w(x,y)$ is a solution of the equation in question. Then the function
$$w_1 = C_1 w(C_2 x + C_3, C_1 C_2 y + \varphi(x)) + C_4,$$
where $\varphi(x)$ is an arbitrary function and $C_1, \ldots, C_5$ are arbitrary constants, is also a solution of the equation.

2°. Solutions involving arbitrary functions:
$$w(x,y) = C_1 y + \varphi(x),$$
$$w(x,y) = C_1 y^2 + \varphi(x) y + \frac{1}{4C_1}\varphi^2(x) + C_2,$$
$$w(x,y) = \frac{6\nu x + C_1}{y + \varphi(x)} + \frac{C_2}{[y + \varphi(x)]^2} + C_3,$$
$$w(x,y) = \varphi(x)\exp(-C_1 y) + \nu C_1 x + C_2,$$
$$w(x,y) = C_1 \exp[-C_2 y - C_2\varphi(x)] + C_3 y + C_3\varphi(x) + \nu C_2 x + C_4,$$
$$w(x,y) = 6\nu C_1 x^{1/3}\tanh\xi + C_2, \qquad \xi = C_1\frac{y}{x^{2/3}} + \varphi(x),$$
$$w(x,y) = -6\nu C_1 x^{1/3}\tan\xi + C_2, \qquad \xi = C_1\frac{y}{x^{2/3}} + \varphi(x),$$
where $C_1, \ldots, C_4$ are arbitrary constants and $\varphi(x)$ is an arbitrary function. The first two solutions are degenerate — they are independent of $\nu$ and correspond to inviscid fluid flows.

TABLE T9.1
Invariant solutions to the hydrodynamic boundary layer equation (the additive constant is omitted)

| No. | Solution structure | Function $F$ or equation for $F$ | Remarks |
|---|---|---|---|
| 1 | $w = F(y) + \nu\lambda x$ | $F(y) = \begin{cases} C_1 \exp(-\lambda y) + C_2 y & \text{if } \lambda \neq 0, \\ C_1 y^2 + C_2 y & \text{if } \lambda = 0 \end{cases}$ | $\lambda$ is any number |
| 2 | $w = F(x)y^{-1}$ | $F(x) = 6\nu x + C_1$ | |
| 3 | $w = x^{\lambda+1}F(z)$, $z = x^\lambda y$ | $(2\lambda + 1)(F'_z)^2 - (\lambda + 1)FF''_{zz} = \nu F'''_{zzz}$ | $\lambda$ is any number |
| 4 | $w = e^{\lambda x}F(z)$, $z = e^{\lambda x}y$ | $2\lambda(F'_z)^2 - \lambda FF''_{zz} = \nu F'''_{zzz}$ | $\lambda$ is any number |
| 5 | $w = F(z) + a\ln|x|$, $z = y/x$ | $-(F'_z)^2 - aF''_{zz} = \nu F'''_{zzz}$ | $a$ is any number |

3°. Table T9.1 lists invariant solutions to the hydrodynamic boundary layer equation. Solution 1 is expressed in additive separable form, solution 2 is in multiplicative separable form, solution 3 is self-similar, and solution 4 is generalized self-similar. Solution 5 degenerates at $a = 0$ into a self-similar solution (see solution 3 with $\lambda = -1$). Equations 3–5 for $F$ are autonomous and generalized homogeneous; hence, their order can be reduced by two.

4°. Generalized separable solution linear in $x$:

$$w(x,y) = xf(y) + g(y), \qquad (1)$$

where the functions $f = f(y)$ and $g = g(y)$ are determined by the autonomous system of ordinary differential equations

$$(f'_y)^2 - ff''_{yy} = \nu f'''_{yyy}, \qquad (2)$$
$$f'_y g'_y - f g''_{yy} = \nu g'''_{yyy}. \qquad (3)$$

Equation (2) has the following particular solutions:

$$f = 6\nu(y + C)^{-1},$$
$$f = Ce^{\lambda y} - \lambda\nu,$$

where $C$ and $\lambda$ are arbitrary constants.

Let $f = f(y)$ be a solution of equation (2) ($f \not\equiv \text{const}$). Then the corresponding general solution of equation (3) can be written in the form

$$g(y) = C_1 + C_2 f + C_3\left(f\int \psi\, dy - \int f\psi\, dy\right), \quad \text{where} \quad \psi = \frac{1}{(f'_y)^2}\exp\left(-\frac{1}{\nu}\int f\, dy\right).$$

5. $\dfrac{\partial w}{\partial y}\dfrac{\partial^2 w}{\partial x \partial y} - \dfrac{\partial w}{\partial x}\dfrac{\partial^2 w}{\partial y^2} = \nu\dfrac{\partial^3 w}{\partial y^3} + f(x).$

This is a hydrodynamic boundary layer equation with pressure gradient.

1°. Suppose $w(x,y)$ is a solution of the equation in question. Then the functions

$$w_1 = \pm w(x, \pm y + \varphi(x)) + C,$$

where $\varphi(x)$ is an arbitrary function and $C$ is an arbitrary constant, are also solutions of the equation.

## T9.5. HIGHER-ORDER EQUATIONS

**TABLE T9.2**
Invariant solutions to the hydrodynamic boundary layer equation
with pressure gradient ($a$, $k$, $m$, and $\beta$ are arbitrary constants)

| No. | Function $f(x)$ | Form of solution $w = w(x, y)$ | Function $u$ or equation for $u$ |
|---|---|---|---|
| 1 | $f(x) = 0$ | See equation 4 | See equation 4 |
| 2 | $f(x) = ax^m$ | $w = x^{\frac{m+3}{4}} u(z)$, $z = x^{\frac{m-1}{4}} y$ | $\frac{m+1}{2}(u'_z)^2 - \frac{m+3}{4} u u''_{zz} = \nu u'''_{zzz} + a$ |
| 3 | $f(x) = ae^{\beta x}$ | $w = e^{\frac{1}{4}\beta x} u(z)$, $z = e^{\frac{1}{4}\beta x} y$ | $\frac{1}{2}\beta(u'_z)^2 - \frac{1}{4}\beta u u''_{zz} = \nu u'''_{zzz} + a$ |
| 4 | $f(x) = a$ | $w = kx + u(y)$ | $u(y) = \begin{cases} C_1 \exp(-\frac{k}{\nu} y) - \frac{a}{2k} y^2 + C_2 y & \text{if } k \neq 0, \\ -\frac{a}{6\nu} y^3 + C_2 y^2 + C_1 y & \text{if } k = 0 \end{cases}$ |
| 5 | $f(x) = ax^{-3}$ | $w = k \ln|x| + u(z)$, $z = y/x$ | $-(u'_z)^2 - k u''_{zz} = \nu u'''_{zzz} + a$ |

2°. Degenerate solutions (linear and quadratic in $y$) for arbitrary $f(x)$:

$$w(x, y) = \pm y \left[ 2 \int f(x)\,dx + C_1 \right]^{1/2} + \varphi(x),$$

$$w(x, y) = C_1 y^2 + \varphi(x) y + \frac{1}{4C_1}\left[ \varphi^2(x) - 2 \int f(x)\,dx \right] + C_2,$$

where $\varphi(x)$ is an arbitrary function and $C_1$ and $C_2$ are arbitrary constants. These solutions are independent of $\nu$ and correspond to inviscid fluid flows.

3°. Table T9.2 lists invariant solutions to the boundary layer equation with pressure gradient.

4°. Generalized separable solution for $f(x) = ax + b$:

$$w(x, y) = xF(y) + G(y),$$

where the functions $F = F(y)$ and $G = G(y)$ are determined by the system of ordinary differential equations

$$(F'_y)^2 - FF''_{yy} = \nu F'''_{yyy} + a, \qquad F'_y G'_y - FG''_{yy} = \nu G'''_{yyy} + b.$$

5°. Solutions for $f(x) = -ax^{-5/3}$:

$$w(x, y) = \frac{6\nu x}{y + \varphi(x)} \pm \frac{\sqrt{3a}}{x^{1/3}}[y + \varphi(x)],$$

where $\varphi(x)$ is an arbitrary function.

6°. Solutions for $f(x) = ax^{-1/3} - bx^{-5/3}$:

$$w(x, y) = \pm\sqrt{3b}\, z + x^{2/3} \theta(z), \qquad z = yx^{-1/3},$$

where the function $\theta = \theta(z)$ is determined by the ordinary differential equation $\frac{1}{3}(\theta'_z)^2 - \frac{2}{3}\theta\theta''_{zz} = \nu \theta'''_{zzz} + a$.

7°. Generalized separable solution for $f(x) = ae^{\beta x}$:

$$w(x, y) = \varphi(x) e^{\lambda y} - \frac{a}{2\beta \lambda^2 \varphi(x)} e^{\beta x - \lambda y} - \nu\lambda x + \frac{2\nu\lambda^2}{\beta} y + \frac{2\nu\lambda}{\beta} \ln|\varphi(x)|,$$

where $\varphi(x)$ is an arbitrary function and $\lambda$ is an arbitrary constant.

## T9.5.2. Fourth-Order Equations

1. $\dfrac{\partial^2 w}{\partial t^2} + \dfrac{\partial}{\partial x}\left(w\dfrac{\partial w}{\partial x}\right) + \dfrac{\partial^4 w}{\partial x^4} = 0.$

*Boussinesq equation.* This equation arises in hydrodynamics and some physical applications.

1°. Suppose $w(x,t)$ is a solution of the equation in question. Then the functions

$$w_1 = C_1^2 w(C_1 x + C_2, \pm C_1^2 t + C_3),$$

where $C_1$, $C_2$, and $C_3$ are arbitrary constants, are also solutions of the equation.

2°. Solutions:

$$w(x,t) = 2C_1 x - 2C_1^2 t^2 + C_2 t + C_3,$$

$$w(x,t) = (C_1 t + C_2)x - \dfrac{1}{12 C_1^2}(C_1 t + C_2)^4 + C_3 t + C_4,$$

$$w(x,t) = -\dfrac{(x+C_1)^2}{(t+C_2)^2} + \dfrac{C_3}{t+C_2} + C_4(t+C_2)^2,$$

$$w(x,t) = -\dfrac{x^2}{t^2} + C_1 t^3 x - \dfrac{C_1^2}{54} t^8 + C_2 t^2 + \dfrac{C_4}{t},$$

$$w(x,t) = -\dfrac{(x+C_1)^2}{(t+C_2)^2} - \dfrac{12}{(x+C_1)^2},$$

$$w(x,t) = -3\lambda^2 \cos^{-2}\!\left[\tfrac{1}{2}\lambda(x \pm \lambda t) + C_1\right],$$

where $C_1, \ldots, C_4$ and $\lambda$ are arbitrary constants.

3°. Traveling-wave solution (generalizes the last solution of Item 1°):

$$w(x,t) = w(\zeta), \quad \zeta = x + \lambda t,$$

where the function $w(\zeta)$ is determined by the second-order ordinary differential equation $w''_{\zeta\zeta} + w^2 + 2\lambda^2 w + C_1 \zeta + C_2 = 0.$

4°. Self-similar solution:

$$w(x,t) = t^{-1} u(z), \quad z = x t^{-1/2},$$

where the function $u = u(z)$ is determined by the ordinary differential equation $u''''_{zzzz} + (uu'_z)'_z + \tfrac{1}{4} z^2 u''_{zz} + \tfrac{7}{4} z u'_z + 2u = 0.$

5°. There are exact solutions of the following forms:

$$w(x,t) = (x+C)^2 F(t) - 12(x+C)^{-2};$$

$$w(x,t) = G(\xi) - 4C_1^2 t^2 - 4C_1 C_2 t, \quad \xi = x - C_1 t^2 - C_2 t;$$

$$w(x,t) = \dfrac{1}{t} H(\eta) - \dfrac{1}{4}\left(\dfrac{x}{t} + Ct\right)^2, \quad \eta = \dfrac{x}{\sqrt{t}} - \dfrac{1}{3} C t^{3/2};$$

$$w(x,t) = (a_1 t + a_0)^2 U(\zeta) - \left(\dfrac{a_1 x + b_1}{a_1 t + a_0}\right)^2, \quad \zeta = x(a_1 t + a_0) + b_1 t + b_0,$$

where $C, C_1, C_2, a_1, a_0, b_1,$ and $b_0$ are arbitrary constants.

6°. The Boussinesq equation is solved by the inverse scattering method. Any rapidly decaying function $F = F(x, y; t)$ as $x \to +\infty$ and satisfying simultaneously the two linear equations

$$\frac{1}{\sqrt{3}}\frac{\partial F}{\partial t} + \frac{\partial^2 F}{\partial x^2} - \frac{\partial^2 F}{\partial y^2} = 0, \qquad \frac{\partial^3 F}{\partial x^3} + \frac{\partial^3 F}{\partial y^3} = 0$$

generates a solution of the Boussinesq equation in the form

$$w = 12\frac{d}{dx}K(x, x; t),$$

where $K(x, y; t)$ is a solution of the linear Gel'fand–Levitan–Marchenko integral equation

$$K(x, y; t) + F(x, y; t) + \int_x^\infty K(x, s; t)F(s, y; t)\,ds = 0.$$

Time $t$ appears here as a parameter.

2. $\quad \dfrac{\partial w}{\partial y}\dfrac{\partial}{\partial x}(\Delta w) - \dfrac{\partial w}{\partial x}\dfrac{\partial}{\partial y}(\Delta w) = \nu\Delta\Delta w, \qquad \Delta w = \dfrac{\partial^2 w}{\partial x^2} + \dfrac{\partial^2 w}{\partial y^2}.$

There is a two-dimensional stationary equation of motion of a viscous incompressible fluid — it is obtained from the Navier–Stokes equation by the introduction of the stream function $w$.

1°. Suppose $w(x, y)$ is a solution of the equation in question. Then the functions

$$w_1 = -w(y, x),$$
$$w_2 = w(C_1 x + C_2, C_1 y + C_3) + C_4,$$
$$w_3 = w(x\cos\alpha + y\sin\alpha, -x\sin\alpha + y\cos\alpha),$$

where $C_1, \ldots, C_4$ and $\alpha$ are arbitrary constants, are also solutions of the equation.

2°. Any solution of the Poisson equation $\Delta w = C$ is also a solution of the original equation (these are "inviscid" solutions).

3°. Solutions in the form of a one-variable function or the sum of functions with different arguments:

$$w(y) = C_1 y^3 + C_2 y^2 + C_3 y + C_4,$$
$$w(x, y) = C_1 x^2 + C_2 x + C_3 y^2 + C_4 y + C_5,$$
$$w(x, y) = C_1 \exp(-\lambda y) + C_2 y^2 + C_3 y + C_4 + \nu\lambda x,$$
$$w(x, y) = C_1 \exp(\lambda x) - \nu\lambda x + C_2 \exp(\lambda y) + \nu\lambda y + C_3,$$
$$w(x, y) = C_1 \exp(\lambda x) + \nu\lambda x + C_2 \exp(-\lambda y) + \nu\lambda y + C_3,$$

where $C_1, \ldots, C_5$ and $\lambda$ are arbitrary constants.

4°. Generalized separable solutions:

$$w(x, y) = A(kx + \lambda y)^3 + B(kx + \lambda y)^2 + C(kx + \lambda y) + D,$$
$$w(x, y) = Ae^{-\lambda(y+kx)} + B(y + kx)^2 + C(y + kx) + \nu\lambda(k^2 + 1)x + D,$$
$$w(x, y) = 6\nu x(y + \lambda)^{-1} + A(y + \lambda)^3 + B(y + \lambda)^{-1} + C(y + \lambda)^{-2} + D,$$

$$w(x,y) = (Ax + B)e^{-\lambda y} + \nu\lambda x + C,$$
$$w(x,y) = \left[A\sinh(\beta x) + B\cosh(\beta x)\right]e^{-\lambda y} + \frac{\nu}{\lambda}(\beta^2 + \lambda^2)x + C,$$
$$w(x,y) = \left[A\sin(\beta x) + B\cos(\beta x)\right]e^{-\lambda y} + \frac{\nu}{\lambda}(\lambda^2 - \beta^2)x + C,$$
$$w(x,y) = Ae^{\lambda y + \beta x} + Be^{\gamma x} + \nu\gamma y + \frac{\nu}{\lambda}\gamma(\beta - \gamma)x + C, \quad \gamma = \pm\sqrt{\lambda^2 + \beta^2},$$

where $A$, $B$, $C$, $D$, $k$, $\beta$, and $\lambda$ are arbitrary constants.

5°. Generalized separable solution linear in $x$:
$$w(x,y) = F(y)x + G(y), \tag{1}$$

where the functions $F = F(y)$ and $G = G(y)$ are determined by the autonomous system of fourth-order ordinary differential equations
$$F'_y F''_{yy} - F F'''_{yyy} = \nu F''''_{yyyy}, \tag{2}$$
$$G'_y F''_{yy} - F G'''_{yyy} = \nu G''''_{yyyy}. \tag{3}$$

Equation (2) has the following particular solutions:
$$F = ay + b,$$
$$F = 6\nu(y + a)^{-1},$$
$$F = ae^{-\lambda y} + \lambda\nu,$$

where $a$, $b$, and $\lambda$ are arbitrary constants.

Let $F = F(y)$ be a solution of equation (2) ($F \not\equiv \text{const}$). Then the corresponding general solution of equation (3) can be written in the form
$$G = \int U\,dy + C_4, \quad U = C_1 U_1 + C_2 U_2 + C_3\left(U_2\int\frac{U_1}{\Phi}\,dy - U_1\int\frac{U_2}{\Phi}\,dy\right),$$

where $C_1$, $C_2$, $C_3$, and $C_4$ are arbitrary constants, and
$$U_1 = \begin{cases} F''_{yy} & \text{if } F''_{yy} \not\equiv 0, \\ F & \text{if } F''_{yy} \equiv 0, \end{cases} \quad U_2 = U_1\int\frac{\Phi\,dy}{U_1^2}, \quad \Phi = \exp\left(-\frac{1}{\nu}\int F\,dy\right).$$

6°. There is an exact solution of the form (generalizes the solution of Item 5°):
$$w(x,y) = F(z)x + G(z), \quad z = y + kx, \quad k \text{ is any number}.$$

7°. Self-similar solution:
$$w = \int F(z)\,dz + C_1, \quad z = \arctan\left(\frac{x}{y}\right),$$

where the function $F$ is determined by the first-order autonomous ordinary differential equation $3\nu(F'_z)^2 - 2F^3 + 12\nu F^2 + C_2 F + C_3 = 0$ ($C_1$, $C_2$, and $C_3$ are arbitrary constants).

8°. There is an exact solution of the form (generalizes the solution of Item 7°):
$$w = C_1\ln|x| + \int V(z)\,dz + C_2, \quad z = \arctan\left(\frac{x}{y}\right).$$

# References for Chapter T9

**Ablowitz, M. J. and Segur, H.,** *Solitons and the Inverse Scattering Transform*, Society for Industrial and Applied Mathematics, Philadelphia, 1981.

**Andreev, V. K., Kaptsov, O. V., Pukhnachov, V. V., and Rodionov, A. A.,** *Applications of Group-Theoretical Methods in Hydrodynamics*, Kluwer Academic, Dordrecht, 1999.

**Bullough, R. K. and Caudrey, P. J. (Editors),** *Solitons*, Springer-Verlag, Berlin, 1980.

**Burde, G. I.,** New similarity reductions of the steady-state boundary-layer equations, *J. Physica A: Math. Gen.*, Vol. 29, No. 8, pp. 1665–1683, 1996.

**Calogero, F. and Degasperis, A.,** *Spectral Transform and Solitons: Tolls to Solve and Investigate Nonlinear Evolution Equations*, North-Holland, Amsterdam, 1982.

**Cariello, F. and Tabor, M.,** Painlevé expansions for nonintegrable evolution equations, *Physica D*, Vol. 39, No. 1, pp. 77–94, 1989.

**Clarkson, P. A. and Kruskal, M. D.,** New similarity reductions of the Boussinesq equation, *J. Math. Phys.*, Vol. 30, No. 10, pp. 2201–2213, 1989.

**Dodd, R. K., Eilbeck, J. C., Gibbon, J. D., and Morris, H. C.,** *Solitons and Nonlinear Wave Equations*, Academic Press, London, 1982.

**Galaktionov, V. A.,** Invariant subspaces and new explicit solutions to evolution equations with quadratic nonlinearities, *Proc. Roy. Soc. Edinburgh*, Vol. 125A, No. 2, pp. 225–448, 1995.

**Gardner, C. S., Greene, J. M., Kruskal, M. D., and Miura, R. M.,** Method for solving the Korteweg–de Vries equation, *Phys. Rev. Lett.*, Vol. 19, No. 19, pp. 1095–1097, 1967.

**Goursat, E.,** *A Course of Mathematical Analysis, Vol. 3, Part 1* [Russian translation], Gostekhizdat, Moscow, 1933.

**Grundland, A. M. and Infeld, E.,** A family of non-linear Klein-Gordon equations and their solutions, *J. Math. Phys.*, Vol. 33, pp. 2498–2503, 1992.

**Hirota, R.,** Exact solution of the Korteweg-de Vries equation for multiple collisions of solutions, *Phys. Rev. Lett.*, Vol. 27, p. 1192, 1971.

**Ibragimov, N. H. (Editor),** *CRC Handbook of Lie Group Analysis of Differential Equations, Vol. 1, Symmetries, Exact Solutions and Conservation Laws*, CRC Press, Boca Raton, 1994.

**Kawahara, T. and Tanaka, M.,** Interactions of traveling fronts: an exact solution of a nonlinear diffusion equation, *Phys. Lett.*, Vol. 97, p. 311, 1983.

**Kersner, R.,** On some properties of weak solutions of quasilinear degenerate parabolic equations, *Acta Math. Academy of Sciences, Hung.*, Vol. 32, No. 3–4, pp. 301–330, 1978.

**Novikov, S. P., Manakov, S. V., Pitaevskii, L. B., and Zakharov, V. E.,** *Theory of Solitons. The Inverse Scattering Method*, Plenum Press, New York, 1984.

**Pavlovskii, Yu. N.,** Investigation of some invariant solutions to the boundary layer equations [in Russian], *Zhurn. Vychisl. Mat. i Mat. Fiziki*, Vol. 1, No. 2, pp. 280–294, 1961.

**Polyanin, A. D. and Zaitsev, V. F.,** *Handbook of Exact Solutions for Ordinary Differential Equations, 2nd Edition*, Chapman & Hall/CRC Press, Boca Raton, 2004.

**Polyanin, A. D. and Zaitsev, V. F.,** *Handbook of Nonlinear Partial Differential Equations*, Chapman & Hall/CRC Press, Boca Raton, 2004.

**Samarskii, A. A., Galaktionov, V. A., Kurdyumov, S. P., and Mikhailov, A. P.,** *Blow-Up in Problems for Quasilinear Parabolic Equations*, Walter de Gruyter, Berlin, 1995.

**Svirshchevskii, S. R.,** Lie–Bäcklund symmetries of linear ODEs and generalized separation of variables in nonlinear equations, *Phys. Lett. A*, Vol. 199, pp. 344–348, 1995.

# Chapter T10

# Systems of Partial Differential Equations

## T10.1. Nonlinear Systems of Two First-Order Equations

**1.** $\dfrac{\partial u}{\partial x} = auw, \quad \dfrac{\partial w}{\partial t} = buw.$

General solution:
$$u = -\frac{\psi'_t(t)}{a\varphi(x) + b\psi(t)}, \quad w = -\frac{\varphi'_x(x)}{a\varphi(x) + b\psi(t)},$$

where $\varphi(x)$ and $\psi(t)$ are arbitrary functions.

**2.** $\dfrac{\partial u}{\partial x} = auw, \quad \dfrac{\partial w}{\partial t} = bu^k.$

General solution:
$$w = \varphi(x) + E(x)\left[\psi(t) - \frac{1}{2}ak \int E(x)\,dx\right]^{-1},$$
$$u = \left(\frac{1}{b}\frac{\partial w}{\partial t}\right)^{1/k}, \quad E(x) = \exp\left[ak \int \varphi(x)\,dx\right],$$

where $\varphi(x)$ and $\psi(t)$ are arbitrary functions.

**3.** $\dfrac{\partial u}{\partial x} = auw^n, \quad \dfrac{\partial w}{\partial t} = bu^k w.$

General solution:
$$u = \left(\frac{-\psi'_t(t)}{bn\psi(t) - ak\varphi(x)}\right)^{1/k}, \quad w = \left(\frac{\varphi'_x(x)}{bn\psi(t) - ak\varphi(x)}\right)^{1/n},$$

where $\varphi(x)$ and $\psi(t)$ are arbitrary functions.

**4.** $\dfrac{\partial u}{\partial x} = uf(w), \quad \dfrac{\partial w}{\partial t} = u^k g(w).$

$1°$. First integral:
$$\frac{\partial w}{\partial x} = kg(w) \int \frac{f(w)}{g(w)}\,dw + \theta(x)g(w), \tag{1}$$

where $\theta(x)$ is an arbitrary function. The first integral (1) may be treated as a first-order ordinary differential equation in $x$. On finding its general solution, one should replace the constant of integration $C$ with an arbitrary function of time $\psi(t)$, since $w$ is dependent on $x$ and $t$.

2°. To the special case $\theta(x) = \text{const}$ in (1) there correspond special solutions of the form
$$w = w(z), \quad u = [\psi'(t)]^{1/k} v(z), \quad z = x + \psi(t) \tag{2}$$
involving one arbitrary function $\psi(t)$, with the prime denoting a derivative. The functions $w(z)$ and $v(z)$ are described by the autonomous system of ordinary differential equations
$$v'_z = f(w)v,$$
$$w'_z = g(w)v^k,$$
the general solution of which can be written in implicit form as
$$\int \frac{dw}{g(w)[kF(w) + C_1]} = z + C_2, \quad v = [kF(w) + C_1]^{1/k}, \quad F(w) = \int \frac{f(w)}{g(w)} dw.$$

5. $\dfrac{\partial u}{\partial x} = f(a_1 u + b_1 w), \quad \dfrac{\partial w}{\partial t} = g(a_2 u + b_2 w).$

Let $\Delta = a_1 b_2 - a_2 b_1 \neq 0$.

Additive separable solution:
$$u = \frac{1}{\Delta}[b_2 \varphi(x) - b_1 \psi(t)], \quad w = \frac{1}{\Delta}[a_1 \psi(t) - a_2 \varphi(x)],$$
where the functions $\varphi(x)$ and $\psi(t)$ are determined by the autonomous ordinary differential equations
$$\frac{b_2}{\Delta} \varphi'_x = f(\varphi), \quad \frac{a_1}{\Delta} \psi'_t = g(\psi).$$

Integrating yields
$$\frac{b_2}{\Delta} \int \frac{d\varphi}{f(\varphi)} = x + C_1, \quad \frac{a_1}{\Delta} \int \frac{d\psi}{g(\psi)} = t + C_2.$$

6. $\dfrac{\partial u}{\partial x} = f(au + bw), \quad \dfrac{\partial w}{\partial t} = g(au + bw).$

Solution:
$$u = b(k_1 x - \lambda_1 t) + y(\xi), \quad w = -a(k_1 x - \lambda_1 t) + z(\xi), \quad \xi = k_2 x - \lambda_2 t,$$
where $k_1$, $k_2$, $\lambda_1$, and $\lambda_2$ are arbitrary constants, and the functions $y(\xi)$ and $z(\xi)$ are determined by the autonomous system of ordinary differential equations
$$k_2 y'_\xi + bk_1 = f(ay + bz),$$
$$-\lambda_2 z'_\xi + a\lambda_1 = g(ay + bz).$$

7. $\dfrac{\partial u}{\partial x} = f(au - bw), \quad \dfrac{\partial w}{\partial t} = ug(au - bw) + wh(au - bw) + r(au - bw).$

Here, $f(z)$, $g(z)$, $h(z)$, and $r(z)$ are arbitrary functions.

Generalized separable solution:
$$u = \varphi(t) + b\theta(t)x, \quad w = \psi(t) + a\theta(t)x.$$

Here, the functions $\varphi = \varphi(t)$, $\psi = \psi(t)$, and $\theta = \theta(t)$ are determined by a system involving one algebraic (transcendental) and two ordinary differential equations:
$$b\theta = f(a\varphi - b\psi),$$
$$a\theta'_t = b\theta g(a\varphi - b\psi) + a\theta h(a\varphi - b\psi),$$
$$\psi'_t = \varphi g(a\varphi - b\psi) + \psi h(a\varphi - b\psi) + r(a\varphi - b\psi).$$

**8.** $\dfrac{\partial u}{\partial x} = f(au - bw) + cw, \quad \dfrac{\partial w}{\partial t} = ug(au - bw) + wh(au - bw) + r(au - bw).$

Here, $f(z)$, $g(z)$, $h(z)$, and $r(z)$ are arbitrary functions.

Generalized separable solution:
$$u = \varphi(t) + b\theta(t)e^{\lambda x}, \quad w = \psi(t) + a\theta(t)e^{\lambda x}, \quad \lambda = \dfrac{ac}{b}.$$

Here, the functions $\varphi = \varphi(t)$, $\psi = \psi(t)$, and $\theta = \theta(t)$ are determined by a system involving one algebraic (transcendental) and two ordinary differential equations:
$$f(a\varphi - b\psi) + c\psi = 0,$$
$$\psi'_t = \varphi g(a\varphi - b\psi) + \psi h(a\varphi - b\psi) + r(a\varphi - b\psi),$$
$$a\theta'_t = b\theta g(a\varphi - b\psi) + a\theta h(a\varphi - b\psi).$$

**9.** $\dfrac{\partial u}{\partial x} = e^{\lambda u} f(\lambda u - \sigma w), \quad \dfrac{\partial w}{\partial t} = e^{\sigma w} g(\lambda u - \sigma w).$

Solutions:
$$u = y(\xi) - \dfrac{1}{\lambda}\ln(C_1 t + C_2), \quad w = z(\xi) - \dfrac{1}{\sigma}\ln(C_1 t + C_2), \quad \xi = \dfrac{x + C_3}{C_1 t + C_2},$$

where the functions $y(\xi)$ and $z(\xi)$ are determined by the system of ordinary differential equations
$$y'_\xi = e^{\lambda y} f(\lambda y - \sigma z),$$
$$-C_1 \xi z'_\xi - \dfrac{C_1}{\sigma} = e^{\sigma z} g(\lambda y - \sigma z).$$

**10.** $\dfrac{\partial u}{\partial x} = u^k f(u^n w^m), \quad \dfrac{\partial w}{\partial t} = w^s g(u^n w^m).$

Self-similar solution with $s \neq 1$ and $n \neq 0$:
$$u = t^{\frac{m}{n(s-1)}} y(\xi), \quad w = t^{-\frac{1}{s-1}} z(\xi), \quad \xi = xt^{\frac{m(k-1)}{n(s-1)}},$$

where the functions $y(\xi)$ and $z(\xi)$ are determined by the system of ordinary differential equations
$$y'_\xi = y^k f(y^n z^m),$$
$$m(k-1)\xi z'_\xi - nz = n(s-1)z^s g(y^n z^m).$$

**11.** $\dfrac{\partial u}{\partial x} = u^k f(u^n w^m), \quad \dfrac{\partial w}{\partial t} = w g(u^n w^m).$

1°. Solution:
$$u = e^{mt} y(\xi), \quad w = e^{-nt} z(\xi), \quad \xi = e^{m(k-1)t} x,$$

where the functions $y(\xi)$ and $z(\xi)$ are determined by the system of ordinary differential equations
$$y'_\xi = y^k f(y^n z^m),$$
$$m(k-1)\xi z'_\xi - nz = z g(y^n z^m).$$

2°. If $k \ne 1$, it is more beneficial to seek solutions in the form
$$u = x^{-\frac{1}{k-1}} \varphi(\zeta), \quad w = x^{\frac{n}{m(k-1)}} \psi(\zeta), \quad \zeta = t + a \ln|x|,$$
where $a$ is an arbitrary constant, and the functions $\varphi(\zeta)$ and $\psi(\zeta)$ are determined by the autonomous system of ordinary differential equations
$$a\varphi'_\zeta + \frac{1}{1-k}\varphi = \varphi^k f(\varphi^n \psi^m),$$
$$\psi'_\zeta = \psi g(\varphi^n \psi^m).$$

12. $\dfrac{\partial u}{\partial x} = uf(u^n w^m), \quad \dfrac{\partial w}{\partial t} = wg(u^n w^m).$

Solution:
$$u = e^{m(kx - \lambda t)} y(\xi), \quad w = e^{-n(kx - \lambda t)} z(\xi), \quad \xi = \alpha x - \beta t,$$
where $k$, $\alpha$, $\beta$, and $\lambda$ are arbitrary constants, and the functions $y(\xi)$ and $z(\xi)$ are determined by the autonomous system of ordinary differential equations
$$\alpha y'_\xi + kmy = yf(y^n z^m),$$
$$-\beta z'_\xi + n\lambda z = zg(y^n z^m).$$

13. $\dfrac{\partial u}{\partial x} = uf(u^n w^m), \quad \dfrac{\partial w}{\partial t} = wg(u^k w^s).$

Let $\Delta = sn - km \ne 0$.

Multiplicative separable solutions:
$$u = [\varphi(x)]^{s/\Delta} [\psi(t)]^{-m/\Delta}, \quad w = [\varphi(x)]^{-k/\Delta} [\psi(t)]^{n/\Delta},$$
where the functions $\varphi(x)$ and $\psi(t)$ are determined by the autonomous ordinary differential equations
$$\frac{s}{\Delta}\varphi'_x = \varphi f(\varphi), \quad \frac{n}{\Delta}\psi'_t = \psi g(\psi).$$

Integrating yields
$$\frac{s}{\Delta} \int \frac{d\varphi}{\varphi f(\varphi)} = x + C_1, \quad \frac{n}{\Delta} \int \frac{d\psi}{\psi g(\psi)} = t + C_2.$$

14. $\dfrac{\partial u}{\partial x} = au \ln u + uf(u^n w^m), \quad \dfrac{\partial w}{\partial t} = wg(u^n w^m).$

Solution:
$$u = \exp(Cme^{ax})y(\xi), \quad w = \exp(-Cne^{ax})z(\xi), \quad \xi = kx - \lambda t,$$
where $C$, $k$, and $\lambda$ are arbitrary constants, and the functions $y(\xi)$ and $z(\xi)$ are determined by the autonomous system of ordinary differential equations
$$ky'_\xi = ay \ln y + yf(y^n z^m),$$
$$-\lambda z'_\xi = zg(y^n z^m).$$

**15.** $\dfrac{\partial u}{\partial x} = uf(au^k + bw), \quad \dfrac{\partial w}{\partial t} = u^k.$

Solution:
$$w = \varphi(x) + C\exp\left[-\lambda t + k\int f(b\varphi(x))\,dx\right], \quad u = \left(\dfrac{\partial w}{\partial t}\right)^{1/k}, \quad \lambda = \dfrac{b}{a},$$
where $\varphi(x)$ is an arbitrary function and $C$ is an arbitrary constant.

**16.** $\dfrac{\partial u}{\partial x} = uf(au^n + bw), \quad \dfrac{\partial w}{\partial t} = u^k g(au^n + bw).$

Solution:
$$u = (C_1 t + C_2)^{\frac{1}{n-k}}\theta(x), \quad w = \varphi(x) - \dfrac{a}{b}(C_1 t + C_2)^{\frac{n}{n-k}}[\theta(x)]^n,$$
where $C_1$ and $C_2$ are arbitrary constants, and the functions $\theta = \theta(x)$ and $\varphi = \varphi(x)$ are determined by the system of differential-algebraic equations
$$\theta'_x = \theta f(b\varphi),$$
$$\theta^{n-k} = \dfrac{b(k-n)}{aC_1 n}g(b\varphi).$$

**17.** $\dfrac{\partial u}{\partial x} = \dfrac{cw^n u}{a + bw^n}, \quad \dfrac{\partial w}{\partial t} = (aw + bw^{n+1})u^k.$

General solution with $b \neq 0$:
$$w = \left[\psi(t)e^{F(x)} - be^{F(x)}\int e^{-F(x)}\varphi(x)\,dx\right]^{-1/n},$$
$$u = \left(\dfrac{w_t}{aw + bw^{n+1}}\right)^{1/k}, \quad F(x) = \dfrac{ck}{b}x - a\int \varphi(x)\,dx,$$
where $\varphi(x)$ and $\psi(t)$ are arbitrary functions.

## T10.2. Linear Systems of Two Second-Order Equations

**1.** $\dfrac{\partial u}{\partial t} = a\dfrac{\partial^2 u}{\partial x^2} + b_1 u + c_1 w, \quad \dfrac{\partial w}{\partial t} = a\dfrac{\partial^2 w}{\partial x^2} + b_2 u + c_2 w.$

*Constant-coefficient second-order linear system of parabolic type.*

Solution:
$$u = \dfrac{b_1 - \lambda_2}{b_2(\lambda_1 - \lambda_2)}e^{\lambda_1 t}\theta_1 - \dfrac{b_1 - \lambda_1}{b_2(\lambda_1 - \lambda_2)}e^{\lambda_2 t}\theta_2,$$
$$w = \dfrac{1}{\lambda_1 - \lambda_2}\left(e^{\lambda_1 t}\theta_1 - e^{\lambda_2 t}\theta_2\right),$$
where $\lambda_1$ and $\lambda_2$ are roots of the quadratic equation
$$\lambda^2 - (b_1 + c_2)\lambda + b_1 c_2 - b_2 c_1 = 0,$$
and the functions $\theta_n = \theta_n(x,t)$ satisfy the independent linear heat equations
$$\dfrac{\partial \theta_1}{\partial t} = a\dfrac{\partial^2 \theta_1}{\partial x^2}, \quad \dfrac{\partial \theta_2}{\partial t} = a\dfrac{\partial^2 \theta_2}{\partial x^2}.$$

2. $\dfrac{\partial u}{\partial t} = a\dfrac{\partial^2 u}{\partial x^2} + f_1(t)u + g_1(t)w, \qquad \dfrac{\partial w}{\partial t} = a\dfrac{\partial^2 w}{\partial x^2} + f_2(t)u + g_2(t)w.$

*Variable-coefficient second-order linear system of parabolic type.*
   Solution:
$$u = \varphi_1(t)U(x,t) + \varphi_2(t)W(x,t),$$
$$w = \psi_1(t)U(x,t) + \psi_2(t)W(x,t),$$

where the pairs of functions $\varphi_1 = \varphi_1(t)$, $\psi_1 = \psi_1(t)$ and $\varphi_2 = \varphi_2(t)$, $\psi_2 = \psi_2(t)$ are linearly independent (fundamental) solutions to the system of linear ordinary differential equations
$$\varphi'_t = f_1(t)\varphi + g_1(t)\psi,$$
$$\psi'_t = f_2(t)\varphi + g_2(t)\psi,$$

and the functions $U = U(x,t)$ and $W = W(x,t)$ satisfy the independent linear heat equations
$$\dfrac{\partial U}{\partial t} = a\dfrac{\partial^2 U}{\partial x^2}, \qquad \dfrac{\partial W}{\partial t} = a\dfrac{\partial^2 W}{\partial x^2}.$$

3. $\dfrac{\partial^2 u}{\partial t^2} = k\dfrac{\partial^2 u}{\partial x^2} + a_1 u + b_1 w, \qquad \dfrac{\partial^2 w}{\partial t^2} = k\dfrac{\partial^2 w}{\partial x^2} + a_2 u + b_2 w.$

*Constant-coefficient second-order linear system of hyperbolic type.*
   Solution:
$$u = \dfrac{a_1 - \lambda_2}{a_2(\lambda_1 - \lambda_2)}\theta_1 - \dfrac{a_1 - \lambda_1}{a_2(\lambda_1 - \lambda_2)}\theta_2, \qquad w = \dfrac{1}{\lambda_1 - \lambda_2}(\theta_1 - \theta_2),$$

where $\lambda_1$ and $\lambda_2$ are roots of the quadratic equation
$$\lambda^2 - (a_1 + b_2)\lambda + a_1 b_2 - a_2 b_1 = 0,$$

and the functions $\theta_n = \theta_n(x,t)$ satisfy the linear Klein–Gordon equations
$$\dfrac{\partial^2 \theta_1}{\partial t^2} = k\dfrac{\partial^2 \theta_1}{\partial x^2} + \lambda_1 \theta_1, \qquad \dfrac{\partial^2 \theta_2}{\partial t^2} = k\dfrac{\partial^2 \theta_2}{\partial x^2} + \lambda_2 \theta_2.$$

4. $\dfrac{\partial^2 u}{\partial x^2} + \dfrac{\partial^2 u}{\partial y^2} = a_1 u + b_1 w, \qquad \dfrac{\partial^2 w}{\partial x^2} + \dfrac{\partial^2 w}{\partial y^2} = a_2 u + b_2 w.$

*Constant-coefficient second-order linear system of elliptic type.*
   Solution:
$$u = \dfrac{a_1 - \lambda_2}{a_2(\lambda_1 - \lambda_2)}\theta_1 - \dfrac{a_1 - \lambda_1}{a_2(\lambda_1 - \lambda_2)}\theta_2, \qquad w = \dfrac{1}{\lambda_1 - \lambda_2}(\theta_1 - \theta_2),$$

where $\lambda_1$ and $\lambda_2$ are roots of the quadratic equation
$$\lambda^2 - (a_1 + b_2)\lambda + a_1 b_2 - a_2 b_1 = 0,$$

and the functions $\theta_n = \theta_n(x,y)$ satisfy the linear Helmholtz equations
$$\dfrac{\partial^2 \theta_1}{\partial x^2} + \dfrac{\partial^2 \theta_1}{\partial y^2} = \lambda_1 \theta_1, \qquad \dfrac{\partial^2 \theta_2}{\partial x^2} + \dfrac{\partial^2 \theta_2}{\partial y^2} = \lambda_2 \theta_2.$$

## T10.3. Nonlinear Systems of Two Second-Order Equations

### T10.3.1. Systems of the Form
$$\frac{\partial u}{\partial t} = a\frac{\partial^2 u}{\partial x^2} + F(u, w), \quad \frac{\partial w}{\partial t} = b\frac{\partial^2 w}{\partial x^2} + G(u, w)$$

**Preliminary remarks.** Systems of this form often arise in the theory of heat and mass transfer in chemically reactive media, theory of chemical reactors, combustion theory, mathematical biology, and biophysics.

Such systems are invariant under translations in the independent variables (and under the change of $x$ to $-x$) and admit traveling-wave solutions $u = u(kx - \lambda t)$, $w = w(kx - \lambda t)$. These solutions as well as those with one of the unknown functions being identically zero are not considered further in this section.

The functions $f(\varphi)$, $g(\varphi)$, $h(\varphi)$ appearing below are arbitrary functions of their argument, $\varphi = \varphi(u, w)$; the equations are arranged in order of complexity of this argument.

---

T10.3.1-1. Arbitrary functions depend on a linear combination of the unknowns.

---

1.  $\dfrac{\partial u}{\partial t} = a\dfrac{\partial^2 u}{\partial x^2} + u \exp\left(k\dfrac{w}{u}\right) f(u), \quad \dfrac{\partial w}{\partial t} = a\dfrac{\partial^2 w}{\partial x^2} + \exp\left(k\dfrac{w}{u}\right)[wf(u) + g(u)].$

Solution:
$$u = y(\xi), \quad w = -\frac{2}{k}\ln|bx|\, y(\xi) + z(\xi), \quad \xi = \frac{x + C_3}{\sqrt{C_1 t + C_2}},$$

where $C_1$, $C_2$, $C_3$, and $b$ are arbitrary constants, and the functions $y = y(\xi)$ and $z = z(\xi)$ are determined by the system of ordinary differential equations

$$ay''_{\xi\xi} + \frac{1}{2}C_1\xi y'_\xi + \frac{1}{b^2\xi^2} y \exp\left(k\frac{z}{y}\right) f(y) = 0,$$

$$az''_{\xi\xi} + \frac{1}{2}C_1\xi z'_\xi - \frac{4a}{k\xi} y'_\xi + \frac{2a}{k\xi^2} y + \frac{1}{b^2\xi^2}\exp\left(k\frac{z}{y}\right)[zf(y) + g(y)] = 0.$$

2.  $\dfrac{\partial u}{\partial t} = a_1\dfrac{\partial^2 u}{\partial x^2} + f(bu + cw), \quad \dfrac{\partial w}{\partial t} = a_2\dfrac{\partial^2 w}{\partial x^2} + g(bu + cw).$

Solution:
$$u = c(\alpha x^2 + \beta x + \gamma t) + y(\xi), \quad w = -b(\alpha x^2 + \beta x + \gamma t) + z(\xi), \quad \xi = kx - \lambda t,$$

where $k$, $\alpha$, $\beta$, $\gamma$, and $\lambda$ are arbitrary constants, and the functions $y(\xi)$ and $z(\xi)$ are determined by the autonomous system of ordinary differential equations

$$a_1 k^2 y''_{\xi\xi} + \lambda y'_\xi + 2a_1 c\alpha - c\gamma + f(by + cz) = 0,$$
$$a_2 k^2 z''_{\xi\xi} + \lambda z'_\xi - 2a_2 b\alpha + b\gamma + g(by + cz) = 0.$$

3. $\dfrac{\partial u}{\partial t} = a\dfrac{\partial^2 u}{\partial x^2} + f(bu + cw), \qquad \dfrac{\partial w}{\partial t} = a\dfrac{\partial^2 w}{\partial x^2} + g(bu + cw).$

Solution:
$$u = c\theta(x,t) + y(\xi), \qquad w = -b\theta(x,t) + z(\xi), \qquad \xi = kx - \lambda t,$$

where the functions $y(\xi)$ and $z(\xi)$ are determined by the autonomous system of ordinary differential equations
$$ak^2 y''_{\xi\xi} + \lambda y'_\xi + f(by + cz) = 0,$$
$$ak^2 z''_{\xi\xi} + \lambda z'_\xi + g(by + cz) = 0,$$

and the function $\theta = \theta(x,t)$ satisfies the linear heat equation
$$\dfrac{\partial \theta}{\partial t} = a\dfrac{\partial^2 \theta}{\partial x^2}.$$

4. $\dfrac{\partial u}{\partial t} = a\dfrac{\partial^2 u}{\partial x^2} + uf(bu - cw) + g(bu - cw),$

$\dfrac{\partial w}{\partial t} = a\dfrac{\partial^2 w}{\partial x^2} + wf(bu - cw) + h(bu - cw).$

1°. Solution:
$$u = \varphi(t) + c\exp\left[\int f(b\varphi - c\psi)\,dt\right]\theta(x,t), \qquad w = \psi(t) + b\exp\left[\int f(b\varphi - c\psi)\,dt\right]\theta(x,t),$$

where $\varphi = \varphi(t)$ and $\psi = \psi(t)$ are determined by the autonomous system of ordinary differential equations
$$\varphi'_t = \varphi f(b\varphi - c\psi) + g(b\varphi - c\psi),$$
$$\psi'_t = \psi f(b\varphi - c\psi) + h(b\varphi - c\psi),$$

and the function $\theta = \theta(x,t)$ satisfies the linear heat equation
$$\dfrac{\partial \theta}{\partial t} = a\dfrac{\partial^2 \theta}{\partial x^2}.$$

2°. Let us multiply the first equation by $b$ and the second one by $-c$ and add the results together to obtain
$$\dfrac{\partial \zeta}{\partial t} = a\dfrac{\partial^2 \zeta}{\partial x^2} + \zeta f(\zeta) + bg(\zeta) - ch(\zeta), \qquad \zeta = bu - cw. \tag{1}$$

This equation will be considered in conjunction with the first equation of the original system
$$\dfrac{\partial u}{\partial t} = a\dfrac{\partial^2 u}{\partial x^2} + uf(\zeta) + g(\zeta). \tag{2}$$

Equation (1) can be treated separately. An extensive list of exact solutions to equations of this form for various kinetic functions $F(\zeta) = \zeta f(\zeta) + bg(\zeta) - ch(\zeta)$ can be found in the book by Polyanin and Zaitsev (2004). Given a solution $\zeta = \zeta(x,t)$ to equation (1), the function $u = u(x,t)$ can be determined by solving the linear equation (2) and the function $w = w(x,t)$ is found as $w = (bu - \zeta)/c$.

Note two important solutions to equation (1):

(i) In the general case, equation (1) admits traveling-wave solutions $\zeta = \zeta(z)$, where $z = kx - \lambda t$. Then the corresponding exact solutions to equation (2) are expressed as $u = u_0(z) + \sum e^{\beta_n t} u_n(z)$.

(ii) If the condition $\zeta f(\zeta) + bg(\zeta) - ch(\zeta) = k_1\zeta + k_0$ holds, equation (1) is linear,

$$\frac{\partial \zeta}{\partial t} = a\frac{\partial^2 \zeta}{\partial x^2} + k_1\zeta + k_0,$$

and, hence, can be reduced to the linear heat equation.

5. $\quad \dfrac{\partial u}{\partial t} = a\dfrac{\partial^2 u}{\partial x^2} + e^{\lambda u}f(\lambda u - \sigma w), \qquad \dfrac{\partial w}{\partial t} = b\dfrac{\partial^2 w}{\partial x^2} + e^{\sigma w}g(\lambda u - \sigma w).$

1°. Solution:

$$u = y(\xi) - \frac{1}{\lambda}\ln(C_1 t + C_2), \qquad w = z(\xi) - \frac{1}{\sigma}\ln(C_1 t + C_2), \qquad \xi = \frac{x + C_3}{\sqrt{C_1 t + C_2}},$$

where $C_1$, $C_2$, and $C_3$ are arbitrary constants, and the functions $y = y(\xi)$ and $z = z(\xi)$ are determined by the system of ordinary differential equations

$$ay''_{\xi\xi} + \frac{1}{2}C_1\xi y'_\xi + \frac{C_1}{\lambda} + e^{\lambda y}f(\lambda y - \sigma z) = 0,$$

$$bz''_{\xi\xi} + \frac{1}{2}C_1\xi z'_\xi + \frac{C_1}{\sigma} + e^{\sigma z}g(\lambda y - \sigma z) = 0.$$

2°. Solution with $b = a$:

$$u = \theta(x,t), \qquad w = \frac{\lambda}{\sigma}\theta(x,t) - \frac{k}{\sigma},$$

where $k$ is a root of the algebraic (transcendental) equation

$$\lambda f(k) = \sigma e^{-k}g(k),$$

and the function $\theta = \theta(x,t)$ is determined by the differential equation

$$\frac{\partial \theta}{\partial t} = a\frac{\partial^2 \theta}{\partial x^2} + f(k)e^{\lambda\theta}.$$

For exact solutions to this equation, see Polyanin and Zaitsev (2004).

---

T10.3.1-2. Arbitrary functions depend on the ratio of the unknowns.

6. $\quad \dfrac{\partial u}{\partial t} = a\dfrac{\partial^2 u}{\partial x^2} + uf\!\left(\dfrac{u}{w}\right), \qquad \dfrac{\partial w}{\partial t} = b\dfrac{\partial^2 w}{\partial x^2} + wg\!\left(\dfrac{u}{w}\right).$

1°. Multiplicative separable solution:

$$u = [C_1\sin(kx) + C_2\cos(kx)]\varphi(t),$$
$$w = [C_1\sin(kx) + C_2\cos(kx)]\psi(t),$$

where $C_1$, $C_2$, and $k$ are arbitrary constants, and the functions $\varphi = \varphi(t)$ and $\psi = \psi(t)$ are determined by the autonomous system of ordinary differential equations

$$\varphi'_t = -ak^2\varphi + \varphi f(\varphi/\psi),$$
$$\psi'_t = -bk^2\psi + \psi g(\varphi/\psi).$$

2°. Multiplicative separable solution:

$$u = [C_1 \exp(kx) + C_2 \exp(-kx)]U(t),$$
$$w = [C_1 \exp(kx) + C_2 \exp(-kx)]W(t),$$

where $C_1$, $C_2$, and $k$ are arbitrary constants, and the functions $U = U(t)$ and $W = W(t)$ are determined by the autonomous system of ordinary differential equations

$$U'_t = ak^2 U + U f(U/W),$$
$$W'_t = bk^2 W + W g(U/W).$$

3°. Degenerate solution:

$$u = (C_1 x + C_2)U(t),$$
$$w = (C_1 x + C_2)W(t),$$

where $C_1$ and $C_2$ are arbitrary constants, and the functions $U = U(t)$ and $W = W(t)$ are determined by the autonomous system of ordinary differential equations

$$U'_t = U f(U/W),$$
$$W'_t = W g(U/W).$$

This autonomous system can be integrated since it is reduced, after eliminating $t$, to a homogeneous first-order equation. The systems presented in Items 1° and 2° can be integrated likewise.

4°. Multiplicative separable solution:

$$u = e^{-\lambda t} y(x), \qquad w = e^{-\lambda t} z(x),$$

where $\lambda$ is an arbitrary constant and the functions $y = y(x)$ and $z = z(x)$ are determined by the autonomous system of ordinary differential equations

$$ay''_{xx} + \lambda y + y f(y/z) = 0,$$
$$bz''_{xx} + \lambda z + z g(y/z) = 0.$$

5°. Solution (generalizes the solution of Item 4°):

$$u = e^{kx-\lambda t} y(\xi), \qquad w = e^{kx-\lambda t} z(\xi), \qquad \xi = \beta x - \gamma t,$$

where $k$, $\lambda$, $\beta$, and $\gamma$ are arbitrary constants, and the functions $y = y(\xi)$ and $z = z(\xi)$ are determined by the autonomous system of ordinary differential equations

$$a\beta^2 y''_{\xi\xi} + (2ak\beta + \gamma)y'_\xi + (ak^2 + \lambda)y + y f(y/z) = 0,$$
$$b\beta^2 z''_{\xi\xi} + (2bk\beta + \gamma)z'_\xi + (bk^2 + \lambda)z + z g(y/z) = 0.$$

To the special case $k = \lambda = 0$ there corresponds a traveling-wave solution. If $k = \gamma = 0$ and $\beta = 1$, we have the solution of Item 4°.

**6a.** $\dfrac{\partial u}{\partial t} = a\dfrac{\partial^2 u}{\partial x^2} + uf\left(\dfrac{u}{w}\right), \quad \dfrac{\partial w}{\partial t} = a\dfrac{\partial^2 w}{\partial x^2} + wg\left(\dfrac{u}{w}\right).$

This system is a special case of system T10.3.1.6 with $b = a$ and hence it admits the above solutions given in Items 1°–5°. In addition, it has some interesting properties and other solutions, which are given below.

Suppose $u = u(x,t)$, $w = w(x,t)$ is a solution of the system. Then the functions

$$u_1 = Au(\pm x + C_1, t + C_2), \quad w_1 = Aw(\pm x + C_1, t + C_2);$$
$$u_2 = \exp(\lambda x + a\lambda^2 t)u(x + 2a\lambda t, t), \quad w_2 = \exp(\lambda x + a\lambda^2 t)w(x + 2a\lambda t, t),$$

where $A$, $C_1$, $C_2$, and $\lambda$ are arbitrary constants, are also solutions of these equations.

6°. Point-source solution:

$$u = \exp\left(-\dfrac{x^2}{4at}\right)\varphi(t), \quad w = \exp\left(-\dfrac{x^2}{4at}\right)\psi(t),$$

where the functions $\varphi = \varphi(t)$ and $\psi = \psi(t)$ are determined by the autonomous system of ordinary differential equations

$$\varphi'_t = -\dfrac{1}{2t}\varphi + \varphi f\left(\dfrac{\varphi}{\psi}\right),$$
$$\psi'_t = -\dfrac{1}{2t}\psi + \psi g\left(\dfrac{\varphi}{\psi}\right).$$

7°. Functional separable solution:

$$u = \exp\left(kxt + \tfrac{2}{3}ak^2 t^3 - \lambda t\right) y(\xi),$$
$$w = \exp\left(kxt + \tfrac{2}{3}ak^2 t^3 - \lambda t\right) z(\xi),\quad \xi = x + akt^2,$$

where $k$ and $\lambda$ are arbitrary constants, and the functions $y = y(\xi)$ and $z = z(\xi)$ are determined by the autonomous system of ordinary differential equations

$$ay''_{\xi\xi} + (\lambda - k\xi)y + yf(y/z) = 0,$$
$$az''_{\xi\xi} + (\lambda - k\xi)z + zg(y/z) = 0.$$

8°. Let $k$ be a root of the algebraic (transcendental) equation

$$f(k) = g(k). \tag{1}$$

Solution:

$$u = ke^{\lambda t}\theta, \quad w = e^{\lambda t}\theta, \quad \lambda = f(k),$$

where the function $\theta = \theta(x,t)$ satisfies the linear heat equation

$$\dfrac{\partial \theta}{\partial t} = a\dfrac{\partial^2 \theta}{\partial x^2}.$$

9°. Periodic solution:

$$u = Ak\exp(-\mu x)\sin(\beta x - 2a\beta\mu t + B),$$
$$w = A\exp(-\mu x)\sin(\beta x - 2a\beta\mu t + B),\quad \beta = \sqrt{\mu^2 + \dfrac{1}{a}f(k)},$$

where $A$, $B$, and $\mu$ are arbitrary constants, and $k$ is a root of the algebraic (transcendental) equation (1).

**10°. Solution:**

$$u = \varphi(t)\exp\left[\int g(\varphi(t))\,dt\right]\theta(x,t), \quad w = \exp\left[\int g(\varphi(t))\,dt\right]\theta(x,t),$$

where the function $\varphi = \varphi(t)$ is described by the separable first-order nonlinear ordinary differential equation

$$\varphi'_t = [f(\varphi) - g(\varphi)]\varphi, \tag{2}$$

and the function $\theta = \theta(x,t)$ satisfies the linear heat equation

$$\frac{\partial \theta}{\partial t} = a\frac{\partial^2 \theta}{\partial x^2}.$$

To the particular solution $\varphi = k = \text{const}$ of equation (2) there corresponds the solution given in Item 8°. The general solution of equation (2) is written out in implicit form as

$$\int \frac{d\varphi}{[f(\varphi) - g(\varphi)]\varphi} = t + C.$$

**11°.** The transformation

$$u = a_1 U + b_1 W, \quad w = a_2 U + b_2 W,$$

where $a_n$ and $b_n$ are arbitrary constants ($n = 1, 2$), leads to an equation of similar form for $U$ and $W$.

**7.** $\dfrac{\partial u}{\partial t} = a\dfrac{\partial^2 u}{\partial x^2} + uf\left(\dfrac{u}{w}\right) + g\left(\dfrac{u}{w}\right), \quad \dfrac{\partial w}{\partial t} = a\dfrac{\partial^2 w}{\partial x^2} + wf\left(\dfrac{u}{w}\right) + h\left(\dfrac{u}{w}\right).$

Let $k$ be a root of the algebraic (transcendental) equation

$$g(k) = kh(k).$$

**1°.** Solution with $f(k) \neq 0$:

$$u(x,t) = k\left(\exp[f(k)t]\theta(x,t) - \frac{h(k)}{f(k)}\right), \quad w(x,t) = \exp[f(k)t]\theta(x,t) - \frac{h(k)}{f(k)},$$

where the function $\theta = \theta(x,t)$ satisfies the linear heat equation

$$\frac{\partial \theta}{\partial t} = a\frac{\partial^2 \theta}{\partial x^2}. \tag{1}$$

**2°.** Solution with $f(k) = 0$:

$$u(x,t) = k[\theta(x,t) + h(k)t], \quad w(x,t) = \theta(x,t) + h(k)t,$$

where the function $\theta = \theta(x,t)$ satisfies the linear heat equation (1).

8. $\dfrac{\partial u}{\partial t} = a\dfrac{\partial^2 u}{\partial x^2} + uf\left(\dfrac{u}{w}\right) + \dfrac{u}{w}h\left(\dfrac{u}{w}\right), \quad \dfrac{\partial w}{\partial t} = a\dfrac{\partial^2 w}{\partial x^2} + wg\left(\dfrac{u}{w}\right) + h\left(\dfrac{u}{w}\right).$

Solution:

$$u = \varphi(t)G(t)\left[\theta(x,t) + \int \dfrac{h(\varphi)}{G(t)}\,dt\right], \quad w = G(t)\left[\theta(x,t) + \int \dfrac{h(\varphi)}{G(t)}\,dt\right], \quad G(t) = \exp\left[\int g(\varphi)\,dt\right],$$

where the function $\varphi = \varphi(t)$ is described by the separable first-order nonlinear ordinary differential equation

$$\varphi'_t = [f(\varphi) - g(\varphi)]\varphi, \tag{1}$$

and the function $\theta = \theta(x,t)$ satisfies the linear heat equation

$$\dfrac{\partial \theta}{\partial t} = a\dfrac{\partial^2 \theta}{\partial x^2}.$$

The general solution of equation (1) is written out in implicit form as

$$\int \dfrac{d\varphi}{[f(\varphi) - g(\varphi)]\varphi} = t + C.$$

9. $\dfrac{\partial u}{\partial t} = a\dfrac{\partial^2 u}{\partial x^2} + uf_1\left(\dfrac{w}{u}\right) + wg_1\left(\dfrac{w}{u}\right), \quad \dfrac{\partial w}{\partial t} = a\dfrac{\partial^2 w}{\partial x^2} + uf_2\left(\dfrac{w}{u}\right) + wg_2\left(\dfrac{w}{u}\right).$

Solution:

$$u = \exp\left\{\int [f_1(\varphi) + \varphi g_1(\varphi)]\,dt\right\}\theta(x,t), \quad w(x,t) = \varphi(t)\exp\left\{\int [f_1(\varphi) + \varphi g_1(\varphi)]\,dt\right\}\theta(x,t),$$

where the function $\varphi = \varphi(t)$ is described by the separable first-order nonlinear ordinary differential equation

$$\varphi'_t = f_2(\varphi) + \varphi g_2(\varphi) - \varphi[f_1(\varphi) + \varphi g_1(\varphi)],$$

and the function $\theta = \theta(x,t)$ satisfies the linear heat equation

$$\dfrac{\partial \theta}{\partial t} = a\dfrac{\partial^2 \theta}{\partial x^2}.$$

10. $\dfrac{\partial u}{\partial t} = a\dfrac{\partial^2 u}{\partial x^2} + u^3 f\left(\dfrac{u}{w}\right), \quad \dfrac{\partial w}{\partial t} = a\dfrac{\partial^2 w}{\partial x^2} + u^3 g\left(\dfrac{u}{w}\right).$

Solution:

$$u = (x+A)\varphi(z), \quad w = (x+A)\psi(z), \quad z = t + \dfrac{1}{6a}(x+A)^2 + B,$$

where $A$ and $B$ are arbitrary constants, and the functions $\varphi = \varphi(z)$ and $\psi = \psi(z)$ are determined by the autonomous system of ordinary differential equations

$$\varphi''_{zz} + 9a\varphi^3 f(\varphi/\psi) = 0,$$
$$\psi''_{zz} + 9a\varphi^3 g(\varphi/\psi) = 0.$$

**11.** $\dfrac{\partial u}{\partial t} = \dfrac{\partial^2 u}{\partial x^2} + au - u^3 f\!\left(\dfrac{u}{w}\right), \qquad \dfrac{\partial w}{\partial t} = \dfrac{\partial^2 w}{\partial x^2} + aw - u^3 g\!\left(\dfrac{u}{w}\right).$

$1°$. Solution with $a > 0$:
$$u = \left[C_1 \exp\!\left(\tfrac{1}{2}\sqrt{2a}\,x + \tfrac{3}{2}at\right) - C_2 \exp\!\left(-\tfrac{1}{2}\sqrt{2a}\,x + \tfrac{3}{2}at\right)\right]\varphi(z),$$
$$w = \left[C_1 \exp\!\left(\tfrac{1}{2}\sqrt{2a}\,x + \tfrac{3}{2}at\right) - C_2 \exp\!\left(-\tfrac{1}{2}\sqrt{2a}\,x + \tfrac{3}{2}at\right)\right]\psi(z),$$
$$z = C_1 \exp\!\left(\tfrac{1}{2}\sqrt{2a}\,x + \tfrac{3}{2}at\right) + C_2 \exp\!\left(-\tfrac{1}{2}\sqrt{2a}\,x + \tfrac{3}{2}at\right) + C_3,$$

where $C_1$, $C_2$, and $C_3$ are arbitrary constants, and the functions $\varphi = \varphi(z)$ and $\psi = \psi(z)$ are determined by the autonomous system of ordinary differential equations
$$a\varphi''_{zz} = 2\varphi^3 f(\varphi/\psi),$$
$$a\psi''_{zz} = 2\varphi^3 g(\varphi/\psi).$$

$2°$. Solution with $a < 0$:
$$u = \exp\!\left(\tfrac{3}{2}at\right) \sin\!\left(\tfrac{1}{2}\sqrt{2|a|}\,x + C_1\right) U(\xi),$$
$$w = \exp\!\left(\tfrac{3}{2}at\right) \sin\!\left(\tfrac{1}{2}\sqrt{2|a|}\,x + C_1\right) W(\xi),$$
$$\xi = \exp\!\left(\tfrac{3}{2}at\right) \cos\!\left(\tfrac{1}{2}\sqrt{2|a|}\,x + C_1\right) + C_2,$$

where $C_1$ and $C_2$ are arbitrary constants, and the functions $U = U(\xi)$ and $W = W(\xi)$ are determined by the autonomous system of ordinary differential equations
$$aU''_{\xi\xi} = -2U^3 f(U/W),$$
$$aW''_{\xi\xi} = -2U^3 g(U/W).$$

**12.** $\dfrac{\partial u}{\partial t} = a\dfrac{\partial^2 u}{\partial x^2} + u^n f\!\left(\dfrac{u}{w}\right), \qquad \dfrac{\partial w}{\partial t} = b\dfrac{\partial^2 w}{\partial x^2} + w^n g\!\left(\dfrac{u}{w}\right).$

If $f(z) = kz^{-m}$ and $g(z) = -kz^{n-m}$, the system describes an $n$th-order chemical reaction (of order $n - m$ in the component $u$ and of order $m$ in the component $w$).

$1°$. Self-similar solution with $n \ne 1$:
$$u = (C_1 t + C_2)^{\frac{1}{1-n}} y(\xi), \qquad w = (C_1 t + C_2)^{\frac{1}{1-n}} z(\xi), \qquad \xi = \dfrac{x + C_3}{\sqrt{C_1 t + C_2}},$$

where $C_1$, $C_2$, and $C_3$ are arbitrary constants, and the functions $y = y(\xi)$ and $z = z(\xi)$ are determined by the system of ordinary differential equations
$$ay''_{\xi\xi} + \tfrac{1}{2}C_1 \xi y'_\xi + \dfrac{C_1}{n-1} y + y^n f\!\left(\dfrac{y}{z}\right) = 0,$$
$$bz''_{\xi\xi} + \tfrac{1}{2}C_1 \xi z'_\xi + \dfrac{C_1}{n-1} z + z^n g\!\left(\dfrac{y}{z}\right) = 0.$$

$2°$. Solution with $b = a$:
$$u(x,t) = k\theta(x,t), \qquad w(x,t) = \theta(x,t),$$

where $k$ is a root of the algebraic (transcendental) equation
$$k^{n-1} f(k) = g(k),$$

and the function $\theta = \theta(x,t)$ satisfies the heat equation with a power-law nonlinearity
$$\dfrac{\partial \theta}{\partial t} = a\dfrac{\partial^2 \theta}{\partial x^2} + g(k)\theta^n.$$

13. $\dfrac{\partial u}{\partial t} = a\dfrac{\partial^2 u}{\partial x^2} + uf\left(\dfrac{u}{w}\right)\ln u + ug\left(\dfrac{u}{w}\right),$

$\dfrac{\partial w}{\partial t} = a\dfrac{\partial^2 w}{\partial x^2} + wf\left(\dfrac{u}{w}\right)\ln w + wh\left(\dfrac{u}{w}\right).$

Solution:
$$u(x,t) = \varphi(t)\psi(t)\theta(x,t), \quad w(x,t) = \psi(t)\theta(x,t),$$

where the functions $\varphi = \varphi(t)$ and $\psi = \psi(t)$ are determined by solving the first-order autonomous ordinary differential equations

$$\varphi'_t = \varphi[g(\varphi) - h(\varphi) + f(\varphi)\ln\varphi], \tag{1}$$
$$\psi'_t = \psi[h(\varphi) + f(\varphi)\ln\psi], \tag{2}$$

and the function $\theta = \theta(x,t)$ is determined by the differential equation

$$\dfrac{\partial \theta}{\partial t} = a\dfrac{\partial^2 \theta}{\partial x^2} + f(\varphi)\theta\ln\theta. \tag{3}$$

The separable equation (1) can be solved to obtain a solution in implicit form. Equation (2) is easy to integrate — with the change of variable $\psi = e^\zeta$, it is reduced to a linear equation. Equation (3) admits exact solutions of the form

$$\theta = \exp\left[\sigma_2(t)x^2 + \sigma_1(t)x + \sigma_0(t)\right],$$

where the functions $\sigma_n(t)$ are described by the equations

$$\sigma'_2 = f(\varphi)\sigma_2 + 4a\sigma_2^2,$$
$$\sigma'_1 = f(\varphi)\sigma_1 + 4a\sigma_1\sigma_2,$$
$$\sigma'_0 = f(\varphi)\sigma_0 + a\sigma_1^2 + 2a\sigma_2.$$

This system can be integrated directly, since the first equation is a Bernoulli equation and the second and third ones are linear in the unknown. Note that the first equation has a particular solution $\sigma_2 = 0$.

Remark. Equation (1) has a special solution $\varphi = k = $ const, where $k$ is a root of the algebraic (transcendental) equation $g(k) - h(k) + f(k)\ln k = 0$.

14. $\dfrac{\partial u}{\partial t} = a\dfrac{\partial^2 u}{\partial x^2} + uf\left(\dfrac{w}{u}\right) - wg\left(\dfrac{w}{u}\right) + \dfrac{u}{\sqrt{u^2 + w^2}}h\left(\dfrac{w}{u}\right),$

$\dfrac{\partial w}{\partial t} = a\dfrac{\partial^2 w}{\partial x^2} + wf\left(\dfrac{w}{u}\right) + ug\left(\dfrac{w}{u}\right) + \dfrac{w}{\sqrt{u^2 + w^2}}h\left(\dfrac{w}{u}\right).$

Solution:
$$u = r(x,t)\cos\varphi(t), \quad w = r(x,t)\sin\varphi(t),$$

where the function $\varphi = \varphi(t)$ is determined from the separable first-order ordinary differential equation

$$\varphi'_t = g(\tan\varphi),$$

and the function $r = r(x,t)$ satisfies the linear equation

$$\dfrac{\partial r}{\partial t} = a\dfrac{\partial^2 r}{\partial x^2} + rf(\tan\varphi) + h(\tan\varphi). \tag{1}$$

The change of variable

$$r = F(t)\left[Z(x,t) + \int \frac{h(\tan\varphi)\,dt}{F(t)}\right], \qquad F(t) = \exp\left[\int f(\tan\varphi)\,dt\right]$$

brings (1) to the linear heat equation

$$\frac{\partial Z}{\partial t} = a\frac{\partial^2 Z}{\partial x^2}.$$

15. $\quad\displaystyle\frac{\partial u}{\partial t} = a\frac{\partial^2 u}{\partial x^2} + uf\!\left(\frac{w}{u}\right) + wg\!\left(\frac{w}{u}\right) + \frac{u}{\sqrt{u^2 - w^2}}h\!\left(\frac{w}{u}\right),$

$\displaystyle\frac{\partial w}{\partial t} = a\frac{\partial^2 w}{\partial x^2} + wf\!\left(\frac{w}{u}\right) + ug\!\left(\frac{w}{u}\right) + \frac{w}{\sqrt{u^2 - w^2}}h\!\left(\frac{w}{u}\right).$

Solution:
$$u = r(x,t)\cosh\varphi(t), \qquad w = r(x,t)\sinh\varphi(t),$$

where the function $\varphi = \varphi(t)$ is determined from the separable first-order ordinary differential equation

$$\varphi'_t = g(\tanh\varphi),$$

and the function $r = r(x,t)$ satisfies the linear equation

$$\frac{\partial r}{\partial t} = a\frac{\partial^2 r}{\partial x^2} + rf(\tanh\varphi) + h(\tanh\varphi). \qquad (1)$$

The change of variable

$$r = F(t)\left[Z(x,t) + \int \frac{h(\tanh\varphi)\,dt}{F(t)}\right], \qquad F(t) = \exp\left[\int f(\tanh\varphi)\,dt\right]$$

brings (1) to the linear heat equation

$$\frac{\partial Z}{\partial t} = a\frac{\partial^2 Z}{\partial x^2}.$$

---

**T10.3.1-3. Arbitrary functions depend on the product of powers of the unknowns.**

16. $\quad\displaystyle\frac{\partial u}{\partial t} = a\frac{\partial^2 u}{\partial x^2} + uf(u^n w^m), \qquad \frac{\partial w}{\partial t} = b\frac{\partial^2 w}{\partial x^2} + wg(u^n w^m).$

Solution:
$$u = e^{m(kx-\lambda t)}y(\xi), \qquad w = e^{-n(kx-\lambda t)}z(\xi), \qquad \xi = \beta x - \gamma t,$$

where $k$, $\lambda$, $\beta$, and $\gamma$ are arbitrary constants, and the functions $y = y(\xi)$ and $z = z(\xi)$ are determined by the autonomous system of ordinary differential equations

$$a\beta^2 y''_{\xi\xi} + (2akm\beta + \gamma)y'_\xi + m(ak^2 m + \lambda)y + yf(y^n z^m) = 0,$$
$$b\beta^2 z''_{\xi\xi} + (-2bkn\beta + \gamma)z'_\xi + n(bk^2 n - \lambda)z + zg(y^n z^m) = 0.$$

To the special case $k = \lambda = 0$ there corresponds a traveling-wave solution.

**17.** $\dfrac{\partial u}{\partial t} = a\dfrac{\partial^2 u}{\partial x^2} + u^{1+kn} f(u^n w^m), \qquad \dfrac{\partial w}{\partial t} = b\dfrac{\partial^2 w}{\partial x^2} + w^{1-km} g(u^n w^m).$

Self-similar solution:
$$u = (C_1 t + C_2)^{-\tfrac{1}{kn}} y(\xi), \qquad w = (C_1 t + C_2)^{\tfrac{1}{km}} z(\xi), \qquad \xi = \dfrac{x + C_3}{\sqrt{C_1 t + C_2}},$$

where $C_1$, $C_2$, and $C_3$ are arbitrary constants, and the functions $y = y(\xi)$ and $z = z(\xi)$ are determined by the system of ordinary differential equations

$$ay''_{\xi\xi} + \tfrac{1}{2}C_1 \xi y'_\xi + \dfrac{C_1}{kn} y + y^{1+kn} f(y^n z^m) = 0,$$
$$bz''_{\xi\xi} + \tfrac{1}{2}C_1 \xi z'_\xi - \dfrac{C_1}{km} z + z^{1-km} g(y^n z^m) = 0.$$

**18.** $\dfrac{\partial u}{\partial t} = a\dfrac{\partial^2 u}{\partial x^2} + cu\ln u + uf(u^n w^m), \qquad \dfrac{\partial w}{\partial t} = b\dfrac{\partial^2 w}{\partial x^2} + cw\ln w + wg(u^n w^m).$

Solution:
$$u = \exp(Ame^{ct})y(\xi), \qquad w = \exp(-Ane^{ct})z(\xi), \qquad \xi = kx - \lambda t,$$

where $A$, $k$, and $\lambda$ are arbitrary constants, and the functions $y = y(\xi)$ and $z = z(\xi)$ are determined by the autonomous system of ordinary differential equations

$$ak^2 y''_{\xi\xi} + \lambda y'_\xi + cy \ln y + yf(y^n z^m) = 0,$$
$$bk^2 z''_{\xi\xi} + \lambda z'_\xi + cz \ln z + zg(y^n z^m) = 0.$$

To the special case $A = 0$ there corresponds a traveling-wave solution. For $\lambda = 0$, we have a solution in the form of the product of two functions dependent on time $t$ and the coordinate $x$.

---

**T10.3.1-4. Arbitrary functions depend on the sum of squares of the unknowns.**

**19.** $\dfrac{\partial u}{\partial t} = a\dfrac{\partial^2 u}{\partial x^2} + uf(u^2 + w^2) - wg(u^2 + w^2),$
$\dfrac{\partial w}{\partial t} = a\dfrac{\partial^2 w}{\partial x^2} + ug(u^2 + w^2) + wf(u^2 + w^2).$

$1^\circ.$ A periodic solution in the spatial coordinate:
$$u = \psi(t)\cos\varphi(x,t), \qquad w = \psi(t)\sin\varphi(x,t), \qquad \varphi(x,t) = C_1 x + \int g(\psi^2)\,dt + C_2,$$

where $C_1$ and $C_2$ are arbitrary constants, and the function $\psi = \psi(t)$ is described by the separable first-order ordinary differential equation

$$\psi'_t = \psi f(\psi^2) - aC_1^2 \psi,$$

whose general solution can be represented in implicit form as

$$\int \dfrac{d\psi}{\psi f(\psi^2) - aC_1^2 \psi} = t + C_3.$$

**2°.** A periodic solution in time:
$$u = r(x)\cos[\theta(x) + C_1 t + C_2], \qquad w = r(x)\sin[\theta(x) + C_1 t + C_2],$$
where $C_1$ and $C_2$ are arbitrary constants, and the functions $r = r(x)$ and $\theta = \theta(x)$ are determined by the autonomous system of ordinary differential equations
$$ar''_{xx} - ar(\theta'_x)^2 + rf(r^2) = 0,$$
$$ar\theta''_{xx} + 2ar'_x\theta'_x - C_1 r + rg(r^2) = 0.$$

**3°.** Solution (generalizes the solution of Item 2°):
$$u = r(z)\cos[\theta(z) + C_1 t + C_2], \qquad w = r(z)\sin[\theta(z) + C_1 t + C_2], \qquad z = x + \lambda t,$$
where $C_1$, $C_2$, and $\lambda$ are arbitrary constants, and the functions $r = r(z)$ and $\theta = \theta(z)$ are determined by the system of ordinary differential equations
$$ar''_{zz} - ar(\theta'_z)^2 - \lambda r'_z + rf(r^2) = 0,$$
$$ar\theta''_{zz} + 2ar'_z\theta'_z - \lambda r\theta'_z - C_1 r + rg(r^2) = 0.$$

**20.**
$$\frac{\partial u}{\partial t} = a\frac{\partial^2 u}{\partial x^2} + uf(u^2 + w^2) - wg(u^2 + w^2) - w\arctan\left(\frac{w}{u}\right)h(u^2 + w^2),$$
$$\frac{\partial w}{\partial t} = a\frac{\partial^2 w}{\partial x^2} + wf(u^2 + w^2) + ug(u^2 + w^2) + u\arctan\left(\frac{w}{u}\right)h(u^2 + w^2).$$

Functional separable solution (for fixed $t$, it defines a structure periodic in $x$):
$$u = r(t)\cos[\varphi(t)x + \psi(t)], \qquad w = r(t)\sin[\varphi(t)x + \psi(t)],$$
where the functions $r = r(t)$, $\varphi = \varphi(t)$, and $\psi = \psi(t)$ are determined by the autonomous system of ordinary differential equations
$$r'_t = -a r\varphi^2 + rf(r^2),$$
$$\varphi'_t = h(r^2)\varphi,$$
$$\psi'_t = h(r^2)\psi + g(r^2).$$

---

**T10.3.1-5. Arbitrary functions depend on the difference of squares of the unknowns.**

**21.**
$$\frac{\partial u}{\partial t} = a\frac{\partial^2 u}{\partial x^2} + uf(u^2 - w^2) + wg(u^2 - w^2),$$
$$\frac{\partial w}{\partial t} = a\frac{\partial^2 w}{\partial x^2} + ug(u^2 - w^2) + wf(u^2 - w^2).$$

**1°.** Solution:
$$u = \psi(t)\cosh\varphi(x,t), \qquad w = \psi(t)\sinh\varphi(x,t), \qquad \varphi(x,t) = C_1 x + \int g(\psi^2)\,dt + C_2,$$
where $C_1$ and $C_2$ are arbitrary constants, and the function $\psi = \psi(t)$ is described by the separable first-order ordinary differential equation
$$\psi'_t = \psi f(\psi^2) + aC_1^2\psi,$$

whose general solution can be represented in implicit form as

$$\int \frac{d\psi}{\psi f(\psi^2) + aC_1^2 \psi} = t + C_3.$$

2°. Solution:

$$u = r(x)\cosh\bigl[\theta(x) + C_1 t + C_2\bigr], \quad w = r(x)\sinh\bigl[\theta(x) + C_1 t + C_2\bigr],$$

where $C_1$ and $C_2$ are arbitrary constants, and the functions $r = r(x)$ and $\theta = \theta(x)$ are determined by the autonomous system of ordinary differential equations

$$ar''_{xx} + ar(\theta'_x)^2 + rf(r^2) = 0,$$
$$ar\theta''_{xx} + 2ar'_x \theta'_x + rg(r^2) - C_1 r = 0.$$

3°. Solution (generalizes the solution of Item 2°):

$$u = r(z)\cosh\bigl[\theta(z) + C_1 t + C_2\bigr], \quad w = r(z)\sinh\bigl[\theta(z) + C_1 t + C_2\bigr], \quad z = x + \lambda t,$$

where $C_1$, $C_2$, and $\lambda$ are arbitrary constants, and the functions $r = r(z)$ and $\theta = \theta(z)$ are determined by the autonomous system of ordinary differential equations

$$ar''_{zz} + ar(\theta'_z)^2 - \lambda r'_z + rf(r^2) = 0,$$
$$ar\theta''_{zz} + 2ar'_z \theta'_z - \lambda r\theta'_z - C_1 r + rg(r^2) = 0.$$

22. $$\frac{\partial u}{\partial t} = a\frac{\partial^2 u}{\partial x^2} + uf(u^2 - w^2) + wg(u^2 - w^2) + w\operatorname{arctanh}\Bigl(\frac{w}{u}\Bigr)h(u^2 - w^2),$$
$$\frac{\partial w}{\partial t} = a\frac{\partial^2 w}{\partial x^2} + wf(u^2 - w^2) + ug(u^2 - w^2) + u\operatorname{arctanh}\Bigl(\frac{w}{u}\Bigr)h(u^2 - w^2).$$

Functional separable solution:

$$u = r(t)\cosh\bigl[\varphi(t)x + \psi(t)\bigr], \quad w = r(t)\sinh\bigl[\varphi(t)x + \psi(t)\bigr],$$

where the functions $r = r(t)$, $\varphi = \varphi(t)$, and $\psi = \psi(t)$ are determined by the autonomous system of ordinary differential equations

$$r'_t = ar\varphi^2 + rf(r^2),$$
$$\varphi'_t = h(r^2)\varphi,$$
$$\psi'_t = h(r^2)\psi + g(r^2).$$

T10.3.1-6. Arbitrary functions depend on the unknowns in a complex way.

23. $$\frac{\partial u}{\partial t} = a\frac{\partial^2 u}{\partial x^2} + u^{k+1} f(\varphi), \quad \varphi = u\exp\Bigl(-\frac{w}{u}\Bigr),$$
$$\frac{\partial w}{\partial t} = a\frac{\partial^2 w}{\partial x^2} + u^{k+1}[f(\varphi)\ln u + g(\varphi)].$$

Solution:

$$u = (C_1 t + C_2)^{-\frac{1}{k}} y(\xi), \quad w = (C_1 t + C_2)^{-\frac{1}{k}} \Bigl[z(\xi) - \frac{1}{k}\ln(C_1 t + C_2) y(\xi)\Bigr], \quad \xi = \frac{x + C_3}{\sqrt{C_1 t + C_2}},$$

where $C_1$, $C_2$, and $C_3$ are arbitrary constants, and the functions $y = y(\xi)$ and $z = z(\xi)$ are determined by the system of ordinary differential equations

$$ay''_{\xi\xi} + \frac{1}{2}C_1\xi y'_\xi + \frac{C_1}{k}y + y^{k+1}f(\varphi) = 0, \qquad \varphi = y\exp\left(-\frac{z}{y}\right),$$

$$az''_{\xi\xi} + \frac{1}{2}C_1\xi z'_\xi + \frac{C_1}{k}z + \frac{C_1}{k}y + y^{k+1}[f(\varphi)\ln y + g(\varphi)] = 0.$$

**24.** $\quad \dfrac{\partial u}{\partial t} = a\dfrac{\partial^2 u}{\partial x^2} + uf(u^2+w^2) - wg\left(\dfrac{w}{u}\right), \qquad \dfrac{\partial w}{\partial t} = a\dfrac{\partial^2 w}{\partial x^2} + ug\left(\dfrac{w}{u}\right) + wf(u^2+w^2).$

Solution:
$$u = r(x,t)\cos\varphi(t), \qquad w = r(x,t)\sin\varphi(t),$$

where the function $\varphi = \varphi(t)$ is determined by the autonomous ordinary differential equation

$$\varphi'_t = g(\tan\varphi), \tag{1}$$

and the function $r = r(x,t)$ is determined by the differential equation

$$\frac{\partial r}{\partial t} = a\frac{\partial^2 r}{\partial x^2} + rf(r^2). \tag{2}$$

The general solution of equation (1) is expressed in implicit form as

$$\int \frac{d\varphi}{g(\tan\varphi)} = t + C.$$

Equation (2) admits an exact, traveling-wave solution $r = r(z)$, where $z = kx - \lambda t$ with arbitrary constants $k$ and $\lambda$, and the function $r(z)$ is determined by the autonomous ordinary differential equation

$$ak^2 r''_{zz} + \lambda r'_z + rf(r^2) = 0.$$

For other exact solutions to equation (2) for various functions $f$, see Polyanin and Zaitsev (2004).

**25.** $\quad \dfrac{\partial u}{\partial t} = a\dfrac{\partial^2 u}{\partial x^2} + uf(u^2-w^2) + wg\left(\dfrac{w}{u}\right), \qquad \dfrac{\partial w}{\partial t} = a\dfrac{\partial^2 w}{\partial x^2} + ug\left(\dfrac{w}{u}\right) + wf(u^2-w^2).$

Solution:
$$u = r(x,t)\cosh\varphi(t), \qquad w = r(x,t)\sinh\varphi(t),$$

where the function $\varphi = \varphi(t)$ is determined by the autonomous ordinary differential equation

$$\varphi'_t = g(\tanh\varphi), \tag{1}$$

and the function $r = r(x,t)$ is determined by the differential equation

$$\frac{\partial r}{\partial t} = a\frac{\partial^2 r}{\partial x^2} + rf(r^2). \tag{2}$$

The general solution of equation (1) is expressed in implicit form as

$$\int \frac{d\varphi}{g(\tanh\varphi)} = t + C.$$

Equation (2) admits an exact, traveling-wave solution $r = r(z)$, where $z = kx - \lambda t$ with arbitrary constants $k$ and $\lambda$, and the function $r(z)$ is determined by the autonomous ordinary differential equation

$$ak^2 r''_{zz} + \lambda r'_z + rf(r^2) = 0.$$

For other exact solutions to equation (2) for various functions $f$, see Polyanin and Zaitsev (2004).

## T10.3.2. Systems of the Form
$$\frac{\partial u}{\partial t} = \frac{a}{x^n}\frac{\partial}{\partial x}\left(x^n\frac{\partial u}{\partial x}\right) + F(u,w),$$
$$\frac{\partial w}{\partial t} = \frac{b}{x^n}\frac{\partial}{\partial x}\left(x^n\frac{\partial w}{\partial x}\right) + G(u,w)$$

T10.3.2-1. Arbitrary functions depend on a linear combination of the unknowns.

1. $\displaystyle\frac{\partial u}{\partial t} = \frac{a}{x^n}\frac{\partial}{\partial x}\left(x^n\frac{\partial u}{\partial x}\right) + uf(bu-cw) + g(bu-cw),$
$\displaystyle\frac{\partial w}{\partial t} = \frac{a}{x^n}\frac{\partial}{\partial x}\left(x^n\frac{\partial w}{\partial x}\right) + wf(bu-cw) + h(bu-cw).$

1°. Solution:
$$u = \varphi(t) + c\exp\left[\int f(b\varphi - c\psi)\,dt\right]\theta(x,t),$$
$$w = \psi(t) + b\exp\left[\int f(b\varphi - c\psi)\,dt\right]\theta(x,t),$$

where $\varphi = \varphi(t)$ and $\psi = \psi(t)$ are determined by the autonomous system of ordinary differential equations
$$\varphi'_t = \varphi f(b\varphi - c\psi) + g(b\varphi - c\psi),$$
$$\psi'_t = \psi f(b\varphi - c\psi) + h(b\varphi - c\psi),$$

and the function $\theta = \theta(x,t)$ satisfies linear heat equation
$$\frac{\partial \theta}{\partial t} = \frac{a}{x^n}\frac{\partial}{\partial x}\left(x^n\frac{\partial \theta}{\partial x}\right). \tag{1}$$

2°. Let us multiply the first equation by $b$ and the second one by $-c$ and add the results together to obtain
$$\frac{\partial \zeta}{\partial t} = \frac{a}{x^n}\frac{\partial}{\partial x}\left(x^n\frac{\partial \zeta}{\partial x}\right) + \zeta f(\zeta) + bg(\zeta) - ch(\zeta), \qquad \zeta = bu - cw. \tag{2}$$

This equation will be considered in conjunction with the first equation of the original system
$$\frac{\partial u}{\partial t} = \frac{a}{x^n}\frac{\partial}{\partial x}\left(x^n\frac{\partial u}{\partial x}\right) + uf(\zeta) + g(\zeta). \tag{3}$$

Equation (2) can be treated separately. Given a solution $\zeta = \zeta(x,t)$ to equation (2), the function $u = u(x,t)$ can be determined by solving the linear equation (3) and the function $w = w(x,t)$ is found as $w = (bu - \zeta)/c$.

Note two important solutions to equation (2):

(i) In the general case, equation (2) admits steady-state solutions $\zeta = \zeta(x)$. The corresponding exact solutions to equation (3) are expressed as $u = u_0(x) + \sum e^{\beta_n t}u_n(x)$.

(ii) If the condition $\zeta f(\zeta) + bg(\zeta) - ch(\zeta) = k_1\zeta + k_0$ holds, equation (2) is linear,
$$\frac{\partial \zeta}{\partial t} = \frac{a}{x^n}\frac{\partial}{\partial x}\left(x^n\frac{\partial \zeta}{\partial x}\right) + k_1\zeta + k_0,$$

and hence can be reduced to the linear heat equation (1) with the substitution $\zeta = e^{k_1 t}\bar\zeta - k_0 k_1^{-1}$.

2. 
$$\frac{\partial u}{\partial t} = \frac{a}{x^n}\frac{\partial}{\partial x}\left(x^n\frac{\partial u}{\partial x}\right) + e^{\lambda u}f(\lambda u - \sigma w),$$
$$\frac{\partial w}{\partial t} = \frac{b}{x^n}\frac{\partial}{\partial x}\left(x^n\frac{\partial w}{\partial x}\right) + e^{\sigma w}g(\lambda u - \sigma w).$$

Solution:
$$u = y(\xi) - \frac{1}{\lambda}\ln(C_1 t + C_2), \quad w = z(\xi) - \frac{1}{\sigma}\ln(C_1 t + C_2), \quad \xi = \frac{x}{\sqrt{C_1 t + C_2}},$$

where $C_1$ and $C_2$ are arbitrary constants, and the functions $y = y(\xi)$ and $z = z(\xi)$ are determined by the system of ordinary differential equations

$$a\xi^{-n}(\xi^n y'_\xi)'_\xi + \frac{1}{2}C_1\xi y'_\xi + \frac{C_1}{\lambda} + e^{\lambda y}f(\lambda y - \sigma z) = 0,$$
$$b\xi^{-n}(\xi^n z'_\xi)'_\xi + \frac{1}{2}C_1\xi z'_\xi + \frac{C_1}{\sigma} + e^{\sigma z}g(\lambda y - \sigma z) = 0.$$

T10.3.2-2. Arbitrary functions depend on the ratio of the unknowns.

3. 
$$\frac{\partial u}{\partial t} = \frac{a}{x^n}\frac{\partial}{\partial x}\left(x^n\frac{\partial u}{\partial x}\right) + uf\left(\frac{u}{w}\right), \quad \frac{\partial w}{\partial t} = \frac{b}{x^n}\frac{\partial}{\partial x}\left(x^n\frac{\partial w}{\partial x}\right) + wg\left(\frac{u}{w}\right).$$

1°. Multiplicative separable solution:
$$u = x^{\frac{1-n}{2}}[C_1 J_\nu(kx) + C_2 Y_\nu(kx)]\varphi(t), \quad \nu = \tfrac{1}{2}|n-1|,$$
$$w = x^{\frac{1-n}{2}}[C_1 J_\nu(kx) + C_2 Y_\nu(kx)]\psi(t),$$

where $C_1$, $C_2$, and $k$ are arbitrary constants, $J_\nu(z)$ and $Y_\nu(z)$ are Bessel functions, and the functions $\varphi = \varphi(t)$ and $\psi = \psi(t)$ are determined by the autonomous system of ordinary differential equations

$$\varphi'_t = -ak^2\varphi + \varphi f(\varphi/\psi),$$
$$\psi'_t = -bk^2\psi + \psi g(\varphi/\psi).$$

2°. Multiplicative separable solution:
$$u = x^{\frac{1-n}{2}}[C_1 I_\nu(kx) + C_2 K_\nu(kx)]\varphi(t), \quad \nu = \tfrac{1}{2}|n-1|,$$
$$w = x^{\frac{1-n}{2}}[C_1 I_\nu(kx) + C_2 K_\nu(kx)]\psi(t),$$

where $C_1$, $C_2$, and $k$ are arbitrary constants, $I_\nu(z)$ and $K_\nu(z)$ are modified Bessel functions, and the functions $\varphi = \varphi(t)$ and $\psi = \psi(t)$ are determined by the autonomous system of ordinary differential equations

$$\varphi'_t = ak^2\varphi + \varphi f(\varphi/\psi),$$
$$\psi'_t = bk^2\psi + \psi g(\varphi/\psi).$$

3°. Multiplicative separable solution:
$$u = e^{-\lambda t}y(x), \quad w = e^{-\lambda t}z(x),$$

where $\lambda$ is an arbitrary constant and the functions $y = y(x)$ and $z = z(x)$ are determined by the system of ordinary differential equations

$$ax^{-n}(x^n y'_x)'_x + \lambda y + y f(y/z) = 0,$$
$$bx^{-n}(x^n z'_x)'_x + \lambda z + z g(y/z) = 0.$$

4°. This is a special case of equation with $b = a$. Let $k$ be a root of the algebraic (transcendental) equation

$$f(k) = g(k).$$

Solution:

$$u = k e^{\lambda t} \theta, \quad w = e^{\lambda t} \theta, \quad \lambda = f(k),$$

where the function $\theta = \theta(x, t)$ satisfies the linear heat equation

$$\frac{\partial \theta}{\partial t} = \frac{a}{x^n} \frac{\partial}{\partial x}\left(x^n \frac{\partial \theta}{\partial x}\right). \tag{1}$$

5°. This is a special case of equation with $b = a$. Solution:

$$u = \varphi(t) \exp\left[\int g(\varphi(t))\,dt\right] \theta(x,t), \quad w = \exp\left[\int g(\varphi(t))\,dt\right] \theta(x,t),$$

where the function $\varphi = \varphi(t)$ is described by the separable first-order nonlinear ordinary differential equation

$$\varphi'_t = [f(\varphi) - g(\varphi)]\varphi, \tag{2}$$

and the function $\theta = \theta(x,t)$ satisfies the linear heat equation (1).

To the particular solution $\varphi = k = \text{const}$ of equation (2), there corresponds the solution presented in Item 4°. The general solution of equation (2) is written out in implicit form as

$$\int \frac{d\varphi}{[f(\varphi) - g(\varphi)]\varphi} = t + C.$$

4.
$$\frac{\partial u}{\partial t} = \frac{a}{x^n} \frac{\partial}{\partial x}\left(x^n \frac{\partial u}{\partial x}\right) + u f\left(\frac{u}{w}\right) + \frac{u}{w} h\left(\frac{u}{w}\right),$$
$$\frac{\partial w}{\partial t} = \frac{a}{x^n} \frac{\partial}{\partial x}\left(x^n \frac{\partial w}{\partial x}\right) + w g\left(\frac{u}{w}\right) + h\left(\frac{u}{w}\right).$$

Solution:

$$u = \varphi(t) G(t)\left[\theta(x,t) + \int \frac{h(\varphi)}{G(t)}\,dt\right], \quad w = G(t)\left[\theta(x,t) + \int \frac{h(\varphi)}{G(t)}\,dt\right], \quad G(t) = \exp\left[\int g(\varphi)\,dt\right],$$

where the function $\varphi = \varphi(t)$ is described by the separable first-order nonlinear ordinary differential equation

$$\varphi'_t = [f(\varphi) - g(\varphi)]\varphi, \tag{1}$$

and the function $\theta = \theta(x,t)$ satisfies the linear heat equation

$$\frac{\partial \theta}{\partial t} = \frac{a}{x^n} \frac{\partial}{\partial x}\left(x^n \frac{\partial \theta}{\partial x}\right).$$

The general solution of equation (1) is written out in implicit form as

$$\int \frac{d\varphi}{[f(\varphi) - g(\varphi)]\varphi} = t + C.$$

5.  $\dfrac{\partial u}{\partial t} = \dfrac{a}{x^n}\dfrac{\partial}{\partial x}\left(x^n\dfrac{\partial u}{\partial x}\right) + u f_1\!\left(\dfrac{w}{u}\right) + w g_1\!\left(\dfrac{w}{u}\right),$

   $\dfrac{\partial w}{\partial t} = \dfrac{a}{x^n}\dfrac{\partial}{\partial x}\left(x^n\dfrac{\partial w}{\partial x}\right) + u f_2\!\left(\dfrac{w}{u}\right) + w g_2\!\left(\dfrac{w}{u}\right).$

Solution:

$$u = \exp\!\left\{\int [f_1(\varphi) + \varphi g_1(\varphi)]\, dt\right\}\theta(x,t), \quad w(x,t) = \varphi(t)\exp\!\left\{\int [f_1(\varphi) + \varphi g_1(\varphi)]\, dt\right\}\theta(x,t),$$

where the function $\varphi = \varphi(t)$ is described by the separable first-order nonlinear ordinary differential equation

$$\varphi'_t = f_2(\varphi) + \varphi g_2(\varphi) - \varphi[f_1(\varphi) + \varphi g_1(\varphi)],$$

and the function $\theta = \theta(x,t)$ satisfies the linear heat equation

$$\dfrac{\partial \theta}{\partial t} = \dfrac{a}{x^n}\dfrac{\partial}{\partial x}\left(x^n\dfrac{\partial \theta}{\partial x}\right).$$

6.  $\dfrac{\partial u}{\partial t} = \dfrac{a}{x^n}\dfrac{\partial}{\partial x}\left(x^n\dfrac{\partial u}{\partial x}\right) + u^k f\!\left(\dfrac{u}{w}\right), \quad \dfrac{\partial w}{\partial t} = \dfrac{b}{x^n}\dfrac{\partial}{\partial x}\left(x^n\dfrac{\partial w}{\partial x}\right) + w^k g\!\left(\dfrac{u}{w}\right).$

Self-similar solution:

$$u = (C_1 t + C_2)^{\frac{1}{1-k}} y(\xi), \quad w = (C_1 t + C_2)^{\frac{1}{1-k}} z(\xi), \quad \xi = \dfrac{x}{\sqrt{C_1 t + C_2}},$$

where $C_1$ and $C_2$ are arbitrary constants, and the functions $y = y(\xi)$ and $z = z(\xi)$ are determined by the system of ordinary differential equations

$$a\xi^{-n}(\xi^n y'_\xi)'_\xi + \tfrac{1}{2} C_1 \xi y'_\xi + \dfrac{C_1}{k-1} y + y^k f(y/z) = 0,$$

$$b\xi^{-n}(\xi^n z'_\xi)'_\xi + \tfrac{1}{2} C_1 \xi z'_\xi + \dfrac{C_1}{k-1} z + z^k g(y/z) = 0.$$

7.  $\dfrac{\partial u}{\partial t} = \dfrac{a}{x^n}\dfrac{\partial}{\partial x}\left(x^n\dfrac{\partial u}{\partial x}\right) + u f\!\left(\dfrac{u}{w}\right)\ln u + u g\!\left(\dfrac{u}{w}\right),$

   $\dfrac{\partial w}{\partial t} = \dfrac{a}{x^n}\dfrac{\partial}{\partial x}\left(x^n\dfrac{\partial w}{\partial x}\right) + w f\!\left(\dfrac{u}{w}\right)\ln w + w h\!\left(\dfrac{u}{w}\right).$

Solution:
$$u = \varphi(t)\psi(t)\theta(x,t), \quad w = \psi(t)\theta(x,t),$$

where the functions $\varphi = \varphi(t)$ and $\psi = \psi(t)$ are determined by solving the autonomous ordinary differential equations

$$\varphi'_t = \varphi[g(\varphi) - h(\varphi) + f(\varphi)\ln\varphi],$$
$$\psi'_t = \psi[h(\varphi) + f(\varphi)\ln\psi], \qquad (1)$$

and the function $\theta = \theta(x,t)$ is determined by the differential equation

$$\dfrac{\partial \theta}{\partial t} = \dfrac{a}{x^n}\dfrac{\partial}{\partial x}\left(x^n\dfrac{\partial \theta}{\partial x}\right) + f(\varphi)\theta\ln\theta. \qquad (2)$$

The first equation in (1) is a separable equation; its solution can be written out in implicit form. The second equation in (1) can be solved using the change of variable $\psi = e^\zeta$ (it is reduced to a linear equation for $\zeta$).

Equation (2) admits exact solutions of the form

$$\theta = \exp\left[\sigma_2(t)x^2 + \sigma_0(t)\right],$$

where the functions $\sigma_n(t)$ are described by the equations

$$\sigma_2' = f(\varphi)\sigma_2 + 4a\sigma_2^2,$$
$$\sigma_0' = f(\varphi)\sigma_0 + 2a(n+1)\sigma_2.$$

This system can be successively integrated, since the first equation is a Bernoulli equation and the second one is linear in the unknown.

If $f = \text{const}$, equation (2) also has a traveling-wave solution $\theta = \theta(kx - \lambda t)$.

---

**T10.3.2-3. Arbitrary functions depend on the product of powers of the unknowns.**

8. $\quad \dfrac{\partial u}{\partial t} = \dfrac{a}{x^n}\dfrac{\partial}{\partial x}\left(x^n \dfrac{\partial u}{\partial x}\right) + uf(x, u^k w^m),$

$\quad \dfrac{\partial w}{\partial t} = \dfrac{b}{x^n}\dfrac{\partial}{\partial x}\left(x^n \dfrac{\partial w}{\partial x}\right) + wg(x, u^k w^m).$

Multiplicative separable solution:

$$u = e^{-m\lambda t} y(x), \quad w = e^{k\lambda t} z(x),$$

where $\lambda$ is an arbitrary constant and the functions $y = y(x)$ and $z = z(x)$ are determined by the system of ordinary differential equations

$$ax^{-n}(x^n y_x')_x' + m\lambda y + yf(x, y^k z^m) = 0,$$
$$bx^{-n}(x^n z_x')_x' - k\lambda z + zg(x, y^k z^m) = 0.$$

9. $\quad \dfrac{\partial u}{\partial t} = \dfrac{a}{x^n}\dfrac{\partial}{\partial x}\left(x^n \dfrac{\partial u}{\partial x}\right) + u^{1+kn} f(u^n w^m),$

$\quad \dfrac{\partial w}{\partial t} = \dfrac{b}{x^n}\dfrac{\partial}{\partial x}\left(x^n \dfrac{\partial w}{\partial x}\right) + w^{1-km} g(u^n w^m).$

Self-similar solution:

$$u = (C_1 t + C_2)^{-\frac{1}{kn}} y(\xi), \quad w = (C_1 t + C_2)^{\frac{1}{km}} z(\xi), \quad \xi = \frac{x}{\sqrt{C_1 t + C_2}},$$

where $C_1$ and $C_2$ are arbitrary constants, and the functions $y = y(\xi)$ and $z = z(\xi)$ are determined by the system of ordinary differential equations

$$a\xi^{-n}(\xi^n y_\xi')_\xi' + \frac{1}{2}C_1 \xi y_\xi' + \frac{C_1}{kn} y + y^{1+kn} f(y^n z^m) = 0,$$
$$b\xi^{-n}(\xi^n z_\xi')_\xi' + \frac{1}{2}C_1 \xi z_\xi' - \frac{C_1}{km} z + z^{1-km} g(y^n z^m) = 0.$$

10. $\dfrac{\partial u}{\partial t} = \dfrac{a}{x^n}\dfrac{\partial}{\partial x}\left(x^n \dfrac{\partial u}{\partial x}\right) + cu\ln u + uf(x, u^k w^m),$

$\dfrac{\partial w}{\partial t} = \dfrac{b}{x^n}\dfrac{\partial}{\partial x}\left(x^n \dfrac{\partial w}{\partial x}\right) + cw\ln w + wg(x, u^k w^m).$

Multiplicative separable solution:
$$u = \exp(Ame^{ct})y(x), \quad w = \exp(-Ake^{ct})z(x),$$

where $A$ is an arbitrary constant, and the functions $y = y(x)$ and $z = z(x)$ are determined by the system of ordinary differential equations

$$ax^{-n}(x^n y'_x)'_x + cy\ln y + yf(x, y^k z^m) = 0,$$
$$bx^{-n}(x^n z'_x)'_x + cz\ln z + zg(x, y^k z^m) = 0.$$

---

**T10.3.2-4. Arbitrary functions depend on $u^2 \pm w^2$.**

11. $\dfrac{\partial u}{\partial t} = \dfrac{a}{x^n}\dfrac{\partial}{\partial x}\left(x^n \dfrac{\partial u}{\partial x}\right) + uf(u^2 + w^2) - wg(u^2 + w^2),$

$\dfrac{\partial w}{\partial t} = \dfrac{a}{x^n}\dfrac{\partial}{\partial x}\left(x^n \dfrac{\partial w}{\partial x}\right) + wf(u^2 + w^2) + ug(u^2 + w^2).$

Time-periodic solution:
$$u = r(x)\cos[\theta(x) + C_1 t + C_2], \quad w = r(x)\sin[\theta(x) + C_1 t + C_2],$$

where $C_1$ and $C_2$ are arbitrary constants, and the functions $r = r(x)$ and $\theta = \theta(x)$ are determined by the system of ordinary differential equations

$$ar''_{xx} - ar(\theta'_x)^2 + \dfrac{an}{x}r'_x + rf(r^2) = 0,$$
$$ar\theta''_{xx} + 2ar'_x \theta'_x + \dfrac{an}{x}r\theta'_x + rg(r^2) - C_1 r = 0.$$

12. $\dfrac{\partial u}{\partial t} = \dfrac{a}{x^n}\dfrac{\partial}{\partial x}\left(x^n \dfrac{\partial u}{\partial x}\right) + uf(u^2 - w^2) + wg(u^2 - w^2),$

$\dfrac{\partial w}{\partial t} = \dfrac{a}{x^n}\dfrac{\partial}{\partial x}\left(x^n \dfrac{\partial w}{\partial x}\right) + wf(u^2 - w^2) + ug(u^2 - w^2).$

Solution:
$$u = r(x)\cosh[\theta(x) + C_1 t + C_2], \quad w = r(x)\sinh[\theta(x) + C_1 t + C_2],$$

where $C_1$ and $C_2$ are arbitrary constants, and the functions $r = r(x)$ and $\theta = \theta(x)$ are determined by the system of ordinary differential equations

$$ar''_{xx} + ar(\theta'_x)^2 + \dfrac{an}{x}r'_x + rf(r^2) = 0,$$
$$ar\theta''_{xx} + 2ar'_x \theta'_x + \dfrac{an}{x}r\theta'_x + rg(r^2) - C_1 r = 0.$$

**T10.3.2-5. Arbitrary functions have different arguments.**

13. 
$$\frac{\partial u}{\partial t} = \frac{a}{x^n}\frac{\partial}{\partial x}\left(x^n\frac{\partial u}{\partial x}\right) + uf(u^2+w^2) - wg\left(\frac{w}{u}\right),$$
$$\frac{\partial w}{\partial t} = \frac{a}{x^n}\frac{\partial}{\partial x}\left(x^n\frac{\partial w}{\partial x}\right) + wf(u^2+w^2) + ug\left(\frac{w}{u}\right).$$

Solution:
$$u = r(x,t)\cos\varphi(t), \quad w = r(x,t)\sin\varphi(t),$$

where the function $\varphi = \varphi(t)$ is determined by the autonomous ordinary differential equation

$$\varphi'_t = g(\tan\varphi), \tag{1}$$

and the function $r = r(x,t)$ is determined by the differential equation

$$\frac{\partial r}{\partial t} = \frac{a}{x^n}\frac{\partial}{\partial x}\left(x^n\frac{\partial r}{\partial x}\right) + rf(r^2). \tag{2}$$

The general solution of equation (1) is expressed in implicit form as

$$\int\frac{d\varphi}{g(\tan\varphi)} = t + C.$$

Equation (2) admits a time-independent exact solution $r = r(x)$.

14. 
$$\frac{\partial u}{\partial t} = \frac{a}{x^n}\frac{\partial}{\partial x}\left(x^n\frac{\partial u}{\partial x}\right) + uf(u^2-w^2) + wg\left(\frac{w}{u}\right),$$
$$\frac{\partial w}{\partial t} = \frac{a}{x^n}\frac{\partial}{\partial x}\left(x^n\frac{\partial w}{\partial x}\right) + wf(u^2-w^2) + ug\left(\frac{w}{u}\right).$$

Solution:
$$u = r(x,t)\cosh\varphi(t), \quad w = r(x,t)\sinh\varphi(t),$$

where the function $\varphi = \varphi(t)$ is determined by the autonomous ordinary differential equation

$$\varphi'_t = g(\tanh\varphi), \tag{1}$$

and the function $r = r(x,t)$ is determined by the differential equation

$$\frac{\partial r}{\partial t} = \frac{a}{x^n}\frac{\partial}{\partial x}\left(x^n\frac{\partial r}{\partial x}\right) + rf(r^2). \tag{2}$$

The general solution of equation (1) is expressed in implicit form as

$$\int\frac{d\varphi}{g(\tanh\varphi)} = t + C.$$

Equation (2) admits a time-independent exact solution $r = r(x)$.

## T10.3.3. Systems of the Form $\Delta u = F(u,w)$, $\Delta w = G(u,w)$

T10.3.3-1. Arbitrary functions depend on a linear combination of the unknowns.

1. 
$$\frac{\partial^2 u}{\partial x^2} + \frac{\partial^2 u}{\partial y^2} = uf(au - bw) + g(au - bw),$$
$$\frac{\partial^2 w}{\partial x^2} + \frac{\partial^2 w}{\partial y^2} = wf(au - bw) + h(au - bw).$$

1°. Solution:
$$u = \varphi(x) + b\theta(x,y), \quad w = \psi(x) + a\theta(x,y),$$

where $\varphi = \varphi(x)$ and $\psi = \psi(x)$ are determined by the autonomous system of ordinary differential equations
$$\varphi''_{xx} = \varphi f(a\varphi - b\psi) + g(a\varphi - b\psi),$$
$$\psi''_{xx} = \psi f(a\varphi - b\psi) + h(a\varphi - b\psi),$$

and the function $\theta = \theta(x,y)$ satisfies the linear Schrödinger equation of the special form
$$\frac{\partial^2 \theta}{\partial x^2} + \frac{\partial^2 \theta}{\partial y^2} = F(x)\theta, \qquad F(x) = f(au - bw).$$

Its solutions are determined by separation of variables.

2°. Let us multiply the first equation by $a$ and the second one by $-b$ and add the results together to obtain
$$\frac{\partial^2 \zeta}{\partial x^2} + \frac{\partial^2 \zeta}{\partial y^2} = \zeta f(\zeta) + ag(\zeta) - bh(\zeta), \qquad \zeta = au - bw. \tag{1}$$

This equation will be considered in conjunction with the first equation of the original system
$$\frac{\partial^2 u}{\partial x^2} + \frac{\partial^2 u}{\partial y^2} = uf(\zeta) + g(\zeta). \tag{2}$$

Equation (1) can be treated separately. An extensive list of exact solutions to equations of this form for various kinetic functions $F(\zeta) = \zeta f(\zeta) + ag(\zeta) - bh(\zeta)$ can be found in the book by Polyanin and Zaitsev (2004).

Note two important solutions to equation (1):

(i) In the general case, equation (1) admits an exact, traveling-wave solution $\zeta = \zeta(z)$, where $z = k_1 x + k_2 y$ with arbitrary constants $k_1$ and $k_2$.

(ii) If the condition $\zeta f(\zeta) + ag(\zeta) - bh(\zeta) = c_1 \zeta + c_0$ holds, equation (1) is a linear Helmholtz equation.

Given a solution $\zeta = \zeta(x,y)$ to equation (1), the function $u = u(x,y)$ can be determined by solving the linear equation (2) and the function $w = w(x,y)$ is found as $w = (bu - \zeta)/c$.

2. 
$$\frac{\partial^2 u}{\partial x^2} + \frac{\partial^2 u}{\partial y^2} = e^{\lambda u} f(\lambda u - \sigma w), \qquad \frac{\partial^2 w}{\partial x^2} + \frac{\partial^2 w}{\partial y^2} = e^{\sigma w} g(\lambda u - \sigma w).$$

1°. Solution:
$$u = U(\xi) - \frac{2}{\lambda} \ln |x + C_1|, \quad w = W(\xi) - \frac{2}{\sigma} \ln |x + C_1|, \quad \xi = \frac{y + C_2}{x + C_1},$$

where $C_1$ and $C_2$ are arbitrary constants, and the functions $U = U(\xi)$ and $W = W(\xi)$ are determined by the system of ordinary differential equations

$$(1 + \xi^2)U''_{\xi\xi} + 2\xi U'_\xi + \frac{2}{\lambda} = e^{\lambda U} f(\lambda U - \sigma W),$$

$$(1 + \xi^2)W''_{\xi\xi} + 2\xi W'_\xi + \frac{2}{\sigma} = e^{\sigma W} g(\lambda U - \sigma W).$$

2°. Solution:

$$u = \theta(x, y), \qquad w = \frac{\lambda}{\sigma}\theta(x, y) - \frac{k}{\sigma},$$

where $k$ is a root of the algebraic (transcendental) equation

$$\lambda f(k) = \sigma e^{-k} g(k),$$

and the function $\theta = \theta(x, y)$ is described by the solvable equation

$$\frac{\partial^2 \theta}{\partial x^2} + \frac{\partial^2 \theta}{\partial y^2} = f(k)e^{\lambda\theta}.$$

This equation is encountered in combustion theory; for its exact solutions, see Polyanin and Zaitsev (2004).

---

**T10.3.3-2. Arbitrary functions depend on the ratio of the unknowns.**

3. $\quad \dfrac{\partial^2 u}{\partial x^2} + \dfrac{\partial^2 u}{\partial y^2} = uf\left(\dfrac{u}{w}\right), \quad \dfrac{\partial^2 w}{\partial x^2} + \dfrac{\partial^2 w}{\partial y^2} = wg\left(\dfrac{u}{w}\right).$

1°. A space-periodic solution in multiplicative form (another solution is obtained by interchanging $x$ and $y$):

$$u = [C_1 \sin(kx) + C_2 \cos(kx)]\varphi(y),$$
$$w = [C_1 \sin(kx) + C_2 \cos(kx)]\psi(y),$$

where $C_1$, $C_2$, and $k$ are arbitrary constants and the functions $\varphi = \varphi(y)$ and $\psi = \psi(y)$ are determined by the autonomous system of ordinary differential equations

$$\varphi''_{yy} = k^2\varphi + \varphi f(\varphi/\psi),$$
$$\psi''_{yy} = k^2\psi + \psi g(\varphi/\psi).$$

2°. Solution in multiplicative form:

$$u = [C_1 \exp(kx) + C_2 \exp(-kx)]U(y),$$
$$w = [C_1 \exp(kx) + C_2 \exp(-kx)]W(y),$$

where $C_1$, $C_2$, and $k$ are arbitrary constants and the functions $U = U(y)$ and $W = W(y)$ are determined by the autonomous system of ordinary differential equations

$$U''_{yy} = -k^2 U + U f(U/W),$$
$$W''_{yy} = -k^2 W + W g(U/W).$$

3°. Degenerate solution in multiplicative form:
$$u = (C_1 x + C_2) U(y),$$
$$w = (C_1 x + C_2) W(y),$$

where $C_1$ and $C_2$ are arbitrary constants and the functions $U = U(y)$ and $W = W(y)$ are determined by the autonomous system of ordinary differential equations
$$U''_{yy} = U f(U/W),$$
$$W''_{yy} = W g(U/W).$$

**Remark.** The functions $f$ and $g$ in Items 1°–3° can be dependent on $y$.

4°. Solution in multiplicative form:
$$u = e^{a_1 x + b_1 y} \xi(z), \qquad w = e^{a_1 x + b_1 y} \eta(z), \qquad z = a_2 x + b_2 y,$$

where $a_1$, $a_2$, $b_1$, and $b_2$ are arbitrary constants, and the functions $\xi = \xi(z)$ and $\eta = \eta(z)$ are determined by the autonomous system of ordinary differential equations
$$(a_2^2 + b_2^2)\xi''_{zz} + 2(a_1 a_2 + b_1 b_2)\xi'_z + (a_1^2 + b_1^2)\xi = \xi f(\xi/\eta),$$
$$(a_2^2 + b_2^2)\eta''_{zz} + 2(a_1 a_2 + b_1 b_2)\eta'_z + (a_1^2 + b_1^2)\eta = \eta g(\xi/\eta).$$

5°. Solution:
$$u = k\theta(x, y), \qquad w = \theta(x, y),$$

where $k$ is a root of the algebraic (transcendental) equation $f(k) = g(k)$, and the function $\theta = \theta(x, y)$ is described by the linear Helmholtz equation
$$\frac{\partial^2 \theta}{\partial x^2} + \frac{\partial^2 \theta}{\partial y^2} = f(k)\theta.$$

For its exact solutions, see Subsection T8.3.3.

4. $\dfrac{\partial^2 u}{\partial x^2} + \dfrac{\partial^2 u}{\partial y^2} = uf\left(\dfrac{u}{w}\right) + \dfrac{u}{w} h\left(\dfrac{u}{w}\right), \qquad \dfrac{\partial^2 w}{\partial x^2} + \dfrac{\partial^2 w}{\partial y^2} = wg\left(\dfrac{u}{w}\right) + h\left(\dfrac{u}{w}\right).$

Solution:
$$u = kw, \qquad w = \theta(x, y) - \frac{h(k)}{f(k)},$$

where $k$ is a root of the algebraic (transcendental) equation
$$f(k) = g(k),$$

and the function $\theta = \theta(x, y)$ satisfies the linear Helmholtz equation
$$\frac{\partial^2 \theta}{\partial x^2} + \frac{\partial^2 \theta}{\partial y^2} = f(k)w.$$

5. $\dfrac{\partial^2 u}{\partial x^2} + \dfrac{\partial^2 u}{\partial y^2} = u^n f\left(\dfrac{u}{w}\right), \quad \dfrac{\partial^2 w}{\partial x^2} + \dfrac{\partial^2 w}{\partial y^2} = w^n g\left(\dfrac{u}{w}\right).$

For $f(z) = kz^{-m}$ and $g(z) = -kz^{n-m}$, the system describes an $n$th-order chemical reaction (of order $n-m$ in the component $u$ and of order $m$ in the component $w$); to $n = 2$ and $m = 1$ there corresponds a second-order reaction, which often occurs in applications.

1°. Solution:

$$u = r^{\frac{2}{1-n}} U(\theta), \quad w = r^{\frac{2}{1-n}} W(\theta), \quad r = \sqrt{(x+C_1)^2 + (y+C_2)^2}, \quad \theta = \dfrac{y+C_2}{x+C_1},$$

where $C_1$ and $C_2$ are arbitrary constants, and the functions $y = y(\xi)$ and $z = z(\xi)$ are determined by the autonomous system of ordinary differential equations

$$U''_{\theta\theta} + \dfrac{4}{(1-n)^2} U = U^n f\left(\dfrac{U}{W}\right),$$

$$W''_{\theta\theta} + \dfrac{4}{(1-n)^2} W = W^n g\left(\dfrac{U}{W}\right).$$

2°. Solution:

$$u = k\zeta(x,y), \quad w = \zeta(x,y),$$

where $k$ is a root of the algebraic (transcendental) equation

$$k^{n-1} f(k) = g(k),$$

and the function $\zeta = \zeta(x,y)$ satisfies the equation with a power-law nonlinearity

$$\dfrac{\partial^2 \zeta}{\partial x^2} + \dfrac{\partial^2 \zeta}{\partial y^2} = g(k)\zeta^n.$$

### T10.3.3-3. Other systems.

6. $\dfrac{\partial^2 u}{\partial x^2} + \dfrac{\partial^2 u}{\partial y^2} = u f(u^n w^m), \quad \dfrac{\partial^2 w}{\partial x^2} + \dfrac{\partial^2 w}{\partial y^2} = w g(u^n w^m).$

Solution in multiplicative form:

$$u = e^{m(a_1 x + b_1 y)} \xi(z), \quad w = e^{-n(a_1 x + b_1 y)} \eta(z), \quad z = a_2 x + b_2 y,$$

where $a_1, a_2, b_1$, and $b_2$ are arbitrary constants, and the functions $\xi = \xi(z)$ and $\eta = \eta(z)$ are determined by the autonomous system of ordinary differential equations

$$(a_2^2 + b_2^2)\xi''_{zz} + 2m(a_1 a_2 + b_1 b_2)\xi'_z + m^2(a_1^2 + b_1^2)\xi = \xi f(\xi^n \eta^m),$$

$$(a_2^2 + b_2^2)\eta''_{zz} - 2n(a_1 a_2 + b_1 b_2)\eta'_z + n^2(a_1^2 + b_1^2)\eta = \eta g(\xi^n \eta^m).$$

7.  $$\frac{\partial^2 u}{\partial x^2} + \frac{\partial^2 u}{\partial y^2} = uf(u^2 + w^2) - wg(u^2 + w^2),$$
    $$\frac{\partial^2 w}{\partial x^2} + \frac{\partial^2 w}{\partial y^2} = wf(u^2 + w^2) + ug(u^2 + w^2).$$

$1°$. A periodic solution in $y$:
$$u = r(x)\cos[\theta(x) + C_1 y + C_2], \quad w = r(x)\sin[\theta(x) + C_1 y + C_2],$$
where $C_1$ and $C_2$ are arbitrary constants, and the functions $r = r(x)$ and $\theta = \theta(x)$ are determined by the autonomous system of ordinary differential equations
$$r''_{xx} = r(\theta'_x)^2 + C_1^2 r + rf(r^2),$$
$$r\theta''_{xx} = -2r'_x \theta'_x + rg(r^2).$$

$2°$. Solution (generalizes the solution of Item $1°$):
$$u = r(z)\cos[\theta(z) + C_1 y + C_2], \quad w = r(z)\sin[\theta(z) + C_1 y + C_2], \quad z = k_1 x + k_2 y,$$
where $C_1$, $C_2$, $k_1$, and $k_2$ are arbitrary constants, and the functions $r = r(z)$ and $\theta = \theta(z)$ are determined by the autonomous system of ordinary differential equations
$$(k_1^2 + k_2^2)r''_{zz} = k_1^2 r(\theta'_z)^2 + r(k_2\theta'_z + C_1)^2 + rf(r^2),$$
$$(k_1^2 + k_2^2)r\theta''_{zz} = -2[(k_1^2 + k_2^2)\theta'_z + C_1 k_2]r'_z + rg(r^2).$$

8.  $$\frac{\partial^2 u}{\partial x^2} + \frac{\partial^2 u}{\partial y^2} = uf(u^2 - w^2) + wg(u^2 - w^2),$$
    $$\frac{\partial^2 w}{\partial x^2} + \frac{\partial^2 w}{\partial y^2} = wf(u^2 - w^2) + ug(u^2 - w^2).$$

Solution:
$$u = r(z)\cosh[\theta(z) + C_1 y + C_2], \quad w = r(z)\sinh[\theta(z) + C_1 y + C_2], \quad z = k_1 x + k_2 y,$$
where $C_1$, $C_2$, $k_1$, and $k_2$ are arbitrary constants, and the functions $r = r(z)$ and $\theta = \theta(z)$ are determined by the autonomous system of ordinary differential equations
$$(k_1^2 + k_2^2)r''_{zz} + k_1^2 r(\theta'_z)^2 + r(k_2\theta'_z + C_1)^2 = rf(r^2),$$
$$(k_1^2 + k_2^2)r\theta''_{zz} + 2[(k_1^2 + k_2^2)\theta'_z + C_1 k_2]r'_z = rg(r^2).$$

## T10.3.4. Systems of the Form
$$\frac{\partial^2 u}{\partial t^2} = \frac{a}{x^n}\frac{\partial}{\partial x}\left(x^n \frac{\partial u}{\partial x}\right) + F(u, w),$$
$$\frac{\partial^2 w}{\partial t^2} = \frac{b}{x^n}\frac{\partial}{\partial x}\left(x^n \frac{\partial w}{\partial x}\right) + G(u, w)$$

T10.3.4-1. Arbitrary functions depend on a linear combination of the unknowns.

1.  $$\frac{\partial^2 u}{\partial t^2} = \frac{a}{x^n}\frac{\partial}{\partial x}\left(x^n \frac{\partial u}{\partial x}\right) + uf(bu - cw) + g(bu - cw),$$
    $$\frac{\partial^2 w}{\partial t^2} = \frac{a}{x^n}\frac{\partial}{\partial x}\left(x^n \frac{\partial w}{\partial x}\right) + wf(bu - cw) + h(bu - cw).$$

$1°$. Solution:
$$u = \varphi(t) + c\theta(x, t), \quad w = \psi(t) + b\theta(x, t),$$

where $\varphi = \varphi(t)$ and $\psi = \psi(t)$ are determined by the autonomous system of ordinary differential equations
$$\varphi''_{tt} = \varphi f(b\varphi - c\psi) + g(b\varphi - c\psi),$$
$$\psi''_{tt} = \psi f(b\varphi - c\psi) + h(b\varphi - c\psi),$$
and the function $\theta = \theta(x, t)$ satisfies linear equation
$$\frac{\partial^2 \theta}{\partial t^2} = \frac{a}{x^n} \frac{\partial}{\partial x}\left(x^n \frac{\partial \theta}{\partial x}\right) + f(b\varphi - c\psi)\theta.$$

For $f = \text{const}$, this equation can be solved by separation of variables.

$2°$. Let us multiply the first equation by $b$ and the second one by $-c$ and add the results together to obtain
$$\frac{\partial^2 \zeta}{\partial t^2} = \frac{a}{x^n} \frac{\partial}{\partial x}\left(x^n \frac{\partial \zeta}{\partial x}\right) + \zeta f(\zeta) + bg(\zeta) - ch(\zeta), \qquad \zeta = bu - cw. \tag{1}$$

This equation will be considered in conjunction with the first equation of the original system
$$\frac{\partial^2 u}{\partial t^2} = \frac{a}{x^n} \frac{\partial}{\partial x}\left(x^n \frac{\partial u}{\partial x}\right) + u f(\zeta) + g(\zeta). \tag{2}$$

Equation (1) can be treated separately. Given a solution $\zeta = \zeta(x, t)$ to equation (1), the function $u = u(x, t)$ can be determined by solving equation (2) and the function $w = w(x, t)$ is found as $w = (bu - \zeta)/c$.

Note three important solutions to equation (1):

(i) In the general case, equation (1) admits a spatially homogeneous solution $\zeta = \zeta(t)$. The corresponding solution to the original system is given in Item $1°$ in another form.

(ii) In the general case, equation (1) admits a steady-state solution $\zeta = \zeta(x)$. The corresponding exact solutions to equation (2) are expressed as $u = u_0(x) + \sum e^{-\beta_n t} u_n(x)$ and $u = u_0(x) + \sum \cos(\beta_n t) u_n^{(1)}(x) + \sum \sin(\beta_n t) u_n^{(2)}(x)$.

(iii) If the condition $\zeta f(\zeta) + bg(\zeta) - ch(\zeta) = k_1\zeta + k_0$ holds, equation (1) is linear,
$$\frac{\partial^2 \zeta}{\partial t^2} = \frac{a}{x^n} \frac{\partial}{\partial x}\left(x^n \frac{\partial \zeta}{\partial x}\right) + k_1\zeta + k_0,$$
and, hence, can be solved by separation of variables.

2. $\dfrac{\partial^2 u}{\partial t^2} = \dfrac{a}{x^n} \dfrac{\partial}{\partial x}\left(x^n \dfrac{\partial u}{\partial x}\right) + e^{\lambda u} f(\lambda u - \sigma w),$

$\dfrac{\partial^2 w}{\partial t^2} = \dfrac{b}{x^n} \dfrac{\partial}{\partial x}\left(x^n \dfrac{\partial w}{\partial x}\right) + e^{\sigma w} g(\lambda u - \sigma w).$

$1°$. Solution:
$$u = y(\xi) - \frac{2}{\lambda} \ln(C_1 t + C_2), \qquad w = z(\xi) - \frac{2}{\sigma} \ln(C_1 t + C_2), \qquad \xi = \frac{x}{C_1 t + C_2},$$
where $C_1$ and $C_2$ are arbitrary constants, and the functions $y = y(\xi)$ and $z = z(\xi)$ are determined by the system of ordinary differential equations
$$C_1^2(\xi^2 y'_\xi)'_\xi + 2C_1^2 \lambda^{-1} = a\xi^{-n}(\xi^n y'_\xi)'_\xi + e^{\lambda y} f(\lambda y - \sigma z),$$
$$C_1^2(\xi^2 z'_\xi)'_\xi + 2C_1^2 \sigma^{-1} = b\xi^{-n}(\xi^n z'_\xi)'_\xi + e^{\sigma z} g(\lambda y - \sigma z).$$

2°. Solution with $b = a$:

$$u = \theta(x,t), \quad w = \frac{\lambda}{\sigma}\theta(x,t) - \frac{k}{\sigma},$$

where $k$ is a root of the algebraic (transcendental) equation

$$\lambda f(k) = \sigma e^{-k} g(k),$$

and the function $\theta = \theta(x,t)$ is described by the equation

$$\frac{\partial^2 \theta}{\partial t^2} = \frac{a}{x^n} \frac{\partial}{\partial x}\left(x^n \frac{\partial \theta}{\partial x}\right) + f(k)e^{\lambda \theta}.$$

This equation is solvable for $n = 0$; for its exact solutions, see Polyanin and Zaitsev (2004).

T10.3.4-2. Arbitrary functions depend on the ratio of the unknowns.

3. $\quad \dfrac{\partial^2 u}{\partial t^2} = \dfrac{a}{x^n} \dfrac{\partial}{\partial x}\left(x^n \dfrac{\partial u}{\partial x}\right) + uf\left(\dfrac{u}{w}\right), \quad \dfrac{\partial^2 w}{\partial t^2} = \dfrac{b}{x^n} \dfrac{\partial}{\partial x}\left(x^n \dfrac{\partial w}{\partial x}\right) + wg\left(\dfrac{u}{w}\right).$

1°. Periodic multiplicative separable solution:

$$u = [C_1 \cos(kt) + C_2 \sin(kt)]y(x), \quad w = [C_1 \cos(kt) + C_2 \sin(kt)]z(x),$$

where $C_1$, $C_2$, and $k$ are arbitrary constants and the functions $y = y(x)$ and $z = z(x)$ are determined by the system of ordinary differential equations

$$ax^{-n}(x^n y'_x)'_x + k^2 y + y f(y/z) = 0,$$
$$bx^{-n}(x^n z'_x)'_x + k^2 z + z g(y/z) = 0.$$

2°. Multiplicative separable solution:

$$u = [C_1 \exp(kt) + C_2 \exp(-kt)]y(x), \quad w = [C_1 \exp(kt) + C_2 \exp(-kt)]z(x),$$

where $C_1$, $C_2$, and $k$ are arbitrary constants and the functions $y = y(x)$ and $z = z(x)$ are determined by the system of ordinary differential equations

$$ax^{-n}(x^n y'_x)'_x - k^2 y + y f(y/z) = 0,$$
$$bx^{-n}(x^n z'_x)'_x - k^2 z + z g(y/z) = 0.$$

3°. Degenerate multiplicative separable solution:

$$u = (C_1 t + C_2)y(x), \quad w = (C_1 t + C_2)z(x),$$

where the functions $y = y(x)$ and $z = z(x)$ are determined by the system of ordinary differential equations

$$ax^{-n}(x^n y'_x)'_x + y f(y/z) = 0,$$
$$bx^{-n}(x^n z'_x)'_x + z g(y/z) = 0.$$

4°. Multiplicative separable solution:

$$u = x^{\frac{1-n}{2}}[C_1 J_\nu(kx) + C_2 Y_\nu(kx)]\varphi(t), \quad \nu = \tfrac{1}{2}|n-1|,$$

$$w = x^{\frac{1-n}{2}}[C_1 J_\nu(kx) + C_2 Y_\nu(kx)]\psi(t),$$

where $C_1$, $C_2$, and $k$ are arbitrary constants, $J_\nu(z)$ and $Y_\nu(z)$ are Bessel functions, and the functions $\varphi = \varphi(t)$ and $\psi = \psi(t)$ are determined by the autonomous system of ordinary differential equations

$$\varphi''_{tt} = -ak^2\varphi + \varphi f(\varphi/\psi),$$
$$\psi''_{tt} = -bk^2\psi + \psi g(\varphi/\psi).$$

5°. Multiplicative separable solution:

$$u = x^{\frac{1-n}{2}}[C_1 I_\nu(kx) + C_2 K_\nu(kx)]\varphi(t), \quad \nu = \tfrac{1}{2}|n-1|,$$

$$w = x^{\frac{1-n}{2}}[C_1 I_\nu(kx) + C_2 K_\nu(kx)]\psi(t),$$

where $C_1$, $C_2$, and $k$ are arbitrary constants, $I_\nu(z)$ and $K_\nu(z)$ are modified Bessel functions, and the functions $\varphi = \varphi(t)$ and $\psi = \psi(t)$ are determined by the autonomous system of ordinary differential equations

$$\varphi''_{tt} = ak^2\varphi + \varphi f(\varphi/\psi),$$
$$\psi''_{tt} = bk^2\psi + \psi g(\varphi/\psi).$$

6°. Solution with $b = a$:

$$u = k\theta(x,t), \quad w = \theta(x,t),$$

where $k$ is a root of the algebraic (transcendental) equation $f(k) = g(k)$, and the function $\theta = \theta(x,t)$ is described by the linear Klein–Gordon equation

$$\frac{\partial^2 \theta}{\partial t^2} = \frac{a}{x^n}\frac{\partial}{\partial x}\left(x^n \frac{\partial \theta}{\partial x}\right) + f(k)\theta.$$

4. $$\frac{\partial^2 u}{\partial t^2} = \frac{a}{x^n}\frac{\partial}{\partial x}\left(x^n \frac{\partial u}{\partial x}\right) + uf\left(\frac{u}{w}\right) + \frac{u}{w}h\left(\frac{u}{w}\right),$$

$$\frac{\partial^2 w}{\partial t^2} = \frac{a}{x^n}\frac{\partial}{\partial x}\left(x^n \frac{\partial u}{\partial w}\right) + wg\left(\frac{u}{w}\right) + h\left(\frac{u}{w}\right).$$

Solution:
$$u = k\theta(x,t), \quad w = \theta(x,t),$$

where $k$ is a root of the algebraic (transcendental) equation $f(k) = g(k)$, and the function $\theta = \theta(x,t)$ is described by the linear equation

$$\frac{\partial^2 \theta}{\partial t^2} = \frac{a}{x^n}\frac{\partial}{\partial x}\left(x^n \frac{\partial \theta}{\partial x}\right) + f(k)\theta + h(k).$$

5. $\dfrac{\partial^2 u}{\partial t^2} = \dfrac{a}{x^n}\dfrac{\partial}{\partial x}\left(x^n \dfrac{\partial u}{\partial x}\right) + u^k f\left(\dfrac{u}{w}\right), \qquad \dfrac{\partial^2 w}{\partial t^2} = \dfrac{b}{x^n}\dfrac{\partial}{\partial x}\left(x^n \dfrac{\partial w}{\partial x}\right) + w^k g\left(\dfrac{u}{w}\right).$

Self-similar solution:
$$u = (C_1 t + C_2)^{\frac{2}{1-k}} y(\xi), \qquad w = (C_1 t + C_2)^{\frac{2}{1-k}} z(\xi), \qquad \xi = \dfrac{x}{C_1 t + C_2},$$
where $C_1$ and $C_2$ are arbitrary constants, and the functions $y = y(\xi)$ and $z = z(\xi)$ are determined by the system of ordinary differential equations
$$C_1^2 \xi^2 y''_{\xi\xi} + \dfrac{2C_1^2(k+1)}{k-1}\xi y'_\xi + \dfrac{C_1^2(k+1)}{(k-1)^2} y = \dfrac{a}{\xi^n}(\xi^n y'_\xi)'_\xi + y^k f\left(\dfrac{y}{z}\right),$$
$$C_1^2 \xi^2 z''_{\xi\xi} + \dfrac{2C_1^2(k+1)}{k-1}\xi z'_\xi + \dfrac{C_1^2(k+1)}{(k-1)^2} z = \dfrac{b}{\xi^n}(\xi^n z'_\xi)'_\xi + z^k g\left(\dfrac{y}{z}\right).$$

---

**T10.3.4-3. Arbitrary functions depend on the product of powers of the unknowns.**

6. $\dfrac{\partial^2 u}{\partial t^2} = \dfrac{a}{x^n}\dfrac{\partial}{\partial x}\left(x^n \dfrac{\partial u}{\partial x}\right) + u f(x, u^k w^m),$

$\dfrac{\partial^2 w}{\partial t^2} = \dfrac{b}{x^n}\dfrac{\partial}{\partial x}\left(x^n \dfrac{\partial w}{\partial x}\right) + w g(x, u^k w^m).$

Multiplicative separable solution:
$$u = e^{-m\lambda t} y(x), \qquad w = e^{k\lambda t} z(x),$$
where $\lambda$ is an arbitrary constant and the functions $y = y(x)$ and $z = z(x)$ are determined by the system of ordinary differential equations
$$ax^{-n}(x^n y'_x)'_x - m^2\lambda^2 y + y f(x, y^k z^m) = 0,$$
$$bx^{-n}(x^n z'_x)'_x - k^2\lambda^2 z + z g(x, y^k z^m) = 0.$$

---

**T10.3.4-4. Arbitrary functions depend on the $u^2 \pm w^2$.**

7. $\dfrac{\partial^2 u}{\partial t^2} = \dfrac{a}{x^n}\dfrac{\partial}{\partial x}\left(x^n \dfrac{\partial u}{\partial x}\right) + u f(u^2 + w^2) - w g(u^2 + w^2),$

$\dfrac{\partial^2 w}{\partial t^2} = \dfrac{a}{x^n}\dfrac{\partial}{\partial x}\left(x^n \dfrac{\partial w}{\partial x}\right) + w f(u^2 + w^2) + u g(u^2 + w^2).$

$1°$. Periodic solution in $t$:
$$u = r(x)\cos\big[\theta(x) + C_1 t + C_2\big], \qquad w = r(x)\sin\big[\theta(x) + C_1 t + C_2\big],$$
where $C_1$ and $C_2$ are arbitrary constants, and the functions $r = r(x)$ and $\theta(x)$ are determined by the system of ordinary differential equations
$$ar''_{xx} - ar(\theta'_x)^2 + \dfrac{an}{x}r'_x + C_1^2 r + r f(r^2) = 0,$$
$$ar\theta''_{xx} + 2ar'_x \theta'_x + \dfrac{an}{x}r\theta'_x + r g(r^2) = 0.$$

$2°$. For $n = 0$, there is an exact solution of the form
$$u = r(z)\cos\big[\theta(z) + C_1 t + C_2\big], \qquad w = r(z)\sin\big[\theta(z) + C_1 t + C_2\big], \qquad z = kx - \lambda t.$$

8.
$$\frac{\partial^2 u}{\partial t^2} = \frac{a}{x^n}\frac{\partial}{\partial x}\left(x^n\frac{\partial u}{\partial x}\right) + uf(u^2 - w^2) + wg(u^2 - w^2),$$
$$\frac{\partial^2 w}{\partial t^2} = \frac{a}{x^n}\frac{\partial}{\partial x}\left(x^n\frac{\partial w}{\partial x}\right) + wf(u^2 - w^2) + ug(u^2 - w^2).$$

1°. Solution:
$$u = r(x)\cosh\bigl[\theta(x) + C_1 t + C_2\bigr], \qquad w = r(x)\sinh\bigl[\theta(x) + C_1 t + C_2\bigr],$$

where $C_1$ and $C_2$ are arbitrary constants, and the functions $r = r(x)$ and $\theta(x)$ are determined by the system of ordinary differential equations
$$ar''_{xx} + ar(\theta'_x)^2 + \frac{an}{x}r'_x - C_1^2 r + rf(r^2) = 0,$$
$$ar\theta''_{xx} + 2ar'_x\theta'_x + \frac{an}{x}r\theta'_x + rg(r^2) = 0.$$

2°. For $n = 0$, there is an exact solution of the form
$$u = r(z)\cosh\bigl[\theta(z) + C_1 t + C_2\bigr], \qquad w = r(z)\sinh\bigl[\theta(z) + C_1 t + C_2\bigr], \qquad z = kx - \lambda t.$$

## T10.3.5. Other Systems

1.
$$ax\frac{\partial u}{\partial x} + ay\frac{\partial u}{\partial y} = \frac{\partial^2 u}{\partial x^2} + \frac{\partial^2 u}{\partial y^2} - f(u, w),$$
$$ax\frac{\partial w}{\partial x} + ay\frac{\partial w}{\partial y} = \frac{\partial^2 w}{\partial x^2} + \frac{\partial^2 w}{\partial y^2} - g(u, w).$$

Solution:
$$u(x, y) = U(z), \qquad w(x, y) = W(z), \qquad z = k_1 x + k_2 y,$$

where $k_1$ and $k_2$ are arbitrary constants, and the functions $U = U(z)$ and $W = W(z)$ are described by the system of ordinary differential equations
$$azU'_z = (k_1^2 + k_2^2)U'' - f(U, W), \qquad azW'_z = (k_1^2 + k_2^2)W'' - g(U, W).$$

2.
$$\frac{\partial u}{\partial t} = \frac{\partial}{\partial x}\left[f\left(t, \frac{u}{w}\right)\frac{\partial u}{\partial x}\right] + ug\left(t, \frac{u}{w}\right),$$
$$\frac{\partial w}{\partial t} = \frac{\partial}{\partial x}\left[f\left(t, \frac{u}{w}\right)\frac{\partial w}{\partial x}\right] + wh\left(t, \frac{u}{w}\right).$$

Solution:
$$u = \varphi(t)\exp\left[\int h(t, \varphi(t))\,dt\right]\theta(x, \tau), \quad w = \exp\left[\int h(t, \varphi(t))\,dt\right]\theta(x, \tau), \quad \tau = \int f(t, \varphi(t))\,dt,$$

where the function $\varphi = \varphi(t)$ is described by the ordinary differential equation
$$\varphi'_t = [g(t, \varphi) - h(t, \varphi)]\varphi,$$

and the function $\theta = \theta(x, \tau)$ satisfies the linear heat equation
$$\frac{\partial \theta}{\partial \tau} = \frac{\partial^2 \theta}{\partial x^2}.$$

## T10.4. Systems of General Form

### T10.4.1. Linear Systems

1. $\dfrac{\partial u}{\partial t} = L[u] + f_1(t)u + g_1(t)w, \qquad \dfrac{\partial w}{\partial t} = L[w] + f_2(t)u + g_2(t)w.$

Here, $L$ is an arbitrary linear differential operator with respect to the coordinates $x_1, \ldots, x_n$ (of any order in derivatives), whose coefficients can be dependent on $x_1, \ldots, x_n, t$. It is assumed that $L[\text{const}] = 0$.

Solution:
$$u = \varphi_1(t)U(x_1, \ldots, x_n, t) + \varphi_2(t)W(x_1, \ldots, x_n, t),$$
$$w = \psi_1(t)U(x_1, \ldots, x_n, t) + \psi_2(t)W(x_1, \ldots, x_n, t),$$

where the two pairs of functions $\varphi_1 = \varphi_1(t)$, $\psi_1 = \psi_1(t)$ and $\varphi_2 = \varphi_2(t)$, $\psi_2 = \psi_2(t)$ are linearly independent (fundamental) solutions to the system of first-order linear ordinary differential equations

$$\varphi'_t = f_1(t)\varphi + g_1(t)\psi,$$
$$\psi'_t = f_2(t)\varphi + g_2(t)\psi,$$

and the functions $U = U(x_1, \ldots, x_n, t)$ and $W = W(x_1, \ldots, x_n, t)$ satisfy the independent linear equations

$$\dfrac{\partial U}{\partial t} = L[U], \qquad \dfrac{\partial W}{\partial t} = L[W].$$

2. $\dfrac{\partial^2 u}{\partial t^2} = L[u] + a_1 u + b_1 w, \qquad \dfrac{\partial^2 w}{\partial t^2} = L[w] + a_2 u + b_2 w.$

Here, $L$ is an arbitrary linear differential operator with respect to the coordinates $x_1, \ldots, x_n$ (of any order in derivatives).

Solution:
$$u = \dfrac{a_1 - \lambda_2}{a_2(\lambda_1 - \lambda_2)}\theta_1 - \dfrac{a_1 - \lambda_1}{a_2(\lambda_1 - \lambda_2)}\theta_2, \qquad w = \dfrac{1}{\lambda_1 - \lambda_2}(\theta_1 - \theta_2),$$

where $\lambda_1$ and $\lambda_2$ are roots of the quadratic equation

$$\lambda^2 - (a_1 + b_2)\lambda + a_1 b_2 - a_2 b_1 = 0,$$

and the functions $\theta_n = \theta_n(x_1, \ldots, x_n, t)$ satisfy the independent linear equations

$$\dfrac{\partial^2 \theta_1}{\partial t^2} = L[\theta_1] + \lambda_1 \theta_1, \qquad \dfrac{\partial^2 \theta_2}{\partial t^2} = L[\theta_2] + \lambda_2 \theta_2.$$

### T10.4.2. Nonlinear Systems of Two Equations Involving the First Derivatives in $t$

1. $\dfrac{\partial u}{\partial t} = L[u] + uf(t, bu - cw) + g(t, bu - cw),$

   $\dfrac{\partial w}{\partial t} = L[w] + wf(t, bu - cw) + h(t, bu - cw).$

Here, $L$ is an arbitrary linear differential operator (of any order) with respect to the spatial variables $x_1, \ldots, x_n$.

**1°. Solution:**
$$u = \varphi(t) + c\exp\left[\int f(t, b\varphi - c\psi)\,dt\right]\theta(x,t), \quad w = \psi(t) + b\exp\left[\int f(t, b\varphi - c\psi)\,dt\right]\theta(x,t),$$

where $\varphi = \varphi(t)$ and $\psi = \psi(t)$ are determined by the system of ordinary differential equations
$$\varphi'_t = \varphi f(t, b\varphi - c\psi) + g(t, b\varphi - c\psi),$$
$$\psi'_t = \psi f(t, b\varphi - c\psi) + h(t, b\varphi - c\psi),$$

and the function $\theta = \theta(x_1, \ldots, x_n, t)$ satisfies linear equation
$$\frac{\partial \theta}{\partial t} = L[\theta].$$

*Remark 1.* The coefficients of the linear differential operator $L$ can be dependent on $x_1, \ldots, x_n, t$.

**2°.** Let us multiply the first equation by $b$ and the second one by $-c$ and add the results together to obtain
$$\frac{\partial \zeta}{\partial t} = L[\zeta] + \zeta f(t, \zeta) + bg(t, \zeta) - ch(t, \zeta), \quad \zeta = bu - cw. \tag{1}$$

This equation will be considered in conjunction with the first equation of the original system
$$\frac{\partial u}{\partial t} = L[u] + uf(t, \zeta) + g(t, \zeta). \tag{2}$$

Equation (1) can be treated separately. Given a solution of this equation, $\zeta = \zeta(x_1, \ldots, x_n, t)$, the function $u = u(x_1, \ldots, x_n, t)$ can be determined by solving the linear equation (2) and the function $w = w(x_1, \ldots, x_n, t)$ is found as $w = (bu - \zeta)/c$.

*Remark 2.* Let $L$ be a constant-coefficient differential operator with respect to the independent variable $x = x_1$ and let the condition
$$\frac{\partial}{\partial t}\left[\zeta f(t, \zeta) + bg(t, \zeta) - ch(t, \zeta)\right] = 0$$
hold true (for example, it is valid if the functions $f, g, h$ are not implicitly dependent on $t$). Then equation (1) admits an exact, traveling-wave solution $\zeta = \zeta(z)$, where $z = kx - \lambda t$ with arbitrary constants $k$ and $\lambda$.

**2.** $\quad \dfrac{\partial u}{\partial t} = L_1[u] + uf\left(\dfrac{u}{w}\right), \quad \dfrac{\partial w}{\partial t} = L_2[w] + wg\left(\dfrac{u}{w}\right).$

Here, $L_1$ and $L_2$ are arbitrary constant-coefficient linear differential operators (of any order) with respect to $x$.

**1°. Solution:**
$$u = e^{kx - \lambda t} y(\xi), \quad w = e^{kx - \lambda t} z(\xi), \quad \xi = \beta x - \gamma t,$$

where $k, \lambda, \beta$, and $\gamma$ are arbitrary constants and the functions $y = y(\xi)$ and $z = z(\xi)$ are determined by the system of ordinary differential equations
$$M_1[y] + \lambda y + yf(y/z) = 0, \quad M_2[z] + \lambda z + zg(y/z) = 0,$$
$$M_1[y] = e^{-kx} L_1[e^{kx} y(\xi)], \quad M_2[z] = e^{-kx} L_2[e^{kx} z(\xi)].$$

To the special case $k = \lambda = 0$ there corresponds a traveling-wave solution.

**2°.** If the operators $L_1$ and $L_2$ contain only even derivatives, there are solutions of the form
$$u = [C_1 \sin(kx) + C_2 \cos(kx)]\varphi(t), \quad w = [C_1 \sin(kx) + C_2 \cos(kx)]\psi(t);$$
$$u = [C_1 \exp(kx) + C_2 \exp(-kx)]\varphi(t), \quad w = [C_1 \exp(kx) + C_2 \exp(-kx)]\psi(t);$$
$$u = (C_1 x + C_2)\varphi(t), \quad w = (C_1 x + C_2)\psi(t),$$

where $C_1, C_2$, and $k$ are arbitrary constants. Note that the third solution is degenerate.

3. $\dfrac{\partial u}{\partial t} = L[u] + uf\left(t, \dfrac{u}{w}\right), \quad \dfrac{\partial w}{\partial t} = L[w] + wg\left(t, \dfrac{u}{w}\right).$

Here, $L$ is an arbitrary linear differential operator with respect to the coordinates $x_1, \ldots, x_n$ (of any order in derivatives), whose coefficients can be dependent on $x_1, \ldots, x_n, t$:

$$L[u] = \sum A_{k_1 \ldots k_n}(x_1, \ldots, x_n, t) \dfrac{\partial^{k_1 + \cdots + k_n} u}{\partial x_1^{k_1} \ldots \partial x_n^{k_n}}. \tag{1}$$

1°. Solution:

$$u = \varphi(t) \exp\left[\int g(t, \varphi(t))\, dt\right] \theta(x_1, \ldots, x_n, t),$$
$$w = \exp\left[\int g(t, \varphi(t))\, dt\right] \theta(x_1, \ldots, x_n, t), \tag{2}$$

where the function $\varphi = \varphi(t)$ is described by the first-order nonlinear ordinary differential equation

$$\varphi'_t = [f(t, \varphi) - g(t, \varphi)]\varphi, \tag{3}$$

and the function $\theta = \theta(x_1, \ldots, x_n, t)$ satisfies the linear equation

$$\dfrac{\partial \theta}{\partial t} = L[\theta].$$

2°. The transformation

$$u = a_1(t) U + b_1(t) W, \quad w = a_2(t) U + b_2(t) W,$$

where $a_n(t)$ and $b_n(t)$ are arbitrary functions ($n = 1, 2$), leads to an equation of the similar form for $U$ and $W$.

*Remark.* The coefficients of the operator (1) can also depend on the ratio of the unknowns $u/w$, $A_{k_1 \ldots k_n} = A_{k_1 \ldots k_n}(x_1, \ldots, x_n, t, u/w)$ (in this case, $L$ will be a quasilinear operator). Then there also exists a solution of the form (2), where $\varphi = \varphi(t)$ is described by the ordinary differential equation (3) and $\theta = \theta(x_1, \ldots, x_n, t)$ satisfies the linear equation

$$\dfrac{\partial \theta}{\partial t} = L^\circ[\theta], \quad L^\circ = L_{u/w = \varphi}.$$

4. $\dfrac{\partial u}{\partial t} = L[u] + uf\left(\dfrac{u}{w}\right) + g\left(\dfrac{u}{w}\right), \quad \dfrac{\partial w}{\partial t} = L[w] + wf\left(\dfrac{u}{w}\right) + h\left(\dfrac{u}{w}\right).$

Here, $L$ is an arbitrary linear differential operator with respect to $x_1, \ldots, x_n$ (of any order in derivatives), whose coefficients can be dependent on $x_1, \ldots, x_n, t$:

$$L[u] = \sum A_{k_1 \ldots k_n}(x_1, \ldots, x_n, t) \dfrac{\partial^{k_1 + \cdots + k_n} u}{\partial x_1^{k_1} \ldots \partial x_n^{k_n}},$$

where $k_1 + \cdots + k_n \geq 1$.

Let $k$ be a root of the algebraic (transcendental) equation

$$g(k) = kh(k). \tag{1}$$

1°. Solution if $f(k) \neq 0$:

$$u(x, t) = k\left(\exp[f(k)t]\theta(x, t) - \dfrac{h(k)}{f(k)}\right), \quad w(x, t) = \exp[f(k)t]\theta(x, t) - \dfrac{h(k)}{f(k)},$$

where the function $\theta = \theta(x_1, \ldots, x_n, t)$ satisfies the linear equation

$$\frac{\partial \theta}{\partial t} = L[\theta]. \tag{2}$$

2°. Solution if $f(k) = 0$:

$$u(x, t) = k[\theta(x, t) + h(k)t], \quad w(x, t) = \theta(x, t) + h(k)t,$$

where the function $\theta = \theta(x_1, \ldots, x_n, t)$ satisfies the linear equation (2).

5. $\dfrac{\partial u}{\partial t} = L[u] + uf\left(t, \dfrac{u}{w}\right) + \dfrac{u}{w}h\left(t, \dfrac{u}{w}\right), \quad \dfrac{\partial w}{\partial t} = L[w] + wg\left(t, \dfrac{u}{w}\right) + h\left(t, \dfrac{u}{w}\right).$

Solution:

$$u = \varphi(t)G(t)\left[\theta(x_1, \ldots, x_n, t) + \int \frac{h(t, \varphi)}{G(t)}\, dt\right], \quad G(t) = \exp\left[\int g(t, \varphi)\, dt\right],$$

$$w = G(t)\left[\theta(x_1, \ldots, x_n, t) + \int \frac{h(t, \varphi)}{G(t)}\, dt\right],$$

where the function $\varphi = \varphi(t)$ is described by the first-order nonlinear ordinary differential equation

$$\varphi'_t = [f(t, \varphi) - g(t, \varphi)]\varphi,$$

and the function $\theta = \theta(x_1, \ldots, x_n, t)$ satisfies the linear equation

$$\frac{\partial \theta}{\partial t} = L[\theta].$$

6. $\dfrac{\partial u}{\partial t} = L[u] + uf\left(t, \dfrac{u}{w}\right) \ln u + ug\left(t, \dfrac{u}{w}\right),$

$\dfrac{\partial w}{\partial t} = L[w] + wf\left(t, \dfrac{u}{w}\right) \ln w + wh\left(t, \dfrac{u}{w}\right).$

Solution:

$$u(x, t) = \varphi(t)\psi(t)\theta(x_1, \ldots, x_n, t), \quad w(x, t) = \psi(t)\theta(x_1, \ldots, x_n, t),$$

where the functions $\varphi = \varphi(t)$ and $\psi = \psi(t)$ are determined by solving the ordinary differential equations

$$\begin{aligned} \varphi'_t &= \varphi[g(t, \varphi) - h(t, \varphi) + f(t, \varphi) \ln \varphi], \\ \psi'_t &= \psi[h(t, \varphi) + f(t, \varphi) \ln \psi], \end{aligned} \tag{1}$$

and the function $\theta = \theta(x_1, \ldots, x_n, t)$ is determined by the differential equation

$$\frac{\partial \theta}{\partial t} = L[\theta] + f(t, \varphi)\theta \ln \theta. \tag{2}$$

Given a solution to the first equation in (1), the second equation can be solved with the change of variable $\psi = e^\zeta$ by reducing it to a linear equation for $\zeta$. If $L$ is a constant-coefficient one-dimensional operator ($n = 1$) and $f = \text{const}$, then equation (2) has a traveling-wave solution $\theta = \theta(kx - \lambda t)$.

7. $F_1\left(w, \dfrac{\partial w}{\partial x}, \ldots, \dfrac{\partial^m w}{\partial x^m}, \dfrac{1}{u^k}\dfrac{\partial w}{\partial t}, \dfrac{1}{u}\dfrac{\partial u}{\partial x}, \ldots, \dfrac{1}{u}\dfrac{\partial^n u}{\partial x^n}\right) = 0,$

$F_2\left(w, \dfrac{\partial w}{\partial x}, \ldots, \dfrac{\partial^m w}{\partial x^m}, \dfrac{1}{u^k}\dfrac{\partial w}{\partial t}, \dfrac{1}{u}\dfrac{\partial u}{\partial x}, \ldots, \dfrac{1}{u}\dfrac{\partial^n u}{\partial x^n}\right) = 0.$

Solution:
$$w = W(z), \quad u = [\varphi'(t)]^{1/k} U(z), \quad z = x + \varphi(t),$$

where $\varphi(t)$ is an arbitrary function, and the functions $W(z)$ and $U(z)$ are determined by the autonomous system of ordinary differential equations

$$F_1\bigl(W, W'_z, \ldots, W^{(m)}_z, W'_z/U^k, U'_z/U, \ldots, U^{(n)}_z/U\bigr) = 0,$$
$$F_2\bigl(W, W'_z, \ldots, W^{(m)}_z, W'_z/U^k, U'_z/U, \ldots, U^{(n)}_z/U\bigr) = 0.$$

### T10.4.3. Nonlinear Systems of Two Equations Involving the Second Derivatives in $t$

1. $\dfrac{\partial^2 u}{\partial t^2} = L[u] + u f(t, au - bw) + g(t, au - bw),$

$\dfrac{\partial^2 w}{\partial t^2} = L[w] + w f(t, au - bw) + h(t, au - bw).$

Here, $L$ is an arbitrary linear differential operator (of any order) with respect to the spatial variables $x_1, \ldots, x_n$, whose coefficients can be dependent on $x_1, \ldots, x_n, t$. It is assumed that $L[\mathrm{const}] = 0$.

1°. Solution:
$$u = \varphi(t) + a\theta(x_1, \ldots, x_n, t), \quad w = \psi(t) + b\theta(x_1, \ldots, x_n, t),$$

where $\varphi = \varphi(t)$ and $\psi = \psi(t)$ are determined by the system of ordinary differential equations

$$\varphi''_{tt} = \varphi f(t, a\varphi - b\psi) + g(t, a\varphi - b\psi),$$
$$\psi''_{tt} = \psi f(t, a\varphi - b\psi) + h(t, a\varphi - b\psi),$$

and the function $\theta = \theta(x_1, \ldots, x_n, t)$ satisfies linear equation

$$\dfrac{\partial^2 \theta}{\partial t^2} = L[\theta] + f(t, a\varphi - b\psi)\theta.$$

2°. Let us multiply the first equation by $a$ and the second one by $-b$ and add the results together to obtain

$$\dfrac{\partial^2 \zeta}{\partial t^2} = L[\zeta] + \zeta f(t, \zeta) + ag(t, \zeta) - bh(t, \zeta), \qquad \zeta = au - bw. \tag{1}$$

This equation will be considered in conjunction with the first equation of the original system

$$\dfrac{\partial^2 u}{\partial t^2} = L[u] + u f(t, \zeta) + g(t, \zeta). \tag{2}$$

Equation (1) can be treated separately. Given a solution $\zeta = \zeta(x,t)$ to equation (1), the function $u = u(x_1,\ldots,x_n,t)$ can be determined by solving the linear equation (2) and the function $w = w(x_1,\ldots,x_n,t)$ is found as $w = (au - \zeta)/b$.

Note three important cases where equation (1) admits exact solutions:

(i) Equation (1) admits a spatially homogeneous solution $\zeta = \zeta(t)$.

(ii) Suppose the coefficients of $L$ and the functions $f$, $g$, $h$ are not implicitly dependent on $t$. Then equation (1) admits a steady-state solution $\zeta = \zeta(x_1,\ldots,x_n)$.

(iii) If the condition $\zeta f(t,\zeta) + bg(t,\zeta) - ch(t,\zeta) = k_1\zeta + k_0$ holds, equation (1) is linear. If the operator $L$ is constant-coefficient, the method of separation of variables can be used to obtain solutions.

2. $\quad \dfrac{\partial^2 u}{\partial t^2} = L_1[u] + uf\left(\dfrac{u}{w}\right), \quad \dfrac{\partial^2 w}{\partial t^2} = L_2[w] + wg\left(\dfrac{u}{w}\right).$

Here, $L_1$ and $L_2$ are arbitrary constant-coefficient linear differential operators (of any order) with respect to $x$. It is assumed that $L_1[\text{const}] = 0$ and $L_2[\text{const}] = 0$.

$1°$. Solution in the form of the product of two waves traveling at different speeds:

$$u = e^{kx-\lambda t}y(\xi), \quad w = e^{kx-\lambda t}z(\xi), \quad \xi = \beta x - \gamma t,$$

where $k$, $\lambda$, $\beta$, and $\gamma$ are arbitrary constants, and the functions $y = y(\xi)$ and $z = z(\xi)$ are determined by the system of ordinary differential equations

$$\gamma^2 y''_{\xi\xi} + 2\lambda\gamma y'_\xi + \lambda^2 y = M_1[y] + yf(y/z), \quad \gamma^2 z''_{\xi\xi} + 2\lambda\gamma z'_\xi + \lambda^2 z = M_2[z] + zg(y/z),$$
$$M_1[y] = e^{-kx}L_1[e^{kx}y(\xi)], \qquad M_2[z] = e^{-kx}L_2[e^{kx}z(\xi)].$$

To the special case $k = \lambda = 0$ there corresponds a traveling-wave solution.

$2°$. Periodic multiplicative separable solution:

$$u = [C_1\sin(kt) + C_2\cos(kt)]\varphi(x), \quad w = [C_1\sin(kt) + C_2\cos(kt)]\psi(x),$$

where $C_1$, $C_2$, and $k$ are arbitrary constants and the functions $\varphi = \varphi(x)$ and $\psi = \psi(x)$ are determined by the system of ordinary differential equations

$$L_1[\varphi] + k^2\varphi + \varphi f(\varphi/\psi) = 0,$$
$$L_2[\psi] + k^2\psi + \psi g(\varphi/\psi) = 0.$$

$3°$. Multiplicative separable solution:

$$u = [C_1\sinh(kt) + C_2\cosh(kt)]\varphi(x), \quad w = [C_1\sinh(kt) + C_2\cosh(kt)]\psi(x),$$

where $C_1$, $C_2$, and $k$ are arbitrary constants and the functions $\varphi = \varphi(x)$ and $\psi = \psi(x)$ are determined by the system of ordinary differential equations

$$L_1[\varphi] - k^2\varphi + \varphi f(\varphi/\psi) = 0,$$
$$L_2[\psi] - k^2\psi + \psi g(\varphi/\psi) = 0.$$

$4°$. Degenerate multiplicative separable solution:

$$u = (C_1 t + C_2)\varphi(x), \quad w = (C_1 t + C_2)\psi(x),$$

where $C_1$ and $C_2$ are arbitrary constants and the functions $\varphi = \varphi(x)$ and $\psi = \psi(x)$ are determined by the system of ordinary differential equations

$$L_1[\varphi] + \varphi f(\varphi/\psi) = 0, \qquad L_2[\psi] + \psi g(\varphi/\psi) = 0.$$

*Remark 1.* The coefficients of $L_1$, $L_2$ and the functions $f$ and $g$ in Items 2°–4° can be dependent on $x$.

*Remark 2.* If $L_1$ and $L_2$ contain only even derivatives, there are solutions of the form

$$u = [C_1 \sin(kx) + C_2 \cos(kx)]U(t), \qquad w = [C_1 \sin(kx) + C_2 \cos(kx)]W(t);$$
$$u = [C_1 \exp(kx) + C_2 \exp(-kx)]U(t), \qquad w = [C_1 \exp(kx) + C_2 \exp(-kx)]W(t);$$
$$u = (C_1 x + C_2)U(t), \qquad w = (C_1 x + C_2)W(t),$$

where $C_1$, $C_2$, and $k$ are arbitrary constants. Note that the third solution is degenerate.

3. $\quad \dfrac{\partial^2 u}{\partial t^2} = L[u] + u f\!\left(t, \dfrac{u}{w}\right), \qquad \dfrac{\partial^2 w}{\partial t^2} = L[w] + w g\!\left(t, \dfrac{u}{w}\right).$

Here, $L$ is an arbitrary linear differential operator with respect to the coordinates $x_1, \ldots, x_n$ (of any order in derivatives), whose coefficients can be dependent on the coordinates.

Solution:
$$u = \varphi(t)\theta(x_1, \ldots, x_n),$$
$$w = \psi(t)\theta(x_1, \ldots, x_n),$$

where the functions $\varphi = \varphi(t)$ and $\psi = \psi(t)$ are described by the nonlinear system of second-order ordinary differential equations

$$\varphi''_{tt} = a\varphi + \varphi f(t, \varphi/\psi),$$
$$\psi''_{tt} = a\psi + \psi g(t, \varphi/\psi),$$

$a$ is an arbitrary constant, and the function $\theta = \theta(x_1, \ldots, x_n)$ satisfies the linear steady-state equation

$$L[\theta] = a\theta.$$

4. $\quad \dfrac{\partial^2 u}{\partial t^2} = L[u] + u f\!\left(\dfrac{u}{w}\right) + g\!\left(\dfrac{u}{w}\right), \qquad \dfrac{\partial^2 w}{\partial t^2} = L[w] + w f\!\left(\dfrac{u}{w}\right) + h\!\left(\dfrac{u}{w}\right).$

Here, $L$ is an arbitrary linear differential operator with respect to the coordinates $x_1, \ldots, x_n$ (of any order in derivatives), whose coefficients can be dependent on $x_1, \ldots, x_n, t$.

Solution:
$$u = k\theta(x_1, \ldots, x_n, t), \qquad w = \theta(x_1, \ldots, x_n, t),$$

where $k$ is a root of the algebraic (transcendental) equation $g(k) = kh(k)$ and the function $\theta = \theta(x, t)$ satisfies the linear equation

$$\dfrac{\partial^2 \theta}{\partial t^2} = L[\theta] + f(k)w + h(k).$$

5. $\quad \dfrac{\partial^2 u}{\partial t^2} = L[u] + au \ln u + u f\!\left(t, \dfrac{u}{w}\right), \qquad \dfrac{\partial^2 w}{\partial t^2} = L[w] + aw \ln w + w g\!\left(t, \dfrac{u}{w}\right).$

Here, $L$ is an arbitrary linear differential operator with respect to the coordinates $x_1, \ldots, x_n$ (of any order in derivatives), whose coefficients can be dependent on the coordinates.

Solution:
$$u = \varphi(t)\theta(x_1, \ldots, x_n),$$
$$w = \psi(t)\theta(x_1, \ldots, x_n),$$

where the functions $\varphi = \varphi(t)$ and $\psi = \psi(t)$ are described by the nonlinear system of second-order ordinary differential equations

$$\varphi''_{tt} = a\varphi \ln \varphi + b\varphi + \varphi f(t, \varphi/\psi),$$
$$\psi''_{tt} = a\psi \ln \psi + b\psi + \psi g(t, \varphi/\psi),$$

$b$ is an arbitrary constant, and the function $\theta = \theta(x_1, \ldots, x_n)$ satisfies the steady-state equation

$$L[\theta] + a\theta \ln \theta - b\theta = 0.$$

## T10.4.4. Nonlinear Systems of Many Equations Involving the First Derivatives in $t$

1. $\dfrac{\partial u_m}{\partial t} = L[u_m] + u_m f(t, u_1 - b_1 u_n, \ldots, u_{n-1} - b_{n-1} u_n)$
$\qquad + g_m(t, u_1 - b_1 u_n, \ldots, u_{n-1} - b_{n-1} u_n), \qquad m = 1, \ldots, n.$

The system involves $n + 1$ arbitrary functions $f, g_1, \ldots, g_n$ that depend on $n$ arguments; $L$ is an arbitrary linear differential operator with respect to the spatial variables $x_1, \ldots, x_n$ (of any order in derivatives), whose coefficients can be dependent on $x_1, \ldots, x_n, t$. It is assumed that $L[\text{const}] = 0$.

Solution:
$$u_m = \varphi_m(t) + \exp\left[\int f(t, \varphi_1 - b_1\varphi_n, \ldots, \varphi_{n-1} - b_{n-1}\varphi_n)\,dt\right]\theta(x_1, \ldots, x_n, t).$$

Here, the functions $\varphi_m = \varphi_m(t)$ are determined by the system of ordinary differential equations

$$\varphi'_m = \varphi_m f(t, \varphi_1 - b_1\varphi_n, \ldots, \varphi_{n-1} - b_{n-1}\varphi_n) + g_m(t, \varphi_1 - b_1\varphi_n, \ldots, \varphi_{n-1} - b_{n-1}\varphi_n),$$

where $m = 1, \ldots, n$, the prime denotes the derivative with respect to $t$, and the function $\theta = \theta(x_1, \ldots, x_n, t)$ satisfies the linear equation

$$\frac{\partial \theta}{\partial t} = L[\theta].$$

2. $\dfrac{\partial u_m}{\partial t} = L[u_m] + u_m f_m\!\left(t, \dfrac{u_1}{u_n}, \ldots, \dfrac{u_{n-1}}{u_n}\right) + \dfrac{u_m}{u_n} g\!\left(t, \dfrac{u_1}{u_n}, \ldots, \dfrac{u_{n-1}}{u_n}\right),$
$\dfrac{\partial u_n}{\partial t} = L[u_n] + u_n f_n\!\left(t, \dfrac{u_1}{u_n}, \ldots, \dfrac{u_{n-1}}{u_n}\right) + g\!\left(t, \dfrac{u_1}{u_n}, \ldots, \dfrac{u_{n-1}}{u_n}\right).$

Here, $m = 1, \ldots, n - 1$ and the system involves $n + 1$ arbitrary functions $f_1, \ldots, f_n, g$ that depend on $n$ arguments; $L$ is an arbitrary linear differential operator with respect to the spatial variables $x_1, \ldots, x_n$ (of any order in derivatives), whose coefficients can be dependent on $x_1, \ldots, x_n, t$. It is assumed that $L[\text{const}] = 0$.

Solution:
$$u_m = \varphi_m(t)F_n(t)\left[\theta(x_1,\dots,x_n,t) + \int \frac{g(t,\varphi_1,\dots,\varphi_{n-1})}{F_n(t)}\,dt\right], \qquad m = 1,\dots,n-1,$$
$$u_n = F_n(t)\left[\theta(x_1,\dots,x_n,t) + \int \frac{g(t,\varphi_1,\dots,\varphi_{n-1})}{F_n(t)}\,dt\right],$$
$$F_n(t) = \exp\left[\int f_n(t,\varphi_1,\dots,\varphi_{n-1})\,dt\right],$$

where the functions $\varphi_m = \varphi_m(t)$ are described by the nonlinear system of first-order ordinary differential equations
$$\varphi'_m = \varphi_m[f_m(t,\varphi_1,\dots,\varphi_{n-1}) - f_n(t,\varphi_1,\dots,\varphi_{n-1})], \qquad m = 1,\dots,n-1,$$

and the function $\theta = \theta(x_1,\dots,x_n,t)$ satisfies the linear equation
$$\frac{\partial\theta}{\partial t} = L[\theta].$$

3. $\displaystyle \frac{\partial u_m}{\partial t} = L[u_m] + \sum_{k=1}^{n} u_k f_{mk}\!\left(t,\frac{u_1}{u_n},\dots,\frac{u_{n-1}}{u_n}\right), \qquad m = 1,\dots,n.$

Here, the system involves $n^2$ arbitrary functions $f_{mk} = f_{mk}(t,z_1,\dots,z_{n-1})$ that depend on $n$ arguments; $L$ is an arbitrary linear differential operator with respect to the spatial variables $x_1,\dots,x_n$ (of any order in derivatives), whose coefficients can be dependent on $x_1,\dots,x_n,t$. It is assumed that $L[\text{const}] = 0$.

Solution:
$$u_m(x_1,\dots,x_n,t) = \varphi_m(t)F(t)\theta(x_1,\dots,x_n,t), \qquad m = 1,\dots,n,$$
$$F(t) = \exp\left[\int \sum_{k=1}^{n}\varphi_k(t)f_{nk}(t,\varphi_1,\dots,\varphi_{n-1})\,dt\right], \qquad \varphi_n(t) = 1,$$

where the functions $\varphi_m = \varphi_m(t)$ are described by the nonlinear system of first-order ordinary differential equations
$$\varphi'_m = \sum_{k=1}^{n}\varphi_k f_{mk}(t,\varphi_1,\dots,\varphi_{n-1}) - \varphi_m\sum_{k=1}^{n}\varphi_k f_{nk}(t,\varphi_1,\dots,\varphi_{n-1}), \qquad m = 1,\dots,n-1,$$

and the function $\theta = \theta(x_1,\dots,x_n,t)$ satisfies the linear equation
$$\frac{\partial\theta}{\partial t} = L[\theta].$$

# References for Chapter T10

**Barannyk, T.,** Symmetry and exact solutions for systems of nonlinear reaction-diffusion equations, *Proc. of Inst. of Mathematics of NAS of Ukraine*, Vol. 43, Part 1, pp. 80–85, 2002.

**Barannyk, T. A. and Nikitin, A. G.,** *Proc. of Inst. of Mathematics of NAS of Ukraine*, Vol. 50, Part 1, pp. 34–39, 2004.

**Cherniha, R. and King, J. R.,** Lie symmetries of nonlinear multidimensional reaction-diffusion systems: I, *J. Phys. A: Math. Gen.*, Vol. 33, pp. 267–282, 7839–7841, 2000.

**Cherniha, R. and King, J. R.,** Lie symmetries of nonlinear multidimensional reaction-diffusion systems: II, *J. Phys. A: Math. Gen.,* Vol. 36, pp. 405–425, 2003.

**Nikitin, A. G.,** Group classification of systems of non-linear reaction-diffusion equations with general diffusion matrix. II. Diagonal diffusion matrix, From Website *arXiv.org* (a service of automated e-print archives of articles), http://arxiv.org/abs/math-ph/0411028.

**Nikitin, A. G. and Wiltshire, R. J.,** Systems of reaction-diffusion equations and their symmetry properties, *J. Math. Phys.,* Vol. 42, No. 4, pp. 1667–1688, 2001.

**Polyanin, A. D.,** Exact solutions of nonlinear sets of equations of the theory of heat and mass transfer in reactive media and mathematical biology, *Theor. Foundations of Chemical Engineering,* Vol. 38, No. 6, pp. 622–635, 2004.

**Polyanin, A. D.,** Exact solutions of nonlinear systems of diffusion equations for reacting media and mathematical biology, *Doklady Mathematics,* Vol. 71, No. 1, pp. 148–154, 2005.

**Polyanin, A. D.,** *Systems of Partial Differential Equations,* From Website *EqWorld – The World of Mathematical Equations,* http://eqworld.ipmnet.ru/en/solutions/syspde.htm.

**Polyanin, A. D. and Vyaz'mina, E. A.,** New classes of exact solutions to nonlinear systems of reaction-diffusion equations, *Doklady Mathematics,* Vol. 74, No. 1, pp. 597–602, 2006.

**Polyanin, A. D. and Zaitsev, V. F.,** *Handbook of Exact Solutions for Ordinary Differential Equations, 2nd Edition,* Chapman & Hall/CRC Press, Boca Raton, 2004.

# Chapter T11
# Integral Equations

## T11.1. Linear Equations of the First Kind with Variable Limit of Integration

1. $\int_a^x (x-t)y(t)\,dt = f(x), \qquad f(a) = f'_x(a) = 0.$

Solution: $y(x) = f''_{xx}(x)$.

2. $\int_a^x (Ax + Bt + C)y(t)\,dt = f(x), \qquad f(a) = 0.$

1°. Solution for $B \neq -A$:
$$y(x) = \frac{d}{dx}\left\{[(A+B)x + C]^{-\frac{A}{A+B}} \int_a^x [(A+B)t + C]^{-\frac{B}{A+B}} f'_t(t)\,dt\right\}.$$

2°. Solution for $B = -A$:
$$y(x) = \frac{1}{C}\frac{d}{dx}\left[\exp\left(-\frac{A}{C}x\right) \int_a^x \exp\left(\frac{A}{C}t\right) f'_t(t)\,dt\right].$$

3. $\int_a^x (x-t)^n y(t)\,dt = f(x), \qquad n = 1, 2, \ldots$

It is assumed that the right-hand side of the equation satisfies the conditions $f(a) = f'_x(a) = \cdots = f_x^{(n)}(a) = 0$.

Solution: $y(x) = \dfrac{1}{n!} f_x^{(n+1)}(x)$.

4. $\int_a^x \sqrt{x-t}\,y(t)\,dt = f(x), \qquad f(a) = 0.$

Solution: $y(x) = \dfrac{2}{\pi}\dfrac{d^2}{dx^2}\displaystyle\int_a^x \dfrac{f(t)\,dt}{\sqrt{x-t}}$.

5. $\int_a^x \dfrac{y(t)\,dt}{\sqrt{x-t}} = f(x).$

*Abel equation.*
Solution:
$$y(x) = \frac{1}{\pi}\frac{d}{dx}\int_a^x \frac{f(t)\,dt}{\sqrt{x-t}} = \frac{f(a)}{\pi\sqrt{x-a}} + \frac{1}{\pi}\int_a^x \frac{f'_t(t)\,dt}{\sqrt{x-t}}.$$

6. $\int_a^x (x-t)^\lambda y(t)\, dt = f(x)$,  $f(a) = 0$,  $0 < \lambda < 1$.

Solution: $y(x) = \dfrac{\sin(\pi\lambda)}{\pi\lambda} \dfrac{d^2}{dx^2} \int_a^x \dfrac{f(t)\, dt}{(x-t)^\lambda}$.

7. $\int_a^x \dfrac{y(t)\, dt}{(x-t)^\lambda} = f(x)$,  $0 < \lambda < 1$.

*Generalized Abel equation.*
   Solution:
$$y(x) = \frac{\sin(\pi\lambda)}{\pi} \frac{d}{dx} \int_a^x \frac{f(t)\, dt}{(x-t)^{1-\lambda}} = \frac{\sin(\pi\lambda)}{\pi} \left[ \frac{f(a)}{(x-a)^{1-\lambda}} + \int_a^x \frac{f'_t(t)\, dt}{(x-t)^{1-\lambda}} \right].$$

8. $\int_a^x e^{\lambda(x-t)} y(t)\, dt = f(x)$,  $f(a) = 0$.

Solution: $y(x) = f'_x(x) - \lambda f(x)$.

9. $\int_a^x e^{\lambda x + \beta t} y(t)\, dt = f(x)$,  $f(a) = 0$.

Solution: $y(x) = e^{-(\lambda+\beta)x} \left[ f'_x(x) - \lambda f(x) \right]$.

10. $\int_a^x \left[ e^{\lambda(x-t)} - 1 \right] y(t)\, dt = f(x)$,  $f(a) = f'_x(a) = 0$.

Solution: $y(x) = \frac{1}{\lambda} f''_{xx}(x) - f'_x(x)$.

11. $\int_a^x \left[ e^{\lambda(x-t)} + b \right] y(t)\, dt = f(x)$,  $f(a) = 0$.

For $b = -1$, see equation T11.1.10.
   Solution for $b \neq -1$:
$$y(x) = \frac{f'_x(x)}{b+1} - \frac{\lambda}{(b+1)^2} \int_a^x \exp\left[ \frac{\lambda b}{b+1}(x-t) \right] f'_t(t)\, dt.$$

12. $\int_a^x \left[ e^{\lambda(x-t)} - e^{\mu(x-t)} \right] y(t)\, dt = f(x)$,  $f(a) = f'_x(a) = 0$.

Solution:
$$y(x) = \frac{1}{\lambda - \mu} \left[ f''_{xx} - (\lambda + \mu) f'_x + \lambda\mu f \right], \qquad f = f(x).$$

13. $\int_a^x \dfrac{y(t)\, dt}{\sqrt{e^{\lambda x} - e^{\lambda t}}} = f(x)$,  $\lambda > 0$.

Solution: $y(x) = \dfrac{\lambda}{\pi} \dfrac{d}{dx} \int_a^x \dfrac{e^{\lambda t} f(t)\, dt}{\sqrt{e^{\lambda x} - e^{\lambda t}}}$.

## T11.1. Linear Equations of the First Kind with Variable Limit of Integration

**14.** $\int_a^x \cosh[\lambda(x-t)]y(t)\,dt = f(x), \qquad f(a) = 0.$

Solution: $y(x) = f_x'(x) - \lambda^2 \int_a^x f(x)\,dx.$

**15.** $\int_a^x \{\cosh[\lambda(x-t)] - 1\}y(t)\,dt = f(x), \qquad f(a) = f_x'(a) = f_{xx}''(x) = 0.$

Solution: $y(x) = \dfrac{1}{\lambda^2} f_{xxx}'''(x) - f_x'(x).$

**16.** $\int_a^x \{\cosh[\lambda(x-t)] + b\}y(t)\,dt = f(x), \qquad f(a) = 0.$

For $b = 0$, see equation T11.1.14. For $b = -1$, see equation T11.1.15.

1°. Solution for $b(b+1) < 0$:

$$y(x) = \frac{f_x'(x)}{b+1} - \frac{\lambda^2}{k(b+1)^2} \int_a^x \sin[k(x-t)]f_t'(t)\,dt, \qquad \text{where} \quad k = \lambda\sqrt{\frac{-b}{b+1}}.$$

2°. Solution for $b(b+1) > 0$:

$$y(x) = \frac{f_x'(x)}{b+1} - \frac{\lambda^2}{k(b+1)^2} \int_a^x \sinh[k(x-t)]f_t'(t)\,dt, \qquad \text{where} \quad k = \lambda\sqrt{\frac{b}{b+1}}.$$

**17.** $\int_a^x \cosh^2[\lambda(x-t)]y(t)\,dt = f(x), \qquad f(a) = 0.$

Solution:

$$y(x) = f_x'(x) - \frac{2\lambda^2}{k}\int_a^x \sinh[k(x-t)]f_t'(t)\,dt, \qquad \text{where} \quad k = \lambda\sqrt{2}.$$

**18.** $\int_a^x \sinh[\lambda(x-t)]y(t)\,dt = f(x), \qquad f(a) = f_x'(a) = 0.$

Solution: $y(x) = \dfrac{1}{\lambda} f_{xx}''(x) - \lambda f(x).$

**19.** $\int_a^x \{\sinh[\lambda(x-t)] + b\}y(t)\,dt = f(x), \qquad f(a) = 0.$

For $b = 0$, see equation T11.1.18.
Solution for $b \neq 0$:

$$y(x) = \frac{1}{b} f_x'(x) + \int_a^x R(x-t)f_t'(t)\,dt,$$

$$R(x) = \frac{\lambda}{b^2}\exp\left(-\frac{\lambda x}{2b}\right)\left[\frac{\lambda}{2bk}\sinh(kx) - \cosh(kx)\right], \qquad k = \frac{\lambda\sqrt{1+4b^2}}{2b}.$$

20. $\int_a^x \sinh(\lambda\sqrt{x-t}\,)y(t)\,dt = f(x),\qquad f(a)=0.$

Solution: $y(x) = \dfrac{2}{\pi\lambda}\dfrac{d^2}{dx^2}\int_a^x \dfrac{\cos(\lambda\sqrt{x-t}\,)}{\sqrt{x-t}}f(t)\,dt.$

21. $\int_0^x \ln(x-t)y(t)\,dt = f(x).$

Solution:

$$y(x) = -\int_0^x f''_{tt}(t)\,dt \int_0^\infty \dfrac{(x-t)^z e^{-Cz}}{\Gamma(z+1)}\,dz - f'_x(0)\int_0^\infty \dfrac{x^z e^{-Cz}}{\Gamma(z+1)}\,dz,$$

where $C = \lim\limits_{k\to\infty}\left(1 + \dfrac{1}{2} + \cdots + \dfrac{1}{k+1} - \ln k\right) = 0.5772\ldots$ is the Euler constant and $\Gamma(z)$ is the gamma function.

22. $\int_a^x [\ln(x-t) + A]y(t)\,dt = f(x).$

Solution:

$$y(x) = -\dfrac{d}{dx}\int_a^x \nu_A(x-t)f(t)\,dt,\qquad \nu_A(x) = \dfrac{d}{dx}\int_0^\infty \dfrac{x^z e^{(A-C)z}}{\Gamma(z+1)}\,dz,$$

where $C = 0.5772\ldots$ is the Euler constant and $\Gamma(z)$ is the gamma function.
For $a = 0$, the solution can be written in the form

$$y(x) = -\int_0^x f''_{tt}(t)\,dt \int_0^\infty \dfrac{(x-t)^z e^{(A-C)z}}{\Gamma(z+1)}\,dz - f'_x(0)\int_0^\infty \dfrac{x^z e^{(A-C)z}}{\Gamma(z+1)}\,dz.$$

23. $\int_a^x (x-t)\bigl[\ln(x-t) + A\bigr]y(t)\,dt = f(x),\qquad f(a) = 0.$

Solution:

$$y(x) = -\dfrac{d^2}{dx^2}\int_a^x \nu_A(x-t)f(t)\,dt,\qquad \nu_A(x) = \dfrac{d}{dx}\int_0^\infty \dfrac{x^z e^{(A-C)z}}{\Gamma(z+1)}\,dz,$$

where $C = 0.5772\ldots$ is the Euler constant and $\Gamma(z)$ is the gamma function.

24. $\int_a^x \cos[\lambda(x-t)]y(t)\,dt = f(x),\qquad f(a) = 0.$

Solution: $y(x) = f'_x(x) + \lambda^2\int_a^x f(x)\,dx.$

25. $\int_a^x \{\cos[\lambda(x-t)] - 1\}y(t)\,dt = f(x),\qquad f(a) = f'_x(a) = f''_{xx}(a) = 0.$

Solution: $y(x) = -\dfrac{1}{\lambda^2}f'''_{xxx}(x) - f'_x(x).$

**26.** $\int_a^x \{\cos[\lambda(x-t)] + b\} y(t)\, dt = f(x), \qquad f(a) = 0.$

For $b = 0$, see equation T11.1.24. For $b = -1$, see equation T11.1.25.

1°. Solution for $b(b+1) > 0$:

$$y(x) = \frac{f'_x(x)}{b+1} + \frac{\lambda^2}{k(b+1)^2} \int_a^x \sin[k(x-t)] f'_t(t)\, dt, \qquad \text{where} \quad k = \lambda\sqrt{\frac{b}{b+1}}.$$

2°. Solution for $b(b+1) < 0$:

$$y(x) = \frac{f'_x(x)}{b+1} + \frac{\lambda^2}{k(b+1)^2} \int_a^x \sinh[k(x-t)] f'_t(t)\, dt, \qquad \text{where} \quad k = \lambda\sqrt{\frac{-b}{b+1}}.$$

**27.** $\int_a^x \sin[\lambda(x-t)] y(t)\, dt = f(x), \qquad f(a) = f'_x(a) = 0.$

Solution: $y(x) = \frac{1}{\lambda} f''_{xx}(x) + \lambda f(x).$

**28.** $\int_a^x \sin(\lambda\sqrt{x-t})\, y(t)\, dt = f(x), \qquad f(a) = 0.$

Solution: $y(x) = \frac{2}{\pi\lambda} \frac{d^2}{dx^2} \int_a^x \frac{\cosh(\lambda\sqrt{x-t})}{\sqrt{x-t}} f(t)\, dt.$

**29.** $\int_a^x J_0(\lambda(x-t))\, y(t)\, dt = f(x).$

Here, $J_\nu(z)$ is the Bessel function of the first kind and $f(a) = 0$.

Solution:
$$y(x) = \frac{1}{\lambda}\left(\frac{d^2}{dx^2} + \lambda^2\right)^2 \int_a^x (x-t)\, J_1(\lambda(x-t))\, f(t)\, dt.$$

**30.** $\int_a^x J_0(\lambda\sqrt{x-t})\, y(t)\, dt = f(x).$

Here, $J_\nu(z)$ is the Bessel function of the first kind and $f(a) = 0$.

Solution:
$$y(x) = \frac{d^2}{dx^2} \int_a^x I_0(\lambda\sqrt{x-t})\, f(t)\, dt.$$

**31.** $\int_a^x I_0(\lambda(x-t))\, y(t)\, dt = f(x).$

Here, $I_\nu(z)$ is the modified Bessel function of the first kind and $f(a) = 0$.

Solution:
$$y(x) = \frac{1}{\lambda}\left(\frac{d^2}{dx^2} - \lambda^2\right)^2 \int_a^x (x-t)\, I_1(\lambda(x-t))\, f(t)\, dt.$$

**32.** $\int_a^x I_0(\lambda\sqrt{x-t})y(t)\,dt = f(x).$

Here, $I_\nu(z)$ is the modified Bessel function of the first kind and $f(a) = 0$.

Solution: $y(x) = \dfrac{d^2}{dx^2}\int_a^x J_0(\lambda\sqrt{x-t})f(t)\,dt.$

**33.** $\int_a^x [g(x) - g(t)]y(t)\,dt = f(x).$

It is assumed that $f(a) = f'_x(a) = 0$ and $f'_x/g'_x \neq \text{const}$.

Solution: $y(x) = \dfrac{d}{dx}\left[\dfrac{f'_x(x)}{g'_x(x)}\right].$

**34.** $\int_a^x [g(x) - g(t) + b]y(t)\,dt = f(x), \qquad f(a) = 0.$

For $b = 0$, see equation T11.1.33.

Solution for $b \neq 0$:

$$y(x) = \frac{1}{b}f'_x(x) - \frac{1}{b^2}g'_x(x)\int_a^x \exp\left[\frac{g(t) - g(x)}{b}\right]f'_t(t)\,dt.$$

**35.** $\int_a^x [g(x) + h(t)]y(t)\,dt = f(x), \qquad f(a) = 0.$

For $h(t) = -g(t)$, see equation T11.1.33.

Solution:

$$y(x) = \frac{d}{dx}\left[\frac{\Phi(x)}{g(x) + h(x)}\int_a^x \frac{f'_t(t)\,dt}{\Phi(t)}\right], \qquad \Phi(x) = \exp\left[\int_a^x \frac{h'_t(t)\,dt}{g(t) + h(t)}\right].$$

**36.** $\int_a^x K(x-t)y(t)\,dt = f(x).$

$1^\circ$. Let $K(0) = 1$ and $f(a) = 0$. Differentiating the equation with respect to $x$ yields a Volterra equation of the second kind:

$$y(x) + \int_a^x K'_x(x-t)y(t)\,dt = f'_x(x).$$

The solution of this equation can be represented in the form

$$y(x) = f'_x(x) + \int_a^x R(x-t)f'_t(t)\,dt. \qquad (1)$$

Here the resolvent $R(x)$ is related to the kernel $K(x)$ of the original equation by

$$R(x) = \mathfrak{L}^{-1}\left[\frac{1}{p\tilde{K}(p)} - 1\right], \qquad \tilde{K}(p) = \mathfrak{L}[K(x)],$$

where $\mathcal{L}$ and $\mathcal{L}^{-1}$ are the operators of the direct and inverse Laplace transforms, respectively.

$$\widetilde{K}(p) = \mathcal{L}[K(x)] = \int_0^\infty e^{-px} K(x)\, dx, \qquad R(x) = \mathcal{L}^{-1}[\widetilde{R}(p)] = \frac{1}{2\pi i} \int_{c-i\infty}^{c+i\infty} e^{px} \widetilde{R}(p)\, dp.$$

2°. Let $K(x)$ have an integrable power-law singularity at $x = 0$. Denote by $w = w(x)$ the solution of the simpler auxiliary equation (compared with the original equation) with $a = 0$ and constant right-hand side $f \equiv 1$,

$$\int_0^x K(x-t) w(t)\, dt = 1. \tag{2}$$

Then the solution of the original integral equation with arbitrary right-hand side is expressed in terms of $w$ as follows:

$$y(x) = \frac{d}{dx} \int_a^x w(x-t) f(t)\, dt = f(a) w(x-a) + \int_a^x w(x-t) f'_t(t)\, dt.$$

37. $\displaystyle\int_a^x \sqrt{g(x) - g(t)}\, y(t)\, dt = f(x), \qquad f(a) = 0, \quad g'_x(x) > 0.$

Solution:

$$y(x) = \frac{2}{\pi} g'_x(x) \left( \frac{1}{g'_x(x)} \frac{d}{dx} \right)^2 \int_a^x \frac{f(t) g'_t(t)\, dt}{\sqrt{g(x) - g(t)}}.$$

38. $\displaystyle\int_a^x \frac{y(t)\, dt}{\sqrt{g(x) - g(t)}} = f(x), \qquad g'_x(x) > 0.$

Solution: $\displaystyle y(x) = \frac{1}{\pi} \frac{d}{dx} \int_a^x \frac{f(t) g'_t(t)\, dt}{\sqrt{g(x) - g(t)}}.$

## T11.2. Linear Equations of the Second Kind with Variable Limit of Integration

1. $\displaystyle y(x) - \lambda \int_a^x y(t)\, dt = f(x).$

Solution: $\displaystyle y(x) = f(x) + \lambda \int_a^x e^{\lambda(x-t)} f(t)\, dt.$

2. $\displaystyle y(x) + \lambda \int_a^x (x-t) y(t)\, dt = f(x).$

1°. Solution for $\lambda > 0$:

$$y(x) = f(x) - k \int_a^x \sin[k(x-t)] f(t)\, dt, \qquad k = \sqrt{\lambda}.$$

2°. Solution for $\lambda < 0$:

$$y(x) = f(x) + k \int_a^x \sinh[k(x-t)] f(t)\, dt, \qquad k = \sqrt{-\lambda}.$$

**3.** $y(x) + \lambda \int_a^x (x-t)^2 y(t)\, dt = f(x)$.

Solution:
$$y(x) = f(x) - \int_a^x R(x-t)f(t)\, dt,$$
$$R(x) = \tfrac{2}{3}ke^{-2kx} - \tfrac{2}{3}ke^{kx}\left[\cos(\sqrt{3}\,kx) - \sqrt{3}\sin(\sqrt{3}\,kx)\right], \qquad k = \left(\tfrac{1}{4}\lambda\right)^{1/3}.$$

**4.** $y(x) + \lambda \int_a^x (x-t)^3 y(t)\, dt = f(x)$.

Solution:
$$y(x) = f(x) - \int_a^x R(x-t)f(t)\, dt,$$
where
$$R(x) = \begin{cases} k\big[\cosh(kx)\sin(kx) - \sinh(kx)\cos(kx)\big], & k = \left(\tfrac{3}{2}\lambda\right)^{1/4} & \text{for } \lambda > 0, \\ \tfrac{1}{2}s\big[\sin(sx) - \sinh(sx)\big], & s = (-6\lambda)^{1/4} & \text{for } \lambda < 0. \end{cases}$$

**5.** $y(x) + A\int_a^x (x-t)^n y(t)\, dt = f(x), \qquad n = 1, 2, \ldots$

1°. Differentiating the equation $n+1$ times with respect to $x$ yields an $(n+1)$st-order linear ordinary differential equation with constant coefficients for $y = y(x)$:
$$y_x^{(n+1)} + An!\, y = f_x^{(n+1)}(x).$$
This equation under the initial conditions $y(a) = f(a)$, $y'_x(a) = f'_x(a)$, $\ldots$, $y_x^{(n)}(a) = f_x^{(n)}(a)$ determines the solution of the original integral equation.

2°. Solution:
$$y(x) = f(x) + \int_a^x R(x-t)f(t)\, dt,$$
$$R(x) = \frac{1}{n+1}\sum_{k=0}^{n} \exp(\sigma_k x)\big[\sigma_k \cos(\beta_k x) - \beta_k \sin(\beta_k x)\big],$$
where the coefficients $\sigma_k$ and $\beta_k$ are given by
$$\sigma_k = |An!|^{\frac{1}{n+1}} \cos\!\left(\frac{2\pi k}{n+1}\right), \quad \beta_k = |An!|^{\frac{1}{n+1}} \sin\!\left(\frac{2\pi k}{n+1}\right) \qquad \text{for } A < 0;$$
$$\sigma_k = |An!|^{\frac{1}{n+1}} \cos\!\left(\frac{2\pi k + \pi}{n+1}\right), \quad \beta_k = |An!|^{\frac{1}{n+1}} \sin\!\left(\frac{2\pi k + \pi}{n+1}\right) \qquad \text{for } A > 0.$$

**6.** $y(x) + \lambda \int_a^x \dfrac{y(t)\, dt}{\sqrt{x-t}} = f(x)$.

*Abel equation of the second kind.* This equation is encountered in problems of heat and mass transfer.

Solution:
$$y(x) = F(x) + \pi\lambda^2 \int_a^x \exp[\pi\lambda^2 (x-t)] F(t)\, dt,$$
where
$$F(x) = f(x) - \lambda \int_a^x \frac{f(t)\, dt}{\sqrt{x-t}}.$$

7. $y(x) - \lambda \int_0^x \dfrac{y(t)\,dt}{(x-t)^\alpha} = f(x), \qquad 0 < \alpha < 1.$

*Generalized Abel equation of the second kind.*

1°. Assume that the number $\alpha$ can be represented in the form

$$\alpha = 1 - \frac{m}{n}, \qquad \text{where} \quad m = 1, 2, \ldots, \quad n = 2, 3, \ldots \quad (m < n).$$

In this case, the solution of the generalized Abel equation of the second kind can be written in closed form (in quadratures):

$$y(x) = f(x) + \int_0^x R(x-t)f(t)\,dt,$$

where

$$R(x) = \sum_{\nu=1}^{n-1} \frac{\lambda^\nu \Gamma^\nu(m/n)}{\Gamma(\nu m/n)} x^{(\nu m/n)-1} + \frac{b}{m}\sum_{\mu=0}^{m-1} \varepsilon_\mu \exp(\varepsilon_\mu b x)$$

$$+ \frac{b}{m}\sum_{\nu=1}^{n-1}\frac{\lambda^\nu \Gamma^\nu(m/n)}{\Gamma(\nu m/n)}\left[\sum_{\mu=0}^{m-1}\varepsilon_\mu \exp(\varepsilon_\mu b x)\int_0^x t^{(\nu m/n)-1}\exp(-\varepsilon_\mu b t)\,dt\right],$$

$$b = \lambda^{n/m}\Gamma^{n/m}(m/n), \quad \varepsilon_\mu = \exp\!\left(\frac{2\pi\mu i}{m}\right), \quad i^2 = -1, \quad \mu = 0,1,\ldots, m-1.$$

2°. Solution for any $\alpha$ from $0 < \alpha < 1$:

$$y(x) = f(x) + \int_0^x R(x-t)f(t)\,dt, \qquad \text{where} \quad R(x) = \sum_{n=1}^\infty \frac{\left[\lambda\Gamma(1-\alpha)x^{1-\alpha}\right]^n}{x\,\Gamma\!\left[n(1-\alpha)\right]}.$$

8. $y(x) + A\displaystyle\int_a^x e^{\lambda(x-t)} y(t)\,dt = f(x).$

Solution: $y(x) = f(x) - A\displaystyle\int_a^x e^{(\lambda-A)(x-t)} f(t)\,dt.$

9. $y(x) + A\displaystyle\int_a^x \left[e^{\lambda(x-t)} - 1\right] y(t)\,dt = f(x).$

1°. Solution for $D \equiv \lambda(\lambda - 4A) > 0$:

$$y(x) = f(x) - \frac{2A\lambda}{\sqrt{D}}\int_a^x R(x-t)f(t)\,dt, \qquad R(x) = \exp\!\left(\tfrac{1}{2}\lambda x\right)\sinh\!\left(\tfrac{1}{2}\sqrt{D}\,x\right).$$

2°. Solution for $D \equiv \lambda(\lambda - 4A) < 0$:

$$y(x) = f(x) - \frac{2A\lambda}{\sqrt{|D|}}\int_a^x R(x-t)f(t)\,dt, \qquad R(x) = \exp\!\left(\tfrac{1}{2}\lambda x\right)\sin\!\left(\tfrac{1}{2}\sqrt{|D|}\,x\right).$$

3°. Solution for $\lambda = 4A$:

$$y(x) = f(x) - 4A^2\int_a^x (x-t)\exp\!\left[2A(x-t)\right] f(t)\,dt.$$

10. $y(x) + A \int_a^x (x-t) e^{\lambda(x-t)} y(t)\, dt = f(x)$.

1°. Solution for $A > 0$:

$$y(x) = f(x) - k \int_a^x e^{\lambda(x-t)} \sin[k(x-t)] f(t)\, dt, \qquad k = \sqrt{A}.$$

2°. Solution for $A < 0$:

$$y(x) = f(x) + k \int_a^x e^{\lambda(x-t)} \sinh[k(x-t)] f(t)\, dt, \qquad k = \sqrt{-A}.$$

11. $y(x) + A \int_a^x \cosh[\lambda(x-t)] y(t)\, dt = f(x)$.

Solution:

$$y(x) = f(x) + \int_a^x R(x-t) f(t)\, dt,$$

$$R(x) = \exp\left(-\tfrac{1}{2} A x\right) \left[ \frac{A^2}{2k} \sinh(kx) - A \cosh(kx) \right], \qquad k = \sqrt{\lambda^2 + \tfrac{1}{4} A^2}.$$

12. $y(x) + A \int_a^x \sinh[\lambda(x-t)] y(t)\, dt = f(x)$.

1°. Solution for $\lambda(A - \lambda) > 0$:

$$y(x) = f(x) - \frac{A\lambda}{k} \int_a^x \sin[k(x-t)] f(t)\, dt, \qquad \text{where} \quad k = \sqrt{\lambda(A-\lambda)}.$$

2°. Solution for $\lambda(A - \lambda) < 0$:

$$y(x) = f(x) - \frac{A\lambda}{k} \int_a^x \sinh[k(x-t)] f(t)\, dt, \qquad \text{where} \quad k = \sqrt{\lambda(\lambda-A)}.$$

3°. Solution for $A = \lambda$:

$$y(x) = f(x) - \lambda^2 \int_a^x (x-t) f(t)\, dt.$$

13. $y(x) - \lambda \int_0^x J_0(x-t) y(t)\, dt = f(x)$.

Here, $J_0(z)$ is the Bessel function of the first kind.
Solution:

$$y(x) = f(x) + \int_0^x R(x-t) f(t)\, dt,$$

where

$$R(x) = \lambda \cos\left(\sqrt{1-\lambda^2}\, x\right) + \frac{\lambda^2}{\sqrt{1-\lambda^2}} \sin\left(\sqrt{1-\lambda^2}\, x\right) + \frac{\lambda}{\sqrt{1-\lambda^2}} \int_0^x \sin\left[\sqrt{1-\lambda^2}\,(x-t)\right] \frac{J_1(t)}{t}\, dt.$$

**14.** $y(x) - \int_a^x g(x)h(t)y(t)\,dt = f(x)$.

Solution:

$$y(x) = f(x) + \int_a^x R(x,t)f(t)\,dt, \quad \text{where} \quad R(x,t) = g(x)h(t)\exp\left[\int_t^x g(s)h(s)\,ds\right].$$

**15.** $y(x) + \int_a^x (x-t)g(x)y(t)\,dt = f(x)$.

1°. Solution:

$$y(x) = f(x) + \frac{1}{W}\int_a^x \left[Y_1(x)Y_2(t) - Y_2(x)Y_1(t)\right]g(x)f(t)\,dt, \tag{1}$$

where $Y_1 = Y_1(x)$ and $Y_2 = Y_2(x)$ are two linearly independent solutions ($Y_1/Y_2 \not\equiv \text{const}$) of the second-order linear homogeneous differential equation $Y''_{xx} + g(x)Y = 0$. In this case, the Wronskian is a constant: $W = Y_1(Y_2)'_x - Y_2(Y_1)'_x \equiv \text{const}$.

2°. Given only one nontrivial solution $Y_1 = Y_1(x)$ of the linear homogeneous differential equation $Y''_{xx} + g(x)Y = 0$, one can obtain the solution of the integral equation by formula (1) with

$$W = 1, \qquad Y_2(x) = Y_1(x)\int_b^x \frac{d\xi}{Y_1^2(\xi)},$$

where $b$ is an arbitrary number.

**16.** $y(x) + \int_a^x (x-t)g(t)y(t)\,dt = f(x)$.

1°. Solution:

$$y(x) = f(x) + \frac{1}{W}\int_a^x \left[Y_1(x)Y_2(t) - Y_2(x)Y_1(t)\right]g(t)f(t)\,dt, \tag{1}$$

where $Y_1 = Y_1(x)$ and $Y_2 = Y_2(x)$ are two linearly independent solutions ($Y_1/Y_2 \not\equiv \text{const}$) of the second-order linear homogeneous differential equation $Y''_{xx} + g(x)Y = 0$. In this case, the Wronskian is a constant: $W = Y_1(Y_2)'_x - Y_2(Y_1)'_x \equiv \text{const}$.

2°. Given only one nontrivial solution $Y_1 = Y_1(x)$ of the linear homogeneous differential equation $Y''_{xx} + g(x)Y = 0$, one can obtain the solution of the integral equation by formula (1) with

$$W = 1, \qquad Y_2(x) = Y_1(x)\int_b^x \frac{d\xi}{Y_1^2(\xi)},$$

where $b$ is an arbitrary number.

**17.** $y(x) + \int_a^x K(x-t)y(t)\,dt = f(x)$.

*Renewal equation.*

$1^\circ$. To solve this integral equation, direct and inverse Laplace transforms are used. The solution can be represented in the form

$$y(x) = f(x) - \int_a^x R(x-t)f(t)\,dt. \qquad (1)$$

Here the resolvent $R(x)$ is expressed via the kernel $K(x)$ of the original equation as follows:

$$R(x) = \frac{1}{2\pi i}\int_{c-i\infty}^{c+i\infty}\widetilde{R}(p)e^{px}\,dp, \quad \widetilde{R}(p) = \frac{\widetilde{K}(p)}{1+\widetilde{K}(p)}, \quad \widetilde{K}(p) = \int_0^\infty K(x)e^{-px}\,dx.$$

$2^\circ$. Let $w = w(x)$ be the solution of the simpler auxiliary equation with $a = 0$ and $f \equiv 1$:

$$w(x) + \int_0^x K(x-t)w(t)\,dt = 1. \qquad (2)$$

Then the solution of the original integral equation with arbitrary $f = f(x)$ is expressed via the solution of the auxiliary equation (2) as

$$y(x) = \frac{d}{dx}\int_a^x w(x-t)f(t)\,dt = f(a)w(x-a) + \int_a^x w(x-t)f'_t(t)\,dt.$$

## T11.3. Linear Equations of the First Kind with Constant Limits of Integration

**1.** $\int_a^b |x-t|\,y(t)\,dt = f(x), \qquad 0 \le a < b < \infty.$

Solution:
$$y(x) = \tfrac{1}{2}f''_{xx}(x).$$

The right-hand side $f(x)$ of the integral equation must satisfy certain relations. The general form of $f(x)$ is as follows:

$$f(x) = F(x) + Ax + B,$$
$$A = -\tfrac{1}{2}\bigl[F'_x(a) + F'_x(b)\bigr], \quad B = \tfrac{1}{2}\bigl[aF'_x(a) + bF'_x(b) - F(a) - F(b)\bigr],$$

where $F(x)$ is an arbitrary bounded twice differentiable function (with bounded first derivative).

**2.** $\int_0^a \dfrac{y(t)}{\sqrt{|x-t|}}\,dt = f(x), \qquad 0 < a \le \infty.$

Solution:
$$y(x) = -\frac{A}{x^{1/4}}\frac{d}{dx}\left[\int_x^a \frac{dt}{(t-x)^{1/4}}\int_0^t \frac{f(s)\,ds}{s^{1/4}(t-s)^{1/4}}\right], \quad A = \frac{1}{\sqrt{8\pi}\,\Gamma^2(3/4)}.$$

3. $\displaystyle\int_a^b \frac{y(t)}{|x-t|^k}\,dt = f(x), \qquad 0 < k < 1.$

It is assumed that $|a| + |b| < \infty$. Solution:

$$y(x) = \frac{1}{2\pi}\cot(\tfrac{1}{2}\pi k)\frac{d}{dx}\int_a^x \frac{f(t)\,dt}{(x-t)^{1-k}} - \frac{1}{\pi^2}\cos^2(\tfrac{1}{2}\pi k)\int_a^x \frac{Z(t)F(t)}{(x-t)^{1-k}}\,dt,$$

where

$$Z(t) = (t-a)^{\frac{1+k}{2}}(b-t)^{\frac{1-k}{2}}, \qquad F(t) = \frac{d}{dt}\left[\int_a^t \frac{d\tau}{(t-\tau)^k}\int_\tau^b \frac{f(s)\,ds}{Z(s)(s-\tau)^{1-k}}\right].$$

4. $\displaystyle\int_0^b \frac{y(t)}{|x^\lambda - t^\lambda|^k}\,dt = f(x), \qquad 0 < k < 1, \quad \lambda > 0.$

Solution:

$$y(x) = -Ax^{\frac{\lambda(k-1)}{2}}\frac{d}{dx}\left[\int_x^b \frac{t^{\frac{\lambda(3-2k)-2}{2}}\,dt}{(t^\lambda - x^\lambda)^{\frac{1-k}{2}}}\int_0^t \frac{s^{\frac{\lambda(k+1)-2}{2}}f(s)\,ds}{(t^\lambda - s^\lambda)^{\frac{1-k}{2}}}\right],$$

$$A = \frac{\lambda^2}{2\pi}\cos\left(\frac{\pi k}{2}\right)\Gamma(k)\left[\Gamma\left(\frac{1+k}{2}\right)\right]^{-2},$$

where $\Gamma(k)$ is the gamma function.

5. $\displaystyle\int_{-\infty}^\infty \frac{y(t)}{|x-t|^{1-\lambda}}\,dt = f(x), \qquad 0 < \lambda < 1.$

Solution:

$$y(x) = \frac{\lambda}{2\pi}\tan\left(\frac{\pi\lambda}{2}\right)\int_{-\infty}^\infty \frac{f(x)-f(t)}{|x-t|^{1+\lambda}}\,dt.$$

It assumed that the condition $\int_{-\infty}^\infty |f(x)|^p\,dx < \infty$ is satisfied for some $p$, $1 < p < 1/\lambda$.

6. $\displaystyle\int_{-\infty}^\infty \frac{\operatorname{sign}(x-t)}{|x-t|^{1-\lambda}}y(t)\,dt = f(x), \qquad 0 < \lambda < 1.$

Solution:

$$y(x) = \frac{\lambda}{2\pi}\cot\left(\frac{\pi\lambda}{2}\right)\int_{-\infty}^\infty \frac{f(x)-f(t)}{|x-t|^{1+\lambda}}\operatorname{sign}(x-t)\,dt.$$

7. $\displaystyle\int_{-\infty}^\infty \frac{a + b\operatorname{sign}(x-t)}{|x-t|^{1-\lambda}}y(t)\,dt = f(x), \qquad 0 < \lambda < 1.$

Solution:

$$y(x) = \frac{\lambda\sin(\pi\lambda)}{4\pi\left[a^2\cos^2(\tfrac{1}{2}\pi\lambda) + b^2\sin^2(\tfrac{1}{2}\pi\lambda)\right]}\int_{-\infty}^\infty \frac{a + b\operatorname{sign}(x-t)}{|x-t|^{1+\lambda}}\left[f(x)-f(t)\right]dt.$$

8. $\int_0^\infty \dfrac{y(x+t) - y(x-t)}{t}\, dt = f(x).$

Solution: $y(x) = -\dfrac{1}{\pi^2} \int_0^\infty \dfrac{f(x+t) - f(x-t)}{t}\, dt.$

▶ *In equations T11.3.9 and T11.3.10 and their solutions, all singular integrals are understood in the sense of the Cauchy principal value.*

9. $\int_{-\infty}^\infty \dfrac{y(t)\, dt}{t - x} = f(x).$

Solution: $y(x) = -\dfrac{1}{\pi^2} \int_{-\infty}^\infty \dfrac{f(t)\, dt}{t - x}.$

The integral equation and its solution form a Hilbert transform pair (in the asymmetric form).

10. $\int_a^b \dfrac{y(t)\, dt}{t - x} = f(x).$

This equation is encountered in hydrodynamics in solving the problem on the flow of an ideal inviscid fluid around a thin profile ($a \leq x \leq b$). It is assumed that $|a| + |b| < \infty$.

1°. The solution bounded at the endpoints is

$$y(x) = -\dfrac{1}{\pi^2} \sqrt{(x-a)(b-x)} \int_a^b \dfrac{f(t)}{\sqrt{(t-a)(b-t)}} \dfrac{dt}{t-x},$$

provided that

$$\int_a^b \dfrac{f(t)\, dt}{\sqrt{(t-a)(b-t)}} = 0.$$

2°. The solution bounded at the endpoint $x = a$ and unbounded at the endpoint $x = b$ is

$$y(x) = -\dfrac{1}{\pi^2} \sqrt{\dfrac{x-a}{b-x}} \int_a^b \sqrt{\dfrac{b-t}{t-a}} \dfrac{f(t)}{t-x}\, dt.$$

3°. The solution unbounded at the endpoints is

$$y(x) = -\dfrac{1}{\pi^2 \sqrt{(x-a)(b-x)}} \left[ \int_a^b \dfrac{\sqrt{(t-a)(b-t)}}{t-x} f(t)\, dt + C \right],$$

where $C$ is an arbitrary constant. The formula $\int_a^b y(t)\, dt = \dfrac{C}{\pi}$ holds.

11. $\int_a^b e^{\lambda|x-t|} y(t)\, dt = f(x), \qquad -\infty < a < b < \infty.$

Solution:
$$y(x) = \dfrac{1}{2\lambda} \left[ f''_{xx}(x) - \lambda^2 f(x) \right].$$

The right-hand side $f(x)$ of the integral equation must satisfy certain relations. The general form of the right-hand side is given by

$$f(x) = F(x) + Ax + B,$$

$A = \dfrac{1}{b\lambda - a\lambda - 2} \left[ F'_x(a) + F'_x(b) + \lambda F(a) - \lambda F(b) \right], \quad B = -\dfrac{1}{\lambda} \left[ F'_x(a) + \lambda F(a) + Aa\lambda + A \right],$

where $F(x)$ is an arbitrary bounded, twice differentiable function.

**12.** $\int_a^b \ln|x - t|\, y(t)\, dt = f(x).$

*Carleman's equation.*

1°. Solution for $b - a \neq 4$:

$$y(x) = \frac{1}{\pi^2 \sqrt{(x-a)(b-x)}} \left[ \int_a^b \frac{\sqrt{(t-a)(b-t)}\, f'_t(t)\, dt}{t-x} + \frac{1}{\ln\left[\frac{1}{4}(b-a)\right]} \int_a^b \frac{f(t)\, dt}{\sqrt{(t-a)(b-t)}} \right].$$

2°. If $b - a = 4$, then for the equation to be solvable, the condition

$$\int_a^b f(t)(t-a)^{-1/2}(b-t)^{-1/2}\, dt = 0$$

must be satisfied. In this case, the solution has the form

$$y(x) = \frac{1}{\pi^2 \sqrt{(x-a)(b-x)}} \left[ \int_a^b \frac{\sqrt{(t-a)(b-t)}\, f'_t(t)\, dt}{t-x} + C \right],$$

where $C$ is an arbitrary constant.

**13.** $\int_a^b \bigl(\ln|x - t| + \beta\bigr) y(t)\, dt = f(x).$

By setting

$$x = e^{-\beta} z, \quad t = e^{-\beta} \tau, \quad y(t) = Y(\tau), \quad f(x) = e^{-\beta} g(z),$$

we arrive at an equation of the form T11.3.12:

$$\int_A^B \ln|z - \tau|\, Y(\tau)\, d\tau = g(z), \qquad A = ae^\beta,\ B = be^\beta.$$

**14.** $\int_{-a}^a \left( \ln \dfrac{A}{|x - t|} \right) y(t)\, dt = f(x), \qquad -a \leq x \leq a.$

Solution for $0 < a < 2A$:

$$y(x) = \frac{1}{2M'(a)} \left[ \frac{d}{da} \int_{-a}^a w(t, a) f(t)\, dt \right] w(x, a)$$

$$- \frac{1}{2} \int_{|x|}^a w(x, \xi) \frac{d}{d\xi} \left[ \frac{1}{M'(\xi)} \frac{d}{d\xi} \int_{-\xi}^\xi w(t, \xi) f(t)\, dt \right] d\xi$$

$$- \frac{1}{2} \frac{d}{dx} \int_{|x|}^a \frac{w(x, \xi)}{M'(\xi)} \left[ \int_{-\xi}^\xi w(t, \xi)\, df(t) \right] d\xi,$$

where the prime stands for the derivative and

$$M(\xi) = \left( \ln \frac{2A}{\xi} \right)^{-1}, \qquad w(x, \xi) = \frac{M(\xi)}{\pi \sqrt{\xi^2 - x^2}}.$$

**15.** $\int_0^a \ln\left|\dfrac{x+t}{x-t}\right| y(t)\, dt = f(x).$

Solution:
$$y(x) = -\dfrac{2}{\pi^2}\dfrac{d}{dx}\int_x^a \dfrac{F(t)\, dt}{\sqrt{t^2 - x^2}}, \qquad F(t) = \dfrac{d}{dt}\int_0^t \dfrac{s f(s)\, ds}{\sqrt{t^2 - s^2}}.$$

**16.** $\int_0^\infty \cos(xt) y(t)\, dt = f(x).$

Solution: $y(x) = \dfrac{2}{\pi}\int_0^\infty \cos(xt) f(t)\, dt.$

Up to constant factors, the function $f(x)$ and the solution $y(t)$ are the Fourier cosine transform pair.

**17.** $\int_0^\infty \sin(xt) y(t)\, dt = f(x).$

Solution: $y(x) = \dfrac{2}{\pi}\int_0^\infty \sin(xt) f(t)\, dt.$

Up to constant factors, the function $f(x)$ and the solution $y(t)$ are the Fourier sine transform pair.

**18.** $\int_0^{\pi/2} y(\xi)\, dt = f(x), \qquad \xi = x\sin t.$

*Schlömilch equation.*
Solution:
$$y(x) = \dfrac{2}{\pi}\left[f(0) + x\int_0^{\pi/2} f'_\xi(\xi)\, dt\right], \qquad \xi = x\sin t.$$

**19.** $\int_0^{2\pi} \cot\!\left(\dfrac{t-x}{2}\right) y(t)\, dt = f(x), \qquad 0 \le x \le 2\pi.$

Here the integral is understood in the sense of the Cauchy principal value and the right-hand side is assumed to satisfy the condition $\int_0^{2\pi} f(t)\, dt = 0.$

Solution:
$$y(x) = -\dfrac{1}{4\pi^2}\int_0^{2\pi} \cot\!\left(\dfrac{t-x}{2}\right) f(t)\, dt + C,$$

where $C$ is an arbitrary constant.

It follows from the solution that $\int_0^{2\pi} y(t)\, dt = 2\pi C.$

The equation and its solution form a Hilbert transform pair (in the asymmetric form).

**20.** $\int_0^\infty t J_\nu(xt) y(t)\, dt = f(x), \qquad \nu > -\tfrac{1}{2}.$

Here, $J_\nu(z)$ is the Bessel function of the first kind.
Solution:
$$y(x) = \int_0^\infty t J_\nu(xt) f(t)\, dt.$$

The function $f(x)$ and the solution $y(t)$ are the Hankel transform pair.

**21.** $\int_{-\infty}^{\infty} K_0(|x-t|) y(t)\, dt = f(x).$

Here, $K_0(z)$ is the modified Bessel function of the second kind.

Solution:
$$y(x) = -\frac{1}{\pi^2}\left(\frac{d^2}{dx^2} - 1\right)\int_{-\infty}^{\infty} K_0(|x-t|) f(t)\, dt.$$

**22.** $\int_{-\infty}^{\infty} K(x-t) y(t)\, dt = f(x).$

The Fourier transform is used to solve this equation.

$1^\circ$. Solution:
$$y(x) = \frac{1}{2\pi}\int_{-\infty}^{\infty} \frac{\tilde{f}(u)}{\tilde{K}(u)} e^{iux}\, du,$$
$$\tilde{f}(u) = \frac{1}{\sqrt{2\pi}}\int_{-\infty}^{\infty} f(x) e^{-iux}\, dx, \quad \tilde{K}(u) = \frac{1}{\sqrt{2\pi}}\int_{-\infty}^{\infty} K(x) e^{-iux}\, dx.$$

The following statement is valid. Let $f(x) \in L_2(-\infty, \infty)$ and $K(x) \in L_1(-\infty, \infty)$. Then for a solution $y(x) \in L_2(-\infty, \infty)$ of the integral equation to exist, it is necessary and sufficient that $\tilde{f}(u)/\tilde{K}(u) \in L_2(-\infty, \infty)$.

$2^\circ$. Let the function $P(s)$ defined by the formula
$$\frac{1}{P(s)} = \int_{-\infty}^{\infty} e^{-st} K(t)\, dt$$
be a polynomial of degree $n$ with real roots of the form
$$P(s) = \left(1 - \frac{s}{a_1}\right)\left(1 - \frac{s}{a_2}\right)\cdots\left(1 - \frac{s}{a_n}\right).$$
Then the solution of the integral equation is given by
$$y(x) = P(D) f(x), \quad D = \frac{d}{dx}.$$

**23.** $\int_0^{\infty} K(x-t) y(t)\, dt = f(x).$

*Wiener–Hopf equation of the first kind.* This equation is discussed in the books by Gakhov and Cherskii (1978), Mikhlin and Prössdorf (1986), Muskhelishvili (1992), and Polyanin and Manzhirov (1998) in detail.

## T11.4. Linear Equations of the Second Kind with Constant Limits of Integration

**1.** $y(x) - \lambda \int_a^b (x-t) y(t)\, dt = f(x).$

Solution:
$$y(x) = f(x) + \lambda(A_1 x + A_2),$$

where

$$A_1 = \frac{12f_1 + 6\lambda(f_1\Delta_2 - 2f_2\Delta_1)}{\lambda^2\Delta_1^4 + 12}, \quad A_2 = \frac{-12f_2 + 2\lambda(3f_2\Delta_2 - 2f_1\Delta_3)}{\lambda^2\Delta_1^4 + 12},$$

$$f_1 = \int_a^b f(x)\,dx, \quad f_2 = \int_a^b xf(x)\,dx, \quad \Delta_n = b^n - a^n.$$

2. $y(x) + A \int_a^b |x-t|\,y(t)\,dt = f(x).$

1°. For $A < 0$, the solution is given by

$$y(x) = C_1 \cosh(kx) + C_2 \sinh(kx) + f(x) + k\int_a^x \sinh[k(x-t)]f(t)\,dt, \quad k = \sqrt{-2A}, \quad (1)$$

where the constants $C_1$ and $C_2$ are determined by the conditions

$$\begin{aligned} y'_x(a) + y'_x(b) &= f'_x(a) + f'_x(b), \\ y(a) + y(b) + (b-a)y'_x(a) &= f(a) + f(b) + (b-a)f'_x(a). \end{aligned} \quad (2)$$

2°. For $A > 0$, the solution is given by

$$y(x) = C_1 \cos(kx) + C_2 \sin(kx) + f(x) - k\int_a^x \sin[k(x-t)]f(t)\,dt, \quad k = \sqrt{2A}, \quad (3)$$

where the constants $C_1$ and $C_2$ are determined by conditions (2).

3°. In the special case $a = 0$ and $A > 0$, the solution of the integral equation is given by formula (3) with

$$C_1 = k\frac{I_s(1+\cos\lambda) - I_c(\lambda + \sin\lambda)}{2 + 2\cos\lambda + \lambda\sin\lambda}, \quad C_2 = k\frac{I_s\sin\lambda + I_c(1+\cos\lambda)}{2 + 2\cos\lambda + \lambda\sin\lambda},$$

$$k = \sqrt{2A}, \quad \lambda = bk, \quad I_s = \int_0^b \sin[k(b-t)]f(t)\,dt, \quad I_c = \int_0^b \cos[k(b-t)]f(t)\,dt.$$

▶ In equations T11.4.3 and T11.4.4 and their solutions, all singular integrals are understood in the sense of the Cauchy principal value.

3. $Ay(x) + \dfrac{B}{\pi}\displaystyle\int_{-1}^{1} \dfrac{y(t)\,dt}{t-x} = f(x), \quad -1 < x < 1.$

Without loss of generality we may assume that $A^2 + B^2 = 1$.

1°. The solution bounded at the endpoints:

$$y(x) = Af(x) - \frac{B}{\pi}\int_{-1}^{1} \frac{g(x)}{g(t)} \frac{f(t)\,dt}{t-x}, \quad g(x) = (1+x)^\alpha(1-x)^{1-\alpha}, \quad (1)$$

where $\alpha$ is the solution of the trigonometric equation

$$A + B\cot(\pi\alpha) = 0 \quad (2)$$

on the interval $0 < \alpha < 1$. This solution $y(x)$ exists if and only if $\displaystyle\int_{-1}^{1} \frac{f(t)}{g(t)}\,dt = 0.$

2°. The solution bounded at the endpoint $x = 1$ and unbounded at the endpoint $x = -1$:

$$y(x) = Af(x) - \frac{B}{\pi} \int_{-1}^{1} \frac{g(x)}{g(t)} \frac{f(t)\,dt}{t - x}, \qquad g(x) = (1+x)^{\alpha}(1-x)^{-\alpha}, \qquad (3)$$

where $\alpha$ is the solution of the trigonometric equation (2) on the interval $-1 < \alpha < 0$.

3°. The solution unbounded at the endpoints:

$$y(x) = Af(x) - \frac{B}{\pi} \int_{-1}^{1} \frac{g(x)}{g(t)} \frac{f(t)\,dt}{t - x} + Cg(x), \qquad g(x) = (1+x)^{\alpha}(1-x)^{-1-\alpha},$$

where $C$ is an arbitrary constant and $\alpha$ is the solution of the trigonometric equation (2) on the interval $-1 < \alpha < 0$.

4. $y(x) - \lambda \int_0^1 \left( \dfrac{1}{t-x} - \dfrac{1}{x+t-2xt} \right) y(t)\,dt = f(x), \qquad 0 < x < 1.$

*Tricomi's equation.*
   Solution:

$$y(x) = \frac{1}{1 + \lambda^2 \pi^2} \left[ f(x) + \int_0^1 \frac{t^{\alpha}(1-x)^{\alpha}}{x^{\alpha}(1-t)^{\alpha}} \left( \frac{1}{t-x} - \frac{1}{x+t-2xt} \right) f(t)\,dt \right] + \frac{C(1-x)^{\beta}}{x^{1+\beta}},$$

$$\alpha = \frac{2}{\pi}\arctan(\lambda\pi)\ (-1 < \alpha < 1), \qquad \tan\frac{\beta\pi}{2} = \lambda\pi\ (-2 < \beta < 0),$$

where $C$ is an arbitrary constant.

5. $y(x) + \lambda \int_0^{\infty} e^{-|x-t|} y(t)\,dt = f(x).$

Solution for $\lambda > -\frac{1}{2}$:

$$y(x) = f(x) - \frac{\lambda}{\sqrt{1 + 2\lambda}} \int_0^{\infty} \exp\bigl(-\sqrt{1 + 2\lambda}\,|x - t|\bigr) f(t)\,dt$$
$$+ \left( 1 - \frac{\lambda + 1}{\sqrt{1 + 2\lambda}} \right) \int_0^{\infty} \exp\bigl[-\sqrt{1 + 2\lambda}\,(x + t)\bigr] f(t)\,dt.$$

6. $y(x) - \lambda \int_{-\infty}^{\infty} e^{-|x-t|} y(t)\,dt = 0, \qquad \lambda > 0.$

*Lalesco–Picard equation.*
   Solution:

$$y(x) = \begin{cases} C_1 \exp\bigl(x\sqrt{1-2\lambda}\bigr) + C_2 \exp\bigl(-x\sqrt{1-2\lambda}\bigr) & \text{for } 0 < \lambda < \tfrac{1}{2}, \\ C_1 + C_2 x & \text{for } \lambda = \tfrac{1}{2}, \\ C_1 \cos\bigl(x\sqrt{2\lambda-1}\bigr) + C_2 \sin\bigl(x\sqrt{2\lambda-1}\bigr) & \text{for } \lambda > \tfrac{1}{2}, \end{cases}$$

where $C_1$ and $C_2$ are arbitrary constants.

**7.** $y(x) + \lambda \int_{-\infty}^{\infty} e^{-|x-t|} y(t)\, dt = f(x)$.

1°. Solution for $\lambda > -\frac{1}{2}$:
$$y(x) = f(x) - \frac{\lambda}{\sqrt{1+2\lambda}} \int_{-\infty}^{\infty} \exp\left(-\sqrt{1+2\lambda}\,|x-t|\right) f(t)\, dt.$$

2°. If $\lambda \leq -\frac{1}{2}$, for the equation to be solvable the conditions
$$\int_{-\infty}^{\infty} f(x)\cos(ax)\, dx = 0, \qquad \int_{-\infty}^{\infty} f(x)\sin(ax)\, dx = 0,$$
where $a = \sqrt{-1-2\lambda}$, must be satisfied. In this case, the solution has the form
$$y(x) = f(x) - \frac{a^2+1}{2a} \int_0^{\infty} \sin(at) f(x+t)\, dt \qquad (-\infty < x < \infty).$$

In the class of solutions not belonging to $L_2(-\infty, \infty)$, the homogeneous equation (with $f(x) \equiv 0$) has a nontrivial solution. In this case, the general solution of the corresponding nonhomogeneous equation with $\lambda \leq -\frac{1}{2}$ has the form
$$y(x) = C_1 \sin(ax) + C_2 \cos(ax) + f(x) - \frac{a^2+1}{4a} \int_{-\infty}^{\infty} \sin(a|x-t|) f(t)\, dt.$$

**8.** $y(x) + A \int_a^b e^{\lambda|x-t|} y(t)\, dt = f(x)$.

1°. The function $y = y(x)$ obeys the following second-order linear nonhomogeneous ordinary differential equation with constant coefficients:
$$y''_{xx} + \lambda(2A - \lambda) y = f''_{xx}(x) - \lambda^2 f(x). \tag{1}$$
The boundary conditions for (1) have the form
$$\begin{aligned} y'_x(a) + \lambda y(a) &= f'_x(a) + \lambda f(a), \\ y'_x(b) - \lambda y(b) &= f'_x(b) - \lambda f(b). \end{aligned} \tag{2}$$

Equation (1) under the boundary conditions (2) determines the solution of the original integral equation.

2°. For $\lambda(2A - \lambda) < 0$, the general solution of equation (1) is given by
$$y(x) = C_1 \cosh(kx) + C_2 \sinh(kx) + f(x) - \frac{2A\lambda}{k} \int_a^x \sinh[k(x-t)]\, f(t)\, dt, \tag{3}$$
$$k = \sqrt{\lambda(\lambda - 2A)},$$
where $C_1$ and $C_2$ are arbitrary constants.

For $\lambda(2A - \lambda) > 0$, the general solution of equation (1) is given by
$$y(x) = C_1 \cos(kx) + C_2 \sin(kx) + f(x) - \frac{2A\lambda}{k} \int_a^x \sin[k(x-t)]\, f(t)\, dt, \tag{4}$$
$$k = \sqrt{\lambda(2A - \lambda)}.$$

For $\lambda = 2A$, the general solution of equation (1) is given by
$$y(x) = C_1 + C_2 x + f(x) - 4A^2 \int_a^x (x-t) f(t)\, dt. \tag{5}$$

The constants $C_1$ and $C_2$ in solutions (3)–(5) are determined by conditions (2).

**3°.** In the special case $a = 0$ and $\lambda(2A - \lambda) > 0$, the solution of the integral equation is given by formula (4) with

$$C_1 = \frac{A(kI_c - \lambda I_s)}{(\lambda - A)\sin\mu - k\cos\mu}, \quad C_2 = -\frac{\lambda}{k}\frac{A(kI_c - \lambda I_s)}{(\lambda - A)\sin\mu - k\cos\mu},$$

$$k = \sqrt{\lambda(2A - \lambda)}, \quad \mu = bk, \quad I_s = \int_0^b \sin[k(b-t)]f(t)\,dt, \quad I_c = \int_0^b \cos[k(b-t)]f(t)\,dt.$$

**9.** $y(x) + \lambda \displaystyle\int_{-\infty}^{\infty} \frac{y(t)\,dt}{\cosh[b(x-t)]} = f(x).$

Solution for $b > \pi|\lambda|$:

$$y(x) = f(x) - \frac{2\lambda b}{\sqrt{b^2 - \pi^2\lambda^2}} \int_{-\infty}^{\infty} \frac{\sinh[2k(x-t)]}{\sinh[2b(x-t)]} f(t)\,dt, \quad k = \frac{b}{\pi}\arccos\left(\frac{\pi\lambda}{b}\right).$$

**10.** $y(x) - \lambda \displaystyle\int_0^{\infty} \cos(xt)y(t)\,dt = f(x).$

Solution:

$$y(x) = \frac{f(x)}{1 - \frac{\pi}{2}\lambda^2} + \frac{\lambda}{1 - \frac{\pi}{2}\lambda^2} \int_0^{\infty} \cos(xt)f(t)\,dt, \quad \lambda \neq \pm\sqrt{2/\pi}.$$

**11.** $y(x) - \lambda \displaystyle\int_0^{\infty} \sin(xt)y(t)\,dt = f(x).$

Solution:

$$y(x) = \frac{f(x)}{1 - \frac{\pi}{2}\lambda^2} + \frac{\lambda}{1 - \frac{\pi}{2}\lambda^2} \int_0^{\infty} \sin(xt)f(t)\,dt, \quad \lambda \neq \pm\sqrt{2/\pi}.$$

**12.** $y(x) - \lambda \displaystyle\int_{-\infty}^{\infty} \frac{\sin(x-t)}{x-t} y(t)\,dt = f(x).$

Solution:

$$y(x) = f(x) + \frac{\lambda}{\sqrt{2\pi} - \pi\lambda} \int_{-\infty}^{\infty} \frac{\sin(x-t)}{x-t} f(t)\,dt, \quad \lambda \neq \sqrt{2/\pi}.$$

**13.** $Ay(x) - \dfrac{B}{2\pi} \displaystyle\int_0^{2\pi} \cot\left(\frac{t-x}{2}\right) y(t)\,dt = f(x), \quad 0 \leq x \leq 2\pi.$

Here the integral is understood in the sense of the Cauchy principal value. Without loss of generality we may assume that $A^2 + B^2 = 1$.

Solution:

$$y(x) = Af(x) + \frac{B}{2\pi}\int_0^{2\pi}\cot\left(\frac{t-x}{2}\right)f(t)\,dt + \frac{B^2}{2\pi A}\int_0^{2\pi} f(t)\,dt.$$

**14.**  $y(x) - \lambda \int_0^\infty e^{\mu(x-t)} \cos(xt) y(t)\, dt = f(x).$

Solution:
$$y(x) = \frac{f(x)}{1 - \frac{\pi}{2}\lambda^2} + \frac{\lambda}{1 - \frac{\pi}{2}\lambda^2} \int_0^\infty e^{\mu(x-t)} \cos(xt) f(t)\, dt, \qquad \lambda \neq \pm\sqrt{2/\pi}.$$

**15.**  $y(x) - \lambda \int_0^\infty e^{\mu(x-t)} \sin(xt) y(t)\, dt = f(x).$

Solution:
$$y(x) = \frac{f(x)}{1 - \frac{\pi}{2}\lambda^2} + \frac{\lambda}{1 - \frac{\pi}{2}\lambda^2} \int_0^\infty e^{\mu(x-t)} \sin(xt) f(t)\, dt, \qquad \lambda \neq \pm\sqrt{2/\pi}.$$

**16.**  $y(x) - \int_{-\infty}^\infty K(x-t) y(t)\, dt = f(x).$

The Fourier transform is used to solve this equation.

Solution:
$$y(x) = f(x) + \int_{-\infty}^\infty R(x-t) f(t)\, dt,$$

where

$$R(x) = \frac{1}{\sqrt{2\pi}} \int_{-\infty}^\infty \widetilde{R}(u) e^{iux}\, du, \quad \widetilde{R}(u) = \frac{\widetilde{K}(u)}{1 - \sqrt{2\pi}\, \widetilde{K}(u)}, \quad \widetilde{K}(u) = \frac{1}{\sqrt{2\pi}} \int_{-\infty}^\infty K(x) e^{-iux}\, dx.$$

**17.**  $y(x) - \int_0^\infty K(x-t) y(t)\, dt = f(x).$

*Wiener–Hopf equation of the second kind*. This equation is discussed in the books by Noble (1958), Gakhov and Cherskii (1978), and Polyanin and Manzhirov (1998) in detail.

## References for Chapter T11

Bitsadze, A. V., *Integral Equation of the First Kind*, World Scientific Publishing Co., Singapore, 1995.

Ditkin, V. A. and Prudnikov, A. P., *Integral Transforms and Operational Calculus*, Pergamon Press, New York, 1965.

Gakhov, F. D., *Boundary Value Problems*, Dover Publications, New York, 1990.

Gakhov, F. D. and Cherskii, Yu. I., *Equations of Convolution Type* [in Russian], Nauka Publishers, Moscow, 1978.

Gorenflo, R. and Vessella, S., *Abel Integral Equations: Analysis and Applications*, Springer-Verlag, Berlin, 1991.

Krasnov, M. L., Kiselev, A. I., and Makarenko, G. I., *Problems and Exercises in Integral Equations*, Mir Publishers, Moscow, 1971.

Lifanov, I. K., Poltavskii, L. N., and Vainikko, G., *Hypersingular Integral Equations and Their Applications*, Chapman & Hall/CRC Press, Boca Raton, 2004.

Mikhlin, S. G. and Prössdorf, S., *Singular Integral Operators*, Springer-Verlag, Berlin, 1986.

Muskhelishvili, N. I., *Singular Integral Equations: Boundary Problems of Function Theory and Their Applications to Mathematical Physics*, Dover Publications, New York, 1992.

Noble, B., *Methods Based on Wiener–Hopf Technique for the Solution of Partial Differential Equations*, Pergamon Press, London, 1958.

Polyanin, A. D. and Manzhirov, A. V., *Handbook of Integral Equations*, CRC Press, Boca Raton, 1998.

**Prudnikov, A. P., Brychkov, Yu. A., and Marichev, O. I.,** *Integrals and Series, Vol. 5, Inverse Laplace Transforms,* Gordon & Breach, New York, 1992.

**Sakhnovich, L. A.,** *Integral Equations with Difference Kernels on Finite Intervals,* Birkhäuser Verlag, Basel, 1996.

**Samko, S. G., Kilbas, A. A., and Marichev, O. I.,** *Fractional Integrals and Derivatives. Theory and Applications,* Gordon & Breach, New York, 1993.

**Zabreyko, P. P., Koshelev, A. I., et al.,** *Integral Equations: A Reference Text,* Noordhoff International Publishing, Leyden, 1975.

# Chapter T12
# Functional Equations

## T12.1. Linear Functional Equations in One Independent Variable

### T12.1.1. Linear Difference and Functional Equations Involving Unknown Function with Two Different Arguments

**T12.1.1-1. First-order linear difference equations involving $y(x)$ and $y(x+a)$.**

**1.** $y(x+1) - y(x) = 0.$

*This functional equation may be treated as a definition of periodic functions* with unit period.

1°. Solution:
$$y(x) = \Theta(x),$$
where $\Theta(x) = \Theta(x+1)$ is an arbitrary periodic function with unit period.

2°. A periodic function $\Theta(x)$ with period 1 that satisfies the Dirichlet conditions can be expanded into a Fourier series:
$$\Theta(x) = \frac{a_0}{2} + \sum_{n=1}^{\infty} \left[ a_n \cos(2\pi n x) + b_n \sin(2\pi n x) \right],$$
where
$$a_n = 2\int_0^1 \Theta(x) \cos(2\pi n x)\, dx, \qquad b_n = 2\int_0^1 \Theta(x) \sin(2\pi n x)\, dx.$$

**2.** $y(x+1) - y(x) = f(x).$

Solution:
$$y(x) = \Theta(x) + \bar{y}(x),$$
where $\Theta(x) = \Theta(x+1)$ is an arbitrary periodic function with period 1, and $\bar{y}(x)$ is any particular solution of the nonhomogeneous equation. Table T12.1 presents particular solutions of the nonhomogeneous equation for some specific $f(x)$.

**3.** $y(x+1) - ay(x) = 0.$

*A homogeneous first-order constant-coefficient linear difference equation.*

1°. Solution for $a > 0$:
$$y(x) = \Theta(x) a^x,$$
where $\Theta(x) = \Theta(x+1)$ is an arbitrary periodic function with period 1.

For $\Theta(x) \equiv \text{const}$, we have a particular solution $y(x) = Ca^x$, where $C$ is an arbitrary constant.

## TABLE T12.1
Particular solutions of the nonhomogeneous functional equation $y(x+1) - y(x) = f(x)$

| No. | Right-hand side of equation, $f(x)$ | Particular solution, $\bar{y}(x)$ |
|---|---|---|
| 1 | $1$ | $x$ |
| 2 | $x$ | $\frac{1}{2}x(x-1)$ |
| 3 | $x^2$ | $\frac{1}{6}x(x-1)(2x-1)$ |
| 4 | $x^n$, $n = 0, 1, 2, \ldots$ | $\frac{1}{n+1}B_n(x)$, where the $B_n(x)$ are Bernoulli polynomials. The generating function: $\dfrac{te^{xt}}{e^t - 1} = \sum_{n=0}^{\infty} B_n(x)\dfrac{t^n}{n!}$ |
| 5 | $\dfrac{1}{x}$ | $\psi(x) = -\mathcal{C} + \int_0^1 \dfrac{1 - t^{x-1}}{1 - t}\,dt$ is the logarithmic derivative of the gamma function, $\mathcal{C} = 0.5772\ldots$ is the Euler constant |
| 6 | $\dfrac{1}{x(x+1)}$ | $-\dfrac{1}{x}$ |
| 7 | $a^{\lambda x}$, $a \ne 1$, $\lambda \ne 0$ | $\dfrac{1}{a^\lambda - 1}a^{\lambda x}$ |
| 8 | $\sinh(a + 2bx)$, $b > 0$ | $\dfrac{\cosh(a - b + 2bx)}{2\sinh b}$ |
| 9 | $\cosh(a + 2bx)$, $b > 0$ | $\dfrac{\sinh(a - b + 2bx)}{2\sinh b}$ |
| 10 | $xa^x$, $a \ne 1$ | $\dfrac{1}{a-1}a^x\left(x - \dfrac{a}{a-1}\right)$ |
| 11 | $\ln x$, $x > 0$ | $\ln \Gamma(x)$, where $\Gamma(x) = \int_0^\infty t^{x-1}e^{-t}\,dt$ is the gamma function |
| 12 | $\sin(2ax)$, $a \ne \pi n$ | $-\dfrac{\cos[a(2x-1)]}{2\sin a}$ |
| 13 | $\sin(2\pi n x)$ | $x\sin(2\pi n x)$ |
| 14 | $\cos(2ax)$, $a \ne \pi n$ | $\dfrac{\sin[a(2x-1)]}{2\sin a}$ |
| 15 | $\cos(2\pi n x)$ | $x\cos(2\pi n x)$ |
| 16 | $\sin^2(ax)$, $a \ne \pi n$ | $\dfrac{x}{2} - \dfrac{\sin[a(2x-1)]}{4\sin a}$ |
| 17 | $\sin^2(\pi n x)$ | $x\sin^2(\pi n x)$ |
| 18 | $\cos^2(ax)$, $a \ne \pi n$ | $\dfrac{x}{2} + \dfrac{\sin[a(2x-1)]}{4\sin a}$ |
| 19 | $\cos^2(\pi n x)$ | $x\cos^2(\pi n x)$ |
| 20 | $x\sin(2ax)$, $a \ne \pi n$ | $\dfrac{\sin(2ax)}{4\sin^2 a} - x\dfrac{\cos[a(2x-1)]}{2\sin a}$ |
| 21 | $x\sin(2\pi n x)$ | $\frac{1}{2}x(x-1)\sin(2\pi n x)$ |
| 22 | $x\cos(2ax)$, $a \ne \pi n$ | $\dfrac{\cos(2ax)}{4\sin^2 a} + x\dfrac{\sin[a(2x-1)]}{2\sin a}$ |
| 23 | $x\cos(2\pi n x)$ | $\frac{1}{2}x(x-1)\cos(2\pi n x)$ |
| 24 | $a^x \sin(bx)$, $a > 0$, $a \ne 1$ | $a^x\dfrac{a\sin[b(x-1)] - \sin(bx)}{a^2 - 2a\cos b + 1}$ |
| 25 | $a^x \cos(bx)$, $a > 0$, $a \ne 1$ | $a^x\dfrac{a\cos[b(x-1)] - \cos(bx)}{a^2 - 2a\cos b + 1}$ |

**2°.** Solution for $a < 0$:

$$y(x) = \Theta_1(x)|a|^x \sin(\pi x) + \Theta_2(x)|a|^x \cos(\pi x),$$

where the $\Theta_k(x) = \Theta_k(x+1)$ are arbitrary periodic functions with period 1 ($k = 1, 2$).

**Remark.** See also Subsection 17.2.1, Example 1.

**4.** $y(x+1) - ay(x) = f(x).$

*A nonhomogeneous first-order constant-coefficient linear difference equation.*

**1°.** Solution:

$$y(x) = \begin{cases} \Theta(x)a^x + \bar{y}(x) & \text{if } a > 0, \\ \Theta_1(x)|a|^x \sin(\pi x) + \Theta_2(x)|a|^x \cos(\pi x) + \bar{y}(x) & \text{if } a < 0, \end{cases}$$

where $\Theta(x)$, $\Theta_1(x)$, and $\Theta_2(x)$ are arbitrary periodic functions with period 1, and $\bar{y}(x)$ is any particular solution of the nonhomogeneous equation.

**2°.** For $f(x) = \sum_{k=0}^{n} b_k x^n$ and $a \neq 1$, the nonhomogeneous equation has a particular solution of the form $\bar{y}(x) = \sum_{k=0}^{n} A_k x^n$; the constants $B_k$ are found by the method of undetermined coefficients.

**3°.** For $f(x) = \sum_{k=1}^{n} b_k e^{\lambda_k x}$, the nonhomogeneous equation has a particular solution of the form

$$\bar{y}(x) = \begin{cases} \displaystyle\sum_{k=1}^{n} \frac{b_k}{e^{\lambda_k} - a} e^{\lambda_k x} & \text{if } a \neq e^{\lambda_m}, \\ \displaystyle b_m x e^{\lambda_m(x-1)} + \sum_{k=1, k \neq m}^{n} \frac{b_k}{e^{\lambda_k} - a} e^{\lambda_k x} & \text{if } a = e^{\lambda_m}, \end{cases}$$

where $m = 1, \ldots, n$.

**4°.** For $f(x) = \sum_{k=1}^{n} b_k \cos(\beta_k x)$, the nonhomogeneous equation has a particular solution of the form

$$\bar{y}(x) = \sum_{k=1}^{n} \frac{b_k}{a^2 + 1 - 2a \cos \beta_k} \left[(\cos \beta_k - a)\cos(\beta_k x) + \sin \beta_k \sin(\beta_k x)\right].$$

**5°.** For $f(x) = \sum_{k=1}^{n} b_k \sin(\beta_k x)$, the nonhomogeneous equation has a particular solution of the form

$$\bar{y}(x) = \sum_{k=1}^{n} \frac{b_k}{a^2 + 1 - 2a \cos \beta_k} \left[(\cos \beta_k - a)\sin(\beta_k x) - \sin \beta_k \cos(\beta_k x)\right].$$

**5.** $y(x+1) - xy(x) = 0.$

Solution:
$$y(x) = \Theta(x)\Gamma(x), \qquad \Gamma(x) = \int_0^\infty t^{x-1} e^{-t}\, dt,$$
where $\Gamma(x)$ is the gamma function, $\Theta(x) = \Theta(x+1)$ is an arbitrary periodic function with period 1.

The simplest particular solution corresponds to $\Theta(x) \equiv 1$.

**6.** $y(x+1) - a(x-b)(x-c)y(x) = 0.$

Solution:
$$y(x) = \Theta(x) a^x \Gamma(x-b)\Gamma(x-c),$$
where $\Gamma(x)$ is the gamma function, $\Theta(x)$ is an arbitrary periodic function with period 1.

**7.** $y(x+1) - R(x)y(x) = 0, \qquad R(x) = a\dfrac{(x-\lambda_1)(x-\lambda_2)\dots(x-\lambda_n)}{(x-\mu_1)(x-\mu_2)\dots(x-\mu_m)}.$

Solution:
$$y(x) = \Theta(x) a^x \frac{\Gamma(x-\lambda_1)\Gamma(x-\lambda_2)\dots\Gamma(x-\lambda_n)}{\Gamma(x-\mu_1)\Gamma(x-\mu_2)\dots\Gamma(x-\mu_m)},$$
where $\Gamma(x)$ is the gamma function, $\Theta(x)$ is an arbitrary periodic function with period 1.

**8.** $y(x+1) - a e^{\lambda x} y(x) = 0.$

Solution:
$$y(x) = \Theta(x) a^x \exp\left(\tfrac{1}{2}\lambda x^2 - \tfrac{1}{2}\lambda x\right),$$
where $\Theta(x)$ is an arbitrary periodic function with period 1.

**9.** $y(x+1) - a e^{\mu x^2 + \lambda x} y(x) = 0.$

Solution:
$$y(x) = \Theta(x) a^x \exp\left[\tfrac{1}{3}\mu x^3 + \tfrac{1}{2}(\lambda - \mu)x^2 + \tfrac{1}{6}(\mu - 3\lambda)x\right],$$
where $\Theta(x)$ is an arbitrary periodic function with period 1.

**10.** $y(x+1) - f(x)y(x) = 0.$

Here, $f(x) = f(x+1)$ is a given periodic function with period 1.

Solution:
$$y(x) = \Theta(x)\bigl[f(x)\bigr]^x,$$
where $\Theta(x) = \Theta(x+1)$ is an arbitrary periodic function with period 1.

For $\Theta(x) \equiv \text{const}$, we have a particular solution $y(x) = C\bigl[f(x)\bigr]^x$, where $C$ is an arbitrary constant.

**11.** $y(x+a) - by(x) = 0.$

Solution:
$$y(x) = \Theta(x) b^{x/a},$$
where $\Theta(x) = \Theta(x+a)$ is an arbitrary periodic function with period $a$.

For $\Theta(x) \equiv \text{const}$, we have particular solution $y(x) = C b^{x/a}$, where $C$ is an arbitrary constant.

**12.** $y(x+a) - by(x) = f(x)$.

1°. Solution:
$$y(x) = \Theta(x) b^{x/a} + \bar{y}(x),$$

where $\Theta(x) = \Theta(x+a)$ is an arbitrary periodic function with period $a$, and $\bar{y}(x)$ is any particular solution of the nonhomogeneous equation.

2°. For $f(x) = \sum_{k=0}^{n} A_k x^n$ and $b \neq 1$, the nonhomogeneous equation has a particular solution of the form $\bar{y}(x) = \sum_{k=0}^{n} B_k x^n$; the constants $B_k$ are found by the method of undetermined coefficients.

3°. For $f(x) = \sum_{k=1}^{n} A_k \exp(\lambda_k x)$, the nonhomogeneous equation has a particular solution of the form $\bar{y}(x) = \sum_{k=1}^{n} B_k \exp(\lambda_k x)$; the constants $B_k$ are found by the method of undetermined coefficients.

4°. For $f(x) = \sum_{k=1}^{n} A_k \cos(\lambda_k x)$, the nonhomogeneous equation has a particular solution of the form $\bar{y}(x) = \sum_{k=1}^{n} B_k \cos(\lambda_k x) + \sum_{k=1}^{n} D_k \sin(\lambda_k x)$; the constants $B_k$ and $D_k$ are found by the method of undetermined coefficients.

5°. For $f(x) = \sum_{k=1}^{n} A_k \sin(\lambda_k x)$, the nonhomogeneous equation has a particular solution of the form $\bar{y}(x) = \sum_{k=1}^{n} B_k \cos(\lambda_k x) + \sum_{k=1}^{n} D_k \sin(\lambda_k x)$; the constants $B_k$ and $D_k$ are found by the method of undetermined coefficients.

**13.** $y(x+a) - bxy(x) = 0$, $\quad a, b > 0$.

Solution:
$$y(x) = \Theta(x) \int_0^\infty t^{(x/a)-1} e^{-t/(ab)}\, dt,$$

where $\Theta(x) = \Theta(x+a)$ is an arbitrary periodic function with period $a$.

**14.** $y(x+a) - f(x)y(x) = 0$.

Here, $f(x) = f(x+a)$ is a given periodic function with period $a$.
Solution:
$$y(x) = \Theta(x)\bigl[f(x)\bigr]^{x/a},$$

where $\Theta(x) = \Theta(x+a)$ is an arbitrary periodic function with period $a$.

For $\Theta(x) \equiv \text{const}$, we have a particular solution $y(x) = C\bigl[f(x)\bigr]^{x/a}$, where $C$ is an arbitrary constant.

**T12.1.1-2. Linear functional equations involving $y(x)$ and $y(ax)$.**

**15.** $y(ax) - by(x) = 0, \qquad a, b > 0.$

Solution for $x > 0$:
$$y(x) = x^\lambda \Theta\left(\frac{\ln x}{\ln a}\right), \qquad \lambda = \frac{\ln b}{\ln a},$$
where $\Theta(z) = \Theta(z+1)$ is an arbitrary periodic function with period 1, $a \neq 1$.

For $\Theta(z) \equiv \text{const}$, we have a particular solution $y(x) = Cx^\lambda$, where $C$ is an arbitrary constant.

**16.** $y(ax) - by(x) = f(x).$

1°. Solution:
$$y(x) = Y(x) + \bar{y}(x),$$
where $Y(x)$ is the general solution of the homogeneous equation $Y(ax) - bY(x) = 0$ (see the previous equation), and $\bar{y}(x)$ is any particular solution of the nonhomogeneous equation.

2°. For $f(x) = \sum_{k=0}^{n} A_k x^n$, the nonhomogeneous equation has a particular solution of the form
$$\bar{y}(x) = \sum_{k=0}^{n} \frac{A_k}{a^k - b} x^k, \qquad a^k - b \neq 0.$$

3°. For $f(x) = \ln x \sum_{k=0}^{n} A_k x^k$, the nonhomogeneous equation has a particular solution of the form
$$\bar{y}(x) = \sum_{k=1}^{n} x^k (B_k \ln x + C_k), \qquad B_k = \frac{A_k}{a^k - b}, \qquad C_k = -\frac{A_k a^k \ln a}{(a^k - b)^2}.$$

**17.** $y(2x) - a \cos x \, y(x) = 0.$

Solution for $a > 0$ and $x > 0$:
$$y(x) = x^{\frac{\ln a}{\ln 2} - 1} \sin x \, \Theta\left(\frac{\ln x}{\ln 2}\right),$$
where $\Theta(x) = \Theta(x+1)$ is an arbitrary periodic function with period 1.

**T12.1.1-3. Linear functional equations involving $y(x)$ and $y(a-x)$.**

**18.** $y(x) = y(-x).$

*This functional equation may be treated as a definition of even functions.*

Solution:
$$y(x) = \frac{\varphi(x) + \varphi(-x)}{2},$$
where $\varphi(x)$ is an arbitrary function.

**19.** $y(x) = -y(-x)$.

*This functional equation may be treated as a definition of odd functions.*
   Solution:
$$y(x) = \frac{\varphi(x) - \varphi(-x)}{2},$$
where $\varphi(x)$ is an arbitrary function.

**20.** $y(x) - y(a - x) = 0$.

1°. Solution:
$$y(x) = \Phi(x, a - x),$$
where $\Phi(x, z) = \Phi(z, x)$ is any symmetric function with two arguments.

2°. Specific particular solutions may be obtained using the formula
$$y(x) = \Psi\big(\varphi(x) + \varphi(a - x)\big)$$
by specifying the functions $\Psi(z)$ and $\varphi(x)$.

**21.** $y(x) + y(a - x) = 0$.

1°. Solution:
$$y(x) = \Phi(x, a - x),$$
where $\Phi(x, z) = -\Phi(z, x)$ is any antisymmetric function with two arguments.

2°. Specific particular solutions may be obtained using the formula
$$y(x) = (2x - a)\Psi\big(\varphi(x) + \varphi(a - x)\big)$$
by specifying $\Psi(z)$ and $\varphi(x)$.

**22.** $y(x) + y(a - x) = b$.

Solution:
$$y(x) = \tfrac{1}{2}b + \Phi(x, a - x),$$
where $\Phi(x, z) = -\Phi(z, x)$ is any antisymmetric function with two arguments.
   Particular solutions: $y(x) = b \sin^2\!\left(\dfrac{\pi x}{2a}\right)$ and $y(x) = b \cos^2\!\left(\dfrac{\pi x}{2a}\right)$.

**23.** $y(x) + y(a - x) = f(x)$.

Here, the function $f(x)$ must satisfy the condition $f(x) = f(a - x)$.
   Solution:
$$y(x) = \tfrac{1}{2}f(x) + \Phi(x, a - x),$$
where $\Phi(x, z) = -\Phi(z, x)$ is any antisymmetric function with two arguments.

**24.** $y(x) - y(a - x) = f(x)$.

Here, the function $f(x)$ must satisfy the condition $f(x) = -f(a - x)$.
   Solution:
$$y(x) = \tfrac{1}{2}f(x) + \Phi(x, a - x),$$
where $\Phi(x, z) = \Phi(z, x)$ is any symmetric function with two arguments.

**25.** $y(x) + g(x)y(a - x) = f(x)$.

Solution:
$$y(x) = \frac{f(x) - g(x)f(a - x)}{1 - g(x)g(a - x)} \qquad [\text{if } g(x)g(a - x) \neq 1].$$

**T12.1.1-4. Linear functional equations involving $y(x)$ and $y(a/x)$.**

**26.** $y(x) - y(a/x) = 0.$

*Babbage equation.*
Solution:
$$y(x) = \Phi(x, a/x),$$
where $\Phi(x, z) = \Phi(z, x)$ is any symmetric function with two arguments.

**27.** $y(x) + y(a/x) = 0.$

Solution:
$$y(x) = \Phi(x, a/x),$$
where $\Phi(x, z) = -\Phi(z, x)$ is any antisymmetric function with two arguments.

**28.** $y(x) + y(a/x) = b.$

Solution:
$$y(x) = \tfrac{1}{2}b + \Phi(x, a/x),$$
where $\Phi(x, z) = -\Phi(z, x)$ is any antisymmetric function with two arguments.

**29.** $y(x) + y(a/x) = f(x).$

The right-hand side must satisfy the condition $f(x) = f(a/x)$.
Solution:
$$y(x) = \tfrac{1}{2}f(x) + \Phi(x, a/x),$$
where $\Phi(x, z) = -\Phi(z, x)$ is any antisymmetric function with two arguments.

**30.** $y(x) - y(a/x) = f(x).$

Here, the function $f(x)$ must satisfy the condition $f(x) = -f(a/x)$.
Solution:
$$y(x) = \tfrac{1}{2}f(x) + \Phi(x, a/x),$$
where $\Phi(x, z) = \Phi(z, x)$ is any symmetric function with two arguments.

**31.** $y(x) + x^a y(1/x) = 0.$

Solution:
$$y(x) = (1 - x^a)\Phi(x, 1/x),$$
where $\Phi(x, z) = \Phi(z, x)$ is any symmetric function with two arguments.

**32.** $y(x) - x^a y(1/x) = 0.$

Solution:
$$y(x) = (1 + x^a)\Phi(x, 1/x),$$
where $\Phi(x, z) = \Phi(z, x)$ is any symmetric function with two arguments.

**33.** $y(x) + g(x)y(a/x) = f(x).$

Solution:
$$y(x) = \frac{f(x) - g(x)f(a/x)}{1 - g(x)g(a/x)} \qquad [\text{if } g(x)g(a/x) \neq 1].$$

**T12.1.1-5. Linear equations involving unknown function with rational argument.**

**34.** $y(x) - y\left(\dfrac{a-x}{1+bx}\right) = 0.$

Solution:
$$y(x) = \Phi\left(x, \dfrac{a-x}{1+bx}\right),$$
where $\Phi(x, z) = \Phi(z, x)$ is any symmetric function with two arguments.

**35.** $y(x) + y\left(\dfrac{a-x}{1+bx}\right) = 0.$

Solution:
$$y(x) = \Phi\left(x, \dfrac{a-x}{1+bx}\right),$$
where $\Phi(x, z) = -\Phi(z, x)$ is any antisymmetric function with two arguments.

**36.** $y(x) + y\left(\dfrac{a-x}{1+bx}\right) = f(x).$

The function $f(x)$ must satisfy the condition $f(x) = f\left(\dfrac{a-x}{1+bx}\right)$.

Solution:
$$y(x) = \tfrac{1}{2}f(x) + \Phi\left(x, \dfrac{a-x}{1+bx}\right),$$
where $\Phi(x, z) = -\Phi(z, x)$ is any antisymmetric function with two arguments.

**37.** $y(x) - y\left(\dfrac{a-x}{1+bx}\right) = f(x).$

Here, the function $f(x)$ must satisfy the condition $f(x) = -f\left(\dfrac{a-x}{1+bx}\right)$.

Solution:
$$y(x) = \tfrac{1}{2}f(x) + \Phi\left(x, \dfrac{a-x}{1+bx}\right),$$
where $\Phi(x, z) = \Phi(z, x)$ is any symmetric function with two arguments.

**38.** $y(x) - cy\left(\dfrac{a-x}{1+bx}\right) = f(x), \qquad c \neq \pm 1.$

Solution:
$$y(x) = \dfrac{1}{1-c^2}f(x) + \dfrac{c}{1-c^2}f\left(\dfrac{a-x}{1+bx}\right).$$

**39.** $y(x) + g(x)y\left(\dfrac{a-x}{1+bx}\right) = f(x).$

Solution:
$$y(x) = \dfrac{f(x) - g(x)f(z)}{1 - g(x)g(z)}, \qquad z = \dfrac{a-x}{1+bx}.$$

**40.** $y(x) + cy\left(\dfrac{ax - \beta}{x + b}\right) = f(x), \qquad \beta = a^2 + ab + b^2.$

A special case of equation T12.1.2.12.

41. $y(x) + cy\left(\dfrac{bx+\beta}{a-x}\right) = f(x)$, $\qquad \beta = a^2 + ab + b^2$.

A special case of equation T12.1.2.12.

42. $y(x) + g(x)y\left(\dfrac{ax-\beta}{x+b}\right) = f(x)$, $\qquad \beta = a^2 + ab + b^2$.

A special case of equation T12.1.2.13.

43. $y(x) + g(x)y\left(\dfrac{bx+\beta}{a-x}\right) = f(x)$, $\qquad \beta = a^2 + ab + b^2$.

A special case of equation T12.1.2.13.

---

**T12.1.1-6. Linear functional equations involving $y(x)$ and $y\left(\sqrt{a^2-x^2}\right)$.**

44. $y(x) - y\left(\sqrt{a^2-x^2}\right) = 0$, $\qquad 0 \le x \le a$.

Solution:
$$y(x) = \Phi\left(x, \sqrt{a^2-x^2}\right),$$
where $\Phi(x,z) = \Phi(z,x)$ is any symmetric function with two arguments.

45. $y(x) + y\left(\sqrt{a^2-x^2}\right) = 0$, $\qquad 0 \le x \le a$.

Solution:
$$y(x) = \Phi\left(x, \sqrt{a^2-x^2}\right),$$
where $\Phi(x,z) = -\Phi(z,x)$ is any antisymmetric function with two arguments.

46. $y(x) + y\left(\sqrt{a^2-x^2}\right) = b$, $\qquad 0 \le x \le a$.

Solution:
$$y(x) = \tfrac{1}{2}b + \Phi\left(x, \sqrt{a^2-x^2}\right),$$
where $\Phi(x,z) = -\Phi(z,x)$ is any antisymmetric function with two arguments.

47. $y(x) + y\left(\sqrt{a^2-x^2}\right) = f(x)$, $\qquad 0 \le x \le a$.

Here, the function $f(x)$ must satisfy the condition $f(x) = f\left(\sqrt{a^2-x^2}\right)$.

Solution:
$$y(x) = \tfrac{1}{2}f(x) + \Phi\left(x, \sqrt{a^2-x^2}\right),$$
where $\Phi(x,z) = -\Phi(z,x)$ is any antisymmetric function with two arguments.

48. $y(x) - y\left(\sqrt{a^2-x^2}\right) = f(x)$, $\qquad 0 \le x \le a$.

Here, the function $f(x)$ must satisfy the condition $f(x) = -f\left(\sqrt{a^2-x^2}\right)$.

Solution:
$$y(x) = \tfrac{1}{2}f(x) + \Phi\left(x, \sqrt{a^2-x^2}\right),$$
where $\Phi(x,z) = \Phi(z,x)$ is any symmetric function with two arguments.

**49.** $y(x) + g(x)y(\sqrt{a^2 - x^2}) = f(x)$, $\quad 0 \le x \le a$.

Solution:
$$y(x) = \frac{f(x) - g(x)f(\sqrt{a^2 - x^2})}{1 - g(x)g(\sqrt{a^2 - x^2})}.$$

---

**T12.1.1-7. Linear functional equations involving $y(\sin x)$ and $y(\cos x)$.**

**50.** $y(\sin x) - y(\cos x) = 0$.

Solution in implicit form:
$$y(\sin x) = \Phi(\sin x, \cos x),$$
where $\Phi(x, z) = \Phi(z, x)$ is any symmetric function with two arguments.

**51.** $y(\sin x) + y(\cos x) = 0$.

Solution in implicit form:
$$y(\sin x) = \Phi(\sin x, \cos x),$$
where $\Phi(x, z) = -\Phi(z, x)$ is any antisymmetric function with two arguments.

**52.** $y(\sin x) + y(\cos x) = a$.

Solution in implicit form:
$$y(\sin x) = \tfrac{1}{2}a + \Phi(\sin x, \cos x),$$
where $\Phi(x, z) = -\Phi(z, x)$ is any antisymmetric function with two arguments.

**53.** $y(\sin x) + y(\cos x) = f(x)$.

Here, the function $f(x)$ must satisfy the condition $f(x) = f(\tfrac{\pi}{2} - x)$.
Solution in implicit form:
$$y(\sin x) = \tfrac{1}{2}f(x) + \Phi(\sin x, \cos x),$$
where $\Phi(x, z) = -\Phi(z, x)$ is any antisymmetric function with two arguments.

**54.** $y(\sin x) - y(\cos x) = f(x)$.

Here, the function $f(x)$ must satisfy the condition $f(x) = -f(\tfrac{\pi}{2} - x)$.
Solution in implicit form:
$$y(\sin x) = \tfrac{1}{2}f(x) + \Phi(\sin x, \cos x),$$
where $\Phi(x, z) = \Phi(z, x)$ is any symmetric function with two arguments.

**55.** $y(\sin x) + g(x)y(\cos x) = f(x)$.

Solution in implicit form:
$$y(\sin x) = \frac{f(x) - g(x)f(\tfrac{\pi}{2} - x)}{1 - g(x)g(\tfrac{\pi}{2} - x)}.$$

T12.1.1-8. Other equations involving unknown function with two different arguments.

**56.** $y(x^a) - by(x) = 0$, $\quad a, b > 0$.

Solution:
$$y(x) = |\ln x|^p \Theta\left(\frac{\ln|\ln x|}{\ln a}\right), \qquad p = \frac{\ln b}{\ln a},$$

where $\Theta(z) = \Theta(z+1)$ is an arbitrary periodic function with period 1.

For $\Theta(z) \equiv \text{const}$, we have a particular solution $y(x) = C|\ln x|^p$, where $C$ is an arbitrary constant.

**57.** $y(x) - y(\omega(x)) = 0$, $\quad$ where $\quad \omega(\omega(x)) = x$.

Solution:
$$y(x) = \Phi(x, \omega(x)),$$

where $\Phi(x, z) = \Phi(z, x)$ is any symmetric function with two arguments.

**58.** $y(x) + y(\omega(x)) = 0$, $\quad$ where $\quad \omega(\omega(x)) = x$.

Solution:
$$y(x) = \Phi(x, \omega(x)),$$

where $\Phi(x, z) = -\Phi(z, x)$ is any antisymmetric function with two arguments.

**59.** $y(x) + y(\omega(x)) = b$, $\quad$ where $\quad \omega(\omega(x)) = x$.

Solution:
$$y(x) = \tfrac{1}{2}b + \Phi(x, \omega(x)),$$

where $\Phi(x, z) = -\Phi(z, x)$ is any antisymmetric function with two arguments.

**60.** $y(x) + y(\omega(x)) = f(x)$, $\quad$ where $\quad \omega(\omega(x)) = x$.

Here, the function $f(x)$ must satisfy the condition $f(x) = f(\omega(x))$.

Solution:
$$y(x) = \tfrac{1}{2}f(x) + \Phi(x, \omega(x)),$$

where $\Phi(x, z) = -\Phi(z, x)$ is any antisymmetric function with two arguments.

**61.** $y(x) - y(\omega(x)) = f(x)$, $\quad$ where $\quad \omega(\omega(x)) = x$.

Here, the function $f(x)$ must satisfy the condition $f(x) = -f(\omega(x))$.

Solution:
$$y(x) = \tfrac{1}{2}f(x) + \Phi(x, \omega(x)),$$

where $\Phi(x, z) = \Phi(z, x)$ is any symmetric function with two arguments.

**62.** $y(x) + g(x)y(\omega(x)) = f(x)$, $\quad$ where $\quad \omega(\omega(x)) = x$.

Solution:
$$y(x) = \frac{f(x) - g(x)f(\omega(x))}{1 - g(x)g(\omega(x))}.$$

## T12.1.2. Other Linear Functional Equations

### T12.1.2-1. Second-order linear difference equations.

**1.** $y_{n+2} + ay_{n+1} + by_n = 0.$

A *homogeneous second-order linear difference equation* defined on a discrete set of points $x = 0, 1, 2, \ldots$ Notation adopted: $y_n = y(n)$.

Let $\lambda_1$ and $\lambda_2$ be roots of the characteristic equation
$$\lambda^2 + a\lambda + b = 0.$$

$1°$. If $\lambda_1 \neq \lambda_2$, the general solution of the difference equation has the form
$$y_n = y_1 \frac{\lambda_1^n - \lambda_2^n}{\lambda_1 - \lambda_2} - y_0 b \frac{\lambda_1^{n-1} - \lambda_2^{n-1}}{\lambda_1 - \lambda_2},$$
where $y_1$ and $y_0$ are arbitrary constants, equal to the values of $y$ at the first two points.

In the case of complex conjugate roots, one should separate the real and imaginary parts in the above solution.

$2°$. If $\lambda_1 = \lambda_2$, the general solution of the difference equation is given by
$$y_n = y_1 n \lambda_1^{n-1} - y_0 b(n-1)\lambda_1^{n-2}.$$

**2.** $y_{n+2} + ay_{n+1} + by_n = f_n.$

A *nonhomogeneous second-order linear difference equation* defined on a discrete set of points $x = 0, 1, 2, \ldots$ Notation adopted: $y_n = y(n)$.

Let $\lambda_1$ and $\lambda_2$ be roots of the characteristic equation
$$\lambda^2 + a\lambda + b = 0.$$

$1°$. If $\lambda_1 \neq \lambda_2$, the general solution of the difference equation has the form
$$y_n = y_1 \frac{\lambda_1^n - \lambda_2^n}{\lambda_1 - \lambda_2} - y_0 b \frac{\lambda_1^{n-1} - \lambda_2^{n-1}}{\lambda_1 - \lambda_2} + \sum_{k=2}^{n} f_{n-k} \frac{\lambda_1^{k-1} - \lambda_2^{k-1}}{\lambda_1 - \lambda_2},$$
where $y_1$ and $y_0$ are arbitrary constants, equal to the values of $y$ at the first two points.

In the case of complex conjugate roots, one should separate the real and imaginary parts in the above solution.

$2°$. If $\lambda_1 = \lambda_2$, the general solution of the difference equation is given by
$$y_n = y_1 n \lambda_1^{n-1} - y_0 b(n-1)\lambda_1^{n-2} + \sum_{k=2}^{n} f_{n-k}(k-1)\lambda_1^{k-2}.$$

$3°$. In boundary value problems, a finite set of points $x = 0, 1, \ldots, N$ is often taken and the initial and final values of the unknown function, $y_0$ and $y_N$, are prescribed. It is required to find the $y_n \equiv y(x)|_{x=n}$ for $1 \leq n \leq N - 1$.

If $\lambda_1 \neq \lambda_2$, the solution is given by
$$y_n = y_0 \frac{\lambda_1^N \lambda_2^n - \lambda_1^n \lambda_2^N}{\lambda_1^N - \lambda_2^N} + y_N \frac{\lambda_1^n - \lambda_2^n}{\lambda_1^N - \lambda_2^N}$$
$$+ \sum_{k=2}^{n} f_{n-k} \frac{\lambda_1^{k-1} - \lambda_2^{k-1}}{\lambda_1 - \lambda_2} - \frac{\lambda_1^n - \lambda_2^n}{\lambda_1^N - \lambda_2^N} \sum_{k=2}^{N} f_{N-k} \frac{\lambda_1^{k-1} - \lambda_2^{k-1}}{\lambda_1 - \lambda_2}.$$

For $n = 1$, the first sum is zero.

3. $y(x + 2) + ay(x + 1) + by(x) = 0$.

*A homogeneous second-order constant-coefficient linear difference equation.*

Let us write out the characteristic equation:
$$\lambda^2 + a\lambda + b = 0. \tag{1}$$

Consider the following cases.

$1°$. The roots $\lambda_1$ and $\lambda_2$ of the quadratic equation (1) are real and distinct. Then the general solution of the original finite-difference equation has the form
$$y(x) = \Theta_1(x)\lambda_1^x + \Theta_2(x)\lambda_2^x, \tag{2}$$
where $\Theta_1(x)$ and $\Theta_2(x)$ are arbitrary periodic functions with period 1, that is, $\Theta_k(x) = \Theta_k(x+1)$, $k = 1, 2$.

For $\Theta_k \equiv \text{const}$, formula (2) gives particular solutions
$$y(x) = C_1\lambda_1^x + C_2\lambda_2^x,$$
where $C_1$ and $C_2$ are arbitrary constants.

$2°$. The quadratic equation (1) has equal roots: $\lambda = \lambda_1 = \lambda_2$. In this case, the general solution of the original functional equation is given by
$$y = \big[\Theta_1(x) + x\Theta_2(x)\big]\lambda^x.$$

$3°$. In the case of complex conjugate roots, $\lambda = \rho(\cos\beta \pm i\sin\beta)$, the general solution of the original functional equation is expressed as
$$y = \Theta_1(x)\rho^x \cos(\beta x) + \Theta_2(x)\rho^x \sin(\beta x),$$
where $\Theta_1(x)$ and $\Theta_2(x)$ are arbitrary periodic functions with period 1.

4. $y(x + 2) + ay(x + 1) + by(x) = f(x)$.

*A nonhomogeneous second-order constant-coefficient linear difference equation.*

$1°$. Solution:
$$y(x) = Y(x) + \bar{y}(x),$$
where $Y(x)$ is the general solution of the corresponding homogeneous equation $Y(x+2) + aY(x+1) + bY(x) = 0$ (see the previous equation), and $\bar{y}(x)$ is any particular solution of the nonhomogeneous equation.

$2°$. For $f(x) = \sum_{k=0}^{n} A_k x^n$ and $a + b + 1 \neq 1$, the nonhomogeneous equation has a particular solution of the form $\bar{y}(x) = \sum_{k=0}^{n} B_k x^n$; the constants $B_k$ are found by the method of undetermined coefficients.

$3°$. For $f(x) = \sum_{k=1}^{n} A_k \exp(\lambda_k x)$, the nonhomogeneous equation has a particular solution of the form $\bar{y}(x) = \sum_{k=1}^{n} B_k \exp(\lambda_k x)$; the constants $B_k$ are found by the method of undetermined coefficients.

$4^\circ$. For $f(x) = \sum_{k=1}^{n} A_k \cos(\lambda_k x)$, the nonhomogeneous equation has a particular solution of the form $\bar{y}(x) = \sum_{k=1}^{n} B_k \cos(\lambda_k x) + \sum_{k=1}^{n} D_k \sin(\lambda_k x)$; the constants $B_k$ and $D_k$ are found by the method of undetermined coefficients.

$5^\circ$. For $f(x) = \sum_{k=1}^{n} A_k \sin(\lambda_k x)$, the nonhomogeneous equation has a particular solution of the form $\bar{y}(x) = \sum_{k=1}^{n} B_k \cos(\lambda_k x) + \sum_{k=1}^{n} D_k \sin(\lambda_k x)$; the constants $B_k$ and $D_k$ are found by the method of undetermined coefficients.

**5.** $y(x+2) + a(x+1)y(x+1) + bx(x+1)y(x) = 0.$

This functional equation has particular solutions of the form

$$y(x; \lambda) = \int_0^\infty t^{x-1} e^{-t/\lambda} \, dt, \tag{1}$$

where $\lambda$ is a root of the square equation

$$\lambda^2 + a\lambda + b = 0. \tag{2}$$

For the integral on the right-hand side of (1) to converge, the roots of equation (2) that satisfy the condition $\operatorname{Re} \lambda > 0$ should be selected. If both roots, $\lambda_1$ and $\lambda_2$, meet this condition, the general solution of the original functional equation is expressed as

$$y(x) = \Theta_1(x) y(x, \lambda_1) + \Theta_2(x) y(x, \lambda_2),$$

where $\Theta_1(x)$ and $\Theta_2(x)$ are arbitrary periodic functions with period 1.

---

**T12.1.2-2. Linear equations involving composite functions $y(y(x))$ or $y(y(y(x)))$.**

**6.** $y(y(x)) = 0.$

Solution:

$$y(x) = \begin{cases} \varphi_1(x) & \text{for } x \leq a, \\ 0 & \text{for } a \leq x \leq b, \\ \varphi_2(x) & \text{for } b \leq x, \end{cases}$$

where $a \leq 0$ and $b \geq 0$ are arbitrary numbers; $\varphi_1(x)$ and $\varphi_2(x)$ are arbitrary continuous functions satisfying the conditions

$$\varphi_1(a) = 0, \quad a \leq \varphi_1(x) \leq b \quad \text{if} \quad x \leq a;$$
$$\varphi_2(b) = 0, \quad a \leq \varphi_2(x) \leq b \quad \text{if} \quad b \leq x.$$

**7.** $y(y(x)) - x = 0.$

*Babbage equation* or the *equation of involutory functions*. It is a special case of equation T12.1.2.21.

$1^\circ$. Particular solutions:

$$y_1(x) = x, \qquad y_2(x) = C - x, \qquad y_3(x) = \frac{C}{x}, \qquad y_4(x) = \frac{C_1 - x}{1 + C_2 x},$$

where $C, C_1, C_2$ are arbitrary constants.

2°. On the interval $x \in (a,b)$, there exists a decreasing solution of the original equation involving an arbitrary function:
$$y(x) = \begin{cases} \varphi(x) & \text{for } x \in (a,c], \\ \varphi^{-1}(x) & \text{for } x \in (c,b), \end{cases}$$
where $c$ is an arbitrary point belonging to the interval $(a,b)$, and $\varphi(x)$ is an arbitrary continuous decreasing function on $(a,c]$ such that
$$\lim_{x \to a+0} \varphi(x) = b, \quad \varphi(c) = c.$$

3°. Solution in parametric form:
$$x = \Theta\left(\frac{t}{2}\right), \quad y = \Theta\left(\frac{t+1}{2}\right),$$
where $\Theta(t) = \Theta(t+1)$ is an arbitrary periodic function with period 1.

4°. Solution in parametric form:
$$x = \Theta_1(t) + \Theta_2(t)\sin(\pi t),$$
$$y = \Theta_1(t) - \Theta_2(t)\sin(\pi t),$$
where $\Theta_1(t)$ and $\Theta_2(t)$ are arbitrary periodic functions with period 1.

5°. The original functional equation has a single increasing solution: $y(x) = x$.

6°. Particular solutions of the equation may be represented in implicit form using the algebraic (or transcendental) equation
$$\Phi(x,y) = 0,$$
where $\Phi(x,y) = \Phi(y,x)$ is some symmetric function with two arguments.

**8.** $y(y(x)) + ay(x) + bx = 0.$

1°. General solution in parametric form:
$$x = \Theta_1(t)\lambda_1^t + \Theta_2(t)\lambda_2^t,$$
$$y = \Theta_1(t)\lambda_1^{t+1} + \Theta_2(t)\lambda_2^{t+1},$$
where $\lambda_1$ and $\lambda_2$ are roots of the quadratic equation
$$\lambda^2 + a\lambda + b = 0$$
and $\Theta_1 = \Theta_1(t)$ and $\Theta_2 = \Theta_2(t)$ are arbitrary periodic functions with period 1.

2°. For $\Theta_1(t) = C_1 = \text{const}$ and $\Theta_2(t) = C_2 = \text{const}$, we have a particular solution in implicit form
$$\frac{\lambda_2 x - y(x)}{\lambda_2 - \lambda_1} = C_1 \left[\frac{\lambda_1 x - y(x)}{C_2(\lambda_1 - \lambda_2)}\right]^k, \quad k = \frac{\ln \lambda_1}{\ln \lambda_2}.$$

**9.** $y(y(y(x))) - x = 0.$

A special case of equation T12.1.2.21.

1°. Particular solutions:
$$y_1(x) = -\frac{C^2}{C+x}, \quad y_2(x) = C - \frac{C^2}{x}, \quad y_3(x) = C_1 - \frac{(C_1+C_2)^2}{C_2+x},$$
where $C, C_1, C_2$ are arbitrary constants.

2°. Solution in parametric form:
$$x = \Theta\left(\frac{t}{3}\right), \quad y = \Theta\left(\frac{t+1}{3}\right),$$
where $\Theta(t) = \Theta(t+1)$ is an arbitrary periodic function with period 1.

**T12.1.2-3. Equations involving unknown function with three different arguments.**

**10.** $Ay(ax) + By(bx) + y(x) = 0.$

This functional equation has particular solutions of the form $y(x) = Cx^\beta$, where $C$ is an arbitrary constant and $\beta$ is a root of the transcendental equation $Aa^\beta + Bb^\beta + 1 = 0$.

**11.** $Ay(x^a) + By(x^b) + y(x) = 0.$

This functional equation has particular solutions of the form $y(x) = C|\ln x|^p$, where $C$ is an arbitrary constant and $p$ is a root of the transcendental equation $A|a|^p + B|b|^p + 1 = 0$.

**12.** $Ay(x) + By\left(\dfrac{ax - \beta}{x + b}\right) + Cy\left(\dfrac{bx + \beta}{a - x}\right) = f(x), \qquad \beta = a^2 + ab + b^2.$

Let us substitute $x$ in the equation first by $\dfrac{ax - \beta}{x + b}$ and then by $\dfrac{bx + \beta}{a - x}$ to obtain two more equations. So we get the system (the original equation is given first)

$$\begin{aligned} Ay(x) + By(u) + Cy(w) &= f(x), \\ Ay(u) + By(w) + Cy(x) &= f(u), \\ Ay(w) + By(x) + Cy(u) &= f(w), \end{aligned} \qquad (1)$$

where $u = \dfrac{ax - \beta}{x + b}$ and $w = \dfrac{bx + \beta}{a - x}$.

Eliminating $y(u)$ and $y(w)$ from the system of linear algebraic equations (1) yields the solution of the original functional equation.

**13.** $f_1(x)y(x) + f_2(x)y\left(\dfrac{ax - \beta}{x + b}\right) + f_3(x)y\left(\dfrac{bx + \beta}{a - x}\right) = g(x), \qquad \beta = a^2 + ab + b^2.$

Let us substitute $x$ in the equation first by $\dfrac{ax - \beta}{x + b}$ and then by $\dfrac{bx + \beta}{a - x}$ to obtain two more equations. So we get the system (the original equation is given first)

$$\begin{aligned} f_1(x)y(x) + f_2(x)y(u) + f_3(x)y(w) &= g(x), \\ f_1(u)y(u) + f_2(u)y(w) + f_3(u)y(x) &= g(u), \\ f_1(w)y(w) + f_2(w)y(x) + f_3(w)y(u) &= g(w), \end{aligned} \qquad (1)$$

where
$$u = \frac{ax - \beta}{x + b}, \qquad w = \frac{bx + \beta}{a - x}.$$

Eliminating $y(u)$ and $y(w)$ from the system of linear algebraic equations (1) yields the solution $y = y(x)$ of the original functional equation.

**T12.1.2-4. Higher-order linear difference equations.**

**14.** $y_{n+m} + a_{m-1}y_{n+m-1} + \cdots + a_1 y_{n+1} + a_0 y_n = 0.$

A *homogeneous $m$th-order linear difference equation* defined on a discrete set of points $x = 0, 1, 2, \ldots$ Notation adopted: $y_n = y(n)$.

Let $\lambda_1, \lambda_2, \ldots, \lambda_m$ be roots of the characteristic equation

$$P(\lambda) \equiv \lambda^m + a_{m-1}\lambda^{m-1} + \cdots + a_1\lambda + a_0 = 0. \tag{1}$$

If all the roots of the characteristic equation (1) are distinct, then the general solution of the original difference equation has the form

$$y_n = \sum_{i=0}^{m-1} y_i \sum_{j=0}^{m-i-1} a_{i+j+1} \sum_{k=1}^{m} \frac{\lambda_k^{n+1}}{P'(\lambda_k)}, \tag{2}$$

where the prime denotes a derivative.

Formula (2) involves the initial values $y_0, y_1, \ldots, y_m$. They can be set arbitrarily.

In the case of complex conjugate roots, one should separate the real and imaginary parts in solution (2).

**15.** $y_{n+m} + a_{m-1}y_{n+m-1} + \cdots + a_1 y_{n+1} + a_0 y_n = f_n.$

*A nonhomogeneous mth-order linear difference equation* defined on a discrete set of points $x = 0, 1, 2, \ldots$ Notation adopted: $y_n = y(n)$.

The general solution of the difference equation has the form $y(x) = Y(x) + \bar{y}(x)$, where $Y(x)$ is the general solution of the corresponding homogeneous equation (with $f_n \equiv 0$) and $\bar{y}(x)$ is any particular solution of the nonhomogeneous equation.

Let $\lambda_1, \lambda_2, \ldots, \lambda_m$ be roots of the characteristic equation

$$P(\lambda) \equiv \lambda^m + a_{m-1}\lambda^{m-1} + \cdots + a_1\lambda + a_0 = 0. \tag{1}$$

If all the roots of the characteristic equation (1) are distinct, then the general solution of the original difference equation has the form

$$y_n = \sum_{i=0}^{m-1} y_i \sum_{j=0}^{m-i-1} a_{i+j+1} \sum_{k=1}^{m} \frac{\lambda_k^{n+1}}{P'(\lambda_k)} + \sum_{\nu=m}^{n} f_{n-\nu} \sum_{k=1}^{m} \frac{\lambda_k^{\nu-1}}{P'(\lambda_k)}, \tag{2}$$

where the prime denotes a derivative.

Formula (2) involves the initial values $y_0, y_1, \ldots, y_m$. They can be set arbitrarily.

In the case of complex conjugate roots, one should separate the real and imaginary parts in solution (2).

**16.** $y(x+n) + a_{n-1}y(x+n-1) + \cdots + a_1 y(x+1) + a_0 y(x) = 0.$

*A homogeneous nth-order constant-coefficient linear difference equation.*

Let us write out the characteristic equation:

$$\lambda^n + a_{n-1}\lambda^{n-1} + \cdots + a_1\lambda + a_0 = 0. \tag{1}$$

Consider the following cases.

$1°$. All roots $\lambda_1, \lambda_2, \ldots, \lambda_n$ of equation (1) are real and distinct. Then the general solution of the original finite-difference equation has the form

$$y(x) = \Theta_1(x)\lambda_1^x + \Theta_2(x)\lambda_2^x + \cdots + \Theta_n(x)\lambda_n^x, \tag{2}$$

where $\Theta_1(x), \Theta_2(x), \ldots, \Theta_n(x)$ are arbitrary periodic functions with period 1, which means that $\Theta_k(x) = \Theta_k(x+1)$, $k = 1, 2, \ldots, n$.

For $\Theta_k(x) \equiv C_k$, formula (2) gives a particular solution

$$y(x) = C_1\lambda_1^x + C_2\lambda_2^x + \cdots + C_n\lambda_n^x,$$

where $C_1, C_2, \ldots, C_n$ are arbitrary constants.

2°. There are $m$ equal real roots, $\lambda_1 = \lambda_2 = \cdots = \lambda_m$ ($m \leq n$), the other roots real and distinct. In this case, the general solution of the functional equation is expressed as

$$y = \left[\Theta_1(x) + x\Theta_2(x) + \cdots + x^{m-1}\Theta_m(x)\right]\lambda_1^x + \Theta_{m+1}(x)\lambda_{m+1}^x + \Theta_{m+2}(x)\lambda_{m+2}^x + \cdots + \Theta_n(x)\lambda_n^x.$$

3°. There are $m$ equal complex conjugate roots, $\lambda = \rho(\cos\beta \pm i\sin\beta)$ ($2m \leq n$), the other roots real and distinct. Then the original functional equation has a solution corresponding to $\Theta_n(x) \equiv \mathrm{const}_k$:

$$\begin{aligned} y &= \rho^x \cos(\beta x)(A_1 + A_2 x + \cdots + A_m x^{m-1}) \\ &\quad + \rho^x \sin(\beta x)(B_1 + B_2 x + \cdots + B_m x^{m-1}) \\ &\quad + C_{m+1}\lambda_{m+1}^x + C_{m+2}\lambda_{m+2}^x + \cdots + C_n \lambda_n^x, \end{aligned}$$

where $A_1, \ldots, A_m, B_1, \ldots, B_m, C_{2m+1}, \ldots, C_n$ are arbitrary constants.

**17.** $y(x+n) + a_{n-1}y(x+n-1) + \cdots + a_1 y(x+1) + a_0 y(x) = f(x)$.

*A nonhomogeneous $n$th-order constant-coefficient linear difference equation.*

1°. Solution:
$$y(x) = Y(x) + \bar{y}(x),$$

where $Y(x)$ is the general solution of the corresponding homogeneous equation

$$Y(x+n) + a_{n-1}Y(x+n-1) + \cdots + a_1 Y(x+1) + a_0 Y(x) = 0$$

(see the previous equation), and $\bar{y}(x)$ is any particular solution of the nonhomogeneous equation.

2°. For $f(x) = \sum_{k=0}^{n} A_k x^n$, the nonhomogeneous equation has a particular solution of the form $\bar{y}(x) = \sum_{k=0}^{n} B_k x^n$; the constants $B_k$ are found by the method of undetermined coefficients.

3°. For $f(x) = \sum_{k=1}^{n} A_k \exp(\lambda_k x)$, the nonhomogeneous equation has a particular solution of the form $\bar{y}(x) = \sum_{k=1}^{n} B_k \exp(\lambda_k x)$; the constants $B_k$ are found by the method of undetermined coefficients.

4°. For $f(x) = \sum_{k=1}^{n} A_k \cos(\lambda_k x)$, the nonhomogeneous equation has a particular solution of the form $\bar{y}(x) = \sum_{k=1}^{n} B_k \cos(\lambda_k x) + \sum_{k=1}^{n} D_k \sin(\lambda_k x)$; the constants $B_k$ and $D_k$ are found by the method of undetermined coefficients.

5°. For $f(x) = \sum_{k=1}^{n} A_k \sin(\lambda_k x)$, the nonhomogeneous equation has a particular solution of the form $\bar{y}(x) = \sum_{k=1}^{n} B_k \cos(\lambda_k x) + \sum_{k=1}^{n} D_k \sin(\lambda_k x)$; the constants $B_k$ and $D_k$ are found by the method of undetermined coefficients.

**18.** $y(x+b_n) + a_{n-1}y(x+b_{n-1}) + \cdots + a_1 y(x+b_1) + a_0 y(x) = 0$.

There are particular solutions of the form $y(x) = \lambda_k^x$, where $\lambda_k$ are roots of the transcendental (or algebraic) equation

$$\lambda^{b_n} + a_{n-1}\lambda^{b_{n-1}} + \cdots + a_1 \lambda^{b_1} + a_0 = 0.$$

> T12.1.2-5. Equations involving unknown function with many different arguments.

**19.** $y(a_n x) + b_{n-1} y(a_{n-1} x) + \cdots + b_1 y(a_1 x) + b_0 y(x) = 0.$

This functional equation has particular solutions of the form $y = Cx^\beta$, where $C$ is an arbitrary constant and $\beta$ is a root of the transcendental equation

$$a_n^\beta + b_{n-1} a_{n-1}^\beta + \cdots + b_1 a_1^\beta + b_0 = 0.$$

**20.** $y(x^{a_n}) + b_{n-1} y(x^{a_{n-1}}) + \cdots + b_1 y(x^{a_1}) + b_0 y(x) = 0.$

This functional equation has particular solutions of the form $y(x) = C|\ln x|^p$, where $C$ is an arbitrary constant and $p$ is a root of the transcendental equation

$$|a_n|^p + b_{n-1}|a_{n-1}|^p + \cdots + b_1 |a_1|^p + b_0 = 0.$$

**21.** $y^{[n]}(x) + a_{n-1} y^{[n-1]}(x) + \cdots + a_1 y(x) + a_0 x = 0.$

Notation used: $y^{[2]}(x) = y(y(x)), \ldots, y^{[n]}(x) = y(y^{[n-1]}(x))$.

1°. Solutions are sought in the parametric form

$$x = w(t), \quad y = w(t+1).$$

With it, the original equation is reduced to an $n$th-order linear finite-difference equation (see equation 16 above):

$$w(t+n) + a_{n-1} w(t+n-1) + \cdots + a_1 w(t+1) + a_0 w(t) = 0.$$

2°. In the special case $a_{n-1} = \ldots = a_1 = 0$ and $a_0 = -1$, we have the following solution in parametric form:

$$x = \Theta\left(\frac{t}{n}\right), \quad y = \Theta\left(\frac{t+1}{n}\right),$$

where $\Theta(t) = \Theta(t+1)$ is an arbitrary periodic function with period 1.

## T12.2. Nonlinear Functional Equations in One Independent Variable

### T12.2.1. Functional Equations with Quadratic Nonlinearity

> T12.2.1-1. Difference equations.

**1.** $y_n y_{n+1} = a_n y_{n+1} + b_n y_n + c_n.$

*Riccati difference equation.* Here, $n = 0, 1, \ldots$ and the constants $a_n, b_n, c_n$ satisfy the condition $a_n b_n + c_n \neq 0$.

1°. The substitution

$$y_n = \frac{u_{n+1}}{u_n} + a_n$$

leads to the linear second-order difference equation of the form T12.1.2.1:

$$u_{n+2} + (a_{n+1} - b_n) u_{n+1} - (a_n b_n + c_n) u_n = 0.$$

**2°**. Let $y_n^*$ be a particular solution of the Riccati difference equation. Then the substitution

$$z_n = \frac{1}{y_n - y_n^*}, \qquad n = 0, 1, \dots$$

reduces this equation to the nonhomogeneous first-order linear difference equation

$$z_{n+1} + \frac{(y_n^* - a_n)^2}{a_n b_n + c_n} z_n + \frac{y_n^* - a_n}{a_n b_n + c_n} = 0.$$

About this equation, see Paragraph 17.1.1-2.

**2.** $y(x+1) - ay^2(x) = f(x)$.

A special case of equation T12.2.3.1.

**3.** $y(x)y(x+1) + a[y(x+1) - y(x)] = 0$.

Solution:

$$y(x) = \frac{a}{x + \Theta(x)},$$

where $\Theta(x) = \Theta(x+1)$ is an arbitrary periodic function with period 1.

**4.** $y(x)y(x+1) = a(x)y(x+1) + b(x)y(x) + c(x)$.

*Riccati difference equation.* Here, the functions $a(x)$, $b(x)$, $c(x)$ satisfy the condition $a(x)b(x) + c(x) \not\equiv 0$.

**1°**. The substitution

$$y(x) = \frac{u(x+1)}{u(x)} + a(x)$$

leads to the linear second-order difference equation

$$u(x+2) + [a(x+1) - b(x)]u(x+1) - [a(x)b(x) + c(x)]u(x) = 0.$$

**2°**. Let $y_0(x)$ be a particular solution of Riccati difference equation. Then the substitution

$$z(x) = \frac{1}{y(x) - y_0(x)}$$

reduces this equation to the nonhomogeneous first-order linear difference equation

$$z(x+1) + \frac{[y_0(x) - a(x)]^2}{a(x)b(x) + c(x)} z(x) + \frac{y_0(x) - a(x)}{a(x)b(x) + c(x)} = 0.$$

### T12.2.1-2. Functional equations involving $y(x)$ and $y(a-x)$.

**5.** $y(x)y(a-x) = b^2$.

Solutions:

$$y(x) = \pm b \exp[\Phi(x, a-x)],$$

where $\Phi(x, z) = -\Phi(z, x)$ is any antisymmetric function with two arguments.

On setting $\Phi(x, z) = C(x - z)$, we arrive at particular solutions of the form

$$y(x) = \pm b e^{C(2x-a)},$$

where $C$ is an arbitrary constant.

**6.** $y(x)y(a-x) = -b^2$.

Discontinuous solutions:

$$\pm y(x) = \begin{cases} b\varphi(x) & \text{if } x \geq a/2, \\ -\dfrac{b}{\varphi(a-x)} & \text{if } x < a/2, \end{cases}$$

where $\varphi(x)$ is an arbitrary function. There are no continuous solutions.

**7.** $y(x)y(a-x) = f^2(x)$.

The function $f(x)$ must satisfy the condition $f(x) = \pm f(a-x)$.

1°. The change of variable $y(x) = f(x)u(x)$ leads to one of the equations of the form T12.2.1.5 or T12.2.1.6.

2°. For $f(x) = f(a-x)$, there are solutions of the form

$$y(x) = \pm f(x)\exp[\Phi(x, a-x)],$$

where $\Phi(x, z) = -\Phi(z, x)$ is any antisymmetric function with two arguments.
On setting $\Phi(x, z) = C(x-z)$, we arrive at particular solutions

$$y(x) = \pm e^{C(2x-a)} f(x),$$

where $C$ is an arbitrary constant.

**8.** $y^2(x) + y^2(a-x) = b^2$.

1°. Solutions:

$$y(x) = \pm\sqrt{\tfrac{1}{2}b^2 + \Phi(x, a-x)},$$

where $\Phi(x, z) = -\Phi(z, x)$ is any antisymmetric function of two arguments.

2°. Particular solutions:

$$y_{1,2}(x) = \pm\frac{b}{\sqrt{2}}, \quad y_{3,4}(x) = \pm b\sin\left(\frac{\pi x}{2a}\right), \quad y_{5,6}(x) = \pm b\cos\left(\frac{\pi x}{2a}\right).$$

**9.** $y^2(x) + Ay(x)y(a-x) + By^2(a-x) + Cy(x) + Dy(a-x) = f(x)$.

A special case of equation T12.2.3.2.
  Solution in parametric form ($w$ is a parameter):

$$y^2 + Ayw + Bw^2 + Cy + Dw = f(x),$$
$$w^2 + Ayw + By^2 + Cw + Dy = f(a-x).$$

Eliminating $w$ gives the solutions in implicit form.

## T12.2.1-3. Functional equations involving $y(x)$ and $y(ax)$.

**10.** $y(2x) - ay^2(x) = 0.$

A special case of equation T12.2.3.3.
    Particular solution:
$$y(x) = \frac{1}{a}e^{Cx},$$
where $C$ is an arbitrary constant.

**11.** $y(2x) - 2y^2(x) + a = 0.$

A special case of equation T12.2.3.3.
    Particular solutions with $a = 1$:
$$y(x) = 0, \quad y(x) = \cos(Cx), \quad y(x) = \cosh(Cx),$$
where $C$ is an arbitrary constant.

**12.** $y(x)y(ax) = f(x).$

A special case of equation T12.2.3.3.

## T12.2.1-4. Functional equations involving $y(x)$ and $y(a/x)$.

**13.** $y(x)y(a/x) = b^2.$

Solution:
$$y(x) = \pm b \exp\bigl[\Phi(x, a/x)\bigr],$$
where $\Phi(x, z) = -\Phi(z, x)$ is any antisymmetric function with two arguments.
    On setting $\Phi(x, z) = C(\ln x - \ln z)$, one arrives at particular solutions of the form
$$y = \pm b a^{-C} x^{2C},$$
where $C$ is an arbitrary constant.

**14.** $y(x)y(a/x) = f^2(x).$

The function $f(x)$ must satisfy the condition $f(x) = \pm f(a/x)$. For definiteness, we take $f(x) = f(a/x)$.
    Solution:
$$y(x) = \pm f(x) \exp\bigl[\Phi(x, a/x)\bigr],$$
where $\Phi(x, z) = -\Phi(z, x)$ is any antisymmetric function with two arguments.
    On setting $\Phi(x, z) = C(\ln x - \ln z)$, one arrives at particular solutions of the form
$$y = \pm a^{-C} x^{2C} f(x),$$
where $C$ is an arbitrary constant.

**15.** $y^2(x) + Ay(x)y(a/x) + By^2(a/x) + Cy(x) + Dy(a/x) = f(x).$

A special case of equation T12.2.3.4.
    Solution in parametric form ($w$ is a parameter):
$$y^2 + Ayw + Bw^2 + Cy + Dw = f(x),$$
$$w^2 + Ayw + By^2 + Cw + Dy = f(a/x).$$

Eliminating $w$ gives the solutions in implicit form.

T12.2.1-5. Other functional equations with quadratic nonlinearity.

**16.** $y(x^2) - ay^2(x) = 0.$

Solution:
$$y(x) = \frac{1}{a}x^C,$$
where $C$ is an arbitrary constant. In addition, $y(x) \equiv 0$ is also a continuous solution.

**17.** $y(x)y(x^a) = f(x), \qquad a > 0.$

A special case of equation T12.2.3.8.

**18.** $y(x)y\left(\dfrac{a-x}{1+bx}\right) = A^2.$

Solutions:
$$y(x) = \pm A \exp\left[\Phi\left(x, \frac{a-x}{1+bx}\right)\right],$$
where $\Phi(x, z) = -\Phi(z, x)$ is any antisymmetric function with two arguments.

**19.** $y(x)y\left(\dfrac{a-x}{1+bx}\right) = f^2(x).$

The right-hand side function must satisfy the condition $f(x) = \pm f\left(\dfrac{a-x}{1+bx}\right)$. For definiteness, we take $f(x) = f\left(\dfrac{a-x}{1+bx}\right).$

Solutions:
$$y(x) = \pm f(x)\exp\left[\Phi\left(x, \frac{a-x}{1+bx}\right)\right],$$
where $\Phi(x, z) = -\Phi(z, x)$ is any antisymmetric function with two arguments.

**20.** $y^2(x) + Ay(x)y\left(\dfrac{a-x}{1+bx}\right) + By(x) = f(x).$

A special case of equation T12.2.3.5.

**21.** $y(x)y(\sqrt{a^2 - x^2}) = b^2, \qquad 0 \le x \le a.$

Solutions:
$$y(x) = \pm b \exp\left[\Phi(x, \sqrt{a^2 - x^2})\right],$$
where $\Phi(x, z) = -\Phi(z, x)$ is any antisymmetric function with two arguments.

**22.** $y(x)y(\sqrt{a^2 - x^2}) = f^2(x), \qquad 0 \le x \le a.$

The right-hand side function must satisfy the condition $f(x) = \pm f(\sqrt{a^2 - x^2})$. For definiteness, we take $f(x) = f(\sqrt{a^2 - x^2}).$

Solutions:
$$y(x) = \pm f(x)\exp\left[\Phi(x, \sqrt{a^2 - x^2})\right],$$
where $\Phi(x, z) = -\Phi(z, x)$ is any antisymmetric function with two arguments.

**23.** $y(\sin x)y(\cos x) = a^2$.

Solutions in implicit form:
$$y(\sin x) = \pm a \exp\bigl[\Phi(\sin x, \cos x)\bigr],$$
where $\Phi(x, z) = -\Phi(z, x)$ is any antisymmetric function with two arguments.

**24.** $y(\sin x)y(\cos x) = f^2(x)$.

The right-hand side function must satisfy the condition $f(x) = \pm f\bigl(\frac{\pi}{2} - x\bigr)$. For definiteness, we take $f(x) = f\bigl(\frac{\pi}{2} - x\bigr)$.

Solutions in implicit form:
$$y(\sin x) = \pm f(x) \exp\bigl[\Phi(\sin x, \cos x)\bigr],$$
where $\Phi(x, z) = -\Phi(z, x)$ is any antisymmetric function with two arguments.

**25.** $y(x)y\bigl(\omega(x)\bigr) = b^2$, where $\omega\bigl(\omega(x)\bigr) = x$.

Solutions:
$$y(x) = \pm b \exp\bigl[\Phi\bigl(x, \omega(x)\bigr)\bigr],$$
where $\Phi(x, z) = -\Phi(z, x)$ is any antisymmetric function with two arguments.

**26.** $y(x)y\bigl(\omega(x)\bigr) = f^2(x)$, where $\omega\bigl(\omega(x)\bigr) = x$.

The right-hand side function must satisfy the condition $f(x) = \pm f\bigl(\omega(x)\bigr)$. For definiteness, we take $f(x) = f\bigl(\omega(x)\bigr)$.

Solutions:
$$y(x) = \pm f(x) \exp\bigl[\Phi\bigl(x, \omega(x)\bigr)\bigr],$$
where $\Phi(x, z) = -\Phi(z, x)$ is any antisymmetric function with two arguments.

### T12.2.2. Functional Equations with Power Nonlinearity

**1.** $y(x + a) - by^\lambda(x) = f(x)$.

A special case of equation T12.2.3.1.

**2.** $y^\lambda(x)y(a - x) = f(x)$.

A special case of equation T12.2.3.2.
   Solution:
$$y(x) = \bigl[f(x)\bigr]^{-\frac{\lambda}{1-\lambda^2}} \bigl[f(a-x)\bigr]^{\frac{1}{1-\lambda^2}}.$$

**3.** $y^{2n+1}(x) + y^{2n+1}(a - x) = b$,   $n = 1, 2, \ldots$

The change of variable $w(x) = y^{2n+1}(x)$ leads to a linear equation of the form T12.1.1.22: $w(x) + w(a - x) = b$.

**4.** $y(ax) = by^k(x)$.

Solution for $x > 0$, $a > 0$, $b > 0$, and $k > 0$ ($a \neq 1$):

$$y(x) = b^{\frac{1}{1-k}} \exp\left[x^{\frac{\ln k}{\ln a}} \Theta\left(\frac{\ln x}{\ln a}\right)\right],$$

where $\Theta(x) = \Theta(x+1)$ is an arbitrary periodic function with period 1.

**5.** $y^\lambda(x) y(a/x) = f(x)$.

A special case of equation T12.2.3.4.
   Solution:

$$y(x) = [f(x)]^{-\frac{\lambda}{1-\lambda^2}} [f(a/x)]^{\frac{1}{1-\lambda^2}}.$$

**6.** $y^\lambda(x) y\left(\dfrac{a-x}{1+bx}\right) = f(x)$.

A special case of equation T12.2.3.5.

**7.** $y^\lambda(x) y\left(\dfrac{ax-\beta}{x+b}\right) = f(x)$, $\quad \beta = a^2 + ab + b^2$.

A special case of equation T12.2.3.13.

**8.** $y^\lambda(x) y\left(\dfrac{bx+\beta}{a-x}\right) = f(x)$, $\quad \beta = a^2 + ab + b^2$.

A special case of equation T12.2.3.13.

**9.** $y^\lambda(x) y(x^a) = f(x)$.

A special case of equation T12.2.3.8.

**10.** $y^\lambda(x) y(\sqrt{a^2 - x^2}) = f(x)$.

A special case of equation T12.2.3.9.

**11.** $y^\lambda(\sin x) y(\cos x) = f(x)$.

A special case of equation T12.2.3.10.

## T12.2.3. Nonlinear Functional Equation of General Form

**1.** $F(x, y(x), y(x+a)) = 0$.

We assume that $a > 0$. Let us solve the equation for $y(x+a)$ to obtain

$$y(x+a) = f(x, y(x)). \tag{1}$$

1°. First, let us assume that the equation is defined on a discrete set of points $x = x_0 + ak$ with integer $k$. Given an initial value $y(x_0)$, one can make use of (1) to find sequentially $y(x_0 + a)$, $y(x_0 + 2a)$, etc.

Solving the original equation for $y(x)$ yields

$$y(x) = g(x, y(x+a)). \tag{2}$$

On setting $x = x_0 - a$ here, one can find $y(x_0 - a)$ and then likewise $y(x_0 - 2a)$ etc.

Thus, given initial data, one can use the equation to find $y(x)$ at all points $x_0 + ak$, where $k = 0, \pm 1, \pm 2, \ldots$

2°. Now assume that $x$ in the equation can vary continuously. Also assume that $y(x)$ is a continuous function defined arbitrarily on the semi-interval $[0, a)$. Setting $x = 0$ in (1), one finds $y(a)$. Then, given $y(x)$ on $[0, a]$, one can use (1) to find $y(x)$ on $x \in [a, 2a]$, then on $x \in [2a, 3a]$, and so on.

*Remark.* The case of $a < 0$ can be reduced, using the change of variable $z = x + a$, to an equation of the form $F(z + b, y(z + b), y(z)) = 0$ with $b = -a > 0$, which was already considered above.

**2.** $F(x, y(x), y(a - x)) = 0.$

Substituting $x$ with $a - x$, one obtains $F(a - x, y(a - x), y(x)) = 0$. Now, eliminating $y(a - x)$ from this equation and the original one, one arrives at an ordinary algebraic (or transcendental) equation of the form $\Psi(x, y(x)) = 0$.

To put it differently, the solution $y = y(x)$ of the original functional equation is determined parametrically with the system of two algebraic (or transcendental) equations

$$F(x, y, t) = 0, \qquad F(a - x, t, y) = 0,$$

where $t$ is a parameter.

**3.** $F(x, y(x), y(ax)) = 0, \qquad a > 0.$

The transformation $z = \ln x$, $w(z) = y(x)$ leads to an equation of the form T12.2.3.1:

$$F(e^z, w(z), w(z + b)) = 0, \qquad b = \ln a.$$

**4.** $F(x, y(x), y(a/x)) = 0.$

Substituting $x$ with $a/x$ yields $F(a/x, y(a/x), y(x)) = 0$. Eliminating $y(a/x)$ from this equation and the original one, we arrive at an ordinary algebraic (or transcendental) equation of the form $\Psi(x, y(x)) = 0$.

To put it differently, the solution $y = y(x)$ of the original functional equation is determined parametrically with the system of two algebraic (or transcendental) equations

$$F(x, y, t) = 0, \quad F(a/x, t, y) = 0,$$

where $t$ is the parameter.

**5.** $F\left(x, y(x), y\left(\dfrac{a - x}{1 + bx}\right)\right) = 0.$

Substituting $x$ with $\dfrac{a - x}{1 + bx}$ yields

$$F\left(\frac{a - x}{1 + bx}, y\left(\frac{a - x}{1 + bx}\right), y(x)\right) = 0.$$

Eliminating $y\left(\dfrac{a - x}{1 + bx}\right)$ from this equation and the original one, we arrive at an ordinary algebraic (or transcendental) equation of the form $\Psi(x, y(x)) = 0$.

In other words, the solution $y = y(x)$ of the original functional equation is determined parametrically with the system of two algebraic (or transcendental) equations

$$F(x, y, t) = 0, \quad F\left(\frac{a - x}{1 + bx}, t, y\right) = 0,$$

where $t$ is the parameter.

6. $F\left(x,\, y(x),\, y\!\left(\dfrac{ax-\beta}{x+b}\right)\right) = 0$,    $\beta = a^2 + ab + b^2$.

A special case of equation T12.2.3.13 below.

7. $F\left(x,\, y(x),\, y\!\left(\dfrac{bx+\beta}{a-x}\right)\right) = 0$,    $\beta = a^2 + ab + b^2$.

A special case of equation T12.2.3.13 below.

8. $F(x,\, y(x),\, y(x^a)) = 0$.

The transformation $\xi = \ln x$, $u(\xi) = y(x)$ leads to an equation of the form 3 above:

$$F(e^\xi,\, u(\xi),\, u(a\xi)) = 0.$$

9. $F(x,\, y(x),\, y(\sqrt{a^2 - x^2})) = 0$,    $0 \le x \le a$.

Substituting $x$ with $\sqrt{a^2 - x^2}$ yields

$$F(\sqrt{a^2 - x^2},\, y(\sqrt{a^2 - x^2}),\, y(x)) = 0.$$

Eliminating $y(\sqrt{a^2 - x^2})$ from this equation and the original one, we arrive at an ordinary algebraic (or transcendental) equation of the form $\Psi(x, y(x)) = 0$.

In other words, the solution $y = y(x)$ of the original functional equation is determined parametrically with the system of two algebraic (or transcendental) equations

$$F(x,\, y,\, t) = 0, \qquad F(\sqrt{a^2 - x^2},\, t,\, y) = 0,$$

where $t$ is the parameter.

10. $F(x,\, y(\sin x),\, y(\cos x)) = 0$.

Substituting $x$ with $\frac{\pi}{2} - x$ yields $F(\frac{\pi}{2} - x,\, y(\cos x),\, y(\sin x)) = 0$. Eliminating $y(\cos x)$ from this equation and the original one, we arrive at an ordinary algebraic (or transcendental) equation of the form $\Psi(x, y(\sin x)) = 0$ for $y(\sin x)$.

11. $F(x,\, y(x),\, y(\omega(x))) = 0$,    where $\omega(\omega(x)) = x$.

Substituting $x$ with $\omega(x)$ yields $F(\omega(x),\, y(\omega(x)),\, y(x)) = 0$. Eliminating $y(\omega(x))$ from this equation and the original one, we arrive at an ordinary algebraic (or transcendental) equation of the form $\Psi(x, y(x)) = 0$.

Thus, the solution $y = y(x)$ of the original functional equation is determined parametrically with the system of two algebraic (or transcendental) equations

$$F(x,\, y,\, t) = 0, \qquad F(\omega(x),\, t,\, y) = 0,$$

where $t$ is the parameter.

12. $F(x,\, y(x),\, y(x+1),\, y(x+2)) = 0$.

A second-order nonlinear difference equation of general form. A special case of equation T12.2.3.14.

13. $F\left(x,\, y(x),\, y\!\left(\dfrac{ax-\beta}{x+b}\right),\, y\!\left(\dfrac{bx+\beta}{a-x}\right)\right) = 0, \qquad \beta = a^2 + ab + b^2.$

Let us substitute $x$ first with $\dfrac{ax-\beta}{x+b}$ and then with $\dfrac{bx+\beta}{a-x}$ to obtain two more equations. As a result, we get the following system (the original equation is given first):

$$\begin{aligned} F\big(x,\, y(x),\, y(u),\, y(w)\big) &= 0, \\ F\big(u,\, y(u),\, y(w),\, y(x)\big) &= 0, \\ F\big(w,\, y(w),\, y(x),\, y(u)\big) &= 0. \end{aligned} \qquad (1)$$

The arguments $u$ and $w$ are expressed in terms of $x$ as

$$u = \frac{ax-\beta}{x+b}, \qquad w = \frac{bx+\beta}{a-x}.$$

Eliminating $y(u)$ and $y(w)$ from the system of algebraic (transcendental) equations (1), we arrive at the solutions $y = y(x)$ of the original functional equation.

14. $F\big(x,\, y(x),\, y(x+1),\, \ldots,\, y(x+n)\big) = 0.$

*An $n$th-order nonlinear difference equation of general form.*

Let us solve the equation for $y(x+n)$ to obtain

$$y(x+n) = f\big(x,\, y(x),\, y(x+1),\, \ldots,\, y(x+n-1)\big). \qquad (1)$$

$1^\circ$. Let us assume that the equation is defined on a discrete set of points $x = x_0 + k$ with integer $k$. Given initial values $y(x_0),\, y(x_0+1),\, \ldots,\, y(x_0+n-1)$, one can make use of (1) to find sequentially $y(x_0+n),\, y(x_0+n+1)$, etc.

Solving the original equation for $y(x)$ gives

$$y(x) = g\big(x,\, y(x+1),\, y(x+2),\, \ldots,\, y(x+n)\big). \qquad (2)$$

On setting $x = x_0 - 1$ here, one can find $y(x_0-1)$, then likewise $y(x_0-2)$ etc.

Thus, given initial data, one can use the equation to find $y(x)$ at all points $x_0 + k$, where $k = 0, \pm 1, \pm 2, \ldots$

$2^\circ$. Now assume that $x$ in the equation can vary continuously. Also assume that $y(x)$ is a continuous function defined arbitrarily on the semi-interval $[0, n)$. Setting $x = 0$ in (1), one finds $y(n)$. Then, given $y(x)$ on $[0, n]$, one can use (1) to find $y(x)$ on $x \in [n, n+1]$, then on $x \in [n+1, n+2]$, and so on.

15. $F\big(x,\, y(x),\, y^{[2]}(x),\, \ldots,\, y^{[n]}(x)\big) = 0.$

Notation: $y^{[2]}(x) = y\big(y(x)\big),\, \ldots,\, y^{[n]}(x) = y\big(y^{[n-1]}(x)\big).$

Solutions are sought in the parametric form

$$x = w(t), \qquad y = w(t+1). \qquad (1)$$

Then the original equation is reduced to an $n$th-order difference equation (see the previous equation):

$$F\big(w(t),\, w(t+1),\, w(t+2),\, \ldots,\, w(t+n)\big) = 0. \qquad (2)$$

The general solution of equation (2) has the structure

$$\begin{aligned} x &= w(t) = \varphi(t;\, C_1, \ldots, C_n), \\ y &= w(t+1) = \varphi(t+1;\, C_1, \ldots, C_n), \end{aligned}$$

where $C_1 = C_1(t),\, \ldots,\, C_n = C_n(t)$ are arbitrary periodic functions with period 1, that is, $C_k(t) = C_k(t+1),\, k = 1, 2, \ldots, n$.

**16.** $F\bigl(x, y(\theta_0(x)), y(\theta_1(x)), \ldots, y(\theta_{n-1}(x))\bigr) = 0.$

Notation: $\theta_k(x) \equiv \theta\bigl(x + \frac{k}{n}T\bigr)$, where $k = 0, 1, \ldots, n-1$. The functions $\theta(x)$ are assumed to be periodic with period $T$, i.e., $\theta(x) = \theta(x+T)$. Furthermore, the left-hand side of the equation is assumed to satisfy the condition $F(x, \ldots) = F(x+T, \ldots)$.

In the original equation, let us substitute $x$ sequentially by $x + \frac{k}{n}T$ with $k = 0, 1, \ldots, n-1$ to obtain the following system (the original equation is given first):

$$\begin{aligned} F\bigl(x, y_0, y_1, \ldots, y_{n-1}\bigr) &= 0, \\ F\bigl(x + \tfrac{1}{n}T, y_1, y_2, \ldots, y_0\bigr) &= 0, \\ &\cdots\cdots\cdots\cdots\cdots\cdots\cdots\cdots, \\ F\bigl(x + \tfrac{n-1}{n}T, y_{n-1}, y_0, \ldots, y_{n-2}\bigr) &= 0, \end{aligned} \qquad (1)$$

where the notation $y_k \equiv y\bigl(\theta_k(x)\bigr)$ is used for brevity.

Eliminating $y_1, y_2, \ldots, y_{n-1}$ from the system of nonlinear algebraic (or transcendental) equations (1), one finds the solutions of the original functional equation in implicit form: $\Psi(x, y_0) = 0$, where $y_0 = y\bigl(\theta(x)\bigr)$.

## T12.3. Functional Equations in Several Independent Variables

### T12.3.1. Linear Functional Equations

T12.3.1-1. Equations involving functions with a single argument.

**1.** $f(x+y) = f(x) + f(y).$

*Cauchy's equation.*
Solution:
$$f(x) = Cx,$$
where $C$ is an arbitrary constant.

**2.** $f\left(\dfrac{x+y}{2}\right) = \dfrac{f(x) + f(y)}{2}.$

*Jensen's equation.*
Solution:
$$f(x) = C_1 x + C_2,$$
where $C_1$ and $C_2$ are arbitrary constants.

**3.** $af(x) + bf(y) = f(ax + by) + c.$

$1^\circ$. Solution:
$$f(x) = \begin{cases} Ax + \dfrac{c}{a+b-1} & \text{if } a+b-1 \ne 0, \\ Ax + B & \text{if } a+b-1 = 0 \text{ and } c = 0, \end{cases}$$
where $A$ and $B$ are arbitrary constants.

$2^\circ$. If $a+b-1 = 0$ and $c \ne 0$, then there is no solution.

**4.** $Af(a_1x + b_1y + c_1) + Bf(a_2x + b_2y + c_2) = Cf(a_3x + b_3y + c_3) + D.$

All continuous solutions of this equation have the form

$$f(x) = \alpha x + \beta,$$

where the constants $\alpha$ and $\beta$ are determined by substituting this expression into the original equation.

**5.** $f(x + y) + f(x - y) = 2f(x) + 2f(y).$

Solution:
$$f(x) = Cx^2,$$

where $C$ is an arbitrary constant.

**6.** $f(x + y) = f(x)e^{ay}.$

Solution:
$$f(x) = Ce^{ax},$$

where $C$ is an arbitrary constant.

**7.** $f(x + y) + f(x - y) = 2f(x)\cosh y.$

Solution:
$$f(x) = C_1 e^x + C_2 e^{-x},$$

where $C_1$ and $C_2$ are arbitrary constants.

**8.** $f(x + y) + f(x - y) = 2f(x)\cosh(ay) + 2f(y).$

Solution:
$$f(x) = C[2 - \cosh(ax)],$$

where $C$ is an arbitrary constant.

**9.** $f(x + y) + f(x - y) = 2f(x)\cos y.$

Solution:
$$f(x) = C_1 \cos x + C_2 \sin x,$$

where $C_1$ and $C_2$ are arbitrary constants.

**10.** $f(x + y) + f(x - y) = 2f(x)\cos(ay) + 2f(y).$

Solution:
$$f(x) = C[2 - \cos(ax)],$$

where $C$ is an arbitrary constant.

**11.** $f(xy) = f(x) + f(y).$

*Cauchy's logarithmic equation.*
Solution:
$$f(x) = C \ln|x|,$$

where $C$ is an arbitrary constant.

12. $f((x^n + y^n)^{1/n}) = f(x) + f(y)$.

Solution:
$$f(x) = Cx^n,$$

where $C$ is an arbitrary constant.

13. $f(x) + f(y) = f\left(\dfrac{x+y}{1-xy}\right)$, $\quad xy < 1$.

Solution:
$$f(x) = C \arctan x,$$

where $C$ is an arbitrary constant.

14. $f(x) + (1-x)f\left(\dfrac{y}{1-x}\right) = f(y) + (1-y)f\left(\dfrac{x}{1-y}\right)$.

*A basic equation of information theory.* The quantities $x$, $y$, and $x+y$ can assume values from 0 to 1.
   Solution:
$$f(x) = C[x \ln x + (1-x)\ln(1-x)],$$

where $C$ is an arbitrary constant.

15. $f(x) + (1-x)^\alpha f\left(\dfrac{y}{1-x}\right) = f(y) + (1-y)^\alpha f\left(\dfrac{x}{1-y}\right)$.

Here, the quantities $x$, $y$, and $x+y$ can assume values from 0 to 1; $\alpha \neq 0, 1, 2$.
   Solution:
$$f(x) = C[x^\alpha + (1-x)^\alpha - 1],$$

where $C$ is an arbitrary constant.

16. $f\left(xy - \sqrt{(1-x^2)(1-y^2)}\right) = f(x) + f(y)$, $\quad |x| \leq 1,\ |y| \leq 1$.

Solution:
$$f(x) = C \arccos x,$$

where $C$ is an arbitrary constant.

17. $f\left(xy + \sqrt{(x^2-1)(y^2-1)}\right) = f(x) + f(y)$, $\quad |x| \geq 1,\ |y| \geq 1$.

Solution:
$$f(x) = C \operatorname{arccosh} x,$$

where $C$ is an arbitrary constant.

18. $f(x) + g(y) = h(x+y)$.

*Pexider's equation.* Here, $f(x)$, $g(y)$, and $h(z)$ are the unknown functions.
   Solution:
$$f(x) = C_1 x + C_2, \quad g(y) = C_1 y + C_3, \quad h(z) = C_1 z + C_2 + C_3,$$

where $C_1$, $C_2$, and $C_3$ are arbitrary constants.

T12.3.1-2. Equations involving functions with two arguments.

**19.** $f(x, y) = f(y, x)$.

*This equation may be treated as a definition of functions symmetric with respect to permutation of the arguments.*

1°. Solution:
$$f(x, y) = \Phi(x, y) + \Phi(y, x),$$
where $\Phi(x, y)$ is an arbitrary function with two arguments.

2°. Particular solutions may be found using the formula
$$f(x, y) = \Psi(\varphi(x) + \varphi(y))$$
by specifying the functions $\Psi(z)$, $\varphi(x)$.

**20.** $f(x, y) = -f(y, x)$.

*This equation may be treated as a definition of functions antisymmetric with respect to permutation of the arguments.*

1°. Solution:
$$f(x, y) = \Phi(x, y) - \Phi(y, x),$$
where $\Phi(x, y)$ is an arbitrary function with two arguments.

2°. Particular solutions may be found using the formulas
$$f(x, y) = \varphi(x) - \varphi(y),$$
$$f(x, y) = (x - y)\Psi(\varphi(x) + \varphi(y)),$$
by specifying the functions $\varphi(x)$ and $\Psi(z)$.

**21.** $f(x, y) = f(x + ak_1, y + ak_2)$.

*Traveling-wave equation.* Here, $a$ is an arbitrary number and $k_1$ and $k_2$ are some constants.
Solution:
$$f(x, y) = \Phi(k_2 x - k_1 y),$$
where $\Phi(z)$ is an arbitrary function.

**22.** $f(ax, ay) = f(x, y)$.

Here, $a$ is an arbitrary number ($a \neq 0$).
Solution:
$$f(x, y) = \Phi(y/x),$$
where $\Phi(z)$ is an arbitrary function.

**23.** $f(ax, ay) = a^\beta f(x, y)$.

*Equation of a homogeneous function.* Here, $a$ is an arbitrary number ($a \neq 0$) and $\beta$ is a fixed number called the order of homogeneity.
Solution:
$$f(x, y) = x^\beta \Phi(y/x),$$
where $\Phi(x)$ is an arbitrary function.

**24.** $f(ax, a^\beta y) = f(x, y)$.

Here, $a$ is an arbitrary number ($a \neq 0$) and $\beta$ is some constant.
Solution:
$$f(x, y) = \Phi(yx^{-\beta}),$$
where $\Phi(x)$ is an arbitrary function.

**25.** $f(ax, a^\beta y) = a^\gamma f(x, y)$.

*Equation of self-similar solutions.* Here, $a$ is an arbitrary number ($a \neq 0$) and $\beta$ and $\gamma$ are some constants.
Solution:
$$f(x, y) = x^\gamma \Phi(yx^{-\beta}),$$
where $\Phi(x)$ is an arbitrary function.

**26.** $f(x, y) = a^n f(x + (1 - a)y, ay)$.

Here, $a$ is an arbitrary number ($a > 0$) and $n$ is some constant.
Solution:
$$f(x, y) = y^{-n} \Phi(x + y),$$
where $\Phi(x)$ is an arbitrary function.

**27.** $f(x, y) = a^n f(a^m x, y + \ln a)$.

Here, $a$ is an arbitrary number ($a > 0$) and $n$ and $m$ are some constants.
Solution:
$$f(x, y) = e^{-ny} \Phi(xe^{-my}),$$
where $\Phi(x)$ is an arbitrary function.

**28.** $f(x, y) + f(y, z) = f(x, z)$.

*Cantor's first equation.*
Solution:
$$f(x, y) = \Phi(x) - \Phi(y),$$
where $\Phi(x)$ is an arbitrary function.

**29.** $f(x + y, z) + f(y + z, x) + f(z + x, y) = 0$.

Solution:
$$f(x, y) = (x - 2y)\varphi(x + y),$$
where $\varphi(x)$ is an arbitrary function.

**30.** $f(xy, z) + f(yz, x) + f(zx, y) = 0$.

Solution:
$$f(x, y) = \varphi(xy) \ln \frac{x}{y^2} \quad \text{if} \quad x > 0, \; y \neq 0;$$
$$f(x, y) = \varphi(xy) \ln \frac{-x}{y^2} \quad \text{if} \quad x < 0, \; y \neq 0;$$
$$f(x, y) = A + B \ln |x| \quad \text{if} \quad x \neq 0, \; y = 0;$$
$$f(x, y) = A + B \ln |y| \quad \text{if} \quad x = 0, \; y \neq 0,$$
where $\varphi(x)$ is an arbitrary function and $A$ and $B$ are arbitrary constants.

## T12.3.2. Nonlinear Functional Equations

T12.3.2-1. Equations involving one unknown function with a single argument.

**1.** $f(x+y) = f(x)f(y)$.

*Cauchy's exponential equation.*
   Solution:
$$f(x) = e^{Cx},$$
where $C$ is an arbitrary constant. In addition, the function $f(x) \equiv 0$ is also a solution.

**2.** $f(x+y) = af(x)f(y)$.

Solution:
$$f(x) = \frac{1}{a}e^{Cx},$$
where $C$ is an arbitrary constant. In addition, the function $f(x) \equiv 0$ is also a solution.

**3.** $f\left(\dfrac{x+y}{2}\right) = \sqrt{f(x)f(y)}$.

Solution:
$$f(x) = Ca^x,$$
where $a$ and $C$ are arbitrary positive constants.

**4.** $f\left(\dfrac{x+y}{n}\right) = [f(x)f(y)]^{1/n}$.

Solution:
$$f(x) = a^x,$$
where $a$ is an arbitrary positive constant.

**5.** $f(y+x) + f(y-x) = 2f(x)f(y)$.

*D'Alembert's equation.*
   Solutions:
$$f(x) = \cos(Cx), \qquad f(x) = \cosh(Cx), \qquad f(x) \equiv 0,$$
where $C$ is an arbitrary constant.

**6.** $f(y+x) + f(y-x) = af(x)f(y)$.

Solutions:
$$f(x) = \frac{2}{a}\cos(Cx), \qquad f(x) = \frac{2}{a}\cosh(Cx), \qquad f(x) \equiv 0,$$
where $C$ is an arbitrary constant.

**7.** $f(x+y) = a^{xy}f(x)f(y)$.

Solution:
$$f(x) = e^{Cx}a^{x^2/2},$$
where $C$ is an arbitrary constant. In addition, the function $f(x) \equiv 0$ is also a solution.

**8.** $f(x+y) = f(x) + f(y) - af(x)f(y)$,   $a \neq 0$.

*For $a = 1$, it is an equation of probability theory.*
   Solution:
$$f(x) = \frac{1}{a}\left(1 - e^{-\beta x}\right),$$
where $\beta$ is an arbitrary constant. In addition, the function $f(x) \equiv 0$ is also a solution.

**9.** $f(x+y)f(x-y) = f^2(x)$.

*Lobachevsky's equation.*
   Solution:
$$f(x) = C_1 \exp(C_2 x),$$
where $C_1$ and $C_2$ are arbitrary constants.

**10.** $f(x+y+a)f(x-y+a) = f^2(x) + f^2(y) - 1$.

Solutions:
$$f(x) = \pm 1, \quad f(x) = \pm \cos\frac{n\pi x}{a},$$
where $n = 1, 2, \ldots$ For trigonometric solutions, $a$ must be nonzero.

**11.** $f(x+y+a)f(x-y+a) = f^2(x) - f^2(y)$.

Solutions:
$$f(x) = 0, \quad f(x) = C\sin\frac{2\pi x}{a}, \quad f(x) = C\sin\frac{(2n+1)\pi x}{a},$$
where $C$ is an arbitrary constant and $n = 0, 1, 2, \ldots$ For trigonometric solutions, $a$ must be nonzero.

**12.** $(x-y)f(x)f(y) = xf(x) - yf(y)$.

Solutions:
$$f(x) \equiv 1, \quad f(x) = \frac{C}{x+C},$$
where $C$ is an arbitrary constant.

**13.** $f(xy) = af(x)f(y)$.

*Cauchy's power equation* (for $a = 1$).
   Solution:
$$f(x) = \frac{1}{a}|x|^C,$$
where $C$ is an arbitrary constant. In addition, the function $f(x) \equiv 0$ is also a solution.

**14.** $f(xy) = [f(x)]^y$.

Solution:
$$f(x) = e^{Cx},$$
where $C$ is an arbitrary constant. In addition, the function $f(x) \equiv 0$ is also a solution for $y > 0$.

**15.** $f\left(\sqrt{x^2 + y^2}\right) = f(x)f(y)$.

*Gauss's equation.*
Solution:
$$f(x) = \exp(Cx^2),$$
where $C$ is an arbitrary constant. In addition, the function $f(x) \equiv 0$ is also a solution.

**16.** $\left(\dfrac{f^2(x) + f^2(y)}{2}\right)^{1/2} = f\left(\left(\dfrac{x^2 + y^2}{2}\right)^{1/2}\right)$.

Solution:
$$f(x) = (ax^2 + b)^{1/2},$$
where $a$ and $b$ are arbitrary positive constants.

**17.** $f\left((x^n + y^n)^{1/n}\right) = af(x)f(y)$, $\quad n$ **is any number.**

Solution:
$$f(x) = \frac{1}{a}\exp(Cx^n),$$
where $C$ is an arbitrary constant. In addition, the function $f(x) \equiv 0$ is also a solution.

**18.** $\left(\dfrac{f^n(x) + f^n(y)}{2}\right)^{1/n} = f\left(\left(\dfrac{x^n + y^n}{2}\right)^{1/n}\right)$, $\quad n$ **is any number.**

Solution:
$$f(x) = (ax^n + b)^{1/n},$$
where $a$ and $b$ are arbitrary positive constants.

**19.** $f\left(x + y\sqrt{f(x)}\right) + f\left(x - y\sqrt{f(x)}\right) = 2f(x)f(y)$.

Solutions:
$$f(x) \equiv 0, \quad f(x) = 1 + Cx^2,$$
where $C$ is an arbitrary constant.

**20.** $f\left(g^{-1}(g(x) + g(y))\right) = af(x)f(y)$.

*Generalized Gauss equation.* Here, $g(x)$ is an arbitrary monotonic function and $g^{-1}(x)$ is the inverse of $g(x)$.
Solution:
$$f(x) = \frac{1}{a}\exp[Cg(x)],$$
where $C$ is an arbitrary constant. The function $f(x) \equiv 0$ is also a solution.

**21.** $M(f(x), f(y)) = f(M(x, y))$.

Here, $M(x, y) = \varphi^{-1}\left(\dfrac{\varphi(x) + \varphi(y)}{2}\right)$ is a quasiarithmetic mean for a continuous strictly monotonic function $\varphi$, with $\varphi^{-1}$ being the inverse of $\varphi$.
Solution:
$$f(x) = \varphi^{-1}(a\varphi(x) + b),$$
where $a$ and $b$ are arbitrary constants.

T12.3.2-2. Equations involving several unknown functions of a single argument.

**22.** $f(x)g(y) = h(x+y)$.

Here, $f(x)$, $g(y)$, and $h(z)$ are unknown functions.
  Solution:
$$f(x) = C_1 \exp(C_3 x), \qquad g(y) = C_2 \exp(C_3 y), \qquad h(z) = C_1 C_2 \exp(C_3 z),$$
where $C_1$, $C_2$, and $C_3$ are arbitrary constants.

**23.** $f(x)g(y) + h(y) = f(x+y)$.

Here, $f(x)$, $g(y)$, and $h(z)$ are unknown functions.
  Solutions:
$$f(x) = C_1 x + C_2, \quad g(x) = 1, \quad h(x) = C_1 x \qquad \text{(first solution)};$$
$$f(x) = C_1 e^{ax} + C_2, \quad g(x) = e^{ax}, \quad h(x) = C_2(1 - e^{ax}) \quad \text{(second solution)},$$
where $a$, $C_1$, and $C_2$ are arbitrary constants.

**24.** $f_1(x)g_1(y) + f_2(x)g_2(y) + f_3(x)g_3(y) = 0$.

*Bilinear functional equation.*
  Two solutions:
$$f_1(x) = C_1 f_3(x), \quad f_2(x) = C_2 f_3(x), \quad g_3(y) = -C_1 g_1(y) - C_2 g_2(y);$$
$$g_1(y) = C_1 g_3(y), \quad g_2(y) = C_2 g_3(y), \quad f_3(x) = -C_1 f_1(x) - C_2 f_2(x),$$
where $C_1$ and $C_2$ are arbitrary constants, the functions on the right-hand sides of the solutions are prescribed arbitrarily.

**25.** $f_1(x)g_1(y) + f_2(x)g_2(y) + f_3(x)g_3(y) + f_4(x)g_4(y) = 0$.

*Bilinear functional equation.*
  Equations of this type often arise in the generalized separation of variables in partial differential equations.

1°. Solution:
$$f_1(x) = C_1 f_3(x) + C_2 f_4(x), \quad f_2(x) = C_3 f_3(x) + C_4 f_4(x),$$
$$g_3(y) = -C_1 g_1(y) - C_3 g_2(y), \quad g_4(y) = -C_2 g_1(y) - C_4 g_2(y).$$

It depends on four arbitrary constants $C_1, \ldots, C_4$. The functions on the right-hand sides of the solution are prescribed arbitrarily.

2°. The equation also has two other solutions,
$$f_1(x) = C_1 f_4(x), \quad f_2(x) = C_2 f_4(x), \quad f_3(x) = C_3 f_4(x), \quad g_4(y) = -C_1 g_1(y) - C_2 g_2(y) - C_3 g_3(y);$$
$$g_1(y) = C_1 g_4(y), \quad g_2(y) = C_2 g_4(y), \quad g_3(y) = C_3 g_4(y), \quad f_4(x) = -C_1 f_1(x) - C_2 f_2(x) - C_3 f_3(x),$$

involving three arbitrary constants.

## 26. $f(x) + g(y) = Q(z)$, where $z = \varphi(x) + \psi(y)$.

Here, one of the two functions $f(x)$ and $\varphi(x)$ is prescribed and the other is assumed unknown, also one of the functions $g(y)$ and $\psi(y)$ is prescribed and the other is unknown, and the function $Q(z)$ is assumed unknown. (In similar equations with a composite argument, it is assumed that $\varphi(x) \not\equiv \text{const}$ and $\psi(y) \not\equiv \text{const}$.)

Solution:
$$f(x) = A\varphi(x) + B, \quad g(y) = A\psi(y) - B + C, \quad Q(z) = Az + C,$$

where $A$, $B$, and $C$ are arbitrary constants.

## 27. $f(x)g(y) = Q(z)$, where $z = \varphi(x) + \psi(y)$.

Here, one of the two functions $f(x)$ and $\varphi(x)$ is prescribed and the other is assumed unknown; also one of the functions $g(y)$ and $\psi(y)$ is prescribed and the other is unknown, and the function $Q(z)$ is assumed unknown. (In similar equations with a composite argument, it is assumed that $\varphi(x) \not\equiv \text{const}$ and $\psi(y) \not\equiv \text{const}$.)

Solution:
$$f(x) = ABe^{\lambda\varphi(x)}, \quad g(y) = \frac{A}{B}e^{\lambda\psi(y)}, \quad Q(z) = Ae^{\lambda z},$$

where $A$, $B$, and $\lambda$ are arbitrary constants.

## 28. $f(x) + g(y) = Q(z)$, where $z = \varphi(x)\psi(y)$.

Here, one of the two functions $f(x)$ and $\varphi(x)$ is prescribed and the other is assumed unknown; also one of the functions $g(y)$ and $\psi(y)$ is prescribed and the other is unknown, and the function $Q(z)$ is assumed unknown. (In similar equations with a composite argument, it is assumed that $\varphi(x) \not\equiv \text{const}$ and $\psi(y) \not\equiv \text{const}$.)

Solution:
$$f(x) = A \ln \varphi(x) + B, \quad g(y) = A \ln \psi(y) - B + C, \quad Q(z) = A \ln z + C,$$

where $A$, $B$, and $C$ are arbitrary constants.

## 29. $f(y) + g(x) + h(x)Q(z) + R(z) = 0$, where $z = \varphi(x) + \psi(y)$.

Equations of this type often arise in the functional separation of variables in partial differential equations.

1°. Solution:
$$f = -\tfrac{1}{2}A_1 A_4 \psi^2 + (A_1 B_1 + A_2 + A_4 B_3)\psi - B_2 - B_1 B_3 - B_4,$$
$$g = \tfrac{1}{2}A_1 A_4 \varphi^2 + (A_1 B_1 + A_2)\varphi + B_2,$$
$$h = A_4 \varphi + B_1,$$
$$Q = -A_1 z + B_3,$$
$$R = \tfrac{1}{2}A_1 A_4 z^2 - (A_2 + A_4 B_3)z + B_4,$$

where the $A_k$ and $B_k$ are arbitrary constants and $\varphi = \varphi(x)$ and $\psi = \psi(y)$ are arbitrary functions.

2°. Solution:
$$f = -B_1 B_3 e^{-A_3 \psi} + \left(A_2 - \frac{A_1 A_4}{A_3}\right)\psi - B_2 - B_4 - \frac{A_1 A_4}{A_3^2},$$
$$g = \frac{A_1 B_1}{A_3} e^{A_3 \varphi} + \left(A_2 - \frac{A_1 A_4}{A_3}\right)\varphi + B_2,$$
$$h = B_1 e^{A_3 \varphi} - \frac{A_4}{A_3},$$
$$Q = B_3 e^{-A_3 z} - \frac{A_1}{A_3},$$
$$R = \frac{A_4 B_3}{A_3} e^{-A_3 z} + \left(\frac{A_1 A_4}{A_3} - A_2\right) z + B_4,$$

where $A_k$ and $B_k$ are arbitrary constants and $\varphi = \varphi(x)$ and $\psi = \psi(y)$ are arbitrary functions.

3°. In addition, the functional equation has two degenerate solutions:
$$f = A_1 \psi + B_1, \quad g = A_1 \varphi + B_2, \quad h = A_2, \quad R = -A_1 z - A_2 Q - B_1 - B_2,$$
where $\varphi = \varphi(x)$, $\psi = \psi(y)$, and $Q = Q(z)$ are arbitrary functions; $A_1$, $A_2$, $B_1$, and $B_2$ are arbitrary constants; and
$$f = A_1 \psi + B_1, \quad g = A_1 \varphi + A_2 h + B_2, \quad Q = -A_2, \quad R = -A_1 z - B_1 - B_2,$$
where $\varphi = \varphi(x)$, $\psi = \psi(y)$, and $h = h(x)$ are arbitrary functions; $A_1$, $A_2$, $B_1$, and $B_2$ are arbitrary constants.

**30.** $f(y) + g(x)Q(z) + h(x)R(z) = 0$, where $z = \varphi(x) + \psi(y)$.

Equations of this type often arise in the functional separation of variables in partial differential equations.

1°. Solution:
$$\begin{aligned} g(x) &= A_2 B_1 e^{k_1 \varphi} + A_2 B_2 e^{k_2 \varphi}, \\ h(x) &= (k_1 - A_1) B_1 e^{k_1 \varphi} + (k_2 - A_1) B_2 e^{k_2 \varphi}, \\ Q(z) &= A_3 B_3 e^{-k_1 z} + A_3 B_4 e^{-k_2 z}, \\ R(z) &= (k_1 - A_1) B_3 e^{-k_1 z} + (k_2 - A_1) B_4 e^{-k_2 z}, \end{aligned} \quad (1)$$

where $B_1, \ldots, B_4$ are arbitrary constants and $k_1$ and $k_2$ are roots of the quadratic equation
$$(k - A_1)(k - A_4) - A_2 A_3 = 0.$$

In the degenerate case $k_1 = k_2$, the terms $e^{k_2 \varphi}$ and $e^{-k_2 z}$ in (1) must be replaced by $\varphi e^{k_1 \varphi}$ and $z e^{-k_1 z}$, respectively. In the case of purely imaginary or complex roots, one should extract the real (or imaginary) part of the roots in solution (1).

The function $f(y)$ is determined by the formulas
$$\begin{aligned} B_2 = B_4 = 0 &\implies f(y) = [A_2 A_3 + (k_1 - A_1)^2] B_1 B_3 e^{-k_1 \psi}, \\ B_1 = B_3 = 0 &\implies f(y) = [A_2 A_3 + (k_2 - A_1)^2] B_2 B_4 e^{-k_2 \psi}, \\ A_1 = 0 &\implies f(y) = (A_2 A_3 + k_1^2) B_1 B_3 e^{-k_1 \psi} + (A_2 A_3 + k_2^2) B_2 B_4 e^{-k_2 \psi}. \end{aligned} \quad (2)$$

Solutions defined by (1) and (2) involve arbitrary functions $\varphi = \varphi(x)$ and $\psi = \psi(y)$.

2°. In addition, the functional equation has two degenerate solutions,

$$f = B_1 B_2 e^{A_1 \psi}, \quad g = A_2 B_1 e^{-A_1 \varphi}, \quad h = B_1 e^{-A_1 \varphi}, \quad R = -B_2 e^{A_1 z} - A_2 Q,$$

where $\varphi = \varphi(x)$, $\psi = \psi(y)$, and $Q = Q(z)$ are arbitrary functions; $A_1$, $A_2$, $B_1$, and $B_2$ are arbitrary constants; and

$$f = B_1 B_2 e^{A_1 \psi}, \quad h = -B_1 e^{-A_1 \varphi} - A_2 g, \quad Q = A_2 B_2 e^{A_1 z}, \quad R = B_2 e^{A_1 z},$$

where $\varphi = \varphi(x)$, $\psi = \psi(y)$, and $g = g(x)$ are arbitrary functions; and $A_1$, $A_2$, $B_1$, and $B_2$ are arbitrary constants.

T12.3.2-3. Equations involving functions of two arguments.

**31.** $f(x, y) f(y, z) = f(x, z).$

*Cantor's second equation.*
  Solution:
$$f(x, y) = \Phi(y)/\Phi(x),$$
where $\Phi(x)$ is an arbitrary function.

**32.** $f(x, y) f(u, v) - f(x, u) f(y, v) + f(x, v) f(y, u) = 0.$
Solution:
$$f(x, y) = \varphi(x)\psi(y) - \varphi(y)\psi(x),$$
where $\varphi(x)$ and $\psi(x)$ are arbitrary functions.

**33.** $f\bigl(f(x, y), z\bigr) = f\bigl(f(x, z), f(y, z)\bigr).$

*Skew self-distributivity equation.*
  Solution:
$$f(x, y) = g^{-1}\bigl(g(x) + g(y)\bigr),$$
where $g(x)$ is an arbitrary continuous strictly increasing function.

**34.** $f\left(\dfrac{x+y}{2}\right) = G\bigl(f(x), f(y)\bigr).$

*Generalized Jensen equation.*

1°. A necessary and sufficient condition for the existence of a continuous strictly increasing solution is the existence of a continuous strictly monotonic function $g(x)$ such that

$$G(x, y) = g^{-1}\left(\frac{g(x) + g(y)}{2}\right),$$

where $g(x)$ is an arbitrary continuous strictly monotonic function and $g^{-1}(x)$ is the inverse of $g(x)$.

2°. If condition 1° is satisfied, the general continuous strictly monotonic solution of the original equation is given by
$$f(x) = \varphi(ax + b),$$
where $\varphi(x)$ is any continuous strictly monotonic solution, and $a$ and $b$ are arbitrary constants.

# References for Chapter T12

Aczél, J., *Functional Equations: History, Applications and Theory*, Kluwer Academic, Dordrecht, 2002.

Aczél, J., *Lectures on Functional Equations and Their Applications*, Dover Publications, New York, 2006.

Aczél, J., Some general methods in the theory of functional equations with a single variable. New applications of functional equations [in Russian], *Uspekhi Mat. Nauk*, Vol. 11, No. 3 (69), pp. 3–68, 1956.

Aczél, J. and Dhombres, J., *Functional Equations in Several Variables*, Cambridge University Press, Cambridge, 1989.

Agarwal, R. P., *Difference Equations and Inequalities, 2nd Edition*, Marcel Dekker, New York, 2000.

Belitskii, G. R. and Tkachenko, V., *One-Dimensional Functional Equations*, Birkhäuser Verlag, Boston, 2003.

Castillo, E. and Ruiz-Cobo, R., *Functional Equations in Science and Engineering*, Marcel Dekker, New York, 1992.

Czerwik, S., *Functional Equations and Inequalities in Several Variables*, World Scientific Publishing Co., Singapore, 2002.

Doetsch, G., *Guide to the Applications of the Laplace and Z-transforms* [in Russian, translation from German], Nauka Publishers, Moscow, 1974, pp. 213, 215, 218.

Elaydi, S., *An Introduction to Difference Equations, 3rd Edition*, Springer-Verlag, New York, 2005.

Fikhtengol'ts, G. M., *A Course of Differential and Integral Calculus, Vol. 1* [in Russian], Nauka Publishers, Moscow, 1969, pp. 157–160.

Goldberg, S., *Introduction to Difference Equations*, Dover Publications, New York, 1986.

Kuczma, M., *An Introduction to the Theory of Functional Equations and Inequalities*, Polish Scientific Publishers, Warsaw, 1985.

Kuczma, M., *Functional Equations in a Single Variable*, Polish Scientific Publishers, Warsaw, 1968.

Kuczma, M., Choczewski, B., and Ger, R., *Iterative Functional Equations*, Cambridge University Press, Cambridge, 1990.

*Mathematical Encyclopedia, Vol. 2* [in Russian], Sovetskaya Entsikolpediya, Moscow, 1979, pp. 1029, 1030.

*Mathematical Encyclopedia, Vol. 5* [in Russian], Sovetskaya Entsikolpediya, Moscow, 1985, pp. 699, 700, 703, 704.

Mirolyubov, A. A. and Soldatov, M. A., *Linear Homogeneous Difference Equations* [in Russian], Nauka Publishers, Moscow, 1981.

Mirolyubov, A. A. and Soldatov, M. A., *Linear Nonhomogeneous Difference Equations* [in Russian], Nauka Publishers, Moscow, 1986.

Nechepurenko, M. I., *Iterations of Real Functions and Functional Equations* [in Russian], Institute of Computational Mathematics and Mathematical Geophysics, Novosibirsk, 1997.

Polyanin, A. D., *Functional Equations*, From Website *EqWorld – The World of Mathematical Equations*, http://eqworld.ipmnet.ru/en/solutions/fe.htm.

Polyanin, A. D. and Manzhirov, A. V., *Handbook of Integral Equations: Exact Solutions (Supplement. Some Functional Equations)* [in Russian], Faktorial, Moscow, 1998.

Polyanin, A. D. and Zaitsev, V. F., *Handbook of Nonlinear Partial Differential Equations (Supplements S.4.4 and S.5.5)*, Chapman & Hall/CRC Press, Boca Raton, 2004.

# Supplement
# Some Useful Electronic Mathematical Resources

**arXiv.org** (http://arxiv.org). A service of automated e-print archives of articles in the fields of mathematics, nonlinear science, computer science, and physics.

**Catalog of Mathematics Resources on the WWW and the Internet** (http://mthwww.uwc.edu/wwwmahes/files/math01.htm).

**CFD Codes List** (http://www.fges.demon.co.uk/cfd/CFD_codes_p.html). Free software.

**CFD Resources Online** (http://www.cfd-online.com/Links). Software, modeling and numerics, etc.

**Computer Handbook of ODEs** (http://www.scg.uwaterloo.ca/ecterrab/handbook_odes.html). An online computer handbook of methods for solving ordinary differential equations.

**Deal.II** (http://www.dealii.org). Finite element differential equations analysis library.

**Dictionary of Algorithms and Data Structures — NIST** (http://www.nist.gov/dads/). The dictionary of algorithms, algorithmic techniques, data structures, archetypical problems, and related definitions.

**DOE ACTS Collection** (http://acts.nersc.gov). The Advanced CompuTational Software (ACTS) Collection is a set of software tools for computation sciences.

**EEVL: Internet Guide to Engineering, Mathematics and Computing** (http://www.eevl.ac.uk). Cross-search 20 databases in engineering, mathematics, and computing.

**EqWorld: World of Mathematical Equations** (http://eqworld.ipmnet.ru). Extensive information on algebraic, ordinary differential, partial differential, integral, functional, and other mathematical equations.

**FOLDOC — Computing Dictionary** (http://foldoc.doc.ic.ac.uk/foldoc/index.html). The free on-line dictionary of computing is a searchable dictionary of terms from computing and related fields.

**Free Software** (http://www.wseas.com/software). Download free software packages for scientific-engineering purposes.

**FSF/UNESCO Free Software Directory** (http://directory.fsf.org).

**GAMS: Guide to Available Mathematical Software** (http://gams.nist.gov). A cross-index and virtual repository of mathematical and statistical software components of use in computational science and engineering.

**Google — Mathematics Websites** (http://directory.google.com/Top/Science/Math/). A directory of more than 11,000 mathematics Websites ordered by type and mathematical subject.

**Google — Software** (http://directory.google.com/Top/Science/Math/Software). A directory of software.

**Mathcom — PDEs** (http://www.mathcom.com/corpdir/techinfo.mdir/scifaq/q260.html). Partial differential equations and finite element modeling.

**Mathematical Atlas** (http://www.math-atlas.org). A collection of short articles designed to provide an introduction to the areas of modern mathematics.

**Mathematical Constants and Numbers** (http://numbers.computation.free.fr/Constants/constants.html). Mathematical, historical, and algorithmic aspects of some classical mathematical constants; easy and fast programs are also included.

**Mathematical WWW Virtual Library** (http://www.math.fsu.edu/Virtual/index.php). A directory of mathematics-related Websites ordered by type and mathematical subject.

**Mathematics Archives** (http://archives.math.utk.edu). Combined archive and directory of mathematics Websites, mailing lists, and teaching materials.

**Mathematics Genealogy Project** (http://www.genealogy.ams.org). A large biographical database of mathematicians.

**Mathematics Websites** (http://www.math.psu.edu/MathLists/Contents.html). A directory of mathematics Websites ordered by type and mathematical subject.

**Math Forum: Internet Mathematics Library** (http://mathforum.org). A directory of mathematics Websites ordered by the mathematical subject.

**MathGuide: SUB Gottingen** (http://www.mathguide.de). An Internet-based subject gateway to mathematics-related Websites.

**MathWorld: World of Mathematics** (http://www.mathworld.com). An online encyclopedia of mathematics, focusing on classical mathematics. The Web's most extensive mathematical resource.

**MGNet** (http://www.mgnet.org/mgnet-codes.html). Free software.

**Netlib** (http://www.netlib.org). A collection of mathematical software, papers, and databases.

**Numerical Solutions** (http://www.numericalmathematics.com/numerical_solutions.htm). A library of mathematical programs.

**PlanetMath.Org** (http://planetmath.org). An online mathematics encyclopedia.

**Probability Web** (http://www.mathcs.carleton.edu/probweb/probweb.html). A collection of probability resources on the World Wide Web; the pages are designed to be especially helpful to researchers and teachers.

**Science Oxygen — Mathematics** (http://www.scienceoxygen.com/math.html). Topics from various sections of mathematics.

**Scilab** (http://scilabsoft.inria.fr). A free scientific software package.

**Software — Differential Equations** (http://www.scicomp.uni-erlangen.de/SW/diffequ.html). General resources and methods for ODEs and PDEs.

**S.O.S. Mathematics** (http://www.sosmath.com). A free resource for math review material from algebra to differential equations.

**Statistics Online Computational Resources** (http://socr.stat.ucla.edu). Interactive distributions, statistical analysis, virtual probability-related experiments and demonstrations, computer games, and others.

**Stat/Math Center** (http://www.indiana.edu/statmath/bysubject/numerics.html). Numerical computing resources on the Internet.

**UW-L Math Calculator** (http://www.compute.uwlax.edu/index.php). Calculus, differential equations, numerical methods, statistics, and others.

**Wikipedia: Free Encyclopedia — List of Open Source Software Packages** (http://en.wikipedia.org/wiki/List_of_open-source_software_packages).

**Wikipedia: Free Encyclopedia — Mathematics** (http://en.wikipedia.org/wiki/Mathematics). A collection of short articles from various sections of mathematics.

**Wolfram Functions Site** (http://functions.wolfram.com). More than 87,000 formulas used by mathematicians, computer scientists, physicists, and engineers; more than 10,000 graphs and animations of the functions.

**Yahoo — Mathematics Websites** (http://dir.yahoo.com/science/mathematics/). A directory of mathematics Websites ordered by type and mathematical subject.

**Yahoo — Software** (http://dir.yahoo.com/Science/Mathematics/Software). A directory of software.

# INDEX

## A

Abel convergence criterion
  functional series, 349
  power series, 341
Abel differential equation
  first kind, 462
  first kind, canonical form, 463
  second kind, 464
  second kind, canonical form, 465
Abel differential equations, 462–465
Abel functional equation, 914
Abel identity, 902
Abel integral equation, 1385
  generalized, 1386
  second kind, 1392
  second kind, generalized, 1393
Abel theorem, 343, 411
  on convergence of power series, 351
Abel transformation, 345
Abel–Dini convergence criterion for series, 341
Abel–Ruffini theorem, 162
abelian group, 226
abscissa axis, 78
abscissa of point, 79, 115
absence of aftereffect, 1075
absolute convergence, 342
absolute convergence of infinite products, 348
absolute extremum, 991
absolute moment, 1044
absolute value of real number, xxx, 8
absolute value theorem, 296
  double integral, 318
  triple integral, 325
absolutely convergent functional series, 348
absolutely convergent infinite product, 348
absolutely convergent series, 342
acceleration of convergence of series, 345
acceleration vector, 381
acceptance region, 1095
acute triangle, 43
addition
  asymptotic series, 364
  hyperbolic functions, 37
  inverse hyperbolic functions, 41
  inverse trigonometric functions, 33
  matrices, 168
  real numbers, 6
  trigonometric functions, 28
addition theorem, 1055
additive inverse of real number, 6
additive separable solution, 571, 596, 597, 678
additively inverse matrix, 169
additivity of conditional expectation, 1063
adjoint matrix, 170
adjoint operator, 206
adjugate matrix, 180
adjusted sample variance, 1088
admissible controlled process, 1010
admissible element, 1009
admissible extremal, 994, 998
admissible function, 993, 1002, 1006
affine coordinate system, 190
affine space, 189
  axioms, 189
  dimension, 190
  point, 189
Airy equation, 955
Airy functions, 477, 955
  first kind, 955
  second kind, 955
Airy stress function, 1295
algebra, 155
algebra of sets, 1032
algebraic branch point, 415
algebraic curve, 87
  on plane, 87
algebraic equation
  arbitrary degree, 161
  binomial, 161
  bounds for roots, 163
  complete cubic, 159
  cubic, 158
  discriminant, 163
  fourth-degree, 159
  incomplete cubic, 158
  linear, 157
  $n$th-degree, 162
  quadratic, 158
  reciprocal, 160, 161
  reciprocal modified, 160
  roots and coefficients, 163
algebraic equations, 157–167
algebraic multiplicity, 209
  eigenvalue, 184
alternative Fourier transform, 444
alternative Fredholm, 834
alternative hypothesis, 1094
altitude
  pyramid, 63
  trapezoid, 54
  triangle, 46
amplitude, 973
analogs Neper, 73
analysis
  sequential, 1097
  variance, 1107
analytic continuation, 418
analytic function, 413
analytic geometry, 77

angle
  between coordinate axes and segments, 82
  between elements, 193
  between lines in space, 135
  between planes, 137
  between straight line and plane, 139
  between two straight lines, 94
  between vectors, 120
  central, 55, 56
  dihedral, 60
  edge, 60
  external, 43
  face, 60
  inscribed, 56
  interior, 43
  interior of polygon, 51
  polar, 80
  polyhedral, 60, 61
  solid, 61
  trihedral, 60
angle bisector of triangle, 45, 46
angle trisectors, 49
angles
  in space, 59
  symmetric polyhedral, 61
angular point, 252
annihilating polynomial, 186
annulus, 58
antagonistic game, 1024
antagonistic two-person zero-sum game, 1024
  minimax theorem, 1026
antiderivative, 273
antihermitian matrix, 168
antiperiodic functions, 885
antisymmetric matrix, 168
apex, 67
apex of pyramid, 63
apothem of regular pyramid, 64
Appell transformation, 1268
applicable surfaces, 395
applicate of point, 115
approximation, best mean-square, 1063, 1066
approximation function, 508
arbitrary curvilinear coordinate systems, 1195
arbitrary series, 341
arc
  geodesic, 70
  length, 390
  length differential, 390
  regular, 379
arccosine, xxx, 32
arccotangent, xxx, 32
arcsine, xxx, 31
arctangent, xxx, 32
area
  polygon, 83
  triangle, 83
argument, 236
  complex number, 400
  increment, 992
  variation, 992

arithmetic, 3
arithmetic mean, 13
  of functions, 297
arithmetic progression, 11
arithmetic root, 8
arithmetic sequence, 11
arithmetic series, 11
arm of hyperbola, 101
arrangement, 1033
  with repetitions, 1034
artificial basic variables, 1016
ascending Landen transformation, 975
associated Legendre functions, 964, 965
  first kind, 965
  second kind, 965
associated linear space of affine space, 189
associated vector, 213
associative composition law, 225
asymmetric form of Fourier cosine transform, 445
asymmetric form of Fourier sine transform, 446
asymmetric form of Fourier transform, 444
asymmetry coefficient, 1045
asymptote, 241, 374
  oblique, 242
  vertical, 242
asymptotes, 241, 374
  of hyperbola, 101
asymptotic cone, 144
asymptotic expansion, 363
asymptotic formulas
  calculation of integrals, 290
  improper integrals with parameter, 307
  Laplace integrals of general form, 292
  Laplace integrals of special form, 291
  power Laplace integral, 294
asymptotic lines, 393
asymptotic point, 373
asymptotic properties of Fourier coefficients, 359
asymptotic series, 363
  addition, 364
  division, 364
  integration, 365
  multiplication, 364
  Poincaré type, 363
  subtraction, 364
asymptotic stability, 546, 549
asymptotic zero, 364
asymptotically stable solution, 546
asymptotically unbiased estimator, 1089
attractive fixed point, 908
augmented matrix, 198
auto-Bäcklund transformations, 663
autocorrelation function, 1072
automorphic function, 912
automorphism of group, 227
autonomous equation, 525, 1207, 1223, 1225
autonomous system
  equilibrium point, 545
  general form, 1240
  ODEs, 545
averaging method, 503

axioms of affine space, 189
axis, 77
    coordinate of point, 77
    coordinate system, 77
    imaginary, 401
    imaginary of hyperbola, 101
    iteration, 876
    ordinate, 78
    polar, 80
    principal, 152
    principal of hyperbola, 101
    principal of second-order curve, 109
    real, 235
    semiminor of ellipse, 98
axisymmetric steady hydrodynamic boundary layer, 712

# B

Babbage equation, 917, 1416, 1423
Bäcklund transformation, 497, 663–665
backward Kolmogorov equation, 1076
backward substitution in Gaussian elimination method, 200
balanced transportation problem, 1020
ball, 68
Banach problem, 1038
Banach space, 196
Bartlett test, 1103
base
    cylinder, 65
    isosceles triangle, 50
    natural (Napierian) logarithms, 9, 22, 23
    pyramid, 63
    spherical segment, 69
    trapezoid, 54
base face of prism, 61
basic columns, 178
basic equation of information theory, 1440
basic inequalities for mean values, 13
basic integrals, 274
basic matrix of linear system of algebraic equations, 197
basic minor, 177, 178
basic properties
    equalities, 5
    inequalities, 5
basic rows, 178
basic variable, 1015
    artificial, 1016
basis
    canonical, 217
    linear space, 189
    orthonormal, 194
Bayes formula, 1036, 1063
Bellman equations, 1028
Bellman optimality principle, 1028
bending, 395
Bernoulli equation, 458, 1207
Bernoulli formula, 1037
Bernoulli numbers, 938
Bernoulli polynomials, 891, 988

Bernoulli process, 1037
Bernoulli scheme, 1037
Bernoulli theorem, 1070
Bertrand convergence criterion for series, 340
Bessel equation, 1216
    modified, 1216
Bessel formula, 950
Bessel function, 947, 1216
    first kind, xxxi, 947, 1216
    second kind, xxxii, 947, 1216
    third kind, 952
best mean-square approximation, 1063, 1066
beta function, 945
biangles, spherical, 71
biased estimator, 1089
bicylindrical coordinates, 1204
biharmonic equation, 1297
bilinear form, 214, 992
    degenerate, 215
    Euclidean space, 219
    finite-dimensional space, 215
    matrix, 215
    nondegenerate, 215
    polar, 216
    rank, 215
    skew-symmetric, 214
    symmetric, 214
bilinear functional, 992
bilinear functional equations, 930
bilinear series of iterated kernels, 838
bimodal distribution, 1045
binomial algebraic equation, 161
binomial coefficients, xxx, 10, 937
binomial distribution, 1047
binomial theorem, 10
binormal, 382
bipolar coordinates, 1204
biquadratic equation, 159
birth–death processes, 1077
bisection method, 261
bisector, perpendicular, 46
bivariate random variable, 1060
    central moment, 1060
    expectation, 1060
block matrix, 174
blow-up regime, 1303
Bochner transform, 449
body
    coordinates of center of mass, 327
    moments of inertia, 327
Bolza functional, 1001
Bolza Lagrangian, 1001
Bolza problem, 1001
Bolza terminal cost function, 1001
Bolzano–Cauchy theorem, 237
Born–Infeld equation, 659
bound
    lower, 235
    upper, 235
boundary
    confidence, 1081

boundary (*continued*)
  domain, 401
  positive sense, 402
boundary conditions, 484, 591, 592
  first kind, 484
  homogeneous, 481, 591, 649
  mixed, 486
  nonhomogeneous, 481, 649
  second kind, 485
  third kind, 486
boundary correspondence principle, 420
boundary correspondence theorem, 419
boundary function, 419
boundary point of domain, 401
boundary value problem, 427
  arbitrary cylindrical domain, 642
  difference equations, 879
  elliptic equations, 631
  first, 480, 593, 627
  Hilbert–Privalov, 432
  hyperbolic equations, 623
  mixed, 480, 593, 627
  nonhomogeneous, 620, 624
  ordinary differential equations, 480
  parabolic equations, 645
  rectangular domains, 640
  second, 480, 593, 627
  third, 480, 593, 627
  with many space variables, 634
bounded function, 236
bounded operator, 204
bounded sequence, 237
bounded set, 263
bounded variation, 247, 313
Boussinesq equation, 709, 1332
Boussinesq solution, 1294
bracket, Jacobi–Mayer, 574
branch of many-valued function, 404
branch point, 414
  function, 404
  infinite order, 415
  logarithmic, 415
  order, 414
Brauer theorem, 877
break point, 373
broken extremal, 997, 999
Bubnov–Galerkin method
  for integral equations, 850
  for ODEs, 509
Budan–Fourier method, 166
Bugaev convergence criterion for series, 341
Bunyakovsky inequality, 296
Burgers equation, 664, 666, 751, 1307
  radial symmetric case, 1307
Bürman–Lagrange formula, 356

# C

calculus of variations, 991
  simplest problem, 993
canonical basis, 217
canonical coefficients, 217

canonical coordinates
  ellipse, 98
  hyperbola, 101
  parabola, 104
canonical equation
  circle, 97
  ellipse, 98
  hyperbola, 101
  noncentral hypersurface, 224
  parabola, 104
  straight line in space, 132
  straight line passing through given point on plane, 90
  straight line passing through two given points on plane, 90
canonical form
  Abel differential equation, of first kind, 463
  Abel differential equation, of second kind, 465
  central surfaces, 143
  elliptic equation, 588
  gas dynamics systems, 786
  hyperbolic equation, 587
  Jordan, of matrix, 182
  linear operators, 213
  matrix, 181
  parabolic equation, 586
  quadric, 148
  Riccati differential equation, 460
  second-order curve, 107
canonical representation of quadratic form, 217
canonical second-order curves, 107
canonical substitutions, 465
Cantor equation
  first, 1442
  second, 1449
cap, spherical, 69
Cardano solution, 158
Carleman equation, 1399
Cartesian coordinate system, 78, 114
  in plane, 78
  in space, 114
  oblique, 79
Casoratti determinant, 901
Casoratti theorem, 902
Cauchy convergence criterion for series, 339
Cauchy criterion, 337
Cauchy distribution, 1054
Cauchy equation, 1438
  exponential, 1443
  logarithmic, 1439
  power, 1444
Cauchy formula
  for multiple integration, 289
  for ODE, 519
Cauchy inequality, 14, 296, 343
  for functions of complex variables, 410
Cauchy integral formula, 409
Cauchy principal value for singular integral, 311, 430

Cauchy problem, 591, 592, 785, 1308
    difference equations, 874, 889, 893
    existence and uniqueness theorems for ODEs, 454, 488, 524, 577
    first-order ODEs, 454
    first-order PDEs, 556, 576
    fundamental solution, 616, 617
    generalized, 627
    Hamilton–Jacobi equation, 577
    hyperbolic equations, 617
    parabolic equations, 615
    second-order ODEs, 488
    system of PDEs, 785
Cauchy series multiplication formula, 343
Cauchy tests, 303
Cauchy theorem, 254, 343
    multiply connected domain, 409
    residue, 416
    simply connected domain, 409
Cauchy-type singular integral, 310, 409, 430
Cauchy–Bunyakovski inequality, 14, 296, 343
Cauchy–Hadamard formula, 411
Cauchy–Riemann conditions, 402
Cauchy–Riemann sum, 286
Cauchy–Schwarz inequality, 121, 193, 1043
Cauchy–Schwarz–Bunyakovski inequality, 14, 314
Cayley–Hamilton theorem, 186, 212
center
    curvature, 375
    ellipse, 99
    equilibrium point, for ODEs, 537
    gravity of triangle, 45
    hyperbola, 102
    pencil, 92
    regular polygon, 55
    second-order curve, 109
    second-order hypersurface, 222
    surface, 143, 151
center of mass
    of body, 327
    of flat plate, 324
centered rarefaction wave, 558
central angle, 55, 56
central cross-correlation function, 1073
central cylinder, 225
central hypersurface, 222
central limit theorem, 1071
central moment, 1043
    $n$-dimensional random variable, 1065
    bivariate random variable, 1060
central second-order curve, 109
central surface, 143
centroid, 45
Ceva theorem, 48
character of group element, 231
characteristic, 654
characteristic equation, 514, 654, 1046, 1066
    difference equation, 881
    linear operator, 211
    linear PDEs, 585

characteristic equation (*continued*)
    matrix, 185
    quadratic form, 150
    second-order curves, 108
    system of ODEs, 529
characteristic function, 1046, 1066
    continuity theorem, 1067
characteristic index, 1221
characteristic matrix, 183
characteristic point, 377
characteristic polynomial
    linear operator, 211
    matrix, 185
characteristic quadratic form, 150
    of quadric, 150
characteristic system for first-order PDEs, 553, 555, 573, 576
characteristic value
    integral equation, 829, 834–836, 840
    linear operator, 209
characteristic velocity, 790
Chebyshev formula, 808
Chebyshev functions of second kind, 985
Chebyshev inequalities, 14, 297, 1070
    generalized, 15
Chebyshev nodes, 523
Chebyshev polynomials
    first kind, 983
    second kind, 984
Chebyshev theorem, 1070
Chetaev theorem of instability, 550
chi-square distribution, 1054
chord, 56, 143
Christoffel symbols
    first kind, 396
    second kind, 396
circle, 56, 97
    canonical equation, 97
    curvature, 375
    diameter, 56
    great of a sphere, 68, 70
    nine-point, 49
    osculating, 375
    parametric equations, 98
    small of a sphere, 71
circular point, 393
circular sector, 58
circulation, 331
circumcenter, 46
circumcircle, 46
    of triangle, 46
circumscribed about circle, polygon, 51
cissoid of Diocles, 372
Clairaut equation, 467, 572, 1262
Clairaut system, 1240
Clarkson–Kruskal direct method, 708–710
classical definition of probability, 1033
classical integral functional, 993
classical method of symmetry reductions, 716
classical solution, 594
classification
    central second-order hypersurfaces, 223

classification (*continued*)
  noncentral second-order hypersurfaces, 224
  points on surface, 394
  quadrics, 148
  second-order linear PDEs, 585
closed domain, 402
closed model of transportation problem, 1020
closed path integral, 331
coalition game, 1024
Cochran statistic, 1104
Cochran test, 1104
coefficient
  asymmetry, 1045
  asymmetry, sample, 1088
  correlation, multiple, 1066
  excess, 1045
  excess, sample, 1088
  Pearson first skewness, 1045
  reflection, 760
  regression, 1063
  variation, 1045
coefficients
  binomial, xxx, 10, 28
  canonical for quadratic form, 217
  Fourier, 358
  Lamé, 1196
  system of linear equations, 197
  undetermined, 847
cofactor
  of matrix entry, 176
  of minor, 177
Cole–Hopf transformation, 751
Cole–Kevorkian scheme (method), 504
collinear vectors, 113
collocation method, 509, 523, 847, 848, 860
  convergence theorem, 524
  integral equations, 847, 860
collocation points, 848
column, pivot, 1016
column expansion of determinant, 176
column vector, 167
column vectors
  linear combination, 171
  linearly dependent, 171
  linearly independent, 171
columns, basic, 178
combinations, 1033
combinations, integrable, 545
  with repetitions, 1034
common difference, 11
common logarithm, 9
common ratio, 11
commutative composition law, 225
commutative group, 226
commutator of operators, 755
commuting matrices, 169
compact subgroup, 229
compatibility conditions
  for PDEs, 738, 740, 742, 744
  for systems of linear equations, 757
  of initial and boundary conditions, 607, 609

compatible size of matrices, 169
complementary error function, xxx, 939, 958
complementary event, 1031
complementary modulus, 971
complete cubic equation, 159
complete elliptic integral, 969
  first kind, 969
  second kind, 969
complete equation of plane, *see* general equation of plane
complete group of events, 1031
complete group of incompatible events, 1032
complete integral, 570
complete space, 196
complete statistic, 1090
complete system of eigenfunctions, 836
completely integrable, 773
completely reducible representation of group, 231
complex conjugate matrix, 170
complex Euclidean space, 195
complex linear space, 188
complex number, 399
  argument, 400
  conjugate, 399
  exponential form, 405
  geometric interpretation, 399
  modulus, 400
  root, 399
  trigonometric form, 400
complex numbers
  difference, 399
  equal, 399
  product, 399
  quotient, 399
  sum, 399
complex plane, 401
  domain, 401
  extended, 402
complex separation of variables, 679
complex stochastic processes, 1073
components, metric tensor, 1195
composite function, 237, 265
composite hypothesis, 1094
composite number, 3
composition, 225
composition law, 225
compound percentage, 7
concave function, 245
condition
  boundary, 484, 591, 592
  boundary, of first kind, 484
  boundary, of second kind, 485
  boundary, of third kind, 486
  first-order necessary for extremum, 994
  Hölder, 310, 430
  initial homogeneous, 591
  invariance, 720, 732
  invariant surface, 732
  Jacobi, 998, 1000, 1004, 1007
  Jacobi, strengthened, 1000, 1004, 1007
  jump, 561

condition (*continued*)
   Lax, 792
   Legendre, 998, 999, 1004, 1007
   Legendre, strengthened, 999, 1004, 1007
   line and plane to be parallel, 140
   line and plane to be perpendicular, 140
   line to be entirely contained in plane, 140
   Lipschitz, 248, 310, 313, 543
   necessary of extremum, 269
   planes to be parallel, 138
   planes to be perpendicular, 138
   planes to coincide, 138
   stability for generalized solution, 568
   straight lines to be parallel on plane, 96
   straight lines to be perpendicular on plane, 95
   straight lines to coincide on plane, 96
   three points to be collinear, 93
   transversality, 577
   two lines in space to meet, 143
   two lines to be parallel in space, 135
   two lines to be perpendicular in space, 136
   two planes to be parallel, 138
   two planes to be perpendicular, 138
   unitarity, 206
   vectors to be parallel, 122
   vectors to be perpendicular, 122
   Weierstrass, 998
conditional convergence, 342
conditional cumulative distribution function, 1062
conditional distribution, 1062
conditional expectation, 1063, 1066
conditional extremum of function, 271
conditional minimum, 270
conditional probability, 1035
conditionally convergent series, 342
conditions
   Cauchy–Riemann, 402
   compatibility for PDEs, 738, 740, 742, 744
   compatibility for systems of linear equations, 757
   compatibility of initial and boundary conditions, 607, 609
   d'Alembert–Euler, 402
   Dirichlet, 357
   evolutionary, 792
   existence of extremum, 258
   extremum of function of two variables, 269
   radiation, 1290
   Sommerfeld, 1290, 1291
   sufficient of extremum, 269
   transversality, 1001
   Weierstrass–Erdmann, 997, 999
cone, 67, 68, 145
   asymptotic, 144
   frustum, 68
   right circular, 68, 145
confidence boundary, 1081
confidence interval, 1093
confidence level, 1093
conformal mapping, 419
congruence transformation, 182
congruent matrices, 182
congruent polyhedral angles, 61
congruent transformation, 182
conic point, 391
   of surface, 391
conical coordinates, 1202
conical surface, 67
conjugate complex number, 399
conjugate diameters, 151
   of ellipse, 100
   of hyperbola, 104
conjugate element of group, 231
conjugate matrix, 170
conjugate point, 998, 1000, 1004, 1007
conjunctive matrices, 183
conjunctive transformation, 183
conservation law, 564, 766, 996
   at discontinuity, 561
   for PDEs, 766
consistency condition for general linear system, 198
consistency of estimator, 1089
consistent system of linear algebraic equations, 197
constant
   Euler, xxx, 944, 949, 958
   Hölder, 310
   separation, 678
constrained maximum, 270
constrained minimum, 270
constraint, 270
   differential, 737, 738, 1009, 1010
   differential, first-order, 739
   differential, secod-order, 744
   isoperimetric, 1002
contact transformation
   for ordinary differential equations, 468
   for partial differential equations, 660
continuation, direct analytic, 418
continuity equation, 784
continuity theorem for characteristic functions, 1067
continuous bivariate random variable, 1059
continuous curve in space, 123
continuous distributions, 1051
continuous function, 243
   of bounded variation, 249
continuous point group, 730
continuous probability distribution, mode, 1045
continuous random variable, 1040, 1064
   expectation, 1042
continuous subgroups, 228
continuous surface in space, 123
contraction ratio, 403
control, 1010
controlled process, admissible, 1010
convergence
   almost sure, 1068
   in distribution, 1069
   in mean, 1068
   in probability, 1068
   mean-square, 315

convergence (*continued*)
   nonuniform, 249
   pointwise, 249
   uniform, 249
   weak, 1069
convergence criteria
   arbitrary series, 341
   infinite products, 347
   series with positive terms, 338
convergence criterion
   Abel, 341
   Abel–Dini, 341
   Bertrand, 340
   Bugaev, 341
   D'Alembert, 338
   Dini, 341
   Dirichlet, 342
   Dirichlet–Jordan, 358, 361
   Ermakov, 340
   Gauss, 339
   generalized Ermakov, 340
   Kummer, 340
   Lipschitz, 358
   Lobachevsky, 341
   Maclaurin–Cauchy, 339
   Raabe, 339
   Sapogov, 340
   Weierstrass, 349
convergence domain of series, 348
convergence radius of power series, 350, 411
convergence theorem, collocation method, 524
convergence with probability 1, 1068
convergent infinite products, 346
convergent sequence, 237, 238
convergent series, 337, 338
convex function, 245
convex polygon, 51
convex polyhedron, 61
convex programming, 1027
convolution theorem, 437, 444
   Fourier transform, 444
   Laplace transform, 437
convolution-type equation, 825, 829
coordinate axes
   in space, 114
   on plane, 78
coordinate functions, 847, 851
coordinate net, 387
coordinate of point on axis, 77
coordinate space, 188
coordinate system
   affine, 190
   in space, 114
   left rectangular Cartesian, 79
   oblique Cartesian, 79
   on axis, 77
   on plane, 78
   polar, 80
   rectangular Cartesian, 78, 114
   right rectangular Cartesian, 79
coordinate vectors, 388

coordinates
   bicylindrical, 326, 1204
   bipolar, 1204
   Cartesian, in plane, 79
   Cartesian, in space, 115
   center of mass of body, 327
   center of mass of flat plate, 324
   conical, 1202
   current, 84
   curvilinear, 387
   cylindrical, xxxi, 118, 326, 1198
   element, 188, 189
   ellipse canonical, 98
   elliptic, 1202
   elliptic cylinder, 1202
   first prolongation, 718
   Gaussian, 387
   general orthogonal curvilinear, 1196
   hyperbola canonical, 101
   line, 77
   oblate ellipsoid of revolution, 1200
   orthogonal, 1196
   parabola canonical, 104
   parabolic, 1203
   parabolic cylinder, 1203
   plane, 78
   point, 79, 115
   polar, 80, 1198
   second prolongation, 718
   spherical, xxxi, 119, 326, 1199
   toroidal, 326, 1205
coplanar vectors, 113
correlated random variables, 1061
correlation
   function, 1072
   matrix, 1066
   random variables, 1061
   sample, 1104
   two random variables, 1061
correspondence, isometric, 395
cosecant, xxx, 24
coset, 227
   left, 227
   right, 227
cosine, 24, 25
cosine integral, 942
cost, reduced, 1015
cotangent, xxx, 24, 26
countable set, 3
counterexample, 17
covariance, 1060, 1061
   matrix, 1065
   of two random variables, 1061
Cramer rule, 199
Cramér–Rao efficient unbiased estimator, 1089
Cramér–Rao inequality, 1089
criterion
   Cauchy convergence for series, 339
   Dini convergence for series, 341
   Dirichlet convergence, 342, 350
   Dirichlet–Jordan convergence for Fourier series, 358, 361

criterion (*continued*)
   Ermakov convergence for series, 340
   first comparison, 338
   Gauss convergence for series, 339
   Kummer convergence for series, 340
   Leibnitz convergence for series, 341
   Lipschitz convergence for series, 358, 361
   Lobachevsky convergence for series, 341
   Maclaurin–Cauchy convergence for series, 339
   Routh–Hurwitz, 167
   Sapogov convergence for series, 340
   second convergence, 338
   special Cauchy, 341
   Sylvester, 219
   Weierstrass convergence, 349
critical region, 1095
cross-correlation function, 1072
cross-covariance function, 1072
cross product
   of vectors, 121
cube, 62
cubic equation, 158
cumulative distribution function, 1039
   conditional, 1062
cumulative frequency, 1084
cumulative relative frequency, 1084
curl, 331
   of vector field, 331
   334
curvature, 376, 384
   center, 375
   circle, 375
   extrinsic, 394
   Gaussian, 394
   geodesic, 392
   lines, 393
   mean, 394
   normal, 392, 394
   of curve on surface, 392
   of plane curves, 376
   of space curves, 384
   radius, 376, 384
   vector, 383
curve
   discriminant, 378
   in space, 120, 123
   intrinsic equations, 388
   on plane, 84
   parabolic type of second-order, 108
   second-order, 107
   in space, 119, 123
   on plane, 84
   on plane algebraic, 87
curves, parametric, 387
curvilinear coordinates, 387
   on surface, 387
curvilinear integral, 329
curvilinear lines, 387
cusp, first kind, 372
cusp, second kind, 372
cycle, 1022

cyclic functional equation, 920
cyclic groups, 229
cylinder, 65, 66
   base, 65
   central, 225
   function, 947
   paraboloidal, 225
   right, 66
   right circular, 66
   round, 66
   segment, 67
   truncated, 66
cylindrical coordinates, xxxi, 118, 1198
cylindrical function, 947, 1214
cylindrical Korteweg–de Vries equation, 1328
cylindrical surface, 65
cylindrical tube, 67

# D

D'Alembert convergence criterion for series, 338
D'Alembert equation, 1443
D'Alembert formula, 1278
D'Alembert (Gauss) formulas, 73
D'Alembert method, 533
D'Alembert solution for wave equation, 600
D'Alembert–Euler conditions, 402
damped vibrations, 1213
Darboux equation, 458
Darboux vector, 386
de Moivre formula, 30, 400
de Morgan laws, 1032
death process, 1077
decadic logarithm, 9
decile, 1045
decomposition
   of matrices, 173, 201
   theorems, 208, 1055
   triangular, 180
decreasing function, 236
decreasing sequence, 238
defect of matrix, 178
definite integral, 286, 1147
   geometric applications, 299
   geometric meaning, 287
   physical applications, 299
   properties, 287
definite integrals
   involving exponential functions, 1150
   involving hyperbolic functions, 1152
   involving logarithmic functions, 1152
   involving trigonometric functions, 1153
degenerate bilinear form, 215
degenerate hypergeometric equation, 1215
degenerate hypergeometric function, xxxii, 956
   Wronskian, 958
degenerate kernel, 801, 802, 810, 811, 825, 829
degenerate matrix, 178
degenerate quadratic form, 216
density, spectral, 1075
density multiplication theorem, 1062
dependent variable, 236

derivation formulas, 396
derivative, 250
   basic elementary functions, 252
   directional, 267, 268
   geometrical meaning, 251
   higher-order, 255
   of function, 402
   of polynomial, 157
   one-sided, 252
   partial, 264
   physical meaning, 251
   total, 265
Descartes, Folium, 374
Descartes theorem, 165
Descartes–Euler solution, 160
descending Landen transformation, 974
determinant, 175
   Casoratti, 901
   column expansion, 176
   Gram, 193
   Jacobian, 321
   linear operator, 209
   order $n$, 176
   order 2, 175
   order 3, 176
   product of matrices, 179
   properties, 176
   row expansion, 176
   sum of matrices, 179
   Vandermonde, 179
   Wronskian, 473, 482, 518, 539
determined system of linear algebraic equations, 197
deviation
   of point from line, 93
   of point from plane, 141
   residual standard, 1063
   root-mean-square, 1045
   sample, 1088
   standard, 1045
diagonal
   main, 168
   secondary, 168
diagonal matrix, 168
diameter, 151
   conjugate to chords, 108
   conjugate to family of planes, 151
   of circle, 56
   of ellipse, 100
   of hyperbola, 104
   of parabola, 106
   of second-order curve, 108
   of set, 317
   of sphere, 69
   partition, 317
   principal, of ellipse, 100
   principal, of hyperbola, 104
   principal, of parabola, 106
diameters
   conjugate, 151
   conjugate, of ellipse, 100
   conjugate, of hyperbola, 104

Dido problem, 1003
diet problem, 1012
difference
   common, 11
   finite, 884
   of complex numbers, 399
   of events, 1031
   of matrices, 169
   of polynomials, 155
   of vectors, 113, 188
difference equation, 889, 1409, 1422, 1425–1426
   characteristic, 881
   general solution, 874, 884
   in integer argument, 873
   linear, constant-coefficient, 873, 877
   linear, higher-order, 1425
   linear, second-order, 877, 1421
   linear homogeneous, 881, 885, 889, 906, 904
   linear homogeneous, first-order, 873
   linear homogeneous, second-order, 877, 1421
   linear nonhomogeneous, 883, 890–893
   linear nonhomogeneous, constant-coefficient, first-order, 1411
   linear nonhomogeneous, constant-coefficient, $n$th-order, 1427
   linear nonhomogeneous, constant-coefficient, second-order, 1422
   linear nonhomogeneous, first-order, 873
   linear nonhomogeneous, $n$th-order, 1426
   linear nonhomogeneous, second-order, 879, 880, 1421
   linearly independent solutions, 882
   logistic, 875
   nonlinear, 884
   nonlinear, first-order, 873
   Riccati, 875, 918
difference kernel, 801, 804, 810, 813, 825, 829, 841
differentiable function, 251
differential, 250, 251, 265
   higher-order, 255
differential constraint, 737, 1009, 1010
   first-order, 739
   second-order, 744
differential constraints method, 737
differential equation
   Abel, of first kind, 462
   Abel, of first kind, canonical form, 463
   Abel, of second kind, 464
   Abel, of second kind, canonical form, 465
   Airy, 955
   autonomous, 1207, 1223, 1225
   axisymmetric steady hydrodynamic boundary layer, 712
   Bernoulli, 458, 1207
   Bessel, 1216
   Bessel, modified, 1216
   biharmonic, 1297
   biharmonic, homogeneous, 1297
   biharmonic, nonhomogeneous, 1298
   Born–Infeld, 659

differential equation (*continued*)
  Boussinesq, 709, 1332
  Burgers, 664, 666, 751, 1307
  Burgers, radial symmetric case, 1307
  Burgers–Korteweg–de Vries, generalized, 709
  characteristic, for system of ODEs, 529
  Clairaut, 467, 572, 1262
  Darboux, 458
  degenerate hypergeometric, 1215
  diffusion, nonlinear, with cubic source, 753
  diffusion boundary layer, 1276
  Duffing, 501
  elliptic, 585, 588, 590, 1284
  Emden–Fowler, 1223
  Ermakov, 1222
  Euler, 473, 520, 994, 998, 1216
  evolution, second-order, 739
  exact, first-order, 458
  exact, second-order, 491
  Fisher, 1299
  Fitzhugh–Nagumo, 1302
  free oscillations, 1213
  gas dynamics, 558
  Gaussian hypergeometric, 960, 1218
  generalized homogeneous, 526
  Hamilton–Jacobi, 576
  heat, linear, 585, 1267
  heat, linear, nonhomogeneous, 1268
  heat, nonlinear, 1318
  heat, stationary anisotropic, 1323
  heat, stationary, with nonlinear source, 1320
  heat, with axial symmetry, 1270
  heat, with central symmetry, 1272
  Helmholtz, 1289
  Helmholtz, three-dimensional, 600
  Helmholtz, two-dimensional, 599
  homogeneous, 526, 591, 592, 810, 829, 834, 1207, 1224
  homogeneous, generalized, 1210, 1224
  hyperbolic, 585, 587, 590
  hyperbolic, first canonical form, 587
  hyperbolic, normal, 590
  hyperbolic, second canonical form, 587
  hypergeometric, 960, 1218
  hypergeometric, degenerate, 1215
  Jacobi, 998, 1004, 1007
  Khokhlov–Zabolotskaya, 658
  Khokhlov–Zabolotskaya, stationary, 1322
  Klein–Gordon, 618, 625, 1280
  Klein–Gordon, linear, 625
  Klein–Gordon, nonlinear, 702, 1315
  Klein–Gordon, two-dimensional, 618
  Korteweg–de Vries, 752, 756, 759, 1327
  Korteweg–de Vries, cylindrical, 1328
  Korteweg–de Vries, modified, 1328
  Lagrange, 467
  Laplace, ODE, 520
  Laplace, three-dimensional, PDE, 643
  Laplace, two-dimensional, PDE, 585, 633, 1284
  Legendre, 1217–1219

differential equation (*continued*)
  Lienard, 1225
  light rays, 1259
  linear, constant-coefficient, second-order, 473, 1213
  linear homogeneous, second order, 1212
  Liouville, 664
  Mathieu, 980, 1221
  Mathieu modified, 982, 1220
  minimal surfaces, 770
  model of gas dynamics, 1254
  Monge–Ampère, 1326
  Monge–Ampère, homogeneous, 668, 1326
  Monge–Ampère, nonhomogeneous, 1326
  Newell–Whitehead, 733
  ordinary, 453–550, 1207–1228
  Painlevé, 495
  parabolic, 585, 586, 590
  parabolic, in narrow sense, 590
  partial, 553–798, 1247–1382
  Pfaffian, 772, 773
  Poisson, 1287
  Poisson, three-dimensional, 599
  Poisson, two-dimensional, 599
  Riccati, 460
  Riccati, general, 461, 462, 1209, 1210
  Riccati, special, 1208
  Schrödinger, linear, 1276
  Schrödinger, nonlinear, 759, 1309
  Schrödinger, of general form, 1311
  Schrödinger, with cubic nonlinearity, 1309
  Schrödinger, with power-law nonlinearity, 1310
  separable, 456, 1207
  separated, 456
  sine-Gordon, 1314
  sinh-Gordon, 1314
  steady transonic gas flow, 1324
  telegraph, 1284
  telegraph, generalized, 626
  thermal boundary layer, 1276
  transonic gas flow, 1324
  ultrahyperbolic, 590
  vibration, with axial symmetry, 1282
  vibration, with central symmetry, 1283
  vibration of string, 1278
  wave, 585, 1278
  wave, nonlinear, 723, 1312–1317
  wave, three-dimensional, 618
  wave, two-dimensional, 618
  Whittaker, 960
differential geometry, 367
differential operator
  linear, 591, 592
  total, 740, 742
differential operators, of field theory, 272
differential substitution, 666
differentiation, 266
  of implicit function, 266
  of stochastic process, 1073
  rules, 253
differentiation method for PDEs, 700
digamma function, 944, 949

dihedral angle, 60
   edge, 60
   face, 60
dilatation ratio, 403
dimension
   of affine space, 190
   of linear space, 189
   of matrix, 167
   of representation, 230
Dini convergence criterion for series, 341
Dini formula, 429
Dini integral, 1287
Diocles cissoid, 372
Dirac delta function, 599
direct analytic continuation, 418
direct method of symmetry reductions, 708
direct scattering problem, 760
direct sum
   of subspaces, 191
   of two matrices, 175
directed segment, 77
direction, principal of second-order curve, 109
direction coefficients of straight line, 131
direction cosine of segment, 116
directional derivative, 267, 268
directrix, 65, 67
   of ellipse, 99
   of hyperbola, 102
   of parabola, 105
Dirichlet conditions, 357
Dirichlet convergence criteria, 341, 342
Dirichlet criterion of uniform convergence of functional series, 350
Dirichlet problem, 427, 428
   for elliptic equations, 593, 1330
Dirichlet theorem, 295, 358
   representation of function by Fourier series, 357
Dirichlet–Jordan convergence criterion for Fourier series, 358, 361
Dirichlet–Mehler integral, 963
discontinuity, conservation law, 561
discontinuous solutions of quasilinear equations, 558
discrete bivariate random variable, 1058
discrete probability distribution, 1045
discrete probability space, 1033
discrete random variable, 1039, 1064
   expectation, 1042, 1063
discrete subgroups, 228
discriminant
   large, 108
   of algebraic equation, 163
   small, 108
discriminant curve, 378
disk, 56, 97
dispersion equation, 532, 596
distance
   between elements, 193, 195
   between lines in space, 142
   between parallel lines, 96

distance (*continued*)
   between points on axis, 78
   between points on plane, 81
   between two parallel planes, 142
   from point to line in space, 142
   from point to plane, 141
   from point to straight line, 93
   shortest, 70
distribution
   bimodal, 1045
   binomial, 1037, 1047
   Cauchy, 1054
   chi-square, 1054
   conditional, 1062
   continuous, 1051
   convergence, 1069
   exponential, 1052
   Fisher–Snedecor, 1086
   geometric, 1048
   hypergeometric, 1049
   Kolmogorov, 1086
   log-normal, 1042
   multimodal, 1045
   negative binomial, 1050
   normal, 1053
   Poisson, 1050
   Student, 1056
   uniform, 1051
   unimodal, 1045
distribution function
   empirical, 1084
   joint, 1064
   marginal, 1064
   multivariate random variable, 1064
   two-dimensional random vector, 1057
distribution polygon, 1085
divergence, 334
divergence theorem, 334
divergent infinite products, 346
divergent integral, 301
divergent series, 337
divisibility tests, 3
division
   of asymptotic series, 364
   of polynomials, 155
   of segment in given ratio, 82
divisor, greatest common, 4
domain
   boundary, 401
   closed, 402
   formula for calculation of volume, 327
   multiply connected, 409
   of attraction of fixed point, 877
   of function, 236
   on complex plane, 401
   simply connected, 409
double factorial, 10
double integral, 317
   change of variables, 321
   estimation theorem, 318
   geometric applications, 321
   geometric meaning, 317

double integral (*continued*)
   mean value theorem, 318
   physical applications, 321
   properties, 317
double point, 371
dual linear programming problem, 1017
duality theorem, second, 1018
Duffing equation, 501
Duhamel integral, 516
Duhamel principle, 646
   first, 646
   second, 648
dynamic programming, 1027

# E

eccentricity
   of ellipse, 98
   of hyperbola, 102
edge
   of dihedral angle, 60
   of polygon, 51
   of polyhedral angle, 61
   of polyhedron, 61
   of trihedral angle, 60
efficient unbiased estimator, 1089
eigenfunction, 482, 604, 605, 829
   integral equation, 835
eigenfunctions
   Fredholm integral equation, 829
   incomplete system, 836
   system, 836
eigenvalue, 183, 482, 605, 829
eigenvalue problem, 604
   for linear ordinary differential equations, 482
eigenvalues
   algebraic multiplicity, 209
   geometric multiplicity, 209
   linear operator, 211, 209
   matrix, 183
   operator, 209
eigenvector
   linear operator, 209
   matrix, 184
elastic rods, transverse vibration, 1294
electronic mathematical resources, 1451
element
   admissible, 1009
   conjugate, of group, 231
   coordinates, 188, 189
   inverse, 225
   length, 193
   neutral, 225
   norm, 193
   probability, 1040
   surface linear, 390
elementary event, 1031
elementary events space, 1031
elementary functions, 19–41
elementary geometry, 43
elementary matrices, 181
elementary symmetric functions, 163

elementary transformation
   of matrix, 180, 199
   of system of linear equations, 199
elements
   linearly dependent, 188
   linearly independent, 188
   of linear space, 187
   orthogonal, 193
ellipse, 98
   canonical equation, 98
   center, 99
   conjugate diameters, 100
   diameter, 100
   directrix, 99
   eccentricity, 98
   focal axis, 99
   focal distance, 98
   focal parameter, 98
   focal property, 99
   focus-directrix property, 100
   gorge, 145
   in polar coordinate system, 101
   left directrix, 99
   left focus, 99
   linear eccentricity, 98
   minor axis, 98
   numerical eccentricity, 98
   optical property, 100
   parameter, 98
   right directrix, 99
   right focus, 99
   semimajor axis, 98
   semiminor axis, 98
   throat, 145
   vertex, 99
ellipsoid, 143, 224
   imaginary, 224
   right, 145
   scalene, 144
   semiaxis, 143
   triaxial, 144
elliptic coordinates, 1202
elliptic equation, 585, 590, 642, 972, 1284
   canonical form, 588
   first boundary value problem, 633, 643, 644
   fundamental solution, 642
   mixed boundary value problem, 644
   second boundary value problem, 633, 643, 644
   third boundary value problem, 643, 644
elliptic equations
   boundary value problems, 631, 643, 644
   linear, 588, 609–611, 631–633, 637, 1284–1298
   nonlinear, 653, 654, 318–1334
elliptic integral, 969
   complete, 969
   first kind, 970
   incomplete, 970
   second kind, 970
   third kind, 971
elliptic modulus, 971

elliptic paraboloid, 145
elliptic point, 394
Emden–Fowler equation, 1223
empirical distribution
  function, 1084
  variance, 1088
energy conservation law, 996
entire axis, 825, 826, 841
entire function, 414
entry, pivot, 1016
envelope, 377
  of family of curves, 377
equal complex numbers, 399
equal matrices, 168
equal polynomials, 155
equation
  Abel differential, first kind, 462
  Abel differential, first kind, canonical
    form, 463
  Abel differential, second kind, 464
  Abel differential, second kind, canonical
    form, 465
  Abel functional, 914
  Abel integral, 1385
  Abel integral, generalized, 806, 1386
  Abel integral, second kind, 1392
  Abel integral, second kind, generalized, 1393
  Airy, 955
  algebraic, 162
  algebraic, binomial, 161
  algebraic, fourth-degree, 159
  algebraic, linear, 157
  algebraic, $n$th-degree, 162
  algebraic, quadratic, 158
  algebraic, reciprocal, 160, 161
  algebraic, reciprocal, modified, 160
  autonomous, 1207, 1223, 1225
  axisymmetric steady hydrodynamic boundary
    layer, 712
  Babbage, 917, 1416, 1423
  Bernoulli, 1207
  Bessel, 1216
  Bessel, modified, 1216
  biharmonic, 1297
  biharmonic, homogeneous, 1297
  biharmonic, nonhomogeneous, 1298
  bilinear functional, 930, 1446
  binomial algebraic, 161
  biquadratic, 159
  Born–Infeld, 659
  Boussinesq, 709, 1332
  Burgers, 664, 666, 751, 1307
  Burgers, radial symmetric case, 1307
  Burgers–Korteweg–de Vries, generalized, 709
  Cantor first, 1442
  Cantor second, 1449
  Carleman, 1399
  Cauchy, 1438
  Cauchy exponential, 1443
  Cauchy logarithmic, 1439
  Cauchy power, 1444

equation (continued)
  characteristic, 514, 654
  characteristic, for difference equation, 881
  characteristic, for system of ODEs, 529
  characteristic, of linear operator, 211
  characteristic, of matrix, 185
  characteristic, of quadratic form, 150
  circle, canonical, 97
  circle, parametric, 98
  Clairaut, 572, 1262
  complete, of plane, 124
  continuity, 784
  convolution-type, of first kind, 825
  convolution-type, of second kind, 829
  cubic, complete, 159
  cubic, incomplete, 158
  curve, on plane, 84
  cyclic functional, 920
  D'Alembert, 1441
  damped vibrations, 1213
  Darboux, 458
  difference, see difference equation
  differential, see differential equation
  diffusion, linear, see heat equation
  diffusion, nonlinear, with cubic source, 753
  diffusion, stationary anisotropic, 1323
  diffusion boundary layer, 1276
  dispersion, 532, 596
  Duffing, 501
  ellipse, canonical, 98
  elliptic, 585, 588, 590, 1284
  Emden–Fowler, 1223
  Ermakov, 1224
  Euler, 473, 994, 1216
  Euler–Darboux, 602
  Euler–Lagrange, 767
  Euler–Poisson, 1006
  evolution, second-order, 739
  exact, first-order ODE, 458
  exact, second-order ODE, 491
  family, 377
  Fisher, 1301
  Fitzhugh–Nagumo, 1302
  Fredholm, of first kind, 825
  Fredholm, of second kind, 829
  free oscillations, 1213
  functional, see functional equation
  functional differential, 682, 684, 688
  gas dynamics model, 558
  Gauss, 1445
  Gauss, generalized, 1445
  Gaussian hypergeometric, 960, 1218
  Gel'fand–Levitan–Marchenko integral, 1328,
    1333
  general, of plane, 124
  general, of second-order hypersurface, 220
  general, of straight line in space, 132
  general, of straight line on plane, 90
  Hamilton–Jacobi, 576
  heat, 585, 672, 680, 698, 728, 1267
  heat, nonhomogeneous, 1268
  heat, stationary, with nonlinear source, 1320

equation (*continued*)
    heat, stationary anisotropic, 1323
    Helmholtz, 1289
    Helmholtz, three-dimensional, 600
    Helmholtz, two-dimensional, 599
    homogeneous, 591, 592, 810, 829, 834, 1207, 1224
    homogeneous, generalized, 1210, 1224
    homogeneous function, 1441
    hyperbola, canonical, 101
    hyperbolic, 585, 587, 590
    hyperbolic, first canonical form, 587
    hyperbolic, normal, 590
    hyperbolic, second canonical form, 587
    hypergeometric, 960, 1218
    hypergeometric, degenerate, 1215
    incomplete, of plane, 124
    information theory, basic, 1440
    integral, *see* integral equation
    intercept, of plane, 125
    intercept-intercept, of straight line, 91
    involutory functions, 1423
    Jacobi, 998, 1004, 1007
    Jensen, 1438
    Jensen, generalized, 1449
    Khokhlov–Zabolotskaya, 658
    Khokhlov–Zabolotskaya, stationary, 1322
    Klein–Gordon, 618, 625, 1280
    Klein–Gordon, linear, 625
    Klein–Gordon, nonlinear, 702, 1315
    Klein–Gordon, two-dimensional, 618
    Kolmogorov, backward, 1076
    Kolmogorov, forward, 1076
    Korteweg–de Vries, 752, 756, 759, 1327
    Korteweg–de Vries, cylindrical, 1328
    Korteweg–de Vries, modified, 1328
    L'Huiller, 74
    Lalesco–Picard, 1403
    Laplace, ODE, 520
    Laplace, three-dimensional, PDE, 643
    Laplace, two-dimensional, PDE, 585, 633, 1284
    Legendre, 1217–1219
    Lienard, 1225
    light rays, 1259
    likelihood, 1092
    line, in projections, 133
    linear, algebraic, 157
    linear, constant-coefficient, ODE, 473, 514, 1213
    linear, constant-coefficient, PDE, 588
    linear, first-order, ODE, 457
    linear, integral, 801, 810, 824, 829
    linear, nonhomogeneous, ODE, 473
    linear, ODE, 472, 518, 1213
    linear, PDE, 585, 591, 592
    Liouville, 664
    Lobachevsky, 1444
    logarithmic Cauchy, 1439
    logistic, 875
    Mathieu, 980, 1221

equation (*continued*)
    Mathieu, modified, 982, 1220
    minimal surfaces, 770
    model, of gas dynamics, 1254
    Monge–Ampère, 1326
    Monge–Ampère, homogeneous, 668, 1326
    Monge–Ampère, nonhomogeneous, 1326
    Newell–Whitehead, 733
    noncentral hypersurface, canonical, 224
    normal, 252
    normal to surface, 153
    normalized, of plane, 126
    normalized, of straight line, 91
    on entire axis, 825, 826, 841
    on semiaxis, 826, 842
    one-dimensional vibration, with central symmetry, 1281
    ordinary differential, *see* ordinary differential equation
    Painlevé, 495
    parabola, canonical, 104
    parabolic, 585, 586, 590
    parabolic, in narrow sense, 590
    parametric, of curve on plane, 87
    parametric, of line in space, 131
    parametric, of plane, 125
    partial differential, *see* partial differential equation
    pencil of planes, 131
    pencil of straight lines, 92
    Pexider, 925, 1440
    Pfaffian, 772, 773
    plane passing through line of intersection of planes, 131
    plane passing through point and parallel to another plane, 127
    plane passing through point and parallel to two straight lines, 129
    plane passing through three points, 127
    plane passing through two points and parallel to line, 128
    Poisson, 1285
    Poisson, three-dimensional, 599
    Poisson, two-dimensional, 599
    quadratic, 158
    reciprocal, algebraic, 160, 161
    reciprocal, algebraic, generalized, 160
    reciprocal, algebraic, modified, 160
    reciprocal, functional, 919
    renewal, 1395
    Riccati, 460
    Riccati, general, differential, 461, 462, 1207, 1210
    Riccati, special, differential, 1208
    Riccati difference, 875, 918, 1428, 1429
    right-hand side, 801, 825, 829
    root, 162
    Schlömilch, 1400
    Schröder–Koenigs, 911
    Schrödinger, linear, 1276
    Schrödinger, nonlinear, 759, 1309
    Schrödinger, of general form, 1311

equation (*continued*)
　Schrödinger, with cubic nonlinearity, 1309
　Schrödinger, with power-law nonlinearity, 1310
　self-similar solutions, 1442
　separable, 456, 1207
　separated, 456
　sine-Gordon, 1314
　sinh-Gordon, 1314
　skew self-distributivity, 1449
　slope-intercept, of straight line, 89
　steady transonic gas flow, 1324
　straight line in polar coordinates, 92
　straight line in projections in space, 133
　straight line in space, canonical, 132
　straight line passing through given point on plane, canonical, 90
　straight line passing through point and perpendicular to plane, 134
　straight line passing through two given points, 90, 134
　straight line passing through two given points on plane, canonical, 90
　surface, 386
　surface in space, 119
　tangent line, 251
　tangent plane, 268
　tangent plane to surface, 152
　telegraph, 1284
　telegraph, generalized, 626
　thermal boundary layer, 1276
　transonic gas flow, 1324
　transposed, 825, 829, 834
　traveling wave, 1441
　Tricomi integral, 1403
　ultrahyperbolic, 590
　Urysohn, with degenerate kernel, 863, 864
　Urysohn integral, nonlinear, 856
　vibration of string, 1278
　vibration with axial symmetry, 1282
　Volterra, with quadratic nonlinearity, 856
　Volterra integral, nonlinear, 856, 858–861
　Volterra integral, of first kind, 801, 807
　Volterra integral, of second kind, 807, 810
　wave, 585, 1278
　wave, nonlinear, 723, 1312–1317
　wave, three-dimensional, 618
　wave, two-dimensional, 618
　Whittaker, 960
　Wiener–Hopf, of first kind, 825, 1401
　Wiener–Hopf, of second kind, 829, 1406
　with difference kernel, integral, 825, 829
　with quadratic nonlinearity, 865
　with weak singularity of second kind, 829
equations
　admitting variational formulation, 767
　autonomous, 525
　Bellman, 1028
　center of hypersurface, 222
　difference, 871–907
　elliptic, linear, 588, 609–611, 631–633, 637, 1284–1298

equations (*continued*)
　elliptic, nonlinear, 653, 654, 318–1334
　Fredholm integral, of second kind, 854
　functional, linear, 907–917
　hyperbolic, linear, 587, 617, 623–630, 636, 647, 648, 1278–1284
　hyperbolic, nonlinear, 653, 654, 1312–1317
　integral, 801–871, 1385–1406
　integral, linear, 801–855, 1385–1406
　integral, linear, of second kind, with constant limits of integration, 824, 829
　integral, nonlinear, 856–871
　intrinsic, of curve, 388
　motion of mass point, 87
　ordinary differential, 453–550, 1207–1228
　Painlevé, 495
　parabolic, linear, 586, 615–616, 618–622, 634, 635, 646, 648, 1267–1278
　parabolic, nonlinear, 653, 654, 1301–1311
　partial differential, 553–798, 1247–1382
　pencil, 92
　reaction-diffusion type, 796
　straight line on plane parametric, 91
　straight lines on plane, 89
equilateral hyperbola, 101
equilateral triangle, 43, 50, 51
equilibrium point of autonomous system, 545
equilibrium points of two-dimensional linear systems, 534
equilibrium solution, 874, 884
equivalence transformation, 180
equivalent events, 1031
equivalent infinitely small functions, 243
equivalent matrices, 180
equivalent representations of group, 230
Ermakov convergence criterion for series, 340
Ermakov equation, 1224
error function, xxxi, 939, 957, 1267
essential singularity, 413
estimation theorem
　for double integral, 318
　for triple integral, 325
estimator, 1088
　asymptotically unbiased, 1089
　consistent, 1089
　interval, 1088, 1093
　maximum likelihood, 1092
　minimality of variance, 1090
　statistical, 1088
　unbiased, 1089
Euclidean space, 192
　isomorphic, 194
　normed, 193
Euler constant, xxx, 944, 949, 958
Euler differential equations, 998
Euler equation, calculus of variations, 994
Euler equation, ODE, 473, 520, 1216
Euler formula, 30, 405, 944
Euler line, 47
Euler numbers, 939
Euler polynomials, 989
Euler substitutions, 280

Euler system of ODEs, 542
Euler theorem, 61, 393
Euler transform of first kind, 449
Euler transform of second kind, 450
Euler transformation, 661
Euler triangles, 71
Euler–Darboux equation, 602
Euler–Fourier formulas, 358
Euler–Lagrange equation, 767
Euler–Lagrange theorem, 1009
Euler–Poisson equation, 1006
even function, 236
even number, 3
even permutation, 178
event, 1031
   complementary, 1031
   elementary, 1031
   favorable, 1031
   frequency, 1083
   impossible, 1031
   random, 1031
   relative frequency, 1083
   sure, 1031
events
   difference, 1031
   equivalent, 1031
   incompatible, 1031
   intersection, 1031
   statistically independent, 1036
   union, 1031
evolute, 378
evolutionary conditions, 792
evolvent, 378, 379
exact differential equation, 458
exact solutions of linear PDEs, 594
example
   Hilbert, 997
   Weierstrass, 997
excenter, 47
excess, spherical, 72
excess coefficient, 1045
excircle of triangle, 47
excircles, 46
existence and uniqueness theorems
   for ODEs, 454, 488, 524, 543
   for PDEs, 557, 577
expansion
   asymptotic, 363
   asymptotic two-term, 478
   of polynomial in powers of linear binomial, 156
expectation, 1042
   conditional, 1066
   of bivariate random variable, 1060
   of continuous random variable, 1042
   of discrete random variable, 1042
   of function of multivariate random variable, 1065
   population, 1087
expected value of random variable, 1042
exponent, Hölder, 310

exponential distribution, 1052
exponential form of complex number, 405
exponential function, 21
   graphs, 21
   properties, 22
exponential integral, 940, 958
exponential self-similar solution, 675
extended complex plane, 402
external angle, 43
   of polygon, 51
extrema, relative, 991
extremal, 994, 998
   admissible, 994, 998
   broken, 997, 999
   nonsingular, 995
   regular, 995
   regular point, 995
extremal points of function, 269
extremal properties of quadratic form, 220
extremum
   absolute, 991
   first-order necessary condition, 994
   necessary condition, 269
   point, 269
   sufficient conditions, 269
extrinsic curvature, 394

# F

face
   of dihedral angle, 60
   of polyhedral angle, 60
   of polyhedron, 61
   of trihedral angle, 60
factor, reflection, 760
factor group, 227
factorial, xxxi, 10
   double, xxxi, 10
faithful representation of group, 230
false negative rate, 1096
false position method, 262
false positive rate, 1095
family
   of curves, 377
   of planes, 151
family equation, 377
favorable event, 1031
feasible region, 1012
feasible solution, 1012
Feller theorem, 1077
Ferrari solution, 161
Feuerbach points, 49
Feuerbach theorem, 49
fifth Painlevé transcendents, 499
finite differences, 884
finite games, 1024
finite matrix, 167
finite series, 12
finite subgroup, 228
finite sums, 1111
finite-dimensional Euclidean space, 193

first boundary value problem
  for elliptic equations, 633
  for hyperbolic equations, 627
first boundary value problem, for ODEs, 480
  for PDEs, 593
first comparison criterion, 338
first duality theorem, 1018
first group of derivation formulas, 396
first integral, 542, 544, 574
  for system of ODEs, 544
first mean value theorem, 295
first noteworthy limit, 241
first Painlevé transcendent, 496
first quadratic form of surface, 390
first variation of functional, 992
first-order differential constraint, 739
first-order hyperbolic system of quasilinear equations, 780
first-order linear difference equations, 1407
first-order necessary condition for extremum, 994
first-order neighborhood, 991
first-order ordinary differential equation
  Bernoulli, 458
  linear, 457
first-order ordinary differential equations, 453–472, 1207–1211
  Cauchy problem, 454
  homogeneous, 456
  homogeneous, generalized, 457
  integrable by quadrature, 453
  not solved for derivative, 454
  singular solutions, 455
  solved for derivative, 453
  uniqueness and existence theorems, 454
  with separated or separable variables, 456
first-order partial differential equations, 553, 1247
  Cauchy problem, 556
  general solution, 553
  linear, 1247
  method of separation of variables, 571
  nonlinear, 1256
  quasilinear, 553, 1252
Fisher equation, 1301
Fisher information, 1089
Fisher test, 1104
Fisher transformation, 1105
Fisher–Snedecor distribution, 1086
Fitzhugh–Nagumo equation, 1302
fixed point of function, 908
flat plate
  coordinates of center of mass, 324
  moments of inertia, 324
flux of vector field, 333
focal axis
  of ellipse, 99
  of hyperbola, 101
focal distance
  of ellipse, 98
  of hyperbola, 102
  of parabola, 105

focal parameter
  of ellipse, 98
  of hyperbola, 102
focal properties
  of hyperbola, 102
  of parabola, 105
focal property of ellipse, 99
focal radius
  of parabola, 105
  right, 99, 102
focus
  of hyperbola, 102
  of parabola, 105
  stable, for ODEs, 537
  unstable, for ODEs, 537
focus-directrix property
  of ellipse, 100
  of hyperbola, 103
  of parabola, 105
Folium of Descartes, 374
form
  bilinear, 214, 992
  bilinear degenerate, 215
  bilinear nondegenerate, 215
  bilinear skew-symmetric, 214
  bilinear symmetric, 214
  canonical, 462, 473
  canonical for central surfaces, 143
  canonical of gas dynamics systems, 786
  canonical of matrix, 181
  canonical of quadric, 148
  canonical of second-order curve, 107
  linear, 213
  multilinear, 215
  multilinear, skew-symmetric, 215
  multilinear, symmetric, 215
  normal, 473
  quadratic, 216, 218, 840
  quadratic, degenerate, 216
  quadratic, negative definite, 216
  quadratic, nondegenerate, 216
  quadratic, nonnegative, 216
  quadratic, nonpositive, 216
  quadratic, positive definite, 216
  quadratic, second of surface, 392
  sesquilinear, 213
formal solution, 594
formula
  Bayes, 1036
  Bernoulli, 1037
  Bessel, 950
  Bürman–Lagrange, 356
  Cauchy for ODE, 519
  Cauchy integral, 409
  Cauchy multiple integration, 289
  Cauchy series multiplication, 343
  Cauchy–Hadamard, 411
  Chebyshev, 808
  D'Alembert, 1278
  de Moivre, 400
  Dini, 429
  Dirichlet, 295

formula (*continued*)
   Euler, 405, 944
   for calculation of volume of domain, 327
   Fourier cosine inversion, 445
   Fourier inversion, 444
   Frénet, 377
   Gauss, 396, 808
   Green, 428
   Heron, 48
   Hopf, 565
   Kontorovich–Lebedev inversion, 448
   Leibnitz, 256
   Liouville, 473, 518
   Liouville for system of ODEs, 540
   Maclaurin, 257
   Meijer inversion, 447
   Newton–Leibniz, 288
   Newton–Leibniz for function of complex variable, 408
   of finite increments, 254
   of integration by parts, 275, 409
   of repeated integration by parts, 275
   Peterson–Codazzi, 396
   Poisson, 949, 1038
   Post–Widder, 441
   quadrature, 808
   Serret–Frénet, 386
   Stirling, 944
   Sturges, 1083
   total probability, 1036
   Watson, 292
   Willier, 74
formulas
   Bayes, 1063
   D'Alembert (Gauss), 73
   Euler–Fourier, 358
   Frénet, 377
   Frobenius, 174
   Mollweide, 44
   Peterson–Codazzi, 397
   Serret–Frénet, 386
   Weingarten, 396
forward Kolmogorov equation, 1076
Fourier coefficients, 358
Fourier cosine inversion formula, 445
Fourier cosine transform, 445, 449
   asymmetric, 445
   tables, 1177
Fourier expansion
   of even functions, 360
   of odd functions, 360
Fourier inversion formula, 444
Fourier method, 602
Fourier series, 357, 358
   criteria of uniform convergence, 361
   differentiation, 359
   Dirichlet–Jordan convergence criteria, 358
   in complex form, 360
   integration, 359
   Lipschitz convergence criteria, 358
   of square-integrable functions, 361
   Riemann localization principle, 358

Fourier sine transform, 445, 446, 449
   asymmetric, 446
   tables, 1182
Fourier transform, 443, 449, 614, 843
   alternative, 444
   asymmetric, 444
   asymmetric form, 444
   inverse, 443
   $n$-dimensional, 445
   solving linear problems of mathematical physics, 615
   standard form, 443
fourth Painlevé transcendent, 498
fourth-degree algebraic equation, 159
fourth-order linear PDEs, 1294
fourth-order nonlinear PDEs, 1332
fraction, 6
fractional integral, 449
fractional power series, 415
Fredholm alternative, 834
   for symmetric integral equations, 839
Fredholm integral equation
   first kind, 824, 825
   second kind, 829, 852
   second kind, with degenerate kernel, 830
   second kind, with symmetric kernel, 835
Fredholm kernel, 825, 829
Fredholm theorems, 834
free oscillations, 1213
free terms of linear system of equations, 197
frequency
   cumulative, 1084
   of event, 1083
   of variate value, 1083
frequency histogram, 1086
frequency polygon, 1085
Frénet formulas, 377
Fresnel integrals, 304, 942
Frobenius formulas, 174
Frullani integral, 307
frustum
   of cone, 67, 68
   of pyramid, 64
Fuchs index, 749
function
   admissible, 993, 1002, 1006
   Airy, 477, 955
   analytic, 413
   analytic at point, 403
   analytic at point at infinity, 403
   analytic in domain, 403
   antiperiodic, 885
   approximation, 508
   arccosine, xxx, 32
   arccotangent, xxx, 32
   arcsine, xxx, 31
   arctangent, xxx, 32
   autocorrelation, 1072
   automorphic, 912
   Bessel, 947, 1216
   Bessel, modified, 953

function (*continued*)
  Bessel, modified, of first kind, xxxi, 953, 1216
  Bessel, modified, of second kind, xxxi, 953, 1216
  Bessel, of first kind, xxxi
  Bessel, of second kind, xxxii
  beta, 945
  boundary, 419
  bounded, 236
  bounded variation, 246, 247, 313
  branch point, 404, 405
  central cross-correlation, 1073
  characteristic, 1046, 1066
  Chebyshev, 985
  complementary error, xxxi, 939, 958
  complex variable, 401, 402, 407, 408
  composite, 237, 265
  concave, 245
  continuous, 243
  convex, 245
  correlation, 1072
  cosine, 24, 25
  cosine, hyperbolic, xxx, 35
  cotangent, xxx, 24, 26
  cotangent, hyperbolic, xxx, 35
  cross-correlation, central, 1073
  cross-correlation, normalized, 1073
  cross-covariance, 1072
  cumulative distribution, 1039
  cylindrical, 947, 1216
  decreasing, 236
  degenerate hypergeometric, xxxii, 956
  differentiable, 251
  differentiable at point, 402
  digamma, 944, 949
  Dirac delta, 599
  distribution of two-dimensional random vector, 1057
  distribution sample, 1084
  domain, 236
  elliptic, 972
  entire, 414
  entire transcendental, 414
  error, xxxi, 939, 957, 1267
  even, 236
  exponential, 21
  gamma, xxxii, 943, 1215
  gamma, incomplete, xxxii, 946, 957, 958
  Gauss error, xxxi
  Gegenbauer, 1218
  generating, 815, 1047
  generating, of numerical sequence, 1047
  Green, 428, 631, 633–637
  Hankel, 952
  harmonic, 404
  Heaviside unit step, 618
  Hermite, 986
  holomorphic at point, 403
  homogeneous, 1441
  Hurwitz, 893
  hyperbolic arccosine, xxx

function (*continued*)
  hyperbolic arccotangent, xxx
  hyperbolic arcsine, xxx
  hyperbolic arctangent, xxx
  hyperbolic cosine, xxx, 35
  hyperbolic cotangent, xxx, 35
  hyperbolic sine, xxxi, 35
  hyperbolic tangent, xxxi, 35
  hypergeometric, 960
  incomplete beta, 947
  incomplete gamma, xxxii, 946
  increasing, 236
  infinitely large, 243
  infinitely small, 242
  integrable, 286, 312, 314
  integrand, 273
  inverse, 237
  inverse hyperbolic cosine, 40
  inverse hyperbolic cotangent, 40
  inverse hyperbolic functions, 39
  inverse hyperbolic sine, 39
  inverse hyperbolic tangent, 40
  involutory, 917, 1423
  Jacobi theta, 978
  joint distribution, 1057, 1064
  joint probability, 1059, 1064
  Kummer, 956
  Lagrange, 270
  Legendre, 962
  Legendre, associated, 964, 965
  Legendre, associated, modified, 966
  Legendre, associated, of first kind, 965
  Legendre, associated, of second kind, 965
  Legendre, of second kind, 1217
  limit, 264
  logarithmic, 22
  Lyapunov, 549
  Macdonald, 953
  marginal distribution, 1064
  marginal probability, 1059
  Mathieu, 980, 1219
  Mathieu, modified, 982
  mean value, 318, 325
  meromorphic, 414
  moment-generating, 1047
  monotone (monotonic), 236
  multi-valued, 236, 402
  multivariate random variable, 1065
  Neumann, 947
  nonincreasing, 236
  normalized cross-correlation, 1073
  odd, 236
  of exponential growth, 905
  of several variables, 263
  of two variables, 263
  Painlevé transcendental, 495
  parabolic cylinder, 967
  penalty, 1027
  periodic, 236
  Pontryagin, 1011
  power, 19
  primitive, 273

function (*continued*)
  probability density, 1040
  probability-generating, of random variable, 1047
  psi, 944
  random variables, 1041
  range, 236
  rational, 276
  real variable, 236
  regression, 1063, 1066
  regular at point, 403
  Riemann, 628, 629, 783
  sample, 1072
  sign, xxxi
  sine, 24, 25
  sine, hyperbolic, xxxi, 35
  single-valued, 236, 402
  spectral of stationary process, 1075
  square integrable, 314
  stationary value, 220
  stream, 1295
  strictly concave, 245
  strictly convex, 245
  Struve, 448, 449
  symmetric elementary, 163
  tangent, xxxi, 24, 26, 56, 368, 380
  tangent, hyperbolic, xxxi, 35
  theoretical distribution, 1082
  theta, 978
  total variation, 247
  Weber parabolic cylinder, 967
  Weierstrass, 998, 1000
  Weierstrass elliptic, 976
  Whittaker, 960
  zero, 413
  Zhukovskii, 404, 424
functional, 991
  bilinear, 992
  Bolza, 1001
  classical integral, 993
  differentiable at point, 992
  first variation, 992
  increment, 992
  linear, 213
  quadratic, 992
  quadratic definite positive, 993
  second variation, 992, 993
  strong minimum, 991
  strongly positive, 993
  variation, 992
  weak minimum, 991
functional differential equations, 682, 684, 688, 931
functional equation
  Abel, 914
  Babbage, 917, 1416, 1423
  bilinear, 930, 1446
  Cantor first, 1442
  Cantor second, 1449
  Cauchy, 1438
  Cauchy exponential, 1443

functional equation (*continued*)
  Cauchy logarithmic, 1439
  Cauchy power, 1444
  cyclic, 920
  D'Alembert, 1443
  Darboux, 458
  difference, 873
  difference, logistic, 875
  Gauss, 1445
  Gauss, generalized, 1445
  homogeneous function, 1441
  information theory, basic, 1440
  involutory functions, 917, 1423
  Jensen, 1438
  Jensen, generalized, 1449
  linear homogeneous, 910
  linear nonhomogeneous, 912
  Lobachevsky, 1444
  logarithmic, Cauchy, 1439
  Pexider, 925, 1440
  reciprocal, 919
  Riccati, difference, 875, 918, 1428, 1429
  Schröder–Koenigs, 911
  self-similar solutions, 1442
  traveling-wave, 1441
functional equations, 705–707, 1409, 1414–1421, 1446
  containing iterations of unknown function, 917
  in several independent variables, 922, 1438
  linear, 1414–1421
functional mean, 13
functional separable solution, 697
  special, 697
functional separation of variables, 697
functional series, 348, 1120
  criteria of uniform convergence, 349
  Dirichlet criterion, 350
  pointwise convergence, 348
  properties, 350
  remainder, 349
  uniform convergence, 349
functional sums, 1116
functions
  coordinate, 847, 851
  dependent, 267
  elementary, 19–41
  independent, 267
  of same order, 243
  small infinitely equivalent, 243
  special, 939–990
fundamental solution, 599, 615
  of Cauchy problem, 616, 617
fundamental solutions for system of ODEs, 539
fundamental system of solutions, 477, 478, 522, 523
  of homogeneous system, 198
fundamental theorem of algebra, 162

# G

Gakhov first theorem, 432
Gakhov second theorem, 433

Galerkin method, 508
Galileo transformation, 718
game, 1024
    antagonistic, 1024
    coalition, 1024
    finite, 1024
    infinite, 1024
    lower price, 1025
    theory, 1024
    two-person, 1024
    upper price, 1025
    with saddle point, 1025
game price in mixed strategies, 1026
game theory, 1024
game value, 1025
"games with nature", 1024
gamma function, xxxii, 943, 1213
    incomplete, xxxii, 946
    logarithmic derivative, 949, 954
gas dynamics, model equation, 1254
Gauss derivation formulas, 396
Gauss equation, 1445
Gauss error function, xxxi
Gauss formula, 396, 808
Gauss theorem, 335
Gauss transform, 449
Gauss transformation, 974
Gauss–Jordan elimination, 201
Gaussian coordinates, 387
Gaussian curvature, 394
Gaussian elimination, 199
Gaussian elimination method, backward substitution, 200
Gaussian hypergeometric equation, 960, 1216
Gegenbauer functions, 1218
Gegenbauer polynomials, 987
Gel'fand–Levitan–Marchenko integral equation, 1328, 1333
general equation
    of plane, 124
    of quadric, 148
    of second-order curve, 107
    of second-order hypersurface, 220
    of straight line, 90
    of straight line in space, 132
    of straight line on plane, 90
general form
    of Laplace integral, 292
    of Schrödinger equation, 1311
    of second-order finite-difference equation, 1436
    of second-order nonlinear difference equation, 1436
general fourth-degree equation, 160
general integral, 570, 571
general linear group, 228
general mathematical programming problem, 1012
general orthogonal curvilinear coordinates, 1196
general Riccati equation, 461, 462, 1209, 1210
general solution
    of difference equation, 874, 884

general solution (*continued*)
    of first-order PDEs, 553
    of linear algebraic system, 203
    of linear hyperbolic equations, 600–602
    of linear ODEs, 472, 514, 518
    of ODEs, 453, 488
general square system of linear algebraic equations, 199
general system of $m$ linear algebraic equations with $n$ unknown quantities, 203
generalized Abel equation, 806
    of second kind, 1393
generalized Burgers–Korteweg–de Vries equation, 709
generalized Cauchy problem, 627
generalized Chebyshev inequalities, 15
generalized classical method of characteristics, 582
generalized Dirichlet problem, 428
generalized Ermakov criterion of convergence of series, 340
generalized Ermakov system, 1242
generalized Gauss equation, 1445
generalized homogeneous equation, 490, 526, 1210, 1224
generalized Jensen equation, 1449
generalized Jensen inequalities, 246
generalized Laguerre polynomials, 983
generalized permutation matrix, 168
generalized reciprocal equation, 160
generalized self-similar solution, 677
generalized separable solution, 681, 682
generalized separation of variables, 681
generalized similarity reduction, 746
generalized solution, 563
    Hopf formula, 565
    of first-order PDE, 563, 568
    regular point, 580
    singular point, 580
    stable, 569
generalized telegraph equation, 626
generalized traveling-wave solution, 697
generalized triangle inequality, 343
generalized variance of $n$-dimensional distribution, 1065
generalized viscosity solutions, 579, 580
generating function, 815, 1047
    of numerical sequence, 1047
generator, 65, 67
    group, 717, 730
    infinitesimal, 717
generic term
    of sequence, 237
    of series, 337
geodesic arc, 70
geodesic curvature, 392
geodesic curvature vector, 392
geometric definition of nondegenerate second-order curve, 111
geometric distribution, 1048
geometric mean, 13
    of functions, 297

geometric multiplicity of eigenvalue, 209
geometric probability, 1034
geometric programming, 1027
geometric progression, 11
geometric sequence, 11
geometric series, 11
geometry
   analytic, 77
   differential, 367
   intrinsic, of surface, 391
   spherical, 70
Gergonne point, 47
Goodness-of-fit tests, 1098, 1101
gorge ellipse, 145
Goursat problem, 592, 629
Gram determinant, 193
Gram–Schmidt orthogonalization, 194
graph of function, 84, 236
gravitational field potential, 328
great circle, 68, 70
   spherical geometry, 70
greatest common divisor, 4
Green first identities, 335
Green formula, 428
Green function, 428, 623, 631, 633–637
   admitting incomplete separation of variables, 640
   for elliptic equations, 631–634, 637
   for hyperbolic equations, 623–625, 636
   for linear ordinary differential equation, 481
   for parabolic equations, 618–622, 635
   in terms of infinite series, 639
Green second identities, 335
Green theorem, 335
group
   abelian, 226
   automorphism, 227
   commutative, 226
   completely reducible representation, 231
   conjugate element, 231
   continuous point, 730
   equivalent representations, 230
   faithful representation, 230
   invariant, 731
   Lie, 730
   linear representation, 230
   of derivation formulas, second, 396
   of events, complete, 1031
   of incompatible events, complete, 1032
   of leading terms of equation, 220
   of linear transformations, 228
   of mirror symmetry, 229
   of modulo 2 residues, 226
   of orthogonal transformations, 229
   orthogonal, 229
   quotient, 227
   special linear, 228
   special orthogonal, 229
   trivial representation, 230
group analysis of differential equations, 716
group element, character, 231

group generator, 717, 730, 732
group invariant, 731
group theory, 225
groups
   basic properties, 225
   cyclic, 229
   isomorphic, 227
   permutation, 226
   unitary, 229
growth exponent, 436

# H

half-plane
   left, 79
   lower, 79
   right, 79
   upper, 79
half-side theorem, 73
Hamilton operator, 272
Hamilton–Jacobi equation, 576
Hamiltonian, 576
Hankel functions, 952
Hankel transform, 446, 449, 1400
Hardy transform, 449
harmonic function, 404
harmonic mean, 13
   of functions, 297
harmonic series, 339
Hartley transform, 449
heat equation
   linear, 585, 1267
   linear, nonhomogeneous, 1268
   nonlinear, 1318
   stationary, anisotropic, 1323
   stationary, with nonlinear source, 1320
   with axial symmetry, 1270
   with central symmetry, 1272
Heaviside unit step function, 618
Helmholtz equation, 1289
Hermite functions, 986
Hermite polynomials, xxxi, 957, 958, 985
Hermitian matrix, 168
Hermitian operator, 206
   negative, 207
   nonnegative, 207
   nonpositive, 207
   positive, 207
Heron formula, 48
higher-order derivatives of elementary functions, 255
higher-order linear difference equations, 1425
Hilbert boundary value problem, 432
Hilbert example, 997
Hilbert space, 196
Hilbert transform, 449, 1396, 1400
Hilbert–Privalov boundary value problem, 432
Hilbert–Schmidt theorem, 837
histogram, 1086
   frequency, 1086
   relative frequency, 1086

hodograph transformation
  for gas dynamic system, 784
  for ODEs, 455
  for PDEs, 657, 658
Hölder condition, 310, 430
Hölder constant, 310
Hölder exponent, 310
Hölder inequality, 14, 297, 343
Hölder–Jensen inequality, 15
homogeneity, order, 1441
homogeneous boundary condition, 591
homogeneous equations
  linear difference, 881, 885, 889, 1421, 1425
  linear difference, constant-coefficient, 1409, 1422, 1426
  linear integral, 810, 829, 834
  ordinary differential, 456, 490, 526, 1207, 1224
  ordinary differential, generalized, 457, 490, 526, 1224
  ordinary differential, linear, 472, 514
  partial differential, elliptic, 592
  partial differential, hyperbolic, 592
  partial differential, parabolic, 591
homogeneous function, 1441
homogeneous initial condition, 591
homogeneous linear algebraic system, 203
  general solution, 203
homogeneous linear integral equation, 810, 829
homogeneous linear ODEs, 472, 514
  of higher-order ODEs, 532
homogeneous Monge–Ampère equation, 668, 1326
homogeneous system
  of linear algebraic equations, 197
  of linear ODEs, 539
  of linear ODEs, superposition principle, 528, 539
homomorphic image, 226
homomorphisms, 226
"hoof", 67
Hopf formula for generalized solution, 565
Hopf–Cole transformation, 666, 1308
Horner scheme, 156
Hurwitz functions, 893
hyperbola, 101
  arm, 101
  asymptotes, 101
  canonical equation, 101
  center, 102
  conjugate diameters, 104
  diameter, 104
  directrix, 102
  eccentricity, 102
  equilateral, 101
  focal axis, 101
  focal distance, 102
  focal parameter, 102
  focal properties, 102
  focus, 102
  focus-directrix property, 103
  imaginary axis, 101

hyperbola (*continued*)
  imaginary semiaxis, 102
  in polar coordinate system, 104
  left arm, 101
  left directrix, 102
  left focus, 102
  linear eccentricity, 102
  numerical eccentricity, 102
  optical property, 103
  parameter, 102
  real semiaxes, 102
  right arm, 101
  right directrix, 102
  right focus, 102
  vertex, 102
hyperbolic arccosine, xxx
hyperbolic arccotangent, xxx
hyperbolic arcsine, xxx
hyperbolic arctangent, xxx
hyperbolic cosine, xxx, 35
hyperbolic cotangent, xxx, 35
hyperbolic equation
  boundary value problems, 623
  canonical form, 587
  first canonical form, 587
  second canonical form, 587
hyperbolic equations, 585, 590, 647, 648, 742
  Cauchy problem, 617
  linear, 587, 617, 623–630, 636, 647, 648, 1278–1284
  nonlinear, 653, 654, 1312–1317
hyperbolic functions, 34
  addition formulas, 36
  differentiation formulas, 38
  expansion in power series, 39
  integration formulas, 38
  inverse, 39
  of half argument, 38
  of multiple argument, 38
  properties, 36
hyperbolic paraboloid, 146
hyperbolic point, 394, 536
hyperbolic sine, xxxi, 35
hyperbolic spiral, 86, 368
hyperbolic system of PDEs, 781, 790
hyperbolic tangent, xxxi, 35
hyperbolic type of second-order curves, 108
hyperboloid, 144, 224
  of revolution, 145
  one-sheeted, 144
  real semiaxes, 144
  two-sheeted, 144
hypergeometric distribution, 1049
hypergeometric equation, 960, 1218
  degenerate, 1215
hypergeometric functions, 960
hypergeometric series, 960, 1218
hypersurface, 220
  center, 222
  central, 222
  noncentral, 224
  semiaxis, 223

hypotenuse, 50, 72
hypothesis, 1094
　alternative, 1094
　composite, 1094
　nonparametric, 1094
　null, 1094
　parametric, 1094
　simple, 1094
　statistical, 1094

# I

identity
　Abel, 902
　Parseval, 361
identity matrix, 168
identity operator, 204
image, homomorphic, 226
imaginary axis, 401
　of hyperbola, 101
imaginary ellipsoid, 224
imaginary part, 399
imaginary semiaxis
　of hyperbola, 102
　of one-sheeted hyperboloid, 144
　of two-sheeted hyperboloid, 144
imaginary unit, 399
impossible event, 1031
improper integral, 301
improper integrals
　involving logarithmic functions, 305
　involving power functions, 304
　involving trigonometric functions, 305
　of unbounded functions, 309
　sufficient conditions for convergence, 302
improper orthogonal transformations, 229
improper subspace, 190
incenter, 47
incenter point, 46
incircle, 46
　of triangle, 46, 47
incompatible events, 1031
incomplete beta function, 947
incomplete cubic equation, 158
incomplete elliptic integral, 970
incomplete equation of plane, 124
incomplete gamma function, xxxii, 946, 957, 958
incomplete separation of variables, 597, 616, 640
incomplete system of eigenfunctions, 836
inconsistent system of linear algebraic equations, 197
increasing function, 236, 238
increment
　of argument, 992
　of functional, 992
　partial, 264
indefinite integrals, 273, 1129
　involving exponential functions, 1137
　involving hyperbolic functions, 1137
　involving inverse trigonometric functions, 1147
　involving irrational functions, 1134
　involving logarithmic functions, 1140

indefinite integrals (*continued*)
　involving rational functions, 1129
　involving trigonometric functions, 1142
　properties, 274
independent random variables, 1060, 1067
independent trials, 1037
independent variable, 236
index
　characteristic, 1221
　Fuchs, 749
　of inertia, 218
　of inertia, negative, 218
　of inertia, positive, 218
　of matrix, 182
　of multiplication, 12
　of summation, 12
inequality
　Bunyakovsky, 296
　Cauchy, 14, 296, 343, 410
　Cauchy–Bunyakovski, 14, 296, 343
　Cauchy–Schwarz, 121, 193, 1043
　Cauchy–Schwarz–Bunyakovski, 14, 314
　Chebyshev, 14, 297, 1070
　Chebyshev, generalized, 15
　Cramér–Rao, 1089
　Hölder, 14, 297, 343
　Hölder–Jensen, 15
　Jensen, 245, 297
　Markov, 1070
　Minkowski, 14, 121, 193, 296, 343
　Steklov, 297
　Sylvester, 178
　triangle, 44, 73, 193, 314
　triangle, generalized, 14, 343
infimum, xxxi
infinite game, 1024
infinite matrix, 167
infinite product, 346, 348
　absolute convergence, 346
　convergent, 346
　divergent, 346
infinite products
　convergence criteria, 347
　with positive factors, 347
infinite series, 1116
infinite-dimensional linear space, 189
infinitely large function, 243
infinitely large quantity, 239
infinitely large sequence, 239
infinitely small function, 242, 243
infinitely small quantity, 239
infinitesimal generator, 717
infinitesimal operator, 716, 717, 730
inflection point, 258
information, Fisher, 1089
information theory, basic equation, 1438
initial condition
　for first-order ODEs, 454
　for parabolic equations, 591
　homogeneous, 591

initial conditions
  for hyperbolic equations, 592
  for ODEs, 488, 524
  for Riemann problem, 785, 793
  homogeneous, 592
initial moment
  of multivariate random variable, 1065
  of two-dimensional random variable, 1060
initial value problem, 580, 785
injective linear operator, 205
inner product, 192, 314, 840
inscribed angle, 56
inscribed polygon, in circle, 51
integer number, 3
integer power
  of complex number, 399
  of linear operator, 205
  of matrix, 172
integrable combinations, 545
integrable function, 286, 312, 314
integral
  absolutely convergent, 303
  Cauchy-type (singular), 310, 409, 430
  closed path, 331
  complete, 570
  curvilinear, 329
  definite, 286
  Dini, 1287
  Dirichlet–Mehler, 963
  divergent, 301
  Duhamel, 516
  elliptic, complete, of first kind, 969
  elliptic, complete, of second kind, 969
  elliptic, of first kind, 970
  elliptic, of second kind, 970
  elliptic, of third kind, 971
  exponential, 958
  fractional, 449
  Fresnel, 304
  Frullani, 307
  general, 571
  improper, 301
  indefinite, 273
  involving power-law functions, 1147
  Laplace, 418, 521, 963
  Laplace, of general form, 292
  Laplace, power, 294
  Laplace, power, multiple, 328
  Laplace, special form, 291
  line, of first kind, 329
  line, of second kind, 330, 331
  logarithmic, 958
  multiple Laplace-type, 328
  of differential binomial, 281
  of function of complex variable, 407
  path, 329
  Poisson, 428, 1287
  Poisson, for half-plane, 429
  potential-type, 291
  iterated, 319
  Riemann, 286

integral (*continued*)
  Schwarz, 428
  Schwarz, for half-plane, 429
  singular (Cauchy-type), 310, 311, 430, 571
  Stieltjes, 312
  surface, of first kind, 332
  surface, of second kind, 333
  with infinite limits, 301
integral equation
  Abel, 1385
  Abel, generalized, 1386
  Abel, of second kind, 1392
  Abel, of second kind, generalized, 1393
  existence and uniqueness of solution, 802
  first kind, with constant integration limit, 824
  first kind, with variable integration limit, 801
  Fredholm, of first kind, 824, 825
  Fredholm, of second kind, 829
  Gel'fand–Levitan–Marchenko, 1328, 1333
  homogeneous, 810, 829
  kernel, 801, 825, 829
  Lalesco–Picard, 1403
  linear, 801, 810, 824, 829
  nonhomogeneous, 810, 829
  nonlinear, 856
  nonlinear, with constant integration limits, 856, 863
  nonlinear, with degenerate kernel, 863
  nonlinear, with variable integration limit, 856
  parameter, 829
  regularization methods, 827
  right-hand side, 810
  Schlömilch, 1400
  second kind, with constant integration limits, 829
  second kind, with variable integration limit, 810
  skew-symmetric, 840
  symmetric, 835
  Tricomi, 1403
  Urysohn, nonlinear, 856
  Urysohn, with degenerate kernel, 863, 864
  Volterra, nonlinear, 856, 858–861
  Volterra, of first kind, 801, 807
  Volterra, of second kind, 807, 810
  Volterra, with quadratic nonlinearity, 856
  Wiener–Hopf, of first kind, 825, 1401
  Wiener–Hopf, of second kind, 829, 1406
  with degenerate kernel, 802, 803, 811, 812
  with difference kernel, 804, 813, 825, 829, 841
  with quadratic nonlinearity, 865
  with weak singularity of second kind, 829
integral equations, 801–871, 1385–1406
  first kind, with constant limits of integration, 824, 1396–1400
  first kind, with variable limit of integration, 801, 1385–1390
  first kind, with weak singularity, 825
  linear, 801–855, 1385–1406
  method of differentiating, 866
  second kind, with constant limits of integration, 829, 1401–1408

integral equations (*continued*)
   second kind, with variable limit of integration, 810, 1391–1395
integral formulas of vector calculus, 334
integral sum, 286, 317
   Stieltjes, 312
integral theorem of de Moivre–Laplace, 1038
integral transform
   Bochner, 449
   Euler, first kind, 449
   Euler, second kind, 450
   Fourier, 443, 449
   Fourier, asymmetric form, 444
   Fourier, inverse, 443
   Fourier, $n$-dimensional, 445
   Fourier, standard form, 443
   Fourier cosine, 445, 449
   Fourier sine, 446, 449
   Gauss, 449
   Hankel, 446, 449
   Hardy, 449
   Hartley, 449
   Hilbert, 449
   kernel, 435
   Kontorovich–Lebedev, 448, 449
   Laplace, 436, 449
   Laplace, inverse, 437
   Laplace, two-sided, 449
   Meijer, 447, 449
   Meler–Fock, 449
   Mellin, 441, 449
   Mellin, inverse, 442
   Weber, 449
   Weierstrass, 449
integral transforms, 435, 1155
   general form, 435
   inversion formulas, 435
   method for linear PDEs, 611
integrals, tables, 1129
integrals of motion, 766
integrand (function), 273
integrating factor, 459
   of Pfaffian equation, 773
integration
   by parts, 275
   by parts, repeated, 275
   of asymptotic series, 365
   of exponential functions, 281
   of hyperbolic functions, 281
   of improper fractions, 278
   of irrational functions, 279
   of proper fraction, 277
   of rational functions, 276
   of trigonometric functions, 282
intercept equation of plane, 125
intercept-intercept equation of straight line, 91
intergroup sample variance, 1108
interior angle, 43
   of polygon, 51
interpolation nodes, 808
interpolation polynomial, 186

intersection
   of events, 1031
   of four planes, 139
   of line and plane, 140
   of subspaces, 190
   of three planes, 139
interval, 235
   closed, xxx, 235
   confidence, 1093
   estimator, 1088, 1093
   half-open, xxx, 234
   of convergence of power series, 351
   open, xxx, 235
   series, 1083
intrinsic equations of curve, 388
intrinsic geometry of surface, 391, 395
invariance condition, 719, 720, 732
invariant, 222, 731
   of group, 731
   of infinitesimal operator, 717
   of linear operator, 211
   of operator, 717, 731
   of quadric, 148
   of second-order curve, 107
   universal, 731
invariant condition, 731
invariant representation, 230
invariant solution, 669, 674, 716, 724
invariant subspace, 205, 230
invariant surface condition, 732
inverse element, 225
inverse Fourier transform, 443, 614
inverse function, 237
inverse hyperbolic cosine, 40
inverse hyperbolic cotangent, 40
inverse hyperbolic functions, 39
   differentiation formulas, 41
   expansion in power series, 42
   integration formulas, 41
   properties, 41
inverse hyperbolic sine, 39
inverse hyperbolic tangent, 40
inverse Laplace transform, 437, 611
   tables, 1162
inverse matrix, 171
   left, 171
   right, 171
inverse Mellin transform, 442
   tables, 1190
inverse of mapping, 226
inverse operator, 205
inverse scattering problem, 760
   method, 755
inverse transform, 435
   Fourier, 443
   Fourier cosine, 445
   Fourier sine, 446
   Hankel, 446
   Kontorovich–Lebedev, 448
   Laplace, 437
   Meijer, 447

inverse trigonometric functions, 30
  arccosine, 32
  arccotangent, 32
  arcsine, 31
  arctangent, 32
  differentiation formulas, 34
  expansion in power series, 34
  integration formulas, 34
inversion, 178, 422
invertible linear operator, 205
invertible mapping, 226
invertible transformation, 267
involutive matrix, 172
involutory function, 917, 1423
irrational numbers, 4, 5
irreducible representation, 231
isolated point, 371
isolated singularity, 413
isometric correspondence, 395
isomorphic Euclidean spaces, 194
isomorphic groups, 227
isomorphic linear spaces, 189
isomorphic mapping of groups, 227
isomorphic unitary spaces, 196
isomorphism, 226, 227
isomorphisms of linear spaces, 188
isoperimetric constraints, 1002
isoperimetric problem, 1002
isosceles trapezoid, 54
isosceles triangle, 43, 50
    base, 50
    leg, 50
iterated integral, 319
iterated kernel, 822, 832, 838
iteration axis, 876
iteration of function, 907
iteration process, 858

# J

Jacobi condition, 998, 1000, 1004, 1007
    strengthened, 1000, 1004, 1007
Jacobi elliptic functions, 979
Jacobi equation, 998, 1004, 1007
Jacobi method, 217
Jacobi polynomials, 986
Jacobi theorem, 180
Jacobi theta functions, 978
Jacobi–Mayer bracket, 574
Jacobian, 267, 321
Jacobian determinant, 321
Jensen equation, 1436
Jensen inequality, 15, 245, 297
joining face of prism, 61
joint distribution function, 1064
    of random variables, 1057
joint probability function, 1059, 1064
Jordan cell, 182
Jordan form, of matrix, canonical, 182
Jordan lemma, 417, 436
jump conditions, 561, 568, 791

# K

$K$-transform, 449
kernel
  degenerate, 801, 802, 810, 811, 825, 829
  difference, 801, 804, 810, 813, 825, 829, 841
  Fredholm, 825, 829
  iterated, 822, 832
  logarithmic, 801
  of integral equation, 801, 825, 829
  of integral transform, 435
  of linear operator, 205
  polar, 801, 825
  positive definite, 837
  symmetric, 825, 829
  with weak singularity, 801, 825, 829
Khokhlov–Zabolotskaya equation, 658
Klein–Gordon equation, 625, 1280
Kolmogorov distribution, 1086
Kolmogorov test, 1099
Kolmogorov theorem, 1070
Kontorovich–Lebedev inversion formula, 448
Kontorovich–Lebedev transform, 448, 449
Korteweg–de Vries equation, 752, 756, 759, 764, 1327
  cylindrical, 1328
  modified, 1328
Kronecker delta, xxxii
Kronecker product of matrices, 175
Kronecker–Capelli theorem, 198
$k$th absolute moment, 1044
$k$th central moment, 1043
$k$th moment, 1043
$k$th sample central moment, 1087
$k$th sample moment, 1087
Kummer convergence criterion for series, 340
Kummer functions, 956
Kummer series, 956, 1215
Kummer transformation, 345, 957
Kummer–Liouville transformation, 475
kurtosis, 1045

# L

$L_2$-norm, 314
L'Hospital rules, 254
L'Huiller equation, 74
Lagrange equation, 467
Lagrange function, 270
Lagrange method, 217
Lagrange multiplier, 270
Lagrange multiplier rule, 1002
Lagrange principle, 1008
Lagrange problem, 1008
Lagrange theorem, 254
Lagrange–Charpit method, 573
Lagrangian, 767, 993, 1002, 1006
  Bolza, 1001
  Noetherian symmetry, 768
Laguerre polynomials, 957, 982
  generalized, 983
Lalesco–Picard equation, 1403

Lamé coefficients, 1196
Landen transformations, 974, 975
Laplace cascade method, 600
Laplace equation
 ODE, 520
 PDE, three-dimensional, 643
 PDE, two-dimensional, 633, 1282
Laplace integral, 418, 521, 963
 general form, 292
 special form, 291
Laplace operator, 272
Laplace theorem, 177
Laplace transform, 436, 548, 611, 804, 813, 843, 856
 convolution theorem, 437
 for derivatives, 611
 inverse, 437
 limit theorems, 440
 properties, 437
 solving linear problems for hyperbolic equations, 614
 solving linear problems for parabolic equations, 612
 tables, 1157–1163
Laplace-type multiple integral, 328
large discriminant, 108
large numbers
 strong law, 1070
 weak law, 1069
largest value of function, 258
lateral face of pyramid, 63
latitude, 119
Laurent series, 410, 412
Lavrentiev regularization method, 827
law
 commutative composition, 225
 composition, 225
 conservation, 766
 conservation at discontinuity, 561
 energy conservation, 996
 of conservation of mass, 784
 of cosines, 44
 of cosines of angles, 73
 of cosines of sides, 73
 of large numbers, 1069, 1070
 of sines, 44, 73
 of tangents, 44
 weak of large numbers, 1069
laws, de Morgan, 1032
Lax condition, 792
Lax pair, 755
least common multiple, 4
least-squared error method, 511
least-squares method, 509, 1092
left arm of hyperbola, 101
left coset, 227
left directrix
 of ellipse, 99
 of hyperbola, 102
left focal radius, 99, 102
left focus
 of ellipse, 99

left focus (*continued*)
 of hyperbola, 102
left half-plane, 79
left inverse matrix, 171
left rectangular Cartesian coordinate system, 79
left-hand limit, 240
leg
 of isosceles triangle, 50
 of right triangle, 50
 of trapezoid, 54
Legendre condition, 998, 999, 1004, 1007
 strengthened, 999, 1004, 1007
Legendre equation, 1217–1219
Legendre function, 962
 associated, 964, 965
 first kind, associated, 965
 modified associated, 966
 second kind, 1217
 second kind, associated, 965
Legendre functions, Wronskians, 967
Legendre polynomial, xxxi, 962, 1217
Legendre theorem, 72
Legendre transformation, 469, 660
 with many variables, 663
Leibnitz convergence criteria, 341
Leibnitz convergence criterion for series, 341
Leibnitz formula, 256
lemma
 Jordan, 417
 Neyman–Pearson, 1097
length
 of element, 193
 of segment, 77
 of vector, 113
level, confidence, 1093
level line, 264, 1014
Levi-Civita symbol, 179
Lévy solution, 914
Lie group, 730
 of point transformations, 730
Lienard equation, 1225
light rays, equation, 1259
likelihood equation, 1092
likelihood function, 1092
likelihood ratio, 1097
likelihood ratio test, 1097
limit
 first noteworthy, 241
 left-hand, 240
 lower of sequence, 240
 noteworthy second, 241
 of function, 240, 264
 of sequence, 237
 partial of sequence, 240
 right-hand, 240
 upper of sequence, 240
limit theorems, 1068, 1069
 for Bernoulli process, 1038
 for Laplace transform, 440
limiting self-similar solution, 675
Lindeberg central limit theorem, 1071

line
  deviation of point from line, 93
  Euler, 47
  in space, 131
  level, 264, 1014
  Simpson, 47
  support, 1014
  tangent, 251
  tangent to surface, 388
line integral
  first kind, 329
  second kind, 330, 331
linear algebraic equation, 157
linear combination
  of column vectors, 171
  of row vectors, 171
  of vectors, 188
linear constant-coefficient partial differential equation, 588
linear difference homogeneous equation with constant coefficients, 906
linear eccentricity
  of ellipse, 98
  of hyperbola, 102
linear equation
  algebraic, 157
  difference, 881, 885, 889, 1421, 1425
  difference, constant-coefficient, 1409, 1422, 1426
  integral, 801, 810, 824, 829
  ordinary differential, 472, 518, 1213
  partial differential, elliptic, 592
  partial differential, hyperbolic, 592
  partial differential, parabolic, 591
linear equations of mathematical physics, 1267
linear first-order ODE, 457
linear form, 213
linear functional, 213
linear functional equations, 907, 1414–1421, 1438
  in one independent variable, 1409
linear group, general, of dimension $n$, 228
linear harmonic oscillator, 1277
linear homogeneous difference equations, 904
  with constant coefficients, 894, 904
  with variable coefficients, 895, 901
linear homogeneous functional equations, 910
linear integral equations
  first kind, with constant limits of integration, 824, 1396–1400
  first kind, with variable limit of integration, 801, 1385–1390
  second kind, with constant limits of integration, 829, 1401–1408
  second kind, with variable limit of integration, 810, 1391–1395
linear nonhomogeneous difference equations, 906
  with constant coefficients, 895, 899
  with variable coefficients, 896, 902
linear nonhomogeneous functional equations, 912
linear operator, 204
  canonical form, 213

linear operator (*continued*)
  characteristic equation, 211
  determinant, 209
  in matrix form, 208
  injective, 205
  invariant, 211
  inverse, 205
  invertible, 205
  kernel, 205
  matrix, 208
  nondegenerate, 205
  norm, 204
  range, 205
  rank, 206
  trace, 209
linear operators
  product, 205
  sum, 204
linear ordinary differential equations, 457, 472–488, 514–524, 1212–1222
  canonical form, 473
linear partial differential equations, 553–650, 1267–1298
  canonical form, 585–590
linear programming problem, 1012
  standard form, 1013
  symmetric, 1013
linear representation of group, 230
linear space, 187, 189
  basis, 189
  complex, 188
  dimension, 189
  elements, 187
  infinite-dimensional, 189
  linear transformation, 204
  of affine space, associated, 189
linear spaces
  isomorphic, 189
  isomorphisms, 188
linear span, 190
linear subspace, 190
  invariant under nonlinear operator, 693, 695
linear superposition principle, 594
linear surface element, 390
linear systems
  of algebraic equations, 197
  of ODEs, 528
  of PDEs of general form, 1374
  of two second-order PDEs, 1341
linear transformation, 656
  of linear space, 204
linear transformations, product, 228
linear-fractional mappings, 422
linearly dependent column vectors, 171
linearly dependent elements, 188
linearly dependent row vectors, 171
linearly independent column vectors, 171
linearly independent elements, 188
linearly independent row vectors, 171

linearly independent solutions of difference
    equation, 882
lines
    asymptotic, 393
    condition for two lines in space to meet, 143
    curvature, 393
    curvilinear, 387
    in space, theorem on arrangement, 136
    parallel, 59, 96
    perpendicular, 95
    skew, 59
Liouville equation, 664
Liouville formula, 473, 518
    for system of ODEs, 540
Liouville theorem, 403, 414
Lipschitz condition, 248, 310, 313, 454, 488, 543, 584
Lipschitz convergence criterion for Fourier series, 358
Lipschitz criterion of uniform convergence of Fourier series, 361
Lobachevsky convergence criterion for series, 341
Lobachevsky equation, 1444
local maximum, 257
    point, 269
local minimum, 257
    point, 269
local theorem of de Moivre–Laplace, 1038
log-normal distribution, 1042
logarithm, xxxi, 9
    common, 9
    decadic, 9
    natural, xxxi
logarithmic branch point, 415
logarithmic derivative of gamma function, 949, 954
logarithmic function, 22
    graphs, 22
    properties, 23
logarithmic identity, 9
logarithmic integral, 941, 958
logarithmic kernel, 801
logarithmic residue, 416
logarithmic spiral, 86
logarithms, natural, 9, 22, 23, 238
logistic difference equation, 875
longitude, 119
Lorentz transformation, 718, 1278
lower bound, 235
lower half-plane, 79
lower limit of sequence, 240
lower price of game, 1025
lower triangular matrix, 168
Lyapunov central limit theorem, 1071
Lyapunov function, 549
Lyapunov stability, 546
Lyapunov stable, 546
Lyapunov theorem
    of asymptotic stability, 549
    of stability, 549

# M

Macdonald function, 953
Maclaurin formula, 257
Maclaurin method, 164
Maclaurin power series, 352
Maclaurin series, 353
Maclaurin–Cauchy integral criterion of convergence of series, 339
main diagonal, 168
major axis of ellipse, 98
majorant series, 349
many-valued function, branch, 404
mapping, 226
    conformal, 419
    groups, 227
    groups, isomorphic, 227
    invertible, 226
    linear-fractional, 422
    onto set, 226
    range, 226
    schlicht, 402
marginal distribution function, 1064
marginal probability functions, 1059
Markov inequality, 1070
Markov process, 1075
Markov property, 1075
mass of body of variable density, 327
mathematical induction, 16
mathematical model of transportation problem, 1019
mathematical physics equations, nonlinear, 1301
mathematical programming, 1012
    general problem, 1012
mathematical statistics, 1081
    problems, 1081
Mathieu equation, 980, 1221
    modified, 982, 1220
Mathieu functions, 980, 1221
    modified, 982
matrices, 167
    commuting, 169
    compatible size, 169
    congruent, 182
    conjunctive, 183
    difference, 169
    direct sum, 175
    elementary, 181
    elementary transformations, 180
    equal, 168
    equivalent, 180
    Kronecker product, 175
    multiplication, 169
    similar, 181
    subtraction, 169
    sum, 168
    unitary transformations, 183
    169
matrix, 167
    additively inverse, 169
    adjoint, 170

matrix (*continued*)
 adjugate, 180
 antihermitian, 168
 antisymmetric, 168
 block, 174
 canonical form, 181
 canonical Jordan form, 182
 characteristic, 183
 characteristic equation, 185
 characteristic polynomial, 185
 complex conjugate, 170
 correlation, 1066
 covariance, 1065
 defect, 177, 178
 determinant, of order $n$, 176
 determinant, of order 2, 175
 determinant, of order 3, 176
 diagonal, 168
 dimension, 167
 eigenvalue, 183
 elementary transformation, 199
 finite, 167
 generalized permutation, 168
 Hermitian, 168
 identity, 168
 index, 182
 infinite, 167
 inverse, 171
 inverse left, 171
 inverse right, 171
 involutive, 172
 multiplication by scalar, 169
 nilpotent, 172
 nondegenerate, 171
 nonsingular, 171
 normal, 171
 normal form, 181
 null, 167
 of bilinear form, 215
 of linear operator, 208
 of linear system of algebraic equations, basic, 197
 of moments, 1065
 of sesquilinear form, 214
 of simple structure, 182
 opposite, 169
 orthogonal, 170
 payoff, 1024
 payoff, saddle point, 1025
 rank, 177, 178
 real, 171
 rectangular, 177
 self-adjoint, 168
 signature, 182
 skew-Hermitian, 168
 skew-symmetric, 168
 spectrum, 183
 square, 168, 176
 subdiagonal, 168
 superdiagonal, 168
 symmetric, 168

matrix (*continued*)
 trace, 171
 transforming, 181
 triangular, lower, 168
 triangular, strictly lower, 168
 triangular, strictly upper, 168
 unitary, 171
 upper triangular, 168
 Vandermonde, 179
 zero, 167
matrix decomposition, 201
matrix entry, 167
 cofactor, 176
maximin principle, 1025
maximum
 conditional, 270
 constrained, 270
 local, 257
 local weak, 993, 1001, 1002, 1006
 of function, xxxi, 257
 strong, 1006
 strong local, 994
 strong of functional, 991
 weak of functional, 991
maximum likelihood estimator, 1092
maximum modulus principle, 403
maximum principle, Pontryagin, 1011
mean
 functional, 13
 geometric, 13
 harmonic, 13
 sample, 1087
mean curvature, 394
mean value of function, 318, 325
mean value theorems, 295
 for double integral, 318
 for triple integral, 325
mean values, general approach, 13
mean-square convergence, 315
mean-square deviation, 1088
mean-square regression, 1063, 1066
median, 1045
 of trapezoid, 54
 of triangle, 45
Mehler–Fock transform, 449
Meijer inversion formula, 447
Meijer transform, 447
Meijer transform ($K$-transform), 449
Mellin transform, 441, 449, 826, 842, 844
 inverse, 442
 properties, 442
 tables, 1187–1190
Menelaus theorem, 49
meromorphic function, 414
Mertens theorem, 343
method
 averaging, 503
 backward substitution in Gaussian elimination, 200
 bisection, 261
 Bubnov–Galerkin, for integral equations, 850
 Bubnov–Galerkin, for ODEs, 509

method (*continued*)
   Budan–Fourier, 166
   characteristics, generalized classical, 582
   Clarkson–Kruskal direct, 708–710
   Cole–Kevorkian, 504
   collocation, 509, 523, 847, 848, 860
   collocation, convergence theorem, 524
   collocation, for integral equations, 847, 848
   collocation, for ODEs, 509, 523
   D'Alembert, 533
   D'Alembert, for nonhomogeneous higher-order linear systems, 533
   differential constraints, 737
   estimation, 1091
   false position, 262
   Fourier, 602
   Galerkin, 508
   Gaussian elimination, backward substitution, 200
   integral transforms, 611
   Jacobi, 217
   Lagrange, 217
   Lagrange–Charpit, 573
   Laplace cascade, 600
   Lavrentiev regularization, 827
   least-squared error, 511
   least squares, 509, 1092
   Lindstedt–Poincaré, 502
   Maclaurin, 164
   mathematical induction, 16
   minimal cost, 1020
   modified Newton–Kantorovich, 859, 868
   moments, 509, 1091
   Newton, 164
   Newton–Kantorovich, 859, 868
   Newton–Raphson, 262
   of accelerated convergence in eigenvalue problems, 513
   of approximating kernel by degenerate one, 844
   of argument elimination, for functional equations, 928
   of differentiation for integral equations, 857, 866
   of differentiation for ODEs, 857
   of differentiation in independent variables, for functional equations, 925
   of differentiation in parameter, for functional equations, 922
   of Euler polygonal lines, 471
   of functional separation of variables, 697
   of generalized separation of variables, 681
   of integral transforms, 825, 856, 865
   of integral transforms, for linear PDEs, 611
   of integral transforms, for nonlinear integral equations, 856, 865
   of integration by differentiation, 465
   of inverse scattering problem, 755
   of least-squared error, 511
   of least squares, 509, 849, 1092
   of least squares, for integral equations, 849

method (*continued*)
   of least squares, normal system, 850
   of matched asymptotic expansions, 506
   of model solutions, 818
   of partitioning domain, 511
   of potentials, 1021
   of quadratures, 808, 823
   of quadratures, for integral equations, 808, 823, 852, 861, 870
   of regular expansion in small parameter, for ODEs, 470, 501
   of replacing kernel by degenerate kernel, 844
   of scaled parameters, 502
   of separation of variables, 602
   of separation of variables, for first-order PDEs, 571
   of separation of variables, for linear PDEs, 602
   of substituting particular values of independent arguments, for functional equations, 926
   of successive approximations, 469, 511, 832, 858
   of symmetry reductions, classical, 732
   of symmetry reductions, direct, 708
   of symmetry reductions, nonclassical, 732, 747
   of Taylor series expansion in independent variable, for ODEs, 470
   of two-scale expansions, 504
   of undetermined coefficients, 17, 277
   of vanishing viscosity, 564, 580
   parametric for defining curve, 123
   parametric for defining surface, 123
   penalty function, 1027
   perturbation, 499
   Picard, 469
   projection, 508
   quadrature, 861
   quadrature, for integral equations, 870
   quadric, 148
   regula falsi, 262
   regularization, for integral equations, 827
   Runge–Kutta, 472, 512
   shooting, 512
   similarity, 669
   simplex, 1014
   single-step, of second-order approximation, 471
   splitting, 704
   step, 884, 885, 889, 893, 919
   Sturm, 166
   successive approximation, 822, 833, 858, 867
   summation of series, 344
   Tikhonov regularization, 828
   Titov–Galaktionov, 693
   Van der Pol–Krylov–Bogolyubov, 503
methods
   numerical for ODEs, 511, 512
   of defining function, 236
   of integral transforms, 841
   perturbation, 499–507
   projection, 508
metric space, 196
metric tensor, 1193
metric tensor components, 1195

Meusnier theorem, 392
midline, 44
minimal cost method, 1020
minimal polynomial, 186
minimal surface, 394, 770
minimality of variance of estimator, 1090
minimax principle, 1025
minimax property, 210
minimax theorem
    for antagonistic two-person zero-sum games, 1026
    von Neumann, 1026
minimum
    conditional, 270
    constrained, 270
    local, 257
    local, strong, 994
    local, weak, 993, 1001, 1002, 1006
    of function, xxxi, 257
    strong, 1006
    strong, of functional, 991
    weak, of functional, 991
Minkowski inequality
    for finite sums, 14
    for integrals, 296
    for series, 343
    for vectors, 121, 193
minor, 177
    basic, 178
    cofactor, 177
    first kind, 177
    of rectangular matrix, 177
    of square matrix, 176
    second kind, 177
minor axis of ellipse, 98
mixed boundary value problem, 480, 593
    for hyperbolic equations, 627
    for PDEs, 593
mixed strategy, 1025
    game price, 1026
mode
    of continuous probability distribution, 1045
    of discrete probability distribution, 1045
model
    of transportation problem, 1020
    of transportation problem, closed, 1020
model equation of gas dynamics, 558, 1254
modified associated Legendre functions, 966
modified Bessel equation, 1216
modified Bessel functions, 953
    first kind, xxxi, 953, 1216
    second kind, xxxi, 953, 1216
modified Korteweg–de Vries equation, 759
modified Mathieu equation, 982, 1220
modified Mathieu functions, 982
modified Newton–Kantorovich method, 859, 868
modified reciprocal equation, 160
modulus
    complementary, 971
    elliptic, 971
    of complex number, xxxii, 400

modulus (*continued*)
    of real number, xxx, 8
    of vector, xxx
modulus set, 907
Mollweide formulas, 44
moment, 1043
    absolute, 1044
    central, 1043
    central, of $n$-dimensional random variable, 1065
    initial, of multivariate random variable, 1065
    initial, of two-dimensional random variable, 1060
    $k$th absolute, 1044
    $k$th central, 1043
    $k$th sample, 1087
    $k$th sample, central, 1087
    of multivariate random variable, 1065
    of two-dimensional random variable, 1060
    sample, 1087
    sample, central, 1087
moment method, 509
moment-generating function, 1047
moments of inertia
    of body, 327
    of flat plate, 324
momentum conservation law, 996
Monge–Ampère equation, 654, 1326
    homogeneous, 668, 1326
monotone function, 236
monotone sequence, 238
Morera theorem, 410
Morley theorem, 49
Morse potential, 1278
motion of mass point, 87
movable critical points, 495
movable singularities of solutions of ordinary differential equations, 494
moving trihedral, 382
    of curve, 382
multidimensional stochastic process, 1072
multifactor analysis of variance, 1108
multilinear form, 215
    skew-symmetric, 215
    symmetric, 215
multimodal distribution, 1045
multiple, common least, 4
multiple correlation coefficient, 1066
multiple integral, 328
    asymptotic formulas, 328
multiple integration, 289
multiplication
    index, 12
    of asymptotic series, 364
    of matrices, 169
    of matrix by scalar, 169
    of polynomials, 155
    of real numbers, 6
    of series, 343
multiplicative inverse of real number, 6
multiplicative separable solution, 571, 596, 678

multiplicity of eigenvalue
   algebraic, 184, 209
   geometric, 209
multiplier, Lagrange, 270
multiply connected domain, 409
multivalued function, 236, 402
multivariate random variable
   initial moment, 1065
   moment, 1065

## N

$n$-dimensional coordinate space, 188
$n$-dimensional distribution, generalized variance, 1065
$n$-dimensional ellipsoid, 224
$n$-dimensional Fourier transform, 445
$n$-dimensional linear space, 189
$n$-dimensional random variable, 1065
$n$-dimensional random vector, 1057
$n$-fold singular point, 371
$n$-person game, 1024
$n$-sided pyramid, 63
$N$-soliton solution, 765
   Korteweg–de Vries equation, 764
Nagel point, 47
Napierian base, 23
Napierian logatithm, *see* natural logarithm
natural logarithm, xxxi, 9, 22, 23, 238
   base, 9, 23
natural number, 3
natural trihedral, 382
necessary condition
   for extremum, 269, 1002, 1006, 1009
   for solvability of Neumann problem, 429
negative binomial distribution, 1050
negative definite quadratic form, 216
negative Hermitian operator, 207
negative index of inertia of quadratic form, 218
negative symmetric operator, 207
neighborhood
   first-order, 991
   of point at infinity, 402
   of point on complex plane, 402
   strong, 991
   weak, 991
   zero-order, 991
Neper analogs, 73
Neper rules, 75
net, coordinate, 387
Neumann function, 947
Neumann problem, 429
   for elliptic equations, 593, 1330
   necessary condition for solvability, 429
Neumann series, 823, 833
neutral element, 225
Newell–Whitehead equation, 733
Newton method, 164
Newton–Kantorovich method, 859, 868
Newton–Leibniz formula, 288
   for function of complex variable, 408
Newton–Raphson method, 262

Neyman–Fisher theorem, 1090
Neyman–Pearson lemma, 1097
nilpotent matrix, 172
nine-point circle, 49
node, 372
   Chebyshev, 523
   interpolation, 808
   quadrature, 808
   stable, 535, 538
   unstable, 536
Noetherian symmetry, 767
   of Lagrangian, 768
nonbasic variable, 1015
nonclassical method of symmetry reductions, 732, 747
nonclassical symmetries, 733
nondecreasing function, 236
nondegenerate bilinear form, 215
nondegenerate linear operator, 205
nondegenerate matrix, 171
nondegenerate quadratic form, 216
nonhomogeneous $n$th-order constant-coefficient linear difference equation, 1427
nonhomogeneous $n$th-order linear difference equation, 1426
nonhomogeneous biharmonic equation, 1298
nonhomogeneous first-order constant-coefficient linear difference equation, 1411
nonhomogeneous higher-order linear systems, D'Alembert method, 533
nonhomogeneous integral equation, 810, 829
nonhomogeneous linear boundary value problems, 620, 624
nonhomogeneous linear difference equations, 882
nonhomogeneous linear ODEs, 474, 515
nonhomogeneous linear PDEs, 598
nonhomogeneous Monge–Ampère equation, 1326
nonhomogeneous second-order constant-coefficient linear difference equation, 1422
nonhomogeneous second-order linear difference equation, 1421
nonhomogeneous system
   of linear algebraic equations, 197
   of linear first-order ODEs, 531, 540
   superposition principle, 541
nonincreasing function, 236
nonlinear $n$th-order difference equation, 919
nonlinear difference equation, 884, 918, 1428
   of general form, 1437
nonlinear diffusion equation with cubic source, 753
nonlinear elliptic equations, 1318
nonlinear first-order partial differential equation with two independent variables, 570
nonlinear functional equation, 1443
   in one independent variable, 1428
   power series solution, 922
nonlinear heat equation, 1301, 1318
   in radial symmetric case, 1307
nonlinear integral equations, 856
   with constant integration limits, 856, 863

nonlinear integral equations (*continued*)
  with degenerate kernels, 863
  with variable integration limit, 856
nonlinear Klein–Gordon equation, 702, 1315
nonlinear mathematical physics equations, 1301
nonlinear operator, 693, 695
nonlinear ordinary differential equations, 453–472, 488–513, 524–527
  systems, 542–550
nonlinear partial differential equations, 570–584, 653–779, 1301–1335
  systems, 780–798, 1343–1382
nonlinear programming, 1027
  problem, 1027
nonlinear Schrödinger equation, 759, 1307
nonlinear systems
  of first-order ODEs, 1239
  of ODEs, 542–550, 1239, 1244
  of PDEs, 780–798, 1343–1382
  of PDEs of general form, 1374–1384
  of second-order ODEs, 1240
  of two first-order PDEs, 1337
  of two second-order PDEs, 1343
nonlinear Urysohn integral equation, 856
nonlinear Volterra integral equation, 856, 858, 860, 861
nonlinear wave equations, 723, 1312–1317
nonnegative Hermitian operator, 207
nonnegative quadratic form, 216
nonnegative symmetric operator, 207
nonorthogonal polynomials, 988
nonparametric hypothesis, 1094
nonpositive Hermitian operator, 207
nonpositive quadratic form, 216
nonpositive symmetric operator, 207
nonseparated by straight line points, 97
nonsingular extremal, 995
nonsingular matrix, 171
nonuniform convergence of functions, 249
norm, 314, 840
  of element, 193
  of linear operator, 204
normal, 91, 252, 368, 380, 389
  positive sense, 369
  principal, 381
  to quadric, 152
  to second-order curve, 111
  to surface, 153
normal curvature, 392
normal curvature vector, 392
normal direction, 389
normal distribution, 1053
normal form, 473
  of matrix, 181
normal fundamental system of solutions of homogeneous system, 198
normal hyperbolic equation, 590
normal matrix, 171
normal operator, 207
normal plane, 380, 382
normal representation, of quadratic form, 218

normal section, 392
normal subgroup, 227
normal system of method of least squares, 850
normal vector, 124
normalized cross-correlation function, 1073
normalized equation
  of plane, 126
  of straight line, 91
normed Euclidean space, 193
normed unitary space, 195
noteworthy limit, second, 241
null hypothesis, 1094
null matrix, 167
null subspace, 190
number
  absolute value, 8
  Bernoulli, 938
  complex, 399
  complex conjugate, 399
  composite, 3
  $e$, base of natural logarithms, xxxi, 9, 22, 23
  Euler, 939
  even, 3
  integer, 3
  irrational, 4, 5
  modulus, 8
  natural, 3
  odd, 3
  of real roots of polynomial, 165
  "pi" ($\pi$), xxxii
  prime, 3
  pure imaginary, 399
  rational, 4, 5
  real, 3, 4
number axis, 77
numerical eccentricity
  of ellipse, 98
  of hyperbola, 102
numerical integration of differential equations, 471
numerical sequence, 237
  generating function, 1047
numerical series, 337, 1118
numerical sum, 1113

# O

obelisk, 63, 65
oblate spheroid, 144
oblique asymptote, 242
oblique Cartesian coordinate system, 79
obtuse triangle, 43
octant, 114
odd function, 236
odd number, 3
odd permutation, 178
ODE, *see* ordinary differential equation
one-dimensional Klein–Gordon equation, 618
one-dimensional stochastic process, 1072
one-parameter family of curves, 377
one-parameter Lie group of point transformations, 730

one-parameter solution, 724
one-parameter transformation, 716
one-sheeted hyperboloid, 144
   imaginary semiaxis, 144
one-sided derivative, 252
one-sided limits, 240
one-soliton solution, 765
one-to-one mapping, 226
open interval, 235
open model of transportation problem, 1020
operations with asymptotic series, 364
operator, 204
   adjoint, 206
   bounded, 204
   characteristic value, 209
   differential, linear, 591, 592
   differential, total, 740, 742
   eigenvalue, 209
   eigenvector, 209
   Hermitian, 206, 207
   Hermitian negative, 207
   Hermitian nonnegative, 207
   Hermitian nonpositive, 207
   Hermitian positive, 207
   identity, 204
   infinitesimal, 716, 717, 730
   invariant, 717, 731
   inverse, 205
   Laplace, xxxii, 272
   linear, 204
   linear invertible, 205
   normal, 207
   opposite, 205
   orthogonal, 207
   reducible, 205
   root, 207
   second prolongation, 720
   self-adjoint, 206
   skew-Hermitian, 206
   skew-symmetric, 207
   symmetric, 207
   symmetric negative, 207
   symmetric nonnegative, 207
   symmetric nonpositive, 207
   symmetric positive, 207
   transpose, 207
   unitary, 206
   zero, 204
operators
   commutator, 755
   differential, of field theory, 272
   orthogonality condition, 207
opposite matrix, 169
opposite of real number, 6
optical property
   of ellipse, 100
   of hyperbola, 103
   of parabola, 105, 106
optimal solution, 1012
optimal strategy, 1024
optimality principle, Bellman, 1028

order
   of associated element, 213
   of branch point, 414
   of homogeneity, 1441
   of pole, 413
   of zero of function, 413
ordinary differential equation
   Abel, of first kind, 462
   Abel, of first kind, canonical form, 463
   Abel, of second kind, 464
   Abel, of second kind, canonical form, 465
   Airy, 955
   asymptotic solutions, 477
   autonomous, 489, 525, 1207, 1223, 1225
   Bernoulli, 458, 1207
   Bessel, 1216
   Bessel, modified, 1216
   boundary value problems, 480
   Clairaut, 467
   Darboux, 458
   degenerate hypergeometric, 1215
   Duffing, 501
   Emden–Fowler, 1223
   Ermakov, 1224
   Euler, 473, 520, 1216
   exact, first-order, 458
   exact, second-order, 491
   first boundary value problem, 480
   Gaussian hypergeometric, 960, 1218
   Green function, 481
   homogeneous, first-order, 456
   homogeneous, first-order, generalized, 456
   hypergeometric, 960, 1218
   hypergeometric, degenerate, 1215
   Lagrange, 467
   Laplace, 520
   Legendre, 1217–1219
   Lienard, 1225
   linear, 472, 518, 1213
   linear, constant-coefficient, 473, 514, 1213
   linear, first-order, 457, 1207
   linear, homogeneous, 472, 1212
   linear, nonhomogeneous, 473
   linear, second-order, 472, 1212
   Mathieu, 980, 1221
   Mathieu, modified, 982, 1220
   method of regular expansion in small
      parameter, 470
   method of Taylor series expansion in
      independent variable, 470
   mixed boundary value problem, 480
   Riccati, 460
   Riccati, general, 461, 462, 1209, 1210
   Riccati, special, 1208
   second boundary value problem, 480
   second-order, linear, 467
   second-order, nonlinear, 488
   self-adjoint form, 481
   separable, 456, 1207
   separated, 456
   third boundary value problem, 480
   Whittaker, 960

ordinary differential equations, 453–550, 1207–1228
  admitting reduction of order, 489, 525
  asymptotic solutions, 522
  boundary value problems, 480, 481, 512
  Cauchy problem, 488, 524
  existence and uniqueness theorem, 488
  first-order, 453–472, 1207–1212
  homogeneous, 490, 526
  homogeneous, generalized, 490, 526
  linear, 457, 472–488, 514–524, 1212–1222
  linear, homogeneous, 514, 518
  linear, nonhomogeneous, 515, 519
  linear, of arbitrary order, 514
  linear, with constant coefficients, 514
  nonlinear, 453–472, 488–513, 524–527, 1223–1228
  not solved for derivative, 465
  numerical methods, 511
  Painlevé, 495
  second-order, 472–514, 1212–1228
  solved for derivative, 488
  systems, nonlinear, 542–550
ordinate axis, 78
ordinate of point, 79, 115
oriented area, 83
oriented surface, 333
origin
  in space, 114
  of affine coordinate system, 190
  on axis, 77
  on plane, 78
orthocenter of triangle, 46
orthogonal complement of subspace, 194
orthogonal coordinates, 1194
orthogonal curvilinear systems of coordinate, 1195
orthogonal elements, 193
  of unitary space, 195
orthogonal group, 229
orthogonal matrix, 170
orthogonal operator, 207
orthogonal polynomials, 982
orthogonal projection, 197
orthogonal trajectory, 379
orthogonal transformation of matrix, 182
orthogonality condition for operators, 207
orthogonalization, Gram–Schmidt, 194
orthonormal basis
  in finite-dimensional Euclidean space, 193
  in finite-dimensional unitary space, 195
osculating circle, 375
osculating plane, 381
osculation, point, 372
Ostrogradsky–Gauss theorem, 334
overdetermined systems of PDEs, 770, 771

# P

Padé approximants, 493
Painlevé equations, 495
Painlevé test for nonlinear PDEs, 748
Painlevé transcendent, 495
  fifth, 499
  first, 496
  fourth, 498
  second, 496
  sixth, 499
  third, 497
Painlevé transcendental functions, 495
pair, Lax, 755
parabola, 104
  canonical equation, 104
  diameters, 106
  directrix, 105
  focal distance, 105
  focal properties, 105
  focal radius, 105
  focus, 105
  focus-directrix property, 105
  in polar coordinates, 107
  optical property, 105, 106
  vertex, 105
  with vertical axis, 106
parabolic coordinates, 1203
parabolic cylinder coordinates, 1203
parabolic cylinder function, 967
parabolic equation, 585, 590
  boundary value problems, 645
  canonical form, 586
  Cauchy problem, 615
  in narrow sense, 590
  linear, 586, 615–616, 618–622, 634, 635, 646, 648, 1267–1278
  nonlinear, 653, 654, 1301–1311
parabolic point, 394
parabolic type of second-order curves, 108
paraboloid, 145, 224
  elliptic, 145
  hyperbolic, 146
  of revolution, 146
paraboloidal cylinder, 225
parallel lines, 59
parallel planes, 59
parallel translation, 220
  of axes, 79
parallel vectors, 113
parallelepiped, 61, 62
  rectangular, 62
parallelogram, 52
parallelogram rule, 113
parameter
  of ellipse, 98
  of hyperbola, 102
  of integral equation, 829
  of plane, polar, 127
  variable, 87
parametric curves, 387
parametric equations
  of circle, 98
  of curve, 87
  of curve on plane, 87
  of plane, 125

parametric equations (*continued*)
   of straight line in space, 131
   of straight line on plane, 91
parametric hypothesis, 1094
parametric method
   for defining curve, 123
   for defining surface, 123
parametrized surface, 387
Parseval identity, 361
part of Laurent series, 412
partial derivative, 264
partial differential equation
   biharmonic, 1297
   biharmonic, homogeneous, 1297
   biharmonic, nonhomogeneous, 1298
   Born–Infeld, 659
   Boussinesq, 709, 1332
   Burgers, 664, 666, 751, 1307
   Burgers, radial symmetric case, 1307
   Burgers–Korteweg–de Vries, generalized, 709
   Clairaut, 1262
   continuity, 784
   diffusion, linear, *see* heat equation, linear
   diffusion, nonlinear, with cubic source, 753
   elliptic, 585, 588, 590, 1284
   evolution, second-order, 739
   first-order, quasilinear, 553
   Fisher, 1301
   Fitzhugh–Nagumo, 1302
   Hamilton–Jacobi, 576
   heat, linear, 585, 1267
   heat, linear, nonhomogeneous, 1268
   heat, nonlinear, 1318
   heat, stationary anisotropic, 1323
   heat, stationary, with nonlinear source, 1320
   heat, with axial symmetry, 1270
   heat, with central symmetry, 1272
   Helmholtz, 1289
   Helmholtz, three-dimensional, 600
   Helmholtz, two-dimensional, 599
   hyperbolic, 585, 587, 590
   hyperbolic, first canonical form, 587
   hyperbolic, normal, 590
   hyperbolic, second canonical form, 587
   Jacobi, 998, 1004, 1007
   Khokhlov–Zabolotskaya, 658
   Khokhlov–Zabolotskaya, stationary, 1322
   Klein–Gordon, 618, 625, 1280
   Klein–Gordon, linear, 625
   Klein–Gordon, nonlinear, 702, 1315
   Klein–Gordon, two-dimensional, 618
   Korteweg–de Vries, 752, 756, 759, 1327
   Korteweg–de Vries, cylindrical, 1328
   Laplace, three-dimensional, 643
   Laplace, two-dimensional, 585, 633, 1284
   linear, constant-coefficient, 588
   Liouville, 664
   model, of gas dynamics, 1254
   Monge–Ampère, 1326
   Monge–Ampère, homogeneous, 668, 1326
   Monge–Ampère, nonhomogeneous, 1326

partial differential equation (*continued*)
   Newell–Whitehead, 733
   nonlinear, first-order, 570
   of diffusion boundary layer, 1276
   of gas dynamics, model, 558
   of minimal surfaces, 770
   of steady transonic gas flow, 1324
   of thermal boundary layer, 1276
   of transonic gas flow, 1324
   of vibration of string, 1278
   parabolic, 585, 586, 590
   parabolic, in narrow sense, 590
   Poisson, 1287
   Poisson, three-dimensional, 599
   Poisson, two-dimensional, 599
   Schrödinger, linear, 1276
   Schrödinger, nonlinear, 759, 1309
   Schrödinger, of general form, 1311
   Schrödinger, with cubic nonlinearity, 1309
   Schrödinger, with power-law nonlinearity, 1310
   second-order, nonlinear, 653
   second-order, quasilinear, 654
   second-order, semilinear, 653
   sine-Gordon, 1314
   sinh-Gordon, 1314
   telegraph, 1284
   telegraph, generalized, 626
   ultrahyperbolic, 590
   vibration, with axial symmetry, 1282
   wave, 585, 1278
   wave, nonlinear, 723, 1312–1317
   wave, three-dimensional, 618
   wave, two-dimensional, 618
partial differential equations
   first-order, 553, 1247
   first-order, linear, 553, 1247
   first-order, nonlinear, 570, 1258
   first-order, quasilinear, 553, 1252
   fourth-order, linear, 1294–1298
   fourth-order, nonlinear, 1332–1334
   nonlinear, 653–769, 1301–1335
   nonlinear systems, 770–798, 1337–1341, 1343–1382
   second-order, linear, 585–650, 1267–1299
   second-order, linear, classification, 585
   second-order, nonlinear, classification, 653, 654
   third-order, nonlinear, 1327–1331
partial fraction decomposition of rational function, 276
partial increments, 264
   of function, 264
partial limit of sequence, 240
partial sum of series, 337
particular solution, 599
partition diameter, 317
path integral, 329
payoff matrix, 1024
   saddle point, 1025
PDE, *see* partial differential equation
Peano existence theorem for system of ODEs, 543
Pearson first skewness coefficient, 1045

Pearson test, 1100
penalty function, 1027
penalty function method, 1027
pencil, 92
   center, 92
   of planes, 131
   of straight lines, 92
percentage, 7
   simple, 7
percentile, 1045
periodic function, 236
permutation, 178
   even, 178
   odd, 178
permutation groups, 226
permutations, 1033, 1034
   with repetitions, 1034
perpendicular bisector, 46
perturbation methods, 499
Peterson–Codazzi formulas, 396, 397
Pexider equation, 925, 1438
Pfaffian equation, 772
   integrating factor, 773
phase plane, 546
phase variable, 546, 1010
physical meaning of line integral of second kind, 331
Picard method, 469
pivot column, 1016
pivot entry, 1016
pivot row, 1016
plane, 124
   complex, 401
   condition for planes to be parallel, 138
   condition for planes to be perpendicular, 138
   condition for planes to coincide, 138
   coordinate axes, 78
   coordinate system, 78
   curve, 84
   deviation of point from plane, 141
   general (complete) equation, 124
   incomplete equation, 124
   intercept equation, 125
   normal, 380, 382
   normalized equation, 126
   osculating, 381
   parametric equation, 125
   phase, 546
   principal, 151
   rectifying, 382
   tangent, 268, 381, 389
   tangent to surface, 152
plane curves, 367
plane geometry, 43
plane in space, 124
plane polygon, 51
plane triangle, 43
planes
   in space, 59
   parallel, 59
player, 1024

Pochhammer symbol, 938
Poincaré type series, asymptotic, 363
point
   applicate, 115
   asymptotic, 373
   at infinity, 402
   at infinity, neighborhood, 402
   boundary of domain, 401
   characteristic, 377
   circular, 393
   conjugate, 998, 1000, 1004, 1007
   coordinates, 79, 115
   derivative of function, 402
   deviation from line, 93
   deviation from plane, 141
   double, 371
   elliptic, 394
   Feuerbach, 49
   fixed, 908
   Gergonne, 47
   hyperbolic, 394, 536
   incenter, 46
   isolated, 371
   movable critical, 495
   Nagel, 47
   of affine space, 189
   of extremum, 257, 269
   of intersection of straight lines, 95
   of local minimum, 269
   of osculation, 372
   of rectification, 382
   of self-intersection, 372
   of tangency, 56
   on complex plane, 402
   on plane, 89
   parabolic, 394
   ramification, 414
   regular, 367, 379, 387
   saddle, 394
   saddle, of payoff matrix, 1025
   singular, 371, 391
   singular of generalized solution, 580
   spiral stable, 537
   spiral unstable, 537
   stationary, 220, 269, 545
   termination, 373
   transcendental branch, 415
   transition, 477
   triple, 371
   umbilical, 393
point estimators, 1088
point transformation, 455
   for ODEs, 455
   for PDEs, 655
point-slope equation of straight line, 89
points
   collocation, 848
   nonseparated by straight line, 97
   separated by straight line, 97
pointwise convergence
   of functional series, 348
   of functions, 249

Poisson distribution, 1050
Poisson equation, 1287
Poisson formula
  for Bessel function, 949
  in probability theory, 1038
Poisson integral, 428, 1287
  for half-plane, 429
Poisson process, 1076
Poisson theorem, 428, 1070
polar, 112
polar angle, 80
polar axis, 80
polar bilinear form, 216
polar coordinate system, 80
polar coordinates, 80, 1198
polar decomposition
  of linear operator, 208
  of matrices, 173
polar kernel, 801, 825
polar parameter of plane, 127
polar radius, 80
pole, 80, 413
  order, 413
  simple, 413
polygon, 51
  center, 55
  circumscribed about circle, 51
  convex, 51
  distribution, 1085
  edge, 51
  external angle, 51
  frequency, 1085
  inscribed in circle, 51
  interior angle, 51
  plane, 51
  relative frequency, 1085
  side, 51
  simple, 51
  vertex, 51
polyhedra, 61
polyhedral angle, 60
  edge, 61
  face, 60
  vertex, 60
polyhedral angles
  congruent, 61
  symmetric, 61
polyhedron, 61
  convex, 61
  edge, 61
  face, 61
  vertex, 61
polynomial, 155
  annihilating, 186
  Bernoulli, 891, 988
  characteristic, of linear operator, 211
  characteristic, of matrix, 185
  Chebyshev of first kind, 983
  Chebyshev of second kind, 984
  derivative, 157
  Euler, 989

polynomial (*continued*)
  expansion in powers of linear binomial, 156
  Gegenbauer, 987
  Hermite, xxxi, 957, 958, 985
  interpolation, 186
  Jacobi, 986
  Laguerre, 982, 957
  Laguerre generalized, 983
  Legendre, xxxi, 962, 1217
  minimal, 186
  number of real roots, 165
  remainder theorem, 156
  ultraspherical, 987
  with matrix argument, 172
  with matrix coefficients, 172
polynomials
  difference, 155
  division, 155
  equal, 155
  multiplication, 155
  orthogonal, 982
  nonorthogonal, 988
  product, 155
  sum, 155
Pontryagin function, 1011
Pontryagin maximum principle, 1011
population, 1081
  expectation, 1087
  variance, 1087
position vector of point, 115
positive definite kernel, 837
positive definite quadratic form, 216
positive definite quadratic functional, 993
positive Hermitian operator, 207
positive index of inertia of quadratic form, 218
positive integer power
  of complex number, 399
  of matrix, 172
positive normal direction, 389
positive operator, 207
positive sense, 367
  of boundary, 402
  of normal, 369
  of tangent, 369
positive sequence, 238
positive symmetric operator, 207
Post–Widder formula, 441
potential, 331, 760, 1021
  of gravitational field, 328
  of vector field, 331
potential-type integrals, 291
power, test, 1096
power function, 19
  graphs, 19
  properties, 19
power Laplace integral, 294
  multiple, 328
power series, 350, 1120
  Abel theorem, 350
  addition, 354
  convergence radius, 350
  division, 354

power series (*continued*)
  interval of convergence, 351
  Maclaurin, 350
  multiplication, 354
  properties, 350
  solution of nonlinear functional equations, 922
  subtraction, 354
  summation formulas, 357
  Taylor, 350
powers, 8
  of hyperbolic functions, 37
  of linear operators, 205
  of matrices, 172
  of trigonometric functions, 28
price, upper of game, 1025
primal linear programming problem, 1017
prime number, 3
primitive function, 273
principal axis, 109, 151, 152
  of hyperbola, 101
  of second-order curve, 109
principal diameter
  of ellipse, 100
  of hyperbola, 104
  of parabola, 106
principal directions of second-order curve, 109
principal normal, 381
principal normal sections, 393
principal part of Laurent series, 412
principal plane, 151
principal value, Cauchy, of singular integral, 311
principle
  boundary correspondence, 420
  Duhamel first, 646
  Duhamel second, 648
  maximin, 1025
  maximum modulus, 403
  minimax, 1025
  of linear superposition, for PDEs, 594
  Rayleigh–Ritz, 484, 513
  Riemann localization, 358
  Schwarz analytic continuation, 426
  superposition, for homogeneous system of linear ODEs, 528, 539
  superposition, for linear functional equations, 880, 883
  superposition, for linear PDEs, 594
  superposition, for nonhomogeneous linear ODE, 519
  superposition, for nonhomogeneous system of linear ODEs, 541
prism, 61
  base face, 61
  joining face, 61
  right, 62
  truncated, 62
Privalov boundary value problem, 432
probabilistic models, 1031
probability, 1032–1034
  classical definition, 1033
  conditional, 1035

probability (*continued*)
  convergence, 1068
  geometric, 1034
  multiplication theorem, 1035
  of event, 1032
  transition, 1075
probability density function, 1040
probability element, 1040, 1059, 1065
probability function
  joint, 1059, 1064
  marginal, 1059
probability integral, 939, 1265
probability multiplication theorem, 1035
probability space, 1032
  discrete, 1033
probability theory, 1031
probability-generating function of random variable, 1047
problem
  Banach, 1038
  Bolza, 1001
  boundary value, first, 480, 593, 627
  boundary value, for difference equations, 879
  boundary value, for elliptic equations, 631
  boundary value, for hyperbolic equations, 623
  boundary value, for ODEs, 480
  boundary value, for parabolic equations, 645
  boundary value, Hilbert–Privalov, 432
  boundary value, mixed, 480, 593, 627
  boundary value, nonhomogeneous, 620, 624
  boundary value, second, 480, 593, 627
  boundary value, third, 480, 593, 627
  Cauchy, for difference equations, 889
  Cauchy, for first-order ODEs, 454
  Cauchy, for first-order PDEs, 556, 576
  Cauchy, for Hamilton–Jacobi equation, 577
  Cauchy, for hyperbolic equations, 617
  Cauchy, for parabolic equations, 615
  Cauchy, for PDEs, 591, 592, 785, 1308
  Cauchy, for second-order ODEs, 488
  Cauchy, for system of PDEs, 785
  Cauchy, generalized, for PDEs, 627
  Dido, 1003
  diet, 1012
  direct scattering, 760
  Dirichlet, 427
  Dirichlet, for elliptic equations, 593, 1330
  Dirichlet, generalized, 428
  dual linear programming, 1017
  eigenvalue, 604
  first boundary value, 593
  general mathematical programming, 1012
  Goursat, 592, 629
  initial value, 580, 785
  inverse scattering, 760
  isoperimetric, 1002
  Lagrange, 1008
  linear programming, 1012
  Neumann, 429
  Neumann, for elliptic equations, 593, 1328
  nonlinear programming, 1027
  of propagation of signal, 566

problem (*continued*)
   primal linear programming, 1017
   Privalov boundary value, 432
   product mix, 1012
   programming linear, 1012
   Riemann, 785
   simplest, of calculus of variations, 993
   Sturm–Liouville, 482, 621, 622, 625, 626, 633
   terminal value, 580, 583, 584
   third boundary value, 593
   transportation, 1019
   transportation balanced, 1020
   two-body, 1258
   with higher derivatives, 1006
problems
   for elliptic equations, 637, 643, 644
   for hyperbolic equations, 607, 636
   for parabolic equations, 605, 634
   of mathematical physics, 590
   of mathematical statistics, 1081
process
   Bernoulli, 1037
   birth-death, 1077
   iteration, 858
   Markov, 1075
   optimal, in strong sense, 1010
   point simplest, 1076
   Poisson, 1076
   pure birth, 1077
   pure death, 1077
   stationary, 1075
   stochastic, 1071, 1073, 1074
   stochastic multidimensional, 1072
   stochastic one-dimensional, 1072
   stochastic stationary in narrow sense, 1074
   stochastic stationary in wide sense, 1074
   stochastic vector, 1072
   stochastic with continuous time, 1072
   stochastic with discrete time, 1072
   Wiener, 1086
product
   inner, 192, 314, 840
   Kronecker of matrices, 175
   mix problem, 1012
   of complex numbers, 399
   of linear operators, 203
   of linear transformations, 227
   of matrices, 157
   of matrix by scalar, 157
   of polynomials, 155
   of real numbers, 6
   of scalar and linear operator, 203
   of scalar and vector, 187
   of subsets of group, 227
   of vector by number, 113
   scalar, 192
   scalar of vectors, 120
   scalar triple of vectors, 123
   triple cross of vectors, 122
products
   of hyperbolic functions, 37
   of trigonometric functions, 28

programming
   convex, 1027
   dynamic, 1027
   geometric, 1027
   quadratic, 1027
progression, 1113, 1118
   arithmetic, 11
   geometric, 11
projection, 212
   of vector onto axis, 114
   on segment, 81
   orthogonal, 197
projection methods, 508
projections, theorem, 44
prolate spheroid, 144
prolongation
   of group generator, 732
   of operator, second, 720
proof by contradiction, 15
proof by counterexample, 17
proper orthogonal transformations, 229
proper subspace, 190
properties of limits, 240
property
   Markov, 1075
   minimax, 210
   optical of ellipse, 100
   optical of hyperbola, 103
   optical of parabola, 106
   stability, 1055
proportions, 7
psi function, 944
Ptolemy theorem, 52
Puiseux power series, 415
pure birth process, 1077
pure imaginary complex number, 399
pyramid, 63
   altitude, 63
   apex, 63
   apothem, 64
   base, 63
   frustum, 64
   lateral face, 63
   triangular, 63
Pythagorean theorem, 50, 193

# Q

quadrant, 79
quadratic equation, 158
quadratic form, 216, 840
   alternating, 216
   canonical representation, 217
   characteristic, 150
   characteristic equation, 150
   degenerate, 216
   extremal properties, 220
   in Euclidean space, 219
   in finite-dimensional linear space, 216
   index of inertia, 218
   negative definite, 216

quadratic form (continued)
  negative index of inertia, 218
  nondegenerate, 216
  nonnegative, 216
  nonpositive, 216
  normal representation, 218
  positive definite, 216
  positive index of inertia, 218
  rank, 216
  second of surface, 392
  signature, 218
quadratic functional, 992
  positive definite, 993
  strongly positive, 993
quadratic mean, 13
  of functions, 297
quadratic nonlinearity, 865
quadratic programming, 1027
quadrature formula, 808, 860
quadrature method, for integral equations, 852, 861, 870
quadrature nodes, 808
quadric
  canonical equations, 143
  canonical form, 143, 148
  classification, 148
  general equation, 148
  general theory, 148
  invariant, 148
  semi-invariant, 148
quadric surfaces, 143
quadrilaterals, 52
quartile, 1045
quasidiagonal canonical form of matrix, 182
quasilinear equations in conservative form, 567
quasilinear partial differential equations, 553
quotient, 155
  of complex numbers, 399
quotient group, 227

# R

Raabe convergence criterion for series, 339
radian, 24
radiation conditions, 1290, 1291
radical, 8
radius
  focal left, 99, 102
  of convergence, 351
  of convergence, of power series, 411
  of curvature, 376, 384
  of torsion, 385
  polar, 80
ramification point, 414
random event, 1031, 1032
random sampling
  with replacement, 1034
  without replacement, 1033
random sequence, 1072
random variable, 1039, 1057
  bivariate, 1060
  central moment, 1065

random variable (continued)
  continuous, 1040, 1042, 1064
  continuous bivariate, 1059
  discrete, 1039, 1042, 1064
  discrete bivariate, 1058
  expected value, 1042
  initial moment, 1060, 1065
  symmetric, 1041
  symmetric about expected value, 1041
  variance, 1044
random variables
  correlated, 1061
  independent, 1060, 1067
  joint distribution function, 1057
  uncorrelated, 1061
random vector, 1057
range
  of function, 236
  of linear operator, 205
  of mapping, 226
  sample, 1082
rank
  of bilinear form, 215
  of linear operator, 206
  of matrix, 178
  of quadratic form, 216
Rankine–Hugoniot jump conditions, 561, 568, 790, 791
rarefaction wave, 559, 785, 1253
rate
  false negative, 1096
  false positive, 1095
ratio
  common, 11
  contraction, 403
  dilatation, 403
  likelihood, 1097
  Romanovskii, 1101
  Student, 1090
rational function, 276
rational numbers, 4, 5
rational polynomial function, 276
Rayleigh–Ritz principle, 484, 513
real axis, 235, 401
  of hyperbola, 101
real Euclidean space, 192
real linear space, 188
real matrix, 171
real number, 3, 4
  multiplicative inverse, 6
  reciprocal, 6
real part, 399
real semiaxes
  of hyperbola, 102
  of one-sheeted hyperboloid, 144
  of two-sheeted hyperboloid, 144
real-valued function of real variable, 236
reciprocal (algebraic) equation, 160, 161
reciprocal functional equation, 919
reciprocal of real number, 6
rectangle rule, 315, 808
  for computation of integrals, 315

rectangular Cartesian coordinate system, 78, 114
rectangular Cartesian coordinates on plane, 78
rectangular matrix, minor, 177
rectangular parallelepiped, 62
rectangular spherical triangle, 72
rectification, point, 382
rectifying plane, 382
reduced cost, 1015
reducible operator, 205
reducible representation, 230
reduction
    of central second-order curves to canonical form, 110
    of noncentral second-order curves to canonical form, 110
    of quadratic form to sum of squares, 219
    of square matrix to diagonal form, 184
    of square matrix to triangular form, 184
reduction formulas, 1145
    for trigonometric functions, 27
reflectance, 760
reflection coefficient, 760
reflection factor, 760
regime
    blow-up, 1303, 1304
    critical, 1095
    feasible, 1012
regression, 1063, 1066
    mean-square, 1063, 1066
regression analysis, 1105
regression coefficient, 1063
regression function, 1063, 1066
regula falsi method, 262
regular arc, 379
regular extremal, 995
regular part of Laurent series, 412
regular point, 367, 379, 387
    of extremal, 995
    of generalized solution, 580
    of plane curve, 367
    of space curve, 379
regular polygon, 55
    center, 55
regular polyhedron, 65
regular prism, 62
regular pyramid, 64
regular triangle, 43, 51
regular value, 829, 834
regularization methods for integral equations, 827
relation
    Parseval for Fourier cosine transform, 445
    Parseval for Fourier sine transform, 446
    Parseval for Hankel transform, 447
relative extrema, 991
relative frequencies, 1083
relative frequency histogram, 1086
relative frequency of event, 1083
relative frequency polygon, 1085
remainder
    in Taylor formula in form of Cauchy, 257

remainder (*continued*)
    in Taylor formula in form of Lagrange, 257
    in Taylor formula in form of Peano, 257
    in Taylor formula in form of Schlömilch and Roche, 257
    in Taylor formula in integral form, 257
    of functional series, 349
    of series, 337
removable singularity, 413
renewal equation, 1395
representation
    dimension, 230
    invariant, 230
    irreducible, 231
    of group, completely reducible, 231
    of quadratic form canonical, 217
    of vector in terms of basis, 189
    reducible, 230
    triangular, 201
representation basis, 230
representation space, 230
residual, 847, 1066
residual standard deviation, 1063
residual variance, 1066, 1108
residue, 415, 435
    logarithmic, 416
resolvent, 810, 823, 829, 833
    of symmetric kernel, 839
resonance, Painlevé test, 749
rhombus, 53
Riccati difference equation, 875, 918, 1428, 1429
Riccati differential equation, 460
    general, 461, 462, 1209, 1210
    polynomial solutions, 460
    reduction to canonical form, 460
    reduction to second-order linear equation, 461
    special, 1208
Riemann function, 628, 629, 783
Riemann integral, 286
Riemann invariant, 783, 786, 787
Riemann localization principle, 358
Riemann problem, 785
    for system of PDEs, 785
Riemann property, 294
Riemann sum, 286
Riemann theorem, 419
Riemann wave, 782
Riemann–Schwarz theorem, 426
Riemannian space, 395
right arm of hyperbola, 101
right circular cone, 68, 145
right circular cylinder, 66
right coset, 227
right cylinder, 66
right directrix
    of ellipse, 99
    of hyperbola, 102
right ellipsoid, 145
right focal radius, 99, 102
right focus of ellipse, 99
right focus of hyperbola, 102

right half-plane, 79
right inverse matrix, 171
right prism, 62
right rectangular Cartesian coordinate system, 79
right triangle, 43, 50
  leg, 50
right-angled triangles, 43, 50
right-hand limit, 240
right-hand side of integral equation, 810
right-handed trihedral, 121
Rolle theorem, 254
Romanovskii ratio, 1101
Romanovskii statistic, 1101
Romanovskii test, 1101
root
  of complex number, 399
  of equation, 162
  of multiplicity $m$, 162
  of operator, 207
root mean square, 13
root-finding algorithms for continuous functions, 260
root-mean-square deviation, 1045
rotation of axes, 79
rotation transformation, 656
Rouché theorem, 413
round cylinder, 66
  segment, 67
Routh–Hurwitz criterion, 167
row, pivot, 1016
row expansion of determinant, 176
row vectors, 167
  linear combination, 171
  linearly dependent, 171
  linearly independent, 171
rows, basic, 165
rule
  Cramer, 199
  Lagrange multiplier, 1002
  of closing chain of vectors, 113
  parallelogram, 113
  rectangle, 315, 808
  sigma, 1054
  Simpson, 315, 808
  trapezoidal, 315, 808
  triangle, 113, 176
  Whitham, "equal areas", 562
ruled surface, 146
rules, Neper, 75
rulings, 146
  of ruled surfaces, 146
Runge–Kutta method, 472, 512

## S

saddle, 536
saddle point, 394
  of payoff matrix, 1025
sample, 1082
  asymmetry coefficient, 1088
  central moment, 1087
  correlation, 1104

sample (*continued*)
  distribution function, 1084
  excess coefficient, 1088
  function, 1072
  mean, 1087
  mean square deviation, 1088
  moment, 1087
  range, 1082
  size, 1082
  space, 1031
  variance, 1088
  variance adjusted, 1088
sampling
  random with replacement, 1034
  random without replacement, 1033
Sapogov convergence criterion for series, 340
Sarrus scheme, 176
scalar, 187
scalar product, 192
  of vectors, 120
scalar square, 120
scalar triple product of vectors, 123
scale segment, 77
scalene ellipsoid, 144
scalene triangle, 43
scaling transformation, 656, 676, 923
scattering data, 761
scheme
  Bernoulli, 1037
  Cole–Kevorkian, 504
  Horner, 156
  Sarrus, 176
  Van der Pol–Krylov–Bogolyubov, 503
schlicht mapping, 402
Schlömilch equation, 1400
Schröder–Koenigs functional equation, 911
Schrödinger equation
  linear, 1276
  nonlinear, of general form, 1311
  with cubic nonlinearity, 1309
  with power-law nonlinearity, 1310
Schwarz analytic continuation principle, 426
Schwarz integral, 428
  for half-plane, 429
Schwarz–Christoffel theorem, 427
secant, xxxi, 24
second boundary value problem, 480, 593
  for elliptic equations, 633
  for hyperbolic equations, 627
  for PDEs, 593
second convergence criterion, 338
second differential, 266
second duality theorem, 1018
second group of derivation formulas, 396
second mean value theorem, 295
second noteworthy limit, 241
second Painlevé transcendent, 496
second partial derivatives, 266
second prolongation of operator, 720
second quadratic form of surface, 392
second variation of functional, 992, 993

second-order constant coefficient linear ODE, 473, 1213
second-order curve, 107
   canonical form, 107
   center, 109
   central, 109
   characteristic equation, 108
   diameter, 108
   elliptic type, 108
   general equation, 107
   hyperbolic type, 108
   invariant, 107
   on plane, 97
   normal, 111
   semi-invariant, 108
   tangent, 111
   vertex, 109
second-order difference equations
   linear homogeneous, 877, 1421
   linear nonhomogeneous, 879, 880, 1421, 1422
   nonlinear, 1436
second-order differential constraint, 744
second-order evolution equation, 739
second-order hypersurface, 220
   center, 222
   centeral, 222
   equation, 223
   general equation, 220, 222
   noncenteral, 224
second-order ordinary differential equations
   linear, 472–488, 1212–1222
   linear, general solution, 472
   nonlinear, 488–514, 1223–1228
second-order partial differential equations
   linear, 585–650, 1267–1298
   nonlinear, 653, 1301–1334
   quasilinear, 654
   semilinear, 653
second-order variation of functional, 993
secondary diagonal, 168
section, normal, 392
sections, principal normal, 393
sector, 58
   circular, 58
   spherical, 69
secular terms, 502
segment, 58, 235
   directed, 77
   direction cosine, 116
   length, 77
   of round cylinder, 67
   on axis, 77
   scale, 77
   spherical, 69
   value, 77
self-adjoint matrix, 168
self-adjoint operator, 206
self-intersection, point, 372
self-similar solution, 669, 1442
   continuous, 781
   exponential, 675

self-similar solution (*continued*)
   generalized, 677
   limiting, 675
semi-invariant
   for probability distribution, 1046
   of quadric, 148
   of second-order curve, 108
semiaxis, 826, 842
   imaginary, of hyperbola, 102
   imaginary, of one-sheeted hyperboloid, 144
   imaginary, of two-sheeted hyperboloid, 144
   of ellipsoid, 143
   of hypersurface, 223
semilinear partial differential equations, 653
semimajor axis of ellipse, 98
semiminor axis of ellipse, 98
separable equation, 1207
separable solution, 596
   additive, 596
   multiplicative, 596
separated by straight line points, 97
separated equation, 456
separation constant, 678
separation of variables, 602
   complex, 679
   functional, 697
   generalized, 681
   incomplete, 597, 616, 640
   simple, 678
sequence
   arithmetic, 11
   bounded, 237
   convergent, 237, 238
   decreasing, 238
   generic term, 237
   geometric, 11
   increasing, 238
   infinitely large, 239
   limit, 237
   lower limit, 240
   monotone, 238
   numerical, 237
   random, 1072
   upper limit, 240
sequential analysis, 1097
series, 337
   Abel convergence criteria, 341
   Abel theorem, 343
   Abel transformation, 344
   absolute convergence, 341
   arbitrary, 341
   arithmetic, 11
   asymptotic, 363
   Cauchy theorem, 343
   conditional convergence, 341
   convergence domain, 348
   convergent, 337
   Dirichlet convergence criteria, 341
   divergent, 337
   Fourier, 357, 358
   fractional power, 415
   functional, 348

series (*continued*)
  generic term, 337
  geometric, 11
  harmonic, 339
  hypergeometric, 960, 1218
  interval, 1083
  interval of convergence, 351
  Kummer, 956, 1215
  Kummer transformation, 344
  Laurent, 410, 412
  Leibnitz convergence criterion, 341
  Maclaurin, 353
  majorant, 349
  Mertens theorem, 343
  Neumann, 823, 833
  numerical, 337
  of order statistics, 1082
  of Poincaré type, asymptotic, 363
  power, 350
  Puiseux power, 415
  remainder, 337
  statistical, 1083
  sum, 337
  summation methods, 344
  Taylor, 352, 410, 411
  uniformly convergent, 349
series solutions
  for hyperbolic equations, 607
  for ODEs, 470, 475, 492
  for parabolic equations, 605
series with positive terms
  convergence criteria, 338
  divergence criteria, 338
Serret–Frénet formulas, 386
sesquilinear form
  in unitary space, 213
  matrix, 214
set
  bounded, 263
  countable, 3
  diameter, 317
  modulus, 907
  submodulus, 907
  unbounded, 263
shock wave, 790–793, 1255
shooting method, 512
shortest distance, 70
side of polygon, 51
sigma rule, 1054
sign function, xxxi
signature
  of matrix, 182
  of quadratic form, 218
similar matrices, 181
similarity method, 669
similarity transformation, 181
simple hypothesis, 1094
simple percentage, 7
simple pole, 413
simple polygon, 51

simple Riemann wave, 782, 783
simple separation of variables, 678
simple zero, 413
simplest point process, 1076
simplest problem of calculus of variations, 993
simplex method, 1014
simply connected domain, 409
Simpson line, 47
Simpson rule, 315, 808
  for computation of integrals, 315
sine, 24, 25
sine integral, 941
sine-Gordon equation, 1314
single-step methods of second-order approximation, 471
single-valued function, 236, 402
singular integral, 310, 311, 570, 571
  in sense of Cauchy principal value, 430
  principal value, 311
singular point, 371, 391
  of generalized solution, 580
  of plane curve, 371
  of solutions, 494
  of surface, 391
singular solutions, 455
singularity
  essential, 413
  isolated, 413
  removable, 413
sinh-Gordon equation, 756, 1314
sink, 535, 538
  spiral, 537
sinusoid, sine curve, 25
sixth Painlevé transcendents, 499
size
  sample, 1082
  test, 1095
skew lines, 59
skew self-distributivity equation, 1449
skew-Hermitian matrix, 168
skew-Hermitian operator, 206
skew-symmetric bilinear form, 214
skew-symmetric integral equation, 840
skew-symmetric matrix, 168
skew-symmetric multilinear form, 215
skew-symmetric operator, 207
skewness, 1045
slope of straight line, 89
slope of tangent, 368
slope-intercept equation of straight line, 89
small circle, of a sphere, 71
small discriminant, 108
smallest value of function, 258
Smirnov test, 1099
Sokhotskii theorem, 431
solid angle, 61
solid geometry, 59
soliton, 1325
soliton solution, 765
  of Korteweg–de Vries equation, 764
soliton theory, 755

solution
    additive separable, 571, 596, 597, 678
    asymptotic for ODEs, 477
    asymptotically stable, 546
    Boussinesq, 1294
    Cardano, 158
    Cauchy problem, fundamental, 616, 617
    classical, 594
    D'Alembert for wave equation, 600
    Descartes–Euler, 160
    discontinuous for quasilinear equations, 558
    equilibrium, 874, 884
    exponential self-similar, 675
    feasible, 1012
    Ferrari, 161
    formal, 594
    functional separable, 697
    functional separable, special, 697
    fundamental, 599, 615
    fundamental for Cauchy problem, 616, 617
    fundamental for elliptic equations, 642
    fundamental for system of ODEs, 539
    general for difference equation, 874, 884
    general for first-order PDEs, 553
    general for ODEs, 453, 472, 488
    generalized, 563
    generalized, for first-order PDEs, 563, 568
    generalized, Hopf formula, 565
    generalized self-similar, 677
    generalized separable, 681, 682
    generalized stable, 569
    generalized traveling-wave, 698
    generalized viscosity, 579, 580
    in form of infinite series, 597
    invariant, 669, 674, 716, 724
    Lévy, 914
    limiting self-similar, 675
    multiplicative separable, 571, 596, 678
    $N$-soliton, 764, 765
    of eigenvalue problems, 604
    of system of linear algebraic equations, 198
    one-parameter, 724
    one-soliton, 765
    optimal, 1012
    particular, 453, 472, 518, 594, 599, 604
    self-similar, 669, 1440
    self-similar, continuous, 781
    self-similar, exponential, 675
    self-similar, generalized, 677
    self-similar, limiting, 675
    separable, 596
    separable, generalized, 681
    separable, multiplicative, 571
    singular for first-order ODEs, 455
    special functional separable, 697
    stable generalized, 569
    traveling-wave, 554, 667, 775, 790, 796
    trivial, 197
    unstable, 546
    viscosity, 564
    weak, 563

solutions
    fundamental system, 477, 478, 522, 523
    linearly independent of difference equation, 882
    of homogeneous system, fundamental system, 198
Sommerfeld conditions, 1288, 1289
source, 536, 538
    spiral, 537
space
    affine, 189, 190
    Banach, 196
    complete, 196
    continuous curve, 123
    continuous surface, 123
    coordinate axes, 114
    coordinate system, 114
    curve, 120
    Euclidean, 192, 193
    Euclidean, complex, 195
    Euclidean, real, 192
    Hilbert, 196
    linear, 187–189
    linear, complex, 188
    linear, infinite-dimensional, 189
    linear, real, 188
    metric, 196
    normed, 195
    of elementary events, 1031
    probability, 1032
    probability, discrete, 1033
    representation, 230
    sample, 1031
    surface, 119, 120
    unitary, 195
    vector, 187
space curves, 379
spaces
    Euclidean isomorphic, 194
    linear isomorphic, 189
    unitary isomorphic, 196
span, linear, 190
special Cauchy criterion, 341
special form, Laplace integral, 291
special functional separable solution, 697
special functions, 939–990
special linear group, 228
special orthogonal group, 229
special Riccati equation, 1208
special right-hand side, 815
specific Euclidean space, 192
specificity test, 1095
spectra of square matrices, 183
spectral decomposition
    of Hermitian operator, 212
    of normal operator, 212
spectral density, 1075
spectral function of stationary process, 1075
spectrum of matrix, 183
sphere, 68, 224
    diameter, 69
    unit, 220

spherical biangles, 71
spherical cap, 69
spherical coordinates, xxxi, 118, 119, 1199
spherical excess, 72
spherical geometry, 70
spherical sector, 69
spherical segment, 69
　base, 69
spherical triangle, 71, 72
　rectangular, 72
　vertices, 71
spherical zone, 69
spheroid, 144
　oblate, 144
　prolate, 144
spiral
　hyperbolic, 86, 368
　logarithmic, 86
　of Archimedes, 85
spiral point
　stable, 537
　unstable, 537
spiral sink, 537
spiral source, 537
splitting in derivatives, 719
splitting method, 704
square, scalar, 120
square integrable function, 314
square matrix, 168
　minor, 176
　trace, 171
square system of linear algebraic equations, 197
stability
　asymptotic, 546
　Lyapunov, 546
stability condition, 568
　for generalized solution, 568
stability property, 1055
stability theory, 546
stable focus, 537
stable generalized solutions, 569
stable node, 535, 538
stable spiral point, 537
standard deviation, 1045
standard form of linear programming problem, 1013
stationary equation
　heat (diffusion), anisotropic, 1323
　heat (diffusion), with nonlinear source, 1320
　Khokhlov–Zabolotskaya, 1322
stationary point, 220, 269, 545
stationary process, spectral function, 1075
stationary stochastic process, 1074
　in narrow sense, 1074
　in wide sense, 1074
stationary value of function, 220
statistic, 1084
　Cochran, 1104
　complete, 1090
　Romanovskii, 1101
　sufficient, 1090

statistical estimation, 1088
statistical estimator, 1088, 1090
statistical hypothesis, 1094
　simple, 1096
statistical series, 1082, 1083
statistical test, 1095
statistical transformations, 1082
statistically independent events, 1036
steady transonic gas flow, 1324
Steklov inequality, 297
step method, 884, 885, 889, 893, 919
Stewart theorem, 48
Stieltjes integral, 312
Stieltjes integral sum, 312
Stirling formula, 944
stochastic process
　differentiation, 1073
　in narrow sense, 1074
　in wide sense, 1074
　multidimensional, 1072
　stationary, 1074
　vector, 1072
　with continuous time, 1072
　with discrete time, 1072
stochastic processes, 1071–1073
　complex, 1073
Stokes theorem, 334
straight line
　direction coefficients, 131
　general equation, 90
　in polar coordinates, 92
　in projections in space, 133
　in space, 59
　in space, general equation, 132
　intercept-intercept equation, 91
　normalized equation, 91
　on plane, 89
　slope, 89
　slope-intercept equation, 89
strategy, 1024
　mixed, 1025
　optimal, 1024
stream function, 1297
strengthened Jacobi condition, 1000, 1004, 1007
strengthened Legendre condition, 999, 1004, 1007
stress function, Airy, 1297
strictly concave function, 245
strictly convex function, 245
strictly hyperbolic system, 781, 790
strictly lower triangular matrix, 168
strictly upper triangular matrix, 168
strong law of large numbers, 1070
strong local minimum, 994
strong maximum, 1006
　of functional, 991
strong minimum, 1006
　of functional, 991
strong neighborhood, 991
Struve function, 448, 449
Student $t$-distribution, 1056
Student distribution, 1056

Student ratio, 1090
Sturges formula, 1083
Sturm comparison theorem on zeros of solutions, 487
Sturm method, 166
Sturm–Liouville problem, 482, 621, 622, 625, 626, 633
subdiagonal matrix, 168
subgroup, 227
   compact, 229
   continuous, 228
   discrete, 228
   finite, 228
   normal, 227
submodulus set, 907
subsequence, 238
subspace
   improper, 190
   invariant, 205
   invariant of representation, 230
   linear, 190
   null, 190
   of linear spaces, 190
   orthogonal complement, 194
   proper, 190
subspaces
   direct sum, 189
   intersection, 190
   sum, 189
substitution
   canonical, 465
   Euler, 280
   trigonometric, 280
   trigonometric versatile, 282
subtraction
   of asymptotic series, 364
   of hyperbolic functions, 37
   of inverse hyperbolic functions, 41
   of inverse trigonometric functions, 33
   of matrices, 169
   of trigonometric functions, 28
successive approximation method
   for integral equations, 822, 833, 858, 867
   for ODEs, 469, 511
sufficient conditions of extremum, 269
sufficient estimators, 1090
sufficient statistic, 1090
sum
   Cauchy–Riemann, 286
   integral, 286, 317
   of complex numbers, 399
   of elements of linear space, 187
   of linear operators, 204
   of matrices, 168
   of polynomials, 155
   of series, 337
   of subspaces, 190
   of vectors, 113
   partial of series, 337
   Stieltjes integral, 312
summation, formulas for trigonometric series, 362

summation, index, 12
sums
   functional, finite, 1116
   involving hyperbolic functions, 1116
   involving trigonometric functions, 1117
   numerical, finite, 1113–1116
superdiagonal matrix, 168
superposition principle, 519, 880, 883
   for homogeneous system, 528, 539
   for nonhomogeneous system, 541
support line, 1014
supremum, xxxi
sure event, 1031
surface
   center, 143, 151
   central, 143
   conical, 67
   continuous in space, 123
   cylindrical, 65
   first quadratic form, 390
   in Cartesian coordinate system, 119
   in space, 119, 120, 123
   intrinsic geometry, 391
   minimal, 394, 770
   normal, 153
   of constant curvature, 394
   oriented, 333
   parametrized, 387
   ruled, 146
   second quadratic form, 392
   tangent line, 388
Sylvester criterion, 219
Sylvester inequality, 178
Sylvester theorem, 187
symbol, Levi-Civita, 179
symbols
   Christoffel, of first kind, 396
   Christoffel, of second kind, 396
   of order $O$ and $o$, 243
symmetric bilinear form, 214
symmetric integral equation, 835
symmetric kernel, 825, 829
symmetric linear programming problem, 1013
symmetric matrix, 168
symmetric multilinear form, 215
symmetric operator, 207
   nonnegative, 207
   nonpositive, 207
symmetric polyhedral angles, 61
symmetric random variable, 1041
symmetries
   differential equations, 716
   nonclassical, 733
   nonlinear second-order equations, 719
symmetry, 422
symmetry principle, 426
symmetry reductions
   based on generalized separation of variables, 712
   classical method, 732
   direct method, 708
   nonclassical method, 732, 747

system
- algebraic equations, linear, 197
- characteristic, 553
- Clairaut, 1238
- Ermakov generalized, 1240
- first-order linear homogeneous ODEs, 528, 529
- first-order ordinary differential equations, 1227
- hyperbolic, 781, 790
- incomplete of eigenfunctions, 836
- characteristic values, 836
- conservation laws, 780
- eigenfunctions, 836
- eigenfunctions, complete, 836
- Fredholm integral equations of second kind, 854
- general form, autonomous, 1238
- linear algebraic equations, 197
- linear algebraic equations, consistent, 197
- linear algebraic equations, determined, 197
- linear algebraic equations, homogeneous, 197
- linear algebraic equations, inconsistent, 197
- linear algebraic equations, nonhomogeneous, 197
- linear algebraic equations, square, 197
- linear algebraic equations, underdetermined, 197
- linear constant-coefficient ODE, 528
- linear constant-coefficient ODE, first-order, 1239
- linear equations, elementary transformation, 199
- ODEs, autonomous, 545
- ODEs, characteristic equation, 529
- ODEs, first integral, 544
- ODEs, fundamental solutions, 539
- ODEs, Liouville formula, 540
- ODEs, Wronskian determinant, 539
- random variables, 1057
- second-order ODEs, 1232
- solved for derivative (for ODEs), 542
- strictly hyperbolic, 781, 790
- two algebraic equations in two unknowns, 198
- two constant-coefficient first-order linear ODEs, 1229
- two constant-coefficient second-order linear ODEs, 1232

systems
- linear ODEs, 1229–1237
- linear PDEs, 1341, 1342
- nonlinear ODEs, 1239–1246
- nonlinear first-order PDEs, 1336–1341
- nonlinear PDEs, of general form, 1374–1382
- nonlinear second-order PDEs, 1343–1373

# T

$t$-distribution, 1056
tangency, point, 56
tangent, xxxi, 24, 26, 56, 368, 380
- positive sense, 369
- slope, 368

tangent line, 251
- to surface, 388
tangent plane, 268, 381, 389
- to quadric, 152
- to surface, 152
tangent to second-order curve, 111
tangential curvature vector, 392
Taylor formula, 257, 266
- remainder, 257
Taylor power series, 352, 410, 411
telegraph equation, 1282
terminal cost function, Bolza, 1001
terminal value problem, 580, 583, 584
termination point, 373
terms, secular, 502
test, 1095
- Bartlett, 1103
- Cauchy, 303
- Cochran, 1104
- Fisher, 1104
- Kolmogorov, 1099
- likelihood ratio, 1097
- Pearson, 1100
- Romanovskii, 1101
- Smirnov, 1099
- statistical, 1095
- Wald, 1097
test power, 1096
test size, 1095
test specificity, 1095
theorem
- Abel, 343, 411
- Abel, on convergence of power series, 351
- Abel–Ruffini, 162
- absolute value, 296
- absolute value, for double integral, 318
- absolute value, for triple integral, 325
- addition, 1055
- algebra, fundamental, 162
- Bernoulli, 1070
- binomial, 10
- Bolzano–Cauchy, 237
- boundary correspondence, 419
- Brauer, 877
- Casoratti, 902
- Cauchy, 254
- Cauchy, multiply connected domain, 409
- Cauchy, residue, 416
- Cauchy, series, 343
- Cauchy, simply connected domain, 409
- Cayley–Hamilton, 186, 212
- central limit, 1071
- Ceva, 48
- Chebyshev, 1070
- Chetaev, of instability, 550
- continuity, for characteristic functions, 1067
- convergence of collocation method, 524
- convolution, for Fourier transform, 444
- convolution, for Laplace transform, 437
- curl, 334
- de Moivre–Laplace, integral, 1038

theorem (*continued*)
  de Moivre–Laplace, limit, 1048
  de Moivre–Laplace, local, 1038
  decomposition, 208, 1055
  density multiplication, 1062
  Descartes, 165
  Dirichlet, 295, 358
  Dirichlet, representation of function by Fourier series, 357
  divergence, 334
  double integral, mean value, 318
  duality, second, 1018
  estimation for double integral, 318
  estimation for triple integral, 325
  Euler, 61, 393
  Euler–Lagrange, 1009
  existence and uniqueness, Cauchy problem for ODEs, 488, 524
  existence and uniqueness, for first-order ODEs, 454
  existence and uniqueness, for first-order PDEs, 557, 577
  existence and uniqueness, for systems of ODEs, 543
  Feller, 1077
  Feuerbach, 49
  first duality, 1018
  first mean value for integration, 295
  Fredholm, 834
  fundamental of algebra, 162
  Gakhov first, 432
  Gakhov second, 433
  Gauss, 335
  Green, 335
  half-side, 73
  Hilbert–Schmidt, 837
  integral of de Moivre–Laplace, 1038
  Jacobi, 180
  Kolmogorov, 1070
  Kronecker–Capelli, 198
  Lagrange, 254
  Laplace, 177
  Laplace transform, convolution, 437
  Laplace transform, limit, 440
  Legendre, 72
  limit, for Bernoulli process, 1038
  limit, for Laplace transform, 440
  Lindeberg central limit, 1071
  Liouville, 403, 414
  local of de Moivre–Laplace, 1038
  Lyapunov, central limit, 1071
  Lyapunov, of asymptotic stability, 549
  Lyapunov, of stability, 549
  mean value, 295
  mean value for double integral, 318
  mean value for triple integral, 325
  Menelaus, 49
  Mertens, series, 343
  Meusnier, 392
  minimax, for antagonistic two-person zero-sum games, 1026

theorem (*continued*)
  minimax, von Neumann, 1026
  Morera, 410
  Morley, 49
  Neyman–Fisher, 1090
  of arrangement of two lines in space, 136
  of boundary correspondence, 419
  of projections, 44
  on basic minor, 178
  Ostrogradsky–Gauss, 334
  Peano existence for system of ODEs, 543
  Poisson, 428, 1070
  polynomial remainder, 156
  probability multiplication, 1035
  projections, 44
  Ptolemy, 52
  Pythagorean, 50, 193
  Riemann, 419
  Riemann–Schwarz, 426
  Rolle, 254
  Rouché, 413
  Schwarz–Christoffel, 427
  second duality, 1018
  second mean value for integration, 295
  Sokhotskii, 431
  Stewart, 48
  Stokes, 334
  Sturm comparison, 487
  Sylvester, 186, 187
  triple integral, mean value, 325
  uniqueness, 413
  Viète, 158, 159, 163
  von Neumann minimax, 1026
theorems
  Fredholm, 834
  of differentiable functions, 254
  of instability, 549
  of stability, 549
theoretical distribution function, 1082
theory
  game, 1024
  stability, 546
theta functions, 978
third boundary value problem
  for elliptic equations, 634
  for hyperbolic equations, 627
  for ODEs, 481
  for PDEs, 593
third Painlevé transcendent, 497
third-order nonlinear PDEs, 1327
three-dimensional Helmholtz equation, 600
three-dimensional Laplace equation, 643, 644
three-dimensional Poisson equation, 599
three-dimensional wave equation, 618
throat ellipse, 145
Tikhonov regularization method, 828
Titov–Galaktionov method, 693
toroidal coordinates, 1205
torsion, 385
  of space curves, 385
torsion radius, 385
torus, 68, 70

total derivative, 265
total increments of function, 264
total probability formula, 1036
total variation of function, 247
trace of linear operator, 209
trace of matrix, 171
trajectory, orthogonal, 379
transcendent, Painlevé, 495
transcendental branch point, 415
transform, 435
    Bochner, 449
    Euler, of first kind, 449
    Euler, of second kind, 450
    Fourier, 443, 449, 614, 843
    Fourier, asymmetric form, 444
    Fourier, $n$-dimensional, 445
    Fourier, standard form, 443
    Fourier cosine, 445, 449
    Fourier inverse, 443, 614
    Fourier sine, 445, 446, 449
    Gauss, 449
    Hankel, 446, 449, 1398
    Hardy, 449
    Hartley, 449
    Hilbert, 449, 1396, 1398
    integral, 435
    inverse, 435
    Kontorovich–Lebedev, 448, 449
    Laplace, 436, 449, 611, 804, 813, 843, 856
    Laplace, inverse, 437
    Laplace, two-sided, 449
    Meijer ($K$-transform), 447, 449
    Mellin, 441, 449, 826, 842, 844
    Mellin, inverse, 442
    Weber, 449
    Weierstrass, 449
transformation
    Abel, 345
    Appell, 1266
    auto-Bäcklund, 663
    Bäcklund, 497, 663–665
    Cole–Hopf, 751
    congruence, 182
    congruent, 182
    conjunctive, 183
    contact, 660
    equivalence, 180
    Euler, 661
    Fisher, 1105
    Galileo, 718
    Gauss, 974
    graphs of functions, 259
    hodograph, 455, 657, 658
    Hopf–Cole, 666, 1308
    in plane, 717
    invertible, 267
    Kummer, 345, 957
    Kummer–Liouville, 475
    Landen, 974, 975
    Landen, ascending, 975
    Landen, descending, 974

transformation (*continued*)
    Legendre, 469, 660, 663
    linear, 656
    Lorentz, 718, 1278
    of equation of quadric to canonical form, 152
    of matrices, orthogonal, 182
    one-parameter, 716
    orthogonal, 182
    orthogonal, improper, 229
    point, 655
    proper orthogonal, 229
    scaling, 656, 676, 923
    similarity, 181
    simplifying boundary conditions, 649
    simplifying initial conditions, 649
    translation, 656, 668
    unitary, 183
    von Mises, 667
transformation group, 228
transforming matrix, 181
transition point, 477
transition probability, 1075
translation
    parallel, 220
    transformation, 656, 668
transportation, unbalanced, 1020
transportation problem, 1019, 1020
    closed model, 1020
    mathematical model, 1019
    open model, 1020
transpose, 170
    of matrix, 170
    of operator, 207
transposed equation, 825, 829, 834
transposition, 178
transversality condition, 577, 1001
transverse vibration of elastic rods, 1294
trapezoid, 54
    altitude, 54
    base, 54
    isosceles, 54
    leg, 54
    median, 54
trapezoidal rule, 315, 808, 810, 824
    computation of integrals, 315
traveling-wave equation, 1439
traveling-wave solution, 554, 667, 775, 790, 796
    generalized, 697
trefoil, 373
trials, independent, 1037
triangle, 43
    altitude, 46
    area, 48
    center of gravity, 45
    circumcircle, 46
    equilateral, 43, 51
    excircle, 47
    incircle, 46, 47
    isosceles, 43, 50
    median, 45
    obtuse, 43
    orthocenter, 46

triangle (*continued*)
  plane, 43
  right, 43, 50
  right-angled, 43
  scalene, 43
  spherical, 71, 72
triangle inequality, 14, 44, 73, 193, 314
triangle rule, 113, 176
triangles
  Euler, 71
  plane, 43
triangular decomposition, 180
triangular pyramid, 63
triangular representation, 201
triaxial ellipsoid, 144
Tricomi equation, 1403
trigonometric circle, 24
trigonometric form of complex number, 400
trigonometric functions, 24
  differentiation formulas, 29
  expansion in power series, 30
  integration formulas, 30
  of half argument, 29
  of multiple arguments, 29
trigonometric series, 1122–1128
trigonometric substitutions, 280
trihedral
  moving, 382
  natural, 382
  right-handed, 121
trihedral angle, 60
  edge, 60
  face, 60
  vertex, 60
triple cross product of vectors, 122
triple integral, 324
  change of variables, 326
  estimation theorem, 325
  geometric applications, 327
  mean value theorem, 325
  physical applications, 327
  properties, 324
triple point, 371
trivial representation of group, 230
trivial solution, 197
truncated cylinder, 66
truncated expansion, 749
truncated prism, 62
tube, cylindrical, 67
two-body problem, 1260
two-dimensional Helmholtz equation, 599
two-dimensional Klein–Gordon equation, 618
two-dimensional Poisson equation, 599
two-dimensional random variable
  initial moment, 1060
  moment, 1060
two-dimensional random vector, distribution function, 1057
two-dimensional wave equation, 618
two-person game, 1024
two-sheeted hyperboloid, imaginary semiaxis, 144
two-sided Laplace transform, 449
two-term asymptotic expansions, 478
types of linear PDEs, 585
types of matrices, 167

## U

ultrahyperbolic equation, 590
ultraspherical polynomials, 987
umbilical point, 393
unbalanced transportation, 1020
unbiased estimator, 1089
unbounded set, 263
uncomputable integrals, 276
uncorrelated random variables, 1061
underdetermined system of linear algebraic equations, 197
undetermined coefficients method, 17, 277
uniform convergence
  of functions, 249
  Weierstrass criterion, 349
uniform distribution, 1051
uniformly convergent functional series, 349
uniformly convergent series, 349
unimodal distribution, 1045
union of events, 1031
uniqueness and existence theorems for first-order ODEs, 454
uniqueness theorem, 171, 413
unit, imaginary, 399
unit normal vector, 389
unit sphere, 220
unit vector, 113
unitarity condition, 206
unitary group, 229
unitary matrix, 171
unitary operator, 206
unitary space, 195
  isomorphic, 196
  normed, 195
  orthogonal elements, 195
unitary transformation, 183
  of matrices, 183
universal invariant, 731
unknown quantities, 199, 201
unstable focus, 536, 537
unstable solution, 546
unstable spiral point, 537
upper bound, 235
upper half-plane, 79
upper limit of sequence, 240
upper price of game, 1025
upper triangular matrix, 168
Urysohn equation with degenerate kernel, 863, 864
Urysohn integral equation, 856

## V

value
  Cauchy principal, for singular integral, 311
  characteristic, 829, 834

value (*continued*)
  characteristic of operator, 209
  game, 1025
  of segment, 77
Van der Pol–Krylov–Bogolyubov scheme (method), 503
Vandermonde determinant, 179
Vandermonde matrix, 179
variable
  artificial, 1016
  basic, 1015
  incomplete separation, 597
  nonbasic, 1015
  phase, 546, 1010
  random, 1039
  symmetric random, 1041
variable density, 327
variable parameter, 87
variables
  dependent, 236
  independent, 236
  random independent, 1060, 1067
variance, 1044
  multifactor analysis, 1108
  of empirical distribution, 1088
  of random variable, 1044
  population, 1087
  residual, 1066, 1108
  sample, 1088
variate value, frequency, 1083
variation
  coefficient, 1045
  of argument, 992
  of functional, 992
  second of functional, 993
  total of function, 247
vector, 113, 167, 187
  associated, 213
  curvature, 383
  Darboux, 386
  geodesic curvature, 392
  issuing from point $A$ with endpoint $B$, 189
  length, 113
  normal, 124
  normal curvature, 392
  normal unit, 389
  position, 115
  tangential curvature, 392
  two-dimensional random, 1057
  unit, 113
  zero, 113
vector field
  flux, 333
  potential, 331
vector opposite to vector, 113
vector space, 187
vector stochastic process, 1072
vectors
  collinear, 113
  coordinate, 388
  coplanar, 113

vectors (*continued*)
  cross product, 121
  difference, 113, 188
  in space, 113
  linear combination, 188
  parallel, 113
  scalar product, 120
  scalar triple product, 123
  sum, 113
  triple cross product, 122
velocity, characteristic, 790
versatile trigonometric substitution, 282
vertex, 43
  of ellipse, 99
  of hyperbola, 102
  of parabola, 105
  of polygon, 51
  of polyhedral angle, 60
  of polyhedron, 61
  of second-order curve, 109
  of trihedral angle, 60
vertical asymptote, 242
vertices of spherical triangle, 71
vibration equation with axial symmetry, 1282
vibration of string, 1276
Viète theorem, 158, 159, 163
viscosity solution, 564
Volterra integral equation
  first kind, 801, 807
  nonlinear, 856
  second kind, 807, 810, 858
  with quadratic nonlinearity, 856
von Mises transformation, 667
von Neumann minimax theorem, 1026

# W

Wald test, 1097
Watson formula, 292
wave
  breaking, 559
  centered rarefaction, 558
  "overturn", 559
  rarefaction, 559, 785, 1255
  Riemann, 782, 783
  shock, 561, 790–793, 1255
wave equation
  linear, 585, 1278
  linear, three-dimensional, 618
  linear, two-dimensional, 618
  nonlinear, 723, 1312–1317
weak convergence, 1069
weak law of large numbers, 1069
weak local maximum, 993, 1001, 1002, 1006
weak maximum of functional, 991
weak neighborhood, 991
weak singularity, 801
  of second kind, 829
weak solution, 563
Weber parabolic cylinder function, 967
Weber transform, 449
wedge, 63, 65

Weierstrass condition, 998
Weierstrass criterion of uniform convergence, 349
Weierstrass example, 997
Weierstrass function, 998, 1000
    elliptic, 976
Weierstrass transform, 449
Weierstrass–Erdmann conditions, 997, 999
    in multidimensional case, 999
weight of variate value, 1083
Weingarten formulas, 396
Whitham rule of "equal areas", 562
Whittaker equation, 960
Whittaker functions, 960
Wiener process, 1086
Wiener–Hopf equation
    first kind, 825, 1401
    second kind, 829, 1406
Willier formula, 74
Wronskian, 473, 482, 518
    for system of ODEs, 539

Wronskian (*continued*)
    of degenerate hypergeometric functions, 958
    of Legendre functions, 967
Wronskian determinant, *see* Wronskian

# Y

$Y$-transform, 448, 449

# Z

zero
    asymptotic, 364
    of function, 413
    simple, 413
zero matrix, 167
zero operator, 204
zero vector, 113
zero-order neighborhood, 991
zeros of Bessel functions, 951
Zhukovskii function, 404, 424
zone, spherical, 69